Second Edition

HANDBOOK OF VIRTUAL ENVIRONMENTS

Design, Implementation, and Applications

T0138999

Human Factors and Ergonomics Series

PUBLISHED TITLES (CONTINUED)

FORTHCOMING TITLES

Second Edition

HANDBOOK OF VIRTUAL ENVIRONMENTS

Design, Implementation, and Applications

Edited by
Kelly S. Hale
Kay M. Stanney

CRC Press
Taylor & Francis Group
Boca Raton London New York

CRC Press is an imprint of the
Taylor & Francis Group, an **informa** business

CRC Press
Taylor & Francis Group
6000 Broken Sound Parkway NW, Suite 300
Boca Raton, FL 33487-2742

First issued in paperback 2017

© 2015 by Taylor & Francis Group, LLC
CRC Press is an imprint of Taylor & Francis Group, an Informa business

No claim to original U.S. Government works

ISBN-13: 978-1-4665-1184-2 (hbk)
ISBN-13: 978-1-138-07463-7 (pbk)

Library of Congress Cataloging-in-Publication Data

Handbook of virtual environments : design, implementation, and applications / editors, Kelly S. Hale, Kay M. Stanney. -- Second edition.
    pages cm -- (Human factors and ergonomics)
  Includes bibliographical references and index.
  ISBN 978-1-4665-1184-2 (hardback)
  1. Computer simulation--Handbooks, manuals, etc. 2. Human-computer interaction--Handbooks, manuals, etc. 3. Virtual reality--Handbooks, manuals, etc. I. Hale, Kelly S. II. Stanney, Kay M.

QA76.9.C65H349 2014
004.01'9--dc23   2014023644

Visit the Taylor & Francis Web site at
http://www.taylorandfrancis.com

and the CRC Press Web site at
http://www.crcpress.com

Contents

SECTION I   Introduction

SECTION II   System Requirements: Hardware

SECTION III   System Requirements: Software

SECTION IV   Design Approaches and Implementation Strategies

SECTION V Health and Safety Issues

SECTION VI Evaluation

SECTION VII   *Selected Applications of Virtual Environments*

SECTION VIII Conclusion

Perspective by Jaron Lanier

The public remains as entranced by a romantic ideal of virtual reality (VR) as ever. VR the ideal, as opposed to the real technology, weds the nerdy thing with the hippie mystic thing; it's high tech and like a dream or an elixir of unbounded experience, all at the same time.

I wish I could fully convey what it was like in the early days. There was a feeling, in the early 1980s, of opening up a new plane of experience. Inhabiting the first immersive avatars, seeing others as avatars, experiencing one's body for the first time as a nonrealistic avatar; these things transfixed us. Everything else in the world seemed dull in comparison.

The irony is that I cannot use VR to share what that experience was like with you. VR, for all it can do, is not a medium of internal states. There might never be such a medium. There is little need for me to make this point in an introduction to a book for VR professionals, but it's a clarification that I'm sure many of the contributors and readers of this book have been called upon to give many times. What the public seems to want from VR is telepathic conjuring of arbitrary reality. It can be difficult to explain that VR is wonderful for what it is, precisely because it isn't magic.

Eventually a new kind of culture, a massive tradition of tricks of the VR trade, might arise and that culture might allow me to convey something to you about how early VR felt, using VR-borne metaphors. I have spent many hours daydreaming about what a mature culture of expression would be like in VR. A cross between cinema, jazz, and programming, I used to say.

Even though we don't know how expressive VR might eventually become, there is always that little core of thrill in the very idea of VR. Arbitrary experience, shared with other people, conversationally, under our control. An approach to a holistic form of expression. Shared lucid dreaming. A way out of the dull persistence of physicality. This thing we seek, it's a way of being that isn't tied just to our circumstances in this world.

Every few years a new wave of public VR mania rises, yet somehow very few people seem to realize how often it's happened before. Long ago, a scrappy little startup called VPL Research seemed for a moment like a much larger company, a real force. The same kind of thing can still happen; it is happening once again at the time of this writing with a startup selling an HMD called Oculus Rift.

The actual experience of high-quality VR remains elusive, however. You really still have to have access to a top lab to experience the good stuff. By "the good stuff" I mean a system good enough to fool you, to engage your whole body, to include others as avatars with you in there, to be usable in the long term, and one that gives you enough to do to outlast the first few demos. There are remarkably few places on Earth today where one can experience anything close to the whole package.

However, such places exist. This is well known to lucky students who work with Jeremy Bailenson at Stanford, with Henry Fuchs or Fred Brooks at the University of North Carolina, Chapel Hill, or with Carolina Cruz-Neira at the University of Louisiana, Lafayette—or at any of the other fine VR research labs. Then there is the even less accessible world of training and design centers that use VR. To my knowledge, no one has ever thought to maintain an atlas of VR systems, but there are still probably only a few thousand decent VR systems in the world as I write this Perspective in 2013.

The VR known to researchers is sometimes a little like what the public dreams of, but there is a colossal difference, because in research labs, VR is a practical, if expensive, tool. Researchers continue to learn more about the human organism by using VR as an experimental device to test hypotheses in cognitive science, biomechanics, and other disciplines. VR research continues to spin off useful technologies, such as motion capture suits, surgical simulators, and vehicle design simulators. From the point of view of pop culture, the realization of the dream remains elusive, while to practitioners, our progress has been satisfying.

Nonetheless, the cost of the parts we use gradually lowers, and a horizon comes into view. Over that horizon lies the long-anticipated popularization of VR, meaning a form of VR that would be complete enough to feel like a protean dream, at least a little like The Matrix, if you squint. The world waits for this.

Much of these could have been written at any time in the last thirty years. In earlier times, we might have mentioned Star Trek's Holodeck instead of The Matrix, but the pattern was similar. We've been at work, while the public frets that what we do hasn't reached its potential yet. The popular circumstances in which we work have remained about the same even as we have learned more, and successive generations of researchers have come to the fore. Journalists periodically write about how VR has "failed to deliver on its promises." But this last year has been different.

Consumer devices related to VR are appearing with enough frequency that they no longer feel exceptional. The last few years have seen Microsoft's Kinect success in the marketplace and Google's Glass public trials, for instance. Neither device is a VR system, but both are consumer gadget versions of components that are cousins to what would be needed in full-on VR systems. Both have changed the public's expectations of information technology.

It does appear that we are about to arrive at the long-anticipated threshold. Even so, it could still take some years for the details to come into alignment. Thirty years ago, my colleagues and I used to guess that a reasonably comprehensive form of VR would be popularized around 2020, based on Moore's Law and related trends, and perhaps we will turn out to have been a little pessimistic.

Those of us who have worked on VR for many years wonder what it will be like once VR becomes both cheap and well-wrought, and the novelty factor has eroded to nothing. Will it be banal—like plumbing or ever new—like books? This is always the question with a new layer of digital capability.

So, a drama of anticipation builds. It is like wondering how a child will turn out. Have we been building something banal or beautiful? Reality can only be known empirically. We don't yet really know what we're building. We don't know to what degree it will be a stream of vital passion and poetry, a banal waste of time, or a cruel gauntlet for manipulation and bullying. Any human medium will be all these things, but in what proportions?

I am currently feeling confident in our hardware and not so confident in our software. Hardware first.

Unsolved problems remain in VR hardware. Some of the problems are shared in common with other technologies. Most approaches to VR involve wearable components, but batteries still weigh too much, for instance. Therefore, VR components tend to either be heavy or need to be charged too often. Trends in battery technology are positive, so we expect this problem to be addressed.

But VR also poses some unique hardware challenges. What is the right haptic feedback design for general popularization? What is the best general scheme for reconciling the specific limited physical space a user can explore at a given time with the unbounded, beckoning virtual spaces that might be presented? Researchers have presented hundreds of good ideas for answers to both questions, but there is no consensus because there haven't yet been adequate, widespread tests of the most promising ideas.

VR intrinsically presents conundrums. The best known one might be this: To fill the human visual field with visual information, it is common to place a source of photons near the eyes, in the form of an HMD. But for people to see each other as avatars, the facial expressions should be measured so as to have an influence on one's avatar. After all, we present ourselves visually using our faces most of all. But HMDs often obscure or interfere with facial expressions. To resolve this difficulty, a clever combination of optics, industrial design, and graphics/machine vision will probably be needed.

VR also intrinsically pushes requirements, for the simple reason that the human organism is remarkable. We are sensitive to minute latencies, and in some cases to tiny inconsistencies in our sensory data. Fortunately, as everyone who works with it knows, VR is based on finding ways to present what should by all rights be inadequate equipment in a way that somehow meets the expectations of the human nervous system. Our field can be considered as the study of highly advanced stage magic without the stage.

Even so, we sometimes find a challenge we must meet head-on. For instance, to obscure items in the real world in order to show virtual content instead (in a mixed- or augmented-reality presentation), presents a healthy challenge to both optics engineers, machine vision scientists, and systems engineers doing battle with latency.

Fortunately, the public's fascination with our work creates an interesting external laboratory that is not available in many other research fields. The marketplace can support widespread testing of hardware components in the form of entertainment products. For instance, real-time human motion sensing in a form that would be suitable for VR systems has also been packaged independently in gaming devices like Matte's Power Glove and Microsoft's Kinect. With each market test we learn more about what a popular, complete VR system might be like. Fortunately, these tests have yielded positive results in many cases.

The hardware side of VR progresses reliably towards fulfilling the collective dream of VR, it seems. Each year, new and better optical and display strategies appear; same for new sensors and actuators. The mega-reorientation of consumer computing towards portable devices has resulted in vast improvements to components like microdisplays and inertial motion units, so VR researchers have lately been showered with hardware gifts.

But no one is helping us with our software, and there are two huge reasons for concern that our software might not turn out as well as our hardware. The first reason is that previous attempts to define software architectures for VR have all turned out to be too specialized to support a broad community. The second is that as we approach our long-anticipated threshold, the world at large is entering into what might be called a surveillance economy that could render our dreams into creepy nightmares.

One problem with VR software architecture is that VR can hypothetically align with a huge variety of human experiences. Software on a personal computer was stuck on the screen, and the person using it was stuck on a desk, with a hand on a mouse. Every program would be reliably circumscribed by those conditions, so it was possible to declare a significant commonality between programs that otherwise had no relationship. Both Mathematica and Angry Birds can be scrolled in a window and pricked by a mouse. More recently, approximately the same kinds of rules have applied to mobile devices; now programs assume multitouch, for instance. But the background in which programs are defined is still the class of device, not the world at large.

VR isn't defined on an artifact. The background is the human body and—if the system is presented for mixed or augmented reality—the physical environment. Neither are human inventions, and both are still under study. We will never have complete knowledge of our physical environment or of our bodies.

So the first job of a VR architecture is to ignore almost everything about reality in order to choose a few hooks that can be used to align virtuality. But what should be the preferred blindness?

Different VR software architectures choose quite different emphases. A major divide can be observed between attempts to interface individuals as well as possible and attempts to interface large numbers of people.

Designs that emphasize increasing the number of participants typically do so at the expense of the interaction model. Examples are military ground combat simulators and popular network sites with a VR flavor, like Second Life. In these systems, users move about, talk to each other, lob projectiles, and so on, but there is little use for detailed hand tracking. You couldn't conduct a dance lesson in such a system.

Then there are designs that take great care to measure and integrate human activity carefully. These include, on a professional level, surgical simulators as well as entertainment systems that use depth cameras or other means to measure human motion comprehensively in real time, like Kinect.

Designers of these systems have their hands full, so to speak, and don't deal with the many issues that come up with massive numbers of users distributed across large distances.

These two schools are not the end of it. There are also designs that emphasize representing the physical world and presenting that to users, but are poor at measuring the users or allowing users to interact. This stream of design might even be the largest. It started long ago with forgotten attempts at standards like VRML and Java3D, and continues today with cloud services that integrate street view, satellite, and other image sources to generate streaming models of the world. It is never elegant to inhabit a virtual world driven by an architecture that is focused obsessively on data rather than on the experience of data.

Here we find another example of how our field can sometimes seem to stand still for decades, even though we're all working very hard. For decades there has been endless talk about combining the different approaches to networked VR software architecture. Wouldn't it be grand to be able to have dance lessons over a service resembling Second Life, using something like Kinect, in a simulation of a historic dance studio, gathered by something like Photosynth?

I have been involved in all three projects, and the simple truth is that just to get those out the door well involved total devotion, total exhaustion, really, for the teams who had to get the jobs done. There was no further energy or time left to figure out dreams of integration or generalization. We don't know how hard it will be to "get it all" in one network VR architecture, but we know it will be hard.

I hope we do it in the next few years. I don't want to have to be carrying four different headsets with me in ten years, each specialized to a different sort of task.

This brings up a correlate to the unbounded nature of VR. Everyone understands that not every Android app will run on an iPhone. But are we ready for a world where certain customers can see certain augmentations of physical reality while others cannot?

Will we enter into a situation in which people walk down the street and see entirely different augmentations of the world depending on which vendor they bought their headsets from? Sherry Turkle wrote a book called "Alone Together" about how people can become oddly insulted by using network technology, which was certainly not the intent of the designers. I worry that mixed reality systems used in public could result in an "Apart Together" situation, in which people are adjacent, but not able to experience much in common, even when they are trying to.

This eventuality could be the result of cloud service rivalries, carrier wars, or—and this becomes our problem—architectural specializations that corral users into disjoint communities. Quests for perfection in computer architectures tend to backfire, but that doesn't mean we have to resign ourselves to worst case scenarios. If the research community can prove that an architecture can come into existence that supports a wide range of scenarios, then the political fight over whether the various companies will play nice with each other can at least have a chance to play out.

Finally, the rise of creepiness must be addressed. VR is the most intimate communication technology. In order to work well, VR instrumentation senses the saccade of the eyes and the turn of the lip. What we do is interface intimately with the human organism.

It happens, however, that as we approach the day when the fruits of our labors become inexpensive enough for everyone to enjoy, another commercial trend is on the rise that would use these data against our users, our customers.

It is now commonplace for people to accept constant surveillance in exchange for so-called bargains or convenience. Most people have accounts on social networks and trust companies to keep track of many of their life memories, for free. We have also recently learned that governments routinely piggyback on these arrangements to maintain their own surveillance. The data derived from surveillance are used commercially to power big data systems that, in turn, micromanage some of the options placed in front of people in the form of paid links. Some feel this is a helpful economic arrangement, while I do not. In any case, everyone should be concerned that the surveillance economy is creating astonishing potential for abuses.

It is one thing for remote corporations or governments to have access to the metadata about who one has been in touch with or one's recent purchases. But it is another entirely for remote powers to compile information about how your body responds to stimuli. There is then a potential for neo-Pavlovian manipulation, for instance, where a person might not realize that a manipulation had even taken place.

Many new consumer products will not function if a user does not "click through" an agreement sending their experiences to the company's servers. In general, all tech companies, and all governments, have come to expect people to click through such agreements and accept surveillance.

I had always thought I could keep politics and VR research separate, but that is no longer true. Our community has to consider how to protect our eventual users. Should sensors of the most important signals in the human body encrypt their results as part of the measurement? Should we be baking in privacy at the lowest level to make man-in-the-middle attacks harder? I'd rather not have to think about such questions, but I do, we all do.

This is a wonderful, but challenging time for the VR community. I have lately become more excited than ever to meet new students who have decided to enter our field because I know their careers will be filled with delightful surprises and thrills. I also worry. Will they come to feel that they were part of the problem instead of a solution to the mounting wave of cyber-creepiness?

The story unfolds. Our time has arrived.

Jaron Lanier
Interdisciplinary Scientist
Microsoft Research

Preface

The first edition of the *Handbook of Virtual Environments* was published in 2000. A lot has changed since that time. Virtual environments (VEs) have always presented infinite possibilities for extending the interactivity and engagement of an ever-expanding variety of applications in the areas of entertainment, education, medicine, and beyond. But now, with the technology advancing at rates exceeding conventional laws, the possibility to create VEs that provide true-to-life experiences has never been more attainable. We thus felt it was time to assemble a second edition of the handbook to summarize the current cutting-edge research and technology advances that are bringing these truly compelling virtual worlds to life.

This second edition includes nine new as well as 41 updated chapters that reflect the progress made in basic and applied research related to the creation, application, and evaluation of VEs. Leading researchers and practitioners from multidisciplinary domains have contributed to this second edition to provide a wealth of both theoretical and practical information, resulting in a complete toolbox of theories and techniques that researchers, designers, and developers of VEs can rely on to develop more captivating and effective virtual worlds.

Section I consists of two chapters. Chapter 1 provides a thorough summary of VEs in the twenty-first century, while Chapter 2 provides an updated glossary of terms to promote common language throughout the community. Section II focuses on system requirements, both hardware and software. Individual chapters are provided for various modalities of presentation, highlighting known techniques for eliciting and manipulating visual (Chapter 3), auditory (Chapter 4), and haptic perception (Chapter 5), as well as perception of body motion (Chapter 7). This second edition also includes a new contribution examining olfactory perception and display design (Chapter 6). In addition, interaction methods that support optimized human–system collaboration are reviewed, including eye tracking (Chapter 8), gesture (Chapter 9), and locomotion systems (Chapter 10). These chapters highlight advances in hardware design and capabilities to support natural interaction that increases presence and human performance within VEs. Four chapters in this section focus on software design of VEs, including development of VE models (Chapter 11), designing effective interaction techniques from the software perspective (Chapter 12), simulating and optimizing VE realism where and when needed (Chapter 13), and design of embodied autonomous agents (Chapter 14).

Section III focuses on design approaches and implementation strategies for VEs. Each chapter in this section has been updated to reflect advances made in the last decade and provides a solid foundation for new and experienced designers. Chapter 15 summarizes the VE development structure, a standardized process for designing and developing VE technology. Design implications for human cognition (Chapter 16), modeling multimodal interaction (Chapter 17), illusory self-motion (Chapter 18), spatial orientation and wayfinding (Chapter 19), technology management for user acceptance (Chapter 20), and product liability (Chapter 21) round out this informative section regarding design of VEs.

Section IV focuses on health and safety issues, which have been observed since the earliest VE systems were created. The chapters in this section examine the direct effects of VEs on users (Chapter 22) and provide a thorough review of motion sickness symptomatology and origins (Chapter 23) and methods to measure motion sickness (Chapter 24) within a VE. The section also includes a review of adaptation effects and their implications (Chapter 25), as well as discussions of visually induced motion sickness (Chapter 26) and the social impact of VE technology (Chapter 27).

Section V focuses on evaluation of VE systems. It begins with a review of usability evaluation techniques for VEs (Chapter 28), summarizing how traditional usability methods may be adapted and extended to evaluate 3D spaces and experiences. Chapter 29 summarizes methods for measuring human performance within VEs, while the addition of Chapter 30 provides guidance on evaluating the effectiveness of VE training systems. VE usage protocols are summarized in Chapter 31, which outlines methods to minimize negative impacts of VE exposure. Methods to evaluate visual aftereffects from VE exposure (Chapter 32), proprioceptive adaption and aftereffects (Chapter 33), presence experienced while in a VE (Chapter 34), and a new chapter highlighting the potential of utilizing an augmented cognition system to evaluate VE systems (Chapter 35) round out this section of the handbook.

Section VI, which focuses on the application of VE technology, has been substantially updated to reflect the tremendous progress made over the last decade in applying VE technology to a growing number of domains. Chapter 36 provides an overview of application areas, while follow-on chapters discuss applications in national defense (Chapter 37) and various training and learning domains, including team training (Chapter 38), perceptual skills training (Chapter 39), conceptual learning (Chapter 40), experimental, STEM, and health sciences learning (Chapter 41), special needs education (Chapter 42), and cultural training (Chapter 46). Additionally, reviews of VEs in the application areas of assisted teleoperation (Chapter 43), human–robot communication (Chapter 44), clinical settings (Chapter 45), geology (Chapter 47), information visualization of big data (Chapter 48), and entertainment (Chapter 49) are provided.

The handbook concludes with Section VII, which provides a review of VE research and advances made since the inception of the technology, as well as a list of references that can provide additional information for VE professionals (Chapter 50).

We are very pleased to have the opportunity to bring together the vast expertise and knowledge provided in this handbook. We trust you will find this compendium a valuable resource in the advancement of VE applications as you take them from the laboratory to the real-world lives of people everywhere.

Kelly S. Hale
Kay M. Stanney

Editors

Dr. Kelly S. Hale is senior vice president of technical operations at Design Interactive, Inc., a woman-owned small business focused on human–systems integration. Her research and development efforts in human–systems integration have examined virtual environment design and evaluation, augmented cognition, multimodal interaction and haptic interfaces, and training sciences.

Over the past 10 years, Dr. Hale has guided Design Interactive, Inc. to be a leader in developing advanced performance metrics and diagnostic capabilities, including behavioral, physiological, and neural metric suites for capturing traditionally *unobservable* behavioral data to evaluate perceptual skills and cognitive processing in real time by synchronizing data with system events/areas of interest dynamically throughout simulated scenarios. She holds a patent for a tactile display language and serves on the editorial boards for the *International Journal of Human–Computer Interaction* and *Theoretical Issues in Ergonomics Science*. Dr. Hale received her BSc in kinesiology (ergonomics option) from the University of Waterloo in Ontario, Canada, and her master's degree and PhD in industrial engineering, with a focus on human factors engineering, from the University of Central Florida.

Dr. Kay Stanney founded Design Interactive, Inc. (DI) in 1998 and serves as president. She provides executive leadership and strategic direction, formulating and driving key business strategies across DI's three business units. Dr. Stanney is recognized as a world leader in virtual environment (VE) technology, especially as it relates to training. During her more than 25 years of carrying out research in the area of VE training, she has led numerous efforts involved in furthering adaptive VE training techniques. She is the recipient of the 2013 Women Who Mean Business Award, 2012 and 2011 Top Simulation & Training Company for innovation, and the IEEE VGTC Virtual Reality Technical Achievement Award.

Dr. Stanney has an MS and PhD in human factors engineering from Purdue University and a BS in industrial engineering from SUNY Buffalo and is a Certified Human Factors Practitioner.

Contributors

David R. Badcock
School of Psychology
University of Western Australia
Crawley, Western Australia, Australia

Daniel Barber
Institute for Simulation and Training
University of Central Florida
Orlando, Florida

Angelos Barmpoutis
Digital Worlds Institute
University of Florida
Gainesville, Florida

Kathleen M. Bartlett
DSCI MESH Solutions, LLC
Orlando, Florida

Cagatay Basdogan
College of Engineering
Koc University
Istanbul, Turkey

Elizabeth Sheldon Biddle
Human Systems Department
Naval Air Warfare Center Training Systems
    Division
Orlando, Florida

Mark Billinghurst
University of Canterbury
Christchurch, New Zealand

Richard A. Blade
IPI Press
Aurora, Colorado

James P. Bliss
Department of Psychology
Old Dominion University
Norfolk, Virginia

Christoph W. Borst
Emerging Markets and Technologies Division
Center for Advanced Computer Studies
University of Louisiana at Lafayette
Lafayette, Louisiana

Doug A. Bowman
Center for Human-Computer Interaction
Virginia Tech
Blacksburg, Virginia

Grigore (Greg) Burdea
Department of Electrical and Computer
    Engineering
Rutgers, The State University of New Jersey
Piscataway, New Jersey

Sandra L. Calvert
Department of Psychology
Georgetown University
Washington, DC

Meredith Carroll
Defensive Solutions Division
Design Interactive, Inc.
Oviedo, Florida

Roberto K. Champney
Medical Innovations Division
Design Interactive, Inc.
Oviedo, Florida

Eric T. Chancey
Department of Psychology
Old Dominion University
Norfolk, Virginia

Bradley Chase
Department of Engineering
University of San Diego
San Diego, California

Aashish Chaudhary
Scientific Computing Division
Kitware, Inc.
Clifton Park, New York

Ryad Chellali
Istituto Italiano di Tecnologia
Genova, Italy

Dustin Chertoff
Intelligent Automation, Inc.
Atlanta, Georgia

Sue Cobb
Virtual Reality Applications Research Team
University of Nottingham
Nottingham, United Kingdom

Joseph V. Cohn
Human and Bioengineered Systems Division
Office of Naval Research
Arlington, Virginia

Philippe Coiffet
National Academy of Technologies of France
Paris, France

Daniel S. Coming
Google, Inc.
Mountain View, California

Jeffrey L. Cowgill, Jr.
Department of Psychology
Wright State University
Dayton, Ohio

Peter R. Culmer
School of Mechanical Engineering
University of Leeds
Leeds, United Kingdom

Rudolph P. Darken
Department of Computer Science
Naval Postgraduate School
Monterey, California

Mirabelle D'Cruz
Virtual Reality Applications Research Team
University of Nottingham
Nottingham, United Kingdom

Sara Dechmerowski
Medical Innovations Division
Design Interactive, Inc.
Oviedo, Florida

Patricia Denbrook
DCS Inc.
Washington, DC

Benjamin DeVane
Learning Sciences
University of Iowa
Iowa City, Iowa

Nuray Dindar
College of Engineering
Koc University
Istanbul, Turkey

Paul DiZio
Ashton Graybiel Spatial Orientation Laboratory
Volen Center for Complex Systems
Brandeis University
Waltham, Massachusetts

Tripp Driskell
Institute for Simulation and Training
University of Central Florida
Orlando, Florida

Richard M. Eastgate
Virtual Reality Applications Research Team
University of Nottingham
Nottingham, United Kingdom

Joseph R. Fanfarelli
Institute for Simulation and Training
University of Central Florida
Orlando, Florida

Andrew Feng
Institute for Creative Technologies
University of Southern California
Playa Vista, California

Neal M. Finkelstein
US Army Research Laboratory
Orlando, Florida

Joseph L. Gabbard
Grado Department of Industrial and Systems
 Engineering
Virginia Tech
Blacksburg, Virginia

Robert H. Gilkey
Department of Psychology
Wright State University
Dayton, Ohio

Adams Greenwood-Ericksen
Department of Game Studies
Full Sail University
Orlando, Florida

Tami Griffith
Simulation and Training Technology Center
US Army Research Laboratory
Orlando, Florida

David Gross
Lockheed Martin
Marietta, Georgia

Kelly S. Hale
Design Interactive, Inc.
Oviedo, Florida

Frank P. Hannigan
Design Interactive, Inc.
Oviedo, Florida

Tessa Hawkins
Virtual Reality Applications Research Team
University of Nottingham
Nottingham, United Kingdom

Heiko Hecht
Department of Psychology
Johannes Gutenberg University
Mainz, Germany

Lawrence J. Hettinger
Center for Behavioral Sciences
Liberty Mutual Research Institute for Safety
Hopkinton, Massachusetts

Charles E. Hughes
Department of Electrical Engineering and
 Computer Science
University of Central Florida
Orlando, Florida

Matthew Johnston
Emerging Markets and Technologies Division
 Center for Advanced Computer Studies
Design Interactive, Inc.
Oviedo, Florida

David L. Jones
Medical Innovations Division
Design Interactive, Inc.
Oviedo, Florida

Brenna Kelly
MESH Solutions, LLC
Orlando, Florida

Kristyne E. Kennedy
CSK Legal
Orlando, Florida

Robert C. Kennedy
Department of Game Studies
Full Sail University
and
RSK Assessments, Inc.
Orlando, Florida

Robert S. Kennedy
RSK Assessments, Inc.
Orlando, Florida

Behrang Keshavarz
Research Department (iDAPT)
Toronto Rehabilitation Institute
Toronto, Ontario, Canada

G. Drew Kessler
SRI International
Princeton, New Jersey

Abderrahmane Kheddar
CNRS-AIST Joint Robotics Laboratory
Centre National de la Recherche Scientifique
National Institute of Advanced Industrial
    Science and Technology
Tsukuba, Japan

and

Interactive Digital Human Group
Centre National de la Recherche Scientifique
Montpellier, France

Gary L. Kinsland
School of Geosciences
University of Louisiana at Lafayette
Lafayette, Louisiana

Sebastian Koenig
Institute for Creative Technologies
University of Southern California
Playa Vista, California

Regis Kopper
Pratt School of Engineering
Duke University
Durham, North Carolina

Stephanie Lackey
Institute for Simulation and Training
University of Central Florida
Orlando, Florida

James R. Lackner
Ashton Graybiel Spatial Orientation Laboratory
Volen Center for Complex Systems
Brandeis University
Waltham, Massachusetts

Belinda Lange
Institute for Creative Technologies
University of Southern California
Playa Vista, California

Ben D. Lawson
US Army Aeromedical Research Laboratory
Fort Rucker, Alabama

Margaux Lhommet
Institute for Creative Technologies
University of Southern California
Playa Vista, California

Albert Y.M. Lin
Qualcomm Institute
University of California, San Diego
La Jolla, California

Robert W. Lindeman
Worcester Polytechnic Institute
Worcester, Massachusetts

Robb Lindgren
College of Education
University of Illinois at Urbana-Champaign
Champaign, Illinois

J. Peter A. Lodge
Department of Hepatobiliary and Transplant
    Surgery
St James's University Hospital
Leeds, United Kingdom

Valerie Lugo
Medical Innovations Division
Design Interactive, Inc.
Oviedo, Florida

Stacy Marsella
Institute for Creative Technologies
University of Southern California
Playa Vista, California

Douglas Maxwell
Simulation and Training Technology Center
US Army Research Laboratory
Orlando, Florida

James G. May
Department of Psychology
University of New Orleans
Lakefront, New Orleans

Ryan P. McMahan
Department of Computer Science
University of Texas at Dallas
Richardson, Texas

Laura Millen
Virtual Reality Applications Research Team
University of Nottingham
Nottingham, United Kingdom

Betty J. Mohler
Max Planck Institute for Biological Cybernetics
Tübingen, Germany

Mark Mon-Williams
School of Psychology
University of Leeds
Leeds, United Kingdom

J. Michael Moshell
School of Visual Arts and Design
University of Central Florida
Orlando, Florida

Allen Munro
University of Southern California
Los Angeles, California

Denise Nicholson
DSCI MESH Solutions, LLC
Orlando, Florida

Margaret Nolan
DSCI MESH Solutions, LLC
Orlando, Florida

Razia Oden
AMTIS, Inc.
Oviedo, Florida

James C. Oliverio
Digital Worlds Institute
University of Florida
Gainesville, Florida

Eric Ortiz
Institute for Simulation and Training
University of Central Florida
Orlando, Florida

Mary Lou Padgett
Life Senior Member, IEEE
Marietta, Georgia

Robert Page
US Naval Research Laboratory
Washington, DC

Stephen Palmisano
School of Psychology
University of Wollongong
Wollongong, News South Wales, Australia

Jim Patrey
Human Systems Department
Naval Air Warfare Center Aircraft Division
Orlando, Florida

Barry Peterson
Department of History
University of Nevada, Reno
Reno, Nevada

William Pike
Simulation and Training Technology Center
US Army Research Laboratory
Orlando, Florida

Nicholas F. Polys
Interdisciplinary Center for Applied
  Mathematics
Virginia Tech
Blacksburg, Virginia

George V. Popescu
National Institute for Laser, Plasma and
  Radiation Physics
Bucharest, Romania

B.J. Price
Department of Engineering
Walt Disney World
Orlando, Florida

Alexandra B. Proaps
Department of Psychology
Old Dominion University
Norfolk, Virginia

Lauren Reinerman-Jones
Institute for Simulation and Training
University of Central Florida
Orlando, Florida

Bernhard E. Riecke
Simon Fraser University
Surrey, British Columbia, Canada

Albert "Skip" Rizzo
Institute for Creative Technologies
University of Southern California
Playa Vista, California

Eduardo Salas
Department of Psychology
Institute for Simulation and Training
University of Central Florida
Orlando, Florida

Sae Schatz
DSCI MESH Solutions, LLC
Orlando, Florida

Tarah N. Schmidt-Daly
Medical Innovations Division
Design Interactive, Inc.
Oviedo, Florida

Dylan Schmorrow
Soar Technologies, Inc.
Vienna, Virginia

Jurgen P. Schulze
Qualcomm Institute
University of California, San Diego
La Jolla, California

Lee W. Sciarini
Operations Research Department
Naval Postgraduate School
Monterey, California

Ari Shapiro
Institute for Creative Technologies
University of Southern California
Playa Vista, California

William R. Sherman
Pervasive Technology Institute
Indiana University
Bloomington, Indiana

Barbara Shinn-Cunningham
Department of Biomedical Engineering
and
Center for Computational Neuroscience and
    Neural Technology
Boston University
Boston, Massachusetts

Brian D. Simpson
Air Force Research Laboratory
Human Effectiveness Directorate
Wright-Patterson Air Force Base, Ohio

Shawn Stafford
Department of Game Studies
Full Sail University
Orlando, Florida

Kay M. Stanney
Design Interactive, Inc.
Oviedo, Florida

Robert J. Stone
School of Electronic, Electrical and Computer
    Engineering
University of Birmingham
Birmingham, United Kingdom

Simon Su
Ball Aerospace & Technologies Corp.
Dayton, Ohio

Glen Surpris
Defensive Solutions Division
Design Interactive, Inc.
Oviedo, Florida

A. Murat Tekalp
College of Engineering
Koc University
Istanbul, Turkey

James Templeman
US Naval Research Laboratory
Washington, DC

Helmuth Trefftz
Virtual Reality Lab
Universidad Escuela de Administración,
    Finanzas y Technologia
Medellín, Colombia

Matthew Turk
Department of Computer Science
University of California, Santa Barbara
Santa Barbara, California

William B. Vessey
Enterprise Advisory Services, Inc.
Johnson Space Center
Houston, Texas

Erik Viirre
Division of Neurosciences
School of Medicine
Perlman Ambulatory Care Center
University of California, San Diego
La Jolla, California

Michael Vorländer
Institute of Technical Acoustics
RWTH Aachen University
Aachen, Germany

Xuezhong Wang
Engineering Division
Design Interactive, Inc.
Oviedo, Florida

Jingjing Wang-Costello
Yahoo, Inc.
Sunnyvale, California

John P. Wann
Department of Psychology
Royal Holloway, University of London
Surrey, United Kingdom

Christine Wasula
CSK Legal
Orlando, Florida

Philip Weber
Qualcomm Institute
University of California, San Diego
La Jolla, California

Janet M. Weisenberger
Department of Speech and Hearing Sciences
The Ohio State University
Columbus, Ohio

Robert B. Welch
Ames Research Center
National Aeronautics and Space Administration
Moffett Field, California

Alan D. White
Department of Hepatobiliary and Transplant
   Surgery
St James's University Hospital
Leeds, United Kingdom

Eric Whiting
Idaho National Laboratory
Idaho Falls, Idaho

Richard M. Wilkie
School of Psychology
University of Leeds
Leeds, United Kingdom

John R. Wilson
Virtual Reality Applications Research Team
University of Nottingham
Nottingham, United Kingdom

Brent Winslow
Design Interactive, Inc.
Oviedo, Florida

Michael Zyda
Department of Computer Science
USC GamePipe Laboratory
Los Angeles, California

Section I

Introduction

1 Virtual Environments in the Twenty-First Century

Kay M. Stanney, Kelly S. Hale, and Michael Zyda

CONTENTS

1.1 INTRODUCTION

> Whatever I have up till now accepted as most true I have acquired either from the senses or through the senses. But from time to time I have found that the senses deceive.
>
> ***Meditations on First Philosophy*, Descartes**

As Descartes suggested, there is no definitive truth; reality emanates from that which is present to our senses, and these senses we trust to distinguish reality from illusion can be deceived. It is this capacity to fool the sensory systems that we attempt to capitalize on when building a virtual world. More than cinema, which can evoke emotion even without relying on identification with psychological characters (Eisenstein, 1987), virtual worlds can provide a direct and egocentric perceptive of an imaginary world within which it is difficult to distinguish the virtual from the real. The goal of this experience is to *deceive* us and represent a *truth* that can educate, train, entertain, and inspire. In their ultimate form, virtual environments (VEs) immerse users in an alternate *reality* that stimulates multiple senses, providing vibrant experiences that are so veridical they fundamentally transform those exposed (e.g., via training, educating, marketing, or entertaining).

When the first edition of the VE Handbook was assembled over a decade ago, visions such as *The Matrix* (1999) IMDb, had elevated the status of VE to the level of pop iconography, and some of those associated with the technology arguably had risen to star status (e.g., Jaron Lanier [2013], who coined the term *virtual reality* and is still influencing the technology of tomorrow):

> Virtual Reality was, by the end of the 20th Century, destined to have helped computer users abandon the keyboard, mouse, joystick and computer display in favour of interfaces exploiting a wide range of natural human skills and sensory characteristics. They would be able to interact intuitively with virtual objects, virtual worlds and virtual actors whilst "immersed" within a multi-sensory, 3D computer-generated world. As is evident today, this brave new world simply did not happen.
>
> **R. Stone (2012, p. 24)**

Technology has been invested in, academic centers of excellence have been stood up, national initiatives have been launched, yet widespread adoption of VE technology has proven elusive. So why embark on the second edition of the VE Handbook? Because the promise is still alive! This promise is evidenced by the fact that the National Academy of Engineering (NAE) has identified the goal of enhancing virtual reality as one of the 14 grand challenges in need of solutions in the twenty-first century (National Academy of Engineering, 2008). As we journey forward, however, rather than being seduced by the technology, we need to take a deliberate look at what has been achieved and which hard problems still stand in the way of providing veridical virtual experiences that can transform those exposed, leading to gains in education, training, entertainment, and more. In this regard, the agenda set forth by Durlach and Mavor (1995) in their seminal National Research Council (NRC) report *Virtual Reality: Scientific and Technological Challenges* still provides an appropriate yardstick by which to judge the current level of maturity of VE technology and associated applications. This report developed a set of recommendations that, if heeded, should assist in realizing the full potential of VE technology. In the first edition, when reviewing this agenda, we found that much progress had been made with regard to improved computer generation of multimodal images and advancements in hardware technologies that support interface devices; however, there was considerable work remaining to realize improvements in the general comfort associated with donning these devices. In revisiting this agenda, it is evident that substantial further progress has been made from a technological perspective over the past decade, yet advances in the areas of psychological considerations and evaluation continue to lag behind (see Table 1.1).

1.2   TECHNOLOGY

The technology used to generate VE systems has rapidly matured over the past decade, driven primarily by the gaming industry, which desires greater realism, more natural and intuitive interaction, and enhanced usability. According to Moore's law, there has been multiple doublings of computer power in this time span and significant advances in massively parallel graphics processing brought about by the widespread adoption of graphical processing units (GPUs) (Mims, 2010) and field-programmable gate arrays (FPGAs) (Inta, Bowman, & Scott, 2012). In fact, GPUs and FPGAs have exceeded traditional Moore's law trends; whereas central processing unit (CPU) have doubled in speed every 18 months, GPUs have increased by a factor of 5 every 18 months (Geer, 2005; Luebke & Humphreys, 2007) and FPGAs promise to be faster than GPUs for some operations (Thomas, Howes, & Luk, 2009), while not for others (Cope, Cheung, Luk, & Witt, 2005). The increased speed, programmability, and parallelism brought about by these specialized circuits allow for high-performance, visually and aurally rich, interactive 3D virtual experiences. Beyond processing capability, technology advances have come in many areas, including human–machine interfaces, the hardware and software used to generate the VE, electromechanical systems used in telerobotics, as well as the communication networks that can be used to transform VE systems into shared virtual worlds.

1.2.1   HUMAN–MACHINE INTERFACE

Human–machine interfaces consist of the multimodal display devices, including visual, auditory, haptic, and olfactory displays used to present information to VE users, and multimodal user input devices used to control movement throughout a virtual world, including the mouse, touch screen, haptic gloves, cameras (to capture gestures, eye gaze, facial expression, etc.), and microphones (to capture voice; see Popescu, Burdea, & Trefftz, 2014, Chapter 17). Advances in this area have been substantial.

A virtual world aims to be as information rich to our eyes as the real world in order to create visual immersion. The state of the art in visual immersion has been driven in large part by the state of the art in real-time computer graphics, which has progressed in terms of detail from flat polygons to smooth shading to texture mapping and on to programmable shaders, the latter of which provide

TABLE 1.1

Status of Durlach and Mavor's (1995) Recommendations for Advancing VE Technology

Area	Recommendation	Status
Technology: human–machine interface	Address issues of information loss due to visual display technology shortcomings (e.g., poor resolution, limited field of view, deficiencies in tracker technology, bulkiness, ruggedness)	S
	Improvements in spatialization of sounds, especially sounds to the front of a listener and outside of the *sweet spot* surrounding a listener's head	S (see Chapter 4)
	Improvements in sound synthesis for environmental sounds	S
	Improvements in real-time sound generation	M
	Better understanding of scene analysis (e.g., temporal sequencing) in the auditory system	M
	Improvements in tactile displays that convey information through the skin	M (see Chapter 5)
	Better understanding of the mechanical properties of skin tissues that come in contact with haptic devices, limits on human kinesthetic sensing and control, and stimulus cues involved in the sensing of contact and object features	M
	Improvements in locomotion devices beyond treadmills and exercise machines	M (see Chapter 10)
	Address fit issues associated with body-based linkage tracking devices; workspace limitations associated with ground-based linkage tracking devices; accuracy, range, latency, and interference issues associated with magnetic trackers; and sensor size and cost associated with inertial trackers	M (see Chapter 7)
	Improvements in sensory, actuator, and transmission technologies for sensing object proximity, object surface properties, and applying force	S (see Chapter 5)
	Improvements in the vocabulary size, speaker independence, speech continuity, interference handling, and quality of speech production for speech communication interfaces	S
	Improvements in olfactory stimulation devices	L (see Chapter 6)
	Improvements in physiological interfaces (e.g., direct stimulation and sensing of neural systems)	M
	Address ergonomic issues associated with interaction devices (e.g., miniaturization, weight, cost, power consumption, and integration methods)	M
	Better understanding of perceptual effects of misregistration of visual images in augmented reality	M
	Better understanding of how multimodal displays influence human performance on diverse types of tasks	M (see Chapter 29)
Technology: computer generation of virtual environments	Improvements in techniques to minimize the load (i.e., polygon flow) on graphics processors	S (see Chapter 11)
	Improvements in data access speeds	S
	Development of operating systems that ensure high-priority processes (e.g., user tracking) receive priority at regular intervals and provide time-critical computing and rendering with graceful degradation	M
	Improvements in rendering photorealistic time-varying visual scenes at high frame rates (i.e., resolving the trade-off between realistic images and realistic interactivity)	S
	Development of navigation aids to prevent users from becoming lost	M (see Chapter 19)
	Improvements in the ability to develop psychological and physical models that *drive* autonomous agents	M (see Chapter 14)
	Improved means of mapping how user control actions update the visual scene	M (see Chapter 12)
	Improvements in active mapping techniques (e.g., scanning-laser range finders, light stripes)	S (see Chapter 7)

(continued)

TABLE 1.1 (continued)
Status of Durlach and Mavor's (1995) Recommendations for Advancing VE Technology

Area	Recommendation	Status
Technology: telerobotics	Improvements in the ability to create and maintain accurate registration between the real and virtual worlds in augmented reality applications	S
	Development of display and control systems that support distributed telerobotics	M (see Chapter 44)
	Improvements in supervisory control and predictive modeling for addressing transport delay issues	M (see Chapter 44)
Technology: networks	Development of network standards that support large-scale distributed VEs	M
	Development of an open VE network	M
	Improvements in the ability to embed hypermedia nodes into VE systems	S
	Development of wide area and local area networks with the capability (e.g., increased bandwidth, speed, reliability, reduced cost) to support the high-performance demands of multimodal VE applications	S
	Development of VE-specific application-level network protocols	M
Psychological consideration	Better understanding of sensorimotor resolution, perceptual illusions, human-information-processing transfer rates, and manual tracking ability	M
	Better understanding of the optimal form of multimodal information presentation for diverse types of tasks	M (see Chapter 13)
	Better understanding of the effect of fixed sensory transformations and distortions on human performance	M
	Better understanding of how VE drives alterations and adaptation in sensorimotor loops and how these processes are affected by magnitude of exposure	M (see Chapters 31 through 33)
	Better understanding of the cognitive and social side effects of VE interaction	M (see Chapters 16, 27)
Evaluation	Establish set of VE testing and evaluation standards	M (see Chapter 28)
	Determine how VE hardware and software can be developed in a cost-effective manner, taking into consideration engineering reliability and efficiency, as well as human perceptual and cognitive features	M (see Chapter 20)
	Identify capabilities and limitations of humans to undergo VE exposure	M (see Chapters 23, 24, 26, 31 through 33)
	Examine medical and psychological side effects of VE exposure, taking into consideration effects on human visual, auditory, and haptic systems, as well as motion sickness and physiological/psychological aftereffects	M (see Chapters 23, 24, 26, 31 through 33)
	Determine if novel aspects of human–VE interaction require new evaluation tools	M (see Chapter 28)
	Conduct studies that can lead to generalizations concerning relationships between types of tasks, task presentation modes, and human performance	L (see Chapter 29)
	Determine areas in which VE applications can lead to significant gains in experience or performance	M (see Chapters 36 through 50)

Note: L, limited to no advancement; M, modest advancement; S, substantial advancement.

unprecedented flexibility (Dionisio, Burns, & Gilbert, 2013). Beyond improved processing power derived through GPUs, improvements in visual displays have also been driven by advances in display hardware, driven today primarily by the gaming industry. In terms of visual displays, the head-mounted display (HMD), which in the past has been looked at as a potential showstopper for virtual worlds, has seen some level of resolution in recent years with regard to its technological hurdles (see Badcock, Palmisano, & May, 2014; Chapter 3). For example, a newcomer to the field targeted for the gaming industry, the Oculus Rift virtual reality headset, has a wide field of view (110°), very little visible optical housing, and low-latency head tracking (due to Oculus' 1000 Hz adjacent reality tracker, which uses a combination of three-axis gyros, accelerometers, and magnetometers), with issues still remaining in terms of resolution (though they are prioritizing improvements in this area; Shilov, 2013), motion sickness, bulkiness, and untested ruggedness (Bruce, 2013; Hester, 2013). To foster widespread adoption, Oculus is aiming for an affordable consumer price point (under $500; Newman, 2013), and with Facebook's recent acquisition of Oculus VR (Zuckerberg, 2014), visually immersive applications are positioned to expand widely. Other HMD options are available, including Carl Zeiss Cinemizer, Sensics zSight, Silicon Micro Display ST1080, and Sony HMZ-T1/HMZ-T2 (Road to VR, 2013). While gains have been made, and low-to-mid cost options are available, remaining issues with HMDs (Boger, 2013) have likely stymied user acceptance. Perhaps for this reason, wearable displays, which may be more opportune for mass adoption than persistently cumbersome headsets, are making their way into virtual reality applications, including Google's Glass (Rivington, 2013), castAR (Lee, 2014), metaglasses (Ramirez, 2013), Atheer Labs' 3D mobile computing technologies (Tweney, 2013), Vuzix's M100 smart glasses (Bohn, 2013). Lumus' optical engine (OE) displays (Wollman, 2012), and Innovega's iOptik contact lens display (Robertson, 2013). Yet, whether it is head-mounted or active goggle displays, none of these solutions create a visual display that arouses the senses in the same manner as one's natural environment. Toward that end, there are attempts to build virtual data displays directly into the physical environment through 3D mapping techniques that are still in their infancy (Jagadeesh, 2013; Lawler, 2013). These nascent technologies, which recently saw a breakthrough in terms of an optical chip based on waveguide-based platform technology (Smalley, Smithwick, Bove, Barabas, & Jolly, 2013), may someday bring holograms in motion (i.e., 3D holographic video displays that are generated by a computer on the fly) and other glass-free 3D displays (Fattal et al., 2013).

A virtual world is not created by visuals alone. The ability to produce ambient audio cues also influences immersion, providing important positional and spatial cues that contribute significantly to one's sense of placement within a VE. There have been gains in the area of virtual auditory displays (see Vorlander & Shinn-Cunningham, 2014; Chapter 4). While early spatialized audio solutions (Blauert, 1997) were expensive to implement, it is currently feasible to include spatialized audio in most VE systems. (For an excellent source on spatialized audio, see Kapralos, Jenkin, and Milios [2008].) On the hardware side, stereo speakers, surround-sound speakers, speaker arrays, and headphone-based systems, along with 16-bit stereo soundcards, can deliver near real-life binaural sound (Chang & Jacobsen, 2013; Fazi & Nelson, 2010; Poletti, Fazi, & Nelson, 2011; Schobben & Aarts, 2005; Seo, Yoo, Kyeongok Kang, & Fazi, 2010). From the software side, many options are available, including Sound Lab (SLAB3D), a software-based real-time virtual acoustic environment (VAE) rendering system (Miller, 2012); SPAT, a real-time modular spatial sound processing software system and an associated semantic and syntactic specification for storing and transmitting spatial audio scene descriptions (Peters, Lossius, & Schacher, 2013; Wozniewski, Settel, Quessy, Matthews, & Courchesne, 2012); SoundScape Renderer (SSR), a tool for real-time spatial audio reproduction (Geier & Spors, 2012); Scatter, a dictionary-based method based on sound diffusion and particle-oriented approaches, which provides an alternative to time–frequency signal representations (e.g., short-term Fourier and wavelet analyses approaches; McLeran, Roads, Sturm, & Shynk, 2008); and Auro-3D, a 3D audio technology suite that provides tools to create sound in three distinct layers—surround, height and overhead (Van Baelen, Bert, Claypool, & Sinnaeve, 2011), and stratified approaches (Peters et al., 2009). These software solutions support the design of complex spatial audio scenarios through control of a variety of signal processing functions, including sound source allocation, specification of reflections, frame-accurate callback configuration, sound output destination selection, and specification of acoustic scene and renderer parameters, as well as allowing for normal head-related transfer function (HRTF) processing (see Vorlander & Shinn-Cunningham, 2014; Chapter 4). Recently, parametric methods have been developed to capture perceptually relevant information from HRTFs, which allow for binaural rendering to be performed at lower complexity compared to conventional HRTF convolution (Breebaart, Nater, & Kohlrausch, 2009; Dal Bó Silva & Götz, 2013). The use of parametric spatial processing provides a more convincing spatial reproduction for conventional stereo signals, and the combined process of spatial decoding and HRTF parameterization reduces processing requirements and increases perceived quality. Given the typically large computational requirements associated with spatialized audio, it is no surprise that the general-purpose GPU is being used to support spatial sound generation and audio processing compiling in VEs (Hamidi & Kapralos, 2009). These approaches can support efficient implementation of complex and computationally costly software-based spatial sound algorithms, although slow data transfer between GPU and CPU remains a bottleneck even with the use of a Peripheral Component Interconnect Express (PCIe) bus (Inta et al., 2012).

Haptic display technology aims to allow those who traverse a virtual world to manipulate and interact with the virtual objects they encounter (Dindar, Tekalp, & Basdogan, 2013, Chapter 5) and some are using haptic displays to communicate via tactile languages (Fuchs, Johnston, Hale, & Axelsson, 2008). The realization of such naturalistic physical interaction is still largely an area of research found in laboratories. Haptic research and development to date has generally been application specific, with many applications focused on the field of medicine (Demain, Metcalf, Merrett, Zheng, & Cunningham, 2013; Kawasaki, Endo, Mouri, Ishigure, & Daniulaitis, 2013). Some work aims to provide means of integrating haptics into existing platforms, such as *Second Life* (de Pascale, Mulatto, & Prattichizzo, 2008). Methods for capturing, modeling, and reproducing touch-related object properties have also been an area of research. To create virtual interaction, haptic interfaces can be used to measure the motion of the human hand, map these movements onto the VE, and trigger haptic device actuators to provide users with appropriate force and vibration feedback, thereby converting virtual contacts into physical interactions. Recently, this mapping has moved from the use of surface models that require extensive and subjective hand tuning via simple parametric relationships to *haptography*, which bases tactile models on haptic data recorded via a handheld stylus that is used to capture haptic properties of items in the real world (Kuchenbecker, 2008). From these data, a *haptograph* is produced, which is a haptic impression of an object or surface patch, including various haptic properties (e.g., shape, stiffness, friction, and texture). Using the haptograph, the feel of an object or surface can then be recreated via a haptic interface. Haptography aims to provide a generalizable approach to designing authentic virtual touch—much as a camera lens provides a means of reproducing the visual world—through the specification of haptic sensations (i.e., determining how to mimic the human's afferent nervous systems), understanding of haptic distillation (i.e., developing algorithms that can automatically reconstruct these haptic impressions), and enhanced haptic rendering (i.e., developing haptic devices that increase the bandwidth of current amplifiers and mechanical linkages). In terms of the latter, considerable research has focused on the design of haptic devices (Bullion & Gurocak 2009; Folgheraiter, Gini, & Vercesi, 2008; Withana et al., 2010). This has evolved into various device options, including electrical tactile systems, vibromechanical systems (both electromechanical [e.g., rotary inertial, linear actuators] and pneumatic tactors), static low-frequency tactors (e.g., pin-based tactile displays, hydraulic), piezoelectric-based devices, and burgeoning approaches such as electroactive polymers and microelectromechanical systems (McGrath et al., 2008). Initially, linear actuators were the most commonly used; however, rotary inertial tactors have become more popular in recent years. Each solution has strengths and limitations and thus the best option is application dependent. Before such technology is embraced by the masses, advances are needed in many areas including miniaturization, weight, cost, power consumption, and integration methods (i.e., means of incorporating tactors into clothing, seats, harnesses, etc.).

Until recently, most virtual worlds did not include stimulation of senses beyond the big three—visual, auditory, and haptic. Advances in olfactory displays and interfaces are making the inclusion of virtual odors a possibility (Nakamoto, 2013; also see Jones, Dechmerowski, Oden, Lugo, Wang-Costello, & Pike, 2014; Chapter 6). Currently, similar to haptic displays, olfactory display techniques have not moved much beyond the research laboratory. There are various scent generation methods and scent delivery methods being explored. In terms of scent generation methods, advances are needed in the areas of vaporization/atomization techniques, scent switching techniques, and formulation techniques (Yanagida & Tomono, 2013). There are various scent delivery methods being explored, including odor releasing vents, aroma chips using functional polymer gels, inkjet printer trigger mechanisms, and projection-based systems (Kim, 2013; Matsukura, Yoneda, & Ishida, 2012; Nakaizumi, Noma, Hosaka, & Yanagida, 2006; Sugimoto & Okada, 2013). Perhaps one day, not too far away, it will be possible to travel through a virtual world with tantalizing smells wafting toward us from various virtual objects.

Since the initial edition of this handbook, touch-based input device technology has expanded in the commercial marketplace beyond conventional control interfaces (e.g., keyboard, mouse, game controller) to realize more natural interaction techniques (e.g., via touch screens, haptic devices, motion tracking, gestures) (Turk, 2014, Chapter 9). The widespread use and adoption of the Nintendo Wii®, released in 2006, and the iPhone®, released in 2007, transformed user expectations regarding acceleration and touch interfaces. Subsequently, gesture (e.g., Microsoft Kinect®, Leap Motion®) and integrated gesture and natural language interfaces (e.g., Creative Senz3D™ Interactive Gesture Camera) have provided *touchless* interactions that are "disrupt[ing] user experience, from the heuristics that guide us, to our design patterns and deliverables" (Pagan, 2012, p. 1). The primary sensing methods for capturing gestural interaction include movement based (e.g., touch interfaces, glove based, and acceleration based) and vision based (e.g., camera systems) (Dardas & Alhaj, 2011; Rautaray & Agrawal, 2012). The gaming industry and interaction researchers have developed gestural libraries for specific needs (e.g., in operating rooms; Ruppert, Reis, Amorim, de Moraes, and Da Silva [2012]) and gesture ontologies for individual users within VEs (Lanier et al., 2013) and multiuser applications (Roman, Lazarov, & Majumder, 2009). There remains a need for gesture-based interface standards, which minimize instances where the same gesture is used to represent differing meanings/actions across platforms or applications, which should, in turn, facilitate learning while minimizing user frustration and errors.

Advances in motion tracking are needed to extend gestural interaction to full-body motion, which would allow virtual images to be calibrated to the head and/or body position of the individual traversing a virtual world. There are several possible approaches to motion tracking including optical, mechanic, magnetic, and inertial. The most popular approach seems to be *outside-in* optical tracking systems. Advances in tracking technology have been realized in terms of 9DOF microelectronic mechanical system (MEMS) sensors, inertial measurement unit (IMU)-enabled GPS devices, emitter tower constellations, and independent acoustic source positioning (Park, 2013; Sun, Ma, Han, Ross, & Wee, 2013; Zou, 2013). The future of tracking technology continues to trend toward hybrid tracking systems, with a hybrid optical–inertial approach recently developed that uses the inside-out concept of the InterSense VisTracker technology (acquired by Thales Visionix), coupled with a tiny high-performance NavChip IMU (Atac & Foxlin, 2013) and several others in research labs (e.g., magnetic inertial, optical magnetic, radio frequency inertial). In addition, ultrawideband radio technology holds promise for an improved method of omnidirectional point-to-point ranging (Venkatesh & Buehrer, 2007).

If those who traverse and interact in virtual worlds are to become truly immersed, they will need to have the ability to seamlessly converse with the virtual agents they encounter. The quality of speech recognition (natural language processing) and synthesis systems is thus of paramount importance. Fluent speech-to-speech applications, fueled by real-time, speaker-independent, automatic speech recognition software and systems, are coming closer to reality (Aggarwal & Dave, 2012; Picheny et al., 2011; Seide, Li, & Yu, 2011). The accuracy of such systems has been substantially improved through recent advances in graphical model–based machine-learning techniques, with computationally tractable training algorithms, and advanced neural-network modeling techniques (e.g., context-dependent deep-neural-network hidden Markov models). Once issues associated with acoustic and language modeling algorithms are fully resolved, the possibility to develop speech recognition systems that can read voice intonation and integrate with body language and facial expression recognition systems so that emotion and intent can be better understood will likely become the holy grail.

Taken together, these display and user input technological advancements, along with those poised for the near future, provide the infrastructure on which to build complex, immersive multimodal VE applications.

1.2.2 COMPUTER GENERATION OF VIRTUAL ENVIRONMENTS

So what exactly is a virtual world? Gilbert (2011) identified five essential characteristics of contemporary state-of-the-art virtual worlds, including the following:

1. A 3D graphical interface and integrated audio (not text based)
2. Immersion derived through spatial, environmental, and multisensory realism that is capable of producing a sensation of presence
3. User-generated activities and goals (not prescripted) with the ability to create content to personalize the VE experience
4. A persistent world that continues to exist even after a user exits it
5. Simultaneous, massively multiuser remote and distributed interactivity

Thus, several elements are involved when generating VEs, including graphics and audio generators, software for effective and flexible content generation, and networks that support online environments. In terms of graphics and audio generators, computer generation of VEs requires very large physical memories, high-speed processors, high-bandwidth mass storage capacity, and high-speed interface ports for input/output devices (Durlach & Mavor, 1995). Remarkable advances in hardware technologies have been realized in the past decade that support generation of simultaneous, massively multiuser virtual worlds. While Moore's law may be nearing its end (theoretical physicist Michio Kaku posits that Moore's law has about 10 years of life remaining; Paul, 2013), the semiconductor industry is working feverishly to solve this problem by introducing trigate (3D) transistor technology, heterogeneous system architectures (HSAs) that use parallel computing schemes (e.g., accelerated processing units [APUs] that couple CPUs with GPUs, FPGAs, or other such specialized processing systems), and a possible future of other-than-silicon-based transistors (e.g., gallium arsenide processors, molecular transistors, quantum computing) sometime between 2018 and 2026 (Chacos, 2013; ITRS, 2011). When using parallel computing schemes, GPU path tracing can be accelerated by an order of magnitude compared to the CPU, which in turn supports near real-time frame rates that enable rendering photorealistic time-varying visual scenes (Rahikainen, 2013). In addition, the first dedicated ray tracing hardware—*Caustic Series2 ray tracing acceleration boards*—has been released, although its current use is not intended for real-time execution but rather for accelerating lighting, *look development*, and design visualization workflows (Imagination, 2013). One caveat: While such advances needed to support real-time generation of virtual worlds will keep making substantial gains, power and resulting heat may become limiting factors. As the NRC has noted, "even as multicore hardware systems are tailored to support software that can exploit multiple computation units, thermal constraints will continue to be a primary concern" (Fuller & Millett, 2011, p. viii).

From a software perspective, over the past decade, it has become possible to rapidly build and render complex virtual worlds. Virtual reality software toolkits (e.g., Vizard, Goblin XNA, Demotride); server platforms (e.g., Open Cobalt, Open Wonderland, OpenSimulator, Solipsis); standard application program interfaces (APIs) (e.g., OpenGL, Direct-3D); cross platform scene graph-based 3D APIs (e.g., Java-3D, H3DAPI, Ardor3D, jMonkeyEngine, Espresso3D, Jreality); 3D modeling languages, toolkits, content generators, and photorealistic rendering tools (e.g., AC3D Modeler, Autodesk 3ds Max, Cobalt, CyberX3D, Java 3D VRML Loader, LightWave 3D, Maya, MMDAgent, Modo, Photosynth, Presagis Creator, Raster3D, Remo 3D, SimVRML, VRML, X3D ToolKit); and real-time image processing libraries (e.g., OpenCV, VXL, IVT) are all accelerating the development process. Using these tools, commercial application developers can build a range of VEs, from the most basic mazes to complex medical simulators and from low-end single-user PC platform applications to massively online collaborative applications supported by client–server environments. Further, today's VE software toolkits support integration with most VE hardware (e.g., HMDs, goggles, 3D projection systems, motion trackers, 3D sound systems, haptic interfaces, datagloves), readily import 3D models and sounds, and provide the ability to build VE applications as executables, and with scripting languages (e.g., Python, OpenSim, Linden, GAML), it is now possible for even nonprogrammers to develop virtual worlds. One aspect of VE content creation that is lagging behind is advances in agent artificial intelligence (AI), especially with regard to techniques to develop emotionally responsive agents and other nonplayer entities (NPEs) (Slater, Moreton, Buckley, & Bridges, 2008). There are tools available to develop *visually* lifelike human avatars (e.g., *CoJACK*, *DI-Guy*; see Feng, Shapiro, Lhommet, & Marsella, 2014; Chapter 14); however, their expressive behavior comes nowhere close to their photorealistic visual fidelity. There have been considerable advances with regard to integration of telerobotic techniques (see Kheddar, Chellali, & Coiffet, 2014; Chapter 43) into autonomous agent design, which is leading to a new breed of social interaction (e.g., virtual humanoid *U-Tsu-Shi-O-Mi*, *Jeeves*, and *MiRA*) that is supported by a combination of robotic and virtual components (Holz, Dragone, & O'Hare, 2009). Interactions with the Kinect®, Wii, and other more naturalistic interfaces are further breaking down the barriers typically found in third person representations of human interactions. The next step may be integrating electroencephalography (EEG) and other neurophysiological measures of human emotional state and rendering these onto virtual agents (Kokini et al., 2012).

When moving about a VE, research has shown that travelers plan ahead, using a spatial map of the environment (Tyson, 2013). Thus, to enable user-generated wayfinding within virtual worlds, navigational techniques should foster the development of spatial maps. Several such techniques have been investigated, including maps, landmarks, trails, and direction finding (see Darken & Peterson, 2014; Chapter 19). Many of these techniques have been perfected by the game industry.

Some informal guidelines to support traversing virtual worlds have evolved. For closed VEs (e.g., buildings), tools that demonstrate the surrounding area (maps, exocentric 3D views) are recommended if training or exposure time is short, while internal landmarks (i.e., along a route) are recommended for longer exposure durations (Stanney, Chen, & Wedell, 2000). For semiopen (e.g., urban areas) and open environments (e.g., sea, sky), demonstrating the surround is appropriate for short exposures, while the use of external landmarks (i.e., outside a route) is recommended for long exposure times. For navigation to far away virtual places, image plane interaction, scaled-world grabbing, steering-based multiscale navigation, target-based multiscale navigation, and World in Miniature (WIM) techniques may prove effective (Kopper, Ni, Bowman, & Pinho, 2006). In addition, Microsoft Kinect–based techniques, such as footpad and multitouch pad controllers, offer an alternative to purely virtual navigation aids (Dam, Braz, & Raposo, 2013). Brain-controlled navigational interfaces provide another alternative to virtual aids, where an individual navigates a virtual world using only their cerebral activity (Lécuyer et al., 2008; Vourvopoulos, Liarokapis, & Petridis, 2012). Any such VE navigational techniques should carefully consider the fact that spatial orientation and navigation in natural environments rely heavily on locomotion and associated activation of motor, vestibular, and proprioceptive systems; thus, the impact of the absence of these motion-based cues in VEs is important to consider when evaluating the efficacy of navigational aids (Taube, Valerio, & Yoder, 2013). More work is needed in the area of navigational aiding because becoming lost or disoriented in a virtual world has been found to be one of the most common usability issues experienced (Sawyerr, Brown, & Hobbs, 2013).

The NRC report (Durlach & Mavor, 1995) indicated the need for a real-time operating system (RTOS) for VEs; however, while it used to be the case that supporting VEs required an RTOS, this no longer holds true. Due to the aforementioned multiple doublings in processing power, the corresponding need for an RTOS has been obviated by the utilization of a common OS (e.g., Windows, Linux), which can be used as a proxy for a realistic VE. This increase in capability of the computer OS is in part due to the increase in popularity of computer gaming, which requires significant processing power in both the processor (to handle relevant computations) and the GPU (to handle rendering of complex VEs). The combination of these technologies has led to the popularity of massively multiplayer online role-playing games (MMORPGs), such as *World of Warcraft*, *Rift*, and *Lord of the Rings* online. These MMORPGs are, for all intents and purposes, the latest incarnation of a VE in that they share the concept of a shared virtual world, a representation of the player as an avatar, and ways to mitigate latency and jitter that were commonplace in older OSs. These have become the de facto standard in deploying this type of software.

1.2.3 TELEROBOTICS

Beyond the advantages to autonomous agent design discussed earlier, there are many areas (e.g., sensing, navigation, object manipulation) in which VE technology can prosper from the application of robotic techniques. Yet, if these techniques are to be adopted, issues of stability and communication time delay (i.e., transport delay), link flexibility, and real-time control architecture design (e.g., object recognition and pose estimation, fusion of vision, tactile, and force control for manipulation) must be resolved (Ambrose et al., 2012; Atashzar, Shahbazi, Talebi, & Patel, 2012). Chapter 44 discusses a number of techniques for addressing these issues.

1.2.4 NETWORKS

The NRC report (Durlach & Mavor, 1995) suggested that with improvements in communications networks, VEs would become shared experiences in which individuals, objects, processes, and autonomous agents from diverse locations interactively collaborate. This has occurred with a number of virtual worlds currently actively populated with online communities (e.g., *Call of Duty*, *Active World*, *Second Life*, *SimCity*, *World of Warcraft*). As these virtual communities grow and online demand increases in general (e.g., crowdsourcing), shortcomings with regard to performance, reliability, scalability, and security are being explored through Future Internet research efforts, which are ongoing around the world (e.g., the United States, Global Environment for Network Innovations [GENI] and Future Internet Network Design [FIND]; Korea, Future of the Internet for Korea [u-IT839]; European Union, FP6 and FP7, Network of Excellence [Euro-NGI, Euro-FGI], and EIFFEL; Japan, Collaborative Overlay Research Environment [CORE]; and Germany, IKT 2020). These efforts are working toward advances in networks (e.g., application-centric multinetwork service and edge-based intelligence [i.e., boundary between providers and users is disappearing; Peer-to-Peer (P2P) content delivery networks]) and service evolution (e.g., overlay and self-organizing networks, changes in user traffic behavior, functional versus stochastic scalability [e.g., Voice over Internet Protocol (VoIP) signaling platform with an overlay network]). Significant challenges remain with regard to transitioning the focus from quality of service or quality of experience and charting the evolutionary path to the Future Internet (e.g., PlanetLab test bed support of GENI) (Tran-Gia, 2007). In addition, commercial successes such as Google Fiber, which aims to provide 1 Gbps networking to residential clients, mean that networked VE applications will become more ubiquitous as time goes on.

1.3 PSYCHOLOGICAL CONSIDERATION

There are a number of psychological considerations associated with the design and use of VE systems. Some of these considerations focus on techniques and concerns that can be used to augment or enhance VE interaction and transfer of training (e.g., perceptual illusions, design based on human-information-processing transfer rates), while others focus on adverse effects due to VE exposure. In terms of the former, we know that perceptual illusions exist, such as auditory–visual cross-modal perception phenomena, yet little is known about how to leverage these phenomena to reduce development costs while enhancing one's experience in a VE. Perhaps, one exception is vection (i.e., the illusion of self-movement), which is known to be related to a number of display factors (see Table 18.1 of Hettinger, 2014; Chapter 18). By manipulating these display factors, designers can provide VE users with a compelling illusion of self-motion throughout a virtual world, thereby enhancing their sense of presence (see Chertoff & Schatz, 2014; Chapter 34) often with the untoward effect of motion sickness (see Keshavarz, Heckt, & Lawson, 2014; Lawson, 2014a, 2014b; Chapters 23, 24, and 26). Other such illusions exist (e.g., visual dominance) and could likewise be leveraged. While current knowledge of how such perceptual illusions occur is limited, it may be sufficient to know that they do occur in order to leverage them to enhance VE system design and reduce development costs. Substantially more research is needed in this area to identify perceptual and cognitive design principles (see Munro, Carroll, Sheldon, & Patrey, 2014; Chapter 16) that can be used to trigger and capitalize on these illusory phenomena.

Another psychological area in need of research is that of transfer of training (see Champney, Carroll, Surpris, & Cohn, 2014; Chapter 30). Stanney, Mourant, and Kennedy (1998, p. 330) suggest that

> To justify the use of VE technology for a given task, when compared to alternative approaches, the use of a VE should improve task performance when transferred to the real-world task because the VE system capitalizes on a fundamental and distinctively human sensory, perceptual, information processing, or cognitive capability.

But what leads to such transfer? VEs provide the ability to reconstruct similar conditions to those in the operational world, which would otherwise be too risky, costly, or cumbersome to reproduce, which imparts them with high face validity. While face validity is important, there are still limited data-grounded best practices that can be used to direct the design of VE training solutions such that they optimize skill acquisition and retention (Burke & Hutchins, 2008). Despite this lack of knowledge, transfer of training from VEs to real-world tasks has been demonstrated across a range of applications from simple sensorimotor tasks (Kenyon & Afenya, 1995 [see refuting evidence in Kozak, Hancock, Arthur, & Chrysler, 1993]; Rose et al., 1998; Rose, Brooks, & Attree, 2002) and procedural tasks to ones that are more complex, both procedurally (Brooks, Rose, Attree, & Elliot-Square, 2002) as well as cognitively and spatially (Foreman, Stanton, Wilson, & Duffy, 2003). While these examples demonstrate the potential of achieving training transfer with VE systems, there is a need for better understanding of the types of tasks or activities for which the unique characteristics of VEs (i.e., egocentric perspective, stereoscopic 3D visualization, real-time interactivity, immersion, and multisensory feedback) can be leveraged to provide significant gains in human performance, knowledge, or experience.

In contrast to the limited knowledge concerning perceptual and cognitive design principles that augment or enhance VE interaction, more is known about identifying and controlling the adverse effects of VE exposure. Adverse effects are of particular concern because they can persist for some time after exposure, potentially predisposing those exposed to harm. These effects are both physiological (see DiZio & Lackner, 2014; Keshavarz, Heckt, & Lawson, 2014; Lawson, 2014a, 2014b; Stanney, Kennedy, & Hale, 2014; Wann & Mon-Williams, 2014; Chapters 23, 24, 26, 31 through 33) and psychological (see Calvert, 2014; Chapter 27), with considerable effort currently focused on the former and less on the latter. With regard to the latter, some are concerned that exposure to VEs that portray violent content, as is often found in entertainment venues, could lead to aggressive, antisocial, or criminal behavior (see Calvert, 2014; Chapter 27). Thus, a proactive approach is needed that weighs the physiological and psychological risks and potential consequences associated with VE exposure against the benefits. Waiting for the onset of harmful consequences should not be tolerated.

Taken together, the research into psychological considerations of VE exposure indicates that more research is needed to derive perceptual and cognitive design strategies that enhance VE interaction and that there are risks associated with VE exposure. However, usage protocols have been developed that, if successfully adopted, can assist in minimizing these risks (see Stanney, Kennedy, & Hale, 2014; Chapter 31). Thus, VE technology is not something to be eschewed as it has many advantages for enticing and didactic experiences; it is rather something to leverage wholly yet vigilantly, taking care to address the associated risks.

1.4 EVALUATION

Most VE user interfaces are fundamentally different from traditional graphical user interfaces, requiring that their design address unique input/output (I/O) devices, perspectives, and physiological interactions (see McMahan, Lopper, & Bowman, 2014; Chapter 12). Thus, when developers and usability practitioners attempt to apply traditional usability engineering methods to the evaluation of VE systems, they find few if any that are particularly well suited to these environments (see Gabbard, 2014; Chapter 28). There is a need to address key characteristics unique to VEs (e.g., perceived presence and real-world fidelity, multidimensional interactivity, immersion, spatial navigation, orientation), for which existing usability methods fall short in their ability to assess VE systems (see Table 1.2). Several have sought to fill this gap.

Stuart (1996) provided basic methods for evaluating general usability components of VEs. Salzman, Dede, and Loftin (1995) developed formative usability evaluation methods for assessing virtual worlds. Bowman, Koller, and Hughes (1998) developed usability techniques specific to the evaluation of various VE travel techniques. Gabbard and Hix (2000; see Gabbard, 2014; Chapter 28) developed a taxonomy of VE usability characteristics that can serve as a foundation for identifying usability criteria that existing evaluation techniques fail to fully characterize. Stanney, Mollaghasemi, and Reeves (2000) used this taxonomy as the foundation on which to develop an automated system, the Multicriteria Assessment of Usability for Virtual Environments (*MAUVE*), which organizes VE usability characteristics into 2 primary usability attributes (VE system usability and VE user considerations), 4 secondary attributes (interaction, multimodal system output, engagement, and side effects), and 11 tertiary attributes (navigation, user movement, object selection and manipulation, visual output, auditory output, haptic output, presence, immersion, comfort, sickness, and aftereffects). Similar to the manner in which traditional heuristic evaluations are conducted, *MAUVE* can be used at various stages in the usability engineering life cycle, from initial storyboard design to final evaluation and testing. It can also be used to compare system design alternatives. The results of a *MAUVE* evaluation not only identify a system's problematic usability components and techniques but also indicate why they are problematic. Such results may be used to remedy critical usability problems as well as to enhance the design for usability of subsequent system development efforts. Sawyerr et al. (2013) developed a hybrid evaluation method that combines a three-cycle (task action, navigation, and system initiative) cognitive walkthrough method and VE-specific usability

---

TABLE 1.2
Limitations of Traditional Usability Methods for Assessing Virtual Environments

Traditional measurement techniques only capture point-and-click interactions, which are not representative of the multidimensional object selection and manipulation characteristics of 3D space.

Quality of multimodal system output (e.g., visual, auditory, haptic) is not comprehensively addressed by traditional evaluation techniques.

Means of assessing sense of presence and aftereffects have not been incorporated into traditional usability methods.

Traditional performance measurements (i.e., time and accuracy) do not comprehensively characterize VE system interaction (e.g., spatial navigation, orientation).

Traditional single-user task-based assessment methods do not consider VE system characteristics in which two or more users interact in the same environment.

heuristics organized into three categories (i.e., design and esthetics, control and navigation, and errors and help) (Munoz, Barcelos, & Chalegre, 2011; Rusu et al., 2011).

Beyond usability, the cost-effectiveness of VE systems should also be evaluated (see Gross, 2014; Chapter 20), as well as the potential for any product liability concerns (see Kennedy, Kennedy, Kennedy, Wasula, & Bartlett, 2014; Chapter 21). With these aspects considered, developers can evaluate if VE technology offers financial advantages as well as acceptable risks over current practices or technologies. This is an essential determination if VE technology is to thrive both commercially and in research domains, which is looking possible as applications have grown in both areas (see Chapters 36 through 50).

1.5 CONCLUSION

With this second coming of virtual reality, we are more hardware and software capable than ever before. The biggest difference this time, though, is that it is the game industry that is getting us the better graphics hardware and low-cost HMDs, as opposed to the mid-1990s Department of Defense (DoD) VR agenda. The 1997 NRC study entitled *Modeling and Simulation—Linking Entertainment and Defense* (Zyda & Sheehan, 1997) predicted this trend. The largest networked VEs today are coming straight out of the game industry—the *Call of Duty* and *World of Warcraft* series of games have networked infrastructures better architected than anything ever conceived or built by the US DoD; DoD may thus be ceding their leadership in the hardware and software base underlying this second coming of VR. These games definitely move toward the ultimate form of VEs, where users are immersed in an alternate *reality* that stimulates at least the visual and aural senses and captures them effectively enough to garner hours and hours of their attention. Imagine when the same powers of immersion are brought to the classroom! As olfaction and gesturing are added, the future is destined to bring a *Matrix-like* experience to the virtual world. Thus, while the VEs to date have yet to meet with expectations, we anticipate the second go around will fulfill many of the early fantasies.

Some of the things we will be looking for as this new VR push goes forward are perhaps solving some of the harder issues with respect to populating our VR worlds—how do we create AI characters or other NPEs that both perceive and display cognition and emotions? We imagine not just a visual display, say from an Oculus Rift HMD, but also a low-cost, hybrid EEG device that reads our emotional state and transmits that to the core virtual world such that our emotionally cognoscent AI characters can appropriately interact with us more subtlety than is currently achieved through the firing of a virtual weapon. The real question then is: who will fund this careful and hard melding of cognition and emotion such that we can achieve these great, next-generation VEs? Our guess is that the entertainment industry will recognize the value in building AI characters that perceive and display emotions as they turn toward generating interactive VR stories imbued with the emotional subtleties currently only found in the cinema.

So, there are lots of great VR research works ahead with this game industry–driven second coming, and it is going to move us toward Descartes' vision—virtual sensory experiences that are so deceiving we take them as truths from which we may have little desire to withdraw.

REFERENCES

Aggarwal, R. K., & Dave, M. (2012). Recent trends in speech recognition systems. In U. S. Tiwary & T. J. Siddiqui (Eds.), *Speech, image, and language processing for human computer interaction: Multi-modal advancements* (pp. 101–127). Hershey, PA: IGI Global.

Ambrose, R., Wilcox, B., Reed, B., Matthies, L., Lavery, D., & Korsmeyer, D. (2012). *Robotics, tele-robotics and autonomous systems roadmap: Technology area 04* [Online]. Available: http://www.nasa.gov/pdf/501622main_TA04-ID_rev6b_NRC_wTASR.pdf.

Atac, R., & Foxlin, E. (2013, May 16). Scorpion hybrid optical-based inertial tracker (HObIT). *Proceedings of SPIE, Head- and Helmet-Mounted Displays XVIII: Design and Applications* (Vol. 8735), 873502, Baltimore, MD. doi:10.1117/12.2012194.

Atashzar, S. F., Shahbazi, M., Talebi, H. A., & Patel, R. V. (2012, October 7–12). Control of time-delayed telerobotic systems with flexible-link slave manipulators. In *International Conference on Intelligent Robots and Systems (IEEE/RSJ)* (pp. 3035–3039), Vilamoura, Algarve, Portugal.

Badcock, D. R., Palmisano, S., & May, J. G. (2014). Vision and virtual environments. In K. S. Hale & K. M. Stanney (Eds.), *Handbook of virtual environments: Design, implementation, and applications* (2nd ed., pp. 39–86). Boca Raton, FL: Taylor & Francis Group, Inc.

Blauert, J. (1997). *Spatial hearing* (Rev. ed.). Cambridge, MA: MIT Press.

Boger, Y. (2013). *Why HMDs sometimes fail (and what to do about it): The 2013 survey on HMD reliability.* Columbia, MD: Sensics, Inc. [Online]. Available: http://sensics.com/wp-content/uploads/2012/03/why-do-hmds-fail-report.pdf.

Bohn, D. (2013, January 6). Vuzix M100 smart glasses: Hands on with the contender trying to beat Google Glass to market. *The Verge* [Online]. Available: http://www.theverge.com/2013/1/6/3843760/vuzix-m100-smart-glasses-hands-on-beating-project-glass.

Bowman, D., Koller, D., & Hughes, L. (1998). A methodology for the evaluation of travel techniques for immersive virtual environments. *Virtual Reality: Journal of the Virtual Reality Society*, 3, 120–131.

Breebaart, J., Nater, F., & Kohlrausch, A. (2009). Parametric binaural synthesis: Background, applications and standards. In M. Boone (Ed.), *Proceedings of the NAG-DAGA International Conference on Acoustics 2009* (pp. 172–175). Rotterdam, the Netherlands: Curran Associates, Inc.

Brooks, B. M., Rose, F. D., Attree, E. A., & Elliot-Square, A. (2002). An evaluation of the efficacy of training people with learning disabilities in a virtual environment. *Disability and Rehabilitation*, 24(11–12), 622–626.

Bruce, J. (2013, May 16). Oculus Rift development kit review and giveaway. *MakeUseOf* [Online]. Available: http://www.makeuseof.com/tag/oculus-rift-development-kit-review-and-giveaway/.

Bullion, C., & Gurocak, H. (2009). Haptic glove with MR brakes for distributed finger force feedback. *Presence: Teleoperators and Virtual Environments*, 18, 421–433.

Burke, L. A., & Hutchins, H. M. (2008). A study of best practices in training transfer and proposed model of transfer. *Human Resource Development Quarterly*, 19(2), 107–128.

Calvert, S. L. (2014). Social impact of virtual environments. In K. S. Hale & K. M. Stanney (Eds.), *Handbook of virtual environments: Design, implementation, and applications* (2nd ed., pp. 699–718). Boca Raton, FL: Taylor & Francis Group, Inc.

Chacos, B. (2013, April 11). Breaking Moore's Law: How chipmakers are pushing PCs to blistering new levels. *PC World* [Online]. Available: http://www.pcworld.com/article/2033671/breaking-moores-law-how-chipmakers-are-pushing-pcs-to-blistering-new-levels.html.

Champney, R. K., Carroll, M., Surpris, G., & Cohn, J. V. (2014). Conducting training transfer studies in virtual environments. In K. S. Hale & K. M. Stanney (Eds.), *Handbook of virtual environments: Design, implementation, and applications* (2nd ed., pp. 781–796). Boca Raton, FL: Taylor & Francis Group, Inc.

Chang, J.-H., & Jacobsen, F. (2013). Experimental validation of sound field control with a circular double-layer array of loudspeakers. *Journal of Acoustical Society of America*, 133(4), 2046–2054.

Chertoff, D., & Schatz, S. (2014). Beyond presence. In K. S. Hale & K. M. Stanney (Eds.), *Handbook of virtual environments: Design, implementation, and applications* (2nd ed., pp. 855–870). Boca Raton, FL: Taylor & Francis Group, Inc.

Cope, B., Cheung, P. Y. K., Luk, W., & Witt, S. (2005). Have GPUs made FPGAs redundant in the field of video processing? In G. J. Brebner, S. Chakraborty, & W.-F. Wong (Eds.), *Proceedings of the IEEE International Conference on Field Programmable Technology* (Vol. 1, pp. 111–118). FPT 2005, December 11–14, 2005, Singapore. IEEE 2005 ISBN 0-7803-9407-0.

Dal Bó Silva, B., & Götz, M. (2013). A structural parametric binaural 3D sound implementation using open hardware. *IFIP Advances in Information and Communication Technology*, 403, 306–317.

Dam, P., Braz, P., & Raposo, A. B. (2013, July 21–26). A study of navigation and selection techniques in virtual environments using Microsoft Kinect®. In *Proceedings of the 15th International Conference on Human-Computer Interaction – HCI International 2013*, Las Vegas, NV.

Dardas, N. H., & Alhaj, M. (2011). Hand gesture interaction with a 3D virtual environment. *The Research Bulletin of Jordan ACM*, 2(3), 186–194. Available: http://ijj.acm.org/volumes/volume2/no3/ijjvol2no3p9.pdf, viewed October 4, 2013.

Darken, R. P., & Peterson, B. (2014). Spatial orientation, wayfinding, and representation. In K. S. Hale & K. M. Stanney (Eds.), *Handbook of virtual environments: Design, implementation, and applications* (2nd ed., pp. 467–492). Boca Raton, FL: Taylor & Francis Group, Inc.

Demain, S., Metcalf, C. D., Merrett, G. V., Zheng, D., & Cunningham, S. (2013). A narrative review on haptic devices: Relating the physiology and psychophysical properties of the hand to devices for rehabilitation in central nervous system disorders. *Disability and Rehabilitation: Assistive Technology, 8*(3), 181–189.

de Pascale, M., Mulatto, S., & Prattichizzo, D. (2008). Bringing haptics to Second Life. In R. Whitaker & B. Liang (Eds.), *Proceedings of the Ambi-Sys Workshop on Haptic User Interfaces in Ambient Media Systems (HAS'08)* (pp. 6:1–6:6). Brussels, Belgium: Institute for Computer Sciences, Social-Informatics and Telecommunications Engineering (ICST).

Dindar, N., Tekalp, A. M., & Basdogan, C. (2013). Dynamic haptic interaction with video. In K. S. Hale & K. M. Stanney (Eds.), *Handbook of virtual environments: Design, implementation, and applications* (2nd ed., pp. 115–130). Boca Raton, FL: Taylor & Francis Group, Inc.

Dionisio, J. D. N., Burns, W. G., & Gilbert, R. (2013). 3D virtual worlds and the metaverse: Current status and future possibilities. *ACM Computing Surveys, 45*(3), Article 34. doi:http://dx.doi.org/10.1145/2480741.2480751.

DiZio, P., Lackner, R., & Champney, R. K. (2014). Proprioceptive adaptation and aftereffects. In K. S. Hale & K. M. Stanney (Eds.), *Handbook of virtual environments: Design, implementation, and applications* (2nd ed., pp. 833–854). Boca Raton, FL: Taylor & Francis Group, Inc.

Durlach, B. N. I., & Mavor, A. S. (1995). *Virtual reality: Scientific and technological challenges.* Washington, DC: National Academy Press.

Eisenstein, S. (1987). *Non-indifferent nature: Film and the structure of things* (Trans. Herbert Marshall). Cambridge, MA: Cambridge University Press.

Fattal, D., Peng, Z., Tran, T., Vo, S., Fiorentino, M., Brug, J., & Beausoleil, R. G. (2013). A multi-directional backlight for a wide-angle, glasses-free three-dimensional display. *Nature, 495*(7441), 348–351.

Fazi, F. M., & Nelson, P. A. (2010, May 26–28). Sound field reproduction with a loudspeaker array. In *Proceedings of the 37th Convegno Nazionale Associazione Italiana Acustica* (p. 4), Siracusa, Italy.

Feng, A., Shapiro, A., Lhommet, M., & Marsella, S. (2014). Embodied autonomous agents. In K. S. Hale & K. M. Stanney (Eds.), *Handbook of virtual environments: Design, implementation, and applications* (2nd ed., pp. 335–350). Boca Raton, FL: Taylor & Francis Group, Inc.

Folgheraiter, M., Gini, G., & Vercesi, D. (2008). A multi-modal haptic interface for virtual reality and robotics. *Journal of Intelligent Robotics Systems, 52*, 465–488.

Foreman, N., Stanton, D., Wilson, P., & Duffy, H. (2003). Spatial knowledge of a real school environment acquired from virtual or physical models by able-bodied children and children with physical disabilities. *Journal of Experimental Psychology: Applied, 9*(2), 67–74.

Fuchs, S., Johnston, M., Hale, K. S., & Axelsson, P. (2008). Results from pilot testing a system for tactile reception of advanced patterns (STRAP). *Proceedings of the HFES 52nd Annual Meeting*, New York, September 2008, pp. 1302–1306.

Fuller, S. H., & Millet, L. I. (2011). *The future of computing performance: Game over or next level?* Washington, DC: The National Academies Press.

Gabbard, J. L., & Hix, D. (2000, July 31). *A taxonomy of usability characteristics in virtual environments* [Online]. Available: http://csgrad.cs.vt.edu/~jgabbard/ve/taxonomy/.

Gabbard, J. L. (2014). Usability engineering of virtual environments. In K. S. Hale & K. M. Stanney (Eds.), *Handbook of virtual environments: Design, implementation, and applications* (2nd ed., pp. 721–748). Boca Raton, FL: Taylor & Francis Group, Inc.

Geer, D. (2005). Taking the graphics processor beyond graphics. *IEEE Computer, 38*(9), 14–16.

Geier, M., & Spors, S. (2012, November 22–25). Spatial audio with the SoundScape Renderer. In *Proceedings of the 27th Tonmeistertagung – VDT International Convention*, Cologne, Germany.

Gilbert, R. L. (2011). The P.R.O.S.E. Project: A program of in-world behavioral research on the Metaverse. *Journal of Virtual Worlds Research, 4*(1), 3–18.

Gross, D. (2014). Technology management and user acceptance of virtual environment technology. In K. S. Hale & K. M. Stanney (Eds.), *Handbook of virtual environments: Design, implementation, and applications* (2nd ed., pp. 493–504). Boca Raton, FL: Taylor & Francis Group, Inc.

Hamidi, F., & Kapralos, B. (2009). A review of spatial sound for virtual environments and games with graphics processing units. *The Open Virtual Reality Journal, 1*, 8–17.

Hester, L. (2013, June 17). The Oculus Rift will make you a virtual believer. *Complex Gaming* [Online]. Available: http://www.complex.com/video-games/2013/06/the-oculus-rift-will-make-you-a-virtual-believer.

Hettinger, L. J., Schmidt, T., Jones, D. L., & Keshavarz, B. (2014). Illusory self-motion in virtual environments. In K. S. Hale & K. M. Stanney (Eds.), *Handbook of virtual environments: Design, implementation, and applications* (2nd ed., pp. 467–492). Boca Raton, FL: Taylor & Francis Group, Inc.

Holz, T., Dragone, M., & O'Hare, G. M. P. (2009). Where robots and virtual agents meet: A survey of social interaction across Milgram's Reality-Virtuality Continuum. *International Journal of Social Robotics*, *1*(1), 83–93.

Imagination. (2013). *Imagination, Caustic* [Online]. Available: https://www.caustic.com/.

Inta, R., Bowman, D. J., & Scott, S. M. (2012). The "Chimera": An off-the-shelf CPU/GPGPU/FPGA hybrid computing platform. *International Journal of Reconfigurable Computing*, *2012*, Article ID 241439, doi:10.1155/2012/241439.

ITRS. (2011). *International technology roadmap for semiconductors (2011 Edition): Emerging research materials* [Online]. Available: http://www.itrs.net/Links/2011ITRS/2011Chapters/2011ERM.pdf.

Jagadeesh, K. (2013, June 13). Virtual race game in the physical setting of your home allows for books and cereal bowls to become obstacles. *PSFK* [Online]. Available: http://www.psfk.com/2013/06/tabletop-augmented-reality-game.html.

Jones, D. L., Dechmerowski, S., Oden, R., Lugo, V., Wang-Costello, J., & Pike, W. (2014). Olfactory interfaces. In K. S. Hale & K. M. Stanney (Eds.), *Handbook of virtual* environments*: Design, implementation, and applications* (2nd ed., pp. 131–162). Boca Raton, FL: Taylor & Francis Group, Inc.

Kapralos, B., Jenkin, M., & Milios, E. (2008). Virtual audio systems. *Presence: Teleoperators and Virtual Environments*, *17*(6), 527–549.

Kawasaki, H., Endo, T., Mouri, T., Ishigure, Y., & Daniulaitis, V. (2013). HIRO: Multi-fingered haptic interface robot and its medical application systems. In I. Galiana & M. Ferre (Eds.), *Multi-finger haptic interaction*, Springer Series on Touch and Haptic Systems (Chapter 5, pp. 85–107). London, U.K.: Springer.

Kennedy, R. S., Kennedy, R. C., Kennedy, K. E., Wasula, C., & Bartlett, K. M. (2014). Virtual environments and product liability. In K. S. Hale & K. M. Stanney (Eds.), *Handbook of virtual environments: Design, implementation, and applications* (2nd ed., pp. 505–518). Boca Raton, FL: Taylor & Francis Group, Inc.

Kenyon, R., & Afenya, M. (1995). Training in virtual and real environments. *Annals of Biomedical Engineering*, *23*(4), 445–455.

Keshavarz, B., Heckt, H., & Lawson, B. D. (2014). Visually induced motion sickness. In K. S. Hale & K. M. Stanney (Eds.), *Handbook of virtual environments: Design, implementation, and applications* (2nd ed., pp. 647–698). Boca Raton, FL: Taylor & Francis Group, Inc.

Kheddar, A., Chellali, R., & Coiffet, P. (2014). Virtual environment–Assisted teleoperation. In K. S. Hale & K. M. Stanney (Eds.), *Handbook of virtual environments: Design, implementation, and applications* (2nd ed., pp. 1107–1142). Boca Raton, FL: Taylor & Francis Group, Inc.

Kim, D. W. (2013). Aroma chip using a functional polymer gel. In T. Nakamoto (Ed.), *Human olfactory displays and interfaces: Odor sensing and presentation* (pp. 384–400). Hershey, PA: IGI Global.

Kokini, C., Carroll, M., Ramirez-Padron, R., Hale, K., Sottilare, R., & Goldberg, B. (2012). Quantification of trainee affective and cognitive state in real time. In *Proceedings of the Interservice/Industry Training, Simulation & Education Conference (I/ITSEC 2012)*, Orlando, FL. Paper # 12064.

Kopper, R., Ni, T., Bowman, D., & Pinho, M. (2006, March 25–29). Design and evaluation of navigation techniques for multiscale virtual environments. In *Proceedings of IEEE Virtual Reality Conference* (*VR 2006*), pp. 175–182, Alexandria, VA, March 25–29, 2009. IEEE Computer Society 2006 ISBN 1-4244-0224-7.

Kozak, J. J., Hancock, P. A., Arthur, E. J., & Chrysler, S. T. (1993). Transfer of training from virtual reality. *Ergonomics*, *36*(7), 777–784.

Kuchenbecker, K. J. (2008). Haptography: Capturing the feel of real objects to enable authentic haptic rendering. In R. Whitaker & B. Liang (Eds.), *Proceedings of the Ambi-Sys Workshop on Haptic User Interfaces in Ambient Media Systems* (*HAS'08*) (pp. 3:1–3:3). Brussels, Belgium: Institute for Computer Sciences, Social-Informatics and Telecommunications Engineering (ICST).

Lanier, J. (2013). *Who owns the future?* New York, NY: Simon & Schuster.

Lanier, R., Bond, A., Johnston, M., LaViola, J., Hale, K., & Nguyen, N. (2013). *Ecological Gesturing Ontology (EGO)*. Final report delivered to AFRL under contract FS8650-13-M-6389.

Lawler, R. (2013, May 9). With $3.2M in funding, 'real-world operating system' startup Dekko refocuses to build its own augmented reality apps. *TechCrunch* [Online]. Available: http://techcrunch.com/2013/05/09/dekko-real-world-os/.

Lawson, B. D. (2014a). Motion sickness symptomatology and origins. In K. S. Hale & K. M. Stanney (Eds.), *Handbook of virtual environments: Design, implementation, and applications* (2nd ed., pp. 531–600). Boca Raton, FL: Taylor & Francis Group, Inc.

Lawson, B. D. (2014b). Motion sickness scaling. In K. S. Hale & K. M. Stanney (Eds.), *Handbook of virtual environments: Design, implementation, and applications* (2nd ed., pp. 601–626). Boca Raton, FL: Taylor & Francis Group, Inc.

Lécuyer, A., Lotte, F., Reilly, R. B., Leeb, R., Hirose, M., & Slater, M. (2008). Brain-computer interfaces, virtual reality, and videogames. *Computer, 41*(10), 66–72.

Lee, N. (2014). castAR's vision of immersive gaming gets closer to final production engadget. http://www.engadget.com/2014/03/20/castar-update-gdc/.

Luebke, D., & Humphreys, G. (2007). How GPUs work. *IEEE Computer, 40*(2), 96–100.

Matsukura, H., Yoneda, T., & Ishida, H. (2012, March 4–8). Smelling screen: Technique to present a virtual odor source at an arbitrary position on a screen. In *Virtual Reality Short Papers and Posters (VRW), 2012 IEEE* (pp. 127–128), Orlando, FL. doi:10.1109/VR.2012.6180915.

McGrath, B., McKinley, A., Duistermaat, M., Carlander, O., Brill, C., Zets, G., & van Erp, J. B. F. (2008). Chapter 4 – Tactile actuator technology. In J. B. F. van Erp & B. P. Self (Eds.), *Tactile displays for orientation, navigation and communication in air, sea and land environments* (pp. 4-1–4-12) (Tech. Rep. No. RTO-TR-HFM-122). Neuilly sur Seine, France: The Research and Technology Organisation (RTO) of the North Atlantic Treaty Organization (NATO).

McLeran, A., Roads, C., Sturm, B. L., & Shynk, J. J. (2008, July 31–August 3). Granular sound spatialization using dictionary-based methods. In *Proceedings of the 5th Sound and Music Computing Conference*, Berlin, Germany.

McMahan, R. P., Kopper, R., & Bowman, D. A. (2014). Principles for designing effective 3D interaction techniques. In K. S. Hale & K. M. Stanney (Eds.), *Handbook of virtual environments: Design, implementation, and applications* (2nd ed., pp. 285–312). Boca Raton, FL: Taylor & Francis Group, Inc.

Miller, J. D. (2012). *SLAB3D User Manual* (v6.7.0) [Online]. Available: http://home.earthlink.net/~slab3d/slab3d_user_manual.pdf.

Mims, C. (2010, October 22). Whatever happened to virtual reality? *MIT Technology Review* [Online]. Available: http://www.technologyreview.com/view/421293/whatever-happened-to-virtual-reality/.

Munoz, R., Barcelos, T., & Chalegre, V. (2011, November). Defining and validating virtual worlds usability heuristics. In *Proceedings of the 30th International Conference of the Chilean Computer Science Society (SCCC)* (pp. 171–178), Curico, Chile.

Munro, A., Patrey, J., Sheldon, E. S., & Carroll, M. (2014). Cognitive aspects of virtual environment design. In K. S. Hale & K. M. Stanney (Eds.), *Handbook of virtual environments: Design, implementation, and applications* (2nd ed., pp. 391–410). Boca Raton, FL: Taylor & Francis Group, Inc.

Nakaizumi, F., Noma, H., Hosaka, K., & Yanagida, Y. (2006, March 25–29). SpotScents: A novel method of natural scent delivery using multiple scent projectors. In *Virtual Reality Conference 2006* (pp. 207–214), Alexandria, VA. doi: 10.1109/VR.2006.122.

Nakamoto, T. (Ed.). (2013). *Human olfactory displays and interfaces: Odor sensing and presentation.* Hershey, PA: IGI Global.

National Academy of Engineering. (2008). *Leading engineers and scientists identify advances that could improve quality of life around the world: 21st century's grand engineering challenges unveiled.* Washington, DC: Author. [online]. Available: http://www8.nationalacademies.org/onpinews/newsitem.aspx?RecordID = 02152008.

Newman, J. (2013). The Oculus Rift commercializes virtual reality. *Desktop Engineering* [Online]. Available: http://www.engineeringontheedge.com/2013/06/the-oculus-rift-commercializes-virtual-reality/.

Pagan, B. (2012, May 7). New design practices for touch-free interactions. *UX Magazine*, Article No. 824. Available: http://uxmag.com/articles/new-design-practices-for-touch-free-interactions, viewed 2013, October 4.

Park, K.-H. (2013). A ubiquitous motion tracking system using sensors in a personal health device. *International Journal of Distributed Sensor Networks*, Article ID 298209. Available: http://dx.doi.org/10.1155/2013/298209.

Paul, I. (2013). The end of Moore's Law is on the horizon, says AMD. *PC World* [Online]. Available: http://www.pcworld.com/article/2032913/the-end-of-moores-law-is-on-the-horizon-says-amd.html.

Peters, N., Lossius, T., & Schacher, J. C. (2013). The spatial sound description interchange format: Principles, specification, and examples. *Computer Music Journal, 37*(1), 11–22.

Peters, N., Lossius, T., Schacher, J., Baltazar, P., Bascou, C., & Place, T. (2009, July 23–25). A stratified approach for sound spatialization. In *Proceedings of the SMC 2009 – 6th Sound and Music Computing Conference* (pp. 219–224), Porto, Portugal.

Picheny, M., Nahamoo, D., Goel, V., Kingsbury, B., Ramabhadran, B., Rennie, S. J., & Saon, G. (2011). Trends and advances in speech recognition. *IBM Journal of Research and Development, 55*(5), 2:1–2:18. doi:10.1147/JRD.2011.2163277.

Poletti, M. A., Fazi, F. M., & Nelson, P. A. (2011). Sound reproduction systems using variable-directivity loudspeakers. *Journal of the Acoustical Society of America, 129*(3), 1429–1438.

Popescu, G. V., Trefftz, H., & Burdea, G. C. (2014). Multimodal interaction modeling. In K. S. Hale & K. M. Stanney (Eds.), *Handbook of virtual environments: Design, implementation, and applications* (2nd ed., pp. 411–434). Boca Raton, FL: Taylor & Francis Group, Inc.

Rahikainen, V. (2013). *On real-time ray tracing* (MSc thesis). University of Tampere School of Information Sciences, Computer Science, Finland [Online]. Available: http://tutkielmat.uta.fi/pdf/gradu06763.pdf.

Ramirez, E. (2013, June 12). Augmented reality, seen through Meta Glasses. *The Wall Street Journal, WSJ Blogs, Digits* [Online]. Available: http://blogs.wsj.com/digits/2013/06/12/augmented-reality-seen-through-meta-glasses/.

Rautaray, S. S., & Agrawal, A. (2012). Real time hand gesture recognition system for dynamic applications. *International Journal of Ubiquitous Computing*, 3(1), 21–31. Available: http://airccse.org/journal/iju/papers/3112iju03.pdf, viewed 2013, October 4.

Rivington, J. (2013). Google Glass: What you need to know: Are Google's glasses more than just a gimmick? *techRadar* [Online]. Available: http://www.techradar.com/us/news/video/google-glass-what-you-need-to-know-1078114.

Road to VR. (2013). *Head mounted display (HMD)/VR headset comparison chart* [Online]. Available: http://www.roadtovr.com/head-mounted-display-hmd-vr-headset-comparison.

Robertson, A. (2013, January 10). Innovega combines glasses and contact lenses for an unusual take on augmented reality. *The Verge* [Online]. Available: http://www.theverge.com/2013/1/10/3863550/innovega-augmented-reality-glasses-contacts-hands-on.

Roman, P., Lazarov, M., & Majumder, A. (2009). A scalable distributed paradigm for multi-user interaction with tiled rear project display walls. *IEEE Transactions on Visualization and Computer Graphics*, 16(6), 1623–1632.

Rose, F. D., Attree, E. A., Brooks, B. M., Parslow, D. M., Penn, P. R., & Ambihaipahan, N. (1998). Transfer of training from virtual to real environments. In P. Sharkey, F. D. Rose, & J-I Lindstrom (Eds.), *Proceedings of the International Conference Series on Disability, Virtual Reality, & Associated Technologies* (pp. 69–75). Berkshire, U.K.: University of Reading.

Rose, F. D., Brooks, B. M., & Attree, E. A. (2002). An exploratory investigation into the usability and usefulness of training people with learning disabilities in a virtual environment. *Disability and Rehabilitation*, 24(11–12), 627–633.

Ruppert, G. C., Reis, L. O., Amorim, P. H., de Moraes, T. F., & Da Silva, J. V. (2012). Touchless gesture user interface for interactive image visualization in urology surgery. *World Journal of Urology*, 30(5), 687–691.

Rusu, C., Munoz, R., Roncagliolo, S., Rudloff, S., Rusu, V., & Figueroa, A. (2011, August). Usability heuristics for virtual worlds. In *Proceedings of the 3rd International Conference Advances in Future Internet (AFIN)* (pp. 16–19), French Riviera, Nice/Saint Laurent du Var, France.

Salzman, M., Dede, C., & Loftin, R. (1995, October). Usability and learning in education virtual realities. In *Proceedings of the Human Factors and Ergonomics Society 39th Annual Meeting* (pp. 486–490), San Diego, CA.

Sawyerr, W., Brown, E., & Hobbs, M. (2013). Using a hybrid method to evaluate the usability of a 3D virtual world user interface. *International Journal of Information Technology & Computer Science*, 8(2), ISSN No: 2091-1610 [Online]. Available: http://www.ijitcs.com/volume%208_No_2/William.pdf.

Schobben, D. W. E., & Aarts, R. M. (2005). Personalized multi-channel headphone sound reproduction based on active noise cancellation. *Acta Acustica*, 91(3), 440–450.

Seide, F., Li, G., & Yu, D. (2011, August 28–31). Conversational speech transcription using context-dependent deep neural networks. In *Interspeech 2011* (pp. 437–440), Florence, Italy.

Seo, J., Yoo, J. H., Kyeongok Kang, K., & Fazi, F. M. (2010, May 22–25). 21-Channel surround system based on physical reconstruction of a three-dimensional target sound field. In *Proceedings of the 128th Convention of the Audio Engineering Society* (Paper 7973, p. 11), London, U.K.

Shilov, A. (2013, June 13). Oculus VR demonstrates full-HD Oculus Rift virtual reality helmet. *Xbit Laboratories* [Online]. Available: http://www.xbitlabs.com/news/multimedia/display/20130613230543_Oculus_VR_Demonstrates_Full_HD_Oculus_Rift_Virtual_Reality_Helmet.html.

Slater, M., Moreton, R., Buckley, K., & Bridges, A. (2008). A review of agent emotion architectures. *Journal for Computer Game Culture*, 2(2), 203–214.

Smalley, D. E., Smithwick, Q. Y. J., Bove, V. M., Barabas, J., & Jolly, S. (2013). Anisotropic leaky-mode modulator for holographic video displays. *Nature*, 498(7454), 313–317.

Stanney, K. M., Chen, J., & Wedell, B. (2000). *Navigational metaphor design* (Final Rep. Contract No. N61339-99-C-0098). Orlando, FL: Naval Air Warfare Center Training Systems Division.

Stanney, K. M., Kennedy, R. S., & Hale, K. S. (2014). Virtual environment usage protocols. In K. S. Hale & K. M. Stanney (Eds.), *Handbook of virtual environments: Design, implementation, and applications* (2nd ed., pp. 796–808). Boca Raton, FL: Taylor & Francis Group, Inc.

Stanney, K. M., Mollaghasemi, M., & Reeves, L. (2000). *Development of MAUVE, the multi-criteria assessment of usability for virtual environments system* (Final Rep. Contract No. N61339-99-C-0098). Orlando, FL: Naval Air Warfare Center Training Systems Division.

Stanney, K. M., Mourant, R., & Kennedy, R. S. (1998). Human factors issues in virtual environments: A review of the literature. *Presence: Teleoperators and Virtual Environments, 7*(4), 327–351.

Stone, R. J. (2012). *Human factors guidance for designers of interactive 3D and games-based training systems.* Birmingham, U.K.: University of Birmingham.

Stuart, R. (1996). *The design of virtual environments.* New York: McGraw-Hill.

Sugimoto, S., & Okada, K. (2013). Olfactory display based on ink jet printer mechanism and its presentation techniques (pp. 401–414). In T. Nakamoto (Ed.), *Human olfactory displays and interfaces: Odor sensing and presentation.* Hershey, PA: IGI Global.

Sun, Y., Ma, T., Han, C. Y., Ross, J., & Wee, W. (2013). A study of tracking system using multiple emitter towers. *Advanced Materials Research, 694–697,* 927–935.

Taube, J. S., Valerio, S., & Yoder, R. M. (2013). Is navigation in virtual reality with fMRI really navigation? *Journal of Cognitive Neuroscience, 25*(7), 1008–1019. doi:10.1162/jocn_a_00386.

Thomas, D. B., Howes, L., & Luk, W. (2009, February 22–24). A comparison of CPUs, GPUs, FPGAs, and massively parallel processor arrays for random number generation. In *Proceedings of the ACM/SIGDA International Symposium on Field Programmable Gate Arrays – FPGA'09* (pp. 63–72), Monterey, CA.

Tran-Gia, P. (2007, May 21–23). Trends towards next generation internet. In *Proceedings of NGI 2007: 3rd EURO-NGI Conference on Next Generation Internet Networks,* Trondheim, Norway. doi:10.1109/NGI.2007.371188.

Turk, M. (2014). Gesture recognition. In K. S. Hale & K. M. Stanney (Eds.), *Handbook of virtual environments* (2nd ed.). Boca Raton, FL: Taylor & Francis Group, Inc.

Tweney, D. (2013, May 30). Atheer Labs demonstrates a 3D, virtual-reality headset technology. *VentureBeat* [Online]. Available: http://venturebeat.com/2013/05/30/atheer-labs-demonstrates-a-3d-virtual-reality-headset-technology.

Tyson, A. (2013, March). What do we know about spatial navigation, and what else could model-based fMRI tell us? *The Einstein Journal of Biology and Medicine,* Online First: E1–E11. http://www.einstein.yu.edu/uploadedFiles/Pulications/EJBM/2013% 20Tyson%20EJBM%20Online%20First.pdf.

Van Baelen, W., Bert, T., Claypool, B., & Sinnaeve, T. (2011, December). A new dimension in cinema sound: Auro-3D. *Cinema Technology, 24*(4), 26–31.

Venkatesh, S., & Buehrer, R. M. (2007). Non-line-of-sight identification in ultra-wideband systems based on received signal statistics. *IET Microwaves, Antennas and Propagation, 1*(6), 1120–1130.

Vorländer, M. & Shinn-Cunningham, B. (2014). Virtual auditory displays. In K. S. Hale & K. M. Stanney, (Eds.), *Handbook of virtual environments: Design, implementation, and applications* (2nd ed., pp. 87–114). Boca Raton, FL: Taylor & Francis Group, Inc.

Vourvopoulos, A., Liarokapis, F., & Petridis, P. (2012, September 2–5). Brain-controlled serious games for cultural heritage. In *Proceedings of the 18th International Conference on Virtual Systems and Multimedia, Virtual Systems in the Information Society (VSMM-2012)* (pp. 291–298), IEEE Computer Society, Milan, Italy.

Wann, J. P., White, A. D., Wilkie, R. M., Culmer, P. R., J. Peter A. Lodge, & Mon-Williams, M. (2014). Measurement of visual aftereffects following virtual environment exposure. In K. S. Hale & K. M. Stanney (Eds.), *Handbook of virtual environments: Design, implementation, and applications* (2nd ed., pp. 809–832). Boca Raton, FL: Taylor & Francis Group, Inc.

Withana, A., Kondo, M., Makino, Y., Kakehi, G., Sugimoto, M., & Inami, M. (2010). ImpAct: Immersive haptic stylus to enable direct touch and manipulation for surface computing. *Computer Entertainment,* 8(9):1–9:16.

Wollman, D. (2012, February 3). Lumus' OE-31 optical engine turns motorcycle helmets, other eyewear into wearable displays. *Engadget* [Online]. Available: http://www.engadget.com/2012/02/23/lumus-oe-31-optical-engine-turns-motorcycle-helmets-other-eyew/.

Wozniewski, M., Settel, Z., Quessy, A., Matthews, T., & Courchesne, L. (2012, September 9–15). SpatOSC: Providing abstraction for the authoring of interactive spatial audio experiences. *International Computer Music Conference (ICMC 2012),* Ljubljana, Slovenia.

Yanagida, Y., & Tomono, A. (2013). Basics for olfactory display. In T. Nakamoto (Ed.), *Human olfactory displays and interfaces: Odor sensing and presentation* (pp. 60–85). Hershey, PA: IGI Global.

Zou, W. (2013). A distributed system for independent acoustic source positioning using magnitude ratios. In Y. Yang & M. Ma (Eds.), *Proceedings of the 2nd International Conference on Green Communications and Networks 2012 (GCN 2012)* (Vol. 2, Lecture Notes in Electrical Engineering, Vol. 224, pp. 109–115). Berlin, Germany: Springer.

Zuckerberg, M. (2014, March 25). https://www.facebook.com/zuck/posts/10101319050523971.

Zyda, M., & Sheehan, J. (Eds.). (1997). *Modeling and simulation: Linking entertainment & defense.* Washington, DC: National Academy Press.

2 Virtual Environments Standards and Terminology

Richard A. Blade and Mary Lou Padgett

CONTENTS

2.1 INTRODUCTION

In just the past decade, virtual environment (VE) applications have emerged in entertainment, training, education, and other areas (see Chapters 37 through 49, this book). In that time, extensive research in VE technology has also been conducted. However, the terminology used to characterize this technology is still evolving. In fact, Durlach and Mavor (1995, p. 2) indicate that "inadequate terminology [is] being used" to describe VE technology and its applications. It is thus important to describe the key terms that are used in this handbook. The objective is not to resolve differences between disparate uses (in fact, often multiple, even conflicting definitions are presented) but rather to provide a coherent set of commonly used terms. While it is customary to present a glossary at the end of handbooks such as this one, this work starts out with a glossary so that readers may develop a common understanding of the terms used throughout the handbook. Paradoxically, the one term that remains particularly elusive is *virtual environment*. Many authors, especially those among the application chapters (see Chapters 36 through 50, this book), have catered the definition of VE to fit the forms of the technology that best suit their needs. Perhaps this definitional multiplicity demonstrates the versatile nature of VE technology and its wide array of potential uses.

While the definitions in this chapter have been presented in a relatively informal manner, the profession continues to work toward a set of standard definitions through the VR Terminology Project. This involves a multiphase effort on the part of the IEEE Computer Society under the auspices of the IEEE Standards Association, of which the first phase is the establishment of a working group on virtual reality (VR) terminology (VI-1392). The first author of this chapter chaired the VI-1392 working group in the very early years of VR and in this chapter attempts to present the most fundamental definitions with the greatest generality.

2.2 IMPORTANCE OF STANDARDS

Standards are critical for systematic and robust development of any emerging technology. *Specification standards* provide for practical descriptions of product characteristics and limitations, critical to an end user. *Interface standards* allow for interchangeability of components developed by different manufacturers, thus permitting specialization and robust competition in the marketplace.

Safety standards ensure the health and safety of product users. *Terminology standards* ensure that technical terminology is used in a consistent and rigorous manner, thus preventing confusion and ambiguity in scientific and technical reports and specifications.

The process of establishing standards is quite possibly as important as the standards ultimately produced for that process establishes a forum for open and systematic dialog between researchers, developers, manufacturers, and end users on the current and future needs and directions of an industry. It often progresses at a *glacial pace*, meaning it could take years to actually reach agreement on a standard, which ensures a systematic, objective, and rigorous examination of all aspects of an issue by all who have interest or involvement.

There are many standards-setting organizations, each having its own rules and procedures. For electrical, electronic, and computer-related standards, the two primary international organizations are the *International Electrotechnical Commission* (IEC) and the *International Organization for Standardization* (ISO). Most standards related to VEs will originate in the IEEE Computer Society, which is a part of the *Institute of Electrical and Electronic Engineers* (IEEE). *The IEEE Standards Association* (IEEE-SA), in turn, belongs to the *American National Standards Institute* (ANSI). Standards organizations can and do work together, but there are no requirement that they do so. Rather, obtaining the broadest possible acceptance of standards is the motivation for cooperation and collaboration. For example, the ISO and IEC have formed the *Joint Technical Committee Number One* (JTC1) to deal with information technology standards. They also cooperate on issues involving safety, electromagnetic radiation, and the environment.

The first step to establishing one or more VE standards (usually a group of related standards) in the IEEE Computer Society is for an individual to prepare a *project authorization request* (PAR) for submission to the *IEEE Standards Board*. If approved, that PAR is given an identifying label (e.g., vi-1392 in the case of VE terminology), and a working group is set up under that label to study the problem and make recommendations. All interested parties are invited to participate and every attempt is made to ensure a broad representation. After much informal discussion and deliberation, the working group prepares draft standards. A *balloting group* is then assembled to vote on the standards and/or modify them if necessary. In contrast to the working group, the process and membership of the balloting group become very formal to ensure a balanced and fair consideration of the issues and proposed standards. The last step in establishing a standard by the IEEE-SA is for the *IEEE Standards Board* to approve the submission of the balloting group. When the standards being proposed overlap between two or more existing groups, a *standards coordinating committee* (SCC) is involved to ensure proper coordination and collaboration. The work of the IEEE-SA is supported by its publication of the standards. Various procedures must be followed throughout the standards process to ensure that the publication rights to each standard produced belong to the IEEE.

2.3 IMPORTANCE OF OFFICIAL TERMINOLOGY

Any standard set by the IEEE or other standards development organizations contains definitions. When multiple uses of a term are found in the literature or are in common use, it should help to have a careful record of the variations, including recommendations and cautions. Consider, for example, the term *megabyte*. It may mean $10^6 = 1,000,000$ bytes, or it may mean $2^{10} = 1,024 \times 1,024 = 1,048,576$ bytes, or even $1,000 \times 1,024 = 1,024,000$ bytes. A consumer needs to be aware of these disparities. One important task of the VR Terminology Project is to provide consumers, including government contractors, reliable definitions so they can properly evaluate product descriptions and proposals. If these definitions are readily available to users and producers, reliable commerce will be enhanced.

A short glossary of terms is given in the next section. Most are technical terms that were provided by the authors of other chapters in this handbook due to their occurrence in those chapters, and many appear in more than one chapter. Although all have definitions in the professional literature, almost none of those definitions have been universally adopted by any standards organization. In preparing this chapter, much discussion took place with the authors, as well as among various authors of this handbook, to ensure a reasonable degree of clarity and consistency. However, in some cases, we found it necessary to provide multiple, sometimes even contradictory, definitions reflecting current usage.

2.4 BASIC GLOSSARY

6DOF: abbreviation for six degrees of freedom, the six being freedom of a three-dimensional object to move in the directions of three perpendicular axes (forward/backward, up/down, and left/right) and rotate about three perpendicular axes (pitch, yaw, and roll).

Accommodation: change in the focal length of the eye's lens to maintain focus on a moving close object.

Active virtual reality: virtual reality where human actions control the model of reality.

Actuator: mechanical means used to provide force or tactile feedback to a user.

Aftereffect: any effect of *VE* exposure that is observed after a participant has returned to the physical world.

Archetypes: prototypes that provide templates to guide learning, development, and the construction of the personality or psyche.

Articulation: objects composed of several parts that are separably moveable.

Artificial reality: same as virtual reality or virtual environment.

Assistive agents: artificial intelligence algorithms developed to guide users through a VR world and to coach the user on available choices within the world.

Attentional inertia: attentional mechanism in which the person's visual system gets locked into engaging, interesting experiences.

Aubert effect or phenomenon: the apparent displacement of an isolated vertical line in the direction opposite to which the observer is tilted. This happens when an observer has a large tilt, for example, 90°.

Audification: an acoustic stimulus involving direct playback of data samples. See also *Sonification.*

Augmented reality: form of virtual reality where the human interacts with a combination of the reality model and true reality, usually through the use of special eyeglasses displaying both data from the model and data from the real world. (An industrial example is Boeing workers constructing wire harnesses using special glasses to display the actual path of each wire in the harness.)

Autonomy: performance or action of the object on the rule of physics, biology, or a virtual world, but not by independent decision of a human operator.

Avatar: an interactive representation of a human in a virtual reality environment.

Back clipping plane: region at a distance beyond which objects are not shown.

Backdrop: stationary background in a virtual world; the boundary of the world that cannot be moved or broken into smaller elements.

Backface removal: elimination of those portions of a displayed object that are facing away from the viewer.

Binocular: displaying a slightly different view to each eye for the purpose of stereographic viewing.

Binocular Omni-Orientation Monitor (BOOM): 3D display device suspended from a weighted boom that can swivel freely so the viewer can use the device by bringing the device up to the eyes and viewing the 3D environment while holding it. The boom's position and orientation communicates the user's point of view to the computer.

Binocular parallax: the means whereby the eyes can judge distance by noticing how closer objects appear to move more than distant ones when the observer moves. See also *Parallax* and *Motion parallax.*

Bi-ocular: displaying the same image to each eye. Sometimes, this is done to conserve computing resources when depth perception is not critical. See also *Stereopsis.*

Biosensors: sensor devices that monitor the state of the body.

Bots: robots or intelligent agents who roam multiuser domains (MUDs) and other virtual environments.

Browser: overviews, such as indexes, lists, or animated maps, which provide a means of navigating through the physical, temporal, and conceptual elements of a virtual world.

CAVE: *cave automatic virtual environment*; virtual images projected on the walls, floor, and ceiling of a room that surrounds a viewer. Oftentimes, a principal viewer's head position is tracked to determine view direction and content, while other viewers *come along for the ride*.

Cognitive map: mental representation of an environment (also referred to as a mental map).

Communication channel: when applied to HCI, it is a pathway between the user and the simulation that allows human–computer interaction.

Computer-assisted teleoperation (CAT): bilateral control of teleoperation through computers, including computer assistance in both robot control and information feedback.

Convergence: occurs in stereoscopic viewing when the left- and right-eye images become fused into a single image.

Convolve: to filter and intertwine signals (e.g., sounds) and render them 3D. Used in VE applications to recreate sounds with directional cues.

Convolvotron: output system for controlling binaural sound production in a VR world.

Coordinates: set of data values that determine the location of a point in a space. The number of coordinates corresponds to the dimensionality of the space.

Coriolis, or cross-coupling, effect: effect resulting from certain kinds of simultaneous multi-axis rotations, especially making head movements while rotating. This illusion is characterized by a feeling of head or self-velocity in a curved path that is roughly orthogonal to the axes of both body and head rotation, which can lead to simulator sickness.

Cue conflict: theory to explain the kind of motion sickness caused when the body tries to interpret conflicting clues being received by the senses. Frequent causes are faulty calibration of eye devices or delay between the sensory inputs and output display.

Culling: removing invisible pieces of geometry and only sending potentially visible geometry to the graphics subsystem. Simple culling rejects entire objects not in the view. More complex systems take into account occlusion of some objects by others, for example, building hiding trees behind it.

Cutaneous senses: skin senses, including light touch, deep pressure, vibration, pain, and temperature.

Cybersickness: sensations of nausea, oculomotor disturbances, disorientation, and other adverse effects associated with VE, typically because the sensory data generated do not properly match reality.

Cyberspace: virtual universe of digital data.

Cyborg: robotic humanoid modeled directly from digital readings of a real human and transformed into a photo realistic, animated character produced via illusionary metamorphosis.

DataGlove: device for sensing hand gestures, which uses fiber-optic flex sensors to track hand orientation and position, as well as finger flexure.

Data sonification: assignment of sounds to digitized data that may involve filtering to give illusion of localized sound.

Data specialization: assignment of orientation (yaw, pitch) and position coordinates (x, y, z) to digital sounds assigned to data.

Deformable object technology (DOT): virtual objects, which bend and deform appropriately when touched.

Depth cueing: use of shading, texture mapping, color, interposition, or other visual characteristics to provide a cue for the distance of an object from the observer.

Desktop virtual systems: virtual experiences that are displayed on a 2D desktop computer; a person can see through the eyes of the character on the screen, but the experience is not 3D.

Dolly shot: display of a scene while moving forward or backward. See also *Pan shot* and *Track shot*.

Dynamic accuracy: system accuracy as a tracker's sensor is moved. See also *Static accuracy.*

Dynamics: rules that govern all actions and behaviors within the environment.

EBAT: *event-based approach to training;* provides the strategies, methods, and tools that are essential for an effective learning environment in a structured and measurable format for training and testing specific knowledge, skills, and abilities.

"E" effect: See *Muller effect.*

Effectors: interfacing devices used in virtual environments for input/output, tactic sensation, and tracking. Examples are gloves, head-mounted displays, headphones, and trackers.

Egocenter: sense of one's own location in a virtual environment.

Embodiment: within the body.

Ergonomics: study of human factors, that is, the interaction between the human and his or her working environment.

Exoskeletal device: flexible interaction devices worn by users (e.g., gloves and suits) or rigid-link interaction systems (i.e., jointed linkages affixed to users).

Exoskeleton: mechanically linked structure for control of feedback from an application.

Eye clearance: most accurate figure of merit used to describe the HMD positioning relative to the eye.

Eye tracking: measurement of the direction of gaze.

Eyeball in the hand: metaphor for visualized tracking where the tracker is held in the hand and is connected to the motion of the projection point of the display.

EyePhone: A VPL Research, Inc., display device consisting of two tiny television monitors (one per eye), earphones, and a sensor for tracking head position and orientation.

Fidelity: degree to which a VE or SE duplicates the appearance and feel of operational equipment (i.e., functional fidelity), sensory stimulation (i.e., physical fidelity), and psychological reactions felt in the real world (i.e., psychological fidelity) of the simulated context.

Field of view (FOV): angle in degrees of the visual field. Since a human's two eyes have overlapping 140° FOV, binocular or total FOV is roughly 180° in most people. A FOV greater than roughly 60°–90° may give rise to a greater sense of immersion.

Fish tank VE: illusion of looking *through* a computer monitor to a virtual outside world using a stereoscopic display system. When looking *out* through the stereo *window*, the observer imagines himself or herself to be in something resembling a fish tank.

Force feedback: output device that transmits pressure, force, or vibrations to provide a VE participant with the sense of resisting force, typically to weight or inertia. This is in contrast to tactile feedback, which simulates sensation to the skin.

Formal features: audiovisual production features that structure, mark, and represent media experiences.

Formative evaluation: assess, refine, and improve user interaction by iteratively placing representative users in task-based scenarios in order to identify usability problems as well as to assess a design's ability to support user exploration, learning, and task performance.

Fractal: self-similar graphical pattern generated by using the same rules at various levels of detail. That is, a graphical pattern that repeats itself on a smaller and smaller scale. Fractals can generate very realistic landscapes with great detail using very simple algorithms.

Frustum of vision: 3D field of view in which all modeled objects are visible.

Functional fidelity: degree VE mimics functional operation or relationship between objects in real world. See also *Fidelity.*

Gesture: hand motion that can be interpreted as a sign, signal, or symbol.

Gouraud shading: shading of polygons smoothly with bilinear interpolation.

GUI: abbreviation for *graphical user interface.*

Gravitoinertial force: resultant force combining gravity and virtual forces created by acceleration.

Guidelines-based evaluation: cost-effective and popular method for evaluating a user interface design; the goal is to find usability problems in a design so that they can be attended to as part of an iterative design process—involves having a small set of evaluators examine the interface to determine its compliance with recognized usability principles (the *heuristics*). See also *Heuristic evaluation.*

Gustatory: pertaining to the sense of taste.

Haptic interface: interface involving physical sensing and manipulation.

HCI: abbreviation for *human–computer interaction.*

Head-mounted display (HMD): a visual display covering the eyes, sometimes having position tracking to provide a computer with the location and orientation of the head.

Heads-up display: display device that allows users to see graphics superimposed on their view of the real world, a form of augmented reality.

Heuristic evaluation: cost-effective and popular method for evaluating a user interface design; the goal is to find usability problems in a design so that they can be attended to as part of an iterative design process—involves having a small set of evaluators examine the interface to determine its compliance with recognized usability principles (the *heuristics*). See also *Guidelines-based evaluation.*

Hidden surface: surface of a graphics object that is occluded from view by intervening objects.

Human–computer interaction (HCI): study of how people work with computers and how computers can be designed to help people effectively use them.

Image distance: perceived distance to the object (in contrast to the real object distance, if there exists a real object).

Immersion: experience of being physically within a VE experience. The term is sometimes subcategorized into external and internal immersion and sensory and perceptual immersion. See also *Presence.*

Intelligent user interface: user interface that is adaptive and has some degree of autonomy.

Interaural amplitude: differences between a person's two ears in the intensity of a sound, typically due to the location of the sound.

Interaural time: differences between a person's two ears in the phase of a sound, typically due to the location of the sound.

Inverse kinematics: specification of the motion of dynamic systems from properties of their joints and extensions.

Kinesthesis: sensations derived from muscles, tendons, and joints and stimulated by movement and tension.

Kinesthetic: all muscle, joint, and tendon senses. Excludes skins' senses, such as touch and vestibular and visual. A subset of somatosensory, which is usually applied to limb position and movement but would also include nonvestibular sensation of head movement (e.g., via neck musculature).

Kinesthetic dissonance: mismatch between feedback and its absence from touch or motion, during VE experiences.

Latency: lag between user motion and tracker system response. Delay between actual change in position and reflection by the program. Delayed response time.

Level of detail (LOD): model of a particular resolution among a series of models of the same object. Greater graphic performance can be obtained by using a lower LOD when the object occupies fewer pixels on the screen or is not in a region of significant interest.

Locomotion: means of travel restricted to self-propulsion. Motion interfaces can be subdivided into those for passive transport (inertial and noninertial displays) and those for active transport (locomotion interfaces).

Magic wand: 3D input device used for pointing and interaction. A sort of 3D mouse.

Master–slave system: teleoperator consisting of a control, called the master, and a remote executing device, called the slave.

Matching: degree to which a VE mimics reality not only in form but also in terms of function and the behaviors the VE elicits from users.

Metaball: surface defined about a point specified by a location, a radius, and an *intensity*. When two metaballs come in contact, their shapes blend together.

Metallic distortion: noise interference or degraded performance in electromagnetic trackers when used near large metallic objects.

Model: simulation of something real.

MOO: object-oriented multiuser domain, a text-based MUD. See also *MUD*.

Motion parallax: the means whereby the eyes can judge distance by noticing how closer objects appear to move more than distant ones when the observer moves. See also *Binocular parallax* and *Parallax*.

Motion platform: platform, often carrying a passenger *pod* resembling an automobile, airplane, or spacecraft that provides the sensations of motion through orientation, vibrations, and jerking and occasionally through spinning.

Move-and-wait strategy: typical teleoperation control mode when time delays occur (with direct feedback).

MUD: *multiuser domain* where users can jointly interact and play games, such as dungeons and dragons.

Muller effect: tendency to perceive an objectively vertical line as slightly tilted in the same direction as the observer when the observer is tilted by only a moderate amount. See also *"E" effect*.

Multimodal command: command issued by the user to a computer simulation using several input communication channels.

Multimodal system: when applied to HCI, a system that allows communication with computers via several modalities, such as voice, gesture, gaze, and touch.

Nanomanipulator: device that allows manipulation on a very microscopic scale.

Navigation: aggregate task of wayfinding and motion (i.e., motoric element of navigation).

Neural interface: ultimate human–computer interface, which connects directly to the human nervous system.

Nystagmus: reflexive eye movements that usually serve to keep the world steady on the retina during self-motion. These eye movements are driven by vestibulo-ocular, cervical-ocular, and optokinetic reflexes. A rapid sideways snap of the eye followed by a slow return to normal fixation or rapid oscillatory movements of the eye, as in following a moving target, in blindness, or after rotation of the body.

Occipital cortex: back of the brain receiving retinotopic projections of visual displays.

Occlusion: hiding an object or a portion of an object from sight by interposition of other objects.

Oculogravic illusion: illusion of visual target displacement that is actually caused by body acceleration. The visual component of the somatogravic illusion (e.g., when an aircraft accelerates and there is a backward rotation of the resultant force vector, in addition to feeling a *pitch-up* sensation, the pilot may sense an apparent upward movement of objects in his or her visual field).

Oculogyral illusion: illusion of visual target displacement that is actually caused by body rotation (e.g., the illusory movement of a faint light in a darkened room following rotation of the body).

Olfactory: pertaining to the sense of smell.

Otoconia: hair cell mechanoreceptors that are embedded in gelatinous membranes containing tiny crystals of calcium carbonate through which the otolith organs detect changes in the magnitude or direction of linear acceleration vectors.

Pan: angular displacement of a view along any axis of direction in a 3D world.

Pan shot: display of a scene while moving about any axis. See also *Dolly shot* and *Track shot*.

Parallax: difference in viewing angle created by having two eyes looking at the same scene from slightly different positions, thereby creating a sense of depth. Also referred to as binocular parallax. See also *Motion parallax*.

Parietal cortex: area of the brain adjacent and above the occipital cortex, which processes spatial location and direction information.

Passive virtual reality: virtual reality in which there is no control of the model by the human. That is, the model only provides information to the human senses.

Persona: public display, or mask, of the self.

Perspective: rules that determine the relative size of objects on a flat-viewing surface to give the perception of depth.

Phase lag: when output from computer-generated images lags the actual position of tracking sensors.

Phi phenomenon: illusion of continuous movement through space induced by a sequence of discontinuous events (e.g., lights flashing in sequence, as on a marquee).

Phong shading: method for calculating the brightness of a surface pixel by linearly interpolating points on a polygon and using the cosine of the viewing angle. Produces realistic shading.

Photo realism: attempt to create realistic appearing images with great detail and texture.

Physical fidelity: degree VE mimics real world in regard to the visual, auditory, and haptic cues present (more similar to sensory simulation). See also *Fidelity*.

Pitch: angular displacement about the horizontal axis perpendicular to the lateral axis. A measure of dipping forward or backward.

Portal: simulated openings that a user can pass through in a virtual space to automatically load a new world or execute a user-defined function.

Position sensor: tracking device that provides information about its location and/or orientation.

Position trigger: hot spot, sensitive spot, or button that causes a change in the computer program when touched in some way.

Presence: illusion of being part of a virtual environment. The more immersive a VE experience, the greater the sense of being part of the experience. See also *Immersion*.

Proprioceptive: internal sense of body position and movement. Includes kinesthetic and vestibular senses but excludes outwardly directed senses such as vision and hearing. This term is not a subset of somatosensory nor is it synonymous with kinesthesia.

Prosocial behavior: socially valued behaviors such as helping, sharing, cooperating, and engaging in constructive imaginary activities.

Pseudo: false.

Radiosity: diffuse illumination calculation system for graphics based on energy balancing that takes into account multiple reflections off many walls.

Ray tracing: technique for displaying a 3D object with shading and shadows by tracing light rays backward from the viewing position to the light source.

Real time: action taking place with no perceptible or significant delay after the input that initiates the action.

Real-time imaging: graphics or images synchronized with real-world time and events.

Reality engine: computer system for generating virtual objects and environments in response to user input, usually in real time.

Refresh rate: frequency with which an image is regenerated on a display surface.

Registration: correspondence between a user's actual position and orientation and that reported by position trackers.

Render: conversion of image data into pixels to be displayed on a screen.

Retinal binocular disparity (RBD): ratio of the convergence angle of an image to the convergence angle of an object.

Roll: angular displacement about the lateral axis.

Tactile display: device that provides tactile and kinesthetic sensations.

Scenes view: virtual display viewed on a large screen or through a terminal window rather than with immersive devices.

Scientific visualization: graphical representation of complex physical phenomena in order to assist scientific investigation and make inferences that are not apparent in numerical form.

Second life: popular website where people construct an avatar that represents themselves and interact socially in an online VE.

Second-person virtual systems: virtual environment experiences in which a person is represented by an on-screen avatar rather than being fully immersed in the environment.

Semantic unification: process of integration and synchronization of information from several input modalities (gesture, speech, gaze, etc.).

Semiocclusion: occlusion to one eye only.

Sensorial redundancy: presenting same or related sensory information to a user using several communication channels.

Sensorial substitution: using a different sensory channel to present information normally presented in the replaced modality.

Shared mental models: overlapping knowledge or understanding a task, team, equipment, and situation between team members.

Shared worlds: virtual environments that are shared by multiple participants.

Shutter glasses: glasses that alternately block out the left- and right-eye views in synchrony with the computer display of left- and right-eye images to provide stereoscopic images on the computer screen.

Side effect: any effect of *VE* exposure that is observed after a participant has returned to the physical world. See also *Aftereffect*.

Simulation overdose: spending too much time in virtual environments.

Simulator sickness: various disturbances, ranging in degree from a feeling of unpleasantness, disorientation, and headaches to extreme nausea, caused by various aspects of a synthetic experience. Possible factors include sensory distortions such as abnormal movement of arms and heads because of the weight of equipment, long delays or lags in feedback, and missing visual cues from convergence and accommodation.

Situational awareness: perception of elements in an environment within a volume of time and space, comprehension of their meaning, and projection of their status in the near future; *up-to-the-minute* cognizance required to operate or maintain the state of a system.

Six degrees of freedom (6DOF): ability to move in three independent directions and rotate about three independent axes passing through the center of the body. Thus, the location and orientation are specified by six coordinates.

Somatogravic illusion: illusions of false attitude occurring mainly in pitch or roll. When lacking appropriate visual cues, the balance organs and brain are unable to sort out a force vector (such as delivered by motion of an aircraft), which differs in magnitude and/or direction from the gravitational vector.

Somatosensory or somesthesia: stimuli or senses arising from the cutaneous, muscle, and joint receptors. Touch and pressure cues. Includes all body senses of the skin, muscle, joint, and internal organs (including Mittelstaedts kidney receptors) but excludes vestibular, visual, auditory, and chemical senses (taste and smell). Forms a superset of kinesthetic, tactile, and haptic stimuli or senses.

Sonification: data are used to control various parameters of a sound generator, thereby providing the listener with information about the controlling data. See also *Audification*.

Sopite syndrome: chronic fatigue, lethargy, drowsiness, nausea, etc.

Spatial navigation: self-orientation and locomotion in virtual worlds.

Spatially immersive display (SID): semisurrounding projected stereo displays.

Static accuracy: ability of a tracker to determine the coordinates of a position in space. See also *Dynamic accuracy*.

Stereopsis: binocular vision of images with different views by the two eyes to distinguish depth.

Summative evaluation: typically performed after a product or design is more or less complete. Its purpose is to statistically compare several different systems, for example, to determine which one is *better*, where better is defined in advance; involves placing representative users in task-based scenarios.

Synthetic experience (SE): experience created through a virtual environment. Some authors include passive virtual environments, such as a movie ride where there is no interaction, in the SE classification, while reserving the term *virtual environment* for active synthetic experiences, where the user interacts with the virtual world.

Tactile: sensory information arising from contact with an object, detected through nerves within the skin. Sometimes restricted to passive information, such as air currents over the skin, in contrast to haptic, which is then applied to active manual exploration. See also *Proprioceptive* and *Kinesthetic*.

Technsplanation: use of VR technology and other communication technology to explain or teach.

Tele-: operating from a remote location.

Tele-existence: virtual reality experienced from remote locations.

Telemanipulation: robotic control of distant objects.

Teleoperation: technology of robotic remote control.

Teleoperation time delay: communication delay necessary to transmit control data to a remote robot or to receive information feedback from the remote location to the operator.

Teleoperator: person doing telemanipulation.

TELEOS(TM): a tool to create Silicon Graphics computer-based real-time interactive environments with *lifelike* deformable objects.

Telepresence: remote control with adequate sensory data to give the illusion of being at that remote location.

Temporal lobe: an area of the brain in front of the occipital cortex and the parietal cortex that is the receiving site for hearing.

Terrain: geographical information and models that can be either randomly generated or based on actual data.

Texture mapping: a bitmap pattern added to an object to increase realism.

Three-dimensional graphics: the presentation of data on a 2D display surface so that it appears to represent a 3D model.

Track shot: rotating display of the same scene. See also *Dolly shot* and *Pan shot*.

Tracker (VR): a device that provides numeric coordinates to identify the current position and/or orientation of an object or user in real space.

Transparency: teleoperation performance measurement (i.e., fidelity of both information feedback and direct remote robot control).

Universe: the collection of all entities and the space they are embedded in for a VR world.

Update rate: tracker's ability to output position and orientation data.

Usability: the effectiveness, intuitiveness, and satisfaction with which specified users can achieve specified goals in particular environments, particularly interactive systems. Effectiveness is the accuracy and completeness of goals achieved in relation to resources expended. Intuitiveness is the learnability and memorability of using a system. Satisfaction is the comfort and acceptability of using a system.

Usability engineering: methodical *engineering* approach to user interface design and evaluation involving practical, systematic approaches to developing requirements, analyzing a usability problem, developing proposed solutions, and testing those solutions.

Usability evaluation: any of a variety of techniques for measuring or comparing the ease of use of a computer system, including usability inspection, user interface critiques, user testing of a wide variety of kinds, safety and stress testing, functional testing, and field testing.

User-centered design: design around the needs and goals of users and with users involved in the design process and design with usability as a primary focus.

User-centered evaluation: approach to evaluating (typically a computer's user interface) that employs representative or actual system users. The evaluation typically involves users performing representative task scenarios in order to reveal how well or poorly the interface supports a user's goals and actions.

User task analysis: process of identifying and decomposing a complete description of tasks, subtasks, and actions required to use a system as well as other resources necessary for user(s) and systems to cooperatively perform tasks.

Vection: illusion of self-motion, usually elicited by viewing a moving image, but also achievable through other sensory modalities (e.g., audition, somesthesia).

Vestibular: sensory structure of the labyrinth of the inner ear that reacts to head and gross bodily movement. The sense organs embedded in the temporal bone on each side of the head are known collectively as the labyrinthine organs or simply the labyrinths. They include the organ of hearing, or cochlea, the three semicircular canals, and the otolith organs, or utricle and saccule.

Vestibular nucleus: a brain stem structure responsible for processing real or apparent motion stimuli and implicated in the maintenance of stable posture and reflexive gaze control. On each side of the brain stem, there are four principal vestibular nuclei: the lateral, medial, superior, and inferior.

Vestibulo-ocular reflex (VOR): when head moves in any direction, the vestibular apparatus senses this movement and sends velocity information directly to the oculomotor system, which responds by driving the eyes (conjunctively) at an approximately equal rate but opposite direction to compensate for head movement and help keep the visual image stabilized on the retina.

Vibratory myesthetic illusions: illusory feeling of limb or body movement that may accompany certain vibrations of the skeletal muscle. For example, vibration of the biceps tendon with the subject's arm fixed and his eyes closed can elicit an illusion of arm extension.

Viewpoints: points from which ray tracing and geometry creation occur. The geometric eye point of the simulation.

Virtual: simulated or artificial.

Virtual environment (VE): model of reality with which a human can interact, getting information from the model by ordinary human senses such as sight, sound, and touch and/or controlling the model using ordinary human actions such as position and/or motion of body parts and voice. Usually *virtual environment* and *virtual reality* are used synonymously, but some authors reserve VE for an artificial environment that the user interacts with.

Virtual heritage: use of computer-based interactive technologies to record, preserve, or recreate artifacts and sites of historic, artistic, religious, and cultural significance and to deliver the results openly to a global audience in such a way as to provide a formative educational experience through electronic manipulations of time and space.

Virtual mechanism: passive artificial modeling structure used in a haptic controller; linked to both a master device (haptic feedback interface) and controlled robot (real or virtual) to allow intuitive and stable (since passive) control and feedback.

Virtual prototype: simulation of an intended design or product to illustrate the characteristics before actual construction. Usually used as an exploratory tool for developers or as a communications prop for persons reviewing proposed designs.

Virtual prototyping: product design that is based on VE technology; provides an alternative concept for the design–test–evaluation cycle by eliminating the fabrication of physical prototypes.

Virtual reality (VR): See *Virtual environment.*

Virtual reality modeling language (VRML): a computer language for creating online virtual reality models.

Virtual reality therapy: any form of psychotherapy using virtual reality. However, the VR is most commonly used to provide desensitization of an anxiety disorder or a phobia by providing a simulation of the situation causing the phobia.

Virtual team members (VTMs): multifunctional autonomous agents or simulated images of humans within a VE that function in the role programmed to them.

Virtual world: entire virtual environment or universe within a given simulation.

Visual suppression of the vestibulo-ocular reflex: use of visual information to suppress the normal eye-beating reflex. An example would be reading text at the same time as undergoing a turning movement while riding on a bus.

Visualization: ability to graphically represent abstract data that would normally appear as text and numbers on a computer.

Voxel: 3D generalization of a pixel, an indivisible small cube that represents a quantum of volume.

Wayfinding: the cognitive element of navigation.

World in the hand: metaphor for visualized tracking where a tracker is held in the hand and is connected to the motion of an object in a display.

Yaw: angular displacement about the vertical axis.

WEBSITES FOR ADDITIONAL INFORMATION ON STANDARDS

American National Standards Institute (ANSI): http://web.ansi.org.

British Standards Institute (BSI): http://www.bsigroup.com.

Human Engineering Design Criteria for Military Systems, Equipment and Facilities (MIL-STD-1472D): http://www.everyspec.com/MIL-STD/MIL-STD-1400-1499/MIL_STD_1472D_1209/.

IEEE Computer Society: http://www.computer.org.

IEEE Standards Association (IEEE-SA): http://standards.ieee.org/. Details of the standards creation process are found at http://standards.ieee.org/resources/index.html.

IEEE Standards Process at a Glance: http://standards.ieee.org/resources/glance.html.

Institute of Electrical and Electronics Engineers (IEEE): http://www.ieee.org.

International Electrotechnical Commission (IEC): http://www.iec.ch/.

International Organization for Standardization (ISO): http://www.iso.ch/.

Italian National Standards Board (Ente Nazionale Italiano di Unificazione, or UNI): http://www.uni.com/.

Joint Technical Committee Number One (JTC1): http://www.iso.org/iso/jtc1_home.html.

Man-Systems Integration Standards Handbook (NASA-STD-3000): http://msis.jsc.nasa.gov/.

ONLINE GLOSSARIES OF VIRTUAL ENVIRONMENTS

http://www.roadtovr.com/virtual-reality-glossary-terminology.
http://www.hitl.washington.edu/scivw/scivw-ftp/other/VR-glossary.
http://inkandvellum.com/blog/2011/04/but-can-you-use-it-in-a-sentence/.
http://www.digitalspace.com/avatars/book/appendix/glossary.htm#glossary.

FURTHER READING

Aukstakainis, S., & Blatner, D. (1992). *Silicon mirage*. Berkeley, CA: Peachpit Press.

Buie, E. (1999). HCI standards: A mixed blessing. *Interactions, VI*(2), 36–42.

Delaney, B. (1994). Glossary and acronyms [Special Ed.]. *CyberEdge Journal* [Online]. Available: http://www.cyberedge.com/4al.html

Durlach, N. I., & Mavor, A. S. (Eds.). (1995). *Virtual reality: Scientific and technological challenges*. Washington, DC: National Academy Press.

Hamit, F. (1993). *Virtual reality and the exploration of cyberspace*. Carmel, IN: Sams.

Hix, D., & Hartson, H. R. (1993). *Developing user interfaces: Ensuring usability through product and process*. New York, NY: John Wiley & Sons.

Kalawsky, R. S. (1993). *The science of virtual reality and virtual environments*. Wokingham, England: Addison-Wesley.

Latham, R. (1991). *The dictionary of computer graphics and virtual reality*. New York, NY: Springer-Verlag.

Manetta, C., & Blade, R. (1995). Glossary of virtual reality terminology. *The International Journal of Virtual Reality*, *1*(2), 35–48.

Nielsen, J. (1994). Heuristic evaluation. In J. Nielsen & R. L. Mack (Eds.), *Usability inspection methods* (pp. 25–62). New York, NY: John Wiley & Sons.

Rada, R., & Ketchel, J. (2000). Standardizing the European information society. *Communications of the ACM*, *43*(3), 21–25.

Rheingold, H. (1991). *Virtual reality*. New York, NY: Simon & Schuster.

Scriven, M. (1967). The methodology of evaluation. In R. E. Stake (Ed.), *Perspectives of curriculum evaluation* [American Educational Research Association Monograph]. Chicago, IL: Rand McNally.

Sheridan, T. (1992). Defining our terms. *Presence*, *1*(2), 212–274.

Steuer, J. (1992). Defining virtual reality: Dimensions determining telepresence. *Journal of Communication*, *42*(4), 73–93. Also available: http://ww.cybertherapy.info/pages/telepresence.pdf

Vince, J. (2004). *Introduction to Virtual Reality*. New York, NY: Springer-Verlag.

Section II

System Requirements: Hardware

3 Vision and Virtual Environments

David R. Badcock, Stephen Palmisano, and James G. May

CONTENTS

3.1   INTRODUCTION

This chapter is intended to provide the reader with knowledge of the pertinent aspects of human visual processing that are relevant to virtual simulation of various environments. Before considering how real-world vision is simulated, it is perhaps prudent to review the kinds of information that are usually extracted by the visual system. It would, of course, be fruitless to provide visual detail that is rarely or never available to the senses, and it may be fatal to the endeavor to omit detail that is crucial. Thus, an overview of normal visual capabilities and idiosyncrasies is provided in Section 3.2. Section 3.3 reviews some of the ways that perceptual systems provide shortcuts to simulating the visual world. The existence of these phenomena allows system developers to compensate for hardware shortcomings with user inferences. One of the most exciting advantages of virtual environment (VE) technology is that it allows a more elaborate and complex interaction between the VE and the observer. Section 3.4 reviews a number of the ways in which we interact with the world and how these mechanisms might augment and detract from virtual simulation. After the discussion of what vision entails, a discussion of techniques that use 2-D renditions of the visual world to simulate normal viewing of the 3-D world is provided in Section 3.5. The emphasis here will be to address the design requirements of VE displays and to determine if existing displays are machine or observer limited. Considering what is optimally required by the user, a review of the adequacy of existing visual displays is also provided in Section 3.6. Suggestions are made as to how existing limitations might be overcome and speculations are made concerning what new technology might allow in Section 3.7.

3.2   WHAT ARE THE LIMITS OF NORMAL VISION? (USER REQUIREMENTS)

Research on human visual capacities is an active field of endeavor. There are still many aspects of visual performance that are not well understood, but knowledge has been accumulating at an accelerating pace throughout the last century. The inquiry continues, but it is important to set forth, from time to time, a compendium of what we think we know. The formal study of vision is perhaps one of the oldest disciplines and the community of visual scientists is larger than for any other sense. Since the early psychophysicists initiated the systematic investigation of the senses, the notion has been to map the subjective sensations against the physical scales of interest and express the relationship as a *psychometric function* that defines the limits of sensitivity and the subjective scaling of the dimension. That approach will be employed here, but it is important to acknowledge that, in practice, we may utilize little information near the limits of visual abilities. It is conservative, in a sense, to employ this approach to ensure that display technologies meet these user capabilities, but we may learn to *cheat* on these limits if the task requirements allow.

A comprehensive attempt to list human performance capabilities was collated in 1986 and published as the *Engineering Data Compendium: Human Perception and Performance*

(Boff & Lincoln, 1988; see also Boff, Kaufman, & Thomas, 1986). In this chapter, we will offer more in the way of explanation of the phenomena described, but the *Engineering Data Compendium* is still an excellent starting location for determining performance limits. More recent extensive reviews on specific topics can be found in *The Visual Neurosciences* (Chalupa & Werner, 2004) and in *The Senses: A Comprehensive Reference*, volumes 1 and 2 (http://dx.doi.org/10.1016/B978-012370880-9.09003-4; Masland & Albright, 2008).

3.2.1 LUMINANCE

The human visual system is sensitive to a broad range of ambient illumination, extending from absolute threshold (~ -6 log cd/m^2) to levels where irreversible damage to the system can result (~ 8 log cd/m^2). Many stimulus factors (e.g., size, wavelength, retinal location) have profound influences on the lower limits of this range (Hood & Finkelstein, 1979). The human visual system contains two types of photoreceptors (rods and cones) with significantly different sensitivities. The lowest absolute threshold is mediated by the more sensitive rods, and the cone threshold is achieved about 2.5 log units above that limit (-3.5 log cd/m^2). At about 1 log cd/m^2, the rod responses begin to saturate, and at higher luminance levels, they provide no significant information because saturation is complete. The cones mediate vision above that point. Thus, the entire range can be broken up into a scotopic region (rods only), a mesopic region (rods and cones), and a photopic region (cones only). Many of the visual abilities reviewed below are different in rod- and cone-mediated vision. They also vary, with luminance level in the ranges subserved by each photoreceptor type. These adjustments take some time and are referred to as light (and dark) adaptation. For example, an observer moving from a sunny street into a dark room may take up to 30 min to reach maximum sensitivity for detecting the presence of a target. However, the reverse process, adapting to an increase in ambient light level, is much more rapid with substantial adjustment in 10–15 s (Hayhoe, Levin, & Koshel, 1992). These adjustments are required to avoid saturation of the neural response and to maintain high sensitivity to contrast changes in a wide range of natural conditions; but it remains the case that sensitivity to luminance contrast is highest when observers are adapted to the ambient illumination level (McCann & Hall, 1980). The same is true for visual acuity (Craik, 1939).

3.2.2 SPATIAL ABILITIES

Much of human vision is concerned with discerning differences in luminance and color across space. The extent of the visual field of view (FOV), visual acuity, contrast sensitivity, and spatial position accuracy is reviewed in this section. The display factors that relate directly to providing adequate stimulation for such abilities are discussed later.

3.2.2.1 Visual Fields

The monocular visual FOV is determined by having a subject fixate a point in the center of a viewing area (preferably approximately 180° H × 150° V) and presenting targets throughout the extent of the area. The limits of the field are defined as the most peripheral points at which target detection achieves some criterion (e.g., 75% correct). These limits vary somewhat with the choice of criterion and also as a result of stimulus factors such as the luminance of the background, the luminance of the target, the size of the target, and the wavelength of the target. In normal individuals with a circular target subtending approximately 20 arcmin, the horizontal extent is approximately 160°, and the vertical extent is approximately 120° (see Figure 3.1). The shape is roughly circular but is limited by the physiognomy of the nose, brows, and cheeks. Most fields show a restriction in the area of the lower nasal quadrant. A smaller target, subtending 1 arcmin, may only be detectable within a 20° diameter area (Borish, 1970), and if the target contains spatial structure, visual fields are larger for lower-spatial-frequency targets (Koenderink, Bouman, Bueno de Mesquita, & Slappendel, 1978). The measured field size is also greatly reduced if the observer needs to name the hue of the light,

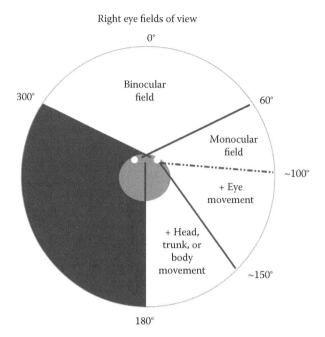

FIGURE 3.1 A depiction of the horizontal monocular and binocular visual fields of the right eye with straight ahead fixation, maximum lateral eye movements, and maximal lateral head movement. The section listing bodily movement can be viewed with maximal eye and head rotation combined, but typically submaximal eye and head movements would be combined with body rotation to view these regions. The black area is not visible by the right eye with clockwise eye and head rotations unless body rotation is included. (The authors thank Professor Paul McGraw and Dr. Andrew Astle for assistance in determining the field limits with eye and head rotations.)

rather than just detecting its presence. This effect varies with the hue to be named. Arakawa (1953) found a 105° horizontal diameter when detecting blue, but only 75° for detecting red and 55° when detecting green. Similar changes occurred vertically. The consequence of this finding suggests that tasks requiring identification of hue should be centered within the visual field.

Careful probing of the area of the visual field about 13°–18° temporal to the fixation point reveals the blind spot, a scotoma formed by the optic nerve head that contains no photoreceptors. It is roughly circular, with a diameter of about 5°. The areas of visibility that are common between the two monocular fields form the binocular visual field. It is roughly circular and is approximately 40° in diameter. Binocular viewing helps to compensate for the blind spot because the visual field location corresponding to the blind spot in one eye coincides with a functional retinal area in the other eye. However, binocular disparity cues for depth (see Section 3.2.3.2) are lost in this region.

3.2.2.2 Visual Acuity

The upper limit of the ability to resolve fine spatial detail is termed visual acuity. It has been measured with numerous techniques (Thomas, 1975), with the normal threshold for spatial resolution ranging between approximately 0.5 and 30 arcsec, depending on the task. The two detection techniques that yield the lowest values are minimum visible and vernier acuity tasks. In the first task, the width of a single line is varied, and the minimally visible width (0.5 arcmin) is determined. In the second, the lateral offset between two vertically oriented line segments is determined (1–2 arcsec). While these are both referred to as acuity tasks, they measure different abilities. The former measures contrast sensitivity (see Section 3.2.2.3) because the change in line width has a greater impact on the retinal luminance difference between the line and background than on the width of the retinal luminance change distribution. This outcome is a result of the line-spread function of the eye. Even in a good optical system, a point of light is spread while passing through the optics. In the human system, this spread covers approximately 1 arcmin. Thus, reducing the width of a small bright line will change the total amount of light passing through the optics, but the spread of light on the retina will change very little. Consequently, a very high acuity for changes can be obtained if those changes create a threshold difference in luminance contrast. Levi, Klein, and Aitsebaomo (1985) used a target composed of five parallel lines in which one of the inner lines was displaced to the left or the right. The displacement varied the width of the gaps on either side of the central line. This produced a contrast difference between the gaps on the two sides of the line, which was detectable with only a 0.5 arcsec line shift—a remarkable spatial discrimination but a normal contrast discrimination.

The vernier acuity task is more appropriately viewed as a position discrimination task. The misalignment of the bars may be detected by discriminating either a difference in the horizontal location of two vertical bars or a difference in the orientation of the overall figure (Watt, Morgan, & Ward, 1983). This task is discussed further later in this chapter.

Resolution tasks require the observer to determine the minimal separation in space of two or more luminance-defined borders. Similar results are obtained with each of two types of commonly employed patterns: gratings and letters. The gratings are high-contrast black lines on a white background with 50% duty cycle, square-wave luminance profiles. The letters are created so that the width of the lines composing the letter is one-fifth of the overall letter size. The acuity measure is calculated from the size of the line width such that the visual angle equals atan (line width/distance between observer and letter).

Since different letters vary in the orientation and position of the lines, it is common to use the Landolt rings instead. The rings have the same line width to overall size ratio but also have a gap the same size as the stroke width of the ring. The observers must resolve the gap and identify its location on the ring. Grating tasks and the Landolt ring task can also be used as discrimination tasks by varying the orientation of the stimuli. These three tasks result in similar thresholds (~30 arcsec in healthy young adult eyes). Numerous variables affect visual acuity performance. Visual acuity increases with luminance, from thresholds of about 1 arcmin at 1 cd/m^2 to 30 arcsec at 100 cd/m^2 (Shlaer, 1937). The recommended standard for illumination for visual acuity measurement is 85 cd/m^2. Under photopic conditions, visual acuity is optimal for targets presented to the center of the visual field (the fovea) and falls off rapidly (see Figure 3.2 for a demonstration). The minimum resolvable detail declines from 30 arcsec at fixation to 20 arcmin at 60° eccentricity for the Landolt ring acuity (Millodot, Johnson, Lamont, & Leibowitz, 1975). At scotopic levels,

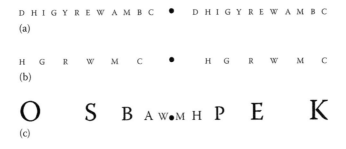

FIGURE 3.2 The reader is invited to fixate the small dot in the center of the displays; while maintaining fixation of the dot, take note of how the visibility of the letters declines as more peripheral letters are considered. (a) Shows the decline in a crowded array, (b) demonstrates the larger range available when crowding is reduced, while (c) attempts to compensate for the decline in visual acuity in the periphery by increasing the size of the letters proportionally. There are individual differences in this rate of decline, but the image indicates the approximate scale of the acuity loss with target eccentricity.

when the cones are not functional, acuity is best about $4°$ from the fixation point. The estimates provided above assume that the targets are presented at the optimal distance. There is a limit to the ability to focus on nearby objects. Targets closer than this near point are blurred and resolution is reduced. The near point recedes with age due to hardening of the lens (presbyopia). This distance may exceed 1–2 m by age 60 and needs to be considered when providing artificial displays because resolution cannot be good if the display is closer than the near point. Optical correction will frequently be needed for older observers.

The identifiability of targets in the visual field is directly influenced by the proximity of surrounding objects. This phenomenon, known as visual crowding, was first reported by Bouma (1970) who noted that the ability to identify a target letter was greatly impaired by adjacent letters. Nandy and Tjan (2012) note that the region in which crowding occurs exhibits several characteristics. First, the size of this crowding zone grows linearly with target eccentricity. Along the axis connecting the fovea to the target, this zone is large, extending to approximately half the target eccentricity. Nandy and Tjan note that this is sometimes referred to as Bouma's law. Second, flanker effects are asymmetric; flankers that are more eccentric than the target have greater crowding effects than flankers at the same distance but on the foveal side of the target. Third, the crowding zone is markedly elongated along the radial, compared to the tangential, axis (see Levi, 2008 and Whitney & Levi, 2011 for reviews). It should also be noted that such crowding effects can impact on the processing of many different stimulus dimensions (van den Berg, Roerdink, & Cornelissen, 2007). Nandy and Tjan have recently presented a physiologically motivated, computational model showing that these critical properties could arise from eye-movement-induced alterations in natural image statistics being reflected in the strength of associations between adjacent cortical receptive fields. For practical purposes, here it is recommended that displays requiring very high acuity should avoid placing objects nearby critical targets.

3.2.2.3 Contrast Sensitivity

While visual acuity has been one of the time-honored descriptors of human spatial ability, it only provides information about the extreme upper limit of the spatial dimension to which humans are sensitive. Much of what humans see and use may not involve spatial detail near these limits. A popular approach to evaluating human sensitivity for objects of different sizes has been to specify the contrast necessary to detect them. In the simplest case, a light bar on a dark background, contrast (C) can be defined as $C = \Delta L/L$, where L is the background luminance and ΔL is luminance increment (or decrement) provided by the bar. Most visual stimuli are more complex, and thus a measure is needed that can provide a sensitivity estimate for these complex patterns. An alternate method takes into account the nature of the neural units early in the visual pathways. At the first stages of processing in the cortex, units are sensitive to short, oriented line segments. Most units give a strong response to a bar of the correct orientation that is surrounded by other bars of the opposite contrast polarity. The underlying weighting function for the cell's response to light can be approximated by a Gabor function that is produced by multiplying a cosinusoidal luminance variation by a Gaussian envelope (Field & Tolhurst, 1986). An image of a Gabor contains a set of bars that alternate between high and low luminance, with contrast that is high in the center of the patch and reducing to zero at the edge of the envelope. When the Gabor is used to represent a weighting function, the bright bars represent excitatory regions and the dark bars inhibitory regions. Individual neurons may vary in the spatial extent of the envelope, the orientation of the cosine, the cosine's spatial frequency, and the cosine's position in the envelope (spatial phase). Since these are the early detectors, one common strategy has been to measure the minimum amount of contrast required for these detectors to reach threshold. Typically, cosinusoidal patterns are employed as stimuli, and with these patterns, contrast is defined as $(L_{MAX} - L_{MIN})/(L_{MAX} + L_{MIN})$, where L_{MAX} and L_{MIN} are the maximum and minimum luminance of the image, respectively. With centrally viewed patches of grating of limited envelope size (e.g., $5°$ of arc) and a large surround (e.g., $30°$) of the same average luminance, the function is band limited from approximately 0.5 to 60 cycles per degree (c/deg).

The peak sensitivity (the reciprocal of the contrast at thresholds of 0.0033 and 0.0025) is approximately 300–400 between 3 and 10 c/deg, but this value varies greatly with display properties.

One of the important motivations for this approach to describing visual sensitivity was the power of the tools of Fourier analysis and now wavelet analysis (Press, Teukolsky, Vetterling, & Flannery, 1992). Using these tools, it is possible to describe any image as a collection of sinusoidal gratings varying in orientation, spatial frequency, phase, and contrast. The similarity between the basis functions employed for this analysis and the weighting functions of some visual neurons led to the suggestion that the visual system may be performing a similar analysis. While this now seems unlikely, it is nevertheless valid to argue that the neurons will only respond to that restricted range of spatial frequencies and orientations within an image to which its receptive field is tuned. Thus, if one could determine the sensitivity of a unit and the amount of contrast in the passband of that unit within an image, it should be possible to predict whether that unit will respond to the particular image. In many cases, such prediction is possible (Campbell & Robson, 1968; Graham, 1980). However, the utility of this approach is limited by the lack of generality of the foveal contrast sensitivity function. In the periphery of the visual field, high spatial frequencies are increasingly poorly resolved, and thus a different contrast sensitivity function is obtained at each eccentricity. Contrast sensitivity declines with luminance (Van Ness & Bouman, 1967) and varies with target motion (Kelly & Burbeck, 1980; Robson, 1966), orientation (Mitchell & Ware, 1974), and chromaticity (Metha, Vingrys, & Badcock, 1993; Mitchell & Ware, 1974; Mullen, 1985). While these limitations are significant, the contrast sensitivity function is a more comprehensive measure of visual sensitivity than spatial acuity alone and is used extensively to characterize the performance of the visual system. The influence of temporal modulation on contrast sensitivity (Kelly & Burbeck, 1980; Robson, 1966) is also relevant when assessing the suitability of different display technologies as discussed below.

3.2.2.4 Spatial Position

The ability to accurately localize image features is an important precursor to object recognition, shape discrimination, and interaction with a cluttered environment. The appreciation of a form presupposes the accurate relative localization of elements of the form, and the perception of peripherally viewed stimuli serves to guide eye movements and locomotion. Approaches to the characterization of such ability in the frontal plane involve relative judgments of the spatial position of two or more objects (Badcock, Hess, & Dobbins, 1996; Westheimer & McKee, 1977), bisection of visual space (Levi et al., 1985), and localization of briefly presented target positions (Solman & May, 1990; Watt, 1987). In central vision, relative spatial position thresholds are on the order of 0.05–1.0 arcsec (see vernier acuity previously mentioned), and once again, the threshold depends on the spatial scale of the pattern. If Gaussian luminance increments or Gabor patches are employed, the threshold is proportional to the standard deviation of the Gaussian envelope (Toet, van-Eekhout, Simons, & Koenderink, 1987). These thresholds can be 10 times higher when targets are presented eccentrically by just 10° (Westheimer, 1982), a decline that is substantially more rapid than that obtained for visual acuity estimates (Levi et al.).

Partitioning studies also indicate that, at eccentricities of 10°, position thresholds vary with the size of the area to be partitioned, increasing from about 0.05° at 0.50° separation to 0.11° at 1.1° separation and remaining constant thereafter (Levi & Klein, 1990). Spatial location difference thresholds for briefly presented, single targets increase with eccentricity from about 0.04° at 2° eccentricity to 3.5° at 12° eccentricity (Solman, Dain, Lim, & May, 1995; Solman & May, 1990). Levi and Klein further clarified that when isoeccentric target element separation approximates target eccentricity, then thresholds are 0.01–0.03 of the eccentricity. It is clear that the relative spatial sense measured with two or more objects, or partitioning, is more precise than is the localization of targets with more limited spatial landmarks (Matin, 1986). Apparent position is also influenced by the spatiotemporal proximity of nearby objects. Badcock and Westheimer (1985) showed that objects closer than 3 arcmin result in an apparent shift of the target toward the distractor's spatial position, whereas larger distances cause repulsion in apparent position (see also Fendick, 1983; Rivest & Cavanagh, 1996). A similar phenomenon occurs over large distances where spatial intervals appear smaller if a larger interval is present nearby and vice versa (Burbeck & Hadden, 1993; Hess & Badcock, 1995).

The apparent location of an object is also greatly affected by motion, with its instantaneous position appearing to be farther along a motion trajectory than its veridical location (DeValois & DeValois, 1991; Edwards & Badcock, 2003; McGraw, Walsh, & Barrett, 2004). This motion position illusion can be induced by either motion in adjacent areas (Whitney, 2002) or motion extracted from the global pooling of local motion signals (even when this global motion differs from dominant local directions [Scarfe & Johnston, 2010]).

3.2.3 Depth

Theoretically, detection of targets in depth is limited by the spatial resolution of the visual system. If the target is large enough (e.g., our moon), an individual can easily see it at a great distance (245,000 miles). However, we will often have difficulty discerning the distance of our moon relative to the sun because they both subtend roughly the same visual angle on our retinas (0.5°) and the binocular disparity between them is too small to reach threshold (Gillam, Palmisano, & Govan, 2011). Their depth order is only unambiguous during an eclipse. At near distances, however, an individual has little difficulty determining the relative distances of two objects because of the presence of numerous depth cues. The cues are traditionally discussed in terms of those available with monocular and binocular viewing.

3.2.3.1 Monocular Cues

Monocular cues are those that would be available to an observer using just one eye.

3.2.3.1.1 Pictorial Cues

Pictorial cues are spatial arrangements that convey relative differences in depth. They are employed by artists adept at conveying 3-D impressions with 2-D depictions. They are the following:

1. Relative size: In the absence of information about the absolute size of an object, its apparent size influences judgments of its distance. The deer on the right-hand side of Figure 3.3 appear to be at increasingly greater distances largely because of their progressively smaller sizes. The moved copies on the left appear larger (upper) and smaller (lower) than the originals (on the right) because of the automatic but inappropriate application of depth scaling.
2. Height relative to the horizon: For objects below the horizon (the identical seagulls indicated by solid arrows in Figure 3.4), objects lower in the visual field appear closer than objects higher in the visual field. For objects above the horizon (the identical clouds in Figure 3.4 indicated by dashed arrows), the reverse is true.
3. Interposition or occlusion: When objects are opaque, those nearby occlude the view of parts of those that are farther away and convey an immediate sense of depth. For example, the boats in Figure 3.4 are interposed between the viewer and the city.
4. Shadows and shading: A directed light source will strike the nearest parts of an object in its path and be prevented from striking other features on the same path. Thus, differences in the intensity of the light reflected from object parts can contribute to the appreciation of depth and 3-D shape (see Figure 3.5a and b). In situations where the position of the light source is not detectable from the image, the visual system resolves ambiguous shading cues by assuming that the light source is above the object (see Figure 3.5c). In addition, objects often cast shadows on surfaces near them, adding additional cues to figure ground configurations by occlusion or interpolation. The degree of occlusion of a shadow

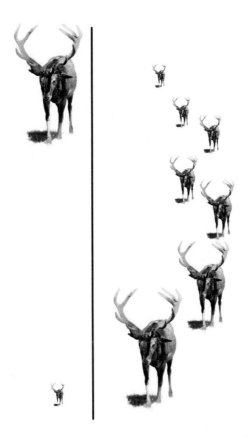

FIGURE 3.3   These images convey differences in apparent depth from decreasing size. The deer on the right are scaled in size so that they vary in apparent distance. The smallest deer appears farther away than the larger ones. The misplaced deer (on the left) appear larger (upper) or smaller (lower) than the identical images on the right (lower and uppermost, respectively) because of inappropriate perceptual depth scaling.

FIGURE 3.4   This scene includes demonstrations of height relative to the horizon and interposition or occlusion cues in determining relative apparent depth (see text for explanation).

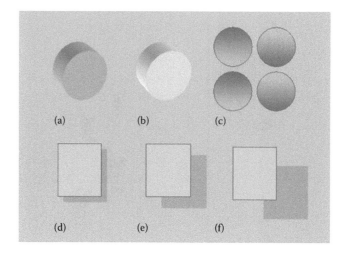

FIGURE 3.5 Brighter objects (b) appear to be somewhat closer than dimmer identical objects (a). (c) Light is assumed to come from above, by default, which means that shadows on the top half of circles produce a percept of concavity while those on the lower half imply convexity. (d–f) show that perceived height of an object, above the picture plane, increases when the object occludes less of its shadow.

by the object casting it can serve as a cue to the distance between an object and the background. These cues are conveyed in 2-D objects by shading (see Figure 3.5d through f).

5. Aerial perspective: The light reflected from objects is both scattered and absorbed by particles in the medium through which it is observed, causing near objects to be perceived as brighter, sharper, and more saturated in color than far objects. Thus, one cue to distance is the brightness of objects (compare objects a and b in Figure 3.5).

6. Linear perspective: When parallel lines recede in distance from an observer (e.g., railroad tracks), their projection on the 2-D retina produces a convergence of the lines as distance increases. The contours do not have to be straight, but merely equally spaced as they become more distant (e.g., a winding roadway). This change in perspective with distance provides a compelling cue for depth and is a consequence of representing 3-D scenes on 2-D surfaces.

7. Texture gradients: Most natural objects are visibly textured. When the image of those objects is projected onto a 2-D surface like the retina, texture elements are distorted and the density of the texture of the surface increases with distance between the surface and the observer. These gradients of texture density are highly salient cues to relative depth. (See Figure 3.6a and c that are warped versions of Figure 3.6b. Figure 3.6b has no texture variation and looks flat. A 180° rotation of Figure 3.6a produces Figure 3.6c. Higher-density regions appear farther away.)

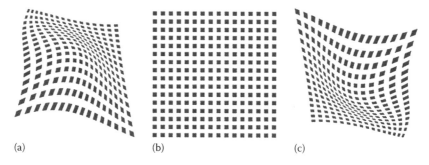

FIGURE 3.6 Texture variation in images can convey a strong impression of relative depth. (b) has no variation and appears flat, whereas the opposite transformations in (a) and (c), which also add a density variation, show that perceived distance increases as density increases.

3.2.3.1.2 *Motion Parallax*

All of the cues so far discussed have concerned static scenes and a stationary observer, but additional depth information is conveyed when the scene is moving relative to the observer or the observer is moving relative to a static scene. The typical example is an observer on a moving train viewing the landscape as it passes (which generates a pattern of visual motion stimulation known as optic flow). If the observer fixates a point midway between the train and the horizon while the train moves forward, two movement-related phenomena can be observed. First, objects beyond the plane of fixation appear to move in the direction that the observer is moving, while objects closer than the fixation plane move in the direction opposite of the observer's movement. Second, the retinal speed of objects is proportional to the distance of the object from the fixation point. Not only do these phenomena provide important information about the depth of objects in the field, but they also tell us about our movement relative to the environment (Rogers & Graham, 1982; see also Section 3.3.5). A special case of motion parallax, known as the kinetic depth effect, occurs when an individual views a moving 3-D object and the relative motion allows the observer to recover the object's 3-D structure. Imagine a sphere constructed of a meshwork of fine wire. When it is stationary, it may be difficult to appreciate its shape, but when it is rotated, the contours near the observer move in one direction, whereas the contours farther away move in the opposite direction. This shape information can even be recovered from a shadow cast by this sphere onto a 2-D screen.

3.2.3.2 Binocular Cues

Although most people view the world through two eyes, they usually see a single unified view of the world. By viewing a scene alternatively with one eye and then the other, an individual can appreciate the differences in the two views. Ignoring the differences in FOV provided, the individual can also notice slight differences in the relative position of objects within the overlapping regions of the two monocular views. These differences are fused to form a singular view when the individual uses two eyes. The horopter (see Figure 3.7) is a hypothetical surface in space determined by the point in depth for which the eyes are converged (and accommodated). For a given depth of focus, all the points on the horopter are associated with homologous pairs of points on the two retinae. The theoretical horopter defines the curved region in which there is no disparity between the two monocular views. The area in front of the horopter contains disparities that are said to be *crossed* (because it would be necessary to cross the eyes to bring the points onto

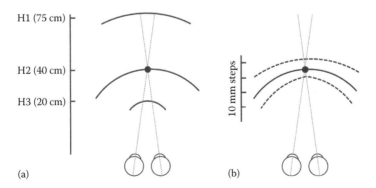

FIGURE 3.7   (a) A depiction of three horopters (H1–H3) associated with fixation in different depth planes. (b) Horopter H2 is replotted with Panum's fusional area, the region of single vision, depicted by dotted lines. (Data replotted from Kalloniatis, M. and Luu, C., Webvision: The organization of the retina and visual system: Perception of depth, 2007. webvision.med.utah.edu/book/part-viii.gabac-receptors/perception-of-depth.)

homologous retinal areas), and the area beyond the horopter contains disparities, which are said to be *uncrossed*. The empirical horopter is shallower in curvature than the theoretical curve, and its curvature decreases with larger viewing distances (see Figure 3.7a). The vertical aspect of the horopter is also tilted, with lower half fields nearer and upper half fields farther away than the point of fixation, a property that may assist when making depth judgments on the ground plane (Schreiber, Hillis, Filippini, Schor, & Banks, 2008).

The region of binocular fusion is a zone approximately centered on the horopter. The reader can appreciate this fact by viewing two objects (e.g., fingers) in different depth planes. If the near object is fixated, attending to the far object reveals double images of that object (diplopia) but a single image of the fixated object. If the far object is fixated, attending to the near object reveals double images of that object (diplopia) but a single image of the fixated object. When fixating 3-D objects, the binocular image contains some areas that fall in front of or behind the horopter and must therefore contain retinal disparities, but these double images are fused and seen as one object. The region in depth over which disparities are fused and singular vision is perceived is termed Panum's fusional area (an example for viewing distance H2 from Figure 3.7a is provided by dotted lines in Figure 3.7b). This area encompasses the horopter and includes areas behind and in front of it.

Studies that have involved dichoptic viewing (presenting separate images to the two eyes) have revealed that subjects will strive to overcome diplopia with vergence eye movements that seek to reduce retinal disparity. If two images can be brought into register with only slight disparity between aspects of the two views, then fusion is achieved and singular vision is experienced. If, however, the disparities are too great, diplopia ensues and binocular rivalry between the two views is often experienced (seeing one view but not the other or seeing part of one view and part of the other). Rivalry can occur at multiple levels in the visual system and can be between the views of different eyes or between different objects, even when part of each object is presented to each eye (Kovacs, Papathomas, Yang, & Feher, 1996; Tong, Meng, & Blake, 2006).

Fusion of disparate images brings with it not only singularity of view but, also more importantly, a vivid sense of depth. For example, imagine viewing dichoptically two separate vertical, square-wave gratings. When fused, they will appear as a single grating in the frontal plane without any depth differences. If the spatial frequency of one of the gratings is shifted slightly lower, the image remains fused but appears to rotate in depth. Retinal disparity is ubiquitously present in our binocular view of the 3-D world, and it may be simulated in 2-D to produce depth information (Howard & Rogers, 1995). The smallest disparity that provides depth information is termed the threshold for stereopsis. This threshold is smallest when the target is viewed foveally and on the horopter (Badcock & Schor, 1985). In the fovea, this threshold may be only a few seconds of arc, but the thresholds increase with the eccentricity of the target to about 300 arcsec at 8°.

In the past, binocular disparity cues were considered useful only for perceiving, and interacting with, very near space (e.g., Arsenault & Ware, 2004; Gregory, 1966). However, recent research conducted in a disused railway tunnel has shown that disparity-based depth is seen well beyond 40 m (Gillam et al., 2011; Palmisano, Gillam, Govan, Allison, & Harris, 2010). With the observers' heads held stationary with complete darkness between the two visible objects in the tunnel, the only cue to depth was binocular disparity. Under these conditions, the estimated depths of objects lying more than 40 m away were still found to increase with their binocular disparity (from 0 to 3 arcmin; corresponding to 0 to 248 m). Interestingly, even when observers were free to move their heads, the contribution of stereopsis to depth appeared to be considerably greater than that of motion parallax.

While stereopsis was long thought to be based simply on binocular disparity cues, this belief has also been challenged by the discovery that stereoscopic depth can still be seen in background regions only visible to one eye (e.g., due to occlusion by a nearer surface). Rather than acting as noise (e.g., by promoting false matches between noncorresponding points in the left and right eyes' images), it is now clear that these types of monocular regions also play important roles in both forming surface representations and binocular depth perception (see Harris & Wilcox, 2009 for a review).

It is also important to recognize that some observers are unable to detect binocular disparity. These stereoanomalous and stereoblind observers, estimated at 3%–5% of the population (Ding & Levi, 2011), with the incidence increasing with age (Rubin et al., 1997), are unable to extract depth signaled purely by disparity differences in modern 3-D display systems and frequently find them unpleasant to view. Studies intending to use 3-D displays should prescreen the observers to ensure their stereoacuity is appropriate for the intended tasks.

3.2.3.3 Accommodation and Vergence

The human visual system contains an elastic lens that can change curvature and refractive index. If an individual looks at an object positioned within about 3 m, the curvature of the lens is adjusted to focus the image sharply on the retina, and the muscular contraction necessary for that adjustment can serve as a cue for the distance of the object. Theoretically, differences in these muscle contractions and differences in sharpness for objects at different distances might also be used as depth cues. With binocular viewing, the two eyes converge when viewing near objects and diverge when viewing objects farther away. Observers are able to use vergence information to assist in judging depth for near objects (Viguier, Clement, & Trotter, 2001), but it is unclear whether the critical information arises from the extrinsic eye muscle tension conveying information about eye position to the brain or, instead, the efferent command to converge the eyes. Accommodation and vergence are normally linked and covary with the depth plane of the target observed. While some doubt the utility of the potential physiological cues to depth, the consensus is that cues associated with accommodation and vergence are viable for near targets (Gillam, 1995) and these variables may be quite important for virtual simulations where depth may be produced by binocular disparity cues presented on screens that are near the eyes. Under these circumstances, the necessary accommodation and convergence may not be congruent with normal visual experience. The consequences of this are discussed in Wann, White, Wilkie, Culmer, Lodge, & Mon-Williams (2014; Chapter 32) and later in this chapter.

3.2.4 Color Vision

In addition to our ability to discern differences in luminance across space and time, humans may also discriminate between the wavelength of light across these dimensions. The history of research on color vision is, perhaps, the most extensive consideration of any human ability and our understanding of the mechanisms by which humans appreciate color differences reflects that impressive effort. The concern here, however, is to define the limits of that human ability and not to explain the mechanisms underlying them. Humans are sensitive to electromagnetic radiation in the range of 370–730 nm (Wandell, 1995). Various light sources (e.g., the sun, tungsten bulbs, fluorescent lights) provide a broad band of radiation across and beyond this visible spectrum. Traditional methods of studying color vision have involved various methods of restricting the spectrum to narrow bandwidths. Transparent devices (e.g., filters, diffraction gratings, prisms) are characterized in terms of spectral transmittance, while opaque media (e.g., paper, paint) are described in terms of their spectral reflectance.

The first question addressed concerning sensitivity to chromaticity involved determining the threshold for detecting the presence of a light composed of only a narrow range of wavelengths. Measures at scotopic levels (rod-mediated vision) revealed that the sensitivity varied with a peak at about 500 nm (the Commission Internationale de l'Eclairage, CIE, V'(λ) function; see upper panel of Figure 3.8). At photopic levels, a similar function was observed, with a peak at 555 nm (the CIE V(λ) function; see upper panel of Figure 3.8). At mesopic levels, the measured functions are a combination of the other two. It must be noted that these curves only reflect the group-averaged sensitivity to the presence of light of different wavelengths. They do not address the ability to identify or discriminate between colors. Color vision is defined as the ability to discriminate between stimuli of equal brightness on the basis of wavelength alone. Under scotopic conditions, this is not possible as all rods have the same spectral sensitivity, but under photopic conditions (where the three different cone types have different spectral sensitivity), the full extent of color vision is measurable.

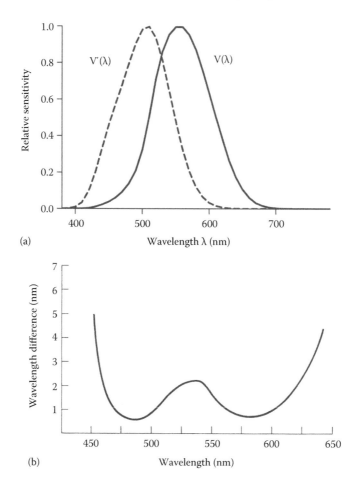

FIGURE 3.8   (a) The upper panel contains the function depicting the spectral sensitivity of the human eye under scotopic [CIE V'(λ)] and photopic [CIE V(λ)] conditions. The two functions have been normalized to remove the overall increased sensitivity of the scotopic system. (b) The lower panel contains the function that describes wavelength discrimination across the spectrum. (From Hurvich, L.M., *Color Vision*, Sunderlund: Sinauer, 1981, p. 210.)

Wavelength discrimination tasks involve asking the observer to vary the wavelength of a comparison stimulus until it can be differentiated from a standard that is of some given wavelength. The increment in wavelength necessary to discern a difference when the stimuli are of equal brightness is plotted against wavelength (see lower panel of Figure 3.8). The wavelength difference varies between 1 and 13 nm, and the function contains two minima in the ranges of 450–500 and 550–600 nm. This kind of discrimination varies with a number of parameters. Chromatic discrimination ability decreases with luminance, especially at lower wavelengths (Brown, 1951; Clarke, 1967; Farnsworth, 1955; Siegal & Siegal, 1972), but remains fairly constant above 100 td. (Trolands are a measure of retinal illuminance that incorporate the effects of changes in pupil diameter with luminance level: I[Trolands] = A [pupil area in mm^2] · L[luminance in cd/m^2].) Discrimination is reduced with field size below 10° (Brown; Wyszecki & Stiles, 1967) but is relatively unaffected by separation between comparison fields (Boynton, Hayhoe, & MacLeod, 1977). If comparison fields are presented successively, discrimination is unaffected until a stimulus onset asynchrony (SOA) of 60 ms but declines at higher intervals (Uchikawa & Ikeda, 1981).

Discrimination for some wavelengths (400 and 580 nm) asymptotes at 100 ms, whereas other wavelengths (480 and 527 nm) require 200 ms for best performance (Regan & Tyler, 1971). Chromatic discrimination is also poor at eccentricities beyond about 8°.

In addition to discrimination of hue, human observers can discern differences in the purity of white light. The threshold for colorimetric purity is defined as the amount of chromatic light that is added to white light to produce a just noticeable difference (JND). Additional JNDs can be observed with increasing steps in the chromatic additive. The number of JNDs observed varies with the wavelength of the additive and is least at about 570 nm. The term *saturation* is used to describe the subjective correlate of colorimetric purity. A highly saturated light appears to have little white light *contamination.*

If the JND is taken as the step size in the dimensions of hue, brightness, and saturation, Gouras (1991) has suggested that there are about 200 JNDs for hue, 500 JNDs for brightness, and 20 JNDs for saturation. This suggests that human color vision capability involves the discrimination of about two million chromatic stimuli ($200 \times 500 \times 20 = 2,000,000$).

The visual subsystem that processes chromatic information differs in a number of significant ways from the subsystem that encodes luminance variation. Critically, for artificial systems where bandwidth limitations are significant, the chromatic system has substantially poorer spatial (Mullen, 1985) and temporal (Cropper & Badcock, 1994) resolutions, and thus chromatic properties can be rendered more coarsely in both space and time. Mullen estimates that the spatial resolution for red-green gratings is three times worse than for achromatic gratings (10–12 c/deg instead of 60 c/deg), whereas Cropper and Badcock reported a twofold reduction in temporal resolution.

This section has dealt with the detectability of light as a function of wavelength and also the ability to discriminate between lights of differing wavelengths, but the hue perceived is greatly affected by the chromaticity of the surround.

3.2.5 MOTION

Humans can see objects moving with respect to themselves, whether they are stationary or moving, and they can detect their own movements through space in a static environment or with objects moving around them. Information about one object's movement relative to another (exocentric motion) is obtained from the image of the objects moving across the retina. An observer may attribute motion to the object that is actually moving or may see movement of a stationary object if the frame around it is displaced (induced movement). Information about object movement relative to an observer's own position or movement (egocentric motion) comes from translations in the retinal image in relation to nonvisual senses (e.g., vestibular and kinesthetic senses) that are involved in body, head, and eye movements. Mere translation of the object on the retina cannot provide information about egocentric information. Humans do not perceive object motion relative to themselves if they move their head or eyes while looking at or away from a stationary object (position constancy).

There are several ways in which one might measure a minimum motion threshold. First, one could measure the minimum distance a feature has to be displaced in order for the direction of motion to be detected. When a visual reference is present, displacements as small as 10 arcsec are sufficient for this judgment (Westheimer, 1978). Second, one could measure the minimum temporal frequency required to detect movement in a temporally extended motion sequence. This threshold depends on both the contrast and the spatial structure of the moving pattern (Derrington & Badcock, 1985). The minimum temporal frequency decreases as contrast increases for low-spatial-frequency periodic sinusoidal grating patterns but is constant for high-spatial-frequency patterns. Third, one could measure the minimum number of dots needed to move in a common direction in a field of randomly moving dots for that direction to be discernible. Edwards and Badcock (1993) have found that as few as 5%–10% of dots moving in a common direction is sufficient either for frontoparallel motion or for expanding/contracting patterns designed to mimic motion in depth, although this threshold does vary with dot step size

(Baker, Hess, & Zihl, 1991). It is also possible to combine signals from dispersed local apertures containing motion into a global rigid solution that may differ from all of the local components (Amano, Edwards, Badcock, & Nishida, 2009). In this latter case, the minimum change in perceived direction or the minimum number of apertures required to be consistent with the rigid solution may be varied, but sensitivity using both measures is high.

While these are impressive abilities, it is also the case that the perception of motion is influenced by a number of stimulus properties. The perceived speed of an object slows if the luminance contrast of the object is reduced (Cavanagh & Favreau, 1985; Thompson, 1982), and observers are unable to see very-high-spatial-frequency repetitive patterns move at all (Badcock & Derrington, 1985). Objects appear to move more slowly through smaller apertures and at greater distances (Rock, 1975).

Finally, even relatively brief periods of exposure to continuous motion (e.g., 30 s to 1 min) produce substantial motion aftereffects, where subsequently viewed stationary patterns appear to drift in the opposite direction to the previously seen physical motion (Mather, Verstraten, & Anstis, 1998). In computer environments, these aftereffects commonly arise due to smooth scrolling of stimulus displays, which can also produce compelling visual illusions of self-motion (see later discussion and Hettinger, 2014; Chapter 18). A common example is observed when trains pull into a station. The continuous visual adaptation to forward motion creates the egocentric impression of rolling backward when the train is stationary.

3.2.6 Motion in Depth

Regan and Beverly have developed the notion that motion in depth is mediated by at least two mechanisms: changing size (Regan & Beverly, 1978a, 1978b) and changing retinal disparity (Regan & Beverly, 1973). Changing size is a monocular mechanism and is characterized by thresholds of about 1 min of arc. The velocity of the size change is proportional to the perceived motion in depth. Changing retinal disparity and inter-ocular velocity differences are stereoscopic cues with approximately the same thresholds for detection as changing size. Figure 3.9 summarizes the movement on the retina caused by monocular and binocular viewing conditions. In Figure 3.9a, the motion of an object toward the viewer results in the edges of the object moving in opposite directions (expansion) on the retina (and vice versa for movement away from the observer, not illustrated). In Figure 3.9b, the motion toward the observer results in both object expansion and

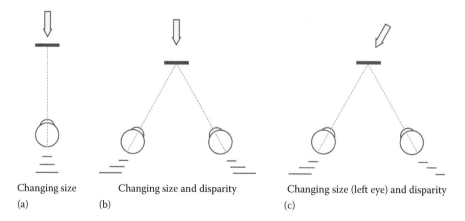

Changing size   Changing size and disparity   Changing size (left eye) and disparity

(a)   (b)   (c)

FIGURE 3.9 (a) Indicates how the edges of an object on a single retina move in opposite directions under expansion and contraction conditions produced by an object moving toward or away from the observer. (b) and (c) Indicate how opposite object motion occurs on the two retinas when it is stereoscopically viewed (see text for elaboration).

movement of the edges of the object in opposite directions (translation) on the two retinas. In Figure 3.9c, the motion of an object toward the left eye will result in only expansion in that eye but translation and less expansion in the right eye. In addition to the differences in direction of movement in the binocular case, differences in velocity occur on the retinas when the object is not moving on a trajectory aimed midway between the eyes. Comparisons of monocular and binocular sensitivities to motion in depth reveal the binocular sense to be slightly more sensitive than the monocular case, as in spatial vision (McKee & Levi, 1987). Both mechanisms appear to provide input to a single motion in depth stage, however, because motion perceived in terms of changing size can be *nulled* with conflicting disparity manipulations. This information concerning motion in depth appears to be uniquely derived because this ability cannot be accounted for by static stereoacuity performance and the visual fields for stereomotion in depth are constricted relative to those observed for static stereopsis. However, receptive fields for motion in depth signaled by optic flow without stereo cues are very large compared to those for stereopsis. Burr, Morrone, and Vaina (1998) have shown signal integration with stimuli up to 70° in diameter in psychophysical experiments, and Duffy and Wurtz (1991) have reported single cells with fields up to 100° in diameter in the medial temporal (MT) cortical area. These mechanisms seem ideal for detecting the optic flow produced by locomotion and appear to be centrally involved in its control. Interestingly though, Edwards and Badcock (1993) have shown that observers are more sensitive to contraction than expansion, which is the kind of flow produced by walking backward.

3.3 WHAT DO WE INFER FROM WHAT WE SEE? (SHORTCUTS)

Previously, the various visual capabilities with regard to detection and discrimination along a number of stimulus dimensions were enumerated. That approach to visual perception assumes a relatively passive observer and defines the abilities in terms of the physical limitations of the visual system. But the human observer is rarely passive and usually interacts with the environment with expectations, goals, and purposes. These psychological aspects of perception have been the focus of numerous lines of research and have provided insight into various phenomena that serve to facilitate and disambiguate information-processing tasks. In this section, we consider these propensities and the implications for perception within the simulated environment.

3.3.1 FIGURE/GROUND ORGANIZATION AND SEGMENTATION

Glance at Figure 3.10a and ask yourself what you see first. You might see lots of Ss or Hs made of Ss or an E made of Hs made of Ss. Whatever it was that you perceived first, it is easily possible to see the other possibilities with further examination. While humans rarely think about it, similar perceptual processes are employed in the analysis of all images. An observer might stare at a forest and see its global outline against the sky or look at a particular tree as an element in the forest or attend to a leaf on one of its branches. In each case, the observer can describe the object of their attention as an object perceived against a background (or ground). Since nothing need change in the scene in order for the observer to experience these different perceptual figure–ground organizations, they are forced to conclude that the experiences are not driven by stimulus change but are determined either physiologically (through adaptation) or psychologically, through directed attention and/or perceptual set. Consider Figure 3.10b, the Necker cube. Although the figure is static, an observer can perceive two possible perceptual organizations involving different depth planes, one with the box protruding to the upper right and the other with the box protruding to the lower left. Much of the visual work accomplished is cognitively driven wherein the observer searches for objects or serially processes objects (e.g., reading) against a background. Early Gestalt psychologists (e.g., Koffka, 1922) proposed a set of *laws* that suggest what stimulus configurations facilitate perceptual organization. Recent work has been attempting to determine the algorithmic and physiological bases for

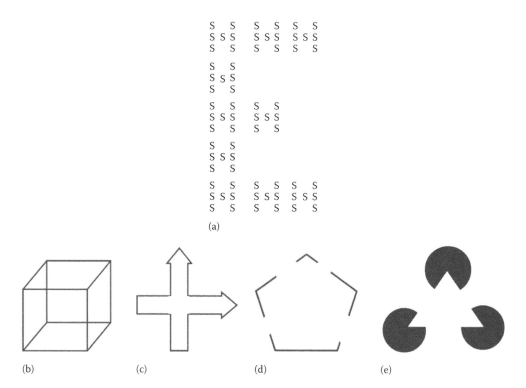

FIGURE 3.10   Stimuli that provide examples of perceptual organization (see text for details). (a) A large letter E composed of mid-level clusters (H), composed of small letters (S). (b) A Necker cube that alternates in 3D interpretation. (c) A compound image interpreted as two arrows crossing because of good continuation. (d) A set of lines form a pentagon through closure. (e) An illusory Kanizsa triangle appears to occlude three circles.

some of these phenomena. They are still poorly understood but have proven to be useful for scene segmentation in computational models. The factors are the following:

1. Proximity: Given an array of elements, those closest together are more likely to be grouped into a global structure. Elements that are close to one another tend to be seen as an object (e.g., the Ss are grouped to form the *H* in Figure 3.10a).
2. Similarity: Similar objects tend to be grouped together to form a global structure (e.g., the *E* in Figure 3.10a).
3. Good continuation: Lines are perceived as continuing with the minimum angular deviation necessary. In Figure 3.10c, the vertical lines of the vertical arrow could deviate sharply along the horizontal arrow but the default grouping avoids this solution. Field, Hayes, and Hess (1993) have described an association field that quantifies the likelihood of binding elements on a smooth path as a function of angular deviation of the path, and the required neural connectivity has been demonstrated in primary visual cortex (Gilbert, Das, Ito, Kapadia, & Westheimer, 1996).
4. Common fate: Elements of an image that undergo a common transformation over time are frequently grouped as one object. The most extensively studied transformation is lateral motion, which produces rapid segmentation of complex scenes. The precise mechanisms are yet to be revealed, but single cells in the MT cortical area have stimulus preferences for dot patterns moving in different directions to their surroundings and thus may form a neural substrate for this segmentation (Allman, Miezin, & McGuinness, 1985). It should be noted that while this segmentation is automatic with moving stimuli, coherent grouping of local texture elements is subject to strong attentional influences (Dickinson, Broderick, & Badcock, 2009).

5. Closure: When a form is drawn with broken or incomplete contours, the perceptual system perceives the form that would result if the contours were complete (Rock, 1975). This is not to say that the breaks in the contours are not detectable, but rather that those breaks do not prevent grouping for object recognition (see Figure 3.10d).

These principles may be applied to images containing a broad array of stimulus attributes. Interestingly, once grouping has occurred, quite distinct boundaries may be formed that are not physically present in the luminance profile of the image. For example, Figure 3.10e appears to be interpreted by the visual system as a white triangle overlapping black circles. An illusory brightness difference is usually perceived between the inferred center of the triangle and the surrounding area. The border between these two zones can appear to be very sharp. Some of these borders may be seen because they stimulate nonlinear texture detection processes within the visual system (Wilson, Ferrera, & Yo, 1992), but no adequate model exists for the full array of illusory contours at this stage. However, there has been substantial progress in understanding how local edge fragments are grouped to define contours (for review, see Loffler, 2008) with demonstrations of explicit global integration processes for both contour (Bell & Badcock, 2008; Loffler, 2008) and texture (Wilson & Wilkinson, 1998), supporting physiological processes in the cortical ventral stream (Connor, 2004) and more detailed computational models (Cadieu et al., 2007).

3.3.2 Perceptual Constancies

Although the 2-D representation of objects on the retina may undergo considerable translations in lightness, color, orientation, size, and position, humans maintain an appreciation for these basic properties of the object. Thus, a white piece of paper (on a gray background) viewed in direct light and in a cast shadow appears to be the same brightness despite the fact that one reflects more light than the other. Likewise, a piece of colored paper (on a gray background) viewed under different colored lights will appear a very similar color despite the differences in hue of the illuminant (although changing the color of just the surround can dramatically alter the perceived hue of the target patch). This page is viewed as rectangular whether viewed from directly above or from a 45° angle, and it is perceived to be the same size whether viewed from 18 in. or 18 ft. The page is seen to be in the same position in space whether an observer looks directly at it or gazes to the side. While the mechanisms whereby these constancies are achieved are still the subject of active research, most explanations emphasize a ratio principle. For lightness and color constancy, it is proposed that the ratio of the lightness or the hue of both the object and the background is extracted, and if the ratio is the same, constancy is achieved. For shape and size constancy, the critical ratio refers to the size and orientation of elements in the object relative to size and orientation in the background under different viewing conditions. See Palmer (1999) for a review of these mechanisms. Position constancy will be elaborated upon later.

3.3.3 Exocentric Motion

Consider a small, stationary point of light in a totally dark room. That object appears to move in a quasirandom fashion, and subjects find it quite difficult to specify its position (by pointing or touching its location). This illusion of exocentric motion is termed the autokinetic effect (Sandström, 1951), and it indicates how difficult it is to determine the motion of an object without a visual frame of reference. The effect is actually caused by implicit eye, head, and body movements in the absence of the stabilizing effect of visual information and is really an egocentric illusion. In a lighted room with referential stimuli, the object is seen as stationary (Mack, 1986). The next question concerns how much that object must be displaced for the observer to detect movement. With a frame of reference, the displacement threshold has been estimated to be 1–2 min, but that value increased tenfold when the reference was removed (Aubert, 1886; Bourdon, 1902; Grim, 1911). With real movement, the moving object is always visible. It moves a given distance (or is visible for a given time), and its movement generally has a fixed velocity (in the simple case). The upper threshold for the detection of movement depends on many factors (luminance, FOV, spatial pattern), but long before an object moves too fast to be seen, the form of the object is obscured. One estimate of this threshold point is termed dynamic visual acuity, and it indicates that form vision as measured by the Landolt C acuity is compromised above about 60°–100°/s (Ludvigh & Miller, 1958). However, very large objects may be detectable up to 3000°/s provided the temporal frequency of the stimuli falls inside the critical flicker-fusion limit (Burr & Ross, 1982). There is a belief that, at moderate velocities, displacements are perceived as movement, but at lower velocities, motion is based on detecting changes in position (Anstis, 1980; Exner, 1875). Some suggest that with brief durations (<0.25 s), only velocity is sensed and only at long durations would velocity thresholds be improved by a framework (Gregory, 1964; Liebowitz, 1955). Velocity thresholds increase in the periphery (Aubert, 1886). There is also a common misconception that the peripheral retina is better at motion detection than the central retina. On the contrary, Tynan and Sekuler (1982) reported minimum movement thresholds to be on the order of 0.03°/s in the fovea, increasing to 0.45°/s at 30° eccentricity, and they become progressively poorer toward the outer limits of the visual field (To, Regan, Wood, & Mollon, 2011).

When two objects are moved in the visual field, the difference threshold for velocity has been estimated to be in the range of 30 arcsec/s (Graham, Baker, Hecht, & Lloyd, 1948) to 2 min/s (Aubert, 1886). But with two objects, induced movement and motion contrast effects come into play. Induced movement is an illusion of motion in which the displacement of some elements results in the attribution of movement in other elements. For example, moving a frame around a stationary object often results in an apparent movement of the object in the direction opposite to frame movement. Motion contrast is a type of induced movement that occurs at the borders of adjacent fields of moving elements. Adjacent elements moving in different directions and/or at different velocities can induce illusory directional and velocity attributes (Levi & Schor, 1984; Whitney, 2002).

With the advent of modern display systems and greater computational power, these findings have now been extended to stimuli covering larger areas of the visual field and more natural scenes. The principles remain, but for comprehensive reviews of recent work, see Nishida (2011) and Burr and Thompson (2011).

3.3.4 EGOCENTRIC MOTION

In most cases, motion in the visual field due to one's own movement is easily discriminated from motion in the visual field due to object motion. The mechanisms that underlie this ability involve a comparison between the movement sensed on the retina and nonvisual signals associated with one's own movements. When movement of an object on the retina is negatively correlated with eye, head, or body movement (later discussed), position constancy holds and objects are seen as stationary. Any mismatch between the visual and nonvisual information signals movement of the object relative to the observer. The limits of position constancy define the precision of these mechanisms. For head movements, position constancy breaks down with displacement ratios (retinal movement/head movement) of 1.5% (Wallach & Kravitz, 1965). While humans are quite accurate at detecting displacements during fixation (2–3 min), displacements as large as 6 min go undetected if they occur during a saccadic eye movement (Stark, Kong, Schwartz, Hendry, & Bridgeman, 1976). Target displacements of less than 1/3 saccadic displacements are rarely seen (Bridgeman, Hendry, & Stark, 1975). Indeed recent work has shown very substantial compression of perceptual space in a time window from approximately 50 ms prior to a saccade to 50–100 ms after the saccade (Burr & Morrone, 2005). Thus, the resolution of the position constancy mechanism during saccades is degraded relative to fixation. When a moving target is tracked with smooth-pursuit eye movements, its perceived speed has been reported to be 63% of that reported when it is moved through the field when the eyes are fixated (Dichgans, Koener, & Voigt, 1969; Turano & Heidenreich, 1999). In addition, stationary objects in the field appear to move (the Filehne illusion) when the eyes track another moving object (Filehne, 1922). Wertheim (1994) argues that reference signals are generated through visual, oculomotor, and vestibular interactions, and it is the comparison of the retinal and reference signal that mediates the perception of self and object motion. He suggests that while oculomotor-generated signals encode how the eyes move in orbits, vestibular signals must be added to form these reference signals to encode how the eyes move in space, and the neural mechanisms supporting this have now been detected in human parietoinsular vestibular cortex (Cardin & Smith, 2010).

3.3.5 Visually Induced Self-Motion

As we move through the real world, either actively (such as walking) or passively (such as sitting on a moving train), multiple sensory systems are responsible for detecting, processing, and perceiving our self-motion. While nonvisual inputs (from the vestibular system of the inner ear, kinesthesia, somatosensation, and even audition—see Lawson et al. [2014; Chapter 7] and Hettinger [2014; Chapter 18]) often contribute significantly to the conscious perception of this self-motion, visual input appears to play a particularly important role. This is clearly demonstrated by the occurrence of visually induced illusions of self-motion, known as vection (Fischer & Kornmüller, 1930). As far back as the time of Helmholtz (1867/1925), it was known that compelling vection could be induced by viewing a quickly moving river from a bridge. Passengers sitting on a stationary train often experience a similar illusion of self-motion when the train on the next track pulls out from the station (Mach, 1875). This vection is a potentially useful tool for enhancing the user experience in many virtual reality applications, such as vehicle simulation, telepresence, and architecture walkthroughs.

Early laboratory research confirmed that large optic flow patterns can generate robust translational and rotational illusions of self-motion in physically stationary observers (Brandt, Dichgans, & Koenig, 1973; Johansson, 1977; Mach, 1875; Tschermak, 1931). In one study, Mach (1922) presented subjects with a large, endless belt (covered with an alternating black-and-white stripe pattern) moving horizontally across two rollers. This induced linear vection (LV) in the opposite direction to the belt's motion. In another study, Mach (1875) placed subjects inside a large rotating drum and found that this induced circular vection (CV) in the opposite direction to the drum's motion. Later research showed that this CV can be subjectively indistinguishable from real self-rotation (e.g., produced by rotating the subject's chair—Dichgans & Brandt, 1978). In fact, tilting one's head during CV produces powerful *pseudo*-Coriolis effects (which include illusions of apparent tilt, dizziness, disorientation, drowsiness, and nausea—Brandt, Wist, & Dichgans, 1971). Interestingly, these symptoms occur even though the complex (Coriolis cross-coupled) vestibular stimulation typically thought to be responsible for them is absent.

Vection is generally perceived in the opposite direction to the motion of the visual background (Ito & Shibata, 2005), except in relatively rare cases of inverted vection where foreground motion dominates the experience (Nakamura & Shimojo, 2000). Typically, there is a 3–12 s delay between the initial exposure to optic flow and vection onset, during which the observer perceives object/scene motion (Brandt et al., 1973; Bubka, Bonato, & Palmisano, 2008; Telford & Frost, 1993). This delay was thought to reflect the time required for the stationary observer to resolve the following sensory conflict: his/her visual input indicates self-motion, but the expected nonvisual inputs for this self-motion are absent (Johnson, Sunahara, & Landolt, 1999; Zacharias & Young, 1981). According to this type of explanation, LV and CV should be reduced/enhanced by maximizing/minimizing this sensory conflict, respectively. However, upright observers still perceive compelling head-over-heels CV when placed inside a fully furnished tumbling room, despite the extreme conflicts between their visual and nonvisual inputs (with the former indicating head-over-heels self-rotation and the latter indicating that the observer is physically stationary—Allison, Howard, & Zacher, 1999; Palmisano, Allison, & Howard, 2006). Another challenge to this explanation has been posed by findings that self-motion displays containing simulated viewpoint jitter/oscillation (thought to generate significant and sustained sensory conflicts) actually induce vection more quickly and intensely than comparable nonjittering self-motion displays (thought to generate only minimal/transient sensory conflict—see Kim, Palmisano, & Bonato, 2012; Palmisano, Allison, Kim, & Bonato, 2011; Palmisano, Gillam, & Blackburn, 2000; Palmisano et al., 2008).

Over the years, parametric research has provided insights into how self-motion might be best simulated with purely visual displays. These studies have shown that LV and CV will generally be more compelling when the optic flow (1) has a faster speed of translation/rotation (up to an optimal velocity; e.g., yaw CV speed is proportional to the stimulus speed up to 90°/s—Allison et al., 1999; Brandt et al., 1973), (2) has more and larger moving elements (e.g., Reason et al., 1982), (3) stimulates a larger retinal area (e.g., Held et al., 1975), and (4) is perceived to be from the background (as opposed to the foreground—Nakamura & Shimojo, 1999; Ohmi & Howard, 1988; Ohmi, Howard, & Landolt, 1987; Seno, Ito, & Sunaga, 2009).

While large fields of view and peripheral motion stimulation are often desirable when simulating self-motion scenarios, they are not necessary to induce vection; for example, Andersen and Braunstein (1985) showed that LV can be induced with as little as 7.5° of central motion stimulation. Both central and peripheral motion stimulation can induce compelling vection (Post, 1988). However, the optimal stimulus spatial frequency for vection appears to depend on the retinal eccentricity of the motion stimulation. Palmisano and Gillam (1998) found that CV was more compelling when higher-spatial-frequency motion stimulation was presented to central vision (where acuity is higher) and low-spatial-frequency motion stimulation was presented peripherally (where the eye's spatial resolution is also lower). Such findings relax the requirement for high-resolution imagery in peripheral display regions (unless the simulation requires the user to look/attend to these regions).

Stereoscopic presentation of the vection-inducing stimulus has also been shown to facilitate CV and LV (Lowther & Ware, 1996; Palmisano, 1996). Consistent stereoscopic cues not only reduce vection onsets and increase vection durations, but they also increase vection speeds and the perceived distances traveled (Palmisano, 2002; Zikovitz, Jenkin, & Harris, 2001). With stereoscopic presentation/display technology becoming increasingly affordable and available, stereoscopic cues can now be used to enhance many self-motion simulations (particularly where the simulated environment is by necessity sparse/basic).

More recent research has also provided evidence that color information plays a role in vection. Bonato and Bubka (2006) found that a chromatic grating (consisting of blue, green, red, yellow, black, and white stripes) is more effective at inducing vection than a gray-shade version of this same grating (consisting of black, white, and four different shades of gray stripes; closely matched in luminance to the stripes in the chromatic condition). However, color manipulations should be used with caution, as research suggests that the use of red backgrounds and perceived changes in scene illumination may inhibit vection (Nakamura, Seno, Sunaga, & Ito, 2010; Seno, Sunaga, & Ito, 2010).

Interestingly, there now appears to be mounting evidence that vection is determined simply not only by physical display parameters but also by how we look at, attend to, perceive, and interpret the optic flow:

1. *Gaze effects*: Palmisano and Kim (2009) have shown that peripheral looking and gaze shifting both significantly improve the LV produced by displays simulating constant-velocity forward self-motion or low frequency self-acceleration (relative to stable central gaze conditions). Displays simulating sideways or rotary self-motions often generate optokinetic nystagmus (OKN—see Section 3.4.1.2), regular eye-movement patterns in which smooth pursuits in one direction are interleaved with rapid return flicks. Suppressing this OKN by fixating a stationary target appears to reduce CV onset and increase CV speed (Becker, Raab, & Jürgens, 2002).

2. *Attentional effects*: Vection appears to be dominated by unattended (as opposed to attended) display motion and is weakened when attentional resources are divided by increasing the task load (Kitazaki & Sato, 2003; Seno, Ito, & Sunaga, 2011a).

3. *Role of perception/interpretation*: In one such study, Seno and his colleagues (2009) used a motion-defined Rubin's vase as a vection stimulus, where the subject's figure/ground perceptions alternated between seeing either two faces (looking at each other) or a vase. Irrespective of which motion region was seen to be the figure, they found vection was always dominated by the motion of the perceived ground.

Recently, other higher-level factors have also been shown to be important in vection induction. As is demonstrated by many theme park rides, the potential for physical observer motion

(Riecke, 2009; Wright, DiZio, & Lackner, 2006), the naturalism (Riecke, Schulte-Pelkum, Avraamides, Heyde, & Bülthoff, 2006), and ecological validity of the display (Bubka & Bonato, 2010) also appear to be important for vection. For example, Bubka and Bonato reported that first-person video taken with a handheld camera when walking through a corridor induces faster vection onsets and longer vection durations than video shot from a rolling cart. They also found that full-color versions of these videos were more effective at inducing vection than their grayscale versions. In another study, Riecke et al. (2006) generated yaw CV using a rotating VE created from either a naturalistic roundshot photo of a city scene or scrambled/inverted versions of this same photograph. Based on the physical parameters of each of these displays (e.g., the numbers of high-contrast edges), the scrambled controls were predicted to produce the best vection. Instead, the naturalistic self-motion stimulus produced superior CV to both the scrambled and inverted controls.

3.4 HOW DO WE LOOK AT AND MANIPULATE OBJECTS? (MOVEMENT OF THE OBSERVER)

Interaction with the real world is rarely passive but instead involves purposive movement of the eyes, head, arms, legs, and body. The way in which an individual accomplishes this involves complex sensory–motor systems that are characterized by reflexive and volitional components. These processes must be considered when simulating the environment, especially when attempting to incorporate simulation of the results of bodily movement within a VE. Some of the systems involved in visual inspection of the world and their implications for man–machine interface are now considered.

3.4.1 EYE MOVEMENTS

There are a variety of eye movements (for a detailed review, see Schutz, Braun, & Gegenfurtner, 2011), and they may be described generally as being conjunctive or disjunctive. Conjunctive eye movements involve the eyes moving in the same direction (OKN, smooth pursuit, and saccades), whereas disjunctive eye movements involve the eyes moving in opposite directions (convergence and divergence). A major consideration in ocular control concerns the ability to control the eyes when examining objects. While this is often done while the head and object are stationary (fixation), it is also done while the object moves (smooth pursuit) or when the head moves (see the vestibulo-ocular reflex—VOR—Lawson et al., 2013; Chapter 7). Visual capabilities are quite different during these oculomotor behaviors.

3.4.1.1 Fixation

When observers look at objects in the environment, they position their eyes so that the object of interest falls on the area of the retina that has the best visual acuity (the central 5° around the optic axis of the eye termed the macula). When this is achieved, the observer is said to have fixated the object. With a stationary target, fixation stability may be defined as the degree to which the eye is stable with reference to some fixation point on the object. Fixation stability is surprisingly poor. The eye may drift as much as a degree without the observer being aware of the drift in fixation. Even with well-controlled fixation, the eye is in constant motion, with microsaccades (movements) of many arcsec (Cherici, Kuang, Poletti, & Rucci, 2012; Riggs, Armington, & Ratliff, 1954). These ocular tremors are necessary for continuous viewing of an object. Artificial conditions that render the stimulus stabilized on the retina result in disappearance of the stimulus (Kelly & Burbeck, 1980). Attempts at maintaining fixation during image movement (see smooth pursuit, Section 3.4.1.3) or head movement involve interactive inputs to and from the vestibular system.

3.4.1.2 Optokinetic Nystagmus

If a contoured visual field is rotated before a stationary observer, OKN is elicited. The eyes drift in the direction of field rotation and then snap back in the opposite direction, and this cyclic pattern of eye movements is repeated in bursts, interrupted by short periods of relative gaze stability.

The initial tracking response (slow phase) is seen as a period of smooth pursuit and the compensatory snap back (fast phase) as a corrective saccade (see Section 3.4.1.4). The existence of this reflex is seen as evidence that there are visual mechanisms that take into account the effect of movement of stationary aspects of the visual field on the retina as the head or the eyes move and that these mechanisms can provide, together with the VOR, additional complementary information about how the eye should move during head movements to provide gaze stability. Unlike the VOR, this response does not abate with continuous stimulation. While the VOR is thought to adapt because the vestibular end organs cease to respond at constant velocities, OKN is seen as a compensatory response for this failing. Another difference between the two responses is that the VOR has a short latency and a slow decay, while OKN has a long latency and slow buildup. While OKN is seen as a reflex, it also can be suppressed with fixation of a stationary target.

3.4.1.3 Smooth Pursuit

Observers are quite capable of tracking moving objects with the head held stationary. This is generally considered a voluntary response engaging the smooth-pursuit system, which must calculate the speed of moving objects to maintain fixation (Turano & Heidenreich, 1999). This system requires a moving object. Smooth-pursuit eye movements cannot be made in total darkness or in a visual field devoid of movement. It is assumed that this system requires volition, because it is also possible to fixate a point when other moving objects are present. The upper limit of this response is about 100°/s.

3.4.1.4 Saccades

Another response, which has generally been considered voluntary, is the saccade. This is a ballistic movement between one fixation and another. Its latency ranges from about 150 to 200 ms, and its velocity is proportional (within limits) to the distance the eye must be moved. Saccades can occur with speeds up to 900°/s. Whereas saccades usually are made from one target to another in visual space, saccades can be executed with high degree of precision to spatial positions defined cognitively (signals in other modalities, verbal commands, memories of spatial locations). Whereas this suggests voluntary control, aspects of saccades (direction, velocity, or amplitude) cannot be changed after they are initiated. However, saccadic latency can be shortened significantly if a presaccadic fixation target is removed shortly before the saccadic target is presented, indicating that disengagement from one target is necessary prior to a saccade (Abrams, Oonk, & Pratt, 1998; Forbes & Klein, 1996). Another characteristic of vision during saccades is that the visibility of achromatic, but not chromatic, stimuli is suppressed throughout the eye movement (Burr, Morrone, & Ross, 1994). Chromatic stimuli are usually not perceived, however, because the retinal velocities exceed the resolution limits of the chromatic pathways.

3.4.1.5 Vergence/Accommodation

All of the eye movements so far considered are conjugate, but vergence movements are also necessary to focus on objects in depth. Accommodation of the crystalline lens is linked to vergence in normal observers, and changes in the refractive power of the eyes are correlated such that objects are maintained in sharp focus and registered with retinal disparities on the order of min of arc. As noted in Section 3.2.3.3, these accommodative and vergence-based responses are not only important for focusing, they may also be important for the perception of the scene's 3-D layout.

3.4.2 Head and Body Movements

One of the most important aspects of contemporary virtual simulations is the ability to provide shifts in the scene contingent on head, hand, or body movement. Position tracking technology provides a major contribution to sense of presence (i.e., the subjective experience of being in the VE—see Chertoff & Schatz, 2013; Chapter 34). The aim of the present section is to review the implications of the visual forcing functions on head and body movement that might be expected to influence the ability to provide realistic scene translations. However, it should be recognized that eye-movement-based adjustments are also needed to fully control scene properties for the impact of observer motions that impact on the visual field.

Postural stability is maintained through the vestibular reflexes acting on the neck and limbs. These reflexes are under the control of three classes of sensory inputs: muscle proprioceptors, vestibular receptors, and visual inputs. As previously discussed, a comparison of the vestibular and visual inputs is necessary to determine if the observer or the environment is moving. Visual inputs that give rise to vection have been shown to result in bodily sway and even falling in young children (Lee, 1980). It is reasonable to expect that such stimulation may result in reflexive neck muscle activity that can lead to head movements. When all or a large part of the visual field is filled with moving contours, standing observers lean in the direction of scene motion (Lee & Lishman, 1975; Lestienne, Soechting, & Berthoz, 1977). There is a compensatory leaning in the opposite direction upon the cessation of stimulation that may last for many seconds (Reason, Wagner, & Dewhurst, 1981). If the virtual scene is driven by head movement, such unintentional reflex head and body movements (and the aftereffects thereof, see Wann, White, Wilkie, Culmer, Lodge, & Mon-Williams, 2014 and DiZio & Lackner, 2013; Chapters 32 and 33) may serve to impede or interfere with scene shifts intended to simulate voluntary attempts at navigation through the VE. In some cases, reflex movements can be suppressed, but this may imply that some learning must occur to achieve efficient and unnatural adjustments that are necessary to facilitate optimal performance.

3.5 HOW DO WE DEPICT THE THREE-DIMENSIONAL WORLD IN TWO-DIMENSIONAL DISPLAYS? (HARDWARE REQUIREMENTS)

Modern VE technology is an extension of all past endeavors to depict renditions of the visual world. There has been an orderly progression in the mastery of monocular cues for depth in 2-D static displays to the movement and disparity cues for depth that are so compelling in dynamic displays. The advent of computer-controlled displays has provided the possibility of greater user interaction with these virtual depictions, and with it comes an illusory sense of presence (see Chertoff & Schatz, 2013; Chapter 34) and some adaptational problems for the user (Welsh & Mohler, 2013; Chapter 25).

3.5.1 STATIC TWO-DIMENSIONAL DISPLAYS

A first approximation to simulating the 3-D world is the traditional 2-D display ubiquitously embodied in drawings, paintings, and photographs. What can be provided relative to the capabilities of the human visual system? Two-dimensional displays can provide fine spatial detail, variations in contrast, differences in hue, the spatial position of objects in the frontal plane, and indications of 3-D using the pictorial cues for depth (relative size, height in the field, occlusion, shading, texture, and linear and aerial perspectives). Using a single image on paper, canvas, or film, one cannot provide the stimulus conditions to provoke motion parallax, kinetic depth, retinal disparity, or, normally, motion (although Akiyoshi Kitaoka provides many examples where illusory motion, induced by oculomotor instability, is perceived in static images. See www.ritsumei.ac.jp/~akitaoka/index-e.html). The FOV is usually restricted and viewed within a 3-D framework provided by the rest of the real world. Some have argued that because of human sensitivity to the actual flatness of the 2-D plane on which 3-D simulations are viewed, the pictorial cues are in conflict with other cues provided simultaneously in the 3-D world and hence are rendered less effective (Ames, 1925; Nakayama, Shimojo, & Silverman, 1989).

As previously noted, in real 3-D views, the ability to focus and converge on different depth planes may provide reafferent information for depth judgments, but these are missing in a depiction of 3-D in only one depth plane. Can these other important visual capabilities be recruited with other manipulations of static 2-D displays? This is possible only if stereoscopic viewing conditions are provided—presenting two versions of the same scene with fusible retinal disparities. This is done in various ways. The first stereo displays were achieved with stereoscopes that allowed viewing of two disparate pictures superimposed by means of prismatic lenses. The two pictures are frequently obtained by simultaneous photography of the same scene with two cameras in slightly different positions and then presented to the appropriate eyes. An anaglyphic approach renders the two members of the stereo pair in different wavelengths (e.g., red and green) and lets the observer view the two disparate scenes with glasses that segregate the visual images with different chromatic filters in front of the eyes. This separation is rarely perfect, with unwanted ghost images appearing in each eye's view that were meant for the opposite eye—referred to as *crosstalk*. However, typically the resulting *left* and *right* eye images can still be fused to provide rather striking depth information and limited motion parallax. Unfortunately, one loses the ability to render color appropriately with this technique (binocular color rivalry is also introduced). A similar technique involves the use of polarizing filters. Two disparate scenes are presented through orthogonally polarized filters. Each eye of the observer views through a polarizer matched to only one of the scenes. The visual system combines the two views and is able to detect the disparity cues. This method allows more natural color rendition; however, it also suffers from crosstalk between the left and right eyes' images (which increases dramatically as the observer rotates his/her head away from vertical). The stereoscopic images so popular in the comics section of Sunday papers use an anaglyphic approach, which does not require chromatic glasses, but require the observer to converge or diverge the eyes to achieve fusion. The motivation to attempt such viewing is to find the hidden figure not visible in the monocular or nonfused binocular view. More recently, with the advent of new 3-D display technology, there has been a revival of interest in displays using either parallax barriers or lenticular prisms. Images are created by alternating left and right eye views in columns of pixels that, after printing, are overlaid with a sheet that either directs the alternate columns of the images to left and right eyes or provides a barrier that alternately blocks the left or right eye view for the appropriate columns. With these displays, disparities are rendered to the two eyes to create compelling 3-D scenes without the need for additional glasses, but in most cases, they must be viewed from a relatively specific angle. Outside these angles, double images may be detectable. In this sense, they are more restrictive than the systems using filter lenses to direct the separate views to each eye.

Modern televisions also support 3-D displays. They may also require glasses, employing a crossed polarization method or more frequently frame interleaving. The polarization method is as described earlier, but frame interleaving requires temporally alternating frames that display left and right eye views, synchronized with glasses that alternately allow only the left or right eye to view the scene. Usually, this is achieved using liquid-crystal display (LCD) technology and is therefore quite slow, but provided frame alternation rates exceed 120 Hz, each eye receives a 60 Hz image sequence and relatively flicker-free rendition of 3-D scenes is achieved. Again, crosstalk is common with these shutter glasses. The polarization reduces the luminance of the images but natural colors can be rendered.

Head-mounted displays (HMDs) can also render 3-D scenes. They provide a separate screen for each eye that can be readily fused to create the retinal disparities needed to perceive depth. This method allows more natural color and perfect separation between the left and right eye images. It does not require frame alternation but usually employs LCD screens with quite slow refresh rates (approximately 60 Hz). While the gain in level of quality of 3-D simulation provided by stereo viewing is impressive, the static 2-D display lacks the vividness that scene motion conveys.

3.5.2 Dynamic Two-Dimensional Displays

In the early part of the last century, considerable amusement was provided to those fortunate enough to own a stereoscope. The excitement generated by such devices paled in comparison to the first moving pictures, however. Movement in nature is an extremely compelling sensation, providing immediate impressions of one's position relative to other objects in the world. The simulation of movement in static displays was first accomplished with a deck of cards, wherein an ordered sequence of pictures denoting differences in scene position was presented in rapid succession by sleight of hand. The deck was bent and individual cards were allowed to slip off the thumb to reveal a simulation of motion. The motion picture projector was developed using the same rudimentary principle, rapidly sequencing a series of still pictures. These pictures were captured on filmstrips with cameras capable of rapid photography. The current standard for film projection is 24 frames/s, and each frame is transilluminated for about 30 ms. A rotary shutter blade is interposed while the film is advanced (taking about 10 ms), yielding a frame rate of about 24 frames/s. This simple arrangement provides a series of still pictures but also produces considerable detectable flicker. By increasing the rate of flicker (to 50–60 Hz), the achromatic critical flicker-fusion limit for the human observer is surpassed and an acceptable sensation of smooth movement is obtained (Cropper & Badcock, 1994).

While these developments added realistic movement to simulations of the real world, stereo-scopic and motion parallax cues were still not available. In the 1950s, the first 3-D movies were screened, and this technology combined the anaglyphic technique for stereo vision with the cin-ematic technique for motion, providing quite compelling perception of motion in depth and some limited motion parallax information. In the last decade, 3-D television has become readily available. Displays using either anaglyphic techniques with passive polarizing glasses, frame alternation with active LCD glasses, or lenticular screens provide ready access to moving 3-D images sequences in quite natural color. Modern cinematography has become an ubiquitous source of entertainment and education in this century, but it does not approach a true simulation of our normal experiences with the 3-D world. In most cases, the observer is quite passive and relatively immobile, with the 3-D point of focus specified by the cinematographer rather than the viewer. That is also the case with standard video simulations, which, although technically more advanced electronically, suf-fer from the same inadequacies. Video, however, affords the possibility to interface with sophisti-cated computer systems that offer the possibility to incorporate user movements into the simulated environment.

3.5.3 ELECTRONIC DISPLAYS AND THE USER INTERFACE

The cathode ray tube (CRT) began as a method of rendering simple electronic activity immediately visible. For example, oscilloscopes are often used to display voltage changes over time. The early versions involved an electron *gun* aimed at a phosphorescent surface. Deflection plates or coils controlled the direction of the gun, and the intensity of fluorescence was varied via accelerating plates that determined the rate of electron flow. Originally, control of the gun was accomplished with *vector scanning* that addressed only the points (pixels) on the screen necessary to *draw* the image in question. This proved adequate for simple geometric shapes, but more complex scenes required the more elaborate *raster scanning* approach wherein every pixel was addressed. In raster scanning, the stream of electrons is swept across the screen horizontally from left to right and then snapped back and down one line in order to paint the next row. This procedure is repeated until the entire screen has been painted and the gun is turned off (blanked) during retrace (snap back). One complete painting of the screen is termed a frame, and frame rates above the human flicker-fusion point (50–60 Hz) result in a flicker-free sequence of pictures. This technology provided the basis for black-and-white television. This relatively simple system was expanded to provide color television by including three independent guns that selectively stimulate colored (red, green, and blue) phosphors laid down as triplets on the screen. One problem with CRT displays is that the tube length increases as the tube width and height increase, resulting in large displays that are quite bulky. For this reason, they have fallen out of favor with television and computer monitor manufacturers in spite of advances in display technology that reduced the size of the apparatus and increased the spatial and temporal resolution. The last decade has seen these displays largely replaced by liquid-crystal, plasma, and electroluminescent digital displays that are far less bulky but have disadvantages associated with resolution, color, or expense; these shortcomings are rapidly being reduced. The switch from the analog technology used in CRTs to digital displays is currently associated with a much poorer ability to render luminous intensity. Advanced graphics systems are available with digital-to-analog converters (DACs) capable of 14-bit intensity resolution (16,384 levels) on each gun that are all available with CRTs, but LCD panels are currently 10 bits (1024) or more frequently 8 (256 levels) and finer gradations of their input signals will not alter the step size between levels (Wang & Nikolic, 2011). The discussion of relevant parameters for meeting user specifications will be undertaken with regard to CRT displays, but the principles will apply to all display technologies, and promising new technologies will be noted in the discussion. The parameters of importance for meeting the user requirements are provided in Table 3.1.

3.5.3.1 Spatial Parameters

The FOV depends on viewing distance (FOV = atan[screen size/viewing distance]), but at a fixed distance, the FOV for modern displays varies widely, from the tiny screens produced for miniature television to the panoramic screens employed in IMAX (Image MAXimum) and movie theaters. The small screens have been incorporated into HMDs, with optics designed to allow sharp focus of images within the near point of vision and to provide a large FOV. Video projection systems can fill the human FOV but often present problems of spatial resolution, require sophisticated warping software to prevent image distortion, and are limited to relatively low luminance levels (although this is rapidly improving). CRT-based HMDs have a limited FOV (<80°) and, essentially now replaced by lighter LCD displays, offer somewhat larger fields of view (105°) and simpler spatial image specification. Although the peripheral retina has poor spatial resolution, wide field of view provides valuable information for visually guided behavior and for producing the illusion of vection (see Hettinger, 2014; Chapter 18).

The spatial resolution of most displays does not approach the limits of human visual acuity (approximately 1 arcmin/pixel) or spatial positioning ability (interpolated displacements of 5–30 arcsec) unless viewing distance is substantial (frequently several meters), which, of course, reduces FOV. This trade-off between FOV and spatial resolution is an important limitation for visual simulation, but there has been steady improvement in spatial resolution over time with all technologies and this is likely to continue. The major cost of less than optimal resolution is a sacrifice of texture and the sudden appearance of objects at simulated distance. This latter characteristic is potentially very serious. Castiello, Badcock, and Bennett (1999) have shown that when observers are reaching to grasp an object, a suddenly illuminated distractor object changes the motor aspects of reaching behavior. This change only occurs if the illumination is sudden; gradual onset has no effect, and if a spotlight suddenly illuminates the peripheral scene but there is no object, the motor movement is not affected. Thus, in VE setups, suddenly occurring peripheral objects would be expected to be the most disruptive of all possible object appearance modes if an observer is trying to make motor movements in the setup. Higher-resolution renditions are needed to overcome this problem.

A second critical aspect with regard to spatial resolution is the uniformity and stability of the pixel matrix over time. In this regard, LCD and the newer organic light emitting diode (OLED) displays are superior to CRTs. The latter use magnetic deflection to spread the image across the screen and this can be influenced by local magnetic effects (e.g., northern vs. southern hemisphere or magnets placed nearby), whereas the LCD and OLED matrices have a physical basis and do not move. CRTs may also perform differently when cold, such as when first turned on (Metha et al., 1993).

To match or exceed the contrast sensitivity of the human visual system, the luminance steps (gray scale) must be small enough to provide differences in luminance to which humans are insensitive. For many natural scenes, 24-bit (8 bits on each of the red, green, and blue guns) graphics systems are adequate. However, if fine contrast discrimination is required, most observers will require either higher-contrast resolution or that the full range of steps be compressed into the near-threshold range of contrasts being presented. The latter has the consequence of restricting the maximum contrast

TABLE 3.1

Relevant Display Factors Relating to Human Visual Abilities

Dimension of Vision	Critical Display Parameters[a]		
Spatial vision	Pixel uniformity: OLED and LCD displays have structured pixel arrays. CRT can be distorted by magnetic interference. This is important for displays that require precise geometry.	Pixel size (screen size ÷ number): finer spatial resolution allows greater positional precision and more accurate shape rendition, particularly for sharp-edged figures that cannot be interpolated across pixels.[b] Optimal positional thresholds can subtend a few seconds of arc. If sharp-edged stimuli are employed, the pixels need to match this resolution. Interpolation allows finer positional adjustment and a tolerance for lower display resolution.	Luminance uniformity: OLED/LCD are more uniform than CRTs, which are brightest in the center. This means object intensity and hue can vary with location when using CRTs, unless corrections are applied when generating the stimuli.	Intensity resolution (bit depth): more bits allow finer control of contrast and larger range and support better antialiasing for rendering sharp edges at small angles to pixel array orientation. CRTs are superior because of analog control. Graphics card sets the limit (best 14 bits, 16,384 levels, per R, G, and B guns). Most LCD and OLED are limited to 8-bit (256 levels) intensity per R, G, or B channel.
Color vision	Display type: OLED has much larger gamut than typical CRT or LCD (Figure 3.11), which allows a bigger range of colors to be represented veridically. The gamut is, however, still smaller than that of normal human vision.	Phosphor types: phosphors, LCD channels, and OLEDs may vary in their exact location in CIE space and thus alter the available gamut (Figure 3.11). It is therefore important to ensure that the desired colors can be represented by the display system.	Gun independence and stability: CRT gun interactions complicate calibration of intensity ranges. High-end CRTs ensure minimal interactions when intensities are high for one gun and low for the others but can require extensive warm-up periods. OLEDs should also be excellent here.	Intensity resolution (bit depth): 8 bits/R, G, or B channel is insufficient for measuring chromatic thresholds unless the intensity range covered by the 256 levels is compressed (only possible for analog displays). If a large color range is required simultaneously with small changes in color, then systems should be chosen with as many bits/gun as possible.

(continued)

TABLE 3.1 (continued)
Relevant Display Factors Relating to Human Visual Abilities

Dimension of Vision	Critical Display Parameters[a]			
Image motion	Display type: LCDs have slow response rates making them poor for displaying moving images veridically; CRTs and OLEDs can both provide the rapid responses needed.	Raster rate and phosphor decay: smooth slow motion requires fine temporal resolution with little temporal smear. LCDs are unsuitable for this purpose with low refresh rates and slow responses. CRTs with >120 Hz refresh rates are preferable.	Pixel size and spacing: moving images are more realistically represented when both small spatial and temporal displacements can be produced. This is particularly important for representing very slow motion. However, as noted in Footnote b, interpolation of smoothly varying luminance profiles can give subpixel precision. The motion of sharp-edged stimuli moving slowly will be quantized by the pixel size.	Intensity resolution (bit depth): the interpolation procedure described in Footnote b is limited by intensity resolution. Very slow motion may require small intensity changes for each pixel. Eight-bit systems are often sufficient for motion of natural scenes, but to measure the limits of motion sensitivity requires finer adjustment of intensity in many cases. Sharp edges can only be moved from one pixel column to the next and are not directly affected by intensity resolution.
Stereopsis	Interlaced frames: flicker-free viewing requires >60 Hz for each eye or >120 Hz if using a single monitor. CRTs can achieve this. Anaglyph methods: >60 Hz sufficient.	Monocular monitors: haploscopic systems using either a different monitor for each eye or different half screens present challenges for matching of intensity ranges. CRT half-screen systems will have intensity gradients falling away from the screen center in opposite directions. Contrast differences can produce distortions in perceived disparity.	Frame disparity: in all displays, large disparities will produce diplopia. Small ranges of disparities[c] within a scene are desirable to avoid this issue.	Intensity resolution (bit depth): as discussed previously, rendering of fine spatial changes of location for the right and left eye image components to produce small disparity differences may require interpolation (or resampling) of the image for a new position relative to the pixel matrix. Fine luminance resolution allows for smaller image changes and more precise interpolation.

Vection (visually simulated self-motion induced in both active and passive observers)	Scene update rate (lag): display lags up to 160 ms impair vection and generate simulator sickness. With longer lags, the visual system appears to override or downplay the visual–vestibular conflict produced by the observer's head motion.	FOV: while it is possible to induce vection with display sizes as small as 7.5° of visual angle, larger displays are required for compelling vection experiences. Recommended display sizes for vection are between 60° and 120° of visual angle.	Frame rate: LCD, DLP[d], OLED, and CRT projectors/displays are all capable of inducing compelling vection. Vection can be induced by displays with refresh rates less than 30 Hz.	Intensity resolution (bit depth): limitations in spatiotemporal resolution generate *jaggies* (or pixel creep) in computer-generated self-motion displays. These artifacts are problematic when vection displays are presented only to the observer's central visual field (where acuity is the highest) and simulate self-motion with respect to a 3-D (as opposed to a 2-D) environment.

[a] This table lists the parameters that are critical for a variety of different tasks. At this point, no display technology is optimal for all applications. The intention here is to indicate what aspects of performance may be limited by the parameters so that the reader can choose the optimal display system for the tasks they intend to implement. In most cases, measuring human performance limits on a particular dimension will require choosing a display that offers the best available resolution on that dimension. Given that is the case, we have not tried to indicate preferred choices of display technology.

[b] The spatial resolution of the pixel array is more critical for sharp-edged stimuli than those with gradual luminance variation at their contours. Sharp edges can only be moved from one pixel location to the next, and thus the pixel size sets a limit on the minimum change but smoothly changing luminance gradients can be resampled at different locations on the pixel array, allowing for positional changes that are smaller than the pixel size, provided the intensity resolution is high enough to adequately sample the luminance waveform. This interpolation procedure is applicable for a wider array of natural images.

[c] There are significant individual differences in sensitivity to diplopia, for example, reported diplopia thresholds have ranged from 2 to 20 arcmin for horizontal disparity in the fovea (Duwaer & Van Den Brink, 1981).

[d] DLP (digital light projector) is a digital micromirror technology produced in two forms. Both are very fast and capable of bright, high-resolution displays with up to 10-bit (1024 levels) intensity resolution per color. However, one method achieves separate color channels by rotating a colored filter through R, G, and B sectors in the otherwise white light path. This presents R, G, and B components of images at separate points in time, and if an observer makes an eye movement during the presentation, objects are seen as different colored replicas smeared across the scene. The second method uses a prismatic process that presents all three color signals simultaneously and should be preferred.

range available at any instant but does allow small contrast steps to be presented. Computers that boast 16-bit DACs for each gun are preferable, but expensive and uncommon. Many commercially available monitors provide adequate luminance (up to 100 cd/m²) and support photopic viewing levels. Thus, the current limitation in intensity resolution in CRTs is usually provided by the DACs employed. However, LCDs and OLED displays are digital, and step sizes in intensity for each R, G, and B component are fixed, as noted previously. Ten-bit (1024 levels, 1920 × 1080 pixels, 60–85 Hz, 100 cd/m²) OLED displays and 12-bit (4096 levels, 1920 × 1200 pixels, 120 Hz refresh, 250 cd/m²) LCD displays are already available.

3.5.3.2 Color

Many modern monitors are capable of providing high spatial resolution (1920 × 1200 pixels or more) and moderate temporal resolution (60–150 Hz frame rates) while also providing a colored image. The CRT monitors contain three electron guns, which are each aimed at a different phosphor type so that in near spatial proximity on the screen, red, green, and blue signals can all be produced, while LCD and OLED displays have individual elements that can be switched on or off to display the color at a particular location. There are a number of important monitor characteristics that are required for high-fidelity color rendition.

The range of achievable colors depends on the chromaticity coordinates of the phosphors used. However, current monitors are only capable of producing a small part of the full range of chromaticities detectable by the human visual system. Figure 3.11 depicts the full-color gamut specified in the CIE 1931 xy chromaticity space for normal human vision, with the more restricted gamuts available for different display systems marked as triangles. The spaces shown are EBU (European Broadcast Union) that corresponds to the PAL television standard, ITU-R (the High Definition Television [HDTV] standard), SMPTE-C (the current American broadcast standard, a variant of NTSC),

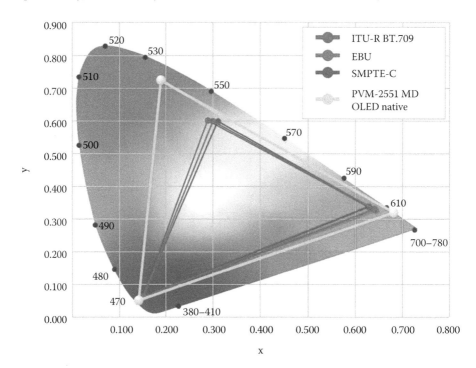

FIGURE 3.11 **(See color insert.)** The CIE (1931) xy chromaticity space for normal human vision, with a 2° field, is depicted as the colored region, with the gamut of color available with the typical RGB monitor broadcast standards depicted as triangles. The perceptual locus of narrowband spectral lights is indicated by numbers specifying their wavelengths. (Adapted from Sony PVM-2551MD Medical OLED monitor catalog, 2011.)

and PVM-2551 MD (an OLED native gamut used in the Sony Medical OLED monitors). The OLED display represents a significant improvement in the available color gamut.

Within the available gamut, accurate color rendition in CRTs requires high gun stability over time and gun independence, but both of these factors need to be rated against the intended use of the display. Gun stability is often a problem during the first hour a monitor is turned on. During this period, the luminance and chromaticity of a display will vary, even in very good monitors. Metha et al. (1993) provide some representative data and a method for evaluating the visual salience of these changes.

Gun interactions are more problematic. On less expensive CRT monitors, the intensity of the output from one gun will vary as the output of the other guns increases. The consequence is that there is not a fixed relationship between a given intensity level specified in the driving software and the output of a particular gun. The output will vary with different images. Observers are quite tolerant of these changes for noncritical applications (e.g., word processing), but the interactions are likely to be destructive when accurate color discriminations are required. Expensive monitors are more likely to avoid this problem, but the only way to be sure is to measure the light output for each gun independently and compare the output for several levels of activation of the other guns (see Metha et al., 1993). This check is critical as many manufacturers of CRT monitors deliberately build in a gun interaction that affects the upper half of the intensity range to minimize the amount of flashing when users move pointers from one window to another in graphical user interfaces.

The final issue of significance for rendering color is the variation in light output across the CRT monitor screen. Some variation is due to nonuniformity of the phosphor deposition during manufacture, but most of it is due to the design characteristics of the shadow mask technology. All such monitors are brightest in the center of the screen and luminance drops considerably toward the periphery. In part, this is due to the smaller effective aperture in the shadow mask at shallower angles of incidence, but the loss is also due to the polarity of the emitted light. The output is much more intense in the direction continuing the path of the electron beam, a direction that points to one side of an observer aligned with the center of the screen. Metha et al. (1993) found this loss to affect luminance more than chromaticity, but ideally stimuli requiring accurate luminance and chrominance discriminations should be small and placed centrally on a CRT screen to minimize spatial variations due to the monitor.

Fortunately, many of these issues are minimized in the OLED displays. The color gamut is larger, and because the pixels are separate elements illuminated in identical manner, screen uniformity is much higher. Stability over time is also claimed to be high but remains to be tested for particular displays.

3.5.3.3 Image Motion

Two factors that have a strong influence on simulated motion when using CRTs are phosphor persistence and frame rate. Phosphor persistence refers to how quickly the phosphorescence fades after it has been energized. Long-persistence phosphors reduce flicker but will also result in image smear when image movement is simulated. Short-persistence phosphors require higher update rates for pixels that convey the presence of static parts of the image, and if the rate is too low, flicker may be a problem. In practice, it is quite easy to determine whether the quantization of motion will be detectable. If the temporal frequency components introduced by the quantization fall within the range detectable by an observer, then the observer will be able to discriminate between smooth and sampled motion sequences (Cropper & Badcock, 1994; Watson, Ahumada, & Farrell, 1986). Other factors may be critical if either speed difference judgments or very slow speeds need to be produced. In both these cases, the achievable steps in speed will depend on a combination of the frame rate of the monitor and the spatial resolution. The quantization in both time and space limits the minimum speed and minimum difference in speed achievable.

The new display technologies differ in their suitability to display moving images. LCDs have slow temporal responses and therefore often produce artifacts, such as motion smear. Manufacturers have tried to remedy this limitation using deblurring algorithms but these alter the specified image sequence and can create detectable quantization artifacts. OLED displays, on the other hand, employ a very fast technology to illuminate a pixel and are therefore able to avoid motion-dependent spatial artifacts and maintain full image intensity for moving images.

3.5.3.4 Stereopsis

Stereoscopic vision is currently enjoying a boom and is supported by either providing separate monitors for each eye (e.g., HMDs and boom-mounted devices), shutter glasses, anaglyphic glasses, or projectors and autostereoscopic displays, including the new arrays of 3-D television systems. Dichoptic presentation using two monitors presents two views of the same scene, one to each eye, and depends on binocular fusion to yield stereopsis. Electronic shutter glasses present alternating views of the display synchronized to the frame rate such that one interleaved frame in each pair is presented to each eye. The display provided in each set of odd or even frames contains a disparity between the two eyes and binocular fusion provides stereopsis. Anaglyphic glasses filter the images provided to the two eyes with chromatic filters, usually red and green in older systems, and the display contains disparate images rendered in the two colors; modern systems usually employ orthogonally polarized filters with full-color images.

Autostereoscopic systems involve lenses imposed between the viewer and display or use a haploscopic arrangement in which dichoptic stimulation is achieved using mirrors to align a different display with each eye. While most would agree that stereoscopic vision is extremely important for the feeling of immersion and presence (see Chertoff & Schatz, 2013; Chapter 34) that the VE seeks to convey, all current techniques involve unwanted attributes. HMDs are bulky, have insufficient FOV, and can cause eyestrain if the optics create competition between the accommodative and vergence systems (see Wann, White, Wilkie, Culmer, Lodge, & Mon-Williams, 2014; Chapter 32). Electronic shutter glasses are less cumbersome, but are also restricted in FOV, suffer from crosstalk, and require higher frame rates to avoid flicker. With flicker-fusion thresholds near 60 Hz at current display luminances, a frame rate of 120 Hz or greater is required for flicker-free vision. The older anaglyphic techniques are compromised because color is used to separate the signals for the two eyes (often not perfectly) and cannot simultaneously render a true-color image. They are also limited by the restrictions of FOV imposed by glasses. Most autostereoscopic systems offer only small display sizes.

However, the modern large screen 3-D televisions (also described in Sections 3.5.1 and 3.5.2) offer improvements in FOV, provided high spatial resolution is not simultaneously required. Stereoscopic capability is desirable because it provides additional binocular depth cues through motion parallax and motion in depth, and this technology is likely to continue to improve given the current resurgence in manufacturer interest. Advances using lenticular displays allow for glasses-free viewing and provide a more natural viewer experience, in addition to providing unobstructed views of the observer that enhances the ability to readily monitor eye movements. This combined with OLED technology would provide a good combination of spatial resolution, color gamut, and temporal response for many research purposes and appears to be achievable with current technology.

3.6 WHAT ARE THE SUCCESSES AND FAILURES OF OUR ATTEMPTS TO SIMULATE THE *REAL WORLD*? (THE STATE OF THE ART AND BEYOND)

3.6.1 Successes

As observed above (3.5.3.1), the spatial aspects of visual simulation seem to be adequate for representing small regions of real-world scenes. HMDs have benefited in the last decade from reduced mass and improved spatial and temporal resolution with less peripheral distortion. Emerging technology promises even higher resolution and more encompassing FOV. As previously discussed, the depiction of color only approximates the limits of human abilities with new OLED technology promising a wider color gamut and more uniformity across displays. The temporal characteristics of most displays are adequate and might be expected to improve with the faster responses and higher intensities of OLEDs that are starting to become available. These improvements will support more realistic depiction of object motion and smooth scene translation. As simulation of motion profiles improves, more pronounced vection and a more realistic sense of presence can be expected. Improving the sense of presence, however, might produce more rather than fewer problems for users.

A significant, recent improvement has been the ability to incorporate eye-movement monitoring in HMDs. Without this, observer-generated changes in fixation within a scene could not be compensated for, precluding the representation of natural changes of the visual field under free viewing.

3.6.2 FAILURES

One of the most problematic areas in VE system design concerns the simulation of environmental motion relative to the observer as their head or body moves in the simulator. VE setups using head-coupled or head-slaved tracking systems all contain unavoidable display lag. *End-to-end* system lag refers to the time it takes to track the observer's physical movements and update them into the graphical display (typically by shifting the observer's virtual viewpoint). This lag depends on a number of factors, including the computational power of the stimulus generation computer, the speed of the particular tracking system, and often the required graphical detail of the scenes to be depicted. A decade ago, the best VE systems boasted 40 ms update rates but achieve this by sacrificing resolution; modern systems are substantially faster but are still unable to render a full field high-resolution display with rapid updates. In most systems, end-to-end system lag ranges between 40 and 250 ms (Moss, Muth, Tyrrell, & Stephens, 2010). The resulting display lag can have detrimental effects on perceptual stability (Allison, Harris, Jenkin, Jasiobedzka, & Zacher, 2001), presence (Meehan, Razzaque, Whitton, & Brookes, 2003), simulator sickness (Draper, Viirre, Furness, & Gawron, 2001), simulator fidelity (Adelstein, Lee, & Ellis, 2003; Mania, Adelstein, Ellis, & Hill, 2004), and virtual task performance (Frank, Casali, & Wierwille, 1988; So & Griffin, 1995—see also Chapters 23 through 26, this book). These detrimental effects are often due to the mismatch between visual and vestibular cues to motion. The lag in updating the scene means that a vestibular cue to head movement is followed by, rather than being coincident with, retinal-image motion. This particular problem can potentially be overcome by increased computational power that may eventually allow the time lag to become imperceptibly small. Estimates of the minimum detectable lag can vary markedly depending on the task. Some studies suggest that display lag has to be at least 150 ms to be detectable (Moss et al., 2010) with as little as 60 ms impairing perceptual stability (Allison et al., 2001). By contrast, other studies suggest that display lags as short as 15 ms are detectable (Mania et al., 2004).

A much more difficult problem is the need to provide vestibular input more generally. When observers are free to navigate around a scene, there is often a mismatch between visually indicated changes and cues from the vestibular system. In the extreme, vision may indicate locomotion, while the vestibular system indicates that the body is stationary. This mismatch frequently produces feelings of nausea. The relationship between visual and vestibular inputs is also under constant recalibration. Thus, periods of mismatch can lead to recalibration, which will maintain feelings of nausea in the natural environment until the normal relationship has been relearned (see Welsh & Mohler, 2013; Chapter 25). Improving visual display technology will not reduce this particular problem; however, providing appropriate vestibular inputs might reduce this (see Lawson et al., 2013; Chapter 7).

While current projection displays do afford FOVs that approach the limits of human observers, HMDs are limited by the size of small monitors relative to the interocular distance and the optics necessary to allow clear vision within the near point. Many optical compensatory arrangements result in eye strain and visual aftereffects. Research on the aftereffects generated with the use of HMDs has indicated another conflict situation in apparent motion situations. Stereoscopic displays that require near viewing often give rise to induced binocular stress. Mon-Williams, Wann, and Rushton (1993) reported loss of visual acuity and eye muscle imbalance after only a 10 min exposure to such displays, with some of the symptoms associated with motion sickness. They suggest that these aftereffects stem from adaptation of the accommodation–vergence system due to the disparity between the stereoscopic depth cues provided and the image focal depth. They later reported that displays that preserve the concordance between vergence and accommodation do not produce these aftereffects with viewing times of 30 min (Rushton, Mon-Williams, & Wann, 1994).

Unlike distance estimation in the real world, which is quite accurate up to 20 m (e.g., measured by having people walk blindfolded to previously seen targets—Loomis, Fujita, Da Silva, & Fukusima, 1992), distances tend to be underestimated in VEs. When viewing VEs through HMDs, helmet weight, low display resolution and level of scene detail, limited FOV and reduction in depth cues have all been identified as possible causes of this distance underestimation. Over the last decade, a number of studies have been conducted to examine the effects of these different factors, both in isolation and in combination (e.g., by having subjects view the real world through mock HMDs with similar weights and FOVs to real HMDs). While their findings have been somewhat mixed, it appears that both the mechanical and optical properties of HMDs contribute to distance underestimation in VEs (Bodenheimer et al., 2007; Knapp & Loomis, 2004; Messing & Durgin, 2005; Thompson et al., 2004; Willemsen, Colton, Creem-Regehr, & Thompson, 2009). This distance underestimation appears to be less likely when using large screen immersive displays (Plumert, Kearney, Cremer, & Recker, 2005). Interestingly, there is evidence that distance estimation in augmented reality (when virtual targets are viewed in the context of a real surrounding environment using a see-through HMD) is similar to that obtained when viewing real targets in a real environment (Jones, Swan, Singh, Kolstad, & Ellis, 2008).

3.7　NEW RESEARCH DIRECTIONS

The development of VE systems has proceeded without an abiding concern for the intricacies of human sensory systems. As these factors are considered in simulator design, the dearth of knowledge of many aspects of human perception is highlighted. In many respects, efforts to create realistic VEs serve to drive and motivate basic research efforts, but they also point to the limitations of classical research settings. One major problem in the approach to the study of human abilities centered on the way in which efforts have been compartmentalized (Kelly & Burbeck, 1984). The use of VE technology has forced, and will continue to force, a broadening in the scope of interests and considerations in the study of sensory motor interactions (e.g., Calvert, Spence, & Stein, 2004). As researchers begin to delve more deeply into these problems, the apparatus necessary to address these issues will require higher resolution on both spatial and temporal dimensions and become more expensive. This may dictate that research goals must be narrowed. It may be laudable, but not practical, to attempt to generate realistically natural displays that are valid in every VE application although improvements in this desire have arisen with the incorporation of augmented reality systems that include cameras mounted on HMDs to combine computer-generated displays with the currently viewed scene (Lee, Oakley, & Ryu, 2012).

One way that research tasks may be limited is to address perceptual problems of the particular device in question. While one can predict, to some extent, how well a display might meet the requirements of the human user, a more empirical approach might be to address the adequacy of the instrument with psychophysical experiments in the simulator at the outset. This approach may also reveal idiosyncratic problems that emerge through the use of the device and belie concerns based on theoretical predictions.

3.7.1　Limitations of Classical Research

Much of the research reviewed previously was generated in laboratories developed for the study of a particular visual capability (e.g., color, motion) without regard to how that ability might be compromised when other sensory and motor systems are in operation. In many experimental situations, the aim was to study the ability in isolation to avoid sources of psychological interference (e.g., arousal, attention, expectation). The aim in VE, however, is often quite different. The user is often actively engaged in tasks where many systems are employed.

Perhaps the greatest research impetus from the use of VE technology has been to emphasize sensory–sensory and sensory–motor interactions. There is a pressing need for increased information with regard to visual–vestibular interactions. In part, this is to alleviate some of the negative effects of using VE technology (e.g., nausea) but also because vestibular inputs can change basic properties of visual neurons, such as their preferred orientation of line stimuli (Kennedy, Magnin, & Jeannerod, 1976). It will be important to determine to what extent vision is identical when vestibular inputs are decorrelated.

Several recent studies have reported that consistent head movements (Ash et al., 2011) and consistent biomechanical cues (generated when walking on a treadmill—Seno, Ito, & Sunaga, 2011b) both appear to facilitate the vection induced by optic flow. Similarly, presenting consistent auditory cues and air flow to the observer's face has also been shown to improve the vection induced by optic flow (Riecke, Väljamäe, & Schulte-Pelkum, 2009; Seno, Ogawa, Ito, & Sunaga, 2011). However, other research suggests that we can be remarkably tolerant to expected Ash, Palmisano, Govan & Kim, 2011 conflicts between visual, vestibular, and nonvisual self-motion inputs (Ash & Palmisano, 2012; Ash, Palmisano, Govan, & Kim, 2011; Kim & Palmisano, 2008). Unfortunately, while treadmills are one of the most widely used (and researched) techniques to allow walking in VEs, they are known to generate problematic visual illusions—decreases in the perceived optic flow speed during treadmill walking (Durgin, Gigone, & Scott, 2005) and increases in perceived real-world walking speeds directly afterward (Pelah & Barlow, 1996).

3.7.1.1 Sensory and Sensory–Motor Interactions

The synthesis of classical theories of color vision into the current view of multistage processing is perhaps a model for how researchers will proceed in elaborating knowledge of visual processing relative to nonvisual sensory and motor processes. While considerable psychophysical and physiological evidence supported both the trichromatic and opponent process views, the controversy cooled when we learned how these mechanisms were staged and integrated. In similar fashion, our understanding of vision during various eye movements (previously discussed) is a step toward understanding how vision is compromised during such activity, but we have yet to fully appreciate how visual scene movement might elicit reflexive eye, head, and body reactions. Becker (1989) reported that head movements are a regular feature of gaze shifts greater than 20°. Some of the studies that have addressed coincident eye, head, and hand movements have done so while the subjects performed a volitional task (e.g., Pelz, Ballard, & Hayhoe, 1993), and they reveal the sequential ordering of gaze, head, and hand movements. With the use of head trackers, incorporating eye-movement monitoring and other pointing devices, it may be the case that these systems are requiring learned modifications to the natural interaction of the sensory and motor systems.

3.7.1.2 Esoteric Devices and Generalization

If it is the case that users are forced to adapt to the idiosyncrasies of a particular VE, then the traditional laboratory work from which researchers hope to predict visual performance may not provide the basis for legitimate generalization. Studies of basic human visual function may not tell about the human capability to cope with unique sensory–motor rearrangements. Investigations designed to study these abilities (e.g., Welsh & Mohler, 2013; Chapter 25) cannot anticipate all the kinds of interactive adaptations that might be required by new technology. For this reason, systematic research on basic visual function and sensory–motor interactions should be incorporated into the design and testing of new VE systems at the outset.

3.7.2 Basic Research in Simulators

Many of the more sophisticated VE systems in use today were developed to address practical problems in training and performance. Various vehicular simulators, for example, are used to train operators initially and to help them maintain their skills. Since this represents the primary mission of the apparatus, little or no time is allotted for investigation of many of the human-interface problems that might exist unless they severely compromise that mission. It is important for the successful development of future systems to document such problems in existing systems and to identify causal variables and usage protocols to prevent them. The work of Kennedy and others (see Stanney et al., 2013; Chapter 31) serves as an exemplary first step in this direction.

3.7.2.1 Machine or Observer Limited

The basic question asked of vision scientists concerning VE systems is: will this system meet or exceed the visual requirements of the human user? This chapter has been concerned with the answer to that question and has attempted to review visual capabilities and relate them to technical specifications. While such an approach has considerable heuristic value, many assumptions are made and conclusions based on such analyses are, in a sense, still quite theoretical. A more direct and empirical approach is to measure visual function within the VE system in question. Some things are readily predictable from extant data. For example, one might ask if the contrast scaling of a display is fine enough to exceed the contrast discrimination capacity of the user. However, the relationship between perceived distance traveled using visual information and the simulated distance traveled is less predictable (Redlick, Harris, & Jenkin, 1999). The most straightforward way to answer that question would be to measure the function of interest inside the VE system and compare the results to measures obtained with more standard methods. If similar findings are obtained with the two methods, the system could be described as observer limited and adequate in that particular dimension. If performance on that measure was poorer in the VE system than by conventional testing, performance could be termed machine limited, and this finding would indicate that the system falls short of the user's capability. It is an open question at the moment as to whether VE systems need always be observer limited to be adequate for a given application. Many visual tasks are not performed with stimuli at the limits of visual capability, but this is sometimes not clear until the tasks are performed with observer-limited and machine-limited systems.

3.7.2.2 Adaptation and Emerging Problems

Many early and continuing studies of human perception indicate that considerable adaptation occurs under conditions of perceptual rearrangement (Webster, 2011). Such adaptation can result in a rescaling of the perceptual systems and to corresponding changes in perceptual–motor responses. With brief exposures (<10 min) to such stimuli, adaptation occurs but recovery is relatively fast. With prolonged exposures (>10 min to several days), such adaptation might be more permanent and require readaptation over an extended period to restore normal perception and performance (Kohler, 1972; see also Welsh & Mohler, 2013; Chapter 25). One notion is that as more realistic visual simulations are developed, less adaptation might occur and this may alleviate the problems associated with VE aftereffects (Stanney et al., 1998). While this seems a reasonable hypothesis for visual adaptation, subtle variations from natural image statistics are still likely to produce adaptation (Falconbridge & Badcock, 2006; Webster, 2011); it may also not hold true for adaptation to mismatches between the senses or the perceptual–motor linkages.

In traditional psychophysical experiments, it is often the case that the variance of the measurements is initially quite high and reduces with continued data collection. It is recognized that subjects often need experience in the paradigm to adopt the best perceptual set and response strategy. Thus, some forms of adaptation to the VE might be expected to occur through perceptual–motor learning and some through changing cognitive problem-solving strategies. These modifications in perceptual set and response strategy would not be expected to result in debilitating aftereffects and might best be considered VE-specific effects. Little research has been carried out on these important but indirect effects on visual processing in VE systems.

REFERENCES

Abrams, R. A., Oonk, H. M., & Pratt, J. (1998). Fixation point offsets facilitate endogenous saccades. *Perception & Psychophysics*, *60*, 201–208.

Adelstein, B. D., Lee, T. G., & Ellis, S. R. (2003, October 13–17). Head tracking latency in virtual environments: Psychophysics and a model. In *Proceedings of the Human Factor Ergonomics Society 47th Annual Meeting* (pp. 2083–2087), HFES, Denver, CO.

Allison, R. S., Harris, L. R., Jenkin, M., Jasiobedzka, U., & Zacher, J. E. (2001, March 13–17). Tolerance of temporal delay in virtual environments. In *Proceedings of the IEEE Virtual Reality Conference* (pp. 247–254), Yokohama, Japan.

Allison, R. S., Howard, I. P., & Zacher, J. E. (1999). Effect of field size, head motion, and rotational velocity on roll vection and illusory self-tilt in a tumbling room. *Perception*, *28*, 299–306.

Allman, J., Miezin, F., & McGuinness, E. (1985). Direction- and velocity-specific responses from beyond the classical receptive field in the middle temporal visual area (MT). *Perception*, *14*, 105–126.

Amano, K., Edwards, M., Badcock, D. R., & Nishida, S. (2009). Adaptive pooling of visual motion signals by the human visual system revealed with a novel multi-element stimulus. *Journal of Vision*, *9*(3), 1–25.

Ames, A. Jr. (1925). Illusion of depth from single pictures. *Journal of the Optical Society of America*, *10*, 137–148.

Andersen, G. J., & Braunstein, M. L. (1985). Induced self-motion in central vision. *Journal of Experimental Psychology: Human Perception and Performance*, *11*(2), 122–132.

Anstis, S. M. (1980). The perception of apparent movement. In H. C. Longuet-Higgins & N. S. Sutherland (Eds.), *The psychology of vision*. London, U.K.: Royal Society.

Arakawa, Y. (1953). Quantitative measurements of visual fields for colors with a direct-current method. *American Journal of Ophthalmology*, *36*, 1594–1601.

Arsenault, R., & Ware, C. (2004). The importance of stereo, eye coupled perspective and touch for eye-hand coordination. *Presence: Teleoperators and Virtual Environments*, *13*, 549–559.

Ash, A., & Palmisano, S. (2012). Vection during conflicting multisensory information about the axis, magnitude and direction of self-motion. *Perception*, *41*, 253–267.

Ash, A., Palmisano, S., Govan, G., & Kim, J. (2011). Effect of visual display lag on vection in active observers. *Aviation, Space and Environmental Medicine*, *82*, 763–769.

Ash, A., Palmisano, S., & Kim, J. (2011). Perceived self-motion induced by consistent and inconsistent multisensory stimulation. *Perception*, *40*, 155–174.

Aubert, H. (1886). Die Bewegungsempfindung. *Archiv fur die Gesamte Physiologie*, *39*, 347–370.

Badcock, D. R., & Derrington, A. M. (1985). Detecting the displacement of periodic patterns. *Vision Research*, *25*, 1253–1258.

Badcock, D. R., Hess, R. F., & Dobbins, K. (1996). Localization of element clusters: Multiple cues. *Vision Research*, *36*, 1467–1472.

Badcock, D. R., & Schor, C. (1985). The depth increment detection function within spatial channels. *Journal of the Optical Society of America*, *A2*, 1211–1216.

Badcock, D. R., & Westheimer, G. (1985). Spatial location and hyperacuity: The centre/surround localization contribution function has two substrates. *Vision Research*, *25*, 1259–1267.

Baker, C. L., Hess, R. F., & Zihl, J. (1991). Residual motion perception in a "motion-blind" patient, assessed with limited-lifetime random dot stimuli. *The Journal of Neuroscience*, *11*(2), 454–461.

Becker, W. (1989). Metrics. In M. E. Goldburg & R. H. Wurtz (Eds.), *The neurobiology of saccadic eye movements*. Amsterdam, the Netherlands: Elsevier Science Publishers.

Becker, W., Raab, S., & Jürgens, R. (2002). Circular vection during voluntary suppression of optokinetic reflex. *Experimental Brain Research*, *144*(4), 554–557.

Bell, J., & Badcock, D. R. (2008). Luminance and contrast cues are integrated in global shape detection with contours. *Vision Research*, *48*(21), 2336–2344.

Bodenheimer, B., Meng, J., Wu, H., Narasimham, G., Rump, B., McNamara, T. P., Rieser, J. J. (2007). Distance estimation in virtual and real environments using bisection. In C. Wallraven & V. Sunstedt (Eds.), *Proceedings of the 4th Symposium on Applied Perception in Graphics and Visualization* (pp. 35–40). Tubingen, Germany: ACM.

Boff, K. R., Kaufman, L., & Thomas, J. P. (Eds.). (1986). *Handbook of human perception* (2 Vols.). New York, NY: John Wiley & Sons.

Boff, K. R., & Lincoln, J. E. (1988). *Engineering data compendium: Human perception and performance*. Wright-Patterson Airforce Base, OH: Armstrong Aerospace Medical Research Laboratory.

Bonato, F., & Bubka, A. (2006). Chromaticity, spatial complexity, and self-motion perception. *Perception, 35*(1), 53–64.

Borish, I. (1970). *Clinical refraction* (3rd ed.). Chicago, IL: Professional Press.

Bouma, H. (1970). Interaction effects in parafoveal letter recognition. *Nature, 226*, 177–178.

Bourdon, B. (1902). *La perception visuelle de l'espace*. Paris, France: Schleicher Freres.

Boynton, R. M., Hayhoe, M. M., & MacLeod, D. I. A. (1977). The gap effect: Chromatic and achromatic visual discrimination as affected by field separation. *Optica Acta, 24*, 159–177.

Brandt, T., Dichgans, J., & Koenig, E. (1973). Differential effects of central versus peripheral vision on egocentric and exocentric motion perception. *Experimental Brain Research, 16*, 476–491.

Brandt, T., Wist, E. R., & Dichgans, J. (1971). Optisch induzierte pseudo-Coriolis-effekte und circularvektion: Ein beitrag zur optisch-vestibulären interaktion. *Archiv Für Psychiatrie und Nervenkrankheiten, 214*, 365–389.

Bridgeman, B., Hendry, D., & Stark, L. (1975). Failure to detect displacement of the visual world during saccadic eye movements. *Vision Research, 15*, 719–722.

Brown, W. R. J. (1951). The influence of luminance level on visual sensitivity to color differences. *Journal of the Optical Society, 41*, 684–688.

Bubka, A., & Bonato, F. (2010). Natural visual-field features enhance vection. *Perception, 39*, 627–635.

Bubka, A., Bonato, F., & Palmisano, S. (2008). Expanding and contracting optical flow patterns and vection. *Perception, 37*, 704–711.

Burbeck, C. A., & Hadden, S. (1993). Scaled position integration areas: Accounting for Weber's law for separation. *Journal of the Optical Society of America, A10*, 5–15.

Burr, D., & Morrone, M. C. (2005). Eye movements: Building a stable world from glance to glance. *Current Biology, 15*(20), R839–R840.

Burr, D. C., Morrone, M. C., & Ross, J. (1994). Selective suppression of the magnocellular visual pathway during saccadic eye movements. *Nature, 371*, 511–513.

Burr, D. C., Morrone, M. C., & Vaina, L. M. (1998). Large receptive fields for optic flow detection in humans. *Vision Research, 38*, 1731–1743.

Burr, D. C., & Ross, J. (1982). Contrast sensitivity at high velocities. *Vision Research, 22*, 479–484.

Burr, D. C., & Thompson, P. (2011). Motion psychophysics: 1985–2010. *Vision Research, 51*(13), 1431–1456.

Cadieu, C., Kouh, M., Pasupathy, A., Connor, C. E., Riesenhuber, M., & Poggio, T. (2007). A model of V4 shape selectivity and invariance. *Journal of Neurophysiology, 98*(3), 1733–1750.

Calvert, G., Spence, C., & Stein, B. E. (Eds.). (2004). *The handbook of multisensory processes*. New York, NY: MIT Press.

Campbell, F. W., & Robson, J. G. (1968). Application of Fourier analysis to the visibility of gratings. *Journal of Physiology [London], 197*, 551–566.

Cardin, V., & Smith, A. T. (2010). Sensitivity of human visual and vestibular cortical regions to egomotion-compatible visual stimulation. *Cerebral Cortex, 20*, 1964–1973.

Castiello, U., Badcock, D. R., & Bennett, K. (1999). Sudden and gradual presentation of distractor objects: Differential interference effects. *Experimental Brain Research, 128*(4), 550–556.

Cavanagh, P., & Favreau, O. E. (1985). Color and luminance share a common motion pathway. *Vision Research, 25*, 1595–1601.

Chalupa, L. M., & Werner, J. S. (2004). *The visual neurosciences* (Vols. 1 & 2). Cambridge, MA: MIT Press.

Cherici, C., Kuang, X., Poletti, M., & Rucci, M. (2012). Precision of sustained fixation in trained and untrained observers. *Journal of Vision, 12*(6:31), 1–16.

CIE (1932). *Commission internationale de l'Eclairage proceedings, 1931*. Cambridge, U.K.: Cambridge University Press, 1932.

Clarke, F. J. J. (1967, April). The effect of field-element size on chromaticity discrimination. In *Proceedings of the Symposium on Colour Measurement in Industry, 1*, The Colour Group, London, U.K.

Connor, C. E. (2004). Shape dimensions and object primitives. In L. M. Chalupa & J. S. Werner (Eds.), *The visual neurosciences* (Vol. 2, pp. 1080–1089). London, U.K.: Bradford.

Craik, K. J. W. (1939). The effect of adaptation upon visual acuity. *British Journal of Psychology, 29*, 252–266.

Cropper, S. J., & Badcock, D. R. (1994). Discriminating smooth from sampled motion: Chromatic and luminance stimuli. *Journal of the Optical Society of America, 11*, 515–530.

Derrington, A. M., & Badcock, D. R. (1985). Separate detectors for simple and complex grating patterns? *Vision Research, 25*, 1869–1878.

DeValois, R. L., & DeValois, K. K. (1991). Vernier acuity with stationary moving Gabors. *Vision Research, 31*, 1619–1626.

Dichgans, J., & Brandt, T. (1978). Visual-vestibular interaction: Effects on self-motion perception and postural control. In *Perception, handbook of sensory physiology* (pp. 756–804). New York, NY: Springer.

Dichgans, J., Koener, F., & Voigt, K. (1969). Verleichende sdalierung des afferenten und efferenten bewegungssehen beim menschen: Lineare funktionen mit verschiedener ansteigssteilheit. *Psychologische Forschung, 32*, 277–295.

Dickinson, J. E., Broderick, C., & Badcock, D. R. (2009). Selective attention contributes to global processing in vision. *Journal of Vision, 9*(2), 1–8.

Ding, J., & Levi, D. M. (2011). Recovery of stereopsis through perceptual learning in human adults with abnormal binocular vision. *Proceedings of the National Academy of Sciences, 108*(37), E733–E741.

Draper, M. H, Viirre, E. S., Furness, T. A., & Gawron, V. J. (2001). Effects of image scale and system time delay on simulator sickness within head-coupled virtual environments. *Human Factors, 43*, 129–145.

Duffy, C. J., & Wurtz, R. H. (1991). Sensitivity of MST neurons to optic flow stimuli. I. A continuum of response selectivity to large-field stimuli. *Journal of Neurophysiology, 65*, 1329–1345.

Durgin, F. H., Gigone, K., & Scott, R. (2005). Perception of visual speed while moving. *Journal of Experimental Psychology: Human Perception and Performance, 31*, 339–353.

Duwaer, A. L., & van den Brink, G. (1981). Diplopia thresholds and the initiation of vergence eye-movements. *Vision Research, 21*(12), 1727–1737.

Edwards, M., & Badcock, D. R. (1993). Asymmetries in the sensitivity to motion in depth: A centripetal bias. *Perception, 22*, 1013–1023.

Edwards, M., & Badcock, D. R. (2003). Motion distorts perceived depth. *Vision Research, 43*, 1799–1804.

Exner, S. (1875). Uber das sehen von bewegungen und die theorie des zusammengesetzen auges. *Sitzungsberichts Akademie Wissen Schaft Wein, 72*, 156–190.

Falconbridge, M. S., & Badcock, D. R. (2006). Implicit exploitation of regularities: Novel correlations in images quickly alter visual perception. *Vision Research, 46*, 1331–1335.

Farnsworth, D. (1955). Tritanomalous vision as a threshold function. *Die Farbe, 4*, 185–197.

Fendick, M. G. (1983). Parameters of the retinal light distribution of a bright line which correspond to its attributed spatial location. *Investigative Ophthalmology and Visual Science, 24*, 92.

Field, D. J., Hayes, A., & Hess, R. F. (1993). Contour integration by the human visual system: Evidence for a local "association field." *Vision Research, 33*, 173–193.

Field, D. J., & Tolhurst, D. J. (1986). The structure and symmetry of simple-cell receptive-field profiles in the cat's visual cortex. *Proceedings of the Royal Society of London B: Biological Sciences, 228*, 379–400.

Filehne, W. (1922). Ueber das optische wahenehmen von bewegungen. *Zeitschrift fur Sinnesphysiologie, 53*, 134–145.

Fischer, M. H., & Kornmüller, A. E. (1930). Optokinetisch ausgelöste bewegungswahrehmungen und optokinetischer nystagmus. *Journal für Psychologie und Neurologie, 41*, 273–308.

Forbes, B., & Klein, R. M. (1996). The magnitude of the fixation offset effect with endogenously and exogenously controlled saccades. *Journal of Cognitive Neuroscience, 8*, 344–352.

Frank, L. H., Casali, J. G., & Wierwille, W. W. (1988). Effect of visual display and motion system delays on operator performance and uneasiness in a driving simulator. *Human Factors, 30*(2), 201–217.

Gilbert, C. D., Das, A., Ito, M., Kapadia, M., & Westheimer, G. (1996). Spatial integration and cortical dynamics. *Proceedings of the National Academy of Science, 93*(2), 615–622.

Gillam, B. (1995). The perception of spatial layout from static optical information. In W. Epstein & S. Rogers (Eds.), *Perception of space and motion* (pp. 23–70). New York, NY: Academic Press.

Gillam, B., Palmisano, S., & Govan, D. G. (2011). Motion parallax and binocular disparity for depth interval estimation beyond interaction space. *Perception, 40*, 39–49.

Gouras, P. (1991). The perception of color. *Vision and visual dysfunction* (Vol. 6). London, U.K.: Macmillan.

Graham, C. H., Baker, K. E., Hecht, M., & Lloyd, V. V. (1948). Factors influencing thresholds for monocular movement parallax. *Journal of Experimental Psychology, 38*, 205–223.

Graham, N. (1980). Spatial frequency channels in human vision: Detecting edges without edge detectors. In C. Harris (Ed.), *Visual coding and adaptability* (pp. 215–262). Hillsdale, NJ: Lawrence Erlbaum Associations.

Gregory, R. L. (1964). Human perception. *British Medical Bulletin, 20*, 21–26.

Gregory, R. L. (1966). *Eye and brain*. London, U.K.: World University Library.

Grim, K. (1911). Uber diie genauigkeit der wahrnehmung und ausfuhrung von augenbeweg unger. *Zeitschrift fur Sinnesphysiologie, 45*, 9–26.

Harris, J. M., & Wilcox, L. M. (2009). The role of monocularly visible regions in depth and surface perception. *Vision Research, 49*, 2666–2685.

Hayhoe, M. M., Levin, M. E., & Koshel, R. J. (1992). Subtractive processes in light adaptation. *Vision Research*, *32*(2), 323–333.

Held, R., Dichgans, J., & Bauer, J. (1975). Characteristics of moving visual areas influencing spatial orientation. *Vision Research*, *15*, 357–365.

Helmholtz, H. (1867/1925). *Physiological optics* (3rd ed., Vol. 3). Menasha, WI: The Optical Society of America.

Hess, R. F., & Badcock, D. R. (1995). Metric for separation discrimination by the human visual system. *Journal of the Optical Society of America*, *A12*, 3–16.

Hettinger, L. G. (2014). Illusory self-motion in virtual environments. In K. S. Hale & K. M. Stanney (Eds.), *Handbook of virtual environments* (2nd ed., pp. 437–468). New York: CRC Press.

Hood, D. C., & Finkelstein, M. A. (1979). Comparison of changes in sensitivity and sensation: Implications for the response-intensity function of the human photopic system. *Journal of Experimental Psychology: Human Perception and Performance*, *5*, 391–405.

Howard, I. P., & Rogers, B. J. (1995). Binocular vision and stereopsis. Oxford, England: Oxford University Press.

Hurvich, L. M. (1981). *Color vision*, p. 210. Sunderland: Sinauer.

Ito, H., & Shibata, I. (2005). Self-motion perception from expanding and contracting optical flows overlapped with binocular disparity. *Vision Research*, *45*, 397–402.

Johansson, G. (1977). Studies on visual-perception of locomotion. *Perception*, *6*, 365–376.

Johnson, W. H., Sunahara, F. A., & Landolt, J. P. (1999). Importance of the vestibular system in visually induced nausea and self-vection. *Journal of Vestibular Research – Equilibrium & Orientation*, *9*, 83–87.

Jones, J. A., Swan, J. E., Singh, G., Kolstad, E., & Ellis, S. R. (2008). The effects of virtual reality, augmented reality, and motion parallax on egocentric depth perception. In S. Creem-Regehr & K. Myszkowski (Eds.), *Proceedings of the 5th Symposium on Applied Perception in Graphics and Visualization* (pp. 9–14). Los Angeles, CA: ACM.

Kalloniatis, M., & Luu, C. (2007). Webvision: The organization of the retina and visual system: Perception of depth. webvision.med.utah.edu/book/part-viii.gabac-receptors/perception-of-depth.

Kelly, D. H., & Burbeck, C. A. (1980). Motion and vision. III. Stabilized pattern adaptation. *Journal of the Optical Society of America*, *70*, 1283–1289.

Kelly, D. H., & Burbeck, C. A. (1984). Critical problems in spatial vision. *Critical Reviews in Biomedical Engineering*, *10*, 125–177.

Kennedy, H., Magnin, M., & Jeannerod, M. (1976). Receptive field response of LGB neurons during vestibular stimulation in awake cats. *Vision Research*, *16*, 119–120.

Kim, J., & Palmisano, S. (2008). Effects of active and passive viewpoint jitter on vection in depth. *Brain Research Bulletin*, *77*, 335–342.

Kim, J., Palmisano, S., & Bonato, F. (2012). Simulated angular head oscillation enhances vection in depth. *Perception*, *41*, 402–414.

Kitazaki, M., & Sato, T. (2003). Attentional modulation of self-motion perception. *Perception*, *32*, 475–484.

Knapp, J. M., & Loomis, J. M. (2004). Limited field of view of head-mounted displays is not the cause of distance underestimation in virtual environments. *Presence: Teleoperators and Virtual Environments*, *13*, 572–577.

Koenderink, J. J., Bouman, M. A., Bueno de Mesquita, A. E., & Slappendel, S. (1978). Perimetry of contrast detection thresholds of moving spatial sine wave patterns II. The far peripheral visual field (eccentricity $0°–50°$). *Journal of the Optical Society of America*, *68*, 850–854.

Koffka, K. (1922). Perception: An introduction to Gestalt theorie. *Psychological Bulletin*, *19*, 531–585.

Kohler, I. (1972). Experiments with goggles. In R. Held & W. Richards (Eds.), *Perception: Mechanisms and models* (p. 390). San Francisco, CA: W. H. Freeman & Co.

Kovacs, I., Papathomas, T. V., Yang, M., & Feher, A. (1996). When the brain changes its mind: Interocular grouping during binocular rivalry. *Proceedings of the National Academy of Sciences*, *93*, 15508–15511.

Lawson, B. D., & Riecke, B. E. (2014). Perception of body motion. In K. S. Hale & K. M. Stanney (Eds.), *Handbook of virtual environments* (2nd ed., pp. 163–196). New York, NY: CRC Press.

Lee, C., Oakley, I., & Ryu, J. (2012). Exploring the impact of visual-haptic registration accuracy in augmented reality. In P. Isokoski & J. Springare (Eds.), *Haptics: Perception, devices, mobility, and communication* (pp. 85–90). Heidelberg, Germany: Springer Verlag.

Lee, D. N. (1980). The optic flow field: The foundation of vision. *Philosophical Transactions of the Royal Society of London*, *290*, 169–179.

Lee, D. N., & Lishman, J. R. (1975). Visual proprioceptive control of stance. *Journal of Human Movement Studies*, *1*, 89–95.

Leibowitz, H. W. (1955). The relation between the rate threshold for the perception of movement and luminance for various durations of exposure. *Journal of Experimental Psychology, 49*, 209–214.

Lestienne, F., Soechting, J., & Berthoz, A. (1977) Postural readjustments induced by linear motion of visual scenes. *Experimental Brain Research, 28*, 363–384.

Levi, D. M. (2008). Crowding—An essential bottleneck for object recognition: A mini-review. *Vision Research, 48*(5), 635–654.

Levi, D. M., & Klein, S. A. (1990). The role of separation and eccentricity in encoding position. *Vision Research, 30*, 557–585.

Levi, D. M., Klein, S. A., & Aitsebaomo, A. P. (1985). Vernier acuity, crowding and cortical magnification. *Vision Research, 25*, 963–977.

Levi, D. M., & Schor, C. M. (1984). Spatial and velocity tuning of processes underlying induced motion. *Vision Research, 24*, 1189–1195.

Loffler, G. (2008). Perception of contours and shapes: Low and intermediate stage mechanisms. *Vision Research, 48*(20), 2106–2127.

Loomis, J. M., Fujita, N., Da Silva, J. A., & Fukusima, S. S. (1992). Visual space perception and visually directed action. *Journal of Experimental Psychology: Human Perception and Performance, 18*, 906–921.

Lowther, K., & Ware, C. (1996). Vection with large screen 3D imagery. In *ACM CHI '96* (pp. 233–234). New York, NY: ACM.

Ludvigh, E., & Miller, J. W. (1958). Study of visual acuity during the ocular pursuit of moving test objects I. Introduction. *Journal of the Optical Society of America, 48*, 799–802.

Mach, E. (1875). *Grundlinien der Lehre von der Bewegungsempfindungen.* Leipzig, Germany: Engelmann.

Mach, E. (1922). *Die Analyse der Empfindungen [The analysis of sensations].* Gena: Gustav Fischer.

Mack, A. (1986). Perceptual aspects of motion in the frontal plane. In K. R. Boff, L. Kaufman, & J. P. Thomas (Eds.), *Handbook of perception and human performance: Sensory processes and perception* (Vol. 1), pp. 17.1–17.38. New York, NY: John Wiley & Sons.

Mania, K., Adelstein, B. D., Ellis, S. R., & Hill, M. I. (2004). Perceptual sensitivity to head tracking latency in virtual environments with varying degrees of scene complexity. In V. Interrante & A. McNamara (Eds.), *Proceedings of the 1st Symposium on Applied Perception in Graphics and Visualization* (pp. 39–48). Los Angeles, CA: ACM.

Masland, R. H., & Albright, T. D. (2008). *The senses: A comprehensive reference* (Vols. 1 and 2). New York, NY: Academic Press. http://dx.doi.org/10.1016/B978-012370880-9.09003-4.

Mather, G., Verstraten, F., & Anstis, S. (Eds.). (1998). *The motion aftereffect: A modern perspective.* London, U.K.: MIT Press.

Matin, L. (1986). Visual localization and eye movements. In K. R. Boff, L. Kaufman, & J. P. Thomas (Eds.), *Handbook of human perception and performance* (Vol. 1, pp. 1–45). New York, NY: John Wiley & Sons.

McCann, J. J., & Hall, J. A. (1980). Effects of average-luminance surrounds on the visibility of sine-wave gratings. *Journal of the Optical Society of America, 70*(2), 212–219.

McGraw, P. V., Walsh, V., & Barrett, B. T. (2004). Motion-sensitive neurones in V5/MT modulate perceived spatial position. *Current Biology, 14*, 1090–1093.

McKee, S. P., & Levi, D. M. (1987). Dichoptic hyperacuity: The precision of nonius alignment. *Journal of the Optical Society of America, A4*, 1104–1108.

Meehan, M., Razzaque, S., Whitton, M. C., & Brookes, F. P. Jr. (2003). Effect of latency on presence in stressful virtual environments. In *VR '03: Proceedings of the IEEE Conference on Virtual Reality* (pp. 141–148). Washington, DC: IEEE Computer Society.

Messing, R., & Durgin, F. H. (2005). Distance perception and the visual horizon in head-mounted displays. *ACM Transactions of Applied Perception, 2*, 234–250.

Metha, A. B., Vingrys, A. J., & Badcock, D. R. (1993). Calibration of a colour monitor for visual psychophysics. *Behavior Research Methods, Instruments, and Computers, 25*(3), 371–383.

Millodot, M., Johnson, C. A., Lamont, A., & Leibowitz, H. A. (1975). Effect of dioptrics on peripheral visual acuity. *Vision Research, 15*, 1357–1362.

Mitchell, D. E., & Ware, C. (1974). Interocular transfer of a visual after-effect in normal and stereoblind humans. *Journal of Physiology [London], 236*, 707–721.

Mon-Williams, M., Wann, J. P., & Rushton, S. (1993). Binocular vision in a virtual world: Visual deficits following the wearing of a head-mounted display. *Ophthalmic and Physiological Optics, 13*, 387–391.

Moss, J. D., Muth, E. R., Tyrrell, R. A., & Stephens, B. R. (2010). Perceptual thresholds for display lag in a real visual environment are not affected by field of view or psychophysical technique. *Displays, 31*, 143–149.

Mullen, K. T. (1985). The contrast sensitivity of human colour vision to red-green and blue-yellow chromatic gratings. *Journal of Physiology, 359*, 381–400.

Nakamura, S., Seno, T., Ito, H., & Sunaga, S. (2010). Coherent modulation of stimulus colour can affect visually induced self-motion perception. *Perception*, *39*, 1579–1590.

Nakamura, S., & Shimojo, S. (1999). Critical role of foreground stimuli in perceiving visually induced self-motion (vection). *Perception*, *28*, 893–902.

Nakamura, S., & Shimojo, S. (2000). A slowly moving foreground can capture an observer's self-motion a report of new motion illusion: Inverted vection. *Vision Research*, *40*, 2915–2923.

Nakayama, K., Shimojo, S., & Silverman, G. H. (1989). Stereoscopic depth: Its relation to image segmentation, grouping and the recognition of occluded objects. *Perception*, *18*, 55–68.

Nandy, A. S., & Tjan, B. S. (2012). Saccade-confounded image statistics explain visual crowding. *Nature Neuroscience*, *15*(3), 463–469.

Nishida, S. (2011). Advancement of motion psychophysics: Review 2001–2010. *Journal of Vision*, *11*(5:11), 1–53.

Ohmi, M., & Howard, I. P. (1988). Effect of stationary objects on illusory forward self-motion induced by a looming display. *Perception*, *17*, 5–11.

Ohmi, M., Howard, I. P., & Landolt, J. P. (1987). Circular vection as a function of foreground-background relationships. *Perception*, *16*, 17–22.

Palmer, S. (1999). *Vision science: Photons to phenomenology*. Cambridge, MA: MIT Press.

Palmisano, S. (1996). Perceiving self-motion in depth: The role of stereoscopic motion and changing-size cues. *Perception & Psychophysics*, *58*, 1168–1176.

Palmisano, S. (2002). Consistent stereoscopic information increases the perceived speed of vection in depth. *Perception*, *31*, 463–480.

Palmisano, S., Allison, R. S., & Howard, I. P. (2006). Illusory scene distortion occurs during perceived self-rotation in roll. *Vision Research*, *46*, 4048–4058.

Palmisano, S., Allison, R. S., Kim, J., & Bonato, F. (2011). Simulated viewpoint jitter shakes sensory conflict accounts of self-motion perception. *Seeing and Perceiving*, *24*, 173–200.

Palmisano, S., Allison, R. S., & Pekin, F. (2008). Accelerating self-motion displays produce more compelling vection in depth. *Perception*, *37*, 22–33.

Palmisano, S., & Gillam, B. J. (1998). Stimulus eccentricity and spatial frequency interact to determine circular vection. *Perception*, *27*, 1067–1078.

Palmisano, S., Gillam, B. J., & Blackburn, S. (2000). Global perspective jitter improves vection in central vision. *Perception*, *29*, 57–67.

Palmisano, S., Gillam, B. J., Govan, D., Allison, R. S., & Harris, J. (2010). Stereoscopic perception of real depths at large distances. *Journal of Vision*, *10*(6), 1–16.

Palmisano, S., & Kim, J. (2009). Effects of gaze on vection from jittering, oscillating and purely radial optic flow. *Attention, Perception & Psychophysics*, *71*, 1842–1853.

Pelah, A., & Barlow, H. B. (1996). Visual illusion from running. *Nature*, *6580*, 381.

Pelz, J. B., Ballard, D. H., & Hayhoe, M. M. (1993). Memory use during the performance of natural visuomotor tasks. *Investigative Ophthalmology and Visual Science [Suppl.]*, *34*(4), 1234.

Plumert, J. M., Kearney, J. K., Cremer, J. F., & Recker, K. (2005). Distance perception in real and virtual environments. *ACM Transactions on Applied Perception*, *2*, 216–233.

Post, R. B. (1988). Circular vection is independent of stimulus eccentricity. *Perception*, *17*, 737–744.

Press, W. H., Teukolsky, S. A., Vetterling, W. T., & Flannery, B. P. (1992). *Numerical recipes in C: The art of scientific computing*. Cambridge, England: Cambridge University Press.

Reason, J. T., Mayes, A. R., & Dewhurst, D. (1982). Evidence for a boundary effect in roll vection. *Perception & Psychophysics*, *31*(2), 139–144.

Reason, J., Wagner, H., & Dewhurst, D. (1981). A visually driven postural aftereffect. *Acta Psychologica*, *48*, 241–251.

Redlick, F. P., Harris, L. R., & Jenkin, M. (1999). Active motion reduces the perceived self-displacement created by optic flow. *Investigative Ophthalmology & Visual Science*, 40, S798–S798.

Regan, D., & Beverley, K. I. (1973). The dissociation of sideways movements from movements in depth: Psychophysics. *Vision Research*, *13*, 2403–2415.

Regan, D., & Beverley, K. I. (1978a). Illusory motion in depth: Aftereffect of adaptation to changing size. *Vision Research*, *18*, 209–212.

Regan, D., & Beverley, K. I. (1978b). Looming detectors in the human visual pathway. *Vision Research*, *18*, 415–421.

Regan, D., & Tyler, C. W. (1971). Temporal summation and its limit for wavelength changes: An analog of Bloch's law for color vision. *Journal of the Optical Society of America*, *61*, 1414–1421.

Riecke, B. E. (2009). Cognitive and higher-level contributions to illusory self-motion perception ("vection"): Does the possibility of actual motion affect vection? *Japanese Journal of Psychonomic Science*, *28*, 135–139.

Riecke, B. E., Schulte-Pelkum, J., Avraamides, M. N., Heyde, M. V. D., & Bülthoff, H. H. (2006). Cognitive factors can influence self-motion perception (vection) in virtual reality. *ACM Transactions on Applied Perception, 3*, 194–216.

Riecke, B. E., Väljamäe, A., & Schulte-Pelkum, J. (2009). Moving sounds enhance the visually-induced self-motion illusion (circular vection) in virtual reality. *ACM Transactions on Applied Perception, 6*, 1–27.

Riggs, L. A., Armington, J. C., & Ratliff, F. (1954). Motions of the retinal image during fixation. *Journal of the Optical Society of America, 44*, 315–321.

Rivest, J., & Cavanagh, P. (1996). Localizing contours defined by more than one attribute. *Vision Research, 36*, 53–66.

Robson, J. G. (1966). Spatial and temporal contrast-sensitivity functions of the visual system. *Journal of the Optical Society of America, 56*, 1141–1142.

Rock, I. (1975). *An introduction to perception*. London, U.K.: Macmillan.

Rogers, B., & Graham, M. (1982). Similarities between motion parallax and stereopsis in human depth perception. *Vision Research, 22*, 261–270.

Rubin, G. S., West, S. K., Munoz, B., Bandeen-Roche, K., Zeger, S., Schein, O., & Fried, L. P. (1997). A comprehensive assessment of visual impairment in a population of older Americans. The SEE study. Salisbury Eye Evaluation Project. *Investigative Ophthalmology & Visual Science, 38*(3), 557–568.

Rushton, S., Mon-Williams, M., & Wann, J. (1994). Binocular vision in a bi-ocular world: New generation head-mounted displays avoid causing visual deficit. *Displays, 15*, 255–260.

Sandström, C. I. (1951). *Orientation in the present space*. Stockholm, Sweden: Almgvist & Wiksell.

Scarfe, P., & Johnston, A. (2010). Motion drag induced by global motion Gabor arrays. *Journal of Vision, 10*(5:14), 1–15.

Schreiber, K. M., Hillis, J. M., Filippini, H. R., Schor, C. M., & Banks, M. S. (2008). The surface of the empirical horopter. *Journal of Vision, 8*(3:7), 1–20.

Schutz, A. C., Braun, D. I., & Gegenfurtner, K. R. (2011). Eye movements and perception: A review. *Journal of Vision, 11*(5:9), 1–30.

Seno, T., Ito, H., & Sunaga, S. (2009). The object and background hypothesis for vection. *Vision Research, 49*, 2973–2982.

Seno, T., Ito, H., & Sunaga, S. (2011a). Attentional load inhibits vection. *Attention, Perception & Psychophysics, 73*, 1467–1476.

Seno, T., Ito, H., & Sunaga, S. (2011b). Inconsistent locomotion inhibits vection. *Perception, 40*, 747–750.

Seno, T., Ogawa, M., Ito, H., & Sunaga, S. (2011). Consistent air flow to the face facilitates vection. *Perception, 40*, 1237–1240.

Seno, T., Sunaga, S., & Ito, H. (2010). Inhibition of vection by red. *Attention, Perception, & Psychophysics, 72*, 1642–1653.

Sony PVM-2551MD Medical OLED monitor (2011). http://pro.sony-asia.com/product/resources/en_BA/images/Brochure/Medical/BP000460_pvm-2551md.pdf

Shlaer, S. (1937). The relation between visual acuity and illumination. *Journal of General Physiology, 21*, 165–188.

Siegal, M. H., & Siegal, A. B. (1972). Hue discrimination as a function of luminance. *Perception & Psychophysics, 12*, 295–299.

So, R. H., & Griffin, M. J. (1995). Head-coupled virtual environment with display lag. In K. Carr & R. England (Eds.), *Simulated and virtual realities: Elements of perception* (pp. 103–111). London, U.K.: Taylor & Francis Group.

Solman, R. T., Dain, S. J., Lim, H. S., & May, J. G. (1995). Reading-related wavelength and spatial frequency effects in visual spatial location. *Ophthalmic and Physiological Optics, 15*, 125–132.

Solman, R. T., & May, J. G. (1990). Spatial localization discrepancies: A visual deficiency in poor readers. *American Journal of Psychology, 103*(2), 243–263.

Stanney, K. M., Salvendy, G., Deisigner, J., DiZio, P., Ellis, S., Ellison, E., … Witmer, B. (1998). Aftereffects and sense of presence in virtual environments: Formulation of a research and development agenda. Report sponsored by the Life Sciences Division at NASA Headquarters. *International Journal of Human-Computer Interaction, 10*(2), 135–187.

Stark, L., Kong, R., Schwartz, S., Hendry, D., & Bridgeman, B. (1976). Saccadic suppression of image displacement. *Vision Research, 16*, 1185–1187.

Telford, L., & Frost, B. J. (1993). Factors affecting the onset and magnitude of linear vection. *Perception & Psychophysics, 53*, 682–692.

Thomas, J. P. (1975). Spatial resolution and spatial interaction. In E. C. Carterette & M. P. Friedman (Eds.), *Handbook of perception* (Vol. 5). New York, NY: Academic Press.

Thompson, P. (1982). Perceived rate of movement depends on contrast. *Vision Research, 22,* 377–380.

Thompson, W. B., Willemsen, P., Gooch, A. A., Creem-Regehr, S. H., Loomis, J. M., & Beall, A. C. (2004). Does the quality of the computer graphics matter when judging the distance in visually immersive environments? *Presence: Teleoperators and Virtual Environments, 13,* 560–571.

To, M. P. S., Regan, B. C., Wood, D., & Mollon, J. D. (2011). Vision out of the corner of the eye. *Vision Research, 51,* 203–214.

Toet, A., van-Eekhout, M. P., Simons, H. L., & Koenderink, J. J. (1987). Scale invariant features of differential spatial displacement discrimination. *Vision Research, 27,* 441–451.

Tong, F., Meng, M., & Blake, R. (2006). Neural bases of binocular rivalry. *Trends in Cognitive Sciences, 10*(11), 502–511.

Tschermak, A. (1931). Optischer raumsinn (optical sense of space). In A. Bethe, G. Bergnann, G. Emden, & A. Ellinger (Eds.), *Handbuch der normalen und pathologischen physiologie* (pp. 824–1000). Leipzig, Germany: Springer.

Turano, K. A., & Heidenreich, S. M. (1999). Eye movements affect the perceived speed of visual motion. *Vision Research, 39,* 1177–1188.

Tynan, P., & Sekuler, R. (1982). Motion processing in peripheral vision: Reaction time and perceived velocity. *Vision Research, 22,* 61–68.

Uchikawa, K., & Ikeda, M. (1981). Temporal deterioration of wavelength discrimination with successive comparison method. *Vision Research, 21,* 591–595.

van den Berg, R., Roerdink, J. B. T. M., & Cornelissen, F. W. (2007). On the generality of crowding: Visual crowding in size, saturation and hue compared to orientation. *Journal of Vision, 7*(2), 1–11.

Van Ness, F. L., & Bouman, M. A. (1967). Variation of contrast sensitivity with luminance. *Journal of the Optical Society of America, 57,* 401–406.

Viguier, A., Clement, G., & Trotter, Y. (2001). Distance perception within near visual space. *Perception, 30*(1), 115–124.

Wallach, H., & Kravitz, J.-H. (1965). Rapid adaptation in the constancy of visual direction with active and passive rotation. *Psychonomic-Science, 3*(4), 165–166.

Wandell, B. A. (1995). *Foundations of vision.* Sunderland, MA: Sinauer Associates.

Wang, P., & Nikolic, D. (2011). An LCD monitor with sufficiently precise timing for research in vision. *Frontiers in Human Neuroscience, 5*(85), 1–10.

Wann, J. P., White, A. D., Wilkie, R. M., Culmer, P. R., Lodge, J. P. A., & Mon-Williams, M. (2014). Measurement of visual aftereffects following virtual environment exposure. In K. S. Hale & K. M. Stanney (Eds.), *Handbook of Virtual Environments* (2nd ed., pp. 809–852). Boca Raton, FL: CRC Press.

Watson, A. B., Ahumada, A. J., & Farrell, J. E. (1986). Window of visibility: A psychophysical theory of fidelity in time-sampled visual motion displays. *Journal of the Optical Society of America A-Optics and Image Science, 3,* 300–307.

Watt, R. J. (1987). Scanning from coarse to fine spatial scales in the human visual system after the onset of a stimulus. *Journal of the Optical Society of America, A4,* 2006–2021.

Watt, R. J., Morgan, M. J., & Ward, R. M. (1983). The use of different cues in vernier acuity. *Vision Research, 23,* 991–995.

Webster, M. A. (2011). Adaptation and visual coding. *Journal of Vision, 11*(5:3), 1–23.

Wertheim, A. H. (1994). Motion perception during self-motion: The direct versus inferential controversy revisited. *Behavioral and Brain Sciences, 17,* 293–355.

Westheimer, G. (1978). Spatial phase sensitivity for sinusoidal grating targets. *Vision Research, 18,* 1073–1074.

Westheimer, G. (1982). The spatial grain of the perifoveal visual field. *Vision Research, 22,* 157–162.

Westheimer, G., & McKee, S. P. (1977). Spatial configurations for visual hyperacuity. *Vision Research, 17,* 941–947.

Whitney, D. (2002). The influence of visual motion on perceived position. *Trends in Cognitive Sciences, 6*(5), 211–216.

Whitney, D., & Levi, D. M. (2011). Visual crowding: A fundamental limit on conscious perception and object recognition. *Trends in Cognitive Sciences, 15*(4), 160–168.

Willemsen, P., Colton, M. B., Creem-Regehr, S. H., & Thompson, W. B. (2009). The effects of head-mounted display mechanical properties and field of view on distance judgments in virtual environments. *ACM Transactions on Applied Perception, 6,* 1–14.

Wilson, H. R., Ferrera, V. P., & Yo, C. (1992). A psychophysically motivated model for two-dimensional motion perception. *Visual Neuroscience, 9,* 79–97.

Wilson, H. R., & Wilkinson, F. (1998). Detection of global structure in Glass patterns: Implications for form vision. *Vision Research, 38,* 2933–2947.

Wright, W. G., DiZio, P., & Lackner, J. R. (2006). Perceived self-motion in two visual contexts: Dissociable mechanisms underlie perception. *Journal of Vestibular Research, 16,* 23–28.

Wyszecki, G., & Stiles, W. S. (1967). *Color science: Concepts and methods, quantitative data and formulae.* New York, NY: John Wiley & Sons.

Zacharias, G. L., & Young, L. R. (1981). Influence of combined visual and vestibular cues on human perception and control of horizontal rotation. *Experimental Brain Research, 41,* 159–171.

Zikovitz, D. C., Jenkin, M., & Harris, L. R. (2001). Comparison of stereoscopic and non-stereoscopic optic flow displays. *Journal of Vision, 1*(3), 317.

4 Virtual Auditory Displays

Michael Vorländer and Barbara Shinn-Cunningham

CONTENTS

4.1   INTRODUCTION

Auditory processing is often given little attention when designing virtual environments (VEs) or simulations. This lack of attention is unfortunate because auditory cues play a crucial role in everyday life. Auditory cues increase awareness of surroundings, cue visual attention, and convey a variety of complex information without taxing the visual system. The entertainment industry has long recognized the importance of sound to create ambience and emotion, aspects that are often lacking in VEs. Placing someone in a virtual world with an improperly designed auditory interface is equivalent to creating a *virtual* hearing impairment for the user, making them less aware of their surroundings, and contributing to feelings of isolation.

Auditory perception, especially localization, is a complex phenomenon affected by physiology, expectation, and even the visual interface. This chapter will consider different methods for creating auditory interfaces. As will be discussed, spatialized audio using headphones or transaural systems is necessary to create compelling sound, and spatialized sound offers the sound engineer the greatest amount of control over the auditory experience of the listener. Multichannel audio systems can produce virtual sound events in 3D to some extent, but they require complex equipment and signal processing. For many applications, especially using projection screens, standard stereo speaker systems may be simpler to implement and provide benefits not available to headphone systems. Properly designed speaker systems, especially using subwoofers, may contribute to emotional context. The positives and negatives associated with each option will be discussed.

It is impossible to include everything that needs to be known about designing auditory interfaces in a single chapter. The current aim is to provide a starting point, laying out the essential theory behind implementing sound in a VE without overwhelming the novice designer. Instead of trying to review all perceptual and technical issues related to creating virtual auditory displays, this chapter focuses on fundamental aspects of spatial auditory perception and the generation of spatial auditory cues in VEs. Specifically, the chapter begins by introducing basic properties of sound and discussing the perception of these sound properties (psychoacoustics), with a special emphasis on spatial hearing. General techniques for producing auditory stimuli, both with and without spatial cues, are then considered (see Letowski et al., 2001, for a lexicon for understanding auditory displays). Unlike the visual channel, very little effort has been put into formulating theories concerning the creation of synthetic sound sources in VEs; the question of how to generate realistic sounds (rather than using sources from some precomputed, stored library of source sounds) is beyond the scope of this chapter. In addition, the technology involved in producing spatialized audio is rapidly changing, with new products introduced all the time as others disappear, so that any specific recommendations would quickly be dated. However, an overview of current technology and solutions is presented at the conclusion of the chapter.

4.1.1   WHY ARE VIRTUAL AUDITORY INTERFACES IMPORTANT?

4.1.1.1   Environmental Realism and Ambience

If it does nothing else, an auditory interface should convey basic information about the VE to the user. For instance, in the real world, pedestrians walking through a shopping area are aware of everything from the sound of their own footsteps to the sounds of other shoppers to the mechanical sounds from cash registers, scanners, escalators, and other machines. In *control room* situations such as nuclear power plants, air traffic control centers, or the bridge of a ship, sounds such as alarms, switches being toggled, and verbal communications with other people in the room (including sounds of stress or uncertainty) provide vital information for participants. The location of these voices, switches, and alarms also provides information concerning their function and importance. In the absence of these basic auditory cues, situational awareness is severely degraded. The same is true in VEs.

The entertainment industry has recognized that sound is a vital aspect of creating ambience and emotion for films. George Lucas, best recognized by the public for stunning visual effects in his movies, has stated that sound is 50% of the movie experience (THX Certified Training Program, 2000). In VEs, the argument is often erroneously made that sound is secondary, since the visual image of a police car chasing down a city street can be compelling on its own. However, without appropriate sounds (squealing tires, a police siren, the tortured breathing of the driver, etc.), the emotional impact of a simulation is muted. The sound quality of footsteps depends on whether you are in grass, on pavement, or in a hallway. Likewise, the sound of one's own voice differs depending on whether you are inside a room or in an open field. These are the types of things that create ambience and feeling in film; the same is true in VEs.

4.1.1.2 Presence/Immersion and Perceived Simulation Quality

Presence (Chertoff & Schatz, 2013; Chapter 34, this book) can be defined as the "sense of being immersed in a simulation or virtual environment." Such a nebulous concept is difficult to quantify. Although definitive evidence is lacking, it is generally believed that the sense of presence is dependent on auditory, visual, and tactile fidelity (Sheridan, 1996). Referring back to the previous section, it can be inferred that as environmental realism increases, the sense of presence increases. However, although realism probably contributes to the sense of presence, it is not necessarily true that an increased sense of presence results in a greater sense of realism. Specifically, although virtual or spatial audio does not necessarily increase the perceived realism of a VE, it does increase the sense of presence (Hendrix & Barfield, 1996). Thus, if implemented properly, appropriately designed audio increases the overall sense of presence in a VE or simulation. Indeed, using medium- and high-quality auditory displays can enhance the perceived quality of visual displays. Inversely, using low-quality auditory displays reduces the perceived quality of visual displays (Storms, 1998).

4.1.1.3 Selective Auditory Attention

In a multisource sound environment, it is easier to segregate, attend to, and comprehend sound sources if they are separated in space, something known as the *cocktail party effect* (Cherry, 1953; Shinn-Cunningham, 2008; Yost, 2006). This ability to direct selective auditory attention, which enables a listener to process whatever sound source is most important at a given moment, is critical in many common situations such as teleconferencing (Begault, 1999) or multichannel radio communications (Begault, 1993; Begault & Wenzel, 1992; Haas, Gainer, Wightman, Couch, & Shilling, 1997). Even when spatial sound cues are imperfect (which can degrade sound localization accuracy), they can improve communication in multichannel situations (Drullman & Bronkhorst, 2000; Shinn-Cunningham, Ihlefeld, & Satyavarta, 2005).

4.1.1.4 Spatial Auditory Displays

While graphical displays are an obvious choice for displaying spatial information to a human operator (particularly after considering the spatial acuity of the visual channel), the visual channel is often overloaded, with operators monitoring a myriad of dials, gauges, and graphic displays. In these cases, spatial auditory cues can provide invaluable information to an operator, particularly when the visual channel is saturated (Begault, 1993; Bronkhorst, Veltman, & van Breda, 1996; Shilling & Letowski, 2000). Spatial auditory displays are also being developed for use in applications for which visual information provides no benefit, for instance, in limited field-of-view (FOV) applications or when presenting information to the blind. In command/control applications, the primary goal is to convey unambiguous information to the human operator. In such situations, realism, per se, is not useful, except to the extent that it makes the operator's task easier (i.e., reduces the workload); however, spatial resolution is critical. In these applications, signal-processing schemes that could enhance the amount of information transferred to the human operator may be useful, even if the result is *unnatural*, as long as the user is able to extract this information (e.g., see Durlach, Shinn-Cunningham, & Held, 1993). It should be noted that when designing spatialized auditory displays for noisy environments such as cockpits, electronic noise cancellation technology should be employed and user's hearing loss taken into account to make certain that the displayed information is perceptible to the user (Begault, 1996). Also, for high-g environments, more work needs to be conducted to discover the contribution of g-forces to displacements in sound localization (e.g., Clark & Graybiel, 1949; DiZio, Held, Lackner, Shinn-Cunningham, & Durlach, 2001).

4.1.1.5 Cross-Modal Interactions

The importance of multimodal interactions involving the auditory system cannot be ignored (Popescu et al., 2014, Chapter 17; Simpson Cowgill, Gilkey, & Weisenberger, 2014, Chapter 13). A whole range of studies show that judgments about one sensory modality are influenced by information in other sensory modalities. For instance, localized auditory cues reduce response times to visual targets (Frens, van Opstal, & van der Willigen, 1995; Perrott, Saberi, Brown, & Strybel, 1990; Perrott, Sadralodabai, Saberi, & Strybel, 1991). Similarly, the number of auditory events affects the perceived number of visual events occurring at that time (e.g., see Shams, Kamitani, & Shimojo, 2004). Even noninformative auditory cues can improve accuracy of perception of visual motion (Kim, Peters, & Sham, 2012), demonstrating the power of cross-modal perceptual effects. Auditory cues also augment or even substitute for tactile and/or visual information about events that are difficult to perceive in these other modalities, such as visual information outside a limited FOV. Through such cross-modal interactions, auditory cues can play an important role in conveying information that may, superficially, seem to be more naturally communicated through some other sensory channel.

4.2 PHYSICAL ACOUSTICS

4.2.1 PROPERTIES OF SOUND

Sound is a pressure wave produced when an object vibrates rapidly back and forth. The diaphragm of a speaker produces sound by pushing against molecules of air, thus creating an area of high pressure (*condensation*). As the speaker's diaphragm returns to its resting position, it creates an area of low pressure (*rarefaction*). This localized disturbance travels through the air as a wave of alternating low pressure and high pressure at approximately 344 m/s or 1128 ft/s (at 70°F), depending on temperature and humidity.

4.2.1.1 Frequency

If the musical note "A" is played as a pure sinusoid, there will be 440 condensations and rarefactions per second. The distance between two adjacent condensations or rarefactions, typically represented by the symbol λ, equals the wavelength of the sound wave. The velocity at which the sound wave is traveling is denoted as c. The time one full oscillatory cycle (condensation through rarefaction) takes is called the frequency (f) and is expressed in Hertz or cycles per second. The relationship between frequency, velocity, and wavelength is given by $f = c/\lambda$.

From a modeling standpoint, this relationship is important when considering the Doppler shift. As a sound source is moving toward a listener, the perceived frequency increases because the wavelength is compressed as a function of the velocity (v) of the moving source. This compression can be explained by the equation $\lambda = (c - v)/f$. For negative velocities (i.e., for sources moving away), this expression describes a relative increase in the wavelength (and a concomitant decrease in frequency).

4.2.1.2 Strength

The amplitude of the waveform determines the intensity of a sound stimulus. It should not be confused, however, with the sound intensity defined in physics as the sound energy propagating per second through a reference area. The meaning of intensity in the context of this chapter is simply the meaning of strength. It is measured in decibels (dB). Decibels give the level of sound (on a logarithmic scale) relative to some reference level. One common reference level is 2×10^{-5} N/m^2. Decibels referenced to this value are commonly used to describe sound intensity expressed in units of dB sound pressure level (SPL). The sound level in dB SPL can be computed by the following equation:

$$\text{dB SPL} = 20 \log_{10}\left(\frac{\text{RMS sound pressure}}{20 \times 10^{-6} \text{ N/m}^2}\right)$$

The threshold of hearing is in the range of 0–10 dB SPL for most sounds, although the actual threshold depends on the spectral content of the sound. When measuring sound strength in the *real world*, it is measured with a sound pressure meter. Most sound pressure meters allow one to collect sound-level information using different scales that weight energy in different frequencies differently in order to approximate the sensitivity of the human auditory system to sound at low-, moderate-, or high-intensity levels. These scales are known as A, B, and C weighted scales, respectively. The B scale is rarely used; however, the C scale (dBC) is useful for evaluating noise levels in high-intensity environments such as traffic noise and ambient cockpit noise. The frequency response of the dBC measurement is closer to an unfiltered (flat) response than dBA. In fact, when conducting *sound surveys* in a complex noise environment, it is prudent to measure sound level in both dBA and flat response (or dBC) to make an accurate assessment of the audio environment.

Frequency, intensity, and complexity are physical properties of an acoustic waveform. The perceptual analogs for frequency, intensity, and complexity are pitch, loudness, and timbre, respectively. Although the distinction between physical and perceptual measures of sound properties is an important one, both physical and perceptual descriptions are important when designing auditory displays.

4.3 PSYCHOPHYSICS

The basic sensitivity of the auditory system is reviewed in detail in a number of textbooks (e.g., see Gelfand, 1998; Moore, 1997; Yost, 2006; Zwicker & Fastl, 2007). This section provides a brief overview of some aspects of human auditory sensitivity that are important to consider when designing auditory VEs.

4.3.1 FREQUENCY ANALYSIS IN THE AUDITORY SYSTEM

In the cochlea, acoustic signals are broken down into constituent frequency components by a mechanical Fourier-like analysis. Along the length of the cochlea, the frequency to which that section of the cochlea responds varies systematically from high to low frequencies. The strength of neural signals carried by the auditory nerve fibers arrayed along the length of the cochlea varies with the mechanical displacement of the corresponding section of the cochlea. As a result, each nerve fiber can be thought of as a frequency channel that conveys information about the energy and timing of the input signal within a restricted frequency region. At all stages of the auditory system, these multiple frequency channels are evident.

Although the bandwidth changes with the level of the input signal and with input frequency, to a crude first-order approximation, one can think of the frequency selectivity of the auditory system as constant on a log-frequency basis (approximately one-third octave wide). Thus, a particular auditory nerve responds to acoustic energy at and near a particular frequency.

Humans are sensitive to acoustic energy at frequencies between about 20 and 22,000 Hz. Absolute sensitivity varies with frequency. Humans are most sensitive to energy at frequencies around 2000 Hz and are less sensitive for frequencies below and above this range.

The fact that input waveforms are deconstructed into constituent frequencies affects all aspects of auditory perception. Many behavioral results are best understood by considering the activity of the auditory nerve fibers, each of which responds to energy within about a third of an octave of its particular *best frequency*. For instance, the ability to detect a sinusoidal signal in a noise background degrades dramatically when the noise spectrum is within a third octave of the sinusoid frequency. When a noise is spectrally remote from a sinusoidal target, it causes much less interference with the detection of the sinusoid. These factors are important when one considers the spectral content of different sounds that are to be used in an auditory VE. For instance, if one must monitor multiple kinds of alerting sounds, choosing signals that are spectrally remote from one another will improve a listener's ability to detect and respond to different signals.

4.3.2 Intensity Perception

Listeners are sensitive to sound intensity on a logarithmic scale. For instance, doubling the level of a sound source causes roughly the same perceived change in the loudness independent of the reference level. This logarithmic sensitivity to sound intensity gives the auditory system a large dynamic range. For instance, the range between just detectable sound levels and sounds that are so loud that they cause pain is roughly 110–120 dB (i.e., an increase in sound pressure by a factor of a million). The majority of the sounds encountered in everyday experience span a dynamic intensity range of 80–90 dB. Typical sound reproduction systems use 16 bits to represent the pressure of the acoustic signal (providing a useful dynamic range of about 90 dB), which is sufficient for most simulations.

While sound intensity (a physical measure) affects the loudness of a sound (a perceptual measure), loudness does not grow linearly with intensity. In addition, the same decibel increase in sound intensity can result in different increments in loudness, depending on the frequency content of the sound. Thus, intensity and loudness, while closely related, are not equivalent descriptions of sound.

4.3.3 Masking Effects

As mentioned earlier, when multiple sources are presented to a listener simultaneously or in rapid succession, the sources interfere with one another in various ways. For instance, a tone that is audible when played in isolation may be inaudible when a loud noise is presented simultaneously. Such effects (known as *masking* effects) arise from a variety of mechanisms, from physical interactions of the separate acoustic waves impinging on the ear to high-level, cognitive factors. For a more complete description of these effects than is given in the following, see Yost (2006, pp. 153–167) or Moore (1997, pp. 111–120).

Simultaneous masking occurs when two sources are played concurrently. However, signals do not have to be played simultaneously for them to interfere with one another perceptually. For instance, both *forward* masking (in which a leading sound interferes with perception of a trailing sound) and *backward* masking (in which a lagging sound interferes with perception of a leading sound) occur. Generally speaking, many simultaneous and forward-masking effects are thought to arise from peripheral interactions that occur at or before the level of the auditory nerve. For instance, the mechanical vibrations of the basilar membrane are nonlinear, so that the response of the membrane to two separate sounds may be less than the sum of the response to the individual sounds. These nonlinear interactions can suppress the response to what would (in isolation) be an audible event.

Other, more central factors influence masking as well. For instance, backward masking may reflect higher-order processing that limits the amount of information extracted from an initial sound in the presence of a second sound. The term *informational masking* refers to all masking that cannot be explained by peripheral interactions in the transduction of sound by the auditory periphery. Most such effects can be traced to problems with segregating a source of interest from other, competing sources, problems with identifying which source in a sound mixture is the most important (*target*) source, or some combination thereof (Shinn-Cunningham, 2008). These failures of selective auditory attention can have significant impact on perception, even for sounds that are clearly audible. For instance, perceptual sensitivity in discrimination and detection tasks is often degraded when there is uncertainty about the characteristics of a target source (e.g., see Yost, 2006, pp. 219–220).

4.3.4 Pitch and Timbre

Just as sound intensity is the physical correlate of the percept of loudness, source frequency is most closely related to the percept of pitch. For sound waves that are periodic (including pure sinusoids, for instance), the perceived *pitch* of a sound is directly related to the inverse of the period of the sound signal. Thus, sounds with low pitch have relative long periods and sounds with high pitch generally have short periods. Many real-world sounds are not strictly periodic in that they have a temporal pattern that repeats over time but has fluctuations from one cycle to the next. Examples of such pseudo-periodic signals include the sound produced by a flute or a vowel sound spoken aloud. The perceived pitch of such sounds is well predicted by the average period of the cyclical variations in the stimulus.

The percept of pitch is not uniformly strong for all sound sources. In fact, nonperiodic sources such as noise do not have a salient pitch associated with them. For relatively narrow sources that are aperiodic, or for band-limited noise, a weak percept of pitch can arise that depends on the center frequency or the cutoff frequency of the spectral energy of the signal, respectively. In fact, perceived pitch is affected by a wide variety of stimulus attributes, including temporal structure, frequency content, harmonicity, and even loudness. Although the pitch of a pure sinusoid is directly related to its frequency, there is no single physical parameter that can predict perceived pitch for more complex sounds. Nonetheless, for many sounds, pitch is a very salient and robust perceptual feature that can be used to convey information to a listener. For instance, in music, pitch conveys melody. In speech, pitch conveys a variety of information (ranging from the gender of a speaker to paralinguistic, emotional content of a speech). Pitch is also a very important cue for segregating competing sound sources and allowing a listener to focus selective auditory attention on a target source (e.g., see Carlyon, 2004).

The percept of timbre is the sound property that enables a listener to distinguish an oboe from a trumpet. Like pitch, the percept of timbre depends on a number of physical parameters of sound, including spectral content and temporal envelope (such as the abruptness of the onset and offset of sound). Like pitch, timbre is an important property for enabling listeners to identify a target source and thus can be used to convey information through an auditory display (e.g., see Brewster, Wright, & Edwards, 1993). However, sounds with different timbres have different perceptual weight, a factor that should be considered in designing discrete sounds for an auditory display (e.g., see Chon & McAdams, 2012). As with pitch, timbre is a feature that allows listeners to direct selective auditory attention to a desired target amidst competing sounds, enabling them to extract desired information from that source despite the presence of interfering information (e.g., Maddox & Shinn-Cunningham, 2011).

4.3.5 Temporal Resolution

The auditory channel is much more sensitive to temporal fluctuations in sensory inputs than either visual (Badcock, Palmisano, & May, 2013, Chapter 3) or proprioceptive (Dindar, Tekalp, & Basdogan, 2013,

Chapter 5; Lawson & Riecke, 2014, Chapter 7) channels. For instance, the auditory system can detect amplitude fluctuations in input signals up to 50 Hz (i.e., a duty cycle of 20 ms) very easily (e.g., see Yost, 2006, pp. 146–149). Sensitivity degrades slowly with increasing modulation rate, so that some sensitivity remains even as the rate approaches 1000 Hz (i.e., temporal fluctuations at a rate of 1 per ms). The system is also sensitive to small fluctuations in the spectral content of an input signal for roughly the same modulation speeds. Listeners not only can detect rapid fluctuations in an input stimulus, but they can react quickly to auditory stimuli. For instance, reaction times to auditory stimuli are faster than visual reaction times by 30–40 ms (an improvement of roughly 20%; e.g., see Welch & Warren, 1986).

4.3.6 SPATIAL HEARING

Spatial acuity of the auditory system is far worse than that of the visual (Badcock et al., 2013, Chapter 3) or proprioceptive (Dindar et al., 2013, Chapter 5; Lawson & Riecke, 2013, Chapter 7) systems (for a review, see Middlebrooks & Green, 1991). For a listener to detect an angular displacement of a source from the median plane, the source must be displaced laterally by about a degree. For a source directly to the side, the listener does not always detect a lateral displacement of 10°. Auditory spatial acuity is even worse in other spatial dimensions. A source in the median plane must be displaced by as much as 15° for the listener to perceive the directional change accurately. While listeners can judge relative changes in source distance, absolute distance judgments are often surprisingly inaccurate even under the best of conditions (e.g., see Zahorik, Brungart, & Bronkhorst, 2005).

Functionally, spatial auditory perception is distinctly different from that of the other *spatial* senses of vision and proprioception. For the other spatial senses, position is neurally encoded at the most peripheral part of the sensory system. For instance, the photoreceptors of the retina are organized topographically so that a source at a particular position (relative to the direction of gaze) excites a distinct set of receptors (Badcock et al., 2013, Chapter 3). In contrast, spatial information in the auditory signals reaching the left and right ears of a listener must be computed from the peripheral neural representations. The way in which spatial information is carried by the acoustic signals reaching the eardrums of a listener has been the subject of much research. This section provides a brief review of how acoustic attributes convey spatial information to a listener and how the perceived position of a sound source is computed in the brain (for more complete reviews, see Blauert, 1997; Middlebrooks & Green, 1991; Mills, 1972; Wightman & Kistler, 1993).

4.3.6.1 Head-Related Transfer Functions

The pairs of spatial filters that describe how sound is transformed as it travels through space to impinge on the left and right ears of a listener are known as head-related transfer functions (HRTFs). HRTFs describe how to simulate the direct sound reaching the listener from a particular position but do not generally include any reverberant energy. Empirically measured HRTFs vary mainly with the direction from head to source but also vary with source distance (particularly for sources within reach of the listener). For sources beyond about a meter away, the main effect of distance is just to change the overall gain of the HRTFs. In the time domain, the HRTF pair for a particular source location provides the pressure waveforms that would arise at the ears if a perfect impulse were presented from the spatial location in question. Often, HRTFs are represented in the frequency domain by taking the Fourier transform of time-domain impulse responses.

HRTFs contain most of the spatial information present in real-world listening situations. In particular, binaural cues are embodied in the relative phase and magnitude (respectively) of the linear filters for the left and right ears. Spectral cues and source intensity are present in the absolute frequency-dependent magnitudes of the two filters.

Figure 4.1 shows two HRTF pairs from a human subject in the time domain (left side of figure) and in the frequency domain (magnitude only, right side of figure). All panels are for a source at

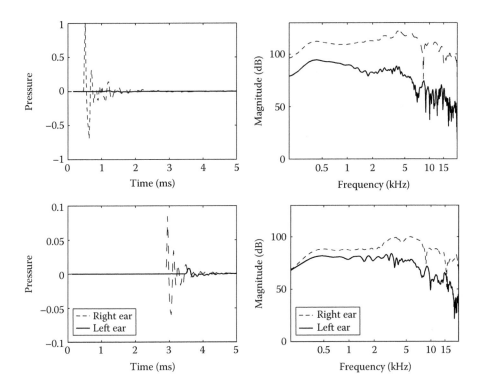

FIGURE 4.1 Time-domain (left panels) and magnitude spectra (right panels) representations of anechoic HRTFs for a human subject. All panels show a source at 90 azimuth and 0 elevation. Top panels are for a source at 15 cm, and bottom panels for a source at 1 m.

azimuth 90 and elevation 0. The top two panels show the HRTF for a source very close to the head (15 cm from the center of the head). The bottom two panels show the HRTF for a source 1 m from the head. In the time domain, it is easy to see the interaural differences in time and intensity, while the frequency-domain representation shows the spectral notches that occur in HRTFs, as well as the frequency-dependent nature of the interaural level difference (ILD). The ILDs are larger in all frequencies for the nearer source (top panels) as expected. In the time domain, the 1 m source must traverse a greater distance to reach the ears than the near source, resulting in additional time delay before the energy reaches the ears (note time-onset differences in the impulse responses in the left top and left bottom panels).

4.3.6.2 Binaural Cues

The most robust cues for source position depend on differences between the signals reaching the left and right ears. Such *interaural* or *binaural* cues are robust specifically because they can be computed by comparing the signals reaching each ear. As a result, binaural cues allow a listener to factor out those acoustic attributes that arise from source content from those attributes that arise from source position.

Depending on the angle between the interaural axis and a sound source, one ear may receive a sound earlier than the other. The resulting interaural time differences (ITDs) are the main cue indicating the laterality (left/right location) of the direct sound. The ITD grows with the angle of the source from the median plane; for instance, a source directly to the right of a listener results in an ITD of 600–800 μs favoring the right ear. ITDs are most salient for sound frequencies below about 2 kHz but occur at all frequencies in a sound. At higher frequencies, listeners use ITDs in signal *envelopes* to help determine source laterality but are insensitive to differences in the interaural phase of the signal.

Listeners can reliably detect ITDs of 10–100 μs (depending on the individual listener), which grossly correspond to ITDs that would result from a source positioned 1°–10° from the median plane. Sensitivity to changes in the ITD deteriorates as the reference ITD gets larger. For instance, the smallest detectable change in ITD around a reference source with an ITD of 600–800 μs (corresponding to the ITD of a source far to the side of the head) can be more than a factor of 2 larger than for a reference with zero ITD.

At the high end of the audible frequency range, the head of the listener reflects and diffracts signals so that less acoustic energy reaches the far side of the head (causing an *acoustic head shadow*). Due to the acoustic head shadow, the relative intensity of a sound at the two ears varies with the lateral location of the source. The resulting ILDs generally increase with source frequency and angle between the source and median plane. ILDs are perceptually important for determining source laterality for frequencies above about 2 kHz.

When a sound source is within reach of a listener, extralarge ILDs (at all frequencies) arise due to differences in the relative distances from source to left and right ears (e.g., see Brungart & Rabinowitz, 1999; Duda & Martens, 1998). These additional ILDs are due to differences in the relative distances from source to left and right ears and help to convey information about the relative distance and direction of the source from the listener (Shinn-Cunningham, Santarelli, & Kopco, 2000). Other low-frequency ILDs that may arise from the torso appear to help determine the elevation of a source (Algazi, Avendano, & Duda, 2001). Most listeners are able to detect ILDs of 0.5–1 dB, independent of source frequency.

The perceived location of a sound source usually is consistent with the ITD and ILD information available. However, there are multiple source locations that cause roughly the same ITD and ILD cues. For sources more than a meter from the head, the locus of such points is approximately a hyperbolic surface of rotation symmetric about the interaural axis that is known as the *cone of confusion* (see left side of Figure 4.2). When a sound is within reach of the listener, extralarge ILDs provide additional robust, binaural information about the source location. For a simple spherical

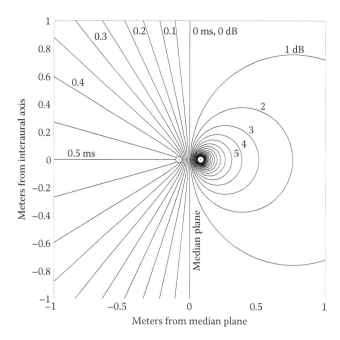

FIGURE 4.2 Iso-ITD (left side of figure) and iso-ILD (right side of figure) contours for sources near the head. On the left, sources at each location along a contour give rise to nearly the same ITD. On the right, sources at each location along a contour give rise to nearly the same unique near-field ILD component.

head model, low-frequency ILDs are constant on spheres centered on the interaural axis (see right side of Figure 4.2). The rate at which the extralarge ILD changes with spatial location decreases as sources move far from the head or near the median plane. In fact, once a source is more than a meter or so from the head, the contribution of this *near-field* ILD is perceptually insignificant (Shinn-Cunningham et al., 2000). In general, positions that give rise to the same binaural cues (i.e., the intersection of constant ITD and ILD contours) form a circle centered on the interaural axis (Shinn-Cunningham et al.). Since ITD and ILD sensitivity is imperfect, the locus of positions that cannot be resolved from binaural cues may be more accurately described as a *torus of confusion* centered on the interaural axis. Such tori of confusion degenerate to the more familiar cones of confusion for sources more than about a meter from a listener.

4.3.6.3 Spectral Cues

The main cue to resolve source location on the torus of confusion is the spectral content of signals reaching the ears. These spectral cues arise due to interactions of the outer ear (pinna) with the impinging sound wave that depend on the relative position of sound source and listener's head (Batteau, 1967). Spectral cues only occur at relatively high frequencies, generally above 6 kHz. Unlike interaural cues for source location, spectral cues can be confused with changes in the spectrum of the source itself. Perhaps because of this ambiguity, listeners are more likely to make localization errors in which responses fall near the correct torus of confusion but are not in the right direction. Individual differences in spectral filtering of the pinnae are large and are important when judging source direction (e.g., see Wenzel, Arruda, Kistler, & Wightman, 1993).

4.3.6.4 Anechoic Distance Cues

In general, the intensity of the direct sound reaching a listener (i.e., sound that does not come off of reflective surfaces like walls) decreases with the distance of the source. In addition, the atmosphere absorbs energy in high, audible frequencies as a sound propagates, causing small changes in the spectrum of the received signal with changes in source distance. If a source is unfamiliar, the intensity and spectrum of the direct sound are not robust cues for distance because they can be confounded with changes in the intensity or spectral content (respectively) of the signal emitted from the source. However, even for unfamiliar sources, overall level and spectral content provide relative distance information (Mershon, 1997).

4.3.6.5 Reverberation

Reverberation (acoustic energy reaching a listener from indirect paths, via walls, floors, etc.) generally has little affect on or degrades the perception of source direction (e.g., see Begault, 1993; Hartmann, 1983; Shinn-Cunningham, 2000b). However, it actually aids in the perception of source distance (e.g., see Mershon, Ballenger, Little, McMurty, & Buchanan, 1989; Shinn-Cunningham, Kopco, & Santarelli, 1999). At least grossly, the intensity of reflected energy received at the ears is independent of the position of the source relative to the listener (although it can vary dramatically from one room to another). As a result, the ratio of direct to reverberant energy provides an absolute measure of source distance for a given listening environment.

Reverberation not only provides a robust cue for source distance, but it also provides information about the size and configuration of a listening environment. For instance, information about the size and *spaciousness* of a room can be extracted from the pattern of reverberation in the signals reaching the ears. While many psychophysical studies of sound localization are performed in anechoic (or simulated anechoic) environments, reverberation is present (in varying degrees) in virtually all everyday listening conditions. Anechoic environments (such as those used in many simulations and experiments) seem subjectively unnatural and strange to naive listeners. Conversely, adding reverberation to a simulation causes all sources to seem more realistic and provides robust information about relative source distance (e.g., see Begault, Wenzel, Lee, & Anderson, 2012; Brungart & D'Angelo, 1995). While reverberation may improve distance perception and improve the realism of a display, it can decrease accuracy in directional perception, albeit slightly, and may interfere with the ability to extract information in a source signal (e.g., degrade speech) and to attend to multiple sources (e.g., see Section 4.3.6.8).

4.3.6.6 Dynamic Cues

In addition to *static* acoustic cues like ITD and ILD, changes in spatial cues with source or listener movement also influence perception of source position and help to resolve torus-of-confusion ambiguities (e.g., see Wallach, 1940; Wightman & Kistler, 1989). For instance, a source either directly in front or directly behind a listener would cause near-zero ITDs and ILDs; however, a leftward rotation of the head results in ITDs and ILDs favoring either the right ear (for a source in front) or the left ear (for a source behind).

While the auditory system generally has good temporal resolution, the temporal resolution of the binaural hearing system is much poorer. For instance, investigations into the perception of moving sound sources imply that binaural information averaged over a time window lasting 100–200 ms results in what has been termed *binaural sluggishness* (e.g., see Grantham, 1997; Kollmeier & Gilkey, 1990).

4.3.6.7 Effects of Stimulus Characteristics on Spatial Perception

Characteristics of a source itself affect auditory spatial perception in a number of ways. For instance, the bandwidth of a stimulus can have a large impact on both the accuracy and precision of sound localization. As a result, one must consider how nonspatial attributes of a source in a VE will impact spatial perception of a signal. In cases where one can design the acoustic signal (i.e., if the signal is a warning signal or some other arbitrary waveform), these factors should be taken into consideration when one selects the source signal.

For instance, the spectral filtering of the pinnae cannot be determined if the sound source does not have sufficient bandwidth. This makes it difficult to unambiguously determine the location of a source on the torus of confusion for a narrowband signal. Similarly, if a source signal does not have energy above about 5 kHz, spectral cues will not be represented in the signals reaching the ears and errors along the torus of confusion are more common (e.g., Gilkey & Anderson, 1995).

Ambiguity in narrowband source locations arises in other situations as well. For instance, narrowband, low-frequency signals in which ITD is the main cue can have ambiguity in their heard location because the auditory system is only sensitive to interaural phase. Thus, a low-frequency sinusoid with an ITD of half cycle favoring the right ear may also be heard far to the left side of the head. However, binaural information is integrated across frequency so that ambiguity in lateral location is resolved when interaural information is available across a range of frequencies (Brainard, Knudsen, & Esterly, 1992; Stern & Trahiotis, 1997; Trahiotis & Stern, 1989). When narrowband sources are presented, the heard location is strongly influenced by the center frequency of the source (Middlebrooks, 1997).

While spectral bandwidth is important, temporal structure of a source signal is also important. In particular, onsets and offsets in a signal make source localization more accurate, particularly when reverberation and echoes are present. A gated or modulated broadband noise will generally be more accurately localized in a reverberant room (or simulation) than a slowly gated broadband noise (e.g., see Rakerd & Hartmann, 1985, 1986).

4.3.6.8 Top-Down Processes in Spatial Perception

Experience with or knowledge of the acoustics of a particular environment also affects auditory localization, and implicit learning and experience affects performance (e.g., see Clifton, Freyman, & Litovsky, 1993; Shinn-Cunningham, 2000b). In other words, spatial auditory perception is not wholly determined by stimulus parameters but also by the state of the listener. Although such effects are not due to conscious decision, they can measurably alter auditory localization and spatial perception. For instance, when localizing a sound followed by a later *echo* of that sound, the influence of the later sound diminishes with repetition of the sound pairing, as if the listener learns to discount the lagging echo (Freyman, Clifton, & Litovsky, 1991).

4.3.6.9   Benefits of Binaural Hearing

Listeners benefit from receiving different signals at the two ears in a number of ways. As discussed earlier, ITD and ILD cues allow listeners to determine the location of sound sources. However, in addition to allowing listeners to locate sound sources in the environment, binaural cues allow the listener to selectively attend to sources coming from a particular direction. This ability is extremely important when there are multiple competing sources in the environment (e.g., see Shinn-Cunningham, 2008).

Imagine a situation in which there is both a speaker (whom the listener is trying to attend) and a competing source (that is interfering with the speaker). If the speaker and competitor are both directly in front of the listener, the competitor degrades speech reception much more than if the competitor is off to one side, spatially separated from the speaker. This *binaural advantage* arises in part because when the competitor is off to one side of the head, the energy from the competitor is attenuated at the far ear. As a result, the signal-to-noise ratio at the far ear is larger than when the competitor is in front. In other words, the listener has access to a cleaner signal in which the speaker is more prominent when the speaker and noise are spatially separated. However, the advantage of the spatial separation is even larger than can be predicted on the basis of energy.

A homologous benefit can be seen under headphones. In particular, if one varies the level of a signal until it is just detectable in the presence of a masker, the necessary signal level is much lower when the ITD of the signal and masker are different than when they are the same. The difference between these thresholds, referred to as the masking level difference (MLD), can be as large as 10–15 dB for some signals (e.g., see Durlach & Colburn, 1978; Zurek, 1993).

The binaural advantage affects both signal detection (e.g., see Gilkey & Good, 1995) and speech reception (e.g., see Bronkhorst & Plomp, 1988). It is one of the main factors contributing to the ability of listeners to monitor and attend multiple sources in complex listening environments (i.e., the *cocktail party effect*; see, e.g., Shinn-Cunningham, 2008; Yost, 1997). Thus, the binaural advantage is important for almost any auditory signal of interest. In order to get these benefits of binaural hearing, signals reaching a listener must have appropriate ITDs and/or ILDs.

4.3.6.10   Adaptation to Distorted Spatial Cues

While a naive listener responds to ITD, ILD, and spectral cues based on their everyday experience, listeners can learn to interpret cues that are not exactly like those that occur naturally. For instance, listeners can learn to adapt to unnatural spectral cues when given sufficient long-term exposure (Hofman, Van Riswick, & Van Opstal, 1998). Short-term training allows listeners to learn how to map responses to spatial cues to different spatial locations than normal (Shinn-Cunningham, Durlach, & Held, 1998). These studies imply that for applications in which listeners can be trained, *perfect* simulations of spatial cues may not be necessary. However, there are limits to the kinds of distortions of spatial cues to which a listener can adapt (e.g., see Shinn-Cunningham, 2000a).

4.3.6.11   Intersensory Integration of Spatial Information

Acoustic spatial information is integrated with spatial information from other sensory channels (particularly vision) to form spatial percepts (e.g., see Welch & Warren, 1986). In particular, auditory spatial information is combined with visual (and/or proprioceptive) spatial information to form the percept of a single, multisensory event, especially when the inputs to the different modalities are correlated in time (e.g., see Popescu et al., 2013; Warren, Welch, & McCarthy, 1981; Chapter 17). When this occurs, visual spatial information is much more potent than that of auditory information, so the perceived location of the event is dominated by the visual spatial information (although auditory information does affect the percept to a lesser degree, e.g., see Pick, Warren, & Hay, 1969;

Welch & Warren, 1980). *Visual capture* refers to the perceptual dominance of visual spatial information, describing how the perceived location of an auditory source is captured by visual cues.

Summarizing these results, it appears that the spatial auditory system computes source location by combining all available acoustic spatial information. Perhaps even more importantly, a priori knowledge and information from other sensory channels can have a pronounced effect on spatial perception of auditory and multisensory events.

4.3.7 AUDITORY SCENE ANALYSIS

Listeners in real-world environments are faced with the difficult problem of listening to many competing sound sources that overlap in both time and/or frequency. The process of separating out the contributions of different sources to the total acoustic signals reaching the ears is known as *auditory scene analysis* (e.g., see Bregman, 1990; Carlyon, 2004).

In general, the problem of grouping sound energy across time and frequency to reconstruct each sound source is governed by a number of basic (often intuitive) principles. For instance, naturally occurring sources are often broadband, but changes in the amplitude or frequency of the various components of a single source are generally correlated over time. Thus, comodulation of sound energy in different frequency bands tends to group these signal elements together and cause them to fuse into a single perceived source. Similarly, temporal and spectral proximity both tend to promote grouping so that signals close in time or frequency are grouped into a single perceptual source (sometimes referred to as a stream). Spatial location can also influence auditory scene analysis such that signals from the same or similar locations are grouped into a single stream. Other factors affecting streaming include (but are not limited to) harmonicity, timbre, and frequency or amplitude modulation (Bregman, 1990; Darwin, 1997).

For the development of auditory displays, these grouping and streaming phenomena are very important because they can directly impact the ability to detect, process, and react to a sound. For instance, if a masker sound is played simultaneously with a target sound, the ability to process the target is significantly worse if the target is heard as just one component of a single sound source comprised of the masker plus the target; when the target is heard as a distinct sound source, a listener is much better at both detecting the target's presence and extracting meaning from the target. Such *grouping* effects cannot be explained solely by peripheral mechanisms, since many times, the target sound is represented faithfully in activity on the auditory nerve. Instead, such effects arise from *central* limitations (e.g., see Shinn-Cunningham, 2008).

4.3.8 SPEECH PERCEPTION

Arguably the most important acoustic signal is that of speech. The amount of information transmitted via speech is larger than any other acoustic signal. For many applications, accurate transmission of speech information is the most critical component of an auditory display.

Speech perception is affected by many of the low-level perceptual issues discussed in previous sections. For instance, speech can be masked by other signals, reducing a listener's ability to determine the content of the speech signal. Speech reception in noisy environments improves if the speaker is located at a different position than the noise source(s), particularly if the speaker and masker are at locations giving rise to different ILDs. Speech reception is also affected by factors that affect the formation of auditory streams, such as comodulation, harmonic structure, and related features. However, speech perception is governed by many high-level, cognitive factors that do not apply to other acoustic signals. For instance, the ability to perceive a spoken word improves dramatically if it is heard within a meaningful sentence rather than in isolation. Speech information is primarily conveyed by sound energy between 200 and 5000 Hz. For systems in which speech communication is critical, it is important to reduce the amount of interference in this range of frequencies or it will impede speech reception.

4.4 SPATIAL SIMULATION

Spatial auditory cues can be simulated using headphone displays or loudspeakers. Headphone displays generally allow more precise control of the spatial cues presented to a listener, both because the signals reaching the two ears can be controlled independently and because there is no indirect sound reaching the listener (i.e., no echoes or reverberation). However, headphone displays are generally more expensive than loudspeaker displays and may be impractical for applications in which the listener does not want to wear a device on the head. While it is more difficult to control the spatial information reaching a listener in a loudspeaker simulation, loudspeaker-based simulations are relatively simple and inexpensive to implement and do not physically interfere with the listener.

Simulations using either headphones or speakers can vary in complexity from providing no spatial information to providing nearly all naturally occurring spatial cues. This section reviews both headphone and speaker approaches to creating spatial auditory cues.

4.4.1 Room Modeling

HRTFs generally do not include reverberation or echoes, although it is possible to measure binaural transfer functions (known as binaural room transfer functions) that incorporate the acoustic effects of a room. While possible, such approaches are generally not practical because such filters vary with listener and source position in the room as well as the relative position of listener and source to produce a combinatorially large number of transfer functions. In addition, such filters can be an order of magnitude longer than traditional HRTFs, increasing both computational and storage requirements of the system.

There has been substantial effort devoted to developing computational models for room reverberation, including high-quality commercial software packages (e.g., see www.odeon.dk or www.catt.se). The required computations are quite intensive; in order to simulate each individual reflection, one must calculate the distance the sound wave has traveled, how the waveform was transformed by every surface on which it impinged, and the direction from which it is arriving at the head. The resulting waveform must then be filtered by the appropriate anechoic HRTF based on the direction of incidence with the head (Vorländer, 2008, pp. 141–146).

If one looks at the resulting reflections as a function of time from the initial sound, the number of reflections in any given time slice increases quadratically with time. At the same time, the level of each individual reflection decreases rapidly, both due to energy absorption at each reflecting surface and increased path length from source to ear. Moreover, those reflections lose their coherence due to surface scattering and edge diffraction. Although second- or third-order reflections may be individually resolvable, higher-order reflections occur so densely in time that the distinct specular content of each *echo* becomes practically irrelevant. Instead, from a certain transition time (which depends on the size of the room and on the surface corrugations), the reflections are smeared in time and heavily overlap to the point that they are well approximated as a so-called *diffuse* sound field. Therefore, many simulations only *spatialize* a relatively small number of the loudest, earliest-arriving reflections (e.g., up to second or third order) and then add random noise that dies off exponentially in time (uncorrelated at the two ears) to simulate later-arriving reflections that are dense in time and arriving from essentially random directions. Even with such simplifications, the computations necessary to generate *realistic* reverberation (particularly in a system that tries to account for movement of a listener) are a challenge; however, with today's computational power, such simulations are feasible and produce plausible results.

Figure 4.3 shows the room impulse response at the right ear for a source located at 45 azimuth, 0 elevation, and distance of 1 m. This impulse response was measured in a moderate-size classroom in which significant reverberant energy persists for as long as 450 ms. The initial few milliseconds of the response are shown in the inset. In the inset, the initial response is that caused by the sound

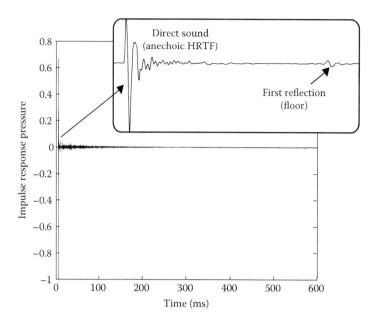

FIGURE 4.3   Impulse response at the right ear for a source at 45 azimuth, 0 elevation, and distance 1 m in a standard classroom. Inset shows first 10 ms of total impulse response.

wave that travels directly from the source to the ear. The first reflection is also evident at the end of the inset, at a much reduced amplitude. In the main figure, the decay of the reverberant energy can be seen with time.

The development of tractable reverberation algorithms for real-time systems has achieved some promising results but is still an ongoing area of research. Most algorithms are based on geometrical acoustics. They are known as hybrid specular and diffuse reflection models and combination of early (mainly specular) and late (mainly diffuse) components (e.g., see Vorländer, 2008, pp. 216–226).

Wave models are also in use in order to compute wave effects such as eigenmodes or diffraction (Aretz, Maier, & Vorländer, 2010; Botteldooren, 1995; Savioja, 2010). Usually this is implemented as a second level into the hybrid models so that the low-frequency range and the mid- and high-frequency range are treated by wave models and geometric models, respectively.

The rendering process using geometrical acoustics is based on a 3D room model consisting of polygons defining the room boundaries. These polygons represent the reflecting surfaces, and accordingly, they are tagged with acoustic absorption and scattering coefficients. For the simulation of reflections, the amplitude and the time of arrival must be calculated. For this, geometrical construction methods are used, which result in identifying the reflection paths between the reflection boundaries of sound propagation in the room.

As soon as the reflection components of the room impulse response are found, they must be convolved with the directional-dependent HRTF and with the sound source stimulus—the so-called *dry* signal.

One crucial part of real-time systems is a rapid calculation of the geometric reflection paths using the polygon model of the room. This task can be highly accelerated by using tree structures in the polygon database (Schröder & Lentz, 2006) or other search algorithms such as spatial hashing (e.g., Schröder, Ryba, & Vorländer, 2010) or frustum tracing (Chandak, Lauterbach, Taylor, Ren, & Manocha, 2008).

Another crucial part of the sound rendering process is real-time convolution of the binaural impulse response with the sound source. Room impulse responses are typically of some seconds in length. The corresponding binaural room impulse response is a finite impulse response (FIR) filter.

For achieving appropriate immersion, the latency of the audio output cannot exceed more than a few milliseconds. Discrete convolution in the time domain and synchronous audio output may be straightforward conceptually but is inefficient in the number of operations required to compute the output samples one by one. Fast Fourier transform (FFT)-based convolution is more efficient, computationally, but requires block processing; this, in turn, leads to block-size-dependent latency in the output. One approach to solving these problems is to use nonuniform segmented block convolution algorithms (Garcia, 2002; Wefers & Vorländer, 2012) or infinite impulse response (IIR) filters.

The binaural filters can be interpreted in the time domain as a temporal series of reflections. Alternatively, they can be interpreted in the frequency domain as a binaural transfer function. The extent to which listeners are sensitive to interaural (temporal) and spectral details in reverberant energy is also not well understood and requires additional research. Nonetheless, there is clear evidence that the inclusion of reverberation can have a dramatic impact on the subjective realism of a virtual auditory display and can aid in perception of source distance (e.g., see Brungart & D'Angelo, 1995).

4.4.2 HEADPHONE SIMULATION

In order to simulate any source somewhere in space or in a room over headphones, one must simply play a stereo headphone signal that recreates at the eardrums the exact acoustic waveforms that would actually arise from a source at the desired location. This is generally accomplished by empirically measuring the HRTF that describes how an acoustic signal at a particular location in space is transformed as it travels to and impinges on the head and ears of a listener. Then, in order to simulate an arbitrary sound source at a particular location, the appropriate transfer functions are used to filter the desired (known) source signal. The resulting stereo signal is then corrected to compensate for transfer characteristics of the display system (for instance, to remove any spectral shaping of the headphones) and presented to the listener. This holds for a single source at a particular direction or for numerous sources or room reflections, provided they have been filtered with each particular directional HRTF.

4.4.2.1 Diotic Displays

The simplest headphone displays present identical signals to both ears (*diotic* signals). With a diotic display, all sources are perceived as inside the head (not *externalized*), at midline. This internal sense of location is known as *lateralization* not *localization* (Plenge, 1974). While a diotic display requires no spatial auditory processing, it also provides no spatial information to a listener. Such displays may be useful if the location of an auditory object is not known or if spatial auditory information is unimportant. However, diotic displays are the least realistic headphone display. In addition, as discussed in Section 4.3.6.9, benefits of spatial hearing can be extremely useful for detection and recognition of auditory information. For instance, when listeners are required to monitor multiple sound sources, spatialized auditory displays are clearly superior to diotic displays (Haas et al., 1997).

4.4.2.2 Dichotic Displays

While normal interaural cues vary with frequency in complex ways, simple frequency-independent ITDs and ILDs affect the perceived lateral position of a sound source (e.g., see Durlach & Colburn, 1978). Stereo signals that only contain a frequency-independent ITD and/or ILD are herein referred to as *dichotic* signals (although the term is sometimes used to refer to any stereo signal in which left and right ears are different).

Generation of a constant ITD or ILD is very simple over headphones since it only requires that the source signal be delayed or scaled (respectively) at one ear. Just as with diotic signals, dichotic signals result in sources that appear to be located on an imaginary line inside the head, connecting the two ears. Varying the ITD or ILD causes the lateral position of the perceived source to move toward the ear receiving the louder and/or earlier-arriving signal. For this reason, such sources are usually referred to as *lateralized* rather than *localized*.

Dichotic headphone displays are simple to implement but are only useful for indicating whether a sound source is located to the left or right of a listener. On the other hand, when multiple sources are lateralized at different locations (using different ITD and/or ILD values), some binaural unmasking can be obtained (see Section 4.3.6.9).

4.4.2.3 Spatialized Audio

Using signal-processing techniques, it is possible to generate stereo signals that contain most of the normal spatial cues available in the real world. In fact, if properly rendered, spatialized audio can be practically indistinguishable from free-field presentation (Langendijk & Bronkhorst, 2000). When coupled with a head-tracking device, spatialized audio provides a true virtual auditory interface. Using a spatialized auditory display, a variety of sound sources can be presented simultaneously at different directions and distances. One of the early criticisms of spatialized audio was that it was expensive to implement; however, as hardware and software solutions have proliferated, it has become feasible to include spatialized audio in most systems. Spatialized audio solutions can be fit into any budget, depending on the desired resolution and number of sound sources required. Most virtual reality (VR) systems are currently outfitted with headphones of sufficient quality to reproduce spatialized audio, making it relatively easy to incorporate spatialized audio in an immersive VE system.

4.4.2.4 Practical Limitations on Spatialized Audio

While in theory, HRTF simulation should yield stimuli that are perceptually indistinguishable from natural experience, a number of practical considerations limit the realism of stimuli simulated using HRTFs. Measurement of HRTFs is a difficult, time-consuming process. In addition, storage requirements for HRTFs can be prohibitive. As a result, HRTFs are typically measured only at a single distance, relatively far from the listener, and at a relatively sparse spatial sampling. Changes in source distance are simulated simply by scaling the overall signal intensity. Because the HRTFs are only measured for a finite number of source directions at this single source distance, HRTFs are interpolated to simulate locations for which HRTFs are not measured. While this approach is probably adequate for sources relatively far from the listener and when some inaccuracy can be tolerated, the resulting simulation cannot perfectly recreate spatial cues for all possible source locations (Wenzel & Foster, 1993). Individual differences in HRTFs are very important for some aspects of sound source localization (particularly for distinguishing front/back and up/down). However, most systems employ a standard set of HRTFs that are not matched to the individual listener. Using these *nonindividualized* HRTFs reduces the accuracy and externalization of auditory images but still results in useful performance increases (Begault & Wenzel, 1993). Researchers have explored a variety of HRTF compression schemes in which individual differences are encoded in a small number of parameters that can be quickly or automatically fit to an individual (e.g., see Kistler & Wightman, 1991; Middlebrooks & Green, 1992). Nonetheless, many *typical* systems cannot simulate source position along a cone of confusion because they do not use individualized HRTFs.

The most sophisticated spatialized audio systems use trackers to measure the movement of a listener and update the HRTFs in real time to produce appropriate dynamic spatial cues. The use of head tracking dramatically increases the accuracy of azimuthal localization (Moldrzyk, Ahnert, Feistel, Lentz, & Weinzierl, 2004; Sorkin, Kistler, & Elvers, 1989). However, time lag in such systems (from measuring listener movement, choosing the new HRTF, and filtering the ongoing source signal) can be greater than 30 ms. While the binaural system is sluggish, the resulting delay can be perceptible. Real-time systems are also too complex and costly for some applications. Instead, systems may compute signals off-line and either ignore or limit the movement of the listener; however, observers may hear sources at locations inside or tethered to the head (i.e., moving with the head) with such systems.

Many simulations do not include any echoes or reverberation in the generated signals. Although reverberation has little impact (or degrades) perception of source direction, it is important for distance perception. In addition, anechoic simulations sound subjectively artificial and less realistic than do simulations with reverberation.

4.4.3 Simulation Using Speakers

The total acoustic signal reaching each ear is simply the sum of the signals reaching that ear from each source in an environment. Using this property, it is possible to vary spatial auditory cues (e.g., ITD, ILD, and spectrum) by controlling the signals played from multiple speakers arrayed around a listener. In contrast with headphone simulations, the signals at the two ears cannot be independently manipulated; that is, changing the signal from any of the speakers changes the signals reaching both ears. As a result, it is difficult to precisely control the interaural differences and spectral cues of the binaural signal reaching the listener to mimic the signals that would occur for a real-world source. However, various methods for specifying the signals played from each loudspeaker exist to simulate spatial auditory cues using loudspeakers.

To reduce the variability of audio signals reaching the ears, careful attention should be given to speaker placement and room acoustics. If speaker systems are not properly placed and installed in a room, even the best sound systems will sound inferior. Improperly placed speakers can reduce speech comprehension, destroy the sense of immersion, and dramatically reduce bass response (Holman, 2000). This is especially true when dealing with small rooms. If the system is installed properly, there will be a uniform (flat) frequency response at the listening area.

One example of a four-speaker system (two front and two surround) is described in an International Telecommunications Union (ITU) recommendation and places speakers at ±30° in front of the listener and at ±110° behind the listener (ITU-R BS. 775-1). It is further recommended that the signals emanating from the two surround channels be decorrelated to increase the sense of spaciousness. Correlated mono signals may give a sense of lateralization rather than localization. If a subwoofer is used, it is usually placed in front of the room. Placing the subwoofer too close to a corner may increase bass response but may result in a muddier sound. The subwoofer should be moved to achieve the best response in the listening area. Unfortunately, speaker placement will vary depending on the dimensions and shape of the room, as well as the number of speakers employed. If the system is mobile, the sound system will have to be readjusted for every new location, unless the simulation incorporates its own enclosure. If the simulation will be housed in different sized rooms, the audio system (amplifiers and speakers) must have enough headroom (power) to accommodate both large enclosures as well as small. When possible, acoustical tile and diffusers should be employed where appropriate to reduce reverberation and echoes.

4.4.3.1 Nonspatial Display

Many systems use free-field speakers in which each speaker presents an identical signal. Such systems are analogous to diotic headphone systems; although simple to develop, these displays (like diotic headphone displays) provide no spatial information to the listener. Such systems can be used when spatial auditory information is unimportant and when segregation of simultaneous auditory signals is not critical. For instance, if the only objects of interest are within the visual field and interference between objects is not a concern, this kind of simplistic display may be adequate.

4.4.3.2 Stereo Display

The analog of the dichotic headphone display presents signals from two speakers simultaneously in order to control the perceived laterality of a *phantom* source. For instance, simply varying the level of otherwise identical signals played from a pair of speakers can alter the perceived laterality of a phantom source. Most commercial stereo recordings are based on variations of this approach.

Imagine a listener sitting equidistant from two loudspeakers positioned symmetrically in front of the listener. When the left speaker is played alone, the listener hears a source in the direction of the left speaker (and ITD and ILD cues are consistent with a source in that leftward direction). When the right speaker is played alone, the listener hears a source in the direction of the right speaker. When identical signals at identical levels are played from both speakers, each ear receives two direct signals, one from each of the symmetrically placed speakers. To the extent that the listener's head is left–right symmetric, the total direct sound in each ear is identical, and the resulting percept will be of a single source at a location that gives rise to zero ITD and zero ILD (e.g., in the listener's median plane). Varying the relative intensity of otherwise identical signals played from the two speakers causes the gross ITD and ILD cues to vary systematically, producing a phantom source whose location between the two speakers varies systematically with the relative speaker levels (e.g., see Bauer, 1961).

This simple *panning* technique produces a robust perception of a source at different lateral locations; however, it is nearly impossible to precisely control the exact location of the phantom image. In particular, the way in which the perceived direction changes with relative speaker level depends upon the location of the listener with respect to the two loudspeakers. As the listener moves outside a restricted area (the *sweet spot*), the simulation degrades rather dramatically. In addition, reverberation can distort the interaural cues, causing biases in the resulting simulation. Nonetheless, such systems provide some information about source laterality and can be very effective when one wishes to simulate sounds from angular positions falling between the loudspeaker positions.

4.4.3.3 Multichannel Loudspeaker Systems

Two-channel stereo can be extended to multichannel panning techniques, so-called vector base amplitude panning (VBAP). The technique can be used to place virtual source in 3D space if the loudspeaker arrays are surrounding the listener (Pulkki, 1997). Loudspeaker triplets are used with particular amplitudes in order to create a sound image in the corresponding triangle. Larger areas around the listener are covered by several of those triplets.

Another technique, actually one of the most popular techniques of spatial audio, is *Ambisonics* (Gerzon, 1976). The mathematical basis is the set of spherical harmonics (SH), a set of orthogonal functions in spherical coordinates. With these, sound fields can be decomposed into their directional components (SH coefficients); these exact coefficients are used to filter the speakers of the reproduction array. The speaker arrays can be freely designed in 3D space, but simple solutions correspond to Platonic bodies such as cubes, dodecahedrons, or icosahedrons. The SH coefficients can be derived from simulation or from recordings with spherical microphone arrays. The first-order approach as defined by Gerzon requires a setup of an omnidirectional and three figure-eight microphones. The spatial resolution and the corresponding details of directional sounds are limited by the low-order SH representation, which creates a kind of spatial smoothing. Higher-order Ambisonics (HOA) allow reproduction of more spatial detail; for this, higher-order microphone arrays must be used (Meyer & Elko, 2002). Wave field synthesis (WFS) technology is another theoretical approach to wave field reconstruction (Berkhout, 1988). In WFS, sound waves are decomposed into plane waves sampled on linear or circular microphone arrays, typically in 2D. If the discrete spatial sampling of the array is sufficiently high, any wave field can be reconstructed with a corresponding large number of loudspeakers in a dense distribution. This goes back to the Huygens principle that explains wave propagation by arrays of numerous point sources, each radiating elementary waves, which interfere to form wave fronts (e.g., see De Vries, 2012). To use WFS, the process of sound recording and mixing is different from usual techniques applied in audio engineering. The spatial decoding of virtual sources is integrated in a flexible way so that position, orientation, and movements of the listener are not restricted in dynamic scenes.

All of the multichannel techniques listed suffer from the fact that a large number of loudspeakers must be used with accordingly a large amount of signal processing, control, and amplifier units. For VR installations with surrounding displays, a practical problem often arises from the conflicting demands of trying to place loudspeaker arrays along with video screens with an undisturbed video image and free line of sight. Acoustically transparent video screens would solve the problem, but the image resolution of such screens is significantly less than for high-quality hard projection screens.

4.4.3.4   Cross-Talk Cancellation and Transaural Simulations

More complex signal-processing schemes can also be used to control spatial cues using a small setup of loudspeakers. In such approaches, the total signal reaching each ear is computed as the sum of the signals reaching that ear from each of the speakers employed. By considering the timing and content of each of these signals, one can try to reproduce the exact signal desired at each ear.

The earliest such approach attempted to recreate the sound field that a listener would have received in a particular setting from stereo recordings taken from spatially separated microphones. In the playback system, two speakers were positioned at the same relative locations as the original microphones. The goal of the playback system was to play signals from the two speakers such that the total signal at each ear was equal to the recorded signal from the nearer microphone. To the extent that the signal from the far speaker was acoustically canceled, the reproduction would be accurate. Relatively simple schemes involving approximations of the acoustic alterations of the signals as they impinged on the head were used to try to accomplish this *cross-talk cancellation*.

As signal-processing approaches have been refined and knowledge of the acoustic properties of HRTFs improves, more sophisticated algorithms have been developed. In particular, it is possible to calculate the contribution of each speaker to the total signal at each ear by considering the HRTF corresponding to the location of the speaker. The total signal at each ear is then the sum of the HRTF-filtered signals coming from each speaker. If one also knows the location and source of the signal that is to be simulated, one can write equations for the desired signals reaching each ear as HRTF-filtered versions of the desired source. Combining these equations yields two frequency-dependent, complex-valued equations that relate the signals played from each speaker to the desired signals at the ears. To the extent that one can find and implement solutions to these equations, it is possible (at least in theory) to recreate the desired binaural signal by appropriate choice of the signals played from each speaker. The problem with such approaches is that the simulation depends critically on the relative location of the speakers and the listener. In particular, if the listener moves his head outside of the sweet spot, the simulation degrades rapidly. Head tracking and dynamic cross-talk filtering is one solution to this problem, which also allows head movements explicitly so that the listener can benefit from a more natural behavior in the virtual scene (Lentz & Behler, 2004). Head trackers can be used in conjunction with multispeaker simulations in order to improve the simulation. However, this requires that computations be performed in real time and significantly increases the cost and complexity of the resulting system. It can be difficult to compute the required loudspeaker signals and the computations are not particularly stable numerically (Masiero & Vorländer, 2012). The technique called *stereo dipole* or *sound bar* is actually more stable. It is an optimized distributed source approach where the high frequencies are radiated from the frontal region while the loudspeakers for the low frequencies span a larger angle (Takeushi, Teschl, & Nelson, 2001). To the extent that reverberation in the listening space further distorts the signals reaching the ears, the derived solutions are less robust than those for headphones. In all cases of binaural reproduction using headphones or cross-talk cancellation, the quality is best if the HRTFs used in the equations are matched to the listener.

4.4.3.5   Lessons from the Entertainment Industry

The ability to generate an accurate spatial simulation using loudspeakers increases dramatically as the number of speakers used in the display increases. With an infinite number of speakers around the listener, one would simply play the desired signal from the speaker at the desired location of the source to achieve a *perfect* reproduction. Surround sound technologies, which are prevalent in the entertainment industry, are implemented via a variety of formats. Surround sound systems find their genesis in a three-channel system created for the movie *Fantasia* in 1939.

Fantasound speakers were located in front of the listener at left, middle, and right. The *surround* speakers consisted of approximately 54 speakers surrounding the audience and carried a mixture of sound from the left and right front speakers (i.e., it was not true stereo; Garity & Hawkins, 1941).

Currently, the most common surround sound format is the 5.1 speaker system, in which speakers are located at the left, middle, and right in front of the listener and left and right behind the listener. The so-called 0.1 speaker is a subwoofer. The middle speaker, located in front of the listener, reproduces most of the speech information to the listener. Typical 5.1 surround formats are Dolby Digital Surround and Digital Theater Systems (DTS).

More complex surround sound formats include Dolby Digital Surround EX, which is a 6.1 speaker system (adding a center speaker behind the listener); a 10.2 system has also been developed (Holman, 2000). In the future, even greater numbers of speakers and more complex processing are likely to become standard. However, it is important to note that adding additional speakers may be detrimental to producing a sense of immersion, especially in a small room. As the number of speakers increases, explicit care must be taken to assure that the sound field in the room is diffuse enough that the speakers themselves are not obtrusive. If the user notices the speakers in the room, the illusion of reality will be destroyed. To negate this possibility, extreme care must be taken into account for room acoustics, speaker design and placement, and the location of the listener in the room (Holman).

In general, it is possible to generate relatively realistic phantom sources using multiple loudspeakers, but such techniques often only simulate sound in the left–right (azimuthal) angle direction of the desired source precisely. More complex simulations in which spectral cues accurately simulate the direction of the desired sound source on the cone of confusion are rarely undertaken as they require much more complex signal processing that is tailored to the individual user (and his or her HRTFs). However, oftentimes, robust left–right angular simulations are sufficient, depending on the goals of an auditory display.

4.5 DESIGN CONSIDERATIONS

4.5.1 Defining the Auditory Environment

When creating the auditory portion of a VE, careful attention should be placed on what is absolutely essential for the task. The adage of motion picture sound designers, *see it*, *hear it* (Holman 1997; Yewdall, 1999), is also valid for designing audio for VEs. The sense of immersion experienced in a movie theater is a carefully orchestrated combination of expertly designed sound effects and skillfully applied auditory ambiences. It is also interesting to note that *realistically* rendered sound is often perceived as emotionally flat in motion pictures. Sound effects are often designed as exaggerated versions of reality to convey emotion or to satisfy the viewers' expectations of reality (Holman). Sound design in VE needs to balance the need for accurate reproduction with the need to make the user emotionally involved in a synthetic environment.

On the hardware side, a simple sound card solution may be adequate if spatialized audio and fidelity are not paramount. However, if there are multiple sound sources and/or a multiuser interface, a special-purpose audio server may be necessary. Just as photos and films of the visual environment are taken during the development process, it is a good idea to make audio recordings, including sound-level measurements, when developing an auditory interface. In addition to cataloguing the different sounds in a real environment, it is also important to systematically measure the intensity of sounds being experienced by the listener. In this manner, the VE developer has a detailed reference with which to compare the real-world auditory environment with the virtual auditory environment. Given the wide dynamic range involved with recording sounds ranging from concussive events to footsteps and the necessity that recordings be absolutely clean and accurate, the best solution may be to rely on professionals for making appropriate recordings that contain the requisite content and ambience. Different combinations of microphones and recording equipment produce vastly differing sound quality. Choosing the appropriate combination for a particular application is still more of an art than a science. In addition, there are many high-quality commercially available sound libraries for obtaining a wide variety of sound effects and ambiences (Holman, 1997; Yewdall, 1999). Given these facts, the use of prerecorded, high-quality sound content is the de facto standard for many current auditory displays.

4.5.2 How Much Realism Is Necessary?

Much of the research devoted to developing and verifying virtual display technology emphasizes the subjective realism of the display (Chertoff & Schatz, 2013; Chapter 34); however, this is not the most important consideration for all applications. In some cases, signal processing that improves realism actually interferes with the amount of information a listener can extract. For instance, inclusion of echoes and reverberation can significantly increase the perceived realism of a display and improve distance perception. However, echoes can degrade perception of source direction. For applications in which information about the direction of a source is more important than the realism of the display or perception of source distance, including echoes and reverberation may be ill advised.

In headphone-based systems, realism is enhanced with the use of individualized HRTFs, particularly in the perception of up/down and front/back positions. Ideally, HRTFs should also be sampled in both distance and direction at a spatial density dictated by human sensitivity. Thus, while the most *realistic* system would use individualized HRTFs that are sampled densely in both direction and distance, most systems use generic HRTFs sampled coarsely in direction and at only one distance.

If a particular application requires a listener to extract 3D spatial information from the auditory display, HRTFs may have to be tailored to the listener to preserve directional information and reverberation may have to be included to encode source distance. On the other hand, if a particular application only makes use of one spatial dimension (for instance, to indicate the direction that a blind user must turn), coarse simulation of ITD and ILD cues (even without detailed HRTF simulation) is probably adequate.

If information transfer is of primary importance, it may be useful to present acoustic spatial cues that are intentionally distorted so that they are perceptually more salient than more *realistic* cues. For instance, it may be useful to exaggerate spatial auditory cues to improve auditory resolution; however, such an approach requires that listeners are appropriately trained with the distorted cues. With such training, listeners can respond accurately with altered cues but also respond accurately to *normal* localization cues with little or no negative aftereffects of training, suggesting that both the new and the old mappings from spatial cues to exocentric location can be maintained simultaneously (e.g., see Hofman et al., 1998; Shinn-Cunningham, 2000a).

The processing power needed to simulate the most realistic virtual auditory environment possible is not always cost-effective. For instance, the amount of computation needed to create realistic reverberation in a VE may not be justifiable when source distance perception and subjective realism are not important. Other acoustic effects are often ignored in order to reduce computational complexity of an acoustic simulation, including the nonuniform radiation pattern of a realistic sound source, spectral changes in a sound due to atmospheric effects, and the Doppler shift of the received spectrum of moving sources. The perceptual significance of many of these effects is not well understood; further work must be done to examine how these factors affect the realism of a display, as well as what perceptual information such cues may convey.

In command-and-control applications, the goal is to maximize information transfer to the human operator; subjective impression (i.e., realism) is unimportant. In these applications, both technological and perceptual issues must be considered to achieve this goal. If nonverbal warnings or alerts are created, the stimuli must be wideband enough to be localizable. In addition, stimuli should be significantly intense to be at least 15 dB above background noise level. Stimulus onset should be fairly gradual so as not to be excessively startling to the user. In many instances, one may want the stimuli to be aesthetically pleasing to the user. As can be imagined, creating acceptable spatialized auditory displays is no trivial chore and should involve formal evaluations to ensure perceptual accuracy and system usability. For applications in which speech is the main signal of interest, basic interaural cues are important for preserving speech intelligibility, particularly in noisy, multisource environments. On the other hand, there is probably little benefit gained from including the detailed frequency dependence of normal HRTFs. In entertainment applications (Greenwood-Ericksen, & Stafford, 2013; Chapter 49), cost is the most important factor; the precision of the display is unimportant as long as the simulation is subjectively satisfactory. For scientific research (Polys, 2013; Chapter 49), high-end systems are necessary in order to allow careful examination of normal spatial auditory cues. In clinical applications, the auditory display must only be able to deliver stimuli that can distinguish listeners with normal spatial hearing from those with impaired spatial hearing. Such systems must be inexpensive and easy to use, but there is no need for a perfect simulation.

REFERENCES

Algazi, V. R., Avendano, C., & Duda, R. O. (2001). Elevation localization and head-related transfer function analysis at low frequencies. *Journal of the Acoustical Society of America, 109*(3), 1110–1122.

Aretz, M., Maier, P., & Vorländer, M. (November 2010). Simulation based auralization of the acoustics in a studio room using a combined wave and ray based approach. VDT Tonmeistertagung, Leipzig, Germany. (8 pages)

Badcock, D. R., Palmisano, S., & May, J. G. (2014). Vision and virtual environments. In K. S. Hale & K. M. Stanney (Eds.), *Handbook of virtual environments* (2nd ed., pp. 39–86). Boca Raton, FL: CRC Press.

Batteau, D. W. (1967). The role of the pinna in human localization. *Proceedings of the Royal Society of London, B168*, 158–180.

Bauer, B. B. (1961). Phasor analysis of some stereophonic phenomena. *Journal of the Acoustical Society of America, 33*(11), 1536–1539.

Begault, D., Wenzel, E., Lee, A., & Anderson, M. (2012, July). *Direct comparison of the impact of head tracking, reverberation, and individualized head-related transfer functions on the spatial perception of a virtual speech source.* In the 108th Audio Engineering Society (AES) Convention, Vol. 5134, Paris, France.

Begault, D. R. (1993a). Head-up auditory displays for traffic collision avoidance system advisories: A preliminary investigation. *Human Factors, 35*(4), 707–717.

Begault, D. R. (1996, November). *Virtual acoustics, aeronautics, and communications.* Presented at the 101st convention of the Audio Engineering Society, Los Angeles, CA.

Begault, D. R. (1999). Virtual acoustic displays for teleconferencing: Intelligibility advantage for "telephone-grade" audio. *Journal of the Audio Engineering Society, 47*(10), 824–828.

Begault, D. R., & Wenzel, E. M. (1992). Techniques and applications for binaural sound manipulation in human–machine interfaces. *The International Journal of Aviation Psychology, 2*(1), 1–22.

Begault, D. R., & Wenzel, E. M. (1993). Headphone localization of speech. *Human Factors, 35*(2), 361–376.

Berkhout, A. J. (1988). A holographic approach to acoustic control. *Journal of the Audio Engineering Society, 36*, 977–995.

Blauert, J. (1997). *Spatial hearing* (2nd ed.). Cambridge, MA: MIT Press.

Botteldooren, D. (1995). Finite-difference time-domain simulation of low-frequency room acoustic problems. *Journal of the Acoustical Society of America, 98*(6), 3302–3308.

Brainard, M. S., Knudsen, E. I., & Esterly, S. D. (1992). Neural derivation of sound source location: Resolution of ambiguities in binaural cues. *Journal of the Acoustical Society of America, 91*, 1015–1027.

Bregman, A. S. (1990). *Auditory scene analysis: The perceptual organization of sound.* Cambridge, MA: MIT Press.

Brewster, S. A., Wright, P. C., & Edwards, A. D. N. (1993). An evaluation of earcons for use in auditory human-computer interfaces. In *Proceedings of the INTERACT'93 and CHI'93 Conference on Human Factors in Computing Systems* (pp. 222–227). New York, NY: ACM.

Bronkhorst, A. W., & Plomp, R. (1988). The effect of head-induced interaural time and level differences on speech intelligibility in noise. *Journal of the Acoustical Society of America, 83*, 1508–1516.

Bronkhorst, A. W., Veltman, J. A., & van Breda, L. (1996). Application of a three-dimensional auditory display in a flight task. *Human Factors, 38*(1), 23–33.

Brungart, D. S., & D'Angelo, W. (1995). Effects of reverberation cues on distance identification in virtual audio displays. *The Journal of the Acoustical Society of America, 97*, 3279.

Brungart, D. S., & Rabinowitz, W. M. (1999). Auditory localization of nearby sources: Near-field head-related transfer functions. *Journal of the Acoustical Society of America, 106*, 1465–1479.

Carlyon, R. P. (2004). How the brain separates sounds. *Trends in Cognitive Science, 8*, 465–471.

Chandak, A., Lauterbach, C., Taylor, M., Ren, Z., & Manocha, D. (2008, October 19–24). AD-frustum: Adaptive frustum tracing for interactive sound propagation. In *Proceedings of the IEEE Visualization*, Columbus, OH.

Cherry, E. C. (1953). Some experiments on the recognition of speech, with one and with two ears. *Journal of the Acoustical Society of America, 25*, 975–979.

Chertoff, D., & Schatz, S. (2014). Beyond presence. In K. S. Hale & K. M. Stanney (Eds.), *Handbook of Virtual Environments* (2nd ed., pp. 855–870). Boca Raton, FL: CRC Press.

Chon, S. H., & McAdams, S. (2012). Investigation of timbre saliency, the attention-capturing quality of timbre. *Journal of the Acoustical Society of America, 13*, 3433–3433.

Clark, B., & Graybiel, A. (1949). The effect of angular acceleration on sound localization: The audiogyral illusion. *The Journal of Psychology, 28*, 235–244.

Clifton, R. K., Freyman, R. L., & Litovsky, R. L. (1993). Listener expectations about echoes can raise or lower echo threshold. *Journal of the Acoustical Society of America, 95*, 1525–1533.

Darwin, C. J. (1997). Auditory grouping. *Trends in Cognitive Sciences, 1*, 327–333.

De Vries, D. (2012). Sound reinforcement by wavefield synthesis: Adaptation of the synthesis operator to the loudspeaker directivity characteristics. In *Proceedings of the Audio Engineering Society Convention* (Paper No. 8661), San Francisco, CA.

Dindar, N., Tekalp, A. M., & Basdogan, C. (2014). Dynamic haptic interaction with video. In K. S. Hale & K. M. Stanney (Eds.), *Handbook of virtual environments* (2nd ed., pp. 117–132). Boca Raton, FL: CRC Press.

DiZio, P., Held, R., Lackner, J. R., Shinn-Cunningham, B. G., & Durlach, N. I. (2001). Gravitoinertial force magnitude and direction influence head-centric auditory localization. *Journal of Neurophysiology, 85*, 2455–2560.

Drullman, R., & Bronkhorst, A. W. (2000). Multichannel speech intelligibility and talker recognition using monaural, binaural, and three-dimensional auditory presentation. *Journal of the Acoustical Society of America, 107*(4), 2224–2235.

Duda, R. O., & Martens, W. L. (1998). Range dependence of the response of a spherical head model. *Journal of the Acoustical Society of America, 104*(5), 3048–3058.

Durlach, N. I., & Colburn, H. S. (1978). Binaural phenomena. In E. C. Carterette & M. P. Friedman (Eds.), *Handbook of perception* (Vol. 4, pp. 365–466). New York, NY: Academic Press.

Durlach, N. I., Shinn-Cunningham, B. G., & Held, R. M. (1993). Supernormal auditory localization. I. General background. *Presence, 2*(2), 89–103.

Frens, M. A., van Opstal, A. J., & van der Willigen, R. F. (1995). Spatial and temporal factors determine auditory-visual interactions in human saccadic eye movements. *Perception & Psychophysics, 57*, 802–816.

Freyman, R. L., Clifton, R. K., & Litovsky, R. Y. (1991). Dynamic processes in the precedence effect. *Journal of the Acoustical Society of America, 90*, 874–884.

García, G. (2002). Optimal filter partition for efficient convolution with short input/output delay. In *Proceedings of Audio Engineering Society Convention* (Paper No. 5660), Los Angeles, CA.

Garity, W. E., & Hawkins, J. A. (1941, August). Fantasound. *Journal of the Society of Motion Picture Engineers 37*, 127–146.

Gelfand, S. A. (1998). *Hearing: An introduction to psychological and physiological acoustics*. New York, NY: Marcel Dekker, Inc.

Gerzon, M. (1976). Multidirectional sound reproduction systems. UK-Patent 3,997,725.

Gilkey, R. H., & Anderson, T. R. (1995). The accuracy of absolute localization judgments for speech stimuli. *Journal of Vestibular Research, 5*(6), 487–497.

Gilkey, R. H., & Good, M. D. (1995). Effects of frequency on free-field masking. *Human Factors, 37*(4), 835–843.

Grantham, D. W. (1997). Auditory motion perception: Snapshots revisited. In R. Gilkey & T. Anderson (Eds.), *Binaural and spatial hearing in real and virtual environments* (pp. 295–314). New York, NY: Lawrence Erlbaum Associates.

Greenwood-Ericksen, A., Kennedy, R. C., & Stafford, S. (2014). Entertainment applications of virtual environments. In K. S. Hale & K. M. Stanney (Eds.), *Handbook of virtual environments* (2nd ed., pp. 1291–1316). Boca Raton, FL: CRC Press.

Haas, E. C., Gainer, C., Wightman, D., Couch, M., & Shilling, R. D. (1997, September). Enhancing system safety with 3-D audio displays. In *Proceedings of the Human Factors and Ergonomics Society 41st Annual Meeting* (pp. 868–872), Albuquerque, NM.

Hartmann, W. M. (1983). Localization of sound in rooms. *Journal of the Acoustical Society of America*, *74*, 1380–1391.

Hendrix, C., & Barfield, W. (1996). The sense of presence in auditory virtual environments. *Presence*, *5*(3), 290–301.

Hofman, P. M., Van Riswick, J. G. A., & Van Opstal, A. J. (1998). Relearning sound localization with new ears. *Nature Neuroscience*, *1*(5), 417–421.

Holman, T. (1997). *Sound for film and television*. Boston, MA: Focal Press.

Holman, T. (2000). *5.1 Surround sound: Up and running*. Boston, MA: Focal Press.

Kim, R., Peters, M. A., & Shams, L. (2012). 0+1 > 1: How adding noninformative sound improves performance on a visual task. *Psychological Science*, *23*(1), 6–12.

Kistler, D. J., & Wightman, F. L. (1991). A model of head-related transfer functions based on principal components analysis and minimum-phase reconstruction. *Journal of the Acoustical Society of America*, *91*, 1637–1647.

Kollmeier, B., & Gilkey, R. H. (1990). Binaural forward and backward masking: Evidence for sluggishness in binaural detection. *Journal of the Acoustical Society of America*, *87*(4), 1709–1719.

Langendijk, E. H., & Bronkhorst, A. W. (2000). Fidelity of three-dimensional-sound reproduction using a virtual auditory display. *Journal of the Acoustical Society of America*, *107*(1), 528–537.

Lawson, B. D., & Riecke, B. E. (2014). Perception of body motion. In K. S. Hale & K. M. Stanney (Eds.), *Handbook of virtual environments* (2nd ed., pp. 115–130). New York, NY: CRC Press.

Lawson, B. D., & Riecke, B. E. (2014). Perception of body motion. In K. S Hale & K. M. Stanney (Eds.), *Handbook of virtual environments* (2nd ed., pp. 163–196). Boca Raton, FL: CRC Press.

Lentz, T., & Behler, G. K. (2004). Dynamic crosstalk cancellation for binaural synthesis in virtual environments. In *Proceedings of the 117th Convention of the Audio Engineering Society* (p. 6315), San Francisco, CA, October 2004.

Letowski, T., Karsh, R., Vause, N., Shilling, R., Ballas, J., Brungart, D., & McKinley, R. (2001). Human factors military lexicon: Auditory displays. ARL Technical Report, ARL-TR-2526; APG (MD).

Maddox, R., & Shinn-Cunningham, B. B. (2011). Influence of task-relevant and task-irrelevant feature continuity on selective auditory attention. *Journal of the Association for Research in Otolaryngology*, *13*, 119–129.

Masiero, B., & Vorländer, M. (2012). A framework for the calculation of dynamic crosstalk cancellation filters. *IEEE Transactions on Audio, Speech and Language Processing* (under review).

McMahan, R. P., Kopper, R., & Bowman, D. A. (2014). Principles for designing effective 3D interaction techniques. In K. S Hale & K. M. Stanney (Eds.), *Handbook of virtual environments* (2nd ed., pp. 285–312). Boca Raton, FL: CRC Press.

Mershon, D. H. (1997). Phenomenal geometry and the measurement of perceived auditory distance. In R. Gilkey & T. Anderson (Eds.), *Binaural and spatial hearing in real and virtual environments* (pp. 251–214). New York, NY: Lawrence Erlbaum Associates.

Mershon, D. H., Ballenger, W. L., Little, A. D., McMurty, P. L., & Buchanan, J. L. (1989). Effects of room reflectance and background noise on perceived auditory distance. *Perception*, *18*, 403–416.

Meyer, J., & Elko, G. W. (2002, May 13–17). A highly scalable spherical microphone array based on an orthonormal decomposition of the soundfield. In *Proceedings of the IEEE International Conference on Acoustics, Speech, Signal Processing* (Vol. II, pp. 1781–1784), Orlando, FL.

Middlebrooks, J. C. (1997). Spectral shape cues for sound localization. In R. Gilkey & T. Anderson (Eds.), *Binaural and spatial hearing in real and virtual environments* (pp. 77–98). New York, NY: Lawrence Erlbaum Associates.

Middlebrooks, J. C., & Green, D. M. (1991). Sound localization by human listeners. *Annual Review of Psychology*, *42*, 135–159.

Middlebrooks, J. C., & Green, D. M. (1992). Observations on a principal components analysis of head-related transfer functions. *Journal of the Acoustical Society of America*, *92*, 597–599.

Mills, A. W. (1972). Auditory localization. In J. V. Tobias (Ed.), *Foundations of modern auditory theory* (pp. 303–348). New York, NY: Academic Press.

Moldrzyk, C., Ahnert, W., Feistel, S., Lentz, T., & Weinzierl, S. (2004). Head-tracked Auralization of acoustical simulation. In *Proceedings of the 117th Convention of the Audio Engineering Society*, San Francisco, CA, October 2004, p. 6275.

Moore, B. C. J. (1997). *An introduction to the psychology of hearing* (4th ed.). San Diego, CA: Academic Press.

Perrott, D. R., Saberi, K., Brown, K., & Strybel, T. (1990). Auditory psychomotor coordination and visual search behavior. *Perception & Psychophysics, 48,* 214–226.

Perrott, D. R., Sadralodabai, T., Saberi, K., & Strybel, T. (1991). Aurally aided visual search in the central visual field: Effects of visual load and visual enhancement of the target. *Human Factors, 33,* 389–400.

Pick, H. L., Warren, D. H., & Hay, J. C. (1969). Sensory conflict in judgements of spatial direction. *Perception & Psychophysics, 6,* 203–205.

Plenge, G. (1974). On the differences between localization and lateralization. *Journal of the Acoustical Society of America, 56*(3), 944–951.

Polys, N. (2014). Information visualization in virtual environments. In K. S. Hale & K. M. Stanney (Eds.), *Handbook of virtual environments* (2nd ed., pp. 1267–1296). New York, NY: CRC Press.

Popescu, G. V., Trefftz, H., & Burdea, G. C. (2014). Multimodal interaction modeling. In K. S. Hale & K. M. Stanney (Eds.), *Handbook of virtual environments* (2nd ed., pp. 411–434). Boca Raton, FL: CRC Press.

Pulkki, V. (1997). Virtual sound source positioning using vector base amplitude panning. *Journal of the Audio Engineering Society, 45*(6), 456–466.

Rakerd, B., & Hartmann, W. M. (1985). Localization of sound in rooms. II. The effects of a single reflecting surface. *Journal of the Acoustical Society of America, 78,* 524–533.

Rakerd, B., & Hartmann, W. M. (1986). Localization of sound in rooms. III. Onset and duration effects. *Journal of the Acoustical Society of America, 80,* 1695–1706.

Savioja, L. (2010). Real-time 3D finite-difference time-domain simulation of mid-frequency room acoustics. In *Proceedings of the 13th International Conference on Digital Audio Effects, DAFx* (p. 43), Graz, Austria.

Schröder, D., & Lentz, T. (2006). Real-time processing of image sources using binary space partitioning. *Journal of the Audio Engineering Society, 54,* 604–619.

Schröder, D., Ryba, A., & Vorländer, M. (2010). Spatial data structures for dynamic acoustic virtual reality. *Proceedings of the International Congress on Acoustics,* Sydney, Australia, August 2010.

Shams, L., Kamitani, Y., & Shimojo, S. (2004). Modulations of visual perception by sound. In G. A. Calvert, C. Spence, & B. E. Stein (Eds.), *Handbook of multisensory processes* (pp. 27–34). Cambridge, MA: MIT Press.

Sheridan, T. B. (1996). Further musings on the psychophysics of presence. *Presence, 5*(2), 241–246.

Shilling, R. D., & Letowski, T. (2000). *Using spatial audio displays to enhance virtual environments and cockpit performance.* NATO Research and Technology Agency Workshop entitled What is essential for Virtual Reality to Meet Military Human Performance Goals, The Hague, The Netherlands.

Shinn-Cunningham, B. G. (2000a). Adapting to remapped auditory localization cues: A decision-theory model. *Perception & Psychophysics, 62*(1), 33–47.

Shinn-Cunningham, B. G. (2000b, April 2–5). Learning reverberation: Implications for spatial auditory displays. In *Proceedings of the International Conference on Auditory Displays* (pp. 126–134), Atlanta, GA.

Shinn-Cunningham, B. G. (2008). Object-based auditory and visual attention. *Trends in Cognitive Sciences, 12,* 182–186.

Shinn-Cunningham, B. G., Durlach, N. I., & Held, R. (1998). Adapting to supernormal auditory localization cues. 1: Bias and resolution. *Journal of the Acoustical Society of America, 103*(6), 3656–3666.

Shinn-Cunningham, B. G., Ihlefeld, A., Satyavarta, & Larson, E. (2005). Bottom-up and top-down influences on spatial unmasking. *Acta Acustica united with Acustica, 91,* 967–979.

Shinn-Cunningham, B. G., Kopco, N., & Santarelli, S. G. (1999). *Computation of acoustic source position in near-field listening.* Paper presented at the Third International Conference on Cognitive and Neural Systems, Boston, MA.

Shinn-Cunningham, B. G., Santarelli, S., & Kopco, N. (2000). Tori of confusion: Binaural localization cues for sources within reach of a listener. *Journal of the Acoustical Society of America, 107*(3), 1627–1636.

Sorkin, R. D., Kistler, D. S., & Elvers, G. C. (1989). An exploratory study of the use of movement-correlated cues in an auditory head-up display. *Human Factors, 31*(2), 161–166.

Stern, R. M., & Trahiotis, C. (1997). Binaural mechanisms that emphasize consistent interaural timing information over frequency. In A. R. Palmer, A. Rees, A. Q. Summerfield, & R. Meddis (Eds.), *Psychophysical and physiological advances in hearing, Proceedings of the XI international symposium on hearing,* August 1997, Grantham, London, U.K.: Whurr Publishers, 1998.

Storms, R. L. (1998). *Auditory-visual cross-modal perception phenomena* (Doctoral dissertation). Naval Postgraduate School, Monterey, CA.

Takeushi, T., Teschl, M., & Nelson, P. A. (2001, June). Subjective evaluation of the optimal source distribution system for virtual acoustic imaging. *Proceedings of the 19th Audio Engineering Society International Conference* (pp. 373–385), Schloss Elmau, Germany, June 21–24, 2001.

THX Certified Training Program. (2000, June). *Presentation materials*. San Rafael, CA.

Trahiotis, C., & Stern, R. M. (1989). Lateralization of bands of noise: Effects of bandwidth and differences of interaural time and phase. *Journal of the Acoustical Society of America*, *86*(4), 1285–1293.

Vorländer, M. (2008). *Auralization—Fundamentals of acoustics, modelling, simulation, algorithms and acoustic virtual reality*. Berlin, Germany: Springer.

Wallach, H. (1940). The role of head movements and vestibular and visual cues in sound localization. *Journal of Experimental Psychology*, *27*, 339–368.

Warren, D. H., Welch, R. B., & McCarthy, T. J. (1981). The role of visual-auditory "compellingness" in the ventriloquism effect: Implications for transitivity among the spatial senses. *Perception & Psychophysics*, *30*, 557–564.

Wefers, F., & Vorländer, M. (2012, October 26–29). Optimal filter partitions for non-uniformly partitioned convolution. In *Proceedings of the 133rd Audio Engineering Society Convention* (pp. 6–4) San Francisco, CA, October 2012.

Welch, R., & Warren, D. H. (1986). Intersensory interactions. In K. R. Boff, L. Kaufman, & J. P. Thomas (Eds.), *Handbook of perception and human performance* (Vol. 2, pp. 25.1–25.36). New York: John Wiley & Sons.

Welch, R. B., & Warren, D. H. (1980). Immediate perceptual response to intersensory discrepancy. *Psychological Bulletin*, *88*, 638–667.

Wenzel, E. M., Arruda, M., Kistler, D. J., & Wightman, F. L. (1993). Localization using nonindividualized head-related transfer functions. *Journal of the Acoustical Society of America* (pp. 102–105), *94*, 111–123.

Wenzel, E. M., & Foster, S. H. (1993, October). Perceptual consequences of interpolating head-related transfer functions during spatial synthesis. In *Proceedings of the 1993 Workshop on the Applications of Signal Processing to Audio and Acoustics*, New York.

Wightman, F. L., & Kistler, D. J. (1989). Headphone simulation of free-field listening. II. Psychophysical validation. *Journal of the Acoustical Society of America*, *85*, 868–878.

Wightman, F. L., & Kistler, D. J. (1993). Sound localization. In W. A. Yost, A. N. Popper, & R. R. Fay (Eds.), *Human psychophysics* (pp. 155–192), New York, NY: Springer Verlag.

Yewdall, D. L. (1999). *Practical art of motion picture sound*. Boston, MA: Focal Press.

Yost, W. A. (1997). The cocktail party problem: Forty years later. In R. Gilkey & T. Anderson (Eds.), *Binaural and spatial hearing in real and virtual environments* (pp. 329–346). New York, NY: Lawrence Erlbaum Associates.

Yost, W. A. (2006). *Fundamentals of hearing: An introduction* (5th ed.). San Diego, CA: Academic Press.

Zahorik, P., Brungart, D. S., & Bronkhorst, A. W. (2005). Auditory distance perception in humans: A summary of past and present research. *Acta Acustica united with Acustica*, *91*, 409–420.

Zurek, P. M. (1993). Binaural advantages and directional effects in speech intelligibility. In G. Studebaker & I. Hochberg (Eds.), *Acoustical factors affecting hearing aid performance* (pp. 255–276). Boston, MA: College-Hill Press.

Zwicker, E., & Fastl, H. (2007). *Psychoacoustics—Facts and models* (3rd ed.). Berlin, Germany: Springer.

5 Dynamic Haptic Interaction with Video

Nuray Dindar, A. Murat Tekalp, and Cagatay Basdogan

CONTENTS

This chapter introduces the notion of passive dynamic haptic interaction with video and describes the computation of force due to relative motion between an object in a video and the haptic interface point (HIP) of a user, given associated pixel-based depth data. While the concept of haptic video, that is, haptic rendering of forces due to geometry and texture of objects in a video from the associated depth data, has already been proposed, passive dynamic haptic interaction with video has not been studied before. It is proposed that in passive dynamic interaction, a user experiences motion of a video object and dynamic forces due to its movement, even though the content of the video shall not be altered by this interaction. To this effect, the acceleration of a video object is estimated using video motion estimation techniques while the acceleration of the HIP is estimated from the HIP position acquired by the encoders of the haptic device. Mass values are assigned to the video object and HIP such that user interaction shall not alter the motion of the video object according to the laws of physics. Then, the dynamic force is computed by using Newton's second law. Finally, it is scaled and displayed to the user through the haptic device in addition to the static forces due to the geometry and texture of the object. Experimental results are provided to demonstrate the difference in rendered forces with and without including the dynamics.

5.1   INTRODUCTION

The next step in digital media technologies will be truly immersive 2D or 3D high-definition media with the help of additional modalities. One of these new modalities will be touch, which is a powerful sense for humans. Haptic video refers to adding a sense of touch to video by preauthoring shape and texture of certain objects, which offers viewers the possibility of passive immersive experiences in entertainment and education applications. With haptic video, viewers will be able to touch these video objects to experience their 3D shape, texture, and motion.

Haptic interaction has been originally proposed for virtual environments (Basdogan, Ho, Srinivasan, & Slater, 2000; Basdogan & Srinivasan, 2001; Magnenat-Thalmann & Bonanni, 2006; Salisbury, Conti, & Barbagli, 2004; Srinivasan & Basdogan, 1997). The potential of adding haptic force feedback into 2D broadcast video has first been discussed by O'Mohdrain and Oakley (2003, 2004). They have defined the term presentation interaction that involves changing the location of the main character in a cartoon movie through the use of a 2-degree-of-freedom (dof) haptic force feedback interface without altering the structure of the narrative. When the user moves the haptic cursor, the character's rendered location in the scene is changed and the force vector derived from the character's displacement from the center of the screen is displayed to the user through the device. In some applications, vibrotactile feedback is also displayed in addition to the force feedback.

Recently, haptic interaction with video has advanced to include rendering structure and texture. Haptic structure refers to touching/feeling the 3D geometry of a video object. The 3D structure can be rendered using full 3D polygon or mesh models or 2.5D pixel-based depth data. A video together with associated depth data for each pixel is called a *2.5D representation* in the computer vision literature and *video-plus-depth representation* in the Motion Picture Experts Group (MPEG) community. Since recent MPEG standards for 3D video rely on such representations, for example, the multiview video-plus-depth (MVD) or layered-depth-video (LDV) representations of 3D video (Müller, Smolic, Dix, Kauff, & Wiegand, 2008; Smolic & Kauff, 2005), it is expected that video together with associated depth images will be widely available in the future and can be used for haptic interaction. Haptic texture refers to rendering surface properties (e.g., roughness) of objects in a video, which can be recorded and encoded as separate video channels. Cha, Kim, Oakley, Ryu, and Lee (2005) and Cha, Kim, Ho, and Ryu (2006) have been the first to propose computing haptic information from depth images rather than full 3D object/scene models. They have modified the proxy graph algorithm (Walker & Salisbury, 2003) for haptic rendering of depth images by constructing a triangular surface model from the depth data and taking the projection of the line segment constructed between the current and previous HIP positions onto the depth image to determine a list of candidate triangles for the collision detection. Then, the collision between the HIP and the video object is detected using this list of candidate triangles. Using this approach, they displayed the rendered force due to the geometrical properties of the object via a 3-dof haptic interface.

More recently, rendering forces to enable the viewer to follow/sense the motion trajectory of a video object has been proposed. Gaw, Morris, and Salisbury (2006) proposed a system for manually annotating haptic motion trajectory information in sync with a movie, though the haptic information due to the geometry has not been considered in rendering. The system proposed by Yamaguchi, Akabane, Murayama, and Sato (2006) generated the haptic effect automatically from 2D graphics using the metadata description of the movement characteristics of the content. Cha, Seo, Kim, and Ryu (2007) developed an authoring/editing framework for haptic broadcasting in the context of passive haptic interaction by extending MPEG binary format for scene (BIFS). They defined two modes of haptic playback, kinesthetic and tactile. In the kinesthetic mode, viewers are forced to follow a predefined motion trajectory with the help of a proportional-derivative (PD) controller, where the applied force is proportional to the distance between the object and the HIP position and the derivative of this distance. The tactile mode involves display of predetermined vibrations to the user while the video plays. The magnitude of the vibration is determined by the grayscale level in the tactile effect channel embedded into the video clip. The user experiences haptic sensation when a selected character in the video interacts with another character or an object or performs an action where the trajectory of the motion can be sensed

(Cha et al., 2007). Rasool and Sourin added haptic information to image and videos as a function-based description where they embedded predefined functional representation of the force fields due to the motion of a video object (Rasool & Sourin, 2010). Lastly, Rydén, Chizeck, Kosari, King, and Hannaford (2011) demonstrated real-time haptic interaction with a moving object recorded by Kinect™. They estimated the force displayed to the user using the contacts between the HIP and the point cloud representation of the object acquired by the camera without considering the relative motion between the object and the HIP (Ryden et al., 2011).

Previous literature on passive haptic interaction with a video deals with either displaying prerecorded vibrations to the user when an object in the video interacts with another object or displaying prerecorded forces due to the trajectory of a moving object in the scene or rendering forces only due to the geometrical and material properties of a video object. There is not any earlier study that considers the computation of dynamic forces due to the relative acceleration of video objects. Two important observations regarding the passive dynamic haptic interaction are as follows: (1) Motion is always observed and defined relative to a frame of reference. For example, if there is camera pan in the direction of motion of a moving object, the apparent motion of the object shall be slower than its actual motion. Furthermore, the HIP, which is used to interact with the object, may or may not be moving. In this chapter, the term *relative motion* refers to the apparent motion of a video object. (2) Unlike in active haptic interaction such as in virtual reality or game applications, in passive interaction, neither the geometry nor the motion of an object in the video can be modified by interacting with it, but the geometry and motion of the video object can be experienced by users simultaneously. To this effect, the HIP should act like a *fly* interacting with a *horse*. Even though a fly may have acceleration, it cannot change the velocity or direction of the motion of a horse. However, a fly moving on the surface of a horse can feel its geometry as well as the dynamic forces due to its motion.

The aim of this chapter is to introduce passive dynamic haptic interaction with a video object, given a 2.5D video representation, considering the dynamic forces due to the relative motion of a video object with respect to the HIP in addition to the static forces due to the geometry of the video object. The main novelties of our approach include (1) a method for computation of dynamic forces for passive interaction considering the relative motion between a video object and the HIP and (2) a new haptic rendering method based on pixel-based depth data without any partial reconstruction of the object surface. The organization of the chapter is as follows: Section 5.2 discusses the computation of static and dynamic forces as well as describing the mapping between graphical/image and haptic workspaces and force interpolation. Section 5.3 describes video processing methods for acquiring more smooth haptic feedback. Section 5.4 elaborates on the experimental setup and results while conclusions are given in Section 5.5.

5.2 COMPUTATION OF FORCE EXPERIENCED BY THE USER

Haptic video requires the computation of forces that a user would experience as if the user actually touches an object in the video. Because an HIP is used to interact with the object, the user would feel as if touching and feeling the object with a fingertip. The total force experienced by the user, $\vec{F}_{user}$, can be modeled as the sum of static force, $\vec{F}_{static}$, based on the geometrical and material properties of an object and dynamic force, $\vec{F}_{dynamic}$, due to the relative motion between an object in the video and the cursor called the HIP:

$$\vec{F}_{user} = \vec{F}_{static} + \vec{F}_{dynamic} \tag{5.1}$$

In computing $\vec{F}_{dynamic}$, the acceleration of a video object is estimated from the video using motion estimation techniques, while the acceleration of HIP is computed from the HIP position data via numerical differentiation.

The force $\vec{F}_{user}(t)$ is calculated when the HIP penetrates into the video object. If $(x_{haptic}(t), y_{haptic}(t), z_{haptic}(t))$ are defined as the current position of the HIP in the 3D haptic space, $\vec{F}_{user}(t)$ can be calculated only if $z_{haptic}(t)$ is inside the object as determined from the object depth map value at $h_{dep}(x_{img}(t), y_{img}(t))$ where $(x_{img}(t), y_{img}(t))$ denotes the HIP position mapped into the image coordinates. This mapping and further implementation details are discussed in Section 5.2.3.

5.2.1 Static Force

This section discusses a modification of the method proposed by Ho, Basdogan, and Srinivasan (1999) for haptic rendering of textures mapped onto a 3D geometrical object. Ho et al. (1999) perturbed the surface normal of the object, where the HIP contacts object, computed from a 3D geometric model, using the gradient of the texture field, which is defined as a texture height for each pixel.

In order to generate the 3D structure of a video object, it can be assumed that the depth values at each pixel, obtained from a given depth image (which is a height value defined for each pixel), are superposed onto a 2D planar surface. Then the surface normal at each pixel is computed using the perturbation method proposed by Ho et al. (1999). The most general form of the expression for calculating the perturbed surface normal, proposed by Ho et al., is given by

$$\vec{M}_{geo} = \vec{N}_s - \vec{\nabla}h_{dep} + \left(\vec{\nabla}h_{dep} \cdot \vec{N}_s\right)\vec{N}_s \tag{5.2}$$

where the perturbed surface normal, $\vec{M}_{geo}$, is defined as the difference between the original surface normal of the object, $\vec{N}_s$, and the local gradient of depth map value, $\vec{\nabla}h_{dep}$, and added to the projection of the local gradient onto the surface normal at the contact point. Since the image plane is initially 2D, the surface normal at any given point is taken as (0, 0, 1) by default. Furthermore, the local gradient of the depth map, which is calculated by the Sobel operator around a given point, has only x and y components. Hence, the dot product of $\vec{\nabla}h_{dep}$ with $\vec{N}_s$ is equal to 0, which simplifies (5.2). Then, for stability reasons as suggested by Ho et al. (1999), the direction and magnitude of the force vector experienced by the user can be calculated by

$$\vec{F}_{static} = \begin{cases} \left(d - Kh\right)\vec{N}_s + Kh\vec{M}_{geo} & \text{if } d \geq Kh \\ d\vec{M}_{geo} & \text{if } d \leq Kh \end{cases} \tag{5.3}$$

$$d = \alpha h_{dep}\left(x_{img}(t), y_{img}(t)\right) - z_{haptic}(t) \tag{5.4}$$

In (5.3), K is a scalar that depends on the surface texture properties, while h is the normalized depth value at the current HIP position. In (5.4), d denotes depth of penetration into the object surface. Penetration is the difference between the depth image (map) value h_{dep} at the HIP position mapped into image coordinates $(x_{img}(t), y_{img}(t))$ and the z component of the HIP in the haptic workspace, $z_{haptic}(t)$. Note that h_{dep} is scaled by α in order to convert the depth code value to the physical units of the haptic device workspace.

5.2.2 Dynamic Force

One of the main novelties of this chapter is introducing the concept of dynamic force experienced by a user due to the relative motion between the video object and the HIP. The motion of a video object actually represents the relative motion between the corresponding scene object and the camera. For example, in the case of a camera pan, the apparent motion of the video object may feel larger or smaller than its actual motion depending on the direction of the pan. Since typically access to camera calibration parameters is limited, force experienced by the user is calculated based on the apparent motion of the video object relative to the HIP. Since the HIP can be stationary, or moving toward or away from the video object, the dynamic component of the force experienced by the user will vary according to the relative motion between the video object and the HIP.

For simplicity, it is assumed that the object has a constant acceleration from one frame to another and its acceleration is calculated using

$$\vec{a}_{object} = \frac{\vec{\Delta v}}{\Delta t} = \frac{25\vec{v}_{flow}}{10^{-s}} \ (pixel/s^2) \tag{5.5}$$

where

$\vec{v}_{flow}$ is velocity of the object found by the optical flow algorithm (defined in Section 5.3.2)
$\vec{a}_{object}$ is the acceleration of the video object

Note that $\vec{v}_{flow}$ is scaled by 25 assuming that the video runs at 25 fps and object acceleration is calculated for each haptic loop; hence, Δt is taken as 1 ms. Then, the dynamic force experienced by the user can be calculated as

$$\vec{F}_{dynamic} = \beta m_{object} \vec{a}_{object} - m_{cursor} \vec{a}_{cursor} \ (Newton) \tag{5.6}$$

where

m_{object} and m_{cursor} are mass values assigned to the video object and cursor, respectively, as discussed in Section 5.2.3
$\vec{a}_{cursor}$ is the acceleration of the cursor
β is a constant to convert the object acceleration from units of image workspace to that of the haptic workspace

5.2.3 IMPLEMENTATION OF DEPTH IMAGE-BASED HAPTIC RENDERING

Another novelty of this chapter is to introduce a new depth-based haptic rendering method using only pixel-based depth data without any partial reconstruction of object surface. The steps of defining a mapping between image and haptic space coordinates and interpolation of depth at the HIP position are described later. It is assumed that the user interacts with video using a 3-dof HIP, which provides force feedback. The x–y coordinates of the point of contact of the HIP with the video are shown as a 3D cone on the image plane. The size of the cone is scaled according to the z coordinate of the HIP to provide the user with a sense of depth on the screen.

5.2.3.1 Mapping between Image Coordinates and Haptic Coordinates

As a first step, the position of the HIP in the physical workspace of the haptic device must be registered with the image coordinates. The x–y coordinates of the HIP position $(x_{haptic}(t), y_{haptic}(t), z_{haptic}(t))$ in the physical workspace are mapped to $(x_{img}(t), y_{img}(t))$ on the image plane using the following linear relations:

$$x_{img}(t) = a_x x_{haptic}(t) + b_x \tag{5.7}$$

$$y_{img}(t) = a_y y_{haptic}(t) + b_y \tag{5.8}$$

where coefficients

$$a_x = \frac{w}{\left(x_{haptic}^{max} - x_{haptic}^{min}\right)} \qquad b_x = -a_x x_{haptic}^{min} \tag{5.9}$$

$$a_y = \frac{h}{\left(y_{haptic}^{max} - y_{haptic}^{min}\right)} \qquad b_y = -a_y y_{haptic}^{min} \qquad (5.10)$$

Here, $\left(x_{haptic}^{min}, y_{haptic}^{max}\right)$ and $\left(x_{haptic}^{max}, y_{haptic}^{min}\right)$ correspond to the top-left and bottom-right corners of the haptic workspace and w and h represent width and height of the image, respectively. The bias terms b_x and b_y are needed since x_{haptic}^{min} and y_{haptic}^{min} can be negative and the pixel indexes of the image must be integers in the range of $[0, w)$ and $[0, h)$, respectively.

5.2.3.2 Interpolation of Depth Map

In order to determine the image depth at the current HIP position $(x_{img}(t), y_{img}(t))$ on the image plane, which is in general a noninteger value, bilinear interpolation is employed from the depth values at the neighboring pixels as depicted in Figure 5.1.

In Figure 5.1, the red dot shows the actual HIP position on the image plane $(x_{img}(t), y_{img}(t))$ and the blue dots show the neighboring pixels. The weights used in the bilinear interpolation are inversely proportional to the distances of the neighboring pixels to $(x_{img}(t), y_{img}(t))$:

$$d_x = \left| x_{img} - \lfloor x_{img} \rfloor \right| \qquad d_y = \left| y_{img} - \lceil y_{img} \rceil \right| \qquad (5.11)$$

The depth value corresponding to the HIP position is then calculated using the bilinear interpolation as

$$\begin{aligned}
h_{dep}\left(x_{img}, y_{img}\right) = {} & d_x d_y h_{dep}\left(\lfloor x_{img} \rfloor, \lceil y_{img} \rceil\right) \\
& + (1 - d_x) d_y h_{dep}\left(\lceil x_{img} \rceil, \lceil y_{img} \rceil\right) \\
& + d_x(1 - d_y) h_{dep}\left(\lfloor x_{img} \rfloor, \lfloor y_{img} \rfloor\right) \\
& + (1 - d_x)(1 - d_y) h_{dep}\left(\lceil x_{img} \rceil, \lfloor y_{img} \rfloor\right) \qquad (5.12)
\end{aligned}$$

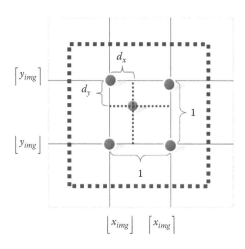

FIGURE 5.1 Bilinear interpolation of the scene/object depth at the HIP position $(x_{img}(t), y_{img}(t))$ on the image plane, shown by the red dot, using the depth map values at the neighboring image pixels, shown by the blue dots.

The penetration, which was defined in (5.4), can then be calculated as the difference between h_{dep} and the z coordinate of the HIP. However, in order to take this difference, the dynamic range of the depth image needs to be mapped to the workspace of the haptic device along the z-axis linearly such that the range of depth values between 0 and 255 corresponds to the range of movements of HIP along the z-axis in the haptic workspace.

5.2.3.3 Selection of Mass for Video Object and Haptic Interface Point

Because in general there is no access to the physical video object, the selection of proper mass values for the HIP and video object should be addressed carefully considering applicable physical laws. Because the velocity of a video object cannot be altered as a result of the user interaction, it is assumed that collision of the HIP with a video object shall only affect the HIP, whereas in real life, collision would affect both parties. Hence, for a real-life-like experience, the mass of the video object should be chosen much larger than that of the HIP so that the velocity of the video object would not be altered by the user interaction. This is in analogy to a fly interacting with a horse as mentioned in the Section 5.1.

In an active interaction, the force experienced by the user should vary according to how tight the user grips the HIP, which should be reflected into the mass of the HIP. However, in our passive dynamic interaction scenario, the effect of the grip of the user is neglected in accordance to the horse–fly analogy. Therefore, the apparent mass of the haptic device at the tip is set to 0.045 kg, as measured by the manufacturer of the device and stated in the data specifications sheet (Sensable Technologies Inc., 2010). Then, the mass of the video object is set equal to at least an order of magnitude larger than this value, such that $m_{object} \gg m_{cursor}$ again in accordance to the horse–fly analogy.

5.2.3.4 Calculation of Accelerations and Forces

For the computation of dynamic force, the acceleration of the video object and the HIP needs to be estimated. To this effect, first HIP velocity is estimated as the difference of the HIP position in time:

$$v_{cursor_x} = x_{haptic}\left(t\right) + x_{haptic}\left(t-1\right) \tag{5.13}$$

$$v_{cursor_y} = y_{haptic}\left(t\right) + y_{haptic}\left(t-1\right) \tag{5.14}$$

Then, the acceleration can be computed similarly as the difference of HIP velocity in time.

The object accelerations are converted to the units of mm/s² using the inverse of linear mapping defined in (5.7), and then $\vec{F}_{dynamic}$ is calculated using (5.6). The sum of $\vec{F}_{dynamic}$ and $\vec{F}_{static}$ is rendered and displayed to the user through the haptic device. For smooth and stable haptic interaction, all forces are interpolated from the video frame rate to the force update rate of the haptic device.

5.2.3.5 Temporal Interpolation of Forces

Smooth and stable force rendering requires the haptic display rate to be significantly higher than the video/graphic frame rate (Salisbury et al., 2004). For example, the haptic display rate is typically 1 kHz, while the video/graphic frame rate is 25 Hz. Hence, interpolated forces calculated at the video frame rate to the haptic display rate for effective haptic experience are needed. To this effect, cubic/spline interpolation can be used assuming that the cursor position is fixed between consecutive video frames. Hence, if the cursor is at the position (x_1, y_1) in frame i and the calculated total force is total $\vec{F}_{total}^{i}(x_1, y_1)$, forces between $\vec{F}_{total}^{i}(x_1, y_1)$ and $\vec{F}_{total}^{i+1}(x_1, y_1)$, which are the total force calculated at the cursor position (x_1, y_1) in frame $i + 1$, are interpolated in the temporal dimension using cubic interpolation.

5.3 VIDEO PROCESSING FOR HAPTIC RENDERING

In order to compute reliable haptic information from 2D video clip and associated depth images, depth images for the video object should be smooth and detailed. The depth images are typically quantized with 8 bits/pixel, and hence, the depth values are in the range [0,255]. If the range of depth in the scene (depth of field) is large, the quantization can cause loss of fine detail regarding the structure/geometry of some objects in the scene. Therefore, postprocessing of depth images may be required to smooth depth data and improve the haptic experience. Furthermore, motion estimation from video and associated depth images is required to estimate the acceleration of video objects in order to render dynamic forces. These postprocessing operations are described in detail in the following.

5.3.1 ENHANCEMENT OF DEPTH IMAGES

The first video processing step is to segment *the touchable object* or the so-called hot object from the rest of the scene at each frame. Hence, the geometrical information of touchable video object would be represented more precisely within the range of the depth map. To segment the object, both the color and depth images are used. The well-known Canny edge detection algorithm is applied to find the edges of the object of interest (Canny, 1986). The color edge images and edges of depth images are fused, in order to create the object segmentation mask as shown in Figure 5.2, where segmentation of the horse from rest of the frame is shown.

After the video object is segmented, depth values outside the object mask are set equal to zero, and depth values within the object mask are rescaled between the minimum and maximum depth values throughout the video clip. Finally, a Gaussian low-pass filter is applied to smooth depth images for a more realistic haptic experience.

5.3.2 MOTION ESTIMATION

In the most general case, precise calculation of the relative motion/acceleration between a video object and the HIP requires estimation of 3D motion of the object from the video and associated depth images,

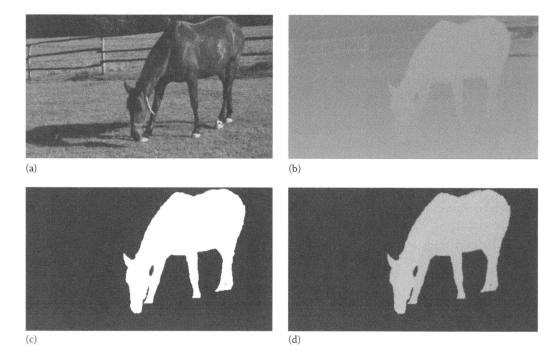

(a)   (b)

(c)   (d)

FIGURE 5.2 Illustration of video object segmentation and depth image enhancement: (a) original frame (Mobile3DTV), (b) corresponding depth image, (c) computed object segmentation mask, and (d) enhanced depth image.

FIGURE 5.3 Sample frames from the synthetic video with $\vec{a}_{object_z} = 0$. Here, a ball moves vertically down in a plane that is perpendicular to the optical axis of the camera.

so that the z component of the object acceleration (in addition to x and y components) can also be computed for dynamic force calculation using (5.6). Estimation of 3D motion and structure from video has been extensively studied in the literature and some widely used methods and procedures exist (Robertson & Cipolla, 2009). These methods can be classified as 3D motion and structure using point correspondences or using dense 2D optical flow. In the former, the first step is to extract common features between successive frames that are registered automatically or manually. The next step requires estimation of the 3D motion and structure parameters using either the essential matrix method (Hartley & Sturm, 1994; Hartley & Zisserman, 2000) or the factorization method proposed by Tomasi and Kanade (1992) that can be formulated for affine or perspective projection models. Since the haptic interaction by the user cannot affect the motion of objects in the video, the computation of 3D motion parameters can be done offline.

In certain special cases, such as 2D motion on a plane perpendicular to the camera or motion purely along the z direction (toward to or away from the camera), there are simpler approaches to estimate acceleration $\vec{a}_{object_x}$, $\vec{a}_{object_y}$, and $\vec{a}_{object_z}$. In the former, $\vec{a}_{object_z} = 0$; in the latter, $\vec{a}_{object_x} = \vec{a}_{object_y} = 0$; and $\vec{a}_{object_z}$ can be estimated by processing depth map images.

Special case $\vec{a}_{object_z} = 0$: If the video object performs 2D planar motion in a plane that is perpendicular to the camera, the dense 2D velocity, and hence acceleration, can be estimated using the Lucas and Kanade (1981) optical flow algorithm over the segmented textured images (see Figure 5.3 for some sample frames of a synthetic video). This algorithm provides us with the x and y components of the 2D object motion with respect to the previous frame. The outlier motion vectors can be filtered using a 2D median filter to obtain a smoother motion field.

Special case $\vec{a}_{object_x} = \vec{a}_{object_y} = 0$: If the motion of the object is purely along the z direction, object motion can be estimated from the variation of the depth observed from the depth map images (see Figure 5.4 for some sample frames of a synthetic video). Since it is assumed that the object structure does not change (rigid object) through the video, the computed differences in depth between the frames will be due to the motion of the object.

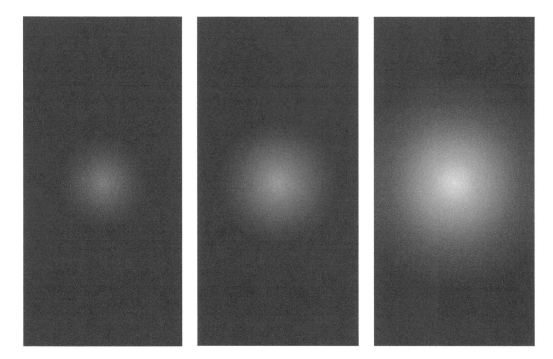

FIGURE 5.4 Sample frames from the synthetic video with $\vec{a}_{object_x} = \vec{a}_{object_y} = 0$. Here a ball moves toward the camera along the optical axis of the camera.

5.4 CASE STUDIES AND RESULTS

The proposed concepts are tested using the Omni haptic device (by Sensable Technologies) having 6-dof sensing and 3-dof force feedback. In our system, the user interacts with video passively through the HIP. Both static and dynamic forces are experienced while interacting with an object moving in the video. The video can be also paused at a particular frame to feel the geometry of the video object alone (i.e., the static force).

In order to demonstrate the passive dynamic haptic interaction concept in a controlled experiment, two synthetic video clips of a bouncing ball are generated. In the first video clip, shown in Figure 5.3, the ball moves vertically up and down perpendicular to the camera, while in the second one, shown in Figure 5.4, it moves back and forth toward the camera along the z-axis. Moreover, some results on a real *horse* video with associated depth data are provided (Mobile3DTV).

5.4.1 SIMULATION RESULTS

Video paused—static force only: A slice of the ball on the y–z plane at a particular frame is shown in Figure 5.5. The depth is zero at the reference plane (background) and increases as the object approaches the camera along the $+z$-axis. The rendered static forces are depicted using the blue arrows while the black curve represents the depth values of the ball at the specified y positions. Note that the HIP moves along the y-axis only; hence, only the y- and z-axes are shown in Figure 5.5.

Case 1: The ball moves perpendicular to the camera axis.

1. *The HIP is stationary* $\left(\left\|a_{cursor} = 0\right\|\right)$: The ball can move along the x-axis, y-axis, or both. In our example (see Figure 5.6), the ball accelerates along the $+y$ direction and the HIP is stationary. Hence, the total force experienced by the user, $\vec{F}_{user}$, shown as red, becomes larger than the static force, $\vec{F}_{static}$ alone, shown as blue. The x and z components of the total

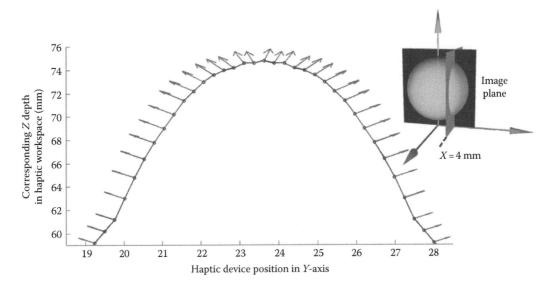

FIGURE 5.5 Static forces rendered for a slice of the ball projected onto the y–z plane in haptic workspace.

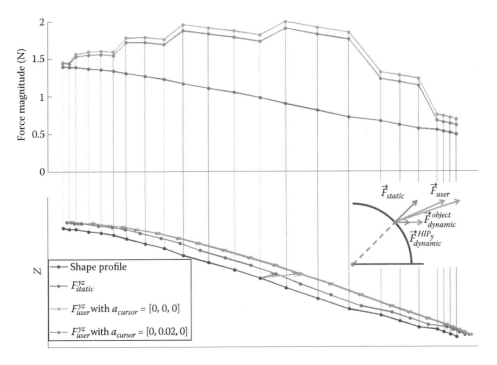

FIGURE 5.6 (a) The force magnitudes and (b) the profile of the force vectors rendered for a half slice of the ball (i) when only the static forces exist (blue curve), (ii) when the cursor is not moving but the object has an acceleration along the +y direction (red curve) and (iii) when the cursor and ball accelerates along the +y direction (purple curve).*

---

* The color versions of the plots and the figures appearing in this chapter are available in the authors' website (http://home. ku.edu.tr/~cbasdogan/).

force are due to the geometry of the video object only, while the y component is due to both the acceleration and the geometry of the object. Note that, since the acceleration of the object is modified intentionally between the frames, the effect of dynamic force is changing.

2. *The HIP has a constant acceleration (a_{cursor} is constant)*: The effect of relative motion of the video object with respect to the HIP motion is also shown in Figure 5.6. The HIP and the object have a constant acceleration along the $+y$ direction. Hence, the total force is perturbed slightly toward the y direction (see the purple curve in Figure 5.6). Note that the acceleration of the HIP is in the same direction as the video object, and hence, the magnitude of the force experienced by the viewers in this case (purple curve) is less than what he or she feels when the HIP is stationary (red curve).

Case 2: The ball moves toward the camera.

1. *The HIP is stationary $\left(\left\|a_{cursor}\right\| = 0\right)$*: It is assumed that the object is rigid and its structure is not changing as it moves toward the camera. Hence, the changes in the size of the object and depth values at each pixel are due to the motion of the object. Since the object moves toward the camera, there is a dynamic force along the $+z$ direction. The effect of this dynamic force is shown by the red curve in Figure 5.7. The blue and red curves again represent the static force and total force including the dynamic force, respectively.

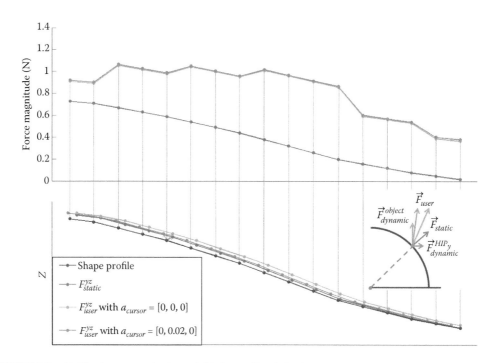

FIGURE 5.7 (a) The force magnitudes and (b) the profile of the force v ectors rendered for a half slice of the ball (i) when only the static forces exist (blue curve), (ii) when the cursor is not moving but the object has an acceleration along the $+z$ direction (red curve) and (iii) when the object accelerates along the $+z$ direction and the cursor accelerates along the $+y$ direction (purple curve). The corresponding force magnitudes for each HIP position are shown in the upper plot.*

---

* The color versions of the plots and the figures appearing in this chapter are available in the authors' website (http://home. ku.edu.tr/~cbasdogan/).

FIGURE 5.8   Sample frames from a real video clip (Mobile3DTV). Here, the horse slowly raises his head.

2. *The HIP has a constant acceleration (a_{cursor} is constant)*: The effect of the relative motion of the ball with respect to the HIP when the object moves toward the camera is shown as a purple curve in Figure 5.7. The HIP has a constant acceleration, having component along the y direction. Since the force due to the acceleration of the ball is in the z direction only, the y component of the total force is due to both the acceleration of the HIP and the geometry of the video object.

5.4.2 RESULTS WITH A REAL VIDEO CLIP

The *horse* video, where the horse raises his head, has been used as a real video clip for testing the proposed approach (see Figure 5.8 for some sample frames). While calculating the acceleration of the head, it is assumed $\vec{a}_{object_z} = 0$ since the movement of the horse's head toward the camera is relatively small compared to its movement perpendicular to the camera. Hence, only $\vec{a}_{object_x}$ and $\vec{a}_{object_y}$ are calculated from the video clip using the motion estimation techniques discussed earlier. The variation in the static and the total forces in the x- and y-axes as a function of the selected frames is shown in Figure 5.9a and b, respectively. In rendering the dynamic forces, the cursor is virtually attached to a fixed spot on the horse's head (i.e., it is stuck to the same spot on the horse as the head moves); hence, the static force is constant since the geometry does not change; but the dynamic force varies due to the acceleration of the horse's head. The red curve in each plot represents the total force due to the *sticking* effect and the acceleration of horse head at the HIP position.

The user can also pause the video and feel the static force alone due to the geometry of the horse, generated from the depth data. The smoothness of the depth data is important for the user experience since it directly affects the gradient calculation and hence the direction of the static force vector displayed to the user through the haptic device. In order to reduce the noise in the depth data and make the haptic experience more enjoyable, a low-pass filter was applied to the depth data.

5.5 CONCLUSION

This chapter introduces the concept of passive dynamic haptic interaction with a video and describes a method for rendering dynamic forces due to the relative motion of the video object. Earlier studies on this subject have focused on the rendering of only static forces due to the geometry of video objects and neglected the dynamic effects. However, in passive haptic interactions with videos involving dynamic scenes such as a ball kicked by a soccer player, a bullet fired by a shooter, or a jet plane making acrobatic movements in the air, displaying the forces due to the inertia of the video object (e.g., the ball, bullet, and jet plane) to the user is important for a more immersive experience. As shown in Figure 5.6, depending on the mass and the acceleration of the video object with respect to those of the HIP, the contribution of the dynamic force to the total force can be significant (e.g., compare the separation distances of the red- and blue-colored force profiles from the black-colored depth profile in Figure 5.6) and may alter the user experience. Note that the masses of the video object and the HIP are set to ensure that the user is a passive observer; that is, the momentum transferred to the video object

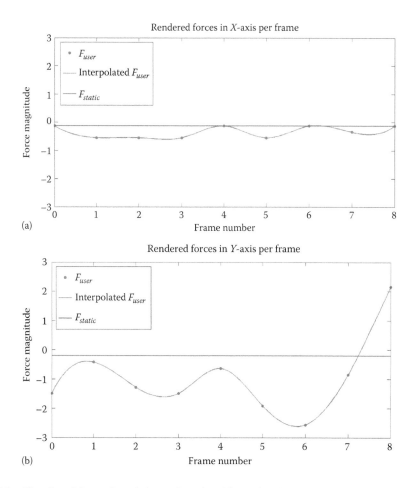

FIGURE 5.9   The plot of the static and the total rendered forces in (a) x-axis and (b) y-axis. A cubic interpolation is used to connect the discrete force values calculated at each video frame and then display them to the user at the haptic update rate of 1 kHz.

from the user (through the HIP) is minimal and does not cause a change in its state. In other words, the user cannot change the state of the object in the video by pushing or pulling it via the HIP. Since the mass of the video object is selected significantly higher than that of the cursor, the dynamic force felt by the user is mainly due to the acceleration of the video object. Experimental results demonstrate that adding dynamic forces results in an observable difference in the total force that leads to more realistic haptic experience.

In conclusion, *immersive* and *interactive* are frequently used adjectives in applications such as virtual reality, computer games, and human–computer interaction. However, their use with real media (both 2D and 3D media) entertainment, broadcast, and communications services is new and in its infancy, since stimuli represented by modalities beyond 3D video and spatial audio, such as touch, heat, and smell, need also be captured, authored, transmitted, rendered, and displayed at the receiving side to even create a fully immersive passive haptic media experience. Furthermore, new modalities of interaction with media, such as active haptic interaction, have not yet been considered. Novel ideas and approaches that would enable haptic experience and haptic interaction with real media open new horizons that will significantly impact next-generation media technologies.

ACKNOWLEDGMENT

A. Murat Tekalp acknowledges support from Turkish Academy of Sciences (TUBA).

REFERENCES

Basdogan, C., Ho, C. H., Srinivasan, M. A., & Slater, M. (2000). An experimental study on the role of touch in shared virtual environments. *ACM Transactions on Computer-Human Interaction (TOCHI)*, 7(4), 443–460.

Basdogan, C., & Srinivasan, M. A. (2001). Haptic rendering in virtual environments. In K. M. Stanney (Ed.), *Handbook of virtual environments: Design, implementation, and applications* (pp. 117–134). Hillsdale, NJ: Lawrence Erlbaum Associates.

Canny, J. (1986). A computational approach to edge detection. *IEEE Transactions on Pattern Analysis and Machine Intelligence, PAMI, 8*(6), 679–698.

Cha, J., Kim, S. Y., Ho, Y. S., & Ryu, J. (2006). 3D video player system with haptic interaction based on depth image-based representation. *IEEE Transactions on Consumer Electronics*, 52(2), 477–484.

Cha, J., Kim, S. M., Oakley, I., Ryu, J., & Lee, K. H. (2005). Haptic interaction with depth video media. In Y. S. Ho & H. J. Kim (Eds.), *Advances in multimedia information processing-PCM 2005* (pp. 420–430). Berlin, Germany: Springer.

Cha, J., Seo, Y., Kim, Y., & Ryu, J. (2007). An authoring/editing framework for haptic broadcasting: Passive haptic interactions using MPEG-4 BIFS. *Second Joint Eurohaptics Conference and Symposium on Haptic Interfaces for Virtual Environment and Teleoperator Systems, World Haptics 2007* (pp. 274–279). Washington, DC: IEEE.

Gaw, D., Morris, D., & Salisbury, K. (2006). Haptically annotated movies: Reaching out and touching the silver screen. *14th Symposium on Haptic Interfaces for Virtual Environment and Teleoperator Systems* (pp. 287–288). Washington, DC: IEEE.

Hartley, R., & Sturm, P. (1994, November). Triangulation. *ARPA Image Understanding Workshop* (pp. 957–966). Monterey, CA.

Hartley, R., & Zisserman, A. (2000). *Multiple view geometry in computer vision* (Vol. 2). New York, NY: Cambridge University Press.

Ho, C. H., Basdogan, C., & Srinivasan, M. A. (1999). Efficient point-based rendering techniques for haptic display of virtual objects. *Presence*, 8(5), 477–491.

Lucas, B. D., & Kanade, T. (1981, August 24–28). An iterative image registration technique with an application to stereo vision. *Proceedings of the 7th International Joint Conference on Artificial Intelligence*. Vancouver, British Columbia, Canada.

Magnenat-Thalmann, N., & Bonanni, U. (2006). Haptics in virtual reality and multimedia. *IEEE Multimedia, 13*(3), 6–11.

Mobile3DTV. (2009). *Video plus depth database*. Retrieved from www.mobile3dtv.eu/video-plus-depth/

Müller, K., Smolic, A., Dix, K., Kauff, P., & Wiegand, T. (2008). Reliability-based generation and view synthesis in layered depth video. *IEEE 10th Workshop on Multimedia Signal Processing* (pp. 34–39). Washington, DC: IEEE.

O'Modhrain, S., & Oakley, I. (2003, December). Touch TV: Adding feeling to broadcast media. In *Proceedings of the European Conference on Interactive Television* (pp. 41–47), Brighton, U.K.

O'Modhrain, S., & Oakley, I. (2004, March). Adding interactivity: Active touch in broadcast media. *Proceedings of the 12th International Symposium on Haptic Interfaces for Virtual Environment and Teleoperator Systems, 2004, HAPTICS'04* (pp. 293–294). Washington, DC: IEEE.

Rasool, S., & Sourin, A. (2010). Towards tangible images and video in cyberworlds—Function-based approach. *2010 International Conference on Cyberworlds (CW)* (pp. 92–96). Washington, DC: IEEE.

Robertson, D. P., & Cipolla, R. (2009). Structure from motion. In Varga, M. (Ed.), *Practical Image Processing and Computer Vision*, New York, NY: John Wiley.

Rydén, F., Chizeck, H. J., Kosari, S. N., King, H., & Hannaford, B. (2011). Using kinect and a haptic interface for implementation of real-time virtual fixtures. *Proceedings of the 2nd Workshop on RGB-D: Advanced Reasoning with Depth Cameras (in conjunction with RSS 2011)*, June 27, Los Angeles, CA.

Salisbury, K., Conti, F., & Barbagli, F. (2004). Haptic rendering: Introductory concepts. *IEEE Computer Graphics and Applications*, 24(2), 24–32.

Sensable Technologies Inc. (2010). PHANTOM Omni datasheet. Wilmington, MA: Sensable Technologies Inc.

Smolic, A., & Kauff, P. (2005). Interactive 3-D video representation and coding technologies. *Proceedings of the IEEE*, 93(1), 98–110.

Srinivasan, M. A., & Basdogan, C. (1997). Haptics in virtual environments: Taxonomy, research status, and challenges. *Computers & Graphics, 21*(4), 393–404.

Tomasi, C., & Kanade, T. (1992). Shape and motion from image streams under orthography: A factorization method. *International Journal of Computer Vision, 9*(2), 137–154.

Walker, S. P., & Salisbury, J. K. (2003, April). Large haptic topographic maps: Marsview and the proxy graph algorithm. *Proceedings of the 2003 Symposium on Interactive 3D Graphics* (pp. 83–92). New York, NY: ACM.

Yamaguchi, T., Akabane, A., Murayama, J., & Sato, M. (2006, July 3–6). Automatic generation of haptic effect into published 2D graphics. *Proceedings of the Eurohaptics*. Paris, France.

6 Olfactory Interfaces

David L. Jones, Sara Dechmerowski, Razia Oden,
Valerie Lugo, Jingjing Wang-Costello, and William Pike

CONTENTS

6.1 INTRODUCTION

The sense of smell provides a continuous stream of critical information about our environment, which we use to drive decisions every day. Smells cue us to potentially dangerous situations, attract us to things that we enjoy, and even help us to identify people from an early age. Because of our evolutionary dependency on scents, we are able to differentiate between thousands of smells. Although these natural uses of smells in the live environment suggest potential applications of scent in virtual environments (VEs), the most powerful applications of scent may lie in eliciting targeted emotional states or modify other user states to prepare them for educational or entertainment experiences.

The potential benefits of integrating scents into VEs have led designers to develop a battery of more realistic simulated scents, although very little guidance is available to support the integration to ensure that the targeted states of VE users are elicited. This chapter is designed to summarize guidance and research gaps associated with integrating olfactory cues into VEs. Section 6.2 outlines the odor sensation and perception process with a focus on conditions and phenomenon that change this process. Section 6.3 builds on this description outlining the effects of scent on human states with a particular focus on those states that are critical to create effective learning environments in VEs. Due to the tight coupling between scents and emotion, a significant portion of that section is devoted to the effects of scent on emotional state and the interaction of resulting emotional states on other critical human cognitive states. Section 6.4 presents technology options for integrating scent within VEs, outlining the advantages and disadvantages of each. Finally, Section 6.5 presents a series of examples of how scents are currently being applied in VEs to support training and improve immersion. At the end of each section, the guidance and research gaps that were introduced within are summarized to support direct application when designing and evaluating VEs.

6.2   HUMAN PROCESSING OF SMELL

Scent detection in humans follows a process of translating chemical molecules into signals that are passed to the primary olfactory cortex. As airborne scent molecules travel through the nasal cavity and over the olfactory epithelium, they are pulled in by receptor cells that activate the olfactory bulbs through a complex structure of reciprocal synapses and neurotransmitters (Bear, Connors, & Paradiso, 2001; Blake & Sekuler, 2006). During this process, the olfactory bulb segregates odorant signals into broad categories that are used to send signals to the primary olfactory cortex, which resides below the anterior portion of the temporal lobe (Blake & Sekuler, 2006; Gottfried, Deichmann, Winston, & Dolan, 2002; Sobel et al., 2000). The primary olfactory cortex communicates with other areas of the brain including the limbic system, which controls emotional responses. The strong emotion-evoking capability of odors that will be discussed later in this chapter is thought to be the product of the two-way link between the olfactory and the limbic system.

This process of scent sensation does not lead to an equal level of sensitivity to all scents (Blake & Sekuler, 2006). For example, mercaptan can be detected at a concentration 10 million times less than that required to smell carbon tetrachloride (Wenger, Jones, & Jones, 1956), leading it to be added to natural gas to warn people for gas leaks (Cain & Turk, 1985). Aside from the scent type, a number of other factors affect detection sensitivity. For example, human characteristics such as gender and age have an effect on odor sensitivity. Older adults are less sensitive to scents (Cain & Gent, 1991), and males are less sensitive than females, especially for musky odors (Koelega & Koster, 1974). Handedness has also been associated with differences in odor sensitivity, where left-handed people are generally less sensitive than right handed (Bensafi et al., 2002). These naturally occurring degradations to an individual's perception of scent can be further compounded by environmental and temporal effects. For example, research has shown that smokers are less sensitive to scents than nonsmokers (Ahlström, Berglund, Berglund, Engen, & Lindvall, 1987) and people are generally able to detect weak odors better in the mornings than in the evenings (Stone & Pryor, 1967). Although there are a great deal of factors that affect the human sensation of odors beyond what is presented in this overview, this demonstrates the complexities that must be taken into account when trying to create a universal olfactory experience in VEs prior to perceiving and applying meaning to scents. The remaining sections focus on describing how individuals perceive scents and the effect of those scents on the physiological and psychological states.

6.2.1   ODOR PERCEPTION

Subjective odor perception is a complex process that is often described using several dimensions. For example, pleasantness (or hedonicity), familiarity, and intensity are three common dimensions that are used to describe the subjective qualities of an odor (Delplanque et al., 2008). Using numeric methods for odor description, research has shown that when a wide range of odors are assessed at similar odor intensity, the hedonic character is the most salient psychological dimension in the perception (Yoshida, 1975). Because of this, hedonic tone is often the most prominent driver of odor experience and has a strong influence on mood, where pleasant odors induce positive moods and unpleasant odors induce negative moods (Rétiveau, Chambers, & Milliken, 2004). While hedonic perception is primarily driven by a learned association with the emotional context that the odorant was first encountered (Herz, Beland, & Hellerstein, 2004), a wide range of factors also have an effect on the human perception of scents.

6.2.1.1 Odor Intensity

From the physiological perspective, odor intensity largely depends on the concentration of the airborne molecules and the amount of odorant that reaches the olfactory receptors in the olfactory epitheliums. Odor intensity is reflected by the firing rate of neurons where weak odors elicit fewer neural impulses, while the strong odors elicit more neural impulses, and this pattern is similar at all levels of the olfactory system from receptor neurons to the cortex (Blake & Sekuler, 2006).

Unlike the physiological process, the perception of odor intensity is influenced by multiple factors. For example, there is a relationship between the perceived intensity and pleasantness of scents (Moskowitz, Dravnieks, & Klarman, 1976). Research suggests that although observers can separate odor intensity and pleasantness most of the time, hedonics are sufficiently pervasive enough to cloud the judgment of odor intensity. Similarly, research has shown that when people were able to identify scents, they were rated as both more familiar and more intense (Distel & Hudson, 2001).

6.2.1.2 Cross-Modality Interference

According to Gottfried and Dolan (2003), human olfactory perception is notoriously unreliable because it can be influenced by sensory input from other modalities (especially visual cues). This research suggests important cross-modal interaction between these primary sensory modalities. The majority of prior research regarding cross-modality interference to olfactory perception has been associated with food (Banks, Ng, & Jones-Gotman, 2012). This research has shown that pairing an odor with a congruent image (e.g., pairing orange smell with a picture of an orange) may facilitate the detection of that odor (Gottfried & Dolan). Similarly, pairing a congruent color (i.e., an orange color patch) or shape can also improve odor identification (Dematte, Sanabria, & Spence, 2009). On the other hand, incongruent images could easily trick one's odor perception. For example, dying orange juice green has led people to identify orange juice as lime juice (Blackwell, 1995). Outside of the realm of food and drink, smell and vision cues can be combined to inform decision making on other topics too. For instance, decisions about the pleasantness of a neutral visual stimulus may be swayed by the presentation of either a pleasant or an unpleasant odor (Van Reekum, Van den Berg, & Frijda, 1999). This effect of cross-modality interference suggests that scents can be used to influence the experience within VEs even when they are not directly applicable to the content being presented in the environment.

6.2.1.3 Personal Experiences/Emotion

One of the strongest influences on the perception of an odor and the emotional response to it is the past experience and current emotional connections associated with the cue. Although this is the case, there is a long-standing debate regarding whether hedonic responses to odors are innate or learned (Herz et al., 2004). The innate view of hedonic perception suggests that people are born with a predisposition to like or dislike certain smells. On the other hand, the learned view suggests that people are merely born with a disposition to *learn* to like or dislike smells. Regardless of the predisposition to like or dislike a smell, a great deal of research has linked emotional association and the hedonic evaluation of scents. Ultimately, the decision to like or dislike a scent is linked to the emotional valance of the experience that the scent is associated with (Bartoshuk, 1991). Herz, Beland, & Hellerstein (2004) suggested that "odor hedonic responses are formed from a learned association combining the sensory percept and the emotional experience when the percept was first encountered" (p. 315). Because of the strength of this association between scent evaluation and emotion, it is critical to tightly control the emotional state of VE users when introducing a novel scent into the environment so the desired emotional state can be triggered in the future. Likewise, it is important to gain an understanding of the emotional effects of scents to each VE user prior to presenting them in the environment. The strong ties between emotion and scent are described in more detail in Section 6.3.

6.2.1.4 Culture

A person's cultural background is tightly linked to their past experiences, and the effects on scent perception are just as strong. Culture and associated background can lead groups to respond to particular scents or categorize scents differently (Candau, 2004). Furthermore, culture can drive the decision of whether a scent is perceived as positive or negative valence, a primary component of both emotional state and scent classification. For instance, the response to the scent of wintergreen has been shown to be affected by cultural background (Herz, 2005). In a study conducted in 2004, US participants rated the scent as very pleasant, while British participants rated the scent as unpleasant. These results were related to cultural differences associated with common applications of the scents. Specifically, British participants associated the scent with medicine because it was a common additive at the time of the study leading to the negative ratings, while US participants associated the scent with candy, leading to positive ratings. These findings are important to take into account and suggest a need to evaluate previous applications of scents across cultures when developing VEs to be applied in a multicultural environment.

6.2.1.5 Gender

The effects of gender differences on olfactory functions are well researched. Neurological studies suggest that the gender has an effect on limbic structures of the brain. More specifically, the volume of amygdala is found to be larger and the bed nucleus of stria terminalis smaller in the females (Swaab, Chung, Kruijver, Hofman, & Ishunina, 2001). Although this difference in makeup has mixed effects on the olfactory perceptual thresholds between men and women (Hummel, Kobal, Gudziol, & Machay-Sim, 2007; Wysocki & Beauchamp, 1991), one of the most consistent findings regarding gender differences is that women tend to outperform men in odor identification tasks. According to Bengtsson, Berglund, Gulyas, Cohen, and Savic (2001), females are better at identifying scents in both free memory recall and recognition tasks, regardless of cultural background. Although these scent identification performance differences between genders are well accepted, the cause remains unclear (Bengtsson et al., 2001), as it could occur at the level of elementary encoding of odorous signals (i.e., at the sensation stage) or at the higher cognitive stages of odor processing (i.e., at the perception stage, which involves odor discrimination, memory, and identification). Regardless of the root cause of the difference, it is important to take this information into account when scent identification is critical in targeted applications.

6.2.1.6 Age

As people age, there is a decline in olfactory sensitivity. Studies have shown that this decline is generally associated with an elevation of threshold and a diminished sensitivity to suprathreshold stimuli (Loo, Youngentob, Kent, & Schwob, 1996). In addition to increased thresholds, elderly also perceive odors as less intense and have more trouble identifying odors than younger people. It is thought that the degradation in olfactory sensitivity is due to degenerative changes that occur with aging, including a reduced epithelium thickness, receptor neuron losses, intermingling of respiratory cells with olfactory neurons, and extensive patchy replacement of olfactory by respiratory epithelium (Naessen, 1971). Age-related diseases such as Alzheimer's disease, Parkinson's disease, cystic fibrosis, and various sinus diseases have been proven to cause olfactory loss as well.

6.2.1.7 Impairments

Although it is less well known than visual blindness, *odor blindness*, or anosmia, can occur in both temporary and permanent forms. A blow to the head is a frequent cause of this inability to detect odor (Varney, 1988), although in most cases the loss is temporary. Anosmia may also be caused by inhaling toxic agents that damage olfactory receptors, such as lead, zinc sulfate, or cocaine.

Neurological diseases such as Alzheimer's disease may also reduce one's odor sensitivity and one's ability to identify an odor (Rezek, 1987). This sensitivity decrement may be a result of neuron degeneration in the olfactory epithelium. Although full loss of scent perception is the most severe, the loss of perception of particular odors is more common. In fact, research has shown that 3% of the US population could not smell the odor of sweat, 12% have limited sensitivity to musky odor, and 47% have difficulty detecting the scent of urine (Amoore, 1991). The loss of scent perception associated with both impairments and age degradation suggests a need to integrate a variety of scents into VEs to ensure that a wide range of users will experience the full multimodal experience and further strengthen the need to follow thorough visual, auditory, and haptic design guidance that is discussed in previous chapters to account for users that cannot perceive the scents presented.

6.2.1.8   Olfactory Adaptation

While the conditions presented in the previous two sections are counterproductive, the human body also regulates the perception of scent to improve the utility of the modality. Specifically, in order to maintain a high level of sensitivity to the wide range of odorants and concentrations that stimulate the olfactory system, humans must adjust their responses to the diverse odorants in an environment. This is a common feature of all sensory modalities and is known as sensory adaptation (Blake & Sekuler, 2006). Adaptation is the decrease in responsiveness during or following repetitive exposure to a stimulus. Sensitization (or recovery), conversely, is the increased responsiveness during or following exposure (Schiffman, 1998). Olfactory adaptation allows the olfactory system to respond appropriately to the appearance of novel odors or changes in odorant concentration while still maintaining a state of equilibrium with environmental odors (Dalton, 2000). Olfactory adaptation is similar to adaptation in other sensory systems, and research on olfactory adaptation leveraging behavioral, psychophysical, and electrophysiological techniques suggests that olfactory adaptation can occur at multiple levels in the olfactory system involving both peripheral (receptor level) and more central (postreceptor) components (Dalton, 2000). Stimulus-specific decreases in olfactory sensitivity occur after repeated or prolonged exposure to an odorant, but sensitivity recovers over time in the absence of further exposure (Zufall & Leinders-Zufall, 2000). The magnitude of the decrease and the time course of adaptation and recovery are dependent on the duration of exposure and the concentration of the odor (Blake & Sekuler, 2006). When compared to other modalities, adaptation in the olfactory system can occur for very long durations (Herz & Engen, 1996) and can occur after short- or long-term exposure periods. Short-term adaptation is a reduction in perceived intensity of an odor during or immediately after exposure, whereas long-term adaptation causes odor insensitivity to be consistently maintained over long periods of time.

A number of factors can impact the olfactory adaptation process, including the intensity of the odor, pairings with other odors, odor type, and presentation method. Adaptation to more intense odors is faster than less intense odors because detection threshold limits change quicker for more intense stimuli (Powers, 2004). In addition, adapting to multiple odors increases the time needed to adapt to those odors, and odor adaptation can occur at different paces for each odor (Powers, 2004). Another factor that affects the rate of adaptation to smells is the perceived consequences of being exposed to the smell. Particularly, it has been shown that individuals adapt significantly less to a smell when it is thought that there is a negative effect of the exposure to that odor (Dalton, 2000). Finally, presenting an odorant continuously decreases adaptation time, whereas intermittent bursts of an odorant prolong the adaptation process (Ho & Spence, 2005). Taking this research into account, when designing olfactory cues into VEs to enhance the experience or to provide critical information to users, it is important to present scents at a subtle intensity level through intermittent bursts.

The adaptation process is not immune to the conditions of previous experience introduced earlier. In fact, adaptation to certain odors can also be impacted by previous exposure to similar odors. This concept, known as cross adaptation, suggests that humans can adapt to a group of similar smelling odors, particularly those odors that share receptor types (Stevenson & Boakes, 2003). Currently, the

TABLE 6.1
Guidelines Associated with Scent Processing

Summary	Guideline
Odors as warnings	Employ an odor that can be detected at very low concentrations (such as mercaptan) when utilizing an odor as a warning stimulus.
Hedonics to differentiate odors	Consider hedonics first when it is important to differentiate between multiple odors. Pleasantness is the most salient psychological dimension in the perception of odors.
Odor intensity and hedonics interaction	When setting odor intensity, consider that it is influenced by hedonics. Malodors will be perceived as substantially more intense than pleasant odors.
Odor intensity and familiarity interaction	When setting odor intensity, consider that it is influenced by familiarity. Familiar odors will be perceived as more intense than unfamiliar odors.
Cross-modality interference	Facilitate odor detection by pairing with congruent stimuli such as a color or image.
Cross-modality interference	Influence appraisals of neutral stimuli by presenting with a pleasant or unpleasant odor.
Increasing discriminability	Facilitate parsing of constituent odorants in a mixture by naming the odorants to increase discriminability.
Cultural differences	Consider cultural differences in perception when selecting odors for training.
Gender differences	Tailor odorant stimuli to gender-specific training populations. Females are more sensitive to and better at identifying odors than males.
Effects of aging	Tailor odorant stimuli to older adults by increasing intensity. An elevated threshold and a diminished sensitivity to suprathreshold stimuli are observed in aging adults.
Countering odor sensitivity decrements	Employ a variety of scents (when feasible) in a training environment to account for trainees that cannot perceive all scents presented.
Adaptation to negative odors	Reduce the saliency of malodors by leveraging olfactory adaptation capabilities through repeated and long-duration odor exposure.
Hastening odor adaptation	Quicken the adaptation process by presenting odors individually and at a high intensity.

results from cross-adaptation studies are inconclusive. Although some similar odor pairs have demonstrated the potential for cross adaptation (Cain & Polak, 1992; Pierce, Wysocki, Aronov, Webb, & Boden, 1996), others do not (Todrank, Wysocki, & Beauchamp, 1991). Moreover, cross adaptation has also been demonstrated in clearly distinct odors as well (Cain & Polak), suggesting that the similarity of odor is not the only characteristic that determines whether this process can occur. These findings suggest a need for further research to determine the odor characteristics that cause cross adaptation.

Although basic research has been conducted to gain an understanding of the olfactory adaptation process, little research has been done to determine how to leverage this process to better prepare trainees to operate in olfactory-rich environments. As scents are integrated more commonly within VEs that are used to entertain and train people, it becomes increasingly important to understand the process that drives cross adaptation to leverage it in VE design. Table 6.1 outlines design guidelines that were introduced in this section to improve VE design based on human scent processing.

6.3 EFFECTS OF SMELLS ON HUMAN STATE

6.3.1 PSYCHOLOGICAL EFFECTS OF ODORS

Odors can have profound positive and negative psychological effects on operators, and extensive research has been conducted on how odor affects the human state and performance. To fully understand the mechanism by which odor affects performance, we must first understand the underlying processes, including how task-critical knowledge is obtained. Albert Bandura has been a pioneer in the science of learning and performance for decades. His social learning theory (Bandura, 1986) provides a framework that provides a basis for information processing and learning. This theory

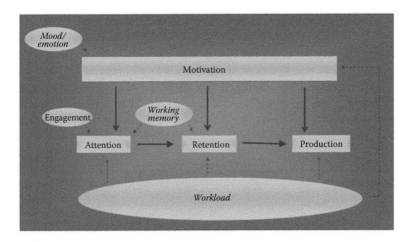

FIGURE 6.1 Expanded social learning theory model. (Modified from Bandura, A., *Social foundations of thought and action: A social cognitive theory*, National Institutes of Mental Health, Rockville, MD, 1986.)

has served as the foundation for work on observation and demonstration-based learning, including research on lower-level motor tasks (Rosen et al., 2010). The social learning theory has four key components—attention, retention, production, and motivation. As Figure 6.1 shows, there are also a number of constructs that positively or negatively impact those four key components. Particularly, four driving factors that are of interest when integrating scents into VEs to optimize learning are mood/emotion, engagement, working memory, and workload.

Attention refers to process of engaging and processing the most critical stimuli. In research of motor tasks, this often equates to spatial and temporal patterns of movement where observation is one way to help learners reproduce unfamiliar movement patterns (Rosen et al., 2010). Attention is affected by workload such that too high or too low amounts decrease attention, while an appropriate amount of workload increases attention. The *ideal* amount of workload is largely dependent on the person and the task (Hancock & Warm, 1989).

Retention takes into account that observations of a task must be stored in memory by moving information from working memory into long-term memory, which involves cognitive rehearsal. The process by which information gets moved from working memory into long-term memory, and ultimately into a form that can be retrieved by the individual at a much later time, is disrupted when workload is too high or too low (Bandura, 1986). Similar to attention, retention is best served by an *ideal* amount of workload. Contrary to popular belief, retention is actually improved by practicing a task first and then being shown a demonstration of the correct way to perform the task because typically the information is stored differently (Bagget, 1987).

Production refers to the process by which stored knowledge is converted into action, which applies to task acquisition as well as transfer (Bandura, 1986). Workload that is too high or low compromises one's ability to make decisions about actions and retrieve information and, at extreme levels, compromises one's ability to perform physical tasks (Hancock & Warm, 1989).

Finally, motivation accounts for the variance in performance that is accompanied by perceived consequences. Motivation is greatly affected by extreme levels of workload and mood/emotion, but the way in which these factors influence motivation is highly dependent on the person, the task, and the initial levels of motivation (Bandura, 1986). For example, viewing demonstrations by high-status individuals (e.g., doctors, lawyers) increases motivation over that of demonstrations by low-status individuals (e.g., peers; McCullagh, 1986). Each of these key and supporting constructs of learning theory (motivation, engagement, attention, workload, memory, and emotional state) can be influenced by odors, and these effects should be taken into account when VEs are designed with the goal of training.

6.3.1.1 Motivation

Situational motivation is characterized by both intrinsic and extrinsic motives for engaging in any particular behavior (Johnson, Prusak, Pennington, & Wilkinson, 2011). Research in other sensory modalities has demonstrated that sensory stimuli serve as both intrinsic and extrinsic motivators (Isbell & Wyer, 2012) and stimuli characteristics such as valence and arousal impact the motivational effect (Zald, 2002). The strong emotional influence of odorant stimuli has also demonstrated a potential connection between odors and motivational behaviors. Studies have shown that the presentation of peppermint has the potential to increase self-reported motivation (Raudenbush, Smith, Graham, & McCune, 2004). Although this research suggests that positive scents can improve motivation, it is important to note when integrating them into VEs that it is unclear whether scents alone directly cause this effect. Instead, scents act as intermediaries to trigger emotions that improve performance and if presented incorrectly during frustrating stages of early training can have negative effects. For example, Herz, Schankler, & Beland (2004) determined that participants who were exposed to an unfamiliar pleasant odor during a frustrating mood manipulation later spent less time working on difficult word puzzle tests when that same odor was present. This is because the negative mood caused by the mood manipulation became associated to the pleasant odor, and elicited frustration, having a detrimental effect on motivation for working on the word puzzle task. This connection between emotion and other constructs critical to the learning process is discussed in more detail within 6.3.1.6 below.

6.3.1.2 Engagement

Engagement is defined as a positive, fulfilling state of mind that is characterized by vigor, dedication, and absorption (Schaufeli, Salanova, González-Romá, & Bakker, 2002). Vigor may be characterized as high levels of energy and mental resilience. Dedication refers to being strongly involved in the task at hand and experiencing a sense of significance, enthusiasm, and challenge. Absorption is characterized by being fully concentrated and happily engrossed in one's task (Bakker & Demerouti, 2008). Odors have the potential to influence task engagement by facilitating vigor, dedication, and absorption. Other sensory modalities have demonstrated that task engagement increases with the presence of pleasant stimuli (Cohen, Davidson, Senulis, Saron, & Weisman, 1992). However, initial research on the topic of increasing task engagement with pleasant odors has hinted at a more complex relationship.

The majority of olfactory research on engagement has focused on the manipulation of time perception as an indicator of task engagement. Research has shown that task engagement varies inversely with time perception, and as engagement increases, time estimation decreases (Chaston & Kingstone, 2004). Some odors have been shown to influence this measure of task engagement. For example, the introduction of peppermint into a testing environment results in decreased estimation of a 60 s time period (Lorig, 1992). In addition, both peppermint and cinnamon reduced the perception of time during a simulated driving task (Raudenbush, Grayhem, Sears, & Wilson, 2009). Initial research investigating the trends in odor characteristics has suggested that the alerting capability of an odor, not necessarily the pleasantness, is the driving factor in the effects on engagement. For example, a coffee odor (high valence/high arousal) was found to significantly decrease time estimation, whereas a baby powder odor (high valence/low arousal) was found to significantly increase time estimation (Hirsch, 2011). Although this research suggests that the perception of time (and therefore task engagement) can be altered by administering an odorant independent of the hedonics of the odor, these findings should be further researched prior to applying guidance during VE design. Specifically, there is a need to further investigate the effects of malodors on the perception of time and potential stress factors.

6.3.1.3 Attention

Models of attention incorporating arousal suggest that low- and high-arousal stimuli exert opposite effects (Gould & Martin, 2001). Administering high-valence, high-arousal odors such as peppermint and cinnamon has demonstrated improvements in attentional process measures such as vigilance or sustained attention and reaction time (Parasuraman, Warm, & Dember, 1992; Raudenbush et al., 2009; Warm, Dember, & Parasuraman, 1991; Zoladz, Raudenbush, & Lilley, 2004). Although the general consensus is that high-valence/high-arousal odors improve attentional processes, there are a few exceptions to this trend. For example, attempting to improve sustained attention with peppermint was not successful when an elevation in mood did not occur simultaneously (Johnson, 2011). This indicates that odor alone may not be sufficient to impact attentional processes but further research is needed to investigate this finding. In contrast to high-valence/high-arousal odors, high-valence/low-arousal odors have demonstrated mixed results in attentional process effects. For example, jasmine and cherry did not affect attentional processes (Zoladz et al.), whereas muguet (relaxing odor) improved sustained attention performance (Warm et al.). Further, bergamot (relaxing odor) demonstrated detrimental effects on sustained attention task performance (Gould & Martin). This research suggests that both high- and low-arousal odors have the potential to impact attentional processes in different ways. Specifically, fragrances that are categorized as alerting have the potential to improve attention, whereas relaxing scents might either reduce the tension and stress that are by-products of performing a vigilance task or create a state of inattentiveness on nonstressful tasks.

Since most of the olfactory research on attention has focused on sustained attention (vigilance), the results do not necessarily transfer to the impact of odors on other types of attention and indicators of attention such as divided attention, reaction time, or attention tunneling. In one investigation into the effects of odors on other types of attention, galaxolide (a musky odor) demonstrated an increase in divided attention as shown by doubling the time necessary to solve a visual search task (Lorig, Huffman, DeMartino, & DeMarco, 1991). Interestingly, this odor was presented at a subthreshold level (remained undetected) and was selected because it is generally considered an unfamiliar odor leading to the increase in divided attention to be attributed to the novelty of the odor. Based on this, it is critical to take into account the types of attention required when completing tasks in VE and past experiences with odors prior to integrating them to support or affect training.

6.3.1.4 Memory

Memory is a core supporting construct to all learning models, and the process of perceiving and processing odors is well situated to affect it. As previously discussed, olfactory signals pass through the neocortex, and this area of the brain, specifically the prefrontal cortex and the lateral intraparietal cortex, controls memory; thus, olfaction and memory are closely linked and invariably influence each other (Bear et al., 2001).

The effects of olfactory cues on memory are varied for different types of memory. While there is research on short-term memory for odors, the effects on working memory are of particular interest when determining how the presentation can persuade the learning process in VEs. While generally alerting odors with high valence and high arousal, such as peppermint, cinnamon, and rosemary, have been shown to improve overall working memory (Moss, Cook, Wesnes, & Duckett, 2003; Zoladz et al., 2004), these results can be mixed depending on the task being performed. For instance, research has shown that peppermint improves accuracy on spatial and numeric working memory tasks but does not improve reaction time for those tasks (Moss, Hewitt, Moss, & Wesnes, 2008). This well-studied scent also did not improve a memorization task that required participants to recreate multimodal stimuli (Barker et al., 2003).

Results are varied for high-valance, low-arousal odors; some of these scents, such as jasmine and cherry, seem to have no effect at all on working memory, yet others, such as ylang-ylang, increase accuracy for spatial and numeric working memory tasks (Moss et al., 2008; Zoladz et al., 2004). Interestingly, unlike peppermint, which also improved accuracy on these tasks but did not affect reaction time, ylang-ylang actually decreased reaction time (Moss et al.), suggesting a difference among odors in the high-valence category. On the other hand, low-valance, high-arousal odors, such as H_2S and NH_3 (which smell like feces and rotten cheese, respectively), consistently have a negative effect on working memory (Danuser, Moser, Vitale-Sethre,

Hirsig, & Krueger, 2003). Taking this into account, these scents should be avoided in VE training platforms when novices are early in the knowledge acquisition process.

Much like results for working memory, results for long-term memory are also varied. For retention of novel stimuli, calming odors with high valence and low arousal, such as jasmine, typically have no effect on memory (Zoladz et al., 2004). However, results get more interesting as we examine specific types of long-term memory measures, such as recognition and recall performance. Certain scents have been shown to improve memory on recognition tasks. Typically, this effect has been found in virtual task performance with high-valence, high-arousal alerting odors, such as peppermint and cinnamon (Moss et al., 2008; Zoladz et al., 2004). Interestingly, some scents that are high valence and low arousal, such as jasmine and cherry, tend not to have an effect at all on recognition tasks (Zoladz et al.). However, other scents in the same high-valence, low-arousal MDS category, such as ylang-ylang, have a negative effect on recognition tasks, and this remains true whether the recognition is immediate or delayed (Moss et al.).

The scent effect pattern for recognition tasks also transfers to recall tasks that require long-term memory, where high-valence and high-arousal scents improve memory (Matlin, 2002; Moss et al., 2008). Also similar to recognition, some calming high-valence, low-arousal scents, such as cherry, tend to have no effect on recall tasks, while others, such as geranium, have a positive effect (Morrin & Ratneshar, 2000; Zoladz et al., 2004).

Although the effects outlined earlier are telling, when designing VEs to support training, additional factors must be taken into account to ensure that learned material can be leveraged in a transfer environment. One such mediating factor is state-dependent learning, which is a potentially negative condition where the recognition or recall of information can only be successfully completed when in the same physiological state as that in which the learning occurred (Galotti, 1999). Since memory is so closely tied to olfaction, there is high potential to create state-dependent learning conditions when odors are present during initial encoding (Cann & Ross, 1989; Parker, Ngu, & Cassaday, 2001; Schab, 1990). It appears that the extent to which learning is state dependent is determined by the subjective pleasantness of the odor, which varies across individuals (Ball, Shoker, & Miles, 2010). Because state dependency may occur without the knowledge of the trainer or trainee, it is necessary to counter this effect if operational performance will not be conducted with the same odors presented during learning. There is currently no research that outlines the best way to do this. Because of this, the authors are in the process of conducting research to specifically address this gap.

Generally, while high-valence odors, both low and high arousal, have been researched with respect to their effects on memory, there is a large gap in the literature concerning the effects of low-valence odors on memory. There is a growing trend to integrate these malodors into military training VEs to follow the "train as you fight" mantra with little objective guidance on how this should be done to avoid negative side effects. Given this, there is a critical need to expand research in this domain to further support training system designers and developers.

6.3.1.5 Workload

As Figure 6.1 shows, workload has the potential to affect all stages of the learning process. Workload can be categorized into two levels—global workload and the subcomponents of workload, which are typically mental workload, physical workload, temporal workload, effort, frustration, and performance, which is a person's appraisal of their own performance on a task (Hart & Staveland, 1988). It is intuitive that scent could affect workload; however, there is a plethora of research in this area detailing how and when this occurs.

When compared to other modalities, olfactory cues result in the highest amount of workload when presented. These findings may be due to the fact that participants are not used to leveraging information being presented via olfactory channels (Warnock, 2011). However, the type of odor presented largely influences how the odor affects workload. Specifically, some odors have a positive effect on workload, promoting levels in an acceptable range (Hancock & Warm, 1989). Several studies have found that the presentation of alerting odors with high valence and high arousal, such as peppermint and cinnamon, reduces global workload as well as the physical, temporal, and mental subcomponents (McCombs, Raudenbush, Bova, & Sappington, 2011; Raudenbush et al., 2009; Raudenbush, Koon, Meyer, Corley, & Flower, 2004; Raudenbush, Koon, Meyer, & Flower, 2002; Wilmes, Harrington, Kohler-Evans, & Sumpter, 2008). Peppermint in particular has also been shown to reduce levels of perceived effort and frustration (Raudenbush, 2000; Raudenbush, Koon et al., 2002). Similarly, calming odors, which have a high valence and low arousal, such as jasmine and vanilla, have the same effect on workload (Raudenbush, Koon et al., 2004; Raudenbush, Koon et al., 2002; Wilmes et al., 2008). As would be expected, malodors tend to have a negative effect on workload. Specifically, odors with a low valence and low arousal, such as ethyl mercaptan (which smells similar to rotting cabbage or sewer gas), have negative effects on workload due to increased feelings of frustration (Rotton, 1983). These effects of odors on workload provide direct guidance to those who are developing VEs to optimize learning. Specifically, in adaptive VE platforms where workload can be evaluated in real time, scents can be used to optimize workload of learners, ensuring that they do not become bored or overwhelmed, two conditions that decrease learning effectiveness.

6.3.1.6   Emotion

Odorants are well identified as having strong emotional correlates. Numerous studies have found that olfactory cues can effectively elicit emotion and have proven successful in shifting a person's preference (Van Reekum et al., 1999), influencing motivation of children (Epple & Herz, 1999) and adults (Herz, Schankler, & Beland, 2004), and eliciting affective responses in those suffering from posttraumatic stress disorder (PTSD; McCaffrey, Lorig, Pendrey, McCutcheon, & Garrett, 1993).

Based on the findings of a growing number of studies, it appears obvious that pleasant odors elicit positive affect with an opposite effect for unpleasant odors, despite the diversity of odors tested and methods used for assessment (Millot & Brand, 2001). For example, high-valence odors such as peppermint, lavender, chamomile, vanillin, and orange have all been successfully used to induce positive affect (Diego et al., 1998; Lehrner, Eckersberger, Walla, Potsch, & Deecke, 2000; Miltner, Matjak, Diekmann, & Brody, 1994; Roberts & Williams, 1992; Zoladz et al., 2004). Conversely, low-valence odors such as dimethyl sulfide (rotting cabbage) and ethyl mercaptan (sewer gas) have successfully been used to induce negative affect (Knasko, 1992; Raudenbush, Meyer, & Eppich, 2002; Rotton, 1983). Research has also shown that odors can be used to mitigate both emotional valence and arousal. For example, peppermint and jasmine have demonstrated decreases in frustration (Raudenbush, Meyer, & Eppich, 2002), and lavender, spiced apple, eucalyptus, and orange have all demonstrated decreases in anxiety and tension (Diego et al., 1998; Lehrner et al., 2000; Lorig & Schwartz, 1987). It appears that odor valence and arousal characteristics illicit the mirrored emotional dimensions of valence, ranging from highly positive to highly negative, and arousal, ranging from calming/soothing to exciting/agitating (see Figure 6.2; Kensinger, 2004; Russell, 1980). However, there are a few exceptions to these findings, for example, a study involving geranium (high-valence odor) found no effects on participant mood or arousal level (Morrin & Ratneshwar, 2000).

It is important to consider that differences in the perception of odor valence can result in drastically different emotions elicited from different individuals exposed to the same odor. This is partially a result of associative learning. Affective responses to odors can be associatively learned and used to impact the emotional state and motivation of individuals when presented at later occasions (Engen, 1991; Herz, 2001). Odor hedonic perception is influenced by the emotional context in which a novel odor is first encountered, thus explaining how certain odors are liked or disliked by different individuals (Herz, 2005). A number of studies have shown that hedonic responses to odors are learned through specific experiences and even cultural learning from indirect experience (Herz, 2005). For example, a fear response was elicited from the smell of eugenol (odor used in dental cement) among patients who are afraid of dental procedures; however, the same odor was rated as pleasant smelling among individuals who are not afraid of

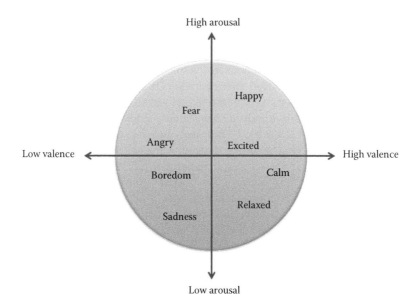

FIGURE 6.2   Valence/arousal dimensional emotion space.

dental procedures (Robin, Alaoui-Ismaili, Dittmar, & Vernet-Mauri, 1998). In addition, studies of both children (Epple & Herz, 1999) and adults (Herz, Schankler, & Beland, 2004) placed in frustrating conditions where odors were present showed a negative impact on future tasks when the same odors were present, including higher levels of frustration and lower motivation to complete tasks. In PTSD studies, simulated scents associated with trauma (e.g., burning flesh) had a significant effect on emotional response, suggesting that smells associated with previously experienced stressful events can cause the onset of affective memories/responses associated with them. Because of this, the army and marines are using scent to prepare soldiers for battle by exposing soldiers to the emotional experiences associated with noxious odors (e.g., melting plastic, rotting flesh) before deployment to Iraq in safe and controlled settings (Vlahos, 2006).

Cultural backgrounds also persuade the emotional response to odors. In addition to the example regarding the effects of the wintergreen smell across groups presented previously, findings consistently show that culture has a great impact on the emotional response to scents. Another example that demonstrates this disconnect in cultural odor hedonic evaluations is the research effort undertaken by the US military to create a universal *stink bomb*. Researchers on this effort were unable to find an odor that was unanimously considered repulsive across various ethnic groups, even when considering the U.S. Army issued latrine scent (Dilks, Dalton, & Beauchamp, 1999). These examples further illustrate that emotional responses to odor stimuli are learned on the basis of previous experiences.

In the discussion of changing the learned emotional responses to odors, it is important to consider the power of initial associations with odors. Familiar odors are necessarily associated with past experiences, and once an association to an odor has been formed, it is difficult to reassociate the odor to be perceived differently (Herz, 2005). The first association made to an odor has been shown to interfere with subsequent learning associations to the same scent (proactive interference), and new pairings to the scent do not override the previous associations made (Lawless & Engen, 1977). Odor novelty prior to associative learning, therefore, is an important implication for introducing scents and attempting to change odor preferences once they have been formed. This tendency to hold on to initially learned emotional responses is of critical importance to VE training designers as it further suggests against haphazardly integrating scents that are expected in live transfer environments early in the learning cycle when trainees are potentially overwhelmed or frustrated as this will have negative connotations during live performance especially when you consider that emotional responses triggered by odors can last a lifetime (Vermetten & Bremner, 2003).

In addition to the initial association with an odor, other factors also affect the process of associative learning. Herz, Beland, & Hellerstein (2004) demonstrated that odor-associative learning time is dependent upon whether the target odor was evaluated as positive or negative prior to presentation and whether the paired emotional experience was positive or negative. Specifically, when an unpleasant odor was associated with a positive emotional experience, a positive association was established and persisted for at least 1 week. However, when a pleasant odor was associated with a negative emotional experience, changes in olfactory perception were not observed until 24 h after the association had been made and were weaker after 7 days (Herz, Beland, & Hellerstein, 2004). Other factors affecting the duration and strength of the emotional association to an odor include the number of associations (i.e., multiple stimuli associated with the odor), level of familiarity or similarity to familiar odors, and the contextual relationship between the odor and stimuli.

There are occasions where it may be beneficial to leverage VEs to mold the emotional experience associated with odors (e.g., in preparation for highly intense performance where malodors will be present). When doing so, there are methods that can be leveraged to improve this process. Specifically, evaluative conditioning research trends show more successful results in establishing associative learning when multiple stimuli are utilized (Houwer, Thomas, & Baeyens, 2001). In one olfactory example, a positive emotional manipulation could not be achieved when the target odor was presented only once with a single stimulus (video game). However, two positive pairing sessions with the same odor and different positive stimuli (video game and film clips) were effective for altering hedonic perception (Herz, Beland, & Hellerstein, 2004). Different conditioning approaches may also need to be considered when attempting to create positive and negative emotional associations. It is well known that negative emotion tends to be more potent and more motivating than positive emotion (Baumeister, Bratslavsky, Finkenauer, & Vohs, 2001; Rozin & Roysman, 2001) and creating a negative emotional association to an odor may require less exposure time than a positive association.

The strong emotional influence in odor perception should be considered when introducing any odors into a VE. Odors may be one of the most effective stimuli for eliciting an emotional response in a VE, particularly for context-relevant odors that have been previously associated with emotional stimuli. Because of this increased potential to affect emotion when integrating scents into VEs, it is important to consider the effects of emotional states on other learner state constructs introduced earlier when VEs are designed to target learning. These interaction effects are introduced in the following section.

6.3.1.6.1 *Effects of Emotion on Other Learner State Constructs*

As the previous section introduced, emotional state plays a particularly complex role when trying to influence learning within VEs, with effects extending across the entirety of the model. What is more, emotional stimuli benefit from an enhanced significance that grants them rapid and often preferential access to our neurocognitive resources. Both functional neuroimaging and neuropsychological studies have demonstrated that emotionally valenced stimuli need not even reach conscious awareness to engage amygdala processing (Vuilleumier & Pourtois, 2007; Zald, 2002). Research further indicates that specific circuits in the brain may serve to amplify neural responses to emotional stimuli (Vuilleumier & Huang, 2009). Such mechanisms display themselves through observed improvements in performance tasks associated with emotional versus neutral stimuli. This suggests that emotional experiences have the potential to enhance learning and, when combined with the direct connection between scents and emotion activation described previously, further suggests the power of scents to shape perception (Li, Moallem, Paller, & Gottfried, 2007) and subsequently the learning process, potentially even prior to perception.

Emotional stimuli have been associated with improvements in numerous functions, among them are motor readiness, autonomic functions, and cognitive processes such as attention and memory (Zald, 2002). They also affect cognition from a lower level, rapidly acting upon the earliest stages of appraisal and commanding the first signs of attentional resources, sometimes leading to attentional narrowing (Vuilleumier, 2005). This suggests that learning could either benefit or suffer from exposure to emotional events (Dolcos, 2010; Dolcos & Denkova, 2008; Dolcos, LaBar, & Cabeza, 2006; McGaugh, 2004; Phelps, 2004; Ritchey, Dolcos, & Cabeza, 2008). Numerous theories have been offered up to explain such phenomena: dedicated brain circuits for emotional attention have been identified (Phelps & LeDoux, 2005), and increased parietal activation has been observed during spatial-orienting tasks when neutral targets are preceded by emotional cues (Armony & Dolan, 2002; Pourtois, Schwartz, Seghier, Lazeyras, & Vuilleumier, 2006; Vuilleumier & Huang, 2009). The modulation hypothesis states that the amygdala and medial temporal lobe memory regions interact more intimately during the encoding of emotional stimuli than during the encoding of neutral stimuli (McGaugh, 2002). Whatever the underlying neurological mechanisms may be, it seems emotional stimuli had an early relevance for survival, benefiting a range of critical behaviors that caused their development during evolution. From a learner state perspective, what is particularly encouraging is the potential emotion-laden stimuli present as a means to enhance cognitive processes involved in perception, attention, and memory (Brefezynski & DeYoe, 1999; Morris et al., 1998; Yeshurun & Carrasco, 1998). Given this connection, there is potential that the potential positive effects of scents on learner states introduced can in part or in whole be due to intermediate changes in emotional states caused by odor presentation.

In terms of learner state, however, the influences of emotion may not always be beneficial. For example, the preoccupation with emotional information has been associated with disruptions in functioning and psychiatric disorders. Impaired executive control and enhanced emotional distractibility are linked to hypofunctioning of the dorsal executives and hyperfunction of the ventral affective systems (Denkova et al., 2010; Dolcos & McCarthy, 2006; Drevets & Raichle, 1998; Mayberg, 1997). This reveals the downside of the priority granted to emotional information: increased distractibility to task-irrelevant emotional stimuli, a potentially highly detrimental drawback. Given this connection, it is critical that VE designers who are targeting learning in VEs ensure that users are highly engaged in the VE content prior to presenting scents when they do not support the improvement of VE immersion.

The complex influence of emotion on the learner state—with its both positive and negative implications—is likely the result of the complexity of this construct. Emotional processing within the human brain is by no means an operation taken up by any kind of single, focused brain structure nor even by any specifiable region of the brain. Though the entirety of the human emotional system is far from well understood by scientists, neurological studies reviewed collectively support the conclusion that emotional processing is a multimodal action involving an extensive list of brain structures, all scattered about different regions of the brain, and all of which have been identified in an even wider range of cognitive functions. Neuroimaging data of these brain structures, by visually depicting the overlapping roles that emotion and cognition play in cerebral processes, reveal that emotions are anatomically intertwined with almost every other cognitive function, implicating an interactive network with distributed activity in time and space (Alter, 2011; Amaral, Behnia, & Kelly, 2003; Anderson, Christoff, Panitz, De Rosa, & Gabrieli, 2003; Anderson & Phelps, 2001; Bechara, 2004; Bechara et al., 1995; Carrasco & McElree, 2001; Cheal & Lyon, 1991; Kosslyn et al., 1996; Liotti et al., 2000; McCarthy et al., 1994; Mitchell, 2011; Morris et al., 1998; Surguladze et al., 2003; Taylor, Liberzon, & Koeppe, 2000; Vuilleumier & Huang, 2009; Vuilleumier & Pourtois, 2007; Whalen, 1998; Winston, Vuilleumier, & Dolan, 2003; Zald, Mattson, & Pardo, 2002). Such a physically explicit connection among brain systems would seem to imply a functional link among them, as well. For example, research has identified regions of the prefrontal cortex to be associated with both behavioral conflict, such as motor response selection or inhibition, and affective conflict, such as emotional representation and awareness. With the same region of the brain playing a role in modulating operant responses (what we do) as well as emotional responses (how we feel), the idea is thus raised that key aspects of these two operations are bound by a common functional objective (Mitchell, 2011; Mitchell et al., 2009). Indeed, patients with lesions of the prefrontal cortex are not only unable to process emotional information normally but are also reported to exhibit impairments in decision making (Bechara; Bechara & Damasio, 2005). Thus, as much as we would like to think our judgments are based purely on our brain's capacity for rational maximization of expected utility, logically evaluating probability and consequences, it appears they are implicitly formed at a gut or emotional level as shown by the somatic marker hypothesis (Bechara, 2004; Bechara & Damasio, 2005; Bechara, Damasio, & Damasio, 2000; Bechara, Tanel, & Damasio, 2002; Ghashghaei & Barbas, 2002; Ghashghaei, Hilgetag, & Barbas, 2007; Greene, Sommerville, Nystrom, Darley, & Cohen, 2001; Hastie & Dawes, 2001; Kahneman & Tversky, 1979; O'Doherty, Critchley, Deichmann, & Dolan, 2003; Rolls, 2004; Rudebeck & Murray, 2008). The entirety of the human emotional system may span at least 90 brain regions, overlapping with such well-known brain systems such as the auditory/language, visuospatial, social, memory/learning, control of action, decision making, and visual systems (Critchley, 2005; Damasio, Bechara, Tranel, & Damasio, 1997; D'Esposito et al., 1995; He & Evans, 2010; Honey, Thivierge, & Sporns, 2010; Mesulam, 1990; Pegna, Khateb, Michel, & Landis, 2004; Petrides, Alivisatos, Meyer, & Evans, 1993; Rolls & Baylis, 1994; Rolls, Hornak, Wade, & McGrath, 1994; Stam, 2010). How and to what degree these allegedly objective cognitive tools are affected by our *gut* are not yet completely understood. Simple anatomy, however, corroborates this empirical evidence that emotion is undoubtedly a powerful and overarching construct, exerting diverse influences on a multitude of functions of the human brain.

Given the extraordinary power emotion wields over most every aspect of cognition, the relationship is, nevertheless, not entirely one sided. Rather, there are reciprocal influences between emotion and cognition (Dolcos, Iordan, & Dolcos, 2011). Studies have shown that people are motivated to actively regulate their emotions through the utilization of high-level executive functions, with the goal of curbing emotion's rampant effects on cognition and subsequent effects on performance (Augustine & Hemenover, 2008; Bishop, Karageorghis, & Loizou, 2007; Davis, Woodman, & Callow, 2010; Hanin, 1993, 1997; Jones, 2003; Karageorghis, Terry, & Lane, 1999; Nieuwenhuys, Hanin, & Bakker, 2008; Niven, Totterdell, & Holman, 2009; Ochsner & Gross, 2005; Robazza, Bortoli, Nocini, Moser, & Arslan, 2000; Woodman & Hardy, 2001; Woodman et al., 2009). Following this approach, under optimal conditions, an individual would be able to utilize emotional information when it is significant and minimize its influence when it is not. Although this is apparently not always possible, the myriad evidence supporting the fact that people do consistently improve performance when they are able to successfully utilize their cognitive resources for emotion regulation clearly implicates the power emotions wield over our performance. From learner state standpoint, the human ability to have some degree of cognitive emotion regulation can be looked at as a behavioral strategy targeted at regulating the underlying states that are critical to improve performance and optimize learning. When taken into account with the aforementioned guidance provided in support of evaluative conditioning of scents, this two-sided relationship between emotion and cognition suggests a need to balance the presentation of emotion-eliciting scents during VE training to ensure that learners can actively control negative emotions while retaining a high level of performance. This guidance along with the other guidance that has been outlined in this section is presented in Table 6.2.

Each of the preceding sections presents the results of the research that has been conducted to determine the effects of olfactory cues on constructs critical to learning. As Figure 6.3 shows, while there is some general consistency across odor types (e.g., high-valance/high-arousal odors improve motivation and low-valance/high-arousal odors reduce motivation), there is a great deal of research that is inconclusive (e.g., the effects of high-valance/low-arousal odors on attention). It is highly likely that the inconclusive nature of research on scents that are generally categorized as positive or negative is due to the strong connection with emotional experiences previously presented and the individualized nature of the appraisal process.

TABLE 6.2

Guidelines Associated with Psychological and Physiological Effects of Scents

Summary	Guideline
Task-congruent odors to influence motivation	Employ task-congruent odors with targeted emotional connections to influence motivation. For example, employ a gasoline odor (suggested fuel leak) to elicit an alerted fear response and motivate trainees to accelerate tasking.
Alerting odors to increase engagement	Employ alerting odors such as peppermint, cinnamon, and coffee to increase engagement and relaxing odors such as baby powder to decrease engagement.
Alerting odors to improve attention	Utilize alerting odors such as peppermint or cinnamon to improve attentional processes.
Peppermint to reduce user effort/frustration	Use peppermint to decrease user effort and frustration during VE use.
Malodors to induce frustration	To induce frustration and increase workload, use malodors such as ethyl mercaptan.
High-valance, high-arousal odors to improve working memory	Choose which high-valence, high-arousal odor depending on the working memory task to be performed, such as using peppermint to improve accuracy though it does not affect reaction time.
Odor valence to elicit emotions	Employ high-valence odors such as peppermint, lavender, chamomile, vanillin, and orange to induce positive affect and low-valence odors such as dimethyl sulfide or ethyl mercaptan to induce negative affect.
Past associations of odors to elicit emotions	Leverage odors that have been associated with past emotional experiences to target those same emotions during training. Ensure odor perception is aligned with the expected valence to achieve a mirrored emotional response.
Leverage odor novelty to create new emotional associations	Consider odor novelty when selecting olfactory cues for training. Leverage previous associations with familiar odors to elicit emotional responses or employ novel odors to condition new associations.
Creating emotional associations to odors	Utilize multiple stimuli to create a stronger emotional association to an odor.
Context relevance of scents to improve emotional conditioning	Employ contextually relevant odor cues during training to improve emotional conditioning.
Control alert states by using scent valence/arousal combinations	To induce an alert state, use odors with high valence and high arousal, such as peppermint; to induce a relaxed state, use odors with high valence and low arousal, such as lavender.

6.3.2 PHYSIOLOGICAL EFFECTS OF ODORS

The previous sections describe the effects of odors on the psychological state with a focus on learner state. The effects of odors go beyond psychological state and have the potential to modify physiological state both negatively and positively. This review classifies the effects on human physiology into three categories: basic physiological processes, advanced physiological processes, and motor tasks. It is important to note that although the effects are reviewed in these three categories, they are highly interrelated and this should be taken into account when integrating scents into VEs.

There are a number of basic physiological processes that are affected by odors. Oxygen saturation and pulse rate can be increased and blood pressure decreased by presenting high-valence odors, such as peppermint and jasmine (Raudenbush, Koon, et al., 2004; Raudenbush, Koon, et al., 2002; Raudenbush & Zoladz, 2003). Some high-valence odors, particularly those with low arousal, have an effect on electrical brain activity. Interestingly, jasmine increases neural activity, while lavender

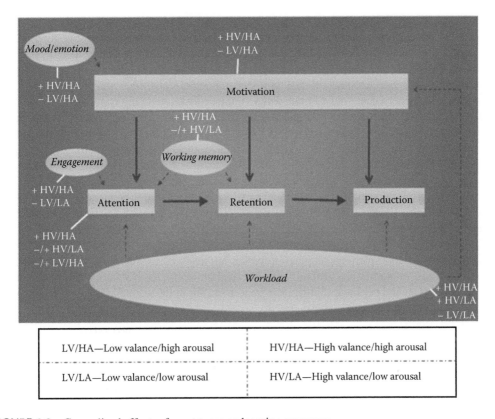

FIGURE 6.3 Generalized effects of scent types on learning processes.

decreases it although the reasons for the difference between the two scents are unknown (Torri et al., 1988). In contrast, malodors, such as methylformate, H_2S, and NH_3, tend to have the opposite effect on physiological processes, decreasing pulse rate, respiratory flow, and visual contrast sensitivity and impairing nerve and muscular functioning (Sethre, Laubli, Berode, Hangartner, & Krueger, 2000).

Perhaps the largest areas of advanced physiological processes for which odors play a role are pain and pain management, particularly because pain is continually the most costly medical problem in the United States (National Institute of Neurological Disorders and Stroke, 2000). High-valence odors, such as peppermint and jasmine, have been shown to significantly decrease pain over time and increase a person's ability to withstand pain (Raudenbash, 2002; Raudenbush, Koon et al., 2004; Raudenbush, Koon et al., 2002). However, combining these types of scents does not always produce the same effect. For instance, peppermint oil combined with eucalyptus oil did not produce any pain relief, though they are both rated as high valence and high arousal (Gobel, Schmidt, & Soyka, 1994). Further, the administration method of scent is important in determining the effectiveness of pain reduction benefits. Research has shown that scent presented through the nose was more effective in reducing pain than scent presented through the mouth (e.g., via chewing gum; Bayley, Matthews, Street, Almeida, & Raudenbush, 2007). Perceived health is a related construct that can also be affected by odors. Remarkably, it seems that high-valence odors, such as lemon, have the same effect as low-valence odors, such as skatole, both leading to reduced perceived health in participants (Knasko, 1992, 1993). These effects should be taken into account when designing VEs for healthcare applications.

Research has also been conducted on how odors affect fatigue and sleep. Odors with high valence and high arousal, such as peppermint and cinnamon, have been shown to decrease fatigue and increase alertness and vigor (Raudenbush, Smith, et al., 2004; Zoladz et al., 2004). As would be expected, calming odors with high valence and low arousal, such as lavender and muguet, induce sleep and promote drowsiness and relaxation (Buchbauer, Jirovetz, Jager, Dietrich, & Plank 1991; Van Toller, 1988; Warm et al., 1991). When applying these findings to guide the design of VEs, it is important to take into account the task and environmental conditions and leverage the scents that optimize the experience. For example, when vigilance tasks are being conducted in VEs, high-valance, low-arousal odors should be avoided and replaced with high-arousal odors when possible.

Finally, motor tasks can also be influenced by odors. High-valence, high-arousal odors, specifically peppermint, can increase basic motor performance, including handgrip strength and speed and accuracy on typing tasks (Barker et al., 2003; Raudenbush, Corley, & Eppich, 2001). Likewise, these odors have the potential to increase gross motor task performance. For example, peppermint increases the number of push-ups an individual can do, speed on a 400 m run, and reaction time and strength for basketball tasks, although it has no effect on task accuracy (Raudenbush et al., 2001; Raudenbush, Smith, et al., 2004).

While the preceding guidance presented and summarized in Table 6.2 shows the potential of leveraging odors to improve physiological processes and tasks, there has been a narrow focus on researching scents to modify physical performance, with most studies evaluating the effects of peppermint. Thus, it should be noted that we, as researchers, are in the infancy stages for determining the nuances of why scents lead to these effects.

6.4 PRESENTATION METHODS OF SCENT IN VIRTUAL ENVIRONMENTS

A primary advantage of VE training systems over alternative training methods (e.g., classroom) is the ability to recreate the fidelity and realism of live training through implementation of affective experiences using multiple modalities for cue presentation. This approach allows for the creation of affective experiences that are more closely matched to the real world, increasing the potential for training transfer. For example, Wilfred et al. (2004) used the sound of random explosions in a search and rescue task to create an affectively intense environment and found that those who trained in an affectively intense environment performed substantially better in the *real* environment than those who trained in an affectively neutral environment. Current VEs have traditionally included advanced visual and audio cues, but olfactory cues are either not incorporated or very limited. Integrating olfactory cues into VEs has the potential to significantly increase immersion by leveraging the strong emotional and memory associations with odors to create an affectively rich VE. However, a fundamental problem common to all studies of olfaction concerns the control and delivery of odor stimuli. There is no standardized or widely accepted method for producing odors or presenting them to test subjects despite the publication of several dozen types of odor generator systems and odor delivery devices (Simon & Nicolelis, 2002). In addition, and in contrast to most other sensory modalities, no standard stimulus against which other stimuli can be compared has been established nor is there general agreement on which metric should be used to measure odor concentration. In part, the absence of such standards is because chemical and vapor stimuli are difficult to control and the equipment to control odor generation and flow and to measure concentration is generally difficult to employ in low-cost solutions. As a result, a wide variety of odor presentation methods and test procedures have evolved.

Two overall categories of odor presentation are personal/specific and ambient methods. Specific or personal presentation methods such as nasal strips, inhalers, masks, or collars are intended to present an odor to an individual trainee. For example, head-mounted displays (HMDs) used in many aircraft simulation environments can be used to disperse odors to a single pilot through the O_2 mask (Washburn & Jones, 2004). A scent collar (fits around a user's neck and holds scent cartridges controlled by a wireless interface) can be used to disperse personal odors to individual trainees during movement in a VE (Washburn & Jones). Such personalized or specific odor presentation devices are useful because of their small size, portability, and the ease of dissipation and control of odors.

Because these technologies aim to present odors to a specific individual, dispersion and apparent concentration are more easily controlled. However, these devices can interfere with user tasks because they often must be worn by the user and these devices do not always represent a realistic environment because odors often appear to have unrealistic concentrations and sources. Thus, it is important to pilot test the scents and concentrations that are in use within VEs if personal presentation methods are selected.

Ambient presentation methods are intended to fill an environment or portion of an environment through the use of larger equipment such as fans, diffusers, hoses, and pumps. For example, some major theme parks have added olfactory cues using theater seating to emit scents throughout the theater (Washburn & Jones, 2004). Also, medical simulation training research manikins can have embedded odor wounds to simulate a contextual odor cue directly from a wound or injury source (Allen, Pike, Lacy, Jung, & Wiederhold, 2010). Ambient presentation methods allow for less precision control of odor stimuli because odors can disperse throughout a large area with different apparent concentrations throughout the volume of the VE space. Therefore, the localization of odors is difficult to control with ambient presentation methods. In addition, ambient presentation methods cannot be utilized to target training for specific individuals as a subset of a group because all odors will be delivered to all trainees. However, ambient presentation technologies are unobtrusive (often even unseen) and are useful for creating a realistic external environment.

The selection of ambient or personal odor presentation methods requires specific considerations for the VE and training/simulation needs. Considerations should include dispersal amount, directionality, portability, operating capacity, and environment modeling. One olfactory simulation challenge is determining the proper amount of odors to disperse. Employing synthetic odors in indoor and outdoor environments requires considerations (in terms of quantity, concentration, time of dispersal, and time dispersed in terms of the training event) to realistically simulate the odor (Allen et al., 2010). Furthermore, ventilation systems in indoor training environments may unintentionally carry odors to other parts of the facility. If the source and/or direction of the odor is critical for training, technologies should be selected to support this requirement. For example, a personal presentation technology might disperse odors based on the trainee's head orientation in an HMD and position tracking to present a localized smell. Portability is essential if scent systems must present scents over large areas (unless numerous ambient presentation technologies are used throughout that space). Since odor dispersion technologies all utilize physical substances to produce odors, size constraints limit the number of odors that can be stored and represented while facilitating unencumbered motion.

Odor presentation timing methods can assist in creating a more salient, realistic odor in a virtual testing or training environment as well as achieve a balance of desired apparent concentration of odors. Of the odor presentation timing methods, there are two main categories: static and dynamic presentation. Static odor presentation is the constant presentation of an odor throughout a session. This is the current best practice method of odor dispersal in military and medical training environments. However, there is some research that suggests the constant presentation of an olfactory stimulus will result in the adaptation to that stimulus and a perceived decrease in intensity over time (Blake & Sekuler, 2006). To maintain saliency of the odor stimulus, some researchers have proposed to use a dynamic presentation method. Dynamic presentation of an odor involves intermittent bursts to prevent adaptation and maintain perceived intensity of the odor, and some research has shown positive performance effects for this method (Knasko, 1993). In other research, intermittent exposure to androstenone has been shown to induce a highly significant increase in odor sensitivity in previously insensitive individuals (Wysocki, Dorries, & Beauchamp, 1989). Although the benefits of reducing adaptation are apparent, further research is needed to evaluate the effects of odor presentation type (constant or intermittent bursts) on task performance and perceived stimulus intensity to determine the best practice odor presentation method.

Once an odor has been introduced into a training environment, that odor then becomes a potential context-/state-dependent cue for later retrieval of the learned material (Schab, 1990).

TABLE 6.3
Guidelines Associated with Scent Presentation Methods

Summary	Guideline
Personal odor presentation	Employ a personal/specific odor presentation method when an odor only needs to be presented to one trainee or to disperse an odor from a specific location.
Group odor presentation	Employ an ambient odor presentation method when attempting to create realistic environmental odor cues for group training and evaluating group performance.
Matching training environment scents with recall environment	The scents that are expected in a recall environment should be leveraged in a training environment.

Because training progression should not necessitate the implementation of an odor for recall, a method of odor removal must be employed during the training to prevent a dependency on that odor. According to the encoding specificity hypothesis (Tulving & Thomson, 1973), the salient elements of the context in which learning occurs are encoded along with the learned material as part of the memory trace. These context-specific elements then function as retrieval cues to the target information during testing or recall (Smith, Glenberg, & Bjork, 1978). The effects of encoding specificity, also known as context- or state-dependent learning, have been demonstrated in odor cue studies. Schab examined the effectiveness of odors as memory retrieval cues and showed that a single ambient odor present on both learning and testing improved recall, confirming that olfactory cues are encoded with learned material and are influential in context-/state-dependent learning. To avoid context-/state-dependent learning and to optimize training transfer in the absence of odor cues, research is needed to determine how best to remove odors from the testing environment. One potential method of odor removal that the authors are currently exploring entails gradually lowering the intensity of the odor during training (Table 6.3).

6.5   SCENT PRESENTATION IN PRACTICE

Presence in a VE occurs when an individual cannot distinguish between sensory inputs from the hardware of the system and inputs from reality (Chertoff, Schatz, McDaniel, & Bowers, 2008). One of the primary reasons that many efforts have integrated olfactory cues into VEs is to increase presence and immersion for either training or entertainment purposes. In one of the most pertinent efforts to the military, the Institute for Creative Technologies (ICT) developed a virtual reality (VR) system for Warfighter that includes olfactory cues provided via a scent collar, a device worn around the participant's neck that emits scents at certain points during the simulation. Subjective feedback with the scent collar suggests the olfactory presentation does increase presence in the VR system (Vlahos, 2006).

Work has also been conducted on the civilian side of presence research. One study incorporated a range of odors to increase presence in a computer cooking game in which participants learn to *cook* a certain dish (Rosenblum et al., 2008). In this application, multiple odor components used were blended in real time to produce unique scents that increased the impact of other cues. In a similar vein, olfactory cues have been used in combination with other sensory data to increase presence in a virtual office viewing task (Dinh, Walker, Song, Kobayashi, & Hodges, 1999) where participants moved through a virtual office environment and answered questions about how they felt about the environment.

Related to presence, immersion is another key characteristic of a successful VE. Immersion refers to the feeling of being physically situated in a VE, the occurrence of which leads to greater presence (Stanney, 2002). Research suggests olfaction can be successfully used to increase immersion. For example, the deep-immersion virtual environment (DIVE) was created specifically to increase the feeling of immersion in a VE (Orr & Carter, 1995). Research with the DIVE system has shown increases in immersion with the use of olfactory (and other) sensory cues. Similarly, scent was used to increase immersion in a commercial system called the Virtual Cocoon (*Science Daily*, 2009), which leverages a helmet with a nasal cannula to deliver relevant olfactory cues to the participant.

In the military, scent is leveraged in a very well-known trainer to increase immersion. The Infantry Immersion Trainer (IIT) at Camp Pendleton, CA, is a mixed-reality trainer that uses olfactory cues to increase the sense of immersion in military ground training. Used primarily as a predeployment trainer, the IIT uses odors in a controlled VE to help marines and soldiers experience smells of a location in context prior to arriving there and being required to operate under potentially stressful conditions. The development of this highly immersive environment demonstrates the military's interest in preparing Warfighters for future experiences by leveraging all modalities in training, including scents. U.S. Army training system designers have embraced the use of simulated malodors in training events, especially for first responders (i.e., combat medics, combat lifesavers) in point-of-injury training. These first responders deal in the "Platinum Ten Minute" arena, rather than the "Golden Hour" arena (Bendall & Parsell, 2005). As such, any impairment to performance could result in the death of a casualty. The modern battlefield is filled with unpleasant odors (Jean, 2007); odors caused by injuries to humans only add to the "stink of the battlefield" (Jean, p. 52). As an example, improvised explosive devices often cause abdominal injuries and the smell of feces, as well as the smell of burnt flesh. Schiffman and Williams (2005) indicated that unpleasant odors can cause stress and a host of physical issues. These can easily be traced to causes of human performance impairments, which could in turn degrade the performance of the first responder. But simply installing simulated odor machines in a training environment may not resolve the issue. More research is needed to determine if previous exposure to malodors can create an adaptation environment sufficient to alleviate, if not eliminate, the potentially harmful effects of malodors. In addition, sometimes malodors can serve to help, not hinder, a medic. For example, smelling feces when a casualty is lying on his stomach could trigger the medic to roll the casualty over and look for an eviscerated bowel. The U.S. Army is exploring how olfactory adaptation may help medics and combat lifesavers alleviate performance issues while still detecting the presence of critical olfactory cues.

In addition to supporting training, olfactory cues can also be used in medical treatment. Several studies have been conducted on clinical applications of scent in VEs, and this seems to be a very promising area (Baus & Bouchard, 2010). Scent has been used in VEs to cause cravings above a baseline level in individuals who use targeted substances. For instance, cannabis cravings can be induced in users when presented with paired visual and olfactory cues in a VE (Bordnick et al., 2009). Similar results have been found for binge drinkers using alcohol-related olfactory cues (Ryan, Kreiner, Chapman, & Stark-Wroblewski, 2010). Though more testing is required, this type of stimulation could be used in treatment to bring patients to a specified level of cravings and then work through them, ultimately resulting in reduced or eliminated cravings. In the military's twist on this approach, they have focused on therapy for PTSD. Specifically, virtual Iraq is a VE that incorporates multimodal cues, including olfactory cues, to treat conditions such as PTSD. Results of a case study show promise of using this multimodal approach, as one veteran showed improvement of PTSD symptoms after treatment in this environment (Gerardi, Rothbaum, Ressler, Heekin, & Rizzo, 2008).

Olfactory cues have also been incorporated in less obvious applications. For example, entertainment venues have used scent for quite some time. Some of the most popular users for scent in entertainment are theme parks and amusement parks. These venues employ olfactory cues to increase presence and immersion in their patrons within select attractions. For example, Walt Disney World and Universal Studios each have several attractions that use timed, relevant scents to increase the feeling of being in a sometimes fantasy environment. Similarly, olfactory cues were included in a VR theater for the World Culture Expo in 2000. Pine scent was spread throughout the theater during

TABLE 6.4

Guidelines Associated with Current Applications of Scent

Summary	Guideline
Scents to improve presence	Scents can be used to improve presence in VE environments.
Scents to improve immersion	Context-relevant scents should be integrated into VEs to increase immersion in both military and civilian applications.
Odors to support medical treatment	Scents should be used within VEs aimed at medical treatment to elicit targeted craving and emotional responses.

certain points of the movie. The cues did increase presence and, along with other sensory cues, helped keep participants engaged (Park, Ahn, Kwon, Kim, & Kim, 2003).

Testing is also being conducted to use scent for advertising. The Advanced Telecommunications Research Institute developed a working prototype of a system that targets specific scents at consumers via air cannons to entice them to buy certain products depending on where they are in a store (Knight, 2004). However, work is still being conducted to refine the specificity of the scent and odor area of this type of system.

As this review shows, scent has been used for a wide range of purposes from increasing immersion and presence in VEs to enticing customers to buy certain products. While the applications may not have much in common, there are similarities throughout this work, namely, identifying the best way to present scent and the specific odor to present, for the desired effect. The guidance included in the previous sections of this chapter provides some direction on how these selections should be made and characteristics to take into account when making decisions to apply scent in VE designs (Table 6.4).

6.6   FUTURE DIRECTION FOR OLFACTORY PRESENTATION IN VIRTUAL ENVIRONMENTS

When compared to the other sensory modalities reviewed in this handbook, applied olfactory research is relatively limited, providing a breadth of opportunities to expand research and further guide the integration of scents into VEs. The research used to drive the development of olfactory presentation guidance should leverage studies to evaluate the cognitive and physical responses to odors in VEs as well as the biological and neurophysiological studies to outline the processes that lead to the responses. Although a number of areas of future research have been introduced throughout this chapter, the following are critical areas that require additional research to support the effective integration of this relatively new information presentation modality into VEs.

Standardization of odor presentation and reporting: There is no standardized or widely accepted method for producing odors or presenting them to participants despite the publication of several dozen types of odor generator systems and odor delivery devices (Simon & Nicolelis, 2002). Likewise, a great deal of research in the olfactory domain does not provide enough information regarding the chemical composition of olfactory stimuli that were evaluated to support the recreation of the scents for future studies. This lack of standardization hinders the scientific progress that can be made in this field and undoubtedly leads to difficult to explain variations in results when evaluating similar constructs across studies. Thus, there is a need to standardize presentation methods that are used in research studies and provide more detailed descriptions of the composition of scents studied.

Connections between odor, emotion, and cognition: Very little research has been conducted to investigate the complex relationships between odor presentation, emotion, and cognition. Although a great deal of research has been conducted to evaluate any one or two of these constructs, given the complex relationship outlined in Section 6.3, it is critical to develop an integrated model of the process that is followed during scent perception that takes into account all three components. A better understanding of these relationships may help to better understand the inconsistent results in olfactory research on basic human state and performance and provide additional guidance on how these cues should be integrated into VEs.

Expand the study of olfactory stressors: Most of the research conducted to date focuses on the effects of presenting scents that are generally positively appraised. Although this research has led to the development of guidance on how these cues should be integrated to improve the experience in VEs, there is a critical need to further expand on basic research conducted with odors that are negatively appraised. Although it is by no means exhaustive, the following list of future research directions provides an example of the breadth of research questions that should be evaluated:

- Determine whether or not malodors trigger an orienting response (OR) following an approach similar to other modalities.
- Determine if novel malodors are more likely to trigger an OR than familiar malodors.
- Determine if psychomotor skills are more resilient to olfactory stressors than cognitive skills as has been shown in other modalities.
- Determine whether olfactory stressors could elicit attention tunneling.
- Determine if olfactory stressors impact decision making similar to other stressors. Specifically, determine if they lead to rigid alternative selection and falsely relying on previous responses.

Expansion of odors evaluated: Similar to the preceding point, it is critical to expand the breadth of odors that are being researched. To date, a significant proportion of the research on olfactory cues has focused on evaluating the effects of peppermint. Given that it is now possible to recreate most any scent to be integrated into VEs and the potential effects of scent integration introduced in this chapter, it is essential to expand the base of scents that are evaluated in future studies.

Leveraging/avoiding adaptation: All of the guidance presented in this chapter is applicable when VE users experience scents that are presented to them. Given the persistence of olfactory adaptation introduced in Section 6.2, it is critical that future applied research focuses on developing methods to avoid this naturally occurring phenomenon when presenting scents during VE use. One specific method that may prove beneficial in reducing this effect is intermittent (vs. constant presentation) presentation of odors.

Develop methods to avoid state-dependent learning when using odors: Given the increased potential to create conditions of state-dependent learning when olfactory cues are present in an environment (Cann & Ross, 1989; Parker et al., 2001; Schab, 1990), it is critical to develop methods to counter these effects when VEs are targeted at creating learning environments.

Evaluate cross-modality interference and determine how to leverage this phenomenon: When integrating scents into VEs, it is important to note that smells are not perceived in isolation. Thus, further research is needed to evaluate the cross-modality interference effects of scent integration. This phenomenon is important to take into account, and methods should be developed to leverage it to persuade the evaluation of scents in a VE as well as persuade the evaluation of other conditions in VEs using scent.

REFERENCES

Ahlström, R., Berglund, R., Berglund, U., Engen, T., & Lindvall, T. (1987). A comparison of odor perception in smokers, nonsmokers, and passive smokers. *American Journal of Otolaryngology, 8*, 1–6.

Allen, C., Pike, W., Lacy, L., Jung, L., & Wiederhold, M. (2010, November 29–December 2). *Using biological and environmental malodors to enhance medical simulation training.* Presented at Interservice/Industry Training, Simulation, and Education Conference (I/ITSEC) 2010, Orlando, FL.

Alter, K. (2011). The essence of human communication—The processing of emotional information. *Physics of Life Reviews, 8,* 406–407.

Amaral, D. G., Behnia, H., & Kelly, J. L. (2003). Topographic organization of projections from the amygdala to the visual cortex in the macaque monkey. *Neuroscience, 118,* 1099–1120.

Amoore, J. E. (1991). Specific anosmias. In T. V. Getchell, R. L. Doty, L. M. Bartoshuk, & J. B. Snow (Eds.), *Smell and taste in health and disease* (pp. 655–664). New York, NY: Raven Press.

Anderson, A. K., Christoff, K., Panitz, D., De Rosa, E., & Gabrieli, J. D. (2003). Neural correlates of the automatic processing of threat facial signals. *Journal of Neuroscience, 23,* 5627–5633.

Anderson, A. K., & Phelps, E. A. (2001). Lesions of the human amygdala impair enhanced perception of emotionally salient events. *Nature Neuroscience, 411,* 305–309.

Armony, J. L., & Dolan, R. J. (2002). Modulation of spatial attention by fear-conditioned stimuli: An event-related fMRI study. *Neuropsychologia, 40,* 817–826.

Augustine, A. A., & Hemenover, S. H. (2008). On the relative effectiveness of affect regulation strategies: A meta-analysis. *Cognition and Emotion, 23,* 1181–1220.

Bagget, P. (1987). Learning a procedure from multimedia instructions: The effect of film and practice. *Applied Cognitive Psychology, 1,* 183–195.

Bakker, B., & Demerouti, E. (2008). Towards a model of work engagement. *Career Development International, 13*(3), 209–223.

Ball, L. J., Shoker, J., & Miles, J. N. V. (2010). Odour-based context reinstatement effects with indirect measures of memory: The curious case of rosemary. *British Journal of Psychology, 101*(4), 655–678.

Bandura, A. (1986). *Social foundations of thought and action: A social cognitive theory.* Rockville, MD: National Institutes of Mental Health.

Banks, S. J., Ng, V., & Jones-Gotman, M. (2012). Does good + good = better? A pilot study on the effect of combining hedonically valenced smells and images. *Neuroscience Letters, 514,* 71–76.

Barker, S., Grayhem, P., Koon, J., Perkins, J., Whalen, A., & Raudenbush, B. (2003). Improved performance on clerical tasks associated with administration of peppermint odor. *Perceptual and Motor Skills, 97,* 1007–1010.

Bartoshuk, L. M. (1991). Taste, smell and pleasure. In R. C. Bolles (Ed.), *The hedonics of taste* (pp. 15–28). Hillsdale, NJ: Erlbaum.

Baumeister, R. F., Bratslavsky, E., Finkenauer, C., & Vohs, K. D. (2001). Bad is stronger than good. *Review of General Psychology, 5,* 323–370.

Baus, O., & Bouchard, S. (2010). The sense of olfaction: Its characteristics and its possible applications in virtual environments. *Journal of Cybertherapy and Rehabilitation, 3,* 31–50.

Bayley, R., Matthews, L., Street, E., Almeida, J., & Raudenbush, B. (2007). Ability of gum flavors to distract participants from painful stimuli: Differential effects of retronasal vs. orthonasal scent administration. *Appetite, 49*(1), 278.

Bear, M. F., Connors, B. W., & Paradiso, M. A. (2001). The neocortex and working memory. In *Neuroscience: Exploring the brain* (pp. 768–772). Philadelphia, PA: Lippincott Williams & Wilkins.

Bechara, A. (2004). The role of emotion in decision-making: Evidence from neurological patients with orbitofrontal damage. *Brain and Cognition, 55,* 30–40.

Bechara, A., & Damasio, A. (2005). The somatic marker hypothesis: A neural theory of economic decision-making. *Games and Economic Behavior, 52,* 336–372.

Bechara, A., Damasio, H., & Damasio, A. R. (2000). Emotion, decision making and the orbitofrontal cortex. *Cerebral Cortex, 10,* 295–307.

Bechara, A., Tanel, D., & Damasio, A. R. (2002). The somatic marker hypothesis and decision-making. In F. Boller & J. Grafman (Eds.), *Handbook of neuropsychology: Frontal lobes* (Vol. 7, 2nd ed., pp. 117–143). Amsterdam, the Netherlands: Elsevier.

Bechara, A., Tranel, D., Damasio, H., Adolphs, R., Rockland, C., & Damasio, A. R. (1995). Double dissociation of conditioning and declarative knowledge relative to the amygdala and hippocampus in humans. *Science, 269,* 1115–1118.

Bendall, J. C., & Parsell, B. (2005). Pre-hospital trauma life support (PHTSL) advanced provider course. *Journal of Emergency Primary Health Care, 3*(1–2). Retrieved from http://www.jephc.com/uploads/Bendall&ParsellReview.pdf

Bengtsson, S., Berglund, H., Gulyas, B., Cohen, E., & Savic, I. (2001). Brain activation during odor perception in males and females. *Chemical Senses, 12*(9), 2027–2033.

Bensafi, M., Rouby, C., Farget, V., Bertrand, B., Vigouroux, M., & Holly, A. (2002). Influence of affective and cognitive judgments on autonomic parameters during inhalation of pleasant and unpleasant odors in humans. *Neuroscience Letters, 319*(3), 162–166.

Bishop, D. T., Karageorghis, C. I., & Loizou, G. (2007). A grounded theory of young tennis players' use of music to manipulate emotional state. *Journal of Sport and Exercise Psychology, 29*, 584–607.

Blackwell, L. (1995). Visual cues and their effects on odour assessment. *Nutrition and Food Science, 95*(5), 24–28.

Blake, R., & Sekuler, R. (2006). *Perception* (5th ed., pp. 523–524). New York, NY: McGraw Hill.

Bordnick, P. S., Copp, H. L., Traylor, A., Graap, K. M., Carter, B. L., Walton, A., & Ferrer, M. (2009). Reactivity to cannabis cues in virtual reality environments. *Journal of Psychoactive Drugs, 41*(2), 105–112.

Brefezynski, J. A., & DeYoe, E. A. (1999). A physiological correlate of the 'spotlight' of visual attention. *Nature Neuroscience, 2*, 370–374.

Buchbauer, G., Jirovetz, L., Jager, W., Dietrich, H., & Plank, C. (1991). Aromatherapy: Evidence for sedative effects of the essential oils of lavender after inhalation. *Zeitschrift fur Naturforschung Teil C, 46*, 1067–1072.

Cain, W., & Polak, E. (1992). Olfactory adaptation as an aspect of odor similarity. *Chemical Senses, 17*, 481–491.

Cain, W. S., & Gent, J. F. (1991). Olfactory sensitivity: Reliability, generality, and association with aging. *Journal of Experimental Psychology: Human Perception and Performance, 17*, 382–391.

Cain, W. S., & Turk, A. (1985). Smell of danger: An analysis of LP-gas odorization. *American Industrial Hygiene Association Journal, 46*, 115–126.

Candau, J. (2004). The olfactory experience: Constants and cultural variables. *Water Science and Technology, 49*(9), 11–17.

Cann, A., & Ross, D. A. (1989). Olfactory stimuli as context cues in human memory. *American Journal of Psychology, 102*(1), 91–102.

Carrasco, M., & McElree, B. (2001). Covert attention accelerates the rate of visual information processing. *Proceedings of the National Academy of Sciences of the United States of America, 98*, 5363–5367.

Chaston, A., & Kingstone, A. (2004). Time estimation: The effect of cortically mediated attention. *Brain and Cognition, 55*(2), 286–289.

Cheal, M., & Lyon, D. (1991). Central and peripheral pre-cuing of forced-choice discrimination. *Quarterly Journal of Experimental Psychology: Human Experimental Psychology, 43A*, 859–880.

Chertoff, D. B., Schatz, S. L., McDaniel, R., & Bowers, C. A. (2008). Improving presence theory through experiential design. *Presence, 17*(4), 405–413.

Cohen, B., Davidson, R., Senulis, J., Saron, C., & Weisman, D. (1992). Muscle tension patterns during auditory attention. *Biological Psychology, 33*, 133–156.

Critchley, H. D. (2005). Neural mechanisms of autonomic, affective, and cognitive interaction. *Journal of Comparative Neurology, 493*, 154–166.

Dalton, P. (2000). Psychophysical and behavioral characteristics of olfactory adaptation. *Chemical Senses, 25*, 487–492.

Damasio, H., Bechara, A., Tranel, D., & Damasio, A. R. (1997). Double dissociation of emotional conditioning and emotional imagery relative to the amygdala and right somatosensory cortex. *Society for Neuroscience Abstracts, 23*, 1318.

Danuser, B., Moser, D., Vitale-Sethre, T., Hirsig, R., & Krueger, H. (2003). Performance in a complex task and breathing under odor exposure. *Human Factors, 45*(4), 549–562.

Davis, P., Woodman, T., & Callow, N. (2010). Better out than in: The influence of anger regulation on physical performance. *Personality and Individual Differences, 49*(5), 457–460.

Delplanque, S., Grandjean, D., Chrea, C., Aymard, L., Cayeux, I., Le Calve, B., ... Sanders, D. (2008). Emotional processing of odors: Evidence of a nonlinear relation between pleasantness and familiarity evaluations. *Chemical Senses, 33*, 469–479.

Dematte, L., Sanabria, D., & Spence, C. (2009). Olfactory discrimination: When vision matters? *Chemical Senses, 34*(2), 103–109.

Denkova, E., Wong, G., Dolcos, S., Sung, K., Wang, L., Coupland, N., & Dolcos, F. (2010). The impact of anxiety inducing distraction on cognitive performance: A combined brain imaging and personality investigation. *PLoS ONE, 5*(11), e141–e150.

D'Esposito, M., Detre, J. A., Alsop, D. C., Shin, R. K., Atlas, S., & Grossman, M. (1995). The neural basis of central execution systems of working memory. *Nature, 378*, 279–281.

Diego, M., Jones, N., Field, T., Hernandez-Reif, M., Schanberg, S., Kuhn, C., ... Galamaga, R. (1998). Aromatherapy positively affects mood, EEG patterns of alertness, and math computations. *International Journal of Neuroscience, 96*, 217–224.

Dilks, D., Dalton, P., & Beauchamp, G. (1999). Cross-cultural variation in responses to malodors. *Chemical Senses, 24*, 599.

Dinh, H. Q., Walker, N., Song, C., Kobayashi, A., & Hodges, L. F. (1999). Evaluating the importance of multi-sensory input on memory and the sense of presence in virtual environments. *Proceedings of IEEE, Virtual Reality '99*, 222–228.

Distel, H., & Hudson, R. (2001). Judgment of odor intensity is influenced by subjects' knowledge of the odor source. *Chemical Senses*, *26*, 247–251.

Dolcos, F. (2010). *The impact of emotion on memory: Evidence from brain imaging studies*. Saarbrucken, Germany: VDM Publishing.

Dolcos, F., & Denkova, E. (2008). Neural correlates of encoding emotional memories: A review of functional neuroimaging evidence. *Cell Science Reviews*, *5*(2), 78–122.

Dolcos, F., Iordan, A. D., & Dolcos, S. (2011). Neural correlates of emotion-cognition interactions: A review of evidence from brain imaging investigations. *Journal of Cognitive Psychology*, *23*(6), 669–694.

Dolcos, F., LaBar, K. S., & Cabeza, R. (2006). The memory-enhancing effect of emotion: Functional neuroimaging evidence. In B. Uttl, N. Ohta, & A. L. Siegenthaler (Eds.), *Memory and emotion: Interdisciplinary perspectives* (pp. 107–133). New York, NY: Wiley.

Dolcos, F., & McCarthy, G. (2006). Brain systems mediating cognitive interference by emotional distraction. *Journal of Neuroscience*, *26*(7), 2072–2079.

Drevets, W. C., & Raichle, M. E. (1998). Reciprocal suppression of regional cerebral blood flow during emotional versus higher cognitive processes: Implications for interactions between emotion and cognition. *Cognition and Emotion*, *12*(3), 353–385.

Engen, T. (1991) *Odor sensation and memory*. New York, NY: Praeger.

Epple, G., & Herz, R. S. (1999). Ambient odors associated to failure influence cognitive performance in children. *Developmental Psychobiology*, *35*, 103–107.

Galotti, K. M. (1999). Retrieving memories from long-term storage. In *Cognitive psychology in and out of the laboratory* (2nd ed., pp. 156–187). Belmont, CA: Wadsworth Publishing Company.

Gerardi, M., Rothbaum, B. O., Ressler, K., Heekin, M., & Rizzo, A. A. (2008). Virtual reality exposure therapy using a virtual Iraq: Case report. *Journal of Traumatic Stress*, *21*(2), 209–213.

Ghashghaei, H. T., & Barbas, H. (2002). Pathways for emotion: Interactions of prefrontal and anterior temporal pathways in the amygdala of the rhesus monkey. *Neuroscience*, *115*, 1261–1279.

Ghashghaei, H. T., Hilgetag, C. C., & Barbas, H. (2007). Sequence of information processing for emotions based on the anatomic dialogue between prefrontal cortex and amygdala. *Neuroimage*, *34*(3), 905–923.

Gobel, H., Schmidt, G., & Soyka, D. (1994). Effect of peppermint and eucalyptus oil preparations on neurophysiological and experimental algesimetric headache parameters. *Cephalalgia*, *14*, 228–234.

Gottfried, J. A., Deichmann, R., Winston, J. S., & Dolan, R. J. (2002). Functional heterogeneity in human olfactory cortex: An event-related functional magnetic resonance imaging study. *Journal of Neuroscience*, *22*, 10819–10828.

Gottfried, J. A., & Dolan, R. J. (2003). The nose smells what the eye sees: Crossmodal visual facilitation of human olfactory perception. *Neuron*, *39*(2), 375–386.

Gould, A., & Martin, N. (2001). A good odor to breathe? The effect of pleasant ambient odor on human visual vigilance. *Applied Cognitive Psychology*, *15*(2), 225–232.

Greene, J. D., Sommerville, R. B., Nystrom, L. E., Darley, J. M., & Cohen, J. D. An fMRI investigation of emotional engagement in moral judgement. *Science*, *293*, 2105–2108.

Hancock, P. A., & Warm, J. S. (1989). A dynamic model of stress and sustained attention. *Human Factors*, *31*(5), 519–537.

Hanin, Y. L. (1993). Optimal performance emotions in top athletes. Sport psychology: An integrated approach. In *Proceedings of the 8th World Congress of Sport Psychology* (pp. 229–232). Lisbon, Portugal: ISSP.

Hanin, Y. L. (1997). Emotions and athletic performance: Individual zones of optimal functioning model. *European Yearbook of Sport Psychology*, *1*, 29–72.

Hart, S. G., & Staveland, L. E. (1988). Development of NASA-TLX (Task Load Index): Results of empirical and theoretical research. In P. A. Hancock & N. Meshkati (Eds.), *Human mental workload* (pp. 139–183). Amsterdam, the Netherlands: North Holland Press.

Hastie, R., & Dawes, R. M. (2001). *Rational choice in an uncertain world*. Thousand Oaks, CA: Sage Publications.

He, Y., & Evans, A. (2010). Graph theoretical modeling of brain connectivity. *Current Opinion in Neurology*, *23*, 341–350.

Herz, R. (2001). Ah, sweet skunk: Why we like or dislike what we smell. *Cerebrum*, *3*, 31–47.

Herz, R. (2005). Odor-associative learning and emotion: Effects on perception and behavior. *Journal of Chemical Senses*, *30*, 250–251.

Herz, R., & Engen, T. (1996). Odor memory: Review and analysis. *Psychonomic Bulletin & Review*, *3*, 300–313.

Herz, R. S., Beland, S. L., & Hellerstein, M. (2004). Changing odor hedonic perception through emotional associations in humans. *International Journal of Comparative Psychology*, *17*(4), 315–338.

Herz, R. S., Schankler, C., & Beland, S. (2004). Olfaction, emotion and associative learning: Effects on motivated behavior. *Motivational Emotion*, *28*, 363–383.

Hirsch, A. (2011). Method of altering perception of time. United States Patent Application Publication. U.S. 2011/0171131 A1.

Ho, C., & Spence, C. (2005). Olfactory facilitation of dual-task performance. *Neuroscience Letters*, *389*, 35–40.

Honey, C. J., Thivierge, J. P., & Sporns, O. (2010). Can structure predict function in the human brain? *Neuroimage*, *52*, 766–776.

Houwer, J., Thomas, S., & Baeyens, F. (2001). Associative learning of likes and dislikes: A review of 25 years of research on human evaluative conditioning. *Psychological Bulletin*, *127*(6), 853–869.

Hummel, T., Kobal, G., Gudziol, H., & Mackay-Sim, A. (2007). Normative data for the "sniffin'sticks" including tests of odor identification, odor discrimination, and olfactory thresholds: An upgrade based on a group of more than 3,000 subjects. *European Archives of Otorhinolaryngol*, *264*, 237–243.

Isbell, L., & Wyer, R. (2012). Correcting for mood-induced bias in the evaluation of political candidates: The roles of intrinsic and extrinsic motivation. *Personality and Social Psychology Bulletin*, *25*, 237–249.

Jean, G. (2007, December). More realism sought in urban combat training. *National Defense*, *92*(649), 50–52.

Johnson, A. (2011).Cognitive facilitation following intentional odor exposure. *Sensors*, *11*, 5469–5488.

Johnson, T., Prusak, K., Pennington, T., & Wilkinson, C. (2011). The effects of the type of skill test, choice, and gender on the situational motivation of physical education students. *Journal of Teaching in Physical Education*, 30, 281–295.

Jones, M. V. (2003). Controlling emotions in sport. *The Sport Psychologist*, *17*, 471–486.

Kahneman, D., & Tversky, A. (1979). Prospect theory: An analysis of decision under risk. *Econometrica*, *47*, 263–291.

Karageorghis, C. I., Terry, P. C., & Lane, A. M. (1999). Development and initial validation of an instrument to assess the motivational qualities of music in exercise and sport: The Brunel Music Rating Inventory. *Journal of Sports Sciences*, *17*, 713–724.

Kensinger, E. (2004). Remembering emotional experiences: The contribution of valence and arousal. *Reviews in the Neurosciences*, *15*, 241–253.

Knasko, S. (1992). Ambient odor's effect on creativity, mood, and perceived health. *Chemical Senses*, *17*(1), 27–35.

Knasko, S. (1993). Performance, mood, and health during exposure to intermittent odors. *Archives of Environmental Health*, *48*(5), 305–308.

Knight, W. (2004). Where's that funny smell coming from? *New Scientist*, *182*(2441), 22.

Koelega, H. S., & Koster, E. P. (1974). Some experiments on sex differences in odor perception. *Annals of the New York Academy of Sciences*, *237*, 234–246.

Kosslyn, S. M., Shin, L. M., Thompson, W. L., McNally, R. J., Rauch, S. L., Pitman, R. K., & Alpert, N. M. (1996). Neural effects of visualizing and perceiving aversive stimuli: A PET investigation. *NeuroReport*, *7*, 1569–1576.

Lawless, H., & Engen, T. (1977). Associations to odors: Interference, mnemonics and verbal labeling. *Journal of Experimental Psychology: Human Learning and Memory*, *3*, 52–59.

Lehrner, J., Eckersberger, C., Walla, P., Potsch, G., & Deecke, L. (2000). Ambient odor of orange in a dental office reduces anxiety and improves mood in female patients. *Physiology & Behavior*, *71*(1), 83–86.

Li, W., Moallem, I., Paller, K. A., & Gottfried, J. A. (2007). Subliminal smells can guide social preferences. *Psychological Science*, *18*, 1044–1049.

Liotti, M., Maybert, H. S., Brannan, S. K., McGinnis, S., Jerabek, P., & Fox, P. T. (2000). Differential limbic-cortical correlates of sadness and anxiety in healthy subjects: Implications for affective disorders. *Biological Psychiatry*, *48*, 30–42.

Loo, A., Youngentob, S. L., Kent, P. F., & Schwob, J. E. (1996). The aging olfactory epithelium: Neurogenesis, response to damage, and odorant-induced activity. *International Journal of Development Neuroscience*, *14*(7/8), 881–900.

Lorig, T. (1992). Cognitive and "non-cognitive" effects of odor exposure: Electrophysiological and behavioral evidence. In S. Van Toller & G. Dodd (Eds.), *The psychology and biology of perfume* (pp. 161–172). Amsterdam, the Netherlands: Elsevier.

Lorig, T., & Schwartz, G. (1987). EEG activity during relaxation and food imagery. *Psychophysiology*, *24*, 599.

Lorig, T. S., Huffman, E., DeMartino, A., & DeMarco, J. (1991). The effects of low concentration odors on EEG activity and behavior. *Journal of Psychophysiology*, *5*, 69–77.

Matlin, M. (2002). Long-term memory. In *Cognition* (5th ed., pp. 111–156). Fort Worth, TX: Harcourt College Publishers.

Mayberg, H. S. (1997). Limbic-cortical dysregulation: A proposed model of depression. *Journal of Neuropsychiatry and Clinical Neurosciences, 9*(3), 471–481.

McCaffrey, R. J., Lorig, T. S., Pendrey, D. L., McCutcheon, N. B., & Garrett, J. C. (1993). Odor-induced EEG changes in PTSD Vietnam veterans. *Journal of Traumatic Stress, 6*, 213–224.

McCarthy, G., Blamire, A. M., Puce, A., Nobre, A. C., Boch, G., Hyder, F., … Shulman, R. G. (1994). Functional magnetic resonance imaging of human prefrontal cortex activation during a spatial working memory task. *Proceedings of the National Academy of Sciences of the United States of America, 91*, 8690–8694.

McCombs, K., Raudenbush, B., Bova, A., & Sappington, M. (2011). Effects of peppermint scent administration on cognitive video game performance. *North American Journal of Psychology, 13*(3), 383–390.

McCullagh, P. (1986). Model status as a determinant of observational learning and performance. *Journal of Sport Psychology, 8*, 319–331.

McGaugh, J. L. (2002). Memory consolidation and the amygdala: A systems perspective. *Trends in Neurosciences, 25*, 456–461.

McGaugh, J. L. (2004). The amygdala modulates the consolidation of memories of emotionally arousing experiences. *Annual Review of Neuroscience, 27*, 1–28.

Mesulam, M. M. (1990). Large-scale neurocognitive networks and distributed processing for attention, language, and memory. *Annals of Neurology, 28*, 597–613.

Millot, J., & Brand, G. (2001). Effects of pleasant and unpleasant ambient odors on human voice pitch. *Neuroscience Letters, 297*, 61–63.

Miltner, W., Matjak, M., Diekmann, H., & Brody, S. (1994). Emotional qualities of odors and their influence on the startle reflex in humans. *Psychophysiology, 31*, 107–110.

Mitchell, D. G., Luo, Q., Avny, S. B., Kasprzycki, T., Gupta, K., Chen, G., … Blair, R. J. R. (2009). Adapting to dynamic stimulus-response values: Differential contributions of inferior frontal, dorsomedial, and dorsolateral regions of prefrontal cortex to decision making. *Journal of Neuroscience, 29*, 10827–10834.

Mitchell, D. G. V. (2011). The nexus between decision making and emotion regulation: A review of convergent neurocognitive substrates. *Behavioral Brain Research, 217*, 215–231.

Morrin, M., & Ratneshwar, S. (2000). The impact of ambient scent on evaluation, attention, and memory for familiar and unfamiliar brands. *Journal of Business Research, 49*, 157–165.

Morris, J. S., Friston, K. J., Buchel, C., Frith, C. D., Young, A. W., Calder, A. J., & Dolan, R. J. (1998). A neuromodulatory role for the human amygdala in processing emotional facial expressions. *Brain, 121*, 47–57.

Moskowitz, H. R., Dravnieks, A., & Klarman, L. A. (1976). Odor intensity and pleasantness for a diverse set of odorants. *Perception & Psychophysics, 19*(2), 122–128.

Moss, M., Cook, J., Wesnes, K., & Duckett, P. (2003). Aromas of rosemary and lavender essential oils differentially affect cognition and mood in healthy adults. *International Journal of Neuroscience, 113*, 15–38.

Moss, M., Hewitt, S., Moss, L., & Wesnes, K. (2008). Modulation of cognitive performance and mood by aromas of peppermint and ylang-ylang. *International Journal of Neuroscience, 118*(1), 59–77.

Naessen, R. (1971). An enquiry on the morphological characteristic and possible changes in age in the olfactory region of man. *Acta Otolaryngologica, 71*, 49–62.

National Institute of Neurological Disorders and Stroke. (2000). *Pain information pamphlet.* Washington, DC: NINDS.

Nieuwenhuys, A., Hanin, Y. L., & Bakker, F. C. (2008). Performance-related experiences and coping during races: A case of an elite sailor. *Psychology of Sport and Exercise, 9*, 61–76.

Niven, K., Totterdell, P., & Holman, D. (2009). A classification of controlled interpersonal affect regulation strategies. *Emotion, 4*, 498–509.

Ochsner, K. N., & Gross, J. J. (2005). The cognitive control of emotion. *Trends in Cognitive Science, 9*, 242–249.

O'Doherty, J., Critchley, H., Deichmann, R., & Dolan, R. J. (2003). Dissociating valence of outcome from behavioral control in human orbital and ventral prefrontal cortices. *Journal of Neuroscience, 23*, 7931–7939.

Orr, J. L., & Carter, J. (1995). Preliminary risk assessment for the olfactory component of the deep-immersion virtual environment (DIVE). *Toxic Substance Mechanisms, 14*(3), 242.

Parasuraman, R., Warm, J. S., & Dember, W. N. (1992). *Effects of olfactory stimulation on skin conductance and event-related potentials during visual sustained attention* (Progress report No. 6). New York: Fragrance research fund Ltd.

Park, C., Ahn, S. C., Kwon, Y.-M., Kim, H.-G., & Kim, T. (2003). Gyeongju VR theater: A journey into the breath of Sorabol. *Presence: Teleoperators and Virtual Environments, 12*(2), 125–139.

Parker, A., Ngu, H., & Cassaday, H. G. (2001). Odour and Proustian memory: Reduction of context-dependent forgetting and multiple forms of memory. *Applied Cognitive Psychology, 15*(2), 159–171.

Pegna, A. J., Khateb, A., Michel, C. M., & Landis, T. (2004). Visual recognition of faces, objects, and words using degraded stimuli: Where and when it occurs. *Human Brain Mapping, 22*(4), 300–311.

Petrides, M., Alivisatos, B., Meyer, E., & Evans, A. C. (1993). Functional activation of the human frontal cortex during the performance of verbal working memory tasks. *Proceedings of the National Academy of Sciences of the United States of America, 90,* 878–882.

Phelps, E. A. (2004). Human emotion and memory: Interactions of the amygdala and hippocampal complex. *Current Opinion in Neurobiology, 14*(2), 198–202.

Phelps, E. A., & LeDoux, J. E. (2005). Contributions of the amygdala to emotion processing: From animal models to human behavior. *Neuron, 48,* 175–187.

Pierce, J., Wysocki, C., Aronov, E., Webb, J., & Boden, R. (1996). The role of perceptual and structural similarity in cross-adaptation. *Chemical Senses, 21,* 223–237.

Pourtois, G., Schwartz, S., Seghier, M. L., Lazeyras, F., & Vuilleumier, P. (2006). Neural systems for orienting attention to the location of threat signals: An event-related fMRI study. *Neuroimage, 31,* 920–933.

Powers, W. (2004). *Odor perception and physiological response.* Ames, IA: Iowa State University.

Raudenbush, B. (2000). The effects of odors on objective and subjective measures of physical performance. *The Aroma-Chology Review, 9,* 1–5.

Raudenbush, B. (2002). Pain threshold and tolerance mediation through the administration of peppermint odor: A preliminary study. *Aromachology, 10,* 1–12.

Raudenbush, B., Corley, N., & Eppich, W. (2001). Enhancing athletic performance through the administration of peppermint odor. *Journal of Sport & Exercises Psychology, 23,* 156–160.

Raudenbush, B., Grayhem, R., Sears, T., & Wilson, I. (2009). Effects of peppermint and cinnamon odor administration on simulated driving alertness, mood and workload. *North American Journal of Psychology, 11*(2), 245–256.

Raudenbush, B., Koon, J., Meyer, B., Corley, N., & Flower, N. (2004). Effects of odorant administration on pain and psychophysiological measures in humans. *North American Journal of Psychology, 6*(3), 361–370.

Raudenbush, B., Koon, J., Meyer, B., & Flower, N. (2002). Effects of ambient odor on pain threshold, pain tolerance, mood, workload, and anxiety. *Psychophysiology, 39,* supplement.

Raudenbush, B., Meyer, B., & Eppich, W. (2002). Effects of odor administration on objective and subjective measures of athletic performance. *International Sports Journal, 6,* 1–15.

Raudenbush, B., Smith, J., Graham, K., & McCune, A. (2004). Effects of peppermint odor administration on augmenting basketball performance during game play. *Chemical Senses, 29,* supplement.

Raudenbush, B., & Zoladz, P. (2003). *The effects of peppermint odor administration on lung capacity and inhalation ability.* Technical Report for HealthCare International, Seattle, WA.

Rétiveau, A. N., Chambers, I. V. E., & Milliken, G. A. (2004). Common and specific effects of fine fragrances on the mood of women. *Journal of Sensory Studies, 19,* 373–394.

Rezek, D. L. (1987). Olfactory deficits as a neurologic sign in dementia of the Alzheimer type. *Archives of Neurology, 44,* 1030–1032.

Ritchey, M., Dolcos, F., & Cabeza, R. (2008). Role of amygdala connectivity in the persistence of emotional memories over time: An event-related fMRI investigation. *Cerebral Cortex, 18*(11), 2494–2504.

Robazza, C., Bortoli, L., Nocini, F., Moser, G., & Arslan, C. (2000). Normative and idiosyncratic measures of positive and negative affect in sport. *Psychology of Sport and Exercise, 1*(2), 103–116.

Roberts, A., & Williams, J. (1992). The effect of olfactory stimulation on fluency, vividness of imagery and associated mood: A preliminary study. *British Journal of Medical Psychology, 65,* 197–199.

Robin, O., Alaoui-Ismaili, O., Dittmar, A., & Vernet-Mauri, E. (1998). Emotional responses evoked by dental odors: An evaluation from autonomic parameters. *Journal of Dental Research, 77,* 1638–1646.

Rolls, E. T. (2004). The functions of the orbitofrontal cortex. *Brain and Cognition, 55,* 11–29.

Rolls, E. T., & Baylis, L. L. (1994). Gustatory, olfactory, and visual convergence within the primate orbitofrontal cortex. *Journal of Neuroscience, 14,* 5437–5452.

Rolls, E. T., Hornak, J., Wade, D., & McGrath, J. (1994). Emotion-related learning in patients with social and emotional changes associated with frontal lobe damage. *Journal of Neurology, Neurosurgery & Psychiatry, 57*(12), 1518–1524.

Rosen, M. A., Salas, E., Pavlas, D., Jensen, R., Fu, D., & Lampton, D. (2010). Demonstration-based training: A review of instructional features. *Human Factors, 52*(5), 596–609.

Rosenblum, L., Julier, S., Nakamoto, T., Otaguro, S., Kinoshita, M., Nagahama, M., ... Ishida, T. (2008). Cooking up an interactive olfactory game display. *IEEE Computer Graphics & Applications, 28*(1), 75–78.

Rotton, J. (1983). Affective and cognitive consequences of malodorous pollution. *Basic and Applied Social Psychology, 4*(2), 171–191.

Rozin, P., & Roysman, E. B. (2001). Negativity bias, negativity dominance, and contagion. *Personality and Social Psychology Review*, 5, 296–320.

Rudebeck, P. H., & Murray, E. A. (2008). Amygdala and orbitofrontal cortex lesions differentially influence choices during object reversal learning. *Journal of Neuroscience*, 28, 8338–8343.

Russell, J. (1980). A circumplex model of affect. *Journal of Personality and Social Psychology*, 39, 1161–1178.

Ryan, J. J., Kreiner, D. S., Chapman, M. D., & Stark-Wroblewski, K. (2010). Virtual reality cues for binge drinking in college students. *Cyberpsychology, Behavior, and Social Networking*, 13, 159–162.

Schab, F. R. (1990). Odors and the remembrance of things past. *Journal of Experimental Psychology: Learning, Memory, and Cognition*, 16(4), 648–655.

Schaufeli, W. B., Salanova, M., González-Romá, V., & Bakker, A. B. (2002). The measurement of engagement and burnout: A confirmative analytic approach. *Journal of Happiness Studies*, 3, 71–92.

Schiffman, S. (1998). Livestock odors: Implications for human health and well-being. *Journal of Animal Science*, 76(5), 1343–1355.

Schiffman, S. S., & Williams, C. M. (2005). Science of odor as a potential health issue. *Journal of Environmental Quality*, 34(1), 581–588.

Science Daily. (2009). First virtual reality technology to let you see, hear, smell, taste, and touch. Retrieved July 16, 2012, from http://www.sciencedaily.com/releases/2009/03/090304091227.htm

Sethre, T., Laubli, T., Berode, M., Hangartner, M., & Krueger, H. (2000). Experimental exposure to methylformate and its neurobehavioral effects. *International Archives of Occupational and Environmental Health*, 73(6), 401–409.

Simon, S. A., & Nicolelis, M. A. L. (2002). *Methods in chemosensory research*. Boca Raton, FL: CRC Press LLC.

Smith, S., Glenberg, A., & Bjork, R. (1978). Environmental context and human memory. *Memory and Cognition*, 6, 342–353.

Sobel, N., Prabhakaran, V., Zhao, Z., Desmond, J. E., Glover, G. H., Sullivan, E. V., & Gabrieli, J. D. E. (2000). Time course of odorant-induced activation in the human primary olfactory cortex. *Journal of Neurophysiology*, 83, 537–551.

Stam, C. J. (2010). Characterization of anatomical and functional connectivity in the brain: A complex networks perspective. *International Journal of Psychophysiology*, 77, 186–194.

Stanney, K. M. (Ed.). (2002). *Handbook of virtual environments: Design, implementation, and applications*. Mahwah, NJ: Lawrence Erlbaum Associates, Inc.

Stevenson, R., & Boakes, R. (2003). A mnemonic theory of odor perception. *Psychological Review*, 110(2), 340–364.

Stone, H., & Pryor, G. (1967). Some properties of the olfactory system of man. *Perception & Psychophysics*, 2, 516–518.

Surguladze, S. A., Brammer, M. J., Young, A. W., Andrew, C., Travis, M. J., Williams, S. C., & Phillips, M. L. (2003). A preferential increase in the extra striate response to signals of danger. *Neuroimage*, 19(4), 1317–1328.

Swaab, D. F., Chung, W. C. J., Kruijver, F. P. M., Hofman, M. A., & Ishunina, T. A. (2001). Structural and functional sex differences in the human hypothalamus. *Hormones and Behavior*, 40, 93–98.

Taylor, S. F., Liberzon, I., & Koeppe, R. A. (2000). The effect of graded aversive stimuli on limbic and visual activation. *Neuropsychologia*, 38(10), 1415–1425.

Todrank, J., Wysocki, C., & Beauchamp, G. (1991). The effects of adaptation on the perception of similar and dissimilar odors. *Chemical Senses*, 16, 467–482.

Torri, S., Fukada, H., Kanemoto, H., Miyanchi, R., Hamauzu, Y., & Kawasaki, M. (1988). Contingent negative variation (CNV) and the psychological effects of odour. In S. Van Toller & G. H. Dodd (Eds.), *Perfumery—The psychology and biology of fragrance* (pp. 107–120). London, U.K.: Chapman & Hall.

Tulving, E., & Thomson, D. (1973). Encoding specificity and retrieval processes in episodic memory. *Psychological Review*, 80, 352–372.

Van Reekum, C. M., Van den Berg, H., & Frijda, N. H. (1999). Cross-modal preference acquisition: Evaluative conditioning of pictures by affective olfactory and auditory cues. *Cognition and Emotion*, 13(6), 831–836.

Van Toller, S. (1988). Emotion and the brain. In S. Van Toller & G. Dodd (Eds.), *Perfumery: The psychology and biology of fragrance* (pp. 121–146). London, U.K.: Chapman & Hall.

Varney, N. R. (1988). The prognostic significance of anosmia in patients with closed-head trauma. *Journal of Clinical and Experimental Neuropsychology*, 10, 250–254.

Vermetten, E., & Bremner, J. D. (2003). Olfaction as a traumatic reminder in posttraumatic stress disorder: Case reports and review. *Journal of Clinical Psychiatry*, 64, 202–207.

Vlahos, J. (2006). The smell of war. *Popular Science*, 8, 73–95.

Vuilleumier, P. (2005). How brains beware: Neural mechanisms of emotional attention. *Trends in Cognitive Science, 9*, 585–594.

Vuilleumier, P., & Huang, Y. M. (2009). Emotional attention- Uncovering the mechanisms of affective biases in perception. *Current Directions in Psychological Science, 18*, 148–152.

Vuilleumier, P., & Pourtois, G. (2007). Distributed and interactive brain mechanisms during emotion face perception: Evidence from functional neuroimaging. *Neuropsychologia, 45*, 174–194.

Warm, J., Dember, W., & Parasuramann, R. (1991). Effects of olfactory stimulation on performance and stress in a visual sustained attention task. *Journal of the Society of Cosmetic Chemists, 42*, 199–210.

Warnock, D. (2011). A subjective evaluation of multimodal notifications. *Proceedings of the 5th International Conference on Pervasive Computing Technologies for Healthcare,* 461–468.

Washburn, D., & Jones, L. (2004). Could olfactory displays improve data visualization? *Computing in Science and Engineering, 6*(6), 80–83.

Wenger, M. A., Jones, F. N., & Jones, M. H. (1956). *Physiological psychology.* New York, NY: Holt, Rinehart & Winston.

Whalen, P. J. (1998). Fear, vigilance, and ambiguity: Initial neuroimaging studies of the human amygdala. *Current Directions in Psychological Science, 7*, 177–188.

Wilfred, L., Hall, R., Hilgers, M., Leu, M., Hortenstine, J., Walker, C., & Reddy, M. (2004). Training in affectively intense virtual environments. In J. Nall & R. Robson (Eds.), *Proceedings of World Conference on E-Learning in Corporate, Government, Health, & Higher Education* (pp. 2233–2240). Chesapeake, VA: AACE.

Wilmes, B., Harrington, L., Kohler-Evans, P., & Sumpter, D. (2008). Coming to our senses: Incorporating brain research findings into classroom instruction. *Education, 128*(4), 659–666.

Winston, J. S., Vuilleumier, P., & Dolan, R. J. (2003). Effects of low-spatial frequency components of fearful faces on fusiform cortex activity. *Current Biology, 13*(20), 1824–1829.

Woodman, T., Davis, P., Hardy, L., Callow, N., Glasscock, I., & Yuill-Proctor, J. (2009) Emotions and sport performance: An exploration of happiness, hope, and anger. *Journal of Sport & Exercise Psychology, 31*(2), 169–188.

Woodman, T., & Hardy, L. (2001). A case study of organizational stress in elite sport. *Journal of Applied Sport Psychology, 13*, 207–238.

Wysocki, C. J., & Beauchamp, G. K. (1991). Individual differences in human olfaction. In C. J. Wysocki & M. R. Kare (Eds.), *Chemical senses: Vol. 3. Genetics of perception and communications* (pp. 353–373). New York, NY: Marcel Dekker, Inc.

Wysocki, C. J., Dorries, K. M., & Beauchamp, G. K. (1989). Ability to perceive androstenone can be acquired by ostensibly anosmic people. *Proceedings of the National Academy of Sciences of the United States of America, 86*, 7976–7978.

Yeshurun, Y., & Carrasco, M. (1998). Attention improves or impairs visual performance by enhancing spatial resolution. *Nature, 396*, 72–75.

Yoshida, M. (1975). Psychometric classification of odors. *Chemical Senses, 1*(4), 443–464.

Zald, D. (2002). The human amygdala and the emotional evaluation of sensory stimuli. *Brain Research Reviews, 41*, 88–123.

Zald, D. H., Mattson, D. L., & Pardo, J. V. (2002). Brain activity in ventromedial prefrontal cortex correlates with individual differences in negative affect. *Proceedings of the National Academy of Sciences of the United States of America, 99*(4), 2450–2454.

Zoladz, P., Raudenbush, B., & Lilley, S. (2004). Impact of the chemical senses on augmenting memory, attention, reaction time, problem solving, and response variability: The differential role of retronasal versus orthonasal odorant administration. *Chemical Senses, 29*, supplement.

Zufall, F., & Leinders-Zufall, T. (2000). The cellular and molecular basis of odor adaptation. *Chemical Senses, 25*, 473–481.

7 Perception of Body Motion

Ben D. Lawson and Bernhard E. Riecke

CONTENTS

7.1   INTRODUCTION AND SCOPE

Coordinated locomotion requires matching efferent commands (e.g., to one's limbs) against afferent visual flow information and vestibular and somatosensory (skin, muscle, and joint) signals concerning changing forces on one's body (Guedry, 1992). If accurate afferent information about forces related to one's own movement was suddenly absent during walking, running, carrying, reaching, climbing, etc., one would fall over. Nevertheless, simulation-corroborating acceleration/motion cues are seldom available to users of simulators or virtual environments (VEs) *moving* through a simulated world, a situation that contributes to incoordination, imbalance, cybersickness, and simulator aftereffects. Even VE users riding *passively* in simulated automobiles or aircraft are forced to control the simulated vehicle via visual flow alone, without being able to match visual flow information to the (actual and expected) afferent information about the forces on their own bodies, as they do during real acceleration, deceleration, and turning in vehicles. Consequently, the virtual experience causes unwanted symptoms and fails to be realistic.

This chapter describes the known techniques for eliciting and manipulating perceptions of body acceleration and motion. The focus of discussion is on the user's requirements for perceiving body acceleration, motion, or tilt in a simulator or VE. Understanding how to create virtual acceleration and motion perceptions should improve the training effectiveness and believability of VEs while avoiding unwanted effects (such as motion sickness). This chapter details various methods for exploiting visual, auditory, vestibular, and somatosensory senses (and intermodal sensory interactions) to elicit perceptions of self-motion, self-tilt, or insertion of oneself into an unusual force environment (such as a nonterrestrial planetary body). The authors compare physical acceleration of the user to other (potentially complementary) methods, such as visually induced illusions of self-motion (vection) or somatosensory illusions of self-motion induced by locomotion without displacement (as occurs on some treadmill-based VE). The authors conclude that the most compelling perceptions of body acceleration in VEs will be achieved through mutually enhancing *combinations* of the vestibular, visual, and somatosensory stimuli such as those described in this report.

It is important at the outset for the reader to understand that a *virtual movement* in a VE does not necessarily have to be caused by an illusion of movement in a stationary user. It would be inaccurate to suppose that an acceleration or motion display is only *virtual* when it employs stimuli other than real acceleration to create an illusion of self-motion. By analogy, no one would argue that a visual display is only virtual when it does not stimulate the retina with light. Similarly, the feeling of virtual acceleration or motion refers not to the means by which the acceleration perception is generated, but rather to the resulting perceptual event experienced. Therefore, a successful acceleration/motion simulation will elicit the desired motion perceptions by whatever means best fits the virtual event being portrayed and the overall goals and limitations of the VE designer. The virtual effect desired may be created by generating an illusion of motion where none exists, by really moving the user but in a way that does not correspond exactly to the event being simulated, or by moving the user exactly according to the event being simulated. This last method is a particularly effective approach that employs externally generated acceleration stimuli and self-generated physical motions of the user to match the real acceleration profiles being simulated. This approach involves the use of isomorphic simulations (Grubb, Schmorrow, & Johnson, 2011; Lawson, Sides, & Hickinbotham, 2002), wherein users walk, drive, or fly in a safe and controlled real environment while perceiving themselves to be in a different (e.g., less controlled, more dangerous) VE, such as assisting with a medical evacuation or moving over or through disputed territory. The advantages and limitations of this approach are described and compared to other methods for simulating self-motion. Acceleration simulations that are designed correctly could be used to create more compelling and realistic experiences outside the user's immediate, veridical experience (e.g., training in a simulator building), but appropriate to the VE (e.g., flying in an aircraft).

Our premise is that the purpose of an acceleration simulation is to elicit perceptions of acceleration appropriate to the overall simulation goal by whatever combination of illusory or real motion is most effective. Assuming simulation of a real event is the purpose of a particular VE (which is not always the case), then to be deemed effective that VE should produce no significant adverse side effects unless they are also caused by the real event being simulated. Most importantly, the VE should provide good transfer of skills learned in VE training to performance in the situation being simulated. Provided side effects are appropriate and transfer of training is satisfactory, it would also be advantageous for the VE to elicit consistent and controllable acceleration perceptions that are difficult to distinguish from the real event being simulated.

Normal movement control relies on more senses than are typically stimulated by a VE. The perception of real body acceleration, motion, and orientation involves vision, audition, vestibular sensation, and somesthesia* (which refers to detection by tactile and kinesthetic somatosensory systems, including cutaneous, muscle, and joint receptors). Similarly, during simulated activities, the aforementioned sensory modalities can be exploited to elicit the perception of acceleration. For example, the motion of large visual fields around the user is known to act upon the vestibular nucleus (and associated structures) and thereby to induce a compelling illusion of self-motion (Dichgans & Brandt, 1978). A visually induced illusion of self-motion is commonly referred to as vection, a term attributed to Tschermak (1931). Vection can be considered an acceleration perception that indirectly implicates vestibular involvement. The use of vection as a means for simulating self-motion in a VE is treated in Section 7.2.1 and in Chapter 18 of this book (Hettinger, Schmidt, Jones, & Keshavarz, 2014).

Just as visual stimuli can provide acceleration/motion cues, so can somatosensory stimuli. For example, some evidence suggests that walking in place on a substrate that moves can be exploited to enhance the perception of naturalistic locomotion through a VE and can induce compensatory eye movements not unlike those elicited by real body motion (Bles, 1979; Brandt, Büchele, & Arnold, 1977; DiZio & Lackner, 2002; Lackner & DiZio, 1988). Like visual vection, walking in place can be considered an acceleration simulation if it leads the user to perceive self-acceleration or self-motion (Section 7.2.3.2; also Appendix A of Hollerbach, 2002; Lawson et al., 2002; Steinicke, Vissell, Campos, & Lecuyer, 2013).

Visual and somatosensory cues are considered in this chapter, but emphasis is placed on the vestibular modality. Since the vestibular organs are specifically designed to sense acceleration, a rudimentary understanding of human vestibular function is necessary in order to employ physical acceleration effectively as part of a VE. This chapter provides an introduction to the *vestibular channel* of the VE stimulus and details different methods for exploiting knowledge about the vestibular modality and cross-modal interactions to elicit perceptions of self-motion, self-tilt, and insertion of oneself into an unusual force environment (such as a nonterrestrial planetary body). These effects can be achieved by the use of moving simulators, centrifuges, and other acceleration

---

* The authors avoid the terms *proprioceptive* or *haptic* when referring collectively to skin, muscle, and joint senses, because *proprioception* includes the vestibular modality, while *haptic* refers to active manual exploration of the world via both cutaneous and kinesthetic senses.

devices (Stanney & Cohn, 2006) or by isomorphic real locomotion (Lawson et al., 2002). All of these *real-motion* methods tend to directly stimulate the vestibular organs. The vestibular modality should be exploited by VE designers to induce feelings of acceleration when they are desirable and to avoid feelings of acceleration when they are undesirable, yet are required because of the logistic constraints of the simulator and the space within which it operates. For example, moving-base simulators often employ a subthreshold motion during relatively quiescent periods in the simulation. This is done to slowly reposition the user and ready the motion platform to achieve the desired acceleration profile the next time a strong acceleration perception is required as part of the simulation. In discussing the vestibular channel in VEs, the essential qualities of the stimulus (acceleration) and the receptor (the vestibular end organ) will be briefly introduced. The authors will introduce ways in which the vestibular, visual, and somatosensory systems (and some of their cross-modal interactions) can be exploited to create, prevent, or modify acceleration/motion perceptions. We first discuss the general approaches for eliciting motion percepts without physical acceleration and then address the approaches that require physical acceleration. Finally, we detail specific ways in which knowledge concerning the vestibular system can be exploited systematically to induce perceptions of body tilt, rotation, translation, absence of tilt, absence of rotation, absence of translation, and changes in the magnitude of gravity.

7.2 ELICITING ACCELERATION/MOTION PERCEPTIONS WITHOUT PHYSICAL ACCELERATION OF THE USER

Physical motion of the user is the most effective and straightforward way to elicit perceptions of acceleration and self-motion. Unfortunately, this approach often requires significant time and money to implement. Large-scale moving-base simulators can cost millions of dollars and require considerable expertise, space, and safety precautions. Even seemingly simple free-space walking areas (Section 7.5.3) can pose considerable challenges due to the need to provide (1) low-latency and accurate tracking across the entire motion envelope (Foxlin, 2002), (2) high-quality naturalistic visuals spanning a large field of view (FOV), and (3) measures to ensure that users do not stray beyond the safe boundaries of the walking area (Sections 7.5.3 and 7.5.5). Thus, different options have been explored for eliciting illusory acceleration/motion perceptions without the need for physical acceleration of the user.

In this section, we will discuss how different stimuli and procedures can be used to cause VE users to experience sensations of self-motion and consider how these can and be combined with other corroborating cues to yield maximum benefit. For recent reviews of the relevance of self-motion illusions to VE, see Chapter 18 (Hettinger, Schmidt, Jones, & Keshavarz, 2014) and also Palmisano, Allison, Kim, and Bonato (2011); Riecke (2009, 2011); Riecke and Schulte-Pelkum (2013); and Väljamäe (2009). Excellent older reviews on visually induced vection are provided by Andersen (1986), Dichgans and Brandt (1978), Howard (1986a), Mergner and Becker (1990), and Warren and Wertheim (1990). The known methods for inducing illusions of self-motion are discussed in the following text.

7.2.1 VISUALLY INDUCED ILLUSIONS OF SELF-MOTION

Self-motion and displacement sensations can be elicited via tilting the frame of reference, moving a visual frame of reference relative to the observer, or the aftereffects of prolonged movement of a visual frame of reference, such as an *optokinetic drum* surrounding an observer (Dichgans & Brandt, 1978). These are promising methods if applied with a full knowledge of the psychophysics of human perception and the factors that give rise to visually induced discomfort (Lawson et al., 2002). Vection can be elicited by nonvisual stimuli, but visual cues have received the most attention in vection research to date, and the first accounts of how large-field visual motion can induce illusory embodied self-motion date back more than a century ago (Helmholtz, 1866; Mach, 1875;

Urbantschitsch, 1897; Warren, 1895; Wood, 1895). Many readers are familiar with the illusion described by Helmholtz (1866), wherein one can experience the illusion that one's train is starting to move when instead the train on the adjacent track has started to move.

Such self-motion illusions can be elicited for translations and rotations and have been termed linear and circular vection, respectively (Berthoz, Pavard, & Young, 1975; Fischer & Kornmüller, 1930; Tschermak, 1931). Adverse symptoms of motion sickness may occur, with at least 20% of people reporting stomach symptoms during prolonged exposure to a provocative optokinetic stimulus (Lawson, 2014, Chapter 23). Roll and pitch rotations of the visual field may be more disturbing than yaw rotations (Bles, Bos, Graaf, Groen, & Wertheim, 1998), since a visual illusion of body motion or position that does not match the actual upright position of a seated person creates a *sensory conflict* between visual and proprioceptive (vestibular and somatosensory) cues (Reason & Brand, 1975). Roll or pitch optokinetics stimuli also can elicit paradoxical illusions of continued body rotation despite a limited sensation of body tilt not usually exceeding approximately 20° (Allison, Howard, & Zacher, 1999; Held, Dichgans, & Bauer, 1975; Young, Oman, & Dichgans, 1975).

For many simulations, it is essential to ensure low incidence and severity of adverse symptoms such as simulator sickness, which sometimes correlates positively with vection, as discussed in Chapter 26 (Keshavarz, Hecht, & Lawson, 2014), Chapter 18 (Hettinger, Schmidt, Jones, & Keshavarz, 2014), and elsewhere (Hettinger, Berbaum, Kennedy, Dunlap, & Nolan, 1990; Kennedy et al., 2003; Lawson, 2005; Palmisano, Bonato, Bubka, & Folder, 2007). The causal relation between vection and sickness is a subject of debate and the evidence is mixed. For example, it has been observed that simulator sickness increases for larger visual–vestibular cue conflicts, whereas vection is enhanced by reduced cue conflicts (Kennedy et al., 2003; Palmisano et al., 2007). A number of other factors have been found to facilitate and strengthen visually induced vection, as detailed by Andersen (1986), Dichgans and Brandt (1978), Howard (1986a), Mergner and Becker (1990), Riecke (2011), Riecke and Schulte-Pelkum (2013), and Warren and Wertheim, 1990). In the following section, we will discuss some of these factors.

Increasing the FOV covered by the moving visual stimulus generally enhances vection to a point where full-field stimulation can yield sensations of self-motion that are difficult to distinguish from actual self-motion (Berthoz et al., 1975; Brandt, Dichgans, & Koenig, 1973; Dichgans & Brandt, 1978; Held et al., 1975; Lawson, 2005). However, small fields can induce vection also when they are radially expanding (Andersen & Braunstein, 1985).

Higher stimulus velocity (but not necessarily higher acceleration to reach that constant dwell velocity) generally enhances vection, up to 120°/s for circular yaw vection (Allison et al., 1999; Brandt et al., 1973; Dichgans & Brandt, 1978; Howard, 1986a). Furthermore, vection can be enhanced by increasing the amount of optical texture (Brandt, Wist, & Dichgans, 1975; Dichgans & Brandt, 1978). For applications where there is little optic flow (e.g., flight or space travel simulations where there are few nearby objects), this could be a challenge, and one may need to carefully add ecologically valid visual elements to provide sufficient optic flow. For example, it may be necessary to add a few clouds to a flight simulation.

Forward linear vection can also be facilitated by adding moderate simulated camera shake (*viewpoint jitter*) to the simulations (Palmisano et al., 2011; Palmisano, Gillam, & Blackburn, 2000). This finding from basic research could be employed in motion simulations by carefully adding (instead of filtering out or ignoring) some naturally occurring viewpoint jitter, for example, from bouncing head motions during walking (Ash, Palmisano, Apthorp, & Allison, 2013; Bubka & Bonato, 2010) or vehicle shaking due to motor vibrations or rough roads.

Illusory self-motion may arise instantaneously but usually requires several seconds. When 45 experiment participants were each completely immersed in a slowly turning optokinetic sphere (Lawson, 2005), 16 reported vection within 3 s of opening their eyes. Therefore, under optimal conditions, only about 36% of simulator/VE users would be expected to get vection without an appreciable delay. In fact, 14 s was the average delay in Lawson. A key challenge when attempting to employ self-motion illusions to enhance VEs is the need to reduce the onset latency to an acceptable value. To this end, it can be useful to consider not only *bottom-up*, visual stimulus parameters but also intermodal facilitations (Section 7.6; Riecke & Schulte-Pelkum, 2013) and higher-level/cognitive or *top-down* factors (Section 7.8; also Riecke, 2009, 2011; Riecke & Schulte-Pelkum, 2013; Riecke, Västfjäll, Larsson, & Schulte-Pelkum, 2005; Seno, Ito, & Sunaga, 2009; and Väljamäe 2009).

7.2.2 AUDITORY ILLUSIONS OF SELF-MOTION

Self-motion sensations can be elicited via motion of auditory surrounds and the aftereffects thereof (Lackner, 1977). Blindfolded listeners can perceive *audiokinetic* illusions of self-motion from moving sound fields (Dodge, 1923; Lackner, 1977; Marme-Karelse & Bles, 1977). While visually induced vection can be induced in all users and can yield strong self-motion illusions that may be indistinguishable from physical self-motion, auditory vection tends to be weaker and more variable. Auditory vection occurs in 20%–75% of blindfolded listeners, depending on various stimulus factors (Väljamäe, 2009). For example, Larsson, Västfjäll, and Kleiner (2004) reported circular vection in 23%–50% of blindfolded listeners when single sound sources were simulated and 28%–66% when multiple sound sources were simulated (depending on the type of sounds used). Adding moving sound fields has been shown to enhance both visually induced vection (Riecke, Schulte-Pelkum, Caniard, & Bülthoff, 2005; Riecke, Väljamäe, & Schulte-Pelkum, 2009) and kinesthetic vection elicited by stepping along a rotating floor platter or *circular treadmill* (Riecke, Feuereissen, Rieser, & McNamara, 2011). Thus, moving auditory cues can support and enhance self-motion illusions. Given the relative ease and affordability with which high-fidelity sound can be simulated, adding spatialized sound should be considered, since it is likely to increase the overall realism of a simulation and also facilitate perceived self-motion through a VE.

7.2.3 SOMATOSENSORY ILLUSIONS OF SELF-MOTION

7.2.3.1 Tactile Illusions due to Contact with a Moving Frame of Reference

Acceleration perceptions can be elicited via moving touch and pressure cues and the aftereffects thereof (DiZio & Lackner, 2002; see also Brandt et al., 1977; Lackner & DiZio, 1984; Lackner & Graybiel, 1977). This is a promising method, especially when used in conjunction with confirmatory information from other sensory modalities. Relatively little is known about this approach to creating VE, compared to visual or auditory methods.

7.2.3.2 Kinesthetic Illusions due to Limb Movement

The literature concerning kinesthetic illusions due to limb movement is complex and the findings vary with the experimental situation. Self-motion perceptions can be elicited by voluntary kinesthetic behaviors such as locomotion without displacement (walking in place), stationary pedaling, or moving one's arm with a rotating surround (as well as the aftereffects of these behaviors) (Brandt et al., 1977; Dizio & Lackner, 2002; Hollerbach, 2002; Lackner & DiZio, 1988, 1993). For example, when people sit stationary inside a slowly rotating optokinetic drum in complete darkness, they may perceive compelling illusory self-rotation (*arthrokinetic circular vection*) when they touch the rotating surrounding cylinder and follow its motion with one hand such that their arm moves about the shoulder joint (Brandt et al., 1977). Although this condition was called *passive*, it was accomplished by reaching out with an unsupported arm and maintaining voluntary contact with the drum as it turned under its own power. Under these conditions, circular vection occurs within 1–3 s and is reportedly indistinguishable from actual self-motion. Arthrokinetic nystagmus and aftereffects for nystagmus and circular vection were observed as well. Similar to kinesthetic circular vection caused by walking in place, illusory self-rotation is in the direction opposite of the rotating cylinder (Bles, 1981; Bles & Kapteyn, 1977; Brandt et al., 1977).

For stimulus rotations greater than 10°/s, Brandt et al.'s participants performed the same voluntary unsupported reaching out and following of the drum's motion but used both arms and placed them alternately on the inner side of the rotating cylinder, similar to a walking motion. Contrary to what has been implied by a few secondary literature sources, this multiarm hand-walking motion did not cause appreciable arthrokinetic vection or nystagmus (albeit no descriptive or inferential statistics were provided). This differs from circular treadmill walking, where stepping along in time with a moving platform elicits vection (Bles, 1981; Bles & Kapteyn, 1977; Riecke et al., 2011). Actively stepping to push the platform backward with both feet alternately also seems to elicit vection (anecdotal observations by Riecke).

A few observations may partially explain the different findings for two-hand walking versus two-foot stepping around: (1) Human locomotion is usually performed with the legs, whose rhythmic motion has a close association with human movement through the world (albeit not by stepping in a circle while seated); (2) The ground surface beneath one's feet provides a strong frame of reference that is usually stationary during locomotion; (3) In the aforementioned studies of seated stepping, the legs moved through a smaller arc than the arc the two arms traversed in Brandt et al., so greater rhythmic limb alternation would occur during stepping, which would corroborate the perception of body locomotion. We recommend the evaluation of these factors in future studies.

Bles and colleagues demonstrated that arthrokinetic vection can occur for linear sideways motion, although only in about 37% of trials (Bles, Jelmorini, Bekkering, & de Graaf, 1995). Blindfolded, seated participants were asked to use hand-walking motions to stay in contact with a flat table-like surface that moved at constant (0.1 m/s^2) acceleration in the frontoparallel plane (left/right). While these findings are promising, there has been little attempt to exploit them for motion simulation in a VE. This may be because of the lack of naturalness of this method and the fact that the hands must be free to carry out tasks. Nevertheless, this could be a useful adjunct to simulations of crawling or climbing in a VE.

While it is not clear how well treadmill walking works for eliciting vection, there is clear evidence that stepping in place along a rotating floor platter can induce or contribute to circular vection (if users walk in place or step around near the center of the rotating floor) or curvilinear vection (if users walk off-center) in the direction opposite of the floor motion (Bles, 1981; Bles & Kapteyn, 1977; Bruggeman, Piuneu, Rieser, & Pick, 2009; Lackner & DiZio, 1988; Riecke et al., 2011). Such apparent stepping around can provide compelling self-motion illusions and is even accompanied by nystagmus and Coriolis-like (pseudo-Coriolis or pseudo-Purkinje) effects when participants perform active head tilts (Bles, 1981; Bles & Kapteyn, 1977).

Feeling like one is actually walking forward through a VE may enhance realism and presence. Given the increasing availability and affordability of linear treadmills, they are becoming more common in VE applications. They allow for fairly natural locomotion, at least in the forward direction (Steinicke et al., 2013). The biomechanics and coordination of walking on a treadmill are more similar to natural walking than walking in place on a stationary floor (another approach that has been used in VEs), especially in advanced treadmills that use force-feedback harnesses (Hollberbach, 2002; Steinicke et al., 2013). Nevertheless, somatosensory cues from walking on even advanced linear treadmills are not identical to real walking, which requires more backward-directed force against the substrate to maintain forward locomotion and requires different forces to start, stop, and turn compared to a treadmill. It is not clear whether linear forward walking on a treadmill is sufficient to elicit a compelling sensation of self-motion. For example, Durgin et al. (2005) pointed out that "during treadmill locomotion, there is rarely any illusion that one is actually moving forward" (p. 401).

Unfortunately, adding velocity-matched treadmill walking to a visual motion simulation does not always enhance self-motion perception (Onimaru, Sato, & Kitazaki, 2010; Seno, Ito, & Sunaga, 2011; but see Lackner & Dizio, 1988). It can sometimes impair self-motion perception, even though it is designed to correspond to our natural (eyes-open) walking experience (Ash et al., 2013; Kitazaki, Onimaru, & Sato, 2010; Onimaru et al., 2010; see also discussion in Riecke & Schulte-Pelkum, 2013). It is possible that the additional head motions associated with walking decrease visual vection (Lackner & Teixiera, 1977; Teixeira & Lackner, 1979; Viirre, Sessoms, & Gotshall, 2012). It is also possible that visual and tactile/kinesthetic cues for self-motion must be very closely matched to be interpreted as synergistic rather than conflicting. However, the pattern of findings is complicated, because Lackner and DiZio (1988) did not observe walking on a circular treadmill to weaken concordant visual vection, although interpretations of body motion were reported, which *captured* visual vection in favor of the kinesthetic cues when the two were put in opposition. Further research is needed to better understand the interactions between visual and tactile/kinesthetic walking cues during circular apparent stepping around and linear treadmill walking. Until then, it seems advisable to carefully evaluate VE walking applications to ensure that the cues from walking provide the intended benefit. A more extensive discussion of walking in VE can be found in Steinicke et al. (2013). It is already clear that rotational, but not necessarily translational, self-motion can be elicited merely by walking in place in the dark. Moreover, recent evidence shows that walking motions on a circular treadmill can enhance visually induced circular vection (Freiberg et al., 2013).

7.2.3.3 Kinesthetic Illusions due to Vibration of Localized Body Regions

Self-motion perceptions can be elicited by vibration of localized body regions. Vibration of skeletal muscles and tendons involved in human postural coordination can elicit a variety of self-motion perceptions (Lackner, 1988; Lackner & Levine, 1979; Levine & Lackner, 1979), most likely because the vibration-induced activity in the muscle spindles is interpreted by the central nervous system as muscle lengthening associated with movement around a body joint. For example, vibration of the Achilles tendons of a standing, restrained observer can trigger an illusion of forward pitch about an ankle-centered axis. Skeletal muscle vibration can elicit apparent motion perceptions of many different types but will not be employed as a primary method of inducing motion perceptions in VE because it requires dark or near-dark conditions as well as precise placement and pressure of the vibrator and appropriate restraint of the limb to achieve an optimal effect. Nevertheless, the technique is inexpensive to implement and should be explored further to determine whether it can strengthen illusions of self-motion elicited by other means.

7.2.4 Methods for Eliciting Vestibular Illusions without Accelerating the User

7.2.4.1 Caloric Stimulation

Caloric stimulation of the ear canal via irrigation with ice water (National Research Council, 1992) causes a change in temperature on one side of the semicircular canal, which triggers convection currents that can elicit a sensation of turning. This is an effective method that can be applied unilaterally, but it is somewhat invasive and quite unpleasant, and the specific motion perception is difficult to alter from moment to moment.

7.2.4.2 Galvanic Vestibular Stimulation

Electrical stimulation of the vestibular nerve fibers is an effective way to elicit movement sensations and balance reflexes (St. George & Fitzpatrick, 2011), but when done invasively (as in the animal model), it is not a promising method for general (nonprosthetic) use in humans. Fortunately, low-intensity galvanic vestibular stimulation (GVS) can be achieved by placing electrodes in the mastoid region just behind the ear, and this may be a method for achieving somewhat controlled velocity perceptions in normal humans (Lenggenhager, Lopez, & Blanke, 2008; Etard, Normand, Pottier, & Denise, 1998). At a neural level, applying a small current bilaterally to the mastoids modulates vestibular firing rates, decreasing them on the anodal side and increasing them on the cathodal side. This presumably is caused by GVS conveying a signal of head motion toward the cathodal side of stimulation (in the absence of an efferent command to move the head), which in turn leads to an automatic adjustment response toward the anodal side (St. George & Fitzpatrick, 2011; Wardman & Fitzpatrick, 2002; Wardman, Taylor, & Fitzpatrick, 2003). This GVS method does not place electrodes inside the skull, but it uses currents that are salient and can be disturbing. Fortunately, even low currents of GVS can be used to affect gait, presumably by altering the perception of self-motion (Bent, McFadyen, French Merkley, Kennedy, & Inglis, 2000). GVS also can modulate the strength of visually induced self-motion illusions (Maeda, Ando, & Sugimoto, 2005) and affect the perceived trajectory of visually induced illusory self-motion (Lepecq et al., 2006).

GVS has wider applications beyond the simple elicitation of a self-motion illusion. GVS can systematically alter the perceived path during active walking and passive wheelchair transport (being pushed by an experimenter; Fitzpatrick, Wardman, & Taylor, 1999). Similarly, GVS can be used to modify the walking direction of blindfolded participants, who will tend to turn toward the side on which the anodal electrode was placed (Fitzpatrick et al., 1999). While dynamically adjusting GVS can be used to *remotely control* blindfolded users' walking trajectories, the effectiveness of this approach depends on one's head orientation with respect to gravity and is maximal when the head is tilted forward by 72° (St. George & Fitzpatrick, 2011). Raising one's head to be upright results in sideways stumbling to regain balance, but no further orientation responses due to GVS. This limits this potential VE application of GVS to situations where head orientation is either fixed or can be dynamically monitored and kept within a useful range.

Responses are fairly predictable for different head orientations. When the head is slightly tilted backward, GVS will induce sideways (lateral) sway toward the anode in standing humans (St. George & Fitzpatrick, 2011). Correspondingly, GVS will yield sagittal (forward/backward) sway when one's head is tilted 90° to the left or right. Finally, sensations of horizontal rotation (but no sway) can be induced by GVS when one's head is tilted forward (by 72°).

7.2.4.3 Drug Effects

Dysfunctions in a given vestibular labyrinth or imbalances in the normal interaction of the respective labyrinths (such as may result from inner ear infections, Ménière's syndrome, or ototoxic drug treatments) can create feelings of movement. It would be unethical to intentionally produce such invasive effects by similar means because of the long-term adverse effects. The vestibular effects of certain drugs such as alcohol (Fetter, Haslwanter, Bork, & Dichgans, 1999) are reversible, but not pleasant or easily controllable. Moreover, such drugs produce unpleasant side effects and raise safety concerns. Overall, current drugs are not a promising approach for eliciting illusions of self-motion.

7.2.4.4 Low-Frequency Sound Effects

Low-frequency (e.g., below 20 Hz) acoustic vibrations of high intensity are not necessarily audible but may act upon the vestibular organs of healthy people (Parker, Tubbs, Ritz, & Wood, 1976) and people with vestibular pathologies (Minor, Solomon, Zinreich, & Zee, 1998), eliciting vertigo. This is currently not a promising method for eliciting acceleration perceptions because loud sounds (125 dB and above; Parker et al., 1976) are required to get only a portion of users vertiginous, because unpleasant symptoms can result, and because long-term consequences to auditory and vestibular organs could be adverse (Lawson & Rupert, 2010).

7.2.4.5 Other *Vestibular* Effects Not Requiring Body Acceleration

Eliciting vertigo via looking over a visual cliff (Brandt, 1999), via postural hypotension (Baloh & Honrubia, 1979), or via atmospheric pressure changes (Benson, 1988) is effective for producing nonspecific feelings of instability, dizziness, or falling but is not conducive to controllable motion perceptions and would give rise to safety concerns for certain individuals. The vestibular or visual–vestibular systems play a role in these effects.

7.3 ELICITING ACCELERATION OR MOTION PERCEPTIONS

The various methods for creating the perception of displacement, tilt, velocity, acceleration, or abnormal G-force upon one's body are enumerated briefly in this section. Not all of these methods are practical, nor do all of them involve imposing real accelerations on the user. Nevertheless, the most obvious way to elicit the perception of dynamic body movement through a VE is by real physical motion of the user that is perceived as such. Physical motion can be achieved passively with moving devices such as moving-base simulators, centrifuges, and real vehicles. Physical motion also can be achieved by allowing the user to move physically through a real environment that serves as an ambient context for the VE. Allowing for real user motion is a very effective and controllable way to create acceleration perceptions in a VE, provided the VE customer can afford it and the simulations are designed by specialists who understand simulators, the psychophysics of human acceleration sensations, and the factors that give rise to motion-induced discomfort.

7.3.1 Changing Perceived Body Weight Using Real Acceleration Stimuli

Physical acceleration of the user is not always perceived as such. Acceleration that is prolonged (e.g., aboard a centrifuge) can lead to the perception of altered body weight instead. Perceived addition or subtraction of body weight during movement through space can be used to enhance the simulation of high-performance vehicles or nonterrestrial environments (Guedry, 1974; Guedry, Richard, Hixson, & Niven, 1973; Lackner & Graybiel, 1980). It is important for the reader to realize that acceleration perceptions one can elicit in a VE are not limited to feelings of movement through space. Since gravity is one type of acceleration (according to Einstein's stated equivalence of inertial and gravitational force; see de Broglie, 1951), the perception of unusual G-forces on one's body qualifies as an acceleration perception also.

It is also possible to cause a temporary illusion of changed body weight (or at least the amount of effort required to move the body) via a simple trampoline. Prolonged jumping on a trampoline causes an altered sensation of effort required to jump (partly because one feels *too heavy*) immediately afterward (Marquez et al., 2010). This is presumably due to a rapid recalibration of the meaning of vestibular and somatosensory in reference to one's own motor commands, similar to that which occurs during knee bends performed during the high-G phases of parabolic flight (Lackner, 1992).

7.3.2 Illusions of Self-Motion Associated with a Change of Velocity

Changing the motion velocity or coming to a complete halt can induce compelling self-motion illusions and motion aftereffects, especially when no visual or other cues are available to disambiguate the situation (Guedry, 1974). For example, when being rotated on center in the earth-vertical z-axis at constant velocity in darkness, vestibular signals from the horizontal semicircular canals will fade over time. When this constant-velocity rotation is suddenly stopped, one can perceive a strong (illusory) rotation in the direction opposite of the original motion, despite being physically stationary. This is a useful adjunct to other methods, especially when combined with confirmatory information from other sensory modalities. For example, a rotation aftereffect indicating that one is turning to the right can be used to hasten and strengthen a desired visual vection illusion of turning to the right, as is frequently done in a military acceleration device used for ground-based disorientation training (Guedry, 1980).

7.3.3 Illusions of Self-Motion Associated with Multiaxis Rotation

Self-rotation perceptions can be produced by prolonged constant-velocity rotation in one axis followed by a head movement in another, nonparallel axis (Guedry & Benson, 1978). Under certain circumstances (e.g., after prolonged constant-velocity rotation without good visual references concerning the outside world), this stimulus elicits the perception of angular head velocity in a third axis orthogonal to the axis of the body rotation and the head rotation. This is a useful approach only when the resulting Coriolis cross-coupling stimulus is relatively mild or the user is resistant to motion sickness. Pronounced discomfort can arise among susceptible individuals making repeated, large, or rapid head movements after prolonged body rotation, especially when body rotation occurs at high velocity. However, veridical earth-referenced cues concerning one's true motion can lessen the unpleasant effects (Lawson, Guedry, Rupert, & Anderson, 1994; Lawson, Rupert, Guedry, Grissett, & Mead, 1997). Unfortunately, these helpful cues are likely to reduce the perceptual effects of illusory rotation as well. Therefore,

they are mostly useful for diminishing sickening effects when passive or voluntary head movement must be performed to accomplish other (e.g., training-related) purposes of the simulation.

7.3.4 ILLUSIONS ASSOCIATED WITH STATIC OR DYNAMIC BODY TILT

Alterations in perceived orientation can be induced via static body tilt and the aftereffects thereof.
In this case, the acceleration stimulus is provided by gravity. This is a useful adjunct to other methods in this section, but simple body tilt by itself elicits illusory effects that are mainly restricted to
the perception of the orientation of visual targets. Dynamic body tilt is, however, frequently used in
moving-base motion simulations, for example, when users are tilted backward and thus pressed
against their seat to simulate physical forward accelerations that cannot be veridically performed
due to the limited motion envelope of most platforms. Since otoliths by themselves may not readily
distinguish between inertial (acceleration) and gravitational forces, the human system readily uses
information from other sensory modalities to disambiguate gravitoinertial forces, a process that can
be modeled by Bayesian sensor fusion (MacNeilage, Banks, Berger, & Bülthoff, 2007). This can
be used to widen the coherence zone in which cross-sensory conflict remains unnoticed or at least
has little detrimental impact (Steen, 1998) and can help to mask imperfections in motion cueing and
motion washout filtering in moving-base simulators.

7.3.5 WHOLE-BODY VIBRATION EFFECTS

Subtle vibrations of the participants' seat may not by themselves be able to reliably induce sensations of self-motion, but they have been found to facilitate visual (Riecke et al., 2005; Schulte-
Pelkum, 2007) and auditory vection (Riecke, Feuereissen, & Rieser, 2009; Riecke et al., 2005;
Väljamäe, Larsson, Västfjäll, & Kleiner, 2006), which are discussed in the following texts. Since
typical vibrations (and to some degree, special infrasound vibrations) can be applied in motion
simulations relatively cheaply and easily, they should be added whenever feasible to enhance the
overall realism of the simulation. This should be done with an awareness of the effects arising from
localized head vibration, per se (Lackner & Graybiel, 1974).

7.4 SPECIFIC CATEGORIES OF ACCELERATION/MOTION PERCEPTIONS INVOLVING ACCELERATION STIMULI TO THE VESTIBULAR MODALITY

To understand how best to elicit the desired perceptions in a simulator or VE, one should start with
a basic knowledge of how perceptions are formed in the natural world. To understand how perceptions are formed, one must begin with the stimulus, that is, the qualities in the physical world that
are detected by the sensory system of interest. If one wishes to meet the user's *visual requirements*
with a display, one manipulates the intensity and wavelength of the light reaching the user's eye so
as to match the stimuli that would be received if the viewer were physically present inside the virtual
world. Similarly, the amplitude and frequency of vibration of air molecules are sensed by the auditory modality and then perceived by the listener as loudness and pitch. Air vibrations of sufficient
magnitude can be sensed by other sensory modalities as well, but the auditory modality is specifically designed to detect them. In the case of motion perception, the usual stimulus is acceleration
(or more properly the forces to the body resulting from acceleration), which is sensed or inferred by
multiple sensory systems, among which the vestibular system is specifically designed with this as
its primary purpose. Therefore, if one wishes to meet the users' *vestibular requirements*, one either
manipulates the acceleration stimulus directly or finds some other way to stimulate the vestibular
system in a way comparable to acceleration.

In any VE, certain system requirements must be met to generate an appropriate stimulus, while
critical user requirements dictated by the functional properties of the user's perceptual apparatus
must be met for the VE to be compelling and effective. Manipulating the quality of the stimulus

itself is only one aspect of a good simulation. It is equally important to understand the functioning and limitations of the sensory modalities one is stimulating to evoke a synthetic experience effectively. Sometimes it may not be possible or desirable in a VE to recreate the stimuli the users would receive if they were in the real world (e.g., harmful forces to the body). In cases where it is desirable but does not appear possible, a solution might be finessed by exploiting known principles of sensory functioning.

Introductory information concerning the functioning of the vestibular system is presented in this section. Such information can be exploited to enhance feelings of acceleration, force, and self-motion. The human vestibular modality is evolved to detect accelerations; it can do so in the absence of visual or somatosensory cues, but these cues are present most of the time during self-generated movements in the natural world. The sensory qualities to which the vestibular modality is sensitive include linear and angular accelerations due to self-motion and gravity. Linear and angular accelerations are detected by the otolith organs and the semicircular canals, respectively. These organs are described briefly as follows. (For detailed review, see the recommended readings at the end of this chapter.)

The otolith organs of the inner ear detect linear accelerations of the head; they also sense angular motion that produces changes in the angle of the perceiver's body relative to the earth's gravity vector. The otolith organs detect changes in the magnitude and direction of linear gravitoinertial force vectors via hair cell mechanoreceptors, which are embedded in gelatinous membranes containing tiny crystals of calcium carbonate called otoconia. The otoconia are more dense than the surrounding medium and hence lag behind motions of the head very slightly due to their inertia, causing a corresponding deformation of the hair cells.

Three semicircular canals located in each inner ear code angular motion in three roughly orthogonal axes. The semicircular canals are fluid-filled rings with a gelatinous structure called the cupula obstructing each ring. The gelatinous cupula is embedded with hair cells that are deformed when the cupula is itself deformed by very small changes in fluid pressure caused by angular acceleration of the semicircular canals. This slight deformation of the cupula is caused by the inertia of the fluid in the canal, which wants to remain at rest when the head is turned.

Collectively, the semicircular canals and otolith organs make up the equilibrium organs of the nonauditory portion of the human labyrinth. The semicircular canals and otolith organs interact with one another (and with other sensory modalities) during the processing of complex stimuli involving simultaneous angular and linear acceleration. The mutual interaction of otoliths and canals provides one explanation for why the perception of self-tilt during off-center rotation (i.e., rotation around a central axis that is located at some radial distance from one's body, as occurs on a merry-go-round) initially lags behind the stimulus given by the rate of change in the direction of the resultant gravitoinertial force vector (Graybiel & Brown, 1950; Lawson, Mead, Chapman, & Guedry, 1996). This lag effect is often attributed to suppression of the otolith system by the angular acceleration signal being processed and transmitted by the semicircular canals (Lawson et al., 1996). Numerous other intralabyrinthine interactions can be identified (Guedry & Benson, 1978; Reason & Brand, 1975) as well as important interactions among the labyrinthine and nonlabyrinthine sensory modalities, including vision (Dichgans & Brandt, 1978), audition (Lackner, 1977), and somesthesia (Bles, 1979; Lackner & Graybiel, 1978). Finally, purely otolithic mechanisms for distinguishing tilt from translation have been proposed (Wood, 2002).

The vestibular system is involved in several aspects of human functioning besides the perception of acceleration, including the reflexive control of gaze stabilization (to maintain clear vision during head movement), head righting, postural equilibrium, coordinated locomotion, certain reaching behaviors, and even wayfinding (Goldberg et al., 2011). Functioning vestibular organs are probably critical to the elicitation of unpleasant symptoms such as motion sickness induced by real or apparent motion (Cheung, Howard, & Money, 1991). A comprehensive explanation of vestibular function, neurovestibular pathways, or the psychophysics of vestibular perception is not necessary for the reader to gain a rudimentary grasp of the practical options available for eliciting acceleration perceptions in VEs. (The Appendix for this chapter provides a list of recommended readings for those who wish to learn more about vestibular function and psychophysics.) The most practical options available for eliciting acceleration perceptions are evaluated in the following text.

The vestibular end organs are stimulated directly by many of the methods of eliciting acceleration/motion perceptions covered in Section 7.3, including physical tilt or motion of the user, rotation aftereffects, caloric stimulation, certain acoustic vibrations, physical vibrations to the cranium or whole body, electrical stimulation of the mastoid region, vestibular diseases, and drug or alcohol effects impacting the end organs. Many other acceleration perceptions arise by stimulating the vestibular nucleus and associated structures (Goldberg et al., 2011) without directly stimulating the vestibular end organs themselves (e.g., via a moving visual field, which indirectly stimulates vestibular brain centers). Several of the aforementioned methods for inducing acceleration perceptions without using a direct stimulus to the vestibular end organ are promising and should be explored further. Even if many of them do not prove to be viable ways of eliciting consistent and controllable acceleration perceptions in all situations, they must be understood if VE designers are to create effective acceleration simulations that are not inadvertently corrupted by interference from competing stimuli from visual or other modalities. Moreover, using such methods can reduce the requirements for physical motion of the user, which can help to reduce overall space requirements, technical complexity, and cost.

Three promising approaches for eliciting acceleration/motion perceptions were compared in detail in Table 7.1 of Lawson et al. (2002): (1) physical motion of the user (either passively or actively produced), (2) visual stimuli that induce illusions of self-motion, and (3) illusory self-motion induced by locomotion without displacement. Lawson et al. (2002) evaluated the relative advantages and limitations of each of these approaches along 12 dimensions, among which were the range of situations that could be simulated, how compelling or realistic the resulting simulation is, the degree to which the user's activities must be restricted for the desired perception to be elicited, the discomfort associated with the stimulus, and logistic considerations such as expense, technical difficulty, and space requirements. For example, real body acceleration/motion of the user (via moving devices or active real locomotion through a VE) offers the greatest range of situations it can simulate and the greatest realism but is generally the most logistically challenging method to employ and the most likely to induce discomfort if not implemented carefully.

Optimal acceleration simulations will coordinate a variety of vestibular and nonvestibular stimuli so that different cues confirm one another and mutually enhance the simulation (Section 7.6). Presently, the authors recommend that the best VEs for simulating self-motion perceptions will entail simulation-appropriate combinations of (1) real physical motion cues (including motion/tilt devices, vibration, and redirected locomotion), (2) visual, (3) auditory frame-of-reference motion, and (4) illusory locomotion (e.g., walking in place, where real locomotion is not feasible). In addition, the first author believes that judicious application of GVS could be useful in some VE applications (e.g., to simulate recognized spatial disorientation in a flight simulator). If virtual displays are eventually to fulfill the promise of their initial hype and become fully immersive and compelling, it will likely be via a careful combination of the methods outlined earlier.

An important step toward developing a good acceleration simulation involving the vestibular modality is to gain some familiarity with the various vestibular acceleration sensations that are possible and how each one is generated. A wide variety of acceleration perceptions can be elicited via stimulation of the vestibular system, some of which are illusory perceptions. In the succeeding text, most of the categories of acceleration/motion/tilt perceptions involving the vestibular modality are listed, along with a brief description of the stimuli that can cause them. Most of the perceptions described in the following are especially salient when veridical information from nonvestibular modalities is absent (e.g., a helpful earth-referenced visual cue from the outside world is lacking that would have diminished the vestibular effect). The following list of illusions should prove useful for helping VE designers know how to elicit the desired illusions. Just as important, the succeeding information will guide VE designers in knowing which stimuli to avoid in order to prevent unwanted illusions from occurring. The reader should note that the somatosensory system is stimulated also in most of the cases that follows, but the discussion centers on the vestibular system because the dynamics of the vestibular system are important to the dynamics of the illusions and perceptions discussed. An important consideration for most of the methods discussed as follows is that the stimuli can be sickening and so should be used with a full understanding of vestibular reactions and careful design of the motion profile and other sensory conditions of the simulation (e.g., to achieve the purposes of the simulation while limiting the magnitude, duration, and frequency of use of the stimuli).

7.4.1 Perception of Tilt

The simplest way of achieving the feeling of body tilt (i.e., misalignment with the earth vertical) in a VE is to physically tilt the user in the real world at the same speed and by the same amount. The effective stimulus is the change in the direction of the gravity vector relative to the user. When this is not possible or the effect needs to be magnified, one can produce or enhance the feeling of tilt in various ways. For example, a feeling of tilt can be produced by the presence of a linear (including centripetal) acceleration vector not aligned with the gravitational acceleration vector. In a laboratory setting, stimuli for eliciting tilt perceptions in this way include off-center rotation about the vertical axis (leading to a feeling of tilting away from the central axis of rotation) and horizontal linear oscillation at low frequency (leading to a feeling of tilting away from the gravity vector during each deceleration that must accompany a reversal in direction).

In aerospace operations, strong accelerations of the aircraft (e.g., forward thrust during jet-assisted takeoff) can induce a false perception of tilt known as the *somatogravic illusion* (Gillingham & Krutz, 1974). Similarly, an overall increase of the net downward head-to-seat (+Gz)* force (e.g., during a banking turn such as occurs in an aircraft) can cause a complex tilt sensation during head movement (known as the *G-excess effect*; Guedry et al., 1973; Guedry & Rupert, 1991; Chelette, Martin, & Albery, 1992). Rolling back to the straight and level after a prolonged banking turn can cause a pilot to feel a false perception of tilt known as *the leans* (Gillingham & Krutz, 1974).[†] The feeling of self-inversion is a special case of perceived tilt. A perception of inversion can be induced in a VE by turning the user upside down, or it can be done by less direct means. Turning a user upside down in a VE could augment the ground-based simulation of the inversion illusion that may occur in real aviation operations when a transition from higher to lower (or negative) G-force occurs (e.g., climbing followed by rapidly leveling off or diving; entry into weightlessness; Graybiel & Kellog, 1967). The same technique could be used to simulate the inversion illusions that can occur during the sudden removal of downward forces that occurs upon transition to microgravity during parabolic flight or entry into space flight (Nicogossian, Huntoon, & Pool, 1989).

7.4.2 Perception of Rotation

The simplest way to achieve a feeling of rotation (or other curvilinear motion) in a VE is to move the user in a rotating (or other curved) path that exactly duplicates the intended simulation. When an

---

* Positive Gz refers to a downward force (caused by upward acceleration) along the longitudinal axis of the head and body. Negative Gz refers to an upward force (caused by downward acceleration) in the longitudinal axis. The three cardinal head axes used in vestibular research are as follows (Hixson et al., 1966): (1) The z-axis is a dorsal–ventral (or cephalocaudal) line traveling along the intersection of the midcoronal and midsagittal planes and is commonly called *yaw* (for angular acceleration about this axis) or *heave* (for linear acceleration within this axis); (2) the x-axis is an anterior–posterior (or naso-occipital) line traveling along the intersection of the midsagittal and midhorizontal planes and commonly called *roll* (for angular acceleration) or *surge* (for linear acceleration); and (3) the y-axis is an interaural line traveling along the intersection of the midcoronal and midhorizontal planes and commonly called *pitch* (for angular acceleration) or *sway/sideslip* (for linear acceleration).

† Related phenomena such as the *graveyard spin*, *graveyard spiral*, and *giant hand phenomenon* were originally described by researchers such as Graybiel, Benson, and Gillingham. A history of this area of research and description of each illusion is provided in Previc and Ercoline (2004).

illusion of rotation is caused by rapid deceleration following prolonged unidirectional rotation in the darkness, it is called after-rotation. The after-rotation sensation involves a feeling of turning opposite to the direction of prior body rotation when the prior body rotation was passively generated, but when the body rotation is actively generated (by stepping around in a circle), the after-rotation illusion is diminished and may be felt in the same direction as prior rotation (Guedry, Mortenson, Nelson, & Correia, 1978).

When an illusion of rotation is caused by a sudden motion of the head in an axis not parallel to the central rotation axis of the body during prolonged rotation in the darkness, it is called a *Coriolis, cross-coupling* effect and is characterized by a feeling of head or self-velocity in a curved path that is predominantly (Lawson, 1995) orthogonal to the axes of both body and head rotation (Guedry & Benson, 1978; Newman, Lawson, Rupert, & McGrath, 2012). The aforementioned (see Section 7.4.1) feeling of tilt vis-à-vis gravity that arises from a purely linear acceleration (usually via low-frequency oscillation) in a horizontal plane is also a perception that has an angular component because the resultant of acceleration due to gravity and due to horizontal movement creates a rotating gravitoinertial acceleration vector that is perceived as an angular change in the direction of *up* when earth-referenced visual cues are absent or fixed relative to the observer (Howard, 1986a,b). Note that these stimuli should be used with caution and careful design of the simulation.

7.4.3 PERCEPTION OF TRANSLATION

When users' actions within a VE require them to move in a straight line, one can move them through the physical world passively or actively (Section 7.5.3) in the same manner. However, given the limited size of most buildings' housing VE displays, often it is desirable to create the illusion of moving in a straight path over a longer time and distance than is physically possible. Such a magnified movement perception can be accomplished via the use of a centrifuge, albeit with the introduction of some angular acceleration. The centrifuge technique is especially useful when the device can minimize the undesirable consequences of angular acceleration by employing the longest feasible rotation radius and allowing for strategic combinations of radial linear translation (toward or away from the center of rotation) and capsule-centered counterrotation (on-axis capsule rotation opposite to the direction of rotation of the central axis of the centrifuge) during the off-center rotation stimulus. (For an explanation of these complex stimuli, see Hixson, Niven, and Correia, 1966.)

It is also possible to elicit translation illusions without a centrifuge. During off-vertical rotation at certain frequencies, participants may experience conical or orbital paths of self-motion of considerable radius, where little or no real translation is occurring (Lackner & Graybiel, 1978; Wood, 2002). However, these stimuli should be used with caution and careful design of the simulation.

7.4.4 ILLUSORY ABSENCE OF TILT

In this case, one has the illusion of remaining in one's former orientation vis-à-vis gravity when one has actually tilted. This can occur when the tilt stimulus is below the threshold of detection, estimates of which vary widely, with some as low as 1.7° of tilt, depending upon the exact stimulus and cues (Lewis, Preiesol, Nicoucar, Lim, & Merfeld, 2011; Mann, Dauterive, & Henry, 1949; Nesti, Masone, Barnett-Cowan et al., 2010; Bringoux, Scherber, Nougier et al., 2002). Tilt can also fail to be perceived when the participant remains aligned with the resultant gravitoinertial force vector during off-center rotation, such as occurs during a banking turn in an airplane (Gillingham & Krutz, 1974).

A failure to detect tilting of the resultant gravitoinertial force vector can occur when the individual remains seated upright while rotating off-center at a very low velocity (wherein the threshold for stimulation given by the oculogravic illusion would suggest that a tilt of the resultant that is less than 1.1° would be undetectable, according to Graybiel and Patterson, 1955). Finally, during off-vertical rotation at frequencies near 0.2–0.3 Hz (Denise, Etard, Zupan, & Darlot, 1996; Golding, Arun, Wotton-Hamrioui, Cousins, & Gresty, 2009), the feeling of rotating in a tilted axis can be replaced by a feeling of moving orbitally in a fixed plane with little or no rotation occurring. For example, subjects rotating about a spine-centered z-axis that is tilted 20 relative to the gravitational vertical may instead feel they are upright and are orbiting in a horizontal plane with their noses pointing at one wall of the room throughout. Similarly, participants lying down and rotating about an earth-horizontal longitudinal z-axis or even seated and pitching end over end about a horizontal (e.g., interaural) y-axis may feel upright (Lackner & Graybiel, 1977; Leger, Landolt, & Money, 1980). Note that these stimuli require caution and careful design of the simulation.

7.4.5 Illusory Absence of Rotation

One can be moving in a curved path but experience the illusion of moving in a straight path or of not moving at all. This effect can occur during off-center rotation below the threshold of the semicircular canals. The minimum threshold for rotation sensation is around $0.44°/s^2$ using verbal reporting of self-rotation but can be as low as $0.11°/s^2$ when detection of the oculogyral illusion is used as the threshold measure (Clark & Stewart, 1968). Even when the rotation stimulus is above threshold, it is possible to fail to perceive a part of the rotation stimulus during off-center angular acceleration with the nose (x-axis) aligned with the resultant of the tangential and centripetal acceleration vectors (Guedry, 1992). A failure to perceive self-rotation can also occur after prolonged constant-velocity rotation because the elastic properties of the cupula of the semicircular canal will cause it to return to resting position during prolonged constant-velocity rotation. The user will feel stationary in this situation if there are no veridical visual (or other nonvestibular) cues to inform him or her that he is still turning. Finally, failure to perceive rotation can occur when rotating with the longitudinal z-axis of the body orthogonal to gravity (e.g., rotating as if on a horizontal barbecue spit). At certain frequencies, this stimulus results in a feeling of orbital motion of the body about a horizontal axis located at a radius from the body, with little or no change in the direction one's face seems to point and hence no negligible felt *rotation* about one's own longitudinal z-axis (Lackner & Graybiel, 1977). Similarly, rotation about an earth-horizontal y-axis at certain frequencies elicits the illusion of orbital motion without rotation, known as the Ferris wheel illusion (Mayne, 1974; Leger et al., 1980). Note that these stimuli require caution and careful design of the simulation to avoid motion sickness.

7.4.6 Illusory Absence of Translation

In this instance, one has the illusion of not moving in a straight line when one is doing so. This effect can be produced by linearly translating an individual (without tilting his or her body out of its former alignment vis-à-vis gravity) below the threshold of the otolith organs, which is about $6 \, cm/s^2$ (Melvill-Jones & Young 1978), but may be slightly lower for participants oriented sideways to the movement (i.e., translating along the y-axis; Melvill et al., 1978; Travis & Dodge 1928). Note that when the linear translation exceeds the threshold of detection, both translation and tilt can be experienced (due to the changing direction and magnitude of the resultant gravitoinertial vector).

7.4.7 Perception of Increased Weight or G-Forces

When driving a real vehicle such as a car or an airplane, one is subjected to G-forces during acceleration in various directions. However, in most current VE, these forces are absent during simulated driving or flying. This absence should be no surprise when one considers the problem of simulating a high-performance jet aircraft. For example, to simulate the forces during a catapult launch takeoff from an aircraft carrier, one would need a simulator and a maneuvering space several hundred feet long. Fortunately, G-force can also be created via centrifugation. One can readily reproduce most of the perceptual aspects of a catapult launch takeoff in a much smaller space by rotating the pilot off-center inside a centrifuge capsule while keeping his nose pointing into (i.e., his x-axis aligned with) the resultant of the tangential and centripetal acceleration vectors (Guedry, 1992). The predominant perception will be of rapid forward and upward acceleration while pitching back. Another way to produce the feeling of changing aircraft G-forces during centrifugation is by exploiting the linear Coriolis acceleration that occurs when a body is moved radially in a rotating environment. Due to the inertia of the body, a force tangential to its motion will act on it during (actual) displacement toward or away from the center of rotation. This method will produce a force that is dependent upon factors such as the speed of rotation of the environment and the mass and speed of the user. Coriolis acceleration should be taken into account whenever simulating aircraft G-forces on a centrifuge that permits radial capsule motion.

7.5 ROLE OF ISOMORPHIC *REAL*-MOTION STIMULI

Even the most careful orchestration of visual displays, locomotion devices, and man-moving machines will not provide an acceleration experience that can pass the virtual reality Turing test (Barberi, 1992; Turing, 1950) for all of the synthetic experiences in a given VE's repertoire. There is a way of simulating real events that has the potential to elicit acceleration perceptions that are indistinguishable from reality, however. This method involves identically duplicating the real acceleration profiles being simulated, but doing so in an alternate real location (e.g., a simulation facility) from the user's virtual location (e.g., a simulated battleground). This isomorphic method is similar to the concept of augmented reality, wherein users see and move through a real environment while being able to view virtual information as well. However, the purpose of the isomorphic simulation presently under consideration is not to provide additional information to a user who is otherwise engaged with the physical world. Rather, the goal is to provide a physical mock-up that serves as the ambient context for a synthetic event that would be difficult or impossible to experience in the physical world. What follows are several potential applications of isomorphic motion stimuli for practical purposes, such as enhancing flight training and law enforcement training.

7.5.1 APPLYING ISOMORPHIC REAL ACCELERATION TO ENHANCE AVIATION MISSION REHEARSAL

During advanced mission rehearsal, military aviators flying real aircraft would execute all the same maneuvers and actions as they would execute during a real combat mission, except they would fly in a designated friendly airspace. The flying would be real, but the targets, threats, and ordnance would be simulated. Aviators would view simulated targets and simulated ground or air threats through virtual displays and neutralize them either through virtual representations of weapons systems or via live fire on real practice targets whose position in a safe location coincides with the virtual image and virtual targeting information represented to the pilot. Opponents could also be real but remote (e.g., aviators and air defense personnel from allied countries coordinating in the exercise from their own friendly airspaces). Safety of the crew and allied ground troops would be ensured by simulating (or deactivating) ordnance in all situations involving close support of ground troops and by close monitoring from the ground and from an onboard safety observer who is not part of the VE. This approach would allow for more realistic, varied, and challenging practice than is possible with current simulators or small-unit training exercises. It would also provide many of the training benefits of full-scale war games on a more frequent basis and without the same amount of preparation time, cost, danger, environmental impact, or political visibility (useful in cases where demonstration of force is not a mission goal). Even as unmanned, remotely controlled systems become increasingly important in military and civilian applications, appropriate motion cues will be desirable to enhance situation awareness concerning the moment-by-moment state of the unmanned vehicle under control (so as to aid tasks requiring mental rotation, avoid a control reversal error, etc.).

7.5.2 APPLYING ISOMORPHIC REAL ACCELERATION TO ENHANCE AUTOMOBILE RACING REHEARSAL

Race car drivers routinely practice alone on tracks to familiarize themselves before a race. Using virtual displays, it would be possible for drivers to do practice runs against virtual (real or simulated) competitors represented on an otherwise transparent (*look through*) display while driving alone on a real track at a speed reduced just enough to keep the margins of exposure to monetary and personnel risk appropriate to the practice session. If the driver makes a driving error while maneuvering against a simulated (or real but remote) competitor, he will receive error feedback but will not suffer a dangerous and expensive car-to-car collision. This type of practice would form a bridge between lone practice on the track and live competition with other drivers. It could also save lives by familiarizing drivers with hazardous episodes of driver disorientation that can be caused by excessive G-forces (Guedry, Raj, & Cowin, 2003).

7.5.3 APPLYING ISOMORPHIC REAL ACCELERATION VIA FREE-SPACE WALKING IN A VE

Allowing for natural walking in real space during VE use is probably the most intuitive and straightforward way for navigating computer-simulated environments and readily provides the appropriate vestibular and somatosensory self-motion cues. In this approach, users wear position and orientation tracking head-mounted displays (HMDs) that allow the system to match their physical locomotion with visual scene cues. When optic flow is displayed on an HMD, it can significantly influence human walking and is readily integrated with vestibular and somatosensory cues (Warren, Kay, Zosh, Duchon, & Sahuc, 2001). Foxlin (2002) provides an overview of the challenges and technological requirements involved in providing low-latency motion tracking over extended free-space walking areas.

Apart from being a natural and intuitive way to move through VEs, allowing users to naturally walk can increase the naturalness and presence of a VE simulation, compared to simpler locomotion methods like walking in place or button-based navigation (Usoh et al., 1999). When rapid and efficient (i.e., naturalistic) navigation is required, actual walking outperforms joystick-based motion metaphors (Suma et al., 2010). Walking in VE can also help to improve user's navigation ability and reduce disorientation (Chance, Gaunet, Beall, & Loomis, 1998; Ruddle & Lessels, 2009), although merely allowing the user to turn naturally can sometimes be sufficient to provide many of these benefits (Riecke et al., 2010).

Walking in a VE typically requires wearing an HMD, which frequently results in an underestimation of visually presented distances (Creem-Regehr, Willemsen, Gooch, & Thompson, 2005; Loomis & Knapp, 2003). While the underlying reasons are not fully understood, recent work suggests that this systematic distance underestimation can be compensated for by artificially increasing the simulated FOV beyond the physical FOV of the HMD (Zhang, Nordman, Walker, & Kuhl, 2012) or by enhancing visual motion (with moving particles, sinus gratings) in the periphery of the stimulus to increase perceived self-motion (Bruder, Steinicke, & Wieland, 2011; Bruder, Steinicke, Wieland, & Lappe, 2012).

While it seems desirable to simply allow users to walk through VEs whenever possible, there are several technical and perceptual challenges (Steinicke et al., 2013). One of the most critical challenges is how to simulate and navigate a sufficiently large virtual space when the physical space in which users can walk around with head-tracked HMD is severely limited. Several approaches have been suggested to tackle this challenge (Steinicke et al.), and this chapter highlights some of them, for example, subtly redirecting users' walking trajectories such that they stay within the physical confounds using redirected walking (Section 7.5.5) or variants of walking in place using linear or circular treadmills (Section 7.2.3). In the following discussion, we briefly present one example of a practical training application where free-space walking in a VE would be beneficial.

7.5.3.1 Applying Real Locomotion through Simulated Environments as Part of Police Training

In traditional video-based training simulations, self-motion was suggested solely by the visual stimulus, and *movement* was scripted in advance, based on options selected by the user. Even in some current devices where movement is not scripted, the user's navigation and visual reconnaissance may be accomplished via nonintuitive methods such as foot switches that initiate visual scene motion, the direction of which is then determined by head pointing. One of the key features distinguishing future individual (or small unit) combat simulations will be that body movement through the VE will be realistic and naturally accomplished.

Law enforcement officers serving on special weapons and tactics (SWAT) teams could go armed into a physical mock-up of a house and proceed to *clear* the armed and hostile criminals resisting within (e.g., as part of hostage crisis training). This simulation would be achieved via voluntary movement through the typical *fun house* or *Hogan's Alley* (a custom-built practice area that looks like a regular building but has bullet-resistant backstops all around to enhance the safety of live-fire practice), coupled with head-mounted *look-through* visual displays wherein virtual criminals and bystanders (either computer simulations or real-but-remote users sharing the VE) appear suddenly and either respond to certain key verbal commands or attempt to attack the officers with various simulated weapons. The performance score of the trainee would be based on his or her speed and skill in moving through the simulation (e.g., skill with *quartering techniques* when rounding corners or entering rooms, use of cover and concealment) and speed in neutralizing threats with appropriate force (e.g., coordination with other officers, skill with nonlethal force alternatives, accuracy and speed of shooting).

Simulation of return fire could be based on tracking technology on the bodies and weapons of two opposing teams of officers moving through one virtual space but physically located in different practice areas. The simplest example would be the case where one officer is locomoting through the Hogan's Alley and directing live fire against the virtual representation of another officer playing the role of a violent criminal in a different Hogan's Alley. The approach described is a marriage of multiplayer tactical shooter or virtual battlefield gaming concepts (e.g., Virtual Battlespace Systems) with training simulations already in use to allow live fire against projected images of prerecorded actors or programmed computer characters portraying criminals (http://teams.drc.com/fast/index.html) and simulated fire against live opponents (www.ais-sim.com; www.meggitttrainingsystems.com). As the most useful features of these existing techniques are enhanced by the new virtual display features previously mentioned, more sophisticated and realistic simulations will emerge. Even though automated virtual avatars are growing more complex (Biron, 2012), having the option for human-to-human engagement will yield clear advantages for the training of law enforcement personnel and for certain national defense applications (Finkelstein, Griffith, & Maxwell, 2014, Chapter 37). The situation is somewhat analogous to the way massively multiplayer online first-person-shooter games replaced games where competition was only possible with the computer.

7.5.4 Advantages and Limitations of Using Isomorphic Acceleration/Motion Stimuli

Real acceleration provides the most direct means for eliciting acceleration perceptions in VE and allows incredible flexibility and control concerning the types of acceleration perceptions that can be elicited at any given moment. Real acceleration that is isomorphic to the psychophysics of the VE should minimize the amount of interference a VE trainer creates regarding the transfer of psychomotor skills to a real situation. For example, if the aircraft-based VE previously mentioned (see Section 7.5.1) is designed and executed properly, it will be able to simulate a hazardous military combat mission over friendly airspace, eliciting only those adaptive effects and aftereffects that occur during a comparable real flight.

The primary disadvantage of using real acceleration in VE is that this approach will require large and often costly practice spaces. In the case of military flight training, this concern is not insurmountable because the infrastructure for real flying (of simulated missions) is already in place (albeit fuel is a significant factor). For police training, the infrastructure of *fun houses* is also in place among many of the larger agencies, although numerous software challenges remain in order to enhance realism and simulate the stress of actual operations. For smaller applications, the introduction of redirected walking techniques (Section 7.5.5; Steinicke, Bruder, Jerald, Frenz, & Lappe, 2010) can reduce the size of the space required, provided the physical mock-up is simple or highly flexible.

The cost associated with introducing real acceleration into VE training for the most popular professional or Olympic sports will probably be less of an impediment than it is for the military and law enforcement. However, the technical ease of implementation and the revenue available for implementation of virtual acceleration or motion will be very different for different sports. For example, implementation of virtual acceleration/motion training could be achieved readily for either automobile racing or whitewater kayaking but would probably be implemented for automobile racing first (despite the fact that a larger practice area is necessary), because automobile racing has greater spectator and commercial support (and thus commands greater revenue). Also, virtual acceleration/motion training during automobile racing can be accomplished on existing automobile tracks without substantive modifications to the track itself, whereas the simulation of whitewater kayaking requires the construction of a special recycling water course with programmable high-speed currents. Some sports will be very difficult to train for via real movement through virtual spaces. A good example would be equitation, which requires the mutual coordination of perception and action on the part of a horse and rider, each of whom has profoundly different user requirements.

7.5.5 Offering Fairly Isomorphic Acceleration Stimuli in a Smaller Space via *Redirected Walking*

How can we naturally walk through large virtual spaces without bumping into the physical walls or leaving the designated area wherein movements can be tracked? To solve this space and safety challenge, various redirected walking techniques have recently been developed and refined. While *repositioning techniques* aim to compress the VE into a more constrained physical space, *reorientation techniques* try to subtly steer users away from the physical boundaries that users might otherwise bump into (see taxonomy by Suma, Bruder, Steinicke, Krum, & Bolas, 2012). When the real space is only slightly smaller than the to-be-simulated virtual space, simple continuous repositioning by using a gain factor between virtual (simulated) and actual (walked) distance might be sufficient; by measuring detection thresholds, Steinicke et al. (2010) showed that walked distances could be upscaled by up to 26% or downscaled by up to 14% without users noticing this discrepancy. When larger simulated environments are needed, there are multiple options. Users could be repositioned by employing real-word metaphors like elevators, escalators, or various vehicle types to reduce disruptions. For longer distances, *teleportation* can provide a possibility, especially if metaphors inspired by science fiction such as portals are employed to provide a consistent motion framework, such as in the Arch-Explore natural user interface for architecture explorations proposed by Bruder, Steinicke, and Hinrichs (2009). Apart from repositioning techniques, various user reorientation techniques can also be used to trick the user into remaining within the physical boundaries, ideally without noticing that they are being redirected. The most promising approaches involve continuous subtle reorientation techniques, which exploit the fact that users are not very accurate in estimating the angle through which they have actually turned. For example, users can be made to believe that they are walking on a straight path despite being redirected on a fairly circular path. While promising, this still requires a curvature radius of a least 22 m and thus will not by itself fit into most current free-space walking areas (Steinicke et al.). When the simulation does not require or afford long straight paths, however, it can become possible to continuously and smoothly adjust the gain factors between virtual and physical rotations and translations to ensure that users stay within the physical boundaries. In fact, our perception of rotations seems to be more malleable than our perception of translations; in a detection threshold study, users did not notice any difference when physical rotations were up to 49% larger or 20% lower than the perceived simulated rotation (Steinicke et al.). This extends earlier findings of often surprisingly large coherence zones where visuo-vestibular cue mismatches remain unnoticed or have only negligible impact (Steen, 1998). Especially when users' attention can be directed away from the locomotion itself, the coherence zone might be further increased. For example, Hodgson, Bachmann, and Waller (2011) steered users to walk arcs with curvature radiuses as low as 8 m without the majority of users noticing any discrepancy. Even those who did notice some discrepancy did not seem to mind or be affected much. This allowed Hodgson et al. (2011) to develop a first generalized redirected walking algorithm that can dynamically steer users away from the physical boundaries while they are exploring a VE. While some users detected discrepancies between real and simulated motions, their spatial memory of the VE remained unaffected, which is promising for future applications.

From an applied perspective, benefits and costs of various redirected walking procedures should be carefully evaluated, since they tend to be specific to the task and goals of the simulation. For some VE goals, walking would not be essential. Naturalistic visual cues in immersive VE can sometimes be sufficient for enabling rapid and effective spatial orientation despite the lack of physical motion cues (Riecke, Cunningham, & Bülthoff, 2007; Riecke, von der Heyde, & Bülthoff, 2005), although there can be individual differences between users (Riecke, Sigurdarson, & Milne, 2012). When physical motion cues are relevant and larger distances need to be covered, walking can be flexibly integrated with redirected driving methods where larger sensory discrepancies can remain unnoticed (Bruder, Interrante, Phillips, & Steinicke, 2012). Methods developed for optimizing redirected walking algorithms can also provide useful guidance for improving driving or even flight simulation, where coherence zones tend to be larger and thus allow for more flexibility.

Finally, when budgetary or spatial constraints are tight, the requirements for physical (loco)motion can be reduced by exploiting multimodal synergistic effects (see also Sections 7.2 and 7.6). In particular, naturalistic full-field visual stimuli can provide compelling self-motion illusions in the absence of any physical motion (Riecke, 2011; Riecke, Schulte-Pelkum, Avraamides, Heyde, & Bülthoff, 2006; Wright, DiZio, & Lackner, 2006). The potential of the visual modality is probably best demonstrated in *tumbling rooms*, where stationary observers may report tumbling sensations of 180° (head down) or greater (full revolutions) when an immersive room is rotated around them (Allison et al., 1999; Howard & Childerson, 1994; Howard, Jenkin, & Hu, 2000; Palmisano, Allison, & Howard, 2006). Thus, the motion paradigm and simulated modalities should be carefully selected on a case-by-case basis to best suit the specific task, context, and overall simulation goal (Riecke & Schulte-Pelkum, 2013).

7.6 BENEFIT OF COMBINING DIFFERENT SENSORY CUES

A controlled and convincing feeling of acceleration and self-motion within a VE can be achieved via a judicious combination of the more promising stimuli listed earlier. Multimodal inputs act in concert to create acceleration perceptions in the natural world. Similarly, perceptions of motion in VE can be enhanced by exploiting coordinated combinations of multimodal stimuli. For example, feelings of acceleration and self-motion can be significantly enhanced when motions of the visual field (see Section 7.2.1) correspond with real motions of the user (Sections 7.4 and 7.5; also Wong & Frost, 1981; Young, Dichgans, Murphy, & Brandt, 1973). When large-scale movements of the user are not feasible or affordable, visually simulated self-motions can be enhanced by small-scale passive physical motions of the body (e.g., simple jerks or more sophisticated motion cueing of a few degrees or centimeters) that corroborate the visually indicated self-motion (Berger, Schulte-Pelkum, & Bülthoff, 2010; Riecke, 2006; Schulte-Pelkum, 2007; Wright, 2009; Wright, DiZio, & Lackner, 2005). Similar benefits can be gained when users themselves initiate and power the motion cueing, for example, by seating people on a modified force-feedback manual wheelchair (Riecke, 2006) or a gaming chair where participants control the virtual locomotion by leaning into the direction they want to travel (Beckhaus, Blom, & Haringer, 2005; Riecke & Feuereissen, 2012). While vibrations (see Section 7.2.3.3) or jerks separately can enhance visually induced percepts of self-motion, combining vibrations and jerks provide additional benefits (Schulte-Pelkum, 2007).

Spatialized sound (Section 7.2.2) can enhance perceptions of self-motion induced by cues from vision (Section 7.2.1; also Riecke et al., 2005; Riecke et al., 2009), somesthesia (Section 7.2.3; also Riecke et al., 2011), or vestibular sensation (Sections 7.2.4, 7.3 through 7.5).

There is mixed evidence concerning whether linear treadmill walking can facilitate visually induced sensations of self-motion (Ash et al., 2013; Kitazaki et al., 2010; Onimaru et al., 2010;

Seno et al., 2011). However, walking on circular treadmills can by itself reliably induce acceleration and self-motion percepts (Bles, 1981; Bles & Kapteyn, 1977; Bruggeman et al., 2009; Riecke et al., 2011) and so should work well with visual cues (Freiberg, Grechkin, & Riecke, 2013), possibly further enhanced by matching moving sound fields (Riecke et al.) as discussed in Section 7.2.3. Moreover, adding matching rotating sound fields can enhance kinesthetic vection in this situation (Riecke et al.) and induce self-motion illusions that are strong enough to facilitate perspective switches (Riecke, Feuereissen, Rieser, & McNamara, 2012). That is, while it is difficult to imagine an orientation change in our immediate environment without physically rotating to the new perspective, the mere illusion of self-rotation was sufficient to facilitate this perspective switch, even in the absence of any supporting visual cues. While further studies are needed to confirm and extend these findings, they suggest that providing compelling sensations of self-motion in VE may help people to remain oriented to the VE during simulated (but not physically executed) self-motions. In our view, the most promising methods for inducing self-motion and acceleration illusion perceptions in a VE will tend to combine three or more of the following stimulus types, in a mutually supporting manner appropriate to the specific purposes of the simulation and the VE designer: (1) visual frame-of-reference motion, (2) auditory surround motion, (3) somatosensory cues, and (4) vestibular cues, including passive body acceleration, active body movement, and/or vestibular stimuli not requiring acceleration (especially electrical stimulation).

7.7   VISUAL CONSEQUENCES OF ACCELERATION

This chapter is concerned with the perception of body acceleration and motion. However, it should be noted that many acceleration stimuli can influence the perception of a visual target or display relative to the user. For example, simple body tilt in darkness can make it hard for an observer to set a line of light to align with the true earth vertical. The visual effects of acceleration stimuli include the following: the Müller (or E) effect, the Aubert (or A) effect (e.g., Bauermeister, 1964; Clark & Graybiel, 1963; Müller, 1916; Passey & Ray, 1950; Wade & Day, 1968), the oculogyral illusion (Graybiel & Hupp, 1946), the oculogravic illusion (Graybiel, 1952), the elevator illusion (Cohen, 1973; Whiteside, 1961), and the degradation of visual acuity that occurs when viewing a head-fixed display while one's body is moving (Guedry, Lentz, Jell, & Norman, 1981). Thus, even when a user is being exposed to a simple tilt, the VE designer must take into account the interactions between visual inputs and vestibular and somatosensory inputs for any task requiring visual estimates of the orientation of objects within the virtual scene or any tasks requiring eye-to-hand coordination. Fortunately, many of these effects will be minimized by the presence of a whole-field visual stimulus (rather than a simple visual target presented in the darkness, as in most of the laboratory studies cited). This will be true especially when the visual stimulus provides veridical information about self-orientation (Lawson et al., 1997).

7.8   COGNITIVE INFLUENCES

Cognitive aspects of the simulation can improve the overall believability and effectiveness of motion simulations (for reviews, see Riecke, 2009, 2011; Riecke and Schulte-Pelkum, 2013). For example, visually simulated self-motions can be facilitated when the moving visual cues are perceived as background motion with respect to an observer-fixed foreground, even when there is no physical depth separation (Brandt et al., 1975; Ito & Shibata, 2005; Kitazaki & Sato, 2003; Nakamura, 2008; Nakamura & Shimojo, 1999; Ohmi & Howard, 1988; Ohmi, Howard, & Landolt, 1987; Seno et al., 2009).

Priming users to believe that physical motion is possible can also enhance visually induced self-motion percepts (Andersen & Braunstein, 1985; Lepecq, Giannopulu, & Baudonniere, 1995; Wright et al., 2006), thus confirming earlier assertions that knowledge that actual motion is impossible might be detrimental to inducing vection (Dodge, 1923). This technique is frequently employed in entertainment applications such as theme park fun rides, where physical setups and narratives are combined to provide a cognitive-perceptual framework that makes actual motion seem possible.

7.9   SUMMARY

The perception of acceleration encompasses feelings of self-motion through space and self-tilt relative to the upright. The feeling that one is being subjected to unusual forces (such as occur during high-performance flight or in nonterrestrial settings) also qualifies as an acceleration perception. The techniques for eliciting acceleration perceptions in future VEs will employ either real motion or illusory motion. Real-motion methods will accelerate the VE user aboard a moving device such as a centrifuge or will allow the user to drive or actively locomote (as appropriate) within a real space that serves as an ambient context for the VE. Illusory motion methods will induce an illusion of body motion by moving a visual, auditory, or somatosensory surround stimulus relative to the user or by having the user locomote (without displacement) on a treadmill or similar device. Each of these methods has advantages and limitations that have been discussed. The methods for inducing acceleration perceptions via real motion will tend to generate a wider range of simulations and be more effective. The methods involving illusory motion will require less money and space to implement. Methods employing real or illusory motion are not mutually exclusive. When a person moves through space while receiving visual, auditory, or somatosensory information that confirms the movement, the resulting movement perception will tend to be stronger than when only one self-movement cue is present in isolation. Thus, combining different cues to elicit acceleration perceptions will be advisable whenever it is feasible. Such multisensory cueing is exactly what takes place whenever the reader goes for a stroll through the real world.

DISCLAIMER

The views expressed in this report are solely those of the authors; they do not represent the views of Simon Fraser University nor of the US government or any of its subordinate agencies or departments. The mention of any agencies, persons, companies, or products in this report does not imply endorsement by the authors of this work or the authors' affiliated agencies. The mention of any persons or agencies in this report does not imply that they endorse the contents of this report.

ACKNOWLEDGMENTS

The authors thank Stephanie Sides and Amy Hickinbotham for their contributions to earlier versions of this manuscript. We thank (in alphabetical order) Paul DiZio, Jim Grissett, Fred Guedry (rest in peace), Jim Lackner, Michael Newman, and Angus Rupert, for many interesting conversations concerning the perception of motion and orientation.

APPENDIX: RECOMMENDED READINGS ON VESTIBULAR FUNCTION

7.A.1   Basic Reviews

For a simple and clear introduction to the perception of body position and movement via kinesthesia and vestibular sensation, the authors recommend the textbook by Ludel (1978). For more detailed explanation of acceleration stimuli and acceleration perceptions, see Nicogossian et al. (1989), Guedry (1992), or Previc and Ercoline (2004). Classic scientific reviews on vestibular sensation were written by Guedry (1974) and Howard (1986a,b). The comprehensive paper by Guedry (some of which is updated in Guedry, 1992) focuses on psychophysics of vestibular sensation during a wide variety of stimulus situations. The 1986b paper by Howard provides a lucid review of vestibular structure, dynamics, neural projections, and psychophysics, ranging afield of the vestibular modality as well.

7.A.2 Reviews on Spatial Orientation in Flight

The most comprehensive recent review of this subject was contributed by Previc and Ercoline (2004). This book covers all the aviation aspects of spatial orientation. The classic introductory papers written on spatial disorientation vis-à-vis aerospace operations were contributed by Benson (1988) and by Gillingham and Krutz (1974). Another classic resource is Guedry's Chapter 8 of the 1968 U.S. Naval Flight Surgeon's Manual. This is a thorough effort with a clinical slant. The most recent online version of this manual is at http://www.operationalmedicine.org/TextbookFiles/USNavalFlightSurgeonsManual.htm. These various works introduce the common illusions associated with spatial disorientation in flight, common vestibular acceleration nomenclature, and human reflexive reactions to unusual accelerations, especially high G. For recent reviews focused on vestibular problems during spaceflight, see Clément (2011) and Buckey (2006).

7.A.3 Detailed Research Compendia or Books Treating Special Topic Areas

Many works on vestibular function are from special edition compilations of papers or chapters by various scholars. Works of note include Cohen, Tomko, and Guedry (1992), Arenberg (1993), and Marcus (1992). The book by Cohen, Tomko, and Guedry covers a variety of topics on vestibular and sensorimotor functions. Arenberg gives a thorough treatment of dizziness and balance disorders. The most comprehensive new scientific review has been contributed by Goldberg et al. (2011). The book by Goldberg et al. covers topics beyond vestibular pathology but has a strong neurophysiological emphasis, rather than a perceptual emphasis geared for VE designers.

7.A.4 Works on Side Effects of Virtual Environments

The latest information on adverse side effects of VE exposure and most other human factor issues related to VE can be found throughout this second edition of the Handbook of Virtual Environments (Hale & Stanney, 2014). Earlier sources for information on unpleasant vestibular side effects relevant to VE include Pausch, Crea, and Conway (1992); Chien and Jenkins (1994); Durlach and Mavor (1995); Kolasinski (1995); Stanney, Mourant, and Kennedy (1998); and Stanney et al. (1998).

7.A.5 General Human Factor Reference Works of Relevance to Orientation

The *Handbook of Human Factors and Ergonomics* (Salvendy, 2006) is a good general reference that briefly touches upon vestibular displays and vestibular effects (in Chapters 40 by Stanney and Cohn or Chapter 23 by Griffin). The older *Engineering Data Compendium: Human Perception and Performance* is a multivolume set edited by Boff and Lincoln (1988). This compendium of human factor findings has succinct sections devoted to orientation, vestibular function, and simulation.

REFERENCES

Allison, R. S., Howard, I. P., & Zacher, J. E. (1999). Effect of field size, head motion, and rotational velocity on roll vection and illusory self-tilt in a tumbling room. *Perception, 28*(3), 299–306.

Andersen, G. J. (1986). Perception of self-motion—Psychophysical and computational approaches. *Psychological Bulletin, 99*(1), 52–65.

Andersen, G. J., & Braunstein, M. L. (1985). Induced self-motion in central vision. *Journal of Experimental Psychology—Human Perception and Performance, 11*(2), 122–132.

Arenberg, I. K. (1993). *Dizziness and balance disorders.* New York, NY: Kugler Publications.

Ash, A., Palmisano, S., Apthorp, D., & Allison, R. S. (2013). Vection in depth during treadmill walking. *Perception, 42*(5), 562–576. doi:10.1068/p7449

Baloh, R. W., & Honrubia, V. (1979). *Clinical neurophysiology of the vestibular system.* Philadelphia, PA: F. A. Davis.

Barberi, D. (1992). *The ultimate turing test* [Online]. Available: http://metalab.unc.edu/dbarberi/vr/ultimate-turing/, info@2meta.com

Bauermeister, M. (1964). Effect of body tilt on apparent verticality, apparent body position, and their relation. *Journal of Experimental Psychology, 67*(2), 142–147.

Beckhaus, S., Blom, K. J., & Haringer, M. (2005). A new gaming device and interaction method for a First-Person-Shooter. *Proceedings of the Computer Science and Magic*, GC Developer Science Track (Vol. 2005). Presented in Leipzig, Germany, 2005.

Benson, A. J. (1988). Motion sickness. In J. Ernsting & P. King (Eds.), *Aviation medicine* (pp. 318–493). London, U.K.: Buttersworth.

Bent, L. R., McFadyen, B. J., French Merkley, V., Kennedy, P. M., & Inglis, J. T. (2000). Magnitude effects of galvanic vestibular stimulation on the trajectory of human gait. *Neuroscience Letters, 279*(3), 157–160. doi:10.1016/S0304-3940(99)00989-1

Berger, D. R., Schulte-Pelkum, J., & Bülthoff, H. H. (2010). Simulating believable forward accelerations on a stewart motion platform. *ACM Transactions on Applied Perceptions, 7*(1), 1–27. doi:10.1145/1658349.1658354

Berthoz, A., Pavard, B., & Young, L. R. (1975). Perception of linear horizontal self-motion induced by peripheral vision (linearvection)—Basic characteristics and visual-vestibular interactions. *Experimental Brain Research, 23*(5), 471–489.

Biron, L. (2012). Virtual humans become more lifelike at USMC trainer. *Defense News*. Retrieved from http://www.defensenews.com/article/20120830/TSJ01/308300004/Virtual-Humans-Become-More-Lifelike-USMC-Trainer.

Bles, W. (1979). *Sensory interactions and human posture: An experimental study*. Amsterdam, the Netherlands: Academische Pers.

Bles, W. (1981). Stepping around: Circular vection and Coriolis effects. In J. Long & A. Baddeley (Eds.), *Attention and performance IX* (pp. 47–61). Hillsdale, NJ: Erlbaum.

Bles, W., Bos, J. E., Graaf, B. de, Groen, E., & Wertheim, A. H. (1998). Motion sickness: Only one provocative conflict? *Brain Research Bulletin, 47*(5), 481–487.

Bles, W., Jelmorini, M., Bekkering, H., & de Graaf, B. (1995). Arthrokinetic information affects linear self-motion perception. *Journal of Vestibular Research: Equilibrium & Orientation, 5*(2), 109–116. doi:10.1016/0957-4271(94)00025-W

Bles, W., & Kapteyn, T. S. (1977). Circular vection and human posture. 1. Does proprioceptive system play a role. *Agressologie, 18*(6), 325–328.

Boff, K. R., & Lincoln, J. E. (Eds.). (1988). *Engineering data compendium: Human perception and performance*. Wright-Patterson Air Force Base, OH: AAMRL.

Brandt, T. (1999). *Vertigo: Its multisensory syndromes*. New York, NY: Springer-Verlag.

Brandt, T., Büchele, W., & Arnold, F. (1977). Arthrokinetic nystagmus and ego-motion sensation. *Experimental Brain Research, 30*(2), 331–338. doi:10.1007/BF00237260

Brandt, T., Dichgans, J., & Koenig, E. (1973). Differential effects of central versus peripheral vision on egocentric and exocentric motion perception. *Experimental Brain Research, 16*, 476–491.

Brandt, T., Wist, E. R., & Dichgans, J. (1975). Foreground and background in dynamic spatial orientation. *Perception & Psychophysics, 17*(5), 497–503.

Bringoux, L., Schmerber, S., Nougier, V., Dumas, G., Barrad, P. A., & Raphel, C. (2002). Perception of slow pitch and roll body tilts in bilateral labryinthine-defective subjects. *Neuropsychologica, 40*(3), 367–372.

Bruder, G., Interrante, V., Phillips, L., & Steinicke, F. (2012). Redirecting walking and driving for natural navigation in immersive virtual environments. *Visualization and Computer Graphics, IEEE Transactions on, 18*(4), 538–545. doi:10.1109/TVCG.2012.55

Bruder, G., Steinicke, F., & Hinrichs, K. H. (2009). Arch-explore: A natural user interface for immersive architectural walkthroughs. *3D User Interfaces, 2009. 3DUI 2009. IEEE Symposium on*, Lafayette, LA (pp. 75–82). doi:10.1109/3DUI.2009.4811208

Bruder, G., Steinicke, F., & Wieland, P. (2011). Self-motion illusions in immersive virtual reality environments. *Virtual Reality Conference (VR), 2011 IEEE*, Singapore (pp. 39–46). doi:10.1109/VR.2011.5759434

Bruder, G., Steinicke, F., Wieland, P., & Lappe, M. (2012). Tuning self-motion perception in virtual reality with visual illusions. *Visualization and Computer Graphics, IEEE Transactions on, 18*(7), 1068–1078. doi:10.1109/TVCG.2011.274

Bruggeman, H., Piuneu, V. S., Rieser, J. J., & Pick, H. L. J. (2009). Biomechanical versus inertial information: Stable individual differences in perception of self-rotation. *Journal of Experimental Psychology: Human Perception and Performance, 35*(5), 1472–1480. doi:10.1037/a0015782

Bubka, A., & Bonato, F. (2010). Natural visual-field features enhance vection. *Perception, 39*(5), 627–635. doi:10.1068/p6315

Buckey, J. C. (2006). *Space physiology*. New York, NY: Oxford University Press.

Chance, S. S., Gaunet, F., Beall, A. C., & Loomis, J. M. (1998). Locomotion mode affects the updating of objects encountered during travel: The contribution of vestibular and proprioceptive inputs to path integration. *Presence—Teleoperators and Virtual Environments, 7*(2), 168–178.

Chelette, T. L., Martin, E. J., & Albery, W. B. (1992). *The nature of the g-excess illusion and its effect on spatial orientation* (Technical Report No. AL-TR-1992-0182). Wright-Patterson Air Force Base, OH: Armstrong Laboratory.

Cheung, B. S., Howard, I. P., & Money, K. E. (1991). Visually-induced sickness in normal and bilaterally labyrinthine-defective subjects. *Aviation, Space, and Environmental Medicine, 62*(6), 527–531.

Chien, Y. T., & Jenkins, J. (1994). *Virtual reality assessment.* Alexandria, VA: Institute for Defense Analyses. (A Report to the Task Group on Virtual Reality to the High Performance Computing and Communications and Information Technology subcommittee of the Information and Communications Research and Development Committee of the National Science and Technology Council.)

Clark, B., & Graybiel, A. (1963). Perception of the postural vertical in normals and subjects with labyrinthine defects. *Journal of Experimental Psychology: General, 65*, 490–494.

Clark, B., & Stewart, J. D. (1968). Comparison of three methods to determine thresholds for perception of angular acceleration. *American Journal of Psychology, 81*, 207–216.

Clément, G. (2011). *Fundamentals of space medicine* (2nd ed.). New York, NY: Springer, Space Technology Library.

Cohen, B., Tomko, D. L., & Guedry, F. (Eds.). (1992). *Sensing and controlling motion: Vestibular and sensorimotor function.* New York, NY: New York Academy of Sciences.

Cohen, M. M. (1973). Elevator illusion: Influences of otolith organ activity and neck proprioception. *Perception and Psychophysics, 14*(3), 401–406.

Creem-Regehr, S. H., Willemsen, P., Gooch, A. A., & Thompson, W. B. (2005). The influence of restricted viewing conditions on egocentric distance perception: Implications for real and virtual indoor environments. *Perception, 34*(2), 191–204. doi:10.1068/p5144

De Broglie, L. (1951). A general survey of the scientific work of Albert Einstein. In P. A. Schilpp (Ed.), *Albert Einstein: Philosopher-Scientist* (pp. 107–127). New York, NY: Tudor.

Denise, P., Etard, O., Zupan, L., & Darlot, C. (1996). Motion sickness during off-vertical axis rotation: Prediction by a model of sensory interactions and correlation with other forms of motion sickness. *Neuroscience Letters, 203*(3), 183–186.

Dichgans, J., & Brandt, T. (1978). Visual–vestibular interaction: Effects on self-motion perception and postural control. In R. Held, H. W. Leibowitz, & H. L. Teuber (Eds.), *Handbook of sensory physiology* (Vol. 8, pp. 755–804). New York, NY: Springer-Verlag.

DiZio, P., & Lackner, J. R. (2002). Proprioceptive adaptation and aftereffects. In Stanney, K. M. (Ed.), *Handbook of virtual environments: Design, implementation, and applications* (pp. 751–771). New York, NY: Lawrence Erlbaum Associates.

Dodge, R. (1923). Thresholds of rotation. *Journal of Experimental Psychology, 6*(2), 107–137. doi:10.1037/h0076105

Durgin, F. H., Pelah, A., Fox, L. F., Lewis, J. Y., Kane, R., & Walley, K. A. (2005). Self-motion perception during locomotor recalibration: More than meets the eye. *Journal of Experimental Psychology: Human Perception and Performance, 31*(3), 398–419.

Durlach, N. I., & Mavor, A. S. (Eds.). (1995). *Virtual reality—Scientific and technological challenges.* Washington, DC: National Academy Press.

Fetter, M., Haslwanter, T., Bork, M., & Dichgans, J. (1999). New insights into positional alcohol nystagmus using three-dimensional eye-movement analysis. *Annals of Neurology, 45*(2), 216–223.

Finkelstein, N. Griffith, T., & Maxwell, D. (2014). National defense. In K. S. Hale and K. M. Stanney (Eds.), *Handbook of virtual environments: Design, implementation, and applications,* (2nd ed., pp. 957–998). New York, NY: CRC Press.

Fischer, M. H., & Kornmüller, A. E. (1930). Optokinetisch ausgelöste Bewegungswahrnehmung und optokinetischer Nystagmus [Optokinetically induced motion perception and optokinetic nystagmus]. *Journal für Psychologie und Neurologie, 41*, 273–308.

Fitzpatrick, R. C., Wardman, D. L., & Taylor, J. L. (1999). Effects of galvanic vestibular stimulation during human walking. *The Journal of Physiology, 517*(3), 931–939. doi:10.1111/j.1469-7793.1999.0931s.x

Foxlin, E. (2002). Motion tracking requirements and technologies. In K. M. Stanney (Ed.), *Handbook of virtual environments: Design, implementation, and applications* (pp. 163–210) New York, NY: Lawrence Erlbaum Associates.

Freiberg, J., Grechkin, T., & Riecke, B. E. (2013). Do walking motions enhance visually induced self-motion illusions in virtual reality? *IEEE Virtual Reality* (pp. 101–102). Lake Buena Vista, FL, USA. doi:10.1109/VR.2013.6549382.

Freiberg, J., Grechkin, T., & Riecke, B. E. (2013). Do walking motions enhance visually induced self-motion illusions in virtual reality? *Proceedings of IEEE Virtual Reality* (pp. 101–102). Orlando, FL.

Gillingham, K. K., & Krutz, R. W. (1974). *Aeromedical review—Effects of the abnormal acceleratory environment of flight* (Technical Report Review 10–74). Brooks Air Force Base, TX: U.S. Air Force School of Aerospace Medicine, Aerospace Medical Division.

Goldberg, M. J., Wilson, V. J., Cullen, K. E., Angelaki, D. E., Broussard, D. M., Buttner-Ennever, J., … Minor, L. B. (2011). *The vestibular system: A sixth sense*. New York, NY: Oxford University Press.

Golding, J. F., Arun, S., Wotton-Hamrioui, K., Cousins, S., & Gresty, M. A. (2009). Off-vertical axis rotation of the visual field and nauseogenicity. *Aviation, Space, and Environmental Medicine*, *80*(6), 516–521.

Graybiel, A. (1952). Oculogravic illusion. *Archives of Ophthalmology*, *48*, 605–615.

Graybiel, A., & Brown, R. H. (1950). The delay in visual reorientation following exposure to a change in direction of resultant force on a human centrifuge. *Journal of General Psychology*, *45*, 143–150.

Graybiel, A., & Hupp, D. I. (1946). The oculo-gyral illusion. A form of apparent motion which can be observed following stimulation of the semicircular canals. *Journal of Aviation Medicine*, *17*, 2–27.

Graybiel, A., & Kellogg, R. S. (1967). The inversion illusion in parabolic flight: Its probably dependence on otolith function. *Aerospace Medicine*, *38*, 1099–1103.

Graybiel, A., & Patterson, J. L., Jr. (1955). Thresholds of stimulation of the otolith organs as indicated by oculogravic illusion. *Journal of Applied Physiology*, *7*, 666–670.

Grubb, J., Schmorrow, D., & Johnson, B. (2011). VIMS challenges in the military. *Visual Image Safety Conference Proceedings*, Las Vegas, NV.

Guedry, F. E. (1968). The nonauditory labyrinth in aerospace medicine. In Naval Aerospace Medical Institute (Eds.), *U.S. naval flight surgeon's manual* (pp. 240–262). Washington, DC: U.S. Government Printing Office.

Guedry, F. E. (1974). Psychophysics of vestibular sensation. In H. H. Kornhuber (Ed.), *Handbook of sensory physiology* (pp. 1–154). New York, NY: Springer-Verlag.

Guedry, F. E. (1980). A multistation spatial disorientation demonstrator. In *Spatial Disorientation in Flight: Current Problems, Advisory Group for Aerospace Research and Development Conference Proceedings No. 287*. London, U.K.: Technical Editing and Reproduction Ltd.

Guedry, F. E. (1992). Perception of motion and position relative to Earth: An overview. In B. Cohen, D. L. Tomko, & F. E. Guedry (Eds.), *Sensing and controlling motion: Vestibular and sensorimotor function* (pp. 315–328). New York, NY: New York Academy of Sciences.

Guedry, F. E., Jr., & Benson, A. J. (1978). Coriolis cross-coupling effects: Disorienting and nauseogenic or not? *Aviation, Space, and Environmental Medicine*, *49*, 29–35.

Guedry, F. E., Lentz, J. M., Jell, R. M., & Norman, J. W. (1981). Visual–vestibular interactions: The directional component of visual background movement. *Aviation, Space, and Environmental Medicine*, *52*(5), 304–309.

Guedry, F. E., Mortenson, C. E., Nelson, J. B., & Correia, M. J. (1978). A comparison of nystagmus and turning sensations generated by active and passive turning. In J. D. Hood (Ed.), *Vestibular mechanisms in health and disease*. New York, NY: Academic Press.

Guedry, F. E., Raj, A. K., & Cowin, T. B. (2003). *Disorientation, dizziness, and postural imbalance in race car drivers, a problem in G-tolerance, spatial orientation or both*. Paper presented at the RTO HFM Symposium on Spatial Disorientation in Military Vehicles: Causes, Consequences and Cures, La Coruna, Spain.

Guedry, F. E., Richard, D. G., Hixson, W. C., & Niven, J. I. (1973). Observation on perceived changes in aircraft attitude attending head movements made in a 2-G bank and turn. *Aerospace Medicine*, *44*, 477–483.

Guedry, F. E., & Rupert, A. H. (1991). Steady state and transient g-excess effects, technical note. *Aviation, Space, and Environmental Medicine*, *62*, 252–253.

Hale, K. S., & Stanney, K. M., (Eds.). (2014). *Handbook of virtual environments: Design, implementation, and applications*, (2nd ed.). New York, NY: CRC Press.

Held, R., Dichgans, J., & Bauer, J. (1975). Characteristics of moving visual scenes influencing spatial orientation. *Vision Research*, *15*(3), 357–365, IN1. doi:10.1016/0042-6989(75)90083-8

Helmholtz, H. von. (1866). *Handbuch der physiologischen Optik*. Leipzig, Germany: Voss.

Hettinger, L. J., Berbaum, K. S., Kennedy, R. S., Dunlap, W. P., & Nolan, M. D. (1990). Vection and simulator sickness. *Military Psychology*, *2*(3), 171–181. doi:10.1207/s15327876mp0203_4

Hettinger, L. J., Schmidt, T., Jones, D. L., & Keshavarz, B. (2014). Illusory self-motion in virtual environments. In K. S. Hale & K. M. Stanney (Eds.), *Handbook of virtual environments: Design, implementation, and applications* (2nd ed., pp. 467–492). Boca Raton, FL: Taylor & Francis Group, Inc.

Hixson, W. C., Niven, J. I., & Correia, M. J. (1966). *Kinematics nomenclature for physiological accelerations: With special reference to vestibular applications* [Monograph No. 14]. Pensacola, FL: Naval Aerospace Medical Institute.

Hodgson, E., Bachmann, E., & Waller, D. (2011). Redirected walking to explore virtual environments: Assessing the potential for spatial interference. *ACM Transactions on Applied Perception*, *8*(4), 22:1–22:22. doi:10.1145/2043603.2043604

Hollerbach, J. M. (2002). Locomotion interfaces. In K. M. Stanney (Ed.), *Handbook of virtual environments: Design, implementation, and applications* (pp. 239–254). Mahwah, NJ: Lawrence Erlbaum Associates.

Howard, I. P. (1986a). The perception of posture, self motion, and the visual vertical. In K. R. Boff, L. Kaufman, & J. P. Thomas (Eds.), *Sensory processes and perception, Handbook of human perception and performance* (Vol. 1, pp. 18.1–18.62). New York, NY: Wiley.

Howard, I. P. (1986b). The vestibular system. In K. R. Boff, L. Kaufman, & J. P. Thomas (Eds.), *Sensory processes and perception, Handbook of human perception and performance* (Vol. 1, pp. 11.1–11.30). New York, NY: Wiley.

Howard, I. P., & Childerson, L. (1994). The contribution of motion, the visual frame, and visual polarity to sensations of body tilt. *Perception*, *23*(7), 753–762. doi:10.1068/p230753

Howard, I. P., Jenkin, H. L., & Hu, G. (2000). Visually-induced reorientation illusions as a function of age. *Aviation, Space, and Environmental Medicine*, *71*(9 Suppl), A87–A91.

Ito, H., & Shibata, I. (2005). Self-motion perception from expanding and contracting optical flows overlapped with binocular disparity. *Vision Research*, *45*(4), 397–402. doi:10.1016/j.visres.2004.11.009

Kennedy, R. S., Drexler, J. M., Compton, D. E., Stanney, K. M., Lanham, D. S., & Harm, D. L. (2003). Configural scoring of simulator sickness, cybersickness, and space adaptation syndrome: Similarities and differences. In Hettinger, L. J., & Haas, M. W. (Eds.), *Virtual and Adaptive Environments: Applications, Implications, and Human Performance Issues* (pp. 247–278). Boca Raton, FL: CRC Press.

Keshavarz, B., Hecht, H., & Lawson, B. D. (2014) Visually-induced motion sickness: Causes, characteristics, and countermeasures. In K. S. Hale & K. M. Stanney (Eds.), *Handbook of virtual environments: Design, implementation, and applications* (2nd ed., pp. 647–697). New York, NY: CRC Press.

Kitazaki, M., Onimaru, S., & Sato, T. (2010). *Vection and action are incompatible*. Presented at the 2nd IEEE VR 2010 Workshop on Perceptual Illusions in Virtual Environments (PIVE) (pp. 22–23), Waltham, MA.

Kitazaki, M., & Sato, T. (2003). Attentional modulation of self-motion perception. *Perception*, *32*(4), 475–484. doi:10.1068/p5037

Kolasinski, E. M. (1995). *Simulator sickness in virtual environments* (Technical Report No. 1027). Alexandria, VA: U.S. Army Research Institute for the Behavioral and Social Sciences.

Lackner, J. R. (1977). Induction of illusory self-rotation and nystagmus by a rotating sound-field. *Aviation, Space, and Environmental Medicine*, *48*, 129–131.

Lackner, J. R. (1988). Some proprioceptive influences on the perceptual representation of body shape and orientation. *Brain*, *111*, 281–297.

Lackner, J. R. (1992). Spatial orientation in weightless environments. *Perception*, *2*, 803–812.

Lackner, J. R., & DiZio, P. (1984). Some efferent and somatosensory influences on body orientation and oculomotor control. In L. Spillman & B. R. Wooten (Eds.), *Sensory experience, adaptation, and perception* (pp. 281–301). Clifton, NJ: Lawrence Erlbaum Associates.

Lackner, J. R., & DiZio, P. (1988). Visual stimulation affects the perception of voluntary leg movements during walking. *Perception*, *17*, 71–80.

Lackner, J. R., & DiZio, P. (1993). Spatial stability, voluntary action and causal attribution during self locomotion. *Journal of Vestibular Research*, *3*, 15–23.

Lackner, J. R., & Graybiel, A. (1974). Elicitation of vestibular side effects by regional vibration of the head. *Aerospace Medicine*, *45*, 1267–1272.

Lackner, J. R., & Graybiel, A. (1977). Somatosensory motion after-effect following Earth-horizontal rotation about the z-axis: A new illusion. *Aviation, Space, and Environmental Medicine*, *48*, 501–502.

Lackner, J. R., & Graybiel, A. (1978). Postural illusions experienced during z-axis recumbent rotation and their dependence on somatosensory stimulation of the body surface. *Aviation, Space, and Environmental Medicine*, *49*, 484–488.

Lackner, J. R., & Graybiel, A. (1980). Visual and postural motion aftereffects following parabolic flight. *Aviation, Space, and Environmental Medicine*, *51*, 230–233.

Lackner, J. R., & Levine, M. S. (1979). Changes in apparent body orientation and sensory localization induced by vibration of postural muscles: Vibratory myesthetic illusions. *Aviation, Space, and Environmental Medicine*, *50*, 346–354.

Lackner, J. R., & Teixeira, R. A. (1977). Optokinetic motion sickness: Continuous head movements attenuate the visual induction of apparent self-rotation and symptoms of MS. *Aviation, Space, and Environmental Science*, *48*(3), 248–253.

Larsson, P., Västfjäll, D., & Kleiner, M. (2004). Perception of self-motion and presence in auditory virtual environments. *Proceedings of the Seventh Annual Workshop of Presence,* Valencia, Spain (pp. 252–258).

Lawson, B. D. (1995). Characterizing the altered perception of self motion induced by a Coriolis, cross-coupling stimulus. *Abstracted Proceedings of the Third International Symposium on the Head/Neck System,* Vail, CO.

Lawson, B. D. (2005). Exploiting the illusion of self-motion (vection) to achieve a feeling of "virtual acceleration" in an immersive display. In C. Stephanidis (Ed.), *Proceedings of the 11th International Conference on Human–Computer Interaction* (pp. 1–10). Las Vegas, NV: Lawrence Erlbaum Associates, Inc.

Lawson, B. D. (2014). Motion sickness symptomatology and origins. In K. S. Hale and K. M. Stanney (Eds.), *Handbook of virtual environments: Design, implementation, and applications* (2nd ed., pp. 533–602). New York, NY: CRC Press.

Lawson, B. D., Guedry, F. E., Rupert, A. H., & Anderson, A. M. (1994). Attenuating the disorienting effects of head movement during whole-body rotation using a visual reference: Further tests of a predictive hypothesis. In *Advisory Group for Aerospace Research and Development: Virtual Interfaces: Research and Applications.* Neuilly-Sur Seine, France: AGARD.

Lawson, B. D., Mead, A. M., Chapman, J. E., & Guedry, F. E. (1996). Perception of orientation and motion during centrifuge rotation. In *Proceedings of the 66th Annual Meeting of the Aerospace Medical Association. Atlanta GA: Aviation, Space, and Environmental Medicine,* Atlanta, GA (Vol. 67, p. 702).

Lawson, B. D., & Rupert, A. H. (2010, June 16–18). Vestibular aspects of head injury and recommendations for evaluation and rehabilitation following exposure to severe changes in head velocity or ambient pressure. In O. Turan, J. Bos, J. Stark, & J. Colwell (Eds.), *Peer-Reviewed Proceedings of the International Conference on Human Performance at Sea (HPAS)* (pp. 367–380). Glasgow, U.K.: University of Strathclyde. ISBN: 978-0-947649-73-9.

Lawson, B. D., Rupert, A. H., Guedry, F. E., Grissett, J. D., & Mead, A. M. (1997). The human–machine interface challenge of using virtual environment (VE) displays aboard centrifuge devices. In M. J. Smith, G. Salvendy, & R. J. Koubek (Eds.), *Design of computing systems: Social and ergonomic considerations* (pp. 945–948). Amsterdam, the Netherlands: Elsevier.

Lawson, B. D, Sides, S. A., & Hickinbotham, K. A. (2002). User requirements for perceiving body acceleration. In K. M. Stanney (Ed.), *Handbook of virtual environments: Design, implementation, and applications* (pp. 135–161). Mahwah, NJ: Lawrence Erlbaum Associates, Inc.

Leger, A., Landolt, J. P., & Money, K. E. (1980). *Illusions of attitude and movement during earth-horizontal rotation* (Report No: PUB-80-P-07). Townto, Ontario, Canada: Defence and Civil Inst of Environmental Medicine.

Lenggenhager, B., Lopez, C., & Blanke, O. (2008). Influence of galvanic vestibular stimulation on egocentric and object-based mental transformations. *Experimental Brain Research, 184*(2), 211–221. doi:10.1007/s00221-007-1095-9

Lepecq, J. C., De Waele, C., Mertz-Josse, S., Teyssedre, C., Huy, P. T. B., Baudonniere, P. M., & Vidal, P. P. (2006). Galvanic vestibular stimulation modifies vection paths in healthy subjects. *Journal of Neurophysiology, 95*(5), 3199–3207. doi:10.1152/jn.00478.2005

Lepecq, J. C., Giannopulu, I., & Baudonniere, P. M. (1995). Cognitive effects on visually induced body motion in children. *Perception, 24*(4), 435–449.

Levine, M. S., & Lackner, J. R. (1979). Some sensory and motor factors influencing the control and appreciation of eye and limb position. *Experimental Brain Research, 36,* 275–283.

Lewis, R. R., Preiesol, A. J., Nicoucar, K., Lim, K., & Merfeld, D. M. (2011). Dynamic tilt thresholds are reduced in vestibular migraine. *Journal of Vestibular Research, 21,* 323–330.

Loomis, J., & Knapp, J. (2003). Visual perception of egocentric distance in real and virtual environments. In L. J. Hettinger & M. W. Haas (Eds.), *Virtual and adaptive environments: Applications, implications, and human performance issues.* (pp. 21–46). Mahwah, NJ: Lawrence Erlbaum.

Ludel, J. (1978). *Introduction to sensory processes.* San Francisco, CA: W. H. Freeman.

Mach, E. (1875). *Grundlinien der Lehre von der Bewegungsempfindung.* Leipzig, Germany: Engelmann.

MacNeilage, P. R., Banks, M. S., Berger, D. R., & Bülthoff, H. H. (2007). A Bayesian model of the disambiguation of gravitoinertial force by visual cues. *Experimental Brain Research, 179*(2), 263–290. doi:10.1007/s00221-006-0792-0

Maeda, T., Ando, H., & Sugimoto, M. (2005). Virtual acceleration with galvanic vestibular stimulation in a virtual reality environment. *Virtual Reality (VR) 2005,* Bonn, Germany (pp. 289–290). IEEE. doi:10.1109/VR.2005.1492799

Mann, C. W., Dauterive, N. H., & Henry, J. (1949). The perception of the vertical: I. Visual and non-labyrinthine cues. *Journal of Experimental Psychology, 39,* 538–547.

Marcus, J. T. (1992). *Vestibulo-ocular responses in man to gravito-inertial forces* (Unpublished doctoral dissertation). University of Utrecht, Utrecht, the Netherlands.

Marme-Karelse, A. M., & Bles, W. (1977). Circular vection and human posture, II. Does the auditory system play a role? *Agressologie, 18*(6), 329–333.

Marquez, G., Aguado, X., Alegre, L. M., Lago, A., Acero, R. M., & Fernandez-del-Olmo, M. (2010). The trampoline aftereffect: The motor and sensory modulations associated with jumping on an elastic surface. *Experimental Brain Research, 204*(4), 575–584.

Mayne, R. (1974). A system concept of the vestibular organ. In: H. Kornhuber (Ed.), *Handbook of sensory physiology, Vol VI/2: The vestibular system* (pp. 493–580). Berlin, Germany/New York, NY: Springer.

Melvill-Jones, G., & Young, L. R. (1978). Subjective detection of vertical acceleration: A velocity-dependent response. Cited in K. R. Boff & J. E. Lincoln (Eds.) (1988). *Engineering data compendium: Human perception and performance*. Wright-Patterson Air Force Base, OH: AAMRL.

Mergner, T., & Becker, W. (1990). Perception of horizontal self-rotation: Multisensory and cognitive aspects. In R. Warren & A. H. Wertheim (Eds.), *Perception & control of self-motion* (pp. 219–263). London, U.K.: Erlbaum.

Minor, L. B., Solomon, D., Zinreich, J. S., & Zee, D. S. (1998). Sound- and/or pressure-induced vertigo due to bone dehiscence of the superior semicircular canal. *Archive Otolaryngologica Head and Neck Surgery, 124*(3), 249–258.

Müller, G. E. (1916). Über das Aubertsche Phänomen. *Zeitschrift für Psychologie und Physiologie der Sinnesorgane, 49*, 109–244.

Nakamura, S. (2008). Effects of stimulus eccentricity on vection reevaluated with a binocularly defined depth. *Japanese Psychological Research, 50*(2), 77–86. doi:10.1111/j.1468-5884.2008.00363.x

Nakamura, S., & Shimojo, S. (1999). Critical role of foreground stimuli in perceiving visually induced self-motion (vection). *Perception, 28*(7), 893–902.

National Research Council. (1992). Evaluation of tests for vestibular function N[2, Suppl.]. *Aviation, Space, and Environmental Medicine, 63*, A1–A34. (Report of the Working Group on Evaluation of Test for Vestibular Function; Committee on Hearing, Bioacoustics, and Biomechanics, and the Commission on Behavioral and Social Sciences and Education).

Nesti, A., Masone, C., Barnett-Cowan, M., Giordano, P. R., Bülthoff, H., & Pretto, P. (2012). Roll rate thresholds and perceived realism in driving simulation. *Actes INRETS*, 23–31.

Newman, M. C., Lawson, B. D., Rupert, A. H., & McGrath, B. J. (2012, August 15). The role of perceptual modeling in the understanding of spatial disorientation during flight and ground-based simulator training. *Proceedings of the American Institute of Aeronautics and Astronautics*, Minneapolis, MN, 14 p.

Nicogossian, A. E., Huntoon, C. L., & Pool, S. L. (Eds.). (1989). *Space physiology and medicine*. Philadelphia: Lea & Febiger.

Ohmi, M., & Howard, I. P. (1988). Effect of stationary objects on illusory forward self-motion induced by a looming display. *Perception, 17*(1), 5–12. doi:10.1068/p170005

Ohmi, M., Howard, I. P., & Landolt, J. P. (1987). Circular vection as a function of foreground-background relationships. *Perception, 16*(1), 17–22.

Onimaru, S., Sato, T., & Kitazaki, M. (2010). Veridical walking inhibits vection perception. *Journal of Vision, 10*(7), 860. doi:10.1167/10.7.860

Palmisano, S., Allison, R. S., & Howard, I. P. (2006). Illusory scene distortion occurs during perceived self-rotation in roll. *Vision Research, 46*(23), 4048–4058. doi:10.1016/j.visres.2006.07.020

Palmisano, S., Allison, R. S., Kim, J., & Bonato, F. (2011). Simulated viewpoint jitter shakes sensory conflict accounts of vection. *Seeing and Perceiving, 24*(2), 173–200. doi:10.1163/187847511X570817

Palmisano, S., Bonato, F., Bubka, A., & Folder, J. (2007). Vertical display oscillation effects on forward vection and simulator sickness. *Aviation, Space, and Environmental Medicine, 78*(10), 951–956.

Palmisano, S., Gillam, B. J., & Blackburn, S. G. (2000). Global-perspective jitter improves vection in central vision. *Perception, 29*(1), 57–67.

Parker, D. E, Tubbs, R. L., Ritz, L. A., & Wood, D. L. (1976). *Effects of sound on the vestibular system*. Technical Report 75-89. Wright-Patterson Air Force Base, OH: Aerospace Medical Research Laboratory.

Passey, G. E., & Ray, J. T. (1950). *The Perception of the vertical: 10. Adaptation effects in the adjustment of the visual vertical*. The Tulane University of Louisiana under Contract N7onr-434. U.S. Naval school of Aviation Medicine, U.S. Navy Publication NM001 063.01.17.

Pausch, R., Crea, T., & Conway, M. (1992). A literature survey for virtual environments: Military flight simulator visual systems and simulator sickness. *Presence, 1*(3), 344–363.

Previc, F. H., & Ercoline, W. R, (2004). Spatial Disorientation in Aviation. *Progress in Aeronautics and Astronautics, 203*, Reston, VA: AIAA, Inc.

Quarck, G., Etard, O., Normand, H., Pottier, M., & Denise, P. (1998). Low-intensity galvanic vestibulo-ocular reflex in normal subjects. *Neurophysiologie Clinique, 28*(5), 413–22.

Reason, J. T., & Brand, J. J. (1975). *Motion sickness*. London, U.K.: Academic Press.

Riecke, B., Bodenheimer, B., McNamara, T., Williams, B., Peng, P., & Feuereissen, D. (2010). Do we need to walk for effective virtual reality navigation? physical rotations alone may suffice. In C. Hölscher, T. Shipley, M. Olivetti Belardinelli, J. Bateman, & N. Newcombe (Eds.), *Spatial Cognition VII, Lecture Notes in Computer Science* (Vol. 6222, pp. 234–247). Berlin/Heidelberg, Germany: Springer. Retrieved from doi: 10.1007/978-3-642-14749-4_21

Riecke, B. E. (2006). Simple user-generated motion cueing can enhance self-motion perception (Vection) in virtual reality. *Proceedings of the ACM Symposium on Virtual Reality Software and Technology (VRST)* (pp. 104–107). Limassol, Cyprus: ACM. doi:10.1145/1180495.1180517

Riecke, B. E. (2009). Cognitive and higher-level contributions to illusory self-motion perception ("vection"): Does the possibility of actual motion affect vection? *Japanese Journal of Psychonomic Science, 28*(1), 135–139.

Riecke, B. E. (2011). Compelling self-motion through virtual environments without actual self-motion—Using self-motion illusions ("Vection") to improve user experience in VR. In J.-J. Kim (Ed.), *Virtual Reality* (pp. 149–176). doi: 10.5772/13150. InTech. Retrieved from http://www.intechopen.com/articles/show/title/compelling-self-motion-through-virtual-environments-without-actual-self-motion-using-self-motion-ill.

Riecke, B. E., Cunningham, D. W., & Bülthoff, H. H. (2007). Spatial updating in virtual reality: The sufficiency of visual information. *Psychological Research, 71*(3), 298–313. doi:http://dx.doi.org/10.1007/s00426-006-0085-z

Riecke, B. E., & Feuereissen, D. (2012). To move or not to move: Can active control and user-driven motion cueing enhance self-motion perception ("Vection") in virtual reality? *ACM Symposium on Applied Perception SAP* (pp. 17–24). Los Angeles, CA: ACM. doi: 10.1145/2338676.2338680

Riecke, B. E., Feuereissen, D., & Rieser, J. J. (2009). Auditory self-motion simulation is facilitated by haptic and vibrational cues suggesting the possibility of actual motion. *ACM Transactions on Applied Perception (TAP), 6*, 20:1–20:22. doi:http://doi.acm.org.proxy.lib.sfu.ca/10.1145/1577755.1577763

Riecke, B. E., Feuereissen, D., Rieser, J. J., & McNamara, T. P. (2011). Spatialized sound enhances biomechanically-induced self-motion illusion (vection). *Proceedings of Conference on Human Factors in Computing Systems* (pp. 1–4). Presented at the (Chi'11), British Columbia, Vancouver, Canada: ACM Press.

Riecke, B. E., Feuereissen, D., Rieser, J. J., & McNamara, T. P. (2012). Self-motion illusions (vection) in VR— Are they good for anything? *IEEE Virtual Reality 2012* (pp. 35–38). Orange County, CA. doi:10.1109/VR.2012.6180875

Riecke, B. E., & Schulte-Pelkum, J. (2013). Perceptual and cognitive factors for self-motion simulation in virtual environments. In F. Steinicke, Y. Vissell, J. L. Campos, & A. Lecuyer (Eds.), *Human walking in virtual environments*. Berlin, Germany: Springer. Retrieved from http://www.springer.com/engineering/robotics/book/978-1-4419-8431-9.

Riecke, B. E., Schulte-Pelkum, J., Avraamides, M. N., Heyde, M. V. D., & Bülthoff, H. H. (2006). Cognitive factors can influence self-motion perception (vection) in virtual reality. *ACM Transactions on Applied Perception (TAP), 3*(3), 194–216. doi:10.1145/1166087.1166091

Riecke, B. E., Schulte-Pelkum, J., Caniard, F., & Bülthoff, H. H. (2005). Towards lean and elegant self-motion simulation in virtual reality. *Proceedings of the 2005 IEEE Conference 2005 on Virtual Reality, VR'05* (pp. 131–138). doi:10.1109/VR.2005.83

Riecke, B. E., Sigurdarson, S., & Milne, A. P. (2012). Moving through virtual reality without moving? *Cognitive Processing*, in print. doi:10.1007/s10339-012-0491-7

Riecke, B. E., Väljamäe, A., & Schulte-Pelkum, J. (2009). Moving sounds enhance the visually-induced self-motion illusion (circular vection) in virtual reality. *ACM Trans. Appl. Percept., 6*(2), 1–27.

Riecke, B. E., Västfjäll, D., Larsson, P., & Schulte-Pelkum, J. (2005). Top-down and multi-modal influences on self-motion perception in virtual reality. *Proceedings of HCI international 2005* (pp. 1–10). Las Vegas, NV. Retrieved from http://en.scientificcommons.org/20596227.

Riecke, B. E., von der Heyde, M., & Bülthoff, H. H. (2005). Visual cues can be sufficient for triggering automatic, reflex-like spatial updating. *ACM Transactions on Applied Perception (TAP), 2*(3), 183–215. doi:http://doi.acm.org/10.1145/1077399.1077401

Ruddle, R. A., & Lessels, S. (2009). The benefits of using a walking interface to navigate virtual environments. *ACM Transactions on Computer-Human Interaction, 16*(1), 1–18. doi:10.1145/1502800.1502805

Salvendy, G., Ed. (2006). *Handbook of human factors and ergonomics* (3rd ed.). Hoboken, NJ: Wiley.

Schulte-Pelkum, J. (2007). *Perception of self-motion: Vection experiments in multi-sensory virtual environments* (PhD thesis). Ruhr-Universität Bochum. Retrieved from http://www-brs.ub.ruhr-uni-bochum.de/netahtml/HSS/Diss/SchultePelkumJoerg/).

Seno, T., Ito, H., & Sunaga, S. (2009). The object and background hypothesis for vection. *Vision Research*, *49*(24), 2973–2982. doi:10.1016/j.visres.2009.09.017

Seno, T., Ito, H., & Sunaga, S. (2011). Inconsistent locomotion inhibits vection. *Perception*, *40*(6), 747.

St. George, R. J., & Fitzpatrick, R. C. (2011). The sense of self-motion, orientation and balance explored by vestibular stimulation. *The Journal of Physiology*, *589*(4), 807–813. doi:10.1113/jphysiol.2010.197665

Stanney, K. M., & Cohn, J. V. (2006). Virtual environments. In G. Salvendy (Ed.), *Handbook of human factors and ergonomics*, (3rd ed., pp. 1079–1096). Hoboken, NJ: John Wiley & Sons.

Stanney, K. M., Mourant, R. R., & Kennedy, R. S. (1998). Human factors issues in virtual environments: A review of the literature. *Presence, 7*(4), 327–351.

Stanney, K. M., Salvendy, G., Deisinger, J., DiZio, P., Ellis, S., Ellison, J., … Witmer, B. (1998). Aftereffects and sense of presence in virtual environments: Formulation of a research and development agenda. *International Journal of Human-Computer Interaction, 10*(2), 135–187.

Steen, F. A. M. van der. (1998). *Self-motion perception* (PhD thesis). Delft University of Technology, Delft, the Netherlands.

Steinicke, F., Bruder, G., Jerald, J., Frenz, H., & Lappe, M. (2010). Estimation of Detection Thresholds for Redirected Walking Techniques. *IEEE Transactions on Visualization and Computer Graphics*, *16*(1), 17–27. doi:10.1109/TVCG.2009.62

Steinicke, F., Vissell, Y., Campos, J. L., & Lecuyer, A. (Eds.). (2013). *Human walking in virtual environments*. Berlin, Germany: Springer. Retrieved from http://www.springer.com/engineering/robotics/book/978-1-4419-8431-9.

Suma, E. A., Bruder, G., Steinicke, F., Krum, D. M., & Bolas, M. (2012). A taxonomy for deploying redirection techniques in immersive virtual environments. *Virtual Reality Workshops (VR), 2012 IEEE* (pp. 43–46). doi:10.1109/VR.2012.6180877

Suma, E. A., Finkelstein, S. L., Reid, M., Babu, S. V., Ulinski, A. C., & Hodges, L. F. (2010). Evaluation of the cognitive effects of travel technique in complex real and virtual environments. *IEEE Transaction on Visualization and Computer Graphics, 16*(4), 690–702. doi:10.1109/TVCG.2009.93

Teixeira, R. A., & Lackner, J. R. (1979). Optokinetic motion sickness: Attenuation of visually-induced apparent rotation by passive head movements. *Aviation, Space, and Environmental Medicine, 50*, 264.

Travis, R. C., & Dodge, R. (1928). Experimental analysis of the sensorimotor consequences of passive oscillation, rotary rectilinear. Cited in K. R. Boff & J. E. Lincoln (Eds.), *Engineering data compendium: Human perception and performance*. Wright-Patterson Air Force Base, OH: AAMRL. (1988).

Tschermak, A. (1931). Optischer Raumsinn. In A. Bethe, G. Bergmann, G. Embden, & A. Ellinger (Eds.), *Handbuch der Normalen und Pathologischen Physiologie* (pp. 834–1000). Berlin, Germany: Springer.

Turing, A. (1950). Computing machinery and intelligence. *Mind, 59*, 433–60.

Urbantschitsch, V. (1897). Über Störungen des Gleichgewichtes und Scheinbewegungen. *Z. Ohrenheilk.*, *31*, 234–294.

Usoh, M., Arthur, K., Whitton, M., Steed, A., Slater, M., & Brooks, F. (1999, August 11–13). Walking: Virtual walking: Flying, in virtual environments. *Proceedings of SIGGRAPH99 Computer Graphics Annual Conference Series*, (pp. 359–364). Los Angeles, CA.

Väljamäe, A. (2009). Auditorily-induced illusory self-motion: A review. *Brain Research Reviews*, *61*(2), 240–255. doi:10.1016/j.brainresrev.2009.07.001

Väljamäe, A., Larsson, P., Västfjäll, D., & Kleiner, M. (2006). Vibrotactile enhancement of auditory induced self-motion and spatial presence. *Journal of the Acoustic Engineering Society*, *54*(10), 954–963.

Viirre, E., Sessoms, P., & Gotshall, K. (2012, June 10–13). Head stabilization during walking as a predictor of performance in persons with amputation and traumatic brain injury: Pilot study [abstract P100]. *27th Meeting of the Bárany Society*, Uppsala, Sweden.

Wade, N. J., & Day, R. H. (1968). Apparent head position as a basis for a visual aftereffect of prolonged head tilt. Cited in K. R. Boff & J. E. Lincoln (Eds.) (1988). *Engineering data compendium: Human perception and performance*. Wright-Patterson Air Force Base, OH: AAMRL.

Wardman, D. L., & Fitzpatrick, R. C. (2002). What does galvanic vestibular stimulation stimulate? *Advances in Experimental Medicine and Biology*, *508*, 119–128.

Wardman, D. L., Taylor, J. L., & Fitzpatrick, R. C. (2003). Effects of galvanic vestibular stimulation on human posture and perception while standing. *The Journal of Physiology*, *551*(3), 1033–1042. doi:10.1111/j.1469-7793.2003.01033.x

Warren, H. C. (1895). Sensations of rotation. *Psychological Review*, *2*(3), 273–276. doi:10.1037/h0074437

Warren, R., & Wertheim, A. H. (Eds.). (1990). *Perception & Control of Self-Motion*. Hillsdale, NJ: Erlbaum.

Warren, W. H., Kay, B. A., Zosh, W. D., Duchon, A. P., & Sahuc, S. (2001). Optic flow is used to control human walking. *Nature Neuroscience*, *4*(2), 213–216.

Whiteside, T. C. D. (1961). Hand–eye coordination in weightlessness. *Aerospace Medicine, 32*, 719–725.

Wong, S. C. P., & Frost, B. J. (1981). The effect of visual-vestibular conflict on the latency of steady-state visually induced subjective rotation. *Perception & Psychophysics*, *30*(3), 228–236.

Wood, R. W. (1895). The "Haunted Swing" illusion. *Psychological Review*, *2*(3), 277–278. doi:10.1037/h0073333

Wood, S. J. (2002). Human otolith-ocular reflexes during off-vertical axis rotation: Effect of frequency on tilt-translation ambiguity and motion sickness. *Neuroscience Letters*, *323*, 41–44.

Wright, W. G. (2009). Linear vection in virtual environments can be strengthened by discordant inertial input. *31st Annual International Conference of the IEEE EMBS (Engineering in Medicine and Biology Society)* (pp. 1157–1160). Minneapolis, MN. doi:10.1109/IEMBS.2009.5333425

Wright, W. G., DiZio, P., & Lackner, J. R. (2005). Vertical linear self-motion perception during visual and inertial motion: More than weighted summation of sensory inputs. *Journal of Vestibular Research - Equilibrium & Orientation*, *15*(4), 185–195.

Wright, W. G., DiZio, P., & Lackner, J. R. (2006). Perceived self-motion in two visual contexts: dissociable mechanisms underlie perception. *Journal of Vestibular Research*, *16*(1–2), 23–28.

Young, L. R., Dichgans, J., Murphy, R., & Brandt, T. (1973). Interaction of optokinetic and vestibular stimuli in motion perception. *Acta Oto-Laryngologica*, *76*(1), 24–31.

Young, L. R., Oman, C. M., & Dichgans, J. M. (1975). Influence of head orientation on visually induced pitch and roll sensation. *Aviation, Space, and Environmental Medicine*, *46*(3), 264–268.

Zhang, R., Nordman, A., Walker, J., & Kuhl, S. A. (2012). Minification affects verbal and action-based distance judgments differently in head-mounted displays. *ACM Transactions on Applied Perception, 9*(3), 1–13.

8 Eye Tracking in Virtual Environments

Xuezhong Wang and Brent Winslow

CONTENTS

8.1 INTRODUCTION

Eye tracking, that encompasses the scientific study and measurement of eye movements, has been evolving for over a century (Delabarre, 1898; Dodge & Cline, 1901). Initially, it was developed for observing basic visual behaviors, but early technology was invasive, painful, and of low accuracy. As the technology advanced, eye tracking has become more accurate, user friendly, and much more affordable (Hiley, Redekopp, & Fazel-Rezai, 2006). It has extended to various fields and developed into a wide range of applications (Mele & Federici, 2012). Eye tracking applications generally fall into two categories: diagnostic and interactive (Duchowski, 2002). Diagnostic applications aim to monitor eye movements and pupil constriction and evaluate user's visual attention and cognitive load (Tokuda, Obinata, Palmer, & Chaparro, 2011). In the medical field, eye tracking has been used to assess and improve surgeon training (Atkins, Tien, Khan, Meneghetti, & Zheng, 2013). Wilson suggested that gaze control parameters extracted from eye movement data can be used to differentiate novices from experts when performing laparoscopic surgery in a virtual reality (VR) simulator (Wilson et al., 2010). Tomizawa, using similar reasoning, found significant differences in eye scan patterns between expert and novice perfusionists when conducting extracorporeal circulation surgery (Tomizawa, Aoki, Suzuki, Matayoshi, & Yozu, 2012). Recent reports have shown the utility of eye tracking in the diagnosis of progressive supranuclear palsy in which decreased ability to carry out vertical saccades is a key symptom (Marx et al., 2012). Eye tracking has also been used to diagnose driver distraction or fatigue in real time and to mitigate driver error or accident (Qiang, Zhiwei, & Lan, 2004). In marketing, eye tracking has also been used to examine the effectiveness of warning labels (Strasser, Tang, Romer, Jepson, & Cappella, 2012).

Interactive eye tracking applications are designed to respond to or interact with the user on the basis of observed eye movements and constitute a promising means of communication for people with severe disabilities (Lin, Chang, & Jain, 2002; Sesin, Adjouadi, Cabrerizo, Ayala, & Barreto, 2008). For the general population, eye tracking has the potential to enhance daily human–computer interfaces (HCI) by functioning as a noncontact interface device (Nouredding, Lawrence, & Man, 2005). However, there are inherent difficulties in using a perceptual organ such as the eye as a control organ, since gaze data only provide information on where the eye is focusing, not necessarily user intent. Outstanding issues remain including the relative accuracy of the eye tracker, measured in visual degrees, and the *Midas touch* problem (Jacob, 1991), named after the character from Greek mythology that was cursed in that anything he touched turned to gold. In eye tracking, this refers to the fact that the eyes are never *off*, and would activate anything that is looked at, thus requiring a need to engage/disengage eye-gaze control. Conventionally, a long dwell time is used to signal user intent, such that a mouse click or button press would occur after the user had fixated on an area of interest for a certain period of time (Jacob).

This chapter first provides an overview of the working theory behind current eye tracking systems and common metrics available for diagnosis and interaction. Then, a brief survey of different types of eye trackers, including commercial and open-source eye tracking systems, is provided. Finally, a case study is presented, demonstrating the integration of a low-cost head-mounted eye tracker within a pair of VR goggles for a flight simulation study.

8.2　STEREOTYPICAL EYE MOVEMENT AND EYE TRACKERS

8.2.1　Eye Movement

Eye movements connect external stimuli with virtual perception during visual observation of the surrounding environment. In addition, eye movements are highly stereotyped in nature compared to other skeletal muscle–based movements. This feature is generally attributed to the fact that the extraocular muscles deal with almost the same mechanical load throughout life (Porter et al., 2001). The reader is referred to several excellent texts for an in-depth analysis of models of eye movements, visual attention, and visual psychophysics (Duchowski, 2007; Enderle, 2010; Hammoud, 2008; Sundstedt, 2012). Briefly, based on functional and neurological differences, major eye movements can be classified into four categories: saccade, pursuit, vergence, and vestibule-ocular (VO) (Robinson, 1968). Such categories may be classified based upon whether the eyes are moving together, termed conjunctive movements, or separately, termed disjunctive movements, based upon their velocity or function.

Saccades are rapid eye movements with velocities as high as $800°\%$ (Abrams, Meyer, & Kornblum, 1989). They occur frequently during reading, looking at a scene, or searching for an object. During saccades, sensitivity to visual input is reduced because the eyes are moving so quickly that only a blurred image would be perceived. The peak velocity and duration of a saccade is mostly determined as a function of the distance covered by the eye movements. Pursuit eye movements occur when the eyes follow a slowly moving target. Vergence is a disjunctive movement in which the eyes rotate in the opposite direction to accommodate a change of visual field depth. VO eye movements occur to compensate for head movement and maintain fixation on an object of interest (Table 8.1).

Additional smaller movements exist, including nystagmus, drifts, and microsaccades. These movements are often categorized with fixational eye movements since they are beyond the threshold of most motion detection systems (Nakayama & Tyler, 1978). Nystagmus is the low-magnitude constant tremor of the eyes (Engelken, Stevens, & Enderle, 1991; Galiana, 1991). Nystagmus is generally considered to be related to perceptual activity in that retinal neural activity is high during nystagmus (Abadi, 2002). Drifts and microsaccades are larger movements in magnitude than the nystagmus. Drift is a small and slow movement caused by pulsation and the less than perfect control of the oculomotor system (Poletti, Listorti, & Rucci, 2010). Microsaccades are small rapid movements to compensate for drift (Rayner, 1998). It should be noted that normal visual perception

TABLE 8.1
Classification of Major Eye Movements

Eye Movement	Directions	Velocity/Frequency	Function
Saccade	Conjunctive	400°–800°/s	Switch focus
Pursuit	Conjunctive	Up to 100°/s	Follow slow-motion target
Vergence	Disjunctive	20°/s	Focus
VO	Conjunctive	300°/s	Compensate head motion
Fixation	Random	About 60 Hz	Maintain eye on fixed location

involves all types of eye movements. Though it is usually not practical to distinguish different types of micro eye movements, fixation serves as a vital metric for both analysis and gaze-based interaction.

8.2.2 CONTACT AND NONCONTACT EYE TRACKERS

Generally, eye tracking systems are divided into three categories: electrooculography (EOG), contact lens, and video-oculography (VOG). EOG requires multiple electrodes to be attached to facial muscles such as corrugator supercilii and orbicularis oculi, to measure the electric potential changes caused by eye movement (Bulling, Ward, Gellersen, & Troster, 2011). EOG has the advantage that it does not interfere with corrective lenses and is relatively inexpensive (Figure 8.1). In addition, since EOG does not depend upon the eyes being open to track eye position, it has been useful in sleep studies and the description of sleep cycles (Aserinsky & Kleitman, 1953).

Contact lens eye tracking is a more invasive approach. A small coil is embedded onto a contact lens that is worn by the subjects in an environment with an electromagnetic field present (Ferman, Collewijn, Jansen, & Van den Berg, 1987; Robinson, 1963). The change in potential induced in the coil is measured to estimate gaze coordinates (Murphy, Duncan, Glennie, & Knox, 2001). Contact lens eye tracking is by far the most reliable approach with accuracy as high as 0.08°. However, the contact lens and associated wires produce ocular discomfort, increased intraocular pressure, and a reduction in visual acuity (Irving, Zacher, Allison, & Callender, 2003).

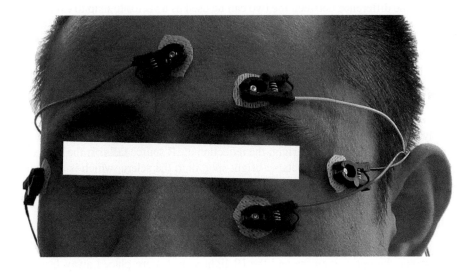

FIGURE 8.1 EOG system.

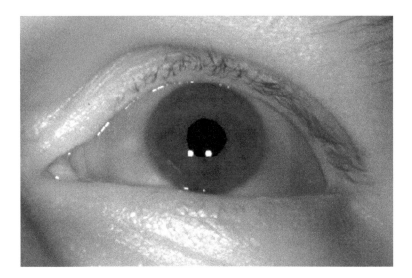

FIGURE 8.2   Pupil and corneal reflections.

VOG is a noninvasive eye tracking approach that depends on processing images of pupil position to extract features for gaze direction estimation. Due to its relatively low-cost and noninvasive nature, most commercial and open-source eye trackers are VOG-based systems. The working theory behind VOG eye trackers is introduced here to help readers better understand common procedures such as the calibration involved in this technology. The hardware usually includes an infrared (IR) illuminator, a camera with an IR pass filter, and a head motion tracker. The camera acquires the images of the eyes illuminated by IR light, and with the IR pass filter, only the face is visible which facilitates image processing. On the eye image acquired, based on the position of the IR illuminator, the pupil appears either bright (if the IR source is close to the camera axis, similar to the red-eye effect) or dark (if IR source is off axis with the camera). This bright/dark pupil effect makes pupil tracking much more robust in response to lighting conditions. Another important eye feature is the corneal reflection, usually seen as a bright dot close to the pupil (Figure 8.2). The corneal reflection is mainly used as the anchor point for compensating head motion in remote eye trackers. A wide range of image processing algorithms can be used to detect the pupil along with the corneal reflection, after which the difference between the two can be used as a vector to map to external screens. The mapping consists of a transformation of gaze position from eye image coordinates to on screen pixel coordinates. This is usually accomplished with a calibration procedure during which multiple calibration points appear on screen, and the user is instructed to fixate on each dot in turn so that the corresponding eye position on the eye image is calculated. The calibration point coordinates with regard to the screen being known; the calculated gaze position on the image is fitted using least squares estimation. Usually, a second-order polynomial is sufficient for mapping. In practice, failure to get a good calibration is one of the most common difficulties for novice eye tracker users. This is usually caused by failure to detect either the pupil center or the corneal reflection accurately. When eyes are rotated to the far left or right, the reflection of IR source falls off the edge of cornea, a smooth surface that results in a detectable bright spot, onto the sclera, which is a much coarser surface. For this reason, the working range of most VOG eye trackers is limited to about 40°.

8.2.3   HEAD-MOUNTED AND REMOTE EYE TRACKERS

VOG eye tracking systems are commonly grouped into two categories: remote and head mounted. In remote systems, the optics (camera and the IR light sources) are placed away from the user, usually right below the monitor mounted on a stand (Figure 8.3). Remote VOG systems have the

FIGURE 8.3 Easy Gaze® remote eye tracker and Tobii head-mounted eye tracker. (Image from www.tobi/eye-tracking-research.com.)

advantage of lower invasiveness, but due to the fact that the optics are at a greater distance from the user, a high-resolution camera with a long focal length lens is required to obtain eye images of sufficient resolution. Typically, the raw image captured will include the whole facial area. To locate the pupils, sophisticated image processing algorithms are used to first detect the eyes, which can be computationally demanding. Head motion is accounted for as long as the eyes are within the field of view of the camera. Remote eye trackers are usually not compatible with VR helmets or head-mounted displays (HMDs), thus limiting the utility of remote eye tracking in immersive virtual environments (VEs).

In head-mounted eye trackers (Figure 8.3), the optics are usually embedded into a wearable mounting device such as glasses or goggles, making it possible to track gaze points within a VE presented to the user. In addition, because the optics are much closer to the eye, less sophisticated cameras or even webcams are sufficient for image acquisition. Since the captured images only include the eyes, simple feature detection is adequate to estimate the center of pupils. However, under certain circumstances, such as flight simulation with surrounding screens outside the cockpit, head-mounted eye trackers require additional motion tracking to compensate for user head movement (Anders, 2001). Depending on the setup of the applications, choosing the appropriate eye tracker is the first and most critical step to apply eye tracking technology successfully.

8.2.4 EYE TRACKING RESEARCH IN ANIMAL MODELS

Animal research in VE is beginning to elucidate the underlying neuronal basis of visual coding, search, and spatial navigation in awake, behaving species. Recently, Harvey developed a method to place head-restrained mice onto a spherical treadmill, which then interacted within a VE to perform spatial movements while neuronal activity was recorded with implanted microelectrodes (Harvey, Collman, Dombeck, & Tank, 2009) or monitored with in vivo microscopy (Dombeck, Harvey, Tian, Looger, & Tank, 2010). These studies began to elucidate the functions of hippocampal place cells and how they encode information but, more importantly, opened a new chapter in VR research by allowing the study of neuronal circuits during activity. It is also feasible to include eye tracking within such a VE. Previous work in rodent eye tracking has used VOG to track pupil movements in head-fixed rodents (Stahl, 2004a, 2004b; Stahl, van Alphen, & DeZeeuw, 2000; Zoccolan, Graham, & Cox, 2010). A combinational approach wherein rodent eye tracking occurs alongside neuronal recording will begin to unravel the function of the brain at a higher level than has been possible previously.

8.3 USE OF EYE TRACKERS FOR HCI AND COGNITIVE STATE MEASUREMENT

Among early studies on utilizing gaze control for HCI, simple mouse clicking and scrolling was attempted with gaze (Jacob, 1991). It turned out that such an interaction scheme was not efficient since eyes move very differently from hand interactions with a mouse. Later studies attempted to mitigate this issue by using eye tracking in tandem with traditional mouse interaction (Hornof & Halverson, 2002), which was expected to reduce manual control stress and fatigue and to decrease performance time for long-distance pointing. However, the benefit gained by using eye tracking technology was not promising.

Aside from the *Midas touch* problem, early studies also suffered from the low accuracy of the available eye tracking devices. Eye tracking is very sensitive to environmental changes, and accuracy tends to deteriorate temporally that requires periodic recalibration. Such effects made eye tracking inefficient for daily HCI (Duchowski, 2002).

Various methods have been proposed to improve the usability of eye tracking technology. Given the fact that eye tracking accuracy is usually measured in visual angles instead of pixels, researchers focused on using a two-stage zoom and then select approach to enable a small icon on-screen be picked up by gaze control (Ashmore, Duchowski, & Shoemaker, 2005; Kumar, Paepcke, & Winograd, 2007; Lankford, 2000). These studies suggested that interfaces developed for computer mouse operation are not suitable for eye tracking techniques. New user interface icons and buttons must be designed to accommodate the dynamics of eye movements. New approaches aimed to integrate gaze input with multimodal interaction are emerging (Bulling et al., 2012). For example, applications combining gaze data with multitouch surfaces, hand gestures, or speech are actively under development (Bardins, Poitschke, & Kohlbecher, 2008; Sundstedt, 2010; Yoo et al., 2010).

Jacob (Jacob et al., 2008) categorized four themes of reality-based interaction involving eye tracking, including naïve physics, body awareness and skills, environmental, and social. Naïve physics refers to commonsense knowledge about the physical world. The information extracted from eye movements may be beneficial to support or improve some visual effects in VEs. For example, it has been proposed that eye tracking techniques can be used to improve camera motions and depth of field blur effects in VEs (Hillaire, Lecuyer, Cozot, & Casiez, 2008). Some researchers also suggest that gaze vergence can drive stereoscopic display disparity (Duchowski, Pelfrey, House, & Wang, 2011) or reduce visual fatigue for immersive stereoscopic displays (Leroy, Fuchs, & Moreau, 2012). Body awareness represents an individual's proprioception and coordination, in which gaze plays an important role (Vercher et al., 1996). Environment awareness denotes a person's sense of surroundings and how to interact with the environment. Applications with gaze control in this field include the use of fixation on a particular item to facilitate selection, the use of gaze to infer intention (Vertegaal, Shell, Chen, & Mamuji, 2006), and entering graphical passwords (Forget, Chiasson, & Biddle, 2010; Luca, Weiss, & Drewes, 2007). Social awareness is a person's understanding of others in the environment and how to interact with them. In this area, expert systems can respond to novice users' gaze when learning search strategies (Vitak, Ingram, Duchowski, Ellis, & Gramopadhye, 2012), and virtual agents can be improved by assessing users' attention through gaze location (Thies, Marc, & Ipke, 2009).

As a vital component of human perception, the visual system performs different perceptual tasks throughout life. These tasks range from simple learned tasks such as reading and visual search to increasingly complicated ones such as medical image diagnosis (Dempere-Marco et al., 2002) and monitoring of information displays, which require a higher level of knowledge. It is reasonable to expect that insights with regard to underlying cognitive load will be obtained by investigating eye movements and patterns. Metrics used for cognitive load measurement ranges from basic eye movements (fixation and saccade) and pupillometry (Hampson, Opris, & Deadwyler, 2010) to scan paths and the sequences of eye movements. When dealing with a challenging cognitive task, individual's pupils tend to dilate. This phenomenon, also referred to as the evoked pupillary response, provides the foundation for using pupillometric measures for cognitive load (Hampson et al., 2010; Iqbal, Zheng, & Bailey, 2004). There are a number of pupillometric cognitive load measures including index of cognitive activity (ICA), mean pupil diameter, and percent change of pupil size. ICA is defined as the frequency of pupil dilations per minute, which is used almost exclusively with head-mounted eye trackers (Marshall, 2002). Mean pupil diameter does not require very high precision to calculate and can be obtained with remote trackers. It is also more noise resistant than the ICA. Recently, it was also proposed to use pupil diameter change rate to estimate rapid cognitive load changes (Palinko, Kun, Shyrokov, & Heeman, 2010). By characterizing the shape of the pupil, it is possible to estimate the degree of eye opening or blinking, which is key to calculating percentage of eye closure over time (PERCLOS) and average eye closure speed. These metrics are good indicators of fatigue or drowsiness and are commonly used for driving simulation (Qiang et al., 2004) and bag screening tasks (Soussou, Rooksby, Forty, Weatherhead, & Marshall, 2012).

Electroencephalography (EEG) is a well-established noninvasive technique for the monitoring of brain activity with high temporal resolution and low cost. As such, EEG has proven to be a critical monitoring and diagnostic tool in the clinic during surgery as well as neurological assessment (Lagerlund, Cascino, Cicora, & Sharbrough, 1996; Mendez & Brenner, 2006). EEG is also a popular research tool among scientists for evaluating somatosensory responses to stimuli, error detection (Davidson, Jones, & Peiris, 2007), and sleep and fatigue monitoring (Colrain, 2011; Landolt, 2011). EEG also has been evaluated as a control or communication device (Wolpaw, Birbaumer, McFarland, Pfurtscheller, & Vaughan, 2002). Eye tracking technology offers the possibility of capturing visual behavior in real time and monitoring locations of fixations within images, and such time-synchronized eye behavior can be used to drive real-time parsing and assessment of EEG recordings, thus providing a better indication of underlying neuronal processes. Recently, several groups have reported the utility of combining EEG and eye tracking to assess simple visual search performance (Hale, Fuchs, Axelsson, Baskin, & Jones, 2007; Kamienkowski, Ison, Quiroga, & Sigman, 2012), correct artifacts in the EEG (Plochl, Ossandon, & Konig, 2012), improve brain–computer interfaces for disabled individuals (Lee, Woo, Kim, Whang, & Park, 2010), and study the process of reading (Dimigen, Sommer, Hohlfeld, Jacobs, & Kliegl, 2011) and in neuromarketing efforts (Khushaba et al., 2013).

8.4 INTEGRATION OF EYE TRACKERS WITH VE SYSTEMS

Depending on the hardware configurations of the VE, users may choose either a remote or head-mounted eye tracker to be integrated within a simulator. For remote eye trackers, the optics are stationary with regard to the display screens so that once calibration is completed, the output gaze point is stable with reference to the screen coordinate system. This is ideal for tasks performed in a sitting position such as driving simulation (Cai & Lin, 2012). One disadvantage is that the tracking range is usually only about 40°, which is not sufficient to cover a wide field of view. For simulators with multiscreen displays, a more advanced eye tracking system involving an array of illuminators and cameras is necessary. Head-mounted eye trackers can also be used for 3D eye tracking provided that an external head motion tracker is integrated. This approach also has the advantage of the optics being much smaller and lightweight, making it suitable to be integrated with HMDs or transparent display for augmented reality (AR).

8.4.1 COMMERCIAL SOLUTIONS

Table 8.2 lists major eye tracker manufacturers and their product lines. Remote eye trackers are more popular due to their noninvasive nature, cost, and suitability for a wider range of applications, such as behavioral research and usability testing. Regarding VE applications, advanced remote

TABLE 8.2
Major Eye Tracker Manufacturers and Models

Manufacturer	Remote	Head Mounted	HMD Integration
Arrington Research	BCU/MCU series	EyeFrame	Yes
Applied Science Laboratories	EYE-TRAC remote	Mobile Eye-XG	Yes
Cambridge Research	LiveTrack FM	None	No
DynaVox	EyeMax	None	No
Ergoneers	None	Dikablis	No
Eye Tech	TM series, VT2	None	No
ISCAN	iScan Labs	OmniView Mobile	Yes
Smart Eye	Smart Eye Pro	None	No
Sensomotoric Instruments	RED series	IVIEW X HED	?
Seeing Machine	faceLAB	None	No
SR Research	EyeLink series	None	No
Tobii	IS-2	None	No

eye trackers can be found in high-end simulators. For example, Smart Eye Pro supports up to eight cameras and covers a 360° field of view. Arrington Research's (AR) 3DWorkSpace provides similar capability and includes a head-mounted eye tracker paired with a fiducial marker-based motion tracking system. Applied Science Laboratories (ASL) also offers head-mounted eye trackers integrated with various motion trackers for multidisplay simulators. For immersive VE applications using HMDs, the options are more limited. AR and ASL are two companies that have experience in integrating eye tracking modules with major HMDs such as NVIS and Sensics. They provide out-of-box products as well as customization service to help users integrating eye tracking module with their own HMD devices. Different commercial eye trackers are generally more robust and under specific conditions (iris color, glasses/contact lens, etc.) but carry a price tag of around $30,000. As an alternative, there have been several low-cost eye tracking projects developed over the past few years, which provide an affordable solution to more users.

8.4.2 Low-Cost Solutions

With recent advances in computing technology, it has become feasible to develop low-cost eye tracking systems with off-the-shelf hardware at significantly lower prices than commercial products. Open-source software has also been developed to allow end users to experience eye tracking techniques for free. For example, ITU Gaze Tracker (Agustin et al., 2010) has exceeded 30,000 downloads since its debut in 2009. With an investment of less than $500 in hardware, a user can quickly set up a rudimentary remote eye tracking system sufficient for many applications. Figure 8.4 shows the prototype of *Raven*, a low-cost remote eye tracker consisting of a webcam and IR

FIGURE 8.4 Raven prototype with Gaze Tracker 2.0.

illuminators developed by Martin Tall with Gaze Tracker version 2.0. Gaze Tracker 2.0 also supports head-mounted eye tracking, without motion tracking support.

There are other open-source eye tracking systems including OpenEyes (Li, Babcock, & Parkhurst, 2006) and EyeWriter (Wiesspeiner, Lileg, & Hutten, 1999) that provide head-mounted tracking capability. The main difficulty faced by most users is the need to assemble eye tracking hardware themselves. Furthermore, without head motion tracking, these head-mounted eye trackers are not suitable to be used within simulators that require gaze direction to be reported with reference to the world. However, once the technical difficulties are resolved, these low-cost eye trackers are very versatile, flexible, and customizable. In the next section, a case study is presented to show the potential of such low-cost eye tracking solutions.

8.5 CASE STUDY OF INTEGRATION OF EYE TRACKERS WITH VE GOGGLES

The Helmet-Mounted Display ASsessment System for the Evaluation of eSsential Skills (HMD ASSESS) development effort is designed to capture performance by integrating eye tracking and EEG to measure perceptual and cognitive processes such as visual scan, attention allocation, and cognitive workload during HMD-based performance. Based on these measures, HMD ASSESS then diagnoses performance deficiencies such as failures in monitoring and detection and inefficiencies such as cognitive overload, distraction, or inefficient scan strategies and provides real-time and after-action review summaries of individualized performance issues. These summaries can be used to support training instructors in identifying skill decrements that need to be effectively remediated to achieve criterion performance, to assist system designers in gaining an understanding of how a pilot is interacting with the system, and to identify specific problem areas within the display. The phase I feasibility demonstration resulted in the development and integration of a low-cost eye tracking solution into the Joint Strike Fighter (JSF) Pilot Training Aid (PTA), a desktop training system for F-35 pilots that allows students to interface with the F-35 touch panel display and representative throttles and sticks. In the PTA, HMD information is represented through the use of Vuzix iWear 920 VR goggles.

Low-cost eye tracking hardware was made in-house with a modified webcam and IR LEDs. A Microsoft LifeCam VX 1000 webcam was modified by replacing the original filter in the webcam with a low-cost IR filter made from developed camera film. This modification was made because the VX 1000 original filter allows visual light to pass and the dark pupil images required by the eye tracking software can only be obtained by IR illumination. Additionally, three IR LEDs were installed as the illuminator, which required a 5 V 50 mA power supply. The power is provided by the LifeCam VX 1000 USB port, which significantly simplified the module. A 22 Ω resistor was connected to the LEDs in serial for power. The modified eye tracker module was mounted to the PTA VR goggles (Vuzix goggles) through an eight-gauge aluminum wire that not only provided the necessary stiffness to hold the module but also was easily adjustable to position the camera in the optimal direction. The module is lightweight, minimizing the impact on the user (Figure 8.5).

ITU Gaze Tracker 2.0 was utilized as the software platform for this project. Like most video-based eye tracking software programs, ITU GT relies on image processing to locate the center of the pupil, which is then mapped to screen coordinates. The eye tracking module streams video at 640×480 pixel resolution. Given that the webcam is situated relatively close to the eye, that resolution is sufficient to locate the gaze position on-screen precisely. With proper positioning, the accuracy from calibration is around $1°$ of visual angle, compared to the $0.5°$ standard of many commercial eye trackers. To prevent the eyes from being blocked by the goggles on the captured image, proper clearance between the eyes and the goggles must be ensured by adjusting the nose pad. It is also important to reduce image distorting by positioning the eye tracker module properly such that the eye is roughly at the center of the captured image. This low-cost eye tracking solution allows the collection of eye tracking metrics such as gaze position and fixation duration.

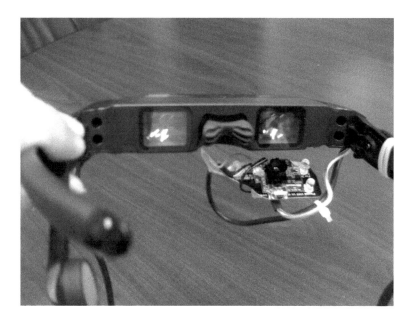

FIGURE 8.5   Low-cost eye tracker integrated with Vuzix VR goggles.

8.6 FUTURE DIRECTIONS

Eye tracking technology is under rapid development with the goal of improving accuracy and reliability and reducing cost. There have been attempts to integrate remote eye tracking with laptops resulting from recent progress in hardware design and implementation. Aside from hardware improvements, software support is also critical toward true gaze-based interaction. Providing a framework that handles low-level eye tracking details and a clean interface for application programmers will be key to promoting gaze-control-enabled applications. Nontraditional input modalities such as phones and tablets combined with eye tracking technique are also becoming more popular. With the success of smartphones, tablets, and game consoles, it is possible that eye tracking will be brought into mainstream products to facilitate interactions (Miluzzo, Wang, & Campbell, 2010).

REFERENCES

Abadi, R. V. (2002). Mechanisms underlying nystagmus. *Journal of the Royal Society of Medicine, 95*(5), 231–234.

Abrams, R. A., Meyer, D. E., & Kornblum, S. (1989). Speed and accuracy of saccadic eye movements: Characteristics of impulse variability in the oculomotor system. *Journal of Experimental Psychology: Human Perception and Performance, 15*(3), 529–543.

Agustin, J. S., Skovsgaard, H., Mollenbach, E., Barret, M., Tall, M., Hansen, D. W., & Hansen, J. P. (2010). *Evaluation of a low-cost open-source gaze tracker.* Paper presented at the Proceedings of the 2010 Symposium on Eye-Tracking Research & Applications, Austin, TX.

Anders, G. (2001). *Pilot's attention allocation during approach and landing—Eye and head-tracking research in an A 330 full flight simulator.* Focusing Attention on Aviation Safety: The 11th International Symposium on Aviation Psychology, Columbus, OH.

Aserinsky, E., & Kleitman, N. (1953). Regularly occurring periods of eye motility, and concomitant phenomena, during sleep. *Science, 118*(3062), 273–274.

Ashmore, M., Duchowski, A. T., & Shoemaker, G. (2005). *Efficient eye pointing with a fisheye lens.* Paper presented at the Proceedings of Graphics Interface 2005, Victoria, British Columbia, Canada.

Atkins, M. S., Tien, G., Khan, R. S., Meneghetti, A., & Zheng, B. (2013). What do surgeons see: Capturing and synchronizing eye gaze for surgery applications. *Surgical Innovation, 20*(3), 241–248.

Bardins, S., Poitschke, T., & Kohlbecher, S. (2008). *Gaze-based interaction in various environments.* Paper presented at the Proceedings of the First ACM Workshop on Vision Networks for Behavior Analysis, Vancouver, British Columbia, Canada.

Bulling, A., Dachselt, R., Duchowski, A., Jacob, R., Stellmach, S., & Sundstedt, V. (2012). *Gaze interaction in the post-WIMP world.* Paper presented at the Proceedings of the 2012 ACM Annual Conference Extended Abstracts on Human Factors in Computing Systems Extended Abstracts, Austin, TX.

Bulling, A., Ward, J. A., Gellersen, H., & Troster, G. (2011). Eye movement analysis for activity recognition using electrooculography. *IEEE Transactions on Pattern Analysis and Machine Intelligence, 33*(4), 741–753.

Cai, H., & Lin, Y. Z. (2012). An integrated head pose and eye gaze tracking approach to non-intrusive visual attention measurement for wide FOV simulators. *Virtual Reality, 16*(1), 25–32.

Colrain, I. M. (2011). Sleep and the brain. *Neuropsychology Review, 21*(1), 1–4.

Davidson, P. R., Jones, R. D., & Peiris, M. T. (2007). EEG-based lapse detection with high temporal resolution. *IEEE Transactions on Biomedical Engineering, 54*(5), 832–839.

Delabarre, E. B. (1898). A method of recording eye-movements. *American Journal of Psychology, 9*(4), 572–574.

Dempere-Marco, L., Hu, X. P., MacDonald, S. L., Ellis, S. M., Hansell, D. M., & Yang, G. Z. (2002). The use of visual search for knowledge gathering in image decision support. *IEEE Transactions on Medical Imaging, 21*(7), 741–754.

Dimigen, O., Sommer, W., Hohlfeld, A., Jacobs, A. M., & Kliegl, R. (2011). Coregistration of eye movements and EEG in natural reading: Analyses and review. *Journal of Experimental Psychology: General, 140*(4), 552–572.

Dodge, R., & Cline, T. S. (1901). The angle velocity of eye movements. *Psychological Review, 8*(2), 145–157.

Dombeck, D. A., Harvey, C. D., Tian, L., Looger, L. L., & Tank, D. W. (2010). Functional imaging of hippocampal place cells at cellular resolution during virtual navigation. *Nature Neuroscience, 13*(11), 1433–1440.

Duchowski, A. T. (2002). A breadth-first survey of eye-tracking applications. *Behavior Research Methods, Instruments, & Computers, 34*(4), 455–470.

Duchowski, A. T. (2007). *Eye tracking methodology—Theory and practice* (Vol. 373). London, U.K.: Springer.

Duchowski, A. T., Pelfrey, B., House, D. H., & Wang, R. (2011). *Measuring gaze depth with an eye tracker during stereoscopic display.* Paper presented at the Proceedings of the ACM SIGGRAPH Symposium on Applied Perception in Graphics and Visualization, Toulouse, France.

Enderle, J. D. (2010). *Models of horizontal eye movements. Part I: Early models of saccades and smooth pursuit* (Vol. 5). San Rafael, CA: Morgan & Claypool Publishers.

Engelken, E. J., Stevens, K. W., & Enderle, J. D. (1991). Optimization of an adaptive nonlinear filter for the analysis of nystagmus. *Biomedical Sciences Instrumentation, 27*, 163–170.

Ferman, L., Collewijn, H., Jansen, T. C., & Van den Berg, A. V. (1987). Human gaze stability in the horizontal, vertical and torsional direction during voluntary head movements, evaluated with a three-dimensional scleral induction coil technique. *Vision Research, 27*(5), 811–828.

Forget, A., Chiasson, S., & Biddle, R. (2010). *Shoulder-surfing resistance with eye-gaze entry in cued-recall graphical passwords.* Paper presented at the Proceedings of the 28th International Conference on Human Factors in Computing Systems, Atlanta, GA.

Galiana, H. L. (1991). A nystagmus strategy to linearize the vestibulo-ocular reflex. *IEEE Transactions on Biomedical Engineering, 38*(6), 532–543.

Hale, K. S., Fuchs, S., Axelsson, P., Baskin, A., & Jones, D. (2007). Determining gaze parameters to guide EEG/ERP evaluation of imagery analysis. In D. D. Schmorrow, D. M. Nicholson, J. M. Drexler, & L. M. Reeves (Eds.), *Foundations of augmented cognition* (4th ed.). Arlington, VA: Strategic Analysis, Inc.

Hammoud, R. (2008). *Passive eye monitoring—Algorithms, applications, and experiments.* Berlin, Germany: Springer.

Hampson, R. E., Opris, I., & Deadwyler, S. A. (2010). Neural correlates of fast pupil dilation in nonhuman primates: Relation to behavioral performance and cognitive workload. *Behavioral Brain Research, 212*(1), 1–11.

Harvey, C. D., Collman, F., Dombeck, D. A., & Tank, D. W. (2009). Intracellular dynamics of hippocampal place cells during virtual navigation. *Nature, 461*(7266), 941–946.

Hiley, J. B., Redekopp, A. H., & Fazel-Rezai, R. (2006). A low cost human computer interface based on eye tracking. *Conference Proceedings: IEEE Engineering in Medicine and Biology Society, 1*, 3226–3229.

Hillaire, S., Lecuyer, A., Cozot, R., & Casiez, G. (2008, March 8–12). *Using an Eye-tracking system to improve camera motions and depth-of-field blur effects in virtual environments.* Paper presented at the Virtual Reality Conference, Reno, NV. VR '08. IEEE.

Hornof, A. J., & Halverson, T. (2002). Cleaning up systematic error in eye-tracking data by using required fixation locations. *Behavior Research Methods, Instruments, & Computers, 34*(4), 592–604.

Iqbal, S. T., Zheng, X. S., & Bailey, B. P. (2004). *Task-evoked pupillary response to mental workload in human-computer interaction.* Paper presented at the CHI '04 Extended Abstracts on Human Factors in Computing Systems, Vienna, Austria.

Irving, E. L., Zacher, J. E., Allison, R. S., & Callender, M. G. (2003). Effects of scleral search coil wear on visual function. *Investigative Ophthalmology & Visual Sciences, 44*(5), 1933–1938.

Jacob, R. J. K. (1991). The use of eye movements in human-computer interaction techniques: What you look at is what you get. *ACM Transactions on Information Systems, 9*(2), 152–169.

Jacob, R. J. K., Girouard, A., Hirshfield, L. M., Horn, M. S., Shaer, O., Solovey, E. T., & Zigelbaum, J. (2008). *Reality-based interaction: A framework for post-WIMP interfaces.* Paper presented at the Proceedings of the 26th Annual SIGCHI Conference on Human Factors in Computing Systems, Florence, Italy.

Kamienkowski, J. E., Ison, M. J., Quiroga, R. Q., & Sigman, M. (2012). Fixation-related potentials in visual search: A combined EEG and eye tracking study. *Journal of Vision, 12*(7), 4.

Khushaba, R. N., Wise, C., Kodagoda, S., Louviere, J., Kahn, B. E., & Townsend, C. (2013). Consumer neuroscience: Assessing the brain response to marketing stimuli using electroencephalogram (EEG) and eye tracking. *Expert Systems with Applications, 40*(9), 3803–3812.

Kumar, M., Paepcke, A., & Winograd, T. (2007). *EyePoint: Practical pointing and selection using gaze and keyboard.* Paper presented at the Proceedings of the SIGCHI Conference on Human Factors in Computing Systems, San Jose, CA.

Lagerlund, T. D., Cascino, G. D., Cicora, K. M., & Sharbrough, F. W. (1996). Long-term electroencephalographic monitoring for diagnosis and management of seizures. *Mayo Clinic Proceedings, 71*(10), 1000–1006.

Landolt, H. P. (2011). Genetic determination of sleep EEG profiles in healthy humans. *Progress in Brain Research, 193*, 51–61.

Lankford, C. (2000). *Effective eye-gaze input into Windows.* Paper presented at the Proceedings of the 2000 Symposium on Eye Tracking Research & Applications, Palm Beach Gardens, FL.

Lee, E. C., Woo, J. C., Kim, J. H., Whang, M., & Park, K. R. (2010). A brain-computer interface method combined with eye tracking for 3D interaction. *Journal of Neuroscience Methods, 190*(2), 289–298.

Leroy, L., Fuchs, P., & Moreau, G. (2012). Visual fatigue reduction for immersive stereoscopic displays by disparity, content, and focus-point adapted blur. *IEEE Transactions on Industrial Electronics, 59*(10), 3998–4004.

Li, D., Babcock, J., & Parkhurst, D. J. (2006). *openEyes: A low-cost head-mounted eye-tracking solution.* Paper presented at the Proceedings of the 2006 Symposium on Eye Tracking Research & Applications, San Diego, CA.

Lin, C. S., Chang, K. C., & Jain, Y. J. (2002). A new data processing and calibration method for an eye-tracking device pronunciation system. *Optics & Laser Technology, 34*(5), 405–413.

Luca, A. D., Weiss, R., & Drewes, H. (2007). *Evaluation of eye-gaze interaction methods for security enhanced PIN-entry.* Paper presented at the Proceedings of the 19th Australasian Conference on Computer-Human Interaction: Entertaining User Interfaces, Adelaide, South Australia, Australia.

Marshall, S. P. (2002). *The Index of Cognitive Activity: Measuring cognitive workload.* Paper presented at the Proceedings of the 2002 IEEE Seventh Conference on Human Factors and Power Plants, Scottsdale, AZ.

Marx, S., Respondek, G., Stamelou, M., Dowiasch, S., Stoll, J., Bremmer, F., … Einhauser, W. (2012). Validation of mobile eye-tracking as novel and efficient means for differentiating progressive supranuclear palsy from Parkinson's disease. *Frontiers in Behavioral Neuroscience, 6*, 88.

Mele, M. L., & Federici, S. (2012). Gaze and eye-tracking solutions for psychological research. *Cognitive Processing, 13*(Suppl 1), S261–S265.

Mendez, O. E., & Brenner, R. P. (2006). Increasing the yield of EEG. *Journal of Clinical Neurophysiology, 23*(4), 282–293.

Miluzzo, E., Wang, T., & Campbell, A. T. (2010). *EyePhone: Activating mobile phones with your eyes.* Paper presented at the Proceedings of the Second ACM SIGCOMM Workshop on Networking, Systems, and Applications on Mobile Handhelds, New Delhi, India.

Murphy, P. J., Duncan, A. L., Glennie, A. J., & Knox, P. C. (2001). The effect of scleral search coil lens wear on the eye. *British Journal of Ophthalmology, 85*(3), 332–335.

Nakayama, K., & Tyler, C. W. (1978). Relative motion induced between stationary lines. *Vision Research, 18*(12), 1663–1668.

Nouredding, B., Lawrence, P. D., & Man, C. F. (2005). A non-contact device for tracking gaze in a human computer interface. *Computer Vision and Image Understanding, 98*(1), 52–82.

Palinko, O., Kun, A. L., Shyrokov, A., & Heeman, P. (2010). *Estimating cognitive load using remote eye tracking in a driving simulator*. Paper presented at the Proceedings of the 2010 Symposium on Eye-Tracking Research & Applications, Austin, TX.

Plochl, M., Ossandon, J. P., & Konig, P. (2012). Combining EEG and eye tracking: Identification, characterization, and correction of eye movement artifacts in electroencephalographic data. *Frontiers in Human Neuroscience, 6*, 278.

Poletti, M., Listorti, C., & Rucci, M. (2010). Stability of the visual world during eye drift. *The Journal of Neuroscience, 30*(33), 11143–11150.

Porter, J. D., Khanna, S., Kaminski, H. J., Rao, J. S., Merriam, A. P., Richmonds, C. R., ... Andrade, F. H. (2001). Extraocular muscle is defined by a fundamentally distinct gene expression profile. *Proceedings of the National Academy of Sciences of the United States of America, 98*(21), 12062–12067.

Qiang, J., Zhiwei, Z., & Lan, P. (2004). Real-time nonintrusive monitoring and prediction of driver fatigue. *IEEE Transactions on Vehicular Technology, 53*(4), 1052–1068.

Rayner, K. (1998). Eye movements in reading and information processing: 20 years of research. *Psychological Bulletin, 124*(3), 372–422.

Robinson, D. A. (1963). A method of measuring eye movement using a scleral search coil in a magnetic field. *IEEE Transactions on Biomedical Engineering, 10*, 137–145.

Robinson, D. A. (1968). Eye movement control in primates. The oculomotor system contains specialized subsystems for acquiring and tracking visual targets. *Science, 161*(3847), 1219–1224.

Sesin, A., Adjouadi, M., Cabrerizo, M., Ayala, M., & Barreto, A. (2008). Adaptive eye-gaze tracking using neural-network-based user profiles to assist people with motor disability. *Journal of Rehabilitation Research and Development, 45*(6), 801–817.

Soussou, W., Rooksby, M., Forty, C., Weatherhead, J., & Marshall, S. (2012). EEG and eye-tracking based measures for enhanced training. *Conference Proceedings: IEEE Engineering in Medicine and Biology Society, 2012*, 1623–1626.

Stahl, J. S. (2004a). Eye movements of the murine P/Q calcium channel mutant rocker, and the impact of aging. *Journal of Neurophysiology, 91*(5), 2066–2078.

Stahl, J. S. (2004b). Using eye movements to assess brain function in mice. *Vision Research, 44*(28), 3401–3410.

Stahl, J. S., van Alphen, A. M., & De Zeeuw, C. I. (2000). A comparison of video and magnetic search coil recordings of mouse eye movements. *Journal of Neuroscience Methods, 99*(1–2), 101–110.

Strasser, A. A., Tang, K. Z., Romer, D., Jepson, C., & Cappella, J. N. (2012). Graphic warning labels in cigarette advertisements: Recall and viewing patterns. *American Journal of Preventive Medicine, 43*(1), 41–47.

Sundstedt, V. (2010). *Gazing at games: Using eye tracking to control virtual characters*. Paper presented at the ACM SIGGRAPH 2010 Courses, Los Angeles, CA.

Sundstedt, V. (2012). Gazing at games—An introduction to eye tracking control. *Synthesis Lectures on Computer Graphics and Animation, 5*(1), 1–113.

Thies, P., Marc, E. L., & Ipke, W. (2009). Evaluation of binocular eye trackers and algorithms for 3d gaze interaction in virtual reality environments. *Journal of Virtual Reality and Broadcasting, 5*(16), 1–14.

Tokuda, S., Obinata, G., Palmer, E., & Chaparro, A. (2011). Estimation of mental workload using saccadic eye movements in a free-viewing task. *Conference Proceedings: IEEE Engineering in Medicine and Biology Society, 2011*, 4523–4529.

Tomizawa, Y., Aoki, H., Suzuki, S., Matayoshi, T., & Yozu, R. (2012). Eye-tracking analysis of skilled performance in clinical extracorporeal circulation. *Journal of Artificial Organs, 15*(2), 146–157.

Vercher, J. L., Gauthier, G. M., Guedon, O., Blouin, J., Cole, J., & Lamarre, Y. (1996). Self-moved target eye tracking in control and deafferented subjects: Roles of arm motor command and proprioception in arm-eye coordination. *Journal of Neurophysiology, 76*(2), 1133–1144.

Vertegaal, R., Shell, J. S., Chen, D., & Mamuji, A. (2006). Designing for augmented attention: Towards a framework for attentive user interfaces. *Computers in Human Behavior, 22*(4), 771–789.

Vitak, S. A., Ingram, J. E., Duchowski, A. T., Ellis, S., & Gramopadhye, A. K. (2012). *Gaze-augmented think-aloud as an aid to learning*. Paper presented at the Proceedings of the 2012 ACM Annual Conference on Human Factors in Computing Systems, Austin, TX.

Wiesspeiner, G., Lileg, E., & Hutten, H. (1999). Eyewriter, a new communication tool for severely handicapped persons. *Medical & Biological Engineering & Computing, 37*, 1346–1347.

Wilson, M., McGrath, J., Vine, S., Brewer, J., Defriend, D., & Masters, R. (2010). Psychomotor control in a virtual laparoscopic surgery training environment: Gaze control parameters differentiate novices from experts. *Surgical Endoscopy, 24*(10), 2458–2464.

Wolpaw, J. R., Birbaumer, N., McFarland, D. J., Pfurtscheller, G., & Vaughan, T. M. (2002). Brain-computer interfaces for communication and control. *Clinical Neurophysiology, 113*(6), 767–791.

Yoo, B., Han, J.-J., Choi, C., Yi, K., Suh, S., Park, D., & Kim, C. (2010). *3D user interface combining gaze and hand gestures for large-scale display.* Paper presented at the Proceedings of the 28th of the International Conference Extended Abstracts on Human Factors in Computing Systems, Atlanta, GA.

Zoccolan, D., Graham, B. J., & Cox, D. D. (2010). A self-calibrating, camera-based eye tracker for the recording of rodent eye movements. *Frontiers in Neuroscience, 4*, 193.

9 Gesture Recognition

Matthew Turk

CONTENTS

9.1 INTRODUCTION

A primary goal of virtual environments (VEs) is to provide natural, efficient, powerful, and flexible interaction for users while navigating, exploring, and communicating. Providing gestural capabilities as an input modality, particularly in the context of multimodal interaction (Jaimes & Sebe, 2005), can help meet these requirements. Human gestures are natural and flexible and may often be efficient and powerful, especially as compared with alternative interaction modes. This chapter will cover automatic gesture recognition, focusing particularly on computer vision–based techniques that do not require the user to wear extra sensors, clothing, or equipment.

The traditional 2D, keyboard- and mouse-oriented graphical user interface (GUI) is not well suited for VEs. Synthetic environments provide the opportunity to utilize several different sensing modalities and technologies and to integrate them into the user experience; these may be alternative methods for utilizing mouse- and/or keyboard-like interactions in VEs or, more generally, natural interaction techniques that move beyond desktop interaction paradigms. Devices that sense body position and orientation, direction of gaze, speech and sound, facial expression, galvanic skin response, and other aspects of human state or behavior can be used to mediate communication between the user and the environment. Combinations of communication modalities and sensing devices can produce a wide range of unimodal and multimodal interface techniques. The potential for these techniques to support natural and powerful interfaces for communication in VEs is compelling, and advances in the past decade are bringing this closer to being a commonplace reality.

If interaction technologies are overly obtrusive, awkward, or constraining, the user's experience with the synthetic environment is severely degraded. If the interaction itself draws attention to the technology rather than the experience or the task at hand or if it imposes a high cognitive load on the user, it becomes a burden and an obstacle to a successful VE experience. It is therefore critical that gesture recognition technologies are unobtrusive and passive, supporting the immersive experience rather than detracting from it.

To support gesture recognition, human position and movement must be tracked and interpreted in order to recognize semantically meaningful gestures. While tracking a user's head position or hand configuration may be quite useful for directly controlling objects or inputting parameters, people naturally express communicative acts through higher-level constructs, as shown schematically in Figure 9.1. The output of position (and other) sensing must be interpreted to allow users to communicate more naturally and effortlessly through gesture. Gesture recognition, then, includes not only low-level, bottom-up processing of image data but also top-down processing that brings context and semantics into the equation.

Gesture has been used for control and navigation in cave automatic virtual environments (CAVEs) (Pavlovic, Sharma, & Huang, 1996) and in other VEs, such as smart rooms, virtual work environments, and performance spaces. More recently, gesture recognition has become a popular consumer technology for use in gaming, driven largely by the introduction of the Microsoft Kinect in 2010 (Xbox, 2012), and in *smart televisions*, with recent commercial offerings by Samsung and others. Sensors providing depth images (or RGB plus depth) at frame rate have made these systems possible.

In addition, gesture may be perceived by the environment in order to be transmitted elsewhere (e.g., as a compression technique to be reconstructed at the receiver). Gesture recognition may also influence—intentionally or unintentionally—a system's model of the user's state. For example, a look of frustration may cause a system to slow down its presentation of information, or the urgency of a gesture may cause the system to speed up. Gesture may also be used as a communication *back channel* (i.e., a visual or verbal behavior such as nodding or saying, "uh-huh," to indicate "I'm with you, continue," or raising a finger to indicate the desire to interrupt) to communicate agreement, participation, attention, conversation turn taking, and so forth.

Given that the human body can express a huge variety of gestures, which are of interest for human–computer interaction (HCI), and what is appropriate to sense and recognize? Clearly, the

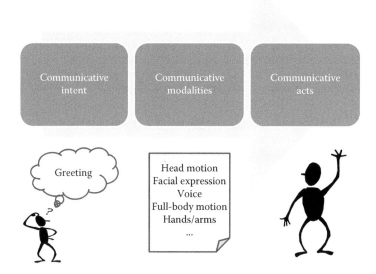

FIGURE 9.1 Observable communications acts (such as gestures) are the result of expressing intent via communication modalities.

position and orientation of each body part—the parameters of an articulated body model—would be useful, as well as features that are derived from those measurements, such as velocity and acceleration. Facial expressions are very expressive and meaningful to people. More subtle cues such as hand tension, overall muscle tension, locations of self-contact, and even pupil dilation may be useful in sensing a person's intention or state.

Dindar, Tekalp, and Basdogan (2014, Chapter 5) cover technologies to track the head, hands, and body. These include instrumented gloves, body suits, and marker-based optical tracking. Most of the gesture recognition work applied to VEs has used these tracking technologies as input. Wang and Winslow (2014, Chapter 8) cover eye-tracking devices and discuss their limitations in tracking gaze direction. The current chapter covers the representation and interpretation of tracked data from such devices in order to recognize gestures. Additional attention is focused on passive sensing from image-based sensors using computer vision techniques. The chapter concludes with suggestions for gesture-recognition system design.

9.2 NATURE OF GESTURE

Gestures are expressive, meaningful body motions—that is, physical movements of the fingers, hands, arms, head, face, or body with the intent to convey information or interact with the environment. Cadoz (1994) described three functional roles of human gesture:

- Semiotic—to communicate meaningful information
- Ergotic—to manipulate the environment
- Epistemic—to discover the environment through tactile experience

In an HCI environment, gesture recognition is the process by which semiotic gestures made by the user are made known to the system. One could argue that in GUI-based systems and standard mouse and keyboard actions used for selecting items and issuing commands are gestures; however, we are interested in less trivial cases. While static position (also referred to as posture, configuration, or pose) is not technically considered gesture, it is included for the purposes of this chapter, as certain poses may be characteristic of the gestures that created them, and a gesture may be considered as a temporal sequence of poses.

In VEs, users need to communicate in a variety of ways, to the system itself and also to other local or remote users. Communication tasks include specifying commands and/or parameters for tasks such as

- Navigating through a space
- Specifying items of interest
- Manipulating objects in the environment
- Changing object values
- Controlling virtual objects
- Issuing task-specific commands

In addition to user-initiated communication, a VE system may benefit from observing a user's behavior for purposes such as

- Analyzing system usability
- Analyzing user behavior
- Monitoring changes in a user's state
- Better understanding a user's intent or emphasis
- Communicating user behavior to other users or environments

Messages can be expressed through gesture in many ways. For example, an emotion such as sadness can be communicated through facial expression, a lowered head position, relaxed muscles, and lethargic movement. Similarly, a gesture to indicate "Stop!" can be simply a raised hand with the palm facing forward or an exaggerated waving of both hands above the head. In general, there exists a many-to-one mapping from concept to gesture (i.e., gestures are ambiguous); there is also a many-to-one mapping from gesture to concept (i.e., gestures are not completely specified). Like speech and handwriting, gestures vary among individuals, from instance to instance for a given individual, and are subject to the effects of coarticulation. As with many recognition tasks, it is important to consider both *within-class* and *between-class* variations in the data.

An interesting real-world example of the use of gestures in visual communications is a US Army field manual (Anonymous, 1987) that serves as a reference and guide to commonly used visual signals, including hand and arm gestures for a variety of situations. The manual describes visual signals used to transmit standardized messages rapidly over short distances. In contrast is the training material for the Microsoft Kinect, where a small number of gestures are imprecisely defined and users are mostly learned by doing.

Despite the richness and complexity of gestural communication, researchers have made progress in beginning to understand and describe the nature of gesture. Kendon (1972) described a *gesture continuum*, defining five different kinds of gestures:

- *Gesticulation*—spontaneous movements of the hands and arms that accompany speech
- *Language-like gestures*—gesticulation that is integrated into a spoken utterance, replacing a particular spoken word or phrase
- *Pantomimes*—gestures that depict objects or actions, with or without accompanying speech
- *Emblems*—familiar gestures such as "V for victory," "thumbs up," and assorted rude gestures (these are often culturally specific)
- *Sign languages*—linguistic systems, such as American Sign Language (ASL), which are well defined

As the list progresses (moving from left to right in Figure 9.2), the association with speech declines, language properties increase, spontaneity decreases, and social regulation increases.

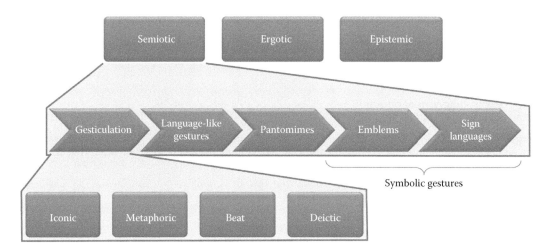

FIGURE 9.2 Top: Cadoz's functional roles of human gesture. Middle: Kendon's gesture continuum. Bottom: McNeill's four gesture types.

Within the first category—spontaneous, speech-associated gesture—McNeill (1992) defined four gesture types:

- *Iconic*—representational gestures depicting some feature of the object, action, or event being described
- *Metaphoric*—gestures that represent a common metaphor, rather than the object or event directly
- *Beat*—small, formless gestures, often associated with word emphasis
- *Deictic*—pointing gestures that refer to people, objects, or events in space or time

Figure 9.2 depicts the relationships among Cadoz's functional roles of human gesture, Kendon's gesture continuum, and McNeill's gesture types.

These types of gesture modify the content of accompanying speech and may often help to disambiguate speech, similar to the role of spoken intonation. Cassell et al. (1994) described a system that models the relationship between speech and gesture and generates interactive dialogs between 3D animated characters that gesture as they speak.

These spontaneous gestures (*gesticulation* in Kendon's continuum) are estimated to make up some 90% of human gestures (McNeill, 2000). People even gesture when they are on the telephone, and blind people regularly gesture when speaking to one another. Across cultures, speech-associated gesture is natural and common. For HCI to be truly natural, the technology to understand both speech and gesture together must be developed.

Despite the importance of this type of gesture in normal human-to-human interaction, most research to date in HCI, and most VE technology, focuses on the lower right side of Figure 9.2, where gestures tend to be less ambiguous, less spontaneous and natural, more learned, and more culture specific. Emblematic gestures and gestural languages, although perhaps less spontaneous and natural, carry more clear semantic meaning and may be more appropriate for the kinds of command-and-control interaction that VEs tend to support. The main exception to this is work in recognizing and integrating deictic (mainly pointing) gestures, beginning with the well-known *Put That There* demonstration by Bolt at MIT in the context of the Media Room project (Bolt, 1980). The remainder of this chapter will focus on *symbolic gestures* (which include emblematic gestures and predefined gesture languages) and *deictic gestures*, keeping our attention largely on the bottom-up approaches.

9.3 REPRESENTATIONS OF POSTURE AND GESTURE

Although most gesture recognition systems share some common approaches, there is no standard way to do gesture recognition—a variety of representations and classification schemes are used. The concept of gesture is not precisely defined, and the interpretation of gestures depends on the context of the interaction. Recognition of natural, continuous gestures requires temporally segmenting gestures, since gestures are fundamentally time-varying events. Automatically segmenting gestures is difficult and has often been finessed or ignored in systems by requiring a starting position in time and/or space, similar to *push to talk* speech recognition systems. Morency, Quattoni, and Darrell (2007) proposed a solution to this problem using latent-dynamic discriminative models, which was also used by Song, Demirdjian, and Davis (2012) based on extracted body and hand features for continuous gesture recognition. Alon, Athitsos, Quan Yuan, and Sclaroff (2009) used spatiotemporal matching and offline learning to incorporate both top-down and bottom-up information flow in gesture segmentation. Similar to this is the problem of distinguishing intentional gestures from other *random* movements.

Gestures can be static, where the user assumes a certain pose, posture, or configuration, or dynamic, defined by movement. McNeill (1992) defined three phases of a dynamic gesture: prestroke, stroke, and poststroke. Some gestures have both static and dynamic elements, where the pose is important in one or more of the gesture phases; this is particularly relevant in sign languages. When gestures are produced continuously, each gesture is affected by the gesture that preceded it and possibly by the gesture that follows it. These *coarticulations* may be taken into account as a system is trained.

There are several aspects of a gesture that may be relevant and therefore may need to be represented explicitly. Hummels and Stappers (1998) described four aspects of a gesture that may be important to its meaning:

- Spatial information—where it occurs, the location a gesture refers to
- Pathic information—the path that a gesture takes
- Symbolic information—the sign that a gesture makes
- Affective information—the emotional quality of a gesture

In order to infer these aspects of gesture, human position, configuration, and movement must be sensed. This can be done directly with sensing devices such as magnetic field trackers, instrumented gloves, inertial sensors, and data suits, which are attached to or held by the user, or indirectly using techniques such as electric field sensing (as in a theremin) or cameras and computer vision techniques. Each sensing technology differs along several dimensions, including accuracy, resolution, mobility, latency, range of motion, user comfort, and cost. The integration of multiple sensors in gesture recognition is a complex task since each sensing technology varies along these dimensions. Although the output from these sensors can be used to directly control parameters such as navigation speed and direction or movement through a virtual space, our interest is primarily in the interpretation of sensor data to recognize gestural information.

The output of initial sensor processing is a time-varying sequence of parameters typically describing positions, velocities, and angles of relevant body parts and features—ideally 3D and view independent, but in some cases 2D, view-dependent features. These should (but often do not) include a representation of uncertainty that indicates limitations of the sensor and processing algorithms. Recognizing gestures from these parameters is a pattern recognition task that typically involves transforming input into the appropriate representation (feature space) and then classifying it from a database of predefined gesture representations, as shown in Figure 9.3. The parameters produced by the sensors may be transformed into a global coordinate space, processed to produce sensor-independent features, or used directly in the classification step.

Because gestures are highly variable from one person to another (interperson variations) and from one example to another within a single person (intraperson variation), it is essential to capture the essence of a gesture—its invariant properties—and use this to represent the gesture. Besides the choice of representation itself, a significant issue in building gesture recognition systems is how to create and update the database of known gestures. Hand-coding gestures only work for trivial systems; in general, a gesture recognition system needs to be trained through some kind of learning procedure. As with speech recognition systems, there is often a trade-off between accuracy and generality—the more accuracy desired, the more user-specific training is required. In addition, systems may be fully trained when in use, or they may adapt over time

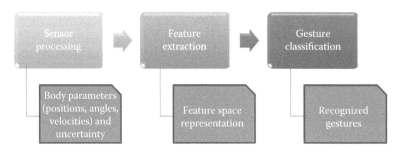

FIGURE 9.3 Pattern recognition systems. (Image from www.tobiiglasses.com.)

to the current user. As should be apparent throughout this chapter, there is no one-size-fits-all solution to problems in gesture recognition.

Static gesture, or *pose*, recognition can be accomplished by a straightforward implementation of Figure 9.3, using template matching, geometric feature classification, neural networks, or other standard pattern recognition techniques to classify pose. Dynamic gesture recognition, however, requires consideration of temporal events. This is typically accomplished through the use of techniques such as time-compressing templates, finite-state machines (Hong, Turk, & Huan, 2000), dynamic time warping (Darrell & Pentland, 1993; Reyes, Dominguez, & Escalera, 2011), hidden Markov models (Lee & Kim, 1999; Wilson & Bobick, 1999), conditional random fields (Wang, Quattoni, Morency, Demirdjian, & Darrell, 2006), and Bayesian networks (Suk, Sin, & Lee, 2008). Elgammal, Shet, Yacoob, and Davi (2003) described learning dynamics for gestures, applied to sequences of body poses.

9.4 APPROACHES FOR RECOGNIZING GESTURE

Gesture recognition is useful in a wide range of application areas (Wachs, Kölsch, Stern, & Edan, 2011), and approaches in most of these are applicable to the use of gesture in VEs. For example, gesture recognition has been used extensively in surface computing (Wobbrock, Morris, & Wilson, 2009), immersive environments (Kehl & Van Gool, 2004), smart home control (Kühnel et al., 2011), interactive gaming (Bleiweiss et al., 2010; Kang, Lee, & Jung, 2004), interactive music (Overholt et al., 2009), and artistic interfaces (Faste, Ghedini, Avizzano, & Bergamasco, 2008). These different scenarios may assume the use of various devices or sensing capabilities.

9.4.1 PEN-BASED AND TOUCH-BASED GESTURE RECOGNITION

Recognizing gestures from a 2D input device such as a pen or mouse has been considered for some time. The early Sketchpad system in 1963 (Johnson, 1963) used light-pen gestures, for example. Some commercial systems have used pen gestures since the 1970s. There are examples of gesture recognition for document editing, for air traffic control, and for design tasks such as editing splines. In the 1990s, systems such as the OGI QuickSet (Cohen et al., 1997) demonstrated the utility of pen-based gesture recognition and speech recognition to control a VE. QuickSet recognized 68 pen gestures, including map symbols, editing gestures, route indicators, area indicators, and taps. Oviatt (1996) demonstrated significant benefits of using speech and pen gestures together in certain tasks. Zeleznik, Herndon, and Hughes (1996) and Landay and Myers (1995) developed interfaces that recognize gestures from pen-based sketching.

The introduction of inexpensive touch screens and multitouch capabilities on mobile phones (with iOS, Android, Windows Phone, and other such devices) in recent years has made touch-based gesture a common experience for smartphone users, expanding the existing vocabulary of mouse- and pen-based gestures. A significant benefit of pen-based and touch-based gestural systems is that sensing and interpretation is relatively straightforward as compared with vision-based techniques. Early pen-based PDAs performed handwriting recognition and allowed users to invoke operations by various, albeit quite limited, pen gestures. Long, Landay, and Rowe (1998) surveyed problems and benefits of these gestural interfaces and provide insight for interfaced designers. A much wider range of more sophisticated gestures are now available with multitouch devices and prototypes of various sizes and configurations (Wilson, 2004; Wobbrock et al., 2009).

Although pen- and touch-based gesture recognition is promising for many HCI environments, it presumes the availability of, and proximity to, an interactive surface or screen. In VEs, this is often too constraining; techniques that allow the user to move around and interact in more natural ways are more compelling. The following sections cover two primary technology approaches for gesture recognition in VEs: instrumented or device-based (active) interfaces and vision-based (passive) interfaces.

9.4.2 Device-Based Gesture Recognition

There are a number of commercially available tracking systems (see Wang & Winslow, 2014, Chapter 8), which can be used as input to gesture recognition, primarily for tracking eye gaze, hand configuration, and overall body position. Each sensor type has its strengths and weaknesses in the context of VE interaction. While eye gaze can be quite useful in a gestural interface, the focus here is on gestures based on input from tracking the hands and body.

9.4.2.1 Instrumented Gloves

People naturally use their hands for a wide variety of manipulation and communication tasks. Besides being quite convenient, hands are extremely dexterous and expressive, with approximately 29 degrees of freedom (including the wrist). In his comprehensive thesis on whole-hand input, Sturman (1992) showed that the hand can be used as a sophisticated input and control device in a wide variety of application domains, providing real-time control of complex tasks with many degrees of freedom. He analyzed task characteristics and requirements, hand action capabilities, and device capabilities and discussed important issues in developing whole-hand input techniques. Sturman suggested a taxonomy of whole-hand input that categorizes input techniques along two dimensions:

- Classes of hand actions: continuous or discrete
- Interpretation of hand actions: direct, mapped, or symbolic

The resulting six categories describe the styles of whole-hand input. A given interaction task can be evaluated as to which style best suits the task. Mulder (1996) presented an overview of hand gestures in HCI, discussing the classification of hand movement, standard hand gestures, and hand-gesture interface design. LaViola (1999) provided a thorough survey on hand posture and gesture recognition techniques at the end of the decade; Dipietro, Sabatini, and Dario (2008) provided a more recent survey of glove-based systems.

For several years, commercial devices have been available that measure, to various degrees of precision, accuracy, and completeness, the position and configuration of the hand. These include *data gloves* and exoskeleton devices (Figure 9.4) mounted on the hand and fingers (the term *instrumented glove* is used to include both types). Some advantages of instrumented gloves include

- Direct measurement of hand and finger parameters (joint angles, 3D spatial information, wrist rotation)
- Provide data at a high sampling frequency
- Ease of use
- No line-of-sight occlusion problems
- Relatively low-cost versions available
- Data that are translation independent (within the range of motion)

Disadvantages of instrumented gloves include the following:

- Calibration can be difficult.
- Tethered gloves reduce range of motion and comfort.
- Data from inexpensive systems can be very noisy.
- Accurate systems are expensive.
- The user is forced to wear a cumbersome device.

Many projects have used hand input from instrumented gloves for *point, reach, and grab* operations or more sophisticated gestural interfaces. Latoschik and Wachsmuth (1997) present a multiagent

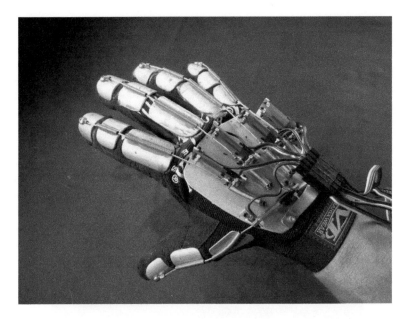

FIGURE 9.4   An exoskeleton data glove. (Courtesy of Christian Ristow.)

architecture for detecting pointing gestures in a multimedia application. Väänänen and Böhm (1993) developed a neural network system that recognized static gestures and allows the user to interactively teach new gestures to the system. Böhm et al. (1994) extended that work to dynamic gestures using a Kohonen feature map (KFM) for data reduction.

Baudel and Beaudouin-Lafon (1993) developed a system to provide gestural input to a computer while giving a presentation. This work included a gesture notation and set of guidelines for designing gestural command sets. Fels and Hinton (1995) used an adaptive neural network interface to translate hand gestures to speech. Kadous (1996) used glove input to recognize Australian sign language as did Takahashi and Kishino (1991) for the Japanese kana manual alphabet. The system of Lee and Xu (1996) could learn and recognize new gestures online.

More recent examples of glove-based gesture research can be found in Parvini et al. (2009) who compare the performance of different approaches; data glove gestures were recently studied in Huang, Monekosso, Wang, and Augusto (2011).

Despite the fact that many or most gestures involve two hands, a large portion of the research efforts in glove-based gesture recognition use only one glove for input. The features that are used for recognition and the degree to which dynamic gestures are considered vary quite a bit in the research literature.

9.4.2.2   Body Suits and Motion Tracking Systems

It is well known that by viewing only a small number of strategically placed dots on the human body, people can easily perceive complex movement patterns such as the activities, gestures, identities, and other aspects of bodies in motion (Johansson, 1973). One way to approach the recognition of human movements and postures is to optically measure the 3D position of several such markers attached to the body (Figure 9.5) and then recover the time-varying articulated structure of the body. The articulated structure may also be measured more directly by sensing joint angles and positions using electromechanical body sensors. Although some of the optical tracking systems require a full body suit to be donned, others only require dots or small balls to be placed on top of a subject's clothing. Both systems are referred to generically here as *motion capture systems.*

Motion capture systems have advantages and disadvantages that are similar to those of instrumented gloves: they can provide reliable data at a high sampling rate (at least for electromagnetic devices),

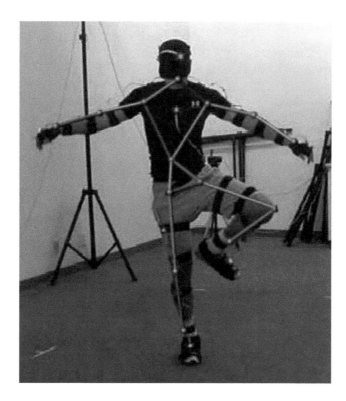

FIGURE 9.5   (See color insert.) An optical motion capture system in use. (From Kirk, A.G. et al., Skeletal parameter estimation from optical motion capture data, *IEEE Conference on Computer Vision and Pattern Recognition*, Silver Spring, MD, 2005. With permission.)

but they are expensive and very cumbersome. Calibration is often nontrivial. While the optical systems typically use several cameras and may have to process the data offline, their major advantage is the lack of wires and a tether.

Motion capture systems have been used, often along with instrumented gloves, in several gesture recognition systems. Wexelblat (1994) implemented a continuous gesture analysis system using a data suit, data gloves, and an eye tracker. In this system, data from the sensors is segmented in time (between movement and inaction), key features are extracted, motion is analyzed, and a set of special-purpose gesture recognizers look for significant changes. Marrin and Picard (1998) developed an instrumented jacket for an orchestral conductor that includes physiological monitoring to study the correlation between affect, gesture, and musical expression.

Although many optical and electromechanical tracking technologies are cumbersome and therefore contrary to the desire for more natural interfaces, advances in sensor technology may well enable a new generation of devices (including stationary field-sensing devices, gloves, watches, and rings) that are just as useful as current trackers but much less obtrusive. Similarly, the currently cumbersome instrumented body suits may be displaced by sensing technologies embedded in belts, shoes, eyeglasses, cell phones, and even shirts and pants. For example, inertial sensors in cell phones have been used to classify human activity, including gestures (Kratz & Rohs, 2010; Liu, Zhong, Wickramasuriya, & Vasudevan, 2009; Pylvänäinen, 2005; Wang, Tarrío, Metola, Bernardos, & Casar, 2012; Wu, Pan, Zhang, Qi, & Li, 2009).

Note that although some of the body tracking methods in this section use cameras and computer vision techniques to track joint or limb positions, they require the user to wear special markers. In the next section, only passive techniques that do not require the user to wear any special markers or equipment are considered.

9.4.3 PASSIVE VISION-BASED GESTURE RECOGNITION

The most significant disadvantage of the tracker-based systems in Section 9.4.2 is that they are cumbersome. This detracts from the immersive nature of a VE by requiring the user to don an unnatural device that cannot easily be ignored and that often requires significant effort to put on and calibrate. Even optical systems with markers applied to the body or clothing suffer from these shortcomings, albeit not as severely. Computer vision techniques have the potential to provide real-time data useful for analyzing and recognizing human motion in a manner that is passive and unobtrusive.

Vision-based interfaces use one or more cameras to capture images, typically at a frame rate of 30 Hz or more, and interpret those images to produce visual features that can be used to represent human activity and recognize gestures (Mitra & Acharya, 2007; Wu & Huang, 1999). Typically, the camera locations are fixed in the environment, although they may also be mounted on moving platforms or on other people. For the past two decades, there has been a significant amount of research in the computer vision community on detecting and recognizing faces, analyzing facial expression, extracting lip and facial motion to aid speech recognition, interpreting human activity, and recognizing particular gestures, most of which is relevant to this topic.

Unlike sensors worn on the body, vision approaches to body tracking have to contend with occlusions. From the point of view of a given camera, there are always parts of the user's body that are occluded and therefore not visible; for example, the backside of the user is not visible when the camera is in front. More significantly, self-occlusion often prevents a full view of the fingers, hands, arms, and body from a single view. Multiple cameras can be used, but at the cost of higher complexity in setup, calibration, and processing.

The occlusion problem makes full-body tracking difficult, if not impossible, without a strong model of body kinematics and perhaps dynamics. However, recovering all the parameters of body motion may not be a prerequisite for gesture recognition. The fact that people can reliably and robustly recognize gestures leads to three possible conclusions: (1) the parameters that cannot be directly observed are inferred, (2) these parameters are not needed to accomplish the task, or (3) some are inferred and others are ignored.

It is a mistake to consider passive vision approaches and direct tracking devices (such as instrumented gloves and body suits) as alternative paths to the same end. Although there is overlap in what they can provide, these technologies in general produce qualitatively and quantitatively different output that enables different analysis and interpretation. For example, tracking devices can in principle detect and measure fast and subtle movements of the fingers while a user is waving his or her hands, whereas human vision in that case may at best get a general sense of the type of finger motion. Similarly, vision can use properties like texture and color in its analysis of gesture, whereas tracking devices do not. From a practical perspective, these observations imply that it may not be an optimal strategy to merely substitute vision at a later date into a system that was developed to use an instrumented glove or a body suit—or vice versa.

Unlike special devices that measure human position and motion, vision uses a multipurpose sensor; the same device used to recognize gestures can be used to recognize other objects in the environment and also to transmit video for teleconferencing, surveillance, and other purposes. On the other hand, advances in miniaturized, low-cost, low-power cameras integrated with processing circuitry on a single chip increasingly make possible a special-purpose *gesture sensor* that outputs motion or gesture parameters to the VE.

Currently, most computer vision systems for recognition look something like Figure 9.3. A digital camera feeds video frames to a computer's memory. There may be a preprocessing step, where images are normalized, enhanced, or transformed in some manner, and then a feature extraction step. The features—which may be any of a variety of 2D or 3D features, statistical properties, or estimated body parameters—are analyzed and classified as a particular gesture if appropriate.

Vision-based systems for gesture recognition vary along a number of dimensions, most notably the following:

- Number of cameras. How many cameras are used? If more than one, is their information combined early (at the pixel or feature level) or late (after recognition of body parts or basic movements)?
- Speed and latency. Is the system real time (i.e., fast enough, with low enough latency, to support interaction)?
- Structured environment. Are there restrictions on the background, lighting, speed of movement, and so forth?
- User requirements. Must the user wear anything special (e.g., markers, gloves, long sleeves)? Is anything disallowed (e.g., glasses, beard, rings)?
- Primary features. What low-level features are computed (edges, regions, silhouettes, moments, histograms, depth maps, etc.)?
- Two- or three-dimensional representation. Does the system construct a 3D model of the body part(s), or is classification done on some other (view based) representation?
- Representation of time. How is the temporal aspect of gesture represented and used in recognition (e.g., via a state machine, dynamic time warping, HMMs, time-compressed template)?

9.4.3.1 Head and Face Gestures

When people interact with one another, they use an assortment of cues from the head and face to convey information. These gestures may be intentional or unintentional, they may be the primary communication mode or back channels, and they can span the range from extremely subtle to highly exaggerated. Some examples of head and face gestures include

- Nodding or shaking the head
- Direction of eye gaze
- Raising the eyebrows
- Opening the mouth to speak
- Winking
- Flaring the nostrils
- Looks of surprise, happiness, disgust, anger, sadness, etc.

People display a wide range of facial expressions. Ekman and Friesen (1978) developed the facial action coding system (FACS) for measuring facial movement and coding expression; this description forms the core representation for many facial expression analysis systems.

A real-time system to recognize actions of the head and facial features was developed by Zelinsky and Heinzmann (1996), who used feature template tracking in a Kalman filter framework to recognize thirteen head and face gestures. Moses, Reynard, and Blake (1995) used fast-contour tracking to determine facial expression from a mouth contour. Essa and Pentland (1997) used optical flow information with a physical muscle model of the face to produce accurate estimates of facial motion. This system was also used to generate spatiotemporal motion-energy templates of the whole face for each different expression. These templates were then used for expression recognition. Oliver, Pentland, and Bérard (1997) described a real-time system for tracking the face and mouth that recognized facial expressions and head movements. Otsuka and Ohya (1998) modeled coarticulation in facial expressions and used an HMM for recognition.

Black and Yacoob (1995) used local parametric motion models to track and recognize both rigid and nonrigid facial motions. Demonstrations of this system show facial expressions being detected from television talk show guests and news anchors (in nonreal time). La Cascia, Isidoro, and Sclaroff (1998) extended this approach using texture-mapped surface models and nonplanar parameterized motion models to better capture the facial motion.

9.4.3.2 Hand and Arm Gestures

Hand and arm gestures receive the most attention among those who study gesture; in fact, many (perhaps most) references to gesture recognition only consider hand and arm gestures. The vast majority of automatic recognition systems are for deictic gestures (pointing), emblematic gestures (isolated signs), and sign languages (with a limited vocabulary and syntax). Some are components of bimodal systems, integrated with speech recognition. Some estimate precise hand and arm configuration, whereas others only coarse motion.

Stark and Kohler (1995) developed the ZYKLOP system for recognizing hand poses and gestures in real time. After segmenting the hand from the background and extracting features such as shape moments and fingertip positions, the hand posture is classified. Temporal gesture recognition is then performed on the sequence of hand poses and their motion trajectory. A small number of hand poses comprise the gesture catalog, whereas a sequence of these makes a gesture. Similarly, Maggioni and Kämmerer (1998) described the GestureComputer, that recognized both hand gestures and head movements. Other early systems that recognize hand postures amidst complex visual backgrounds are reported by Weng and Cui (1998) and Triesch and von der Malsburg (1996). Oka, Sato, and Koike (2002) tracked fingertips; Bretzner, Laptev, and Lindeberg (2002) used multiscale skin color features and particle filtering; and Yang, Ahuja, and Tabb (2002) used motion trajectories to represent and recognize hand gestures.

There has been a great deal of interest in creating devices to automatically interpret various sign languages to aid the deaf community (Ong and Ranganath, 2005). One of the first to use computer vision without requiring the user to wear specialized devices on the hands was built by Starner and Pentland (1995), who used HMMs to recognize a limited vocabulary of ASL sentences. A similar effort, which uses HMMs to recognize sign language of the Netherlands, was described by Assan and Grobel (1997). Vogler and Metaxas (2001) dealt with recognizing simultaneous aspects of complex sign language; Wang, Chen, Zhang, Wang, and Gao (2007) focused on viewpoint-invariant recognition, while Zaki and Shaheen (2011) proposed a combination of new image features for sign language recognition.

The recognition of hand and arm gestures has been applied to entertainment applications. Freeman, Tanaka, Ohta, and Kyuma (1996) developed a real-time system to recognize hand poses using image moments and orientation histograms and applied it to interactive video games. Cutler and Turk (1998) described a system for children to play virtual instruments and interact with lifelike characters by classifying measurements based on optical flow; more recent optical flow-based gesture recognition is presented in Holte, Moeslund, and Fihl (2010). A nice overview of work-up to 1995 in hand-gesture modeling, analysis, and synthesis is presented by Huang and Pavlovic (1995).

The Leap (Leap Motion, 2012), a gesture control system introduced in 2012 initially focused on desktop use, is a small device that, according to early reports, accurately tracks the 3D positions of fingertips, hands, and lower arms, capturing subtle movements in an interaction space of about eight cubic feet. The device senses individual hand and finger movements independently, providing touch-free motion sensing and control. While there are few details available for this device as of Q1 2013, the demonstrations are impressive, and there is great interest among researchers and developers to try this device in a variety of interactive environments.

9.4.3.3 Body Gestures

Full-body motion is needed to recognize large-scale gestures and to recognize or analyze human activity (Gavrila, 1999; Moeslund, Hilton, Krüger, & Sigal, 2011). Activity may be defined over a much longer period of time than what is normally considered a gesture; for example, two people meeting in an open area, stopping to talk and then continuing on their way, may be considered a recognizable activity. A different view of these terms was proposed by Bobick (1997) in his hierarchy of motion perception:

- Movement—the atomic elements of motion
- Activity—a sequence of movements or static configurations
- Action—high-level description of what is happening in context

Most research to date has focused on the first two levels.

The Pfinder system (Wren, Azarbayejani, Darrell, & Pentland, 1996) developed at the MIT Media Lab has been used by a number of groups to do body tracking and gesture recognition. It formed a 2D representation of the body using statistical models of color and shape. The body model provided an effective interface for applications such as video games, interpretive dance, navigation, and interaction with virtual characters. Lucente, Zwart, and George (1998) combined Pfinder with speech recognition in an interactive environment called visualization space, allowing a user to manipulate virtual objects and navigate through virtual worlds. Paradiso and Sparacino (1997) used Pfinder to create an interactive performance space where a dancer can generate music and graphics through their body movements; for example, hand and body gestures can trigger rhythmic and melodic changes in the music.

Systems that analyze human motion in VEs may be quite useful in medical rehabilitation and athletic and military training (Maxwell et al., 2014, Chapter 37). For example, a system like the one developed by Boyd and Little (1998) to recognize human gaits could potentially be used to evaluate rehabilitation progress. Yamamoto, Kondo, and Yamagiwa (1998) described a system that used computer vision to analyze body motion in order to evaluate the performance of skiers.

Davis and Bobick (1997) used a view-based approach by representing and recognizing human action based on *temporal templates*, where a single image template captures the recent history of motion. This technique was used in the KidsRoom system (Bobick et al., 1999), an interactive, immersive, narrative environment for children.

Video surveillance and monitoring of human activity has received significant attention in recent years. For example, the W⁴ system developed at the University of Maryland (Haritaoglu, Harwood, & Davis, 1998) tracked people and detected patterns of their activity. Although partly relevant to VEs—especially in understanding context and in multiperson environments—the topic is beyond the scope of this chapter.

9.4.4 Depth Cameras

There has been a recent proliferation in the use of depth sensors to detect, track, and recognize the gestures and activity of people, largely due to the introduction and rapid popularity of the Microsoft Kinect device (Figure 9.6), which is based on the PrimeSense depth sensor. Others have used stereo camera rigs, time-of-flight depth sensors, and other technologies to obtain 3D information about the scene for use in gesture recognition. Many of these current approaches use both RGB and depth data, which are available from the Kinect and other similar RGBD cameras.

Breuer, Eckes, and Müller (2007) and Hackenberg, McCall, and Broll (2011) explored hand-gesture recognition using a time-of-flight range camera. Girshick, Shotton, Kohli, Criminisi, and Fitzgibbon (2011) have used the Kinect for modeling body pose and determining human activity; Ren, Meng, Yuan, and Zhang (2011) and Van den Bergh and Van Gool (2011) have used RGB and depth data for robustly recognizing hand gestures. Zafrulla, Brashear, Starner, Hamilton, and Presti

FIGURE 9.6   The Microsoft Kinect in action (depth image, skeleton model, RGB image.)

(2011) used a Kinect to recognize ASL. Doliotis, Stefan, McMurrough, Eckhard, and Athitsos (2011) have studied gesture recognition accuracy using an RGBD sensor.

Although current depth sensors have significant limitations for general use (such as a limited range of distances from the sensor and mostly indoor lighting conditions), improvements are underway to improve on these, and there is great promise for continued progress in real-time, robust gesture recognition in VEs using such devices. Table 9.1 summarizes some of the main benefits and drawbacks of various categories of gesture recognition devices.

TABLE 9.1
Pros and Cons of Gesture Recognition Devices

Device	Pros	Cons
Pen- or stylus-based	High resolution and precision; inexpensive; high familiarity	Inconvenience of additional device (pen or stylus) and surface; proximity to surface; limited 2D recognition space
Touch-based	Ease of use; very high familiarity; multitouch; inexpensive	Inconvenience of additional surface; proximity to surface; limited 2D recognition space
Haptic device (e.g., Phantom)	Relatively high resolution and precision; 3D gestures	Expensive; proximity to devices
Instrumented glove	High-input dimensionality; whole-hand input; natural use; no line-of-sight occlusion; two-hand gesture possible	Expensive; need to don extra equipment; limited range if tethered; complex processing/interpretation
Motion tracking	No direct contact required; relatively wide range of motion possible; full-body gestures	Very expensive; calibration needed; obtrusive to wear; fixed area of use
Vision-based tracking and recognition	No direct contact required; wide range of motion (esp. with mobile sensors); high resolution possible; 3D gestures possible; full-body gestures; ease of use; may be inexpensive; synergy with gaming and other commercial markets	Must be aware of camera occlusion, field of view; calibration may be needed; IR illumination may interfere with other sensors; complex processing

9.5 GUIDELINES FOR GESTURE RECOGNITION SYSTEMS

There has been surprising little work in evaluating the utility and usability of gesture recognition systems in realistic settings. However, those developing gestural systems have learned a number of lessons along the way. Here, a few guidelines are presented in the form of *dos and don'ts* for gestural interface designers:

- *Do inform the user.* As discussed in Section 9.2, people use different kinds of gestures for many purposes, from spontaneous gesticulation associated with speech to structured sign languages. Similarly, gesture may play a number of different roles in a VE. To make compelling use of gesture, the types of gestures allowed and their effects must be clear to the user or easily discoverable.
- *Do give the user feedback.* Feedback is essential to let the user know when a gesture has been recognized (or has been detected but not recognized). This could be inferred from the action taken by the system, when that action is obvious, or by more subtle visual or audible confirmation methods.
- *Do take advantage of the uniqueness of gesture.* Gesture is not just a substitute for a mouse or keyboard. It may not be as useful for 2D pointing or text entry but great for more expressive input.
- *Do understand the benefits and limits of the particular technology.* For example, precise finger positions are better suited to instrumented gloves than vision-based techniques. Tethers from gloves or body suits may constrain the user's movement.
- *Do usability testing on the system.* Do not just rely on the designer's intuition (Gabbard & Hix, 2014; Chapter 28).
- *Do avoid temporal segmentation if feasible.* At least with the current state of the art, an initial, bottom-up segmentation of gestures can be quite difficult and error prone.
- *Don't tire the user.* Gesture is seldom the primary mode of communication. When a user is forced to make frequent, awkward, or precise gestures, the user can become fatigued quickly. For example, holding one's arm in the air to make repeated hand gestures becomes tiring very rapidly.
- *Don't make the gestures to be recognized too similar.* This is for ease of classification and to help the user more easily make distinguishable gestures.
- *Don't use gesture as a gimmick.* If something is better done with a mouse, keyboard, speech, or some other device or mode, use it—extraneous use of gesture should be avoided.
- *Don't unduly increase the user's cognitive load.* Having to remember the whats, wheres, and hows of a gestural interface can be a burden to the user. The system's gestures should be as intuitive and simple as possible. The learning curve for a gestural interface is more difficult than for a mouse and menu interface because it requires recall rather than just recognition among a visible list of options.
- *Don't require precise motion.* Especially when motioning in space with no tactile feedback, it is difficult to make highly accurate or repeatable gestures.
- *Don't create new, unnatural gestural languages.* If it is necessary to devise a new gesture language, make it as intuitive as possible.

9.6 CONCLUSIONS AND FUTURE DIRECTIONS IN GESTURE RECOGNITION

Although several research efforts have been referenced in this chapter, these are just a sampling; many more have been omitted for the sake of brevity. Good sources for much of the work in gesture recognition can be found in the *Proceedings of the IEEE International Conference on Automatic Face and Gesture Recognition* (FG) and in a number of workshops devoted to various aspects of gesture recognition.

There is still much to be done before rich and fully natural gestural interfaces that track and recognize local human activities become pervasive and cost-effective for the masses. However, much progress has been made in the past decade, and with the continuing march toward computers and sensors that are faster, smaller, and more ubiquitous, there is cause for optimism. As touch-based and pen-based computing continues to proliferate, 2D gestures have become more common, and some of the technology will transfer to 3D hand, head, and body gestural interfaces. Gaming and entertainment is, of course, driving much of the progress in gesture recognition. Similarly, technology developed in surveillance and security areas will also find uses in gesture recognition for VEs.

There are many open questions in this area. There has been little activity in evaluating usability (Gabbard & Hix, 2013; Chapter 28) and understanding performance requirements and limitations of gestural interaction. Error rates are reported from 1% to 50% depending on the difficulty and generality of the scenario. There are currently few common databases or metrics with which to compare research results. Can gesture recognition systems adapt to variations among individuals, or will extensive individual training be required? What about individual variation due to fatigue and other factors? How good do gesture recognition systems need to be to become truly useful in mass applications beyond simple games?

Each technology discussed in this chapter has its benefits and limitations. Devices that are worn or held—pens, gloves, body suits—are currently more advanced, as evidenced by the fact that there are many commercial products available. However, passive sensing (using cameras or other sensors) promises to be more powerful, more general, and less obtrusive than other technologies. It is likely that both camps will continue to improve and coexist and that new sensing technologies will arise to give even more choice to VE developers.

REFERENCES

Alon, J., Athitsos, V., Quan Yuan, Q., & Sclaroff, S. (2009, September). A unified framework for gesture recognition and spatiotemporal gesture segmentation. *IEEE Transactions on Pattern Analysis and Machine Intelligence*, *31*(9), 1685–1699.

Anonymous (1987). Visual Signals. U.S. Army Field Manual FM-2160 [online]. Available: http://armypubs. army.mil/doctrine/DR_pubs/dr_a/pdf/fm21_60.pdf. Accessed on March 19, 2014.

Assan, M., & Grobel, K. (1997). Video-based sign language recognition using hidden Markov models. In I. Wachsmuth & M. Fröhlich (Eds.), *Gesture and sign language in human–computer interaction* (*Proceedings of the International Gesture Workshop*). Bielefeld/Berlin, Germany: Springer-Verlag.

Baudel, T., & Beaudouin-Lafon, M. (1993). CHARADE: Remote control of objects using free-hand gestures. *Communications of the ACM*, *36*(7), 28–35.

Black, M., & Yacoob, Y. (1995). Tracking and recognizing rigid and non-rigid facial motions using local parametric models of image motion. *International Conference on Computer Vision* (pp. 374–381).Cambridge, MA.

Bleiweiss, A., Eshar, D., Kutliroff, G., Lerner, A., Oshrat, Y., & Yanai, Y. (2010). Enhanced interactive gaming by blending full-body tracking and gesture animation. *ACM SIGGRAPH ASIA 2010 Sketches*, Article 34, 2pp.

Bobick, A. (1997). Movement, activity, and action: The role of knowledge in the perception of motion. *Royal Society Workshop on Knowledge-Based Vision in Man and Machine*. London, U.K.

Bobick, A. F., Intille, S. S., Davis, J. W., Baird, F., Pinhanez, C. S., Campbell, L. W., … Wilson, A.. (1999, August). The KidsRoom: A perceptually-based interactive and immersive story environment. *Presence: Teleoperators and Virtual Environments*, *8*(4), 367–391.

Böhm, K., Broll, W., & Solokewicz, M. (1994). Dynamic gesture recognition using neural networks: A fundament for advanced interaction construction. In S. Fisher, J. Merrit, & M. Bolan (Eds.), *Stereoscopic displays and virtual reality systems*. (*SPIE Conference on Electronic Imaging Science and Technology*, Vol. 2177). San Jose, CA.

Bolt, R. A. (1980). Put-That-There: Voice and gesture at the graphics interface. *Computer Graphics*, *14*(3), 262–270.

Boyd, J., & Little, J. (1998). Shape of motion and the perception of human gaits. *IEEE Workshop on Empirical Evaluation Methods in Computer Vision*. Santa Barbara, CA.

Bretzner, L., Laptev, I., & Lindeberg, T. (2002, May 21). Hand gesture recognition using multi-scale colour features, hierarchical models and particle filtering. *IEEE International Conference on Automatic Face and Gesture Recognition,* Washington, DC (pp. 423–428).

Breuer, P., Eckes, C., & Müller, S. (2007). Hand gesture recognition with a novel IR time-of-flight range camera—A pilot study. In A. Gagalowicz & W. Philips (Eds.), *MIRAGE* 2007, *LNCS 4418* (pp. 247–260).

Cadoz, C. (1994). Les réalités virtuelles. Dominos, Flammarion, Paris. Rocquencourt, France: Springer.

Cassell, J., Steedman, M., Badler, N., Pelachaud, C., Stone, M., Douville, B., … Achorn, B. (1994). Modeling the interaction between speech and gesture. *Proceedings of the 16th Conference of the Cognitive Science Society*. Atlanta, GA.

Cohen, P. R., Johnston, M., McGee, D., Oviatt, S., Pittman, J., Smith, I., … Clow, J.. (1997). QuickSet: Multimodal interaction for distributed applications. *In Proceedings of the Fifth Annual International Multimodal Conference* (pp. 31–40). Seattle, WA.

Cutler, R., & Turk, M. (1998). View-based interpretation of real-time optical flow for gesture recognition. *International Conference on Automatic Face and Gesture Recognition*. Nara, Japan.

Darrell, T., & Pentland, A. (1993, June). Space-time gestures. *IEEE Conference on Computer Vision and Pattern Recognition* (pp. 335–340). Cambridge, MA.

Davis, J., & Bobick, A. (1997). The representation and recognition of human movement using temporal trajectories. *IEEE Conference on Computer Vision and Pattern Recognition*. San Juan, Puerto Rico.

Dindar, N., Tekalp, A. M., & Basdogan, C. (2014). Haptic rendering and associated depth data. In K. S. Hale & K. M. Stanney (Eds.), *Handbook of Virtual Environments* (2nd ed., pp. 115–130). New York, NY: CRC Press.

Dipietro, L., Sabatini, A., & Dario, P. (2008). A survey of glove-based systems and their applications. *IEEE Transactions on Systems, Man and Cybernetics*, *38*(4), 461–482.

Doliotis, P., Stefan, A., McMurrough, C., Eckhard, D., & Athitsos, V. (2011). Comparing gesture recognition accuracy using color and depth information. *ACM International Conference on Pervasive Technologies Related to Assistive Environments (PETRA '11)*, Hearaklion, Crete. Article 20, 7pp.

Ekman, P., & Friesen, W. V. (1978). *Facial action coding system: A technique for the measurement of facial movement.* Palo Alto, CA: Consulting Psychologists Press.

Elgammal, A., Shet, V., Yacoob, Y., & Davis, L. S. (2003, June). Learning dynamics for exemplar-based gesture recognition. *IEEE Conference on Computer Vision and Pattern Recognition,* Madison, WI (pp. I-571–I-578).

Essa, I., & Pentland, A. (1997). Coding, analysis, interpretation and recognition of facial expressions. *IEEE Transactions on Pattern Analysis and Machine Intelligence*, *19*(7), 757–763.

Faste, H., Ghedini, F., Avizzano, C. A., & Bergamasco, M. (2008, April). Passages: An artistic 3D interface. *CHI 2008*. Florence, Italy.

Fels, S., & Hinton, G. (1995). Glove-Talk II: An adaptive gesture-to-formant interface. *CHI'95*. Denver, CO.

Freeman, W., Tanaka, K., Ohta, J., & Kyuma, K. (1996). Computer vision for computer games. *International Conference on Automatic Face and Gesture Recognition*. Killington, VT.

Gabbard, J. L. (2014). Usability engineering of virtual environments. In K. S. Hale & K. M. Stanney (Eds.), *Handbook of virtual environments* (2nd ed., pp. 721–748). New York, NY: CRC Press.

Gavrila, D. M. (1999, January). The visual analysis of human movement: A survey. *Computer Vision and Image Understanding*, *73*(1), 82–98.

Girshick, R., Shotton, J., Kohli, P., Criminisi, A., & Fitzgibbon, A. (2011, October). Efficient regression of general-activity human poses from depth images. *IEEE International Conference on Computer Vision*, *2011* (pp. 415–422).

Hackenberg, G., McCall, R., & Broll, W. (2011, March 19–23). Lightweight palm and finger tracking for real-time 3D gesture control. *IEEE Virtual Reality Conference*, Singapore: IEEE (VR 2011, pp.19–26).

Haritaoglu, I., Harwood, D., & Davis, L. (1998). W4: Who? When? Where? What? A real-time system for detecting and tracking people. *International Conference on Automatic Face and Gesture Recognition*. Nara, Japan.

Holte, M. B., Moeslund, T. B., & Fihl, P. (2010, December). View-invariant gesture recognition using 3D optical flow and harmonic motion context. *Computer Vision and Image Understanding*, *114*(12), 1353–1361.

Hong, P., Turk, M., & Huang, T. S. (2000). Gesture modeling and recognition using finite state machines. *IEEE International Conference on Automatic Face and Gesture Recognition,* Grenoble, France (pp. 410–415).

Huang, T., & Pavlovic, V. (1995). Hand-gesture modeling, analysis, and synthesis. *International Workshop on Automatic Face- and Gesture-Recognition*. Zurich, Switzerland.

Huang, Y., Monekosso, D., Wang, H., & Augusto, J. C. (2011). A concept grounding approach for glove-based gesture recognition. *International Conference on Intelligent Environments* (pp. 358–361). Nottingham, U.K.

Hummels, C., & Stappers, P. (1998). Meaningful gestures for human-computer interaction: Beyond hand gestures. *International Conference on Automatic Face and Gesture Recognition*. Nara, Japan.

Jaimes, A., & Sebe, N. (2007). Multimodal human-computer interaction: A survey. *Computer Vision and Image Understanding*, 108, 1–2 (October 2007), 116–134. DOI=10.1016/j.cviu.2006.10.019. http://dx.doi.org/10.1016/j.cviu.2006.10.019

Johansson, G. (1973). Visual perception of biological motion and a model for its analysis. *Perception and Psychophysics*, *14*, 201–211.

Johnson, T. (1963). Sketchpad III: Three-dimensional graphical communication with a digital computer. *AFIPS Spring Joint Computer Conference*, *23*, 347–353.

Kadous, W. (1996). Computer recognition of Auslan signs with PowerGloves. *Workshop on the Integration of Gesture in Language and Speech*. Wilmington, DE.

Kang, H., Lee C. W., & Jung, K. (2004, November). Recognition-based gesture spotting in video games. *Pattern Recognition Letters*, *25*(15), 1701–1714.

Kehl, R., & Van Gool, L. (2004, May). Real-time pointing gesture recognition for an immersive environment. *IEEE Conference on Automatic Face and Gesture Recognition*. Seoul, Korea.

Kendon, A. (1972). Some relationships between body motion and speech. In A. W. Siegman & B. Pope (Eds.), *Studies in dyadic communication*. New York, NY: Pergamon Press.

Kirk, A. G., O'Brien, J. F., & Forsyth, D. A. (2005). Skeletal parameter estimation from optical motion capture data. *IEEE Conference on Computer Vision and Pattern Recognition*. Silver Spring, MD.

Kratz, S., & Rohs, M. (2010). A $3 gesture recognizer: Simple gesture recognition for devices equipped with 3D acceleration sensors. *International Conference on Intelligent User Interfaces (IUI '10)* (pp. 341–344). New York, NY: ACM.

Kühnel, C., Westermann, T., Hemmert, F., Kratz, S., Müller, A., & Möller, S. (2011, October). I'm home: Defining and evaluating a gesture set for smart-home control. *International Journal of Human-Computer Studies*, *69*(11), 693–704.

La Cascia, M., Isidoro, J., & Sclaroff, S. (1998). Head tracking via robust registration in texture map images. *IEEE Conference on Computer Vision and Pattern Recognition*. Santa Barbara, CA.

Landay, J. A., & Myers, B. A. (1995). Interactive sketching for the early stages of user interface design. *ACM CHI'95,* Denver, CO (pp. 43–50).

Latoschik, M., & Wachsmuth, I. (1997). Exploiting distant pointing gestures for object selection in a virtual environment. In I. Wachsmuth & M. Fröhlich (Eds.), *International gesture workshop: Gesture and sign language in human–computer interaction*. Bielefeld, Germany.

LaViola, J. J. Jr. (1999). *A survey of hand posture and gesture recognition techniques and technology* (Technical Report CS-99-11). Providence, IL: Department of Computer Science, Brown University.

Leap Motion (2012). http://www.leapmotion.com. Accessed on March 19, 2014.

Lee, C., & Xu, Y. (1996). Online, interactive learning of gestures for human/robot interfaces. *IEEE International Conference on Robotics and Automation*, *4*, 2982–2987, Minneapolis, MN.

Lee, H.-K., & Kim, J.-H. (1999, October). An HMM-based threshold model approach for gesture recognition. *IEEE Transactions on Pattern Analysis and Machine Intelligence*, *21*(10), 961–973.

Liu, J., Zhong, L., Wickramasuriya, J., & Vasudevan, V. (2009, December). uWave: Accelerometer-based personalized gesture recognition and its applications. *Pervasive and Mobile Computing*, *5*(6), 657–675.

Long, A., Landay, J., & Rowe, L. (1998). *PDA and gesture uses in practice: Insights for designers of pen-based user interfaces* (Report CSD-97-976). Berkeley, CA: University of California, Berkeley, CS Division, EECS Department.

Lucente, M., Zwart, G., & George, A. (1998). Visualization space: A testbed for deviceless multimodal user interface. *Intelligent Environments Symposium*. Stanford, CA: AAAI Spring Symposium Series.

Maggioni, C., & Kämmerer, B. (1998). GestureComputer—History, design and applications. In R. Cipolla & A. Pentland (Eds.), *Computer vision for human–machine interaction*. Cambridge, U.K.: Cambridge University Press .

Marrin, T., & Picard, R. (1998). The conductor's jacket: A testbed for research on gestural and affective expression. *Twelfth Colloquium for Musical Informatics*. Gorizia, Italy.

Maxwell, D., Griffith, T., and Finkelstein, N. M., (2014). Use of virtual worlds in the military services as part of a blended learning strategy. In K. S. Hale & K. M. Stanney (Eds.), *Handbook of virtual environments* (2nd ed., pp. 961–1002). New York, NY: CRC Press.

McNeill, D. (1992). *Hand and mind: What gestures reveal about thought*. Chicago, IL: University of Chicago Press.

McNeill, D. (Ed.) (2000). *Language and gesture*. Cambridge, U.K./New York, NY/Melbourne, Australia/ Madrid, Spain: Cambridge University Press.

Mitra, S., & Acharya, T. (2007, May). Gesture recognition: A survey. *IEEE Transactions on Systems, Man, and Cybernetics, Part C: Applications and Reviews*, *37*(3), 311–324.

Moeslund, Th. B., Hilton, A., Krüger, V., & Sigal, L. (Eds.) (2011). *Visual analysis of humans: Looking at people.* New York, NY: Springer.

Morency, L.-P., Quattoni, A., & Darrell, T. (2007, June). Latent-dynamic discriminative models for continuous gesture recognition. *IEEE Conference on Computer Vision and Pattern Recognition.* Minneapolis, MN.

Moses, Y., Reynard, D., & Blake, A. (1995). Determining facial expressions in real time. *International Conference on Computer Vision.* Cambridge, MA.

Mulder, A. (1996). *Hand gestures for HCI* (Technical Report No. 96-1). Ann Arbor, MI: Simon Fraser University, School of Kinesiology.

Oka, K., Sato, Y., & Koike, H. (2002, November/December). Real-time fingertip tracking and gesture recognition. *IEEE Computer Graphics and Applications*, *22*(6), 64–71.

Oliver, N., Pentland, A., & Bérard, F. (1997). LAFTER: Lips and face real-time tracker. *IEEE Conference on Computer Vision and Pattern Recognition.* San Juan, Puerto Rico.

Ong, S. C. W., & Ranganath, S. (2005, June). Automatic sign language analysis: A survey and the future beyond lexical meaning. *IEEE Transactions on Pattern Analysis and Machine Intelligence*, *27*(6), 873–891.

Otsuka, T., & Ohya, J. (1998). Recognizing abruptly changing facial expressions from time-sequential face images. *IEEE Conference on Computer Vision and Pattern Recognition.* Santa Barbara, CA.

Overholt, D., Thompson, J., Putnam, L., Bell, B., Kleban, J., Sturm, B., & Kuchera-Morin, J. (2009, December). A multimodal system for gesture recognition in interactive music performance. *Computer Music Journal 33*(4), 69–82.

Oviatt, S. L. (1996). Multimodal interfaces for dynamic interactive maps. *ACM CHI: Human Factors in Computing Systems* (pp. 95–102). New York, NY, NY.

Paradiso, J., & Sparacino, F. (1997). Optical tracking for music and dance Performance. *Fourth Conference on Optical 3-D Measurement Techniques.* Zurich, Switzerland.

Parvini, F., Mcleod, D., Shahabi, C., Navai, B., Zali, B., & Ghandeharizadeh, S. (2009). An approach to glove-based gesture recognition. In J. A. Jacko (Ed.), *International Conference on Human-Computer Interaction. Part II: Novel Interaction Methods and Techniques* (pp. 236–245). Berlin/Heidelberg, Germany: Springer-Verlag.

Pavlovic, V., Sharma, R., & Huang, T. (1996). Gestural interface to a visual computing environment for molecular biologists. *International Conference on Automatic Face and Gesture Recognition.* Killington, VT.

Pylvänäinen, T. (2005). Accelerometer based gesture recognition using continuous HMMs. *LNCS: Pattern Recognition and Image Analysis*, *3522*, 413–430, Berlin, Germany: Springer.

Ren, Z., Meng, J., Yuan, J., & Zhang, Z. (2011). Robust hand gesture recognition with kinect sensor. *ACM International Conference on Multimedia* (pp. 759–760). New York, NY.

Reyes, M., Dominguez, G., & Escalera, S. (2011, November). Feature weighting in dynamic timewarping for gesture recognition in depth data. *IEEE Workshop on Consumer Depth Cameras for Computer Vision.* Barcelona, Spain.

Song, Y., Demirdjian, D., & Davis, R. (2012, March). Continuous body and hand gesture recognition for natural human-computer interaction. *ACM Transactions on Interactive and Intelligent Systems 2*(1), Article 5, 28.

Stark, M., & Kohler, M. (1995). Video-based gesture recognition for human-computer interaction. In W. D. Fellner (Ed.), *Modeling—Virtual Worlds—Distributed graphics.*

Starner, T., & Pentland, A. (1995). Visual recognition of American Sign Language using hidden Markov models. *International Workshop on Automatic Face- and Gesture-Recognition.* Zurich, Switzerland.

Sturman, J. (1992). *Whole-hand input* (Unpublished doctoral dissertation). MIT Media Laboratory, Cambridge, MA.

Suk, H., Sin, B., & Lee, S. (2008). Recognizing hand gestures using dynamic Bayesian network. *IEEE Conference on Automatic Face and Gesture Recognition.* Amsterdam, the Netherlands.

Takahashi, T., & Kishino, F. (1991). Gesture coding based in experiments with a hand-gesture interface device. *SIGCHI Bulletin*, *23*(2), 67–73.

Triesch, J., & von der Malsburg, C. (1996). Robust classification of hand postures against complex backgrounds. *International Conference on Automatic Face and Gesture Recognition.* Killington, VT.

Väänänen, K., & Böhm, K. (1993). *Virtual Reality Systems*, R. Earnshaw, M. Gigante, & H. Jones (Eds.), Chapter 7, pp. 93–106. Academic Press.

Van den Bergh, M., & Van Gool, L., (2011, January 5–7). Combining RGB and ToF cameras for real-time 3D hand gesture interaction. *IEEE Workshop on Applications of Computer Vision (WACV 2011),* Kona, HI (pp. 66–72).

Vogler, C., & Metaxas, D. (2001, March). A framework for recognizing the simultaneous aspects of American sign language. *Computer Vision and Image Understanding*, *81*(3), 358–384.

Wachs, J. P., Kölsch, M., Stern, H., & Edan, Y. (2011, February). Vision-based hand-gesture applications. *Communications of the ACM, 54*(2), 60–71.

Wang, Q., Chen, X., Zhang, L.-G., Wang, C., & Gao, W. (2007, October–November). Viewpoint invariant sign language recognition. *Computer Vision and Image Understanding, 108*(1–2), 87–97.

Wang, S. B., Quattoni, A., Morency, L.-P., Demirdjian, D., & Darrell, T. (2006). Hidden conditional random fields for gesture recognition. *IEEE Conference on Computer Vision and Pattern Recognition,* New York, NY (pp. 1521–1527).

Wang, X., Tarrío, P., Metola, E., Bernardos, A. M., & Casar, J. R. (2012). Gesture recognition using mobile phone's inertial sensors. In S. Omatu et al. (Eds.), *Distributed computing and artificial intelligence* (AISC 151, pp. 173–184). Berlin, Germany: Springer-Verlag.

Wang, X., & Winslow, B. (2014). Eye tracking in virtual environments. In K. S. Hale & K. M. Stanney (Eds.), *Handbook of virtual environments* (2nd ed., pp. 197–210). Boca Raton, FL: CRC Press.

Weng, J., & Cui, Y. (1998). Recognition of hand signs from complex backgrounds. In R. Cipolla & A. Pentland (Eds.), *Computer vision for human-machine interaction.* New York, NY: Cambridge University Press.

Wexelblat, A. (1994). A feature-based approach to continuous-gesture analysis (Unpublished master's thesis). Cambridge, MA: MIT Media Labortory.

Wilson, A., & Bobick, A. (1999, September). Parametric hidden Markov models for gesture recognition. *IEEE Transactions on Pattern Analysis and Machine Intelligence, 21*(9), 884–900.

Wilson, A. D. (2004, October). TouchLight: An imaging touch screen and display for gesture-based interaction. *International Conference on Multimodal Interfaces.* State College, PA.

Wobbrock, J. O., Morris, M. R., & Wilson, A. D. (2009, April 4–9). User-defined gestures for surface computing. *International Conference on Human Factors in Computing Systems* (pp. 1083–1092). Boston, MA.

Wren, C., Azarbayejani, A., Darrell, T., & Pentland, A. (1996). Pfinder: Real-time tracking of the human body. *International Conference on Automatic Face and Gesture Recognition.* Killington, VT.

Wu, J., Pan, G., Zhang, D., Qi, G., & Li, S. (2009). Gesture recognition with a 3-D accelerometer. In D. Zhang et al. (Ed.), *UIC 2009, LNCS 5585,* Brisbane, Australia (pp. 25–38).

Wu, Y., & Huang, T. S. (1999). Vision-based gesture recognition: A review. In A. Braffort et al. (Eds.), *Gesture Workshop (GW'99), LNAI 1739,* Gif-sur-Yvette, France (pp. 103–115). Springer-Verlag.

Xbox (2012, September). http://support.xbox.com/en-US/xbox-360/kinect/body-controller. Accessed on March 19, 2014.

Yamamoto, J., Kondo, T., Yamagiwa, T., & Yamanaka, K. (1998). Skill recognition. *International Conference on Automatic Face and Gesture Recognition.* Nara, Japan.

Yang, M.-H., Ahuja, N., & Tabb, M. (2002, August). Extraction of 2D motion trajectories and its application to hand gesture recognition. *IEEE Transactions on Pattern Analysis and Machine Intelligence, 24*(8), 1061–1074.

Zafrulla, Z., Brashear, H., Starner, T., Hamilton, H., & Presti, P. (2011). American sign language recognition with the kinect. *ACM International Conference on Multimodal Interfaces* (pp. 279–286). New York, NY.

Zaki, M. M., & Shaheen, S. I. (2011, March 1). Sign language recognition using a combination of new vision based features. *Pattern Recognition Letters, 32*(4), 572–577.

Zeleznik, R. C., Herndon, K. P., & Hughes J. F. (1996). Sketch: An interface for sketching 3D scenes. *ACM SIGGRAPH,* 163–170.

Zelinsky, A., & Heinzmann, J. (1996). Real-time visual recognition of facial gestures for human–computer interaction. *International Conference on Automatic Face and Gesture Recognition.* Killington, VT.

10 Avatar Control in Virtual Environments

James Templeman, Robert Page, and Patricia Denbrook

CONTENTS

10.1   INTRODUCTION

A *virtual environment* (VE) may be generally defined as an array of sensory cues generated in response to a user's actions that give the impression of dealing directly with a 3D model. These sensory cues rely on a display system that generates a clear, high-resolution image of the virtual scene and a realistic, low-latency rendering engine to drive it. Quality 3D audio and haptic feedback also contributes to a user's sense of immersion within the VE. The components of the 3D model may include inanimate objects (roads, buildings, vehicles), embodied entities representing human players (avatars), and embodied entities representing computer-driven nonplayer characters (NPCs).

A VE is considered *realistic* when all of its components interact in a realistic manner. Natural laws of physics are simulated and embodied entities exhibit realistic behavior. Within many VEs, the user interacts with the VE through his *avatar*, which represents his body in the VE. The user picks up sensory information from the vantage point of his avatar and performs tasks within VE by animating his avatar. From the user's perspective, the realism of the VE is determined by how realistically he senses the environment and controls his avatar to act in the environment and how realistically the VE responds to his actions. A user interface (UI) is used to support this human to VE interaction and consists of one or more *interaction techniques*. These are the means by which a person applies devices to perform tasks (Foley, Wallace, & Chan, 1984). In virtual simulation, these tasks include navigating through and interacting with objects in the VE.

Two classes of interaction techniques enable the user to interact realistically with the VE through his avatar: *stimulus substitution* and *motor substitution*. These address the sensory and control aspects, respectively, of the immersive UI. *Stimulus substitution* involves the substitution of one stimulus for another, within the same sensory modality. It presents the sensory cues by which the user senses the VE. Stimulus substitution is at the heart of VE: a computer-generated image displayed as an array of picture elements substitutes for the ambient array of light a person sees in a natural environment. The substituted stimulus, however, is almost always an approximation to the real one, due to differences in resolution, latency, alignment, and other factors. *Motor substitution* involves the substitution of a range of artificial gestures for the range of natural motions. It employs *naturalistic* gestures for controlling virtual motion, such as stepping-in-place or sliding foot pedals to move one's avatar through VE. As with stimulus substitution, motor substitution is never perfect, but some actions work better for controlling certain simulated actions than others.

The user interacts with the VE in a constantly recurring *cycle of sensing and acting*. The quality of stimulus substitution impacts the user's ability to sense the VE, while the quality of motor substitution impacts the user's ability to act in VE. Stimulus and motor substitution must work together for the user to interact with the VE in a realistic manner. What you sense alters your course of action, and how you move alters the vantage from which you sense.

Our group at the Naval Research Laboratory's Immersive Simulation Lab first started developing UIs for VEs back in the mid-1990s. We began by addressing the problem of navigating over large distances in the virtual world and eventually expanded our research to address the full range of motions required to allow the user to interact with the virtual world in a demanding application: dismounted infantry simulation. Early on in our research, we realized that we needed to develop a framework for characterizing and evaluating avatar interfaces for VE, focusing on the crucial aspects of stimulus and motor substitution. Using this framework, we sought to develop UIs to enable the user to interact with a VE in a behaviorally realistic manner. The remainder of this chapter will show how our group approached this challenge and, over a number of years, arrived at a compelling solution.

10.2   FIRST STEPS TOWARD VIRTUAL LOCOMOTION

Our first entry into the realm of UIs for VE was to develop a more natural means of moving and navigating through virtual worlds. It had long been observed that people tend to get disoriented in VE more than they do in the real world (Darken & Sibert, 1996; Mills & Noyes, 1999; Ruddle, Payne, & Jones, 1998). In these earlier studies, users were also more likely to collide with walls, doorframes, and other virtual surfaces in VE. We set out to develop a more realistic control for moving through VE to enable the user to navigate more easily and move around virtual obstacles.

We define *virtual locomotion* as an interaction technique that enables a person to move in a natural way over long distances in the VE, while remaining within a relatively small physical space (Templeman, Denbrook, & Sibert, 1999). Ideally, it should allow users to perform the same coordinated actions and move with the same agility in VE as in the real world. This goal turned out to be more daunting than we expected.

The approach we took was to develop a high-fidelity virtual locomotion control based on stepping-in-place. We postulated that stepping-in-place would be a natural means of expressing one's intention to move. Using stepping motions would allocate control over locomotion to the legs, the part of the body that normally performs ambulation, and would avoid overloading the user's hands (i.e., via a mouse or game controller) to control locomotion. Our goal was to achieve a natural coordination between the way the user moved his head, arms, and legs to perform tasks in VE. The resultant stepping-in-place locomotion control was called *Gaiter.*

While a number of other stepping-in-place controls were developed over the past decade (Feasel, Whitton, & Wendt, 2008; Grant & Magee, 1998; Iwata & Fujii, 1996; Lampton & Parsons, 2001; Lane, Marshall, & Roberts, 2006; Wendt, Whitton, & Brooks, 2010), what set our approach apart was that we sought to carry over as many of the expressive characteristics of stepping-in-place that users could make *naturally.* We wanted to convey the extension and cadence of the user's in-place steps, as the steps were being taken. We also wanted the user's leg motions to control the direction of locomotion, in terms of walking to the side, backward, or diagonally. It is important to distinguish between *heading* (the direction in which the body faces) and *course* (the direction relative to the heading in which the body translates by stepping). One turns his body to change his heading. Although one may twist one's neck, torso, or pelvis to turn momentarily, most of the time, people step-to-turn their body. One steps forward, sideways, backward, or diagonally to set the course. We coupled the extent and direction of the horizontal excursion of the swing leg away from its previous support position to the horizontal translation of the user's avatar. When the swing leg moved forward, or to the side, locomotion was coupled to the excursion of the knee; when the leg swung backward, locomotion was coupled to the excursion of the foot. Gaiter discriminated between stepping-in-place and taking actual steps, and the user could apply either form of stepping to move through VE, although actual steps were constrained by the available tracking space or the length of the cable on a tethered head-mounted display (HMD).

One aspect of stepping-in-place that matches actual stepping is that there is a transition point as the rate of stepping increases, at which both feet become airborne at the same time (Thorstensson & Roberthson, 1987). One goes from walking to jogging in place just as one transitions from walking to running. This helped us match the relationship between stride length and cadence (Templeman, Sibert, Page, & Denbrook, 2006). With Gaiter, we adjusted virtual steps to account for a natural increase in stride length as a function of cadence (Inman, Ralston, & Todd, 1981).

The initial Gaiter system had magnetic trackers attached to the user's lower legs to track leg motion and another to the HMD to control the viewpoint and viewing direction (Templeman et al., 1999). Due to the accuracy limitations of magnetic tracking, we had to insert a pair of pressure sensors on insoles inside the user's shoes to detect when each foot was in contact with the floor. Later iterations of Gaiter utilized an optical tracking system to avoid issues with magnetic tracker limitations (e.g., limited utility in buildings with much metal rebar in the floor). We tracked the user's legs by attaching markers to the calves and found that the optical inputs were accurate enough to discern when each foot rested on the floor. This allowed Gaiter to dispense with the insole pressure sensors.

10.3 USER INTERFACE DESIGN PRINCIPLES FOR AVATAR CONTROL

We developed a framework for describing and analyzing methods for controlling one's avatar's motion over the course of our research in virtual locomotion. This framework focuses on the class of interaction techniques we refer to as *motor substitution*, substituting a range of artificial gestures for the range of natural motions. Our goal was to discern the best ways of mapping the user's natural and control actions to drive his avatar's movements in VE.

We refer to the motion a person would normally make in the real world as the *natural action* and the motion used to substitute for that motion the *control action*. The results obtained by performing these actions are called the *natural effect* and *control effect*, respectively. The natural action of taking actual steps produces the natural effect of translating one's body through space (natural locomotion), while the control action of walking in place produces the control effect of translating one's avatar through VE (virtual locomotion).

To aid the design process, we developed a list of attributes to characterize actions and effects (Templeman, Sibert, Page, & Denbrook, 2008):

1. *Body segments*: Which parts of the body are involved in performing the action?
2. *Effort*: How much effort is required to perform the action?
3. *Coordination*: How do different kinds of action interact with each other?
4. *Degrees of freedom and range of motion*: In what directions and to what extent do the body segments involved in executing the action move?
5. *Rate and accuracy*: What is the trade-off between speed and accuracy when performing the action?
6. *Open- or closed-loop control*: Open-loop control can be performed without external sensory feedback. Closed-loop control requires external feedback.

We posited that a closer match between control and natural actions (and control and natural effects) would provide better motor substitution, making it more natural for the user to perform and resulting in a more realistic simulation. In general, the most complex attribute to address is *coordination*. The control action should interact with all other actions the user might make in the same manner that the natural action being simulated interacts with those other actions. That is a tall order.

Although all six attributes apply to both actions and effects, the first three have a greater bearing on actions, while the last three are more pertinent to effects. Consider how the first three attributes describe the natural action of stepping with the legs. *Body segments:* One's legs alternatively swing in the direction of motion. *Effort:* It takes more effort to go faster, further, and uphill. *Coordination:* The segments of the upper body can be applied fairly independently of the lower body when stepping. The upper body must remain balanced above the pelvis, and the sensory and mechanical design of the body favors forward movement. The legs are used to both translate and turn the body by stepping, so locomotion and turning are closely interrelated.

The natural effect of stepping with the legs is to translate the pelvis and with it the entire upper body. The latter attributes pertain to this effect. *Degrees of freedom (DoFs):* The pelvis moves horizontally above the surface of support being walked upon. Since one can move forward, backward, and sideways with respect to one's heading, virtual locomotion provides 2 DoFs of control. *Range of motion:* Steps are delimited by stride length. *Rate and accuracy:* Although people have fine control over their movement, there is a trade-off between speed and precision. *Open- or closed-loop control:* People are quite good at visually picking out a destination and then walking directly to it with their eyes closed (Loomis, Da Silva, Fujita, & Fukusima, 1992).

Now consider how these attributes pertain to the control action of stepping-in-place. *Body segments:* The stepping-in-place motions of the legs are used to translate the avatar's body. *Effort:* Although walking-in-place requires less effort than actual walking, it still takes more effort to raise the legs higher, step faster, and keep stepping for a longer time. *Coordination:* Walking-in-place coordinates naturally with movements of the upper body. Stepping-in-place allows the user to turn the body naturally, by stepping-to-turn. The knees are not lifted very far up when simply stepping-to-turn. Stepping-to-turn can be combined with walking-in-place by raising the knees further while stepping-to-turn.

Gaiter supports translating the avatar's body in any direction by any realistic stride length (2 DoF translational motion). The rate and accuracy with which users can move were empirically studied and reported (Sibert, Templeman, Page, Barron, & McCune, 2004). Gaiter is open loop in the sense that users can count the number of steps taken, although the user's intuitive perception of distance traveled is another matter. One cannot consider the control effect in isolation; the mapping between the control action and effect is critical for developing an effective interaction technique used to simulate natural locomotion.

10.4 VIRTE: DISMOUNTED INFANTRY SIMULATION

Demonstration of Gaiter sparked interest by the Marine Corps representatives at the Office of Naval Research (ONR). This coincided with ONR's establishment of the Virtual Technology and Environments (VIRTE) Future Naval Capability program. The demonstration II component of VIRTE addressed simulation-based training of close quarters battle (CQB) for military operations in urban terrain (MOUT). Beginning in 2002, VIRTE funded development to turn Gaiter into a high-fidelity dismounted infantry simulator. Shifting gears from using Gaiter to control virtual locomotion through a VE to fully controlling one's avatar in an infantry combat simulator introduced a host of challenges.

Vehicle simulators used by the military have a long and successful track record. UIs for vehicle simulators are relatively straightforward. In the real world, the user is sequestered in a cockpit and drives using a set of vehicular controls. This allows a realistic set of control devices to be used in simulation. The near-field environment that the user physically contacts and manipulates moves with him and can simply remain in place in a simulator. The far-field environment is viewed through a window, so the user does not come in direct physical contact with it.

Dismounted infantry simulation poses a greater challenge. An infantryman makes direct contact with his environment and moves through it, propelled by his own limbs. If he keeps moving toward objects in the distance, they will come within reach. People move differently than vehicles and change their posture in complex ways. Rather than controlling a vehicle, the user of a dismounted simulation controls an *avatar* that represents his body in the virtual world. The primary role of the UI is to enable the user to control his avatar and sense the virtual world from the avatar's perspective. The avatar exists within the simulation, and the user interacts with the virtual world through his avatar.

UIs that alter or restrict the way one's avatar acts and reacts in VE make poor training simulators. The true measures of an infantry simulator are how well it allows the user to (1) fully control the motion of his avatar and (2) receive the sensory input that would be available in the real world (Templeman & Denbrook, 2012).

Further, perception is an active process, based on how the eyes, head, and body are directed, so the sensory input received depends on the user's control over the avatar's movements. The actions involved with infantry tactics are often divided along the lines of *look–move–shoot* (Templeman et al., 2008). Looking entails directing the eyes and ears to pick up information. An infantryman must freely look around to detect and engage threats in a 360°/180° battlefield. Movement involves

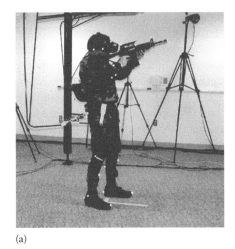

(a)   (b)

FIGURE 10.1   Gaiter infantry simulator (1999–2004). (a) Physical setup. (b) Third-person view of avatar.

turning the body and stepping or crawling to translate the body in any direction. *Shooting* encompasses weapon manipulations ranging from holding the weapon during movement to firing it. Looking and shooting overlap in terms of aiming and interact with movement in terms of the different shooting positions adopted (standing, kneeling, sitting, and prone). The look–move–shoot paradigm also applies to the use of cover and concealment. An infantryman is not heavily armored, so he must rely on available cover and concealment for protection. It is not good enough to simply hide behind cover. He must be able to peer out and shoot from behind it and move from cover to cover while minimizing exposure and maximizing situational awareness. Tactical movements such as pieing a corner involve moving with respect to cover and concealment. In addition to the specialized tactical actions required for infantry simulation, there is also a need to simulate the direct physical interaction between a person's body and the surfaces it makes physical contact with. We refer to the simulation of this in VE as *virtual contact.*

To overcome these challenges using Gaiter, we sought to create a high-fidelity infantry simulator to allow the user to control his avatar in as natural a manner as possible (Figure 10.1). We tracked the rotation and translation of the user's major body segments, an instrumented rifle prop, and an HMD. Since the avatar's posture constantly reflected the user's posture, the user simply turned his head and body and lined up the rifle, to change the view, heading, and aim. To move through the VE, the user walked or jogged in place.

With the full-body-tracked approach that maps stepping-in-place into virtual locomotion, we had to address two forms of motor substitution: virtual locomotion and virtual contact. An advantage of the stepping-in-place locomotion technique is that it allows the user to adopt natural motor control over looking and shooting. When the user's movements are mapped one for one to the movements of the avatar, no motor substitution is applied. Stimulus substitution, however, is still involved in supporting the two natural actions of looking and shooting. The quality of the simulator depended on just how well these essential actions—look, move, shoot, and virtual contact—could be made to work together (coordinate) in the UI.

10.4.1 STIMULUS SUBSTITUTION

10.4.1.1 Looking Using an HMD

A UI for dismounted infantry simulation must allow the user to look in any direction in the virtual world. We decided to continue using HMDs to keep the footprint of the system to a manageable size and make the system more deployable. A completely immersive surround screen display would have to surround the user on all sides, including the ceiling and floor. This would require a large, dedicated commitment of space for each user. We wanted users to focus all of their attention on the VE. HMDs can provide full or partial immersion: Full immersion is achieved by covering the rim of the display with a cowl to block the view of the surrounding physical environment. Partial immersion is achieved by leaving the rim of the display uncovered so the user has a partial view of the surroundings. Full immersion makes for a more realistic simulation because the user is exposed to a single visual environment. Gaiter employed a cowled HMD to present a fully immersive view.

10.4.1.2 Shooting Using a Rifle Prop

We also had to address natural control over aiming when the user could see only a rendered image of the weapon he was holding, displayed in the tracked HMD. With carefully placed markers attached to the ridged portions of the HMD and rifle prop, with a good spacing between markers to optimize for triangulation, the optical motion capture system was just precise enough to track them well enough to aim a one-for-one rendition of the rifle in VE, without undue jitter. The orientation readings of the HMD and rifle prop were stable within 3 minutes of arc.

10.4.2 MOTOR SUBSTITUTION

10.4.2.1 Virtual Locomotion

For a simulator intended to represent people moving about on foot, *virtual locomotion creates virtual space.* The user needs to move through a virtual terrain that does not match the physical terrain he is in. If the user tried to simply walk around in the physical environment, problems arise due to the mismatch of the two spaces.

The virtual locomotion technique adopted goes a long way toward determining the user's level of control over his avatar. The rate and accuracy with which the avatar walks and runs through the virtual world affect the user's sense of timing and scale. Many games allow unrealistically fast movement at sustained speeds (Clarke & Duimering, 2006). It would be impossible to obtain correct timing estimates when rehearsing a mission. If an action in the real world can be achieved with open-loop control using only internal sensory feedback (without external cues), the interaction technique should also make that possible.

Full immersion in a standing interface comes at a price. People tend to drift off center when they simply walk in place with their eyes closed. This also occurs when users step-in-place wearing fully immersive HMDs. This drift accumulates over time and leads to bumping into obstacles or other immersed users. We developed a series of *centering harnesses* to overcome this problem and to manage cables from the HMD so the user would not get caught up on them. A *go-prone* centering harness was built (Kaufman, 2007) to allow users to turn in place and to go between standing, kneeling, and prone while being held at the center of the tracking space. The frame holding the harness is large and heavy, reducing portability. Using a partially immersive HMD would eliminate the need for a centering harness by making the user responsible for avoiding moving off center, but at the cost of splitting the user's attention between the virtual and physical environments.

10.4.2.2 Virtual Contact

At a minimum, virtual contact is applied to determine what happens when the user's avatar runs into a wall and the user keeps stepping. In addition to simply redirecting the overall motion of the avatar's body to stop or slide along walls that the avatar would otherwise pass through, the simulator had to handle collisions of the avatar's limbs and virtual rifle to keep them from going through walls. Once the user is given full control over his avatar's limbs, he will want to use them to manipulate objects in the VE. A physics simulation engine was used to allow the Gaiter avatar to interact with virtual objects by pushing and kicking doors open. A good deal of additional R&D would be required to realize the potential of using virtual contact to enhance the realism of the UI.

10.5 PROBLEMS WITH THE FULL-BODY-TRACKED INFANTRY SIMULATOR

10.5.1 SENSORY–MOTOR MISMATCH

Sensory–motor mismatch occurs when the user performs an action and the sensory feedback derived from that action fails to match its real-world counterpart (McCabe, Haigh, Halligan, & Blake, 2005). In virtual simulation, this results from both stimulus and motor mismatches. *Stimulus mismatch* occurs when the synthetic stimulus fails to match what would be sensed in the real world. *Motor mismatch* occurs due to inadequacies in the motor substitution techniques applied. Sensory–motor mismatch should be avoided in VEs, as it can produce simulator sickness (Cobb, Nichols, Ramsey, & Wilson, 1999; Kolasinski, 1995; Reason & Brand, 1975), degrade performance, and produce negative training.

10.5.2 STIMULUS MISMATCH

Even when the UI attempts to make the user's avatar follow the user's physical actions one for one, natural interaction can still be disrupted by stimulus mismatch. The Gaiter infantry simulator suffered due to inadequacies of stimulus substitution, caused by spatial and temporal misalignments, and limitations of the display.

10.5.2.1 Limiting the Field of View and Visual Acuity

Human vision provides a wide visual field of view (FoV), consisting of two partially overlapping binocular regions. Each eye has an oval FoV measuring about 160°h × 120°v (Boff & Lincoln, 1988). They overlap by about 120°. A view of the virtual world approximating the eyes' natural FoV is essential for realistic infantry simulation. Currently available HMDs suitable for this application limit the user's view to less than a quarter of the natural FoV. Interacting with a VE with a narrow FoV display is like wearing blinders that block one's view. It requires extra head movement to compensate for wearing blinders. The narrower the FoV, the longer it takes to search for threats. A narrow FoV makes it easier to get caught on doorframes and other obstacles when moving through buildings and cluttered terrain. A narrow FoV greatly reduces one's ability to move as part of a team. The simple task of walking down the sidewalk next to someone requires a wide FoV to see both the person and the path ahead.

For well-lit objects directly in front of the eyes, a person with good visual acuity can discern detail at about 1 minute of arc (ibid.). Viewing a low-acuity display is like looking through a fogged pair of goggles. Distant targets blend into the background, making them difficult to discriminate. High-acuity vision is essential for recognizing different characters and discerning body language over the range of distances.

In our work with Gaiter, we used an NVIS nVisor SX HMD with a 47°h × 37°v physical FoV, which provided a visual acuity of 2.2 min of arc. We found that users tended to get hung up on doorframes because the frame fell outside their FoV as they approached a doorway. We found that rendering a larger geometric FoV made it easier for users to move through virtual buildings without getting caught on doorframes. Kuhl, Thompson, and Creem-Regehr (2006) recommend scaling the geometric FoV up in this way by 20% to improve depth perception as gauged by blind walking. Unfortunately, scaling the geometric FoV trades off a lower level of visual acuity to obtain a wider geometric FoV and distorts the visual feedback obtained from head, arm, and rifle movements.

10.5.2.2 Aiming a Rifle

Within the Gaiter system, the user holds and manipulates a physical rifle prop but sees only a rendered image of the rifle in the HMD. It is not technically possible to fully align the image of the rifle with the physical prop. There is inherent delay (latency) between when the tracker detects the positions of the HMD and rifle and when the rifle's image can be depicted at that position against the current background of the simulated environment. We noticed in a pilot study of rapid target engagement in VE that latency first caused the shooter to *overshoot* the target, but with practice, the user accommodated for this delay and exhibited a fairly balanced mix of undershooting and overshooting the target (i.e., firing before or after the rifle passed over the target). The user altered his shooting technique to optimize his performance in the VE.

Another mismatch is due to the failure to align the rendered image on the optical display with the position of the user's eyes, in order to present a correct perspective view of the virtual world. People wear HMDs in different ways due to the shape of their head, what feels comfortable, and how their movements jostle the HMD. Misalignment between how the physical rifle is held and how the virtual rifle appears to be held is problematic, since the user is forced to compensate by misaligning the physical rifle aim in order to line the virtual rifle's sights on the target. HMD-based immersive simulation creates a sensory–motor illusion powerful enough to recalibrate hand–eye coordination (Biocca & Rolland, 1998; Swan & Gabbard, 2005). Rapid target engagement relies on the practiced ability to snap the rifle from a ready position into an aim with the sights nearly aligned on the target, yet these skills acquired on a shooting range cannot be applied in the virtual simulator. Worse yet, repeated practice at misaligning the physical rifle to get the virtual gunsights on target will likely undermine the vital skill of engaging actual threats.

Why employ a realistic rifle prop for training if the user's real-world skills cannot be brought to bear in simulation and practicing with it degrades the user's ability to rapidly engage real-world targets? Negative training that would compromise a shooter's ability to engage threats must be eliminated.

10.5.3 MOTOR MISMATCH

10.5.3.1 Reflexes Tend to Override Gestural Substitutes

When we ran experienced Marines through simulated combat engagements in Gaiter, it was not uncommon for a marine to attempt to physically move for cover during the scenario. This was most likely to occur when the Marine was near virtual cover and was surprised by incoming fire. We were often concerned about them moving so forcibly that they would tear the centering harness apart. Only with constant coaching were we able to avert this catastrophe. The take-home point is that although users readily learn to apply motor substitution to maneuver through the environment, they are likely to revert to reflexive responses under stressful conditions.

10.5.3.2 Phase Mismatch between In-Place and Natural Stepping

There is an underlying problem with mapping the user's stepping-in-place control actions into *corresponding* natural translational steps. This is due to the fact that in-place stepping is inherently out of phase with actual stepping. It would be desirable to associate in-place steps with natural steps at the point at which both feet are down as well as the point at which the feet come together, but this cannot be done. If we define a step to start at the moment the swing foot lifts off, then with stepping-in-place, the legs are furthest apart in the middle of the step when the motion of the swing leg stops and reverses direction. In a natural step, however, the legs are closest together in the middle of the step when the swing leg passes by the support leg and the speed of the swing leg is at its peak. One may stop walking in a stable stance in the middle of a natural step (with legs apart and both feet on the ground), but the user does not want to stop moving in the middle of an in-place step with one leg raised, for any length of time. Another way of saying this is that stepping-in-place does not exhibit the same kind of reciprocal leg motion as natural stepping. Stepping-in-place better matches a series of half steps, since the feet start and end up next to each other in both cases. This kept us from achieving a strong correspondence between the way the user's and avatar's legs move.

This mismatch was further highlighted in one of our experimental studies. During a blind walking assessment at gauging distance using Gaiter (Sibert et al., 2008), we found that users regularly overshot the distance needed to move to virtual targets by a factor of about two. Comments from the subjects suggested they misinterpreted an in-place step as a half step rather than a full step and

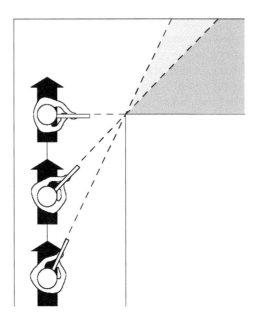

FIGURE 10.2   A rifleman *pies a corner.*

traveled twice as far as expected. The user's underlying sense of what a walking-in-place step felt like (proprioceptively) overrode the 10 min of practice they received at walking through the VE using Gaiter.

10.5.3.3   Difficulty Executing Tactical Movements

Tactical movements made with respect to cover and concealment often require changing the heading while maintaining the course. When walking down an L-shaped hallway, a rifleman *pies the corner* (Figure 10.2) by turning his body to keep his view and rifle directed at the corner he is approaching. This allows him to focus on the area behind the corner that comes into view as he moves down the hall. Moving along a path while turning the body is easy to do in the real world. The momentum of the body and the vestibular sense allow you to smoothly and subconsciously coordinate the motion of the legs. With Gaiter, it is possible, but awkward, to change the heading while maintaining the course. The user must step-in-place to turn the body while swinging the raised leg in a fixed direction to maintain the course. Although one can learn to execute this motion sequence, it remains an awkward movement to make.

10.5.3.4   Poorly Simulated Virtual Contact

Motor mismatch also resulted from rudimentary collision handling and the lack of haptic feedback. These deficiencies impeded the avatar's movement in close proximity to obstacles. The avatar was frequently redirected when moving near walls and obstructed when trying to move through tight spaces, especially when carrying a rifle. The lack of haptics also eliminated the UI's ability to support touch as a means of communication. A basic technique used in room clearing is for the members of the entry team to *stack* (gather) outside of a room and relay the signal to move by tapping the person in front of you on the back.

10.6   TOWARD A MORE ABSTRACT USER INTERFACE FOR AVATAR CONTROL

We initially assumed that a fully immersive, full-body-tracked approach would provide the highest fidelity of control over one's avatar in a dismounted infantry simulation. When we applied our Gaiter-driven avatar interface to this application, however, we found that problems arose when the user's interaction with the virtual world resembles natural interaction, but is not close enough for a person to apply their real-world skills and reflexes, as outlined previously. Using natural actions to control their avatar without providing truly realistic stimulus substitution disrupts the user's ability to perform tasks. Also, when control action used for motor substitution comes close but does not match natural action, real-world skills and reflexes interfere with the interaction.

Our group invested a lot of time and effort in designing, implementing, and testing a body-tracked simulator. There was a great deal that we liked about the naturalness of that UI, but the vital abilities of properly aiming a rifle, gracefully pie-ing a corner, or rapidly diving into cover were not among them. This led us back to the drawing board. We took a fresh look at existing UIs to first-person shooter (FPS) games and considered how we could extend them to include the advantages of body-tracked simulators such as Gaiter. Perhaps, a more abstract interface would help avoid negative training and free us to adopt a better way of simulating human locomotion.

Conventional desktop and console game controllers provide control over look–move–shoot, albeit at a minimal level. They have a limited number and range of input channels, due to being operated exclusively by the hands. Over the years, the operation of the mouse–keyboard and dual thumb stick gamepad controllers for FPS games has evolved and converged to those found in any of today's popular game titles (e.g., *Call of Duty*). The mouse or the right thumb stick of a gamepad turns the avatar's body and tilts the head up and down. Keys or the left thumb stick directs the course and speed of virtual locomotion. Keys or buttons trigger animations for looking to the side, leaning the upper body right or left, shifting between firing positions, and raising or lowering a weapon. These conventional input devices provide a means of controlling the avatar's motion and posture, and through them, the user directs his avatar to perform the combat-related actions described earlier.

We sought to increase the users' effective control over their avatar. Our overriding goal was to enable a Marine to control his avatar to act in the virtual world in as similar a manner as possible to the way he would act in a corresponding real-world situation. We refer to this as *behavioral realism* (refer to Templeman & Denbrook, 2012, for an in-depth discussion of behavioral realism and its importance for training small unit decision making and team coordination). We also wanted it to be easy to use, both in terms of being quick to learn and compatible with users' basic reactions. That meant adopting controls that the users were already familiar with and only adding controls that are intuitive to use.

Our most basic criticism of the interfaces to FPS games is that the user controls his entire avatar with just his hands. The hands are *overloaded* and thus constrained to control the avatar in a cursory manner. This is literally the case for the mouse–keyboard control: the only continuous input device used to control the avatar is the same as that used to control the cursor's movement in a conventional window–icon–menu–pointer UI. We wanted to leverage a person's abilities to apply and coordinate the motions of different parts of their body, in a fairly natural way. Moreover, we had to do that in order to provide a (behaviorally) realistic dismounted infantry simulation.

We learned from working with Gaiter that body-tracked UIs can enhance the user's control over the posture and motion of the user's avatar, and that an infantryman's posture is critical to executing tactics, techniques, and procedures (TTPs; as described, e.g., in MCWP 3-35.3, 1998). Our solution was to add input devices to the gamepad controller to provide a similar level of control in a seated interface.

10.6.1 MOTOR SUBSTITUTION: CORRESPONDENCE AND ABSTRACTION

We were willing to relinquish a number of actions that could be performed in a natural manner using the full-body-tracked system in order to achieve more balanced control over one's avatar in a dismounted infantry simulator. Once seated, the user can no longer step-to-turn. Once we dispensed with naturally turning the body, looking and shooting could no longer be performed in a fully natural manner. Motor substitution must then be applied to enable the user to look, move, and shoot in the VE. This was a trade-off that we were willing to make, based on our experiences using Gaiter. It allowed us to adopt a more effective means of controlling virtual locomotion in terms of quickly moving to a desired position without fighting one's instincts, and perform tactical movements like pie-ing a corner without having to execute awkward gestural movements. It also allowed for the adoption of a less realistic means of shooting that reduced the negative impact of the sensory–motor mismatch.

There are advantages to adopting more *abstract* forms of stimulus and motor substitution in a UI. The primary reason was to avoid having real-world skills and reflexes interfere when control actions come close but do not match the corresponding natural actions. In the case of simulated stepping, abstraction allowed us to match some aspects of natural action better than the stepping-in-place technique. In the case of controlling the avatar's upper body posture, abstraction did little to detract from allowing the user to control the avatar since the user is able to move his upper body in a natural, largely reflexive manner even while seated.

We adopted the term *correspondence* to denote effective motor substitution in the face of abstraction. That meant finding the maximum level of similarity between clearly distinct actions in order to close the gap between the user's and avatar's actions. As each interface element was designed and integrated, we sought to maximize correspondence, both in terms of matching the characteristic properties (listed earlier), when possible, and also in terms of continuously tying the avatar's posture to the way the input device was operated.

This freed us to alter the way the user controls the movement of his avatar and manipulates his virtual weapon. It allowed us to adopt body-tracked control over some aspects of the avatar's motion and device-driven control over other aspects of motion.

10.6.2 Pointman™: A Seated Interface Using the Head, Hands, and Feet

We developed a new desktop UI for dismounted infantry simulation, called *Pointman*. It uses a set of three input devices: a head tracker, gamepad, and rudder pedals (Figure 10.3). The additional input from the head and feet is designed to off-load the hands from having to control the entire avatar and allows for a more natural assignment of control. Together, the three devices offer twelve independent channels of control over the avatar's posture.

10.6.2.1 Slide Pedals to Step

The CH Products rudder pedals slide back and forth and also move up and down like accelerator pedals. The user slides the pedals back and forth to move the avatar by stepping (Figure 10.4). The pedals are mechanically coupled so when one pedal slides forward, the other slides back.

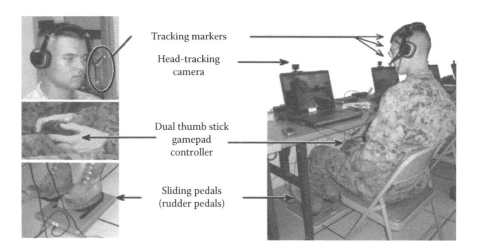

FIGURE 10.3 Setup of Pointman input devices.

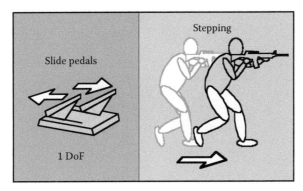

1 DoF: Displacement

FIGURE 10.4 Slide pedals to step.

A key advantage of using the sliding foot pedals is that their motion exhibits the same kind of reciprocal motion seen in natural stepping. This allowed us to set up a one-to-one mapping between the position of the foot pedals and the positions of the avatar's feet when stepping or when stationary (with the legs together or apart by any extent). When stepping forward, the avatar's right leg swings forward when the user moves the right pedal forward. When stepping backward, the avatar's right leg swings backward when the user moves the right pedal backward. This provides a stronger form of correspondence than could be achieved with Gaiter using the gestural motions of stepping-in-place. Work was needed to stabilize the optical flow associated with Pointman's foot-driven control. This led to a real-time model of how the legs flex to keep the motion of the pelvis more constant.

Sliding pedals allow the user to take measured steps to precisely control his motion. Precise control is critical when moving into and around cover. The rate at which the pedals are moved back and forth determines the speed and cadence with which the avatar steps. This can range anywhere from a slow, careful movement around cover to running at full speed between cover. The continuous reciprocal sliding input provides more nuanced control and a wider range of stepping speeds than a thumb stick rate control. Many computer games couple a *turbo speed* button with the thumb stick to increase the dynamic range of the control. The sliding pedals enable the user to control stepping speed in a way that mirrors natural locomotion.

The adoption of the seemingly artificial sliding foot pedal device allowed us to obtain this stronger form of correspondence. It also modified the relationship between the stepping control action and its effect on locomotion. In Gaiter, the control action of stepping-in-place produces the effect of translating the avatar's body, whereas in Pointman, the control action of sliding the foot pedals produces the effect of moving the avatar's feet, which in turn determines the translation of the avatar's body. Rather than acting *on* the avatar, the user acts *through* the avatar.

10.6.2.2 Depress Pedals to Lower Postural Height

When the user presses down on the pedals, tilting the pedals down, the avatar bends its legs to lower its postural height (Figure 10.5). The avatar's stance goes from standing tall to a low squat. If he pushes the pedals down with the pedals apart (i.e., one forward and one back), the avatar takes a kneeling position. A button press is used to transition from standing to prone; there is a distinct shift in balance in physically going between the two. When prone, the user crawls by sliding the pedals and can move continuously from a high to a low prone (from hands and knees to belly on the ground) by depressing the pedals.

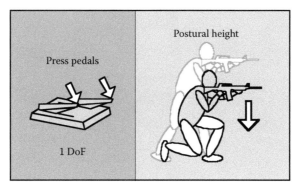

1 DoF: Height

FIGURE 10.5 Depress pedals to lower postural height.

10.6.2.3 Thumb Sticks Direct Heading and Course

We adopted a dual thumb stick gamepad controller when we started developing a desktop UI for infantry simulation, as we needed a means of directing the course and heading while remaining seated and applying foot pedals. While both the gamepad and the mouse–keyboard are commonly used for this purpose in FPS games, we favored the dual thumb stick gamepad since it offers two continuous controls over course and heading, compared with the mouse–keyboard's single continuous control over heading. Also, we originally planned on using an HMD as we had done with Gaiter, and unlike the mouse–keyboard, a gamepad can be used effectively without having to see it directly (an open-loop control device).

Conventional gamepad controls used in FPS games apply the right thumb stick as a rate control over the direction of view. Pushing the right stick to the side turns the view to the side by rotating the avatar's body (directing the heading). Pushing the right stick back or forward turns the view up or down by tilting the avatar's head. The left stick controls translational motion. The direction in which the left stick is pushed determines the direction of motion (directing the course). The degree to which the left stick is deflected determines the speed at which the avatar moves; it is a rate control over translation. The course is coupled to the heading, since it is redirected as the heading changes.

Pointman retains the use of the thumb sticks for directing the avatar's course and heading (Figure 10.6). Unlike conventional gamepad controls, however, the right thumb stick is used exclusively for turning the avatar's body, and the left stick is used exclusively for directing the avatar's course.

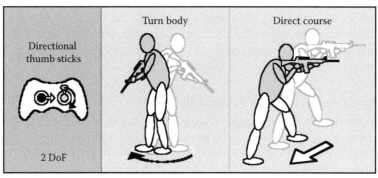

1 DoF: Yaw   1 DoF: Yaw course

FIGURE 10.6 Thumb sticks direct heading and course.

Pointman off-loads control over translating the avatar's body to the sliding foot pedals, and changing the avatar's view to the user's tracked head.

Early on, we experimented with other ways of applying the right stick to control heading (Templeman, Sibert, Page, & Denbrook, 2007). The right stick functioned as a positional control over turning, since there was a direct correspondence between how far the thumb stick was revolved around the rim and how far the avatar's body turned. We also uncoupled the course from the heading to make it easy to vary the heading without altering the course. The combination of directional control over heading and uncoupling control over the course works very well for executing the tactical movements involved in infantry combat. In the real world, pie-ing a corner is performed by turning the heading while maintaining the course. While the new control mapping looked very promising, young Marine's familiarity with the coupled/rate-based turning control used in console games led us to adopt the more conventional approach, to expedite learning to use Pointman.

10.6.2.4   Tilt Gamepad to Vary Weapon Hold

The gamepad used with Pointman, the Sony DualShock 3, includes a pair of tilt sensors. The tilt of the gamepad controls how the virtual rifle is held (Figure 10.7). The user tilts the gamepad down to lower the rifle and tilts the gamepad up to continuously raise the rifle up through a low ready, into an aim, and then to a high ready. This allows users to practice muzzle discipline. The user lowers the virtual weapon to avoid muzzle sweeping friendly or civilian characters, minimize collisions when moving through tight spaces, and avoid leading with the rifle when moving around cover. Once the rifle is raised into an aim, the user's head motion aligns the sight picture. Rolling the gamepad (tilting it side to side) cants the weapon.

10.6.2.5   Head Tracker Controls Upper Body

Pointman couples the orientation of the user's head to that of his avatar to provide a strong correspondence between the user's and the avatar's movements and support head-coupled viewing and head-coupled aiming. The NaturalPoint TrackIR 5 head tracker registers the translation and rotation of the user's head (Figure 10.8). The avatar's upper body posture directly reflects the tracked movement of the user's head. The rotation (yaw, pitch, roll) of the avatar's head follows that of the user's head. The view changes accordingly so that the user simply turns his head to look around (as further described later, under the heading of *Head-Coupled View Control*). When aiming a weapon, its sights remain centered in the FoV, so that turning the head also adjusts the aim. This mimics

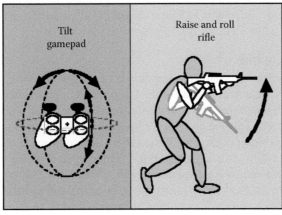

2 DoF: Pitch roll

FIGURE 10.7   Tilt gamepad to vary weapon hold.

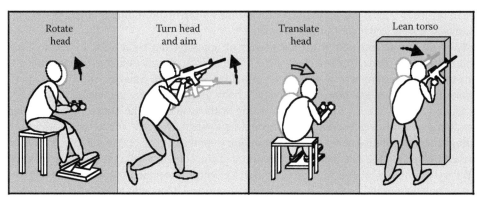

3 DoF: Yaw-pitch-roll   3 DoF: Roll-pitch-hunch

FIGURE 10.8   Head tracker based upper body control.

physical aiming in the sense that a rifleman moves his head and rifle together as a unit by maintaining a consistent cheek-weld to index the weapon. The user can aim as precisely as he can hold his head on target. The translational (x, y, z) readings of the head tracker allow the Pointman to fully couple the motions of the avatar's head and torso to those of the user. The head translates as the torso leans forward or to the side. Hunching the head down by flexing the spine is also registered by the head's translation, and the avatar adopts a matching posture. Leaning forward and hunching are used to duck behind cover. Rising up and leaning to the side are used to look out and shoot from behind cover.

10.6.3   APPLICATION OF INTERFACE DESIGN PRINCIPLES

Pointman achieves a good match between control and natural actions. Compared to conventional FPS interfaces, the more natural assignment of control actions to the different *body segments* affords more natural *correspondence* between the user's and the avatar's movements: the head turns to look, the torso leans to duck or lean around corners, the reciprocal motion of the sliding foot pedals mimics stepping, and the hands tilt the gamepad to raise and lower the weapon. This correspondence allows us to tap into the user's natural ability to perform *coordinated* actions by employing different parts of the body. Although the user does not exert a realistic level of *effort* involved in natural locomotion or in holding a rifle, the user is at least aware of having to keep moving his feet and raising and lowering the gamepad to wield a virtual weapon.

10.6.3.1   Positional Controls

The additional input devices allowed us to upgrade both the quality and quantity of the controls available to enhance the behavioral realism of the UI. As each interface element was designed and integrated, we sought to maximize correspondence by tying the avatar's posture to the way each input device was employed. This led us to opt, whenever feasible, for positional controls. A *positional control* uses the position of an input device or tracked segment of the user's body to determine the position of an output element (Zhai, 1995). For our purposes, the output element will be some aspect of the avatar's body or a virtual object held by the avatar. The term *position* may refer to translation or rotation. In contrast, a *rate control* uses the position of an input element to determine the rate of change in position (speed of motion) of an output element, and a *discrete control* uses the binary state of an input element (usually a button or key) to determine the fixed position (e.g., pose or stepping velocity) of an output element.

Positional control allows the user to move the avatar by continuously varying its posture. The user animates the avatar by moving the parts of his body. To implement this level of control, we had to depart from the way conventional game control software works, where the game plays out an animation in response to an input. Instead, we directly couple the position of the input to the avatar's posture. It feels more like controlling a puppet or inhabiting the avatar than like driving the avatar through a vehicular control interface.

This is fundamentally different from either discrete control over the avatar's posture or rate control over its motion. With discrete or rate controls, the user presses a button or pushes a stick and the simulation plays out a predefined motion sequence. Some body-tracked simulators also limit which aspects of the user's motion are reflected by the avatar. They gather just enough information to know when and how to transition the avatar from one posture or motion to another. These simulators provide the same sorts of stereotyped avatar motion control found in desktop and console interfaces.

Positional controls are well suited for controlling the avatar's posture; they provide direct, continuous control over the different postural elements. The user controls the avatar's posture with the same dexterity and precision as he can physically position his head, hands, and feet. This incorporates a natural means of trading off the *rate and accuracy* of movement in the simulation, allowing users to attain more realistic levels of performance. The user constantly feels the avatar's posture through his body. He can stop moving at any point along the arc of motion and can sense how his avatar is posed without the need of visual feedback (*open-loop control*).

The use of the pedals to simulate stepping further illustrates the effectiveness of positional control; the sliding pedals capture the reciprocal nature of leg motion, in that sliding the right pedal forward swings the avatar's right leg forward, past its left leg. The avatar moves by continuously varying the postures of its legs. The user can stop moving with the legs apart or together and feels their separation at every moment. The user can vary the stride length and cadence of stepping to control the speed of motion. The user can also make fine adjustments in the avatar's position by sliding small amounts. This allows for precise movement in and around cover.

Pointman provides continuous control over the following *DoFs* of motion, typically supported in conventional FPS games:

1. Turning the body (view and aim)
2. Pitching the head (view and aim)
3. Directing the course
4. Moving the body by stepping
5. Leaning the upper body to the side
6. Lowering the body
7. Raising the rifle

Pointman provides rate control over turning the body and positional control over all the other motions. In FPS games, gamepads typically provide positional control over only the avatar's course, and mouse–keyboard interfaces provide positional control over only the heading and the pitch of the avatar's head.

Point man also provides positional control over the following DoFs of motion, rarely supported in FPS games:

8. Yawing the head independently of the body
9. Canting the head
10. Leaning the upper body forward and back
11. Hunching the torso
12. Canting the rifle

This gives the user 12 continuous DoFs of control over the avatar's posture, eleven of which are positional. The controls work together to afford smooth, coordinated motion. This is quite unique, in that the posture of the user's avatar reflects the positions of the various inputs at every moment in time. There is no delay involved (as there would be with cueing a canned animation). The user can fluidly adopt and hold a wide array of postures, applying them when and however he sees fit, enhancing the behavioral realism provided by the UI.

The head tracker provides 6 DoFs of input, which are mapped to control 6 DoFs describing the avatar's head and upper body posture. The pedals provide 3 DoFs of input, which control the avatar's leg posture in 2 DoFs. While their sliding motion is physically coupled, the pedals move up and down independently. The up and down movements of the pedals are aggregated and used to control a single output DoF (effect). While each thumb stick provides 2 DoFs of input, only 1 DoF is applied to control the avatar's motion. This was done to off-load control from the hands to other parts of the body. The tilt sensors in the gamepad provide 2 DoFs of input, which are mapped to control the weapon hold in 2 DoFs. Thus, a total of 15 DoFs of input are mapped to 12 DoFs of control over the avatar's posture.

10.6.4 STIMULUS SUBSTITUTION: MORE ABSTRACTIONS

We have seen that switching from a full-body-tracked interface to a seated desktop interface required us to adopt more abstract forms of motor substitution, such as device-driven control over stepping, turning, and weapon handling, and that, surprisingly, these abstractions served to enhance the user's overall control of his avatar. Advantages can also be gained by adopting more abstract forms of stimulus substitution.

10.6.4.1 Head-Coupled View Control

Head-coupled viewing is commonly used with HMDs to allow the user to look in any direction. The HMD is tracked, and an image of the virtual world as seen from the perspective of the user's avatar is rendered on the display.

Desktop interfaces can also use head tracking to control the view. When the user's head turns, his avatar's head turns in the same direction, and an image of what the avatar would see looking in that direction is displayed on the monitor. Figure 10.9 shows the nominal case when the user looks straight ahead, while Figure 10.10 shows what happens in the virtual world and on the screen when the user tilts his head to look up.

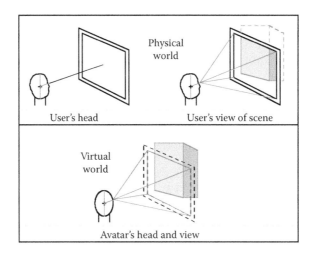

FIGURE 10.9 Head-coupled viewing—fixed display: looking straight ahead.

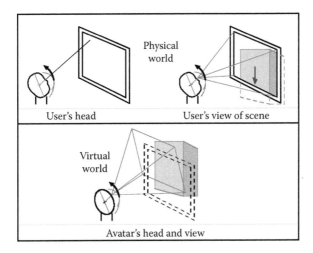

FIGURE 10.10 Head-coupled viewing—fixed display: looking upward.

This approach works well in practice, as can be seen in the popular use of the NaturalPoint TrackIR head tracker for controlling viewing in a wide variety of desktop simulators (Loviscach, 2009; Yim, Qiu, & Graha, 2008). Head-coupled view control is taken even further in the Pointman interface. Pointman uses the 6 DoF tracking of the head to lean and hunch the avatar's upper body independently of the head's orientation. It is able to accomplish this because the user's pelvis remains fixed to the chair. When a desktop display is used to perform head-coupled viewing, the pitch of the user's head (tilting up and down) is typically amplified to allow the user to look straight up or down in the virtual world while still being able to see the screen. Combining this with turning the avatar's body using a thumb stick allows the user to look in any direction (360°/180°: around, above, and below).

10.6.4.2 Advantages of Using a Fixed Screen Display

Although Pointman can be used with an HMD, we generally prefer a desktop display. Using a fixed screen display relieves expectations of realistic views, which adds flexibility. For example, it is far more acceptable to vary the geometric FoV presented in a desktop display than in an HMD (contrast Neale, 1997, with Kuhl et al., 2006). What matters most is having a higher-resolution display available to present high-acuity visual information, and even midlevel fixed screen displays offer better resolution and brightness and a wider physical FoV than currently available high-end HMDs.

10.6.4.3 Advantages of Being Seated

Being seated resolves a host of issues relating to virtual locomotion. A simple stationary chair serves the same role as the turning harness used in Gaiter: it holds the user *centered* so that he need not be concerned with the physical surroundings and can focus exclusively on his interaction with the VE. The chair supports the user in making reciprocal leg movements and makes it easy for him to transition between virtually walking and driving a vehicle. A seated desktop interface takes up much less space than a standing interface in which the user is allowed to turn 360° while holding a rifle prop out in front of him.

Since infantrymen operate in small unit teams, an infantry simulator must support a team of users working together. Marines typically train team tactics at the squad level (thirteen men) and above. A desktop interface fits neatly into a classroom setting, albeit for Pointman accommodations need to be made to ensure that each tracking camera only sees one set of tracking markers. The simplest layout for training a squad of marines is to set up all the Pointman stations around a long row of tables with all cameras facing outward.

10.6.4.4 Advantages of Virtual Weapons

There are many advantages to working with purely virtual weapons. They avoid the possibility of negative training by eliminating the sensory–motor mismatch that causes existing skills and reflexes to interfere with performance in simulation. There is no need to maintain an armory containing all the props needed to represent every weapon and instrument a squad of Marines might use. Since the weapon exists only within the VE, it can be set down or passed to another avatar at will.

10.6.5 Abstraction Can Enhance Behavioral Realism

Perhaps the most surprising advantage of abstracting is that it can be applied to overcome the difficulties associated with sensory–motor mismatch. If coming close but not matching real-world interaction causes difficulties, then one solution is to move virtual interaction further from natural interaction. FPS interfaces demonstrate how readily people can control an avatar just by using their hands. We found that adding input devices to engage more of the user's body, while applying the principles of correspondence, allowed us to enhance the behavioral realism of the UI for simulation.

The forms of stimulus and motor substitution used in Pointman enabled the following actions to be performed with greater behavioral realism than in conventional desktop interfaces:

- *Direct the view*: knowing where you are looking and turning your head to search and access
- *Aim*: as precisely as you can hold your head directed at the target
- *Adjust the rifle hold*: to continuously vary the rifle hold from down, up through a low ready, into an aim, and then to a high ready
- *Step*: to move precisely in tight spaces and with respect to cover and concealment, varying speed continuously over a wide dynamic range, and the ability to immediately stop at any moment with feet together or apart
- *Lean and hunch*: to duck behind cover and raise or lean out from cover or concealment to look and shoot

10.6.6 Training to Use a New Interface

The time required to learn a new UI is a major concern, especially if the system is not used on a regular basis. Ease of learning often goes hand in hand with ease of use; interfaces that require users to associate button presses with a large set of essential actions are not well suited for use by infrequent users who are apt to forget which button invokes which action. Pointman employs distinct motor actions that resemble their real-world counterpart. It is obvious which actions are controlled by head or foot movements. Users have a good idea of how the interface works when they return to it and progressively gain confidence in using it.

We experimented with making the turning thumb stick a positional control and changing the way the course changes as the avatar turns to provide better support for pie-ing corners (Templeman et al., 2007). We found that experienced console gamers required a great deal of practice to unlearn the rate-controlled turning that they were familiar with and proficient at using. It is far easier to learn a new interface when the user does not have to overcome highly engrained habits, as is required to use a familiar device in a new way.

10.6.7 Integration with Virtual Battlespace 2

Virtual Battlespace 2 (VBS2) is a combined arms training simulator used by the US Marine Corps, US Army, and North Atlantic Treaty Organization (NATO) allies. VBS2 is a central component of the United States Marine Corps Deployable Virtual Training Environment (DVTE) program. Bohemia Interactive Systems, the creators of VBS2, worked closely with the US Naval Research

Laboratory (NRL) to tightly integrate the Pointman interface with VBS2 and to allow Pointman to fully and continuously control the posture of the user's avatar. The detailed articulation of the user's avatar is made visible to other squad members running in a networked simulation. Pointman-enhanced VBS2 (VBS-Pointman) also supports a wide range of small arms and additional forms of mobility, including climbing, swimming, and mounted roles (driver, passenger, and gunner) using the full complement of manned vehicles.

10.6.8 Testing and Assessment

VBS-Pointman was first shown at the Future Immersive Training Environment (FITE) Joint Capability Technology Demonstration (JCTD) as an excursion demonstration in December 2009, at Camp Pendleton and in March 2010, at Camp Lejeune. Since then, it has undergone extensive testing to evaluate and refine enhancements to the UI. Testing was performed at the Simulation Center in Camp Lejeune in June 2010 and at The Basic School in Quantico in March 2011.

10.6.8.1 Military Utility Assessment

ONR Code 30's demonstration and assessment team performed a full Military Utility Assessment (MUA) of the Pointman UI in September 2011. The assessment was held at the 3-MEF Simulation Center, Marine Corps Base Hawaii. A squad of Marines with combat experience from Golf Company, 2nd Battalion, 3rd Marine Regiment participated in the study.

The MUA focused on the ability of the Pointman UI to replicate real-world movement in a VE and to enhance the virtual training experience for Marine infantrymen. The primary data used to inform the assessment were subjective user feedback solicited from the Marines. This was collected using automated surveys and interviews. Objective data were gathered by the assessment team throughout the MUA and documented via logs used to capture system performance times and events.

The data were compiled and analyzed by the Demonstration & Assessment team to arrive at a conclusion regarding Pointman's realism and training value. A description of the study, the analysis of the collected data, and its conclusions are included in the final report (Office of Naval Research, Code 30, 2012). The study found that the Marines adapted easily to the Pointman input devices. They gave the head tracker a 97% approval rating for the ability it provided to precisely control the movements of their avatar's head and torso. They liked that the gamepad gave them a familiar means of controlling their course and heading, that it enabled them to fluidly control the level of the weapon, and that it integrated well with the head tracker for aiming, scanning, and engaging targets.

The Marines' reception to the use of foot pedals was met with guarded skepticism at the start of Pointman training sessions, but after completing the first training session, their feedback was overwhelmingly in favor of the pedals (92% either strongly agreed or agreed they were easy to use). In characterizing Pointman, the Marine squad leader observed that the "foot pedals allowed better, smaller and more precise avatar movement." The foot pedals also proved to be durable over the course of the MUA, withstanding five days of continuous use by Marines in combat boots with no breaks or failures.

The MUA report concluded that Pointman provided realistic movement and utility in enhancing the training system. The Pointman UI allowed the Marines to control their avatars to move realistically and enhanced their ability to utilize cover more effectively. Training utility to individual and small Marine units was also demonstrated. The report recommended transitioning Pointman into VBS2 to enhance the training of Marine infantry units up to the squad level. NRL is continuing to work with ONR and the US Marine Corps toward this goal.

10.7 CONCLUSION

When first introduced to an immersive virtual simulation, people marvel at the new means of delivering synthetic sensory information. Often, we get so caught up in surrounding the user with high-fidelity sensory streams that we neglect the importance of representing the user's actions in the VE.

All too often, the user's experience is akin to *steering* his avatar through the virtual terrain, and other immersed users see him as a rather stilted animated figure that moves through a limited set of stereotypical actions.

We have come to the conclusion that the ability to exert greater control over one's avatar is essential to making an interactive simulation feel immersive. If the user is really going to inhabit the VE, he needs to have true freedom of action and a sense of embodiment in that world. You cannot really judge the realism of a virtual simulation until you have walked a mile in your avatar.

Based on our firsthand experience with Pointman and Gaiter, it is far easier to quickly direct and redirect motion using the seated desktop control than with the high-end stepping-in-place control, where the user's natural body motions are engaged as fully as possible, but not in a fully natural way. This is especially true when the user needs to move tactically, as when scanning for threats while moving along a path and using cover effectively.

Pointman allows users to control their avatars to act in a more realistic way but makes no attempt to train real-world motor skills. It abstracts how the user controls the avatar; the user's actions do not match every motion the avatar makes in detail. Pointman is a more realistic simulation interface in the sense of being able to control the avatar to perform a wider range of combat-relevant actions. This enhanced behavioral realism supports the training of cognitive skills, including tactical decision making and team coordination.

ACKNOWLEDGMENTS

We would like to thank the following: NRL for the 6.2 base funding to initially develop Pointman; ONR's rapid technology transition program office for supporting the integration of point man with VBS2; Bohemia Interactive Systems for their assistance in integrating with VBS2; and ONR's Human Performance, Training, and Education Science and Technology (S&T) thrust area managers for their support in refining, demonstrating, and assessing Pointman.

REFERENCES

Biocca, F. A., & Rolland, J. P. (1998). Virtual eyes can rearrange your body: Adaptation to visual displacement in see-through, head-mounted displays. *Presence*, 7(3), 262–277.
Boff, R. B., & Lincoln, J. E. (Eds.). (1988). *Engineering data compendium* (Vol. 1). Wright Patterson Air Force Base, OH: Harry G. Armstrong Aerospace Medical Research Laboratory.
Clarke, D., & Duimering, P. R. (2006). How computer gamers experience the game situation: A behavioral study. *Computers in Entertainment (CIE)*, 4(3), 6.
Cobb, S. V., Nichols, S., Ramsey, A., & Wilson, J. R. (1999). Virtual reality-induced symptoms and effects (VRISE). *Presence*, 8(2), 169–186.
Darken, R. P., & Sibert, J. L. (1996). Wayfinding strategies and behaviors in large virtual worlds. *Proceedings of the SIGCHI Conference on Human Factors in Computing Systems: Common Ground*, Vancouver, British Columbia, Canada. (pp. 142–149). ACM.
Feasel, J., Whitton, M. C., & Wendt, J. D. (2008). LLCM-WIP: Low-latency, continuous-motion walking-in-place. *3D User Interfaces, 2008. 3DUI 2008. IEEE Symposium on*, Reno, NV (pp. 97–104). IEEE.
Foley, J. D., Wallace, V. L., & Chan, P. (1984). Human factors of computer graphics interaction techniques. *IEEE Computer Graphics Applications*, 4(11), 13–48.
Grant, S. C., & Magee, L. E. (1998). Contributions of proprioception to navigation in virtual environments. *Human Factors: The Journal of the Human Factors and Ergonomics Society*, 40(3), 489–497.
Inman, V. T., Ralston, H. J., & Todd, F. (1981). *Human walking*. Baltimore, MD: Williams & Wilkins.
Iwata, H., & Fujii, T. (1996). Virtual perambulator: A novel interface device for locomotion in virtual environment. *Proceedings of the IEEE 1996 Virtual Reality Annual International Symposium, 1996*, Santa Clara, CA (pp. 60–65). IEEE.
Kaufman, R. E. (2007). A family of new ergonomic harness mechanisms for full-body natural constrained motions in virtual environments. *IEEE Symposium on 3D User Interfaces, 2007. 3DUI'07*. Charlotte, NC. IEEE.

Kolasinski, E. M. (1995). *Simulator sickness in virtual environments* (No. ARI-TR-1027). Alexandria, VA: Army Research Inst for the Behavioral and Social Sciences.

Kuhl, S. A., Thompson, W. B., & Creem-Regehr, S. H. (2006). Minification influences spatial judgments in virtual environments. *Proceedings of the Third symposium on Applied Perception in Graphics and Visualization*, Boston, MA (pp. 15–19). ACM.

Lampton, D. R., & Parsons, J. B. (2001). The fully immersive team training (FITT) research system: Design and implementation. *Presence*, *10*(2), 129–141.

Lane, S. H., Marshall, H., & Roberts, T. (2006). *Control interface for driving interactive characters in immersive virtual environments.* Princeton Junction, NJ: SOVOZ.

Loomis, J. M., Da Silva, J. A., Fujita, N., & Fukusima, S. S. (1992). Visual space perception and visually directed action. *Journal of Experimental Psychology: Human Perception and Performance*, *18*(4), 906.

Loviscach, J. (2009). Playing with all senses: Human–Computer interface devices for games. *Advances in Computers*, *77*, 79–115.

McCabe, C. S., Haigh, R. C., Halligan, P. W., & Blake, D. R. (2005). Simulating sensory–motor incongruence in healthy volunteers: Implications for a cortical model of pain. *Rheumatology*, *44*(4), 509–516.

MCWP 3-35.3. (1998). *Military operations on urbanized terrain (MOUT).* Washington, DC: Department of the Navy.

Mills, S., & Noyes, J. (1999). Virtual reality: An overview of user-related design Issues: Revised paper for special issue on "Virtual reality: User Issues" in interacting with computers, May 1998. *Interacting with Computers*, *11*(4), 375–386.

Neale, D. C. (1997). Factors influencing spatial awareness and orientation in desktop virtual environments. *Proceedings of the Human Factors and Ergonomics Society Annual Meeting*, *41*(2), 1278–1282. Albuquerque, NM: SAGE Publications.

Office of Naval Research, Code 30: Expeditionary Maneuver Warfare & Combating Terrorism Department. (2012). *Pointman Dismounted Infantry Simulation Interface Military Utility Assessment Report* (FOUO).

Reason, J. T., & Brand, J. J. (1975). *Motion sickness.* London, U.K.: Academic Press.

Ruddle, R. A., Payne, S. J., & Jones, D. M. (1998). Navigating large-scale "desk-top" virtual buildings: Effects of orientation aids and familiarity. *Presence*, *7*(2), 179–192.

Sibert, L. E., Templeman, J. N., Page, R. C., Barron, J. T., & McCune, J. A. (2004). *Initial assessment of human performance using the Gaiter interaction technique to control locomotion in fully immersive virtual environments* (No. NRL/FR/5510—04-10). Washington, DC: Naval Research Lab.

Sibert, L. E., Templeman, J. N., Stripling, R. M., Coyne, J. T., Page, R. C., La Budde, Z., & and Afergan, D. (2008). Comparison of locomotion interfaces for immersive training systems. *Proceedings of the Human Factors and Ergonomics Society Annual Meeting*, *52*(27), 2097–2101. New York, NY: SAGE Publications.

Swan, J. E. & Gabbard, J. L. (2005). *Survey of user-based experimentation in augmented reality.* Proceedings of First International Conference on Virtual Reality. Las Vegas, NV, pp. 1–9.

Templeman, J. & Denbrook, P. (2012). *The Interservice/Industry Training, Simulation & Education Conference (I/ITSEC)* (Vol. 2012, No. 1), Orlando, FL. National Training Systems Association.

Templeman, J. N., Denbrook, P. S., & Sibert, L. E. (1999). Virtual locomotion: Walking in place through virtual environments. *Presence*, *8*(6), 598–617.

Templeman, J. N., Sibert, L. E., Page, R. C., & Denbrook, P. S. (2006). *Immersive simulation to train urban infantry combat.* Washington, DC: Naval Research Lab Information Technology Div.

Templeman, J. N., Sibert, L. E., Page, R. C., & Denbrook, P. S. (2007). Pointman-A device-based control for realistic tactical movement. *IEEE Symposium on 3D User Interfaces, 2007. 3DUI'07*, Charlotte, NC. IEEE.

Templeman, J. N., Sibert, L. E., Page, R. C., & Denbrook, P. S. (2008). Designing user interfaces for training dismounted infantry. In D. Nicholson, D. Schmorrow, & J. Cohn (Eds.), *The PSI handbook of virtual environments for training and education.* Westport, CT: Greenwood Publishing Group.

Thorstensson, A., & Roberthson, H. (1987). Adaptations to changing speed in human locomotion: speed of transition between walking and running. *Acta Physiologica Scandinavica*, *131*(2), 211–214.

Wendt, J. D., Whitton, M. C., & Brooks, F. P. (2010, March). Gud wip: Gait-understanding-driven walking-in-place. *2010 IEEE Virtual Reality Conference (VR)*, Waltham, MA (pp. 51–58). IEEE.

Yim, J., Qiu, E., & Graham, T. C. (2008). Experience in the design and development of a game based on head-tracking input. *Proceedings of the 2008 Conference on Future Play: Research, Play, Share*, Toronto, ON (pp. 236–239). ACM.

Zhai, S. (1995). *Human performance in six degree of freedom input control* (Doctoral dissertation). University of Toronto, Toronto, Ontario, Canada. http://etclab.mie.utoronto.ca/people/shumin_dir/papers/PhD_Thesis/Chapter2/Chapter23.html.

Section III

System Requirements: Software

11 Virtual Environment Models

G. Drew Kessler

CONTENTS

11.1 INTRODUCTION

Many of the design decisions of virtual environment (VE) applications are driven by the capabilities and limitations of users. These considerations affect both the design of the hardware and the design of the software. To be interactive, a VE software application must constantly present the current *view* of a computer-generated world and have that world quickly react to the user's actions. To be convincing, the presentation must provide enough objects and sufficient details to make objects easily recognizable to give the user the sense of being in the world. To be useful, the environment must respond to the user. The user's location in the world should change when a navigation action is performed. Objects that the user can grab or nudge should move as expected. Three-dimensional interface elements, such as floating buttons, tabs, and sliders, should allow users to take actions that change the environment in some way.

An environment can be described as a set of geometry, the spatial relationship of the geometry to the user or users, and the change in the geometry in response to user actions and as time progresses. The set of geometry can be decomposed into geometric objects that may have a visual appearance, make a sound, or have an odor, taste, or texture to be felt. A geometric object can also be simply a point in space. This geometric environment is presented to each user using output devices. Virtual environment systems usually provide a visual display, and many provide audio, although output for all of the far senses—sight, sound, and odor—has been shown to be effective (Dinh, Walker, Hodges, Song, & Kobayashi, 1999). Tactile and force feedback devices can also present an aspect of the world to the user and are currently in development. It is much more difficult, however, to mimic the touch sensation of arbitrary geometry than it is to provide a visual representation.

Although display technology has a significant impact on the fidelity of the virtual environment being displayed, the technology used for user input can have a much more significant impact. If the user's perspective of the environment is being driven by a head tracker, the presentation will need to be refreshed much more frequently than if the view is driven by mouse control. The display latency, the time between a user movement and the presentation of the view, will need to be shorter. Rendering speed, therefore, will need to be orders of magnitude faster. Using a glove input device with force feedback implies a much more sophisticated collision detection and response computation than a point-and-click interface. An augmented or mixed reality display, where the virtual environment is combined with the real world, has even more strict requirements for display latency to have the virtual and real world appear in sync.

Given that a 3D environment may occupy a large space, providing a convincing illusion that the user is in a realistic world requires many geometric objects that constantly react to other objects and the users. However, even though new technologies are constantly being developed to improve the ability to present a realistic world, there are still significant limits to the amount of environment detail that can be handled for an interactive world. In addition, creating all of the pieces of the world at such a level of detail would be a significant effort. Luckily, many applications can be created with VE technology that does not require a full illusion, and a somewhat realistic world can be quite convincing. It is still a challenge, though, to create and process the description of an environment, the environment's model.

An environment model has at least three representations. One representation describes the model in a stored file. Another representation describes the model to a running application, which can be accessed and modified to produce complex behaviors that could not be described in the model itself. A final representation stores the model internally. This representation is optimized for the task of constantly supplying information to the output devices to render a view of the model, providing 3D sound, and so on. At some point, the model is translated from one representation to another. For example, a number of file formats provide a stored model, including the FBX (Filmbox) format (Autodesk FBX, 2013), used by many 3D modeling tools; the COLLADA (Collaborative design activity) format (COLLADA, 2013), used as a common format to interchange 3D models between tools; and the XML-based X3D language (ISO/IEC 19775-1, 2008), a follow-on to the Virtual Reality Modeling Language (VRML) (Ames, Nadeau, & Moreland, 1997). The Java 3D package (Sowizral, Rushforth, & Deering, 2000) provides a run-time model for Java programs. Parts of the Java 3D run-time model can be explicitly compiled to produce an optimized internal model, at the cost of making modification of the environment model more difficult.

Some file representations are designed to translate easily to the internal representation, but cannot be easily translated to a model that can be modified at run-time. Other file representations may translate quickly to a model that is easy to modify at run-time but require more processing to obtain an internal model that is optimized for interactive use. Over long execution times, a translation step from a file to an internal representation is cheap, as it occurs only once. On the other hand, if the run-time model is modified often, the translation from the run-time model to the internal model can become a critical performance bottleneck.

In this chapter, a discussion of how an environment model is described in terms of geometric primitives and behaviors is provided. This model is often used as the run-time representation. Next, this representation is related to common file formats used to describe a virtual environment and to an internal representation that provides optimizations for efficient presentation of the model. Finally, a description of how the user can be incorporated into the model to provide an interface between the user and environment is given.

11.2   GEOMETRY

For the most part, virtual environments are spatial in nature. A virtual environment consists of properties associated with locations in three spaces, such as sounds and appearance. A VE application, therefore, maintains a representation of spatial information and attempts to organize that information so that arbitrary views can be produced quickly and repetitively. Since visual perception is the primary channel used by sighted people to obtain information about the world around them, VE systems have focused on representing geometric models of visual objects and their placement in the environment.

A virtual environment is usually described by low-level primitives such as planar polygons, lines, and text and by higher-level primitives such as spheres, boxes, cones, polygonal meshes, polyhedra, curves, and curved surfaces. Current graphics hardware renders low-level primitives, as well as polygonal meshes, very efficiently. However, many of the higher-level primitives, such as spheres, curves, and curved surfaces, cannot be precisely rendered fast enough. Instead, they are converted to a polygonal representation when they are added to the environment. Recent advances in graphics architectures allow applications to define programs, called shaders, which can be loaded into and executed by the graphics hardware. These do allow an application to implement, in the graphics hardware, translations of higher-level representations into what is drawn on screen, if they can be expressed within the limitations of the shader context (which is stateless and receives limited data from the application).

Although the visual appearance of an environment can be completely described, albeit approximately in some cases, by a set of polygons, lines, and text images, it is more useful to organize the geometry into separate groups that are related to each other. This network of geometries and relationships is known as a *scene graph*. The scene graph contains nodes that define geometric primitives and properties, relationships between geometry, and perhaps behaviors of the geometry. In this section, we will describe the geometric primitives used in VE applications, how these primitives are organized into a scene graph, what relationships the scene graph provides between groups of primitives, and what properties may be stored in the scene graph that determine the appearance of primitives and the environment, such as materials and light sources.

11.2.1   GEOMETRIC PRIMITIVES

Current computer graphics systems are optimized to display convex polygons and lines and are capable of rendering millions of polygons per second to the screen. Therefore, geometry is decomposed into, or approximated by, a set of planar polygons or straight lines. Text is often treated as a special primitive, although it can be crudely represented by a set of lines and a set of polygons or more nicely represented by an image with transparent areas. Geometric primitives are described by point coordinates in a 3D coordinate system. By convention, the coordinate system is right handed (when curling the fingers of the right hand from one positive axis to another, the outstretched thumb points in the direction of the third positive axis). Whenever a group of primitives describe an object that has a particular orientation (e.g., a tree generally grows upward), convention assigns the Y coordinate axis to represent the *up* direction, the X coordinate axis to represent the *right* direction, and the Z-axis to represent the *back* direction. This coordinate system is a remnant of models developed for display on a computer monitor, where the X-axis is the horizontal axis of the monitor and the Y-axis is the vertical axis of the monitor. (If only monitors had been placed face up on or in a desk, then the coordinate system convention might match the Z-up convention of mathematicians and physicists!)

Polygons, also known as faces, are described by a list of coplanar points in three spaces. Each pair of points on the list (and the pair containing the first and last points) defines an edge of the polygon. Graphics systems, however, may only be able to render convex polygons and may just render triangles, as they are simple to process. Concave polygons or polygons with four or more edges are usually decomposed into triangles before they are rendered (often, a graphics system needs to be told to spend the time to convert concave polygons correctly). Polygons are shown filled with some color or pattern but are considered infinitely thin, meaning that if they are viewed from the side, they will be invisible. Lines, however, are of a given width regardless of the viewing angle.

In addition to allowing for polygons and lines, modeling systems will generally allow users to create higher-level geometries, such as boxes, spheres, cones, and cylinders. These shapes may be converted to polygons by the modeling system or may be stored as their shapes and must be converted to a polygonal representation by the rendering system. Unfortunately, a polygonal representation is often a rough approximation of the shape. This is especially true of shapes with curved lines or surfaces. Since a computer image is an approximation in and of itself, the error in the shape representation may be hidden by the error introduced in using discrete pixels to define an image. If the rendering system knows the complete shape description, it can translate the shape to a representation that is as accurate as possible given its visibility, at the cost of extra processing time. In general, a more accurate representation requires more polygons or lines. Current modeling packages, such as 3D Studio Max (2013) and Maya (2012), can produce polygonal representations of their models. The primary application for modeling packages is to produce high-quality animation frames or still images, but they are increasingly providing capabilities for creating 3D models that can be used for interactive applications (e.g., generating low-polygon count models). Thus, they often produce many more polygons than can be handled by an interactive rendering system unless configured otherwise.

11.2.2 Surfaces

An environment generally consists of a set of geometric objects, where each object is described by a set of solids or lines and each solid is described by a set of surfaces, which, in turn, are described by a set of edge-connected polygons (see Figure 11.1). Geometry in the world could be described by a higher-level representation, such as in a technique called pose space deformation or skinning, where surface points are tied to a virtual skeleton for articulated figures and moved whenever the skeleton bones are moved (Lewis, Cordner, & Fong, 2000). Or geometry could be generated from patches or other input data to add detail to the world, either by the application or within the rendering pipeline. In either case, the geometry is ultimately transformed into 3D polygons or points that are used to draw the geometry on the screen.

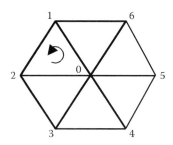

FIGURE 11.1 Polygonal surface.

When drawing the surface of an object, the viewpoint is assumed to always be outside of a solid. With this assumption, the rendering system can limit the polygons they draw to those on surfaces that face the viewer, since surfaces that do not face the viewer are on the other side of a solid and are obscured by the front surface of the solid. This technique is known as back-face culling. However, when the assumption fails (the viewpoint moves inside of a solid), then the effect of back-face culling is that the surfaces disappear altogether (all surfaces are facing away from the viewpoint). At the cost of deciding if the viewpoint is inside an object, a rendering system may display an inside color instead. However, this technique is expensive for arbitrary solids and large numbers of objects.

In order for the back-face culling technique to work, polygons of a surface must have a front and a *back*. By convention, the order of the points that describe the polygon determines the front and back sides of the polygon; if the viewer is on the front side, then the points are given in counterclockwise order. If the points are given in clockwise order from the point of view of the viewer, the viewer is on the back side of the polygon (as shown in Figure 11.1).

Surfaces can be approximated as a set of polygons, but surface representation methods that require less information can be loaded into the application faster and transmitted faster from the application to the graphics rendering system. One such method represents a surface that is defined by a set of height values along a regular grid. This method can represent terrain that contains no caves or outcroppings (all bumps go up, all holes go down). This method allow a surface to be described by the coordinates of the corners, the distance between grid points, and a height value for each grid point, roughly 1/3 the information needed if only points were used. In addition, the representation can be shrunk further using algorithms that can quickly approximate these surfaces with fewer polygons when the viewpoint is far away (Lindstrom et al., 1996).

A polygon mesh also describes a surface using less information than the same surface given as a set of distinct polygons. Graphics systems generally support meshes of triangles or quadrangles. A mesh is described by a list of polygons, where each successive polygon shares one edge completely with the previous polygon and therefore shares two points. For example, a triangle mesh can be described by three points for the first triangle, then one more point for the next triangle, another point for the third triangle, and so on. A surface described in this manner uses roughly 1/3 fewer points to describe a surface and does not have the restrictions of the elevation map. In fact, an elevation map will likely be translated to a triangle mesh before it is rendered.

When a polyhedron is specified by a set of face polygons, each being described by a set of coordinates in three spaces, many of the same coordinates are given more than once. This is a result of the surface polygons' common edges. To avoid storing redundant information, the geometric model is very often given as a set of labeled coordinates and a set of faces that list the labels of the coordinates that describe the face's edges. This storage scheme is shown in Figure 11.2 for a $2 \times 2 \times 2$ cube centered at the origin. Note that the order of the coordinates for each face is given in counterclockwise order from the point of view of being outside of the cube. A similar strategy can be used to store coordinates of a list of line segments.

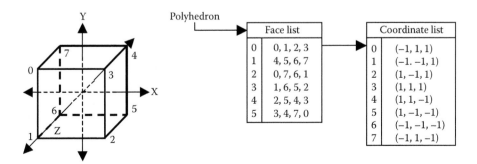

FIGURE 11.2   Face and coordinate lists for a cube.

11.2.3 Designing for Rendering Efficiency

Graphics hardware venders tend to describe the performance of their display system in terms of how many polygons per second can be drawn. However, this measurement does not tell the full story. Depending on the hardware architecture, a graphics system may be polygon bound or fill bound. If a system is fill bound, then the number of polygons that can be displayed to the screen is heavily dependent on how much screen space each polygon takes when rendered. In other words, the time taken to fill each polygon dominates the rendering time. If a system is polygon bound, then the changes to the number of polygons have a more significant effect to rendering time than the size of the polygons. This difference in architectures has a profound effect on how models should be designed and built for VE applications. If the VE application will use a polygon-bound system, then the model should contain as few polygons as possible, no matter how large they are individually. Mostly flat surfaces could be replaced by single polygons, for example. If the system used is fill bound, however, this change in the model will not have much of an effect on the rendering time. In fact, a system may be fill bound because it is designed to handle many polygons in parallel and works best when polygons take about the same amount of time to fill because the parallel processing is balanced. If this is the case, it would be a good idea to break up a large polygon (or, more precisely, one that will appear prominently on the display) into a set of smaller, coplanar polygons.

11.2.4 Scene Graph

One could represent the geometric model of a virtual environment with just a set of polygons, lines, and text. However, when a particular object, such as a chair, table, or vase, moved in the environment, then all of the coordinates of the polygons that represent the moving object would need to be changed, or transformed, in the same way. In addition, if the vase was conceptually on top of the table, the coordinates of the vase should move when the table moves. A more efficient representation of the model would group geometric primitives together into geometric objects and would define *attachment* relationships between the objects.

An environment model, therefore, can be represented by a coordinate system graph, which is a directed acyclic graph. Each node of a coordinate system graph contains a set of geometric primitives, which usually represent a conceptual object or object part. The geometry of a node is described in terms of a local coordinate system. The edges of the graph define a transformation between the coordinate system of one node and the coordinate system of another node. Conceptually, the coordinate system of one node (the child) is contained in the coordinate system of another (the parent). If the parent's coordinate system is transformed, the child's coordinate system is transformed, as well (unbeknownst to the child), but if the child's coordinate system is transformed, the parent's coordinate system is unaffected, and the child's transformation is visible in the parent's coordinate system.

A coordinate system graph does not contain cycles because it is unclear how a change in the relationship between objects in the cycle should change the relationships between the other objects in the cycle. Should all of the transformations be changed equally, or should just one other edge change? If just one edge changes, which one should it be? For example, if five marbles represented in a coordinate system graph were placed next to each other and the graph contained edges between adjacent marbles to keep them spaced evenly and an edge between the outer marbles to maintain a certain length for the marble line, how should the marbles move if the distance between the outer marbles changes? Or if the distance between two adjacent marbles changes? Should the marbles adjust to space out evenly again, or should just one other distance change?

Coordinate system transformations are described by a 4×4 matrix that operates on homogeneous coordinates, where a location in space is given by the four-tuple $(x, y, z, 1)$. A set of locations is transformed for one coordinate system to another by multiplying each location by the matrix transformation between the coordinate systems. Matrix transformations can be combined by multiplying them together, but matrix multiplications are not commutative. A change in the order in which the transformations are given will change the total transformation (unless the transformations are all the identity matrix, which does nothing).

A matrix transformation in a coordinate system graph is often described as combination of scale changes along the major axes, X, Y, and Z, rotations about the major axes, and translations along the major axes. However, the transformation may include shears, mirrors, or other affine transformations. Standard 3D matrix transformations and the use of these transformations in a coordinate system graph are described in more detail by Foley, van Dam, Feiner, and Hughes (1996). Some systems or file formats (such as X3D) may describe the rotation component as a certain rotation around a given vector, known as an angle-axis rotation. Another representation of rotations is a *quaternion*, which defines a rotation of θ degrees around a unit vector $\left[n_x, n_y, n_z \right]$ as $\left(\cos(\theta/2), \sin(\theta/2) n_x, \sin(\theta/2) n_y, \sin(\theta/2) n_z \right)$. These representations describe a rotation more exactly than a set rotations around major axes and more efficiently than a rotation transformation matrix. The quaternion, although more difficult to interpret visually, can be more easily composed mathematically with other quaternion rotations and is better for interpolating between two rotations (smoothly changing from one rotation to another without experiencing gimbal lock). However, an extra step is required to incorporate a rotation in this form into a 4×4 matrix transformation (Watt & Watt, 1992).

A coordinate system graph is usually represented as a tree, where the root node contains geometry in a world coordinate system and the child nodes have local coordinate systems that related to the world coordinate system. Therefore, points in the geometry of a child node can be transformed to the world coordinate system. Points further down a subtree of the root node can be transformed to their parent's coordinate system, then to its grandparent's coordinate system, and so on until it is in the world coordinate system. Similarly, a point in the world coordinate system can be transformed to the local coordinate system of a descendant node by applying the inverse of each transformation on the path to that node. When rendering a scene for the display, geometry must be transformed to the world coordinate system, then transformed to a descendant node representing the coordinate system of the viewer, where the origin is at the eye point.

The coordinate system graph is a type of *scene graph*, as it describes the scene for an environment. Other types of scene graphs have nodes that do not represent geometry and do not associate a transformation for every edge in the graph. For example, the X3D file format (ISO/IEC 19775-1, 2008) and the Cosmo3D run-time library (Eckel, 1998) provide a scene graph with many types of nodes, one of which is the *transform* node that defines a spatial relationship between geometry attached to it and the geometry it is attached to. Edges are still used to group geometry by defining, directly or indirectly, rigid attachments. Other nodes in the graph define light sources, appearance properties, and different types of grouping.

11.2.5 Material Properties

Surfaces in the real world have many properties that affect how they appear beyond their position and orientation. They may be rough or smooth. They may be dyed or painted with different colors, or they may contain grains. They may have fine detail that is too expensive to model geometrically, such as the texture of skin or hair. At a great distance, even features like brick grooves, windowsills, or door handles might be considered fine detail. These features can be described by their contribution to the color of a pixel on the display, if the feature is visible at all. In addition to the various properties of the face, this contribution will be affected by the light sources and global lighting model that is in effect for the environment.

The appearance of a face under a given lighting configuration is generally described by properties of the face as a whole and properties of the points on the face. Properties given to the entire face include the color of the face under no light, the color response to a diffuse light, the color response to a specular (mirror) reflection, and the transparency of the face. Properties of individual points on a face can be given by an image or texture map, which contains color and transparency information

Shaded using single directional light source (behind viewer)   Shaded using point light sources and shadows   Polygon edges

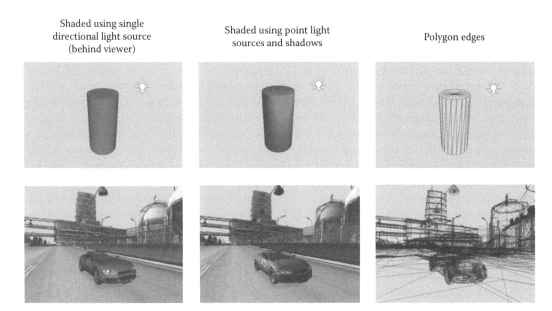

FIGURE 11.3 **(See color insert.)** The effects of light sources and shading. (From screenshots of the Unity Car Tutorial in the Unity Editor, reproduced with permission from Unity Technologies ApS.)

for points on a 2D grid. The color information of the image map may be combined with the face properties or may override them (the scene in Figure 11.3 uses texture maps for most surfaces, including the road and building sides). As a whole, the material properties define a contribution of certain values of red, green, and blue to a pixel on the display, if the face is visible at that point on the display.

11.2.6 LIGHTING AND SHADING

The appearance of a geometric primitive depends not only on its material properties but also how light sources in the environment are positioned. An environment is assumed to have a certain amount of low-level ambient light, and the material definition given to a face describes the color to use for the interior of the face in the absence of any light sources. For each light sources that does exists, a proportional amount of diffuse color, from the material definition, is added to particular pixel colors for parts of the face that are visible. The amount of diffuse color is dependent on the incident angle of the ray from the light source onto the face. Light sources can be given at infinity, where all light rays are parallel and travel in a certain direction, or can be given a position from which the rays emanate. There are *shadow mapping* techniques to project shadows from each light source onto the environment's geometry, although the generation of shadows requires significant computation and memory resources. Shadow generation can require a complete rendering of the scene for each light source, which can be too time consuming to execute for each frame rendered to the screen. However, GPU-based shadow mapping techniques (Fernando, 2004) can provide the performance necessary to produce shadows for each rendered view and follow the moving objects in the environment. The left two screenshots in Figure 11.3, taken from the Unity (2013) game development application, demonstrate the effect of light sources. The location of the light source for the view of the cylinder in the top left image is shown in the top right corner of the image.

Light sources can be included as a node or with geometry in the scene graph, but the rendering system must only use the location and direction of the light in the coordinates in which objects are rendered. The coordinate system transform that is usually used to define a node's position in the world or relative to another node may include nonrigid transformations, such as scales and shears, which do not preserve angles or distances.

Many surfaces in the real world are curved rather than flat. In general, a curved surface is represented by a collection of flat polygons, called a polygon mesh, which approximate the surface. The rendering of the polygon mesh, however, can appear curved by combining a shading model with a lighting model to determining the color of the interior pixels of the polygons. The Phong lighting model (Foley et al., 1996) describes how the color of a surface can be computed from the collection of light sources, material properties of the surface such as diffuse and specular reflection parameters, and the normal vector to the surface. When applying a shading model to a set of polygons, therefore, the difference between a flat polygon and a polygon with *smooth shading* is how the surface normal is defined for the vertices and interior pixels of a polygon. The cylinder figure in Figure 11.3 demonstrates how a surface whose vertex normal all points in the same direction, as is the case of the cylinder top, appears flat, while a surface whose vertex normal points out as if from a curved surface, as is the case of the cylinder side, appears curved. Note that because a vertex may be at the boundary between different curves of flat surfaces, it may have different normal vectors assigned to it.

The simplest technique for smooth shading, called Gouraud shading , uses the normal to the surface at each vertex of the polygons that make up the surface to compute the vertex colors. The colors of the interior pixels are then interpolated from the vertex colors (the value is a weighted average of the vertex colors, based on the relative distance from each vertex). This technique was commonly used in fixed *function* graphics pipelines, as the lighting calculation was fixed to be computed before a polygon was *filled* on the drawing area. In the modern shader graphics architecture, the Phong shading model or the Blinn–Phong shading model variant (Blinn, 1977) is more commonly used and is shown in shaded geometry in Figure 11.3. In these models, the surface normal for the interior pixels of the polygon is interpolated from the normal vectors at the corners, and the pixel color is computed using the Phong lighting equation.

11.2.7 SURFACE DETAIL

Surfaces with solid colors (such as a brand new car) are fairly rare in the real world. Although a real-time rendering system could not handle geometric surfaces with all of the nooks, crannies, and blemishes that make a surface appear more realistic, there are techniques that can introduce theses details into a virtual environment. Three commonly used techniques are texture mapping, environment mapping, and bump mapping. In general, each of these techniques modifies or replaces the color that is used for an interior point of a polygon being drawn to the screen. A number of other techniques have been developed to produce different visual effects, enabled by the introduction of shader programming in modern graphics libraries, where a pixel color can be generated programmatically, rather than using one of a set of selected techniques supported by the graphics library (see, e.g., the *GPU Gems* book series, Fernando, 2004).

The simplest surface detail technique maps a pixel on the screen to an area on a polygon and then that area on the polygon to an area on a 2D image (whose image points are called *texels*). The image usually represents a *texture* to be applied to the polygon, such as wood grain for a hardwood floor or rough brick faces for a brick wall, but can also be used to apply a graphic, such as a poster to appear on a wall. Using a *multitexturing* technique, texture images can be combined to add even more realism. For example, the poster image could be combined with a dirty texture image to make the poster appear weathered. This is accomplished by *alpha blending* between a current color and a new color for a pixel, where the percentage of the color that should come from the new color is represented by an alpha value. When the number of layers increases beyond two or three, though, the combination of texel colors can become too light or too dark. More complex combinations can be handled by a shader program that can sample and combine texel colors for geometry that is about to be drawn with more control.

A surface that is represented by a set of polygons can be drawn with a given texture by mapping polygon corners to interior points in the texture image, selecting the portions of the image that together present a cohesive texture or image. When images just represent a surface texture, a texture image is often tiled by defining the texture coordinates in the range of 0 and 1 and wrapping the coordinates. That is, coordinates 1.5, 2.5, 3.5, and so on are interpreted as 0.5. Tiled texture image should be designed (like flooring tiles) so that copies can be placed adjacent to each other without breaking the pattern. The building walls in the scene shown in Figure 11.3 are an example of tiling a texture across a large area.

The general approach of texture mapping, where a pixel may map to a large area on an image, would often be too expensive for a real-time rendering system. Imagine, for example, a grass texture for a meadow that recedes into the distance. As pixels approach the horizon, they will be representing larger numbers of texels from the grass texture, and combining them into a single color value becomes very expensive. A technique that is used to address this problem is called MIP mapping (MIP is short for multum in parvo, or "many things in a small place"). In this technique, a texture image is represented by the original image and a series of copies that are each a quarter the size of the previous copy, increasing the storage needs by a third of the original. When a color is needed for a particular area, the image whose texel size most closely matches the area is selected to provide the color. Even with this technique, there may be multiple texels that are contained in the area mapping to a pixel. In this case, real-time rendering systems generally select the nearest texel to the center of the area or average the nearest four texels to the center of the area, at the expense of introducing aliasing effects. The top two images in Figure 11.4 demonstrate the visual artifacts of choosing the nearest texel. Notice that the white squares do not appear to be of the same size when the checkerboard is not viewed straight on. The bottom two images in Figure 11.4 demonstrate that this artifact

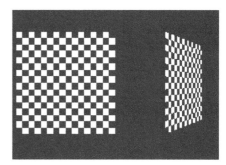

Texture mapping using the nearest texel

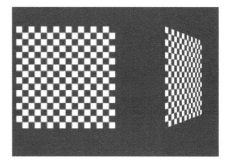

Texture mapping using the average of the four nearest texels

FIGURE 11.4 Comparison of texel selection approaches for texture mapping.

can be alleviated by using an average of the nearest four texels, at the cost of blurring the edges of the checkerboard squares.

11.2.8 RENDERING PIPELINE

The tools for creating and programming virtual environments range from high-level tools that allow creators to describe the geometry and behaviors of the environment and let the tool store the run-time model and generate the user's view of the environment to low-level programming libraries that allow creators to directly generate the view of the world from the run-time model. Although the high-level tools generally make programming the graphical presentation at a low level unnecessary, a knowledge of the capabilities and function of the low-level programming libraries can be very helpful in understanding how to use the higher-level tools to maximize rendering efficiency and visual realism for a run-time rendering system.

There are two low-level programming libraries that commonly used to write applications that render views in 3D in real time: OpenGL (2012) and Direct3D (2012). In recent years, the design architectures of how these libraries generate the views shown on screen, known as the rendering pipeline, have converged into a shader-based design, where programmers provide snip-its of code written in a C-like shader language (the GL shader language [GLSL] for OpenGL and the high-level shader language [HLSL] for Direct3D) for most stages in the pipeline. Many higher-level programming libraries, such as Vizard (2012), and file-based environment descriptions, such as X3D (ISO/IEC 19775-1, 2008), allow for shader programs to be given to handle application-specific rendering techniques. Many development tools for game engines, such as Unity (2013), also provide the ability to plug in shader programs to produce different rendering effects.

Figure 11.5 diagrams a generic shader-based 3D rendering pipeline. In this diagram, vertices are provided as parts of geometric primitives to the vertex processor, where a user-provided vertex shader transforms the vertices into camera clipping coordinates. These vertices may be sent to one or both of the tessellator and primitive generator optional stages. The tessellator uses two user-provided tessellation shader programs and a provided tessellation function to tessellate given patches into a set of polygons at a selected a level of detail. The primitive generator uses a user-defined geometry shader to transform selected primitives into 0 or more new primitives. The tessellation and primitive generation stages both provide an opportunity to change the number of primitives (lines and polygons) to process on the fly, where the tessellator provides faster generation and the primitive generator allows for a more general approach. The user-provided shaders for these

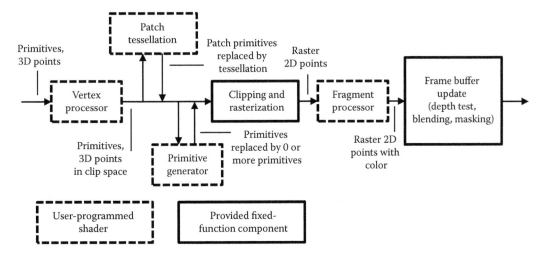

FIGURE 11.5   Generic shader-based 3D rendering pipeline.

stages have the ability to modify an incoming set of primitives into another set of primitives that are more or less detailed to provide better visual realism or more efficient rendering.

Given the output of the vertex and primitive-based shaders, the primitives are clipped to the viewing volume and projected to the 2D coordinates of the drawing raster. The primitives are *rasterized* by interpolating the attributes of the points (3D location, surface normal, color, etc.) between the primitive vertices into a set of point fragments. It is the job of the user-provided fragment shader to generate a color for the fragment location on the frame buffer, using a lighting model and attributes of the vertices that contributed to the fragment. A fragment shader could also modify the lighting model to make the surface look bumpy (known as *bump mapping*) or take the color from an image representing the reflection off of the surface (known as *reflection mapping*). Once a color has been generated, then the rendering system will perform a set of tests, including a depth test and any screen-based masks, to determine if the color should be used for the screen location and may overwrite or blend it with the color that exists at the location already.

The 3D rendering library architectures have also returned to more of a *retained mode* design, where the model is stored with the rendering hardware (on the GPU) rather than provided piece by piece during the rendering, as is done with an *immediate mode* design. The implication of this design is that processing and drawing geometry that is stored with the rendering hardware can be done very efficiently and quickly but that changing the geometry can be relatively slow, as the new geometry must be copied to the rendering hardware and reprocessed. However, the shader program snip-its can be used to make a number of changes to the geometry or how the geometry is used to create the view in real time without requiring a change to the internal model in the rendering hardware. For example, the vertices of the geometry could be perturbed based on a mathematical or data-sampled model in a vertex shader to produce a 3D terrain, ocean surface, or body shapes; and collections of geometries can be repositioned relative to each other using coordinate system transformations controlled by a small number of parameters provided by the application.

11.3   BEHAVIOR

A dynamic environment can be represented by a set of objects (often nodes in a scene graph) that have particular behaviors. These behaviors may be given as part of the stored model of the object, may be defined by the execution of the application code, or may be some combination of the two. An object may have a set of behaviors that it follows given its current context. These behaviors can be roughly categorized as *environment-independent behaviors*, which do not consider the current state of the other objects in the environment, and *environment-dependent behaviors*, which do consider other objects, as shown in Figure 11.6. Environment-independent behaviors include changes to the object that are solely based on the passing of time, or *time-based* behaviors, and changes that occur in a particular sequence, or *scripted* behavior. Environment-dependent behaviors include *event-driven* behaviors, which respond to events initiated by users or other objects, and *constraint maintenance* behaviors, which react to changes of other objects to maintain defined constraints, such as relative placement, gravity, or penetration limits. The following subsections describe each

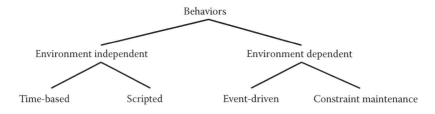

FIGURE 11.6   Categories of behavior specification in virtual environments.

type of behavior in more detail, concluding with a subsection on interactive behaviors encapsulated into *interactor* objects.

11.3.1 Time Based

An object's behavior may describe a change in one of the object's properties over time or at a certain point in time. Object properties that may be changed include the object's position, orientation, color, and visibility. A circling airplane may follow a particular path as it flies around a town, where its position and orientation are determined by the amount of time that has passed from when it began flying. A distant mountain may change from a detailed model displayed during daylight hours to a simpler model for dawn and dusk to being invisible at night.

Time-based behaviors may be defined to use wall-clock or absolute time, which is the time frame of the user or a time frame that is transformed from wall-clock time. Behaviors usually have a start and end time. A behavior may change a property continually over the time between the start and end times. For convenience, the time values used to define such a change usually are zero at the start time and one at the end time. If a behavior should last, say, 5 min, then the behavior can be scaled appropriately.

Time frames can be used like 1D coordinate system frames. A time frame can define a translation (time 0 to a start time) and a scale (time 1 to an end time). Time frames can be related to the wall-clock time, or they can be organized hierarchically, like a coordinate system graph. For example, in the Mirage system (Tarlton & Tarlton, 1992), an activity consists of the change of an object's property from a $t = 0$ to a $t = 1$ value. A single activity can be set to begin and end at certain times or can be combined with other activities by becoming siblings of a parent activity. The relationship between a parent activity and its children may specify that the child activities occur simultaneously or one after the other. The parent activity will have a relationship with wall-clock time that will have it occur at a certain time and pace. With this structure, an activity could flow at various speeds and could even be paused or reversed!

For example, Figure 11.7 shows an activity hierarchy where a *rolling wheel* activity consists of a *rotating wheel* activity and a *moving wheel* activity. These child activities may be combined for a coordinated motion by making no transformation from the parent time frame to the time frame of each child, as is shown on the diagram on the left. Alternatively, the motions can be performed one after the other by transforming the 0–0.5 time values of the rolling wheel activity to the 0–1 time values of the rotating wheel activity and transforming the 0.5 time values of the rolling wheel

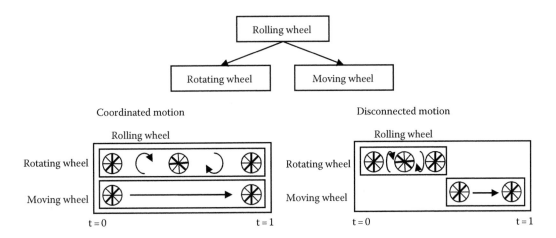

FIGURE 11.7 Activity combinations. (Adapted from Tarlton, M.A. and Tarlton, P.N., A framework for dynamic visual applications. *Proceedings of the 1992 Symposium on Interactive 3D Graphics*, ACM, Cambridge, MA, pp. 161–164, 1992. With permission.)

activity to the 0–1 time values of the moving wheel activity. This is shown on the diagram to the right of Figure 11.7. The left structure of Figure 11.7 produces a rolling wheel, and the right structure produces a wheel that rotates and then slides.

Individual activities may specify an instantaneous change in an object's property at the activity's start time or may specify a continuous change of a property based on the time value. The continuous change may be given as a function of the time value or may simply be specified as an interpolation between starting and ending property values. Different interpolation methods may be used depending on the property type. For example, a linear interpolation is often used for position changes, while a color change may be a linear interpolation of colors in hue, saturation, value (HSV) space, having been transformed from red, green, blue (RGB) space. Other interpolation methods can be used to produce a different transition between values. For example, an ease-in, ease-out method slowly increases the rate a value changes until the halfway point, then slowly decreases the rate back to zero, at which point the new value should be obtained. A common animation technique is to define particular property values, such as position and orientation, for objects at certain moments in time. A set of these values are known as a *key frame*, which defines the start and ending values of interpolation activities that follow one another. See Watt and Watt (1992) for a detailed description of how interpolation is used in animating object movements.

11.3.2 SCRIPTED

A scripted behavior defines a set of steps to be taken in order. The run-time system may make a new step every time it renders an image to the display or one step every so many image frames. The speed of the behavior, therefore, is directly tied to the frame rate of the application. Scripted behaviors may not appear smooth, since they are not tied to a constant passage of time. However, a scripted behavior can provide a transition or animation that shows every step to the user while completing as fast as possible. If the behavior is time based, there is a chance that a step will not be seen when it occurs between the times an image frame is rendered.

11.3.3 EVENT-DRIVEN

The behavior of an object may be specified as a reaction to an event that occurs in the environment. An event may represent a user's action, such as pressing a button, or it may represent an incidence in the environment, such as the collision of two objects, or it may represent a change to the property of another object, such as its appearance or mass.

Events can be stored in a global event queue (which may actually be many queues) and dispatched to objects that have expressed an interest in them or can be stored locally in an object's out field and routed to the in field of other objects when such a route has been set up. Generally, systems that are designed to support event response in application code use the event queue model, while systems that use only behaviors defined in a file description, such as VRML (Ames et al., 1997), will use event routes. However, the Java 3D package (Sowizral et al., 2000) uses an event route network, which eases the incorporation of VRML models into its run-time system. The event route model is not as flexible as the event queue model, as a route must have a single source, while an event can come from anywhere. It is not difficult for an object to listen for events from an event queue that comes from any object, but it is difficult for an object to set up routes from many objects. File descriptions can support a more general approach to handling events if they allow script code to be specified that can observe and change the run-time state of the environment, as is provided by X3D (ISO/IEC 19775-1, 2008).

11.3.4 CONSTRAINT MAINTENANCE

It is sometimes easier to describe a behavior declaratively, stating a set of relationships that should be maintained, rather than defining a procedure to enforce the relationships. This is akin to saying what result you want, but not how to get the result. The positional relationship between a parent and a child in a coordinate system graph is an example of a declarative constraint: the child should have a certain position in relation to the parent. Of course, the simplicity of describing a behavior with constraints comes at a cost. The run-time must ensure that the constraints are maintained and must deal with situations where the behavior is overconstrained or underconstrained. In addition, it is difficult to design a system of constraints of any complexity, as one change can have far-reaching, and sometimes puzzling, effects.

Constraints can be specified for any set of properties of the objects in the environment. For example, the TBAG toolkit (Elliot, Schechter, Yeung, & Abi-Ezzi, 1994) is a graphics library and run-time system that provides constrainable program objects that describe relationships between values that should be maintained. The TBAG toolkit uses the SkyBlue constraint satisfaction algorithm (Sannela, 1992) to generate the values that satisfy the given constraints as best as possible. However, two types of constraints are found more often than others in VE run-time support systems: geometric constraint maintenance and physics-based motion.

A geometric constraint uses properties such as attachment and spatial relationships with other objects. Using the coordinate system graph as part of a scene graph, a run-time system automatically maintains a spatial relationship between child and parent objects in the graph. The run-time system may also maintain a direct spatial relationship between two objects that are not a child or a parent of the other in the scene graph. For example, objects that represent 3D interface components (buttons, menus, etc.) may be automatically arranged so that they are adjacent to each other, so that they do not overlap even if components are moved or scaled. In another example, the WALKEDIT system (Bukowski & Sequin, 1995) allows users to quickly position objects to construct rooms with furniture, books, and other items. This task is facilitated by the enforcement of spatial constraints that come from abstract relationships or associations. Picture frames are constrained to vertical surfaces, cups sit on horizontal surfaces such as desks or floors, and so on.

A run-time system can reasonably maintain a list of one-way constraints, where a change in one property results in a change in another property, but not vice versa. For example, a change in the position of a parent node in the scene graph affects the position of the child node in the environment, but a change in the child node's position does not affect the parent node's position. It is much more difficult to support two-way constraints or constraint dependences that contain cycles, as a solution must be found by solving a system of simultaneous equations and it may be over- or underconstrained (a solution does not exist, or many possible solutions exist). A run-time system that needs to enforce two-way constraints can implement constraint satisfaction algorithms such as SkyBlue.

11.3.5 REINTRODUCING PHYSICS

A virtual environment can look more realistic if the objects in the environment follow the physical laws we are used to in the real world. Real-world objects have mass, may be rigid or flexible and solid or fluid, will fall at a constant acceleration when not supported, will move and twist when pushed by some force, and so on. A software module that simulates these behaviors is called the *physics engine*. Physics engines are complex enough that they are often provided by a third-party software toolkit, such as PhysX (2013) and Havok (2013). Although the capability of computers at simulating the real world continues to increase, with some computation being off-loaded onto the GPU, there are still limits to how many potential interactions between objects and parts of objects can be handled at interactive rates. For the purpose of interaction of objects, an object's geometry is generally approximated by a simple shape, such as a sphere, box, or plane. As such, a physics engine generally operates on a separate collection of geometric shapes, which must be kept in sync with the rendered geometries that the shape approximates. More complex polyhedral may be supported, but fewer in number and with some limitations. For example, the Vizard toolkit (Vizard, 2012) allows for polyhedral collision shapes, but these objects do not react to other objects but are only reacted to. A physics engine generally moves objects using given properties such as mass, velocity, angular velocity, acceleration, and angular momentum as time passes (perhaps faster or slower than wall-clock time). For objects with collision shapes, the penetration of one object into another is detected, and a change in object movement may be computed for use in the next time step. In addition, the collision may be reported so that a different response can be executed.

11.3.6 Interactors

As is true of 2D graphical applications, a 3D environment may contain geometric objects that belong to a class that interact with the user and other objects in a well-defined way. In 2D interfaces, these objects are called widgets, components, or interactors and include buttons, menus, scrollbars, and pointers. Each class of object encapsulates an interactive behavior, parameterized to allow for multiple uses. Buttons, for example, have the same behavior when pressed by the user pointer but have different labels and reactions to that user action. Since the number of ways a user interacts with a virtual environment can be larger when going beyond the mouse point-and-click interface and the interaction between 3D objects is more complex, the implementation of such interactor classes is important to facilitate the development of interactive VE applications.

A 3D interactor is generally a geometric object with a defined behavior when certain events occur and certain properties of the environment and the objects in it change. For example, a 3D button might react to the collision of a ray or hand object with it. Figure 11.8 shows a VE application with several 3D interactors. An interactor can be defined as a specialized node in the scene graph, as is done in the X3D file format using PROTO nodes (ISO/IEC 19775-1, 2008). These specialized nodes can be given fields to store state and in and out connection points for event routes. A system, such as the Virtuality Builder II (VB2) (Gobbetti & Balaguer, 1993), which uses a constraint satisfaction algorithm, can enforce constraints between fields internal to an interactor node and between fields of different interactors. In addition, one-way constraints can constrain field values to user input, such as hand position.

An interactor framework that uses event routes to connect interactors together can be difficult to wire in, especially if the interactor reacts to a property for a set of objects (e.g., the hand position of any of a group of users). In the Simple Virtual Interactor Framework and Toolkit (SVIFT), an extension to the simple virtual environment (SVE) library, interactor classes can themselves identify the set of events and property changes that they are interested in, based on the parameters given at instantiation and the behavior desired (Kessler, 1999). For example, a *narrator* interactor may look for any *description* property that is stored with any object added to the scene graph and observe the gaze direction of the user. If the user looks in the direction of an object with a description for a set amount of time, the narrator object would display or read the description. Figure 11.8 contains a set of SVIFT interactors, including buttons, menus, and tabs that can be grabbed and moved (or bestow the grab

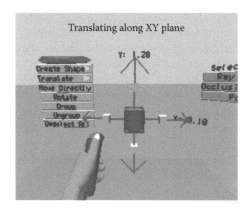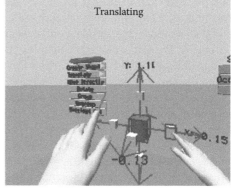

FIGURE 11.8   Interactors in a shape manipulation task.

and move ability to another object), constraint maintainers (e.g., keeping the manipulated object on a particular plane), and selectors. The figure on the left is using a ray selector interactor, and the figure on the right is using a poking selector. Since the other interactors in the environment are defined to respond to any *selection* event that has them as the selected object, they can work with either selection mechanism. This sort of capability is supported by VE libraries that can define behaviors with an interpreted language, as is done in Vizard with the Python programming language (Vizard, 2012). Python allows new fields to be assigned to software objects at run-time and has support for an event dispatcher mechanism to allow interested observers to react to changes in fields.

11.4   FILE FORMATS

In the previous sections, we have defined the vocabulary with which virtual environments are described. The geometry and behavior of a VE application may be given procedurally, using this vocabulary as a basis. A VE application may contain code that constructs geometry using a low-level graphics library, such as OpenGL, or a high-level library such as Vizard or Java 3D. The application may also contain a step-by-step procedure for how geometry changes over time or as a reaction to the user's actions.

A geometric model can generally be described more efficiently by its geometric primitives and properties, such as vertices and colors, than giving a procedure for building the geometry. A file format can be designed to describe geometry and spatial relationships between geometries and therefore allow for clear and concise descriptions of a geometric model. It becomes the task of a file loader and the run-time system to translate the model from the stored representation to the run-time model and ultimately to the internal model for efficient rendering.

A file format might also allow simple, object-based behaviors to be defined. However, complex behaviors and behaviors involving many objects are usually best left to a programming language to describe the behavior procedurally. For example, the Vizard (2012) and Alice systems (Pausch et al., 1995) use geometries that are described by model files, but object and environment behaviors are described using Python scripts.

Files that describe geometric models may be intended to be read and edited directly by developers, such as those defined by X3D (ISO/IEC 19775-1, 2008), or be intended to be created and modified only by modeling applications, such those defined by the 3D Studio Max (2013), Maya (2012), or Multigen "OpenFlight" (MultiGen, Inc., 1998). A format that will be edited by a developer must define an organization of geometric primitives and components that match the developer's conceptual model of the environment's construction. In addition, it should use descriptive key words and allow components to be named. A model format that is not intended to be edited by a person can be more compact, even if it can still be read. For example, the Wavefront *obj* format defines a single geometric object using vertices, vertex normals, texture coordinates, faces, and face materials, each of which is defined using a short key word (*v* for a vertex, *vn* for a vertex normal, etc.). Numbers are represented as a series of numeric characters, rather than being stored as a binary value, allowing the file to be read (and edited if necessary). A Wavefront material file supplements an obj file to provide the definition of face materials.

11.5   MAKING IT REAL TIME

The internal model representation of the environment is designed to provide the rendering performance required for VE applications. The application-level representation of a geometric model could be passed directly to the low-level graphics library and audio libraries to present to the user, but the application representation is not likely to be the most efficient representation to use, especially for visual components. An internal representation, therefore, can be created that provides a more efficient transfer to the graphics hardware and therefore allows for faster performance of the system.

The goal of many rendering optimizations is to reduce the amount of information passed from a software application to the rendering hardware. These optimizations can be automatic or application-assisted. One type of automatic optimization generates additional representations of geometry from an application or stored model that uses less information to define the geometry. Another computes extra information that allows for run-time selection of parts of the geometry that can be ignored. Automatic optimizations include primitive sorting for reduced state changes, display list generation, and culling faces for the render list based on the user's viewpoint. Application-assisted optimizations include switching between different levels of detail and using *billboarded* geometries, described later. This section describes these optimizations in more detail. Many of these are implemented, for example, by the IRIS Performer toolkit, described by Rohlf and Helman (1994).

11.5.1 RENDERING PIPELINE EFFICIENCY

A graphics rendering system uses a current *graphics state* to determine how primitives will appear when they are rendered. The current graphics state includes the coordinate system, material properties, current texture image, and global properties such as fog and light sources. When describing each primitive of a model to be rendered, an application could specify all properties of the graphics state required to render the primitive correctly. However, more often than not, a primitive is part of a greater whole and shares most of its properties with other primitives. For example, a set of triangles that describe the surface of a hood for a car model will share the same material and lighting properties and will share the same lighting properties with the triangles that describe the hubcap. Repeating the current state for every polygon on a particular surface would be quite inefficient.

Systems that generate an internal model for rendering efficiency, such as the Performer toolkit, provide a framework that allows primitives to be sorted by their graphics state. Primitives that have the same material and global properties will appear together, while those with similar properties will be close to each other, and so on. This modified list will be used to describe the model to the renderer, and state changes will only be given when necessary. The result of this optimization is that much of the redundancy of the model is removed.

The most optimal list of primitives would be a sorted list of all of the primitives. However, this list would need to be updated whenever one of an object's properties changes. For example, since all of the primitives would need to have a common coordinate system, all of the primitives of an object would need to be transformed to a common, world coordinate system, taking into account the coordinate system transformations between the object and its parent and ancestors. If the object were to move, however, through the change of one of these transformations, all of the primitives in the list that are part of the object would need to be updated with a new position. This would be a bookkeeping nightmare. Instead, lists of primitives that are sorted are usually limited to individual objects, as described in the next section. However, Durbin, Gossweiler, and Pausch (1995) describe a system that provides a list of primitives for each object that is either just the primitives of the object or one that includes the primitives of the object and its children, if the children have not changed their state in a while. In this system, when a child changes its state and its primitives are part of a parent, the parent returns to its individual list until its children do not change for a while. Therefore, object trees that do not change often are rendered more efficiently (perhaps dramatically so, if the objects share many properties).

Another method to reduce the transfer of geometry and properties changes through the graphics system is to, instead, create *display lists* representing the graphics commands required to create the geometry and to make property changes. A display list can be created and transferred once through the graphics system, then referred to by a unique ID each time the commands should be executed by the graphics system. The graphics system may perform the sorting optimization. Display lists can contain the commands to create a geometric object or may simply consist of the commands to set up a particular material with a texture image to be used for a variety of primitives.

Like a sorted primitive list, a display list is only useful as long as the properties of the primitives do not change. When one property of a single primitive changes, the list needs to be recreated. Since this can cause delays, the Performer library does not use display lists for most primitives. Systems that do use display lists and sort primitives must catch changes to properties and perform a regeneration of the list or simply prevent the change.

11.5.2   FACE AND OBJECT CULLING

Two common methods to reducing the amount of information transferred to the rendering system involve omitting faces that cannot be seen by the user. One method, back-face culling, makes the assumption that face primitives define the surface of a solid object and that the face will only be viewed from outside of the object. Therefore, the face is only rendered when the user's eye point is on the front side of the primitive, which is generally defined to be the side of the face for which the list of vertices is given in a counterclockwise manner. If the viewer is on the back side of the face, it is not rendered. If the object is one that the viewer can be inside of, such as a room, the faces must be made to be two sided, usually by giving the same list of face vertices twice, in opposite orders. The back-face removal technique is usually a feature of the low-level graphics library.

The other method to reduce the amount of faces rendered involves the higher-level representation of the objects in the environment. Objects are checked for an intersection or containment in the user's view volume frustum, which defines the volume of the environment that the user would see if there were no obstructing objects. This volume is defined by the position of the user's viewpoint and viewing window in the environment and the near and far clipping planes. If an object is found to be outside of the viewing frustum, it is simply not rendered at all, saving the rendering system the effort of determining that each face of the object falls outside of the rendered display.

This method depends on having a simple geometric representation of each object that is a good approximation of the volume the object occupies. Common representations include spheres and boxes. It is easy to check for intersections between spheres and other spheres or plane-based volumes, but sphere boundaries often contain much more empty space than space occupied by the object. In addition, to make intersection checking fast, spheres must be defined in a common coordinate system, as nonuniform scale and shear transformations will change the spheres into nonaligned ellipsoids that are much more difficult to work with. A box volume can all be aligned to a single world axis for easier intersection testing or aligned with the local coordinate system of the object it represents, for a more accurate approximation (objects are often defined so that its widest length is parallel to a major axis).

Since a scene graph defines a hierarchy of objects, this hierarchy can be exploited for more effective view volume culling. The parent of group of child objects might have a bounding volume that contains the geometry of the parent and the geometries of the group. Therefore, if the parent object's bounding volume is outside of the viewing volume, so are all of its children and that whole subtree can be omitted from the rendering of the current frame. If the environment is sparsely populated by objects, then a more effective bounding volume hierarchy may be constructed separately from the scene graph based on grouping the closest objects, then the closest groups together, until one group contains all objects.

11.5.3   LEVEL OF DETAIL SWITCHING

The amount of detail necessary for an object's description depends on how much detail the user can possibly notice. If the object is far away or in a foggy or low-light environment, the user will not notice much detail, while if the object is nearby on a clear day, the most detail will be perceived. A method to reducing the amount of information transferred to the rendering system involves selecting the object geometry from a list of progressively less-detailed representations that has the least detail but will appear very similar to the object at its full detail. That object is used, potentially saving the wasted effort of rendering detail geometry that would not be noticed. This method of omitting detail can work for all presentation mediums involving the distance senses. Sights, sounds, and smells diminish in intensity and detail based on distance or other factors. In general, the application must supply the different levels of detail for an object and give some measure of how it compares to a full detail model. A common approach, as seen in the LOD node of the X3D file format, is to associate a representation with a range of distances from the user's eye point for which that representation should be used. A system, such as Performer, can also automatically determine when an object's geometry is so small that it does not occupy enough of the screen to be noticed and therefore can be omitted completely.

11.5.4 Billboarded Geometry

A common technique that is used to reduce the amount of detail needed for a geometric model is to use an image containing a picture of the detail. This technique works well for distant objects, as the absence of depth to parts of the model is not noticed. However, this technique alone will not work for objects that can be viewed from all directions, such as trees or clouds. Another technique can be used, in addition to an image, to give the appearance of depth to a model. This technique, known as billboarding, can be used at the cost of some additional computation if the model can be assumed to appear the same from all directions. The image is placed on a flat polygon, but the polygon automatically always faces the viewer or always orients to the viewer while keeping one axis aligned (such as the *up* axis for standing objects). For example, a tree image will remain upright but will appear to have a trunk and crown from any point of view (except from directly above or below). The tree will appear to be 3D, if a bit too symmetrical. This technique can be combined with changing the image based on the rotated angle to give an approximation of an object that is not symmetric around one or more axes.

11.6   INCLUDING THE USER

When creating 2D images for interface and content, computer applications use a standard location for the viewer some distance in front of the computer screen. In fact, the viewer is mathematically assumed to be at a great distance from the screen, as raised buttons appear the same whether they are on the left, right, top, or bottom. The geometric model of these applications is 2D or an ordered composite of 2D image layers, all of which are in front of the viewer. However, when the geometric model becomes truly 3D, we can place the viewer in the model itself, giving the user a presence in the 3D environment. This allows users to view a 3D model from any perspective, to get closer and even fly through the model. Including a representation of the user in the 3D model itself clearly specifies the point of view for the image rendered for the user, and it allows the behavior of the model to affect the view; when the user is attached to a moving airplane object, the viewpoint moves with it. In addition, geometry can be attached to the user so that it moves when the view changes (through mouse manipulations or tracking device movements), as in a head's up display. Finally, the user model can describe the type of display being used and specify the window in the 3D world that matches the display image edges. The image may be glasses into the virtual world, moving as if attached to the user's head (an HMD or head-mounted display), or a window into the world that moves independently of the user's head (a fish-tank, workbench, or handheld display). Figure 11.9 shows three different representations of the user in a scene graph for a X3D world file, a Java 3D application, and an SVE application (Kessler, Bowman, & Hodges, 2000).

Placing a representation of the user in the 3D model of the virtual environment not only provides a convenient way to specify the user's view of the world but also allows for a unified method for handling the user's visible representation in the world and interaction with the world. Geometric objects can be given for important parts of the user's body, such as the arms and hands, which may be controlled by any positional input device (e.g., mouse, joystick, or six degree-of-freedom tracker)

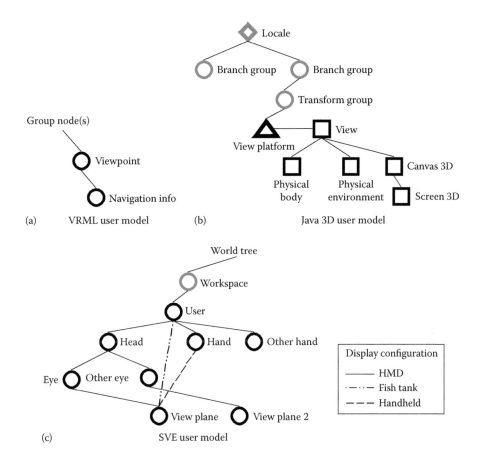

FIGURE 11.9   User model in the scene graph for (a) X3D and (b) the Java 3D and (c) SVE libraries.

or scripted animations. The direct interaction of the user with the world can be handled just as the interaction between any two or more objects is handled. The following sections describe how the user can be represented in the environment model, how tracking input devices can control the user model, and, finally, how the user model can be used to generate the viewing parameters for the various display devices that the user may be using.

11.6.1   USER MODEL

The simplest representation of a user in a 3D environment is simply a location, gaze direction, and *up* direction for the view. However, this viewpoint method conceptually places the user outside of the environment. A better method, common to most VE systems, is to represent the user by one entity or a group of entities in the environment model, which can be attached to other entities of the model and be manipulated like other entities. This method allows the user's viewpoint follow an object that she is in, such as an airplane or car. The X3D file format provides an example, shown in Figure 11.9a, where the user is represented by a Viewpoint node (ISO/IEC 19775-1, 2008). The Viewpoint node can be attached to a Transform or other Group node, which means that the user's viewpoint will change as the positions of the Transform nodes further up the tree change. The NavigationInfo node describes how the user's viewpoint can change within the coordinate system of the Viewpoint node, including the speed, and constraints on movement.

The viewpoint representation of the user also fails to include other important information, such as the position of two eyes for stereo viewing; the position of hands, feet, the torso, and other body parts; and the movement of digits. Information about the user is used to help determine the view presented, to provide a visible representation of the user in the environment, to allow direction interaction with the geometric objects in the environment, and to provide input to gesture recognition algorithms. Given that the information is spatial in nature and that it is all about one subject (the user, of which there may be more than one in the environment), it is natural to include the user data in the environment model.

Figure 11.9 shows the user representation in three different 3D environment models: the X3D file format (ISO/IEC 19775-1, 2008), the Java 3D scene graph (Sowizral et al., 2000), and the SVE coordinate system graph (Kessler et al., 2000). In addition to the viewpoint being associated with a Viewpoint node in the X3D file format, an attached NavigationInfo node describes the size of the user's avatar in the environment. The X3D avatar is described as having a width, given as a radius from the user's position, a height, and a maximum step size. The width is used to determine collisions between the user and geometry in the world, the height is used to describe the distance the user should be above the ground if following a terrain, and the step size is used to determine if a height difference in the terrain is too large to allow the user to pass over it.

The representation of the user in the Java 3D scene graph is similar to the X3D format. The location of the user in the 3D environment is defined by the ViewPlatform node, which can be placed in the scene graph so that the user automatically moves with objects. However, Java 3D has two additional program objects: the PhysicalBody and PhysicalEnvironment objects that define the positions of the user's head, eye(s), and hands in the environment. These objects encapsulate the transformation from the ViewPlatform coordinate system to a *coexistence* coordinate system and describe the locations of user body parts in that coexistence coordinate system. The application programmer can use the calculated positions of the various body parts in Behavior nodes to define interactions between the user and the environment and to have separate geometric objects follow the user's movements and provide a visible representation of the user in the environment.

In contrast to Java 3D, the SVE coordinate system graph associates a separate geometric object in the coordinate system graph for each important body part, as is shown in Figure 11.9c. A User object represents the user as a whole and can be attached to the environment model at any place, like the Viewpoint and ViewPlatform nodes. Attached to the User are a Head object and two Hand objects. Attached to the Head object are two Eye objects. Other user objects, such as feet, palms, and finger joints, can be added to the representation as needed. The next section describes how the relationship between the physical locations of the user's body parts, if tracked, and location of the geometric object in the environment is maintained. These objects may not have a geometry defined for it, simply representing a location in the world, or may have a geometry or child objects with geometry to give a visible representation of the user in the environment. An advantage to this user model is that the interaction of the user with the world can be handled in the same way as interactions between any set of objects in the environment are handled.

11.6.2 Incorporating Tracking Information

Tracking devices generally report the position and orientation of a *tracker* in relationship to a particular reference frame. If the tracker is the end of a set of mechanical linkages, then the reference frame is the base of the linkages. If the tracker is an electromagnetic device, then the reference frame is the transmitter part of the device (often placed on the desk or hung above the user).

Information from the tracking device can be used in raw form or adjusted by a given offset to relate the tracking reference frame to the world coordinate system of the environment. This information can be used to determine the position of the viewer or be used to determine how the user interacts with the scene. However, if the tracking information is to be used in a coordinate system other than the world coordinate system, then the reference frame offset must be carefully calculated and updated when the scene graph changes. In fact, two offsets might need to be maintained: a transformation between the tracker reference frame and the environment and a transformation between the tracking device and the reference point of the associated user's body part. For example, the transmitter may be mounted above the user's head, and a head tracker may be mounted on a head-mounted display such that it is a few centimeters above and behind the top of the head and tilted.

The user model of Java 3D defines a set of coordinate systems between the ViewPlatform node in the scene graph representation of the user's position (or, more precisely, the position of the user's feet, screen, or eye) and the user's head, eyes, and hands. The ViewPlatform node defines where its origin is in coexistence coordinates, based on whether it represents the user's feet, screen, or eye. The PhysicalEnvironment object defines a transformation from the coexistence coordinates to the tracker reference frame, the tracker base (the first offset discussed). Finally, the PhysicalBody object defines a transformation from the head position to the head tracker position (the second offset discussed). As we will see in the next section, this model supports a variety of display configurations. The particular display configuration used can be defined separately from the Java 3D scene graph, since the necessary parameters are in the View object and associated objects (shown as squares in Figure 11.9b).

In the SVE library, tracking information is introduced into the user model to control a body part (or any other object in the environment) by adding two new geometric objects. One object is used to represent the reference frame itself, and it is set to be the child of the controlled object's parent. The other new object represents the tracker and is placed as a child of the reference frame object. The controlled object is then set to be the child of the tracker object. The tracker object's coordinate system transformation is set to the raw tracking information of each frame. By initializing the coordinate system transform of the reference frame object correctly, the controlled object will move around its previous parent correctly. An added benefit to having a new reference frame object for each tracker is that different tracking devices can be used simultaneously, even if their tracker bases are at different locations and the tracker reference frames do not match. If a correction needs to be made for how the tracker is attached to the body part it is tracking (it is 90° off, perhaps), then a correction can be placed in the transformation from the tracker to the controlled object. The SVE model provides the application with spatial information such as tracker reference frame and tracker position in the same way it provides spatial information about any other object, simplifying the implementation of user interaction and environment behaviors.

11.6.3 VIEWING MODEL

One of the main tasks of the user model is to define the parameters necessary to generate the user's current view, so that it can be rendered to the user's display. The X3D Viewpoint node provides a common representation for the user's view, that of a camera in the environment. The Viewpoint node provides a location, orientation, and a field of view for a camera model, which defines how geometry is projected onto the display toward an eye point. Two viewpoints can be used with a camera model when generating the images for a head-mounted display. In this case, however, precision requires that the optics of the display be accounted for, as described by Robinett and Holloway (1995). For example, the standard assumptions that each eye is centered on the image and that the display plane is perpendicular to the gaze direction are generally not quite correct.

The main assumption of the camera model is that the eye point is related to the display image by a constant offset. When the eye moves relative to the display, then the camera model no longer applies. This is the case for *fish-tank* display configurations (Deering, 1992), which include desktop monitors and projected images on horizontal workbenches or on walls. For these configurations, the user model provides the position of the user's eye (or eyes, if a stereo image is displayed) and the location in the environment of the window that defines the view seen from each eye point. For example, the Java 3D scene graph supports both HMD and fish-tank display configurations by storing transformations in the PhysicalEnvironment, ViewPlatform, and Scene3D objects that relate the tracker reference frame (the tracker base) to the displayed image. For HMD configurations, this relationship changes constantly. For fish-tank configurations, it is constant. Regardless, the view is determined by the relationship between the screen and the eye point (which is controlled by the tracker that moves in the user model's tracker reference frame).

Following the philosophy that spatial objects in the user's environment should be represented by geometric objects in the coordinate system graph, the SVE user model includes a Viewplane object that represents the outline of the display. In SVE, a view is defined by one or two Eye objects, a Viewplane object, and the edges of the display, given in the coordinate system of the Viewplane object. Since the Eye objects and the Viewplane object are in the same coordinate system graph, they can be related spatially and can determine the viewing parameters. Obviously, if the Viewplane and Eye are placed in the graph so that they can move independently, then this model can support fish-tank display configurations as well as HMD configurations. In fact, since the Viewplane and the Eye can be moved independently of the tracker reference frame, the SVE user model supports a third display configuration, handheld displays (Rekimoto, 1997). In this configuration, the window to the virtual world is rendered to a handheld display that moves with the user's hand motions.

The Viewplane object is the representation of the projection plane for the default view of the environment and is placed in the coordinate system graph differently depending on the display configuration, as shown in Figure 11.9c. For HMD configurations, the Viewplane object is positioned to correspond to the optical projection plane of the HMD (with window extent values that provide the correct vertical and horizontal field of view angles for the HMD) and is attached to the Head object. For fish-tank displays, the view plane represents the actual monitor screen in the real-world configuration. Therefore, the view plane does not move with the user's head but stays stationary in the user's reference frame. Thus, the Viewplane object is attached to the User or Workspace object. For fish-tank displays, the view plane must be positioned carefully so that the position difference, or disparity, between left and right views of an object is correct (objects at the view plane should have no disparity, while objects further in front of or behind the view plane should have larger disparities).

11.7 SUMMARY

Virtual environment applications present 3D environments to a user that is located in the environment. In this chapter, we have described how the environment is defined in terms of three representations. One representation defines the model in a file or set of files. The format of these files is designed to provide a concise, geometrically organized description of geometric primitives and simple behaviors. Another representation is used by the application at run-time to modify the environment, to effect behavior of the objects in the environment, and to make the environment react to user interactions. The final environment representation is an internal representation that, when possible, distills the description of the environment to the information that is strictly necessary for processing and provides that information to a rendering system to produce the image for the display. Many of the techniques described in this chapter are given in terms of a visual result, but they could also be adapted to the presentation for other senses, such as sound and smells.

REFERENCES

3D Studio Max. (2013, February 11). *Autodesk 3D studio max* [Online]. Available: http://usa.autodesk.com/3ds-max/.

Ames, A. L., Nadeau, D. R., & Moreland, J. L. (1997). *VRML 2.0 sourcebook* (2nd ed.). New York, NY: John Wiley & Sons, Inc.

Autodesk FBX. (2013, February 11). *Autodesk FBX SDK* [Online]. Available: http://usa.autodesk.com/fbx/.

Blinn, J. F. (1977, July). Models of light reflection for computer synthesized pictures. *Proceedings of ACM SIGGRAPH'97* (pp. 192–198). New York, NY: ACM.

Bukowski, R. W., & Sequin, C. H. (1995). Object associations. *Proceedings of the 1995 Symposium on Interactive 3D Graphics* (pp. 131–138). Monterey, CA: ACM.

COLLADA. (2013, February 11). *COLLADA digital asset and FX exchange schema* [Online]. Available: https://collada.org/mediawiki/index.php/COLLADA_-_Digital_Asset_and_FX_Exchange_Schema.

Deering, M. (1992, July). High resolution virtual reality. *Proceedings of ACM SIGGRAPH 92* (pp. 195–202). Chicago, IL: ACM.

Dinh, H. Q., Walker, N., Hodges, L. F., Song, C., & Kobayashi, A. (1999, March). Evaluating the importance of multi-sensory input on memory and the sense of presence in virtual environments. *Proceedings of IEEE Virtual Reality '99* (pp. 222–228). Houston, TX: IEEE Press.

Direct3D. (2012, October 1). *Direct3D 11 graphics* [Online]. Available: http://msdn.microsoft.com/en-us/library/windows/desktop/ff476080%28v=vs.85%29.aspx : Microsoft.

Durbin, J., Gossweiler, R., & Pausch, R. (1995). Amortizing 3D graphics optimization across multiple frames. *Proceedings of the ACM Symposium of User Interface Software and Technology (UIST'95).* New York, NY: ACM.

Eckel, G. (1998). *Cosmo 3D™ Programmer's Guide* (Document 007-3445-002). Mountain View, CA: Silicon Graphics, Inc.

Elliot, C., Schechter, G., Yeung R., & Abi-Ezzi, S. (1994, August). TBAG: A high level framework for interactive, animated, 3D graphics applications. *Proceedings of ACM SIGGRAPH'94* (pp. 421–434). Orlando, FL: ACM.

Fernando, R. (Ed.). (2004). *GPU gems: Programming techniques, tips and tricks for real-time graphics.* Reading, MA: Addison-Wesley.

Foley, J., van Dam, A., Feiner, S., & Hughes, J. (1996). *Computer graphics: Principles and practice.* (2nd ed. in C, pp. 222–226). Reading, MA: Addison-Wesley.

Gobbetti, E., & Balaguer, J. (1993, November). VB2: An architecture for interaction in synthetic worlds. *Proceedings of the ACM Symposium on User Interface Software and Technology (UIST'93)* (pp. 167–178). Atlanta, GA: ACM.

Havok. (2013, February 21). *Havok physics* [Online]. Available: http://www.havok.com/products/physics.

ISO/IEC 19775-1. (2008). *Information technology—Computer graphics and image processing—Extensible 3D* (X3D).

Kessler, G. D. (1999, March). A framework for interactors in immersive virtual environments. *Proceedings of IEEE Virtual Reality'99* (pp. 190–197). Houston, TX: IEEE Press.

Kessler, G. D., Bowman, D. A., & Hodges, L. F. (2000). The simple virtual environment library: An extensible framework for building VE applications. *Presence, 9*(2), 187–208.

Lewis, J., Cordner, M., & Fong, N. (2000, July). Pose space deformation: A unified approach to shape interpolation and skeleton-driven deformation. *Proceedings of ACM SIGGRAPH 2000* (pp. 165–172). New Orleans, LA: ACM SIGGRAPH.

Lindstrom, P., Koller, D., Ribarsky, W., Hodges, L. F., Faust, N., & Turner, G. A. (1996, August). Real-time, continuous level of detail rendering of height fields. *Proceedings of ACM SIGGRAPH 96* (pp. 109–118). New Orleans, LA: ACM SIGGRAPH.

Maya. (2012, October 1). *Autodesk maya* [Online]. Available: http://usa.autodesk.com/maya/.

MultiGen Inc. (1998, July 31). *Openflight specification* (v15.6.0).

OpenGL. (2012, October 1). *OpenGL* [Online]. Beaverton, OR: The Kronos Group. Available: http://www.opengl.org/.

Pausch, R., Burnette, T., Capehart, A. C., Conway, M., Cosgrove, D., DeLine, R., ... White J. (1995, May). Alice: Rapid prototyping for virtual reality. *IEEE Computer Graphics & Applications, 15*(3), pp. 8–11.

PhysX. (2013, February 21). *NVIDIA geForce physX* [Online]. Available: http://www.geforce.com/hardware/technology/physx.

Rekimoto, J. (1997). NaviCam: A magnifying glass approach to augmented reality. *Presence, 6*(4), 399–412.

Robinett, W., & Holloway, R. (1995). The visual display transformation for virtual reality. *Presence, 4*(1), 1–23.

Rohlf, J., & Helman, J. (1994, August). IRIS performer: A high performance multiprocessing toolkit for real-time 3D graphics. *Proceedings of ACM SIGGRAPH '94* (pp. 381–394). Orlando, FL: ACM.

Sannela, M. (1992). *The skyblue constraint solver* (Technical Report TR-92-07-02). Seattle, WA: Department of Computer Science, University of Washington.

Sowizral, H., Rushforth, K., & Deering, M. (2000, April). *The Java 3D™ API Specification* (Version 1.2). Santa Clara, CA: Sun Microsystems, Inc.

Tarlton, M. A., & Tarlton, P. N. (1992). A framework for dynamic visual applications. *Proceedings of the 1992 Symposium on Interactive 3D Graphics* (pp. 161–164). Cambridge, MA: ACM.

Unity. (2013, February 21). *Unity game engine, tools, and multiplatform* [Online]. Available: http://unity3d.com/unity/.

Vizard. (2012, October 1). *Worldviz vizard virtual reality toolkit* [Online]. Available: http://www.worldviz.com/products/vizard/.

Watt, A. H., & Watt, M. (1992). *Advanced animation and rendering techniques: Theory and practice.* Reading, MA: Addison-Wesley Pub. Co.

12 Principles for Designing Effective 3D Interaction Techniques

Ryan P. McMahan, Regis Kopper, and Doug A. Bowman

CONTENTS

12.1 INTRODUCTION

Applications of virtual environments (VEs) have become increasingly interactive, allowing the user not only to look around a 3D world but also to navigate the space, select and manipulate virtual objects, and give commands to the system. Thus, it is crucial that researchers, developers, and designers understand the issues related to 3D interfaces and interaction techniques. In this chapter, the space of possible interaction techniques for several common tasks is explored, and guidelines for their use in VE applications are offered. These guidelines are drawn largely from empirical research results.

12.1.1 MOTIVATION

Interaction (communication between users and systems) in a 3D VE can be extremely complex. Users must often control six degrees of freedom (DOFs) simultaneously, move in three dimensions, and give a wide array of commands to the system. To make matters worse, the standard and familiar input devices such as mice and keyboards are usually not present, especially in immersive VEs.

Meanwhile, VE applications are themselves becoming increasingly complicated. Once a technology only for interactively simple systems (those in which interaction is infrequent or lacks complexity), such as architectural walk-through (Brooks et al., 1992) or phobia treatment (Hodges et al., 1995), VEs are now used in domains such as manufacturing, design, medicine, education, and entertainment. Due to several interactions beyond simple navigation, these domains require a much more active user and therefore a more complex user interface.

One of the main concerns with the advent of these complex applications is the broadly defined *user experience*. Three factors that contribute to the user experience are considered in this chapter: general usability, task performance, and naturalism. General usability refers to the general qualitative experience of the user during interaction, including ease of use, ease of learning, user comfort, and affordance. Task performance refers to the objective quality of task completion, such as speed, accuracy, and precision. Finally, naturalism describes how natural the application is as well as how natural it is to interact with the application. Naturalism is important when presence, training transfer, or appropriate physical exertion is desired.

The focus of this chapter is how to design effective 3D interaction techniques (3D ITs) that positively impact the user experience in terms of usability, task performance, or naturalism. This is an extremely important topic. Since VEs support user tasks, it is essential that VE developers

TABLE 12.1

Summary of High-Level Guidelines for the Design of Effective 3D Interaction Techniques

Based on Existing HCI Guidelines	Based on Interacting in 3D Space	Based on Hardware Considerations
Practice user-centered design and follow well-known general principles from HCI research.	Take advantage of the user's proprioceptive sense for precise and natural 3D interactions.	In semisurrounding SIDs, design the application to minimize virtual rotations.
Ensure all interaction tasks and techniques integrate well together.	Use well-known 2D interaction metaphors if the interaction task is inherently 1D or 2D.	For narrow-FOV HMDs, use amplified head rotations when visual search tasks are important.
	Allow two-handed interaction for more precise input.	Use an input device with the appropriate number of DOFs for the task.
	Provide redundant interaction techniques for a single task.	Use physical props to constrain and disambiguate complex spatial tasks.
		Use absolute devices for positioning tasks and relative devices for tasks to control the rate of movement.

and designers show concern for user-experience issues when selecting interaction techniques and metaphors for their systems. Here, techniques for the most common VE tasks (selection, manipulation, travel, and system control) are presented. Perhaps, more importantly, a large number of design guidelines are provided. These guidelines are taken, where possible, from published empirical evaluations of 3D ITs (and in many other cases, from personal experience and observation of user interaction within VE applications) and are meant to give VE developers practical and specific ways to improve the user experience of their applications. High-level guidelines are summarized in Table 12.1, and task-specific guidelines are summarized in Table 12.2.

12.1.2 Universal Interaction Tasks

At first glance, there appear to be an extremely large number of possible user tasks in immersive VEs—too many, in fact, to think about scientific design and evaluation for all of them. However, as Foley (1979) has argued for 2D interaction, there is also a set of *universal* tasks (simple tasks that are present in most applications, which can be combined to form more complex tasks) for 3D interfaces. These universal tasks include *selection*, *manipulation*, and *navigation*.

Selection refers to the specification of one or more objects from a set. Selection may be used on its own to specify an object to which a command will be applied (e.g., *delete the selected object*), or it might denote the beginning of a manipulation task. Selection may also be used to specify a target toward which one desires to travel. Manipulation refers to the modification of various object attributes (including position and orientation and possibly scale, shape, color, texture, or other properties). Finally, navigation refers to the task of moving the viewpoint within a 3D space and includes both a cognitive component (wayfinding) and a motor component (travel, also called viewpoint motion control; Darken & Peterson, 2014; Chapter 19).

These simple tasks are the building blocks from which more complex interactions arise. For example, the user of a surgery simulator may have the task of making an incision. This task might involve approaching the operating table (navigation), picking up a virtual scalpel (selection), and moving the scalpel slowly along the desired incision line (manipulation). One class of complex tasks, *system control*, involves the user giving commands to the system. For example, this might be accomplished by bringing a virtual menu into view (manipulation) and then choosing a menu item

TABLE 12.2

Summary of Task-Specific Guidelines for the Design of Effective 3D Interaction Techniques

	Selection	Manipulation	Travel	System Control
General usability guidelines	Use the natural simple virtual hand technique if all selection is within arm's reach.	Use general or application-specific constraints for manipulation.	Use smooth transitional motions between locations. Use physical movement techniques for maneuvering tasks. Train users in exploration strategies.	Use object-first system control task sequences. Use an appropriate spatial reference frame. Structure the functions in an application.
Performance guidelines	Use a pointing technique for high-performance selection. Increase precision by increasing the CD ratio. Use progressive refinement if accuracy is more important than speed.	Increase precision by reducing the DOFs. Use indirect manipulation if accuracy is more important than speed.	Use a steering technique for high-performance exploration and search. Use a target-based technique for goal-oriented travel.	Use a 2D or 1D selection method when interacting with a graphical menu. Verify commands if accuracy is more important than speed.
Naturalism guidelines	Use a virtual hand technique for more-natural selection.	Use a virtual hand technique for more-natural manipulation. Increase the probability of training transfer by using a direct mapping.	Use a physical locomotion technique for more-natural travel. Use redirected walking for more-natural exploration and search.	Use redundant, intuitive utterances when using voice commands. Use gestures that have high biomechanical symmetry.
Special guidelines	Consider designing the environment to maximize the perceived size of objects. Consider using object snapping for environments with limited selectable objects.	Consider using nonisomorphic rotations to reduce clutching.	Consider using wayfinding aids to help the user decide where to move. Consider using manual manipulation techniques for manipulation-oriented travel.	Consider using multimodal input. Consider reducing the number of commands when using voice or gestural commands.

(selection). However, system control is so ubiquitous in VE applications that the design of system control techniques can be considered separately.

This chapter is targeted at developers, researchers, and designers who produce complete VE applications. It provides background information, a large set of potential techniques for universal interaction tasks, and guidelines to help in the choice of an existing technique or the design of a new technique for a particular system. The use of these guidelines should lead to more usable, useful, efficient, and effective VEs.

12.1.3   USER-EXPERIENCE REQUIREMENTS

In order to determine whether or not a 3D IT positively affects the user experience, metrics must be defined that capture its effects. Metrics allow the effectiveness of a technique to be quantified, the performance of competing techniques compared, and the interaction requirements of an application specified. Listed later are some (but certainly not a complete set) of the most common user-experience metrics for 3D interaction, including metrics for general usability, task performance, naturalism, and some special requirements. For each individual interaction task, the metrics may have slightly different meanings.

12.1.3.1   General Usability Requirements

1. *Ease of learning.* This is commonly discussed in the human–computer interaction (HCI) community and refers to the ease with which a novice user can comprehend and begin to use a technique. It may be measured by subjective ratings, or the time for a novice to reach some level of performance, or by characterizing the performance gains by a novice as exposure time to the technique increases.
2. *Ease of use.* This is another HCI concept that may be difficult to quantify. It refers to the simplicity of a technique from the user's point of view. In psychological terms, this may relate to the amount of mental workload induced upon the user of the technique. This metric is usually obtained through subjective self-reports, but measures of mental workload may also indicate ease of use.
3. *User comfort.* Most of the interaction techniques discussed herein require activity on the part of the user (e.g., moving the arm, turning the head). It is important in systems that require a moderate to long exposure time that these motions do not cause discomfort in the user. Discomfort can range from classic simulator sickness to eyestrain to hand fatigue and so on. Although VEs in general may induce some level of discomfort, this may be increased or decreased depending on the interaction techniques and input devices chosen. Comfort measures are usually self-reported (Kennedy, Lane, Berbaum, & Lilienthal, 1993).
4. *Affordance.* Finally, a technique's usability can be described by the affordances that it presents for a task. An affordance (Norman, 1990) is simply a characteristic of a technique or tool that helps the user understand what the technique is to be used for and how it is to be used. For example, voice commands in general have little affordance because the user must know what the commands are. Listing the available commands on screen is an affordance that aids the user. Affordance is an innate characteristic of a technique that is not easily measured. Nonetheless, it must be taken into consideration when designing 3D ITs.

12.1.3.2   Performance Requirements

1. *Speed.* This refers to the classic quantifier of performance: task completion time. This efficiency metric will undoubtedly be important for many tasks, but should not be the only measurement considered.
2. *Accuracy.* Accuracy is a measurement of the exactness with which a task is performed. For travel or manipulation tasks, this will likely be measured by the distance of the user or object from the desired position or path. For selection, one might measure the number of errors that were made.
3. *Precision.* Precision refers to the degree of refinement required to complete a task with respect to the task difficulty. The precision of a 3D IT can be measured by analyzing the amount of fine-grained control afforded by the technique. For example, a selection technique is precise if it allows consistent accurate selection of small objects, and a travel technique is precise if it allows en route navigation of narrow paths without collisions.

12.1.3.3 Naturalism Requirements

1. *Presence.* Another goal of VEs is to induce a feeling of presence (or *being there*) in users (Chertoff & Schatz, 2014; Chapter 34). This quality is purported to lend more realism to a VE system, which may be desirable in systems for entertainment, education, or simulation. Presence can be affected by the interaction techniques in a system (McMahan, Bowman, Zielinski, & Brady, 2012). It is usually measured by subjective reports and questionnaires (Slater, Usoh, & Steed, 1994; Witmer & Singer, 1998).

2. *Training transfer.* This refers to how effectively knowledge and skills obtained in VEs transfer to the real world. In general, the more natural a VE, and the more natural it is to interact with the VE, the more likely training transfer will occur. Natural interaction techniques are especially important for enabling the training transfer of psychomotor skills (Dede, 2009), in addition to the transfer of cognitive skills (e.g., Gutiérrez et al., 2007).

3. *Physical exertion.* Another requirement of some VEs is to induce physical exertion from the user. This usually is to provide the user with exercise for health purposes, particularly in physical rehabilitation (e.g., Rand, Kizony, & Weiss, 2008) or entertainment applications. Interaction techniques that involve large muscle groups, like walking, are more likely to cause physical exertion in users.

12.1.3.4 Special Requirements

1. *Spatial orientation.* A user's spatial orientation is related to his knowledge of the layout of a space and his own position within it. This may be an important performance requirement in large, highly occluded, or complex VEs. Most often, movement within the space (travel) affects spatial orientation, but other interaction tasks may also affect this metric.

2. *Information gathering.* One of the goals of many immersive VEs is for the user to obtain information from or about the environment while in it. The choice of interaction techniques may affect the user's ability to gather information, and so the measurement of this ability can be seen as an aspect of technique usability.

3. *Engagement.* This requirement refers to the user's involvement with a VE during a particular experience. Since interacting with one's surroundings is part of being involved, engagement and presence are often entwined. In fact, some presence questionnaires touch upon the concept of engagement (Witmer & Singer, 1998), though engagement can be measured independently (McMahan et al., 2012).

4. *Enjoyment.* For some VEs, enjoyment is the most important requirement. This is particularly true for entertainment, since enjoyment is the goal and an indicator of success. Enjoyment is usually measured by subjective reports and questionnaires.

12.2 NATURALISM VERSUS MAGIC

In the early days of 3D interaction research, two design approaches emerged: naturalism and *magic*. Naturalism attempts to design interaction techniques to work exactly the way the real world works or at least as close as is practically possible. The very term *virtual reality* promotes this approach—that virtual reality should be the same as *real reality*. On the other hand, magic techniques try to enhance usability and performance by giving the user new abilities and nonnatural methods for performing tasks (Smith, 1987). Allowing the user to fly like a bird is an example of a magic technique. One way to describe the two design approaches is to say that they differ in terms of interaction fidelity.

Interaction fidelity is the objective degree of exactness with which real-world actions are reproduced in an interactive system (McMahan et al., 2012). Hence, swinging one's arms to hit a virtual golf ball clearly provides greater interaction fidelity than pressing a sequence of buttons to accomplish the same task. Naturalism essentially strives to provide the highest level of interaction fidelity possible. Conversely, magic techniques strive to enhance the user experience by reducing interaction fidelity and circumventing the limitations of the real world.

By examining the level of interaction fidelity magic techniques provide, a third design approach becomes evident: *hypernatural*. Hypernatural techniques are magic techniques that use natural movements but make them more powerful by giving the user new abilities or intelligent guidance (Bowman, McMahan, & Ragan, 2012). Increased interaction fidelity is what distinguishes hypernatural techniques from standard magic techniques. An example of a hypernatural technique is the go-go technique (Poupyrev, Billinghurst, Weghorst, & Ichikawa, 1996), in which a natural arm extension causes the user's virtual hand to extend far into the environment, allowing the user to select objects from a great distance.

A key component to distinguishing hypernatural techniques from traditional magic ones and in terms of naturalism is to objectively identify their level of interaction fidelity. The Framework for Interaction Fidelity Analysis (FIFA—McMahan, 2011) presents interaction fidelity as a composite of three broad concepts—biomechanical symmetry, control symmetry, and system appropriateness—described in the following.

12.2.1 BIOMECHANICAL SYMMETRY

Perhaps, the most distinct feature of immersive VEs is the prevalent use of gestures. Instead of using joysticks and buttons, gestures allow users to interact with VEs using natural postures and body movements. The greater these gestures correspond to real-world actions, the greater the level of interaction fidelity provided to the user. The term *biomechanical symmetry* refers to the objective degree of exactness with which real-world body movements are reproduced for a task during a successful interaction. For example, redirected walking, which subtly rotates the VE to allow for walking when the virtual space is larger than the physical space (Razzaque, 2005), has a high biomechanical symmetry to real walking considering the natural body movements involved. Conversely, a walking-in-place technique, which interprets in-place steps for virtual locomotion (Slater Usoh, & Steed, 1995) would have a lower biomechanical symmetry due to its less-natural movements.

12.2.2 CONTROL SYMMETRY

Another component of interaction fidelity is related to the amount of control provided to the user. Techniques with lower interaction fidelity can be frustrating due to the need to switch between modes of interaction to control objects in 3D space. High-fidelity techniques, however, may provide the same control without the need for modes of interaction. The term *control symmetry* refers to the objective degree of exactness with which control in a real-world task is provided by an interaction technique to complete the task virtually. For instance, directly manipulating an object with a 6-DOF controller provides greater control symmetry than manipulating the same object with a 2-DOF mouse while using multiple planar and rotational modes.

12.2.3 SYSTEM APPROPRIATENESS

Aside from biomechanical symmetry and control symmetry, there are numerous factors (such as input device accuracy, latency, and form factor) that can influence the overall fidelity of an interaction. Most of these factors characterize how suitable a system is for implementing a particular aspect of interaction. Therefore, these factors are considered components of system appropriateness, which constitutes part of the overall interaction fidelity provided by a particular implementation of a technique. For example, a technique in which the user swings a virtual baseball bat by swinging an input device would have lower interaction fidelity if the input device has high latency. This might result in the user swinging the virtual bat sooner than he would in the real world, hence promoting negative training transfer.

12.3 HIGH-LEVEL GUIDELINES FOR 3D INTERACTION TECHNIQUES

When attempting to develop guidelines that will produce highly effective 3D interaction, both generic guidelines that inform interaction design at a high level and specific guidelines for common tasks (i.e., selection, manipulation, travel, and system control) should be considered. The next two sections will cover these areas. The high-level guidelines presented in this section are not intended to be exhaustive, but are limited to those that are especially relevant to enhancing the user experience and those that have been verified through formal evaluation. A large number of VE-specific usability guidelines can be found in Gabbard and Hix (1998).

12.3.1 EXISTING HCI GUIDELINES

The first thing to remember when developing interaction for VEs is that interaction is not new. The field of HCI has its roots in many areas, including perceptual and cognitive psychology, graphic design, and computer science, and has a long history of design and evaluation of 2D computer interfaces. Through this process, a large number of general-purpose guidelines have been developed that have wide applicability to interactive systems, and not just the standard desktop computer applications with which everyone is familiar. This existing knowledge and experience can be leveraged in interaction design of VEs. If VE design does not meet these most basic requirements, then VE systems are sure to be unusable. Furthermore, the application of HCI principles to VEs may lead to VE-specific guidelines as well. These guidelines are well known, if not always widely practiced, so they are not reviewed in detail here.

Practice user-centered design and follow well-known general principles from HCI research.

Two important sources for such general guidelines are Donald Norman's *The Design of Everyday Things* (Norman, 1990) and Jakob Nielsen's usability heuristics (Nielsen & Molich, 1992). These guidelines focus on high-level and abstract concepts such as making information visible (how to use the system, what the state of the system is, etc.), providing affordances and constraints, using precise and unambiguous language in labeling, designing for both novice and expert users, and designing for prevention of and recovery from errors.

Following such guidelines should lead to a more understandable, efficient, and usable system. However, because of their abstract nature, applying these principles is not always straightforward. Nevertheless, they must be considered as the first step toward a usable system.

Ensure all interaction tasks and techniques integrate well together.

Interaction techniques can be quite effective when considered by themselves, but the overall usability of a system depends on the seamless integration of the various tasks and techniques provided by the application. Selection is most often used to begin object manipulation, and so there must be a smooth transition between the selection and manipulation techniques to be used in an application. The ability to select or manipulate objects while traveling is a powerful integration that decreases time-wasted stopping travel to manipulate an object. Finally, system control should integrate well enough with the other techniques to avoid unintended interactions due to mode errors and technique switching.

12.3.2 INTERACTING IN 3D SPACE

By its very nature, 3D interaction is qualitatively and quantitatively different than standard 2D interaction. Here are some general principles related to the way 3D ITs are implemented.

Take advantage of the user's proprioceptive sense for precise and natural 3D interaction.

Proprioception is a person's sense of the location of the parts of his body no matter how the body is positioned. For example, a driver can easily change gears without looking because of his knowledge of his body and hand position relative to the gearshift. Mine, Brooks, and Sequin (1997) discuss how to take advantage of this sense in VEs by providing body-centered interactions. One possibility is to give users a virtual tool belt on which various tools (e.g., pointer, cutting plane, spray paint) can be hung. Because users know where the various tools are located on their body, they can interact and choose tools much more efficiently and easily without looking away from their work.

Use well-known 2D interaction metaphors if the interaction task is inherently 1D or 2D.

There are many 2D interaction metaphors that can be used directly in or adapted for use in VEs. Pull-down or pop-up menus, 2D buttons, and 2D drag-and-drop manipulation have all been implemented in VEs with success (e.g., Bowman, Hodges, & Bolter, 1998). With these interaction techniques, issues related to reducing the number of DOFs the user must control often arise. When 2D interaction metaphors are used, the provision of a 2D surface for interaction (such as the pen-and-tablet metaphor discussed in Section 12.3.2) can increase precision and efficiency.

Allow two-handed interaction for more precise input.

Most VE interfaces *tie one hand behind the user's back*, allowing only input from a single hand. This severely limits the flexibility and expressiveness of input. By using two hands in a natural manner, the user can specify arbitrary spatial relationships, not just absolute positions in space. For example, the distance between the hands can be used to control the length of a flexible virtual pointer, while twisting the hands can cause the pointer to curve and indicate an obscured object (Olwal & Feiner, 2003). However, it should not be assumed that both hands will be used in parallel to increase efficiency. Rather, the most effective two-handed interfaces are those in which the nondominant hand provides a frame of reference in which the dominant hand can do precise work (Hinckley, Pausch, Profitt, Patten, & Kassell, 1997).

Provide redundant interaction techniques for a single task.

One of the biggest problems facing evaluators of VE interaction is that the individual differences in user performance seem to be quite large relative to 2D interfaces. Some users seem to comprehend complex techniques easily and intuitively, whereas others may never become fully comfortable. Work on discovering the human characteristics that cause these differences is ongoing, but one way to mitigate this problem is to provide multiple interaction techniques for the same task. For example, in a system that provides both steering and automated travel techniques, one user may prefer the steering technique while another user may be more comfortable with automated travel. Of course, the addition of techniques also increases the complexity of the system, and so this must be done with care and only when there is a clear benefit.

12.3.3   Hardware Considerations

A basic question one must ask when designing a VE system regards the choice of input and output devices to be used. Recently, a number of studies have investigated the effects of display and input devices on the user experience (Bowman & McMahan, 2007; Bowman et al., 2012). The following guidelines reflect some of the empirical results.

12.3.3.1   Display Devices

Three common VE display devices, as described in Badcock, Palmisano, and May (2014, Chapter 3, this book), are head-mounted displays (HMDs), spatially immersive displays (SIDs, fully or semisurrounding projected stereo displays, such as the CAVE™), and single-screen stereo displays, such as the Responsive Workbench. These display types have very different characteristics, and interaction with these displays is likely to be extremely different as well.

In semisurrounding SIDs, design the application to minimize virtual rotations.

HMDs and fully surrounding SIDs provide the ability to view the entire VE by physically turning. This is useful because physical turning has been demonstrated to afford faster searching (McMahan, 2011) and better spatial orientation (Bakker, Werkhoven, & Passenier, 1998) than using virtual rotations. Semisurrounding SIDs require the use of virtual rotations, so applications intended for such displays should be designed in a manner to minimize these less-desirable rotations. One way to alleviate this problem is to adopt a vehicle metaphor for navigation so that the user is always facing the center of the display.

For narrow-FOV HMDs, use amplified head rotations when visual search tasks are important.

HMDs that have a narrow FOV require extensive head rotation in order for the user to see the entire environment. This process can be particularly straining in visual search tasks, which require full environment exploration. To reduce fatigue, designers should consider using a nonlinear mapping between the physical rotation of the head and the corresponding environment rotation. A study by Jay and Hubbold (2003) showed that a doubling (2× amplification) of head rotations may lead to better search performance. A more recent evaluation suggests that a tripling (3× amplification) of head rotations may not provide any additional benefits over the smaller 2× amplification (Kopper, Stinson, & Bowman, 2011), although results are inconclusive.

12.3.3.2   Input Devices

Common VE input devices include 6-DOF trackers, continuous posture-recognition gloves, discrete event gloves, pen-like devices, simple button devices, and special-purpose devices such as the Spaceball or force-feedback joysticks. In recent years, modern game consoles have made 3D input devices available to and popular with the public. This movement started with Nintendo's Wii Remote, which used accelerometers integrated with an infrared camera to afford limited 3D tracking, and was furthered by Microsoft's Kinect, which has made controller-less tracking widely available.

Use an input device with the appropriate number of DOFs for the task.

Many inherently simple tasks become more complex if an improper choice of input device is made. For example, toggling a switch is inherently a 1-DOF task (the switch is on or off). Using an interaction technique that requires the user to place a tracked hand within a virtual button (a 3-DOF task) makes it overly complex. A simple discrete event device, such as a pinch glove, makes the task simpler. Of course, one must trade off the reduced DOFs with the arbitrary nature of the various actions the user must learn when using a pinch glove or other such device. In general, designers should strive to reduce unnecessary DOFs when it is practical (Hinckley, Pausch, Goble, & Kassell, 1994). If only a single input device is available, software constraints can be introduced to reduce the number of DOFs the user must control (see Section 12.2.2).

Use physical props to constrain and disambiguate complex spatial tasks.

This guideline is related to the previous discussion about DOFs. Physical props can help to reduce the number of DOFs that the user must control. For example, the pen-and-tablet interaction metaphor (Bowman, Wineman, Hodges, & Allison, 1998) uses a physical tablet (a 2D surface) and a tracked pen. A 2D interface is virtually displayed on the surface of the tablet, for tasks such as button presses, menu selection, and 2D drag and drop (see Figure 12.1). The physical props allow the user to do these tasks precisely, because the tablet surface guides and constrains the interaction to two dimensions.

Physical props can also make complex spatial visualization easier. For example, in the Netra system for neurosurgical planning (Goble, Hinckley, Pausch, Snell, & Kassell, 1995), it was found that

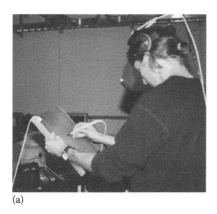

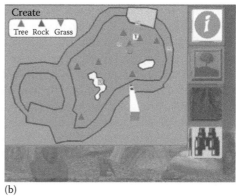

(a)   (b)

FIGURE 12.1 **(See color insert.)** Physical (a) and virtual (b) views of a pen-and-tablet system.

surgeons had difficulty rotating the displayed brain data to the correct orientation when a simple tracker was used to control rotation. However, when the tracker was embedded within a doll's head, the task became much easier because the prop gave orientation cues to the user.

Use absolute devices for positioning tasks and relative devices for tasks to control the rate of movement.

This guideline is well known in desktop computing, but not always followed in VEs. Absolute positioning devices such as trackers will work best when their position is mapped to the position of a virtual object. Relative devices (devices whose positional output is relative to a center position that can be changed), such as joysticks, excel when their movement from the center point is mapped to the rate of change (velocity) of an object, usually the viewpoint. Interaction techniques that use absolute devices for velocity control or relative devices for position control will perform less efficiently. Zhai (1993) extended this idea by comparing isometric and isotonic devices in a 3D manipulation task.

12.4 TECHNIQUES AND GUIDELINES FOR COMMON VE TASKS

12.4.1 SELECTION

Selection is simply the task of specifying an object or set of objects for some action. Most often, selection precedes object manipulation (see Section 12.2), determines the target to travel to (see Section 12.3), or specifies the object of some command (see Section 12.4), such as *delete the selected object*. In interactively complex VEs, selection tasks occur quite often and therefore, efficiency and ease of use are important requirements for this task.

12.4.1.1 Categories of Selection Techniques

Most selection techniques can be categorized by their method of indicating of an object, whether done by touching it with a virtual hand, pointing at it, occluding it, encapsulating it in a volume, or indirectly selecting it. These categories of selection are discussed in the following.

1. *Virtual Hand*
   The most obvious VE selection technique is again the one that mimics real-world interaction—simply touching the desired object with a virtual hand controlled directly by the user's real hand. Within this general metaphor, there are several possible implementations. The virtual hand could simply be a rigid object controlled by a single 6-DOF tracker, or it could be controlled by a glove that recognizes a multitude of hand postures. No matter how implemented, the simple virtual hand metaphor suffers from a serious problem: the user can only select objects in the VE that are actually within

arm's reach. In many large-scale VEs, especially in the design or prototyping application domains, the user will wish to select remote objects—those outside the local area
surrounding the user. An efficient solution is to give the virtual hand a greater range
than the user's physical hand. The simplest example is a technique that uses a linear
mapping between the physical hand movements and the virtual hand movements, so
that for each unit the physical hand moves away from the body, the virtual hand moves
away N units. The go-go technique (Poupyrev et al., 1996) takes a more thoughtful
approach. It defines a radius around the user within which the physical hand is mapped
directly to the virtual hand. Outside that radius, a nonlinear mapping is applied to
allow the virtual hand to reach quite far into the environment, although still only a
finite distance (see Figure 12.2).

2. *Pointing*
Other selection techniques move away from the object-touching metaphor and instead
adopt a pointing metaphor. The fundamental concept of pointing is ray casting, in which a
ray emanates from the user and the first object the ray intersects may be selected. The most
common implementation of ray casting is to attach the ray to the user's virtual hand so that
simple wrist movements allow pointing in any direction (Mine, 1995). Another class of
pointing techniques uses gaze direction for ray casting so that an object can be selected by
placing it in the center of one's field of view.

Pointing techniques generally do not offer high precision, due to natural hand tremor and
tracker jitter (Bowman, Johnson, & Hodges, 1999). Some techniques deal with this issue by
providing an increased control–display (CD) ratio of the pointing ray (e.g., Gallo & Minutolo,
2012; see Section 12.1.3).

3. *Occlusion*
Some techniques use a combination of eye position and hand position for selection (Pierce
et al., 1997). These techniques still use a form of pointing, but the ray emanates from the
eyepoint and passes through the virtual hand position (see Figure 12.3). This is often called
occlusion, framing, or image-plane selection. Occlusion-based selection techniques are
more precise than standard ray casting, as the ray's origin is fixed, and the user adjusts its
orientation by *aiming* with the hand. However, they may also be more fatiguing, since the
user must keep the hand in view (Bowman et al., 1999).

4. *Volume-Based*
In some cases, it may be difficult to point precisely at far away or very small objects.
For such cases, volume-based selection techniques can be used. These techniques involve

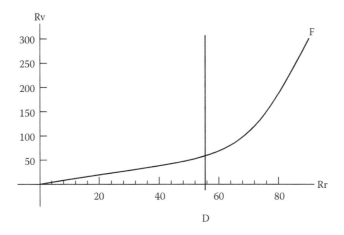

FIGURE 12.2   Mapping function for go-go technique.

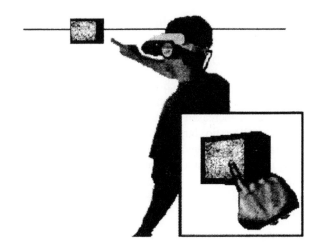

FIGURE 12.3 Occlusion selection.

encapsulating the target object with a volume (e.g., a cone or a box). One challenge with these techniques is how to disambiguate when more than one object is selected. The cone-casting flashlight technique (Liang & Green, 1994) addresses this challenge by selecting the object closest to the centerline of the cone or selecting the object closer to the user if two or more objects are equally distant from the centerline. A more advanced technique is the aperture technique (Forsberg, Herndon, & Zeleznik, 1996), which modifies the flash-light technique to allow the user to control the spread angle of the selection volume by bringing the hand sensor closer or farther away. Another way to determine which object to select from the volume is to perform the selection in multiple stages, through *progressive refinement*. The SQUAD technique (Kopper, Bacim, & Bowman, 2011) is one example of a progressive refinement technique. After a selection volume is used to indicate candidates for selection, an out-of-context quad menu is used to make coarse indications of subsets of the candidates until the remaining candidates are refined to one desired object. In some cases, disambiguation is not necessary, such as using volume-based selection for multiple object selection (Lucas, 2005).

5. *Indirect*

A final selection category includes techniques that achieve selection through an indirect action, such as using two buttons to extend and retract a virtual hand or specifying attributes of the desired object. Indirect selection can be useful when selection is limited to a small set of selectable objects, but generally it is less natural and induces more mental workload.

12.4.1.2 General Guidelines for Designing Selection Techniques

Use the simple virtual hand technique if all selection is within arm's reach.
The simple virtual hand metaphor is natural and works well in systems where all interaction with objects is local. This usually includes VE applications implemented on a tabletop display, where most objects lie on or above the surface of the table.

12.4.1.3 Guidelines for High-Performance Selection Techniques

Use a pointing technique for high-performance selection.
Evaluation (Bowman & Hodges, 1999) has shown that pointing techniques perform more efficiently than virtual hand techniques over a wide range of possible object distances, sizes, and densities.

This is due to the fact that ray casting is essentially 2D (in the most common implementation, the user simply changes the pitch and yaw of the wrist).

Increase precision by increasing the CD ratio.

Selection tasks that require high precision, such as ones involving objects of very small visual size, can be almost impossible to achieve using standard ray casting (Kopper, Bowman, Silva, & McMahan, 2010). Precision can be increased by varying the CD ratio of the selection technique. For example, by increasing the CD ratio, large rotations of a pointing device will lead to small movements of the selection ray within the environment. Increasing the CD ratio can be achieved in a discrete manner by having the user request a high-precision mode (Kopper, 2011) or, dynamically, based on the velocity of the hand rotation (e.g., Gallo & Minutolo, 2012). When using an increased CD ratio to increase precision, designers should keep in mind that an offset between the input device and the ray is introduced.

Use progressive refinement if accuracy is more important than speed.

Though pointing techniques provide for fast selections, hand tremor and tracking jitter can reduce the accuracy of these techniques. When accuracy is the most important performance requirement (e.g., when permanently deleting an object), progressive refinement (Kopper, Bacim, & Bowman, 2011) should be used. Progressive refinement has been shown to allow for the accurate selection of a desired object from a set of candidate objects, even when there are many small or tightly packed objects.

12.4.1.4 Guidelines for More-Natural Selection Techniques

Use a virtual hand technique for more-natural selection.
Virtual hand techniques provide a high level of interaction fidelity due to high biomechanical and control symmetries. This high interaction fidelity makes the virtual hand techniques the most natural for selection.

12.4.1.5 Special Guidelines for Selection Techniques

Consider designing the environment to maximize the perceived size of objects.
Selection errors are affected by both the size and distance of objects, using either ray casting or virtual hand techniques (Bowman & Hodges, 1999). These two characteristics can be combined in the single attribute of visual angle or the perceived size of an object in an image (Kopper et al., 2010). Unless the application requires precise replication of a real-world environment, manipulating the perceived size of objects will allow more efficient selection (Poupyrev, Weghorst, Billinghurst, & Ichikawa, 1997).

Consider using object snapping for environments with limited selectable objects.

As discussed earlier, selection by pointing is very fast and often effective, but it suffers from a lack of precision. If a VE has a small number of selectable objects, pointing accuracy can be greatly increased if the ray *hops* from one object to the next. This style of interaction is called object snapping (de Haan, Koutek, & Post, 2005).

12.4.2 MANIPULATION

Manipulation goes hand in hand with selection. Manipulation refers broadly to the modification of attributes of the selected object. Attributes may include position, orientation, scale, shape, color, or texture. For the most part, research has mainly considered the manipulation of the position and orientation of rigid objects, although some special-purpose applications include object deformation or scaling. Object manipulation tasks have importance in such applications as design, prototyping, simulation, and entertainment, all of which may require environments that can be modified by the user.

12.4.2.1 Categories of Manipulation Techniques

The design space for manipulation techniques is quite large. To provide a simple overview of the techniques already developed in the design space, three categories are presented in the following—virtual hand, proxy, and indirect.

1. *Virtual Hand*

   The most common object manipulation technique is a natural one, in which the selected object is rigidly attached to the virtual hand and moves along with it until some signal is given to release the object. This technique is simple and intuitive, but certain object orientations may require the user to twist the arm or wrist to uncomfortable positions, and it does not use the inherent dexterity of the user's fingers. Some research has focused on more precise and dexterous object manipulation using fingertip control (Kitamura, Yee, & Kishino, 1998). This can be simulated to a degree using a rigid virtual hand if a clutching mechanism is provided.

   As with selection, virtual hand techniques suffer from limitations of reach. Therefore, techniques for remote manipulation are also important. A simple solution is to use the same physical-to-virtual hand mapping used for remote selections. The go-go technique is such a solution (Poupyrev et al., 1996). Another solution is to select using ray casting and then move the virtual hand to the selected object for direct manipulation (Bowman & Hodges, 1997). Yet another solution to remote manipulation is to scale the user or the environment so that the virtual hand, which was originally far from the selected object, is actually touching this object so that it can be manipulated directly (Pierce et al., 1997).

2. *Proxy*

   Another category of techniques solves the remote object manipulation problem by giving the user local proxies of objects to directly manipulate instead of the virtual objects. One example of a proxy technique is called world in miniature (WIM—Stoakley, Conway, & Pausch, 1995). This technique gives the user a small handheld copy of the environment. Direct manipulation of the WIM objects causes the larger environment objects to move as well. This technique is usually implemented using a small 3D model of the space, but a 2D interactive map can also be considered a type of WIM. Another proxy technique is voodoo dolls (Pierce, Stearns, & Pausch, 1999) in which users create handheld proxies of target objects (or groups of objects) called *dolls*. Pinch gloves and image-plane selection are used for creating dolls in users' dominant and nondominant hands. The object corresponding to the doll in a user's nondominant hand provides a stationary frame of reference and does not move while the doll in the dominant hand defines the position and orientation of its corresponding object relative to the stationary frame of reference.

3. *Indirect*

   A final category involves techniques that achieve manipulation through an indirect action, such as using numerical input to specify the exact position of a selected object. Another example of indirect manipulation is the fishing-reel technique (Bowman & Hodges, 1999), which uses two buttons to move selected objects nearer or farther away from the user. Though indirect manipulation techniques are usually unnatural and may require some training to use, they can provide more accuracy than directly manipulating objects in 3D space with a virtual hand or surrogate technique.

12.4.2.2 General Guidelines for Designing Manipulation Techniques

Use general or application-specific constraints for manipulation.
Adding general or application-specific constraints to generic 3D manipulation techniques can help users be more efficient and precise by reducing the complexity of interaction. For example, in an

interior design task, the furniture should remain on the floor. Hence, instead of the user needing to accurately manipulate furniture in 3D space to appear on the floor, vertical positions can be constrained to the plane of the floor. This reduces the complexity of manipulations by an entire dimension and requires less mental workload from the user.

12.4.2.3   Guidelines for High-Performance Manipulation Techniques

Increase precision by reducing the DOFs.
Like application constraints, reducing the DOFs reduces the complexity of manipulations by entire dimensions and requires less mental workload from the user. Instead of leveraging general or application-specific characteristics, however, reducing the DOFs can be achieved by providing widgets that allow the manipulation of one or more related DOFs (Conner et al., 1992; Mine, 1997).

Use indirect manipulation if accuracy is more important than speed.

Unlike virtual hand techniques, indirect manipulations are not based on continuous hand tracking in 3D space. Instead, discrete increments or values can be applied to objects, which provides for better accuracy. For example, the fishing-reel technique, which uses joystick buttons to move objects nearer or farther away from the user, has been shown to provide more accurate object placement, especially if the target is far from the user (Bowman & Hodges, 1999). Moreover, indirect manipulation techniques normally do not require much physical exertion from the user, allowing for extended sessions of use.

12.4.2.4   Guidelines for More-Natural Manipulation Techniques

Use a virtual hand technique for more-natural manipulation.
In addition to providing high levels of interaction fidelity, manipulation techniques that allow the direct positioning and orientation of virtual objects with the user's hand have been shown empirically to perform more efficiently and to provide greater user satisfaction than other manipulation techniques (Bowman & Hodges, 1999).

Increase the probability of training transfer by using a direct mapping.

The training of psychomotor skills usually occurs when direct, one-to-one mappings are used for manipulations since those same mappings will be used in the real world (Taffinder, Sutton, Fishwick, McManus, & Darzi, 1998). In the medical field, laparoscopic simulators that use direct mappings have been demonstrated to provide positive training transfer for surgical psychomotor skills (Grantcharov et al., 2004).

12.4.2.5   Special Guidelines for Manipulation Techniques

Consider using nonisomorphic rotations to reduce clutching.
In some applications, it might be useful to use nonisomorphic rotations (Poupyrev, Weghorst, & Fels, 2000) rather than direct, one-to-one rotation mappings. For example, the user can be allowed to control large ranges of 3D rotations with small rotations of the wrist. These amplified virtual rotations reduce clutching, which has been found to significantly hinder user performance in rotation tasks (Zhai, Milgram, & Buxton, 1996).

12.4.3   Travel

Travel, also called viewpoint motion control, is the most ubiquitous and common VE interaction task—simply the movement of the user within the environment. Travel and wayfinding (the cognitive process of determining one's location within a space and how to move to a desired location—see Darken and Peterson (2014, Chapter 19, this book) make up the task of navigation.

There are three primary tasks for which travel is used within a VE. *Exploration* is travel that has no specific target, but which is used to build knowledge of the environment or browse the space.

Search tasks have a specific target, whose location may be completely unknown (naive search) or previously seen (primed search). Finally, *maneuvering tasks* refer to short, precise movements with the goal of positioning the viewpoint for another type of task, such as object manipulation. Each of these three types of tasks may require different travel techniques to be most effective, depending on the application.

12.4.3.1 Categories of Travel Techniques

Because travel is so universal, a multitude of techniques have been proposed (see Mine, 1995, for a survey of early techniques). Most travel techniques can be categorized as physical locomotion, steering, automated, or manual manipulation. Here, these categories are presented in detail:

1. *Physical Locomotion*
   Physical locomotion techniques leverage physical exertion from the user to travel through the environment. These techniques often mimic real-world locomotion and therefore usually provide a high degree of interaction fidelity. The most direct and obvious physical locomotion is real walking, which requires head tracking to directly map the viewpoint of the VE to the location and orientation of the user's head. This technique is extremely natural and provides vestibular cues, which help the user understand the size of the environment. However, real walking is not always feasible due to space limitations.

   Redirected walking is a type of physical locomotion technique that circumvents space limitations. Redirected walking applies gains to the user's tracked motions to keep the immediately accessible areas of the larger VE within the smaller physical space of the real world (Razzaque, 2005). An alternate, physical locomotion that requires even less space is walking in place (Slater et al., 1995), in which users move their feet to simulate walking without actually translating their bodies. The *human joystick* technique is yet another physical locomotion requiring a small physical space. This technique utilizes the user's position relative to a central zone to create a 2D vector that defines the horizontal direction and velocity of virtual travel (McMahan et al., 2012; Wells, Peterson, & Aten, 1996). Some physical locomotion techniques are enabled by locomotion interface devices, which are discussed in Chapter 11 of this book.

2. *Steering*
   While physical locomotion techniques are concerned with active movements, steering techniques focus on passive movement in which the user remains physically stationary while virtually moving through the VE. Steering involves the continuous control of the direction of motion by the user. That direction can be either a 3D vector for virtual flying or a horizontal direction for virtual walking. Though steering techniques provide less biomechanical symmetry than physical locomotion techniques, they do allow users to virtually travel greater distances with much less (if any) physical exertion. One of the most common steering techniques is gaze-directed steering, which allows users to move in the direction toward which they are looking (Mine, 1995). To avoid coupling gaze direction and travel direction, some designers have chosen to use pointing to indicate steering directions (e.g., Kopper, Ni, Bowman, & Pinho, 2006). Torso-directed steering is another technique that avoids coupling gaze direction and travel direction, but also allows for object selection while traveling.

3. *Automated*
   While physical locomotion and steering techniques give users constant control over travel, automated travel techniques allow users to indicate locations or paths before taking control of their movement through the environment. The simplest automated technique is target-based travel, which allows the user to specify the goal or final destination before the system determines how the viewpoint is moved from the current location to the target

(Bowman, Koller, & Hodges, 1997). In a similar fashion, route-planning techniques allow users to specify a path between the current location and the goal before the system executes that path (Bowman, Davis, Hodges, & Badre, 1999). This might be implemented by drawing a path on a map of the environment or placing markers that the system uses to interpolate a path.

4. *Manual Manipulation*

The final category of travel techniques presented here use manual manipulation to specify viewpoint motion. Ware and Osborne (1990) developed the *camera-in-hand* technique, where the user's hand motion above a map or model of the space specifies the viewpoint from which the scene will be rendered, and the *scene-in-hand* technique, in which the environment itself is attached to the user's hand position. Both of these techniques are exocentric in nature, but manual viewpoint manipulation can also be done from a first-person perspective. Any direct object manipulation technique (see Section 12.2) can be modified so that the user's hand movements affect the viewpoint instead of the selected object. The selected object remains fixed in the environment, and the user moves relative to that object using hand motions. Such a technique can be extremely useful for maneuvering tasks where the user is constantly switching between travel and manipulation.

12.4.3.2   General Guidelines for Designing Travel Techniques

Use smooth transitional motions between locations.

Teleportation, or *jumping*, refers to a target-based travel technique in which the user is moved immediately from the starting position to the target. Such a technique seems very attractive from the perspective of efficiency. However, empirical results (Bowman et al., 1997) have shown that disorientation results from teleportation techniques. Interestingly, all techniques that used continuous smooth motion between the starting position and the target caused little disorientation in this experiment, even when the velocity was relatively high.

Use physical movement techniques for maneuvering tasks.

Almost all immersive VE systems use head tracking to render the scene from the user's point of view. However, in certain applications, especially those in which the user is seated, it might be tempting to specify viewpoint orientation indirectly, for example, by using a joystick. It has been shown that such indirect orientation control, which does not take advantage of proprioception, has a damaging effect on the spatial orientation of the user (Chance, Gaunet, Beall, & Loomis, 1998). Therefore, physical head motion should almost always be used for maneuvering.

Train users in exploration strategies.

Users can easily be trained to perform strategies that will help them obtain spatial knowledge in unfamiliar environments (Bowman et al., 1999). These strategies include flying up to get a bird's-eye view of the environment, traversing the environment in a structured fashion, retracing paths to see the same part of the environment from the opposite perspective, and stopping to look around during travel. These strategies are especially important if spatial orientation is an application requirement.

12.4.3.3   Guidelines for High-Performance Travel Techniques

Use a steering technique for high-performance exploration and search.

Exploring and searching a large-scale VE can require a lot of travel. While physical locomotion techniques are better for maneuvering (as discussed in the guideline earlier), these high-fidelity techniques require more time and can leave the user fatigued when a lot of travel is required. Instead, steering techniques, which require less physical exertion and have been shown to improve spatial orientation (Bowman et al., 1999), should be used for exploration and search.

Use a target-based technique for goal-oriented travel.

If the goal of travel is simply to move to a known location, such as moving to the location of another task, target-based techniques provide the simplest metaphor for the user to accomplish this task. In many cases, the exact path of travel itself is not important; only the goal is important. In such situations, target-based techniques make intuitive sense and leave the user's cognitive and motor resources free to perform other tasks. The use of target-based techniques assumes that the desired goal locations are known in advance or will always coincide with a selectable position in the environment. If this is not true (e.g., the user wishes to obtain a bird's-eye view of a building model), target-based techniques will not be appropriate. Users of these techniques should also pay attention to the guideline regarding smooth transitional motions earlier.

12.4.3.4   Guidelines for More-Natural Travel Techniques

Use a physical locomotion technique for more-natural travel.
As discussed in Section 12.3.1, physical locomotion techniques usually provide high levels of interaction fidelity due to mimicking real-world locomotion. Studies (McMahan et al., 2012; Usoh et al., 1999) have shown that physical locomotion techniques also provide users with a greater sense of presence and are more engaging. Real walking is the most-natural travel technique due to its high biomechanical and control symmetries and should be used when possible.

Use redirected walking for more-natural exploration and search.

While real walking is the most-natural technique for travel, it requires enough physical space to cover the area of the VE. When space limitations are an issue, redirected walking provides the ability to explore and search the areas of the VE beyond the boundaries of the physical space, as discussed in Section 12.3.1. Due to its subtle differences from real walking, redirected walking provides biomechanical and control symmetries comparable to real walking's high degree of interaction fidelity, making it a good choice for more-natural exploration and search.

12.4.3.5   Special Guidelines for Travel Techniques

Consider using wayfinding aids to help the user decide where to move.
Wayfinding aids (e.g., Darken, 1996) may be needed, especially in large-scale VEs where the user is expected to build survey knowledge of the space. Such aids can include maps, signposts, compass markings, and paths. Similarly, for multilevel or multiscale VEs, wayfinding aids can help users avoid becoming disoriented and losing track of the path they should take. Previous studies have shown that a miniature of a higher-level structure, displayed with correct orientation, can improve the user's spatial orientation while inside a smaller-scale space (e.g., Kopper et al., 2006).

Consider using manual manipulation techniques for manipulation-oriented travel.

Manual viewpoint manipulation techniques use object manipulation metaphors (see Section 12.3.1) to specify viewpoint position. Such techniques have been shown experimentally to perform poorly on general travel tasks such as exploration and search (Bowman et al., 1999). However, such techniques allow the use of the same metaphor for both travel and object manipulation tasks, which reduce the mental workload of the user.

12.4.4   System Control

Many of the other interactions found in VE applications fall under the heading of system control. This category includes commands, mode changes, and other modifications of system state. Often, system control tasks are composites of the other universal tasks. For example, choosing a menu item is a selection task, whereas dragging an object to a trash can for deletion is a manipulation task.

12.4.4.1 Categories of System Control Techniques

In this section, several categories of system control techniques are presented including graphical menus, voice commands, gestures, and tools:

1. *Graphical Menus*

   Graphical menus are the most common forms of system control found in VEs, and many of the menu systems that have been developed are simple adaptations of menus from 2D desktop systems. The most simple menu system is a series of labeled buttons that appears in the VE. These may be at a specific location in the environment, or they may be attached to the user for greater availability from any location. Slightly more complex are pull-down menus, which appear only as a label and whose items are revealed when the label is selected (Jacoby & Ellis, 1992). Pop-up menus have also been implemented so that the menu appears at the location of the user's hand for easy access. Other implementations use menus on a virtual surface, such as in the pen-and-tablet metaphor or on the surface of a workbench display. Mine (1997) developed a rotary menu system in which items are chosen by rotating the wrist. This takes advantage of the fact that menu selection is essentially a 1D task, requiring one DOF. TULIP menus (Bowman & Wingrave, 2001) map menu items to fingers, which are selected using pinch gloves. Figure 12.4 shows three example graphical menu systems.

   Many graphical menu systems have faced a set of common problems. One is that the resolution of some VE displays is low, so menus and labels must contain few items and take up considerable display space due to large font sizes. Also, using 3D input is imprecise, so menu items must be large and few submenus can be used. For a command-intensive application such as immersive design, these problems force designers to think of creative ways to issue commands.

2. *Voice Commands*

   The use of voice as a command input has many advantages, including its simple input device (a microphone), freedom to use the hands for other operations, and flexibility of voice input to specify complex commands. Voice also has disadvantages, including limited recognition capability, forcing the user to remember arbitrary command utterances, inappropriateness for specifying continuous quantities, and the distraction to other users in the same room.

   Voice has most often been used to implement simple, discrete commands such as "Save," "Delete," or "Quit," but it has also been used in more complex menu hierarchies. Darken (1994) combined voice input with visual menus so that recall of command names was eliminated. Voice research has also focused on multimodal interaction (Bolt, 1980), where voice and gestures or other input modalities are combined to form richer interactions.

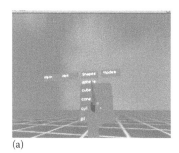 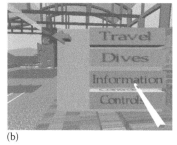 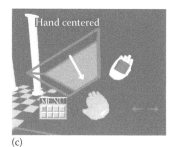

(a) (b) (c)

FIGURE 12.4 Virtual pull-down menu (a), pen-and-tablet menu (b), and rotary menu (c).

3. *Gestures*

   Many of the earliest VE systems used gloves as input and glove gestures to indicate commands. Gesture recognition is covered in detail in Chapter 10 (this book). Advantages of using gestures include the flexibility and number of DOFs of the human hand, the lack of a need for traditional input devices, and the ability to use two hands. However, gestures suffer from many of the same problems as voice, including forced user recall and poor recognition rates. Regardless, gestures have become mainstream with their use in video game consoles.

4. *Virtual Tools*

   Since having a large command space proves problematic for VE applications, developers have looked for ways to reduce the number of commands needed. One way to do this is through the use of virtual tools. Tools are common in many desktop applications as a more direct way to indicate an action to the system. For example, instead of selecting an area of the screen and choosing an *erase* command, the user can simply select an eraser tool and directly erase the desired areas. There have been similar efforts in VEs. One example is Mine's ISAAC system (Mine, 1997), which makes use of a wide variety of tools to perform interactive modifications to a geometric scene. The virtual tricorder (Wloka & Greenfield, 1995) is a multipurpose tool that changes appearance and function depending on system mode and state.

12.4.4.2 General Guidelines for Designing System Control Techniques

Use object-first system control task sequences.

Most system control techniques, with the exception of virtual tools, provide the designer the option of determining the order of tasks to be used for system control. The two most common task sequences used in VEs are action-object and object-action. With action-object task sequences, the user first indicates the action or command to issue and then selects the object to apply that action to. Object-action task sequences function more like context menus in which the user selects an object and then selects an action to apply to it. McMahan and Bowman (2007) have demonstrated that object-first task sequences require less mental workload from users and should be used instead of action-first sequences.

Use an appropriate spatial reference frame.

The placement of graphical menus and virtual tools in the *right* positions can make a big difference in the usability of a system control interface. If system control elements are difficult or impossible to see due to being positioned far away or not oriented toward the user, the user will waste time attempting to access the system control interface. On the other hand, if system control elements are placed in the central focus of the user's viewpoint, occlusion will occur, and the user will be forced to work around the interface.

Structure the functions in an application.

Adopted from desktop applications, hierarchical and context-sensitive menus are good methods for structuring the functionality of an application. Hierarchical menus allow the designer to logically group system commands into high-level menu items that afford users an overview of the application's capabilities. On the other hand, context-sensitive menus avoid overwhelming users with commands by only providing the currently relevant choices.

12.4.4.3 Guidelines for High-Performance System Control Techniques

Use a 2D or 1D selection method when interacting with a graphical menu.

Input to VEs (usually based on 6-DOF trackers) can be quite imprecise. Requiring users to navigate more than one level of menus can lead to frustration and inefficiency. The same holds for menu systems that require 3D input. The use of a 2D physical surface such as the pen-and-tablet metaphor can alleviate some of these problems, as can a 1D solution such as rotary menus.

Verify commands if accuracy is more important than speed.

Like deleting files on a desktop computer, commands should be verified or confirmed when accuracy is the most important requirement. Though this applies to all system control techniques, it is especially true of voice commands and gestures, which are prone to recognition errors. If commands will be verified often, the designer should choose a confirmation method that is quick to execute and requires little precision, such as clicking a button or uttering "Confirm."

12.4.4.4 Guidelines for More-Natural System Control Techniques

Use redundant, intuitive utterances when using voice commands.

Voice commands are often meant to semantically represent real-world actions. For example, a verbal *remove* can indicate the system should remove the currently selected object from the environment. Sometimes, however, users may forget the exact utterance that represents the remove command and instead will attempt to use semantically similar utterances, such as *delete* or *erase*. Because of this, redundant and intuitive utterances should be provided for each system control task when using voice commands.

Use gestures that have high biomechanical symmetry.

Gestures on their own do not give users a reminder of available commands. Instead, users are forced to recall gestural commands. Sometimes, a user may have an intended command in mind but not know what gesture to use to issue the command. Because gestures are often used to mimic real-world actions, it is probable that gestures with high biomechanical symmetries are more likely to be remembered (or even discovered) by users.

12.4.4.5 Special Guidelines for System Control Techniques

Consider using multimodal input.

When selection is used in combination with system control tasks, it may be more efficient and natural to use multimodal interaction (Bolt, 1980; Chapter 21, this book). For example, one may point at an object and then give the voice command "Delete."

Consider reducing the number of commands when using voice or gestural commands.

In many cases, system control in VEs is awkward and distracts from the actual task at hand. If users are struggling with the menu interface, for example, their domain-specific goals will take much longer to realize. This seems to affect immersive VEs more than traditional desktop applications because of the narrow field of view (see Chapter 3, this book), fatigue from spatial interaction (see Chapter 41, this book), and adverse effects of extended VE use (see Chapters 29 through 32, this book). Developers need to look closely at what commands are absolutely necessary and what actions can be done using direct manipulation or virtual tools instead.

12.5 GOAL-BASED DESIGN TRADE-OFFS

In this section, common design goals and their trade-offs are discussed with regard to the preceding guidelines.

12.5.1 PERFORMANCE-BASED DESIGN

Performance-based design focuses on how effectively a task can be completed, with particular emphasis on speed, accuracy, and precision. Unfortunately, these performance aspects are usually improved at the cost of sacrificing naturalism, whether potentially breaking the user's sense of presence, decreasing training transfer due to indirect mappings, or requiring less physical exertion from the user. These trade-offs can be seen when using progressive refinement for more accurate selection, reducing the DOFs to increase the precision of manipulation, or using a target-based technique for goal-oriented travel.

12.5.2 Naturalism-Based Design

While performance-based design regards user performance as the most important design factor, naturalism-based design focuses on creating an application and interaction techniques that mimic the real world. Usually, this approach is used for achieving high levels of presence or training transfer. In turn, this approach often requires more concentration and effort from the user for completing tasks, as opposed to performance-based design, which facilitates tasks to reduce user effort. These trade-offs are apparent with using direct mappings for manipulation, requiring physical locomotion for travel, and using gestures that have a high biomechanical symmetry.

12.5.3 Entertainment-Based Design

For some VEs, the design goal is to simply create a fun application. As seen with modern video games, this approach is usually achieved by creating interaction techniques that have a high level of naturalism, but requires less effort from the user to successfully create a task. For example, consider the baseball minigame in Wii Sports. Instead of requiring the user to accurately swing the Wii Remote in all six DOFs and with a large amount of force to hit the virtual baseball out of the park, the minigame simply requires a large acceleration along one axis at the correct time. Entertainment-based design is normally best accomplished by using an interaction technique with a high biomechanical symmetry, like the virtual hand technique for grabbing objects or a punching gesture for a boxing task, and reducing the DOFs required to successfully complete the task. Essentially, entertainment-based design is a hybrid of the performance-based and naturalism-based design approaches.

12.6 EVALUATION AND APPLICATION

This chapter has focused on the design of VE interaction techniques with high performance, but design is not the end of the story. Evaluation and application are also necessary components to the entire process of determining the appropriate interaction techniques for a VE application.

Following the guidelines and principles in this chapter should lead to interaction techniques that are well suited for the performance requirements of VE applications. However, because 3D interaction is still nonstandardized and diverse, one cannot guarantee performance levels no matter how many guidelines are followed. For this reason, evaluation and assessment of VE interaction is essential. Chapter 34 (this book) covers the topic of usability evaluation in detail. Application designers need to test their systems with members of the intended user population to ensure that required performance levels are met.

For those who are designing new VE interaction techniques and performing basic research, the aspect of application must be kept in mind. Future research should focus on developing techniques that meet the performance requirements of current and proposed real-world VE applications. For examples of such applications, see Chapters 36 through 49 (this book).

12.7 CONCLUSIONS

This chapter has been intended as a practical guide, both for researchers and developers of interactive VE applications. It is essential to remember that most of the guidelines and principles presented here are the results of formal evaluation and that further guidelines should also be the product of careful testing.

This chapter emphasized the concept of effective 3D ITs, with a broad definition of effectiveness. In this area, user comfort, enjoyment, ease of use, and other subjective parameters are in some cases more important than more traditional performance measures such as speed and accuracy. Application developers must carefully consider the requirements of their systems before choosing interaction techniques.

Another consideration is the required naturalism of the interface. Certain applications require high levels of interaction fidelity. However, other applications with different requirements may benefit from the use of magic and hypernatural techniques that extend the user's real-world capabilities.

Finally, the issue of standards should be addressed. There has been a great deal of discussion about the possibility of a standard interface or set of interaction techniques for VEs, much like the desktop metaphor for personal computers. Time may prove the authors wrong, but the current situation appears to suggest that it will not be fruitful to pursue a standardized interface for VE interaction. Most of the types of VE applications that are currently in use or that have been proposed are highly specialized and domain specific, and they exhibit a wide range of interaction performance requirements. There is currently no killer application for VEs. Furthermore, evaluation has shown time and time again that the optimal interaction technique is not absolute but is instead application and task dependent. For these reasons, the authors advocate the development of VE interaction techniques that are optimized for particular tasks and applied to particular systems.

ACKNOWLEDGMENTS

The authors would like to thank the following people for their help, support, and discussion regarding the issues in this chapter: Eric Ragan, Felipe Bacim, Tao Ni, Chad Wingrave, Nicholas Polys, Jian Chen, Larry Hodges, Donald Allison, Drew Kessler, Rob Kooper, Donald Johnson, Albert Badre, Elizabeth Davis, Ivan Poupyrev, Joseph LaViola, Ernst Kruijff, Mark Mine, Matthew Conway, Jeffrey Pierce, Andrew Forsberg, Ken Hinckley, and other members of the 3DUI mailing list.

REFERENCES

Badcock, D. R., Palmisano, S., & May, J. G. (2014). Vision and virtual environments. In K. S. Hale & K. M. Stanney (Eds.), *Handbook of Virtual Environments* (2nd ed., pp. 39–86). New York, NY: CRC Press.

Bakker, N., Werkhoven, P., & Passenier, P. (1998). Aiding orientation performance in virtual environments with proprioceptive feedback. In *Proceedings of the Virtual Reality Annual International Symposium* (pp. 28–33). Atlanta, GA: IEEE Computer Society Press.

Bolt, R. (1980). "Put-that-there": Voice and gesture at the graphics interface. In *Proceedings of SIGGRAPH* (pp. 262–270). Los Angeles, CA: ACM Press.

Bowman, D., Davis, E., Hodges, L., & Badre, A. (1999). Maintaining spatial orientation during travel in an immersive virtual environment. *Presence: Teleoperators and Virtual Environments, 8*(6), 618–631.

Bowman, D., & Hodges, L. (1997). An evaluation of techniques for grabbing and manipulating remote objects in immersive virtual environments. In *Proceedings of the ACM Symposium on Interactive 3D Graphics* (pp. 35–38). Providence, RI: ACM Press.

Bowman, D., & Hodges, L. (1999). Formalizing the design, evaluation, and application of interaction techniques for immersive virtual environments. *Journal of Visual Languages and Computing, 10*(1), 37–53.

Bowman, D., Hodges, L., & Bolter, J. (1998). The virtual venue: User-computer interaction in information-rich virtual environments. *Presence: Teleoperators and Virtual Environments, 7*(5), 478–493.

Bowman, D., Johnson, D., & Hodges, L. (1999). Testbed evaluation of VE interaction techniques. In *Proceedings of the ACM Symposium on Virtual Reality Software and Technology* (pp. 26–33). London, U.K.: ACM Press.

Bowman, D., Koller, D., & Hodges, L. (1997). Travel in immersive virtual environments: An evaluation of viewpoint motion control techniques. In *Proceedings of the Virtual Reality Annual International Symposium* (pp. 45–52). Albuquerque, NM: IEEE Computer Society Press.

Bowman, D., & McMahan, R. (2007). Virtual reality: How much immersion is enough? *IEEE Computer, 40*(7), 36–43.

Bowman, D., McMahan, R., & Ragan, E. (2012). Questioning naturalism in 3D user interfaces. *Communications of the ACM, 55*(9), 78–88.

Bowman, D., Wineman, J., Hodges, L., & Allison, D. (1998). Designing animal habitats within an immersive VE. *IEEE Computer Graphics and Applications, 18*(5), 9–13.

Bowman, D., & Wingrave, C. (2001). Design and evaluation of menu systems for immersive virtual environments. In *Proceedings of IEEE Virtual Reality* (pp. 149–156). Yokohana, Japan: IEEE Computer Society Press.

Brooks, F., Airey, J., Alspaugh, J., Bell, A., Brown, R., Hill, C., Nimscheck, U., Rheingans, Rohlf, J., Smith, D., Turner, D., Varshney, A., Wang, Y., Weber, H., & Yuan, X. (1992). Final technical report: Walkthrough project. National Science Foundation, TR92-026.

Chance, S., Gaunet, F., Beall, A., & Loomis, J. (1998). Locomotion mode affects the updating of objects encountered during travel. *Presence: Teleoperators and Virtual Environments, 7*(2), 168–178.

Chertoff, D., & Schatz, S. (2014). Beyond Presence. In K. S. Hale & K. M. Stanney (Eds.), *Handbook of Virtual Environments* (2nd ed., pp. 855–870). New York, NY: CRC Press.

Conner, B., Snibbe, S., Herndon, K., Robbins, D., Zeleznik, R., & van Dam, A. (1992). Three-dimensional widgets. In *Proceedings of the ACM Symposium on Interactive 3D Graphics* (pp. 183–188). Los Angeles, CA: ACM Press.

Darken, R. (1994). Hands-off interaction with menus in virtual spaces. In *Proceedings of SPIE, Stereoscopic Displays and Virtual Reality Systems, 2177* (pp. 365–371). Bellingham, WA: SPIE.

Darken, R. (1996). Wayfinding behaviors and strategies in large virtual worlds. In *Proceedings of CHI* (pp. 142–149). Los Angeles, CA: ACM Press.

Darken, R. P., & Peterson, B. (2014). Spatial orientation, wayfinding, and representation. In K. S. Hale & K. M. Stanney (Eds.), *Handbook of Virtual Environments*. (2nd ed., pp. 467–492). New York, NY: CRC Press.

Dede, C. (2009). Immersive interfaces for engagement and learning. *Science, 323*(5910), 66–69.

de Haan, G., Koutek, M., & Post, F. (2005). IntenSelect: Using dynamic object rating for assisting 3D object selection. In *Proceedings of 9th Immersive Projection Technology (IPT) Workshop and 11th Eurographics VE (EGVE) Workshop* (pp. 201–209). Aalborg, Denmark.

Foley, J. (1979). A standard computer graphics subroutine package. *Computers and Structures, 10*, 141–147.

Forsberg, A., Herndon, K., & Zeleznik, R. (1996). Aperture based selection for immersive virtual environments. In *Proceedings of ACM Symposium on User Interface Software and Technology* (pp. 95–96). Los Angeles, CA: ACM Press.

Gabbard, J., & Hix, D. (1998). *Usability engineering for virtual environments through a taxonomy of usability characteristics* (Unpublished masters thesis). Virginia Tech, Blacksburg, VA.

Gallo, L., & Minutolo, A. (2012). Design and comparative evaluation of smoothed pointing: A velocity-oriented remote pointing enhancement technique. *International Journal of Human-Computer Studies, 70*(4), 287–300.

Goble, J., Hinckley, K., Pausch, R., Snell, J., & Kassell, N. (1995, July). Two-handed spatial interface tools for neurosurgical planning. *IEEE Computer, 28*(7), 20–26.

Grantcharov, T., Kristiansen, V., Bendix, J., Bardram, L., Rosenberg, J., & Funch-Jensen, P. (2004). Randomized clinical trial of virtual reality simulation for laparoscopic skills training. *British Journal of Surgery, 91*(2), 146–150.

Gutiérrez, F., Pierce, J., Vergara, V., Coulter, R., Saland, L., Caudell, T., … Alverson, D. (2007) The effect of degree of immersion upon learning performance in virtual reality simulations for medical education. *Studies in Health Technology and Informatics, 125*, 155–160.

Hinckley, K., Pausch, R., Goble, J., & Kassell, N. (1994). Design hints for spatial input. In *Proceedings of the ACM Symposium on User Interface Software and Technology* (pp. 213–222). Los Angeles, CA: ACM Press.

Hinckley, K., Pausch, R., Profitt, D., Patten, J., & Kassell, N. (1997). Cooperative bimanual action. In *Proceedings of CHI* (pp. 27–34). Los Angeles, CA: ACM Press.

Hodges, L., Rothbaum, B., Kooper, R., Opdyke, D., Meyer, T., North, M., … Williford, J. (1995). Virtual environments for treating the fear of heights. *IEEE Computer, 28*(7), 27–34.

Jacoby, R., & Ellis, S. (1992). Using virtual menus in a virtual environment. In *Proceedings of SPIE, Visual Data Interpretation, 1668* (pp. 39–48). Bellingham, WA: SPIE.

Kennedy, R., Lane, N., Berbaum, K., & Lilienthal, M. (1993). A simulator sickness questionnaire (SSQ): A new model for quantifying simulator sickness. *International Journal of Aviation Psychology, 3*(3), 203–220.

Kitamura, Y., Yee, A., & Kishino, F. (1998). A sophisticated manipulation aid in a virtual environment using dynamic constraints among object faces. *Presence: Teleoperators and Virtual Environments, 7*(5), 460–477.

Jay, C., & Hubbold, R. (2003). Amplifying head movements with head-mounted displays. *Presence: Teleoperators and Virtual Environments, 12*(3), 268–276.

Kopper, R. (2011). *Understanding and improving distal pointing interaction* (PhD dissertation). Virginia Tech, Blacksburg, VA.

Kopper, R., Bacim, F., & Bowman, D. (2011). Rapid and accurate 3D selection by progressive refinement. In *Proceedings of IEEE Symposium on 3D User Interfaces* (pp. 67–74). Piscataway, NJ: IEEE Computer Society Press.

Kopper, R., Bowman, D., Silva, M., & McMahan, R. (2010). A human motor behavior model for distal pointing tasks. *International Journal of Human-Computer Studies*, *68*(10), 603–615.

Kopper, R., Ni, T., Bowman, D., & Pinho, M. (2006). Design and evaluation of navigation techniques for multiscale virtual environments. In *Proceedings of IEEE Virtual Reality* (pp. 175–182). Piscataway, NJ: IEEE Computer Society Press.

Kopper, R., Stinson, C., and Bowman, D. (2011). Towards an understanding of the effects of amplified head rotations. In *Proceedings of the Workshop on Perceptual Illusions in Virtual Environments,* Singapore, (pp. 10–15).

Liang, J., & Green, M. (1994). JDCAD: A highly interactive 3D modeling system. *Computers and Graphics*, *18*(4), 499–506.

Lucas, J. (2005). *Design and evaluation of 3D multiple object selection techniques* (Master's thesis). Virginia Tech, Blacksburg, VA.

McMahan, R. (2011). *Exploring the effects of higher-fidelity display and interaction for virtual reality games* (PhD dissertation). Virginia Tech, Blacksburg, VA.

McMahan, R., & Bowman, D. (2007). An empirical comparison of task sequences for immersive virtual environments. In *Proceedings of IEEE Symposium on 3D User Interfaces* (pp. 25–32). Piscataway, NJ: IEEE Computer Society Press.

McMahan, R., Bowman, D., Zielinski, D., & Brady, R. (2012). Evaluating display fidelity and interaction fidelity in a virtual reality game. *IEEE Transactions on Visualization and Computer Graphics*, *18*(4), 626–633.

Mine, M. (1995). *Virtual environment interaction techniques* (Technical Report No. TR95-018). Chapel Hill, NC: University of North Carolina.

Mine, M. (1997). ISAAC: A meta-CAD system for virtual environments. *Computer-Aided Design*, *29*(8), 547–553.

Mine, M., Brooks, F., & Sequin, C. (1997). Moving objects in space: Exploiting proprioception in virtual-environment interaction. In *Proceedings of SIGGRAPH* (pp. 19–26). Los Angeles, CA: ACM Press.

Nielsen, J., & Molich, R. (1992). Heuristic evaluation of user interfaces. In *Proceedings of CHI* (pp. 249–256). Los Angeles, CA: ACM Press.

Norman, D. (1990). *The design of everyday things*. New York, NY: Doubleday.

Olwal, A., & Feiner, S. (2003). The flexible pointer: An interaction technique for augmented and virtual reality. In *User Interface Software and Technology Conference Supplement* (pp. 81–82). Los Angeles, CA: ACM Press.

Pierce, J., Forsberg, A., Conway, M., Hong, S., Zeleznik, R., & Mine, M. (1997). Image plane interaction techniques in 3D immersive environments. In *Proceedings of the ACM Symposium on Interactive 3D Graphic* (pp. 39–44). Providence, RI: ACM Press.

Pierce, J., Stearns, B., & Pausch, R. (1999). Voodoo dolls: Seamless interaction at multiple scales in virtual environments. In *Proceedings of ACM Symposium on Interactive 3D Graphics* (pp. 141–145). Los Angeles, CA: ACM Press.

Poupyrev, I., Billinghurst, M., Weghorst, S., & Ichikawa, T. (1996). The go-go interaction technique: Nonlinear mapping for direct manipulation in VR. In *Proceedings of the ACM Symposium on User Interface Software and Technology* (pp. 79–80). Los Angeles, CA: ACM Press.

Poupyrev, I., Weghorst, S., Billinghurst, M., & Ichikawa, T. (1997). A framework and testbed for studying manipulation techniques for immersive VR. In *Proceedings of the ACM Symposium on Virtual Reality Software and Technology* (pp. 21–28). Los Angeles, CA: ACM Press.

Poupyrev, I., Weghorst, S., & Fels, S. (2000). Non-isomorphic 3D rotational interaction techniques. In *Proceedings of CHI* (pp. 540–547). Los Angeles, CA: ACM Press.

Rand, D., Kizony, R., & Weiss, P. L. (2008). The sony playstation II eye toy: Low cost virtual reality for use in rehabilitation. *Journal of Neurologic Physical Therapy*, *32*, 155–163.

Razzaque, S. (2005). *Redirected walking* (PhD dissertation). University of North Carolina, Chapel Hill, NC.

Slater, M., Usoh, M., & Steed, A. (1994). Depth of presence in virtual environments. *Presence: Teleoperators and Virtual Environments, 3*(2), 130–144.

Slater, M., Usoh, M., & Steed, A. (1995). Taking steps: The influence of a walking technique on presence in virtual reality. *ACM Transactions on Computer-Human Interaction*, *2*(3), 201–219.

Smith, R. (1987). Experiences with the alternate reality kit: An example of the tension between literalism and magic. In *Proceedings of ACM CHI+GI* (pp. 61–67). Los Angeles, CA: ACM Press.

Stoakley, R., Conway, M., & Pausch, R. (1995). Virtual reality on a WIM: Interactive worlds in miniature. In *Proceedings of CHI* (pp. 265–272). Los Angeles, CA: ACM Press.

Taffinder, N., Sutton, C., Fishwick, R., McManus, I., & Darzi, A. (1998). Validation of virtual reality to teach and assess psychomotor skills in laparoscopic surgery: Results from randomised controlled studies using the MIST VR laparoscopic simulator. *Studies in Health Technology and Informatics, 50*, 124–130.

Usoh, M., Arthur, K., Whitton, M., Bastos, R., Steed, A., Slater, M., & Brooks, F. (1999). Walking > Walking-in-place > Flying in virtual environments. In *Proceedings of SIGGRAPH* (pp. 359–364). Los Angeles, CA: ACM Press.

Ware, C., & Osborne, S. (1990). Exploration and virtual camera control in virtual three-dimensional environments. In *Proceedings of the ACM Symposium on Interactive 3D Graphics. Computer Graphics, 24*(2), 175–183.

Wells, M., Peterson, B., & Aten, J. (1996). The virtual motion controller: A sufficient-motion walking simulator. In *Proceedings of IEEE Virtual Reality Annual International Symposium* (pp. 1–8). Washington, DC: IEEE Computer Society Press.

Witmer, B., & Singer, M. (1998). Measuring presence in virtual environments: A presence questionnaire. *Presence: Teleoperators and Virtual Environments, 7*(3), 225–240.

Wloka, M., & Greenfield, E. (1995). The virtual tricorder: A unified interface for virtual reality. In *Proceedings of the ACM Symposium on User Interface Software and Technology* (pp. 39–40). Pittsburgh, PN: ACM Press.

Zhai, S. (1993, October). Investigation of feel for 6-DOF inputs: Isometric and elastic rate control for manipulation in 3D environments. In *Proceedings of the Human Factors and Ergonomics Society 37th Annual Meeting*. Seattle, WA.

Zhai, S., Milgram, P., & Buxton, W. (1996). The influence of muscle groups on performance of multiple degree-of-freedom input. In *Proceedings of CHI* (pp. 308–315). Los Angeles, CA: ACM Press.

13 Technological Considerations in the Design of Multisensory Virtual Environments
How Real Does It Need to Be?

Brian D. Simpson, Jeffrey L. Cowgill, Jr.,
Robert H. Gilkey, and Janet M. Weisenberger

CONTENTS

13.1 INTRODUCTION

What makes a virtual environment (VE) feel real? How real does the VE need to be? These are questions that designers of VEs must ask either explicitly or implicitly. The answers to these questions will depend to a very large degree on the intended use of the environment (e.g., basic research, test and evaluation, training, entertainment). Moreover, each real environment and each intended use will likely suggest different sensory, aesthetic, emotional, and cognitive requirements for the corresponding VE. The designer's ability to meet these requirements will be limited by the available technology and its cost and also by the simple fact that the users know they are participating in a simulation. This chapter broadly considers technological and practical limits on multisensory VEs and the implications of these limits. The current discussion is framed in terms of the requirements for simulating the environment that an Air Force Pararescueman, or Pararescue Jumper (PJ), might encounter while performing a combat search and rescue (CSAR) mission.

We chose this domain in part because the CSAR mission has been one focus of our research in recent years and in part because the PJ's task is inherently complex, dynamic, multisensory, physically and psychologically demanding, and thereby difficult to simulate. The capabilities and limitations of current and foreseeable technology for capturing key elements of this environment can be readily understood and are representative of the issues that will be encountered in other similarly complex domains.

13.2 COMBAT SEARCH AND RESCUE MISSION

The PJ is a part of an elite group of United States Air Force Special Operations Forces known as the Guardian Angel Weapons System, whose primary mission is focused on the rescue and recovery of personnel from dangerous, enemy-controlled, and isolated environments. The diversity of situations in which they must operate requires them to complete broad training in parachuting, mountaineering, deep-water dive and rescue operations, close-quarters combat, and emergency medical treatment. They are required to carry up to 150 lb of equipment to support rescue, recovery, survival, battle, extraction, and medical care. When not in direct combat operations, PJs may be deployed in civil search and rescue, disaster response, medical evacuation, and humanitarian relief. They are among the most elite of all military teams, working alongside US Navy SEALS, Army Rangers, and Marine Reconnaissance teams, and have the skills to survive in the most severe and hazardous of environments.

Although the PJs are involved in a variety of missions, to focus our discussion, consider the following fictional CSAR mission:

> Sergeant Johnson has been a PJ for years, and the sights, sounds, and smells of the HH-60G Pave Hawk helicopter bring back memories of previous CSAR missions and raise his adrenaline as he anticipates the rescue of the downed Black Hawk helicopter crew. As his helicopter hovers low to the ground above the drop site, the signal is given, the doors open, and the team fast ropes down to the street below, in near-complete darkness. Johnson and his team rely on their night-vision goggles to see, but with the brownout from the rotor downwash, they still struggle with situation awareness trying to relate the little they can see with the maps and images they have studied. The helicopter quickly flies away after the last of the team members reach the ground. The low pitch of the receding helicopter always emphasizes the conflicting feelings of excitement and isolation.
>
> As the dust clears, Johnson quickly assesses the location of his teammates. He sees the clock tower he had planned to use as a landmark and quickly moves his team in that direction. Moments later, he hears the unmistakable sound of bullets flying by his head and runs for cover behind a nearby ox cart. The sidearm holstered to his leg slams into a rock as he lands. He hears bullets riddle the cart and crawls behind an adjacent retaining wall for better cover. As bullets continue to hit the cart, he carefully peers above the wall and sees the muzzle flashes in a window across the plaza. He raises his weapon, aims, and fires, providing covering fire as the rest of the team moves out. The firing stops and there is an eerie silence except for a few dogs barking in the distance.
>
> The team reforms into their stack, moving in tactical formation, each man with a hand on the man in front, squeezing to provide hand signals as needed. As the point man turns at the first intersection, he sees a blockade of debris, wrecked cars, and old tires; two hostiles stand guard staring out into the darkness. The team quickly retreats back around the corner, pressing their bodies against the building. The guards don't seem to have seen them. Johnson faintly hears the voices of women and children, most likely civilians—not a good place for a firefight. A new path must be found. The combat rescue officer (CRO) in the helicopter above reports that an alley back 100 m should do the trick. A UAV reports the rooftops clear all the way to the crash site except for one obstacle, a solid concrete wall at the end of the alley. They quickly move through the narrow alley and climb over the wall. They smell the burning helicopter. Rounding the last corner, they see the site, the downed Black Hawk helicopter, still smoking.
>
> The team spreads out into two groups, one securing the crash site. Johnson and a teammate enter the downed helicopter, assessing the injured men inside. Two are dead, and two wounded, one of the wounded is able to walk, and the other is bleeding badly. A head-to-toe inspection discovers a severed artery on the left arm; Sergeant Johnson quickly takes vital signs and starts an IV. The other PJ holds

the wounded soldier as Johnson probes with his index finger to find the severed artery and clamp it. After stabilizing the wounded and packaging the dead, the PJs quickly construct an extraction plan. The soldier with the severed artery is moved out first via a portable stretcher, the other is helped out under his own power. It should be safe to extract the bodies from the wreckage now, but a piece of hot metal touches Johnson's arm, and he recoils in pain. Once everyone is clear, they blow up the Black Hawk wreckage and the team moves toward the extraction point, a clear intersection 100 m away. The higher pitch of the approaching helicopters signals that the end is near. Their helicopters swoop in, and once everyone is on board, they fly away to safety.

The PJ on a CSAR mission faces a wide range of sensory stimulation, physical and cognitive demands, and emotional challenges that a multisensory VE should ideally represent. Because the display choices made for one sensory modality will have consequences for the other modalities, we begin this chapter by separately considering some of the display options, technical problems, and possible solutions available in visual, auditory, haptic, and olfactory modalities and also some of the technologies available for controlling and representing motion in VEs. We then consider the challenges of integrating these displays and technologies into an immersive multisensory VE, which produces a meaningful and compelling experience of presence.

13.3 VISUAL ENVIRONMENT

Although VEs are conceived as multisensory display systems, the visual component has typically had a level of primacy in thought, research, and expenditure. The visual component of our CSAR scenario is critical to the PJ's immersion and sense of presence, and faithfully reproducing the visual cues that support his job in the real world is fundamental to the design process. The dim starlit cityscape, the geometry of the physical objects, the movements of the vehicles and people, the view through night-vision goggles, the condition of the downed aircraft, the appearance of the wounds, and the facial expressions of the wounded are all a part of the visual environment that are needed to accurately simulate the real visual world of the PJ.

When simulating a visual environment in a VE, the developer must consider the capabilities and limitations of the available technology and of the human visual system, the fidelity required to support the specific application, and the broader issues of immersion and presence. Ideally, a VE display technology should meet or exceed the limits of the human visual system: that is, ~210° horizontal by 125° vertical field of view (FoV), *normal* visual acuity of 1 min of arc, color sensitivity of 300 chromaticities per luminance level, and photopic daylight luminance levels of 0.314–3.14 × 10^6 cd/m^2 (Boff & Lincoln, 1988). Display technologies for VEs have largely followed three major tracks of development: large projection arrays (domes, cubes, cave automatic virtual environments [CAVEs®], etc.), head-mounted displays (HMDs), and tiled flat-panel displays (Schmidt, Staadt, Livingston, Ball, & May, 2006). For a large projection VE display system that is 10 ft × 10 ft per screen, the resolution in pixels needed to meet *normal* visual acuity is approximately 4900 × 4900 (~24 million) pixels at 5 ft. An HMD with a 1 in. × 1 in. display and an eye distance of 0.7086 in. would also need approximately 4900 × 4900 (~24 million) pixels. The current state of the art is exceeding 16 million pixels per wall in multiwalled CAVE-like systems (10 ft × 10 ft), 100 million pixels in tiled display walls with 50 or more tiles, and 2.3 million pixels per eye in HMDs. That said, the CSAR scenario includes objects occurring at multiple distances from the PJ that must be displayed with a high degree of fidelity so that, for example, the PJ can accurately identify the sniper in the window at a distance of 50 m as well as accurately treat the wound of a downed airman at 0.5 m. The best display choice for one distance is unlikely to be best at the other, with substantial conflicts between vergence and accommodation cues at some particular distances.

Field of view, color reproduction, and brightness are also fundamental to providing immersive virtual experiences, and the different display technologies each have their own weaknesses and strengths and should be evaluated based on the specific needs of the application. Tiled displays have become more prevalent in recent years because they can be built with very high pixel densities and produce high-fidelity visual scenes that exceed the resolution of the human visual system. Despite this capability, research has shown that gaps between tiles resulting from the bezels on each flat panel can degrade performance on visual tasks (Bi, Bae, & Balakrishnan, 2010). Moreover, it is not clear how these gaps might affect immersion and the sense of presence in VEs. CAVE-like systems can fully surround the user in a visual scene, while HMDs are limited in field of view to approximately 120° horizontal and 50° vertical but are very portable and less costly.

The frame rate determines the temporal resolution of a visual display. For perspective, many modern 35 mm movie projectors display images at 24 frames per second, displaying each frame three times, producing a total of 72 images per second. A high frame rate (>20 Hz; Watson, Spaulding, Walker, & Ribarsky, 1997) is necessary for producing a visual scene in which rapid motion appears smooth and natural, for example, when simulating a scene in which the PJ is flying low to the ground in the helicopter. Display technologies such as projectors and CRT monitors also have an internal hardware refresh rate known as the vertical refresh rate. This rate need not be the same as the frame rate generated by the VE software, but the two processes are frequently synchronized to prevent tearing of the image on screen. A sufficient vertical refresh rate (a minimum of 60 Hz mono, 100 Hz stereo) is necessary for minimizing image flicker. Software design, 3D model size, and operating system setup all play roles in limiting a VE system's frame rates, and system memory and I/O bandwidth influence the degree to which the system can maintain these high rates consistently. The hardware employed to generate the scene also plays a critical role in determining frame rates, and rapid advancements in graphics processing unit (GPU) hardware, in particular, have pushed real-time graphics to levels approaching real-world visuals in some applications. Current generation graphics cards capable of generating over 1.9 trillion triangles per second can be configured as a series of multiple cards in one workstation powered by commercial or open-source operating systems and cost several times less than only 10 years ago (e.g., Advanced Micro Devices, 2013). The 3D stereographic capability (NVidia 3D Vision Pro and AMD HD3D Pro), 30-bit color fidelity, digital video connections (DVI and DisplayPort), and multiple synchronized GPUs (NVIDIA SLI, AMD CrossFire, Genlock, and Framelock) are state-of-the-art technologies found in current systems. These technologies allow for lower-cost image generation (IG) computers with multiple GPUs in each IG and multiple IGs linked together to power large numbers of displays all synchronized and displaying 3D stereographic images.

Modern stereoscopic display technologies typically use HMDs, polarized glasses, or active shutter glasses. They take advantage of the ability of humans to fuse different images presented to the two eyes into a single 3D image. Polarized glasses have a cost advantage and no power requirements; however, active shutter glasses let more light through and provide better picture quality (Wickens, 1990). Although not all VE applications require 3D stereographic images, they tend to increase immersion (Hendrix & Barfield, 1995) and are a key part of most VE systems. Many tasks required of the PJ during the CSAR mission will require accurate depth perception (e.g., jumping for cover, inserting an IV). Choosing the best stereographic solution is problem dependent, and the designer of the VE should consider the impact of color and light filtration, field of view, and cost. Moreover, the choice of display technology (HMD, projection, or panel display) will dictate how stereographics are ultimately implemented in the VE. However, some stereographic solutions (e.g., glasses and HMDs) require the user of the VE to be encumbered by additional equipment that may interfere with the use of props incorporated into the scenario (e.g., the PJ's helmet), can impede user movement within the VE, and may generally reduce the degree to which the PJ feels immersed in the simulation. As stereographic technologies that do not require the user to wear glasses or an HMD continue to mature, more advanced approaches, such as multiview displays and light-field techniques, will become more tenable (Lawton, 2011). Multiview technologies render a 3D image on a 2D display using vertical slices and include lenticular lenses and parallax barriers that block one set of views from each eye. Light-field displays recreate the 3D image by producing the actual pattern of light coming from all parts of a 3D object; examples include volumetric light-field displays and holographic prism technologies.

Although we have not yet met the goal of surpassing the human visual system's perceptual limits in all areas, we now have display technologies available that allow us to present very robust and immersive visual environments. Nevertheless, a number of challenges persist, and design decisions must be made to determine the display capabilities (e.g., resolution, luminance, field of view, frame rate) most critical for achieving the visual environment needed to support the performance requirements of the specific application. Finally, the images presented in most VEs are not *photorealistic*. They typically lack an adequate representation of the illumination of objects. Such a representation requires models of the surface properties of objects and of the light sources (which may themselves be combinations of other reflections) and the ability to produce photopic daylight luminance levels. In a dynamic, user-controlled environment like the CSAR scenario, this is extremely computationally challenging (see Mortensen et al., 2008).

13.4 AUDITORY ENVIRONMENT

Whether for team communication, target detection, threat avoidance, or merely monitoring the battlespace, the PJ relies on information from the auditory environment to establish, enhance, and maintain situation awareness. With a 360° FOV, audition naturally functions as an early warning system, alerting the PJ to the presence and location of critical events, regardless of their position in the environment. Moreover, we have argued that the auditory environment may be the critical element that determines the degree to which the user of a VE has a compelling experience in the virtual world (Gilkey & Weisenberger, 1995). In the following section, we briefly describe the methods for the generation of an auditory world for VEs and review some of the issues and challenges in generating a virtual auditory environment for the CSAR mission (for a more thorough review, see Vorlander & Shinn-Cunningham, 2014; Chapter 4).

The rendering of auditory VEs once required extensive dedicated signal processing hardware but can now be accomplished through a number of software architectures such as OpenAL, Microsoft DirectX, Sound Lab (Wenzel, Miller, & Abel, 2000), and other custom-built applications, with varying degrees of capability. Nevertheless, the basic approach to creating an auditory VE has changed little since the first edition of this handbook was published. In general, there are two broad classes of technology that have been employed: loudspeaker-based systems and headphone-based systems. In each case, the intent of the system is to provide the user with a realistic perception of a 3D acoustic world. This is achieved by trying to generate a sound field at the eardrum that is *equivalent* to the sound field that would result from a real sound in the external environment.

In most loudspeaker-based systems, an auditory VE is generated by varying the relative level and/or timing of the signals reaching the individual loudspeakers in order to generate the illusion of sounds (i.e., phantom images) at locations between the actual sound sources. Simple systems containing a few sources on a horizontal plane (simulating only sounds on that plane) have been extended to larger configurations containing loudspeakers at different vertical positions, allowing for the rendering of images at varying elevations, as employed in vector-based amplitude panning (Pulkki, 1997) and ambisonics (Gerzon, 1985), among other approaches. Notably, these systems are valid over only a limited region of space (the so-called auditory *sweet spot*), the size of which varies with the number and spatial layout of the loudspeakers. Newer technologies have been developed to steer this sweet spot in real time (e.g., Song, Zhang, Florencio, & Kang, 2010), but the techniques to extend this capability to large-area, multiloudspeaker arrays are still under development. Wave field synthesis (Berkhout, 1988) is a technique used to recreate the physical characteristics of a sound field by reconstructing the elements of an actual wavefront using a large number of loudspeakers. This approach produces an auditory environment that is valid over a large region of space and for multiple simultaneous users, thus allowing for a scenario in which the PJ can physically move about a large area and interact intuitively with teammates in the shared virtual space. Currently, however, the computational and hardware requirements limit the viability of this approach for most current interactive VE applications.

The other general approach, headphone-based virtual audio, employs miniature microphones placed in the ears of a listener (or a manikin), to record, from multiple surrounding locations, the direction-specific, frequency-dependent transformations imposed on a free-field sound by the torso, head, and pinna. Those recordings (head-related transfer functions [HRTFs]) are then used to generate filters that are convolved with monophonic stimuli to produce virtual sounds over headphones that are perceived to originate from appropriate locations in virtual space. These systems, when implemented properly, can generate virtual sounds that are indistinguishable from real-world sounds (e.g., Kulkarni & Colburn, 1998; Langendijk & Bronkhorst, 2000) and can support localization performance comparable to that found for free-field sources (Martin, McAnally, & Senova, 2001; Romigh & Brungart, 2009), providing the precision required by the PJ to accurately localize threats and targets (e.g., the sniper), remain spatially oriented in low-visibility situations, and monitor and track the position of teammates and support vehicles.

However, several issues must be considered by the VE developer in order for this quality of display to be obtained. First, great care must be taken during HRTF measurement so that the influence of the recording environment is minimized and the response characteristics of the loudspeakers, microphones, headphones, and the rest of the signal chain are compensated for; indeed, even the simple removal and replacement of headphones on a listener can alter certain characteristics of the stimulus and thus influence the spatial accuracy of the virtual auditory environment. Research has demonstrated that listening through HRTFs different from one's own results in degraded localization performance (Wenzel, Arruda, Kistler, & Wightman, 1993) and a reduced sense of presence in the virtual world (Väljamäe, Larsson, Västfjäll, & Kleiner, 2004). This may prove to be a formidable obstacle for HRTF-based virtual audio environment generation because it is generally impractical to collect HRTFs for each potential user of a VE. However, a number of techniques have recently been developed to overcome the need for recording individualized HRTFs, including HRTF modification based on anthropometric measures (Middlebrooks,1999; Zotkin, Hwang, Duraiswami, & Davis, 2003), artificial enhancement of the spectral features present in generic HRTFs (Brungart & Romigh, 2009), and perceptual matching from a large database of HRTFs (Seeber & Fastl, 2003). Although promising, none can as yet support the same level of performance obtained with individualized HRTFs. Moreover, these techniques can generate artifacts that cause virtual stimuli to sound unnatural.

In addition to the cues required for achieving accurate sound source directionality, it is important to ensure that the relevant cues for source distance are provided. Distance information is critical for maintaining an understanding of the relationships among objects and events present in the CSAR scenario, which include sound sources over a wide range of distances, from the explosions and gunfire at 50 m or more to the whisper of a teammate a few centimeters from the PJ's head. Unfortunately, headphone-based virtual auditory display systems sometimes produce sounds that are heard as inside the head and routinely produce sounds that are perceived to be closer than their intended distance. Cues for simulating source distance include distance-based amplitude variations and changes in the ratio of the direct to reverberant sound (Mershon & King, 1975; Zahorik, Brungart, & Bronkhorst, 2005). Events at distances greater than about 15 m should also include spectral filtering that would arise from atmospheric absorption. For sounds within about 1 m of the PJ, the HRTFs vary with distance (Brungart & Rabinowitz, 1999), and the ability to generate a compelling auditory VE in this region may require the inclusion of multiple sets of HRTFs or the implementation of range-based interpolation algorithms (Duraiswami, Zotkin, & Gumerov, 2004). Finally, in order to provide a realistic simulation of a dynamic acoustic environment, the auditory scene must be updated in real time to account for changes in the position of sound sources (e.g., the Pave Hawk as it flies overhead) as well as changes in the position and orientation of the PJ's head (e.g., as he turns to visually acquire the source of the gunfire). Low-latency head tracking (Brungart, Simpson, & Kordik, 2005) and effective interpolation algorithms are required for smooth, natural motion, and for rapidly moving objects, effects such as Doppler shift should be included to enhance realism.

The discussion thus far has been focused on the rendering of individual sounds, but it is important to also consider the acoustic impact of the environment (e.g., walls, ground, objects) in which these sounds occur. Several methods based on sound field decomposition (image source, beam tracing, etc.; Svensson, 2002) have been developed for real-time modeling of acoustic spaces. In this general approach, each reflection from a surface is modeled as an additional sound source, with properties that define the direction, intensity, and frequency content of those reflections. Because the requirement is to generate, in real time, a complex auditory VE, conservation of computational resources should be a consideration, and prioritizing the detail in which each reflection (sound source) is rendered based on psychoacoustic relevance and perceptual constraints (Min & Funkhouser, 2000; Wakefield, Roginska, & Santoro, 2010) is important. In addition, the use of statistical reverberation estimation techniques (Taylor, Chandak, Antani, & Manocha, 2009) can help to reduce computational requirements, enhance realism, and allow for a greater number of sources in a real-time rendered scene. Most of these efforts have focused on the modeling of indoor spaces; somewhat less attention has been given to modeling the acoustics of outdoor spaces, where much of the CSAR mission takes place. Relatively simple outdoor environments can be simulated using models of propagation and dispersion over terrain, but city environments contain buildings and other large objects, and often, the direct sound path between the source and the listener is occluded. Advanced techniques that account for diffraction effects as well as direct and reflected paths provide more accurate models of the environment but add computational complexity. Finally, more research and development is needed to design models of the acoustic characteristics that exist at the boundary between two environments and the characteristics of sound traveling through different media (e.g., walls, doorways).

Despite what is achievable today in terms of audio fidelity, spatial precision, and acoustic room modeling, auditory VE generation has yet to achieve the level of realism and immersion that has been found with simple recordings made with a binaural manikin. Even without the benefit of individualized HRTFs, head motion, or interactivity, these recordings are so striking that listeners (with eyes closed) routinely experience powerful cross-modal illusions: the sound of a virtual haircut gives the tactile sensation that someone is touching their hair and they *see* a shadow when they hear a virtual person walk by. Conversely, although it appears that current systems have the ability to recreate, in an auditory VE, all aspects of a real-world physical stimulus accurately, and performance on most low-level auditory tasks can come close to what is possible in the real world, we seem to have hit a *realism wall* that cannot be surmounted. Technological advancements have led to faster systems and more powerful, efficient tools for rendering auditory VEs, but there is clearly more to address concerning the perceptual and cognitive requirements. And, for the mission at hand, the goal must be more than just providing the cues for best performance—we must provide the appropriate immersive environment to support a sufficient level of presence. Ultimately, we must strive to render an auditory VE that the PJ feels he is listening *in* rather than listening *to*.

13.5 OLFACTORY ENVIRONMENT

Although typically ignored, olfaction can be a key component of the multisensory virtual experience (see Jones et al., 2014; Chapter 6)—the noxious exhaust fumes from the helicopter, the cloud of dust and sand kicked up by the rotor downwash, and the sweat- and dirt-soaked battle gear. Perhaps more than any other sensory information, olfactory stimulation has an immediacy that transports an individual, in a very visceral way, to the place and time when these smells were first experienced. Olfactory cues can trigger specific memories of objects, people, and events (Herz & Cupchik, 1995), particularly when the associations are with highly charged and traumatic past situations (Vermetten & Bremner, 2003). For the PJ, even the faintest of odors can serve as critical information indicating recent weapon activity, his proximity to an outdoor market, or the prognosis for his patient.

Although olfaction is rarely considered when simulating an environment, the strong link between olfaction and cognitive and emotional responses suggests that the sense of smell may in fact be a crucial component of a VE for recreating the intensity of a CSAR mission. It has been argued that the introduction of smells associated with the actions taken and the objects encountered on a military mission enhances the sense of presence in VEs (Vlahos, 2006), supports the learning of tasks, improves the efficiency of technical training (Washburn, Jones, Vijaya Satya, Bowers, & Cortes, 2003), and can lead to a user having greater spatial awareness in the world, as well as a more complete memory of the environment encountered (Dinh, Walker, Song, Kobayashi, & Hodges, 1999). As a result, the smells associated with battle have been incorporated into simulator-based posttraumatic stress disorder treatment regimens for soldiers returning from the Iraq and Afghanistan wars (Burch, 2012).

However, designers of olfactory displays for VEs face a number of obstacles that stem from unique characteristics of the basic olfactory stimulus and the sensory system itself. In order to simulate the complex olfactory environment of the CSAR mission, which is likely to contain a large number of different odors, it is necessary that the VE system be capable of presenting the required number of different scents into the VE. The difficulty here is that, unlike color, there is no small set of primary chemicals from which a large number of odors may be generated, and so each individual odor that will be presented in the virtual world must exist in a library of individual, premade chemical solutions. This poses a practical limit on the total number of scents that may be experienced by the user in any given scenario. For some situations, having a small number of scents may be sufficient, and for the PJ, there are certain odors that are likely to be encountered in any battlespace and thus should be included as part of a *basic* stimulus set (e.g., the smell of gunpowder and the scent of his own battle gear). However, there are other odors that may be more specific to certain environments and certain missions—smells such as cooking spices, fire, burning tires, and sanitation are more likely to be present in an urban battlespace; odors of pine and decaying vegetation are more common in forested and/or mountainous terrain. Moreover, central to the PJ's mission is the provision of medical care for the wounded, and the scene of the wreckage in the scenario we have described would be filled with odors all too familiar to the PJ—those associated with blood, flesh, infection, and even death, mixed with the smell of burning fuel and hot metal emanating from the downed helicopter. Many of the odors in an olfactory VE would be passively experienced by the PJ, but his training, like any medical training (Spencer, 2006), requires him to learn to use olfactory cues for diagnostic purposes. Thus, the PJ would likely be keenly attuned to the olfactory environment in those critical situations and the implementation of these odors would be central to the simulation. In general, the limited capacity of existing scent delivery systems for VEs necessitates the establishment of a prioritization scheme based on the predicted experiences of the PJ so that the odors assumed to contribute most substantially to the scenario remain, and those that contribute less are eliminated.

Many rudimentary olfactory display systems that have been employed simply diffuse chemical odors into the air within the laboratory space containing the VE. The problem with this approach is that it is difficult to vary the olfactory environment in a way that is consistent with the environment rendered to the other senses, and the smells of the current location will still be present when the user transitions to another point in the virtual scenario. In order for an olfactory display to be convincing, one must have control of the spatial and temporal properties of the display to effectively support the simulation of a CSAR scenario. A number of current approaches are employed for generating odors in a VE, including the use of compressed air to inject liquid scents into spaces, fans to diffuse scents stored in aerosol bottles or from heated scented oils, or, more recently, inkjet-type technologies to generate scents from a set of ingredients. These systems vary in their practicality as well as in the degree to which they can control the quantity and quality of scent provided.

Throughout the scenario, smells in the environment provide the PJ with information regarding objects and activities of which he should be aware and help to guide him along his mission. He knows he must get to the downed helicopter and crew, but they are not in sight; rather, it is the smell of burning fuel and hot metal that direct him to the crash location. The ability to localize a smell is based in part on differences in scent concentration at his two nostrils as an odorant drifts from a particular direction (Kobal, Van Toller, & Hummel, 1989). In the VE, technologies must be in place to support the diffusion of localized odors that will create a cross-lateral odor concentration gradient in order to reach the PJ with perceivable directionality. One approach to controlling the environment involves a system of tubes worn by the user through which chemical odors are released under the user's nose. While this provides some spatial and temporal control, it does mean the user will be tethered. One prototype system that addresses this issue, developed by Yanagida, Kawato, Noma, Tomono, and Tetsutani (2004), tracks the position of a user's nose throughout a VE and uses an air cannon system to present a punctate concentration of odorant particles precisely into the area of the nose at the appropriate time in the scenario. One advantage of this system is that the user of the VE is not encumbered by a worn display; however, this system may not provide the desired control over the general concentration, and dissipation, of odorant, as well as control over the relative concentration of odorant across nostrils, all of which are critical for positioning a scent in a desired location. Controlling the overall concentration may be a critical tool for providing information about the distance of a source from the perceiver, as everyday experiences demonstrate that odors become stronger as one draws nearer to the source. Researchers have demonstrated that users in a VE can find the source of an odor through exploration of the virtual space by following changes in the intensity of the odor (Yamada, Yokoyama, Tanikawa, Hirota, & Hirose, 2006). In the CSAR scenario, the intensity of the many odors associated with the scene of the downed helicopter should increase as a PJ nears the wreckage, and he could use this intensity change to determine the proximity of the wreckage to his current position.

Although the air cannon may not provide the control to vary the type and concentration of odors continuously throughout the VE, another display, known as the Scent Collar (US Patent 7,484,716 B2), may provide this level of control. The Scent Collar is a lightweight, wireless device, which is worn around the neck and contains a small number of canisters (currently four), each of which holds a different chemical scent. Release of the chemicals is controlled so that different scents can be presented to an individual at specified times and intensities, for specified durations, throughout the simulation, and, because the location of the delivery point of the stimulus is always very near the user's nose, very precise spatial and temporal control over the delivery of the stimulus is possible.

The unique demands associated with simulating a dynamic, complex environment for the PJ in a CSAR scenario require a scent delivery system that can efficiently provide a wide variety of olfactory stimuli, with the spatial and temporal control to accurately recreate an olfactory world consistent with the rich multisensory environment required to simulate a CSAR scenario. Although the impact of a realistic olfactory environment may be particularly effective when the simulation involves very emotionally charged situations such as those in the current scenario, such a high-fidelity solution seems more remote at this time than for the other modalities.

13.6   HAPTIC ENVIRONMENT

Substantial advances in haptic technology in the past decade, as well as the development of numerous applications of virtual haptics in research devices and products, suggest a positive future for the use of haptic displays in multisensory VEs. Improvements in computational speed and network bandwidth, together with the development of new algorithms, also suggest an optimistic future for the complexities of haptic rendering of virtual objects and interactions. In this section, requirements for both haptic interfaces and haptic rendering are considered in the context of the CSAR task, together with a very brief assessment of the current state of haptic technology in regard to these requirements (see also Dindar, Tekalp, & Basdogan, 2014; Chapter 5). Specific to medical tasks, Coles, Meglan, and John (2011) provide an excellent review of the current state of technology, comparing the capabilities of numerous force feedback and tactile displays in a wide variety of medical tasks, including palpation, needle insertion, laparoscopy, endovascular procedures, and arthroscopy.

The CSAR task constraints, as described in Section 13.2, indicate that tactile and thermal stimulation would be likely at many points on the body surface as the task was performed. This suggests that the optimal configuration for a haptic display would be a whole-body, wearable haptic interface that could present force, vibratory, and possibly thermal stimulation. There is also a need to track the PJ's movements, both in distance and direction of locomotion and in direction and extent of movements of hands, fingers, and limbs. The optimal haptic interface would also need to incorporate sensors that could provide information about these movements back to a central controller, such that the PJ's environment could be updated appropriately. Because of the requirement of motion, both locomotion and limb, hand, and finger motion in performing different aspects of the CSAR task, the optimal haptic interface would need to be lightweight and flexible in design, and any exoskeleton components should not interfere with user movements. Such an interface is a tall order in the face of present technology. But a number of new developments point the way to a more comprehensive solution to providing haptic feedback for at least some parts of the CSAR scenario. A few examples of relevant new technologies are provided in the following to convey a sense of what is currently possible.

Advances in the field of wearable haptics have been fueled by the development of new, lightweight materials for actuators and networking, as well as by the increasing sophistication of haptics in medical and surgical simulation. Some of the most exciting innovations have arisen through the entrepreneurial development of new commercial products incorporating haptic technology. Whole-body or partial-body displays are of three basic types: exoskeletons, which typically include an armature that permits articulation of limbs and can provide position sensing and appropriate oppositional forces in the VE scenario; actuator arrays built into clothing, belts, vests, or gloves, typically used to provide pressure or vibratory stimulation to users to cue navigation or simulate other sensory experiences; and thin-film microactuator displays that can wrap around fingers or limbs, which can provide mechanoreceptive cues for a variety of virtual tasks.

Recent designs for exoskeletons include those developed at the Technical University of Berlin's Robotics and Biology Laboratory (Kossyk, Dorr, & Kondak, 2010) and at the University of Washington (Rosen & Perry, 2007). Each of these devices provides a lightweight exoskeleton for the arm, offering 3- or 7-degree-of-freedom (DOF) force feedback. Each also permits the user to operate in a relatively large workspace. Rosen and Perry's device places control of the device (the human–machine interface) at the neural level, by implementing a Hill-based muscle model, called a myoprocessor. Stored representations from the model allow prediction of the intended muscle movement, minimizing the delay in actuating the exoskeleton. The use of predictive models can speed processing time substantially, making it more possible to cover the complex and rapidly changing movements that a rescuer in the CSAR task might initiate.

A number of hand exoskeleton interfaces have been developed as well. One example of a hand exoskeleton is the CyberGrasp (Immersion Corporation). The CyberGrasp was designed to fit over a CyberGlove and provide resistive force to each finger. In this way, users could feel the size and shape of virtual objects. A backpack can be used to carry the necessary actuator module, making the system portable and wearable. The grasp forces can be specified differently for each fingertip, making multifinger haptics a possibility. The ability to sense with multiple fingers simultaneously provides a much better simulation of real object manipulation. A similar and potentially more wearable effort was reported by Chinello, Malvezzi, Pacchierotti, and Prattichizzo (2012). Other technologies have been adapted for use in haptic gloves, including pneumatic actuators in the Rutgers II-ND glove, which provides up to 16 N of force at each fingertip (Bouzit, Burdea, Popescu, & Boian, 2002). Its direct drive actuators make cables and pulleys unnecessary and thus make the interface much lighter in weight than other exoskeleton systems. Wu et al. (2012) have used pneumatic technology as well in a next-generation glove, while Cassar and Saliba (2010) implemented magnetorheological actuators in a recent haptic glove. Glove technologies are well suited to the CSAR task, given their lightweight design, wearability, and potential to be used in a variety of task situations across a broad workspace. A glove technology that incorporates both force and vibratory feedback could provide significant tactile stimulation in a VE and create a more realistic and immersive experience.

The use of a haptic glove or exoskeleton display could be coupled with a whole-body display, which could provide stimulation to limbs or torso via arrays of tactors. Developments in the commercial realm for haptic garments have considerable potential for use in VEs. The concept of *generative wardrobe* (Talk2myShirt, 2010) is a *haptic-synth* shirt embedded with haptic-synth cells, a type of compression touch sensor coupled with a miniature haptic actuator. Other haptic wardrobe items include the haptic hug shirt (Rosella & Genz, 2005), a vest that provides force and thermal stimulation. The Touch Sensitive (Tangible Media Group, MIT) is a matrix of clothing elements that employs heat sensors, mechanically driven textural display, and liquid diffusion (Vaucelle & Abbas, 2007; Vaucelle, Bonanni, & Ishii, 2009).

New flexible polymer actuators incorporating an ionic liquid polymer electrolyte can present lightweight and nonintrusive haptic stimulation (Ferre et al., 2006), while Koo et al. (2008) described a thin-film display consisting of eight layers of electroactive polymer actuator films sprayed with electrodes arranged in particular patterns. This display is flexible enough to wrap around the fingertip, extremely power efficient, and relatively easy to manufacture. Such a technology holds immense promise for haptic interfaces in VEs, given its nonintrusive and flexible design. It could be incorporated into many interfaces to add a fingertip stimulation component. Again, the need for large-scale movement of the rescuer in the CSAR task argues for the use of such minimal displays.

As is evident from just these few examples, there is keen interest in the use of arrays of haptic actuators embedded in articles of clothing to convey information and stimulation to users in a variety of contexts. In the CSAR task, the likelihood of PJs coming into contact with objects or other actors in the VE with parts of the body other than the fingertips strongly argues for the use of such articles of clothing. If thermal sensors could be included with pressure or vibratory actuators, a very convincing simulation could potentially be created for many elements of the CSAR task.

The presentation of thermal stimulation is also an important component in many aspects of the CSAR task. In addition to being a basic property of most touched objects that aids in the identification and manipulation of objects, thermal sensations are an integral part of medical care. Further, the realism of encountering objects in the VE that are normally very hot or very cold is considerably reduced if the thermal properties of these objects are not displayed. Finally, aspects of the ambient environment in many CSAR scenarios (e.g., extreme heat or extreme cold) must be managed by the PJs in achieving a successful rescue outcome and would thus ideally be part of the training paradigm. Haptic thermal devices must be able to provide realistic thermal sensations based on the prediction of thermal interactions between the skin and an object during contact, and thus, appropriate models of thermal interaction need to be incorporated into the device hardware and software.

Jones and Ho (2008) summarize recent research on thermal displays, describing both the modeling aspect and the hardware most typically used. Peltier devices have been the most widely used in these displays, typically constructed in a layered design of two ceramic substrates interspersed with a layer of semiconductors. Thermal sensors in the display provide feedback that can be used to control the magnitude of the dc to maintain the desired temperature. Using thermal actuators to present extreme heat could pose ethical difficulties for training in the CSAR task due to the potential for injury. However, possible future devices could take advantage of a particular haptic illusion, known as the thermal grill illusion (Craig & Bushnell, 1994). This effect is created by presentation of interlaced warm and cool bars to the skin, resulting in a sensation of extreme heat but without any associated tissue damage. Incorporation of thermal grill technology into a haptic interface could effectively simulate the sense of extreme heat as might be experienced when the PJ in our CSAR scenario burns his arm.

Concomitant with the need for lightweight, wearable, flexible haptic displays that permit a large range of motion in a virtual space for optimal simulation of the CSAR task is the need for rapid, real-time rendering of the properties of virtual objects that the PJ might encounter. Consensus in the haptic research community suggests that hardware update rates of 1 kHz or greater are necessary for a smooth and realistic virtual haptic experience; thus, the haptic display of virtual information requires fast rendering of the complex environment. This can pose a daunting problem, particularly as the potential objects in the PJ's reach change with locomotion, as the object is contacted with one or more fingers or body parts, and as the virtual scenario and other actors in the environment modify objects within it. However, researchers in haptic rendering techniques have taken advantage of a variety of potential *shortcuts* to make progress in achieving the goal of a seamless representation of a rapidly changing environment. Rapid increases in computational power and networking bandwidth, coupled with the use of predictive models of motion and interaction and with algorithmic approaches like those of Duriez, Dubois, Khedder, and Andriot (2006), hold considerable promise for addressing the challenge of haptic rendering in a dynamic scenario such as the CSAR task.

13.7 LOCOMOTION

The ability to move through, and interact with, a VE in a natural way is central to the user's experience (for a thorough review, see Templeman, Page, & Denbrook, 2014; Chapter 10). In the CSAR scenario, ideally the user would be permitted to control movement including walking, running, jumping, crawling, and/or climbing. In addition, the PJ needs to be able to aim and fire a weapon, use medical devices to treat virtual patients, and interact realistically with the various props and virtual objects incorporated into the environment. In VEs, ease of interaction is an important goal for virtual travel/locomotion techniques and contributes to a user's sense of presence (Welch, Blackmon, Liu, Mellers, & Stark, 1996).

Several different tracking technologies are available for use in VEs. Optical tracking systems like the Advanced Realtime Tracking (ARTTRACK) System and InterSense IS-1200 are scalable 6-DOF systems that can track users in very large spaces. Small space tracking (less than ~10 ft × 10 ft) has also become cheaper and more accurate with advances in acoustic, magnetic, and optical tracking technologies providing accurate position measurements at 120 Hz with more than 5 mm of accuracy. More mature, magnetic technologies such as the Polhemus G^4 have also continued to improve resolution and accuracy. The Microsoft Kinect and other low-cost optical systems have greatly facilitated the implementation of skeletal tracking for virtual travel techniques. In the CSAR task, having skeletal tracking would allow for hand signals to be captured and used as input or transmitted to other users in a distributed VE. With the wide array of options in tracking, choosing the best solution should take into account the resolution/accuracy needed, the ease of calibration, the setup requirements, and the cost. In simulating our CSAR task, there are several key items that need to be tracked in both position and orientation to properly present the VE, including the operator's head, hand, and any real guns, litters, or medical devices that are used as interactive props. The choice of what to track and how to track them also greatly depends on the choice of both the visual display system and the haptic feedback system. An HMD solution within a large open space would require a different tracking system, as compared to a CAVE display system that has a limited real movement space. In many cases, haptic hardware devices, discussed in the previous section, are also integrated into the travel techniques in the simulation. An exoskeleton, as described in the haptic section of this chapter, would provide position information along with haptic feedback and would change what the tracking system would need to track. As discussed in the auditory section of this chapter, tracking systems also serve as inputs to most spatial audio solutions and this requirement must also be considered.

A diverse selection of other input devices can be used in VEs, including joysticks, button boxes, speech recognition, touch surfaces, real object props, and large locomotion devices. At the lower end of cost are basic game controllers like the Sony PlayStation DUALSHOCK3 wireless controller, containing high-quality analog and digital joysticks and buttons. When attached to a 6-DOF tracking system, they can be very robust input devices for virtual travel techniques like *moving where pointing* locomotion models. While not fast enough for traditional walking locomotion models, speech recognition can still be used as an input for some types of virtual travel techniques, particularly useful for teleportation-based travel or in data visualization where the user's hands might be better left open to do other tasks. Touch surfaces have had a huge impact on computing in the last few years since the release of the Apple iPhone. While tablets and smartphones are another option for input, touch surfaces do not need to be limited to handheld devices; rather, they can be entire floors that sense the users' feet positions (Kim, Gračanin, Matković, & Quek, 2009; Visell, Law, Ip, Smith, & Cooperstock, 2010).

When using CAVE-like display systems, it is frequently useful to use real object props as input devices. In helicopter simulations, a helicopter cockpit with real controls can be integrated into the VE, creating a mixed reality. However, large props like entire cockpits and cars are usually very specific to the task and can create challenges when using a VE system for different applications due to weight and storage constraints. Smaller props like guns can be more easily integrated and can create powerful bridges between the real and virtual. The VirTra 300 immersive training platform uses real guns converted to simulate recoil with CO_2 and wired into control computers to present police and military users with very realistic training environments (VirTra, 2013). At the top end of cost are large motion platforms like treadmills and moving floors. Developed by Virtual Space Devices, Inc., the ODT (Darken, Cockayne, & Carmein, 1997) is an omnidirectional treadmill. The US Army research in treadmill technology has made progress in creating an input device that allows for some of the motions required by the CSAR scenario, but not all. Crowell, Faughn, Tran, and Wiley (2006) have also shown that soldiers prefer a foot pedal interface to real walking on an ODT, with no real degradation in task performance. This presents an interesting example of how striving for exact realism might not be necessary or needed to create an immersive and effective VE.

The algorithms and hardware interface drivers are just as important in creating useful virtual travel techniques as the hardware itself. Not only are the software algorithms critical in translating the user's actions into virtual movements in a realistic, controllable way, but they can frequently make up for hardware weaknesses. Update rates (how often is the location of the entity computed) are critical technical considerations when working with locomotion models. This will limit how often the displays will be updated from the point of view of the entity and where the entity will appear to look, sound, and feel from the point of view of other entities. Task demands will dictate the necessary minimum update rate. For example, a low-flying fighter jet traveling at 340 m/s would require an update rate much higher than a walking user traveling at 2 m/s. Choosing the software language and/or APIs in which to implement the virtual travel techniques is also important for optimal performance; programming languages like C++ should be considered over interpreted languages like Java. In addition, there are many companies that provide solutions for virtual travel in VEs, ranging from basic walking models up through flight simulation software based on real aircraft data. The same rules apply for choosing a COTS solution or writing your own; the update rate and how the hardware interfaces with the other software used are important, and both should be considered carefully.

Current locomotion technologies have reached a fairly robust state that can simulate many complex motions in VEs, but there are still some challenges and limitations to overcome. When a single control device is used for multiple subtasks in the CSAR application (walking, running, shooting, etc.), the user may have difficulty focusing on the correct operations. A person has a limited number of fingers, toes, and limbs to control these devices. As designers, we are challenged with creating a set of input devices that feel natural and comfortable to use while allowing the control of all of types of virtual travel required for accomplishing the task. The action of moving through the environment can interact with the action of looking, aiming, and firing a gun. So, if an actual physical gun is used, it will occupy the user's hands, interfering with *moving where pointing* locomotion models. Pedals could possibly be added for the feet, or an ODT might be implemented to control walking. Another option to consider is the use of a large tracked space where the user could physically walk around, interacting with a VE seen through a portable HMD. Although allowing for normal movement is generally desirable, it is not without its challenges. For example, how might one aim a gun while wearing an HMD, how might one deal with virtual spaces larger than the user's physically tracked space, and how should haptic feedback be provided to prevent walking through virtual objects (Bruder et al., 2013; Interrante, Ries, & Anderson, 2007; Peck, Fuchs, & Whitton, 2010).

13.8 IMPLEMENTING A MULTISENSORY VIRTUAL ENVIRONMENT

Our experiences in the real world are inherently multisensory. As such, VE systems should strive to recreate these rich sensory experiences through the use of multimodal display technologies. A number of advanced technologies are employed to immerse users in simultaneous, high-fidelity, multisensory stimulation that replaces/blocks stimulation from the actual environment. Thus far, we have reviewed key issues to consider in the individual modalities: vision, audition, olfaction, haptics, and locomotion. However, when creating a VE, the technological challenges do not exist solely in modal silos, independent of the other modalities. Thus, although the technology available to stimulate the individual senses is often quite effective, combining those technologies to create a fully immersive multisensory VE is not trivial. A viable solution to provide stimulation in one modality may interact with viable solutions for the other senses. Moreover, for maximum realism, multisensory stimulation needs to be spatially and temporally aligned and semantically coherent.

13.8.1 DISPLAY COMPATIBILITY

For our CSAR task, the choice of a visual display system has immediate consequences for the other sensory modalities. A projected display system (e.g., a CAVE or multiwall tiled display) has a number of advantages. It typically provides a very wide FOV, which can be useful for depicting many environments. We have found, for example, that aerial views (e.g., looking out the open side door of a helicopter) can be extremely compelling in our CAVE. In general, CAVEs seem to be relatively unlikely to produce simulator sickness. Real or mocked-up physical objects can often be successfully incorporated into the environment (e.g., a medical manikin for our CSAR task or a cockpit mock-up for a flight simulation), reducing the need to render these elements through visual and other sensory displays. On the other hand, a multisensory VE in a CAVE can create a number of problems. For example, most high-end projectors are quite noisy and can interfere with the presentation of auditory stimuli. If sound is to be delivered through a loudspeaker array, rather than headphones, ridged projection screens may limit sound transmission. Optimal loudspeaker placement may be limited by the need to keep the speakers out of the light path and the projectors out of the sound path. Haptic stimulators, such as exoskeletons, will be clearly visible and may draw the user's attention and/or reduce the sense of presence. A typical CAVE allows normal walking but over a very limited range (~10 ft × 10 ft); it does not allow normal running, jumping, climbing, etc., over substantial distances, as in our CSAR scenario. Treadmills such as the ODT have capabilities to support *natural* walking and running (although it is not obvious how such a design could support motions such as climbing and jumping) but would clearly be visible and have a significant acoustic signature (>82 dB).

HMDs, on the other hand, have the advantage of largely or completely blocking undesirable visual stimuli (i.e., other sensory stimulators like an exoskeleton) typically at the cost of reduced field of view and some additional weight (although the weight of these devices is progressively diminishing). HMDs also allow natural motion in a large tracked space as discussed in Section 13.7. Walking and running are readily supported, but some strategy is needed to keep the subject within the tracked space when the virtual space is much larger, as would be the case in our CSAR task (Peck et al., 2010). Other motions such as jumping, diving for cover, and climbing could be supported to some degree, although general-purpose, on-demand solutions are not obvious.

Body-worn devices such as HMDs, exoskeletons, headphones, and olfactory delivery tubes may have unintended haptic consequences with the very active/physical users in our CSAR simulation and may make certain movements difficult or impossible, particularly if the devices are anchored or tethered. Even with lightweight, untethered devices, motions like diving for cover may damage the device and/or hurt the wearer.

Potentially, one could transition from one type of display to another based on the needs of the scenario, but how best to manage such transitions is not readily apparent. Putting on or removing an HMD will certainly reduce immersion. Similarly, as mentioned, in a CAVE, a physical mock-up like a medical manikin might be a useful and powerful display while the PJ is treating a wounded soldier. But the removal of such a physical mock-up for portions of the scenario when it is not appropriate, such as when the PJ is running through the street, offers challenges. Having a manikin *lying around* that is incongruous with the current scene and that would appear to move through the VE with the PJ would definitely be expected to disrupt immersion.

13.8.2 TEMPORAL SYNCHRONY AND SPATIAL ALIGNMENT

The displays in complex multimodal VE applications like our CSAR scenario are controlled by an array of computers that must communicate and synchronize effectively in order to present a coherent virtual world to the user. If the temporal and spatial attributes of objects and entities are not aligned across modalities and do not respond quickly and *naturally* to the actions of the user, we can expect negative consequences for both performance and presence. While it might seem straightforward simply to choose a single rapid update rate for all modalities, different display systems have different temporal capabilities and the individual senses have different temporal sensitivity/ requirements. For example, a haptic display requires an update rate of about 1 kHz for smooth and realistic texture perception, but an update rate of 20 Hz is adequate for most visual displays. So, ideally, update rates are chosen for the individual modalities that match the specific demands and capabilities but are all integer multiples of each other. Synchronization may be more difficult when delays are introduced, either because the VE is distributed or because complex computational models of physics or psychology operate at slower than real-time rates. If a virtual teammate's synthetic speech and lip movements do not match, we can expect a jarring effect similar to watching bad lip-syncing. Similarly, if the PJ feels the haptic feedback of his arm hitting a wall before or after he sees it hitting the wall, his coordination and sense of presence may both be degraded. In general, failure to take these synchronization and temporal issues into account can give rise to unwanted effects, including simulator sickness and loss of immersion.

13.8.3 PERCEPTUAL ILLUSIONS AND MODAL SYNERGIES

On some level, everything in a VE is an *illusion*, using perspective, stereopsis, and other cues to create a 3D perception from a 2D image, using filtering and time delays to create the perception of a remote sound at a specific location, using force applied through an exoskeleton to creating a perception of heaviness for a virtual object, etc. The VE designer exploits these illusions in order to create a compelling experience of the illusory environment. Multisensory illusions can play a similar role. The illusion of self-motion, or vection, is a fundamental and powerful experience in VEs. Vection is typically achieved by moving the visual scene past the user, eliciting the feeling that the user himself is moving rather than the visual scene. Auditory stimulation alone can also induce illusory self-motion in rotation (Lackner, 1977) and translation (Väljamäe et al., 2004). When combined with the visual information into a multisensory display, the illusion is even stronger. Vibrotactile stimulation on the user's seat and footrest can further enhance the illusion of circular motion generated by a rotating visual scene (Riecke, Schulte-Pelkum, Caniard, & Bulthoff, 2005), and haptic stimulation of the feet has been shown to enhance the illusion of self-motion in the vertical dimension when presented concurrently with a vertically dynamic visual scene (Nordahl, Nilsson, Turchet, & Serafin, 2012). On the other hand, there can be unintended consequences of multisensory stimulation. For example, the ventriloquist effect, in which a visual image of a sound source's location overwhelms auditory information indicating a different source location (e.g., Alais & Burr, 2004), could help to ensure the desired auditory perception, despite a poorly rendered audio display. Similarly, a misaligned (or unintended) visual target will still dominate a veridical audio signal creating a misperception. Providing correct lipreading cues can aid speech communication, but poorly rendered cues can lead to misperceptions (e.g., the McGurk effect, McGurk & MacDonald, 1976). Similarly, if the auditory and visual self-motion information is inconsistent, the sense of vection will be reduced (Väljamäe et al., 2004).

13.8.4 Adaptive Environments

Although the focus in this chapter has been on the quality and capabilities of the multiple sensory displays used to present the VE, it is also important that the information presented through those displays describe a meaningful, consistent, and coherent world. The VE should react realistically to the user's actions. Does the building fall down when the user shoots it with a rocket? Can the user shoot holes in a door and then look through the holes to see what is on the other side? What happens when the user interacts with a virtual person in the world? Is there a sense of relative personal space, will the virtual person say "hello" or get annoyed? In a complex, dynamic, and unpredictable environment like the battlespace, this is likely to be a major issue. As unexpected situations develop, the PJ must adapt and find creative solutions to problems. In order for the virtual environment to appear realistic, it must respond appropriately to these solutions. For example, rather than fight his way through a situation, the PJ might choose to create a distraction by making a Molotov cocktail out of some gauze and a bottle of alcohol in his pack and then throwing it at a pile of rubble a few meters away. If the programmer anticipates these actions, the appropriate scripts can be developed. If not, appropriate models of the physical objects involved and how they interact must have been developed. In this case, that would include a large number of models: a model of the pack and its contents, a model of the bottle and its contents, a model of how the flammable liquid can be ignited and how it burns once ignited, a model of the gauze and how it burns, a model of how the fire spreads when it hits the rubble, etc. Modeling human behavior is potentially even more difficult. A virtual teammate might discuss options for the rescue, respond to the user's facial expressions, behave based on a knowledge of the PJ's likely reactions *learned* from a *shared history* of previous missions, etc. The adaptability of a VE is very important to the user's experience and sense of presence in the VE. A virtual *sandbox* (Onoue & Nishita, 2003) with virtual objects and people who react and interact with both each other and real users in a completely natural way is ideal but presently unattainable. Even approximating this goal is likely to be cost prohibitive in terms of both time and money. In practice, the VE designer will need to make compromises based on the resources available and the needs of the project. Depending on the emphasis of a CSAR study, one might choose to not support deformable terrain and destroyable buildings and instead emphasize better AI to simulate team interactions or better physiological models to support realistic battlefield medicine.

13.9 REALISM AND PRESENCE

Since the first edition of this book, there have been dramatic advances in technology in terms of speed, resolution, accuracy, wearability/comfort, etc. When properly utilized, this technology can produce stunning virtual experiences. Nevertheless, creating a completely realistic simulation, which is indistinguishable from the real-world analog, is a very high bar. For each modality, there remain significant challenges. Combining displays to create an immersive and well-synchronized multisensory VE that describes a believable world is even more difficult. The good news is that solving *all* of these seemingly unsolvable problems is rarely, if ever, needed. A completely realistic virtual world may not be critical for many applications.

13.9.1 Importance of Realism

As noted by Sadowski and Stanney (2002), complete realism may not be required to generate a sense of presence on the part of the user. As long as the user perceives himself or herself as included in the VE and can perceive the virtual self moving as the real self moves and interacts with objects in the environment, a sense of presence is possible (Witmer & Singer, 1998). Perhaps more importantly, there is at least some evidence that training in a VE does transfer to real-world tasks, even if the VE is not totally realistic (Knerr, 2003). This last point provides a strong argument for implementing training scenarios, such as a CSAR task, even in VEs that do not provide a completely realistic simulation. In fact, it has long been acknowledged that some aspects of training might be more effectively realized if the simulation is *not* completely realistic (e.g., Durlach & Mavor, 1995). Rather, critical aspects of the scenario for training effectiveness might be emphasized in size, color, or other characteristics that make them more salient to the trainee, with the idea that when training transfers to the real environment, those aspects would be well trained even if the real environmental signals were less salient.

Some researchers (e.g., Cooke & Shope, 2004; Elliot, Dalrymple, Schiflett, & Miller, 2004; Schiflett & Elliot, 2000) have argued that the most important issue when developing what they call *synthetic task environments* is *psychological* fidelity. Although physical fidelity can be an important component of psychological fidelity, it is often not required. Other aspects of psychological fidelity, which are often more important, include cognitive fidelity (does the task environment require the same underlying processes as the real operational environment and create the same overall cognitive demand?), functional fidelity (are the roles of the participants and the goals of the task similar to the real environment?), and construct fidelity (do the measurements taken accurately reflect the underlying processes that govern performance in the operational environment?). Training specific operator skills may require a great deal of physical fidelity (e.g., an accurate representation of displays and controls), but higher-level processes, such as teamwork, can be examined and/or trained with lower physical fidelity. Moreover, one of the most desirable aspects of conducting research in simulated environments is the high level of experiment control and internal validity that is possible relative to field studies. A simulation that is too realistic or too flexible would almost certainly compromise this advantage.

13.9.2   PRESENCE AND THE CSAR TASK

It seems that the question of how real is real enough must be addressed on a case-by-case basis. The answer will depend on both the purpose of the environment and the resources available. Ideally, the need for realism for effective transfer of training or for the external validity of research findings would be verified experimentally. In practice, the answer will most often depend on the informed judgment of the VE designer.

That said, presence seems likely to be particularly important in CSAR research. Obviously, the PJ's mission is extremely stressful; fear, heroism, duty, and exhilaration are central aspects of his job. We are interested in how PJs perform while doing their job in the real world and whether and how they will benefit from the wearable displays (mainly audio displays) we are designing. The PJ's ability to monitor and utilize our displays is likely to be affected by the *distracting* impact of task stress and the emotions it generates. Similarly, exposure to these emotions before a PJ's first battle experience would seemingly be a useful component of any training scenario. However, fully recreating these emotions outside of the battlespace will be difficult or impossible. If we are able to generate a sense of presence in the user, these emotions are likely to emerge to some degree. *Willing suspension of disbelief* will, of course, be a major contributor to this sense of presence. To the degree that the users think their participation is important (they want the training, they hope that the research will lead to fieldable devices that will benefit themselves or their buddies, etc.) and can *get into* the scenario, small problems with breaks in immersion may not pose a problem. If the users have combat experience, the depicted scenario may take them back to a previously experienced scene and engender past emotions. Some real cost of poor performance (e.g., loss of bragging rights, pay, or duty opportunities) could also add to the level of stress and exhilaration. But in the end, the user knows it is a simulation: no one will die, and no one will get hurt. Real danger, such as from live fire, is practically and ethically unavailable. In the final analysis, this is probably the boundary between simulation and reality—it can never get that real.

13.10   CONCLUSION

In summarizing the capability of VE technology available at that time to simulate a baseball game, Nelson and Bolia (2002) state that "The virtual field of dreams will have to wait" (p. 301). Despite significant advances during the last decade, our conclusion is largely the same. However, this conclusion depends on what your dreams are. Today's VE technologies allow rich multisensory simulations that are likely to be real enough for many purposes. In practice, most VE designs will consider a smaller range of circumstances (e.g., just ground navigation or just medical treatment). They will more selectively strive for fidelity (e.g., just medical instruments or just auditory stimulation). The task will be tailored in part to support the level of internal and external validity the application demands and in part to the budget and technology available. Indeed, these limitations are true of our own simulations of the CSAR mission. The virtual field of dreams continues to be a goal to strive for, but some useful dreams can be realized today.

REFERENCES

Advanced Micro Devices, I. (2013). AMD firePro W9000. Retrieved from http://www.amd.com/us/products/workstation/graphics/ati-firepro-3d/w9000/pages/w9000.aspx#1. Accessed on January 28, 2013.

Berkhout, A. J. (1988). A holographic approach to acoustic control. *Journal of the Audio Engineering Society*, *36*(12), 977–995.

Bi, X., Bae, S.-H., & Balakrishnan, R. (2010). Effects of interior bezels of tiled-monitor large displays on visual search, tunnel steering, and target selection. *Proceedings of the 28th International Conference on Human Factors in Computing Systems—CHI'10*, pp. 65–74, Atlanta, GA, doi:10.1145/1753326.1753337

Boff, K. R., & Lincoln, J. E. (1988). *Engineering data compendium. Human perception and performance* (Vol. 1). Wright-Patterson AFB, OH: Harry G. Armstrong Aerospace Medical Research Laboratory.

Bouzit, M., Burdea, G., Popescu, G., & Boian, R. (2002). The Rutgers Master II-new design force-feedback glove. *IEEE/ASME Transactions on Mechatronics*, *7*(2), 256–263. doi:10.1109/TMECH.2002.1011262

Bruder, G., Steinicke, F., Bolte, B., Wieland, P., Frenz, H., & Lappe, M. (2013). Exploiting perceptual limitations and illusions to support walking through virtual environments in confined physical spaces. *Displays*, *34*, 132–141. doi:10.1016/j.displa.2012.10.007

Brungart, D., & Romigh, G. (2009). *Spectral HRTF enhancement for improved vertical-polar auditory localization*. IEEE Workshop on Applications of Signal Processing to Audio and Acoustics (WASPAA'09). New Paltz, NY, pp. 305–308. doi:10.1109/aspaa.2009.5346479

Brungart, D., Simpson, B., & Kordik, A. (2005). The detectability of headtracker latency in virtual audio displays. *Proceedings of the International Conference on Auditory Display (ICAD05)* (pp. 37–42). Limerick, Ireland.

Brungart, D. S., & Rabinowitz, W. M. (1999). Auditory localization of nearby sources. Head-related transfer functions. *The Journal of the Acoustical Society of America*, *106*(3), 1465–1479. doi:10.1121/1.427180

Burch, A. D. S. (2012). Bringing the smells of war home, via virtual reality. Miami Herald. Retrieved from http://www.miamiherald.com/2012/09/09/2994320/bringing-the-smells-of-war-home.html. Accessed on February 6, 2013.

Cassar, D. J., & Saliba, M. A. (2010, April 26–28). A force feedback glove based on magnetorheological fluid: Preliminary design issues. *Melecon 2010—2010 15th IEEE Mediterranean Electrotechnical Conference* (pp. 618–623). Valletta, Malta. doi:10.1109/MELCON.2010.5476012

Chinello, F., Malvezzi, M., Pacchierotti, C., & Prattichizzo, D. (2012). A three DoFs wearable tactile display for exploration and manipulation of virtual objects. *2012 IEEE Haptics Symposium (HAPTICS)*. Vancouver, British Columbia, Canada. doi:10.1109/HAPTIC.2012.6183772

Coles, T. R., Meglan, D., & John, N. W. (2011). The role of haptics in medical training simulators: A survey of the state of the art. *IEEE Transactions on Haptics*, *4*(1), 51–66. doi:10.1109/TOH.2010.19

Cooke, N., & Shope, S. (2004). Designing a synthetic task environment. In S. Schiflett, L. Elliott, E. Salas, & M. Coovert (Eds.), *Scaled worlds: Development, validation, and applications* (pp. 263–278). Surrey, U.K.: Ashgate.

Craig, A. D., & Bushnell, M. C. (1994). The thermal grill illusion: Unmasking the burn of cold pain. *Science*, *265*(5169), 252–255.

Crowell, H. P. III, Faughn, J., Tran, P., & Wiley, P. (2006). Improvements in the omni-directional treadmill: Summary report and recommendations for future development. Retrieved from http://oai.dtic.mil/oai/oai?verb = getRecord&metadataPrefix = html&identifier = ADA456606. Accessed on February 9, 2013.

Darken, R., Cockayne, W., & Carmein, D. (1997). The omni-directional treadmill: A locomotion device for virtual worlds. *Proceedings of the 10th Annual ACM Symposium on User Interface Software and Technology* (pp. 213–221). Banff, Alberta, Canada.

Dindar, N., Tekalp, A. M., & Basdogan, C. (2014). Haptic rendering and associated depth data. In K. S. Hale & K. M. Stanney (Eds.), *Handbook of virtual environments, design, implementation, and applications* (2nd ed., pp. 115–130). New York, NY: CRC Press.

Dinh, H. Q., Walker, N., Hodges, L., Song, C., and Kobayashi, A. (1999). Evaluating the importance of Multisensory Input on Memory and the Sense of Presence in Virtual Environments. *Proceedings of IEEE Virtual Reality 1999*, Houston, 222–228.

Duraiswami, R., Zotkin, D. N., & Gumerov, N. A. (2004). Interpolation and range extrapolation of HRTFs [head related transfer functions]. *2004 IEEE International Conference on Acoustics, Speech, and Signal Processing* (Vol. 4, pp. iv-45–iv-48). Montreal, Quebec, Canada, doi:10.1109/ICASSP.2004.1326759

Duriez, C., Dubois, F., Kheddar, A., & Andriot, C. (2006). Realistic haptic rendering of interacting deformable objects in virtual environments. *IEEE Transactions on Visualization and Computer Graphics, 12*(1), 36–47. doi:10.1109/TVCG.2006.13

Durlach, N. I., & Maver, A. (1995). *Virtual reality: Scientific and technological challenges*. Washington, DC: National Academies Press.

Elliott, L. R., Dalrymple, M. A., Schiflett, S. G., & Miller, G. C. (2004). Scaling scenarios: Development and application to C4ISR sustained operations research. In S. Schiflett, L. Elliott, E. Salas, & M. Coovert (Eds.), *Scaled worlds: Development, validation, and applications* (pp. 119–133). Surrey, U.K.: Ashgate.

Ferre, M., Barrio, J., Monroy, M., Hölldampf, J., Wang, Z., Juger, J., Vidal, F., Kheddar, A., Chevrot, C., Teyssié, D., Sgambelluri, N., & Scilingo, E. (2006). Immersence Project D3.2.1: Integrated haptic display system prototypes. http://citeseerx.ist.psu.edu/viewdoc/download?doi=10.1.1.107.2602&rep=rep1&type=pdf. Retrieved 1/3/2013.

Gerzon, M. A. (1985). Ambisonics in multichannel broadcasting and video. *Journal of the Audio Engineering Society, 33*(11), 859–871.

Gilkey, R. H., & Weisenberger, J. M. (1995). The sense of presence for the suddenly deafened adult—Implications for virtual environments. *Presence: Teleoperators and Virtual Environments, 4*(4), 357–363.

Hendrix, C., & Barfield, W. (1995). *Presence in virtual environments as a function of visual and auditory cues*. Proceedings of Virtual Reality Annual International Symposium'95. Research Triangle Park, NC, pp. 74–82. doi:10.1109/VRAIS.1995.512482

Herz, R. S., & Cupchik, G. C. (1995). The emotional distinctiveness of odor-evoked memories. *Chemical Senses, 20*(5), 517–528.

Interrante, V., Ries, B., & Anderson, L. (2007). *Seven league boots: A new metaphor for augmented locomotion through moderately large scale immersive virtual environments*. 2007 IEEE Symposium on 3D User Interfaces. Charlotte, NC. doi:10.1109/3DUI.2007.340791

Jones, D. L., Dechmerowski, S., Oden, R., Lugo, V., Wang-Costello, J., & Pike, W. (2014). Olfactory interfaces. In K. S. Hale & K. M. Stanney (Eds.), *Handbook of virtual environments, design, implementation, and applications* (2nd ed., pp. 133–164). New York, NY: CRC Press.

Jones, L. A., & Ho, H.-N. (2008). Warm or cool, large or small? The challenge of thermal displays. *IEEE Transactions on Haptics, 1*(1), 53–70. doi:10.1109/TOH.2008.2

Kim, J., Gračanin, D., Matković, K., & Quek, F. (2009). *iPhone/iPod touch as input devices for navigation in immersive virtual environments*. IEEE Virtual Reality 2009. Lafayette, LA, pp. 261–262. doi: 10.1109/VR.2009.4811045

Knerr, B. W. (2003). Virtual environments for dismounted soldier simulation, training, and mission rehearsal: Results of the FY 2002 culminating event. Retrieved from http://oai.dtic.mil/oai/oai?verb = getRecord&metadataPrefix = html&identifier = ADA417360. Accessed on February 5, 2013.

Koo, I. M., Jung, K., Koo, J. C., Nam, J., Lee, Y. K., & Choi, H. R. (2008). Development of soft-actuator-based wearable tactile display. *IEEE Transactions on Robotics, 24*(3), 549–558. doi:10.1109/TRO.2008.921561

Kossyk, I., Dorr, J., & Kondak, K. (2010). *Design and evaluation of a wearable haptic interface for large workspaces*. 2010 IEEE/RSJ International Conference on Intelligent Robots and Systems. Taipei, Taiwan, pp. 4674–4679. doi:10.1109/IROS.2010.5650890

Kulkarni, A., & Colburn, H. S. (1998). Role of spectral detail in sound-source localization. *Nature, 396*(6713), 747–749.

Lackner, J. R. (1977). Induction of illusory self-rotation and nystagmus by a rotating sound- field. *Aviation Space and Environmental Medicine, 48*, 129–131.

Langendijk, E. H. A., & Bronkhorst, A. W. (2000). Fidelity of three-dimensional-sound reproduction using a virtual auditory display. *The Journal of the Acoustical Society of America, 107*(1), 528–537. doi:10.1121/1.428321

Lawton, G. (2011). 3D displays without glasses: Coming to a screen near you. *Computer*, *44*(1), 17–19. doi: 10.1109/MC.2011.3

Martin, R. L., McAnally, K. I., & Senova, M. A. (2001). Free-field equivalent localization of virtual audio. *Journal of the Audio Engineering Society*, *49*(1/2), 14–22.

McGurk, H., & MacDonald, J. (1976). Hearing lips and seeing voices. *Nature, 264*, 746–748.

Mershon, D. H., & King, L. E. (1975). Intensity and reverberation as factors in the auditory perception of egocentric distance. *Attention, Perception, & Psychophysics, 18*, 409–415.

Middlebrooks, J. C. (1999). Individual differences in external-ear transfer functions reduced by scaling in frequency. *The Journal of the Acoustical Society of America*, *106*(3), 1480–1492. doi:10.1121/1.427176

Min, P., & Funkhouser, T. (2000). Priority-driven acoustic modeling for virtual environments. *Computer Graphics Forum*, *19*(3), 179–188. doi:10.1111/1467–8659.00410

Mortensen, J., Yu, I., Khanna, P., Tecchia, F., Spanlang, B., Marino, G., & Slater, M. (2008). Real-time global illumination for VR applications. *IEEE Computer Graphics and Applications*, *28*(6), 56–64. doi:10.1109/MCG.2008.121

Nelson, W., & Bolia, R. (2002). Technological considerations in the design of multisensory virtual environments: The virtual field of dreams will have to wait. In K. Stanney (Ed.), *Handbook of virtual environments: Design, implementation, and applications* (pp. 301–312). Mahwah, NJ: Lawrence Erlbaum Associates.

Nordahl, R., Nilsson, N. C., Turchet, L., & Serafin, S. (2012). Vertical illusory self-motion through haptic stimulation of the feet. *IEEE VR Workshop on Perceptual Illusions in Virtual Environments* (pp. 21–26). Orange County, CA.

Onoue, K., & Nishita, T. (2003). Virtual sandbox. *11th Pacific Conference on Computer Graphics and Applications, 2003* (pp. 252–259). Proceedings. Canmore, Alberta, Canada. doi:10.1109/PCCGA.2003.1238267

Peck, T. C., Fuchs, H., & Whitton, M. C. (2010). Improved redirection with distractors: A large-scale-real-walking locomotion interface and its effect on navigation in virtual environments. *2010 IEEE Virtual Reality Conference (VR)* (pp. 35–38). Waltham, MA. doi:10.1109/VR.2010.5444816

Pulkki, V. (1997). Virtual sound source positioning using vector base amplitude panning. *Journal of the Audio Engineering Society*, *45*(6), 456–466.

Riecke, B., Schulte-Pelkum, J., Caniard, F., & Bulthoff, H. (2005). Influence of auditory cues on the visually-induced self-motion illusion (circular vection) in virtual reality. *Proceedings of Eighth Annual Workshop on Presence* (pp. 49–57). London, U.K.

Romigh, G. D., & Brungart, D. S. (2009). Real-virtual equivalent auditory localization with head motion. *The Journal of the Acoustical Society of America*, *125*(4), 2690.

Rosella, F., & Genz, R. (2005). Hug shirt: Wearable haptic telecommunication devices for sharing emotional and physical closeness over distance. *ISWC'05 International Symposium on Wearable Computers*. Osaka, Japan.

Rosen, J., & Perry, J. C. (2007). Upper limb powered exoskeleton. *International Journal of Humanoid Robotics*, *04*(03), 529–548. doi:10.1142/S021984360700114X

Sadowski, W., & Stanney, K. (2002). Presence in virtual environments. In K. Stanney (Ed.), *Handbook of virtual environments: Design, implementation, and applications* (pp. 791–806). Mahwah, NJ: Lawrence Erlbaum Associates.

Schiflett, S., & Elliott, L. (2000). Synthetic team training environments: Application to command and control aircrews. In H. F. O'Neill, Jr & D. Andrews (Eds.), *Aircrew training and assessment: Methods, technologies, and assessment* (pp. 313–338). Mahwah, NJ: Lawrence Erlbaum Associates.

Schmidt, G. S., Staadt, O. G., Livingston, M. A., Ball, R., & May, R. (2006). A survey of large high-resolution display technologies, techniques, and applications. *IEEE Virtual Reality Conference (VR 2006)* (pp. 223–236). Alexandria, VA. doi:10.1109/VR.2006.20

Seeber, B., & Fastl, H. (2003). Subjective selection of non-individual head-related transfer functions. *Proceedings of the 2003 International Conference on Auditory Display (ICAD03)* (pp. 259–262). Boston, MA.

Song, M., Zhang, C., Florencio, D., & Kang, H. (2010). Personal 3D audio system with loudspeakers. *2010 IEEE International Conference on Multimedia and Expo (ICME)* (pp. 1600–1605). Singapore. doi:10.1.1.170.9860

Spencer, B. S. (2006). Incorporating the sense of smell into patient and haptic surgical simulators. *IEEE Transactions on Information Technology in Biomedicine, 10*(1), 168–173.

Svensson, U. (2002). Modeling acoustic spaces for audio virtual reality. *Proceedings of the IEEE Benelux Workshop on Model Based Processing and Coding of Audio* (pp. 109–116). Leuven, Belgium.

Talk2myShirt. (2010). Generative wardrobe. *talk2myShirt*. Retrieved February 5, 2013, from http://www.talk-2myshirt.com/blog/archives/4471.

Taylor, M. T., Chandak, A., Antani, L., & Manocha, D. (2009). RESound. *Proceedings of the 17th ACM International Conference on Multimedia—MM'09* (p. 271). New York, NY: ACM Press. doi:10.1145/1631272.1631311

Templeman, J., Page, R., & Denbrook, P. (2013). Locomotion interfaces. In K. S. Hale & K. M. Stanney (Eds.), *Handbook of virtual environments, design, implementation, and applications* (2nd ed., pp. 235–258). New York, NY: CRC Press.

Väljamäe, A., Larsson, P., Västfjäll, D., & Kleiner, M. (2004). Auditory presence, individualized head-related transfer functions, and illusory ego-motion in virtual environments. In M. A. Raya & B. R. Solaz (Eds.), *Proceedings of Presence 2004, 7th International Workshop on Presence*, Valencia, Spain (pp. 252–258).

Vaucelle, C., & Abbas, Y. (2007). Touch sensitive apparel. *Proceedings of CHI 2007* (p. 2723). New York, NY: ACM Press. doi:10.1145/1240866.1241069

Vaucelle, C., Bonanni, L., & Ishii, H. (2009). Design of haptic interfaces for therapy. *Proceedings of the 27th International Conference on Human Factors in Computing Systems—CHI 09* (p. 467). Boston, MA: ACM Press. doi:10.1145/1518701.1518776

Vermetten, E., & Bremner, J. D. (2003). Olfaction as a traumatic reminder in posttraumatic stress disorder: case reports and review. *Journal of Clinical Psychiatry, 64*, 202–207.

VirTra. (2013). VirTra 300 | Virtra. *www.virtra.com*. Retrieved from http://www.virtra.com/virtra-300/. Accessed on January 28, 2013.

Visell, Y., Law, A., Ip, J., Smith, S., & Cooperstock, J. (2010). Interaction capture in immersive virtual environments via an intelligent floor surface. *Proceedings of the IEEE Virtual Reality Conference (VR)* (pp. 313–314). Waltham, MA.

Vlahos, J. (2006). The smell of war. *Popular Science, 269*(2), 72–95.

Vorlander, M. & Shinn-Cunningham, B. (2014). Virtual auditory displays. In K. S. Hale & K. M. Stanney (Eds.), *Handbook of virtual environments, design, implementation, and applications* (2nd ed., pp. 89–116). New York, NY: CRC Press.

Wakefield, G. H., Roginska, A., & Santoro, T. S. (2010). Human-constrained design of spatial audio systems. *Proceedings of the Undersea Human System Integration Symposium*. Providence, RI.

Washburn, D. A., Jones, L. M.,Vijaya Satya, R., Bowers, C. A., & Cortes, A. (2003). Olfactory use in virtual environment training. *Modeling and Simulation, 2*(3), 19–25.

Watson, B., Spaulding, V., Walker, N., & Ribarsky, W. (1997). Evaluation of the effects of frame time variation on VR task performance. *Proceedings of the IEEE 1997 Annual International Symposium on Virtual Reality* (pp. 38–44). Albuquerque, NM. doi:10.1109/VRAIS.1997.583042

Welch, R. B., Blackmon, T. T., Liu, A., Mellers, B. A., & Stark, L. W. (1996). The effects of pictorial realism, delay of visual feedback, and observer interactivity on the subjective sense of presence. *Presence: Teleoperators and Virtual, 5*(3), 263–273.

Wenzel, E. M., Arruda, M., Kistler, D. J., & Wightman, F. L. (1993). Localization using nonindividualized head-related transfer functions. *The Journal of the Acoustical Society of America, 94*(1), 111–123. doi:10.1121/1.407089

Wenzel, E. M., Miller, J. D., & Abel, J. S. (2000). A software-based system for interactive spatial sound synthesis. *Proceedings of the Sixth International Conference on Auditory Display (ICAD2000)* (pp. 151–156). Atlanta, GA.

Wickens, C. (1990). Three-dimensional stereoscopic display implementation: Guidelines derived from human visual capabilities. In J. O. Merritt & S. S. Fisher (Eds.), *Stereoscopic displays and applications: Proceedings of SPIE, 1256*. Santa Clara, CA (pp. 2–11).

Witmer, B. G., & Singer, M. J. (1998). Measuring presence in virtual environments: A presence questionnaire. *Presence: Teleoperators and Virtual Environments, 7*, 225–240.

Wu, Y., Schmidt, L., Parker, M., Strong, J., Bruns, M., & Ramani, V. K. (2012). Active-hand: Automatic configurable tactile interaction in virtual environment. *Proceedings of the ASME International Design Engineering Technical Conferences and Computers and Information in Engineering Conference* (pp. 1–10). Chicago, IL.

Zahorik, P., Brungart, D. S., & Bronkhorst, A. W. (2005). Auditory distance perception in humans: A summary of past and present research. *Acta Acustica United with Acustica, 91*(3), 409–420.

Zotkin, D., Hwang, J., Duraiswami, R., & Davis, L. (2003). HRTF personalization using anthropometric measurements. *2003 IEEE Workshop on Applications of Signal Processing to Audio and Acoustics (WASPAA'03)* (pp. 157–160). New Paltz, NY.

14 Embodied Autonomous Agents

Andrew Feng, Ari Shapiro, Margaux Lhommet, and Stacy Marsella

CONTENTS

14.1 INTRODUCTION

Since the last decade, virtual environments have been extensively used for a wide range of applications, from training systems to video games. Virtual humans are animated characters that are designed to populate these environments and to interact with the objects of the world as well as with the user. A virtual agent must perceive the world in which it exists, reason about those perceptions, and decide on how to act on them in pursuit of its own agenda.

The work on virtual humans has become especially concerned with the bold challenge of realizing naturalistic face-to-face interactions between virtual (and often humanlike) agents and human participants, typically seeking to create virtual agents that interact with people using the same verbal and nonverbal behavior that people use to interact with each other. These virtual agents have gone by a variety of names, most notably embodied conversational agents (Cassell, 2000) and virtual humans (Gratch et al., 2002; Rickel et al., 2002).

Virtual humans have been proposed for a wide range of educational, social, medical, and training applications, where the virtual humans operate as peers, mentors, subordinates, patients, adversaries, or background characters.

The design and evaluation of such sophisticated artifacts is a multidisciplinary effort, requiring the integration of research spanning artificial intelligence, social psychology, linguistics, and computer graphics and animation.

Designers of virtual humans aspire to satisfy multiple requirements. Foremost, virtual humans must be *responsive*; that is, they must respond flexibly to the human user and to the events in the virtual environment. Second, they must be *believable*; that is, they must provide a sufficient illusion of humanlike behavior so that the human user will be drawn into the interaction. Finally, they must be *interpretable*; the user must be able to interpret their response to situations, including their dynamic cognitive and emotional state, using the same verbal and nonverbal behaviors that people use to understand one another.

Whereas virtual human research draws on work in artificial intelligence, graphics, and dialogue systems, it also faces unique challenges that stem from virtual humans being autonomous, embodied facsimiles of people that must be responsive, believable, and interpretable. In this chapter, we therefore touch on those challenges that are more unique to the realization of virtual humans capable of interacting with human users.

In particular, the burden of realizing believable, responsive, and interpretable behavior falls, in part, on the virtual humans' outward behavior in the virtual world, for it is that behavior that a user perceives. The virtual humans must be responsive to events in the virtual world, acting and reacting. They must express realistic emotions in their facial expressions, their gestures, their postures, etc. They must be able to carry on spoken dialogues with humans and other agents, including all the nonverbal communication that accompanies human speech (e.g., eye contact and gaze aversion, postural shifts, facial displays, and gestures). Finally, they must be able to take physical actions in the world such as walking or grasping an object.

For that reason, we begin our discussion of virtual humans with a discussion of realizing its outward appearance and behavior. We then proceed to discuss the mental capabilities required and conclude with a discussion of how the daunting challenges of realizing a virtual human are often addressed.

14.2 APPEARANCE OF A VIRTUAL CHARACTER

It is important to decide which aspects of an autonomous character will be modeled, which will depend on the circumstances in which it is used. Applications that require physically close, verbally rich interactions between a single agent and a user differ, such as those used for conversational agents (Heloir & Kipp, 2009; Poggi, Pelachaud, de Rosis, Carofiglio, & De Carolis, 2005; Thiebaux, Marsella, Marshall, & Kallmann, 2008; van Welbergen, Reidsma, Ruttkay, & Zwiers, 2009), from those used for personal or small group interaction (Shapiro, 2011) and from those that require interactions at a distance with crowds (Shoulson, Marshak, Kapadia, & Badler, 2013) or groups, such as a city-scale event simulations (Yersin, Maïm, Pettré, & Thalmann, 2009).

14.2.1 TRADITIONAL 3D CHARACTER REPRESENTATION

A common method to model virtual characters is to use geometry-based 3D meshes that are controlled through a hierarchical set of nodes representing the character's joints and other moving parts. A 3D mesh is constructed either manually by digital artists using 3D modeling software or automatically via a 3D scanning process and later adjusted manually. Such models are effective at representing surface features, such as the outward appearance of a character's face, skin, or clothing. In addition to the 3D mesh, surface features such as colors, textures, light reflectivity properties, and some subsurface structures can be visually modeled by adding different kinds of maps, including texture maps (2D images transferred onto the 3D geometry), and other 2D maps that create complex interactions with lighting and other external objects, such as bump, normal, specular, diffuse, and related maps. Specialized surface features can also be mimicked by creating special textures for features such as wrinkles and blushing. Such methods rely on the illusion of an outward appearance as a representation of the internal structures of a virtual character. For example, subcutaneous structures such as muscles and bones are indirectly shown through their effects on the surface features rather than explicitly modeled (Ng-Thow-Hing & Fiume, 1997). Modeling such internal structures explicitly can yield greater realism at the expense of complexity and computation time. Internal structures are often modeled for effect rather than for realism, for example, a muscle layer that does not contain the anatomically correct number of the type of muscles, but rather a representative muscular layer that contributes bulging and stretching to the skin layer above (Lee, Sifakis, & Terzopoulos, 2009). See Section 14.2.5, *Skinning,* for details of the skeleton to mesh binding algorithms.

14.2.2 GPU-BASED SHADING

In recent years, the advent of GPU-based shaders has allowed the incorporation of many high-quality techniques suitable for offline rendering available for interactive simulations. By running the algorithms on the hardware of a video card, as opposed to on software in the computer's CPU, techniques for shadowing, deformation, simulation, and display can be run in real time on commodity hardware. For example, GPU-based physics simulation enables the rapid physical simulation of rigid or soft bodies (Yeh, Faloutsos, & Reinman, 2006), and GPU-based shadowing algorithms (Kim & Neumann, 2001) allow for detailed lighting effects on volumetric surfaces. Shaders can also be designed to model complex anatomical structures such as the reflectivity and behavior of the human eye (Pamplona, Oliveira, & Baranoski, 2009) or skin (Jimenez, Sundstedt, & Gutierrez, 2009).

14.2.3 IMAGE-BASED CHARACTER REPRESENTATION

Image-based techniques for character representation show promise for character modeling. Rather than modeling a character using 3D geometry explicitly, numerous images are taken from video and then reconstructed in 3D to form a character (Casas, Tejera, Guillemaut, & Hilton, 2011; Starck, Miller, & Hilton, 2005). The advantage of such techniques is the ability to capture the dynamic aspects of character without explicit modeling. For example, the movement of a clothing as it deforms in response to a person's bending and moving can be captured automatically by an image-based technique without the need to understand the physical characteristics of the clothing material and its interactions with the person wearing it (Starck & Hilton, 2007). The advantages of image-based techniques also lend themselves to a similar disadvantage; without an understanding of the underlying system, it is easy and common to make errors during the reconstruction of a character. For example, during a particular movement, parts of the body may momentarily appear as if they are connected to each other, when they are not. These techniques represent interesting approaches that are useful in situations where characters cannot be explicitly modeled but are relatively new and are not widespread in their use.

Image-based character representation may include video-based representations. By recording and playing back a video of a live actor, a virtual character can be synthesized using a set of carefully controlled video clips (Bregler, Covell, & Slaney, 1997). This, of course, affords limited interaction and viewing perspectives but affords the possibility of photorealistic characters, which is difficult to do using traditional 3D techniques due to the resolution and fidelity requirements of photorealism.

14.2.4 FORWARD KINEMATICS/INVERSE KINEMATICS

With a virtual character's skeleton hierarchy, we are able to define new postures for the character by changing the skeletal configuration. There are mainly two different ways to adjust the skeletal configuration—forward kinematics (FK) and inverse kinematics (IK).

14.2.4.1 Forward Kinematics

The goal of FK is to obtain the position of a joint based on input joint parameters such as joint angles. It works by using both the kinematic equation and the skeleton hierarchy to recursively update the joint positions given a new set of joint parameters. This computation is simple to implement and is the most straightforward way to animate a character. However, when the hierarchy is very complicated and contains a large number of joints, FK update may become too slow to be computed in real time. This is one of the reasons that the skeleton hierarchy for interactive virtual character is usually much simpler than the one used in feature animation films; real-time performance is crucial for interactive application while animation quality is the key competence for feature animation production. To allow interactive editing and update of highly complex skeleton hierarchy, DreamWorks Animation utilized the multicore CPU by segmenting the hierarchy into multiple dependency graphs (Watt et al., 2012). Each graph can be updated independently in parallel. This helps accelerate the FK process to allow interactive update in their production pipeline. Similar technique could be applied for an interactive virtual character should the complexity of the character increase to the level for animation production.

14.2.4.2 Inverse Kinematics

IK, as its name suggests, is an inverse problem to FK. It takes the desired position(s) for some target joint(s) as constraints and computes joint parameters that will move the target joints to those positions. This can be useful when we want the virtual character to accurately execute some actions. For example, when a character is reaching for an object, it is desirable for his hand to touch the object. However, the provided joint angles from motion capture may not accurately put his hand on the right place due to various capturing errors. Thus, IK can be applied to adjust the joint angles to ensure the hand constraint is satisfied.

In general, this is a more difficult problem than FK since there are multiple sets of the feasible joint parameters that could satisfy an input constraint. This problem can be regarded as the nonlinear optimization problem that minimizes the error between the desired joint constraint and actual joint positions by adjusting joint parameters over the IK chain. Here, the IK chain is defined as the bone segments from target joint to root joint. The problem is not trivial since the mapping from joint angles to joint positions is nonlinear. There are several different methods to compute IK for a given constraint. They differ by their simplicity, performance, and robustness.

Cyclic coordinate descent (CCD) is a popular IK method widely adapted in the video game industry (Canutescu & Dunbrack, 2003; Lander & Content, 1998). It is based on a heuristic by iterating through each joint in the IK chain and rotating the bone segment to move target joints toward the constraint positions. It is simple to implement and is very efficient to compute. However, since its computation is based on a simple heuristic, it is difficult to guarantee the consistency across different animation frames. Depending on the traverse order, it may be possible that two similar constraint positions give rise to vastly different joint parameters and thus causes discontinuity in the resulting animations. This problem is even more obvious when the skeleton hierarchy is deep. In practice, this method is more suitable for real-time animation with relatively few joints involved.

Analytical method derives a direct solution for the IK problem (Badler & Tolani, 1996; Kallmann, 2008; Tolani, Goswami, & Badler, 2000). It requires the whole IK chain to consist of no more degrees of freedom (DOF) than the target joint constraints to uniquely determine the solution. Although this restriction reduces the applicability for this method, it is usually enough to use only 7-DOF to model an arm or a leg for a virtual character. For example, a typical human arm consists of 3-DOF for shoulder, 2-DOF for elbow, and 2-DOF for the wrist. A similar analogy can be made for a leg with hip, knee, and ankle joints. Therefore, a typical virtual character can be modeled by four separate IK chains to model both arms and legs. Notice that the joint parameters are not unique given the target joint position since target constraints only have 6-DOF (3 for position and 3 for rotation). Therefore, the user needs to provide a swivel angle for elbows or knees to obtain a unique solution. The analytical method is very efficient to compute and could be suitable for virtual characters with simple skeleton hierarchy.

Jacobian method approximates the solution by solving a numerical optimization problem directly through linearization (Buss & Kim, 2005; Yamane & Nakamura, 2003). It works by computing a Jacobian matrix, which encodes the partial derivative from target joint position to joint parameters. This reduces the nonlinear optimization problem into a series of linear least square problem and can be solved via a numerical linear solver. Since the problem is linearized, the solution is varying smoothly based on initial skeletal configurations and target joint constraints. Moreover, it can integrate the desired joint angles as the secondary target. Thus, the solution will match the target joint constraints while satisfying the desired posture as much as possible. This provides a nice IK solution to adjust a character's final posture while staying faithful to the original posture. However, the method has two drawbacks. The first is that it suffers from the singularity when the target position is out of reach from the designated IK chain. In this situation, the Jacobian matrix becomes singular since there are no angle adjustments that can move the target joint closer to the constraint. Thus, the solution becomes unstable and will introduce jerky motions as a result. The second issue is that the solution requires solving a dense linear system during each iteration. For complex skeleton hierarchy, this computation becomes very expensive for real-time application since the computational complexity is cubic to the number of joints in the system. Therefore, the method is more applicable when the IK chain is relatively simple.

Particle-based method is a variation of CCD based on similar heuristics (Hecker et al., 2008). It also recursively updates the bone segments to orient the target joint toward the constraint position. The difference is that it operates directly on the joint positions instead of rotations. The method adjusts the bone segments in a two-pass manner. During the first pass, the end joint of a bone segment is simply moved toward the desired position without considering the length constraint of its bone segment. The length constraint is then enforced during the second pass by moving both end joints of that bone segment. The method then iterates this two-pass process over each bone segment until converges. The analogy of the process is like attaching a spring to each bone segment. Thus, the end joint is free to move by extending its bone segment, but the bone segment must eventually return to rest length for equilibrium. Overall, this method has similar property as CCD but provides smoother solutions due to the fact that it only operates on the position domain instead of joint angles.

14.2.5 SKINNING

Skinning is the process that transforms the skeletal animation to character mesh. In general, it animates a mesh by blending joint transformations and applying them on each mesh vertex. Here, the blending weights for transformations need to be defined on each vertex and are usually manually created by the animator. With skinning, the motions only need to be defined at the skeleton level to animate the whole mesh. Thus, the effort for animating a mesh is greatly simplified. There are several different techniques involved in mesh skinning that address issues that arise during the skinning process.

Linear blend skinning, sometimes also called skeletal subspace deformation (SSD), is the simplest method for skinning. It represents the joint transformations, which are usually rigid transformations, as 4×4 matrices and computes the weighted sum of the matrices to transform each vertex (Lander, 1998; Thalmann, 1989). This is a straightforward process that has been widely used in many interactive applications such as video games due to its simplicity (Lander). However, this simple method comes with a drawback since combining matrices by weighted average will not necessarily preserve the desired quality. For example, the average of rotation matrices will not result in a rotation matrix. Therefore, the resulting mesh deformation may have a collapsed shape and lose volume. This problem is especially obvious when the joint is bent more than 90° and twisted, which results in the notorious *candy wrap* artifacts (Mohr & Gleicher, 2003).

Pose space deformation (PSD) was introduced to alleviate the artifacts from SSD (Lewis, Cordner, & Fong, 2000). The main idea is to create a set of example shapes for some important poses and compute the residue differences between these shapes and the shapes produced from SSD. Vertex offset is used to represent these residues and it encodes the errors caused by the SSD method. To compensate these errors, the residues are interpolated for a new pose and added to the new mesh shape generated by SSD. The technique can effectively improve the artifacts common in SSD skinning. However, it requires additional efforts from artists to sculpt the example mesh shapes for various poses and adds more computation during run-time animation.

Dual quaternion skinning (DQS) is an alternative to SSD (Kavan, Collins, Žára, & O'Sullivan, 2007; Kavan, McDonnell, Dobbyn, Žára, & O'Sullivan, 2007). Similar to SSD, it blends joint transformations together to produce animated mesh. Instead of representing rigid transformations as matrices, it uses dual quaternion (DQ) to encode the transformation. DQ is a compact representation for both rotation and translation. Its advantage is that one can blend different DQs together to still produce a DQ, which is always a valid rigid transformation. Therefore, it does not suffer from the collapsing and candy wrap artifacts that plaque the SSD technique. Overall, it is a better skinning method than traditional SSD in terms of quality, though it requires more efforts to correctly implement dual DQ algebra so the conversions between DQ and rigid transformation are handled correctly.

Although skinning techniques can efficiently represent the mesh animations of a virtual character, they also have some limitations. One limitation is that it is difficult to model some deformation effects such as muscle bulges with rigid transformations from skeletal joints. PSD can partially alleviate this problem with example mesh shapes. However, for deformations with large rotations such as 180° bending or twisting, PSD may not produce high-quality results since its residues are represented as vertex offsets. Recent research works addressed this issue by adding the shear and scale components to model the muscle bulge effects in the joint transformation in addition to rigid transformation. The correct shear and scale parameters can be predicted based on skeletal pose by precomputing a regression model from example mesh shapes (Feng, Kim, & Yu, 2008; Kim, Feng, & Yu, 2010; Wang, Pulli, & Popović, 2007). The quality of results would depend on how well the learned regression model captures the muscle deformations from the example shapes and whether it can generalize to new input skeletal poses.

Another limitation is that self-collisions could happen in the resulting mesh deformations. Since skeleton is a simplified representation for the virtual character, it does not have the underlying knowledge about its corresponding character mesh. Therefore, it would be difficult to resolve potential self-collisions in the animated mesh from simply checking the skeleton configurations. One possible solution for this problem is to attach a bounding volume such as box or cylinder to each bone segment as an approximation to the corresponding mesh. These bounding volumes can then be used to detect self-collisions given a new skeleton configuration. A new set of joint parameters can then be inferred to resolve these collisions between bounding volumes. This method can work as an approximate solution to collision problems, though it cannot model more subtle deformations such as compressed skins due to collisions. This issue is still an open problem in the research. More accurate and efficient methods are desired to resolve the collisions while maintaining these subtle skin deformations.

14.2.6 Cloth

Cloth is an important part of the appearance for a virtual character. Appropriate clothing can improve the realism and produce distinct visual styles for each character. Based on the type of clothes and the available computation resources, a cloth can be either animated from kinematic body movements or simulated physically.

Since most types of clothes we wear such as shirts or jeans fit tightly on a character, their deformations will roughly follow the deformation of body segments. Therefore, one can approximate the cloth shapes by computing its deformations in a similar manner like one computes them for the character. This is done by assigning a suitable skin blend weights for each vertex on the cloth and deforming them based on joint transformations of the character. This provides a simple method to quickly produce cloth animations with lower quality. The problem with this method is that the secondary motions due to dynamics and other fine wrinkle details are totally ignored. Therefore, it can only be used to approximate tightly fit cloth and is not suitable to model highly dynamic cloth such as skirts.

Physical simulation, on the other hand, can produce highly detailed cloth deformations with interesting dynamic effects according to character movements (Baraff & Witkin,1998; Choi & Ko, 2005). For example, a dancer with a long skirt can demonstrate highly dynamic body movements. These body movements can be visually enhanced by simulating the skirt to highlight some interesting actions such as spinning or speed changes.

The most common way of cloth simulation is to model the cloth as a set of mass particles connected by springs (Baraff & Witkin,1998; Bridson, Marino, & Fedkiw, 2003; Choi & Ko, 2005). During each simulation step, the particles are affected by both external force such as gravity and internal spring forces. The spring forces keep the particles within a reasonable distance to each other and thus model the inextensible property of a cloth sheet. In addition to mass-spring model, the properties of different textiles can also be measured from the real-world material and simulated using a finite element method (Etzmuss, Keckeisen, & Strasser, 2003). This requires significant setup to correctly measure the desired material. The run-time simulation using finite element method is also more expensive than a mass-spring model. However, this method produces more realistic and accurate simulation results that are closer to real textile material. This makes it is suitable for applications that require high-quality simulations such as virtual fashion design.

With recent advances in graphics hardware, now the cloth simulation with moderate complexity can be computed in real time (Cordier & Magnenat-Thalmann, 2002). It is done by reformulating the cloth simulation steps so they can be computed in parallel on modern graphics hardware. Specifically, it divides a simulation step into two stages, the particle simulation and the constraint limiting (Müller, Heidelberger, Hennix, & Ratcliff, 2007; Zeller 2005). The particle simulation stage treats each cloth particle as unconstrained. Thus, each particle can be simulated independently. The constraint limiting stage checks each pair of particles with a spring connection and adjusts the particle positions if their distance is too far from or too close to each other. Both stages can be run in parallel, and thus, the whole simulation can be executed efficiently on hardware. This makes it possible for an interactive virtual character to be dressed with dynamic cloth to produce interesting effects.

14.3   MENTAL PROCESSES

To create a socially responsive embodied facsimile of a human capable of interacting with a person face to face requires the modeling of an array of capabilities in a virtual human.

First, the virtual human must *perceive* the environment in which it is embodied, including any humans it interacts with. It must *reason* about how events in that environment impact its goals and react to those events appropriately. More specifically, for the virtual human to seem humanlike, it should *react emotionally* appropriately.

We may additionally want it to support spoken language interaction with humans that cohabit in the virtual environment, including the use of nonverbal behaviors such as facial expressions, head movements, gaze, and gestures that play such a central role in human face-to-face interaction. In this section, we discuss the challenges that underlie such capabilities and requirements.

14.3.1   PERCEPTION, ATTENTION, AND PERCEPTUAL UNDERSTANDING

For a virtual human to be responsive to what happens in the environment, it must perceive and understand the situation, that is, to what extent it impacts its goals and what action can be made to change the situation.

14.3.1.1　Environment

One effort here is to ensure that the virtual human's perceptual and attentional capabilities are consistent with human limitations. Thus, the virtual human should not be omniscient, not hear or look through walls, and show semblance of a focus of attention. Computational models that drive the virtual human perceptual attention and perception ensure behavior consistent with human capabilities. For example, the level of detail at which a virtual human will perceive objects and their properties in the virtual world can be predicted (Hill, 1999; Kim, Hill, & Traum, 2006). The types of visual attention required for several basic tasks (such as locomotion, object manipulation, or visual search), as well as the mechanisms for dividing attention among multiple tasks can be determined (Khullar & Badler, 2001).

Two methods can be used for the agent to understand the perceived situation.

In informed environments, the virtual human can directly read inside the objects what actions can be realized with or upon it. Such objects are called *smart objects* (Kallmann & Thalmann, 1999) and they have been extensively used in virtual environments.

More compatible with the reality, some virtual humans have their own model of the world that they maintain by using their perceptions. For example, STEVE has a symbolic model of the world that he updates whenever he perceives changes of objects and attributes (Rickel & Johnson, 1997, 2000).

14.3.1.2　Human Interactant

Another key effort here is to ensure the virtual human can perceive the behavior of humans immersed in the virtual environment. Most notably, there is considerable effort being undertaken to give virtual humans the ability to perceive and recognize the nonverbal behavior of human interactant including facial expressions, postural shifts, gestures, head movements, and gaze. As we noted earlier, such behavior plays a critical role in human face-to-face interaction. A complete review of the recent work in this area can be found in Scherer et al. (2012).

14.3.2　Reasoning and Representation

The basis that allows virtual humans to act as autonomous agents and to be perceived by human users as possessing agency is the goal of a virtual human. Virtual humans do not usually react only to the environment, but they also are proactive. To achieve these goals, virtual humans must be able to generate plans or intentions, decide on appropriate actions, and react to unexpected events. They need to represent their beliefs about past events, present circumstances, and future expectations, especially in terms of how those impact their goals. Additionally, to interact effectively with other agents and humans, a virtual human may maintain beliefs about them as well, including what others believe and what their goals are.

To achieve this functionality, work in virtual human research has explored a variety of techniques drawn from research in artificial intelligence, cognitive science, and robotics. This includes planning and decision-making, cognitive architectures, work on belief, desire and intention frameworks for agent design, and robotics work on path planning.

14.3.3　Emotion

Emotion has a central, powerful effect on human behavior. It affects how people perceive the world, think, act, and speak. Studies have identified its critical role in human decision-making, influencing the subjective value of alternative choices (Busemeyer, Dimperio, & Jessup, 2007) and guiding decision-making (Bechara, Damasio, Damasio, & Lee, 1999). Work by Simon (1967) argued that emotions serve a critical function in human behavior, interrupting cognition when unattended goals need to be addressed and therefore providing a means for a person to balance competing goals as well as supporting reactions to unexpected events. Research has also argued how social emotions such as anger and guilt may reflect a mechanism that improves group utility (Frank, 1988). Collectively, these findings underline that in many respects, it is obvious to us all that in everyday life, emotions have important influences in human behavior.

Emotion expression plays a powerful role in shaping human behavior and social interaction. From emotional displays, observers can form interpretations of a person's beliefs (e.g., frowning at an assertion may indicate disagreement), desires (e.g., joy gives information that a person values an outcome), and intentions/action tendencies (e.g., fear suggests flight). With such a powerful signal, it is not surprising that emotions can be a means of social influence and control (Campos, Thein, & Owen, 2003; Fridlund, 1997; Waal, 2003). For example, anger can coerce reactions in others and enforce social norms; displays of guilt can elicit reconciliation after some transgression, distress can be seen as a way of recruiting social support, and displays of joy or pity are a way of signaling such support to others. Other emotion displays seem to exert control indirectly by inducing emotional states in others and thereby influencing an observer's behavior. Specific examples of this are empathy and emotional contagion that can lead individuals to *catch* the emotions of those around them (Hatfield, Cacioppo, & Rapson, 1994).

These findings on the functional, often adaptive, role that emotions play in human behavior have led researchers to incorporate models of human emotion and emotional expression as core capabilities in virtual human systems in order to realize more humanlike behavior (Dias & Paiva, 2005; Rickel & Johnson, 1997). Emotional displays can make the virtual human seem human or lifelike and thereby influence the user to respond to, and interact with, it as if it were a person. In that people utilize these behaviors in their everyday interpersonal interactions, modeling the function of these behaviors is essential for any application that hopes to faithfully mimic face-to-face human interaction. More importantly, however, the ability of emotional behaviors to influence a person's emotional and motivational states could potentially, if exploited effectively, guide a user toward more effective interactions.

Since emotions are so pervasive and influence all cognitive processes, incorporating them into virtual human systems poses certain constraints. Marsella, Gratch, and Petta (2010) present an overview of the existing computational models of emotions and their applications. For example, researchers have looked at emotion and emotional expression in characters as a means to engender empathy and bonding between users and virtual characters (Marsella & Gratch, 2003; Paiva et al., 2005).

14.3.4 Natural Language Dialogue

The ability of a virtual human to engage in natural language interactions with a human user must address a range of challenges common to work in dialogue systems generally. This includes speech recognition to determine what words are being uttered by a human user, natural language understanding to comprehend the meaning of the utterance, dialogue management to determine the role in the ongoing conversation and how to respond, natural language generation to transform the virtual human's response into utterance text, and speech synthesis to transform that text into spoken language. In practice, virtual humans differ dramatically in how these functions are realized. For example, at one simpler extreme, the natural language understanding and dialogue management could be realized by a system that simply maps words recognized by a speech recognizer to an appropriate response. A more sophisticated approach may employ understanding and dialogue management techniques that attempt to match the interpretation of the utterance against the virtual human's representation of the context, including their beliefs about the past, present, and future as well as the state of the conversation and their goals (see Traum, Swartout, Gratch, and Marsella, 2008; Traum & Larsson, 2003; Larsson, Staffan, & Traum, 2000). Such sophistication allows the agent to make determinations, for example, about whether to listen, take the dialogue turn, seek clarification on an issue, change topics, or take nondialogue actions instead, overall providing for more flexible, natural interactions. Similarly, speech synthesis may employ more flexible generative text-to-speech synthesis techniques for generating the spoken language or simpler techniques of having a fixed prespecified set of utterances recorded by a voice actor and simply playing back that recorded audio to generate the virtual human's spoken language.

14.3.5 NONVERBAL BEHAVIOR

The flip of a hand, a raising of an eyebrow, a gaze shift: the physical, nonverbal behaviors convey a wide variety of information that powerfully influences face-to-face interactions. The relation between nonverbal behavior and speech is complex. Nonverbals can stand in different, critical relations to the verbal content, providing information that embellishes, substitutes for, and even contradicts the information provided verbally (Ekman & Friesen, 1969; Kendon, 2000). For example, a nod can convey agreement, and a beat gesture emphasizes a point. Nonverbal behaviors also serve a variety of rhetorical functions. Shifts in topic, for example, can be cued by shifts in posture or shifts in head pose. In addition, a wide range of mental states and character traits can be conveyed: gaze reveals thought processes, blushing suggests shyness, and facial expressions intentionally or unintentionally convey emotions and attitudes. Finally, nonverbal behavior helps manage conversation, for example, by signaling the desire to hold onto, get, or hand over the dialogue turn (Argyle & Cook, 1976; Bavelas, 1994). A speaker's aversion of gaze reflects they are thinking, in essence, regulating cognitive load as they consider what to say next while also signaling they want to hold onto the dialogue turn.

Nonverbal behaviors are so pervasive in every moment of face-to-face interaction that their absence also signals information—that something is wrong, for example, about the physical health or mental state of the person. Integrating nonverbal behaviors is therefore important to improve the quality of interaction. However, the issue encountered is the absence of a computational model of nonverbal behaviors that would answer the question of what behaviors to exhibit and when to exhibit them. Creating such a model faces several challenges.

First, the context in which the behavior occurs can transform the interpretation, as can even subtle changes in the dynamics of the behavior: head nods signaling affirmation versus emphasis typically have different dynamics. Behaviors can also be composed with each other, further transforming their interpretation. The generation of the behaviors must additionally take into account that the behaviors are synchronized, often tightly, with the dialogue, and changes in this synchronization can lead to significant changes in what is conveyed to a listener. For instance, the stroke of a hand gesture, a nod, and eyebrow raise performed individually or together are often used to emphasize the significance of a word or phrase in the speech. To achieve that emphasis, the behavior must be closely synchronized with the utterance of the associated words or phrases being emphasized. Alteration of the timing will change what words are being emphasized and consequently change what is conveyed to a listener. Achieving such synchronization in a virtual character can be difficult, especially in the case of behaviors such as hand gestures that involve relatively large-scale motion and preparatory phases to bring the hand into position to perform the gesture.

Such challenges make the pattern and timing of the behavior animations that accompany utterances unique to the utterance and the state of the character.

14.4 GROUP AND CROWDS

Some virtual environments integrate crowds or large groups of characters. Although aforementioned techniques can indeed be applied to multiple characters, several aspects need to be considered. In particular, animating a large number of high-quality intelligent virtual humans requires significant computation resources.

14.4.1 LEVEL OF DETAIL

Animating a large number of high-quality virtual characters requires significant computation resources, so it may not be feasible to animate and render all characters with full resolution.

The common technique is to apply level-of-detail control to dedicate more computation resources to characters that are closer to the user (Di Giacomo, Moccozet, Kim, & Magnenat-Thalmann, 2007; Luebke, Watson, Cohen, Reddy, & Varshney, 2002). This would require a multiresolution mesh for the virtual character; the mesh will be simplified or refined based on the current animation and camera view (Feng, Kim, Yu, Peng, & Hart, 2010; Kircher & Garland 2005).

14.4.2 GENERATING DIFFERENT APPEARANCES

Finally, it is desirable to create variations in both appearance and behavior when simulating crowds. The appearance variation can be achieved using a multiple of different texture images, face geometry, body sizes (McDonnell, Larkin, Hernández, Rudomin, & O'Sullivan, 2009). Behavior variations can be produced by applying a different personality behavior for each character during steering simulation (Guy, Kim, Lin, & Manocha, 2011) or synthesizing motions variations from original motions (Lau, Bar-Joseph, & Kuffner, 2009).

14.4.3 PATHFINDING AND STEERING

The next is crowd movement control. Naively animating the crowds to navigate in the environment will result in a monotonous behavior for each character. Moreover, intersections between characters cannot be prevented or resolved since the character is not aware of the environment. Therefore, a steering system is needed to coordinate the movements of each character to generate desired crowd behaviors (Reynolds, 1987, 1999). In general, the steering method needs to plan the valid path for each character based on the environment. The environment would include static obstacles as well as dynamically moving objects and other moving characters. For navigation in static environment, it is possible to precompute the maneuverable space and store them in a structure such as navigation mesh (Kallmann, 2010) or probabilistic roadmap (Kavraki, Svestka, Latombe, & Overmars, 1996). These structures can be utilized at run time by steering system for efficient path planning. On the other hand, dynamic obstacles, such as other moving characters, can only be resolved at run time. Reactive methods have the character check the adjacent environment periodically and try to move away from obstacles when close to an obstacle (Reynolds, 1999). Predictive method such as reciprocal velocity obstacles (RVO) (Van den Berg, Lin, & Manocha, 2008) controls each character to avoid each other before collision by anticipating the movements from incoming characters. For more complex situation such as resolving deadlocks inside a tunnel, more expensive space-time planning techniques are required (Levine, Lee, Koltun, & Popović, 2011). Crowd behavior can be simplified by organizing members into smaller units of crowds, groups and individuals (Musse & Thalmann, 2001).

In addition to environment navigation, the steering method can also be extended to simulate crowd movement behavior to adapt for stressful situation (Kim, Guy, Manocha, & Lin, 2012).

14.5 CONCLUSION: MAKING COMPROMISES

As we have noted, the creation of virtual humans faces a set of imposing challenges. What has made the realization of virtual humans feasible is the bounded context of the environment and task in which a human user interacts with them. Open-ended face-to-face interaction with a virtual human is currently considerably beyond the state of the art, especially in terms of representing the wealth of human experience and dialogue capabilities that a person brings to an open-ended interaction. However, if the user is performing a specific task with the virtual human, then the interaction becomes more constrained. For example, the virtual human Max can play with a user (Kopp, Sowa, & Wachsmuth, 2004). STEVE instructs a user how to operate a machine (Rickel & Johnson, 1997). Similarly, if the virtual environment's setting elicits well-defined associations, then the human user's behavior becomes more predictable. For example, the Gunslinger system (Hartholt, Gratch, Weiss, & The Gunslinger Team 2009) places a human user in a scenario from the American Old West, where the user plays a sheriff facing a gunslinger in a bar. The scenario draws on people's common experiences watching Hollywood films about the Old West to help constrain the way they interact with the various virtual humans they encounter.

Nevertheless, the design of these artifacts for specific tasks and scenarios requires an iterative design process. For example, early versions of the system will be used to explore how users interact and what they say and do. That may include the so-called Wizard of Oz studies where the virtual human is no more than a puppet controlled by a hidden person. Results of such explorations lead to refinements in the world and dialogue knowledge encoded in the virtual human or changes in the scenario that subtly constrain how users tend to interact with the system. Assuming sufficient data are acquired from these studies, machine learning techniques can be employed to create the dialogue models used by the virtual human.

REFERENCES

Argyle, M., & Cook, M. (1976). *Gaze and mutual gaze.* Cambridge, U.K.: Cambridge University Press.

Badler, N. I., & Tolani, D. (1996). Real-time inverse kinematics of the human arm. *Presence, 5*(4), 393–401.

Baraff, D., & Witkin, A. (1998). Large steps in cloth simulation. *Proceedings of the 25th Annual Conference on Computer Graphics and Interactive Techniques* (pp. 43–54). Los Angeles, CA: ACM.

Bavelas, J. B. (1994). Gestures as part of speech: Methodological implications. *Research on Language and Social Interaction, 27*(3), 201–221.

Bechara, A., Damasio, H., Damasio, A., & Lee, G. (1999). Different contributions of the human amygdala and ventromedial prefrontal cortex to decision-making. *The Journal of Neuroscience, 19*(13), 5473–5481.

Bregler, C., Covell, M., & Slaney, M. (1997). Video rewrite: Driving visual speech with audio. *Proceedings of the 24th Annual Conference on Computer Graphics and Interactive Techniques.* Los Angeles, CA: ACM Press.

Bridson, R., Marino, S., & Fedkiw, R. (2003). Simulation of clothing with folds and wrinkles. *Proceedings of the 2003 ACM SIGGRAPH/Eurographics Symposium on Computer Animation* (pp. 28–36). Granada, Spain: Eurographics Association.

Busemeyer, J., Dimperio, E., & Jessup, R. (2007). Integrating emotional processes into decision-making models. In W. Gray (Ed.), *Integrated models of cognitive systems* (pp. 213–229). Cambridge, MA: MIT Press.

Buss, S. R., & Kim, J. S. (2005). Selectively damped least squares for inverse kinematics. *Journal of Graphics, GPU, and Game Tools, 10*(3), 37–49.

Campos, J. J., Thein, S., & Owen, D. (2003). A darwinian legacy to understanding human infancy. *Annals of the New York Academy of Sciences, 1000*(1), 110–134.

Canutescu, A. A., & Dunbrack, R. L. (2003). Cyclic coordinate descent: A robotics algorithm for protein loop closure. *Protein Science, 12*(5), 963–972.

Casas, D., Tejera, M., Guillemaut, J., & Hilton, A. (2011). Parametric control of captured mesh sequences for real-time animation. *Proceedings of the Fourth International Conference on Motion in Games* (pp. 242–253). Edinburgh, U.K.: Springer.

Cassell, J. (2000). *Embodied conversational agents.* Cambridge, MA: MIT Press.

Choi, K. J., & Ko, H. S. (2005). Stable but responsive cloth. *ACM SIGGRAPH 2005 Courses* (p. 1). Los Angeles, CA: ACM.

Cordier, F., & Magnenat-Thalmann, N. (2002). Real-time animation of dressed virtual humans. *Computer Graphics Forum, 21*(3), 327–335.

Di Giacomo, T., Moccozet, L., Kim, H., & Magnenat-Thalmann, N. (2007). State-of-the-art on level-of-detail for virtual human animation and representation. In L. de Floriani & M. Spagnuolo (Eds.), *In shape analysis and structuring.* Berlin, Germany: Springer.

Dias, J., & Paiva, A. (2005). Feeling and reasoning: A computational model for emotional characters. *Progress in Artificial Intelligence, 3808*, 127–140.

Ekman, P., & Friesen, W. V. (1969). The repertoire of nonverbal behavior: Categories, origins, usage, and coding. *Semiotica, 1*, 49–98.

Etzmuss, O., Keckeisen, M., & Strasser, W. (2003). A fast finite element solution for cloth modelling. *Proceedings of the 11th Pacific Conference on Computer Graphics and Applications* (pp. 244–251). Canmore, Alberta, Canada: IEEE.

Feng, W. W., Kim, B. U., & Yu, Y. (2008). Real-time data driven deformation using kernel canonical correlation analysis. *ACM Transactions on Graphics (TOG), 27*(3), 91.

Feng, W. W., Kim, B. U., Yu, Y., Peng, L., & Hart, J. (2010). Feature-preserving triangular geometry images for level-of-detail representation of static and skinned meshes. *ACM Transactions on Graphics (TOG)*, *29*(2), 11.

Frank, R. (1988). *Passions within reason*. New York, NY: W. W. Norton & Company.

Fridlund, A. J. (1997). The new ethology of human facial expressions. In J. A. Russell & J. M. Fernandez-Dols (Eds.), *The psychology of facial expression* (p. 103). Cambridge, U.K.: Cambridge University Press.

Gratch, J., Rickel, J., André, E., Cassell, J., Petajan, E., & Badler, N. (2002). Creating interactive virtual humans: Some assembly required. *Intelligent Systems, IEEE*, *17*(4), 54–63.

Guy, S. J., Kim, S., Lin, M. C., & Manocha, D. (2011). Simulating heterogeneous crowd behaviors using personality trait theory. *Proceedings of the 2011 ACM SIGGRAPH/Eurographics Symposium on Computer Animation* (pp. 43–55). ACM, Wales, U.K.

Hartholt, A., Gratch, J., Weiss, L., & The Gunslinger Team. (2009). At the virtual frontier: Introducing gunslinger, a multi-character, mixed-reality, story-driven experience. In *Intelligent virtual agents, Vol: 5773, Lecture Notes in Computer Science* (pp. 500–501). Berlin, Germany: Springer.

Hatfield, E., Cacioppo, J. T., & Rapson, R. L. (1994). *Emotional contagion*. Cambridge, MA: Cambridge University Press.

Hecker, C., Raabe, B., Enslow, R. W., DeWeese, J., Maynard, J., & van Prooijen, K. (2008). Real-time motion retargeting to highly varied user-created morphologies. *ACM Transactions on Graphics (TOG)*, *27*(3), 27.

Heloir, A., & Kipp, M. (2009). EMBR—A realtime animation engine for interactive embodied agents. In *Intelligent Virtual Agents* (pp. 393–404). Berlin, Germany: Springer.

Hill, R. (1999). Modeling perceptual attention in virtual humans. *Proceedings of the 8th Conference on Computer Generated Forces and Behavioral Representation* (pp. 563–573). Orlando, FL.

Jimenez, J., Sundstedt, V., & Gutierrez, D. (2009). Screen-space perceptual rendering of human skin. *ACM Transactions on Applied Perception (TAP)*, *6*(4), 3–17.

Kallmann, M. (2008). Analytical inverse kinematics with body posture control. *Computer Animation and Virtual Worlds*, *19*(2), 79–91.

Kallmann, M. (2010). Navigation queries from triangular meshes. *Proceedings of the Third International Conference on Motion in Games* (pp. 230–241). Zeist, the Netherlands: Springer.

Kallmann, M., & Thalmann, D. (1999). A behavioral interface to simulate agent-object interactions in real time. *Computer Animation, Proceedings* (pp. 138–146). Geneva, Switzerland: IEEE.

Kavan, L., Collins, S., Žára, J., & O'Sullivan, C. (2007). Skinning with dual quaternions. *Proceedings of the 2007 Symposium on Interactive 3D Graphics and Games* (pp. 39–46). Seattle, WA: ACM.

Kavan, L., McDonnell, R., Dobbyn, S., Žára, J., & O'Sullivan, C. (2007). Skinning arbitrary deformations. *Proceedings of the 2007 Symposium on Interactive 3D Graphics and Games* (pp. 53–60). Seattle, WA: ACM.

Kavraki, L. E., Svestka, P., Latombe, J. C., & Overmars, M. H. (1996). Probabilistic roadmaps for path planning in high-dimensional configuration spaces. *IEEE Transactions on Robotics and Automation*, *12*(4), 566–580.

Kendon, A. (2000). Language and gesture: Unity or duality. In D. McNeill (Ed.), *Language and gesture* (pp. 47–63). Cambridge, MA: Cambridge University Press.

Khullar, S. C., & Badler, N. I. (2001, March 1). Where to look? Automating attending behaviors of virtual human characters. *Autonomous Agents and Multi-Agent Systems*, *4*(1–2), 9–23. doi:10.1023/A:1010010528443

Kim, B. U., Feng, W. W., & Yu, Y. (2010). Real-time data driven deformation with affine bones. *The Visual Computer*, *26*(6–8), 487–495.

Kim, S., Guy, S. J., Manocha, D., & Lin, M. C. (2012). Interactive simulation of dynamic crowd behaviors using general adaptation syndrome theory. *Proceedings of the ACM SIGGRAPH Symposium on Interactive 3D Graphics and Games* (pp. 55–62). Costa Mesa, CA: ACM.

Kim, T., & Neumann, U. (2001). Opacity shadow maps. *Rendering techniques 2001*. Vienna, Austria: Springer.

Kim, Y., Hill Jr., R., & Traum, D. R. (2006). *A computational model of dynamic perceptual attention for virtual humans*. Los Angeles, CA: University of Southern California.

Kircher, S., & Garland, M. (2005). Progressive multiresolution meshes for deforming surfaces. *Proceedings of the 2005 ACM SIGGRAPH/Eurographics Symposium on Computer Animation* (pp. 191–200). Dublin, Ireland: ACM.

Kopp, S., Timo S., & Ipke, W. (2004). Imitation games with an artificial agent: From mimicking to understanding shape-related iconic gestures. *Gesture-based communication in human-computer interaction* (pp. 436–447). Berlin Heidelberg: Springer.

Lander, J. (1998). Making kine more flexible. *Game Developer Magazine*, *1*, 15–22.

Lander, J. (1999). Over my dead, polygonal body. *Game Developer Magazine, 17*(1), 1–4.

Larsson, S., & Traum, D. R. (2000). Information state and dialogue management in the TRINDI dialogue move engine toolkit. *Natural Language Engineering, 6*(3 & 4), 323–340.

Lau, M., Bar-Joseph, Z., & Kuffner, J. (2009). Modeling spatial and temporal variation in motion data. *ACM Transactions on Graphics (TOG), 28*(5), 171.

Lee, S., Sifakis, E., & Terzopoulos, D. (2009). Comprehensive biomechanical modeling and simulation of the upper body. *ACM Transactions on Graphics (TOG), 28*(4), 99.

Levine, S., Lee, Y., Koltun, V., & Popović, Z. (2011). Space-time planning with parameterized locomotion controllers. *ACM Transactions on Graphics (TOG), 30*(3), 23.

Lewis, J. P., Cordner, M., & Fong, N. (2000). Pose space deformation: A unified approach to shape interpolation and skeleton-driven deformation. *Proceedings of the 27th Annual Conference on Computer Graphics and Interactive Techniques* (pp. 165–172). New Orleans, LA: ACM Press.

Luebke, D., Watson, B., Cohen, J. D., Reddy, M., & Varshney, A. (2002). *Level of detail for 3D graphics.* New York, NY: Elsevier Science Inc.

Marsella, S., & Gratch, J. (2003). Modeling coping behavior in virtual humans: Don't worry, be happy. *Proceedings of the Second International Joint Conference on Autonomous Agents and Multiagent Systems. AAMAS'03* (pp. 313–320). New York, NY: ACM.

Marsella, S., Gratch, J., & Petta, P. (2010). Computational models of emotion. In K. R. Scherer, T. Bänziger, and E. B. Roesch (Eds.), *A blueprint for a affective computing: A sourcebook and manual.* Oxford, U.K.: Oxford University Press.

McDonnell, R., Larkin, M., Hernández, B., Rudomin, I., & O'Sullivan, C. (2009). Eye-catching crowds: Saliency based selective variation. *ACM Transactions on Graphics (TOG), 28*(3), 55.

Mohr, A., & Gleicher, M. (2003). Building efficient, accurate character skins from examples. *ACM Transactions on Graphics (TOG), 22*(3), 562–568.

Müller, M., Heidelberger, B., Hennix, M., & Ratcliff, J. (2007). Position based dynamics. *Journal of Visual Communication and Image Representation, 18*(2), 109–118.

Musse, S. R., & Thalmann, D. (2001). Hierarchical model for real time simulation of virtual human crowds. *IEEE Transactions on Visualization and Computer Graphics, 7*(2), 152–164.

Ng-Thow-Hing, V., & Fiume, E. (1997). Interactive display and animation of B-spline solids as muscle shape primitives. *Computer Animation and Simulation'97.* Vienna, Austria: Springer.

Pamplona, V. F., Oliveira, M. M., & Baranoski, G. V. (2009). Photorealistic models for pupil light reflex and iridal pattern deformation. *ACM Transactions on Graphics (TOG), 28*(4), 106.

Paiva, A., Dias, J., Sobral, D., Aylett, R., Woods, S., Hall, L., & Zoll, C. (2005). Learning by feeling: Evoking empathy with synthetic characters. *Applied Artificial Intelligence, 19*(3–4), 235–266.

Poggi, I., Pelachaud, C., de Rosis, F., Carofiglio, V., & De Carolis, B. (2005). Greta: A believable embodied conversational agent. In O. Stock, & M. Zancarano (Eds.), *Multimodal intelligent information Presentation* (pp. 3–25). Dordrecht, the Netherlands: Springer.

Reynolds, C. W. (1987). Flocks, herds and schools: A distributed behavioral model. *ACM SIGGRAPH Computer Graphics, 21*, 25–34.

Reynolds, C. W. (1999). Steering behaviors for autonomous characters. *Game Developers Conference.* San Francisco, CA. http://www. Red3d. Com/cwr/steer/gdc99

Rickel, J., & Johnson, W. L. (1997). Integrating pedagogical capabilities in a virtual environment agent. *Proceedings of the First International Conference on Autonomous Agents* (pp. 30–38). New York, NY.

Rickel, J., & Johnson, W. L. (2000). Task-oriented collaboration with embodied agents in virtual worlds. In J. Cassell, J. Sullivan, S. Prevost, & E. Churchill (Eds.), *Embodied conversational agents* (pp. 95–122). Cambridge, MA: MIT Press.

Rickel, J., Marsella, S., Gratch, J., Hill, R., Traum, D., & Swartout, W. (2002). Toward a new generation of virtual humans for interactive experiences. *Intelligent Systems, IEEE, 17*(4): 32–38.

Scherer, S., Glodek, M., Layher, G., Schels, M., Schmidt, M., Brosch, T., Tschechne, S., … Palm, G. (2012). A generic framework for the inference of user states in human computer interaction: How patterns of low level communicational cues support complex affective states. *Journal on Multimodal User Interfaces, Special Issue on: Conceptual Frameworks for Multimodal Social Signal Processing, 6*(3), 117–141. doi:DOI 10.1007/s12193-012-0093-9

Shapiro, A. (2011). Building a character animation system. *Proceedings of the 4th International Conference on Motion in Games* (pp. 98–109). Edinburgh, U.K.: Springer.

Shoulson, A., Marshak, N., Kapadia, M., & Badler, N. I. (2013). ADAPT: The agent development and prototyping testbed. *Proceedings of the Symposium on Interactive 3D Graphics and Games, I3D.* Orlando, FL: ACM.

Simon, H. A. (1967). Motivational and emotional controls of cognition. *Psychological Review, 74*(1), 29–39.

Starck, J., & Hilton, A. (2007). Surface capture for performance-based animation. *Computer Graphics and Applications, IEEE, 27*(3), 21–31.

Starck, J., Miller, G., & Hilton, A. (2005). Video-based character animation. *Proceedings of the 2005 ACM SIGGRAPH/Eurographics Symposium on Computer Animation (SCA '05)* (pp. 49–58). New York, NY: ACM. doi:10.1145/1073368.1073375 http://doi.acm.org/10.1145/1073368.1073375

Thalmann, D. (1990). Robotics methods for task-level and behavioral animation. *Scientific Visualization and Graphics Simulation* (pp. 129–147). Chichester, UK: John Wiley.

Thiebaux, M., Marsella, S., Marshall, A. N., & Kallmann, M. (2008). Smartbody: Behavior realization for embodied conversational agents. *Proceedings of the 7th International Joint Conference on Autonomous Agents and Multiagent Systems* (Vol. 1, pp. 151–158). Estoril, Portugal: International Foundation for Autonomous Agents and Multiagent Systems.

Tolani, D., Goswami, A., & Badler, N. I. (2000). Real-time inverse kinematics techniques for anthropomorphic limbs. *Graphical Models, 62*(5), 353–388.

Traum, D., Swartout, W., Gratch, J., & Marsella, S. (2008). A virtual human dialogue model for non-team interaction. In L. Dybkjoer & W. Minker (Eds.), *Recent trends in discourse and dialogue* (pp. 45–67). Dordrecht, the Netherlands: Springer. http://link.springer.com/chapter/10.1007/978-1-4020-6821-8_3

Traum, D. R., & Larsson, S. (2003). The information state approach to dialogue management. *Current and New Directions in Discourse and Dialogue, 212*, 325–353.

Van den Berg, J., Lin, M., & Manocha, D. (2008). Reciprocal velocity obstacles for real-time multi-agent navigation. *IEEE International Conference on Robotics and Automation, 2008. ICRA 2008* (pp. 1928–1935). Pasadena, CA: IEEE.

van Welbergen, H., Reidsma, D., Ruttkay, Z. M., & Zwiers, J. (2009). Elckerlyc. *Journal on Multimodal User Interfaces, 3*(4), 271–284.

Waal, F. (2003). Darwin's legacy and the study of primate visual communication. *Annals of the New York Academy of Sciences, 1000*(1), 7–31.

Wang, R. Y., Pulli, K., & Popović, J. (2007). Real-time enveloping with rotational regression. *ACM Transactions on Graphics (TOG), 26*(3), 73.

Watt, M., Cutler, L. D., Powell, A., Duncan, B., Hutchinson, M., & Ochs, K. (2012). LibEE: A multi-threaded dependency graph for character animation. *Proceedings of the Digital Production Symposium* (pp. 59–66). Glendale, CA: ACM.

Yamane, K., & Nakamura, Y. (2003). Natural motion animation through constraining and deconstraining at will. *IEEE Transactions on Visualization and Computer Graphics, 9*(3), 352–360.

Yeh, T. Y., Faloutsos, P., & Reinman, G. (2006). Enabling real-time physics simulation in future interactive entertainment. *Proceedings of the 2006 ACM SIGGRAPH Symposium on Videogames* (pp. 71–81). Boston, MA: ACM.

Yersin, B., Maïm, J., Pettré, J., & Thalmann, D. (2009). Crowd patches: Populating large-scale virtual environments for real-time applications. *Proceedings of the 2009 Symposium on Interactive 3D Graphics and Games* (pp. 207–214). Boston, MA: ACM.

Zeller, C. (2005). Cloth simulation on the GPU. *ACM SIGGRAPH 2005 Sketches* (p. 39). Los Angeles, CA: ACM.

Section IV

Design Approaches and Implementation Strategies

15 Structured Development of Virtual Environments

Richard M. Eastgate, John R. Wilson[†], and Mirabelle D'Cruz

CONTENTS

[†] Deceased.

15.1 INTRODUCTION

15.1.1 BACKGROUND

In the early 1990s, little systematic attention was paid to the way in which virtual environments (VEs) were or should be designed. This was for a number of reasons. First, many early VEs were *proof-of-concept* demonstrations (with fairly crude object representations), produced at great speed often by trial and error to sell the idea of VEs to potential user organizations. Little time or thought was given to defining the development process. Second, several early VEs and development processes were commercially or militarily confidential and therefore publication was not an option. Third, VE games were often based on other computer games, and in any case, their developers had no scientific reason to publish anything on their design and development process and may have had good commercial reasons not to; no published accounts therefore exist. The fourth category of early VEs was simple usability test prototypes, to allow ergonomists and psychologists to experiment with various aspects of the virtual experience and with factors of VE usability. Again, there would be little need for, or thought given to, the VE design process.

In more recent times, especially with the transfer of VE technology into manufacturing and construction (for instance, in the automotive industry or public space architecture), there has been much greater need for formal specification of the VE development process and thus for documentation of this.

Early accounts or proposals relevant to VE development were produced, inter alia, by the 3D interaction group at Virginia Tech (e.g., Bowman, Gabbard and Hix, 2002; Bowman, Kruijff, LaViola, & Poupyrev, 2004; Gabbard, Hix, & Swan, 1999; Hix et al., 1999) and University of Nottingham (e.g., D'Cruz, 1999; D'Cruz, Stedmon, Wilson, Modern, & Sharples, 2003; Eastgate, 2001; Neale, 2001; Sharples et al., 2007; Tromp, 2001; Tromp, Sharples, & Patel, 2006). More recently, POSTECH has looked at making the VE development process more efficient but concentrates on the modeling aspects of VE development (Seo & Kim, 2002). Chen and Bowman (2009) looks at interaction techniques with a view to finding a useful compromise between generic reusable interactions and those designed for a specific application. Mansouri, Kleinermann, and De Troyer (2009) describe a tool for applying semantic rules to VE development so as to reduce the development iterations, but only reported on its use with static walkthrough VEs. Again, concentrating on modeling, Trescak, Esteva, and Rodriguez (2010) present a Virtual World Grammar that, in combination with their Virtual World Builder Toolkit (VWBT), can automatically generate a 3D world from a multiagent system specification.

Notwithstanding the somewhat anarchic nature of the PC games software industry, and especially the unstructured nature of their development alluded to the aforementioned, there are moves to better understand best practice in games development. The past decade has seen an unprecedented growth in 3D interactive gaming and in *serious games* or edutainment, even receiving funding from the European Framework Seven (FP7) work program for a Network of Excellence called Games and Learning Alliance for Serious Games (GALA). Serious games are those designed for another purpose (usually some form of education or training), and not just for entertainment. As such, the gaming community is looking toward the scientific community for educational models and 3D user interface techniques (Ferdig & de Freitas, 2012). Even recognizing the contributions identified earlier, there is still a consensus that much work of VE specification, building, and testing is characterized by a lack of guidance across the whole process (Molina, García, López, & González, 2005; Seidel, Gartner, Froschauer, Berger, & Merkl, 2010). Systematic guidance for VE development can reap dividends in terms of a more efficient design process and more functional and usable VEs. The allocation of limited processing resources, the nonintuitive nature of (most of) the physical interfaces, and the limits that must be applied in deciding what aspects of the real world to model are all issues that must be addressed within a structured VE development framework.

To some extent, VEs could be developed following design process guidance for any human–machine system or human–computer interaction (HCI), of which there are vast numbers (e.g., Stanton, Hedge, Brookhuis, Salas, & Hendrick, 2005; Sutcliffe, 1995; Wickens, Gordan, & Liu, 2004). However, the effort required to model objects and actions of the real world and the nature of the cognitive interface (the participant being *within* the database) mean that VE development also requires guidance specific to its own special characteristics and needs. Further, the experience of the authors is that guidelines and models within human factors and *conventional* HCI communities are notorious for not always being useful or usable by designers, engineers, or even ergonomists, as they are often too general to be of use or too specific for transfer to other applications. Because VE design has sufficiently different attributes and purpose, it is recommended that a standard, VE-specific structured development framework and associated guidelines be developed.

This chapter first reviews past research in VE design standards and remaining challenges. A framework—Virtual Environment Development Structure (VEDS) for VE design—is then presented. Subsequently, this chapter expands upon the framework's major stages of project definition, requirements analysis, specification, overall design, resource acquisition, detail design, building, verification, deployment, and validation. Throughout the chapter, the focus is on design *of* VEs, in contrast to design *with* VEs (e.g., solving design problems) or design *for* VEs (VE system improvements). As Bolas (1994, p. 53) asserted "design of virtual environments is perhaps the most exciting and challenging VR design task."

15.1.2 Barriers to VE Application

From our previous industrial work and supported by other sources (DTI, 1996; Sharples et al., 2007; Wilson, Cobb, D'Cruz, & Eastgate, 1996; Wilson & D'Cruz, 2006), four barriers to wide development and acceptance of VE applications have been identified, which has meant that most companies were not expecting take-up of the technology until at least 2001. The barriers were defined as

- *Technical*—particularly integration with other technologies and the implicit technical trade-offs in data processing
- *Applications*—the paucity of serious applications working in an operational environment, and available for inspection
- *Usability*—the nonintuitive nature of the interfaces, both hardware and software, and concerns about possible side and aftereffects, particularly with head-mounted displays (HMDs) (Gross, 2014; see Chapter 20)
- *Evaluation*—the poor evidence of added value from VE application (for some notable exceptions, see Chapters 28 through 35)

These four barriers to VE application still exist today to varying degrees. The technical barrier is still regarded as the main obstacle to widespread VE use—developing a VE that works in the way an end customer wants it to is still very time consuming, less so because of the technical limitations that have largely been overcome but more so as customer expectations are raised by big-budget films and computer games that can take hundreds of thousands of person hours to program. There are still limitations in that sharing of data and models with other systems is often not possible because of the specific requirements of the VEs. Often, even sharing models produced on the same VR platform is not possible because the models were developed to simulate different aspects of an object's behavior. When the added complication of converting models between different file formats is added, this can often make it more economic to build models from scratch.

The paucity of applications has in some ways been alleviated by the embedding of VR-type interfaces in domestic technologies such as satellite navigation systems and 3D TV and in more professional systems such as those used in computer-aided design (CAD). However, while these systems serve to familiarize people with the possibilities of 3D, they do little to increase the understanding of what a VR system is capable of and what it can be applied to.

Usability is still an issue in VE development for the simple reason that many developers think that it can be bolted on afterward so do not give it much thought until it is too late. Developers can also be somewhat blinkered in that they believe that if they can use the software, everyone else should be able to as well. The result of this is that many VEs reach the validation stage with nonintuitive, inconsistent, and often overcomplicated interaction requirements that make the VE unusable except to the developers themselves. This situation is often exacerbated by inadequate evaluation whereby the failings are not recognized as such and the VEs are released in an immature state. This reduces the users' confidence in the technology and will prevent them from using that technology again or recommending it to other users. For methods to address these limitations, see Chapter 29, this book.

To the four original limitations or barriers outlined earlier, one can add a fifth, resulting from and leading to some of the others: the lack of a structured methodology by which organizations might establish a potential need for a VE solution and then specify, develop, employ, and evaluate VEs in real applications. All five barriers have been ongoing challenges in more recent and current research projects. For example, the aim of Virtual and Interactive Environments for Workplaces of the Future (VIEW of the Future) (IST-2000-26089) was to develop best practices for appropriate design, implementation, and use of VEs in industrial settings, particularly for the purposes of product development and testing and of industrial training. The outcome was a large number of new methodologies, technologies, and concepts related to analysis of user needs and requirements (User Requirements Document [URD]), application platform development (BAF, PI-casso, V€-VIEW), new interaction concepts and devices (multiple decoupled interaction, Bug, Dragonfly, Virtual Prints), development of guidelines and interactive design tool (I-doVE, VIEW-IT, NAÏVE), and evaluation criteria and methodologies (usability test battery [UTB]) (Wilson & D'Cruz, 2006). This initiated the European-funded network of excellence INTUITION (virtual reality and virtual environment [VR/VE] applications for future workspaces), aimed at systematically acquiring and clustering knowledge on VR concepts, methodologies, and guidelines to provide an understanding of the state of the art and provide a reference point for future projects (www.intuition-eunetwork.org). This effort ran for 4 years (2004–2008) and consisted of 58 organizations around Europe. The outcome was a large number of consolidation reports related to users, applications, technologies, and methodologies, which were integrated into a public wiki knowledge base to enable the information to continue to evolve and be current. In light of the barriers given earlier, the knowledge base provided links to technology available and more specifically VR technology developers so that potential users could discuss with experts their technical issues. Regarding the lack of applications, usability issues, and real evidence, INTUITION had 11 working groups—aerospace, automotive and transport, constructions and energy, entertainment and culture, medicine and neuroscience, education and training, engineering and design, haptic interaction, augmented reality, evaluation and testing, and VR/VE technologies—dedicated to exchanging best practices, eliciting needs and requirements, and supporting all relevant stakeholders in the process. In addition, it provided reports and guidelines for design and evaluation (Patel, D'Cruz, Cobb, & Wilson, 2006). INTUITION led to the founding of the European Association for Virtual and Augmented Reality (euroVR—http:// www.euro-association.org) whose goal is to support industry in their design, development, and evaluation of the technologies by providing a forum to meet and exchange needs, requirements, and developments. Therefore, the work of overcoming these barriers will continue.

Other industrial projects such as DiFac (Digital Factory for Human-Oriented Production System), ManuVAR (Manual work support throughout system lifecycle by exploiting virtual and augmented reality), and IMOSHION (Improving Occupational safety and health in European SMEs with help of simulation and virtual reality) have all focused on aspects of the technical limitations. The aim of DiFac (IST-5-035079), working with 12 partners (five industrial), was to develop a collaborative manufacturing environment based on a framework to support group work for product design, prototyping, and manufacturing as well as worker training (Redaelli, Lawson, Sacco, & D'Cruz, 2009). The underlying foundation of the framework was based on three pillars—presence, ergonomics, and collaboration. These pillars supported the design of the technology and performance of the tasks; however, the difficulty in integrating different software and interfaces into a shared collaborative VE (CVE) was the greatest challenge. ManuVAR (CP-IP-211548) had 18 partners (10 industrial) and aimed to support highly skilled, high-value manual workers with human factors methods supported by VR and AR technologies (Krassi, D'Cruz, & Vink, 2010). In a similar way to DiFac, the aim was to develop an innovative technology platform and a framework to create a shared collaborative space for bidirectional information exchange throughout the product life cycle. Five industrial case studies represented the stages of procedural design, workplace design, training, maintenance, and product review. Again, the technical issues posed the most challenges, in particular interoperability between technologies, integration within existing company legacy systems, and robustness of AR technologies on the factory floor. At the end of the project (April 2012), a roundtable involving senior representatives from the industrial partners was held in Turin. The most positive result was the eagerness of all the organizations to continue to develop their applications and to use the structured approach examined in ManuVAR. Finally, the ongoing IMOSHION project has four research partners and four small-to-medium-sized enterprise (SME) associations providing access to over 200 SMEs in Spain, Bulgaria, and Germany, to develop tools including VEs to support them in their training of health and safety issues. The barriers to implementing such high-end technologies in SMEs differ from those in large industries. Unlike large industries who may have legacy systems that make it difficult to introduce new technologies, SMEs can often adopt new technologies in a shorter time frame. However, within the company, there needs to be someone who has information and communication technology (ICT) capabilities. It is often difficult to create bespoke systems as still now there are little off-the-shelf VE applications. The challenges of IMOSHION are to provide tools that SMEs can use to learn to create applications and that can be maintained beyond the end of the project.

15.1.3   Understanding Human Performance in VEs

The goal of VE development is to create effective, safe, usable VEs that enable participants to achieve goals in a better, more motivational, and cost-effective fashion than by using alternative technologies or by using the real situation. Producing VEs should thus ideally be based on an understanding of how people will behave and carry out tasks within the VEs. In practice, though, these considerations are often ignored or forgotten during the development process and, because most (real world) environments have largely intuitive navigation and interactions, developers assume that this will also be true of their VE. However, the previously mentioned limited and nonintuitive VE physical interfaces and clumsy efforts to compensate for these often result in VEs that are far from easy to use. Developers need general guidance on human performance within VEs and specific guidance on enhancing VE usability. Clarity about what determines usability of VEs will be improved with better understanding of participant performance.

There are two distinct fundamental approaches to account for the user in VE development that relate to the philosophy of the VE itself. These are *VE as an alternative real world* and *VE as a computer interface*. VEs can be treated as alternative representations of *real worlds*, and existing knowledge, models, criteria, and methods can be applied as appropriate. There are a number of challenges with this approach. First, there is little agreement about the meaning of such concepts as spatial awareness and perception in the real world (e.g., Flach & Holden, 1998), let alone in a virtual one. Second, even if the software could provide a strong representation of equivalent perceptions in the real world, the peripherals that people need to experience the VE currently interfere with this parallel. Third, there is still considerable confusion about the underlying psychological mechanisms and processes involved in the use of VEs, and whether and how these might be similar to those in the real world. For instance, what is a sense of presence and is it equivalent to or related to attention

(Zahorik & Jenison, 1998)? Although presence has been subject to much debate over time and now even groups involved in much of the early definition work are choosing to consider different terminology such as *place illusion* for the type of presence that provides "a strong illusion of being in a place in spite of the sure knowledge that you are not there" and plausibility illusion that refers to "the illusion that the scenario being depicted is actually occurring" (Slater, 2009, p. 3549). This is to distinguish the type of presence felt in VEs from that potentially produced by other types of media such as an interesting book or interesting film.

The other approach is to treat the VE as simply another, if potentially richer, computer interface and apply existing HCI knowledge, models, criteria, and methods. Again, there are a number of challenges with this approach. The state of knowledge and coherence of the HCI field has changed markedly, and HCI as a discipline has already had to redefine itself, first with the almost universal move from command line to graphical user interfaces and then with the growth of various multimedia distributed and ubiquitous interfaces. Is HCI knowledge ready to be applied to VEs, especially those where a participant is navigating across the interface and within the VE itself? There are a large number of technical variables that are specific to VEs and less relevant to work elsewhere in HCI, including the impact of temporal lags or problems with the optics and field of view of HMDs (see Chapter 3, this book), and that leave developers with trade-offs between potential attributes of complexity, pictorial realism, and speed of update. Therefore, although VE developers and researchers might build on the 30 or more years' research in HCI, any models, criteria, standards, or knowledge must be used with care. Their applicability needs testing by careful research.

To overcome some of these challenges, methods have been developed to systematically capture user and task needs, and utilize this knowledge to build system requirements and specifications. For example, task and user analyses are useful methods in identifying and categorizing participants' requirements, and storyboarding can outline how the VE should be specified and designed to meet those requirements. A more detailed examination, for example, using personas, scenarios, and/or use cases, should be made of how participants will respond to different elements of the VE, utilize all its functionality, and be able to comprehend and use interface elements to meet application goals. As part of this, developers need to know how to encourage participants to explore the VE and enable them to understand which elements can be interacted with, identify how this interaction might be achieved, and minimize dysfunctional participant behavior and serious errors. In other words, an improved understanding of human behavior and performance within a VE is needed.

There are a number of candidate theories and approaches that might help embed understanding of human behavior within the designed VE. First, behavior of groups of workers in complex technical and organizational environments can be interpreted within a framework of distributed cognition (e.g., Hutchins, 1995; Perry, 2003). In such a view, the handling of information and decision making that characterize work in transport control, for instance, is distributed in time and space across teams of people and various computer and telecommunication systems. An ethnographic approach may then be taken to study and interpret work behavior and skills. This may be a framework within which to study performance in networked VEs or CVEs.

Second, we can build an understanding of participation in VEs and thus guidance for their design, around the notion of situated action (Suchman, 1987) whereby the user responds to the situation as they perceive it utilizing the resources (skills, experience, etc.) they have at their disposal. In this view, work cannot be understood merely in terms of individual skills and institutional norms, but is the product of continually changing people, technology, information, and space. Further, people do not approach tasks with a clear set of plans in mind, but build and modify their plans in the course of situated action.

A third possibility for understanding performance in VEs is to utilize the notion of situation awareness. People perceive cues from their environment and use them to make sense of the current state of the world and to project this understanding into the future (Endsley, 1995). Various measurement methods are available to assess situation awareness that might have value in understanding and measuring performance in VEs. Also, there is a strong connection between the role played by situation awareness and the participant's mental model (Endsley & Jones, 1997; Sarter & Woods, 1991). Guidance on design usability then could concentrate on those VE elements expected to help build a strong and *correct* mental model for the participant or to match their existing ones and to contribute to high levels of situation awareness.

Finally, developers could borrow from a large variety of models and frameworks used to understand, predict, and reduce human error, generally in safety critical systems. A particular possibility here is Rasmussen's (1986) well-known SRK model of skill-, rule-, and knowledge-based behavior. If the types of tasks in the VE can be defined as to whether they are skill, rule, or knowledge based, one might be able to distinguish between the types of design guidance offered for each. More usefully, characterization of real-world tasks as skill, rule, or knowledge based, and particular variants of these, could allow association of different ways to deliver their representations in the VE, with rules and knowledge using largely cognitive interactions and skills being more likely to require specific physical interactions. The SRK model is often used in conjunction with Reason's (1990) typology of human error—slips, mistakes, and violations—and the underlying psychological error mechanisms involved. Again, adaptation of these to the VE domain could allow a better understanding of the types of behavior exhibited by VE participants and, subsequently, lead to production of useful development guidance.

15.1.4 Usability of VEs

The notion of usability of VEs and usability engineering is explored in depth elsewhere in this book (see Gabbard, 2014; Chapter 28). Included here are only the major issues of usability that must be addressed in the VE development process. Some aspects of usability are covered later in this chapter in the relevant sections of VEDS, but a number of usability issues have implications for across many parts of the VE development process and so these are introduced here.

A North Atlantic Treaty Organization (NATO) sponsored workshop on VE research requirements started with a brief that the most important issues to discuss were aftereffects and sense of presence. Twenty-five specialists from Germany, the United States, and United Kingdom prioritized problems and research needs and established interconnections between issues. Many concerns were with systems design, for instance, display design, latency, and real-time interaction. Beyond these, the workshop participants agreed with the importance of minimizing aftereffects, the need for better understanding of participants' sense of presence, and on the need for a substantial human factors research agenda to address VE usability concerns (Stanney et al., 1998).

One of the barriers to the proliferation of VE technology identified earlier was widespread concern over the usability of both the VE equipment and of the VE itself. The equipment concerns will limit the freedom of systems developers who, for example, will not want to specify or deliver a system that is hard to set up, calibrate, or maintain. These concerns also manifest themselves in terms of the design, comfort and fit of HMDs and handheld devices, temporal and spatial resolution limitations, field of view restrictions, and generally the limitations implicit in communication using visual and auditory information only, with none or at best primitive use of other real-world information that can be acquired, for example, from vibrations, forces, smells, and temperatures. Some of these systems concerns translate into potential problems of usability within the VE itself, for example, if the field of view is narrow resulting in the user missing information off to one side or having to move their head excessively to view a process. Others might be minimized by careful design of the VE.

There are a large number of usability issues related to successful implementation and use of VE technology (see Barfield & Furness, 1995; Draper, Kaber, & Usher, 1998; Sharples et al., 2007; Wann & MonWilliams, 1996; Wilson, 1997, 1999). Few have been satisfactorily addressed as yet, although the work of Gabbard (2014; see Chapter 28, this book), Kaur (1998), Tromp (2000), and Sutcliffe and Gault (2004) is helping to address this. For each usability issue, one must ask: Is it a significant issue for VE development and for its safe and effective use? Are the usability issues unique? Can we utilize data and criteria from other domains and from knowledge of performance of related tasks? Broad usability issues to be considered during VE development include

- *Forms of representation of the participant within the VE*: None, limbs, full lifelike; these affect how the user understands where they are, what they are doing, and what the possibilities for interaction are in the VE
- *Modeling of avatars and mannequins*: Sizes, shapes, appearance, movements, facial expressions; how does the user relate to them and how do they affect presence (Marsella & Shapiro, 2014; see Chapter 14)
- *Supporting navigation and orientation within VEs*: Interface tools and other aids, shortcuts, familiarization routines, optimum world sizes (Darken & Peterson, 2014; see Chapter 19)
- *Understanding and enhancing presence and involvement in VEs*: Balancing pictorial realism, size and complexity of the VE, update rules, use of sound and shadowing; enhancing interest (Schatz & Chertoff, 2013; see Chapter 34)
- *Requirements for cues and feedback to assist the participant (see later discussion)*
- *Minimizing any side and aftereffects*: Sickness, performance decrement, and physiological change (Keshavarz et al., 2014; see Chapter 26)
- *Providing interface support and tools for interactivity*: Mixed-reality design, metaphors, interface elements, tool kits, and so forth

15.2 VEDS: THE VE DEVELOPMENT STRUCTURE

Development of any product or system, including setting appropriate requirements and specifications, will be aided by guidance from a clear, flexible framework that defines what issues are relevant, what data and information are needed, what decisions must be made at what stages, and how these stages interconnect, including feedback loops and iterations. The human factors/ergonomics community is clear on the need for a process or framework to be established. Only in this way can early design decisions be taken to improve the usability, likeability, and general acceptability of a product or system in enough detail in order to influence the final outcomes. VEs are no different than any other product, system, or human–computer interface, in that their development needs clear goal setting and constraints, requirements analysis, task and user analyses or models, appropriate interface guidelines, predictions of task performance, an iterative design/test cycle, and a clear evaluation process.

Over a number of years, we have worked with a wide range of potential users of VEs in aerospace, automotive, manufacturing, medicine, and education to produce a structured framework to guide VE development—the VEDS. Various generations of this framework and additional detail can be found in Wilson et al. (1996), Wilson (1997, 1999), Eastgate (2001), D'Cruz et al. (2003), and Lawson, Hermawati, D'Cruz, Cobb, and Shalloe (2012). A top level and simplified outline of the main stages of VEDS is shown in Figure 15.1.

As drawn, it appears as if only a top-down approach is suggested. Although a structured development process is perhaps best suited to top-down working, it is recognized that certain aspects of development will need to take place bottom-up. Bryson (1995) terms these as *design for the task and interface metaphor* (top-down) and *design for performance* (bottom-up). Also, only the main feedback loops are shown; this is for clarity in the diagram, the boldness of the line representing the frequency of use rather than the importance of the link. The whole process is iterative, embracing formative and summative evaluation and a design morphology as the VE (and test methods) becomes more sophisticated through design cycles within the process.

The key stages of VEDS are described in the rest of this chapter, but the whole process is summarized here. At the outset, a clear view of the attributes of VR/VEs and knowledge of the domain are needed to obtain agreement on goals for the application and to identify likely priorities and constraints.

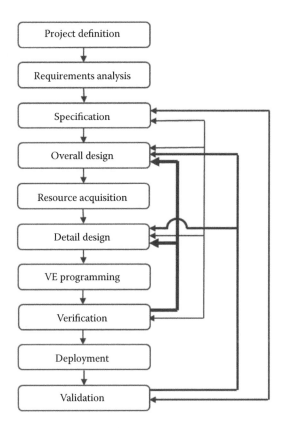

FIGURE 15.1 VEDS outline.

Considerable effort in requirements analysis and in function allocation (for instance, when it is preferable for the VE or intelligent agents within it to control a participant's options or actions) will pay off later in improved design phases. Technical and systems limitations can be reduced in their effects on the application by appropriate choice of technology in the light of a task analysis and by goal-oriented and user-need-based VE programming. Design must be rational, realistic, and parsimonious, rejecting any elements not important to the core application goals and to support user performance. The number of objects modeled, and their level of detail and degree of associated intelligence, must also be considered carefully before including them within the VE. As with any product or system, the participants' performance within the VE will be determined by the environment itself, personal characteristics, tasks to be performed in or with the VE, and the circumstances under which those tasks are carried out (for instance, without training, under time pressures, or with incentives). A virtual task analysis, which may build on the real-world task analysis carried out earlier, can support decisions within VE design and development. In addition to participants' performance within the VE and behavior with the VE system interface, subsequent outcomes might be manifest in the transfer of the participants' effects upon the real world (for instance, measured as design process effectiveness, better product designs, or training program efficiency), their reported and observable experiences, and any health-and-safety-related side effects. At the outset, when defining the application and activities, alternative solutions that do not employ VEs should be identified, and evaluation methods, measures, and criteria set. Although the validation may be somewhat crude at times, it must be carried out to assess the success of the VE in meeting its goals, examining both individual and organizational effects against the evaluation criteria.

In the rest of this chapter, the different stages of VEDS are discussed with particular emphasis on structured analysis, specification and building.

15.2.1 Project Definition

The process begins with defining the project by describing the background, identifying relevant stakeholders, and understanding the potential priorities and constraints (Figure 15.2). This can be achieved by talking to the client (or other persons proposing the project) and any other parties the client deems relevant. It is important to understand the background of the project as this provides some context and some understanding as to why this effort has been initiated. This can be used to drive the purpose and direction of the work and should include the ultimate goal (e.g., time and cost savings) and expected outcomes (e.g., decrease in time taken to perform task, 20% cost savings) but will also include other aspects such as whether this is entirely new or replaces an existing system, the client's vision of how the end product might work, and any specific problems they might foresee. This will enable you to assess how important the project is, whether and why it is needed, and how this can be measured.

This in turn will also help to identify the relevant stakeholders. It is necessary to define all the people who will have some influence and/or be influenced by the solution to ensure requirements meet all stakeholders' needs. This includes not just the end users but could also include the client, sponsor, customer, subject matter experts, members of the public, users of a current system, marketing experts, legal experts, domain experts, usability experts, representatives of external associations, business analysts, designers, developers, testers, system engineers, software engineers, technology experts, and system designers (Robertson & Robertson, 2006a).

With the relevant stakeholders, it is then necessary to define the priorities and constraints of the project. It is unlikely that everything can be developed or that the solution can address all issues. Therefore, it is necessary to agree upon the main aims and objectives and thus the priorities of the project. In addition, at this stage, it is necessary to also consider the likely constraints, the most obvious being the budget. Money will have a major influence on the whole project as well as the rest of the resources available for the project. There is no doubt that the cost of VEs is presently high in terms of the money required for the development time and personnel. Also, the limitations on the time and number of the stakeholders available for the design and evaluation process need to be considered in the scheduling of the development. Therefore, to have some control over the costs of developing a VE application, reasonable constraints must be placed on the project.

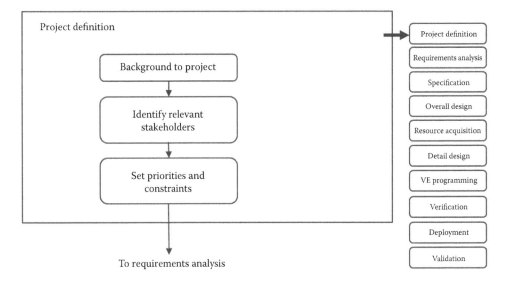

FIGURE 15.2 An expanded view of the project definition stage of VEDS.

15.2.2 Requirements Gathering and Analysis

An important stage in requirements gathering is the selection of appropriate methods (Figure 15.3). There are several different methods for requirements elicitation (e.g., interviews, questionnaires, observation, ethnography) as well as several good guides on their use and the types of data they provide (Champney, Carrol, Milham, & Hale, 2008; Kirwan & Ainsworth, 1992; Stanton et al., 2004; Stanton, Salmon, Walker, Baber, & Jenkins, 2005; Milham, Carroll, Stanney, & Becker, 2009; Wilson, 2005). When working with industrial partners, in particular with those from SMEs, it is important to base selection of the methods not only on the type of participant or data required but also on the time constraints and availability of key stakeholders. As it is unlikely that most stakeholders will be engaged full time in the design of the VE, it is necessary to consider methods that can minimize their time input or methods that can be conducted while they continue to do their job (Lawson et al., 2012). The classic starting point for application of ergonomics in systems design, following the agreement of aims, objectives, constraints, and stakeholders, is the task or function analysis and user analysis (Kirwan & Ainsworth, 1992).

The early process of VE development to meet the needs of users differs little from that normally followed in system design, namely, to (1) assess what tasks must be completed in the VE and in the real world subsequently via a task or function analysis; (2) understand the context for this use and the environmental, organizational, cultural, and other constraints; (3) determine all the stakeholders and their roles and needs; and (4) define the end-user characteristics and needs relevant to the systems design and especially the interface design. Such a process should allow the client and the developer of the VE, together with the ergonomist or psychologist, to specify the goals to be achieved by the VE. Again, these goals should be prioritized, because only rarely can all be attained. In developing VEs, if the intention really is to enable cognitive (if not always physical) replication of interaction with the real world, then the task analysis will be carried out twice, firstly for the actual tasks being modeled, the application task analysis, and secondly for the tasks to be carried out within the VE, the virtual task analysis. The first of these is axiomatic for human–machine systems development, to understand what tasks must be carried out within the domain where the system will be used and the requirements or constraints this will impose upon the people involved. This should be followed by a user analysis

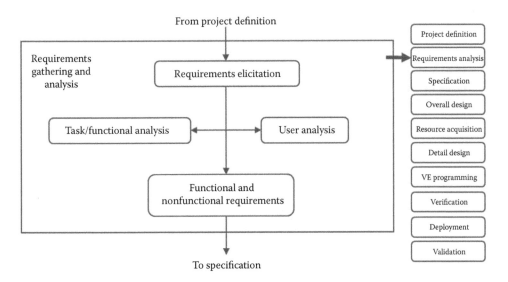

FIGURE 15.3 An expanded view of the requirements gathering and analysis stage of VEDS.

to define the role, characteristics, and activities of the user in performing the tasks. Such early task and user analysis will then enable decisions to be made about

- Whether VE technology even has a role for the particular application and what alternative technologies there are
- Which functions within the application will be supported by using VE technology (in many applications, a mixed setup, comprising virtual and hard elements, may be appropriate)
- How VE technology will be integrated with other technologies if there is a mixed environment
- How eventual use of the VE will be evaluated in terms of direct and indirect outcomes and effects
- Which VE system (hardware and software) is most appropriate for the tasks, task environment, and user group
- What elements of the real world and tasks should be included in early specification of the VE design

Some of the development decisions that might emanate from the task and user analysis are shown in Table 15.1 (see also Gabbard & Hix, 1997, 2013; Chapter 28).

Some of the decisions implied in Table 15.1 may not be possible to make at this early stage and their resolution may depend upon the virtual task analysis that is carried out later. This analysis phase of development provides information, ideas, and a set of requirements on which to base conceptual design and VE specification covered in the next section. There are several types of requirements—user, functional, and nonfunctional requirements. User requirements should be clear, complete, and verifiable while still providing flexibility to enable the developer to provide some innovation. Functional requirements should specify what the system must enable the user to achieve, and nonfunctional requirements are other considerations that may place restrictions or constraints on the solution, such as performance or development constraints (e.g., usability, aesthetics, security, reliability, maintainability). There are a number of ways of documenting user requirements, such as the Volere Requirements Specification Template (Robertson & Robertson, 2006b). An example of a requirements gathering form as used in the CoSpaces project (IST-5-034245) is shown in Table 15.2.

TABLE 15.1

Table of Decisions That Need to Be Made on the Basis of Application Task Analysis and Virtual Task Analysis

Decision	Choices
Degree of *reality* desirable	Real-world replication → highly abstract
Type of presence	Egocentric, exocentric
Participants	One, few, many
Temporal collaboration	None, asynchronous, synchronous
View of scale	Close-up, arm's length, room sized, far horizon
Vision	Monocular, biocular, stereoscopic
Manipulation	None, click/drag, one hand, both hands, upper limb, feet, whole body
Virtual tool provision	None, separate palette, within the VE
Mobility	Stationary, ground anchored, flying
Viewpoint control	None, restricted, six degrees of freedom (DOF)
Sensory cues	Visual, auditory, haptic, olfactory

TABLE 15.2

Excerpt from a Requirements Gathering Form, as Used in the CoSpaces Project

Scenario Heading		Detail of Current Situation	Vision for Future; Comments
1. Company(ies)—type, etc. *Which is the company concerned, or which grouping or network of companies? What is their market position?*			
2. Area/Dept(s) *For example, design engineering, structural testing, architect, client, etc.*			
3. Function(s) and process(es) *What is carried out in the focus area/dept.—for example, testing wind resistance of body profile or assessing service systems access to the building. The process could usefully be described with diagrams and a timeline based on an actual example.* *(a) What is the approximate time frame of this activity/stage?*			
4. Goals *These are the goals for the current functions and processes.*	Business *Market position, structuring or financial, etc., goals.*		
	Operational *Goals of the function or process focused on.*		
	Human *Goals in terms of the type, contribution, and support for the people involved.*		
5. Evaluation of current functions and processes *This is probably the most important part (together with the process/function description) of the scenario and more detail is better here than less. The problems and current good points will determine where and how CoSpaces collaborative work environments may bring improvements.*	Needs/problems *In the current situation—these probably explain why this area, function, process, or activity is the one selected as relevant to potential CWEs.*		
	Successes *In the current situation what works well, what does not want to be lost in any change implementation—in terms of performance, technical setups, human factors, etc.*		

Source: Wilson, J.R. et al., Human factors and development of next generation collaborative engineering, in *Proceedings of the Ergonomics Society Annual Conference*, Taylor & Francis, London, U.K., April 22–23, 2009.

15.2.3 SPECIFICATION

The requirements analysis phase will leave the development team in a position to prioritize the goals for the VE implied by the application goals (Figure 15.4). Out of the analyses, the target VE can be specified in terms of its goals, the expected user tasks, complexity, and the balance between interactivity and exploration afforded. This specification must then be agreed by the VE development team and client using techniques such as storyboards, personas, and scenarios.

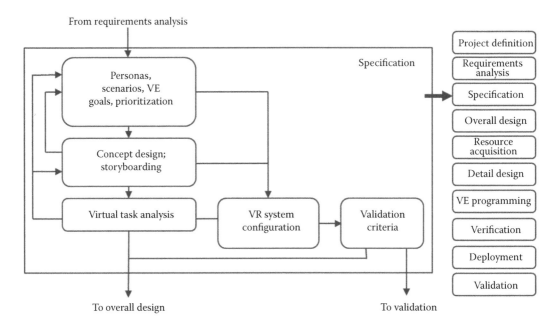

FIGURE 15.4　An expanded view of the specification stage of VEDS.

15.2.3.1　Personas and Scenarios

The requirements generated will often be specific to certain groups of stakeholders and these can be used to generate personas. Personas are a useful method for helping developers to understand more about the typical users of the system (Cooper, 1995). They are descriptions that often include an example name, image, and user characteristics (e.g., job role, experience, attitudes, competences).

These personas and requirements can be used to generate scenarios or concepts of operations that are short stories or other descriptions involving the various users, performing tasks in the real environment. Initial *as is* scenarios can be produced describing the current situation based on the analysis phase so that all the stakeholders and in particular the development team understand the context of the project. These *as is* scenarios can be used to generate a number of *to be* scenarios. These are descriptions of the future vision based on the expectations and desires of all the stakeholders. These *to be* scenarios can be used to identify the goals and specification of the VE and to identify some priorities.

15.2.3.2　VE Goals and Prioritization

The goals for the VE can be derived from the application goals. Some basic choices as to the structure and functionality of the VE might be made, in the light of the computer processing limitations for candidate VE systems and on the basis of the application task analysis and the virtual task analysis. (Note: Virtual task analysis is defined in Figure 15.4 as occurring at the end of the VE specification phase. In fact, it can occur at any time during specification or even during the design phase, much as application task analysis can take place at any time in the early phase of development.) For instance, VE developers and their clients might choose to look for added value from interaction via *manipulation* of objects or via *exploration* around the VE. The concept design will need to be robust enough to manage both VEs built to provide the participant with the ability to manipulate (e.g., switch on, switch off, open, turn) elements or objects in the VE and also those to walk through or around a large, rich, complex, but essentially unchanging VE (see Figure 15.5a and b).

From the scenarios, the target VE will be specified in terms of its goals, the expected user tasks, complexity, and the balance between interactivity and exploration afforded. This specification must then be agreed to by the VE development team and client, the most common means being storyboard development, which is another form of scenario.

(a)   (b)

FIGURE 15.5 **(See color insert.)** (a and b) An example of a VE that gives the participant the ability to manipulate elements within the VE as well as navigate through them. This shows a virtual ATM that can be *racked out* to practice day-to-day replenishment procedures by selecting buttons, catches, and switches.

15.2.3.3 Concept Design and Storyboarding

A storyboard is a set of frames, each using pictures and words to describe a stage in the experience offered by a VE. It can differ from a storyboard for a film in that a VE may not be a straightforward narrative sequence, so the links and relationships between frames may be more complex. On the other hand, it may not be necessary to show the links between the frames if, for example, the sequence of events does not affect the outcome. In that case, the frames will contain descriptions of interactive elements within that part of the VE.

VEs are often designed by more than one person, frequently by a group whose members work for different organizations and come from different backgrounds. Some of these people will have a good understanding of the possibilities offered by a VE. Others may be experts in their own field and have a good idea of what they want the VE to do, but little idea of what is feasible. A storyboard is a method by which the different parties involved can describe, design, outline, and agree on the form the VE should take. The developer can use the storyboarding process to work out how to match the requirements of the client to the limiting factors of the VE system. The storyboard can then be used by the developer to build the VE, ensuring during the process that the design closely matches the requirements of the client.

It is important to get all stakeholders involved at the storyboarding stage. Experience has shown that if this is not done, a VE may be developed, which does not meet the requirements of the one person who was not consulted, resulting in expensive and time-consuming redevelopment. Generally, the stakeholders in the storyboarding process are the project manager (to oversee the process and keep it within budget); the VE developer(s) (to establish exactly what the client wants and to explain what is technically feasible and how long it will take); the client (the person or people who specified the VE to make sure the VE will fulfill the specified requirements); representatives of the user population or participants (to make sure the VE will be suited to their characteristics and needs); and a usability or HCI expert—who is possibly one of the developers (to guide discussion from a user perspective, instantiate this into the design, and incorporate best practice in interface design). It is often the case that one person may fill more than one of these roles, in which case the size of the storyboarding group will be reduced. Typically, between two and five people collaborate to develop a storyboard. A storyboard can be graphical or textual (see Figures 15.6 and 15.7), and its content will depend on the type of VE and size of project. A set of potential contents is shown in Figure 15.8.

Once developers start building the VE, they will be confronted with many unexpected and previously unforeseen problems at every stage. It may not be possible for developers to make some decisions, and

4.3 Screen layout
Resolution 1024 x 768

4.3.1 VE window
- How big? 800 x 600
- Where? Top (centre?)

4.3.2 Icons
- Company logos? VIRART, XXXX, Uni of Nott
- Any buttons? Reset View (not world), automated walk through
- Where? Lower screen

4.3.3 Help boxes
Temporary over VE window (possibly only visible whilst right mouse button is held down?)

4.3.4 Instructions
Automated walkthrough (in the form of an AVI?) shows users what to do

4.4 Initial scenario
4.4.1 Room layout
4.4.1.1 Doors
Door to EPA room

4.4.1.2 Furniture
- Shelf
- Mat for HG testing
- Panel for WB testing
- Pictures on walls (XXXX, no smoking, ESD signs, etc.)

4.4.2 Clothing / equipment
4.4.2.1 Wrist band (WB)
On shelf

4.4.2.2 Heel grounder (HG)
On shelf

4.4.2.3 Overalls
On shelf

4.4.3 Actions
- No entry to EPA room without tested HGs and WBs and wearing overalls
- If attempt is made without them message tells user how to get them
- Click on HGs and WBs to select
- Feedback? Icon changes from bare wrist to wrist with WB

4.5 EPA room scenario
On entering the room a message appears;

There are 10 ESD hazards in this room. Use the ESD meter to find them. Click on the ESD meter with the left mouse button to pick it up. Use the left mouse button to click on items which you think are an ESD hazard.

4.5.1 Room layout
As built so far from electronic image provided by Neil

4.5.1.1 Doors
One entrance door

4.5.1.2 Furniture
- Benches
- Stuff on benches, e.g., soldering iron
- Shelves
- Swivel chairs
- Trolley

4.5.2 Hazards
4.5.2.1 Plastic Cup
Likely ESD voltage level (for ESD meter)?

4.5.2.2 Jacket on back of chair
Likely ESD voltage level?

FIGURE 15.6 Extract from a text-based storyboard.

- A wire-frame model or skeleton of the structure of the VE
- The initial scene presented when the participant first enters the environment
- The expected layout of the scene at subsequent points
- Any narration or instructions within the VE, whether aural or textual
- Any implicit or explicit signs or cues given to the participant
- Images to describe the activities in which the participant is expected to take part
- The broad class of control and input devices by which participants perform activities
- Images to show the consequences of various actions (both correct and incorrect)
- The approximate sequence of events given an "ideal" use (e.g., for training); where there is no sequence required (i.e., the user can perform the tasks in any order) this should be stated
- Any links with other software (e.g., multimedia clips)
- A "formal end" to the virtual experience if this is needed (for instance when a VE lesson has been completed or a full escape route navigated)

FIGURE 15.7   An example of a picture-based storyboard.

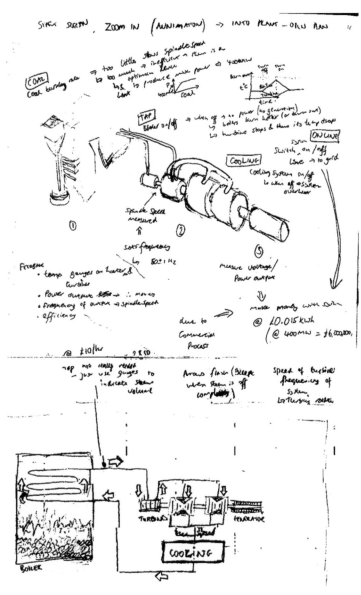

FIGURE 15.8   Potential elements in a storyboard.

they will have to be referred back to the relevant person in the storyboarding group, which can delay the development process. It is therefore important to establish as much detail as possible at the group storyboarding stage and to use the conceptualization that storyboarding allows to obtain consensus. Developers have to drive this process by asking questions about the design of the VEs.

The storyboard can be used for the basis of the virtual task analysis, which provides the developer more detailed specifications.

15.2.3.4   Virtual Task Analysis

During or after specification of the VE, there is the second application of task analysis: the analysis of the tasks to be performed by the participant within the VE. For example, if the activity is hammering a nail, the task analysis could break this down into picking up the hammer, picking up the nail, placing the nail over the location where it is to be hammered, raising the hammer, swinging it down onto the nail, etc. The virtual task analysis, however, may be much simplified in that it may only require mouse-clicking on the hammer, mouse-clicking on the nail, and a final mouse click on the target location. This virtual task analysis is a critical component of the VE development process for several reasons. First, there are trade-offs across the technical capabilities of the systems. For any particular level of system sophistication and cost, there must be prioritization among the variables of VE complexity (and thus sensory richness; update rate) related to sense of presence and any disturbing effects and interactivity (numbers of objects that can be *manipulated* in real time and how this is to be done). Analysis of the tasks that the participant must perform within the VE and any consequences and support needs can assist this prioritization. Second, a critical part of VE specification, assisted by the task analysis, is to define the minimum set of objects capable of interactivity and how cues to this interactivity might be provided (Eastgate, Nichols, & D'Cruz, 1997; Milham et al., 2009). Third, by extension, there is the production (through prediction and observation) of task analyses for collaboration and task performance within CVEs (Tromp et al., 1998a, p. 61). Finally, task analysis of what must be done within the VE to achieve set goals can be used to define behaviors that are expected (or not) to underpin successful use of the VE—for instance, in achieving learning objectives (e.g., Neale, Brown, Cobb, & Wilson, 1999)—and to give a basis to evaluation of participation within a VE.

Many of the decisions, which could have been made on the basis of the application task analysis (see Figure 15.4), may, in fact, be made after the virtual task analysis or be revisited at this stage in the process. Some assistance for the process of virtual task analysis may be obtained by use of taxonomies (e.g., Gabbard & Hix, 1997) or cognitive models, such as the resources model reported by Smith, Duke, and Wright (1999) or Kaur, Maiden, and Sutcliffe (1999) model of interaction.

15.2.3.5   VR System Configuration: Hardware and Software

The virtual task analysis will also influence the system configuration. It is likely that many of the priorities have already been set based on the existing system configuration. However, it is necessary to review to ensure that the technologies selected provide the optimal support for the user. Currently, there are a vast array of display technologies (e.g., desktop, HMDs, passive and active stereo/mono glasses, see-through glasses, flat and curved projected displays, powerwalls, 3–6-sided cave automatic virtual environments [CAVEs]) and interaction devices (e.g., 3–6 DoF mice, joystick, wand, keyboard, gesture, speech). There is no standard VE system. Hardware choice is largely based on affordability and availability as well as tasks to be supported by the environment. Software selection is more difficult as there are few commercial off-the-shelf (COTS) programs and all require a certain level of computer competence to develop good VEs. The key is to ensure that the technology enhances performance and is not a barrier; the more intuitive the better. The virtual task analysis will identify the various interactions required and these should be the basis of selecting the appropriate display and device.

15.2.3.6   Validation Criteria

Furthermore, it is possible to use the outcome of the virtual task analysis to set the validation criteria. The virtual task analysis will list the expected interactions, and these can be used to measure the success of the VE.

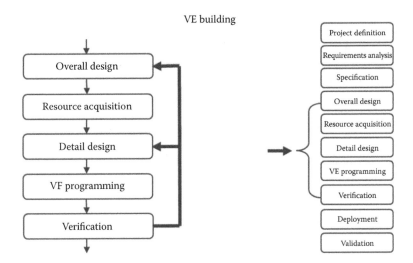

FIGURE 15.9 A section of the VEDS outline showing the VE building stages.

15.2.4 VE Building

The next five stages of VEDS can all be said to be part of VE building (Figure 15.9). This process has feedback loops; some of this iteration will be to correct problems found during verification, but the majority takes place as each object is designed in detail, built, and tested before moving on to the next object.

15.2.4.1 Basic Choices

After VE specification is complete, construction of the VE involves a series of hard choices over what can be done to meet the user and application needs. A fundamental choice facing the world developer is where to concentrate development effort (in time and creativity) and computer processing resources. Crudely, VEs may be designed to maximize the participant's ability to interact with them through exploration (i.e., visualization and walkthroughs) and manipulation (i.e., alterations in the state of objects) or to compromise by trying to meet needs for both. Technical limitations of VE technology present in differing degrees in all systems, from the simplest desktop to a CAVE, and usually result in the developer having to make a number of trade-offs. The aim is to render recognizable 3D representations in real time, updating synchronously with a participant's control inputs or movements in the VE. The trade-offs will often involve choosing between providing the user with functionality, including a high degree and speed of interactivity, and producing a higher specification for the complexity and appearance of the environment itself. The total number of objects to be modeled and the level of detail and degree of associated intelligence for each must be considered carefully before including any within the VE (see Eastgate & Wilson, 1994; Richard et al., 1996; Smets & Overbeeke, 1995). The application and virtual task analyses and prioritization exercises will be used for this. Some decisions over what to model, and how, can be made through modeling tools, for instance, extraction of geometry from photographs and drawings.

15.2.4.2 Choices over Object "Intelligence"

One way in which VEs may not match the real world, at least in their underlying construction, is in the selection of which objects have the capability to be *intelligent*, to change configuration or else determine the form of the total VE. For example, materials or objects being worked on or manipulated during a manufacturing process are generally *dumb* in that they have no influence over the process. It is the manufacturing machines or systems that can influence the position, orientation, deformation, and joining of the materials or objects. Although veridical representation means we should attach code to give intelligence to the machines, it is sometimes simpler to attach intelligence to the objects being manipulated without affecting the visual representation of the process. In this way, the real world can be represented in the most efficient way possible, rather than exactly simulating it. For example, where a robot has to move components from a conveyor onto an assembly, in principle, it is one *intelligent* object working with an unknown number of *dumb* objects of known type. It is often simpler to create one component that has a program associated with it, such that it can react correctly to the robot, and then to duplicate that component (and associated program) as necessary, than it is to program the robot to cope correctly with an unknown number of components. This effectively takes the *intelligence* from the robot and assigns it to the component.

This approach, however, has disadvantages. If an object has been programmed to sense when a robot is picking it up and to behave accordingly, it may be given code only to be able to recognize one robot or type of robot. Also the robot will be incapable of manipulating other objects for which the user would expect it to be suitable but which do not have the required program associated with them.

15.2.4.3 PROGRAMMING INTERACTIVITY

One vital aspect of building VEs, and a major component of usability, is interactivity. VEs have, of course, been typically defined in terms of possessing presence, autonomy, and interactivity (Zeltzer, 1992). Of these, interactivity is arguably the most critical attribute as it clearly defines the difference between VE and other 3D modeling systems (Eastgate, 2001). Here, the term interactivity is used broadly to mean any action on the part of the participant that results in a change in the VE. An obvious case is selecting one object to initiate change in another object or in the scene, but interactivity can also include navigation where a user moves through a VE while it updates and changes according to one's movements and position. There are many computer applications that offer varying amounts of interactivity, but it is reasonable to suppose that VE technology can offer more representations of the types of interactivity commonly encountered in real life.

As with all facets of VE development, the technical limitations of any particular system and user interface mean that there will be limits on the amount and quality of interactivity available. In an ideal world, the participant would feel that they are interacting directly with the virtual objects in the VE while they are actually interacting with peripherals in the interface. However, due to the limitations of current VE technology, the user interface is more likely to form a barrier between the user and the VE, whatever the nature of the sensors and effectors and whichever of the user's senses are engaged. One of the jobs of the VE developer is to make this barrier seem minimal within the constraints of the application.

There are two extremes in VE interactivity. The first is to aim to accurately model as many of the characteristics of the real application as possible. The second approach is to deliberately create an approximation of reality, where the result is a crude representation of salient characteristics rather than an accurate simulation. The first approach may end up with a VE that is too complex for the processor to update as fast as required for exploration and manipulation interactivity, whereas the second approach may result in an impoverished or very simple VE, insufficient for the requirements of the application user. Therefore, a compromise between the two extremes of modeling interactivity will generally be used to achieve the best solution for any given application. The model will be built to mimic the behavior of the real world, behaving *correctly* in a variety of circumstances. Object-oriented design, maximizing object autonomy, will make the VE easier to adapt or expand in the future if the need arises. Where accurate models of the structure or dynamics of an object or process cannot be justified on the grounds of programming time or processing overhead, they can be replaced by, for example, a texture, an animation, or a model based on a simplified algorithm.

15.2.4.4 Implementation of Physical Properties

VE software usually gives developers limited ability to assign some physical properties to virtual objects, particularly dynamic characteristics approximating those expected in the real world. One factor limiting the *realism* from doing this is a lack of temporal consistency in some 3D visualization systems. Any factor that decreases the rendering speed, such as visual complexity, can result in a slowing down of the dynamic properties of moving objects. To get around this, it is usually possible to link the dynamic properties to the processor's real-time clock. However, it is not possible to increase the rendering speed using this method; instead, the step size of the movement of the objects is altered, so if the rendering speed becomes very slow, a fast-moving object may appear to jump across the screen. This can become particularly pronounced in CVEs, where the collaboration is taking place over a network with its inherent potential for lag. Local activity will render smoothly, but remote activity will jump about as packets of information arrive over the network. Thus, to achieve smooth but temporally accurate dynamics, the developer has to be aware of the potential bottlenecks in the processing and rendering of the virtual world on particular VE systems.

Some VE development systems provide facilities to give objects physical properties, for example, gravity, coefficient of friction, coefficient of restitution, and velocity, both linear and angular. These are not generally designed to simulate the laws of physics but rather to give VE developers shortcuts to making objects appear to obey the laws of physics. Coefficients of friction and restitution are determined by the properties both of the materials that are coming into contact and cannot be allocated to an object in isolation. One physical property that is not generally available is center of gravity; this is because it is not easy to calculate even if the object is of uniform density, but most objects do not possess uniform density. Without this property, it is not possible to make an object topple realistically as the center of gravity needs to be known in order to predict when an object will become unstable and start to fall.

Dynamic characteristics not provided by a VE development system can sometimes be programmed by developers using a relevant programming or scripting language. By assigning a piece of program code, any virtual object can be programmed to behave in a particular way independent of other virtual objects in the same environment. Alternatively, objects can have their separate codes linked (e.g., via common variables) such that one object's behavior depends upon another's.

Any use of simplified physical constraints can make an interaction more realistic without sacrificing real-time updating and can provide a more intuitive interface (Papper & Gigante, 1993).

15.2.4.5 Overcoming Limiting Factors in VE Development

The combination of VE complexity and finite system processing power will limit the rendering speed of a VE presentation. It is important that rendering speeds are sufficient to avoid causing usability, likeability, and utility problems. When faced with restrictions, there are various possibilities available to the developer. More than one of these techniques would normally be used in combination. The facilities offered by the VE system software and the developer's ability and preferences are most likely to influence the choice (see Grinstein & Southard, 1996).

15.2.5 Overall Design

VE overall design is the process of designing the application as it is going to be built in VR (Figure 15.10). This is done with reference to the characteristics of the software and hardware of the target VR system. The process is one of working out how the concepts established in the storyboard can be realized using the tools available. During this phase, the VE developer will need to plan out the development process so that the finished VE can be delivered in the time available, the resources needed can be prepared, and the different parts of the VE and the user interface will work together when assembled.

When planning the layout of a VE, the first thing to think about is the extent of the model. Whether the model is on the microscale (e.g., subatomic) or macroscale (e.g., an entire city), there will be a

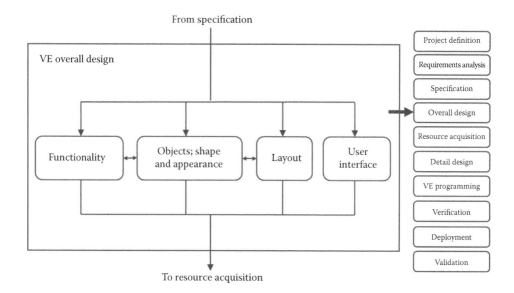

FIGURE 15.10   An expanded view of the VE overall design stage of VEDS.

system boundary, beyond which no more features will be modeled. There may also be a second boundary within this, beyond which the user is not able to navigate. The purpose of this second boundary is to prevent the user navigating to a point where they can see outside the modeled area.

The next decision for the VE developer is which way to orient the world with respect to the VR system axes alignment. In a VE consisting largely of orthogonal cuboids, this choice will be an easy one. In a less structured VE, making a good decision at this point could prevent rendering or dynamics problems later on. Often linked to orientation is the planning of the hierarchical structure of the objects. Grouping the objects into hierarchies speeds up the rendering process and allows multiple objects to be manipulated easily.

The objects that are to be included in the VE need to be prioritized with respect to the overall purpose of the VE, and the resource requirements (dimensions, textures, etc.) need to be specified.

The final area for planning the overall design of the VE is the user interface. With reference to the task and virtual task analysis, the VE developer needs to decide at this stage what viewpoints are going to be available, what the screen layout will look like, and what methods of interaction will be possible. This will prevent inconsistencies in the user interface or unnecessary duplication of points of interaction (e.g., an overlaid button with the same function as an interaction point within the VE).

15.2.6   Resource Acquisition

During the VE overall design phase, the resource requirements for the development of the VE should have been specified (Figure 15.11). The types of resources required in order to model an object depend on what role that object will play in the finished VE. With reference to resource acquisition, objects can be split into four types:

1. Background objects are there to act as scenery at the periphery of the modeled VE. An example of a background objects would be 2D mountain scenery at the edge of a VE.
2. Context objects are not individually fundamental to the VE, but collectively help the user to establish what type of environment they are in. They need to be distinct and recognizable. An example of a visual context object would be a filing cabinet in an office. This category also includes navigational aids such as landmarks and paths.

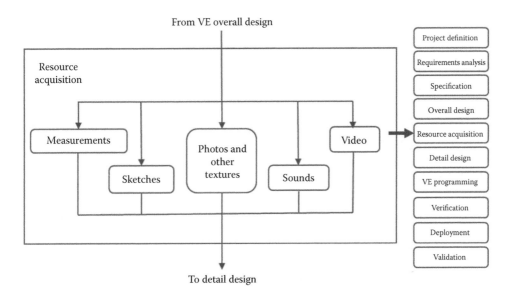

FIGURE 15.11   An expanded view of the resource acquisition stage of VEDS.

3. Visual primary features are those that are fundamental to the purpose of a VE but the user cannot interact with them. As such, they only need to look correct. Examples of a visual primary features are a sculpture in a virtual art gallery or a direction sign.
4. Functional primary features may need to be both functionally and visually correct. Examples of a functional primary feature are a component in a virtual assembly task, an on–off switch, or a car steering wheel.

Some objects will fall into more than one of the aforementioned categories. For example, a wall may be part of the background, it may also provide context, and it may also perform a function in that it is a boundary that the user cannot cross. In these cases the resources required are those of the highest category that the object falls into.

A background object has to look right from a distance. This can normally be achieved using a texture applied to a large 2D object that cannot be viewed close up or from an acute angle by the user. A photograph of the real location being modeled or a typical location of the right type can be used to create the texture.

A context object does not need to accurately represent an actual object of that type, but rather it must be recognizable for what it is and should be accurate enough not to look wrong. If it is being used to help the user judge the scale or size of a virtual space, then it needs to be scaled appropriately itself. For example, if a chair is placed in a room to give the user an idea of the size of the room, the chair should be scaled within the range of sizes that one would find similar chairs in the real world. As many visual context objects exist at the edges of areas navigable by the user, these can be replaced by 2D textures where this is easier than modeling the object in 3D.

The accuracy required when modeling primary features of a VE will depend on the exact role that the feature plays in the VE, but they would normally need to be visually accurate. Where the VE is representing objects that exist in the real world, this would require dimensions measured from the real object being used, along with textures to add realism where necessary. If a primary feature is functional, this can increase the dimensional accuracy required (i.e., so that the object fits accurately into the same space as it would occupy in the real world). Figure 15.12 shows the accuracy requirements during resource acquisition for various object types.

Type of Object	Model Accuracy		
	Approximate	Visually Accurate	Mechanically Accurate
Background	2D texture		
Context	2D texture(s)	AND/OR 3D dimensions	
Primary feature— visual		3D dimensions and textures	
Primary feature— functional		3D dimensions and textures	AND/OR 3D dimensions (+sound? +video?)

FIGURE 15.12 Types and accuracy of resource information required when acquiring resources for different object types.

As well as dimensional measurements and photographs for the creation of textures, it may be necessary to collect other forms of information about the objects to be modeled. Sketches of the objects and their relative positions (with dimensions added where necessary) will be easier to understand than written descriptions. A video of a place or a piece of equipment in action will store a wealth of information that can be referred to repeatedly as necessary. Functional objects may have sounds associated with their function that will need to be recorded (or extracted from video footage) so that they can be added to the VE.

It is common for the location at which the acquisition of the resources takes place to be different from that in which the VE development takes place. In these circumstances, it is important to maximize the amount acquired in each visit; once VE building starts, it is better to have extra redundant information than not enough information.

Shortcuts can be taken in resource acquisition. One obvious source is CAD and other solid model (3D) data. Translators may be required here and some precision in geometry may be lost, but this can allow much more rapid development. Other sources may be image and mapping agencies or the Internet for models or animations.

15.2.7 DETAIL DESIGN

The detail design, development, and verification phases of VE development are so closely interwoven that it is sometimes hard to distinguish between them (Figure 15.13). It is common for developers to work on one object at a time, designing, building, importing it into the VE, adding behaviors and interactivity where required, and testing and refining it before moving on to the next object. Where objects are interdependent, several objects may be worked on at once. Thus, this period of the development of a VE is a repeating pattern of detail design, development, and testing, at the end of which an initial version of the finished VE will have been created. Occasionally, during this process, omissions or flaws will be found that require a brief return to the specification or overall design phases. The order in which objects are designed, built, and tested for the VE will be determined by the prioritizations defined in the specification, the nature of the VE itself, and the preferences of the VE developer. As well as virtual object creation, the developer will also be setting up viewpoints that afford appropriate navigation and additional user interface features such as text boxes and buttons overlaid on the screen. This user interface programming will be integral to the process of object creation; as new objects are created, changes to the methods of navigation available, buttons to control the behavior of those objects, and text boxes giving relevant information will be developed throughout the process to align with object development processes.

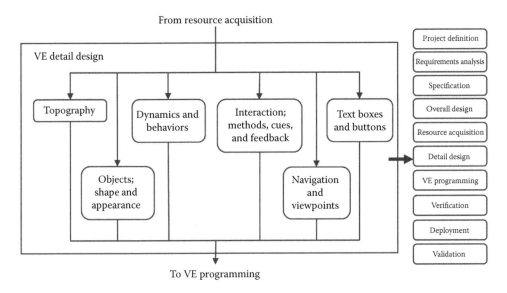

FIGURE 15.13 Expanded view of the VE detail design stage of VEDS.

Of these three phases, the VE detail design phase is concerned with taking each detail in turn and seeing how it can best be made to contribute to the overall requirements of the VE. These details can be object geography, shape or appearance, dynamics or behaviors, and interaction methods, or they can be the user interface issues of navigation viewpoints, on-screen buttons, and text boxes. It is during this detail design phase that the developer is faced with a number of decisions. Each decision has to be made with reference to the overall goal of the VE and the specific requirements of the expected user population. Having come up with a design, it must be considered in terms of its impact on the rendering speed and frame update rate of the finished VE. The number of decisions required at this stage in the VE development process demonstrates a need for tools to enable the VE developer to devise appropriate design solutions. It may be possible for the VE developer to work in close collaboration with the other stakeholders throughout this process. If this is not possible, it is important that the input from the specification and overall design stages is adhered to and that any areas of doubt are cleared up to avoid VE development effort being wasted.

15.2.7.1 Cues and Feedback

One particular aspect of usability that is intimately linked to design choices made by developers, and to the notion of interactivity addressed earlier, is the provision of cues and feedback to the participant. In order that users can interact successfully, the VE should offer some cues that tell participants what types of activity are afforded. In this sense, VEs are no different to human–computer interfaces or to consumer products generally in terms of the concept and importance of affordances (e.g., Norman, 1988). Cues to assist the participant could be in the form of navigational signs to aid their movement through the VE and to help them recognize their current location. Alternatively, cues could suggest that a certain type of control action would afford a particular response. The participant will benefit from knowing which objects afford interaction, whether they are available for interaction at that moment, what method of interaction is appropriate, and what effect interaction with that object might have on the VE. Then, having interacted with a chosen object, they will want to know whether that interaction was successful. Figure 15.14 shows guidance being produced by our group at VIRART to support VE developers in improving the interactivity and feedback in their VEs.

The design choice outlined earlier—of whether to design to model realism or some abstraction of this or even to model a deliberate distortion of reality—is as valid for interactivity as for visual

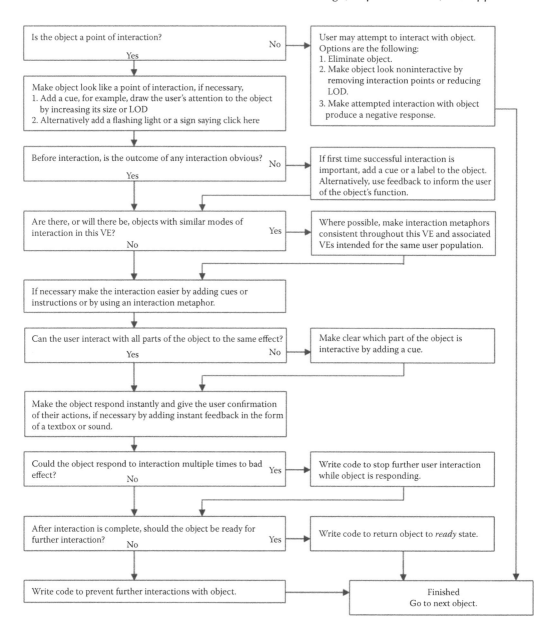

FIGURE 15.14 A flowchart taking the developer through the steps required to ensure that an object can be successfully interacted with where interaction is afforded.

appearance. The points of interaction, the cues for these points, and the feedback given might all interfere with, or detract from, any visual fidelity in the scene and objects. Points of interaction may be of a size, shape, and color very different from their real-world counterparts; feedback and cues may need to be explicit in the form of text or audio prompts and messages. Such elements may strongly influence any usability/presence trade-off in the VE.

15.2.7.1.1 Texture Mapping

A large amount of 3D detail can be replaced by a bitmap texture pasted onto a single facet in a VE. Depending on the subject matter, textures can increase rendering speeds and will generally be quicker to create than 3D geometry. Whereas textures are a good way of adding realism to a VE, they can only be used in certain circumstances. A new development that offers a compromise between the increased speed from textures and the interactivity available with modeled objects are algorithms that automatically replace textures with modeled objects according to the participant's apparent distance away from an object (e.g., Rosenblum, Burdea, & Tachi, 1998).

15.2.7.1.2 Level-of-Detail Management

Level-of-detail (LOD) management, or distancing, is a widely used technique, whereby if a virtual object is more than a specified distance from the viewpoint of the participant, it is replaced by a simpler model or else disappears altogether, thus reducing the amount of processing time required to render that object. Extensions of this technique can be found in layered depth images, and, for one specific problem, use of portal culling and portal textures (e.g., Rafferty, Aliaga, Popescu, & Lastra, 1998).

15.2.7.1.3 Selectively Including Objects

Reducing the number of objects in an environment will increase rendering speed and decrease development time. The decisions as to which objects to include should be based on the task analyses and rational thought about which objects are really needed, but tend also to be made on the basis of the size of objects, ease of programming, and individual programmers' preferences. In the worst case, such decisions are made arbitrarily with little understanding of how the choice will affect participants.

A technique that can both decrease development time and increase rendering speeds is to selectively include objects. It is applicable across all hardware and software platforms and will be compatible with future VE systems. The resulting VE should not be significantly inferior (at achieving its specified purpose) to a hypothetical one in which all objects are modeled. Indeed, it may be found that omitting some objects may increase the effectiveness of a VE. A parallel is the use of a stylized graphic drawing in a book of instructions rather than a photograph to avoid confusion through unnecessary detail. What is required is a set of guidelines to help developers through the process of deciding which objects to include and which to omit. This guidance will have to be developed with reference to the specified purpose of a VE and user behavior within the VE.

15.2.8 VE PROGRAMING

To a certain extent, VE development has to be done in a fixed order. A 3D shape has to be created before it can be imported into the VE (Figure 15.15). Once it is in the VE, it can be assembled with other shapes, have colors and textures mapped onto it, and be given dynamics and behaviors and associated sounds where appropriate. Viewpoints, on-screen buttons, and text boxes can be added at any time. There are many toolkits and some domain-specific languages available for the actual scripting of VEs. Many of these are open source and freely available to download from the Internet. The exact scripting process carried out will depend on the language/toolkit being used. The purpose of this chapter is to describe the approach rather than specific programming technique that is covered elsewhere in this book.

15.2.9 VERIFICATION

The purpose of verification is to ensure that the product is internally consistent and that it meets the specification and behaves as expected/specified (Figure 15.16). It is not intended to make any subjective assessment of the product at this stage. For a VE, this process is needed to find out whether the components of a VE or the entire VE have the specified interactivity, navigation, and fidelity. If the specification has been written carefully, the developer should be able to decide definitively whether the VE and its components meet that specification. However, ambiguities in the specification are often uncovered at this stage, which will need to be addressed:

- *Smoke testing*—This is a quick and simple test to confirm that the system being tested works in its primary function. In a VE, this will require entering the environment and

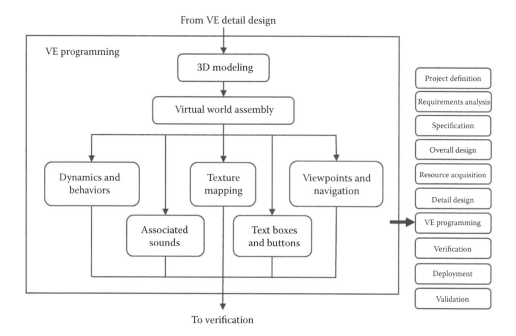

FIGURE 15.15 An expanded view of the VE programming section of VEDS.

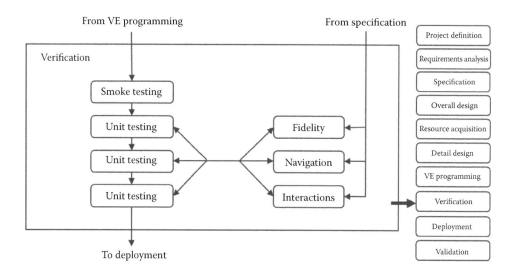

FIGURE 15.16 An expanded view of the verification section of VEDS.

viewing, navigating through, and/or interacting with the component to be tested. A system that fails a smoke test is automatically referred back to the previous stage before more work is carried out.

- *Unit testing*—Individual VE components are more thoroughly tested before being assembled or combined to make larger components of the VE. This testing is done with reference to the specification to make sure the component is fully compliant.
- *Integration testing*—Smaller components are integrated into larger assemblies and larger assemblies into the overall VE. This differs from unit testing in that components are no longer tested independently but in groups, the focus shifting from the individual units to

the interaction between them and any unforeseen consequences of combining 3D objects and their code. Again, this is done with reference to the specification.

- *System testing*—The overall test on the final version of a VE. System testing is important because it is only at this stage that the full complexity of the environment is tested. The focus in systems testing is typically to ensure that the VE responds correctly to all possible input conditions and that it handles unanticipated interactions in an acceptable fashion.

15.2.10 DEPLOYMENT

While they may not be present at the time, the VE developer needs to understand the process of deployment of a VE. Once a VE has been delivered to a client, there are four factors that may determine the success of its implementation and the achievement of the application goals (Figure 15.17). Firstly, there are the characteristics of the actual users, which may or may not match those of the anticipated user population, particularly in the areas of aptitude and motivation. The user's motivation may be further affected by the circumstances of use. Factors such as time of day, what they would normally be doing (instead of using a VE), as well as environmental factors such as ambient temperature and background noise could affect the user's motivation and overall performance in the VE. From user requirements and specifications, the VE developer should have fully considered usability throughout the development process. Despite this, there may be aspects of system usability that need improvement and these can best be established through validation. Finally, the fidelity and validity of the VE itself, resulting from the earlier specification, design, and development phases, will certainly have an effect on the overall success of the application

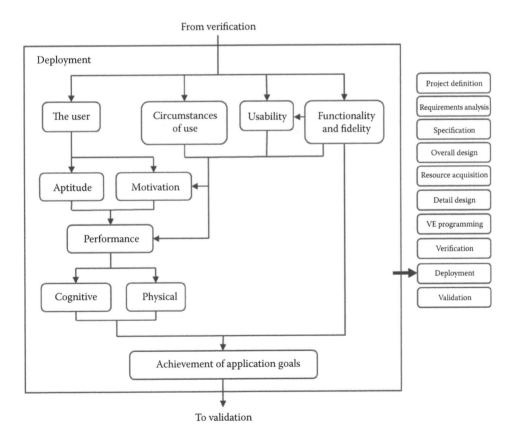

FIGURE 15.17 An expanded view of the deployment section of VEDS.

15.2.11 VALIDATION

For all VE applications, evaluations should be made of both the fidelity of the environments themselves and also their use and usefulness (Figure 15.18). Such evaluations can be divided into examinations of validity, outcomes, user experience, and process (see also Tromp, Istance, Hand, Steed, & Kaur, 1998b; Chapters 28 through 35).

Developers should seek to address the validity of VEs in every project, at the very least by *walking through* them with the client or other experts. The agile cycles of development within a participatory design approach adopted in the applied European project work (e.g., DiFac, ManuVAR, IMOSHION, VISTRA) has meant that several versions of early prototypes are developed during the formative stage of development. These cycles enable the developers to receive feedback as early as possible so that larger, more complicated changes are less likely at a later, more difficult stage. The process involves hands-on workshops and focus groups directly with the developers and various stakeholders, specifically the anticipated end users. The developers often find that users have difficulties with unexpected parts of the VR technology and VE. Sometimes the choice of interaction device hinders performance of tasks. It is a good way of seeing firsthand the design limitations and capturing more detail on a set of requirements directly with users. From the user point of view, it helps with their understanding and eventual use of the technology if they are part of the design process, giving a sense of ownership.

The potential usefulness of VEs has been somewhat harder to evaluate sensibly since few, if any, are sufficiently developed in a form whereby potential users can directly see how they might be applied and carry out test applications. Often in applied work, there is little time or resources available to conduct any longitudinal studies so most structured evaluation work has been of the immediate outcomes of use of a VE (Crosier, Cobb, & Wilson 2002; D'Cruz, 1999; Karaseitanidis et al., 2006). Not surprisingly—given the better developed nature of measures in this area—evaluation studies have concentrated on education or training applications in aerospace, automotive, and medical domains (e.g., Darken & Banker, 1998; Gurusamy, Aggarwal, Palanivelu, & Davidson, 2009; Hall, Stiles, & Horwitz, 1998; Haritos & Macchiarella, 2005; Regenbrecht, Baratoff, &Wilke, 2005).

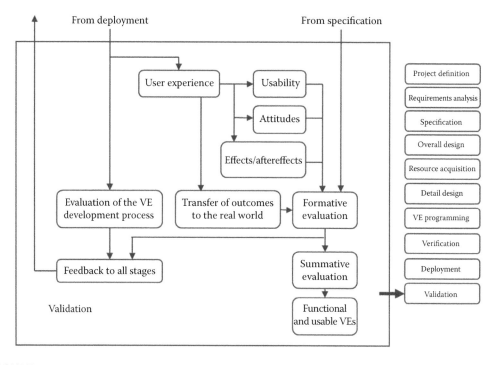

FIGURE 15.18 An expanded view of the validation section of VEDS.

The user experience can be evaluated in terms of performance measures within the VE (reaction times, navigation, or selection tasks such as in the VEPAB from Lampton et al., 1994; Darken & Peterson, 2014; Chapter 19, Bliss, 2014; Chapter 29), side and aftereffects (e.g., Cobb, Nichols, Ramsey, & Wilson, 1999; Nichols, Cobb, & Wilson, 1997; Nichols, Haldane, & Wilson, 1998, as well as Chapters 22 through 27), and participant attitudes and usability of the interface and VE (Gabbard, 2014; see Chapter 28). Finally, there is evaluation of the process of both developing and using VEs. One example of this is evaluation of how potential users will help specify, build, and evaluate VEs within a participatory design process (Neale, Brown, Cobb, & Wilson, 1999); another is the evaluation of the process whereby VEs are used in participatory work redesign in control rooms (Wilson, 1999).

15.3 VEDS IN PRACTICE

In practice, the way VEDS is applied will be determined by a number of factors. These factors include the size of the project and the domain that it is being developed for. Figures 15.19 and 15.20 show how VEDS was applied in the development process of a VE for education (Crosier et al., 2002). Figure 15.20 shows how the development structure was envisaged at the start of the project. The proposed framework is on the right of the diagram and the outline of VEDS is on the left showing how this was used to create the framework. Figure 15.20 has a similar layout but this time shows the

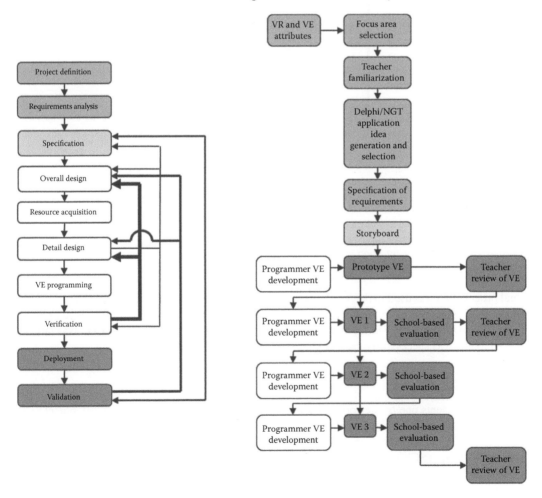

FIGURE 15.19 How it was planned that VEDS would support the development of a VE. (From Crosier, J.K. et al., *Comp. Educ.*, 38, 77, 2002.)

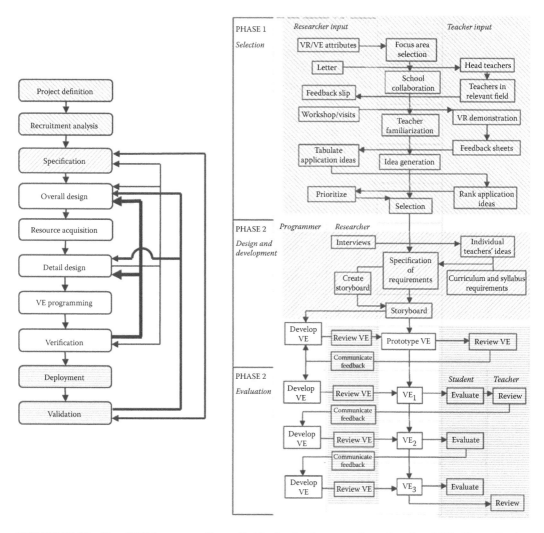

FIGURE 15.20   How VEDS was actually applied in the development of an educational VE. (From Crosier, J.K. et al., *Comp. Educ.*, 38, 77, 2002.)

actual process that was used in the VE development. Once the project got underway, it became clear that the process needed many more iterations, both before and after the VE building stage in order to specify and validate the VE. Despite these adaptations, the basic structure of VEDS was largely adhered to and was successfully used to guide the process of VE development.

15.4   CONCLUSIONS

This chapter presents one framework to support a systematic and structured VE development process, VEDS. This framework has been used to support the development of VEs in industrial use (e.g., Wilson et al., 1996), industrial training (D'Cruz, 1999), and education (Barmpoutis et al., 2014, Cobb et al., 2014; see Chapters 41 and 42).

By careful, but flexible, use of such a framework, the consequences of some of the barriers to VE application can be minimized. Technical limitations can be circumvented or their effects reduced by appropriate choice of technology and careful specification of the minimum VE size and least complexity required to meet user goals. Trade-offs, such as choices between visual fidelity and interactivity, can more easily be resolved. Usability will be enhanced through selection of appropriate hardware and development of cognitive interface tools and aids, all on the basis of the task and user analysis and storyboarding. Real working applications will only be achieved through selection of those domains to be modeled that make the most of the attributes of VE, using a thorough understanding of what constraints will apply and of how to overcome them in specifying and building the VE. Finally, these applications will best show added value if a rational evaluation process is established at the outset on the basis of realistic and achievable targets for use of VEs and for their incorporation into industrial and commercial practices, education curricula, or training programs.

The chapter is largely focused on VEs that model and give experience of interacting with a device or that allow exploration of a space such as a factory or dwelling. Therefore, the sorts of VE used for stress analysis or as a virtual wind tunnel, for instance, are not covered in terms of the engineering models on which they build. Nor are very large geographical representations addressed, for instance, as used in landscaping projects, harbor navigation, or city planning. Furthermore, the support of groups of participants in CVEs is not explicitly addressed. Many of these issues are taken up elsewhere throughout this book (see Chapters 12, 14, 25, 34, and 41).

Since the first edition of this book in 2002, VEDS has been applied by the authors in a number of research projects including VIEW of the Future (IST-2000-26089) (Wilson and D'Cruz 2006), INTUITION (IST-507248-2), DiFac (FP6-2005-IST-5-035079) (Redaelli et al. 2009), ManuVAR (CP-IP-211548) (Krassi, D'Cruz, and Vink 2010), IMOSHION, Tempus VR-Lab (JEP-34012-2006), and Tempus eLab (JPCR-517102-2011). As mentioned previously, there have been continuing attempts during this period to provide structure to the VE development process, and these attempts have been reviewed along the way. Molina et al. (2005) repeat the assertion that "Virtual Reality is not easy" and reiterate how the technology is not as simple as 2D HCI, before reporting on how they have built on other VE development research, including VEDS, to produce their TRES-D methodology. Sanders and Silvanus (2010) have extensively drawn on VEDS and Shewchuk's model (Shewchuk, Chung, & Williges, 2002) in the development of their virtual environment information provision (VEIP) framework. VEDS is embedded within the framework that is aimed at training in manufacturing and manages the provision of information for virtual training. Seidel et al. (2010) express their opinion that there is still no general agreement on process for developing VEs and review a number of contenders including VEDS before presenting their own methodology that echoes much of what is in VEDS while adding more detail to the project definition and specification stages.

ACKNOWLEDGMENT

This chapter is dedicated to the memory of Professor John R. Wilson, our inspiration, mentor, motivator and friend.

REFERENCES

Barfield, W., Zeltzer, D., Sheridan, T., & Slater, M. (1995). Presence and performance within virtual environments. In W. Barfield & Design, implementation, and applications (T. Furness III (Eds.), *Virtual environments and advanced interface design* (pp. 473–513). New York, NY: Oxford University Press.

Barmpoutis, A., DeVane, B., & Oliverio, J. C. (2014). Applications of Virtual Environments in Experiential, STEM, and Health Science Education. In K. S. Hale & K. M. Stanney (Eds.), *Handbook of virtual environments: Design, implementation, and applications* (2nd ed., pp. 1055–1072). Boca Raton, FL: Taylor & Francis Group, Inc.

Bliss, J. P., Proaps, A., & Chancey, E. T. (2014). Virtual environment usage protocols. In K. S. Hale & K. M. Stanney (Eds.), *Handbook of virtual environments: Design, implementation, and applications* (2nd ed., pp. 749–780). Boca Raton, FL: Taylor & Francis Group, Inc.

Bolas, M. (1994). Designing virtual environments. In C. Loeffler & T. Anderson (Eds.), *The virtual reality casebook* (pp. 49–55). New York, NY: Van Nostrand Rheingold.

Bowman, D., Gabbard, J. L., & Hix, D. (2002). A survey of usability evaluation in virtual environments: Classification and comparison of methods. *Presence: Teleoperators and Virtual Environments*, *11*(4), 404–424.

Bowman, D., Kruijff, E., LaViola, J., & Poupyrev, I. (2004). *3D user interfaces: Theory and practice*. Boston, MA: Addison-Wesley.

Bryson, S. (1995). Approaches to the successful design and implementation of VR applications. In R. Earnshaw, J. Vince, & H. Jones (Eds.), *Virtual reality applications* (pp. 3–15). London, U.K.: Academic Press.

Champney, R. K., Carrol, M. B., Milham, L. M., & Hale, K. (2008). Sensory-perceptual objective task (SPOT) taxonomy: A task analysis tool. *Proceedings of the Human Factors and Ergonomics Society 52nd Annual Meeting*. New York, NY, September 22–26.

Chen, J., & Bowman, D. (2009). Domain-specific design of 3d interaction techniques: An approach for designing useful virtual environment applications. *Presence: Teleoperators and Virtual Environments*, *18*(5), 370–386.

Chertoff, D., & Schatz, S. (2014). Beyond presence. In K. S. Hale & K. M. Stanney (Eds.), *Handbook of virtual environments: Design, implementation, and applications* (2nd ed., pp. 855–870). Boca Raton, FL: Taylor & Francis Group, Inc.

Cobb, S. V. G., Nichols, S. C., Ramsey, A. D., & Wilson, J. R. (1999). Virtual reality-induced symptoms and effects (VRISE). *Presence: Teleoperators and Virtual Environments*, 8(2), 169–186.

Cobb, S., Hawkins, T., Millen, L., & Wilson, J. R. (2014). Design and development of 3D interactive environments for special educational needs. In K. S. Hale & K. M. Stanney (Eds.), *Handbook of virtual environments: Design, implementation, and applications* (2nd ed., pp. 1073–1106). Boca Raton, FL: Taylor & Francis Group, Inc.

Cooper, A. (1995). *About face—The essentials of user interface design*. New York, NY: Hungry Minds.

Crosier, J. K., Cobb, S. V. G., & Wilson, J. R. (2002). Key lessons for the design and integration of virtual environments in secondary science. *Computers and Education*, *38*, 77–94.

Darken, R. P., & Banker, W. P. (1998). Navigating in natural environments: A virtual environment training transfer study. *Proceedings of the IEEE Virtual Reality Annual International Symposium* (pp. 12–19). Los Alamitos, CA.

Darken, R. P., & Peterson, B. (2014). Spatial orientation, wayfinding, and representation. In K. S. Hale & K. M. Stanney (Eds.), *Handbook of virtual environments*: Design, implementation, and applications (2nd ed., pp. 467–492). Boca Raton, FL: Taylor & Francis Group, Inc.

Deol, K. K., Sutcliffe, A., & Maiden, N. (1999 September 15–21). A design advice tool presenting usability guidance for virtual environments. *Proceedings of Workshop on User Centred Design and Implementation of Virtual Environments, at York, UK*. London, U.K.: British HCI Group.

Draper, J. V., Kaber, D. B., & Usher, J. M. (1998) Telepresence. *Human Factors*, *40*, 354–375.

DTI. (1996). *A study of the virtual reality market. Information society instructive report*. Department of Trade and Industry, London, U.K.: HMSO.

D'Cruz, M. D. (1999). *Structured evaluation of training in virtual environments* (PhD thesis). Virtual Reality Applications Research Team, University of Nottingham, Nottingham, U.K.

D'Cruz, M., Stedmon, A. W., Wilson, J. R., Modern, P. J., & Sharples, S. C. (2003). Building virtual environments using the virtual environment development structure: A case study. In C. Stephanidis (Ed.), *HCI International '03. Proceedings of the 10th International Conference on Human-Computer Interaction*. Mahwah, NJ: Lawrence Erlbaum Associates.

Eastgate, R., Nichols, S., & D'Cruz, M. (1997). Application of human performance theory to virtual environment development. In D. Harris (Ed.), *Engineering psychology & cognitive ergonomics* (Vol. 2: Job Design and Product Design, pp. 467–475). Aldershot, U.K.: Ashgate.

Eastgate, R. M. (2001). *The structured development of virtual environments: Enhancing functionality and interactivity* (PhD thesis). University of Nottingham, Nottingham, U.K.

Eastgate, R. M., & Wilson, J. R. (1994). Virtual worlds. *EXE: The Software Developers' Magazine*, *9*, 14–16.

Endsley, M. (1995). Toward a theory of situation awareness in dynamic systems. *Human Factors*, *37*, 32–64.

Endsley, M. R., & Jones, W. M. (1997). *Situation awareness, information dominance, and information warfare* (No. AL/CF-TR-1997-0156). Wright-Patterson AFB, OH: United States Air Force Armstrong Laboratory.

Feng, A., Shapiro, A., Lhommet, M., & Marsella, S. (2014). Embodied autonomous agents. In K. S. Hale & K. M. Stanney (Eds.), *Handbook of virtual environments: Design, implementation, and applications* (2nd ed., pp. 335–350). Boca Raton, FL: Taylor & Francis Group, Inc.

Ferdig, R., & de Freitas, S. (Eds.) (2012). *Interdisciplinary advancements in gaming, simulations and virtual environments: Emerging trends*. Hersey, PA: IGI Global.

Flach, J. M., & Holden, J. G. (1998). The reality of experience: Gibson's Way. *Presence: Teleoperators and Virtual Environments*, *7*, 90–95.

Gabbard, J. L., & Hix, D. (1997). *A taxonomy of usability characteristics in virtual environments*. Virginia Polytechnic Institute and State Univ. Technical Report; http://csgrad.cs.vt.edu/~jgabbard/ve/taxonomy. Accessed on January 2001.

Gabbard, J. L., Hix, D., & Swan, J. E. (1999). User-centred design and evaluation of virtual environments. *IEEE Computer Graphics and Applications*, *19*, 51–59.

Grinstein G., & Southard, D. A. (1996). *Rapid Modeling and Design in Virtual Environments*, PRESENCE, 5(1), Winter 1996, 146–158.

Gross, D. (2014). Technology management and user acceptance of virtual environment technology. In K. S. Hale & K. M. Stanney (Eds.), *Handbook of virtual environments: Design, implementation, and applications* (2nd ed., pp. 493–504). Boca Raton, FL: Taylor & Francis Group, Inc.

Gurusamy, K. S., Aggarwal, R., Palanivelu, L., & Davidson, B. R. (2009). Virtual reality training for surgical trainees in laparoscopic surgery. *Cochrane Database of Systematic Reviews*, (1), CD006575. doi:10.1002/14651858.CD006575.pub2

Hall, C. R., Stiles, R. J., & Horwitz, C. D. (1998). Virtual reality for training for evaluating knowledge retention. *Proceedings of the IEEE Virtual Reality Annual International Symposium* (pp. 184–189). Atlanta, GA.

Haritos, T., & Macchiarella, N. D. (2005). A mobile application of augmented reality for aerospace maintenance training. *AIAA/IEEE Digital Avionics Systems Conference—Proceedings*. Vol. 1. Washington, DC.

Hix, D., Swan, J. E., Gabbard, J. L., McGee, M., Durbin, J., & King, T. (1999). User-centered design and evaluation of a real-time battlefield visualisation virtual environment. *Proceedings of VR'99 Conference* (pp. 96–103), IEEE. Houston, TX.

Hutchins, E. (1995). *Cognition in the wild*. Cambridge, MA: MIT Press.

Karaseitanidis, I., Amditis, A., Patel, H., Sharples, S., Bekiaris, E., Bullinger, A., & Tromp, J. (2006).Evaluation of virtual reality products and applications from individual, organisational and societal perspectives—The VIEW case study. *International Journal of Human-Computer Studies*, *64*(3), 251–266.

Kaur, K. (1998). *Designing virtual environments for usability* (PhD thesis). Centre for HCI Design, City University, London, U.K.

Kaur, K., Maiden, N., & Sutcliffe, A. (1999). Interacting with virtual environments: An evaluation of a model of interaction. *Interacting with Computers*, *11*, 403–426.

Keshavarz, B., Heckt, H., & Lawson, B. D. (2014). Visually induced motion sickness. In K. S. Hale & K. M. Stanney (Eds.), *Handbook of virtual environments: Design, implementation, and applications* (2nd ed., pp. 647–698). Boca Raton, FL: Taylor & Francis Group, Inc.

Kirwan, B., & Ainsworth, L. K. (Eds.) (1992). *A guide to task analysis*. London, U.K.: Taylor & Francis.

Krassi, B., D'Cruz, M., & Vink, P. (2010). ManuVAR: A framework for improving manual work through virtual and augmented reality. *Conference on Applied Human Factors and Ergonomics—AHFE 2010, USA*. Miami, FL.

Lampton, D. R., Knerr, B. W., Goldberg, S. L. Bliss, J. P., Moshell, J. M., & Blau, B. S. (1994). The virtual environment performance assessment battery (VEPAB): Development and evaluation. *Presence: Teleoperators and Virtual Environments*, *3*, 145–157.

Lawson, G., Hermawati, S., D'Cruz, M., Cobb, S., & Shalloe, S. (2012). Human Factors research methods in the design and evaluation of applied virtual environments. *Tijdschrift voor Ergonomie 2012–3. (Dutch Journal of Ergonomics)*, *37*(3), 5–10.

Mansouri, H., Kleinermann, F., & De Troyer, O. (2009). Detecting inconsistencies in the design of virtual environments over the web using domain specific rules. *Proceedings of the 14th International Symposium on 3D Web Technology (Web3D 2009)* (pp. 31–38). New York, NY: ACM Press.

Milham, L. M., Carroll, M. B., Stanney, K. M., & Becker, W. (2009). Training system requirements analysis. In: D. Schmorrow, J. Cohn, & D. Nicholson (Eds.), *The handbook of virtual environment training: Understanding, predicting and implementing effective training solutions for accelerated and experiential learning*. Aldershot, U.K.: Ashgate Publishing.

Molina, J. P., García, A. S., López, V., & González, P. (2005). Developing VR applications: The TRES-D methodology. *Proceedings of the First International Workshop on Methods and Tools for Designing VR Applications*. Ghent, Belgium.

Neale, H. (2001). *Virtual environments in special education: Considering users in design* (Unpublished doctoral dissertation). VIRART, University of Nottingham, Nottingham, U.K.

Neale, H., Brown, D. J., Cobb, S. V. G., & Wilson, J. R. (1999).Structured evaluation of virtual environments for special needs education. *Presence: Teleoperators and Virtual Environments*, *8*(3), 264–282. MIT Press.

Nichols, S. C., Cobb, S. V. G., & Wilson, J. R. (1997) Health and safety implications of virtual environments: Measurement issues. *Presence: Teleoperators and Virtual Environments*, 6(6), 667–675.

Nichols, S. C., Haldane, C., & Wilson, J. R. (1998). Measurement of presence and its consequences in virtual environments. *International Journal of Human-Computer Studies*, 52(3), 471–491.

Norman, D. A. (1988). *The psychology of everyday things*. New York, NY: Basic Books.

Papper, M. J., & Gigante, M. A. (1993). Using physical constraints in a virtual environment. In R. Earnshaw, M. Gigante, & H. Jones (Eds.), *Virtual reality systems* (pp. 107–118). London, U.K.: Academic Press.

Patel, H., D'Cruz, M., Cobb, S., & Wilson, J. R. (2006). Initial report on existing evaluation methodologies. INTUITION Deliverable D1.D_1 (former D1.10_1), www.intuition-eunetwork,org. Accessed on March 2014.

Perry, M. (2003). Distributed cognition. In J. M. Carroll (Ed.), *HCI models, theories, and frameworks: Toward an interdisciplinary science* (pp. 193–223). San Francisco, CA: Morgan Kaufmann.

Rafferty, M. M., Aliaga, D. G., Popescu, V., & Lastra, A. A. (1998). Images for accelerating architectural walk-throughs. *IEEE Computer Graphics and Applications*, 18(6), 21–23.

Rasmussen, J. (1986). *Information processing and human-machine interaction: An approach to cognitive engineering*. Wiley.

Reason, J. (1990). *Human error*. New York, NY: Cambridge University Press.

Redaelli, C., Lawson, G., Sacco, M., & D'Cruz, M. (2009). DiFac: Digital factory for human oriented production system. In I. Maurtua (Eds.), *Human Computer Interaction* (pp. 339–354). Vukovar, Croatia: In-Tech.

Regenbrecht, H., Baratoff, G., & Wilke, W. (2005). Augmented reality projects in the automotive and aerospace industries. *IEEE Computer Graphics and Applications*, 25(6), 48–56.

Richard, P., Birebent, G., Coiffet, P., Burden, G., Gomez, D., & Langrama, N. (1996). Effect of frame rate and force feedback on virtual object manipulation. *Presence: Teleoperators and Virtual Environments*, 5, 95–108.

Robertson, J., & Robertson, S. (2006b). Volere: Requirements specification template. *The Atlantic Systems Guild*, http://systemsguild.com/GuildSite/Robs/Template.html. Accessed on March 2014.

Robertson, S., & Robertson, J. (2006a). *Mastering the requirements process* (2nd ed.). London, U.K.: Addison-Wesley.

Rosenblum, L., Burdea, G., & Tachi, S. (1998). VR reborn. *IEEE Computer Graphics and Applications*, 18(6), 21–23.

Sanders, J., & Silvanus, J. (2010). An information provision framework for performance-based interactive elearning application for manufacturing. *Simulation and Gaming*, 41(4), 511–536.

Sarter, N. B., & Woods, D. D. (1991). Situation awareness: A critical but ill-defined phenomenon. *International Journal of Aviation Psychology*, 1, 45–57.

Seidel, I., Gartner, M., Froschauer, J., Berger, H., & Merkl, D. (2010). Towards a holistic methodology for engineering 3D virtual world applications. *2010 International Conference on Information Society (i-Society)* (pp. 224–229). London, U.K: IEEE.

Seo, J., & Kim, G. J. (2002). Design for presence: A structured approach to virtual reality system design. *Presence: Teleoperators and Virtual Environments*, 11(4), 378–403.

Sharples, S., Stedmon, A. W., D'Cruz, M., Patel, H., Cobb, S., Yates, T., Saikayasit, R., & Wilson, J. R. (2007). Human factors of virtual reality—Where are we now? In R. Pikaar, E. Koningsveld, & P. Settels (Eds.), *Meeting diversity in ergonomics*. Amsterdam, the Netherlands: Elsevier Science.

Shewchuk, J. P., Chung, K. H., & Williges, R. C. (2002). Virtual environments in manufacturing. In K. Stanney (Ed.), *Handbook of virtual environments* (pp. 1119–1141). Mahwah, NJ: Lawrence Erlbaum.

Slater, M. (2009). Place illusion and plausibility can lead to realistic behaviour in immersive virtual environments. *Philosophical Transactions of the Royal Society, Biological Sciences*, 364(1535), 3549–3557.

Smets, G. J. F., & Overbeeke, K. J. (1995). Trade-off between resolution and interactivity in spatial task performance. *IEEE Computer Graphics and Applications*, 15(5), 46–51.

Smith, S., Duke, D., & Wright, P. (1999). Using the resources model in virtual environment design. In *User centred design and implementation of virtual environments, Proceedings of Workshop at York, UK* (pp. 57–72). London, U.K.: British HCI Group.

Stanney, K. M., Salvendy, G., Deisigner, J., DiZio, P., Ellis, S., Ellison, E., … Witmer, B. (1998). Aftereffects and sense of presence in virtual environments: Formulation of a research and development agenda. *International Journal of Human-Computer Interaction*, 10(2), 135–187.

Stanton, N., Hedge, A., Brookhuis, K., Salas, E., & Hendrick, H. (2004). *Handbook of human factors and ergonomics methods*. Boca Raton, FL: CRC Press.

Stanton, N., Salmon, P. M., Walker, G. H., Baber, C., & Jenkins, D. P. (2005). *Human factors methods: A practical guide for engineering and design*. Farnham, U.K.: Ashgate Publishing Limited.

Suchman, L. A. (1987). *Plans and situated actions*. New York, NY: Cambridge University Press.

Sutcliffe, A. G. (1995). *Human-computer interface design* (2nd ed.). London, U.K.: McMillan.

Sutcliffe, A. G., & Gault, B. (2004). Heuristic evaluation of virtual reality applications. *Interacting with Computers*, *16*(4), 831–849.

Trescak, T., Esteva, M., & Rodriguez, I. (2010). A virtual world grammar for automatic generation of virtual worlds. *The Visual Computer*, *26*(6), 521–531.

Tromp, J. G. (2000). *Systematic design for usability in collaborative virtual environments* (PhD thesis). Communications Research Group, University of Nottingham, Nottingham, U.K.

Tromp, J. G. (2001). *Systematic design for usability in collaborative virtual environments* (Unpublished doctoral dissertation). Communications Research Group, University of Nottingham, Nottingham, U.K.

Tromp, J., Bullock, A., Steed, A., Sadagic, A., Slater, M., & Frécon, E. (1998a). Small group behaviour experiments in the COVEN project. *IEEE Computer Graphics and Applications*, *18*(6), 53–63.

Tromp, J., Sharples, S., & Patel, H. (2006). Special issue: VR design and usability workshop—Guest editors' introduction. *Presence, Teleoperators and Virtual Environments*, *15*(6), iii–iv.

Tromp, J., Istance, H., Hand, C., Steed, A., & Kaur, K. (1998b). *Proceedings of the First International Workshop on Usability Evaluation for Virtual Environments*. Leicester, U.K.

Wann, J., & Mon-Williams, M. (1996). What does virtual reality really NEED?: Human factors issues in the design of three-dimensional computer environments. *International Journal of Human-Computer Studies*, *44*, 829–847.

Wickens, C. D., Gordan, S. E., & Liu, Y. (2004). *An introduction to human factors engineering*. New York, NY: Longman.

Wilson, J. R. (1997). Virtual environments and ergonomics: Needs and opportunities. *Ergonomics*, *40*(10), 1057–1077. London, U.K.: Taylor & Francis Ltd.

Wilson, J. R. (1999). Virtual environments applications and applied ergonomics. *Applied Ergonomics*, *30*(1), 3–9.

Wilson, J. R. (2005). Methods in the understanding of human factors. In J. R. Wilson & N. Corlett (Eds.), *Evaluation of human work*. (3rd ed., pp. 1–31). London, U.K.: Taylor & Francis, pp. 1–31.

Wilson, J. R., & D'Cruz, M. D. (2006). Virtual and interactive environments for work of the future. *International Journal of Hum-Computer Studies*, *64*(3), 158–169.

Wilson, J. R., Cobb, S. V. G., D'Cruz, M. D., & Eastgate, R. M. (1996). *Virtual reality for industrial application: Opportunities and limitations*. Nottingham, U.K.: Nottingham University Press.

Zahorik, P., & Jenison, R. L. (1998). Presence as being-in-the-world. *Presence: Teleoperators and Virtual Environments*, *7*(1), 78–89.

Zeltzer, D. (1992). Autonomy, interaction and presence. *Presence: Teleoperators and Virtual Environments*, *1*, 127–132.

16 Cognitive Aspects of Virtual Environment Design

Allen Munro, Jim Patrey, Elizabeth Sheldon Biddle, and Meredith Carroll

CONTENTS

16.1 INTRODUCTION: COGNITIVE ISSUES FOR VIRTUAL ENVIRONMENTS

The uses of virtual environments (VEs) raise a variety of cognitive issues. VEs provide information to users, and in many cases, they provide users with the ability to interact with the environment in response to the information that the environment conveys. Many different cognitive factors play roles in the use of VE applications. These include issues related to perception, attention, learning and memory, problem solving and decision making, and motor cognition. Research and experimental applications of VEs have clarified some cognitive aspects of VE usage, but many questions have been raised for further research.

16.1.1 PERCEPTION

Cognitive research has revealed that perception is an active process, not merely an automatic bottom-up *conveyance* of sensory data such as visual images to higher-order cognitive centers. Expectations and experience exert a top-down contribution to perception, which results from the active interpretation of sensations in the context of these expectations and experience. In the past, the relatively low visual fidelity of VEs raised issues for perception due to the very poor resolution of displays (e.g., in terms of number of pixels presented per degree of visual angle), the frequent problems with alignment and convergence, and the often-primitive 3D models and texture maps presented on the displays. These characteristics place substantive constraints on the role of bottom-up data processing in perception in VEs. However, incredible advancements have been made in VE technology over the past decade resulting in high-resolution displays and complex 3D models that many times appear to accurately mimic reality. These advancements have shifted the importance of the top-down contributions of expectations and experience as VEs are now able to more accurately represent the cues that impact bottom-up processing. Despite this, the question remains: "To what extent must the perceptual *experience* with VE representations replicate real-world experience to ensure accurate perceptions?" The answer is not a simple one as it depends on the goals of the VE. In today's high-technology culture, people are able to interpret the graphical representations on television and on computer screens. However, when it is necessary for these interpretations to elicit emotional response, oftentimes, they fall short. To what extent must people learn to interpret—learn to *see*—what is presented in a VE and to what extent must these VEs be better designed to elicit accurate perceptions?

16.1.2 ATTENTION

There are several aspects of VEs that have potentially important consequences for the role of attention processes during VE experiences. A VE system has access to clues about the focus of a user's attention that are not available in an ordinary screen-presented graphics system. The field of view (FOV) provided by current VE systems is often so limited that a user must often carry out navigation and orientation actions so that the visual area of interest is brought into view. This has implications for both task performance and performance measurement. It can be incredibly disruptive to task performance when cues that are typically available instantly through peripheral vision in the real world now require the user to take time and cognitive resources to bring these cues into view. However, these actions constitute an announcement—a sort of nonverbal protocol—about the range of visual cues to which the user is currently attending. Additional precision can be obtained by leveraging eye-tracking technology to determine information about a person's visual attention via gaze, scan path, and fixation data (Hyona, Radach, & Deubel, 2003). Visual attention can provide important insights into the information used in task performance, such as the importance of various features or cues (Raab & Johnson, 2007).

16.1.3 LEARNING AND MEMORY

A number of research projects have studied the application of VE technology to learning environments, including advanced technical training and intelligent tutoring systems. The results of these projects provide useful guidance for the developers of VE-based learning environments. For instance, how can student attention most effectively be directed in VE? Highlighting and decluttering are techniques that have been used to orient visual attention (Chapman, 2002 cited by Underwood, 2007; Hagemann, Strauss, & Cañal-Bruland, 2006; John, Smallman, & Manes, 2005). How can a tutoring system know whether a student has perceived a feature of an environment? Eye-tracking and EEG-based diagnosis methods have been developed that can provide insight into a performer's cognitive processing (Carroll, 2010; Hale, Carpenter, & Wang-Costello, 2012; Hale, Fuchs, Axelsson, Berka, & Cowell, 2008). How much fidelity is required to produce desired levels of learning transference to a real-world context such as on-the-job task performance? Stanney, Milham, Champney, and Carroll (2007) assessed transfer of skills learned in low- and high-fidelity VEs to live exercises and found that the higher-fidelity environment facilitated greater live trial savings for skills such as visual search/scan. However, a number of important questions remain to be answered. Some of these questions relate closely to issues of perception and attention, such as to what extents do the special cognitive demands of the VE experience interfere with learning? Does VE-based learning offer special advantages for conveying some types of knowledge, but not for others?

16.1.4 Problem Solving and Decision Making

The same concerns about the cognitive demands of VE experiences of concern for learning are also very relevant to problem solving and decision making in the context of VEs. A VE provides the opportunity for decision makers to effectively and naturally collect data necessary to make decisions, such as observations of critical cues in an environment that would impact decision alternatives considered. Further, VEs make it possible for a user to quite naturally find new perspectives for a graphically represented problem or set of data; this offers the potential for VE-based aiding systems for analysts and decision makers. VEs also allow for a decision maker to observe the consequences of their decisions play out, a natural form of feedback that can be quite powerful and help to develop recognition-primed decision-making expertise (Klein, 1993). On the other hand, the additional cognitive processing required for perception and for the VE navigation actions could tie up cognitive processing resources, reducing the effective deployment of higher-level cognitive processing functions in some contexts.

16.1.5 Motor Cognition

In many real-world tasks, people must move about and change their physical orientation in order to make observations and to carry out actions. Infants and children spend many years acquiring the motor skills that make it possible, in most contexts, to carry out these navigation and orientation actions in a largely or even completely unconscious manner. In many VEs, the full set of motor skills used to navigate and orient in the real world is not wholly transferable to a VE. This could limit the utility of these VEs as it is unknown to what extent the nonautomatic nature of navigation and orientation in some VEs interferes with other cognitive processes. To overcome this, many emerging VEs are moving toward more natural interactions. For instance, the Naval Research Lab is developing point man, a novel virtual locomotion control for use in first-person shooter VEs that leverages a dual joystick gamepad in hand with head tracking and sliding foot pedals to facilitate more effective control over a wide range of tactical motions (Templeman & Denbrook, 2012). However, as up and coming generations spend more and more of their time interacting with video games and other virtual technology, this limitation may become obsolete.

Much of what is known about the cognitive aspects of VE design is a result of using VE for learning environments and for training. The next section of this chapter discusses cognitive issues in the context of VE training. VE technology has implications for human information processing, mental models, and metaphors/analogies for the design of training systems that make use of immersion. We will discuss cognitive issues in training and go on to describe recent VE systems used by the Navy and Marine Corps to teach cognitive tasks. The overall intent of this chapter is to give the reader some insight into the design of VE that lends itself to teaching cognitive skills. The work is in progress. There are many unanswered questions in this area. However, we have also made great progress with the systems that have been tested so far. We will discuss a series of applications that have been sponsored by the Office of Naval Research. These include VEs for maintenance training, harbor navigation, complex-maneuvering skills in open ocean, and infantry military operations in urban terrain (MOUT) room clearing. These efforts led to the development of objective performance measures and VEs that integrate visual, audio, and haptic cues.

16.2 COGNITIVE ISSUES IN VIRTUAL ENVIRONMENT TRAINING

16.2.1 Knowledge Type and the Acquisition of Knowledge in VEs

The cognitive factors listed earlier must be considered with respect to the types of information that are being conveyed in a VE. Consider these major types of information that may be presented in a VE: location knowledge, structural knowledge, behavioral knowledge, and procedural knowledge.

For each type of information that can be presented in a VE, there are potential issues to be addressed in determining how to best support the cognitive processes—perceptual, memory, decision making, and motor—that are brought into play in interacting with or learning the information. We can view these types of knowledge as four broad categories of knowledge with subtypes of knowledge that may each benefit from different learning affordances in VEs, as discussed in the following four sections.

16.2.1.1 Location Knowledge

Several types of knowledge about location may be particularly amenable to conveying with VEs rather than through more conventional means. Many VEs give their participants the facility to change their location and their orientation while observing objects in the environment. For instance, VE-based training systems such as Virtual Battlespace 2 (VBS2) provide a God's-eye view of the simulated world. This type of experience can give the participant a richer set of information about the relative locations of simulated objects, about how to get to a location, about how to bring objects into view, and about how to access and to manipulate them.

Relative position knowledge. To teach the relative location of objects in a 2D plane, it is enough to present a map or a picture that shows the layout of the objects in two dimensions. Drill and practice or mnemonic techniques can be used to fix the information in the student's memory, but his or her basic understanding of the position relationships in 2D space is an immediate apperception when the map or image is presented and the relevant relationships are pointed out. When objects have more complex relative locations in a 3D space layout, a single visual map will not suffice ordinarily. Students often require a variety of visual and/or positional experiences in 3D space in order to understand relative positional relationships. Two-dimensional presentations are particularly insufficient when the parts of two complex objects overlap because they have complex relationships to each other in the third dimension. However, in a VE, a student can move about while looking at such *intertwined* objects and can, thereby, quickly develop an accurate understanding of their relative positions. VEs do have significant limitations with respect to supporting identification of the relative positions of objects within a 3D environment, such as when a forward air controller (FAC) must determine if an incoming aircraft is pointed at the correct target. In the real world, this task may be accomplished quickly and easily with a slight turn of the head; however, in the virtual world, there is the added task requirement of panning back and forth that requires additional time and cognitive resources and often results in difficulty understanding the orientation of one's FOV (e.g., Where am I looking now?). Some VEs have attempted to overcome this with the incorporation of interfaces that give information regarding FOV orientation. For instance, the Marine Corps Deployed Virtual Training Environment–Combined Arms Network (DVTE-CAN; Muller, Schaffer, & McDonnough, 2008) presents a compass-like display that provides the range of the visual angle being displayed.

Navigation knowledge. Regian and Yadrick (1994) have shown, under controlled experimental conditions, that VEs promote more rapid acquisition of personal navigation knowledge in a complex environment of corridors and stairways than can be accomplished with an otherwise similar 2D graphical trainer. Results suggest that accurate textural detail in such environments may enhance personal navigation learning.

"How to view" knowledge. Many real-world tasks require that workers know where to position themselves and how to orient in order to make a necessary observation. Viewing a conventional 2D representation of the environment may not provide a learner with enough information about what can actually be seen from where. A VE gives the student an opportunity to find the available lines of sight for viewing indicators. For instance, when a fire support team (FiST) is emplacing in order to direct close air support and indirect fires to support a company maneuver, they must ensure they have a clear line of sight of the objective as well as the maneuvering company. A VE allows the verification of this line of sight while taking into account terrain that could obstruct the view.

"How to use" knowledge. Knowledge about how to use objects in an environment includes two characteristics. First, a student must know how the object can be accessed. Is it necessary to open a door? Must one reach through an opening or between two other objects? Second, the student must know how the object should be manipulated. Should it be pushed away, or grasped and twisted, or pulled toward one, or pushed down or pulled up? Is some more complex manipulation required? VEs offer good potential for teaching how to access objects that are to be manipulated. For instance, the medic vehicle environment/task trainer (MVETT) was designed to train the location of/spatial recognition of equipment within the high-armored ground ambulance utilizing infrared sensor technology (HAGA; Jones, Del Guidice, Moss, & Hale, 2010). Given the limitations of current technologies for manipulation in VE, the potential for teaching the details of fine motor manipulation is somewhat limited. Many VE systems have only one or two types of possible fine motor manipulation action (such as *pinch* and *release*). In these VEs, the simulation software typically interprets the action that is taken as being the one appropriate to the device that is being touched. With the emergence of sensor-based technology such as the Wii and Kinect, performance of many gross motor movements such as swinging a tennis racket or golf club is now feasible. Of course, there is every reason to expect that finer motor performance will be possible in future VE systems.

16.2.1.2 Structural Knowledge

Structural knowledge is the knowledge of how concepts in a domain are interrelated (Diekhoff, 1983). VEs have the potential to convey a rich variety of structural information, although only a subset of those possibilities have thus far been exploited.

Part–whole knowledge. Participants in VE rely on their real-world knowledge of the represented world to perceive part–whole relationships. If an environment includes a display panel with many lights and labels, an observer is likely to conclude that the labels and lights are part of the panel. When less familiar complex objects are presented in an environment, it may be necessary to give the participant means for exploring part–whole relationships. One such means is to give the participant the ability to move objects. A set of objects that move together may reasonably be construed as the parts of a whole.

Support–depend knowledge. If object A is above and flush with object B, then a participant can draw upon ordinary pragmatic knowledge to conclude that object B supports object A, at least in VEs that simulate conventional gravity conditions. If the environment allows the participant to move object B out from beneath object A, then the participant can test the hypothesis that B supports A. Such a VE would allow the exploration of structural relationships in complex structures such as buildings much like a child utilizes blocks to understand basic physics principles such as gravity.

Containment knowledge. Those VEs that permit a participant to open one object and manipulate or move another object within the first one provide a means for learning containment relationships. For instance, the MVETT VE referenced earlier (Jones et al., 2010) also supports the development of an understanding of what medical tools reside in different compartments within the vehicle, including an understanding of the need to extract all relevant tools from their respective locations before being ready to perform the task.

16.2.1.3 Behavioral Knowledge

There are several types of behavioral knowledge that can be effectively conveyed in a VE. By *behavioral knowledge*, we mean knowledge about how objects in the VE interact with each other and with participants.

Cause-and-effect knowledge. Cause-and-effect knowledge lets a participant predict that if one state change takes effect, then another state change will also occur. For example, "if the power switch is depressed when the system if functioning normally, then the *power on* light will be lit." Cause-and-effect knowledge can be acquired by observing a system as it changes. Such knowledge is more likely to be acquired, however, if the participant has the ability to effect state changes directly. This makes it possible for participants to observe cascades of state changes and to develop cause-and-effect knowledge as a result.

This does not mean that exploratory environments are enough in themselves to assure learning of cause-and-effect knowledge. For most students, it is necessary to guide exploration and to point out explicitly the causal relationships.

Function knowledge. Function knowledge is related to cause-and-effect knowledge. It has to do with the use of an object, that is, with what the object is *for*. Given the actions that can be taken on an object, the sets of intended effects of those actions constitute the function of the object. Again, a VE with interactive behavior can offer participants the opportunity to learn by experimenting, but guidance and tutorial exposition will typically be required to ensure effective learning for many students.

Systemic-behavior knowledge. Some VEs are designed to help students to come to appropriate generalizations about behavior. Simulations of Newtonian mechanics systems, for example, can be constructed so as to work with tutors that help students to understand Newton's model for the behavior of masses subjected to accelerations. In these VEs, students are expected not to learn simply the behaviors of particular objects but rather the principles that explain the behaviors of whole classes of objects.

16.2.1.4 Procedural Knowledge

VEs may also be appropriate for conveying knowledge about how to carry out procedures. Behavioral knowledge alone is not enough to determine how procedures should be carried out. In many cases, there are many action sequences that will result in a desired goal, but some will be more desirable than others for reasons of cost-effectiveness, safety, or speed.

Task-prerequisite knowledge. Before a procedure can be undertaken, an initial state must often be achieved. This state has characteristics that are the prerequisites for successfully carrying out the task. A VE is a particularly apt context for teaching task prerequisites based on the positions and orientations of objects or of the participants who are to carry out the task.

Goal-hierarchy knowledge. Task knowledge can be viewed as knowledge of a goal hierarchy and sequences of actions that can achieve transitions to subgoal states. Again, VEs are particularly useful for teaching procedures in which there are subgoals that prescribe positions and orientations or states of relative movement.

Action-sequence knowledge. The simplest procedures can be represented simply by a sequence of actions and a final goal. The subgoals of more complex procedures also have associated action sequences that are capable of bringing about the subgoal states. When these action sequences include movements and/or orientation changes, whether of the participant or objects under the participants control, a VE may be an appropriate learning context.

Hall, Stiles, and Horwitz (1998) conducted a study that can serve as a standard for evaluating the effectiveness of competing designs for learning how to move about in a complex environment in order to carry out a procedure. Two sets of students were trained to carry out a moderately complex task that involved moving among electrohydraulic machines on board a ship to make observations and to set controls. Students had to learn both the sequence of actions (observations and manipulations) that were to be carried out and where to go to carry out the actions. One group learned the task in the virtual environment technology (VET) facility (Johnson, Rickel, Stiles, & Munro, 1998; Stiles et al., 1996), while the other learned the task in a 2D graphical simulation environment. In both cases, the simulation was under the control of the same simulation, which was developed with the Virtual Interactive Intelligent Tutoring System (ITS) Development Shell (VIVIDS) simulation development and delivery system (Munro & Pizzini, 1998). Students from both groups were evaluated by asking them to perform the task in a real-life mockup environment modeled after the actual shipboard systems. There were no significant performance differences between the two groups, suggesting that, at least for this procedure, simulation training in a 2D graphical environment may be as effective as VE-based training.

16.2.2 Tutor Architecture

Figure 16.1 presents an architecture for VE tutoring systems (Munro, Surmon, Johnsson, Pizzini, & Walker, 1999). On the left side are those components responsible for the user interface and for the interactive behavior of the environment. On the right are the components that manage the course of instruction, that maintain a model of the student, and that conduct the moment-by-moment details of tutorial interactions.

The user view has subcomponents that are the responsibility of the VE software. An example of these components in use is shown in Figure 16.2.

A user view is responsible for everything that can be seen (or heard, or touched, etc.) by a student during simulation learning. Its model view component is responsible for rendering the simulated world prescribed by the behavior model and for detecting user events and passing them on to the behavior model for semantic processing. Presentation channels may include text, speech, audio, HTML, and video presenters, to name a few. Entries are user interface components that students can use to answer questions that cannot be answered by actions in the

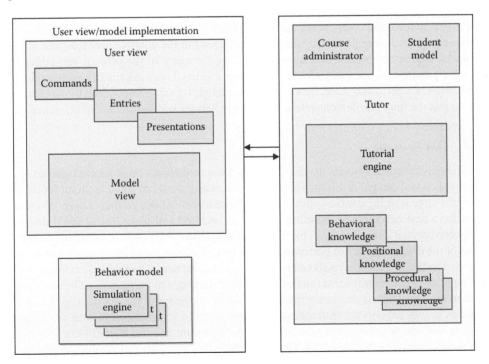

FIGURE 16.1   A generic architecture for simulation-centered tutors.

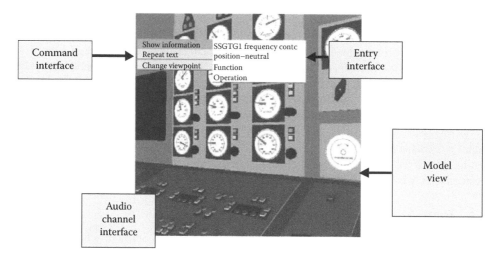

Command interface

Entry interface

Show information
Repeat text
Change viewpoint

SSGTG1 frequency contc
position–neutral
Function
Operation

Model view

Audio channel interface

FIGURE 16.2   A user view in a VE.

simulated model view. Commands are interfaces for higher-level student interactions with the tutorial system, such as pacing instruction and asking for help.

A variant of this architecture, shown in Figure 16.3, is capable of supporting implementations in which multiple students collaborate with each other in a simulation environment. Each student has a user view, including a model view that is under the control of a central behavior model. In this way, other students can view the cascading effects of actions taken by one student. The model views shown to the different students need not depict the same scenes. Team members often work with different parts of a complex interconnected system and are able to see only one part of it at a time. The single behavior model determines the values of simulation attributes, and some of these value changes cause changes in what is displayed in the model views.

The instruction control components are responsible for the configuration of the complete learning environment, including the availability and positioning of the possible user interface elements. The tutorial engine component is a service requester for many of the other major components. It can ask a presentation channel to fetch and present a video. It can ask the behavior model to put the simulation in a particular state. It can ask the model view to stop responding to user actions for a time, so that the tutor can demonstrate a sequence of actions without student interference.

16.2.3 Task Analysis

Large-scale instructional systems developed for sophisticated clients must have a rigorous foundation in instructional design. A military or large business unit would no more contract for the development of a large training course without a detailed instructional design than a corporation would contract for a new headquarters building without an architectural plan. Instructional designs are typically constructed so as to realize objectives. In many cases, these objectives are best understood in terms of the desired quality of performance of a task. Students are to learn enough as a result of the planned instruction that they will be able to carry out some task at a specified performance level. This type of objective requires that the task the trained student is to perform must be analyzed well enough so that the analysis can guide the design and development of the instruction. A number of approaches to task analysis for instructional design have been summarized in such works as those of Leshin, Pollock, and Reigeluth (1992); Jonassen, Hannum, and Tessmer (1989); and Jonassen (1999). A key approach in the design of VEs is the conduct of a sensory task analysis (STA; Milham, Carroll, Stanney, & Becker, 2008) to ensure that multimodal cues are instantiated at the appropriate fidelity to ensure training objectives can be achieved without unnecessary expense in time and

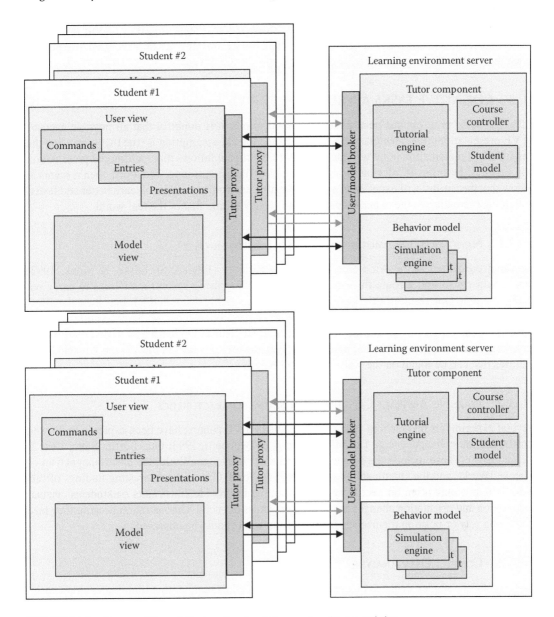

FIGURE 16.3   One possible architecture for simulation-centered team training.

processing power. Task analysis tools such as the sensory perceptual objective task (SPOT) tax-onomy have been developed to aid in the STA process of task decomposition (Champney, Carroll, Milham, & Hale, 2008).

Several characteristics of VE training require special attention to certain task characteristics dur-ing the task analysis process. VEs may be especially well suited to teaching students where to move and where to direct their gaze during a task. When combined with eye-tracking-based performance assessment and feedback, VE-based training has been shown to lead to significant gains in search performance during both training (Carroll, 2010) and transfer to the real-world application (Carroll, Kokini, & Moss, 2013). To facilitate this, however, it is important that an analysis of a task should pay attention to elements of the task that require moving or orienting actions as well as observation of critical cues, so that those steps in the task will be effectively taught in the VE training system.

If a task requires either that the student learn to move and/or orient himself or herself or that the student must learn to move and/or orient objects, then the VE training can be structured to explicitly teach the student to carry out these steps.

16.3 COGNITIVE TASKS APPROPRIATE FOR VE

Although some progress has been made with applications, it is doubtful that all possible avenues of VE presentation have been explored. That is, if the logical assumption is true that increasing the bandwidth of information input will facilitate learning, then it follows that multimodal presentation will facilitate learning. If information can be received via vision, sound, and touch, then it seems to follow that the learner will experience a richer learning environment and will learn better and faster, with longer retention as the experience will more closely mimic that of the real world.

16.3.1 NAVIGATION/LOCOMOTION IN COMPLEX ENVIRONMENTS

Research conducted at the Air Force Research Laboratory (Regian, Shebilske, & Monk, 1992) taught subjects to find a route through a complex building using several modalities to carry out the instruction. One result of this work is that providing a richly decorated VE may have the effect of improving how well a student learns to navigate in the environment as the decorations provide intermittent landmarks by which they can navigate. If students are to learn how to find their way in a building with many corridors and many turns, then the texture maps in the VE should include the pictures that can be found on the walls of the real building.

16.3.2 LEARNING ABSTRACT CONCEPTS WITH SPATIAL CHARACTERISTICS

Several experiments on teaching scientific concepts in a VE context have been explored by Bowen Loftin and his colleagues (e.g., Loftin, Engelberg, & Benedetti, 1993; Ota, Loftin, Saito, Lea, & Keller, 1995). These include the Newtonian mechanics of NewtonWorld, the electromagnetism of MaxwellWorld, and the chemical bonding of PaulingWorld. One of the interesting findings of this work is that it may often be necessary in VE training to provide constraints on students' virtual locomotion and orientation changes as students may miss a crucial demonstration because they face the wrong way or position themselves far away from what should be observed.

16.3.3 COMPLEX DATA ANALYSIS

Some types of complex data can be presented in 3D space in a manner that illuminates characteristics of the data that would not be evident in conventional 2D presentations. For instance, many VEs support the viewing of objects within the VE from both a first-person view and a God's-eye view to support an understanding of the geospatial relation as well as object proximities and size differentials. Furthermore, users can be given the opportunity to manipulate the representation and to view it from any angle. For instance, in training battlespace planning, it is useful for a learner to be able to view targets, friendly locations, weapons trajectories, and ammunition impact zones from different angles in order to understand the impact of weapon deployment as well as potential effectiveness and risk of alternative courses of action.

16.3.4 MANIPULATION OF COMPLEX OBJECTS AND DEVICES IN 3D SPACE

When there is a requirement to train someone on the physical manipulation of a complex object in 3D space, training that makes use of the actual object under the conditions that prevail on the job offers the highest probability for good transfer to the actual task. In certain cases, however, the use of the actual object for training is inappropriate or impossible. If the training involves the handling of an extremely delicate and valuable object that can be damaged through inappropriate handling, it may be advantageous to train with a VE simulator. In other cases, danger to the operator or to others may argue for training in a simulation. For example, an inadequately trained crane operator on a construction site could carry out actions that would cause his crane to topple, possibly landing on other workers or bystanders and pulling the operator's cab off its pedestal.

16.3.5 DECISION MAKING

Where the domain of decision involves relationships that must be understood in 3D and, especially, when spatial relationships are changing over time, VEs may prove extremely valuable. Shiphandling tasks require the conning officer to develop the *seaman's eye* (Crenshaw, 1975). Part of that skill appears to be recognition of relative motion of objects in the path of own ship. Another part of that skill appears to be maneuvering own ship to avoid obstacles. Now, wind and sea currents effect the direction the ship must point in order to achieve a desired course through the water. As they exert force upon the ship to push it in one direction and the ship's crew desires it to go in another, the heading of the ship must be adjusted slightly to compensate. It appears that one method to teach this compensation angle computation would be a mathematical formula. It also appears that one way of augmenting the concept might be to actually experience the forces of wind and current. There are a couple of ways to do that that have been considered in work completed by the Navy. One method would be to use a VE to enlarge the user to the size of a giant so that they could actually hold the ship in their hand and feel the force of the wind and current acting on the force of the ship. This could be achieved both visually and with a haptic device for the hand. Other laws of force and motion might be taught by having one feel the forces with a haptic device through force feedback, a technique being used by laparoscopic surgery VEs (e.g., Rosen, Brown, Chang, Sinanan, & Hannaford, 2006). Current haptic devices do not appear to be robust enough for sophisticated forces acting in multiple dimensions. Thus, this area requires further work.

VEs also facilitate training complex decision-making skills needed to operate successfully in unpredictable environments with potentially innumerable situations and decision alternatives. VEs provide the opportunity to present the learner with a large array of variations in the situation in order to both enhance their library of experiences on which to draw and hone the skills necessary for effective decision making such as situation assessment and evaluation of decision alternatives. Further, VEs provide the opportunity to allow performance of decision-making skills under conditions that mimic the stressors they will experience on the battlefield. For instance, the Marine Corps Infantry Immersion Trainer (IIT) is a mixed-reality training facility that has integrated high-fidelity VE technology that mimics not only the sights and sounds of the battlefield but also the smell in an attempt to enhance the ability to make effective decisions under the stress of combat (Dean et al., 2008).

16.4 APPLICATIONS OF COGNITIVE TRAINING USING VE

Martin and his colleagues (Breaux, Martin, & Jones, 1998) sought to lay a foundation for application to VE of a technique whereby the learner codes information both visually and verbally (see Paivio, 1991). This was intended to explain why VE immersion could be effective in developing a mental representation of the environment. Consider the VE that contains visual, haptic, and audio components to teach shiphandling skills. The intent of traditional training is to apply the concepts described by the teacher verbally and with mathematics in the classroom and augment with static depictions of course tracks. The real-world application is, instead, a dynamic, interactive environment of ships at sea maneuvering to avoid one another and to achieve a desired course to reach their destination. The intent of VE-based cognitive training is to solidify a relationship between the verbal concepts of the classroom and the dynamic visual environment of the simulation. VE is intended as a method to coalesce the cognitive model from components of dynamic interactive objects of the visual scene with static linguistic concepts from the classroom. The model thereby developed by the learner is expected to be a richer, more robust model that generalizes to novel ship maneuvering situations. A more detailed explanation of specific tasks will follow.

Another challenge posed for VE by the work of Breaux et al. (1998) was the potential for VE to shape mental behavior. Let us say we seek to teach a complex concept of adjusting the heading of the ship to correct its path through the water because it has been disturbed by wind and sea current forces acting to push the ship off course. If the student adjusts the course too much or too little, we may want to understand what it is about the student's mental concept that results in the errors. If we can create a model of the student's behavior of adjusting too much or too little, then we can claim to understand the misperception of the student. That model may be computationally intensive. However, if VE can project the path of the ship with the error introduced by the student, we may say we have depicted the current mental model being used by the student. The instructional intervention that VE allows is then to demonstrate visually and haptically how the forces are, in fact, acting upon the ship's path. Next, VE can dynamically transition the immersed student from the misperception to the correct perception by the use of visual and haptic displays to adjust what the student is experiencing visually and haptically. Consider the analogy of the teacher taking the hand of the student and moving it through the motions required in drawing an image on paper. Now, extend that analogy to the VE adjusting the image and forces felt so that the student sees and feels the proper forces acting on the ship. This is what was intended by the concept espoused by Breaux et al. of morphing the student's misperception into the proper perception.

Let us move now to some specific applications of VE to training. We consider applications of submarine navigation, shiphandling, electronic maintenance, and infantry MOUT room clearing.

16.4.1 VESUB Harbor Navigation

Hays, Vincenzi, Seamon, and Bradley (1998) provide a description of a VE training test bed, the VE for SUBmarine (VESUB) training that evaluated the effectiveness of teaching harbor navigation using an immersive visual system with audio. The officer of the deck (OOD) stands in the open air while a submarine is on the surface and gives commands to the crew for speed and direction of the submarine (see Figure 16.4). The surfaced submarine must navigate its way through the harbor, avoiding obstacles, and tie up at its pier. Current training is typically on the job, so this VE system was designed to provide concentrated training for a variety of situations, including various harbors, weather conditions, traffic, and time of day. Because of the expense of running aground or collision of the submarine with other objects, it is seldom that inexperienced OOD has the opportunity to train in a real submarine. Therefore, the system can provide much needed training for inexperienced officers as well as refresher training for the *old hands*.

FIGURE 16.4 **(See color insert.)** VESUB harbor navigation training.

To make the system look like a real submarine, the visual system depicted a very detailed harbor and displayed various ship instruments on the bridge. A computer voice recognition system captured the commands from the student and translated them for use by the computer mathematical models driving the visual display. The computer models generated sounds of the water, foghorns, and other relevant audio cues as well. An instructor was present to create and observe a particular scenario for the student and to evaluate performance.

One of the features of this system is team communication. The computer expects the student officer to issue proper commands using proper terminology. Also, the instructor is available to provide feedback and manipulate the scenario to challenge the student's knowledge and skills. Unlike the real task where the commanding officer, acting as instructor, must be ever vigilant to the danger of an error that can be very expensive or even life threatening, the VE system can provide the instructor a less-intensive work pace with concentration on teaching the student.

Hays et al. (1998) report that the evaluation of VESUB was quite successful. Two Navy training facilities spent 3 weeks each evaluating the system, using 41 participants. Of the 15 shiphandling variables tested, improvement ranged from 13% to 57% on 11 of them. It was proposed that in order to perform the task of navigating the submarine through the harbor, one must develop a mental model of not only the task at hand but also the relative motion of other water traffic and the effect of tides, current, and wind upon the submarine. Some of the cognitive components of the task include the relationship of what one sees in the environment to its representation on the harbor chart, the relative size and height of objects with the angle on the bow of the submarine, the relative motion of the objects, the prioritization of other water traffic that may conflict with the submarine's course, and how to maneuver the submarine out of danger.

16.4.2 COVE Underway Replenishment

Following the apparent success of the VESUB project, the Navy undertook application of VE to more complex cognitive skills. The Conning Officer Virtual Environment (COVE) project sought to combine VE with computer-generated performance assessment. In his thesis, Norris (1998) describes some of the cognitive components of the task of commanding a ship. The challenge for the COVE project was to ascertain and develop methods by which the VE could provide measures of the student's performance then use those to deliver diagnosis, feedback, and prescriptive tasks to the student. COVE sought to divide the task into cognitive and perceptual components. The VE not only presented the environment to the student but also allowed measures of the objects in the environment, indications of where the student looks, and computer speech recognition of the verbal commands that are given.

The task most studied in preparation for the COVE project was that of underway replenishment. In this task, a supply ship is steaming a set course, and the ship to be replenished steams to an alongside position. Both ships then continue to maintain their course and speed while supplies are transferred across approximately 120 ft separating the two ships. The perceptual components include distance, speed, and relative position of the ships. The cognitive components include issuing verbal commands to guide the ship through four different phases of the task. Each of the four phases requires slightly different skills.

The following were proposed as phases of the UNREP task:

- Approach
- Slide-in
- Alongside
- Station
- Breakaway

The approach is defined as being from an arbitrary starting position behind the replenishment ship to a point where the officer must order an initial deceleration to slow to a speed matching that of the replenishment ship. Slide-in is from the initial deceleration to a point where the bow of the receiving ship crosses the stern of the replenishment ship. The alongside phase is from the bow–stern crossing to a point where the two ships are even with one another and at the same speed. The station phase is maintaining the position so that replenishment can occur. Breakaway phase is from separation of the two ships to the end of the scenario. The slide-in and alongside are transitional phases, the slide-in being the *end* of the approach and the alongside representing the *beginning* of station. It was proposed that the phases require different combinations of component skills, some of which include

- Ship control
- Perception
- Decision making

Ship control refers to the ability to effectively control the speed and direction of the ship. Some implicit parameters include ship mass, hydrodynamics, and acceleration. Perceptual skill includes relative motion, the sensation of the visual scene flowing by the conning officer, and distance estimation (both range and lateral separation estimation). Decision making entails the processes involved in monitoring speed and position as well as the ability to *stay ahead* so as to prevent errors and to identify critical incidents. It was proposed that decision making, perception, and, most particularly, ship dynamics best lend themselves to training in a VE.

These component skills were proposed to vary across the different phases. The approach and slide-in required more distal perceptual cues with larger, ballistic-type ship movements (and commands) and simpler, slower decisions. The alongside and station phases required more proximate perceptual cues with small, iterative ship movements (and commands) and quicker decisions with foresight. Finally, the breakaway was proposed to involve transitioning from proximate cues and small, iterative movements to distal cues and large, ballistic movements. Decision making appeared to require analysis of *errors* and *critical incidents*, but unfortunately Norris (1998) did not provide operational definitions for *errors* and *critical incidents*.

The benefits that were expected from a COVE system included

- Training perceptual skills through multiple modalities in VE (visual and auditory)
- Tutor software to teach complex hydrodynamic concepts
- Techniques to teach spatial and temporal orientation of ship movements relative to one another

The preliminary work on planning the COVE project included the development of measures of the cognitive components of the task. These will be discussed next.

16.4.3 Performance Measures

The underlying cognitive components of the COVE task were discussed earlier and include issuing verbal commands to guide the ship through four different phases of the task. In order to arrive at more specific components that would lend themselves to being useable for feedback, diagnosis, and remedial tasks, the preliminary COVE work sought to determine specific measures of cognitive shiphandling skills. Although Norris (1998) provided a task analysis, what was needed was specific measures that correlated with a *good UNREP*.

Patrey, Sheldon, Breaux, and Mead (2001) proposed the application of a perceptual–action task model. This requires one to acquire and use spatial knowledge in order to be successful at shiphandling. Actions the officer of the receiving vessel is able to do include giving heading and speed commands for the ship. This means that the officer must be aware of water traffic, some of which may be on a collision course and of a hypothetical position that would put the receiving ship in the proper position with the replenishment ship. Patrey et al. (2001) defined four characteristics of a ship's dynamics, position, heading, velocity, and acceleration as its *moments*.

In their study, subjects spent about 1.5 h to perform two UNREPs. They were evaluated by six expert shiphandlers, and the average rating was correlated with 13 moments that were measurable in the VE. Four of the moments proved significant: lateral separation of the two ships, bearing, speed, and acceleration. In addition, it appeared that 83% of the performance of the UNREP could be explained by these four moments at the point when the bow of the receiving ship passed the stern of the replenishment ship. This is the transition point between slide-in and alongside phases of the UNREP.

The conclusions drawn in this study contradicted current thinking about UNREP performance. Experts typically suggest that there is no single best way to UNREP. There are many variations that are just as good. However, Patrey et al. (2001) suggest that certain moments correlate so highly with performance rating that it is possible to quantify what is traditionally thought of as a qualitative task. Further, typical training for UNREP done at sea is a trial and error proposition. This study suggested that there are in fact measures that may indicate the strength of the cognitive model used by students to perform the UNREP. Further, it was suggested the VE training, by its ability to measure the entire environment, will allow more structured training of the cognitive model than will occur by trial and error training. One approach would be that the moments would be used to develop remedial exercises or part-task exercises. For example, it may be that UNREP skills develop quicker if a student practices just the slide-in and approach phases until their performance is high, then moves on to approach, station, and breakaway.

16.4.4 OPERATIONS TRAINING WITH VE

In the Virtual Environments for Training Project (Johnson, Rickle, Stiles, & Munro, 1999; Stiles et al., 1996), there was a concerted effort to generalize features of the training environment to exploit cognitive characteristics of operations training in a VE. This project made use of Virtual Interactive ITS Development Shell (VIVIDS) (Munro et al., 1998; Munro & Pizzini, 1998), a tool for authoring interactive graphical simulations and tutorials that are delivered in the context of interactive graphical simulations. VIVIDS includes a deep representation of the notions of effectuating actions and observations. As a simulation author builds a simulation, the author designates those attributes that store the immediate results of effectuating actions and those attributes that store directly observable values in the simulation. Because the simulation is coded with these cognitively relevant concepts, it is possible to use the VIVIDS tool to rapidly author procedural lessons simply by carrying out the procedures that students are to learn.

VIVIDS provides a set of direct manipulation interfaces for building simulations and tutorials. A number of features support productive simulation authoring. Authors can select from libraries of behaving objects to add interactive components to a simulation. Alternatively, authors can draw new objects or import object images and then write rules that control the interactive behavior of the objects. A simulation engine that supports constraint-based effect propagation makes it possible to author complex simulation behaviors without attending to details of the flow of control of simulation effects. In the right portion of Figure 16.5, a portion of the behavior authoring interface is displayed. An object data view displays the attributes of a simulated shipboard throttle. Authors can add and delete attributes, change their values, and write constraint rules that prescribe the values of given attributes. Such rules specify values in terms of the values of other attributes and in terms of student actions. In this figure, the author has marked an attribute named *mode* as a *control attribute*, that is, an attribute whose value is directly determined by user actions. During instructional authoring and delivery, VIVIDS knows that it can make use of the values shown in control attributes to talk about student actions.

In the left portion of the figure is a snapshot of a portion of the VE that exhibits the authored behavior. The behavior of simulated devices such as the throttle shown here is determined by the behavior rules created in the VIVIDS authoring system.

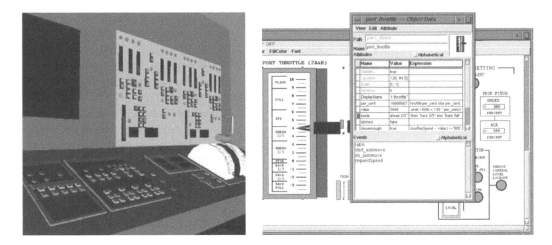

FIGURE 16.5   VIVIDS behavior authoring interface.

VIVIDS supports productivity in tutorial development by exploiting the simulation author's designations of the cognitively relevant attributes for actions and observations. In Figure 16.6, for example, a simple procedural tutorial is being authored by carrying out a task and then pointing to the objects in the simulation that indicates that the task has been achieved. As the instructor carries out the steps in the procedure, the tutorial authoring system notes the actions and the associated changes in attribute values. It generates brief action descriptions, which appear in a list in the tutorial authoring view. When it is time to present the tutorial to the student, VIVIDS will be able to talk to the student in terms of these cognitively relevant action and indicator values.

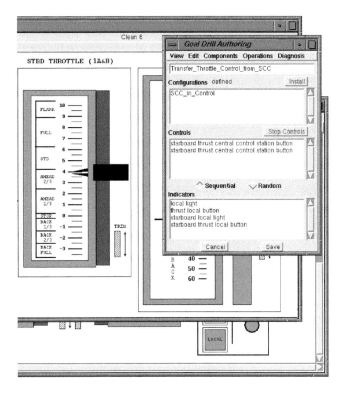

FIGURE 16.6   VIVIDS tutorial authoring system.

FIGURE 16.7   Steve demonstrating how to carry out a step in a procedure.

The VET project supported more than one approach to tutoring procedures. An intelligent agent called *Steve*, developed by the Information Sciences Institute, could be used in conjunction with VIVIDS (Johnson et al., 1998). At the cost of some expert knowledge engineering, *Steve* could be given a robust representation of the structure of as task in terms of plans and subgoals. During instruction, *Steve* could watch student actions and monitor changes in VIVIDS simulation attributes to determine where the student might be in the plan structure. Because *Steve* has a richer representation of the decision process, it is capable of generating richer tutorial dialogs during instruction than is the easily authored native VIVIDS procedure lesson. For example, *Steve* can describe the reasons for actions. In Figure 16.7, *Steve* is shown demonstrating to a student how to carry out a step in a procedure.

16.4.5   VE MOUT

Under ONR's Virtual Technologies and Environments (VIRTE) program, the VE MOUT trainer was developed to train team-based infantry room clearing skills. The goal of the MOUT team is to search and eliminate enemy threats in an urban environment (e.g., a building). To accomplish this, a unit team enters a building and moves rapidly along a hallway while covering the entire area with their weapons to maintain security. If people are encountered, they are quickly evaluated and engaged if hostile. The team clears from one end of the hallway to the other, clearing rooms as they are reached. As team members cross the threshold of a doorway into a room, they perform immediate target engagement of any enemies detected. Each team member clears a designated area of responsibility, such as the left, right, or overhead areas of the room. Next, team members search behind furniture or other obstacles to ensure that enemies are not hiding behind them, avoiding dangerous areas in the environment such as windows in order to minimize exposure to enemy units.

There are several key knowledge and skill sets required for successful performance that are targeted with these VE prototype trainers, including relational knowledge (e.g., distances between objects, speed required to move effectively), perceptual skills (e.g., ability to effectively search for and recognize relevant cues in the environment such as how an enemy looks or the feel of a teammate giving them a *bump* to set them in motion), decision making (e.g., shoot/no shoot), procedural knowledge/skill (e.g., steps to complete a buttonhook room entry), and team coordination and communication (e.g., communication and physical coordination for fluid entry into a room).

Two VE MOUT prototypes were developed, both low and high fidelity to evaluate their effectiveness in training the earlier referenced skills. The low-fidelity training system consisted of a desktop trainer with a monitor, 3D audio (low-fidelity generalized head-related transfer functions [HRTF]), and a game *rumble* pad haptic interface. The high-fidelity training system consisted of a fully immersive system with a head-mounted display (HMD), head and body tracking system, 3D audio (generalized HRTF), vibrotactile vest, and an airsoft haptic gun interface. The high-fidelity environment provided several key multimodal cue and interaction capabilities hypothesized to improve training effectiveness such as haptic vibration that provided information such as when/where the trainee was in contact with a teammate and/or wall as they stacked outside a room entry, when/where on the body the trainee had been shot, and the ability to more naturally maneuver through hallways and peer around/into doorways.

A training effectiveness experiment was conducted in which trainees trained on either low- or high-fidelity VE MOUT systems and transferred to a live MOUT shoothouse in which performance improvements and time to train to criterion were evaluated. Results indicated that pretraining in the high-fidelity prototype led to greater increases in overall performance levels in the live environment, specifically for physical tasks that required search and scanning and for outcome measures. These differences may be due to the high-fidelity system's ability to support the development of perceptual skills such as search and scan skills (due to the HMD and head and body tracking), which allowed trainees to physically scan and to peer around corners using their head and upper body. For full description of research findings, see Stanney et al. (in preparation).

16.5   CONCLUSIONS

Cognitive issues in the fields of perception, attention, learning and memory, motor cognition, and decision making are all relevant to VEs. Research on and prototype developments of VEs for learning have dealt with these issues in ways that may help to inform future VE projects of many types.

A number of research and practical issues remain to be addressed in future research. These include such questions as "How should a user's attention be drawn to objects or areas that are currently outside of the student's FOV?" and "Under what conditions is it appropriate to assume that a user has perceived something that is in his FOV?"

Validation studies also remain to be conducted. Under what conditions does procedural expertise in a VE translate to competence in a corresponding real-world task?

As for training applications, where are VEs best suited? VE-based training may be appropriate for concepts that require mental images in conjunction with decisions about location, about relative location of things over time, or about the timing of location and direction changes. VE can help students to recognize, or visually imagine, patterns of time and location in tasks done before, so that the student can see a familiar pattern in a new situation and then do the job in the new situation.

Why should VE work well for training? The tutor can control the mental images, timing, and location not only of objects but also of the location of the learner. This lets the learner explore relationships from different perceptual viewpoints so that the learner can later recognize the same kind of situation. Even if the situation looks like it is new, the learner can see that it is really the old situation presented at just a little different angle. The use of VEs encourages the presentation of many perceptual viewpoints to the learner so that an appropriate generalization of the concept can be rapidly acquired.

Another area of research to be explored is the further development of tools that encourage the production of VEs that take advantage of cognitive factors to ensure that such an environment will achieve its purpose in an effective and efficient manner. A key issue may be the development of familiar metaphors for VE that facilitate the learner's association of novel situations with more easily understood ones.

REFERENCES

Breaux, R., Martin, M., & Jones, S. (1998). Cognitive precursors to virtual reality applications. *Proceedings of SimTecT'98* (pp. 21–26). Adelaide, South Australia.

Carroll, M. B. (2010). *Empirical evaluation of the effectiveness of eye tracking-based search performance diagnosis and feedback methods* (Unpublished doctoral dissertation). University of Central Florida, Orlando, FL.

Carroll, M., Kokini, C., & Moss, J. (2013). Training effectiveness of eye tracking-based feedback at improving visual search skills. *International Journal of Learning Technology*, 8(2), 147–168.

Champney, R. K., Carroll, M. B., Milham, L., & Hale, K. (2008). Sensory-perceptual objective task (SPOT) taxonomy: A task analysis tool. *Proceedings of the Human Factors and Ergonomics Society Conference 52nd Annual*. New York, NY. September 22–26, 2008.

Chapman, P., Underwood, G., & Roberts, K. (2002). Visual search patterns in trained and untrained novice drivers. *Transportation Research Part F: Psychology and Behaviour*, 5(2), 157–167.

Crenshaw, R. S. (1975). *Naval shiphandling*. Annapolis, MD: Naval Institute Press.

Dean, S., Milham, L., Carroll, M., Schaeffer, R., Alker, M., & Buscemi, T. (2008). Challenges of scenario design in a mixed-reality environment. *Proceedings of the Interservice/Industry Training, Simulation, and Education Conference (I/ITSEC) Annual Meeting*. Orlando, FL. December 1–4, 2008.

Diekhoff, G. M. (1983). Relationship judgments in the evaluation of structural understanding. *Journal of Educational Psychology*, 75, 227–233.

Hagemann, N., Strauss, B., & Cañal-Bruland, R. (2006). Training perceptual skill by orienting visual attention. *Journal of Sport & Exercise Psychology*, 28(2), 143–158.

Hale, K. S., Carpenter, A., & Wang-Costello, J. (2012). *Screener's auto diagnostic adaptive precision training (ScreenADAPT) System. Year 1 research and development technical status report. Delivered to Department of Homeland Security and T* (Human Factors Behavioral Division under contract D11PC20053). Washington, DC: Department of Homeland Security.

Hale, K. S., Fuchs, S., Axelsson, P., Berka, C., & Cowell, A. J. (2008). Using physiological measures to discriminate signal detection outcome during imagery analysis. *Human Factors and Ergonomics Society Annual Meeting Proceedings*, 52(3), 182–186.

Hall, C., Stiles, R., & Horwitz, C. D. (1998). *Virtual reality for training: Evaluating knowledge retention*. Paper presented at the Proceedings of the Virtual Reality Annual International Symposium, Atlanta, GA. March 14–18, 2008.

Hays, R. T., Vincenzi, D. A., Seamon, A. G., & Bradley, S. K. (1998). *Training effectiveness evaluation of the VESUB technology demonstration system* (Technical Report 98-003). Naval Air Warfare Center Training Systems Division. Orlando, FL.

Hyona, J., Radach, R., & Deubel, H. (Eds.). (2003). *The mind's eye: Cognitive and applied aspects of eye movement research*. Oxford, U.K.: Elsevier.

John, M. S., Smallman, H. S., & Manes, D. I. (2005). Heuristic automation for decluttering tactical displays. *Human Factors*, 47(3), 509–525.

Johnson, W. L., Rickel, J., Stiles, R., & Munro, A. (1998). Integrating pedagogical agents into virtual environments. *Presence*, 7(6), 523–546.

Johnson, W. L., Rickle, J., Stiles, R., & Munro, A. (1999). Integrating pedagogical agents into virtual environments. *Presence: Teleoperators and Virtual Environments*, 7(6), 523–546.

Jonassen, D. H. (1999). *Task analysis methods for instructional design*. Mahwah, NJ: Lawrence Erlbaum Associates.

Jonassen, D. H., Hannum, W. H., & Tessmer, M. (1989). *Handbook of task analysis procedures*. New York, NY: Praeger.

Jones, D. L., Del Guidice, K., Moss, J., & Hale, K. (2010). *Medic vehicle environment/task trainer (M-VETT): Phase I process and results* (Final Technical Report submitted to the Office of the Secretary of Defense under contract W81XWH-10-C-0198). Washington, DC: Office of the Secretary of Defense.

Klein, G. (1993). A recognition primed decision (RPD) model of rapid decision making. In G. A. Klein, J. Orasanu, R. Calderwood, & C. Zsambok (Eds.), *Decision making in action: Models and methods* (pp. 138–147). Norwood, NJ: Ablex.

Leshin, C. B., Pollock, J., & Reigeluth, C. M. (1992). *Instructional design strategies and tactics*. Englewood Cliffs, NJ: Educational Technology Publications, Inc.

Loftin, R. B., Engelberg, M., & Benedetti, R. (1993). Virtual environments in education: A virtual physics laboratory. *Proceedings of Society for Information Display Conference*, May 16–21, 1993, Seattle, WA.

Milham, L. M., Carroll, M. B., Stanney, K. M., & Becker, W. (2008). Training requirements analysis. In D. Schmorrow, J. Cohn, & D. Nicholson (Eds.), *The handbook of virtual environment training: Understanding, predicting and implementing effective training solutions for accelerated and experiential learning.* Aldershot, U.K.: Ashgate Publishing.

Muller, P., Schaffer, R., & McDonough, J. (2008). U.S. marine corps deployable virtual training environment. In J. Cohn, D. Nicholson, & D. Schmorrow (Eds.), *The PSI handbook of virtual environments for training and education.* Westport, CT: Praeger Security International.

Munro, A., & Pizzini, Q. A. (1998). *VIVIDS reference manual.* Los Angeles, CA: Behavioral Technology Laboratories, University of Southern California.

Munro, A., Johnson, M. C., Pizzini, Q. A., Surmon, D. S., Towne, D. M., & Wogulis, J. L. (1998). Authoring simulation-centered tutors with RIDES. *International Journal of Artificial Intelligence in Education*, *8*(3–4), 284–316.

Munro, A., Surmon, D. S., Pizzini, Q. A., & Johnson, M. C. (1997). Collaborative authored simulation-centered tutor components. In C. L. Redfield (Ed.), *Intelligent tutoring system authoring tools* (Technical Report FS-97-01). Menlo Park, CA: AAAI Press.

Munro, A., Surmon, D., Johnson, M., Pizzini, Q., & Walker, J. (1999). An open architecture for simulation-centered tutors. In Lajoie, S. P. & Vivet, M. (Eds.), *Artificial Intelligence in Education: Open Learning Environments: New Computational Technologies to Support Learning, Exploration, and Collaboration* (pp. 360–367). Amsterdam: IOS Press.

Norris, S. D. (1998). *A task analysis of underway replenishment for virtual environment ship-handling simulator scenario development* (Master's thesis). Naval Postgraduate School, Monterey, CA.

Ota, D., Loftin, R. B., Saito, T., Lea, R., & Keller, J. (1995). Virtual reality in surgical education. *Computers in Biology and Medicine*, *25*(2), 127–137.

Paivio, A. (1991). Dual coding theory: Retrospect and current status. *Canadian Journal of Psychology*, 45, 255–287.

Patrey, J., Sheldon, B., Breaux, R., & Mead, A. (2001, April 13–15). *Performance measurement in VR.* Paper presented at the RIO HFM Workshop on "What Is Essential for Virtual Reality to Meet Military Performance Goals?" (RTO MP-058) The Hague, the Netherlands.

Raab, M., & Johnson, J. G. (2007). Expertise-based differences in search and option-generation strategies. *Journal of Experimental Psychology: Applied*, *13*(3), 158–170.

Regian, J. W., & Yadrick, R. (1994). Assessment of configurational knowledge of naturally and artificially-acquired large-scale space. *Journal of Environmental Psychology*, *14*, 211–223.

Regian, J. W., Shebilske, W., & Monk, J. (1992). A preliminary empirical evaluation of virtual reality as an instructional medium for visual-spatial tasks. *Journal of Communication*, *42*(4), 136–149.

Rosen, J., Brown, J. D., Chang, L., Sinanan, M. N., & Hannaford, B. (2006). Generalized approach for modeling minimally invasive surgery as a stochastic process using a discrete Markov model. *Biomedical Engineering, IEEE Transactions*, *53*(3), 399–413.

Stanney, K. M., Milham, L. M., Champney, R., & Carroll, M. B. (2007). *Transfer of training for two virtual environments (VEs) of differing fidelity* (Final Technical Report, Contract No. N0001404C0024). Arlington, VA: Office of Naval Research.

Stiles, R., McCarthy, L., Munro, A., Pizzini, Q., Johnson, L., & Rickel, J. (1996). *Virtual Environments for Shipboard Training,* Intelligent ship symposium, American Society of Naval Engineers, Pittsburgh, PA.

Templeman, J., & Denbrook, P. (2012). Enhancing realism in desktop interfaces for dismounted infantry simulation. *Proceedings of the Interservice/Industry Training, Simulation, and Education Conference (I/ITSEC) Annual Meeting.* Orlando, FL, December 3–6, 2012.

Underwood, G. (2007). Visual attention and the transition from novice to advanced driver. *Ergonomics, 50*(8), 1235–1249.

17 Multimodal Interaction Modeling

George V. Popescu, Helmuth Trefftz,
and Grigore (Greg) Burdea

CONTENTS

17.1 INTRODUCTION

Virtual environments (VEs) represent advanced, immersive, human–computer interaction (HCI) systems. Such interaction occurs through communication over several sensorial channels. Since the communication pathway between users and the simulation system groups several distinct channels, it is termed *multimodal*. The modalities used are primarily the visual, auditory, and haptic ones. The number, quality, and interaction between such modalities are key to the realism of the simulation and eventually to its usefulness. Thus, the need for increased immersion and interaction motivates system designers to explore the integration of additional modalities in VE systems and to take advantage of cross-modal effects. The downside, of course, is increased system complexity, cost, and possible integration/synchronization problems. In addition, cross-modal interaction can lead to perceptual illusions that must be considered by system developers.

Finding the best compromise between multimodal simulation realism and its cost and drawbacks requires a good understanding of each modality individually and in combination. The first step is to have a model of the communication loop between the user and the computer(s) running the simulation, as illustrated in Figure 17.1 (Schomaker et al., 1995). The human output channels and computer input modalities define the input flow into the simulation. The computer output modalities and human input channels define the feedback flow by which the user receives feedback from the VE. The two communication channels form the human–VE interaction loop.

The processes involved at the human side of the interaction loop are perception, cognition, and control. *Perception* is the process of transforming sensorial information to higher-level representations (cognitive and associative processes; Schomaker et al., 1995). In the model adopted here, perception involves the machine–human communication flow. The corresponding perception channels are visual, auditory, haptic (kinesthetic, tactile), olfactory, gustatory, and vestibular. Interface devices that need to be matched to a user's input characteristics mediate the user's perception of the computer output. In order to guide the design of such interfaces, numerous studies have been dedicated to human sensory and motor physiology. A comprehensive comparison of human sensory capabilities with feedback device specifications can be found in Barfield, Hendrix, Bjorneseth, Kaczmarek, and Lotens (1995).

The *cognition* process takes place both at the user's side and at the computer side (as part of *computational intelligence*) of the HCI system. Human cognition processes have a far greater complexity, however, and can only be approximated by simplistic computer *cognition* algorithms. As with the perception modalities, present limitations lay at the computer side of the interaction loop.

The *control* process, in the context of VE systems, represents the translation of human actions into task-related information for the computer. Thus, control involves the human output channels and computer input modalities. The control modalities of gesture, speech, gaze, and touch are used in conjunction with hardware interfaces. Such devices convert a user's output to computer digital

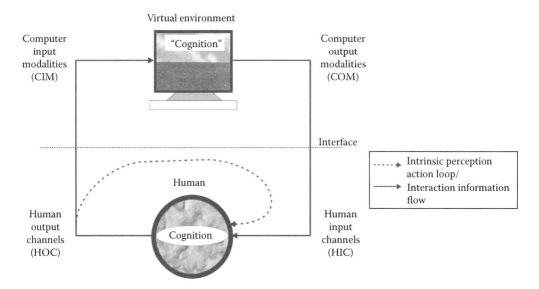

FIGURE 17.1 Human–VE interaction model. (Adapted from Schomaker, L. et al., *A Taxonomy of Multimodal Interaction in the Human Information Processing System* [Online], Multimodal Integration for Advanced Multimedia Interfaces (MIAMI), ESPRIT III, Basic Research Project 8579, Available at http://hwr.nici.kun. nl/~miami/, 1995. Reprinted by permission.)

input data as accurately and as fast as possible. Control and perception are not totally independent processes. The dotted line in Figure 17.1 accounts for input and output (I/O) channel coupling of the human information processing system, which will be discussed later in this chapter.

The capacity and bandwidth requirements for the VE system communication channels vary with human information processing capacity (visual, auditory, haptic, etc.) and with human motor performance (eye motion, body motion). Other communication parameters, such as time delays, interchannel lags, and channel interference, are also important for integrating all of the previously mentioned modalities. There are no models available, however, for describing multimodal integration due to the complexity of human information processing channels. The design guidelines for multimodal integration are mostly based on experimental studies (human factors). Typically, these studies focus on either input or output modalities and do not evaluate integrated systems. Therefore, the following sections will treat separately the two components of the VE loop (I/O). The discussion of input–output coupling is limited to a short final section.

17.2 MULTIMODAL INPUT FROM USER TO THE COMPUTER

Human input in VE is mostly achieved through gesture-controlled input devices. Even though many of these devices are intuitive and easy to use, they cannot provide a fluent dialogue between the user and VE. Integrating multiple input modalities (gesture plus speech and gaze) achieves a high level of interactivity while increasing the naturalness of the human–VE interface (Oviatt, 2002). Development of multimodal input for VE systems is essential due to the increased user control and interactivity required by recent VE applications (e.g., medical VE, VRCAD). Natural interfaces extract more information from human output (speech, gestures, gaze) and use *cognitive* models to respond intelligently to a user's input. These models integrate parallel streams of information from speech, gesture, and gaze and provide real-time semantic interpretation. The present discussion on multimodal input starts with a presentation of performance and limitations of input modalities. This highlights the need for integration of input modalities in order to improve the interface. Performance-oriented interaction techniques (Bowman & Hodges, 1999; McMahan, Kopper & Bowman, 2013, Chapter 12) increase human–computer communication in VE by proposing gesture-based metaphors, which appeal to user imagination (*magic* interaction methods).

17.2.1 INPUT MODALITIES

The most prevalent input modality used in present VE systems is through tracking of the movement of the user's body. Typically, input is provided through hand motions: wrist motions (when a joystick or spaceball is used) and fingers motion (e.g., when sensing gloves are used); however, whole-body input devices (when the user wears a sensing suit or interacts with systems such as Wii® or KINECT®) are becoming available. Joysticks, mice, or trackballs are simpler and cheaper but constrain the user's freedom of motion, a very important ingredient for increased simulation realism.

Sensing gloves and sensing suits preserve the user's freedom of motion (within the range of 3D trackers) and are more appropriate for modern large-volume VE. Furthermore, sensing gloves allow dexterous interactions, overcoming the limited menu of actions of simpler devices (button clicks). They are, of course, more expensive and more complex to program (Kinect 2013; Wii 2013).

Hand gesture has been extensively studied as an efficient input modality. Gesture languages with different levels of complexity (starting from grasp–release commands to American Sign Language) have been proposed for various applications (Billinghurst & Wang, 1999; Kjeldsen & Kender, 1997; Pavlovic, Sharma, & Huang, 1997; Starner & Pentland, 1995). Large vocabularies necessary for a fluent human–VE dialogue cannot be implemented through gesture only because this requires extensive user training and memorization.

The speech channel has been used extensively as an input channel and has great potential for control in VEs. Being the dominant channel of human–human communication, speech can convey levels of abstraction inaccessible to other input modalities. Speech input relies on speech recognition capabilities of the computer(s) running the simulation. Current speech recognition algorithms use real-time hidden Markov models (HMMs): Dragon NaturallySpeaking (2013); WebSphere Voice (2014); Microsoft SAPI (2013). Natural language understanding technologies, using large databases of application-dependent vocabularies and powerful inference methods (IBM Watson, 2013), may be used in the future to deepen the sense of immersion in VE. However, speech input is less powerful in many task-oriented human–VE interactions. Thus, speech input should be combined with other modalities, such as gestures, to provide better spatiotemporal information (Oviatt, DeAngeli, & Kuhn, 1997). Gesture can also help disambiguate speech sentences.

Another approach to augment speech input is to add gaze input. During control tasks, gaze automatically focuses on the VE region of interest (Jacob, 1995). This can be exploited for navigation or object selection tasks. More elaborate tasks, such as moving objects from one location in the VE to another, require user training in order to exercise eye movement control and to minimize eyestrain (Wang & Winslow, 2013, Chapter 8). Extracting the semantics of eye gestures is still a difficult problem. Eye movements represent a very fast way to communicate user intentions, but they also carry unintended (subconscious) content. These limitations make gaze input useful only in combination with speech and/or gesture.

The previous discussion suggests that multimodal input may be preferred to unimodal speech input. Experiments done by Oviatt et al. (1997), Oviatt and Clow (1998), and Oviatt (1999) have demonstrated that users have a strong preference for multimodal as compared to unimodal interactions, especially in the spatial domain. In experiments with a 2D map application, users were more likely to express multimodal commands when describing spatial information (location, number, size, orientation, or shape of objects); user commands carrying spatial information were best sent to the computer through gestures. The studies concluded that simple speak-and-point interfaces were too rigid and of limited practical use.

In a review of multimodal interface design presented by Oviatt (2002), the author stated that innovative, well-integrated, and robust multimodal system designs are needed. The author proposed that such designs will be successful if guidance is used from cognitive science on the coordinated human perception and production of natural modalities. Finally, the author advocated for multimodal interface designs that incorporate input from multiple heterogeneous information sources in order to gain robustness and discriminative power to understand user input. In a later article, Oviatt et al. (2003) made the first steps toward a theory of organized multimodal integration patterns during HCI. As an example, quantitative modeling on the organization of users' speech and pen multimodal integration patterns was presented. Reidsma, Nijholt, Tschacher, and Ramseyer (2010) surveyed multimodal synchronization as well as a quantitative method for measuring the level of nonverbal synchrony in an interaction, focusing on the interaction between humans and the virtual humans that inhabit cyberworlds.

There is a vast literature on multimodal user input for HCI, which cannot be entirely covered in this chapter; for additional references, interested readers should also consult Blattner and Glinert (1996) or Oviatt (2002) or Dumas et al. (2009). The following sections focus on analyzing three aspects of multimodal input: software architecture, multimodal integration, and multimodal input applications.

17.2.2 MULTIMODAL INPUT ARCHITECTURE

VEs need to minimize time delays in system response, whether these delays are due to software processing, communication bottlenecks, or slow graphics. Real-time speech recognition, gesture recognition, and haptic rendering represent a significant computation load. Early multimodal systems were implemented as a distributed computing platform, with modalities running on separate computers. The various input devices were attached to their own computation platforms running a driver and data processing software and were communicating through a network interface. At present, multicore CPUs and large amounts of memory make desktop multimodal computing a reality.

The bottleneck is still the availability of programming tools to help developers integrate the different hardware interfaces. Even in cases when the input device has an associated programming interface, integration is not easy. Most of the programming interfaces were not developed for multimodal input and have limitations when used jointly in advanced computing environments, such as difficulties in input channels synchronization and integration. In order to disambiguate input signals, robust multimodal architectures integrate input modes synergistically by taking advantage of modality-specific information content, redundancy, and complementary input modes.

Typically, multimodal input software uses a hierarchical architecture, with input devices at the lowest level and the application at the highest. Intermediate layers composed of functional modules that describe how input modalities are used and what tasks need to be accomplished. This modularity is important because it allows a better separation of application functionality from the user interface.

A layered model for modal input and interaction in VEs was presented in Schomaker et al. (1995) and is summarized in Table 17.1. The lower levels of the model contain the input devices and the associated events. The user–VE interaction is realized through gesture, speech, and gaze commands. When modalities are used in parallel, the interaction is multimodal, based on uniform access to the pool of events received from all input devices. The last two levels specify the application and the interaction tasks implemented by the command language. Tasks are modality independent, even though some modalities are better suited for certain tasks (e.g., gestures for object manipulation).

A layered multimodal input platform was implemented as part of the STIMULATE NSF program and is illustrated in Figure 17.2 (Marsic, Medl, & Flanagan, 2000; Sletterink, 1999; Flagnan et al., 1999). The first layer consists of modal software interfaces, which receive data from input devices and detect input events (gestures, spoken units, mouse clicks, etc.). These events are then transformed into text units and sent to the next layer of the architecture. The second layer contains the central component of the software architecture, namely, the *multimodal input manager* (MIM). The MIM receives streams of data from several modal interfaces and issues commands to the current application. The MIM in turn consists of three units, namely, the *modality handler*, the *customizer*, and the *fusion agent*. The *modality handler* allows adaptive use of available input modalities. It detects active input modalities, decides which modalities are relevant for the current application, and creates input streams from the corresponding modal interfaces. The *fusion agent* handles automatic context by keeping track of a focused application (the selected window). Context switching increases the reliability of speech and gesture recognition modules by reloading the grammar and gesture sets through the *customizer*. Instead of implementing a large grammar that will reduce recognition performance, each application has its own grammar and a corresponding parser. This intelligent multimodal input platform implements the integration (fusion) and synchronization of input modalities, as described in the following.

TABLE 17.1
Layered Model for Interaction in VE

Application	Virtual Reality, CAD, Visualization, Architectural Design
Interaction tasks	Navigation, manipulation (move, rotate, scale), identification, selection, VE editing
Interaction techniques	Gesture language (grabbing, releasing, pointing), 3D menu, speech commands, gaze commands, multimodal commands
Events	Hand and body gestures, 3D motion, button click, force, 2D motion, torque, spoken units, eye motion, gaze direction
Input devices	Sensing gloves, trackballs, 3D mouse, 6D tracker, eye tracker, joystick, microphone, tactile gloves

Source: Adapted from Schomaker, L. et al., *A Taxonomy of Multimodal Interaction in the Human Information Processing System* [Online], Multimodal Integration for Advanced Multimedia Interfaces (MIAMI), ESPRIT III, Basic Research Project 8579. Available at http://hwr.nici.kun.nl/~miami/, 1995. Reprinted by permission.

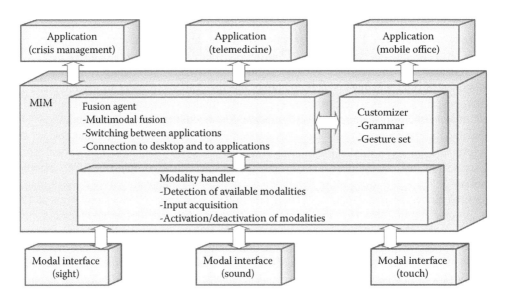

FIGURE 17.2 Structure of the MIM. (From Marsic, I. et al., Natural communication with information systems, *Proceeding of the IEEE*, 88(8), 1354, Copyright 2000, IEEE. Reprinted by permission.)

17.2.3 MULTIMODAL INPUT INTEGRATION AND SYNCHRONIZATION

Multimodal integration performs the mapping of modal inputs into a semantic interpretation. The main elements of multimodal integration are (1) the *unified data representation*, (2) *input synchronization*, and (3) *semantic data fusion*. *Unification of data representation* is the first step after capturing the data from user input. To obtain a unified representation, each input data stream is translated into a sequence of tokens contributing to the semantic of HCI. An example of unified data representation tokens is time-stamped text items. The second step of multimodal integration is the *synchronization of the input streams*. The input streams are buffered, segmented (in order to create a context for semantic analysis), and aligned on the temporal axis—as synchronization will take into account the temporal precedence of modes and intermodal lags. The last step is the *semantic data fusion*. Here, the temporally aligned input tokens are translated into control sentences. This can be achieved using natural language processing methods. The steps described previously are typically performed in sequence, but feedback loops from synchronization and semantic analysis can modify some input tokens during the integration process.

Several mechanisms have been proposed for implementing multimodal integration, including frames, neural nets, and agents. The frame–slot integration mechanism applies the well-known artificial intelligence method (slot–filler) for multimodal fusion. In this method, information necessary for command or language understanding is encoded in structures called *frames*. For the multimodal input model shown in Figure 17.2, a frame stores the entire information about multimodal input commands. Command frames are composed of slots, which are lexical units provided by the multimodal input. As an example, the *move* frame is composed of two slots: *object*, to identify the object, and *where*, to specify the final position.

A frame–slot-based method, with predefined and application-dependent frames, was used in Medl et al. (1998). The multimodal architecture is illustrated in Figure 17.3. The parser extracts lexical units from different input modalities and fills the appropriate positions in the slot buffer. The concurrency of input modalities is handled on a first-come, first-served basis. The central component of the fusion agent is the slot buffer, which stores the information inputted by the user. This allows back referencing of past lexical units (e.g., *it* could be used to reference the previously

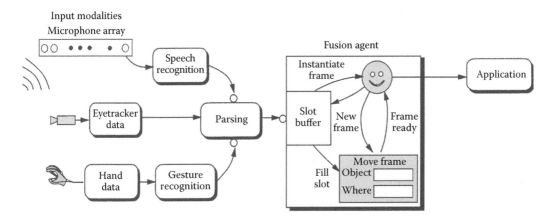

FIGURE 17.3 Frame-based multimodal input integration. (Adapted from Medl, A. et al., Multimodal man–machine interface for mission planning, Paper presented at *Intelligent Environments-AAAI Spring Symposium*, Stanford, CA, 1998. Reprinted by permission.)

selected object), increasing the fluency and naturalness of the command language. Data fusion uses lexical units drawn uniformly from input channels. The fusion agent continuously monitors the slot buffer, trying to fill predefined frames. Once a frame is filled (i.e., there is enough information to generate a command), the fusion agent forwards it to the application. The method demonstrated an increased capacity for human–computer communication mainly due to the uniform access of the input modes and its innovative context switching and modality fusion features.

Another less structured approach to multimodal integration uses neural networks. Neural networks are adaptive systems designed to model the way in which the brain performs cognitive tasks. Their main quality is the capacity to learn complex relations between inputs and outputs, which makes them suited for multimodal input integration. Vo and Waibel (1993) were among the first to use neural networks for multimodal input systems. In their initial architecture, each input modality uses time-delay neural nets (TDNNs) for temporal alignment, allowing for simultaneous processing of data received over a period of time. The design includes temporal synchronization rules, such as precedence of hand gesture (writing) over speech. Later designs added more structure, using connectionist networks for multimodal semantic integration (Vo & Waibel, 1997). In this architecture, the input modality channels were considered parallel streams that can be aligned in action frames and jointly segmented in parameter slots. Temporal alignment was based on time stamps associated with input tokens. The integration mechanism used a mutual information network architecture and a dynamic programming algorithm to generate the input segmentation. After obtaining the frames and the corresponding slot information, grammar-based algorithms were applied to capture syntactic elements and extract their meaning.

Another approach to multimodal input integration is the use of a set of hierarchically organized entities (agent-based algorithms). An agent implements a certain function (task) of the multimodal input integration. The agent has a specified set of sensors (inputs), a model to process the information, and a set of effectors (outputs). A typical agent-based architecture contains three layers. The first layer consists of modal agents processing the data from input devices. The next layer contains language recognition agents. A language recognition agent gets input from a modal interface (speech, gesture, etc.) and generates a potential interpretation based on the specified language model. This layer is responsible also for unifying the modal data representation. The third and last layer contains a multimodal fusion agent, which receives language units from recognition agents and integrates them to produce the best possible interpretation of the user's input.

An agent-based fusion of speech and gesture modalities was presented in Johnston et al. (1997). Their multimodal architecture used six agents: interface client, speech recognition, natural language processing, gesture recognition, multimodal integration, and bridge (a command execution agent). The fusion agent used a probabilistic model: probabilities were associated with each modal input, with the highest unified score interpretation being selected. This allowed grammar-based techniques employed in natural language processing to be used in the semantic integration process as well.

New designs of multimodal integration are emerging, centered on flexible rather rigid frame-based input. Lin and Makedon (2010) proposed a hierarchical multimodal planning for pervasive interaction, composed of Markov decision processes for activity planning and multimodal partially observable Markov decision processes for action planning and executing.

17.2.4 SUMMARY OF MULTIMODAL INPUT

The data available from human factors evaluations suggest users have a clear subjective preference for multimodality versus unimodal input. From a user's perspective, multimodal input allows better control of VE simulation through integration of several communication channels. By using voice, gaze, and gesture, simulation interactivity as well as a user's feeling of immersion (see Chertoff & Schatz, 2013, Chapter 34) in the VE increases. Multimodal input helps alleviate some limitations of unimodal, voice-only input: commands that otherwise may be difficult or impossible to understand are disambiguated, objects are placed in an environment more precisely, and navigation is faster. The most challenging aspects of multimodal input implementation are multimodal integration and semantic analysis. Several approaches have been developed for integrating input modalities, namely, application-dependent frames, neural networks, and hierarchically organized agents.

17.3 MACHINE–HUMAN COMMUNICATION

Machine–human communication, or multimodal feedback, represents a computer response displayed to the user as a result of his interaction with the simulated environment. Early VE systems had unimodal visual feedback through stereo or monoscopic graphics displayed on CRTs or head-mounted displays (HMDs). As research interest for other communication channels grew, bimodal VE display systems (visual–auditory, visual–haptic feedback) have been developed and evaluated. The most frequently used feedback channels are visual, auditory, and haptic (force–tactile feedback). Additional channels such as olfactory feedback may significantly enhance a user's immersion (see Jones et al., 2013, Chapter 6) but are less developed at present.

The visual, auditory, and haptic feedback channels differ strongly from each other in many respects. The information flow received through the visual channel is much larger than that corresponding to the haptic or auditory channels. Thus, visual feedback takes the dominant role among the three primary VE feedback modalities. Adding stereoscopic vision allows 3D perception from images rendered to the left and right eyes. When the visual feedback modality is coupled with body posture input (head motion), the user takes control of the field of view. This is thought to result in increased immersion and simulation realism (see Chertoff & Schatz, 2013, Chapters 34; Simpson, Cowgill, Gilkey, & Weisenberger, 2013, Chapter 13). Earlier HMDs were heavy and had low image resolution, problems that were compounded by low-accuracy tracking of a user's head position (Thorpe & Hodges, 1995). In recent years, problems with HMD weight and image resolution have largely been solved, while more accurate and robust trackers are being developed.

Sounds are ubiquitous in the real world and therefore their presence in VEs should be beneficial to the simulation realism. Auditory displays can be classified as nonlocalized and localized (see Vorlander & Shinn-Cunningham, 2013, Chapter 4). Nonlocalized audio feedback is provided by multimedia-quality stereo sound, meant to increase user interaction. Localized (or spatialized) sound uses head tracking and head-related transfer functions (HRTFs) to display 3D soundscapes. A localized sound source remains fixed in space regardless of the user's change in head position, similar to what happens in the real world (Barfield et al., 1995). Thus, 3D sound technology has a significant potential to increase immersion in VE. A current limitation is related to the inability to customize the HRTF to the individual user's pine (external ear) characteristics, which results in less accurate mapping of spatialized sound information in VEs than in the real world. Other limitations of earlier commercial products were the high computation load required and the high hardware cost. These problems have been largely addressed in recent years, with the introduction of significantly faster hardware, at lower costs. The computation of 3D sound remains an area of active research.

Haptic feedback is a critical modality for VE interactions involving active virtual object manipulation (Burdea, 1996; Burdea, 1999; Burdea et al., 1999; Dindar, Tekalp, & Basdogan, 2013, Chapter 5). Adding haptics increases user immersion and interactivity by allowing an object's physical characteristics (weight, compliance, inertia, surface smoothness, slippage, temperature, etc.) to be displayed. Haptic feedback is also needed when the visual channel is degraded or nonexistent. Current commercial haptic feedback interfaces suffer from large weight, reduced maneuverability, and large costs. Additionally, modeling physical interactions is complex and computationally intensive, typically requiring multiprocessor computers. Forces applied by the computer through the haptic interface are real, which makes user's safety a concern (see Dindar et al., 2013, Chapter 5).

17.3.1 MULTIMODAL FEEDBACK INTEGRATION

Each type of feedback interface mentioned earlier is the subject of other chapters in this handbook (see Chapters 3 through 6). By contrast, the focus here is on the integration of the separate feedback modalities, including such topics as sensorial redundancy and sensorial transposition. A user's perception of VEs spans the spatiotemporal and physical domains. The spatiotemporal domain refers to the perception of 3D space and time. The physical domain describes such virtual object properties as weight, compliance, inertia, and surface temperature. In homogeneous environments, these domains are orthogonal, simplifying the implementation of multimodal feedback.

The feedback channel interaction consists of channel complementarity and modal redundancy. *Complementarity* refers to nonoverlapping information received through separate feedback channels. This allows the VE designer to increase simulation realism by simply adding more feedback modalities. *Modal redundancy* comes from overlapping the same information in the representation domain, fed back to the user through several sensorial channels. The VE designer can use such redundancy to reinforce perception of certain features of a VE as long as coherence constrains are imposed. Maintaining spatiotemporal and physical model coherence requires that feedback modalities be synchronized. Nonsynchronized events or objects affect immersion drastically and reduce user's participation (e.g., loss of spatial and/or temporal synchronization between haptic and visual channels reduces user immersion [Richard, Burdea, Gomez, & Coiffet, 1994]). Thus, it is important to analyze complementarity and redundancy of the visual, auditory, and haptic feedback modalities.

17.3.1.1 Visual–Auditory Feedback Interaction

Sound is frequently used in combination with the visual channel because it provides additional simulation cues and increased user interactivity. The acoustic characteristics of virtual objects may include their spatial location, audio-temporal properties, timbre, pitch, intensity, and rhythm. Sounds can be mapped to physical objects or be independent of them. The visual and auditory channels are similar and complementary, as they can both represent spatiotemporal information. However, their spatiotemporal resolution differs. Sound spatial localization of objects is poor relative to visual presentation, whereas time accuracy associated with auditory feedback is superior to that of the visual channel (Barfield et al., 1995). Therefore, sound feedback will greatly enhance applications where a user's reaction time is critical.

The overlapping of visual and auditory information on the spatiotemporal domain results in sensorial redundancy as long as the two information flows are synchronized. This augments the perception of both modalities. For example, a graphics scene that has 3D sound cues that are well mapped to graphical objects appears visually sharper to the user than the same scene without the associated sound.

Sound can convey complementary spatial information when the object of interest is not in the field of view or when objects closer to the user occlude more distant ones. In the time domain, sound can provide very precise temporal stamps when needed. Sounds can be used to convey physical properties of objects as well, such as consistency (discriminate between an empty and a full container), weight (a heavy ball sounds different from a light ball bouncing off the floor), and elasticity (vibration of a string is correlated to its pitch). Sound can also act as a gaze direction guide by focusing the user's attention on a certain spatial event in the VE. Three-dimensional sound cues can point the user to an object location (a telephone that rings) or help navigation tasks by suggesting a new direction to be explored. As a consequence of redundancy and complementarity, 3D sound integrated with the visual display provides a more consistent spatiotemporal representation of a simulated environment.

17.3.1.2 Visual–Haptic Feedback Interaction

The haptic channel can represent information related to the physical properties of virtual objects in addition to their spatiotemporal properties. Haptics is rarely used for spatial discrimination by itself (except when visual feedback is missing). In most cases, haptic feedback is combined with information received on the visual channel as a redundant modality. The benefits of such redundancy were mentioned earlier, but a quantitative assessment can only be based on experimental data.

Human factors experiments were conducted by Richard et al. (1994) to assess the benefits of adding haptics to a partially immersive environment in direct manipulation tasks. The experiments studied the influence of haptic feedback on dexterous manipulation of a plastically deformable ball. The performance measure was the amount of ball deformation (which had to be kept less than 10% of the ball radius). Experimental results showed increased spatial resolution (inversely proportional to ball deformation in excess of 10%) for VEs with haptic feedback. When the haptic modality was added to the visual feedback, the spatial resolution increased almost three times. Force resolution can therefore be used to increase spatial resolution when the relationship between space and force can be mentally evaluated by the user (e.g., a linear relationship—Hooke's law). This supports the idea that haptic feedback is an essential ingredient for direct manipulation tasks in VEs.

Coherence of spatiotemporal representation should be imposed for tactile and kinesthetic channels. For example, the *roughness* of a surface evaluated through visual inspection should be matched by the rugosity information provided by the tactile feedback interface. Large time lags between the graphics and haptic loops can confuse the user and may result in control instabilities. Information related to the physical properties of VE displayed on the haptic and visual channels needs to be coherent as well. Object surface deformation should be synchronized with force calculation to provide increased immersion in VEs. A *soft* ball (small forces applied to the user's finger when squeezing) should also be highly deformable. Virtual walls should resist with very high force when being pushed and should have no visual surface deformation. Plastically deformed objects should present a hysteresis behavior both in shape deformation (their surface remains deformed after the interaction) and in the associated force feedback profile produced by the haptic interface. In summary, physical behavior of objects should be implemented both in graphics and haptics domains and displayed synchronized to the user.

17.3.1.3 Haptic Channels Coupling

The haptic feedback channel constitutes a complex coupled system. There is a very tight coupling between force and touch feedback components. Minsky, Ouh-young, Steele, Brooks, and Behensky (1990) demonstrated that high-bandwidth (500–1000 Hz) force feedback displays could be used to render tactile information as well. Their method used the texture surface gradient and real-time physics (spring–damper model) to calculate the forces to be displayed to the user. The simulation integrated a high-bandwidth two-degree-of-freedom force feedback joystick that rendered various textured surfaces like sandpaper and elastic bumps. This research suggests that haptic displays can be classified as a function of bandwidth: low bandwidth corresponding to force feedback and high bandwidth to tactile feedback.

The benefits of superimposing vibratory feedback over force feedback for manipulation tasks were illustrated in Kotarinis and Howe (1995). This study identified the kind of tasks where high-frequency vibratory feedback is important (inspection, haptic exploration, direct manipulation). Additionally, it explored the use of tactile display for conveying task-related vibratory information. Their results showed that adding vibratory feedback to a force feedback system was very beneficial, resulting in increased performance in manipulation tasks, such as peg-in-hole insertion.

17.3.2 SENSORIAL TRANSPOSITION

Sensorial transposition is the provision of feedback to the user through a different channel than the expected one. Sensorial transposition is typically used to substitute unavailable communication channels required by an application. For example, force feedback required in direct manipulation tasks could be substituted with visual, auditory, or tactile feedback. Sensorial transposition can be *simple*, when one modality is replaced by another, or *complex*, when one modality is substituted by multiple other types of feedback.

Sensorial transposition requires user adaptation; however, the level of user adaptation needed in the mappings involved in sensorial transpositions varies. Some mappings feel *natural*, whereas others require more training. For easy user adaptation, the mapping should use the strongest representation domains (visual → spatial domain, auditory → temporal, tactile → temporal, etc.) of the transposed channel. Several examples of sensorial transposition and corresponding mapping domains are shown in Table 17.2.

Sensorial transposition schemes can also provide *sensorial redundancy* in VEs. The same feedback information is mapped through a different channel in addition to the one normally used to communicate that type of feedback. This is done in order to reinforce the original message and can be used to increase user performance in complex tasks.

When using sensorial substitution to obtain redundancy, the substitution scheme should be chosen carefully in order to avoid sensorial contradictions or sensorial overload. Otherwise, instead of reinforcing the original signal, such methods may confuse the user and induce reaction delays as the user copes with unexpected sources of information. Sensorial overload resulting from too much feedback data may also decrease human performance in VEs.

Research on human psychology has resulted in several studies describing sensorial transposition effects (Kaczmarek & Bach-y-Rita, 1995). The VE literature focuses mostly on sensorial transposition and redundant feedback effects related to the haptics channel (Fukui & Shimojo, 1992; Massimo & Sheridan, 1993; Richard et al., 1994). This was motivated by the limitations of current haptic feedback devices, which are difficult and expensive to use. The tactile channel was used in many sensorial substitution schemes to display visual and auditory feedback information. For example, tactile–auditory substitution remaps of the frequency analysis performed by the ear to intensities of electrotactile stimulation of the skin have been used for the hearing impaired. Sensorial

TABLE 17.2
Examples of Sensorial Substitution Schemes

Initial Channel	Input Domain	Transposed Channel	Mapping Domain
Visual	Spatial	Tactile	Spatial
Auditory	Temporal (frequency)	Tactile	Sensorial intensity
Force feedback	Sensorial intensity	Auditory	Temporal (frequency)
Force feedback	Sensorial intensity	Auditory, tactile	Temporal (frequency), sensorial intensity
Force feedback	Sensorial intensity	Auditory, tactile	Temporal (frequency), sensorial intensity
		Visual	Spatial
Force feedback	Sensorial intensity	Auditory, tactile	Temporal, spatial

transposition has been used to discriminate phonemes by lip reading (Kaczmarek & Bach-y-Rita, 1995). The resolution of the electrotactile stimulus was much lower than the frequency discrimination capability of the ear. Therefore, the approach was found not beneficial in the discrimination of connected text, making it difficult to develop high-performance systems. Kaczmarek and Bach-y-Rita (1995) also cite another sensorial transposition scheme, namely, tactile substitution of visual data. This consists of mapping of spatiotemporal information normally received through the visual channel to the tactile display. The mapping represented pixel intensity as vibrotactile and electrotactile pulse width and amplitude. Two-dimensional images (patterns of lines) were *projected* on the skin rather than on the retina. With practice, users were able to identify more complicated objects, such as faces. However, the limited spatial resolution and narrow dynamic range of the tactile display prevented the rendering of complex scenes.

Another tactile–visual sensorial transposition is described in Fritz and Barner (1996). Here, proprioceptive information was combined with the tactile feedback produced by a PHANToM interface to substitute for visual information. Finger position information provided by the device was coupled with haptic rendering to display textures in 3D, as shown in Figure 17.4. A stochastic modeling technique was used to generate random texture patterns in order to spatially display text data. High-resolution position detection, as well as the resolution of randomly generated haptic textures (256 for this application), was key in supporting a pattern discrimination task.

Redundant tactile feedback to reinforce the visual channel was described in Bouzit and Burdea (1998). The tactile feedback was used to increase pilot performance and reduce spatial disorientation when the visual feedback data were degraded. Airplane attitude orientation was mapped to pressure applied by pairs of air bellows on an elbow joint to resist flexion and extension movements. Tactile feedback was selected instead of force feedback due to its quicker response and unobtrusiveness of the interface. When asked to track a random attitude trajectory with degraded visual feedback, the user registered a 27.4-degree error without tactile feedback. When the tactile feedback was added to the simulation, the performance was significantly increased. This underscores the advantage of tapping into the underutilized haptic channel to convey spatial information when the visual channel is saturated. A similar example is the tactile mapping of aircraft instrumentation on the pilot's torso using a tactile feedback vest developed at the Chaiasson, McGrath, & Rupert (2002).

Another area where sensorial substitution was shown to be beneficial is teleoperation with long time delays. Under such conditions, the force feedback system that normally is beneficial becomes detrimental and may lead to system instabilities. Massimo and Sheridan (1993) studied the efficacy of using tactile and auditory substitution of the delayed teleoperation force feedback signal. Their experimental setup consisted of master and slave robotic manipulators, visual feedback of the remote site, and additional tactile and auditory feedback. In force–auditory feedback substitution, the intensity of the sound mapped the force magnitude. In force–vibrotactile feedback substitution, the vibration amplitude was proportional to the magnitude of force feedback. The experimental data for contact detection (taps) are illustrated in Figure 17.5. Sensorial substitution schemes outperformed the visual display (worst) and the force feedback scheme. This experiment showed that

(a)

(b)

FIGURE 17.4   (a) Original 2D image; (b) spatial textured pattern. (Reprinted from Fritz, J.P. and Barner, K.E., Stochastic models for haptic textures, in M.R. Stein, Ed., *Proceedings of SPIE Vol. 2901, Telemanipulator and Telepresence Technologies III*, pp. 34–44, SPIE, Boston, MA, 1996. With permission from SPIE.)

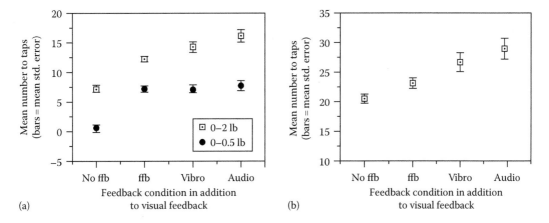

FIGURE 17.5 Sensorial substitution of force feedback: (a) contact force detection; (b) magnitude of contact force. (Reprinted from Massimo, M. and Sheridan, T., Sensory substitution for force feedback in teleoperation, *Presence*, 2(4), 344–352, Copyright 1993. With permission from MIT Press Journals.)

for remote manipulation tasks, the user could take advantage of the fast human response to auditory and tactile stimuli to reduce reaction time. In the second experiment, a peg-in-hole insertion task was executed with either zero or very large visual time delay. The auditory channel was used this time to convey spatial information in addition to force magnitude. Collisions with the left or right side of the hole were conveyed to the corresponding ear. Tactile mapping was similarly enhanced to display spatial information on the lower and upper part of the palm. Results showed that sensory substitution schemes allow decreased manipulation time when no time delay was present. Auditory and vibratory substitution of forces could function independent of visual information and insure task completion for as much as 3 s time delay. Under such large time delays, sensorial substitution allowed teleoperation, which would otherwise have been impossible to perform.

Richard et al. (1994) and later Fabiani, Burdea, Langrana, and Gomez (1996) studied the influence of sensorial substitution and sensorial redundancy in a task involving dexterous virtual object manipulation. The experiment required the user to pick and place a plastically deformable virtual ball, as previously discussed. The substitution of force feedback through visual feedback used LED bar graphs (group VI) or bar graphs overimposed on the VE scene (V2). These were proportional with the contact forces that should have been applied to the user's fingertips by the Rutgers Master I (RMI) glove. Ball deformation was also mapped on the auditory channel proportional to the sound intensity (group A). Experimental results showed that redundant feedback (haptics H + another modality) provided through different mapping methods increases user performance (groups H–V and H–A; see Figure 17.6). In the absence of sensorial redundancy, force feedback was better than methods using sensorial substitution approaches. Users were able to quickly adapt to the additional mappings available (LED and auditory) and use these successfully during the dexterous manipulation task. The overall relative performance structure remained the same in subsequent experiments with an improved haptic glove (Rutgers Master II [RMII]). However, absolute performance in each category improved due to the larger work envelope and dynamic range of RMII versus the older RMI.

17.3.3 I/O Channel Coupling

The previous discussion treated multimodal input separate from multimodal feedback for ease of understanding. In reality, the human I/O systems are coupled (illustrated in Figure 17.1 by the dashed feedback line). The VE designer should not overlook this coupling that closes the feedback loop *internally* as opposed to going through the simulator.

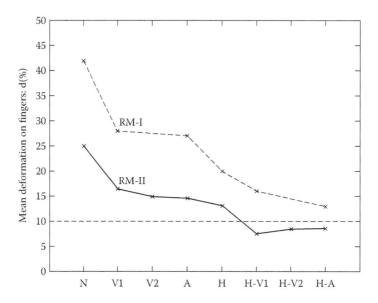

FIGURE 17.6 Ball deformation for different force feedback modalities: N, graphics only; V, visual; A, auditory; H, haptic. The target was 10% deformation. (Reprinted from Fabiani, L., Burdea, G., Langrana, N., and Gomez, D., Human performance using the Rutgers Master II force feedback interface, in *IEEE International Symposium on Virtual Reality and Applications (VRAIS '96)*, pp. 54–59, IEEE Press, Santa Clara, CA, Copyright 1996. With permission from IEEE.)

Visuomotor coupling is a well-known example; data from head-position trackers are used to compute the stereoscopic display images in immersive VEs. Gaze tracking can be used to further improve the design of visual displays. The resolution of displayed images can be decreased as pixels get further away from gaze direction because the resolution of the retina decreases toward the periphery of field of view (the so-called foveating effect). Head orientation influences hearing, as described in the binaural model. Additional dynamic effects (acceleration, speed) should also be considered when the user moves in the VE (Cohen & Wenzel, 1995).

The strongest form of integrated perception/control occurs in haptic perception. In direct manipulation tasks, tactile feedback is accompanied by muscle stimulation (force feedback). In order to increase immersion, tactile interfaces need to include force reflection devices. Otherwise, the functionality of the tactual display is limited to a simple signaling mechanism (events in the VE can be associated with tactile feedback response). An example involving tactile, force feedback, and user control is grasping. An accurate model of object grasping in VE should include haptic feedback information in the hand-motion control loop. Otherwise, the grasping is artificial (objects snap to the palm and force feedback is applied suddenly) and may result in control instabilities.

Hand–eye coordination is important in direct manipulation tasks. The mapping of the user's hand to a virtual hand depends on the position and orientation accuracy of the wrist tracking system. Misalignments between virtual and real hands (sensed by the user's proprioceptive system) are resolved through users' adaptation mechanism (Groen & Werkhoven, 1998), but aftereffects can be observed (see Chapters 22 through 26). In addition to mapping, hand–eye coordination is influenced by the response time of the system. Early studies have shown that delay of hand movement in response to visual changes takes several hundred milliseconds. This limits the minimum refresh rate of the graphic scene (see Badcock, Palmisano, & May, 2013, Chapter 3) required for preserving the same order of magnitude for the response time for hand–eye coordination in VEs.

Other forms of coordination are mentioned in Burdea and Coiffet (1994). Hand–hand coordination has three forms: coordination between hand sensing and hand force feedback, coordination between hand input and force feedback to other parts of the body, and two-hand manipulation.

The delay between sensing and feedback should be less than 100 ms in order to maintain coordination. Two-hand manipulation requires hand–hand coordination; one hand has a leading role while the other is playing an assisting role. Time delays between hand-relative movements, as well as hand misalignment, are important factors of two-hand coordination. The previous discussion on hand–hand coordination applies to other parts of the body involving sensing and feedback.

Another important aspect of VE simulation is the hand–ear coordination. This coordination depends on hand movement time delay in response to a sound event and on 3D localization of the sound relative to hand position. Voice–ear and voice–eye are less important coordination modalities for a simulation as voice is frequently used as an independent human communication channel. Voice synthesis is sometimes used to give feedback to users in response to their verbal commands. The lag between command and feedback can be on the order of seconds. Changes in the visual feedback as result of a voice command can have a response time of the same order of magnitude (seconds).

17.3.4 SUMMARY OF MULTIMODAL FEEDBACK

Looking again at Figure 17.1 depicting the VE simulation loop, one realizes that multimodal feedback is key to simulation realism and eventually to application usefulness. Multimodal feedback allows users immersed in a VE to not only see the graphics scene but also to hear or feel virtual objects, in case such objects are manipulated. Multimodal feedback in immersive VEs thus should include visual, 3D sound, and haptic (touch and force) feedback. These modalities can be used simultaneously to transmit modality-specific data independent of each other. Alternately, one feedback modality can be used to transmit data from another channel in a sensorial substitution arrangement. Such sensorial substitution may be needed when interfaces for another modality are expensive or difficult to use. Finally, several modalities can be used to convey a single event/data, such as contact between objects transmitted visually, through sound as well as haptically. Such sensorial redundancy is beneficial in overcoming interface limitations (friction, limited dynamic range, etc.) but requires good synchronization of the redundant sensorial data. Otherwise, sensorial feedback redundancy can become detrimental to the simulation. Human sensing and control are coupled such that the feedback loop has an internal component as well. This coupling involves hand–eye coordination, the proprioception component of the haptic loop, and hand–hand and hand–ear coordination. The sensorial coupling is another reason why synchronization of modalities and reduction in overall time delays are key to effective VE system design.

17.4 MULTIMODAL VE DESIGN

17.4.1 ADVANCES IN VE DESIGN USING MULTIMODAL INPUT

Several applications have benefited substantially from advances in multimodal interfaces design. Computer-aided design (CAD) systems have a great need to replace the windows–icons–menu–pointer (WIMP) paradigm with more-natural user interfaces. This is especially true in the early *concept* phase of the CAD process. The main limitations are related to the inability to create 3D shapes intuitively and provide interactive 3D visualization. Chu, Dani, and Gadh (1997) investigated the potential use of multimodal input for COVIRDS (conceptual virtual design)—a VE-based CAD system. Candidate input modalities were gaze, voice commands, and hand gestures. Multimodal interface requirements were analyzed for three stages of product design: part and assembly generation, part and assembly modification, and design review. Sets of interaction tasks (commands) were produced according to application requirements. Subjective evaluation tests showed that voice and hand gestures were the preferred modes of interaction, whereas gaze input was found useful for *select* commands. No comments referring to sensory input fusion or modalities concurrency were given in this study.

In another example, speech and gesture inputs were used to develop an intuitive interface for concept shape creation (Chu, Dani, & Gadh, 1998). A series of tasks were implemented using different modalities (e.g., zoom-in, viewpoint translation and rotation, selection, resizing, translation). Evaluation of the interface was based on user questionnaires. Voice was found to be intuitive to use in abstract commands like viewpoint zooming and object creation and deletion. Hand gestures were effective in spatial tasks (resizing, moving). Some tasks (resizing, zoom in a particular direction) were performed better when combining voice and hand input. The command language was very simple and the integration of modalities was implemented at syntax level. Therefore, in some cases, users showed preference for a simple input device (a wand with five buttons) rather than for multimodal input.

Another early multimodal application was MDScope—a VE test bed for visualization and interactive manipulation of complex molecular structures, developed by Beckman Institute. Multimodal input (speech, hand gesture, and gaze) was integrated in the VE in order to improve virtual object manipulation. The multimodal framework for object manipulation (Sharma, Huang, & Pavlovic, 1996) used an HMM speech recognizer. The hand gesture input module used a pair of cameras and HMM-based recognition software. Speech and gesture were integrated using a fixed syntax: <action>; <object>; and <modifier>. This early system has several drawbacks in the integration of modalities. The user command language was too rigid to allow easier synchronization of input modalities. In addition, the synchronization process assumed modality overlapping (the lag between speech and gesture input was considered to be at most one word), leading to difficulties in interpreting input sequences. Gaze input was used only to provide complementary information for gesture recognition; for example, the direction of gaze was exploited in order to disambiguate object selection. Despite these limitations, the experimental work with the aforementioned test bed concluded that multimodal interfaces allow much better interactivity and user control compared to unimodal, joystick-based input.

One of the first studies to analyze the relationship between speech and gesture in VE was presented in Cassell et al. (1994). Their test-bed application involved animated human figures capable of conversation in a VE. The underlying assumption was that gesture and speech are representations of a single mental representation. The set of rules to predict the location and type of gestures was inferred from the analysis of the correlation between speech and gesture. The system provided therefore a test bed where predictive gestures were computed during the dialogue. The implementation used the Jack virtual human toolkit, which was animated using PaT-Nets (Dauville, Levison, & Badler, 1996). Hand, wrist, beat (a specialized form of wrist motion), and arm motion were driven by computation modules that used information from the output of the speech synthesis module. Using this predictive approach, VEs were populated with autonomous animated conversational agents, increasing user's immersion in the VE.

Another application area for multimodal input is military planning, where users perform very fast control tasks in a stressful and noisy environment. Rosenblum and colleagues at the Navy Research Laboratory developed Dragon 2, a VE research platform for battlefield planning and control (Rosenblum, Durbin, Doyle, & Tate, 1997). The multimodal input was provided by QuickSet (Cohen et al., 1997, 1999), a pen-and-voice-based system that ran on handheld PCs and communicated via a wireless local area network (LAN) with the host computer running the simulation. The multimodal input for Dragon 2 consisted of voice commands and a *flight stick*, a commercial joystick modified to incorporate a Polhemus tracker. The flight stick was used for navigation, selection, and drawing on a 2D or 3D map. Users' speech and gesture input were recognized, parsed, and then fused via the QuickSet multimodal integration agent.

A second military planning application—3D Mission Control—was developed on top of the multimodal architecture shown in Figure 17.3 (Medl et al., 1998). The system used a commercial speech recognition engine (Microsoft WHISPER), a haptic glove (the RMII), and a gaze tracker (ISCAN). The data fusion mechanism used the slot–filler method described earlier in this section. The user saw a 3D map of the area targeted by the planned mission (shown in Figure 17.7, Marsic et al., 2000).

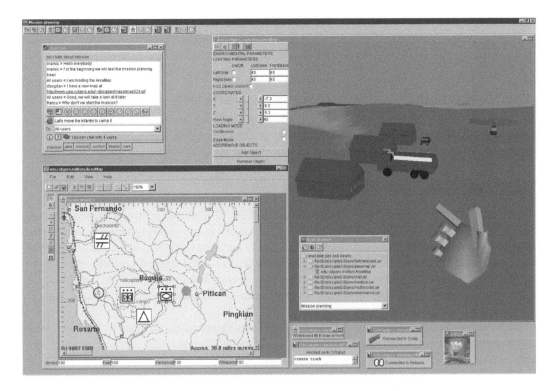

FIGURE 17.7 "3D mission control." (Reprinted from Marsic, I., Medl, A., and Flanagan, J., Natural communication with information systems, *Proceeding of the IEEE*, 88(8), 1354–1366, Copyright 2000. With permission from IEEE.)

The scene was composed of a 3D terrain created from a digital terrain elevation data (DTED) map file, a virtual hand, and several 3D objects representing *assets* (base camps, jeeps, trucks, helicopters, etc.). The user could add, select, move, and delete these objects in the 3D environment. The manipulation of virtual objects relied on gesture input, whereas speech was most useful for object creation and deletion and for viewpoint change. Feedback was provided to the user through visual, auditory, and haptic channels. Besides reinforcing the visual feedback, force feedback was used to map the physical properties of the objects in the VE. For instance, the hardness of a transport vehicle indicated whether it was loaded or not. Synthesized voice feedback was used to confirm successful completion of a command, identify specific objects, and warn the user of errors.

17.4.2 Using Multiple Modalities in CVEs

Collaborative VEs (CVEs) allow multiple users to cooperate in a virtual space in which they seek a common goal. The common goal may be entertainment, training, education, or work, among others (Churchill, Snowdon, & Munro, 2001). Despite the fact that users may be dispersed geographically, distributed systems try to create the illusion of a shared time and space (Singhal & Zyda, 1999). Since CVEs are implemented on computer networks, time delays can have a large impact on the immersion and the shared state impression as perceived by the user.

While early CVEs included only visual feedback, a growing number of CVEs use modalities other than visual due to advances in networking and computer I/O devices. In field surgery training simulations, for instance, haptic feedback is added to visual feedback, providing additional control information to the trainee. Voice input is typically added to CVE in order to allow better communication between users; synthesized voice is frequently used as feedback for players on massively multiuser online games or MMOGs (Rozak, 2007). Auditory and/or the haptic information, however, are very sensitive to time delays and require new simulation strategies in order to accommodate different types of events while achieving effective collaboration and a shared sense of space and time.

17.4.2.1   Effect of Multiple Modalities on Collaboration in CVEs

Adding haptic and auditory feedback to a CVE increases not only task performance but also a user's sense of presence. Nam, Shu, and Chung (2008) created a collaborative air hockey game and provided users with the following interfaces: (1) only visual, (2) visual and haptic, or (3) visual, haptic, and audio feedback. The study found that users could perform better in playing the game as more modalities were added. In addition, the sense of presence, togetherness, and collaboration was increased as more modalities were added (Nam et al., 2008). This is consistent with a previous result by Basdogan and Ho (2000), where haptic feedback effect was evaluated in a CVE task requesting two remote users to move a ring around a wire. The comparison was realized between visual feedback only and visual and haptic feedback (allowing one user to feel the movements applied by the remote user). The study found that the time to complete the task was reduced when haptic feedback was added. It was also concluded that the subjective measurement of *togetherness* (as reported in users' questionnaires) was increased in the presence of haptic feedback.

In the case of multimodal CVEs that are used for training purposes, the objective of the system is to develop skills and dexterities that can be transferred to the real world. In these types of systems, evaluating the time to complete the task in the VE only is not very meaningful. Stevenson and Hutchins (2006) proposed several ways to evaluate the effectiveness of a multimodal system for training users to perform a surgical procedure. They used (1) subjective evaluation through questionnaires (to evaluate the users' subjective perceptions on the system), (2) performance evaluation through transfer trials (by measuring the performance of the users in the real world before and after the virtual training in order to determine the effectiveness of the training), and (3) dialogue analysis (to study how the users took advantage of the multimodal aspects of the system).

17.4.2.2   Architectures for Multimodal CVEs

Several architectures have addressed the bandwidth and computing power limitations in order to provide the best performance in multimodal CVEs. Trefftz, Marsic, and Zyda (2003) proposed the switchboard architecture in order to accommodate users with heterogeneous computing systems in a CVE. In this architecture, a server receives updates from various modalities from all participants and performs resolution subsampling, allowing users with less powerful computers or less bandwidth to subscribe to a channel with lesser resolution. The initial implementation used a static measure of the client's computer capabilities obtained with a benchmark. In a later implementation, an adaptive system that performs a search for optimal performance during the execution of the simulation was used (Quiroz & Trefftz, 2004).

One of the factors that affect collaboration the most is lag. Lag arises due to the physical distance that messages need to travel between remotely located users. Glencross et al. (2007) created a CVE that allowed users to perform a collaborative haptic task in the presence of tans-oceanic lag (140 ms). Their system was based on the concept of virtual time, which is related to, but independent from, the actual clocks of the two participants, and is calculated at each client based on the received messages time stamps. A prediction scheme is used to anticipate the remote user's movements at the local client. When messages describing the remote movement arrive, they are inserted into the local queue, and a correction is performed. Corrections might be perceived as small glitches that can be filtered out so as to not affect performance (Glencross et al., 2007).

As the number of users in the CVE increases, it may be necessary to add servers, since a single server cannot replicate the multimodal events to all participants. This creates the need to define a method to assign clients to servers, in order to achieve load balance. Morillo, Rueda,

Orduña, and Duato (2007) proposed a heuristic load balancing method, based on genetic algorithms, that dynamically assigned clients to servers while providing a predefined quality of service (QoS) for the system as a whole.

17.4.2.3 Multimodal Interaction in Distributed Surgical Simulators

One particular application that benefits from multimodal distributed interaction is distributed surgical training. Surgical tasks imply the use of haptic feedback. In real surgeries, either traditional or laparoscopic, the surgeon controls his movements in part based on the sense of touch. But the sense of touch is particularly problematic in simulators, since it requires high bandwidth and frequency of updates.

Boulanger, Wu, Bischof, and Yang (2006) created a system that allowed a student to see and feel every movement that a teacher performed while undergoing a real ophthalmologic surgery. Additionally, the student had access to a sound channel, allowing him/her to hear the conversations in the operating room. Using prediction algorithms and a very high-speed connection, the system was tested in various surgeries from Ottawa to Edmonton (over a distance of 4000 km). They reported that a lag of 55 ms was acceptable for their system.

Distributed surgical simulators are highly complex systems in which consistency has to be maintained in several domains: physical simulation, related to haptic feedback, and graphical simulation, related to visual feedback. Additionally, network conditions (available bandwidth, jitter, package drop count) can change in time, introducing variables that an adaptive system needs to take into account (Cote et al., 2008). Several simulators have been designed to adapt to the previously mentioned conditions.

Diaz, Trefftz, Quintero, Acosta, and Srivastava (2013) report a study showing the effects of network conditions on a collaborative task to perform a cholecystectomy in a multimodal collaborative surgical simulator (Figure 17.8). They report that processor speed and memory affect the visual feedback (measured in frames per second), whereas the network conditions (delay, jitter, and packet loss) affect the collaboration task (measured as time to complete the task). Tang, Rong, Guo, and Prabhakaran (2010) report the creation and evaluation of a middleware that manages the issues related to the network connection and maintains the consistency of the collaborative surgical VE.

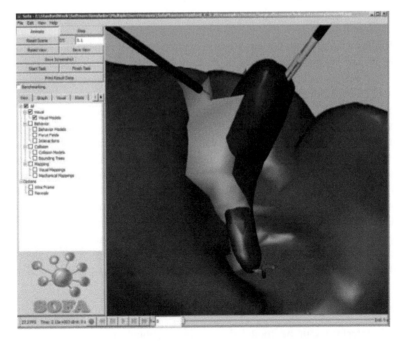

FIGURE 17.8 **(See color insert.)** Multimodal interaction in distributed surgical simulators: cholecystectomy procedure.

Qin, Choi, and Heng (2009) designed an integrated framework using cluster-based hybrid network architecture to support collaborative virtual surgery. Multicast transmission was employed to transmit updated information among participants in order to reduce network latencies while an administrative server maintained system consistency. Reliable multicast was implemented using a distributed message acknowledgment scheme.

17.4.3 Distributed Web-Based Multimodal VEs

Among the early multimodal distributed VEs were those designed for telemedical applications and teleoperation. Popescu, Burdea, Bouzit, Girone, and Hentz (2000) and Burdea et al. (2000) described a VE system using haptic feedback for home telerehabilitation while allowing remote monitoring from the clinic. A hand interface is used to sample a patient's hand positions (Popescu et al., 1999) and to provide resistive forces using the RMII glove (Bouzit, Burdea, Popescu, & Boian, 2002). Popescu, Burdea, and Boian (2002) later developed a two-user shared VE for hand telerehabilitation. Each site has an RMII force feedback glove (with haptic feedback and hand input) and video and audio input. The shared VE application was used to measure and assess issues in controlling a virtual hand and interact haptically with virtual objects over a broadband network and to simulate physical interactions (using hand force feedback) between two users (therapist and patient).

A predicted increase of broadband access is expected to accelerate the deployment of network VE over the Internet. With this, new multimodal issues arise in the design of distributed multimodal VEs. Among these are the network-related issues such as synchronization of content, group communication, and intelligent methods for multimodal user interaction. Popescu & Codella (2002) proposed a QoS architecture for addressing synchronization issues in network VEs. QoS is achieved by predicting the load and adapting to network traffic variations. Data are prefetched at the client based on network traffic estimates and viewpoint navigation prediction. QoS negotiation allows the server to control the network resources allocated per client. Such QoS framework is appealing from the multimodal input design perspective as well. Performance limited systems (either standalone or part of distributed system) should define measures of multimodal interaction and continuously optimize human–computer communication performance. Chang, Popescu, and Codella (2002) addressed the problem of efficient and scalable update dissemination in an environment where client interests can change dynamically and the number of multicast channels available for update dissemination is limited. A new algorithm that can group objects and clients in a way that it handles limited bandwidth resources was designed and analyzed.

Among the key design points are communication protocols that improve group communication and intelligent CVE partitioning methods. Popescu and Liu (2004) designed an application-level (or end-system) multicast and a stateless group communication mechanism together with its tree building algorithms. Stateless multicast—a multicast protocol that does not maintain a distributed state of communicating groups—reduces the control signaling of dynamic multicast groups linearly with the group size. To support interactive applications involving a large number of dynamic multicast groups, the application-level multicast uses stateless forwarding within clusters of network nodes and hierarchical aggregation of multicast group membership. Popescu and Liu (2006) proposed efficient algorithms for clustering network nodes dynamically based on their communication interest. The control algorithms were designed for large-scale collaborative systems optimized for scalability as well as end-to-end data dissemination. The proposed control algorithms use proximity-based clustering of network nodes and hierarchical communication interest aggregation. The new network overlay control algorithms achieve scalability and low overhead with a controlled degradation of end-to-end data path performance.

17.5 CONCLUSIONS AND FUTURE DIRECTIONS

The human–VE interface can be modeled as a bidirectional pathway containing perceptual, cognitive, and control processes. These processes are not independent but interact with each other following complex models. These interactions manifest themselves through sensorial substitution and I/O coupling effects. The design of human–VE interfaces need not be limited to the use of a single I/O modality but should exploit multimodal input and feedback.

Gesture and voice are the most popular inputs in VE, but they are typically not used together. Voice input is unreliable when using a large vocabulary. Multimodality allows erroneous speech recognition to be screened out and ambiguous gestures to be resolved, increasing application control and naturalness of human–machine dialogue. The increasing demand for interactivity and immersion will drive further development of multimodal input for human–VE interfaces. More experimental work is necessary to understand speech–gesture interaction. The studies should lead to guidelines for more efficient implementation of VE control.

Users' sensorial experience in a VE is a fusion of information received through the human perception channels (visual, auditory, somatic, etc.). While feedback in a VE is provided to users predominantly through the visual channel, additional modalities can increase the efficiency of VE simulations. Three-dimensional sound and haptic feedback increase the sense of presence in a VE with additional spatiotemporal and physical information. User immersion in a VE simulation results from rich sensorial interaction coupled with real-time simulation response. Multimodal feedback is not just the sum of visual, auditory, and somatic feedback, because there is redundancy and transposition in the human sensorial processes. Thus, to increase immersion and interactivity, VE designers can exploit sensorial redundancy and sensorial transposition effects.

Looking at the future, it is clear that the present state of multimodal interaction is still rudimentary. Large-volume distributed simulations pose additional challenges to system designers, because sensing needs to be done at larger distances, and feedback interfaces need to be powerful yet light and portable. New technologies are thus needed not only for user's multimodal input but also for multimodal feedback. The current human factors effort to quantify the usability of multimodal VEs (see Gabbard, 2013, Chapter 28) needs to be accelerated and its scope broadened to cover the needs of large-volume VEs. On the theoretical front, more complex models of an intrinsically more complex system need to be developed. Such models will eventually lead to design guidelines and make system optimization efforts easier to undertake.

ACKNOWLEDGMENTS

The authors' research reported here was supported by grants from the National Science Foundation (NSF grant IRI-9618854—STIMULATE and BES-9708020) and from the CAIP Center at Rutgers University with funds provided by the New Jersey Commission on Science and Technology and by CAIP industrial members. George V. Popescu acknowledges the support of EU project POSDRU 86/1.2/S/61756.

REFERENCES

Barfield, W., Hendrix, C., Bjorneseth, O., Kaczmarek, K., & Lotens, W. (1995). Comparison of human sensory capabilities with technical specifications of virtual environment equipment. *Presence, 4*(4), 329–356.

Basdogan, C., & Ho, C. (2000). An experimental study on the role of touch in shared virtual environments. *ACM Transactions on Computer-Human Interaction, 7*(4), 443–460.

Billinghurst, M., & Wang, X. (1999). *GloveGRASP* [Online]. Available at http://www.hitl.washington.edu/people/grof/GestureGRASP.html [September 20, 1999].

Blattner, M., & Glinert, E. (Winter 1996). Multimodal integration. *IEEE Multimedia, 3*, 14–24.

Boulanger, P., Wu, G., Bischof, W. F., & Yang, X. D. (2006). Hapto-audio-visual environments for collaborative training of ophthalmic surgery over optical network. *HAVE 2006—IEEE International Workshop on Haptic Audio Visual Environments and Their Applications* (pp. 21–26). Ottawa, Ontario, Canada.

Bouzit, M., & Burdea, G. (1998). Force feedback interface to reduce pilot's spatial disorientation. *Third Annual Symposium and Exhibition on Situational Awareness in the Tactical Air Environment* (pp. 69–76). Piney Point, MD.

Bouzit, M., Burdea, G., Popescu, G., & Boian, R. (2002). The Rutgers Master II-new design force-feedback glove. *IEEE/ASME Transactions on Mechatronics*, *7*(2), 256–263.

Bowman, D., & Hodges, L. (1999). Formalizing the design, evaluation, and application of interaction techniques for immersive virtual environments. *The Journal of Visual Languages and Computing*, *10*(1), 37–53.

Burdea, G. (1996). *Force and touch feedback for virtual reality*. New York, NY: John Wiley & Sons.

Burdea, G. (1999). *Haptic interfaces for virtual reality*. Keynote address in Proceedings of International Workshop. Laval, France.

Burdea, G., & Coiffet, P. (1994). *Virtual reality technology*. New York, NY: John Wiley & Sons.

Burdea, G., Patounakis, G., Popescu, G. V., & Weiss, R. E. (1999). Virtual reality-based training for the diagnosis of prostate cancer. *IEEE Transactions on Biomedical Engineering*, *46*(10), 1253–1260.

Burdea, G., Popescu, V., Hentz, V., & Colbert, K. (2000). Virtual reality-based orthopedic tele-rehabilitation. *IEEE Transactions on Rehabilitation Engineering*, *8*(3), 429–432.

Cassell, J., Steedman, M., Badler, N., Pelachaud, C., Stone, M., Douville, B., … Achron, B. (1994). Modeling interaction between speech and gesture. *Proceedings of the 16th Annual Conference of the Cognitive Science Society*. Atlanta, GA: Georgia Institute of Technology.

Chaiasson, J. E., McGrath, B. J., & Rupert, A. H. (2002). Enhanced situation awareness in sea, air, and land environments. *NATO Conference Proceedings, Spatial Disorientation in Military Vehicles: Causes, Consequences and Cures*, A Coruna, Spain.

Chang, T., Popescu, G. V., & Codella, C. (2002). Scalable and efficient update dissemination for interactive distributed applications. *Proceedings of ICDCS 2002* (pp. 143–151). Vienna, Austria.

Chu, C. P., Dani, T. H., & Gadh, R. (1997). Multimodal interface for a virtual reality-based computer-aided design system. *Proceedings of the 1997 IEEE International Conference on Robotics and Automation* (pp. 1329–1334). Albuquerque, NM: IEEE Press.

Chu, C. P., Dani, T. H., & Gadh, R. (1998). Evaluation of a virtual reality interface for product shape design. *IIE Transactions*, *30*, 629–643.

Churchill, E. F., Snowdon, D. N., & Munro, A. J. (2001). *Collaborative virtual environments: Digital places and spaces for interaction*. London, U.K.: Springer Verlag.

Cohen, M., & Wenzel, E. (1995). The design of multidimensional sound interfaces. In W. Barfield, & T. A. Furness, III (Eds.), *Virtual environments and advance interface design* (pp. 291–346). New York, NY: Oxford University Press.

Cohen, P. R., Johnston, M., McGee, D., Smith, I., Oviatt, S., Pittman, J., … Clow, J. (1997). QuickSet: Multimodal interaction for simulation set-up and control. *Proceedings of the Fifth Applied Natural Language Processing Meeting*. Washington, DC: Association for Computational Linguistics.

Cohen, P., McGee, D., Oviatt, S., Wu, L., Clow, J., King, R., … Rosenblum L. (1999). Multimodal interaction for 2D and 3D environments. *IEEE Computer Graphics and Applications*, *19*, 10–13.

Cote, M., Boulay, J.-A., Ozell, B., Labelle, H., & Aubin, C.-E. (2008, October 18–19). Virtual reality simulator for scoliosis surgery training: Transatlantic collaborative tests. *IEEE International Workshop on Haptic, Audio and Visual Environments and Their Applications*. Ottawa, Ontario, Canada, October 18–19.

Dauville, B., Levinson, L., & Badler, N. (1996) Task-level object grasping for simulated agents. *Presence*, *5*(4), 416–30, Cambridge, U.K.: MIT Press.

Diaz, C., Trefftz, H., Quintero, L., Acosta, D., & Srivastava, S. (2013). Collaborative networked virtual surgical simulators (CNVSS): Factors affecting collaborative performance. *Presence: Teleoperators and Virtual Environments*, *22*(1), 54–66.

Dragon Naturally Speaking (2014). *Dragon Naturally Speaking Developer Suite* [Online]. Available: http://www.nuance.com/support/dragon-naturallyspeaking/index.htm. Accessed on April 2, 2014.

Dragon Systems. (1999). *Dragon naturally speaking developer suite* [Online]. Available: http://www.dragonsystems.com/products/developer/naturallyspeaking/index.html [September 20, 1999].

Dumas, B., Lalanne, L., & Oviatt, S. L. (2009). Multimodal interfaces: A survey of principles, models and frameworks. *Human Machine Interaction*, *5440*, 3–26.

Fabiani, L., Burdea, G., Langrana, N., & Gomez, D. (1996). Human performance using the Rutgers Master II force feedback interface. *IEEE International Symposium on Virtual Reality and Applications (VRAIS '96)* (pp. 54–59). Santa Clara, CA: IEEE Press.

Flanagan, J., Kulikovski, C., Marsic, I., Burdea, G., Wilder, J., & Meer, P. (1999). *Synergistic multimodal communication in collaborative multiuser environments* (Ann. Rep.): National Science Foundation.

Fritz, J. P., & Barner, K. E. (1996). Stochastic models for haptic textures. In M. R. Stein (Ed.), *Proceedings of SPIE Vol. 2901, Telemanipulator and Telepresence Technologies III* (pp. 34–44). Boston, MA: SPIE.

Fukui, Y., & Shimojo, M. (1992). Differences in recognition of optical illusion using visual and tactual sense. *Journal of Robotics and Mechatronics, 4910*, 58–62.

Glencross, M., Jay, C., Feasel, J., Kohli, L., Whitton, M., & Hubbold, R. (2007). Effective cooperative haptic interaction over the internet. *IEEE Virtual Reality Conference 2007* (pp. 115–122). Charlote, NC.

Groen, J., & Werkhoven, P. (1998). Visuomotor adaptation to virtual hand position in interactive virtual environments. *Presence, 7*(5), 429–446.

IBM Watson. (2014). *The deepQA project* [Online]. Available: http://researcher.ibm.com/researcher/view_grouppubs.php?grp=2099. Accessed on April 2, 2014.

Jacob, L. (1995). Eye tracking in advance interface design. In W. Barfield & T. A. Furness, III (Eds.), *Virtual environments and advance interface design* (pp. 258–90). New York, NY: Oxford University Press.

Johnston, M., Cohen, P. R., McGee, D., Oviatt, S. L., Pittman, J. A., & Smith, I. (1997). Unification-based multimodal integration. *Proceedings of the 35th Annual Meeting of the Association for Computational Linguistics*. Stroudsburg, PA: Association for Computational Linguistics Press.

Kaczmarek, K. A., & Bach-y-Rita, P. (1995). Tactile displays. In W. Barfield & T. A. Furness, III (Eds.), *Virtual environments and advance interface design* (pp. 349–414). New York, NY: Oxford University Press.

Kinect. (2014). *Kinect for Windows* [Online]. Available: http://www.microsoft.com/en-us/kinectforwindows/ Accessed on April 2, 2014.

Kjeldsen, R., & Kender, J. (1997). Interaction with on-screen objects using visual gesture recognition. *Proceedings of the IEEE Computer Vision and Pattern Recognition* (pp. 788–793). Son Juan, Puerto Rico: IEEE Press.

Kotarinis, D., & Howe, R. (1995). Tactile display of vibratory information in teleoperation and virtual environments. *Presence, 4*(4), 387–402.

Lin, Y., & Makedon, F. Hierarchical multimodal planning for pervasive interaction. *AAAI Fall Symposium Series*, North America, November 2010. Available at: http://aaai.org/ocs/index.php/FSS/FSS10/paper/view/2211. Accessed on April 2, 2014.

Marsic, I., Medl, A., & Flanagan, J. (2000). Natural communication with information systems. *Proceeding of the IEEE, 88*(8), 1354–1366. IEEE Press.

Massimo, M., & Sheridan, T. (1993). Sensory substitution for force feedback in teleoperation. *Presence, 2*(4), 344–352.

Medl, A., Marsic, I., Andre, M., Liang, Y., Shaikh, A., Burdea, G., … Flanagan, J. (1998). *Multimodal man–machine interface for mission planning*. Paper presented at Intelligent Environments-AAAI Spring Symposium. Stanford, CA.

Microsoft SAPI. (2014). *Microsoft speech API (2013)* [Online]. Available: http://msdn.microsoft.com/en-us/library/ee125663(VS.85).aspx. Accessed on April 2, 2014.

Minsky, M., Ouh-young, M., Steele, O., Brooks, F., Jr., & Behensky, M. (1990). Feeling and seeing: Issues in force display. *Computer Graphics, 24*(2), 235–43.

Morillo, P., Rueda, S., Orduña, J. M., & Duato, J. (2007). A latency-aware partitioning method for distributed virtual environment systems. *IEEE Transactions on Parallel and Distributed Systems, 18*(9), 1–12.

Nam, C. S., Shu, J., & Chung, D. (2008). The roles of sensory modalities in collaborative virtual environments (CVEs). *Computers in Human Behavior, 24*(4), 1404–1417. doi:10.1016/j.chb.2007.07.014.

Oviatt, S. (2002). Multimodal interfaces. In J. A. Jacko & A. Sears (Eds.), *The human-computer interaction handbook: Fundamentals, evolving technologies and emerging applications* (pp. 286–304). Hillsdale, NJ: L. Erlbaum Assoc.

Oviatt, S. L. (1999). Ten myths of multimodal interaction, *Communications of the ACM* [Online]. Available: http://church.cse.ogi.edu/CHCC/Personnel/oviatt.html [September 20, 1999].

Oviatt, S. L., & Clow, J. (1998). An automated tool for analysis of multimodal system performance. *Proceedings of the International Conference on Spoken Language Processing*. Sydney, New South Wales, Australia.

Oviatt, S. L., DeAngeli, A., & Kuhn, K. (1997). Integration and synchronization of input modes during multimodal human–computer interaction. *Proceedings of Conference on Human Factors in Computing Systems: CHI '97*. New York, NY: ACM Press.

Oviatt, S., Coulston, R., Tomko, S., Xiao, B., Lunsford, R., … Carmichael, L. (2003). Toward a theory of organized multimodal integration patterns during human-computer interaction. *Proceedings of the Fifth International Conference on Multimodal Interfaces (ICMI '03)* (pp. 44–51). New York, NY: ACM.

Pavlovic, V., Sharma, R., & Huang, T. S. (1997). Visual interpretation of hand gestures for human–computer interaction: A review. *IEEE Transactions on Pattern Analysis and Machine Intelligence, 19*(7), 677–695.

Popescu, G. V., & Codella, C. F. (2002). An architecture for QoS data replication in network virtual environments. *VR2002* (pp. 41–48).

Popescu, G. V., & Liu, Z. (2004). Stateless application-level multicast for dynamic group communication. *DS-RT2004* (pp. 20–28).

Popescu, G. V., & Liu, Z. (2006). Network overlays for efficient control of large scale dynamic groups. *DS-RT2006* (pp. 135–142).

Popescu, G. V., Burdea, G., & Boian, R. (2002, January 23–26). Shared virtual environments for telerehabilitation. *Proceedings of Medicine Meets Virtual Reality 2002* (pp. 362–368). Newport Beach CA: IOS Press.

Popescu, G. V., Burdea, G., Bouzit, M., Girone, M., & Hentz, V. (2000). A virtual-reality-based telerehabilitation system with force feedback. *IEEE Transactions on Information Technology in Biomedicine*, *4*(1), 45–51.

Popescu, V., Burdea, G., & Bouzit, M. (1999). Virtual reality modeling for a haptic glove. *Proceedings of Computer Animation '99* (pp. 195–200). Geneva, Switzerland: IEEE Press.

Qin, J., Choi, K. S., & Heng, P. A. (2009). Collaborative simulation of soft-tissue deformation for virtual surgery applications. *Journal of Medical Systems*, *34*(3), 367–378. doi:10.1007/s10916-008-9249-2.

Quiroz, A., & Trefftz, H. (2004). Combinatory multicast for differentiated data transmission in distributed virtual environments. *Computer Graphics and Imaging (CGIM 2004)* (pp. 158–163). Anaheim, CA: Acta Press.

Reidsma, D., Nijholt, A., Tschacher, W., & Ramseyer F. (2010). Measuring multimodal synchrony for human-computer interaction. *Proceedings of the 2010 International Conference on Cyberworlds (CW '10)* (pp. 67–71). Washington, DC: IEEE Computer Society.

Richard, P., Burdea, G., Gomez, D., & Coiffet, P. (1994). A comparison of haptic, visual and auditive force feedback for deformable virtual objects. *Proceedings of ICAT 94 Conference* (pp. 49–62). Tokyo, Japan.

Rosenblum, L., Durbin, J., Doyle, R., & Tate, D. (1997). Situational awareness using the VR responsive workbench. *IEEE Computer Graphics and Applications*, *16*(4), 12–13.

Rozak, M. (2007). Text-to-speech designed for a massively multiplayer online role-playing game (MMORPG). *Proceedings of the IEEE Speech Synthesis Workshop* (pp. 1–6). Boston, MA.

Schomaker, L., Nijtmans, J., Camurri, A., Lavagetto, F., Morasso, P., Benoit, C., Guiard-Marigny, T., Le Goff, B., Robert-Ribes, J., Adjoudani, A., Defee, I., Munch, S., Hartung, K., & Blauert, J. (1995). *A taxonomy of multimodal interaction in the human information processing system* [Online]. Multimodal Integration for Advanced Multimedia Interfaces (MIAMI). ESPRIT III, Basic Research Project 8579. Available: http://www.ai.rug.nl/~lambert/projects/miami/reports/taxrep-300dpi.pdf. Accessed on April 2, 2014.

Sharma, R., Huang, T. S., & Pavlovic, V. I. (1996). A multimodal framework for interacting with virtual environments. In C. A. Ntuen & E. H. Park (Eds.), *Human interaction with complex systems: Conceptual principles and design practice* (pp. 53–71). Boston, MA: Kluwer Academic Publishers.

Singhal, S., & Zyda, M. (1999). *Networked virtual environments: Design and implementation*. New York, NY: ACM Press.

Sletterink, B. (1999). *A managing agent for sharing multiple modalities* (Unpublished master's thesis). Delft University of Technology, Delft, the Netherlands.

Starner, T., & Pentland, A. (1995). Visual recognition of american sign language using hidden Markov models. *Proceedings of the International Workshop on Automatic Face and Gesture Recognition* (pp. 189–194). Zurich, Switzerland.

Stevenson, D., & Hutchins, M. (2006). Multiple approaches to evaluating multi-modal collaborative systems. *Workshop on Effective Multimodal Dialogue Interfaces, International Conference on Intelligent User Interfaces (IUI 2006)* (pp. 2–4). Sydney, New South Wales, Australia.

Tang, Z., Rong, G., Guo, X., & Prabhakaran, B. (2010). Streaming 3D shape deformations in collaborative virtual environment. *2010 IEEE Virtual Reality Conference (VR)* (pp. 183–186). Waltham, MA: IEEE.

Thorpe D., & Hodges, L. (1995). Human stereopsis, fusion, and stereoscopic virtual environments. In W. Barfield & T. A. Furness, III, (Eds.), *Virtual environments and advance interface design* (pp. 145–74). New York, NY: Oxford University Press.

Trefftz, H., Marsic, I., & Zyda, M. (2003). Handling heterogeneity in networked virtual environments. *Presence*, *12*(1), 7–14.

Vo, M., & Waibel, A. (1993). A multimodal human–computer interface: Combination of gesture and speech recognition. *Adjunct Proceedings of InterCHI '93*. Amsterdam, the Netherlands.

Vo, M., & Waibel, A. (1997). *Modeling and interpreting multimodal input: A semantic integration approach* (Technical Report No. CMU-CS-97-192).

WebSphere Voice. (2014). *IBM webSphere voice products* [Online]. Available at http://www-01.ibm.com/software/voice/. Accessed on April 2, 2014.

Wii. (2014). *Nintendo integrated research and development* [Online]. Available at: http://nintendo.wikia.com/wiki/Nintendo_Integrated_Research_%26_Development. Accessed on April 2, 2014.

18 Illusory Self-Motion in Virtual Environments

Lawrence J. Hettinger, Tarah N. Schmidt-Daly,
David L. Jones, and Behrang Keshavarz

CONTENTS

18.1 INTRODUCTION

Current virtual environment (VE) systems can often induce compelling illusions of self-motion in users. In many instances, the effects of visually depicted motion in a VE can be strong enough to create an overwhelming sense of actual self-motion (*vection*) in physically stationary observers, along with pronounced postural adjustment and/or a strong sense of disequilibrium and, occasionally, motion sickness. Indeed, the ability to generate realistic sensations of self-motion within VEs might be considered an important element in the overall sense of *presence* (see Chertoff & Schatz, 2014, Chapter 34) imparted by a given system.

In all likelihood, future VE systems will be able to produce stronger and more durable illusions of self-motion. In some cases, there may be compelling reasons for doing so, particularly if the attainment of larger human–VE system objectives is thereby aided. (Note: A *human–VE system objective* is defined as the intended, often measurable, outcome of the VE system with respect to its human user. For instance, the objective of a VE-based training system is to promote the acquisition of a set of general or well-defined skills in its users. The experience of illusory self-motion, or any other component of presence, may or may not assist in the attainment of overall human–VE system objectives. Indeed, it may in some cases interfere with them [cf. Hettinger & Haas, in press].) In other cases, the benefits of these illusory experiences may be dramatically offset by the occurrence of negative side effects such as motion sickness (see Keshavarz, Hecht, & Lawson, 2014, Chapter 26; Lawson, 2014a, Chapter 23; Welch & Mohler, 2014, Chapter 25) and persistent perceptual–motor disruptions. Long-standing assumptions point to sensory conflicts that exacerbate motion sickness (Palmisano, Allison, Kim, & Bonato, 2011; Prothero & Parker, 2003). In any event, rational design and usage decisions will need to be based on knowledge of the factors that promote (or inhibit) the occurrence of these illusions, their specific benefits with respect to the system user, and the potential risks that they entail.

The term *vection* is used to describe a broad class of illusory self-motion phenomena that are primarily induced by dynamic, large field-of-view optic flow patterns (Dichgans & Brandt, 1978; Howard, 1986a), but which have also been noted to occur with relatively small fields of view (Andersen & Braunstein, 1985; Andersen & Dyer, 1987). Although most commonly a visual phenomenon, vection illusions have also been reported as a result of exposure to specific classes of auditory stimulation (*audiokinetic vection*—e.g., Dodge, 1923; Lackner, 1977; Larsson, Västfjäll, & Kleinder, 2004; Riecke, Väljamäe, & Schulte-Pelkum, 2009; see also Väljamäe, 2009 for a review) and electrical vestibular stimulation (e.g., Cress et al., 1997; Dzendolet, 1963). Dichgans and Brandt (1978) describe a haptic form of vection (*haptokinetic vection*—e.g., Nordahl, Nilsson, Turchet, & Serafin, 2012; Roll et al., 2012) that produces a weak sensation of illusory self-motion induced through tactile motion stimulation of a large part of the body. Systematic examination of each of these nonvisual forms of vection has been very limited, and they will not be described in any further detail.

This chapter will examine the implications of the considerable experimental literature in this area for the design and use of VE systems. It provides a general summary of the experimental findings from the primary areas of research in this area (e.g., optical flow pattern characteristics of effective vection displays, neurophysiological correlates of vection, vection-induced motion sickness and disorientation) and describes their relevance to VE design and usage issues. One of the major concerns of the chapter is to heighten the reader's sensitivity to questions of whether the user's experience of vection in a given VE system in any way promotes its overall system objectives, under what circumstances and/or within which applications such benefits arise, and at what potential cost to the user's safety and well-being. With regard to the latter point, particular attention will be paid to the literature that has examined the negative side effects that have occasionally been associated with the vection illusion, particularly those involving motion sickness and disrupted perceptual–motor behavior or aftereffects (see Keshavarz et al., 2014, Chapter 26; Lawson, 2014a, Chapter 23; Welch & Mohler, 2014, Chapter 25).

The earlier questions are closely related to a major debate that has existed within the VE community almost since its inception centering around the potential benefits and costs of *presence* or *telepresence* in VE systems (cf. Held & Durlach, 1992; Slater & Steed, 2000; Zahorik & Jenison, 1998; Zeltzer, 1992). This debate echoes an essentially identical controversy that existed in the flight simulation community for many years, that of the role of *realism* in the design and use of these devices. Although difficult to define and even more so to measure (see Chapter 40, this book), presence and realism are too often assumed to be necessary in order for VE systems to be successful and/or accepted by users. Although realism pertains to the fidelity of a simulator or VE system, it is not a necessary component for training and educational purposes. Presence, on the other hand, should be a characteristic that developers of a VE system should strive for. Feeling present and being able to interact in and with the VE provide users with presence (Steuer, 1992). The potential validity of this largely unexamined assumption may vary widely from one application to the next, but it is certain that it entails important cost and design implications for VE systems. In addition, in the case of vection, it also involves important issues of user safety and well-being.

While this chapter will not examine the notion of presence in depth, we do wish to emphasize that vection, as a key component of presence, may or may not meaningfully contribute to the system objectives of a human–VE system (von der Heyde & Riecke, 2002). It will, however, occasionally have undesired consequences for users. Therefore, it is important to understand the nature of the vection illusion and to approach its inclusion as a VE system design feature with something more than the simplistic assumption that *if it looks and feels real, then it must be good*.

The strength and duration of the vection illusion, its role within and effect on the overall design goals of any given VE system, and the nature, strength, and duration of potential negative side effects are all dependent on a variety of factors to be discussed in this chapter. This chapter is intended to provide readers with a basic understanding of these phenomena, the factors that underlie their occurrence, and the degree to which they may help or hinder specific applications of VE technology.

This chapter will not provide an encyclopedic examination of all aspects of the vection illusion, nor will it be a complete and exhaustive review of the considerable experimental literature in the area. For excellent in-depth presentations of these aspects of vection, readers are referred to Dichgans and Brandt (1978), Howard (1982, 1986a), Prothero and Parker (2003), and Väljamäe (2009). This chapter will deal solely with summarizing the research to date on vection phenomena and describe its relevance for the effective design and use of VE systems.

18.2 GENERAL CHARACTERISTICS OF VECTION

Actual self-motion through the real world is visually specified by reliable and highly structured transformations in the pattern of optical information—the optic array or flow—that directly correspond to the dynamics of specific forms of self-motion (e.g., running, crawling, driving in a car) through specific types of structured environments (Cutting, 1986; Gibson, 1958, 1961, 1979; Owen, 1990; Warren, 1976). Generally speaking, these patterns of optical transformations take place over a wide field of view (corresponding to one's normal field of vision) and primarily occur when one is in fact moving. Therefore, when found in a situation in which large-field optical transformations of the sort that normally accompany and specify actual self-motion are made available, it is not surprising that one experiences a sense of being physically displaced even when in fact stationary.

VEs, particularly those encompassing large visual fields, are particularly well suited to produce vection illusions. But how are these illusions produced? What controls their onset, compellingness, and duration? Why do these illusions appear to lead to motion sickness and perceptual–motor disturbances in some individuals? And, most importantly, of what potential benefit (if any) are vection illusions to the larger objectives of individual human–VE systems, and what are their potential costs?

Experimental psychologists and neurophysiologists have studied vection for well over a century. Mach (1875) was the first to systematically study vection in the laboratory. Tschermak

(1931, cited in Dichgans & Brandt, 1978) appears to have been the first to refer to illusory self-motion as *vection*. Although the empirical investigation of this phenomenon has primarily occurred within traditional laboratory settings using various types of specially designed devices such as vection cylinder *drums* and *moving rooms*, the advent of high-fidelity flight simulation systems in the 1970s and 1980s extended its occurrence, as well as its study, into more *applied* domains (e.g., Frigon & Delorme, 1992; Hettinger, Berbaum, Kennedy, Dunlap, & Nolan, 1990; van der Steen, and Brockhoff, 2000). More recently, the increased development and use of VE systems for multisensory integration in wayfinding (Darken & Peterson, 2014, Chapter 19), posture rehabilitation, prevention of cortical disruption, vehicular simulation, for entertainment (see Greenwood-Ericksen, Kennedy, & Stafford, 2014, Chapter 50), and many other applications has produced an even greater incidence of vection illusions (Roll et al., 2012; Seno, Ito, & Sunaga, 2009; Väljamäe, 2009). Indeed, the tremendous popularity of *virtual reality rides* at many of the world's most popular amusement and entertainment parks is based in large part on the reliable production of vection in paying customers and has exposed very large segments of the population to its intended effects and occasionally to its unintended negative side effects.

One of the earliest (and certainly most literary) accounts of vection is found in a description by Wood (1895) of the *Haunted Swing*—a popular attraction at the 1895 San Francisco Midwinter Fair and clearly a predecessor of today's VE-based entertainment devices. There is little doubt that the experience of illusory self-motion underlies the exhilarating and, apparently for some, slightly disturbing experience described by Wood. To experience this attraction, fairgoers entered a large room furnished in a more or less normal way, with the exception of a very large swing in the center of the room capable of holding up to 40 people and suspended from a large iron rod that passed through the center of the room. Visitors took their seats on the swing, and Wood describes what happened next as follows:

> We took our seats and the swing was set in motion, the arc gradually increasing in amplitude until each oscillation carried us apparently into the upper corners of the room. Each vibration of the swing caused those peculiar 'empty' sensations within which one feels in an elevator; and as we rushed backwards toward the top of the room there was a distinct feeling of "leaning forward," if I can so describe it—such as one always experiences in a backward swing, and an involuntary clutching at the seats to keep from being pitched out. We were then told to hold on tightly as the swing was going clear over, and sure enough, so it did. … (p. 277)

The Haunted Swing did not, of course, traverse a 360° arc. In fact, it did not move at all (with the exception of *merely being joggled a trifle*). The entire effect was based on the presence of a vection illusion—the room itself (with its furnishings securely fastened to the floor and walls) moving back and forth around the astonished, yet physically stationary, observers.

Two further comments by Wood are particularly interesting in light of their relevance for the current discussion of vection: "The curious and interesting feature … was that even though the action was fully understood, as it was in my case, it was impossible to quench the feeling of 'goneness within' with each apparent rush of the swing" (p. 277). As is common of contemporary vection illusions, even objective knowledge of the true state of affairs is insufficient to suppress the illusory sensation of motion. Finally, and somewhat more ominously, "Many persons were actually made sick by the illusion. I have met a number of gentlemen who said they could scarcely walk out of the building from dizziness and nausea" (pp. 277–278). Indeed, people are still being made dizzy and sick by similar phenomena.

Vection illusions occasionally occur in everyday life, outside of laboratories, simulators, and VE devices. One of the most frequently cited of these ordinary circumstances involves sitting in a stationary railroad car with another stationary train directly alongside. When the neighboring train begins to accelerate from a dead stop, it is quite common for passengers in the stationary railcar to experience a sudden and somewhat arresting experience of self-motion, an experience that is particularly compelling when viewed in the periphery of the visual field. This experience manifests itself as a strong phenomenological sense of one's own motion relative to the adjacent train (which is experienced as stationary), accompanied by a distinct postural control motion as the observer *adjusts* to the experienced motion. The same phenomenon occurs in stationary cars, busses, airplanes, and other vehicles. Research has shown vection onset and intensity to be exacerbated by the notion that observers have the potential to move or that actual self-motion is possible. In line with this example, being on a train, a passenger would expect that *eventually* they will be in motion (Nordahl et al., 2012; Riecke, Feuereissen, Rieser, & McNamara, 2012; Seno, Ito, & Sunaga, 2012). Another common setting for self-motion illusions is the cinema, particularly the wide field-of-view setting of *cinemax* and I-Max theaters. Large, hemispheric dome displays, an attraction at several of the world's largest entertainment parks, on which compelling patterns of motion are projected (e.g., fighter jets flying in formation), are designed to produce strong vection illusions. Wisely, railings are included for customers to hang onto and thereby avoid pitching over en masse in response to pronounced twists and turns of the motion pattern.

In most, if not all, VE applications, there is a strong perceived need among designers to provide very-high-fidelity synthetic replications of the real (or, in some cases, imaginary) world (Riecke et al., 2012). A corollary of this emphasis on design realism is the desire to provide high-fidelity representations of the multisensory consequences of users' motor behaviors and interactions with virtual entities. In other words, there is a perceived need in the VE design community to provide a highly realistic simulation of the *response* of the perceptual world to each significant activity undertaken by the user of the system.

For example, when a user initiates optical (and other sensory) transformations specifying self-motion in a VE (either by physically walking on a treadmill or by means of some sort of intermediary device such as a dataglove or joystick), the pattern of visual (see Badcock, Palmisano, & May, 2014, Chapter 3), auditory (see Vorlander & Shinn-Cunningham, 2014, Chapter 4), and in some cases haptic (see Dindar, Tekalp, & Basdogan, 2014, Chapter 5) and vestibular (see Lawson & Riecke, 2014, Chapter 7) information made available to the user should correspond to the intended self-motion profile (Revol et al., 2009; Riecke et al., 2012; Riecke & Wiener, 2007). In most cases, VE users will not in fact be physically displaced. However, the entire pattern of multisensory stimulation to which they are exposed will specify self-motion, and in many of these cases, users will experience strong vection illusions. As seen from Woods' description earlier, this experience can be compelling and exhilarating but also occasionally disturbing and even dangerous (depending on the activities the user engages in during exposure to the VE and immediately upon leaving it). It is incumbent on designers of VE systems to understand the nature of this illusion in order to assure that the user's experience is safe and accomplishes its intended goals.

Before embarking on specific discussions of how an understanding of the vection illusion can substantially impact the design and use of VE systems, the major findings in vection research that are of most relevance to these applications will be reviewed. These findings are classified into the following general areas: (1) the nature of the effects experienced by observers and methods used to measure those effects, (2) the nature of the optical transformations that underlie the illusion, (3) the neurophysiological correlates of vection, and (4) motion sickness and perceptual–motor adaptation phenomena. Prior to discussing these areas, the various types of vection illusions that have received the most research attention will be reviewed.

18.2.1 TYPES OF VECTION

Vection can be induced across all six degrees of freedom of body motion (e.g., roll, pitch, yaw, and linear translations) and may be experienced as rotational motion, linear motion, or some combination of the two. In this section, the major forms of vection that have been identified and examined as a means of describing the conditions under which vection has traditionally been investigated and the general nature of the experimental findings that have been obtained will be discussed.

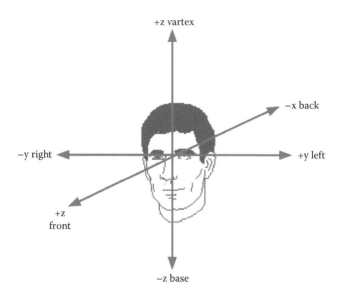

FIGURE 18.1   Three body-centric axes used to classify forms of vection.

In describing these forms of vection, the following nomenclature will be used to describe the three principal physiocentric axes of real and illusory self-motion (see Figure 18.1):

- Z-axis—refers to the vertical, spinal, or yaw axis of the body, or an imaginary line passing directly through the head, down through the body and out the soles of the feet
- X-axis—refers to the roll axis of the body corresponding to the forward line of sight, or an imaginary line passing through the middle of the body from front to back
- Y-axis—refers to the pitch axis of the body, or an imaginary line passing through the body from its right side to its left side

18.2.1.1   Circular Vection

Dichgans and Brandt (1978) define circular vection as illusory rotation about one or more of the three body axes. Generally speaking, however, the term *circular vection* has primarily been applied to illusory rotation about the upright body's z-axis, while rotation about the body's x- and y-axes is more commonly referred to as roll and pitch vection, respectively. Roll and pitch vection are described separately below.

Circular vection is perhaps the most thoroughly investigated of all illusory self-motion phenomena. Ernst Mach first studied it in the nineteenth century, using a rotating, striped cylinder that completely surrounded his subjects (Mach, 1875). He later used a large, effectively endless belt that moved across two rollers to induce the illusion (Mach, 1922, cited in Dichgans & Brandt, 1978). The former apparatus, frequently referred to as a *vection drum*, has been the most common means of inducing and examining the illusion in laboratory settings over the years.

Though certain aspects of the experience of circular vection vary somewhat from one observer to the next (notably its latency of onset), the experience itself is quite uniform across individuals. Once the motion stimulus has been initiated (i.e., the vection drum is turned on and begins rotating about the observer), the common initial impression is of object or *surround* motion—that is, the observer perceives the drum as moving and not himself. However, shortly thereafter, a distinct transformation in the observer's phenomenological experience occurs as he or she begins to perceive himself as moving in the direction opposite to that of drum rotation, while the drum itself appears more or less stationary. *Saturated vection* refers to the condition in which the motion stimulus is perceived as completely stationary, with all motion experienced as self-motion.

As described in more detail below, the latency and strength of the circular vection illusion are dependent on a number of key optical flow parameters. These include the velocity of the motion display, spatial and temporal frequency of the visual pattern, presence of foreground and background visual information, and size of the visual field of view. As Woods noted in reference to the *Haunted Swing* illusion, it is nearly impossible to suppress the illusion of illusory rotation given an optimal combination of these parameters. Objective knowledge of one's circumstances appears to have very little influence on any aspect of the user's experience of this, or any other form of vection, although the knowledge that one *could* move tends to increase measures of illusory self-motion (onset latency is decreased) and intensity, duration, and saturation are increased. As noted by Brandt, Dichgans, and Koenig (1973), it is common for the experience of illusory circular self-motion to persist beyond termination of the optical motion, a phenomenon that may partially underlie problems with disrupted perceptual–motor behavior observed in some users of flight simulation and VE systems (Kennedy & Fowlkes, 1992).

18.2.1.2 Roll Vection

Roll vection is defined as illusory self-motion about the x- or frontal axis of the body (Dichgans & Brandt, 1978). It has been studied using devices such as circular disks with a patterned surface that are positioned immediately in front of an observer or hollow spheres with a patterned inner surface that are set into motion around an observer (e.g., Dichgans, Held, Young, & Brandt, 1972; Held, Dichgans, & Bauer, 1975; Howard, Cheung, & Landolt, 1987). In each case, the surface rotates (or in some cases oscillates) around the observer's x-axis. Recently, Allison, Howard, and Zacher (1999) examined roll vection and illusory self-tilt in a *tumbling room* very reminiscent of the *Haunted Swing*, capable of a full 360° translation in the roll axis. When observers view such displays, they experience an illusory rotation and tilt in the direction opposite that of the motion of the display; the strength of the illusion is dependent on characteristics of the display described below.

The experience of vection in the roll axis is often quite paradoxical. As noted by Howard (1986a), it often seems as if the body is rotating continuously but remaining at a more or less constant and fairly limited angle of tilt. As described below, this paradoxical phenomenon appears to be due to the influence of the vestibular otoliths (see Lawson & Riecke, 2014, Chapter 7).

The same factors that influence the latency and strength of circular vection also appear to affect roll vection. For instance, Held et al. (1975) observed that the illusory self-tilt component of roll vection increased along with the size and retinal eccentricity of the visual display. Allison et al. (1999) observed that subjects were far more likely to report experiences of *tumbling sensations* (i.e., complete 360° apparent self-rotation) as a function of increased field-of-view size. This effect was also greater at the faster of the two rotational velocities tested (30°/s vs. 15°/s). For smaller fields of view, observers reported sensations of constant tilt rather than full, 360° roll.

18.2.1.3 Pitch Vection

Pitch vection occurs as a result of viewing motion patterns that rotate about the body's y-axis (Dichgans & Brandt, 1978; Fushiki, Kobayashi, Asai, & Watanabe, 2005; Kitazaki & Sato, 2003; Young, Oman, & Dichgans, 1975) and has been studied using devices similar to those employed in the study of roll vection, such as rotating disks or hollow spheres within which observers view patterns of moving visual stimuli (e.g., Howard et al., 1987).

The general characteristics of pitch vection are very similar in most respects to those of roll vection. That is, observers can be made to experience constant illusory self-motion, but only a limited degree of displacement or self-tilt (which tends to be less than that experienced in roll vection). Unlike roll vection, however, there appears to be a body-centric asymmetry in the degree of tilt experienced in pitch vection—specifically, smaller magnitude estimates of tilt are observed for apparent pitch-up conditions as compared to apparent pitch-down conditions (Young et al., 1975). Additionally, the stimulus parameters that underlie the experience of pitch vection are very similar to those of roll vection—in particular, larger visual fields of view tend to produce more compelling pitch illusions.

18.2.1.4 Linear Vection

Linear vection can be produced along any of the three body axes described in the preceding text (or as a vector sum of two or more of these axes), but has most often been examined as illusory self-motion along the body's x-axis. It has been studied using a broad array of devices, including moving rooms (e.g., Lee & Lishman, 1975), devices that incorporate projection of linear optical flow patterns onto the walls of a stationary room in which an observer is standing or seated (e.g., Berthoz, Lacour, Soechting, & Vidal, 1978; Lestienne, Soechting, & Berthoz, 1977), and frontal stereoscopic presentation of motion patterns (e.g., Ohmi & Howard, 1988). Most recently, linear vection has been examined in studies that implicate haptic stimulation in the feet (Nordahl et al., 2012) and *elevator* simulation (Bringoux, Lepecq, & Danion, 2012). Flight simulators and numerous VE devices, particularly those with wide field-of-view display capabilities, are also capable of inducing compelling sensations of linear vection.

Postural sway responses to linear vection stimuli are often quite profound. Generally, observers will make a postural adjustment toward the direction of perceived self-motion in an apparent attempt to compensate for a perceived shift in the direction of gravitoinertial force vector. However, observers tend to show greater amplitude in backward as opposed to forward postural adjustments (Lestienne et al., 1977). It is not difficult to create conditions in a linear vection study that will cause observers to fall over, a finding of clear relevance for the safe design and operation of VE devices that incorporate such a capability. Indeed, a standing, unsupported observer in any compelling vection situation is at risk of falling. In addition, linear vection can induce sensations of disequilibrium that are at least as strong as those produced in other forms of the illusion (described in more detail below). For example, Lestienne et al. (1977) reported that of the 30 subjects in their study, 3 (10%) became so disoriented that they fainted while viewing a linear vection display.

18.2.2 PHENOMENOLOGICAL AND BEHAVIORAL ASPECTS OF VECTION

The most common phenomenological and behavioral manifestations of vection include (1) a compelling illusory sense of self-motion in the absence of true inertial displacement and (2) distinct adjustments in postural control activity that are closely coupled to the nature of the inducing stimulus. The majority of research described in this chapter employs metrics tied to these two classes of response, with particular emphasis placed on the former.

As aforementioned, when observers are first exposed to a vection-inducing display or apparatus, such as a rotating cylindrical drum in a circular vection setting, they typically first perceive the drum itself as being in motion. It is not unusual for a similar first impression to occur within a VE setting—that is, users perceive themselves as immersed in a computer-generated environment within which motion is being depicted, but the motion is not experienced as being their own. Shortly thereafter, however, the perception in most laboratory vection settings changes dramatically from one of object or surround motion to one of self-motion (assuming that conditions for inducing the experience of the illusion are appropriately configured)—the direction of self-motion being opposite in direction to that of the surround motion. Coincident with the onset of self-motion, surround motion may appear to effectively cease.

In the case of circular vection, the latency of this effect is on the order of 5–30 s (e.g., Dichgans & Brandt, 1973, 1978). This range is representative of all forms of vection, and specific values are influenced by a host of factors described in the following section. For example, onset of the vection illusion tends to occur sooner at lower optical accelerations (Howard, 1986a; Melcher & Henn, 1981). In addition, its latency tends to decrease as a direct function of increases in the size of the visual field of view (Dichgans & Brandt, 1978). In some cases, the experience of vection may remain fairly constant once it has begun—that is, it will remain present until the motion stimulus is discontinued and may persist for a while afterward. However, in other cases, the sensation of vection may come and go and may also increase and decrease in intensity.

In measuring the phenomenological component of vection, two key questions that are typically addressed are as follows: (1) Is vection present (i.e., is the subject in a vection experiment or the user of a VE system currently experiencing the vection illusion?), and (2) if present, how powerful is the phenomenological experience of the illusion? A number of approaches have been used over the years in experimental studies of vection, and they may be useful to VE system designers as they attempt to either enhance or diminish the presence and/or strength of this illusion.

As is often the case with phenomena that are largely subjective in nature (e.g., presence), the measurement of the phenomenological component of vection is of necessity largely an indirect process. A number of approaches derived from psychophysical measurement techniques have been used, including magnitude estimation of surround and/or self-motion, magnitude estimation of degree of perceived self-tilt, and subjective scaling of the strength of the illusion (cf. Diener, Wist, Dichgans, & Brandt, 1976; Kennedy, Hettinger, Harm, Ordy, & Dunlap, 1996; Ohmi & Howard, 1988). These measures are widely accepted in the vection research community and have been shown to have a high degree of psychometric reliability (Kennedy et al., 1996). However, new methods for objectively measuring vection are being developed. Physiological measurements are unobtrusive and provide researchers with individual differences on vection research. These include body posture (sway and postural shifts), muscle tension, and eye tracking (Bringoux et al., 2012; Fushiki et al., 2005; Ishida, Fushiki, Nishida, & Watanabe, 2008; Prothero, Parker, Furness, & Wells, 1995). Use of neurological measures has also been promising as well (Ishida et al., 2008; Roll et al., 2012; Väljamäe, 2009).

Magnitude estimation techniques often require subjects to make estimates of the velocity of apparent self-motion in the case of continuous motion (e.g., in degrees/second), displacement (e.g., in degrees) in the case of discrete motion, or in estimating the illusory tilt so commonly perceived in roll and pitch vection. Subjects have occasionally been asked to provide magnitude estimates of their own apparent motion in addition to the perceived motion of the stimulus (e.g., Wong & Frost, 1978). In the latter case, the strength of the vection illusion can be inferred by greater magnitude estimates of self-motion relative to stimulus motion.

One of the more common methods of assessing the strength of the vection illusion involves the use of comparatively simple subjective scaling techniques. Typically, these scales range from zero, corresponding to a rating of no illusory self-motion, to some upper limit meant to reflect a sense of *saturated* vection (cf. Hettinger et al., 1990; Howard et al., 1987).

As noted earlier, postural readjustment responses are also commonly observed in vection settings, and their characteristics can be used to assess the degree of vection experienced by an observer in a given situation. These types of metrics are derived from quantitative measures of the observer's reflex-like reactions to perceived motion in which body sway and/or other forms of postural control activity are the most common response (e.g., Bles & DeWit, 1976; Dichgans & Brandt, 1978; Stoffregen, Hettinger, Haas, Roe, & Smart, 2000). Under conditions of sufficient stimulus magnitude and/or duration, it is not unusual to observe subjects losing functional postural control and eventually falling or collapsing (Lee & Aronson, 1974; Lestienne et al., 1977). As Kennedy and his colleagues (e.g., Kennedy, Drexler, Compton, Stanney, and Harm, in press) have noted, it is also not unusual for postural control disturbances to persist well beyond the cessation of the self-motion experience.

The precise, objective measurement of postural control responses during exposure to illusory self-motion stimulation is a more complex and equipment-intensive procedure than is the use of psychophysical magnitude estimation or scaling techniques; however, as technology continues to improve, video camera capture, gyroscopic measurements, and 3D rendering are all becoming more feasible and require little procedural setup. Equipment that does require moving platforms or chairs is still time and labor intensive, but these techniques will improve with time as well (see Wann & Mon-Williams, 2014, Chapter 32). Typically, devices such as force platforms, body-mounted accelerometers, and calibrated photographic/video methods have been used to assess postural sway, and the data are then reduced to quantitative profiles of postural control activity. While this level of precision may require excessive instrumentation for most VE applications outside the laboratory, the simple observation of users' postural control behaviors in the presence of self-motion stimuli may be a useful converging measure for assessing the presence of vection if used in conjunction with the psychophysical techniques described earlier and may also provide information regarding early onset of the illusion. For example, Stoffregen (1985) found compensatory sway in a moving room to have a lower threshold than reports of motion.

One of the least understood aspects of vection involves the nature of differences between individuals in terms of their phenomenological and postural responses. In perhaps the only study of this aspect of the illusion, Kennedy and his colleagues (Kennedy et al., 1996) examined the psychometric reliability (within individuals) and variability (between individuals) of various measures of circular vection, such as latency and magnitude estimation of perceived self-motion. Results indicated that while the classes of measures used exhibited strong intraobserver reliability (i.e., observers were quite consistent in their responses from one session to the next), there were strong and consistent differences between individuals. This finding suggests that key aspects of the vection illusion may not be as uniformly experienced across individuals as previously believed and/or that observers may respond to aspects of vection-inducing situations differently. In addition, the authors hypothesize that those observers who exhibit greater sensitivity to vection may be among those most likely to experience occasional side effects and/or problems in readapting to normal perceptual–motor relationships once out of the vection environment. However, these hypotheses have yet to be tested.

In a similar vein, von der Heyde and Riecke (2002) have introduced their framework on reference frames and users' expectations, imagination, reality checks, spatial learning, and cognition as they relate to spatial orientation. Not only are physical parameters of vection-inducing stimuli able to affect the illusion of self-motion, but the way in which the stimuli are perceived and interpreted could modulate the phenomenon, hence leading to individual differences in vection perception (Riecke, 2011; Riecke et al., 2012).

18.2.3 OPTICAL DETERMINANTS OF VECTION

In order to attempt to control (or perhaps, in some cases, prevent) the onset of vection illusions in VE devices, it is important to understand the nature of the optical information that influences its onset and strength. Most of these factors have been alluded to earlier but are summarized in more detail in this section. It is interesting to note that many aspects of features that appear to enhance the vection illusion are the very factors that are widely sought in current VE systems and are generally thought to support the experience of presence. In other words, the visual display features that enhance vection, such as the size of the field of view, spatial and temporal frequency of the optical pattern (i.e., *scene detail*), and others described below, are widely emphasized in the design of contemporary VE systems. Therefore, vection in VEs should become increasingly common as systems continue to more closely attain the engineering and display parameters that promote its occurrence.

Vection is strongly influenced by a number of visual display factors (see Table 18.1). The influence of most of these is summarized in Dichgans and Brandt (1978), Howard (1986a), and Riecke (2011).

TABLE 18.1
Visual Display Factors Affecting Vection

1. Size of the visual field of view
2. Optical velocity (or temporal frequency) of the visual stimulus
3. Spatial frequency of the stimulus pattern
4. Presence of background and foreground information

The vast majority of the vection literature is characterized by studies that focus on one form of vection at a time, and little work has been done to attempt to generalize findings across the various forms of the illusion (Seno et al., 2009). Recognizing that the critical elements underlying specific forms of vection may vary according to the form of the illusion under examination, it is nevertheless possible to make a number of general statements concerning visual display factors.

Despite several findings indicating that it is possible to induce sensations of vection with relatively small (less than 10° of visual angle), centrally presented motion patterns (e.g., Andersen & Braunstein, 1985), there appears to be little doubt that peripheral visual stimulation is far more effective in inducing the illusion (and corresponding postural adjustments) than central visual stimulation. Simply put, wide fields of view are more effective for inducing vection than are comparatively narrow fields of view. This is almost certainly due to the greater connectivity between the vestibular system and peripheral retina in comparison to the central retina (see later discussion of neurophysiological correlates of vection).

Many studies have examined field-of-view effects on vection. The findings of Lestienne et al. (1977) and Brandt et al. (1973) are illustrative of this work in general. In their examination of linear vection, Lestienne and his colleagues observed that the amplitude of the postural sway response observed in their subjects depended primarily on the size of the motion pattern. Specifically, increases in postural sway amplitude were linearly related to field-of-view size—the larger the field of view, the more pronounced the postural sway response to the linear vection stimulus. This is consistent with the findings of Brandt et al. that the intensity (i.e., strength of the illusion) and velocity of illusory self-rotation in circular vection are strongly dependent on the size of the available field of view, with intensity being more adversely affected than apparent velocity at smaller field-of-view sizes. A centrally located field of view of 30° (the lowest value tested) produced very low estimates of intensity compared to full-field stimulation. In a compelling illustration of the impact of the visual periphery on vection, Brandt et al. masked the central 120° of the visual field so that subjects could only view the rotating motion stimulus in the far periphery. The intensity of the vection illusion was scarcely affected by this manipulation.

The relevance of these findings for VE system design is very clear. Larger fields of view will generally be more effective for eliciting vection illusions than will smaller fields of view. In some cases, this may be advantageous (e.g., in helping to make a VE-based entertainment ride more exhilarating) but in other cases may result in excessive problems with motion sickness and/or postural disturbances. The addition of accurate representation of motion-in-depth information (e.g., radially expanding optical flow patterns characteristic of forward, linear self-motion), which Andersen and Braunstein (1985) assert to be the critical factor in supporting illusions of self-motion with small, centralized displays, should serve to enhance the field-of-view effects.

Optical velocity is another key visual display factor, influencing both the intensity and velocity of illusory self-motion. Howard (1986a) reports the general finding that the apparent velocity of illusory self-rotation in circular vection is directly proportional to optical velocity up to values of approximately 90°/s, although this relationship is influenced by the spatial frequency (optical texture density) of the stimulus pattern. Similarly, Brandt et al. (1973) and Dichgans and Brandt (1973) observed that perceived velocity of illusory self-motion in circular vection is linearly related to stimulus velocity up to about 90°s–120°s, beyond which perceived illusory self-motion velocity lags behind stimulus velocity.

The spatial frequency of the visual stimulus, or what can generally be considered to be equivalent to the amount of *detail* presented in the display, also affects the perception of illusory self-motion. Diener et al. (1976) observed that subjects' perceptions of the velocity of illusory self-rotation more closely matched the true velocity of the stimulus with increases in the spatial frequency of the visual display. These findings correspond to those obtained by Owen and his colleagues (e.g., Owen, Wolpert, & Warren, 1983; Warren, Owen, & Hettinger, 1982) who observed that *edge rate* information was more salient than global optical flow in affecting observers' judgments of self-motion.

One of the earliest and most common findings in the vection literature is that the perception of illusory self-motion is greatly enhanced when observers fixate a stationary target located in front of the visual motion pattern (e.g., Mach, 1875). By implication, this finding suggests that the inclusion of foreground visual information in VEs that requires sustained attention by the user will result in more pronounced sensations of vection, if a wide field-of-view pattern of motion is made available. A frame that remains stationary will more than likely result in an individual choosing that frame as their reference, with which to compare other frames to and help delineate foreground versus background perceptions (Prothero & Parker, 2003; Riecke, 2011).

When there are multiple displays in view, each depicting patterns of motion and each at different distances from the observer, the most distant display has been observed to control the direction of illusory self-motion (Brandt, Wist, & Dichgans, 1975). Additionally, vection is suppressed by stationary objects seen beyond the moving display but not by stationary objects in the foreground (Howard & Howard, 1994; see also Seno et al. (2009), for a review of the Object-Background Hypothesis).

18.2.4 Neurophysiological Correlates of Vection

Significant research attention has been devoted to the identification and description of the neurophysiological correlates of vection—that is, the nature of the activity that occurs in the central nervous system (CNS) in conjunction with exposure to vection-inducing displays. This work strongly suggests that the neurophysiological basis of the illusion lies in the activity of the vestibular nuclei (Daunton & Thomsen, 1979; Dichgans & Brandt, 1978; Waespe & Henn, 1977a), a CNS site whose activity is also strongly tied to actual self-motion. Such a convergence makes sense not only because vection and actual self-motion are very similar in a phenomenological sense but also because they share many observable behavioral similarities (e.g., postural sway, motion sickness) and physiological similarities, such as the occurrence of optokinetic afternystagmus (the tendency of the eyes to exhibit a pattern of nystagmic motion following exposure to real or illusory self-motion), a phenomenon known to be directly affected by the activity of the vestibular system (e.g., Waespe & Henn, 1977b, 1978).

Interactions between the visual and vestibular system are central to the vection illusion. Research in visual–vestibular interactions has its own long history (see Henn, Cohen, & Young, 1980; Howard, 1982, 1986b, 1993, for reviews of this literature), and much of the work either overlaps with research on vection or has direct relevance for it. Indeed, the desire to understand and explain the neurophysiology of the visual–vestibular system has always been a prime motivator of vection research (e.g., Dichgans & Brandt, 1978; Straube & Brandt, 1987).

One method used to examine the nature of visual–vestibular interactions in vection involved attempts to isolate components of this ordinarily highly interactive system. For instance, Howard et al. (1987) demonstrated that the paradoxical combination of continuous illusory self-rotation with limited, constant tilt in upright observers is a function of activity from the vestibular otoliths. Specifically, when observers lying on their back (effectively nulling inputs from the otoliths) view a roll vection stimulus, the experience is simply one of continuous self-rotation. This clearly suggests a strong role of the vestibular system in vection illusions.

A different approach for isolating visual and vestibular inputs from one another involves the examination of vection and other visual–vestibular phenomena in microgravity, where vestibular inputs are greatly reduced. For example, Young et al. (1986) studied roll vection in four crew members under microgravity conditions during a space shuttle mission, as well as during the microgravity phases of parabolic flight. Their findings indicated that perception of roll vection was enhanced in the early phases of weightlessness (i.e., early in the mission), possibly due to sharply reduced functionality of nonvisual cues for gravity and orientation. Results obtained during later phases of the mission were somewhat more variable, perhaps due to effective incorporation of nonvestibular cues for orientation.

As noted by Howard (1986a), it was originally thought that vection could be ascribed to the effects of optokinetic nystagmus (OKN)—the eye's reflexive tendency to alternative slow pursuit motions with rapid, saccadic return movements when viewing a scene or object (such as a train) moving in one direction (see Badcock et al., 2014, Chapter 3). However, Brandt et al. (1973) demonstrated that the phenomenological, vection-inducing effects of a large moving field are unaltered even upon adding a small central display of motion in the opposite direction. Whereas the latter changes the direction of OKN as long as the subject fixates on it, it has no effect on the vection experience. On the other hand, the duration of positive and negative aftereffects of illusory self-motion appears to be directly related to the time course of optokinetic afternystagmus, the tendency of the eye to show nystagmic patterns of motion following the cessation of a motion stimulus. (Note: *Positive aftereffects* are illusory motion sensations experienced as being in the same direction as the immediately preceding, stimulus-driven illusion. *Negative aftereffects* are illusory sensations experienced as being in the opposite direction.)

18.2.5 MOTION SICKNESS AND ADAPTATION PHENOMENA

One of the most widely studied areas in vection research, and one with tremendous relevance to VE design and usage issues, concerns the relation between vection, motion sickness, postural disturbances, and other negative perceptual–motor side effects. Circular vection has been shown to induce symptoms of motion sickness in approximately 60% of healthy human subjects (Stern et al., 1985; Stern, Koch, Stewart, & Lindblad, 1987), and, as previously noted, Lestienne et al. (1977) observed that 3 of the 30 subjects in their linear vection study were disoriented by the experience to the extent that they lost consciousness. Anyone who has been around flight simulators or vehicular simulators for very long is familiar with *simulator sickness* (Kennedy, Hettinger, & Lilienthal, 1990; see Lawson, 2014a, Chapter 23; Keshavarz et al., 2014, Chapter 26), a phenomenon that appears to be spreading to the larger domain of VE systems in general.

Motion sickness resulting from vection has been discussed as a type of *visually induced motion sickness* (Hettinger & Riccio, 1992; McCauley & Sharkey, 1992). The underlying cause of sickness in these circumstances has traditionally been attributed to the existence of conflict between sources of sensory information specifying self-motion and orientation. The *sensory conflict theory* (Reason & Brand, 1975) is the most widely accepted theoretical explanation of motion sickness in real and virtual environments. However, over the years, many concerns have been raised with the theory, including why only certain types of conflicts appear to be nauseogenic, why the same conflict might not reliably produce sickness across different individuals (or, indeed, even within the same individual across time), and how to attempt to quantify the amount of conflict present in a given situation and relate it to the frequency and severity of motion sickness. Stoffregen, Riccio, and their colleagues have proposed an alternative explanation, arguing that motion sickness, whether it occurs in virtual or real environments, results from prolonged disruptions in normal postural control activities (Riccio & Stoffregen, 1991; Stoffregen & Riccio, 1991). Results do suggest that this is a promising theoretical approach (Smart, Pagulayan, & Stoffregen, 1998; Stoffregen et al., 2000; Stoffregen & Smart, 1998). However, in recent years, other researchers have developed methodological explanations for motion sickness and possible attenuation methods. Prothero and Parker (2003) suggest that motion sickness occurs when there are too many rest (reference) frames to choose from and the individual becomes confused and conflicted. A possible way to combat motion sickness in this case would be to introduce one rest frame, an independent visual background (IVB), salient enough to attenuate focus from other competing and conflicting frames.

Another possible explanation for motion sickness is the inability to adopt an appropriate reference frame as it related to perceptions about the spatiotemporal surrounds and specific spatial behaviors (Prothero & Parker, 2003; von der Heyde & Riecke, 2002). Given this explanation, training programs that teach individuals how to reliably switch to a more appropriate reference frame will alleviate motion sickness effects.

As suggested earlier, cognition plays an important role in the illusion of self-motion. Similarly, cognition plays a role in the experience of motion sickness as well. If the spatiotemporal expectations of an environment (real or virtual) do not match the sensory cues or perceived experiences, then motion sickness may occur (von der Heyde & Riecke, 2002). Cognitive conflicts also occur when space–time is perceived as being inconsistent and discontinuous, which occurs more often in VE systems. Cognitive resolutions can occur when a VE system allows an individual to update their spatial surrounds in a reliable manner.

However, the notion of sensory conflict still seems to be of value in explaining many of the situations that lead to prolonged postural disruptions and/or motion sickness. It may be that all views are to some extent correct and that motion sickness, including that induced by vection stimuli, ultimately results from unusually provocative disruptions in the normal cycle of perceiving and acting.

One of the most vivid similarities between motion sickness observed in real and illusory motion conditions involves the presence of *Coriolis effects* in the former and *pseudo–Coriolis effects* in the latter. Coriolis effects are powerful and well-known stimuli for motion sickness and occur as a result of the cross-coupled angular accelerations that are produced within the vestibular system when the head is tilted out of the body's axis of rotation (cf. Benson, 1990). This is often done by placing subjects in a rotating *Barany chair*—a device that rotates the body about its z-axis like a barber chair might and then asking the participant to tilt the head slightly out of the axis of that rotation. Intense nausea is quickly produced (cf. Melville-Jones, 1970) and vomiting will nearly always follow shortly afterward if the motion and/or cross-coupled stimulation is not ceased. Dichgans and Brandt (1973) demonstrated the existence of a pseudo–Coriolis effect through what one might today refer to as a virtual simulation of the Coriolis situation. By exposing subjects to a circular vection stimulus (a rotating cylindrical drum) and then asking them to tilt the head outside the axis of apparent rotation, they were able to produce the same sort of motion sickness symptomatology observed in actual Coriolis situations.

Stern, Koch, and their colleagues have pioneered the study of vection-induced motion sickness. In much of their research (e.g., Stern, Hu, Vasey, & Koch 1989; Stern et al., 1985), they have studied sickness by recording subjects' gastric myoelectric activity using an electrogastrogram (EGG). The EGG provides a very sensitive and reliable index of gastric motility associated with subjective symptoms of nausea and is often reflected by a shift from the normal gastric rhythm of 3 cpm to 4 to 9 cpm, referred to as *tachygastria* (Stern et al., 1985; see also Chapter 32, this book). The work of Stern, Koch, and their colleagues has demonstrated that motion sickness is a common side effect of the experience of vection, which appears to diminish in severity as a result of adaptation to repeated exposures (Hu, Grant, Stern, & Koch, 1991), concentrated focus on deep breathing (Jokerst, Gatto, Fazio, Stern, & Koch, 1999), and treatment with acupressure techniques (Hu, Stritzel, Chandler, & Stern, 1995).

There is nearly unanimous agreement among VE system designers, and especially users, that motion sickness is a distinctly unpleasant experience, the occurrence of which should be minimized to the greatest degree possible. Of note: An interesting exception is the NASA Preflight Adaptation Trainer, a VE-based system whose intent was to preadapt astronauts to conditions considered to be highly evocative of space motion sickness. This unique system's objective was, in fact, to induce feelings of queasiness and disorientation in its users, thereby inoculating them to such conditions in the microgravity environment of space (cf. Harm, Zogofros, Skinner, & Parker, 1993; Parker, 1991). However, the most urgent risk to users might not involve the feelings of discomfort that occasionally accompany use of the system. Rather the greatest risk might involve the disrupted perceptual–motor control that is often observed after use of the system (see DiZio & Lackner, 2014, Chapter 33). A point made repeatedly by Kennedy and his colleagues in their discussions of sickness in flight simulators and VE systems is that negative side effects experienced after exposure to a VE are of greater safety concern than those that occur during exposure (e.g., Kennedy & Fowlkes, 1992; Kennedy et al., 1990).

Aftereffects of exposure to vection-inducing events have been frequently noted in the experimental literature. For example, in his review of vection phenomena, Howard (1986a) notes that aftereffects can be readily demonstrated to occur in circular vection settings: "If the lights are put out after a moving scene has been observed for some time, the illusory self rotation continues as *a positive aftereffect* followed by a *negative aftereffect* in the opposite direction" (p. 28). Some form of adaptation to vection-inducing events may underlie these aftereffect phenomena, suggesting that on leaving a vection-inducing situation such as a VE system, flight simulator, or laboratory device, observers may need to undergo a process of readaptation and recalibration, during which time their normally effective perceptual–motor control might be disrupted.

18.3 FRAMEWORKS FOR VECTION

The section above introduced characteristics of vection that have been evaluated for decades. In fact, vection has been studied since the late nineteenth century. Since that time, researchers have been trying to explain the experience of illusory self-motion and its behavioral, psychological, and physiological influences on the observer. Although vection has been described in the literature for over 100 years, the past 20 years have led to the development of a variety of frameworks to explain vection in terms of best practices, both for inducing and suppressing vection. A brief review of the five most referenced frameworks is outlined in the following.

18.3.1 PERCEPTUAL DIFFERENCE MODEL/DEPTH-ORDER EFFECT

The long-standing view on the characteristics that induce vection has been described by the depth-order effect, or the perceptual differences of objects in both the foreground and background. Since the mid-1970s, this perceptual difference viewpoint has dominated the literature (Brandt et al., 1975; Howard & Heckmann, 1989; Ito & Shibata, 2005; Nakamura, 2008; Ohmi & Howard, 1988; Ohmi, Howard, & Landolt, 1987; Telford, Spratley, & Frost, 1992). Vection is dominated by motion characteristics of the more distant display, or motion in the perceived background. Recently, studies using 3D depth effects were able to show background-motion-determined vection direction over 2D effects (Ito & Shibata, 2005; Nakamura, 2008). Similarly, when depth-order information was presented visually, vection direction was modulated by the farther of the displays and attentional effects were attenuated (Kitazaki & Sato, 2003). Brandt and colleagues concluded that the retinal periphery and depth of visual stimuli are main modulators of vection perception and intensity, and this view has persisted into present studies of vection.

18.3.2 REST FRAME HYPOTHESIS

Extensively studied by Prothero and colleagues, the rest frame hypothesis (RFH) posits that individuals select a particular reference frame within their environment and surroundings to which they will compare other motion and spatiotemporal frames (Prothero, 1998; Prothero & Parker, 2003). The idea that certain objects or scenes remain stationary is the typical viewpoint of humans' sensory systems. Instead of perceiving the room through which we walk to be moving, rather we tend to perceive that our body is moving through the stationary room, but according to the RFH, either the stationary self or stationary room is possible. The frame that we perceive to remain stationary is the *rest frame*. The comparator frame is typically the ground or background of a visual scene. Other frames compared to the rest frame include the following:

- Egocentric frames—frames that are defined by the person. An egocentric frame can be further divided into oculocentric (eye-based), head-centric (head-based), and body-centric (body-based) frames.
- Allocentric frames—frames that are defined by environmental features.
- Geocentric frames–frames that take into account the forces of gravity.

Only recently, illusory self-motion studies have included the use of VEs to induce and manipulate vection. Presence in a VE, the vivid feeling of being in, or ability to interact with the VE, can be enhanced when the user perceives their rest frame to be a part of the virtual world instead of the real world (Prothero & Parker, 2003; Schubert, Regenbracht, & Friedmann, 2000; Steuer, 1992). The rest frame must remain congruent with inertial and visual cues. If this congruence is misaligned, the user will experience motion sickness.

Motion sickness occurs when there are too many rest frames to choose from and the individual is forced to switch between multiple frames. To reduce motion sickness caused by rest frame inconsistency, an IVB should be utilized as a stable rest frame outside of the VE, which should reduce the number of conflicting rest frames within the visual scene. Prothero and Parker (2003) have shown that the IVB did not affect the vection illusion but did reduce self-reported simulator sickness.

18.3.3 REFERENCE FRAME MODEL

The reference frame model relates presence, spatial orientation, and spatial updating into a framework that can be used as a base for future research (von der Heyde & Riecke, 2002). According to this straightforward framework, spatial presence is described as "intuitively and spontaneously knowing where one is with respect to the immediate surround" (von der Heyde & Riecke, p. 1), as it relates to either the real world or a VE.

Humans perceive the world as being both consistent and continuous in terms of spatiotemporal properties. In other words, we cannot be in two places at once (inconsistency) nor can we teleport (discontinuity) to other locales (von der Heyde & Riecke, 2002). As such, our mental model, or our egocentric reference frame, is seen from the first-person perspective in a consistent and continuous way. Additionally, because we have multiple sensory modalities, we can perceive multiple egocentric reference frames whether they be visual, haptic, kinetic, auditory, or olfactory (although vection induction has not been previously studied using olfactory stimuli).

As we move through the real world or a VE, we perceive transformations to our egocentric reference frames. These transformations are based on orienting information we receive from our spatial surroundings. This process of orienting ourselves in our surroundings is known as continuous spatial updating and/or instantaneous spatial updating. The reference frame model posits that both continuous and instantaneous spatial updating are necessary for spatial behavior, which is how we behave in the environment in which we perceive ourselves to be. Similar to the real world, intuitive VE systems should allow users to accurately spatially update, both continuously (gathering information in a linearly temporal fashion) and instantaneously (immediately gathering data when you become aware of it in order to realign your mental reference frame).

According to von der Heyde and Riecke (2002), motion sickness and, in a similar vein, simulator sickness can be described when VEs do not allow us to reliably update our spatial reference frames. Additionally, motion sickness can also occur when our expectations about our surroundings compared with what we actually perceive do not line up. The comparison between expectations and what is actually perceived is known as the reality check. When the reality check is in line with our expectations, we believe the system to be fine. When the reality check is misaligned, our attention is required to rectify the situation or cognitively resolve the conflict that could result in motion sickness.

In terms of vection induction, VEs must allow for users to experience presence. Users must be able to update spatial information in a meaningful and intuitive way, they must be able to react with appropriate spatial behavior, and vection perceptions must not conflict with users' expectations.

18.3.4 ATTENTIONAL MODULATION FRAMEWORK

According to Kitazaki and Sato (2003), attentional effects modulate the illusion of self-motion and vection intensity. They posit that nonattended visual stimuli will dominate the vection illusion.

Built upon the perceptual depth-order effects, attended stimuli are most often the objects of a scene, while the nonattended stimuli are considered the background. These nonattended background stimuli determine vection direction and intensity, depending on stimuli properties.

After using red and green dot fields in their study on attentional modulation, Kitazaki and Sato (2003) found that the more salient stimuli, those with characteristics that attract attention, could be perceived as objects or *foreground* and those that do not attract attention may be perceived as background. Once participants in the study were told to attend to just one color, most participants voluntarily attended to the more salient of the two colors presented. As vection direction matched that of the red dot field (i.e., illusion of forward self-motion) over a significant period of time and across trials, it can be concluded that vection direction was modulated by the unattended stimuli, the green dot field, which moved in a downward motion.

In the same study, Kitazaki and Sato (2003) separated the dot fields through stereoscopic 3D vision, which introduced physical depth to the stimuli. When physical depth cues were available, attentional modulation of vection was weakened. They argue that even though actual physical depth is a stronger modulator of vection, this is only because attended stimuli are in the physical foreground and unattended stimuli are in the physical background. One potential weakness of this research on attentional modulation is the argument for figure–ground versus perceived depth order as they covary in most situations regarding attentional effects.

18.3.5 OBJECT–BACKGROUND HYPOTHESIS

In 2009, Seno, Ito, and Sunaga proposed a new hypothesis to explain the vection illusion. The object–background hypothesis provides a unified approach for the study of vection and vection stimuli, taking into account other frameworks and models. The unique experimental design used in the investigation of object versus background helped parse out stimuli characteristics that have confounded researchers in the past. As an example, in the previous section, the point was made that the authors could not definitively explain their results based on the covarying nature of depth-order effects and figure–ground effects on the attentional modulation of vection. Seno et al. (2009) employed Rubin's vase (1915, cited in Seno et al., 2009), the optical illusion with reversible foreground–background stimuli. They included motion-defined properties into Rubin's vase to study the vection illusion (see Figure 18.2).

They concluded object properties weaken vection, while background properties enhance vection. The use of the optical illusion helped isolate component properties of object versus background,

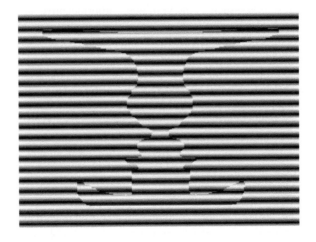

FIGURE 18.2 Motion-defined Rubin's vase. (Adapted from Seno, T. et al., *Vision Res.*, 49(2009), 2973, 2009.)

motion direction (left to right or up and down), and depth order (stereoscopic vision was employed in one study). When the 3D stimuli field was incorporated with motion on the farthest plane, if stimuli field contained object properties, vection was weak or the illusion did not occur at all. Furthermore, when the separation of object–background was salient, there was a more effective determinant of vection than each of the individual component properties (i.e., attention, depth— whether physical or perceived).

Seno et al. (2009) also argue that attention affects the possibility of *object likelihood*. If one is to attend to certain stimuli, most often the attended stimuli contain object-like properties compared with those unattended stimuli that would take on characteristics of the background.

18.3.6 Summary

There has been a marked increase in the study of vection and illusory self-motion since the early 2000s. The addition of five new frameworks of vection clearly denotes interest in the phenomenon. New guidelines on how to induce or suppress vection have been developed and can be used as *best practices* for eliciting self-motion in VEs in a safe, effective manner. There does not seem to be a one-size-fits-all approach to vection, and the goals of the human–VE system will determine which framework will be adopted. It is important to note, the common thread to all frameworks presented in this section is our ability to perceive self-motion even if stationary. Cognition, as it relates to these frameworks, is briefly touched upon in each of the studies outlined. Cognition can introduce bias into the experimental processes based on researcher instruction aid in the perception of self-motion even in the absence of motion stimuli or influence behavior in light of vivid illusory self-motion irrespective of perceptual stimuli (Bringoux et al., 2012; Seno et al., 2009; Seno et al., 2012).

New experimental findings have helped shape our understanding of the illusory self-motion experience, but with exponentially better VE systems and new frameworks for the study of vection, more questions have arisen than have been answered. A promising future for vection research has been mapped, and VE systems will provide the tools to answer these new questions. Additionally, the findings pertaining to vection properties and vection-inducing stimuli will in turn create robust VE systems for research into human systems.

18.4   MULTISENSORY VECTION ILLUSIONS

So far, we have outlined the most prominent factors contributing to visually induced vection. However, other nonvisual senses—such as the vestibular, proprioceptive, auditory, haptic, and tactile systems—are also involved in maintaining and stabilizing the human's posture during locomotion (e.g., Sakellari & Soames, 1996). Klemen and Chambers (2012) provided a comprehensive overview of essential neural processes involved in multisensory integration, including the colliculus superiores, thalamus, and other cortical and subcortical areas. Although the importance of cross-modal integration is doubtlessly proven for real motion, vision is by far the most frequently discussed modality in vection research. The neglect of other perceptual senses is quite surprising for two reasons. First, vection induced by nonvisual senses was first documented more than a century ago. For instance, auditory vection was reported in the late nineteenth and early twentieth century (Dodge, 1923; Urbantschitsch, 1897). Despite this early discovery, auditory vection has not been paid much attention during the last decades. Only recently have researchers started to reestablish the role of acoustics in the genesis of vection. Second, nonvisual aspects of vection are becoming more and more important for the design of VEs, simulators, and entertainment systems (Brooks, 1999; Mourant & Refsland, 2003). The focus of interest has shifted toward multisensory stimulation (e.g., surround sound, gamepad and feedback), in the quest for a highly realistic and exciting user experience. Thus, the following section will briefly describe and discuss the role of auditory and tactile cues during multisensory vection illusions.

18.4.1 Auditory Vection

Auditory cues contribute to vection in two ways. On the one hand, acoustic stimuli can produce vection in the absence of additional sensory information (i.e., without additional visual or tactile information). On the other hand, sound can have a cumulative effect regarding vection in addition to other simultaneously presented sensory cues. Both aspects will be described next, starting with a description of pure auditory vection, followed by a discussion of the role of acoustics in multisensory vection illusions. A comprehensive review of auditory vection can be found by Väljamäe (2009).

18.4.1.1 Pure Auditory Vection

As previously mentioned, the sensation of pure auditory vection is a rather ancient phenomenon and describes the illusion of self-motion solely based on auditory stimulation. Systematic auditory vection research has been left largely unattended (except for Lackner, 1977) and has only recently become a topic of interest (Kapralos, Zikovitz, Jenkin, & Harris, 2004; Larsson et al., 2004; Riecke, Schulte-Pelkum, Caniard, & Bülthoff, 2005; Sakamoto, Suzuki, & Gyoba, 2008. Typically, two conventional methods exist to simulate auditory vection, including an array of speakers that are positioned around (or along) a stationary observer or stereo headphones that present prerecorded sounds. The latter technique offers the advantage of minimizing other disturbing noises, such as acoustic reflections of environmental objects.

Compared to visually induced vection, the illusion of auditory self-motion is less frequent and less strong. Riecke, Feuereissen, and Rieser (2009) estimate that 20%–60% of all participants experience auditory vection, which is in accordance with other findings (Lackner, 1977; Väljamäe, 2009). To create a compelling sensation of auditory vection, conflicting information delivered from other modalities, especially from the visual system, needs to be excluded or reduced to a minimum. Thus, blindfolding participants is a prerequisite to create and maintain auditory vection, as visual cues that are not in concordance with the acoustic stimulus would quickly terminate the sensation of auditory vection. Providing the possibility of real motion, for instance, by applying a rotatable chair or a moving platform, enhances the chance to elicit auditory vection (Dodge, 1923; Wright, Dizio, & Lackner, 2006). Riecke et al. (2009) analyzed the role of participants' awareness of *movability*. In this study, two groups of participants were seated in a Hammock chair, one group with their feet rested on the ground (movement impossible) and one group with their feet positioned on a footrest, which was connected to the chair (movement possible). Results showed that placing the feet on the ground significantly decreased the strength and duration of auditory vection.

Just like visually induced self-motion, auditory vection can be divided into (1) circular and (2) linear vection. Circular auditory vection describes illusory *rotation* along one of the three body axes (i.e., roll, pitch, and yaw axis). Several factors have been proven to facilitate the occurrence of circular auditory vection, such as the velocity of a rotating auditory source. Among others, Larsson et al. (2004) exposed their participants to rotational velocities of 20°/s, 40°/s, or 60°/s and found significantly enhanced auditory vection when the stimulus rotated with faster speed (see also Lackner, 1977). Furthermore, the nature of the sound source plays a crucial role in auditory vection. Early research typically used rather artificial sounds, such as pink or white noise, whereas recent studies focused on more realistic and ecologically valid sounds. Larsson et al. chose three categories of sound sources, including artificial sounds (e.g., pink noise), movable objects (e.g., birds), or fixed objects (e.g., fountain). Fixed objects (so-called landmarks) turned out to significantly increase the level of auditory vection. Väljamäe, Larsson, Vastfjall, and Kleiner (2009) found similar results and reported stronger vection ratings when acoustic stimuli were generated from a nonmovable object. In addition, increasing the number of sound sources (from one to three) also enhanced the level of vection significantly (Larsson et al.).

Linear vection, in contrast, contains *translational* motion along one of the three body axes. Research regarding linear auditory vection is rather sparse (see Väljamäe, 2009); only a handful of studies have assessed components of such vection. For instance, Väljamäe, Larsson, Västfjäll, and Kleiner (2008) analyzed the role of contextual variables associated with linear auditory vection. Participants were seated in a stationary mock car and were exposed to different sounds that passed the car, thereby creating the sensation of linear auditory vection. They found that if the car's engine sound was added to the virtual scene, the subjective sensation of auditory vection was significantly increased. In another study, Väljamäe, Larsson, Vastfjall, and Kleiner (2005) noted that forward linear auditory vection was more easily achieved than backward vection.

18.4.1.2 Auditory Cues in Multimodal Vection Illusions

Audiovisual cross-modal integration (e.g., the ventriloquism effect) is a well-known phenomenon in the current literature (see Storms, 2002). However, the role of auditory cues in addition to other modalities has only been vaguely discussed with respect to vection. Nichols, Haldane, and Wilson (2000) reported that corresponding background sound, presented along with a visual stimulus, enhanced the feeling of vection. Similar results were reported by Seno, Hasuo, Ito, and Nakajima (2012) as well as Riecke et al. (2005). In the latter study, the authors used either mono or spatialized sound in addition to a visual stimulus that elicited visually induced vection; spatialized sound turned out to enhance vection strength and to decrease its onset time, whereas mono sound had no effect. Although the impact of sound on visually induced vection was rather weak, the results of Riecke et al.'s study emphasize the importance of multisensory integration in the genesis of vection (see also Marme-Karelse & Bles, 1977). In contrast, Keshavarz and Hecht (2012) found that adding corresponding background sounds to a visual stimulus (i.e., recorded bicycle ride) did not significantly enhance the level of perceived vection.

18.4.2 HAPTIC AND TACTILE CUES

Nonvisual and nonauditory cues, such as haptic and tactile information, have been shown to affect the perception of vection. Dichgans and Brandt (1978) reported that blindfolded participants who were seated inside a rotating drum experienced vection when touching the surface of the drum's interior even in the absence of any other stimuli. Passive rotation of a participant's arm in darkness elicited the feeling of vection as well (Brandt, Büchele, & Arnold, 1977). Besides pure haptic or pure tactile induction of vection, there is some evidence that vection can be facilitated by adding nonvisual and nonauditory cues. Wong and Frost (1981) seated their participants in a rotatable chair that was surrounded by a vection drum, creating circular vection along the yaw axis. Half of their participants received an additional short initial haptic stimulation of the rotating chair when the vection drum started to move. Although the vestibular stimulation elicited by the rotation of the chair was rather weak and lasted for only a very short period, the onset time of vection was significantly reduced. In another study, Seno, Ogawa, Ito, and Sunaga (2011) exposed their participants to a visual stimulus that elicited vection. Additionally, half of their participants received a consistent airflow that was produced by a stationary fan and matched the visual stimulus. Introducing the airflow significantly increased the participants' vection ratings.

Enhanced vection has also been reported when vibrotactile information were presented along with visual or acoustic stimuli (Higashiyama & Koga, 2002; Riecke et al., 2009; Sakurai, Kubodera, Grove, Sakamoto, & Suzuki, 2010; Seno et al., 2012; Väljamäe, Larsson, Vastfjall, & Kleiner, 2006). For instance, Schulte-Pelkum (2008) successfully increased vection by adding jittering motion to a participant's chair. Similar results were obtained by Riecke et al. (2005) and Schulte-Pelkum, Riecke, and Bülthoff (2004). The same holds for auditory vection, as Riecke et al. (2009) reported stronger vection ratings when vibrotactile cues were simultaneously presented with auditory stimuli.

Lately, Ash, Palmisano, and Kim (2011; see also Ash & Palmisano, 2012) analyzed the role of head oscillation with respect to visually induced vection. In their experiment, participants had to perform active head movements (e.g., turning right–left or back–forth) while watching a visual stimulus that indicated self-motion (expanding radial display of randomly placed blue squares on a black background). The authors tested two separate groups of participants: In the first group, participants performed active head movements that were simultaneously transferred to the visual stimulus, resulting in a matching visual–vestibular scene. In the second condition, the participants' head movements were incongruent with the presented visual stimulus; in fact, the visual stimulus responded in the opposite direction of the head movement (e.g., turning the head to the right resulted in a shift of the optic display to the left). The authors found increased feelings of vection not only when the visual and vestibular information was congruent but also when they contradicted each other. In contrast, Seno, Ito, and Sunaga (2011) compared two different groups of participants that walked on a treadmill in a VE. The subjects' locomotion either matched or was inconsistent with the visual stimulus. Congruent information turned out to facilitate and enhance vection, whereas the incongruent locomotion did not.

Taken together, vection is not exclusively a phenomenon produced by the visual system. Several other senses—including the auditory, haptic, and tactile senses—can influence the genesis and severity of vection. Although the level of vection induced by these alternative senses is not as strong as visually induced self-motion, further attention needs to be paid toward multisensory integration during the perception of vection. To date, little is known about the mechanisms and factors contributing to auditory or haptic vection; thus, further studies are mandatory to understand the phenomenon of vection as a whole. The multisensory approach assumes that vection is meant to be more complex than a simple bottom-up perception with respect to an adequate stimulus. Instead, recent findings accentuate the importance of cognitive factors in the genesis of vection and highlight the role of potentially top-down-driven mechanisms (Harris et al., 2002; Palmisano & Chan, 2004; Riecke et al., 2006; Wright et al., 2005).

18.5 RELEVANCE OF VECTION FOR VE SYSTEM DESIGN AND USE

An understanding of vection phenomena is important for the design of safe and effective VE technology for a number of reasons:

- Vection appears to be one of the key elements underlying the broader illusion of *presence* (see Chertoff & Schatz, 2014, Chapter 34) in a VE. To the extent that presence is a desirable design feature of a particular VE device, a proper understanding of the design elements that underlie vection is an important engineering issue.
- Vection may or may not contribute to the achievement of a VE device's *behavioral goals.* To the extent that vection in fact facilitates the achievement of various desired outcomes with respect to the user of a device, it is important to understand the factors that promote its occurrence.
- Vection has frequently been identified as a correlate of, and perhaps a causal element in, the phenomenon of motion sickness in VEs—or *cybersickness*. Additionally, exposure to a compelling vection stimulus may be a crucial factor in a larger process of perceptual–motor adaptation (see DiZio & Lackner, 2014, Chapter 33) to VEs. To the extent that adaptation does occur, a period of readaptation upon return to the normal spatiotemporal constraints of the actual world is necessarily required. These periods of adaptation and readaptation may expose users to a number of risks. To control these risks, it is necessary to understand the factors that create vection.
- Auditory and haptic vection research has forged the way for new practical uses of VE systems. According to Väljamäe (2009), these uses include posture rehabilitation, navigation in microgravity or other unusual environments (i.e., underwater), nonvisual navigation, and multisensory self-motion.

All of these concerns and potentials are directly related and cannot be considered in isolation from one another. There is little doubt that vection can play a key role in helping to produce a sense of presence in a VE system. An illusion of *being in* a computer-generated environment is clearly enhanced by a sense of illusory self-motion so compelling that it can hardly be distinguished from actual self-motion. Furthermore, the research literature provides relatively firm guidance, summarized in the preceding sections, concerning system design and usage guidelines for producing the illusion and enhancing its strength. However, much of this literature also describes the risks that system users may face as a result of exposure to self-motion illusions, most notably motion sickness and perceptual–motor disturbances.

The fundamental questions that VE system designers must address with respect to vection are as follows:

- In light of the potential risks to system users posed by vection, does the presence of illusory self-motion meaningfully contribute to the overall objectives of the human–VE system? In other words, are the overarching goals that the system is intended to accomplish being supported by producing compelling sensations of self-motion in users, are they being interfered with, or does the experience of vection have little or no impact on them? Will there be risks if the vection is executed correctly within the VE?
- If a reasonable case can be made that vection is a key factor in promoting the overall objectives of the human–VE system, then the question becomes one of how to safeguard users from the risks associated with its negative side effects.

With regard to the first of these two points, there is little evidence to date that vection plays a critical role in the attainment of human–VE system objectives when executed improperly. However, there are situations when it is likely to play a key role, notably in the successful application of VE-based entertainment devices (see Greenwood-Ericksen et al., 2014, Chapter 50) and potentially in certain VE-based training and education applications (see Rizzo, Lang, & Koenig, 2013, Chapter 45; Bartlett, Nicholson, Nolan, & Kelly, 2014, Chapter 46) in which skill in the perception and control of self-motion is critical (Bringoux et al., 2012; Roll et al., 2012). A major potential problem exists in the former application area, however, in that VE-based entertainment devices are intended to expose as many people as possible to the illusory perceptual features characteristic of this technology. This, of course, raises the risks of significant numbers of users being made ill and/or suffering postexposure side effects.

In many cases, the occurrence of vection may actually interfere with the objectives of VE systems. *Self-motion artifacts* might arise in situations where no perception of self-motion is desired, but where the stimulus information is such that momentary (or longer) vection illusions arise. For instance, this could occasionally occur if motion of virtual objects in the periphery of the visual field gives rise to transient vection illusions, causing disruptions in both the phenomenological experience of the VE and psychomotor disturbances associated with postural readjustments.

Certainly, the occurrence of vection-induced motion sickness and perceptual–motor disturbances can be expected to interfere dramatically with the objectives of most human–VE systems. Any benefit to be derived from the inclusion of vection-inducing stimuli in a VE system has to be carefully weighed against such potential risks. And if there is no benefit to be gained from the presence of self-motion (or, indeed, if there is significant risk associated with the occurrence of its side effects), then design and usage guidelines to minimize its occurrence should be adhered to. Familiarity with the stimulus factors that underlie the illusion, summarized earlier and described in detail in the literature cited in this chapter, can help to promote effective and safe system design. In addition, simulator and VE system usage guidelines developed by Kennedy and his colleagues (e.g., Kennedy et al., 1990) should be used to further protect users from the occasional negative side effects of exposure to simulated self-motion.

18.6 RESEARCH ISSUES

This section addresses research issues that may be of potential value in further illuminating our understanding of the relation between vection and VEs. These issues fall into two broad categories: research on vection and related phenomena that would be useful in enhancing the future safe and effective design and use of VE systems and research on vection using VE systems as research tools.

As anyone who has ever experienced or worked with VEs knows, they are fascinating systems from many perspectives. As this chapter has discussed, they raise many important psychological, physiological, and behavioral issues as developers attempt to maximize their safe and effective use. However, they not only raise important research issues, they also provide new means for examining these issues. In other words, because of their compelling multisensory, interactive nature and because they are programmable and permit the acquisition of very detailed information about user behavior (based, for instance, on data retrievable from body motion sensors—see Lawson & Riecke, 2014, Chapter 7), they afford the means to examine important psychological and physiological issues in ways that have not been possible to date (cf. Durlach & Mavor, 1996). Even in restricting the present discussion to matters concerned with vection and self-motion perception, it is clear that VE technology offers many exciting possibilities for enhancing our knowledge.

For example, Oman, Howard, and their colleagues have been engaged in research on circular and linear vection in the microgravity environment of outer space (Oman et al., 2000) using a helmet-mounted display and computer-generated VE as a motion stimulus. As described previously, the microgravity environment is useful for the study of visual–vestibular interactions because it allows researchers to isolate the influence of the two systems. However, until the introduction of VE technology, it was very difficult to examine vection phenomena in microgravity because of the severe weight and size limitations of the equipment involved. Oman's and Howard's work has provided important information on the relative roles of visual and proprioceptive sources of information for orientation in microgravity. Their results, indicating dramatically increased circular and linear vection after several days of exposure to microgravity, also provide a very useful source of converging information to those found by Young et al. (1986) for roll vection.

18.6.1 Effect of Complex Motion Patterns on Vection

Research conducted to date on vection has concentrated almost exclusively on relatively simple patterns of motion, specifically motion along or around a single axis. However, many current and future VE applications will require the depiction of far more complex patterns of self-motion, such as that involved in depicting the intricate and aggressive motion patterns of aircraft engaged in air combat (for training such skills) and any number of possible complex motion profiles for entertainment purposes.

Very little is known about illusory self-motion under such conditions, primarily because until very recently the technology for examining such issues has not existed. There was, however, one such study in which Seno et al. (2009) employed the Rubin's vase optical illusion with motion-defined face–vase areas. Manipulation of interchangeable background and foreground areas along with left–right, up–down vection allowed the researchers to isolate the effects of complex motion patterns. Additionally, Seno, Yamada, and Palmisano (2012) recently discovered a new vection phenomenon they described as directionless vection. Simultaneous motion planes with expanding and contraction stimuli were able to induce illusory self-motion that participants described as *directionless*. Although there has been significant progress made in the past 5 years, there is further need to apply VE systems to research a number of important questions regarding illusionary self-motion.

18.6.2 Multisensory Patterns of Information Specifying Self-Motion

VEs are fundamentally multisensory devices. Their intent is to take advantage of the information acquisition capabilities of as many of the sensory systems as possible to replicate the perceptual experience of the real world (or create compelling multisensory experiences of imagined or synthetic worlds). Research to date on the vection illusion has concentrated almost exclusively on the visual modality, with limited research attention having been devoted to audition (e.g., Lackner, 1977; Larsson et al., 2004; Väljamäe, 2009). In addition, several studies on electrical stimulation of the vestibular system have shown that it is possible to produce compelling vection illusions (primarily of roll) using such an approach (e.g., Cress et al., 1997; Dzendolet, 1963; Revol et al., 2009). However, there has been very little research to date examining the effects of multisensory information on vection. Cress et al. (1997) contrasted the effects of visually specified roll vection, electrical stimulation of the vestibular (producing illusions of roll motion), and combined visual and electrical vestibular stimulation and found that the latter condition produced sensations of motion that were rated as significantly more realistic by observers than either of the two former conditions alone. Visually induced vection has been paired with auditory stimuli and haptic stimulation, which has shown that the illusion of self-motion can be enhanced with the addition of multiple sensory modalities (Larsson et al., 2004; Nordahl et al., 2012; Prothero et al., 1995; Revol et al., 2009; Riecke et al., 2009, 2012; Väljamäe, 2009; von der Heyde & Riecke, 2002).

Additional research of this type would not only help to provide design and usage guidance for VE systems but may even help to illuminate the role that sensory conflict plays (or does not play) in creating vection-induced motion sickness.

18.6.3 Adaptation, Readaptation, Transfer of Adaptation, and Virtual Environments

The problems that users occasionally experience when adapting to the rearranged perceptual–motor relationships of VEs and/or when readapting to normal actual environment spatiotemporal constraints after having spent time immersed in a VE have been described in the preceding text and in Welch and Mohler (2014), Chapter 25, and Dizio and Lackner (2014), Chapter 33. Not all of these problems are due to vection illusions of course, but motion sickness and adverse postexposure symptomatology have been repeatedly observed in vection studies and related settings with sufficient frequency to identify it as a probable key contributor to this problem.

As VE systems become more widely distributed and more commonly used in everyday work and entertainment settings, these problems can be expected to increase in frequency. In addition to being more widely available, the fidelity and resulting strength of vection illusions (indeed, the overall sense of presence) afforded by VE systems will almost certainly continue to increase for the foreseeable future. Each of these factors will combine to increase problems with undesired side effects unless research on their cause and amelioration is aggressively pursued.

One avenue of research that may help to alleviate the severity of these problems involves the examination of factors that promote functional adaptation across environments (actual and virtual) that vary in their spatiotemporal characteristics (cf. Hu, Stern, & Koch, 1991; Welch, in press). The identification of system design and usage features that promote the safe transition of users between the differing spatiotemporal arrangements of actual and VEs should remain a high priority in order to assure a safe and successful future for this technology.

18.6.4 Effect of Visual Frames of Reference on Vection and Orientation

Many sources of information interact to produce one's sense of static and dynamic orientation in the world, including the potentially powerful effect of *visual frames of reference. Visual frames*, as described by Howard and his colleagues, are sets of distinct horizontal lines and surfaces (e.g., elements of the optic array corresponding to walls, ceilings, pillars, etc.) whose orientation is highly linked to that of the gravitoinertial force vector (Allison et al., 1999; Howard & Childerson, 1994). Familiar objects (e.g., trees, furniture) within these frames of reference also exhibit an *intrinsic visual polarity* (Allison et al., p. 299). When the visual frame of reference and all objects contained within it are tilted with respect to gravity, the result is a compelling experience of self-tilt.

Prothero and Parker (2003), in examining their *RFH*, have identified strong interactive effects between observers' visual frames of reference, the perception of vection, and subsequent occurrences of motion sickness and disorientation in a unified approach.

As described in Section 18.3, a great deal of research has been conducted over the past decade leading to an evolution of frameworks regarding visual frames and the effects on vection. It is essential to continue to apply VE technologies to continue to study this core characteristic of vection in order to determine precisely what characteristics of the visual scene will consistently affect the phenomenon. By continuing to apply this technology to research frames of reference, design guidance can be elicited to further enhance VE design.

18.7 CONCLUSIONS

The purpose of this chapter has been to provide a summary of research performed to date on the vection illusion, and in so doing demonstrate its relevance, and the relevance of vection phenomena in general, to the design and use of effective VE systems. The illusion of self-motion is an increasingly ubiquitous aspect of modern VE systems and it is likely to become even more characteristic of future systems. While there is little doubt that a compelling sense of vection can be a powerful contributor to an overall sense of presence in a VE system, it is by no means clear that it will always be of benefit in achieving the overall system objectives of a given device. Given the *cost* of the vection illusion—occasional, and occasionally severe, problems with motion sickness and perceptual–motor aftereffects—there is a clear need to understand the nature of this phenomenon and to incorporate knowledge of its stimulus characteristics and human performance implications into the design and use of VE systems.

ACKNOWLEDGMENTS

The very helpful comments of Robert S. Kennedy and Dean H. Owen on an earlier version of this chapter are gratefully acknowledged.

REFERENCES

Allison, R. S., Howard, I. P., & Zacher, J. E. (1999). Effect of field size, head motion, and rotational velocity on roll vection and illusory self-tilt in a tumbling room. *Perception, 28,* 299–306.

Andersen, G. J., & Braunstein, M. L. (1985). Induced self-motion in central vision. *Journal of Experimental Psychology: Human Perception and Performance, 11,* 122–132.

Andersen, G. J., & Dyer, B. P. (1987). Induced roll vection from stimulation of the central visual field. *Proceedings of the Human Factors Conference 31st Annual Meeting*, New York, NY (pp. 263–265).

Ash, A., & Palmisano, S. (2012). Vection during conflicting multisensory information about the axis, magnitude, and direction of self-motion. *Perception, 41*(3), 253–267. doi:10.1068/p7129

Ash, A., Palmisano, S., & Kim, J. (2011). Vection in depth during consistent and inconsistent multisensory stimulation. *Perception, 40*(2), 155–174. doi:10.1068/p6837

Badcock, D. R., Palmisano, S., & May, J. G. (2014). Vision and virtual environments. In K. S. Hale & K. M. Stanney (Eds.), *Handbook of virtual environments: Design, implementation, and applications* (2nd ed., pp. 39–86). Boca Raton, FL: Taylor & Francis Group, Inc.

Bartlett, K. M., Nicholson, D., Nolan, M., & Kelly, B. (2014). Modeling and simulation for cultural training. In K. S. Hale & K. M. Stanney (Eds.), *Handbook of virtual environments: Design, implementation, and applications* (2nd ed., pp. 1201–1226). Boca Raton, FL: Taylor & Francis Group, Inc.

Benson, A. J. (1990). Sensory functions and limitations of the vestibular system. In R. Warren & A. H. Wertheim (Eds.), *Perception and control of self-motion*. Hillsdale, NJ: Lawrence Erlbaum Associates.

Berthoz, A., Lacour, M., Soechting, J. F., & Vidal, P. P. (1978). The role of vision in the control of posture during linear motion. In R. Granit & O. Pompeiano (Eds.), *Reflex control of posture and movement*. Amsterdam, the Netherlands: Elsevier/North-Holland Biomedical Press.

Bles, W., & DeWit, G. (1976). Study of the effects of optic stimuli on standing. *Agressologie, 17,* 1–5.

Brandt, T., Büchele, W., & Arnold, F. (1977). Arthrokinetic nystagmus and ego-motion sensation. *Experimental Brain Research*, *30*(2–3), 331–338.

Brandt, T., Dichgans, J., & Koenig, E. (1973). Differential effects of central versus peripheral vision on egocentric and exocentric motion perception. *Experimental Brain Research*, *16*, 476–491.

Brandt, T., Wist, E. R., & Dichgans, J. (1975). Foreground and background in dynamic spatial orientation. *Perception and Psychophysics*, *17*, 497–503.

Bringoux, L., Lepecq, J., & Danion, F. (2012). Does visually induced self-motion affect grip force when holding an object? *Journal of Neurophysiology,* *108*(2012), 1685–1694.

Brooks, F. P. (1999). What's real about virtual reality? *IEEE Computer Graphics and Applications*, *19*(6), 16–27.

Chertoff, D., & Schatz, S. (2014). Beyond presence. In K. S. Hale & K. M. Stanney (Eds.), *Handbook of virtual environments: Design, implementation, and applications* (2nd ed., pp. 855–870). Boca Raton, FL: Taylor & Francis Group, Inc.

Cress, J. D., Hettinger, L. J., Cunningham, J. A., Riccio, G. E., Haas, M. W., & McMillan, G. R. (1997). Integrating vestibular displays for VE and airborne applications. *IEEE Computer Graphics and Applications*, *17*(6), 46–52.

Cutting, J. E. (1986). *Perception with an eye for motion*. Cambridge, MA: MIT Press.

Darken, R. P., & Peterson, B. (2014). Spatial orientation, wayfinding, and representation. In K. S. Hale & K. M. Stanney (Eds.), *Handbook of virtual environments: Design, implementation, and applications* (2nd ed., pp. 467–492). Boca Raton, FL: Taylor & Francis Group, Inc.

Daunton, N., & Thomsen, D. (1979). Visual modulation of otolith-dependent units in cat vestibular nuclei. *Experimental Brain Research*, *37*, 173–176.

Dichgans, J., & Brandt, T. (1973). Optokinetic motion sickness and pseudo-coriolis effects induced by moving visual stimuli. *Acta Otolaryngologica*, *76*, 339–348.

Dichgans, J., & Brandt, T. (1978). Visual-vestibular interaction: Effects on self-motion perception and postural control. In R. Held, H. Leibowitz, & H. L. Teuber (Eds.), *Handbook of sensory physiology: Vol 8. Perception*. Berlin, Germany: Springer-Verlag.

Dichgans, J., Held, R., Young, L., & Brandt, T. (1972). Moving visual scenes influence the apparent direction of gravity. *Science*, *178*, 1217–1219.

Diener, H. C., Wist, W. R., Dichgans, J., & Brandt, T. (1976). The spatial frequency effect on perceived velocity. *Vision Research*, *16*, 169–176.

Dindar, N., Tekalp, A. M., & Basdogan, C. (2013). Dynamic haptic interaction with video. In K. S. Hale & K. M. Stanney (Eds.), *Handbook of virtual environments: Design, implementation, and applications* (2nd ed., pp. 115–130). Boca Raton, FL: Taylor & Francis Group, Inc.

DiZio, P., Lackner, R., & Champney, R. K. (2014). Proprioceptive adaptation and aftereffects. In K. S. Hale & K. M. Stanney (Eds.), *Handbook of virtual environments: Design, implementation, and applications* (2nd ed., pp. 833–854). Boca Raton, FL: Taylor & Francis Group, Inc.

Dodge, R. (1923). Thresholds of rotation. *Journal of Experimental Psychology*, *6*, 107–137.

Durlach, N. I., & Mavor, A. S. (1996). *Virtual reality: Scientific and technical challenges*. Washington, DC: National Academy Press.

Dzendolet, E. (1963). Sinusoidal electrical stimulation of the human vestibular system. *Perceptual and Motor Skills*, *17*, 171–185.

Frigon, J. Y., & Delorme, A. (1992). Roll, pitch, longitudinal and yaw vection induced by optical flow in flight simulation conditions. *Perceptual and Motor Skills*, *74*, 935–955.

Fuskiki, H., Kobayashi, K., Asai, M., & Watanabe, Y. (2005). Influence of visually induced self-motion on postural stability. *Acta Oto-Laryngologica*, *125*(2005), 60–64.

Gibson, J. J. (1958). Visually controlled locomotion and visual orientation in animals. *British Journal of Psychology*, *49*, 182–194.

Gibson, J. J. (1961). Ecological optics. *Vision Research*, *1*, 253–262.

Gibson, J. J. (1979). *The ecological approach to visual perception*. Boston, MA: Houghton-Mifflin.

Greenwood-Ericksen, A., Kennedy, R. C., & Stafford, S. (2014). Entertainment applications of virtual environments. In K. S. Hale & K. M. Stanney (Eds.), *Handbook of virtual environments: Design, implementation, and applications* (2nd ed., pp. 1291–1316). Boca Raton, FL: Taylor & Francis Group, Inc.

Harm, D. L., Zogofros, L. M., Skinner, N. C., & Parker, D. E. (1993). Changes in compensatory eye movements associated with simulated conditions of space flight. *Aviation, Space, and Environmental Medicine*, *64*, 820–826.

Harris, L. R., Jenkin, M. R., Zikovitz, D., Redlick, F., Jaekl, P., & Jasiobedzka, U. T. (2002). Simulating self-motion I: Cues for the perception of motion. *Virtual Reality*, *6*, 75–85.

Held, R. M., & Durlach, N. I. (1992). Telepresence. *Presence*, *1*, 109–112.

Held, R. M., Dichgans, J., & Bauer, J. (1975). Characteristics of moving visual scenes influencing spatial orientation. *Vision Research, 15*, 357–365.

Henn, V., Cohen, B., & Young, L. R. (1980). Visual-vestibular interaction in motion perception and the generation of nystagmus. *Neurosciences Research Progress Bulletin, 18*, 459–651.

Hettinger, L. J., & Haas, M. W. (in press). Psychological issues in the design and use of virtual and adaptive environments. In L. J. Hettinger & M. W. Haas (Eds.), *Psychological issues in the design and use of virtual and adaptive environments*. Mahwah, NJ: Lawrence Erlbaum Associates.

Hettinger, L. J., & Riccio, G. E. (1992). Visually induced motion sickness in virtual environments. *Presence, 1*, 306–310.

Hettinger, L. J., Berbaum, K. S., Kennedy, R. S., Dunlap, W. P., & Nolan, M. D. (1990). Vection and simulator sickness. *Military Psychology, 2*, 171–181.

Higashiyama, A., & Koga, K. (2002). Integration of visual and vestibulo-tactile inputs affecting apparent self-motion around the line of sight. *Perception & Psychophysics, 64*(6), 981–995.

Howard, I. P. (1982). *Human visual orientation*. London, U.K.: John Wiley & Sons.

Howard, I. P. (1986a). The perception of posture, self motion, and the visual vertical. In K. R. Boff, L. Kaufman, & J. P. Thomas (Eds.), *Handbook of perception and human performance: Vol. 1. Sensory processes and perception*. New York, NY: John Wiley & Sons.

Howard, I. P. (1986b). The vestibular system. In K. R. Boff, L. Kaufman, & J. P. Thomas (Eds.), *Handbook of perception and human performance: Vol. I. Sensory processes and perception*. New York, NY: John Wiley & Sons.

Howard, I. P. (1993). The optokinetic system. In J. A. Sharpe & H. O. Barber (Eds.), *The vestibulo-ocular reflex and vertigo*. New York, NY: Raven Press.

Howard, I. P., & Childerson, L. (1994). The contribution of motion, the visual frame, and visual polarity to sensations of body tilt. *Perception, 23*, 753–762.

Howard, I. P., & Heckmann, T. (1989). Circular vection as a function of the relative sizes, distances, and positions of two competing visual displays. *Perception, 18*(5), 657–665.

Howard, I. P., & Howard, A. (1994). Vection: The contributions of absolute and relative visual motion. *Perception, 23*,745–751.

Howard, I. P., Cheung, B., & Landolt, J. (1987). Influence of vection axis and body posture on visually induced self-rotation and tilt. In *Motion cues in flight simulation and simulator induced sickness* (AGARD Conference Proceedings No. 433). Neulliy-Sur-Seine, France: North Atlantic Treaty Organization, Advisory Group for Aerospace Research and Development.

Hu, S., Grant, W. R., Stern, R. M., & Koch, K. L. (1991). Motion sickness severity and physiological correlates during exposures to a rotating optokinetic drum. *Aviation, Space, and Environmental Medicine, 62*, 308–314.

Hu, S., Stern, R. M., & Koch, K. L. (1991). Effects of pre-exposures to a rotating optokinetic drum on adaptation to motion sickness. *Aviation, Space, and Environmental Medicine, 62*, 53–56.

Hu, S., Stritzel, R., Chandler, A., & Stern, R. M. (1995). P6 acupressure reduces symptoms of vection-induced motion sickness. *Aviation, Space, and Environmental Medicine, 66*, 631–634.

Ishida, M., Fushiki, H., Nishida, H., & Watanabe, Y. (2008). Self-motion perception during conflicting visual-vestibular acceleration. *Journal of Vestibular Research, 18*(2008), 267–272.

Ito, H., & Shibata, I. (2005). Self-motion perception from expanding and contracting optical flows overlapped with binocular disparity. *Vision Research, 45*(2005), 397–402.

Jokerst, M. D., Gatto, M., Fazio, R., Stern, R. M., & Koch, K. L. (1999). Slow deep breathing prevents the development of tachygastria and symptoms of motion sickness. *Aviation, Space, and Environmental Medicine, 70*, 1189–1192.

Kapralos, B., Zikovitz, D., Jenkin, M., & Harris, L. R. (2004). Auditory cues in the perception of self-motion. *Journal of the Audio Engineering Society (Abstract), 52*, 801–802.

Kennedy, R. S., & Fowlkes, J. E. (1992). Simulator sickness is polygenic and poly symptomatic: Implications for research. *International Journal of Aviation Psychology, 2*(1), 23–38.

Kennedy, R. S., Drexler, J. M., Compton, D. E., Stanney, K. M., & Harm, D. L. (in press). Configural scoring of simulator sickness, cybersickness, and space adaptation syndrome: Similarities and differences. In L. J. Hettinger & M. W. Haas (Eds.), *Psychological issues in the design and use of virtual and adaptive environments*. Mahwah, NJ: Lawrence Erlbaum Associates.

Kennedy, R. S., Hettinger, L. J., & Lilienthal, M. G. (1990). Simulator sickness. In G. H. Crampton (Ed.), *Motion and space sickness* (pp. 179–215). Boca Raton, FL: CRC Press.

Kennedy, R. S., Hettinger, L. J., Harm, D. L., Ordy, J. M., & Dunlap, W. P. (1996). Psychophysical scaling of circular vection (CV) produced by optokinetic (OKN) motion: Individual differences and effects of practice. *Journal of Vestibular Research, 6*, 331–341.

Keshavarz, B., & Hecht, H. (2012). Stereoscopic viewing enhances visually induced motion sickness but sound does not. *Presence: Teleoperators and Virtual Environments*, *21*(2), 213–228.

Keshavarz, B., Heckt, H., & Lawson, B. D. (2014). Visually induced motion sickness. In K. S. Hale & K. M. Stanney (Eds.), *Handbook of virtual environments: Design, implementation, and applications* (2nd ed., pp. 647–698). Boca Raton, FL: Taylor & Francis Group, Inc.

Kitazaki, M., & Sato, T. (2003). Attentional modulation of self-motion perception. *Perception*, *32*(4), 475–484.

Klemen, J., & Chambers, C. D. (2012). Current perspectives and methods in studying neural mechanisms of multisensory interactions. *Neuroscience and Biobehavioral Reviews*, *36*(1), 111–133. doi:10.1016/j.neubiorev.2011.04.015

Lackner, J. R. (1977). Induction of illusory self motion and nystagmus by a rotating sound field. *Aviation, Space, and Environmental Medicine*, *44*, 129–131.

Larsson, P., Västfjäll D., & Kleiner M. (2004). Perception of self-motion and presence in auditory virtual environments. *Proceedings of the Seventh Annual Workshop of Presence* (pp. 252–258). Valencia, Spain.

Lawson, B. D., & Riecke, B. E. (2014). Perception of body motion. In K. S. Hale & K. M. Stanney (Eds.), *Handbook of virtual environments: Design, implementation, and applications* (2nd ed., pp. 163–196). Boca Raton, FL: Taylor & Francis Group, Inc.

Lawson, B. D. (2014a). Motion sickness symptomatology and origins. In K. S. Hale & K. M. Stanney (Eds.), *Handbook of virtual environments: Design, implementation, and applications* (2nd ed., pp. 531–600). Boca Raton, FL: Taylor & Francis Group, Inc.

Lee, D. N., & Aronson, E. (1974). Visual proprioceptive control of standing in human infants. *Perception and Psychophysics*, *15*, 529–532.

Lee, D. N., & Lishman, J. R. (1975). Visual proprioceptive control of stance. *Journal of Human Movement Studies*, *1*, 87–95.

Lestienne, F., Soechting, J., & Berthoz, A. (1977). Postural readjustments induced by linear motion of visual scenes. *Experimental Brain Research*, *28*, 363–384.

Mach, E. (1875). *Grundlinien der Lehre von der Bewegungsempfindungen*. Leipzig, Germany: Engelmann.

Mach, E. (1922). *Die Analyse der Empfindugnen*. Jena, Germany: G. Fischer.

Marme-Karelse, A., & Bles, W. (1977). Circular vection and human posture II: Does the auditory-system play a role? *Agressologie*, *18*(6).

McCauley, M. E., & Sharkey, T. J. (1992). Cybersickness: Perception of self-motion in virtual environments. *Presence*, *17*, 311–318.

Melcher, G. A., & Henn, V. (1981). The latency of circular vection during different accelerations of the optokinetic stimulus. *Perception and Psychophysics*, *30*, 552–556.

Melville-Jones, G. (1970). Origin, significance and amelioration of Coriolis illusions from semicircular canals: A non-mathematical appraisal. *Aerospace Medicine*, *40*, 482–490.

Mourant, R. M., & Refsland D. (2003). *Developing a 3D sound environment for a driving simulator*. Paper presented at the Ninth International Conference on Virtual Systems and Multi-media (VSMM) (pp. 711–719).

Nakamura, S. (2008). Effects of stimulus eccentricity on vection reevaluated with a binocularly defined depth. *Japanese Psychological Research*, *50*(2), 77–86.

Nichols, S., Haldane, C., & Wilson, J. R. (2000). Measurement of presence and its consequences in virtual environments. *International Journal of Human-Computer Studies*, *52*(3), 471–491. doi:10.1006/ijhc.1999.0343

Nordahl, R., Nilsson, N. C., Turchet, L., & Serafin, S. (2012, March 5). *Vertical illusory self-motion through haptic stimulation of the feet*. Paper presented at the 2012 IEEE Virtual Reality Workshop on Perceptual Illusions in Virtual Environments, Orange Country, CA.

Ohmi, M., & Howard, I. P. (1988). Effect of stationary objects on illusory forward self-motion induced by a looming display. *Perception*, *17*, 5–12.

Ohmi, M., Howard, I. P., & Landolt, J. P. (1987). Circular vection as a function of foreground-background relationships. *Perception*, *16*(1), 17–22.

Oman, C. M., Howard, I. P., Carpenter-Smith, T., Beall, A. C., Natapoff, A., Zacher, J. E., & Jenkin, H. L. (2000). Neurolab experiments on the role of visual cues in microgravity spatial orientation. *Aviation, Space, and Environmental Medicine*, *71*, 293.

Owen, D. H. (1990). Lexicon of terms for the perception and control of self-motion and orientation. In R. Warren and A. H. Wertheim (Eds.), *Perception and control of self-motion*. Hillsdale, NJ: Lawrence Erlbaum Associates.

Owen, D. H., Wolpert, L., & Warren, R. (1983). Effects of optical flow acceleration, edge acceleration, and viewing time on the perception of egospeed acceleration. In D. H. Owen (Ed.), *Optical flow and texture variables useful in detecting decelerating and accelerating self-motion* (AFHRL-TP-84-4). Williams AFB, AZ: Air Force Human Resources Laboratory. (NTIS No. AD-A148 718)

Palmisano, S., & Chan, A. Y. C. (2004). Jitter and size effects on vection are immune to experimental instructions and demands. *Perception, 33*(8), 987–1000.

Palmisano, S., Allison, R. S., Kim, J., & Bonato, F. (2011). Simulated viewpoint jitter shakes sensory conflict accounts of vection. *Seeing and Perceiving, 24*(2), 173–200. doi:10.1163/187847511X570817

Parker, D. E. (1991). Human vestibular function and weightlessness. *The Journal of Clinical Pharmacology, 31*, 904–910.

Prothero, J. D. (1998). *The role of rest frames in vection, presence, and motion sickness* (Doctoral dissertation). University of Washington, Seattle, WA. Retrieved from ProQuest Dissertations and Theses.

Prothero, J. D., & Parker, D. E. (2003). A unified approach to presence and motion sickness. In L. J. Hettinger and M. W. Haas (Eds.), *Virtual and adaptive environments: Applications, implications, and human performance issues* (pp. 47–66). Mahwah, NJ: Lawrence Erlbaum.

Prothero, J. D., Parker, D. E., Furness, T. A. III, & Wells, M. J. (1995). Towards a robust, quantitative measure for presence. *Proceedings of the Conference on Experimental Analysis and Measurement of Situation Awareness* (pp. 359–366). Daytona Beach, FL: Embry-Riddle Aeronautical University Press.

Reason, J. T., & Brand, J. J. (1975). *Motion sickness.* London, U.K.: Academic Press.

Revol, P., Farné, A., Pisella, L., Holmes, N. P., Imai, A., Susami, K., … Rossetti, Y. (2009). Optokinetic stimulation induces illusory movement of both out-of-the-body and on-the-body hand-held visual objects. *Experimental Brain Research, 193*(2009), 633–638.

Riccio, G. E., & Stoffregen, T. A. (1991). An ecological theory of motion sickness and postural instability. *Ecological Psychology, 3*, 195–240.

Riecke, B. E. (2011). Compelling self motion through virtual environments without actual self-motion—Using self-motion illusions ("Vection") to improve user experience in VR. In J. J. Kim (Ed.), *Virtual reality* (pp. 149–176). Retrieved from http://www.intechopen.com/books/virtual-reality/compelling-self-motion-through-virtual-environments-without-actual-self-motion-using-self-motion-ill.

Riecke, B. E., & Wiener, J. M. (2007, March 10–14). *Can people not tell left from right in VR? Point-to-origin studies revealed qualitative errors in visual path integration.* Paper presented at the 2007 IEEE Virtual Reality Conference, Charlotte, NC.

Riecke, B. E., Feuereissen, D., & Rieser, J. J. (2009). Auditory self-motion simulation is facilitated by haptic and vibrational cues suggesting the possibility of actual motion. *ACM Transactions on Applied Perception, 6*(3), 1–22. doi:10.1145/1577755.1577763

Riecke, B. E., Feuereissen, D., Rieser, J. J., and McNamara, T. P. (2012, March 4–8). *Self-motion illusions (vection) in VR—Are they good for anything?* Paper presented at the 2012 IEEE Virtual Reality Conference, Orange Country, CA.

Riecke, B. E., Schulte-Pelkum, J., Avraamides, M. N., Heyde, M. von der, & Bülthoff, H. H. (2006). Cognitive factors can influence self-motion perception (vection) in virtual reality. *ACM Transactions on Applied Perception, 3*(3), 194–216. doi:10.1145/1166087.1166091

Riecke, B. E., Schulte-Pelkum, J., Caniard, F., & Bülthoff, H. H. (2005). Influence of auditory cues on the visually-induced self-motion illusion (circular vection) in virtual reality. *Proceedings of the Eighth International Workshop on Presence 2005* (pp. 49–57).

Riecke, B. E., Väljamäe, A., & Schulte-Pelkum, J. (2009). Moving sounds enhance the visually-induced self-motion illusion (circular vection) in virtual reality. *ACM Transactions on Applied Perception, 6*(2), 1–27. doi:10.1145/1498700.1498701

Rizzo, A. "Skip", Lange, B., & Koenig, S. (2014). Clinical virtual reality. In K. S. Hale & K. M. Stanney (Eds.), *Handbook of virtual environments: Design, implementation, and applications* (2nd ed., pp. 1157–1200). Boca Raton, FL: Taylor & Francis Group, Inc.

Roll, R., Kavoundoudias, A., Albert, F., Legré, R., Gay, A., Fabre, B., & Roll, J. P. (2012). Illusory movements prevent cortical disruption caused by immobilization. *NeuroImage, 62*(2012), 510–519.

Sakamoto, S., Suzuki, F., Suzuki, Y., & Gyoba, J. (2008). The effect of linearly moving sound image on perceived self-motion with vestibular information. *Acoustical Science and Technology, 29*(6), 391–393. doi:10.1250/ast.29.391

Sakellari, V., & Soames, R. W. (1996). Auditory and visual interactions in postural stabilization. *Ergonomics, 39*(4), 634–648. doi:10.1080/00140139608964486

Sakurai, K., Kubodera, T., Grove, P., Sakamoto, S., & Suzuki, Y. (2010). Multi-modally perceived direction of self-motion from orthogonally directed visual and vestibular stimulation. *Journal of Vision, 10*(7), 866. doi:10.1167/10.7.866

Schubert, T., Regenbrecht, H., & Friedmann, F. (2000, March 27–28). *Real and illusory interaction enhance presence in virtual environments*. Paper presented at the Third International Workshop on Presence, University of Delft, Delft, the Netherlands.

Schulte-Pelkum, J. (2008). *Perception of self-motion: Vection experiments in multi-sensory virtual environments* (PhD thesis). Ruhr Universität, Bochum, Germany.

Schulte-Pelkum, J., Riecke, B. E., & Bülthoff, H. H. (2004). Vibrational cues enhance believability of ego-motion simulation. *Fifth International Multisensory Research Forum (IMRF 2004)*. Barcelona, Spain.

Seno, T., Hasuo, E., Ito, H., & Nakajima, Y. (2012). Perceptually plausible sounds facilitate visually induced self-motion perception (vection). *Perception, 41*(5), 577–593. doi:10.1068/p7184

Seno, T., Ito, H., & Sunaga, S. (2009). The object and background hypothesis for vection. *Vision Research, 49*(2009), 2973–2982.

Seno, T., Ito, H., & Sunaga, S. (2011). Inconsistent locomotion inhibits vection. *Perception, 40*(6), 747–750. doi:10.1068/p7018

Seno, T., Ito, H., & Sunaga, S. (2012). Vection can be induced in the absence of explicit motion stimuli. *Experimental Brain Research, 219*(2), 235–244. doi:10.1007/s00221-012-3083-y

Seno, T., Ogawa, M., Ito, H., & Sunaga, S. (2011). Consistent air flow to the face facilitates vection. *Perception, 40*(10), 1237–1240. doi:10.1068/p7055

Seno, T., Yamada, Y., & Palmisano, S. (2012). Directionless vection: A new illusory self-motion perception. *i-Perception, 3*(2012), 775–777.

Slater, M., & Steed, A. (2000). A virtual presence counter. *Presence, 9*, 413–134.

Smart, L. J., Pagulayan, R. J., & Stoffregen, T. A. (1998). Self-induced motion sickness in unperturbed stance. *Brain Research Bulletin, 47*, 449–457.

Stern, R. M., Hu, S., Vasey, M. W., & Koch, K. L. (1989). Adaptation to vection-induced symptoms of motion sickness. *Aviation, Space, and Environmental Medicine, 60*, 566–572.

Stern, R. M., Koch, K. L., Leibowitz, H. W., Lindblad, I. M., Shupert, C. L., & Stewart, W R. (1985). Tachygastria and motion sickness. *Aviation, Space, and Environmental Medicine, 56*, 1074–1077.

Stern, R. M., Koch, K. L., Stewart, W. R., & Lindblad, I. M. (1987). Spectral analysis of tachygastria recorded during motion sickness. *Gastroenterology, 92*, 92–97.

Steuer, J. (1992). Defining virtual reality: Dimensions determining telepresence. *Journal of Communication, 42*(4), 73–93.

Stoffregen, T. A. (1985). Flow structure versus retinal location in the optical control of stance. *Journal of Experimental Psychology: Human Perception and Performance, 11*, 554–565.

Stoffregen, T. A., & Riccio, G. E. (1991). An ecological critique of the sensory conflict theory of motion sickness. *Ecological Psychology, 3*, 159–194.

Stoffregen, T. A., & Smart, L. J. (1998). Postural instability precedes motion sickness. *Brain Research Bulletin, 47*, 437–448.

Stoffregen, T. A., Hettinger, L. J., Haas, M. W., Roe, M. M., & Smart, L. J. (2000). Postural instability and motion sickness in a fixed-base flight simulator. *Human Factors, 42*, 458–469.

Storms, R. L. (2002). Auditory-visual cross-modality interaction and illusions. In K. M. Stanney (Ed.), *Handbook of virtual environments. Design, implementation, and applications* (pp. 589–618). Mahwah, NJ: Lawrence Erlbaum Associates.

Straube, A., & Brandt, T. (1987). Importance of the visual and vestibular cortex for self-motion perception in man (circularvection). *Human Neurobiology, 6*, 211–218.

Telford, L., Spratley, J., & Frost, B. J. (1992). Linear vection in the central visual field facilitated by kinetic depth cues. *Perception, 21*(3), 337–349.

Tschermak, A. (1931). Optischer raumsinn. In A. Bethe, G. Bergmann, G. Emden, & A. Ellinger (Eds.), *Handbuch 'der normalen und pathologischen Physiologie*. Berlin, Germany: Springer.

Urbantschitsch, V. (1897). Über Störungen des Gleichgewichtes und Scheinbewegungen [On disturbances of the equilibrium and illusory motions]. *Zeitschrift für Ohrenheilkunde, 31*, 234–294.

Väljamäe, A. (2009). Auditorily-induced illusory self-motion: A review. *Brain Research Reviews, 61*(2), 240–255. doi:10.1016/j.brainresrev.2009.07.001

Väljamäe, A., Larsson, P., Vastfjall, D., & Kleiner, M. (2005). Travelling without moving: Auditory scene cues for translational self-motion. *Proceedings of the 11th International Conference on Auditory Display* (ICAD2005) (pp. 9–16). Limerick, Ireland.

Väljamäe, A., Larsson, P., Vastfjall, D., & Kleiner, M. (2006). Vibrotactile enhancement of auditory-induced self-motion and spatial presence. *Journal of the Audio Engineering Society, 54*(10), 954–963.

Väljamäe, A., Larsson, P., Vastfjall, D., & Kleiner, M. (2009). Auditory landmarks enhance circular vection in multimodal virtual reality. *Journal of the Audio Engineering Society, 57*(3), 111–120.

Väljamäe, A., Larsson, P., Västfjäll, D., & Kleiner, M. (2008). Sound representing self-motion in virtual environments enhances linear vection. *Presence: Teleoperators and Virtual Environments, 17*(1), 43–56. doi:10.1162/pres.17.1.43

Van der Steen, F. A. M., & Brockhoff, P. T. M. (2000). Induction and impairment of saturated yaw and surge vection. *Perception and Psychophysics, 62*, 89–99.

von der Heyde, M., & Riecke, B. E. (2002). Embedding presence-related terminology in a logical and functional model. In F. Gouveia (Ed.), *Presence 2002* (pp. 37–52). Porto, Portugal: Universidare Fernando Pessoa.

Vorländer, M., & Shinn-Cunningham, B. (2014). Virtual auditory displays. In K. S. Hale & K. M. Stanney, (Eds.), *Handbook of virtual environments: Design, implementation, and applications* (2nd ed., pp. 87–114). Boca Raton, FL: Taylor & Francis Group, Inc.

Waespe, W., & Henn, V (1977a). Neuronal activity in the vestibular nuclei of the alert monkey during vestibular and optokinetic stimulation. *Experimental Brain Research, 27*, 523–538.

Waespe, W., & Henn, V. (1977b). Vestibular nuclei activity during optokinetic after-nystagmus (OKAN) in the alert monkey. *Experimental Brain Research, 30*, 323–330.

Waespe, W., & Henn, V. (1978). Reciprocal changes in primary and secondary optokinetic afternystagmus (OKAN) produced by repetitive optokinetic stimulation in the monkey. *Archiv fur Psychiatrie und Nervenkrankheiten, 225*, 23–30.

Wann, J. P., White, A. D., Wilkie, R. M., Culmer, Peter R. J., Lodge, P. A., & Mon-Williams, M. (2014). Measurement of visual aftereffects following virtual environment exposure. In K. S. Hale & K. M. Stanney (Eds.), *Handbook of virtual environments: Design, implementation, and applications* (2nd ed., pp. 809–832). Boca Raton, FL: Taylor & Francis Group, Inc.

Warren, R. (1976). The perception of egomotion. *Journal of Experimental Psychology: Human Perception and Performance, 2*, 448–456.

Warren, R., Owen, D. H., & Hettinger, L. J. (1982). Separation of the contributions of optical flow rate and edge rate on the perception of egospeed acceleration. In D. H. Owen (Ed.), *Optical flow and texture variables useful in simulating self-motion (I)* (Interim Technical Report for Grant No. AFOSH-81–0078). Columbus, OH: The Ohio State University, Department of Psychology, Aviation Psychology Laboratory. (NTIS No. AD-A117 016).

Welch, R. B., & Mohler, B. J. (2014). Adapting to virtual environments. In K. S. Hale & K. M. Stanney (Eds.), *Handbook of virtual environments: Design, implementation, and applications* (2nd ed., pp. 627–646). Boca Raton, FL: Taylor & Francis Group, Inc.

Welch, R. B. (in press). Adapting to telesystems. In L. J. Hettinger & M. W Haas (Eds.), *Psychological issues in the design and use of virtual and adaptive environments*. Mahwah, NJ: Lawrence Erlbaum Associates.

Wong, S. C. P., & Frost, B. J. (1978). Subjective motion and acceleration induced by the movement of the observer's entire visual field. *Perception and Psychophysics, 24*, 115–120.

Wong, S. C., & Frost, B. J. (1981). The effect of visual-vestibular conflict on the latency of steady-state visually induced subjective rotation. *Perception & Psychophysics, 30*(3), 228–236.

Wood, R. W. (1895). The "haunted swing" illusion. *The Psychological Review, 2*, 277–278.

Wright, W. G., Dizio, P., & Lackner, J. R. (2005). Vertical linear self-motion perception during visual and inertial motion: More than weighted summation of sensory inputs. *Journal of Vestibular Research: Equilibrium & Orientation, 15*(4), 185–195.

Wright, W., Dizio, P., & Lackner, J. (2006). Perceived self-motion in two visual contexts: Dissociable mechanisms underlie perception. *Journal of Vestibular Research: Equilibrium & Orientation, 16*(1–2), 23–28.

Young, L. R., Oman, C. M., & Dichgans, J. M. (1975). Influence of head orientation on visually induced pitch and roll sensations. *Aviation, Space, and Environmental Medicine, 46*, 264–269.

Young, L. R., Shelhamer, M., & Modestino, O. (1986). MIT/Canadian vestibular experiments on the Spacelab-1 mission: 2. Visual vestibular interaction in weightlessness. *Experimental Brain Research, 64*, 299–307.

Zahorik, P., & Jenison, R. L. (1998). Presence as being-in-the-world. *Presence, 7*, 78–89.

Zeltzer, D. (1992). Autonomy, interaction, and presence. *Presence, 1*, 127–132.

19 Spatial Orientation, Wayfinding, and Representation

Rudolph P. Darken and Barry Peterson

CONTENTS

19.1 INTRODUCTION

Everyone has been disoriented at one time or another. It is an uncomfortable, unsettling feeling to be unfamiliar with your immediate surroundings and unable to determine how to correct the situation. Accordingly, we might think that the goal of navigation research in virtual environments (VEs) is to create a situation where everyone is oriented properly all the time and knows exactly where everything is and how to get there. This, however, may not be absolutely correct. Much is gained from the navigation process beyond just spatial knowledge. The path of discovery rarely lies on a known road. The experience of serendipitous discovery is an important part of human navigation and should be preserved. But how do we resolve the conflicts between this and the not-so-pleasant experience of lostness?

Navigation tasks are essential to any environment that demands movement over large spaces. However, navigation is rarely, if ever, the primary task. It just tends to get in the way of what you really want to do. Our goal is to make the execution of navigation tasks as transparent and trivial as possible, but not to preclude the elements of exploration and discovery. Disoriented people are anxious, uncomfortable, and generally unhappy. If these conditions can be avoided, exploration and discovery can take place.

This chapter is about navigation in VEs—understanding how people navigate and how this affects the design of VE applications. The literature on navigation in VEs extends back to the earliest applications of the technology. One of the best examples of how VE research has impacted our world is here—methods, tools, and techniques for navigation that were developed to aid virtual world navigators have emerged in the real world as navigation tools in the mobile era.

We begin with a clarification of terms and some theoretical background on navigation. A discussion of methods for navigation performance enhancement will follow. This is about how to improve performance in a VE. This is different from the next topic concerning the use of VEs as training tools where we are interested in performance in the real world. Lastly, the chapter concludes with a summary of principles for the design of navigable VEs.

19.1.1 DEFINITION OF TERMS

One of the problems we find in the literature is confusion over terms involving navigation. It is difficult to compare two studies that use different terms in different ways. We may be unknowingly comparing apples to oranges. We use specific terms with specific definitions and encourage the research community to adopt these.

Wayfinding is the cognitive element of navigation. It does not involve movement of any kind but only the thoughtful parts that guide movement. As we will see later in the chapter, wayfinding is not merely a planning stage that precedes motion. Wayfinding and motion are intimately tied together in a complex negotiation that is navigation. An essential part of wayfinding is the development and use of a *cognitive map*, also referred to as a *mental map*. Still not well understood, a cognitive map is a mental representation of an environment. It has been called a *picture in the head*, although there is significant evidence that it is not purely based on imagery but rather has a symbolic quality. The representation of spatial knowledge in human memory that constitutes a cognitive map will be an important part of this chapter.

Motion is the motoric element of navigation. A reasonable synonym for motion is travel as used by Bowman, Koller, and Hodges (1997). Durlach and Mavor (1995) subdivide this further into *passive transport* such as a *point-and-fly* interface or other abstraction and *active transport* such as the omnidirectional treadmill or other literal motion interfaces that replicate human bipedal motion (Darken, Cockayne, & Carmein, 1997). Active transport interfaces are often referred to as *locomotion* interfaces. *Maneuvering* is a subset of motion involving smaller movements that may not necessarily be a part of getting from *here* to *there* but rather adjusting the orientation of perspective, as in rotating the body or sidestepping. This is an important distinction to make for the development of active transport interfaces for locomotion such as Gaiter or Pointman (Templeman, Denbrook, & Sibert, 2000; Templeman, Sibert, Page, & Denbrook, 2006). In this chapter, we will address locomotion issues as they pertain to overall navigation performance. For a more detailed discussion of locomotion devices and techniques, see *Locomotion Interfaces* (Templeman, Page, & Denbrook, 2014, Chapter 10).

Navigation is the aggregate task of wayfinding and motion. It inherently must have both the cognitive element (wayfinding) and the motoric element (motion). Consequently, we use this term only when we mean to imply the aggregate task and not merely a part. The literature is replete with references to *navigation* that are only interested in novel motion techniques. We find this to be confusing and counterproductive to the discussion.

It is also useful to define what is implied by *navigation performance*, as this is the metric we need to use to determine the relative effectiveness of specific navigation tools and techniques. This is entirely dependent on the navigation task in question. Are we studying the ability of a person to find an unknown object in a complex space? Then search time might be an appropriate measure. Are we interested in the ability to find a known location in a complex space? Then route following might be appropriate. Are we interested in a person's overall knowledge of the configuration of a space? Then a map drawing exercise might be appropriate. These issues will be discussed in more detail in Section 19.2.

19.1.2 TRAINING TRANSFER OR PERFORMANCE ENHANCEMENT?

There are two primary classes of applications having to do with navigation in VEs. All VEs that simulate a large volume of space will have navigation problems of one sort or another. Typically, any space that cannot be viewed from a single vantage point will exhibit these problems as users move from one location to another. The need to maintain a concept of the space and the relative locations between objects and places is essential to navigation. This is called *spatial comprehension* and, like verbal comprehension, involves the ability to perceive, understand, remember, and recall for future use.

In applications where we find that users tend to become disoriented or are unable to relocate previously visited points of interest, it is desirable to either redesign these applications so these problems do not appear or provide tools or mediators to help alleviate these problems. This class of application involves a need to enhance performance within the VE. This distinction differentiates these applications from those where improved performance is required outside of the VE in the real world. These are training transfer applications that we will discuss next.

The second class of applications that involves navigation has to do with the use of VEs as training aids for real-world navigation tasks. The fact that we can construct virtual representations of real environments has led many to consider the use of VEs for much the same purposes that a conventional paper map might be used. Indeed, we will show examples where techniques initially developed for use in VEs, video games, and visual simulation have evolved for use on mobile devices for real navigation tasks.

While there are certainly similarities between these two classes of applications, the validation process is entirely different. If we want to show that a visualization scheme in a virtual building walkthrough, for example, can be used to lessen the severity of disorientation, we need only show that users of the VE with this visualization scheme perform better on navigation tasks than users of an identical VE without the visualization scheme. However, if we want to show that this same VE can be used as a training aid for navigation tasks in the actual building, navigation performance within the VE, while interesting, does not prove our point. A training transfer study or comparative analysis must be completed to show that users who trained on the VE navigate the building better than users who received some other form of training or possibly no training at all. This is an example of the use of a VE as a training aid for specific environments. Navigation performance is expected to improve in one specific real environment and nowhere else.

Another issue is the use of a VE as a training aid for general navigation tasks. If it could be shown that a VE could help people use paper maps more efficiently, or select landmarks in an environment effectively, that performance increase would be expected to exist across physical spaces, possibly assuming some spatial similarities* with that of the training environments. This is beyond the scope of our discussion here but is an important use of VEs that has not been explored as yet.

19.2 BACKGROUND

Whether or not a VE attempts to simulate the real world, we have to consider the fact that people are accustomed to navigation in physical spaces. Certainly, there are differences between regions of physical space that alter how navigation works, such as navigating in a forest versus navigating in a city, but there are assumptions that can be made based on past experience in real environments that are useful in any real space. While this is not a satisfactory reason to blindly copy the real world in every way we can, we certainly have to learn everything possible about how people relate to the physical world so we can understand how to build better VEs.

19.2.1 SPATIAL KNOWLEDGE ACQUISITION

There are many ways to acquire spatial knowledge of an environment. The fundamental distinction between sources of spatial knowledge is whether the information comes directly from the environment (primary) or from some other source (secondary) such as a map or mobile device. An issue specific to secondary sources has to do with whether or not the source is used inside or outside of the actual environment and whether it is static, like a paper map, or dynamic, like Google Maps™ or Apple Maps™.

19.2.1.1 Direct Environmental Exposure

When you navigate in an environment, you extract information both for use on whatever navigation task currently is being executed and for any subsequent navigation task. Exactly what information is useful for navigation? We cannot possibly attend to every stimulus and make use of it. Much of it is irrelevant or at least of lesser importance. If we knew what information was the most important, this could be useful in designing VEs. Designers would know what to put where to help people find their way around. While there is no clear answer to this question, Kevin Lynch presents the most compelling, environment-independent answer to date (Lynch, 1960).

In studying urban environments (Lynch was an urban planner), he found that there are certain similarities that cross cities. There are in fact *elements* of urban environments, building blocks that can be used to construct or decompose any city. He starts with *landmarks* that are the most salient cues in any environment. They are also directional, meaning that a particular building might be a landmark from one side but not another. Then there are *routes* (or *paths* if you prefer) that connect the landmarks. They do not necessarily connect them directly but they move you through the city such that the spatial relationships between landmarks become known. Cities tend to have complex road and rail structures. Interchanges or junctions between routes are called *nodes*. These are important because they are fundamental to the structure of the routes. This structure must be understood before proficient navigation can take place. Most cities have specific regions that are explicitly or implicitly separated from the rest of the city. These are *districts*. Landmarks and nodes typically live in districts. Routes pass through districts and connect them. Finally, regions of the city, and in fact the city itself, are bounded by *edges*. Edges prevent or deter travel. A typical edge is a river, lake, or a fence line.

---

* For example, a virtual environment that could be used to train a person to effectively navigate in a generic city would probably have little impact on that person's ability to navigate in a generic forest.

An interesting fact is that classifying some city object as one element or another does not preclude it being classified as something else in another context. To a pedestrian, a walking path is a route while highways and railroads are edges. In aviation navigation, linear elements (paths and edges) are often used as *catching features* that pilots use to indicate when they have gone too far or have otherwise missed a checkpoint. To a driver, the roads are routes and everything else is an edge. It depends on the mode of travel. Furthermore, the mode of travel also effects *what* gets encoded, not just *how* it gets encoded (Goldin & Thorndyke, 1982).

19.2.1.2   Map Usage

There are a number of secondary sources that have been used for spatial knowledge acquisition. These include maps, photographs, videotape, verbal directions, and recently VEs. The most common of these is the map. We will begin this discussion with static maps because much of what we have learned about how to design paper maps has directly impacted the design of dynamic maps that are ubiquitous today on mobile devices. For any static map, we need to know when the map is used or, more appropriately, what tasks the map is to be used for. The critical issue is in whether or not the map is to be used preceding navigation or concurrent with navigation. This is important because maps that are used concurrently with navigation involve the placement of oneself on the map. The first part of any task of this nature is, "Where am I? What direction am I facing?" A transformation is required from the egocentric perspective to the geocentric perspective. If the map is used as a precursor to navigation, it is used only for planning and familiarization. No perspective transformation is required. Planning a trip is one example of such a geocentric navigation task. The planning is done outside of the environment so there is no perspective transformation needed. However, when such a transformation is required, as in using a map during navigation, the problem is more complicated.

When a perspective transformation must be performed, the rotation of the map can have a great effect on performance. Aretz and Wickens (1992) showed that maps used during navigation tasks (egocentric tasks) should be oriented *forward-up* (the top of the map shows the environment in front of the viewer), while maps used for planning or other geocentric tasks should be *north-up*. This same concept is reinforced by Rossano and Warren (1989) who showed that judgments of direction are adversely affected by misaligned maps. Levine, Marchon, and Hanley (1984) goes on to provide basic principles of map presentation including aligning the map with the environment, always showing two concurrent points on the map so the viewer can triangulate position, and avoiding symmetry. Aviators (especially rotary wing pilots because they travel slower and lower to the ground) are taught to rotate their maps in the direction of flight and to track their current position. Péruch, Pailhous, and Deutsch (1986) showed that redundancy of information is crucial to resolve conflicts between different frames of reference.

The key to map use for navigation is resolving the egocentric to geocentric perspective transformation. This certainly is not the entire problem, but it is the biggest part of it. This involves the ability to perform a mental rotation. The easier this rotation is, the easier the task is. Unfortunately, mental rotation is not a level playing field. Some of us are better at it than others, so much so that it affects the way individuals perform navigation tasks (McGee, 1979; Thorndyke & Stasz, 1980). We all know individuals who maintain a high level of proficiency in navigating environments, including environments they have not ever been in before. Alternatively, there are individuals who will get lost on the way home from work if they are detoured from their usual route.

19.2.1.3   Other Techniques and Tools

Today, dynamic maps are commonplace, so much so that we often take it for granted that we will not become disoriented as long as we have a smartphone in our pocket. However, cellular coverage, noisy signals, and the *deniability* of mobile navigation services (e.g., they can be jammed) suggest that there is still a need for basic navigation skills. Furthermore, as we will see later in the chapter, what we take for granted in mobile mapping services follows the same design principles that we apply to VEs.

FIGURE 19.1 Google StreetView™. The user remains oriented via several well-known tools. There is a compass in the upper left corner and a moving map in the lower right corner. The map not only shows where the viewer is located but the arrow shows which way they are looking Figure 19.1. (From Google, Inc., Street view, 2014, https://www.google.com/maps/views/streetview?gl=us, viewed on May 9, 2014.)

Beyond maps, there are other media that have utility for navigation. Google StreetView™, for example, is a pseudo-video interface that is visually correlated with a map view. It follows Levine's principles for user orientation discussed earlier by showing the position and direction of view of the user (Figure 19.1).

The next extension of this technique is Google Glass™, which will take the next step of melding virtual and real by using an eye-worn display with projected spatially relevant information. By aligning the map to the real world, users instantly understand their relative location and direction. The precursor to this technique can be seen in earlier work by Höllerer, Feiner, Terauchi, Rashid, and Hallaway (1999). Voice control will be an issue since this is an eye-worn device used while in motion. Principles from navigation systems for drivers will likely be adapted for pedestrian navigation (Figure 19.2).

Lastly, the use of video and still images is also a means to acquire spatial information about a space. While it may appear that video is the dynamic counterpart to still images, their uses are quite different and somewhat nonobvious. Still images are useful for landmarks but not the routes between them. Video could be used to capture routes that connect landmarks; however, a cautionary note is that viewing a video of someone else's movement through a space is an extremely difficult way to gain spatial knowledge. Partly due to the lack of vestibular feedback and peripheral view, landmarks can be identified but it is difficult to remain oriented, and consequently, spatial layout can be incomprehensible if only video is used (Darken, Kempster, & Peterson, 2001).

19.2.2 Representations of Spatial Knowledge

Arguably the most important part of this puzzle is the part we understand the least. Once spatial knowledge is acquired, how is it organized in the brain for future use? Spatial knowledge must be organized in some way such that it can be used during navigation tasks. The term

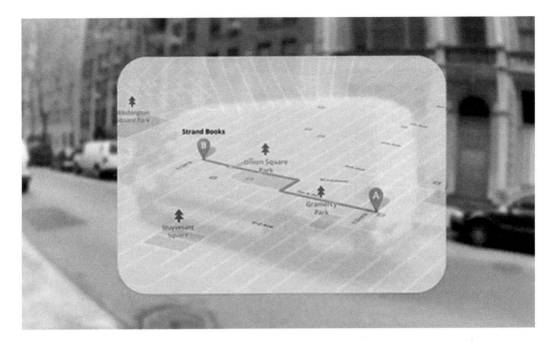

FIGURE 19.2 Google Glass™ concept. Using an eye-worn display, the Google Map™ is presented to the user in context. The selected route is clearly identified and aligned with the environment. (From Santhosh, K.D., Google glass goes online, 2014, http://www.digitash.com/2012/04/google-glass-goes-online.html, viewed on May 28, 2014.)

cognitive map was first used by Tolman to describe a mental representation of spatial information used for navigation (Tolman, 1948). However, 50 years later, we still do not have any hard answers about the neurological structure of spatial knowledge.

The representation of spatial knowledge is affected by the method used to acquire it. Knowledge acquired from direct navigation is different from knowledge acquired from maps. After studying a map of San Francisco before an initial visit there, information gleaned from the map is in a north-up orientation, just like the map. Consequently, when entering the city from the north, the mental representation of the city is upside down from what is being seen. This requires a 180° mental rotation before the cognitive map can be used to direct movement. Spatial knowledge acquired from maps tends to be orientation specific (McNamara, Ratcliff & McKoon, 1984; Presson & Hazelrigg, 1984; Presson, DeLange, & Hazelrigg, 1989). This implies that accessing information from a misaligned cognitive map is more difficult and error prone than if it were aligned (Boer, 1991; Rieser, 1989).

The most common model of spatial knowledge representation is the Landmark, Route, Survey (or LRS) model described by Seigel and White (1975) and Thorndyke and Goldin (1983). This model addresses not only spatial knowledge but also the development process. The theory states that we first extract landmarks from an environment. These are salient cues but are static, orientation dependent, and disconnected from one another. Landmark knowledge is like viewing a series of photographs. Later, route knowledge develops as landmarks are connected by paths. Route knowledge can be thought of as a graph of nodes and edges that is constantly growing as more nodes and edges are added. Finally, survey knowledge (or configurational knowledge) develops as the graph becomes complete. At this point, even if I have not traversed every path through my environment, I can generate a path on the fly since I have the ability to estimate relative distances and directions between any two points. This model directly fits the elements of urban environments described by Lynch (1960).

The most important caveat to this development process has to do with the use of maps. Maps allow us to jump over the route knowledge level and proceed directly to survey knowledge since they afford a picture of the completed graph all at one time. However, there is no free lunch. Survey knowledge attained from maps is inferior to survey knowledge developed from route knowledge and direct navigation because of the orientation specificity issue described earlier. What is currently unknown is what the impact is of modern mobile mapping tools like Google Maps™. Does survey knowledge develop faster using dynamic maps or slower? Is it more or less robust? Many of the classic experiments in the spatial knowledge acquisition literature bear repeating with modern tools to help us better understand the impact these devices have on learning and performance.

A modification to the LRS model is the hierarchical model (Colle & Reid, 1998; Stevens & Coupe, 1978). In some cases, direct exposure to an environment for extremely long durations never results in survey knowledge (Chase, 1983). In other cases, survey knowledge develops almost immediately. The model proposed by Colle and Reid suggests a dual mode, whereby survey knowledge can be acquired quickly for local regions and slowly for remote regions. The *room effect* comes from the ability to develop survey knowledge of a room rather quickly, but survey knowledge of a series of rooms develops relatively slowly and with more errors.

19.2.3 MODELS OF NAVIGATION

Understanding how navigation tasks are constructed is useful in determining how best to improve performance. If it were possible to decompose navigation tasks in a general way, we might be able to determine where assistance is needed or where training can occur. Several attempts have been made at such a model (Chen & Stanney, 2000; Darken, 1996; Downs & Stea, 1977; Neisser, 1976; Passini, 1984; Spence, 1998), but most are either too specific to one type of environment or they do not capture the intricacies of the entire task.

The model proposed by Jul and Furnas (1997) is relatively complete in that it incorporates the motion component into the process in a way not attempted before (see Figure 19.3). The model works like this. I am at the shopping mall and decide I need a pair of shoes. I have just formulated a goal. How should I go about finding shoes? I decide to try the department stores. Department stores are typically on the far points of the mall. I have just formulated a strategy. The next step is to gather information so I do not walk off in a random direction. I decide to seek out a map of the mall. I am acquiring information and scanning (perceiving) my environment. This is the tight wayfinding/motion loop referred to earlier. I view my surroundings, assess my progress toward my

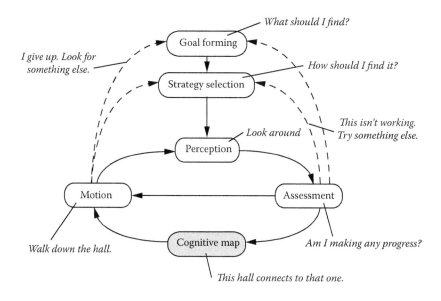

FIGURE 19.3 A model of navigation. (Adapted from Jul, S. and Furnas, G.W., *SIGCHI Bull.*, *29*(4), 44, 1997.)

goal, and make judgments as to how to guide my movement. At any time in this loop, I may decide to stop looking for shoes and look for books instead. This is a change in goal. I could also decide to look for a small shoe store instead of a department store. This is a change of strategy. In any case, the task continues, shifting focus and process as necessary.

An important point is that navigation is a situated action (Suchman, 1987). Planning and task execution are not serial events but rather are intertwined in the context of the situation. It is not possible nor practical to consider the task, the environment, and the navigator as separate from each other. Observable and measurable behavior is a product of these factors, yet the relationships between them are, as yet, poorly understood.

In the real world, this process is performed so often that it is typically automatic. When we know where we want to go, we go there. When we do not know where to go, we ask someone or look for some other source of assistance. VEs are an entirely different animal. Movement is not typically so intuitive. We have to think about it. Knowing where to go is problematic. While display resolution is improving, application designers are reluctant to surrender pixels for navigation aids. Who are you going to ask for help? How do you get unlost? These are the issues we will discuss next.

19.3   NAVIGATION PERFORMANCE ENHANCEMENT

This section will look at ways to help people navigate a virtual space, without regard to whether or not their improved performance transfers outside of the VE. The most obvious way to address navigation problems in complex VEs is to provide some sort of tool or mediator that can be used directly on the task at hand. This, of course, is the history of the map, the compass, the sextant, and the chronometer, to name a few real-world tools. Used together, these tools are able to help their users determine their position in the environment, their direction of travel, and the relative position of other objects or places in the environment.

An alternative to mediators is the organization of the space itself. There are a few real-world vocations related to this including architectural design and urban planning, each of which has extensively studied the relationship between people and their environment. They are interested in much more than just navigation, of course, but there is much to learn from these disciplines about how to construct space in a meaningful way in which people can comprehend and operate effectively. In VEs where the contents of the environment are not designed, such as in a flight simulator, there is not much that can be done in the way of spatial organization. However, in other applications, such as many scientific visualizations, it is possible to organize the data expressly for the purpose of navigability.

19.3.1   NAVIGATION TOOLS AND MEDIATORS

This section discusses a number of navigation tools and mediators that we have investigated in our laboratory. While they are similar to others in the literature, it should be noted that our intent is to investigate principles for the design of navigable VEs, not merely new techniques. It is unlikely that one of these tools as described here will perfectly match the needs of a real application. But by explaining how certain types of information affect behavior and its related performance, designers can mix and match the techniques described here and elsewhere to construct custom-built tools specifically for the demands of their application.

There is an extensive literature regarding tactile and haptic cues in navigation that we will not discuss here (but see Dindar, Tekalp, & Basdogan, 2013; Chapter 5 for more detail on these technologies and their general application to VEs). Much of this work has been intended to assist the visually impaired with surrogate nonvisual navigational cues (e.g., Caddeo, Fornara, Nenci, & Piroddi, 2006). It is important to note that the principles remain the same—providing relative directional cues to the in situ navigator and exocentric, global views (even via tactile maps) for planning or explicitly learning the layout of a space.

19.3.1.1 Maps

The navigation mediator people are most familiar with is the map. Maps are extremely powerful tools for navigation because of the wealth of information they can provide and the rate at which people can digest this information. However, their use in any VE application is not to be taken lightly. There are right ways and wrong ways to use maps. Maps come in a variety of forms, the differences usually being in terms of symbology or projection. However, VEs have certain qualities the real world does not have that make the use of maps in VEs different.

It is possible to navigate directly on a map in a VE. Rather than use the map to determine where to go in the VE, why not just point on the map to where you want to go? This has been attempted several times. The worlds-in-miniature (Stoakley, Conway, & Pausch, 1995) metaphor was one such implementation. In this case, a scaled-down version of the world, a virtual map, was held in the hand. Movement could be specified directly on the virtual map. Similarly, the use of maps can be a moded or unmoded task in a VE, meaning that its use can be in lieu of or concurrent with motion. In the real world, it is typically unmoded. Under certain conditions, you might want to stop moving to read a map but it is not required. Some games and VEs mode the map so that map use precludes motion.

For maps of very large VEs, there is a scaling problem. How do I view the map such that I can see the detail I need to navigate but still maintain a sense of the overall space? This is a classic problem of navigation in any problem domain (e.g., Donelson, 1978; Furnas, 1986). There are ways to zoom into a map or otherwise scale it to a usable level. The familiar *pinch zoom* popularized by the Apple iPhone™ would be one such technique.

The last issue we will discuss about maps has to do with the orientation of the map. As we discussed earlier, the orientation of a map with respect to the viewer has a strong effect on the viewer's ability to perform the mental rotation required to use a map during navigation. However, if we already know that forward-up maps are best for egocentric tasks and north-up maps are best for geocentric tasks, is this not a moot point? The fact that many video games continue to use north-up maps is enough to warrant an investigation. Is it possible that we can put enough redundant information on a north-up map to make it equivalent to a forward-up map?

We created two very large VEs: one was a sparse environment and the other was an urban environment (Darken & Cevik, 1999). Participants were asked to locate objects in the space. Sometimes they were shown the targets on the map (a targeted search), sometimes they had to return to known targets that were not shown on the map (a primed search), and other times they had to locate a target not shown on the map and not seen before (a naïve search). We timed their performance and marked wrong turns where participants clearly moved away from a target rather than toward it. At the conclusion of each trial, participants were asked to sketch a map of the environment from memory.

We created two virtual maps: one was in a north-up configuration and the other was in a forward-up configuration (see Figure 19.4). Both maps had a *you-are-here* marker that dynamically moved across the map as the user moved through the environment. The only difference on the maps, besides their orientation, was that the you-are-here indicator on the north-up map was a cone while the you-are-here indicator on the forward-up map was a small sphere. This was necessary because a north-up map does not indicate direction implicitly while the forward-up map does.

What we found was comparable to the results of Aretz and Wickens (1992) but to a lesser degree. The forward-up configuration seems to indeed be best for egocentric navigation tasks while the north-up map is best for geocentric tasks. We also found that individuals with high spatial abilities (as measured with the Guilford–Zimmerman standardized tests) are able to use either type of map better than participants with low spatial abilities on similar tasks. Furthermore, we found that these principles apply across different types of environments with vastly different spatial characteristics, but sparse environments seem to exhibit less of a performance difference than dense environments. Some of the same principles were applied in a subsequent helicopter navigation study that focused on map-to-ground correlation (Sullivan, Darken, & McLean, 1998) confirming that the mode of travel does not substantially alter the utility of maps when oriented optimally.

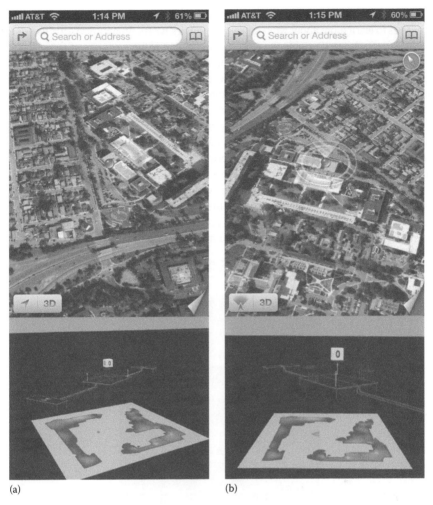

(a)   (b)

FIGURE 19.4 The *north-up* configuration of Apple Maps™ (a) and the *forward-up* configuration (b). Notice that the forward-up map includes a compass in the upper right corner. Both of these configurations were tested in a VE long before mobile technology enabled the types of maps we all use today. (From Apple, Inc., iOS 7 maps, 2014, http://www.apple.com/ios/maps/, viewed on May 9, 2014.)

VE designers should make virtual map decisions by carefully weighing the priorities of navigation tasks versus the spatial ability of their users.

Our results would suggest one improvement to the Apple Maps™ design shown in Figure 19.4. The forward-up configuration centers on the user's position and renders a cone indicating the direction of view even though the direction of view is obvious—it is forward. The north-up configuration displays the user as a blue sphere that can be problematic because users do not know what direction they are facing. Imagine using your Apple Map™ in north-up mode as you come out of a subway exit to street level. How would you know which way to go? Getting these design details right is important because another factor our work identified is that regardless of how they perform, people like what they are comfortable with. Therefore, our job is to provide the best tools no matter what the user selects.

19.3.1.2 Landmarks

We learned earlier that landmarks are extremely important to spatial knowledge acquisition and representation. In the real world, landmarks are critical to all modes of travel—ground (on foot or in a vehicle), air, or sea. For example, the use of coastal features as landmarks in nautical navigation is a fundamental technique that all sailors utilize. As such, it might be useful to allow users to affect the placement of landmarks themselves to learn more about what makes a landmark useful. If landmarks are useful as *anchors* on which to relatively place other objects, what would happen if the locations of the objects remained constant but we allowed users to insert highly salient landmarks on which to anchor their locations?

Participants were given a set of 10 different colored markers to place on the surface of the environment at any position. These markers were visible from a far greater distance than the objects themselves. We wondered where they might be placed and how this might affect performance. Markers tended to be placed in between objects such that as one marker would disappear from view, another would appear on the horizon. Moving between markers and objects, the participant could always have something to *hold on to* much like a *handrail*[*] is used in sport orienteering. We learned that for most individuals, being in a void, even a partial void like this environment, is very uncomfortable. Most people need regular reassurance that they are not lost. Only the most advanced navigators do not seem to need this kind of assurance.

We also tried this same condition but combined it with a forward-up map (see Figure 19.5). As participants would place a marker, it would appear both in the environment and on the map. Would the strategy remain the same or would the map override the utility of the markers?

FIGURE 19.5 *Dropped pins* in Google and Apple Maps are very similar to the landmark + map treatment with very similar resulting use strategies and behaviors. (From Apple, Inc., iOS 7 maps, 2014, http://www. apple.com/ios/maps/, viewed on May 9, 2014.)

---

[*] A handrail is a linear feature, such as a stream or road, which land navigators use to guide movement. They typically use it to constrain movement, keeping it to one side and traveling along it for some distance.

This turned out to be unquestioningly the favorite condition of everything we tried. The reason is simple. We still did not give them any help in locating the target objects. But as they found them, they no longer had to remember anything about them. They simply placed a marker at each object. The markers in the environment were not used at all. But the markers on the map were like colored *pushpins*. All they had to do was remember what color coincided with what target and the posttrial map drawing exercise was trivial. What are *pushpins* on Apple and Google Maps if not user-placed landmarks that appear on the map but not in the environment? The strategies and behaviors we observed in the VE are indeed similar to what we see today using maps on mobile devices.

19.3.1.3 Trails and Directional Aids

Since participants used the markers to create a sort of *trail* connecting objects with markers and other objects, we decided to revisit an old idea. If a trail was left behind such that it showed where the participant had been, this might be an even better tool than the markers (see Figure 19.6). We called this the *Hansel and Gretel* technique (Darken & Sibert, 1993). A better analogy is that of footprints since footprints are directional and breadcrumbs are not.

FIGURE 19.6 The *trail* technique looks very much like the route following display on most navigation aids, but it operates in a reverse fashion. We dropped a trail as the user moved through the environment while the route on a navigation aid shows the user where to go. (From Apple, Inc., iOS 7 maps, 2014, http://www.apple.com/ios/maps/, viewed on May 9, 2014.)

This technique is useful for scanning space in an exhaustive search (e.g., a naïve search). One of the problems in an exhaustive search is knowing if you have been in some place before. An optimal exhaustive search never revisits the same place twice. However, with this technique, the environment becomes cluttered with footprints after a time. It is more useful to identify *regions searched* than *paths followed*.

Thus far, we have discussed a number of tools and mediators that deal with absolute position information. The maps showed exactly where the user was at all times. The markers specify an exact location in the environment. Each footprint designates an exact location as well. What about orientation? How important is it to know absolute direction versus absolute position?

To study this, we tested two simple tools, both of which show only direction—a virtual compass and a virtual sun. We found that neither tool was particularly useful when used by itself, but when used with other techniques, it does add value. The compass in the upper right corner of an Apple Map™ (see Figure 19.5) is a good example. The compass by itself is marginally useful, but when paired with a forward-up map in this case, it provides the necessary information to execute a wide variety of navigation tasks.

19.3.2 Organizational Remedies

It could be said that the tools and mediators described in the previous section should be the last recourse of the designer if all else fails. The fact that modern mobile navigation aids have developed many of the same techniques we were studying for VEs several years ago suggests that when you cannot redesign your environment, these are effective tools to assist the navigator. But what if you could redesign the environment? If we were constructing a VE visualization of the stock market, for example, we would have complete control over how objects looked, where they were located, and how they were organized.

19.3.2.1 Environmental Design

Passini (1984) talks about the use of an *organizational principle* in architectural design. If a space has an understandable structure and that structure can be made known to the navigator, it will have a great influence on the strategies employed and resulting performance on navigation tasks. For example, knowing that Manhattan is generally a rectangular grid is of great benefit to any navigator. Given the knowledge of the grid and its orientation, words like *uptown* and *downtown* instantly have meaning. But while there is great power in using such an organizational principle, there is a danger that goes along with it. Violations to that principle will have a much greater negative effect than they might otherwise have. For example, we know that Manhattan is generally a grid, but Broadway cuts through on an angle, violating the grid principle. A naïve tourist thinking that the grid principle held throughout may be misled. It is important to develop a clear organizational principle and stick to it throughout the environment. If it must be violated, it should be made clear where the violation occurs and that it is a violation; otherwise, the navigator may attempt to erroneously fit it into a cognitive representation.

Organizational principles can also add meaning to cues that might be seen during navigation. Again using our Manhattan example, if we are looking for 57th St., street signs for 44th St., 45th St., etc. tell us that we are not only going in the right direction but they also give us a rough estimate of distance as well.

Game designers have become highly proficient at building virtual worlds that are highly navigable through good design. They typically use *channels* that give the player the feeling of an open environment when it actually is not very open at all. This was a particularly important design factor in early high-fidelity gaming, but recently the rendering performance of game engines has improved to the point that open environments are now commonplace (e.g., most of the Rockstar Games titles such as *Grand Theft Auto* are all open environments). In these worlds, the player is mostly unconstrained and can move at will throughout the space. *Playability* of these games is very much tied to navigability. *Liberty City*[*] is a well-designed virtual city that illustrates Lynch's principles of landmarks, routes, nodes, edges, and districts. A player who finds the need to constantly refer to the map in GTA will not perform well. They must learn the space, and to learn the space, it must be designed well.

---

[*] Liberty City is the fictional city used in *Grand Theft Auto* developed by Rockstar Games.

It is common to see urban metaphors applied to unstructured environments such as websites because the structure of the city can add meaning to information that may be viewed as otherwise amorphous. This works in some cases, even when there is no obvious semantic connection between the information presented and the city metaphor. In other cases, this fails because the constraints implied by an urban landscape do not coincide with the constraints of the information presented. This also applies to most persistent online virtual worlds like Second Life™ where users often cannot connect spatial layout to meaning. Consequently, this should be used judiciously with an understanding that there is no free lunch. Using a metaphor to simplify navigation may unknowingly inhibit some other piece of functionality. It is important when studying user performance in these types of applications to study the whole task, not just the navigation component. Again, navigation is not an isolated action but is situated in some other higher-order task.

19.3.2.2 Visualization

There have been attempts to apply Lynch's elements of the urban landscape to abstract VEs (Darken, 1996; Ingram & Benford, 1995). In principle, this will work, but implementations seem contrived as if we are forcing an inappropriate structure on abstract data. It may be that the very generic nature of VEs might demand a different set of elements similar to Lynch's but not necessarily identical.

Darken and Sibert (1996) commented that people inherently dislike a lack of structure. A person who is in an environment that is nearly void of useful cues and that does not suggest ways to move through it is generally uncomfortable. Users will grasp at anything they can view as structure whether or not the designer intended it to be that way. In earlier experiments, we observed participants using coastlines and the edges of the world as paths even though this was not a particularly effective strategy toward completion of the task. People do this because it is all they have to work with. If even a simple bit of structure is added such as a rectangular or radial grid, performance immediately improves. A path should suggest to the navigator that it leads to somewhere interesting and useful.

In that sense, basic environmental design principles do apply to abstract visualizations. The example in Figure 19.7 has no obvious structure to it, but the way in which it is displayed to

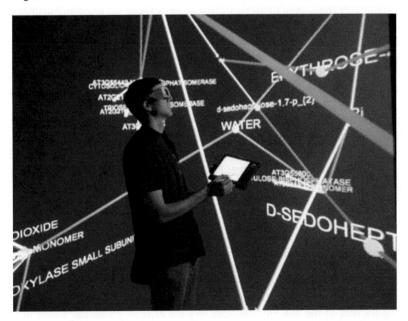

FIGURE 19.7 This is a visualization of hurricane and ocean surges for educating science teachers and the public through interactive displays in aquariums and ocean science centers. It has no obvious structure but could be organized using Lynch's principles. (Image courtesy of Carolina Cruz-Neira, CREATE, University of Louisiana at Lafayette, ©2013.)

the user could follow a simple organizational principle or sectioning (Lynch would call this a *district*) to aid the user's spatial comprehension.

19.4   ENVIRONMENTAL FAMILIARIZATION

Imagine visiting places using a VE before you actually got there. If it were a vacation spot or resort, you might like to see if you like the views, the beaches, or anything else that might catch your interest. If you were a soldier about to enter a hostile environment, you might use a VE to rehearse a planned route and familiarize yourself with the area.

In these examples, the VE in question is not a mere abstraction but is a representation of a real environment. In each case, we are not interested in how well the user navigates the VE but rather in how they navigate the real world after exposure to the VE. This section will examine several empirical studies involving the use of VEs for environmental familiarization. Studies of this kind are central to the topic of this chapter for two primary reasons:

1. Such investigations highlight the differences between navigation in the real world and in VEs, whereas studying VE navigation alone does not. Studying behavioral differences between virtual and real-world navigation may provide insights to help us better understand and model human navigation: virtual or real. People know how to navigate in the real world. They bring this knowledge with them into VEs. While this does not mean that VE navigation must replicate real-world navigation in every way, it makes sense that we need to understand how humans navigate in the real world before we can optimize methods for navigation in the VE.

2. While the potential use of VEs to enhance training for real-world performance of many tasks has been touted from the inception of the technology, few application domains have clearly demonstrated a significant enhancement. Environmental familiarization (or mission rehearsal) is among these applications. In fact, a close inspection of the literature will lead to the conclusion that there is much confusion over whether or not VEs offer a significant enhancement over traditional methods of spatial knowledge acquisition. Furthermore, even if we knew they were useful, we still do not know exactly how they should be used to optimize the positive effects we want while minimizing the negative effects (e.g., reverse training) we do not want. Transfer of spatial information might be a near-term training domain that is within the reach of current technology. So, environmental familiarization may represent a microcosm of the issues involved in virtual training of any knowledge domain.

A key aspect of the studies presented in this section is that performance measures are made in the real world to evaluate how much spatial information was acquired in the VE or how the VE may have affected behavior. This is not to say that measurements are never made in the VE, only that without real-world measurements, we cannot know what affect, positive or negative, the VE tool may have had on the participants.

19.4.1   SPATIAL KNOWLEDGE TRANSFER STUDIES

To ground this discussion, we compare four studies. There are many more studies of this type in the literature but we have selected a subset for presentation here. We are particularly interested in the methodologies behind these studies, not the technology or applications themselves. Each of these is a transfer study implying that there is a training phase involving a VE of some type and a testing phase involving transfer to the real environment. We are also only interested here in studies about environmental familiarization rather than skill development so the VE must replicate a specific real environment rather than some generic real environment such as learning to navigate in a generic city.

The purpose of this comparison is to systematically point out the similarities and differences in the studies so that we can make some statement about what is currently known about the use of

VEs for environmental familiarization. An issue that will become very clear is that there is little consistency in the literature about what to study or how to study it. Consequently, we see a variety of experiments controlling a variety of parameters but in a way that it is difficult, if not impossible, to leverage off of what was done previously.

We will briefly discuss four experiments and conclude with a discussion of environmental familiarization and how VEs might be used for these tasks. The experiments are as follows:

- Witmer, Bailey, and Knerr: *Virtual Spaces and Real World Places: Transfer of Route Knowledge*
- Darken and Banker: *Navigating in Natural Environments: A Virtual Environment Training Transfer Study*
- Koh, von Wiegand, Garnett, Durlach, and Shinn-Cunningham: *Use of Virtual Environments for Acquiring Configurational Knowledge About Specific Real-World Spaces: Preliminary Experiment*
- Waller, Hunt, and Knapp: *The Transfer of Spatial Knowledge in Virtual Environment Training*

19.4.1.1 Basis for Comparison

In an attempt to make a meaningful comparison between these experiments, we will look at each in terms of a structured set of parameters. The key elements for this comparison are (1) the characteristics of the human participants, (2) the characteristics of the environment, (3) the characteristics of the tasks to be performed, (4) the characteristics of the human–computer interface, and (5) the characteristics of the experimental design to include dependent measures. Table 19.1 is a summary of the four experiments in terms of these elements.

While each experiment investigated human navigation performance, the characteristics of the participant sample must be taken into account. Traditionally, important issues such as the quantity, age, and gender of the participants are considered. However, navigation is a specialized task, so we suggest that two specific differences may be quite critical. First, some of the experiments situated the navigation task within a higher-level task context. In those cases, participants' experience in the respective domain could be expected to influence both motivation to participate and task competence level. Second, individuals enter the experiment with a given level of innate spatial ability. Although measures

---

TABLE 19.1

Comparison of Training Transfer Studies

Experiment	Participants	Environment	Tasks	Interface	Measures
Darken and Banker	Domain expertise	Natural, unstructured	Maps used, 60 min exposure, route planning and execution	Keyboard, mouse, desktop display	Deviation from route
Witmer, Bailey, and Knerr	No domain expertise	Architectural, structured	No maps used, 15 min repeated exposure, route replication	Immersive display, buttons	Wrong turns, bearing/range estimation, time
Koh, Durlach, and von Wiegand	No domain expertise	Architectural, structured	No maps used, 10 min exposure, survey knowledge	Both immersive and desktop displays, joystick	Bearing/range estimation
Waller, Hunt, and Knapp	No domain expertise	Structured	No maps used, 2 or 5 min exposure, route replication, path integration	Both immersive and desktop displays, joystick	Time, bumps into walls

Sources: Darken, R.P. and Banker, W.P., Navigating in natural environments: A virtual environment training transfer study, Paper presented at the *IEEE Virtual Reality Annual International Symposium*, Atlanta, GA; Witmer, B.G. et al., Training dismounted soldiers in virtual environments: Route learning and transfer, Technical Report 1022, U.S. Army Research Institute for the Behavioral and Social Sciences, 1995; Waller, D. et al., *Presence Teleop. Virt. Environ.*, 7(2), 129, 1998.

of individual spatial and navigation ability may provide ambiguous results, attempts to quantify and categorize participants based upon individual ability can help to explain differences in performance.

The real-world environment that is modeled in the VE influences navigation behavior. Just as individuals possess differing navigation ability, so do various environments afford different navigation experiences. Some real-world environments simply provide more navigation cues than others. Furthermore, some environments lend themselves to a higher model fidelity level than others. How closely a VE matches its real counterpart is referred to as *environmental fidelity* (Waller, Hunt, & Knapp, 1998). However, do not assume that higher environmental fidelity must correlate with higher performance. There are other issues that are equally important.

With so many varieties of VEs and their associated interfaces, it is necessary to be more descriptive in terms of the specific differences between them. The devices and interaction styles used by the system provide differing levels of sensory stimulation to the user. How closely a VE interface matches the interface to the real world (e.g., walking, driving) is referred to as *interface fidelity* (Waller et al., 1998). Again, do not assume that higher interface fidelity equates to higher performance or training transfer. This has not yet been established and it is unclear if we will eventually find that to be the case.

Desktop VEs channel visual output to a computer monitor that rests on the desktop. Immersive VEs use a head-mounted unit or projection system to display the world to the user. Still, within both the desktop and immersive categories, the mix of input and output devices requires a more detailed description. The power of the system to deliver high-fidelity multimodal output and monitor user input commands in real time is a critical discriminator. The primary issue with interface fidelity in the studies we are concerned with has to do with the motion technique; specifically, what interaction method is used to control speed and direction of travel? Finally, some interfaces provide the user with special abilities and computer-generated tools that further differentiate one from another.

Experimental tasks, conditions, and standards differ across experiments. Since transfer studies consider both the training task and the testing task, we must consider both cases. The experimenter's instructions to the participant will constrain task performance. So, two items of interest are the procedures and the dependent measures. Other items of consideration include the use of maps and exposure duration to the VE.

While the experiments investigate a wide range of issues related to knowledge transfer, the central issue is what is learned and how is it applied to the real environment. The level of spatial knowledge acquired is of particular interest. A study that develops route knowledge but then tests survey knowledge may mislead the reader to believe that some other factor was the cause of low performance. Even if a system had the right interface and the right level of fidelity for a given rehearsal task, it can be used incorrectly resulting in poor performance on the transfer task in the real world.

19.4.1.2 Experiments

Witmer et al. (1995) were among the first to show that a VE could be useful for spatial knowledge acquisition. They compared a VE to the real world in an architectural walkthrough application. They used an immersive display and their population was a random sampling without any expertise on the task. Motion was controlled by gaze-directed movement using buttons on the display.

The experimental protocol divided the session into four stages: individual assessment, study, rehearsal, and testing. During the individual assessment stage, participants responded to numerous questionnaires, some of which probed their sense of direction and navigation experience level. Next, regardless of experimental treatment condition, every participant was given 15 min to study written step-by-step route directions and color photographs of landmarks. Half of the participants in each treatment condition were also provided a map of the building as a third study aid. Following the 15-min study stage, each participant rehearsed the route three times, either in the VE or the real world depending on their group. All participants were required to identify six landmarks on the route, and researchers provided immediate correction if a participant made a wrong turn or misidentified a landmark. Finally, participants were tested in the real building. They were asked to replicate the route they had learned and to identify the six landmarks. The route replication measures included attempted wrong turns, route traversal time, route traversal distance, and misidentified landmarks. Configurational knowledge was tested by requiring participants to draw a line on a map from their known location to an unseen target.

This study effectively showed how landmark knowledge can become route knowledge but survey knowledge was not given the opportunity to develop. Survey knowledge takes time to develop by primary exposure to an environment. Exposure times for this experiment were not long enough for this to occur. Nevertheless, this study effectively provided optimism that the technology could work for this purpose—but not how well or under what conditions.

Darken and Banker (1998) studied how a VE might be used as an augmentation to traditional familiarization methods. They compared performance of three groups: a map-only group, a VE group that also had the use of the map, and a real-world group that also had the use of the map. The interface to the VE was a desktop display controlled with a keyboard. The environment used was a natural region of central California with a few man-made structures but largely vegetation and rough paths (see Figure 19.8). They used a participant population with specialized knowledge of the task, specifically sport orienteers and experienced military land navigators.

The experimental session was comprised of two phases: planning/rehearsal and testing. During the planning/rehearsal portion of the session, which lasted 60 min, participants studied and created a personal route from the starting point to nine successive control points. By the end of the planning phase, participants were required to draw their planned route on the map. Testing involved execution of the planned route in the real environment without the aid of the map or compass. As the participant navigated the real-world course, the researcher followed, videotaping participant behavior with a head-mounted camera. A GPS unit worn by the participant recorded their position. This information was used to measure the quantity of unplanned deviations from the route and the total distance traveled. In addition, the frequency of map/compass checks was recorded as a dependent measure.

FIGURE 19.8 The top image is a photograph from the testing area in the former Ft. Ord, California. The bottom image is a snapshot from virtual Ft. Ord at that exact location.

This study is unique in many respects. It required the participants to plan their own routes rather than practice a given one. This serves to develop survey knowledge since alternative routes must be explored. They also attempted to introduce individual experience as a factor in addition to spatial ability. Since the task was specific to a particular domain, experience on these types of tasks should have an effect.

The results show that only intermediate participants seem to improve with the use of the VE. Beginners have not yet developed enough skill at the task to be able to make use of the added information the VE offers, and advanced participants are so highly proficient at map usage that the VE simply does not add much information they cannot gain in other ways. On a complex environment such as this, even the hour of exposure provided does not seem to be enough. Given very short train-up times, maps still seem to be the best alternative for spatial knowledge acquisition.

Koh, von Wiegand, Garnett, Durlach, and Shinn-Cunningham (2000) expanded previous work by specifically looking toward the development of configurational knowledge in architectural environments. In their experiment, they compared a real-world group, an immersive VE group, a desktop VE group, and what they call a *virtual model* group that is similar to a noninteractive world in miniature that is held in the hand. They used a general population sample and varied the interface to the VE as described by the group (immersive or desktop). In the two VE conditions, participants controlled their motion direction and speed with a joystick. The desktop group viewed a typical computer monitor. The immersive group members wore a head-mounted display. Their heads were tracked, and head rotation updated the visual scene although gaze direction was not linked to the direction of travel.

The experimental sessions were split into three phases: administration, training, and testing. During the administrative phase, the participants were informed about the bearing and distance estimation task, although the specific stations and targets were not disclosed. Participants in the three non-real-world groups then were provided a period of time to familiarize themselves with the interface. Training consisted of 10 min of free exploration under the assigned treatment conditions. The members of the real-world group explored the real building and members of the VE groups explored the synthetic building. Testing was specific to configurational knowledge. Participants were asked to estimate distances and bearings to unseen targets in the real world while being transported from place to place while blindfolded in a wheelchair.

The purpose of this study was to specifically focus on the development of configurational knowledge. The experimenters were not interested in landmark or route knowledge, although they do reference the hierarchical landmark, route, and survey model. Furthermore, the researchers pronounced a bias that higher-fidelity experiences may not necessarily lead to better transfer of knowledge, hence the use of three different fidelities in different configurations. Their results show that the VE conditions do develop configurational knowledge at a comparable level to the real world.

Waller et al. (1998) used six different conditions based on the practice method. The conditions were real-world, map, no-study, and three different virtual groups—desktop, short-duration immersive, and long-duration immersive. The environment was a maze constructed of full-length curtains with targets placed at selected locations. They also used a general population sample with no specific experience in these tasks.

All participants in VE conditions controlled their motion using a joystick, and the immersive groups both used a head-mounted display. The short-duration immersive group spent a total of 12 min in practice, while the long-duration immersive group spent a total of 30 min in practice over repeated trials.

The experimental protocol is comprised of four phases. The first was administrative and included proctoring of the Guilford–Zimmerman spatial abilities test. The second phase tested route knowledge by interleaving practice and testing six times. The participant would make a practice run followed immediately by a testing run. The third phase tested survey knowledge. The experimenter altered the configuration of the maze so that portions of the learned route were now blocked. The blindfolded participant then had to find a new route from one location to another.

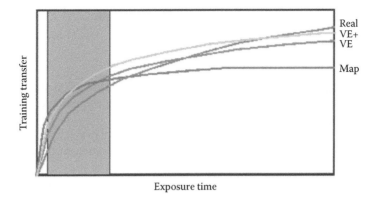

FIGURE 19.9   This graph shows a hypothetical picture of how spatial knowledge might be acquired over time depending on the apparatus used. Maps are best for short-term events, but the real world or VEs with training interventions (VE+) are best over time.

Dependent measures for both phases two and three included time to traverse the route and quantity of times the participant bumped into a wall. Finally, in the last phase, participants completed a pencil-and-paper test of their configurational knowledge.

This study showed that maps were just as effective as short durations of training, but that if enough time was given, the VE did prove to be more effective—even to the point of surpassing the real-world condition.

We attempt to explain all these results in the graph shown in Figure 19.9. Note that these curves are hypothetical since we cannot directly compare all the training transfer studies in the literature, and even if we could, there are far too few data points to establish the shape of the curves. But based on our research, we believe that given a short exposure duration, maps are better than any VE alternative simply because they do not overload the user with information that cannot be absorbed. However, the map is only so useful. Given enough time, the added information a VE may provide will increase performance. The dark vertical bar represents the time slice that encompasses most of the studies in the literature. Very rarely are participants given enough time to develop survey knowledge in any meaningful way. We also differentiate a general VE that is assumed to be only a virtual replication of the real space, from a specialized VE (VE+) that might have added training features such as aerial views, transparent walls, or other features that may enhance the training effect. However, it is important to note that current research in this area is attempting to determine what those features are. We only suggest here that we will eventually know what tools to use under what conditions in such a way that we can create VE systems for environmental familiarization that compare or even surpass the real world.

19.5   PRINCIPLES FOR THE DESIGN OF NAVIGABLE VIRTUAL ENVIRONMENTS

This final section will serve to summarize the chapter into a series of principles we have discussed in earlier sections. This is not to be interpreted as a design *cookbook* where designers can look to see how to select navigation aids for their application. We again stress that navigation is a highly aggregate task involving people, tasks, and environments. It is not possible to develop a design solution without addressing all three elements as a whole system and not as a set of parts. What we provide here is a set of guidelines based on the literature and on our experiments as a starting point for designers to address the important issues they may face. In some cases, a technique we have discussed may fit perfectly; in others, an adaptation may be needed. The key is in providing enough spatial information so that users can execute navigation tasks as demanded by the application without overconstraining the interface, thereby eliminating exploration and discovery.

19.6 PERFORMANCE ENHANCEMENT (SECTION 19.3)

19.6.1 TOOLS AND MEDIATORS (SECTION 19.3.1)

19.6.1.1 Map Usage (Section 19.3.1.1)

1. Maintain orientation but match map orientation to the task—predominantly egocentric tasks like searching should use a forward-up map while geocentric tasks such as exploration should use a north-up map.
2. Always show the user's position and view direction on the map and update dynamically.
3. The orientation problem will be more severe with a user population that includes individuals with low spatial abilities such as mental rotation—be aware of who your users are.
4. Video game play may have an effect on the selection of an appropriate map for a VE. If the user population is largely a gaming community, their spatial abilities are likely to be high.
5. Use moded maps (e.g., where the use of the map precludes motion) only where appropriate. The default method should be unmoded.

19.6.1.2 Landmarks (Section 19.3.1.2)

1. Allowing users to annotate the environment in some way to *personalize* the spatial cues they wish to use can be effective for complex spaces and is easily adaptable to a wide variety of navigation tasks.
2. Beware of giving the user the ability to clutter the space with excessive information. What they think will be helpful can become distracting noise.
3. Provide enough obvious landmarks (typically dependent on the context of the application and the task) so that the navigator has *reassurance* cues along a route to know that they are on the right path.
4. Make sure that landmarks are directional as well as salient—they should help provide orientation information to the navigator as well as position information.

19.6.1.3 Trails or Footprints (Section 19.3.1.3)

1. Simply leaving a trail is marginally useful since it tends to clutter the space. Making the trail such that it disappears over time is better but can be confusing since it no longer tells the navigator that this place has been visited before, only that it has not been visited lately.
2. Trails can be particularly effective for exhaustive searches. It may be appropriate to use them in the context of a specific exhaustive search but then turn them off afterward.

19.6.1.4 Directional Cues (Section 19.3.1.3)

1. Directional cues (e.g., sun, compass) alone will not be satisfactory as a navigation aid. They should be used with other techniques since they do not provide positional information.
2. Directional cues can be effective when moded.
3. Directional cues, when used with directional landmarks, are highly effective since they place landmarks in a global coordinate system.

19.6.2 ORGANIZATIONAL REMEDIES (SECTION 19.3.2)

19.6.2.1 Environmental Design (Section 19.3.2.1)

1. Use an organizational principle wherever possible and do not violate it.
2. If the organizational principle must be violated, make it obvious where and why the violation occurred so the navigator does not attempt to resolve the violation into the organization principle.
3. Match landmarks to the organizational principle whenever possible. They can be used to reinforce the shape of the space.

4. Do not blindly try to use the elements of urban design in any VE. They might not be appropriate. Keep in mind the key concepts: provide useful paths, observable edges, and usable landmarks and divide big, complex spaces into a number of smaller navigable spaces that are connected in some clear, understandable way.

5. Use an urban metaphor for abstract data judiciously. Make it clear where the metaphor fits and where it does not.

19.6.2.2 Visualization (Section 19.3.2.2)

1. Use explicit sectioning, particularly when implicit reorganization does not work for a particular environment.

2. Again, use an organizational principle and make it obvious.

3. Select a scheme for organizing your space based on what tasks people are likely to do there. If they are doing a lot of naïve searches, for example, make sure there is a way for them to easily and systematically explore the entire space without repetition.

19.7 ENVIRONMENTAL FAMILIARIZATION (SECTION 19.4)

1. Do not assume that because someone can efficiently navigate a VE that they can navigate the real world as well. This is simply not the case.

2. Beware of creating performance *crutches* by adding features in the VE that enhance performance there but that are not available in the real world.

3. Given a short amount of time with which to familiarize someone with an environment, use a map and maybe some photos if they are available.

4. Beware of developing orientation-specific survey knowledge if only maps are used. Given enough time, a VE can be used to develop orientation-independent spatial knowledge, but it takes time.

5. Be careful when deciding how to use a VE for environmental familiarization because it is extremely difficult to compare studies in the literature. Make a decision based on the whole problem—the people you are training, the tasks they are doing, and the environment they are navigating in.

REFERENCES

Apple, Inc. (2014). iOS 7 maps. http://www.apple.com/ios/maps/ (viewed on May 9, 2014).

Aretz, A. J., & Wickens, C. D. (1992). The mental rotation of map displays. *Human Performance, 5*(4), 303–328.

Boer, L. C. (1991). Mental rotation in perspective problems. *Acta Psychologica, 76*, 1–9.

Bowman, D., Koller, D., & Hodges, L. (1997). *Travel in immersive virtual environments: An evaluation of viewpoint motion control techniques.* Paper presented at the Virtual Reality Annual Internation Symposium (VRAIS), Albuquerque, NM.

Caddeo, P., Fornara, F., Nenci, A., & Piroddi, A. (2006). Wayfinding tasks in visually impaired people: The role of tactile maps. Dynamics in spatial interactions, *International Conference on Spatial Cognition.* Bremen, Germany.

Chase, W. G. (1983). Spatial representations of taxi drivers. In D. R. Rogers & J. A. Sloboda (Eds.), *Acquisition of symbolic skills.* New York, NY: Plenum.

Chen, J. L., & Stanney, K. M. (2000). A theoretical model of wayfinding in virtual environments: Proposed strategies for navigational aiding. *Presence: Teleoperators and Virtual Environments, 8*(6), 671–685.

Colle, H. A., & Reid, G. B. (1998). The room effect: Metric spatial knowledge of local and separated regions. *Presence: Teleoperators and Virtual Environments, 7*(2), 116–128.

Darken, R., Kempster, K., & Peterson, B. (2001). Effects of streaming video quality of service on spatial comprehension in a reconnaissance task. *Proceedings of I/ITSEC 2001.* Orlando, FL.

Darken, R. P. (1996). *Wayfinding in large-scale virtual worlds* (Unpublished Doctoral dissertation). The George Washington University, Washington, DC.

Darken, R. P., & Banker, W. P. (1998). *Navigating in natural environments: A virtual environment training transfer study.* Paper presented at the IEEE Virtual Reality Annual International Symposium, Atlanta, GA.

Darken, R. P., & Cevik, H. (1999). *Map usage in virtual environments: Orientation issues.* Paper presented at the IEEE Virtual Reality 99, Houston, TX.

Darken, R. P., Cockayne, W. R., & Carmein, D. (1997). *The omni-directional treadmill: A locomotion device for virtual worlds.* Paper presented at the ACM UIST '97, Banff, Alberta, Canada.

Darken, R. P., & Sibert, J. L. (1993). *A toolset for navigation in virtual environments.* Paper presented at the ACM Symposium on User Interface Software and Technology, Atlanta, GA.

Darken, R. P., & Sibert, J. L. (1996). Wayfinding strategies and behaviors in large virtual worlds. *ACM SIGCHI 96,* Vancouver, BC (pp. 142–149).

Donelson, W. C. (1978). Spatial management of information. *Proceedings of ACM SIGGRAPH '78,* Atlanta, GA (pp. 203–209).

Downs, R. M., & Stea, D. (1977). *Maps in minds: Reflections on cognitive mapping.* New York, NY: Harper & Row.

Durlach, N., & Mavor, A. (Eds.). (1995). *Virtual reality: Scientific and technological challenges.* Washington, DC: National Academy Press.

Furnas, G. W. (1986). Generalized fisheye views. *Proceedings of ACM SIGCHI '86,* Boston, MA (pp. 16–23).

Goldin, S. E., & Thorndyke, P. W. (1982). Simulating navigation for spatial knowledge acquisition. *Human Factors, 24*(4), 457–471.

Google, Inc. (2014). Street view. https://www.google.com/maps/views/streetview?gl=us (viewed on May 9, 2014).

Höllerer, T., Feiner, S., Terauchi, T., Rashid, G., & Hallaway, D. (1999). Exploring MARS: Developing indoor and outdoor user interfaces to a mobile augmented reality system. *Computers and Graphics, 23*(6), 779–785.

Ingram, R., & Benford, S. (1995). *Legibility enhancement for information visualisation.* Paper presented at the Visualization 1995, Atlanta, GA.

Jul, S., & Furnas, G. W. (1997). Navigation in electronic worlds: A CHI 97 workshop. *SIGCHI Bulletin, 29*(4), 44–49.

Koh, G., von Wiegand, T., Garnett, R., Durlach, N., & Shinn-Cunningham, B., (2000). Use of virtual environments for acquiring configurational knowledge about specific real-world spaces: Preliminary experiment. *Presence: Teleoperators and Virtual Environments, 8*(6), 632–656.

Levine, M., Marchon, I., & Hanley, G. (1984). The placement and misplacement of you-are-here maps. *Environment and Behavior, 16*(2), 139–157.

Lynch, K. (1960). *The image of the city.* Cambridge, MA: MIT Press.

McGee, M. G. (1979). Human spatial abilities: Psychometric studies and environmental, genetic, hormonal, and neurological influences. *Psychological Bulletin, 86*(5), 889–918.

McNamara, T. P., Ratcliff, R., & McKoon, G. (1984). The mental representation of knowledge acquired from maps. *Journal of Experimental Psychology: Learning, Memory, and Cognition, 10*(4), 723–732.

Neisser, U. (1976). *Cognition and reality: Principles and implications of cognitive psychology.* New York, NY: W. H. Freeman & Company.

Passini, R. (1984). *Wayfinding in architecture.* New York, NY: Van Nostrand Reinhold Company Inc.

Péruch, P., Pailhous, J., & Deutsch, C. (1986). How do we locate ourselves on a map: A method for analyzing self-location processes. *Acta Psychologica, 61,* 71–88.

Presson, C. C., DeLange, N., & Hazelrigg, M. D. (1989). Orientation specificity in spatial memory: What makes a path different from a map of the path? *Journal of Experimental Psychology: Learning, Memory, and Cognition, 15*(5), 887–897.

Presson, C. K., & Hazelrigg, M. D. (1984). Building spatial representations through primary and secondary learning. *Journal of Experimental Psychology: Learning, Memory, and Cognition, 10*(4), 716–722.

Rieser, J. J. (1989). Access to knowledge of spatial structure at novel points of observation. *Journal of Experimental Psychology: Learning, Memory, and Cognition, 15*(6), 1157–1165.

Rossano, M. J., & Warren, D. H. (1989). Misaligned maps lead to predictable errors. *Perception, 18,* 215–229.

Siegel, A. W., & White, S. H. (1975). The development of spatial representations of large-scale environments. In H. Reese (Ed.), *Advances in child development and behavior* (Vol. 10). New York, NY: Academic Press.

Spence, R. (1998). *A Framework for navigation* (Technical report 98/2). London, U.K.: Imperial College of Science, Technology and Medicine.

Stevens, A., & Coupe, P. (1978). Distortions in judged spatial relations. *Cognitive Psychology, 10,* 422–437.

Stoakley, R., Conway, M. J., & Pausch, R. (1995). *Virtual reality on a WIM: Interactive worlds in miniature.* Paper presented at the Proceedings of ACM SIGCHI 95, Denver, CO.

Suchman, L. A. (1987). *Plans and situated actions: The problem of human machine communication.* Cambridge, U.K.: Cambridge University Press.

Sullivan, J., Darken, R., & McLean, T. (1998, June 2–3). Terrain navigation training for helicopter pilots using a virtual environment. *Third Annual Symposium on Situational Awareness in the Tactical Air Environment.* Piney Point, MD.

Templeman, J., Denbrook, P. S., & Sibert, L. E. (2000). Virtual locomotion: Walking-in-place through virtual environments. *Presence: Teleoperators and Virtual Environments, 8*(6), 598–617.

Templeman, J., Page, R., & Denbrook, P. (2014). Avatar control in virtual environments. In K. S. Hale & K. M. Stanney (Eds.), *Handbook of virtual environments: Design, implementation, and applications* (2nd ed., pp. 233–256). Boca Raton, FL: Taylor & Francis Group, Inc.

Templeman, J. N., Sibert, L. E., Page, R. C., & Denbrook, P. S. (2006). Immersive simulation to train urban infantry combat. In *Virtual Media for Military Applications. Meeting Proceedings RTO-MP-HFM-136* (pp. 23-1–23-16). Paper 23, RTO, Neuilly-sur-Seine, France. Available from: http://www.rto.nato.int/abstracts.asp.

Thorndyke, P. W., & Goldin, S. E. (1983). Spatial Learning and Reasoning Skill. In H. L. Pick & L. P. Acredolo (Eds.), *Spatial orientation: Theory, research, and application* (pp. 195–217). New York, NY: Plenum Press.

Thorndyke, P. W., & Stasz, C. (1980). Individual differences in procedures for knowledge acquisition from maps. *Cognitive Psychology, 12*, 137–175.

Tolman, E. C. (1948). Cognitive maps in rats and men. *Psychological Review, 55*(4), 189–208.

Waller, D., Hunt, E., & Knapp, D. (1998). The transfer of spatial knowledge in virtual environment training. *Presence: Teleoperators and Virtual Environments, 7*(2), 129–143.

Witmer, B. G., Bailey, J. H., & Knerr, B. W. (1995). *Training dismounted soldiers in virtual environments: Route learning and transfer* (Technical Report 1022): U.S. Army Research Institute for the Behavioral and Social Sciences. Orlando, FL.

20 Technology Management and User Acceptance of Virtual Environment Technology

David Gross

CONTENTS

20.1 INTRODUCTION

Technology deployment, or transitioning innovation into use, is the natural objective of anyone who develops an innovation. Despite the old saw "build a better mousetrap, and the world will beat a path to your door," the particular power or utility of a technological innovation is just one of the criteria for successful technology deployment. Technology deployment takes place within a society of (potential) users, which may (or may not) adopt the innovation. Innovators seek to deploy their innovation, but success is in the control of users who may adopt it. The economic circumstances of that society of potential users, as well as its mores (i.e., formal or informal moral attitudes), also play important roles in the success (or failure) of a technological deployment. Furthermore, the ongoing advance of technology, independent of the technology in question, as well as evolving demands from consumers in society, creates forces that affect a society's adoption choice.

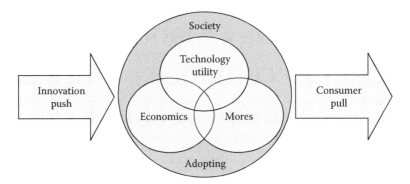

FIGURE 20.1   Multiple effects on a technology's diffusion in society.

If adoption of virtual environment (VE) technology occurs within a society, then there are many forces external to the technology itself that affect the success (or failure) of deploying a technological innovation (see Figure 20.1). Some of these forces tend to encourage technology deployment, but some discourage it because the cross direction of these forces creates societal inertia. This resistance to change is a phenomenon so pervasive and widely recognized that it scarcely requires documentation (Mackie & Wylie, 1988). The forces resisting change can slow and stunt or accelerate and advance the adoption of new technologies beyond the intrinsic attributes of the technology, a classic example being the deployment of nuclear power technology.

VE technology, like other innovations before it, has great potential for improving the way people work and live; therefore, VE technologists seek to deploy it. However, fulfilling this potential will certainly require equally great changes in the way these things are done now. Every significant technology deployment has changed the adopting society's way of life. Consider, for example, the deployment of automobile technology and the wide-ranging impact it has had on every aspect of modern life. Some of those changes were anticipated and appreciated; some were not. Whether desirable or not, change is a certainty with a new technology. The resistance to deploying VE technology can be reduced by understanding how organizations within a society come to understand and adopt technological innovations.

While adopting technology changes the society, the change is most often one of addition rather than replacement. Humanity has needed transportation as long as there have been humans: the automobile in this light is just an addition of a powered machine–based mode to existing modes. Likewise, the Internet appeals to basic human needs for information previously and still met with libraries, social groups, and even talking to one's neighbor. Blascovich and Bailenson (2011) argue that VE technology likewise appeals to basic human needs to escape, fantasize, and imagine. They place VE technology in a context of simply another mode of virtual media, adding to storytelling, graphics, sculpture, theater, photography, cinematography, etc.

20.1.1 TECHNOLOGY DEPLOYMENT AND ADOPTION

The rate of technology deployment and adoption can be very frustrating to the technology developer, who feels that the relative advantage of this technological innovation over present approaches is so overwhelming that only the naive would not immediately adopt it. Figure 20.2 suggests an expected time frame for technology innovation adoption in software on the order of a decade or more (Utz & Walter, 1992). The reluctance of adopters to easily adopt innovation has led to what Rogers (2003) calls the *innovativeness–needs paradox*, in which individuals or other units in a system who most need the benefits of an innovation are generally the last to adopt it.

However, potential adopters see the technological innovation differently than developers. Adopters (in contrast with innovators) are concerned with issues such as an innovation's compatibility

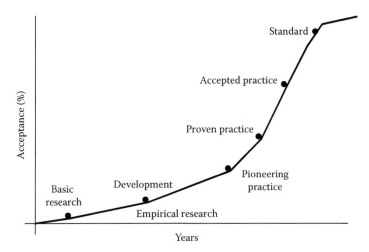

FIGURE 20.2   Typical technology transition cycle.

with the rest of an organization and with the complexity of using and maintaining the innovation. Technologists can increase the rate of adoption if they understand the obstacles and concerns an adopting organization faces.

20.1.2   Toward an Understanding of Technology Adoption

One approach to understanding technology adoption is *innovation diffusion theory*, especially as described by Rogers (2003). Technology adoption cannot occur unless the innovation is distributed or diffused through a society. The essence of the technology diffusion process is human interaction in which one person communicates a new idea to another person. Thus, at the most elemental level of conceptualization, the diffusion process consists of (1) a new idea, (2) individual "A" who knows about the innovation, and (3) individual "B" who does not yet know about the innovation. The innovation diffusion model suggests a four-step process for technology diffusion: innovation, uncertainty, diffusion, and adoption/rejection. Although the technology itself plays an important role in this process, social relationships have a great deal to say about the conditions under which A will tell B about the innovation and the results of this telling (Rogers). In this light, the factors that most affect a particular organization's decision to adopt a particular technology are internal to the organization, specifically factors describing the social relationships in the organization. These factors can be grouped into three categories: (1) an organization's past experience with technology, (2) the organization's characteristics, and (3) the organization's pursued strategy (Lefebvre & Lefebvre, 1996).

   While innovation diffusion has much explanatory power, the complaint has been made that it focuses too much on internal factors. Innovation diffusion theory models technology adoption based on a company's communication patterns in which prior adopters inform potential adopters and persuade them. One alternative is the theory of network externality, which argues that adopter behavior is strongly influenced by relevant parties outside the company (e.g., vendors, consultants, standards bodies). These external factors can be grouped into the following categories: (1) the industry level, (2) the macroeconomic environment, and (3) national policies (Lefebvre & Lefebvre, 1996). Konana and Balasubramanian (2005) studied the factors influencing the adoption of online financial investing and concluded: "that to achieve a comprehensive understanding of online investing, and more generally, of other technology-intensive domains where consumers of a product or service adopt an active role, a multifaceted view of the domain that jointly accommodates social, economic, and psychological perspectives is called for" [sic].

If classical innovation diffusion theory focuses on internal factors, and network externality theory focuses on external factors, perhaps an approach that balances these factors is needed. Bhattacherjee and Gerlach (1998) suggest that while these approaches have explanatory power for relatively simply technology adoptions, they fail to explain more complicated pervasive technology adoptions such as software's object-oriented technology (OOT). The potential of VE technology certainly is as pervasive as the introduction of OOT was, so it is worth considering a more holistic approach. An emerging approach that balances the internal and external influences on adoption decisions is *organizational learning theory*. Organizational learning theory suggests that organizations learn by a collective process very similar to how individuals learn; that is, organizational learning occurs when we observe, act, and reflect with others. Shibley (1998) divides an organization's learning into two parts, observation and reflection. Observation involves collection of the raw information of daily events and the discovery of patterns in those events. Reflection involves the attempt by organizations to place their observations in a useful context, either system structures, mental models, or corporate vision. Most often, observations are forced into preexisting structures, but valuable learning takes place when the raw data forces a change in these structures (Senge, 1990). An Organization Learning Theory approach to understanding technology adoption depicts an organization as attempting to integrate both the internal and external factors into a corporate view of a technological innovation, which dictates an adoption decision.

20.2 MANAGEMENT CHALLENGES

Managers of organizations considering adoption of any technology face real challenges in realizing the opportunities that the innovation affords them, while avoiding their pitfalls. This has led to difficulty in explaining experiences in technology adoption. For example, despite the widespread awareness of OOT's benefits for software development, and the commercial availability of reusable libraries, databases, methodologies, programming languages, and packaged applications, many businesses remain cautious about deploying OOT for major system development (Bhattacherjee & Gerlach, 1998). This experience may prove instructive for technologist concerned about deploying VE technologies in many of the same businesses that have been slow to adopt OOT.

20.2.1 Integration into Corporate Culture

Chief among management's concerns of about VE technological innovations is the difficulty of integrating new technologies into existing corporate culture. Corporate cultures clash with proposed innovations at many levels, including even the language used. While the VE technology community has a clear definition of what it means by *virtual environment* consistently used throughout this book, there are other common definitions that confuse potential adopters such as "A computer that is running in a virtual machine environment, which is the combination of virtual machine monitor and hardware platform" (*PC Magazine Encyclopedia,* 2013).

The innovation may clash with the culture by challenging established organization. At many companies, the introduction of personal computers capable of supporting technical documentation maintenance was slowed by concerns about what would happen to existing central word processing organizations and personnel. Of course, all adopting organizations are not the same; some will express innovation rapidly and easily, and some will not. Table 20.1 shows Roger's classification of different kinds of adopter organizations.

Even within these classifications, an organization may choose to have a more or less aggressive stance toward a particular technology, choosing, for example, to lead the market in one technology but simply defending its competitive position in another. Table 20.2 illustrates some objective organization characteristics and whether or not they were significant factors affecting the adoption of a particular technology, namely, integrated services digital network (ISDN). The data are analyses of formal corporate surveys.

TABLE 20.1
Classification of Innovation Adopters

Adopter Categories	Share of Whole	Description
Innovators	2.5%	Venturesome, cosmopolite, networked with other innovators, financial resources, understand complex technical knowledge, cope with uncertainty
Early adopters	13.5%	Respectable, more local than innovators, strong opinion leadership
Early majority	34%	Interact frequently with peers, seldom hold positions of opinion leadership, interconnectedness to system's interpersonal networks, long period of deliberation before making an adoption decision
Late majority	34%	Adoption might result from economic or social necessity due to the diffusion effect, skeptical and cautious, relatively scarce resources
Laggards	16%	Most localized, point of reference is the past, suspicious of change agents and innovations, few resources

Source: Adapted from Rogers, E.M., *Diffusion of Innovations,* 5th ed., Free Press, New York, 2003.

TABLE 20.2
Significance of Organization Factors in the Adoption of ISDN

Organization Characteristic	Significant	Not Significant
Adoption encouragement	X	
Open vs. closed organizations	X	
Slack (uncommitted) resources	X	
Large vs. small organization	X	
Norms encouraging change		X
Organization centralization		X
Organization formalization		X
Organization complexity		X

Source: Adapted from Lai, V.S. and Guynes J.L., *IEEE Trans. Eng. Manage.,* 44(2), 46, 1997.

Understanding what kind of organization is considering adoption can help the technology developer plan for and accommodate the particular problems of an organization, thereby increasing the likelihood of success. For example, a VE technologist proposing a first VE project should realize from Tables 20.1 and 20.2 that organizations that commit significant resources toward a VE project and are at or near the forefront of their industrial domain are more likely to experience success. In contrast, the VE technologist should probably not be as concerned about the impact of the organization's degree of formality (rank, office, titles, etc.) or complexity (layers or styles of management) on success. In addition, a VE project proposal for a laggard organization will need plenty of support in the form of experience by other similar organizations.

Rogers (2003) cites numerous studies that show that larger organizations are more innovative; in fact, size is consistently shown as one of the best predictors of organization innovativeness. This may also indicate that they are also more likely to adopt innovations. Joskow and Rose (1990) find strong evidence that large firms tend to lead the electricity industry in adoption of new technologies. However, Rogers points out that while size is easily measured, it encompasses a number of effects (such as total resources, slack resources, and range of employee expertise) and their contribution to either innovativeness or adoption is not known.

20.2.2 Need for Credible Productivity Gains

The second major challenge facing VE technology adopters' management is the need to address the *bottom line*. Sometimes simplified as just showing a profit, the more accurate concern is to increase the process' productivity (return for resources invested). Figure 20.3 illustrates this desire in terms of moving an example process capability curve to a greater capability at a lower cost. VE technology innovators must consider the cost of adoption as well as the potential benefit if they are to understand the adopter's challenges. A legitimate claim that a VE project will improve productivity would make it easier for an organization to adopt the innovation. Can VE technology make this claim? The capability of business processes is a function of the knowledge, techniques, and tools that their employees apply. Human actions determine whether production is done efficiently and whether new and improved methods of production are introduced (Ostberg & Chapman, 1988). VE technology, integrating visualization and artificial intelligence, can legitimately claim to impact the productivity of business processes because it offers powerful opportunities for improvements in the way humans perceive and utilize knowledge (see Figure 20.4).

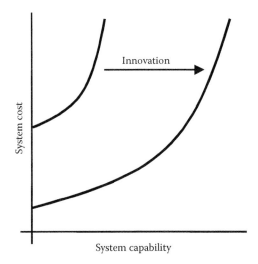

FIGURE 20.3 Innovation increasing process productivity.

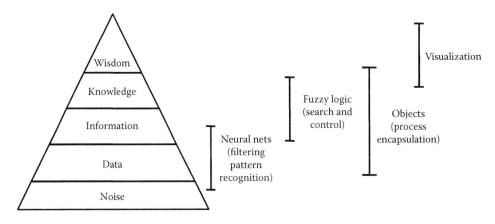

FIGURE 20.4 Relevant merits of technologies to perceive and utilize knowledge. (From Hoffman, M. and Wright, A., Technology enablers for the intelligent enterprise, Technical Report No. D97-2065, Menlo Park, CA, SRI Consulting, 1997.)

20.2.3   Fear of the Operations and Sustainability Tail

The second major challenge facing VE technology adopters' management is the potential operations and sustainability tail. Major manufacturers have seen operations expense resulting from introduction of the personal computer far outstrip their acquisition cost as well as their productivity gain, particularly from earlier installations. Management has legitimate concerns about VE technology becoming another logistics nightmare, particularly considering the fragile nature of some of the equipment. When this is combined with the rapid evolution/revolution in VE devices and the rapidly changing nature of VE applications, this challenge can be a show-stopper for deploying VE technology. Any VE technology deployment will have to address this challenge to survive scrutiny. Despite marketing hype from manufactures, adopters have not acted in concert with a belief that VE equipment with life cycle costs that produces sufficient compensatory benefit for widespread adoption exists. As a result, VE technology today has been adopted primarily in research and highly specialized design applications (see Section 20.4 in this chapter).

20.3   USER CHALLENGES

Management can be convinced, if the rank and file have powerful applications for the proposed technological innovation. The potential users of VE technology have challenges to overcome in their own right about discovering and explaining these applications. These challenges arise from the interaction between the user's context and the innovation's attributes. Rogers (2003) identified five variables that the adopter's perception, and can influence his or her rate of adoption:

1. Relative advantage
2. Compatibility
3. Complexity
4. Trailability
5. Observability

The following discusses the challenges faced by VE adopters as they consider the variables that affect rate of innovation adoption.

20.3.1   Killer Application

Worldwide, the market for business virtual reality services, software, hardware, and peripherals is seen as rising from 1995's level of 134.9 million to a little over $1 billion no later than 2001 (Boman, 1995). However, the marketplace is still hampered by the lack of a *killer application* (Stanney et al., 1998). The concept of a killer application dates from the development of the business spreadsheet, which motivated the purchase of millions of personal computers. Two years ago, 3D and virtual technology on the Internet identified potential key applications: games, e-commerce, collaboration, education, training, and visualization. However, significant applications for VE in these fields are still just potential (Leavitt, 1999).

20.3.2   Physical Ergonomics

The VE technology to date has unresolved issues with physical ergonomics, which creates an adoption challenge. Cybersickness and related syndromes continue to be significant obstacles to widespread use of VE technology (see Lawson, 2014a,b, Chapters 23 and 24). At least for some users, cybersickness is a significant problem, and many organizations are reluctant to make significant commitment to technologies that clearly bypass significant portions of their population.

For example, wearers of head-mounted displays (HMDs) have been known to suffer stress because they are unable to utilize eye convergence and accommodation when viewing virtual objects at different distances. In general, some people are very sensitive to unfamiliar motion stimuli, such as that generated in a VE, and these individuals may be discomforted by VE exposure (Vince, 1995).

Beyond cybersickness, much VE technology is uncomfortable or unpleasant to use (see Viirre, Price, & Chase, 2014, Chapter 22, this book). Stuart (1996) notes some current VE technology limitations for users:

1. Position trackers with small working volumes, inadequate robustness, and problems of latency and poor registration
2. HMDs with very poor resolution, somewhat limited field of view, and encumbering form factors (i.e., weight and tethers)
3. Virtual acoustic displays that require a great deal of computational resources in order to simulate a small number of sources
4. Force and tactile displays, still in their infancies, with limited functionality
5. Image generators that cannot provide low-latency rendering of head-tracked complex scenes, requiring severe trade-offs between performance and scene quality

In addition to these limitations for users, Stuart notes developers of VEs experience limitations with the real-time capabilities of operating systems and inadequate software development tools.

20.3.3 Cognitive Ergonomics

As discussed, one of the primary benefits of VE systems is their ability to assist users in developing knowledge about subject matters portrayed in the virtual world. However, VEs developed to date deliver little of this promise. Stuart (1996) notes that in order for VEs to fulfill their promise, they will have to deliver

1. Sociability: the capability to permit multiple users to share an environment (see Driskell, Salas, & Vessey, 2014, Chapter 38)
2. Veridicality: the accuracy with which a VE reproduces its modeled environment (fidelity) (see Simpson, Cowgill, Gilkey, & Weisenberger, 2014, Chapter 13)
3. Resolution: the level of precision and detail provided by a VE system (see Chapters 3 through 6)
4. Immersion presence: the extent to which users believe that they are *in* a VE (see Chertoff & Schatz, 2014, Chapter 34)
5. Engagement: the extent to which users are deeply involved and occupied with what they are doing in the VE (see McMahan, Kopper, & Bowman, 2014, Chapter 12 and Popescu, Burdea, & Trefftz, 2014, Chapter 17)
6. Reconfigurability: the capability to modify a VE model
7. Responsiveness: the ability of a VE to respond to user inputs (lag, latency; see McMahan, Kopper, & Bowman, 2014, Chapter 12)
8. Stability: lack of jitter or oscillation in the VE display (see Badcock, Palmisano, & May, 2014, Chapter 3)
9. Robustness: the ability to function correctly across all operational ranges

20.4 DOMAIN-SPECIFIC CHALLENGES

As discussed, technology adoption happens within a society of potential users. Obviously, each individual situation is different and therefore cannot be addressed, but there are similarities in the challenges faced by potential adopters in similar domains. The following discusses the challenge in several domains of interest for VE technology.

20.4.1 Design and Testing

One of the first and still most important applications of VEs has been for architectural design. The success of this application has led to experimentation with VE in other kinds of product design and testing. The use of VE systems for designing and testing new products and systems is gaining acceptance, particularly as a complement to computer-aided design technology (Boman, 1995). The Boeing Company's 777 commercial airliner marked a significant milestone in the fielding of *paperless* design techniques, centered particularly on virtual representations of the aircraft. Less well known is the application of these techniques to projects such as the international space station and the B-2 aircraft. Virtual representation of product design has been particularly helpful in the automobile industry in shortening product design cycle times.

20.4.2 Manufacturing

VEs offer much potential for improving manufacturing productivity, particularly in the emerging *mass-customization* marketplace. Most applications to date have been with experiments in laying custom assembly instructions over a standardized work area. Breakthroughs in this domain await improvements in the physical ergonomics challenges discussed.

20.4.3 Operations

Experiments have been successfully conducted in using VE technology to support operations, such as feeding firefighters real-time maps of buildings as they fight fires (Bliss, Tidwell, & Guest, 1997). Breakthrough applications in this domain await more robust (i.e., not fragile) VE equipment to support the harsh demands of varied operating environments.

20.4.4 Information Management

Successful VEs have been developed and fielded for information management problems, such as database maintenance. This domain is limited by the cognitive ergonomics problem discussed earlier.

20.4.5 Entertainment

The broadest and most successful application of VE technologies has been for entertainment, both home-based and location-based systems (see Greenwood-Ericksen, Kennedy, & Stafford, 2014, Chapter 50). The Disney Corporation has demonstrated its commitment to exploring the advantages of VE-based entertainment with its Aladdin VE ride at Epcot Center and DisneyQuest arcade. However, the fact that 5%–10% of Aladdin visitors experience dizziness or nausea is a reminder of the problems that may hinder mass appeal of VE systems (Boman, 1995).

20.4.6 Medicine/Health Care

Extensive experimentation continues in the use of VEs for medicine, particularly in the specialized field of medical training (see Rizzo, Lang, & Koenig, 2014, Chapter 46). Beyond medical training, researchers are investigating the use of VE for aiding people with disabilities, cognitive rehabilitation, and telepresence for remote surgery (Boman, 1995). The National Institutes of Health's Visible Human Project is creating complete, anatomically detailed, 3D representations of the male and female human body (Banvard, 2002). The project's data sets are designed to serve as a common reference point for the study of human anatomy. The data sets are being applied to a wide range of educational, diagnostic, treatment planning, virtual reality, artistic, mathematical, and industrial uses by over 1000 licensees in 41 countries. But key issues remain in the development of methods to link such image data to text-based data. Standards do not currently exist for such linkages. Basic research is needed in the description and representation of image-based structures and to connect image-based structural–anatomical data to text-based functional–physiological data (Card, Mackinlay, & Shneiderman, 1999).

20.4.7 EDUCATION AND TRAINING

Many observers believe that education and training will be a major application domain for VE technologies (see Barmpoutis, DeVane, & Oliverio, 2014, Chapter 41). Some specific application opportunities include astronaut, military, and sports training (Boman, 1995). VE technology has been used effectively in persuading users to adopt other technologies, for example, the successful use of a virtual reality design support tool for breaking down barriers in adoption of hybrid concrete innovations (Goulding et al., 2007). VE technology is in use in many education endeavors, but there are significant variables affecting success (Raaji & Schepers, 2006; Selim, 2003).

20.4.8 MARKETING AND SALES

VE technology has been successfully deployed and adopted as a means for marketing products to potential buyers and for closing sales. Kim and Foreythe (2008), for example, discuss the use of VE technology for delivering product information that is similar to the information obtained from direct product examination, with focus on apparel.

20.5 CONCLUSIONS

VE technology offers exciting potential for revolutionizing the way work is performed in many domains. However, VE technology will fail to deliver on its potential if system developers do not begin to understand and grapple with the problems facing organizations desiring to adopt this technology. Chief among these challenges is the need for *killer* applications in varied domains, supported by robust and stable equipment. It is tempting to project the adoption of VE technology based on our experience with related technologies, such as the personal computer. However, any such projection would be conjecture as sufficient information about the degree to which VE is actually similar to such technologies, and ubiquitous applications do not yet exist.

REFERENCES

Badcock, D. R., Palmisano, S., & May, J. G. (2014). Vision and virtual environments. In K. S. Hale & K. M. Stanney (Eds.), *Handbook of virtual environments: Design, implementation, and applications* (2nd ed., pp. 39–86). Boca Raton, FL: Taylor & Francis Group, Inc.

Banvard, R. A. (2002). The visible human project® image data set from inception to completion and beyond. *Proceedings CODATA 2002: Frontiers of Scientific and Technical Data*. Montréal, Quebec Canada, Track I-D-2: Medical and Health Data.

Barmpoutis, A., DeVane, B., & Oliverio, J. C. (2014). Applications of virtual environments in experiential, STEM, and health science education. In K. S. Hale & K. M. Stanney (Eds.), *Handbook of virtual environments: Design, implementation, and applications* (2nd ed., pp. 1055–1072). Boca Raton, FL: Taylor & Francis Group, Inc.

Bhattacherjee, A., & Gerlach, J. (1998). Understanding and managing OOT adoption. *IEEE Software, 15*(3), 91–96.

Blascovich, J., & Bailenson, J. (2011). *Infinite reality: Avatars, eternal life, new worlds, and the dawn of the virtual revolution* (Kindle ed.) (P.S.) (Kindle Location 4527).New York, NY: HarperCollins.

Bliss, J. P., Tidwell, P. D., & Guest, M. A. (1997). The effectiveness of virtual reality for administering spatial navigation training to firefighters. *Presence, 6*(1), 73–86.

Boman, D. (1995). *Virtual environment applications* (Technical Report No. D95-1917). Menlo Park, CA: SRI Consulting.

Card, S. K., Mackinlay, J. D., & Shneiderman, B. (1999). *Readings in information visualization: Using vision to think.* San Francisco, CA: Morgan Kaufmann Publishers.

Chertoff, D., & Schatz, S. (2014). Beyond presence. In K. S. Hale & K. M. Stanney (Eds.), *Handbook of virtual environments: Design, implementation, and applications* (2nd ed., pp. 855–870). Boca Raton, FL: Taylor & Francis Group, Inc.

Driskell, T., Salas, E., & Vessey, W. B. (2014). Team training in virtual environments. In K. S. Hale & K. M. Stanney (Eds.), *Handbook of virtual environments: Design, implementation, and applications* (2nd ed., pp. 999–1026). Boca Raton, FL: Taylor & Francis Group, Inc.

Goulding, J. S., Kagioglou, M., Sexton, M. G., Zhang, X., Aouad, G., & Barrett, P. S. (2007). Technology adoption: Breaking down barriers using a virtual reality design support tool for hybrid concrete. *Construction Management and Economics, 25*(12), 1239–1250.

Hoffmann, M., & Wright, A. (1997). *Technology enablers for the intelligent enterprise* (Technical Report No. D97-2065). Menlo Park, CA: SRI Consulting.

Joskow, P. L., & Nancy L. Rose. The Diffusion of New Technologies: Evidence from the Electric Utility Industry. *The RAND Journal of Economics*, 1990, *21*(3), 354–373.

Kim, J., & Foreythe, S. (2008). Adoption of virtual try-on technology for online apparel shopping. *Journal of Interactive Marketing, 22*(2), 45–59.

Konana, P., & Balasubramanian, S. (2005). The social–economic–psychological model of technology adoption and usage: An application to online investing. *Decision Support Systems, 39*(3), 505–524.

Lai, V. S., & Guynes, J. L. (1997). An assessment of the influence of organizational characteristics information technology adoption decision: A discriminative approach. *IEEE Transactions on Engineering Management, 44*(2), 46–157.

Leavitt, N. (1999, July). Online 3D: Still waiting after all these years. *Computer, 32*(7), 4–7.

Lefebvre, É., & Lefebvre, L. A. (1996). *Information and telecommunication technologies: The impact of their adoption on small and medium-sized enterprises.* Ottawa, Ontario, Canada: International Development Research Centre.

Mackie, R. R., & Wylie, C. D. (1988). Factors influencing acceptance of computer-based innovations. In M. Helander (Ed.), *Handbook of human–computer interaction* (pp. 1081–1106). Amsterdam, the Netherlands: Elsevier Science Publishers.

McMahan, R. P., Kopper, R., & Bowman, D. A. (2014). Principles for designing effective 3D interaction techniques. In K. S. Hale & K. M. Stanney (Eds.), *Handbook of virtual environments: Design, implementation, and applications* (2nd ed., pp. 285–312). Boca Raton, FL: Taylor & Francis Group, Inc.

Ostberg, O., & Chapman, L. (1988). Social aspects of computer use. In M. Helander (Ed.), *Handbook of human–computer interaction* (pp. 1033–1049). Amsterdam, the Netherlands: Elsevier Science Publishers.

PC Magazine Encyclopedia: Definition of virtual environment. (2013, May 10). Retrieved from http://www.pcmag.com/encyclopedia/term/58705/virtual-environment

Popescu, G. V., Trefftz, H., & Burdea, G. C. (2014). Multimodal interaction modeling. In K. S. Hale & K. M. Stanney (Eds.), *Handbook of virtual environments: Design, implementation, and applications* (2nd ed., pp. 411–434). Boca Raton, FL: Taylor & Francis Group, Inc.

Raaji, E. M., & Schepers, J. L. (2006). The acceptance and use of a virtual learning environment in china. *Computers and Education, 50*(3), 838–852.

Rogers, E. M. (2003). *Diffusion of innovations* (5th ed.). New York, NY: Free Press.

Selim, H. M. (2003, May). An empirical investigation of student acceptance of course websites. *Computers and Education, 40*(4), 343–360.

Senge, P. M. (1990). *The fifth discipline.* New York, NY: Doubleday.

Shibley, J. J. (1998). A primer on systems thinking and organizational learning [Portland Learning Organization Group, Online]. Available: http://www.systemsprimer.com

Simpson, B. D., Cowgill, J. L., Gilkey, R. H., & Weisenberger, J. M. (2014). Technological considerations in the design of multisensory virtual environments: How real does it need to be? In K. S. Hale & K. M. Stanney (Eds.), *Handbook of virtual environments: Design, implementation, and applications* (2nd ed., pp. 313–334). Boca Raton, FL: Taylor & Francis Group, Inc.

Stanney, K. M., Salvendy, G., Deisinger, J., DiZio, P., Ellis, S., Ellison, J., … Witmer, B. (1998). Aftereffects and sense of presence in virtual environments: Formulation of a research and development agenda. *International Journal of Human–Computer Interaction, 10*(2), 135–187.

Stuart, R. (1996). *The design of virtual environments*. New York, NY: McGraw-Hill.

Utz, Jr., & Walter, J. (1992). *Software technology transitions: Making the transition to software engineering*. Englewood Cliffs, NJ: Prentice-Hall.

Viirre, E., Price, B. J., & Chase, B. (2014). Direct effects of virtual environments on users. In K. S. Hale & K. M. Stanney (Eds.), *Handbook of virtual environments: Design, implementation, and applications* (2nd ed., pp. 521–530). Boca Raton, FL: Taylor & Francis Group, Inc.

Vince, J. (1995). *Virtual reality systems*. Workingham, England: Addison-Wesley.

21 Virtual Environments and Product Liability

Robert S. Kennedy, Robert C. Kennedy,
Kristyne E. Kennedy, Christine Wasula,
and Kathleen M. Bartlett

CONTENTS

21.1 INTRODUCTION

Simulators and other virtual environments (VEs) utilize dynamic images and motion stimuli to produce the vicarious experience of immersion and presence for training and entertainment. However, these exposures often produce unwanted motion sickness–like side effects including balance disturbances, profound drowsiness, and coordination problems. Not everyone appears to be affected to the same extent, but such changes, when they occur, imply that the human nervous system has been temporarily recalibrated, or adapted, to the virtual world. This adaptation involves fundamental, significant changes in the basic function of an individual that may not be noticed when the individual exits the VE (see Chapters 24 through 27 and 32 through 33). As individuals subsequently perform routine tasks, unaware of any lingering effects of VE exposure, negative consequences in the form of accidents may occur; thus, these aftereffects may have implications for safety and health for both consumers of VE technology and persons and property surrounding those consumers. If a VE product occasions problems, the liability of VE developers could range from simple accountability (i.e., reporting what happened) to full legal liability (i.e., paying compensation for damages). In order to minimize their liability, manufacturers and corporate users of VE devices (i.e., companies) can proactively assess the potential risks associated with human factors by using a comprehensive systems safety approach. Human factors systems safety actions undertaken include the following: (1) systems should be properly designed; or (2) aftereffects should be removed, guarded against, or warned against; or (3) adaptation methods should be developed; or (4) users should be certified to be at their preexposure levels; or (5) users should be monitored and debriefed.

21.2 AFTEREFFECTS FOLLOWING USAGE OF A PRODUCT

While most VE product manufacturers have given only slight, if any, consideration to issues of litigation involving aftereffects of VE exposure, the inevitable risks incurred by placing these products into wide circulation must be considered. It has been established, for example, that exposure of humans to simulated environments, for entertainment or education, produces motion sickness–like discomfort and aftereffects such as balance disturbances (Stanney, Mourant, & Kennedy, 1998; see Chapters 24 through 27 and 33 through 34). The military flight simulators of the 1960s, which were the VEs of that day, first called our attention to this problem (Miller & Goodson, 1960). In those early days, fielded devices had equipment limitations such as excessive transport delays, optical distortions, and flickering imagery; these equipment problems were considered to be the agents of the discomfort. Many of these equipment problems have been remedied in modern simulator and VE systems. At the same time, technological advances in simulation engineering have enabled newer systems to portray even more compellingly realistic dynamic scenarios. Yet varying degrees of motion sickness–like symptoms persist and have been reported in nearly every flight simulator fielded by the military services.

Although carefully controlled studies that employ comparable experimental design, kinematics (i.e., velocity and spatial frequency of a visual scene), equipment features, participants, durations of exposure, and others have not yet been conducted, it is known from survey work (Kennedy, 1998; Kennedy, Stanney, Drexler, Compton, & Jones, 1999b) that the overall incidence of VE sickness (cybersickness, McCauley & Sharkey, 1992) appears to be greater than it ever was with flight simulators. In this chapter, some of these reports of motion sickness–like symptoms occasioned by VE exposures are described for purposes of illustration only. The reader is referred elsewhere in this book for more complete treatments dealing with the cause and incidence of cybersickness (see Chapters 25 and 27). Here, we merely wish to suggest that if sickness occurs, accidents could result, and if so there may be an issue of product liability. Specifically, it is known that VEs and simulators are products intended for use by adults and children and that they are designed to be manufactured and sold. Sickness, which may occur in connection with usage, may imply a hidden defect in the product, or, alternatively, it may be an expected outcome of the product's usage. Also, in addition to the obvious symptoms of sickness, other adverse consequences of exposure may occur, and the symptoms of discomfort may only be a harbinger of other outcomes.

For the purposes of this discussion, the established definition (McCauley, 1984) that describes simulator sickness as the experience of symptomatology during and after the use of a VE that would not ordinarily be experienced if the same activity were carried out in the real world is followed. Obviously, when a real environment ordinarily occasions sickness, if a replicated VE also produces sickness, then the simulator actually provides a *good* simulation.

Outcomes known to exist following exposure to VEs and simulators include nausea (up to 30% of exposures), eyestrain (up to 40% of exposures), drowsiness, salivation, sweating, headache, and dizziness/vertigo, as well as loss of postural stability (Kennedy, Fowlkes, & Lilienthal, 1993). Vomiting is infrequent (<1%) and when it occurs, it is usually more than 1 h after exposure. Long-term aftereffects, including visual flashbacks and disorientation, have been routinely reported to have occurred up to 12 h after simulator exposure (Baltzley, Kennedy, Berbaum, Lilienthal, & Gower, 1989; Lilienthal, Kennedy, Berbaum, & Hooper, 1987), but such data are only now being collected experimentally. Sleepiness and fatigue are symptoms reported by large numbers, but the significance of this for subsequent performance is not well appreciated (see Chapter 25). The incidence of all of these long-term aftereffects is generally low (<10% of all exposures) and is correlated with the strength of aftereffects immediately after exposure and the duration of exposure. Although less than 10% with long-term aftereffects may seem like a low proportion, the huge number of persons who are exposed each day to such devices is staggering and may be expected to increase; 10% of a growing number of exposures represents a serious number of persons who may be expected to experience long-term aftereffects.

Visual and vestibular scientists have previously experimented with ancillary aftereffects when they were examining distortion due to tilted figures (Ebenholtz, 1988, 1992) and altered lenses (Wallach & Kravitz, 1965, 1968, 1969). These same distortions can be found in today's head mounted displays (HMDs). The vestibulo-ocular reflex (Leigh, 1996; O'Leary, 1992) produces involuntary rotation of the eyes in the direction opposite to head movement in order to maintain fixation and reduce slippage of images across the retina (see Chapter 3). The human internal calibration of this reflex seems to depend upon the recent experience of a relationship of head movement and the resulting image motion (Gonshor & Melville-Jones, 1976). If a new and consistent relationship between head movement and scene displacement is introduced, such as via an HMD, the scenery will at first appear to move because eye rotation will fail to stabilize the image (cf. DiZio & Lackner, 1997; see Chapter 3). Apparent shifts in the scene with head movement may produce impressions of actual individual movement (vection), motion sickness, and eyestrain (Kennedy, Hettinger, Harm, Ordy, & Dunlap, 1996a). Eventually, the individual will adapt so that the new relationship elicits eye rotation that yields a stable retinal image (see Chapter 26). However, the adapted individual will experience aftereffects; once adjusted to distortion, a head movement that produces the ordinary displacement of the visual scene will again appear to shift. Readaptation to the natural environment will be required. From the dynamics and kinematics of the helmet-mounted VE displays, one can predict that such a recalibration is occurring in users, although this problem has only infrequently been studied (Prothero & Parker, 1999). Considerable research needs to be conducted in order to determine the size, time course, and duration of such aftereffects.

This peculiar kind of adaptation can be expected to occur with increasing frequency as VE systems are employed for longer exposure periods, the natural consequence of using them as training devices. This effect, when it occurs, can produce significant disability when entering the real world; indeed, initial exposure to VE would probably be a lot like getting used to a new pair of glasses and eventually could be no more troublesome than changing glasses. Nevertheless, this adaptation involves fundamental, significant changes in the function of the individual at a level at which he or she may be unaware. If developers take steps to address and prepare for the potential safety hazards involved with VE system interaction, they may be able to minimize harm. Companies that take such proactive measures and perform extensive preliminary safety analyses may circumvent product liability issues.

Other aftereffects are worth noting, particularly balance problems, which imply that the human nervous system has been temporarily recalibrated to the virtual world, and time may be required before the individual is ready to navigate safely in the *stable world*. Sea legs or *le mal de debarquement* (Gordon, Spitzer, Shupak, & Doweck, 1992) is a related malady that also results in perturbed posture, but such perceptual and motor aftereffects have long been known to occur. See Dolezal (1982) for a review of studies on human perceptual distortion in connection with mirror and prism use. They have also been reported following weightlessness. One could argue that if such effects outlast the period of time an individual is under the control of the VE device owner/manager, care for well-being should extend beyond the owner's property line. More is said about this later.

In one particularly interesting simulator study (Gower, Lilienthal, Kennedy, & Fowlkes, 1987), US Army pilots who were exposed to the flight simulator daily for 2 weeks reported less sickness, as would be expected, with each subsequent exposure. But ironically, and perhaps with implications for safety, they also demonstrated progressively more balance problems over this same time frame and as adaptation ensued (Kennedy, 1993).

Similarly, in a remarkable case following VE exposure, posteffects were reported to have occurred in a middle-age man after he spent approximately 2 h in a three-story VE entertainment facility (Kennedy, Stanney, & Fernandez, 1999c). After exposure, the individual, who historically had not been judged suprasensitive to motion sickness, exhibited the characteristic signs and symptoms (viz., nausea, imbalance, dizziness, vertigo, sleep difficulties) of motion sickness and had to remain in bed for his first 48 h at home. Subsequently, in medical examinations (neurologic and otolaryngologic), no organic cause for his symptoms could be found. The symptoms persisted for weeks and then months, and he was still experiencing significant distress as much as 6 months after the initial exposure. Fortunately, by the ninth month, he recovered, and now, several months later, he is considered to be symptom-free. If, during a bout of aftereffects, he were to become injured in an auto accident related to the persistent illusory perception of motion, it could be argued that the VE exposure caused the accident.

Note that the legal issue that surfaced here is not simply whether one can recover from cybersickness or any other form of motion sickness. The problem concerns the individual's ability to safely perform routine tasks. If his mental and physical functioning appears to have been perturbed because of his exposure-related distress, were this individual to fall from a roof, have an automobile accident, or participate in some other activity that he had not been warned about, then entities responsible for placing the VE device "in the stream of commerce" may be liable. For example, suppose that a driver of an automobile were to have a balance disturbance following a VE exposure, and suppose that his balance performance were equivalent to the disturbances seen with a 10% blood alcohol concentration level (Kennedy, Turnage, Rugotzke, & Dunlap, 1994). In this case, it could be argued that if postural equilibrium is disrupted, steering ability is also likely to be disrupted, and the cause of the degraded performance could be associated with a VE exposure. If that individual had an auto accident and had had no alcohol, he could argue that his imbalance might relate to the aftereffect experienced in the VE exposure.

Finally, profound drowsiness, sleepiness, and the experience of fatigue, frequently reported byproducts of VE and simulator exposure, may occur during that period after a user is released from the VE device and is occupied with walking or driving home. These aftereffects can represent real hazards in the operation of motor vehicles and in certain types of work and recreational activities. Time may be required before the individual is ready to navigate safely in the *real world*. Therefore, developers of VE systems should take steps to warn VE users about potential aftereffects and their implications for safety. As discussed in the following, the adequacy of warnings about potential aftereffects is likely to be the key factor in determining whether the manufacturer of a VE product will be held liable for injuries to the user of the product.

How long aftereffects may last after individuals leave the VE is an important question for which there are scant data. In one survey, Baltzley et al. (1989) have shown that with 2–4 h exposures in flight simulators, 1%–10% of the operators had one or more symptoms that lasted a period of time after exposure and persisted after subjects left the simulator building. Careful analysis of problems associated with VE may help developers take proactive steps (such as the use of warnings, certification tests, and/or checklists) to assure that people who are exposed to VE may safely reenter the real world and to minimize developers' legal liability.

21.3   PRODUCT STANDARDS

With this potential for negative aftereffects from a product that may achieve wide usage in the private sector for education and training, as well as for entertainment, developers of VE systems should have concern over product liability issues. The law of product liability has its origins in tangible property but has been extended beyond tangible goods to include intangibles such as electricity after it has been delivered to the consumer (Phillips, 1998). It has also been applied to other items such as natural products and real estate fixtures such as a house (Phillips). However, the concept of a *product* is not limitless, and one court recently held that a virtual reality video game (Mortal Kombat), which allegedly "uses sophisticated technology to make players physically feel as if they are killing the characters in the game, and rewards players when they tap their 'killer responses,'" was not a *product* for purposes of the Connecticut Products Liability Act, as required to support a parent's claim for failure to warn of the inappropriate level of violent content and mentally addictive nature of the game and for defective design, which were filed by a mother whose son was murdered by a friend who was allegedly addicted to and obsessed with the game (Wilson v. Midway Games, Inc., 2002, p. 169).

The rationale for imposing strict product liability rests on the following ideas: (1) that the manufacturer is in the best position to reduce the risk and insure against the risk, (2) that the manufacturer is the one responsible for the product being on the market, and (3) that the consumer lacks the means and skill to investigate the soundness of a product for herself. This section will deal with product liability issues and definitions of ordinary defects, unknowable defects, and product and consumer expectation, among others. The relevance of defective warnings and negligence on causation will be discussed.

If a commercial product is placed in the stream of commerce and injury occurs, then the product may be adjudged defective because it occasions injury and the responsible person or entity that knew or should have known it was defective can be held liable (Phillips, 1998). This liability can be attached to any individual or business that profited from the defective technology. There are several grounds for imposing legal liability. First, legal liability may be found if a product *defect* is "unreasonably dangerous to the user" (Phillips, p. 53). Second, liability may be imposed for inadequate design. Third, liability may be imposed for inadequate warnings of foreseeable risks of harm that could have been avoided by the adoption of a reasonable alternative design or by reasonable instructions or warnings (American Law Institute, 1998). In the present case, the VE system must be considered "dangerous to the extent beyond that which would be contemplated by the ordinary consumer" (p. 11). If VE systems have this known propensity for aftereffects documented in published scientific literature, the ordinary consumer may not be aware of any of the effects or of all of the effects. Furthermore, if they do not use reasonable care, the potentially adverse consequences of product liability may accrue to persons involved in research and development of these systems for purposes of bringing them to consumer use.

Because of the postural unsteadiness, drowsiness, and the discomfort of vomiting and retching, it would appear likely that aftereffects could result in accidents after usage of a VE system (Kennedy & Stanney, 1996a). This is a likely scenario since while many users of VE technology are aware of the possible side effects of cybersickness and, when afflicted, these individuals may restrict their own movements, predictably few are aware of the potential for loss of balance and eye–hand incoordination effects (Kennedy, Stanney, Compton, Drexler, & Jones, 1999a) as well as long-term aftereffects. As an extreme example, there are reports of work and sickness following exposure to immersive video games at commercial amusement parks where vomiting occurred less than 15 h later; our own research has shown that vomiting can occur with VE users long after the experience of a 60-min exposure to a VE display (Kennedy, Stanney, Dunlap, & Jones, 1996b). If a manufacturer or seller of a VE product is aware of these potential aftereffects but fails to provide adequate warnings to consumers of the product, it may be held liable under theories of strict liability or negligence, as discussed in the following. Incidentally, a similar situation occurs with the posteffects following long durations in microgravity (Reschke et al., 1994), although the number of persons exposed to 4 h extended durations in space is far smaller than the anticipated number that will be exposed under training regimes planned for the military (see Chapter 37) and other federal services and those exposed in educational (see Chapters 40 through 42) or recreational settings (see Chapter 50). In our research, we advocate the use of a procedure to certify that an individual has recovered to at least preexposure levels. This process, if in force, can protect users, as well as vendors, manufacturers, and other purveyors of VE systems from potentially damaging lawsuits.

Manufacturers might wish to follow the trend in modern law to avoid costly, time-consuming litigation all together by anticipating and responding to any potentially compromising possibilities. Thus, in order to minimize legal liability, VE systems may need to be properly prepared and accompanied by warning labels and appropriate directions (i.e., debriefing protocols) to raise the level of consumer awareness of the potential for harm. In order to provide effective debriefing protocols, objective measures of adaptation resulting from VE exposure are needed. Specifically with respect to VE, rather than waiting for issues to arise, it might be prudent to address safety concerns before they result in crisis or harm. A proactive rather than reactive approach may allow researchers to identify and address potentially harmful side effects related to the use of VE technology. Standards or criteria to be developed are needed to guide VE developers' decisions, such as the principle of harm minimization. Under this principle, an activity should minimize harm to all exposed, which requires recognizing the concerns and interests of the various individuals involved (Kallman, 1993). More principles need to be established to guide and direct the safe development of virtual technology.

21.4   PRODUCT WARNINGS

Under a product liability theory, a manufacturer or seller of products may be liable for failure to provide adequate warnings as to dangers associated with its products. Generally, a manufacturer has a duty to warn against *latent* dangers resulting from *foreseeable* uses of its products that it knew or should have known. Thus, a manufacturer has no duty to warn against dangers that are open and obvious. In Hartman v. Cedar Fair (2009), the attraction was found not liable for a patron's injuries that occurred when he was disembarking a water ride by attempting to climb over a railing that was open and obvious. In another case (Durmon v. Billings, 2004), it was ruled that there was no duty to warn or protect the patron of a *haunted* cornfield maze from a reaction to an actor in costume who was carrying a running chainsaw. In this case, it was obvious to the patron that the purpose of the attraction was to scare, frighten, and/or surprise and she knew that the chainsaw was meant to scare and haunt the patrons since she even paid an additional fee to enter the portion of the attraction in which the characters were present.

In addition, a manufacturer has no duty to warn about injuries that are not *foreseeable*. In a case against distributors of violent movies and video games, it was held that they were not liable for a violent shooting spree by teenage students since the shooting was not a foreseeable result of playing the games or watching the movies (Sanders v. Acclaim Entertainment, Inc., 2002). A key factor in determining the adequacy of a warning is often whether there were prior injuries or accidents of which the manufacturer was aware. In Colon v. Mountain Creek Waterpark (2012), a patron fractured his lower leg while boarding a ride. In this case, the park was not held liable because there were no previous injuries so the warning signs were deemed adequate. Conversely, prior complaints or injury tends to support liability, as in the case of a child who had a seizure while playing a Nintendo video game. Since others had previously complained of seizures resulting from playing their game, the warnings were deemed to be inadequate for informing players of the risk for seizures (Roccaforte v. Nintendo of America, Inc., 2001).

An adequate warning is one that is reasonable under the circumstances (Lehto & Miller, 1986). To be adequate, a warning must describe the nature and extent of the danger involved (Cunitz, 1981). A key factor in determining whether a warning is adequate is whether a better warning would have been more effective.

Warning cases often involve issues of providing adequate instructions for safe use of a product. A warning is distinguished from an instruction in that instructions are calculated primarily to secure the efficient use of a product, whereas warnings are designed to insure safe use (Phillips, 1998). A product distributed without adequate warnings or instructions is sometimes treated as one with a design defect. Therefore, a product may be faultlessly manufactured and designed but may be defective when placed in the consumers' hands without an adequate warning concerning the manner in which to use the product safely.

It is important to note that, ultimately, a manufacturer's liability for breach of duty to warn will depend on the plaintiff proving causation. This means the plaintiff must prove that he or she would not have suffered the harm in question if adequate warnings and instructions had been provided. This is often proven through expert testimony, with the expert explaining how a more detailed or expansive warning would have resulted in a different outcome. If the plaintiff cannot show that he would have avoided using a product if a more detailed warning had been given, he will not be able to recover against the manufacturer of the product. Although this element may be difficult to prove in many cases, it is of course in the manufacturer's best interest to protect itself from liability by giving adequate warnings. A warning must also be effectively communicated, and if it is, then failure to read or comply with the warning as given may bar recovery for a plaintiff (Lehto & Miller, 1986). One example is the case of Desai v. Silver Dollar City, Inc., (1997) in which a patron in an amusement park was injured when she was getting off of the raft at the end of the ride. She was struck by another raft and was injured, and since there were clearly posted warning signs and also an audible recorded announcement instructing patrons to remain on the raft until they are assisted by park staff, it was ruled that she disregarded adequate warnings and she was unable to recover damages.

While there exists nothing in the case law addressing the situation of a consumer who is ultrasensitive to motion sickness, we do know that such persons exist in many environments (cf. Crampton, 1990). Furthermore, an analogy of hypersusceptibility can also be made to the allergic consumer where the duty normally owed to allergic users is one of warning and then only when the plaintiff is a member of a substantial or appreciable number of persons subject to allergy, where the defendant should have known of this risk (Phillips, 1998). A seller may be required to warn that his product contains an ingredient that is a known allergen and may have to warn of the symptoms associated with an allergic reaction (Phillips). The American Law Institute's influential section on torts states that

> in order to prevent the product from being unreasonably dangerous, the seller may be required to give directions or warning, on the container, as to its use. The seller may reasonably assume that those with common allergies, as for example to eggs or strawberries, will be aware of them, and he is not required to warn against them. Where, however, the product contains an ingredient to which a substantial number of the population are allergic, and the ingredient is one whose danger is not generally known, or if known is not one which the consumer would reasonably not expect to find in the product, the seller is required to give warning against it. (p. 347, 1965)

In the science community, there have been similar calls for such warnings for video games and other video-based systems, which may produce epileptiform symptoms (e.g., Harding & Harding, 2010; Harding & Hodgetts, 2009), and there is precedent for this action in that certain video games have been sold with labels that warn of the prospect of epileptiform seizures by susceptible individuals (Solodar, 2010).

Furthermore, a *postsale* or *continuing* duty to warn generally exists where a manufacturer is responsible for warning of a defective product at any time after it has been manufactured and sold if the manufacturer becomes aware of the defect. This means that the manufacturer is under a duty to issue warnings or instructions as they later become known to the manufacturer, or even as they should become known, and that the manufacturer is expected to stay informed about advancements in the state of the art and/or of later accidents involving dangers in the product. For example, a drug manufacturer is treated as an expert and is under a continuing duty to keep abreast of scientific developments in regard to a product and to notify the medical profession of any additional side effects discovered from its use. This was held in a case between Lance v. Wyeth (2010) ruling that the manufacturer had a postsale duty to warn consumers of any dangerous side effects as long as the product is on the market. Thus, because VE systems arguably affect human safety, this idea could be applicable in the case of VE systems already in existence, imposing a continuing duty on the manufacturers to keep up with developments and to provide proper warnings (Lawyers Co-Operative Publishing Company, 1988). In the case between Adams v. Genie Industries, Inc. (2008), a dangerous defect in the design of a product was discovered after the product had been sold and the manufacturer was held responsible to either remedy the defect or, at a minimum, provide the user with adequate warning of the defect as to how to safely use the product and minimize risk to the user and others.

21.5   PRODUCT LIABILITY ISSUES

The increase in product liability claims and settlements witnessed in the last half of the twentieth century may be linked to the development of more sophisticated products. The sheer complexity of modern products is a source of danger because consumers may not understand safe use; risks may not be readily apparent. Additionally, consumers may not possess the knowledge to judge the quality, efficacy, and safety of a product; thus, assurance of minimum standards of performance is demanded from manufacturers (Howells, 1993).

In recent years, product liability litigation has been extended into many areas, including recreational equipment such as pool tables, trampolines, amusement rides, and attractions.

Cases have arisen involving injuries on amusement park rides, such as water slides and roller coasters, where the amusement park and/or manufacturer was sued under a product liability theory. In some cases, a higher standard of care is imposed on amusement park operators as well as manufacturer designers of amusement park rides because of the risk created to thousands of passengers, many of them children, for monetary gain (this is similar to the higher standard of care imposed on a common carrier, who is required to exercise great care to protect passengers).

However, courts have been reluctant to impose liability on the manufacturers and operators of amusement park rides when (1) there are adequate warnings, (2) the dangers associated with the ride are open and obvious, and (3) the injuries in question are not foreseeable. In a recent case between Nilson and Hershey Entertainment and Resorts Company (2009), the parents of an 11-year-old boy filed suit against an amusement park (Hersheypark) and the manufacturer of a roller coaster called the *Wildcat* alleging that their son lost hearing in his left ear as a result of riding the roller coaster. The crux of the parties' dispute was whether the warnings provided by the manufacturer and the park adequately advised riders of the latent risks associated with riding the roller coaster. The parents asserted that the warnings failed to advise their son of how violent and fast this ride would be. Had they done so, according to the parents, their son would have heeded the appropriate warnings, not ridden the *Wildcat*, and would still be able to hear from his left ear. The parents asserted that the defendants were liable for their failure to warn under theories of negligence and strict product liability.

After considering these arguments, the court ruled in favor of the defendants, finding that the warnings posted on the ride were adequate and that the warning suggested by plaintiffs' expert would not have prevented the child's injury. The warnings posted on the ride warned plaintiffs' son and other riders that

THROUGHOUT THE RIDE RIDERS WILL EXPERIENCE
SPEED CHANGES AND UNEXPECTED FORCES.

The parents' expert opined that this warning was insufficient because it failed to provide adequate warning and information, which expressed the severity of the forces associated with the *Wildcat* roller coaster. However, the court found that the warning in place was not significantly different from the warning suggested by plaintiffs' expert, which was as follows:

WARNING!!!
THIS IS A HIGH SPEED RIDE WITH RAPID JOLTS IN SHARP TURN
AND HARD DIPS. DO NOT RIDE UNLESS YOU ARE PREPARED
FOR THE MOST SEVERE LEVEL OF PHYSICAL IMPACT AND THRILL.

The court found that there was no evidence that the alternative warning suggested by the parents' expert would have deterred the child from riding the roller coaster. Thus, the court held that parents failed to demonstrate that the *Wildcat*'s warning was defective. The court also found that the parents had failed to demonstrate that the defects complained of—the speed and force exerted by the ride—were not obvious. The court found that even an 11-year-old should know that roller coasters go up and down at a high rate of speed and that they exert strong forces on the body.

Finally, the court found that any violent shaking and resultant hearing loss that the parents' son allegedly experienced was not reasonably foreseeable in light of minimal prior injuries experienced by other riders and the lack of any prior published cases or claims of roller coaster–induced hearing loss. Thus, the parents' claims for strict liability and negligence failed.

In contrast, in Beroutsos v. Six Flags Theme Park, Inc. (2000), the plaintiff boarded a roller coaster ride at an amusement park and subsequently suffered neck and back injuries allegedly caused by the ride. In considering the defendant's argument that the plaintiff had assumed the risk of injury by voluntarily riding the coaster, the court ruled that ordinary people assume risks on roller coasters, which include dizziness, vomiting, and nausea. Plaintiff's evidence included an expert affidavit from an engineer who inspected the subject roller coaster and opined that its defective design and operation, including insufficient head cushioning and lack of head restraint, proximately caused the plaintiff's injuries and that the ride was unsafe and its problems were not open and obvious to an ordinary patron. On this state of evidence, the court held that the defendant was not entitled to summary judgment in its favor because regardless of whether dizziness and vomiting should be considered normal, the patron should not expect lasting physical injury.

In the case of VE sickness, a duty to warn will likely exist because the dangers associated with a VE system are usually not open and obvious. However, the manufacturer of a VE system will only have a duty to warn about dangers that are reasonably foreseeable. As a result, the adequacy of the warnings provided will likely depend in large on how many complaints or reports of injury the manufacturer has received in the past. If a manufacturer provides adequate warnings concerning foreseeable risks, based on prior incidents and complaints, it may be able to avoid liability for any injuries.

Product liability law involves four basic types of liability standards: (1) contractual or warranty standards, which involve products that fail to meet the promised standard; (2) negligence standards, which judge the conduct of the defendant in light of risks and benefits that result from his conduct; (3) strict liability standards, which do not judge the conduct of the producer/supplier but rather assess the product; and (4) absolute liability standards, which are based on proof of damage caused by the product (Howells, 1993). Although any of these standards might be the basis for litigation in a case involving VE sickness and manufacturers of VE systems, negligence standards, in particular, merit careful consideration.

Negligence has been defined as conduct that involves an unreasonably great risk of causing damage (Terry, 1915) and, alternatively, as conduct that "falls below the standard established by law for the protection of others against unreasonably great harm" (American Law Institute, 1965, Sec. 282, para. 1). To make out a cause of action in negligence, the plaintiff does not need to establish that the defendant either intended harm or acted recklessly. Negligence focuses on the conduct of the various participants who are responsible for the product's creation and entry into the marketplace. The judge balances the risks and benefits of the defendant's conduct; risks include those that the defendant knew of, or ought to have known of, at that time. Risks that become known at a later date are irrelevant unless they could have been discovered by the defendant via testing or research that he or she could be expected to have undertaken (Howells, 1993).

The reasonable manufacturer is expected to be aware of current industry-wide standards and technology. Although evidence that a manufacturer has met self-imposed standards, voluntary standards, or even statutory standards is significant, such evidence alone does not establish that the manufacturer is not negligent, as in the case with the defective gas can (Pfeiffer v. Eagle Mfg. Co., 1991). An entire industry can be found negligent for continuing to apply unacceptable standards of manufacture (Hooper, 1932) or for failing to make currently available technological improvements. However, although conformity with industry practice does not demonstrate conclusively that a product is safe, an industry member will only rarely be held liable for failing to do what no other member has done before (Lawyers Co-Operative Publishing Co., 1988). The flexibility of the negligence standard is achieved by permitting interplay among three factors. In every negligence action, the court must consider (1) the probability that harm would occur, (2) the gravity of the harm, and (3) the burden of taking precautions to protect against the harm—what the manufacturer might have done to minimize or eliminate the risk of harm.

Because all human activity involves an element of risk, a defendant's conduct is deemed negligent only when it falls below what a *reasonable* person would have done under similar circumstances (Brown, 1991). Risks to health justify more precautions than risks to property, and the more serious the risk, the greater the precautions required. Although this assessment is made at the time of the alleged negligent act, for example, the time of manufacture of a defective product, there may be a separate postmarketing duty to monitor the product. When risks are discovered, it may be necessary to either order the recall of the product or take other steps to avert any danger (Howells, 1993). However, in several recent cases, courts have declined to hold manufacturers liable for failing to recall a product. In Jablonski v. Ford Motor Co. (2011), the court found no statutory obligation that the manufacturer had no duty to recall the product because the defect was not discovered until after the product left the manufacturer's control (see also Ford Motor Co. v. Reese, 2009).

Even though such safety principles are developed, unintended outcomes may occur. Thus, after performing a safety analysis to minimize harm, the issue of who is accountable (i.e., the appropriate person to respond when something undesirable occurs), responsible, and strictly liable when harm occurs must be resolved (Johnson, 1993). Responsibility is broken into role responsibility (i.e., responsibility linked to an individual's duties or by virtue of occupying a certain role), causal responsibility (i.e., a person who does something or fails to do something and this causes something else to happen), blameworthy responsibility (i.e., a person who did something wrong that led to an event or circumstance), and legal liability responsibility (i.e., one is legally liable when it is claimed that the individual must pay damages or otherwise compensate those harmed). If a VE product malfunctions, the consequences for the VE developers could range from simple accountability (i.e., reporting what happened) to full legal liability (i.e., paying compensation for damages). When legal liability is imposed, the potential damages include out-of-pocket expenses, loss of earnings, loss or impairment of future earning capacity, lost profits, medical expenses, loss of household services, loss of business or employment opportunities, pain and suffering, mental anguish, emotional distress, loss of consortium, attorney's fees, litigation costs, prejudgment interest, and, in extreme cases, punitive damages (Lawyers Co-Operative Publishing Co., 1988). If VE products are being employed for training purposes, whether or not the training is internal to the company or carried out under contract may also influence liability issues. In addition, a company's position on this liability continuum may be related to how proactive that company was in terms of product standards and how extensive a preliminary safety analysis was performed.

21.6   SYSTEMS SAFETY APPROACH

In order to minimize their liability, manufacturers and corporate users of VE devices (i.e., companies) can proactively assess the potential risks associated with human factors by using a comprehensive systems safety approach. The human factors systems safety approach, as originally put forward by Christensen (1993), has five elements, including (1) design, (2) remove, (3) guard, (4) warn, and (5) train, in general order of preference of application. We follow the Christensen model and, for purposes of the following VE product safety rendition offered, provide two new approaches that can usefully be added to the case for VE systems: (6) certification and (7) monitoring/debriefing (Kennedy & Stanney, 1996b).

1. *Design*. Design new products to be as free of hazard as possible. Obviously, the first method of choice would be to eliminate any known hazard, which requires knowledge of the exact causes of safety concerns. Unfortunately, the exact causes of simulator sickness associated with VE use are not currently well understood, which underscores the need for cautious introduction of new products.
2. *Remove*. Remove hazards from existing systems. The ability to fix existing systems is limited by the same lack of knowledge that hinders the ability to design safe new systems. Making changes in existing systems and observing the impact on sickness incidence represents the best methodology for gaining knowledge to deal with the problem.
3. *Guard*. Where hazards cannot be eliminated or removed, users should be prevented from coming into inadvertent contact with the hazard. Guards must be convenient and obviously essential and must not impair operations. If a guard is not effective in eliminating a problem or interferes with training, it should be eliminated.
4. *Warn*. Warn users of remaining hazards. A simulator sickness field manual has been developed by the Department of Defense (NTSC, 1989) and is distributed at simulator sites. The scientific basis of this field manual is described in a *guideline* report (Kennedy et al., 1987).

The *field manual* and *guidelines* are used to teach instructors and trainees to recognize simulator sickness symptoms and have been distributed to the three military services and the U.S. Coast Guard; they are also available in NATO countries. The *field manual* warns trainees using military simulators about potential risks involved when resuming activities, including flying, immediately after a bout of simulator sickness. Similar manuals and guidelines might be provided to users of VE systems. It is important to note, however, that warning is a less preferable alternative than the first three approaches (i.e., effective design, removing hazards, and guarding; Christensen, 1993).

5. *Train.* Even the best products may require training for safe usage. Were humans not so adaptable to altered sensory environments, no one would be able to tolerate simulator training or VE devices. Much of the advice contained in the *field manual* described earlier has to do with how to facilitate sensory-motor adaptation to the simulator and readaptation once outside it.

   Teaching people to adapt rapidly to the peculiarities of VEs will be the primary means of enhancing the benefits of these devices while reducing the discomfort and risk that they may produce. If the first four methods (i.e., design, remove, guard, warn) are not sufficiently effective, a program of training can be prepared so that users can be taught to avoid hazards. In the case of VE sickness, perceptual adaptation is likely to occur (cf. Kennedy, Smith, & Jones, 1991; Uliano, Kennedy, & Lambert, 1986; see Chapter 26) and can be employed to reduce symptoms if proper schedules are followed.

6. *Certification.* In connection with the five main human factors systems safety approaches outlined by Christensen (1993), we believe two additional approaches should be added (Kennedy & Stanney, 1996b). The first concerns measurement of the effectiveness of the freedom from hazard. This is called process certification. Should human factors problems be suspected, VE systems should be measured to determine the level of hazard they are expected to represent. Until recently, self-report was the chief method for certifying that a simulator was safe. Currently, objective measures and tests of human balance are being developed that could be used for certification purposes (Kennedy & Lilienthal, 1994).

7. *Monitoring/debriefing.* The second additional approach involves measuring individuals to determine if they have been adversely affected by exposure to VE systems (Kennedy & Stanney, 1996b). Even if a system *passes* its certification procedure, human variability is such that some individuals may still be adversely affected by VE system exposure. In order to limit one's liability, it may be prudent to establish monitoring procedures, such as a postexposure balance test. The same or similar tests of human balance used for certification could also be used for monitoring. For those individuals who *fail* or marginally *pass* the balance test, debriefing protocols could be developed that provide information to those affected so they can act accordingly to minimize ham.

21.7 CONCLUSIONS

VE systems, while holding much promise to educate, train, and entertain, also have the potential to harm. Individuals exposed to VE systems may experience malaise during exposure and adverse aftereffects that could linger days or even months after exposure. There has been substantial case law that addresses the liability of products and manufacturers, but there is still ambiguity and it is incumbent upon the sponsors of VE training and entertainment systems to identify potential risks for symptoms and long-term aftereffects in order to provide a safe and effective experience, which includes any subsequent behaviors that may be interfered with by aftereffects. System developers can follow the systems safety certification approach discussed in this chapter to ensure they have exercised due care in protecting their patrons from harm.

REFERENCES

Adams v. Genie Industries, Inc., 53 A.D.3d 415 (N.Y.A.D. 1st Department. 2008). Retrieved from LexisNexis Academic database.

American Law Institute. (1965). *Restatement of the Law, Second of Torts*, 291, 395, 402a, St. Paul, MI: American Law Institute Publishers.

American Law Institute. (1998). *Restatement of the law, torts-products liability.* St. Paul, MN: American Law Institute Publishers.

Baltzley, D. R., Kennedy. R. S., Berbaum. K. S., Lilienthal, M. G., & Gower. D. W. (1989). The time course of post-flight simulator sickness symptoms. *Aviation, Space, and Environmental Medicine, 60*(11), 1043–1048.

Beroutsos v. Six Flags Theme Park, Inc., 185 Misc. 2d 557 (Sup. Ct. 2000). Retrieved from LexisNexis Academic database.

Brown, S. (Ed.). (1991). *The product liability handbook: Prevention, risk, consequence and forensics of product failure.* New York, NY: Van Nostrand Reinhold.

Christensen, J. M. (1993). Forensic human factors psychology: Part 2. A model for the development of safer products. *CSERIAC Gateway* (Vol. 4(3), pp. 1–5). Dayton, OH: Crew System Ergonomics Information Analysis Center.

Colon v. Mountain Creek Waterpark, 465 Fed. Appx. 186 (3d. Cir. 2012). Retrieved from LexisNexis Academic database.

Crampton, G. (Ed.). (1990). *Motion and space sickness.* Boca Raton, FL: ERC Press.

Cunitz, R. J. (1981, May/June). Psychologically effective warnings. *Hazard Prevention, 17*, 5–7.

Desai v. Silver Dollar City, Inc., 493 S. E. 2d 540 (Ga. Ct. App. 1997). Retrieved from LexisNexis Academic database.

DiZio, P., & Lackner. J. R. (1997). Circumventing side effects of immersive virtual environments. In M. J. Smith, G. Salvendy, & R. J. Koubek (Eds.), *Design of computing systems: Social and ergonomic considerations* (pp. 893–896). Amsterdam, the Netherlands: Elsevier Science Publishers.

Dolezal, H. (1982). *Living in a world transformed: Perceptual and performatory adaptation to visual distortion.* New York, NY: Academic Press.

Durmon v. Billings, 873 So. 2d 872 (La. Ct. App. 2d Cir. 2004). Retrieved from LexisNexis Academic database.

Ebenholtz, S. M. (1988). *Sources of asthenopia in navy flight simulators.* Alexandria, VA: Defense Logistics Agency, Defense Technical Information Center. (Accession No. AD~A212699)

Ebenholtz, S. M. (1992). Motion sickness and oculomotor systems in virtual environments. *Presence, 1*(3), 302–305.

Ford Motor Co. v. Reese, 684 S. E.2d 279 (Ga. App. 2009). Retrieved from LexisNexis Academic database.

Gonshor, A., & Melville-Jones, G. (1976). Extreme vestibulo-ocular adaptation induced by prolonged optical reversal of vision. *Journal of Physiology, 256*, 381–414.

Gordon, C. R., Spitzer, O., Shupak, A., & Doweck, H. (1992). Survey of mal de debarquement. *BMJ, 304*, 544.

Gower, D. W., Lilienthal, M. G., Kennedy. R. S., & Fowlkes, J. E. (1987, September). Simulator sickness in U.S. Army and Navy fixed and rotary wing flight simulators. *Proceedings of the Conference AGARD Medical Panel Symposium on Motion Cues in Flight Simulation and Simulator-Induced Sickness No. 433* (pp. 8.1–8.20). Brussels, Belgium.

Harding, G. F. A., & Harding, P. F. (2010) Photosensitive epilepsy and image safety. *Applied Ergonomics, 41*(4), 504–508.

Harding, G. F. A., & Hodgetts, M. A. (2009). The application of PSE guidelines to electronic screen games. *Proceedings of the VIMS2009 Conference.* Utrecht, the Netherlands. Available at: http://www.vims2009.org/Abstracts/Harding-paper.pdf.

Hartman v. Cedar Fair, L. P. (2009). Ohio App. LEXIS 3337 (Ohio App. 6th Dist. 2009). Retrieved from LexisNexis Academic Database.

Howells, G. (1993). *Comparative product liability.* Aldershot, England: Dartmouth Publishing.

Jablonski v. Ford Motor Co., 955 N.E.2d 1138 (Ill. 2011). Retrieved from LexisNexis Academic database.

Johnson, D. (1993), *Computer ethics* (2nd ed.). Englewood Cliffs, NJ: Prentice-Hall.

Kallman, E. A. (1993). Ethical evaluation: A necessary element in virtual environment research. *Presence, 2*(2), 143–146.

Kennedy, R. S. (1993). *Device for measuring head position as a measure of postural stability* (Final Report No. 9260166). Washington, DC: National Science Foundation.

Kennedy, R. S. (1998, September). *Gaps in our knowledge about motion sickness and cybersickness.* Paper presented at the Motion Sickness, Simulator Sickness, Balance Disorders and Sopite Syndrome Conference, New Orleans, LA.

Kennedy, R. S., Berbaum, K. S., Lilienthal, M. G., Dunlap, W. P., Mulligan, B. E., & Funaro, I. F. (1987). *Guidelines for alleviation of simulator sickness symptomatology* (NAVTRASYSCEN TR-87-OOJ). Orlando, FL: Naval Training Systems Center.

Kennedy, R. S., Fowlkes, J. F., & Lilienthal, M. G. (1993). Postural and performance changes in Navy and Marine Corps pilots following flight simulators. *Aviation, Space. and Environmental Medicine, 64*, 912–920.

Kennedy, R. S., Hettinger, L. J., Harm, D. L., Ordy, J. M., & Dunlap, W. P. (1996). Psychophysical scaling of circular vection (CV) produced by optokinetic (OKN) motion: Individual differences and effects of practice. *Journal of Vestibular Research, 6*(5), 331–341.

Kennedy, R. S., & Lilienthal, M. G. (1994). Measurement and control of motion sickness aftereffects. *The Official Conference Proceedings of Virtual Reality and Medicine, The Cutting Edge* (pp. 111–119). New York, NY: SIG-Advanced Applications.

Kennedy, R. S., Smith, M. G., & Jones, S. A. (1991). Variables affecting simulator sickness: Report of a semi-automatic scoring system. *Proceedings of the Sixth International Symposium on Aviation Psychology* (pp. 851–856). Columbus, OH.

Kennedy, R. S., & Stanney, K. M. (1996a). Postural instability induced by virtual reality exposure: Development of a certification protocol. *International Journal of Human-Computer Interaction, 8*(1), 25–47.

Kennedy, R. S., & Stanney, K. M. (1996b). Virtual reality systems and products liability. *Journal of Medicine and Virtual Reality, 1*(2), 60–64.

Kennedy, R. S., Stanney, K. M., Compton, D. E., Drexler, J. M., & Jones, M. B. (1999). *Virtual environment adaptation assessment test battery* (Phase II Final Report, Contract No. NAS9·91022). Houston, TX: NASA Lyndon B. Johnson Space Center.

Kennedy, R. S., Stanney, K. M., Drexler, J. M., Compton, D. E., & Jones, M. B. (1999). Computerized methods to evaluate virtual environment aftereffects. *Proceedings of the Driving Simulation Conference "DSC '99"* (pp. 273–287). Paris, France: French Ministry of Equipment, Transport, and Housing.

Kennedy, R. S., Stanney, K. M., Dunlap, W. P., and Jones, M B. (1996). *Virtual environment adaptation assessment test battery* (Final Report No. NASA1-%-1. Contract No. NAS9–19453). Houston, TX: NASA Lyndon B. Johnson Space Center.

Kennedy, R. S., Stanney, K. M., & Fernandez, E. (1999). Six months residual after effects from a virtual reality entertainment system. Unpublished manuscript.

Kennedy, R. S., Turnage, J. J., Rugotzke, G. G., & Dunlap, W. P. (1994). Indexing cognitive tests to alcohol dosage and comparison to standardized field sobriety tests. *Journal of Studies on Alcohol, 55*(5), 615–628.

Lance v. Wyeth, 4 A.3d 160 (Pa. Super. Ct. 2010). Retrieved from LexisNexis Academic database.

Lawyers Co-Operative Publishing Co. (1988. with 1999 supplement). *American law of products liability* (3rd ed.). Rochester, NY: Author.

Lehto, M. R., & Miller, J. M. (1986). *Warnings, Vol. 1: Fundamentals, design, and evaluation methodologies* (1st ed.). Ann Arbor, MI: Fuller Technical Publications.

Leigh, R J. (1996). What is the vestibulo-ocular reflex and why do we need it? In R W. Baloh & G. M. Halmagyi (Eds.), *Disorders of the vestibular system* (pp. 12–19), New York, NY: Oxford University Press.

Lilienthal, M. G., Kennedy, R. S., Berbaum, K. S., & Hooper. J. (1987, November). *Vision-motion-induced sickness in navy flight simulators: Guidelines.* Paper presented at the Ninth Interservice Industry Training Systems Conference, Washington, DC.

McCauley, M. E. (Ed.). (1984). *Research issues in simulator sickness: Proceedings of a workshop.* Washington, DC: National Academy Press.

McCauley, M. E., & Sharkey, T. J. (1992). Cybersickness: Perception of self-motion in virtual environments. *Presence, 1*(3), 311–318.

Miller, J. W., & Goodson, J. E. (1960). Motion sickness in a helicopter simulator. *Aerospace Medicine, 31*, 204–212.

Nilson v. Hershey Entertainment and Resorts Company, 649 F. Supp. 2d 378 (M.D. Pa. 2009). Retrieved from LexisNexis Academic database.

NTSC. (1989, November). *Simulator sickness field manual MOD 4.* Orlando, FL: Naval Training Systems Center, Human Factors Laboratory.

O'Leary, D. P. (1992). Physiological bases and a technique for testing the full range of vestibular function. *Revue de Laryngologie, 113*, 407–412.

Pfeiffer v. Eagle Mfg. Co., 771 F. Supp. 1133 (D. Kan, 1991). Retrieved from LexisNexis Academic database.

Phillips, J. J. (1998). *Products liability in a nutshell.* St. Paul, MN: West Group.

Prothero, J. D., & Parker. D. E. (1999). A unified approach to presence and motion sickness. In L. J. Hettinger & M. W. Haas (Eds.), *Virtual and adaptive environments: Psychological and human performance issues.* Mahwah, NJ: Lawrence Erlbaum Associates.

Reschke, M. F., Hann, D. L., Parker, D. E., Sandoz, G. R., Homick, J. L., & Vanderploeg, J. M. (1994). Neurophysiologic aspects: Space and motion sickness. In A. E. Nicogossian, C. L. Huntoon, & S. L. Pool (Eds.), *Space physiology and medicine* (3rd ed., pp. 228–260). Philadelphia, PA: Lee & Febiger.

Roccaforte v. Nintendo of America, Inc., 802 So. 2d 764 (La. App. 5th Cir. 2001). Retrieved from LexisNexis Academic database.

Sanders v. Acclaim Entertainment, Inc., 188 F. Supp. 2d 1264 (D. Colo. 2002). Retrieved from LexisNexis Academic database.

Solodar, J. (2010). Seizures triggered by video games: Underestimated and underdiagnosed. *Epilepsy: Insights & Strategies, 3*, 7–15.

Stanney, K. M., Mourant, R. R., & Kennedy, R. S. (1998). Human factors issues in virtual environments: A review of the literature. *Presence, 7*(4), 327–351.

Terry, H. T. (1915). Negligence. *Harvard Law Review, 29*, 40–54.

T. J. Hooper, 60 F 2d 737 1932 U.S. App. LEXIS 2592, cert. denied, 287 U.S. 662 (2d Cir. 1932). Retrieved from LexisNexis Academic Database.

Uliano, K. C., Kennedy, R. S., & Lambert, E. Y. (1986). Asynchronous visual delays and the development of simulator sickness. *Proceedings of the Human Factors Society 30th Annual Meeting* (pp. 422–426). Dayton, OH: Human Factors Society.

Wallach, H., & Kravitz, J. H. (1965). Rapid adaptation in the constancy of visual direction with active and passive rotation. *Psychonomic Science, 3*, 165–166.

Wallach, H., & Kravitz, J. H. (1968).Adaptation in the constancy of visual direction tested by measuring the constancy of auditory direction. *Perception and Psychophysics, 4*(5), 299–303.

Wallach, H., & Kravitz, J. H. (1969). Adaptation to vertical displacement of the visual direction. *Perception and Psychophysics, 6*(2), 111–112.

Wilson v. Midway Games, Inc., 198 F. Supp. 2d 167 (D. Conn. 2002). Retrieved from LexisNexis Academic database.

Section V

Health and Safety Issues

22 Direct Effects of Virtual Environments on Users

Erik Viirre, B.J. Price, and Bradley Chase

CONTENTS

22.1   INTRODUCTION

To deal with the effects that virtual environment (VE) technologies might have on users, the categorization of direct versus indirect effects is used herein. *Indirect effects* affect the user at a high functional level. These include psychological effects, such as modification of phobias and enhancement or repression of emotions, as well as neurological effects on the visual system (eyestrain, modification of stereoscopic vision, and visual acuity). VEs can also affect the motion detection system (i.e., vestibular system) and may result in imbalance, nausea, and motion sickness. Research and recommendations into the indirect effects of VEs including eyestrain and motion sickness are reported in Chapters 23 through 26 and 32 through 33. *Direct effects* of VEs are less complex to study and control but are potentially as great a hazard to users as indirect effects. The direct effects of a VE system are those that act at a direct tissue level, as opposed to the body systems level of the indirect effects. Direct effects are the influence of energy on body tissues from the technology and the risk of trauma because of encumbrances (e.g., weight of a helmet-mounted display [HMD]). The incorporation of motion platforms and other heavy equipment in the home, vehicle simulators, and industry applications of VR increases physical trauma risks. Trauma is a particularly important issue, because the essential features of VEs are *interactivity* and *immersion*, where the user interacts in three dimensions with computer graphics. Irrational exuberance by users totally absorbed in the virtual experience might occur in some situations and result in injuries. Fortunately, dealing with direct effects of VEs is mostly a matter of common sense and awareness. Well-established standards based on extensive research provide clear guidance on how to implement safety programs.

22.2 DIRECT TISSUE EFFECTS

22.2.1 VISUAL SYSTEM

In the evolution of VE systems, considerable development efforts have gone into construction of visual displays. In the visible, ultraviolet (UV), and infrared (IR) wavelengths of light, there are hazards to the eye. See Table 22.1 for a delineation of biological hazards to the eye from these various light wavelengths.

Light energy can be concentrated enough to damage ocular structures directly; however, such light levels are unlikely to be used in VE systems as they would be uncomfortable to look at and not serve any purpose in the display. There are a few possible exceptions. Note that eyestrain from visual displays will be discussed in Wann and Mon-Williams (2014, Chapter 32). This chapter includes a discussion of seizures and migraine headaches induced by flashing lights and moving images.

22.2.1.1 Visible Light

Visual display systems for virtual reality (VR) and for augmented reality (AR) now have high levels of resolution but also have a high incident light output. Indeed, for AR systems to be used in ambient daylight, the light level in direct daylight has to be high. Anyone who has attempted to read his or her smartphone screen in daylight understands the viewability issue. With HMD systems, uncollimated light will be spread broadly through the pupil of the eye. Much of this light may not be incident on the retina in the viewable area of the display image but can still be powerful, uncomfortable, and theoretically a risk to the retina. The ANSI Standard Z87-2010 provides the light intensity standards for industrial eye safety.

A novel approach to development of visual displays is the use of scanned laser light directly onto the retina. In practice, as with the conventional displays described earlier, there should be little hazard. However, high-power light could potentially be introduced to the eye, so care must be taken to ensure safety (see Table 22.2). These light scanning systems use high-frequency vertical and horizontal scanners to scan tightly focused noncoherent light or coherent laser light directly onto the retina. The virtual retinal display (VRD) (TM, Microvision Incorporated, Seattle, Washington) is one such display. The VRD was invented at the University of Washington in Seattle and uses laser light to scan images onto the retina. Practical use of the VRD in room light conditions has shown that images that appear very bright can be generated with very-low-power light levels, typically on the order of 100–500 nW. The power limit for a Class 1 laser as defined by the

---

TABLE 22.1

Summary of Basic Biological Effects of Light on the Eye

Photobiological Spectral Domain	Eye Effects
UV C (0.200–0.280 μm)	Photokeratitis
UV B (0.280–0.315 μm)	Photokeratitis
UV A (0.315–0.400 μm)	Photochemical UV cataract
Visible (0.400–0.780 μm)	Photochemical and thermal retinal injury
IR A (0.780–1.400 μm)	Cataract, retinal burns
IR B (1.400–3.00 μm)	Corneal burn, aqueous flare, IR cataract
IR C (3.00–1000 μm)	Corneal burn only

Source: U.S. Department of Labor, Occupational Safety and Health Administration, (November 2000), *OSHA Technical Manual* [online], U.S. Department of Labor, Occupational Safety and Health Administration, Washington, DC, Available at: http://wwwosha-slc.gov/dts/osta.html.

TABLE 22.2
Laser Classifications: Summary of Hazards

	Applies to Wavelength Ranges				Hazards		
Class	UV	VIS	NIR	IR	Direct Ocular	Diffuse Ocular	Fire
I	X	X	X	X	No	No	No
IA	—	X[a]	—	—	Only after 1000 s	No	No
II	—	X	—	—	Only after 0.25 s	No	No
IIIA	X	X[b]	X	X	Yes	No	No
IIIB	X	X	X	X	Yes	Only when laser output is near Class IIIB limit of 0.5 W	No
IV	X	X	X	X	Yes	Yes	Yes

Note: X indicates class applied in wavelength range.

[a] Class IA applicable to lasers *not intended for viewing* only.

[b] CDRH Standard assigns Class IIIA to visible wavelengths only. ANSI Z 136.1 assigns Class IIIA to all wavelength ranges.

American National Standards Institute (ANSI) is 400 nW (ANSI Z136.1, 2007). Lasers with wavelengths only in the visible range that emit less than this power level can be viewed continuously without damage to the retina (ANSI Z136.1, 2007). Beyond this power level (Class 2 and above), damage can occur to the retina or the optics of the eye, depending on the wavelength, the duration of exposure, whether the exposure is pulsed or continuous, and the intensity. The ANSI standard explains how the risks can be determined.

The VRD has a potential risk when it is used as an AR display, that is, where the computer-generated image is viewed superimposed on the ambient visual scene. If the VRD were used in very-high-altitude aircraft, superimposition of the images on the ambient scene might require very-high light power levels to provide a perceived image of adequate brightness and contrast. These power levels could damage the eye. It should be noted, though, that such images would be uncomfortably bright and probably would not be usable without attenuation of the light from the external scene and from the VRD.

Safety of laser light can be analyzed through straightforward means and calculations using the ANSI standard. Interestingly, the same intensity limits might well be applied to noncoherent (nonlaser) light. Such analyses would be most relevant for displays using scanners.

22.2.1.2 Infrared and Ultraviolet Light

There are a few circumstances in which IR and UV light could be encountered in VE systems. The most likely reason would be incidental output of these wavelengths from the particular display technology. As with visible light, safety of IR and UV can be readily determined with simple measurements and calculations. These can also be found in the ANSI standard or in other sources. If there is no need for these wavelengths, inexpensive and effective optical coatings can be used to prevent their transmission into the eye.

Occasionally, UV or particularly IR can be useful. IR systems can be used to measure eye movements or body movements with special body-worn systems. The technique of video-oculography (VOG) can measure the eye movements of a VE system user (Wang, Lawson, & Winslow, 2014, Chapter 8). To obtain these measures, the user wears a head-mounted video camera, usually based on a micro CCD chip. Good illumination of the eye is necessary in order for the video image recognition technology to work. In order to avoid interfering with viewing, the illumination and video imaging are done in IR. Cameras for capturing the images of the moving eye are mounted on the head. In a VE system, they can be mounted inside an HMD. The cameras are generally fitted with filters to block light above about 800 nm. The eye is illuminated with an IR light-emitting diode (LED), whose illumination is not visible to the eye. High-output IR LEDs can be a hazard to the eye, as they can lead to cataracts. Furthermore, as the eye cannot see the illumination, the user is not aware of the hazard. Class 1 light emission device limits are generally a good place to start. The FDA regulation Part 1040—Performance Standards for Light-Emitting Devices—describes the calculations. The most stringent standard is for continuous viewing: defined as longer than 10^4 s. For example, for IR light of 850 nm wavelength, the permissible radiant power is 7.6×10^{-5} W. The regulations and basis for calculations can be found at www.fda.gov/cdrh/radhlth/cfr/21cfrl040.10. pdf or through the Food and Drug Administration. Further, the ANSI Standard publication Z87-2010 publishes light intensity standards for IR and UV light for industrial eyewear and thus is a very useful guideline for light safety.

22.2.1.3 Photic Seizures

Statistics show that 1.1 in 10,000 people are prone to photic seizures (Quirk et al., 1995; Shoja et al., 2007) A photic seizure is an epileptic event that is provoked by flashing lights. A bright light pulsing at the rate of 1–10 Hz can drive repetitive firing of neural cells in large groups. If enough of these cells begin firing in synchrony, the activity causes a chain reaction spreading throughout reasonably large portions of the brain. Individuals who experience such neural activity generally show a brief period of *absence* where they are awake but do not respond. Rarely, photic seizures will become generalized convulsions. In standard neurological tests for epilepsy or the tendency for epilepsy, pulsed bright lights are used to see if seizure activity can be induced and recorded on an electroencephalograph. These types of events are at times triggered accidentally in a seizure-prone driver passing a picket fence with the sun behind it or with other sources of repetitive light flashes. Video images in the 50–60 Hz range showing striped patterns have been used to induce seizures (Kasteleijn-Nolst Trenite et al., 1999). Repeated seizures can lead to brain injury and a lower threshold for future episodes.

A well-publicized incident of large numbers of induced photic seizures occurred in Japan, where a cartoon television program featured a scene in which red and blue frames alternated at 12 Hz (Takada et al., 1999). Hundreds of children viewing this program had absence attacks. A similar event could occur, inadvertently or deliberately, if pulsing light was shown to a VE user. A plausible scenario involves transport delay: the processing time from the movement of a user's head to the movement of the scene in an HMD. If a complex scene is in the VE system, the computer processor may slow in the real-time reconstruction of the scene. If the slowing went to several frames per second, the potential for flashing in the range that provokes photic seizures might occur. As with most safety problems, this circumstance could be detected by an individual dedicated to safety monitoring from the company producing the images. Producers of VE display technologies and content would do well to have a reviewer monitor their products for prolonged flashing imagery. The National Society for Epilepsy website states that photic seizures are triggered by flashing images or lights in the range of 5–30 Hz (5–30 flashes per second) (http://www.epilepsynse.org.uk/pages/info/leaflets/photo.html.). It follows that designers and reviewers of VEs should avoid flashing images in the range of 5–30 Hz. The National Society for Epilepsy website also suggests that if using a television, utilize one that possesses a 100 Hz or greater rate of refresh. The Job Accommodation Network website recommends providing a high-resolution VGA monitor/flicker-free screen and glare guards or tinted computer glasses for photic seizure-prone computer users (http://www.jan.wvu.edu).

22.2.1.4 Migraines

Migraines are a common phenomenon in the general population, occurring in 15%–20% of women and approximately 10% of men (Stewart & Lipton, 1992). Migraine sufferers have a constellation of possible symptoms including the well-known headache, aurae (i.e., an ocular sensory phenomena immediately preceding the migraine attack), and dizzy spells. Migraine sufferers are prone to motion sensitivity, as well as light and sound sensitivity, particularly during an attack. Thus, the

TABLE 22.3
Minimizing Migraine Stimulation in VEs

Sensory Stimulus	Control Measures
Visual imagery	Minimize brightness, give brightness control to subjects
Visual motion	Minimize visual motion, especially point-of-view image motion(view from a head or body mounted camera viewpoint)
Auditory	Minimize loud pulsing sounds
Duration	Minimize duration, provide breaks, allow user control

sensory experiences created in VE systems are the very phenomena that might make a migraine sufferer uncomfortable, if not ill. Fortunately, there are few serious consequences, but a bad experience could influence migraine sufferers to avoid future VE exposures. VEs should provide explanations of the experiences that they create. Migraine sufferers are often aware of their sensitivities, and so many would avoid experiencing something described as *the ultimate roller coaster* or *a sound and light explosion.* As with the problems with photic seizures described earlier, a technology or content producer might want to have a reviewer consider the amount of visual motion, brightness and contrast, and loudness of sounds that their system produces. Particularly for systems intended for training or other general purposes where migraineurs might be required to work in a VE, control of the amount and intensity of motion, sound, and light contours would be prudent. Control of the duration and repetition of intense sound and light experiences is also recommended. Some useful guidelines are provided in Table 22.3.

22.2.2 AUDITORY SYSTEM

Sound is an essential part of VE experiences, not only for making more realistic sensations, but also for providing cues to actions in the environment (see Chapter 4). Sounds for VEs can be presented through speaker systems or through headsets worn over or in the ears. The realism of some VE experiences might depend on loud sounds (e.g., an aircraft) presented through a headset. Prolonged exposure to such sounds can result in noise levels that exceed the US Federal Department of Labor—Occupational Safety and Health Administration's (OSHA) permissible exposure limit (PEL; OSHA, Technical Manual). The OSHA PEL is 90 dB, averaged over an 8 h period on the A scale of a standard sound-level meter set on slow response (OSHA, Technical Manual). Noise levels exceeding OSHA's PEL will cause hearing damage (OSHA, Technical Manual). While no guidelines exist for VE systems, user guides for a typical Walkman-style audio player headset suggest avoiding continuous loud sounds and that the headset should be removed every 30 min. Unfortunately, few users pay attention to such warnings. Manufacturers might thus consider limits on the intensity of sounds produced by their systems.

OSHA has established safety standards for sound exposures (OSHA regulations [Standards-29CFR], occupational noise exposure—1910.95). The standards establish sound-level exposures that require monitoring or head protection for workers. The louder the average noise level, the lower the amount of time of exposure allowed. For example, an OSHA *action level* is 85 dBA averaged over an 8 h workday. Exposure to noise of this level requires a hearing conservation program consisting mainly of regular hearing tests. If the sound level is over 90 dBA averaged over 8 h, hearing protection is required. A level of 100 dBA exposure over 2 h/day also requires hearing protection. These sound-level standards are for continuous sounds. It should be noted that repeated high-intensity pulsatile sounds can cause hearing loss as well. Industrial users of VE systems would do well to review the sound levels produced by their experiences and determine the amount and duration of exposures for users. These could then be compared to the guidelines set by OSHA to determine if the VE system should be used in moderation to prevent hearing damage (see Table 22.4).

TABLE 22.4

Comparison Table of Duration per Day in Hours to Allowable Sound Level in Decibels, Measured on the Scale of a Standard Sound-Level Meter Set on Slow Response

Duration per Day, Hours	Sound Level, dBA, Slow Response
8	90
6	92
4	95
2	100
1	105
0.5	110
0.25	115

Source: 29 CFR 1910.95, Table G-16, OSHA technical manual.

Of note in VE systems is *spatialized* or 3D sound systems (see Vorlander, 2014, Chapter 4). The same standards presented earlier apply to 3D systems, but the variation in intensity over the virtual sound field should be considered. If loud sounds occur within a VE system, users might intentionally or unintentionally get close to the loud sound source. Controls to reduce prolonged exposure to such intense sounds may be needed. If a user is receiving a prolonged exposure to a loud sound, a safety circuit to automatically reduce the sound levels might be a desirable design feature.

22.2.3 SKIN AND TISSUE EFFECTS

The use of VE systems can injure the skin via body-worn technology. Sensible reviews of the technology should take place. For example, if the system produces x-rays (e.g., some cathode ray tube), there is a risk of inducing carcinomatous changes in the skin (Cade, 1957). Similarly, systems producing high-radiofrequency (rF) fields at other frequencies (such as magnetic position trackers or IR emitters) should be monitored for heat generation and other tissue effects. The IEEE standard C95.1-2005 outlines the standards for rF fields from 3 kHz up to 300 GHz. There are extensive discussions on the carcinogenicity, genetic damage, and other tissue-disrupting effects of this wide range of VR in body-worn technology; these tissue effects need to be considered. As so much of our modern technology is closely applied to the body, such as mobile phones, there is substantial attention and research in these areas.

Finally, there is a potential to transmit infective agents through objects worn on the skin of one user and transferred to the skin of another, such as bacteria, viruses, and fungi. In the absence of specific governmental guidelines in the workplace, contamination of shared devices will be reduced by the use of materials resistant to infective agents (e.g., metals and plastics, as opposed to porous materials) and materials that can be disinfected.

The skin is sensitive to prolonged physical pressure, which can result in skin irritation or breakdown. Furthermore, there is a risk of passing on skin contaminants such as bacteria between multiple users, as previously noted. Fortunately, there are numerous other types of devices with similar patterns of use as public VE systems already in practice (e.g., goggles for 3D movies, public telephones), and there appears to be minimal risk associated with them. Developers of haptic feedback devices (see Dindar, Tekalp, & Basdogan, 2014, Chapter 5) and applications using them should be on the lookout for feedback or other unusual situations that could produce high pressure from their devices and cause injury. The devices for force generation should have limits on the maximal forces generated in order to minimize risk to hands and fingers.

22.3 TRAUMA

The encumbrances of many VE systems present a risk to users for physical injury. By its nature, VEs have compelling visual imagery that obliterates or obscures the visual world. Furthermore, the scenes and sequences that occur in VEs are often intended to induce the user to move, rapidly at times. Sound systems can produce disorienting sensations and the myriad of cables, weights, and tethers may cause a user to lose balance. For these reasons, system designers should watch ill effects of VE systems carefully. These are not *set and forget* kinds of systems. A *spotter* should watch users and be available to interact with them should ill effects arise. Finally, motion sickness does occur in VE systems (see Keshavarz, Hecht, & Lawson, 2014, Chapter 26). Users should thus be monitored for imbalance or attendant symptoms of motion sickness, as its occurrence makes a fall more likely.

22.4 MOTION BASE PLATFORMS

VR systems now incorporate sophisticated motion platform technology. These systems, which have been adapted from the flight-simulator industry, have become so inexpensive that they can be adapted to computer assisted virtual environment (CAVE) systems. For example, the computer assisted rehabilitation environment (CAREN) by Motek Medical of Amsterdam, Netherlands, has a six degree-of-freedom motion platform with a built-in treadmill. This technology allows operators to design VEs that incorporate moving surfaces, such as riding on the ocean; uneven terrain, such as climbing mountains; and even virtual *obstacles*. The obstacles are created by altering the motion of the treadmill to mimic a soft piece of ground or even a protruding item, such as a tree root. These virtual items are literally intended to induce tripping and fall recovery. Of course, an actual fall to the ground, especially in a compromised individual such as an amputee whose prosthesis function is being tested, is risky by definition. Safety systems and procedures must be incorporated into these platforms. The safety systems should include harnesses suspended from gantries, safety railings, and potentially body-worn equipment such as padding. The procedures will have to include line-of-sight direct observation of users, emergency stops, and professional, dedicated, and trained observers in control of the system operations. Automatic stops, detecting stumbles, strain on safety lines, or other warning features should also be incorporated. Of note, remote operators also need to be protected. Motion platforms are designed to have violent actions that will be unpredictable to bystanders. Further, other hazards, such as pits where the motion platform is mounted may also be present and present risk of severe injury or death to users and operators.

Safety standards for motion platforms are described in ASTM F2291-11, Standard Practice for Design of Amusement Rides and Devices. The ASTM Standard addresses user safety considerations including restraint systems, the size of the ride envelope, acceleration forces on the user, structural considerations for the motion base system, and considerations for operator controls. Motion base platforms in VEs should adhere to the same strict standards that apply to amusement rides. Restraint systems are designed to prevent the user from exiting the attraction unsafely. Restraint systems also restrict the size of the ride envelope, which is the area around them that the user can reach. For example, a user in a seated position restrained by a lap bar can move more in the horizontal and forward directions than a user restrained by an over-the-head harness system. Restraint systems are chosen based on the magnitude and direction of g-forces likely to be experienced by the user. The design of the structure and the materials used should be able to accommodate the largest users under the most extreme forces expected by the ride system design. Operator controls should be designed to be in easy reach of the operator and to avoid unintended activation of the ride system.

The CAREN systems use a harness system attached to a single point above the user to catch the user should they trip or fall during the VR environment experience. The user is positioned such that they cannot easily grasp the structural system designed to hold the harness system aloft. The design of the motion base should be such that the user's fingers cannot get trapped in the mechanical systems and belts that are part of the motion base design. The easiest way to accommodate this is to ensure that when the harness is supporting most of the weight of the user, the motion base should automatically stop motion. The materials and design of the structure used to support the harness system should be able to accommodate the largest expected user under the largest expected g-force. Operator controls appear to be in close proximity to the operator. It is unclear if the design of these controls prohibits inadvertent activation of the system.

22.5   AVOIDANCE OF INJURY

Fortunately, the use of VE systems is not inherently unsafe. Most circumstances that may involve risk should be mitigated with a little forethought. An individual from VE system development teams should be dedicated to the task of monitoring risks. One important component of the monitoring should be reviews with developers of any hazards or sensations they notice. For example, if developers find they are experiencing motion sickness in the development stage, users should be expected to have the same sensation. Similarly, if developers repeatedly trip over a cable, a reconfiguration should occur to fix the problem. If there is dedicated responsibility for testing and monitoring risks, the possibility of injury will be reduced. The simple framework of this chapter provides an approach for risk review, summarized in Table 22.5.

The body has its own feedback for injury risk. If a light is uncomfortably bright, sound uncomfortably loud, gear uncomfortably tight, or system uncomfortably encumbering in the short term, the risk for long-term injury should be considered. Thus, a straightforward review by potential users should reveal a number of the risks involved with the use of VE technology. Some issues, however, such as those associated with repetitive use or those to which only a portion of the user population will be susceptible (e.g., seizures) should be monitored by safety engineers. In addition, users might need to be screened for their ability to wear and appropriately interact with VE systems prior to allowing unlimited exposure. Warnings and restrictions might be needed in some situations (see Kennedy, Kennedy, Wasula, & Bartlett, 2014, Chapter 21).

The type of experience being created for users should also be considered. There are potential indirect or psychological effects from being in VEs. The possibility of producing emotional and cognitive changes in users must be considered (see Calvert, 2014, Chapter 27). Indeed, the compelling nature of VE experiences is being used to affect the nervous systems and psyches of patients with a variety of disorders (see Rizzo, Lang, & Koenig, 2014, Chapter 46). Thus, the effects of battlefields or dramatic flight sequences on users should be considered. Indirect effects can combine with direct effects to produce injury. An anxious user might pull on cables or become uncomfortable with headgear. The best method of prevention is close monitoring of users by developers and manufacturers, especially naive users of an environment.

TABLE 22.5
Steps to Minimize Risk in Development of VE Systems

1. Assign responsibility to someone for risk review.
2. Monitor and review user and developer comments for discomfort.
3. Review visual stimuli for brightness and visual motion.
4. Review auditory levels.
5. Consider emissions from electronics (e.g., heat, ionizing radiation).
6. Review hazards from cables, tethers, and other devices.
7. Review the experience being provided.

22.6 CONCLUSION

VE technology promises to provide a variety of experiences to users: bright visual displays, compelling sound environments, and dramatic interactivity. Further, the influences of VE interaction often result in changes in users' mind/brain via training or therapy. VE systems can potentially be used to improve the balance systems of patients with disturbed vestibular systems; reduce the phobias of people afraid of heights, flying, or spiders; and teach concepts of art or science to induce an emotional state or a level of education. Yet along with the promise of VE technology, one can postulate the potential negatives associated with VE exposure: bright lights, loud sounds, motion sickness, phobia induction, negative emotions, and destructive tendencies. While harnessing the potential of VE technology, developers cannot neglect to invoke common sense and research into new observations while determining the intent of technology creators to minimize the adverse effects of VE interaction.

REFERENCES

American National Standards Institute. (2007). *American national standard on the safe use of lasers* (ANSI Z136.1-2007). Orlando, FL: Laser Institute of America.

ASTM F2291-11. (2003). *Standard practice for design of amusement rides and devices*. ASTM International, West Conshohocken, PA. doi:10.1520/C0033-03, www.astm.org.

Cade, S. (1957). Radiation induced cancer in man. *British Journal of Radiology, 30*, 3939.

Calvert, S. L. (2013). The social impact of virtual environment technology. In K. S. Hale & K. M. Stanney (Eds.), *Handbook of virtual environments, design, implementation, and applications* (2nd ed., pp. 699–718). New York, NY: CRC Press.

Dindar, N., Tekalp, A. M., & Basdogan, C. (2013). Haptic rendering and associated depth data. In K. S. Hale, & K. M. Stanney (Eds.), *Handbook of virtual environments, design, implementation, and applications* (2nd ed., pp. 115–130). New York, NY: CRC Press.

IEEE International Committee on Electromagnetic Safety (SCC39). (2006, April 19). *IEEE standard for safety levels with respect to human exposure to radio frequency electromagnetic fields, 3 kHz to 300 GHz IEEE Std C95.1™-2005 IEEE*. New York, NY.

Kasteleijn-Nolst Trenite, D. G., da Silva, A. M., Ricci, S., Binnie, C. D., Rubboli, G., Tassinari, C. A., & Segers, J. P. (1999). Video-game epilepsy. *Epilepsia, 40*(4 Suppl.), 70–74.

Kennedy, R. S., Kennedy, R. C., Kennedy, K. E., Wasula, C., & Bartlett, K. M. (2013). Virtual environments and product liability. In K. S. Hale & K. M. Stanney (Eds.), *Handbook of virtual environments, design, implementation, and applications* (2nd ed., pp. 505–518). New York, NY: CRC Press.

Keshavarz, B., Hecht, H., & Lawson, B. (2013). Visually-induced motion sickness: Causes, characteristics, and countermeasures. In K. S. Hale & K. M. Stanney (Eds.), *Handbook of virtual environments, design, implementation, and applications* (2nd ed., pp. 647–698). New York, NY: CRC Press.

Quirk, J. A., Fish, D. R., Smith, S. J., Sander, J. W., Shorvon, S. D., & Allen, P. J. (1995). Incidence of photosensitive epilepsy: A prospective national study. *Electroencephalography Clinical Neurophysiology, 95*(4), 260–267.

Shoja, M. M., Tubbs, R. S., Malekian, A., Jafari Rouhi, A. H., Barzgar, M., & Oakes, W. J. (2007). Video game epilepsy in the twentieth century: A review. *Child's Nervous System, 23*(3), 265–267.

Stewart, W. F., & Lipton, R. B. (1992). Prevalence of migraine headache in the United States: Relation to age, income range and other sociodemographic factors. *Journal of American Medical Association, 267*, 64–69.

Takada, H., Aso, K., Watanabe, K., Okumura, A., Negoro, T., & Ishikawa, T. (1999). Epileptic seizures induced by animated cartoon, "Pocket Monster." *Epilepsia, 40*(7), 997–1002.

U.S. Department of Labor, Occupational Safety and Health Administration. (2000, November). *OSHA technical manual* [online]. Washington, DC: U.S. Department of Labor, Occupational Safety and Health Administration. Available: http://wwwosha-slc.gov/dts/osta.html

Wang, X., & Winslow, B. (2013). Eye tracking in virtual environments. In K. S. Hale & K. M. Stanney (Eds.), *Handbook of virtual environments, design, implementation, and applications* (2nd ed., pp. 197–210). New York, NY: CRC Press.

Wann, J. P., & Mon-Williams, M. (2013). Measurement of visual aftereffects following virtual environment exposure. In K. S. Hale & K. M. Stanney (Eds.), *Handbook of virtual environments, design, implementation, and applications* (2nd ed., pp. 809–832). New York, NY: CRC Press.

23 Motion Sickness Symptomatology and Origins

Ben D. Lawson

CONTENTS

23.1   INTRODUCTION

This chapter describes the *symptomatology* of the maladaptation syndrome commonly called *motion sickness* (MS) (Irwin, 1881), that is, the adverse symptoms and readily observable signs that are associated with exposure to real and/or apparent (e.g., visual) motion. MS incidence, observable clinical signs, and subjective symptoms of MS are described in detail (Sections 23.2 and 23.3), as well as the typical temporal progression of signs and symptoms (Section 23.4). Related syndromes (in Section 23.5) that do not necessarily involve nausea and vomiting but may pose a health or safety risk are discussed: these include the *sopite syndrome*, loss of dynamic visual acuity during head or body motion, and postural disequilibrium caused by real or apparent motion and their aftereffects. Situations conducive of these various problems are discussed as well. For example, the evidence to date indicates that many *virtual environments* (VEs) are capable of producing significant decrements in well-being (Sections 23.2.5 through 23.2.7; Sections 23.6 and 23.7). Careful assessment and correction of human factors problems in VEs is recommended to minimize user concerns about comfort, health, and safety.

MS is caused by real motion on land, at sea, and in the air. It is also caused by various *synthetic experiences* (SEs, Lawson, Graeber, Mead, & Muth, 2002), such as fixed- or moving-base simulators, VEs, teleoperators, augmented reality (Durlach & Mavor, 1995), and large-screen or 3D movies (Robinett, 1992). The term *SE* also applies to the wearing of optical prisms that alter the visual stimulus (Welch, 1978), and some of the principles of MS adaptation* to prism goggles will be similar to those that apply to modern head-mounted displays (HMDs; Mohler & Welch, 2014). Finally, the term *SE* could be applied to simulations that allow people to experience nonterrestrial forces or other feelings of body acceleration (Bowman, 2014; DiZio & Lackner, 2014; Lawson & Riecke, 2014, Chapter 7; Templeman & Denbrook, 2014).

The type of SE known as VE (Durlach & Mavor, 1995; Fisher, 1990) is of special interest. The clinical symptomatology of VE exposure is often referred to as *cybersickness* (McCauley & Sharkey, 1992). In cases where cybersickness findings are limited, conjectures must be based on related syndromes such as *simulator sickness* (the effects of which were first observed by Havron & Butler, 1957), *regular* motion-induced sickness, and visually induced motion discomfort (Crampton & Young, 1953).

---

* *Adaptation* (rather than *habituation*) is preferred in this discussion of MS to reflect the following four points: (1) exposure to a wide range of stimuli (Section 23.2) causes complex central and physiological changes that (2) cannot usually be reversed immediately by attending to the stimulus; (3) multiple sensory systems are important to the MS response, as well as sensorimotor interactions (e.g., reafference during voluntary movement); and (4) adaptation can occur under conditions of continuous or repeated stimulation.

23.1.1 SCOPE OF THIS REVIEW

The following pages describe the incidence of MS and related syndromes associated with vehicle motion and SEs and the nature of these syndromes (characteristic signs and symptoms). Methods for quantifying the signs and symptoms of these syndromes are treated elsewhere (Lawson, 2014), as are MS countermeasures (Keshavarz, Hecht, & Lawson, 2014). The present chapter is intended to be a practical guide to MS symptoms; thus, MS-related symptoms and syndromes will be treated mainly from the standpoint of empirically observed contributory factors, rather than viewed strictly through the lens of one or more of the hypotheses concerning MS etiology.

There is no comprehensive and universally accepted theory of MS etiology or the evolutionary origins of the MS response. A discussion concerning some of the key requirements a complete MS theory would need to meet is included (Section 23.9.1). Also, hypotheses concerning the evolutionary origins of the MS response are discussed. Treisman's (1977) indirect *poison hypothesis* of MS evolution is evaluated, and an alternative *direct evolutionary hypothesis* is considered (Section 23.9.2), as well as a modification of Treisman's hypothesis, called the *direct poison hypothesis* (Section 23.9.3), is considered.* Finally, a *MS amplification hypothesis* is offered (Section 23.9.4) to explain a potentially important contributor to the wide individual variability that has been observed in MS susceptibility and the fact that some people become more sensitive to MS over time (rather than more adapted and resistant). For additional review of MS hypotheses (especially concerning MS causation), see Keshavarz et al. (2014). Earlier reviews of MS hypotheses were contributed by Money (1970), Reason and Brand (1975), McCauley and Kennedy (1976), Kennedy and Frank (1985), Crampton (1990), Kolasinski (1995), Jones (1998), and Panos (2005).

This chapter emphasizes subjective symptoms and observable signs rather than physiological responses. The physiological pathogenesis of MS-related syndromes is discussed by Keshavarz et al. (2014), and more extensive coverage of physiological correlates of MS can be found in the earlier works of Crampton (1990); Money, Lackner, and Cheung (1996); and Harm (2002). Sensorimotor effects relevant to VEs are discussed by DiZio and Lackner (2014); Champney et al. (2007); Stanney, Kennedy, Drexler, and Harm (1999); N. I. Durlach and Mavor (1995); and Keshavarz et al.

The present review will aid the reader in comparing human syndromes elicited across different studies by itemizing specific signs and symptoms whenever feasible, rather than simply offering numerical summary scores of *overall sickness*. Individual symptoms are discussed because MS researchers do not use the same MS scales (see Lawson 2014), and even when they do, the computation and interpretation of summary scores vary from study to study (Lawson, 1993), such that the same sickness score can mean different things.

While the present chapter will focus on group responses that characterize MS, individual differences in susceptibility and ability to adapt to SEs are important topics reviewed by Kennedy, Dunlap, and Fowlkes (1990), Guedry (1991), Kolasinski (1995), and Mohler and Welch (2014). This chapter will not discuss individual variability in MS susceptibility extensively, other than in relation to the question of relative susceptibility of women versus men. The male versus female question receives special attention (Section 23.8.1) because the author disagrees with the prevailing sentiment that women have been proven to be more susceptible to MS than men.

This chapter strives to be up-to-date, but does not restrict discussion to the most recent work. In the field of MS research, some of the finest studies were carried out from 1960 to 1980. Therefore, older research will be given full consideration. In addition, this chapter strives to give the reader a more complete perspective than a space-limited journal article, for example, by relating problems in MS research to those in other fields of inquiry, by incorporating negative findings and gray literature (e.g., theses, technical reports), and by providing the reader background on the history and terminology of the field.

---

* The word *hypothesis* (rather than *theory*) is preferred throughout because the MS explanations in this chapter (including the author's) only account for limited aspects of MS and are not fully established, unifying explanations.

23.1.2 TERMINOLOGY

The term *cybersickness* was coined by McCauley and Sharkey (1992) to describe the motion-sickness-like symptoms associated with VE characterized by "far applications involving distant objects, that is, terrain, self-motion (travel) through the environment, and the illusion of self motion (*vection*)" (p. 313). McCauley and Sharkey chose cybersickness as a more general term than *simulator sickness* (Havron & Butler, 1957) because they considered moving- and fixed-base simulators to be a subset of VEs. The word *cybersickness* does not denote a diseased or pathological state, but rather a normal physiological response to an unusual stimulus. Similarly, in the case of *classical* MS (e.g., nausea during high sea states), failure to be susceptible to nauseogenic motion stimuli sometimes indicates an abnormal clinical vestibular state (e.g., Graybiel, 1967).

Cybersickness and simulator sickness share much in common with MS, but it is more semantically accurate to describe these *sicknesses* as *syndromes* (Kennedy & Frank, 1985). For example, Benson (1988) noted that *motion maladaptation syndrome* was a more accurate phrase than *MS*, a view that has been adopted by Probst and Schmidt (1998) and Finch and Howarth (1996). For convenience, Benson's apt phrase *motion maladaptation syndrome* could be shortened to *motion adaptation syndrome* (Lawson et al., 2002), roughly in accordance with the precedent of Nicogossian and Parker (1984), who recommended the phrase *space adaptation syndrome* in lieu of *space sickness*. Nevertheless, these semantic considerations are of most interest to experts, while lay persons would not readily understand what was meant by these various adaptation syndromes. Moreover, any researchers who want their MS paper to be found by colleagues in a topic search are obliged to include the phrase *motion sickness* in the report. Therefore, despite the drawbacks of the phrase *motion sickness*, it will be used in this chapter.

23.2 MS INCIDENCE AND SYMPTOMS IN VARIOUS ENVIRONMENTS

23.2.1 ACCELERATION AND UNUSUAL FORCE ENVIRONMENTS

It is difficult to estimate the proportion of humans who are susceptible to unusual forces created by passive acceleration within a moving device or by travel into space. The incidence estimate will vary with the stimulus (e.g., frequency, duration), the individual exposed to the stimulus (e.g., susceptibility, experience), and the experimenter's measurement criteria. The author estimates that about 10% of the population has not experienced significant nausea during transportation, whereas about 1% of the population will vomit or be made nauseated by vehicles as mild and ubiquitous as the automobile (Birren, 1949; Reason, 1975). Pethybridge (1982) found that 70% of naval personnel experience episodes of seasickness, with the greatest problems occurring in rougher seas and on smaller ships. Seasickness seems to decrease work efficiency and/or motivation. Among the sufferers, 80% report experiencing some difficulty working when they feel ill. Schwab's (1943) clinical assessment indicated that the worst-affected naval personnel were only capable of about 5%–10% of their normal land-based efficiency when they were serving on medium or small vessels.

Up to 5% of those who go to sea fail to adapt fully to the motion of the sea throughout the duration of the voyage, and these *chronically MS* individuals may show effects weeks after the end of a voyage (from Reason & Brand, 1975; Schwab, 1943; Tyler & Bard, 1949). Particularly long-lasting effects have been referred to as *mal de débarquement* (Hain, Hanna, & Rheinberger, 1999; Nachum et al., 2004). Although the biochemical changes associated with repeated vomiting and fluid loss confound the interpretation of the lasting effects of ship motion, Gordon, Spitzer, Doweck, Melamed, and Shupak (1995) have found that *land sickness* can strike persons of widely varying susceptibility after a sea voyage, resulting in feelings of postural instability and perceived instability of the visual field during self-movement (even among some individuals for whom fluid loss due to vomiting is not a concern). Similar lasting aftereffects have been noted following space flight (Clément, 2007; Nicogossian, Huntoon, & Pool, 1989) and simulator use (Gower, Lilienthal, Kennedy, & Fowlkes, 1988). According to Kennedy, Berbaum, and Lilienthal (1997). There is reason to suspect that the postural disequilibrium aftereffects of simulator exposure implicate disorientation mechanisms, not just those mechanisms involved in the nausea response to real or apparent motion.

The most provocative laboratory tests of MS yield very high incidence estimates. Using the *Pensacola Diagnostic Criteria* for MS,* Miller and Graybiel (1970b) found that 90%–96% of participants will suffer from stomach symptoms by the time they reach the maximum number of head movements called for during a rotation protocol. Specifically, in E. F. Miller and Graybiel (1970b), 90% of 250 subjects tested to a prevomiting endpoint experienced at least *minimal nausea, epigastric discomfort*, or *epigastric awareness*, whereas another 96% of 25 subjects tested all the way to vomiting experienced one of the aforementioned symptoms of nausea syndrome. The maximum possible number of head movements in the protocol was a total of 204 carried out across 10 steps of increasing chair rotation velocity.

Over the years, laboratory tests and some very provocative field tests (e.g., Kennedy, Graybiel, McDonough, & Beckwith, 1968) have shown that persons with normal vestibular function can be made nauseated by motion, whereas participants without vestibular function cannot be made nauseated by motion (Money, 1970). Similarly, participants without vestibular function cannot be made nauseated by alternating periods of weightlessness and high-G (Graybiel, 1967), although at least 67% of shuttle astronauts feel some unpleasant symptoms during their first trip into space (Davis, Vanderploeg, Santy, Jennings, & Stewart, 1988), with approximately 33% of them experiencing moderate to severe symptoms (Wood, Reschke, Harm, Paloski, & Bloomberg, 2012). According to Reschke (from Crampton, 1990), the most common signs/symptoms of space adaptation syndrome include *headache* (45% of crew members), *malaise* (43%), *vomiting* (42%), *lethargy* (40%), and *anorexia* (40%).

23.2.2 Moving Visual Surrounds

Like motion stimuli, moving visual fields can cause significant discomfort in stationary observers who are healthy, but not in individuals without vestibular function (Cheung, Howard, & Money, 1991, but see Johnson, Sunahara, & Landolt, 1999). As early as 1895, Wood (from Reason & Brand, 1975) described the aftereffects of immersing nearly stationary observers into a swinging room: "I have met a number of gentlemen who said they could scarcely walk out of the building from dizziness and nausea." (Reason & Brand, 1975, p. 110). Such adverse symptoms were confirmed by later experimentation. Using a visual surround that oscillated about the earth-vertical axis, Crampton and Young (1953) found that 46% of participants (n = 26) experienced *dizziness* (without nausea), whereas 35% experienced *nausea*. Benfari (1964) found that all nine of his participants experienced *dizziness* while viewing a wide-screen film involving vehicle motion, two of whom experienced *extreme dizziness and nausea*. (Note that the experimental conditions also involved varying amounts of peripheral flicker.)

Using an optokinetic drum, Lackner and Teixeira (1977) found that 40% of participants (n = 10) viewing a moving visual surround terminated the experimental trial because of aversive symptoms. Over all experimental conditions, the most common symptom during or after exposure was *dizziness* (mean 50%), followed by *epigastric disturbance* (mean 23%), *headache* or *eyestrain* (mean 20%), and *drowsiness* (mean 10%). In the head-fixed condition of Teixeira and Lackner (1979), *dizziness* was observed in 50% of participants (n = 8) and *eyestrain* in 13%. Using the same optokinetic drum as Teixeira and Lackner but a different velocity profile, Lawson (1993) found that 57% of participants (n = 14) experienced stomach symptoms (*stomach awareness* or *stomach discomfort* or *nausea*) during the viewing of a rotating visual surround.

---

* MS symptom assessment methods such as the *Pensacola Diagnostic Criteria*, the *Simulator Sickness Questionnaire*, the *Motion Sickness Questionnaire*, the *Motion Sickness Assessment Questionnaire*, the *Misery Scale*, and others, are discussed in the next chapter of this book (Lawson, 2014).

Using a smaller optokinetic drum (that surrounded the upper body) and summarizing across four previous studies, Stern, Hu, Anderson, Leibowitz, and Koch (1990) concluded that about 60% of people viewing a moving visual surround will report symptoms. The most common symptoms found in the experiment described in Stern et al. were *dizziness*, *warmth*, and *nausea* (during unrestricted viewing without fixation).

Several optokinetic drum studies employing similar stimuli were reviewed by Rose (2004), with MS incidence estimates obtained from six. Optokinetic exposure ranged from 12 to 30 min. Participants were defined as motion sick by Rose if they achieved a score of ≥ 6 on the *Pensacola Diagnostic Criteria* (Miller & Graybiel, 1970a,b) or reported *nausea*. Inferring from the data in Table 1 of Rose, MS incidence ranged from 45% to 75%, with an average of 57%. Longer exposures tended to produce more MS.

When MS incidence is quantified by the number of subjects deciding to terminate the experiment due to severe symptoms, the estimates are lower, between 35% and 42%. Dahlman (2009) reports four studies carried out on MS, two of which (Study II and III) are relevant here. In the two studies, participants (38 in Study II and 40 in Study III) were exposed to an optokinetic drum rotating at 60°/s for a maximum of 25 min. In Study II, 16/38 (42%) terminated early due to MS, while 22/38 (58%) endured the entire 25 min; In Study III, 14/40 (35%) terminated early, while 21/40 (53%) either endured all 40 min or were excluded from analysis due to insufficient MS (5/40 or 13% of participants had a Borg rating below 2, which would indicate no MS or very weak MS). Figure 23.1 shows the sickness ratings for those who terminated Study II due to MS versus those who did not.

Looking at the data from Dahlman, several inferences can be made that are hopeful for VE applications: (1) Approximately 60% of Dahlman's participants could endure a full 25 min of stimulation, which is longer than usual time required to achieve the goals of many simulation applications where the designers may wish to induce *vection* (Tschermak, 1931, cited in Jones, 1998) in the participants; (2) Even among the 40% of people who could not endure 25 min of optokinetic stimulation, the average discomfort rating did not exceed *weak* (2 on a scale from 0 to 10) until minute 4, which may be a sufficient duration of exposure to achieve the vection goals of some simulation applications.

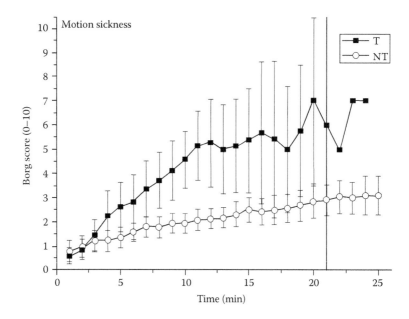

FIGURE 23.1 Means and 95% confidence intervals of 0–10 Borg sickness ratings among optokinetic drum participants who terminated (T) versus those who did not (NT). (From Dahlman, J., *Psychophysiological and performance aspects on motion sickness*, Doctoral dissertation, retrieved from http://urn.kb.se/resolve?urn=urn:nbn:se:liu:diva-15919, 2009.)

TABLE 23.1

Most Common (Non-Retching/Vomiting) Signs and Symptoms Observed in Vision-Centric Situations (Described in Section 23.2)

Situations	Most Common Symptoms Mentioned in Literature
Optokinetic tests	Eyestrain
	Headache
	Dizziness
Simulators	Drowsiness/fatigue
	Eyestrain
	Headache
	Sweating
	Difficulty focusing
	Nausea
	Difficulty concentrating
Virtual environments	Dizziness/vertigo
	Stomach symptoms
	Headache
	Fullness of head
	Blurred vision

Investigators have noticed that symptoms induced by moving visual fields seem to linger for some time after the stimulus has terminated (Crampton & Young, 1953; Havron & Butler, 1957; Lackner & Teixeria, 1977; Teixeira & Lackner, 1979). Similarly, lingering effects have been associated with visual *fixed-base simulators* (see Section 23.2.3) and the wearing of scene-altering prism goggles (Dolezal, 1982; Welch, 1978). However, it should be noted that lingering symptoms are not limited to situations involving visual motion but can also occur at sea (Reason & Brand, 1975).

Generalizing across the different visual motion experiments mentioned earlier, it appears that while a few subjects are unaffected by moving visual fields, 50%–100% will experience dizziness and 20%–60% will experience stomach symptoms of some kind. The frequency of other symptoms cannot be estimated, but it seems that oculomotor symptoms such as *eyestrain* are a prominent feature of human responses to moving visual fields (Ebenholtz, 1992; McCauley & Kennedy, 1976; Teixeira & Lackner, 1979). It is possible that moving visual fields induce more *head* (rather than *gut*) symptoms (Lackner, from national research council (NRC) Report BRL-CR-629, 1990; Reason & Brand, 1975), such as *eyestrain*, *headache*, and *dizziness*. Such nongastric effects have been observed extensively in flight simulators, which have a strong visual component (e.g., Kennedy, Berbaum, et al., 1997). Sections 23.2.3 through 23.2.6 describe MS incidence and symptoms in response to several vision-centric stimuli besides visual surround motion (Table 23.1).

23.2.3 SIMULATORS

Havron and Butler (1957) observed 36 trainees using a U.S. Navy helicopter visual simulator and noticed problems in 78% of the participants. The most frequently reported effects were (in descending order) *nausea, dizziness, vertigo, blurred vision,* and *headache*. Some effects lingered long after participants left the simulator, particularly *nausea, dizziness,* and *drowsiness*. More than half of the participants reported sickness lasting an hour or longer, and 13.9% said the effects lasted overnight. (Symptoms that are immediate or lingering are detailed on pages 5 and 8 of Appendix F from Havron & Butler, 1957).

Soon after the Havron and Butler study, Miller and Goodson (1958, 1960) reported similar effects in 12% of student pilots and 60% of pilot instructors (n = 36), using the same model of simulator. Effects included *disorientation, nausea, dizziness, vertigo, visual distortions,* and *headache.* Miller and Goodson (1958) also noted that "even those individuals who did not become ill reported that they usually felt very tired after a run. This fatigued feeling lasted frequently throughout the day" (p. 208), but they noted that longer training exposures led to more fatigue, so both common fatigue and cumulative MS effects may have been involved. Many of the symptoms lasted for hours after simulator exposure.

Kellog, Castore, and Coward (1984) found that 88% of pilots reported adverse symptoms when exposed to a particularly disturbing simulator, experiencing *visual flashbacks* (sensations of climbing, turning, or visual inversion) as much as 8–10 h after exposure (thus confirming the observations of Miller & Goodson, 1958, 1960). In reviewing selected simulator findings between the years 1957 and 1982, McCauley (1984) noted an overall incidence ranging from 10% to 88% across a number of studies. Additional effects mentioned by McCauley during these years included a feeling of tilt called *the leans* (McGuinness, Bouwman, & Forbes, 1981, in McCauley 1984) and postural disequilibrium (Crosby & Kennedy, 1982). Crowley's subsequent (1987) helicopter simulator review yielded an MS estimate of 40%, which is near the middle of the range described by McCauley. Helicopter and other flight simulators vary widely in their stimulus characteristics, so a wide range of incidence estimates are to be expected. This is still true today: Hicks and Durbin (2011) studied MS in 14 helicopter simulators and found two they considered *problem simulators*, as indicated by a score of >20 on the *Simulator Sickness Questionnaire* or SSQ (Kennedy, Lane, Berbaum, & Lilienthal, 1993; Lane & Kennedy, 1988). At the other extreme, one simulator produced negligible symptoms (SSQ score < 5). (No MS incidence data were presented.)

Magee, Kantor, and Sweeney (1988, in AGARD CP-433) found that 95% of their participants (n = 42) reported at least one symptom following simulator exposure, with 83% of the symptoms classified as *simulator sickness* (according to the criteria of Kennedy, Dutton, Ricard, & Frank, 1984). The most common symptoms were *eyestrain* (27%), *after-sensations of motion* (25%), *mental fatigue* (22%), *physical fatigue* (21%), and *drowsiness* (17%). For 81% of the participants, symptoms lingered for a median of 2.5 h postexposure, with the most common delayed symptoms being *physical fatigue* (19%), *eyestrain* (17%), and *mental fatigue* (16%). Chapelow (1988) also observed that physical and mental fatigue were the predominant symptoms of prolonged exposure to two different simulators. Gower et al. (1988) estimated an overall *simulator sickness* of 44% across a number of simulator exposures (n = 434), with the most common symptoms being *drowsiness/fatigue* (43%), *sweating* (30%), *eyestrain* (29%), *headache* (20%), and *difficulty concentrating* (11%).

Kennedy, Lilienthal, Berbaum, Baltzley, and McCauley (1989) reviewed the data from 1186 U.S. Navy flight simulator training sessions involving both fixed- and moving-base simulators. (Note that their database partly overlapped with Gower et al., 1988.) *Drowsiness* or *fatigue* occurred in 26% of the hops, *eyestrain* in 25%, *headache* in 18%, *sweating* in 16%, *difficulty focusing* in 11%, *nausea* in 10%, and *difficulty concentrating* in 10%. Only 0.2% of the hops induced *retching* or *vomiting*. The raw frequency with which one or more symptoms characteristic of motion discomfort (i.e., *vomiting, retching, increased salivation, nausea, pallor, sweating,* or *drowsiness*) were reported ranged from 10% to 60% across several (9) different simulators, with an average of 34.3%. In another analysis looking for lasting aftereffects, Baltzley, Kennedy, Berbaum, Lilienthal, and Gower (1989) found that of 700 pilots queried, 45% felt some symptoms after exposure. These effects lasted longer than 1 h in 25% of the cases and longer than 6 h in 8% of the cases.

Studies by Ungs (1988) and Kennedy, Massey, and Lilienthal (1995) indicate that the most frequently reported symptoms of simulator exposure are *drowsiness* and *fatigue*. In general, about 30%–50% of people report *fatigue* or *drowsiness* (or both) upon exiting flight simulators (Kennedy et al., 1995). Ungs found that *fatigue* was not only the most common symptom after one's first simulator flight (occurring in 34% of participants) but also one of the most severe symptoms (along with *sweating* and *nausea*). Fatigue and sleeping problems were still present after several simulator flights, albeit reduced in frequency and severity. These effects seem to occur in both fixed- and moving-base simulators (Kennedy et al., 1995) and may indicate the presence of the *sopite syndrome* (Graybiel & Knepton, 1976; Lawson & Mead, 1998), which is discussed later in this chapter (Section 23.5.1).

A driving simulator study was done recently by Brooks et al. (2010). They reported individual symptoms from which incidence can be inferred. Data from 114 participants were combined from previous studies involving the first author and using up to 30 min of exposure to the stimulus. The three most common signs/symptoms among 19 participants who quit the experiment were *annoyed/irritated* (10/19 or 53%), *clammy/cold sweat* (10/19 or 53%), and *as if I might vomit* (10/19 or 53%). Among the 95 subjects who did not terminate the study, the three most common items were *annoyed/irritated* (10/95 or 11%), *uneasy* (10/95 or 11%), and *sick to my stomach* (8/95 or 8%). It should be noted that the only symptoms above which are usually classified as MS by all researchers are those relating to cold sweating, stomach symptoms, or imminent vomiting; however, *annoyed/irritated* may be relevant to the *sopite syndrome* (Graybiel & Knepton, 1976) if other causes for irritation can be ruled out.

A 180° field-of-view (FOV) desktop flight simulator was evaluated by Cevette et al. (2012). Among 8 symptoms of MS assessed, there were 43 reports of symptoms among 11 participants exposed to the control condition, with the most common symptoms being *dizziness* (9 cases), *nausea* (8 cases), and *drowsiness* or *warmth* (6 cases each).

A recent study was conducted by Chu, Li, Huang, and Lee (2013), using a moving-base flight simulator during spatial disorientation training. The training scenarios produced mild simulator sickness, with 80% of the participants (*n* = 15) reporting at least one symptom, usually of slight severity. The most common symptoms were *increased salivation* (reported by 80% of participants) and *vertigo* (53%), with 47% of participants reporting each of the following symptoms: *fullness of head*, *eyestrain*, or *dizziness* (*eyes closed*). These symptom incidence estimates are based on mostly passive exposure to simulator profiles (under control by outside operators) and do not include the experimental condition where transcutaneous electrical nerve stimulation was evaluated as MS countermeasure during simulator exposure (Dr. Hsin Chu, personal communication, May 20, 2013).

In summary, the incidence and severity of the adverse effects associated with simulator exposure vary widely from one simulator study to the next, which is to be expected considering how different the stimuli are in each simulator employed. Adverse signs and symptoms are experienced by at least 10% and sometimes as many as 90% of simulator trainees. Effects noted in the largest simulator sickness dataset (Kennedy, Lilienthal et al., 1989) indicate that the most common effects are (in descending order of prevalence): *drowsiness/fatigue*, *eyestrain*, *headache*, *sweating*, *difficulty focusing*, *nausea*, and *difficulty concentrating*. Of these symptoms, *sweating* and *nausea* were confirmed as common by a recent driving simulator study (Brooks et al., 2010), while *eyestrain* was confirmed as common by a recent flight simulator study (Chu et al., 2013).

The distribution of susceptibility to such effects is skewed, with approximately 40% of simulator pilots reporting no symptoms and 25% reporting only mild symptoms. Approximately 5% of participants may be so severely disturbed that they should restrict subsequent activities until symptoms subside (Kennedy, 1996).

23.2.4 SIMULATORS VERSUS OTHER SYNTHETIC ENVIRONMENTS

As with visually induced motion discomfort, simulator side effects can bother the sufferer long after simulator exposure ends, in which case the most common symptoms are *fatigue* and *eyestrain*. It is interesting to note that the proportion (about 10%) of individuals who do not report symptoms even in the most provocative simulator mentioned earlier is about the same as the proportion of people highly resistant to real-motion stimuli. Similarly, the proportion of individuals who report profound and lasting symptoms (about 1%–5%) in simulators is roughly the same as the proportion of the population extremely susceptible to real motion (Section 23.2.1). Finally, the distributions for classical motion discomfort (during transportation) and for simulator discomfort appear to be skewed, with both syndromes showing more individuals with susceptibility below the middle of the range.

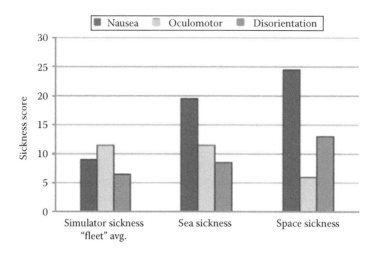

FIGURE 23.2 SSQ scores along three dimensions for simulators, sea travel, and space travel. (From Kennedy, R.S. et al., *Appl. Ergon.*, 41(4), 494, 2010.)

Thus, it can be inferred that the incidence and nature of simulator syndromes is comparable to other real or apparent motion stimuli, even though different stimuli show different subfactor profiles as measured by the SSQ (Kennedy & Stanney, 1997; Lane & Kennedy, 1988; Stanney & Kennedy, 1997), as shown in Figure 23.2.

Differences exist in the SSQ symptom clusters for simulator discomfort versus VE discomfort. It is necessary to explain the SSQ symptom clusters before discussing the findings. The three symptom clusters of the SSQ include *Nausea, Disorientation*, and *Oculomotor*. The *Nausea* cluster is composed of *general discomfort, increased salivation, sweating, nausea, difficulty concentrating, stomach awareness*, and *burping*. The *Disorientation* cluster is composed of *difficulty focusing, nausea, fullness of head, blurred vision, dizzy* (rated with eyes open and with eyes closed), and *vertigo*. The *Oculomotor* cluster is composed of *general discomfort, fatigue, headache, eyestrain, difficulty, focusing, difficulty concentrating*, and *blurred vision*. The four descriptors that appear across more than one symptom cluster are *nausea, difficulty concentrating, difficulty focusing*, and *blurred vision*.

As for the findings, Kennedy and Stanney (1997; Stanney & Kennedy, 1997) have shown that immediate postexposure profiles usually indicate that VEs produce more of the *Disorientation* symptom cluster, whereas simulators produce more of the *Oculomotor* symptom cluster. Ranking the relative magnitude of each symptom cluster, they summarized the SSQ profile for VEs as follows: *Disorientation > Nausea > Oculomotor*, whereas the profile for simulator discomfort was *Oculomotor > Nausea > Disorientation*. For comparison, the profile of space adaptation syndrome was *Oculomotor > Disorientation > Nausea*, whereas the profile aboard ships and planes was *Nausea > Disorientation > Oculomotor* (Kennedy, Lanham, Drexler, Massey, & Lilienthal, 1997; Kennedy & Stanney, 1997; Stanney & Kennedy, 1997). The distinct VE symptom profile (wherein the *Disorientation* cluster predominates) has been replicated with five different VE display systems (Stanney, Mourant, & Kennedy, 1998).

As can be seen in Figure 23.2, simulator sickness is associated with a lower score in the *Nausea* symptom cluster of the SSQ compared with seasickness or space sickness. Also, the *Oculomotor* symptom cluster contributes more to the overall sickness rating for simulators than for the other two environments (albeit, the *Oculomotor* scores appear to be about 12–13 for both simulator and seasickness, so there is little absolute difference between these two situations).*

---

* Since seasickness involves real motion, the full 28-item *Motion Sickness Questionnaire* (not just the 16-item *SSQ*) typically would be scored, but in this case, a comparison of SSQ subscale scores for *Nausea, Oculomotor*, and *Disorientation* was of interest.

23.2.5 Virtual Environments and Head-Mounted Displays

The technological aspects of a VE display play an important role in the experience of the user. The fact that displays vary widely among VE devices (and within one manufacturer's device over time) may account for the range of findings reported in the research performed to date. Specific signs and symptoms were detailed in one of the earliest published studies of the side effects of VEs, conducted by Regan and Price (1994). They reported that about 5% of their participants (*n* = 146) withdrew from the 20 min experiment due to severe *nausea* or severe *dizziness*, whereas 61% reported adverse symptoms at some time during exposure, (28% of whom had *nausea*). Among those participants who reported some symptoms but no *nausea*, the most frequently reported symptoms were *dizziness*, *stomach awareness*, *headache*, and *eyestrain*. Regan and Ramsey (1996) subsequently conducted a drug study during VE exposure using a similar protocol. They found that none of the 39 participants (among the 19 in the drug condition and 20 in the placebo condition) withdrew from this experiment due to severe discomfort, but 15 of the 20 participants (75%) in the placebo group reported symptoms at some point during or immediately after VE immersion, and 5 of whom (25%) reported *nausea*. The most common symptoms other than nausea were *dizziness*, *stomach awareness*, *disorientation*, and *headache*.

Wilson (1997) reviewed 12 previous experiments by his research group using 233 participants immersed in one of three HMDs for exposure times between 20 min and 2 h. He found that approximately 80% of the 233 participants experienced some symptoms (type not specified). For most participants, the symptoms were mild and short-lived, but for 5%, the symptoms were so severe that the participants terminated the experiment.

Howarth and Finch (1999) asked participants to report symptoms every minute during VE use, up to 20 min. Of the original 17 participants, 3 (18%) chose to terminate their participation, 1 of whom continued to feel symptoms similar to car sickness for several hours after immersion. All 14 participants who completed both conditions of the experiment reported an increase in *nausea*, and 43% of the participants reported *moderate nausea* within the 20 min immersion period. Howarth and Finch note that an unspecified number of participants who did not terminate their participation also experienced symptoms persisting for hours. The persisting symptoms mentioned included *general discomfort*, *severe hangover*, and *feeling vacant*.

Summarizing the MS incidence estimates established in the VE literature, it appears that 61%–80% of participants will experience adverse symptoms, 25%–43% of whom will experience *nausea*, and 0%–17% of whom will withdraw from the experiment due to symptoms. Next, we consider how MS develops over time in HMD-based VEs.

MS can build during continued exposure to a VE/HMD. Moss and Muth (2011) used the SSQ to assess MS among 80 participants (73 of whom completed all trials). Participants were asked to perform head movements to locate various dispersed objects viewed through live video on an HMD (under various fields of view and image update delays). They performed 40 head movements per trial over the course of five 2 min trials. The reader will note that despite the many differences between this stimulus and an optokinetic drum, the head movements led to a steady rise in overall MS (Figure 23.3) similar to that caused by the optokinetic stimulus in Figure 23.1.

23.2.6 Virtual Environments versus Other Synthetic Environments

Reviewing a number of VE studies, Stanney, Mourant, et al. (1998) noted that 80%–95% of participants exposed to a VE (for 15–60 min) reported adverse symptoms, whereas 5%–30% experienced symptoms severe enough to end participation. Stanney and colleagues observed cybersickness rates higher than had been observed in flight simulators, where they estimated 60%–70% of users report side effects (Stanney et al., 1999). They cautioned that most simulator users are military aviators or aviator candidates, who may be less susceptible to VE than the college students typically used in VE experiments. Of relevance to this assertion is a study by Regan and Price (1994), who exposed

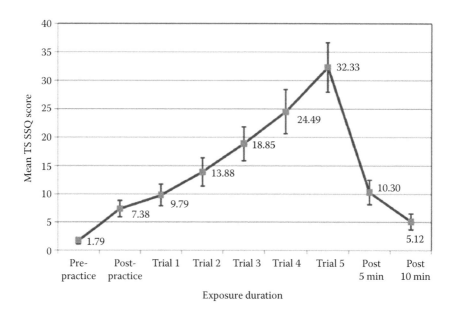

FIGURE 23.3 SSQ mean ratings (and standard error bars) for total sickness score over the trials of exposure to an HMD. (From Moss, J.D. and Muth, E.R., *Hum. Factors*, 53, 308, 2011.)

150 participants to VE, 20 of whom were in the military. Eight participants quit prior to the end of the experiment, none of whom were military personnel (however, it should be noted that malaise ratings were not significantly different between civilian and military participants). Such preliminary findings can be interpreted in a number of ways. There may be no real difference at all, or the pattern of findings could be attributable to the greater level of experience of a military service member with various SEs (Kennedy, Hettinger, & Lilienthal, 1990). It is not known whether stimulus experience rather than job description is the key factor.

If there proves to be a real difference between military personnel and civilians, it could be because people who are highly susceptible to motion are not as likely to join the military, where they may be exposed to a wide variety of challenging vehicle motions. It is also possible that military personnel acquire greater than average adaptation to challenging motions after joining. Military people may even have a higher level of commitment to finishing the experiment. Further comparisons of different participant groups are needed to corroborate and understand the apparent difference in prevalence of adverse effects in VE versus other devices (such as simulators). Further device comparisons are also needed, because the latest VE technology may yield a different estimate than was obtained with the VE technology of past decades.

Projected moving images represent synthetic environments relevant to the study of VE. Bos, de Vries, Emmerick, and Groen (2010) completed a recent study of MS on 20 participants viewing a rear-projected virtual tour of coast and city environments (50 min maximum exposure; 3 days of repeated exposure to study adaptation; at least 1 day off between sessions). Although the maximum (Day 1) MS ratings averaged only a bit more than 3 out of 10 on the *Misery Scale* (see Figure 2 of Bos et al.), the authors report that 8 out of 20 subjects (40%) got "fairly nauseated (thus terminating the trial)" (p. 519). It is implied earlier in the paper that a rating of *slight nausea* (6 out of 10) would constitute a reason to terminate the experiment. Dr. Bos clarified (Bos, personal communication, October 15, 2012) that the stated 40% means that 8/20 subjects reached a rating of at least 6 on any of the 3 days.

In summary, at least 60% of VE users will report adverse symptoms during their first exposure, with the MS incidence estimates ranging from 40% to 100%, depending upon the VE and the MS criteria. Approximately 5% of users will not be able to tolerate the prolonged use of VE, whereas at least 5% of users will remain symptom free. The withdrawal rate from VE experiments ranges widely, from 0% to 40%. The number of people highly susceptible to VE side effects during their first exposure appears to be about the same as the number of people highly susceptible to motion stimuli, moving visual fields, and simulators. The number of people immune to VE effects may be lower than for some of these aforementioned stimuli, but it is too early to be certain. VE stimuli seem to share in common with moving visual fields and simulators the tendency to elicit lasting symptoms following the cessation of the stimulus. Finally, both VE and optokinetic drums appear to produce discomfort during visual scene oscillation in pitch, roll, or yaw (Cheung et al., 1991; So & Lo, 1999). However, VE syndromes appear to be distinguishable from other syndromes via the ranking of symptom clusters from the SSQ (Kennedy & Stanney, 1997; Stanney & Kennedy, 1997), with the *Disorientation* symptom cluster most in evidence in VE. Note that three of the five descriptors within the *Disorientation* cluster (viz., *difficulty focusing, nausea,* and *blurred vision*) are shared with the other two symptom clusters of the SSQ (*Oculomotor* and *Nausea*), while three descriptors within the *Disorientation* cluster are unique (*fullness of head, dizzy,* and *vertigo*). While *dizziness, stomach symptoms,* and *headache* are among the most common symptoms in the MS literature, it is possible that symptoms such as *vertigo* or *fullness of head* are somewhat more uniquely characteristic of reactions to VE exposure (e.g., vs. *seasickness*). It would be useful to determine how individual VE symptoms (rather than clusters) compare to the symptoms generated by other real or apparent motion challenges.

23.2.7 Using VE in a Moving Environment: Uncoupled Motion

As VEs and other displays proliferate, they will be embedded within a wide range of motion environments, including moving cars, bus, trains, ships, aircraft, etc. Since VEs can be sickening and real vehicle motion can also be sickening, it would be advantageous to know what happens when both situations are paired, especially when the information from the visual display does not match the motions of the vehicle. This is an issue of practical importance, since simulation training is becoming more prevalent on board ships at sea. Very little is known about this issue. It is known that reading in a moving vehicle can be sickening and that reading a visual display during low-frequency rotatory oscillation about the vertical axis can be sickening (Guedry, Lentz, & Jell, 1978). An operational study of MS during uncoupled motion (i.e., two asynchronous real or virtual motions presented simultaneously) was performed by Muth and Lawson (2003). They evaluated 26 military subjects (12 of whom had experience as pilots) flying a virtual aircraft via a multimonitor desktop simulator while onboard a U.S. Navy yard patrol boat at sea. They found negligible additional effects on MS due to travel by sea, but they encountered very mild seas on the days of the experiment. They did observe some physiological aftereffects from the uncoupled motion. A second uncoupled motion study by Muth, Walker, and Fiorello (2006) involved a simulated driving task during real transportation in a car. In this case, performance was degraded, but it was difficult to separate MS from the other effects of motion on the task, for example, *motion-induced interruption* (Baitis, Applebee, & McNamara, 1984) or *sopite syndrome* (Graybiel & Knepton, 1976). Walker, Gomer, and Muth (2007) subsequently determined that both specific MS factors and general motion-related factors contributed to performance degradation. Muth (2009) tested 11 participants on an HMD-based flight simulator during a mild vertical oscillation stimulus (1.22 m total amplitude at 0.2 Hz) and found significant cognitive and physiological aftereffects.

23.3 SIGNS AND SYMPTOMS OF MS

23.3.1 Consensus regarding Symptomatology

The clinical techniques for measuring dizziness, nausea, and similar effects associated with MS derive from controlled laboratory studies. As pointed out by Durlach and Mavor (1995), motion discomfort is relatively easy to identify in the laboratory because fairly provocative stimulation is usually employed, multiple and systematic measurements of symptomatology are taken, and trained observers are present.

Experts tended to agree about the chief characteristics of motion discomfort by the middle of the twentieth century. For example, Birren (1949), Tyler and Bard (1949), and Chinn and Smith (1955) all agreed that motion can elicit *nausea, vomiting* (or *retching*), *(cold) sweating*, and *pallor*. Tyler and Bard and Chinn and Smith additionally mentioned *drowsiness, (increased) salivation* or *swallowing, epigastric awareness* or *epigastric discomfort* (usually defined as a sensation just short of *nausea*), *headache*, and *dizziness*. In a comprehensive review, Money (1970) noted that the four most frequently reported side effects of motion are *pallor, cold sweating, nausea*, and *vomiting*. (However, Money enumerated dozens of other characteristics that had been mentioned in the literature.)

The majority of MS researchers define the *cardinal* signs or symptoms of MS as *nausea* (possibly leading to *retching/vomiting*), *increased salivation, pallor, cold sweating*, and *drowsiness*, based on Wood, Graybiel, and McDonough (1966), Miller and Graybiel (1970a,b), and other studies by Graybiel and colleagues. Some investigators (e.g., Harm, 1990; Reason & Brand, 1975) tend to restrict the preceding list of cardinal signs and symptoms to just those four mentioned most frequently in the motion discomfort literature (according to Money, 1970), namely, *pallor, cold sweating, nausea,* and *vomiting*. On the other hand, it should be remembered that *pallor* and *cold sweating* were not more heavily weighted in the original Wood et al. scoring procedure than were *increased salivation* or *drowsiness*. Moreover, Miller and Graybiel (1970b) found that *flushing* and feelings of *warmth* were more commonly reported than was *increased salivation*. Therefore, it seems prudent not to restrict the list of cardinal MS symptoms to just four.

MS criteria such as *vomiting* and *retching* are easily quantifiable and yield parametric data; furthermore, they may be the only obvious evidence of discomfort in some cases (Lackner & Graybiel, 1986). When stimuli do not typically produce vomiting, the cardinal symptom of *nausea* is likely to heavily influence estimates of overall well-being (Reason & Diaz, 1971) and willingness to continue as a participant in a sickenin experiment (Balk, Bertola, and Inman, 2013). The picture changes when the feeling of overall well-being is not the paramount criterion, but rather the likelihood of occurrence of a given symptom and the possible deleterious effects that symptom might have on the performance of tasks. For example, *drowsiness* and *fatigue* are the most commonly reported symptoms of simulator exposure (Kennedy et al., 1995; Ungs, 1988, both from Lawson & Mead, 1998) and armored vehicle operations (Cowings, Toscano, DeRoshia, & Tauson, 1999), and such symptoms have been implicated in transportation accidents (Lawson & Mead, 1998). In general, the author of the present chapter recommends that multisymptom MS assessments be made by MS investigators and that such assessments include, at minimum, the following list of signs and symptoms (Table 23.2): *nausea, retching/vomiting, increased salivation, cold sweating, pallor* (when visual inspection can be made), *drowsiness, headache, flushing/warmth*, and *dizziness*. (MS assessment methods are described by Lawson, 2014.)

23.3.1.1 Descriptions of Signs/Symptoms

There is fairly good agreement among researchers concerning the signs and symptoms that comprise MS, but specific definitions of symptoms vary. A list of definitions that has been used

---

TABLE 23.2

Common Signs and Symptoms of Motion Sickness (Section 23.3.1)

Nausea (sometimes preceded by epigastric awareness or discomfort)

Retching/vomiting

Increased salivation

Cold sweating

Drowsiness

Pallor (if visual inspection can be made)

Dizziness

Headache

Warmth/flushing

---

by the author for prebriefing the subjects (before MS experiments) is provided in the following. Explaining these definitions to subjects should improve the accuracy and consistency of data during MS severity scaling, which is treated separately by Lawson (2014). The following definitions are adapted from Lawson (1993), which was based on sources such as Wood et al. (1966), Miller and Graybiel (1970a,b), and especially Oman, Rague, and Rege (1987). It should be noted that subjective sensations such as *nausea* are usually considered symptoms, while readily observable characteristics such as *pallor* of the skin are considered a sign of MS. However, researchers generally mean to refer to signs and symptoms whenever they use words such as *symptoms* or *symptomatology*. In fact, distinguishing the signs and symptoms of MS too carefully can be artificial and misleading. As the following symptom descriptions demonstrate, subjective symptoms such as *increased salivation* are often accompanied by observable signs such as swallowing. Similarly, *drowsiness* is considered a symptom, and yet it is often accompanied by yawning and drooping of the eyelids. Likewise, *dizziness* may be accompanied by visible eye *nystagmus*. Conversely, *cold sweating* may be detectable by the subject before it is observable to the experimenter. While the subject ultimately determines the severity rating of every MS descriptor except *pallor*, the experimenter should be attentive to the many observable effects of MS during symptom questioning. Also, while the experimenter should be cautious about using change scores, it is a good idea to determine whether any of the following symptoms are present before the experiment. For example, when assessing *headache*, it is important to know whether the symptom is attributable to the stimulus or was present beforehand. In the following, we define some of the main signs and symptoms of MS (Table 23.2) and describe a few special considerations the experimenter should keep in mind during the assessment of participants:

- *Low-level stomach symptoms. Epigastric awareness (or stomach awareness)* is a sensation that draws attention to the stomach, upper abdomen, esophagus, and/or throat. It is best described as the awareness that one possesses a stomach/epigastric/esophageal region—an awareness that is not usually present in daily life unless something is amiss. The word *epigastric* is more accurate in this context than *stomach*, but the more common word is more widely understood, and as long as the experimenter ensures the participants read and understand the full scope of this definition, common words such as *stomach* can be employed. The symptom is barely noticeable and not uncomfortable.

In contrast, *epigastric discomfort (or stomach discomfort)* is a sensation in the same regions mentioned earlier that has become uncomfortable. The severity of the sensation lies somewhere between the awareness of one's epigastric region and the first feeling of slight nausea. It is roughly comparable (but not identical) to feelings of *butterflies in the stomach* or pangs of hunger. It is difficult for some people to distinguish this sensation from minimal nausea. If *epigastric/stomach discomfort* worsens or persists enough to cause doubt, the participant should consider calling the feeling *minimal nausea*.* Participants should be encouraged to keep their criteria consistent during repeated assessments. Also, they should be warned that they may not pass through *stomach awareness or stomach discomfort*, but move straight into *minimal nausea*. In fact, it is not mandatory to assess *stomach awareness/discomfort* and some experimenters only assess *nausea*. The pros and cons of this choice during MS scaling are discussed by Lawson (2014).

- *Nausea. Minimal nausea* is the first genuine recognition of detectable *true* nausea. Unpleasant sensations are experienced in the stomach, upper abdomen, esophagus, and/ or throat that the subject unequivocally recognizes as being associated with the earliest feelings that accompany the need to vomit. However, vomiting is not imminent. *Moderate*

---

* When only stomach awareness or stomach discomfort (but not both) are used in an MS scale (as occurs in the SSQ of Kennedy, Lane, Berbaum, & Lilienthal, 1993), then the higher severity level should be used in the definition to indicate a feeling of discomfort just short of nausea.

nausea is roughly halfway to imminent vomiting; specifically, it is associated with the middle third of the range of the nausea continuum, which runs from the threshold of nausea up to intense nausea that usually precedes vomiting. *Major nausea* is associated with the intense level experienced prior to vomiting. It is a signal by the participant that unless the sickening stimulation ceases, vomiting will occur soon.

- *Increased salivation.* This symptom is relatively self-explanatory; it starts with *minimally increased salivation*, or the first noticeable increase (above normal) in the amount of saliva accumulating in the mouth, leading to the need to swallow. *Moderately increased salivation* is characterized by a pronounced increase in excess saliva and a clear need to swallow more frequently. *Majorly increased salivation* indicates copious amounts of saliva, leading to the desire to spit and frequent swallowing. (Drooling may occur but is not required for a rating of *majorly increased salivation.*)

- *Cold sweating.* This symptom is defined as sweating not entirely attributable to the ambient temperature; for example, the participant may not feel hot but may still become clammy and sweaty suddenly. It is not required for the participant to feel cool to have *cold sweating*; rather, the phrase *cold sweating* is employed to emphasize that the experimenter's intent is not to measure thermal sweating. *Minimal cold sweating* is usually experienced as slight wetness of the forehead, upper torso, or underarms. The skin may also feel cooler due to evaporation. Sweating may not be visible yet to the experimenter. Note that sweating of the palms is not as reliable a sign of MS. *Moderate cold sweating* is defined as distinct and visible beads of perspiration on the forehead or face, or slight dampening of the clothing, often followed by coolness or clamminess. Major *cold sweating* is defined as profuse whole-body sweating. Sheets of sweat appear on the face and neck. The clothing becomes quite damp, particularly on the chest, underarms, and back. *Cold sweating* is obviously easier to measure in a temperature-controlled environment where thermal sweating is not likely.

- *Drowsiness.* This symptom is relatively self-explanatory. *Minimal drowsiness* is defined as a slight decrease in alertness or feeling slightly sleepy. *Moderate drowsiness* is defined as being roughly halfway to falling asleep (i.e., in the middle third of the range). By this point, drowsiness is noticeable to the subject and yawning is noticeable to the experimenter. The participant may be slower to respond. *Major drowsiness* is defined as feeling like one could fall asleep and having trouble staying awake or concentrating on experimental tasks. If the experiment does not end soon, the subject may start dozing off. Yawning may be frequent or the participant's eyes may droop or close.

- *Pallor. Minimal–moderate–major pallor* ranges from the first noticeable whitening or loss of normal color of the skin of the face (especially the forehead and lips) to more loss of normal color (including ears, neck, and chest) and finally to severe ashen white, gray, or green tint to the head and upper body. This sign is usually assessed by the experimenter based upon how the subject's skin looked before the experiment. If the experiment allows, it may be beneficial to have the subject participate before and during the experiment by looking into a mirror. However, this important sign of MS tends to be assessed less commonly in MS experiments than the other signs and symptoms of MS for practical reasons, such as the requirement for darkness during the experiment or the impossibility of frequent access to the experimental chamber (because the device must continue to move and the subject's face is therefore only observable by camera). Whenever feasible, good lighting and direct observation should be used, or a high-quality color camera that faithfully reproduces the subject's skin tone.

- *Headache.* This symptom is relatively self-explanatory and refers to a *minimal, moderate, or major headache* that appears after the onset of stimulation and seems attributable to the same.

- *Dizziness.* The author (Lawson, 1993) uses a fairly liberal definition in accord with common English usage, wherein *dizziness* refers to a sensation of movement, spinning, tumbling, drifting, disorientation, or *wooziness* of the head. *Minimal, moderate, or major dizziness* is determined by how strong the sensation is (e.g., how disorienting; how much the sensation feels like real movement) and by whether it is transient or more than transient. Investigators should be aware that major dizziness or vertigo is often startling and may be accompanied by visible eye nystagmus, an exclamation from the participant, a stiffening of posture, or other behaviors such as grabbing the arms of the chair.

Dizziness and *vertigo* are lumped together under the term *dizziness* by the present author. The author could find no recommendation to distinguish these symptoms from the authors of the two most influential books on MS (Crampton, 1990; Reason & Brand, 1975), nor are they distinguished in the MS scales reviewed by Lawson (2014), with the important exception of the SSQ of Kennedy, Lane, Berbaum, & Lilienthal (1993). The SSQ does not define *dizziness* in the explanatory footnotes to the scale, but does define vertigo as a loss of orientation with respect to the vertical upright. By contrast, Oman, Lichtenburg, and Money (in Crampton, 1990) defined *dizziness* during spaceflight as "any uncertainty in orientation which outlasted head movement," while they defined *vertigo* as "a feeling of spinning or movement in a definable direction" (p. 222).

Dizziness is sometimes distinguished from *vertigo* in vestibular clinical medicine as well, with *dizziness* referring to a nonspecific sense of disorientation and *vertigo* referring to an illusion of self-motion suggestive of vestibular function (Luxon, 1996). In everyday life, people tend to use the terms interchangeably, with *vertigo* being the less commonly understood term. When descriptors are created by participants and used to generate a *Motion Sickness Assessment Questionnaire* (MSAQ, Gianaros, Muth, Mordkoff, Levine, & Stern, 2001), words such as *dizzy, spinning, light-headed,* and *disoriented* are generated, which appear to encompass both *dizziness* and *vertigo* without explicitly employing the term *vertigo*.

Distinguishing the terms *dizziness* and *vertigo* may yield additional useful information but may be difficult to do in some MS experiments and in the vestibular clinical setting as well. In a review of clinical vestibular effects of head injury, Luxon (1996) pointed out that suffering patients are not likely to make subtle distinctions between *dizziness* and *vertigo*, so both symptoms should be considered as potentially related to posttraumatic vertigo. Nevertheless, with careful explanation of the symptoms prior to the experiment, this problem can be overcome, and additional information can be obtained by assessing *dizziness* and *vertigo* separately, if time allows. The present author is usually in favor of preserving individual symptom descriptors and assessing them separately. If time does not allow for this, *dizziness* is the more common term for the subject to employ and more commonly used in the MS literature and MS scales and therefore should fill the needs of most MS research well. The key point for the investigator to remember is that if only the word *dizziness* is employed, the experimental definition of that term probably should include experiences of self-motion often associated with the term *vertigo*.

- *Warmth/flushing.* This is a mixture of signs and symptoms that usually occur together. Typically, it takes the form of a sudden sensation of increased warmth, especially of the face, neck, chest, back, underarms, or thighs. The experience is similar to a wave of fever. This sensation may be accompanied by a rush of blood to the face, causing a reddish face. (Sweating often commences as well.) The key point to remember for rating this item is that the participant must feel increased warmth not entirely attributable to the ambient temperature, which is in keeping with the phrase *subjective warmth or flushing* from the original *Pensacola Diagnostic Criteria* (Miller & Graybiel, 1970a,b). Like *cold sweating*, the measurement of this symptom benefits from a comfortable ambient temperature in the experimental chamber.

23.3.2 IMPORTANCE OF SUBJECTIVE MS SYMPTOMATOLOGY

A great deal can be learned by careful inquiry into the subjective aspects of motion discomfort. Researchers have yet to develop a complete description of the relation between the characteristics of different stimuli and the particular signs and symptoms elicited by those stimuli. Important steps in this direction have been taken by Kennedy and his colleagues (e.g., Kennedy, Berbaum, & Lilienthal, 1997; Kennedy, Lanham et al., 1997). To date, much inquiry has been restricted to a consideration of acute MS during short-term exposure to highly provocative stimuli. Little is known about certain aspects of motion discomfort, such as the rate and retention of adaptation, the reactions occurring during and after prolonged or repeated exposures, or the distinct symptoms of low-grade exposure (Graybiel & Knepton, 1976).

Graybiel and Lackner (1980) suggested that the aspect of motion discomfort that is most operationally relevant will depend upon the stimulus situation and the goals of the operation; for example, rate of adaptation to flying discomfort and length of retention of that adaptation may prove more important than an aviator's inherent susceptibility on initial exposure. A better understanding of these aspects of motion-related symptomatology would be immensely useful in the future. Ultimately, the best approach to the quantification of signs and symptoms associated with MS will not rely entirely on any single category of measure (subjective, physiological, performance) but will employ multiple cross-validated measures in order to get the best look at MS and its related syndromes (see book Sections II, IV, and V of Hale & Stanney, 2014). Moreover, within the domain of subjective estimates of MS, it is best not to rely entirely upon a single overall MS measure taken once following stimulation, for the reasons outlined in the next section.

23.4 SYMPTOM ONSET, EARLY SYMPTOMS, AND INITIAL SYMPTOM PROGRESSION

Most published accounts of MS, simulator sickness, and cybersickness present quantitative summary scores as estimates of total symptom severity before and after exposure to a nauseogenic stimulus, without specifying the temporal progression of individual signs and symptoms during MS. As a consequence, it is difficult to determine the onset and progression of individual symptoms in the literature. This is unfortunate, since most studies that have done repeated assessments of MS (e.g., once per minute) during stimulation using a multisymptom checklist will possess the needed information in their original participant databases, and there are many practical and scientific benefits to be derived from reporting onset and progression information concerning individual symptoms in one's final publication. For example, some symptoms are more likely to predict who will quit a simulator study prematurely (Balk, Bertola, & Inman, 2013). Other benefits are described below.

23.4.1 PRACTICAL BENEFITS OF UNDERSTANDING SYMPTOM ONSET AND PROGRESSION

- *Identifying early warning signs of MS.* The laboratory science of MS was helped immensely by the development of experimental symptom criteria that relied on a prevoming endpoint such as *moderate nausea*. Similarly, the future study of MS should be helped by the identification of the most likely pre*nausea* symptoms, that is, the symptoms that are likely to appear prior to the emergence of true *minimal nausea*. For example, MS adaptation has been recognized as a key problem to study and understand in the last few decades, but most adaptation protocols identify prenausea criteria for adaptation arbitrarily, that is, without the benefit of knowing which prenausea criteria most frequently appear as the early warning symptoms of MS progression.
- *Helping users avoid MS.* Knowing the earliest symptoms of MS may prevent a naïve person from pushing himself or herself too far. It is important in many operational simulation or transportation settings to know when to *back off* while there is still time, especially when one can exert some control over the situation (e.g., by asking the driver to pull over on a winding road to take a break, then taking control of the car). Similarly, military pilots are trained to be familiar with their earliest symptoms of hypoxia, but not their earliest symptoms of MS. It would be useful for them to learn (at least didactically via their aeromedical training personnel) the most common early symptoms of MS. This could also aid their

general awareness of their state of readiness during flying and help them determine the likely origin and hazard posed by any sudden malaise they experience. Knowledge gained in the classroom and by experience could give aviators the tools to distinguish the early symptoms of hypoxia, G-induced loss of consciousness, MS, and fatigue. This knowledge could also aid them in taking corrective actions quickly.

- *Facilitating the development of automated MS warning devices.* Any physiological correlate of MS that seeks to predict the temporal onset of MS in advance must perform better (in terms of reliability, validity, etc.) than simply asking people how they feel. For example, a physiological measure that predicts MS onset would be required to set off a warning indicator earlier than the subject's first few symptoms are recognized consciously. This difficult process of technology development will be nearly impossible unless the early symptoms of MS are studied and reported in the literature. Similarly, increased knowledge about the early symptoms of MS would help investigators determine which objective predictors to study next. For example, if one intends to find an objective measure temporally predictive of MS onset, one might initially rule out the study of measures that seem more directly related to symptoms that only appear late in the symptom progression.

23.4.2 Scientific Benefits of Understanding Symptom Onset and Progression

- *Reconciling differences in the literature.* Researchers seeking to learn more about MS by reading general reviews of the subject can read nearly as many descriptions of symptom order in the literature as there are reviews of MS. Many published reviews of MS postulate different temporal orderings of the progression of individual symptoms of MS and tend to lack sufficient cited evidence to corroborate the symptom order being postulated. Further attention to symptom onset and progression during publication would correct this problem, giving MS reviewers and their readers better evidence-based information.
- *Facilitating focused research efforts.* Some researchers are not interested in MS as a whole, but rather, in specific aspects of MS, for example, those symptoms that may relate to postural control (such as *dizziness*), those that may relate to visual acuity (such as *visual blurring*), or those that may relate to loss of concentration (such as *drowsiness*). Such researchers can be helped greatly by being able to select out individual symptoms of interest from the literature.
- *Facilitating comparison across studies.* Preserving and reporting individual symptom findings would make it easier to compare studies conducted across different laboratories (or over time in a single laboratory). There are many different MS scales, but the individual symptoms of MS are an area of much better consensus. There have been at least 70 years of widespread and systematic laboratory experiments on MS without our community having reached a universal agreement that we should all adopt one MS scale. In fact, the number of different MS scales being advocated in the literature is greater now than it was a few decades ago, despite the fact that some existing scales are established and reliable (Lawson, 2014). There is no reason to suppose this situation will change in the future, despite the fact that multilab collaborations and meta-analyses have grown in popularity and importance. While meta-analysis can handle disparate data types via indirect methods of overall comparison, being able to compare data directly across studies allows for the exploration of a greater number of specific questions and greater confidence concerning one's conclusions. Lacking a universally adopted scale, the best chance for cross-experiment comparison is represented by the reporting and preservation of individual symptom data. If researchers agree to preserve whatever individual symptom data they gather, some degree of comparison will be possible in the future concerning the core set of symptoms that overlap for most investigators.
- *Informing the refinement of evolutionary hypotheses of MS.* Symptom onset and progression evidence could be important to evolutionary hypotheses concerning why the symptoms of

MS developed as a response to unusual real or apparent motion. For example, one could ask whether the most common early indicator of MS is *stomach discomfort* (Miller & Graybiel, 1969), *dizziness* (Reason & Graybiel, 1970b), or *decreased concentration/arousal* (Bos, MacKinnon, & Patterson, 2005). One's answer to this question is very important to the type of functional explanation for the origins of MS one devises. The potential importance of the first warning signs of MS to evolutionary explanations of MS has not received much attention in the literature.

- *Informing MS scaling.* Accurate knowledge of average symptom onset and progression can aid investigators in knowing which scales to use and how to order the questions for an MS study (Lawson, 2014).

23.4.3 FINDINGS CONCERNING AVERAGE SYMPTOM ONSET AND PROGRESSION

A published report could not be located in the literature that specifically focused on the onset of individual symptoms and their complete time course of development throughout a period of exposure to a VE. There are published accounts of the time course of postflight simulator symptoms (e.g., Baltzey et al., 1989), and there is some evidence from McGuinness, Bouwman, and Forbes (1981 in McCauley, 1984) that shorter periods of simulator exposure (45 min or less) are associated most commonly with reports of *dizziness*, *vertigo*, and *fatigue* or *drowsiness*. (The most common symptom was *dizziness*, reported by 17% of participants.) Such studies do not directly address the question of what symptoms, on average, are likely to occur first, second, third, etc., during the temporal progression of MS. The most thorough elucidation of temporal onset and progression of MS symptoms has come from studies involving a real-body motion, so these findings will be summarized in the following.

One of the earliest mentions of initial symptoms during the onset of MS is by Bennett (1928), who informally observed that seasickness typically begins with *yawning* or *sighing*. In a later review, Tyler and Bard (1949) asserted that the typical order of appearance of MS symptoms would be *drowsiness* first, followed by *pallor, cold sweating, salivation*, and *stomach awareness*. On the other hand, de Wit (1953) stated that *dizziness* and *visual world movement* (loss of visual constancy due to nystagmus) would be among the earliest symptoms of MS.

Some of the first direct empirical evidence concerning onset and progression of MS symptoms comes from a survey of performance at sea by Nieuwenhujsen in 1958. During a voyage in a 149 m long ship traveling from New York to Rotterdam, 193 passengers encountered rough seas during the part of the voyage, 62% of whom became sick at sea. The first clear symptoms reported by the sick passengers were (from most to least common): *general weakness or malaise, headache, slight nausea, dizziness/disorientation, salivation*, and *cold sweating*. Nieuwenhujsen also said that these symptoms usually preceded vomiting in the order just listed. Of the specific symptoms of MS in Nieuwenhujsen that are recognized by the widely used *Pensacola Diagnostic Criteria* (Miller & Graybiel, 1970a,b; see Lawson, 2014 for history and application), the most common one reported early on was *headache*, which was experienced by 23% of the MS-affected passengers in Nieuwenhujsen's study.

One of first controlled laboratory studies of MS from which some information about early symptom progression can be inferred was performed by Crampton (1955). Crampton subjected 22 participants to up to 1 h of vertical oscillation at 7.6 ft (2.3 m) of amplitude,* at a peak G of 0.25, and a frequency of 15.6 cycles per minute (0.26 Hz). In 21/22 subjects, *pallor* preceded *nausea*, and in the remaining person, they arose simultaneously. Progression of other symptoms is not reported.

---

* Crampton does not specify if this is peak or peak-to-peak amplitude, but Dr. Robert Kennedy (personal communication, June 22, 2013), who worked at the same institution, confirms that the research elevators there were capable of 15–16 ft of peak-to-peak amplitude. Also, since 15/22 (68%) of the subjects experienced nausea or vomiting, it is likely Crampton is referring to a peak-to-peak amplitude of 15 ft (4.6 m).

Some information relevant to symptom progression can be inferred from a laboratory study by Miller and Graybiel (1969). They subjected 250 participants to a Coriolis cross-coupling test that entailed making paced head and torso movements while seated and rotating (at ever increasing speeds) about an earth-vertical axis in the dark. Coriolis cross-coupling is probably the most common method for eliciting MS in the laboratory and has proven to be a controllable and reliable test (Calkins, Reschke, Kennedy, & Dunlap, 1987; Kennedy, Dunlap, & Fowlkes, in Crampton, 1990; Lawson et al., 2007; Miller & Graybiel, 1970b). This highly sickening test caused 90% of the participants to report stomach symptoms by the end of testing. During the early phase of testing (*malaise I* or *mild sickness*), the most common symptom was *stomach awareness/discomfort* (i.e., 56% of the MS-affected subjects reported prenausea stomach symptoms), followed by *minimal sweating* and *subjective warmth/flushing*.

Reason and Graybiel (1970a) also employed a variant of the Coriolis cross-coupling test where 10 subjects actively/voluntarily moved their heads and torsos with their eyes open while seated inside a rotating enclosure. The purpose of this study was not to elicit MS rapidly, but rather to study the gradual acquisition of adaptation during incremental exposure to increasing rotation velocities. The most common early effects reported by the seven susceptible subjects were *stomach awareness* (71%), followed by *dizziness* and *pallor*.

The study that is most directly relevant to the question of onset and progression of individual MS symptoms was done by Reason and Graybiel (1970b) who exposed 41 participants to the dial test (Kennedy & Graybiel, 1965) on board a slow rotation room. The dial test involves actively moving the head, eyes, and arms in order to adjust five dials in various positions around one's body during whole-body rotation in a moving room (inside a rotating enclosure, i.e., with a participant-fixed visual surround). The basic protocol is to complete 20 sequences of five-dial settings (per sequence) at 7.5 RPM, with 6 s intervals between each setting. Participants rated their individual symptoms and also gave concurrent *overall well-being* ratings of 0–10 (0 = *I feel fine* to 10 = *I feel awful, just like I am about to vomit*). Looking at the overall *well-being* ratings of 1–1.5 (the initial onset of effects), one finds that the earliest signs and symptoms during the dial test in the slow rotation room are usually *minimal dizziness* and *visible nystagmus* (Figure 23.4). Next, at overall *well-being* ratings of about 1.6–5.0, a number of symptoms appear, including (roughly in temporal order) *bodily warmth, moderate dizziness, headache, minimal pallor, minimal cold sweating, stomach awareness,* and *minimal(ly) increased salivation*. The symptoms overlap quite a bit, but at overall ratings greater than 5.5, *sweating* and *pallor* tend to increase to moderate levels while *stomach discomfort* and unequivocal *nausea* build. This study comes fairly close to answering the question "which symptom comes first, second, etc., on average, as a group of people get more MS?"

Reschke (from Crampton, 1990) reports the progression of symptoms of *space adaptation syndrome* among 14 astronauts. One can infer from Figure 3 of his report that one of the first symptoms to appear (and to subside) is *headache*, which is also the most common symptom, reported by 45% of crew members. (It should be noted that *stomach symptoms, nausea,* and even *vomiting* are seen very early in *space adaptation syndrome*, also.)

The majority of studies from which symptom progression can be inferred are rotation studies. In a study that was relevant to VEs, Takahashi et al. (1991) asked 10 participants to wear horizontal-reversing prism goggles for up to 2 h while voluntarily locomoting and interacting with their environment. The most common initial symptoms included *discomfort, headache, cold sweating,* and *stomach ache*. The most common subsequent effects were *thirst, nausea, pallor,* and *salivation*. Of these, the most common symptom widely recognized as a distinct symptom of MS was *headache*, reported by 40% of participants.

The most recent study from which symptom progression can be inferred was performed by Bos et al. (2005). Participants (*N* = 24) were exposed to a ship-motion simulator under three 30 min viewing conditions (inside, outside, and no view). After each simulator session, participants were given a list of sickness-related symptoms and asked to recall the temporal order of individual symptom appearance. An overall group ranking of symptom order was then calculated for each condition.

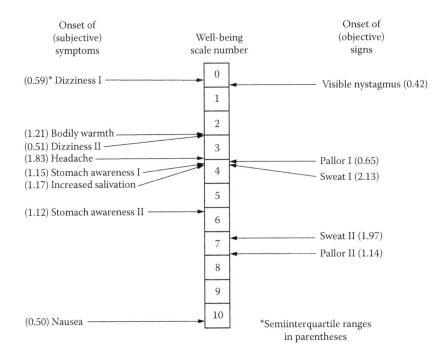

FIGURE 23.4   Median ratings and interquartile ranges associated with various points of well-being. (From Reason, J.T. and Graybiel, A., *Aerosp. Med.*, 41, 166, 1970b; This study is also described in Reason, J.T. and Brand, J.J., *Motion sickness*, Academic Press Inc., New York, pp. 74–79.)

The earliest symptoms reported were related to *concentration, drowsiness, stomach symptoms,* and *headache*. Across all conditions, *concentration* and *drowsiness* were reported before *stomach symptoms*, and *stomach symptoms* were reported before *headache*. A correlation of 0.59 was observed between symptom order and symptom severity. An additional symptom endorsed fairly early in the symptom order was *dizziness* (inferred from Figure 3 of Bos et al., 2005).

23.4.4   Summary and Gaps in Knowledge concerning MS Onset and Progression

Great individual variability exists in the order of symptoms among participants. Nevertheless, trained individuals are reliable within themselves regarding their responses to repeated provocative tests (DiZio & Lackner, 1991; Graybiel & Lackner, 1980). Ignoring the informal observations or the literature reviews discussed in Section 23.4.3 (Bennett, 1928; Tyler & Bard, 1949; de Wit, 1953), the space study (Reschke, from Crampton, 1990), and the early laboratory study where only nausea and pallor onset were tracked (Crampton, 1955), we can summarize the trends among the six remaining studies tracking multiple symptoms of *regular* MS. First, the studies involved one ship survey (Nieuwenhujsen, 1958) and five laboratory experiments (viz., Bos et al., 2005; Miller & Graybiel, 1969; Reason & Graybiel, 1970a,b; Takahashi et al., 1991). Among them, the trend (Table 23.3) was for mild stomach symptoms to be likely in the early stages of MS during all six studies. This is interesting, since the literature often describes stomach symptoms or slight nausea as being temporally in the middle or at the end of the chain of symptom progression, whereas it would appear that prenausea stomach symptoms appear very early. Other common early effects mentioned by these studies included *dizziness* (which appeared early in 4/6 studies), *headache* (3/6 studies), thermoregulatory effects (*warmth, flushing, and/or sweating*) (4/6), and *pallor* (4/6).

While the literature permits a glimpse at the early onset and progression of MS, caution should be exercised in attempting to establish the temporal order of symptoms later during exposure.

TABLE 23.3

Likely Early Symptoms and Signs during Initial Onset and Early Progression of Motion Sickness (Section 23.4)

Mild stomach symptoms (of any kind)

Dizziness

Headache

Thermoregulatory effects (warmth/flushing or sweating)

Pallor

It should come as no surprise that the final stages of MS reported in the majority of the six studies reviewed included at least *minimal nausea* as an endorsed symptom. Nausea is the signature symptom of MS and is probably the single-most important symptom to one's overall well-being report (Keshavarz & Hecht, 2011; Reason & Diaz, 1971). It should be noted that the experiment endpoint criteria during acutely sickening experiments usually require a report of at least minimal (and more commonly moderate) nausea for any normal run where the experimenter calls the halt. Clearly, nausea is to be expected as a prominent feature of laboratory studies designed to elicit MS quickly and terminate upon a report of nausea.

A potential limitation concerning the stated inferences about early symptoms is that since 3/6 studies cited earlier (viz., Miller & Graybiel, 1969; Reason & Graybiel, 1970a,b) involved some variant of Coriolis cross-coupling, *dizziness* may have featured more prominently in the data than it otherwise would. This is because rotational velocity sensations in the head are a hallmark of the Coriolis cross-coupling stimulus. Fortunately, *dizziness* was also reported as an early symptom in the survey at sea (Nieuwenhujsen, 1958) and the ship-motion simulator study (Bos et al., 2005), implying that *dizziness* is not solely an early indicator of adverse Coriolis cross-coupling effects.

It should be mentioned that while the three studies by Graybiel and his colleagues (E. F. Miller & Graybiel, 1969; Reason & Graybiel, 1970a,b) were well controlled, yielded systematic symptom reports, and generated concurrent information about minute-by-minute symptom development during testing, they were not without some potential biases. For example, the methodology of Reason and Graybiel (1970a,b) probably biased participants to focus on nausea at higher (worse) point levels of *well-being*, to the exclusion of other symptoms, because the word *vomiting* was used to anchor the definition for a *well-being* rating of 10. This is a common demand characteristic present in the symptom definitions offered earlier in this report and in most MS scales (Bos et al., 2010; Lawson, 2014), going back to the beginning of MS scaling (e.g., Hemingway, 1946, in Table 8 from Reason & Brand, 1975). Moreover, Miller and Graybiel used a weighted scoring system wherein some symptoms (especially *nausea*) contributed more than others to their overall *malaise* rating. Certainly, *nausea* and *vomiting* are very important aspects of motion discomfort, but certain stimuli produce vomiting without marked *nausea* (Lackner & Graybiel, 1986), and *nausea* is a construct that incorporates more than just descriptors beyond those associated with purely stomach symptoms leading to *vomiting* (Muth, Stern, Thayer, & Koch, 1996).

A final limitation to the current state of knowledge (about MS progression) concerns how early *drowsiness* typically appears during the progression of MS. The Bos et al. (2005) experiment is the only one of the six studies included in this summary section that specifically mentioned *drowsiness* and/or *difficulty concentrating* among the earliest symptoms of MS.* Four differences in

---

* Note, however, that the two early reviews (by Bennett, 1928; Tyler & Bard, 1949) not included in the six most important studies in this summary corroborate the symptom order in Bos et al.

methodology can be observed between the study of Bos et al. and the majority of the other studies, one or more of which may explain the different findings:

- First, the Bos et al. (2005) study did not require the participant to be as active as many of the remaining studies. In the studies by Miller and Graybiel (1969), Reason and Graybiel (1970a,b), and Takahashi et al. (1991), a key contributor to MS was the voluntary head movements made by the participants. In the Bos et al. study, while head movement would have exacerbated MS, it was not required by the procedure in order to elicit MS. The requirement for active head movement tends to keep the participant more alert (at least initially) via the need to attend to head movement commands and execute voluntary muscle activity. Also, startling or disturbing sensations will accompany each head movement in a Coriolis cross-coupling experiment, regardless of whether the head movement is voluntary or passive, causing this stimulus to be inherently dynamic, regardless of whether the head movement is voluntary and active.
- Second, Bos et al.'s (2005) study represented one of the milder stimuli in the group of six studies reviewed in this summary section, with overall MS ratings (see Figure 2 of Bos et al.) peaking *around 3 on a 0–10 scale (3 = non-nausea symptoms of slight severity)*. It is possible for *drowsiness symptoms* consistent with *early sopite syndrome* to be reported during certain mild stimuli (Graybiel & Knepton, 1976; Lawson & Mead, 1998). It should be noted that in what was probably the second-mildest of the six MS studies in this summary (viz., Reason & Graybiel, 1970a), *drowsiness* was only reported by 1/10 participants. Nevertheless, some symptoms of MS were reported by 7/10 subjects in the Reason and Graybiel (1970b) study, with 2/10 reporting *nausea* and 4/10 participants choosing to terminate the experiment before the planned end. Therefore, it is likely that the Reason and Graybiel (1970b) study was not mild enough to be comparable to the Bos et al. study.
- Third, the symptom order data in Bos et al. (2005) were based on retrospective reports given by participants 5 min after leaving the device. The use of retrospective reporting is sometimes superior for vestibular psychophysics (Mead & Lawson, 1997), but when a variety of symptoms must be recalled in an MS study, retrospective estimates may introduce an opportunity for recency effects (later symptoms being remembered as earlier symptoms) or other recall errors or demand characteristics.
- Fourth, the overall *Misery Scale* employed by Bos et al. during the stimulation involved certain assumptions and limitations, for example, concerning the contribution of different individual symptoms to overall MS as it grows more severe (Keshavarz & Hecht, 2011; Lawson, 2014, Section 24.4). It is not known whether or how the *Misery Scale* influences the findings obtained after testing when subjects are asked to retrospectively order the appearance of 25–27 individual symptoms (see Table III and Figure 3 of Bos et al.).

Assuming the potential confounds are ruled out, the Bos et al. study may reveal an important aspect of MS symptom onset and progression, that is, that the earliest symptoms one sees depend upon the type or intensity of the stimulus one uses to elicit MS. This possibility should be explored further, since it may relate to the assertion that early *sopite syndrome* is more likely to emerge with mild stimuli (Graybiel & Knepton, 1976) or that differing symptom clusters emerge with different stimuli (Kennedy, Drexler, & Kennedy, 2010; see Sections 23.2.4 and 23.2.7 of this chapter). It would be instructive to employ a similar stimulus and participant's task as in Bos et al., but to gather concurrent, minute-by-minute reports of individual symptoms (rather than using an overall MS scale) and to do so during (rather than after) stimulation.

23.4.5 Conclusions and Recommendations for Future Study

It would be useful in the future studies of symptom onset and progression to determine not only which symptoms appear first but, more importantly, which symptoms specific to MS appear first. For example, it should be determined whether *stomach awareness/discomfort* is the first symptom of MS because it is the most sensitive conscious indicator of motion discomfort or, rather, because it is present during initial entry into any novel or noxious situation, for example, because of anticipatory arousal or anxiety. It is commonly known that anxiety alone (e.g., stage fright) can cause nausea and vomiting.

To explore this question further for this chapter, the author and a colleague* looked at data from a previous Coriolis cross-coupling study by Lawson et al. (2007). Their findings corroborate 5/6 of the studies mentioned earlier (except for Bos et al.) in finding that *stomach awareness/discomfort* is the most common symptom endorsed first by participants. Unfortunately, the Lawson et al. data also indicate that *stomach awareness/discomfort* is the most common symptom endorsed immediately prior to the start of rotation. Although *stomach awareness/discomfort* increases further during the early period of rotation (compared to its baseline level), the appearance of this symptom prior to rotation in a naïve sample of participants (who are more likely to be anxious) raises questions about its specificity as an early warning sign of MS. Researchers should explore various explanations, including the possibility that *stomach awareness/discomfort* is an indicator of the earliest conscious awareness of the onset of MS, that it is an indicator of arousal/anxiety associated with the anticipation of an uncertain and noxious situation, and that it is a response deriving from past experiences of challenging motion (i.e., more akin to classical conditioning than anticipatory anxiety) or combinations of these potential factors. The answer to such questions can inform the theories of MS and the practical development of temporal predictors for MS onset. For now, all that can be said is that the most common first symptom of MS from the Lawson et al. dataset that is not also common prior to stimulation is *dizziness*, which is also noted as one of the first symptoms observed by Reason and Graybiel (1970b) (another Coriolis cross-coupling experiment and the published study deemed most relevant to the question of symptom onset in Section 23.4.4).

Traditionally, the study of symptom progression has tended to follow a group-level or *nomothetic* approach, in which the focus is on overall group results based on aggregated participant data measured at relatively few time points. Rather than relying exclusively on group-level statistics, a better approach to understanding symptom progression is to study single cases (i.e., take an *idiographic* approach) to quantify individual symptom patterns and individual differences (Barlow, Nock, & Hersen, 2008; Salvatore, Valsiner, Strout-Yagodzynski, & Clegg, 2010; Velicer, 2010). Quantitative methods that focus on the individual (e.g., via time series analysis) have the potential to reveal unique symptom patterns for an individual over time, while subgroup-level analysis (e.g., cluster analysis) may help identify homogeneous subgroups of individuals who share similar patterns in the temporal progression of different symptoms. In addition to providing a more comprehensive view of symptom progression, the inclusion of individual and subgroup-level analysis could lead to novel treatment approaches specifically tailored to individuals or homogenous subgroups.

Attempts should be made to evaluate temporal symptom progression in VE thoroughly to see how it compares to the motion responses summarized earlier for real motion. This would help identify similarities and differences between VE symptoms and symptoms associated with other SEs or with motion. It might also aid the development of health and safety warning labels for VEs (Kennedy, Kennedy, & Bartlett, 2002; Kennedy, Stanney, & Hale, 2013). Since individuals are likely to be fairly consistent in how they respond to a given VE, it is a realistic goal for persons whose jobs will require repeated VE training (e.g., future military pilots) to learn their own unique progression of signs and symptoms so they can pace themselves accordingly. Of course, persons exposed to VE frequently should also avail themselves of dual adaptation strategies that emerge from research (Mohler & Welch, 2014) as long as the strategies have been proven to enhance the performance of the person's job duties.

---

* Ms. Heather McGee, a PhD candidate at the University of Rhode Island.

TABLE 23.4
Additional Adverse Effects of Motion Maladaptation (Section 23.5)

Sopite syndrome
Degraded dynamic visual acuity
Postural disequilibrium

23.5 OTHER LESS OBVIOUS SIDE EFFECTS OF REAL OR APPARENT MOTION

Nausea and vomiting are the obvious manifestations of motion discomfort. Nausea at sea or during space flight can be lessened using appropriate oral drugs or, in extreme cases, drug injections (Graybiel & Lackner, 1987; Harm, 2002). While current interactive SEs (such as VE) may elicit the nausea syndrome, VE users are not as likely to seek drug solutions as they would for a sea voyage because they may not anticipate that the VE stimulus is a sickening experience for which drug remedies are appropriate. Even experienced VE users may choose to avoid drug therapies because (unlike the occasional sea cruise) an experienced VE user is likely to subject himself to the stimulus more frequently (than a sea voyage) and because the magnitude of the stimulus is largely under the user's control (e.g., by restricting head movements, eye movements, or duration of exposure to the VE).

People who cannot adapt readily to the nausea created by a VE are likely to limit themselves to brief exposures. However, the more subtle side effects of VE syndromes will not deter such persons (as strongly as nausea will), so they may contribute disproportionately to the future health and safety problems that arise among VE users. Some of these other side effects include the *sopite syndrome*, degraded dynamic visual acuity, and postural disequilibrium (Table 23.4). Of these three, the least overtly noxious is the *sopite syndrome*, although it may have significant implications for safety.

23.5.1 SOPITE SYNDROME

Graybiel and Knepton (1976) coined the phrase *sopite syndrome* to describe a sometimes sole manifestation of motion discomfort characterized by such symptoms as motion-induced *drowsiness*, *difficulty concentrating*, and *apathy*. Evidence that the *sopite syndrome* is somewhat distinct from the nausea syndrome and from fatigue due to nonmotion stimuli are discussed by Lawson and Mead (1998), who review what is known about the scope of the problem and its potentially insidious character. Evidence that the fatigue associated with motion is not entirely attributable to increased energy expenditure (e.g., due to the additional work performed by a person repeatedly traversing the deck of a moving ship) is to be found in Crossland and Lloyd (1993) and Lewis and Griffin (1998). An introduction to *sopite syndrome* will be provided in this section, adding some information not previously presented by Lawson and Mead and expanding upon Lawson et al. (2002).

The term *sopite* derives from the same Latin root as the more common English words *sopor* and *soporific*, namely, the Latin verb *sopire*, meaning *to lay to rest or put to sleep* (Neilsen, Knott, & Carhart, 1956). Since before recorded history, rocking back and forth has helped adults to relieve anxiety and tension (Watson, Wells, & Cox, 1998) and infants to fall asleep (Ter Vrugt & Peterson, 1973, from Leslie, Stickgold, DiZio, Lackner, & Hobson, 1997). Erasmus Darwin (1795) was probably the first person to publish the observation that motion induces drowsiness; he described a remarkable method that was related to him for procuring sleep by lying on a rotating mill wheel. The soporific effects of motion were recognized by a number of other researchers prior to Graybiel and Knepton. In 1894, DeZouche (according to Reason & Brand, 1975, p. 53) described the effects of *chronic seasickness* as "great exhaustion … [and] … heavy sleepiness." In 1922, Quix (cited by Graybiel & Knepton, 1976, p. 879) said the effects of motion discomfort include "slow ideation, lack of inclination to work, abulia [slow reaction], weakness, fatigue, feeling of uneasiness, and apathy that can lead to melancholy." In 1928, Bennet observed that MS usually begins with *yawning* or *sighing*. Similarly, Money (1970) listed *sleepiness* and *apathy* among the symptoms of MS. A number of early studies are described by Wiker, Pepper, and McCauley (1980), who note that symptoms such as *apathy* and *fatigue* have been reported by de Wit (1953), Abrams, Earl, Baker, and Buckner (1971), and Wiker and Pepper (1978) (cited in Wiker et al., 1980), the last two of which (Abrams et al.; Wiker & Pepper) saw more evidence of a fatigue response due to motion than any other mood change from a mood adjective checklist of more than 30 items (Abrams et al.), although in the case of Abrams, the findings may have been influenced by time-of-day (circadian) variations.

According to Guignard and McCauley (1990), MS can elicit *lassitude*, *yawning*, and *disinclination to be active*. They state that such effects (which can be serious in operational situations) occasionally lead to the abandonment of performance of critical tasks. They also point out that continuous oscillatory motions during rough seas can impair cognitive performance in a cumulative way and affect the quality of sleep and wakefulness, even among the persons who are not seasick. The many factors contributing to fatigue at sea can progressively degrade and delay the work of the ship's departments, which can, in turn, reduce morale (Guignard & McCauley, 1990).

Just as MS susceptibility varies widely, it appears that a few people are profoundly affected by *sopite syndrome*. Such people could be at special risk in certain operational or transportation situations. For example, a senior chief petty officer in the U.S. Navy (confidential personal communication to the author, 1998) with extensive experience at sea reported that despite being essentially immune to nausea at sea, he had often experienced overwhelming drowsiness and fatigue during the first few days after leaving the harbor, provided he had been on land for some time before the voyage. He related a disturbing event where he was standing watch aboard ship during Operation Desert Shield. The sea was fairly mild, but he found the motion overwhelmingly sedating. He felt that in this instance, his fitness for duty may have been compromised.

Similarly, a U.S. Navy flight surgeon related the case (confidential personal communication to the author, 1998) of a patient he had seen who finds it nearly impossible to stay awake when riding in a car. The person often falls asleep, even while driving, and has wrecked two cars in this manner. This person does not seem to suffer from overwhelming drowsiness or sleeping spells at any time when he is not in a moving vehicle.

Some of the first systematic studies of *sopite syndrome* were performed by Graybiel and his colleagues. In one of many rotating room studies carried out at Pensacola Naval Air Station, participants lived aboard a 20 ft diameter rotating room for 12 days (Graybiel et al., 1965). The participants suffered few episodes of overt nausea or vomiting during the study; only one out of four participants vomited as late as day 2 and none vomited after that. Nevertheless, the participants (and some onboard experimenters) yawned frequently and complained of strong fatigue and drowsiness throughout the experiment, despite taking frequent naps and sometimes sleeping longer than usual. Participants showed little motivation for mental or physical work until the fifth day of adaptation and had not fully recovered from the fatiguing effects of rotation by the end of the rotation period. A possible vestibular etiology for the *sopite syndrome* was inferred by Graybiel et al. because the participants restricted their head movements intentionally even after the cessation of nausea. Also, normal participants exposed to 2 days in the rotating room showed evidence of drowsiness and apathy at rotation rates as low as 1.71–3.82 RPM, while a control participant who had lost vestibular function was free of such symptoms under the same conditions (Graybiel, Clark, & Zarriello, 1960). Finally, it is interesting to note that a paper on the vestibular malady known as Menière's disease (Eklund, 1999) explicitly compares the aftereffects of an attack of rotational vertigo to *sopite syndrome* symptoms, which can also appear following motion (described as *late sopite* by Graybiel & Knepton, 1976).

While many researchers are aware of the earliest *sopite syndrome* paper written by Graybiel and Knepton (1976), fewer researchers are aware of the sopite-relevant studies that were carried out after that time. For example, Wright, Bose, and Stiles (1994) observed worse digit-span test performance among nauseated individuals ($n = 26$) following helicopter flight. Interestingly, the experimenters also observed worse digit-span test performance when participants had symptoms indicative of *sopite syndrome* and nausea was not prominent. In this latter case, degradation in performance would be attributable mostly to factors other than nausea, such as degraded attention and wakefulness. This notion is supported by studies that indicate low-frequency electroencephalographic (EEG) activity is associated with real or apparent motion. For example, preliminary studies by Miller (1995) and Yano et al. (1997) reported increased EEG alpha activity soon after entry into a driving situation, as well as longer gaze time and slower eye movements. Miller observed EEG (and other) evidence of drowsiness in 80 truck drivers driving late at night. Of potential relevance to the *sopite syndrome* is Figure 1 of J. C. Miller, which shows the alteration of alpha and theta activity occurring within only about 30–40 min of transferring from a stationary rest break to a moving state. However, it is possible that the restorative effect of a rest break late at night is short-lived regardless of whether motion is present. Nevertheless, the pilot data of Yano et al. (1997) indicated EEG alpha activity and eye movements associated with decreased arousal in three automobile drivers after only 30 min of daytime driving.

Chelen, Kabrisky, and Rogers (1993), and Hu et al. (1999) noted increased EEG delta activity during real or apparent motion. Chelen et al. measured the EEG of 10 subjects during real motion, whereas Hu et al. measured the EEG of 52 subjects observing a moving optokinetic visual surround.

Woodward, Taubar, Spielmann, and Thorpy (1990) tested participants ($n = 8$) on two separate days using a parallel swing and found EEG evidence of altered sleep functioning on nights following the motion stimulus. Using a relatively mild stimulus, Leslie et al. (1997) found that after 10 min of optokinetic stimulation, participants ($n = 14$) became significantly more drowsy than after a control condition (reading), but the participants ($n = 13$) from whom good EEG data could be obtained did not show a significant decrease in sleep latency onset. Specifically, using an optokinetic drum (the same as in Teixeira & Lackner, 1979) and the common 10 min period of exposure to a 60°/s constant velocity optokinetic stimulus, Leslie et al. found that scores on the *Median Sleep Latency Test* dropped from 13.6 to 12.7 min; $F(1–12) = 3.53$, $p = 0.085$.*

Kiniorski et al. (2004) also employed an optokinetic stimulus to study *sopite syndrome* ($n = 48$, independent measures, 24 per condition), but did not detect much evidence of it in their study. Most of the dimensions of their MS scale (the MSAQ) were greater for the treatment than control group (stationary drum) conditions, but the increase in the sopite-related dimension was not significant, and both groups showed similar changes in a mood scale. One limitation of the study (noted by its authors) was that the treatment group was usually tested in the afternoon while the control group was usually tested in the early evening, leading to detectable baseline differences in mood. Also, it could be argued that the control condition for a moving optokinetic drum is not a stationary optokinetic drum, but rather a stationary optokinetic drum with some other dynamic cue present suggesting a comparable lack of static state (e.g., an auditory beat stimulus).

A general issue for consideration in sopite research is the optimal use of optokinetic drums. While optokinetic drums are inexpensive and simple to set up, it should be noted that even if such indirect vestibular stimuli can cause drowsiness, the drowsiness is less likely to emerge and be detectable under conditions where the participants sit upright in a well-lit and highly novel situation (a laboratory experiment inside an optokinetic drum) while viewing a dynamically moving visual stimulus for a brief period (e.g., 12 min in Kiniorski et al.). Longer exposures should be attempted and aftereffects of stimulation tracked at the end of the experiment with participants sitting in the dark.

When feasible, milder stimuli and more naturalistic working environments should be studied so that the novelty and arousal/anxiety associated with the situation is not so great. Also, real-motion stimuli should be included in the study of sopite, because this is the setting from which most of the limited evidence for sopite syndrome derives. Real-motion studies allow investigators to determine the influence of normal lighting, low lighting, and darkness during motion.

Finally, the author strongly recommends that any fatigue study done in the transportation setting where the acceleration is known (e.g., via a black box) should include acceleration as a covariate of fatigue (during analysis). This would contribute to understanding the role of *sopite syndrome* during transportation operations while possibly filling an important knowledge gap in fatigue research as well.

---

* The sample size of $n = 13$ is deduced from the degrees of freedom of the Analysis of Variance.

Sleep deprivation studies do not generally occur in highly stimulating situations in which nausea is a worry, but this is the case for much of the sopite research. Just as sleep deprivation studies attempt to study the effects of sleep deprivation under conditions where sleepiness can emerge and be detected *if* it is present (e.g., comfortable participants executing prolonged, boring vigilance tasks), so such approaches are recommended for the studies of *sopite syndrome*. If *sopite syndrome* proves to be a fairly subtle and insidious effect for the majority of people, then it is not likely that sopite will emerge readily during the circadian peak in a well-rested person viewing a visual stimulus in a strange laboratory experiment. However, it is quite possible that a tired person driving at night during the circadian nadir will find the combination of visual flow and changes in vehicle direction and velocity associated with driving to be the *last straw*, pushing him or her over into an unsafe condition of drowsiness. Answering this question experimentally is difficult, but of great practical significance. For example, federal transportation agencies (e.g., the U.S. Department of Transportation) should have an interest in knowing whether real and apparent motion stimuli during driving or flying contribute to drowsiness-related accidents. In fact, it could be said that almost every possible contributor to drowsiness in the transportation setting (sleep deprivation, time of day, time on task, type of task, time-zone change, age, medication state, etc.) has been studied *except for the fact that the vehicle is moving*.

There have been some attempts to study the brain correlates of arousal that are relevant to the *sopite syndrome*. Wood et al. (1990) observed a slowing of EEG alpha activity and emergence of some theta and delta waves in participants with symptoms of motion discomfort and identified some drugs that alleviate nausea but do not appear to alleviate symptoms of the *sopite syndrome*. This finding may have implications for the effort to develop drugs to alleviate motion discomfort, because the relief of motion discomfort clearly entails more than alleviating nausea. Earlier, Brandt, Dichgans, and Wagner (1974) had found that either dimenhydrinate or scopolamine was effective in reducing the incidence of vomiting during real or apparent motion stimuli, but that scopolamine subjects still exhibited *fatigue*, *drowsiness*, and *loss of concentration* (Harm, 2002). It is possible that some anti-motion-sickness drugs that are sedating are very effective in preventing emesis but may mimic or exacerbate the symptoms associated with *sopite syndrome*. However, the performance effects of MS medications can sometimes be ruled out after careful testing and consideration (Lawson et al., 2007).

Other EEG studies relevant to *sopite syndrome* were performed after C. D. Wood et al. (1990). An evoked potential study by Dornhoffer, Mamiya, Bray, Skinner, and Garcia-Rill (2002) found that a rotation stimulus affected the habituation of the auditory-evoked P50 potential of the EEG, which is a measure of arousal and presumably involved in filtering of information. One of the few MS studies of EEG response in a VE was carried out by Chen et al. (2010). They measured EEG responses among 24 participants in a moving-base immersive driving simulator. They reported significant MS-related power increases in mainly the theta and delta bands in the occipital components of the EEG. These and many other findings by site and frequency band are discussed by Chen et al.

An important nondrug therapy for alleviating motion discomfort is adaption to the disturbing stimulus (Mohler & Welch, 2014). The U.S. Navy maintains a Self-Paced Airsickness Desensitization (SPAD) program wherein airsick flight students adapt to a rotation stimulus. This has proven to be effective in helping the students get over their nausea and return to flight status. Unfortunately, it appears that whether or not a student returns to flight status is unrelated to how much *sopite syndrome* he exhibits during desensitization training (Flaherty, 1998). Since *sopite syndrome* effects may linger (Graybiel & Knepton, 1976), it is possible that a few of the students who have successfully adapted to motion-induced nausea are returned to further flight training despite their tendency to exhibit greater drowsiness during and after motion stimuli (compared with their fellow students). This possibility should be investigated, because flight training itself is known to produce symptoms indicative of *sopite syndrome*. For example, Kay, Lawson, and Clarke (1998) found that two of the common effects that naval flight officers in primary flight training experienced during flight days (vs. nonflight days) were *drowsiness despite adequate rest* and *persistent unexplained fatigue*.

Similar effects have been observed at sea. For example, Colwell (2000) completed an extensive performance assessment questionnaire of individuals engaged in North Atlantic Treaty

Organisation (NATO) Atlantic Fleet Operations. Of the dozens of subjective variables surveyed, two of the most commonly reported items were *fatigue* and *difficulty sleeping* (the night before), which correlated with the amount of ship motion (as a function of sea state). Similarly, Holmes, Robertson, and Crossland (2002) found that greater ship motion and greater MS susceptibility were correlated with poorer sleep and subsequent fatigue, as revealed by surveys and wrist-worn activity monitors (n = 12). (Also see Haward, Lewis, & Griffin, 2000; Wertheim, 1998). It should be noted that fatigue at sea could be caused by *sopite syndrome* but also by being awakened at night due to vessel motions. Holmes et al. suggested sopite may have been a minor factor in their study, since they found no significant relationship between daytime vessel motion and reported fatigue during duty. However, the majority of the participants in this study had taken a nap between shifts and were well adapted to sea motion already, so being rested and adapted may have played a role in the pattern of findings observed.

It is possible that *sopite syndrome* occurs in other settings as well, such as aboard flight simulators, in VEs, and during spaceflight. Kennedy, Massey et al. (1995) found ample reason to suspect that *sopite syndrome* is a feature of simulator exposure. The possibility for *sopite syndrome* as a result of VE exposure was first mentioned in an NRC report (Lackner, in BRL-CR-629, 1990) and has been recommended as a research priority (Durlach & Mavor, 1995; Stanney, Salvendy et al., 1998), but to date, little is known about *sopite syndrome* in VEs. There are several interesting issues that could be explored. For example, *sopite syndrome* sometimes causes the sufferer to feel detached, distant, less communicative, and less willing to engage in group behavior (Graybiel & Knepton, 1976). Such an affective state would adversely affect the VE user's sense of presence (Sadowski & Stanney, 2002; Schatz, 2014) and degree of performance in shared VEs (Durlach & Slater, 1998; Kaber, Draper, & Usher, 2002).

Sopite symptoms might also compromise crew coordination during some military operations; for example, teleoperation of unmanned aerial vehicles, highly mobile amphibious landings, or maneuvers aboard a crew transport vehicle (Cowings et al., 1999). Finally, *space adaptation syndrome* has been suggested as a factor hindering communication between ground and space crews (Kelly & Kanas, 1993). Hettinger, Kennedy, and McCauley (1990) noted that while good communication is required in almost all situations where people work together aboard moving vehicles and is a vital aspect of teamwork (Lawson, Kelley, & Athy, 2012), this remains an aspect of MS that has received little attention. It is known that sopite-like symptoms occur in operational settings such as space flight (Kanas & Manzey, 2008; Lawson & Mead, 1998) and military flight training (Flaherty, 1998). In space flight, such nicknames as *space stupids, mental viscosity,* and *space fog* have been used to describe the soporific effects of an altered force environment (Boyle, 2007; Kanas & Manzey, 2008).

Kennedy et al. (2010) explored the hormonal correlates of visually-induced-MS, partially due to an interest in the *sopite syndrome.* They measured saliva cortisol and melatonin during visually induced MS elicited by an optokinetic surround or a VE. They observed an elevation of cortisol during exposure to either type of stimulus. Melatonin results were mixed. Nine of the twelve participants showed increased melatonin levels after exposure to the visual stimulus, two of whom were dramatically elevated. However, this trend was not significant overall. Kennedy et al. note that not all subjects will experience a given symptom (e.g., drowsiness) but that melatonin may prove a useful indicator among those who respond to motion with symptoms of *sopite syndrome.* Larger studies with more comprehensive measures were recommended. In an earlier study, Kennedy and colleagues (Kennedy, French, Drexler, & Compton, 2005) recruited 25 participants, 13 of whom were given 3 mg of melatonin and 12 of whom were given placebo. Thirty minutes later, the participants were exposed to either an optokinetic stimulus or a VE. Oral melatonin had no discernible effect (compared with placebo) on *vection*, cognitive performance, fatigue ratings, or MS ratings. It would be interesting to carry out a study of this type with a greater number of participants per condition and a larger dose of melatonin. The potential challenges associated with studying *sopite syndrome* in novel, highly interactive visual motion situations have already been discussed earlier in this Section 23.5.1.

The mechanisms of *sopite syndrome* are not known, but it is known that the vestibular system is highly integrated with the arousal systems of the brain. In a series of studies on vestibular influences during sleep, Pompeiano and colleagues (e.g., Morrison & Pompeiano, 1970) described several neural connections between the vestibular brainstem, the reticular activating arousal control systems, and associated structures. More recently, Horowitz, Blanchard, and Morin (2005) identified projections from the vestibular nucleus to the intergeniculate leaflet and the suprachiasmatic nucleus of the circadian rhythm system. It is reasonable to hypothesize that *vestibular signals might influence circadian rhythmicity*, which could explain why sleep disturbances are associated with exposure to motion (Graybiel & Knepton, 1976). McCandless and Balaban (2010) have argued that the parabrachial nucleus is an important center for the processing of head acceleration signals and note that it also has a role in the processing and mediation of autonomic and affective aspects of the pain response. They hypothesize that this center may contribute to affective responses associated with falling (which is accompanied by heightened arousal), MS (accompanied by a decreased sense of well-being), and comorbid balance and anxiety disorders (Lawson, Rupert, & Kelley, 2013). It has been explained in this section, the *sopite syndrome* also is accompanied by arousal and mood effects (Graybiel & Knepton, 1976; Lawson & Mead, 1998).

Why should the brain be organized in this way? At present, only broad functional conjectures can be made. In closing our consideration of *sopite syndrome*, we will briefly consider whether any purpose could possibly be served by responding to motion by entering a quiescent state. Several possible benefits are considered in the following, wherein entering a quiescent state tends to minimize sickening motion and foster adaptation during or after sickening stimuli:

- Lying down maximizes somatosensory input, which could aid orientation, reduce sensory conflict, and reduce postural control demands.
- Lying still decreases the sickening vestibular stimulus, improving the probability that one's adaptive capacity will not be exceeded.
- Lying on one's back may put the otolith organs in a less sensitive position, somewhat reducing the troubling input in certain situations.
- Sleeping takes the vestibulo-ocular reflex mostly *off-line*, reducing MS contributions from oculomotor mechanisms, retinal slip, and visual suppression of vestibulo-ocular reflex.
- Sleeping allows consolidation of learning during rapid-eye-movement sleep, and adaptation is a form of learning. Rapid eye movements are controlled by the vestibular nucleus and associated structures (Morrison & Pompeiano, 1970), making the circumstantial link tighter. It is possible that the brain reprograms itself while sleeping to recalibrate to the new sensorimotor/force environment.

The hypothesized protective effects of *sopite syndrome* should not be considered a given in every situation. Nature may need to strike a delicate balance in each situation between the functional advantages of minimizing severe nausea/vomiting and improving adaptation versus the functional disadvantages of lying down and/or sleeping. For example, it makes sense for an infant in the mother's womb or a child not yet weaned to respond to motion by becoming quiescent, since a strong vomiting response to ingested toxins (Treisman, 1977) would not be needed yet and vomiting due to motion may expose one to detection by (or decreased vigilance for) predators. By contrast, it makes sense for an adult to be able to adopt either response, or both, depending on the strength and duration of the stimulus. If a stimulus is mild, the reaction would be less similar to the effect of ingested toxins and quiescence (without vomiting) may be sufficient. Conversely, if a stimulus is very strong and vomiting has not sufficiently rendered enough immunity, it makes sense for the organism to fall back to quiescence as another line of defense instead of cycling indefinitely through endless episodes of vomiting and retching.

23.5.2 LOSS OF VISUAL ACUITY DURING HEAD OR BODY MOTION

Another likely side effect for VE users is the loss of visual acuity during head or body motion. Most whole-body motions will elicit a vestibulo-ocular reflex that helps gaze remain stable as the head or body moves. This natural coordination of visual and vestibular function is disrupted when one dons prism goggles that alter the amount of visual-field movement corresponding to a given head movement (Gonshor & Melville-Jones, 1976); it can also be altered when the user attempts to read a head-fixed display during prolonged whole-body motion. In this latter situation, interpreting the display requires a visual suppression of the ongoing vestibulo-ocular reflex. Anyone who has attempted to read fine print during turbulent flight aboard a commercial airplane will appreciate the difficulty and unpleasantness of the effort. Those who design VE will need to understand the conditions in which a viewer's ability to suppress his vestibulo-ocular reflex is enhanced without creating unpleasant symptoms (Lawson, Rupert, Guedry, Grissett, & Mead, 1997).

23.5.3 POSTURAL DISEQUILIBRIUM

Postexposure disruptions of balance have been associated with motion and unusual force environments (see Baltzley et al., 1989). People can adapt to unusual gravitoinertial force environments such as microgravity (Nicogossian et al., 1989), high G-force, ship motion, and rotating rooms (reviewed in Crampton, 1990). Unfortunately, the acquired adaptation often creates a negative aftereffect that disrupts balance and coordination on returning to the stationary world. A familiar example is the way in which sailors must regain their *land legs* after returning to shore (Gordon et al., 1995).

Postural disruption can also be caused by abnormally integrated visual–vestibular signals. Lee and colleagues (Lee & Aronson, 1974; Lee & Lishman, 1975) showed that a moving visual surround can influence postural sway in toddlers standing on a regular floor or in adults standing on compliant surfaces (which degrade ankle kinesthesia). A study by Brandt, Wenzel, and Dichgans (1976) implies that the postural influence of a moving visual field is greatest from ages 2 to 5. The possible causal and temporal relations among *vection*, postural sway, and feelings of discomfort are viewed differently by different researchers (see Jones, 1998, Keshavarz et al., 2014). Regardless of the relations involved, it is clear from data gathered in ground-based flight simulators that significant ataxic aftereffects occur after short exposures to certain visual-field motions (Kennedy, Berbaum et al., 1997; Kennedy, Lilienthal et al., 1989). Research by Kennedy and Stanney (1996) suggests that simple measures of postural instability can reveal the aftereffects of VE exposure. These various findings point to a potential health and safety risk associated with the prolonged use of *inertial* and *noninertial* virtual displays (Durlach & Mavor, 1995; Lawson et al., 1997; Stanney, Salvendy et al., 1998). Of particular concern is the fact that soon, the majority of the population will be routinely exposed to VEs and *augmented reality*, while the elderly segment of the population that is most vulnerable to falls will be larger than it is now (Fregly, 1974; Grabowski, 1996).

23.6 FACTORS RELATED TO MOTION SICKNESS IN VIRTUAL ENVIRONMENTS

23.6.1 MOVING VISUAL FIELDS

23.6.1.1 Relation between Vection and Feelings of Discomfort

The role of *visually induced illusions of self-motion (vection)* in enhancing the experience of VE is discussed by Storms (2002) and Lawson and Riecke (2014). This section briefly summarizes the evidence concerning a possible relation between *vection* and MS. Hettinger, Berbaum, Kennedy, Dunlap, and Nolan (1990) hypothesized that simulator discomfort and VE discomfort should be correlated with the presence of a visually induced illusion of self-motion, known as *vection* (Fischer & Kornmüller, 1930). Hettinger et al. is the most cited study on this topic in the VE literature, and thus, it merits a brief summary. Hettinger et al. tested 15 participants; they found that among the 10/15 participants reporting *vection*, 8 of those 10 also experienced MS symptoms, while 2/10 experienced *vection* without MS. Of the 5/15 subjects who did not report any *vection*, only 1/5 became sick. When sickness incidence was evaluated as a function of *vection* incidence, the observed Chi-square = 5.00, with Phi coefficient = 0.557 and $p = 0.25$. In short, a nonsignificant trend was observed. Jones (1998) later tested a large sample ($n = 78$) with a driving simulation task and detected a small but significant correlation between the magnitude of *vection* during the stimulus and sickness severity score immediately after the stimulus (Pearson $r = 0.25$, $p < 0.03$, $n = 78$). Similarly, when Hu et al. (1997) observed 100 participants, they found that their optokinetic stripe condition that elicited the most *vection* also elicited the most sickness (vs. the other stripe frequencies tested). Hu et al. did not report the direct correlation value obtained between *vection* and MS, only the difference test results for *vection* and sickness separately across different stripe frequency conditions, since stripe frequency was their variable of interest. One of the strongest correlations between *vection* and MS was found in an HMD study by Muth and Moss (2009) involving 30 men and 50 women. Muth and Moss observed a moderate correlation between peak MS and total *vection* ($r = 0.53$, $p < 0.05$). They concluded that feelings of *vection* seem to be related to MS, but not for all subjects. Personal communication with Dr. Jason Moss (October 18, 2012) confirms that 31 subjects had at least slight MS (SSQ scores greater than 7.48) without reporting vection.

The relation between *vection* and sickness was less convincing in several other studies. For example, in one of the earliest studies from which an inference can be drawn about the relation between *vection* and MS, Crampton and Young (1953) found that nine of their participants (i.e., 35% of $n = 26$) experienced nausea during visual-field motion, five of whom perceived themselves to be moving (no inferential statistics presented). Similarly, Webb and Griffin (2002) found that ratings of sickness and *vection* were not significantly correlated with each other in 16 participants tested across two visual conditions (Spearman Rho ≥ 0.184, $p > 0.479$). In his dissertation, Webb (2000) evaluated *vection* and MS in a number of experiments (n ranging from 13 to 20 per study) involving an optokinetic or a VE and did not find *vection* to correlate with MS in any of them. Finally, Kleinschmidt et al. (2002) found that 9/9 of their participants experienced *vection*, *none* of whom experienced nausea. Chen, Chow, and So (2011) obtained similar subjective velocity ratings of *vection* under optokinetic stimulus conditions that differed in terms of MS.

Lawson (2005) concludes that the literature yields the estimates of *sickness-free vection* ranging widely (from 20% to 100% of subjects) and that a significant correlation between *vection* and sickness was usually observed only when the sample size was large. Lawson sought to determine whether *vection* could be experimentally dissociated from visually induced MS using limited exposures to a completely immersive optokinetic stimulus (Figure 23.5). He tested 45 participants, eliciting nearly maximal *vection* ratings (on a scale of 1–5 where 1 = *not moving* and 5 = *as if I am actually rotating*) using a brief stimulus exposure that would be more than sufficient for many VE applications (two 5 min exposures with a 1 min break in between). The high *vection* ratings are shown in Figure 23.6. Despite the near-maximal *vection* ratings obtained in this study, the stimulus produced negligible visually induced MS, with no subjects reporting minimal nausea or greater and only 4/45 (9%) of the subjects reporting even prenausea stomach symptoms such as *stomach awareness/discomfort*, which were transient (gone in less than 1 min) for all but one of the four participants experiencing them. Therefore, Lawson demonstrated that strong vection can be elicited even when MS is negligible.

In summary, only 3/10 studies reviewed earlier found a significant relation between vection and MS (Table 23.5). The author concludes that the relation between *vection* and sickness is not strong enough to preclude exploiting *vection* as a helpful cue in VE (Lawson & Riecke, 2014) and that, under the right conditions, strong *vection* can be elicited readily with negligible sickness (Lawson, 2005). The converse observation that visually induced MS can arise in the absence of *vection* has also been reported, most recently by Ji, So, and Cheung (2009), who employed a mixture of central and peripheral motions in opposition to one another. Therefore, the earlier view that *vection* may

FIGURE 23.5 (**See color insert.**) The immersive optokinetic sphere: outside (left) and inside (right) views. (From Lawson, B.D., Exploiting the illusion of self-motion [vection] to achieve a feeling of "virtual acceleration" in an immersive display, in Stephanidis, C., Ed., *Proceedings of the 11th International Conference on Human-Computer Interaction*, Las Vegas, NV, pp. 1–10, 2005.)

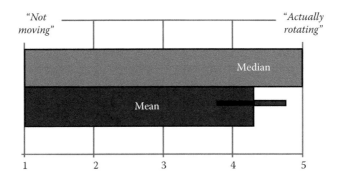

FIGURE 23.6 Maximal median ratings of compellingness of vection (mean and SD shown for reference).

TABLE 23.5
Findings from the Literature concerning the Relation between Vection and Motion Sickness

Studies	Findings
Crampton and Young (1953)	Some participants got nauseated without reporting vection.
Hettinger et al. (1990)	Nonsignificant correlation between vection and MS ($n = 15$).
Hu et al. (1997)	Most MS in condition that elicited most vection ($n = 100$)[a].
Jones (1998)	Small but significant correlation ($n = 78$)[a].
Webb (2000)	No significant correlation across several studies ($n = 13–20$).
Kleinschmidt et al. (2002)	All participants reported vection; none reported nausea.
Webb and Griffin (2002)	Nonsignificant correlation between vection and MS ($n = 16$).
Muth and Moss (2009)	Positive, significant correlation ($n = 80$)[a].
Lawson (2005)	Maximal vection ratings with little or no MS.
Chen, Chow, and So (2011)	Different MS levels at the same vection velocity rating.
	Conclusion: 3/10 studies found a relation between vection and MS.

[a] Significant findings obtained.

---

be correlated with visually induced MS but not necessarily be a precursor of it (Ebenholtz, 1992; Hettinger & Riccio, 1992) can now be expanded. The current evidence implies that *vection* is not consistently correlated with MS and that conditions can be devised for which *vection* is neither a necessary nor a sufficient condition for visually induced MS. It is recommended that any future studies attempting to study *vection* versus MS should employ a large sample of participants viewing a moving surround for a prolonged period of time at various velocities (e.g., below and above the velocity limit of the visual following or optokinetic reflex). Conversely, VE users or designers seeking to elicit strong vection without provoking MS should limit exposure duration, frequency, and velocity of movement of the visual scene.

Two other studies may bear indirectly on the question of *vection* versus MS. First, individuals without functional vestibular systems cannot be made sick by motion, yet can experience *vection* when viewing visual stimuli (Cheung et al., 1991). Second, Witmer and Singer (1998) found a negative correlation between the measures of *presence* and simulator discomfort, such that greater *presence* resulted in less discomfort. The Cheung et al. study poses a further challenge to the hypothesis that *vection* is strongly related to MS, while the Wittmer and Singer study may be of future interest to VE designers seeking to increase *presence* by eliciting *vection*.

A related topic of interest is the potential link between *vection* and the adverse effects of head movement during optokinetic drum rotation. The case here seems just as mixed. Studies by Dichgans and Brandt (1978) suggested that if strong *vection* was being experienced by the participant, head movements could elicit discomfort. However, studies by Lackner and Teixeira (1977) and Teixeira and Lackner (1979) indicated that if head movements were performed prior to the onset of *vection*, the movement suppressed the *vection* illusion and prevented discomfort.

This section (23.6.1) discusses adverse effects of moving visual fields. However, a simple view of cybersickness as a subset of visually induced MS is unlikely to prove adequate in the long term because sensations of self-motion can be elicited by other than visual modalities (Bles, 1979; Lackner & DiZio, 1988), and future VE will not be limited to visual and auditory displays but will incorporate the cutaneous, kinesthetic (Basdogan & Srinivasan, 2002; Biggs & Srinivasan, 2002; DiZio & Lackner, 2014; Templeman & Denbrook, 2014), and vestibular stimuli (Lawson & Riecke, 2014) as well.

Side effects involving real or apparent motion stimuli are the focus of this chapter. Nevertheless, certain VE display characteristics can cause discomfort without requiring whole-field visual motion or other apparent motion stimuli. For example, some of the earliest or least expensive head-mounted VEs use low-resolution liquid crystal displays placed close to the eyes without ideal clarity, contrast, or illumination. Also, stereoscopic VE displays may require the user to accommodate to a particular depth plane while requiring vergence movements to a range of depths. As a result of such display factors, Wann and Mon-Williams (1997) observed changes in visual functioning that were correlated with reports of discomfort and have replicated these results with various HMDs (Wann & Mon-Williams, 2002).

23.6.1.2 Field of View (FOV)

Throughout the literature on simulators, display FOV has been implicated in a trade-off between performance and side effects (e.g., Kennedy, Fowlkes, & Hettinger, 1989; Van Cott, 1990; Westra, Sheppard, Jones, & Hettinger, 1987). Although a larger FOV often enhances performance, it also increases the probability of experiencing side effects. DiZio and Lackner (1997) showed that by decreasing the FOV in a VE (and thus the delay or lag in visual display update), they were able to decrease discomfort ratings.

Symptoms elicited by moving visual fields generated by rotating optokinetic drums are worse if one is viewing a large FOV display (Dichgans & Brandt, 1978); however, Andersen and Braunstein (1985) found that 17 of their participants (i.e., 31% of $n = 55$) reported adverse symptoms (13 felt dizzy, 5 had a headache, 5 felt nauseated, and 4 felt warm).*

---

* These were the four most common symptoms. Subjects could report more than one symptom each.

MS can also be elicited while viewing radially expanding patterns subtending only 7.5° to 21.2° of visual angle. One can conclude that although the Andersen and Braunstein stimulus elicited a lower incidence of symptoms than previous experiments that had used a wider FOV, it was a stimulus sufficient to induce 10% of the participants to terminate their participation.

23.6.1.3 Frequency of Visual Scene Motion

In simulators, the amount of movement of the visual scene accounts for 20% or more of the variance in simulator discomfort (Kennedy, Berbaum, Dunlap, & Hettinger, 1996). The velocity and spatial frequency of the visual display contributes to this effect (Dichgans & Brandt, 1978). The predominant frequency of oscillation of the visual display (Kennedy et al., 1996) is important to the resulting level of discomfort. It is well known that real motion is most disturbing when it involves low frequencies of oscillation approximately in the 0.2 Hz range of motion or 1 cycle every 5 s (e.g., McCauley & Kennedy, 1984); it is believed that the same generalization applies to moving visual scenes (Hettinger et al., 1990; Stoffregen & Smart, 1998). Interestingly, in the case of very-low-frequency oscillation (0.02 Hz or 1 cycle every 50 s) of the visual scene, the visual stimulus can sometimes be more disturbing than real-body motion at the same frequency (Dichgans & Brandt, 1978).

23.6.1.4 Intervals between Exposure and Exposure Duration

Kennedy, Lane, Berbaum, and Lilienthal (1993) found that repeated exposure in simulators was effective in promoting adaptation and reducing discomfort when the interval between exposures was 2–5 days. Watson (1998) obtained similar results in a driving simulator, noting that a 2–3 day interval between exposures was best.

It has been estimated that 20%–50% of the variance in VE-related discomfort can be accounted for by the amount of time a person spends in a VE and the intersession interval (Kennedy, Stanney, & Dunlap, 2000). Kennedy, Jones, Stanney, Ritter, and Drexler (1996) found marked reductions in discomfort in the second of two 40 min VE exposures. Cobb, Nichols, Ramsey, and Wilson (1999) looked at repeated 20 min exposures to a passive VE. They observed a reduction in self-reported discomfort over the three exposures spaced a week apart. (They noted that the greatest reduction occurred for the Disorientation symptom cluster of the SSQ.) Cobb et al. reported a preliminary study of four participants interacting with a VE for up to 2 h. All four participants reported increased discomfort up to 1 h, at which time two of the participants withdrew. However, the two participants who continued with the experiment reported decreasing levels of discomfort at 75 min. At the end of the 2 h session, their reports of discomfort were back to preexposure levels.

23.6.1.5 Lack of Correspondence between Visual Scene and Head Movement

The literature on simulators indicates that an asynchrony between inertial and visual motion will cause discomfort (Frank, Casali, & Wierwille, 1988; Uliano, Lambert, Kennedy, & Sheppard, 1986); this has been suggested as a problem for VE users also (Hettinger & Riccio, 1992). VE users were observed to have higher levels of disorientation, discomfort, and postural instability as lag increased in the update rate of the visual display (Cobb et al., 1999; DiZio & Lackner, 1997). Specific problems stemming from a temporal discordance between head movement and visual-field movement are discussed in Chapter 6 of Durlach and Mavor (1995). Improving the correspondence between head movement and visual-field movement is an important area of VE development and dissemination (Stanney, Salvendy, et al., 1998). Since head-tracking, headset-mounted displays are well known for lag and have been central to the concept of VEs used in many applications, the advantages and limitations of using an HMD will be considered in the following, expanding the consideration beyond just lag.

23.6.1.6 Control and Navigation Factors

Individuals who have little control over the movements of a simulator seem to experience greater discomfort than individuals who are controlling the simulator (Casali & Wierwille, 1998; Reason & Diaz, 1971). Similarly Stanney and Hash (1998) found that giving participants control over their actions will reduce their VE side effects. Stanney and Hash examined three conditions: (1) *active* (using a joystick to maneuver with six degrees of freedom), (2) *active–passive* (using a joystick to move forward and backward, side to side, up and down, and in specific circumstances yaw and pitch), and (3) *passive* (passively observing scripted movements). Stanney and Hash found that the average discomfort scores were highest for the passive condition. The active condition did not reduce the severity of the symptoms as much as the active–passive condition. Stanney and Hash suggest that in the active condition, the participants might not have been able to adapt as quickly as in the active–passive condition, since more abundant sensory information was available during unrestricted movements made in the active condition.

Howarth and Finch (1999) found that head-controlled navigation produced greater severity and sooner onset of nausea than hand-controlled navigation, most likely due to greater system lag. Anecdotal evidence has suggested that as users gain experience with a VE, they tend to reduce the number and magnitude of their head movements to avoid discomfort (Cobb et al., 1999; Howarth & Finch, 1999). These findings suggest that exploring VEs via head movements is more provocative than using alternative input devices (see also Bowman, 2002; Durlach & Mavor, 1995).

23.7 ADVANTAGES AND CHALLENGES OF HEAD-MOUNTED DISPLAYS

23.7.1 ADVANTAGES OF HMDs FOR DISPLAYING VEs

VEs based on HMDs have several strengths, including low cost, small footprint, privacy, and effective 360° FOV. This can be a way to administer training that would be too difficult, dangerous, time-consuming, or expensive to administer in a real-world environment. Unfortunately, HMD-based VEs have not been disseminated widely yet because of negative side effects and the potential for aftereffects to last hours after exposure (Muth, 2010).

23.7.2 CHALLENGES ASSOCIATED WITH HMDs

Undesirable MS side effects of HMDs can lead to the cessation of training due to trainee discomfort. These side effects can also lead to negative transfer of training because (in an effort to minimize MS) the users may perform behaviors differently in the VE than they would in the real-world environment (Walker, Muth, Switzer, & Hoover, 2010). Furthermore, in the consumer market, negative side effects can lead to decreased sales and dissatisfied customers. Therefore, it is pertinent to identify and minimize the causes of the negative side effects.

There are several unique features of HMDs that may contribute to their side effects. First, HMD systems have a small actual display FOV in terms of what the user can see on the display at any given movement, but this is partially compensated by an effective 360° FOV (what the user can see by moving his or her head). Second, HMDs have an image scaling factor that corresponds to the ratio of the actual display FOV to the geometric FOV or the simulated FOV of the scene within the HMD (Draper, Viirre, Furness, & Gawron, 2001). Third, HMDs often use eyecups to occlude peripheral vision. Finally, because the effective FOV is created by using a sensor to track head movement and determine visual point of view within the simulated scenes, there is a delay between head movement and visual scene movement to the new visual point of view. These various potential contributors to HMD side effects are explored in the following.

Early work investigating MS associated with HMD considered actual display FOV and system latency as causal factors (DiZio & Lackner, 1997). Actual FOV is still thought to be a factor, but it should be noted that the variability in range of FOVs of HMDs reaching a broad market is rather limited. On the related issue of FOV image scaling factor, a recent study did not reveal an effect on MS (Moss & Muth, 2011).

In terms of system latency, human perceptual threshold for latency is approximately 150 ms ± 85 SD (Moss, Muth, Tyrrell, & Stephens, 2010), a figure just at or above most current system latencies, and would therefore not seem to account for the high degree of MS caused by HMD use. Furthermore, systematically exposing individuals to increasing latency does not appear to increase MS (Moss et al., 2011). Occluding peripheral vision does appear to increase sickness but may represent only a minor contribution to overall sickness (Moss & Muth, 2011).

Recent investigation into why HMDs make people sick found that system latency is actually not a single constant number but rather varies over time (Wu, 2011). For the system components examined, two types of latency were found: (1) latency variability in the 15–50 Hz range with a magnitude of 10–20 ms (2) latency variability in the 0.5–1.0 Hz range with a magnitude of 20–100 ms. The higher frequency variability was associated with the system clocks and buffer times, and the lower frequency variability was associated with head tracker sensor error. Both of these frequencies are of interest from a human standpoint, since flicker vertigo occurs at a frequency of 4–20 Hz (Rash, 2004), and both horizontal and vertical translations seem to result in peak sickness at a frequency of 0.2 Hz (Golding, Phil, Mueller, & Gresty, 2001; O'Hanlon & McCauley, 1974). Further work is needed to determine if this variability in latency is a causal factor in MS in HMDs.

It is important to note that *augmented reality* systems are poised to become more important and ubiquitous tools (e.g., consider *Google Glasses*). These systems may involve a *see-through* version of an HMD in which the wearer can see the real environment he or she is in with virtual images presented in the display appearing as if they are in the real world. These augmented reality systems are subject to the same latency issues that HMDs are subject to, since they must sense and track a user's visual point of view. Therefore, research into the occurrence of variable latency in augmented reality systems is recommended.

A challenge with HMDs, augmented reality systems, and VEs in general is that there is a lack of standardization by which the components (e.g., orientation sensors) are evaluated by manufacturers. This results in specifications being published that may only be accurate for a limited set of operating circumstances. These inaccuracies are often discovered only during system integration after an end user has invested in the technology. Standardized evaluation processes need to be developed for individual components as well as complete systems. Having accurate measurements of sensor error, total system latency, actual FOV, and image scaling factor would help in evaluating and improving human safety and comfort when using these systems.

Some VE systems are presented as a surround display that is not mounted on the head. This tends to give the user a wider FOV when the head is stationary and has the advantage of smaller lag associated with rapid head movement but requires a larger setup and presents a more limited FOV when the head scans the VE. Kennedy et al. (2010) observed that the symptom clusters of the SSQ were somewhat different for HMD versus non-HMD VEs, with HMD systems causing more sickness in the *Nausea* cluster but less in the *Oculomotor* cluster.

23.8 PREDICTION OF MS SUSCEPTIBILITY

One of the most useful and well-supported descriptions of MS prediction was written by Kennedy, Dunlap, and Fowlkes (in Crampton, 1990). They evaluated the predictive validity of a wide range of factors from the literature. They concluded that the measurement of past MS in the same operational situation one is trying to predict was the best approach (accounting for up to 67% of the variance, after correction for attenuation). Controlled laboratory MS testing was second-best for predicting MS in an operational setting (38% of variance after correction). Among those factors that did not require MS testing and related to the question of individual traits, Kennedy et al. found that the two best predictors of MS in a given situation were as follows: (1) a past history of susceptibility to MS in a number of different situations (estimated via an MS history questionnaire, accounted for up to 34% of corrected variance) and (2) perceptual style, especially as reflected by field dependency scores (up to 26% of corrected variance).* Factors that were not shown to be adequate predictors

---

* The reader should note that these separate predictors share overlapping sources of variance.

of MS included *autonomic nervous system tendency* (p. 186, i.e., certain baseline physiological measures of predisposition to MS) and *function of sensory end organs* (p. 182, i.e., clinical and psychophysical tests of vestibular function and sensation).

A number of other individual factors possibly predictive of MS susceptibility have been mentioned frequently in the literature, including age, race, personality type, and sex (Crampton, 1990; Keshavarz, Hecht, & Lawson, 2014; Reason & Brand, 1975). It is generally thought that susceptibility peaks at around ages 8–12, that Chinese people are more susceptible than some other ethnic groups, that there is a weak tendency for neurotics to be more susceptible, and that women are more susceptible. While such factors as race may account for some variance in susceptibility that is of interest to researchers, transportation personnel wishing to quickly identify susceptible individuals should concentrate on whether persons of interest have been susceptible to past exposures to the same stimulus, whether they report a past history of susceptibility to various motions (Kennedy, Dunlap, & Fowlkes, in Crampton, 1990), and whether their parents are highly susceptible to MS (Reavley, Golding, Cherkas, Spector, & MacGregor, 2006).

23.8.1 Susceptibility of Men versus Women: An Alternative View

The majority of MS researchers would probably say that women are more susceptible to MS than men, but alternative explanations for this trend have never been sufficiently controlled to allow a definitive statement on the issue. The author can identify up to 19 possible confounds to interpretation (Lawson, Kass, Lambert, & Smith, 2004) that would need to be controlled before it can be concluded for certain that sex differences reported in MS studies are attributable to biological sex, per se, and not some other mediating factor. A few of the potential confounds during vehicle transport include the amount of unusual motion experience/adaptation among men versus women, usual extent of reafferent information derived from vehicle control, amount of head movement typically caused by vehicle motion (due to body size and trunk/neck muscle mass/stiffness relative to head mass), likelihood of needing to move one's head around more voluntarily during transportation (to spot signs and landmarks, to assist children and other passengers), and likelihood of needing to be *head down* reading in-vehicle displays or visually interacting with passengers. Psychological factors play a mediating role in apparent biological sex differences in reported MS as well, for example, enculturated willingness to report adverse symptoms, proficiency at introspecting about body state, field dependency, etc.

In a preliminary study, Lawson et al. (2004) reviewed 46 experiments presented in 44 studies published from 1940 to 2001, all of which were relevant to the question of sex differences in MS. They found that while the majority (26/46 = 56.5%) of the experiments found females to be more susceptible, the proportion falls far short of the percentage one would expect to observe in support of a conclusive statement regarding a clear effect that has been widely replicated under different circumstances. Moreover, Lawson found several disturbing (and probably overlapping) trends in the literature: (1) studies that controlled for fewer confounds were more likely to conclude that females are more susceptible; (2) studies done before 1980 were approximately twice as likely to conclude women are more susceptible than studies done after 1980; and (3) studies involving surveys of one's MS history were approximately three times more likely to conclude women are more susceptible than studies involving direct observation of MS during a controlled laboratory experiment. Clearly, more careful laboratory studies are needed, especially those that control or assess many of the potential confounds to interpretation. Such studies would probably need to employ double-blind laboratory MS tests on men and women who were matched in as many other respects as feasible (experience with motion, age, field dependency, race, aerobic fitness, FOV, body size, absence of migraine, absence of vestibular pathology, absence of phobia, etc.). For now, the author will hold to the view that within-sex variability in response to MS probably is just as wide as between-sex variation, a view that is in accord with the literature on sex differences in reported pain sensitivity (Edwards, Augustson, & Fillingim, 2000; Pennebaker, 1999; Robinson et al., 2001).

Since Lawson et al. (2004) questioned whether the case for increased susceptibility of women has been proven by the literature from 1940 to 2001, at least 10 additional studies on the topic have been published*, and the findings continue to be mixed, with the majority of the new studies (8/10) failing to find significant differences. In agreement with the overall survey-versus-lab study findings from a 1999 study by Park & Hu, two newer studies (B. Cheung & Hofer, 2003; Klosterhalfen et al., 2005) observed higher susceptibility of women in an MS history but not during laboratory testing.

Graeber and Stanney (2002) did not find a difference between men and women in an MS history survey when susceptibility to MS was balanced within sexes and treatments. Bruder, Steinicke, Hinrichs, Frenz, and Lappe (2009) studied how seven men and six women distinguished between real and virtual stimuli. They incidentally noted pre- versus postexposure SSQ scores as follows: rising from 13.4 to 26.6 for men and from 8.1 to 26.2 for women (no inferential statistics provided). This result can be interpreted in different ways, but it cannot be interpreted as demonstrating a strong difference.

Bos et al. (2005) evaluated MS in 14 women and 10 men exposed to 30 min of motion (under various visual conditions) in a ship-motion simulator. They found no significant difference for their overall pooled *Misery Scale* scores during the stimulus or for retrospective estimates of their past history of MS.

Bledsoe, Brill, Zak, and Li (2007) reviewed the medical logs of 1057 passengers on a cruise ship traveling to Antarctica, 47.4% (501) of whom were male. The most common complaint among the passengers was MS, which was reported by 40 men and 56 women. Thus, it can be inferred that the MS rate by sex was 8% for men and 10.1% for women. Bledsoe et al. concluded that there were no strong sex differences for the rate of reporting the various medical complaints in the study. Vibert et al. (2006) found a nonsignificant trend for the history of MS susceptibility, with 52 women yielding a Motion Sickness Susceptibility Questionnaire (Golding, 1998, 2006) score of 48 ± 36 and 47 men yielding a score of 36 ± 32 ($p = 0.08$). Meissner, Enck, Muth, Kellerman, and Klosterhalfen (2009) did not see a significant difference between 16 men and 16 women in their sickness ratings over 5 days of optokinetic drum testing, although physiological responses differed.

In contrast to the preceding eight studies, two recent studies since 2001 have reported a sex difference in MS susceptibility. First, Flanagan (2005; Flanagan, May, & Dobie, 2005) found some sex differences in a situation of visually induced MS involving 6 men and 18 women. Second, Garcia, Baldwin, and Dworsky (2010) observed lower levels of driving simulator sickness among 8 men versus 8 women experiencing various and moving-base simulations.

Potential sex differences in the *sopite syndrome* have not been studied extensively, but at least two surveys have been conducted, and much like the rest of the MS literature on sex differences, they have reported mixed findings. Smith, Kass, and Lawson (2003) observed a difference between men and women in a retrospective survey of *sopite syndrome* involving 213 (mostly military) men and 44 women, with women reporting stronger sopite effects, while Neilson and Brill (2012) did not observe significant sex differences in a recent sopite survey of 249 female and 97 male college students. Interestingly, military women in the Smith et al. study tended to have slightly *lower* average sopite scares than nonmilitary men who were survey respondents in the same study. This does not explain the difference between Smith et al. and Neilson and Brill, but it does serve as a note of caution concerning making overall conclusions from different samples. For now, the author of the present chapter concludes that, like sex differences in MS, there is not enough evidence to conclude that there are sex differences in *sopite syndrome*.

Overall, 28/56 (50%) of the studies the author has reviewed conclude that women are more susceptible to MS than men. This suggests that the case for increased MS susceptibility of women is not yet proven. In addition, a number of important confounds have not been addressed by the research, and the tendency remains for fewer differences to be found in lab studies than in surveys.

It appears that the odds of finding women to be more susceptible than men by reading studies in the MS literature are equal to the odds of winning a coin toss by selecting *heads*. Nevertheless, it should be noted that while a failure to find women more susceptible is common in the literature, one seldom sees reports of the opposite outcome, that is, men being found to be significantly more

---

* Based on a literature search from 2002 to Spring of 2013 when this manuscript was submitted.

susceptible than women, which should be the case, assuming the studies to date are not biased or confounded and their findings can be attributed entirely to chance (i.e., assuming there is no true underlying sex difference). Unfortunately, the studies have not controlled for enough of the confounds to be sure of how to interpret this lack of a converse trend.

What is needed is a comprehensive meta-analysis in the short term and some adequately powered and carefully controlled laboratory experiments in the long term. Future research should not only seek to detect a sex difference but also to determine whether the sex difference is a function of genetically determined sex throughout the entire life span, hormonal fluctuations leading to a transient monthly increase in MS susceptibility during the portion of the life span involving menstruation, enculturated gender role influences, or other factors. It should also be established whether the effect is large enough to be functionally, clinically, and operationally significant (e.g., in terms of human performance). Finally, very little is known about the differences between men and women in terms of specific MS symptoms (rather than overall MS score). For example, a study by Reason and Diaz (1971) reported 10 different symptoms among men and women during a sickening driving simulation but noted that the only marked difference in symptoms between the sexes was in pallor. An interesting side question is whether the pattern of response to stressful motion by women may actually be healthier in some respects (Rohleder et al., 2006). Currently, there are more questions than answers concerning potential sex differences in susceptibility to MS.

23.9   THEORETICAL CONSIDERATIONS

23.9.1   What Should a Complete Theory of Motion Sickness Be Able to Explain?

As stated in Section 23.1, the purpose of this chapter is not to review each of the theories of MS causation and response but to offer practical information concerning the likelihood (incidence) of MS caused by different stimuli, the typical symptoms to be expected, the suitable approaches for assessing those symptoms, and the main factors that influence susceptibility and symptomatology. The various theories of MS have been reviewed by Keshavarz, Hecht, and Lawson (2014), Panos (2005), Jones (1998), Kolasinski (1995), Crampton (1990), Kennedy and Frank (1985), McCauley (1984), and Reason and Brand (1975). The search for a completely satisfying theory of MS has been in effect at least since the days of Hippocrates, but it appears there is still no single theory that is universally accepted. Fortunately, we are nearing a point where we at least understand the many and sometimes difficult requirements such a theory must meet. This section outlines some of the broad characteristics a strong theory of MS should possess and the empirical observations it should be able to explain (Table 23.6). This is not meant to be a comprehensive list of every feature an MS theory should have (e.g., it does not list all the known sickening stimuli that should be explainable by an optimal theory of MS); rather, the list tends to focus on the most troublesome observations for current theories (e.g., sickening stimuli that seem to defy explanation by a given theory). It is hoped that the following list assists the search for a theory of MS by focusing the current theoretical debate

---

TABLE 23.6

Some Key Features of a Complete Theory of Motion Sickness (Section 23.9.1)

Should enable quantitative predictions of MS

Should account for findings on *space adaptation syndrome*

Should account for findings concerning average susceptibility across the life span

Should account for postural equilibrium in relation to MS

Should account for importance of vestibular system to MS

Should account for findings in dispute concerning the *sensory conflict hypothesis*

Should explain the evolutionary origin of the MS response

---

and promoting needed modifications of current hypotheses or research efforts in order to integrate their most useful features into a new theory.

An optimal, fully validated theory of MS should have the following features:

1. *It should enable quantitative predictions.* If the acceleration and visual stimuli are known, it should be possible to make and evaluate quantitative predictions concerning MS and disorientation. These predictions should match the average responses of normal, healthy persons. Some general steps in this direction have been made by various researchers (e.g., Bos & Bles, 2002; Guedry & Benson, 1979; Matsangas, 2004; Newman, 2009; Newman, Lawson, Rupert, & McGrath, 2012; Oman, 1990).

2. *It should account for space adaptation syndrome*, which is most readily elicited by head movements within a microgravity environment. In this circumstance, passive vehicle motion per se is not the proximal cause (except in its widest definition as motion due to vehicle free fall), nor is there an obvious demand on postural equilibrium and a concomitant need to avoid falling (although the postural mechanisms do not necessarily go completely off-line as a result).

3. It should be able to explain average empirically observed age-related changes in MS susceptibility across the life span, including the following:
   a. MS susceptibility is low early in the life span when motor control programs are developing rapidly and effective strategies for maintenance of postural stability have not fully developed. This observation has implications for motor development or postural control theories.
   b. MS susceptibility during *regular* motion (e.g., seasickness) tends to decrease much later in the life span.
   c. Visually induced MS susceptibility also appears to vary across the life span. This has implications for the assertion (Johnson, 2005) that *sensory conflict* and *postural instability* hypotheses would make opposite predictions concerning age-related susceptibility (an alternative view is presented by Keshavarz et al., 2014).

4. It should be able to explain the pattern of findings in the literature regarding MS and postural equilibrium, including the following:
   a. The fact that postural instability is often related to MS (Kennedy, Berbaum, et al., 1997) and VE exposure (Champney et al., 2007).
   b. The fact that postural instability is not always related to MS nor always the most important contributor (Jones, 1998 dissertation; Kolasinski, 1995; Kolasinski & Gilson, 1996; Warwick-Evans & Beaumont, 1995; Warwick-Evans, Symons, Fitch, & Burrows, 1998).
   c. The fact that visually induced MS severity peaks early during repeated simulator exposure and postural instability is greatest later during exposure (Kennedy, Lanham, Drexler, & Lilienthal, 1995).
   d. The fact that MS can still be elicited under the conditions of passive restraint (Faugloire, Bonnet, Riley, Bardy, & Stoffregen, 2007).
   e. The fact that some of the same head and body motions that may precede or co-occur with MS (and therefore would be relevant to the confirmation of Riccio & Stoffregen's, 1991, *ecological hypothesis* concerning the role of postural instability in MS) would also be expected to increase sensory conflict.
   f. Conversely, the fact that many of the same head and body motions that may increase sensory conflict would also be relevant to the confirmation of the postural instability theory. These potential confounds can be dealt with by testing experimental conditions that produce only postural instability without sensory conflict and vice versa.
   g. The fact that being inside a statically tilted room represents a sensory conflict and a potential challenge to the maintenance of the subjective vertical and/or standing posture but does not generally lead to MS.
   h. It has been widely reported that the predominant frequencies of sway overlap with the typical frequencies of MS. A complete theory of MS should account for this finding.

Also any MS experiment that estimates the predominate frequencies of body sway as being around 0.1–0.3 Hz should measure respiration as a covariate, since respiration overlaps this same frequency range and will affect postural sway. While the relation between MS and respiration rate is dubious, respiration can be affected by changes in arousal or anxiety regardless of whether a sickening stimulus is present. Does respiration change always accompany changes in postural sway frequency in MS studies? If so, why?

5. A complete MS theory should be able to account for the observed importance of the vestibular system to the MS response, including the following:

   a. The fact that the majority of evidence indicates that the total chronic bilateral absence of vestibular function renders a person immune to MS.

   b. The fact that some evidence (e.g., Money et al., 1996) suggests that people who lack vestibular function are immune to visually induced MS.

   c. The fact that surgical removal of the vestibular apparatus in animals reduces their vomiting response in reaction to some poisons (Money & Cheung, 1983).

   d. The fact that unresolved vestibular pathology can cause decreased postural equilibrium during baseline conditions and increased susceptibility to MS during the conditions of real or apparent motion challenge (Gottshall, Moore, & Kopke, 2002; Markham & Diamond, 1992). These two observations are important because they may support *sensory conflict, postural instability*, or some third hypothesis in which vestibular pathology acts as a contributor to MS independent of sensory conflict or postural instability.

      i. For the *postural instability hypothesis*, it would be important to rule out any latent or undiagnosed vestibular problems among subjects so that any positive findings can be confidently related to the normal control of postural equilibrium and potential sensory conflicts arising from unreliable vestibular afference can be ruled out (e.g., in the case of Smart, Pagulayan, & Stoffregen, 1998).

      ii. Conversely, for *sensory conflict hypothesis*, it would be important to know the exact vestibular medical status of all subjects in order to determine the extent to which the occurrence of sickness under conditions where it has been argued there is no sensory conflict can be explained by a type of conflict arising from unreliable vestibular afference.

6. A complete MS theory should be able to resolve or explain certain potential questions concerning *sensory conflict hypothesis* and its offshoots (such as subjective vertical mismatch):

   a. Why is there apparent sensory conflict without MS?

      i. A complete theory should incorporate the fact that novel sensory inputs may occur whenever any new movement is engaged in for the first time, but do not always lead to MS. Note that this is not the same point as stating that an expectation violation will occur during any novel movement (Riccio & Stoffregen, 1991), for example, because invariant properties of certain classes of stimuli experienced in the past may generalize to similar but novel stimuli.

   b. Why is there MS without apparent sensory conflict?

      i. A complete theory should explain the fact that MS is reported during the circumstances of pure vertical oscillation at low frequency, even with the eyes open (O'Hanlon & McCauley, 1974).

         A. A possible explanation is offered in Wood (2002) and discussed in Keshavarz et al. (2014).*

         B. Mittelstaedt (e.g., 1996) and Yates, Miller, and Lucot (1998) describe possible conflicts between vestibular and visceral graviceptors.

---

* Also, a relevant discussion concerning hypothesized biomechanical influences on tilt versus (horizontal) translation perception is offered by Gresty, Golding, Gresty, Powar, and Darwood (2011).

c. A complete theory should explain the fact that visually induced MS is reported in the absence of head movements during exposure to optokinetic drums putatively rotating about an earth-vertical axis.

d. A complete theory should account for the fact that MS is reported in the absence of head movements during exposure to sudden-stop rotation and sudden-stop visual-vestibular stimulation, both of which involve rotations putatively limited to the earth-vertical axis.

e. A complete theory should account for the fact that MS is reported during low-frequency vertical axis oscillatory rotation without an outside world view, especially while attempting to read a head-fixed visual display (Guedry et al., 1978).

   i. The theory should account for the fact that the most sickening frequency of MS across a wide range of situations is ~0.2 Hz (Gresty et al., 2011; O'Hanlon & McCauley, 1974).

7. *A complete MS theory should be able to explain why the main symptoms of MS evolved* as a response to certain real or apparent motion stimuli. An evolutionary hypothesis has been offered by Treisman (1977),* which views MS as a by-product from what was originally a poison response. Treisman hypothesizes that nausea and vomiting occur during motion because normal sensorimotor relations have been disrupted in a way that emulates what happens when the organism has ingested poisons. In essence, the body reacts to motion as if it has been poisoned and is seeking to rid itself of toxins. The unpleasant symptoms triggered by motion are, in essence, an indirect and accidental by-product of the way the poison response system works. This hypothesis offers an explanation for the evolutionary origin of MS and for the particular symptoms triggered by motion, albeit Oman (2012) recently has pointed out incongruities in how well the symptoms of poisoning match the symptoms of MS and has questioned some of the other evidence in support of Treisman's poison hypothesis.[†]

   a. It should be noted that regardless of whether one accepts Treisman's poison hypothesis, any alternative theory positing a functional purpose for MS *without resorting to MS as an accidental poison response* (e.g., Bowins, 2010; Guedry, Rupert, & Reschke, 1998; Watt, 1992) must explain how MS would evolve directly. What the author proposes to call the *direct evolutionary hypothesis* for MS (Section 23.9.2) could be considered viable if one could explain how the predominant symptoms of MS (such as nausea) would be subject to preservation by natural selection *during the types of aversive motions that would have been part of human experience for tens of thousands of years* or more (and possibly inherited from our earlier primate ancestors). This modification of Treisman's indirect *poison hypothesis* would need to explain the origins of MS by reference to the evolutionary benefits of an organism finding certain ancient but unusual motions in nature to be noxious, for example, because they threaten survival or reproduction directly in some way (rather than being noxious via the indirect mechanism of a co-opted poison response). The possibility for direct evolution of an aversive response to motion is explored in Section 23.9.2. Such a hypothesis may account for direct evolution of aversive reactions to certain motions, but like other hypothesis, without poisoning as a feature (Bowins, 2010; Guedry et al., 1998), it does not necessarily account for the nausea and vomiting that characterize the MS response. Therefore, a more direct role for the poisoning response in MS is forwarded in Section 23.9.3. (A summary of the questions raised in Sections 23.9.2 and 23.9.3 is shown in Table 23.7.)

---

* This hypothesis is called (1) *Treisman's hypothesis*, (2) *poison theory*, (3) *toxin detector hypothesis* or (4) the *theory of toxins*. Name #1 is not adequately descriptive. In names #2 and 4, "theory" is premature. "Poison" is more recognizable than "toxin," and in general usage, comes closest to incorporating all poisons, venoms, toxins, and intoxicants that may be relevant to Treisman's hypothesis. For these reasons, the author will use *poison hypothesis* throughout this report when referring to the idea proposed by Treisman (1977).

† Moreover, the idea that vomiting mainly functions to rid the body of poisons has been questioned by Johnson, Hill, and Cooper (2007).

TABLE 23.7

Questions Raised concerning Treisman's Poison Hypothesis of Motion Sickness Evolution (Sections 23.9.2 and 23.9.3)

Q1: Is Treisman's *poison hypothesis* necessary to explain the evolution of aversive reactions to real or apparent motion?

• Answer (A): No (Section 23.9.2)

Q2: Are aversive reactions to real or apparent motion an indirect by-product of an ancient poison response system accidentally triggered by modern challenging stimuli?

• A: Not necessarily (Section 23.9.3)

Q3: Are there plausible means by which ancient forms of real or apparent motion could have contributed directly to the evolution of aversive reactions (without the need for a co-opted poison response)?

• A: Yes, possibly (*direct evolutionary hypothesis*, Section 23.9.2)

Q4: Is it plausible that a poison response system may still have played a role in shaping the evolution of the signature symptoms and signs that characterize MS (viz., nausea and vomiting)?

• A: Yes, possibly (*direct poison hypothesis*, Section 23.9.3)

23.9.2 Is a *Poison Hypothesis* Necessary to Explain Aversive Reactions to Motion, or Is a *Direct Evolutionary Hypothesis* Possible without a Role for Toxins?

To determine whether the *poison hypothesis* is critical for explaining the evolution of MS or whether the direct evolutionary hypothesis of MS is viable, it is necessary to start by considering two questions:

• Are there any motion stimuli likely to have been aversive or sickening prior to the invention of motorized means of vehicle travel?
• If so, are any of them likely to be sources of evolutionary selection pressure in a way that would directly lead to an aversive response to motion, that is, without the need to resort to a co-opted *poison* response *hypothesis?*

Examples of noxious motions known in the present day that might reasonably be expected to have been feasible long before the beginning of civilization would include spinning in place, self-generated torso rotation in the vertical axis (e.g., Bouyer & Watt, 1999), prolonged head shaking (extrapolating from Gresty, Golding, Le, & Nightingale, 2008), and repeated nodding of the head while supine (Cloutier & Watt, 2006, 2007). Another partially self-generated activity logically and anecdotally linked to MS is swimming in the sea. While it is possible to envision ways in which many of these bizarre motions would cause sensory conflict or challenge postural control, the explanatory link is weaker when one requires each to have been aspects of human motion that would occur regularly in the course of the daily activities of survival, and which were important enough to have impacted evolutionary fitness to the extent that engaging in them would harm fitness over the course of many tens of thousands of years. Certainly, in the case of prolonged or repeated bouts of spinning, there is little evidence that children and adults avoid this stimulus to the degree that would be expected by an evolutionary theory. One only needs to imagine the myriad ways children or adults vigorously pursue leisure or sport activities involving spinning as part of party games, ice skating, roller skating, gymnastics, discus/hammer throwing, martial arts, dancing (e.g., ballet or break-dancing), skateboarding, snowboarding, water-park activities (e.g., bumper boats, spinning inner tubes on a water slide), amusement rides, and playground activities.

Nevertheless, a clear example where selection pressure could significantly impact a self-generated activity over a sufficient period of time would be in the case of swimming at sea or, more likely, floating with the assistance of a buoyant object. Such an activity would be likely to benefit food procurement and to be engaged in regularly for as long as humans occupied coastal areas of sufficient warmth. Such an activity may have preceded the existence of modern *Homo sapiens sapiens,* but it could conservatively be expected to take place well within the 40,000–50,000 year period that has been conjectured as necessary for human changes to occur, such as the evolution of food preferences conferring minor evolutionary fitness advantages (Kock, 2010). Moreover, engaging in this activity only during calmer seas would make spotting food or predators easier and drowning less likely, so a potentially hoxions aspect can be argued.

Ranging further afield in our reasoning, we can identify several fairly ancient forms of relatively passive travel that are sickening also, including riding aboard camels, elephants, and sea-worthy boats (Benson, 1988). Travel by camel and elephant is limited in extent and has only been possible since circa 2600 BC and 200 BC, respectively (Gascoigne, 2001; Hirst, n.d.), so will be excluded from an evolutionary argument. The oldest and most widespread of these forms of travel is sea travel, and it can be argued that sea travel also has had the greatest impact on the evolutionary fitness of humans able to accomplish it. Sea travel by boat can easily be argued as a logical extension of floating or swimming at sea. While reliable evidence of sea travel goes back only 12,000 years, it is possible that the first sea travel by humans occurred as early as 45,000 years ago (Balter, 2011). If so, then sufficient generations may have passed for natural selection to have taken place concerning individual reactions to sea motion. Overall, swimming, floating, or boating at sea are viable examples of activities that could have contributed directly to the evolution of an aversive response to the unusual low-frequency motion associated with rough seas.

Another situation that has largely been ignored in the MS literature but might be expected to produce MS would be sitting on a tree branch that is swaying in the wind (Money, in Crampton, 1990). Data are available for conifers, which have a natural frequency of sway as low as 0.2–0.5 Hz (see Figure 1 from Moore & Maguire, 2005), although it is not known whether the typical amplitude or peak acceleration of conifers would be sufficient or what the natural frequencies of other tree species are. What can be said with confidence is that there has been sufficient time and inpetus for the larger primates (including human ancestors) to have evolved any traits favorable to evolutionary fitness in an arboreal setting. Moreover, the case for a direct link between postural equilibrium and control versus evolutionary fitness can be made readily for the case of a swaying tree, since failure of balance and postural coordination in a tree leads, at minimum, to significant energy expenditure for recovery and, at maximum, to death by falling.* It is known that vestibular function is necessary for the initial fast muscle response to falling in baboons (Lacour, Xerri, & Hugon, 1978), which implies that vestibular cues are important to the detection of falling and control of landing posture.

A final motion challenge that would be present throughout the history of humans and other terrestrial animals is the motion of the earth and visual swaying of associated objects (such as trees) due to earthquakes. The author has experienced unsteadiness during mild earthquakes, while disturbing sensorimotor aftereffects of earthquakes have been noted by Nomura et al. (2012). Interestingly, during their poster presentation, Nomura and colleagues said that only 2% of the cases involved vomiting and very few involved nausea, despite the fact that 97% of the people had sensations of dizziness, primarily characterized by illusory sensations of body sway. It is not known whether there is any logical relation between earthquake-related feelings of unsteadiness and the potential role of evolution in the various theories of MS. It is likely that earthquakes are too infrequent, short in duration, and geographically localized to explain the fact that most animals can be made motion sick regardless of their natural geographic distribution. Even a single natural catastrophe can be a significant driver of evolution if it is severe and widespread, but overall, this particular explanation does not seem as convincing as the preceding sea-based or tree-based examples.

Consideration of the preceding list of alternative explanations supporting the *direct evolutionary hypothesis* of MS evolution (without the need to resort to MS as a co-opted poison response) leads the author to make some inferences about Treisman's indirect *poison hypothesis*. First, the fact that several ancient motions that were likely to be harmful or noxious can be identified suggests that the

---

* Balance can be affected not only by motion of the tree but also by the sight of trees moving all around the organism. A moving tree also presents a whole-field visual motion stimulus. Whole-field visual motion can elicit illusions of self-motion in humans and cause postural reactions (Dichgans & Brandt, 1978). It is known that the vestibulospinal reflex responses of baboons during falling are modulated by the visual cues available (Lacour, Vidal, & Xerri, 1981).

hypothesis of an indirectly co-opted poison response is not the only viable evolutionary hypothesis for the origin of the MS response.* Second, among the alternative explanations, natural selection to avoid sitting in severely swaying trees or floating on rough seas appear to be the two possibilities that logically merit consideration.† Third, while direct experimental evidence is lacking, initial and indirect exploration of the alternative hypotheses (viz., the *direct evolutionary hypothesis* of MS) can be carried out comparatively by observing whether the many terrestrial species that are made sick by motions of the correct frequency range (for their species‡) also avoid swaying trees or floating in rough seas. The *direct hypothesis* of MS evolution would not be fully supported if it was discovered that many terrestrial species that suffer MS do not avoid floating in rough seas or sitting in trees swaying significantly. To a lesser extent, the hypothesis would be challenged if many terrestrial species do not experience MS but still avoid swaying trees or floating in the sea. Animal studies are recommended to gather preliminary evidence concerning whether the alternative *direct evolutionary hypothesis* warrants full experimental evaluation versus the *poison hypothesis.*

The answers to the preceding questions are not yet known and may prove difficult to obtain. Nevertheless, there is some observational evidence of possible relevance to the question of whether primates avoid trees that are swaying. First, the author has already mentioned that some large trees sway at the right frequencies for elicitation of MS in humans. Second, anecdotal observations have been reported by deer hunters, confirming that they can be made sick by sitting in a deer tree stand when the tree is swaying.§ It is likely, therefore, that experimental observations will confirm that humans can be made motion sick by sitting in large swaying trees.

It seems reasonable to ask whether tree-induced MS could be one of the oldest forms of MS among humans and possibly their ancestors. It is known that many mammals are susceptible to MS during laboratory testing, including nonhuman primates such as monkeys and chimpanzees (Daunton, in Crampton, 1990); however, no studies could be found documenting whether nonhuman primates get MS in swaying trees. There is evidence that falling from trees is a significant source of selection pressure among nonhuman primates and that nonhuman primates modify their behavior when a tree starts swaying, so it is logical to suppose that primates find swaying trees aversive. These issues are discussed below.

Evidence suggests that falling from trees would be subject to natural selection during primate evolution. It has been established in field studies that primates from prosimians to apes and humans fall from trees and that young primates fall from trees more often than mature primates, sometimes with fatal consequences (Young, Fernandez, & Fleagle, 2010). According to Young et al., this may be the reason that modern infant monkeys who are primarily arboreal have strong bones relative to their body size (vs. nonarboreal primates), making it more likely for them to avoid fatal injury.

It is not known whether primates find swaying trees aversive, although adult primates do tend to avoid unstable perches in zoos.¶ Primates also tend to avoid sitting on thin swaying branches in the wild and to modify their behavior when wind and storms make trees sway (e.g., by moving to sheltered places on the ground); however, they are also known to shake tree branches on which they are perched, perhaps as part of play or communication. (Author's inferences based on personal communications with several animal behavior specialists** from April 10–15, 2013.)

In summary, it is possible that *tree sickness* is one of the earliest forms of MS among humans and that it evolved as an early aversive response to the energy and survival costs of sitting in a tree that is swaying excessively. Insufficient evidence exists to determine whether this *direct evolutionary*

---

* This does not mean the alternative MS hypotheses are necessarily superior to the *poison hypotheses,* only that the *poison hypothesis* is not critical for explaining aversion to unusual motion.

† It should also be noted that Treisman states that animals who spend much of their time in trees or water should not have evolved MS response unless there was a positive selection pressure for its presence.

‡ MS occurs readily at frequencies around 0.2 Hz for humans and at higher frequencies of oscillation for smaller mammals, e.g., the Asian house shrew gets MS at 1–2 Hz of oscillation (Javid & Naylor, 1999).

§ http://randygodwin.hubpages.com; www.deeranddeerhunting.com; www.homesteadingtoday.com.

¶ www.awionline.org/lab_animals/biblio/enri3.html

** The experts are listed by name in the Acknowledgments. They were identified individually online via a Google search or *en masse* via the Primate-Science discussion forum, www.primate.wisc.edu/mailman/listinfo/primate-science.

hypothesis is a viable alternative to Treisman's indirect *poison hypothesis*. All that can be confirmed presently is that large trees can sway in frequencies that are sickening to humans, that humans have anecdotally reported being made sick by sitting in swaying trees, that trees are an attractive source of refuge and food for many otherwise terrestrial primates (possibly even including early hominids, according to Choi, 2012), and that humans and nonhuman primates have been known to fall from trees. Finally, it can be concluded that it is not critical to resort to the *poison hypothesis* to explain how failing to avoid nonadaptive motion (such as excessively swaying tree limbs) could harm the evolutionary fitness of those organisms (by death or injury from falling) who lack an early aversion reaction of some kind.

A critical limitation of the *direct evolutionary hypothesis* is that it only explains why motion would be noxious and cause aversion, not why it would be *sickening*. The *direct evolutionary hypothesis* concerning MS origins still needs further development to explain why the specific symptoms associated with MS evolved (e.g., nausea and vomiting), rather than some other aversive reaction, such as perceived pain. In fact, it has been argued many times in the MS literature that vomiting is an evolutionarily costly reaction to aversive motion (in terms of a loss of calories) and so would emerge only under strong positive selection pressure.* A hypothesis that attempts to account for specific MS reactions such as vomiting is explored in the following.

23.9.3 NAUSEA AND VOMITING AS DIRECT RATHER THAN INDIRECT EVOLUTIONARY RESPONSES TO MOTION: A *DIRECT POISON HYPOTHESIS*

While Treisman's *poison hypothesis* may not be critical to explain the evolutionary origins of an aversive response to excessive motion, *a co-opted poison response may still have played an important role in shaping the specific symptoms that characterize MS*. Moreover, it is possible that even if the emergence of symptoms associated with poisoning during exposure to motion is not an accidental and indirect by-product (Treisman, 1977), poison response mechanisms may still play a role in the evolution of MS. It could be that *generalization of the poison response to the MS response is a direct consequence of important and related selection pressures (for each), which were encountered by human ancestors during their evolution* (rather as an indirect by-product). In contrast to Treisman's indirect *poison hypothesis* as originally stated, this modification of Treisman's view could be called a *direct poison hypothesis* concerning the evolution of MS. The *direct poison hypothesis* is similar to the possibility raised (but deemed unlikely) by Money (in Crampton, 1990); namely, that the stimulus for MS could have played a direct role in the evolution of MS *if poisonous foods became available during exposure to motion* (such as wave motion or tree motion). In the following, we present preliminary evidence for the *direct poison hypothesis* and summarize some of the additional evidence that would need to be obtained to determine the viability of the hypothesis.

The *direct poison hypothesis* argues that *the human MS response evolved because the circumstances under which our primate ancestors were poisoned or intoxicated also harmed their evolutionary fitness directly during the natural conditions of real-body motion and/or visual-field motion*. For example, if it can be shown that species with an arboreal heritage are likely to encounter poisons, venoms, toxins, and/or intoxicants during their natural foraging for tree-grown foods and that such unwanted substances affect their balance, then this relation would suggest a direct contributor to the evolution of the characteristic aversive poisoning responses (e.g., nausea, vomiting) triggered by certain real or apparent motions.

---

* While positive selection pressure would be necessary, it would be an overstatement to argue that very strong selection pressure would be necessary. Many animals vomit fairly frequently (ask any dog owner), and there is only a partial loss of calories when this happens. That is because some absorption of calories has already occurred and because many animals eat their own (and other animals') vomit. Examples of vomit eating from the primate world include the proboscis monkey and bonnet macaque. One of our closest relatives, the gorilla, does this also, but since this behavior has been observed in captivity in the case of the gorilla, it is not known if it is an aberration (Gould & Bres, 1986) that would not occur in the wild. Answering this question should interest MS researchers and animal rights activists alike.

Indeed, preliminary evidence indicates that primates do ingest poisons or intoxicants, that these substances have been known to affect balance and coordination, and that ingestion of such substances sometimes contributes to falling from trees (Glander, 1977, cited in Kricher, 1999). Kricher describes the case of a disoriented-looking howler monkey and her clutching juvenile observed to survive a 35 ft fall from a tree in an area where several dead howler monkeys were found on the ground. The cause of these events was traced to toxic defense compounds produced by the leaves of the tree. Such defense toxins are thought by Kricher to be a major selection pressure in the selective feeding preferences of howler monkeys and their ability to remember which trees are safe. Kricher points out that one of the common ways in which plants discourage predation is by the production of L-dopa, which can be hallucinogenic. Therefore, it can be argued that plant defense compounds do not always exert selection pressure by poisoning plant predators to death but rather, by making them more likely to fall or fail to escape predators. A final way poisonous plants can exert selection pressure on plant predators without killing them is via the development of learned (or conditioned) taste aversion, which is discussed further in Section 23.9.4.

There is a distinct possibility that the earliest emergence of the generalization of a poison response to a MS response began with our primate ancestors. It is not known whether poisoned or intoxicated primates exhibit all the clinical signs that are logically akin to the MS response (Oman, 2012). The MS-like signs or behaviors interested primatologists should look for in the wild in the cases of poisoning or intoxication include nystagmus, clutching tightly to a limb, unsteadiness (all evidence of dizziness or vertigo), drooling, retching/vomiting, holding the head still, anorexia, and evidence of drowsiness such as drooping eyelids or yawning.

Why would the poison response get generalized into an aversion response to unusual motion? Treisman explains the change as being due to the fact that intersensory integration and sensorimotor integration are disrupted by poisons in the same way as they are by motion stimuli that disrupt normal sensory integration. Thus, certain unusual motions of the body that occur during modern travel accidentally simulate the internal conditions that prompt the body to react as if it has been poisoned. A more *direct poison hypothesis* is conceivable, namely, that *the evolutionary circumstances that caused the ingestion of toxins were the same circumstances in which unusual, maladaptive motions of ancient origin frequently were present.* For example, it is logical to suppose that the circumstances that tend to cause fruit to be produced rapidly, to drop from trees, and to ferment on the ground or sometimes in the trees (viz., the fruiting season and warmer temperatures) are often associated with the times of rapid tree growth (causing more thin branches to appear) and greater likelihood of wind and storms (causing the branches to move more). These conditions (swaying trees and fruit on the ground) could cause primates to move to the ground to forage for fallen fruits, some of which are fermented. Fruit-induced intoxication would make death or injury more likely if a somewhat intoxicated primate subsequently returned to the trees (e.g., to rest, sleep, evade a competitor, or respond to a warning call concerning a threat), especially if the primate is compelled to move to higher, smaller, or more mobile branches of the tree for the purposes of evasion or to procure more fruit from the tree.

Is there any evidence supporting this idea? The author has already mentioned evidence that primates sometimes ingest plant-defensive poisons while foraging for food and subsequently fall from trees, sometimes to their deaths. There also is evidence that some primates ingest fermented fruits (Brenner, 2013; Stephens & Dudley, 2004) containing alcohol, which is known to be intoxicating and to disrupt balance and coordination. In the following, the author will briefly discuss the pertinent implications of alcohol intoxication.*

Stephens and Dudley explain that the smell of ethanol is an important cue for locating the ripest fruit, which must be identified and consumed rapidly due to competition. They also present evidence concerning their own observations of "alcohol-driven fruit binges" (p. 42), which they

---

* Not discussed are the potential evolutionary implications of the fact that primates also encounter venomous animals (such as snakes and spiders) whose bites or stings are known to affect balance. Interestingly, fear of snakes, spiders, and heights are among the most common phobias of humans.

associate with reckless behavior (e.g., inadequate precaution at heights to reach more fruit) but do not explicitly link to observations of disequilibrium or falling from trees. Rather, the purpose of their *drunken monkey hypothesis* is to explain human attraction to ethanol as arising during primate evolution because of the importance of ethanol as a source of energy and an indicator of ripe fruit. Dr. Dudley (personal communication, April 30, 2013) confirmed that the time of year when the palm fruits ripen was probably the second-windiest time of the year but was not able to confirm or disconfirm whether intoxicated primates fall from trees due to disequilibrium. He ruled out the possibility that swaying horizontal limbs were a potential hazard in the cases described by Stephens and Dudley (2004) because the fruiting palm trees they observed do not have such woody branches. However, the present author has confirmed that the fruits of the Astrocaryum tree hang from the end of a thick vine or stalk in a massed bunch that would be expected to behave like a pendulum if the tree or the fruit bunch were moved by monkeys, wind, or other perturbations.

There is some disagreement concerning how early in evolution the attraction to fermenting fruit may have occurred. Pen-tailed tree shrews consume fermented nectar from the flower buds of the Bertram palm plant (Zielinski, 2011). Stephens and Dudley describe the consumption of fruit with significant ethanol by howler monkeys, while colobus and proboscis monkeys digest leaves via stomach and foregut fermentation (MapOfLife.org, 2010).* By contrast, some other primate researchers have not observed several other primate species preferring to eat fermented fruit (Graber, 2008). A recent paleogenetic presentation by Benner (2013) reconstructs when (in evolutionary history) the key enzyme needed to metabolize ethanol would have become efficient at doing so. Benner deduces that this would have occurred with the common ancestor of the chimpanzee, gorilla, and human. He argues that the ability to metabolize ethanol in the fallen fruit would have been of advantage to these large, mainly terrestrial apes. Dr. Brenner (personal communication April 30, 2013) does not agree with the idea that monkeys in the wild also eat fermented fruit and become intoxicated. This debate is mentioned here so that the reader knows there are recent questions about how widely and recently the *drunken monkey hypothesis* can be applied. From the standpoint of the issue at hand in this chapter (viz., consideration of the viability of a *direct evolutionary hypothesis* for MS), the resolution of this question may not be critical, so long as fermentation-induced (or even poison-induced) intoxication has been demonstrated. It could be argued that the later emergence of such intoxication during primate evolution may strengthen the argument for its influence on human evolution, however.

23.9.3.1 Summary

The author's thoughts concerning poisoning by plant-defensive compounds (Kricher, 1999) and alcohol intoxication in the wild are summarized as follows:

1. There is preliminary evidence that accidental poisoning has occurred in primates and has contributed to their deaths by falling, in a manner consistent with the *direct poison hypothesis* of MS.
2. There is evidence that intoxication has occurred in primates and may constitute a second possible means by which the *direct poison hypothesis* could be supported:
   a. It is widely known that some fruit can ferment on the tree and much of it can ferment on the ground and that ripe fruit falls to the ground more readily when the fruiting tree is moved by wind or animals.
   b. There is some evidence that the ancestors of humans were able to metabolize fermented fruit and probably ate fermented fruit, which is known to be intoxicating and to cause disequilibrium and ataxia.
   c. Many primate and other mammal species have long been known to become intoxicated after consuming human alcoholic beverages, which some do quite readily.

---

* *Gut fermentation syndrome* occurs rarely in humans also.

It should be stressed that the potential link between fermented fruit consumption in the wild by primates and their likelihood of falling from trees (aided by tree motion) cannot be established without further observation. Even if it is established, it would constitute comparative evidence that would need to be corroborated (if feasible) by genetic and paleontological evidence. It is known that a number of bird species become intoxicated in the wild and fall from trees, but presently, it is not possible to confirm or disconfirm the hypothesis that the evolutionary circumstances that caused the ingestion of poisons or intoxicants were directly linked to the circumstances in which unusual, maladaptive motion frequently was present. Further research is recommended to determine whether MS is an accidentally co-opted poison response and, if so, whether it was first triggered indirectly by humankind's exposure to unusual vehicle motion or, rather, has more ancient origins. Over the coming decades, it is hoped that such investigations will determine which hypothesis concerning the evolution of MS has optimal parsimony balanced with explanatory power: Treisman's *poison hypothesis*, the *direct poison hypothesis*, the *direct evolutionary hypothes* (not requiring poisons), or some other idea. (Additional hypotheses concerning the long-term evolution and immediate causes of MS are discussed in Keshavarz, Hecht, & Lawson, 2014.) While the immediate causes of MS are not simple to study, the hypotheses concerning the evolution of MS require an even more daunting effort to evaluate fully, via sources of evidence derived from fields such as genetics, developmental biology, paleontology, comparative animal studies, and even artificial selection.

23.9.4 HYPOTHESES CONCERNING SOURCES OF INDIVIDUAL VARIABILITY IN MS SUSCEPTIBILITY

It has been said throughout the literature that one limitation of the *sensory conflict hypothesis* of MS (Oman, 1990; Reason & Brand, 1975) is revealed by the observation that susceptibility varies widely among people even in situations involving approximately the same amount of sensory conflict. Keshavarz, Hecht, and Lawson (2014) argue that individual variability is not a critical problem for *sensory conflict* or most other MS hypotheses, since these hypotheses only need to explain why the stimulus is disturbing, not why the population varies in its response to the stimulus. Population variability* is the fundamental material for natural selection. By analogy, clear vision could be argued to be a critical factor for evolutionary fitness among predominantly visual animals, yet the existence of variability among the members of a particular species in their ability to perceive the same visual target or threat cannot be accepted as disproof of the importance of vision, nor is natural selection expected to lead inexorably to a circumstance wherein all members of a species eventually have perfect and identical visual acuity.

Not only is individual variability to be expected in general, but *there may be an evolutionary reason why MS varies widely among people* without such variability necessarily disproving *sensory conflict*. Specifically, if one considers MS to be derived from what was originally a response to ingesting poisons (Treisman, 1977), variability could arise because *vomiting is not the only mechanism by which humans avoid poisoning*. Treisman alludes to other mechanisms of response to poisons, such as taste aversion, which he argues is a possible explanation for the presence of nausea (i.e., nausea during MS contributes to aversion to the harmful substance in the future). Poststimulus taste aversion can develop when a particular taste is followed by nauseating motion (Stockhorst, Enck, & Klosterhalfen, 2007); however, nausea is just one symptom of the MS response (and the poison response); nausea does not always precede conditioned taste aversion for other types of stimuli such as sickening drugs (Fox, in Crampton, 1990).

Taste aversion has evolved to be one of the fastest and most robust learning mechanisms in nature, and the learned association can readily extend to the environmental context beyond the simple pairing of taste with later nausea (Klosterhalfen et al., 2000; Stockhorst et al., 2007). Human studies suggest that the development of anticipatory nausea and vomiting (e.g., associated with chemotherapy) is facilitated by the presence of anxiety as well (Burish & Carey, 1986). It is possible

---

* For example, due to random mutations in the genome leading to differing individual traits, some of which are not neutral in their effect upon reproductive or survival fitness.

that similar mechanisms of learning facilitation evolved in nonhuman primates, which would further strengthen an already strong form of learning linked closely to survival. For example, it would be beneficial for evolutionary fitness if the association between a near fall from a tree and the disequilibrium related to eating a poison plant was readily burned into the animal's memory by the twin brands of nausea associated with eating the plant and anxiety associated with the near fall. In this case, the learning would occur rapidly during the animal's lifetime, but the proclivity to learn so readily would be a product of natural selection.

Benson, Hooker, Koch, and Weinberg (2012) report that people who are more sensitive to bitter tastes report less severe MS symptoms during optokinetic stimulation. They hypothesize that sensitivity to bitter taste and susceptibility to nausea and vomiting are both protective mechanisms against toxin ingestion, which evolved together. They predict that increased nausea susceptibility would have evolved in environments where not being as sensitive to bitter tastes would confer an evolutionary advantage by enabling the intake of bitter fruits and vegetables containing useful nutrients or providing other advantages. For example, Elert (2012) conjectures that people from malaria-infested parts of the world tend to carry a gene that makes them less sensitive to bitter plants containing low levels of cyanide, presumably because a small amount of cyanide is detrimental to the malarial parasite but not to the host organism.

If these two very different types of responses (vomiting and taste aversion) to the presence of poisons in food exist, then it seems logical to hypothesize that an equally challenging real or apparent motion stimulus could be presented to two persons and only one of them would readily exhibit vomiting as a result, partly because he or she has greater reliance on mechanisms designed to eject poisons rather than taste-related mechanisms of protection. It would be interesting to compare reported racial differences in MS susceptibility (Klosterhalfen, Pan, Kellermann, & Enck, 2006; Stern et al., 1996) to racial differences in taste sensitivity.

There are many other potential contributors to the wide individual differences in response to the same motion stimulus (besides varying reliance on taste aversion). For example, latent vestibular pathologies can cause the same stimulus to be more *conflicting* for one person than another (if one has a pathology and the other does not). Such problems can be caused by hereditary lateral asymmetries, vestibular disease, concussion, or aging.

Something akin to classical conditioning is likely to play a profound role in nausea. Golding and Stott (1995) have observed that some participants appear to sensitize instead of desensitize during repeated exposure to high sickness levels; they have likened this phenomenon to the anticipatory nausea that develops during the courses of chemotherapy (Burish & Carey, 1986; Morrow & Dobkin, 1988). Subsequent sensitization to motion appears to be related to the magnitude of the initial response in multiple studies, with sicker subjects more likely to develop anticipatory nausea later (Morrow & Dobkin, 1988).

Why is it that some people become sensitized to motion while others adapt? The author hypothesizes that *conditioned aversion could be one of the important contributors to the wide individual variability seen in MS susceptibility and the emergence of sensitization to MS among some people.* If two people start out with moderately different levels of susceptibility to a motion challenge (or slightly different ability to adapt), it is reasonable to suppose that their responses to challenging stimuli could diverge further from one another as they are each exposed to motion, because the more-susceptible person would be far more likely to acquire conditioned aversion, pairing certain motions (and associated vibration, noise, smells, etc.) with nausea and/or vomiting. Conversely, the moderately less-susceptible person would encounter a greater number of motion stimuli within his or her adaptive capacity and not pair them with nausea. Importantly, the more-susceptible person would also tend to avoid a greater number of subsequent opportunities to experience motion than the less-susceptible person, leading to less maintenance of protective motion adaptation. This *downward spiral* could occur despite the fact that some of the motion opportunities would have been within the adaptive capacity of the more-susceptible person and would have helped his or her resistance to motion. The essential notion behind this proposed contributor to MS variability and sensitization can be stated as the following *MS amplification* hypothesis*: *moderate individual differences in MS susceptibility should become larger over time if a more-susceptible person acquires conditioned aversion to motion that alters his or her subsequent response (increased symptoms) and behavior (avoidance of motion)*. In other words, the traits that render a person more susceptible initially[†] may also trigger conditioning that increases his or her response to subsequent motion and causes avoidance behavior that prevents the building of future resistance. Conversely, the person who starts out less susceptible learns that motion is benign and builds further resistance. The reader should note that the person who adapts to MS and the person who becomes more sensitive are each responding according to the learned importance of the real or apparent motion challenges they have encountered.

The MS amplification effect would not necessarily cause a bimodal curve of overall population susceptibility to MS, because MS amplification of individual differences may take years to affect the many young people born into the population regularly, while the amplification of differences may diminish as elderly people travel less and experience more vestibular maladies. Nevertheless, the amplification effect should increase the differences over time between two young-to-middle-aged adults of sufficiently differing initial susceptibility (but the same age and same initial level motion experience). Also, if MS amplification occurs, it would tend to ensure there is a small group of highly susceptible people in the population which appears to be the case.

Even if MS amplification by conditioned aversion and motion avoidance does not prove sufficient to explain much of the variance in interindividual variability, it is likely that the *MS amplification hypothesis* should be applicable to changes in susceptibility *within an individual over time*. Although the majority of people will adapt over time during repeated exposure to motion (thus groups statistics will show adaptation on average), some will become more sensitive. It is hypothesized that *MS amplification can occur in some moderately susceptible people because they become more sensitive to MS due to conditioned aversion and subsequent avoidance of motion adaptation opportunities* (some of which would have been within their adaptive capacity).

The reader should note that the *MS amplification hypothesis* concerning how animals avoid poisoning by taste sensitivity or nausea susceptibility is not dependent upon the two-mechanism poison protection hypothesis of Benson et al. (2012). Nevertheless, whether an individual has a greater tendency for one reaction (vomiting) versus another (taste aversion) could be a factor contributing to his or her initial MS susceptibility and subsequent increase in susceptibility via MS amplification (or failure to increase in susceptibility due to adaptation).

There are many questions concerning the *MS amplification hypothesis*, especially in regard to interindividual differences. For example, how large must the initial difference in susceptibility be for a divergence in susceptibility to occur? How susceptible must a person be before conditioned aversion and motion avoidance occurs? Why are some people able to adapt to subsequent motion even when they have become quite sick (e.g., vomiting at sea) while others become sensitized? How dependent is MS amplification on the initial stimulus conditions? This last question could prove especially important, for several reasons. First, it is less likely that MS amplification would occur if the initial stimuli experienced by a susceptible person were very mild. Second, it is known that anticipatory nausea and vomiting (due to conditioned aversion) are lessened when the person first experiences the challenging situation in a benign way. For example, it has been recommended that chemotherapy patients benefit from becoming familiar with the clinic well before the treatment commences (Hall, Symonds, & Rodriguez, 2009). Similarly, it is likely that a susceptible person would benefit from having grown up around the sights and smells of the sea and of boats (and building up positive associations with each) long before taking his or her first voyage at sea. For the reasons mentioned in this paragraph, the *MS amplification hypothesis* at present must be considered tentative.

---

* The word *amplification* is used instead of *sensitization* because the hypothesis encompasses a concept that is not identical to the relevant MS literature on sensitization, desensitization, conditioned aversion, or behavioral therapy for conditioned aversion (e.g., desensitization therapy).

† Such as inherited motion susceptibility, latent vestibular asymmetry, latent vestibular pathology, etc.

23.10　CONCLUSIONS AND RECOMMENDATIONS

Humans tend to experience MS during and after real or apparent (usually visual) motion stimuli. MS is typically measured via symptom checklists, the most popular of which are described in Lawson (2014). The most common side effects of exposure to real-motion stimuli, moving visual fields, or flight simulators include the following: *dizziness*, epigastric symptoms (*stomach aware-ness, stomach discomfort*, or *nausea*), *flushing/subjective warmth, eyestrain, fatigue, drowsiness, cold sweating, headache*, or *difficulty concentrating*. Other classically defined cardinal signs or symptoms include *facial pallor, increased salivation, retching*, and *vomiting*. The side effects of VE exposure are not as well studied, but they seem to include many of the symptoms produced by other situations involving real or apparent motion. However, it is likely that different real or apparent motion challenges produce somewhat different clusters of predominant symptoms.

The most conservative prediction that can be made from the limited VE research is that at least 5% of all users will not be able to tolerate the prolonged use of current VEs; this is about the same as the proportion of people highly susceptible to motion stimuli, moving visual fields, and simulators. VE stimuli also seem to share in common with many other SEs the tendency to elicit symptoms that persist after exposure. Side effects such as *stomach symptoms, dizziness, eyestrain*, and *headache* among susceptible VE users will present an obstacle to widespread acceptance. This may be one of the reasons that the early excitement about VEs has not survived the *hype cycle* and the extent of dissemination has fallen short of early expectations. In addition to the stated side effects, less-obvious side effects may pose a significant risk to VE users who do not believe themselves to be susceptible; these include *sopite syndrome*, postural disequilibrium, and loss of visual acuity during head or body motion.

Since the observable signs and subjective symptoms of real or apparent motion discomfort have valid and reliable measures of a person's state (Lawson, 2014), they will continue to be important criteria for interpreting physiological and performance data regarding VE side effects. However, there are gaps in knowledge about MS symptomatology everywhere one turns. For example, the majority of techniques used to measure real or apparent motion discomfort have been developed and applied in the situations of acute and highly provocative exposure. Much less is known about response to mild, chronic, or repeated exposure situations. Also, careful evaluations of the tempo-ral progression of VE side effects should be made in order to identify the early warning signs of an adverse reaction to a VE and to understand the time course of VE adaptation and postexposure recovery.

Some general advice is offered in parting, in the hope it will facilitate future discoveries in MS research. The MS literature has for too long been plagued by difficulties with the interpretability of reports, often due to large individual variability. There are several approaches to MS research that should improve interpretability but have not been implemented sufficiently in the past research (Table 23.8). These ideas are presented in the following, starting with the simplest and progressing to the more complex:

1. Researchers should err on the side of caution and add a healthy margin above their sample size estimate before proceeding. MS is not an area of study where it pays to be optimistic concerning the sample.
2. Greater standardization is needed in symptom measurement. While it would help MS research immensely if all researchers used the same MS scale, this is not likely to hap-pen, partly because scientific questions of interest vary and partly because the personal incentives are too great for researchers to devise new scales or to use only their own scale. Nevertheless, it should be feasible for almost every researcher to agree (regardless of their favored choice of MS scale) to administer the most widespread scale (viz., the full list of symptoms in the SSQ/MSQ of Kennedy, Lane, Berbaum & Lilienthal, 1993) as soon as possible after the cessation of a motion challenge, which would allow greatly improved

---

TABLE 23.8

Improvements Recommended for Future Motion Sickness Studies (Section 23.10)

Larger sample sizes

Greater standardization in symptom measurement across studies

Greater reliance on controlled experiments employing established methods of eliciting
    MS and ensuring sufficient MS is elicited

Permitting more time to elapse between trials of exposure in repeated-measures studies

More careful control over between-subjects variability

- Via focused experiments on specific populations
- Via analyzing the sources of individual variability as covariates
- Via idiographic research methods

---

cross-study comparison. Doing so would partly ameliorate the cross-study comparison problem in MS research, which is analogous to the situation in human civilization before the invention of standard weights and measures. (Further recommendations regarding MS scales and standardization are elaborated in Lawson, 2014.)

a. A related issue that is important to MS multisymptom measurement within and across studies relates to the participants' understanding of the symptoms. It is critical for participants to understand the established symptoms of MS so they do not become confused concerning thermal sweating versus *cold sweating*, prenausea epigastric symptoms versus *nausea*, the criteria for different severity ratings of a given symptom, etc. (Symptom definitions are provided in Lawson, 2014).

3. It is important to rely upon the use of controlled, reliable, and well-established methods of evoking MS (e.g., Miller & Graybiel, 1970b), rather than using unique, *home-grown* stimuli whose reliability and level of sickness provocation are unknown. Also, unless very mild MS or the initial onset of MS is the main topic of research, it is imperative to ensure that an established MS challenge is used that causes the participants to become sufficiently motion sick. Far too many studies over the decades have detected statistical differences in MS between experimental conditions *where hardly any sickness was experienced* in either condition. Obviously, less confidence can be placed in such studies than those in which the participants were required to experience unequivocal MS. Many studies use a nausea scale of *none–minimal–moderate–major* and require *moderate nausea* to be reported for the experimenter to halt the experiment (when the participant is still willing to continue). Regrettably, in many published studies, only a small proportion of participants report even *minimal nausea*. It is difficult to draw confident inferences about the functional or clinical significance of the findings of such studies.

4. Repeated-measures designs can be very useful for dealing with the problem of between-subjects variability in MS susceptibility. Such studies should be considered in cases where the number of repeated trials or return sessions is relatively few, because it is not realistic to expect subjects to volunteer initially or return repeatedly when they can expect to experience unequivocal MS. In fact, it is probable that the more times subjects must experience the stimulus, the more likely it is that the subjects who complete all experimental conditions are the least susceptible subjects, which would introduce a systematic bias in the experiment.

The anticipated number of experimental conditions is one of the most important considerations when the investigator is deciding whether to employ repeated measures of MS or an independent design. Lawson, McGee, Castaneda, Golding, Kass, and McGrath (2009) discuss the strengths and limitations of repeated measures of MS and estimate that attempting four repeated evaluations of MS is likely to cause approximately half of the participants to drop out of the study. Based on their guidelines, it is recommended that repeated measures generally should be preferred for two-condition MS studies and independent designs for studies of four or more conditions, with three-condition studies representing a *gray area* where either design can be considered, depending upon the characteristics of the participants and the recruitment plan (how many people are available to be solicited, how willing past subjects have been, how many weeks the participants are likely to be available, how long each session will last, the investigators' allowed recruitment incentives, etc.).

    a. Adaptation to motion is an important challenge for repeated-measures studies of MS, at least when adaptation is not the topic of study. The carryover effect of motion adaptation from one session to the next should be distributed equally by presenting the conditions in balanced order, but the investigators should be aware that a confounding error is still present in the data and can wash out the overall findings. To minimize this problem, it is recommended that investigators allow at least 1 week to elapse between each motion exposure. This procedure has the added advantage of enhancing control over time-of-day (circadian) and day-of-week sources of variability. That is because a subject can be scheduled for the same time of day and the same day of the week on two successive weeks. Far too many repeated-measures studies in the literature have not allowed a sufficient number of days to elapse between motion exposures or attempted to control for variability related to the hour of the day or day of the week.

5. In simple group MS experiments not intended to study individual traits, interpretability of findings should be weighted more heavily in the investigators' considerations than generalizability of their MS research, at least until some of the fundamental MS questions can be settled unequivocally. For example, if a group of investigators is interested in the aggregate nauseogenic effects of a specific stimulus, they should seek to restrict the individual characteristics of their sample (to a certain age range, a certain range of reported MS susceptibility on a questionnaire, a certain level of experience with motion, etc.). This will require additional experiments to generalize new discoveries, but it will increase the odds of obtaining a clear result with each experiment.

6. Conversely, in more complicated experimental designs, investigators can benefit from monitoring the key sources of individual variability (e.g., reported MS susceptibility) and including them as covariat during analysis.

7. Finally, given the sizable limitations of aggregate analyses of MS, investigators should make greater use of individual and subgroup analysis (i.e., the idiographic approach discussed in Section 23.4.5) when studying the development of symptoms over time.

The author hopes that the information and recommendations in this report will aid the small but dedicated group of researchers worldwide who are engaged in one of the most thorny and yet addicting of research topics—the problem of motion maladaptation. For those who are considering initiating MS research but have not yet been *bitten by the bug*, I will conclude this report with a caveat. The late Dr. Frederick Guedry (1921–2011) was one of the most knowledgeable and accomplished scientists on the planet concerning spatial orientation and vestibular function, yet he once said, "Don't do motion sickness research—it's too hard!"

DISCLAIMER

ACKNOWLEDGMENTS

The author thanks Laura Stanley for making the writing of this chapter possible. The author thanks the following colleagues (in alphabetical order) for their very helpful contributions to this report: Chris Brill, Joakim Dahlman, David Graeber, Deahndra Grigley, Shauna Legan, Heather McGee, Andrew Mead, Jason Moss, Eric Muth, Jill Parker, Bethany Ranes, Linda-Brooke Thompson, and Catherine Grandizio. The author also thanks Bob Cheung, John Golding, Fred Guedry (rest in peace), Bob Kennedy, Mike McCauley, and Angus Rupert for many insightful discussions (and debates) over the years concerning MS. Finally, the author thanks the following animal behavior specialists for considering questions concerning the behavior of nonhuman primates and providing tentative replies or leads relevant to some of the author's conjectures in Sections 23.9.2 and 23.9.3: Steve Brenner, John Capitano, Frans de Waal, Robert Dudley, Frank Ervin, Paul Houghton, Sue Howell, Sarah Johns, and Irene Pepperberg. The author thanks the sponsors of his past research reported in this chapter, including the U.S. Navy (e.g., Bureau of Medicine, Office of Naval Research), the U.S. Special Operations Command (the Biomedical Initiatives Steering Committee), and the National Aeronautics and Space Administration.

REFERENCES

Abrams, C., Earl, W. K., Baker, C. H., & Buckner, D. N. (1971). *Studies of the effects of sea motion on human performance* (Technical Report 798-1). Goleta, CA: Human Factors Research.

Andersen, G. J., & Braunstein, M. L. (1985). Induced self-motion in central vision. *Journal of Experimental Psychology: Human Perception and Performance, 11*, 122–132.

Baitis, A. E., Applebee, T. R., & McNamara, T. M. (1984). Human factors considerations applied to operations of the FFG-8 and LAMPS MK III. *Naval Engineers Journal, 96*(3), 191–199.

Balk, S. A., Bertola, A., & Inman, V. W. (2013). Simulator Sickness Questionnaire: Twenty years later. In *Proceedings of the 7th International Driving Symposium on Human Factors in Driver Assessment, Training, and Vehicle Design*, 257–263.

Balter, M. (2011). *When humans first plied the deep blue sea*. Retrieved from http://news.sciencemag.org/sciencenow/2011/11/when-humans-first-plied-the.html

Baltzley, D. R., Kennedy, R. S., Berbaum, K. S., Lilienthal, M. G., & Gower, D. W. (1989). The time of post-flight simulator sickness symptoms. *Aviation, Space, and Environmental Medicine, 60*(11), 1043–1048.

Barlow, D. H., Nock, M. K., & Hersen, M. (2008). *Single case experimental designs: Strategies for studying behavior change* (3rd ed.). Boston, MA: Allyn & Bacon.

Basdogan, C., & Srinivasan, M. A. (2002). Haptic rendering in virtual environments. In K. M. Stanney (Ed.), *Handbook of virtual environments: Design, implementation, and applications* (2nd ed., pp. 117–134). New York, NY: CRC Press.

Benfari, R. C. (1964). Perceptual vertigo: A dimensional study. *Perceptual and Motor Skills, 18*, 633–639.

Bennett, R. A. (1928). Sea-sickness and its treatment. *British Medical Journal, 1*(3513), 752.

Benson, A. J. (1988). Motion sickness. In J. Ernsting & P. King (Eds.), *Aviation medicine* (pp. 318–493). London, U.K.: Butterworth.

Benson, P. W., Hooker, J. B., Koch, K. L., & Weinberg, R. B. (2012). Bitter taster status predicts susceptibility to vection-induced motion sickness and nausea. *Neurogastroenterology & Motility, 24*(2), 134–140.

Biggs, S. J., & Srinivasan, M. A. (2002). Haptic interfaces. In K. M. Stanney (Ed.), *Handbook of virtual environments: Design, implementation, and applications* (2nd ed., pp. 93–116). New York, NY: CRC Press.

Birren, J. E. (1949). MS: Its psychophysiological aspects. In *A survey report on human factors in undersea warfare* (pp. 375–398). Washington, DC: Committee on Undersea Warfare.

Bledsoe, G. H., Brill, J. D., Zak, D., & Li, G. (2007). Injury and illness aboard an Antarctic cruise ship. *Wilderness & Environmental Medicine, 18*(1), 36–40.

Bles, W. (1979). *Sensory interactions and human posture: An experimental study*. Amsterdam, the Netherlands: Academic Press.

Bos, J. E., & Bles, W. (2002). Theoretical considerations on canal–otolith interaction and an observer model. *Biological Cybernetics, 86*(3), 191–207. doi: 10.1007/s00422-001-0289-7

Bos, J. E., de Vries, S. C., van Emmerik, M. L., & Groen, E. L. (2010). The effect of internal and external fields of view on visually induced motion sickness. *Applied Ergonomics, 41*(4), 516–521.

Bos, J. E., MacKinnon, S. N., & Patterson, A. (2005). Motion sickness symptoms in a ship motion simulator: Effects of inside, outside, and no view. *Aviation, Space, and Environmental Medicine, 76*(12), 1111–1118.

Bouyer, L. J. G., & Watt, D. G. D. (1999). "Torso Rotation" experiments. 4: The role of vision and the cervico-ocular reflex in compensation for a deficient VOR. *Journal of Vestibular Research, 9*(2), 89–102.

Bowins, B. (2010). Motion sickness: A negative reinforcement model. *Brain Research Bulletin, 81*(1), 7–11.

Bowman, D. (2014). Principles for designing effective 3D interaction techniques. In K. S. Hale & K. M. Stanney (Eds.), *Handbook of virtual environments: Design, implementation, and applications* (2nd ed., pp. 285–312). New York, NY: CRC Press.

Bowman, D. A. (2002). Principles for the design of performance-oriented interaction techniques. In K. S. Hale & K. M. Stanney (Eds.), *Handbook of virtual environments: Design, implementation, and applications* (pp. 277–300). New York, NY: CRC Press.

Boyle, A. (2007, May 31). Scares in space. *Cosmic Log-NBC.* Retrieved from http://cosmiclog.nbcnews.com/_news/2007/05/31/4350247-scares-in-space?lite

Brandt, T., Wenzel, D., & Dichgans, J. (1976). Visual stabilization of free stance in infants: A sign of maturity. *Archiv für Psychiatrie und Nervenkrankheiten, 223*(1), 1–13.

Brandt, T. H., Dichgans, J., & Wagner, W. (1974). Drug effectiveness on experimental optokinetic and vestibular MS. *Aerospace Medicine, 45*(11), 1291–1297.

Brenner, S. (2013, Feb). *Paleogenetics and the history of alcohol in primates.* Abstract from Advancing Science, Serving Society Annual Meeting, Boston, MA. Retrieved from: http://aaas.confex.com/aaas/2013/webprogram/Paper8851.html

Brooks, J. O., Goodenough, R. R., Crisler, M. C., Klein, N. D., Alley, R. L., Koon, B. L., … Wills, R. F. (2010). Simulator sickness during driving simulation studies. *Accident Analysis & Prevention, 42*(3), 788–796.

Bruder, G., Steinicke, F., Hinrichs, K. H., Frenz, H., & Lappe, M. (2009, March 14). Impact of gender on discrimination between real and virtual stimuli. In *Workshop on perceptual illusions in virtual environments* (pp. 10–15). Lafayette, LA.

Burish, T. G., & Carey, M. P. (1986). Conditioned aversive responses in cancer chemotherapy patients: Theoretical and developmental analysis. *Journal of Consulting and Clinical Psychology, 54*(5), 593–600.

Calkins, D. S., Reschke, M. F., Kennedy, R. S., & Dunlap, W. P. (1987). Reliability of provocative tests of MS susceptibility. *Aviation, Space, and Environmental Medicine, 58*(9 Suppl.), A50–A54.

Casali, J. G., & Wierwille, W. (1998). Vehicular simulator-induced sickness (Technical Report No. NTSC-TR86-012, AD-A173 266, Vol. 3, p. 155). Arlington, VA: Naval Training Systems Center. In K. M. Stanney, R. R. Mourant, & R. S. Kennedy. Human factors issues in virtual environments: A review of the literature. *Presence: Teleoperators and Virtual Environments*, 7(4), 327–351.

Cevette, M. J., Stepanek, J., Cocco, D., Galea, A. M., Pradhan, G. N., Wagner, L. S., ... & Brookler, K. H. (2012). Oculo-vestibular recoupling using galvanic vestibular stimulation to mitigate simulator sickness. *Aviation, space, and environmental medicine, 83*(6), 549–555.

Champney, R. K., Stanney, K. M., Hash, P. A., Malone, L. C., Kennedy, R. S., & Compton, D. E. (2007). Recovery from virtual environment exposure: Expected time course of symptoms and potential readaptation strategies. *Human Factors: The Journal of the Human Factors and Ergonomics Society, 49*(3), 491–506.

Chapelow, J. W. (1988). Simulator sickness in the Royal Air Force: A survey. In *Advisory Group for Aerospace Research and Development, Motion cues in flight simulation and simulator-induced sickness* (AGARD Conference Proceedings No. 433, pp. 6-1–6-11). NATO, Neuilly-sur-Seine, France.

Chelen, W. E., Kabrisky, M., & Rogers, S. K. (1993). Spectral analysis of the electroencephalographic response to MS. *Aviation, Space, and Environmental Medicine, 64*, 24–29.

Chen, D. J., Chow, E. H., & So, R. H. (2011). The relationship between spatial velocity, vection, and visually induced motion sickness: An experimental study. *i-Perception, 2*(4), 415.

Chen, Y. C., Duann, J. R., Chuang, S. W., Lin, C. L., Ko, L. W., Jung, T. P., & Lin, C. T. (2010). Spatial and temporal EEG dynamics of motion sickness. *Neuroimage, 49*(3), 2862–2870.

Cheung, B., & Hofer, K. (2003). Lack of gender difference in motion sickness induced by vestibular Coriolis cross-coupling. *Journal of Vestibular Research, 12*(4), 191–200.

Cheung, B. S., Howard, I. P., & Money, K. E. (1991). Visually induced sickness in normal and bilaterally labyrinthine-defective subjects. *Aviation, Space, and Environmental Medicine, 62*, 527–531.

Chinn, H. I., & Smith, P. K. (1955). Motion sickness. *Pharmacological Review, 7*, 33–82.

Choi, C. (2012). Early human 'Lucy': Swung from the trees. *LiveScience.* Retrieved from http://www.livescience.com/24297-early-human-lucy-swung-from-trees.html

Chu, H., Li, M. H., Huang, Y. C., & Lee, S. Y. (2013). Simultaneous transcutaneous electrical nerve stimulation mitigates simulator sickness symptoms in healthy adults: A crossover study. *BMC Complementary and Alternative Medicine, 13*(1), 84.

Clément, G. (2007). *Fundamentals of space medicine.* Dordrecht, the Netherlands: Springer Verlag.

Cloutier, A., & Watt, D. G. (2006). Motion sickness provoked by torso rotation predicts that caused by head nodding. *Aviation, Space, and Environmental Medicine, 77*(9), 909–914.

Cloutier, A., & Watt, D. G. (2007). Adaptation to motion sickness from torso rotation affects symptoms from supine head nodding. *Aviation, Space, and Environmental Medicine, 78*(8), 764–769.

Cobb, S., Nichols, S., Ramsey, A., & Wilson, J. R. (1999). Virtual reality–induced symptoms and effects. *Presence: Teleoperators and Virtual Environments, 8*, 169–186.

Colwell, J. L. (2000, September). NATO questionnaire: Correlation between ship motions, fatigue, sea sickness, and naval task performance. In *International Conference: Human Factors in Ship Design and Operation*. London, U.K.: Royal Institution of Naval Architects.

Cowings, P. S., Toscano, W. B., DeRoshia, C., & Tauson, R. (1999). *Effects of command and control vehicle (C2V) operational environment on soldier health and performance* (Technical Report No. NASA TM-1999-208786). Moffett Field, CA: National Aeronautics and Space Administration.

Crampton, G. H. (1955). Studies of motion sickness: XVII. Physiological changes accompanying sickness in man. *Journal of Applied Physiology, 7*(5), 501–507.

Crampton, G. H. (Ed.). (1990). *Motion and space sickness*. Boca Raton, FL: CRC Press.

Crampton, G. H., & Young, F. A. (1953). The differential effects of a rotary visual filed on susceptibles and nonsusceptibles to MS. *Journal of Comparative and Physiological Psychology, 46*, 451–453.

Crosby, T. N., & Kennedy, R. S. (1982). Presented at the 53rd Annual Scientific Meeting of the Aerospace Medical Association, Bal Harbor, FL. Cited in M. E. McCauley & T. J. Sharkey (1992). Cybersickness: Perception of self-motion in virtual environments. *Presence: Teleoperators and Virtual Environments, 1*(3), 311–318.

Crossland, P., & Lloyd, A. R. J. M. (1993). *Experiments to quantify the effects of ship motion on crew task performance—Phase 1, motion-induced interruptions and motion-induced fatigue* (Technical Report No. DRA/AWMH/TR/93025). Farnborough, England: Defence Research Agency.

Crowley, J. S. (1987). Simulator sickness: A problem for Army aviation. *Aviation, Space, and Environmental Medicine, 58*(4), 355.

Dahlman, J. (2009). *Psychophysiological and performance aspects on motion sickness* (Doctoral dissertation). Retrieved from http://urn.kb.se/resolve?urn=urn:nbn:se:liu:diva-15919

Darwin, E. (1795). *Zoonomia*. London, U.K.: J. Johnson. Cited in Chapter 1 of W. J. White. (1964). *A history of the centrifuge in aerospace medicine*. Santa Monica, CA: Douglas Aircraft Company.

Daunton, N. G. (1990). Animal models in motion sickness research. In G. H. Crampton (Ed.), *Motion and space sickness* (pp. 87–104). Boca Raton, FL: CRC Press.

Davis, J. R., Vanderploeg, J. M., Santy, P. A., Jennings, R. T., & Stewart, D. F. (1988). Space MS during 24 flights of the space shuttle. *Aviation, Space, and Environmental Medicine, 59*, 1185–1189.

de Wit, G. (1953). Seasickness (motion sickness) a labyrinthological study. *Acta Otolaryngologica, Supplementum, 108*, 7.

Dichgans, J., & Brandt, T. (1978). Visual-vestibular interaction: Effects on self-motion perception and postural control. In R. Held, H. W. Leibowitz, & H. L. Teuber (Eds.), *Handbook of sensory physiology* (Vol. 8, pp. 755–804). New York, NY: Springer-Verlag.

DiZio, P., & Lackner, J. R. (1991). MS susceptibility in parabolic flight and velocity storage activity. *Aviation, Space, and Environmental Medicine, 62*, 300–307.

DiZio, P., & Lackner, J. R. (1997). Circumventing side effects of immersive virtual environments. In M. Smith, G. Salvendy, & R. Koubek (Eds.), *Design of computing systems: Social and ergonomic considerations* (pp. 893–896). Amsterdam, the Netherlands: Elsevier Science Publishers.

DiZio, P., & Lackner, J. (2014). Proprioceptive adaptation and aftereffects. In K. S. Hale & K. M. Stanney (Eds.), *Handbook of virtual environments: Design, implementation, and applications* (2nd ed., pp. 833–854). New York, NY: CRC Press.

Dolezal, H. (1982). *Living in a world transformed: Perceptual and performatory adaptation to visual distortion*. New York, NY: Academic Press.

Dornhoffer, J. L., Mamiya, N., Bray, P., Skinner, R. D., & Garcia-Rill, E. (2002). Effects of rotation on the sleep state-dependent midlatency auditory evoked P50 potential in the human. *Journal of Vestibular Research-Equilibrium & Orientation, 12*(5–6), 205–209.

Draper, M. H., Viirre, E. S., Furness, T. A., & Gawron, V. J. (2001). Effects of image scale and system time delay on simulator sickness within head-coupled virtual environments. *Human Factors, 43*, 129–146.

Durlach, N., & Slater, M. (1998). *Presence in shared virtual environments and virtual togetherness* [Online]. Paper presented at the Presence in Shared Virtual Environments Workshop, Ipswich, England. Available from http://www.cs.ucl.ac.uk/staff/m.slater/BTWorkshop/durlach.html [1999, September 1]

Durlach, N. I., & Mavor, A. S. (Eds.). (1995). *Virtual reality: Scientific and technological challenges.* Washington, DC: National Academy Press.

Ebenholtz, S. M. (1992). MS and oculomotor systems in virtual environments. *Presence, 1*(3), 302–305.

Edwards, R., Augustson, E. M., & Fillingim, R. (2000). Sex-specific effects of pain-related anxiety on adjustment to chronic pain. *The Clinical Journal of Pain, 16*(1), 46.

Eklund, S. (1999). *Gentamicin treatment and headache in Menière's disease* (Unpublished doctoral dissertation). Helsinki University Central Hospital, Helsinki, Finland.

Elert, E. (2012). FYI: Why does some food taste bad to some people and good to others? *POPSCI.* Retrieved from http://www.popsci.com/science/article/2012-03/fyi-why-does-some-food-taste-bad-some-people-and-good-others

Faugloire, E., Bonnet, C. T., Riley, M. A., Bardy, B. G., & Stoffregen, T. A. (2007). Motion sickness, body movement, and claustrophobia during passive restraint. *Experimental Brain Research, 177*(4), 520–532.

Finch, M., & Howarth, P. A. (1996). *A comparison between two methods of controlling movement within a virtual environment* (Technical Report No. VISERG 9606). Leicestershire, England: Loughborough University.

Fischer, M. H., & Kornmüller, A. E. (1930). Perception of motion based on the optokinetic sense and optokinetic nystagmus. *Journal fuer Psychologie und Neurologie, 41*, 273–308.

Fisher, S. (1990). Virtual interface environments. In B. Laurel (Ed.), *The art of human–computer interface design.* Menlo Park, CA: Addison-Wesley.

Flaherty, D. E. (1998). *Sopite syndrome in operational flight training* (Master's thesis). Naval Postgraduate School, Monterey, CA.

Flanagan, M. (2005). *Sex and virtual reality: Posture and motion sickness.* University of New Orleans Dissertation 302.

Flanagan, M. B., May, J. G., & Dobie, T. G. (2005). Sex differences in tolerance to visually-induced motion sickness. *Aviation, Space, and Environmental Medicine, 76*(7), 642–646.

Frank, L. H., Casali, J. G., & Wierwille, W. W. (1988). Effects of visual display and motion system delays on operator performance and uneasiness in a driving simulator. *Human Factors, 30*, 201–217.

Fregly, A. R. (1974). Vestibular ataxia and its measurement in man. In H. H. Kornhuber (Ed.), *Handbook of sensory physiology* (Vol. 6, No. 2, pp. 321–360). New York, NY: Springer-Verlag.

Garcia, A., Baldwin, C., & Dworsky, M. (2010, September). Gender differences in simulator sickness in fixed- versus rotating-base driving simulator. In *Proceedings of the Human Factors and Ergonomics Society Annual Meeting* (Vol. 54, No. 19, pp. 1551–1555). New York, NY: SAGE Publications.

Gascoigne, B. (2001). *History of the domestication of animals. HistoryWorld. From 2001, ongoing.* Retrieved from http://www.historyworld.net/wrldhis/PlainTextHistories.asp?ParagraphID = ayt

Gianaros, P. J., Muth, E. R., Mordkoff, J. T., Levine, M. E., & Stern, R. M. (2001). A questionnaire for the assessment of the multiple dimensions of motion sickness. *Aviation, space, and environmental medicine, 72*(2), 115–119.

Golding, J. F. (1998). Motion sickness susceptibility questionnaire revised and its relationship to other forms of sickness. *Brain Research Bulletin, 47*(5), 507–516.

Golding, J. F. (2006). Motion sickness susceptibility. *Autonomic Neuroscience, 129*(1), 67–76.

Golding, J. F., Phil, D., Mueller, A. G., & Gresty, M. A. (2001). A MS maximum around the 0.2 Hz frequency range of horizontal translational oscillation. *Aviation Space, and Environmental Medicine, 72*(3), 188–192.

Golding, J. F., & Stott, J. R. R. (1995). Effect of sickness severity on habituation to repeated motion challenges in aircrew referred for airsickness treatment. *Aviation, Space, and Environmental Medicine, 66*(7), 625–630.

Gonshor, A., & Melville-Jones, J. G. (1976). Extreme vestibular-oculomotor adaptation induced by prolonged optical reversal of vision. *Journal of Physiology, 256*, 381–414.

Gordon, C. R., Spitzer, O., Doweck, I., Melamed, Y., & Shupak, A. (1995). Clinical features of mal de débarquement: Adaptation and habituation to sea conditions. *Journal of Vestibular Research, 5*, 363–369.

Gottshall, K. R., Moore, R. J., & Kopke, R. D. (2002). *Exercise induced motion intolerance: Role in operational environments.* Paper presented at RTO HFM Symposium on "Spatial Disorientation in Military Vehicles: Causes, Consequences and Cures", La Coruna, Spain.

Gould, E., & Bres, M. (1986). Regurgitation in gorillas: Possible model for human eating disorders (ruminations/bulimia). *Journal of Developmental & Behavioral Pediatrics, 7*(5), 314–319.

Gower, D. W., Lilienthal, M. G., Kennedy, R. S., & Fowlkes, J. E. (1988, September). Simulator sickness in U.S. Army and Navy fixed- and rotary-wing flight simulators. In *Advisory Group for Aerospace Research and Development, motion cues in flight simulation and simulator induced sickness* (AGARD Conference Proceedings No. 433, pp. 8.1–8.20). NATO, Neuilly-sur-Seine, France.

Graber, C. (2008). *Fact or fiction?: Animals like to get drunk*. Retrieved from http://www.scientificamerican. com/article.cfm?id = animals-like-to-get-drunk

Grabowski, L. S. (1996). *Falls among the geriatric population* [Online]. Available from www.sansumcorn/ highlite/1996/13_2_2html

Graeber, D. A., & Stanney, K. M. (2002, September). Gender differences in visually induced motion sickness. In *Proceedings of the Human Factors and Ergonomics Society Annual Meeting* (Vol. 46, No. 26, pp. 2109–2113). New York, NY: SAGE Publications.

Graybiel, A. (1967). Functional disturbances of vestibular origin of significance in space flight. In *Second International Symposium on Man in Space, Paris* (pp. B8–B32). Wien, Austria/New York, NY: Springer-Verlag.

Graybiel, A., Clark, B., & Zarriello, J. J. (1960). Observations on human subjects living in a "slow-rotation room" for periods of two days. *AMA Archives of Neurology, 3*, 55–73.

Graybiel, A., Kennedy, R. S., Knoblock, E. C., Guedry, F. E., Mertz, W., McLead, M. E., … Fregly, A. R. (1965). Effects of exposure to a rotating environment (10 RPM) on four aviators for a period of twelve days. *Aerospace Medicine, 36*, 733–754.

Graybiel, A., & Knepton, J. (1976). Sopite syndrome: A sometimes sole manifestation of MS. *Aviation, Space, and Environmental Medicine, 47*, 873–882.

Graybiel, A., & Lackner, J. R. (1980). A sudden-stop vestibulovisual test for rapid assessment of MS manifestations. *Aviation, Space, and Environmental Medicine, 51*, 21–23.

Graybiel, A., & Lackner, J. R. (1987). Treatment of severe motion sickness with antimotion sickness drug injections. *Aviation, Space, and Environmental Medicine, 58*, 773–776.

Gresty, M. A., Golding, J. F., Gresty, J. M., Powar, J., & Darwood, A. (2011). The movement frequency turning of motion sickness is determined by biomechanical constraints on locomotion. *Aviation, Space, and Environmental Medicine, 82*, 242.

Gresty, M. A., Golding, J. F., Le, H., & Nightingale, K. (2008). Cognitive impairment by spatial disorientation. *Aviation, Space, and Environmental Medicine, 79*(2), 105–111.

Guedry, F. E. (1991). Factors influencing susceptibility: Individual differences and human factors. In *Proceedings of Motion Sickness: Significance in Aerospace Operations and Prophylaxis* (AGARD LS-175:5/1-5/18). NATO, Neuilly-sur-Seine, France.

Guedry, F. E., & Benson, A. J. (1979). Coriolis cross-coupling effects: Disorienting and nauseogenic or not? *Aviation, Space, and Environmental Medicine, 49*(1), 29–35.

Guedry, F. E., Lentz, J. M., & Jell, R. M. (1978). *Visual-vestibular interactions. I. Influence of peripheral vision on suppression of the vestibulo-ocular reflex and visual acuity* (No. NAMRL-1246). Pensacola FL: Naval Aerospace Medical Research Laboratory.

Guedry, F. E., Rupert, A. R., & Reschke, M. F. (1998). Motion sickness and development of synergy within the spatial orientation system. A hypothetical unifying concept. *Brain Research Bulletin, 47*(5), 475–480.

Guignard, J. C., & McCauley, M. E. (1990). The accelerative stimulus for motion sickness. In G. H. Crampton (Ed.), *Motion and space sickness* (pp. 123–152). Boca Raton, FL: CRC Press.

Hain, T. C., Hanna, P. A., & Rheinberger, M. A. (1999). Mal de debarquement. *Archives of Otolaryngology—Head & Neck Surgery, 125*(6), 615.

Hale, K. S., & Stanney, K. M. (Eds.). (2014). *Handbook of virtual environments: Design, implementation, and applications* (2nd ed.,). New York, NY: CRC Press.

Hall, G., Symonds, M., & Rodriguez, M. (2009). Enhanced latent inhibition in context aversion conditioning. *Learning and Motivation, 40*(1), 62–73.

Harm, D. H. (1990). Physiology of MS symptoms. In G. H. Crampton (Ed.), *Motion and space sickness* (pp. 153–177). Boca Raton, FL: CRC Press.

Harm, D. L. (2002). Motion sickness neurophysiology, physiological correlates, and treatment. In K. M. Stanney (Ed.), *Handbook of virtual environments: Design, implementation, and applications* (2nd ed., pp. 637–661). New York, NY: CRC Press.

Havron, M. D., & Butler, L. F. (1957). *Evaluation of training effectiveness of the 2-FH-2 helicopter flight training research tool* (Technical Report No. NAVTRADEVCEN 20-OS-16, Contract 1915). Arlington, VA: U.S. Naval Training Device Center.

Haward, B. M., Lewis, C. H., & Griffin, M. J. (2000, September). Crew response to motions of an offshore oil production and storage vessel. In *International Conference: Human Factors in Ship Design and Operation*. London, U.K.: Royal Institution of Naval Architects.

Hemingway, A. (1946). The relationship of air sickness to other types of motion sickness. *J. Aviat. Med.* 17: 80–85. In Reason, J. T., & Brand, J. J. (1975). *MS*. London: Academic Press.

Hettinger, L. J., Berbaum, K. S., Kennedy, R. S., Dunlap, W. P., & Nolan, M. D. (1990). Vection and simulator sickness. *Military Psychology, 2*(3), 171–181.

Hettinger, L., Kennedy, R., & McCauley, M. (1990). Motion and human performance. In G. H. Crampton (Ed.), *Motion and space sickness* (pp. 411–441). Boca Raton, FL: CRC Press, Inc.

Hettinger, L. J., & Riccio, G. E. (1992). Visually induced MS in virtual environments. *Presence, 1*(3), 306–310.

Hicks, J. S., & Durbin, D. B. (2011). *A summary of simulator sickness ratings for US army aviation engineering simulators* (No. ARL-TR-5573). Aberdeen Proving Ground, MD: Army Research Laboratory.

Hirst, K. (n.d.). *About.com*. When and where were camels domesticated? Retrieved from http://archaeology.about.com/od/cterms/g/camels.htm

Holmes, S. R., Robertson, K. A., & Crossland, P. (2002). *Ship motion adversely affects sleep quality and fatigue*. Paper presented to Conference on Human Factors in Ship Design and Operation, Royal Institute of Naval Architects, London, U.K.

Horowitz, S. S., Blanchard, J., & Morin, L. P. (2005). Medial vestibular connections with the hypocretin (orexin) system. *Journal of Comparative Neurology, 487*(2), 127–146.

Howarth, P. A., & Finch, M. (1999). The nauseogenicity of two methods of navigating within a virtual environment. *Applied Ergonomics, 30*, 39–45.

Hu, S., Davis, M. S., Klose, A. H., Zabinsky, E. M., Meux, S. P, Jacobsen, H. A., … Gruber, M. B. (1997). Effects of spatial frequency of a vertically striped rotating drum on vection-induced motion sickness. *Aviation, Space, and Environmental Medicine, 68*(4), 306–311.

Hu, S., McChesney, K. A., Player, K. A., Bahl, A. A., Buchanan, J. B., & Scozzafava, J. E. (1999). Systematic investigation of physiological correlates of MS induced by viewing an optokinetic rotating drum. *Aviation, Space, and Environmental Medicine, 70*, 759–765.

Irwin, J. A. (1881). The pathology of sea-sickness. *The Lancet, 118*(3039), 907–909.

Javid, F. A., & Naylor, R. J. (1999). Variables of movement amplitude and frequency in the development of motion sickness in *Suncus murinus. Pharmacology, Biochemistry and Behavior, 64*(1), 115–122.

Ji, J. T., So, R. H., & Cheung, R. T. (2009). Isolating the effects of vection and optokinetic nystagmus on optokinetic rotation-induced motion sickness. *Human Factors: The Journal of the Human Factors and Ergonomics Society, 51*(5), 739–751.

Johnson, D. M. (2005). *Introduction to and review of simulator sickness research*. Fort Rucker, AL: Army Research Institute Field Unit.

Johnson, E. C., Hill, E., & Cooper, M. A. (2007). Vomiting in wild bonnet macaques. *International Journal of Primatology, 28*(1), 245–256.

Johnson, W. H., Sunahara, F. A., & Landolt, J. P. (1999). Importance of the vestibular system in visually induced nausea and self-vection. *Journal of Vestibular Research, 9*(2), 83–88.

Jones, S. A. (1998). *Effects of restraint on vection and simulator sickness* (Unpublished doctoral dissertation). University of Central Florida, Orlando, FL.

Kaber, D. B., Draper, J. V., & Usher, J. M. (2002).Influence of individual differences on virtual reality application design for individual and collaborative virtual environments. In K. M. Stanney (Ed.), *Handbook of virtual environments: Design implementation and applications* (Chapter 18, pp. 379–402). Mahwah, MH: Lawrence Erlbaum & Associates.

Kanas, N., & Manzey, D. (2008). *Space psychology and psychiatry: Vol. 22. The space technology library*. Berlin, Germany: Published jointly by Microcosm Press and Springer.

Kay, D. L., Lawson, B. D., & Clarke, J. E. (1998). Airsickness and lowered arousal during flight training. In *Proceedings of the 69th Annual Meeting of the Aerospace Medical Association*, Seattle, WA. *Aviation, Space and Environmental Medicine, 69*(3), 236.

Kellog, R. S., Castore, C. H., & Coward, R. E. (1984). Psychological effects of training in a full vision simulator. In M. E. McCauley (Ed.), *Research Issues in Simulator Sickness: Proceedings of a Workshop* (pp. 2, 6). Washington, DC: National Academy Press. (Original work published in 1980)

Kelly, A. D., & Kanas, N. (1993). Communication between space crews and ground personnel: A survey of astronauts and cosmonauts. *Aviation, Space, and Environmental Medicine, 64*, 795–800.

Kennedy, R. S. (1996). *Analysis of simulator sickness data* (Technical Report under Contract No. N61339-91-D-0004 with Enzian Technology, Inc.). Orlando, FL: Naval Air Warfare Center, Training Systems Division.

Kennedy, R. S., Berbaum, K. S., Dunlap, W. P., & Hettinger, L. J. (1996). Developing automated methods to quantify the visual stimulus for cybersickness. In *Proceedings of the Human Factors and Ergonomics Society 40th Annual Meeting* (pp. 1126–1130). Santa Monica, CA: Human Factors and Ergonomics Society.

Kennedy, R. S., Berbaum, K. S., & Lilienthal, M. G. (1997). Disorientation and postural ataxia following flight simulation. *Aviation, Space, and Environmental Medicine, 68*(1), 13–17.

Kennedy, R. S., Drexler, J., & Kennedy, R. C. (2010). Research in visually induced motion sickness. *Applied Ergonomics, 41*(4), 494–503.

Kennedy, R. S., Dunlap, W. P., & Fowlkes, J. E. (1990). Prediction of MS susceptibility. In G. H. Crampton (Ed.), *Motion and space sickness.* (pp. 179–215) Boca Raton, FL: CRC Press.

Kennedy, R. S., Dutton, B., Ricard, G. L., & Frank, L. H. (1984). *Simulator sickness: A survey of flight simulators for the navy* (SAE Technical Paper Series No. 811597). Warrendale, PA. Society for Automotive Engineering.

Kennedy, R. S., Fowlkes, J. E., & Hettinger, L. J. (1989). *Review of simulator sickness literature* (Technical Report No. NTSC TR89-024). Orlando, FL: Naval Training Systems Center.

Kennedy, R. S., French, J., Drexler, J. M., & Compton, D. E. (2005, July). *Effects of motion exposure on physiological, cognitive, and subjective measures of sickness and fatigue.* Paper presented at 1st Virtual Reality International Conference, Las Vegas, NV.

Kennedy, R. S., & Frank, L. H. (1985). *A review of MS with special reference to simulator sickness* (Technical Report No. NAVTRAEQUIPCEN 81-C-0105-16). Orlando, FL: Naval Training Equipment Center.

Kennedy, R. S., & Graybiel, A. (1965). *The dial test: A standardized procedure for the experimental production of canal sickness symptomatology in a rotating environment* (Technical Report No. NSAM-930). Pensacola, FL: Naval School of Aviation Medicine.

Kennedy, R. S., Graybiel, A., McDonough, R. G., & Beckwith, F. D. (1968). Symptomatology under storm conditions in the North Atlantic in control subjects and in persons with bilateral labyrinthine defects. *Acta Oto-Laryngologica, 66*, 533–540.

Kennedy, R. S., Hettinger, L. J., & Lilienthal, M. G. (1990). Simulator sickness. In G. H. Crampton (Ed.), *Motion and space sickness* (pp. 317–341). Boca Raton, FL: CRC Press.

Kennedy, R. S., Jones, M. B., Stanney, K. M., Ritter, A., & Drexler, J. M. (1996). *Human factors safety testing for virtual environment mission-operations training* (Technical Report No. NASA1-96-2, Contract No. NAS9-19482). Houston, TX: NASA Lyndon B, Johnson Space Center.

Kennedy, R. S., Kennedy, K. E., & Bartlett, K. M. (2002). Virtual environments and product liability. In K. M. Stanney (Ed.), *Handbook of virtual environments: Design, implementation, and applications* (2nd ed., pp. 534–554). New York, NY: CRC Press.

Kennedy, R. S., Lane, N. E., Berbaum, K. S., & Lilienthal, M. G. (1993). Simulator sickness questionnaire: An enhanced method for quantifying simulator sickness. *International Journal of Aviation Psychology, 3*, 203–220.

Kennedy, R. S., Lane, N. E., Lilienthal, M. G., Berbaum, K. S., & Hettinger, L. J. (1992). Profile analysis of simulator sickness symptoms: Application to virtual environment systems. *Presence: Teleoperators and Virtual Environments 1*(3), 295–301.

Kennedy, R. S., Lanham, D. S., Drexler, J. M., & Lilienthal, M. G. (1995). A method for certification that aftereffects of virtual reality exposures have dissipated: Preliminary findings. In A. C. Bittner & P. C. Champney (Eds.), *Advances in industrial ergonomics safety VII* (pp. 263–270). London, U.K.: Taylor & Francis Group.

Kennedy, R. S., Lanham, D. S., Drexler, J. M., Massey, C. J., & Lilienthal, M. G. (1997). A comparison of cybersickness incidences, symptom profiles, measurement techniques, and suggestions for further research. *Presence, 6*, 638–644.

Kennedy, R. S., Lilienthal, M. G., Berbaum, K. S., Baltzley, D. R., & McCauley, M. E. (1989). Simulator sickness in U.S. Navy flight simulators. *Aviation, Space, and Environmental Medicine, 60*, 10–16.

Kennedy, R. S., Massey, C. J., & Lilienthal, M. G. (1995, July). *Incidences of fatigue and drowsiness reports from three dozen simulators: Relevance for the sopite syndrome.* Paper presented at the First Workshop on Simulation and Interaction in Virtual Environments (SIVE '95), Iowa City, IA.

Kennedy, R. S., & Stanney, K. M. (1996). Postural instability induced by virtual reality exposure: Development of a certification protocol. *International Journal of Human-Computer Interaction, 8*(1), 25–47.

Kennedy, R. S., & Stanney, K. M. (1997). Aftereffects of virtual environment exposure: Psychometric issues. In M. J. Smith, G. Salvendy, & R. J. Koubek (Eds.), *Design of computing systems: Social and ergonomic considerations* (pp. 897–900). Amsterdam, the Netherlands: Elsevier Science Publishers.

Kennedy, R. S., Stanney, K. M., & Dunlap, W. P. (2000). Duration and exposure to virtual environments: Sickness curves during and across sessions. *Presence, 9*(5), 466–475.

Kennedy, R. S., Stanney, K. M., & Hale, K. (2014). Virtual environment usage protocols. In K. S. Hale & K. M. Stanney (Eds.), *Handbook of virtual environments: Design, implementation, and applications* (2nd ed., Chapter 31, pp. 508–518). New York, NY: CRC Press.

Keshavarz, B., & Hecht, H. (2011). Validating an efficient method to quantify motion sickness. *Human Factors: The Journal of the Human Factors and Ergonomics Society, 53*(4), 415–426.

Keshavarz, B., Hecht, H., & Lawson, B. D. (2014). Visually induced motion sickness-causes, characteristics, and counter measures. In K. S. Hale & K. M. Stanney (Eds.), *Handbook of virtual environments: Design, implementation, and applications* (2nd ed., Chapter 27, pp. 647–697). New York, NY: CRC Press.

Kiniorski, E. T., Weider, S. K., Finley, J. R., Fitzgerald, E. M., Howard, J. C., Di Nardo, P. A., & Guzy, L. T. (2004). Sopite symptoms in the optokinetic drum. *Aviation, Space, and Environmental Medicine, 75*(10), 872–875.

Kleinschmidt, A., Thilo, K. V., Büchel, C., Gresty, M. A., Bronstein, A. M., & Frackowiak, R. S. (2002). Neural correlates of visual-motion perception as object-or self-motion. *Neuroimage, 16*(4), 873–882.

Klosterhalfen, S., Kellermann, S., Pan, F., Stockhorst, U., Hall, G., & Enck, P. (2005). Effects of ethnicity and gender on motion sickness susceptibility. *Aviation, Space, and Environmental Medicine, 76*(11), 1051–1057.

Klosterhalfen, S., Pan, F., Kellermann, S., & Enck, P. (2006). Gender and race as determinants of nausea induced by circular vection. *Gender Medicine, 3*(3), 236–242.

Klosterhalfen, S., Rüttgers, A., Krumrey, E., Otto, B., Stockhorst, U., Riepl, R. L., Probst, T., & Enck, P. (2000). Pavlovian conditioning of taste aversion using a motion sickness paradigm. *Psychosomatic Medicine, 62*(5), 671–677.

Kock, N. (2010, January 20). *How long does it take for a food-related trait to evolve?* [Web log post]. Retrieved from http://healthcorrelator.blogspot.com/2010/how-long-does-it-take-for-food-related.html

Kolasinski, E. M. (1995). *Simulator sickness in virtual environments* (Technical Report No. 1027). Alexandria, VA: United States Army Research Institute for the Behavioral and Social Sciences.

Kolasinski, E. M., & Gilson, R. D. (1999). Ataxia following exposure to a virtual environment. *Aviation, Space, and Environmental Medicine, 70*(3), 264–269.

Kricher, J. C. (1999). *A neotropical companion: An introduction to the animals, plants, and ecosystems of the New World tropics.* Princeton, NJ: Princeton University Press.

Lackner, J. R. (1990). Human orientation, adaptation, and movement control, In National Research Council Committee on Vision and Working Group on Wraparound Visual Displays, *Proceedings of a Conference on Wraparound Visual Displays Held in Waltham, Massachusetts on 14–15 January 1988 (Motion Sickness, Visual Displays, and Armored Vehicle Design)* (p. 29–51). NRC Report BRL-CR-629. Aberdeen Proving Ground, MD: Army Ballistic Research Laboratory.

Lackner, J. R., & DiZio, P. (1988). Visual stimulation affects the perception of voluntary leg movements during walking. *Perception, 17,* 71–80.

Lackner, J. R., & Graybiel, A. (1986). Sudden emesis following parabolic flight maneuvers: Implications for space MS. *Aviation, Space, and Environmental Medicine, 57,* 343–347.

Lackner, J. R., & Teixeira, R. A. (1977). Optokinetic MS: Continuous head movements attenuate the visual induction of apparent self-rotation and symptoms of MS. *Aviation, Space, and Environmental Science, 48*(3), 248–253.

Lacour, M., Vidal, P. P., & Xerri, C. (1981). Visual influences on vestibulospinal reflexes during vertical linear motion in normal and hemilabyrinthectomized. *Experimental Brain Research, 43,* 3–4.

Lacour, M., Xerri, C., & Hugon, M. (1978). Muscle responses and monosynaptic reflexes in falling monkey. Role of the vestibular system. *Journal de Physiologie, 74*(4), 427–438.

Lane, N., & Kennedy, R. S. (1988). *A new method for quantifying simulator sickness: Development and application of the simulator sickness questionnaire* (SSQ), (EOTR 88-7). Orlando, FL: Essex Corporation.

Lawson, B. D. (1993). *Human physiological and subjective responses during motion sickness induced by unusual visual and vestibular stimulation* (Doctoral dissertation). Brandeis University, Waltham, MA. University Microfilms International, Report #9322354, Ann Arbor, MI.

Lawson, B. D. (2005). Exploiting the illusion of self-motion (vection) to achieve a feeling of "virtual acceleration" in an immersive display. In Stephanidis, C. (Ed.), *Proceedings of the 11th International Conference on Human-Computer Interaction* (pp. 1–10). Las Vegas, NV.

Lawson, B. D. (2014). Motion sickness scaling. In K. S. Hale & K. M. Stanney (Eds.), *Handbook of virtual environments: Design, implementation, and applications* (2nd ed., Chapter 24, pp. 601–626). New York, NY: CRC Press.

Lawson, B. D., Graeber, D. A., Mead, A. M., & Muth, E. R. (2002). Signs and symptoms of human syndromes associated with synthetic experiences. In K. M. Stanney (Ed.), *Handbook of virtual environments: Design, implementation, and applications* (pp. 589–618). New York, NY: CRC Press.

Lawson, B. D., Kass, S., Harmon-Horton, H., Decoteau, P., Simmons, R., McGrath, C. M., & Vacchiano, C. (2007). A test of the usefulness of an aromatic nausea remedy for alleviating symptoms of motion sickness [Abstract]. *Aviation, Space, and Environmental Medicine, 78*(3), 117.

Lawson, B. D., Kass, S. J., Lambert, C., & Smith, S. (2004). Survey and review concerning evidence for gender differences in motion susceptibility. Abstracts of the 75th Aerospace Medical Association Annual Scientific Meeting, Anchorage, AK. *Aviation, Space and Environmental Medicine, 75*(4), (Suppl. 2), B105.

Lawson, B. D., Kelley, A. M., & Athy, J. R. (2012). *A review of computerized team performance measures to identify military-relevant, low-to-medium fidelity tests of small group effectiveness during shared information processing* (No. USAARL-2012-11). Fort Rucker, AL: Army Aeromedical Research Laboratory.

Lawson, B. D., & Mead, A. M. (1998). The sopite syndrome revisited: Drowsiness and mood changes during real or apparent motion. *Acta Astronautica, 43*(3–6), 181–192.

Lawson, B. D., McGee, H. A., Castaneda, M. A., Golding, J. F., Kass, S. J., & McGrath, C. M. (2009). *Evaluation of several common anti-motion sickness medications and recommendations concerning their potential usefulness during special operations* (No. NAMRL-09-15). Pensacola, FL: Naval Aerospace Medicine Research Laboratory.

Lawson, B. D., & Riecke, B. E. (2014). Perception of body motion. In K. S. Hale & K. M. Stanney (Eds.), *Handbook of virtual environments: Design, implementation, and applications* (2nd ed., Chapter 7, pp. 163–195). New York, NY: CRC Press.

Lawson, B. D., Rupert, A. H., Guedry, F. E., Grissett, J. D., & Mead, A. M. (1997). The human–machine interface challenge of using virtual environment (VE) displays aboard centrifuge devices. In: M. J. Smith, G. Salvendy, & R. J. Koubek (Eds.), *Design of computing systems: Social and ergonomic considerations* (pp. 945–948). Amsterdam, the Netherlands: Elsevier Science Publishers.

Lawson, B. D., Rupert, A. H., Kelley, A. M. (2013). Mental disorders comorbid with vestibular pathology. *Psychiatric Annals, 43*(7), 324–327.

Lee, D. N., & Aronson, E. (1974). Visual proprioceptive control of standing in human infants. *Perception & Psychophysics, 15*, 529–532.

Lee, D. N., & Lishman, J. R. (1975). Visual proprioceptive control of stance. *Journal of Human Movement Studies, 1*, 87–95.

Leslie, K. R., Stickgold, R., DiZio, P., Lackner, J. R., & Hobson, J. A. (1997). The effect of optokinetic stimulation on daytime sleepiness. *Archives Italiennes de Biologie, 135*, 219–228.

Lewis, C. H., & Griffin, M. J. (1998). *Modelling the effects of deck motion on energy expenditure and motion-induced fatigue* (ISVR Contract Report No. 98/05). Southampton, England: University of Southampton.

Luxon, L. M. (1996). Posttraumatic vertigo. In R. W. Baloh & G. M. Halmagyi (Eds.), *Disorders of the vestibular system* (pp. 381–395). New York, NY: Oxford Press.

Magee, L. E., Kantor, L., & Sweeney, D. M. (1988). Simulator-induced sickness among Hercules aircrew. In *Advisory Group for Aerospace Research and Development, Motion cues in flight simulation and simulator-induced sickness* (AGARD Conference Proceedings No 433, pp. 5-1–5-8). NATO, Neuilly-Sur-Seine, France.

Map of Life Org. (2010). *Foregut fermentation in mammals*. Retrieved from http://mapoflife.org/topics/topic_573_Foregut-fermentation-in-mammals/

Markham, C. H., & Diamond, S. G. (1992). Further evidence to support disconjugate eye torsion as a predictor of space motion sickness. *Aviation, Space, and Environmental Medicine, 63*(2), 118.

Matsangas, P. (2004). *A linear physiological visual-vestibular interaction model for the prediction of motion sickness incidence* (Doctoral dissertation). Naval Postgraduate School, Monterey, CA.

McCandless, C. H., & Balaban, C. D. (2010). Parabrachial nucleus neuronal responses to off-vertical axis rotation in macaques. *Experimental Brain Research, 202*(2), 271–290.

McCauley, M. E. (Ed.). (1984). *Research issues in simulator sickness: Proceedings of a Workshop*. Washington, DC: National Academy Press.

McCauley, M. E., & Kennedy, R. S. (1976). Recommended human exposure limits for very-low-frequency vibration (TP-76-36). Pacific Missile Test Center, Point Magu, CA. In M. E. McCauley (Ed.), *Research issues in simulator sickness: Proceedings of a workshop*. Washington, DC: National Academy Press, 1984.

McCauley, M. E., & Sharkey, T. J. (1992). Cybersickness: Perception of self-motion in virtual environments. *Presence, 1*(3), 311–318.

McGuinness, J., Bouwman, J. H., & Forbes, J. M. (1984). Simulator sickness occurrence in the 2E6 air combat maneuvering simulator (ACMS). In M. E. McCauley (Ed.), *Research issues in simulator sickness: Proceedings of a workshop* (pp. 3–7). Washington, DC: National Academy Press. (Original work published in 1981.)

Mead, A. M., & Lawson, B. L. (1997, June 8–13). Psychophysical measures of motion and orientation: Implications for human interface design. Abstracted presentation to the *NASA 12th Man in Space Conference*, Washington, DC.

Meissner, K., Enck, P., Muth, E. R., Kellermann, S., & Klosterhalfen, S. (2009). Cortisol levels predict motion sickness tolerance in women but not in men. *Physiology & Behavior, 97*(1), 102–106.

Miller, E. F., & Graybiel, A. (1969). *A standardized laboratory means of determining susceptibility to Coriolis (motion) sickness* (Technical Report No. NAMI-1058). Pensacola, FL: Naval Aerospace Medical Institute.

Miller, E. F., & Graybiel, A. (1970a). *Comparison of five levels of MS severity as the basis for grading susceptibility* (Bureau of Medicine and Surgery, NASA Order R-93). Pensacola, FL: Naval Aerospace Medical Institute.

Miller, E. F., & Graybiel, A. (1970b). A provocative test for grading susceptibility to MS yielding a single numerical score. *Acta Otolarygologica, 274*(Suppl), 1–20.

Miller, J. C. (1995). Batch processing of 10,000 h of trucker driver EEG data. *Biological Psychology, 40*(1–2), 209–222.

Miller, J. W., & Goodson, J. E. (1958). *A note concerning "MS" in the 2-FH-2 Hover Trainer* (Project No. 1 17 01 11, Subtask 3, Report No. 1). Pensacola, FL: Naval School of Aviation Medicine.

Miller, J. W., & Goodson, J. E. (1960). MS in a helicopter simulator. *Aerospace Medicine, 31*, 204–212.

Mittelstaedt, H. (1996). Somatic graviception. *Biological Psychology, 42*(1), 53–74.

Mohler, B., & Welch, R. B. (2014). Adapting to virtual environments. In K. S. Hale & K. M. Stanney (Eds.), *Handbook of virtual environments: Design, implementation, and applications* (2nd ed., Chapter 26, pp. 627–646). New York, NY: CRC Press.

Money, K., Lackner, J., & Cheung, R. (1996). MS and the autonomic nervous system. In A. Miller & W. Yates (Eds.), *Vestibular autonomic regulation* (pp. 147–173). New York, NY: CRC Press.

Money, K. E. (1970). Motion sickness. *Physiological Reviews, 50*, 1–39.

Money, K. E., & Cheung, B. S. (1983). Another function of the inner ear: Facilitation of the emetic response to poisons. *Aviation, Space, and Environmental Medicine, 54*(3), 208–211.

Moore, J. R., & Maguire, D. A. (2005). Natural sway frequencies and damping ratios of trees: Influence of crown structure. *Trees-Structure and Function, 19*(4), 363–373.

Morrison, A. R., & Pompeiano, O. (1970). Vestibular influences during sleep. VI. Vestibular control of autonomic functions during the rapid eye movements of desynchronized sleep. *Archives Italiennes de Biologie, 108*(1), 154.

Morrow, G. R., & Dobkin, P. L. (1988). Anticipatory nausea and vomiting in cancer patients undergoing chemotherapy treatment: Prevalence, etiology, and behavioral interventions. *Clinical Psychology Review, 8*, 517–556.

Moss, J. D., Austin, J., Salley, J., Coats, J., Williams, K., & Muth, E. R. (2011). The effects of display delay on simulator sickness. *Displays, 32*(4), 159–168.

Moss, J. D., & Muth, E. R. (2011). Characteristics of head-mounted displays and their effects on simulator sickness. *Human Factors, 53*, 308–319.

Moss, J. D., Muth, E. R., Tyrrell, R. A., & Stephens, B. R. (2010). Perceptual thresholds for display lag in a real visual environment are not affected by field of view or psychophysical technique. *Displays, 31*, 143–149.

Muth, E. R. (2009). The challenge of uncoupled motion: Duration of cognitive and physiological aftereffects. *Human Factors: The Journal of the Human Factors and Ergonomics Society, 51*(5), 752–761.

Muth, E. R. (2010). The challenge of uncoupled motion: Duration of cognitive and physiological aftereffects. *Human Factors, 51*, 752–761.

Muth, E. R., & Lawson, B. (2003). Using flight simulators aboard ships: Human side effects of an optimal scenario with smooth seas. *Aviation, Space, and Environmental Medicine, 74*(5), 497–505.

Muth, E. R., & Moss, J. D. (2009). Feelings of self motion under varying helmet mounted display conditions and its relation to simulator sickness [PowerPoint Slides]. In *Proceedings of the Second International Symposium on Visual IMage Safety (VIMS)*, Utrecht, Netherlands. Retrieved from http://www.vims2009.org/Abstracts/Muth-ppt.pdf

Muth, E. R., Stern, R. M., Thayer, J. F., & Koch, K. L. (1996). Assessment of the multiple dimensions of nausea: The Nausea Profile. *Journal of Psychosomatic Research, 40*, 511–520.

Muth, E. R., Walker, A. D., & Fiorello, M. (2006). Effects of uncoupled motion on performance. *Human Factors: The Journal of the Human Factors and Ergonomics Society, 48*(3), 600–607.

Nachum, Z., Shupak, A., Letichevsky, V., Ben-David, J., Tal, D., & Tamir, A., (2004). Mal de debarquement and posture: Reduced reliance on vestibular and visual cues. *The Laryngoscope, 114*(3), 581–586. doi: 10.1097/00005537-200403000-00036

Neilson, B. N., & Brill, J. C. (2012, May). *Developing a latent trait model to predict sopite syndrome in aerospace systems*. Poster session presented at the *83rd Annual Scientific Meeting of the Aerospace Medical Association*, Atlanta, GA.

Neilson, W. A., Knott, T. A., & Carhart, P. W. (1956). *Webster's New International Dictionary of the English Language* (2nd ed.). Springfield, MA: G. & C. Merriam Company.

Newman, M. C. (2009). *A multisensory observer model for human spatial orientation perception* (Doctoral dissertation). Massachusetts Institute of Technology, Cambridge, MA.

Newman, M. C., Lawson, B. D., Rupert, A. H., & McGrath, B. J. (2012, August 15). The role of perceptual modeling in the understanding of spatial disorientation during flight and ground-based simulator training. In *Proceedings of the American Institute of Aeronautics and Astronautics* (14 pp.), Minneapolis, MN.

Nicogossian, A. E., Huntoon, C. L., & Pool, S. L. (1989). *Space physiology and medicine*. Philadelphia, PA: Lea & Febiger.

Nicogossian, A. E., & Parker, J. F. (Eds.). (1984). Space and physiology medicine. In M. E. McCauley (Ed.), *Research Issues in Simulator Sickness: Proceedings of a Workshop* (p. 1). Washington, DC: National Academy Press. (Original work published in 1982.)

Nieuwenhujsen, J. H. (1958). *Experimental investigations on seasickness* (MD thesis). University of Utrecht, Drukkerij Schotanus & Jens, Utrecht, the Netherlands.

Nomura, Y., Toi, T., Kaneita, Y., Masuda, T., Shighara, S., & Ikeda, M., (2012, June). Factors of post earthquake dizziness syndrome in Japan. In *27th Barany Society Meeting*, Uppsala, Sweden. Abstract retrieved from http://www.barany2012.se.pdf/barany-abstracts-web.pdf

O'Hanlon, J. F., &McCauley, M. E. (1974). MS incidence as a function of the frequency and acceleration of vertical sinusoidal motion. *Aviation, Space, and Environmental Medicine, 45*, 366–369.

Oman, C. M. (1990). Motion sickness: A synthesis and evaluation of the sensory conflict theory. *Canadian Journal of Physiology and Pharmacology, 68*(2), 294–303.

Oman, C. M. (2012). Are evolutionary hypotheses for motion sickness "just-so" stories? *Journal of Vestibular Research, 22*(2), 117–127.

Oman, C. M., Rague, B. W., & Rege, O. U. (1987). *Standard definitions for scoring acute motion sickness using the Pensacola Diagnostic Index Method. Appendix B, Symptom scoring definitions* (Man-Vehicle Laboratory Report, pp. 120–128). Cambridge, MA: Massachusetts Institute of Technology.

Panos, K. (2005). *The theories of motion sickness—A review*. Retrieved from http://airforcemedicine.a.fms.mil/idc/groups/public/documents/afms/ctb_080202.pdf

Pennebaker, J. W. (1999). Psychological factors influencing the reporting of physical symptoms. In A. A. Stone, J. S. Turkkan, C. A. Bachrach, J. B. Jobe, H. S. Kurtzman, & V. S. Cain (Eds.), *The science of self-report: Implications for research and practice* (pp. 299–316). Mahwah, NJ: Erlbaum Publishers. A summary of research on how self-reports of physical symptoms and emotions are affected by perceptual, cognitive, biological, and personality factors.

Pethybridge, A. D. (1982). Study of association in unsymmetrical electrolytes by conductance measurements. *Zeitschrift für Physikalische Chemie, 133*(2), 143–158.

Probst, T. H., & Schmidt, U. (1998). The sensory conflict concept for the generation of nausea. *Journal of Psychophysiology, 12*, 34–49.

Quix, F. H. (1922). Le mal de mer, le mal des aviateurs. *Mono. Oto-Rhino-Laryngol. Internat.* 8: 825–950. In A. Graybiel, & J. Knepton (1976). Sopite Syndrome: A sometimes sole manifestation of motion sickness (pp. 879). *Aviation, Space, and Environmental Medicine, 47*(8), 873–882.

Rash, C. E. (2004). Awareness of causes and symptoms of flicker vertigo can limit ill effects: *Human Factors and Aviation Medicine, 51*(2), March–April 2004: Flight Safety Foundation (online).

Reason, J. T. (1975). An investigation of some factors contributing to individual variation to MS susceptibility. In J. T. Reason & J. J. Brand (Eds.), *Motion sickness* (p. 7). New York, NY: Academic Press Inc. (Original work published 1967.)

Reason, J. T., & Brand, J. J. (1975). *Motion sickness*. London, U.K.: Academic Press.

Reason, J. T., & Diaz, E. (1971). *Simulator sickness in passive observers* (Technical Report No. 1310). London, U.K.: Flying Personnel Research Committee.

Reason, J. T., & Graybiel, A. (1970a). Progressive adaptation to Coriolis accelerations associated with 1-RPM increments in the velocity of the slow rotation room. *Aerospace Medicine, 41*(1), 73–79.

Reason, J. T., & Graybiel, A. (1970b). Changes in subjective estimates of well-being during the onset and remission of MS symptomatology in the slow rotation room. *Aerospace Medicine, 41*, 166–171. (This study is also described in Reason, J. T. & Brand, J. J., *Motion sickness* (pp. 74–79). New York: Academic Press Inc.)

Reavley, C. M., Golding, J. F., Cherkas, L. F., Spector, T. D., & MacGregor, A. J. (2006). Genetic influences on motion sickness susceptibility in adult women: A classical twin study. *Aviation, Space, and Environmental Medicine, 77*(11), 1148–1152.

Regan, E. C., & Price, K. R. (1994). The frequency of occurrence and severity of side-effects of immersion virtual reality. *Aviation, Space, and Environmental Medicine, 65*, 527–530.

Regan, E. C., & Ramsey, A. D. (1996). The efficacy of hyoscine hydrobromide in reducing side-effects induced during immersion in virtual reality. *Aviation, Space, and Environmental Medicine, 67*(3), 222–226.

Riccio, G. E., & Stoffregen, T. A. (1991). An ecological theory of motion sickness and postural instability. *Ecological Psychology, 3*(3), 195–240.

Robinett, W. (1992). Synthetic experience: A proposed taxonomy. *Presence, 1*(2), 229–247.

Robinson, M. E., Riley, J. L., Myers, C. D., Papas, R. K., Wise, E. A., Waxenberg, L. B., & Fillingim, R. B. (2001). Gender role expectations of pain: relationship to sex differences in pain. *The Journal of Pain, 2*(5), 251–257.

Rohleder, N., Otto, B. R., Wolf, J. M., Klose, J., Kirschbaum, C., Enck, P., & Klosterhalfen, S. (2006). Sex-specific adaptation of endocrine and inflammatory responses to repeated nauseogenic body rotation. *Psychoneuroendocrinology, 31*(2), 226–236.

Rose, P. N. (2004). *Motion sickness: A brief review and a proposed methodology for a research program* (No. ARL-TN-0232). Aberdeen Proving Ground, MD: Army Research Laboratory.

Salvatore, S., Valsiner, J., Strout-Yagodzinski, S., & Clegg, J. (Eds.). (2010). YIS: Yearbook of Idiographic Science, Volume 1/2008. Rome, Italy: Firera & Liuzzo Publishing.

Sadowski, W., & Stanney, K. (2002). Presence in virtual environments. In K. M. Stanney (Ed.), *Handbook of virtual environments: Design, implementation and applications* (pp. 791–806). Mahwah, NJ: Lawrence Erlbaum Associates, Inc.

Schatz, S. (2014). Presence in virtual environments. In K. S. Hale & K. M. Stanney (Eds.), *Handbook of virtual environments: Design, implementation, and applications* (2nd ed., Chapter 34, pp. 855–870). New York, NY: CRC Press.

Schwab, R. S. (1943). Chronic seasickness. *Annals of Internal Medicine, 19*, 28–35. In J. T. Reason, & J. J. Brand (1975), *Motion sickness*. London, U. K.: Academic Press.

Smart, L. J., Pagulayan, R. J., & Stoffregen, T. A. (1998). Self-induced motion sickness in unperturbed stance. *Brain Research Bulletin, 47*(5), 449–457.

Smith, S. A., Kass, S. J., & Lawson, B. (2003, March). *Gender differences in motion sickness and sopite syndrome*. Paper presented at the 49th Annual Meeting of the Southeastern Psychological Association, New Orleans, LA.

So, R. H. Y., & Lo, W. T. (1999). Cybersickness: An experimental study to isolate the effects of rotational scene oscillations. In *Proceedings of the IEEE Virtual Reality Conference* (pp. 237–241). Los Alamitos, CA: IEEE Computer Society.

Stanney, K. M., & Hash, P. (1998). Locus of user-initiated control in virtual environments: Influences on cybersickness. *Presence, 7*(5), 447–459.

Stanney, K. M., & Kennedy, R. S. (1997). Cybersickness is not simulator sickness. In *Proceedings of the Human Factors and Ergonomics Society 41st Annual Meeting* (pp. 1138–1142). Santa Monica, CA: Human Factors and Ergonomics Society.

Stanney, K. M., Kennedy, R. S., Drexler, J. M., & Harm, D. H. (1999). MS and proprioceptive aftereffects following virtual environment exposure. *Applied Ergonomics, 30*, 27–38.

Stanney, K. M., Mourant, R. R., & Kennedy, R. S. (1998). Human factors issues in virtual environments: A review of the literature. *Presence, 7*(4), 327–351.

Stanney, K. M., Salvendy, G., Deisinger, J., DiZio, P., Ellis, S., Ellison, J., …Witmer, B. (1998). Aftereffects and sense of presence in virtual environments: Formulation of a research and development agenda. *International Journal of Human–Computer Interaction, 10*(2), 135–187.

Stephens, D., & Dudley, T. R. (2004). The drunken monkey hypothesis. *Natural History, 40*, 44.

Stern, R. M., Hu, S., Anderson, R. B., Leibowitz, H. W., & Koch, K. L. (1990). The effects of fixation and restricted visual field on vection-induced MS. *Aviation, Space, and Environmental Medicine, 61*, 712–715.

Stern, R. M., Hu, S. E. N. Q. I., Uijedehaage, S. H. J., Muth, E. R., Xu, L. H. & Kock, K. L. (1996). Asian hypersusceptibility to motion sickness. *Human Heredity, 46*(1), 7–14.

Stockhorst, U., Enck, P., & Klosterhalfen, S. (2007). Role of classical conditioning in learning gastrointestinal symptoms. *World Journal of Gastroenterology, 13*(25), 3430–3437.

Stoffregen, T. A., & Smart, L. J. (1998). Postural instability precedes MS. *Brain Research Bulletin, 47*(5), 437–448.

Storms, R. L. (2002). Auditory-visual cross-modality interaction and illusions. In K. M. Stanney (Ed.), *Handbook of virtual environments: Design, implementation, and applications* (pp. 455–470). New York, NY: CRC Press.

Takahashi, M., Saito, A., Okada, Y., Takei, Y., Tomizawa, I., Uyama, K., & Kanazaki, J. (1991). Locomotion and motion sickness during horizontally and vertically reversed vision. *Aviation, Space, and Environmental Medicine, 62*, 136–140.

Teixeira, R. A., & Lackner, J. R. (1979). Optokinetic MS: Attenuation of visually induced apparent self-rotation by passive head movements. *Aviation, Space, and Environmental Medicine, 50*(3), 264–266.

Templeman, J., & Denbrook, P. (2014). Locomotion interfaces. In K. S. Hale & K. M. Stanney (Eds.), *Handbook of virtual environments: Design, implementation, and applications* (2nd ed., Chapter 15, pp. 233–256). New York, NY: CRC Press.

Ter Vrugt, D., & Peterson, D. R. (1973). The effects of vertical rocking frequencies on the arousal level in two-month-old infants. *Child Development, 44*, 205–209.

Treisman, A. (1977). Focused attention in the perception and retrieval of multidimensional stimuli. *Attention, Perception, & Psychophysics, 22*(1), 1–11.

Tschermak, A. (1931). Optischer raumsinn [Optical sense of space]. In A. Bethe, G. Bergmann, G. Emden, & A. Ellinger (Eds.), *Handbuch der normalen undpathologischen physiologic* (pp. 834–1000). Berlin, Germany: Springer-Verlag.

Tyler, D. B., & Bard, P. (1949). Motion sickness. *Physiological Reviews, 29*(4), 311–369.

Uliano, K. C., Lambert, E. Y., Kennedy, R. S., & Sheppard, D. J. (1986). *The effects of asynchronous visual delays on simulator flight performance and the development of simulator sickness symptomatology* (Technical Report No. NAVTRASYSCEN 85-D-0026-1). Orlando, FL: Naval Training Systems Command.

Ungs, T. J. (1988). Simulator induced syndrome in Coast Guard aviators. *Aviation, Space, and Environmental Medicine, 59*, 267–272.

Van Cott (1990). Lessons from simulator sickness studies, In National Research Council Committee on Vision and Working Group on Wraparound Visual Displays, *Proceedings of a Conference on Wraparound Visual Displays Held in Waltham, Massachusetts on 14–15 January 1988 (Motion Sickness, Visual Displays, and Armored Vehicle Design)* (p. 77–85). NRC Report BRL-CR-629. Aberdeen Proving Ground, MD: Army Ballistic Research Laboratory.

Velicer, W. (2010). Applying idiographic research methods: Two examples. In *Proceedings of the 8th International Conference on Teaching Statistics*, Ljubljana, Slovenia. Retrieved from www.stat.auckland.ac.nz/~iase/publications/.../ICOTS8_4F3_VELICER.php

Vibert, N., Hoang, T., Gilchrist, D. P. D., MacDougall, H. G., Burgess, A. M., Roberts, R. D., … Curthoys, I. S. (2006). Psychophysiological correlates of the inter-individual variability of head movement control in seated humans. *Gait & Posture, 23*(3), 355–363.

Walker, A., Muth, E. R., Switzer, F. S., & Hoover, A. W. (2010). The role of head movements in simulator sickness generated by a virtual environment. *Aviation, Space, and Environmental Medicine, 81*, 929–934.

Walker, A. D., Gomer, J. A., & Muth, E. R. (2007, October). The effect of input device on performance of a driving task in an uncoupled motion environment. In *Proceedings of the Human Factors and Ergonomics Society Annual Meeting* (Vol. 51, No. 27, pp. 1627–1630). New York, NY: SAGE Publications.

Wann, J. P., & Mon-Williams, M. (1997, May). Health issues with virtual reality displays: What we do know and what we don't. *ACMSIEERAPH Computer Graphics, 31*(2), 53–57.

Wann, J. P., & Mon-Williams, M. (2002). Measurement of visual aftereffects following virtual environment exposure. In K. M. Stanney (Ed.), *Handbook of virtual environments: Design, implementation, and applications* (pp. 731–749). Hillsdale, NJ: Lawrence Erlbaum Associates.

Warwick-Evans, L., & Beaumont, S. (1995). An experimental evaluation of sensory conflict versus postural control theories of motion sickness. *Ecological Psychology, 7*(3), 163–179.

Warwick-Evans, L. A., Symons, N., Fitch, T., & Burrows, L. (1998). Evaluating sensory conflict and postural instability. Theories of motion sickness. *Brain Research Bulletin, 47*(5), 465–469.

Watson, G. S. (1998). The effectiveness of a simulator screening session to facilitate simulator sickness adaptation for high-intensity driving scenarios. In *Proceedings of the 1998 IMAGE Conference*. Chandler, AZ: IMAGE Society.

Watson, N. W., Wells, T. J., & Cox, C. (1998). Rocking chair therapy for dementia patients: Its effect on psychosocial well-being and balance. *American Journal of Alzheimer's Disease, 13*(6), 296–308.

Watt, D. G. D. (1992). What is motion sickness? *Annals of the New York Academy of Science, 656*, 660–667.

Webb, N. A., & Griffin, M. J. (2002). Optokinetic stimuli: motion sickness, visual acuity, and eye movements. *Aviation, space, and environmental medicine, 73*(4), 351–358.

Webb, N. A. (2000). *Visual acuity, eye movements, motion sickness and the illusion of motion, with optokinetic stimuli* (Doctoral dissertation). University of Southampton, Southampton, UK.

Welch, R. B. (1978). *Perceptual modification: Adapting to altered sensory environments*. New York, NY: Academic Press.

Wertheim, A. H. (1998). Working in a moving environment. *Ergonomics, 41*(12), 1845–1858.

Westra, D. P., Sheppard, D. J., Jones, S. A., & Hettinger, L. J. (1987). *Simulator design features for helicopter shipboard landings* (Technical Report No. TR-87-041 AD-A203 992, p. 61). Orlando, FL: Naval Training Systems Center.

Wiker, S. F., & Pepper, R. L. (1978). Changes in crew performance, physiology and affective state due to motion aboard a small monohull vessel (No USCG-D-75-78). In S. F. Wiker, R. L. Pepper, & M.E. McCauley. (1980). *A vessel class comparison of physiological, affective state, and psychomotor performance changes in men at sea* (No. USCG-D-07-81). Washington, DC: Office of Research and Development.

Wiker, S. F., Pepper, R. L., & McCauley, M. E. (1980). *A vessel class comparison of physiological, affective state and psychomotor performance changes in men at sea* (No. USCG-D-07-81). Washington, DC: Office of Research and Development.

Wilson, J. R. (1997). Virtual environments and ergonomics: Needs and opportunities. *Ergonomics, 40*, 1057–1077.

Witmer, B. G., & Singer, M. J. (1998). Measuring presence in virtual environments: A presence questionnaire. *Presence: Teleoperators and Virtual Environments, 7*(3), 225–240.

Wood, C. D., Graybiel, A., & McDonough, R. G. (1966). Clinical effectiveness of anti-motion-sickness drugs. *Journal of the American Medical Association*, February, 187–190.

Wood, C. D., Stewart, J. J., Wood, M. J., Manno, J. E., Manno, B. R., & Mims, M. E. (1990). Therapeutic effects of antiMS medications on the secondary symptoms of motions sickness. *Aviation, Space, and Environmental Medicine, 61*, 157–161.

Wood, R. W. (1975). The "haunted swing" illusion. In J. T. Reason & J. J. Brand (Eds.), *Motion sickness* (p. 110). New York, NY: Academic Press Inc. (Original work published 1895.)

Wood, S. J. (2002). Human otolith–ocular reflexes during off-vertical axis rotation: Effect of frequency on tilt-translation ambiguity and motion sickness. *Neuroscience Letters, 323*(1), 41–44.

Wood, S. J., Reschke, M. F., Harm, D. L., Paloski, W. H., & Bloomberg, J. J. (2012). Which way is up? Lessons learned from space shuttle sensorimotor research. Abstracted Presentation, *83rd Annual Scientific Meeting of the Aerospace Medical Association*, Atlanta, GA, 13–17.

Woodward, S., Tauber, E. S., Spielmann, A. J., & Thorpy, M. J. (1990). Effects of otolithic vestibular stimulation on sleep. *Sleep, 13*(6), 533–537.

Wright, M. S., Bose, C. L., & Stiles, A. D. (1994). The incidence and effects of MS among medical attendants during transport. *Journal of Emergency Medicine, 13*, 15–20.

Wu, W. (2011). Measuring digital system latency from sensing to actuation at continuous 1 millisecond resolution (Master's thesis). Clemson, SC: Clemson University.

Yano, F., Yazu, Y., Suziki, S., Kasamatsu, K., Idogawa, K., & Ninomija, S. P. (1997). Analysis of driver's eye directions at vehicle driving on the highway. In M. J. Smith & G. Salvendy (Eds.), *Proceedings from the 7th Annual Conference on Human–Computer Interaction* (p. 14). San Francisco, CA: HCI International.

Yates, B. J., Miller, A. D., & Lucot, J. B. (1998). Physiological basis and pharmacology of motion sickness: An update. *Brain Research Bulletin, 47*(5), 395–406.

Young, J. W., Fernandez, D., & Fleagle, J. G. (2010). Ontogeny of long bone geometry in capuchin monkeys (*Cebus albifrons* and *Cebus apella*): Implications for locomotor development and life history. *Biology Letters, 6*(2), 197–200.

Zielinski, S. (2011). *The alcoholics of the animal world*. Retrieved from http://blogs.smithsonianmag.com/science/2011/09/the-alcoholics-of-the-animal-world/

24 Motion Sickness Scaling

Ben D. Lawson

CONTENTS

24.1 INTRODUCTION

This chapter describes *motion sickness* (*MS*) scaling, that is, to the quantification of MS signs and symptoms via questionnaire rating scales. The early development of MS scales is described, along with common currently used scales for assessing past MS history or current state. Methods are discussed for obtaining overall single-answer ratings of discomfort or ratings comprised of multiple clinical signs and subjective symptoms of MS. Some scaling comments are provided that are specific to *simulator sickness*, *cybersickness*, or *sopite syndrome* (Lawson & Mead, 1998; McCauley, 1984; McCauley & Sharkey, 1992). The purpose of this chapter is to assist researchers seeking to choose an existing scale most appropriate to their needs, rather than to offer complete coverage of theoretical scaling issues or scale development.

24.2 HISTORY AND DEVELOPMENT OF THE MOST WIDELY USED MS RATING SCALES

24.2.1 CONSENSUS REGARDING THE NEED FOR A PREVOMITING ENDPOINT

An important aspect of a MS study is the criterion that determines how sick a (still willing) participant should be before the investigator ceases testing. Consideration of this issue was an important factor in the creation of the first MS rating scales. Early researchers struggled with the identification of a symptom-based prevomiting endpoint for the termination of nauseogenic testing. Vomiting has the advantage of being a clearly observable endpoint that permits ratio scaling. However, vomiting is not necessarily a good predictor of a participant's relative stress level, since some participants can be in great distress and still not vomit, while others vomit early on in the testing before nausea is well developed. When vomiting is used as an endpoint, systemic side effects (such as the fluid loss) can confound efforts at continued measurement of physiological responses or repeated testing of MS susceptibility. Moreover, the unpleasantness of vomiting complicates human-use approval, participant recruitment, etc. Some of the early attempts to devise a list of prevomiting signs and symptoms and scale them collectively were made by Hemingway (1943) and Alexander, Cotzin, Hill, Ricciuti, and Wendt (1945a). The authors obtained numerical scores for an individual's overall discomfort based on a number of signs and symptoms in addition to vomiting, but they did not employ a multisymptom checklist wherein data concerning each sign or symptom could be tabulated and analyzed separately before a final score was tabulated. Also, it was not the main purpose of these investigators to develop a prevomiting experimental endpoint.

To determine a prevomiting endpoint, the experimenter must select the symptom severity at which the experiment will be terminated even if the subject is willing to continue. Multisymptom scores have been used to determine the endpoint (Miller & Graybiel, 1970b); currently, it is more common to terminate at the first report of *moderate nausea* of a nontransient duration, (for example, lasting for 1 min) so that the experiment is not be stopped for a mere *wave* of nausea lasting a few seconds (Lawson et al., 2009). There are obvious practical reasons for terminating before *severe nausea* is reported. Experimenters often terminate testing upon the appearance of minimal or moderate nausea because an *avalanche phenomenon* (Reason & Graybiel, 1970, p. 168) can occur where vomiting happens before the experiment can be stopped. Retching or vomiting is quite unpleasant for the participant (and the investigator) and can lead to data loss during the run, difficulty gathering sufficient data after the run (e.g., for postrun performance tests), difficulty getting the participant to return for repeated measures, or difficulty recruiting subsequent participants by word of mouth.

It should be noted that vomiting can be a problem for resistant as well as susceptible participants; in Reason and Graybiel (1970), participants who exhibited greater resistance to the stimulus (by tolerating more head movements and longer testing) also tended to proceed very suddenly from overall *well-being* ratings of 2–4 all the way up to 10 (the maximum discomfort rating). Another reason to not test beyond moderate nausea is that the very act of stopping a rotating chair (the most common device used for controlled study of MS) at the end of testing is in itself a provocative stimulus. Therefore, chair deceleration should commence while the participant reports no-worse-than-moderate nausea, and deceleration should be very gradual to avoid making the participant feel worse.

24.2.2 ORIGINAL AND MOST WIDELY USED MULTIPLE-SYMPTOM CHECKLISTS

The first diagnostic categorization of multiple motion-related symptoms that saw widespread use was initiated by Ashton Graybiel and developed by Drs. Kennedy, Clark, Johnson, Miller, and a number of other researchers working at the Pensacola Naval Air Station, FL, mostly as part of studies of MS and performance during exposure to a rotating chair or a rotating room. These efforts started with a rudimentary checklist that recorded the presence (*yes or no*) of a number of distinct signs or symptoms, including general *malaise, sweating, nausea, vomiting, dizziness, headache, apathy*, and the *number of hours slept* during the day (Graybiel, Clark, & Zarriello, 1960; Kennedy & Graybiel, 1962). Soon after, the list of signs and symptoms was expanded and rated on a 0–3 scale (Graybiel & Johnson, 1963). Many related forms of this diagnostic categorization system were produced during these prolific years. For example, the *Pensacola Motion Sickness Questionnaire*, or MSQ (so named in Kennedy & Graybiel, 1965), consisted of more than 20 signs and symptoms initially presented in Hardacre and Kennedy (1963), which expanded to 33 MSQ items in Kennedy and Graybiel (1965, cited in Kennedy, Drexler, & Kennedy, 2010). The MSQ reached a relatively mature form in Kennedy, Tolhurst, and Graybiel (1965), which is one of the best of the early guides to the MSQ. Kennedy et al. (1965) categorized minor, major, and pathognomic* diagnostic criteria, which were scored variously and tabulated as a single malaise score.

A shorter table was published around the same time by Wood, Graybiel, and McDonough (1966); it employed seven criteria, including the pathognomic sign of *vomiting*, the major sign of *retching*

---

* Also spelled pathognomonic, referring to characteristic and unequivocal signs or symptoms of MS.

(later upgraded to a pathognomic sign in Miller & Graybiel, 1970b), and a number of other signs and symptoms whose classification as minor or major depended on their severity. It is likely that the majority of MS investigators active today first encountered these diagnostic criteria not in the original sources but in the Graybiel, Wood, Miller, and Cramer (1968) criteria reproduced on p. 81 of Reason and Brand's well-known book, *Motion Sickness* (1975). The mature MS criteria in Miller and Graybiel (1970b) include what are now considered by most researchers as the *cardinal signs or symptoms of MS*, namely, *nausea, increased salivation, pallor, cold sweating*, and *drowsiness*. These five criteria are believed to contribute strongly to the participant's overall malaise score and are considered the cardinal characteristics of motion discomfort. Such symptoms have also been accepted as criteria for *simulator sickness* (Kennedy, Fowlkes, & Hettinger, 1989).

Eventually, several mature forms of the diagnostic criteria for MS emerged. One mature set of criteria was produced by Miller and Graybiel (1970b), who presented a diagnostic categorization table along with an accompanying sheet for scoring signs and symptoms of motion discomfort. The mature form of the criteria included some signs and symptoms of presumed lesser importance, including *epigastric awareness, epigastric discomfort, flushing/subjective warmth, headache*, and *dizziness* with eyes closed or open.

The diagnostic criteria, that culminated in the Miller and Graybiel (1970b) paper, have been referred to by various names, including the *Pensacola diagnostic criteria* (PDC), the *Pensacola diagnostic index*, the *Pensacola diagnostic categorization*, the *Pensacola diagnostic rating scale*, the *Graybiel scale*, and the *Miller and Graybiel diagnostic criteria*. The original publications referred to *diagnostic criteria*, but other appellations are provided here for readers who wish to obtain comprehensive search terms for locating published applications of the criteria after 1970. Since the symptoms, that culminated in Miller and Graybiel's (1970b) report, do not yield a formal scale in the modern sense but instead should be considered reliable and established criteria of MS, the present report will refer to them using the name PDC.

Since the Miller and Graybiel (1970b) PDC can be difficult to apply correctly when presented in their original form, a simplified representation is presented in Table 24.1, rendered as an experimenter's

TABLE 24.1

Simplified Pensacola Diagnostic Criteria (SPDC) of Motion Sickness

	None		Minimal I	Minor II	Major III	Pathognomic
		Epigastric Awareness	Epigastric Discomfort	Nausea 1 (Minimal)	Nausea 2–3 (Moderate to Severe)	Vomiting, Retching
	0	1	2	4	8	16
Nausea	None	(NA)	Minimal	Minor	Major	(NA)
Flushing, warmth	0		0	1	1	
Dizziness (eyes closed)	0		0	1	1	
Dizziness (eyes open)	0		0	0	1	
Headache	0		0	1	1	
Drowsiness	0		2	4	8	
Cold sweating	0		2	4	8	
Pallor	0		2	4	8	
Increased salivation	0		2	4	8	

Sources: Adapted from Lawson, B.D. et al., Signs and symptoms of human syndromes associated with synthetic experiences, *Handbook of Virtual Environments: Design, Implementation, and Applications*, pp. 589–618, 2002.
Direct questions to benton.d.lawson.cin@mail.mil.

checklist for the convenient scoring of a single sickness assessment (e.g., one rating of discomfort experienced during one sudden stop following rotation or one Coriolis cross-coupling head movement while rotating). The *simplified Pensacola diagnostic criteria* (SPDC) in Table 24.1 are derived from Lawson, Graeber, Mead, and Muth (2002) and Miller and Graybiel (1970a, 1970b), with earlier contributions from Graybiel et al. (1968), Graybiel (1969), and others. Graybiel, Miller, and their colleagues devised a weighted scoring procedure that recognizes six distinct levels of severity for symptoms indicative of the nausea syndrome and two to four distinct levels of severity for nonnausea symptoms. The original table of criteria is difficult to use as a scoring checklist, and the present author has seen them applied incorrectly at times, so they are simplified in Table 24.1 to facilitate the accurate scoring of a single MS assessment event. It can be seen in Table 24.1 that *nausea, drowsiness, cold sweating, pallor,* and *increased salivation* (the aforementioned cardinal signs/symptoms) are assigned greater weights than the remaining symptoms. Cumulative points are scored for all symptoms and then added to distinguish between different levels of general *malaise*. For example, a cumulative score of eight could be reached by a participant reporting moderate or severe nausea, but an eight would also be assigned to a participant whose only symptom was severe drowsiness. A total score of ≥ 16 points = *frank sickness*, 8–15 points = *severe malaise*, 5–7 points = *moderate malaise A*, 3–4 points = *moderate malaise B*, and 1–2 points = *slight malaise*. These are overall sickness descriptions; individual symptoms are rated *none, minimal, minor, or major*.

Versions of the PDC have been used extensively over the years. The criteria have been used by various researchers at the Pensacola Naval Air Station, Defense and Civil Institute of Environmental Medicine (Canada), National Aeronautics and Space Administration, Massachusetts Institute of Technology, Pennsylvania State University, Brandeis University, and various European laboratories.

The PDC has changed little over the years, although the way it is scored has varied (Lawson, 1993). By contrast, the afore mentioned Kennedy et al. (1965) scale continued to undergo a series of more formal and significant modifications over the years. A notable offshoot of the original Kennedy et al. (1965) scale was the MS severity scale of Wiker, Kennedy, McCauley, and Pepper (1979), which added several more signs and symptoms and offered one of the more convincing attempts to establish the reliability and validity of subjective motion discomfort reporting. The scales derived from Kennedy et al. (1965) have varied from 20 to 33 items over the years (Kennedy et al., 2010), eventually emerging in the form of the *Simulator Sickness Questionnaire* (SSQ) (Lane & Kennedy, 1988), which has seen widespread use in military simulator studies and has become the most popular measurement technique for the assessment of signs and symptoms associated with simulators and other synthetic experiences (e.g., Kennedy, Lane, Berbaum, & Lilienthal, 1993). Perhaps the final version of the 16-descriptor SSQ is described in detail by Kennedy et al. (1993). Kennedy et al. (2010) explain the rationale of the SSQ, which is to identify the 16 items from the MSQ that are the best indicators of *simulator sickness*, while still retaining 27 (28 if one counts dizziness eyes open/closed as two items) of the original MSQ items to preserve the psychometrics of the original factor analysis. While only the first 16 items are scored for simulator studies, the full list of items can be exploited for other types of MS studies.

24.2.3 MODIFICATIONS OF THE ORIGINAL MS SCALES

There is little reason to think that the MS criteria of Kennedy et al. (1965) or Miller and Graybiel (1970b) could be made much more reliable (Calkins, Reschke, Kennedy, & Dunlap, 1987; Kennedy, Dunlap, & Fowlkes, 1990; Miller & Graybiel 1970a, 1970b; Reason & Graybiel, 1970; Wiker et al., 1979). Nevertheless, there is some question as to the metric assumptions made in obtaining the numerical summary scores of overall motion discomfort yielded by these methods (Bock & Oman, 1982). Reason and Diaz (1971) were among the first researchers to publish a *symptom checklist* and scoring procedure that used criteria nearly identical to those in Miller and Graybiel (1970b) but did not assume ratio scale data. During the development of the SSQ, Lane and Kennedy (1988) also dropped many of the metric assumptions in the original Kennedy et al. (1965) scale and opted for a simple 0–3 scale of severity for all signs and symptoms.

Lawson and colleagues (Lawson, 1993; Lawson et al., 2002) similarly created a 0–3 version of the Miller and Graybiel PDC that minimizes several logical but arbitrary and not-fully-proven aspects of the original rating scale. For example, the unequal weighting of signs and symptoms of equally rated severity is eliminated, as is the original doubling of points with increases in severity of certain symptoms (but not others). The latest version of these *modified Pensacola diagnostic criteria* (MPDC) is shown in Table 24.2. The minimal rating of zero is kept to make the bottom anchor comparable to the original scale. It could be asked whether including a zero in past MS scales is optimal (or instead presents too strong an anchor for the scale). The present author thinks that a logical argument for a zero anchor can be made for a MS scale where 0 = no symptom (e.g., the SSQ), eventhough starting with "1" is generally preferable for Likert-type scales of agreement or endorsement of a statement. Practical considerations figure into scale modifications since changing established scales may limit future comparison to previous data.

The MPDC do not use the general *malaise* categories to characterize total sickness or to decide upon a MS endpoint for testing. The peak total sickness score is simply the sum of the severity ratings obtained at the subject's peak point of sickness, while the testing endpoint is usually based on the subject having reported *moderate nausea* (Lawson et al., 2009).

Note that in the SPDC (Table 24.1) and the original PDC severity ratings of Miller and Graybiel (1970b), 0 = *none*, 1 = *minimal*, and 3 = *major*; however, in the MPDC (Table 24.2), *moderate* replaces *minimal* on the rationale that *moderate* represents a semantic label more likely to be equidistant from *minimal* and *major*, that is, more appropriate as a semantic choice for the midpoint between scale extremes. Equidistant labels are desirable for Likert-type scales, and *moderately* is known to be equidistant from *slightly* and *very* (Dobson & Mothersill, 1979), the most comparable labels in Dobson and Mothersill to *minimal* and *major*, respectively.

A balanced scale with equal intervals is generally desirable, but it is sometimes logical to intentionally employ semantically similar descriptors near the low (or high) end of a scale, for example, when the entire scale is not likely to be used otherwise because most responses would group at the

---

TABLE 24.2
Modified Pensacola Diagnostic Criteria (MPDC) of Motion Sickness

Stomach Awareness or Discomfort[a]	No	Yes		
	None	**Minimal**	**Moderate**	**Major**
Nausea	0	1	2	3
Increased salivation	0	1	2	3
Cold sweating	0	1	2	3
Pallor	0	1	2	3
Drowsiness	0	1	2	3
Headache	0	1	2	3
Flushing/warmth	0	1	2	3
Dizziness (eyes: closed/open?)	0	1	2	3
(Circle which or rate both[b])	0	1	2	3
Retching or vomiting?	Tally of discrete events and their clock times:			
Number of minutes of stimulation tolerated (termination time):				

Source: Adapted from Lawson, B.D. et al., Signs and symptoms of human syndromes associated with synthetic experiences, *Handbook of Virtual Environments: Design, Implementation, and Applications*, pp. 589–618, 2002. Direct questions to ben.lawson@amedd.army.mil.

[a] It is not mandatory to measure stomach awareness/discomfort, but it is recommended if time permits for the reasons outlined in Section 24.2.4 of this chapter (Lawson, 2014a).

[b] Most experiments will only require one rating of dizziness based on whether the protocol involves eyes open or closed instructions throughout.

low or high end. In such a case, sensitivity of the scale may be harmed unless an unbalanced scale is considered (Parasuraman, Grewel, & Krishnan, 2006). Such an approach could perhaps be justified for *simulator sickness*, which yields lower average ratings than other forms of MS (Lawson, 2014); however, the aforementioned rotation-based MS studies used to develop the PDC were very sickening, and there is no evidence in these studies that the similarity between *minimal* and *minor* in the PDC was an intentional choice intended to correct low average ratings. In fact, such an approach does not appear to have been chosen with the SSQ either, which (like the MPDC) uses *moderate* as the label between *slight* and *severe*. It would be interesting in future research to explore the utility of unbalanced MS scales, for example, semantically similar labels at the low end of scales (especially those that do not already employ other low-severity descriptors such as *stomach awareness* or *stomach discomfort*).

Stomach symptoms and nausea are assessed first in the current version of the MPDC (Table 24.2). These are probably the most important symptoms to record when time is limited for assessment or the participant is starting to feel sick, since they usually determine termination of the experiment by the researcher. Nausea, salivation, sweating, pallor, and drowsiness are assessed before the remaining symptoms in the list, because these were defined as the cardinal symptoms of MS by Miller and Graybiel (1970b) and are mentioned widely in the literature. Experimenters who are more interested in studying the initial onset of MS may wish to change the order in which symptoms are assessed so that it matches the typical temporal order of early symptom progression (Lawson, 2014).

It is not essential to assess *stomach awareness/discomfort* in all MS studies nor to do so exactly as the MPDC demonstrate in Table 24.2. By assessing stomach awareness/discomfort as in Table 24.2, the MPDC fails to eliminate some metric problems from the original PDC and the SPDC. Nevertheless, the MPDC do tend to relax the implicit PDC/SPDC assumption that stomach awareness/discomfort lies on a continuum denoting gastric discomfort, *per se* (Muth, Stern, Thayer, & Koch, 1996).

In general, it is recommended that investigators should assess stomach awareness/discomfort if time permits, especially when one or more of the following conditions are met:

- The researcher wishes to stay closer to the original PDC (which includes *epigastric awareness* and *epigastric discomfort*, Table 24.1).
- The provocative stimulus is one that allows MS to develop gradually enough for different severity levels to be experienced and reported consecutively; for example, >10 min (at least 10 symptom assessments) was required for the majority of subjects to reach *moderate nausea* (the experimental endpoint) in four Coriolis cross-coupling studies by Lawson and colleagues (summarized in Figure 15 of Lawson et. al., 2009).
- The researcher wishes to study the earliest onset and progression of consciously recognized symptoms (it can be useful in such a case to know that stomach awareness/discomfort was reported at 12 min of stimulation, minimal nausea at 15 min, etc.).
- The researcher is studying MS in a setting where participants may be less likely to recognize the initial onset of true nausea and using a more neutral descriptor would foster subsequent introspection and responding. This may be the case in situations that create a strong demand against reporting MS susceptibility, such as certain group testing situations (e.g., operational or field settings) or when using participants for whom experiencing strong MS susceptibility may create cognitive dissonance (e.g., aviators, sailors, amphibious personnel). For example, the author occasionally receives comments from military aviator candidates who say they do not realize they have true nausea until they realize that their *stomach discomfort is getting much stronger now*. In such cases, asking participants about stomach awareness/discomfort immediately before asking them about nausea helps them decide whether to gastric aspects of the overall nausea construct to be warrant reporting *minimal nausea*. It would be interesting to establish whether such demand characteristics exist and, if so, whether they are beneficial to the reliability and validity of MS reporting.

The approach to MS scaling that is most similar to the MPDC (Table 24.2; Lawson et al., 2002) is the aforementioned *symptom checklist* (Reason and Diaz, 1971), which is represented in its mature form on page 82 of Reason and Brand (1975). Both approaches reduce the number of metric assumptions inherent in the original PDC. The main differences between the two approaches are as follows: (1) Reason and Diaz* assess fewer symptoms than Lawson et al. (Miller & Graybiel 1970b); (2) *drowsiness, stomach awareness/discomfort,* and *retching/vomiting* are not explicitly assessed by Reason and Diaz, while *dry mouth* is not explicitly assessed by Lawson et al. (Miller & Graybiel, 1970b); (3) Reason and Diaz assess *bodily warmth,* while Lawson et al. asses *flushing/warmth* (Miller & Graybiel asses *subjective warmth/flushing*); and (4) the definitions of severity levels are slightly different between the two scales[†]. The main implication of these differences is that the Lawson et al. criteria are more similar to the original PDC (Miller & Graybiel, 1970b) on points #1–4 but Lawson et al. continue to reduce metric assumptions (vs. the PDC), similarly to Reason and Diaz (1971). On the other hand, Reason and Diaz's symptom checklist (Reason & Brand, 1975) seems slightly simpler and faster to administer and does not require as many metric assumptions concerning *stomach awareness/discomfort* versus *nausea*.[‡]

The SSQ and MPDC show good test–retest reliability and correlate well with the earlier PDC scales that make more assumptions. The SSQ is designed for the relatively mild (low vomiting incidence) situation of simulator exposures and is probably the most useful of the *Pensacola school* techniques in that it allows for a three-factor assessment of underlying symptom clusters. The *MPDC* of Table 24.2 is recommended for those investigators already using the original Miller and Graybiel symptom criteria but wishing to limit their metric assumptions to decrease the odds of making erroneous conclusions about the null hypothesis. The Lawson adaptation should also be considered for use if the investigator needs to administer a multisymptom checklist in less time than would be required for the MSQ or SSQ (e.g., when there are brief intervals available for symptom assessment during rotation).

The SSQ/MSQ (see Table 24.3) is recommended for investigators who have sufficient time for administration and are not already committed to the Miller and Graybiel (1970b) PDC. As administered in questionnaires, the SSQ/MSQ descriptors are usually preceded by a number of biographical questions. Table 24.3 shows only the Likert-type items from the SSQ and MSQ, suitable for administration immediately before and after a real or apparent motion stimulus. The first 16 descriptors listed in Table 24.3 are used in the SSQ to assess symptoms in situations that elicit low vomiting rates and tend to include a visual display, with or without a moving base. The entire list of 28 descriptors is used in the MSQ to assess situations where vomiting is more common and motion tends to be present, for example, in ships at sea (SSQ vs. MSQ descriptors are displayed separately in Table 24.3 but are mixed in the actual MSQ/SSQ). Key publications describing the development, administration, and scoring of the SSQ and MSQ were mentioned earlier.

Since its inception, some modifications of the SSQ have been considered. For example, Nichols, Cobb, and Wilson (1997) describe an abbreviated six-item version of the SSQ called the *short symptom checklist* (SSC) that employs a five-point rating system (instead of the original four-point ratings in the SSQ), ranging from *not at all* to *severe.* The scale was subsequently modified to a 10-point rating system of each symptom, ranging from *not at all* to *unbearable,* according to a personal communication with Dr. Sarah Sharples (March 4, 2013, citing her dissertation: Nichols, 1999).

Nichols et al. (1997) do not itemize the six descriptors they selected nor could they be found in other reports mentioning the SSC (Cobb, Nichols, & Wilson, 1995; Cobb, Nichols, Wilson, &

---

* The four points refer to the scale shown in Reason and Brand (1975) and cited as Reason and Diaz (1971). However, the original report by Reason and Diaz (1971) states that *stomach awareness* was measured and gives no mention of *stomach discomfort.* Also, the original report indicates that *drowsiness* and *dry mouth* were symptoms volunteered by subjects after the car simulator runs but not planned *a priori* as questions to ask during the runs. The version in Reason and Brand is discussed because it is the later version and the version with which more researchers are familiar.

[†] The four severity categories are *none–minimal–moderate–major* in Lawson et al. and *none–mild–moderate–severe* in Reason and Diaz. (Graybiel & Miller [1970b] use *none–minimal–minor–major,* with exceptions shown in Table 26.2.)

[‡] This advantage is not as clear in the original Reason and Diaz (1971) source paper, which includes *stomach awareness.*

TABLE 24.3

SSQ (Items 1–16) and MSQ (Items 1–28)[a]

Circle How Much Each Symptom in the Following Is Affecting You Right Now

1. General discomfort	None	slight	moderate	severe
2. Fatigue	None	slight	moderate	severe
3. Headache	None	slight	moderate	severe
4. Eyestrain	None	slight	moderate	severe
5. Difficulty focusing	None	slight	moderate	severe
6. Increased salivation	None	slight	moderate	severe
7. Sweating	None	slight	moderate	severe
8. Nausea	None	slight	moderate	severe
9. Difficulty concentrating	None	slight	moderate	severe
10. Fullness of head	None	slight	moderate	severe
11. Blurred vision	None	slight	moderate	severe
12. Dizziness (eyes open)	None	slight	moderate	severe
13. Dizziness (eyes closed)	None	slight	moderate	severe
14. Vertigo[b]	None	slight	moderate	severe
15. Stomach awareness[c]	None	slight	moderate	severe
16. Burping	None	slight	moderate	severe
17. Boredom	None	slight	moderate	severe
18. Drowsiness	None	slight	moderate	severe
19. Decreased salivation	None	slight	moderate	severe
20. Depression	None	slight	moderate	severe
21. Visual illusions[d]	None	slight	moderate	severe
22. Faintness	None	slight	moderate	severe
23. Awareness of breathing	None	slight	moderate	severe
24. Decreased appetite	None	slight	moderate	severe
25. Increased appetite	None	slight	moderate	severe
26. Desire to move bowels	None	slight	moderate	severe
27. Confusion	None	slight	moderate	severe
28. Vomiting	None	slight	moderate	severe

Table adapted from one obtained from Dr. Robert S. Kennedy.

[a] The entire questionnaire, with background questions, items in original order, and scoring instructions, can be obtained directly from Robert Kennedy, Ph.D. The most pertinent publication is Kennedy et al. (1993). Direct questions to 6kennedy@bellsouth.net.

[b] Vertigo is experienced as loss of orientation with respect to vertical upright.

[c] Stomach awareness is usually used to indicate a feeling of discomfort that is just short of nausea.

[d] Visual illusion of movement or false sensations of movement when not in the simulator, car, or aircraft.

Ramsey, 1996; Sharples, Cobb, Moody, & Wilson, 2008). Personal communication with Dr. Sharples (October 25, 2012) revealed that the SSC descriptors are *headache*, *eyestrain*, *blurred vision*, *dizziness eyes open*, *dizziness eyes closed*, and *sickness*. *Sickness* is used in the SSC in place of the original term *nausea* in the SSQ, under the rationale that *nausea* was less commonly used to indicate gastric aspects of MS in Great Britain (where the scale was developed), at least at the time (Cobb et al., 1995) when the descriptors were chosen (Sharples, personal communication, March 4, 2013, citing Nichols, 1999).

The six descriptors of the SSC are not easy to find in the literature, and the scale properties are not known as of this writing, so further evaluation may be necessary before the SSC is used for widespread evaluation of MS. Nevertheless, it seems that a short version of the SSQ could be useful for situations where time for symptom assessment is very limited due to the constraints of the experiment, provided the shortened scale properties are acceptable and represent a valid shortening of the SSQ. For example, assuming each item can be rated in 1–2 s, the full list of 28 items on the MSQ/SSQ may require 28–56 s, while rating just the 16 scored items of the SSQ may require 16–32 s. Under these assumptions, a six-item scale would require only 6–12 s, making it feasible to use as a during-experiment probe for some experiments where the participant is not fully engaged in a performance task. These time estimates seem realistic, given that (1) Dr. Robert Kennedy has estimated the full 28 symptoms of the MSQ/SSQ can usually be assessed in under 1 min following a motion challenge if the scale is properly explained during the pretest screening administration of the MSQ/SSQ (personal communication, February 13, 2013), and (2) the participants in Lawson et al. (2009) were required to respond verbally to a MS checklist of nine symptoms in ~12 s and could do so with proper training beforehand.

As seems inevitable in the scaling literature, other researchers have also expanded or reduced the number of factors of the SSQ. The original SSQ has a three-factor solution. Kennedy et al. (1995) pointed out that *sopite syndrome* is not explicitly measured in the three-factor solution of the SSQ and may require considering greater-than-three-factor solutions of the SSQ. The present author notes that the *fatigue* symptom is placed in the *oculomotor* factor of the SSQ and the *difficulty concentrating* symptom is in the *nausea* factor. The grouping of these two symptoms into different factors under the stated factor names is not straightforward from the theoretical standpoint, but Dr. Kennedy confirms (personal communication, March 3, 2013) that the groupings were driven by how the factor analysis of simulator data worked out empirically.

While Kennedy has not yet published a four-factor solution for the purpose of distinguishing *sopite syndrome*, Bruck and Watters (2011) developed a four-factor solution for other reasons. They performed factor analysis on a set of mixed variables assessing the responses of 28 participants to three 2-min exposures to a 3D virtual roller-coaster rider. The variables included the SSQ, respiration, cardiac, and anxiety data. Bruck and Watters developed a four-factor solution of the SSQ designed for assessing *cybersickness*, with factors called *general cybersickness*, *vision*, *arousal*, and *fatigue*. While this scale incorporates four factors and is relevant to *sopite syndrome*, it cannot be considered a straightforward four-factor solution of the SSQ. Also, the scale is unusual compared to much past MS research, which would tend to view anxiety as a confound to be controlled or evaluated and a separate pathway to nausea and vomiting, rather than an integral part of the MS response itself, to be included in one's scale and be used as a criterion for MS. Note, however, that Gianaros, Muth, Mordkoff, Levine, and Stern (2001) incorporated some symptoms known to be elicited separately by anxiety within their multidimensional nausea construct.

Bruck and Watters (2011) offer an explanation for a possible link between anxiety and MS symptoms, hypothesizing that disturbing *virtual environments* (VEs) cause increased anxiety, which causes increased respiration and reduced carbon dioxide in cerebral blood flow. This change, in turn, causes lightheadedness, which triggers symptoms such as dizziness and fatigue. Unfortunately, the literature is quite mixed concerning the strength of the relation between MS and objectively measured respiration. However, there is evidence that some people exhibit physiological reactions consistent with fainting, even during low-G motion (Cheung & Hofer, 2001). Concerning the general question of the relation between MS and anxiety, much of the literature sees these as two separate pathways leading to some similar outputs, such as vomiting. For example, it is known that anxiety in the absence of motion can lead to nausea and vomiting, but also that decerebrated individuals unlikely to experience motion-related anxiety still become motion sick (Reason & Brand, 1975). It is recommended that arousal and anxiety should be assessed as covariates whenever an investigator wishes to know if reported MS symptoms are correlated with any physiological responses known to be sensitive to arousal changes also (e.g., heart rate, blood pressure, respiration rate). There are readily administered scales of arousal (e.g., Shapiro et al., 2006) and anxiety (e.g., Marteau & Bekker, 1992) that could be employed in future studies of physiological correlates of MS.

In addition to Bruck & Walters four-factor solution that was loosely based on the SSQ, a two-factor French-language solution more tightly based on the SSQ was developed by Bouchard, Robillard, and Reneaud (2007), with the factors called *nausea* and *oculomotor*. It should be noted that Bouchard and Reneaud used a heterogeneous sample of 371 participants, 164 of whom had been diagnosed with an anxiety disorder. The stimuli varied as well, with different head-mounted displays or immersive projections and different tasks, including exposure to phobic trigger stimuli. To determine the potential advantages of a two-factor solution of the SSQ, it would be necessary to perform a study on the original version using the same stimulus on a more homogenous sample of participants.

24.3   OTHER MULTISYMPTOM CHECKLISTS

Several motion discomfort scaling efforts were initiated outside of Pensacola, which are discussed in this section and Section 24.4. The *Motion Sickness Assessment Questionnaire* (MSAQ) was developed by Gianaros et al. (2001), based on descriptors generated by 67 participants. The 34 top-ranked descriptors were then administered to 747 participants. Participants rated how well each descriptor matched how they feel when motion sick, using a four-point Likert-type scale (where $0 = $ *not at all*, $1 = $ *slightly*, $2 = $ *moderately*, and $3 = $ *very*). Factor analysis revealed four dimensions of MS, called *gastrointestinal* (e.g., "I felt sick to my stomach"), *central* (e.g., "I felt dizzy"), *peripheral* (e.g., "I felt sweaty"), and *sopite related* (e.g., "I felt drowsy"). The MSAQ and the 16 descriptors that make up its four factors are shown in Table 24.4. Confirmatory factor analysis was completed along with some construct validity work versus other questionnaires and a lab test where 21 participants viewed a rotating optokinetic drum. The MSAQ shows good properties and allows assessment of some symptoms relevant to MS and some relevant to *sopite syndrome*. The *drowsy* and *tired/fatigued* items of the scale seem highly relevant to sopite, but more research may be needed to confirm whether the *uneasy* item is sopite specific.

TABLE 24.4

The Motion Sickness Assessment Questionnaire (MSAQ)

Instructions. Using the scale below, please rate how accurately the following statements describe your experience

Not at All   Severely

1—2—3—4—5—6—7—8—9

1. I felt sick to my stomach (G)	9. I felt disoriented (C)
2. I felt faint-like (C)	10. I felt tired/fatigued (S)
3. I felt annoyed/irritated (S)	11. I felt nauseated (G)
4. I felt sweaty (P)	12. I felt hot/warm (P)
5. I felt queasy (G)	13. I felt dizzy (C)
6. I felt lightheaded (C)	14. I felt like I was spinning (C)
7. I felt drowsy (S)	15. I felt as if I may vomit (G)
8. I felt clammy/cold sweat (P)	16. I felt uneasy (S)

Source: Appendix A of Gianaros et al., *Aviat. Space Environ. Med.*, 72(2), 115, 2001. Reproduced with the knowledge of Dr. Eric Muth and permission from the journal *Aviation, Space, and Environmental Medicine*.

Notes: G: gastrointestinal; C: central; P: peripheral; S: sopite related. The overall MS score is obtained by calculating the percentage of total points scored: (sum of points from all items/144) × 100. Subscale scores are obtained by calculating the percent of points scored within each factor: (sum of gastrointestinal items/36) × 100; (sum of central items/45) × 100; (sum of peripheral items/27) × 100; (sum of sopite-related items/36) ×100. Direct questions to Dr. Eric Muth: Muth@clemson.edu.

The *Nausea Profile* of Muth et al. (1996) is not a MS checklist, *per se*. It is the only procedure mentioned in this chapter that measures the construct of *nausea* by itself, regardless of whether the nauseogenic stimulus derives from motion or from some other cause (e.g., illness). The *Nausea Profile* yields a multifactor solution that is likely to be applicable to any provocative stimulus strong enough to trigger the nausea syndrome. It has proven useful for demonstrating that the subjective sensation of nausea includes factors other than the experience of gastric discomfort. An especially interesting feature of the *Nausea Profile* (and the MSAQ) is that the descriptors were generated by laymen rather than experts, thus enhancing the content validity of the scale and making it easier for nonexperts to administer. The *Nausea Profile* is not the first choice for researchers who wish to study MS exclusively, but it could be very helpful to those seeking to study nausea in a number of different situations including motion and nonmotion scenarios (e.g., postoperative nausea and vomiting; see also Wood, Chapman, & Eilers [2011] for a review of 24 other approaches). It is important for MS researchers to be familiar with Muth et al.'s report in order to be aware that when subjects report *nausea*, they are referring to a construct that has a couple of additional dimensions beyond the obvious *gastrointestinal distress* that lies along a continuum leading eventually to vomiting. These other dimensions were defined by Muth et al. as *somatic distress* (shaky, lightheaded, sweaty, tired/fatigued, weak, warmth) and *emotional distress* (upset, worried, hopeless, panicked, nervous, scared/afraid) (Gianaros et al., 2001). The *conditional distress* dimension is especially interesting since it implies that a person who experiences sufficient nausea is also likely to experience arousal or anxiety, which obviously complicates the study of psychophysiological correlates of MS. Thus, it could be argued that nausea is a multidimensional construct, which is itself part of the even wider, multidimensional construct of MS, although perhaps the symptom nausea has a narrower meaning when only real or apparent motion is the cause. Gianaros et al. (2001) discuss the possible relation between the dimensions of nausea and the dimensions of MS, as revealed by the *Nausea Profile* and the MSAQ, respectively.

A questionnaire was developed by Ames, Wolffsohn, and McBrien (2005) to further emphasize oculomotor symptoms associated with head-mounted VEs. The authors assembled 47 items from past VEs questionnaires not necessarily specific to MS but included questions from the SSQ as well. They then narrowed down the list to the 13 items with the most endorsement and/ or the best item-total correlations. The final questionnaire included the following symptoms: *Nonocular* symptoms were *general discomfort, fatigue, boredom, drowsiness, headache, dizziness, difficulty concentrating,* and *nausea*. *Ocular* symptoms included *tired eyes, sore/aching eyes, eyestrain, blurred vision,* and *difficulty focusing*. These symptoms were assembled in the *Virtual Reality Symptom Questionnaire* (VRSQ), a 0–6 point rating system shown in Table 24.5. Note that some potential limitations of the VRSQ have been described by Pesudovs (2005), however.

A new questionnaire has been designed recently to assess visually induced MS caused by viewing 3D movies. Solimini, Mannocci, and Thiene (2011) pilot tested a 20-item questionnaire on 38 participants. The questionnaire contains a mixture of questions relating to the participant, the stimulus, and the participant's symptoms. It does not appear that a multiple-factor solution has been generated from the questionnaire yet. The Solimini et al. questionnaire may be considered for settings where large numbers of moviegoers must be surveyed quickly (in 2–5 min). Solimini et al. intentionally exclude *fatigue, drowsiness,* and *difficulty* concentrating from the symptom checklist, so this questionnaire would not be suitable for those wishing to study *sopite syndrome* (Graybiel & Knepton, 1976) and MS together with one instrument. Moreover, *palpitation* is included as a symptom, which differs from Miller and Graybiel's MS criteria (1970). In fact, Muth et al. (1996) use *increased heart rate* and *aware of my heart* as distractor items because they were not generated by the subjects as being part of nausea during the descriptor development phase of the *Nausea Profile* and thus were meant to serve as a control descriptor, that is, a reference point for the judgment of which descriptors add value during factor analysis and which do not.

TABLE 24.5

Virtual Reality Symptoms Questionnaire (VRSQ)

Subject Code....................   Correction: CL   Specs   None

Date.....................

	None	Slight		Moderate		Severe		None	Slight		Moderate		Severe	
General Body Symptoms														
General discomfort	0	1	2	3	4	5	6	0	1	2	3	4	5	6
Fatigue	0	1	2	3	4	5	6	0	1	2	3	4	5	6
Boredom	0	1	2	3	4	5	6	0	1	2	3	4	5	6
Drowsiness	0	1	2	3	4	5	6	0	1	2	3	4	5	6
Headache	0	1	2	3	4	5	6	0	1	2	3	4	5	6
Dizziness	0	1	2	3	4	5	6	0	1	2	3	4	5	6
Difficulty concentrating	0	1	2	3	4	5	6	0	1	2	3	4	5	6
Nausea	0	1	2	3	4	5	6	0	1	2	3	4	5	6
Eye-Related Symptoms														
Tired eyes	0	1	2	3	4	5	6	0	1	2	3	4	5	6
Sore/aching eyes	0	1	2	3	4	5	6	0	1	2	3	4	5	6
Eyestrain	0	1	2	3	4	5	6	0	1	2	3	4	5	6
Blurred vision	0	1	2	3	4	5	6	0	1	2	3	4	5	6
Difficulty focusing	0	1	2	3	4	5	6	0	1	2	3	4	5	6

Other symptoms/feelings

Source: Reproduced from Ames, S.L. et al. *Optom. Vis. Sci.*, 82(3), 168, 2005. With permission.

Some users take a long time to recover from simulator or VE-related symptoms. A scale presently in development is the *Long Term Aftereffect Questionnaire* (LTAQ; Kennedy, 2009). This scale is designed to assess lasting effects of simulator or VE-related MS. Kennedy has reviewed the literature for reports of lasting symptoms, which will be used to generate a checklist for subsequent scale development.

24.4 APPROACHES THAT SOLICIT A SINGLE MS ANSWER FROM THE PARTICIPANT

While the original Miller and Graybiel (1970b) criteria and the SSQ each allow the investigator to generate a single number corresponding to total sickness severity, this number is derived by the experimenter from multiple ratings made by the participant. There are a number of MS approaches that ask the subject to participate in the generation of a single answer concerning his or her MS state. This is generally done by asking the participant to choose a single number each time an inquiry is made by the experimenter. Most of these techniques do not preserve data concerning the individual symptoms of MS. The main advantage of these techniques is that they are faster to administer than multisymptom checklists and easier to analyze. There are dozens of such approaches, from simple *yes/no* decisions by subjects concerning whether they are experiencing MS (Chen, Dong, Hagstrom, & Stoffregen, 2011) to involve magnitude estimation procedures, a few of which are highlighted in the following.

An early single-answer approach to MS was devised by Hemingway (1943). Hemingway's (1946) swing sickness grading system is summarized on p. 176 of Reason and Brand (1975). The grading system entails a 10-point scale where 10 = *no effects* (in 20 min of swinging) and 1 = *vomiting* in the first 5 min of swinging. The intermediate ratings are obtained via a fairly complicated mixture of criteria involving symptom/sign type (*pallor, nausea, sweating*, or *vomiting*), symptom/sign severity (e.g., *slight pallor, pallor, marked pallor*), and time-of-symptom/sign onset (e.g., 5 = *vomiting* in 14–17 min of stimulation).

Many single-answer MS scales have been used since Hemingway and are in use today. The MS measurement procedure of Bock and Oman (1982) involves magnitude estimation, which has been argued to yield metrically superior data (Stevens, 1946). The Bock and Oman procedure is difficult to implement by the experimenter, but it can be explained to subjects easily and the estimates can be gathered from them rapidly once they have experienced the reference stimulus.

Bock and Oman used left–right reversing prism goggles to elicit symptoms in eight subjects via locomotion and head movements. After experiencing a few different levels of MS, the subjects assigned a "10" to a specific level and rated all subsequent experiences relative to that number, based on overall discomfort rather than any single symptom. If the participant felt half as sick as the reference, he or she would assign that a "5" and so forth. No fixed range was specified and only one anchor was employed to avoid distortion (Bock & Oman, 1982). In its original form, the Bock and Oman procedure called for previous exposure to the stimulus of interest, but this requirement was relaxed by Eagon (1988).

Any single-number scale of this type does not generate multiple-factor solutions for exploring the sort of psychometric questions addressed by scales like the SSQ and MSAQ. Moreover, the general assumption in the literature that magnitude estimation is superior to Likert-type rating has been questioned (Sprouse, 2008; Wills & Moore, 1994). Nevertheless, while magnitude estimation is not the most popular method of MS assessment, it has been found useful in a study of seasickness by Bittner and Guignard (1986a, 1986b).

Dahlman (2009) adapted a non-MS scale originally designed to have ratio properties (Borg & Borg, 2001) for use in rating overall MS. The 0–10 scale used by Dahlman employs five verbal anchors to define ratings from 0 to 3 and three verbal anchors to define ratings from 4 to 10. The definitions are as follows: 0 = *nothing at all*, 0.5 = *extremely weak*, 1 = *very weak*, 2 = *weak*,

3 = *moderate*, 5 = *strong*, 7 = *very strong*, and 10 = *extremely strong*. If this is truly a ratio scale, then some implicit assumptions seem to be present, as follows: (1) *Weak* refers to a state of MS 4× stronger than *extremely weak*, and (2) the scale is nonlinear, since *weak* changes to *moderate* three-tenths of the way to the high end of the scale, *moderate* changes to *strong* five-tenths of the way, and then various versions of *strong* are retained for the upper half of the scale (from 5 to 10). Despite these unproven metric assumptions, the scale should be relatively easy for the investigator to implement and should be able to estimate overall MS state without making unwarranted assumptions about where individual symptoms should appear. Nevertheless, the number of choices and definitions would make this a difficult scale for subjects to apply from memory in order to make verbal symptom reports without a visual aid or coaching from the investigator.

A rating system called the *misery scale* (MISC) was developed by Wertheim, Bos, and Krul (2001) and adapted recently to study visually induced MS and *cybersickness* (Bos, de Vries, van Emmerik, & Groen; 2010; Emmerick, de Vries, & Bos, 2011). The original Wertheim et al. rating system is a 0–10 scale where 0 = *no problems*, 1 = *stuffy/uneasy feeling in head*, 3 = *stomach awareness*, 5 = *nausea*, 7 = *very nauseous*, 9 = *retching*, and 10 = *vomiting*. This is somewhat similar to the earlier *well-being rating scale* (Reason & Graybiel, 1970; Reason & Diaz, 1971), where 0 = *I feel fine* and 10 = *I feel awful, just like I am about to vomit*. The main difference is that the earlier scale assesses overall *well-being* rather than the existence or severity of any given symptom, although there is a general bias toward any symptoms usually associated with vomiting in both scales.

The 2011 version of the MISC is shown in Table 24.6 (from Emmerick et al., 2011). It is identical to the 2010 version (Bos et al., 2010) but is shown here because Emmerick studies VE-related responses and because they present the nonnausea symptoms more completely than the 2010 paper. The Emmerick et al. MISC is a 0–10 scale where 0 = *no problems*; 1 = *uneasiness (no typical symptoms)*; 2 = *vague dizziness, warmth, headache, stomach awareness, [or] sweating*; 3 = *slight dizziness, warmth, headache [etc.]*; 4 = *fairly dizziness [dizzy, etc.]*; 5 = *severe dizziness*; 6 = *slight nausea*; 7 = *fairly nausea [ted]*; 8 = *severe nausea*; 9 = *retching*; and 10 = *vomiting*. Thus, intermediate ratings from 2 to 5 refer to four levels of severity for nonnausea symptoms and ratings from 6 to 8 refer to three levels of severity for nausea symptoms. The advantages and limitations of this rating strategy are discussed by Keshavarz and Hecht (2011). This MISC is quick to

TABLE 24.6
The Misery Scale (MISC)

Symptom		Score
No problems		0
Uneasiness (no typical symptoms)		1
Dizziness, warmth, headache,	Vague	2
stomach awareness, sweating	Slight	3
	Fairly	4
	Severe	5
Nausea	Slight	6
	Fairly	7
	Severe	8
	Retching	9
Vomiting		10

Source: Van Emmerik, M.L. et al., *Displays*, 32(4), 169, 2011. Reproduced with the knowledge of Dr. Jelte Bos and permission from the journal Displays. Available in multiple publications by Dr. Bos and his colleagues, direct questions to: Jelte.bos@tno.nl.

administer and will assess the entire range of MS, so it could be applied in highly sickening situations. Potential limitations of the scale derive from the heterogeneous use of different individual symptoms and symptom severity levels to correspond with different point values along a single 0–10 misery continuum. This method requires several implicit assumptions: (1) that *dizziness, warmth, headache, stomach awareness,* and *sweating* are likely to emerge after other *nontypical* prenausea signs/symptoms (*uneasiness* is mentioned, but *pallor* or *drowsiness* would also be included); (2) that nonnausea symptoms precede *nausea*; (3) that *nausea* precedes *vomiting*; (4) that *slight nausea* elicits more misery than *severe dizziness, severe headache,* etc.; (5) that *retching* involves less misery than *vomiting*; (6) that the word *vague* best defines the lowest severity of nonnausea symptoms, while *slight* best defines the lowest severity of nausea (this is probably because *stomach awareness* was included as a nonnausea symptom); and (7) that *stomach discomfort, pallor, salivation,* and *drowsiness* are not critical to include as explicitly mentioned symptoms. While many of these assumptions are logical *a priori*, there is evidence of possible violations of assumptions 1 and 3 (Lawson, 2014). Furthermore, Lawson has received anecdotal comments from research participants and MS researchers implying that potential violations of assumptions 4 and 5 have occurred in the laboratory or field setting. Finally, the practice among many MS researchers runs counter to assumption 7 (especially in regard to salivation), although a discussion with Dr. Jelte Bos (personal communication, October 21, 2012) indicates that the participants know that scores in the 2–5 range may include other nonnausea symptoms as well symptoms are highly variable. As the MISC matures, it would be worthwhile to carry out controlled experiments to determine the strength of empirical support for the implicit assumptions it contains. Such research would benefit many other MS scales as well, such as the approach described in the next paragraph.

Griffin and Newman (2004) employ a MS rating system adapted from Golding and Kerguelen (1992). Griffin and Newman use a 0–6 scale where 0 = *no symptoms*; 1 = *any symptoms, however slight*; 2 = *mild symptoms, for example, stomach awareness but no nausea*; 3 = *mild nausea*; 4 = *mild to moderate nausea*; 5 = *moderate nausea but can continue*; and 6 = *moderate nausea and want to stop.* The original version by Golding and Kerguelen was nearly identical, but a 1–7 rating was employed. In either version, the rating system is not designed for experiments where severe nausea and retching/vomiting are likely. The system appears somewhat less complicated to explain to participants and to have them implement (without coaching or visual aids) than is the MISC (Emmerick et al., 2011) or the scale used by Dahlman (2009) but more complicated than the 0–10 *well-being* rating scale (Reason & Graybiel, 1970) mentioned earlier. The scale can be administered rapidly but appears to require the implicit assumption that all nonnausea/nonstomach symptoms are to be rated, since they must, by definition, score lower than mild nausea, which is assigned a level of 2. The hypothetical question arises: should a participant reporting *dizziness, headache,* and *stomach awareness* be considered less sick than a person with *mild nausea* but no other symptoms? Such questions could be answered empirically by comparing how well the cases in question correlate with the participant's overall *well-being* score assessed at the same time as the individual symptoms.

The *fast motion sickness scale* (FMS) (Keshavarz & Hecht, 2011) is a verbal rating scale that ranges from 0 = *no sickness at all* to 20 = *frank sickness. Frank sickness* is not defined by Keshavarz and Hecht, but Dr. Behrang Keshavarz confirms (personal communication, October 15, 2012) that it corresponds to extreme sickness, that is, imminent vomiting. This semantic usage should not be confused with the original meaning of *frank sickness* as used by Graybiel (1969) and Miller and Graybiel (1970a, 1970b), since the original usage requires reaching a score of at least 16 points on a multisymptom checklist (e.g., by *retching* or *vomiting*, by experiencing *major nausea* plus *major sweating*, by experiencing *major drowsiness* plus *major pallor*, or by several other possible symptom combinations not necessarily implying imminent vomiting).

The Keshavarz and Hecht scale was developed using participants from two experiments. Experiment 1 involved watching an in-car video of an automobile driving on a racetrack for

18 min (n = 65, 21 of whom were biased to expect a sickening experience, 23 of whom were biased to expect a fun experience, and 21 of whom were not biased concerning what to expect). Experiment 2 involved 15 min of viewing a repeating loop of a computer-simulated roller-coaster ride (n = 61). Participants were instructed to focus on general discomfort, nausea, and stomach problems and to ignore symptoms such as fatigue. Keshavarz and Hecht found that the FMS correlated well with the SSQ. They cautioned that the data gathered with the FMS are not normally distributed. Keshavarz, Hecht, and Lawson (2014) recommend that this nausea-centric scale is not appropriate for sole use if the soporific or oculomotor dimensions of MS are of particular interest. In fact, it is generally true that most of the single-number rating systems in this section are nausea-centric.

There are several other single-answer MS rating systems. For example, Gal (1975) describes an unusual peer-evaluation method called the *Seasickness Questionnaire*. In this assessment method, crew members assign one another a ranking and then a five-point severity-of-sickness score based on their perceptions of their fellows' *pallor, sweating, nausea,* and *vomiting.* The authors assume that the listed symptoms appear and worsen in the order just listed.

The seven approaches reviewed here should suffice to give the reader a fairly complete look at the options available for single-answer MS assessment. The next topics for consideration will be the scientific and practical issues to be considered when soliciting a single answer from the participant (i.e., a single participant-generated number rating per administration) versus applying multisymptom checklists. First, since time saving is the main advantage of the single-number scales, it would be useful to determine how long it takes on average to administer the various published single-number scales. The time required for administration is seldom mentioned in the literature. For example, the FMS was administered once per minute in Keshavarz and Hecht, but average administration time was not reported. This is important to know because a multisymptom checklist can be administered once per minute also. Presumably, the FMS could be administered in 2–6 s. By comparison, a variant of the MPDC (adapted from Table 24.2) was employed in Lawson et al. (2009) wherein subjects were asked to rate nine symptoms once per minute and 12 s were available to make the ratings before the next cycle of Coriolis cross-coupling began. This was a bit rushed but worked well because careful explanation before the study allowed the scale to be implemented successfully during the study. Moreover, the participants only had to consider one symptom at a time and to report only on those symptoms that increased since the last assessment, assigning them a new rating value (*minimal, moderate, or major*). They could say the *same* (as mentioned previously) in answer to any symptoms that had not changed since the last assessment. Thus, in this hypothetical comparison, the FMS would provide a time saving of 6–10 s per administration, but data concerning multiple symptoms would not be obtained. Whether this trade-off is desirable will depend on the experiment. For example, when MS is the most important variable, the researcher may wish to obtain data on multiple symptoms. On the other hand, when the participant is heavily tasked and the main measure is performance, a single-number scale of MS may suffice. The SSQ/MSQ requires 16–28 symptoms to be rated, and so a greater time savings can be obtained by using the FMS versus the SSQ/MSQ would be true when comparing the FMS to the PDC (e.g., as the MPDC was applied in Lawson et al., 2009). The SSQ/MSQ has some unique benefits discussed in Section 24.2, however.

24.5 CONSIDERATION OF THE NUMBER OF CATEGORIES USED TO OBTAIN RATINGS OF MS SEVERITY

The scales discussed in Sections 24.4 and 24.5 vary in format, particularly with regard to the number of categories used to obtain ratings of MS severity. A question of interest is whether there are relative advantages to rating MS via the 0–20, 0–10, 0–4, 1–5, or 1–7 rating schemes used in past scales. Certainly, 1–5 and 1–7 Likert-type scales are the standard, and there are potential limitations to the traditional 0–3 MS ratings used by many multisymptom checklists, such as restricting the range in the data that are available for variation and analysis. However, changing such features at this late stage of scale development complicates future comparison to the large body of previous data from the MSQ/SSQ and the PDC. Moreover, it is possible that participants can respond to a multisymptom checklist slightly faster when they do not have to ponder as many rating choices per question. The decision concerning the optimal number of categories will vary depending on multiple factors. For the MS scales that ask the participant to generate a single answer, it seems appropriate to offer more than a four-point scale to allow for a wider response range and improve reliability. Ideally, the number of choices would be the least number sufficient for obtaining good data, since the key reasons to employ a single-number scale are to get a fairly general answer as quickly as possible and with minimal misunderstanding or need for explanation. Therefore, anything that slows administration or complicates the single-answer scale should be avoided.

The literature has long established that 5–7 points allow an adequate number of choices for many research applications concerning human attitudes (Dobson & Mothersill, 1979; Preston & Colman, 2000). Also, rating ranges that avoid presenting too many choices should be easier to administer verbally in cases where the participant must be in the dark or have his or her eyes engaged in a task and not be able to look at a visual representation of a scale for reference. It could be argued that a 5-point or 10-point scale would be intuitive for situations requiring verbal or gestural (number of fingers) responses, since humans think in base 10. It remains to be proven whether a scale with more points than 10 would increase precision for MS research or merely increase administration time and variability of responding. For Likert scales, in general, it has been established that scale reliability tends to peak at around seven categories and decline as categories increase to 11, with greater midpoint response bias also observed in the higher end of the range (Kieruj & Moors, 2010). However, even seven categories can be too many for certain types of research (Pesdudovs, 2005). For example, Rasch analysis of the seven-level Functional Independence Measure (a rehabilitation outcome measure of how much help a patient needs) by Nilsson, Sunnerhage, and Grimby (2005) indicates that reducing the scale to a four-category scale is more accurate. Nilsson et al. explain this would reduce the disordering of thresholds they observed in the seven-category scale, which implied that more categories exist in that scale than are needed to describe the construct and that problems would arise with the comparison of raw sums. Ultimately, the number of categories chosen will depend upon the particular construct being studied and whether the number of categories chosen is ideal for adding to trait variance but not systematic or random error variance (Chang, 1994). Further discussion of scaling issues is offered by Friedman and Amoo (1999). The summary recommendation of the present author is that MS scales should employ four to seven categories for each question in a multisymptom checklist. No more than 11 categories should be employed in a single-answer MS scale, with fewer categories being acceptable (but no fewer than four).

24.6 RECOMMENDATIONS CONCERNING THE USE OF MS SCALES

Given the wide variety of multisymptom and single-answer scales, it can be difficult for a researcher new to MS to know how to proceed. A good strategy would be to use a rapid rating system (e.g., the MPDC) once per minute during testing and a multisymptom checklist (e.g., the SSQ) immediately after testing (e.g., once every 15 min during the defined period of posttest recovery). Suitable rapid strategies to apply during testing include rating the nausea symptom only (requires 1–3 s), rating overall well-being or general discomfort without requiring particular symptoms to be present (item #1 of the SSQ of Kennedy et al., 1993; Reason & Graybiel, 1970), (1–3 s) or rating the nine signs/symptoms of the MPDC (at least 12 s, Table 24.2). More detailed guidelines are offered in the following to aid investigators in making the best choice for their specific research needs:

- Researchers who already have a program of research and associated databases already committed to the Miller and Graybiel (1970) PDC and who regularly study MS in real-motion laboratory or operational settings (rather than studying only optokinetic and

VE stimuli) should consider using the MPDC shown in Table 24.2. This checklist has relatively fewer metric assumptions than the original criteria (Miller & Graybiel, 1970) and is easier to use, but the results are fairly comparable to the values of the Miller and Graybiel scale if comparison to old data is desired.

- Researchers who have a program of research and databases already committed to the SSQ of Kennedy et al. (1993) should stay with the SSQ, but consider the following issues:
  - Does the research require data collection and analysis only on the first 16 SSQ items? This may be the case if the research limited to visually induced MS research in fixed-base simulators and VE *and* the protocol is under tight time constraints.
  - Would the research benefit from use of the entire list, including the MSQ items 17–28? This will be the case if the research program includes real-motion or space sickness. (Note that Kennedy advises administering the entire 28-item scale even if only the first 16 items are scored.)
  - Does the research need to employ the correlated descriptors in three-factor weighted scoring system of the SSQ or simply add raw ratings to achieve a score? If the researcher wishes to compare SSQ symptom clusters, the original scoring should be used. If the researcher wishes to focus more on individual symptoms and avoid scoring the appearance of any symptom multiple times in final sickness score, then a simple count may be sufficient (Hoyt, Lawson, McGee, Strompolis, & McClellan, 2009; Lawson et al., 2009).
- Researchers who have no committed program of research or MS database should adopt the SSQ of Kennedy et al. (1993) but consider the SSQ questions just posed immediately earlier.
- The MSAQ should be considered by researchers who are not committed solely to simulator/VE studies and wish to employ a single multisymptom scale, which takes roughly the same time to fill out as the first 16 items of the SSQ but captures certain aspects of *sopite syndrome* as well as MS.
- The following alternative strategies should be considered by those researchers who must assess MS *during* testing in < 10 s:
  - Simply ask the participants how nauseated they are (e.g., on a four-point scale). This is very fast, is likely to capture a fair amount of variance in overall comfort or well-being using just one symptom (Reason & Diaz, 1971; Balk, Bertola & Inman, 2013), and allows for direct comparison to the existing nausea item from the SSQ or the Miller and Graybiel PDC, either of which can be administered in full immediately after testing.
  - Administer a simple overall *well-being rating scale* (Reason & Graybiel, 1970) or ask only about *general* discomfort (item #1 of the SSQ in Table 24.3) to get a single number quickly with one answer per administration and relatively few metric assumptions.
- Either of the two approaches just described will minimize some unproven features that tend to muddy other one-answer scales in the literature, such as injecting unproven assumptions about which prenausea symptoms must be present at different point values in the rating.
  - If a single-answer MS response assessment method is used besides the two just recommended, the researcher is advised to limit the number of MS severity rating choices the participant must make from four categories or points minimum (e.g., *none, minimal, moderate, major*) to 11 points maximum and to be aware that metric advantages of more categories will tend to peak by seven categories.
- Regardless of which single-answer or multisymptom scale the investigator employs during testing, he or she should consider administering the entire 28-item MSQ/SSQ immediately after termination of the stimulus so that some degree of comparison is possible across studies, especially to Kennedy's large database of past studies. Dr. Kennedy advises not analyzing pre- versus poststimulus MSQ/SSQ responses for statistical reasons but rather recommends that researchers exclude any volunteers who score higher than 7.48 on the SSQ under baseline conditions (personal communication, March 5, 2013). The rationale

for his advice (Kennedy et al., 2010) seems well founded, relating back to Chronbach and Furby (1970) who advise that change or gain measures derived from "subtracting pretest scores from posttest scores lead to fallacious conclusions, primarily because such scores are systematically related to any random error of measurement" (p. 68): they conclude that "there seems to be no occasion to estimate true gain scores" (p. 80).

- Researchers wishing to study *sopite syndrome* in isolation (or to employ two separate, dedicated, multidimensional scales for MS and *sopite syndrome*) should consider the new sopite instruments under development, discussed in the following.

24.7   *SOPITE SYNDROME*: RELEVANT SCALES IN DEVELOPMENT

Pensacola researchers Graybiel and Knepton (1976) coined the phrase *sopite syndrome* to describe *a sometimes sole manifestation of motion discomfort characterized by such symptoms as motion-induced drowsiness, difficulty concentrating, and apathy.* Evidence that the sopite syndrome is somewhat distinct from the nausea syndrome and from regular fatigue (due to nonmotion challenges such as sleep deprivation) is discussed by Lawson and Mead (1998). Lawson, Kass, Muth, Sommers, and Guzy (2001) report preliminary findings concerning a *Mild Motion Questionnaire* (MMQ) designed to assess *sopite syndrome*. The questionnaire subsequently became the subject of initial construct validation research (Wallace, Kass, & Lawson, 2002; Brill, Kass, & Lawson, 2004).

Unlike MS scales, which ask subjects to rate various items reflecting how they feel during MS, a scale designed to assess sopite cannot ask how subjects feel during soporific motion without creating a large demand characteristic concerning what constitutes *sopite syndrome*. In other words, to ask subjects about *sopite syndrome* requires telling them what to say. For this reason, Lawson et al. 2001 adopted an indirect approach, asking 212 subjects (in Phase 1 of the effort) to generate an exhaustive list of (230 distinct) words to describe how they feel in moving situations that are *mild or not sickening*. The rationale was that less traditional MS and more sopite may be detected under such conditions. The 212 subjects were comprised of 17 MS experts retrospecting about past motion situations, 59 laypersons retrospecting about past motion situations, and 136 laypersons who had just emerged from a brief and mild motion stimulus known to have a low vomiting rate of 0.8%–1.6% (Askins, Mead, Lawson, & Bratley, 1997; Bratley, Lawson, & Mead, 1997; Lawson & Mead, 1998). Each participant generated up to 10 descriptors and ranked them in descending order of preference. The ranking and frequency of occurrence of each descriptor were combined (using the method of Muth et al., 1996) to generate a narrowed-down list of 42 items (plus one *dummy* variable we added as a manipulation check; see Muth et al.).

In the next phase of research, Lawson et al. (2001) administered the 42 + 1 item list to 456 participants. The participants rated how accurately each item described their past experiences with mild or nonsickening motion, using five categories of choice, including *not at all, a little, moderately, fairly strongly, and very strongly.* Thus, the MMQ was initially developed as a trait scale to test whether consistent factors describing *sopite syndrome* could be observed.

Based on an exploratory factor analysis, three items were removed for failing to load on any factor higher than the dummy item (at 0.43). Once these and the dummy item were dropped, the resulting scale had 39 items (see the Appendix of Wallace et al., 2002). High internal consistency (coefficient alpha = 0.92) was observed for the total set of items and for each subset of items in the four-factor solution chosen by the Lawson et al. The four factors of *sopite syndrome* are provisionally called *head/body symptoms* (e.g., *stomach awareness*), *relaxed/content* (e.g., *relaxed*), *drowsy/fatigued* (e.g., *drowsy*), and *poor concentration/motivation* (e.g., *fuzz/foggy headed, lazy/unmotivated*).

Despite the necessity for an indirect inquiry method, these factors fit key aspects of the clinical description of *sopite syndrome* in the literature. The *head/body* factor seems to reflect symptoms consistent with a very mild version of *regular MS*. The *relaxed/content* factor seems to reflect a state of relaxation without necessarily being drowsy. This factor is unusual in that it contains some items reflecting positive effect. The *drowsy/fatigued* factor seems to reflect the core working definition of sopite and is exactly what would be expected from the literature. The *poor concentration/motivation* factor explains less variance than the other three but is preserved because it is operationally relevant to transportation situations and because it matches the literature on *sopite syndrome* well. What was not observed in the MMQ study were many descriptors or factors reflecting strong negative moods of the kind described by Graybiel and Knepton (1976). However, Lawson and colleagues (Lawson & Mead, 1998; Lawson, Mead, & Clark, 1997; Mead & Lawson, 1997) have observed some limited evidence for negative affect in two preliminary case reports concerning aviators.

There are several reasons strong negative affect could emerge more readily in Graybiel and Knepton's (1976) original observations of the *sopite syndrome*. It could be that the MMQ, which inquires indirectly about *sopite syndrome*, reflects a slightly different aspect of sopite from what was being observed during the prolonged, live-in rotation studies reviewed by Graybiel and Knepton. For example, the MMQ may reflect more *early sopite* during the initial phases of a mild stimulus (before negative feelings have emerged), while the rotating room studies also observed *late sopite* effects after adaptation to nausea had been accomplished for most subjects. It is also possible that in Graybiel and Knepton's report, the rotating room situation of close confinement without privacy, the multiday, live-in circumstances of the studies, and the initial phase of inescapable sickness contributed to later negative affects in ways not directly related to the soporific effects of vestibular stimulation. Many questions concerning *sopite syndrome* have yet to be answered.

Brill and colleagues have been conducting research concerned with measuring *sopite syndrome* also. Brill and Neilson (2011) conducted a study using the MMQ (Lawson et al., 2001) with the goal of developing an abbreviated questionnaire for rapid assessments of symptoms. They administered the MMQ to 422 college students, prompting them to rate its 39 items on how they usually respond to mild, nonsickening motion. They performed a factor analysis and determined which items were most representative of each factor. Brill and Neilson adopted a very conservative criterion for item inclusion (factor loadings ≥ .60). As a result, their prospective short-form MMQ was comprised of only 25 items loading on two factors. The first factor represented adverse effects of motion exposure, such as *fatigue* or *irritability*, while the second factor represented more positive symptoms arising from motion, such as feelings of *relaxation* and *contentment*.

24.8   MS SCALES MEASURING PAST SUSCEPTIBILITY OR HISTORY OF MS

One of the first studies demonstrating the usefulness of a MS history questionnaire in predicting subsequent MS induced in the laboratory was contributed by Alexander, Cotzin, Hill, Ricciuti, and Wendt (1945b). Kennedy, Dunlap, and Fowlkes (in Crampton, 1990) sought to determine exactly how useful MS histories are. They estimated the predictive validity of a wide range of factors from the literature, finding that no approach was as predictive as actual testing in a MS situation. Nevertheless, past history of susceptibility to MS (recollected for a number of situations) was one of the most predictive of the factors not requiring MS testing, accounting for up to 34% of corrected variance. Considering this approach only requires filling out a questionnaire, this is a very good result. The main MS history questionnaires will be introduced briefly, as follows.

The instrument Kennedy et al. described for measuring history is the *Pensacola Motion History Questionnaire* (Kennedy & Graybiel, 1965; Kennedy et al., 1990). A similar motion history was developed by Reason (Reason, 1968; Reason & Brand, 1975). Later contributions have been made by Griffin and Howarth (2000) and Golding (1998, 2006). The most recent and convenient among the history questionnaires in this series is the *Motion Sickness Susceptibility Questionnaire Short version* (MSSQ-Short, Golding, 2006, see Table 24.7). Suitability of the MSSQ-Short for VE-specific research is not obvious initially, since items relating to VEs, video games, and Cinerama movies were excluded when the MSSQ was shortened. They were dropped because they accounted for little variance, but Golding (2006) acknowledges that VE responses could become more important to the MSSQ-Short in the future. Currently, it can be said that the MSSQ-Short shows good predictive validity for a stimulus involving conflicting real and visual motion (Table 24.4, Golding 2006). Dr. John Golding

TABLE 24.7

The Motion Sickness Susceptibility Questionnaire (MSSQ-Short)

1. Please state your age. .... years
2. Please state your sex (tick box).   Male   Female   [ ]   [ ]

This questionnaire is designed to find out how susceptible to motion sickness you are and what sorts of motion are most effective in causing that sickness. Sickness here means feeling queasy or nauseated or actually vomiting.

Your *childhood* experience only (before 12 years of age), for each of the following types of transport or entertainment, please indicate:

3. As a *child* (before age 12), how often you felt sick or nauseated (tick boxes):

	Not Applicable—Never Traveled	Never Felt Sick	Rarely Felt Sick	Sometimes Felt Sick	Frequently Felt Sick
Cars					
Buses or coaches					
Trains					
Aircraft					
Small boats					
Ships, e.g., channel ferries					
Swings in playgrounds					
Roundabouts in playgrounds					
Big Dippers, funfair rides					
	t	0	1	2	3

Your experience over the *last 10 years* (approximately), for each of the following types of transport or entertainment, please indicate:

4. Over the *last 10 years*, how often you felt sick or nauseated (tick boxes):

	Not Applicable—Never Traveled	Never Felt Sick	Rarely Felt Sick	Sometimes Felt Sick	Frequently Felt Sick
Cars					
Buses or coaches					
Trains					
Aircraft					
Small boats					
Ships, e.g., channel ferries					
Swings in playgrounds					
Roundabouts in playgrounds					
Big Dippers, funfair rides					
	t	0	1	2	3

Note: Dr. John Golding's version, sent for reproduction in this chapter. The table is also shown in Golding (2006), in the journal Autonomic Neuroscience. Scoring is described in Golding (2006) as well. Direct questions to goldinj@westminster.ac.uk.

confirms that the MSSQ-Short can be applied to visually induced MS studies (personal communication, October 25, 2012). Researcher outside Britain may need to explain to their subjects some MSSQ terms before administration (e.g., "Channel ferries", "Round abouts", "Bigg Dippers").

24.9 SUMMARY

This chapter describes existing MS rating scales for assessing MS history or current MS state. It is worthwhile for MS researchers to consider how MS signs and symptoms will be gathered and scored prior to experimentation. Too often, this aspect of MS research is given insufficient thought,

TABLE 24.8
Procedures for Assessing Motion Sickness (Section 24.6)

	Minimal Procedures	Optimal Additional Procedures
Before stimulus	Give symptom definitions (Lawson 2014, Section 23.3.1) Screen out sick volunteers	Assess history of MS (Table 24.7) Assess *anxiety* (Section 24.2.3)
During stimulus (once/min)	Assess *nausea* and *general discomfort*	Administer MPDC (Table 24.2) Assess *anxiety*
After stimulus	Administer SSQ/MSQ (Table 24.3)	Repeat SSQ/MSQ every 15 min until symptoms disappear Assess *anxiety*

despite the fact that reported sickness will be the criterion against which other (e.g., physiological, performance-based) MS measures are judged.

Methods are discussed for obtaining overall (single-answer) ratings of MS or multi-item ratings of various individual symptoms known to be associated with MS. The author recommends that multisymptom checklists (such as the SSQ/MSQ) should be employed whenever feasible. When the time available for a single MS assessment is too short to fill out a multisymptom questionnaire, a single-answer method can be used (such as rating *nausea*, *general discomfort*, or *well-being*; see Section 24.6) but should be backed up by a multisymptom questionnaire as soon as time allows (e.g., immediately upon termination of the sickening stimulus). Specific recommendations for MS scaling before, during, and after a real or apparent motion challenge are summarized in Table 24.8. The recommendations apply to experiments in which acute MS response is the focus of study, rather than being a secondary variable versus non-MS variables or MS-related variables (e.g., *sopite syndrome*).

There are several issues that are not fully resolved in MS scaling, such as the ideal number of categories to use for rating MS severity, the optimal semantic label to apply to each category, the costs and benefits of a numerical midpoint in the ratings (e.g., by using five or seven categories instead of four or six), the costs and benefits of a zero value at the low end of the scale (vs. a "1"), and the costs and benefits of explicitly numbered ratings (e.g., the SSQ has four labeled categories, but they are not usually numbered). Some of these issues should be explored via Rasch analysis, which may lead to discoveries that would promote further improvements in MS scaling in the future.

DISCLAIMER

The author thanks Lavia Stanley for making the writing of this chapter possible.

The views expressed in this chapter are solely those of the author; they do not represent the views of the US Government or any of its subordinate agencies or departments. The mention of any agencies, persons, companies, or products in this chapter does not imply endorsement by the author or any agency with which he is affiliated, nor does it imply their endorsement of the views expressed in this chapter.

ACKNOWLEDGMENTS

The author thanks the following people for their contributions to this report: (in alphabetical order) Chris Brill, Shauna Legan, Eric Muth, and Jill Parker. The author thanks Justin Campbell, John Golding, Steve Kass, Amanda Kelly, Bob Kennedy, Heather McGee, Eric Muth, and Joel Ventura for offering a wide range of interesting perspectives concerning the quantification of symptoms and the art of scaling.

REFERENCES

Alexander, S. J., Cotzin, M., Hill, C. J., Ricciuti, E. A., & Wendt, G. R. (1945a). Wesleyan University studies of motion sickness: IV. The effects of waves containing two acceleration levels upon sickness. *Journal of Psychology, 20*, 9–18.

Alexander, S. J., Cotzin, M., Hill, C. J., Ricciuti, E. A., & Wendt, G. R. (1945b). Wesleyan University studies of motion sickness: VI. Prediction of sickness on a vertical accelerator by means of a motion sickness history questionnaire. *Journal of Psychology, 20*, 25–30.

Ames, S. L., Wolffsohn, J. S., & McBrien, N. A. (2005). The development of a symptom questionnaire for assessing virtual reality viewing using a head-mounted display. *Optometry and Vision Science, 82*(3), 168–176.

Askins, K., Mead, A. M., Lawson, B. D., & Bratley, M. C. (1997, May). Sopite syndrome study I: Isolated sopite symptoms detected post hoc from a preliminary open-ended survey of subjective responses to a short-duration visual-vestibular stimulus. *Aerospace Medical Association 68th Annual Scientific Meeting*, Chicago, IL. *Aviation, Space, and Environmental Medicine, 68*(7), 648.

Balk, S. A., Bertola, A., & Inman, V. W. (2013). Simulator sickness questionnaire: Twenty years later. In *Proceedings of the 7th International Driving Symposium on Human Factors in Driver Assessment, Training, and Vehicle Design*, 257–263.

Bittner, A. C., & Guignard, J. C. (1986a). Magnitude estimation of motion sickness in an operational environment. Abstracts from the *115th Meeting of the Acoustical Society of America*. *The Journal of the Acoustical Society of America, 79*(S1), S86–S87.

Bittner, A. C., Jr., & Guignard, J. C., (1986b). *Motion sickness evaluations in an at-sea environment: Seakeeping trials of a USCG cutter (WMEC 901)* (Report No. NBDL-86R002). New Orleans, LA: Naval Biodynamics Laboratory.

Bock, O. L., & Oman, C. M. (1982). Dynamics of subjective discomfort in motion sickness as measured with a magnitude estimation method. *Aviation, Space, and Environmental Medicine, 53*, 773–777.

Borg, G., & Borg, E. (2001). A new generation of scaling methods: Level-anchored ratio scaling. *Psychologica, 28*, 15–45.

Bos, J. E., de Vries, S. C., van Emmerik, M. L., & Groen, E. L. (2010). The effect of internal and external fields of view on visually induced motion sickness. *Applied Ergonomics, 41*(4), 516–521.

Bouchard, S., Robillard, G., & Renaud, P. (2007). Revising the factor structure of the simulator sickness questionnaire. *Annual Review of CyberTherapy and Telemedicine, 5*, 117–122.

Bratley, M. C., Lawson, B. D., & Mead, A. M. (1997, May). Sopite syndrome study II: Further evidence of sopite syndrome among aviation students indoctrinated aboard the multi-station disorientation demonstrator (MSDD). *Aerospace Medical Association 68th Annual Scientific Meeting*, Chicago, IL. *Aviation, Space, and Environmental Medicine, 68*(7), 648.

Brill, J. C., Kass, S. J., & Lawson, B. D. (2004, September). Mild motion questionnaire (MMQ): Further evidence of construct validity. In *Proceedings of the Human Factors and Ergonomics Society Annual Meeting* (Vol. 48, No. 21, pp. 2503–2507). SAGE Publications.

Brill, J. C., & Neilson, B. (2011). A factor-analysis perspective of the measurement of sopite syndrome in aerospace system. *International Symposium on Aviation Psychology* (pp. 369–374). Red Hook, NY: Curran Associates, Inc..

Bruck, S., & Watters, P. A. (2011). The factor structure of cybersickness. *Displays, 32*(4), 153–158.

Calkins, D. S., Reschke, M. F., Kennedy, R. S., & Dunlap, W. P. (1987). Reliability of provocative tests of motion sickness susceptibility. *Aviation, Space, and Environmental Medicine, 58*(9 Suppl.), A50–A54.

Chang, L. A. (1994). A psychometric evaluation of 4-point and 6-point Likert-type scales in relation to reliability and validity. *Applied Psychological Measurement, 18*(3), 205–215.

Chen, Y., Dong, X., Hagstrom, J., & Stoffregen, T. A. (2011). Control of a virtual ambulation influences body movement and motion sickness. *BIO Web of Conferences, 1, 00016*. Retrieved from: http://dx.doi.org/10.1051/bioconf/20110100016

Cheung, B., & Hofer, K. (2001). Coriolis-induced cutaneous blood flow increase in the forearm and calf. *Brain Research Bulletin, 54*(6), 609–618.

Chronbach, L. J., & Furby, L. (1970). How should we measure "change"—Or should we? *Psychological Bulletin, 74*(1), 68–80.

Cobb, S., Nichols, S., Ramsey, A., & Wilson, J. (1996). Health and safety implications of virtual reality: Results and conclusions from an experimental programme. *Proceedings of 96: Framework for Immersive Working Environments, the 2nd International Conference* (pp. 154–162). Pisa, Italy.

Cobb, S., Nichols, S., & Wilson, J. R. (1995, December). Health and safety implications of virtual reality: In search of an experimental methodology. *Proceedings of FIVEi95* (*Framework for Immersive Virtual Environments*) (pp. 227–242). London, U.K.: University of London.

Crampton, G. H. (Ed.). (1990). *Motion and space sickness*. Boca Raton, FL: CRC Press.

Dahlman, J. (2009). *Psychophysiological and performance aspects on motion sickness* (No. 1071, Linköping University Medical Dissertations). Linköping University, Linköping, Sweden.

Dobson, K. S., & Mothersill, K. J. (1979). Equidistant categorical labels for construction of Likert-type scales. *Perceptual and Motor Skills, 49*, 575–580.

Eagon, J. C. (1988). *Quantitative frequency analysis of the electrogastrogram during prolonged motion sickness* (Unpublished Master's thesis). Massachusetts Institute of Technology, Cambridge, MA.

Emmerik, M. L., de Vries, S. C., & Bos, J. E. (2011). Internal and external fields of view affect cybersickness. *Displays, 32*(4), 169–174.

Friedman, H. H., & Amoo, T. (1999). Rating the rating scales. *Journal of Marketing Management, 9*(3), 114–123.

Gal, R. (1975). Assessment of seasickness and its consequences by a method of peer evaluation. *Aviation Space and Environmental Medicine, 46*, 836–039.

Gianaros, P. J., Muth, E. R., Mordkoff, J. T., Levine, M. E., & Stern, R. M. (2001). A questionnaire for the assessment of the multiple dimensions of motion sickness. *Aviation, Space, and Environmental Medicine, 72*(2), 115.

Golding, J. F. (1998). Motion sickness susceptibility questionnaire revised and its relationship to other forms of sickness. *Brain Research Bulletin, 47*(5), 507–516.

Golding, J. F. (2006). Motion sickness susceptibility. *Autonomic Neuroscience, 129*(1), 67–76.

Golding, J. F., & Kerguelen, M. (1992). A comparison of the nauseogenic potential of low-frequency vertical versus horizontal linear oscillation. *Aerospace Medical Association, 63*, 491–497.

Graybiel, A. (1969). Structural elements in the concept of motion sickness. *Aerospace Medicine, 40*, 351–367.

Graybiel, A., Clark, B., & Zarriello, J. J. (1960). Observations on human subjects living in a "slow-rotation room" for periods of two days. *AMA Archives of Neurology, 3*, 55–73.

Graybiel, A., & Johnson, W. H. (1963). A comparison of the symptomatology experienced by healthy persons and subjects with loss of labyrinthine function when exposed to unusual patterns of centripetal force in a counter-rotating room. *Annals of Otology, Rhinology, and Laryngology, 72*, 1–17.

Graybiel, A., & Knepton, J. (1976). Sopite syndrome: A sometimes sole manifestation of motion sickness. *Aviation, Space and Environmental Medicine, 47*(8), 873–882.

Graybiel, A., Wood, C. D., Miller, E. F., & Cramer, D. B. (1968). Diagnostic criteria for grading the severity of acute motion sickness. *Aerospace Medicine, 39*, 4453–4455.

Griffin, M. J., & Howarth, H. V. C. (2000). *Motion sickness history questionnaire* (Technical Report No. 283). Hampshire, U.K.: Institute of Sound and Vibration Research (ISVR). Retrieved from: http://resource.isvr.soton.ac.uk/staff/pubs/PubPDFs/Pub1228.pdf

Griffin, M. J., & Newman, M. M. (2004). Visual field effects on motion sickness in cars. *Aviation, Space, and Environmental Medicine, 75*(9), 739–748.

Hardacre, L. E., & Kennedy, R. S. (1963). Some issues in the development of a motion sickness questionnaire for flight students. *Aerospace Medicine, 34*, 401–402.

Hemingway, A. (1943). *Adaptation to flying motion by airsick aviation students* (Research Report Project 170, Technical Report No. 4). Randolph Field, TX: School of Aviation Medicine.

Hoyt, R. E., Lawson, B. D., McGee, H. A., Strompolis, M. L., & McClellan, M. A. (2009). Modafinil as a potential motion sickness countermeasure. *Aviation, Space, and Environmental Medicine, 80*(8), 709–715.

Kennedy, R. S. (2009, June). *Development of a long term aftereffect questionnaire (LTAQ) for simulator and virtual reality sickness*. Paper Presented at International Symposium on Visual IMage Safety, Utrecht, the Netherlands.

Kennedy, R. S., Drexler, J., & Kennedy, R. C. (2010). Research in visually induced motion sickness. *Applied Ergonomics, 41*(4), 494–503.

Kennedy, R. S., Dunlap, W. P., & Fowlkes, J. E. (1990). Prediction of motion sickness susceptibility. In G. H. Crampton (Ed.), *Motion and space sickness* (pp. 179–215). Boca Raton, FL: CRC Press.

Kennedy, R. S., Fowlkes, J. E., & Hettinger, L. J. (1989). *Review of simulator sickness literature* (Technical Report No. NTSC TR89-024). Orlando, FL: Naval Training Systems Center.

Kennedy, R. S., & Graybiel, A. (1962). Symptomatology during prolonged exposure in a constantly rotating environment at a velocity of one revolution per minute. *Aerospace Medicine, 33*, 817–825.

Kennedy, R. S., & Graybiel, A. (1965). *The dial test: A standardized procedure for the experimental production of canal sickness symptomatology in a rotating environment* (Technical Report No. NSAM-930). Pensacola, FL: Naval School of Aviation Medicine.

Kennedy, R. S., Lane, N. E., Berbaum, K. S., & Lilienthal, M. G. (1993). Simulator sickness questionnaire: An enhanced method for quantifying simulator sickness. *International Journal of Aviation Psychology, 3*, 203–220.

Kennedy, R. S., Massey, C. J., & Lilienthal, M. G. (1995, July). Incidences of fatigue and drowsiness reports from three dozen simulators: Relevance for the Sopite Syndrome. In *First Workshop on Simulation and Interaction in Virtual Environments (SIVE'95)*.

Kennedy, R. S., Tolhurst, G. C., & Graybiel, A. (1965). *The effects of visual deprivation on adaptation to a rotating room*. Pensacola, FL: NSAM-918.

Keshavarz, B., & Hecht, H. (2011). Validating an efficient method to quantify motion sickness. *Human Factors: The Journal of the Human Factors and Ergonomics Society, 53*(4), 415–426.

Keshavarz, B., Hecht, H., & Lawson, B. D. (2014). Visually induced motion sickness—Causes, Characteristics, and Counter measures. In K. M. Stanney & K. Hale (Eds.), *Handbook of virtual environments: Design, implementation, and applications* (2nd ed., pp. 647–697). New York, NY: CRC Press, and imprint of Taylor & Francis Group, LLC.

Kieruj, N. D., & Moors, G. (2010). Variations in response style behavior by response scale format in attitude research. *International Journal of Public Opinion Research, 22*(3), 320–342.

Lane, N., & Kennedy, R. S. (1988). *A new method for quantifying simulator sickness: Development and application of the simulator sickness questionnaire (SSQ)* (EOTR 88-7). Orlando, FL: Essex Corporation.

Lawson, B. D. (1993). *Human physiological and subjective responses during motion sickness induced by unusual visual and vestibular stimulation* (Doctoral dissertation, Report #9322354).Brandeis University, Waltham, MA, University Microfilms International, Ann Arbor, MI.

Lawson, B. D. (2014). Motion sickness symptomatology and origins. In K. M. Stanney & K. Hale (Eds.), *Handbook of virtual environments: Design, implementation, and applications* (2nd ed., pp. 531–599). New York, NY: CRC Press, and imprint of Taylor & Francis Group, LLC.

Lawson, B. D., Graeber, D. A., Mead, A. M., & Muth, E. R. (2002). Signs and symptoms of human syndromes associated with synthetic experiences. In K. M. Stanney (Ed.), *Handbook of virtual environments: Design, implementation, and applications* (pp. 589–618). Mahwah, NJ: Lawrence Erlbaum Associates.

Lawson, B. D., Kass, S., Muth, E., Sommers, J., & Guzy, L. (2001). Development of a scale to assess signs and symptoms of sopite syndrome in response to mild or nonsickening motion stimuli. *Annual Meeting of the Aerospace Medical Association*. In *Aviation, Space, and Environmental Medicine, 73*(3), 255.

Lawson, B. D., McGee, H. A., Castaneda, M. A., Golding, J. F., Kass, S. J., & McGrath, C. M. (2009). *Evaluation of several common antimotion sickness medications and recommendations concerning their potential usefulness during special operations* (Technical Report No NAMRL 09-15). Pensacola, FL: Naval Aerospace Medical Research Laboratory.

Lawson, B. D., & Mead, A. M. (1998). The sopite syndrome revisited: Drowsiness and mood changes during real or apparent motion. *Acta Astronautica, 43*(3–6), 181–192.

Lawson, B. D., Mead, A. M., & Clark, J. B. (1997). Sopite syndrome case report II: Debilitating drowsiness, mood changes, and borderline neuro-vestibular performance in an aviator referred for air sickness. *Annual Scientific Meeting of the Aerospace Medical Association*. In *Aviation, Space, and Environmental Medicine, 68*(3), 648.

Marteau, T. M., & Bekker, H. (1992). The development of a six-item short-form of the state scale of the Spielberger State-Trait Anxiety Inventory (STAI). *British Journal of Clinical Psychology, 31*, 301–306.

McCauley, M. E. (Ed.). (1984). *Research Issues in Simulator Sickness: Proceedings of a Workshop*. Washington, DC: National Academy Press.

McCauley, M. E., & Sharkey, T. J. (1992). Cybersickness: Perception of self-motion in virtual environments. *Presence, 1*(3), 311–318.

Mead, A. M., & Lawson, B. D. (1997). Sopite syndrome case report I: Motion-induced drowsiness and mood changes in an individual with no other motion sickness symptoms—A case of "pure" sopite syndrome. *68th Annual Scientific Meeting of the Aerospace Medical Association*. In *Aviation, Space, and Environmental Medicine, 68*(7), 648.

Miller, E. F., & Graybiel, A. (1970a). *Comparison of five levels of motion sickness severity as the basis for grading susceptibility* (Bureau of Medicine and Surgery, NASA Order R-93). Pensacola, FL: Naval Aerospace Medical Institute.

Miller, E. F., & Graybiel, A. (1970b). A provocative test for grading susceptibility to motion sickness yielding a single numerical score. *Acta Otolaryngologica, 274*(Suppl), 1–20.

Muth, E. R., Stern, R. M., Thayer, J. F., & Koch, K. L. (1996). Assessment of the multiple dimensions of nausea: The nausea profile. *Journal of Psychosomatic Research, 40*, 511–520.

Nichols, S. (1999) *Virtual reality induced symptoms and effects: Theoretical and methodological issues* (PhD thesis). University of Nottingham, Nottingham, U.K.

Nichols, S., Cobb, S., & Wilson, J. R. (1997). Health and safety implications of virtual environments-Measurement issues. *Presence: Teleoperators and Virtual Environments*, *6*(6), 667–675.

Nilsson, A. L., Sunnerhagen, K. S., & Grimby, G. (2005). Scoring alternatives for FIM in neurological disorders applying Rasch analysis. *Acta Neurologica Scandinavica*, *111*, 264–273.

Parasuraman, A., Grewel, D., & Krishnan, R. (Eds.) (2006). Measurement and scaling in data collection. In *Marketing research* (2nd ed., Chapter 9). Cincinnati, OH: South-Western College Publishing.

Pesudovs, K. (2005). The development of a symptom questionnaire for assessing virtual reality viewing using a head-mounted display. *Optometry & Vision Science*, *82*(7), 571.

Preston, C. C., & Colman, A. M. (2000). Optimal number of response categories in rating scales: reliability, validity, discriminating power, and respondent preferences. *Acta Psychologica, 104*, 1–15.

Reason, J. T. (1968). Relations between motion sickness susceptibility, the spiral after-effect and loudness estimation. *British Journal of Psychology*, *59*(4), 385–393.

Reason, J. T., & Brand, J. J. (1975). *Motion sickness*. London, U.K.: Academic Press.

Reason, J. T., & Diaz, E. (1971). *Simulator sickness in passive observers* (Technical Report No. FPRC/1310, NTIS No. AD-753 560). Flying Personnel Research Committee. London, U.K.: Ministry of Defence.

Reason, J. T., & Graybiel, A. (1970). Changes in subjective estimates of well-being during the onset and remission of motion sickness symptomatology in the Slow Rotation Room. *Aerospace Medicine, 41*, 166–171.

Shapiro, C. M., Auch, C., Reimer, M., Kayumov, L., Heslegrave, R., Huterer, N., Driver, H., & Devins, G. M. (2006). A new approach to the construct of alertness. *Journal of Psychosomatic Research, 60*(6), 595–603.

Sharples, S., Cobb, S., Moody, A., & Wilson, J. R. (2008). Virtual reality induced symptoms and effects (VRISE): Comparison of head mounted display (HMD), desktop and projection display systems. *Displays, 29*(2), 58–69.

Solimini, A. G., Mannocci, A., & Di Thiene, D. (2011). A pilot application of a questionnaire to evaluate visually induced motion sickness in spectators of tri-dimensional (3D) movies. *Italian Journal of Public Health, 8*(2), 197.

Sprouse, J. (2008). Magnitude estimation and the non-linearity of acceptability judgments. In N. Abner & J. Bishop (Eds.), *Proceedings of the 27th West Coast Conference on Formal Linguistics* (pp. 397–403). Somerville, MA: Cascadilla Proceedings Project.

Stevens, S. S. (1946). On the theory of scales of measurement. *Science, 103*(2684), 677–680. doi:10.1126/science.103.2684.677

Van Emmerik, M. L., de Vries, S. C., & Bos, J. E. (2011). Internal and external fields of view affect cybersickness. *Displays, 32*(4), 169–174.

Wallace, J. C., Kass, S. J., & Lawson, B. D. (2002). *Assessing a specific measure of sopite syndrome: The mild motion questionnaire* (Technical Report No NAMRL-1414). Pensacola, FL: Naval Aerospace Medical Research Laboratory.

Wertheim, A. H., Bos, J. E., & Krul, A. J. (2001). *Predicting motion induced vomiting from subjective misery (MISC) ratings obtained in 12 experimental studies* (Report TNO-TM-01-A066). Soesterbery, NL: TNO Human Factors Research Institute.

Wiker, S. F., Kennedy, R., McCauley, M. E., & Pepper, R. L. (1979). Susceptibility to seasickness: Influence of hull design and steaming direction. *Aviation, Space, and Environmental Medicine, 50*(10), 1046–1051.

Wills, C. E., & Moore, C. F. (1994). A controversy in scaling of subjective states: Magnitude estimation versus category rating methods. *Research in Nursing and Health, 17*, 231–237.

Wood, C. D., Graybiel, A., & McDonough, R. G. (1966). Clinical effectiveness of anti-motion-sickness drugs. *Journal of the American Medical Association*, *198*(11), 1155–1158.

Wood, J. M., Chapman, K., & Eilers, J. (2011). Tools for assessing nausea, vomiting, and retching: A literature review. *Cancer Nursing*, *34*(1), E14.

25 Adapting to Virtual Environments

Robert B. Welch and Betty J. Mohler

CONTENTS

25.1 OVERVIEW

This chapter describes and illustrates a variety of procedures for reducing or eliminating the deleterious effects of the *sensory rearrangements* that plague virtual environment (VE) technology. A sensory rearrangement exists when the sensory array and/or the relationship between sensory systems differs from normal. A well-known example is viewing one's hand through a light-displacing wedge prism (e.g., Redding & Wallace, 1997). Initial exposure to a sensory rearrangement causes misperception, the surprise of violated expectations, and sensory–motor disruption. Due to limitations in many current VEs, users are often exposed to such rearrangements, with predictable consequences for perception and behavior. For example, a mismatch between a hand tracker and the visual image of the user's hand can be expected to produce the same kind of sensory rearrangement as prismatic displacement. Because human beings can adapt to a wide range of sensory rearrangements, it is reasonable to assume they are equally able to adapt to the sensory rearrangements that occur in VEs. Likewise, the variables found to influence adaptation to sensory rearrangements in general, for example, active interaction, should apply as well to VE adaptation. Thus, the VE training procedures described in this chapter are based on the variables demonstrated by previous research to control or influence adaptation to rearranged sensory environments. Research indicates that the problematical postexposure aftereffects that inevitably result from VE adaptation can be reduced or eliminated by means of systematic readaptation procedures and *dual adaptation* training (i.e., repeated alternation between adaptation and readaptation). Finally, the ubiquity of individual differences in adaptation to sensory rearrangements is acknowledged and implications for the individualization of VE adaptation-training procedures considered.

25.2 ADAPTATION AS A SOLUTION TO VE LIMITATIONS

Despite tremendous advances in the technology of so-called virtual reality (e.g., Durlach & Mavor, 1995; Slater, Perez-Marcos, Ehrsson, & Sanchez-Vives, 2009), it is rarely mistaken for reality. Thus, most VEs, including *augmented realities* (e.g., see-through displays), continue to be plagued by such deficiencies as poor lighting, unrealistic graphics, cumbersome wearable displays or interactive devices (see Willemsen, Colton, Creem-Regehr, & Thompson, 2009), and a multitude of sensory rearrangements that render it a poor imitation of everyday experience. More importantly, these shortcomings are the source of a variety of sensory/perceptual, behavioral, and even physical complaints, many of which afflict nearly all VE users to some degree (see Table 25.1). The last malady listed in Table 25.1, motion-sickness symptoms (variously referred to as simulator sickness, VE sickness, and cybersickness), is dealt with in length in Lawson (2014a, 2014b, Chapters 23 and 24). Thus, although the occurrence of simulator sickness is typically considered indirect evidence of the presence of intersensory conflicts and its recovery a sign that adaptation has taken place, these topics will, for the most part, be ignored here.

TABLE 25.1

Sensory/Perceptual, Behavioral, and Physical Complaints Afflicting VE Users

Sensory/perceptual problems reported by VE users

- Momentary reduction in binocular acuity (e.g., Mon-Williams, Rushton, & Wann, 1993)
- Misperception of depth and distance (e.g., Kunz et al., 2009; Loomis & Knapp, 2003; Roscoe, 1993; Thompson et al., 2004; Willemsen et al., 2009)
- Changes in dark accommodative focus (Fowlkes, Kennedy, Hettinger, & Harm, 1993)
- Potentially dangerous *delayed flashbacks* (e.g., illusory experiences of climbing, turning, and inversion) that may not surface until several hours after user has left an airplane simulator or similar VE (e.g., Kennedy, Fowlkes, & Lilienthal, 1993)

Disruptive behavioral effects of VEs

- Disturbed perceptual–motor (e.g., hand–eye) coordination (e.g., Biocca & Rolland, 1998)
- Locomotory and postural instability (e.g., DiZio & Lackner, 1997)
- Degraded task performance (e.g., Fowlkes et al., 1993)

Physical complaints reported by VE users

- Eyestrain, or *asthenopia* (e.g., Mon-Williams et al., 1993), which may be symptomatic of underlying distress of or conflict between oculomotor subsystems (e.g., Ebenholtz, 1992)
- Headaches (e.g., Mon-Williams, Rushton, & Wann et al., 1995)
- Cardiovascular, respiratory, or biochemical changes (e.g., Calvert & Tan, 1994; Mohler et al. 2007a)
- Motion-sickness symptoms (e.g., pallor, sweating, fatigue, and drowsiness, although rarely vomiting; e.g., Gower et al., 1987)
- Weight of HMD is too heavy (e.g., Willemsen et al., 2009)

Clearly, the adverse effects of VEs pose a serious obstacle to optimal task performance and training with these devices. Furthermore, as Biocca (1992) has noted, unpleasant experiences in VEs, especially if well publicized, are likely to have a chilling effect on the general diffusion of VE technology. In short, there is ample motivation to do whatever is necessary to overcome the problems that frequently disrupt the human–VE interface. There are two very different, albeit complementary, ways in which to respond to this imperative. The first is to modify (typically, improve) the VE to accommodate the user, and the second is to modify the user to accommodate the VE. Designers and implementers of VEs have typically chosen the first of these strategies. That is, they tend to seek engineering solutions (e.g., more powerful computers, faster tracking devices, and improved displays) for eliminating the deficiencies of their devices. According to the second and much less common tactic, the VE is left as is, and users are provided with systematic training and/or deliberate strategies aimed at reducing, or perhaps merely circumventing, its inadequacies. It is generally agreed that the first of these two approaches will ultimately prevail as computer technology is perfected and becomes more generally affordable. However, the second strategy has the great advantage that it can be implemented now, without having to wait for those improvements to be realized. Some of the limitations of current VE technology, such as poor lighting, small field of view (FOV), and heavy headgear, can only be overcome by technological improvements. However, to technically achieve a replication of human sensory inputs both in resolution and in relative timing to each other is an extremely difficult technical task, one not likely to be achieved in the near future. The human body is adapted to the specific timing offsets of the sensory systems, and even a slight difference in this timing or any other aspects of the sensory input would require at least a brief period of adaptation, as has been shown by scientists who study multisensory integration (Ernst & Banks, 2002; Ernst & Buelthoff, 2004; Hillis, Ernst, Banks, & Landy, 2002). Additionally, shortcomings such as discrepancies between felt and seen limb position, delays of visual feedback, and distortions of perceived depth represent the kinds of sensory rearrangements that have proven amenable to user-training procedures, in particular, those that promote adaptation. It is these procedures that represent the focus of this chapter.

Adaptation to sensory rearrangement is defined here as a "change of perception and/or sensory-motor coordination that serves to reduce or eliminate a registered discrepancy between or within sensory modalities or the errors in behavior induced by this discrepancy" (Welch, 1978, p. 8). Adaptation is measured in two ways: (1) the reduction of observers' perceptual and/or sensory–motor errors during exposure to the sensory rearrangement (the *reduction of effect*) and (2) postexposure *negative aftereffects* (errors in the direction opposite those initially elicited by the sensory rearrangement). Adaptation is automatic and *unconscious*. For example, the negative aftereffects of hand–eye coordination occur even when the subject is aware that the sensory rearrangement has been removed. Sensory–motor adaptation can occur even when perceptual adaptation has not, as illustrated by the pioneering study of Snyder and Pronko (1952) in which for an entire month a subject wore goggles that inverted his visual field. Despite the fact that by the end of this extended period his sensory–motor coordination was essentially error-free, the world still appeared to be upside-down to him. Thus, it may be concluded that perceptual adaptation (how things look) predicts behavioral adaptation (the ability to correctly interact with it) but not vice versa. Before examining how research on adaptation can assist the design of procedures for training VE users to overcome the sensory rearrangements to which they may be exposed, it will be useful to specify what is currently known about these sensory rearrangements.

25.3 SENSORY REARRANGEMENTS FOUND IN SOME VEs

There are five major deficiencies of VEs for which the processes of perceptual– and sensory–motor adaptation are relevant. These are (1) intersensory conflicts, (2) distortions of depth and distance, (3) distortions of form and size, (4) the loss of stability of the visual field due to visual feedback delays and asynchronies, and (5) sensory inconsistencies in the form of randomly varying distortions. This section will discuss each of these in turn.

25.3.1 Intersensory Conflicts

This category entails mismatches between paired spatial modalities—vision, audition, touch, proprioception, and the haptic and vestibular senses. Thus, the computer-generated surrogate of the observer's hand may be seen in one location, while the hand itself is felt in another, an object may look rough but feel smooth, and a sound source may be seen in one place but heard elsewhere. Fortunately, VE users readily accept *ownership* of surrogate hands or other limbs that do not necessarily look identical to the actual body parts, especially if movements and stimulation of the real and surrogate limbs are synchronous (e.g., Pusch et al., 2011; Slater, 2009).

Some intersensory discrepancies entail the *absence* of one member of a pair of normally correlated sensory impressions. For example, in fixed-base aircraft simulators, visual motion is often displayed without the vestibular and other inertial cues (e.g., tactual and somatosensory stimuli) that usually accompany the visuals. Likewise, many VEs allow observers to manipulate a virtual object with their virtual hands while providing no tactual or force feedback about these *contacts*. (An exception is a study by Lok, Naik, Whitton, and Brooks (2003), in which real objects were used in a spatial-cognitive task.)

The type and severity of the effects of intersensory conflicts depend, in part, on which sensory modalities are involved. For example, conflicts between seen and felt positions of the limb will merely produce reaching errors (e.g., Welch, 1986), whereas conflicts between visual and vestibular–inertial cues' characteristic of many flight simulators are likely to cause disorientation, postural instability (ataxia), and motion sickness–like symptoms (e.g., Oman, 1991). Conflicts between spatial vision and vestibular sensations during movement of the observer's head or entire body are particularly problematic, both because of their unpleasant behavioral and gastrointestinal consequences and the fact that, at least at present, they are very difficult or impossible to avoid. Even very sophisticated VE technology may be incapable of preventing serious intersensory discrepancies of this sort. For example, it has been argued (e.g., DiZio & Lackner, 1992) that the ability of VE technology to simulate inertial motion may never improve to the point where some adaptation will not be necessary. Thus, motion-based aircraft simulators may always be afflicted by significant discrepancies between the ranges and amplitudes of the applied inertial forces and the visual motions with which they are paired. Even the most sophisticated current simulators, such as the MPI CyberMotion simulator (Barnett-Cowan, Meilinger, Vidal, Teufel, & Bülthoff, 2012), the SIMONA Research Simulator (Berkouwer, Stroosma, Van Paassen, Mulder, & Mulder, 2005), and the Toyota research driving simulator (Slob, 2008), are likely to be afflicted by these problems. For example, mimicking the takeoff of a fighter plane from an aircraft carrier or a violently maneuvering airplane will almost certainly produce substantial conflicts between the two senses because the appropriate gravitational forces exceed the capacity of any existing simulator. The fact that motion-based aircraft simulators tend to be more nauseating than their fixed-based counterparts (e.g., Gower, Lilienthal, Kennedy, & Fowlkes, 1987) may be due to the presence of such sensory conflicts. Further limitations in attempting to simulate gravitational–inertial forces stem from the fact that sustained hypogravity (i.e., G < 1.0) cannot be simulated by earthbound VEs, whereas hypergravity, whether brief or prolonged, requires a human centrifuge (e.g., Cohen, Crosbie, & Blackburn, 1973), a device unavailable or unaffordable to most investigators.

A common instance of sensory spatial conflict in a VE is when a normally correlated sensory modality is simply not stimulated. Many display arrangements still use only visuals and head tracking, even though devices involving haptic, olfactory, gustatory, and the vestibular systems do exist (see Basdogan & Loftin, 2008, for a summary of existing multimodal displays). Here, too, the behavioral and physical effects are diverse. The absence of tactual or force feedback when attempting to grasp a virtual object, although unlikely to disturb gross reaching behavior, may interfere with fine manual control. On the other hand, when visual motion occurs without the normally attendant vestibular and other inertial cues, as in fixed-base aircraft simulators, the operator may experience a dramatic sense of virtual body motion known as *vection*. This illusion is nauseating for some individuals, as is the *freezing* of an aircraft simulator (or other VE) in an unusual orientation, which can occur when the user has engaged in an extraordinary maneuver (e.g., McCauley & Sharkey, 1992).

25.3.2 DEPTH AND DISTANCE DISTORTIONS

One form of static depth distortion occurs in some VEs when visual objects that are being depicted as closer to the observer's viewpoint fail to occlude the supposedly more distant objects. Another example is when visual objects, whose edges are fuzzy due to poor resolution and which, on the basis of *aerial perspective*, cues should thus appear far away, are instead presented with the saturated hue and optical sharpness of a much nearer object.

Moving the head or entire body while viewing a VE through a stereoscopic head-mounted display (HMD) frequently causes virtual objects to appear to change their relative positions (as well as their sizes and shapes) in the distance dimension. This form of dynamic depth distortion is caused because the VE system is unable to calculate correctly and instantly the images presented to the HMD display during the bodily movement. The same is true if HMD wearers are stationary and the objects are moved around them. This movement-contingent depth distortion, referred to by perceptual psychologists as the *kinetic depth effect*, is likely to disrupt the user's visual–motor interactions (e.g., pointing or reaching) with respect to these objects and may cause simulator sickness. Fortunately, this form of sensory rearrangement is subject to a certain amount of visual adaptation (e.g., Wallach & Karsh, 1963).

Besides perceptual distortions of depth (i.e., relative distance), the apparent absolute distance of objects is underestimated, as measured, for example, by errors in reaching without visual feedback. Studies using indirect and direct walking and pointing measures (Loomis & Knapp, 2003; Thompson et al., 2004), verbal estimates (Mohler, Creem-Regehr, & Thompson, 2006), and throwing (Sahm, Creem-Regehr, Thompson, & Willemsen, 2005) have found that people underestimate egocentric distances in the virtual world as compared to real-world performance. It has also been shown that while the quality of computer graphics influences action responses to distances, verbal estimates of these distances are not influenced (Kunz, Wouters, Smith, Thompson, & Creem-Regehr, 2009). Recently, scientists have demonstrated that seeing a virtual self-avatar enables people to make veridical estimations of egocentric distances in HMD VEs (Mohler, Creem-Regehr, Thompson, & Bülthoff, 2010; Ries, Interrante, Kaeding, & Anderson, 2008). Finally, in situations that entail very few depth cues, distance perception may be ambiguous, which is likely to result in substantial moment-to-moment variability in hand–eye coordination.

A common discrepancy observed with stereoscopic VE displays is between ocular vergence and accommodation of the lens. Under everyday circumstances, fixating an object in the distance dimension entails convergence of the eyes to produce a single image, together with the appropriate shaping of the lens for fine focusing. However, in many stereoscopic VEs, the convergence and accommodation appropriate for the visual display are placed in conflict. Thus, the stereoscopic stimuli may be set for a distance that differs from the one for which optimal focusing will occur. Besides leading to reaching errors, this discordance may cause diplopia, asthenopia, and even nausea (e.g., Ebenholtz, 1992). Unfortunately, it may be difficult or even impossible to overcome this problem by means of adaptation since, as Wann and Mon-Williams (1997) have argued, many current stereoscopic VE displays present the operator with a *range* of binocular targets to which they may verge. Clearly, such a situation fails to provide the consistent *rule* required for adaptation.

Roscoe (1993) has demonstrated that pilots who are wearing see-through HUDs and viewing collimated virtual images do not focus their eyes on infinity, but instead toward their resting accommodation (approximately 1 m away). The results of this accommodative error are that (1) the actual visual scene beyond the heads-up display (HUD) appears minified, (2) a terrain or airport runway appears to be observed from a higher altitude than is actually the case, and (3) familiar objects look farther away than they really are. The behavioral consequences of these misperceptions are obvious and, according to Roscoe, potentially *lethal*. Unfortunately, magnifying the images in an attempt to compensate for these illusions is not feasible because of the limitation this hardware modification places on display size (Roscoe). Fortunately, humans are able to adapt to distance and size distortions. Much of the evidence for this claim comes from studies of adaptation to underwater distortions caused by the glass–water interface of the diver's mask (e.g., Ross, 1971). More recent studies have shown that using various types of sensory feedback about absolute target distances, observers are able to adapt their spatial estimates of distance (Mohler et al., 2006; Richardson & Waller, 2005, 2007; Richardson 2008; Waller et al., 1998).

25.3.3 Shape and Size Distortions

An artifact of many VE systems is the illusory curvature of contours, most notably the so-called pincushion effect, in which the sides of a rectilinear shape (especially one that occupies most or all of the FOV) appear to be bowed inwardly. Magnification or minification of the visual scene or of objects within it can also occur with some VEs, especially those that distort perceived distance. Thus, in the example of the see-through HUD described previously (Roscoe, 1993), overestimation of the distance of familiar objects caused them to appear too large, presumably based on the mechanism of *misapplied size constancy scaling*, used to explain the *moon illusion* (e.g., Rock & Kaufman, 1962). In a study by Kuhl, Thompson, and Creem-Regehr (2009), pincushion distortion was accounted for in the calibration of an HMD system, and it was discovered that while this distortion is undesirable, it does not in and of itself account for the underestimation of egocentric distances in HMDs.

The few published studies on optically induced size distortions have revealed only a modest amount of visual adaptation (Welch, 1978, Chapter 8). This same resistance to adaptation applies to prismatically induced curvature, as seen most convincingly in an experiment by Hay and Pick (1966) in which subjects who wore prism goggles for heroic seven consecutive weeks underwent adaptation of only about 30% of the optical curvature to which they were exposed. Head movements in the presence of a magnified or minified visual field cause oculomotor disruption (and loss of visual position constancy) from the increased or reduced gain of the vestibulo-ocular reflex (VOR). Although this is likely to be a sickening experience, it can apparently be entirely eliminated by adaptation (e.g., Collewijn, Martins, & Steinman, 1983). Even the drastic loss of visual position constancy due to head movements while wearing right–left reversing goggles eventually disappears (e.g., Stratton, 1897), although it is interesting to note that the resulting loss of VOR under these circumstances resists recovery (e.g., Gonshor & Mevill Jones, 1976).

25.3.4 Delays of Sensory Feedback

Perhaps the most serious and currently intractable flaw of many VEs is the presence of significant delays (lags) between the operator's movements and resulting visual, auditory, and/or somatosensory feedback. Such delays can occur because of insufficient refresh rates and relatively slow position trackers. Studies of flight simulation have revealed that lags of as little as 50 ms can have a measurable effect on performance, whereas longer delays cause serious behavioral oscillations (e.g., Wickens, 1986). Faster computers are reducing these lags, and an easy and inexpensive method of measuring latency in VE systems involving photodiodes and an audio card has recently been proposed (Di Luca, 2010).

An especially disruptive and unpleasant effect of visual delay is the interference of the VOR and concomitant illusory visual motion that occurs during head or entire body movements. Normally, a head movement in one direction causes an approximately equal eye movement in the opposite direction, effectively nulling motion (both real and perceived) of the visual field relative to the head.

However, visual feedback delay causes the visual field to lag behind the head movement, initially making the VOR inappropriate and causing the visual world to appear to move in the same direction as the head. This loss of visual position constancy (or *oscillopsia*) is both behaviorally disruptive and nauseating. Furthermore, it has been suggested that these delays can cause simulator sickness and other physical problems, even when they are below the observer's sensory threshold (Ebenholtz, 1992).

Visual feedback delays from motor movements that do not involve vestibular stimulation, such as reaching or pointing, rarely cause motion-sickness symptoms. However, they can lead to other problems, ranging from simple distraction to serious discoordination. Auditory feedback delays also cause difficulties, particularly with speech (e.g., Smith & Smith, 1962). Fortunately, as with intersensory discrepancies of spatial perception, human beings can adapt, at least partially, to both visual–motor and visual–auditory delays (Cunningham, Billock, & Tsou, 2001; Di Luca, 2010).

Variable delays are more problematic than fixed ones. The same holds for asynchronies between the onsets of two (or more) sensory systems, as, for example, with some motion-based aircraft simulators involving mismatches between the onset times of visual and inertial stimuli. Kennedy (personal communication, May 15, 1994) believes that these problems may be an especially important factor in the etiology of simulator sickness.

25.3.5 SENSORY *DISARRANGEMENT*

When the size and/or direction of a VE-induced sensory rearrangement varies from moment to moment (rather than remaining constant over an extended time period), a condition of sensory *disarrangement* is said to exist (e.g., Cohen & Held, 1960). Examples with respect to VEs include (1) so-called jitter (jiggling of the visual image due to electronic noise in the position tracker and/or the image-generator system) and (2) moment-to-moment variability in the absolute and relative accuracy of position trackers in monitoring the operator's limbs and body transport. These limitations have been discussed by Meyer, Applewhite, and Biocca (1992). Such noise can cause the seen and felt positions of the hand to shift randomly relative to each other. Not surprisingly, such unpredictably changing sensory conditions resist adaptation since there is no constant compensatory *rule* on which such an adaptive process can be based. However, as discussed in a later section, although no change in *average* perceptual or perceptual–motor performance is likely to take place in the presence of a disarrangement, an increase in moment-to-moment behavioral variability (e.g., Cohen & Held, 1960) will occur, which represents a potentially serious degradation of the user's performance.

It is apparent that many of the sensory distortions found in current VEs are similar or identical to those deliberately created by investigators by means of lenses, prisms, and mirrors (see Redding & Wallace, 1997; Welch, 1978, for extensive reviews of this literature). The sensory rearrangements examined include optically induced (1) lateral rotation (yaw) of the visual field (e.g., Held & Hein, 1958), (2) visual tilt (roll, e.g., Ebenholtz, 1969; horizontal, e.g., Adams, Banks, & van Ee, 2001), (3) displacement in the distance dimension (e.g., Held & Schlank, 1959), (4) curvature (e.g., Hay & Pick, 1966), (5) right–left reversal (e.g., Kohler, 1964), (6) inversion (e.g., Stratton, 1897), (7) depth and distance distortion (e.g., Wallach, Moore, & Davidson, 1963), (8) changes in visual size (e.g., Rock, 1965), and (9) altered cues for visual motion during locomotion (e.g., Durgin, Fox, & Kim, 2003; Mohler et al., 2007b; Philbeck, Woods, Arthur, & Todd, 2008; Rieser, Pick, Ashmead, & Garing, 1995). Acoustical rearrangements have entailed (1) small (e.g., 10°) lateral rotations of the auditory field (e.g., Held, 1955), (2) right–left reversal (e.g., Young, 1928), and (3) functional increases in the length of the interaural axis (e.g., Durlach & Pang, 1986). Another atypical sensory environment to which humans have been exposed and adapted is altered gravitational–inertial force—hypergravity (e.g., Welch, Cohen, & DeRoshia, 1996), hypogravity (e.g., Lackner & Graybiel, 1983), and alternating hyper- and hypogravity (e.g., Cohen, 1992). Finally, underwater distortions caused by the air–glass interface of the diver's mask have been the subject of extensive investigation (e.g., Ross, 1971).

On the basis of this vast literature on adaptation to sensory rearrangement, one conclusion is certain: human beings (and other mammals) are able to adjust their behavior and, to a lesser extent, their perception to any sensory rearrangement to which they are actively exposed, as long as this rearrangement remains essentially constant over the time interval in which sensory feedback is provided. It is important to note that adaptation is not a unitary process, but rather varies greatly in both acquisition rate and magnitude as a function of type of sensory rearrangement and the specific adaptive component (visual, proprioceptive, vestibulo-ocular, etc.) under consideration. Furthermore, as mentioned previously, while human subjects are capable of adapting their *behavior* nearly completely to even the most dramatic visual rearrangements, such as 180° rotation of the visual field, adaptive changes in *visual perception* often fail to occur (e.g., Snyder & Pronko, 1952). On the other hand, a certain amount of visual adaptation does result from exposure to lesser distortions, such as prismatic displacement (e.g., Craske & Crawshaw, 1978) and curvature of the visual field (e.g., Hay & Pick, 1966).

Since the technological limitations of many VEs create many of the same sensory rearrangements imposed in laboratory experiments examining adaptation, it is reasonable to conclude that procedures that incorporate the variables demonstrated by these experiments to control or facilitate adaptation will be especially useful for the systematic training of VE users. Expediting this adaptive process should, in turn, facilitate task performance and transfer of training to the real-world task (if any) for which the VE is designed. Let us now look at the variables shown to influence or control adaptation to sensory rearrangement and consider how they might be integrated into VE adaptation-training procedures.

25.4 ADAPTING TO SENSORY REARRANGEMENTS AND APPLICABILITY TO VE TRAINING PROCEDURES

25.4.1 STABLE REARRANGEMENT

As indicated previously, exposure to a sensory rearrangement that is continuously changing in magnitude and/or direction will fail to produce adaptation. This condition is referred to as *disarrangement* and has been shown to have a degrading effect on hand–eye coordination, as seen by an increase in moment-to-moment variability of open-loop target pointing. This conclusion is illustrated by a pioneering experiment by Cohen and Held (1960) in which subjects viewed their actively moving hands through prisms whose strength varied continuously from 22° leftward through no displacement to 22° rightward and the reverse, at a rate of one cycle every 2 min. Although this experience had no effect on average target-pointing accuracy along the lateral dimension, (i.e., no adaptive shift), it greatly increased the variability of subjects' open-loop target-pointing behavior over repeated attempts. An analogous result has been reported for auditory disarrangement (e.g., Freedman & Pfaff, 1962) as measured by auditory localization errors in the right–left dimension. In the *real world* and in VEs, organisms are presented with both variable and constant errors, and it is critical to distinguish between them since it is only with respect to the latter error that adaptation is a useful solution. A recent study by Wei and Körding (2009) demonstrated that human observers are capable of this distinction between irrelevant and relevant errors, adapting only to the latter.

Another example of disarrangement is the situation in which delays of visual feedback from bodily movements vary from moment to moment, as may occur in some VEs. Visual delays are difficult enough to handle when they are constant over time but represent a much more serious problem when they are inconsistent or variably asynchronous (e.g., Uliano, Kennedy, & Lambert, 1986).

In summary, sensory disarrangement represents a distortion whose effects cannot be ameliorated by adaptation-training procedures because the basic premise upon which adaptation is based (i.e., the presence of fixed sensory rearrangement) does not hold. It would appear, therefore, that VEs that suffer from this problem must await engineering solutions such as more stable position trackers.

25.4.2 ACTIVE INTERACTION

Given the presence of a stable sensory rearrangement, one can single out five major variables that control or facilitate adaptation. These are (1) active interaction, (2) error-corrective feedback, (3) immediate feedback, (4) incremental exposure to the rearrangement, and (5) the use of distributed practice.

It is generally agreed that the most powerful of all the controlling conditions for adaptation to sensory rearrangement is active interaction with the altered sensory environment, coupled with the visual consequences of these actions (*reafference*). In short, passive exposure to a sensory rearrangement produces little or no adaptation (e.g., Held & Hein, 1958). Exactly why passive exposure is so ineffectual is, however, subject to serious debate (e.g., Welch, 1978, pp. 28–29). One possibility is that active interaction provides the observer with unmistakable information about the sensory rearrangement, which, in turn, initiates and/or catalyzes the adaptive process. For example, because felt limb position is more precise when bodily movement is self-initiated than when controlled by an external force (e.g., Paillard & Brouchon, 1968), the discrepancy between seen and felt limb positions can be assumed to be particularly pronounced during active movement, thereby facilitating the adaptive process.

One caveat to the endorsement of active visual–motor behavior as a facilitator of adaptation is that it is necessary that the motor intentions (*efference*) generating these bodily movements actually be in conflict with the resulting reafference. For example, when wearing prism goggles, subjects' active hand or head movements initially lead to visual consequences (e.g., errors in localizing a visual target) that conflict with those that usually occur. However, their eye movements will not be in error. Thus, although subjects who are wearing prism goggles and instructed to rapidly turn and face a visually perceived object will err in the direction of the prismatic displacement, they will have no difficulty turning their eyes to fixate the perceived locus of the object. The reason that the latter is true is that the prism goggles do not alter the relationship between retinal locus of stimulation and the appropriate fixating eye movements. On the other hand, instead of being attached to goggles, the prisms are affixed to contact lenses or are otherwise controlled by eye movements (e.g., White, Shuman, Krantz, Woods, & Kuntz, 1990); the situation is quite different. With this optical arrangement, not only will head movements be in error but eye movements as well. The results of the few relevant studies using this unusual arrangement (e.g., Festinger, Burnham, Ono, & Bamber, 1967) suggest that observers who are so accoutered are not only able to correct their eye movements but also undergo much more visual adaptation to prismatically induced curvature of contours than is the case with traditional prism goggles. Perhaps adaptation to other discrepancies such as altered size and displacement would also be enhanced by such means. Therefore, for VEs designed to be controlled by the operator's eyes, it is possible that any visual discordance present will undergo stronger adaptation than that found with VEs controlled by head movements (e.g., those involving an HMD).

A relatively recent method of providing active interaction with a VE is via a self-avatar (or parts thereof). This procedure is supported by the popular theory of grounded cognition, which argues that the body is especially important for perception of the surrounding world (e.g., Barsalou, 2008). Scientists have found that visual–motor coupling, as well as visual–haptic stimulation, is sufficient to feel ownership of the visual body part or whole (Sanchez-Vives, Spanlang, Frisoli, Bergamasco, & Slater, 2010; Sanchez-Vives & Slater, 2006; Slater et al., 2010).

Recent research (e.g., Slater, 2009) investigating the experience of presence in VEs suggests that two factors are important for creating a truly realistic virtual world: *being there*, or the place illusion, and plausibility. Slater argues that these two experiences can lead to realistic behavior in immersive VEs. This is likely due to the idea that if the expectations of what actions can be performed are known and their relationship to the surrounding spatial environment known, less adaptation is necessary.

It is clear from the preceding discussion that to produce substantial sensory and sensory–motor adaptation, VE users must be given the opportunity to actively interact with the sensory environment being displayed. This can be a two-edged sword, however, because before active movement can have this beneficial effect on adaptation, it may first cause some users to get sick. For example, VEs that create a sense of gravito-inertial force by means of passive transport in a centrifuge or visual flow field are very likely to induce motion-sickness symptoms as soon as the user engages in active bodily movements, particularly of the head (e.g., DiZio & Lackner, 1992). In such situations, those participants who cannot tolerate such unpleasant effects may abandon the VE before much adaptation has had a chance to occur.

25.4.3 ERROR-CORRECTIVE FEEDBACK

The combination of making errors and then correcting for them when reaching for objects in a rearranged visual environment facilitates adaptation beyond the effect attributable to visual–motor activity alone. For example, Welch (1969) showed that when subjects were allowed to make error-corrective (visually closed-loop) target-pointing responses, they adapted significantly more than when they actively moved the visible hand in the same manner but without visual targets. Although the facilitating effect of error-corrective feedback on adaptation is now well established, it is not sufficient merely that targets are available in the environment. Rather to be most effective, target-pointing responses must be made either ballistically (i.e., so rapidly that their trajectories cannot be altered en route to the target) or in such a manner that the outcome of a given attempt is not observable until it has reached its goal (often referred to as *terminal exposure*). The latter condition was used in the Welch (1969) experiment. If instead observers are allowed to make slow, visually guided reaching movements (*concurrent exposure*), they will almost certainly *zero in* on the target on each attempt and thus experience little or no error when the hand finally reaches its goal. The fact that concurrent exposure is thus less informative than terminal exposure is almost certainly the reason why adaptation is much greater in the latter condition (e.g., Welch, 1978, pp. 29–31). In sum, it can be concluded that VE exposure should, where possible, entail numerous targets with which to actively interact, together with unambiguous error-corrective feedback from these interactions, preferably at the termination of each response.

25.4.4 IMMEDIATE SENSORY FEEDBACK

Delays of motor-sensory feedback represent one of the most serious limitations of current VEs. Even with quite short delays, severe behavioral disruption occurs and is not amenable to adaptation as defined here (e.g., Smith & Smith, 1962). Such a state of affairs would seem to argue against the utility of adaptation-training procedures as a means of eliminating this problem. Fortunately, as will be seen later in this section, there are some situations in which the effects of delayed feedback in VEs are not quite so dire.

The initial effects of visual feedback delays are quite different when the vestibular system and other sources of inertial information are involved than when they are not. An example of the first of these situations is when an HMD-wearing subject views the visual field while turning the head whose movements are being monitored by a relatively slow tracking device. Thus, in this situation, the opposite sweep of the visual field begins slightly after head movement has begun and continues briefly after it is completed. An example of the second is rapidly moving the hand and seeing its virtual image lagging behind its true position because the refresh rate of the VE cannot keep pace. While the former condition is characterized by visual–spatial instability and possibly symptoms of motion sickness, the latter seriously interferes with hand–eye coordination but does not appear to be nauseating (e.g., Held, Efstathiou, & Greene, 1966).

A further complication involves the situation in which a sensory feedback delay is superimposed on a second intersensory rearrangement, such as a discrepancy between felt and seen limb positions. In this case, the ability to adapt to the latter condition may be severely impeded. For example,

Held et al. (1966) showed that visual–motor adaptation to prismatic displacement is greatly reduced or even prevented by delays of visual feedback as short as 120 ms. Because many current VEs are subject to much greater delays than this when visually rendering a limb movement, it might seem, as Held and Durlach (1993) have argued, that users will be unable to use adaptation to overcome any spatial (or other) discordance that may also be present. Fortunately, however, this pessimistic conclusion appears to be limited to the type of exposure condition typically preferred by Held and his colleagues in which subjects do not reach for specific targets but merely view their hands as they move them from side to side before a visually homogeneous background. In contrast, it has been demonstrated that when subjects are provided with unambiguous error-corrective feedback from discrete target-pointing responses, substantial adaptation is possible despite the presence of significant visual feedback delays. For example, in an experiment by Rhoades (described by Welch, 1978, p. 105), subjects revealed significant hand–eye adaptation to prismatic displacement even when error-corrective feedback was delayed by as much as 8 s. Furthermore, considerable adaptation has been reported in a number of published studies (e.g., Welch, 1972) in which for procedural reasons the investigators were forced to institute visual error-corrective feedback delays of up to approximately 1 s. Fortunately, most current VEs, perhaps with the exception of teleoperation VE arrangements, entail head tracking and visual scene rendering with latencies well below 120 m. Representing the human body movements of a virtual avatar typically involves greater latencies but is now also much below 1 s.

What, then, can we conclude about the role of sensory feedback delays for VE adaptation? First, visual feedback delays of a second or more seriously disrupt hand–eye coordination, a problem that VE users may, at best, only be able to circumvent by the use of conscious strategies such as deliberately moving very slowly (e.g., Smith & Smith, 1962). Second, if the feedback delay is superimposed on another discordance, such as a discrepancy between seen and felt limb positions, and the observer's task does not entail target-pointing responses, not only will visual–motor performance be disturbed, but adaptation to the second discordance is likely to be prevented, even for very short visual delays. Finally, if observers are allowed to engage in discrete sensory–motor responses accompanied by error-corrective feedback, at least some adaptation to the discordance will occur even with feedback delays as long as several seconds.

25.4.5 INCREMENTAL EXPOSURE

Another important variable for adapting to sensory rearrangement, and presumably to VEs, is the provision of incremental (rather than *all at once*) exposure. It has been shown that if exposure to prismatic displacement (e.g., Lackner & Lobovits, 1978), optical tilt (e.g., Ebenholtz & Mayer, 1968), and slow-rotating rooms (e.g., Graybiel & Wood, 1969) entails gradual increases in the strength of the discordance, adaptation and/or the elimination of motion-sickness symptoms is much greater than if the observer is forced to confront the full-blown sensory rearrangement from the start.

There are several ways in which this variable could be applied to VE adaptation. First, it has been advocated (e.g., Kennedy et al., 1987, p. 48) that aircraft simulator operators begin their training with short *hops* before attempting longer ones. Second, since intense, rapid actions are likely to cause greater delays of visual feedback than mild, slow ones, training should probably begin with the latter. Thus, perhaps by working up slowly to the longer delays, VE users will be better able to acquire the deliberate strategies (e.g., *wait and move*) needed to make such delays more manageable. A third potentially useful form of incremental training would be to gradually increase the size of the FOV for HMD devices, rather than beginning with their maximum size. That is, because it is known that VEs with large FOVs produce better task performance than those with small FOVs but are also more nauseating (e.g., Pausch, Crea, & Conway, 1992), gradually increasing the size of the FOV may serve to reduce the simulator sickness symptoms users would otherwise experience and simultaneously improve their performance.

25.4.6 Distributed Practice

It has been demonstrated that providing periodic rest breaks facilitates adaptation to prismatic displacement (e.g., Cohen, 1974) and/or results in greater retention of prism adaptation (Dewar, 1970). Presumably, then, the same would be true for adaptation to VEs. Indeed, Kennedy, Lane, Berbaum, and Lilienthal (1993) determined that the ideal *interhop* interval for *inoculating* users against simulator sickness with respect to the aircraft simulators they examined was two to five days; shorter or longer intervals resulted in less adaptation. Thus, it appears that VE training should entail some sort of distributed practice regimen, although the ideal profile of *on* and *off* periods will almost certainly vary from one VE to another.

25.5 OPTIMAL ADAPTATION-TRAINING PROCEDURE AND HOW TO COUNTERACT THE DRAWBACKS OF MAXIMAL ADAPTATION

25.5.1 Optimal Procedure

Awareness of the critical roles the preceding variables play in perceptual and sensory–motor adaptation to sensory rearrangement provides a critical basis for the design and implementation of effective VE adaptation-training procedures. Thus, assuming a VE display whose discordance is constant rather than transient or variable, the ideal procedure would appear to entail the following: (1) providing operators with many opportunities to interact actively with the displayed environment; (2) providing error-corrective feedback and other salient sensory information about the discordance; (3) avoiding, when possible, significant delays of feedback, especially for conditions of continuous exposure that include no error-corrective feedback; (4) gradually incrementing the length of exposure and/or the magnitude of the discordance; and (5) using a practice regimen that entails frequent rest breaks distributed in an optimal manner.

25.5.2 Simulations of Naturally Rearranged Sensory Environments

Some VEs are designed (or could be designed) for the express purpose of simulating real-life sensory environments that are by their nature rearranged and to which individuals must adapt if they are to perform appropriately. In other words, for these VEs, the presence of a particular intersensory discordance is deliberate rather than an unintended consequence of inadequate technology. An example is a VE designed to simulate the visual–inertial conflicts of hypogravity caused by the absence of the otolithic cues for gravity and used as a means of assisting astronaut trainees to adapt to this environment before actually entering it. This procedure, sometimes referred to as *preflight adaptation training* (PAT), has shown promising results (e.g., Harm & Parker, 1994; Shelhamer & Beaton, 2012). Presumably, PAT procedures could be used to generate adaptive immunity to other disruptive sensory environments such as rocking ocean vessels or the underwater visual world as viewed through a diving mask (e.g., Welch, 1978). Another example of a VE that has been deliberately designed to include a sensory rearrangement is that of Viirre, Draper, Gailey, Miller, and Furness (1998). These investigators proposed rehabilitating people who are suffering from chronically low VOR gains by using a VE to expose and adapt them to gradually incremented gain demands. VE training has also been used for the rehabilitation of stroke and cerebral palsy patients by deliberately imposing discrepancies between their visual and haptic–kinesthetic feedback to which they must adapt (see Adamovich, Fluet, Tunik, & Merians, 2009, for a review). Obviously, all of the facilitating procedures and conditions for adaptation proposed in this chapter should prove useful in helping users adapt to this and any of the other intentionally inflicted VE sensory rearrangements.

Alternatively, a VE might be designed to exaggerate an intersensory discordance on the belief that such a *super* conflict will serve a useful purpose. A good example is the magnification of the interaural axes as a potential means of enhancing the localizability of auditory objects (e.g., Durlach & Pang, 1986). It is important to note that with this arrangement, one must hope that auditory adaptation does *not* occur because this would mean that observers' capacity to localize auditory objects had gone from supernormal back to normal, contrary to the aim of the device. On the other hand, it would probably be advantageous if the observers' *behaviors* (e.g., hand–ear coordination) recovered, presumably by means of adaptation. Otherwise, they might misreach for auditory objects, assuming that these responses were open loop or ballistic. Whether it is possible in this situation to have things both ways remains to be seen.

25.5.3 POTENTIAL DRAWBACKS OF MAXIMAL ADAPTATION

25.5.3.1 Negative Aftereffects and Negative Transfer

Implementation of the optimal training procedures prescribed by the principles of adaptation to sensory rearrangement should maximize adaptation to VEs and thereby minimize the behavioral difficulties frequently experienced while using these devices. Naturally, in the case of VEs that serve as training devices (e.g., aircraft simulators, surgery simulators), it is presumed, or at least hoped, that such adaptation training will also result in improved subsequent performance of the real-world task for which these devices have been designed. Strong support for this assumption has been found for the use of VEs as a training tool for surgeons (e.g., Seymour, 2008). However, assuming that in real life this task does not include the intersensory discordances experienced in the VE, it is possible to imagine circumstances in which adaptation will actually make things *worse* for the user. Thus, although adaptation to a VE as evidenced by improved behavior and the reduction or elimination of simulator sickness indicates that users have become more comfortable with the device, it is still possible for this adaptation to cause them to be less capable of controlling the real-life devices they are being trained to use due to the presence of negative aftereffects and/or negative transfer. Obviously, only if the positive effects of training (i.e., generalization to the real-world task) are found to outweigh these potential deleterious consequences should one advocate and use the adaptation-training regimens described here. These are, of course, empirical issues whose answers are likely to vary from one VE to another.

The unavoidable price for maximal adaptation is maximal aftereffects. Note that it is important to distinguish between the aftereffects of exposure to a VE (or to any sensory rearrangement) and the mere *perseveration* of effects. Only if a particular effect (motion-sickness symptom, hand–eye miscoordination, etc.) disappears by the end of the VE exposure period (or perhaps fails to occur in the first place) and then reappears at some time in the postexposure period is it correct to say that an aftereffect has occurred. It is not accurate, for example, to refer to postexposure malaise as an aftereffect if the subject was experiencing malaise just prior to leaving the VE. The reason why this distinction is important is the assumption, held here, that the presence of an aftereffect is unequivocal evidence that adaptation occurred to the VE during the exposure period. Probably the most dramatic example of a VE-induced aftereffect is the so-called delayed flashback (e.g., vertigo) that can occur many hours after a session of simulator training (e.g., Crampton, 1990). A plausible real-life example of a disruptive aftereffect from a VE was proposed by Biocca and Rolland (1998). They imagined a surgeon who, after using and adapting to a see-through HUD while operating on a human patient, removes the device to continue the operation and makes a serious mistake due to the negative aftereffect. Is there a way to retain the advantages of optimal VE adaptation while simultaneously avoiding its disruptive and potentially dangerous aftermath? Two solutions for ameliorating postexposure aftereffects may be proposed: (1) institute *readaptation* procedures immediately after the user has exited the VE and (2) create *dual* (or *contingent*) adaptation.

25.5.3.2 Eliminating the Aftereffects

25.5.3.2.1 Readaptation Procedures

It is reasonable to presume that the unlearning of adaptation (i.e., readaptation to the normal sensory world) is controlled and facilitated by the same variables that operate during adaptation to the rearranged sensory world. Given this premise, it follows that the ideal way to abolish postexposure aftereffects is to have users engage in the same activities after exiting the VE as they did when in it. Studies of adaptation to the *traditional* forms of sensory rearrangement (e.g., prismatic displacement) have shown that postexposure aftereffects dissipate much more rapidly if subjects are allowed to interact actively with the normal sensory environment and receive feedback from this activity than if they simply sit immobile in the dark or even in a lighted environment (e.g., Welch, 1978). Consider a simulator designed to train people to pilot an oil tanker and which therefore deliberately duplicates the large delay between turning the wheel and the response of the simulated ship that exists for real oil tankers. Clearly, it would be foolhardy for well-trained users to operate an automobile or other vehicle immediately after leaving such a simulator. Rather, one would want to be certain that their presumed training aftereffects had been completely and permanently abolished before allowing them to leave the premises or, if not, to prohibit them from driving vehicles for a prescribed period of time. Reliance on the latter course of action is based on the assumption (or hope) that, as a simple function of time away from the simulator, the aftereffects of training will disappear, from either *decay* or *unlearning/relearning* from random and unspecified sensory–motor activities. In contrast, the present proposal would have operators again pilot the virtual ship immediately after the simulator training session, but this time with the device arranged to eliminate the delay between their actions and the visual consequences. Of course, for VEs with intersensory conflicts that cannot be eliminated, the preceding procedure is inappropriate. In the latter case, if subjects can at least be presented with a real situation that is similar to that provided by the VE and then required to make the same visual–motor actions (e.g., hand–eye, head–eye) in which they engaged during VE exposure, readaptation should be accelerated.

It is unclear what causes *delayed flashbacks* from aircraft simulator training. However, it is reasonable to assume that they are triggered when the individual happens to encounter sensory cues in the environment (e.g., the feel of the automobile steering wheel, the lines on the highway) that are similar or identical to those with which certain perceptions were associated during the simulator training session. The great advantage of the unlearning/relearning procedures suggested here is that they are virtually guaranteed to decondition all such cues, whether or not they can be specifically identified. It can be argued, therefore, that this proposed tactic represents a significant improvement over the current strategy for dealing with these flashbacks in which pilots are grounded for 12–24 h after an aircraft simulator training session.

25.5.3.2.2 Creating Dual (or Context-Specific) Adaptation

Adaptation, as evidenced by either its reduction of effects during exposure or its postexposure aftereffects, does not start from scratch at each transition between one sensory environment and another. Rather, both anecdotal and experimental observations indicate that frequent alternations between adapting to a rearranged sensory environment and readapting to the normal environment lead to decreased interference (e.g., aftereffects) at the point of changeover between the two conditions and/or the more rapid reacquisition of the appropriate perceptions and behavior. This phenomenon of adapting separately to two (and perhaps more) mutually conflicting sensory environments has been referred to as *dual adaptation* (e.g., Welch, Bridgeman, Anand, & Browman, 1993) or *context-specific adaptation* (e.g., Shelhamer, Robinson, & Tan, 1992). An everyday example is adjusting to new prescription lenses in which, after repeatedly donning and doffing their spectacles, wearers report that the depth distortions, illusory visual motion, and behavioral difficulties experienced at the outset have now largely disappeared. Thus, as the result of this alternating experience, the presence or absence of the tactual sensations of the spectacles has become the discriminative cue for turning adaptation on or off.

Empirical evidence of dual or context-specific adaptation comes from a variety of sources. For example, it has been observed that those fortunate astronauts and cosmonauts who have gone into space and returned to Earth several times (thus repeatedly adapting to weightlessness and readapting to Earth's gravity) experience progressively less initial perceptual and perceptual interference when they first enter microgravity (or reenter 1 G) and/or are able to regain normal neurovestibular functioning more rapidly (e.g., Bloomberg, Peters, Smith, Heubner, & Reschke, 1997). Laboratory evidence of dual adaptation for both optically and gravitationally rearranged environments has been reported by, among others, Bingham, Muchisky, and Romack (1991), Cunningham and Welch (1994), Dumontheil et al. (2006), Flook and McGonigle (1977), McGonigle and Flook (1978), and Welch, Bridgeman, Williams, and Semmler (1998). It is important to understand that individuals who have acquired a dual adaptation do not actually remain adapted during the period between exposures to a given rearrangement but rather maintain a *readiness* to adapt (or readapt). Thus, this proclivity (or *immunity*) becomes manifest only in the presence of the sensory cues that signal entry into or departure from the sensory environment in question.

The importance of dual adaptation for adaptation to VEs is this: if users are systematically alternated between adapting to a VE and readapting to the normal environment (or to another VE), they should eventually be able to shift from one sensory environment to the other with little or no interference of perception, performance, or physical well-being. It is conceivable that, by means of such alternation training, operators can eventually acquire the ability to interact with a given VE without serious initial difficulty and then return to the normal world with little or no aftereffect. Presumably, the ideal dual adaptation-training regimen would use as its means of eliminating aftereffects the unlearning/relearning procedures advocated in the preceding section. Finally, the brain's *decision* about the form of adaptation to invoke in a given situation clearly requires the presence of one or more discriminative cues that reliably differentiate the two conditions. Therefore, it will be important for investigators to deliberately provide such cues for the observer and to make them both salient and redundant. Evidence that may outwardly appear to contradict the dual adaptation hypothesis has been reported by Gower et al. (1988), whose subjects revealed increasing, rather than the expected decreasing, postural aftereffects (ataxia) from repeated exposures to an aircraft simulator. Before accepting these data as evidence against the notion of dual adaptation, however, it should be determined if, despite the systematic increase in postexposure ataxia, the rate of readaptation to the normal environment changed. An increase in the latter from one simulator exposure to the next would support the dual adaptation hypothesis. Alternatively, the report by Stanney and Kennedy (1996) of progressively increasing aftereffects might signify that subjects' adaptation was accumulating from one adaptation session to the next, thereby causing the aftereffects to increase as well. According to the dual adaptation hypothesis, however, once adaptation has finally reached asymptote such that further exposures provide little or no increment, the size of the postexposure aftereffects should begin to decline on subsequent exposures, while the rate of readaptation should increase.

25.6 ROLE OF INDIVIDUAL DIFFERENCES

As with all measures of human perception and performance, the acquisition rate and magnitude of adaptation to sensory rearrangement vary widely from person to person (e.g., Welch, 1978, Chapter 11). Therefore, it is reasonable to assume that VE users will differ reliably from one another with respect to such things as (1) the detectability of a given discordance, (2) the degree of interference the discordance causes them, and (3) how adaptable they are to it. Little is known about the causes or correlates of individual traits of adaptability.

Clearly, however, the existence of such individual differences in the response to current VEs means that some users will require more adaptation training than others to attain a given level of adaptation. Indeed, it is possible that those individuals who are particularly prone to the deleterious behavioral and perceptual effects of VEs will be the most benefited from adaptation-training procedures. Furthermore, because of such individual differences, it seems likely that, even after the major problems of current VEs have been overcome by engineering advances, at least some users will continue to register and be reactive to the small sensory and sensory–motor rearrangements that will almost certainly remain. Thus, adaptation-training procedures of the sort advocated here will continue to have a place in VE use. However, even when all that is left are these idiosyncrasies, the likely response of the human-factors engineer would be to create a device that modified itself to conform to them, rather than employing the adaptive fine-tuning suggested here. A potential problem with this solution, however, is that it entails confronting the user with a changing stimulus condition, which might cause problems of its own (M. Draper, personal communication, May 27, 1998).

25.7 SUMMARY AND CONCLUSIONS

Until the sensory rearrangements that characterize many current VEs are corrected by means of improved design and technology, users will continue to experience disruptive perceptual, behavioral, and physiological effects during and after exposure to these devices. One important means of assisting users to deal with these problems is the application of training procedures based on well-established principles of perceptual and perceptual–motor adaptation to sensory rearrangement. The flipside of maximal adaptation is maximal aftereffects. However, the latter can be quickly, completely, and perhaps permanently abolished by means of the same procedures and conditions used to produce adaptation in the first place and by the induction of dual (or contingent) adaptation from repeated experience with the VE in question.

There is one serious caveat to the whole undertaking of training VE users to overcome the limitations of their devices. Namely, because it is possible for adaptation to inhibit transfer of training to the subsequent real-world task for which it is being used, it is important to balance carefully the benefits and potential disadvantages of adaptation for transfer of training. Even when engineering advances ultimately eliminate the major sensory and sensory–motor limitations of current VEs, it is likely that the adaptation-training procedures proposed here will continue to be necessary as a means of accommodating the idiosyncrasies of the individual user. Finally, it goes without saying that many of the arguments and training procedures proposed here are open to question and thus represent topics for empirical investigation with important implications for the VE community.

REFERENCES

Adamovich, S. V., Fluet, G. G., Tunik, E., & Merians, A. S. (2009). Sensorimotor training in virtual reality: A review. *NeuroRehabilitation, 25*(1), 29–44.

Adams, W. J., Banks, M. S., & van Ee, R. (2001). Adaptation to three-dimensional distortions in human vision. *Nature Neuroscience, 4*(11), 1063–1064.

Barnett-Cowan, M., Meilinger, T., Vidal, M., Teufel, H., & Bülthoff, H. H. (2012). MPI CyberMotion Simulator: Implementation of a novel motion simulator to investigate multisensory path integration in three dimensions. *Journal of Visualized Experiments*, (63), e3436. DOI: 10.3791/3436.

Barsalou, L. W. (2008). Grounded cognition. *Annual Review Psychology, 59*, 617–645.

Basdogan, C., & Loftin, R. B. (2008). Multimodal display systems: Haptic, olfactory, gustatory, and vestibular. *The PSI Handbook of Virtual Environments for Training and Education–Developments for the Military and Beyond: VE Components and Training Technologies, 2*, 116–135.

Berkouwer, W. R., Stroosma, O., Van Paassen, M. M., Mulder, M., & Mulder, J. A. (2005, August 15). Measuring the performance of the SIMONA research simulator's motion system. In *Proceedings of the AIAA Modelling and Simulation Technologies Conference* (Vol. 18), San Francisco, CA.

Bingham, G. P., Muchisky, M., & Romack, J. L. (1991, November). *"Adaptation" to displacement prisms is skill acquisition*. In Paper presented at the meetings of the Psychonomic Society, San Francisco, CA.

Biocca, F. (1992). Will simulator sickness slow down the diffusion of virtual environment technology? *Presence: Teleoperators and Virtual Environments, 1*, 334–343.

Biocca, F., & Rolland, J. (1998). Virtual eyes can rearrange your body: Adaptation to visual displacement in see-through, head-mounted displays. *Presence: Teleoperators and Virtual Environments, 7*, 262–277.

Bloomberg, J. J., Peters, B. T., Smith, S. L., Heubner, W. P., & Reschke, M. F. (1997). Locomotor head-trunk coordination strategies following space flight. *Journal of Vestibular Research, 7*, 161–177.

Calvert, S. L., & Tan, S. L. (1994). Impact of virtual reality on young adults' physiological arousal and aggressive thoughts: Interaction vs. observation. *Journal of Applied Developmental Psychology, 15*, 125–139.

Cohen, M. M. (1974). Visual feedback, distribution of practice, and intermanual transfer of prism aftereffects. *Perceptual and Motor Skills, 37*, 599–609.

Cohen, M. M. (1992). Perception and action in altered gravity. *Annals of the New York Academy of Sciences, 656,* 354–362.

Cohen, M. M., Crosbie, R. J., & Blackburn, L. H. (1973). Disorienting effects of aircraft catapult launchings. *Aerospace Medicine, 44,* 37–39.

Cohen, M. M., & Held, R. (1960, April). *Degrading visual-motor coordination by exposure to disordered re-afferent stimulation.* Paper presented at the Eastern Psychological Association, New York, NY.

Collewijn, H., Martins, A. J., & Steinman, R. M. (1983). Compensatory eye movements during active and passive head movements: Fast adaptation to changes in visual magnification. *Journal of Physiology, 340,* 259–286.

Crampton, G. H. (1990). *Motion and space sickness.* Boca Raton, FL: CRC Press.

Craske, B., & Crawshaw, M. (1978). Spatial discordance is a sufficient condition for oculomotor adaptation to prisms: Eye muscle potentiation need not be a factor. *Perception and Psychophysics, 23,* 75–79.

Cunningham, D. W., Billock, V. A., & Tsou, B. H. (2001). Sensorimotor adaptation to violations of temporal contiguity. *Psychological Science, 12*(6), 532–535.

Cunningham, H. A., & Welch, R. B. (1994). Multiple concurrent visual-motor mappings: Implications for models of adaptation. *Journal of Experimental Psychology: Human Perception and Performance, 20,* 987–999.

Dewar, R. (1970). Adaptation to displaced vision: The influence of distribution of practice on retention. *Perception and Psychophysics, 8,* 33–34.

Di Luca, M., Machulla, T. K., & Ernst, M. O. (2009). Recalibration of multisensory simultaneity: Cross-modal transfer coincides with a change in perceptual latency.

Di Luca, M. (2010). New method to measure end-to-end delay of virtual reality. *Presence: Teleoperators and Virtual Environments, 19*(6), 569–584.

DiZio, P., & Lackner, J. R. (1992). Spatial orientation, adaptation, and motion sickness in real and virtual environments. *Presence: Teleoperators and Virtual Environments, 3,* 319–328.

DiZio, P., & Lackner, J. R. (1997). Circumventing side effects of immersive virtual environments. In M. Smith, G. Salvendy, & R. Koubek (Eds.), *Design of computing systems: Social and ergonomic considerations* (pp. 893–896). Amsterdam, the Netherlands: Elsevier Science Publishers.

Dumontheil, I., Panagiotaki, P., & Berthoz, A. (2006). Dual adaptation to sensory conflicts during whole-body rotations. *Brain Research, 1072*(1), 119–132.

Durgin, F. H., Fox, L. F., & Kim, D. H. (2003). Not letting the left leg know what the right leg is doing: Limb-specific locomotor adaptation to sensory-cue conflict. *Psychological Science, 14*(6), 567–572.

Durlach, N. I., & Mavor, A. S. (1995). *Virtual reality: Scientific and technological challenges.* Washington, DC: National Academy Press.

Durlach, N. I., & Pang, X. D. (1986). Interaural magnification. *Journal of the Acoustical Society of America, 80,* 1849–1850.

Ebenholtz, S. M. (1969). Transfer and decay functions in adaptation to optical tilt. *Journal of Experimental Psychology, 81,* 170–173.

Ebenholtz, S. M. (1992). Motion sickness and oculomotor systems in virtual environments. *Presence: Teleoperators and Virtual Environments, 1,* 302–305.

Ebenholtz, S. M., & Mayer, D. (1968). Rate of adaptation under constant and varied optical tilt. *Perceptual and Motor Skills, 26,* 507–509.

Ernst, M. O., & Banks, M. S. (2002). Humans integrate visual and haptic information in a statistically optimal fashion. *Nature, 415*(6870), 429–433.

Ernst, M. O., & Bülthoff, H. H. (2004). Merging the senses into a robust percept. *Trends in Cognitive Sciences, 8*(4), 162–169.

Festinger, L., Burnham, C. A., Ono, H., & Bamber, D. (1967). Efference and the conscious experience of perception. *Journal of Experimental Psychology Monograph, 74*(4, Whole No. 637), 1–36.

Flook, J. P., & McGonigle, B. O. (1977). Serial adaptation to conflicting prismatic rearrangement effects in monkey and man. *Perception, 6,* 15–29.

Fowlkes, J. E., Kennedy, R. S., Hettinger, L. J., & Harm, D. L. (1993). Changes in the dark focus of accommodation associated with simulator sickness. *Aviation, Space, and Environmental Medicine, 64,* 612–618.

Freedman, S. J., & Pfaff, D. W. (1962). The effect of dichotic noise on auditory localization. *Journal of Auditory Research, 2,* 305–310.

Gonshor, A., & Mevill Jones, G. (1976). Extreme vestibulo-ocular adaptation induced by prolonged optical reversal of vision. *Journal of Physiology (London), 256,* 381–414.

Gower, D. W., Lilienthal, M. G., Kennedy, R. S., & Fowlkes, J. E. (1988). Simulator sickness in U.S. Army and Navy fixed- and rotary-wing flight simulators. In *AGARD Conference Proceedings 433. Motion Cues in Flight Simulation and Simulator Induced Sickness.* Brussels, Belgium: AGARD.

Graybiel, A., & Wood, C. D. (1969). Rapid vestibular adaptation in a rotating environment by means of controlled head movements. *Aerospace Medicine, 40*, 638–643.

Harm, D. L., & Parker, D. E. (1994). Preflight adaptation training for spatial orientation and space motion sickness. *The Journal of Clinical Pharmacology, 34*, 618–627.

Hay, J. C., & Pick, H. L., Jr. (1966). Visual and proprioceptive adaptation to optical displacement of the visual stimulus. *Journal of Experimental Psychology, 71*, 150–158.

Held, R. (1955). Shifts in binaural localization after prolonged exposure to atypical combinations of stimuli. *American Journal of Psychology, 68*, 526–548.

Held, R., & Durlach, N. (1993). Telepresence, time delay and adaptation. In S. R. Ellis, M. K. Kaiser, & A. J. Grunwald (Eds.), *Pictorial communication in virtual and real environments* (2nd ed., pp. 232–246). London, U.K.: Taylor & Francis.

Held, R., Efstathiou, A., & Greene, M. (1966). Adaptation to displaced and delayed visual feedback from the hand. *Journal of Experimental Psychology, 72*, 887–891.

Held, R., & Hein, A. (1958). Adaptation of disarranged hand-eye coordination contingent upon re-afferent stimulation. *Perceptual and Motor Skills, 8*, 87–90.

Held, R., & Schlank, M. (1959). Adaptation to disarranged eye-hand coordination in the distance dimension. *American Journal of Psychology, 72*, 603–605.

Hillis, J. M., Ernst, M. O., Banks, M. S., & Landy, M. S. (2002). Combining sensory information: Mandatory fusion within, but not between, senses. *Science, 298*(5598), 1627–1630.

Kennedy, R. S., Berbaum, K. S., Dunlap, M. P., Mulligan, B. E., Lilienthal, M. G., & Funaro, J. F. (1987). *Guidelines for alleviation of simulator sickness symptomatology* (NAVTRA SYSCEN TR-87-007). Orlando, FL: Navy Training Systems Center.

Kennedy, R. S., Fowlkes, J. E., & Lilienthal, M. G. (1993). Postural and performance changes following exposures to flight simulators. *Aviation, Space, and Environmental Medicine, 64*, 912–920.

Kennedy, R. S., Lane, N. E., Berbaum, K. S., & Lilienthal, M. G. (1993). Simulator sickness questionnaire: An enhanced method for quantifying simulator sickness. *The International Journal of Aviation Psychology, 3*, 203–220.

Kohler, I. (1964). The formation and transformation of the perceptual world. *Psychological Issues, 3*, 1–173.

Kuhl, S. A., Thompson, W. B., & Creem-Regehr, S. H. (2009). HMD calibration and its effects on distance judgments. *ACM Transactions on Applied Perception (TAP), 6*(3), 19.

Kunz, B. R., Wouters, L., Smith, D., Thompson, W. B., & Creem-Regehr, S. H. (2009). Revisiting the effect of quality of graphics on distance judgments in virtual environments: A comparison of verbal reports and blind walking. *Attention, Perception, & Psychophysics, 71*(6), 1284–1293.

Lackner, J. R., & Graybiel, A. (1983). Perceived orientation in free-fall depends on visual, postural, and architectural factors. *Aviation, Space, and Environmental Medicine, 54*, 47–51.

Lackner, J. R., & Lobovits, D. (1978). Incremental exposure facilitates adaptation to sensory rearrangement. *Aviation, Space and Environmental Medicine, 49*, 362–264.

Lawson, B. D. (2014a). Motion sickness symptomatology and origins. In K. S. Hale & K. M. Stanney (Eds.), *Handbook of virtual environments: Design, Implementation, and Applications* (2nd ed., pp. 531–600), Boca Raton, FL: Taylor & Francis Group, Inc.

Lawson, B. D. (2014b). Motion sickness scaling. In K. S. Hale & K. M. Stanney (Eds.), *Handbook of virtual environments: Design, Implementation, and Applications* (2nd ed., pp. 601–626), Boca Raton, FL: Taylor & Francis Group, Inc.

Lok, B., Naik, S., Whitton, M., & Brooks, Jr. F. (2003). Effects of handling real objects and self-avatar fidelity on cognitive task performance and sense of presence in virtual environments. *Presence: Teleoperators and Virtual Environments, 12*(6), 615–628.

Loomis, J. M., & Knapp, J. M. (2003). Visual perception of egocentric distances in real and virtual environments. *Virtual and Adaptive Environments, 11*, 21–46.

McCauley, M. E., & Sharkey, T. J. (1992). Cybersickness: Perception of self-motion in virtual environments. *Presence: Teleoperators and Virtual Environments, 1*(3), 311–318.

McGonigle, B. O., & Flook, J. P. (1978). Long-term retention of single and multistate prismatic adaptation by humans. *Nature, 272*, 364–366.

Meyer, K., Applewhite, H., & Biocca, F. (1992). A survey of position trackers. *Presence: Teleoperators and Virtual Environments, 1*, 173–201.

Mohler, B. J., Campos, J. L., Weyel, M., & Bülthoff, H. H. (2007a). Gait parameters while walking in a head-mounted display virtual environment and the real world. In *Proceedings of the 13th Eurographics Symposium on Virtual Environments*, Paris, France (pp. 85–88).

Mohler, B. J., Creem-Regehr, S. H., & Thompson, W. B. (2006, July). The influence of feedback on egocentric distance judgments in real and virtual environments. In *Proceedings of the 3rd Symposium on Applied Perception in Graphics and Visualization*, Boston, MA (pp. 9–14). ACM.

Mohler, B. J., Creem-Regehr, S. H., Thompson, W. B., & Bülthoff, H. H. (2010). The effect of viewing a self-avatar on distance judgments in an HMD-based virtual environment. *Presence: Teleoperators and Virtual Environments, 19*(3), 230–242.

Mohler, B. J., Thompson, W. B., Creem-Regehr, S. H., Willemsen, P., Pick, Jr. H. L., & Rieser, J. J. (2007b). Calibration of locomotion resulting from visual motion in a treadmill-based virtual environment. *ACM Transactions on Applied Perception (TAP), 4*(1), 4.

Mon-Williams, M., Rushton, S., & Wann, J. P. (1995). Binocular vision in stereoscopic virtual-reality systems. *Society for Information Display International Symposium Digest of Technical Papers, 25*, 361–363.

Mon-Williams, M., Wann, J. P., & Rushton, S. (1993). Binocular vision in a virtual world: Visual deficits following the wearing of a head-mounted display. *Ophthalmic and Physiological Optics, 13*, 387–391.

Oman, C. M. (1991). Sensory conflict in motion sickness: An Observer Theory approach. In S. R. Ellis (Ed.), *Pictorial communication in virtual and real environments* (pp. 362–376). New York, NY: Taylor & Francis Group.

Paillard, J., & Brouchon, M. (1968). Active and passive movements in the calibration of position sense. In S. J. Freedman (Ed.), *The neuropsychology of spatially oriented behavior*. Homewood, IL: Dorsey Press.

Pausch, R., Crea, T., & Conway, M. (1992). A literature survey for virtual environments: Military flight simulator visual systems and simulator sickness. *Presence: Teleoperators and Virtual Environments, 1*(3), 344–363.

Philbeck, J. W., Woods, A. J., Arthur, J., & Todd, J. (2008). Progressive locomotor recalibration during blind walking. *Perception & Psychophysics, 70*(8), 1459–1470.

Pusch, A., Martin, O., & Coquillart, S. (2011, March). Effects of hand feedback fidelity on near space pointing performance and user acceptance. In *2011 IEEE International Symposium on VR Innovation (ISVRI)*, Singapore (pp. 97–102).

Redding, G. M. & Wallace, B. (1997). *Adaptive spatial alignment*. Hillsdale, NJ: Lawrence Erlbaum Associates.

Richardson, A. R., & Waller, D. (2005). The effect of feedback training on distance estimation in virtual environments. *Applied Cognitive Psychology, 19*(8), 1089–1108.

Richardson, A. R., & Waller, D. (2007). Interaction with an immersive virtual environment corrects users' distance estimates. *Human Factors: The Journal of the Human Factors and Ergonomics Society, 49*(3), 507–517.

Ries, B., Interrante, V., Kaeding, M., & Anderson, L. (2008, October). The effect of self-embodiment on distance perception in immersive virtual environments. In *Proceedings of the 2008 ACM Symposium on Virtual Reality Software and Technology*, Bordeaux, France (pp. 167–170). ACM.

Rieser, J. J., Pick, H. L., Ashmead, D. H., & Garing, A. E. (1995). Calibration of human locomotion and models of perceptual-motor organization. *Journal of Experimental Psychology-Human Perception and Performance, 21*(3), 480–497.

Rock, I. (1965). Adaptation to a minified image. *Psychonomic Science, 2*, 105–106.

Rock, I., & Kauffman, L. (1962). The moon illusion, II. *Science, 136*, 1023–1031.

Roscoe, S. N. (1993). The eyes prefer real images. In S. R. Ellis, M. K. Kaiser, & A. J. Grunwald (Eds.), *Pictorial communication in virtual and real environments* (2nd ed., pp. 577–585). London, U.K.: Taylor & Francis Group.

Ross, H. E. (1971). Spatial perception underwater. In J. D. Woods, & J. N. Lythgoe (Eds.), *Underwater science* (pp. 69–101). London, U.K.: Oxford University Press.

Sahm, C. S., Creem-Regehr, S. H., Thompson, W. B., and Willemsen, P. (2005). Throwing versus walking as indicators of distance perception in similar real and virtual environments. *ACM Transactions on Applied Perception (TAP), 2*(1), 35–45.

Sanchez-Vives, M., & Slater, M. (2006). The Virtual Arm Illusion: Displacement of sensation of ownership to a virtual arm in virtual reality. In: The 5th Forum of European Neuroscience.

Sanchez-Vives, M. V., Spanlang, B., Frisoli, A., Bergamasco, M., & Slater, M. (2010). Virtual hand illusion induced by visuomotor correlations. *PLoS One, 5*(4), e10381.

Seymour, N. E. (2008). VR to OR: A review of the evidence that virtual reality simulation improves operating room performance. *World Journal of Surgery, 32*(2), 182–188.

Shelhamer, M., & Beaton, K. (2012). Pre-flight sensorimotor adaptation protocols for suborbital flight. *Journal of Vestibular Research, 22*(2), 139–144.

Shelhamer, M., Robinson, D. A., & Tan, H. S. (1992). Context-specific adaptation of the gain of the vestibulo-ocular reflex in humans. *Journal of Vestibular Research, 2*, 89–96.

Slater, M. (2009). Place illusion and plausibility can lead to realistic behaviour in immersive virtual environments. *Philosophical Transactions of the Royal Society B: Biological Sciences, 364*(1535), 3549–3557.

Slater, M. (2010). Virtual hand illusion induced by visuomotor correlations. *PLoS One, 5*(4), e10381.

Slater, M., Perez-Marcos, D., Ehrsson, H. H., & Sanchez-Vives, M. V. (2009). Inducing illusory ownership of a virtual body. *Frontiers in Neuroscience, 3*(2), 214.

Slater, M., Spanlang, B., Sanchez-Vives, M. V., & Blanke, O. (2010). First person experience of body transfer in virtual reality. *PLoS One, 5*(5), e10564.

Slob, J. J. (2008). *State-of-the-art driving simulators, a literature survey.* Eindhoven, the Netherlands: Technische Universiteit.

Smith, K. U., & Smith, W. K. (1962). *Perception and motion.* Philadelphia, PA: Saunders.

Snyder, F. W., & Pronko, N. H. (1952). *Vision with spatial inversion.* Wichita, KS: University of Wichita Press.

Stanney, K. M., & Kennedy, R. S. (1996). Human factors issues with virtual environments technology. *Proceedings of the 40th Annual Human Factors and Ergonomics Society Meeting,* (pp. 298–300). Philadelphia, PA, September 2–6.

Stratton, G. (1897). Vision without inversion of the retinal image. *Psychological Review, 4*(341–460), 463–481.

Thompson, W. B., Willemsen, P., Gooch, A. A., Creem-Regehr, S. H., Loomis, J. M., & Beall, A. C. (2004). Does the quality of the computer graphics matter when judging distances in visually immersive environments? *Presence: Teleoperators & Virtual Environments, 13*(5), 560–571.

Uliano, K. C., Kennedy, R. S., & Lambert, E. Y. (1986). Asynchronous visual delays and the development of simulator sickness. In *Proceedings of the Human Factors Society 30th Annual Meeting* (pp. 422–426). Dayton, OH: Human Factors Society.

Viirre, E. S., Draper, M. H., Gailey, C., Miller, D., & Furness, T. A. (1998). Adaptation of the VOR in patients with low VOR gains. *Journal of Vestibular Research, 8,* 331–334.

Wallach, H., & Karsh, E. B. (1963). Why the modification of stereoscopic depth-perception is so rapid. *American Journal of Psychology, 76,* 413–420.

Wallach, H., Moore, M. E., & Davidson, L. (1963). Modification of stereoscopic depth perception. *American Journal of Psychology, 76,* 191–204.

Waller, D., & Richardson, A. R. (2008). Correcting distance estimates by interacting with immersive virtual environments: Effects of task and available sensory information. *Journal of Experimental Psychology: Applied, 14,* 1, 61–72.

Waller, D., Hunt, E., & Knapp, D. 1998. The transfer of spatial knowledge in virtual environment training. *Presence: Teleoperators and Virtual Environments, 7,* 2, 129–143.

Wann, J. P., & Mon-Williams, M. (1997, May). Health issues with virtual reality displays: What we do know and what we don't. *Computer Graphics, 31*(2), 53–57.

Wei, K., & Körding, K. (2009). Relevance of error: What drives motor adaptation? *Journal of Neurophysiology, 101*(2), 655–664.

Welch, R. B. (1969). Adaptation to prism-displaced vision: The importance of target pointing. *Perception & Psychophysics, 5,* 305–309.

Welch, R. B. (1972). The effect of experienced limb identity upon adaptation to simulated displacement of the visual field. *Perception & Psychophysics, 12,* 453–456.

Welch, R. B. (1978). *Perceptual modification: Adapting to altered sensory environments.* New York, NY: Academic Press.

Welch, R. B. (1986). Adaptation of space perception. In K. R. Boff, L. Kaufman, & J. P. Thomas (Eds.), *Handbook of perception and human performance.* New York, NY: John Wiley & Sons.

Welch, R. B., Bridgeman, B., Anand, S., & Browman, K. E. (1993). Alternating prism exposure causes dual adaptation and generalization to a novel displacement. *Perception and Psychophysics, 54,* 195–204.

Welch, R. B., Bridgeman, B., Williams, J. A., & Semmler, R. (1998). Dual adaptation and adaptive generalization of the human vestibulo-ocular reflex *Perception and Psychophysics, 60,* 1415–1425.

Welch, R. B., Cohen, M. M., & DeRoshia, C. W. (1996). Reduction of the elevator illusion from continued hypergravity exposure and visual error-corrective feedback. *Perception and Psychophysics, 58,* 22–30.

White, K. D., Shuman, D., Krantz, J. H., Woods, C. B., & Kuntz, L. A. (1990, March). Destabilizing effects of visual environment motions simulating eye movements or head movements. In N. I. Durlach & S. R. Ellis (Eds.), *Human–machine interfaces for teleoperators and virtual environments, (NASA Conference Publication 10071)* Santa Barbara, CA: NASA.

Wickens, C. D. (1986). The effects of control dynamics on performance. In K. R. Boff, L. Kaufman, & J. P. Thomas (Eds.), *Handbook of perception and human performance* (pp. 39–60). New York, NY: John Wiley & Sons.

Willemsen, P., Colton, M. B., Creem-Regehr, S. H., &Thompson, W. B. (2009). The effects of head-mounted display mechanical properties and field of view on distance judgments in virtual environments. *ACM Transactions on Applied Perception (TAP), 6*(2), 8.

Young, P. T. (1928). Auditory localization with acoustical transposition of the ears. *Journal of Experimental Psychology, 11,* 399–429.

26 Visually Induced Motion Sickness

Causes, Characteristics, and Countermeasures

Behrang Keshavarz, Heiko Hecht, and Ben D. Lawson

CONTENTS

26.1 INTRODUCTION

Motion sickness (MS) is a familiar and ancient malady: "Sailing the ocean was known to be highly nauseating in ancient Greece" (Hippocrates, as cited in Reason & Brand, 1975). Almost all of us have suffered at least once from MS while sailing on a boat, riding in a bus or car, flying in an airplane, or riding the rollercoaster. Exposure to highly sickening motions initially elicits sudden warmth (sometimes with flushing), cold sweat, dizziness, mild stomach symptoms, headache, and pallor (Lawson, 2001). As sickness worsens, the sufferer may experience severe nausea, increased salivation, and retching or vomiting. Drowsiness may occur with these other symptoms or in isolation (Graybiel & Knepton, 1976; Lawson & Mead, 1998). Although the phrase *motion sickness* implies a clinical malady, every healthy person—with the possible exception of people suffering from vestibular loss (e.g., bilateral hypofunction)—can become motion sick when exposed to certain stimuli. Once severe MS symptoms have set in, it may take hours to fully recover. Only a small percentage of the population suffers from severe MS during stimuli that would not disturb others; such persons also tend to be poor adapters and less responsive to medications (Birren, 1949; Reason, 1978b).

This review focuses on an aspect of MS characterized by the occurrence of MS-like symptoms without real physical motion necessarily being present. This phenomenon is referred to as *visually induced motion sickness* (VIMS) and is known to appear under visual stimulation such as occurs in driving simulators, flight simulators, movie theaters, or video games (Hettinger & Riccio, 1992; Kennedy, Drexler, & Kennedy, 2010; McCauley & Sharkey, 1992). This review will consider theories of VIMS to see whether they help with the interpretation of the empirical evidence. Given the competition among several hypotheses that all have some supporting and some counter evidence, it seems wise to evaluate some studies in terms of whether they could serve as building blocks for a more definitive theory. Therefore, we first consider the nature of VIMS and characterize the most common symptoms, contrasting classical MS from its visually induced equivalent. Next, we describe the major theories of MS. We then discuss historical and recent attempts to reliably measure VIMS, and finally, we present a number of studies relevant to important knowledge gaps, including aspects of vection, habituation, field of view, body axis of the rotation stimulus, and long-term aftereffects. At this point, there is no unified theory of VIMS; however, we hope that this synopsis will help researchers to clarify issues that must be accounted for by a unified theory.

Depending on the study, VIMS has been referred to as *cybersickness* (Kennedy, Lanham, Drexler, Massey, & Lilienthal, 1997; LaViola, 2000; McCauley & Sharkey, 1992), *virtual reality sickness* (Cobb, Nichols, Ramsey, & Wilson, 1999; Nichols & Patel, 2002), *gaming sickness* (Frey, Hartig, Ketzel, Zinkernagel, & Moosbrugger, 2007; Merhi, Faugloire, Flanagan, & Stoffregen, 2007), *cinerama sickness*, or *simulator sickness* (Johnson, 2005; Kennedy, Lilienthal, Berbaum, Baltzley, & McCauley, 1989). The list of symptoms during acute MS or VIMS is long and varies among individuals. Strictly speaking, every deviation from a normal and healthy state of well-being that can be traced back to the exposure to virtual or real motion could qualify as a MS symptom. However, nausea is usually the most disturbing symptom and is considered a cardinal sign or cardinal symptom along with pallor, increased salivation, cold sweating, drowsiness, retching, or vomiting (Miller & Graybiel, 1970). Other symptoms that commonly occur in virtual environments (VEs) include dizziness or vertigo, headache or fullness of head, and eyestrain or blurred vision (Lawson, 2001). A challenge to assessing VIMS is the variable nature of some of these symptoms.

None of them is strictly necessary, and certainly none of them is sufficient for MS, not even vomiting. For example, nausea can be caused by motion without vomiting occurring, or vomiting can be caused by motion without nausea (or any other symptom) being present (Lackner & Graybiel, 1984). Moreover, vomiting can occur for reasons unrelated to MS. Stanney, Kennedy, and Drexler (1997) observed differences between the symptom clusters of simulator sickness and cybersickness, with more disorientation (including focusing problems, vertigo, fullness of head, and blurred vision) and less nausea (including general discomfort, increased salivation, sweating, and difficulty concentrating) in the cybersickness users.

Although the symptom cluster of VIMS is similar to that reported for the classical MS syndrome, there are some differences. In broad comparison to classical MS, VIMS in optokinetic drums (i.e., moving cylindrical surrounds with striped or polka-dotted interiors) and fixed-base flight simulators is associated with more *head* and *oculomotor* symptoms, such as headache, dizziness, and eyestrain (Lawson, 2013; Stanney & Kennedy, 1997—see Chapter 23). Also, the severity of MS tends to be lower for purely optokinetic visual stimuli versus purely rotational stimuli in the dark (Dichgans & Brandt, 1973), and retching or vomiting are rarer during VIMS (Kennedy et al., 1989; Kingdon, Stanney, & Kennedy, 2001). A special cluster of motion-related symptoms has been referred to as the *sopite syndrome* (DiZio & Lackner, 1991; Graybiel & Knepton, 1976; Lawson & Mead, 1998). The sopite syndrome includes symptoms related to motion-based degradation of arousal, including drowsiness, relaxation, or difficulty concentrating, and it is also known to occur during the use of VEs (e.g., Leslie, Stickgold, DiZio, Lackner, & Hobson, 1997; Wright, Bose, & Stiles, 1995). It remains unclear, however, whether the sopite syndrome qualifies as a construct in its own right. On the one hand, drowsiness is a cardinal symptom of classical MS (Miller & Graybiel, 1970). On the other hand, sopite syndrome may be the sole manifestation of MS (Graybiel & Knepton, 1976; Lawson & Mead, 1998). Moreover, the descriptors associated with mild or nonsickening motion have good internal consistency and many of them clearly describe dearousal (Lawson, Kass, Muth, Sommers, & Guzy, 2001), indicating that sopite syndrome has sufficient validity as a construct that is related to MS but not synonymous with it (Brill, Kass, & Lawson, 2004; Wallace, Kass, & Lawson, 2002).

The incidence of VIMS can vary widely, but it is certainly not a rare phenomenon. Crampton and Young (1953) noted that 35% of their subjects reported nausea when exposed to visual surround oscillations along the earth-vertical axis. Stanney, Kennedy, Drexler, and Harm (1999) estimate that 60%–70% of all simulator users suffer at least slightly from VIMS. Cobb et al. (1999) and Stanney, Mourant, and Kennedy (1998) report that 80%–95% of users exposed to VEs show some MS-like symptoms. Similarly, Regan and Price (1994) found that 5% of the VE users in their study aborted a 20 min long experiment because of adverse symptoms. According to Stanney et al. (1998), the dropout quota among VE users can reach 30%.

Now that we have laid out the characteristics of VIMS, we turn to the question of *why* VIMS is generated. The answer is not straightforward, and various competing hypotheses offer rather different answers.

26.2 THEORIES OF MS

Several hypotheses seek to explain the genesis of VIMS in different stimulus situations, but there is no single universally accepted theory. The majority of current hypotheses focus on explaining MS in the presence of real motion and vestibular stimulation. Since MS and VIMS share many characteristics and mechanisms, classic MS theories that incorporate visual components should be discussed in the context of a review of VIMS. Over the past decades, the most accepted and cited MS theory has been commonly called the *sensory conflict theory* (Claremont, 1931; Reason & Brand, 1975). There have been several modifications or offshoots of the theory to render it more quantitative and precise (Oman, 1982, 1990) or to further specify or explain the essential nature of the conflicts and responses (Bles, Bos, Graaf, Groen, & Wertheim, 1998; Dai, Kunin, Raphan, & Cohen, 2003; Guedry,

Rupert, & Reschke, 1998; Khalid, Turan, Bos, & Incecik, 2011; Prothero, Hoffman, Parker, Furness, & Wells, 1995; Sheehan, Oman, & Duda, 2011). In contrast, a theory has been forwarded that explicitly rejects sensory conflict and instead emphasizes the role of postural equilibrium in the causation of MS (Riccio & Stoffregen, 1991; Stoffregen & Riccio, 1991). Some theoretical research efforts fit within a general conceptual framework of sensory conflict but emphasize the role of gaze or ocular mechanisms in MS (Dai et al., 2003; Ebenholtz, 1992; Ebenholtz, Cohen, & Linder, 1994; Reschke, Somers, & Ford, 2006). A new theory seeks to explain MS as a function of the vestibular detection of stimuli that would be disruptive to digestion (Rupert, 2010). Finally, there are several theories that try to determine the evolutionary origins of MS, rather than the immediate cause of MS during a particular motion challenge (Bowins, 2010; Lawson, 2013; Treisman, 1977*). A few of the main theories are listed in Table 26.1 and described in detail later.

26.2.1 Contribution of Sensory Conflict to the Elicitation of MS

The vestibular organs play a key role in the genesis of MS (see Howard, 1986, for an overview of the physiology of the vestibular organs). Intact vestibular organs (namely the semicircular canals and the otoliths) are believed to be an essential requirement to experience MS (Cheung, Howard, & Money, 1991; Graybiel & Johnson, 1963; Johnson, Meek, & Graybiel, 1962; Kellogg, Kennedy, & Graybiel, 1965; Kennedy, Graybiel, McDonough, & Beckwith, 1968; Money, 1970; Reason & Brand, 1975). Patients with damaged or destroyed vestibular end organ function, so-called labyrinthine defectives, are not susceptible to MS. Interestingly, although whole-field visual motion stimulates vestibular brain centers indirectly (e.g., Dichgans & Brandt, 1973; Klinke, 1970), labyrinthine-defective patients are also immune to VIMS (Cheung et al., 1991; Cheung, Howard, Nedzelski, & Landolt, 1989; but see Johnson, Sunahara, & Landolt, 1999). The visual system, in contrast, is not necessary for the genesis of MS, as blind people become motion sick as well (Graybiel, 1970).

In essence, the sensory conflict theory by Reason and Brand (also called sensory rearrangement, neural mismatch, or cue conflict; Guedry, 1991; Reason, 1978a, 1978b; Reason & Brand, 1975) proposes that two crucial processes take part in the genesis of MS: (1) the orientation information transmitted by the eyes, the vestibular organs, and/or the somatosensory senses (nonvestibular proprioceptive senses of skin, muscles, and joints) do not match; and/or (2) the sensed motion pattern among them is not congruent with what is expected based on past experiences. These two processes will be specified in the following section, including their role in MS adaptation. Finally, we will introduce an extension of the classic conflict approach by discussing the subjective vertical conflict theory (Bles et al., 1998; Khalid et al., 2011).

The idea of sensory conflict was first proposed by Irwin (1881). Sensory conflict was developed into a formal theory based on relevant work by many other authors (Claremont, 1931; Graybiel & Johnson, 1963; Groen, 1961; Guedry, 1964; Held, 1961; Holst, 1954; Lansberg, 1963; Oman, 1982, 1990; Reason & Brand, 1975; Steele, 1961; Young, Meiry, & Li, 1966). According to this approach, a sensory conflict happens when the visual, vestibular, and/or somatosensory systems are at variance with each other, that is, they send nonconcordant afferent information to the central nervous system (CNS), or one system fails to send the expected information. For instance, reading a book or a map while sitting in the back of a moving car represents a classic scenario that provokes MS in susceptible persons. In this case, a signal mismatch between the vestibular and the visual system exists, as the vestibular organs (and somatosensory receptors) signal the physical motion of the car

---

* Lawson (2013) considers whether poisoning is a necessary aspect of a theory concerning the evolution of aversion responses to unusual motion or a more *direct evolutionary hypothesis* is viable. Lawson also considers whether Treisman's hypothesis concerning the origin of MS (as an indirect by-product of a poison response system being triggered by motion) can be reframed as a *direct poison hypothesis*, wherein a more direct correlation existed between the circumstances that gave rise to poisoning and unusual real or apparent motion. These hypotheses are not reviewed in this chapter since they are discussed already elsewhere in this same book.

TABLE 26.1
Listing of Some of the Main Theories of MS

	Theories of Immediate Cause			Theories of Origin
	Theories that mainly seek to explain why MS arises in certain sensorimotor circumstances			Theories that mainly seek to explain the evolutionary origins of the responses that constitute MS
Emphasis	**Sensory Conflict**	**Postural Instability**	**Eye Movement**	**Evolutionary**
Strengths	Highly cited approach with good explanatory power; additional variants and elaborations	Improvement of ecological validity and relation to functional outcomes	Most bounded in scope; aided identification of potential neural mechanisms	Explanation of *why* some of the common signs and symptoms of MS would be triggered (instead of other effects)
Limitations	Does not fully predict average incidence and severity of MS during a given conflict (see Section 26.2.1.6)	Variable findings in the literature No consensus on how theory accounts for decreased MS during restraint and some adaptation-related observations (see Section 26.2.2.1)	Generalization to various conditions not fully understood, e.g., in stimuli without a visual component, reduced outside view, blind passengers (see Section 26.2.4)	Difficult to test Can become overly adaptationist (see Section 26.2.5)
		Controlled testing for falsification difficult: • Wide range of conflicts that can be construed • Wide range of sway measures and findings construed to support postural instability theory • Difficulty devising experiments that all researchers agree challenge postural stability without creating conflict and vice versa	Not trivial to disentangle the effects of vection, retinal slip, optokinetic nystagmus, accommodation, and vergence	
		Neural centers and mechanisms not fully understood (e.g., centers of conflict, relation of posture to MS mechanisms) Particular signs or symptoms (across and within individuals; during prolonged or repeated stimuli) are not fully explained by the theory		

(continued)

TABLE 26.1 (continued)
Listing of Some of the Main Theories of MS

	Theories of Immediate Cause			Theories of Origin
Variants:	Classical sensory conflict theory (Section 26.2.1)	NA	NA	Treisman's hypothesis (theory of toxins; Section 26.2.5. See Lawson, 2014, for further variants and alternatives)
	• Elaboration of uncorrelated input in sensory conflict theory (Section 26.2.1.1)			
	• Mathematical elaboration (Section 26.2.1.2)			Bowins' variant (negative reinforcement theory; Section 26.2.1.5)
	• Variant emphasizing the perceived vertical (Section 26.2.1.3)			*Note*: some aspects of theories of immediate MS causation have been considered in relation to evolution also, e.g., the developmental theory (Section 26.2.1.5) and the postural instability theory (Section 26.2.2)
	• Variant emphasizing frames of reference (Section 26.2.1.4)			
	• Variants emphasizing ontological development (Section 26.2.1.5)			

(e.g., accelerating, breaking, or turning), while the eyes signal a comparatively stable visual world. A visual–vestibular conflict is also apparent in VIMS (see Bos, Bles, & Groen, 2008). If the visual stimulus indicates self-motion, and self-motion is experienced by a stationary observer (see Mach, 1875), this vection is often in conflict with the vestibular end organs (and somatosensory receptors), for instance, because they are not also signaling body motion (Dichgans & Brandt, 1973, 1978). Again, VIMS occurs as a result of the intersensory conflict. Reason and Brand (1975) describe six categories of sensory conflict (in Table 6 of their book, p. 102), which have been reiterated and discussed widely over the years. The categories result from the possible combinations of visual–vestibular rearrangements (both signals present but in conflict, only visual motion stimulation, only vestibular motion stimulation), and the possible combinations of canal–otolith rearrangements (both signals present but in conflict, only canals signal movement, only otoliths signal movement). Note that Reason and Brand acknowledge that inertial information includes "both the vestibular and non-vestibular proprioceptors" (p. 106). For example, somatosensory (skin, muscles, and joints) systems are capable of sensing forces applied to the body during movement. Various visual–inertial and vestibular–somatosensory mismatches have since been described extensively in the literature (e.g., Crampton, 1990; Guedry, 1991).

Sensory conflict theory does not attribute MS to a simple mismatch among the senses. The afferent sensory information is often stored, extrapolated, and/or anticipated such that the representation of afferent information may contribute to or even constitute the nauseogenic sensory conflict. In particular, vestibulo–ocular coordination involves such information storage or representation. Reason (1978b) introduced an internal model, the *neural store*, to accommodate this storage component into the sensory conflict model. The neural store is hypothesized to be an (yet undefined) area of the CNS, which contains every sensory pattern of movement experienced in the past. Whenever an active movement is planned, an expectation about the forthcoming sensory pattern of this particular motion is provided by the neural store. The sensory pattern most likely to fit the intended motion is selected from the store and is compared with the new afferent information provided by the eyes, the inner ears, and the somatosensory senses. If the new pattern matches the previous one, then MS should fail to appear. In contrast, if there is a discrepancy between what was expected and what is sensed, a neural mismatch arises that may provoke MS (see Figure 26.1).

The risk of stronger discrepancy between the sensed and the expected pattern is highly increased during passive motion, which is in line with the common observations that a driver controlling the steering wheel of a vehicle usually is less affected by MS than the passengers. In sum, the crucial factor in the genesis of MS is not only the existence of a sensory conflict, but also the discrepancy between the actual and the expected sensory pattern.

So far, we have defined the sensory conflict and demonstrated how this conflict generates MS. However, the idea of the neural mismatch and the concept of neural store also offer an explanation of how people *adapt* to MS and learn to cope with previously sickening scenarios. As described earlier, MS occurs whenever a sensory conflict is present and the performed motion does not match previous experiences. This can be due to the fact that an adequate representation is simply missing (because the motion is new) or that it is very weak and/or not easily accessible. The neural store is updated during the mismatch signal—an adaptation phase is initiated (see Figure 26.1), that is, a modified representation (containing information from the eyes, the inner ears, and the somatosensory senses) is established and saved to the neural store. During this *sensory rearrangement* process, MS is a typical side effect, but when the recalibration of the neural store is finished and the new pattern is available, MS will quickly diminish. Thus, repetitive exposure to the same MS-provoking stimuli should lead to habituation and minimize the occurrence of MS. As we will discuss later, training is indeed a very effective method to reduce and prevent MS and VIMS.

After a sea voyage, a need to readapt to land conditions (*getting your land legs*) is often observed. Occasionally, these disturbing aftereffects persist for weeks or months and are referred to as *mal de débarquement* (Nachum et al., 2004). This phenomenon can again be traced back to the neural

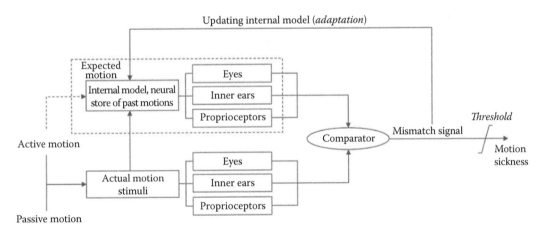

FIGURE 26.1 Schematic illustration of the sensory conflict theory. (Modified from Benson, A.J., Motion sickness, in Pandolf, K.B. and Burr, R.E., [Eds.], *Medical Aspects of Harsh Environments*, Vol. 2, United States Army Medical Department Center and School, Washington, DC, 2002, pp. 1048–1083.) The dotted area represents the internal model (neural store), which is activated in advance by intended active movement, but not by passive motion. The comparator integrates the sensed and expected motion patterns and in cases of a discrepancy (and if the discrepancy is stronger than a threshold), MS results. Note that the mismatch signal steadily updates the internal store to achieve adaptation.

store: when going offshore, the neural store needs time to recalibrate to the moving frame of reference, during which MS may occur. After several days aboard the ship, the organism adapts to the wave motion stimulus. Upon return to solid ground, a negative aftereffect of adaptation is exhibited, and in some highly sensitive individuals, *mal de débarquement* occurs. The process of readaptation is not limited to sea travel; it can also be found after prolonged use of VEs. It is characterized by disturbed locomotor and postural control after finishing the VE experience (Biocca, 1992; Kennedy, Lane, Lilienthal, Berbaum, & Hettinger, 1992; Pausch, Crea, & Conway, 1992). Additionally, such adverse symptoms are frequently reported by astronauts after spaceflight (see Clément, 2007).

26.2.1.1 Role of Uncorrelated Input in Sensory Conflict Theory

Kennedy, Hettinger, and Lilienthal (1990) presented an elaboration of sensory conflict theory intended to help with predictions concerning the magnitude of conflict in a given situation based upon how well the sensory inputs correlate in the spatial and temporal dimensions. Sensory inputs lead to more MS as their correlations approach zero, with differing weights given to differing sensory modalities. The authors present other ideas about how the size of a conflict could be estimated.

26.2.1.2 Mathematical Elaboration of Sensory Conflict Theory

One criticism toward sensory conflict theory is that it does not provide predictions that are sufficiently quantitative. To remedy this situation and as an extension of the classic MS theory by Reason and Brand (1975), Oman (1982, 1990) integrated the crucial components (signals of the senses, neural store, and comparator) into a heuristic and mathematical model of MS. The link between each component is described as a vector force, which makes more specific predictions about the occurrence and strength of MS symptoms. According to Oman's model, the severity of MS is defined as the discrepancy between the true and the expected sensory pattern of the motion. Strictly speaking, the more the true motion differs from the expected one, the stronger the MS. Other mathematical models of MS have also been entertained (Ji, So, & Cheung, 2010; Matsangas, 2004).

26.2.1.3 Contribution of Conflicts with the Perceived Vertical

There are a number of explanations of MS or spatial orientation that emphasize the importance of appreciating one's orientation and motion relative to the true earth-vertical (Bos & Bles, 1998; Dai et al., 2003; Guedry, 1978; Guedry & Benson, 1978; Mittelstaedt, 1983). Such explanations provide some testable assertions. One criticism of sensory conflict theory is that the range of conceivable conflicts is so wide that it is difficult to devise experiments whose results would falsify the theory. An interesting extension of Oman's model was introduced by Bles, Bos, and their colleagues, which restricts the conflicts predicted to elicit MS to those affecting the appreciation of the vertical (Bles et al., 1998; Bles, Bos, & Kruit, 2000; Bos & Bles, 1998, 2002; Bos et al., 2008; Graaf, Bles, & Bos, 1998). The authors offer a *subjective vertical theory*, which states that the key sensory conflict in MS is the conflict between the ongoing sensation of the vertical (given by the various sensory modalities) and the organism's internal model of subjective vertical based on past experience. Based on Oman's model (1982, 1990), a desired body position is achieved via body dynamics and muscle activity. The motion signals are detected by the senses and result in a sensory pattern of information. The internal model includes the same components and results in the expected sensory information. The difference between the real and the expected sensory patterns is fed back into the system to provide an update of the internal model. Bles et al. (1998) extended this model by adding a network that represents the subjective vertical as an intervening variable. Thus, the discrepancy between the sensed vertical and the expected vertical is the main factor in the update process of the internal model and determines the occurrence of MS. (Note that recently the role of horizontal accelerations has also been considered in the *subjective vertical–horizontal* conflict theory of Khalid et al., 2011.)

According to the subjective vertical theory, rotation of a visual surround in the earth-vertical yaw axis (e.g., via an optokinetic drum) should not provoke VIMS, unless head movements are performed. Support for this view is given by Bles (1998), who reported very low VIMS scores when the head was fixed during optokinetic rotations limited to the yaw axis. However, several other studies have revealed significant MS in such rotating drums under situations not involving intentional head movement (Hu, Stern, Vasey, & Koch, 1989; Hu et al., 1999; Muth, Koch, Stern, & Thayer, 1999). Note that the participants' heads were not fixed rigidly in these studies. Bos and Bles (2004) argued that even minimal motions of the head (e.g., while talking) could produce a subjective vertical mismatch and lead to MS, as could small deviations of the optokinetic drum rotation axis from true vertical, thus explaining the MS reported in most optokinetic drum studies. It is true that optokinetic drums elicit more VIMS when they are tilted (Golding, Arun, Wortley, Wotton-Hamrioui, & Gresty, 2007), but it is not known how small the tilt must be before the MS difference becomes negligible.

Also, Bubka, Bonato, and colleagues (Bubka & Bonato, 2003; Bubka, Bonato, Urmey, & Mycewicz, 2006; Bonato, Bubka, & Story, 2005) have observed that an optokinetic drum elicits more VIMS not only when it is tilted, but also when it is rotating about a vertical axis and changes either direction or speed. These findings can be interpreted as evidence that VIMS requires an explanation beyond a simple conflict with the subjective vertical; however, note that Bos (2011) asserts that motion and tilt cues are both subsumed within any theory explaining VIMS in reference to tilt.

It is also true that the head can move slightly when the head restraint is minimal (e.g., a chin rest). Significant head movement is less likely in laboratories where the optokinetic apparatus employs a rigid head restraint (e.g., the Graybiel Laboratory at Brandeis University) that does not touch the jaw, yet VIMS still has been observed (Lackner & Teixeira, 1977; Teixeira & Lackner, 1979). In fact, the frequency of subjects classified as having VIMS was 66% in chin-rest experiments by Stern and colleagues (Stern et al., 1985; Stern, Koch, Stewart, & Lindblad, 1987), but 79%–93% (depending upon whether or not the MS criteria of Stern et al. 1985 or 1987 were applied) in a head-fixed experiment by Lawson (1993). There also have been animal studies where the stimulus rotation axis was vertical and the head was rigidly fixed, yet MS and/or VIMS was still evident, although reduced (Daunton, 1985; Wilpizeski, Lowry, Contrucci, Green, & Goldman, 1985). It is certainly true that under certain conditions, head movements during optokinetic stimulation can be sickening, but it should be remembered that most of these *pseudo-Coriolis* experiments typically involved much larger head movements (which also were not as likely to return immediately to the rest position) than would occur for a subject's jaw movements when talking while using a chin rest (Dichgans & Brandt, 1978). Again, it is not known how minor the head movements must be before the effect on VIMS is negligible. This issue should be explored as subjective vertical theory matures.

A final point concerning head movements during optokinetic stimulation is that the relationship between head movements and VIMS is not always positive. Under certain conditions involving paced head movements commencing early during optokinetic stimulation, reduced VIMS (and vection) have been observed by Lackner and Teixeira (1977) and Teixeira and Lackner (1979). DiZio and Lackner (1986) also qualitatively observed no VIMS during moderate head movements while subjects viewed optokinetic stimuli, even though some of the stimuli also had a tilted rotation axis. These findings should be incorporated into subjective vertical theory.

Other sickening stimuli that should be considered in the context of the subjective vertical theory include low-frequency body rotation about the earth-vertical axis during reading of head-fixed displays (Guedry, Lentz, & Jell, 1979; Guedry, Lentz, Jell, & Norman, 1981) and repeated sudden stops after rotation in the earth-vertical axis (Graybiel & Lackner, 1980). As the subjective vertical theory develops, it should incorporate the sickening effects of these two stimuli along with the findings of the aforementioned head-fixed optokinetic drum experiments. Some potentially relevant limitations of subjective vertical theory are discussed by Bles et al. (1998) and Lackner and DiZio (2006).

Another line of research relevant to the processing of the orientation vector has been forwarded by Dai et al. (2003) and Dai, Raphan, and Cohen (2007). They propose that MS is elicited when the subject's orientation vector relative to gravity is misaligned with axis of eye motion during head motion. The duration of this misalignment is determined by the time constant of the velocity storage mechanism and contributes to the buildup of MS. The authors present some experimental evidence consistent with their theory. (For a review of the wider role of velocity storage in estimates of orientation and motion, see Goldberg et al., 2012.)

26.2.1.4 Contribution of Conflicts with the Perceived Stationary Frame of Reference

Another criticism of sensory conflict theory is that there are some sources of conflict that do not provoke sickness or where an obvious source of conflict has not been identified. Prothero et al. (1995) hypothesize that MS is not caused by conflicting motion sensations per se; rather, MS occurs when motion signals result in conflicting sensations concerning what is stationary. They call the stationary frame of reference the *rest frame*. They present evidence showing that when subjects view a moving visual field through which *independent visual background* cues are provided (i.e., an inertially stable visual reference), MS is diminished. This theory should allow for the existence of sensory conflicts that do not provoke MS, provided they are irrelevant to the determination of the rest frame. See Steele (1961) for an early description of the role of sensory conflict versus reference frames in MS and Feenstra, Bos, and Gent (2009) for a recent experiment that seems consistent with Steele (1961) and Prothero et al. (1995).

26.2.1.5 Contribution of Development of Movement Control

Some researchers (Guedry et al., 1998; Watt, Bouyer, Nevo, Smith, & Tiande, 1992) emphasize the importance of sensorimotor development in relation to the maintenance of spatial orientation. Guedry et al. reason that the MS symptoms that arise during adaptation to unusual gravitoinertial force environments (such as rotating rooms) provide clues to the understanding of the process of normal ontological development. They assume that elimination of inefficient motor programs is important during development and that disturbing symptoms (including but not limited to MS) discourage the development of inefficient programs. Similarly, Bowins (2010) hypothesizes that MS developed as a negative reinforcement system for avoidance of sensory conflict or postural instability.* Bowins contrasts this *negative reinforcement theory* with Guedry et al.'s explanation based on development of efficient motor programs, pointing out that there are already positive reinforcement systems in place to reward moving to a desired goal and because movement development is active early in life when MS is relatively rare (therefore, any additional reinforcement systems hypothesized by Guedry would be redundant). Bowins also contrasts the role of negative reinforcement in his theory versus punishment in Guedry et al.'s earlier theory.†

Several observations can be made about Bowins' (2010) criticisms of Guedry et al. (1998):

1. Regarding the issue Bowins raises concerning redundancy, we think that the avoidance of inefficient motor coordination may be important enough to survival for the allowance of partially redundant systems, as seems to be the case for food aversion (see later in this chapter) and many other responses. Even if this is not the case, redundancy alone would not be a reason to reject the potential existence of multiple processes serving one survival goal. For example, visual, vestibular, and somatosensory systems all assist with balance and with the coordination of head and gaze movement. As Goldberg et al. (2012) have pointed out, these different sensory systems do not have entirely unique roles in the maintenance of balance. Rather they overlap in the information each system provides, as would be expected for a system serving such a critical goal as postural control. This partial redundancy of system design enhances reliability and aids recovery of balance following a serious injury to one of the sensory systems (Goldberg et al., 2012).

---

* Mayne (1974) presented a more general and somewhat complementary concept involving the rejection of new inputs that cannot be integrated in order to maintain well-established responses to sensory information.
† For a discussion of the nondichotomous, interrelated, and context-dependent nature of reward and aversion, the reader is referred to Umberg and Pothos (2011).

2. Considering Bowins' criticism of punishment in Guedry et al., it would seem that since nausea and associated highly aversive symptoms are triggered by self-movement in both theories (Bowins' and Guedry's), the distinction drawn by Bowins between punishment and negative reinforcement is moot. Moreover, Guedry et al. allude to some fairly mild effects preliminary to MS that would not be all that *punishing*, including affects such as displeasure, which could be logically triggered by vestibular reactions (Balaban, 1999; Guedry & Oman, 1990; McCandless & Balaban, 2010). Bowins, on the other hand, appears to rely on MS (which can be highly noxious) as the main source of reinforcement correcting motions that do not contribute to survival. He does not discuss other symptoms, but neither does he reject them.

3. Considering the criticism regarding movement development early in life, Golding (2006a) has offered a possible explanation for why MS does not occur at earlier ages that, in our view, seems to fit within the general framework of Guedry et al. Golding states that the delay in peak susceptibility to MS may be related to the fact that the brain remains highly plastic at an early age and the perceptual–motor map that influences the identification of sensory mismatches is not fully formed until around age seven.

These points are raised not to reject Bowins' (2010) hypothesis, but rather to demonstrate that Guedry's hypothesis has not been rejected adequately by Bowins' critique. In fact, the reader should note that there are unresolved questions that have yet to be addressed by either theory since each attempts to logically relate the origin of MS to natural selection without relying upon poison control theory—see a discussion of this point in Section 26.2.5 (concerning evolutionary theories).

26.2.1.6 Consideration of Sensory Conflict Theory

Although the general conceptual framework of sensory conflict is widely accepted, there are some significant gaps in theory and knowledge. These have been treated extensively in the last 35 years of the literature and will be discussed briefly here. First, the cortical centers for the processing of sensory conflict and the underlying physiological processes have not been fully identified. To do so would firmly establish the biological substrate for sensory conflict (see Oman & Cullen, 2012, for a proposed explanation). Second, it has been argued that since the sensory conflict theory can provide post hoc explanations for almost every stimulus that causes MS, a falsification of this theory is difficult, which makes for few testable predictions (Ebenholtz et al., 1994). Nevertheless, it has been asserted frequently that sensory conflict cannot explain why severe MS can be provoked by low-frequency vertical oscillation. This potential weakness was identified by Reason & Brand (1975, p. 265). Subsequent research has revealed evidence that may be consistent with the presence of sensory conflict during low-frequency vertical oscillation due to differences in response phase between otolithic and abdominal proprioceptors (von Gierke & Parker, 1994; Yates, Miller, & Lucot, 1998) or between otolithic and general somatosensory receptors (Benson, 1988). Analogous arguments concerning MS during horizontal linear acceleration have been traced all the way back to Mach (Mach, 1875, p. 85f; see also Guedry, 1974).

Another source of sensory conflict during vertical oscillation may arise due to the need to distinguish gravity from self-induced acceleration (Mayne, 1974). Wood (2002) employed a laboratory analogue of oscillation (constant velocity off-vertical rotation) and hypothesized that MS in this situation may be related to the fact that normal frequency segregation of ambiguous otolith information (which helps to distinguish tilt versus translation) breaks down near the sickening frequency. It is interesting to note that Reason and Brand (1975) had specifically mentioned the need for more information about otolith dynamics in settling the question of whether there is a sensory conflict during vertical oscillation. Whether the evidence from Wood and from Yates et al. (1998) helps to support sensory conflict theory or to further demonstrate the difficulty of falsifying it is open to debate. The optimal interpretation of translation findings and theories is an active area of discussion (e.g., see Goldberg et al., 2012), but it is clear that the observation that low-frequency vertical oscillation elicits MS should no longer be considered a convenient example to cite against sensory conflict theory. This has been the case for quite some time.

Draper, Viirre, Furness, and Gawron (2001), and many others have criticized the low predictive validity of the sensory conflict idea. While the theory explains why a given stimulus would be disturbing, it encounters difficulties forecasting the incidence and severity of MS for a given stimulus applied to a group or to a single person. Certainly, a unified and universally accepted theory of MS should be able to make accurate, quantitative predictions of the average group response provided the stimulus conditions are known.

While we can expect a MS theory to explain why a given stimulus is disturbing, it may be asking too much to expect an MS theory to predict the susceptibility of a given person since individual variability of responses among people (and within people over time) is a fundamental trait of human and animal physiology and behavior. In fact, variability (e.g., via small genetic differences) is the material for natural selection. By analogy, the fact that pain responses between people vary dramatically (Fillingim, 2005) cannot be used to argue that a theory concerning the cause of pain is wrong because it cannot predict the range of human responses to a given painful stimulus or the pain response of a given person in a particular instance. Rather, it is more reasonable to try to predict the reactions of groups of people of similar susceptibility to pain (or MS) first, and then seek to understand why people have evolved or been enculturated to vary so widely in their responses to such an apparently survival-relevant sensation.

Patterns of age-related susceptibility to VIMS have been proposed as a possible weakness of sensory conflict theory (Johnson, 2005). It has been found that some types of VIMS are more pronounced among older adults (Classen, Bewernitz, & Shechtman, 2011; Stanney & Kennedy, 2009) rather than less pronounced, as is the case for regular MS (Gahlinger, 2000; Reason & Brand, 1975). It has been asserted that if MS is indeed lower among older people whereas VIMS is higher, this pattern of results has implications for sensory conflict versus postural instability theories (Johnson, 2005). Sensory conflict theory should predict that VIMS declines with old age, whereas postural instability theory should predict that VIMS increases with age.

Unfortunately, the situation is more complicated than described by Johnson (2005). It is possible for both of the contrasted theories to be true if older subjects are better adapted to regular motion than younger subjects, but less adapted to visual stimuli that elicit VIMS. Age usually comes with more experience of motion, but experience leads to adaptation, which is specific to the stimulus. If this were not the case, then experienced pilots would report less simulator sickness. Instead, they report more VIMS (Kennedy et al., 1990) because they are highly adapted to *real* flying and thus more likely to exhibit maladaptation responses when encountering a stimulus that does not possess the same pattern of afference and reafference. Johnson acknowledged that the most experienced aviators studied in flight simulators also tend to be the older aviators and proposed a simulator sickness test involving nonaviators across a range of ages to control for simulator experience. This idea is interesting, but it may be necessary to consider extending experimental controls beyond experience with simulators to other types of relevant experience, such as frequent exposure to fast-moving video games, entertainment programs with highly dynamic visual content, wide screens or 3D displays, arcade simulators, VEs, and possibly reading or sending text messages in a moving car. Finally, the assertion that older people are less susceptible to regular MS has not been conclusively proven, according to Daunton (1990) and Golding (2006a). In fact, Golding (2006a) has suggested that MS susceptibility may show a second increase in the elderly years. Currently, there is not sufficiently clear evidence concerning the pattern and the causes of age-related trends in MS susceptibility to cite such trends as falsifications of a particular theory of MS.

Some confounds for a test of any MS theory included the following: whether the subjects are blind to the experimenter's theory and prediction or to their condition assignment, whether the experimenters are blind to the conditions during data collection or analysis, and whether the confounding effects of arousal or anxiety-related symptoms have been controlled or monitored during sickening testing. These are general concerns rather than specific criticisms of sensory conflict theory. They are mentioned in this section because many of the publications concerning the role of sensory conflict in MS and physiological correlates of sensory conflict have not controlled these variables. Fortunately, the range of possible sensory conflicts during MS is wide and not always predictable by laypersons, so blinding to condition would not be hard to accomplish in most sensory conflict experiments.

26.2.2 CONTRIBUTION OF POSTURAL MECHANISMS TO THE ELICITATION OF MS

Other potential limitations of sensory conflict theory not addressed earlier can be found in the writings of Stoffregen and Riccio (1991) and Yardley (1992). Stoffregen and Riccio raise many potential scientific, philosophical, and semantic limitations of sensory conflict theory and forward an alternative explanation for the genesis of MS or VIMS, called the *postural instability theory* (also known as the ecological theory of MS). They reason that sickness arises in situations that challenge one's ability to maintain postural stability. Postural stability is defined as "the state in which uncontrolled movements of the perception and action system are minimized" (Stoffregen & Riccio, 1991, p. 202).

Several studies revealed that low-frequency movement in the range of 0.06–0.4 Hz induces the most severe MS (provided the peak acceleration is sufficient). This is true for a stationary observer confronted with sinusoidal real or visual field motion (Duh, Parker, & Furness, 2004; Golding, Bles, Bos, Haynes, & Gresty, 2003; Golding & Gresty, 2005; Golding, Mueller, & Gresty, 2001; Griffin & Newman, 2004a; Guignard & McCauley, 1982; Lawther & Griffin, 1987, 1988; O'Hanlon & McCauley, 1974; Turner & Griffin, 1999a). However, the findings are mixed for studies attempting to correlate postural instability with MS (e.g., by measuring sway before versus after a challenging motion) (Anderson, Reschke, Homick, & Werness, 1986; Cobb, 1999; Cobb & Nichols, 1998; Hamilton, Kantor, & Magee, 1989; Jones, 1998; Kennedy & Stanney, 1996; Kenyon & Young, 1986; Reed-Jones, Vallis, Reed-Jones, & Trick, 2008).

Cars, boats, and airplanes tend to produce low-frequency movements that fall within this MS-provoking band of frequencies. Interestingly, the lower ranges of natural spontaneous human body sway (spontaneous motion of the body) fall within a similar frequency range (Golding et al., 2003; Golding & Gresty, 2005). Considering the frequency similarity of sway versus sickening passive motions, Riccio and Stoffregen (1991) proposed the idea that sickness arises when there is an overlap of the frequencies produced by natural body sway and by imposed motion. This interference within the same frequency band results in increased challenge to the maintenance of postural stability and, hence, leads to MS. More generally, it would be interesting to make a variable-by-variable comparison of the most posturally destabilizing external motion perturbations to the most sickening motions identified in the literature (in terms of frequency, amplitude, peak velocity, peak acceleration, and direction of greatest instability or MS, respectively). This would help to determine whether the most destabilizing perturbations are also the most sickening.

Several studies have shown that postural instability is correlated with MS (Akiduki et al., 2003; Baltzley, Kennedy, Berbaum, Lilienthal, & Gower, 1989; Flanagan, May, & Dobie, 2004; Kellogg & Gillingham, 1986; Reed-Jones et al., 2008; Smart, Stoffregen, & Bardy, 2002; Stoffregen, Hettinger, Haas, Roe, & Smart, 2000; Stoffregen & Smart, 1998; Tanahashi, Ujike, Kozawa, & Ukai, 2007; Villard, Flanagan, Albanese, & Stoffregen, 2008). For instance, Merhi et al. (2007) report that observers who performed more head movements while playing video games suffered more from VIMS. In another study by Bonnet, Faugloire, Riley, Bardy, and Stoffregen (2006), standing observers were positioned in a moving room and exposed to low-frequency motion. Data analyses revealed larger changes in postural sway in the group with strong MS symptoms compared to the group with weak symptoms. The authors concluded that increased challenges to postural stability precede and are the cause of MS.

26.2.2.1 Consideration of Postural Instability Theory

Postural instability theory (Riccio & Stoffregen, 1991) posits that postural instability is "necessary and sufficient for the occurrence of MS" (Stoffregen, Faugloire, Yoshida, Flanagan, & Merhi, 2008, p. 322). On first blush, this would seem to imply that increased sway would always be associated with MS, and that MS would always be associated with increased sway. However, Stoffregen and colleagues (2008) clarify that in their view postural instability theory can be confirmed by findings other than an increased magnitude of sway. As an example, the findings for the standing condition of their study showed that—unlike some of their group's previous studies (e.g., Faugloire, Bonnet, Riley, Bardy, & Stoffregen, 2007; Merhi et al., 2007)—vertical movement tended to increase for the well group but not for the sick group (both of which were playing a console video game). Nevertheless, they stressed that the effects were consistent with the hypothesis that movement should differ between the sick and well groups. Therefore, it would appear that the authors view postural instability theory as consistent with an increase in postural sway prior to MS, a decrease in postural sway prior to MS, or an increase in the variability of postural sway prior to MS.

It is not reported by Stoffregen et al. (2008) whether subjects were blind to the postural instability theory or the experimenter's predictions, whether the experimenters were blind to the experimental condition during data collection or data analysis, or whether arousal level differed between the groups. These are general concerns for any MS study rather than specific criticisms of postural instability theory. However, controlling or evaluating such factors as arousal or anxiety could be of specific value to tests of postural instability theory. For example, Merhi (2009) reported greater variability in head movement in the vertical axis for sick versus not sick participants playing a video game. It would be instructive in future experiments of this type to determine whether the sick group has an elevated level or arousal or anxiety (e.g., based on past experiences of MS) since it is known that increased arousal causes increased fidgeting apart from whether one is in a sickening situation or whether the increased arousal was precipitated by past experiences of postural instability, sensory conflict, or both. Maki and McIlroy (1996) suggest that the level or arousal or anxiety should be controlled or monitored when studying postural responses, especially when one is testing the effect of an attention-demanding task.

Studies monitoring postural stability throughout the MS testing have claimed greater success in establishing a relation, although not for all conditions or measures of interest (Merhi, 2009; Otten, 2008). Sometimes, sick versus not sick subject groups can be distinguished before stimulation commences and sometimes during the period of stimulation. Certain measures of sway work better in some studies than in others. Summarizing the findings for anterior–posterior sway from Merhi (2009) and nine other experiments reviewed by Otten (2008), it appears that 3/10 of the studies have shown a change in at least one of the following three measures of anterior–posterior sway: range, velocity, or variability. Note, however, that this estimate does not include all the studies done by this research group, nor all the measures attempted by them.

Golding and Gresty (2005) pointed out apparent inconsistencies in the postural instability theory. For example, the theory seems to imply that restraining the observer's body should decrease postural sway and MS to a minimum; however, the data regarding passive restriction and MS are inconsistent. Several studies showed that passive restraint is effective in reducing or even preventing MS (Bonnet, Faugloire, Riley, Bardy, & Stoffregen, 2008; Lackner & DiZio, 1991; Mills & Griffin, 2000). In contrast, other studies did not succeed in reducing MS by restraining the subjects (Graybiel & Miller, 1970; Warwick-Evans, Symons, Fitch, & Burrows, 1998).

Riccio and Stoffregen (1991) point out that postural sway cannot be completely eliminated by restraining the observer's body. Faugloire et al. (2007) had subjects stand on a force plate in a moving room. They were strapped to a vertical surface to restrain their head, torso, and feet. The authors compared the number of participants who became sick with the results of a similar experiment that used the same experimental design but did not restrain the observers (Bonnet et al., 2006). The number of participants who reported sickness during or after stimulus exposure did not differ between the two studies, but Faugloire et al. (2007) found that center of pressure data from their restrained participants predicted their MS. These results are partially in line with the postural instability hypothesis, but restraint of the body should have decreased MS to some extent compared to the unrestrained control group.

Other research questioning the postural instability theory has been carried out by Lubeck, van Geest, Bos, and Stins (2011). The authors have stated that the preponderance of evidence favors the subjective vertical theory, which has been discussed earlier (pp. 5–7). In regard to their criticisms of postural instability theory, the main objections are listed in Bos (2011). This paper lists seven possible problems with the postural instability theory, including the inferior account (as compared to sensory conflict) for the experimental findings of Flanagan et al. (2004) and the failure to adequately account for certain adaptation-related observations (Kennedy & Stanney, 1996; Lubeck et al., 2011).

Some attempts have been made to manipulate the factors that would be deemed important to postural control. For example, Otten (2005) reasoned that more complex visual stimuli would be less predictable and thus more likely to challenge postural control and cause MS. He did not detect a difference in sickness incidence with complex versus simple motion (but see his discussion section for possible explanations of this finding). Otten noted a trend toward a slight, nonsignificant baseline difference in the anterior–posterior velocity of postural sway between groups who later became sick versus those who did not (specifically, the group who later became sick showed 0.03 cm/s faster baseline sway, $p = 0.067$). In follow-up research, Otten (2008) observed more sickness in the complex condition but failed to detect a difference in anterior–posterior sway variability, velocity, or range of motion, regardless of the stimulus condition (simple or complex) or whether the participants reported sickness.

The overlap between the lower frequencies of sway and the frequencies of sickening motions is an interesting observation, which should be explored further. The possible relation is not a simple one since sway occurs at a much lower amplitude and peak acceleration within the frequency of interest for MS than would be employed in MS experiments involving vertical oscillation or would be experienced aboard a ship at high seas. Also, greater sway reactions seem to peak in some studies well outside the highly sickening 0.2 Hz range, but the pattern of findings is complicated. So and Chow (2009) exposed their subjects to scene movement frequencies from 0.05 to 0.8 Hz and found the greatest sway and vection at the lowest frequencies tested. However, with visual cues present, fore–aft translation of a support surface is related to greater center-of-mass sway at lower frequencies, whereas head–trunk anterior–posterior motion decreases at higher frequencies (Buchanan & Horak, 2001). Moreover, when elderly balance patients stand on a platform that perturbs their sway by translating forward and backward, the 0.5 Hz frequency range appears to be more useful in detecting the patients with a history of falling than the lower 0.25 Hz (Ghulyan & Paolino, 2005), although 0.25 Hz is closer to the most sickening frequency used in MS experiments. Finally, Buchanan and Horak (2001) observed less stable patterns of sway emerge in normal subjects at and above 0.5 Hz of fore–aft platform translation when visual cues were absent. Obviously, explanations of human sway control are limited when they depend solely upon reference to a single number representing sway frequency. Sway is a complicated behavior involving different reactions of the ankles, knees, hips, arms, and head at different amplitudes, frequencies, and peak velocities of sway perturbation.

26.2.3 Attempts to Evaluate Multiple Theories

A direct comparison between sensory conflict theory and postural instability theory has been attempted by Warwick-Evans et al. (1998). The authors compared the amount of sickness for two groups, one lying supine (restrained group) and one standing freely (unrestrained group). Their results showed no difference between the groups, although presumably less sickness would have been predicted for the restrained group by the postural instability theory. Based on their results, the authors argue that postural instability theory cannot explain the genesis of MS. However, Warwick-Evans and colleagues did not control or record the participants' body movements. Although lying in a supine posture should reduce sway, they cannot completely exclude possible body movements in their participants. All that can be said is that demands on the maintenance of postural control would be dramatically reduced compared to upright standing, although that reduction should still influence MS.

It is difficult to completely control a test of postural stability theory versus sensory conflict theory, and the problem cuts both ways. Specifically, in many of the published studies by Stoffregen and colleagues, the presence of sensory conflict cannot be ruled out (e.g., when the head moves more in the sickening condition, the head movement can contribute to a sensory conflict). Similarly, postural mechanisms are not always adequately controlled in tests of sensory conflict theory.

Flanagan et al. (2004) compared the predictions of sensory conflict, postural instability, and eye movement theory (described immediately next), concluding that all three theories partially succeed in explaining factors that contribute to MS. Their findings are described further at the end of the following section.

26.2.4 CONTRIBUTION OF EYE MOVEMENT MECHANISMS TO THE ELICITATION OF MS

Ebenholtz (1992) and his colleagues (Ebenholtz et al., 1994) highlighted the contribution of eye movement afferences to MS and VIMS. The authors propose that MS and VIMS are mainly "based on the premise of a specific routing between vestibular and the vagal nuclei, mediated by eye movements" (Ebenholtz et al., 1994, p. 1032). What we will call the *eye movement theory* relies on the fact that the visual and the vestibular systems are strongly connected as indicated by the visual–vestibular–ocular reflex and visual suppression of the vestibulo–ocular reflex. Based on earlier animal and human work by others, Ebenholtz and colleagues propose that optokinetic nystagmus (OKN) evoked by moving visual patterns innervates the vagal nerve (by stimulating cells within the vestibular nuclei) and such innervations lead to MS-typical outcomes such as nausea or emesis. Hence, nystagmus is not simply related to but rather a cause of MS. In a study by Houchin, Dunbar, and Lingua (1992, as cited in Ebenholtz et al., 1994), two groups of strabismus surgery patients were compared: the first group received only general anesthesia during surgery, whereas the second group was additionally treated with a local retrobulbar anesthetic to eliminate eye muscle afferences that frequently occur during strabismus surgery. The second group showed significantly less nausea and emesis in the postoperative 24 h phase.

Since the paper by Ebenholtz et al. (1994), some studies have emphasized the contribution eye movements may have to MS and VIMS (Flanagan, May, & Dobie, 2002; Gupta, 2005; Webb & Griffin, 2002). For instance, Hu et al. (1989) used different velocities of an optokinetic drum, varying the amount of OKN in their participants. Their results showed increased MS severity when the drum rotated with a faster velocity. Other studies introduced a fixation point to reduce eye movement during stimulus presentation and succeeded in decreasing VIMS (Flanagan et al., 2004; Webb & Griffin, 2002). Thus, the reduction of eye movements may ease VIMS, although other possible interpretations are discussed later.

Quarck, Etard, Oreel, and Denise (2000) analyzed the characteristics of OKN for participants either suffering strong MS symptoms or showing no MS at all. Comparisons of gain and time constants of the participants' nystagmus revealed no significant differences between the groups. Furthermore, precise measurement and manipulation of eye movements can be complicated by factors such as vection (Flanagan et al., 2002) or retinal slip (Ji et al., 2010; Reschke et al., 2006). Reducing the velocity of optokinetic stimuli could account for a reduction of the sensation of vection, which by itself might explain the results of Hu et al. (1989). Additionally, Yang and colleagues (Yang, Guo, So, & Cheung, 2011) noted that a fixation point may reduce OKN while increasing the amount of retinal slip. Therefore, it is imperative to disentangle effects of vection, retinal slip, and OKN.

One attempt to resolve some of these confounds was made by Ji et al. (2010). They tested two groups, separating vection from OKN while keeping the amount of retinal slip constant. The first group was exposed to a virtually presented optokinetic rotating drum that included a fixation point in the middle of the screen. Eye fixation has been successfully applied to reduce OKN in previous studies, but it was also known to leave the feeling of illusory self-motion unimpaired (Brandt, Dichgans, & Koenig, 1973; Stern, Hu, Anderson, Leibowitz, & Koch, 1990; Webb & Griffin, 2002). In the second group, vection was divorced from OKN. To do this, the authors used central and peripheral patterns that rotated in opposite directions. The idea behind this manipulation is that central vision is crucial for OKN, but vection relies both on central and peripheral vision (Brandt et al., 1973). As a result, participants looking at the moving central pattern should exhibit OKN but not report vection. The electrooculogram and vection data indicated the authors succeeded in separating vection from OKN. However, both groups showed comparable MS scores. Based on these findings, vection and OKN both seem to be involved in the genesis of MS, but neither component is likely to be a necessary prerequisite for MS.

There are some possible limitations of the eye movement theory. First, it is not clear how this approach explains why congenitally blind people can suffer from MS (Graybiel, 1970) even though they do not exhibit convincing vestibulo–ocular reflex. How can this be explained by afferent signals from the extraocular muscles? Also, people below deck of an ocean liner likely make fewer eye movements compared to those on deck looking at the horizon, but they typically show more severe MS. The same holds for passengers looking at a point inside a car. It would seem that the mechanisms involved in the visual suppression of the vestibulo–ocular reflex would figure prominently in the explanation, but the relation to extraocular afference is not fully resolved. The reader should note that conflicting demands on accommodation versus convergence also have been identified as a disturbing factor in VEs (e.g., Hoffman, Girshick, Akeley, & Banks, 2008; Howarth, 2011).

Flanagan et al. (2004) directly compared the predictions of eye movement theory, sensory conflict, and postural instability theory, using a $2 \times 2 \times 2$ within-subjects design. Each factor of their experiment manipulated a major aspect of each respective theory: sensory conflict was varied by showing moving or still images, postural sway was varied by staying on a moving platform or on the ground, and eye movements were varied by showing a fixation mark or not. Several interactions were found, prohibiting a decision as to the superiority of one single theory. Rather, several additive aspects best accounted for the occurrence of MS. Strongest MS scores were reported when participants were exposed to moving stimuli standing on a moving platform with no fixation point. The authors propose that all three theories partially succeed in explaining factors that contribute to MS.

26.2.5 EVOLUTIONARY EXPLANATIONS FOR THE ORIGIN OF MS

An evolutionary approach to explain MS was described by Treisman (1977) and is commonly called *Treisman's hypothesis*, the *poison theory*, or the *theory of toxins*. This approach primarily focuses on explaining why MS arises, instead of trying to explain the process of how MS develops during a given sickening stimulus. Therefore, this approach cannot be seen as a competitor for the aforementioned theories. Rather, the theory of toxins provides an interesting explanation for the particular signs and symptoms that characterize MS, such as nausea, retching, and vomiting.

According to Treisman's hypothesis (1977), emesis is the organism's natural reaction to an intoxication of the body by poisonous substances. By emptying the gastrointestinal tract, the organism tries to purge the harmful substances. Vomiting and retching appear in most mammals, and MS may have its evolutionary roots in this purging reaction. Treisman hypothesizes that a sensory conflict between the vestibular, the visual, and the somatosensory systems is associated with real intoxication and thereby triggers nausea and emesis during MS. Thus, the sensory conflict activates a similar pattern of reaction as does a real intoxication, including symptoms of stomach awareness, vertigo, or dizziness (see also Balaban, 1999). Similar to actual poisoning, the organism induces vomiting and retching. Reid and collaborators (Reid, Grundy, Khan, & Read, 1995) proposed a direct link between vection-induced nausea and the process of emesis, indicating that gastric motility is a key variable in nausea and vomiting. This assumption is in concordance with the toxin theory and supports Treisman's idea. Additionally, Money and Cheung (1983) showed that the emetic response toward several poisons was inhibited in dogs whose inner ears had been surgically removed. Taken together, the evolutionary need to detect and react to toxins appears to provide a logical argument for why sensory conflict produces nausea.

Bowins (2010) offers another evolutionary theory of MS, named the negative reinforcement model.* Bowins hypothesizes that MS evolved as a negative reinforcement system to motivate the organism to avoid or terminate motion involving sensory conflict or postural instability. Bowins believes this explanation accounts better for the absence of MS in infants and toddlers in comparison to Treisman's evolutionary hypothesis (1977). According to Bowins and others, one weakness of Treisman's theory of toxins is that infants and toddlers should be most susceptible to ingesting toxins yet are relatively immune to MS. While it is true that infants are more vulnerable to ingested toxins, this would not be a serious concern until after weaning, which may take 2 years or more in the *primitive* or *traditional* tribal societies comprising the majority of the history of the *Hominina* (i.e., Genus *Homo*). Moreover, in hunter-gatherer cultures, young children are often carried or swaddled, further decreasing the odds of exposure to toxins. In all societies, young children are closely supervised by adults.

Of course, young children do not exercise much judgment about what they will put into their mouths for the purpose of tactile exploration, but this is a different behavior from eating and swallowing. Nevertheless, it seems logical to suppose that young children would benefit from instinctive mechanisms to prevent them from ingesting harmful foods or to forcefully eject them. This is certainly the case since younger children are much more finicky about what they will eat, find unusual smells and textures more noxious, have greater acuity of taste and smell, and have much more easily triggered gag and vomiting responses than adults. Clearly, younger children are quite willing to reject novel foods (Chatoor & Ganiban, 2003). Interestingly, the willingness to reject foods (e.g., by spitting or vomiting) increases from age 4 to 24 months (Chatoor, 2009), which is the time period during which the child becomes more mobile and less reliant on breast milk. Why then are young children (e.g., age 1–3 years) not also more susceptible to MS than older children (e.g., age 8–12 years)? One possible explanation was discussed earlier in reference to Golding (2006a), who conjectured that the delay in peak susceptibility to MS may be related to the fact that the perceptual–motor map that influences the identification of sensory mismatches is not fully formed until around age seven. Brandt (1991; also Brandt, Wenzel, & Dichgans, 1976) has argued that young children are more dependent on vision for orientation and less on other systems; therefore, young children do not suffer to the same degree from visual–vestibular conflicts that are sickening to adults. Regardless of the explanation for this age-related trend in MS susceptibility, the observation remains that childhood aversion to novel foods does not have a 1:1 relationship with childhood susceptibility to MS, and this has potential implications for Treisman's (1977) theory.

Bowins (2010) points out another possible weakness of Treisman's theory (1977). Vomiting in response to poisons may be redundant since many animals are prepared to avoid toxins in at least two other ways: (1) the senses of taste and smell help to prevent the intake of harmful (e.g., spoiled) food and (2) the liver filters out many harmful substances. This argument makes the case that vomiting probably is not the only poison-protective response under selection pressure, but it does not convincingly establish that adaptive responses such as vomiting will evolve only if they are unique. In fact, avoidance of poisoning is so important that evolutionary fitness would benefit from multiple defenses against it. Be this as it may, if the ancient evolutionary mechanisms for toxin removal are also being triggered by challenging motions, then there is a need to explain why vomiting should not occur sooner and more frequently than it actually does during most challenging motions.

Any evolutionary theory positing a direct functional evolutionary purpose for MS without resorting to the notion that MS is an indirectly co-opted poison response (Treisman, 1977) needs to explain how the predominant symptoms of MS (such as nausea) would be subject to preservation by

---

* *Note*: Bowins' hypothesis also has developmental implications that were discussed in Section 26.2.1.5.

natural selection during the types of presently sickening motions that would have been a common aspect of human experience tens of thousands of years ago (e.g., sitting in a swaying tree, swimming or floating at sea when the waves are high). Such direct functional arguments can be made but are largely lacking so far. In support of his negative reinforcement idea, Bowins (2010) states that vomiting due to MS prompts the organism to seek to avoid or terminate the motion. He does not explain, however, why vomiting would be a necessary outcome of a negative reinforcement against MS when any noxious sensation (for negative reinforcement), coupled with quiescence (to minimize the motion challenge), would be sufficient. In fact, it could be argued that many other noxious sensations establish negative reinforcement without making the organism as vulnerable as do nausea and vomiting. In contrast, Treisman's hypothesis does not need to make as tight a link between prehistoric forms of motion and the evolution of motion-induced nausea and vomiting since part of the explanatory load in Treisman's hypothesis is carried by the co-opted poison ejection reflex.

It should be noted that not every trait or response has a clear and specific evolutionary function and that theories of MS must be careful to avoid becoming too narrowly adaptationist (Gould, 1997; Oman, 2012). In fact, the possibility has been raised that MS may simply represent an aberrant activation of homeostatic vestibuloautonomic pathways (Yates et al., 1998; see also Money, 1990). Of course, when it comes to responses as noxious, distracting, and potentially wasteful of calories as nausea and vomiting, the temptation to seek a foundation for these responses in some kind of functional explanation is strong, especially since nausea and vomiting are so strongly linked to rapid learning (Horn, 2008).

26.2.6 SUMMARY

In summary, there is still no universally accepted, quantitative theory able to explain the phenomenon of MS in a satisfying manner. The sensory conflict theory is the most widely accepted approach, but potential limitations to the theory have been discussed. The role of eye movements and the theory of toxins still need further attention as the data are still rather incomplete. Although all theories emphasize the importance of the visual and vestibular systems, there is unfortunately still no comprehensive approach that integrates the established aspects of the competing theories.

We shall now leave theoretical considerations and take a look at the practical side of MS and VIMS that continues to pose serious problems in research, rehabilitation, and entertainment. In what follows, we discuss the measurement of MS and VIMS, the factors that influence the severity of VIMS, and some promising countermeasures to minimize sickness.

26.3 MS MEASUREMENT

MS is undoubtedly a subjective experience. Some visible signs such as pallor, cold sweating, or vomiting may occur during episodes of MS. However, vomiting has not been a favorite dependent variable among experimenters or subjects because of subject discomfort, subsequent difficulty with recruitment of participants, and the potential for a mess. Also, measuring episodes of vomiting does not capture the nuances of MS, which is a syndrome that includes manifold symptoms that can vary interindividually. Not everybody who suffers from MS has to vomit. A finer differentiation between the shades of MS is impossible with this measure. There are several other techniques that have been used instead to capture MS, namely, questionnaires, self-reports, and psychophysiological measurements. We will discuss these techniques in turn.

26.3.1 MOTION SICKNESS SYMPTOM CHECKLIST QUESTIONNAIRES

Questionnaires are the most popular tools to gather data on MS. One of the early multiple-symptom checklists is the Motion Sickness Questionnaire (MSQ; Kennedy, Tolhurst, & Graybiel, 1965). The MSQ gathers data regarding the current MS state and contains more than 20 symptoms (e.g., general discomfort, fatigue, boredom, sweating, faintness, nausea), which are rated by the participant. Some symptoms such as blurred vision, stomach awareness, or confusion are rated on a dichotomous scale (yes versus no), while others (including general discomfort, sweating, or nausea) are rated on a four-point Likert scale ranging from *none* to *severe*. The MSQ was adopted in several different studies and served as a foundation for further MS questionnaires.

The most common tool for measuring VIMS is the Simulator Sickness Questionnaire (SSQ) by Kennedy, Lane, Berbaum, and Lilienthal (1993), which was derived from the MSQ. Since the experience of simulator sickness differs somewhat from classical MS, the SSQ includes symptoms typical for VIMS scenarios. The SSQ contains 16 scored items that are rated by the participants on 4-point Likert scales. The items are assigned to three subscales with overlapping descriptors: nausea (including general discomfort, increased salivation, sweating, nausea, difficulty concentrating, stomach awareness, and burping), oculomotor (including general discomfort, fatigue, headache, eyestrain, difficulty focusing, difficulty concentrating, and blurred vision), and disorientation (including difficulty focusing, nausea, fullness of head, blurred vision, dizziness, and vertigo). A weighting technique is used to calculate a single score for each subscale as well as a total SSQ score. Note that some adaptations of the SSQ have been made to measure VIMS with respect to special applications (such as VE displays); for example, by modifying some of the items from the symptom checklist and integrating them with items of other scales (e.g., Ames, Wolffsohn, & McBrien, 2005).

Another multisymptom questionnaire of MS is the Motion Sickness Assessment Questionnaire (Gianaros, Muth, Mordkoff, Levine, & Stern, 2001). Besides measuring nausea, the Motion Sickness Assessment Questionnaire also delivers data regarding symptoms germane to other aspects of MS. The questionnaire includes 16 items, which are classified in 4 subscales based on factor analysis. The subscales are gastrointestinal feelings (e.g., sick to the stomach, queasy, nauseated), central issues (e.g., lightheaded, disorientated), peripheral issues (e.g., sweaty, clammy, or cold sweat), and sopite-related issues (e.g., tired or fatigued, drowsy). Each symptom is rated by the participant using a 10-point scale ranging from *not at all* to *severely*. Again, a total score and a score for each subscale can be calculated.

A limitation of many multisymptom checklist questionnaires is the time required for administration. It may be impossible to collect questionnaire data during the stimulus portion of an experiment due to the length of the questionnaire itself, especially when the subject must concentrate on experimental tasks. Therefore, MS symptoms are often assessed at the end of the experiment. Consequently, it is difficult with multisymptom questionnaires to capture the moment-by-moment development of MS over the period of stimulus presentation or recovery from MS afterward. The following section will introduce various self-report procedures that can measure MS more quickly.

26.3.2 RAPID SELF-REPORT QUESTIONNAIRES

When using rapid self-report methods, participants verbally judge their overall state of well-being or the severity of VIMS via a single global rating scale. These scales benefit from simplicity and robustness and make it easier to capture MS data during stimulus presentation. Several studies have made use of such ratings, which varied dramatically from study to study. For instance, a simple and rapid *yes* or *no* dichotomization regarding MS can be explained and memorized easily and has been employed by Stoffregen et al. (2008). Since researchers usually desire to capture finer gradations of VIMS, Likert-type MS scales have been used more often, including four-point scales (Bagshaw & Stott, 1985; Draper et al., 2001), five-point scales (McCauley, Royal, Wylie, O'Hanlon, & Mackie, 1976), and 7-point scales (Griffin & Newman, 2004b).

Bos, Vries, van Emmerik, and Groen (2010) introduced the Misery Scale Index (MISC). This scale combines qualitative and quantitative aspects. It has 10 gradations ranging from 0 (*no problems*) to 10 (*vomiting*), and every gradation represents the presence of particular MS symptoms. The scale starts with uneasiness (score 1), continues with dizziness, warmth, headache, and stomach awareness (vague, slight, fairly, or severe corresponding to scores 2–5), and then reaches nausea (slight, fairly, or severe corresponding to scores 6–8), finally ending in retching (score 9) and vomiting (score 10). Training is required to ensure that participants apply the scores correctly.

The explicit attachment of scores to symptoms includes some assumptions. For example, extreme dizziness would receive a lower score than slight nausea, which is only justified if nausea is always preceded by feelings of dizziness, headache, and warmth. This is not always the case.

Another method for quantifying VIMS during immersion has been introduced by Nichols, Cobb, and Wilson (1997) under the label of Short Symptom Checklist (SSC). The SSC represents a short version of the SSQ by Kennedy et al. (1993) and contains six items in total (two for each subscale). The rating scale—in contrast to the SSQ ratings—ranges from *not at all* to *severe* in five steps. The authors state that the SSC can be used during VR exposure and is able to reflect the participant's sickness profile during stimulus presentation.

The Fast Motion Sickness Scale (FMS) by Keshavarz and Hecht (2011b) is one of the few MS scales that has been tested for concurrent validity versus other scales. The FMS is a verbal rating scale that ranges from 0 (*no sickness at all*) to 20 (*severe sickness*) and is partially based on the concept of classical magnitude estimation (Bock & Oman, 1982; Stevens, 1946), although the FMS does not base its numerical estimates of sickness severity upon a previous standard (or reference) stimulus presented to each subject. The FMS asks respondents to focus on general discomfort, nausea, and stomach problems and to ignore factors such as fatigue. The authors gathered FMS data from 126 participants once a minute and compared these results with the SSQ (Kennedy et al., 1993). High correlations were found among the peak FMS score and the SSQ-subscales nausea ($r = 0.828$), oculomotor ($r = 0.608$), disorientation ($r = 0.795$), and the total score ($r = 0.785$). This indicates that MS can be captured rather well with a single verbal rating. It also indicates that there is considerable overlap between the SSQ subscales. The FMS is an easy to administer tool that is capable of capturing the time course of MS. Note that this scale is not without drawbacks: The data gathered with it are not normally distributed, which complicates statistical analyzes. The scale would not be appropriate for sole use if data concerning the individual symptoms of MS must be preserved or the soporific or oculomotor dimensions of MS are of particular interest.

26.3.3 Psychophysiological Measurements

MS may be linked to several physiological changes during the acute phase, which can be traced to the vestibular apparatus, the CNS, and the autonomic nervous system (ANS). Although the involvement of the CNS in the genesis of MS is far from being understood, the vestibular nuclei, the archicerebellum, and some other brainstem, autonomic, and hypothalamic areas have been proposed (see Golding, 2006a). The activity of the ANS is of particular interest since ANS activity is (at least partly) visible to the experimenter: pallor, sweating, and vomiting are observable parameters (see Cowings, Suter, Toscano, Kamiya, & Naifeh, 1986; Harm, 2002). Sensitive and specific physiological measures of MS are desirable because they would permit real-time measurement of state without requiring the subject to pause during an experimental task to report his or her state. Also, they should be less susceptible to the potential response biases of the subject (e.g., to understate or exaggerate symptoms). Unfortunately, inconsistent findings exist for a host of measures, including heart rate, heart rate variability, blood pressure, respiration rate, pupillary changes, peripheral blood flow, skin temperature, body temperature, electrogastrography, ocular–vestibular evoked myogenic potentials, and various hormone levels (Bertin, Collet, Espié, & Graf, 2005; Blanford, 1990; Blanford & Oman, 1990; Cheung & Vaitkus, 1998; Crampton, 1990; Dahlman, Sjors, Lindstrom, Ledin, & Falkmer, 2009; Eagon, 1988; Harm, 1990; Harvey & Howarth, 2007; Lawson, 1993; Money, 1970; Ohyama et al., 2007; Otto, Riepl, Klosterhalfen, & Enck, 2006; Rague & Oman, 1987; Reason & Brand, 1975; Sheehan et al., 2011; Westmoreland, Krell, & Self, 2007). Explanations have been ventured to reconcile some of these mixed findings (e.g., Lin, Lin, Chiu, Duann, & Jung, 2011; Sheehan et al., 2011). Also, some promising measures may emerge from recent advances in functional brain imaging or cerebral blood flow measurement techniques (Kowalski, Rapps, & Enck, 2006; Serrador, Schlegel, Black, & Wood, 2005). For now, the situation is much the same as it was when Shupak and Gordon (2006) pointed out that no single physiological measure has yet been found with good enough properties for the diagnosis or prediction of MS in a given individual.

Although there is no single physiological measure that captures MS as a whole, approaches involving the combination of multiple measures may be beneficial in certain cases (Harm & Schlegel, 2002). Also, it may be possible to establish single measures that capture the severity of specific symptoms (see Shupak & Gordon, 2006). Some techniques to measure the cardinal symptoms already exist. For instance, pallor can be measured by visual observation (Crampton, 1955), infrared reflectance techniques (e.g., Oman, 1987), or white light reflectance techniques (e.g., Holmes, King, Stott, & Clemes, 2002). Sweating can be measured by recording the electrodermal activity (Dawson, Schell, & Filion, 1990). Gastric pacesetter activity (as a possible correlate of nausea) can be measured by electrogastrography (Hu, Stern, & Koch, 1992; Muth, Stern, & Koch, 1996). However, all of these parameters can also be altered by triggers other than MS.

The ANS responses associated with VIMS contain a mixture of sympathetic and parasympathetic responses. An attempt to explain this mixed set of responses to MS has been put forth by Sheehan et al. (2011). They hypothesize that a cholinomimetic agent is being released as a result of sensory conflict and explain possible ways in which such an agent could stimulate sympathetic and parasympathetic ganglia. There is some evidence that the sympathetic part of the ANS shows increased activity during an acute phase of MS while the parasympathetic part is reduced (Hu, Grant, Stern, & Koch, 1991). An increased release of neurohumoral substances (e.g., cortisol, acetylcholine, epinephrine, norepinephrine, vasopressin) has been reported by Koch et al. (1990). Additionally, some researchers have observed heart rate acceleration (Cowings et al., 1986; Hu et al., 1991), increased skin conductance (Golding, 1992; Hu et al., 1991; Warwick-Evans et al., 1987), and a shift in the frequency of electrogastrogram activity (changing from 3 to 4–9 cycles per minute) (Hu et al., 1991; Muth et al., 1996). Kim, Kim, Kim, Ko, and Kim (2005) measured heart rate, skin conductance, respiration rate, skin temperature, photoplethysmography, and gastric tachyarrhythmia for 61 participants during a 10-min VE exposure. They found significantly increased electrogastrogram frequency, increased electrodermal activity, heart rate acceleration, decreased skin temperature, and a trend for decreased respiration rate. When Kim et al. correlated these results to the severity of VIMS measured via the SSQ (Kennedy et al., 1993), analyses showed significant but only slight to moderate correlations between the SSQ total score and gastric tachyarrhythmia ($r = 0.317$), eye blink rate ($r = 0.267$), respiration rate ($r = 0.392$), and heart rate ($r = 0.373$). Similar results were also found in other studies (Hu et al., 1999; Miller, Sharkey, Graham, & McCauley, 1993) that included skin conductance, increased heart rate, and gastric tachyarrhythmia as measures.

Taken together, an acute phase of MS appears to be linked to physiological changes in the ANS. Nevertheless, it is problematic to relate physiological indicators to MS severity. Although objective ways to capture MS (such as gathering MS data by physiological parameters) are highly desirable, the attempts to measure MS based entirely on physiological parameters have not been conclusive or robust in the face of even minor variations in stimulus or approach. There are some limitations that make current psychophysiological parameters less than ideal for the measurement of MS severity. First, the usually weak or inconsistent correlations between the physiological data and the MS self-reports are unsatisfactory. Second, changes in the ANS responses may be traced to factors other than changes in MS per se. For instance, simple stress responses may lead to increased sympathetic activity and result in a physiological pattern typical for MS (and even some of the same symptoms), but sickness is not necessarily a by-product of stress responses nor is motion required for them to occur. For these reasons, it is important to control and monitor arousal and anxiety during any MS experiment where physiological measures are taken. This is especially important when seeking to identify a temporal predictor of MS. A final concern is that the physiological effects of movement or exercise are not accounted for during some MS tests involving head movement. Overall, physiological correlates of VIMS are unfortunately still lacking.

26.4   RELATED CONCEPTS

Now that we have covered the main theories of VIMS and familiarized ourselves with the available measurement tools, we will take a closer look at other constructs that are related to VIMS. They range from immediate but illusory sensations of self-motion (vection) to stimulus characteristics to trait-oriented viewer characteristics (e.g., susceptibility).

26.4.1   Illusory Self-Motion (Vection)

Harris, Jenkin, and Zikovitz (2000) exposed their participants either to real or to simulated acceleration and observed that subjects tended to rely more on vestibular and somatosensory information than on visual information. Nevertheless, whenever the vestibular and somatosensory systems indicated that the observer was stationary, whole-field visual cues indicating movement exerted a strong influence. Gibson (1958, 1961) proposed that optic flow provides information about the observer's motion direction and speed. In some cases (e.g., nonaccelerated movement), retinal information is the most reliable indicator of self-motion since the vestibular system is only sensitive to accelerated movements. Visual and vestibular information are congruent during locomotion, but when one is seated inside a fixed-base simulator, the eyes indicate motion and override the vestibular information to the contrary. The resulting illusion of self-motion is referred to as *vection* (Dichgans & Brandt, 1973, 1978; Fischer & Kornmüller, 1930). A classic example is the situation of sitting in a stationary train waiting for departure from the station. If a neighboring train starts to move, the passengers sitting in the stationary train frequently get the feeling of self-motion. Vection is the illusory self-motion in the opposite direction of the moving train. Vection has been studied systematically since the nineteenth century (e.g., Mach, 1875; Wood, 1895). This section focuses on some basic characteristics of vection that are relevant to VIMS. A full review of vection and its components was compiled by Hettinger and Keshavarz (2013; see Chapter 18).

Vection can be experienced along all six degrees of freedom of body motion, including circular and linear vection about the pitch, roll, and yaw axes. To study circular vection in the laboratory, researchers frequently made use of optokinetic rotating drums (Bubka & Bonato, 2003; Dichgans & Brandt, 1973; Hu et al., 1989, 1999). Optokinetic drums typically consist of a cylinder with alternating black and white vertical stripes that surround the observer. As the drum starts to rotate, the stationary observer perceives vection in the direction opposite to the drum's motion. These drums are capable of provoking VIMS under special circumstances (e.g., Bonato et al., 2005; Bubka & Bonato, 2003; Bubka et al., 2006; Hu et al., 1989; Klosterhalfen, Muth, Kellermann, Meissner, & Enck, 2008; Stern et al., 1985, 1987).

Some experiments have employed physically swinging rooms (e.g., Allison, Howard, & Zacher, 1999; Bonnet et al., 2006; Howard & Hu, 2001; Stoffregen, Yoshida, Villard, Scibora, & Bardy, 2010) or virtually swinging rooms (Villard, Flanagan, Albanese, & Stoffregen, 2008). The optic flow created by the moving room induces the illusion of self-motion in the stationary participant. In a swinging room, participants are positioned inside a (sometimes furnished) room, which often oscillates along the viewer's line of sight.

Linear vection (especially along the fore–aft axis) is more common than circular vection in VEs such as flight or driving simulators and is related to VIMS (Diels & Howarth, 2006; Lee, Yoo, & Jones, 1997). Nevertheless, the relationship between vection and VIMS is not yet fully understood. Some researchers argue that vection is necessary for VIMS (e.g., Hettinger, Berbaum, Kennedy, Dunlap, & Nolan, 1990) or for pseudo-Coriolis effects (Dichgans & Brandt, 1973). By this reasoning, people who do not get the feeling of illusory self-motion should not experience VIMS (Flanagan et al., 2004; Hettinger et al., 1990; Hettinger & Riccio, 1992; Smart et al., 2002). Hettinger et al. (1990) found that 88% of subjects who reported VIMS also reported vection, whereas only 20% of the observers who failed to get the feeling of self-motion became sick. Note, however, that not every observer who experienced vection suffered from sickness. Moreover, Webb and Griffin (2002) failed to find a correlation between vection and sickness, and Lawson (2005) successfully demonstrated that under the right conditions (brief exposures to an immersive stimulus), vection and sickness could be dissociated. Lawson observed very strong vection ratings among 45 participants, none of whom reported minimal (or greater than minimal) nausea.

Other studies are indirectly pertinent to the relationship between vection and sickness. Cheung et al. (1991) showed that participants without functioning vestibular organs readily experienced vection but did not become motion sick. Finally, Palmisano, Allison, and Pekin (2008) demonstrated that a visual flow field that expanded and contracted could elicit more VIMS but less vection than a steadily expanding flow pattern. Although vection is relevant to VIMS, further knowledge is needed to understand how and why vection may generate VIMS.

26.4.2 Presence

Another construct possibly relevant to VIMS is *presence* (Heeter, 1992; Lombard & Ditton, 1997; Sheridan, 1992; Steuer, 1992; Witmer & Singer, 1998). Presence describes the feeling of *being there* in a VE. In the most extreme case, users of virtual reality systems are entirely engaged in a virtual world and ignore the real world surrounding them (Barfield & Hendrix, 1995). A high degree of presence should accentuate an existing visual–vestibular conflict unless presence is most readily experienced when conflict is minimal.

The degree of presence depends on numerous influences (see Chertoff & Schatz, 2014—see Chapter 34), the most important of which are the ease of interaction, user-initiated control (Sheridan, 1992), pictorial realism (Welch, Blackmon, Liu, Mellers, & Stark, 1996; Witmer & Singer, 1994), exposure duration (Kennedy, Stanney, & Dunlap, 2000), social factors (Heeter, 1992), internal factors (Slater & Usoh, 1993), and system factors (Hendrix & Barfield, 1996; Welch et al., 1996). Furthermore, vection has been considered a contributor to presence (see Hettinger, Schmidt, Jones, & Keshavarz, 2014—see Chapter 18).

Only a handful of studies have investigated whether presence and MS are positively related. In one study, Witmer and Singer (1998) found a negative correlation between presence and VIMS ($r = -0.426$), that is, stronger symptoms of MS when presence was low. Data gathered by Nichols, Haldane, and Wilson (2000), Jerome and Witmer (2004), and Keshavarz and Hecht (2012b) confirmed this negative correlation. Therefore, the relationship between presence, vection, and VIMS is not straightforward. Future studies should go beyond a correlational approach and aim at ruling out the direct influence of presence and immersion on the severity of VIMS. Upcoming studies should compare groups with different levels of presence and immersion regarding the severity of VIMS.

26.4.3 Display Technology Effects

There are four major display technologies that vary with regard to the degree of immersion they create in the user: head-mounted displays (HMDs), multidirectional projection screens (CAVEs), straight large-field projection screens, and computer or television screens. The display type can influence the occurrence and severity of VIMS (Frank, Casali, & Wierwille, 1987; Keshavarz, Hecht, & Zschutschke, 2011; Sharples, Cobb, Moody, & Wilson, 2008). HMDs are generally known to be problematic in several ways. They may introduce processing lags associated with head movement or they may influence the observer because of their weight and the restrictions imposed on the user's mobility (Howarth & Costello, 1997; Howarth & Finch, 1999; Mon-Williams, Warm, & Rushton, 1993; Patterson, Winterbottom, & Pierce, 2006; Peli, 1998). The discomfort of wearing an HMD has been linked to MS. Most recently, Moss and Muth (2011) presented a study on the characteristics of HMDs, including the degree of peripheral vision of the external world (using the HMD with or without the rubber eyecups), the image scale factor related to the geometric field of view, and the latency between head movements and the accompanying change in the visual scenery.

They found that if peripheral vision of the external world was excluded by the HMD eyecups, MS increased significantly. Their results did not reveal an MS effect of image scale factor or of increasing the latency (up to 200 ms) between the real and simulated head movements. Buker, Vincenzi, and Deaton (2012), however, implicate reduced MS when latencies are shorter.

Only a few studies have made direct comparisons among the different display types. Sharples et al. (2008) compared an HMD with a CAVE, a large projection screen, and a smaller television. HMDs were reported to be more provocative than the other displays and resulted in increased sickness ratings. Stoffregen et al. (2008) and Merhi et al. (2007) tested two groups of video game players, one group using an HMD and one group playing on a large television. Sickness ratings did not differ between the two groups. Based on these findings, Keshavarz et al. (2011) attempted to make a comparison between an HMD and a projection screen and found increased VIMS ratings when stimuli were presented on the projection screen.

A key difference between HMDs and other devices is that HMDs (except for see-through augmented reality HMDs) completely occlude the real visual surroundings and limit the display angle to the size of the internal HMD monitors. In contrast, laboratory surroundings are usually visible to the observer when viewing typical monitors and projection screens. Besides the display type, there are several other technological aspects that should be relevant to immersion and influence the severity and occurrence of VIMS, including the field of view. The distance between the observer and the display is linked to VIMS (e.g., Bos et al., 2010; Duh, Lin, Kenyon, Parker, & Furness, 2001; Emoto, Sugawara, & Nojiri, 2008; Keshavarz et al., 2011; Lin, Duh, Parker, Abi-Rached, & Furness, 2002; Stern et al., 1990). Results usually indicate that participants who are exposed to a smaller field of view report significantly less sickness. Additionally, the observer's relative position toward the display screen seems to be of particular interest. Ujike and colleagues (Ujike, Ukai, & Nihei, 2008) gathered post hoc questionnaire data at a Japanese high school after several pupils got sick during a movie presentation in an auditorium. The authors found that the children who were seated directly in front of the cinema screen suffered significantly more from MS than those seated at the edge of the screen.

The amount of VIMS also seems to be related to the velocity and the optic flow of the presented stimuli. Min et al. (2006) reported that users of a driving simulator showed significantly increased autonomic responses and MS at faster driving speeds (30 versus 120 km/h). This result is only partially in line with a study by So, Lo, and Ho (2001), who varied the navigation speed in a VE from 3 m/s (10.8 km/h) to 59 m/s (212 km/h). Increased vection and sickness ratings in their participants were obtained when the speed increased up to 10 m/s (36 km/h); however, further acceleration beyond 10 m/s did not exacerbate MS. Although both studies emphasize a link between navigation speed and the level of VIMS, the nature of this relationship still remains vague.

More recently, the role of stereopsis has been considered in VIMS. Keshavarz and Hecht (2012a) presented video footage (a rollercoaster ride or a bicycle ride) to their participants either stereoscopically (3D) or binocularly (2D). The results indicated that the 3D videos provoked significantly more VIMS than the 2D videos. In a similar manner, Ujike and Watanabe (2011) found higher sickness scores in a driving simulator when their participants were exposed to stereoscopic stimuli. However, in a study by Häkkinen, Pölönen, Takatalo, and Nyman (2006), two groups of participants played a video game either in 3D or 2D, but no significant differences between the two groups arose. Therefore, stereopsis did not consistently cause a detectable increase in the level of VIMS in each of the mentioned studies.

26.4.4 Fundamental Stimulus Characteristics

We have considered some display parameters that affect VIMS, but now we focus on the fundamental stimulus itself. The magnitude and the quality of optic flow obviously affect MS; however, the relationship is complex. VIMS is positively correlated with the amount of optic flow, for instance, the total size of texture influences VIMS (Bubka, Bonato, & Palmisano, 2007). Hu et al. (1989) varied the rotational speed of an optokinetic drum and found significantly more VIMS when the rotational speed was faster. Still-frame pictures are not known to be provocative for VIMS (IJsselsteijn, Ridder, Freeman, Avons, & Bouwhuis, 2001). Kennedy and colleagues (Kennedy, Berbaum, Dunlap, & Hettinger, 1996) varied the amount of optic flow and proposed that the visual motion of the stimuli account for at least 20% of the sickness variance. Diels, Ukai, and Howarth (2007) confirmed the role of optic flow in the genesis of VIMS but found that optic flow has largely different effects as a function of where on the retina it occurs. When their participants were forced to shift their gaze fixation away from the center of motion, sickness significantly increased.

When it comes to cyclical motion, physical movement sensed vestibularly within a visual world that does not indicate motion (e.g., being below deck on an ocean liner) is particularly provocative at low frequencies. Real motion around 0.06–0.4 Hz is most effective in the genesis of MS. The same frequency is less provocative when the body is stationary (barring small oscillations), and instead the visual motion is pronounced and cyclical at low frequencies (Groen & Bos, 2008; Hettinger et al., 1990; Stoffregen & Smart, 1998). Several studies showed that increasing the optic flow of the stimulus increases VIMS in stationary observers, for instance, by adding rotating visual motion along a second body axis (Bonato, Bubka, & Palmisano, 2008, 2009; Lo & So, 2001) or by adding display oscillations to a forward visual motion flow (Palmisano, Bonato, Bubka, & Folder, 2007).

Joseph and Griffin (2008) exposed their participants to actual low-frequency body oscillation (0.2 Hz) by using a tilting motion cabin. Roll and pitch motion of the cabin were both varied between 1.8°, 3.7°, and 7.3°, and sickness was measured via the SSQ (Kennedy et al., 1993). Results showed that MS increased almost linearly up to rotations of 3.7°, whereas motion beyond that level did not increase MS any further. Similar results were obtained by Keshavarz and Hecht (2011a), who compared visual motion along multiple body axes (i.e., either along the pitch, the pitch and roll, or the pitch, roll, and yaw axis). Results revealed lowest VIMS scores in the single-axis condition, whereas adding a third rotational axis did not exceed the VIMS scores reported in the dual-axis condition. The authors discuss these findings with respect to the relationship between the amount of optic flow and the magnitude of VIMS. They hypothesize that the relationship between those two factors follows a stepwise function rather than a linear one.

26.5 SUSCEPTIBILITY TO VIMS

Individual factors such as age, gender, or personality appear to affect the genesis of MS (for a good review, see Golding, 2006a). In the following section, we give a short overview of these factors among others. Almost every healthy participant can be made motion sick when the presented stimuli are severe and provocative enough. The only reliable prevention of VIMS is a complete bilateral loss of the inner ear's labyrinth system, as some studies have revealed that these people are mostly immune toward MS (Cheung et al., 1991; Kellogg et al., 1965; Kennedy et al., 1968). Note that even this evidence for a necessary involvement of the vestibular system has not gone unchallenged. Johnson et al. (1999) noticed sickness in people without functioning vestibular systems under special circumstances (such as head movements under visually induced pseudo-Coriolis).

The literature on the possible causes of susceptibility to VIMS is rather scant. Most research focuses on factors that influence classic or *real* MS instead of its visually induced complement. However, we assume that the factors that correlate with the genesis of real MS are quite similar to the ones that may be causative for VIMS. A reasonable approach to forecast an individual's susceptibility to MS is to use the subject's past episodes of MS to predict future ones. One way to collect data on past MS experiences is via MS history questionnaires (MSHQs) (Griffin & Howarth, 2000; Kennedy, Fowlkes, Berbaum, & Lilienthal, 1992; Reason & Brand, 1975). The MSHQ captures the history of the participant and makes a susceptibility prediction based on previous sickness experiences. Golding (2006b) presented a short version of the MSHQ (called the MS susceptibility short form or MSSQ-short) with high reliability and sufficient internal consistency. Participants rate the frequency of past episodes during travel (never traveled, never felt sick, sometimes felt sick, frequently felt sick, always felt sick) with respect to various transport forms including cars, ships, trains, etc. Participants fill in the short form twice, once for the memory of occurrences during their childhood (prior to the age of 12) and once for the past decade (last 10 years). Golding tested 257 participants and found high correlations (ranging from 0.51 to 0.74) between the MSSQ-short results and reported sickness ratings during stimulus exposure. Interestingly, women report significantly higher MS history scores than men in the MSSQ. The MSSQ was designed and validated using real motion stimuli. Although there should be congruent results between real MS and VIMS, further studies should ascertain the relationship between MS history and VIMS (e.g., see Bijveld, Bronstein, Golding, & Gresty, 2008).

Susceptibility to MS varies within the life span. In the very early stage (0–24 months), children may not be susceptible to MS, then susceptibility increases with age and reaches its peak around the age of 10–12 (Brooks et al., 2010; Reason & Brand, 1975; Turner & Griffin, 1999b). After a slow but steady decrease, MS susceptibility may show a second peak in the elderly years (Golding, 2006a), but further research is needed to confirm the responses of older adults (Daunton, 1990; Golding, 2006b). The reasons for these age-related increases are not yet fully understood. The infantile immunity could be attributed to the incomplete early development of neural and muscular capacities. After children learn to walk, their sensitivity for other movement maneuvers (e.g., riding a bike or travelling in cars, ships, or airplanes) may develop further. With increasing experience in early adulthood, proneness to MS subsides because of coping and adaptation processes. Most people recover from early proneness to MS. Birren (1949) estimates that only 1% of the adult population experience nausea due to moderate vehicle movements such as during typical automobile rides. The second peak along the life span might be influenced by the age-related changes in the sensory (e.g., vestibular), hormonal, and/or endocrine systems.

Another focus of attention regarding the susceptibility to MS has been sex differences. Women report higher sickness scores in several different MS scenarios (e.g., Flanagan, May, & Dobie, 2005; Kaplan, 1964; Klosterhalfen et al., 2005; Lentz & Collins, 1977; Stanney, Hale, Nahmens, & Kennedy, 2003). Most findings rely on self-reports rather than on objective measurements. Thus, it cannot be completely ruled out that men merely report less MS for reasons of social desirability rather than being less affected by MS (Ladwig, Marten-Mittag, Formanek, & Dammann, 2000). Lawther and Griffin (1988) tracked the incidence of vomiting during a cruise. Women had an increased risk ratio (4:3) of vomiting, which suggests that willingness to report prevomiting MS symptoms is not the only factor involved in past findings of greater susceptibility among women. Of course, self-report surveys in an uncontrolled group setting are less definitive than direct observation of isolated individuals in a controlled laboratory setting.

Women may be more susceptible to nausea in general, as in the case of postoperative nausea and chemotherapy-induced vomiting (Golding, 1998; Morrow, 1985), rather than to MS specifically. A complete explanation for possible gender differences in MS has not been found yet. One logical explanation is based on hormone differences, as some evidence suggests that women are most susceptible to MS around their menstruation (Clemes & Howarth, 2005; Golding, Kadzere, & Gresty, 2005; Grunfeld & Gresty, 1998). From an evolutionary point of view, women might be more sensitive toward nausea to prevent damage to the unborn child during pregnancy. According to the theory of toxins, by vomiting early, the female organism would better protect the fetus from harmful toxins (Golding, 2006a).

Susceptibility to MS has also been linked to ethnic origin (Klosterhalfen et al., 2005; Klosterhalfen, Pan, Kellermann, & Enck, 2006). Stern, Hu, LeBlanc, and Koch (1993) found a significantly higher chance to suffer from MS in Chinese compared with American or African subjects. The authors propose a hypersusceptibility for the Asian population, probably due to genetic differences (Stern et al., 1996; Yoshida et al., 2003).

Personality factors such as neuroticism or extraversion have sometimes been found to correlate with reported MS susceptibility. Several studies indicate the relationship between personality factors and MS is rather weak. Bick (1983) correlated several personality factors for male and female participants with MS and found only a single significant correlation ($r = 0.62$) for neuroticism and

MS in women but not in men. On the other hand, the only significant correlation for male participants was between field dependency and MS ($r = 0.83$). Other studies revealed a similar pattern of results, implying that personality factors affect the severity of reported MS, but not in a prominent fashion. In contrast, the individual state of well-being and the subject's health seem to influence MS susceptibility. For instance, Ujike et al. (2008) found an increased risk for MS when the participants had a slight cold. Similarly, MS has quite frequently been related to migraine, indicating that the susceptibility to MS is increased with augmented migraine (Bijveld et al., 2008; Golding, 1998; Grunfeld & Gresty, 1998; Shupak & Gordon, 2006). Almost two-thirds of adults with migraine do suffer from MS as well (Kuritzky, Ziegler, & Hassanein, 1981). The reason for this is not yet fully understood but may relate to past labyrinthine concussion or to low levels of serotonin implicated in both migraine and MS.

Recently, biological and genetic makeup has been suggested as a cause for the occurrence of MS. Finley et al. (2004) found that participants who were more susceptible to autonomic stress reactions caused by motion-induced sickness showed higher α_2-adrenergic reactions than the nonsusceptible group. Increased α_2-adrenergic activity is known to be encoded by a gene located on chromosome 10.

Lastly, athleticism is related to MS. Cheung, Money, and Jacobs (1990) studied the effect of aerobic fitness on MS by subjecting a group of untrained subjects to 8 weeks of stationary bicycle workouts (plus three additional weeks of training during MS testing). MS susceptibility was measured before and after the training period and indicated significantly increased susceptibility to MS after the training interval, especially among more MS-susceptible individuals. However, aerobic fitness and athletic skill are not identical. Some athletes, such as ballet dancers or figure skaters, require an exquisite sense of balance and are highly experienced in the maintenance of their postural balance during unusual movements. Extensive movement training might indeed protect them from MS. In a large questionnaire survey, Caillet et al. (2006) found significantly less reported susceptibility to MS in people who engaged in sports regularly before the age of 18. Of course, it is possible that people who are less susceptible to MS are more likely to start playing a sport initially. In a study by Hecht and Brendel (2008), two groups of experts—gymnasts and pilots of small airplanes—were subjected to a Coriolis cross-coupling stimulus. Their eye movement patterns were more affected than those of control subjects, but they reported smaller degrees of illusory motion of a stimulus in the dark (autokinetic effect) as they attempted to fixate. This suggests that experts have developed strategies to avoid those sensory consequences of heightened sensitivity that should produce MS. They certainly did not report getting motion sick as a consequence of their challenging maneuvers. In another study by Tanguy, Quarck, Etard, Gauthier, and Denise (2008), figure skaters showed a significantly lower gain of vestibulo–ocular reflex and reported less MS during off-vertical axis rotations. These results suggest that experts—although they have a fine-tuned sense of balance—are less susceptible to MS due to adaptation or behavioral coping strategies such as reducing eye movements. It remains to be seen whether self-selection for reduced susceptibility to MS plays a role prior to entry into sports. Also, it remains to be determined how generalizable sports-related adaptation is to motions that differ from those practiced in the sport, whether some sports confer significant resistance to MS while others do not, and how aerobic fitness interacts with sports-related movement experience. Finally, the general MS responses in sports that regularly induce MS during participation should be investigated (Perrin, Perrot, Deviterme, Ragaru, & Kingma, 2000).

26.6 AFTEREFFECTS OF VIMS

Given that there is large individual variability in susceptibility and severity of MS, the time that is needed to completely recover from MS symptoms is likely to vary as well. In some cases, VIMS can be a rather long-lasting phenomenon. It has been reported that aftereffects can persist hours or even days after stimulus presentation has been terminated (Baltzley et al., 1989; Muth, 1996, 2010; Stanney et al., 2003; Stanney & Kennedy, 1998; Stanney & Salvendy, 1998; Ungs, 1987). There is no magic cure to quickly eliminate MS once the offending stimulus has been removed. The usual way to recover from MS is just to wait for the symptoms to subside, which is referred to as natural decay (McCauley & Sharkey, 1992). A few techniques to accelerate the recovery process have been discussed in the current literature (Champney et al., 2007).

Two main measurement techniques have been established over the past years to measure recovery and aftereffects, including postural stability measures (Cobb, 1999; Kennedy & Stanney, 1996) and self-report questionnaires (Kolasinski, 1995; Stanney & Hash, 1998). Similar to the various types of symptoms during stimulus presentation, the recovery process is also influenced by several aspects, such as the exposure duration (Baltzley et al., 1989) or stimulus characteristics. Individual factors may change the recovery speed, but there is no empirical evidence indicative of which personality traits may facilitate the recovery process. Until we know more, we can only suspect the same factors that influence the susceptibility to MS.

Stanney et al. (1999) studied postexperimental sickness recovery using the SSQ (Kennedy et al., 1993) four times, immediately after stimulus offset, 30 min afterward, between 2 and 4 h after stimulus offset, and the next morning. Results showed that immediately after the experiment, sickness was significantly higher than prior to stimulus onset. Participants had not completely recovered 30 min after the cessation of the stimulus, but sickness scores were notably reduced. Keshavarz and Hecht (2011a) gathered sickness data using the FMS (Keshavarz & Hecht, 2011b) and the SSQ up to 5 h after stimulus presentation. Usually, sickness decreased quite rapidly after stimulus offset, but in several cases MS symptoms lasted for the entire postperiod of 5 h.

26.7 COUNTERMEASURES

So far we have discussed how VIMS is thought to arise, which concepts and factors are closely linked to VIMS, and how we can best measure them. But most importantly, research regarding VIMS should allow us to derive ways to reduce VIMS in the real world. One main reason why we study the characteristics of VIMS is to find solutions to prevent observers from getting sick. Thus, we will introduce the most common and promising countermeasures in the following section, starting with classic medical countermeasures, followed by behavioral ways to cope with VIMS. A summary of the most popular countermeasures is given in Table 26.2.

26.7.1 MEDICATION COUNTERMEASURES

Over the past decades, a variety of drugs have been developed and introduced to minimize MS during travel (for an overview, see Sherman, 2002; Shupak & Gordon, 2006). Unfortunately, many of these medications lead to serious side effects, including dizziness and fatigue. The side effects dramatically restrict their field of application (see Golding, 2006a). For instance, fatigue may be a tolerable side effect for travelers during long flights who can easily close their eyes and rest for a longer period of time, but not so for aircrew members. They must have unimpaired mental faculties and fatigue and sleepiness are not acceptable. Medical drugs are still the method of choice to reduce MS in travelers due to their easy accessibility and rapid action. The list of substances known to reduce MS is long, including betahistine, benzodiazepines, and barbiturates, antipsychotics, neurokinin-1 antagonists, opioids, phenytoin, and dopamine antagonists (Golding, 2006a), all of them with only slight or moderate success. Two drug classes have surfaced to be the most popular in the prevention of MS, namely, antihistamines and anticholinergics (Hoyt, Lawson, McGee, Strompolis, & McClellan, 2009). We will take a closer look at these two types of medications and discuss them in more detail in the following section. As the physiological mechanisms of MS genesis involve the vestibular, visual, and somatosensory systems—including the vestibular nuclei, the archicerebellum, and other brainstem, autonomic, and hypothalamic areas—it is exceedingly difficult to target the root of MS. Although the pathway of MS is not yet fully understood, Shupak and Gordon (2006) report that relevant vestibular, visual, and somatosensory signals are sent to the vestibular nuclei, then sent via the cerebellum to the vomiting center in the parvicellular reticular

TABLE 26.2
Prominent Countermeasures for MS and VIMS

Category	Subgroup	Examples	Application	Efficiency	Side Effects and Disadvantages
Medical	Antihistamines	Diphenhydramine, dimenhydrinate, meclozine, cyclizine, promethazine	MS	Medium–high	Very frequent: drowsiness, fatigue, impaired cognitive performance
		Cinnarizine	MS	High	Frequent: fatigue, stomach awareness, drowsiness
	Anticholinergics	Scopolamine	MS/VIMS	High	Frequent: fatigue, sedation
		Scopolamine plus stimulants	MS	Very high	Moderate: drug abuse
	Others	Betahistine, benzodiazepines, barbiturates, antipsychotics, neurokinin-1 antagonists, phenytoin, opioids, dopamine antagonists	MS	Low or variable	Frequent: drug abuse, drowsiness, fatigue, impaired cognitive performance
Behavioral	Active strategies	Training or habituation	MS/VIMS	Very high	No side effects. Time-consuming, unpleasant for patient, readaptation costs
		Active control of movement (e.g., steering the car)	MS/VIMS	Very high	No side effects. Limited applicability
		Minimizing head movements	MS/VIMS	Medium–high	No side effects. Limited applicability
		Antisickness music	VIMS	Low–medium	No side effects. Limited applicability
		Controlled breathing, cognitive-behavioral training	VIMS/MS	Medium—high	No side effects. Limited applicability
	Technical manipulations	Smaller field of view, minimizing the time lag, reduced use of HMDs, independent visual background	VIMS/MS	High	No side effects. Limited applicability
		View of the road ahead	MS	Medium	No side effects. Limited applicability

formation of the medulla oblongata. Finally, various areas of the CNS are also involved, resulting in different MS-typical symptoms. The neural transmission paths from the vestibular nuclei to the vomiting center are known to be histaminergic and cholinergic, which might explain the success of antihistamines and anticholinergic agents in the prevention of MS.

Antihistamines inhibit histaminergic H-receptors and include diphenhydramine, dimen-hydrinate, meclozine, cyclizine, and promethazine. Histamine is known to play a role in the vestibular system with a high density of H1 receptors in the vestibular nuclei (Matsuoka, Ito,

Takahashi, Sasa, & Takaori, 1984; Waele, Mühlethaler, & Vidal, 1995). Although anti-MS drugs based on histaminergic substances work, the exact mechanism is not yet understood. Antihistamines are thought to block H1 receptors in the vestibular nuclei and to minimize the firing rate of afferent nerves in the ampullae of the vestibular organs. Antihistamines that do not cross the blood–brain barrier are not effective in the prevention of MS. Unfortunately, antihistamines often produce strong side effects in the consumer. Drowsiness and fatigue are especially common side effects, another is impaired cognitive performance. For instance, a study by Vuurman, Rikken, Muntjewerff, Halleux, and Ramaekers (2004) showed that reaction times and psychomotor performance were impaired under diphenhydramine medication. Another antihistamine, cinnarizine, is widely accepted in Europe but not available in the United States (Hargreaves, 1980). The efficacy of cinnarizine without adverse side effects has been shown in several studies (Doweck, Gordon, Spitzer, Melamed, & Shupak, 1994; Gordon et al., 2001).

Anticholinergic drugs effective against MS block muscarinic receptors in the brain. Again, the exact procedure and the mechanism of action are not entirely known, but anticholinergics likely inhibit the input from the vestibular organs to the vestibular nuclei. A direct influence of anticholinergics on the vomiting center in the medulla oblongata is also conceivable (Matsuoka et al., 1984; Waele et al., 1995). Scopolamine is currently the most common anticholinergic medication used to prevent and reduce MS (Murray, 1997; Sherman, 2002) and is also effective in the prevention of MS in VEs (Regan & Ramsey, 1996). Wood and Graybiel (1968; also Wood, Graybiel, & Kennedy, 1966) compared 16 medications and found scopolamine to be the most effective single drug against MS. However, scopolamine is sedating when taken in high doses. Thus, anticholinergics have been frequently combined with stimulants, some of which (such as amphetamine) may have anti-MS actions in their own right. Unfortunately such stimulants (including amphetamine or ephedrine) bear the risk of abuse. Therefore, the intake of anticholinergics and stimulants should be done with caution. Alternatively, Hoyt et al. (2009) introduced modafinil—a medication to treat narcolepsy—as an alternate stimulant posing a lower risk of abuse (Lyons & French, 1991). Hoyt et al. tested the effectiveness of scopolamine in combination with modafinil and found significantly reduced MS compared to placebo or modafinil alone.

Depending on the dose of scopolamine and its route of intake, the antinausea effect might last up to several hours. Scopolamine can also be used as a nasal spray (Klöcker, Hanschke, Toussaint, & Verse, 2001) or in transdermal form (Gordon et al., 2001; Gordon, Binah, Attias, & Rolnick, 1986). Transdermal scopolamine intake can provide up to 72 h of MS prophylaxis, but its repeated usage might again lead to the familiar sedating side effects and blurred vision in some individuals. Several potential drawbacks to the use of transdermal scopolamine in the military setting are discussed by Lawson et al. (2009).

26.7.2 NONDRUG AGENTS

In addition to medications, a number of ingestible, inhalable, or alternative medicine approaches not classified as drugs have been tried against MS. Ginger is widely mentioned on the Internet as an anti-MS agent, but evidence for its effectiveness during controlled experimentation is limited and findings have been mixed (Estrada, LeDuc, Curry, Phelps, & Fuller, 2007; Lien et al., 2003). Inhalation of supplemental oxygen may be effective for reducing MS in patients (during ambulance transport), but not in individuals who are otherwise healthy. Limited inhalation of isopropyl alcohol vapors has been reported to help with postoperative nausea, but did not have a significant effect on MS (Lawson et al., 2007). Variants of acupuncture and acupressure have also been evaluated with mixed results (Chu, Li, Juan, & Chiou, 2012, p. 4). Chu et al. tried transcutaneous electrical nerve stimulation as an MS countermeasure, but the results were inconclusive. Judging from their results section, the treatment allowed subjects to tolerate nine additional seconds compared to the control group ($p = 0.059$), and 12 subjects were able to complete the 300 s rotation challenge versus 10 in the control condition.

26.7.3 Behavioral Countermeasures

Because of the undesirable side effects of medications, nonmedication countermeasures would be desirable. The most efficient method to prevent VIMS and real MS is adaptation or habituation. Repeated exposure to the same offending stimulus causes MS symptoms to subside (Guedry et al., 1998), especially when the stimulus starts out mildly and gradually increases in intensity over days or weeks of repeated exposure (Reason & Graybiel, 1969). When this positive effect is retained after a longer time interval, the organism has adapted. Several studies found that habituation reduces MS (e.g., Bagshaw & Stott, 1985; Cheung & Hofer, 2005; Cowings & Toscano, 2000; Golding & Stott, 1995; Hecht, Brown, & Young, 2002; Smither, Mouloua, & Kennedy, 2008; Sugita et al., 2007; Young, Sienko, Le Lyne, Hecht, & Natapoff, 2003). For instance, Hill and Howarth (2000) tested the severity of VIMS during a 20 min video game presented over the course of five consecutive days. Their participants adapted to the nauseating stimulus and reported significantly less VIMS after the fifth experimental day. The most widespread use of MS adaptation protocols has been in the military, where airsick aviator trainees may be referred for a course of ground-based motion adaptation before being returned to flight (Crampton, 1990). This approach avoids the side effects associated with medications and has been quite successful (although Smith & Qureshi, 2012, point out that definitions of success vary); however, it requires time and dedication. It is important that any MS adaptation program be designed and monitored by a vestibular expert because otherwise the possibility exists to do more harm than good. The third author has often had to caution young military aviators to restrain their natural tendency to push as hard as possible when adapting to a rotational stimulus and instead decrease the stimulus intensity whenever they detect the slightest stomach discomfort. The severity of MS may be less important to adaptation than the number of sessions experienced (Golding & Stott, 1995). In fact, it is likely that pushing oneself too hard could trigger aversive conditioning wherein the motion stimulus becomes inextricably linked to the nausea response.

A case study vividly illustrating the adaptation process in the civilian setting can be found in a study by Rine, Schubert, and Balkany (1999). The authors describe the malady of a woman who suffered from severe VIMS. Simple car maneuvers, flickering lights, standing in an elevator, or any form of visual motion evoked visually induced vertigo and VIMS for her. The patient even lost her capability to work, as standing on a floating dock or diving into the ocean (both of which were necessary during her work as marine biologist) had become almost impossible for her. Despite this extreme case of susceptibility to VIMS, a 10-week training program in a simulator and at home helped cure her of her symptoms. Simulators can also be used to prevent real MS, as a study by Stroud, Harm, and Klaus (2005) indicates. The authors report the benefit of space simulator training to prevent space sickness in aircraft members. However, there seem to be some people who are resistant to habituation and continue to suffer from MS even after numerous training sessions (Birren, 1949).

Another very popular recommendation to prevent MS during travels by car or coach is to sit in the first row of the vehicle and to look at the road ahead. This suggestion is backed by findings that a view of the road ahead limits the severity of MS (Rolnick & Lubow, 1991; Turner & Griffin, 1999a). Active participation is also known to reduce MS. The driver of the vehicle usually has milder symptoms than the passengers (Rolnick & Lubow, 1991). The same holds for motion in a VE, as participants who passively watch a movie of a virtual ride report significantly more VIMS than participants who actively steer the virtual vehicle (Stanney & Hash, 1998). As practical as these measures may be during car rides, they cannot be applied during travel by ship, train, or airplane. The front view of the road ahead is not available to all passengers on planes or ships (at least not below deck). Griffin and Newman (2004b) presented a real-time view of the road from the driver's seat to their participants during a car ride. The subjects sat in the back of the car and had no external view but only the real-time video. Surprisingly, the video did not reduce MS significantly in the passengers (also see Butler & Griffin, 2006). In contrast, Lin, Parker, Lahav, and Furness (2005) found that VIMS was reduced when the participants were able to foresee the upcoming motion in virtual reality displays (e.g., by adding a pathway track).

An alternate approach to reduce MS exploits the role of head movements in the genesis of MS (Golding, Markey, & Stott, 1995; Lackner & Graybiel, 1984). Golding et al. (2003) emphasized that reducing head movements during car transport decreased MS. While sitting upright and almost stable is not a guarantee for MS-free travel, it can reduce or delay MS symptoms. Similarly, Howarth and Finch (1999) reported that participants suffered greater sickness when they exhibited increased head movements during a virtual reality scenario while wearing an HMD. In another study, Vogel, Kohlhaas, and Baumgarten (1982) reported that lying supine also reduces the risk of suffering from MS. In their study, participants were transported in the back of an ambulance with no view of the outside world. They were seated or placed supine on a stretcher either facing or with their back to the direction of travel. Sickness was lowest when the participants were lying on the stretcher aligned longitudinally with the direction of travel.

Several countermeasures against VIMS exist. For instance, a smaller field of view (Bos et al., 2010; Keshavarz et al., 2011), minimizing the time lag in VEs (Akizuki et al., 2005; Draper et al., 2001), or reducing the use of HMDs (Moss & Muth, 2011; Patterson et al., 2006) all decrease VIMS. Duh et al. (2004) introduced an independent visual background representing "... a visual scene component that provides visual motion and orientation cues that match those from the vestibular receptors" (p. 578). The independent visual background led to significantly less MS. Note that the presentation of congruent background sound did not decrease the amount of MS in different studies (Dahlman, Sjörs, Ledin, & Falkmer, 2008; Keshavarz & Hecht, 2012a, 2012b). Other countermeasures against VIMS are being developed as well (Cevette et al., 2012).

Based on anecdotal suggestions, standing on deck and focusing the horizon during sea travel is thought to prevent or reduce MS. The effectiveness of a stable (real or artificial) horizon has been proven to alleviate MS (Bos, MacKinnon, & Patterson, 2005; Rolnick & Bles, 1989) and to limit the amount of postural sway (Mayo, Wade, & Stoffregen, 2011). Additionally, other techniques to reduce MS were introduced recently. Yen Pik Sang, Billar, Golding, and Gresty (2003) assessed special antisickness music and controlled breathing as countermeasures, showing that both techniques reduced MS to some extent. The positive effect of controlled breathing on MS was confirmed by several authors (Denise, Vouriot, Normand, Golding, & Gresty, 2009; Jokerst, Gatto, Fazio, Stern, & Koch, 1999; Yen Pik Sang, Billar, Gresty, & Golding, 2005; Yen Pik Sang, Golding, & Gresty, 2003). Additionally, cognitive-behavioral training was also found to reduce MS (Dobie & May, 1994). However, this method is quite time-consuming and elaborate. Finally, some sufferers, such as yachtsmen, believe in acupressure or electrical stimulation of a pressure point on the inner forearm about 2 in. (5.08 cm) above the wrist. However, a controlled study found no significant effects for this procedure (Miller & Muth, 2004) and findings are mixed in the literature.

Although there are many factors known to be effective in reducing MS and VIMS, many people suffer severely from MS, and the perfect way to prevent MS in simulators and VEs has not yet been found. Until then, habituation seems to be the best way to reduce MS in recurring scenarios.

26.8 CONCLUSION

The purpose of this article has been to offer an overview of MS theories, causes, characteristics, and countermeasures. Based on our review, we draw three conclusions regarding the state of the field and the knowledge gaps that should be addressed in future research.

26.8.1 THEORETICAL APPROACHES

We do not have a single, unified, universally accepted theory of MS. The debate of how and why MS is generated is far from being settled. We have discussed several modern theories that try to explain MS. The main theories are partially successful. For example, the sensory conflict approach offers a plausible explanation of how VIMS occurs within simulators or VEs, and the findings generally are in line with the theory. On the other hand, the postural instability account offers an explanation for how increased sway is often correlated with VIMS. The majority of research conducted on VIMS thus far was either couched within the language of sensory conflict or postural instability. A quick evaluation by the authors of this chapter reveals that these two theories are cited the most, followed by the subjective visual vertical offshoot of the sensory conflict theory (Bos et al., 2008). Of course, simple citation counts may not reflect the current or future importance of a theory and may include *self-citations* (citations made of one's own past work), *friendly citations* (citations made by one's coauthors, students, or close colleagues), and *negative citations* (citations made for the purpose of disagreement with a given theory).

There are only a handful of studies that discuss experimental results from multiple theoretical frameworks (Flanagan et al., 2004; Warwick-Evans et al., 1998). Until more such studies are available, it will be impossible to decide which theory is better able to explain the existing data on MS and VIMS. To meet the requirements of such complex studies, the refinement and standardization of different measurement protocols is needed, including agreement on the important confounds, comparable estimates of postural sway (e.g., head tracking, posturography), subjective symptoms (e.g., SSQ, FMS), cognitive performance, and physiological functioning. When different types of measures all provide a consistent answer concerning a construct, one can be more confident concerning construct validity. Similarly, when multiple laboratories use a given measure and get the same result, confidence is increased. We believe that such a comprehensive approach would ultimately foster the development of a universally acceptable theory. It is conceivable that such a superior theory would integrate aspects of the various competing theories. For instance, visual–vestibular conflict, postural sway, and eye movements may all contribute to VIMS once certain thresholds are passed or certain requirements met.

26.8.2 Countermeasures

Knowledge about the origin of VIMS would aid in finding ways to decrease or prevent VIMS. A future successful theory will not only have to address the role of visual–vestibular conflict and postural sway, it will also have to integrate our knowledge of possibly contributing factors such as vection.

Medical countermeasures are fairly effective in reducing MS, but they have negative side effects. We have discussed several behavioral countermeasures, but—with the exception of adaptation—they do not usually suffice to reduce nausea and sickness to acceptable levels. VIMS continues to be a major problem with the use and proliferation of driving simulators and VEs. One primary goal of future research should be the prevention of VIMS. We believe that once the origin of MS and VIMS is understood, more efficient countermeasures can be designed.

26.8.3 Measurement Techniques

We have reviewed and compared several measurement techniques that have been chosen to study VIMS. The differences among them are staggering, which makes it difficult to compare studies. The SSQ (Kennedy et al., 1993) has become the most popular questionnaire to measure VIMS. Many researchers make use of this questionnaire, which aids comparison across studies. The SSQ is very useful for obtaining individual symptom data known to be relevant to sickening situations with a visual component. It is also helpful for exploring psychometric questions such as the relative contribution of different symptom clusters to MS under different circumstances (Stanney et al., 1997). However, the SSQ is limited as a moment-by-moment assessment tool when time is very confined because it can take up to 1 min to fill out. Typically, the SSQ is administered once during screening for inclusion in the study and one or more times after exposure to an offending stimulus. There are alternatives to the SSQ that may prove useful in cases where a moment-by-moment measure is needed and individual symptom data do not need to be preserved (e.g., Bock & Oman, 1982). Recently, the FMS has been shown to produce results that closely match SSQ data while being faster to administer (Keshavarz & Hecht, 2011b). We believe that it is of foremost importance that researchers agree on how VIMS is best measured. Much is to be gained from converging on a smaller set of measures across MS studies. For example, it may be advantageous to use the SSQ before and after the stimulus while using the FMS repeatedly during the stimulus.

In summary, VIMS is a complex phenomenon that is still not completely understood. Although many studies have been conducted over the past decades, a solution that entirely prevents or eliminates this unpleasant side effect of simulators and VEs has not yet been found. Much remains to be done before an entirely convincing theory can emerge.

DISCLAIMER

The views expressed in this report are solely those of the authors; they do not represent the views of the Johannes Gutenberg-University, the US government, or any subordinate agency of the US government. Our mention of agencies or persons in this report does not imply their endorsement by us or their endorsement of our views. Our mention of products does not imply commercial endorsement of those products.

ACKNOWLEDGMENTS

We thank (in alphabetical order) Rene Bertin, Bernie Cohen, Paul DiZio, John Golding, Larry Hettinger, Bob Kennedy, Jim Lackner, Eric Muth, Chuck Oman, and Larry Young for insightful discussions over the years relating to some of the topics we have presented in this chapter. We also thank Eric Muth and John Golding for reading earlier sections of the manuscript and recommending improvements, and Shauna Legan and Heather McGee for assistance with the references.

REFERENCES

Akiduki, H., Nishiike, S., Watanabe, H., Matsuoka, K., Kubo, T., & Takeda, N. (2003). Visual–vestibular conflict induced by virtual reality in humans. *Neuroscience Letters, 340*(3), 197–200. doi:10.1016/S0304-3940(03)00098-3

Akizuki, H., Uno, A., Arai, K., Morioka, S., Ohyama, S., Nishiike, S., ... Takeda, N. (2005). Effects of immersion in virtual reality on postural control. *Neuroscience Letters, 379*(1), 23–26. doi:10.1016/j.neulet.2004.12.041

Allison, R. S., Howard, I. P., & Zacher, J. E. (1999). Effect of field size, head motion, and rotational velocity on roll vection and illusory self-tilt in a tumbling room. *Perception, 28*(3), 299–306. doi:10.1068/p2891

Ames, S. L., Wolffsohn, J. S., & McBrien, N. A. (2005). The development of a symptom questionnaire for assessing virtual reality viewing using a head-mounted display. *Optometry and Vision Science, 82*(3), 168–176.

Anderson, D. J., Reschke, M., Homick, J., & Werness, S. A. S. (1986). Dynamic posture analysis of Spacelab-1 crew members. *Experimental Brain Research, 64*(2), 380–391. doi:10.1007/BF00237754

Bagshaw, M., & Stott, J. R. (1985). The desensitisation of chronically motion sick aircrew in the Royal Air Force. *Aviation, Space, and Environmental Medicine, 56*(12), 1144–1151.

Balaban, C. D. (1999). Vestibular autonomic regulation (including motion sickness and the mechanism of vomiting). *Current Opinion in Neurology, 12*(1), 29–33. doi:10.1097/00019052-199902000-00005

Baltzley, D. R., Kennedy, R. S., Berbaum, K. S., Lilienthal, M. G., & Gower, D. W. (1989). The time course of postflight simulator sickness symptoms. *Aviation, Space, and Environmental Medicine, 60*(11), 1043–1048.

Barfield, W., & Hendrix, C. (1995). The effect of update rate on the sense of presence within virtual environments. *Virtual Reality, 1*(1), 3–15. doi:10.1007/BF02009709

Benson, A. J. (1988). Motion sickness and spatial disorientation. In J. Ernsting & P. King (Eds.), *Aviation medicine* (pp. 318–493). London, U.K.: Buttersworth.

Benson, A. J. (2002). Motion sickness. In K. B. Pandolf & R. E. Burr (Eds.), *Medical aspects of harsh environments* (Vol. 2, pp. 1048–1083). Washington, DC: United States Army Medical Department Center and School.

Bertin, R. J. V., Collet, C., Espié, S., & Graf, W. (2005). Objective measurement of simulator sickness and the role of visual–vestibular conflict situations. In D. L. Fischer, M. Rizzo, J. Caird, & J. D. Lee (Eds.), *Proceedings of the Driving Simulation Conference North America 2005* (pp. 280–293).

Bick, P. A. (1983). Physiological and psychological correlates of motion sickness. *The British Journal of Medical Psychology, 56*(Pt 2), 189–196.

Bijveld, M. M. C., Bronstein, A. M., Golding, J. F., & Gresty, M. A. (2008). Nauseogenicity of off-vertical axis rotation vs. equivalent visual motion. *Aviation, Space, and Environmental Medicine, 79*(7), 661–665. doi:10.3357/ASEM.2241.2008.

Biocca, F. A. (1992). Will simulation sickness slow down the diffusion of virtual environment technology? *Presence: Teleoperators and Virtual Environments, 1*(3), 334–343.

Birren, J. E. (1949). Motion sickness: Its psychophysiological aspects. In National Research Council (U.S.). Committee on Undersea Warfare: United States Office of Naval Research (Ed.), *A survey report on human factors in undersea warfare* (pp. 375–398). Retrieved from http://archive.org/stream/surveyreportonhu00nati#page/n4/mode/1up

Blanford, C. L. (1990). *Frequency analysis and diagnostic classification of changes in the human electrogastrogram during motion sickness* (Master's thesis). Massachusetts Institute of Technology, Boston, MA.

Blanford, C. L., & Oman, C. M. (1990, May). *Diagnostic classification of changes in the human electrogastrogram during motion sickness*. Paper presented at the Annual Scientific Meeting of the Aerospace Medical Association, New Orleans, LA.

Bles, W. (1998). Coriolis effects and motion sickness modelling. *Brain Research Bulletin, 47*(5), 543–549. doi:10.1016/S0361-9230(98)00089-6

Bles, W., Bos, J. E., de Graaf, B., Groen, E. L., & Wertheim, A. H. (1998). Motion sickness: Only one provocative conflict? *Brain Research Bulletin, 47*(5), 481–487. doi:10.1016/S0361-9230(98)00115-4

Bles, W., Bos, J. E., & Kruit, H. (2000). Motion sickness. *Current Opinion in Neurology, 13*(1), 19–25. doi:10.1097/00019052-200002000-00005

Bock, O. L., & Oman, C. M. (1982). Dynamics of subjective discomfort in motion sickness as measured with a magnitude estimation method. *Aviation, Space, and Environmental Medicine, 53*(8), 773–777.

Bonato, F., Bubka, A., & Palmisano, S. (2008). Multiple axis rotation and cybersickness in a virtual environment. *Aviation, Space, and Environmental Medicine, 79*(3), 309–309.

Bonato, F., Bubka, A., & Palmisano, S. (2009). Combined pitch and roll and cybersickness in a virtual environment. *Aviation, Space, and Environmental Medicine, 80*(11), 941–945. doi:10.3357/ASEM.2394.2009

Bonato, F., Bubka, A., & Story, M. (2005). Rotation direction change hastens motion sickness onset in an optokinetic drum. *Aviation, Space, and Environmental Medicine, 76*(9), 823–827.

Bonnet, C. T., Faugloire, E., Riley, M. A., Bardy, B. G., & Stoffregen, T. A. (2006). Motion sickness preceded by unstable displacements of the center of pressure. *Human Movement Science, 25*(6), 800–820. doi:10.1016/j.humov.2006.03.001

Bonnet, C. T., Faugloire, E., Riley, M. A., Bardy, B. G., & Stoffregen, T. A. (2008). Self-induced motion sickness and body movement during passive restraint. *Ecological Psychology, 20*(2), 121–145. doi:10.1080/10407410801949289

Bos, J. E. (2011). Nuancing the relationship between motion sickness and postural stability. *Displays, 32*(4), 189–193. doi:10.1016/j.displa.2010.09.005

Bos, J. E., & Bles, W. (1998). Modelling motion sickness and subjective vertical mismatch detailed for vertical motions. *Brain Research Bulletin, 47*(5), 537–542. doi:10.1016/S0361-9230(98)00088-4

Bos, J. E., & Bles, W. (2002). Theoretical considerations on canal–otolith interaction and an observer model. *Biological Cybernetics, 86*(3), 191–207. doi:10.1007/s00422-001-0289-7

Bos, J. E., & Bles, W. (2004). Motion sickness induced by optokinetic drums. *Aviation, Space, and Environmental Medicine, 75*(2), 172–174.

Bos, J. E., Bles, W., & Groen, E. L. (2008). A theory on visually induced motion sickness. *Displays, 29*(2), 47–57. doi:10.1016/j.displa.2007.09.002

Bos, J. E., MacKinnon, S. N., & Patterson, A. (2005). Motion sickness symptoms in a ship motion simulator: Effects of inside, outside, and no view. *Aviation, Space, and Environmental Medicine, 76*(12), 1111–1118.

Bos, J. E., de Vries, S. C., van Emmerik, M. L., & Groen, E. L. (2010). The effect of internal and external fields of view on visually induced motion sickness. *Applied Ergonomics, 41*(4), 516–521. doi:10.1016/j.apergo.2009.11.007

Bowins, B. (2010). Motion sickness: A negative reinforcement model. *Brain Research Bulletin, 81*(1), 7–11. doi:10.1016/j.brainresbull.2009.09.017

Brandt, T. (1991). *Vertigo: Its multisensory syndromes*. London, U.K.: Springer Verlag.

Brandt, T., Dichgans, J., & Koenig, E. (1973). Differential effects of central verses peripheral vision on egocentric and exocentric motion perception. *Experimental Brain Research, 16*(5), 476–491.

Brandt, T., Wenzel, D., & Dichgans, J. (1976). Die Entwicklung der visuellen Stabilisation des aufrechten Standes beim Kind: Ein Reifezeichen in der Kinderneurologie [Visual stabilization of free stance in infants: A sign of maturity]. *Archiv für Psychiatrie und Nervenkrankheiten, 223*(1), 1–13.

Brill, J. C., Kass, S. J., & Lawson, B. D. (2004). Mild Motion Questionnaire (MMQ): Further evidence of construct validity. *Proceedings of the Human Factors and Ergonomics Society Annual Meeting, 48*(21), 2503–2507. doi:10.1177/154193120404802113

Brooks, J. O., Goodenough, R. R., Crisler, M. C., Klein, N. D., Alley, R. L., Koon, B. L., … Wills, R. F. (2010). Simulator sickness during driving simulation studies. *Accident Analysis & Prevention, 42*(3), 788–796. doi:10.1016/j.aap.2009.04.013

Bubka, A., & Bonato, F. (2003). Optokinetic drum tilt hastens the onset of vection-induced motion sickness. *Aviation, Space, and Environmental Medicine, 74*(4), 315–319.

Bubka, A., Bonato, F., & Palmisano, S. (2007). Expanding and contracting optical flow patterns and simulator sickness. *Aviation, Space, and Environmental Medicine, 78*(4), 383–386.

Bubka, A., Bonato, F., Urmey, S., & Mycewicz, D. (2006). Rotation velocity change and motion sickness in an optokinetic drum. *Aviation, Space, and Environmental Medicine, 77*(8), 811–815.

Buchanan, J. B., & Horak, F. (2001). Transitions in a postural task: Do the recruitment and suppression of degrees of freedom stabilize posture? *Experimental Brain Research, 139*(4), 482–494. doi:10.1007/s002210100798

Buker, T. J., Vincenzi, D. A., & Deaton, J. E. (2012). The effect of apparent latency on simulator sickness while using a see-through helmet-mounted display: Reducing apparent latency with predictive compensation. *Human Factors: The Journal of the Human Factors and Ergonomics Society, 54*(2), 235–249. doi:10.1177/0018720811428734

Butler, C. A., & Griffin, M. J. (2006). Motion sickness during fore-and-aft oscillation: Effect of the visual scene. *Aviation, Space, and Environmental Medicine, 77*(12), 1236–1243.

Caillet, G., Bosser, G., Gauchard, G. C., Chau, N., Benamghar, L., & Perrin, P. P. (2006). Effect of sporting activity practice on susceptibility to motion sickness. *Brain Research Bulletin, 69*(3), 288–293. doi:10.1016/j.brainresbull.2006.01.001

Cevette, M. J., Stepanek, J., Cocco, D., Galea, A. M., Pradhan, G. N., Wagner, L. S., … Brookler, K. H. (2012). Oculo-vestibular recoupling using galvanic vestibular stimulation to mitigate simulator sickness. *Aviation, Space, and Environmental Medicine, 83*(6), 549–555. doi: 10.3357/ASEM.3239.2012

Champney, R. K., Stanney, K. M., Hash, P. A. K., Malone, L. C., Kennedy, R. S., & Compton, D. E. (2007). Recovery from virtual environment exposure: Expected time course of symptoms and potential readaptation strategies. *Human Factors: The Journal of the Human Factors and Ergonomics Society, 49*(3), 491–506. doi:10.1518/001872007X200120

Chatoor, I. (2009). Sensory food aversions in infants and toddlers. *Zero to Three, 29*(3), 44–49.

Chatoor, I., & Ganiban, J. (2003). Food refusal by infants and young children: Diagnosis and treatment. *Cognitive and Behavioral Practice, 10*(2), 138–146. doi:10.1016/S1077-7229(03)80022-6

Chertoff, D., & Schatz, S. (2014). Presence in virtual environments: Augmented cognition for virtual environment evaluation. In K. S. Hale & K. M. Stanney (Eds.), *Handbook of virtual environments: Design, implementation, and applications* (2nd ed., pp. 855–870). New York, NY: CRC Press.

Cheung, B. S., & Hofer, K. (2005). Desensitization to strong vestibular stimuli improves tolerance to simulated aircraft motion. *Aviation, Space, and Environmental Medicine, 76*(12), 1099–1104.

Cheung, B. S., Howard, I. P., & Money, K. E. (1991). Visually-induced sickness in normal and bilaterally labyrinthine-defective subjects. *Aviation, Space, and Environmental Medicine, 62*(6), 527–531.

Cheung, B. S., Howard, I. P., Nedzelski, J. M., & Landolt, J. P. (1989). Circularvection about earth-horizontal axes in bilateral labyrinthine-defective subjects. *Acta Oto-laryngologica, 108*(5–6), 336–344.

Cheung, B. S., Money, K. E., & Jacobs, I. (1990). Motion sickness susceptibility and aerobic fitness: A longitudinal study. *Aviation, Space, and Environmental Medicine, 61*(3), 201–204.

Cheung, B., & Vaitkus, P. (1998). Perspectives of electrogastrography and motion sickness. *Brain Research Bulletin, 47*(5), 421–431.

Chu, H., Li, M.-H., Juan, S.-H., & Chiou, W.-Y. (2012). Effects of transcutaneous electrical nerve stimulation on motion sickness induced by rotary chair: A crossover study. *The Journal of Alternative and Complementary Medicine, 18*(5), 494–500. doi:10.1089/acm.2011.0366

Claremont, C. A. (1931). The psychology of seasickness. *Psyche, 11*, 86–90.

Classen, S., Bewernitz, M., & Shechtman, O. (2011). Driving simulator sickness: An evidence-based review of the literature. *American Journal of Occupational Therapy, 65*(2), 179–188. doi:10.5014/ajot.2011.000802

Clément, G. (2007). *Fundamentals of space medicine*. Dordrecht, the Netherlands: Springer Verlag.

Clemes, S. A., & Howarth, P. A. (2005). The menstrual cycle and susceptibility to virtual simulation sickness. *Journal of Biological Rhythms, 20*(1), 71–82. doi:10.1177/0748730404272567

Cobb, S. V. G. (1999). Measurement of postural stability before and after immersion in a virtual environment. *Applied Ergonomics, 30*(1), 47–57. doi:10.1016/S0003-6870(98)00038-6

Cobb, S. V. G., & Nichols, S. C. (1998). Static posture tests for the assessment of postural instability after virtual environment use. *Brain Research Bulletin, 47*(5), 459–464. doi:10.1016/S0361-9230(98)00104-X

Cobb, S. V. G., Nichols, S. C., Ramsey, A. D., & Wilson, J. R. (1999). Virtual reality-induced symptoms and effects (VRISE). *Presence: Teleoperators and Virtual Environments, 8*(2), 169–186. doi:10.1162/105474699566152

Cowings, P. S., Suter, S., Toscano, W. B., Kamiya, J., & Naifeh, K. (1986). General autonomic components of motion sickness. *Psychophysiology, 23*(5), 542–551. doi:10.1111/j.1469-8986.1986.tb00671.x

Cowings, P. S., & Toscano, W. B. (2000). Autogenic-feedback training exercise is superior to promethazine for control of motion sickness symptoms. *Journal of Clinical Pharmacology, 40*(10), 1154–1165.

Crampton, G. H. (1955). Studies of motion sickness: XVII. Physiological changes accompanying sickness in man. *Journal of Applied Physiology, 7*(5), 501–507.

Crampton, G. H. (Ed.). (1990). *Motion and space sickness*. Boca Raton, FL: CRC Press.

Crampton, G. H., & Young, F. A. (1953). The differential effect of a rotary visual field on susceptibles and non-susceptibles to motion sickness. *Journal of Comparative and Physiological Psychology, 46*(6), 451–453. doi:10.1037/h0058423

Dahlman, J., Sjörs, A., Ledin, T., & Falkmer, T. (2008). Could sound be used as a strategy for reducing symptoms of perceived motion sickness? *Journal of NeuroEngineering and Rehabilitation, 5*(1), 35. doi:10.1186/1743-0003-5-35

Dahlman, J., Sjors, A., Lindstrom, J., Ledin, T., & Falkmer, T. (2009). Performance and autonomic responses during motion sickness. *Human Factors: The Journal of the Human Factors and Ergonomics Society, 51*(1), 56–66. doi:10.1177/0018720809332848

Dai, M., Kunin, M., Raphan, T., & Cohen, B. (2003). The relation of motion sickness to the spatial–temporal properties of velocity storage. *Experimental Brain Research, 151*(2), 173–189. doi:10.1007/s00221-003-1479-4

Dai, M., Raphan, T., & Cohen, B. (2007). Labyrinthine lesions and motion sickness susceptibility. *Experimental Brain Research, 178*(4), 477–487. doi:10.1007/s00221-006-0759-1

Daunton, N. G. (1985). Motion sickness elicited by passive rotation in squirrel monkeys. In M. Igarashi & F. O. Black (Eds.), *Vestibular and visual control on posture and locomotor equilibrium: 7th International symposium* (pp. 164–169). Basel, Switzerland: Karger.

Daunton, N. G. (1990). Animal models in motion sickness research. In G. H. Crampton (Ed.), *Motion and space sickness* (pp. 87–104). Boca Raton, FL: CRC Press.

Dawson, M. E., Schell, A. M., & Filion, D. L. (1990). The electrodermal system. In J. T. Cacioppo & L. G. Tassinary (Eds.), *Principles of psychophysiology: Physical, social, and inferential elements* (pp. 195–324). Cambridge, NY: Cambridge University Press.

Denise, P., Vouriot, A., Normand, H., Golding, J. F., & Gresty, M. A. (2009). Effect of temporal relationship between respiration and body motion on motion sickness. *Autonomic Neuroscience, 151*(2), 142–146. doi:10.1016/j.autneu.2009.06.007

de Waele, C., Mühlethaler, M., & Vidal, P. P. (1995). Neurochemistry of the central vestibular pathways. Brain research. *Brain Research Reviews, 20*(1), 24–46.

Dichgans, J., & Brandt, T. (1973). Optokinetic motion sickness and pseudo-Coriolis effects induced by moving visual stimuli. *Acta Oto-laryngologica, 76*(1–6), 339–348. doi:10.3109/00016487309121519

Dichgans, J., & Brandt, T. (1978). Visual-vestibular interaction: Effects on self-motion perception and postural control. In S. M. Anstis, R. Held, H. W. Leibowitz, & H. L. Teuber (Eds.), *Handbook of sensory physiology: Perception* (Vol. VIII, pp. 756–795). Berlin, Germany: Springer Verlag.

Diels, C., & Howarth, P. A. (2006). Frequency dependence of visually-induced motion sickness in the fore-and-aft direction. *Aviation, Space, and Environmental Medicine, 77*(3), 346–346.

Diels, C., Ukai, K., & Howarth, P. A. (2007). Visually induced motion sickness with radial displays: Effects of gaze angle and fixation. *Aviation, Space, and Environmental Medicine, 78*(7), 659–665.

DiZio, P., & Lackner, J. R. (1986). Perceived orientation, motion, and configuration of the body during viewing of an off-vertical, rotating surface. *Perception & Psychophysics, 39*(1), 39–46. doi:10.3758/BF03207582

DiZio, P., & Lackner, J. R. (1991). Motion sickness susceptibility in parabolic flight and velocity storage activity. *Aviation, Space, and Environmental Medicine, 62*(4), 300–307.

Dobie, T. G., & May, J. G. (1994). Cognitive-behavioral management of motion sickness. *Aviation, Space, and Environmental Medicine, 65*(10 Pt 2), C1–C2.

Doweck, I., Gordon, C. R., Spitzer, O., Melamed, Y., & Shupak, A. (1994). Effect of cinnarizine in the prevention of seasickness. *Aviation, Space, and Environmental Medicine, 65*(7), 606–609.

Draper, M. H., Viirre, E. S., Furness, T. A., & Gawron, V. J. (2001). Effects of image scale and system time delay on simulator sickness within head-coupled virtual environments. *Human Factors: The Journal of the Human Factors and Ergonomics Society, 43*(1), 129–146. doi:10.1518/001872001775992552

Duh, H.-L., Lin, J.-W., Kenyon, R., Parker, D. E., & Furness, T. A. (2001). Effects of field of view on balance in an immersive environment. In H. Takemura & K. Kiyokawa (Eds.), *Proceedings of the IEEE Virtual Reality Conference, Yokohama, Japan*, (pp. 235–240). doi:10.1109/VR.2001.913791

Duh, H.-L., Parker, D. E., & Furness, T. A. (2004). An independent visual background reduced simulator sickness in a driving simulator. *Presence: Teleoperators and Virtual Environments, 13*(5), 578–588. doi:10.1162/1054746042545283

Eagon, J. C. (1988). *Quantitative frequency analysis of the electrogastrogram during prolonged motion sickness* (MD thesis). Massachusetts Institute of Technology, Cambridge, MA.

Ebenholtz, S. M. (1992). Motion sickness and oculomotor systems in virtual environments. *Presence: Teleoperators and Virtual Environments, 1*(3), 302–305.

Ebenholtz, S. M., Cohen, M. M., & Linder, B. J. (1994). The possible role of nystagmus in motion sickness: A hypothesis. *Aviation, Space, and Environmental Medicine, 65*(11), 1032–1035.

Emoto, M., Sugawara, M., & Nojiri, Y. (2008). Viewing angle dependency of visually-induced motion sickness in viewing wide-field images by subjective and autonomic nervous indices. *Displays, 29*(2), 90–99.

Estrada, A., LeDuc, P. A., Curry, I. P., Phelps, S. E., & Fuller, D. R. (2007). Airsickness prevention in helicopter passengers. *Aviation, Space, and Environmental Medicine, 78*(4), 408–413.

Faugloire, E., Bonnet, C. T., Riley, M. A., Bardy, B. G., & Stoffregen, T. A. (2007). Motion sickness, body movement, and claustrophobia during passive restraint. *Experimental Brain Research, 177*(4), 520–532. doi:10.1007/s00221-006-0700-7

Feenstra, P. J., Bos, J. E., & van Gent, R. N. H. W. (2009). *Visual motion counteracting sickness—A simulator study on comfort in air transport.* Paper presented at the 2nd International Symposium on Visual IMage Safety (VIMS), Utrecht, the Netherlands.

Fillingim, R. B. (2005). Individual differences in pain responses. *Current Rheumatology Reports, 7*(5), 342–347. doi:10.1007/s11926-005-0018-7

Finley, J. C., O'Leary, M., Wester, D., MacKenzie, S., Shepard, N., Farrow, S., & Lockette, W. (2004). A genetic polymorphism of the α_2-adrenergic receptor increases autonomic responses to stress. *Journal of Applied Psychology, 96*(6), 2231–2239.

Fischer, M., & Kornmüller, A. (1930). Optokinetisch ausgelöste Bewegungswahrnehmungen und optokinetischer Nystagmus. *Journal für Psychologie und Neurolgie, 41*, 273–308.

Flanagan, M. B., May, J. G., & Dobie, T. G. (2002). Optokinetic nystagmus, vection, and motion sickness. *Aviation, Space, and Environmental Medicine, 73*(11), 1067–1073.

Flanagan, M. B., May, J. G., & Dobie, T. G. (2004). The role of vection, eye movements and postural instability in the etiology of motion sickness. *Journal of Vestibular Research: Equilibrium and Orientation, 14*(4), 335–346.

Flanagan, M. B., May, J. G., & Dobie, T. G. (2005). Sex differences in tolerance to visually-induced motion sickness. *Aviation, Space, and Environmental Medicine, 76*(7), 642–646.

Frank, L. H., Casali, J. G., & Wierwille, W. W. (1987). Effects of visual display and motion system delays on operator performance and uneasiness in a driving simulator. *Proceedings of the Human Factors and Ergonomics Society Annual Meeting, New York, NY, 31*(5), 492–496. doi:10.1177/154193128703100502

Frey, A., Hartig, J., Ketzel, A., Zinkernagel, A., & Moosbrugger, H. (2007). The use of virtual environments based on a modification of the computergame Quake III Arena (R) in psychological experimenting. *Computers in Human Behavior, 23*(4), 2026–2039.

Gahlinger, P. M. (2000). Cabin location and the likelihood of motion sickness in cruise ship passengers. *Journal of Travel Medicine, 7*(3), 120–124. doi:10.2310/7060.2000.00042

Ghulyan, V., & Paolino, M. (2005). Comparative study of dynamic balance in fallers and non-fallers. *Fr ORL, 88*, 89–96.

Gianaros, P. J., Muth, E. R., Mordkoff, J. T., Levine, M. E., & Stern, R. M. (2001). A questionnaire for the assessment of the multiple dimensions of motion sickness. *Aviation, Space, and Environmental Medicine, 72*(2), 115–119.

Gibson, J. J. (1958). Visually controlled locomotion and visual orientation in animals. *British Journal of Psychology, 49*(3), 182–194. doi:10.1111/j.2044-8295.1958.tb00656.x

Gibson, J. J. (1961). Ecological optics. *Vision Research, 1*(3–4), 253–262. doi:10.1016/0042-6989(61)90005-0.

Goldberg, J. M., Wilson, V. R., Cullen, K. E., Angelaki, D. E., Broussard, D. M., Buttner-Ennever, J., … Minor, L. B. (Eds.). (2012). *The vestibular system: A sixth sense.* Oxford, NY: Oxford University Press.

Golding, J. F. (1992). Phasic skin conductance activity and motion sickness. *Aviation, Space, and Environmental Medicine, 63*(3), 165–171.

Golding, J. F. (1998). Motion sickness susceptibility questionnaire revised and its relationship to other forms of sickness. *Brain Research Bulletin, 47*(5), 507–516. doi:10.1016/S0361-9230(98)00091-4

Golding, J. F. (2006a). Motion sickness susceptibility. *Autonomic Neuroscience, 129*(1–2), 67–76. doi:10.1016/j.autneu.2006.07.019

Golding, J. F. (2006b). Predicting individual differences in motion sickness susceptibility by questionnaire. *Personality and Individual Differences, 41*(2), 237–248. doi:10.1016/j.paid.2006.01.012

Golding, J. F., Arun, S., Wortley, E., Wotton-Hamrioui, K., & Gresty, M. A. (2007, December). *The effects of frequency and tilt on motion sickness induced by optokinetic stimuli.* Paper presented at the 1st International Symposium on Visually Induced Motion Sickness, Fatigue, and Photosensitive Epileptic Seizures (VIMS 2007), Hong Kong, People's Republic of China.

Golding, J. F., Bles, W., Bos, J. E., Haynes, T., & Gresty, M. A. (2003). Motion sickness and tilts of the inertial force environment: Active suspension systems vs. active passengers. *Aviation, Space, and Environmental Medicine, 74*(3), 220–227.

Golding, J. F., & Gresty, M. A. (2005). Motion sickness. *Current Opinion in Neurology, 18*(1), 29–34. doi:10.1097/00019052-200502000-00007

Golding, J. F., Kadzere, P., & Gresty, M. A. (2005). Motion sickness susceptibility fluctuates through the menstrual cycle. *Aviation, Space, and Environmental Medicine, 76*(10), 970–973.

Golding, J. F., Markey, H. M., & Stott, J. R. (1995). The effects of motion direction, body axis, and posture on motion sickness induced by low frequency linear oscillation. *Aviation, Space, and Environmental Medicine, 66*(11), 1046–1051.

Golding, J. F., Mueller, A. G., & Gresty, M. A. (2001). A motion sickness maximum around the 0.2 Hz frequency range of horizontal translational oscillation. *Aviation, Space, and Environmental Medicine, 72*(3), 188–192.

Golding, J. F., & Stott, J. R. (1995). Effect of sickness severity on habituation to repeated motion challenges in aircrew referred for airsickness treatment. *Aviation, Space, and Environmental Medicine, 66*(7), 625–630.

Gordon, C. R., Binah, O., Attias, J., & Rolnick, A. (1986). Transdermal scopolamine: Human performance and side effects. *Aviation, Space, and Environmental Medicine, 57*(3), 236–240.

Gordon, C. R., Gonen, A., Nachum, Z., Doweck, I., Spitzer, O., & Shupak, A. (2001). The effects of dimenhydrinate, cinnarizine, and transdermal scopolamine on performance. *Journal of Psychopharmacology, 15*(3), 167–172. doi:10.1177/026988110101500311

Gould, S. J. (1997). Evolution: The pleasures of pluralism. *New York Review of Books, 44*, 47–52.

de Graaf, B., Bles, W., & Bos, J. E. (1998). Roll motion stimuli: Sensory conflict, perceptual weighting and motion sickness. *Brain Research Bulletin, 47*(5), 489–495. doi:10.1016/S0361-9230(98)00116-6

Graybiel, A. (1970). Susceptibility to acute motion sickness in blind persons. *Aerospace Medicine, 41*(6), 650–653.

Graybiel, A., & Johnson, W. H. (1963). A comparison of the symptomatology experienced by healthy persons and subjects with loss of labyrinthine function when exposed to unusual patterns of centripetal force in a counter-rotating room. *The Annals of Otology, Rhinology, and Laryngology, 72*, 357–373.

Graybiel, A., & Knepton, J. (1976). Sopite syndrome: A sometimes sole manifestation of motion sickness. *Aviation, Space, and Environmental Medicine, 47*(8), 873–882.

Graybiel, A., & Lackner, J. R. (1980). A sudden-stop vestibulovisual test for rapid assessment of motion sickness manifestations. *Aviation, Space, and Environmental Medicine, 51*(1), 21–23.

Graybiel, A., & Miller, E. F. (1970). Off-vertical rotation: A convenient precise means of exposing the passive human subject to a rotating linear acceleration vector. *Aerospace Medicine, 41*(4), 407–410.

Griffin, M. J., & Howarth, H. V. C. (2000). *Motion sickness history questionnaire* (ISVR Technical Report No. 283). Retrieved from the University of Southampton, Institute of Sound and Vibration Research website: http://resource.isvr.soton.ac.uk/staff/pubs/PubPDFs/Pub1228.pdf

Griffin, M. J., & Newman, M. M. (2004a). An experimental study of low-frequency motion in cars. *Proceedings of the Institution of Mechanical Engineers, Part D: Journal of Automobile Engineering, 218*(11), 1231–1238. doi:10.1243/0954407042580093

Griffin, M. J., & Newman, M. M. (2004b). Visual field effects on motion sickness in cars. *Aviation, Space, and Environmental Medicine, 75*(9), 739–748.

Groen, E. L., & Bos, J. E. (2008). Simulator sickness depends on frequency of the simulator motion mismatch: An observation. *Presence: Teleoperators and Virtual Environments, 17*(6), 584–593. doi:10.1162/pres.17.6.584

Groen, J. J. (1961). Problems of the semicircular canal from a mechanico-physiological point of view. *Acta Oto-laryngologica. Supplementum, 163*, 59–67.

Grunfeld, E., & Gresty, M. A. (1998). Relationship between motion sickness, migraine and menstruation in crew members of a "round the world" yacht race. *Brain Research Bulletin, 47*(5), 433–436. doi:10.1016/S0361-9230(98)00099-9

Guedry, F. E. (1964). Visual control of habituation to complex vestibular stimulation in man. *Acta Oto-laryngologica, 58*(1–6), 377–389. doi:10.3109/00016486409121398

Guedry, F. E. (1974). Psychophysics of vestibular sensation. In H. H. Kornhuber (Ed.), *Handbook of sensory physiology: Vol. VI. Vestibular system, Part 2* (pp. 1–54). Berlin, Germany: Springer-Verlag.

Guedry, F. E. (1978). Visual counteraction on nauseogenic and disorienting effects of some whole-body motions: A proposed mechanism. *Aviation, Space, and Environmental Medicine, 49*(1 Pt 1), 36–41.

Guedry, F. E. (1991). Motion sickness and its relation to some forms of spatial orientation: Mechanisms and theory. In Various (eds), *Motion sickness: Significance in aerospace operations and prophylaxis* (pp. 1–30). Neuilly sur Siene, France: North Atlantic Treaty Organization, Advisory Group for Aerospace Research & Development.

Guedry, F. E., & Benson, A. J. (1978). Coriolis cross-coupling effects: Disorienting and nauseogenic or not? *Aviation, Space, and Environmental Medicine, 49*(1 Pt 1), 29–35.

Guedry, F. E., Lentz, J. M., & Jell, R. M. (1979). Visual–vestibular interactions: I. Influence of peripheral vision on suppression of the vestibulo–ocular reflex and visual acuity. *Aviation, Space, and Environmental Medicine, 50*(3), 205–211.

Guedry, F. E., Lentz, J. M., Jell, R. M., & Norman, J. W. (1981). Visual–vestibular interactions: The directional component of visual background movement. *Aviation, Space, and Environmental Medicine, 52*(5), 304–309.

Guedry, F. E., & Oman, C. M. (1990). *Vestibular stimulation during a simple centrifuge run* (Report No. NAMRL-1353). Pensacola, FL: Naval Aerospace Medical Research Laboratory. Retrieved from the Defense Technical Information Center website: http://www.dtic.mil/dtic/tr/fulltext/u2/a227285.pdf

Guedry, F. E., Rupert, A., & Reschke, M. (1998). Motion sickness and development of synergy within the spatial orientation system. A hypothetical unifying concept. *Brain Research Bulletin, 47*(5), 475–480. doi:10.1016/S0361-9230(98)00087-2

Guignard, J. C., & McCauley, M. E. (1982). Motion sickness incidence induced by complex periodic waveforms. *Aviation, Space, and Environmental Medicine, 53*(6), 554–563.

Gupta, V. K. (2005). Motion sickness is linked to nystagmus-related trigeminal brain stem input: A new hypothesis. *Medical Hypotheses, 64*(6), 1177–1181. doi:10.1016/j.mehy.2004.11.031

Häkkinen, J., Pölönen, M., Takatalo, J., & Nyman, G. (2006). Simulator sickness in virtual display gaming. In M. Nieminen & M. Röykkee (Eds.), *Proceedings of the 8th Conference on Human–Computer Interaction with Mobile Devices and Services* (p. 227). New York, NY: ACM Press.

Hamilton, K. M., Kantor, L., & Magee, L. E. (1989). Limitations of postural equilibrium tests for examining simulator sickness. *Aviation, Space, and Environmental Medicine, 60*(3), 246–251.

Hargreaves, J. (1980). A double-blind placebo controlled study of cinnarizine in the prophylaxis of seasickness. *The Practitioner, 224*(1343), 547–550.

Harm, D. L. (1990). Physiology of motion sickness symptoms. In G. H. Crampton (Ed.), *Motion and space sickness* (pp. 153–177). Boca Raton, FL: CRC Press.

Harm, D. L. (2002). Motion sickness neurophysiology, physiological correlates, and treatment. In K. M. Stanney (Ed.), *Handbook of virtual environments: Design, implementation, and applications* (pp. 637–661). Mahwah, NJ: Lawrence Erlbaum Associates.

Harm, D. L., & Schlegel, T. T. (2002). Predicting motion sickness during parabolic flight. *Autonomic Neuroscience, 97*(2), 116–121. doi:10.1016/S1566-0702(02)00043-7

Harris, L. R., Jenkin, M., & Zikovitz, D. C. (2000). Visual and non-visual cues in the perception of linear self motion. *Experimental Brain Research, 135*(1), 12–21. doi:10.1007/s002210000504

Harvey, C., & Howarth, P. A. (2007, December). *The effect of display size on visually induced motion sickness (VIMS) and skin temperature.* Paper presented at the 1st International Symposium on Visually Induced Motion Sickness, Fatigue, and Photosensitive Epileptic Seizures (VIMS 2007), Hong Kong, People's Republic of China.

Hecht, H., & Brendel, E. (2008). Künstliche Schwerkraft—Die schwellennahe Adaptation des vestibulären Systems [Artificial gravity—Near-threshold adaptation of the vestibular system]. *Flugmedizin, Tropenmedizin, Reisemedizin, 15*(2), 63–67.

Hecht, H., Brown, E. L., & Young, L. R. (2002). Adapting to artificial gravity (AG) at high rotational speeds. *Journal of Gravitational Physiology: A Journal of the International Society for Gravitational Physiology, 9*(1), P1–P5.

Heeter, C. (1992). Being there: The subjective experience of presence. *Presence: Teleoperators and Virtual Environments, 1*(2), 262–271.

Held, R. (1961). Exposure-history as a factor in maintaining stability of perception and coordination. *The Journal of Nervous and Mental Disease, 132*(1), 26–32. doi:10.1097/00005053-196113210-00005.

Hendrix, C., & Barfield, W. (1996). Presence within virtual environments as a function of visual display parameters. *Presence: Teleoperators and Virtual Environments, 5*(3), 274–289.

Hettinger, L. J., Berbaum, K. S., Kennedy, R. S., Dunlap, W. P., & Nolan, M. D. (1990). Vection and simulator sickness. *Military Psychology, 2*(3), 171–181. doi:10.1207/s15327876mp0203_4.

Hettinger, L. J., Schmidt-Daly, T. N., Jones, D. L., & Keshavarz, B. (2014). Illusory self-motion in virtual environments. In K. S. Hale & K. M. Stanney (Eds.), *Handbook of virtual environments: Design, implementation, and applications* (2nd ed., pp. 435–466). New York, NY: CRC Press.

Hettinger, L. J., & Riccio, G. E. (1992). Visually induced motion sickness in virtual environments. *Presence: Teleoperators and Virtual Environments, 1*(3), 306–310.

Hill, K., & Howarth, P. A. (2000). Habituation to the side effects of immersion in a virtual environment. *Displays, 21*(1), 25–30. doi:10.1016/S0141-9382(00)00029-9.

Hoffman, D. M., Girshick, A. R., Akeley, K., & Banks, M. S. (2008). Vergence–accommodation conflicts hinder visual performance and cause visual fatigue. *Journal of Vision, 8*(3), 33. doi:10.1167/8.3.33

Holmes, S. R., King, S., Stott, R., & Clemes, S. (2002). Facial skin pallor increases during motion sickness. *Journal of Psychophysiology, 16*(3), 150–157.

Horn, C. C. (2008). Why is the neurobiology of nausea and vomiting so important? *Appetite, 50*(2–3), 430–434.

Houchin, K. W., Dunbar, J. A., & Lingua, R. W. (1992). Reduction of postoperative emesis in medial rectus muscle surgery. *Investigative Ophthalmology and Visual Science, 33*(4), 1336.

Howard, I. P. (1986). The perception of posture, self motion, and the visual vertical. In K. R. Boff, L. Kaufman, & J. P. Thomas (Eds.), *Handbook of perception and human performance: Vol. 1. Sensory processes and perception* (pp. 18.1–18.62). New York, NY: Wiley.

Howard, I. P., & Hu, G. (2001). Visually induced reorientation illusions. *Perception, 30*(5), 583–600. doi:10.1068/p3106

Howarth, P. A. (2011). Potential hazards of viewing 3-D stereoscopic television, cinema and computer games: A review. *Ophthalmic and Physiological Optics, 31*(2), 111–122. doi:10.1111/j.1475-1313.2011.00822.x

Howarth, P. A., & Costello, P. (1997). The occurrence of virtual simulation sickness symptoms when an HMD was used as a personal viewing system. *Displays, 18*(2), 107–116. doi:10.1016/S0141-9382(97)00011-5

Howarth, P. A., & Finch, M. (1999). The nauseogenicity of two methods of navigating within a virtual environment. *Applied Ergonomics, 30*(1), 39–45. doi:10.1016/S0003-6870(98)00041-6

Hoyt, R. E., Lawson, B. D., McGee, H. A., Strompolis, M. L., & McClellan, M. A. (2009). Modafinil as a potential motion sickness countermeasure. *Aviation, Space, and Environmental Medicine, 80*(8), 709–715. doi:10.3357/ASEM.2477.2009

Hu, S., Grant, W. F., Stern, R. M., & Koch, K. L. (1991). Motion sickness severity and physiological correlates during repeated exposures to a rotating optokinetic drum. *Aviation, Space, and Environmental Medicine, 62*(4), 308–314.

Hu, S., McChesney, K. A., Player, K. A., Bahl, A. M., Buchanan, J. B., & Scozzafava, J. E. (1999). Systematic investigation of physiological correlates of motion sickness induced by viewing an optokinetic rotating drum. *Aviation, Space, and Environmental Medicine, 70*(8), 759–765.

Hu, S., Stern, R. M., & Koch, K. L. (1992). Electrical acustimulation relieves vection-induced motion sickness. *Gastroenterology, 102*(6), 1854–1858.

Hu, S., Stern, R. M., Vasey, M. W., & Koch, K. L. (1989). Motion sickness and gastric myoelectric activity as a function of speed of rotation of a circular vection drum. *Aviation, Space, and Environmental Medicine, 60*(5), 411–414.

IJsselsteijn, W., Ridder, H. de, Freeman, J., Avons, S. E., & Bouwhuis, D. (2001). Effects of stereoscopic presentation, image motion, and screen size on subjective and objective corroborative measures of presence. *Presence: Teleoperators and Virtual Environments, 10*(3), 298–311. doi:10.1162/105474601300343621

Irwin, J. (1881). The pathology of sea-sickness. *The Lancet, 118*(3039), 907–909. doi:10.1016/S0140-6736(02)38129-7

Jerome, C. J., & Witmer, B. G. (2004). Human performance in virtual environments: Effects of presence, immersive tendency, and simulator sickness. *Proceedings of the Human Factors and Ergonomics Society Annual Meeting, 48*(23), 2613–2617. doi:10.1177/154193120404802302

Ji, J. T. T., So, R. H. Y., & Cheung, R. T. F. (2010). Isolating the effects of vection and optokinetic nystagmus on optokinetic rotation-induced motion sickness. *Human Factors: The Journal of the Human Factors and Ergonomics Society, 51*(5), 739–751. doi:10.1177/0018720809349708

Johnson, D. M. (2005). *Introduction to and review of simulator sickness research.* Fort Rucker, AL: U.S. Army Research Institute for the Behavioral and Social Sciences.

Johnson, W. H., Meek, J., & Graybiel, A. (1962). Effects of labyrinthectomy on canal sickness in squirrel monkey. *The Annals of Otology, Rhinology, and Laryngology, 71*, 289–298.

Johnson, W. H., Sunahara, F. A., & Landolt, J. P. (1999). Importance of the vestibular system in visually induced nausea and self-vection. *Journal of Vestibular Research: Equilibrium and Orientation, 9*(2), 83–87.

Jokerst, M. D., Gatto, M., Fazio, R., Stern, R. M., & Koch, K. L. (1999). Slow deep breathing prevents the development of tachygastria and symptoms of motion sickness. *Aviation, Space, and Environmental Medicine, 70*(12), 1189–1192.

Jones, S. A. (1998). *Effects of restraint on vection and simulator sickness* (Unpublished doctoral dissertation). University of Central Florida, Orlando, FL.

Joseph, J. A., & Griffin, M. J. (2008). Motion sickness: Effect of the magnitude of roll and pitch oscillation. *Aviation, Space, and Environmental Medicine, 79*(4), 390–396.

Kaplan, I. (1964). Motion sickness in railroads. *Industrial Medicine & Surgery, 33*, 648–651.

Kellogg, R. S., & Gillingham, K. K. (1986). United States Air Force experience with simulator sickness, research and training. *Proceedings of the Human Factors and Ergonomics Society Annual Meeting, Dayton, OH, 30*(5), 427–429. doi:10.1177/154193128603000503

Kellogg, R. S., Kennedy, R. S., & Graybiel, A. (1965). Motion sickness symptomatology of labyrinthine defective and normal subjects during zero gravity maneuvers. *Aerospace Medicine, 36*, 315–318.

Kennedy, R. S., Berbaum, K. S., Dunlap, W. P., & Hettinger, L. J. (1996). Developing automated methods to quantify the visual stimulus for cybersickness. *Proceedings of the Human Factors and Ergonomics Society Annual Meeting, Philadelphia, PA, 40*(22), 1126–1130. doi:10.1177/154193129604002204

Kennedy, R. S., Drexler, J. M., & Kennedy, R. C. (2010). Research in visually induced motion sickness. *Applied Ergonomics, 41*(4), 494–503. doi:10.1016/j.apergo.2009.11.006

Kennedy, R. S., Fowlkes, J. E., Berbaum, K. S., & Lilienthal, M. G. (1992). Use of a motion sickness history questionnaire for prediction of simulator sickness. *Aviation, Space, and Environmental Medicine, 63*(7), 588–593.

Kennedy, R. S., Graybiel, A., McDonough, R. C., & Beckwith, D. (1968). Symptomatology under storm conditions in the north atlantic in control subjects and in persons with bilateral labyrinthine defects. *Acta Oto-laryngologica, 66*(1–6), 533–540. doi:10.3109/00016486809126317

Kennedy, R. S., Hettinger, L. J., & Lilienthal, M. G. (1990). Simulator sickness. In G. H. Crampton (Ed.), *Motion and space sickness*. Boca Raton, FL: CRC Press.

Kennedy, R. S., Lane, N. E., Berbaum, K. S., & Lilienthal, M. G. (1993). Simulator sickness questionnaire: An enhanced method for quantifying simulator sickness. *The International Journal of Aviation Psychology, 3*(3), 203–220. doi:10.1207/s15327108ijap0303_3

Kennedy, R. S., Lane, N. E., Lilienthal, M. G., Berbaum, K. S., & Hettinger, L. J. (1992). Profile analysis of simulator sickness symptoms: Application to virtual environment systems. *Presence: Teleoperators and Virtual Environments, 1*(3), 295–301.

Kennedy, R. S., Lanham, D. S., Drexler, J. M., Massey, C. J., & Lilienthal, M. G. (1997). A comparison of cybersickness incidences, symptom profiles, measurement techniques, and suggestions for further research. *Presence: Teleoperators and Virtual Environments, 6*(6), 638–644.

Kennedy, R. S., Lilienthal, M. G., Berbaum, K. S., Baltzley, D. R., & McCauley, M. E. (1989). Simulator sickness in U.S. Navy flight simulators. *Aviation, Space, and Environmental Medicine, 60*(1), 10–16.

Kennedy, R. S., & Stanney, K. M. (1996). Postural instability induced by virtual reality exposure: Development of a certification protocol. *International Journal of Human–Computer Interaction, 8*(1), 25–47. doi:10.1080/10447319609526139

Kennedy, R. S., Stanney, K. M., & Dunlap, W. P. (2000). Duration and exposure to virtual environments: Sickness curves during and across sessions. *Presence: Teleoperators and Virtual Environments, 9*(5), 466–475. doi:10.1162/105474600566952

Kennedy, R. S., Tolhurst, G. C., & Graybiel, A. (1965). *The effects of visual deprivation on adaptation to a rotating environment* (Report No. NASA-CR-69359). Pensacola, FL: U.S. Naval School of Aviation Medicine. Retrieved from the National Aeronautics and Space Administration website: http://ntrs.nasa.gov/archive/nasa/casi.ntrs.nasa.gov/19650026830_1965026830.pdf

Kenyon, R., & Young, L. R. (1986). M.I.T./Canadian vestibular experiments on the Spacelab-1 mission: 5. Postural responses following exposure to weightlessness. *Experimental Brain Research, 64*(2), 335–346. doi:10.1007/BF00237750

Keshavarz, B., & Hecht, H. (2011a). Axis rotation and visually induced motion sickness: The role of combined roll, pitch, and yaw motion. *Aviation, Space, and Environmental Medicine, 82*(11), 1023–1029. doi:10.3357/ASEM.3078.2011

Keshavarz, B., & Hecht, H. (2011b). Validating an efficient method to quantify motion sickness. *Human Factors: The Journal of the Human Factors and Ergonomics Society, 53*(4), 415–426. doi:10.1177/0018720811403736

Keshavarz, B., & Hecht, H. (2012a). Stereoscopic viewing enhances visually induced motion sickness but sound does not. *Presence: Teleoperators and Virtual Environments, 21*(2), 213–228.

Keshavarz, B., & Hecht, H. (2012b). Visually induced motion sickness and presence in videogames: The role of sound. *Proceedings of the Human Factors and Ergonomics Society Annual Meeting, 56*(1), 1763–1767. doi:10.1177/1071181312561354

Keshavarz, B., Hecht, H., & Zschutschke, L. (2011). Intra-visual conflict in visually induced motion sickness. *Displays, 32*(4), 181–188. doi:10.1016/j.displa.2011.05.009

Khalid, H., Turan, O., Bos, J. E., & Incecik, A. (2011). Application of the subjective vertical–horizontal-conflict physiological motion sickness model to the field trials of contemporary vessels. *Ocean Engineering, 38*(1), 22–33. doi:10.1016/j.oceaneng.2010.09.008

Kim, Y. Y., Kim, H. J., Kim, E. N., Ko, H. D., & Kim, H. T. (2005). Characteristic changes in the physiological components of cybersickness. *Psychophysiology, 42*(5), 616–625. doi:10.1111/j.1469-8986.2005.00349.x

Kingdon, K. S., Stanney, K. M., & Kennedy, R. S. (2001). Extreme responses to virtual environment exposure. *Proceedings of the Human Factors and Ergonomics Society Annual Meeting, Minneapolis, MN, 45*(27), 1906–1910. doi:10.1177/154193120104502711

Klinke, R. (1970). Efferent influence on the vestibular organ during active movements of the body. *Pflügers Archiv European Journal of Physiology, 318*(4), 325–332. doi:10.1007/BF00586972

Klöcker, N., Hanschke, W., Toussaint, S., & Verse, T. (2001). Scopolamine nasal spray in motion sickness: A randomised, controlled, and crossover study for the comparison of two scopolamine nasal sprays with oral dimenhydrinate and placebo. *European Journal of Pharmaceutical Sciences, 13*(2), 227–232. doi:10.1016/S0928-0987(01)00107-5

Klosterhalfen, S., Kellermann, S., Pan, F., Stockhorst, U., Hall, G., & Enck, P. (2005). Effects of ethnicity and gender on motion sickness susceptibility. *Aviation, Space, and Environmental Medicine, 76*(11), 1051–1057.

Klosterhalfen, S., Muth, E. R., Kellermann, S., Meissner, K., & Enck, P. (2008). Nausea induced by vection drum: Contributions of body position, visual pattern, and gender. *Aviation, Space, and Environmental Medicine, 79*(4), 384–389. doi:10.3357/ASEM.2187.2008

Klosterhalfen, S., Pan, F., Kellermann, S., & Enck, P. (2006). Gender and race as determinants of nausea induced by circular vection. *Gender Medicine, 3*(3), 236–242. doi:10.1016/S1550-8579(06)80211-1

Koch, K. L., Stern, R. M., Vasey, M. W., Seaton, J. F., Demers, L. M., & Harrison, T. S. (1990). Neuroendocrine and gastric myoelectrical responses to illusory self-motion in humans. *The American Journal of Physiology, 258*(2 Pt 1), E304–E310.

Kolasinski, E. M. (1995). *Prediction of simulator sickness in a virtual environment* (Doctoral dissertation). University of Central Florida, Orlando, FL. Retrieved from http://www.hitl.washington.edu/scivw/kolasinski/

Kowalski, A., Rapps, N., & Enck, P. (2006). Functional cortical imaging of nausea and vomiting: A possible approach. *Autonomic Neuroscience, 129*(1–2), 28–35. doi:10.1016/j.autneu.2006.07.021

Kuritzky, A., Ziegler, D. K., & Hassanein, R. (1981). Vertigo, motion sickness and migraine. *Headache: The Journal of Head and Face Pain, 21*(5), 227–231. doi:10.1111/j.1526-4610.1981.hed2105227.x

Lackner, J. R., & DiZio, P. (1991). Decreased susceptibility to motion sickness during exposure to visual inversion in microgravity. *Aviation, Space, and Environmental Medicine, 62*(3), 206–211.

Lackner, J. R., & DiZio, P. (2006). Space motion sickness. *Experimental Brain Research, 175*(3), 377–399. doi:10.1007/s00221-006-0697-y

Lackner, J. R., & Graybiel, A. (1984). Elicitation of motion sickness by head movements in the microgravity phase of parabolic flight maneuvers. *Aviation, Space, and Environmental Medicine, 55*(6), 513–520.

Lackner, J. R., & Teixeira, R. A. (1977). Optokinetic motion sickness: Continuous head movements attenuate the visual induction of apparent self-rotation and symptoms of motion sickness. *Aviation, Space, and Environmental Medicine, 48*(3), 248–253.

Ladwig, K. H., Marten-Mittag, B., Formanek, B., & Dammann, G. (2000). Gender differences of symptom reporting and medical health care utilization in the German population. *European Journal of Epidemiology, 16*(6), 511–518.

Lansberg, M. P. (1963). Canal-sickness: Fact or fiction? *Industrial Medicine & Surgery, 32*, 21–24.

LaViola, J. J. (2000). A discussion of cybersickness in virtual environments. *ACM SIGCHI Bulletin, 32*(1), 47–56. doi:10.1145/333329.333344

Lawson, B. D. (1993). *Human physiological and subjective responses during motion sickness induced by unusual visual and vestibular stimulation* (Doctoral dissertation). Brandeis University, Waltham, MA. University Microfilms International, Report #9322354, Ann Arbor, MI.

Lawson, B. D. (2001). Changes in subjective well-being associated with exposure to virtual environments. In M. J. Smith, G. Salvendy, D. Harris, & R. J. Koubek (Eds.), *Usability evaluation and interface design: Vol. 1. Cognitive engineering, intelligent agents and virtual reality.* Mahwah, NJ: Lawrence Erlbaum Associates.

Lawson, B. D. (2005). Exploiting the illusion of self-motion (vection) to achieve a feeling of "virtual acceleration" in an immersive display. In C. Stephanidis (Ed.), *Proceedings of the 11th International Conference on Human–Computer Interaction,* Las Vegas, NV, *8,* 1–10.

Lawson, B. D. (2014). Motion sickness symptomatology and origins. In K. S. Hale & K. M. Stanney (Eds.), *Handbook of virtual environments: Design, implementation, and applications* (2nd ed., pp. 531–539). New York, NY: CRC Press.

Lawson, B. D., Kass, S., Harmon-Horton, H., Decoteau, P., Simmons, R., McGrath, C. M., & Vacchiano, C. (2007). A test of the usefulness of an aromatic nausea remedy for alleviating symptoms of motion sickness [Abstract]. *Aviation, Space, and Environmental Medicine, 78*(3), 117.

Lawson, B. D., Kass, S., Muth, E. R., Sommers, J., & Guzy, L. (2001). Development of a scale to assess signs and symptoms of sopite syndrome in response to mild or nonsickening motion stimuli [Abstract]. *Aviation, Space, and Environmental Medicine, 72*(3), 255.

Lawson, B. D., McGee, H. A., Castenada, M. E., Golding, J. F., Kass, S. J., & McGrath, C. M. (2009). *Evaluation of common antimotion sickness medications and recommendations concerning their potential usefulness during special operations* (Technical Report No. NAMRL 09-15). Pensacola, FL: Naval Aerospace Medical Research Laboratory. Retrieved from the Defense Technical Information Center website: http://www.dtic.mil/dtic/tr/fulltext/u2/a511823.pdf

Lawson, B. D., & Mead, A. M. (1998). The sopite syndrome revisited: Drowsiness and mood changes during real or apparent motion. *Acta Astronautica, 43*(3–6), 181–192. doi:10.1016/S0094-5765(98)00153-2

Lawther, A., & Griffin, M. J. (1987). Prediction of the incidence of motion sickness from the magnitude, frequency, and duration of vertical oscillation. *The Journal of the Acoustical Society of America, 82*(3), 957. doi:10.1121/1.395295

Lawther, A., & Griffin, M. J. (1988). A survey of the occurrence of motion sickness amongst passengers at sea. *Aviation, Space, and Environmental Medicine, 59*(5), 399–406.

Lee, G. C., Yoo, Y., & Jones, S. (1997). Investigation of driving performance, vection, postural sway, and simulator sickness in a fixed-based driving simulator. *Computers & Industrial Engineering, 33*(3–4), 533–536. doi:10.1016/S0360-8352(97)00186-1

Lentz, J. M., & Collins, W. E. (1977). Motion sickness susceptibility and related behavioral characteristics in men and women. *Aviation, Space, and Environmental Medicine, 48*(4), 316–322.

Leslie, K. R., Stickgold, R., DiZio, P., Lackner, J. R., & Hobson, J. A. (1997). The effect of optokinetic stimulation on daytime sleepiness. *Archives Italiennes de Biologie, 135*(3), 219–228.

Lien, H.-C., Sun, W. M., Chen, Y.-H., Kim, H., Hasler, W., & Owyang, C. (2003). Effects of ginger on motion sickness and gastric slow-wave dysrhythmias induced by circular vection. *American Journal of Physiology. Gastrointestinal and Liver Physiology, 284*(3), G481–G489. doi:10.1152/ajpgi.00164.2002

Lin, C.-T., Lin, C.-L., Chiu, T.-W., Duann, J.-R., & Jung, T.-P. (2011). Effect of respiratory modulation on relationship between heart rate variability and motion sickness. *Proceedings of the International Conference of the IEEE Engineering in Medicine and Biology Society, Boston, MA* (pp. 1921–1924). doi:10.1109/IEMBS.2011.6090543

Lin, J.-W., Duh, H.-L., Parker, D. E., Abi-Rached, H., & Furness, T. A. (2002). Effects of field of view on presence, enjoyment, memory, and simulator sickness in a virtual environment. In B. Loftin, J. X. Chen, S. Rizzo, M. Goebel, & M. Hirose (Eds.), *Proceedings of the IEEE Virtual Reality Conference, Orlando, FL* (pp. 164–171). doi:10.1109/VR.2002.996519

Lin, J.-W., Parker, D. E., Lahav, M., & Furness, T. A. (2005). Unobtrusive vehicle motion prediction cues reduced simulator sickness during passive travel in a driving simulator. *Ergonomics, 48*(6), 608–624.

Lo, W. T., & So, R. H. Y. (2001). Cybersickness in the presence of scene rotational movements along different axes. *Applied Ergonomics, 32*(1), 1–14. doi:10.1016/S0003-6870(00)00059-4

Lombard, M., & Ditton, T. (1997). At the heart of it all: The concept of presence. *Journal of Computer-Mediated Communication, 3*(2), 0. doi:10.1111/j.1083-6101.1997.tb00072.x, http://onlinelibrary.wiley.com/doi/10.1111/j.1083–6101.1997.tb00072.x/abstract

Lubeck, A. J. A., van Geest, L., Bos, J. E., & Stins, J. F. (2011, September). *Prolonged postural instability after watching 2D motion pictures.* Paper presented at the 3rd International Symposium on Visual IMage Safety (VIMS 2011), Las Vegas, NV.

Lyons, T. J., & French, J. (1991). Modafinil: The unique properties of a new stimulant. *Aviation, Space, and Environmental Medicine, 62*(5), 432–435.

Mach, E. (1875). *Grundlinien der Lehre von den Bewegungsempfindungen.* Leipzig, Germany: Engelmann.

Maki, B. E., & McIlroy, W. E. (1996). Influence of arousal and attention on the control of postural sway. *Journal of Vestibular Research: Equilibrium & Orientation, 6*(1), 53–59.

Matsangas, P. (2004). *A linear physiological visual–vestibular interaction model for prediction of motion sickness incidence* (Master's thesis). Naval Postgraduate School, Monterrey, CA. Retrieved from http://calhoun.nps.edu/public/handle/10945/1382

Matsuoka, I., Ito, J., Takahashi, H., Sasa, M., & Takaori, S. (1984). Experimental vestibular pharmacology: A minireview with special reference to neuroactive substances and antivertigo drugs. *Acta Otolaryngologica. Supplementum, 419*, 62–70.

Mayne, R. (1974). A system concept of the vestibular organ. In H. H. Kornhuber (Ed.), *Handbook of sensory physiology: Vol. VI. Vestibular system, Part 2* (pp. 493–580). Berlin, Germany: Springer-Verlag.

Mayo, A. M., Wade, M. G., & Stoffregen, T. A. (2011). Postural effects of the horizon on land and at sea. *Psychological Science, 22*(1), 118–124. doi:10.1177/0956797610392927

McCandless, C. H., & Balaban, C. D. (2010). Parabrachial nucleus neuronal responses to off-vertical axis rotation in macaques. *Experimental Brain Research, 202*(2), 271–290. doi:10.1007/s00221-009-2130-9

McCauley, M. E., Royal, J. W., Wylie, C. D., O'Hanlon, J. F., & Mackie, R. R. (1976). *Motion sickness incidence: Exploratory studies of habituation, pitch and roll, and the refinement of a mathematical model* (Technical Report No. 1733-2). Goleta, CA: Human Factors Research, Inc. Retrieved from the Naval Postgraduate School website: http://faculty.nps.edu/memccaul/docs/McCauley%20et%20al%201976.pdf

McCauley, M. E., & Sharkey, T. J. (1992). Cybersickness: Perception of self-motion in virtual environments. *Presence: Teleoperators and Virtual Environments, 1*(3), 311–318.

Merhi, O. (2009). *Motion sickness, virtual reality, and postural stability* (Doctoral dissertation). University of Minnesota, Minneapolis, MN. Retrieved from http://conservancy.umn.edu/bitstream/58646/1/Merhi_umn_0130E_10796.pdf

Merhi, O., Faugloire, E., Flanagan, M. B., & Stoffregen, T. A. (2007). Motion sickness, console video games, and head-mounted displays. *Human Factors: The Journal of the Human Factors and Ergonomics Society, 49*(5), 920–934. doi:10.1518/001872007X230262

Miller, E. F., & Graybiel, A. (1970). *Comparison of five levels of motion sickness severity as the basis for grading susceptibility* (Report No. NAMI-1098). Pensacola, FL: Naval Aerospace Medical Institute. Retrieved from the National Aeronautics and Space Administration website: http://ntrs.nasa.gov/archive/nasa/casi.ntrs.nasa.gov/19700023589_1970023589.pdf

Miller, J. C., Sharkey, T. J., Graham, G. A., & McCauley, M. E. (1993). Autonomic physiological data associated with simulator discomfort. *Aviation, Space, and Environmental Medicine, 64*(9 Pt 1), 813–819.

Miller, K. E., & Muth, E. R. (2004). Efficacy of acupressure and acustimulation bands for the prevention of motion sickness. *Aviation, Space, and Environmental Medicine, 75*(3), 227–234.

Mills, K. L., & Griffin, M. J. (2000). Effect of seating, vision and direction of horizontal oscillation on motion sickness. *Aviation, Space, and Environmental Medicine, 71*(10), 996–1002.

Min, Y.-K., Chung, S.-C., You, J.-H., Yi, J.-H., Lee, B., Tack, G.-R., … Min, B. C. (2006). Young adult drivers' sensitivity to changes in speed and driving mode in a simple vehicle simulator. *Perceptual and Motor Skills, 103*(1), 197–209. doi:10.2466/PMS.103.5.197–209

Mittelstaedt, H. (1983). A new solution to the problem of the subjective vertical. *Die Naturwissenschaften, 70*(6), 272–281.

Money, K. E. (1990). Motion sickness and evolution. In: G. H. Crampton (ed), *Motion and space sickness* (pp. 1–8). Boca Raton, FL, CRC Press.

Money, K. E. (1970). Motion sickness. *Canadian Journal of Physiology and Pharmacology, 50*(1), 1–39.

Money, K. E., & Cheung, B. S. (1983). Another function of the inner ear: Facilitation of the emetic response to poisons. *Aviation, Space, and Environmental Medicine, 54*(3), 208–211.

Mon-Williams, M., Warm, J. P., & Rushton, S. (1993). Binocular vision in a virtual world: Visual deficits following the wearing of a head-mounted display. *Ophthalmic and Physiological Optics, 13*(4), 387–391. doi:10.1111/j.1475-1313.1993.tb00496.x

Morrow, G. R. (1985). The effect of a susceptibility to motion sickness on the side effects of cancer chemotherapy. *Cancer, 55*(12), 2766–2770.

Moss, J. D., & Muth, E. R. (2011). Characteristics of head-mounted displays and their effects on simulator sickness. *Human Factors: The Journal of the Human Factors and Ergonomics Society, 53*(3), 308–319. doi:10.1177/0018720811405196

Murray, J. B. (1997). Psychophysiological aspects of motion sickness. *Perceptual and Motor Skills, 85,* 1163–1167. doi:10.2466/pms.1997.85.3f.1163

Muth, E. R. (1996). Assessment of the multiple dimensions of nausea: The Nausea Profile (NP). *Journal of Psychosomatic Research, 40*(5), 511–520. doi:10.1016/0022-3999(95)00638-9

Muth, E. R. (2010). The challenge of uncoupled motion: Duration of cognitive and physiological aftereffects. *Human Factors: The Journal of the Human Factors and Ergonomics Society, 51*(5), 752–761. doi:10.1177/0018720809353320

Muth, E. R., Koch, K. L., Stern, R. M., & Thayer, J. F. (1999). Effect of autonomic nervous system manipulations on gastric myoelectrical activity and emotional responses in healthy human subjects. *Psychosomatic Medicine, 61*(3), 297–303.

Muth, E. R., Stern, R. M., & Koch, K. L. (1996). Effects of vection-induced motion sickness on gastric myoelectric activity and oral–cecal transit time. *Digestive Diseases and Sciences, 41*(2), 330–334. doi:10.1007/BF02093824

Nachum, Z., Shupak, A., Letichevsky, V., Ben-David, J., Tal, D., & Tamir, A. (2004). Mal de debarquement and posture: Reduced reliance on vestibular and visual cues. *The Laryngoscope, 114*(3), 581–586. doi:10.1097/00005537-200403000-00036

Nichols, S. C., Cobb, S. V. G., and Wilson, J. R. (1997). Health and safety implications of virtual environments: Measurement issues. *Presence: Teleoperators and Virtual Environments, 6*(6), 667–675.

Nichols, S. C., Haldane, C., & Wilson, J. R. (2000). Measurement of presence and its consequences in virtual environments. *International Journal of Human–Computer Studies, 52*(3), 471–491. doi:10.1006/ijhc.1999.0343

Nichols, S. C., & Patel, H. (2002). Health and safety implications of virtual reality: A review of empirical evidence. *Applied Ergonomics, 33*(3), 251–271. doi:10.1016/S0003-6870(02)00020-0

O'Hanlon, J. F., & McCauley, M. E. (1974). Motion sickness incidence as a function of the frequency and acceleration of vertical sinusoidal motion. *Aerospace Medicine, 45*(4), 366–369.

Ohyama, S., Nishiike, S., Watanabe, H., Matsuoka, K., Akizuki, H., Takeda, N., & Harada, T. (2007). Autonomic responses during motion sickness induced by virtual reality. *Auris Nasus Larynx, 34*(3), 303–306. doi:10.1016/j.anl.2007.01.002

Oman, C. M. (1982). A heuristic mathematical model for the dynamics of sensory conflict and motion sickness hearing in classical musicians. *Acta Oto-laryngologica, 94*(s392), 4–44. doi:10.3109/00016488209108197

Oman, C. M. (1987). Spacelab experiments on space motion sickness. *Acta Astrocautica, 15*(1), 55–66.

Oman, C. M. (1990). Motion sickness: A synthesis and evaluation of the sensory conflict theory. *Canadian Journal of Physiology and Pharmacology, 68*(2), 294–303. doi:10.1139/y90-044

Oman, C. M. (2012). Are evolutionary hypotheses for motion sickness "just-so" stories? *Journal of Vestibular Research: Equilibrium and Orientation, 22*(2), 117–127. doi:10.3233/VES-2011-0432

Oman, C. M., & Cullen, K. E. (2012). Brainstem processing of vestibular sensory reafference: Implications for motion sickness etiology [Abstract]. *Aviation, Space, and Environmental Medicine, 83*(3), 273.

Otten, E. W. (2005). *Effect of predictability of imposed visual motion on the occurrence of motion sickness* (Master's thesis). Miami University, Oxford, OH.

Otten, E. W. (2008). *The influence of stimulus complexity and perception-action coupling on postural sway* (Doctoral dissertation). Miami University, Oxford, OH.

Otto, B., Riepl, R. L., Klosterhalfen, S., & Enck, P. (2006). Endocrine correlates of acute nausea and vomiting. *Autonomic Neuroscience, 129*(1–2), 17–21. doi:10.1016/j.autneu.2006.07.010

Palmisano, S., Allison, R. S., & Pekin, F. (2008). Accelerating self-motion displays produce more compelling vection in depth. *Perception, 37*(1), 22–33. doi:10.1068/p5806

Palmisano, S., Bonato, F., Bubka, A., & Folder, J. (2007). Vertical display oscillation effects on forward vection and simulator sickness. *Aviation, Space, and Environmental Medicine, 78*(10), 951–956. doi:10.3357/ASEM.2079.2007

Patterson, R., Winterbottom, M. D., & Pierce, B. J. (2006). Perceptual issues in the use of head-mounted visual displays. *Human Factors: The Journal of the Human Factors and Ergonomics Society, 48*(3), 555–573. doi:10.1518/001872006778606877

Pausch, R., Crea, T., & Conway, M. (1992). A literature survey for virtual environments: Military flight simulator systems and simulator sickness. *Presence: Teleoperators and Virtual Environments, 1*(3), 344–363.

Peli, E. (1998). The visual effects of head-mounted display (HMD) are not distinguishable from those of desktop computer display. *Vision Research, 38*(13), 2053–2066. doi:10.1016/S0042-6989(97)00397-0

Perrin, P., Perrot, C., Deviterme, D., Ragaru, B., & Kingma, H. (2000). Dizziness in discus throwers is related to motion sickness generated while spinning. *Acta Oto-laryngologica, 120*(3), 390–395. doi:10.1080/000164800750000621

Prothero, J. D., Hoffman, H. G., Parker, D. E., Furness, T. A., & Wells, M. J. (1995). Foreground/background manipulations affect presence. *Proceedings of the Human Factors and Ergonomics Society Annual Meeting, 39*(21), 1410–1414. doi:10.1177/154193129503902111

Quarck, G., Etard, O., Oreel, M., & Denise, P. (2000). Motion sickness occurrence does not correlate with nystagmus characteristics. *Neuroscience Letters, 287*(1), 49–52. doi:10.1016/S0304-3940(00)01140-X

Rague, B., & Oman, C. M. (1987). Use of a microcomputer system for running spectral analysis of EGGs to predict the onset of motion sickness. *Proceedings of 9th Annual Conference of IEEE Engineering in Medicine & Biology Society*, Boston, MA.

Reason, J. T. (1978a). Motion sickness adaptation: A neural mismatch model. *Journal of the Royal Society of Medicine, 71*(11), 819–829.

Reason, J. T. (1978b). Motion sickness: Some theoretical and practical considerations. *Applied Ergonomics, 9*(3), 163–167. doi:10.1016/0003-6870(78)90008-X

Reason, J. T., & Brand, J. J. (1975). *Motion sickness*. London, U.K.: Academic Press.

Reason, J. T., & Graybiel, A. (1969). *Progressive adaptation to Coriolis accelerations associated with 1-RPM increments in the velocity of the slow rotation room* (Technical Report No. NAMI-1081). Pensacola, FL: Naval Aerospace Medical Institute. Retrieved from the National Aeronautics and Space Academy http://ntrs.nasa.gov/archive/nasa/casi.ntrs.nasa.gov/19700005540_1970005540.pdf

Reed-Jones, R. J., Vallis, L. A., Reed-Jones, J. G., & Trick, L. M. (2008). The relationship between postural stability and virtual environment adaptation. *Neuroscience Letters, 435*(3), 204–209. doi:10.1016/j.neulet.2008.02.047

Regan, E. C., & Price, K. R. (1994). The frequency of occurrence and severity of side-effects of immersion virtual reality. *Aviation, Space, and Environmental Medicine, 65*(6), 527–530.

Regan, E. C., & Ramsey, A. D. (1996). The efficacy of hyoscine hydrobromide in reducing side-effects induced during immersion in virtual reality. *Aviation, Space, and Environmental Medicine, 67*(3), 222–226.

Reid, K., Grundy, D., Khan, M. I., & Read, N. W. (1995). Gastric emptying and the symptoms of vection-induced nausea. *European Journal of Gastroenterology & Hepatology, 7*(2), 103–108.

Reschke, M. F., Somers, J. T., & Ford, G. (2006). Stroboscopic vision as a treatment for motion sickness: Strobe lighting vs. shutter glasses. *Aviation, Space, and Environmental Medicine, 77*(1), 2–7.

Riccio, G. E., & Stoffregen, T. A. (1991). An ecological theory of motion sickness and postural instability. *Ecological Psychology, 3*(3), 195–240.

Rine, R. M., Schubert, M. C., & Balkany, T. J. (1999). Visual–vestibular habituation and balance training for motion sickness. *Physical Therapy, 79*(10), 949–957.

Rolnick, A., & Bles, W. (1989). Performance and well-being under tilting conditions: The effects of visual reference and artificial horizon. *Aviation, Space, and Environmental Medicine, 60*(8), 779–785.

Rolnick, A., & Lubow, R. E. (1991). Why is the driver rarely motion sick? The role of controllability in motion sickness. *Ergonomics, 34*(7), 867–879. doi:10.1080/00140139108964831

Rupert, A. H. (2010). *Motion sickness etiology: An alternative to Treisman's evolutionary theory*. Paper presented at the Spatial Orientation Symposium in Honor of Fred Guedry, Institute of Human and Machine Cognition, Pensacola, FL.

Serrador, J. M., Schlegel, T. T., Black, F. O., & Wood, S. J. (2005). Cerebral hypoperfusion precedes nausea during centrifugation. *Aviation, Space, and Environmental Medicine, 76*(2), 91–96.

Sharples, S., Cobb, S. V. G., Moody, A., & Wilson, J. R. (2008). Virtual reality induced symptoms and effects (VRISE): Comparison of head mounted display (HMD), desktop and projection display systems. *Displays, 29*(2), 58–69. doi:10.1016/j.displa.2007.09.005

Sheehan, S. E., Oman, C. M., & Duda, K. R. (2011). Motion sickness: A cholinomimetic agent hypothesis. *Journal of Vestibular Research: Equilibrium and Orientation, 21*(4), 209–217. doi:10.3233/VES-2011-0417

Sheridan, T. (1992). Musing on telepresence and virtual presence. *Presence: Teleoperators and Virtual Environments, 1*(1), 120–125.

Sherman, C. R. (2002). Motion sickness: Review of causes and preventive strategies. *Journal of Travel Medicine, 9*(5), 251–256. doi:10.2310/7060.2002.24145

Shupak, A., & Gordon, C. R. (2006). Motion sickness: Advances in pathogenesis, prediction, prevention, and treatment. *Aviation, Space, and Environmental Medicine, 77*(12), 1213–1223.

Slater, M., & Usoh, M. (1993). Representation systems, perceptual position, and presence in virtual environments. *Presence: Teleoperators and Virtual Environments, 2*(3), 221–233.

Smart, L. J., Stoffregen, T. A., & Bardy, B. G. (2002). Visually induced motion sickness predicted by postural instability. *Human Factors: The Journal of the Human Factors and Ergonomics Society, 44*(3), 451–465. doi:10.1518/0018720024497745

Smith, A. V., & Qureshi, S. (2012). Motion sickness desensitization: A review of ADF experience of "success" [Abstract]. *Aviation, Space, and Environmental Medicine, 83*(3), 24.

Smither, J. A.-A., Mouloua, M., & Kennedy, R. S. (2008). Reducing symptoms of visually induced motion sickness through perceptual training. *The International Journal of Aviation Psychology, 18*(4), 326–339. doi:10.1080/10508410802346921

So, R., & Chow, E. (2009). *Postural disturbance and vection when viewing visual stimulus oscillating in roll and fore-and-aft directions: Effects of frequency.* Paper presented at the 2nd International Symposium on Visual Image Safety (VIMS), Utrecht, the Netherlands.

So, R. H. Y., Lo, W. T., & Ho, A. T. (2001). Effects of navigation speed on motion sickness caused by an immersive virtual environment. *Human Factors: The Journal of the Human Factors and Ergonomics Society, 43*(3), 452–461. doi:10.1518/001872001775898223

Stanney, K. M., Hale, K. S., Nahmens, I., & Kennedy, R. S. (2003). What to expect from immersive virtual environment exposure: Influences of gender, body mass index, and past experience. *Human Factors: The Journal of the Human Factors and Ergonomics Society, 45*(3), 504–520. doi:10.1518/hfes.45.3.504.27254

Stanney, K. M., & Hash, P. (1998). Locus of user-initiated control in virtual environments: Influences on cybersickness. *Presence: Teleoperators and Virtual Environments, 7*(5), 447–459. doi:10.1162/105474698565848

Stanney, K. M., & Kennedy, R. S. (1997). The psychometrics of cybersickness. *Communications of the ACM, 40*(8), 66–68. doi:10.1145/257874.257889

Stanney, K. M., & Kennedy, R. S. (1998). Aftereffects from virtual environment exposure: How long do they last? *Proceedings of the Human Factors and Ergonomics Society Annual Meeting, 42*(21), 1476–1480. doi:10.1177/154193129804202103

Stanney, K. M., & Kennedy, R. S. (2009). Simulation sickness. In D. A. Vincenzi, J. A. Wise, M. Mouloua, & P. A. Hancock (Eds.), *Human factors in simulation and training* (pp. 117–127). Boca Raton, FL: CRC Press.

Stanney, K. M., Kennedy, R. S., & Drexler, J. M. (1997). Cybersickness is not simulator sickness. *Proceedings of the Human Factors and Ergonomics Society Annual Meeting, 48*(2), 1138–1142.

Stanney, K. M., Kennedy, R. S., Drexler, J. M., & Harm, D. L. (1999). Motion sickness and proprioceptive aftereffects following virtual environment exposure. *Applied Ergonomics, 30*(1), 27–38. doi:10.1016/S0003-6870(98)00039-8

Stanney, K. M., Mourant, R. R., & Kennedy, R. S. (1998). Human factors issues in virtual environments: A review of the literature. *Presence: Teleoperators and Virtual Environments, 7*(4), 327–351. doi:10.1162/105474698565767

Stanney, K. M., & Salvendy, G. (1998). Aftereffects and sense of presence in virtual environments: Formulation of a research and development agenda. *International Journal of Human–Computer Interaction, 10*(2), 135–187. doi:10.1207/s15327590ijhc1002_3

Steele, J. E. (1961). *Motion sickness and spatial perception: A theoretical study* (ASD Technical Report No. 61-530). Wright-Patterson Air Force Base, OH: Aeronautical Systems Division, U.S. Air Force. Retrieved from the Defense Technical Information Center website: http://www.dtic.mil/dtic/tr/fulltext/u2/273602.pdf

Stern, R. M., Hu, S., Anderson, R. B., Leibowitz, H. W., & Koch, K. L. (1990). The effects of fixation and restricted visual field on vection-induced motion sickness. *Aviation, Space, and Environmental Medicine, 61*(8), 712–715.

Stern, R. M., Hu, S., LeBlanc, R., & Koch, K. L. (1993). Chinese hypersusceptibility to vection-induced motion sickness. *Aviation, Space, and Environmental Medicine, 64*(9), 827–830.

Stern, R. M., Hu, S., Uijtdehaage, S. H., Muth, E. R., Xu, L. H., & Koch, K. L. (1996). Asian hypersusceptibility to motion sickness. *Human Heredity, 46*(1), 7–14. doi:10.1159/000154318

Stern, R. M., Koch, K. L., Leibowitz, H. W., Lindblad, I. M., Shupert, C. L., & Stewart, W. R. (1985). Tachygastria and motion sickness. *Aviation, Space, and Environmental Medicine, 56*(11), 1074–1077.

Stern, R. M., Koch, K. L., Stewart, W. R., & Lindblad, I. M. (1987). Spectral analysis of tachygastria recorded during motion sickness. *Gastroenterology, 92*(1), 92–97.

Steuer, J. (1992). Defining virtual reality: Dimensions determining telepresence. *Journal of Communication, 42*(4), 73–93. doi:10.1111/j.1460-2466.1992.tb00812.x

Stevens, S. S. (1946). On the theory of scales of measurement. *Science, 103*(2684), 677–680. doi:10.1126/science.103.2684.677

Stoffregen, T. A., Faugloire, E., Yoshida, K., Flanagan, M. B., & Merhi, O. (2008). Motion sickness and postural sway in console video games. *Human Factors: The Journal of the Human Factors and Ergonomics Society, 50*(2), 322–331. doi:10.1518/001872008X250755

Stoffregen, T. A., Hettinger, L. J., Haas, M. W., Roe, M. M., & Smart, L. J. (2000). Postural instability and motion sickness in a fixed-base flight simulator. *Human Factors: The Journal of the Human Factors and Ergonomics Society, 42*(3), 458–469. doi:10.1518/0018720007779698097

Stoffregen, T. A., & Riccio, G. E. (1991). An ecological critique of the sensory conflict theory of motion sickness. *Ecological Psychology, 3*(3), 159–194. doi:10.1207/s15326969eco0303_1

Stoffregen, T. A., & Smart, L. J. (1998). Postural instability precedes motion sickness. *Brain Research Bulletin, 47*(5), 437–448. doi:10.1016/S0361-9230(98)00102-6

Stoffregen, T. A., Yoshida, K., Villard, S., Scibora, L., & Bardy, B. G. (2010). Stance width influences postural stability and motion sickness. *Ecological Psychology, 22*(3), 169–191. doi:10.1080/10407413.2010.496645

Stroud, K. J., Harm, D. L., & Klaus, D. M. (2005). Preflight virtual reality training as a countermeasure for space motion sickness and disorientation. *Aviation, Space, and Environmental Medicine, 76*(4), 352–356.

Sugita, N., Yoshizawa, M., Abe, M., Tanaka, A., Watanabe, T., Chiba, S., … Nitta, S. (2007). Evaluation of adaptation to visually induced motion sickness based on the maximum cross-correlation between pulse transmission time and heart rate. *Journal of NeuroEngineering and Rehabilitation, 4*(1), 35. doi:10.1186/1743-0003-4-35

Tanahashi, S., Ujike, H., Kozawa, R., & Ukai, K. (2007). Effects of visually simulated roll motion on vection and postural stabilization. *Journal of NeuroEngineering and Rehabilitation, 4*(1) 39. doi:10.1186/1743-0003-4-39

Tanguy, S., Quarck, G., Etard, O., Gauthier, A., & Denise, P. (2008). Vestibulo–ocular reflex and motion sickness in figure skaters. *European Journal of Applied Physiology, 104*(6), 1031–1037. doi:10.1007/s00421-008-0859-7

Teixeira, R. A., & Lackner, J. R. (1979). Optokinetic motion sickness: Attenuation of visually-induced apparent self-rotation by passive head movements. *Aviation, Space, and Environmental Medicine, 50*(3), 264–266.

Treisman, M. (1977). Motion sickness: An evolutionary hypothesis. *Science, 197*(4302), 493–495. doi:10.1126/science.301659

Turner, M., & Griffin, M. J. (1999a). Motion sickness in public road transport: The effect of driver, route and vehicle. *Ergonomics, 42*(12), 1646–1664. doi:10.1080/001401399184730

Turner, M., & Griffin, M. J. (1999b). Motion sickness in public road transport: Passenger behaviour and susceptibility. *Ergonomics, 42*(3), 444–461. doi:10.1080/001401399185586

Ujike, H., Ukai, K., & Nihei, K. (2008). Survey on motion sickness-like symptoms provoked by viewing a video movie during junior high school class. *Displays, 29*(2), 81–89. doi:10.1016/j.displa.2007.09.003

Ujike, H., & Watanabe, H. (2011). Effects of stereoscopic presentation on visually induced motion sickness. *Proceedings of SPIE-IS&T Electronic Imaging, 7863*, 786314. doi:10.1117/12.873500

Umberg, E. N., & Pothos, E. N. (2011). Neurobiology of aversive states. *Physiology & Behavior, 104*, 69–75.

Ungs, T. J. (1987). Simulator induced syndrome: Evidence for long term simulator aftereffects. *Proceedings of the Human Factors and Ergonomics Society Annual Meeting, 31*(5), 505–509. doi:10.1177/154193128703100505

Villard, S. J., Flanagan, M. B., Albanese, G. M., & Stoffregen, T. A. (2008). Postural instability and motion sickness in a virtual moving room. *Human Factors: The Journal of the Human Factors and Ergonomics Society, 50*(2), 332–345. doi:10.1518/001872008X250728

Vogel, H., Kohlhaas, R., & Baumgarten, R. J. (1982). Dependence of motion sickness in automobiles on the direction of linear acceleration. *European Journal of Applied Physiology and Occupational Physiology, 48*(3), 399–405. doi:10.1007/BF00430230

Von Gierke, H. E., & Parker, D. E. (1994). Differences in otolith and abdominal viscera graviceptor dynamics: Implications for motion sickness and perceived body position. *Aviation, Space, and Environmental Medicine, 65*(8), 747–751.

von Holst, E. (1954). Relations between the central nervous system and the peripheral organs. *The British Journal of Animal Behaviour, 2*(3), 89–94. doi:10.1016/S0950-5601(54)80044-X

Vuurman, E. F. P. M., Rikken, G., Muntjewerff, N., Halleux, F., & Ramaekers, J. (2004). Effects of desloratadine, diphenhydramine, and placebo on driving performance and psychomotor performance measurements. *European Journal of Clinical Pharmacology, 60*(5), 307–313. doi:10.1007/s00228-004-0757-9

Wallace, J. C., Kass, S. J., & Lawson, B. D. (2002). *Assessing a specific measure of sopite syndrome: The mild motion questionnaire* (Technical Report No. NAMRL-1414). Pensacola, FL: Naval Aerospace Medical Research Laboratory.

Warwick-Evans, L. A., Church, R. E., Hancock, C., Jochim, D., Morris, P. H., & Ward, F. (1987). Electrodermal activity as an index of motion sickness. *Aviation, Space, and Environmental Medicine, 58*(5), 417–423.

Warwick-Evans, L. A., Symons, N., Fitch, T., & Burrows, L. (1998). Evaluating sensory conflict and postural instability. Theories of motion sickness. *Brain Research Bulletin, 47*(5), 465–469. doi:10.1016/S0361-9230(98)00090-2

Watt, D. G. D., Bouyer, L. J. G., Nevo, I. T., Smith, A. V., & Tiande, Y. (1992). What is motion sickness? *Annals of the New York Academy of Sciences, 656*(1), 660–667. doi:10.1111/j.1749-6632.1992.tb25243.x

Webb, N. A., & Griffin, M. J. (2002). Optokinetic stimuli: Motion sickness, visual acuity, and eye movements. *Aviation, Space, and Environmental Medicine, 73*(4), 351–358.

Welch, R., Blackmon, T., Liu, A., Mellers, B., & Stark, L. (1996). The effects of pictorial realism, delay of visual feedback, and observer interactivity on the subjective sense of presence. *Presence: Teleoperators and Virtual Environments, 5*(3), 263–273.

Westmoreland, D., Krell, R. W., & Self, B. P. (2007). Physiological responses to the Coriolis illusion: Effects of head position and vision. *Aviation, Space, and Environmental Medicine, 78*(10), 985–989.

Wilpizeski, C. R., Lowry, L. D., Contrucci, R. B., Green, S. J., & Goldman, W. S. (1985). Effects of head and body restraint on experimental motion-induced sickness in squirrel monkeys. *Aviation, Space, and Environmental Medicine, 56*(11), 1070–1073.

Witmer, B. G., & Singer, M. J. (1994). *Measuring immersion in virtual environments.* (Technical Report No. 1014). Alexandria, VA: U.S. Army Research Institute for the Behavioral and Social Sciences.

Witmer, B. G., & Singer, M. J. (1998). Measuring presence in virtual environments: A presence questionnaire. *Presence: Teleoperators and Virtual Environments, 7*(3), 225–240. doi:10.1162/105474698565686

Wood, C. D., & Graybiel, A. (1968). Evaluation of sixteen anti-motion sickness drugs under controlled laboratory conditions. *Aerospace Medicine, 39*(12), 1341–1344.

Wood, C. D., Graybiel, A., & Kennedy, R. S. (1966). Comparison of effectiveness of some antimotion sickness drugs using recommended and larger than recommended doses as tested in the slow rotation room. *Aerospace Medicine, 37*(3), 259–262.

Wood, R. W. (1895). The 'Haunted Swing' illusion. *Psychological Review, 2*(3), 277–278. doi:10.1037/h0073333

Wood, S. J. (2002). Human otolith–ocular reflexes during off-vertical axis rotation: Effect of frequency on tilt-translation ambiguity and motion sickness. *Neuroscience Letters, 323*(1), 41–44.

Wright, M. S., Bose, C. L., & Stiles, A. D. (1995). The incidence and effects of motion sickness among medical attendants during transport. *The Journal of Emergency Medicine, 13*(1), 15–20. doi:10.1016/0736-4679(94)00106-5

Yang, J., Guo, C., So, R. H. Y., & Cheung, B. S. (2011, September). *Effects of pursuit eye movements on visually induced motion sickness.* Paper presented at the 3rd International Symposium on Visual IMage Safety (VIMS 2011), Las Vegas, NV.

Yardley, L. (1992). Motion sickness and perception: A reappraisal of the sensory conflict approach. *British Journal of Psychology, 83*(Pt 4), 449–471.

Yates, B. J., Miller, A. D., & Lucot, J. B. (1998). Physiological basis and pharmacology of motion sickness: An update. *Brain Research Bulletin, 47*(5), 395–406.

Yen Pik Sang, F. D., Billar, J. P., Golding, J. F., & Gresty, M. A. (2003). Behavioral methods of alleviating motion sickness: Effectiveness of controlled breathing and a music audiotape. *Journal of Travel Medicine, 10*(2), 108–111.

Yen Pik Sang, F. D., Billar, J. P., Gresty, M. A., & Golding, J. F. (2005). Effect of a novel motion desensitization training regime and controlled breathing on habituation to motion sickness. *Perceptual and Motor Skills, 101*(1), 244–256.

Yen Pik Sang, F. D., Golding, J. F., & Gresty, M. A. (2003). Suppression of sickness by controlled breathing during mildly nauseogenic motion. *Aviation, Space, and Environmental Medicine, 74*(9), 998–1002.

Yoshida, K., Naito, S., Takahashi, H., Sato, K., Ito, K., Kamata, M., … Ohkubo, T. (2003). Monoamine oxidase a gene polymorphism, 5-HT2A receptor gene polymorphism and incidence of nausea induced by fluvoxamine. *Neuropsychobiology, 48*(1), 10–13. doi:10.1159/000071822

Young, L. R., Meiry, J. L., & Li, Y. T. (1966). Control engineering approaches to human dynamic space orientation. In J. Huertas & A. Graybiel (Eds.), *Second Symposium on the Role of the Vestibular Organs in Space Exploration* (pp. 217–227).

Young, L. R., Sienko, K. H., Le, L., Hecht, H., & Natapoff, A. (2003). Adaptation of the vestibulo–ocular reflex, subjective tilt, and motion sickness to head movements during short-radius centrifugation. *Journal of Vestibular Research: Equilibrium and Orientation, 13*(2–3), 65–77.

27 Social Impact of Virtual Environments

Sandra L. Calvert

CONTENTS

27.1 INTRODUCTION

> Sam tells a story about going to ride a big horse named Star, but the horse is nowhere to be found. Sam then goes for the Sheriff, fearing that Star has been stolen. The Sheriff tells Sam not to worry, that a kind lady found Star in the forest and that all was well. Sam is happy as she ends her story in the magical castle where she is interacting with a preschool child, who will then tell her story.
>
> **Ryokai, Vaucelle, and Cassell (2003)**

Storytelling has long been a part of oral culture, but this oral exchange occurs in a special kind of imaginary setting, a toy castle that has an embodied autonomous agent—Sam—projected on a screen behind the castle. Sam is programmed to promote literacy skills by interacting in humanlike ways with children by taking turns as they tell stories. Autonomous agents like Sam allow realistic interactions in 3-D virtual environments (VEs) that can promote literacy skills through social interaction.

How will the time spent in such VEs impact children's social skills? Will something innately human be lost if people choose to interact with virtual rather than with real characters? Or will online virtual characters supplement face-to-face friendships? Will people feel lonely and isolated when they interact with embodied virtual agents or with peers that are not in their immediate environments, or will life on the screen become a part of everyday experiences? How will children construct their identities in these VEs, in which they can create their own avatars and yet are disembodied (Calvert, 2002)?

The very definition of a VE is one that is somewhat ambiguous. While VEs are often conceptualized as 3-D immersive environments (Roussou, 2004), others argue that VEs should include 2-D VEs in which players have a sense of presence, which involves how real an environment feels to people (e.g., see Lee, 2004; Schroeder, 2008). More specifically, Schroeder defined VEs as those that "people experience as ongoing over time and that have large populations which they experience together with others as a world for social interaction" (p. 1). VEs, then, are forums for social experiences that are perceived as real or at least as realistic. In this chapter, I will examine the social influences of VEs using a broad definition of what a VE is.

27.2 FORM AND CONTENT OF INFORMATION TECHNOLOGIES

Knowledge is transmitted and must be decoded in all media experiences (Calvert, 2004) including VEs. Like other technologies before it, VE simulations are comprised of both content and form. Content has to do with domains such as aggression, sexuality, prosocial behavior, and fantasy. Form has to do with the unique representational codes that are used to present content (Huston & Wright, 1998). More specifically, form refers to audiovisual production features, such as action, sound effects, and dialogue, that structure, mark, and represent various kinds of content (Calvert, 1999a). VEs add interactive capabilities and, potentially, a first-person perspective to media experiences.

27.2.1 FORM OF INFORMATION TECHNOLOGIES

Our world provides a rich stream of changing visual, auditory, tactile, and olfactory sensations. The form of VE simulations reflects an ongoing evolution in the symbolic codes that are used to represent media experiences. More specifically, communication was once based on auditory, spoken language. With television, video games, and most recently the Internet, users can interact with visual and audio messages. Although embodied VE simulations will expand these representational experiences into olfactory and tactile modes of expression, VE currently relies on the visual system more than other sensory systems (Schroeder, 1996). Auditory dimensions are not that different from previous applications in the film, television, and video game areas. Some VEs are unique in that they allow users a first-person interaction with 3-D visual images in immersive environments.

In examining the impact of media, McLuhan (1964) argued that television was unique in its form, not its content. This statement is true of other media as well. Take, for instance, the incident where two Colorado adolescent boys killed their classmates, teachers, and themselves. Playing the video game *Doom* was implicated as one source of their antisocial acts. Apparently, they learned a particular style of killing by playing this game (Bushman et al., 2013). Perhaps they could have learned similar behaviors from reading books, newspapers, or watching television programs or films, but *how* the behavior was displayed as they pulled the triggers of their guns was similar to the form of how they fired the guns of their video game.

Embodiment, interactivity, perceptually salient production features, and immersion combine to make some VE simulations seem realistic. Embodiment involves the sensation that you are the character. In VE simulations, a player looks out of the eyes of a character and the world looks similar to how it looks when a person looks out of their own body. Embodiment leads to the sensation of *presence* (Barfield, Zeltzer, Sheridan, & Slater, 1995; Chertoff & Schatz, 2014, Chapter 34;

Stuart, 1996). Virtual embodiment lends itself to a first-person perspective to media experiences, as many video game experiences do.

Presence is augmented by interactivity. In VE games, people interact with and control images much more so than when they are spectators viewing a prepackaged television program or even when they control a video game player but are not embodied in it (Calvert & Tan, 1994). When a person directly experiences a media form and is able to control it, there is a one-to-one correspondence between what is done and what happens. Indeed, control is an important aspect of interactivity that increases children's attention to computer content (Calvert, Strong, & Gallagher, 2005). This increased sense of personal involvement and presence in the VE game may also increase player identification with characters, their perceptions of the vividness of events, and their perceptions of self-efficacy, the belief that they are in personal control of the events that are being experienced (Bandura, 1997).

The realism of virtual images may be augmented further by the use of perceptually salient formal features. *Perceptual salience*, a term coined by Berlyne (1960), involves features that contain high levels of movement, contrast, incongruity, surprise, complexity, novelty, and the like. Perceptually salient television and computer features involve action (movement), sound effects (surprise, incongruity), and loud music (contrast, surprise; Calvert, 2004). Immersion uses perceptual cues that can trick the senses into feelings of presence in a computer-generated environment (Stuart, 1996). Because of primitive orienting responses that ensure our survival, perceptually salient features are likely to draw attention to information, thereby increasing the probability that the content will be processed (Calvert, 1999a).

When perceptually salient images are combined with interactivity, an embodied user feels immersed in these symbolic worlds, often making the virtual world very compelling. This personal experience of interacting with 3-D moving images is why some users want to run away when they are attacked in a VE game. The player personally feels attacked. As virtual interfaces are improved, these audiovisual representational interactions will provide a lifelike quality that is unparalleled in media experiences (Calvert, 1999a). This evolution of technology means that users can create and construct shared realities that are limited only by their imaginations and by the latest technological breakthroughs (Schroeder, 1994).

Although form and content are theoretically separate dimensions, production practices link perceptually salient formal features with certain kinds of content. For instance, rapid action, sound effects, and loud music are often associated with action–adventure television programs containing high levels of violent content (Huston & Wright, 1998). These same perceptually salient features are used extensively in VE games (Calvert, 1999b), making these experiences attention getting and arousing to players. Even so, perceptually salient forms can be used to present prosocial content. When interaction, embodiment, and perceptual salience are used to present content, the potential for that content to be attention getting and acted on increases, be it antisocial or prosocial.

27.2.2 CONTENT OF INFORMATION TECHNOLOGIES

Formal features deliver content. VEs offer the opportunity for users to experience any type of content that interests them. Content can range from realistic simulations where surgeons perform heart catheterizations to fantasy simulations such as slink world. Slinks are yellow-faced characters with cylindrical bodies and distinct personalities who autonomously move, avoid collisions with other objects, and react to other slinks, the latter activity being based upon the slink's happy, angry, or uncaring moods (Green & Sun, 1995). VE games, models of buildings, artistic environments, visualization models, and training environments are the most commonly portrayed content in virtual worlds (Schroeder, 1996, see Chapters 36 through 50, this book).

Obviously, the applications of VEs are diverse, and the content of various applications varies to suit the particular goal of the simulation. Nonetheless, VE content reflects market trends, making games one of the most widespread applications for consumer use (Greenwood-Ericksen,

Kennedy, & Stafford, 2014, Chapter 50; Schroeder, 1996). The content of VE games partly reflects earlier media content, though its roots are in military simulation (Herz, 1997). Just as the entertainment industry relies on action and violence to attract viewers to television, film, and video game content, so too do many VE games (Calvert, 1999a). Because the most common experience of people is currently VE games, games are a focal point here for understanding the social impact of VE technology.

Game content involves a player or players who at times assume various identities and perform certain activities within a virtual simulation. On the one hand, these games provide an avenue for aggressive interactions with other real or imaginary characters. On the other hand, these games provide an avenue for characters to create cognitive strategies and to work together collaboratively to win the game (Pearce, 1998). Therefore, both aggressive and prosocial interactions can occur within these imaginary scenarios.

The company Virtuality was one of the first producers of commercially available VE systems (Schroeder, 1996). They created adventure games such as *Dactyl Nightmare* and *Legend Quest*. *Legend Quest* uses archetypal fantasy images, such as the wizard, the warrior, and the elf. Narrative elements, comparable to those found in film, are central to this game (Schroeder). In *Dactyl Nightmare*, two cartoonlike characters gain points by shooting each other and a pterodactyl as they move through a maze of platforms suspended in space (Calvert & Tan, 1994). The goal is to get as many points as possible. *Virtual Worlds Battle Tech Center, Magic Edge*, and *Fighter Town* also provided early combat-based entertainment. Mainstream commercial virtual entertainment programs were offered early on by Disney Quest and by Sony Metreon.

Immersive VE often connotes a space that arises when a person interacts with a computer with particular hardware, such as gloves and helmets, or in a cave automatic virtual environment (CAVE), where users move through a shared virtual space. Multiuser domains, dungeons, or dimensions (MUDs) are yet another type of VE. MUDs, originally constructed around the idea of a real-life game called *Dungeons and Dragons* (Schroeder, 1997), are now found on the Internet.

One initial form of MUD was a text-based social VE (Turkle, 1995). There were no particular goals or tasks that participants were expected to do in this kind of MUD. Instead, MUD participants assumed complex personae, such as a wizard or a warrior, and embarked on various adventures in relatively unstructured situations. Multiuser object-oriented environments (MOOs) and multiuser simulated environments (MUSEs) are similar to MUDs except they are based on different kinds of software (Turkle).

MUDs evolved into visual, rather than solely textual, virtual spaces (Calvert, 2002). These MUD applications, which were more sophisticated than text-based MUDs but less sophisticated than truly immersive VE simulations, include desktop VE and second-person VE systems (Schroeder, 1997). In desktop VE systems, a person can look out of a character's perspective into a virtual world, but they are not fully immersed, as they would be when using head-mounted displays (HMDs) or in a CAVE. Second-person VE systems are even less immersive than desktop VE systems. In second-person VE systems, the player is represented by an onscreen avatar (i.e., a figure or character). Users communicate through written text that appears on the screen. The player sees a computer-generated environment, but not through their own eyes (Schroeder). The experience is much like playing a video game, but it is often much less structured than a game.

In second-person VE systems, players represented through avatars roam visual spaces through VE worlds such as Alphaworld and Cybergate (Schroeder, 1997). Social interactions change substantially when visual images and personal characters are included in the simulation, rather than having to rely solely on text for communication. For example, verbal aggression expressed in words becomes physical aggression directed at avatars, sexual language becomes sexual interactions between avatars, and belonging to the group via text-based exchanges is expressed by physical proximity to other avatars (Schroeder). The additional visual information provided in these virtual spaces represented a fundamental leap in the experiences of players.

MUDs include visual adventure games on the Internet where players slay beasts and triumph over evil (Sleek, 1998b) as well as social MUDs where players interact in relatively open spaces to construct whatever is of interest to them (Schroder, 1997; Turkle, 1995). The hundreds of thousands of early MUD users were primarily adolescent and adult males in their 20s, but there were also children playing Barbie and the Power Rangers online (Turkle). Through MUDs, players can recreate themselves, their identities, and be anyone that they want to be (Turkle). Social experiences have also been examined in visual and textual MUDs developed for children (Calvert, Mahler, Zehnder, Jenkins, & Lee, 2003; Calvert, Strouse, Strong, Huffaker, & Lai, 2009). Millions of users spent approximately 22 h per week as avatars playing massive-multiplayer online role-playing games (MMORPGs) in VEs, such as the *Second Life* (Yee, 2006).

Immersion influences the relative power of people who interact in VEs. For instance, small groups of young adults solved problems in a VE followed by a continuation of the task in the real world. Only one person was actually immersed. The immersed person was consistently the leader in the VE, but this role did not generalize to the real world (Slater, Sadagic, Usoh, & Schroeder, 1999).

27.3 THEORETICAL PERSPECTIVES AND PREDICTIONS IN RELATION TO VIRTUAL ENVIRONMENT CONTENT

VEs allow people to engage in various kinds of representational experiences in both realistic and imaginary simulations (Calvert, 1999b). Representational thought allows individuals to transcend the present reality, thereby extending the perceptual field in both time and space. Through representational experiences, memories of the past and fantasies about the future become part of current reality.

The impact of media on peoples' lives has been examined from various theories in the fields of psychology and communication. Various predictions are made about the impact of these experiences, depending on the particular theoretical perspective brought to bear on the information. These predictions are summarized next.

27.3.1 SOCIAL COGNITIVE THEORY

Social cognitive theory (Bandura, 1997) focuses on the role that models play as influences on social behaviors. These models can be real people, or they can be symbolic media characters.

As a model displays behaviors, observers attend to and encode relevant information. If rewards are available for imitating the behavior, the observer will often do so. Imitation can occur for all kinds of behaviors, be they prosocial or antisocial (Calvert, 1999a). Models can also increase the probability that another person will exhibit antisocial actions through a process called disinhibition. That is, people often inhibit behaviors that they would like to do because the behaviors are not socially acceptable. However, if a model performs antisocial behaviors and gets rewarded for them (or at least not punished), then the observer's internal controls are undermined or disinhibited (Bandura, 1997). Models can also increase the probability of socially desirable behaviors through a process called response facilitation. In response facilitation, infrequent yet desirable behaviors, such as buying breast cancer stamps at the post office, can be increased when a person observes a model perform them.

Over time, Bandura (1997) increasingly focused on the concept of self-efficacy, the belief in personal control over life events. The subprocesses of attention, retention, production, and motivation occur as a person observes and later imitates a behavior. Self-efficacy regulates these subprocesses, particularly when a person fails and motivation to continue a task is weakened. If a person perceives that they have control over events, then they are more likely to continue to attend to, retain, and produce a behavior.

VEs allow user control, a component of self-efficacy, via interaction. Experienced VE users take control of these environments more than inexperienced, novice users. For instance, novices observe more in simulations, whereas experts are more likely to act (Schroeder, 1997). Also, role-playing, an important rehearsal mechanism for performing a behavior, is integrated into VE simulations through interaction. This means that the behaviors enacted (e.g., pulling the trigger of a gun) will be readily accessible to players after the VE interaction is over.

27.3.2 AROUSAL THEORY

Autonomic arousal is often measured by heart rate, skin conductance, and blood pressure, whereas an electroencephalogram measures cortical arousal. Both kinds of arousal are related, but researchers in the media area typically study autonomic arousal (Calvert, 1999a). People who are exposed to aggressive, sexual, or scary content typically become more aroused and direct that arousal in diverse ways, depending on what is cued by the environment. That is, arousal can be channeled into aggressive, sexual, or prosocial behaviors, regardless of the content, depending upon situational cues. After repeated exposure to aggressive or sexual content, the individual may become desensitized to that content and to the plights of others.

Certainly, VE content is not always meant to be arousing. For instance, behavioral therapeutic applications utilizing systematic desensitization procedures, such as a program to help a person overcome a fear of heights, focus on how to get a person to relax in what is typically an arousing, stressful situation. By contrast, entertainment games (Greenwood-Ericksen et al., 2014, Chapter 50) focus on getting users to be excited, much like an amusement park ride. The arousing qualities of entertainment-driven VEs may increase attentiveness. Interaction and presence should increase users' heart rate and other indices of physiological arousal. The content that is being experienced in VE will be the immediate trigger for releasing arousal, but real-life events just after VE interactions can also trigger behavior. With repeated interaction, users should habituate to the content as they have done in other exciting media experiences. That means that the content will have to be more vivid for arousal levels to return to prior levels (Calvert, 1999a).

27.3.3 PSYCHOANALYTIC THEORY

Psychoanalytic theory focuses on human drives, particularly sex and aggression, as the reasons for our behaviors (Miller, 2009). Most of the reasons for actions occur at an unconscious level, reflecting these basic drives. Although drives are a normal part of human experience, society often disapproves of, and therefore regulates, their expression. Consequently, drive reduction is accomplished in indirect ways. One example is catharsis. Instead of killing someone or having sex with someone, drives are released vicariously through a fantasy experience. Theoretically, VE games should yield the same cathartic releases that television, film, and video game experiences were expected to offer viewers or players.

In an extension of Freud's ideas, Jung (1959) proposed that humans share a collective unconscious, a repository of shared racial experiences that have been inherited. Archetypes, or primordial images, reside in the collective unconscious and are developed by individual experiences. These archetypes include mother, father, hero, trickster, persona, anima, animus, shadow, and self (Hall & Nordby, 1973). On the Internet, the persona, or mask, is a term that is used to describe a character's identity (Turkle, 1995). Moreover, men sometimes create and play a female persona online, thereby developing their anima or feminine side of the self, whereas women often create and play a male persona online, thereby developing their animus, or male side of the self. The shadow, or primitive side of human nature, protects the self but can overreact in very aggressive ways when threatened. The developmental task is to synthesize the various archetypes into one's self, the overarching archetype.

Most fantasies, including virtual ones, include many archetypal images. Playing out personae, or the archetypes, in VE games should lead to the development and differentiation of that archetype and the inclusion of it in the self-archetype. Put another way, virtual fantasy interactions may provide experiences that players can use to construct their identities.

Erikson (1968) extended psychoanalytic theory by focusing on the ego. In particular, identity was conceptualized as the development of the ego, the planful, organizing part of the personality (Miller, 2009). Identity was thought to emerge only after a period of searching, which is provided today, in part, by online experiences (Calvert, 2002).

27.3.4 CULTIVATION THEORY

In cultivation theory, media present messages that form the basis for a shared, constructed reality (Calvert, 1999a). One cultivation effect, called mainstreaming, focuses on the dominant media images that people share and, therefore, come to believe to be true. For example, those who frequently view aggressive content on television come to believe that the world is a more aggressive place than it really is. These heavy viewers buy more guns, watchdogs, and locks than light viewers do (Gerbner, Gross, Morgan, & Signorielli, 1994). Another cultivation effect, called resonance, amplifies real-life experiences that overlap with media experiences.

Mainstreaming effects should become common for those who share the same virtual spaces day after day because they create a reality based on shared experiences. Resonance will occur when these experiences occur in real life as well as in the virtual world. The implication is that experiences in the virtual world will influence people's views of reality, just as television viewing has.

27.3.5 COGNITIVE THEORIES

Information processing theory focuses on humans as information processing devices. Based on their experiences, people construct schemas, that is, cognitive structures that influence perception, attention, and memory (Calvert, 1999a). These schemas are then used to interpret future incoming information, thereby influencing what will be learned.

Other cognitive approaches include priming, which makes recent experiences more likely to be acted upon (e.g., priming the pump for action; Bushman et al., 2013). Scaffolds, in which a more mature person provides bridges to more advanced concepts, are another cognitive approach that influences learning outcomes (Vygotsky, 1978). Finally, constructivist approaches such as those described by Piaget are often used in media experiences to describe play as a mechanism to facilitate learning (Roussou, 2004).

27.3.6 BEHAVIORISM

Behaviorism involves reinforcements and punishments that come to control actions due to the consequences of one's behaviors (Miller, 2009). Behaviorism is often used in therapeutic settings. For example, a person may avoid settings because of fears associated with a traumatic experience. One technique that is used to treat phobias is systematic desensitization, in which the person is gradually introduced to a feared stimulus. VE has successfully been used to create realistic settings that are then used in systematic desensitization interventions (Difede & Hoffman, 2002).

27.3.7 PERCEPTUAL DEVELOPMENTAL THEORY

In perceptual developmental theory, the focus is on how perceptual systems evolved to ensure survival (Miller, 2009). People are active perceivers whose sensory systems evolved to pick out the salient and important environmental stimuli. The visual system is particularly important in evolution. Perceptually salient stimuli such as movement, contrast, incongruity, complexity, surprise, and novelty are features that are likely to elicit attention as well as improve the odds of survival (Berlyne, 1960). Perceptually salient production techniques, such as action and sound effects, are used to create media environments that simulate realistic, symbolic experiences (Calvert, 2004).

In VE simulations, users are tricked into perceiving and seeing events that are not really there. VEs accomplish this task in part by creating natural interactions where characters move and interact with each other without the use of artificial keyboards or textual commands (Bricken & Coco, 1995; Templeman, Page, & Denbrook, 2014, Chapter 10; Turk, 2014, Chapter 9). Perceptual laws are obeyed in many of these simulations. For instance, in *Slink World*, slinks are programmed to avoid collisions with other moving objects. When rapidly approached, slinks pull back and act scared. But when slowly approached, slinks are likely to stay in their current position (Green & Sun, 1995).

27.3.8 Uses and Gratification Theory

Communication media are used by people to satisfy or gratify their needs (Rubin, 1994). Needs are diverse, ranging from sexual stimulation to relaxing to information seeking. In VE, users may play games to entertain themselves, or they may interact with a clinical simulation to reduce phobias. Even so, many people use VE to satisfy entertainment needs, just like they used other media before it.

27.3.9 Parasocial Relationships

Children create emotionally tinged relationships with favorite media characters, known as parasocial relationships (Hoffner, 2008). These characters are often treated as if they are real, coming to serve important social functions in children's lives, that, in turn, influence what children learn (Calvert, Richards & Kent, in press; Gola, Richards, Lauricella, & Calvert, 2013; Lauricella, Gola, & Calvert, 2011). VEs provide yet another forum in which these parasocial relationships take place.

27.4 COGNITIVE, SOCIAL, AND BEHAVIORAL IMPACT OF VIRTUAL ENVIRONMENTS

According to uses and gratification theory, media fulfill certain needs that people have (Calvert, 1999a). Freud theorized that those needs involve sex and aggression, and the Nielsen ratings attest to the popularity of violent content. However, people have diverse needs for entertainment and for learning about others and themselves that extend well beyond violence and sex. Social interactions provide opportunities for constructive prosocial activity as well as for antisocial activity. Media influences on antisocial and prosocial interactions as well as identity construction are explored in this section.

27.4.1 Aggressive Interactions

In 2008, the US incarceration rate was 506 per 100,000 Americans, which was almost triple the rate of incarceration in 1986 (Clear, Cole, & Reisig, 2013). Data from 2005 to 2006 indicate a rate of 457 US juvenile crime suspects per 100,000 population, making the US rate 3rd of 60 nations (UNODC, 2009). From 1953 until the early 1990s, violent crime rose 600% in the United States, and homicide was the main cause of death for US urban males who were 15–24 years of age (Coie & Dodge, 1998). Exposure to violent media content, specifically violent television viewing and aggressive video game play, has been implicated as one reason for aggressiveness by US youth, in particular, and US society, in general (Anderson, Gentile, & Buckley, 2007; Coie & Dodge, 1998; Huston & Wright, 1998). Heavy exposure to violent content also cultivates a view that the world is a mean and scary place (Gerbner, Gross, Signorelli, & Morgan, 1986).

In an early examination of VE and aggression, Calvert and Tan (1994) examined the arousal levels and aggressive thoughts of college students in one of three conditions: (1) an immersive VE condition where they interacted in an aggressive game, (2) an observational condition where they viewed another person's aggressive VE game, and (3) a simulation experience where the players moved as if they were in a VE game, but in which no aggression was displayed (the control group). The game, *Dactyl Nightmare*, involved shooting an opponent and a pterodactyl that both periodically attack the player.

The results best supported arousal theory. Participants who played the game were more aroused, as measured by pulse rate, than those who observed another play the game or those who simulated game movements. Social cognitive theory received mixed support. Specifically, game players reported more aggressive thoughts than those who observed another person play the game or who simulated game movements; however, contrary to prediction, the observers reported no more aggressive thoughts than the game simulators. The psychoanalytic view that catharsis would release aggressive impulses safely via fantasy was not supported because players were more aroused and had more aggressive thoughts, not less. The results suggest that interactive VE experiences lead to stronger effects than simply observing another person perform an antisocial act.

According to Pearce (1998), the visceral kill or be killed formula created by game makers plays on primitive human instincts that are necessary for survival. Violence, as part of the human psyche, has been prevalent throughout time. War is expressed in simulations that were originally played out on board games that became computer games as technology evolved. Thus, VE and video games, more generally, had their roots in military operations. In fact, the first VE games were designed to train military personnel. Through role-playing, war strategies could be perfected. VE entertainment platforms adopted these earlier military game simulations, which can influence aggressive outcomes. For example, players who were primed to be more aggressive by having their avatars wear black rather than white cloaks in a VE subsequently reported more aggressive attitudes and behavioral intentions (Pena, Hancock, & Merloa, 2009). Playing aggressive video games with a personalized avatar also resulted in more aggressive behavior and more arousal than playing with a nonpersonalized avatar (Fischer, Kastenmuller, & Greitemeyer, 2010).

A key question is the extent to which these aggressive encounters will generalize to real-life experiences. Those who view aggressive television content tend to be more aggressive in real life, and research in video games and initial VE games finds similar results for aggressive behavior and aggressive thoughts, respectively (Calvert, 1999a). Estimates of 3%–9% of unique variance are found in studies that link children's aggressive conduct to their exposure to television violence (Huston & Wright, 1998). Millions of dollars were invested in virtual and video game military simulations to impact real fighting skills, not just pretend ones (Herz, 1997).

There are also those who argue that violent media impact people in different ways. That is correct. Some people become more aggressive, but those who see themselves as victims become more afraid, a cultivation effect built in part on the aggressive schemas developed through differential experiences as well as personal characteristics, such as being male or female (Gerbner et al., 1994). Moreover, everyone who sees violent media learns about aggressive behavior, even if they do not act on it immediately, and they can act on it later if motivational incentives are provided (Bandura, 1965). If only a small percent of individuals spontaneously act more aggressively after exposure to media violence, the quality of life is impacted for all. The evidence suggests that a portion of the population does become more aggressive after viewing or interacting with violent media content (Huston & Wright, 1998), as a result of observational learning, arousal, and priming of aggressive responses.

27.4.2 Identity Construction

Constructing a personal identity is a lifelong experience accomplished by day-to-day encounters, particularly those involving other people (Erikson, 1968). VEs are a source of experiences for creating parts of identity. As individuals move between virtual worlds and real worlds, their sense of identity may become increasingly malleable. They can be a male, a female, or both, moving fluidly across varying MUDs (Calvert, 2002). How will virtual construction impact the real-life construction of the self? Who will people be?

Historically, a debate involves whether identity is comprised of a single entity or multiple entities (Turkle, 1995). That is, does each individual have one and only one true self, or do individuals display multiple selves, depending upon the situation? Physical embodiment is a defining characteristic of the self (Harre, 1983), but people are free to use different bodies in VEs. What implication does the lack of real-life embodiment have for self-construction? Will people develop alternate and multiple selves that express hidden aspects of their character?

When children identify with other people, they selectively take in information about them and incorporate that person's characteristics and qualities into their own sense of self (Calvert, 2002). According to social cognitive theory (Bandura, 1997), identification with televised and filmed models is one reason that viewers imitate certain characters; they are making that character's behaviors their own. The VE genre allows players to be, not just to observe, the characters in these narratives. Being an embodied character may allow players to incorporate personality characteristics and behaviors directly into their own personal repertoire (Calvert & Tan, 1994).

The opportunity to create different selves expands greatly in VEs. People are not constrained by their physical body in defining who they are. They can construct the specific body that they want, and they can be as many people as they want to be (Calvert, 2002). They can also be animals, robots, or other nonhuman entities. Although they no longer have to use their own physical body in these virtual worlds, they do have a virtual body. Moreover, presence (Chertoff & Schatz, 2013, Chapter 34) should increase their beliefs about the realism of these embodied entities. These personae may then become incorporated into their self-constructions.

For children, the role-playing that takes place in imaginative play episodes is one way that they try out various perspectives, thereby coming to understand how and why others feel and act as they do (Calvert, 2002). Similarly, VE immersions provide opportunities for both children and adults to play different roles in a fantasy context, a facet of Piagetian theory, thereby mastering a role and understanding the perspective of that person in a safe environment.

In Jungian psychoanalytic theory, key archetypal images recur throughout history. Many of these images, embedded in the collective unconscious and displayed in fiction, involve the roles of hero, father, mother, child, wizard, sorceress, sorcerer, king, queen, emperor, princess, prince, wise old man, wise old woman (or crone), innocent, healer, trickster, witch, magician, persona, anima or animus, shadow, and self (Hall & Nordby, 1973). The self is a meeting place where various archetypes are integrated into one's personality (Turkle, 1995).

Archetypal images are passed from one generation to the next through the collective unconscious, a shared repository of ancestral images that are like the twentieth century negatives that used to be on a roll of undeveloped film (Hall & Nordby, 1973). Individual experiences then develop those latent images. The specific image that emerges is a function of a person's unique experiences. These experiences can be real, or they can be realistic or even imaginary as is the case of VE immersions.

According to Jungian psychoanalytic theory, the persona is the public display, or the mask, of the self that is presented in public (Hall & Nordby, 1973). The Internet is a place where personae are already being constructed in MUDs (Turkle, 1995). Acting on information makes it your own (Calvert, 2002). Thus, as people act on these archetypal images, they should be integrated within their own identities, thereby making their own sense of self much richer and more complex. The implication is that identity can potentially be altered as players enact various roles via these Internet personae, allowing the real person to integrate the archetypes that are being played out into his or her real personality. Personal views of a hero, or of any other archetypal image, should be influenced by role-playing heroes in VE encounters. The role of hero, which has often been reserved for men, may become more available to women through VE games.

One problem with these immersive identities is that an individual may lose track of who he or she really is. Fantasy can sometimes be compelling and even more appealing than reality. The person may choose to withdraw from real life to a more controlled imaginary persona and reality (Turkle, 1995). A person can be important in fantasy, with all the embellishments made possible by being present in an electronically generated environment of that person's choosing. The virtual identity option may be particularly appealing to young adolescents who are struggling with an identity crisis. Negative identities can also be constructed. If a person repeatedly pretends to be a shooter in VEs, antisocial behaviors are being practiced and aggressive schemas are being cultivated. Having such images and behaviors readily available can potentially increase antisocial behaviors and accentuate character flaws through social learning and priming processes, potentially playing a factor in rampage killings such as those that took place in communities like Littleton, Colorado (Bushman et al., 2013).

In online symbolic encounters, people can also deceive others about who they really are (Calvert, 2002). The use of temporary and fictitious identities is widespread in electronic communities (Mantovani, 1995). Is it ethical for a man to present himself as a woman, thereby gaining access to information that would not ordinarily be available to him? One man did, assuming an Internet persona of a woman who was helping other women with their emotional problems. Many women were upset that he deceived them (Turkle, 1995). Men also present themselves as women because they are more likely to be approached, and women present themselves as men to avoid being approached, particularly by men who are looking for online sexual experiences (Turkle). These gender switches are not uncommon in the online experiences of adults but are less likely for children and adolescents (Calvert et al., 2003, 2009; Huffaker & Calvert, 2005; Subramanyam & Greenfield, 2004), perhaps because children are still in the process of initially defining their sexual identities and, hence, more rigid in how they express their gender.

In one MUD, Calvert and colleagues (Calvert et al., 2003, 2009) examined how pairs of 10–12-year-old children interacted in same-gender or opposite-gender peers who they knew each other or who were strangers. Gender differences were found in both studies. Specifically, boys created avatars with fantasy names (e.g., Frotto), whereas girls created avatars with more realistic names (e.g., Julia). Boy pairs were also more playful, particularly when interacting with boys. By contrast, girls were more likely to chat with one another. Although boys and girls altered their preferred interaction styles in opposite-gender pairs when they did not know each other (Calvert et al.), a follow-up study revealed considerable difficulty in cross-gender communication patterns for those who did know one another (Calvert et al.). For instance, a boy tried to get his female peer to play *hide-and-seek* with him by moving their avatars across virtual scenes, but the girl complained via the text box and wanted him to chat with her instead. He eventually complied, but the overall interaction lacked harmony. Consistent with these studies, males more so than females maintain more interpersonal distance and engage in less eye gaze when immersed in *Second Life*, a 3-D virtual community populated by millions of players who interact with one another via their avatars (Yee, Bailenson, Urbanek, Chang, & Merget, 2007). Taken together, these studies reveal that players bring much of who they are to online experiences, with VEs providing another forum for players to express and act upon who they are, including typical gender norms.

Youth also stay close to their ethnic identities in chat rooms, where they express themselves in a stream of verbal interactions with a group of their peers. For example, youth in chat rooms that lacked adult supervision were more likely to be harassed by other youth about their race or ethnicity. In particular, youth in nonmonitored chat rooms were more likely to be pressured to tell others their race and were then sometimes stereotyped and harassed for their backgrounds than were those who participated in monitored chat rooms (Tynes, Reynold, & Greenfield, 2004).

Players can also confuse bots, which are online robots, with a real person. Or players can create a persona where they appear to be a bot, thereby gaining access to private conversations as cyberspace voyeurs (Turkle, 1995). One player created a passive character called Nibbet who was a rabbit. Others forgot he was there or assumed he was a bot, allowing him to listen to conversations where he would probably not be welcome.

In summary, VEs allow people to try out different facets of their identities and to rehearse behaviors in a safe place. Because players can be anyone they want to be, they can try out new ways of being in cyberspace. However, they can deceive others and themselves about their real selves.

Identities can be enacted that are negative and that reinforce character flaws. In spite of the color-blind, ageless, and nongendered opportunities provided by VEs, many players stay close to the truth, choosing to express who they are when interacting with others. Taken together, the research suggests that rather than providing a web of deception, VEs instead appear to provide a forum where online youth often express their actual identities.

27.4.3 Sexual Interactions

Men have long expressed curiosity about having virtual sex (Epley, 1993). Currently, these possibilities are limited because of tactile interface issues. However, once these telehaptic issues are resolved, there is clearly a paying audience for sexual virtual experiences. Computer software and Internet interfaces are already widely used for pornography, particularly by boys and men, which is consistent with arousal theory (Calvert, 1999a).

Opportunities for sexual enjoyment via VE offer positive as well as negative experiences for users. In text-based Internet MUD applications, many characters meet online and engage in virtual sex. Some even get married in virtual ceremonies (Turkle, 1995). These fantasy relationships provide an opportunity for safe sex because there is no danger of contracting or spreading a sexually transmitted disease. Users also are engaged in an experience with another person, allowing them to participate within the boundaries of a shared sexual fantasy rather than an individual one. By knowing how a partner feels and what a partner enjoys, a player may become better able to interact with real partners by understanding their needs. Players may even meet future real-life partners by exploring intimate experiences on the Internet. Sometimes, people really marry a person that they were initially involved with in an Internet relationship, but other times, real meetings are disappointing (Turkle).

However, sexually inappropriate acts and violations of another person's rights can also occur in virtual sexual experiences. For instance, some Internet sites are policed to prevent users from acting in sexually deviant ways on sites that are not meant to be sexual. MUD administrators for the Palace Internet site dealt with issues such as pornography and profanity (Sleek, 1998b). Reports of cyber-rape have also occurred (Wallace, 1999). In one instance, a more sophisticated MUD user took control of another person's character and he forced that character to have a violent sexual encounter (Turkle, 1995). As these sexual encounters become more realistic with the addition of immersive technology, how will the law deal with such verbal actions and symbolic violations? Words are deeds in virtual experiences (Turkle), making ethical issues a pressing concern. The anonymity afforded by cyberspace currently allows sexual deviants to act out with impunity (Wallace). Issues of imitation, disinhibition, and desensitization may become serious issues as sexual activity becomes an immersive, online option.

Ethical issues, such as marital fidelity, will also be experienced in virtual spaces. How will a person feel if their partner has virtual sex with an imaginary character, or with an avatar of a real person? Will betrayal and infidelity be experienced? What will happen when a man pretends to be a woman or a woman pretends to be a man? Do partners have a right to know the real sex of that person, particularly if they are seeking an online sexual encounter? Is this kind of deception permissible, or is it unethical? Will the online partner care if they are being deceived? Bots, that is, cyberspace robots, are also mistaken for real people. In one instance, a man made repeated sexual advances toward Julia, an intelligent computer agent (Turkle, 1995). As these agents become more humanlike in their interactions with people, the line between the robot and the real person will be increasingly difficult for people to discern.

Finally, online sexual predators have a new home to prey on unsuspecting people, particularly children (Thornburg & Linn, 2002). Parents are particularly worried about these potential online dangers as Internet interactions can be used to create avenues for real-life meetings or for virtual online meetings (Turow, 1999).

27.4.4 Social Interaction or Social Isolation?

Humans are social beings who often seek the contact of others. Will computer simulations and online interactions promote friendships and close ties with others, or will simulations lead to social isolation and addiction? Although many parents are concerned that Internet interactions will lead to social isolation and potentially disrupt family interactions (Turow, 1999), the impact of online Internet interactions is mixed.

Kraut and colleagues (1998) examined the social and psychological impact of initial Internet use for 73 households. People used the Internet primarily to communicate. As Internet use increased, people reported that they interacted less with their families, had fewer people in their social circles, and felt more depressed and lonely. Similarly, Slater et al. (1999) found that young adults enjoyed working in groups, felt less isolated, felt more comfortable, and felt other group members were more cooperative after participating in a real, rather than a VE, interaction; however, the design was flawed in that the real interaction always occurred after the virtual one. Sleek (1998a) argues that too much reliance on impersonal, weak, online interactions through e-mail correspondence and chat groups cannot replace the social support of real-life face-to-face interactions with family members and friends.

Subsequent research, however, revealed more positive effects of Internet use on the social well-being of adolescents and adults. Kraut and colleagues (2002), for instance, found increases in the size of social circles, increases in face-to-face interactions, increases in positive feelings, and increases in trust in other people for Internet users as time passed and the novelty effect of Internet experiences decreased. Consistent with these findings, adolescents demonstrated strong preferences for online applications that allowed them to interact with their friends (Gross, Juvonen, & Gable, 2002). Presumably, users became more skilled in how they used the Internet over time, gravitating toward communications with others over solitary online website explorations, which was aided by having applications such as instant messenger emerge that increasingly supported interactions with their friends (Subramanyam, Greenfield, Kraut, & Gross, 2002).

As a participant observer, Schroeder (1997) explored two virtual worlds, Alphaworld and Cybergate, as a human and a cartoon avatar, respectively. Schroeder found that players created social stratification within these worlds. Insiders often developed a distinctive persona such as being a prankster, a bully, or a helpful guide, whereas outsiders acted more like a tourist, roles that fit nicely with Jung's psychoanalytic theoretical approach. In Alphaworld, visitors or tourists literally wore a camera around their avatar's neck so that they were readily identifiable as outsiders. In these VEs, insiders tended to explore and to create spaces with their friends in the far reaches of these worlds, whereas outsiders were less engaged with others, observed more, and stayed in central locations where most of the avatars tended to be located. Overall, these two different groups had a different social status, marked by the distinctive roles and behaviors that separated them, and differential status emerged with varying levels of involvement and expertise in the virtual world.

Online VE interactions can eliminate the geographical constraints that disrupt human interactions by making friends and family members in faraway places accessible in everyday social life. Players can meet and interact in virtual simulations, enacting fantasies when their real worlds impede actual physical contact. The loneliness associated with a move to a new place may decrease when close friends are brought into one's symbolic space, even if it is not a face-to-face interaction. While the experience is not the same as being face-to-face in the real world, having your characters face-to-face in an imaginary setting may also strengthen ties. Texting on mobile phones also allows ongoing interactions with friends.

Online relationships can also lead to new friendships. For example, Parks and Roberts (1998) reported that 94% of a sample of 235 MOO participants, ranging in age from 13 to 74, had developed an ongoing personal relationship on a MOO. These included close friendships, Casual friendships, and romantic relationships. Although offline relationships were better developed than MOO relationships, MOO relationships offered the possibility of creating new face-to-face friendships. One third of these MOO relationships had led to real-life meetings.

In a survey of 2500 respondents, Katz and Aspden (1997) found that the Internet complements social connections. Friendships cultivated online often led to real-life meetings, family contact was enhanced, and a sense of community was fostered. There was no evidence that Internet users dropped out of real-life activities, such as religious, leisure, and community organizations. Eighty-eight percent of the respondents reported no change in the amount of time spent in face-to-face interactions or phone conversations with family and friends. Six percent of respondents spent more time interacting with or talking to family members, and 6% spent less time. Only 14% of the sample reported friendships created online, and 60% of those people had met at least one of their online friends. The authors estimate that about two million face-to-face meetings had taken place because of initial Internet interactions.

A major shift occurred in online interactions with the introduction of social networking sites like Facebook, MySpace, and LinkedIn. These sites emerged, in part, because of faster Internet connections and related applications that allowed youth to interact with their friends (Gola & Calvert, 2012). Rapidly adopted by adolescents and emerging adults, more than 73% of online US teens (ages 12–17) reported involvement in social networking sites (Lenhart, Purcell, Smith, & Zickuhr, 2010). A main reason that college students reported for using social networking sites was to stay in touch with friends, with females reporting 170 and males reporting 123 friends (Pempek, Yermolayeva, & Calvert, 2009). Connections were used to sustain friendships from the high school years as well as more recent friends from college (Pempek et al., 2009). Overall, social networking sites enabled interactions among friends, thereby enhancing friendship circles and peer relationships, including face-to-face meetings as well as following what friends who lived in other locations were doing.

Addiction to computer interaction is one reason that people may have fewer face-to-face interactions (Turkle, 1995). For example, some players choose to interact with text-based VEs from the time that they wake up until they fall asleep (Turkle). Windows on computers allow users to participate in work and in fantasy text-based MUD worlds simultaneously, moving between various applications throughout the day. Similarly, some college students described Facebook experiences as addictive (Pempek et al., 2009).

In certain instances, however, online experiences may be engaging rather than addictive. Consider a lesson from television. Addiction has often been used to describe those who cannot seem to quit looking at the television screen. Anderson and Lorch (1983), however, documented that such experiences actually involve an easy and engaged attentional state, known as attentional inertia, in which users get increasingly involved in the activity that they are doing. These activities can range from watching television to playing to reading a book. Moreover, attentional inertia has been documented from childhood through adulthood, making it a skill that all age groups use to sustain attention in an activity. Therefore, while some may describe rapt attention as addictive, it may just be a normal facet of attention: people become involved in interesting, responsive activities and environments. VEs are certainly one kind of intrinsically interesting environment for many users.

The answer to the complex questions about social isolation and addiction ultimately depends on how a player treats online experiences (Turkle, 1995). The most extreme case involves those who spend most of their time in computer simulations or online computer communications such as social networking sites that can now be accessed by mobile phones. If those computer simulations become a permanent substitute for face-to-face interactions and the person pursues dysfunctional behaviors, then addiction to the computer world and isolation from real social relationships can become serious issues. By contrast, if computer experiences are a temporary working space to deal with adverse circumstances and a supplementary way of interacting with friends—much like the use of telephones has historically served—the person may emerge healthier and more functional (Turkle).

In summary, while some studies found that users felt more lonely and isolated after engaging in Internet interactions, others found that the Internet does not substantially alter the amount of time spent in face-to-face interactions. In many instances, real face-to-face interactions occur because of initial or ongoing Internet contact. For most people, VEs become a recreational and social activity, taking up part, but not all, of their leisure time experiences. In the latter situation, VEs have the potential to satisfy needs by supplementing the everyday social experiences of real life, just as other media experiences have done in the past. Nevertheless, heavy use of online computer technologies has the potential to disrupt social relationships, just as heavy television use can.

27.4.5 PROSOCIAL INTERACTIONS

Just as media exposure can lead to antisocial activity, so too can exposure lead to constructive prosocial responses. Prosocial behavior involves socially constructive activities such as helping, sharing, cooperating, and acting creatively (Calvert, 1999a). There are numerous television studies demonstrating that prosocial television can enhance children's helpfulness, cooperation, empathic responsiveness, and delay of gratification (Huston & Wright, 1998). The same impact may occur for VE interactions.

Many VE games provide opportunities for players to interact with virtual actors, who can be either real or imaginary players. These games are often fantasies where the player assumes an archetypal role (e.g., hero or wizard) and then enters a VE simulation where other players assume the identities of other game characters. These games provide the opportunity for players to share an adventure where they work together and cooperate to overcome evil forces (i.e., the shadow). Characters can talk to each other and plan group strategies to win the game. Prosocial and antisocial (e.g., aggressive) behaviors coexist in these VEs as they sometimes do in real life. Through role-play, characters enact the behaviors of others, thereby helping the player to control and to incorporate those qualities into their own identities (Calvert, 2002). Self-efficacy should increase as users role-play these ancient and powerful images.

Even those who engage in aggressive activities in MUDs can demonstrate and be recipients of prosocial behavior. For instance, a seemingly incompetent, short, squatty avatar named Argyle, who was always in search of his socks, was befriended by more experienced MUD players who came to his aid by giving him weapons as well as tips on who the most vulnerable monsters were (Turkle, 1997).

Pearce (1998), who finds that women enjoy social interaction in computer games, created a non-violent, story-based VE game called *Virtual Adventures*. Players competed against other teams and cooperated within teams. Prosocial actions, such as rescuing the Loch Ness Monster's eggs from bounty hunters, were incorporated in the game. The players became story characters that worked together to solve a common pressing goal. Women rated this software more favorably than men did. Pearce's findings suggest that prosocial immersive programs can be entertaining to certain users.

Trust is an important aspect of offline friendships as it provides a basis for social relationships. With the increased use of online environments, researchers began to examine how they might create embodied conversational agents that players trust and find to be credible. In one such application, Ryokai and colleagues (2003) created the virtual character Sam and had her perform nonverbal (e.g., eye contact when talking to children) and verbal behaviors (e.g., telling slightly advanced stories with children) that are consistent with real-life actions with people. Preschool-aged girls interacted with Sam by telling stories in a toy castle. These interactions resulted in children treating Sam as if she were a real person. For instance, children took turns with Sam, talked to Sam, asked Sam questions, gave Sam advice about her stories, and looked directly at Sam as they took turns telling stories. Those who interacted with Sam told more advanced stories. The results supported Vygotsky's sociocultural theory, though in this instance, a virtual character provided the scaffolds for children through more advanced storytelling, which improved children's literacy skills (Ryokai et al., 2003). Similarly, when an embodied conversational real estate agent named Rea was programmed to engage in small talk during interactions, users found the character to be more reliable, competent, and intelligent, that is, more credible and trustworthy, than when she did not use small talk (Bickmore & Cassell, 2001). Put another way, social relationships with media characters can result in more trust, which can also lead to learning since children do not learn much from those that are not trustworthy and credible to them (Cassell & Bickmore, 2000).

The suspension of belief in which children treat embodied conversational agents as if they are real is consistent with children's interactions and enhanced learning from media characters like *Dora the Explorer* (Calvert, Strong, Jacobs, & Conger, 2007), a pattern that is also found in toddlers' emotionally tinged parasocial relationships and subsequent learning of mathematical skills from onscreen media characters (Calvert et al., in press; Gola et al., 2013). Taken together, the findings suggest that at least through the preschool years, children are somewhat forgiving of media characters' limited interaction skills as they treat these characters as if they are real peers who have intent and feelings.

VEs have also been effective in clinical arenas. For instance, those who have autism spectrum disorder have social impairments, including a poor understanding of a theory of mind, which includes inferences about the emotional feelings of others (Cheng, Moore, McGrath, & Fan, 2005). Those with autism are also uncomfortable looking into the eyes of another person, a cue to how others feel as well as a practice that leads to effective social interactions. Using avatars who displayed emotions in a VE, 8 of 10 males (ages 8–18 years) with autism were able to recognize the emotional displays of avatars in a collaborative VE (Cheng et al., 2005). Using a behavioral approach involving systematic desensitization in a VE, Difede and Hoffman (2002) were also successfully able to treat a woman who had posttraumatic distress by gradually introducing her to a realistic simulation of the feared environment that had traumatized her in real life.

A potential problem for VE players is that they may come to rely on others' fantasies rather than constructing their own. A constructed reality leaves little room for imagination (Mantovani, 1995). Heavy television viewing is associated with a decline in imaginative play and in creativity, but with increases in daydreaming (Valkenburg & van der Voort, 1994; van der Voort & Valkenburg, 1994). However, the kind of program being viewed is crucial in understanding how television impacts fantasy, imagination, and creativity. For instance, television programs that encourage imaginative activities lead to increases in creativity (Anderson, Huston, Schmitt, Linebarger, & Wright, 2001). VEs can also provide a creative space where children can play and experiment. For instance, children can become active participants within VEs created for use in museums and entertainment settings, thereby enhancing interactivity and learning through play (Rousseau, 2004).

Interactive VE experiences may impact a player's imaginative world differently than observational media have, with daydreams taking on a new life as they are developed and enacted in online computer experiences. Unstructured situations tend to enhance imaginative play more so than structured situations (Carpenter, Huston, & Holt, 1986). Therefore, there may be a difference in imaginative outcomes for those who enter relatively unstructured virtual spaces and create parts of that site when compared to those who just visit an unstructured site or for those who play structured VE games.

27.5 LIMITATIONS OF VIRTUAL ENVIRONMENTS

There are always limitations to any innovation and VE is no exception. Recurring media concerns such as the displacement of other activities have been documented. For example, some players spent 40 or more hours each week in text-based MUDs, an experience known as simulation overdose (Turkle, 1995). People who spend too much time in VEs may head back to real life because they decide that the real world offers more than a simulated one. For instance, a real pet may ultimately be more fulfilling than a virtual one, and a real peer may be more engaging than a virtual one. Media research in other areas indicates that displacement usually occurs when a new technology is introduced and is sustained only for activities that fill a similar need for people, such as entertaining them (Calvert, 1999a). Thus, it is likely that a novelty effect for VE experiences will eventually give way to a balanced use of the technology in daily life.

The distinction between what is real and what is pretend is another issue that will be faced. In the everyday world, children already have difficulty in telling the difference between a realistic, but fictional, event versus a real event portrayed on television (Huston & Wright, 1998). These issues are partly resolved though developmental changes in the way that children think, but the roles that television characters play can even confuse adults. For instance, people often expected actors like John Wayne to be like the characters that they played. Perceptions of fantasy and reality should be explored, particularly as virtual simulations become even more realistic in the future, as the enhanced perceptual qualities of VEs may increasingly blur the line between what is real and what is pretend.

27.6 SOCIAL IMPLICATIONS FOR THE FUTURE OF VIRTUAL ENVIRONMENTS

The information age continues to hold new options and challenges for the people who participate. What will social life be like in this coming age of realistic VE immersions and communications?

In VEs, the social behaviors that are played out in real life simply move to a new symbolic plane. The long history of media research can be used to predict that people will use VE to meet their needs, including their social ones, as they do with other media. Virtual experiences have become integrated into the daily fabric of life, just as the earlier experiences with technologies were. People will imitate the characters that they play and become desensitized to situations in which they overindulge themselves. But each new technology also adds its own unique imprint to how representational experiences will affect users.

VEs will allow people to participate in, and to exercise control over, online social interactions. Friends can be virtual as well as real, but the clearest trends favor preferences for social networking sites in which real friendships can be sustained. People can assume alternate personae, both realistic and fantasy driven, yet typically prefer to be themselves. As is true of the technologies that came before it, people can use VEs to enhance their real social lives, or they can use VEs to escape or to act out antisocial tendencies, augmenting the problems that they already have.

McLuhan (1964) told us that the medium was the message. VEs represent a major step forward, bringing people one step closer to a social world where the lines between the symbolic and the real are merged. Overall, people's actions in VEs parallel what they do in *real* environments, in part because they bring an ongoing thread of who they are to all experiences.

REFERENCES

Anderson, C. A., Gentile, D. A., & Buckley, K. E. (2007). *Violent video games effects on children and adolescents: Theory, research, and public policy.* New York, NY: Oxford University Press.

Anderson, D. R., Huston, A. C., Schmitt, K. L., Linebarger, D. L., & Wright, J. C. (2001). Early childhood television viewing and adolescent behavior. *Monographs of the Society for Research in Child Development, 68*(1), Serial No. 264, 36–67 & 119–134.

Anderson, D. R., & Lorch, E. P. (1983). Looking at television: Action or reaction? In J. Bryant & D. R. Anderson (Eds.), *Children's understanding of television: Research on attention and comprehension.* New York, NY: Academic Press.

Bandura, A. (1965). Influence of models' reinforcement contingencies on the acquisition and performance of imitative responses. *Journal of Personality and Social Psychology, 1,* 589–595.

Bandura, A. (1997). *Self-efficacy: The exercise of control.* New York, NY: W. H. Freeman.

Barfield, W., Zeltzer, D., Sheridan, T., & Slater, M. (1995). Presence and performance within virtual environments. In W. Barfield & T. A. Furness, III (Eds.), *Virtual environments and advanced interface design* (pp. 473–513). New York, NY: Oxford University Press.

Berlyne, D. (1960). *Conflict, arousal, and curiosity.* New York, NY: McGraw-Hill.

Bickmore, T., & Cassell, J. (2001). *Relational agents: A model and implementation of building user trust.* Proceedings of the SIGCHI Conference on Human Computer Interaction, pp. 393–403, Seattle, WA.

Bricken, W., & Coco, G. (1995). VEOS: The virtual environment operating shell. In W. Barfield & T. A. Furness, III (Eds.), *Virtual environments and advanced interface design* (pp. 102–142). New York, NY: Oxford University Press.

Bushman, B., Newman, K., Calvert, S., Downey, G., Dredze, M., Gottfredson, M., ... Webster, D. (2013, February). Youth Violence: What we need to know. *Report of the Subcommittee on Youth Violence of the Advisory Committee to the Social, Behavioral and Economic Sciences Directorate*, National Science Foundation, Arlington, VA.

Calvert, S. L. (1999a). *Children's journeys through the information age*. Boston, MA: McGraw-Hill.

Calvert, S. L. (1999b). The form of thought. In I. Sigel (Ed.), *Theoretical perspectives in the concept of representation* (pp. 453–470). Hillsdale, NJ: Lawrence Erlbaum Associates.

Calvert, S. L. (2002). Identity on the Internet. In S. L. Calvert, A. B. Jordan, & R. R. Cocking (Eds.), *Children in the digital age: Influences of electronic media on development*. Westport, CT: Praeger.

Calvert, S. L. (2004). Cognitive effects of video games. In J. Goldstein & J. Raessens (Eds.), *Handbook of computer game studies* (pp. 125-131). Cambridge, MA: MIT Press.

Calvert, S. L., Mahler, B. A., Zehnder, S. M., Jenkins, A., & Lee, M. (2003). Gender differences in preadolescent children's online interactions: Symbolic modes of self-presentation and self-expression. *Journal of Applied Developmental Psychology, 24*, 627–644.

Calvert, S. L., Richards, M., & Kent, C. (in press). Personalized interactive characters for toddlers' learning of seration from a video presentation. *Journal of Applied Developmental Psychology*.

Calvert, S. L., Strong, B., & Gallagher, L. (2005). Control as an engagement feature for young children's attention to, and learning of, computer content. *American Behavioral Scientist, 48*, 578–589.

Calvert, S. L., Strong, B. L., Jacobs, E. L., & Conger, E. E. (2007). Interaction and participation for young Hispanic and Caucasian children's learning of media content. *Media Psychology, 9*(2), 431–445.

Calvert, S. L., Strouse, G. A., Strong, B., Huffaker, D. A., & Lai, S. (2009). Preadolescent boys' and girls' virtual MUD play. *Journal of Applied Developmental Psychology, 30*, 250–264.

Calvert, S. L., & Tan, S. L. (1994). Impact of virtual reality on young adults' physiological arousal and aggressive thoughts: Interaction versus observation. *Journal of Applied Developmental Psychology, 15*, 125–139.

Carpenter, C. J., Huston, A. C, & Holt, W. (1986). Modification of preschool sex-typed behaviors by participation in adult-structured activities. *Sex Roles, 14*, 603–615.

Cassell, J., & Bickmore, T. (2000). External manifestations of trustworthiness in the interface. *Communications of the ACM, 43*(12), 50–56.

Cheng, Y., Moore, D., McGrath, P., & Fan, Y. (2005). Collaborative virtual environment technology for people with autism. *Proceedings of the Fifth International Conference on Advanced Learning Technologies*, Kaohsiung, Taiwan.

Chertoff, D., & Schatz, S. (2014). Beyond presence. In K. S. Hale & K. M. Stanney (Eds.), *Handbook of virtual environments: Design, implementation, and applications* (2nd ed., pp. 855–870). Boca Raton, FL: Taylor & Francis Group, Inc.

Clear, T., Cole, G., & Reisig, M. (2013). *American corrections* (10th ed.). Belmont, CA: Wadsworth, Cengage Learning.

Coie, J., & Dodge, K. (1998). Aggression and antisocial behavior. In W. Damon & N. Eisenberg (Eds.), *Handbook of child psychology: Vol. 3. Social, emotional, and personality development* (5th ed.). New York, NY: John Wiley & Sons.

Difede, J., & Hoffman, H. G. (2002). Virtual reality exposure therapy for World Trade Center post-traumatic stress disorder: A case report. *CyberPsychology and Behavior, 5*, 529–535.

Erikson, E. H. (1968). *Identity, youth, and crisis*. New York, NY: Norton.

Epley, S. (1993, May/June). A female's view: Sex in virtual reality. *CyberEdge, 15*.

Fischer, P., Kastenmuller, A., & Greitemeyer, T. (2010). Media violence and the self: The impact of personalized gaming characters in aggressive video games on aggressive behavior. *Journal of Experimental Child Psychology, 46*, 192–195.

Gerbner, G., Gross, L., Morgan, M., & Signorielli, N. (1994). Growing up with television: The cultivation perspective. In J. Bryant & D. Zillmann (Eds.), *Media effects: Advances in theory and research*. Hillsdale, NJ: Lawrence Erlbaum Associates.

Gerbner, G., Gross, L., Signorielli, N., & Morgan, M. (1986). *Television's mean world: Violence profile no. 14-15*. Philadelphia, PA: Annenberg School of Communications.

Gola, A. A., & Calvert, S. L. (2012). Children's and Adolescents' Internet Access, Use, and Online Behaviors. In Y. Zheng (Ed.), *Encyclopedia of cyber behavior*. Hershey, PA: IGI Global.

Gola, A. A., Richards, M., Lauricella, A., & Calvert, S. L. (2013). Building meaningful relationships between toddlers and media characters to teach early mathematical skills. *Media Psychology, 16*, 390–411.

Green, M., & Sun, H. (1995). Computer graphics modeling for virtual environments. In W. Barfield & T. A. Furness, III (Eds.), *Virtual environments and advanced design interface* (pp. 63–101). New York, NY: Oxford University Press.

Greenwood-Ericksen, A., Kennedy, R. C., & Stafford, S. (2014). Entertainment applications of virtual environments. In K. S. Hale & K. M. Stanney (Eds.), *Handbook of virtual environments: Design, Implementation, and applications* (pp. 1291–1318), Boca Raton, FL: Taylor & Francis Group, Inc.

Gross, E., Juvonen, J., & Gable, S. (2002). Online communication and well-being in early adolescence. *Journal of Social Issues, 58*, 75–90.

Hall, C, & Nordby, V. (1973). *A primer of Jungian psychology*. New York, NY: Mentor Books.

Harre, R. (1983). *Personal being*. Oxford, England: Blackwell.

Herz, J. C. (1997). *Joystick nation*. Boston, MA: Little, Brown.

Hoffner, C. (2008). Parasocial and online social relationships. In S. L. Calvert & B. J. Wilson, (Eds.), *The handbook of children, media, and development* (pp. 309–333). Malden, MA: Blackwell.

Huffaker, D. A., & Calvert, S. L. (2005). Gender, identity, and language use in teenage blogs. *Journal of Computer-Mediated Communication, 10*(2).

Huston, A. C., & Wright, J. C. (1998). Mass media and children's development. In W. Damon, I. Sigel, & K. A. Renninger (Eds.), *Handbook of child psychology: Vol. 4. Child psychology in practice* (5th ed.). New York, NY: John Wiley & Sons.

Jung, C. (1959). *The basic writings of C. G. Jung*. New York, NY: Modern Library.

Katz, J., & Aspden, P. (1997). A nation of strangers? *Communications of the ACM, 40*, 81–87.

Kraut, R., Kiesler, S., Boneva, B., Cummings, J., Helgeson, V., & Crawford, A. (2002). Internet paradox revisited. *Journal of Social Issues, 58*, 49–74.

Kraut, R., Patterson, M., Lundmark, V., Kiesler, S., Mukophadhya, T, & Scherlis, W. (1998). Internet paradox: A social technology that reduces social involvement and psychological well-being. *American Psychologist, 53*, 1017–1031.

Lauricella, A., Gola, A. A. H., & Calvert, S. L. (2011). Meaningful characters for toddlers learning from video. *Media Psychology, 14*, 216–232.

Lee, K. (2004). Presence, explicated. *Communication Theory, 14*, 27–50.

Lenhart, A., Purcell, K., Smith, A., & Zickuhr, K. (2010). Social media and mobile Internet use among teens and young adults. *Pew Internet and American Life Project, February 2010*. Retrieved from http://www.pewinternet.org/Reports/2010/Social-Media-and-Young-Adults.aspx

Mantovani, G. (1995). Virtual reality as a communication environment: Consensual hallucination, fiction, and possible selves. *Human Relations, 48*, 669–683.

McLuhan, H. M. (1964). *Understanding media: The extensions of man*. New York, NY: McGraw-Hill.

Miller, P. (2009). *Theories of development* (5th ed.). New York, NY: Worth Publishers.

Parks, M., & Roberts, L. (1998). Making MOOsic: The development of personal relationships online and a comparison to their offline counterparts. *Journal of Social and Personal Relationships, 15*, 517–537.

Pearce, C. (1998). Beyond shoot your friends: A call to arms to battle against violence. In C. Dodsworth (Ed.), *Digital illusion: Entertaining the future*. Reading, MA: Addison-Wesley.

Pempek, T., Yermolayeva, Y., & Calvert, S. L. (2009). College students social networking experiences on Facebook. *Journal of Applied Developmental Psychology, 30*, 227–238.

Pena, J., Hancock, J., & Merola, N. (2009). The priming effects of avatars in virtual settings. *Communication Research, 36*, 838–856.

Roussou, M. (2004). Learning by doing and learning through play: An exploration of interactivity in virtual environments for children. *ACM Computers in Entertainment, 2*, 1–23.

Rubin, A. (1994). Media uses and effects: A uses-and-gratifications perspective. In J. Bryant & D. Zillmann (Eds.), *Media effects: Advances in theory and research*. Hillsdale, NJ: Lawrence Erlbaum Associates.

Ryokai, K., Vaucelle, C., & Cassell, J. (2003). Virtual peers in storytelling and literacy learning. *Journal of Computer Assisted Learning, 19*(2), 195–208.

Schroeder, R. (1994b, June). *Worlds in cyberspace: A typology of virtual realities and their social contexts*. Paper presented at the BCS Displays Group Conference Applications of Virtual Reality, University of Leeds, Leeds, England.

Schroeder, R. (1996). *Possible worlds: The social dynamic of virtual reality technology*. Boulder, CO: Westview Press.

Schroeder, R. (1997). *Networked worlds: Social aspects of multi-user virtual reality technology [Online]. Sociological Research Online, 2*. Available from http://www.socresonline.org.uk/socresonline/2/4/5.html

Schroeder, R. (2008). Defining virtual worlds and virtual environments. *Journal of Virtual Worlds Research.* Accessed at http://jvwr-ojs-utexas.tdl.org/jvwr/index.php/jvwr/article/viewFile/294/248

Slater, M., Sadagic, M., Usoh, M., & Schroeder, R. (1999). *Small group behavior in a virtual and real environment: A comparative study [Online].* Available from http://www.cs.ucl.ac.uk/staff/m.slater/BTWorkshop

Sleek, S. (1998a). Isolation increases with Internet use. *American Psychological Association Monitor, 29,* 30–31.

Sleek, S. (1998b). New cyber toast: Here's MUD in your psyche. *American Psychological Association Monitor, 29,* 30.

Stuart, R. (1996). *The design of virtual environments.* New York, NY: McGraw-Hill.

Subramanyam, K., & Greenfield, P. M. (2004). Constructing sexuality and identity in an online chat room. *Journal of Applied Developmental Psychology, 25,* 651–666.

Subramanyam, K., Greenfield, P. M., Kraut, R., & Gross, E. (2002). The impact of computer use on children's and adolescent's development. In S. L. Calvert, A. B. Jordan, & R. R. Cocking (Eds.), *Children in the digital age: Influences of electronic media on development.* Westport, CT: Praeger.

Templeman, J., Page, R., & Denbrook, P. (2014). Avator control in virtual environments. In K. S. Hale & K. M. Stanney (Eds.), *Handbook of virtual environments: Design, implementation, and applications* (2nd ed., pp. 233–256). Boca Raton, FL: Taylor & Francis Group Inc.

Thornburgh, D., & Lin, H. S. (Eds.) and the Committee to Study Tools and Strategies for Protecting Kids from Pornography and their Applicability to Other Internet Content. (2002). *Youth, pornography, and the Internet.* Washington, DC: National Academy Press.

Turk, M. (2014). Gesture recognition. In K. S. Hale & K. M. Stanney (Eds.), *Handbook of virtual environments: Design, implementation, and applications* (2nd ed., pp. 211–232). Boca Raton, FL: Taylor & Francis Group, Inc.

Turkle, S. (1995). *Life on the screen: Identity in the age of the Internet.* New York, NY: Simon & Schuster.

Turkle, S. (1997). Constructions and reconstructions of self in Virtual Reality: Playing in MUDs. In S. Kiesler (Ed.), *Culture of the Internet.* Mahwah, NJ: Erlbaum.

Turow, J. (1999, May). *The Internet and the family: The view from parents, the view from the press.* Paper presented at the Annenberg Public Policy Center Conference on the Family and the Internet, National Press Club, Washington, DC.

Tynes, B., Reynolds, L., & Greenfield, P. M. (2004). Adolescence, race, and ethnicity on the Internet: A comparison of discourse in monitored vs unmonitored chat rooms. *Journal of Applied Developmental Psychology, 25,* 667–684.

UNODC (2009). The Tenth United Nations survey of crime trends and operations of criminal justice systems (Tenth CTS, 2005–2006). http://www.unodc.org/unodc/en/data-and-analysis/Tenth-United-Nations-Survey-on-Crime-Trends-and-the-Operations-of-Criminal-Justice-Systems.html

Valkenburg, P. M., & van der Voort, T. H. (1994). Influence of TV on daydreaming and creative imagination: A review of research. *Psychological Bulletin, 116,* 316–339.

Van der Voort, T. H., & Valkenburg, P. M. (1994). Television's impact on fantasy play: A review of research. *Developmental Review, 14,* 27–51.

Vygotsky, L. (1978). *Mind in society. The development of psychological processes.* Cambridge, MA: Harvard University Press.

Wallace, P. (1999). *The psychology of the Internet.* Cambridge, U.K.: Cambridge University Press.

Yee, N. (2006). The demographics, motivations, and derived experiences of users of massively-multiplayer online graphical environments. *Presence: Teleoperators and Virtual Environments, 15,* 309–329.

Yee, N., Bailenson, J., Urbanek, M., Chang, F., & Merget, D. (2007). The unbearable likeness of being digital: The persistence of nonverbal social norms in online virtual environments. *CyberPsychology and Behavior, 10,* 115–121. doi:10.1089/cpb.2006.9984.

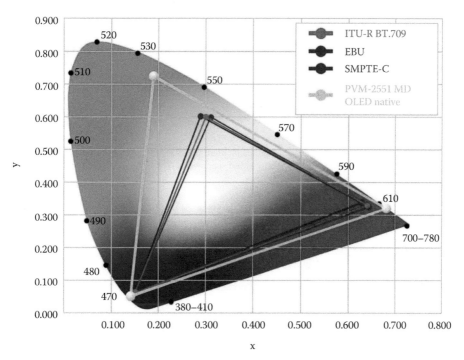

FIGURE 3.11 The CIE (1931) xy chromaticity space for normal human vision, with a 2° field, is depicted as the colored region, with the gamut of color available with the typical RGB monitor broadcast standards depicted as triangles. The perceptual locus of narrowband spectral lights is indicated by numbers specifying their wavelengths. (Adapted from Sony PVM-2551MD Medical OLED monitor catalog, 2011.)

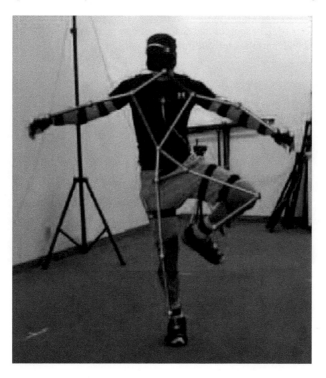

FIGURE 9.5 An optical motion capture system in use. (From Kirk, A.G. et al., Skeletal parameter estimation from optical motion capture data, *IEEE Conference on Computer Vision and Pattern Recognition*, Silver Spring, MD, 2005. With permission.)

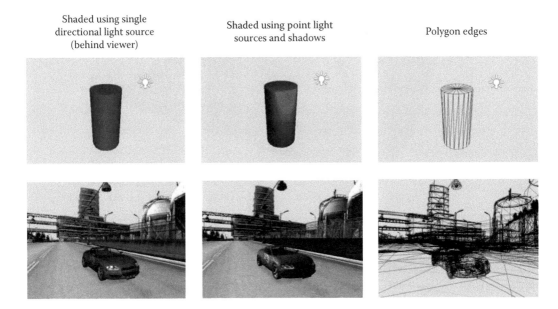

Shaded using single directional light source (behind viewer)

Shaded using point light sources and shadows

Polygon edges

FIGURE 11.3 The effects of light sources and shading. (From screenshots of the Unity Car Tutorial in the Unity Editor, reproduced with permission from Unity Technologies ApS.)

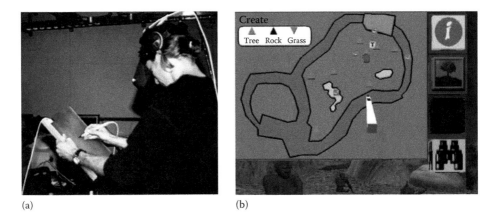

(a)

(b)

Create

Tree   Rock   Grass

FIGURE 12.1 Physical (a) and virtual (b) views of a pen-and-tablet system.

(a) (b)

FIGURE 15.5 (a and b) An example of a VE that gives the participant the ability to manipulate elements within the VE as well as navigate through them. This shows a virtual ATM that can be *racked out* to practice day-to-day replenishment procedures by selecting buttons, catches, and switches.

FIGURE 16.4 VESUB harbor navigation training.

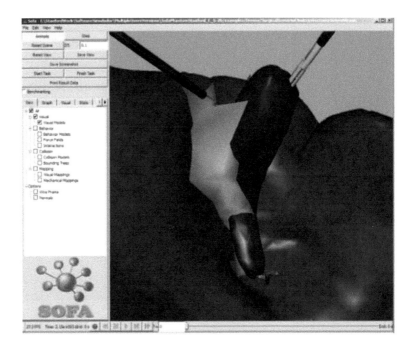

FIGURE 17.8 Multimodal interaction in distributed surgical simulators: cholecystectomy procedure.

FIGURE 23.5 The immersive optokinetic sphere: outside (left) and inside (right) views. (From Lawson, B.D., Exploiting the illusion of self-motion [vection] to achieve a feeling of "virtual acceleration" in an immersive display, in Stephanidis, C., Ed., *Proceedings of the 11th International Conference on Human-Computer Interaction*, Las Vegas, NV, pp. 1–10, 2005.)

FIGURE 28.4 Prototype of the BARS (see Livingston et al., 2002).

FIGURE 39.1 Expert versus trainee fixations.

(a)   (b)

FIGURE 41.2 (a) Middle school students exploring various textures and forces using virtual environment with haptic simulation. (b) A screenshot showing the simulated virtual environment.

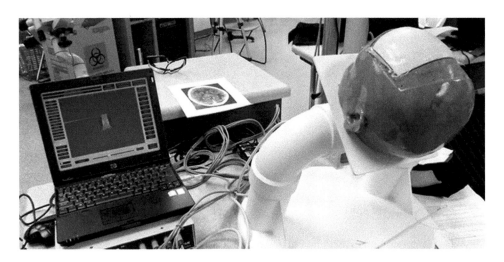

FIGURE 41.7 Picture of the UF Department of Neurosurgery experiential learning mixed-reality ventriculostomy simulator with haptic feedback.

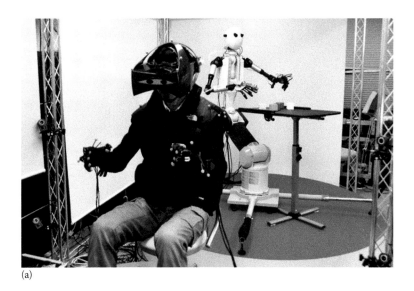

(a)

FIGURE 43.11 (a) TELESAR (tele-existence slave anthropomorphic robot) system. (Courtesy of Professor Tachi, Keio University, Tokyo, Japan.)

FIGURE 47.1 Immersive technologies (aka VR interfaces) enable thecreation of new methods of interaction useful for geologists to inspect and interact with their data. Here Gary Kinsland and a student from the University of Louisiana at Lafayette collaboratively explore the Chicxulub crater on a large immersive projection system. (See Section 47.4.1 for more details.)

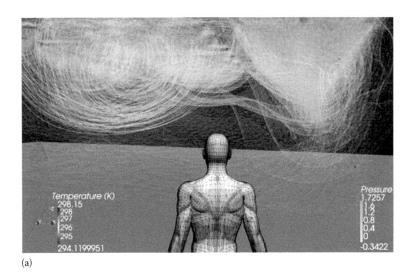

(a)

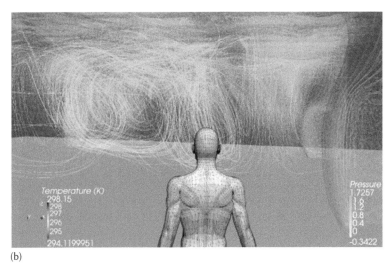

(b)

FIGURE 48.1 HVAC simulation for a green energy retrofit composed and rendered with ParaView, at (a) time step = 15 and (b) time step = 81. (Image: Drs. Burns, Borggaard, Herdman, Cliff, Polys [2012]; Courtesy of US Department of Energy, Washington, DC.)

Section VI

Evaluation

28 Usability Engineering of Virtual Environments

Joseph L. Gabbard

CONTENTS

28.1 INTRODUCTION

Usability engineering, described rather simplistically, is the process by which usability is ensured for an interactive application, at all phases in the development process. These phases include user task analysis, user class analysis, user interaction (UI) design, rapid prototyping, user-centered evaluation, and iterative redesign based on evaluation results. Usability engineering employs representative users and their interests *in all UI life cycle activities*, from early analyses to design, through various evaluations; it is not just applicable at the evaluation phase.

Prior to the first edition of this book, the terms *usability engineering* and *virtual environments* (VEs) were rarely, if ever, used in the same sentence or context. VE developers were at the time focusing largely on producing new hardware and gadgets, developing hybrid tracking methods, or designing *out of the box* UI techniques with too little attention given to how users will benefit (or not) from those gadgets and techniques. Admittedly, such exploration in a new realm such as VEs was and still is necessary. In the past decade or so, there has been growing interest in human–computer interaction (HCI) and usability engineering within the VE community, mainly due to

(1) increased awareness and acknowledgment of the value of user-centered approaches and (2) increased development of successful VE applications (afforded in part by mature hardware and VE application development environments) that needed good usability. These two trends (along with concurrent UX trends in the web development domain) underscore the need and demand for cost-effective usability engineering methods that are easy to understand and conduct and, most importantly, produce intuitive user interfaces that are usable by today's (increasingly growing) expectations of user experience.

Most extant usability engineering methods widely in current use were spawned by the development of traditional 2D graphical user interfaces (GUIs). So even when VE developers attempt to apply these usability engineering methods, most VE user interfaces are so radically different that well-proven techniques that have produced usable GUIs may be neither particularly appropriate nor effective for VEs without modification. In some cases, the general approach for methods may be followed, but the specific manner in which the methods are applied, or the logistics involved in successfully applying the methods, must be modified to avoid interference or degradation of data quality. As a result, applying extant usability engineering methods to VEs often presents HCI researchers and practitioners with interesting challenges. For example, how can we create accurate shared stereoscopic views for evaluators when only the user's head is tracked? How can an evaluator conduct a rich, formative evaluation in a fully immersive VE, such as the CAVE automatic virtual environment (CAVE) or VisCube™ without breaking a users' sense of presence? How do we employ the think-aloud protocol in VEs that uses voice recognition as an interaction technique?

Moreover, in terms of user-centered design (an important phase of usability engineering), few principles for design of VE user interfaces exist or have been empirically derived or validated. When designing interaction techniques to meet specific requirements, windows-based GUI, tablet app and web designers have a *toolbox* of well-known and established guidelines in which to leverage. For example, web designers' toolbox contains a back button, main navigation motifs, right-click content-sensitive menus, left-hand context-sensitive filters, popup windows, and so forth. This toolbox not only makes the design effort more straightforward but also generally affords a more usable interface, since users are familiar with these design elements and have preexisting expectations about how they work. In VE interfaces, the toolbox is not as well established, and in many cases, designers develop their own interaction technique by attempting to extend known techniques from 2D applications (in which cases extending can be tricky) or, in cases where there is no obvious mapping, create their own. The result is interaction techniques that may not be obvious to users (as they cannot leverage past experiences), may have a significant learning curve, and generally do not translate from one VE application experience to another. These challenges force evaluators to address goals at two levels of abstraction: improving the usability of the product and improving the processes by which we improve the products. This chapter focuses mainly on improving the product (i.e., the VE user interface) with a mention of process improvement where applicable. Ultimately researchers and developers of VEs should seek to improve VE applications, *from a user's perspective*—ensuring their usability—by following a systematic approach to VE development such as offered from both extant and emerging usability engineering methods.

To this end, we present several usability engineering methods, mostly adapted from GUI development, that have been successfully applied to VE development. These methods include user task analysis, expert guidelines-based evaluation (also sometimes called heuristic evaluation or usability inspection), and formative usability evaluation. We also present summative evaluation because it is an important aspect of making comparative assessments of VEs from a user's perspective (e.g., comparing a desktop-based VE to a CAVE-based VE). Further, we postulate that—like GUI development—there is no single method for VE usability engineering, and we address how each of these methodologies supports focused, specialized design, measurement, management, and assessment techniques such as those presented in other chapters of this handbook. We include our experiences with usability engineering of two very different VEs: a command-and-control map-based virtual reality (VR) application fielded on a variety of display platforms (ranging from a Responsive Workbench to a CAVE) and a mobile, outdoor augmented reality application designed to give urban warfighter real-time situational awareness.

28.2   SETTING THE CONTEXT FOR VE USABILITY ENGINEERING

Before we present specific usability engineering methods and give examples of their application, it is important to set the context for usability engineering. Developers of interactive systems—systems with user interfaces—often confuse the boundaries between the software engineering process and the usability engineering process. This is due at least in part to a lack of understanding of techniques for usability engineering, as well as which of these techniques is appropriate for use at which stages in the development process. Software engineering has a goal to improve software quality, but this goal, in and of itself, has little impact on *usability* of the resulting interactive system—in this case, a VE. For example, well-established *v&v* (validation and verification) techniques focus on software correctness, robustness, and so on, from a software developer's view, with little or no consideration of whether that software serves its users' needs. Thus, quality of the *user interface*—the usability—of an interactive system is largely independent of quality of the *software* for that system. Usability of the user interface is ensured by a user-centered focus on developing the *UI component*—the look and feel and behavior as a user interacts with an application. The UI component includes all 3D visually rendered items (i.e., models, textures, lighting), text, other graphics, audio, video, and devices through which a user communicates with an interactive system. The UI component is conceived, designed, and evaluated (with the help of others) by usability engineers—sometimes under the pseudo name *interaction designer*, *user experience designer*, and *usability evaluator*. Software engineers and systems engineers develop the *software component* of an interactive system—that is, the back-end components that support application performance—both from a user's perspective and a system performance perspective.

Cooperation between usability engineers, software engineers, and system engineers is essential if VEs are to mature toward a truly user-centric work and entertainment experience. Thus, both the interaction component and the software component are necessary for producing any interactive system, including a VE, but the *component that ensures usability is the UI component.*

28.3   CURRENT USABILITY ENGINEERING METHODS

Current research at Virginia Tech aims to address the applicability of established usability engineering methods in conjunction with development of new usability engineering methods. Simply put, our goal is to provide a methodology—or set of methodologies—to ensure usable and useful VE interfaces.

Usability engineering methods typically consist of one or more usability evaluation, or inspection, methods. Usability inspection is a process that aims to identify usability problems in UI design and, hence, end use (Mack & Nielsen, 1994).

As already mentioned, many of the techniques we present have been derived from usability engineering methods for GUIs. But from our own studies, as well as from collaboration with and experiences of other VE researchers and developers, we have adapted GUI methods and produced some new methods for usability engineering of VE UI design. We have made adaptations and enhancements to existing methods at two levels to evolve a usability engineering methodology applicable to VEs: specific methods themselves had to be altered and extended to account for the complex interactions inherent in multimodal VEs, and various methods had to be applied in a meaningful sequence to both streamline the usability engineering process as well as provide sufficient coverage of the usability space.

To better understand the strengths and applicability of each individual GUI usability engineering method, we present a basic discussion of each method to provide a brief overview of each. Following discussion of individual methods, we present benefits and insights gained from sequential application of these adapted methods for VE UI development.

28.3.1 User Task Analysis

A *user task analysis* is the process of identifying a complete description of tasks, subtasks, and actions required to use a system as well as other resources necessary for user(s) and the system to cooperatively perform tasks (Hartson & Pyla, 2012; Hix & Hartson, 1993).

User task analyses follow a formal methodology, describing and assessing performance demands of UI and application objects. These demands are, in turn, compared with known human cognitive and physical capabilities and limitations, resulting in an understanding of the performance requirements of end users. User task analysis (Hackos & Redish, 1998) may be derived from several components of early systems analysis and, at the highest level, rely on an understanding of several physical and cognitive components. User task analyses are the culmination of insights gained through an understanding of user, organizational, and social workflow; needs analysis; and user modeling.

There are four generally accepted techniques for performing user task analysis: documentation review, questionnaire survey, interviewing, and observation (Eberts, 1999). Documentation review seeks to identify task characteristics as derived from technical specifications, existing components, or previous legacy systems. Questionnaire surveys are generally used to help evaluate interfaces that are already in use or have some operational component. In these cases, task-related information can be obtained by having domain experts such as existing users, trainers, or designers complete carefully designed surveys. Interviewing an existing or identified user base, along with domain experts and application *visionaries*, provides very useful insight into what users need and expect from an application. Observation-based analysis, on the other hand, requires a UI prototype, resembling more the formative evaluation process than development of user task analysis, and, as such, is used as a last resort. A combination of early analysis of application documentation and domain-expert and user interviewing typically provides the most useful and plentiful task analysis.

While user task analyses are typically performed early in the development process, it should be noted that—like all aspects of UI development—task analyses also need to be flexible and potentially iterative, allowing for modifications to performance and UI requirements during any stage of development. However, major changes to user task analysis during late stages of development can derail an otherwise effective development effort and, as such, should only be considered under dire circumstances.

User task analysis generates critical information used throughout all stages of the application development life cycle. One such result is a top-down decomposition of detailed task descriptions. These descriptions serve, among other things, as an enumeration of desired functionality for designers and evaluators. Equally revealing results of task analysis include an understanding of required task sequences as well as sequence semantics. Thus, results of task analysis include not only identification and description of tasks but ordering, relationships, and interdependencies among user tasks. These structured analytical results set the stage for other products of task analysis, including an understanding of information flow as users work through various task structures.

Another useful result of task analysis is indications of where and how users contribute information to, and are required to make decisions that influence, user task sequencing. This information, in turn, can help designers identify what part(s) of the tasking process can be automated by computer (one of the original and still popular services a computer may provide) affording a more productive and useful work environment.

Without a clear understanding of user task requirements, both evaluators and developers are forced to *best guess* or interpret desired functionality that inevitably leads to poor interaction design. Indeed, both UI and user interface software developers claim that poor, incomplete, or missing task analysis is one of the most common causes of both poor software and product design.

28.3.2 Expert Guidelines-Based Evaluation

Expert guidelines-based evaluation or *heuristic evaluation* or *usability inspection* aims to identify potential usability problems by comparing a UI design—either existing or evolving—to established usability design guidelines. The identified problems are then used to derive recommendations for improving interaction design. The method is used by usability experts to identify critical usability problems early in the development cycle so that design issues can be addressed as part of the iterative design process (Nielsen, 1994).

Expert guidelines-based evaluations rely on established usability guidelines to establish whether a UI design supports intuitive user task performance (i.e., usability). Nielsen (1994) recommends three to five evaluators for a heuristic evaluation since fewer evaluators generally cannot identify enough problems to warrant the expense and more evaluators produce diminishing results at higher costs. It is not clear whether this recommendation is cost-effective for VEs since more complex VE interaction designs may require more evaluators than GUIs. Each evaluator first inspects the design alone, independently of other evaluators' findings. Results are then combined, documented, and assessed as evaluators communicate and analyze both common and conflicting usability findings.

A heuristic evaluation session may last 1–2 h or even more depending upon the complexity of the design. Again, VE interaction designs may require more time to fully explore. Further, heuristic evaluation can be done using a simple pencil and paper design (assuming that the design is mature enough to represent a reasonable amount of interaction components and interactions). This allows assessment to begin very early in the application development life cycle.

The output from an expert guidelines-based evaluation should not only identify problematic interaction components and interaction techniques but should also indicate *why* a particular component or technique is problematic. Results from heuristic evaluation are subsequently used to remedy obvious and critical usability problems as well as to shape the design of subsequent formative evaluations. Evaluation results further serve as both a working instructional document for user interface software developers and, more importantly, a fundamentally sound, research-backed, design rationale.

Given that expert guidelines-based evaluations are based largely on a set of usability heuristics, it can be argued that the evaluations are only as effective and reliable as the guidelines themselves. Nielsen (1993) and Nielsen (1994b), respectively, present the original and revised set of usability heuristics for traditional GUIs. While these heuristics are considered to be the de facto standard for GUIs, we have found that they are too general, ambiguous, and high level for effective and practical heuristic evaluation of VEs. Effectiveness is questioned on the simple fact that 3D, immersive, VE interfaces are much more complex than traditional GUIs. The original heuristics (implicitly) assume traditional input/output devices such as keyboard, mouse, and monitor and do not address the appropriateness of VE devices such as CAVEs, HMDs, haptic and tactile gloves, and various force feedback devices. Determining appropriate VE devices for a specific application and its user tasks is critical to designing usable VEs. Sutcliffe and Gault (2004) present an extended version of Nielsen's heuristics that are applicable for VEs that provide applicable guidelines to assist in rapid and reliable expert guidelines-based VE evaluations.

It is well recognized that VE interfaces and VE UI are immature and currently emerging technologies for which standard sets of design, much less usability guidelines, do not yet exist. However, our recent research at Virginia Tech has produced a set of VE usability design guidelines, contained within a taxonomy of usability characteristics (Gabbard, 1997). This taxonomy document (see http://filebox.vt.edu/users/jgabbard/Gabbard-Taxonomy.pdf) provides a reasonable starting point for heuristic evaluation of VEs. The complete document contains several associated usability resources including specific usability guidelines, detailed context-driven discussion of the numerous guidelines, and citations of additional references.

The taxonomy organizes VE UI design guidelines and the related context-driven discussion into four major areas: users and user tasks, input mechanisms, virtual model, and presentation mechanisms. The taxonomy categorizes 195 guidelines covering many aspects of VEs that affect usability including navigation, object selection and manipulation, user goals, fidelity of imagery, input device modes and usage, and interaction metaphors. The guidelines presented within the taxonomy document are well suited for performing heuristic evaluation of VE UI, since they both provide broad coverage of VE interaction/interfaces and are specific enough for practical application. For example, with respect to navigation within VEs, one guideline reads "provide information so that users can always answer the questions: Where am I now? What is my current attitude and orientation? Where do I want to go? How do I travel there?" Another guideline addresses methods to aid in usable object selection techniques stating "use transparency to avoid occlusion during selection."

We have successfully used these guidelines within context as a basis for expert evaluation of several VEs—we describe in detail two examples in Sections 28.5 and 28.6.

28.3.3 FORMATIVE USABILITY EVALUATION

The term *formative evaluation* was coined by Scriven (1967) to define a type of evaluation that is applied during evolving or formative stages of design. Scriven used this in the educational domain for instructional design. Williges (1984) and Hix and Hartson (1993) extended and refined the concept of formative evaluation for the HCI domain. In the past 10 years, numerous VE researchers have embraced the formative evaluation approach as a means of gathering critical empirical evidence regarding the usability of a VE. Some examples can be found in Rivera-Gutierrez et al. (2012); Dalgarno, Bishop, and Bedgood (2012); Komlodi, Józsa, Hercegfi, Kucsora, and Borics (2011); Seidel, Gartner, Froschauer, Berger, and Merkl (2010); Lewis, Deutsch, and Burdea (2006); Wang and Dunston (2006); Wingrave, Haciahmetogl, and Bowman (2006); and Swan, Gabbard, Hix, Schulman, and Ki (2003).

The goal of formative evaluation is to assess, refine, and improve UI by iteratively placing representative users in task-based scenarios in order to identify usability problems as well as to assess the design's ability to support user exploration, learning, and task performance (Hartson & Pyla, 2012). Formative usability evaluation is an observational evaluation method, which ensures usability of interactive systems by including users early and continually throughout user interface development. The method relies heavily on usage context (e.g., user task, user motivation) as well as a solid understanding of HCI (and in the case of VEs, human–VE interaction) and, as such, requires the use of usability experts.

While the formative evaluation process was initially intended to support iterative development of instructional materials, it has proven itself to be a useful tool for evaluation of traditional GUIs. Moreover, in the past few years, we have seen first-hand evidence indicating that the formative evaluation process is also an efficient and effective method of improving the usability of VE interfaces (Hix, Swan, Gabbard, McGee, Durbin, & King, 1999a).

The steps of a typical formative evaluation cycle (Figure 28.1) begin with development of user task scenarios (1) and are specifically designed to exploit and explore all identified task, information, and work flows. Representative users perform these tasks as they explicitly verbalize their thoughts, actions, goals, etc., using the think-aloud protocol (2). Evaluators collect (3) both qualitative and quantitative data. These data are then analyzed to identify UI components or features that both support and detract from user task performance. These observations are in turn used to suggest UI design changes (4) as well as formative evaluation scenario and observation (re)design (5).

The formative evaluation process itself is iterative, allowing evaluators to continually refine user task scenarios in order to fine-tune both the UI *and* the evaluation process. Contrary to popular belief, the formative evolution process produces both qualitative and quantitative results collected from representative users during their performance of task scenarios (del Galdo, Williges, Williges, & Wixon, 1986). One type of qualitative data collected is termed critical incidents (del Galdo et al., 1986; Hartson & Pyla, 2012; Hix & Hartson, 1993). A critical incident is a user event that has a significant impact, either positive or negative, on users' task performance and/or satisfaction (e.g., a system crash or error, being unable to complete a task scenario, user confusion). Critical incidents that have a negative effect on users' work flow can drastically impede usability and may even have a dramatic effect on users' perceptions of application quality, usefulness, and reputation. As such, any obvious critical incidents are best discovered during formative evaluation phases as opposed to consumers' desktops.

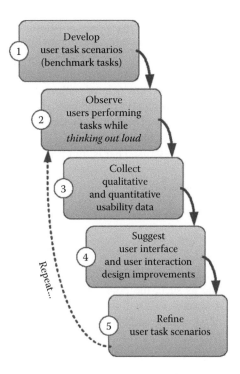

FIGURE 28.1 Basic components of a formative evaluation used for VEs.

Equally important are the quantitative data collected during formative evaluation. These data include measures such as how long it takes a user to perform a given task and the number of errors encountered during task performance. Collected quantitative data are then compared to appropriate baseline metrics, sometimes initially redefining or altering evaluators' perceptions of what should be considered baseline. Both qualitative and quantitative data are equally important since they each provide unique insight into an interaction design's strengths and weaknesses.

28.3.4 SUMMATIVE EVALUATION

Summative evaluation, in contrast to formative evaluation, is typically performed after a product or design is more or less complete; its purpose is to statistically compare several different systems, for example, to determine which one is *better*—where better is defined in advance. Another goal of summative evaluation is to measure and subsequently compare the productivity and cost benefits associated with various designs. In this fashion, evaluators are simply comparing the best of a few refined designs to determine which of the *finalists* is best suited for delivery.

The term *summative evaluation* was also coined by Scriven (1967), again for use in the instructional design field. As with the formative evaluation process, HCI researchers (e.g., Williges, 1984) have applied the theory and practice of summative evaluation to interaction design with surpassingly successful results.

In practice, summative evaluation can take on many forms. The most common are the comparative, field trial and, more recently, the expert review (Stevens, Frances, & Sharp, 1997). While both the field trial and expert review methods are well suited for instructional content and design assessment, they typically involve assessment of single prototypes or field-delivered designs. In the context of VE design, we are mostly interested in assessing the quality of two or more UI designs and, as such, have focused on the comparative approach. Our experiences have found that this approach is very effective for analyzing the strengths and weaknesses of various well-formed, completed designs using representative user scenarios.

Stevens et al. (1997) present a short list of questions that summative evaluation should address. We have modified these questions to address summative, comparative evaluation of VE user interfaces. The questions include the following:

- What are the strengths and weaknesses associated with each UI design?
- To what extent does each UI design support overall user and system goal(s)?
- Did users perceive increased utility and benefit from each design? In what ways?
- What components of each design were most effective?
- Was the UI evaluation effort successful? That is, did it provide a cost-effective means of improving design and usability?
- Were the results worth the program's cost?

28.3.5 Successful Progression

As previously mentioned, one of our long-term research goals is to produce methodologies to improve the usability of VE UI designs. More specifically, the goal is to develop, modify, and fine-tune usability engineering techniques specifically for VEs. Techniques to aid in usability engineering of VEs will mature, and as such, the potential for conducting effective evaluations and delivering subsequent usable and useful VEs will increase.

Our current efforts are focusing on the combination of usability evaluation techniques described in Sections 28.3.1 through 28.3.4. As depicted in Figure 28.2, our applied research over the past several years has shown that, at a high level, progressing from user task analysis to expert guidelines-based evaluation to formative evaluation to summative evaluations is an efficient and cost-effective strategy for designing, assessing, and improving the UI—usability engineering—of a VE.

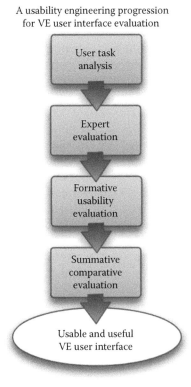

FIGURE 28.2 Successful progression of evaluation methods that are iteratively used to refine and improve a VE user interface.

One of the strengths of this progression is the fact that it exploits a natural ordering or evolution of interaction design and prototyping, with each method generating a streamlined set of information the next method utilizes. In this sense, each method is able to generate a much better starting point for subsequent methods than when applied in a stand-alone fashion. Moreover, simply applying more than one usability engineering method ensures more complete coverage of an interaction design's *usability space*, each revealing its niche of particular usability problems, collectively shaping a more usable VE. And finally, the progression of methods also produces a *paper trail* of persistent documentation that may serve as documented design rationale.

This progression is very cost-effective for assessing and improving VEs. For example, summative studies are often performed on VE interaction designs that have had little or no task analysis or expert guidelines-based or formative evaluation. This may result in a situation where the expensive summative evaluation is essentially comparing *good apples* to *bad oranges*. Specifically, a summative study of two different VEs may be comparing one design that is inherently better, in terms of usability, than the other one. When all designs in a summative study have been developed following our suggested progression of usability engineering, then the comparison is more valid. Experimenters will then know that the interaction designs are basically equivalent in terms of their usability, and any differences found among compared designs are, in fact, due to variations in the fundamental nature of the designs, and not their usability.

28.4 APPLICATION OF USABILITY ENGINEERING METHODS

Research and development performed both in our own labs and in other VE labs has shown the usability engineering techniques just described to be effective for ensuring usability. The following case studies present the application of these methods to interaction design of two different VE applications. The first, called Dragon, is a true VR/VE military command-and-control application developed for the Responsive Workbench—which is a table-based stereoscopic immersive display. The second, called Battlefield Augmented Reality System (BARS), is a wearable, outdoor mobile augmented reality system designed to give real-time situational awareness to individuals in dynamic urban settings. We have chosen depth over breadth in these case studies to give a detailed discussion of how our work progressed from an initial task analysis through various forms of evaluations. These case studies are different from each other in that they employ (for the most part) very different technologies and afford very different user experiences in very different usage contexts. While applying the same usability engineering techniques, we encountered very different challenges. The results for both were improved UIs and lessons learned on the process.

28.5 CASE STUDY 1: DRAGON

Personnel at the Naval Research Laboratory's VR Laboratory, again in collaboration with Virginia Tech researchers, have developed a VE for battlefield visualization, called Dragon (Figure 28.3; Hix et al., 1999a). Implemented on a Responsive Workbench, Dragon's metaphor for visualizing and interacting with 3D computer-generated scenery uses a familiar tabletop environment. Applications in which several users collaborate around a work area, such as a table, are excellent candidates for the Workbench. This metaphor is especially familiar to naval personnel and marines, who have, for decades, accomplished traditional battlefield visualization on a tabletop. Paper maps of a battlespace are placed under sheets of acetate. As intelligence reports arrive from the field, technicians use grease pencils to mark new information on the acetate. Commanders then draw on the acetate to plan and direct various battlefield situations. Historically, before high-resolution paper maps, these operations were performed on a sand table, literally a box filled with sand shaped to replicate battlespace terrain. Commanders moved around small physical replicas of battlefield objects to direct battlefield maneuvers. The fast-changing modern battlefield

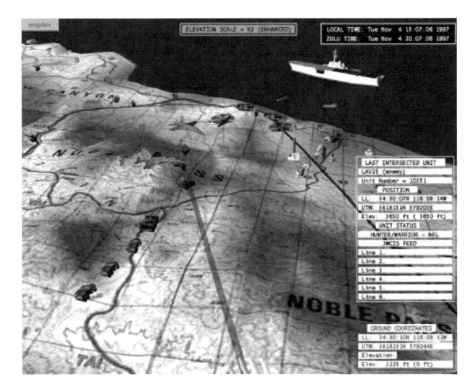

FIGURE 28.3 A screen capture from the Dragon command-and-control VE. In this view, users see a 3D terrain map; models of various air, land, and sea assets; and text-based information overlaid in moveable 2D overlay boxes. The line emanating from the avatar hand (bottom center of image) aids users in selecting objects using a ray casting selection technique.

produces so much time-critical information that these cumbersome, time-consuming methods are inadequate for effectively visualizing and commanding a modern-day battlespace.

Dragon was developed on the Responsive Workbench to give a 3D display for observing and managing battlefield information shared among technicians and commanders. Visualized information includes a high-resolution terrain map; entities representing friendly, enemy, unknown, and neutral units; and symbology representing other features such as obstructions or key map points. Users can navigate to observe the map and entities from any angle and orientation around the Workbench.

Dragon's early development was based on an admittedly cursory user task analysis, which drove early design. This was followed, however, by numerous cycles of expert guidelines-based evaluation as well as formative evaluations. Early in Dragon design, we produced and assessed three general interaction methods for the Workbench, any of which could have been used to interact with Dragon: hand gestures using a pinch glove (Obeysekare et al., 1996), speech recognition, and a handheld flight stick. Although it was an interesting possibility for VE interaction, our formative evaluations found that speech recognition is still too immature for battlefield visualization. We further found the pinch glove to be too fragile and time consuming to pass from user to user around the Workbench. It also worked best for right-handed users whose hands were approximately the same size. In contrast, our formative evaluations revealed that the flight stick was robust, easily handed from user to user, and worked for both right- and left-handed users.

Based on these formative evaluations, we modified a three-button game flight stick by removing its base and placing a six-degree-of-freedom position sensor inside. Our initial designs used a laser pointer metaphor in which a laser beam appeared to come out the *head* of the flight stick as a user pointed it toward the VE map. When a beam intersected terrain or an object, a highlight marker appeared.

In early demonstrations of initial versions of Dragon in real military exercises for battlefield planning, users indicated they found Dragon's accurate and current visualization of the battlespace to be more effective and efficient than the traditional method of maps, acetate, and grease pencils. Following these successful demonstrations and positive feedback, we began intensive usability engineering of Dragon's UI design.

Also during our early demonstrations and evaluations, we observed that navigation—how users manipulate their viewpoint to move from place to place in a virtual world (in this case, the battlefield map)—profoundly affects all other user tasks. If a user cannot successfully navigate to move about in a virtual world, then other user tasks such as selecting an object or grouping objects cannot be performed. A user cannot query an object if the user cannot navigate through the virtual world to get to that object. Although we performed a user task analysis before our guidelines-based and formative evaluations, these evaluations supported our expectations of the importance of navigation.

28.5.1 DRAGON EXPERT EVALUATION

Expert guidelines-based evaluation was done extensively for Dragon, prior to much formative evaluation. However, we were developing the taxonomy of usability characteristics of VEs, and at the same time, we were performing the guidelines-based evaluation, and so the guidelines we used were much more ad hoc and less structured than those that eventually became the taxonomy. Nonetheless, even our informal inspection of the evolving Dragon interaction design provided tremendous feedback in what worked and what did not, especially for use of the wand and other aspects of navigation in Workbench VE applications in general and Dragon in particular. During these evaluations, VE UI design experts worked alone or collectively to assess the Dragon interaction design. In our earliest evaluations, the experts did not follow specific user task scenarios per se but engaged simply in *free play* with Dragon using the wand. All experts knew enough about the purpose of Dragon as a battlefield visualization VE to explore the kinds of tasks that would be more important for Dragon users. During each session, one person was typically *the driver*, holding the flight stick and generally deciding what and how to explore in the application. One and sometimes two other experts were observing and commenting as the *driver* worked, with much discussion during each session.

While typically used in formative and other types of user evaluations, we employed the think-aloud protocol (Hartson & Pyla, 2012) during these expert evaluations. This common and effective technique for generating important qualitative data during usability evaluation sessions encourages participants (or, in this case, evaluators) to talk about their actions, goals, expectations, strategies, biases, and general thoughts regarding the VE interface while they are performing tasks. For Dragon, we used the codiscovery variant of the think-aloud protocol (Kennedy, 1989; O'Malley, Draper, & Riley, 1984), which allows multiple participants to interact with the VE while thinking aloud and affords a more natural conversations between participants (evaluators) resulting in more complex and thorough verbalizations of usability issues.

Care should be taken when employing think aloud since some VEs (but not Dragon) use voice recognition as an interaction technique, which can render the think-aloud protocol much more difficult if not impossible to employ. In these cases, postexperiment interviews can be used to explore information that would have been otherwise obtained via the think-aloud protocol.

Major design problems that were uncovered in our expert guidelines-based evaluation of Dragon included poor mapping of navigation tasks (i.e., pan, zoom, pitch, heading) to flight stick buttons, missing functionality (e.g., exocentric rotate, terrain following), problems with damping of map movement in response to flight stick movement, and graphical and textual feedback to the user about the current navigation task. After these evaluations had revealed and remedied as many design flaws as possible, we moved on to formative evaluation.

We used the basic Dragon application to perform extensive evaluations, using anywhere from one to three users for each cycle of evaluation. From a single evaluation session, we often uncovered design problems so serious that it was pointless to have a different user attempt to perform scenarios with the same design. So we would iterate the design, based on our observations, and begin a new cycle of evaluation. We went through four major cycles of iteration in all.

28.5.2   DRAGON FORMATIVE EVALUATION

In designing our scenarios for formative evaluation, we carefully considered coverage of specific usability issues related to navigation. For example, some of the tasks exploited an egocentric (user moves through the virtual world) navigation metaphor, while others exploited an exocentric (user moved the world) navigation metaphor. Some scenarios exercised various navigation tasks (e.g., degrees of freedom: pan, zoom, rotate, heading, pitch, roll) in the virtual map world. Other scenarios served as primed exploration or nontargeted searches for specific features or objects in the virtual world. Still others were design to evaluation rate control versus position control.

During each of six formative evaluation sessions, we first asked the participant to play with the flight stick to figure out which button activated which navigation task. We timed each user as they attempted to determine this and took notes on comments and any critical incidents that occurred. Once a user had successfully figured out how to use the flight stick, we began having them formally perform the task scenarios. Only one user was unable to figure out the flight stick in less than 15 min; we told this user details they had not yet discovered and proceeded with the scenarios.

Time to perform the set of scenarios ranged from about 20 min to more than 1 h. We timed user performance of individual tasks and scenarios and counted errors made during task performance. A typical error was moving the flight stick in the wrong direction for the particular navigation metaphor (exocentric or egocentric) that was currently in use. Other errors involved simply not being able to maneuver the map (e.g., to rotate it) and persistent problems with mapping navigation tasks (degrees of freedom) to flight stick buttons, despite our extensive prior evaluations to minimize this issue. During each formative evaluation session, we had at least two and often three evaluators present. One served as the facilitator to interact with the participant and keep the session moving; the other one or two evaluators recorded times, counted errors, and collected critical incidents and other qualitative data. While these sessions seem personnel intensive, with two or three evaluators involved, we found that the quality and quantity of data collected by multiple evaluators greatly outweighed the cost of those evaluators. A surprising amount of our effort was spent on mapping flight stick buttons to navigation tasks (pan, zoom, rotate, heading, pitch, roll), but we found it paid off with more effective, intuitive mappings.

As mentioned earlier, we went through four major iterations of the Dragon interaction design, based on our evaluations. The first iteration, the *virtual sand table*, was an egocentric navigation metaphor based on the sand table concept briefly described earlier. This was the version demonstrated in the military exercises also mentioned previously. A key finding of this iteration was that users wanted a terrain-following capability, allowing them to *fly* over the map—an egocentric design. Map-based navigation worked well when globally manipulating the environment and conducting operations on large-scale units. However, for small-scale operations, users wanted this *fly* capability to visually size up terrain features, entity placement, fields of fire, lines of sight, etc.

The second iteration, *point and go*, used the taxonomy of usability characteristics of VEs to suggest various possibilities for an egocentric navigation metaphor design. This metaphor attempted to avoid having different modes (and flight sick buttons) for different navigation tasks because of known usability problems with moded interaction. Further, we based this decision on how a person often navigates to an object or location in the real world; namely, they point (or look) and go (move) there. Our reasoning was that adopting this same idea to egocentric navigation would simplify the design and at least loosely mimic the real world. So in this design, a user simply pointed the flight stick toward a location or object of interest and pressed the trigger to fly there. We found that the single gesture to move about was not powerful enough to support the diverse, complicated variety of navigation tasks inherent in Dragon. Furthermore, a single gesture meant that all degrees of freedom were controlled by that single gesture. This resulted in, for example, unintentional rolling when a user only wanted to pan or zoom. Essentially we observed a control versus convenience trade-off. Many navigation tasks (modes) were active simultaneously, which were convenient but difficult to physically control for a user. With separate tasks (modes), there was less convenience, but physical control was easier because degrees of freedom were more limited in each mode. In addition to these serious problems, we found that users wanted to rotate around an object, such as to move completely around a tank. This indicated that Dragon needed an exocentric rotate ability, which we added. This interesting finding showed that neither a pure egocentric nor a pure exocentric metaphor was desirable; each metaphor has aspects that are more or less useful depending on user goals. While this may seem obvious, it was confirmed by users through our formative evaluation approach. Further, the somewhat poor performance of what we thought was the natural *point and go* metaphor was rather counterintuitive and further demonstrates that mimicking VE design after the real world is not always successful from the VE user's perspective.

The third iteration, *modal*, went from the extreme of all navigation tasks coupled on a single button as in the previous iteration to a rather opposite design in which each navigation task was a separate mode. Specifically, as a user clicked the left or right flight stick button, Dragon cycled successively through the tasks of pan, zoom, pitch, heading, and exocentric rotate. A small textual indicator was displayed on the Workbench to show the current mode. Once a user had cycled to the desired task, the user simply moved the flight stick, and that task was enabled without the need to push any flight stick button. We observed that, as we expected, it was very cumbersome for users to always have to cycle between modes, and it was obvious that we still had not achieved a compromise between convenience and control.

In our fourth iteration of the Dragon interaction design, *integrated navigation*, based on the taxonomy of usability characteristics of VEs and our own user observations of what degrees of freedom could be logically coupled in the Dragon application, we produced a hybrid design of the modeless/moded designs of prior iterations. Specifically, we coupled pan and zoom onto the flight stick trigger, pitch and heading onto the left flight stick button, and exocentric rotate and zoom onto the right flight stick button. This fourth generation interaction design for Dragon finally achieved the desired convenience versus control compromise. In our final formative evaluation studies, we found that at last, we had a design for navigation that seemed to work well. The only usability problem we observed was minor: damping of map movement was too great and needed some adjustment, which we made.

28.5.3 DRAGON SUMMATIVE EVALUATION

As the formative evaluation cycles neared completion for Dragon, we began planning summative studies for our navigation design. Design parameters that affect usability of the VE navigation metaphor to be studied in our summative evaluations are those that, in general, could not be decided from the formative evaluations. We identified 27 such design parameters, shown in Table 28.1, that potentially affect the usability of user navigation. The organization of these parameters is based on the taxonomy of usability characteristics of VEs. The four main areas of this taxonomy are the bold headings in Table 28.1, and our design parameters for navigation are grouped based on these areas.

Based on the taxonomy of usability characteristics, observations during our evaluations, extensive literature review, and our expertise in VE interaction design, we narrowed the numerous variables to five that we feel are most critical for navigation and therefore most important for the first cycle of summative evaluations. These five candidate variables and their level of treatment for a summative study are

- Navigation metaphor (ego- vs. exocentric)
- Gesture control (rate vs. position of hand movement)
- Visual presentation device (Workbench, desktop, CAVE)
- Head tracking (present vs. absent)
- Stereopsis (present vs. absent)

TABLE 28.1

Design Parameters for Navigation, Organized by Taxonomy of Usability Characteristics for VEs

User Tasks	Input Devices	Virtual Model	Presentation Devices
User scenarios	Navigation viewpoint	Mode switching	Visual presentation device
Navigation presets	Navigation degrees of freedom	Mode feedback	Stereopsis
	Gestures to trigger actions	Number of modes	
	Speech input	Visual navigation aids	
	Number of flight stick buttons	Dataset characteristics	
	Input device type	Visual terrain representation	
	Movement dead space	Visual (battlefield) object representation	
	Movement damping	Visual input device representation	
	User gesture work volume	Size of (battlefield) objects	
	Gesture mapping	Object relationship representation	
	Button mapping	Map constrained vs. floating	
	Head tracking		

Source: Gabbard, J.L., A taxonomy of usability characteristics for virtual environments, Master's Thesis, Department of Computer Science, Virginia Tech., Blacksburg, VA, http://filebox.vt.edu/users/jgabbard/Gabbard-Taxonomy.pdf, 1997.

We conducted a summative evaluation by first narrowing the list candidate variables (Table 28.1) to a smaller set of variables, based on the taxonomy of usability characteristics, several of our observations during heuristic and formative evaluations, and our expertise in VE interaction design. We felt these four variables have the greatest effect on locomotion and are therefore the most important for summative evaluations: locomotion metaphor (ego- vs. exocentric), gesture control (controls rate vs. controls position), visual presentation device (workbench, desktop, CAVE), and stereopsis (present vs. not present).

Thirty-two subjects performed a series of 17 tasks, each requiring the subject to navigate to a certain location, manipulate the map, and/or answer a specific question based on the map. We called the series of 17 tasks a *task set*, and because platform was a within-subjects variable, we created four task sets (A through D) and designed questions in each task set to be semantically parallel and therefore functionally equivalent so that users were performing essentially similar, but *not* identical, tasks on all four platforms. We even made small changes to the map (e.g., adding nonexistent towns or bays) to help with this. Table 28.2 shows the parallel wording of three questions from task sets A and C.

TABLE 28.2

Three Sample Tasks (Text, Map, and Geometric) Out of 17 Total Tasks, Shown for Two (Sets A and C) of the Four Task Sets

Task Set A	Task Set C
Text task. Identify the highway number of the long road running east–west in the upper northeast area of the map.	*Text task.* Identify the highway number of the long road running east–west in the lower southwest area of the map.
Map task. Tilt the map so that both the northern horizon and Peru are visible (on the front screen).	*Map task.* Tilt the map so that both the western horizon and Fulcher Landing are visible (on the front screen).
Geometric object task. Remaining on the white cube, look around in all directions and indicate which blue object appears farthest from you.	*Geometric object task.* Remaining on the white cube, look around in all directions and indicate which red object appears farthest from you.

Task set questions fell into three categories (Table 28.2 contains an example question from each category): (1) *text tasks*, which involved searching for named items on the map—the subject was either searching for a terrain object to determine its name or looking for a terrain object when given its name; (2) *map tasks*, which asked the subject to place the map in a given position; and (3) *geometric object tasks*, which asked the subject to navigate relative to geometric solids, such as cubes, towers, and pyramids. Subjects began with a training task set, performed on a different map containing similar geographic features. The training task set comprised seven tasks similar to those in the main task sets.

Our summative evaluation yielded interesting results (Swan et al., 2003). A striking finding of our results was that the desktop had the best overall user performance time of all display platforms. Many user tasks required finding, identifying, and/or reading text or objects labeled with text. While all displays were set to 1024×768 pixels, the size of the projection surface varied enough to conjecture that pixel density is more critical than field of view or display size. Our observations and qualitative data support this claim. This research suggests we should further research user task performance using high-resolution displays. Also interestingly, we found no effect of platform at all in map tasks and geometric object tasks. This begs examination of the important question: "Why are we building large display VEs and incurring the resulting expense if the user benefit is not there?"

28.5.4 Dragon: Lessons Learned

Designing usability evaluation studies within an application context presents a number of challenges that are not present in generic, basic-research, human-factors-type user studies. For example, the tasks users performed in usability evaluations often necessarily are often goal oriented, rely on specific domain knowledge and expertise, and may be achieved through multiple means, whereas basic human-factors studies often examine low-level perceptual or attention tasks. Moreover, application software is often very complex, containing multiple software layers such as user interface and interaction code, middleware layers, network and database connection code, and other back-end components. In basic research studies, software (or testbed) components are often minimized to the smallest amount of code needed to effectively measure human performance and generally exclude these unnecessary software layers in order to reduce variability not associated with human performance and thus eliminate unneeded sources that may confound data.

Throughout many of these studies, as we considered a particular set of statistical results, we noted considerable variance that may be hiding otherwise notable findings. We suggest that much of this variance comes from uncontrollable unknowns associated with the increased complexity of application software, relative to abstract testbed software that is developed only for research and experimentation.

The types of user tasks needed to support summative studies within application contexts are very different than those appropriate in generic, human-factors user studies. We suggest that designing user tasks for an application-based study forces a trade-off between the need to obtain qualitative usability results and the need to truly represent end-user tasking versus the need to obtain clean, powerful, extensive statistical findings.

For example, abstract testbed user studies typically employ atomic tasks that have clear starting points and ending points and that are designed specifically to reduce variance and thereby obtain precise statistical results. In application settings, atomic tasks are strung together to establish higher-level real-world tasking and thus create a high number of task dependencies. In creating scenarios representative of real-world user tasking, it is difficult to neatly divide and constrain singleton tasks, at least in a meaningful fashion. The result is user tasks that may be, for example, hard to precisely time. Results can then vary due to user strategy and previous system state rather than because of specific independent variables manipulated in the study.

This trade-off can be best managed by identifying task sequences and dependencies in advance so that the type of data collected and strategies for collecting data can be designed accordingly, for example, design tasks and types of task responses to maximize quantitative data collected within the constraints of typical application-specific task sets. It is also useful to design task sets that have clear closure (end of task) for both evaluator and user, but without compromising the representative application-level task flow. This can be done by designing and timing small task sets rather than singleton tasks. Another strategy is to design tasks where users have to achieve a certain level of competence before proceeding. While this may help ensure a constant starting point for subsequent tasks, timing strategies will need to take into account the additional time needed to establish the level of competence.

Our application-specific approach to usability evaluation may indicate why we do not have clean, expected, widespread statistical results—the complexity of the application and user tasks introduced variance. Our user tasks were necessarily higher-level tasks, perhaps more appropriate for a qualitative analysis. The alternative was to use focused, atomic tasks that might lead to strong statistical results, but those results may not be widely applicable to real-world application domains.

28.6 CASE STUDY 2: BARS

The battlefield augmented reality system (BARS) was conceived to assist mobile, urban warfighters. Urban terrain is one of the most important and challenging environments for current and future peacekeepers and warfighters. Because of the increased concentration of military operations in urban areas, many future police and military operations will occur in cities. However, urban terrain is also one of the most demanding environments, with complicated 3D infrastructure potentially harboring many types of risks (Dewar, 1992).

BARS was developed at the Naval Research Lab in Washington, DC (Livingston et al., 2002), in conjunction with usability engineers/HCI researchers from Virginia Tech, researcher collaborators from Columbia University, and others. During that time, we worked closely with the BARS software development team to apply usability engineering techniques and conduct user-based studies to advance state-of-the art research in outdoor augmented reality (AR) user interfaces, as well as the actual BARS user interface itself.

The hardware for the implementation of BARS is shown in Figure 28.4. BARS is built using the following commercial off-the-shelf products: an optical see-through head-mounted display,

FIGURE 28.4 **(See color insert.)** Prototype of the BARS (see Livingston et al., 2002).

a GPS-based position tracker (which tells where the user is located), an orientation tracker (which tells the direction the user is looking), a wireless network, interaction devices (e.g., a wrist-mounted keyboard and wireless handheld mouse), a wearable computer with 3D graphics accelerated hardware, and a rechargeable battery.

28.6.1 BARS DOMAIN ANALYSIS

We based the BARS user task analysis on both interaction with a subject matter expert and on extensive bibliographical research, including military history books and military doctrine manuals. We considered both types of sources equally important and complementary. Military history books provide a general overview of urban combat, describe associated problems, and outline tactics that have been successfully and unsuccessfully followed. They also provide a context in which we can discover the importance of information requirements such as situational awareness. Official training manuals are much more specific and enumerate a variety of tactics and rules of engagement. However, because they give no historical context, it is very difficult to identify which tasks are the most important ones.

Table 28.3 shows a snapshot from the resulting user task analysis document, which was 14 pages in length. From the military history books, we assembled a list of general tasks performed during armed conflict in urban areas. Next, we used military doctrine manuals to create a detailed breakdown of the subtasks that each task requires. Table 28.3 shows the high-level task *moving through open areas* and two subsequent subtasks. In total, we found several hundred different tasks. However, we realized that for many of the tasks, AR and other visualization/cognitive systems are just not relevant; an example is the task of *grenade throwing*. Based on this observation, we reduced the list to around 100 relevant AR-appropriate tasks. Each task that could be enhanced by an AR system has two main components: perceptual and cognitive. The perceptual component of the task is the type of perceptual action that the soldier needs to perform, while the cognitive component describes the necessary mental processes. For each task, we analyzed how an AR system could support both the perceptual and cognitive components. Table 28.3 demonstrates this analysis.

TABLE 28.3

A Snapshot from the BARS Task Analysis Document

Task Analysis Components	Task: *Moving through Open Areas*—Representative User Subtasks	
	Task: Identify Open Areas	**Task: Conduct Movement inside the Building**
Perception	The soldier must have a clear view of the area around him. Combination between reading a map and visual observation.	The soldier must be able to identify the buildings around him, their internal geometry and distribution, the access to them (windows, doors, holes), and their distribution in space.
Cognition	The soldier must be able to identify open from closed areas; maps can be difficult to understand in this sense.	The soldier must be able to identify the ways he can use the building's internal geometry and distribution to move (or advance) toward his destination. This includes identifying the location of hidden and occluded objects.
AR support	The soldier using the AR system will have a 3D view of the terrain around him. He can invoke a 3D map that can be manipulated at will. He will see which areas are open and which ones are interiors. He can identify an open area like a park or the internal yard of a building.	The soldier using the AR system will have a 3D view of the terrain around him. He can invoke a 3D map that can be manipulated at will. He can explore the interior of the 3D models of each building and understand more clearly their geometry and distribution.

28.6.2 BARS EXPERT EVALUATION

We analyzed the task analyses to produce user-centered requirements. In doing so, we realized that our user-centered requirements identified a list of features that could not be easily delivered by any current AR system. For example, one BARS user-centered requirement said that the system must be able to display the location of hidden and occluded objects (e.g., personnel or vehicles located somewhere behind a visible building). This raised numerous user interface design questions related to occluded objects and how they should be presented graphically to a user. To address such issues, we began expert evaluations on an evolving BARS user interface design. While we conducted several expert evaluations on BARS, this chapter describes the expert evaluations we conducted related to visually representing occlusion.

During six cycles of expert evaluation over a 2-month period, summarized in Table 28.4, we designed approximately 100 mock-ups depicting various potential designs for representing occlusion, systematically varying drawing parameters such as *lines* (intensity, style, thickness), *shading* (intensity, style, fill, transparency), and *hybrid* techniques employing combinations of lines and shadings.

We were specifically examining several aspects of occlusion, including how best to visually represent occluded information and objects, the number of discriminable levels (layers) of occlusion, and variations on the drawing parameters listed earlier. In each cycle of expert evaluation, team members individually examined a set of occlusion representations (set size ranged from 5 to 30 mock-ups in a cycle), which were created using Adobe Photoshop and Microsoft PowerPoint employing video to capture real-world scenes as background images. Team members each independently performed an expert evaluation of electronically shared mock-ups in advance of extensive teleconference calls. During the calls, we shared our individual expert evaluation results, compiled our assessments, and collaboratively determined how to design the next set of mock-up representations, informed by results of the current cycle. Because the mock-ups supported a very quick turnaround, we were able to evaluate many more designs than could have been implemented *live* in BARS. In fact, this use of mock-ups was extremely cost-effective, allowing the team to begin substantive usability evaluation work even before many BARS features were implemented.

Cycle 1 (see Table 28.4) served to indicate that, in fact, the mock-ups were an effective way of performing expert evaluations. In cycles 2–4, we specifically studied line-based encodings, and our results showed that line intensity appeared to be the most powerful (i.e., consistently recognizable) line-only drawing parameter, followed by line style. Further, line-based representations were discriminable at only three or four levels of occlusion. Interestingly, we found a few instances when color and intensity created misleading cues when used in combination as the encoding scheme. In cycle 5, we studied distance estimation and shading-based representations. Results indicated that shading alone may not be enough to indicate distances; user-controllable overlaying of distance cues onto the ground may be necessary. Again, we found that shading-based representations were also discriminable at only three of four levels of occlusion. In cycle 6, we combined both line- and shading-based representations into some hybrid designs, hoping to maximize the best characteristics of each type of representation. In particular, we found that a hybrid of shaded regions and line width, both with varying intensity, appeared to be the most powerful, discriminable representation for representing occluded objects.

Further, at this point, based on the relatively small changes we were making to the mock-ups, we felt we had iterated to an optimal set of representations for occlusion, so we chose to move on to formative evaluations using them. However, in retrospect (and as part of continually evolving and improving our cost-effective progression of usability evaluation), we realized it would have been scientifically advantageous to have run the user-based statistical evaluations next, to evolve empirically derived user interface designs for our BARS formative evaluation.

TABLE 28.4

Summary of Expert Evaluations to Evolve BARS User Interface Designs for Occlusion

Cycle No.	Purpose of This Evaluation Cycle	Medium for This Evaluation Cycle	Results/Findings
1	Initial expert evaluation and overview of BARS	BARS system.	Focus usability engineering efforts on 1. Tracking and registration 2. Occlusion 3. Distance estimation
2	First cut at representing occlusion in MOUT (military operations in urban terrain)	Five interface mock-ups including line-based building outlines and personnel representations.	• Tracking study will require time to build cage (see Figure 28.5) and focus on occlusion in the interim.
3	Examine exhaustive set of mock-ups that redundantly encode occlusion using various line attributes	25 interface mock-ups systematically varying different types of line width, intensity, and style.	• Line intensity and thickness appear to be the most powerful encoding mechanisms, followed by line style. • Color and intensity of the scene can create misleading cues when using color and intensity together as an encoding scheme.
4	Continue to examine previous set of representations	25 interface mock-ups systematically varying different types of line width, intensity, and style.	• Number of occluded layers that can be discriminably (effectively) represented by line-based encoding is three or four.
5	Examine additional visual cues to aid in distance estimation; examine use of filled polygons to represent occlusion in interior spaces	14 interface mock-ups using shading of occluded objects to show distance estimation as well as occlusion in interior spaces.	• Labeling occluded objects is a hard problem. • Distance cues should be overlaid onto the ground and should be easily turned off and on by the user. • Motion parallax may help resolve some problems. • Number of occluded layers that can be discriminably (effectively) represented by shading-based encoding is three or four.
6	Examine shaded polygon representations in a complex outdoor environment (Columbia campus), as well as hybrid designs employing lines; examine effects of motion parallax on encodings	30 interface mock-ups (5 mock-ups per set, 6 sets) systematically varying representations of occlusions employing filled polygons, transparency, and lines. Mock-ups also simulated motion parallax by paging between images in a set.	• A combination of shaded polygons and line width is most powerful encoding. • Distance encoding may be more powerful than simple occlusion. • Users should be able to push and pull the three to four levels of representation into and out of their real-world scene.

28.6.3 BARS FORMATIVE EVALUATION

We used the results of our expert evaluations to drive the early set of real-time user interface prototypes used for formative evaluations. While the expert evaluation described earlier addressed the issue of occlusion, our formative evaluations addressed a larger set of application-specific issues.

Having anticipated the challenge of working in an outdoor, mobile, highly dynamic environment, BARS team members had to consider innovative approaches to usability evaluation. Our solution was to design and build a specially constructed motion tracking cage so that BARS could accurately track the evaluation participant and accurately register graphics onto the real world. The cage provided a mounting platform for InterSense IS-900 tracking rails, which were and still are currently commonly used for AR tracking. While clearly not usable in a final, fielded outdoor AR system, mounting the tracking rails on top of the cage gave us adequate tracking performance to meet our user task requirements, without waiting for BARS engineers to complete the development of mobile AR tracking devices with the required performance. The main trade-off was that the participant was not able to freely walk large distances, as envisioned for the final BARS. We therefore focused on tasks related to scanning the urban environment from the area covered by the tracking cage. Our setup also included auxiliary evaluator's monitors to provide evaluators an accurate display of a participant's view.

We created a formal set of user tasks and had five individual participants perform the set of tasks. Three participants were marines, and two were user interface/AR experts. The tasks were militarily relevant, inspired by our task analysis work. In the tasks, participants were asked to find explicit information from the augmenting graphics that they could see. Some simple examples included answering questions such as the following: Which enemy platoon is nearest you? Where are the restricted fire areas? Where are other friendly forces? Estimate the distance between the enemy squad and yourself. What direction is the enemy tank traveling?

Participants wore a head-mounted, Sony Glasstron optical see-through display (Figure 28.5). Participants were asked to perform simple perceptual tasks and answer questions about what they perceived using BARS. For example, we wanted to determine whether the system could present information to the participant about relative locations and sizes of objects that may be hidden from actual view but visible using the *x-ray vision* metaphor enabled through AR. We also assessed participants' perception of depth, by having them estimate distances from themselves to both real and virtual objects. We derived some of these measurements from certain operating parameters of the system, such as the accuracy with which various subsystems for determining the participant's location (e.g., GPS and inertial navigation systems or other passive measurement devices) must report the participant's location in order for the participant to answer such questions with a desired confidence or accuracy level. Finally, we asked participants subjective questions about the system: likes and dislikes, potential features to be considered in revised BARS prototypes, etc.

FIGURE 28.5   One of our marine participants employing the *think-aloud* protocol during the BARS formative evaluation.

Researchers and software developers observed participants while the study was performed and recorded data about participant's behavior (e.g., if the participant moved and answered questions quickly or deliberately or if/how the participant gestured while answering questions). For some tasks, the participant was asked to press keys on a miniature wrist-worn keyboard or point using a gyroscopic computer mouse. From these observations, we were able to produce subjective measurements of user performance, such as comfort and ease of learning.

Each participant was present for no more than 2 h, including the orientation, all experimental tasks, intermediate reconfigurations of the system, all questions, and debriefing. The total time actually wearing the AR display was approximately 30–45 min, in spans of approximately 10 min with breaks as needed or requested by participants. Following is the specific list of procedures (protocol) used for this evaluation:

1. Welcomed the participant and introduced him or her to all researchers present.
2. Introduced the study, its objective, and the equipment being used.
3. Had the participant put on the AR display device and look at the real world through it. At this stage, no graphics were displayed.
4. Introduced the computer-generated (i.e., AR) imagery and had the participant survey the surrounding real-world environment in combination with computer-generated imagery. The participant was able to see line-based drawings of objects in the field of view (which may be occluded) and was able to look in any direction.
5. Had the participant—while wearing the display—face a specific direction. A software developer assisting with the evaluation then triggered the system to display or highlight selected objects (buildings, vehicles, or individuals), some of which were real, synthetic, occluded, and/or visible.
6. Researchers then asked a series of questions about what the participant saw and how he or she was interpreting what they saw. Sample questions included whether the participant saw certain colors in use in the display, which item in a set of objects was closest to the participant, identify when a moving object became visible, describe how the participant might navigate to a highlighted location (even though the participant did not physically move to the location), or locate a particular object (e.g., a friendly unit or a mission objective). Steps 5 and 6 were repeated several times with different objects.
7. The participant (at various times during this series of questions) was allowed to press keys on a miniature keyboard to change the parameters of the display methods. These parameters included the distance at which objects were highlighted and the type of method used to highlight objects.
8. At various times, researchers introduced different methods of drawing styles to the participant. For example, objects that were drawn with lines were redrawn with filled shapes, or lines changed from solid to dashed or dotted or changed in intensity or color.
9. Some virtual objects were in motion during some of the questions. No real objects physically approach the participant, nor did any imagery depict an object coming close to the participant.
10. At the end of the study, we handed each participant a written statement and asked him or her to read the statement. We then verbally reiterated the main points of the debriefing statement and encouraged the participant to ask any questions. During this time, we observed the participant carefully for any signs of simulator after-effects. We also asked the participant if they felt dizzy or light-headed, as we asked them to walk around in the space nearby.

Our overall evaluation results showed that participants performed approximately 85% of the tasks correctly and efficiently with less than 10 min of training using BARS. Participants liked having multiple views of various graphical augmentations and liked being able to develop strategies to manipulate the scene and understand how BARS works. They stated that they were able to gain situational awareness from using BARS. Participants disliked use of wireframes (lines) as the main augmentation representation, saying that they made the scene too cluttered. They also disliked some of the controls for manipulating augmentations (e.g., making them appear/disappear), but these controls are temporary, only for our evaluation studies, and are not intended to be included in a deployable BARS. Many scientific results supported findings from our previous observations, such as no more than three or four levels of occlusion are discriminable, and that objects must be perceived by a user as 3D. We made new findings such as the fact that three-dimensionality of occluded objects was easier to perceive in shaded objects than in line-drawn objects. All participants had a very positive, enthusiastic reaction to BARS and its capabilities. Our experience during the evaluation led us to determine that the problem of representing occluded objects in AR required more research and specifically required us to design studies to determine what visual design factors for occluded objects were most effective, independent of other user interface components (e.g., text labels).

Through mostly empirical observation, we identified several usability issues regarding the actual BARS prototype, with these centered around the visual user interface. These usability issues, along with potential solutions or next steps, are presented in Table 28.5.

28.6.4 BARS: Lessons Learned

The most compelling lesson learned from our BARS evaluation work was the simple fact that software developers were designing outdoor AR user interfaces with absolutely no guidance from the literature or standards, as there is a scarcity of both (Swan & Gabbard, 2005). Thus, as a team of engineers and usability experts, we felt that we had reached a critical juncture in the overall BARS program both in terms of product (e.g., how to effectively design and develop a quality BARS UI) and process (i.e., how to modify/extend our existing usability engineering plan to address difficult AR design activities).

> The system continually has to make this choice: it can either continue to exploit a known process and make it more productive, or it can explore a new process at the cost of being less efficient.
>
> **Kevin Kelly**

We started by identifying a set of core scientific issues that needed to be addressed before cost-effective AR application development could continue. We then modified our existing usability engineering process to help inform the BARS application development process; this process includes the addition of statistical user-based studies and is described in detail in Gabbard & Swan (2008).

We made a decision to *take a step back* and catalogue, as a team, the core outdoor AR issues that needed to be resolved before meaningful user interface/interaction design and evaluation activities could take place. Concurrently with domain analysis and the BARS team, we determined a list of over 25 core scientific issues in user interface design of ARs. This list came from an extensive literature review, team brainstorming sessions, our own expertise, and discussion with other AR researchers. This catalogue of core scientific issues that posed design and evaluation challenges for outdoor AR systems forced us to a critical junction where we had to decide whether usability engineering efforts should address domain/scenario-level issues or address core scientific-level AR issues.

We concluded that we would examine core scientific AR issues within the context of our current urban warfighting domain. Specifically, we chose to emphasize five issues (listed in the following in rough chronological order in which they were researched) most likely to improve user performance within an AR in an urban warfighting setting:

- Representing occlusion in far-field AR (occlusion)
- Supporting accurate distance and depth perception in far-field AR (distance perception/ estimation and stereopsis)

TABLE 28.5

Usability Issues and Potential Solutions/Next Steps Gleaned from Our Formative Evaluation of the BARS User Interface

Usability Issue	Potential Solution(s) or Next Step
Participants found far (distant) text difficult to read. Participants could not associate virtual text labels with specific real-world buildings.	• Research methods to increase text legibility. • Use tethers, indicating which labels belong to which buildings. • Employ better registration between text and real-world objects. • Research AR HMDs with autoadjustable focal depths. • Put labels on side of buildings, pasted like billboards.
Participants had difficulty associating text size with distance (e.g., smaller text registered to real-world objects that were farther away than larger text).	• Increasing/decreasing label sizes should be exaggerated and well pronounced (well above the just-noticeable-difference threshold). • Place labels in the vertical direction to redundantly encode distance. • Present actual distances (e.g., in meters) as part of the label (would want this to be user toggleable).
Participants could not discriminate buildings when using wireframe and/or dashed lines as visual cues. Participants found small objects at a distance hard to see.	• Continue to examine the use of transparent fill patterns to represent occluded objects. • Use redundant techniques to visually encode objects, such as shading, transparency, and outlines to indicate front/visible/hidden surfaces. • Use three to four encodings to cardinally encode buildings as opposed to using specific measured distances to visually encode. • Consider integrating MIL-STD 25/25 or MIL-STD 50 to increase recognition of objects. • Display motion vectors for moving objects.

Participants had difficulty determining direction of occluded moving vehicles.
Participants could not determine whether personnel (shown as virtual representations) were inside or outside a specific building.
Participants were not sure which areas were *off limits* to engagement.
Participants had difficulty perceiving real-world threat when there was too much visual clutter. Moreover, the visual clutter was difficult to avoid by the system, since BARS did not have any filtering mechanisms in place.
	• Use subtle visual encoding enhancement to indicate whether personnel is inside or outside building.
	
	• Use subtle encoding enhancement to indicate restricted fire areas and friendly areas of refuge.
	• Legend (if needed) should be used at the bottom of display.
	• Employ a hardware button or voice command to instantly toggle all graphics on and off.
	• Support three levels of database use:
	  1. Whole scene, whole database—supported via map mode
	  2. General areas, three- to five-building deep—supported via optional map and soldier's eye modes
	  3. Nearby areas—the next building or two, who can harm me now—supported via soldier's eye view

Participants liked the overhead map feature but found it lacked rich functionality.	• Extend overhead map feature to include
	  1. Support for route planning
	  2. Distance estimation (e.g., marking points along a path)
	  3. A distance scale in the corner

AR display was heavy on participants' head. Participants found the wrist keyboard and gyro mouse cumbersome.	• Consider alternate displays (e.g., lightweight monocular displays that mount to helmet). • Examine other hardware for truly mobile yet intuitive UI devices. • Examine devices currently used by military in the field that could be leveraged.
The virtual compass feature was well received by participants and was considered vital for performing domain-specific tasks. Participants had no visible indications of nausea at any time during or after their evaluation session.	• Continue development of virtual compass, supporting both the soldier's eye and map mode views. • Consider employing Kennedy's simulator sickness questionnaire (Kennedy et al., 1993) in future evaluations to quantify nausea or lack thereof.

- Designing augmenting graphics for display on real-world backgrounds (text legibility)
- Determining acceptable ranges of registration error and tracking lag (tracking and registration)
- Managing information clutter and user filtering of 3D objects and labels (i.e., information filtering)

Once this catalogue of core scientific issues was created, we reassessed how we could best apply a usability engineering approach to a development effort in which there was much uncertainty. As a result, we produced an early conceptual model for our modified usability engineering approach (Figure 28.6) that would iteratively refine and improve user interface components associated with the select core scientific issues listed previously to produce an integrated BARS user interface that embodied the best component designs.

The results of these evaluations, in time, led to research that produced over 20 publications by members of the BARS team (e.g., Baillot et al., 2003; Gabbard, Swan, & Hix, 2006; Gabbard, Swan II, Hix, Kim, & Fitch, 2007; Julier & LaViola, 2004; Julier, 2003; Julier & Uhlmann, 2003; Julier & Uhlmann, 2004; Julier et al., 2002; Julier et al., 2000; Livingston et al., 2003; Swan et al., 2006; van der Merwe et al., 2004). Indeed, many of the results of these atomic studies were folded back into a mature BARS user interface. We planned to perform a traditional summative evaluation of the integrated BARS user interface; however, even though much energy and iteration was put into the BARS UI, resources did not afford a final summative evaluation.

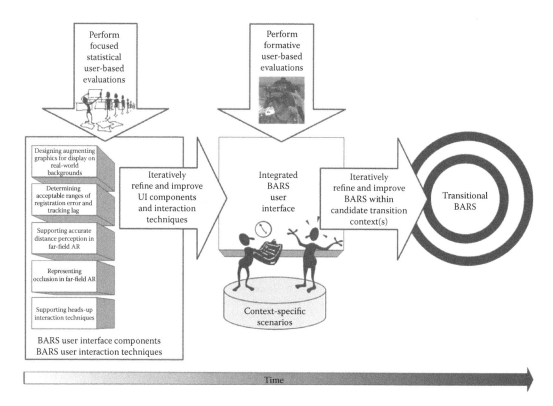

FIGURE 28.6 Early conceptualization of a modified usability engineering approach that employed both focused user-based evaluations on core scientific issues as well as whole-UI formative evaluations in route to a fieldable BARS product.

28.7 CONCLUSION

Despite improvement in efforts to improve usability engineering of VEs, there still is not nearly enough usability engineering applied during development of VEs. And there is still a need more cost-effective, high-impact methods. As VEs have become more mature, established interaction techniques and generic tasks for VEs are starting to emerge and eventually have the potential—as they did with GUIs—to improve usability. But the design space and options for VEs are enormously greater than that for GUIs, and VE applications tend to be much more complex than many GUIS. So even *standard* interaction techniques, devices, and generic tasks for VEs will help improve usability only by a small fraction. Usability engineering will continue to be a necessary process if new and exciting VEs that are, in fact, usable and useful for their users are to be created.

ACKNOWLEDGMENTS

We thank Dr. Deborah Hix for her integral conceptual and applied work described herein. Dr. Hix's contributions in this area have been a guiding light for myself and others. Many present and past members of the VR Lab of the Naval Research Lab in Washington, DC, that were integral in the support of both the Dragon VE and the BARS mobile augmented reality platform are thanked including Dr. Larry Rosenblum (currently at NSF), J. Edward Swan (currently at Mississippi State University), Simon Julier (currently at University College of London), and Mark Livingston (currently at NRL). Dr. Larry Rosenblum and Dr. Helen Gigley, of the Office of Naval Research, funded the Dragon work described herein, as well as development of the taxonomy of usability characteristics of VEs. Dr. Rudy Darken, at the Naval Postgraduate School in Monterey, California, has been a strong proponent of our work for several years; his support and encouragement are greatly appreciated. We are grateful to all these contributors, without whom this large body of work would not have been possible.

REFERENCES

Baillot, Y., Julier, S., Brown, D., & Livingston, M. (2003). A tracker alignment framework for augmented reality. *Proceedings of the International Symposium on Mixed and Augmented Reality (ISMAR .03)*, pp. 142–150, Tokyo, Japan, October 7–10.

Dalgarno, B., Bishop, A. G., & Bedgood, Jr., D. R. (2012). The potential of virtual laboratories for distance education science teaching: Reflections from the development and evaluation of a virtual chemistry laboratory. *Proceedings of the Australian Conference on Science and Mathematics Education,* Vol. 9 (formerly UniServe Science Conference). Sydney, New South Wales, Australia.

del Galdo, E. M., Williges, R. C., Williges, B. H., & Wixon, D. R. (1986). An evaluation of critical incidents for software documentation design. *Proceedings of Thirtieth Annual Human Factors Society Conference*. Anaheim, CA.

Dewar, M. (1992). *War in the Streets: The Story of Urban Combat from Calais to Khafji*. Newton Abbot: David and Charles.

Eberts, R. E. (1999). Unpublished lecture on task analysis: IE486 Work Design and Analysis II, Purdue University Presentation is online at http://palette.ecn.purdue.edu/~ie486/Class/Lecture/lect14/sld001.htm, Accessed on February 2002.

Gabbard, J. L. (1997). *A taxonomy of usability characteristics for virtual environments* (Master's Thesis). Department of Computer Science, Virginia Tech., Blacksburg, VA. http://filebox.vt.edu/users/jgabbard/Gabbard-Taxonomy.pdf.

Gabbard, J. L., & Swan II, E. J. (2008). Usability engineering for augmented reality: employing user-based studies to inform design. *IEEE Transactions on Visualization and Computer Graphics, 14*(3), 513–525.

Gabbard, J. L., Swan II, J. E., & Hix, D. (2006). The effects of text drawing styles, background textures, and natural lighting on text legibility in outdoor augmented reality. *Presence: Teleoperators and Virtual Environments, 15*(1), 16–32, Spring 2006.

Gabbard, J. L., Swan II, J. E., Hix, D., Kim, S. J., & Fitch, G. (2007, March 10–14). *Active text drawing styles for outdoor augmented reality: A user-based study and design implications.* Technical Papers, Proceedings of IEEE Virtual Reality 2007 (pp. 34–42). Charlotte, NC.

Hackos, J. T., & Redish, J. C. (1998). *User and task analysis for interface design.* New York, NY: John Wiley & Sons.

Hartson, R., & Pyla, P. (2012). *The UX book: Process and guidelines for ensuring a quality user experience.* Waltham, MA: Morgan Kaufmann/Elsevier.

Hix, D., & Hartson, H. R. (1993). *Developing user interfaces: Ensuring usability through product and process.* New York, NY: John Wiley & Sons.

Hix, D., Swan, E. J., Gabbard, J. L., McGee, M., Durbin, J., & King, T. (1999a). *User-centered design and evaluation of a real-time battlefield visualization virtual environment.* Proceedings of the IEEE VR'99 Conference. Houston, TX. (NOTE: This paper was awarded Best Technical Paper at the VR'99 Conference.)

Julier, S. J., Baillot, Y., Brown, D., & Lanzagorta, M. (2002). Information filtering for mobile augmented reality. *IEEE Computer Graphics and Applications, 22*(5), pp. 12–15, September/October, 2002.

Julier, S. J., Langzagorta, M., Balliot, Y., Rosenblum, L., Feiner, S., & Höllerer, T. (2000). Information filtering for mobile augmented reality. *Proceedings of IEEE and ACM International Symposium on Augmented Reality (ISAR)*, pp. 3–11.

Julier, S., & LaViola, J. An empirical study into the robustness of split covariance addition (SCA) for human motion tracking. *American Control Conference*, Boston, MA, USA, June 30–July 2, 2004.

Julier, S. J. (2003). The stability of covariance inflation methods for SLAM. *Proceedings of the 2003 IEEE IROS Conference*, pp. 2749–2754, Las Vegas NV, USA, October 27–31, 2003.

Julier, S. J., & Uhlmann, J. K. (2004). Unscented filtering and nonlinear estimation. *IEEE Review, 92*(3), pp. 401–422, March, 2004.

Julier, S. J., & Uhlmann, J. K. (2003). Using multiple SLAM algorithms. *Proceedings of the 2003 IEEE IROS Conference*, pp. 200–205, Las Vegas NV, USA, October 27–31, 2003.

Kennedy, R. S., Lane, N. E., Berbaum, K. S., & Lilienthal, M. G. (1993). Simulator sickness questionnaire: an enhanced method for quantifying simulator sickness. *The International Journal of Aviation Psychology, 3*(3), pp. 203–220.

Kennedy, S. (1989). Using video in the BNR usability lab. *SIGCHI Bulletin, 21*(2), 92–95. doi:10.1145/70609.70624. http://doi.acm.org/10.1145/70609.70624.

Komlodi, A., Józsa, E., Hercegfi, K., Kucsora, S., & Borics, D. (2011). *Empirical usability evaluation of the Wii controller as an input device for the VirCA immersive virtual space.* Second International Conference on Cognitive Infocommunications (CogInfoCom). Budapest, Hungary, pp. 1–6. IEEE.

Lewis, J. A., Deutsch, J. E., & Burdea, G. (2006). Usability of the remote console for virtual reality telerehabilitation: Formative evaluation, *CyberPsychology & Behavior, 9*(2), 142–147. doi:10.1089/cpb.2006.9.142.

Livingston, M. A., Rosenblum, L. J., Julier, S. J., Brown, D., Baillot, Y., Swan II, J. E., Gabbard, J. L., & Hix, D. (2002). An augmented reality system for military operations in urban terrain., *Proceedings of Interservice / Industry Training, Simulation & Education Conference (I/ITSEC)*, December 2–5, Orlando, FL, pp. 89 (abstract only).

Livingston, M., Swan II, J. E., Gabbard, J. L., Höllerer, T. H. Hix, D., Julier, S. J., Baillot, Y., & Brown, D. (2003). Resolving multiple occluded layers in augmented reality. *Proceedings of the Second International Symposium on Mixed and Augmented Reality (ISMAR)*, Tokyo, Japan, October 7–10, 2003.

Mack, R. L., & Nielsen, J. (1994). Executive summary. In *Usability inspection methods* (Chapter 1, pp. 1–23). New York, NY: John Wiley & Sons.

Nielsen, J. (1993). Usability Engineering, Academic Press, Boston.

Nielsen, J. (1994). Heuristic evaluation. In *Usability inspection methods* (Chapter 2, pp. 25–62). New York, NY: John Wiley & Sons.

Obeysekare, U., Williams, C., Durbin, J., Rosenblum, L., Rosenberg, R., Grinstein, F., … Sandberg, W. (1996). Virtual workbench: A non-immersive virtual environment for visualizing and interacting with 3d objects for scientific visualization. *Proceedings of IEEE Visualization'96* (pp. 345–349). Los Alamitos, CA: IEEE Computer Society Press.

O'Malley, C. E., Draper, S. W., & Riley, M. S. (1984). Constructive interaction: A method for studying human-computer-human interaction. *Proceedings of INTERACT 84* (pp. 269–274). Amsterdam: the Netherlands: North-Holland.

Rivera-Gutierrez, D., Welch, G., Lincoln, P., Whitton, M., Cendan, J., Chesnutt, D. A., … Lok, B. (2012). Shader lamps virtual patients: The physical manifestation of virtual patients. *Medicine Meets Virtual Reality 19: NextMed, 173*, 372.

Scriven, M. (1967). The methodology of evaluation. In R. E. Stake (Ed.), *Perspectives of curriculum evaluation, American educational research association monograph*. Chicago, IL: Rand McNally.

Seidel, I., Gartner, M., Froschauer, J., Berger, H., & Merkl, D. (2010). Towards a holistic methodology for engineering 3D Virtual World applications. *International Conference on Information Society (i-Society) 2010*, (pp. 224–229). IEEE, 28–30 June, 2010, London, UK.

Stevens, F., Frances, L., & Sharp, L. (1997). *User-friendly handbook for project evaluation: Science, mathematics, engineering, and technology education* (pp. 93–152). Arlington, VA: National Science Foundation.

Sutcliffe, A., & Gault, B. (2004) Heuristic evaluation of virtual reality applications. *Interacting with Computers, 16*(4), 831–849.

Swan II, J. E., Gabbard, J. L., Hix, D., Schulman, R. S., & Kim, K. P. (2003). A comparative study of user performance in a map-based virtual environment. *Proceedings of the IEEE Virtual Reality 2003* (pp. 259–266). IEEE, Los Alamitos, CA.

Swan II, J. E., & Gabbard, J. L. (2005). *Survey of User-based Experimentation in Augmented Reality*. Paper presented at *the 1st International Conference on Virtual Reality*, HCI International 2005, Las Vegas, NV.

Swan II, J. E., Livingston, M. A., Smallman, H. S., Brown, D., Baillot, Y., Gabbard, J. L. & Hix, D. (2006, March 25–29). A perceptual matching technique for depth judgments in optical, see-through augmented reality. Technical Papers, *Proceedings of IEEE Virtual Reality 2006* (pp. 19–26). Alexandria, Virginia.

van der Merwe, R. Wan, E., Julier, S., Bogdanov, A., Harvey, G., & Hunt, J. (2004). Sigma-point kalman filters for nonlinear estimation and sensor fusion: applications to integrated navigation, *Proceedings of the AIAA Guidance Navigation & Control Conference*, (Providence, RI), August 2004.

Wang, X., & Dunston, P. S. (2006, July). Usability evaluation of a mixed reality collaborative tool for design review. *International Conference on Computer Graphics, Imaging and Visualisation 2006*. Los Alamitos, CA, (pp. 448–451). IEEE.

Williges, R. C. (1984). Evaluating human-computer software interfaces. *Proceedings of the International Conference on Occupational Ergonomics*. Toronto, Ontario, Canada.

Wingrave, C. A., Haciahmetoglu, Y., & Bowman, D. A. (2006). Overcoming world in miniature limitations by a scaled and scrolling WIM. *IEEE Symposium on 3D User Interfaces 2006*. Los Alamitos, CA, (pp. 11–16). IEEE.

29 Human Performance Measurement in Virtual Environments

James P. Bliss, Alexandra B. Proaps, and Eric T. Chancey

CONTENTS

29.1   INTRODUCTION

> When description gives way to measurement, calculation replaces debate.

Stevens (1951)

This chapter includes material that pertains to the measurement of human performance in virtual environments (VEs). First, the general process of measuring human performance is discussed, including challenges faced by all researchers. After that, the special case of measuring performance of humans in VEs is discussed. The chapter presents a variety of issues related to the reasons for measuring performance in VEs, the definition and specification of measurable constructs, and the particular methods that could and should be used when measuring performance in VEs. Our goal in this discussion is to include recommendations about the optimal times to measure performance as well as the preferred methods to achieve specific goals. We emphasize psychophysiological measures because of their potential to complement other measures of performance in VEs and because of the recent advancements that have been made in that area. We also emphasize psychometric properties of measurements and factors that affect data integrity.

The American National Standards Institute's guide to human performance measurement (ANSI, 1993) lists several problems that underlie human performance measurement in the context of scientific research and test and evaluation:

- Lack of a general theory to guide performance measurement
- The inverse relationship between operational control and realism
- The multiple dimensions of behavior
- The ambiguous relationship between objective and subjective data
- Difficulty of measuring cognitive tasks
- Lack of objective performance criteria for most tasks
- Difficulty of generalizing results to the real world

All of these problems apply to human performance measurement in VEs; in fact, some are more pronounced. For example, generalization of performance results from the virtual to the real world is a continual concern that could jeopardize the usefulness of VE research. Whether the VE sufficiently represents the real world, whether the VE system interface allows a recognized expert to demonstrate expertise in the VE, and whether interfaces allow for realistic interaction with the VE are concerns faced often by practitioners and researchers.

Measuring performance in VEs can be advantageous for many reasons. First, flexibility of environmental design enables simulation of situations or environments quickly and with reasonable fidelity. For example, the Microsoft Kinect system is now being used as a low-cost method for quickly scanning and recreating environments that resemble an original environment (Fenlon, 2011). This capability, combined with the potential to keep certain parameters constant, has enabled researchers to complete rigorous research cheaply. As a result, researchers and practitioners have tracked performance in VEs to facilitate training, skill-level certification, and personnel selection. A second advantage is the power to record behavior with detail and precision. For example, advances in optical tracking systems have enhanced the ability to record user behaviors precisely and with great detail. This allows expert raters and automated scoring systems to create meaningful performance profiles. It also facilitates the real-time modification of environments so that they adapt to demonstrated user behaviors.

29.2 WHY MEASURE PERFORMANCE IN VEs?

Pioneers in the design of VEs have stressed their use for education, training, visualization, and entertainment. In all such situations, the interaction between the technology and the human user is paramount. Therefore, it is important to measure performance of the human user to determine whether the VE has been well designed and employed. Within education applications, human performance represents an index of concept mastery. For example, different users may search for items or information within the VE that reveal their relative lack of knowledge about a topic or task. Within skill training applications, measuring human performance serves as a barometer to isolate task steps that are unclear, confusing, or missing from the task learner's repertoire. Similarly, measuring the performance of users who rely on a VE for visualization provides a window into the accuracy of their mental models. For example, users may perform tasks in a way that shows incorrect knowledge of spatial relationships or the relative object size. In entertainment applications, speed, accuracy and efficiency of user performance allows game designers to judge how compelling, challenging, or engaging the VE is.

To ensure that a VE system reflects human performance honestly and faithfully, designers should

- Take a systematic approach to design
- Upgrade system aspects according to user feedback
- Judge user performance against group norms and predetermined performance criteria
- Provide users with meaningful feedback during and after performance

It is important that the environment be designed from the start with human performance measurement in mind. The alternative—modifying a VE to suit human performance measurement after initial design work has ended—is inefficient and will ultimately result in a substandard tool that yields substandard results. Consideration of human performance should be central throughout the life of the VE.

Because a key reason to measure human performance is to compare users, task elements, and environmental variables, it is important to follow a systematic approach to design. Designers should carefully replicate critical elements of the target environment. Doing so will ensure that meaningful comparisons may be made between the VE and a non-VE comparator. Along with cost-effectiveness and safety, human performance measurement should be an early consideration when evaluating a VE system.

Performance measurement is needed to support comparisons across system configurations. Performance measures provide a way to determine that system changes and upgrades produce desired effects. In cases where budgetary concerns mandate design trade-off decisions, performance measurement may support the selection of particular hardware or software options. When components of a system are upgraded, experienced users of the previous configuration may believe that their performance was better with the old system. For example, objective measurement provides a way to address claims from individuals that they preferred system performance with the old visual display system (or tracking system, rendering system, etc.) after an expensive upgrade has been implemented.

Performance measures allow objective comparisons of an individual against other individuals, against group norms, and against predetermined performance criteria. Objective measurement is a critical requirement for testing, skill-level certification, and personnel selection and assignment applications.

One of the most important reasons to measure performance is to provide meaningful feedback and knowledge of results to individual users. Holding (1965) pointed out that knowledge of results could guide users as they perform a task and could improve subsequent task performance. Knowledge of results can also increase user motivation. It can influence motivation by allowing learners to evaluate their progress, understand their level of competence, and sustain their efforts to reach realistic goals.

For VE applications in which measuring detailed user performance is less critical (e.g., entertainment; see Greenwood-Ericksen, Kennedy, & Stafford, 2014, Chapter 49), it is possible that performance measures could reflect cybersickness (see Lawson, 2014, Chapter 23). Therefore, for many situations involving immersion in VEs, there would be safety and probably legal (see Kennedy, Kennedy, Kennedy, Wasula, & Bartlett, 2014, Chapter 21) reasons to measure performance.

29.3 WHAT CRITERIA TO MEASURE

Because VEs allow for the measurement of a large variety of information about users and their actions, the following paragraphs provide a discussion of the concept of measurement in a general sense and as it specifically applies to VEs. We begin by introducing appropriate levels of performance measurement and associated exemplars. We then focus on psychomotor and physical measures of performance as useful reflections of user cognition. Next, we describe procedural task measures in terms of desirable and undesirable critical incidents followed by a discussion of cognitive tests implemented in VEs. Physical and cognitive errors are then discussed, with particular consideration of error taxonomies. At the end of the section, we pay particular attention to collective performance measurement, reflecting the recent emphasis on this area in VE design and application.

29.3.1 Taxonomies of Measurement

The National Research Council identified several early areas for VE application: design, manufacturing, and marketing; medicine and health care; teleoperation for hazardous tasks; training; education; information visualization; telecommunications and teletravel; art; and entertainment (Durlach & Mavor, 1995). Obviously, the performance measures of primary interest may differ greatly across these applications. The following paragraphs include a discussion of levels of measures, the comparative benefits of using established and novel measures, the importance of taxonomies, the use of primary and secondary measures, and the practice of relying on user characteristics as measures.

When devising the performance measurement system for a VE application, two levels of measures should be considered. The primary level will reflect the specific task domain that the VE system is designed to address. The second level of measures will support the interpretation of primary measures. Primary measures should indicate outcome, what the user accomplished in the VE system. Secondary measures suggest why performance was or was not successful.

For some application areas, well-established primary measures exist. For these areas, the challenge is to develop comparable measures within the VE system. For novel application areas, the need to approximate traditional approaches within the VE may reveal inadequacies in the traditional measures. Some new VE technologies may lead to new tasks for which there are no established measures or new tasks that can be performed only in VEs.

For example, some aspects of architectural design performed in a VE could be much easier to measure than in the real world. Interim processes could be more easily observed and recreated in VEs. Time to complete a design or time to complete a number of alternative designs could be easily and precisely determined. Raters or judges of interim designs and the final design could walk through and interact with the designed structure, annotating their observations. However, the ultimate performance measure of utility, style, and structural acceptability would still be defined and judged relative to real-world construction by highly trained and experienced human observers.

Although the primary performance measures of most interest will vary by specific VE application, there are common criteria to meet. For example, certain statistical or psychometric properties are desirable for all measures. Task taxonomies may provide a starting place in the identification of categories and dimensions of performance measures relevant to a specific VE application. Fleishman and Quaintance (1984) presented one example of a comprehensive taxonomy for tasks. Companion and Corso (1982) offered a general review of approaches to creating taxonomies. Fineberg (1995) described a taxonomy specifically designed to address performance in VEs. ANSI (1993) created a useful guide to performance measure taxonomies that included 10 categories:

- Time
- Accuracy
- Amount achieved or accomplished
- Frequency of occurrence
- Behavior categorization by observers
- Consumption or quantity used
- Workload
- Probability
- Space/distance
- Errors

Many widely used global measures (such as task success rate, assembly sequence errors, or magnitude of production) have at their basis speed or accuracy. However, for most areas of human performance, it is possible to trade off speed and accuracy. For example, if a VE user is instructed to *perform as fast and accurate as possible,* he or she will decide the relative importance of those two dimensions of performance. Setting a time limit has the advantage of keeping a constant duration of immersion, a consideration in controlling or measuring the incidence and severity of cybersickness (see Lawson, 2014, Chapter 32; Keshavarz, Hecht, & Lawson, 2014, Chapter 26).

Regardless of the VE task, interpreting primary measures and identifying ways to improve performance may be facilitated by secondary measures. Secondary measures include performance on subtasks and processes needed to complete the primary task, user characteristics, and conditions under which performance was measured.

Process measures are foundational to primary measures. They include abilities such as sensory acuity or basic locomotion that underlie the primary measures. Process measures are not limited to simple divisions or subtasks of the primary task. Cannon-Bowers and Salas (1997) extend the concept of process measures to include personal qualities such as assertiveness and flexibility. The distinction between outcome (primary or secondary) measures and process measures is at times a function of context; for example, manual dexterity may be a primary measure of interest for an assembly task, but may be a process ability that enables superior navigation performance. Process measures may be administered before, during, or after immersion in a VE.

There are many candidate user characteristics that may aid in interpreting performance measures. Examples are user age, sex, handedness, subject matter expertise, and experience computers or specific VE systems. It is common to see individuals who have extensive experience with VE-based gaming systems perform well in immersive training. Finally, measures of the conditions under which performance occurred might aid in the interpretation of primary measures. Examples are personal threat or environmental stress such as temperature extremes.

29.3.2 SENSATION/PSYCHOMOTOR MEASURES

Process performance measures discussed earlier may include measures of sensation and perception in the appropriate sensory modalities, various levels of consideration of movement through the VE (locomotion, travel, and navigation; see Templeman, Page, & Denbrook, 2014, Chapter 10; Darken & Peterson, 2014, Chapter 19), and interaction with objects (interrogation, selection, and manipulation). Gabbard and Hix (1997) presented a useful taxonomy of these VE functions. Lampton et al. (1994) developed a VE test battery that assessed several of these measures and showed them to be sensitive to control device. Many of these measures should be addressed during usability testing of a VE system (see Gabbard, 2013, Chapter 28). For some VE systems, it may be advisable to maintain the ability to collect *usability* measures throughout the life cycle of the VE system.

29.3.3 Physical Behaviors

Behavioral process measures such as navigation and locomotion discussed earlier may reflect real-world actions by the user and VE actions apparent by the user's representation. The former are increasingly important to consider because of advances in optical motion tracking. The latter are also important because of increases in the sophistication of avatar and nonplayer character (NPC) modeling and rendering. This section will detail important aspects of recording physical behaviors: the use of motion capture devices, the increasing popularity of low-cost motion capture systems, and the need for peripheral equipment to facilitate physical recording.

Measurement of physical behaviors such as limb and head movement can provide estimates or predictions of future physical movements within a VE. Motion capture devices have been used to assess virtual work environments for potential physical stressors that can result in musculoskeletal disorders (Ma, Bennis, Chablat, & Zhang, 2008). Motion capture devices can also be useful sources of information for after-action reviews in applied VEs such as medical and military simulations. Apart from tracking the physical motion of the limbs or head of the user, sensors may be placed on a handheld device. This type of motion capture has the ability to provide a more detailed account of response time associated with lifting the object, choosing among options, and then subsequent object-oriented actions.

Industry grade motion capture devices, although highly precise, can be extremely expensive. However, recent off-the-shelf motion sensing technologies and associated software development kits (SDKs) may provide an inexpensive alternative for motion data collection. The Microsoft Kinect has the capability of assessing limb and head movement and has been implemented in several research studies (Stone & Skubic, 2011). Considerations must be made, however, if fine motor movements are being assessed (however, see Oikonomidis, Kyriazis, & Argyros, 2011) or if the research participant is too close to the Kinect camera. If this is the case, extra equipment such as a focusing lens may be required. Another off-the-shelf device, the Nintendo Wii, has also shown some promise to be used as a low-cost alternative for a VE-based physical assessment (Alankus, Proffitt, Kelleher, & Engsberg, 2011).

29.3.4 Critical Incidents

Task domains for VE simulation range from mundane to critical. In the real world, the very nature of emergency situations often precludes careful documentation of what happened and why. In VEs, a structured data collection instrument should be made available to document what happened and the context in which the incident occurred. For example, it is fortuitous to approach simulated emergency situations in a VE system by putting in place a system designed to capture critical incident events in detail (Flanagan, 1954). In such applications, the term *critical incident* should not be limited to indicating an unwanted event, but any notable performance, whether good or bad, by the user or the VE system itself.

29.3.5 Knowledge Tests

As part of the effort to differentiate the effects of the VE system per se on performance, knowledge tests can be administered to users. A common practice during real-world assessment and training, knowledge tests can capture procedural and declarative knowledge. One particularly useful aspect of knowledge tests may be to determine whether a user or group of users has the prerequisite knowledge to benefit from VE training.

29.3.6 Error Documentation

As noted earlier, errors (physical or cognitive) are often included as primary measures for VE performance. The primary challenge with including errors is classifying them in a way that makes sense for the researcher or practitioner. Cognitive errors may reflect specific stages of mental processing.

Therefore, researchers have devoted significant effort to classify them. Norman (1988) identified two fundamental categories of errors: slips and mistakes. Slips are errors in the execution of otherwise correct plans. A mistake involves the execution of a plan as intended, but the plan was inappropriate to accomplish the overall goal. Comprehensive approaches to error analysis are presented by Reason (1990) and Senders and Moray (1991).

As VE user interfaces improve, detailed error analysis may assume greater importance relative to simpler measures of speed and accuracy. VE users often demonstrate calculated risks and may intentionally violate rules or procedures. Occasionally, a VE task may encourage greater risk taking by users. Some performance measures involving judgment within VEs may completely defy clearcut scoring as right or wrong, but rather will involve evaluating idioms, which are styles of artistic expression typical of a particular medium.

29.3.7 COLLECTIVE PERFORMANCE

The increasing sophistication of VEs has led to expanded use by teams. A review of existing team literature reveals complex issues that form the foundation of team study. Those issues include the focus of this section: the definition of teams, the makeup of team task elements, and the distinction between taskwork and teamwork as it relates to measurable behaviors.

Teams are groups of two or more individuals who share a common purpose, interact to accomplish a shared goal, and share responsibility for team performance and outcomes (Smith-Jentsch, Johnston, & Payne, 1998). Many organizations rely on work teams to complete critical missions and tasks. VEs allow geographically dispersed teams (GDTs), distributed teams, or virtual teams to interact across multiple locations with an increasing amount of flexibility in time, communication, and cost (Lipnack & Stamps, 1997).

Team performance consists of taskwork and teamwork (Smith-Jentsch et al., 1998). Taskwork includes the cognitive and technical skills needed to perform a task regardless of how the team accomplishes the task. Teamwork includes the social and communication skills needed to function within the team. Team processes are individual cognitive and behavioral activities that are combined to achieve collective goals (Marks, Mathieu, & Zaccaro, 2001).

A team's ability to communicate and synthesize visual, verbal, and contextual cues varies as a function of media richness (e.g., face-to-face, audio–video, audio-only, textual, immersive graphical representations) (Driskell, Radtke, & Salas, 2003). Distributed team coordination efforts are especially hindered by the difficulty associated with learning, conveying, and processing information critical to performance and by a lack of contextual cues and team or task artifacts (Hollingshead, McGrath, & O'Connor, 1993). Therefore, it is critical to measure differences in taskwork and teamwork processes associated with team effectiveness in distributed environments. Doing so allows knowledge of results and feedback to individuals.

29.3.7.1 Team Outputs

Considering taskwork requires instructors and researchers to address primary and secondary measures of speed and accuracy at the team level. Early VE systems that demonstrated effective taskwork performance measurement capabilities were the US Army's fully immersive team training (FITT) system (Parsons et al., 1998) and the virtual environment performance assessment battery (VEPAB) (Lampton et al., 1994).

The dismounted infantry visual after-action review system (DIVAARS) was another military system that focused specifically on after-action review. It provided dismounted soldiers with opportunities to comprehend their actions immediately following training by automating data recording and analysis (Goldberg, Knerr, & Grosse, 2003). DIVAARS offered playback controls, multiple viewing modes, movement tracks and bullet lines, audio and video recording, and the ability for leaders and researchers to view actions taking place inside buildings.

Proaps and Bliss (2010) measured VE-based search speed, target identification accuracy, and route efficiency by teams using the game-distributed interactive simulation (G-DIS), a modified version of Half-Life 2™ with a database that resembled the Fort McKenna military operations in urban terrain (MOUT) site at Fort Benning, Georgia (Lampton, Bliss, Orvis, Kring, & Martin, 2009).

A number of researchers have studied behavioral, motivational, and cognitive competencies or skills that constitute teamwork. Diversity among teamwork models makes it difficult to select the most appropriate team performance measurements, but researchers must select team performance measurements based on sound theory. The following competencies are often included in teamwork models: adaptability or adaptation, shared situation awareness (SA), initiative and leadership, feedback and monitoring, communication, coordination, giving and accepting suggestions, interpersonal cooperation, assertiveness, decision making, orientation, and shared mental models (e.g., Brannick, Prince, Prince, & Salas, 1995; Dickinson & McIntyre, 1997; Entin & Serfaty, 1999; Fiore, Salas, Cuevas, & Bowers, 2003; Salas, Sims, & Burke, 2005). Fiore et al. (2003) also mention trust, cohesion, collective efficacy, and collective orientation as important for teams.

Methods to assess such teamwork competencies may include observations, interviews, after-action reviews, task analyses, checklists, surveys, card sorting, process tracing, queries or freeze techniques, eye tracking, communication analyses, and experimental manipulations (Cooke et al., 2003). Each of these measurement techniques should be used within its appropriate theoretical framework. Many individual-level measures or team-level measures may be adapted to the team level if constructed around a team task analysis approach.

Finally, it is important to measure team composition factors that may impact colocated and distributed team performance. Kozlowski and Bell (2003) indicated that researchers should make note of team size, demographics, personality, KSAOs, and cognitive ability data. The possibility of automating such measure collection within VEs may improve the possibility of assessing team performance in an integrative fashion.

29.3.7.2 When to Measure Team Performance

Clearly from the previous discussion, a team's attitudes, behaviors, and cognitions can vary across assessments collected before, during, or after immersion in a VE. It is important to measure primary and secondary measures at various points during performance within a VE. Researchers and designers who desire comprehensive performance measurement in VEs should strive to incorporate both perspectives. Additionally, it is generally important to assess predictor (personality, performance, and aptitudes) prior to immersion in a VE. Doing so will enable realistic assessments of performance improvement, particularly during training scenarios.

29.3.7.3 Team Performance Measurement Requirements

There are specific team measurement requirements beyond those discussed at the individual level. For example, Jones and Schilling (2000) describe eight principles for measuring team performance: (1) capture team strategy, (2) align the strategy with the organization, (3) stimulate problem solving that results in improved performance, (4) use measurement to focus team meeting, (5) measure critical items, (6) ensure team members understand measures, (7) involve the customer in measurement development, and (8) address the work of each member (Jones & Schilling, 2000). Salas, Burke, and Fowlkes (2006) state that team measures should assess relevant team competencies as well as team, individual, and subteam performances, processes, and outcomes. Team performance measurement should also control task content, facilitate observation, and be as unobtrusive to team processes as possible. Last, team-level measures should incorporate multiple method approaches to prevent source bias and to ensure researchers measure a full range of performance indicators (Salas et al., 2006).

It is also important to measure reaction outcomes, learning outcomes (cognitive and skill based), affective changes, and behavioral outcomes within team training contexts (Salas & Priest, 2005). For an extensive review of team performance measures in scenario-based training

(SBT) and event-based training and assessment techniques (EBAT), readers are directed to Driskell, Salas, and Vessey (2013) discussion in Chapter 38 of this book.

29.3.7.4 Limitations of Team Measures

There are three methods for operationalizing team performance discussed by researchers. One method accounts for individual variance in scores. The second method uses the highest or lowest team member score to generalize to the entire team. The third, most common, method involves using individual scores to calculate a mean score for the group (Barrick et al., 1998). It is important to emphasize that aggregating data may not be appropriate for heterogenous teams. In such cases, it may be better to focus on the process more than the outcomes or to use a rank-order assessment method (Cooke et al., 2003).

With regard to communication analyses, there is little agreement as to which specific communication behaviors align with the most effective teamwork processes. Therefore, the validity and reliability of communication measures of team behaviors and cognitions are lacking. It is important to create practical, unobtrusive, reliable, valid, and diagnostic measurement approaches that can assess team-level performance in real time.

29.3.8 Reactions to Nonvisual, Nonauditory Sensory Stimuli

In the past, VEs have typically relied on the sense of vision to present stimuli to users. In some limited cases, designers explored the auditory modality for presenting stimuli as well; however, stimuli were typically presented in a limited way that lacked realism. Though 3D audio was explored in the early 1990s (see Begault, 1991), VEs in the 1990s did not usually include such advances. For this reason, options for measuring user performance were limited to those that reflected detection and reactions to visual or simple auditory stimuli.

Recently, researchers have begun to expand the use of nonvisual, nonauditory stimuli in VEs. Madrigal (2009) discussed early attempts to incorporate smell and taste into VEs. An example was the Sensorama, developed by Morton Heilig in the 1960s as a technology to expand theater experiences. It presented users with an enclosed viewing experience that included exposure to smells and tastes. Though somewhat successful at engaging the senses of users, Heilig failed to attract funding, and the Sensorama is now little more than a footnote in VE history (Robinett, 1994). Other technological solutions for chemical senses have included nasal inserts and mouthpieces. Madrigal discusses work underway by Chalmers and his colleagues at the University of Warwick Digital Laboratory to create a *virtual cocoon* for simultaneously stimulating all five senses. In 2003, Hiroo Iwata at the University of Tsukuba created a way to stimulate the sense of taste by injecting taste chemicals into the mouth (Ananthaswamy, 2003). Though most attempts to incorporate taste and smell have been tried for the sake of increasing the subjective feeling of presence, some efforts have specifically explored training benefits of doing so. As an example, Cater (1994) created a method for olfactory presentation for the purpose of training emergency responders. The advanced virtual environment real-time fire trainer (ADVERT-FT) presented odors along with radiant heat to train firefighters (Cater).

More recently, efforts have been made to incorporate taste and smell into the virtual experience; readers may find discussions of these by Kortum (2008) and Basdogan and Loftin (2008) as well as Chapter 6 of this book. Similarly, researchers have been very interested to incorporate the sense of touch into VEs. Broadly, touch can be considered as including aspects of tactile (cutaneous stimulation) and haptic (kinesthetic consideration of forces associated with holding and grasping) stimulation as well as the vestibular (balance) and proprioceptive (body orientation and position) senses. Though the terms *tactile* and *haptic* are often used interchangeably, it is useful to distinguish them, especially when discussing VE applications (see Loomis & Lederman, 1986). Much of the advancement in tactile and haptic stimulation for VEs has resulted from the design of devices to facilitate teleoperation. Solutions have included gloves, suits, joysticks, batons, and other contraptions specific to task domains such as medicine. Vestibular and proprioceptive displays often concern the interaction of the entire body with the environment.

Consideration of human perception and performance suggests that response time differs in reaction to stimuli in various modalities, with auditory reaction time quickest (Shelton & Kumar, 2010). Accuracy, too, may be affected, particularly when redundant or conflicting sources of information are presented to a VE user. Recent work by the military has explored the use of touch-enabled displays to promote SA and to facilitate communication. Investigation of communication modalities in VE contexts has shown differences in performance (Bliss, Liebman, & Brill, 2012).

One concern is the potential for reaction artifacts as more sources of information are implemented into VEs. For example, researchers have for many years documented the startle effect that degrades reaction time to loud or unexpected stimuli (Muhlberger, Wieser, & Pauli, 2006). As more diverse stimuli are incorporated into VEs, such effects may become more frequent and more pronounced. A more practical problem concerns the plethora of performance data that will accompany the incorporation of multimodal displays. Because performance data are generally event driven, researchers will necessarily need increasingly complex, capable sorting and filtering algorithms to organize generated events and their associated response behaviors. The challenge is already evident with optical tracking systems used to record physical movement as a function of task requirements or VE events.

29.4   PSYCHOPHYSIOLOGICAL MEASUREMENT IN VEs

The field of psychophysiology focuses on the relationship between psychological functioning and physiological responses (e.g., vascular and cardiac activity, ocular activity, skin activity, and brain activity; Andreassi, 2006). Psychophysiological measurement techniques have been used and interpreted as physiological correlates of underlying processes and constructs for decades (e.g., Kramer, 1991; Fitts, Jones, & Milton, 1950; Kahneman, Tursky, Shapiro, & Crider, 1969). Moreover, these techniques are becoming prevalent in studying interactions between human and technology. This has led to the discipline of *Engineering Psychophysiology* (Backs & Boucsein, 2000), the interest group *Psychophysiology in Ergonomics* (PIE), and *Neuroergonomics* (Parasuraman & Rizzo, 2007). Neuroergonomics has been referred to as the study of brain and behavior at work (Parasuraman, 2003). Although similar to psychophysiological-based studies, neuroergonomics is primarily concerned with those aspects associated with brain activity and not psychophysical measurement specifically (see Parasuraman, 2003, for distinctions between psychophysiology and neuroergonomics).

With advances in VE realism, these emerging fields, and their associated measurement techniques, are becoming a highly relevant topic for research using VEs. In the real world, many of the physiological correlates of task pertinent cognitive constructs may be impractical or impossible to measure during task performance. However, the controlled nature of VEs can allow measurement of electroencephalographic (EEG) activity, eye movements, blink rate, cardiovascular activity, and galvanic skin responses. The following section will provide a brief overview of several physiological measurement techniques and associated cognitive constructs, paying particular attention to workload, sustained attention, and SA (for a more in-depth review, the reader is referred to Backs & Boucsein, 2000 and Parasuraman & Rizzo, 2007).

Workload: Frequently, researchers are interested in the measurement of workload, as this can have a direct impact on task performance. Various physiological measurement devices have shown promise in evaluating workload and could provide useful input in VE settings. Research has shown that pupil diameter varies according to changes in task demands with a range of cognitively based actions that index perception, reasoning, and memory (Beatty, 1982; Kramer, 1991; Matthews, Davies, Westerman, & Stammers, 2000). However, although pupil diameter is highly sensitive, responses may occur for other reasons such as ambient illumination and emotional arousal (Matthews et al., 2000). Dwell time may also be an indication that the individual is extracting information from certain areas. Dwell times can differ by level of expertise, where novices dwell longer than experts as a function of workload. Finally, in multitask environments, fixation points can indicate the task that is the greatest source of resource allocation (Wickens & Hollands, 2000, p. 466).

EEG devices may also be used to measure workload via event-related potentials (ERP), more specifically the P300. The P300 signal occurs in response to the investment of cognitive and perceptual processing resources (Wickens, 2002). Researchers have shown that as the primary task difficulty increases, the P300 response rate for a secondary task decreases, suggesting an investment of resources (Matthews et al., 2000). However, traditional multielectrode devices can be expensive and require a trained individual to properly place electrodes, ensure the removal of hair at electrode sites, and then apply electrode paste to provide adequate connectivity.

Sustained attention: Many of the tasks once carried out exclusively by human operators are now being automated to some degree. Because of the increasing prevalence of automation, in many cases, the role of the human has transitioned to that of a system monitor (Parasuraman & Riley, 1997). One of the potential performance issues associated with sustained attention is the vigilance decrement. Sustained attention tasks have traditionally been studied under conditions in which an operator must detect scarcely issued signals over long periods of time. Generally, the vigilance level, the ability to maintain a steady state of vigilance performance, is lower than desirable and decreases rapidly after 30 min (Wickens & Hollands, 2000, pp. 34–44). Psychophysiological measurement techniques have been used to study this phenomenon. Warm and Parasuraman (2007) describe a series of studies that monitored cerebral blood flow velocity and oxygenation during vigilance task paradigms. The authors suggest that as a watch progresses, a decline in signal detection is accompanied by a similar decline in blood flow as time progresses. Transcranial Doppler sonography (TCD) is a technique that can be used to measure cerebral blood flow velocity. TCD is a relatively unobtrusive technique, which offers real-time and continuous measurement of cerebral blood flow.

Situation awareness: In addition to workload and attention, many applied domains are concerned with operator SA. Endsley (1995) defines SA as "the perception of the elements in the environment within a volume of time and space, the comprehension of their meaning and the projection of their status in the near future" (p. 36). From this perspective, SA has been conceptualized as a product, rather than the process used to attain a particular degree of SA. However, an index of this process may be measured via eye-tracking technologies (Salmon, Stanton, Walker, & Green, 2006; Smolensky, 1993). Although it may seem reasonable to associate ocular measures, such as visual scanning, with levels of SA, research using eye trackers has found that even though individuals have fixated on an object, they are still unable to later recall seeing the object (Strayer & Drews, 2007). This suggests that users may not necessarily *notice* what it is they have viewed and researchers should be cautious when drawing SA-related inferences from these data (Salmon et al., 2006). For this reason, eye tracking may have more applicability in the detection of task difficulty and levels of workload than SA.

29.4.1 Psychophysiological Approaches to Measurement

As suggested earlier, psychophysiological measurement may in some cases bolster measurement of human performance in VEs and can provide insight to the underlying mechanisms of performance. However, important issues surround strategies for data collection.

Baseline measurement refers to collecting psychophysiological data before any target experimental stimulus of interest is introduced. It is the initial assessment of the physical and psychological state of the individual, including daily changes to psychophysiological characteristics such as fatigue, concentration, and anxiety as well as long-term psychopathologies and medical problems. Both state and trait variables are essential to recognize for two main purposes: to separate the state and trait characteristics from the performance variables related to the VE and to categorize the participant.

Categorically, baseline measurement can separate individuals according to a performance variable of interest. Advances in technology and research have provided the ability for researchers to use psychophysiological baseline data to predict individuals' performance in areas such as task functional state, task performance, workload, adaptability, accuracy, and anxiety. This approach is used not only to predict future performance but also to separate performance-related variables after the experiment.

A correlation approach is often used to determine the specific psychophysiology markers characteristic of human-system variables assessed by subjective measures. The measures of interest here would be those that measure human-system variables such as sense of presence (see Chertoff & Schatz, 2013, Chapter 34) or simulator sickness (see Keshavarz et al., 2013, Chapter 26). Correlating these measures with psychophysiological data clarifies human characteristics underlying the reports and facilitates subsequent analyses to identify desired and relevant evoked responses to various virtual stimuli.

Measuring evoked responses is helpful when an experimenter wishes to determine information about the influence of environmental variables on cognition and performance. The experimenter can separate and analyze various aspects of a condition or system that may induce immediate psychophysiological change such as arousal, attention, and distractibility. As designers create compelling and sensitive VEs, importance is placed on human responses to stimuli that are increasingly real in appearance and VE components.

During real-time psychophysiological monitoring, researchers may choose to identify unwanted cognitive or physiological states and then either provide indicators such as tones or alter the environment and virtual stimuli to return the operator's psychophysiological performance to its desired state. Such an approach is common with therapists who employ VEs and psychophysiological monitoring with biofeedback to reduce acrophobia or agoraphobia. Also, feedback techniques have been used with pilots to maintain a necessary cognitive state of awareness.

29.4.2 Types of Measures and Performance Indicators

Most current psychophysiological techniques involve placing and securing electrodes over the areas of the body appropriate to the types of performance indicators (e.g., concentration, anxiety, and workload) to be measured.

Electroencephalography (EEG) is the recording of electrocortical brainwaves using an electroencephalograph (EEG). This area of applied psychophysiology has rapidly become a commonly used tool for assessing human performance primarily because of its ability to specify internal cognitive functions (e.g., visualization, attention, workload) as well as most physical processing of external stimuli. Unlike other psychophysiological measures, EEG has been shown to predict and indicate a wide range of performance-related variables. The brainwave recording procedure requires that electrodes be placed on the scalp to compare and measure various regions of brain activity.

Analysis of brainwave activity can provide highly sensitive measures of changes in cognitive task difficulty and working memory. When brainwave variables are analyzed in various ways, they can accurately discriminate and predict specific cognitive states. The raw EEG data are broken up into specific bands, usually ranging from 0.5 to 32 Hz, each with specialized contribution to the various performance indices.

A working-memory increase due to task load results in an increase in the frontal theta band (4–7 Hz) and a decrease in alpha band (8–12 Hz; Gevins et al., 1998). A decrease in alpha is usually indicative of fatigue and lack of attention or vigilance and has been used to reflect increased workload. Crawford, Knebel, Vendemia, and Ratcliff (1995) point out that the lower part of alpha band (7.5–9 Hz) can actually differentiate low and high sustained attention. Frontal midline (FM) theta is a useful indicator of increase in concentration and expertise (Yamada, 1998). An increase in task complexity can be determined by relative frontal and central beta (13–32 Hz) activity increase (Wilson, Swain, & Brookings, 1995). The task engagement index has been able to reflect overall task engagement best and is derived by dividing beta by alpha plus theta ($\beta/(\alpha + \theta)$). The sensory–motor reflex (SMR, 12–15 Hz) is used to determine a lateralized readiness potential (LRP). This is a type of event-related potential, useful in determining the processes that prepare the motor system for action (the preparation to respond) and that initiate but inhibit the response before it takes place (Coles, 1998).

Electromyography (EMG) is the process of recording the firing of muscle units as they relate to other muscle units. The data can determine tension and even the slightest of muscle activity without any visible muscle movement. EMG can be used in combination with EEG to weed out cognitive activity from signals related to muscle movement. Tension in the forehead and eye movements can produce EMG artifact signals in EEG recording, but with the use of recorded EMG on facial locations, these artifacts can be easily recognized and removed.

EMG is a measure of tension and thus can provide insight on patterns of bodily reactions, such as startle responses, tension and movement, and breathing effort. An increase in task difficulty can be identified by an increase in EMG from the neck and shoulder muscles (Hanson, Schellekens, Veldman, & Mulder, 1993).

Electrocardiogram (ECG or EKG) measures the electrical activity that spreads across the muscle of the heart. The spread of excitement through the muscle of the ventricle produces the QRS complex, a waveform in which the intervals between the R waves indicate heart rate (HR). HR deceleration is indicative of such activities as orienting to neutral or pleasant stimuli, appropriate responding, and quicker reaction-time responses. HR acceleration is a characteristic response to stimuli that induce a defensive reaction or to unpleasant stimuli. Recent research suggests that cardiovascular indices can indicate individual coping strategies to external noise in order to maintain task performance (Hanson et al., 1993). Research also indicates that respiration rate, HR, and electrodermal activity (EDA) correlate with degree of immersion or presence in the VE (Wiederhold, Davis, & Wiederhold, 1998).

Blood pressure (BP) is a measure of systolic (contraction) over diastolic (relaxation) cardiovascular activity. An increase in BP can indicate an arousal reaction occurring in performance situations that are mentally or physically strenuous (e.g., Rose & Fogg, 1993; Seibt, Boucsein, & Scheuch, 1998; Sharma et al., 1994). Blood volume is a measure of slow changes in blood volume in any limb, whereas *pulse volume* reflects rapid change related to heart pumping and vascular dilation and constriction. Human performance variables that include orienting responses to stimuli can be determined by a general increase or decrease in blood volume. An increase is indicative of improved perceptual ability; a decrease suggests a defensive response to unpleasant stimuli. Measures of finger pulse volume have been shown to be sensitive to stress manipulations and predictive of anxiety (Smith, Houston, & Zurawski, 1984). Recently, Schmidt-Daffy (2013) used pulse volume amplitude as a measure of anxiety-related arousal in a driving simulator study. Results showed that when participants were prescribed a speed at which to travel, pulse volume amplitude was less affected by task demand components than other physiological measures (EDA and HR).

EDA, also known as skin conductance level, is a measure of surface skin electrical activity. The higher the sweat rises in the eccrine sweat glands (located in palm and foot soles), the lower the resistance. EDA is of interest in human performance research because it responds to both external and internal tonic activity. Eccrine glands respond to psychological stimulation instead of body temperature change and can indicate many of the same cognitive variables as the EEG. The number of skin conductance changes in a given period of time is useful in indicating the preparation of physical motor activity. A change in skin conductance and number of spontaneous skin responses are both associated with faster reaction time (Wilson, 1987). An overall increase in EDA is indicative of better task performance, signal detection, and learning (see review by Kramer, 1991). These variations in performance, however, may be due to individual differences in response biases (see Parasuraman, 1975). A decrease in EDA is associated with increased task complexity and difficulty.

29.4.3 Data Collection Considerations

To obtain useful and reliable data, recordings must be taken in highly controlled settings considering both the individual and environmental factors that influence human physiology such as age, gender, neurological damage, noise, and temperature. Moreover, research suggests that electromagnetic field exposure from equipment may change the psychophysiology of participants

(Morris, Shirkey, & Mouloua, 2000). Andreassi (2006) discusses in-depth psychophysiology in reference to human-system performance. Fahrenberg and Wientjes (2000) also provide an overview of methodological issues that should be considered before using physiological measures in VEs and applied environments. Additionally, prior to data collection, researchers should acknowledge specific methodological considerations of each physiological response used in an investigation. Methodological guidelines and considerations have been provided for the following physiological measures:

- ERPs: Picton et al. (2000)
- Heart rate variability (HRV): Berntson et al. (1997) and Jorna (1992)
- Respiratory sinus arrhythmia (RSA): Denver, Reed, and Porges (2007)
- BP: Shapiro et al. (1996)
- Eyeblink startle responses with electromyogram (EMG): Blumenthal et al. (2005)
- Respiration: Ritz et al. (2002)
- HR, HRV, and respiration: Mulder (1992) and Porges and Byrne (1992)

29.4.4 Advances in Low-Cost Psychophysiological Measurement

Behavioral scientists have been committed to the idea of measuring brain and related body activity as directly as possible for many years. In addition to recording electrical activity from the brain, heart, or muscles as described earlier, researchers have also devoted considerable effort to record specific localized behaviors. For example, eye movement recording has long been acknowledged as a way to capture attention. Specific methods used have included dark-pupil tracking and light-pupil tracking (Richardson & Spivey, 2004), as well as electrooculography (recording electrical potential changes in muscles near the eye; Rohner, 2002). More broadly, technologies have been refined to track movements of body parts in space. Historically, methods were developed to track body movements by inertial, mechanical, or magnetic means. More recently, optical motion tracking systems have been developed and refined so that body parts may be tracked faster and with greater precision.

The psychophysiological and behavioral recording methods discussed previously have traditionally required significant effort and cost to integrate into VEs. However, in the last few decades, there have been tremendous advances in low-cost physical and physiological recording. For EEG, new systems such as the Mindset by Neurosky and the Emotiv 14-channel recorder have been explored by researchers. Similarly, low-cost EKG recording devices by companies like Polaris and Zephyr have enabled unobtrusive, convenient, and powerful heart activity measurement. Since the development of the Microsoft Wii and Kinect, optical tracking of body parts has been rendered cheap and increasingly capable. Even eye tracking, once possible only with expensive and cumbersome equipment, has been shown to be possible with low-cost systems such as the iGaze tracker and eye-tracking-enabled laptop computers.

Ultimately, such advances are presenting new possibilities for data collection and performance measurement in VEs. However, because of the complexity of physiological recording and the challenges associated with interpreting physiological data, researchers are urged to exercise caution when implementing low-cost solutions for data capture. As an example, though the Neurosky EEG system has been shown to approximate the recording capabilities of more expensive systems (Grierson & Kiefer, 2011), the conditions under which it operates (ambient light, sound, and distractions) must be tightly controlled for accurate measurement. Such conditions are difficult to achieve in a typical VE. Furthermore, researchers are encouraged to carefully consider the implications of movement recording for the task of interest and in light of overall VE task parameters.

29.5 LOGISTICS OF MEASURING VE TASK PERFORMANCE

Measuring task performance in VEs is not always a straightforward goal. As discussed later, the flexibility of a VE provides certain strengths, but also certain challenges, for measurement. To illustrate these, this section presents examples of VE task measurement from the last three decades, focusing on the importance of measure perspective and application to training.

One of the main strengths of performance measurement in VE is the ability to provide a human observer with flexible, comprehensive views of a VE system user's actions in the VE. For many systems, it is possible to provide a simultaneous view of multiple users' actions, including interface manipulations as well as environmental changes. Some VE systems allow automatic capture of detailed user performances, such as operation of the input control devices, as well as comprehensive perspectives of the user's actions in the VE.

An example of automated performance measurement might be assessing the number of collisions with walls or objects in the VE. Minimum real-time processing of such factors should occur to conserve processing power. Similarly, more detailed processing should be delayed to avoid degrading system performance while the user is on the system. During performance, observers or experimenters could maintain a checklist to record desired measures that cannot be automated. Such activity should be minimized, however, so attention can be paid to controlling the session and safety. In some cases, several raters may score playback independently and replay of a session could help resolve interrater differences.

Beginning in the late 1980s, the US Army's simulation networking (SIMNET) system demonstrated the powerful performance measurement advantages that VE systems can provide for training applications. SIMNET employed a stealth vehicle, invisible to training exercise participants, that allowed trainers and exercise controllers to view training exercises from any perspective within the VE. In addition, a training mission could be replayed and viewed from any angle. Such a tool is a useful addition to many training VEs.

In the 1990s, the FITT research system allowed the observer/mission controller to view the unfolding training exercise from any perspective while trainees wearing head-mounted displays (HMDs) practiced urban search and rescue missions within VEs. The view presented to the observer was intentionally modified from the view of the participants to aid the observer in scoring trainee performance. For example, the observer could view the unfolding training exercise from any angle and zoom in or out. In addition, equipment in use by the trainees was represented to the observer at a much larger scale, so it is easier for the observer to identify. The FITT system featured a replay function that allowed normal, slow, and fast playback. Technical aspects of data sampling rate, filtering, capture, and storage are described in Parsons et al. (1998) (Figure 29.1).

In the 2000s, the US Army Research Institute developed the DIVAARS. This system featured many of the same aspects as FITT, such as the replay function and multiple points of view. In addition, live audio and chat communication was implemented as well as detailed troop movement traces and extensive performance measurement. This allowed improved ability to measure SA (see Lampton et al., 2006). At about this time, the Army began to embrace the use of immersive game-based systems for training critical skills. One important aspect of such systems was the potential to practice distributed exercises. This facilitated the study and refinement of team leadership skills (Proaps & Bliss, 2010).

With modern immersive systems, the scoring playback can be conducted with or without the user. In addition, the user may be interviewed and provided questionnaires to help interpret what occurred and why. Interviews are in general more flexible and easier for the user than questionnaires. Questionnaires are easier to administer and provide for more private responses than interviews.

FIGURE 29.1   The FITT research system replay station.

29.6   WHEN TO MEASURE PERFORMANCE IN VEs

An individual's attitudes, behaviors, and cognitions can vary across assessments collected before, during, or after immersion in the VE. Some measures such as severe susceptibility to motion illness or seizures should be determined before the participant is immersed in the VE. These measures may be used for screening out potential users of the VE system or at least providing a cautionary alert. However, when feasible, other background measures may be taken after the participant has completed the session in the VE. Scheduling paperwork for this time provides a safety period between use of the VE system and activities such as driving. Even though it may be feasible to have an automated means of collecting PC-based or even VE-based information before or after immersion, from the standpoint of mitigating cybersickness, it may be preferable to have the participant complete paper-and-pencil questionnaires rather than prolong the amount of time spent looking at electronic displays. Because there are so many potentially important variables, when feasible, a take-home package approach may be used in which the VE system user completes and returns questionnaires hours or days after using the system.

Suggested time lines for performance measurement activities before, during, and after immersion are presented in this order in Tables 29.1 through 29.3. These tables integrate recommendations on what, how, and when to measure performance in VEs.

TABLE 29.1

Example Time Line of Preimmersion Performance Measurement Activities

Research Participant or System User Recruitment

- Off-site telephone screening to confirm user has at least minimally appropriate sensory and psychomotor skills and to identify safety issues
- Preimmersion health status questionnaire
- Preimmersion indices of cybersickness symptoms
- Preimmersion baseline physiological measures
- Review of participant's previous performance on the system (when appropriate)

TABLE 29.2

Example Time Line of Performance Measurement Activities during Immersion

- System data capture
  - Real-time or frequently updated summary statistics of performance measures
  - Displayed to participant and controller
  - Stored for replay and subsequent data analysis
- Physiological measures
- Observer/controller checklists
- Participant comments/dialog
- Critical incident intervention (e.g., when appropriate, a safety intervention)

TABLE 29.3

Example Time Line of Performance Measurement Activities after Immersion

- Indices of cybersickness
- Postexposure physiological measures
- Replay
  - Participant's comments
  - Controller's comments
- Open interview
- Structured interview
- Questionnaires
- Take-home packet
- Follow-up interview or questionnaire
- Off-line data analyses (e.g., summary statistics across immersion sessions)
- Scoring of session replay and videotape by different scorers or raters
- Session to reconcile difference in scoring
- Archive performance measures in system database

29.7 PSYCHOMETRIC PROPERTIES OF MEASUREMENTS

29.7.1 Psychometric Criteria for Performance Measures

VE researchers have the ability to measure a myriad of performance variables in an automated fashion, but face unique psychometric obstacles. In traditional (non-VE) experimental contexts, it can be a challenge to get enough human performance data, so researchers often resort to using surveys, questionnaires, or other measurement tools. In contrast, researchers using VEs often must make a conscious decision to limit the data generated by the computer system or to filter it after collection.

Because of the many possibilities for data capture in VEs, researchers are encouraged to carefully select performance measures that are psychometrically sound. Experts generally agree on two criteria by which to judge performance measures: reliability and validity (Whitley, 1996). However, sensitivity or discriminability of performance measures is also important (see Leavitt, 1991). In the following paragraphs, each of these criteria is defined and discussed as they pertain to VE performance measurement. Where possible, we include examples of VE research that features optimization of the criteria.

29.7.1.1 Reliability

Performance measures should reflect behavior in a stable fashion. When measuring human performance in VEs, it is especially important to employ reliable measures because of the large and changing variety of VE hardware, software, tasks, and testing paradigms available.

Conventional explanations of reliability center around two issues: repeatability and internal consistency. Repeatability concerns whether the measure reflects performance consistently across time and testing administrations. Researchers who use VEs for training are especially interested in repeatability because performance measures that fluctuate across time or testing sessions are not comparable. Therefore, it becomes difficult to chart a trainee's progress or expertise relative to other trainees. Determining the repeatability of a performance measure is typically done by correlating measures taken during one experimental session with those taken during a subsequent session.

Internal consistency reliability indicates how well a performance measure reflects a unitary construct. For example, it is important that a measure of accuracy in a VE pick-and-place task not include an aspect of timeliness. Otherwise, it would be difficult to isolate the cause of observed variability. Most researchers who study human performance calculate test–retest reliability as a substitute for internal consistency because it is more common and interpretable and is roughly comparable to internal consistency (Whitley, 1996).

Researchers and task designers often desire to know what constitutes acceptable reliability. Although most experts have difficulty agreeing on a precise value, Whitley (1996) suggests that test–retest correlation coefficients should approximate 0.5. However, this value generally applies to paper-and-pencil tests. It is reasonable to assume that acceptable coefficients for other task performance data might be somewhat lower, given the relative complexity of such measures and the variability of conditions across which performance may be observed.

Researchers frequently compare task performances in VEs to those in the *real world*. In some respects, data collected from immersive task performances may be more reliable than data from an actual task performance. Just as with conventional simulation, researchers using VEs have the capability to standardize aspects of the immersive experience and to hold certain task parameters (e.g., criteria for success, difficulty, and duration) constant.

29.7.1.2 Validity

Although repeatability and consistency are necessary qualities for performance measures, researchers must also examine the validity of measures. It is crucial that performance measures be faithful indicators of the constructs they presume to reflect. As noted by Salvendy and Carayon (1997), reliability and validity must be examined together, because a measure may consistently reflect a construct of little relevance.

Evidence of performance measure precision may be reflected by content validity (whether the measure adequately measures the various facets of the target construct), construct validity (how well the measured construct relates to other well-established measures of the same underlying construct), or criterion-related validity (whether measures correlate with performance on some criterion task).

All techniques for establishing validity are inferential. Whereas it is relatively easy to calculate a reliability coefficient directly from available performance data, determining the validity of measures often requires researchers to rely on evidence obtained in an indirect fashion. Frequently, such evidence includes expert judgments or correlations with other marginally validated measures. Historically, validating measures taken in an immersive setting has been difficult to accomplish; however, Parsons and Rizzo (2008) have successfully used a multitrait-multimethod approach to demonstrate convergent and discriminative validity of their virtual reality cognitive performance assessment test (VRCPAT). Other successful validation demonstrations have occurred in specific domains such as medicine (Kundhal & Grantcharov, 2009) and for broader manual skills applications such as sports (Huegel, Celik, Israr, & O'Malley, 2009). It is encouraging to see such efforts; as technology improves, researchers should continue to include measure validation as an objective within their work.

It may be feasible to estimate construct validity of some VE performance measures by comparing them to measures in other contexts such as the real-world (where practical) or well-established task simulations (if such simulations exist). For example, estimating the construct validity of procedural errors in a virtual mine-clearing task may be possible by comparing those errors to performance in more conventional (real world) mine-clearing training programs. However, such validation is not practical for many immersive tasks, because there is no usable standard for comparison or because the task is too dangerous, costly, or complex to be trained in a nonimmersive fashion. Promising research has been reported by Bittner, Mellinger, Imam, Schade, and Macfadyen (2010) who demonstrated construct validity of measures for VE training of a pancreatic visualization procedure, endoscopic retrograde cholangiopancreatography (ERCP). Their measures included times taken to complete various procedure elements, success of the simulated procedure, and number of complications.

As an alternative to construct validity, researchers may wish to determine criterion validity. This may be more feasible than establishing construct validity, because measures of concurrent or future performance may be readily available. For example, data concerning mine-clearing expertise would presumably be available from other test scores, rankings, or safety records. Unfortunately, using future performance as a criterion requires the researcher to wait some time before obtaining comparison data. Using current data is quicker but often forces the researcher to sacrifice comparability. For example, it would be more meaningful to compare VE mine-clearing task measures with future success in mine-clearing operations. However, researchers may opt to use currently available measures (such as performance on a test of mine-clearing equipment). In the medical realm, Oropesa et al. (2011) provide a useful discussion of validity as it applies to metrics for VE simulators of laparoscopic surgery, including a differentiation of validity types and their applicability to medical simulation data.

To maximize convenience, researchers often judge precision by consulting with subject matter experts. Content validity procedures make intuitive sense for situations where there are no current or future data available or where there are no other well-established measures of the construct. The disadvantage is reliance on experts who may or may not have adequate knowledge to judge performance measure acceptability.

For traditional paper-and-pencil tests, acceptable validity coefficients typically approximate 0.3–0.4 (Nunnally & Bernstein, 1994, p. 100). As with reliability, however, the difficulties associated with validation of VE performance measures could drive this figure lower. Yet, given comprehensive knowledge of the target task, researchers have the capability to maximize performance measure validity by exercising creativity. For example, researchers and practitioners have successfully used VEs to treat posttraumatic stress disorder in returning war fighters (Biron, 2012).

29.7.1.3 Sensitivity

Instructional VEs should be created so that good performers may be distinguished from poor performers. Although sensitivity of data is not always identified as a psychometric criterion, it is arguably as important as reliability and validity, particularly in performance measurement situations. According to Leavitt (1991), sensitivity refers to the capability of a performance measure to adequately discriminate between experts and novices. It is conceptually separate from reliability and validity because performance measures available in VEs may be consistent and may truly reflect the construct of interest. However, they may not reflect sufficient variability to be of use.

In cases where dependent measures lack sensitivity, the data may reflect ceiling or floor effects, where the majority of participants score at the high or low end of the measurement range, respectively (Nunnally & Bernstein, 1994). As an example, in some of the early VE research concerning spatial navigation training, a majority of participants had problems maintaining spatial awareness and easily got lost within VEs. As a result, the data did not reveal much, because few people could master virtual wayfinding tasks (see Darken & Peterson, 2013; Chapter 19).

A common problem associated with low sensitivity is that the experimental data lack variability. Because many statistical techniques rely on variability, it may be difficult for experimenters to draw meaningful conclusions from their statistical analyses.

Unfortunately, there are no established guidelines for sensitivity measurement. Generally, it is the researcher's responsibility to examine the raw data to determine whether performance measures show acceptable levels of variability. Provided that researchers are aware of the need for sensitive measures, VEs offer the potential to optimize sensitivity. Zeltzer and Pioch (1996) provide an example of the effort necessary to ensure good sensitivity of performance measures. They describe in detail the steps they followed to construct their VE system for training submarine officers and point out the importance of consulting with subject matter experts throughout the VE development process to determine appropriate task elements to model and the degree of fidelity with which to represent them.

29.7.2 Issues Impacting Data Integrity in VEs

Although immersive technology may hold promise for reliability, validity, and sensitivity of performance measures, researchers should acknowledge VE-related issues that threaten measurement rigor. The following sections identify specific problems researchers may face when collecting data in VEs.

29.7.2.1 Hardware Issues

During the advent of VE technology, it was not uncommon to see processors running at a mere 66 MHz driving to update visual displays. One consequence of slow processors was commensurate delays in screen refresh rate. As a result, participants often had to perform tasks at unnaturally slow speeds as they waited for the system to *catch up*. Kenyon and Afenya (1995) noted that excessive lag led to task performance idiosyncrasies in a pick-and-place task, such as physical movements that were overly deliberate. Unfortunately, lag may also have a negative effect on performance measure sensitivity by masking performance differences between experts and novices. Though improvements in processor speed and use of multiple processors have improved the situation, the problem is still evident, particularly in some game-based or distributed applications. For example, it has been suggested that the Kinect system may introduce enough lag to negatively affect some users (Perna, 2010). However, recent work to develop the new Oculus Rift HMD suggests that developers have worked to minimize the effect of lag while striving to maintain low cost (Morris, 2012).

In other cases, performance measures such as reaction time may not reflect true human reaction time at all, but human reaction time filtered through an overburdened processing interface. Today, the situation has improved dramatically because graphics processors are much faster. However, even the fastest processors may be bogged down by calculations required by complex environments, position tracking, and peripheral device use. Watson, Walker, Ribarsky, and Spalding (1998) discuss how task performance may be influenced by reduced system responsiveness.

In some cases, computing resources used to generate and update VEs and to collect performance measures may be accessible by multiple users on a network. Unfortunately, additional users may increase the processing load on a given processor so that VE rendering occurs at variable rates. Variability in processing loads may decrease performance measure reliability within and among participants.

Performance measures taken from first-generation VE systems often reflected sporadic equipment problems such as a visual channel that malfunctioned from time to time or position trackers that drifted in and out of calibration. When equipment is unreliable, validity may also be compromised because measures may not reflect human performance on the task so much as human adaptation to equipment idiosyncrasies. Sensitivity may also be in jeopardy because of the indiscriminant nature of some equipment problems unpredictably impacting the performances of experts and novices.

Early VE equipment configurations were often uncomfortable as well as unreliable. According to Durlach and Mavor (1995), the discomfort associated with early HMDs led some researchers to consider using off-head displays. Additionally, Stanney, Mourant, and Kennedy (1998) suggested that use of some peripherals may lead to repetitive motion injuries such as carpal tunnel syndrome or tenosynovitis. Such discomfort may negatively impact performance measure reliability because greater performance decrements are likely with prolonged peripheral use. Validity also may suffer because performance data may partly reflect frustration with peripheral devices. Sensitivity may be compromised because participants of all expertise levels may react differently to uncomfortable peripherals.

A related challenge faced by participants is learning to use peripheral devices to navigate and manipulate objects. As shown in early VE research by Lampton et al. (1994), such learning is not a trivial matter. Unfortunately, the reliability and stability of measures often suffer because the normal learning curve is extended to account for control device unfamiliarity. Also, participants may show reluctance to exploit the advantages of immersion (such as head rotation). Some researchers have harnessed voice to interact with elements of the virtual world (Cohen, 2002). If successful, voice interaction may provide a way to overcome the performance limitations of uncomfortable, unreliable peripheral devices. Additionally, there has been work by researchers to specifically improve ergonomic aspects of immersive input devices (Stefani et al., 2005).

Given the plethora of equipment manufacturers and the transient popularity of VE equipment, research participants may navigate and manipulate using many different devices. As a result, performance measures may not be comparable from one experiment to another. Sensitivity likely suffers because experts learn to perform well using one equipment configuration but not another. Additionally, there is the potential for performance measures drawn from low-fidelity systems to have poor validity when compared with measures taken from higher-fidelity VE configurations or from real-world performance conditions. Such differences have implications for training transfer. Consequently, there have been many investigations of training transfer from virtual to real environments (see Hamblin, 2005). Most conclude that transfer is influenced by a myriad of factors, including psychological fidelity, physical fidelity, functional fidelity, cognitive aspects, task elements, and technology. Researchers and developers alike should keep transfer in mind, especially as more use is made of modern immersive systems for rehabilitation and training.

29.7.2.2 Software Issues

VEs that are used for research vary widely in terms of their complexity. Some researchers use simple models consisting of 20,000 polygons or less with elementary object dynamics, and others use models that are extremely complex (over 1,000,000 polygons) and place heavy computing demands on processing resources. Not only does model complexity differ among environments, it may differ within the same environment, making measures of performance appear unreliable. Reaction-time estimates may fluctuate widely between simple and complex environments or between locations within the same environment. Furthermore, the difference between novice and expert performance may vary with environment complexity, decreasing measurement sensitivity.

Another issue concerns fidelity or how closely a constructed environment approximates an operational environment. A distinction may be made among physical, functional, and psychological or cognitive fidelity. Physical fidelity refers to the physical similarity of the VE to the actual environment (e.g., visual, equipment, auditory, motion features, and components of a task) (Hays & Singer, 1989). Functional fidelity refers to the capacity of the VE to provide a realistic interaction with the user (e.g., providing feedback and reacting to the user's inputs) (Hays & Singer, 1989). Psychological or cognitive fidelity refers to the extent to which underlying psychological constructs and processes required for effective performance are present in the VE (e.g., immersion and presence) (Bell, Kanar, & Kozlowski, 2008). It is plausible to assume that a lack of fidelity will compromise measurement validity. Performance measures may not reflect the constructs they were designed to reflect because participants will not interact with the environment in a natural, spontaneous fashion. There is research to suggest that learning and performance may be enhanced in well-designed VEs that are high in fidelity because participants experience higher levels of enthusiasm, motivation, presence, and SA (Mantovani & Castelnuovo, 2003).

29.7.2.3 Task Issues

In some cases, the structure of the tasks that participants are required to perform while immersed may alter the reliability, validity, or sensitivity of the performance measures. Some immersive training situations may in effect be part-task simulation exercises because the training task is simplified in the virtual world. Such simplification may occur for a variety of reasons, including costs associated with full-task programming or the desire to conserve processing power. However, if researchers create VEs without conducting a task analysis (such as that advocated by Stanney, 1995), the result may be measures with low validity. In addition, sensitivity may suffer because of the potential for ceiling effects.

Another potential threat to sensitivity concerns the fact that some research participants may be confused or overwhelmed by the immersive experience. Bliss, Tidwell, and Guest (1997) noted occasional confusion by older participants who were not familiar with computers or VEs. In cases such as these, the performance measures chosen may lose sensitivity due to floor effects, where most of the participants' cognitive resources are focused on adjusting to the novelty of the VE.

Because of modeling and rendering limitations, some simulations of operational tasks may lack realism. For example, in many operational environments, time is a critical constraint. Other factors difficult to replicate include environmental influences such as heat, cold, vibration, or noise. If aspects of the operational task are not represented faithfully, then validity of the performance measures may suffer because they reflect skill at accomplishing a diluted task. Similarly, performance measure sensitivity may be limited if true expertise depends upon the ability to perform in a variety of environmental conditions.

Similarly, most real-world operational tasks include consequences for poor performance. In some cases, these consequences are quite dramatic, such as injury or death. As with most laboratory research in the behavioral sciences, it is difficult to adequately represent in VE consequences for poor performance. In many task simulations, consequences for poor performance generally resemble those of a video game. As a result, the dedication and effort of the participants may be questionable, compromising performance measure validity.

29.7.2.4 User Characteristics

Researchers have found certain participants may perform better in VEs than others because of their demographic makeup or past technological experience. In addition, Howe and Sharkey (1998) created a method for identifying successful users of VEs based on competence and temperament. This underscores the need to select research participants with care, as noted by Whitley (1996). When using VEs to simulate task conditions or training scenarios, users must reflect the target population. Yet behavioral researchers often rely on a sample of convenience, usually college students. If college students are used for testing in a VE, ceiling effects may occur because of students' greater familiarity with computers and VE-based games. On the other hand, if a specialized sample is used, there may be a danger of a floor effect because of limited experience with technology.

Witmer and Singer (1994) showed that users vary with regard to how deeply they may become psychologically immersed in a VE. Some people find it quite easy to accept the VE as a distinct, separate world that they can influence. Others may find it difficult to suspend their disbelief when experiencing the same environment. Because susceptibility for presence often dictates how participants will act when immersed, researchers should assume that such differences will adversely impact performance measure reliability.

Researchers have at times noted performance differences in VEs with regard to sex. In some cases, the differences that arise are due to related factors (i.e., males typically use computers for recreation more often and are generally less susceptible to motion sickness). Researchers should be cautious to maintain participant sex ratios that are ecologically valid. However, they should also be aware that threats to data validity and sensitivity might exist when testing sex-diverse samples.

Speed and accuracy of virtual task performances have been shown to vary as a function of both computer and video game familiarity. Therefore, it is wise to employ caution when investigating data trends. Although participant samples may echo the operational environment demographically, their experience with technology may introduce unexpected variability in performance measures, lowering measure validity and sensitivity.

Finally, of particular interest to researchers has been the influence of cybersickness on human performance in VEs (see Barrett, 2004). Accurate measurement of performance in VEs may be jeopardized by participants' varying levels of susceptibility to cybersickness and the variety of ways such sickness can manifest itself. For example, performance measures may be rendered less reliable because of the growth of nausea, eyestrain, or disorientation during a testing session. Validity may also suffer because participants become incapacitated and cannot respond quickly or accurately. In some cases, measures may lose sensitivity because experts, in contrast to novices, may be disproportionately vulnerable to cybersickness.

29.7.2.5 Combining Performance Metrics into Global Assessment Scores

Many modern systems to track user motions and record control input information yield an abundance of detailed information in real time. For example, the Kinect system tracks spatial coordinates of multiple points within the body frame several times per second. Consequently, the potential exists to record thousands of lines of data during even a short VE exposure. Data overload is compounded by the possibility of tracking not just the user but the objects he or she is using. Clearly, there is a need for researchers and designers to consider implementing summary scores for human performance. Such scores may be generated on the basis of a particular object, a certain task, an interpersonal interaction of interest, or any number of events occurring during VE use.

Psychometrically, there are challenges to be overcome when combining performance scores. The pragmatic approach to doing so stresses the importance of using z-scores to ensure that individual score components are standardized. Though z-scores may facilitate statistical analysis, combining such scores may very well impact validity, reliability, and sensitivity (Wiesen, 2006). For example, when assessing the validity of an intricate object manipulation (e.g., joystick movement), combining position data across time may well mask important information about central tendency or variability that is specific to the task of interest. Likewise, the impact of an environmental event, such as an auditory signal, may change depending on the focus of analysis (e.g., finger movement or combined movement of the finger, arm, and shoulder).

Another common practice is to weight score components to meet the goals of a particular analysis. Ideally, this should be done according to theory-driven expectations and applied goals of the research. Special care should be given to the underlying distributions of the score components as well as the relationships between combined score elements. The best practice is to use factor analysis or a similar correlative method to assess the relation among the scores to be combined. If scores are not related, it is possible that combining them may overshadow underlying data trends. Of particular importance is the weighting of combined scores. A statistical method or program should be used that allows complete control over the weighting coefficients for each combined element. As noted earlier, such combinatorial weighting should be attempted only after judiciously standardizing score components.

29.7.2.6 Extended Use of Avatars and NPCs

Within VEs and gaming contexts, there are two types of computer-generated characters: avatars and agents. This dichotomy is based on the source of agency and autonomy (i.e., who controls the characters). Player characters (PCs) are game avatars representing human players. Sometimes players interact with an NPC or PC that is programmed into the game or VE (i.e., a computer agent, embodied agent, synthetic teammate, or synthetic learning agent). Feng, Shapiro, Lhommet, and Marsella (2013) provide a thorough review of embodied autonomous agent characteristics (see Chapter 14).

Technological advances enable designers to create computer-generated characters that mimic the appearance and behaviors of humans (i.e., anthropomorphism) (Veletsianos, Miller, & Doering, 2010).

Veletsianos et al. (2010) outline a useful design framework to enhance agent–learner interactions. For example, they recommend that designers program agents with attentiveness and sensitivity to learner's needs and wants through responsiveness and reaction, feedback, and balancing on/off task communications. Designers should consider intricacies of the messages the agents will send and receive. Designers should also establish role and relationship to user/task, use a level of anthropomorphism that is appropriate to the context and the learner, and create agents that are polite and positive (e.g., encouraging, motivating) and expressive (Veletsianos et al., 2010). It is important to create avatars and agents with socially appropriate emotional expressions (i.e., well-timed head movements, gestures, nods, eye contact, and expressive faces) (Veletsianos et al., 2010).

Interaction with agents is often mediated by individual characteristics like age, gaming experience, and sex as well as by affective states such as feelings of power, anonymity, and authority. In a training or learning context, individuals interacting with agents may report frustration, anger, mistrust, and apprehension, each of which may decrease or negatively impact interaction with the agent, thereby impeding learning (Veletsianos et al., 2010).

Research suggests that users interact with NPCs differently than they do with other humans using avatars. For example, Weibel, Wissmath, Habegger, Steiner, and Groner (2008) found that participants who believed they were interacting with another human reported more presence, flow, and enjoyment than those who believed that they were interacting with a computer agent (Weibel et al., 2008). Lim and Reeves (2010) found that users became more physiologically aroused when interacting with avatars versus agents, especially when the game required more competition than cooperation. An increase in arousal may negatively impact performance, especially in terms of accuracy.

A number of research programs have explored the development of speech interactive synthetic agents that allow users to develop communication, problem solving, and decision-making skills in team scenarios (e.g., virtual interactive pattern environment and radiocomms simulator [VIPERS] and synthetic cognition for operational team training [SCOTT]) (Bell, Ryder, & Pratt, 2009). Eitelman, Ryder, Szczepkowski, and Santarelli (2006) recently developed an automated performance assessment method for team training: cognitive agent-based real-time measurement and assessment (CARMA). The DARWARS tactical language training system (TLTS) assists learners in the acquisition of communication skills in foreign languages and cultures (Johnson, Marsella, & Vilhjálmsson, 2004).

Speech recognition errors (e.g., recognition failures or misrecognition) occur for the following reasons: improper hardware calibration, trainee accent or dialect, training software algorithms that are too strict, instructors and researchers that overestimate the reliability of current technology, and lack of time to develop speech recognition calibration (Eitelman et al., 2006; Wachowicz & Scott, 1999). Therefore, spoken language interaction through the use of synthetic agents is not accepted or marketed by everyone due to the perception that it can be unrealistic, unconvincing, risky, *brittle*, and expensive (Bell et al., 2009).

29.7.2.7 Performance Measurement with Augmented Reality

Augmented reality (AR) integrates real and virtual or computer-generated items such that computer-generated information is projected into the real world from the user's perspective (Azuma et al., 2001). This computer-generated information is presented in a number of ways. It may be integrated such that the user may not be able to distinguish it from the real world, it may be invisible without augmentation, or it may not be part of the environment and still retain AR system characteristics (e.g., head-up displays). Other AR uses involve annotations and visualizations in advertising and marketing (Goldiez, Saptoka, & Aedunuthula, 2006). The diversity of AR information presentation in and of itself is a potential threat to generalizability and performance measure reliability.

Problems still exist despite recent advancements in the size, weight, accessibility, and cost of AR systems (e.g., Google's Project Glass, augmented video games, and augmented smartphones). See-through displays can lack the brightness, resolution, field of view, and contrast required to integrate real and virtual information in a seamless way. Adaptation and aftereffects can occur due to a parallax error caused by cameras not being placed directly at the eye (Biocca & Rolland, 1998).

Implementation errors (e.g., miscalibration), technological problems (e.g., vertical mismatch in image frames of a stereo display, reliability and efficiency of sensors), and fundamental limitations in the design of HMDs (e.g., accommodation–vergence conflict, eye fatigue and strain with prolonged use) also create perceptual issues (Azuma et al., 2001). There is also some concern about the excessive initial cognitive demands placed on users with AR systems (e.g., Stedmon, Hill, Kalawsky, & Cook, 1999).

Conversely, research does indicate some positive cognitive benefits of AR for human performance. Individuals can create meaningful patterns from stimuli, integrate information across sensory modalities, match new information to a standard or guideline, and make judgments or solve problems within AR systems. AR can allow users to minimize information search between the real world and training materials, can alter reality in a way that improves SA (e.g., removing occlusions), can provide examples of correct states or standards, can provide cues for attention direction, and can be flexible in its knowledge of results presentation (see Macchiarella, Liu, & Vincenzi, 2009).

AR may allow users to use more elaboration, especially during psychomotor tasks due to the sequencing of events, than other forms of VR. AR also allows for dynamic visuospatial cognition that might increase encoding new information into memory and thus improve recall (Macchiarella et al., 2009). AR training allows learners to practice in an environment that is identical to the real-world environment. Retrieval and recall are the most effective if the training environment and real-world environment are similar. Rose et al. (2000) found that this type of training does provide a training of transfer advantage.

AR might level the playing field for learners because AR systems can include information throughout training and during the real task that might normally only be found during the training portion. There is less need for interaction devices like a mouse, keyboard, or joystick that often influence user's abilities in VEs (Macchiarella et al., 2009). However, it is difficult to compare across different augmented information displays and interface interaction styles thereby decreasing the generalizability, sensitivity, and reliability of performance measures in AR.

Goldiez et al. (2006) discussed some differences in performance data based on the type of AR system employed during a task. For example, egocentric displays result in better performance for making Euclidean distance and directional estimations, and exocentric displays lead to better estimation of Cardinal direction estimations. Displays that are only available on demand may increase performance if compared with displays that are available continuously. Latency is a common problem with AR systems. Holloway (1997) suggested that a 1 ms delay for a close range task can lead to 1 mm of error. Ellis, Breant, Manges, Jacoby, and Adelstein (1997) found delays within 10 ms can statistically significant changes in performance. Differences in perceptual calibration across AR platforms decrease the reliability of performance measures. In addition, many studies are domain and task specific and may involve different perceptual requirements.

Within the domain of AR in surgery applications, many performance outcomes are not measured quantitatively. Subjective statements and self-report data may not be validated through empirical research. In addition, performance is often based solely on user-friendly features, accuracy of targeting tissues, and cost–benefit analyses. There are also issues with wearable systems being distracting or uncomfortable, or not being feasible additions to the ergonomics of an operating room (Shuhaiber, 2004). There is also variability in the clinical definition of accuracy in each medical domain, as well as image acquisition, registration techniques, computers and software interfaces, iterative localization devices and intraoperative use, integration of real-time data, tissue displacement, robotics, and clinical experience (Tang, Kwoh, Teo, Sing, & Ling, 1998).

29.8 SUMMARY

The measurement of human performance should be a consideration throughout the design and life cycle of a VE system. An integrated approach to performance measurement should take into account the what, how, and when of performance measurement and consideration of psychometric properties of data. For most VE systems, performance measurement will involve a combination of automated data capture and observation by human raters. Other relevant chapters are as follows: Chapter 12, *Principles for Designing Effective 3D Interaction Techniques*; Chapter 23, *Motion Sickness Symptomatology and Origins*; Chapter 28, *Usability Engineering of Virtual Environments*; and Chapter 38, *Team Training in Virtual Environments: A Dual Approach*. The motion sickness chapter is relevant in that sickness can affect performance and performance can influence motion sickness. The chapter on team training not only presents an excellent discussion of team performance measurement but also describes how a proactive approach to VE design can aid performance measurement.

ACKNOWLEDGMENTS

The authors gratefully acknowledge the efforts and contributions of Donald R. Lampton and Christina Morris, authors on the original version of this chapter.

REFERENCES

Alankus, G., Proffitt, R., Kelleher, C., & Engsberg, J. (2011). Stroke therapy through motion-based games: A case study. *ACM Transactions on Accessible Computing, 4*(1), 1–35.

American National Standards Institute [ANSI]. (1993). *Guide to human performance measurements.* Washington, DC: American Institute of Aeronautics.

Ananthaswamy, A. (2003, July 31). Virtual reality conquers sense of taste. *New Scientist, 14*, 49.

Andreassi, J. L. (2006). *Psychophysiology: Human behavior and physiological response* (5th ed.). Mahwah, NJ: Lawrence Erlbaum Associates.

Azuma, R. T., Baillot, Y., Behringer, R., Feiner, S., Julier, S., & MacIntyre, B. (2001). Recent advances in augmented reality. *Computer Graphics and Applications, IEEE, 21*(6), 34–47.

Backs, R. W., & Boucsein, W. (Eds.) (2000). *Engineering psychophysiology: Issues and applications.* Mahwah, NJ: Lawrence Erlbaum Associates.

Barrett, J. (2004). *Side effects of virtual environments: A review of the literature.* (Technical Report No. DSTO-TR-1419). Edinburgh, South Australia: DSTO Information Sciences Laboratory.

Barrick, M. R., Stewart, G. L., Neubert, M. J., & Mount, M. K. (1998). Relating member ability and personality to work-team processes and team effectiveness. *Journal of Applied Psychology, 83*, 377–391.

Basdogan, C., & Loftin, R. B. (2008). Multimodal display systems: Haptic, olfactory, gustatory, and vestibular. In D. Schmorrow, J. Cohn, & D. Nicholson (Eds.), *PSI handbook of virtual environments for training and education: Developments for the military and beyond.* Westport, CT: Praeger Security International.

Beatty, J. (1982). Task-evoked pupillary responses, processing load, and the structure of processing resources. *Psychological Bulletin, 91*(2), 276–292.

Begault, D. R. (1991). Challenges to the successful implementation of 3-D sound. *Journal of the Audio Engineering Society, 39*(11), 864–870.

Bell, B., Ryder, J. M., & Pratt, S. N. (2009). Communications and coordination training with speech-interactive synthetic teammates: A design and evaluation case study. In D. A. Vincenzi, J. A. Wise, M. Mouloua, & P. A. Hancock (Eds.), *Human factors in simulation and training* (pp. 387–414). Boca Raton, FL: Taylor & Francis Group.

Bell, B. S., Kanar, A. M., & Kozlowski, S. W. J. (2008). Current issues and future directions in simulation-based training in North America. *The International Journal of Human Resource Management, 19*(8), 1416–1434.

Berntson, G. G., Bigger, J. T., Eckberg, D. L., Grossman, P., Kaufmann, P. G., Malik, M., … van der Molen, M. W (1997). Heart rate variability: Origins, methods, and interpretive caveats. *Psychophysiology, 34*, 623–648.

Biocca, F. A., & Rolland, J. P. (1998). Virtual eyes can rearrange your body: Adaptation to visual displacement in see through, head-mounted displays. *Presence: Teleoperators and Virtual Environments, 7*(3), 262–277.

Biron, L. (2012, June 11). Virtual reality helps service members deal with PTSD. *Defense News*, http://www.defensenews.com/article/20120611/TSJ01/306110003/Virtual-Reality-Helps-Service-Members-Deal-PTSD

Bittner, J. G., Mellinger, J. D., Imam, T., Schade, R. R., & Macfadyen, B. V., Jr. (2010). Face and construct validity of a computer-based virtual reality simulator for ERCP. *Gastrointestinal Endoscopy, 71*(2), 357–364.

Bliss, J. P., Liebman, R., & Brill, J. C. (2012). Alert characteristics and identification of avatars on a virtual battlefield. *Intelligent Decision Technologies, 6*(2), 151–159.

Bliss, J. P., Tidwell, P. D., & Guest, M. A. (1997). The effectiveness of virtual reality for administering spatial navigation training to firefighters. *Presence, 6*(1), 73–86.

Blumenthal, T. D., Cuthbert, B. N., Filion, D. L., Hackley, S., Lipp, O. V., & Van Boxtel, A. (2005). Committee report: Guidelines for human startle eyeblink electromyographic studies. *Psychophysiology, 42*, 1–15.

Brannick, M. T., Prince, A., Prince, C., & Salas, E. (1995). The measurement of team process. *Human Factors, 37*(3), 641–651.

Cannon-Bowers, J. A., & Salas, E. (1997). A framework for developing team performance measures in training. In M. T. Brannick, E. Salas, & C. Prince (Eds.), *Team performance assessment and measurement: Theory, methods, and applications.* Mahwah, NJ: Lawrence Erlbaum Associates.

Cater, J. P. (1994). Approximating the senses. Smell/taste: Odors in virtual reality. *Proceedings of the IEEE International Conference on Systems, Man and Cybernetics* (Vol. 2, p. 1781). New York, NY: IEEE Computer Society.

Chertoff, D., & Schatz, S. (2014). Beyond presence. In K. S. Hale & K. M. Stanney (Eds.), *Handbook of virtual environments: Design, implementation, and applications* (2nd ed., pp. 855–870). Boca Raton, FL: Taylor & Francis Group, Inc.

Cohen, P. R. (2002). *Multimodal interaction for virtual environments* (Technical Report, Grant No. N00014-99-1-0380). Washington, DC: Office of Naval Research.

Coles, M. (1998). Preparation, sensory–motor interaction, response evaluation, and event related brain potentials. *Current Psychology of Cognition, 17*(4–5), 737–748.

Companion, M. A. & Corso, G. M. (1982). Task taxonomies: A general review and evaluation. *International Journal of Man–Machine Studies, 77*(8), 459–472.

Cooke, N., Kiekel, P., Salas, E., Stout, R., Bowers, C., & Cannon-Bowers, J. (2003). Measuring team knowledge: A window to the cognitive underpinnings of team performance. *Group Dynamics: Theory, Research and Practice, 7*, 179–199.

Crawford, H. J., Knebel, T. E., Vendemia, L. K., & Ratcliff, B. (1995). EEG activation during tracking and decision-making tasks: Differences between low- and high-sustained attention adults. *Proceedings of the Eighth International Symposium on Aviation Psychology* (Vol. 2, pp. 886–890). Columbus, OH.

Darken, R. P., & Peterson, B. (2014). Spatial orientation, wayfinding, and representation. In K. S. Hale & K. M. Stanney (Eds.), *Handbook of virtual environments: Design, implementation, and applications* (2nd ed., pp. 467–492). Boca Raton, FL: Taylor & Francis Group, Inc.

Denver, J. W., Reed, S. F., & Porges, S. W. (2007). Methodological issues in the quantification of respiratory sinus arrhythmia. *Biological Psychology, 74*, 286–294.

Dickinson, T. L., & McIntyre, R. M. (1997). A conceptual framework for teamwork measurement. In M. T. Brannick, E. Salas, & C. Prince (Eds.), *Team performance and measurement: Theory, methods, and applications* (pp. 19–43). Mahwah, NJ: Lawrence Erlbaum.

Driskell, J. E., Radtke, P. H., & Salas, E. (2003). Virtual teams: Effects of technological mediation on team performance. *Group Dynamics: Theory, Research, and Practice, 7*(4), 297–323.

Driskell, T., Salas, E., & Vessey, W. B. (2014). Team training in virtual environments. In K. S. Hale & K. M. Stanney (Eds.), *Handbook of virtual environments: Design, implementation, and applications* (2nd ed., pp. 999–1026). Boca Raton, FL: Taylor & Francis Group, Inc.

Durlach, N., & Mavor, A. (1995). *Virtual reality: Scientific and technological challenges.* Washington, DC: National Academy Press.

Eitelman, R., Ryder, J., Szczepkowski, & Santarelli, (2006, May). Using agents to enhance performance assessment of team communications. *Proceedings of Behavioral Representation in Modeling and Simulation (BRIMS).* Baltimore, MD.

Ellis, S. R., Breant, F., Manges, B., Jacoby, R., & Adelstein, B. D. (1997). Factors influencing operator interaction with virtual objects viewed via head-mounted see-through displays: Viewing conditions and rendering latency. *Proceedings of the 1997 Virtual Reality Annual International Symposium (VRAIS'97)* (p. 138). Washington, DC: IEEE Computer Society.

Endsley, M. R. (1995). Measurement of situation awareness in dynamic systems. *Human Factors, 37*(1), 65–84.

Entin, E. E., & Serfaty, D. (1999). Adaptive team coordination. *Human Factors, 41*, 312–325.

Fahrenberg, J., & Wientjes, C. J. (2000). Recording methods in applied environments. In R. B. Backs, & W. Boucsein (Eds.), *Engineering psychophysiology: Issues and applications* (pp. 111–136). Mahwah, NJ: Lawrence Erlbaum Associates.

Feng, A., Shapiro, A., Lhommet, M., & Marsella, S. (2014). Embodied autonomous agents. In K. S. Hale & K. M. Stanney (Eds.), *Handbook of virtual environments: Design, implementation, and applications* (2nd ed., pp. 335–350). Boca Raton, FL: Taylor & Francis Group, Inc.

Fenlon, W. (2011, October 3). Microsoft's KinectFusion project scans the world in real time. *Tested News*, http://www.tested.com/tech/photography/2956-microsofts-kinectfusion-project-scans-the-world-in-real-time-video/

Fineberg, M. (1995). *A comprehensive taxonomy of human behaviors for synthetic forces.* (IDA Paper No. P-3155). Alexandria, VA: Institute for Defense Analyses.

Fiore, S. M., Salas, E., Cuevas, H. M., & Bowers, C. A. (2003). Distributed coordination space: Toward a theory of distributed team process and performance. *Theoretical Issues in Ergonomics Science, 4*, 340–364.

Fitts, P. M., Jones, R. E., & Milton, J. L. (1950). Eye movements of aircraft pilots during instrument landing approaches. *Aeronautical Engineering Review, 9*, 1–16.

Flanagan, J. C. (1954). The critical incident technique. *Psychological Bulletin, 51*(4), 327–358.

Fleishman, E. A., & Quaintance, M. K. (1984). *Taxonomies of human performance: The description of human tasks.* New York, NY: Academic Press.

Gabbard, J. L. (2014). Usability engineering of virtual environments. In K. S. Hale & K. M. Stanney (Eds.), *Handbook of virtual environments: Design, implementation, and applications* (2nd ed., pp. 721–748). Boca Raton, FL: Taylor & Francis Group, Inc.

Gabbard, J., & Hix, D. (1997). *A Taxonomy of usability characteristics in virtual environments* (Master's thesis). Virginia Polytechnic Institute and State University, Blacksburg, U.K.

Gevins, A., Smith, M. E., Leong, H., McEvory, L., Whitfield, S., Du, R., & Rush, G. (1998). Monitoring working memory load during computer-based tasks with EEG pattern recognition. *Human Factors, 40*(1), 79–91.

Goldberg, S. L., Knerr, B. W., & Grosse, J. (2003). Training dismounted combatants in virtual environments. *RTO HFM Symposium on "Advanced Technologies for Military Training"*, held Genoa, Italy, October 13–15, 2003, and published in RTO-MP-HFM-101.

Goldiez, B. F., Saptoka, N., & Aedunuthula, P. (2006) Human performance assessments when using augmented reality for navigation. *Proceedings of the Virtual Media for Military Applications Conference.* RTO, Neuilly-sur-Seine, France.

Greenwood-Ericksen, A., Kennedy, R. C., & Stafford, S. (2014). Entertainment application of virtual environments. In K. S. Hale & K. M. Stanney (Eds.), *Handbook of virtual environments: Design, implementation, and applications* (2nd ed., pp. 1291–1316). Boca Raton, FL: Taylor & Francis Group, Inc.

Grierson, M., & Kiefer, C. (2011, May 7–12). Better brain interfacing for the masses: Progress in event-related potential detection using commercial brain computer interfaces. *Proceedings of the Computer-Human Interaction Conference*, Vancouver, British of Columbia, Canada.

Hamblin, C. J. (2005). *Transfer of training from virtual reality environments* (Unpublished Ph.D. Dissertation). Wichita State University, Wichita, KS.

Hanson, E. K.S., Schellekens, J. M. H., Veldman, J. B. P., & Mulder, L. J. M. (1993). Psychomotor and cardiovascular consequences of mental effort and noise. *Human Movement Science, 12*(6), 607–626.

Hays, R. T., & Singer, M. J. (1989). *Simulator fidelity in training system design.* New York, NY: Springer-Verlag.

Holding, D. H. (1965). *Principles of Training.* London, U.K.: Pergamon Press.

Hollingshead, A. B., McGrath, J. E., & O'Connor, K. M. (1993). Group task performance and communication technology: A longitudinal study of computer-mediated versus face-to-face work groups. *Small Group Research, 24*(3), 307–333.

Holloway, R. L. (1997). Registration error analysis for augmented reality, *Presence: Teleoperators and Virtual Environments, 6*(2), 413–432.

Howe, T., & Sharkey, P. M. (1998). Identifying likely successful users of virtual reality systems. *Presence, 7*(3), 308–316.

Huegel, J. C., Celik, O., Israr, A., & O'Malley, M. K. (2009). Expertise-based performance measures in a virtual training environment. *Presence, 18*(6), 449–467.

Johnson, W. L., Marsella, S., & Vilhjálmsson, H. (2004). The DARWARS tactical language training system. *Proceedings of the Interservice/Industry Training, Simulation, and Education Conference (I/ITSEC).* Orlando, FL.

Jones, S. D., & Schilling, D. J. (2000). *Measuring team performance.* San Francisco, CA: Jossey-Bass.

Jorna, P. M. (1992). Spectral analysis of heart rate and psychological state: A review of validity as a workload index. *Biological Psychology, 34*(2–3), 237–257.

Kahneman, D., Tursky, B., Shapiro, D., & Crider, A. (1969). Pupillary, heart rate, and skin resistance changes during a mental task. *Journal of Experimental Psychology, 79*(1), 164–167.

Kenyon, R. V., & Afenya, M. B. (1995). Training in virtual and real environments. *Annals of Biomedical Engineering, 23*, 445–455.

Kennedy, R. S., Kennedy, R. C., Kennedy, K. E., Wasula, C., & Bartlett, K. M. (2014). Virtual environments and product liability. In K. S. Hale & K. M. Stanney (Eds.), *Handbook of virtual environments: Design, implementation, and applications* (2nd ed., pp. 505–518). Boca Raton, FL: Taylor & Francis Group, Inc.

Keshavarz, B., Heckt, H., & Lawson, B. D. (2014). Visually induced motion sickness. In K. S. Hale & K. M. Stanney (Eds.), *Handbook of virtual environments: Design, implementation, and applications* (2nd ed., pp. 647–698). Boca Raton, FL: Taylor & Francis Group, Inc.

Kortum, P. (Ed.) (2008). *CHI beyond the GUI: Design for haptic, speech, olfactory, and other non-traditional interfaces.* Burlington, MA: Morgan Kaufmann (Elsevier).

Kozlowski, S. W. J., & Bell, B. S. (2003). Work groups and teams in organizations. In W. C. Borman, D. R. Ilgen, & R. J. Klimoski (Eds.), *Handbook of psychology,* (Vol. 12, pp. 333–375). Hoboken, NJ: John Wiley & Sons, Inc.

Kramer, A. F. (1991). Physiological metrics of mental workload: A review of recent progress. In D. L. Damos (Ed.), *Multiple-task performance* (pp. 279–328). Bristol, PA: Taylor and Francis.

Kundahl, P. S., & Grantcharov, T. P. (2009). Psychomotor performance measured in a virtual environment correlates with technical skills in the operating room. *Surgical Endoscopy, 23*(3), 645–649.

Lampton, D. R., Bliss, J. P., Orvis, K., Kring, J., & Martin, G. (2009). A distributed game-based simulation training research testbed. *Proceedings of the 53rd Annual Human Factors and Ergonomics Society Conference.* October 19–23, San Antonio, TX.

Lampton, D. R., Knerr, B. W., Goldberg, S. L., Bliss, J. P., Moshell, J. M., & Blau, B. S. (1994). The virtual environment performance assessment battery (VEPAB): Development and evaluation. *Presence: Teleoperators and Virtual Environments, 3*(2), 145–157.

Lampton, D. R., Riley, J. M., Kaber, D. B., Sheik-Nainar, M. A., & Endsley, M. R. (2006). *Use of immersive virtual environments for measuring and training situation awareness* (Final Report). Alexandria, VA: U.S. Army Research Institute for the Behavioral and Social Sciences.

Lawson, B. D. (2014). Motion sickness symptomatology and origins. In K. S. Hale & K. M. Stanney (Eds.), *Handbook of virtual environments: Design, implementation, and applications* (2nd ed., pp. 531–600). Boca Raton, FL: Taylor & Francis Group, Inc.

Leavitt, F. (1991). *Research methods for behavioral scientists.* Dubuque, IA: William C. Brown.

Lim, S., & Reeves, B. (2010). Computer agents versus avatars: Responses to interactive game characters controlled by a computer or other player. *International Journal of Human-Computer Studies, 68*, 57–68.

Lipnack, J., & Stamps, J. (1997). *Virtual teams: Reaching across space, time and organizations with technology.* New York, NY: John Wiley & Sons.

Loomis, J. M., & Lederman, S. J. (1986). Tactual perception. In K. Boff, L. Kaufman, & J. Thomas (Eds.), *Handbook of perception and human performance* (pp. 31-1–31-41). New York, NY: Wiley.

Ma, L., Bennis, F., Chablat, D., & Zhang, W. (2008). Framework for dynamic evaluation of muscle fatigue in manual handling work. *Proceedings of the IEEE International Conference on Industrial Technology.* Sichuan University, Chengdu, China, April 21–24, 2008.

Macchiarella, N. D., Liu, D., & Vincenzi, D. A. (2009). Augmented reality as a means of job task training in aviation. In D. A. Vincenzi, J. A. Wise, M. Mouloua, & P. A. Hancock (Eds.), *Human factors in simulation and training* (pp. 333–348). Boca Raton, FL: Taylor & Francis Group.

Madrigal, A. (2009, March 4). Researchers want to add touch, taste, and smell to virtual reality. *Wired Magazine.*

Mantovani, F., & Castelnuovo, G. (2003). Sense of presence in virtual training: Enhancing skills acquisition and transfer of knowledge through learning experience in virtual environments. In G. Riva, F. Davide, & W. A. Ijsselsteijn (Eds.), *Being there: Concepts, effects and measurement of user presence in synthetic environments.* Amsterdam, the Netherlands: IOS Press.

Marks, M. A., Mathieu, J. E., & Zaccaro, S. J. (2001). A temporally based framework and taxonomy of team processes, *The Academy of Management Review, 26*(3), 356–376.

Matthews, G., Davies, D. R., Westerman, S. J., & Stammers, R. B. (2000). Human performance cognition, stress and individual differences, London: Psychology Press.

Morris, C. (2012, August 6). Virtual reality makes a comeback with Oculus Rift. *Plugged In.*

Morris, C., Shirkey. E. C., & Mouloua, M. (2000). QEEG characteristics of cordless vs. corded phone users (Abstract). *Proceedings of the Human Factors and Ergonomics Society/International Ergonomics Association 44th Annual Meeting.* Santa Monica, CA: Human Factors and Ergonomics Society.

Muhlberger, A., Wieser, A. J., & Pauli, P. (2006). Darkness-enhanced startle responses in ecologically valid environments: A virtual tunnel driving experiment. *Biological Psychology, 77*(1), 47–52.

Mulder, L. M. (1992). Measurement and analysis methods of heart rate and respiration for use in applied environments. *Biological Psychology, 34*, 205–236.

Norman, D. A. (1988). *The psychology of everyday things*. New York, NY: Basic Books.

Nunnally, J. C., & Bernstein, I. H. (1994). *Psychometric theory* (3rd ed., pp. 99–101). New York, NY: McGraw-Hill.

Oikonomidis, I., Kyriazis, N., & Argyros, A. A. (2011, August 29–September 1). Efficient model-based 3D tracking of hand articulations using Kinect. *Proceedings of the 22nd British Machine Vision Conference*, University of Dundee, Dundee, U.K.

Oropesa, I., Lamata, P., Sanchez-Gonzalez, P., Pagador, J. B., Garcia, M. E., Sanchez-Margallo, F. M., & Gomez, E. J. (2011). Virtual reality simulators for objective evaluation on laparoscopic surgery: Current trends and benefits. In J.J. Kim (Ed.), *Virtual reality*. doi:10.5772/13178. Available from: http://www.intechopen.com/books/virtual-reality/virtual-reality-simulators-for-objective-evaluation-on-laparoscopic-surgery-current-trends-and-benef. Rijeka, Croatia.

Parasuraman, R. (1975). Response bias and physiological reactivity. *The Journal of Psychology, 91*, 309–313.

Parasuraman, R. (2003). Neuroergonomics: Research and practice. *Theoretical Issues in Ergonomics Science, 4*, 5–20.

Parasuraman, R., & Riley, V. (1997). Humans and automation: Use, misuse, disuse, abuse. *Human Factors, 39*(2), 230–253.

Parasuraman, R., & Rizzo, M. (Eds.) (2007). *Neuroergonomics: The brain at work*. New York, NY: Oxford University Press.

Parsons, J., Lampton, D. R., Parsons, K. A., Knerr, B. W., Russell, D., Martin, G., … Weaver, M. (1998). Fully immersive team training: A networked testbed for ground-based training missions. *Proceedings of the Interservice/Industry Training Systems and Education Conference*. Orlando, FL.

Parsons, T. D., & Rizzo, A. A. (2008). Initial validation of a virtual environment for assessment of memory functioning: Virtual reality cognitive performance assessment test. *Cyberpsychological Behavior, 11*(1), 17–25.

Perna, G. (2010, December 20). Could the Kinect cause motion sickness? *International Business Times*, http://www.ibtimes.com/could-kinect-cause-motion-sickness-251213.

Picton, T. W., Bentin, S., Berg, P., Donchin, E., Hillyard, S. A., Johnson, R., …Taylor, M.J. (2000). Guidelines for using human event-related potentials to study cognition: Recording standards and publication criteria. *Psychophysiology, 37*, 127–152.

Porges, S. W., & Byrne, E. A. (1992). Research methods for measurement of heart rate and respiration. *Biological Psychology, 34*, 93–130.

Proaps, A. B., & Bliss, J. P. (2010). Team performance as a function of task difficulty in a computer game. *Proceedings of the Human Factors and Ergonomics Society Annual Meeting*. Human Factors and Ergonomics Society, Santa Monica, CA.

Reason, J. (1990). *Human error*. Cambridge, England: Cambridge University Press.

Richardson, D. C., & Spivey, M. J. (2004). Eye tracking: Characteristics and methods. In G. Wnekm, & G. Bowlin (Eds.), *Encyclopedia of biomaterials and biomedical engineering* (pp. 568–572). New York, NY: Marcel Dekker, Inc.

Ritz, T., Dahme, B., Dubois, A. B., Folgering, H., Fritz, G. K., Harver, A., … van de Woestijne, K. P. (2002). Guidelines for mechanical lung function measurements in psychophysiology. *Psychophysiology, 39*, 546–567.

Robinett, W. (1994). Interactivity and individual viewpoint in shared virtual worlds: The big screen vs. networked personal displays. *Computer Graphics, 28*(2), 127.

Rohner, J. C. (2002). The time-course of visual threat processing: High trait anxious individuals eventually avert their gaze from angry faces. *Cognition and Emotion, 16*, 837–844.

Rose, F. D., Attree, E. A. Brooks, B. M. Parslow, D. M., Penn, P. R., & Ambihaipahan, N. (2000). Training in virtual environments: Transfer to real world tasks and equivalence to real task training. *Ergonomics, 43*(4), 494–511.

Rose, R. M., & Fogg, L. F. (1993). Definition of a responder: Analysis of behavioral, cardiovascular, and endocrine responses to varied workload in air traffic controllers. *Psychosomatic Medicine, 55*, 325–338.

Salas, E., Burke, C. S., & Fowlkes, J. E. (2006). Measuring team performance "in the wild": Challenges and tips. In W. Bennet, C. E. Lance, & D. J. Woehr (Eds.), *Performance measurement: Current perspectives and future challenges*. Mahwah, NJ: LEA.

Salas, E., & Priest, H. (2005). Team training. In N. Stanton, A. Hedge, K. Brookhuis, E. Salas, & H. Hendrick (Eds.), *Handbook of human factors and ergonomics methods* (pp. 44-1–44-7). Washington, DC: CRC Press.

Salas, E., Sims, D. E., & Burke, C. S. (2005) Is there a Big Five in teamwork? *Small Group Research, 36*, 555–599.

Salmon, P., Stanton, N., Walker, G., & Green, D. (2006). Situation awareness measurement: A review of applicability for C4i environments. *Applied Ergonomics, 37*(2), 225–238.

Salvendy, G., & Carayon, P. (1997). Data collection and evaluation of outcome measures. In G. Salvendy (Ed.), *Handbook of human factors and ergonomics* (pp. 1451–1470). New York, NY: John Wiley & Sons.

Schmidt-Daffy, M. (2013). Fear and anxiety while driving: Differential impact of task demands, speed, and motivation. *Transportation Research Part F, 16*, 14–28.

Seibt, R., Boucsein, W., & Scheuch, K. (1998). Effects of different stress settings on cardiovascular parameters and their relationship to daily life blood pressure in normotensives, borderline hypertensives and hypertensives. *Ergonomics, 41*(5), 634–648.

Senders, J. W., & Moray, N. P. (1991). *Human error*. Hillsdale, NJ: Lawrence Erlbaum Associates.

Shapiro, D., Jamner, L. D., Lane, J. D., Light, K. C., Myrtek, M., Sawada, Y., & Steptoe, A. (1996). Blood pressure publication guidelines. *Psychophysiology, 33*, 1–12.

Sharma, V. M., Sridharan, K., Selvamurthy, W., Mukherjee, A. K., Kumaria, M. M., Upadhyay, T. N., ... Dimri, G. P. (1994). Personality traits and performance of military parachutist trainees. *Ergonomics, 37*(7), 1145–1155.

Shelton, J., & Kumar, G. (2010). Comparison between auditory and visual simple reaction times. *Neuroscience and Medicine, 1*(1), 30–32.

Shuhaiber, J. H. (2004). Augmented reality in surgery FREE. *Archives of Surgery, 139*(2), 170–174.

Smith, T. W., Houston, B. K., & Zurawski, R. M. (1984). Finger pulse volume as a measure of anxiety in response to evaluative threat. *Psychophysiology, 21*(3), 260–264.

Smith-Jentsch, K. A., Johnston, J. H., and Payne, S. C. (1998). Measuring team-related expertise in complex environments. In J. A. Cannon-Bowers, & E. Salas (Eds.), *Making decisions under stress: Implications for individual and team training* (pp. 61–87). Washington, DC: American Psychological Association.

Smolensky, M. W. (1993). Toward a physiological measurement of situation awareness: The case for eye movement measures. *Proceedings of the Human Factors and Ergonomics Society 37th Annual Meeting* (p. 41). Santa Monica, CA.

Stanney, K. M. (1995). Realizing the full potential of virtual reality: Human factors issues that could stand in the way. *Proceedings of the 1995 VRAIS Conference* (pp. 28–34). RTP, North Carolina, NC.

Stanney, K. M., Mourant, R. R., & Kennedy, R. S. (1998). Human factors issues in virtual environments: A review of the literature. *Presence, 7*(4), 327–351.

Stedmon, A. W., Hill, K., Kalawsky, R. S., & Cook, C. A. (1999). Old theories, new technologies: Comprehension and retention issues in augmented reality systems. *Proceedings of the Human Factors and Ergonomics Society 43rd Annual Meeting*. Santa Monica, CA: Human Factors and Ergonomics Society.

Stefani, O., Mager, R., Mueller-Spahn, F., Sulzenbacher, H., Bekiaris, E., Wiederhold, B., ... Bullinger, A. (2005). Cognitive ergonomics in virtual environments: Development of an intuitive and appropriate input device for navigating in a virtual maze. *Applied Psychophysiology and Biofeedback, 30*(3), 259–269.

Stevens, S. S. (1951). Mathematics, measurement, and psychophysics. In S. S. Stevens (Ed.), *Handbook of experimental psychology* (pp. 1–49). New York, NY: John Wiley & Sons.

Stone, E. E., & Skubic, M. (2011). Evaluation of an inexpensive depth camera for passive in-home fall risk assessment. *Proceedings of the 5th International Conference on Pervasive Computing Technologies for Healthcare* (pp. 71–77). Dublin, Ireland.

Strayer, D. L., & Drews, F. A. (2007). Cell-phone-induced driver distraction. *Current Directions in Psychological Science, 16*(3), 128–131.

Tang, S. L., Kwoh, C. K., Teo, M. Y., Sing, N. W., & Ling, K. V. (1998). Augmented reality systems for medical applications. *IEEE Engineering Medicine Biology Magazine, 17*(3), 49–58.

Templeman, J., Page, R., & Denbrook, P. (2014). Avatar control in virtual environments. In K. S. Hale & K. M. Stanney (Eds.), *Handbook of virtual environments: Design, implementation, and applications* (2nd ed., pp. 233–256). Boca Raton, FL: Taylor & Francis Group, Inc.

Veletsianos, G., Miller, C., & Doering, A. (2010). EnALI: A research and design framework for virtual characters and pedagogical agents. *Journal of Educational Computing Research, 41*(2), 171–194.

Wachowicz, K. A., & Scott, B. (1999). Software that listens: It's not a question of whether, it's a question of how. *CALICO Journal, 16*(3), 253–276.

Warm, J. S., & Parasuraman, R. (2007). Cerebral hemodynamics and vigilance. In R. Parasuraman, & M. Rizzo (Eds.), *Neuroergonomics: The brain at work* (pp. 146–158). New York, NY: Oxford University Press.

Watson, B., Walker, N., Ribarsky, W., & Spaulding, V. (1998). Effects of variation in system responsiveness on user performance in virtual environments. *Human Factors, 40*(3), 403–414.

Weibel, D., Wissmath, B., Habegger, S., Steiner, R., & Groner, Y. (2008). Playing online games against computer- vs. human-controlled opponents: Effects on presence, flow, and enjoyment. *Computers in Human Behavior, 24*(5), 2274–2291.

Whitley, B. E., Jr. (1996). *Principles of research in behavioral science* (pp. 97–128). Mountain View, CA: Mayfield.

Wickens, C. D. (2002). Multiple resources and performance prediction. *Theoretical issues in Ergonomic Science, 3*(2), 159–177.

Wickens, C. D., & Hollands, J. G. (2000). *Engineering psychology and human performance* (3rd ed.). Upper Saddle River, NJ: Prentice Hall.

Wiederhold, B. K., Davis, R., & Wiederhold, M. D. (1998). The effects of immersiveness on physiology. In G. Riva, & B. K. Wiederhold (Eds.), *Virtual environments in clinical psychology and neuroscience: Methods and techniques in advanced patient-therapist interaction: Vol. 58. Studies in health technology and informatics* (pp. 52–60). Amsterdam, the Netherlands: IOS Press.

Wiesen, J. P. (2006, October 27). Benefits, drawbacks and pitfalls of z-score weighting. *Proceedings of the 30th Annual IPMAAC Conference*. Las Vegas, NV.

Wilson, G., Swain, C., & Brookings, J. B. (1995). The effects of simulated air traffic control workload manipulations on EEC. *Proceedings of the Eighth International Symposium on Aviation Psychology* (Vol. 2, pp. 1025–1030). Columbus, OH.

Wilson, K. G. (1987). Electrodermal lability and simple reaction time. *Biological Psychology, 24*, 275–289.

Witmer, B. G., & Singer, M. J. (1994). *Measuring immersion in virtual environments* (Technical Report No. 1014). Alexandria, VA: U.S. Army Research Institute for the Behavioral and Social Sciences.

Yamada, F. (1998). Frontal midline theta rhythm and eye blinking activity during a VDT task and a video game: Useful tools for psychophysiology in ergonomics. *Ergonomics, 41*(5), 678–688.

Zeltzer, D., & Pioch, N. J. (1996). Validation and verification of virtual environment training systems. *Proceedings of the 1996 VRAIS Conference* (pp. 123–130). Los Alamitos, CA.

30 Conducting Training Transfer Studies in Virtual Environments

Roberto K. Champney, Meredith Carroll,
Glen Surpris, and Joseph V. Cohn

CONTENTS

30.1 INTRODUCTION

The methods used to train practitioners in a broad range of occupations as distinct from each other as manufacturing to warfighting are changing. With reduced training budgets, access to live training has been substantially reduced (Erwin, 2013). Unless other highly effective training options are put in place, it is only a matter of time before losses in knowledge and skill proficiency are experienced. Maintaining a high level of proficiency will require training packages to be systematically designed such that they optimize skill acquisition and retention. The key to accomplishing this is identifying those training options that maximize training transfer. Unfortunately, the data-grounded best practices that can be used to direct the design of training packages to meet this goal are still limited (Burke & Hutchins, 2008). In general, a training package typically commences with classroom training to impart a foundation of declarative knowledge (i.e., general facts, principles, rules, and concepts) and basic skills (Cohn, Stanney, Milham, Jones, Hale, Darken & Sullivan, 2007). Next, trainees are typically provided with a means to apply their newly acquired knowledge and skills and practice to proficiency. This is often done through live training exercises. With the reduction in access to live training, computer-based alternatives to live training can be used to fill this gap. Virtual environments (VE) provide one such alternative.

VEs provide a means to experience a variety of simulated environments and conditions in a safe and efficient manner. Their ability to reconstruct similar conditions to those in the operational world, which would otherwise be too risky, costly, or cumbersome to reproduce, imparts VE trainers with high face validity. These qualities, coupled with the ability to re-create scenario conditions multiple times with ease and allow for repetitive learning opportunities, may make VEs a viable venue for replacing some portion of live training. Nonetheless, this does not mean that simply developing a VE implies that it is good enough to achieve training gains. VE training could lead to positive transfer (i.e., improved performance in operational environments), negative transfer (i.e., degraded real-world performance), or have no effect (Stone, 2012). The ultimate test to a VE's training capability is thus the evidence that comes from seeing performance gains in the operational environment following training in the VE (i.e., positive transfer; Baldwin & Ford, 1988).

So the question becomes—what brings about positive training transfer? In its most basic form, training transfer is defined as something learned (knowledge, skills, or attitudes [KSAs]) under a training condition that is retained and applied in the operational condition (Pennington, Nicolich, & Rahm, 1995; Thorndike & Woodworth, 1901). Early efforts at achieving such transfer of training from VEs focused on the *identical element* principle, in which the simulated task environment was designed to have many elements in common with real-world tasks (Thorndike, 1906). While this foundational concept has been key to many types of training systems, the implied focus on physical fidelity often led to overly expensive solutions (McCauley, 2006; Singley & Anderson, 1989). An alternative approach focused on *deep structure*, with the goal being to emulate underlying system functions in the simulated environment (i.e., high functional fidelity, where the simulation acts like the operational equipment in reacting to tasks executed by trainees; Allen, Hays, & Buffordi, 1986; Gick & Holyoak, 1987; Lehman, Lempert, & Nisbett, 1988). Deep structure was, in turn, a precursor to psychological fidelity that focused on ensuring that trainees perceive and act on the training environment as they would on the operational environment (Kozlowski et al., 2004). Together, these three perspectives illustrate the different components of the situation or experience that must be recreated within a VE trainer. It could be said then that the value of VE trainers lies in their ability to recreate situations.

The situated learning theory proposes that learning occurs as a function of three components: activity, context, and culture (Lave & Wenger, 1990; Bossard & Kermarrec, 2006). VE training can thus be said to derive its efficacy from its ability to re-create these components of the real world that are critical for learning (Stanney, Hale, & Cohn, 2012). It is this focus on learning within authentic operational conditions where the value of VEs is posited; because VEs have the capability to present credible physical, functional, and psychological cues, immerse trainees into these synthetic environments, and present situational conditions much like those experienced in the real world. It is proposed that as a result of these characteristics VE training can lead to positive training transfer (Stanney, Hale, Carroll, & Champney, 2012).

Yet, face validity is not enough. Considering the effort and cost involved in the development and acquisition of VE trainers and the risks involved with negative transfer, it is necessary to evaluate VE systems for their training utility and efficacy. The decision to utilize VEs for training should be determined via a comprehensive review of the training progression for a given domain and consider aspects of performance (i.e., transfer), cost, and schedule to understand the return on training investment (ROTI) (Cohn & Fletcher, 2010). This is of particular importance as evaluation has often been a neglected component of the design and development of training systems (Bassi & van Buren, 1999; Carnevale & Shultz, 1990; Champney et al., 2008; Eseryel, 2002; Thompson, Koon, Woodwell, & Beauvais, 2002). This chapter provides an overview of factors to take into consideration when evaluating VE trainers. In particular, such evaluations can be used to (1) make decisions regarding the adoption of new training systems by quantifying their training value, (2) determine the level of training offered by different training systems or components in an instructional program (e.g., classroom, simulated environments, live training), and (3) determine an appropriate mix of training solutions across the training progression from novice to expert (Champney et al., 2008).

30.2 TRAINING TRANSFER

Gick and Holyoak (1987) suggested that transfer is a consequence of learning and the ability of a trainee to draw relationships between two tasks. Thus, the extent of what is transferred will depend on how the trainee is able to utilize newly acquired KSAs in task repetitions, similar tasks, or very different tasks and conditions. This then influences the amount and types of observed transfer.

30.2.1 Types of Transfer

It is important to recognize that there are varying types of transfer, each of which can be related to the training goals of a VE trainer. The type of transfer assessed should match the intended goals of the given training solution (Bossard, Kermarrec, Buche, & Tisseau, 2008; Gick & Holyoak, 1987; Leberman, Mcdonald, & Doyle, 2006; Schunk, 2004).

Positive and Negative Transfer. The distinction between positive and negative transfer relates to whether the trained KSAs promote or hinder performance in the transfer environment (i.e., the operational domain) that was targeted during training (Baldwin & Ford, 1988). A third possibility occurs when the training does not have an impact in operational performance and thus no transfer occurs (i.e., neither positive nor negative performance impacts are observed).

Near and Far Transfer. The distinction between near and far transfer relates to the conceptual similarities between what was trained and the expected application domain (Perkins & Solomon, 1994). Near transfer refers to the application of learned KSAs to similar conditions to those that occurred during training (similar context and tasks). Far transfer refers to the application of learned KSAs to conditions dissimilar to those that occurred in training (different context and tasks). Given the vast possibilities of domains and conditions, the differentiating factor between the two is an abstract similarity scale.

Vertical and Horizontal Transfer. The distinction between vertical and horizontal transfer relates to how learned KSAs influence further learning. This perspective assumes a hierarchical organization of KSAs, where some lower KSAs are necessary prerequisites to more advanced, higher ones (Gagne, 1965). Vertical transfer refers to the direct use of prior learned KSAs to build new KSAs, theorizing that new learning is facilitated only when the same elements are reutilized in the new KSAs, such as when learning to count numbers facilitates addition or subtraction (Bossard et al., 2008). Horizontal transfer refers to the application of learned KSAs to other similar ones within the same level in the hierarchy of KSAs.

High Road and Low Road Transfer. The distinction between low and high road transfer relates to the mechanism by which the transfer occurs (Perkins & Salomon, 1992). Low road transfer is reflexive and involves the activation of highly practiced activities by cues in the transfer environment that are very similar to those in the training environment. High road transfer is conscious and requires a deliberate effort to abstract and find the relationship between the transfer environment and the training environment.

Understanding the different forms in which learned KSAs may be retained transferred to the operational environment is important as it provides an overview of the potential implications of the training provided. For instance, identifying the nature of the KSAs will not only allow one to design and assess near transfer but also understand the potential implications of how and where this same KSAs will have an impact (e.g., support execution of similar tasks [near/far transfer], support learning of new tasks [vertical transfer]).

30.3 EVALUATING TRAINING TRANSFER

… good-quality [transfer] studies are currently few and far between, due in part to the fact that many defence establishments are (unsurprisingly) not receptive to their training régimes being compromised, or their classes being split up into control and treatment groups. A related concern is the possible negative impact on the performance and attitudes of control group participants who discover that they may, as a result of the random condition allocation process, have been "deprived" of access to a new and exciting learning technology. Furthermore, measuring the outcomes of [computer]-based training in naturalistic settings–the military classroom and the training or operational field–is far from straightforward.

Stone (2012, p. 3)

Evaluating training transfer involves three broad challenges (Cohn et al., 2008): (1) developing meaningful measures of skill transfer and selecting associated experimental designs, (2) identifying an operational system or suitable substitute within which to conduct the empirical transfer evaluation, and (3) resolving multiple logistical constraints that surround field studies (accessibility to trainees, scheduling, etc.).

30.3.1 CHALLENGE ONE: DEVELOPING MEANINGFUL MEASURES OF TRANSFER AND SELECTING AN APPROPRIATE EXPERIMENTAL DESIGN

The selection of appropriate measures of transfer of training should first be approached by under-standing the desired evaluation goals. The list provided in Table 30.1 summarizes some of the evaluation questions that one may seek to address when assessing the efficacy of a VE training system (Ahlberg et al., 2002; Alexander, Brunye, Sidman, & Weil, 2005; Campbell & Stanley, 1966; Champney et al., 2008; Koonce, Lintern, & Taylor, 1993; Rantenen & Talleur, 2005; Simpson & Defense Manpower Data Center, 1999; Steensma & Groeneveld, 2010; Stewart, Dohme, & Nullmeyer, 2001).

The answer to each of these evaluation questions implies a different measure of transfer, as well as different assessment methodologies (i.e., experimental designs).

Measures of Transfer. There are multiple options as to how to assess transfer, and each has its own set of benefits and challenges. Those approaches that assess not only training impact but also efficiency are the most robust yet the most resource intensive to collect (Alexander et al., 2005; Champney et al., 2008; Fletcher, 1998; Roscoe, 1972; Taylor, Talleur, Emanuel, & Rantanen, 2005; Taylor et al., 1999). For instance, two different VE trainers may have the ability to produce a compa-rable level of transfer, yet one may do so with less training time. A review of the literature identifies several potential measures of transfer, as summarized in Table 30.2.

Each of the measures in Table 30.2 provides varying degrees of information regarding the train-ing capabilities of the evaluated system via different levels of scrutiny. The decision of which mea-sures of transfer to use largely depends on the desired evaluation goals and resources available. For example, one could systematically evaluate the most optimal amount of training needed by evalu-ating increments of training (i.e., using the Incremental Training Effectiveness Ratio [ITER]) and evaluate transfer for each amount. Alternatively, a less resource–intensive option is to select a single training amount (i.e., identify the point at which transfer should be evaluated) selected by using a learning curve analysis to (Champney et al., 2008) and utilize a measure such as the Training Effectiveness Ratio (TER) (this method is explained further in Section 30.3.3).

Experimental Designs. To address the goals of the various evaluation questions summarized in Table 30.1, it is necessary to identify not only a suitable transfer measure from Table 30.2 but also an

---

TABLE 30.1

Evaluation Questions to Address When Assessing the Efficacy of VE Training Systems

1. Do trainees perform up to criterion after training in the VE simulator?
2. What is the change in performance that a VE simulator can produce?
3. What is the performance difference between trained and untrained individuals?
4. Which VE simulator/intervention is more effective at improving performance?
5. How long does it take to reach criterion?
6. How many live training trials can be saved by using a pretraining in a VE simulator?
7. How effective are successive increments of training in a given VE platform?
8. What is the optimal number of training sessions for a VE platform to produce the best performance?
9. Which order of different training platforms is the most effective at improving performance?
10. What effect does a VE training intervention have on performance at different time periods after training?
11. How can maturation be controlled for as the longitudinal effect of VE training is measured?

TABLE 30.2

Transfer of Training Measures

- *Percent transfer* (Alexander et al., 2005; Roscoe & Williges, 1980; Taylor et al., 1999; Taylor et al., 2001):
  This measure identifies the percent of savings achieved by a training treatment (e.g., VE trainer) compared
  to a baseline (e.g., no VE trainer).
- *TER* (Alexander et al., 2005; Koonce et al., 1993; Roscoe, 1971): This measure identifies the live training
  savings (e.g., trails saved) that results from a given amount of pretraining in a VE system.
- *ITER* (Roscoe, 1972; Taylor et al., 2005): This measure identifies the effectiveness of multiple training
  regimes (e.g., incremental number or training trials) and identifies the impact of training gained for each
  additional increment in training. Given the need for multiple training treatment groups, it requires
  considerable effort and resources, which limit its feasibility (c.f. Roscoe & Williges, 1980).
- *Cumulative transfer effectiveness ratio* (*CTER*) (Roscoe, 1970; Stewart et al., 2001): This measure
  identifies the efficacy of a VE trainer compared to training in the real world (where a ratio of 1.0 indicates
  the VE is as efficient as the real world, and less than 1.0 indicates it is less efficient; Holman, 1979).

---

appropriate evaluation approach (i.e., experimental design). Depending on the evaluation questions
selected from Table 30.1 and amount of resources available, a suitable experimental design may be
utilized as shown in Table 30.3.

30.3.2 CHALLENGE TWO: IDENTIFYING AN OPERATIONAL SYSTEM OR SUITABLE SUBSTITUTE TO SUPPORT EVALUATION

An additional challenge with transfer studies is determining what constitutes the transfer environment for test purposes. For various reasons (e.g., no access to flight time, safety), it may be impossible to find a real-world transfer environment within which to conduct the transfer evaluation. One
option available is to utilize *substitute test transfer environments* when the real operational environment is not accessible (c.f., Stanney et al., in press; Wiederhold & Wiederhold 2006). Substitute
environments should, of course, resemble the operational environment as much as possible, which
begs the question "What defines the operational environment?"

Given that an operational environment represents a human experience, it can be decomposed into
its sensory, cognitive, affective, and functional components. As such, one can define experiential
requirements that the substitute environment must comprise in order to qualify as a suitable alternative. An approach to achieve this is via an experiential task analysis (ETA), which seeks to characterize the operational environment by identifying and defining target operational tasks in terms of
experiential cues (Champney & Carroll, 2013). This process is grounded in social science methodologies and expands on traditional approaches, such as task analysis (Diaper, 2004), by incorporating
further levels of decomposition (i.e., via definition of the sensory, cognitive, affective, and functional
experiences that take place during the task). The sensory component refers to those elements of a task
that are experienced directly by the senses, such as visual, auditory, haptic, proprioceptive, and olfactory cues. These cues represent what the trainee would feel if she or he were executing the task in the
operational environment. The cognitive component refers to those elements of the experience that
impact the cognitive state of the individual, such as workload and attention requirements. Similarly,
the affective component refers to those elements that impact the affective state of the trainee, such
as visceral, behavioral, and symbolic cues. These cues represent how the trainee would be feeling
as she or he experiences tasks under operational conditions. The functional component refers to the
elements in the environment with which the human operator interacts with and how this environment
reacts and behaves to inputs from the human or other actors (i.e., cues a trainee would experience via
reactions from the world as she or he interacts or observes the operational environment).

By decomposing an operational domain and characterizing it in terms of its experiential cues, it
is possible to develop experiential requirements that must exist in order to qualify an environment

TABLE 30.3
Training Effectiveness Evaluation Experimental Designs

Design	Description	Table 30.1 Evaluation Questions Addressed	Advantage(s)	Disadvantage(s)	Formula	Outputs
Solomon 4 group design (Steensma and Groeneveld [2010])	There are two experimental groups that receive training. One receives pretesting and one doesn't. There are two control groups. One receives pretesting and one doesn't. All receive posttesting.	1, 2, 3, 4	High internal and external validity. Accounts for nontraining variability between groups. Accounts for learning effects of a pretest	Resource intensive. Requires a higher N relative to other experimental designs.	N/A	Ability to infer learning improvement in trainees as a result of a training intervention and taking account of priming effects (i.e., performance gains due to a pretest)
Comparison to baseline with pre- and posttests (Cambell and Stanley [1966]; Simpson [1999])	There is a control (baseline) and an experimental group. Both receive pretesting and posttesting.	1, 2, 3, 4	High internal and external validity. Accounts for nontraining variability between groups	Resource intensive. Unable to account for priming effects (i.e., performance gains due to a pretest).	N/A	Ability to infer learning improvement in trainees as a result of a training intervention
Comparison to baseline with only posttests (Campbell and Stanley [1966]; Simpson [1999])	There is an experimental and a control group (baseline). They both receive only posttesting.	1, 2, 3	Resource efficient. Lower complexity design	Randomization helps to control for variability between groups, but lack of pretest doesn't allow for assessment of trainee's initial state, thus unable to quantify skill improvement.	N/A	Ability to compare learning outcomes as a result of type of training intervention
Single-group training improvement (pretest and posttest, but no control group) (Campbell and Stanley [1966]; Simpson [1999])	There is only an experimental group that receives pre- and posttesting (no baseline for comparison).	1, 2	Measures change in performance before versus after training. Uses fewer participants and thus less costly and less complex than experiments that include control condition	Lack of randomization and control introduces confounds (e.g., determining whether changes in learning resulted from different trainings or from characteristics inherent to the groups) that must be accounted for. Cannot be used to compare multiple interventions.	N/A	Limited ability to infer that the difference in performance resulted from the training intervention

Posttraining performance (posttest only, no control group) (Campbell and Stanley [1966]; Simpson [1999])	There is an experimental group that only receives posttesting.	1	Resource efficient. Measures whether performance criterion is met after training	No direct correlation or diagnosis can be made between training and performance.	N/A	Limited to only allowing verification that trainees perform to standard after training
Training impact over time (single-group time–series experiment) (Campbell and Stanley [1966]; Simpson [1999])	There is one group that receives multiple evenly spaced pretests and posttests.	10	Allows for greater insight as to how a training intervention impacts performance over time	Cannot compare two systems, resource intensive due to volume of testing.	N/A	Ability to make inferences on training impact over time
Training impact over time compared to baseline (multiple group time–series design) (Campbell and Stanley [1966]; Simpson [1999])	A control group (baseline) and experimental group receive multiple evenly spaced pretests and posttests.	11	Controls for maturation effects. Allows for training to be measured over time	Many assessments must be conducted for the control and experimental groups. Very resource intensive.	N/A	Ability to make inferences on training impact over time compared to a baseline condition
Training impact of order of training interventions (counterbalanced design) (Campbell and Stanley [1966]; Simpson [1999])	Multiple experimental groups in which every group experiences every treatment in a different order.	9	Allows for the measurement of order effects across a combination of training platforms. Can use a partial design to cut down on resources	Very resource intensive depending upon the number of training systems being evaluated. Many participants required, as well as many training interventions.	N/A	Ability to make inferences about order effects of training interventions

(continued)

TABLE 30.3 (continued)
Training Effectiveness Evaluation Experimental Designs

Design	Description	Table 30.1 Evaluation Questions Addressed	Advantage(s)	Disadvantage(s)	Formula	Outputs
Percent transfer study (Taylor et al. [1999]; Alexander et al. [2005]; Rosco and Williges [1980]; Taylor et al. [2001])	Measures the amount of KSAs learned during a training intervention that carried over and were observed in the operational environment.	6	Allows for measurement of the amount of training that made observable differences in performance in the operational environment	Resource intensive. Performance in operational environment may be hard or impossible to assess due to risks or resource constraints. Pseudo-domain environment may be used as an option, yet may lack some elements of real environment and operational conditions.	$\dfrac{Y_c - Y_x}{Y_c} * 100$ Y_c: time/trials by control group in the prime mission system (PMS) (e.g., aircraft) Y_x: time/trials by experimental group in PMS	Percent transfer—summarizes amount of performance that was observed in operational environment
TER study (Stanney et al. [in press]; Alexander et al. [2005]; Koonce et al. [1993])	Measures the number of live training trials saved per unit of time trained in a simulator.	5,6	Incorporates prior training in simulation while determining how much time was saved in live training. Identifies simulations that are ineffective at producing transfer	May draw conclusions about training transfer at points along the learning curve that have yet to stabilize. Gives a snapshot at one point in time (a single training regimen is used to make the assessment).	$\dfrac{Y_c - Y_x}{X}$ Y_c: time/trials by control group in PMS Y_x: time/trials by experimental group in PMS X: time/trials by experimental group in simulation	TER—ratio of pretraining in simulation to live trial savings, evaluates how effective a simulation is at a given point in time

	Description		Benefits	Limitations		Metric
ITER study (Champney et al. [2008]; Taylor et al. [2005])	Measures the number of trials at which point simulation training is no longer cost effective.	4, 7, 8	Accounts for diminishing learning returns that inevitably occur over time and allows for identification of optimal simulation regimen to use to obtain maximum training gains	Very highly resource intensive. Requires multiple trials of testing.		ITER—incremental point at which simulation is no longer cost effective
Learning curve analysis (Champney et al. [2008])	Identifies the asymptote at which learning improvements reach a plateau.	6 (part), 8	Identifies point at which TER should be performed. Helps identify plateau in learning effectiveness without having to run thousands of trials	Can be resource intensive depending on how much testing is needed to identify a plateau in learning gains.	N/A	Number of trials at which TER should be performed
Cumulative transfer of training effectiveness ratio (CTER) study (Stewart et al. [2001]; Roscoe [1971])	Evaluates the cumulative effects and savings in transfer over progressive simulation training trials.	10	More dynamic than TER; can determine how TER behaves over successive trials. Can identify trends and overall behavior of TER	Requires multiple rounds of testing during the training intervention.		CTER—provides an assessment of training effectiveness over time.

TABLE 30.4
Cue Impact Rating Scale

0—Cue not necessary: The cue is not necessary for effective execution of the task.

1—Cue enhances: The cue enhances effective execution of the task.

2—Cue supports: The cue supports effective execution of the task.

3—Cue required: The cue is required for effective execution of the task.

as representative of the transfer environment. During an ETA, the cues identified should be rated for their criticality to task performance using a scale such as the one provided in Table 30.4. This rating mechanism can be used to categorize cues and allow for a deeper understanding of the necessity of these cues. These ratings should always be determined with respect to the training goals of the system. For instance, if the goal of the system is to train shoot/no shoot decision making, although the trigger is a required haptic cue for shooting, since the goal is not to train marksmanship, it may not be a required cue for training the objective task. It is not expected that all cues will be reproducible or feasible to represent within the constraints surrounding the evaluation study. Therefore, the use of this rating scheme can also support trade-off analyses for determining which cues would provide the greatest return on investment with respect to training impact.

These cues can then be transformed into requirements and the ratings used to judge the substitute environment for its ability to reproduce such conditions and thus determine whether it is a good option to conduct a transfer study. Stanney et al. (2013) describe a substitute test transfer environment developed in such a fashion. In their study, which sought to evaluate the efficacy of two VE trainers for military operations in urban terrain, a substitute test environment was fashioned to provide an environment as realistic as possible, yet providing safety for trainees and access to data collection. Their test environment was designed to replicate the interior of a building containing specific types of features (e.g., room shapes, sizes, windows, doorways) to replicate the conditions expected in real operational urban environments. In addition, the substitute environment was also populated with combatants and noncombatants who were paid *actors* instructed to behave in specific ways, while both trainees and role players utilized laser-based weapon replicas.

30.3.3 CHALLENGE THREE: OVERCOMING LOGISTICAL CONSTRAINTS

Two logistical challenges that are faced by all transfer of training studies are (1) the target population and (2) the training regime. The *target population* challenge refers to the requirement that every transfer of training study must ensure that the target population of trainees is represented in the evaluation. Yet it can be difficult to obtain access to individuals from the target population. The *training regime* challenge refers to the fact that the effects observed from training are dependent not only on the technology used (i.e., the VE trainer) but also on the number of opportunities for learning afforded to the trainees (i.e., amount of time or number of trials trained). Yet determining the right amount of training to provide may be challenging and there may be resource constraints (e.g., access to enough participants, costs, schedules).

Target Population. In applied fields, it is common to encounter difficulties in obtaining access to a sample of representative trainees to conduct transfer of training studies. Due to limited availability, restricted access, willingness to participate, and resource constraints, among others, representative samples are often hard to come by. In these instances, an alternative is to recruit from substitute populations (Ward, 1993). Nonetheless, given the potential differences between the surrogate sample and target population, it is necessary to ensure that the surrogate sample is comparable in terms of the prerequisite KSAs of the target population (i.e., both possess sufficient KSAs regarding the target domain to ensure validity of results; Wintre, North, & Sugar, 2001).

TABLE 30.5

Approach to Developing a Surrogate Sample

1. *Initial classroom instruction*: to introduce participants to the domain and teach desired constructs
2. *Practical instruction and evaluation*: to train skills required in task performance with support of an instructor-evaluator to provide feedback
3. *Rehearsal*: for continued review of the constructs and practice of the skills in the time between the pretraining and the study
4. *Review*: just prior to the study to mitigate any memory decay
5. *Scenario-based feedback*: integrated with the practical instruction to guide appropriate development of target KSAs

For example, one common practice is to recruit college students as experimental participants due to their relative ease of recruiting and availability. In addition to evaluating KSAs, there may be other potential differences between the surrogate population and the target population that may limit generalizability, such as lack of context (e.g., task meaningfulness) and artificiality of experimental conditions (Gordon, Schmitt, & Schneider, 1984). To address these issues, Champney, Milham, Bell Carroll, Stanney, and Cohn (2006) have proposed an approach by which a surrogate population sample is brought up to speed by putting them through a pretraining boot camp. The pretraining focuses on both increasing knowledge of target KSAs and providing a contextual understanding of the target domain. Instructional materials for the boot camp can be developed through a task analysis that provides a thorough domain understanding and identifies training objectives. In the boot camp developed by Champney et al. (2006), the pretraining course curriculum consisted of five elements (see Table 30.5).

Putting a surrogate sample through a pretraining program such as this can help mitigate issues with sample validity that arise when there are limitations in obtaining access to the target domain population.

Training Regime. The effects of a VE trainer are not just due to training with the technology but also due to the number/amount of opportunities to learn (e.g., number of trials). Given this relationship, it is challenging to determine the true effectiveness of a VE trainer without first knowing what is an adequate amount of training necessary to produce the desired training effects. This issue is compounded by the fact that testing multiple amounts of training has a direct impact on the time and resources needed in an already resource-intensive transfer of training study. What is known is that learning has been shown to follow a *learning curve*, a pattern of rapid improvement followed by ever diminishing further improvements with practice (Ebbinghaus, 1885 [in Wozniak, 1999]; Ritter & Schooler, 2001). This implies an inherent reduction in the rate of return for every additional practice opportunity up to a point where it no longer is cost effective to conduct more training (i.e., a plateau in learning). One can use an operational approach to identify this area of limited training returns, such as a learning curve analysis (Champney et al., 2008). While analyzing a graphical learning curve by simple visual inspection may seem feasible, such an approach lacks objectivity and reliability (e.g., the scale used to define the graph of the curve may create false-positive identification of a plateau; also, slopes may appear smaller in larger scales); thus, a more systematic approach is needed. One such approach is to identify the location at which further training reaches a critical point of diminished return by conducting a small-scale pilot study (Champney et al.). This approach is discussed next.

Learning Curve Analysis. The main premise of this approach is to objectively identify the point at which further training will no longer result in sizeable gains in performance observed in a VE trainer. This approach begins with gathering performance data from a series of training sessions. Next, three parameters are identified: (A) percent variability, (B) period of variability stabilization,

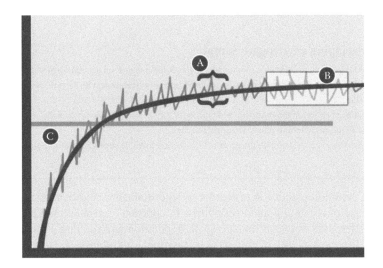

FIGURE 30.1   Learning curve analysis. (Adapted from Champney et al. 2008.)

and (C) general location of plateau, as illustrated in Figure 30.1. The asymptote is then localized at some point along the shaded area. The parameters are discussed next:

A. *Percent variability* is defined as the percentage in the variability of task performance as observed in the cumulative average between subsequent trials. This percentage is used to define the acceptable variability and minimum threshold for gains (or losses) in performance across trials that warrant further training (e.g., 10% improvement).

B. *Period of variability stabilization* is defined as how long performance must remain relatively unchanged within the boundaries of parameter A to be considered a plateau. Three to five trials have been used with adequate success (e.g., Champney et al., 2006).

C. *General location of plateau* serves as safety parameters to mitigate for the potential of temporary plateaus (i.e., periods where performance stabilizes and later increases significantly) and serves as a visual inspection rule to prescribe the likely location of the plateau (e.g., tail end of the curve). While this might be determined through a visual inspection, specifying a rule ensures consistency across multiple measures (e.g., on the second half of the curve).

This learning curve analysis approach results in the identification of an amount of VE training at which further training gains are minimal. It is at this point (i.e., with this amount of training) that one is likely to have the most success in a full-scale transfer of training evaluations. This approach should prevent evaluators in drawing conclusions about training transfer at points along the learning curve that have yet to stabilize.

30.4   CONCLUSIONS

With reduced training budgets, issues with stagnant proficiency can be overcome if appropriate training alternatives to live training are systematically designed such that they optimize skill acquisition and retention. The recommendations provided in this chapter, if appropriately applied, can provide an assessment of the utility and efficacy of a VE training solution, including its potential for replacing some portion of live trials. By first identifying the goals of the VE trainer (see Table 30.1), one can identify the type of transfer sought. This can then lead to the identification of appropriate measures of transfer of training (see Table 30.2) and a suitable evaluation approach (experimental design; see Table 30.3) given the identified training objectives. Within these items identified, it is then possible to address the remaining challenges regarding test environment (i.e., designing a substitute test transfer environment; see Table 30.4), sample population (i.e., pretraining a surrogate test population; see Table 30.5), and training regime (i.e., identifying the appropriate number of trials to train before evaluating transfer; see Figure 30.1), all of which are necessary to ensure an objective and effective transfer evaluation. Together, these recommendations are anticipated to result in successful evaluation of the transfer of training potential for any given VE trainer.

ACKNOWLEDGMENTS

This material is based upon work supported in part by the Office of Naval Research (ONR) under contact No. N00014-04-C-0024, and by the Air Force Research Laboratory (AFRL) under contract FA8650-12-M-6319. Any opinions, findings, and conclusions or recommendations expressed in this material are those of the authors and neither necessarily reflect the views of, nor are endorsed by, ONR or AFRL.

REFERENCES

Ahlberg, G., Heikkinen, T., Iselius, L., Leijonmark, C. E., Rutqvist, J., & Arvidsson, D. (2002). Does training in a virtual reality simulator improve surgical performance? *Surgical Endoscopy, 16*, 126–129.

Alexander, A., Brunye, T., Sidman, J., & Weil, S. (2005). *From gaming to training: A review of studies on fidelity: Immersion, presence, and buy-in and their effects on transfer in PC-based simulations and games.* Alexandria, VA: The American Society for Training and Development Retrieved 28 June, 2007 from http://www.darwars.com/downloads/DARWARS%20Paper%2012205.pdf.

Allen, J. A., Hays, R. T., & Buffordi, L. C. (1986). Maintenance training, simulator fidelity, and individual differences in transfer of training. *Human Factors, 28*, 497–509.

Baldwin, T. T., & Ford, J. K. (1988). Transfer of training: A review and directions for future research. *Personnel Psychology, 41*, 63–105.

Bassi, L. J., & van Buren, M. E. (1999). ASTD state of the industry report. *Training and Development, 52*, 21–43.

Bossard, C., & Kermarrec, G. (2006). Conditions that facilitate transfer of learning in virtual environments. *Proceedings of the 2nd International Conference on Information and Communication Technologies (ICTTA'06)*, (Vol. 1, pp. 604–609). Damascus, Syria.

Bossard, C., Kermarrec, G., Buche, C., & Tisseau, J. (2008). Transfer of learning in virtual environments: A new challenge? *Virtual Reality, 12*, 151–161.

Burke, L. A., & Hutchins, H. M. (2008). A study of best practices in training transfer and proposed model of transfer. *Human Resource Development Quarterly, 19*(2), 107–128.

Campbell, D. T., & Stanley, J. C. (1966). *Experimental and quasi-experimental designs for research.* Chicago, IL: Rand McNally.

Carnevale, A. P., & Schulz, E. R. (1990). Economic accountability for training: Demands and responses. *Training and Development Journal Supplement, 44*(7), s2–s4.

Champney, R., & Carroll, M. B. (2013). *The experiential task analysis: A training system design and analysis tool.* (Technical Report) Oviedo, FL: Design Interactive, Inc.

Champney, R., Milham, L. M., Carroll, M. B., Ahmad, A., Stanney, K. M., Cohn, J., & Muth, E. (2008). Conducting training transfer studies in complex operational environments. In D. Schmorrow, J. Cohn, & D. Nicholson (Eds.), *The handbook of virtual environment training: Understanding, predicting and implementing effective training solutions for accelerated and experiential learning* (Vol. 3, Section 2, pp. 243–253). Westport, CN: Praeger Security International.

Champney, R. K., Milham, L., Bell Carroll, M., Stanney, K. M., & Cohn, J. (2006). A method to determine optimal simulator training time: Examining performance improvement across the learning curve. *Proceedings of the Human Factors and Ergonomics Society 50th Annual Meeting.* San Francisco, CA, October 16–20, 2006.

Cohn, J., & Fletcher, J. (2010). What is a pound of training worth? Frameworks and practical examples for assessing return on investment in training. *Proceedings of the Interservice/Industry Training, Simulation, and Education Conference (I/ITSEC) Annual Meeting* (Paper No. 100113). Orlando, FL.

Cohn, J., Stanney, K. M., Milham, L., Bell Carroll, M., Jones, D., Sullivan, J., & Darken, R. (2008). Training effectiveness evaluation: From theory to practice. In D. Schmorrow, J. Cohn, & D. Nicholson (Eds.), *The handbook of virtual environment training: Understanding, predicting and implementing effective training solutions for accelerated and experiential learning* (Vol. 3, Section 2, pp. 157–172). Westport, CN: Praeger Security International.

Cohn, J. V., Stanney, K. M., Milham, L. M., Jones, D. L., Hale, K. S., Darken, R. P., & Sullivan, J. A. (2007). Training evaluation of virtual environments. In E. L. Baker, J. Dickieson, W. Wulfeck, & H. O'Neil (Eds.), *Assessment of problem solving using simulations* (pp. 81–105). Mahwah, NJ: Lawrence Erlbaum.

Diaper, D. (2004) Understanding task analysis for human-computer interaction. In D. Diaper, & N. Stanton, *The handbook of task analysis for human-computer interaction.* Mahwah, NJ: Lawrence Erlbaum Associates, Inc. Publishers.

Erwin, S. I. (2013, June 4). With warplanes grounded, can flight simulators fill air force training gap? *National Defense Magazine.* Retrieved June 11, 2013: http://www.nationaldefensemagazine.org/blog/Lists/Posts/Post.aspx?ID = 1164.

Eseryel, D. (2002). Approaches to evaluation of training: Theory and practice. *Journal of Educational Technology and Society, Special Issue: Integrating Technology into Learning and Working*, 5(2), 93–98.

Fletcher, J. D. (1998). Measuring the cost, effectiveness, and value of simulation used for military training. *Proceedings of SimTecT 1998 Conference* (pp. 47–52). Adelaide, South Australia, Australia.

Gagne, R. M. (1965). *The conditions of learning.* New York, NY: Holt Rinehart & Winston.

Gick, M. L., & Holyoak, K. J. (1987). The cognitive basis of knowledge transfer. In S. M. Cormier, & J. D. Hagman (Eds.), *Transfer of training: Contemporary research and applications* (pp. 9–46). New York, NY: Academic Press.

Gordon, M. E., Schmitt, N., & Schneider, W. (1984). An evaluation of laboratory research bargaining and negotiations. *Industrial Relations, 23,* 218–233.

Koonce, J. M., Lintern, G., Taylor, H. L. (1993). Quasi-transfer as a predictor of transfer from simulator to airplane. *The Journal of General Psychology, 120*(3), 257–276.

Kozlowski, S. W. J., DeShon, R. P., Schifflett, S. G., Elliott, L. R., Salas, E., & Coovert, M. D. (2004). A psychological fidelity approach to simulation-based training: Theory, research, and principles. In E. Salas, L. R. Elliott, S. G. Schflett, & M. D. Coovert (Eds.), *Scaled worlds: Development, validation, and applications* (pp. 75–99). Burlington, VT: Ashgate Publishing.

Lave, J., & Wenger, E. 1990. *Situated learning: Legitimate peripheral participation.* Cambridge, U.K.: Cambridge University Press.

Leberman, S. I., McDonald, L., & Doyle, S. (2006). *The transfer of learning: Participants' perspectives of adult education and training.* Hampshire, U.K.: Gower Publishing.

Lehman, D. R., Lempert, R. O., & Nisbett, R. E. (1988). The effects of graduate training on reasoning: Formal discipline and thinking about everyday-life events. *American Psychologist, 43,* 431–442.

McCauley, M. E. (2006). *Do army helicopter training simulators need motion bases?* (Technical Report 1176). Arlington, VA: United States Army Research Institute for the Behavioral and Social Sciences.

Pennington, N., Nicolich, R., & Rahm, J. (1995). Transfer of training between cognitive subskills: Is knowledge use specific? *Cognitive Psychology, 28,* 175–224.

Perkins, D. & Salomon, G. (1994). Transfer of learning. In T. Husen, & T. Postelwhite (Eds.), *International encyclopedia of education* (2nd ed., Vol. 11, pp. 6452–6457). Oxford: Elsevier Science Ltd., England: Pergamon Press.

Rantanen, E. M., & Talleur, D. A. (2005). Incremental transfer and cost effectiveness of groundbased flight trainers in university aviation programs. *Proceedings of the Human Factors and Ergonomics Society 49th Annual Meeting* (pp. 764–768). Orlando, FL. Retrieved from http://pro.sagepub.com/content/49/7/764.short.

Ritter, F. E., & Schooler, L. J. (2001). The learning curve. In N. J. Smelser & P. B. Baltes (Eds.) *International encyclopedia of the social and behavioral sciences* (pp. 8602–8605). Amsterdam, the Netherlands: Pergamon Press.

Roscoe, S. N. (1970). *Incremental transfer effectiveness* (Technical Report ARL-70-5/AFOSR-70-1). Arlington VA: United States Air Force Office of Scientific Research.

Roscoe, S. N. (1971) Incremental transfer effectiveness. *Human Factors, 13,* 561–567.

Roscoe, S. N. (1972). A little more on incremental transfer effectiveness. *Human Factors, 14*(4), 363–364.

Roscoe, S. N., & Williges, B. H. (1980). Measurement of transfer of training. In S. N. Roscoe (Ed.), *Aviation psychology* (pp. 182–193). Ames, IA: Iowa State University Press.

Schunk, D. H. (2004). *Learning theories: An educational perspective* (4th ed.). Upper Saddle River, NJ: Merrill & Prentice Hall.

Simpson, H., & Defense Manpower Data Center. (1999). *Evaluating large-scale training simulations, volume I: Reference manual.* (DMDC Technical Report 99-05). Retrieved from http://www.dtic.mil/cgi-bin/GetTRDoc?AD = ADA383851.

Singley, M. K., & Anderson, J. R. (1989). *The transfer of cognitive skill.* Cambridge, MA: Harvard University Press.

Stanney, K. M., Hale, K. S., Carroll, M., & Champney, R. (2012, In Press). Applied perception and virtual environment training systems. In J. L. Szalma, M. Scerbo, R. Parasuraman, P. A. Hancock, & R. R. Hoffman (Eds.), *The handbook of applied perception research*. Cambridge, U.K.: Cambridge University Press.

Stanney, K. M., Hale, K. S., & Cohn, J. V. (2012). Virtual environments. In G. Salvendy (Ed.), *Handbook of human factors and ergonomics* (4th ed., pp. 1031–1056). New York, NY: John Wiley.

Stanney, K. M., Milham, L. M., Champney, R. K., Carroll, M. B., Cohn, J., & Muth, E. (2013). *Can virtual environment training save live training trials?* (Technical Report). Oviedo, FL: Design Interactive, Inc.

Steensma, H., & Groeneveld, K. (2010). Evaluating a training using the "four levels model". *Journal of Workplace and Learning, 22*(5), 319–331.

Stewart, J. E., II, Dohme, J. A., & Nullmeyer, R. T. (2001). U.S. army initial entry rotary-wing transfer of training research. *The International Journal of Aviation Psychology, 12*(4), 359–375.

Stone, R. J. (2012). *Human factors guidance for designers of interactive 3D and games-based training systems*. Birmingham, U.K.: Human Factors Integration Defence Technology, Centre University of Birmingham.

Taylor, H. L., Lintern, G., Hulin, C. L., Talleur, D. A., Emanuel, T. W., Jr., & Phillips, S. I. (1999). Transfer of training effectiveness of a personal computer aviation training device. *The International Journal of Aviation Psychology, 9*(4), 319–335.

Taylor, H. L., Talleur, D. A., Emanuel, T. W., & Rantanen, E. M. (2005). *Effectiveness of flight training devices used for instrument training* (Final Technical Report AHFD-05- 9/FAA-05-4). Retrieved from https://www.hf.faa.gov/docs/508/docs/GAFTDfinal.pdf.

Taylor, H. L., Talleur, D. A., Emanuel, T. W., Jr., Rantanen, E. M., Bradshaw, G., & Phillips, S. I. (2001). *Incremental training effectiveness of personal computers used for instrument training*. Paper presented at the 11th International Symposium on Aviation Psychology, Columbus, OH.

Thompson, C., Koon, E., Woodwell, W. H., & Beauvais, J. (2002). *Training for the next economy: An ASTD state of the industry report* (Report No. 790201). Alexandria, VA: The American Society for Training and Development.

Thorndike, E. L. (1906). *Principles of teaching*. New York, NY: A.G. Seiler.

Thorndike, E. L., & Woodworth, R. S. (1901). Psychological review. In S. Leberman, L. McDonald, & S. Doyle, (Eds.), *Transfer of learning* (No. 8, pp. 247–261). Hampshire, U.K.: Gower Publishing Company.

Ward, E. A. (1993). Generalizability of psychological research from undergraduates to employed adults. *Journal of Social Psychology, 133*(4), 513–519.

Wiederhold, B. K., & Wiederhold, M. D. (2006). *Virtual reality training transfer: A DARWARS study. Physiological monitoring during simulation training and testing*. San Diego, CA: The Virtual Reality Medical Center.

Wintre, M. G., North, C., & Sugar, L. A. (2001). Psychologists' response to criticisms about research based on undergraduate participants: A developmental perspective. *Canadian Psychology, 22*, 216–225.

Wozniak, R. H. (1999). Introduction to memory: Hermann Ebbinghaus (1885/1913). In *Classics in psychology, 1855–1914: Historical essays*. Bristol, U.K.: Thoemmes Press.

31 Virtual Environment Usage Protocols

Kay M. Stanney, Robert S. Kennedy, and Kelly S. Hale

CONTENTS

31.1 INTRODUCTION

While investment in virtual environment (VE) technology has been expanding over the past decade, concomitant with advances in this field has been awareness of the potential consequences of venturing into a virtual world. This technology provides the potential to enhance training, education, and design among other uses. It could help to prevent job mishaps through better procedural safety training, medical mistakes through simulated surgical practice, and poor communication through virtual business connections. If improperly used, however, a whole host of adverse effects can be experienced both during and after exposure. Stone (2012, p. 24) suggests that "the biggest mistake made by the VR community was that it ignored (these) human factor" issues, and thus truly widespread adoption of the technology has yet to be seen. The problems associated with VE technology are real. VE exposure can cause people to discontinue use (13% overall; 6.3% within 15 min condition, 16.9% with 30 min condition, 31.0% with 45 min condition, 45.8% for 60 min condition), experience visual flashbacks (15%), drowsiness (44%), and experience vomiting (about 1%) (Stanney, Hale, Nahmens, & Kennedy, 2003). Approximately 80%–95% of those exposed to a VE report some level of symptomatology postexposure, which may be as minor as a headache or as severe as vomiting or intense vertigo (Bos, 2011a; Stanney et al., 1998; Stern & Koch, 1996). More troubling, the problems do not stop immediately upon cessation of exposure. In fact, VE exposure is associated with aftereffects (see Welch & Mohler, 2014, Chapter 25; Wann & Mon-Williams, 2014, Chapter 32; Dizio & Lackner, 2014, Chapter 33), which can render the exposed individual ill equipped to operate in their normative environment for a period of time after exposure. Such aftereffects bring about product liability concerns for those who use the technology without appropriate safeguards (see Kennedy, Kennedy, Kennedy, Wasula, & Bartlett, 2014, Chapter 21). In addition, beyond the workforce, these problems may be particularly troublesome for other venues, such as entertainment-based applications (Greenwood-Ericksen, Kennedy, & Stafford, 2014, Chapter 49), which are generally less supervised than workplace applications, such as those used in training (Driskell, Salas & Vessey, 2014, Chapter 38; Johnston, Carroll & Hale, 2014, Chapter 39), education (Barmpoutis, DeVane, & Oliverio, 2014, Chapter 41; Cobb, Heale, Crosier, & Wilson, 2014, Chapter 42), and medicine (Vozenilek & Pribaz, 2014, Chapter 43).

The response to VE exposure varies directly with the dose (i.e., VE stimulus intensity), capacity of the individual exposed (e.g., susceptibility, experience), and exposure duration (Lawson, 2014,

Chapter 23). This indicates that through effective usage protocols that address the strength of the VE stimulus, screening of individuals and usage instructions the problems associated with VE technology can be minimized.

31.2 STRENGTH OF THE VIRTUAL ENVIRONMENT STIMULUS

VEs can impose what is called visually induced motion sickness (Bos, Bles, & Groen, 2008; Kennedy, Drexler, & Kennedy, 2010; Sharples, Cobb, Moody, & Wilson, 2008). The *dose* of a VE stimulus, aka its sickness-inducing potential, is determined by technological factors that in some way disturb the internal state of the individual exposed. In particular, sickness and other adverse effects are generally thought to be due to sensory conflicts between the design of the VE stimulus and what users expect due to their previous experiences in the real world; the greater the sensory conflict, the stronger the dose (Lawson, 2014, Chapter 23; Welch & Mohler, 2014, Chapter 25). Within VEs, this conflict can come from a conflict among the visual, vestibular, and proprioceptive systems. Normally, these sensory systems provide concomitant information about body position and movement; however, when *traversing* through a VE, one (or more) of these systems can imply a body position or movement that is not corroborated by the other systems. This sensory conflict mechanism normally is the impetus for vomiting in response to toxin ingestion (Claremont, 1930; Money, 1990; Triesman, 1977), and thus, the body presumes it is being poisoned and sickness can occur. Different VE systems have different levels of sensory conflict, and thus system developers and administrators need to assess the strength of their VE stimulus (i.e., how likely it is to cause adverse effects during and after exposure). Once the stimulus strength is known and the capacity of those exposed has been determined, appropriate usage protocols can be established.

In establishing the strength of a VE stimulus, it is beneficial to know which technological factors lead to a more intense stimulus. Measures can then be taken to reduce technological issues associated with the adverse effects of VE exposure. In general, system factors thought to influence stimulus strength include the following: system consistency (Uliano, Kennedy, & Lambert, 1986), lag (So & Griffin, 1995; Stiller & Cutmore, 2003), update rate (Badcock, Palmisano, & May, 2014, Chapter 3; So & Griffin, 1995), mismatched interpupillary distance (IPDs) (Mon-Williams, Rushton, & Wann, 1995; Mon-Williams, Wann, & Rushton, 1993, 1995; Wann & Mon-Williams, 2014, Chapter 32), large field of view (FOV) (Kennedy & Fowlkes, 1992; Badcock et al., Chapter 3), spatial frequency content (Dichgans & Brandt, 1978, cf. Figure 6, p. 770; note: frequencies around 0.2–0.4 Hz should be avoided; Bos, Bles, & Groen, 2008; Diels & Howarth, 2013; Golding et al., 2009), visual simulation of action motion (i.e., vection; Flanagan, May, & Dobie, 2004; Kennedy, Berbaum, Dunlap, & Hettinger, 1996; Hettinger et al., 2014, Chapter 18), and unimodal and intersensorial distortions (both temporal and spatial). While the technological factors within any given system can produce varying adverse effects, system developers should ensure that the guidelines in Table 31.1 are heeded to minimize stimulus strength.

Focusing on the parameters in Table 31.1, system developers should identify the primary factors that are inducing adverse effects in their system. For example, in the domain of the visual system, the mismatch between accommodation and vergence demands has been highlighted (Howarth & Costello, 1996; Mon-Williams et al., 1993; Wann, Rushton, & Mon-Williams, 1995). Although IPD settings may not be critical (Howarth, 1999), it can be demonstrated that a mismatch between the orientation of the optical axes of a display and the axes assumed for software generation may produce large errors (Wann, Rushton, & Mon-Williams, 1995). DiZio and Lackner (1997) identified large end-to-end visual update delays and a large FOV as significant etiologic factors in a VE that used a head-mounted display (HMD). Although FOV and display resolution may affect the usability of a display system and lead to motion sickness, they have not been shown (with current displays) to be critical factors in producing visual stress. It is pertinent to note that both common and unique stimulus factors can arise from HMD, spatially immersive displays (SIDs or cave automatic virtual environments [CAVEs] [see Cruz-Neira, Sandin, & DeFanti, 1993]), and even large-screen

TABLE 31.1

Addressing System Factors That Influence the Strength of a VE Stimulus

- Ensure any system lags/latencies are stable; variable lags/latencies can be debilitating.
- Minimize display/phase lags (i.e., end-to-end tracking latency between head motion and resulting update of the display).
- Optimize frame rates.
- Provide adjustable IPD.
- When large FOVs are used, determine if it drives high levels of vection (i.e., perceived self-motion) (Hettinger et al., 2014, Chapter 18).
- If high levels of vection are found and they lead to high levels of sickness, then reduce the spatial frequency content of visual scenes, with frequencies around 0.2–0.4 Hz being avoided altogether.
- Provide multimodal feedback that minimizes sensory conflicts (i.e., provide visual, auditory, and haptic/kinesthetic feedback appropriate for situation being simulated).

TABLE 31.2

Steps to Quantifying VE Stimulus Intensity

1. Get an initial estimate. Talk with target users (not developers) of the system and determine the level of adverse effects they experience.
2. Observe. Watch users during and after exposure and note comments and behaviors.
3. Try the system yourself. Particularly if you are susceptible to motion sickness, obtain a firsthand assessment of the adverse effects.
4. Measure the dropout rate. If most people can stay in for an hour without symptoms, then the system is likely benign; if most people drop out within 10 min, then the system is probably in need of redesign.
5. Measure. Use simple rating scales to assess sickness (Lawson, 2014, Chapter 24) and visual, proprioceptive, and postural measures to assess aftereffects (Wann et al., 2014, Chapter 32; DiZio et al., 2014, Chapter 33).
6. Compare. Use Table 31.3 to determine how the system under evaluation compares to other VE systems.
7. Report. Summarize the severity of the problem, specify required interventions (e.g., warnings, instructions), and set expectations for use (e.g., target exposure duration, intersession intervals).
8. Expect dropouts. With a high-intensity VE stimulus, dropout rates can be high.

implementations (Stone, 2012). A unique factor that is present in an HMD is lag between head movement and update of the visual display. Much like an HMD, a stereoscopic SID may also present a discord between accommodation and vergence stimuli (Wann & Mon-Williams, 1997). In addition to this, a SID may, from some viewing positions, require a small degree of aniso-accommodation (unequal focus across the two eyes). It can be demonstrated that under intensive viewing conditions, even a desktop stereoscopic display can produce aftereffects for the binocular visual system, whereas a nonstereoscopic display will not (Mon-Williams & Wann, 1998). As can be seen from the various studies reported here (see Stanney et al., 1998), the manner in which individual software and hardware components can be integrated varies greatly; thus, to establish the stimulus strength of any given system, the steps in Table 31.2 should be taken.

The comparative intensity of a VE system can be determined via reference to three subscales of sickness, which provide quartiles of sickness symptoms based on 29 VE studies (Cobb, Nichols, Ramsey, & Wilson, 1999; Kennedy, 2001; Kennedy, Jones, Stanney, Ritter, & Drexler, 1996; Stanney, 2001; Stanney & Hash, 1998) that evaluated sickness via the simulator sickness questionnaire (SSQ) (Kennedy, Lane, Berbaum, & Lilienthal, 1993). The SSQ contains 16 items, rated on a four-level Likert-type scale, which assesses total sickness as well as three subscales of sickness: disorientation (e.g., vertigo), nausea (e.g., general discomfort, nausea), and oculomotor issues (e.g., eyestrain). If a given VE system is of medium to high intensity (say the 50th or higher percentile, with a total SSQ score of 20 or higher), significant dropouts can be expected. In VE studies, dropout rates of 20% or more are common, with about 50% of the attrition occurring within the first 20 min of exposure (Stanney, Lanham, Kennedy, & Breaux, 1999; Reed, Diels, & Parkes, 2007).

Beyond the SSQ, other survey tools are available to assess the strength of a VE stimulus. The nausea profile (Muth, Stern, Thayer, & Koch, 1996) measures three subscales of nausea, including somatic distress (e.g., fatigue, weakness, perceived elevated body temperature), gastrointestinal distress (e.g., stomach sickness), and emotional distress (e.g., nervousness). The motion sickness assessment questionnaire (Gianaros, Muth, Mordkoff, Levine, & Stern, 2001) measures multiple dimensions of motion sickness, including gastrointestinal feelings (e.g., stomach sickness), central issues (e.g., disorientation), peripheral issues (e.g., sweatiness), and sopite-related issues (e.g., tiredness). All of these measurement scales take a considerable amount of time to administer and thus are not conducive to capturing the experience of symptoms during VE stimulus exposure. A new measurement tool—the fast motion sickness scale (FMS) (Keshavarz & Hecht, 2011)—provides a quick way to capture sickness data during exposure through a verbal rating of motion sickness along a 20-point scale, ranging from zero (no sickness) to 20 (frank sickness). Other verbal reporting scales have been used in the past (Bagshaw & Stott, 1985; Bos, de Vries, van Emmerik, & Groen, 2010; Draper, Viirre, Furness, & Gawron, 2001; Griffin & Newman, 2004; McCauley, Royal, Wylie, O'Hanlon, & Mackie, 1976; Nichols, Cobb, & Wilson, 1997); however, the FMS promises to resolve some of the issues with these earlier scales (e.g., requiring training to correctly interpret the scale, conflating symptom-based criteria with magnitudes, lack of sensitivity, low correlation with the SSQ).

Another concern with regard to VE stimulus strength is that strong-dose VEs may lead to persistent aftereffects (see Table 31.4), which may be unbeknownst to users. Aftereffects can manifest in several different forms, including degraded hand–eye coordination (Kennedy, Stanney, Ordy, & Dunlap, 1997), postural instability (Stoffregen, Yoshida, Villard, Scibora, & Bardy, 2010), and changes in vestibuloocular reflex (VOR) gain (Di Girolamo, Picciotti, Sergi, Di Nardo, Paludetti, & Ottaviani, 2001). Means of assessing such aftereffects have been developed (c.f. Stanney & Kennedy, 1997); however, currently, there is a lack of valid and reliable measures, although informal techniques could be used (e.g., an approach similar to field sobriety tests). Strong VE stimuli may lead to more prolonged aftereffects and thus, aftereffects should be measured. For those systems with prolonged aftereffects, users should be appropriately warned and readaptation mechanisms should be employed (Champney & Stanney, 2005).

31.3 INDIVIDUAL CAPACITY TO RESIST ADVERSE EFFECTS OF VE EXPOSURE

Given the current state of knowledge, it is difficult to differentiate the interpretation of adverse effects of VE exposure according to influencing individual factors, such as age, gender, anthropometrics, drug and/or alcohol consumption, and health status. However, sufficient knowledge is available to set general guidelines regarding the capacity of those exposed (see Table 31.5).

Capacity can be defined as the capability of an individual to resist adverse effects resulting from VE exposure. The capacity of an individual to undergo VE exposure varies greatly both within an individual and between individuals. Within an individual, capacity can be reduced (e.g., via high VE stimulus intensity, extended exposure duration, consumption of drugs and/or alcohol) or enhanced (e.g., via repeated exposures and related adaptation, see Welsh & Mohler, 2014, Chapter 25). Previous exposure to a provocative environment influences susceptibility to motion sickness. Kennedy et al. (1993) found that for repeated exposures in simulators to be effective in desensitizing individuals to sickness, the intersession interval should be short (1 week or less). This finding is consistent with other reports of motion sickness, where (1) some adaptation was retained during 7-day intervals between slow rotation room exposures but not during a 30-day interval (Kennedy, Tolhurst, & Graybiel, 1965) and (2) intersession intervals in Navy flight trainers of 1 day or less showed no evidence of increased tolerance, nor did those greater than 6 days apart, whereas those 2–5 days apart appeared to be optimum (Kennedy et al., 1993).

Between individuals, there are several individual factors related to oculomotor and disorientation effects of VE exposure, including age, gender, height, and weight. (Note: Interestingly, in the data set examined, nausea was not correlated to any individual factors, but solely to technical factors and exposure duration.) Before the age of 2, children appear to be immune to motion sickness, after which time susceptibility increases until about the age of 12, at which point it declines again. In a large study ($N > 4500$), Lawther and Griffin (1986) make this point quite clearly. Those over 25 are about half as susceptible as they were at 18 years of age (Mirabile, 1990). Whether or not this trend will hold in VE systems is not currently known, but Paige (1994) has shown increased susceptibility to vection in older adults and Liu (2012) has shown that the elderly are particularly vulnerable to increasing navigational rotating speed and duration of exposure.

Some studies have found that females generally experience greater motion sickness than males (Kennedy, Lanham, Drexler, & Lilienthal, 1995). In one VE study, a similar gender contrast has been reported (Kennedy, Stanney, Dunlap, & Jones, 1996). In this study, the mean symptomatology for females after VE exposure was 3.4 times greater than for males. This gender difference is thought to be hormonally related (Schwab, 1954). It is possible that at certain phases of the menstrual cycle, a woman is more susceptible to motion sickness than at others (Stiller & Cutmore, 2003). In studies of postoperative nausea and vomiting (PONV), the severity of sickness has been found to be dependent on the stage in the menstrual cycle and if a woman is on the contraceptive pill (Beattie, Lindblad, Buckley, & Forrest, 1991; Honkavaara, Lehtinen, Hovorka, & Korttila, 1991; Ramsay, McDonald, & Faragher, 1994).

As with other poisons, the same dosage appears to be more detrimental to those of smaller stature. Thus, the relation to height, weight, and even gender may indicate that VE stimulus intensity should be given in proportion to body weight. It must be noted, however, that there are great individual differences in susceptibility to toxins (Konz, 1997); thus, it is beneficial to use body stature in conjunction with motion sickness susceptibility to set exposure guidelines.

Susceptibility to motion sickness is thought to be influenced by several factors (Kennedy, Dunlap, & Fowlkes, 1990). Physiological predisposition has been associated with hormonal and cardiovascular factors as well as hypersensitivity of the vestibular system. Psychological predisposition is generally attributed to personality type (e.g., propensity to report adverse effects, neuroticism, and anxiety) and cognitive style (e.g., field articulation and perceptual rigidity). Regardless of the factors that predispose one to motion sickness, an individual is generally very aware of their personal level of susceptibility; thus, self-report can be used to gauge this factor.

The motion history questionnaire (MHQ), which was developed over 45 years ago to study airsickness and disorientation due to Coriolis stimulation, is often used to assess susceptibility to motion sickness (Kennedy & Graybiel, 1965). The MHQ assesses susceptibility based on past occurrences of sickness in inertial environments. Using this multi-item paper-and-pencil questionnaire, individuals list past experiences in and preferences for provocative motion environments. Scores on the MHQ are generally predictive of an individual's susceptibility to motion sickness in physically moving environments. Susceptibility to motion sickness may also be gauged by an individual's physiological response to a VE stimulus. From a physiological perspective, those susceptible to motion sickness generally experience an increased sympathetic nervous system, which decreases stomach activity, and decreased parasympathetic nervous system, which is needed to return the electrical activity in the stomach to normal, which taken together lead to a dysrhythmic frequency in the gut (i.e., increased gastric tachyarrhythmia; Stern & Koch, 1996). There are also neurohormonal responses associated with stress, including increased blood levels of cortisol, norepinephrine, epinephrine, and vasopressin, found in those who are susceptible. Further, Stiller and Cutmore (2003) demonstrated that change in heart rate was a sensitive measure of sickness susceptibility (see additional evidence from Yokota, Aoki, Mizuta, Ito, & Isu, 2005, who also found postural instability in susceptibles, the latter of which was corroborated by Stoffregen et al. (2010); however, Bos (2011b) suggests that while postural stability and motion sickness may be related via a common mechanism, this relation does not imply causality). Similarly, Kim, Kim, Kim, and Ko (2005) found a positive correlation between susceptibility and heart period, as well as gastric tachyarrhythmia, eyeblink rate, and EEG delta wave, and a negative correlation with EEG beta wave. Sugita et al. (2007) suggest that a physiological index based on the maximum cross-correlation between pulse transmission time and heart rate could provide an effective means of assessing adaptation to visually induced motion sickness. However, currently, there is no single valid and reliable physiological indicator of sickness susceptibility (Shupak & Gordon, 2006); this should be pursued in future work as real-time physiological measures become more sensitive and robust.

Except in some clinical cases (e.g., clinical therapy), VE exposure should be limited to those individuals who are free from drug or alcohol consumption. (Note: In clinical user groups, informed sensitivity to the vulnerabilities of these groups [e.g., unique psychological, cognitive, and functional characteristics] should be obtained.) Akin to drug and/or alcohol consumption, VE exposure has been shown to degrade hand–eye coordination, postural stability, and visual functioning (Di Girolamo et al., 2001; Kennedy et al., 1997; Stoffregen et al., 2010). While limited research has been conducted on the combined stress of drug and/or alcohol use and VE exposure, such biodynamics (Hettinger, Kennedy, & McCauley, 1990) have been studied in the context of other environments that cause illusory self-motion (Gilson, Schroeder, Collins, & Guedry, 1972; Schroeder & Collins, 1974). If one applies the findings of these studies to VE exposure, the expectation is that adverse effects of VE exposure will be exacerbated by drug and alcohol use.

VE exposure is also known to lead to motion-induced drowsiness, known as the sopite syndrome (Lawson, 2014, Chapter 23). Thus, individuals should generally be well rested before commencing VE exposure. If the immune system is already compromised by ill health, this may intensify the adverse effects of VE exposure. Thus, those with cold, flu, or other ailments should be discouraged from participating.

There are some individuals who should probably not participate in VE exposure. This includes but is not limited to those susceptible to photic seizures and migraines (Evans & Furman, 2007; although see refuting evidence by Drummond, 2005); those displaying comorbid features of various psychotic, bipolar, paranoid, substance abuse, claustrophobic, or other disorders where reality testing and identity problems are evident (Rizzo, Lang, & Koenig, 2014, Chapter 45); and those with preexisting binocular anomalies (Wann et al., 2014, Chapter 32). Binocular vision develops early and is relatively stable by the early school years, but the age at which children should be allowed unconstrained access to HMD-based systems is still open to debate. Similarly, adults vary in the robustness of their visual systems. It can be predicted that someone with unstable binocular vision may experience stronger postexposure effects if there are stimuli that place some stress on either the accommodation (focal) system, vergence system, or the cross-links between them (Wann et al., 1995).

31.4   VE USAGE PROTOCOL

A systematic VE usage protocol can be developed that minimizes risks to users. A comprehensive VE usage protocol will consider the following factors:

1. Following the guidelines in Table 31.1, design VE stimulus to minimize adverse effects (see Section 31.2).
2. Following the guidelines in Table 31.2, quantify VE stimulus intensity (i.e., dose) of target system and compare to quartiles in Table 31.3.
3. Following the guidelines in Table 31.2, quantify VE aftereffects of target system and compare to percentiles in Table 31.4.
4. Following the guidelines in Table 31.5, identify individual capacity of target user population to resist adverse effects of VE exposure.
5. Provide warnings for those with severe susceptibility to motion sickness, seizures, migraines, cold, flu, or other ailments (see Section 31.3).

TABLE 31.3
VE Stimulus Dose Based on SSQ Quartiles (*n* = 29)

VE Stimulus Dose	Quartile	SSQ Score
Low	25th	15.5
Moderate	50th	20.1
Medium	75th	27.9
High	95th	33.3
Extreme	99th	53.1

[a] Statistics derived from a database of 785 participants exposed to VE for 15–60 min (Stanney, 2001).

TABLE 31.4
Persistence of Aftereffects from VE Exposure

Post-VE Session Time Period	Percentage of Sample with Aftereffects (%)
30 min–1 h	14.6
>4 h	6
>6 h	4
Spontaneously occurring effects	1

Source: Stone, R. J., *Human Factors Guidance for Designers of Interactive 3d and Games-Based Training Systems*, Human Factors Integration Defence Technology, Centre University of Birmingham, Birmingham, U.K., 2012.

TABLE 31.5
Factors Affecting Individual Capacity to Resist Adverse Effects of VE Exposure

1. *Adaptation*. Set intersession exposure intervals 2–5 days apart to enhance individual adaptability.
2. *Age*. Expect little motion sickness for those under age 2; expect greatest susceptibility to motion sickness between the ages of 2 and 12; expect motion sickness to decline after 12, with those over 25 being about half as susceptible as they were at 18 years of age until they reach their elderly year, at which time susceptibility increases once again.
3. *Gender*. Expect females to be more susceptible than males (perhaps as great as three times more susceptible).
4. *Anthropometrics*. Consider setting VE stimulus intensity in proportion to body weight/stature.
5. *Individual susceptibility*. Expect individuals to differ greatly in motion sickness susceptibility, and use the MHQ or another instrument to gauge the susceptibility of the target user population.
6. *Drug and alcohol consumption*. Limit VE exposure to those individuals who are free from drug or alcohol consumption.
7. *Rest*. Encourage individuals to be well rested before commencing VE exposure.
8. *Ailments*. Discourage those with cold, flu, or other ailments (e.g., ear infections/headache, diplopia, blurred vision, sore eyes, eyestrain, conjunctivitis, corneal ulcers/disease, *dry eye*, iritis, cataracts, glaucoma, pregnancy, respiratory ailments) from participating in VE exposure; encourage those susceptible to photic seizures and migraines as well as individuals with preexisting binocular anomalies to avoid exposure. Observe closely those experiencing extreme fatigue, emotional stress, digestive problems, significant sleep loss, or anxiety.
9. *Clinical user groups*. Obtain informed sensitivity to the vulnerabilities of these user groups (e.g., unique psychological, cognitive, and functional characteristics). Encourage those displaying comorbid features of various psychotic, bipolar, paranoid, substance abuse, claustrophobic, or other disorders where reality testing and identity problems are evident to avoid exposure.

6. Educate users as to the potential risks of VE exposure (see Chapters 22 through 27). Inform users of the insidious effects they may experience during exposure, including nausea, malaise, disorientation, headache, dizziness, vertigo, eyestrain, drowsiness, fatigue, pallor, sweating, increased salivation, and vomiting. Depending on VE content, potential adverse psychological effects may also need to be considered (Calvert, 2014, Chapter 27).

7. Educate users as to the potential adverse aftereffects of VE exposure (DiZio & Lackner, 2014, Chapter 33; Wann & Mon-Williams, 2014, Chapter 32). Inform users that they may experience disturbed visual functioning, visual flashbacks, and unstable locomotor and postural control for prolonged periods following exposure. Relating these experiences to excessive alcohol consumption may prove instructional.

8. Inform users that if they start to feel ill, they should terminate their VE interaction because extended exposure is known to exacerbate adverse effects (Kennedy, Stanney, & Dunlap, 2000).

9. Avoid the use of an HMD if possible. If not possible, avoid fully face-enclosing HMDs unless the user can be seated or their stance somehow supported; it is best to provide the latter for nonface-enclosing HMDs as well (Stone, 2012). Never use full-face HMDs for applications that require the user to stand unaided or move around freely.

10. Screen for stereoblindness and recommend those who test positive to avoid exposure (Stone, 2012).

11. Always couple an HMD with an appropriate head tracking system to minimize mismatches between visual and head movement-based cues (Stone, 2012).

12. Donning an HMD is a jarring experience (Pierce, Pausch, Sturgill, & Christiansen, 1999). Depending on the complexity of the virtual world, it can take 30–60 s to adjust to the new space (Brooks, 1988). Prepare users for this transition by informing them that there will be an adjustment period.

13. Adjust environmental conditions. Provide adequate airflow and comfortable thermal conditions (Konz, 1997). Sweating often precedes an emetic response; thus, proper airflow can enhance user comfort. In addition, extraneous noise should be eliminated, as it can exacerbate ill effects.

14. Fatigue can exacerbate the adverse effects of VE exposure. To minimize fatigue, ensure all equipment is comfortable and properly adjusted for fit. HMDs should fit snugly and be evenly weighted about a user's head, stay in place when unsupported, and avoid uneven loading to neck and shoulder muscles. Many HMDs have adjustable head straps, IPDs, and viewing distance between the system's eyepieces and user's eyes. Ensure users optimize these adjustments to obtain proper fit. Tethers should not obstruct movements of users. DataGloves and other effectors should not induce excessive static loads via prolonged unnatural positioning of the arms or other extremities.

15. For low- to medium-dose VE stimuli (see Table 31.3), initial exposure should be of *long* duration (i.e., 15–30 min; however, avoid exposures >30 min) and allow an intersession recovery period of 2–3 days (Stone, 2012).

16. For strong-dose VE stimuli (see Table 31.3), limit initial exposures to a short duration (i.e., 10 min or less) and allow an intersession recovery period of 2–5 days.

17. For strong-dose VE stimuli (see Table 31.3), warn users to avoid movements requiring high rates of linear or rotational acceleration and extraordinary maneuvers (e.g., flying backward) during initial interaction (McCauley & Sharkey, 1992).

18. Throughout VE exposure, an attendant should be available at all times to monitor users' behavior and ensure their well-being. The attendant may also have to assist users if they become stuck or lost within the virtual world, as often happens (Darken & Peterson, 2014, Chapter 19).

19. Indicators of impending trouble include excessive sweating, verbal frustration, lack of movement within the environment for a significant amount of time, and less overall movement (e.g., restricting head movement). Look for red flags. Users demonstrating any of

these behaviors should be observed closely, as they may experience an emetic response. Extra care should be taken with these individuals postexposure. Note: It is beneficial to have a bag or garbage can located near users in the event of an abrupt emetic response.

20. Set criteria for terminating exposure. Exposure should be terminated immediately if users verbally complain of symptoms and acknowledge they are no longer able to continue. Also, to avoid an emetic response, if telltale signs are observed (i.e., sweating, increased salivation), exposure should be terminated. Some individuals may be unsteady upon postexposure. These individuals may need assistance when initially standing up after exposure.

21. After exposure, the well-being of users should be assessed. Measurements of their hand–eye coordination and postural stability should be taken. Similar to field sobriety tests, these can include measures of balance (e.g., standing on one foot, walking an imaginary line, leaning backward with eyes closed), coordination. (e.g., alternate hand clapping and finger-to-nose touch while the eyes are closed), and eye nystagmus (e.g., follow a light pen with the eyes without moving the head). Do not allow individuals who fail these tests to conduct high-risk activities until they have recovered (e.g., have someone drive them home). Consider the use of postexposure recalibration exercises (Champney & Stanney, 2005).

22. Set criteria for releasing users. Specify the amount of time after exposure that users must remain on premises before driving or participating in other such high-risk activities. In our lab, a 2:1 ratio is used; postexposure users must remain in the laboratory twice the amount of exposure time to allow recovery.

23. Call users the next day or have them call to report any prolonged adverse effects.

31.5 CONCLUSIONS

The risks associated with VE exposure are real and include ill effects during exposure as well as the potential for prolonged aftereffects. To minimize these risks, VE system developers should quantify and minimize VE stimulus intensity, identify the capacity of the target user population to resist the adverse effects of VE exposure, and then follow a systematic usage protocol. This protocol should focus on warning, educating, and preparing users, setting appropriate environmental and equipment conditions, limiting initial exposure duration and user movements, monitoring users and looking for red flags, and setting criteria for terminating exposure, debriefing, and release. Adopting such a protocol can minimize the risk factors associated with VE exposure, thereby enhancing the safety of users, and limiting the liability (Kennedy et al., 2014, Chapter 21) of system developers and administrators.

ACKNOWLEDGMENTS

This material is based on work supported in part by the Office of Naval Research (ONR) under grant No. N000149810642, the National Science Foundation (NSF) under grants No. DMI9561266 and IRI-9624968, and the National Aeronautics and Space Administration (NASA) under grants No. NAS9-19482 and NAS9-19453. Any opinions, findings, and conclusions or recommendations expressed in this material are those of the authors and do not necessarily reflect the views or the endorsement of the ONR, NSF, or NASA.

REFERENCES

Badcock, D. R., Palmisano, S., & May, J. G. (2014). Vision and virtual environments. In K. S. Hale & K. M. Stanney (Eds.), *Handbook of virtual environments: Design, implementation, and applications* (2nd ed., pp. 39–86). Boca Raton, FL: Taylor & Francis Group, Inc.

Barmpoutis, A., DeVane, B., & Oliverio, J. C. (2014). Applications of virtual environments in experiential, STEM, and health science education. In K. S. Hale & K. M. Stanney (Eds.), *Handbook of virtual environments: Design, implementation, and applications* (2nd ed., pp. 1055–1072). Boca Raton, FL: Taylor & Francis Group, Inc.

Bagshaw, M., & Stott, J. R. (1985). The desensitization of chronically motion sick aircrew in the royal air force. *Aviation, Space, and Environmental Medicine, 56*, 1144–1151.

Beattie, W. S., Lindblad, T., Buckley, D. N., & Forrest, J. B. (1991). The incidence of postoperative nausea and vomiting in women undergoing laparoscopy is influenced by day of menstrual cycle. *Canadian Journal of Anesthesia, 38*(3), 298–302.

Bos, J. E. (2011a). Visual image safety. *Displays, 32*, 151–152.

Bos, J. E. (2011b). Nuancing the relationship between motion sickness and postural stability. *Displays, 32*, 189–193.

Bos, J. E., Bles, W., & Groen, E. L. (2008). A theory on visually induced motion sickness. *Displays, 29*, 47–57.

Bos, J. E., de Vries, S. C., van Emmerik, M. L., & Groen, E. L. (2010). The effect of internal and external fields of view on visually induced motion sickness. *Applied Ergonomics, 41*(4), 516–521.

Brooks, F. P., Jr. (1988). Grasping reality through illusion: Interactive graphics serving science. *ACM SIGCHI'88 Conference Proceedings* (pp. 1–11). Washington, DC.

Calvert, S. L. (2014). Social impact of virtual environments. In K. S. Hale & K. M. Stanney (Eds.), *Handbook of virtual environments: Design, implementation, and applications* (2nd ed., pp. 699–718). Boca Raton, FL: Taylor & Francis Group, Inc.

Champney, R., & Stanney, K. (2005, July 22–27). Virtual reality aftereffects and potential readaptation exercises. *1st International Conference on Virtual Reality*. Las Vegas, NV.

Claremont, C. A. (1930). The psychology of seasickness. *Psyche, 11*, 86–90.

Cobb, S., Hawkins, T., Millen, L., & Wilson, J. R. (2014). Design and development of 3D interactive environments for special educational needs. In K. S. Hale & K. M. Stanney (Eds.), *Handbook of virtual environments: Design, implementation, and applications* (2nd ed., pp. 1073–1106). Boca Raton, FL: Taylor & Francis Group, Inc.

Cobb, S. V. G., Nichols, S., Ramsey, A., & Wilson, J. R. (1999). Virtual reality–induced symptoms and effects (VRISE). *Presence: Teleoperators and Virtual Environments, 8*(2), 169–186.

Cruz-Neira, C., Sandin, D. J., & DeFanti, T. A. (1993, July). Surround-screen projection-based virtual reality: The design and implementation of the CAVE. In *ACM SIGGRAPH'93 Conference Proceedings* (pp. 135–142). Prentice Hall, New York, NY.

Darken, R. P., & Peterson, B. (2014). Spatial orientation, wayfinding, and representation. In K. S. Hale & K. M. Stanney (Eds.), *Handbook of virtual environments: Design, implementation, and applications* (2nd ed., pp. 467–492). Boca Raton, FL: Taylor & Francis Group, Inc.

Di Girolamo, S., Picciotti, P., Sergi, B., Di Nardo, W., Paludetti, G., & Ottaviani, F. (2001). Vestibulo-ocular reflex modification after virtual environment exposure. *Acta Otolaryngol, 121*(2), 211–215.

Dichgans, J., & Brandt, T. (1978). Visual-vestibular interaction: Effects on self-motion perception and postural control. In R. Held, H. W. Leibowitz, & H. L. Teuber (Eds.), *Handbook of sensory physiology: Vol. 8. Perception* (pp. 756–804). Heidelberg, Germany: Springer-Verlag.

Diels, C., & Howarth, P. A. (2013). Frequency characteristics of visually induced motion sickness. *Human Factors, 55*, 595–604.

DiZio, P., Lackner, R., & Champney, R. K. (2014). Proprioceptive adaptation and aftereffects. In K. S. Hale & K. M. Stanney (Eds.), *Handbook of virtual environments: Design, implementation, and applications* (2nd ed., pp. 833–854). Boca Raton, FL: Taylor & Francis Group, Inc.

DiZio, P., & Lackner, J. R. (1997). Circumventing side effects of immersive virtual environments. In M. Smith, G. Salvendy, & R. Koubek (Eds.), *Design of computing systems: Social and ergonomic considerations* (pp. 893–896). Amsterdam, the Netherlands: Elsevier Science Publishers.

Draper, M. H., Viirre, E. S., Furness, T. A., & Gawron, V. J. (2001). Effects of image scale and system time delay on simulator sickness within head-coupled virtual environments. *Human Factors, 43*, 129–146.

Driskell, T., Salas, E., & Vessey, W. B. (2014). Team training in virtual environments. In K. S. Hale & K. M. Stanney (Eds.), *Handbook of virtual environments: Design, implementation, and applications* (2nd ed., pp. 999–1026). Boca Raton, FL: Taylor & Francis Group, Inc.

Drummond, P. D. (2005). Triggers of motion sickness in migraine sufferers. *Headache, 45*, 653–656.

Evans, R. W., & Furman, J.M. (2007). Motion sickness and migraine. *Headache, 47*, 607–610.

Flanagan, M.B., May, J. G., & Dobie, T. G. (2004). The role of vection, eye movements and postural instability in the etiology of motion sickness. *Journal of Vestibular Research—Equilibrium and Orientation, 14*, 335–346.

Gianaros, P. J., Muth, E. R., Mordkoff, J. T., Levine, M. E., & Stern, R. M. (2001). A questionnaire for the assessment of the multiple dimensions of motion sickness. *Aviation, Space, and Environmental Medicine, 72*, 115–119.

Gilson, R. D., Schroeder, D. J., Collins, W. E., & Guedry, F. E. (1972). Effects of different alcohol dosages and display illumination on tracking performance during vestibular stimulation. *Aerospace Medicine, 43*, 656.

Golding, J. F., Arun, S., Wortley, E., Wotton-Hamrioui, K., Cousins, S., & Gresty, M. A. (2009). Off-vertical axis rotation of the visual field and nauseogenicity. *Aviation, Space and Environmental Medicine, 80*, 516–521.

Greenwood-Ericksen, A., Kennedy, R. C., & Stafford, S. (2014). Entertainment applications of virtual environments. In K. S. Hale & K. M. Stanney (Eds.), *Handbook of virtual environments: Design, implementation, and applications* (2nd ed., pp. 1291–1316). Boca Raton, FL: Taylor & Francis Group, Inc.

Griffin, M. J., & Newman, M. M. (2004). Visual field effects on motion sickness in cars. *Aviation Space and Environmental Medicine, 75*, 739–748.

Hettinger, L. J., Kennedy, R. S., & McCauley, M. E. (1990). Motion and human performance. In G. H. Crampton (Ed.), *Motion and space sickness* (pp. 411–441). Boca Raton, FL: CRC Press.

Hettinger, L. J., Schmidt, T., Jones, D. L., & Keshavarz, B. (2014). Illusory self-motion in virtual environments. In K. S. Hale & K. M. Stanney (Eds.), *Handbook of virtual environments: Design, implementation, and applications* (2nd ed., pp. 467–492). Boca Raton, FL: Taylor & Francis Group, Inc.

Honkavaara, P., Lehtinen, A. M., Hovorka, J., & Korttila, K. (1991). Nausea and vomiting after gynecological laparoscopy depends upon the phase of menstrual cycle. *Canadian Journal of Anesthesia, 38*(7), 876–879.

Howarth, P. A. (1999). Oculomotor changes within virtual environments. *Applied Ergonomics, 30*, 59–67.

Howarth, P. A., & Costello, P. J. (1996). Visual effects of immersion in virtual environments: Interim results from the U.K. Health and Safety Executive Study. *Digest of the Society for Information Display, 27*, 885–888.

Kennedy, R. S. (2001). Unpublished research data, RSL Assessments, Inc., Orlando, FL.

Kennedy, R. S., Berbaum, K. S., Dunlap, W. P., & Hettinger, L. J. (1996). Developing automated methods to quantify the visual stimulus for cybersickness. *Proceedings of the Human Factors and Ergonomics Society 40th Annual Meeting* (pp. 1126–1130). Human Factors and Ergonomics Society, Santa Monica, CA.

Kennedy, R. S., Drexler, J. M., & Kennedy, R. C. (2010). Research in visually induced motion sickness. *Applied Ergonomics, 41*, 494–503.

Kennedy, R. S., Dunlap, W. P., & Fowlkes, J. E. (1990). Prediction of motion sickness susceptibility. In G. H. Crampton (Ed.), *Motion and space sickness* (pp. 179–215). Boca Raton, FL: CRC Press.

Kennedy, R. S., & Fowlkes, J. E. (1992). Simulator sickness is polygenic and poly symptomatic: Implications for research. *International Journal of Aviation Psychology, 2*(1), 23–38.

Kennedy, R. S., & Graybiel, A. (1965). *The Dial test: A standardized procedure for the experimental production of canal sickness symptomatology in a rotating environment* (Report No. 113, NSAM 930). Pensacola, FL: Naval School of Aerospace Medicine.

Kennedy, R. S., Jones, M. B., Stanney, K. M., Ritter, A. D., & Drexler, J. M. (1996). *Human factors safety testing for virtual environment mission-operation training* (Final Report, Contract No. NAS9-19482). Houston, TX: Lyndon B. Johnson Space Center.

Kennedy, R. S., Kennedy, R. C., Kennedy, K. E., Wasula, C., & Bartlett, K. M. (2014). Virtual environments and product liability. In K. S. Hale & K. M. Stanney (Eds.), *Handbook of virtual environments: Design, implementation, and applications* (2nd ed., pp. 505–518). Boca Raton, FL: Taylor & Francis Group, Inc.

Kennedy, R. S., Lane, N. E., Berbaum, K. S., & Lilienthal, M. G. (1993). Simulator sickness questionnaire: An enhanced method for quantifying simulator sickness. *International Journal of Aviation Psychology, 3*(3), 203–220.

Kennedy, R. S., Lanham, D. S., Drexler, J. M., & Lilienthal, M. G. (1995). A method for certification that after effects of virtual reality exposures have dissipated: Preliminary findings. In A. C. Bittner, & P. C. Champney (Eds.), *Advances in industrial safety VII* (pp. 263–270). London, U.K.: Taylor & Francis.

Kennedy, R. S., Stanney, K. M., & Dunlap, W. P. (2000). Duration and exposure to virtual environments: Sickness curves during and across sessions. *Presence: Teleoperators and Virtual Environments, 9*(5), 466–475.

Kennedy, R. S., Stanney, K. M., Dunlap, W. P., & Jones, M. B. (1996). *Virtual environment adaptation assessment test battery* (Final Report, Contract No. NAS9-19453). Houston, TX: Lyndon B. Johnson Space Center.

Kennedy, R. S., Stanney, K. M., Ordy, J. M., & Dunlap, W. P. (1997). Virtual reality effects produced by head-mounted display (HMD) on human eye-hand coordination, postural equilibrium, and symptoms of cybersickness. *Society for Neuroscience Abstracts, 23*, 772.

Kennedy, R. S., Tolhurst, G. C., & Graybiel, A. (1965). *The effects of visual deprivation on adaptation to a rotating environment* (Technical Report No. NSAM 918). Pensacola, FL: Naval School of Aviation Medicine.

Keshavarz, B., & Hecht, H. (2011). Validating an efficient method to quantify motion sickness. *Human Factors, 53*(4), 415–426.

Kim, Y. Y., Kim, H. J., Kim, E.N., & Ko, H. D. (2005). Characteristic changes in the physiological components of cybersickness. *Psychophysiology, 42*, 616–625.

Konz, S. (1997). Toxicology and thermal discomfort. In G. Salvendy (Ed.), *Handbook of human factors and ergonomics* (2nd ed., pp. 891–908). New York, NY: John Wiley & Sons.

Lawson, B. D. (2014). Motion sickness symptomatology and origins. In K. S. Hale & K. M. Stanney (Eds.), *Handbook of virtual environments: Design, implementation, and applications* (2nd ed., pp. 531–600). Boca Raton, FL: Taylor & Francis Group, Inc.

Lawther, A., & Griffin, M. J. (1986). The motion of a sea ship and the consequent motion sickness amongst passengers. *Ergonomics, 29*(4), 535–552.

Liu, C. L. (2012). A study of sickness induced by perceptual conflict in the elderly within a 3D virtual store and avoidance. In J. Wang, G. G. Yen, & M. M. Polycarpou (Eds.), *Proceedings of the 9th International Conference on Advances in Neural Networks—Volume Part I* (ISNN'12), Vol. Part I (pp. 422–430). Berlin, Heidelberg: Springer-Verlag.

McCauley, M. E., & Sharkey, T. J. (1992). Cybersickness: Perception of self-motion in virtual environments. *Presence: Teleoperators and Virtual Environments, 1*(3), 311–318.

McCauley, M. E., Royal, J. W., Wylie, C. D., O'Hanlon, J. F., & Mackie, R. R. (1976). *Motion sickness incidence: Exploratory studies of habituation, pitch and roll, and the refinement of a mathematical model* (Technical Report HFR 1733-2). Santa Barbara, CA: Human Factors Research.

Mirabile, C. S. (1990). Motion sickness susceptibility and behavior. In G.H. Crampton (Ed.), *Motion and space sickness* (pp. 391–410). Boca Raton, FL: CRC Press.

Mon-Williams, M., Rushton, S., & Wann, J. P. (1995). Binocular vision in stereoscopic virtual-reality systems. *Society for Information Display International Symposium Digest of Technical Papers, 25*, 361–363.

Mon-Williams, M., & Wann, J. P. (1998). Binocular virtual reality displays: When problems occur and when they don't. *Human Factors, 40*(1), 42–49.

Mon-Williams, M., Wann, J. P., & Rushton, S. (1993). Binocular vision in a virtual world: Visual deficits following the wearing of a head-mounted display. *Ophthalmic and Physiological Optics, 13*, 387–391.

Mon-Williams, M., Wann, J. P., & Rushton, S. (1995). Binocular vision in stereoscopic virtual reality systems. *Digest of the Society for Information Display, 1*, 361–363.

Money, K. E. (1990). Motion sickness and evolution. In G. H. Crampton (Ed.), *Motion and space sickness* (pp. 1–8), Boca Raton, FL: CRC Press.

Muth, E. R., Stern, R. M., Thayer, J. F., & Koch, K. L. (1996). Assessment of the multiple dimensions of nausea: The Nausea Profile (NP). *Journal of Psychosomatic Research, 40*, 511–520.

Nichols, S., Cobb, S., & Wilson, J. R. (1997). Health and safety implications of virtual environments: Measurement issues. *Presence, 6*, 667–675.

Occupational Safety and Health Administration. (2000). Ergonomics Program, Federal Registration No. 65:68261-68870, CFR Title 29, Part 1910, RIN 1218-AB36 [Docket No. S-777, Online]. Available: http://www.osha.gov/ergo-temp/FED20001114.html

Paige, G. D. (1994). Senescence of human visual–vestibular interactions: Smooth pursuit, optokinetic, and vestibular control of eye movements with aging. *Experimental Brain Research, 98*, 355–372.

Pierce, J. S., Pausch, R., Sturgill, C. B., & Christiansen, K. D. (1999). Designing a successful HMD-based experience. *Presence: Teleoperators and Virtual Environments, 8*(4), 469–473.

Ramsay, T. M., McDonald, P. F., & Faragher, E. B. (1994). The menstrual cycle and nausea or vomiting after wisdom teeth extraction. *Canadian Journal of Anesthesia, 41*(9), 789–801.

Reed, N., Diels, C., & Parkes, A. M. (2007). Simulator sickness management: Enhanced familiarisation and screening processes. *Proceedings of the First International Symposium on Visually Induced Motion Sickness, Fatigue, and Photosensitive Epileptic Seizures* (VIMS2007) (pp. 156–162). VIMS, Hong Kong, People's Republic of China, 2007.

Rich, C. J., & Braun, C. C. (1996). Assessing the impact of control and sensory compatibility on sickness in virtual environments. *Proceedings of the Human Factors and Ergonomics Society 40th Annual Meeting* (pp. 1122–1125). Human Factors and Ergonomics Society, Santa Monica, CA.

Rizzo, A. "Skip", Lange, B., & Koenig, S. (2014). Clinical virtual reality. In K. S. Hale & K. M. Stanney (Eds.), *Handbook of virtual environments: Design, implementation, and applications* (2nd ed., pp. 1157–1200). Boca Raton, FL: Taylor & Francis Group, Inc.

Schroeder, D. J., & Collins, W. E. (1974). Effects of secobarbital and d-amphetamine on tracking performance during angular acceleration. *Ergonomics, 17*(5), 613.

Schwab, R. S. (1954). The nonlabyrinthine causes of motion sickness. *International Record Medicine, 167*, 631–637.

Sharples, S., Cobb, S., Moody, A., & Wilson, J. R. (2008). Virtual reality induced symptoms and effects (VRISE): Comparison of head mounted display (HMD), desktop and projection display systems. *Displays, 29*, 58–69.

Shupak, A., & Gordon, C. R. (2006). Motion sickness: Advances in pathogenesis, prediction, prevention, and treatment. *Aviation, Space, and Environmental Medicine, 77,* 1213–1223.

So, R. H., & Griffin, M. J. (1995). Effects of lags on human operator transfer functions with head-coupled systems. *Aviation, Space, and Environmental Medicine, 66,* 550–556.

Stanney, K. M. (2001). Unpublished research data, University of Central Florida, Orlando.

Stanney, K. M., Hale, K. S., Nahmens, I., & Kennedy, R. S. (2003). What to expect from immersive virtual environment exposure: influences of gender, body mass index, and past experience. *Human Factors, 45*(3), 504–518.

Stanney, K. M., & Hash, P. (1998). Locus of user-initiated control in virtual environments: Influences on cybersickness. *Presence: Teleoperators and Virtual Environments, 7*(5), 447–459.

Stanney, K. M., & Kennedy, R. S. (1997). Development and testing of a measure of the kinesthetic position sense used to assess the aftereffects from virtual environment exposure. *Virtual Reality Annual International Symposium'97* (pp. 87–94). IEEE Computer Society Press, Los Alamitos, CA.

Stanney, K. M., Lanham, S., Kennedy, R. S., & Breaux, R. B. (1999). Virtual environment exposure dropout thresholds. *Proceedings of the 43rd Annual Human Factors and Ergonomics Society Meeting* (pp. 1223–1227). Human Factors and Ergonomics Society, Santa Monica, CA.

Stanney, K. M., Salvendy, G., Deisinger, J., DiZio, P., Ellis, S., Ellison, E.,…Witmer, B. (1998). Aftereffects and sense of presence in virtual environments: Formulation of a research and development agenda. *International Journal of Human–Computer Interaction, 10*(2), 135–187.

Stern, R. M., & Koch, K. L. (1996). Motion sickness and differential susceptibility. *Current Directions in Psychological Science, 5*(4), 115–120.

Stiller, A., & Cutmore, T. (2003, Supplement). Individual and environmental factors influencing the experience of sickness in virtual environment systems. *Combined Abstracts of 2003 Psychology Conferences. Australian Journal of Psychology, 55,* 28.

Stoffregen, T. A., Yoshida, K., Villard, S., Scibora, L., & Bardy, B. G. (2010). Stance width influences postural stability and motion sickness. *Ecological Psychology, 22,* 169–191.

Stone, R. J. (2012). *Human factors guidance for designers of interactive 3D and games-based training systems.* Birmingham, U.K.: Human Factors Integration Defence Technology, Centre University of Birmingham.

Sugita, N., Yoshizawa, M., Abe, M., Tanaka, A., Watanabe, T., Chiba, S.,… Nitta, S. (2007). Evaluation of adaptation to visually induced motion sickness based on the maximum cross-correlation between pulse transmission time and heart rate. *Journal of NeuroEngineering and Rehabilitation, 4,* 34. Available from: http://www.jneuroengrehab.com/content/4/1/35

Treisman, M. (1977). Motion sickness. An evolutionary hypothesis. *Science, 197,* 493–549.

Uliano, K. C., Kennedy, R. S., & Lambert, E. Y. (1986). Asynchronous visual delays and the development of simulator sickness. *Proceedings of the Human Factors Society 30th Annual Meeting* (pp. 422–426). Human Factors Society, Santa Monica, CA.

Wann, J. P., & Mon-Williams, M. (1997, May). Health issues with virtual reality displays: What we do know and what we don't. *Computer Graphics, 31,* 53–57.

Wann, J. P., Rushton, S. K., & Mon-Williams, M. (1995). Natural problems for stereoscopic depth perception in virtual environments. *Vision Research, 19,* 2731–2736.

Wann, J. P., White, A. D., Wilkie, R. M., Culmer, Peter, R. J., Lodge, P. A., & Mon-Williams, M. (2014). Measurement of visual aftereffects following virtual environment exposure. In K. S. Hale & K. M. Stanney (Eds.), *Handbook of virtual environments: Design, implementation, and applications* (2nd ed., pp. 809–832). Boca Raton, FL: Taylor & Francis Group, Inc.

Welch, R. B., & Mohler, B. J. (2014). Adapting to virtual environments. In K. S. Hale & K. M. Stanney (Eds.), *Handbook of virtual environments: Design, implementation, and applications* (2nd ed., pp. 627–646). Boca Raton, FL: Taylor & Francis Group, Inc.

Yokota, Y., Aoki, M., Mizuta, K., Ito, Y., & Isu, N. (2005). Motion sickness susceptibility associated with visually induced postural instability and cardiac autonomic responses in healthy subjects. *Acta Oto-Laryngologica, 125,* 280–285.

32 Measurement of Visual Aftereffects following Virtual Environment Exposure

Implications for Minimally Invasive Surgery

John P. Wann, Alan D. White, Richard M. Wilkie,
Peter R. Culmer, J. Peter A. Lodge, and Mark Mon-Williams

CONTENTS

32.1 INTRODUCTION

Virtual environment (VE) visual displays can result in unusual pressures being placed on the human visual system (Wann, Rushton, & Mon-Williams, 1995). These pressures have the potential to influence the performance in skilled tasks conducted within such environments. The implications of impaired performance are obvious in the case of minimally invasive surgery (MIS). MIS has revolutionized medicine with greatly improved patient outcome postoperatively. MIS is recommended by the UK's National Institute for Health and Clinical Excellence (NICE) and is particularly beneficial in the cases of upper and lower gastrointestinal cancers and bariatric surgery. MIS is associated with many benefits to the patient such as reduced pain, shorter (if any) hospital stays, reduced wound infection rates, and a decreased risk of operative and 30-day mortality (Cuschieri, 1995). There is also a lower risk of longer-term complications such as incisional hernias (where an abdominal viscus protrudes through a weakness in the muscle or surrounding tissue wall). There are several forms of MIS, but the primary techniques practiced in hospitals around the world are laparoscopic *key-hole* surgery (LS) and robotic-assisted surgery (RAS). One primary focus within the development of these technologies is how to present the visual information relayed by cameras within the abdominal cavity—that is, how best to create the VE with which the surgeon must interact. Thus, the topic of visual aftereffects with VE exposure remains an important area even after two decades of VE research.

This chapter details how visual aftereffects can be measured (so that MIS technologies can be tested to ensure they do not create problems) and is organized in the following manner. First, a detailed consideration of the ocular-motor system is provided. Second, a series of clinical tests are detailed that are useful in indicating whether the ocular-motor system has been placed under stress. These tests are used generally to explore ocular-motor status before and after using a particular VE system for a given period of time. We additionally describe a different approach to investigating the impact of VE systems on the ocular-motor system. This section describes experimental results that highlight some interesting features of the ocular-motor system. Finally, we provide an overview of recent advances in MIS and give a consideration of the issues relating to visual aftereffects raised by this technology.

32.2 BINOCULAR ORGANIZATION

32.2.1 REFRACTION

It is possible to think of the eye as an optical system where the cornea and crystalline lens have a certain refractive power with a corresponding focal length (the distance from the optical system to a principal focal point) as illustrated in Figure 32.1. The length of the eye can be considered as being independent of the power of the optical system. If the length of the eye is such that the retina is at the focal length of the optical system, the eye is described as being *emmetropic*. If the retina is closer than the focal length, the eye is described as *hypermetropic*, and if the retina is beyond the focal length, the eye is described as being *myopic*. It is possible for the optical system of the eye to cause two separate focal planes, of which both can fall behind, both can fall in front, or one can fall on either side of the retina. Such eyes are described as *astigmatic* (it is easiest to think of the astigmatic optical system being shaped like an egg instead of a sphere). Eyes that are myopic, hypermetropic, or astigmatic are

Emmetropia

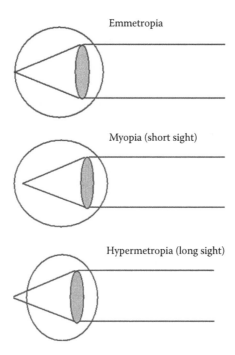

Myopia (short sight)

Hypermetropia (long sight)

FIGURE 32.1 It is possible to think of the eye as an optical system where the cornea and crystalline lens have a certain refractive power with a corresponding focal length (the distance from the optical system to a principal focal point) as illustrated. If the length of the eye is such that the retina is at the focal length of the optical system, the eye is described as being emmetropic (*upper plot*). If the retina is beyond the focal length, the eye is described as being myopic (*middle plot*), and if the retina is closer than the focal length, the eye is described as hypermetropic (*lower plot*).

described as having *refractive error*. The refractive error refers to the extent to which the eye is too long or too short for the power of the optical system and is described in terms of diopters (D): the reciprocal of the refractive error in meters. It is normal to express the refractive error in terms of the lens that needs to be placed in front of the eye to correct for the refractive error—myopia requires a concave (negative) lens, whereas hypermetropia requires a convex (positive) lens. Lens power is specified in diopters, which is equivalent to the reciprocal of focal distance specified in meters.

32.2.2 Eye Movements

Horizontal movements of the two eyes may be considered as either conjugate, when the eyes move in the same direction, or disconjugate when the eyes move in opposite directions. Conjugate and disconjugate eye movements are also commonly referred to as version and vergence, respectively, but because the phonetic similarity leads to confusion, the term *conjugate* is used herein rather than version. Conjugate eye movements are responsible for the maintenance of static fixation (gaze holding), rapidly changing eye fixation position (saccadic movement), or continuously maintaining fixation on a moving object (pursuit). The majority of humans are binocular, and thus, conjugate eye movements require coordinated movement of both eyes. Hering (1868/1977) originally suggested that the nervous system controls the two eyes as if they were a single organ, and this idea has subsequently received a large amount of empirical support (see Mon-Williams & Tresilian, 1998). Hering's law suggests that the same command (e.g., move 10° to the left) is sent to both eyes. In order for this control arrangement to succeed, it is necessary for the peripheral ocular-motor system to adapt in response to changes between the two eyes. Thus, it appears as if the nervous system has a large degree of peripheral adaptability in order that the central command structure can remain reasonably invariant.

The adaptability allows the system to respond to changes induced by neuromuscular fatigue or disease as well as respond to changes in the viewing environment. This suggests that a crucial issue in the design of VE is the extent to which a VE system places adaptive pressures on the ocular-motor system and the manner in which the nervous system responds to those pressures. It seems reasonable to suggest that the primary adaptive pressures placed on the ocular-motor system relate to binocular vision. It is the case that a poorly designed VE headset could induce adaptation in reflexive-type eye movements (e.g., the vestibular ocular reflex; see Chapter 39) by introducing temporal lags between sensed changes in head movements and the resulting visual display. Nonetheless, it is difficult to see a well-designed system placing pressure on monocular viewing. Thus, the fundamental problem faced by the ocular-motor system relates to the binocular coordination between the eyes. The binocular system is, therefore, herein considered in some depth. It is important at this stage to differentiate two different types of VE display technologies—*biocular displays* and *stereoscopic binocular displays*. Biocular displays present identical images to the two eyes, whereas stereoscopic binocular displays present different images to the eyes (the differences, or disparities, can provide a powerful phenomenological impression of three-dimensional [3D] depth). For example, in LS, the surgical image is projected onto a flat screen monitor that can be viewed by all members of the surgical team (biocular display). In RAS, the operative field is viewed by the operating surgeon through a stereoscopic binocular display generating an impression of a 3D image, aiding the surgeon gauging the depth of tools and body structures, which is particularly useful when suturing tissue. As demonstrated in the following discussion, the potential for visual after-effects is far higher in stereoscopic binocular displays than in biocular displays. Indeed, a well-designed biocular display is likely to place few adaptive demands on the ocular-motor system (see Hoffman, Girshick, Akeley, & Banks, 2008; Lambooij, Fortuin, Heynderickx, & IJsselsteijn, 2009).

32.2.3 VERGENCE AND ACCOMMODATION

In this section, the components of the ocular-motor system, which are responsible for providing clear and single vision (accommodation and vergence eye movements, respectively), are briefly described. It is assumed initially that the eye is emmetropic. It is possible to represent the key features of accommodation and vergence in a heuristic model of the vergence and accommodation control system (see Figure 32.2, modified from Schor & Kotulak, 1986). If an observer wishes to change fixation from a distant object to a near one (or vice versa), the retinal image of the target object is initially defocused (blur describes this error of focus), and there is a fixation error between the image of the target and the fovea (disparity refers to this error of fixation). In order to bring clarity to the retinal image, the eye must focus in a process known as accommodation, and to overcome disparity, the eyes must change vergence angle to maintain fixation within corresponding retinal areas (if noncorresponding points of the retina are stimulated, then double vision will result).

Although blur and disparity are extremely effective *retinotopic* feedback signals, they are only effective over small distances so that larger changes are driven by *spatiotopic* stimuli (Schor, Alexander, Cormack, & Stevenson, 1992). Spatiotopic stimuli can be defined as the cues (sources of information) that indicate a target's position within the environment. The spatiotopically driven changes get the accommodation and vergence systems into the right *ballpark*, and then retinotopically driven responses achieve precisely located fixation. Despite the importance of the spatiotopic response, it remains unclear what information provides the necessary stimuli: it is an issue of distance perception, which is still a topic of considerable debate (Mon-Williams & Tresilian, 1999).

In retinotopic conditions, the systems rely on negative feedback, with a phasic element in the feed-forward pathway rapidly eliminating blur and disparity. The feedback loop to accommodation may be negated (opened) by removing blur information (e.g., by viewing through a pinhole), and the vergence feedback loop may be opened by removing disparity (e.g., by covering one eye). These procedures allow measurement of the bias within the system. The bias reflects the resting position of the system and can be modified through visual demand.

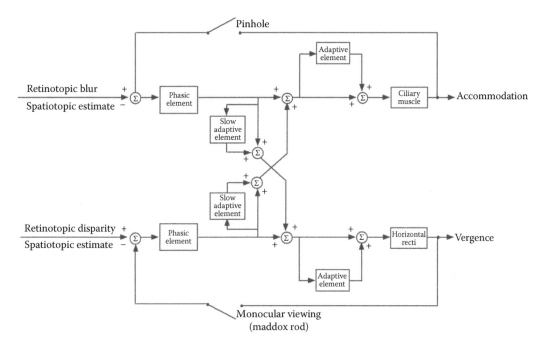

FIGURE 32.2 A heuristic model of the vergence and accommodation control system. (Modified from Schor, C.M. and Kotulak, J.C., *Vision Res.*, 26, 927, 1986.) In normal situations, the systems rely on negative feedback, with a phasic element in the feed-forward pathway rapidly eliminating blur and disparity. A tonic controller is also present in both systems, and this adapts to reduce any steady-state demands placed on the phasic response component. The tonic controller ensures that the accommodation and vergence systems are kept in the middle of their functional range. In order to further maximize system efficiency, the accommodation and vergence responses are neurally cross-linked so that accommodation produces vergence eye movements, whereas vergence causes accommodation. The negative feedback loop to accommodation may be opened by removing blur information (e.g., by viewing through a pinhole), and the vergence feedback loop may be opened by removing disparity (e.g., by covering one eye). The open-loop vergence bias is known as heterophoria (see text for details).

32.2.3.1   Phasic Elements

Figure 32.2 illustrates that accommodation is driven (retinotopically) through blur, whereas vergence responds to disparity (see Schor, 1983, 1986, for a comprehensive overview). An initial change in vergence angle or accommodative state is initiated by a phasic element within the vergence and accommodation system, respectively. The phasic controller acts to rapidly eliminate blur and disparity in order that a clear and single image is achieved. The disparity that drives vergence is the absolute disparity of the target object (i.e., the disparity with respect to the horopter). The horizontal disparities that provide for stereopsis are derived from relative disparities. It follows that the disparity signal for vergence is not the same as that used for stereopsis.

32.2.3.2   Tonic Elements

A tonic controller in the vergence and accommodation system can adapt to reduce any steady-state demands placed on the phasic response component (e.g., Carter, 1965; Schor, 1979). The tonic controller ensures that the accommodation and vergence systems are kept in the middle of their functional range.

32.2.3.3 Cross-Links

In order to maximize system efficiency, the accommodation and vergence responses are neurally cross-linked so that accommodation produces vergence eye movements (accommodative vergence [AV]), whereas vergence causes accommodation (vergence accommodation [VA]). AV is normally expressed in terms of its ratio with accommodation (the AV/A ratio), and VA is expressed in terms of its ratio with vergence (the VA/V ratio). The cross-coupling interactions between accommodation and vergence are stimulated by the phasic but not by the tonic control elements (Schor). Thus, the cross-links have been shown to originate after the phasic element but before the site of the tonic controller.

32.2.4 Vergence (or Prism) Adaptation

In the initial consideration of the vergence system, it was emphasized that a tonic element exists in the feed-forward pathway and that this component acts to minimize any demands placed on the phasic controller. It follows that if the eyes maintain a convergent position for a prolonged period of time, the resting position of the eyes will shift inward, and, reciprocally, after prolonged divergence, the resting position of the eyes will shift outward.

The tonic element means that a change in the demands placed on the vergence system will not necessarily produce visual problems; it is possible that any demands will be accommodated by the adaptable element within the vergence system. On the other hand, it has been established that symptomatic individuals have reduced tonic adaptability (Fisher, Ciuffreda, Levine, & Wolf-Kelly, 1987). A reduced ability to adapt to steady-state viewing together with large vergence demands may cause some individuals to suffer from visual fatigue. Vergence adaptation is often described as *prism adaptation* as the easiest way of artificially inducing the adaptive behavior is to place unyoked prisms in front of the eyes (prisms with the bases pointing in opposite directions relative to the nose). A poorly designed VE system can induce prism through a misalignment of the viewing optics in situations where the lenses are not collimated to infinity (Mon-Williams, Wann, & Rushton, 1993). In a biocular VE system, the screens can be set at any viewing distance from optical infinity to very close to the observer. The closer the viewing distance, the greater the vergence demands. It is necessary to realize that what constitutes too large a vergence demand varies from individual to individual (North & Henson, 1981) and between age groups (Winn, Gilmartin, Sculfor, & Bamford, 1994). Importantly, it has been shown that individuals with binocular vision anomalies lack the ability to adapt to induced changes in vergence bias (Henson & Dharamshi, 1982), and this has been hypothesized as being one of the causes of binocular vision problems (Schor, 1979). A well-designed biocular VE system can avoid the vergence system needing to adapt to the VE.

It is worth drawing attention to another factor that can place demands on the adaptable element of the vergence system. As vertical gaze angle (the vertical orientation of the eyes with respect to the head) is changed, the effort required of the extraocular muscles becomes modified (Heuer, Bruwer, Romer, Kroger, & Knapp, 1991; Heuer & Owens, 1989). This may be readily demonstrated by fixating the tip of a pen held close (e.g., 10 cm) to the face and raising and lowering the pen. It will be noted that it is considerably more comfortable to fixate the pen when it is approximately in line with the mouth and considerably less comfortable to fixate when at eye level. The reason for these changes in comfort relates to the fact that the muscles used to lower the eyes also aid convergence, whereas the muscles used to raise the eyes aid divergence (Mon-Williams, Burgess-Limerick, Plooy, & Wann, 1999). Thus, the vertical position of VE screens can alter the vergence demands placed on the ocular-motor system. It should not be assumed that the user will adjust a VE headset if the gaze angle is inappropriate. The major determinant of a head-mounted display's (HMD's) position on the head is the design of the headset together with the muscular–skeletal position of greatest comfort (see Chapter 41). Once the HMD is in place, it is highly unlikely that a user will equate eyestrain with the vertical location of the headset. It should also be noted that the majority of HMDs allow for initial adjustment of head position, after which time, the headset is fixed in place (normally by means of a headband that is tightened).

We have previously used a software calibration routine that allows for the measurement of heterophoria within a VE headset (essentially, this routine displays a tangent screen on one screen and a vertical line on the other). The use of this routine allows a user to adjust a VE headset until the heterophoria measures reach a minimum value. Furthermore, we have described a routine that enables any demands placed on the vergence system by a VE system to be minimized (Wann et al., 1995). Such software may be extremely useful when adjusting an HMD's position on the head and ensuring that the VE system is not placing large demands on the adaptable component of vergence.

32.2.5 ACCOMMODATIVE ADAPTATION

In the same manner as the vergence system, the accommodation system has a tonic element in the feed-forward pathway in order to minimize any demands placed on the phasic controller. The easiest way of inducing accommodative adaptation is by placing refractive lenses in front of the two eyes. It follows that VE systems that use refractive lenses to present images to the human observer may induce accommodative adaptation. In general, such adaptive demands are met readily by the ocular-motor system unless the viewing distance is very close. One additional issue of accommodative adaptation that arises with VE systems is the extent to which the eyes are capable of *aniso-accommodation* (Marran & Schor, 1998). Aniso-accommodation describes the response of the ocular-motor system to unequal accommodative demands between the two eyes. Situations can arise naturally when the eyes have unequal refractive error and thus have different focusing demands. In general, large interocular refractive discrepancies lead to strabismus and amblyopia (see Section 32.2.7), but it has been reported that the ocular system can respond to small discrepancies by adjusting the tonic accommodative resting position of the eyes independently (Marran & Schor, 1997, 1998). It is possible for a poorly designed VE system to introduce unequal focusing demands if the optical focus is not perfect for the two eyes. It is likely that such a system would place adaptive pressures on both the accommodation and vergence systems.

32.2.6 CROSS-LINK ADAPTATION

Helmholtz (1924) argued that developmental change (such as increasing interocular distance and sclerosis of the crystalline lens with age) requires that the cross-links should be open to adaptation. Although some studies have provided indirect evidence for plasticity (see Schor & Tsuetaki, 1987), there is no conclusive evidence that environmental pressures will directly change the normal cross-link relationship. The VA/V ratio has received relatively little attention owing to difficulties in its measurement, but it is known that the AV/A cross-link may be temporarily modified through cycloplegia, miosis, visual fatigue, and orthoptics (Christoferson & Ogle, 1956; Flom, 1960; Manas, 1955; Parks, 1958). On the other hand, it is not certain whether these transient modifications actually reflect plasticity within the cross-links or arise from an alteration in other components of the motor systems. It has been demonstrated that fatigue in the adaptable tonic component can produce predictable changes in cross-link behavior for individuals with abnormal binocular vision. Schor and Tsuetaki (1987) showed that reducing adaptability of tonic accommodation increases the AV/A and decreases the VA/V, whereas reducing adaptability of tonic vergence has the opposite effect.

What has not been determined unambiguously is whether it is possible to directly modify the cross-links in individuals with normal binocular vision. One technique for placing pressure on the cross-links is to use telestereoscopes to optically increase interocular separation or cyclopean spectacles to negate interocular separation (see Figure 32.3). In telestereoscopes, the observer must increase vergence relative to the focal depth of a target; hence, pressure is placed on the AV/A to increase and the VA/V to decrease. In contrast, cyclopean spectacles require a decrease in the AV/A and an increase in VA/V. An appropriate increase in the AV/A has been observed following the

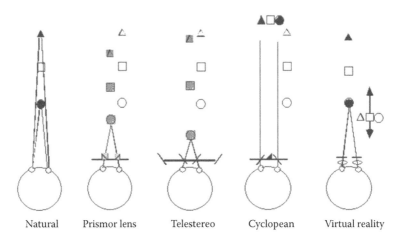

FIGURE 32.3 Stimuli to accommodation and vergence in natural conditions, when a constant vergence or accommodation bias is introduced, and when viewing in optical devices. In natural conditions, the stimuli to accommodation (empty symbols) and vergence (shaded symbols) are in accordance. If a lens or prism power is induced, a constant offset (bias) is introduced. It is possible for a poorly designed VE system to induce such biases (see text for details). Telestereoscopes necessitate an increase in vergence per unit change in natural unaltered accommodation. Cyclopean spectacles require normal accommodation, whereas vergence remains fixed at optical infinity. In a stereoscopic VE display, accommodation and vergence stimuli are disassociated, and the degree of disassociation depends on viewing distance (thus requiring a change in the gain of the AV/A and VA/V ratios) even when the VE system is optically perfect.

use of telestereoscopes for 30 min, but in conflict, a corresponding decrease in the AV/A ratio was not found after the use of cyclopean spectacles (Judge & Miles, 1985; Miles & Judge, 1982; Miles, Judge, & Optican, 1987). Cyclopean spectacles require accommodation to vary normally, whereas vergence is fixed (see Figure 32.3), which is difficult, and hence, single vision is often lost. Miles et al. (1987, p. 2585) reported that "subjects had difficulty in obtaining single, sharp images when viewing nearby objects through the cyclopean spectacles and were never able to overcome this problem entirely." To achieve adaptation, comparison is required between the ocular-motor response and the perceptual goal (a clear, single target image). If the goal of single vision is consistently disrupted due to other neuromuscular factors, the cyclopean paradigm fails to provide the requisite conditions for adaptation to occur. In summary, telestereoscopes have been used to successfully increase the AV/A ratio, but attempts to reduce the AV/A ratio with cyclopean spectacles have been unsuccessful. Schor and Tsuetaki (1987) have argued convincingly that because the AV/A always rose together with a concomitant fall in the VA/V in studies using telestereoscopes, the response to the telestereoscopes might have been due to fatigue of adaptable tonic accommodation rather than direct cross-link modification.

The issue of cross-link adaptation is extremely important with regard to VE systems. Biocular VE systems will not place any adaptive demands on the cross-links. In contrast, a stereoscopic binocular system requires individuals to maintain a constant level of accommodation while changing vergence angle when changing fixation between near and far objects. The stereoscopic binocular systems thus place large adaptive demands on the ocular-motor system and raise the crucial issue of whether the cross-links themselves are open to direct adaptation.

32.2.7 STRABISMUS

Binocular fixation of a stationary object requires that both visual axes are aligned with an object of interest in order to eliminate double vision. Examination of the muscular and neurological systems involved in the coordination of the two eyes can be carried out using a *cover test*. The cover test consists of the covering and uncovering of each eye in turn while the fellow eye maintains fixation on a target. Any movement of the eye when its fellow is being covered indicates the presence of heterotropia (commonly referred to as strabismus or squint). Strabismus can either be convergent (esotropia) or divergent (exotropia) and can affect either eye (e.g., right esotropia). The primary sensory effect of strabismus is amblyopia in the nonfixating eye. One of the consequences of amblyopia is an increased threshold for stereopsis. Strabismus can result from adaptive pressures being placed on a compromised binocular system. Participants with long-standing strabismus will be less subject to difficulties with VE displays because their binocular status is already compromised. Nevertheless, the reduced stereoacuity might be highly problematic in some situations (such as tying off a suture). The emergence of 3D displays for MIS raises questions about the level of stereoacuity that a surgeon should possess. This is an empirical question that will require further research.

32.2.8 HETEROPHORIA

The majority of human observers do not have strabismus (i.e., no bias is present in closed-loop situations) but do show an open-loop vergence bias. It is possible to measure the constant resting point (or bias) that exists within the vergence system by opening the normal feedback loop to vergence (e.g., by removing any disparity information). Open-loop vergence bias is known as heterophoria and may be defined as a slight deviation from perfect binocular positioning that is apparent only when the eyes are dissociated. Esophoria refers to convergent visual axes when the eyes are dissociated and exophoria to divergent axes (see Figure 32.4). Note that the gain of the cross-links is less than one. This means that individuals are generally exophoric when fixating close targets.

It should also be noted that it is possible to observe vertical or rotational biases in the vergence system. The ocular-motor system shows a markedly lower ability to adapt to vertical or rotational deviations from perfect binocular positioning, and thus, such deviations will normally produce double vision. Nevertheless, it is possible for poorly aligned VE systems to require small amounts of vertical or rotational adaptation, and thus, it is necessary to test for such deviations when measuring visual aftereffects following VE exposure. In the interest of brevity, the following discussion restricts the consideration of heterophoria to the more commonly observed horizontal deviations.

32.2.9 BINOCULAR VISION ANOMALIES

It is not uncommon for binocular vision to break down or fail to develop properly. Binocular vision anomalies can be classified according to the scheme in Figure 32.5. Incomitant deviations are far rarer than comitant deviations. Incomitant deviations describe deviations that vary with the direction of gaze and arise due to direct trauma of the neuromuscular system. It is comitant deviations that are important with regard to VE systems. Heterophoria is often a precursor to strabismus (strabismus representing the situation when the system can no longer maintain binocular vision and normal binocular vision breaks down). Heterophoria can be classified in the following manner.

32.2.9.1 Esophoria

1. Divergence weakness esophoria. Esophoria at distance > esophoria near.
2. Convergence excess esophoria. Esophoria at distance < esophoria near.
3. Basic esophoria. Esophoria at distance = esophoria near.

32.2.9.2 Exophoria

1. Convergence weakness exophoria. Exophoria at distance < exophoria near.
2. Divergence excess exophoria. Exophoria at distance > exophoria near. This often breaks down into strabismus for distance viewing.
3. Basic exophoria. Exophoria distance = exophoria near.
4. Convergence insufficiency.

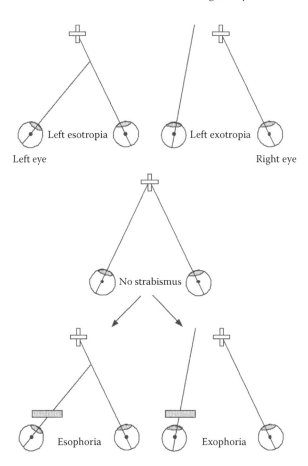

FIGURE 32.4 In normal binocular viewing, the visual axes of both eyes are in alignment with a fixated target (*middle plot*). If the axes are not in alignment, the system is described as strabismic or heterotropic. The left upper plot illustrates esotropia (convergent visual axes), and right upper plot illustrates exotropia (divergent visual axes). It is possible to measure the constant resting point (or bias) that exists within the vergence system in people without strabismus by opening the normal feedback loop to vergence (e.g., by covering one eye). Open-loop vergence bias is known as heterophoria. Esophoria refers to convergent visual axes (*left lower plot*) when the eyes are dissociated and exophoria to divergent axes (*right lower plot*).

A consideration of the organization of the accommodation and vergence systems provides insight into the etiology of heterophoria and strabismus. It has been well established that abnormalities of the cross-link ratios in childhood are associated with strabismus. The need to exert high levels of accommodation to produce clear vision creates a high level of vergence because of the accommodation–vergence cross-link. These demands ultimately result in esotropia. Note that poorly designed VE systems (e.g., systems with poorly designed optics) can induce the same high levels of accommodation as hypermetropia.

An alternative mechanism of strabismus can occur when the cross-links have too high or too low a gain. For example, a high accommodative convergence to accommodation (AC/A) ratio results in convergence excess, and a low AC/A ratio results in convergence insufficiency. Recall that high adaptation of accommodation and low adaptation of vergence results in a low AC/A ratio and vice versa. One possible treatment of high and low AC/A ratios is thus to fatigue the highly adaptable element. Schor and Tsuetaki (1987) have shown that such treatment can restore normal cross-link function. Quick, Newbern, and Boothe (1994) have found that monkeys with esotropia show either a reduced or an abnormally high AV/A ratio. Moreover, Parks (1958) has reported a high incidence of abnormal AV in adult humans with strabismus. The fact that the cross-links are placed under particular adaptive pressure by binocular stereoscopic VE systems has already been discussed.

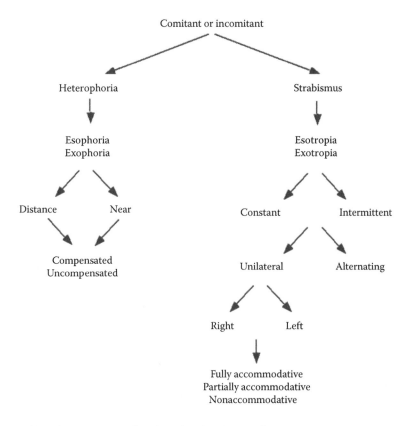

FIGURE 32.5 Classification scheme for binocular visual anomalies.

The importance of considering the etiology of heterophoria when evaluating VE systems relates to the pressures potentially placed by VE on the cross-links and adaptable elements of vergence and accommodation. It seems reasonable to suggest that these issues are of paramount importance when considering the use of VE systems in surgery where the consequences of unstable binocular vision could be profound.

32.2.9.3 Position-Specific Heterophoria Adaptation

Heterophoria describes the difference in position between the two eyes when one eye is occluded. In situations where a prism is placed in front of one eye (see Section 32.2.4), the stimulus for adaptation is reasonably constant at all directions of gaze. In situations where a magnifying lens, for example, is placed in front of one eye, the amount of adaptation required changes as the eye moves eccentrically from the optical center of the lens (Maxwell & Schor, 1994). The ocular-motor system is known to be able to adapt in a position-specific manner, but the time constants of such adaptive behavior are far greater than those observed in uniform adaptation. Indeed, the initial response of the ocular-motor system to position-specific vergence pressure is to adapt uniformly in order to decrease or eliminate the demand on at least one part of the visual field (Maxwell & Schor, 1994). The initial global adaptation process is followed by a slower-acting position-specific alignment. It is possible for position-specific heterophoria to be induced by a poorly designed VE system (once more through optical errors). It should also be noted that the representation of true 3D space with two-dimensional (2D) image planes has the potential to introduce spatial biases between the two eyes, with the biases becoming larger as gaze becomes increasingly eccentric (Wann et al., 1995). It is possible that these biases will place position-specific adaptation pressures on the ocular-motor system. It is not clear whether these small distortions have practical implications with prolonged exposure to VE systems that use flat screen projections of 3D images.

32.2.9.4 Summary of Some Key Points

Although it is possible for a biocular VE display to promote changes in the user's visual system, adaptive pressures are only likely to be found in a system with very poor optical quality or poorly integrated tracking. In contrast, the basic design of all current stereoscopic binocular systems means that they will place adaptive pressures on the user because the binocular stimuli are not commensurate with natural viewing (Figure 32.3). In some cases, the adaptive pressure may be minor, but sustained usage of a binocular display that presents large depth intervals is likely to produce transient aftereffects. The consequence of such effects depends on the individual's visual status, and it cannot be predicted without a full clinical appraisal. Where sustained usage is planned with binocular displays that present large depth intervals, pre and post (see Section 32.3) should be undertaken with candidate observers prior to unsupervised use, and any observers with preexisting binocular anomalies should be excluded or carefully monitored. Symptoms (headache and difficulty focusing) should be used as indicators that display usage should cease, but symptoms cannot be used as the sole criteria for evaluating potential problems; hence, questionnaire studies are imprecise tools for assessing visual aftereffects.

32.3 CLINICAL MEASURES OF VISUAL AFTEREFFECTS FOLLOWING VE EXPOSURE

There are two methods of measuring visual aftereffects following VE exposure. One method is to take a series of measures before and after VE use and explore whether there are any statistically reliable differences between these measures. Additionally, it is possible to compare any changes with a control condition in which the same measures are taken before and after the same period of time viewing, for example, a standard computer monitor. The alternative method is to take a series of clinical measures before and after VE use and determine whether any of these measures has shown a clinically significant change. We would argue that this technique provides a far more powerful tool and avoids the problems associated with testing the null hypothesis (a single person developing strabismus might not produce a statistically reliable change in the population means but would unarguably represent a significant aftereffect). We will outline a series of useful clinical measures that should be evaluated with regard to VE systems. The determination of refractive error can take even an expert 10 min, but the other measures can be incorporated into a battery that can be completed within 10 min. Thus, these clinical measures provide a useful tool for assessing the impact of VE exposure (see Mon-Williams et al., 1993).

32.3.1 HISTORY AND SYMPTOMS

A general medical and ocular history should be recorded together with any symptomatic complaints. Participants should be asked to report if they are suffering from any adverse symptoms, including headache, diplopia, blurred vision, sore eyes, or eyestrain. Participants should be asked to report verbally any such symptoms during or following the use of the VE system.

32.3.2 VISION/VISUAL ACUITY

The term *eyesight* is herein used to describe the clarity of the retinal image. The reason for using this terminology is because the terms *vision* and *visual acuity* have specialist meanings within the ophthalmic literature: *Vision* refers to uncorrected eyesight, whereas *visual acuity* describes the level of eyesight obtained with optimum refractive correction. Participants should be assessed in their normal situation so that vision is recorded for participants who do not normally wear spectacles or contact lenses and visual acuity for those who have (and wear) a refractive correction. It is most important to assess binocular eyesight, as disturbances of the ocular-motor system can cause a reduction in the normal advantage of binocular viewing. Eyesight is typically assessed with some form of letter chart and expressed as a fraction where the numerator refers to the distance of the chart and the denominator refers to the distance at which the smallest visible letter subtends 5 min of arc. The standard clinical chart is placed at 6 m (or 20 ft), and a person with *good* eyesight is expected to read letters that subtend 5 min of arc at that distance. If someone can read the small letters near the bottom of a chart, then they will typically have eyesight equal to 6/6 or better (20/20 in imperial measures), but if only the large letter on top of the chart could be read, then the eyesight would be equal to 6/60 (6/12 occurs about halfway down). Traditional (Snellen) eyesight charts have major disadvantages for scientific studies. These disadvantages relate to the following: (1) the lines changing in unequal step sizes and (2) the lines having an unequal number of letters. These two facts make it impossible to report valid statistics for Snellen eyesight measures. For this reason, we propose the use of logarithmic charts when evaluating eyesight. For example, we have previously used the Glasgow Acuity Cards designed for clinical screening (McGraw & Winn, 1993). The Glasgow Acuity Cards use a progression of letter sizes in a geometrical series with a ratio of 0.1 log units per card. Each card contains four letters of equal legibility, and equivalent letter-row spacing is maintained throughout. Vision is recorded either as the log of the minimum angle of resolution or as a decimal value from 0.0 (6/60 Snellen) to 1.0 (6/6 Snellen). The use of the Glasgow Acuity Cards allows for accurate measurement and better analysis of the vision/visual acuity than the traditional Snellen chart.

It should be noted that ophthalmic charts typically measure high-contrast eyesight (i.e., a well-defined black shape or letter against a bright white background). The problem with such charts is that they do not indicate how well the visual system can process retinal information at different spatial frequencies (e.g., how well someone can see when contrast is low). A number of methods do exist, however, that allow eyesight to be assessed over a range of spatial frequencies. It is possible (in principle) for a VE system to cause a decrease in the midrange spatial frequency sensitivity. The reason for this is that the visual system responds to an amelioration of high spatial frequencies by depressing the midrange spatial frequencies in order to unmask the higher-frequency channels (Mon-Williams, Tresilian, Strang, Kochar, & Wann, 1998). A VE system that has poor image quality effectively filters the high-spatial-frequency information from a display and thus might cause the visual system to adapt. There are a number of commercially available contrast sensitivity charts for use in a clinical setting. One such chart is the Pelli–Robson chart (Pelli, Robson, & Wilkins, 1988). This chart is wall mounted and consists of 16-letter triplets. The triplets are of constant contrast, but the contrast reduces between triplets from the top to the bottom of the chart. The Pelli–Robson chart and other available alternatives allow for easy measurement of contrast sensitivity. In all methods of eyesight measurement, it is important that the recommended level of illumination is used.

32.3.3 REFRACTIVE ERROR

It is worthwhile measuring the refractive state of the eye before and after the use of a VE system to ensure that the refraction has not changed because of accommodative adjustment. The most common clinical method of assessing refractive status involves a *retinoscope*. The standard retinoscopy technique involves shining a light into the eye while viewing the resultant retinal reflection (the *red eye* observed in flash photography) along the visual axis through a semisilvered mirror. The practitioner then rotates the retinoscope and observes the resulting movement of the beam. If there is no resultant movement, there is no refractive error; if the reflex moves in the same direction as the retinoscope (with it), then the eye is hypermetropic, and if the reflex moves in the opposite direction, the eye is myopic.

32.3.4 OCULAR-MOTOR BALANCE

Examination of the muscular and neurological systems involved in the coordination and integration of the two eyes can be carried out using a battery of clinical tests. The cover test is described in Section 32.2.7, which consists of the covering and uncovering of each eye in turn while the other eye maintains fixation on the smallest target visible to both eyes. Any movement of the eye when its fellow is being covered indicates the presence of strabismus, whereas any movement of the eye when it is being uncovered shows the presence of heterophoria. Further assessment of heterophoria can be carried out using a Maddox rod at 6 m and a Maddox wing for use at 33 cm in good illumination (500 lux). A *flashed* approach is the optimum method of obtaining the heterophoria reading for distance and for near. It is important to assess the horizontal, vertical, and rotational components of binocular alignment.

One useful index of binocular function is provided by *fixation disparity*: small errors of binocular fixation that do not disrupt binocular vision, as the errors do not exceed Panum's fusional areas (the corresponding retinal errors in the two eyes). It is straightforward to measure fixation disparity, but we suggest that associated heterophoria provides a much better indicator of binocular stress. *Associated heterophoria* is the term used to describe the amount of prism required to reduce fixation disparity to zero and is readily measured using a Mallett unit (Mallett, 1974). Mallett units use a binocular lock consisting of the letters OXO and monocular markers in line with the X to demonstrate the presence of decompensated heterophoria, because viewed through Polaroid filters, the monocular markers will remain in line only in the absence of decompensated heterophoria. The existence of decompensated heterophoria provides a useful indication of binocular stress. Standard Mallett distance and near units can be used with loose trial case prisms within a trial frame in order to produce absolute alignment of the nonius lines using the minimal value of prism in 0.5 prism diopter steps for the distance and 1 prism diopter steps for near. Participants should be instructed to look for even very small movements of the nonius lines. A full review of binocular vision tests is presented by Pickwell (1984). An example of a Mallet unit adapted for use within a VE display is presented by Wann et al. (1995).

32.3.5 NEAR POINT OF CONVERGENCE

Clinical measurement of pursuit convergence should be taken with a vertical line target brought toward the eyes on the median plane until diplopia (double vision) is reported and/or one eye can be seen to diverge. It is also possible to assess *jump* vergence by asking participants to rapidly change fixation from a near target to a far one and vice versa (see Pickwell, 1984).

32.3.6 AMPLITUDE OF ACCOMMODATION

The dynamic recording of accommodation is technically demanding (in contrast to the relatively straightforward techniques available for recording eye movements). It is possible, however, to use the retinoscope (see Section 32.3.3) to determine the level of accommodation that occurs in response to a proximal target. The eye's ability to change focus from the far point to the near point of vision should be measured in diopters using subjective report of near text becoming illegible. Amplitudes of accommodation should be checked for any abnormality. Accommodative facility can be assessed by placing ophthalmic lenses in front of the eyes and evaluating the time it takes for the eyes to relax or increase accommodation.

32.3.7 STEREOPSIS

There are a large number of clinical tests of stereopsis (including the TNO test, the Frisby test, the Stereo random dot E test, the Titmus random dot test, and the Lang test). Stereopsis provides a good indication of binocular status, because stereoscopic sensitivity drops rapidly when binocular vision is compromised. It is important to ensure that the tests are conducted at a standardized distance, as disparities are inversely proportional to the square of the distance. It is also important to ensure that the test is carried out under good illumination (e.g., 500 lux).

32.4 BEYOND CLINICAL MEASURES

The measures described in Section 32.3 are proposed as the basic requirements in an applied setting for appraising any new or substantially modified display system or to screen any group of individuals who may be required to use a binocular display for sustained or repetitive periods. In addition to these measures, it is possible to undertake a more analytical approach in an experimental setting to evaluate the impact of different adaptive pressures. In this section, an example of undertaking such a study is outlined using an infrared (IR) optometer and eye tracking to appraise the response of the AV/A cross-links (Section 32.2.3.3) to sustained and demanding usage of a conventional binocular VE system. We originally conducted this work for the first edition of this book. There has been great interest in our results subsequently, but unfortunately, there has been little subsequent work conducted into the adaptability of the AV/A and VA/V ratios (see Maxwell, Tong, & Schor, 2012). We would suggest there is an urgent need for such studies given the rapidly increasing importance of VE displays in MIS.

32.4.1 STUDY OF AV/A AND VA/V ADAPTATION

The AV/A and VA/V ratios (Section 32.2.3.3) were objectively measured before and after the usage of a VE system. Two volunteer participants (AS and MB) took part in the experiments. Both were 25 years of age, emmetropic, did not take any medication, had no ophthalmic abnormalities, and were naive to the purpose of the experiment. The participants gave written consent, and the experiments met the approval of the university ethics committee. Clinical measures of accommodation amplitude, stereoacuity thresholds, and AV/A and VA/V ratios were all normal. LCD shutter goggles were used to present field-sequential computer-generated images at 120 Hz on a high-resolution computer monitor. The participants viewed a high-contrast cross for 60 min in the dark oscillating from optical infinity to either 33 or 16 cm with accommodative demand set at the closest distance. The angular size of the target cross was kept constant as it moved in depth, creating the percept of a single cross of unchanging size located at a constant egocentric distance. In the early stages of adaptation, participants had to exert voluntary vergence to eliminate double vision. At first, participants had problems consistently fusing the target, but after the first 10 min, they reported that it was easy to keep a clear single target though both complained of sore eyes and frontal headaches. Hence, this presented a highly demanding task that is unlikely to occur in many occupational setting, but similar stimuli may occur if close visual monitoring is required of targets moving in depth over a sustained period (e.g., some military applications or some games). The experimental issue was what changes resulted from the sustained usage.

Dynamic accommodation was measured using a modified Canon Autoref R-1 IR objective optometer. The autorefractor determines the position of focus in three meridians, from which the spherocylindrical refractive error can be computed to an accuracy of ±0.12D per second. Vergence was recorded with a purposely built differential IR limbal tracker, accurate to 10 arcmin under the employed experimental conditions. The IR sources were chopped at greater than 1000 Hz, with the detectors set to only recognize this frequency (thus avoiding direct current drift, reducing 1/f noise, and eliminating cross-talk between the photodiodes and the IR optometer). Cross-axis sensitivity was found to be ±10%. The analog output of the eye trackers was fed into a digital storage oscilloscope, which was controlled by an IBM-clone computer via an IEEE-488 interface. This system allowed inspection of the analog signals to be made in real time so that artifacts (e.g., blinks) could be identified. Comprehensive calibration routines were carried out before preadaptation measurements and after postadaptation recordings. The ambient room illumination was constant at approximately 250 lux. Natural pupil size was monitored on an external 9 in. video monitor (Sony PVM 97) using the magnified (8.2×) image provided by the IR optometer to ensure that pupillary changes were not a confound (pupil size always exceeded 0.5 cm).

AV was measured by covering the left eye with an occluder. Participants fixated a high-contrast Maltese cross through a 5 D Badal lens, whereas accommodative demand was altered via ophthalmic trial lenses so that the only change in the stimulus was due to accommodation. Targets were presented along the optical axis of the right eye to ensure that eye movements did not interfere with the accommodative recording. VA was measured using Difference of Gaussian (DoG) stimuli. The DoG presented a blurred vertical bar of 0.4 cm, which acted as a stimulus to vergence but not to accommodation. Vergence demand was altered via ophthalmic prisms so that the only change in the stimulus was due to disparity vergence.

Accommodation and vergence responses were expressed in diopters and meter angles (MA), respectively: both are the reciprocal of distance in meters (MA is the angle through which each eye has rotated from the primary position in order to fixate an object located 1 m away), so 1 D or 1 MA corresponds to 100 cm, 2 D, or 2 MA corresponds to 50 cm, etc. The normal AV/A ratio is approximately 0.60 ± 0.2 MA/D, and the VA/V ratio is around 0.5 ± 0.2 D/MA. The cross-link response was assessed using five separate measurements of accommodation and vergence for seven different stimuli conditions presented in a randomized order (to control for hysteresis). The stimuli were in unitary D or MA steps from optical infinity to six. Measurement of the AV/A and VA/V was alternated pre- and postadaptation and between participants.

32.4.2 RESULTS

Figure 32.6 shows that cross-link interaction changed following adaptation. Regression lines are plotted and correlation coefficients are provided, although the nonlinear nature of the postadaptation data weakens any regression analysis (the postadaptation data for AS showed a significant, $p < 0.05$, quadratic component). Dunn's procedure (Keppel, 1982, p. 146) was carried out on the data points of interest.

The target distance did not produce exact accommodation or vergence, emphasizing the importance of measuring ocular response when computing cross-link ratios. The pre- and postadaptation vergence responses were identical for MB but showed a mean decrease of 0.28 MA for AS (maximum of 0.6), whereas the accommodation responses decreased by a mean of 0.36 D for MB and 0.54 for AS (maximum of 0.6 and 0.98, respectively). The maximum shift was always found in the stimuli where the ocular demands were greatest. These decreases suggest that some plant fatigue is occurring (tiredness of the muscles; see Schor & Tsuetaki, 1987), but the shifts cannot explain the cross-link changes.

In participant MB, it is clear that the AV/A and VA/V ratios have been reduced, thereby allowing accommodation and vergence to independently respond to blur and disparity cues. The variability in the AV/A response is typical of an uncoupled system, whereas the accommodation produced by the VA/V stabilizes around the bias found in the absence of accommodative stimuli.

The data from participant AS are less compelling than those of MB. On the other hand, it is clear that the cross-link responses have changed. The change in the VA/V is very similar to the clear lowering of the ratio observed in participant MB. The change in the gain of the AV/A ratio is less straightforward as the postadaptation AV/A ratio showed a significant quadratic component. It therefore appears that part of the slope has increased its gain, whereas the other part has flattened.

The data from participant MB are very compelling, suggesting that the AV/A and VA/V ratios have both decreased following exposure to the VE display. The data from participant AS are less impressive, but detailed examination shows a similar picture to the data from participant MB. The data suggest that both participants have reduced the normal action of AV and VA. The physiological response will reduce the pressure on the visuomotor system, but these changes will impact perception and may have implications for performance (see Hoffman et al., 2008).

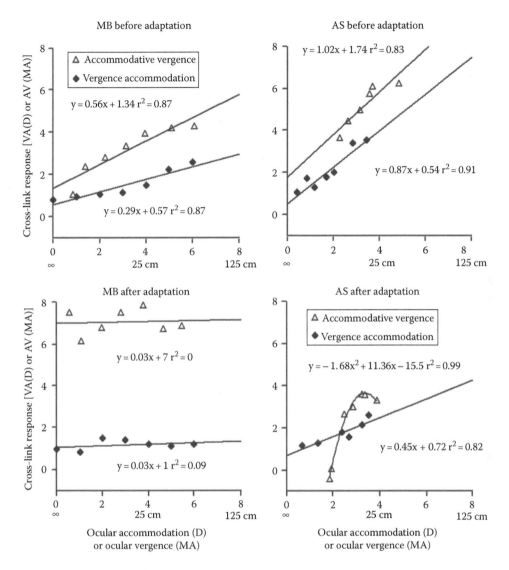

FIGURE 32.6 The cross-links between accommodation and vergence before (*above*) and after (*below*) adaptation for MB (*left*) and AS (*right*). Each point is the mean of five randomly organized measurements. The standard deviation of the measurements was smaller than the plot symbol. The primary response (ocular accommodation [A in diopters] or vergence [V in MA]) is plotted on the abscissa with the secondary (cross-link) response (AV in MA or VA in D, respectively) shown on the ordinate. The normal ratios may clearly be seen to change following adaptation.

32.5 VISUAL AFTEREFFECTS AND MIS

32.5.1 Laparoscopic Surgery

LS has been around since the beginning of the twentieth century. However, it was the development of the computer chip television camera that led to the field progressing exponentially. Prior to this, a surgeon could only use one instrument as he or she would use the other hand to move the laparoscope and visualize the instrument. This meant that LS had limited applications. The advent of the digital camera allowed the surgical image to be projected on a monitor that could be seen by both the surgeon and the assistant and enabled the surgeon to use both hands to operate. Thus, MIS began in earnest in 1987 (Lanfranco, Castellanos, Desai, & Meyers, 2004).

In LS, the patient is positioned on the operating table, and an incision is made in the patient's skin—usually above or below the belly button. A port is inserted either by dissecting through the abdominal wall or under direct vision. A gas supply is attached to fill the abdominal cavity with CO_2, establishing a pneumoperitoneum. The purpose of this is to create the space within the abdomen within which to perform the operation. A laparoscopic camera is inserted through this central port, and further ports are inserted under direct vision to allow access for the laparoscopic instruments.

In contrast to open surgery, MIS creates a variety of constraints on the surgeon, such as restricted movement, compromised dexterity, degradation or loss of haptic feedback, reduced visual depth perception, amplification of hand tremor, and the fulcrum effect (where the hand needs to move in the opposite direction to that in which the tip of the instrument needs to move; Stefanidis, Korndorffer, Markley, Sierra, & Scott, 2006). These constraints mean that during MIS, surgeons need to learn new complex and challenging mappings between the visual input and the movement output. Display of the laparoscopic tools and the surgical site are usually via a 2D monitor positioned at the surgeon's discretion. Figure 32.7 depicts a common operating room setup for LS. The position of the camera, assistant, and monitor(s) varies with each surgeon's particular preference.

Monitors are typically between 21 and 26 in. in size. In 2012, high-definition cameras have 1920 × 1080p resolution, but lower-resolution cameras are often used depending upon the port

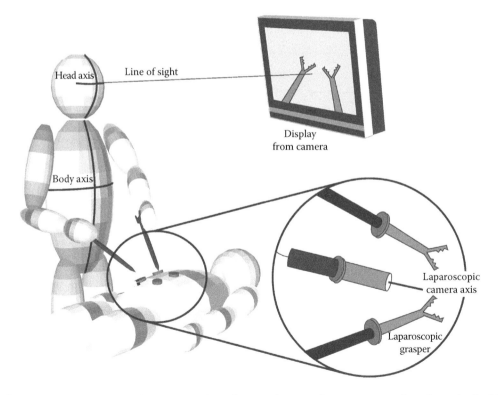

FIGURE 32.7 An example of the setup during LS. Here, the operating surgeon stands at the patient's side, and the viewing monitor is suspended over the patient at the cranial end. The surgeon performs the operation using two instruments inserted through separate incisions made in the patient's abdominal wall. The camera is inserted through a third incision (usually in the region of the umbilicus) and controlled by an assistant, allowing the surgeon to use both hands to perform the operation. Three axes are demonstrated; the surgeon's head and body axes and the axis of the laparoscopic camera.

size being used and equipment available. Different laparoscopic cameras with angulation at the tip (30° and 45°) can assist the surgeon to *look round corners* (i.e., around tissue or an organ that is obscuring the view). This does, however, add further distortion to the mapping between motor action and visual movement of the tools displayed on the screen. While remote viewing of the laparoscopic tool may impair the absolute level of performance, it has been shown that there are some benefits to not looking at your actual hand when learning to compensate for visual distortions (Wilkie, Johnson, Culmer, Allen, & Mon-Williams, 2012).

An additional possible source of variability is the degree to which the camera moves during MIS, which will depend largely on the surgery being performed. An operation such as a laparoscopic cholecystectomy (removal of the gallbladder) or robotic prostatectomy (removal of part/all of the prostate gland) requires a relatively static visual field, and therefore, there will be limited movement of the camera around the abdomen. In contrast, during a gastric bypass or colectomy (removal of a section of bowel), a variety of locations within the abdomen need to be visualized throughout the procedure. It would be expected that a static camera view would provide a useful frame of reference when adapting to the other visual-motor distortions present during MIS. When the image is no longer a fixed reference, then continuous recalibration of the visual-motor mapping may be required.

32.5.2 ROBOTIC-ASSISTED SURGERY

The origins of today's RAS systems can be traced to when the US Army became interested in a system to provide medical assistance on the battlefield from a remote location. One of the early areas where RAS gained prominence was in urological surgery, specifically in prostate resection for cancer. This type of surgery takes place deep within the pelvis where there is little room, making delicate movements difficult, and vision can be obscured by larger laparoscopic instruments or an assistant surgeon's hands. However, in recent years, surgeons have recognized the advantages and begun to practice RAS in fields such as pediatric surgery, antireflux surgery, cardiac procedures, and obstetrics and gynecology.

The main commercial system currently in widespread use is the da Vinci system®. As in LS, the patient is positioned on the operating table, a pneumoperitoneum is established, and several ports are inserted into the patient's abdomen. The camera and surgical instruments are inserted through these ports and connected to the robot's *arms*. The surgeon sits in a console that controls the movements of the instruments and camera using the hands and feet. The console is usually in the same operating room as the patient or in an adjacent room.

RAS has the potential to overcome several of the limitations of conventional LS. In particular, it can offer a 3D view, instrumentation with seven degrees of freedom, reduction of physiological tremor, and elimination of the fulcrum effect. It allows the surgeon to sit in an ergonomic position and offers improved visualization (Behan et al., 2011). However, RAS systems are very expensive with added start-up costs. They may require additional staff to operate the system, lack the sensation of touch, and are currently of unproven benefit (Lanfranco et al., 2004). In addition, the 3D displays have the potential to create additional issues as specifically addressed in Section 32.6.4.

32.5.3 TRAINING SIMULATORS FOR MIS

In order to facilitate surgical training in an increasing litigious society where training opportunities are limited, several virtual reality simulators have been developed. Such simulators allow the practice of a particular procedure without the need for an actual patient. Simulators exist for laparoscopic, arthroscopic, and endovascular surgeries as well for invasive procedures such as upper and lower gastrointestinal endoscopy and bronchoscopy. Simulators have also been produced for practicing RAS. There is some literature that suggests such simulators are beneficial for training (e.g., Grantcharov et al., 2003; Seymour et al., 2002) but the supporting evidence is weak. The lack of proper evaluations is particularly disappointing given the enormous financial investment associated with these simulators. There is a desperate need for proper research to consider how these systems might be used effectively (Gallagher et al., 2005). One of the fundamental problems is that these systems have been designed with little understanding of how humans learn complex skills and what perceptual information is required when learning control strategies. For example, the Simodont has been developed to train dentists in a 3D VE. There is no control, however, over the fixation distance, meaning that the perceptual information regarding the depth of drilling will vary according to where the trainee dentists position themselves. It is thus unsurprising that our colleagues in dental education have expressed concerns to us: they are worried that this system might cause huge problems if the trainees acquire skills specific to this system and then attempt to transfer these skills to a real-world setting (i.e., drilling real teeth). It is clearly important that training simulators prepare the trainee for the real-world situation and that the requisite perceptual information needs to be available: this means that the visual information within the VE must be commensurate with the visual information within the real-world setting. These issues will be considered with specific regard to visual information in MIS in the following sections.

32.5.4 Visual Considerations within MIS

One of the great challenges in MIS is that often the surgeon must interact with poor-quality images of 3D structures presented on 2D screens. One possible technological advance is to provide a stereoscopic view to the surgeon by having two camera viewpoints from within the abdominal cavity. Indeed, systems such as the da Vinci surgical system allow the surgeon to gain a stereoscopic view. The provision of 3D information is clearly advantageous for surgeons trying to determine the depth of structures (and in tasks such as tying off sutures). Nevertheless, the fact that the stereoscopic images are displayed on flat screens creates two difficulties. First, the mismatch between accommodation and vergence can place pressures on the visual system, and these pressures might cause surgical errors as outlined previously. Second, it is extremely likely that the disparities and vergence angle recreated within the virtual display will not match the physical geometry of stereoscopic vision outside the display (see Fukushima, Torii, Ukai, Wolffsohn, & Gilmartin, 2009). This may cause the surgeon to misperceive the depth and distance of anatomical structures (see Priot et al., 2012). These problems will be exacerbated if the viewpoint is not consistent from one operation to the next—a situation that seems probable given the inherent variability in surgical procedures. Prolonged viewing of stereoscopic displays has also been shown to cause a decline in visual functions (see Emoto, Niida, & Okano, 2005; Ukai, 2007; Ukai & Howarth, 2008). The optimal development of MIS technology thus requires a consideration of these issues if the full potential of the technology is to be realized. One method to overcome these disparities may be to utilize the technique of depth filtering in future technologies (see Akeley, Watt, Girshick, & Banks, 2004). Here, each eye is presented with a sum of images at multiple image planes placed at different focal distances. This has been shown to reduce many of the problems caused by accommodation–vergence conflicts in conventional stereoscopic displays and to precisely match variations in accommodation and vergence demand just as in natural vision (see Akeley et al., 2004; Hoffman et al., 2008; Mackenzie, Hoffman, & Watt, 2010; Watt, Akeley, Ernst, & Banks, 2005; Watt, Akeley, Girshick, & Banks, 2005).

32.6 CONCLUSIONS

Binocular visual displays can result in unusual visual pressures being placed on accommodation and vergence. These may be partially reduced by a change in tonic accommodation and tonic vergence, but where objects are presented with a wide range of depths, tonic shifts cannot provide a solution (Figure 32.3). Global shifts in bias can only decrease the ocular-motor demands of uniform disparities or blur. We propose that nonuniform demands (like those produced in VE systems) require a response from the cross-links systems. The results presented in Section 32.4 suggest that the cross-links between accommodation and vergence are directly open to adaptive pressures. These results highlight the importance of addressing the fundamental issues of binocular coordination when considering the pressures placed on the ocular-motor system by VE systems. The simultaneous decrease observed in the AV/A and VA/V ratios cannot be explained (in a parsimonious manner) by an alteration in the adaptability of the respective tonic components or by plant fatigue. This may be viewed as a positive adaptation to the VE setting, but the negative consequences, highlighted by the symptoms of headache and sore eyes, should not be overlooked. Adaptation of this type is not stress free, and we would take the conservative position that it is not desirable if it can be avoided through the design of VE tasks (not just display design, but the rendered stimuli) or limiting exposure periods. Moreover, the adaptation processes are likely to interfere with the stable perceptual platform required for skilled eye–hand coordination. This has particular issues for successfully performing MIS.

Our previous suggestion that binocular VE systems may place pressure on the accommodation and vergence systems was initially controversial (Mon-Williams et al., 1993; Wann et al., 1995). There is now ample evidence from a number of studies to confirm our original hypothesis (see Watt, Akeley, Ernst et al., 2005; Watt, Akeley, Girshick et al., 2005). It is important therefore that such effects are monitored and minimized in any setting where sustained usage of a binocular system is likely to occur or the consequences of an unstable visual-motor mapping are profound. MIS is one such setting, and it is imperative that these systems are designed to minimize adverse visual effects.

ACKNOWLEDGMENTS

Production of this manuscript was supported by grants from the EPSRC Bridging the Gaps fund, The Leeds Teaching Hospitals Charitable Foundation (Registered Charity No. 1075308), and the Medical Research Council, UK.

REFERENCES

Akeley, K., Watt, S. J., Girshick, A. R., & Banks, M. S. (2004). A stereo display prototype with multiple focal distances. *ACM Transactions on Graphics, 23*, 804–813.

Behan, J. W., Kim, S. S., Dorey, F., DeFilippo, R. E., Chang, A. Y., Hardy, B. E., & Koh, C. J. (2011, October). Human capital gains associated with robotic assisted laparoscopic pyeloplasty in children compared to open pyeloplasty. *Journal of Urology, 186*(4 Suppl.), 1663–1667.

Carter, D. B. (1965). Fixation disparity and heterophoria following prolonged wearing of prisms. *American Journal of Optometry and Archives of American Academy of Optometry, 42*, 141–151.

Christoferson, K. W., & Ogle, K. N. (1956). The effect of homatropine on the accommodation-convergence association. *AMA Archives of Ophthalmology, 55*, 779–791.

Cuschieri, A. (1995). Laparoscopic management of cancer patients. *Journal of the Royal College of Surgeons of Edinburgh, 49*, 1–9.

Emoto, M., Niida, T., & Okano, F. (2005). Repeated vergence adaptation causes the decline of visual functions in watching stereoscopic television. *Journal of Display Technology, 1*, 328–340.

Fisher, S. K., Cuiffreda, K. J., Levine, S., & Wolf-Kelly, S. (1987). Tonic adaptation in symptomatic and asymptomatic subjects. *American Journal of Optometry and Physiological Optics, 64*, 333–343.

Flom, M. C. (1960). On the relationship between accommodation and accommodative convergence. I. Linearity. *American Journal of Optometry and Archives of American Academy of Optometry, 37*, 474–482.

Fukushima, T., Torii, M., Ukai, K., Wolffsohn, J. S., & Gilmartin, B. (2009). The relationship between CA/C ratio and individual differences in dynamic accommodative responses while viewing stereoscopic images. *Journal of Vision, 9*(13), 21.1–21.13.

Gallagher, A. G., Ritter, E. M., Champion, H., Higgins, G., Fried, M. P., Moses, G., & Satava, R. M. (2005). Virtual reality simulation for the operating room: Proficiency-based training as a paradigm shift in surgical skills training. *Annals of Surgery, 241*, 364–372.

Grantcharov, T. P., Kristiansen, V. B., Bendix, J., Bardram, L., Rosenberg, J., & Funch-Jensen, P. (2003). Randomized clinical trial of virtual reality simulation for laparoscopic skills training. *British Journal of Surgery, 91*, 146–150.

Henson, D. B., & Dharamshi, B. G. (1982). Oculomotor adaptation to induced heterophoria and anisometropia. *Investigative Ophthalmology & Visual Science, 22*, 234–240.

Hering, E. (1868/1977). *The theory of binocular vision*. New York, NY: Plenum Press.

Heuer, H., Bruwer, M., Romer, T., Kroger, H., & Knapp, H. (1991). Preferred vertical gaze direction and observation distance. *Ergonomics, 34*, 379–392.

Heuer, H., & Owens, D. A. (1989). Vertical gaze direction and the resting posture of the eyes. *Perception, 18*, 363–377.

Hoffman, D. M., Girshick, A. R., Akeley, K., & Banks, M. S. (2008). Vergence–accommodation conflicts hinder visual performance and cause visual fatigue. *Journal of Vision, 8*(3), 33, 1–30.

Judge, S. J., & Miles, F. A. (1985). Changes in the coupling between accommodation and vergence eye movements induced in human subjects by altering the effective interocular separation. *Perception, 14*, 617–629.

Keppel, G. (1982). *Design and analysis: A researcher's handbook*. Englewood Cliffs, NJ: Prentice-Hall.

Lambooij, M., Fortuin, M., Heynderickx, I., & IJsselsteijn, W. (2009). Visual discomfort and visual fatigue of stereoscopic displays: A review. *Journal of Imaging Science, 53*, 30201–30201.

Lanfranco, A. R., Castellanos, A. E., Desai, J. P., & Meyers, W. (2004). Robotic surgery a current perspective. *Annals of Surgery, 239*, 14–21.

MacKenzie, K. J., Hoffman, D. M., & Watt, S. J. (2010, July 1). Accommodation to multiple-focal-plane displays: Implications for improving stereoscopic displays and for accommodation control. *Journal of Vision, 10*(8), 22.

Mallett, R. M. J. (1974). Fixation disparity—Its genesis in relation to asthenopia. *Ophthalmic Optician, 14*, 1159–1168.

Manas, L. (1955). The inconsistency of the AC/A ratio. *American Journal of Optometry and Archives of American Academy of Optometry, 32*, 304–315.

Marran, L., & Schor, C. M. (1997). Multiaccommodative stimuli in VR systems: Problems and solutions. *Human Factors, 39*, 382–388.

Marran, L., & Schor, C. M. (1998). Lens induced aniso-accommodation. *Vision Research, 38*, 3601–3619.

Maxwell, J., Tong, J., & Schor, C. M. (2012). Short-term adaptation of accommodation, accommodative vergence and disparity vergence facility. *Vision Research, 62*, 93–101.

Maxwell, J. S., & Schor, C. M. (1994). Mechanisms of vertical phoria adaptation revealed by time-course and two-dimensional spatiotopic maps. *Vision Research, 34*, 241–251.

McGraw, P. V., & Winn, B. (1993). Glasgow acuity cards: A new test for the measurement of letter acuity in children. *Ophthalmic and Physiological Optics, 13*, 400–403.

McLin, L. N., & Schor, C. M. (1988). Voluntary effort as a stimulus to accommodation and vergence. *Investigative Ophthalmology and Visual Science, 29*, 1739–1746.

Miles, F. A., & Judge, S. J. (1982). Optically induced changes in the neural coupling between vergence eye movements and accommodation in human subjects. In G. Lennerstrand, D. S. Zee, & E. L. Keller (Eds.), *Functional basis of ocular motility disorders* (pp. 93–96). Oxford, U.K.: Pergamon Press.

Miles, F. A., Judge, S. J., & Optican, L. M. (1987). Optically induced changes in the couplings between vergence and accommodation. *Journal of Neuroscience, 7*, 2576–2589.

Mon-Williams, M., Burgess-Limerick, R., Plooy, A., & Wann, J. (1999). Vertical gaze direction and postural adjustment: An extension of the Heuer model. *Journal of Experimental Psychology: Applied, 5*, 35–53.

Mon-Williams, M., & Tresilian, J. R. (1998). A framework for considering the role of afferent and efferent signals in the control and perception of ocular position. *Biological Cybernetics, 79*, 175–189.

Mon-Williams, M., & Tresilian, J. R. (1999). A review of some recent studies on the extra-retinal contribution to distance perception. *Perception, 28*, 167–181.

Mon-Williams, M., Tresilian, J. R., Strang, N., Kochar, P., & Wann, J. P. (1998). Improving vision: Neural compensation for optical defocus. *Proceedings of the Royal Society, B265*, 71–77.

Mon-Williams, M., Wann, J. P., & Rushton, S. K. (1993). Binocular vision in a virtual world: Visual deficits following the wearing of a head-mounted display. *Ophthalmic and Physiological Optics, 13*, 387–391.

North, R. V., & Henson, D. B. (1981). Adaptation to prism-induced heterophoria in subjects with abnormal binocular vision or asthenopia. *American Journal of Optometry and Physiological Optics, 58*, 746–752.

Parks, M. M. (1958). Abnormal accommodative convergence in squint. *AMA Archives of Opthalmology, 59*, 364–380.

Pelli, D., Robson, J., & Wilkins, A. (1988). The design of a new letter chart for measuring contrast sensitivity. *Clinical and Vision Science, 2*, 187–199.

Pickwell, D. (1984). *Binocular vision anomalies: Investigation and treatment*. London, U.K.: Butterworth.

Priot, A. E., Neveu, P., Sillan, O., Plantier, J., Roumes, C., & Prablanc, C. (2012). How perceived egocentric distance varies with changes in tonic vergence. *Experimental Brain Research, 219*, 457–465.

Quick, M. W., Newbern, J. D., & Boothe, R. G. (1994). Natural strabismus in monkeys: Accommodative errors assessed by photorefraction and their relationship to convergence errors. *Investigative Ophthalmology and Visual Science, 35*, 4069–4079.

Schor, C. M. (1979). The influence of rapid prism adaptation upon fixation disparity. *Vision Research, 19*, 757–765.

Schor, C. M. (1983). The Glenn A. Fry Award Lecture: Analysis of tonic and accommodative vergence disorders of binocular vision. *American Journal of Optometry and Physiological Optics, 60*, 1–14.

Schor, C. M., Alexander, J., Cormack, L., & Stevenson, S. (1992). Negative feedback control model of proximal convergence and accommodation. *Ophthalmological and Physiological Optics, 12*, 307–318.

Schor, C. M., & Kotulak, J. C. (1986). Dynamic interactions between accommodation and vergence are velocity sensitive. *Vision Research, 26*, 927–942.

Schor, C. M., & Tsuetaki, T. K. (1987). Fatigue of accommodation and vergence modifies their mutual interactions. *Investigative Ophthalmology and Visual Science, 28*, 1250–1259.

Seymour, N. E., Gallagher, A. G., Roman, S. A., O'Brien, M. K., Bansal, V. K., Andersen, D. K., & Satava, R. M. (2002). Virtual reality training improves operating room performance: Results of a randomized, double-blinded study. *Annals of Surgery, 236*, 458–464.

Stefanidis, D., Korndorffer, J. R. Jr, Markley, S., Sierra, R., & Scott, D. J. (2006). Proficiency maintenance: Impact of ongoing simulator training on laparoscopic skill retention. *Journal of the American College of Surgeons, 202*, 599–603.

Ukai, K. (2007). Visual fatigue caused by viewing stereoscopic images and mechanism of accommodation. In *Proceedings of the First International Symposium on University Communication* (Vol. 1, pp. 176–179). Kyoto, Japan.

Ukai, K., & Howarth, P. A. (2008). Visual fatigue caused by viewing stereoscopic motion images: Background, theories and observations. *Displays, 29*, 106–116.

von Helmholtz, H. (1924). *Physiological optics* (Vol. 3, pp. 191–192). New York, NY: Dover, 1962. English translation by J.P.C. Southall for the Optical Society of America from the third German edition of *Handbuch der physiologiscen optik*, Hamburg, Voss, 1909. (Original work published 1894)

Wann, J. P., Rushton, S., & Mon-Williams, M. (1995). Natural problems for stereoscopic depth perception in virtual environments. *Vision Research, ACM 19*, 2731–2736.

Watt, S. J., Akeley, K., Ernst, M. O., & Banks, M. S. (2005a). Focus cues affect perceived depth. *Journal of Vision, 5*(10), 834–862.

Watt, S. J., Akeley, K., Girshick, A. R., & Banks, M. S. (2005b). Achieving near-correct focus cues in a 3d display using multiple image planes. In B. E. Rogowitz, T. N. Pappas, & S. J. Daly (Eds.), *Human vision and electronic imaging X* (Vol. 5666, pp. 393–401). San Jose, CA: SPIE.

Wilkie, R. M., Johnson, R. J., Culmer, P. R., Allen, R. J., & Mon-Williams, M. (2012). Looking at the task in hand impairs motor learning. *Journal of Neurophysiology, 108*(11), 3043–3048. Published ahead of print September 19, 2012, doi:10.1152/jn.00440.2012

Winn, B., Gilmartin, B., Sculfor, D. L., & Bamford, J. C. (1994). Vergence adaptation and senescence. *Optometry and Vision Science, 71*, 1–4.

33 Proprioceptive Adaptation and Aftereffects

Paul DiZio, James R. Lackner, and Roberto K. Champney

CONTENTS

33.1 INTRODUCTION

Until relatively direct interfaces with brain signals become widely available, users will have to play a physically active part in controlling the virtual environment (VE). This involves moving the head, eyes, limbs, or whole body. Control and perception of movement depend heavily on proprioception, which is traditionally defined as the sensation of limb and whole body position and movement derived from somatic mechanoreceptors. This chapter presents evidence that proprioception actually is computed from somatic (muscle, joint, tendon, skin, vestibular, visceral) sensory signals, motor command signals, vision, and audition. Experimental manipulation of these signals can alter the perceived spatial position and movement of a body part, attributions about the source and magnitude of forces applied to the body, representation of body dimensions and topography, and the localization of objects and support surfaces. The effortless and unified way these qualities are perceived in normal environments depends on multiple, interdependent adaptation mechanisms that continuously update internal models of the sensory and motor signals associated with the position, motion and form of the body, the support surface, the force background, and properties of movable objects in relation to intended movements.

An understanding of these mechanisms is important to VE users because VEs will sometimes inadvertently and sometimes purposely expose active users to never before encountered combinations of sensory and motor signals and environmental constraints. In many cases, this will lead to perceptual and motor errors in the VE until internal adaptation has been updated. When a user who has adapted to a VE carries the new adaptive state back into the normal environment, he or she will experience aftereffects, usually in the form of mirror image errors to those initially made in the VE (see Welch & Mohler, 2014; Chapter 25). These can include deviated execution of limb and whole body movements, proprioceptive errors, misestimates of externally imposed forces, and visual and auditory mislocalizations.

This chapter will use concepts derived from laboratory studies as a basis for interpreting measurements of proprioceptive side effects and aftereffects in VEs. It discusses what VE conditions will alter the state of sensory and motor adaptation, what components of the movement and orientation control system will adapt, what side effects and aftereffects will result from adaptive modification of these subsystems, and what contexts will evoke aftereffects. This survey emphasizes facets of proprioception relevant to VEs involving manual performance and whole body motion tasks. A balanced presentation is attempted of both the value and the limitations of laboratory-based conceptual distinctions for making the best predictions in practice.

33.2 PROPRIOCEPTION, MOTOR CONTROL, AND SPATIAL ORIENTATION

33.2.1 Multisensory and Motor Factors

In his analysis of sensation and motor control, Sherrington (1906) coined the term *proprioceptors* for the mechanoreceptors located inside body tissues that are primarily responsive to changes within the animal itself. Examples of these receptors are spindle organs in the muscles, Golgi tendon organs, unmyelinated Ruffini endings of the joints, and various sensor types in the viscera and vasculature, which were thought to be inaccessible to external energy but well attuned to body orientation, configuration, and movement. In contrast, systems whose sensors are located on the body surface and receive external stimuli were categorized as exteroceptive. The skin mechanoreceptors are located on the boundary of the environment and the body and have overlapping proprioceptive and exteroceptive roles. They are involved in discriminating form, texture, pain, temperature, and other complex properties of the external world as well as body configuration, motion, and orientation (especially for the fingers, lips, and tongue). Vision, audition, and olfaction are exteroceptive systems, but they can also be used to monitor motion and orientation of the limbs and the whole body. Sherrington noted that the vestibular labyrinth of the inner ear (semicircular canals and otolith organs) is developmentally derived from the exteroceptive system, but it acts in concert with the proprioceptive system because it is stimulated by pressure changes and shear forces in the sensory end organs usually brought about by head and body motion and changes in orientation. His analysis acknowledges that proprioception is multisensory and overlaps with other experiential domains.

Proprioception is also directly related to motor control. Internal signals corresponding to movement commands (or the command not to move) have been posited as contributing to proprioception and spatial awareness. They have been called *the desire* (Aristotle, 1978), *the perceived effort of will* (Helmholtz, 1925), *efference copy* (Von Holst & Mittelsteadt, 1950), or *corollary discharge* (Sperry, 1950). The idea that such efference copy signals are sufficient for conscious perception of movement or position lacks strong empirical support (see McCloskey & Torda, 1975, for a review).

For example, when humans are subjected to local ischemic nerve block or systemic neuromuscular block and attempt to move one of their paralyzed limbs, they do not perceive any illusory movement but instead have a sense of great heaviness (Lazlo, 1963; Melzack & Bromage, 1973). However, internal motor signals seem to interact with afferent signals in visual perception and oculomotor control. For example, fruitless attempts to move a paralyzed eye are accompanied by a great sense of effort and apparent motion and displacement of visual targets (Matin et al., 1972). In Von Holst and Mittelsteadt's (1950) terms, motor control and perception mechanisms must discriminate between *reafferent* sensory signals brought about by self-initiated movement and *exafferent* signals evoked by external events. One way for this to happen would be for an efference copy signal to cancel only the reafferent portion of the total afferent signal. This requires an internal calibration in which the afferent and efferent signals are represented in comparable units and in the appropriate proportions.

33.2.2 Sensorimotor Calibration

The correlations between afferent and efferent signals, however, are very complex. For example, there is no unique pattern of motor command and sensory feedback signals associated with even a simple single-joint arm movement. Raising the forearm 45° vertically or flexing it 45° in a horizontal plane involves very different efferent commands to and sensory feedback from the biceps brachii and brachialis muscles because the gravity torques are so different. The muscle command and feedback signals also differ when weights are wielded or machines are manipulated, even for the same movement amplitude and orientation of the arm. A constant efferent command will not produce the same force in fatigued and rested muscle. Self-locomotion or vehicular transport can generate accelerative loads requiring unique muscle forces for accomplishing an arm movement, relative to when the body is stationary. To complicate matters further, body acceleration activates vestibular afferent discharge, which through vestibulospinal pathways innervates skeletal muscles and modulates muscle spindle sensitivity (cf. Wilson & Melville-Jones, 1976). In order for the central nervous system to use an efference copy signal to cancel just the reafferent portion of the total afferent signal, it must first parcel the sensory signals into components due to intended movement and anticipated external loads and the efferent signals into voluntary and involuntary components.

In other words, calibration of perception and motor performance involves internalization of the relationship between combinations of sensory and motor signals in relation to body movement and orientation, external loads, environmental constraints, and internal conditions. Recalibration is required when conditions change. For example, alterations in body dimensions, strength, and sensory capacity throughout the life span, loss or distortion of motor or sensory function due to illness or injury, and migration to a novel environmental medium (land to water) all require adaptive accommodations. There is ample evidence that given sufficient time, humans can adapt to a wide range of stable sensory–motor rearrangements produced in the laboratory or in applied technological environments such as aerospace flight, traditional simulators, teleoperation, and VEs (Held & Durlach, 1991; Kennedy, Fowlkes, & Lilienthal, 1993; Lackner, 1976, 1981; Rock, 1966; Wallach, 1976; Welch, 1978; Chapter 31). Adaptation is often accompanied by aftereffects that degrade performance on return to the normal environment. Such aftereffects raise questions that must be addressed by VE system designers (see Table 33.1).

33.2.3 Muscle Spindles

Muscle spindle organs provide signals correlated with the full natural range of muscle length (limb position) and velocity (Harvey & Matthews, 1961). Spindles are connected in parallel with the muscle body so that they can be unloaded when the muscle contracts on receiving input from alpha motor neurons and loaded when the muscle lengthens. Stretch receptors embedded in each spindle's viscoelastic central region project two types of sensory fibers signaling both the spindle's length and rate of change of length (primary fibers) or only its length (secondary fibers). The polar

TABLE 33.1

Questions concerning the Nature of Aftereffects of VE Exposure

- What conditions will trigger adaptation?
- How does the internal calibration change?
- Can multiple states of calibration be maintained?
- In what contexts will different calibration states dominate?

These theoretical questions translate into practical ones for the VE context:

- Which VEs will elicit adaptation of proprioception, motor control, and perceived orientation?
- What kinds of aftereffects will be expressed, and under what conditions?

regions of each spindle are composed of contractile fibers innervated by gamma motor neurons that provide independent central regulation of their afferent sensitivity (see Figure 33.1). Heightened gamma innervation shortens the polar regions and thereby stretches the central region and produces a greater afferent discharge rate for the same overall muscle length. In contrast, quiescent gamma efferent activity relaxes the spindle poles, unloads the central region, and decreases the afferent activity while muscle length remains constant.

Until relatively recently, muscle spindle organs were thought to be important in reflexive motor control but not to influence the conscious awareness of limb position. Position sense was thought to arise from receptors in the joint capsules, but this view was weakened by evidence from humans with artificial hips. Although capsular receptors are absent in artificial joints, patients postsurgery were found nevertheless to have normal position sense accuracy of their hip (reviewed by Burgess & Wei, 1982). Muscle spindle activity is still present and is implicated in their hip position sensitivity.

In relaxed mammals, the muscle spindle firing rate is very low, about 0–4 impulses/s. Hagbarth and Eklund (1966) found that low-amplitude mechanical vibration applied to the skin over a muscle

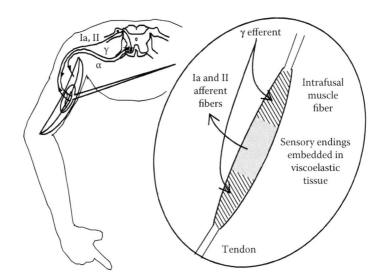

FIGURE 33.1 Muscle spindles reside in parallel with extrafusal muscle fibers, which are excited by α-motor neurons and provide the contractile force for movements. In this simplified diagram, the primary (Ia) and secondary (II) afferent fibers are shown with sensory endings on the central viscoelastic portion (gray area) of a single spindle, but in reality, type Ia and II fibers innervate different types of spindles. Primary fibers discharge in proportion to muscle length and rate of change in length, and secondary fibers discharge in relation to length. Afferent discharge can be modulated by γ efferent activity innervating intrafusal contractile fibers (striped areas) located at the poles of each spindle, which loads or unloads the central region containing the sensory endings but does not contribute to overall muscle force.

causes reflexive contraction of that muscle by entraining spindle afferent discharge to the vibration frequency and activating the stretch reflex. This phenomenon is known as the tonic vibration reflex (TVR). The vibration technique was extended by Goodwin, McCloskey, and Matthews (1972) to demonstrate that spindle afferent signals influence conscious sense of limb position and motion. In these experiments, the biceps brachii was vibrated and the participant's unseen forearm was restrained so that reflexive contraction would not shorten the muscle and unload the spindles. All participants reported feeling an increase in the elbow joint angle. The illusion could be quantified by having participants match the perceived angle of the elbow of their vibrated arm with the other arm and measuring the discrepancy in forearm positions (see Figure 33.2a). Similarly, triceps vibration elicited an illusory flexion of the forearm. Thus, an abnormally high discharge rate in the vibrated muscle is interpreted as lengthening of the vibrated muscle and this *lengthening* is referred to the joint(s) controlled by the muscle. If the same arm vibration experiment is done in 0 G, weightless conditions, the magnitude of the illusion is smaller than in normal 1 G conditions, and in an increased gravitoinertial

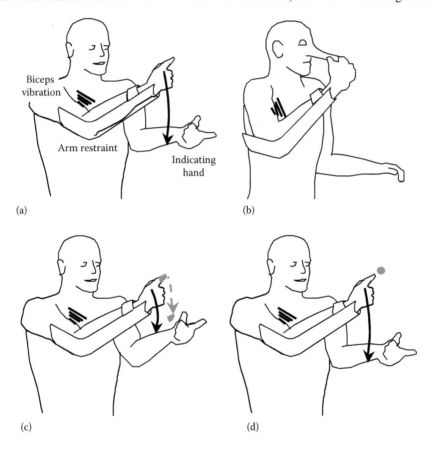

FIGURE 33.2 Illusions experienced during vibration of the restrained right biceps in darkness, with different visual cues and tactile contact conditions. (a) In complete darkness, biceps vibration elicits illusory extension of the restrained forearm (*solid arrow*), which can be measured if the participant points to the felt position of the hand of the vibrated arm with the other unrestrained, nonvibrated arm. (b) In complete darkness, the Pinocchio illusion, elongation of the nose, is experienced when it is grasped by the hand of the arm undergoing illusory elbow extension due to biceps vibration. (c) When just the index finger of the vibrated right arm is made visible with phosphorescent paint (*gray*) in an otherwise dark room, the participant sees the stationary finger move down (*gray, broken arrow*) and feels the forearm displace less than in complete darkness (*compare black arrow with panel A*). (d) A dim target light (*gray dot*) located 1 mm away from the fingertip of the restrained vibrated arm looks stationary, while the concurrent apparent downward displacement of the arm is as great as in complete darkness during biceps vibration.

force background, the illusion is larger (Lackner, DiZio, & Fisk, 1992). (Note: Gravitoinertial force is the resultant of gravitational and inertial acceleration on a mass. One G equals the force due to acceleration due to gravity at sea level.) The novel force background affects the otolith organs, which influence muscle tone by activating gamma motor neurons in the spinal cord, heightening the spindle afferent signal in high-force backgrounds, and decreasing it in 0 G. This mechanism is part of the vestibulospinal regulation of the antigravity musculature of the body. Clearly, muscle spindle afferent and efferent signals influence proprioception and motor control.

33.2.4 Role of Body Schema and Spatial Orientation

Vibration of the proper muscles can produce apparent displacement and motion of the head, arm, leg, and trunk (Lackner, 1988) in a dark room. The character of these vibratory myesthethic illusions demonstrates that perceptual interpretation of the afferent signal takes into account the anatomy and dimensions of the body. For example, a vibration-induced increase in biceps brachii spindle afferent discharge (1) affects the perceived angle of the elbow joint, which is spanned by the biceps brachii, (2) alters it in a direction (extension) that agrees with the bicep's role as a flexor, and (3) evokes apparent spatial hand motion in an arc consistent with the elbow joint's hinge-like motion and the length of the forearm. Some participants experience anatomically impossible hyperextension of the forearm during biceps vibration, whereas others after reaching the normal limits of extension perceive a paradoxical illusion of continuous motion without further displacement (Craske, 1977). Even nonmotile appendages can undergo apparent distortion. For example, Lackner showed that if the participant holds the tip of the nose while the biceps brachii is vibrated, he or she will perceive the nose to grow in length as the forearm extends away from the face, as illustrated in Figure 33.2b. The relatively stable, but modifiable, cortical representation of the body schema (topography and dimensions of the body) interacts with the moment-to-moment representation of body configuration based on peripheral muscle spindle and tactile signals. This conclusion is consistent with phantom limb experiences (Henderson & Smyth, 1948).

Achilles tendon vibration gives rise to a whole body movement backward relative to the support surface because of the TVR activation of the calf muscles (Eklund, 1972). When the Achilles tendons of participants restrained in a standing position are vibrated, they perceive forward body pitch, sometimes in full 360° circles, and even though they are physically stationary, they also show nystagmoid eye movements similar to what would be evoked by vestibulo-ocular reflexes during real body tumbling. Tactile cues from environmental surfaces can modify postural vibratory myesthetic illusions. Achilles tendon vibration normally leads to apparent pivoting about the ankles, but if the participant bites a rigid dental mold, then it may become the pivot point (Lackner & Levine, 1979). If a head-fixed target light is present during such illusions, it will be perceived as moving spatially relative to the current axis of rotation, either the feet or the mouth. Thus, limb position, body orientation, and object localization are overlapping, interdependent representations, each derived from different combinations of muscle spindle information about a particular appendage and tactile and vestibular signals about spatial orientation. Exposure to VEs will generally not alter muscle spindle signals directly, but often will affect the user's spatial orientation, which will influence the central interpretation of proprioceptive signals, resulting in perceptual and motor errors until adaptation occurs.

33.2.5 Bidirectional Interactions of Visual and Muscle Spindle Influences

Most VEs will have prominent visual displays so research on the registration of seen versus felt body position and orientation is very relevant. Lackner and Taublieb (1984) assessed this in experiments where they made either the whole hand or a single finger visible in a dark room by application of phosphorescent paint. When the biceps brachii was vibrated, participants felt the unseen forearm move and saw their finger or hand move as well but through a smaller distance (see Figure 33.2c). The magnitude of felt motion of the forearm was less with the finger visible than in complete darkness, and it was least with the hand visible. In other words, *visual capture* (Hay, Pick, & Ikeda, 1965) was incomplete because the physically stationary visual finger did not prevent participants from seeing and feeling motion of their finger. Proprioceptive capture was substantial; the finger was seen to move nearly as much as the forearm was felt to displace. If participants attempted to fixate their unseen hand as it underwent illusory downward motion, their eyes moved down. However, when they fixated on a visible finger that they perceived as moving down during vibration, their gaze remained spatially constant. The character and magnitude of vibration illusions changed in normal illumination conditions where participants could see their hand or finger in relation to the contours of the room, with the remainder of their arm being hidden below a screen. In this case, when the biceps brachii was vibrated, participants felt their unseen forearm move down but did not see the unoccluded finger or hand move. Felt motion of the arm was about 30% of what it had been in the dark room. These results indicate that multiple, interdependent, body-centered, and spatially centered representations of hand and arm position exist, which are influenced by the visual and muscle spindle inputs. Seen and felt hand positions are dissociable from each other and from motor responses. The strength of the bidirectional visual–proprioceptive influence relates to the amount of the limb that is represented. In VEs, the spatial correspondence of visual and real body positions, the level of detailed visual representation of the body and of the remainder of the virtual visual context, will determine the magnitude and nature of proprioceptive and motor errors.

33.2.6 Role of Tactile Cues in Unifying Muscle Spindle and Other Sensory Influences

When the fingertip is portrayed only by a single point of light or a punctuate sound source, there are still bidirectional influences and the role of tactile contact cues is paramount. For example, if in darkness a small target light is taped to the index finger of a participant's spatially fixed arm, the stationary light will appear to displace when the biceps brachii is vibrated, but the felt motion of the arm will be less than if no target were present (Levine & Lackner, 1979). If the two forearms are restrained in a horizontal plane and vibration is applied to the right and left biceps brachii, then the apparent distance between the fingertips will increase. If target lights are attached to the opposing index fingers, participants will see them get further apart as they feel their fingers move apart (DiZio, Lackner, & Lathan, 1993). During vibration, if physical contact of a target light and finger is broken by moving the light a millimeter or more away from the finger, then the illusory visual target movement and displacement will be abolished and the illusory felt movement of the unseen limb augmented (see Figure 33.2d). In this case, the lights seem to represent external objects instead of the tips of the fingers. During vibration eliciting apparent forearm extension or flexion, a participant will hear a sound source attached to their hand move and change spatial position in keeping with the change in apparent hand position. Breaking tactile contact with the auditory target abolishes its illusory movement during vibration. Taken together with the experiments in which various amounts of the hand were visible, it is clear that the strength of interaction among multimodal representations of self and target position depends on tactile contact cues, visual configural cues, and visual context cues.

These results have important implications for whether VEs should represent body parts with simple visual icons, high-fidelity visual representations, or haptic interfaces. In the natural world, perception and motor control are unified and accurate because there are adequate contextual cues defining how to group multisensory representations and adequate internal calibrations of how these signals should be combined. By contrast, VEs create alternate physical worlds that may require novel combinatorial rules. For example, if adequate visual form cues and tactile contact cues are present for the visual image to be perceptually grouped with the muscle spindle representation of hand position, then the visual and felt locations will influence each other. If either the visual or proprioceptive signal is inaccurate, then both perceptual representations will be biased. However, if only a visual icon is present and there is not a contact cue with the body, then the visual object will likely be perceived as an external target rather than a representation of the hand, and the spatial perception of each may be independent.

33.3 PROPRIOCEPTIVE ADAPTATION OF THE ARM TO VISUAL DISPLACEMENT

33.3.1 SENSORY REARRANGEMENT

Visual distortion is a likely scenario in VEs involving manual tasks, and exposures will likely be prolonged and require users to be active. The retinal image may be degraded or augmented by many VE factors such as the computer model of environmental objects, the technique for tracking the user's visual perspective, the graphical rendering system, and the optics of the display system. Retinal image rearrangement may be caused by inadvertent technological limitations or deliberate augmentation of reality. A typical unplanned distortion is where the visually displayed virtual position of an object does not correspond to its haptic or auditory virtual position because one or more of the display devices is inaccurate. Improper initial alignment and random slippage of a head-mounted display's (HMD's) focal axis relative to the user's optic axis are sources of such unintentional visual inaccuracy (see Chapter 3, this book). A system latency in rendering a user's moving hand in an HMD that occludes the real hand results in a dynamic spatial dissociation of the seen and real hand positions (Ellis, Young, Adelstein, & Ehrlich, 1999; Held & Durlach, 1991). A deliberate distortion is introduced, for example, when the visual image is magnified, as in the case of virtual microsurgery systems (Hunter et al., 1994). The laboratory studies reviewed below will illustrate that the type of rearrangement, activities performed, and opportunity for reafferent and exafferent sensory stimulation govern the nature, internal form, and specificity of adaptation that occurs with exposure to visual rearrangements. Measuring and understanding the side effects and aftereffects of sensory–motor adaptation require observing baseline performance before the rearrangement is introduced, the initial and subsequent performance during exposure, and the initial postexposure performance when normal conditions are restored (cf. Held, 1965).

Lateral displacement of the visual world by wedge prisms is an experimental analog of a possible visual rearrangement in a VE. The vast literature on this topic (cf. Howard, 1966; Rock, 1966; Welch, 1978, for reviews) originates from Helmholtz's (1925) demonstration that small azimuthal displacements by prisms of the visual field can be adapted to in a matter of minutes. If a participant looks through base-left wedge prisms without sight of the hand, any object chosen as the target for a reaching movement will appear to the right of its true position, and a movement will be directed accordingly to the right. The prisms will make the hand as it comes into view look like it is moving more to the right than intended, away from the target, and typically participants will steer the hand nearer to the target in midcourse, but the movement end point will tend to be right of the target. Participants will see this gap and may also describe with surprise not seeing their arm where they feel it. It takes only 10–20 reaches to the same target under these conditions for participants to hit it and to move straight again, but when the prisms are removed, they will reach too far to the left initially. Figure 33.3a illustrates typical reaching paths before, during, and after exposure to wedge prisms. Improvement of performance while the world is viewed through prisms and the aftereffect when they are removed are measures of adaptation. Similar improvement and aftereffects are seen when lateral displacements are purposely introduced in a VE (Groen & Werkhoven, 1998).

33.3.2 INTERNAL FORM OF ADAPTATION

Harris (1963) argued that when one can see only his or her arm and external targets in a featureless background, adaptation is achieved by an internally modified position sense of the arm. The participant eventually feels the arm where he or she sees it. For example, after adaptation to base-right prisms, the participant will reach to the target's true position but see and feel the arm to the right of that position. The felt arm position then does not correspond to the real position, but visual and proprioceptive perceptions are unified and reaching movements are directed accurately to the target.

Harris recognized but rejected other possible forms of adaptation under these conditions. For example, he rejected the possibility that the adaptation is simply a conscious process of reaching to

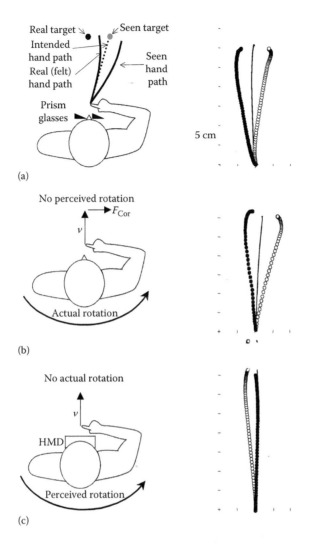

FIGURE 33.3 The panels on the left are schematics of different sensory–motor rearrangements that evoke reaching errors and with continued exposure different forms of adaptation. The panels on the right are plots (*top view*) of finger trajectory before the rearrangement (*solid line*), during the initial reach attempted in the presence of the rearrangement (*open symbols*), and during the initial reach attempted on return to normal conditions (*filled symbols*). (a) Wearing prisms that shift a visual target to the right causes the initial reach to deviate to the right and to curve slightly due to corrections made when the hand comes into view. With many attempts, participants learn to make straight accurate reaches (not shown) by modifying their internal representation of felt arm position (see text). Persistence of the adaptation when the prisms are removed causes leftward end point and curvature errors, mirror symmetric to the initial reaches made during prism exposure. (b) When participants are unaware they are rotating, they do not compensate for the Coriolis forces (F_{Cor}) their movements (v) generate and initially make reaching errors in the direction of the Coriolis forces (open symbols). With practice, they modify their motor commands as necessary to restore their baseline trajectory. When rotation ceases and they again feel stationary, they make leftward end point and curvature errors (filled symbols). The pre- and postexposure reaching errors are the same in (a) and (b), although the internal form of adaptation differs. (c) Stationary participants feel like they are rotating when viewing a moving visual scene presented in an HMD. Their reaching movements deviate in the direction opposite to the Coriolis force, which would be present if they were actually rotating (open symbols). With practice, they regain straight accurate movement paths by learning not to compensate for anticipated Coriolis forces. This motor adaptation is specific to the self-rotation context simulated by the HMD and does not carry over to a normal stationary context; consequently, there are no aftereffects when the HMD scene is again stationary (filled symbols).

the left of where the target appears because when the prisms were removed, his participants showed negative aftereffects instead of reverting to their baseline performance. Second, the adaptation is not an internal change in visual perception. Such a change would mean that a midline target optically shifted right would in the course of adaptation come to be perceived as straight ahead, and when the prisms are removed, it would be registered to the left of its true position and any movement aimed at it should deviate to the left. Contrary to this, Harris found that when prisms were removed, an adapted participant pointed accurately with the unadapted hand; there was no intermanual transfer. Participants positioned their adapted hand to the left of the unadapted one when asked to align the two hands in azimuth in darkness. Also contradicting a visual shift was the finding that after the prisms were removed, participants made comparable errors pointing to a visual target or to an unseen sound source. An alternate explanation to Harris's is motor adaptation. Participants could, for example, change the direction in which they reach to targets without a concurrent change in the felt sense of limb position. Motor adaptation will be discussed in Section 33.4.

The internal form of adaptation is influenced by activities performed and feedback obtained while wearing prisms and how much of one's body and the world is visible. An interesting case is walking around but keeping the arm out of sight while wearing rightward-displacing prisms (Held & Bossom, 1961). At first, participants bump into things but are soon able to get around. When the prisms are removed, the participants initially point incorrectly, leftward, with both arms to visual targets. They also adopt a resting head posture that is deviated in the direction of the optical displacement. These facts have been interpreted as meaning that adaptation in this context is achieved by an internal recalibration of perceived head position on the torso (Harris, 1965). In the paradigm where participants just sat and reached with one arm, only that arm had to be remapped to compensate. This illustrates that the nervous system tends to adapt in the way most specific to the conditions encountered during exposure to rearrangement. More extensive reviews on this topic are available (Dolezal, 1982; Lackner, 1981).

In the two cases presented earlier, the sensory rearrangement is the same but the resulting form of internal adaptive shift differs. There can be an arm–torso recalibration when the participant just sits and points to a single target in a featureless field and a head–torso recalibration when the participant walks about in the normal world. This means that detecting and understanding VE side effects and aftereffects must take into account the physical distortions introduced by the VE, the nature of the visual field, and the tasks required of the user. There is enough evidence to indicate that the form of adaptation to a sensorimotor rearrangement depends on fine details of task characteristics, but not enough to make accurate predictions for every case imaginable in practical VEs. Several examples not explicitly covered by the reviewed studies are VEs where the visual position of the user's hand and the whole visual scene are displaced, VEs where the virtual visual objects have real or virtual counterparts, and VEs where the hand is represented as a symbolic visual icon instead of realistically.

33.3.3 CONDITIONS NECESSARY FOR ADAPTATION TO OCCUR

Understanding what factors are necessary for adaptation to take place is important for predicting proprioceptive side effects and aftereffects from VE exposure. Held and colleagues (Held, 1965; Held & Bossom, 1961) showed that active reaching or locomotory movements during visual rearrangement could elicit adaptation, but a laterally displaced view of one's passively moved arm or sight of the world during passive transport did not generate adaptation. This led to the idea that efference copy signals are required for adaptation. Held developed a model in which efference copy signals were stored along with their correlated reafferent sensory signals so that any active movement would call up the normally associated visual reafferent signal for comparison with the actual one. In the reaching and locomotion paradigms, discrepancies between the reactivated and current reafferent visual patterns were thought necessary for adaptation to occur. A practical implication is that greater activity within a VE should speed up adaptation.

An alternative point of view is that active movements are only superior because they enhance proprioception (Paillard & Brouchon, 1968). Consistent with the idea that active movement is not necessary, Wallach, Kravitz, and Lindauer (1963) demonstrated that partial adaptation of reaching movements to visually displaced targets occurs if an immobile observer simply views the rest of his body through displacing prisms for 10 min. Lackner (1974) went further, showing that a discrepancy between actual visual feedback and the usual feedback associated with a voluntary movement is not sufficient for adaptation to occur, but a visual–proprioceptive discrepancy is necessary and sufficient for eliciting adaptation to laterally displacing prisms. When participants reached without sight of their arm to visually displaced dowels and contacted hidden, vertically aligned, similarly shaped extensions of these targets, there was no visual feedback about the arm to compare with the visual feedback normally associated with the executed movement. Adaptation still occurred in this condition because there was a discrepancy between the visual and felt target position. Adaptation did not occur if the visual–proprioceptive discrepancy was eliminated by having participants point to the same array of prism-displaced dowels whose hidden lower halves were laterally offset to match the optical displacement. The results emphasize the importance for adaptation of sensory discordance as well as active control. They also demonstrate that adaptation is enhanced by establishing through fingertip contact an association between the unseen arm and visual targets. In VE systems, finger contact can contribute to the fidelity of the synthetic experience if vision and haptic interfaces are in register or create side effects and aftereffects if they are discrepant.

33.3.4 RETENTION AND SPECIFICITY OF ADAPTATION

To assess retention of adaptation of reaching errors caused by laterally displacing prisms, Lackner and Lebovitz (1977) had participants participate in two adaptation sessions 24 h apart. Each session had pre- and postexposure periods of reaching to virtual visual targets without sight of the arm and a prism exposure period in which participants reached to the same loci with their arm in view. A surprising finding was that in the preexposure period of the second session, reaches were deviated in the direction of the aftereffects from the previous day, although none of the participants reported any difficulty with visuomotor control between sessions outside the laboratory. This demonstrated a long-term aftereffect and raised the possibility that it was a context-specific aftereffect. To further evaluate retention and context specificity, Yachzel and Lackner (1977) gave participants six adaptation sessions over a 4-week period. The sessions were similar in that participants pointed to visual targets with their arm under a screen before and after exposure to visual displacement in which they reached with sight of their arm. Tests were conducted in two different sets of apparatus requiring different arm movements. Retention of adaptation from session to session was seen in the form of preexposure baseline shifts across the first five sessions, which were spaced 2 or 3 days apart. The size of the long-term aftereffect did not depend on what apparatus the participant was tested in. Aftereffects appeared when the participants were tested without being able to see their arm and never were noticed during daily activities. Diminished aftereffects were still evident in the sixth session, which was delayed for 2 weeks relative to the fifth. These results indicate that proprioceptive adaptation to displaced vision can be retained for long periods and are not necessarily context specific. The adaptation retained does not reveal itself in daily activities unless visual cues about the arm are reduced. Long-term aftereffects from simulator and VE exposure are likely caused by long-term retention of adaptation, but more research is required to understand what contexts will evoke aftereffects and what sensory cues will hold them at bay.

33.4 ADAPTATION TO ALTERED FORCE BACKGROUNDS

The notion of motor adaptation introduced in Section 33.3.2 is crucial for understanding side effects and aftereffects in real and virtual environments involving novel force backgrounds. Motion-coupled VEs rearrange the external force environment. For example, in VEs involving real body motion or visual portrayal of body motion, the background gravitoinertial forces on the body are different from the forces that would be present if the body were physically moving in the experienced fashion. This is a situation where motor adaptation occurs and individuals need to reprogram their limb movements to achieve accurate control for the current force background. This can produce aftereffects when the individual leaves the VE. The situation is very different when a novel real or virtual tool or machine is introduced. In this case, individuals make errors until they learn the properties of the manipulated device. Such tasks do not lead to aftereffects outside the specific context of that device.

33.4.1 Motor Adaptation to Coriolis Force Perturbations in a Rotating Room

Traditional vehicle motion simulators often have motion bases to try to mimic features of the gravitoinertial force backgrounds of moving vehicles. For example, a rotating room generates centripetal acceleration that simulates artificial gravity, but it also generates a Coriolis force on any object moving nonparallel to the spin axis of the room. The Coriolis force is only present when an object is moving in relation to the rotating room and acts perpendicular to the direction of object motion, according to the cross-product rule: $F_{cor} = -2m(\omega \times \nu)$, where m and ν are the mass and linear velocity, respectively, of the object and ω the angular velocity of the room.

When an occupant of a room rotating counterclockwise reaches forward, a rightward Coriolis force is generated on the arm deviating it rightward. Both the movement's path and end point are displaced relative to the prerotation pattern of straight movements directly to the target. Subsequent reaches during rotation return quasiexponentially, within 10–20 trials, to prerotation straightness and accuracy. A new set of motor commands is being issued in order to move the hand straight to the target. When rotation stops and Coriolis forces are absent, reaching movements show curvature and end point deviations in the direction opposite those of the initial perturbed movements (Lackner & DiZio, 1994).

This pattern of pre-, per-, and postrotation movements resembles the pre-, per-, and postexposure phases of a prism displacement experiment (compare Figures 33.3a and b), but there are important differences. Adaptation to rotation involves motor remapping instead of the proprioceptive shifts that underlie some forms of adaptation to prism spectacles. Adaptation to rightward prism displacement can make movements that go straight ahead to a midline target feel like they are going rightward, and it may cause participants to make errors aligning the adapted hand with the unadapted one if it is hidden. By contrast, the true trajectory of the arm is experienced throughout the pre-, per-, and postexposure phases of adaptation to rotation. Participants who adapt to reaching with one arm during rotation also can accurately align their left and right fingertips.

33.4.2 Motor Adaptation and Force Perception

Another special feature of motor adaptation to rotation is the alteration of force perception that accompanies it. Participants report their first movement during rotation as being deviated by a magnetic-like pull in the direction of the Coriolis force, whereas no unusual force is perceived during movements whose visual paths are perturbed by prisms. When participants adapt fully to rotation, they no longer can feel the Coriolis forces that are still present during their movements, even if their attention is called to them. They report that whatever had initially perturbed their arm is now gone and they can produce the desired movement with the same effort as before the perturbation. When rotation stops and there are no Coriolis forces during movements, participants report feeling a force on their arm that is the mirror image of the Coriolis force they had adapted to. The relationship of actual to perceived force is modified during adaptation because of an internal calibration mechanism that interrelates force feedback signals, position, and velocity signals and motor commands. Informal observations clearly show that adaptation to rotation alters the feel and control of objects and surfaces that are handled following return to a normal stationary environment. VEs that alter the force environment or that involve visual simulation of body acceleration without the normal concomitant gravitoinertial forces will cause illusions in the feel of tools and affect the manual control of a vehicle or other machines.

33.4.3 CONTEXT-SPECIFIC MOTOR ADAPTATION AND AFTEREFFECTS IN DIFFERENT FORCE ENVIRONMENTS

Participants who sit quietly in an enclosed rotating room turning at a constant velocity feel after about 60 s like they are in a completely normal, stationary environment. This is because the angular velocity-sensitive semicircular canals have had time to return to their resting discharge level after acceleration to constant velocity, and the room is fully enclosed so there are no visual flow cues about rotation. Thus, participants who adapt their reaching movements in this situation learn to associate making Coriolis force compensations and receiving Coriolis force feedback with an internally registered nonrotating context. These recalibrations are carried over to the postrotation period, which also is internally registered as being a normal stationary environment, and cause aftereffects. If after adaptation to rotation the rotating room instead of being stopped is accelerated to twice the initial speed, then after a minute, the participant will again feel stationary and when reaching for the first time will make renewed end point and curvature errors of the same magnitude and in the same direction as the first movements at the previous speed (Lackner & DiZio, 1995). The Coriolis forces are greater at the new speed, but the motor mapping from the lower speed is carried over because the internally registered nonrotating state is the same.

Understanding whether adaptation to one rotation speed will carry over to a different speed requires the converse of the conditions provided by the rotating room where participants are rotating but feel stationary. Participants who are actually stationary can experience virtual rotation by viewing a moving visual scene in an HMD. The first reaches of participants experiencing constant velocity counterclockwise rotation and displacement will be deviated leftward in path and end point (see Figure 33.3c; Cohn, DiZio, & Lackner, 2000). The magnitude of the reaching errors is proportional to the perceived speed of self-rotation. If the participants had actually been rotating counterclockwise, there would have been a rightward Coriolis force when they reached. Their leftward errors show that they had anticipated and generated muscle forces to resist the expected but absent Coriolis forces. These participants also report feeling a phantom leftward force on their arm, which has the characteristic bell-shaped profile of a Coriolis force. When repeated reaches are made during virtual rotation, the reaches become straighter and progressively more accurate, and the *force* perceived to be deviating the arm vanishes. When the visual scene is again stationary and participants feel stationary, their first postexposure reaches feel normal and go straight to a target. That is, there is no aftereffect.

These results show that the motor plan for a forthcoming reaching movement compensates for the Coriolis forces normally generated at the currently registered speed of self-rotation. In other words, we normally maintain multiple motor adaptation states that are context specific for registered body motion. Experimental alteration of the relationship between Coriolis forces and registered body speed alters the motor compensation only for the body rotation speed at which movements are practiced and is not carried over to other speeds. Thus, motor adaptation is specific to the registered context of body motion where aberrant forces were experienced. Not enough is known to allow prediction of how faithfully a moving base training simulator must reproduce an operational environment in order for training to transfer. Virtual training environments that simulate body motion with just a dynamic visual scene are unlikely to cause motor aftereffects in everyday life, but they may well produce adaptation in which participants cease to compensate or anticompensate in terms of postural and movement control for inertial forces that will actually be present in the operational moving environment. That is, they could maladapt participants to the operational context.

33.4.4 MOTOR ADAPTATION TO ENVIRONMENTS VERSUS LOCAL CONTEXTS

The rotating room and virtual rotation provide environmental contexts in which individuals carry out all their actions. Real and virtual tools and machines create force-reflected feedback or force fields that are local contexts a user may interact with in a limited fashion. For example, humans can use a planar robotic linkage to control a cursor on a video screen in the presence of external forces generated by torque motors on the handle. Shadmehr, Mussa-Ivaldi, and Bizzi (1993) created a force field that resembles the Coriolis force field in a rotating room, such that the manipulandum pushed the hand perpendicular to its current velocity. Their participants' movements were initially perturbed, but after many hundreds of trials, straight paths ending at the target characteristic of movements with a null force field were regained. This contrasts sharply with the complete adaptation participants achieve in 10–20 movements to Coriolis force perturbations of their reaching movements in a rotating room. Thus, learning an internal model of an electromechanical device takes much longer than recalibrating one's own unfettered movements. The nature of force perception also differs sharply during self-calibration versus internalization of a machine's force properties. As described earlier, perturbing Coriolis forces become perceptually transparent when motor adaptation of free reaching movements to rotation is complete. However, participants interacting with a manipulandum can still detect and describe the forces it applies even after they have learned to resist them and to move their arm in the desired path to the target. The implication of this is that learning a local force field by interacting with a real or virtual machine will not cause aftereffects in any context outside the confines of the machine. By contrast, learning to move one's arm will affect perception and performance in any local context embedded in an environment that provides the same stimuli about self-motion as the environment in which adaptation was acquired.

33.4.5 ROLE OF TOUCH CUES IN MOTOR AND PROPRIOCEPTIVE ADAPTATION

Cutaneous contact cues during a movement and at the terminus of the movement contribute to perception and adaptation of limb position and force. Continuous cutaneous contact cues throughout movements are prominent differences between Coriolis force perturbations in the rotating room and force field perturbations produced by a manipulandum. When a manipulandum perturbs a movement, the muscle spindles, Golgi tendon organs, and vision can signal that the movement deviated from the intended path, and the cutaneous mechanoreceptors of the hand signal the continuous presence of a local external force. There is a systematic correlation of the cutaneous force profile and the compensatory muscular forces that need to be learned. The nervous system, following the principle of making the most parsimonious change possible, can simply learn the dynamics of the perturbing object. In adaptation to Coriolis force perturbations, there is no local cutaneous stimulation because the Coriolis force is a noncontacting inertial force applied to every moving particle of the limb. In the absence of an external agent contacting the limb, the most specific form of adaptation is an alteration in motor control of the exposed limb. This is because the central nervous system recognizes a situation requiring motor recalibration when it detects an error in movement path without an external obstruction while the movement is in progress.

The transient cutaneous stimuli occurring when the finger lands on a surface at the end of a reaching movement also influence proprioception and motor adaptation in real and virtual environments. Coriolis force–induced end point and curvature deviations of visually open-loop reaching movements in a rotating room are eliminated by adaptation within 10–15 movements if the reaches end on a smooth surface. However, only 50% of the initial end point error is eliminated before performance reaches a steady asymptote if the reaches end with the finger in the air. In other words, if terminal contact is denied, participants will not fully adapt. This pattern shows that information obtained from finger touchdown provides information about limb position errors, which is critical for adaptation to rotation. The source of this information is the direction of shear forces generated in the first 30 ms after a reaching movement contacts a smooth surface. These shear forces are systematically mapped to the location of the finger and code where the finger is relative to the body (DiZio, Landman, & Lackner, 1999). These shear forces are about 1 N in magnitude. When participants reach in the air to objects on a virtual instrument panel, their end point variability is greater than when reaching to a similar real panel (DiZio & Lackner, in development). The variability is greatest if the virtual panel is presented only in the visual mode and the reaches end in midair. It is greatly reduced by the addition of a real contourless surface in the same spatial plane as the visual

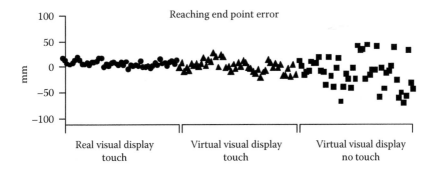

FIGURE 33.4 Plot of end point errors of repeated reaches made without sight of the arm to a single target on a real or virtual horizontal work surface. Variability is least in the first set of movements, where the target is a light-emitting diode (LED) embedded in a smooth sheet of Plexiglas with no distinguishing marks indicating the target location. Variability is greater in the second set of movements, which are aimed at a virtual target on a virtual surface programmed to coincide with the spatial location of a real Plexiglas work surface that the participant contacts at the end of each movement. The same virtual target and surface are presented without the real surface in the last set of reaches, and the variability increases markedly in the absence of physical contact. Reintroducing the real surface (not shown) restores low variability immediately.

virtual panel (see Figure 33.4). Thus, minimal haptic cues, just a flat surface in the proper plane, can improve proprioception, reduce movement errors, and enhance the usability of VE visual interfaces.

33.5 PERCEPTION OF LIMB AND BODY MOTION DURING POSTURE AND LOCOMOTION

Control of posture and movement are highly integrative, multimodal processes, requiring sensitivity to complex environmental constraints. Touch cues play a major role in control and perception of whole body movement, the apparent stability of the environment, and attributions of causality.

33.5.1 TOUCH STABILIZATION OF POSTURE

Fingertip contact with environmental surfaces has a stabilizing influence on standing posture. If participants standing on one foot or heel-to-toe in darkness hold their index finger on a stable surface with a force of about 0.4 N (about 41 g), their body sway amplitude is cut in half relative to not touching (Holden, Ventura, & Lackner, 1994; Jeka & Lackner, 1994). This level of force is too low to provide mechanical stabilization, but it corresponds to the maximum sensitivity range of fingertip sensory receptors (Westling & Johansson, 1987). The horizontal and vertical forces at the fingertip fluctuate with a correlation of 0.6 to both body sway and ankle electromyography (EMG) activity, but lead EMG by about 150 ms and body sway by 250–300 ms. Light touch even attenuates body sway when the eyes are open. The stabilizing influence of touch is lost if sensory–motor control of the arm is disrupted by vibration of brachial muscles, but with the arm functioning normally, touch can stabilize the body even when excessive sway is induced by ankle muscle vibration (Lackner, Rabin, & DiZio, 2000). Patients with no vestibular function who cannot stand in darkness for more than a few seconds can maintain stance for as long as desired and are as stable as control participants with eyes closed if allowed light touch of the finger (Lackner et al., 1999). These findings indicate that the finger–arm system functions as an active proprioceptive probe providing sensory information about body position and velocity. This stabilization system works with nested sensory–motor loops. A finger to brachial sensory–motor loop stabilizes the finger relative to the surface and minimizes the force changes at the fingertip. These residual fingertip force fluctuations activate the leg muscle to stabilize posture better than is possible with ankle proprioception, vision, or vestibular signals alone.

Light contact cues with a simple surface are an effective way to stabilize body sway in VEs where participants are free standing. An HMD and head-tracking system that introduces temporal distortions of visual feedback when head movements are made induces severe postural instability, and light touch suppresses this side effect (DiZio & Lackner, 1997). Postural aftereffects also occur upon return to a normal environment resulting in further postural disruption. Light touch stabilizes posture and suppresses aftereffects in such situations.

33.5.2 Touch and Locomotion

A seated individual holding his or her hands or feet in contact with a rotating floor or railing will experience self-rotation in the opposite direction (Brandt, Buchele, & Arnold, 1977; Lackner & DiZio, 1984). Actively pedaling the freewheeling floor while seated or turning the railing with a hand-over-hand motion makes the experience very powerful. Pedaling movements made without surface contact do not elicit any experience of self-motion. These demonstrations illustrate that in normal bipedal locomotion, contact with the floor is very important.

The pattern of ground reaction forces during locomotion is not related to body motion through space in a univariant way. For example, walking down a hill, one is progressing forward but pushing backward. An experimental situation used by Lackner and DiZio (1988) to assess the interactions among surface contact, leg movements, and whole body motion is illustrated in Figure 33.5. It is basically a VE in which the visual display and the substrate of support can be independently manipulated. A participant holding onto the world-fixed handlebar and walking in place on the backward rotating floor experiences forward motion through space. The experience is compelling and immediate if the chamber walls also are rotated backward, but it occurs in darkness (Bles & Kapteyn, 1977) and even if the stationary walls are in plain view. In the latter case, the motor and somatosensory signals from walking in place capture the visual scene, which will appear to be dragged along (Lackner & DiZio, 1988).

If the visual scene is moved backward twice as fast as the floor, a participant who is walking in place will feel the handlebar pulling him forward and sense an increase in body velocity through space. Some participants report an illusory elongation of each leg at the toe-off phase of the step cycle,

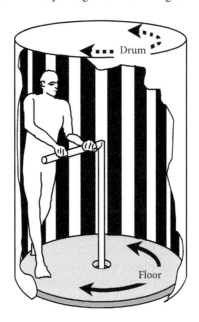

FIGURE 33.5 Schematic of the apparatus used to uncouple visual motion, inertial motion, and overground motion during treadmill locomotion.

and others sense that the floor has become rubbery and bounces them forward. If the visual scene motion is reversed, participants stepping forward in place will report either that they are moving backward through space and making backward stepping movements or the steps, which usually propel them forward, are now propelling them backward. These effects powerfully demonstrate that afferent signals about limb movements, the body schema representation, internal calibration of body motion, and apparent environmental stability are all interrelated.

33.6 AFTEREFFECTS AND READAPTATION TO THE REAL WORLD

It is clear that proprioceptive and vestibular aftereffects may be present after interacting with a VE. Two central issues with the management of these aftereffects are understanding the time course of readaptation and the potential for speeding the return to normal functioning following exposure. As evidenced by the preceding discussion, there has been a significant effort dedicated to identifying and quantifying these aftereffects (e.g., DiZio & Lackner, 1997; 2002; Draper, 1996; Kennedy & Stanney, 1996; Stanney & Kennedy, 1997; Stoffregen, Draper, Kennedy, & Compton, 2002; Wann & Mon-Williams, 2002). Yet few if any efforts have sought to address the exploration and development of protocols to speed a return to normal functioning.

When exiting a VE, the incidence of aftereffects implies that some form of adaptation has occurred to that environment. In essence, the plasticity of the human body has allowed the individual to adapt to the VE and upon exit carry with him those adaptations. In this case, the *real world* may be considered the *novel* environment to which the individual must now adapt to. Thus, a period of readaptation is required to calibrate the afferent signals, which may create issues as an individual reengages normal activities. If this readaptation is to take place while one engages in activities requiring normal proprioceptive or vestibular functioning (e.g., surgery, driving, flying aircraft), it may result in degraded performance or lead to potentially hazardous conditions. Two approaches are possible to address this concern: natural decay and active readaptation exercises.

Natural decay involves refraining from activity, particularly from potentially hazardous activities. Yet this approach does run the risk of prolonging the readaptation period as the sensory systems do not have the opportunity to actively readjust. The natural decay approach involves the utilization of a natural decay process by which given sufficient amount of recovery time an individual returns to normal functioning (McCauley & Sharkley, 1992). Yet the amount of time necessary for this natural decay process to take place may vary widely. It has been suggested that this process may be proportional to the amount of exposure (Baltzley, Kennedy, Berbaum, Lilienthal, & Gower, 1989), yet it has been observed that maladaptations in simulators and VEs have lasted for longer than the duration of the exposure (Baltzley et al., 1989; Champney et al., 2007; Gower & Fowlkes, 1989; Gower, Lilienthal, Kennedy, Fowlkes, & Baltzley, 1987; Kellogg, Castore, & Coward, 1980). In fact, such prolonged aftereffect durations have prompted the creation of regulations regarding the use of immersive simulators (OPNAVINST 3710.7T [Section 8.3.2.17] Navy, 2004).

Active readaptation exercises involve the use of targeted activities specifically aimed at recalibrating the sensory systems affected by the VE exposure. For instance, Kinney, McKay, Luria, and Gratto (1970) showed how participants actively engaged in a hand–eye coordination task (chess game) showed greater adaptation to an underwater environment than those who swam freely or engaged in target pointing. Fregley (1974) evaluated the possibility of improving postural stability on vestibularly impaired individuals by using a rail walking exercise. It was hypothesized that by walking and attempting to maintain their balance, participants would receive visual, vestibular, and hand–eye coordination signals from their surrounding environment. The results suggested that this task may support adaptation. Similarly, Gordon, Fletcher, Melville-Jones, and Block (1995) have shown that the relationship between afferent signals about limb movements, the body schema representation, internal calibration of body motion, and apparent environmental stability can be recalibrated by walking in place on a circular treadmill for an extended period. After doing so for an hour in a visually normal environment, participants walk in a curved arc when trying to maintain a straight line on solid ground in darkness. When passively pushed in a wheelchair, eyes closed, they can differentiate straight and curved paths as well as they can before the treadmill exposure, indicating vestibular function is not changed. These investigators concluded that the aftereffects are due to adaptive recalibration of a *podokinetic system*, which through ground contact, leg proprioception, and motor copy signals provides a representation of trunk rotation relative to the stance foot. When forced not to turn by guide rails after prolonged stepping around in place, participants feel they are turning in the other direction, and they exhibit nystagmoid eye movements consistent with their perceived direction and rate of turning, not their actual body motion (Weber, Fletcher, Melville-Jones, & Block, 1997). Allowed visual feedback, the participants can walk in their desired path with no ocular nystagmus. This is an example where potentially dangerous aftereffects due to internal recalibration of proprioception in the exposure environment can be masked by the appropriate sensory information.

More recently, Champney et al. (2007) conducted a study under which over 900 participants were exposed to an HMD-based VE under a variety of conditions: VE exposure durations (15, 30, 45, and 60 min), user control (3 or 6 degrees of freedom), scene complexly (low or high texture and ceiling height), and readaptation exercises (natural decay, rail walking, or peg-in-hole). Both eye–hand coordination and postural stability were measured via a target pointing task and a tandem Romberg position (see Kennedy & Stanney, 1996) with computerized instruments. Their results showed that regardless of VE exposure duration, only about half of participants showed aftereffects (59.4% for hand–eye coordination, 48.6% for posture). Further, they showed that aftereffects were not proportional to the amount of exposure and present for longer than their amount of exposure. In their study, Champney et al. (2007) also evaluated the efficacy of readaptation exercises to facilitate a return to normalcy (natural decay, rail walking, and peg-in-hole). Their results showed that participants engaged in a peg-in-hole task for 5 min postexposure had reduced aftereffects in eye–hand coordination, yet these gains were neither sustained nor sufficient to return them to their pre-VE state; the rail exercise failed to show significant gains in postural stability as compared to natural decay. This is of interest because it implies that individuals are affected differently. Aftereffect durations are not predictable by exposure duration and thus must be monitored to ensure readaptation has occurred. Further, the sensory adaptations that occur in VEs may require different or more prolonged readaptation exercises to sufficiently and more efficiently return individuals to a state of normalcy after VE exposure.

33.7　CONCLUSIONS

Many VEs will intentionally or inadvertently introduce users to sensorimotor rearrangements that will result in side effects, followed by adaptation and aftereffects or maladaptive transfer on return to a normal environment. Proprioceptive adaptation is one form of adaptation associated with potential VE side effects and aftereffects. Two important facets of proprioception that can undergo adaptive modification are the sense of position and orientation and the sense of force or effort. Predicting and measuring the manifestations and underlying form of adaptation require understanding that it is not a unitary phenomenon. Proprioception involves the interplay of afferent and efferent signals about limb and body position and motion, internal representations of the body schema, and spatial orientation and representations of environmental constraints. All of these variables are labile and can undergo long-term adaptive changes in relation to the others. Reaching errors in a VE that look similar on the surface can have different causes that lead to diverse forms of adaptation and aftereffects. For example, virtual visual displacements, mismatches of the virtual gravitoinertial force environment, and novel contact force fields cause adaptation of proprioception, of motor control, and of internal models of objects, respectively. Understanding the cause and form of adaptation in a specific VE context will help predict ways of enhancing adaptation; for example, when more user activity versus sharpening, the sensory discordance will help. Principles governing the specificity of adaptation are key for understanding what aftereffects will occur when a user leaves the VE and either goes about daily life or engages in the real-world operational task that was simulated. Such principles are scarce, but one important factor is feedback from contact cues during movement. Continuous and terminal cutaneous signals contribute to neural computations that partition the net force environment into functionally relevant components. For example, in adaptation of reaching movements, cutaneous signals are critical for determining whether the motor system will recalibrate or the representation of a tool's properties will be updated. Motor adaptation in a VE will carry over to any task performed with the exposed limb, but learning an internal model of a device will not deleteriously affect performance on dissimilar devices. Touch cues and visual cues can also mask potentially dangerous long-term aftereffects that appear in impoverished conditions or specific contexts.

ACKNOWLEDGMENTS

This material is based upon work supported in part by the National Science Foundation (NSF) under Grant No. IRI-9624968, the Office of Naval Research (ONR) under Grant No. N00014-98-1-0642, NAWCTSD contracts N61339-96-C-0026, and the National Aeronautics and Space Administration (NASA) under Grants No. NAG9-1037, NAG9-1038, and NAS9-19453. Any opinions, findings, and conclusions or recommendations expressed in this material are those of the authors and do not necessarily reflect the views or the endorsement of the NSF, NAWCTSD, ONR, or NASA.

REFERENCES

Aristotle. (1978). *De motu animalium* (pp. 38–42, M. C. Nussbaum, Trans.). Princeton, NS: Princeton University Press.

Baltzley, B. A., Kennedy, R. S., Berbaum, K. S., Lilienthal, M. G., & Gower, D. W. (1989). The time course of postflight simulator sickness symptoms. *Aviation, Space, and Environmental Medicine, 60*(11), 1043–1048.

Bles, W., & Kapteyn, T. (1977). Circular vection and human posture: 1. Does the proprioceptive system play a role? *Agressologie, 18*, 325–328.

Brandt, T., Buchele, W., & Arnold, F. (1977). Arthrokinetic nystagmus and ego-motion sensation. *Experimental Brain Research, 30*, 331–338.

Burgess, P. R., & Wei, J. Y. (1982). Signaling of kinesthetic information by peripheral sensory receptors. *Annual Review of Neuroscience, 5*, 171–187.

Champney, R. K., Stanney, K. M., Kennedy, R. S., Hash, P., Malone, L., & Compton, D. (2007). Recovery from virtual environment exposure: Expected time-course of symptoms and potential readaptation strategies. *Human Factors, 49*(3), 491–506.

Cohn, J., DiZio, P., & Lackner, J. R. (2000). Reaching during virtual rotation: Context-specific compensation for expected Coriolis forces. *Journal of Neurophysiology, 83*, 3230–3240.

Craske, B. (1977). Perception of impossible limb positions induced by tendon vibration. *Science, 196*, 71–73.

DiZio, P., & Lackner, J. R. (1997). Circumventing side effects of immersive virtual environments. In M. J. Smith, G. Salvendy, & R. J. Koubek (Eds.), *Advances in human factors/ergonomics: Vol. 21. Design of computing systems* (pp. 893–397). Amsterdam, the Netherlands: Elsevier Science Publishers.

DiZio, P., & Lackner, J. R. (2002). Proprioceptive adaptation and aftereffects. In K. M. Stanney (Ed.), *Handbook of virtual environments: Design, implementation, and applications* (pp. 751–771). Mahwah, NJ: Lawrence Erlbaum Associates.

DiZio, P., & Lackner, J. R. (in development). Minimal haptic cues enable finer motor resolution during interaction with visual virtual objects.

DiZio, P., Lackner, J. R., & Lathan, C. E. (1993). The role of brachial muscle spindle signals in assignment of visual direction. *Journal of Neurophysiology, 70*(4), 1578–1584.

DiZio, P., Landman, N., & Lackner, J. R. (1999). Fingertip contact forces map reaching endpoint. *Society for Neuroscience Abstracts, 25*, 760.15.

Dolezal, H. (1982). *Living in a world transformed: Perceptual and performatory adaptation to visual distortion.* New York, NY: Academic Press.

Draper, M. H. (1996). *Can your eyes make you sick?: Investigating the Relationship between the vestibulo-ocular reflex and virtual reality.* Seattle, WA: Human Interface Technology Laboratory, University of Washington.

Eklund, G. (1972). Position sense and state of contraction: The effects of vibration. *Journal of Neurology Neurosurgical Psychiatry*, *35*, 606–611.

Ellis, S. R., Young, M. J., Adelstein B. D., & Ehrlich, S. M. (1999). Discrimination of changes of latency during voluntary hand movement of virtual objects. In *Proceedings of the Human Factors and Ergonomics Society*. Houston, TX: Human Factors and Ergonomics Society.

Fregley, A. R. (1974). Vestibular Ataxia and its Measurement in Man. In *Handbook of sensory physiology* (pp. 321–359). New York, NY: Springer-Verlag.

Goodwin, G. M., McCloskey, D. I., & Matthews, P. B. C. (1972). The contribution of muscle afferents to kinesthesia shown by vibration induced illusions of movement and by the effects of paralysing joint afferents. *Brain*, *95*, 705–748.

Gordon, C. R., Fletcher, W. A., Melvill Jones, G., & Block, E. W. (1995). Adaptive plasticity in the control of locomotor trajectory. *Experimental Brain Research*, *102*, 540–545.

Gower, D. W., & Fowlkes, J. E. (1989, September). *Simulator sickness in the UH-60 (Black Hawk) flight simulator* (USAARL 89-20 (AD-A214 434), p. 74).

Gower, D. W., Lilienthal, M. G., Kennedy, R. S., Fowlkes, J. E., & Baltzley, D. R. (1987, November). *Simulator sickness in the AH-64 Apache combat mission simulator* (USAARL 88-1 (AD-A193 419), p. 49).

Groen, J., & Werkhoven, P. J. (1998). Visuomotor adaptation to virtual hand position in interactive virtual environments. *Presence*, *7*, 429–446.

Hagbarth, K. E., & Eklund, G. (1966). Motor effects of vibratory muscle stimuli in man. In R. Granit (Ed.), *Muscular afferents and motor control* (pp. 177–186). Stockholm, Sweden: Almqvist & Wiksell.

Harris, C. S. (1963). Adaptation to displaced vision: Visual, motor, or proprioceptive change? *Science*, *140*, 812–813.

Harris, C. S. (1965). Perceptual adaptation to inverted, reversed, and displaced vision. *Psychological Review*, *72*, 419–444.

Harvey, R. J., & Matthews, P. B. C. (1961). The response of de-efferented muscle spindle endings in the cat's soleus to slow extension in the muscle. *Journal of Physiology*, *157*, 370–392.

Hay, J. C., Pick, H. L., & Ikeda, K. (1965). Visual capture produced by prism spectacles. *Psychonomic Science*, *2*, 215–216.

Held, R. (1965). Plasticity in sensory-motor systems. *Scientific American*, *213*, 84–94.

Held, R., & Bossom, J. (1961). Neonatal deprivation and adult rearrangement: Complementary techniques for analyzing plastic sensory-motor coordinations. *Journal of Comparative and Physiological Psychology*, *54*, 33–37.

Held, R., & Durlach, N. (1991). Telepresence, time delay and adaptation. In S. R. Ellis, M. K. Kaiser, & A. J. Grunwald (Eds.), *Pictorial communication in virtual and real environments* (2nd ed., pp. 232–246). London, U.K.: Taylor & Francis.

Helmholtz, H. (1925). In J. P. C. Southall (Ed.), *Helmholtz's treatise on physiological optics* (3rd ed.). Menasha, WI: Optical Society of America.

Henderson, W. R., & Smyth, G. E. (1948). Phantom limbs. *Journal or Neurology, Neurosurgery and Psychiatry*, *11*, 88–112.

Holden, M., Ventura, J., & Lackner, J. R. (1994). Stabilization of posture by precision contact of the index finger. *Journal of Vestibular Research*, *4*(4), 285–301.

Holst, E. von, & Mittelsteadt, H. (1950). Das reaffferenz-prinzip. *Die Naturwissenschaften*, *20*, 464–467.

Howard, I. P. (1966). *Human spatial orientation*. London, U.K.: John Wiley & Sons.

Hunter, I., Doukoglou, T. D., Lafontaine, S. R., Charrette, P. G., Jones, L. A., Sagar, M. A., … & Hunter, P. J. (1994). A teleoperated microsurgical robot and associated virtual environment for eye surgery. *Presence*, *2*, 265–280.

Jeka, J. J., & Lackner, J. R. (1994). Fingertip contact influences human postural control. *Experimental Brain Research*, *100*(3), 495–502.

Kellogg, R. S., Castore, C., & Coward, R. (1980). Psycho-physiological effects of training in a full vision simulator. *Annual Scientific Meeting of the Aerospace Medical Association*, Anaheim, California.

Kennedy, R. S., Fowlkes, J. E., & Lilienthal, M. G. (1993). Postural and performance changes following exposures to flight simulators. *Aviation, Space, and Environmental Medicine*, *64*, 912–920.

Kennedy, R. S., & Stanney, K. M. (1996). Postural instability induced by virtual reality exposure: Development of a certification protocol. *International Journal of Human-Computer Interaction*, *8*(1), 25–47.

Kinney, J. A. S., McKay, C. L., Luria, S. M., & Gratto C. L. (1970). *The improvement of divers' compensation for underwater distortions* (Technical Report NSMRL 633). Groton, CT: U.S. Naval Submarine Medical Center.

Lackner, J. R. (1974). Adaptation to displaced vision: Role of proprioception. *Perceptual and Motor Skills*, *38*, 1251–1256.

Lackner, J. R. (1976). Influence of abnormal postural and sensory conditions on human sensory motor localization. *Environmental Biology and Medicine*, 2, 139–177.

Lackner, J. R. (1981). Some aspects of sensory-motor control and adaptation in man. In R. D. Walk & H. L. Pick (Eds.), *Intersensory perception and sensory integration* (pp. 143–173). New York, NY: Plenum Press.

Lackner, J. R. (1988). Some proprioceptive influences on the perceptual representation of body shape and orientation. *Brain*, *111*, 281–297.

Lackner, J. R., Dizo, P., & Fisk, J. (1992). Tonic vibration reflexes and background force level. *Acta Astronautica*, 26, 133–136.

Lackner, J. R., & DiZio, P. (1984). Some efferent and somatosensory influences on body orientation and oculomotor control. In L. Spillman & B. R. Wooten (Eds.), *Sensory experience, adaptation, and perception* (pp. 281–301). Clifton, NJ: Lawrence Erlbaum Associates.

Lackner, J. R., & DiZio, P. (1988). Visual stimulation affects the perception of voluntary leg movements during walking. *Perception*, *17*, 71–80.

Lackner, J. R., & DiZio, P. (1994). Rapid adaptation to Coriolis force perturbations of arm trajectory. *Journal of Neurophysiology*, *72*, 99–313.

Lackner, J. R., & DiZio, P. (1995, November 11–16). Generalization of adaptation to Coriolis force perturbations of reaching movements. *Neuroscience 1995 Abstracts*, San Diego, CA: Society for Neuroscience (Online).

Lackner, J. R., DiZio, P., Jeka, J. J., Horak, F., Krebs, D., & Rabin, E. (1999). Precision contact of the fingertip reduces postural sway of individuals with bilateral vestibular loss. *Experimental Brain Research*, *126*, 459–466.

Lackner, J. R., & Levine, M. S. (1979). Changes in apparent body orientation and sensory localization induced by vibration of postural muscles: Vibratory myesthetic illusions. *Aviation Space and Environmental Medicine*, *50*, 346–354.

Lackner, J. R., & Lobovits, D. (1977). Adaptation to displaced vision: Evidence for prolonged aftereffects. *Quarterly Journal of Experimental Psychology*, *29*, 65–69.

Lackner, J. R., Rabin, E., & DiZio, P. (2000). Fingertip contact suppresses the destabilizing influence of tonic vibration reflexes in postural muscles. *Journal of Neurophysiology*, *84*, 2217–2224.

Lackner, J. R., & Taublieb, A. B. (1984). Influence of vision on vibration-induced illusions of limb movement. *Experimental Neurology*, *85*, 97–106.

Lazlo, J. I. (1963). The performance of a simple motor task with kinaesthetic loss. *Quarterly Journal of Experimental Psychology*, *18*, 1–8.

Levine, M. S., & Lackner, J. R. (1979). Some sensory and motor factors influencing the control and appreciation of eye and limb position. *Experimental Brain Research*, *36*, 275–283.

Matin, L., Picoult, E., Stevens, J. K., Edwards, M. W., Young, D., & MacArthur, R. (1972). Oculoparalytic illusion: Visual-field dependent spatial mislocalizations by humans partially paralyzed with curare. *Science*, *216*, 198–201.

McCauley, M. E., & Sharkley, T. J. (1992). Cybersickness: Perception of self-motion in virtual environments. *Presence*, *1*(3), 311–318.

McCloskey, D. I., & Torda T. A. G. (1975). Corollary motor discharges and kinesthesia. *Brain Research*, *100*, 467–470.

Melzack, R., & Bromage, P. R. (1973). Experimental phantom limbs. *Experimental Neurology*, *39*, 261–269.

Navy (2004). Aeromedical and survival. Available online at: http://www.netc.navy.mil/nascweb/sas/files/3710.7t_ch8.pdf. Viewed May 20, 2014.

Paillard, J., & Brouchon, M. (1968). Active and passive movements in the calibration of position sense. In S. J. Freedman (Ed.), *The neuropsychology of spatially oriented behavior*. Homewood, IL: Dorsey Press.

Rock, I. (1966). *The nature of perceptual adaptation*. New York, NY: Basic Books.

Shadmehr, R., Mussa-Ivaldi, F. A., & Bizzi, E. (1993). Postural force fields of the human arm and their role in generating multi-joint movements. *Journal of Neuroscience*, *13*, 45–62.

Sherrington, C. S. (1906). *The integrative action of the nervous system*. London, U.K.: Charles Scribner's Sons. (Reprinted 1948, Cambridge: University Press).

Sperry, R. W. (1950). Neural basis of the spontaneous optokinetic response produced by visual neural inversion. *Journal of Comparative and Physiological Psychology*, *43*, 482–489.

Stanney, K. M., & Kennedy, R. S. (1997). Development and testing of a measure of kinesthetic position sense used to assess the aftereffects from virtual environment exposure. *Proceedings of the 1997 IEEE Virtual Reality Annual International Symposium* (p. 87). Albuquerque, NM: IEEE.

Stoffregen, T. A., Draper, M. H., Kennedy, R. S., & Compton, D. (2002). Vestibular adaptation and aftereffects. In K. M. Stanney (Ed.), *Handbook of virtual environments: Design, implementation, and applications* (pp. 773–790). Mahwah, NJ: Lawrence Erlbaum Associates.

von Holst, E., & Mittelsteadt, H. (1950). Das reaffferenz-prinzip. *Die Naturwissenschaften*, 20, 464–467.

Wallach, H. (1976). *On perception*. New York, NY: Quadrangle.

Wallach, H., Kravitz, J. H., & Lindauer, J. (1963). A passive condition for rapid adaptation to displaced visual direction. *American Journal of Psychology*, 76, 568–578.

Wann, J. P., & Mon-Williams, M. (2002). Measurement of visual aftereffects following virtual environment exposure. In K. M. Stanney (Ed.), *Handbook of virtual environments: Design, implementation, and applications* (pp. 731–749). Mahwah, NJ: Lawrence Erlbaum Associates.

Weber, K. D., Fletcher, W. A., Melvill Jones, G., & Block, E. W. (1997). Oculomotor response to rotational podokinetic (PK) stimulation and its interaction with concurrent VOR in normal and PK adapted states. *Society for Neuroscience Abstracts*, *23*(1), 470.

Welch, R. B. & Mohler, B. J. (2014). Adapting to virtual environments. In K. S. Hale & K. M. Stanney (Eds.), *Handbook of virtual environments: Design, implementation, and applications* (2nd ed., pp. 627–646). Boca Raton, FL: Taylor & Francis Group, Inc.

Welch, R. B. (1978). *Perceptual modification: Adapting to altered sensory environments*. New York, NY: Academic Press.

Westling, G., & Johansson, R. S. (1987). Responses in glabrous skin mechanoreceptors during precision grip in humans. *Experimental Brain Research*, *66*, 128–140.

34 Beyond Presence
How Holistic Experience Drives Training and Education

Dustin Chertoff and Sae Schatz

CONTENTS

34.1 INTRODUCTION

One of the often-cited benefits of virtual environments (VEs) is that they give participants a means to experience places they would never be able to visit and to perform tasks that would otherwise be too dangerous or expensive. Being *transported* to such places and feeling the sensation of *being there* is called *presence* (Barfield, Zeltzer, Sheridan, & Slater, 1995, p. 475; Heeter, 1992).

To many, the concept of presence and the benefits of improving VE participants' senses of presence seem intuitive. However, upon deeper inspection of the academic literature, one finds that this topic is not so clear-cut. For years, portions of the academic community have been debating the definition of presence, arguing about which constructs correlate with it, and contemplating the causal effects of it. The resulting literature provides the intrepid researcher with the impression that presence is a labyrinthine construct related to everything from image resolution (Held & Durlach, 1992) to mood (Apter, 1992), and yet only applicable in specific, hardware-based contexts.

This chapter, however, argues that this debate on semantics misses the mark and that real-world practitioners are less interested in the theoretical nuances of presence and more interested in its practical implications. As such, in this chapter, we encourage readers to look beyond academic definitions of presence and instead evaluate participants' holistic experiences within VEs and related systems.

This chapter introduces readers to presence and briefly describes some of the conceptual deliberations associated with it. After that, we outline the construct of *holistic experience* and describe a tool, the Virtual Experience Test (VET), that can support investigations about the VE experience design. We then demonstrate the VET's applicability by describing its use in several exemplar VEs.

34.2 MANY CONCEPTUALIZATIONS OF PRESENCE

A renowned artificial intelligence researcher, Marvin Minsky (1980), first coined the term *telepresence* to refer to the feeling of *being there* at a remote location by way of remote-access technology, such as video-teleconferencing or a remotely operated vehicle interface. Since telepresence technically refers to the sense of being in a different real-world site, other researchers have used the term *virtual presence* to represent the sense of being present in a VE (Barfield et al., 1995). Still others partition the construct further, attempting to differentiate among concepts such as spatial presence (Böcking et al., 2004); mediated presence (Biocca, Harms, & Gregg, 2001); copresence (Nowak & Biocca, 2003); transactional presence (Shin, 2002); teaching presence (Garrison, Anderson, & Archer, 2001); environmental, social, and personal presence (Heeter, 1992); simple, cybernetic, and experiential presence (Draper, Kaber, & Usher, 1998); or physical, subjective, and objective presence (Schloerb, 1995). As a result, the contemporary academic conceptualization of presence has become rather fragmented, broken into numerous subconstructs that do not necessarily share similar levels of granularity of parallelism.

There is also ongoing debate about the manifestation of presence. Some researchers emphasize the role of hardware- and software-created stimuli, that is, factors exogenous to human participants (e.g., Sheridan, 1992; Steuer, 1992), while others highlight the influence of participants' unique characteristics and experiences and how these shape their subjective interpretation of a medium, that is, factors endogenous to human participants (e.g., Barfield & Weghorst, 1993; Heeter, 1992; Kalawsky, 2000). Some theorists emphasize the role of task performance, suggesting that presence is "… tantamount to successfully supported action in the environment" (Zahoric & Jenison, 1998, p. 30). Still others accentuate social factors, focusing on the interactions participants have with each other, virtual entities, and the mediated environment (e.g., Lemish, 1982; Lombard, 1995; Mantovani & Riva, 1999). In fact, in an extensive review of the literature at the time, Lombard and Ditton (1997) identified six distinct factions of presence definitions, that is, social richness, realism, transportation, immersion, social interaction, and interaction with the medium. In other words, there are many differing opinions within the presence community, and after looking across the literature, it becomes clear that presence cannot be readily defined as a simple, uniform phenomenon.

In their attempts to better understand the multifaceted nature of presence, some researchers have created integrated models of it. These compound models generally begin from a common-ground definition of presence (i.e., being there), but they diverge into different subcomponents and typically represent some subset of the six factors identified by Lombard and Ditton (1997). The various models tend to be nonorthogonal (i.e., they overlap in some ways), yet their components are not wholly congruent. Witmer and Singer (1998), for instance, hypothesize that four clusters of variables influence presence: control, sensory, distraction, and realism factors. Meanwhile, Lee and Nass (2001) offer a three-factor model, including technology, user, and social factors, and Lombard and Ditton (1997) argue that presence comprises only two facets (i.e., physical and social factors). See Table 34.1 for a summary of these models' features, as well as some of the other often-cited presence conceptualizations.

Some reports (usually in an attempt to rapidly summarize the literature) try to abstract these disparate models and then amalgamate their factors into a general list. For instance, in a recent article, Lee (2004) lists 25 factors, ranging from participants' moods to the image resolution of a simulator, which are "either empirically identified or theoretically argued" to be "closely associated with" presence (p. 495). Similarly, Lessiter, Freeman, Keogh, and Davidoff (2001), and

TABLE 34.1
Often-Cited Presence Conceptualizations and Their Theoretical Components

Author(s)	Components of Presence
Sheridan (1992)	• *Sensory information*—amount and fidelity of sensations provided to user. • *Control*—amount of control the user has over the sensor mechanisms. • *Environment modification*—amount of modification the user can make to the environment.
Heeter (1992)	• *Environmental*—amount the environment appears to respond to the user's existence within the world. • *Social*—amount of support received by a user from other users that they are in a VE. • *Personal*—various perceptual factors relating to how and why the user might feel that they are in the environment.
Lombard and Ditton (1997)	• *Physical*—the user's sense of being physically at some location. • *Social*—the user's feeling of being together or communicating with another person.
Zahoric and Jenison (1998)	• *Environment support*—presence emerges based on successful support of user action by the environment.
Mantovani and Riva (1999)	• *Social construction*—view presence as the relationship between actors and their environments. User interaction toward the environment is important as well.
Witmer and Singer (1998)	• *Control*—extent to which a user can interact with and manipulate the VE. • *Sensory*—the number, types, richness, and consistency of sensations a user will feel. • *Distraction*—amount the hardware and surrounding external environment affect the user's ability to focus on the VE. • *Realism*—how connected and consistent VE information is to the real world and how well the user relates to the information.
IJsselsteijn, de Ridder, Freeman, and Avons (2000)	• *Sensory*—extent and fidelity of sensory information. • *Sensory-motor contingencies*—how well a user's actions match the spatiotemporal effects of those actions. • *Content factors*—contains ways in which the user can interact with and modify the environment. • *User characteristics*—user perceptual, cognitive, and motor abilities, previous experience, susceptibility to simulator sickness, and a willingness to suspend disbelief.
Lee and Nass (2001)	• *Technology factors*—objective quality of technology. • *User factors*—individual differences. • *Social factors*—social characteristics of technology.

Lessiter, Freeman, Keogh, and Davidoff (2000) sketch out 15 different relevant *content areas* (i.e., variables and categories that moderate, mediate, or otherwise correlate with presence); these cover a broad conceptual space, including topics such as perception of time, realness, personal relevance, and a sense of social interaction (2001, p. 287).

Bundling together the subcomponents of different models may appear like a practical way to construct a comprehensive articulation of presence; however, not all theorists agree with the viability of this method. For instance, Böcking et al. (2004) point out that "if one examines these components more deeply, most of them turn out not to be part of the presence concept, i.e. the feeling of being in a mediated environment" (p. 226); the variables can be considered "components enhancing presence or effects of presence, but not the experience of being present itself" (p. 226). Ellis (1996) explains that "… catalogs like the above property list for presence are incomplete" because their selection and combination is not "aided by a theory of presence which would describe how these components combine" (p. 249). Ellis cautions researchers to remain mindful of good psychometric principles and be more judicious about identifying which variables, in which ways, comprise presence. That is, clearly delineating, based upon theory and empirical observation, whether various factors cause presence, merely correlate with it, or are subcomponents of it (Barfield et al., 1995; Draper et al., 1998).

Not surprisingly, the uncertainties regarding the decomposition of presence create problems for measurement, and there have subsequently been diverse attempts to create measurement tools. Measures may address presence as a whole or target-specific types of it, use primarily subjective or mainly objective data, or focus on endogenous or exogenous factors (for more detailed reviews of these measurement approaches, see Chertoff, 2009; Insko, 2003).

As one might expect, presence assessments frequently involve self-report questionnaires, which VE participants may complete during or after their mediated experiences (i.e., in situ or post hoc). Three of the most widely used include the Witmer-Singer Presence Questionnaire (PQ) (Witmer & Singer, 1998), Slater-Usoh-Steed (SUS) (Usoh, Catena, Arman, & Slater, 2000), and the ITC-Sense of Presence Inventory (ITC-SOPI) (Lessiter et al., 2001). These surveys, however, fail to pass the *reality* test, which suggests that participants should rate an experience in the real-world more highly than they do in a mediated, virtual one (Usoh et al., 2000; see also, Villania, Repettoa, Cipressob, & Rivaa, 2012). Failing the reality test implies a lack of validity, and other studies have found that these apparatus have low statistical reliabilities as well (Youngblut & Perin, 2002, specifically discussing the PQ and SUS). The PQ, SUS, and ITC-SOPI are not unique in these limitations, though. Presence surveys are frequently described as *unstable* or *biased* (Freeman, Avons, Pearson, & IJsselsteijn, 1999) and "their utility is doubtful for the comparison of experiences across environments" (Usoh et al., 2000, p. 1). All this suggests that such subjective-response questionnaires are, perhaps, not an ideal means of measuring a VE participant's sense of presence, and they are most likely poor mechanisms for rating a VE's inherent potential to engender a certain level of presence.

In an attempt to enhance the robustness of measures, researchers have turned to more objective metrics, including behavioral or physiological assessments. For instance, one group of researchers assessed participants' involvement in a car racing simulation based upon how much they tilted their body postures with the motion depicted in the VE (Freeman, Avons, Meddis, Pearson, & IJsselsteijn, 2000). In another example, researchers studying a 20 ft virtual reality visual drop-off gauged presence by monitoring participants' pit-avoidance behaviors, such as leaning away from the edge or adjusting their gait near the pit (e.g., Meehan, 2001; Phillips, Interrante, Kaeding, Ries, & Anderson, 2012). Other presence evaluation methods include evaluating participants' facial expressions, either scored manually or through the use of automatic pattern recognition software (Huang & Alessi, 1999), evaluating their reflex responses, such as startle effects (Held & Durlach, 1991; Loomis, 1992; Malbos, Rapee, & Kavakli, 2012), and measuring their physiological indices, such as skin conductance, skin temperature, or changes in heart rate (see Insko, 2003). In general, these methods appear to have more reliability and sensitivity than subjective-response presence surveys. However, under certain conditions, their validity may be limited because of their responsiveness to confounding influences that, theoretically, should not significantly affect presence, such as a system's frame rate or the specific experimental instructions given to research participants (Insko, 2003; discussing Meehan, 2001). Behavioral assessments also tend to be highly specialized to a particular VE, limiting their utility to inform comparisons across environments and preventing their use in certain scenarios (i.e., those lacking specific behavior-inducing stimuli or well-defined expected responses).

Finally, to add another complication to the use of presence, many academic presence discussions involve nuanced semantic questions that may appear to lack practical relevance. For instance, researchers frequently debate whether experiences must be mediated by technology to elicit a sense of presence (e.g., Lee, 2004) and what exactly *mediated* means (Lombard & Ditton, 1997; Lombard et al., 2000). Whether presence occurs naturally or whether it requires a willing suspension of disbelief is also under consideration (e.g., Nass & Moon, 2000; Reeves & Nass, 1996; see also Lee, 2004). Such conversations regularly delve into more esoteric topics, too, such as what is a *real experience*, how do perception and reality relate to one another (e.g., Lee, 2004), and other *ontological questions about the nature of existence* (Usoh et al., 2000, p. 497).

In summary, there is no suitable, agreed-upon framework with which to conceptualize presence. The presence community has defined many different subconstructs and developed an array of competing presence models. Theorists continue to debate the variety of potential correlates to presence, and while presence psychometricians attempt to capture these variables and their relationships, their efforts often lack validity, reliability, or practical utility. In sum, although attempts have been made to realign the academic field (e.g., Slater, 2009), considerable confusion and disagreement persist. While the debate and even the profound philosophical discussions that often arise within it are healthy for the academic community, they can make presence a more difficult concept for those outside of the academic sphere to interpret. Practitioners may struggle to appreciate the fragmentation of the presence construct or the nuanced debates surrounding it, and VE system designers may be unsure of how to apply the superfluity of facets in practical ways. Developers may also become frustrated, looking for clearer guidance on which of the subcomponents they should emphasize, how the subcomponents may affect one another, or how these different components may affect their systems. Thus, overall, the academic turmoil surrounding the presence creates challenges for practitioners, and it can limit their ability to readily use presence in applied settings.

34.3 WHY CARE ABOUT PRESENCE, ANYWAY?

Historically, interest in presence was motivated by practical concerns in teleoperation, simulation, and telecommunication. For instance, Minsky (1980) originally hypothesized that higher degrees of telepresence would better facilitate task performance, contributing to greater levels of safety and efficiency. Others have since reinforced the idea that presence enhances performance in both teleoperated and virtual environments (e.g., Bystrom, Barfield, & Hendrix, 1999; Loomis, 1992). In training simulations, presence theoretically enhances training transfer (see Alexander, Brunyé, Sidman, & Weil, 2005, for a review), and in therapeutic simulations, such as those designed to treat phobias or enhance one's body image, presence may improve the efficacy of treatment (e.g., Riva, Bacchetta, Baruffi, & Molinari, 2001; Riva, Bacchetta, Cesa, Conti, & Molinari, 2001; Rothbaum & Hodges, 1999). In simulations designed for entertainment, presence contributes to higher levels of enjoyment (Barfield & Weghorst, 1993), and in telecommunications, higher levels of presence are believed to affect the feeling of being together, which, in turn, influences the feeling of satisfaction and the effectiveness of collaborations (e.g., Gunawardenaa & Zittleb, 1997; Lowenthal, 2010; Shin, 2003; Short, Williams, & Christie, 1976).

In other words, presence is that human–computer je ne sais quoi that theoretically enhances the intended outcome of a system. It has practical relevance for system operators or evaluators, who may instinctively refer to this factor when they describe or appraise a system, and it is relevant for system developers, who may seek to build systems that better engender it (in order to achieve some of the related benefits discussed earlier). The notion of presence—of *being there* or, at least, being in-sync with or in the flow of the mediated experience—is colloquially intuitive. Even without extensive training in modeling and simulation, many VE participants will naturally arrive at some version of the concept when describing their mediated experiences. Thus, there is an inherent reason to name and define the construct.

The academic discussions surrounding the presence concept, however, have made the construct so complex and idiomatic that it does not readily lend itself to practical application. Further, the academic conceptualizations of presence do not align with colloquial or common usage of the term. As a result, many people outside of the academic presence community misinterpret the notion of presence or confuse it with related constructs (e.g., Bordnick et al., 2008). Practitioners need an interpretable framework of presence that they can use to readily discuss, evaluate, and build systems. This approach should be based on a well-defined theory that allows the model to be extended to unique systems and new technologies. It should emphasize the *gestalt*, or overall whole, of associated system interactions, rather than be concerned with the countless subcomponents of the experience (which tend to individually lead to trivial downstream effects). Finally, the concept should facilitate practical discussions about the design, development, operation, and testing of VEs and similar systems, helping those involved with a system achieve their usage goals. As presented in the following, one approach involves the idea of *experience*.

34.4 EXPERIENCE

Most readers likely have an intuitive sense of what *an experience* is. Formally, an experience is something that can be articulated, named, and schematized within a person's memory. Experiences of this type have beginnings and endings, but the anticipation of, and reflection on, the experience may take place before or after the event (for a detailed definition, see Forlizzi & Battarbee, 2004). The art and science of strategically designing the environments, interactions, and activities of an experience is commonly known within the business and marketing fields as experiential design (for reviews, see Leppiman & Same, 2011; Schmitt, 2003; Suri, 2003; Voss & Zomerdijk, 2007).

34.4.1 Experiential Design

In commercial settings, the goal of experiential design is to provide customers with rewarding, holistic experiences—which, naturally, they pay a premium for. For example, a coffee shop might offer live music, provide a lounge area with Internet access, and be known for selling organic food. Compare this with a nonexperiential deli, where a patron can simply purchase a cup of coffee in a Styrofoam cup. By selling a product in the coffee shop manner, the customer receives both the product and the experience surrounding the product, and customers pay for the feelings of engaging in the experience—over and above the costs associated with the goods and services alone (Schmitt, 2003). Experiential design is thus concerned with interactions that result in compelling experiences. This design approach leads customers to have a broader connection with a product or brand, which triggers stronger consumer responses (Battarbee & Mattelmaki, 2002).

Experiential design consists of five dimensions that designers manipulate to produce engaging, holistic experiences: sensory, cognitive, affective, active, and relational (Pine & Gilmore, 1999). The sensory dimension includes all sensory input (visual, aural, haptic, etc.) as well as perception of those stimuli. The cognitive dimension encompasses all mental engagement with an experience, such as anticipating outcomes and solving mysteries, and the affective dimension refers to a participant's emotional state. Next, the active dimension relates to the degree of personal connection a person feels to an experience. For instance, does he incorporate the experience into his personal narrative; does she form meaningful associations via the experience? Finally, the relational dimension is comprised of the social aspects of an experience. This can be operationalized as coexperience, that is, creating and reinforcing meaning through collaborative experiences (Battarbee, 2003; Forlizzi & Battarbee, 2004), which further enables individuals to develop personal and memorable narratives (Battarbee, 2003).

34.4.2 Experiential Design and Virtual Environments

Previously, researchers have explored how each of these dimensions could be used to describe virtual experiences (Chertoff, Schatz, McDaniel, & Bowers, 2008), and subsequent publications have explored the relationship of the experiential dimensions to existing presence conceptualization (Chertoff, 2009; Chertoff, Goldiez, & LaViola, 2010). For VEs, the sensory dimension is represented through hardware, such as monitors or head-mounted displays, and software, such as image generation engines. It encompasses a range of elements, including the aesthetic appeal of stimuli and, when appropriate, their physical fidelity. Much of the cognitive dimension within a VE can be interpreted as task engagement, which is related to the intrinsic motivation, meaningfulness, and continuity (actions yielding expected responses) of an activity. In a broad sense, successful design of the cognitive dimension elicits *flow*, the psychological state of complete and optimal focus on a task (Czikszentmihalyi & Czikszentmihalyi, 1992). Flow may also be referred to as *optimal experience* or, commonly, as *being in the zone*. It occurs when there is an ideal balance between the difficulty of a task and the skills of the person trying to complete that task. The affective component in VEs is linked to the degree to which a person's emotions in the virtual experience would accurately mimic his/her emotional state in the same real-world situation. For example, does a participant feel the same degree of stress and anxiety in a dismounted infantry simulation as he/she would in the real-life equivalent? (In the case where there is no corresponding real-world situation, one could ask whether the participant feels an emotional state close to what the VE designer intended.) Essentially, all emotional factors fall under this domain.

The active dimension can be associated with the degree of empathy, identification, and personal relation a participant feels with the VE's avatars, environment, and scenario. Similarly, the relational component, or coexperience, within VEs involves collaboration with fellow (human) participants; although, realistic virtual characters that influence psychosocial factors also fall under the relational dimension.

Each of the five domains outlined can contribute meaningfully to the experiential design of a VE; however, they should not be considered in isolation. The integration of the domains, that is, their collective emergent properties, also affects the essence of an experience. Accordingly, experiential design involves both the individual domains *and* their integrated sum, and together these elements define a *holistic experience*.

Marketing uses these holistic experiences to encourage consumers to develop deeper connections with a product or brand. However, experiential designers are interested not only in their customers' immediate experiences but also in how their audience develops memories of an experience, associates emotions with those memories, and recalls those memories in the future. Stated more formally, experiential designers attempt to influence consumers' schemata, the learned cognitive structures that people use to process, store, and manipulate patterns of information (Neisser, 1976; Schank & Abelson, 1977). For example, if a customer hears a coffee shop's jingle on the radio, experiential design theory suggests that the consumer may recall positive memories associated with that brand and, subsequently, feel a craving to buy more of its coffee. Thus, holistic experiences involve immediate and referential (i.e., later remembered) elements. In summary, experiential design involves the five experiential domains, the emergent properties that manifest from a combination of the domains (integration) and the experiential components that help foster consumers' memories (persistence). Although experiential design was originally conceived to support marketing efforts in real-world environments, its principles also apply to virtual contexts. In both cases, the five domains of experiential design define and help reinforce the broader system of factors that comprise an experience. Experiential design's emphasis on cross-domain integration forces designers to consider the entire abstract experience and not simply its isolated component parts, and experiential design encourages designers to take a longer-term view, considering not only the immediate experience but also the memories and recall of it.

For VEs, there are several theoretical benefits of using experiential design methodologies. First, the five domains impose a structure on the phenomenon of participating in a VE, and the breadth of the domains encourages VE designers to more systematically consider the range of factors that influence participants' experiences. Second, experiential design helps reinforce the emergent properties of a VE, in contrast to some conceptualizations of presence that may, for instance, overly emphasize the more granular, less impactful subcomponents of the construct. Third, its attention to memory and recall encourages experiential designers to more intentionally consider participants' cognitive schemata, which, in turn, reinforces participants' experiences and helps make the experiences more interpretable.

Fourth, and perhaps most relevant for practitioners, experiential design meaningfully informs system design and development *a priori*. This is in contrast to many presence theories and measurement apparatus, which support postdesign analysis but provide fewer insights for applied requirements and specification authoring. In other words, while presence theories and measures may describe a developed system, they do not always predict how factors of a new system will contribute to an intended user experience.

34.4.3 Virtual Experience Test

Theories of experiential design suggest various guidelines for applying experiential processes to marketing endeavors (e.g., Garg, Rahman, Kumar, & Qureshi, 2011; Pullman & Gross, 2004; Schmitt, 2003, 2010; Voss, Roth, & Chase, 2009). However, because VEs possess unique features and goals as compared with typical consumer commodities, existing experiential design strategies should be updated and validated for VE designers' use. To fill this gap, researchers have devised the VET (Chertoff, 2009), an adaptable Likert-style questionnaire that developers can use to better gauge each dimension of an experience within a VE. Version 1.0 of the VET has been refined through empirical testing and factor analysis (Chertoff et al., 2010), but additional items, iterative psychometric improvements, and more validation testing are still required.

The VET is intended to be an efficient heuristic that VE practitioners can use to support iterative system design efforts. Ultimately, the goal is to develop a pool of validated VET items (addressing different aspects of each experiential design component) that VE practitioners can adapt and, as necessary, extend to facilitate their design efforts. Thus, VET items support the planning and testing of system factors that influence experiences, rather than measuring participants' outcome experiences, directly. In other words, the VET is intended to serve as a design aid, evaluating the *potential* experiences of participants in various VEs, and when used as part of an iterative process, the VET should help system designers better plan and assess a system's potential to deliver compelling, holistic experiences.

Initial versions of the VET built upon theoretical and empirical findings involving presence and flow, as well as the theories associated with experiential design (for initial validation studies, see Chertoff, 2009; Chertoff et al., 2010). The structure of VET 1.0 currently follows the five dimensions of experiential design. Each subscale is calculated by rating a set of subjective-response items, and the overall score is a summation of each subdivision. New subscales involving the emergent properties of experiences (integration) and the longer-term schematic influences of those experiences (persistence) are being developed, and through ongoing empirical testing, meaningful outcome ranges are being identified. It is hypothesized that, given sufficient analysis, value ranges for each subscale can be correlated to representative participant reactions. See Table 34.2 for a summary.

34.5 EXAMPLES OF EXPERIENCE IN VIRTUAL ENVIRONMENTS

To demonstrate the practicality of experiential design within VEs, four example applications are presented in the following from the medical, military, education, and entertainment domains. The examples also show how the VET could assist in evaluating a VE's potential to engender meaningful, holistic experiences.

34.5.1 Medical: Virtual Therapy Environments

Virtual therapy involves the application of virtual reality technologies in clinical treatments, such as the use of an imaginary environment to address a patient's phobia or general anxiety (Cardenas, Munoz, Gonzalez, & Uribarren, 2006; Gorini & Riva, 2008; Krijn, Emmelkamp, Olafsson, & Biemond, 2004), posttraumatic stress disorder (Rizzo et al., 2005; Spira et al., 2006), or addiction (Saladin, Brady, Graap, & Rothbaum, 2006). As patients can have varied reactions to the condition requiring therapy, a similarly varying level of sensation, cognition, affect, activity, and relational interaction is needed for virtual therapy VEs as well. Typically, the created world is tailored toward a patient's condition. For example, a patient with a fear of heights could be brought onto virtual roofs of various heights. Alternatively, a military veteran suffering from posttraumatic stress disorder could relive battlefield events with the guidance of a therapist. In the case of therapy for such a disorder, a therapist might want to gradually introduce new elements to the virtual experience to slowly bring the patient back to the *real* (or more realistically simulated) experience. This could be

TABLE 34.2

Summary of Experiential Design Components and Exemplar VET Items[a]

Experiential Component	Summary	Exemplar VET Items
Sensory domain	Participants' reactions to all of the sensory stimuli within a given experience	• The sensory information of the VE is, as applicable, consistent with reality. For example, the sound of two metal objects colliding sounds metallic. A visually smooth object feels smooth.
Cognitive domain	The extent to which participants mentally engage with an experience	• The content of the VE helps inform (contextualize) participants' tasks in the VE.
Affective domain	The intensity and appropriateness of participants' emotional states in response to an experience	• Elements are designed to evoke emotional reactions while participants work on the environment's tasks.
Active domain	The degree of personal connection participants feel to an experience	• Participants can continuously reuse techniques that they learned on previous tasks on their later tasks.
Relational domain	The extent to which experiences are collaboratively created and socially reinforced	• The experience involves a high level of interaction with computer agents in the VE.
Integration	The degree to which elements of the five domains are intentionally blended to mutually support a consistent message	• A cohesive scheme of visual iconography, signs, and symbols is used throughout the VE to facilitate and reinforce VE tasks. For example, the color orange might be consistently used to signal the presence of a key object.
Persistence	The inclusion of factors that help engender and reinforce coherent mental schemata related to an experience	• The VE explicitly reminds participants of previous activities (e.g., prior tasks or narrative elements) from the environment, in order to link new activities to prior actions.

[a] The VET was originally developed in 2009 and is currently undergoing iterative theoretical and empirical refinement (see Chertoff, 2009; Chertoff et al., 2010). The exemplar VET items included in this table have not been fully validated for their psychometric rigor.

achieved by systematically introducing new sensations, behavior patterns of agents, and tasks that more and more closely resemble a particular troubling event for the patient.

By planning the system's design using experiential design theories, and then monitoring its development using an adapted VET, a virtual therapy simulation developer could more systematically design different levels of *realism* (or, in this case, psychological fidelity) and then give therapists dynamic, rheostatic control of their patents' experiences (e.g., a dial they can use to adjust the poignancy of the experience). When software developers think about realism, however, they may focus myopically on the sensory fidelity of a system. Using experiential design as a guide, and the VET as an index of it, system developers can better account for a wider range of experiential components, their integration, and their influence on memory formation. This more methodical, inclusive definition of the virtual therapy experience should, theoretically, better support therapeutic outcomes by providing therapists with a more sensitive lever for virtual treatment.

34.5.2 MILITARY: INFANTRY TRAINING SIMULATIONS

A wide variety of vehicle simulators, including flight or tank trainers, have been developed for military application. More recently, researchers have also created simulation-based training systems that support dismounted infantry operations (i.e., ground troops performing functions outside of a vehicle).

These simulators present unique challenges, because the real-life environments that they simulate are innately unmediated; this is in contrast to vehicles, for instance, that naturally moderate their occupants' experiences and, therefore, are typically more tractable modeling challenges (e.g., Muller, 2010).

The US Marine Corps (USMC), for example, is investing in various infantry simulators that use Virtual Battlespace 2 (VBS2), a three-dimensional first person shooter-style VE engine. The USMC uses these platforms to support a diverse array of training objectives, such as developing culture and foreign language proficiency (Johnson, 2010), building tactical decision-making prowess (Muller, 2010), or learning sustained observation and sociocultural sensemaking skills (Schatz & Nicholson, 2012). A common property of all these VE simulations is that many subtle factors influence actions in the real-life equivalent contexts. For instance, in a cultural trainer, the minute body language of another person may suggest different conversational approaches. Detailing all of these components in a simulation is a challenging, but ultimately manageable, task. However, without a guiding framework to ensure that the components are integrated across domains, the resultant experiences may lack realism, cohesion, and coherence. Emphasizing the integration of experiential domains, and using a VET to guide those thoughtful integration efforts, could help infantry simulators better convey holistic experiences to their training audiences, theoretically enhancing the relevance and appeal of the simulations, as well as their transfer of training to real-world contexts.

34.5.3 EDUCATION: CASUAL SERIOUS GAMES

Casual games, sometimes also called *microgames*, are designed for wide, diverse audiences; they include fairly straightforward rules, do not expect players to have specialized video game-playing skills, and do not require extensive time commitments (Juul, 2010). In addition to amusement, casual games can be used for training and education. Casual *serious* games, or casual games used for purposes other than entertainment (Susi, Johannesson, & Backlund, 2007), offer unique advantages. Corporations, for instance, appreciate their low development costs, as well as their flexibility to support short, 10 or 15 min increments of training as employees' schedules permit (White, 2007). Schools also appreciate casual games' ability to enhance retention (Brom, Preuss, & Klement, 2011).

Typical measures of presence do not effectively support (typical or serious) casual game design, however, because they emphasize the feeling of *being there* in more immersive virtual worlds. Thus, traditional theories of presence do not adequately capture—let alone suggest strategies for—design features in less physically immersive contexts.

Even though casual games are not physically immersive, they can still engross their players; consequently, theories of experiential design still apply. Even if a game is essentially nonrepresentational, it can still conjure a holistic experience if all experiential factors are present. In fact, using best practices of experiential design may have a particularly salient impact on simpler systems, such as casual games, since they feature fewer elements that can contribute to an experience and, therefore, each element plays a more important role.

For example, serious casual game designers may benefit from experiential design's emphasis on persistence. People are far more likely to recall new information when it can be positively related to their existing experiences (Bellezza, 1992; Bower & Gilligan, 1979; Ganellen & Carver, 1985; Mills, 1983). Holistic experiences increase the opportunity for environment triggers to result in a response from potentially larger networks of schema. Subsequently, new experiences can become associated with a wider range of past experiences, which may theoretically engender performance increases (Barfield et al., 1995). When serious casual games offer more opportunity for the user to relate to that experience, they can enhance a variety of factors, including retention, recall, and training transfer.

34.5.4 ENTERTAINMENT: VIDEO GAMES

Already, the entertainment industry attempts to address overall user experiences. This can be seen through both video games and locative works, such as those found at a theme park. In the case of video games, there has been a steady trend to increase the sensory component of games, such as the quality of graphics or sophistication of auditory effects (sensory), as well as the pathos of stories (affective), depth of characters (relational), and creativity of game-based tasks (cognitive). Although game developers may not intentionally employ an experiential design framework, they successfully make use of its domains to attract players to their games.

Brown and Cairns (2004) interviewed gamers to determine what factors most engaged them in their gaming experiences. They identified three distinct stages of video game experiences, ranging from engagement to engrossment, and, finally, to total immersion. As gamers progress from engagement to total immersion, they demonstrate greater levels of time investment, emotionality, and personal attachment to the game. A wide number of factors contribute toward keeping someone involved in a game: affect, scenario design, task difficulty, and feedback to name just a few. When all of these factors combine, they produce a state of deep involvement, that is, flow (Czikszentmihalyi & Czikszentmihalyi, 1992). The set of factors that engender increased video game engagement can be reframed as experiential design dimensions (e.g., emotionality as the affective dimension of personal attachment as the active dimension). Not surprisingly, higher levels of these factors are correlated with higher levels of enjoyment (Barfield & Weghorst, 1993; Pinchbeck, 2005).

Experiential design, and a comprehensive VET, can help game developers continue to reliably develop compelling game experiences by guiding the iterative design and evaluation process. A game-centric VET may also help developers assess, post facto, why certain games may fail to achieve popularity by drawing attention to the weaker elements of the experience. Most notably, however, a VET can help translate associated academic principles (e.g., from flow theory or cognitive psychology) into an actionable format for game designers.

34.6 CONCLUSION

The modeling and simulation community lacks an interpretable and extensible framework with which to apply the theoretical principles of presence toward VE design. While there exists a theoretical understanding of what contributes to the emergence of presence, there is no practitioner's guide or evaluation technique for VE design to promote the emergence of presence beyond the sensory dimension. Experiential design, however, has the potential to improve the design process of immersive virtual technologies. Further, the concept of *holistic experience* offers a more actionable construct for use with design, development, and applied evaluation because it better accounts for emergent properties, is divided into clearer levels associated with more actionable recommendations for developers, and is associated with less semantic confusion.

In experiential design, the primary goal is to create a holistic experience for participants through the incorporation of five experiential dimensions. Through this approach, designers can more intentionally control the connection participants' form between themselves, the various contributing factors of the experience and the content of the experience. In other words, the aim is to integrate various elements of experience—sensory, cognitive, affective, active (personal), and relational (social), as well as their integration and persistence—to construct a framework that is capable of eliciting an enhanced sense of *presence* and that will create a potential situation for developing accurate, memorable, and stable schema.

There are a variety of avenues open for future studies in experiential design. First, the VET, or other measures of VE experiential design, must be formalized and validated across a series of environments. In addition, the relationship between various individual differences and VE experiential evaluations should be more thoroughly explored. Variables such as age, education, and personality are all likely to affect how different participants individually experience a virtual holistic experience.

At a higher level, an exploration of whether, and to what extent, holistic environments lead to superior outcomes, such as greater training transfer, is needed to quantify the impact of experiential design in VEs. Environments designed to have superior holistic experiences should encourage participants to become more engrossed in their experiences, making it harder to break them away from their experiences, and well-designed holistic experiences should better support cognitive schemata formation, engendering various downstream benefits, from better retention to increased contextualization. The results of such a study could then fuel design guidelines for what types of designs are most relevant for an expected level of possibly competing stimuli.

REFERENCES

Alexander, A. L., Brunyé, T., Sidman, J., & Weil, S. A. (2005). From gaming to training: A review of studies on fidelity, immersion, presence, and buy-in and their effects on transfer in pc-based simulations and games. In *Proceedings of the Interservice/Industry Training, Simulation, and Education Conference (I/ITSEC)*. Washington, DC: NTSA.

Apter, M. J. (1992). *The dangerous edge: The psychology of excitement.* New York, NY: Free Press.

Barfield, W., & Weghorst, S. (1993). The sense of presence within virtual environments: A conceptual framework. In G. Salvendy & M. J. Smith (Eds.), *Human-computer interaction: Software and hardware interfaces* (pp. 669–704). Amsterdam, the Netherlands: Elsevier.

Barfield, W., Zeltzer, D., Sheridan, T., & Slater, M. (1995). Presence and performance within virtual environments. In W. Barfield & T. A. Furness (Eds.), *Virtual environments and advanced interface design.* New York, NY: Oxford University Press.

Battarbee, K. (2003). Defining co-experience. In *Proceedings of the 2003 International Conference on Designing Pleasurable Products and Interfaces* (pp. 109–113). Pittsburgh: ACM, PA.

Battarbee, K., & Mattelmaki, T. (2002). Meaningful product relationships. *Proceedings of the 3rd Conference of Design and Emotion*, Loughborough, England.

Bellezza, F. S. (1992). Recall of congruent information in the self-reference task. *Bulletin of the Psychonomic Society, 30*, 275–578.

Biocca, F., Harms, C., & Gregg, J. (2001). The networked minds measure of social presence: Pilot test of the factor structure and concurrent validity. *Proceedings of the 4th Annual International Workshop on Presence*, Philadelphia, PA.

Böcking, S., Gysbers, A., Wirth, W., Klimmt, C., Hartmann, T., Schramm, H., Laarni, J., Sacau, A., & Vorderer, P. (2004). Theoretical and empirical support for distinctions between components and conditions of spatial presence. In *Proceedings of the Seventh Annual International Workshop Presence 2004*, Mariano Alcaniz Raya, Beatriz Rey Solaz (Ed.), (pp. 224–231), Universidad Politécnica de Valencia, 2004.

Bordnick, P. S., Traylor, A., Copp, H. L., Graap, K. M., Carter, B., Ferrer, M., & Walton, A. P. (2008). Assessing reactivity to virtual reality alcohol based cues. *Addictive Behaviors, 33*(6), 743–756.

Bower, G. H., & Gilligan, S. G. (1979). Remembering information related to one's self. *Journal of Research in Personality, 13*, 420–432.

Brom, C., Preuss, M., & Klement, D. (2011). Are educational computer micro-games engaging and effective for knowledge acquisition at high-schools? A quasi-experimental study. *Computers & Education, 57*(3), 1971–1988.

Brown, E., & Cairns, P. (2004). A grounded investigation of game immersion. In *Proceedings of CHI 2004* (pp. 1297–1300). Vienna: ACM, Austria.

Bystrom, K. E., Barfield, W., & Hendrix, C. (1999). A conceptual model of the sense of presence in virtual environments. *Presence: Teleoperators and Virtual Environments, 8*(2), 241–244.

Cardenas, G., Munoz, S., Gonzalez, M., & Uribarren, G. (2006). Virtual reality applications to agoraphobia: A protocol. *CyberPsychology and Behavior, 9*(2), 248–250.

Chertoff, D. (2009). *Exploring additional factors of presence* (Doctoral dissertation). University of Central Florida, Orlando, FL.

Chertoff, D., Schatz, S., McDaniel, R., & Bowers, C. (2008). Improving presence theory through experiential design. *Presence: Teleoperators and Virtual Environments, 17*(4), 405–413.

Chertoff, D. B., Goldiez, B., & LaViola, J. J. (2010, March). Virtual experience test: A virtual environment evaluation questionnaire. In *Proceedings of the Virtual Reality Conference (VR), 2010* (pp. 103–110). Piscataway, NJ: IEEE.

Czikszentmihalyi, M., & Czikszentmihalyi, I. S. (1992). *Optimal experience: Psychological studies of flow in consciousness.* New York, NY: Cambridge University Press.

Draper, V. D., Kaber, D. B., & Usher, J. M. (1998). Telepresence. *Human Factors, 40*(3), 354–375.

Ellis, S. R. (1996). Presence of mind… a reaction to Thomas Sheridan's musings on telepresence. *Presence: Teleoperators and Virtual Environments, 5*(2), 247–259.

Forlizzi, J., & Battarbee, K. (2004). Understanding experience in interactive systems. In *Proceedings of the 5th ACM Conference on Designing Interactive Systems* (pp. 261–268). Cambridge: ACM, MA.

Freeman, J., Avons, S. E., Meddis, R., Pearson, D. E., & IJsselsteijn, W. A. (2000). Using behavioral realism to estimate presence: A study of the utility of postural response to motion stimuli. *Presence: Teleoperators and Virtual Environments, 9*, 149–164.

Freeman, J., Avons, S. E., Pearson, D. E., & IJsselsteijn, W. A. (1999). Effects of sensory information and prior experience on direct subjective ratings of presence. *Presence: Teleoperators and Virtual Environments, 8*, 1–13.

Ganellen, R. J., & Carver, C. S. (1985). Why does self-reference promote incidental encoding? *Journal of Experimental Social Psychology, 21*, 284–300.

Garg, R., Rahman, Z., Kumar, I., & Qureshi, M. N. (2011). Identifying and modelling the factors of customer experience towards customers' satisfaction. *International Journal of Modelling in Operations Management, 1*(4), 359–381.

Garrison, D. R., Anderson, T., & Archer, W. (2001). Critical thinking, cognitive presence, and computer conferencing in distance education. *American Journal of Distance Education, 15*(1), 7–23.

Gorini, A., & Riva, G. (2008). Virtual reality in anxiety disorders: The past and future. *Expert Review of Neurotherapeutics, 8*(2), 215–233.

Gunawardena, C. N., & Zittle, F. J. (1997). Social presence as a predictor of satisfaction within a computer-mediated conferencing environment. *American Journal of Distance Education, 11*(3), 8–26.

Heeter, C. (1992). Being there: The subjective experience of presence. *Presence: Teleoperators and Virtual Environments, 1*, 262–271.

Held, R., & Durlach, N. I. (1991). Telepresence, time delay, and adaptation. In S. R. Ellis (Ed.), *Pictorial communication in virtual and real environments*. New York, NY: Taylor & Francis.

Held, R. M., &Durlach, N. I. (1992). Telepresence. *Presence: Teleoperators and Virtual Environments, 1*(1), 109–112.

Huang, M., & Alessi, N. (1999). Presence as an emotional experience. In J. D. Westwood, H. M. Hoffman, R. A. Robb, & D. Stredney (Eds.), *Medicine meets virtual reality: The convergence of physical and informational technologies options for a new era in healthcare*. Amsterdam, the Netherlands: IOS Press.

IJsselsteijn, W. A., de Ridder, H., Freeman, J., & Avons, S. E. (2000). Presence: Concept, determinants and measurement. *Proceedings of the SPIE, 3959*, 520–529.

Insko, B. E. (2003). Measuring presence: Subjective, behavioral and physiological methods. In G. Riva, F. Davide, & W. A. IJsselsteijn, (Eds.), *Being there: Concepts, effects and measurement of user presence in synthetic environments* (pp. 109–119). Amsterdam, the Netherlands: IOS Press.

Johnson, W. L. (2010, July). Serious use of a serious game for language learning. *International Journal of Artificial Intelligence in Education, 20*(2), 175–195.

Kalawsky, R. S. (2000). The validity of presence as a reliable human performance metric in immersive environments. *Proceedings of the 3rd International Workshop on Presence*, Delft, the Netherlands.

Krijn, M., Emmelkamp, P. G., Olafsson, R. P., & Biemond, R. (2004). Virtual reality exposure therapy of anxiety disorders: A review. *Clinical Psychology Review, 24*(3), 259–281.

Lee, K. M. (2004). Why presence occurs: Evolutionary psychology, media equation, and presence. *Presence: Teleoperators and Virtual Environments, 13*, 494–505.

Lee, K. M., & Nass, C. (2001). Social presence of social actors: Creating social presence with machine-generated voices. *Proceedings of the 4th Annual International Workshop on Presence*, Philadelphia, PA.

Lemish, D. (1982). The rules of viewing television in public places. *Journal of Broadcasting and Electronic Media, 26*(4), 757–781.

Leppiman, A., & Same, S. (2011). Experience marketing: Conceptual insights and the difference from experiential marketing. In Prause, G., & Venesaar, U. (Eds.), *Regional business and socio-economic development 5: University Business Cooperation, Berliner Wissenschafts-Verlag*, (Vol. 5, pp. 240–258).

Lessiter, J., Freeman, J., Keogh, E., & Davidoff, J. (2000, March 27–28). Development of a cross-media presence questionnaire: The ITC-sense of presence questionnaire. *Proceedings of Presence 2000: The Third International Workshop on Presence*, Delft, the Netherlands.

Lessiter, J., Freeman, J., Keogh, E., & Davidoff, J. (2001). A cross-media presence questionnaire: The ITC-sense of presence inventory. *Presence: Teleoperators and Virtual Environments, 10*, 282–298.

Lombard, M. (1995). Direct responses to people on the screen television and personal space. *Communication Research, 22*(3), 288–324.

Lombard, M., & Ditton, T. (1997). At the heart of it all: The concept of presence. *Journal of Computer Mediated-Communication, 3*(2).

Lombard, M., Ditton, T. B., Crane, D., Davis, B., Gil-Egui, G., Horvath, K., Rossman, J., & Park, S. (2000, March). Measuring presence: A literature-based approach to the development of a standardized paper-and-pencil instrument. *Third International Workshop on Presence* (Vol. 240). Delft, the Netherlands.

Loomis, J. M. (1992). Presence and distal attribution: Phenomenology, determinants, and assessment. *In SPIE/IS&T 1992 Symposium on Electronic Imaging: Science and Technology* (pp. 590–595). International Society for Optics and Photonics.

Lowenthal, J. N. (2010). Using mobile learning: Determinates impacting behavioral intention. *American Journal of Distance Education, 24*(4), 195–206.

Malbos, E., Rapee, R. M., & Kavakli, M. (2012). Behavioral presence test in threatening virtual environments. *Presence: Teleoperators and Virtual Environments, 21*(3), 268–280.

Mantovani, G., & Riva, G. (1999). Real presence: How different ontologies generate different criteria for presence, telepresence, and virtual presence. *Presence: Teleoperators and Virtual Environments, 8*, 540–550.

Meehan, M. (2001). *Physiological reaction as an objective measure of presence in virtual environments* (Doctoral dissertation). University of North Carolina at Chapel Hill, Chapel Hill, NC.

Mills, C. J. (1983). Sex-typing and self-schemata effects on memory and response latency. *Journal of Personality and Social Psychology, 45*, 163–172.

Minsky, M. (1980). Telepresence. *Omni, 2*(9), 45–52.

Muller, P. (2010). The future immersive training environment (FITE) JCTD: Improving readiness through innovation. In *Proceedings of the Interservice/Industry Training, Simulation & Education Conference (I/ITSEC)* (Vol. 2010, pp. 3158–3164), Washington, DC: NTSA.

Nass, C., & Moon, Y. (2000). Machines and mindlessness: Social responses to computer. *Journal of Social Issues, 56*, 81–103.

Neisser, U. (1976). *Cognition and reality*. San Francisco, CA: Freeman.

Nowak, K. L., & Biocca, F. (2003). The effect of the agency and anthropomorphism on users' sense of telepresence, copresence, and social presence in virtual environments. *Presence: Teleoperators and Virtual Environments, 12*(5), 481–494.

Phillips, L., Interrante, V., Kaeding, M., Ries, B., & Anderson, L. (2012). Correlations between physiological response, gait, personality, and presence in immersive virtual environments. *Presence: Teleoperators and Virtual Environments, 21*(2), 119–141.

Pinchbeck, D. M. (2005). Is presence a relevant or useful construct in designing game environments? *Proceedings of the 3rd International Conference in Computer Game Design and Technology*, Liverpool, U.K.

Pine, J., & Gilmore, J. (1999). *The experience economy*. Boston, MA: Harvard Business School Press.

Pullman, M. E., & Gross, M. A. (2004). Ability of experience design elements to elicit emotions and loyalty behaviors. *Decision Sciences, 35*(3), 551–578.

Reeves, B., & Nass, C. I. (1996). *The media equation: How people treat computers, television, and new media like real people and places*. Chicago, IL: Center for the Study of Language and Information.

Riva, G. M., Bacchetta, M., Baruffi, M., & E. Molinari. (2001). Virtual reality-based multidimensional therapy for the treatment of body image disturbances in obesity: A controlled study. *Cyberpsychology and Behavior, 4*, 511–526.

Riva, G. M., Bacchetta, M., Cesa, G., Conti, S., & Molinari, E. (2001). Virtual reality and telemedicine based experiential cognitive therapy: Rationale and clinical protocol. In G. Riva, & C. Galimberti (Eds.), *Towards cyberpsychology: Mind, cognition and society in the internet age* (pp. 273–308). Amsterdam, the Netherlands: IOS Press.

Rizzo, A. A., Pair, J., McNerney, P. J., Eastlund, E., Manson, B., Gratch, J., … Swartout, B. (2005). Design and development of a VR therapy application for Iraq war veterans with PTSD. In J. Westwood (Ed.), *Technology and informatics* (pp. 407–413). Amsterdam, the Netherlands: IOS Press.

Rothbaum, B. O., & L. F. Hodges. (1999). The use of virtual reality exposure in the treatment of anxiety disorders. *Behavior Modification, 23*, 507–525.

Saladin, M. E., Brady, K. T., Graap, K., & Rothbaum, B. O. (2006). A preliminary report on the use of virtual reality technology to elicit craving and cue reactivity in cocaine dependent individuals. *Addictive Behaviors, 31*(10), 1881–1894.

Schank, D. A., & Abelson, T. (1977). *Scripts, plans, goals, and understanding*. Hillsdale, NJ: Lawrence Erlbaum.

Schatz, S., & Nicholson, D. (2012). Perceptual training for cross cultural decision making (session overview). In D. M. Nicholson & D. D. Schmorrow (Eds.), *Advances in design for cross-cultural activities part I* (Chapter 1, pp. 3–12). San Francisco, CA: CRC Press.

Schloerb, D. W. (1995). A quantitative measure of telepresence. *Presence: Teleoperators and Virtual Environments, 4*(1), 64–80.

Schmitt, B. (2003). *Customer experience management.* New York, NY: The Free Press.

Schmitt, B. (2010). Experience marketing: Concepts, frameworks and consumer insights. *Foundations and Trends in Marketing, 5*(2), 55–112.

Sheridan, T. B. (1992). Musings on telepresence and virtual presence. *Presence: Teleoperators and Virtual Environments, 1*, 120–125.

Shin, N. (2002). Beyond interaction: The relational construct of "transactional presence." *Open Learning, 17*(2), 121–137.

Shin, N. (2003). Transactional presence as a critical predictor of success in distance learning. *Distance Education, 24*(1), 69–86.

Short, J., Williams, E., & Christie, B. (1976). *The social psychology of telecommunications.* London, U.K.: Wiley.

Slater, M. (2009). Place illusion and plausibility can lead to realistic behaviour in immersive virtual environments. *Philosophical Transactions of the Royal Society B: Biological Sciences, 364*(1535), 3549–3557.

Spira, J., Pyne, J. M., Wiederhold, B., Wiederhold, M., Graap, K., & Rizzo, A. A. (2006). Virtual reality and other experiential therapies for combat-related PTSD. *Primary Psychiatry, 13*(3), 58–64.

Steuer, E. (1992). Assessment center simulation a university training program for business graduates. *Simulation & Gaming, 23*(3), 354–369.

Suri, J. F. (2003). The experience of evolution: Developments in design practice. *The Design Journal, 6*(2), 39–48.

Susi, T., Johannesson, M., & Backlund, P. (2007). *Serious games: An overview* (HS-IKI-TR-07-001). Skövde, Sweden: School of Humanities and Informatics, University of Skövde.

Usoh, M., Catena, E., Arman, S., & Slater, M. (2000). Using presence questionnaires in reality. *Presence: Teleoperators and Virtual Environments, 9*, 497–503.

Villani, D., Repetto, C., Cipresso, P., & Riva, G. (2012). May I experience more presence in doing the same thing in virtual reality than in reality? An answer from a simulated job interview. *Interacting with Computers, 24*(4), 265–272.

Voss, C., Roth, A. V., & Chase, R. B. (2009). Experience, service operations strategy, and services as destinations: Foundations and exploratory investigation. *Production and Operations Management, 17*(3), 247–266.

Voss, C., & Zomerdijk, L. (2007). Innovation in experiential services–An empirical view. In DTI (Ed.), *Innovation in services* (pp. 97–134). London, U.K.: DTI.

White, K. (2007). Casual games get serious. *Casual Games Quarterly, 2*(3). http://archives.igda.org/casual/quarterly/2_3/index.php?id=2

Witmer, B. G., & Singer, M. J. (1998). Measuring presence in virtual environments: A presence questionnaire. *Presence: Teleoperators and Virtual Environments, 7*, 225–240.

Youngblut, C., & Perrin, B. M. (2002). Investigating the relationship between presence and performance in virtual environments. *IMAGE 2002 Conference*, Scottsdale, AZ.

Zahoric, P., & Jenison, R. L. (1998). Presence as being-in-the-world. *Presence: Teleoperators and Virtual Environments, 7*, 78–89.

35 Augmented Cognition for Virtual Environment Evaluation

Kelly S. Hale, Kay M. Stanney,
Dylan Schmorrow, and Lee W. Sciarini

CONTENTS

35.1 INTRODUCTION

Augmented cognition seeks to revolutionize human–computer interactions by creating a closed loop between the user and the system, one in which the real-time cognitive state of a user is captured via neuroscientific tools and used to modify or adapt the system to optimize human performance. While the notion of better coupling humans and systems together is not new (Licklider, 1960), advances made during the *decade of the brain* (1990–1999) provided neurological and physiological science and technology advances that allow for real-time capture and understanding of the human cognitive state using noninvasive means, thereby providing a key component to the realization of augmented cognition systems. The formal research community stemmed from the Defense Advanced Research Projects Agency's (DARPA's) Augmented Cognition program*, which was initiated in 2000. This initial research into the field of augmented cognition focused on identifying and mitigating human limitations in cognitive processes, particularly attention and working memory, which place a ceiling on the capacity of the brain to process and store information with the goal of optimizing human performance. Since that time, great strides in sensor technology, software-based algorithms to classify a variety of human cognitive states in real time, and software design solutions that incorporate dynamic interfaces to support real-time updates to optimize human performance given a variety of states have been made. While such augmented cognition sensors and systems are beginning to emerge on the market (Libelium, 2014; Lomas, 2014), ongoing research continues to stretch the bounds of science and technology as areas of application are beginning to unfold. One area in which augmented cognition could have value is that of evaluating a virtual environment (VE).

VEs, a display medium that can be used to interact, engage, and inform a user within a closed-loop fashion, are proposed to be alternative realities that can stimulate our senses to generate desired cognitive, affective, and physical responses. While initial VE designs focused on *matching reality* as close as possible, it has since been realized that the desired user responses from VE exposure may not require *real-world mapping* across all modalities, and the extent to which physical, functional, and psychological fidelity are required is dependent upon the targeted use

---

* Augmented Cognition program later became Improving Warfighter Information Intake Under Stress program.

case. Thus, there is a need to determine how to get the amount of *reality* in virtual reality right. While there has been research focused on quantifying the adverse effects associated with VE experiences (e.g., disorientation, instability, sickness; see Keshavarz, Hecht, & Lawson, 2014, Chapter 26; Lawson, 2014, Chapter 23; Lawson, 2014, Chapter 24; Viirre, Price, & Chase, 2014, Chapter 22; Welch & Mohler, 2014, Chapter 25), less focus has been placed on identifying reliable and valid metrics that can be used to prescribe the appropriate level of fidelity to optimize the user's VE experience. Augmented cognition systems, where real-time sensors are used to evaluate and quantify the user state, could be used to address this gap in two ways. The first is in enabling more effective design and development of VEs, while the second is in the assessment of how well the designed and developed VEs achieve their desired goals.

Designing and developing effective VEs—systems that are easy and intuitive to use and that do not incur excessive cognitive costs—requires a deep understanding of users' needs and the degree to which proposed VE systems enable, or disable, them. In the past, this understanding relied on the representations of a user's cognitive and affective states inferred from behavioral measures. With augmented cognition, what was previously inferred can now be directly measured. Augmented cognition tools provide a unique *window* into the human mind where high-resolution representations of a user's cognitive and affective states can be developed. These data can, in turn, be used to specify the fidelity levels that optimize the human response to a VE system design. In terms of evaluation, many of the uses for which VEs are developed focus on enhancing cognitive skills. Augmented cognition provides a unique set of metrics that allow designers and developers to represent underlying changes in user cognition resulting from exposure to VE, providing a means to quantify the impact of VE on a user's performance. These data can be used to assess if a VE system is indeed meeting its training or other experience-based goals. In short, not only can augmented cognition be used in the design of VEs to reduce costs by providing VE designers valuable information on what are the critical elements of their VE, thereby specifying the *appropriate* level of fidelity, it can also be used to improve the quality of the experience by providing a more personalized interaction (e.g., tailored training) based on real-time cognitive and affective state assessment.

The use of augmented cognition in the design and evaluation of systems requiring human interaction with technology is not new. At the turn of the century, the DARPA augmented cognition program funded many efforts applying the augmented cognition technology in a wide range of applications. One of these efforts was a partnership with Lockheed Martin Advanced Technology Laboratory, where Tomahawk cruise missile operators were instrumented with augmented cognition neurocognitive activity monitors for which specialized algorithms were developed to drive a range of cognitive-inspired gauges (Poythress et al., 2006). These gauges provided designers and evaluators of VE training solutions with the real-time assessment of neurocognitive activity, such as visual and spatial working memory, attention, and executive functioning. This use of augmented cognition in the design, and more importantly the evaluation of the design, provided designers unique insight into the VE command and control system they had developed. Instead of having to interrupt tasks and ongoing experiments to ask users what they liked, preferred, or were struggling with, developers could simply monitor the activity and assess their designs. This led to rapid redesign of components that users struggled with or quick adoption of designs that were shown to be highly effective. This early use of augmented cognition in the design and evaluation of VE systems demonstrated the great potential attainable through the union of these two technologies. This chapter will highlight the technologies, processes, methods, and roles of augmented cognition in the design and evaluation of VE systems. First, a look at the traditional VE design and evaluation approaches.

35.2 TRADITIONAL VE DESIGN AND EVALUATION APPROACHES

VEs present a challenge to human–system interaction designers and evaluators across educational, entertainment, and training domains. Specifically, the varying types and levels of interaction that can be supported by VEs are a complexity multiplier. Not limited to the visual domain, VE interaction modalities include tactile, gestural, olfactory, auditory, and speech. If the utilization, coordination, and application of multimodal VE system interaction capabilities are not carefully considered in their design and evaluation, fielded systems may be rendered ineffective and difficult to use, if not altogether useless. There are many ways this complexity can become nettlesome. For example, interactivity involves the ability to navigate and get hopelessly lost (Darken & Peterson, 2014, Chapter 19). It allows objects to be manipulated, but only if one can figure out how to reorient and relocate them with the haptic interfaces that often accompany virtual worlds (Dindar, Tekalp, & Basdogan, 2014, Chapter 5). It provides multimodal sensory cues that can produce a sense of presence (Chertoff & Schatz, 2014, Chapter 34) if well designed, but lead to issues such as the startle effect, frustration, and confusion if ill-designed (Viirre et al., 2014, Chapter 22), while potentially inflicting an individual with sensorial mismatches that can lead to an emetic response (Lawson, 2014, Chapter 23), as well as irksome aftereffects. This multitude of potential shortcomings demonstrates that the cost of getting the design wrong with VEs can be very high. For this reason, there have been efforts to extend traditional usability evaluation approaches to meet the specific needs of VE assessment (Gross, 2014, Chapter 20). These efforts have sought to address the key characteristics unique to VEs, while at the same time identifying the limitations of existing usability methods for assessing VE systems (see Table 35.1; Stanney, Mollaghasemi, Reeves, Breaux, & Graeber, 2003).

The list of limitations presented in Table 35.1 demonstrates the need to extend traditional usability methods to consider VE system interface usability (i.e., issues associated with interaction—to include wayfinding, navigation, and object manipulation, and multimodal system output—to include visual, auditory, haptic, olfactory, and gustatory cues), as well as VE user interface usability (i.e., issues associated with engagement—to include immersion and presence, and side effects—to include difficulties related to comfort, sickness, and aftereffect). Taking these two dimensions of VE usability into account, one can then use a multicriteria assessment approach to VE usability, which is exactly what the Multicriteria Assessment of Usability for Virtual Environments (MAUVE) system does (Stanney et al., 2003). Specifically, MAUVE supports a heuristic evaluation, where expert evaluators can review a VE system and predict its usability based on fit or lack of fit with a detailed list of design considerations that aid in assessing VE usability criteria. The MAUVE methodology uses a two-stage evaluation process, with the first stage focused on evaluating traditional usability heuristics (Nielsen, 1993) and the second phase focused on evaluating and prioritizing VE-specific usability criteria according to the needs of a particular application. This latter evaluation requires evaluators to assess VE-specific usability criteria via either user testing (e.g., interaction, multimodal system output), questionnaire responses (e.g., engagement, sickness, side effects), or a combination of both. While user testing results and subjective reports are informative, they are limited in their ability to assess the unobservable aspects of VE interaction, such as levels of confusion while *orienting* during object manipulation or wayfinding, which multimodal cues were attended to, the level of physiological engagement (e.g., arousal), the precursors to an emetic response, or lingering proprioceptive or visual aftereffects. Heuristic-based design and assessment techniques, such as those supported by the MAUVE system, do not provide a means of assessing such

---

TABLE 35.1
Limitations of Traditional Usability Methods for Assessing VEs

- Multidimensional object selection and manipulation characteristics of 3D space are not captured by traditional point and click interaction evaluation approaches.
- Quality of multimodal system output (e.g., visual, auditory, haptic) is not fully characterized by current assessment methods.
- Sense of presence/immersion and aftereffects present new dimensions to usability evaluation.
- Collaborative VEs are not well addressed by traditional single-user task-based assessment methods.

---

unobservable aspects of VE interaction, which if poorly designed can prevent a VE system from being adopted. To address this gap in the VE system design and evaluation, there is a need for more real-time assessment. Real-time measures of behavior (e.g., interactions, eye tracking) and physiological response (e.g., engagement measures that can be provided by electroencephalography [EEG], arousal measures provided by galvanic skin response [GSR], or heart rate) can uncover those unobservable aspects of VE interaction and multimodal system outputs that lead to positive effects (e.g., high presence and comfort) and those that lead to negative effects (e.g., disorientation, sickness, aftereffects). As aforementioned, augmented cognition, which involves near real-time closed-loop symbiosis between human and computer, is particularly well-suited to filling this measurement gap (Hale, Stanney, & Schmorrow, 2012). With augmented cognition, human state can be captured and analyzed in real time and used to assess the effectiveness of a VE system design.

35.3 AUGMENTED COGNITION DESIGN AND EVALUATION TECHNIQUES

Traditional design and evaluation techniques focus on overt behavioral observations (often outcome-based measures, indicating task success or failure), or self-report of cognitive or affective state. Such measures are limited in that they do not provide continuous data (i.e., they are often intermittent and may be detached from the VE), are subjective (in the case of self-reports), and require conscious awareness (Parsons & Courtney, 2011). Augmented cognition design and evaluation techniques, in comparison, use sensors to capture data on a continual basis, do not *interrupt* an interaction with a VE during task performance, can be captured without conscious awareness, and are objective measures of human cognitive and affective states. Such sensor-based measures can be used to evaluate the utility and usability of a VE design and quantify a VE experience. Further, such measures can be used to evaluate the degree of system fidelity required to optimize human responses or identify whether a user's response to an event in a VE matches their response to the same event in the real world (Jones, Greenwood-Ericksen, Hale, & Johnston, 2008; Parsons & Courtney, 2011).

Multiple sensor technologies are available to capture signals related to the central nervous system (brain and spinal cord), somatic nervous system (muscle activity), and autonomic nervous system (major glands and organs; Allanson & Fairclough, 2004), which may be used in isolation and/or combination to capture cognitive and affective state in real time. Sensors from across these categories have been utilized within augmented cognition systems as outlined in Table 35.2. Such sensors have been used to evaluate psychophysiological responses to VEs, including examining the impact of avatar customization levels on children's electrodermal response (Bailey, Wise, & Bolls, 2009), physiological indicators of presence (Fairclough & Venables, 2006; Meehan, Insko, Whitton, & Brooks, 2002), responses associated with virtual rehabilitation settings (Fidopiastis, Hughes, Smith, & Nicholson, 2007), and the impact of scenario immersion on startle blink response and heart rate interbeat intervals (IBIs) (Parsons et al., 2009). In most cases, a combination of sensors has been utilized to provide an indication of a given cognitive or affective state. These data are then related to the task at hand for interpretation, as any one physiological signal can be influenced by a number of other factors in addition to cognitive or affective state, including physical fitness, ingested stimulants (e.g., caffeine), physical activity, and daily circadian rhythms (Hale et al., 2012).

35.4 CURRENT CHALLENGES/LIMITATIONS

The implementation of augmented cognition technology in the design and evaluation of VEs faces multiple challenges. Sensor technology available today can be cumbersome, invasive, and costly. However, continued advancements in this area promise to provide cost-effective, reliable solutions in the near future. For example, a sensor suite, which included posture sensors, EEG, and a heart rate monitor, was developed for under $600 and was used to quantify cognitive and affective states in real time while users observed virtual combat scenarios (Carroll et al., 2011; Kokini et al., 2012). The form

TABLE 35.2
Augmented Cognition Sensors

Sensor Technology	Measure	Description	States Measured
Electroencephalography (EEG)	Alpha, beta, theta, gamma, sigma waveforms	Records electrical activity produced by the brain via sensors placed on the scalp with high temporal resolution; limited spatial resolution	Sensory memory, working memory, attention, executive function, alertness/vigilance, engagement, cognitive load, workload
	Event-related potentials (ERPs)	Records activity in response to an event (e.g., stimulus presentation)	Detection (hit, miss, false alarm, correct rejection), visual interest
Functional near-infrared imaging (fNIR)	Blood-oxygen-level-dependent (BOLD) signal	Measures blood oxygenation and volume changes in brain relative to where optical sensors are placed on the head	Spatial and verbal working memory, workload, loss of concentration
Eye-tracking camera	Eye/gaze tracking	Measures visual attention allocation	Cognitive load, attention, task difficulty
	Pupillometry	Tracks change in pupil dilation/constriction	Measure of mental workload, cognitive resource taxation, increases in perceptual, cognitive, and response-related processing demands
	Blink rate	Counts reflexive blinks during a trial	Mental workload, engagement in decision making, and memory tasks
Electromyography (EMG)	Muscular microtremors	Activates skeletal muscles measured by sensors on the skin surface	Attention, effort, stress
Body position/posture-tracking systems	Pressure, position, and velocity	Shifts in posture as measured via pressure-sensing chair or head tracker	Attention
Chest band	Respiration	Changes in breathing pattern (rate, depth)	Workload, emotional state (calm–excited; relaxed–tense)
Electrocardiogram (ECG)	Heart rate	Beats per minute	General indicator of cognitive workload
	Heart rate variability	Varied duration of time between heartbeats, the interbeat interval (IBI), or mean heart period	Arousal, engagement, workload
Electrodermal conductance sensor	Skin conductance level	Measures changes of the electrical impedance of the skin	Attention, working memory, cognition, attention, emotion, engagement, anxiety, stress

Source: Adapted from Hale, K.S. et al., *Handbook of Human-Computer Interaction*, 3rd ed., Taylor & Francis/CRC Press, Boca Raton, FL, 2012; Sciarini, L.W., Noninvasive physiological measures and workload transitions: An investigation of thresholds using multiple synchronized sensors (Unpublished doctoral dissertation), University of Central Florida, Orlando, FL, 2009.

factor of these sensors is also making great gains, from cumbersome wires and gel-adhered sensors to tattoos and adhesive bandages (epidermal electronics) placed directly on the skin (Kim et al., 2011).

Additional challenges with augmented cognition design and evaluation solutions include data synchronization and classification. Synchronization of many sensor data streams is critical for the accurate assessment of human state, particularly when sensors with high sampling frequencies (e.g., EEG and ECG) are combined with sensors that may have lower sampling frequencies as well as differing response time windows. For example, an event-based pattern may be evident within an EEG signal within 300 ms, while the same event may be related to changes in pupillometry or heart rate data within a 2–6 s window. Such challenges with different sampling frequencies, local timing information, and timing accuracy all need to be resolved to achieve real-time data synchronization, logging, and analysis (Barber & Hudson, 2011). Frameworks (Barber & Hudson, 2011) and passive algorithms (Olson, 2010) have been developed to address data synchronization issues; however, more research is needed particularly in the areas of classification and accounting for individual differences. In terms of classification, the algorithms derived from neurophysiological sensor data must be of adequate sensitivity to appropriately capture and reliably represent the psychological state of interest as opposed to other related processes, such as behavioral movement or environmental conditions (Allanson & Fairclough, 2004). Continued classification research is needed to ensure that sensor data patterns indicative of specific responses of interest (e.g., cognitive workload, engagement) can be reliably and accurately identified from the noise in the data.

Another challenge is individual differences. Individuals differ between one another in their physiological and neural responses as well as within themselves based on time of day or on successive days caused by changes in diet, sleep, physical exertion, and/or variations in mental state. The quality of data collected and analyzed within an augmented cognition system will thus be dependent on the quality of baseline or calibration procedures that have taken place to account for individual differences. Baseline procedures most often collect state sensor data from individuals during a *rest* period prior to insertion into the closed-loop augmented cognition system (e.g., Berka et al., 2004). Alternatively, baseline procedures can be assessed during initial integration into the augmented cognition system (Fishel, Muth, & Hoover, 2007) using techniques such as orthogonal signal correction (Alamudun, Choi, Gutierrez-Osuna, Khan, & Ahmed, 2012). Practitioners are encouraged to implement baseline techniques to reduce signal bias associated with individual differences and improve the quality of measures that are captured by an augmented cognition design and evaluation system.

35.5 IMPLICATIONS FOR VE SYSTEM DESIGN AND EVALUATION

Utilizing augmented cognition solutions to design and evaluate VEs provides the opportunity to more fully characterize real-time user reactions and interactions than is typically achieved today. With such technology, system designers, evaluators, and acquisition personnel can not only understand how well a VE supports users in their tasks—whether users are able to achieve their mission (outcome performance)—but can also capture how effectively they are able to accomplish their goals (process outcomes). In other words, the underlying causes of performance inefficiencies/deficiencies can be captured by coupling behavioral (observable) outcome-based data (i.e., data tied to specific events within the VE) with real-time (unobservable) cognitive state (e.g., workload, overload, drowsiness, attention) and affective state (e.g., boredom, fear, anger, frustration) data. These data can provide root-cause diagnostics that relate performance breakdowns—an error—to human state breakdowns: periods of high workload, inattention, or frustration. This detailed understanding can be used to diagnose design improvement opportunities within VE systems. This detailed understanding can be used to complete root cause evaluation processes, resulting in diagnosis of design improvement opportunties within VE systems (e.g., TEE-FAST; Jones et al., 2008).

35.6 CONCLUSIONS

While augmented cognition focuses on enabling systems to understand a user's intended actions and cognitive needs in real time, VEs provide a medium for translating this understanding into actionable interactions and appropriate information presentation based on a user's current cognitive and affective state. Consequently, augmented cognition tools are not simply interesting options or features for VE developers and users to consider, but rather they represent new capabilities that can significantly enhance the use of VE—and should be on the critical path for future VE development and use. Arguably, operational use of VEs in any domain, ranging from entertainment to science exploration to operation of autonomy via a VE, will be dramatically improved by the real-time assessment made realizable by augmented cognition technology. It is postulated that only once augmented cognition technologies, methods, and practices are seamlessly integrated into virtual worlds that the true promise of this technology in our society will be realized.

REFERENCES

Alamudun, F., Choi, J., Gutierrez-Osuna, R., Khan, H., & Ahmed, B. (2012). Removal of subject-dependent and activity-dependent variation in physiological measures of stress. In *Proceedings of the 6th International Conference on Pervasive Computing Technologies for Healthcare (PervasiveHealth)* (pp. 115–122). Brussels, Belgium: Institute for Computer Sciences, Social Informatics and Telecommunications Engineering. http://research.cse.tamu.edu/prism/publications/alamudun2012activityNoiseRemoval.pdf

Allanson, J., & Fairclough, S. H. (2004). A research agenda for physiological computing. *Interacting with Computers, 16*, 857–878.

Bailey, R., Wise, K., & Bolls, P. (2009). How avatar customizability affects children's arousal and subjective presence during junk food-sponsored online video games. *Cyberpsychology & Behavior, 12*(3), 277–283.

Barber, D., & Hudson, I. (2011). Distributed logging and synchronization of physiological and performance measures to support adaptive automation strategies. In D. D. Schmorrow & C. M. Fidopiastis (Eds.), *Foundations of augmented cognition, Lecture notes in computer science* (Vol. 6780, pp. 559–566). Berlin, Germany: Springer.

Berka, C., Levendowski, D. J., Cvetinovic, M. M., Petrovic, M. M., Davis, G., Lumicao, M. N., … Olmstead, R. (2004). Real-time analysis of EEG indices of alertness, cognition and memory with a wireless EEG headset. *International Journal of Human-Computer Interaction, 17*(2), 151–170.

Carroll, M., Kokini, C., Champney, R., Fuchs, S., Sottilare, R., & Goldberg, B. (2011). Modeling trainee affective and cognitive state using low cost sensors. *Proceedings of the Interservice/Industry Training, Simulation and Education Conference (I/ITSEC) 2011*, Orlando, FL. Paper #11215.

Chertoff, D., & Schatz, S. (2014). Presence in virtual environments. In K. S. Hale & K. M. Stanney (Eds.), *Handbook of virtual environments: Design, implementation, and applications* (2nd ed., pp. 855–870). Boca Raton, FL: CRC Press.

Darken, R. P., & Peterson, B. (2014). Spatial orientation, wayfinding, and representation. In K. S. Hale & K. M. Stanney (Eds.), *Handbook of virtual environments: Design, implementation, and applications* (2nd ed., pp. 467–492). Boca Raton, FL: CRC Press.

Dindar, N., Tekalp, A. M., & Basdogan, C. (2014). Haptic rendering and associated depth data. In K. S. Hale and K. M. Stanney (Eds.), *Handbook of virtual environments, design, implementation, and applications* (2nd ed., pp. 115–130). New York, NY: CRC Press.

Fairclough, S. H., & Venables, L. (2006). Prediction of subjective states from psychophysiology: A multivariate approach. *Biological Psychology, 71*, 100–110.

Fidopiastis, C. M., Hughes, C. E., Smith, E. M., & Nicholson, D. M. (2007, September 27–29). Assessing virtual rehabilitation design with biophysiological metrics. *Proceedings of Virtual Rehabilitation*, pp. 7–11, Venice, Italy.

Fishel, S. R., Muth, E. R., & Hoover, A. W. (2007). Establishing appropriate physiological baseline procedures for real-time physiological measurement. *Journal of Cognitive Engineering and Decision Making, 1*(3), 286–308.

Gross, D. (2014). Technology management and user acceptance of virtual environments. In K. S. Hale & K. M. Stanney (Eds.), *Handbook of virtual environments: Design, implementation, and applications* (2nd ed., pp. 493–504). Boca Raton, FL: CRC Press.

Hale, K. S., Stanney, K. M., & Schmorrow, D. D. (2012). Augmenting cognition in HCI: 21st century adaptive system science and technology. In J. Jacko & A. Sears (Eds.), *Handbook of human–computer interaction* (3rd ed., pp. 1343–1358). Boca Raton, FL: Taylor & Francis/CRC Press.

Jones, D. L., Greenwood-Ericksen, A., Hale, K., & Johnston, M. (2008). The physiological assessment of VE training system fidelity. *Proceedings of the Annual Conference of the Human Factors and Ergonomics Society, 52*(27), 2107–2111.

Keshavarz, B., Heckt, H., & Lawson, B. D. (2014). Visually induced motion sickness. In K. S. Hale, & K. M. Stanney (Eds.), *Handbook of virtual environments: Design, implementation, and applications* (2nd ed., pp. 647–698). Boca Raton, FL: Taylor & Francis Group, Inc.

Kim, D.-H., Lu, N., Ma, R., Kim, Y., Kim, R., Wang, S.,… Rogers, J. A. (2011). Epidermal electronics. *Science, 333*, 838–843.

Kokini, C., Carroll, M., Ramirez-Padron, R., Hale, K., Sottilare, R., & Goldberg, B. (2012). Quantification of trainee affective and cognitive state in real-time. *Proceedings of the Interservice/Industry Training, Simulation and Education Conference (I/ITSEC) 2012*, Orlando, FL. Paper #12064.

Lawson, B. (2014). Motion sickness symptomatology. In K. S. Hale & K. M. Stanney (Eds.), *Handbook of virtual environments: Design, implementation, and applications* (2nd ed., pp. 531–600). Boca Raton, FL: CRC Press.

Libelium (2014). *e-Health sensor platform for biometrics and medical applications*. Available from http://www.libelium.com/130220224710/. Viewed August 14, 2013.

Licklider, J.C.R. (1960). Man-computer symbiosis. IRE Transactions on Human Factors in Electronics, March, pp. 4–11.

Lomas, N. (2014). PIP is a bluetooth biosensor that aims to use your phone to gamify beating stress. *TechCrunch*. Available from http://techcrunch.com/2013/06/17/pip/. Viewed August 13, 2013.

Meehan, M., Insko, B., Whitton, M., & Brooks, J. (2002). Physiological measures of presence in stressful virtual environments. *ACM Transactions on Graphics—SIGGRAPH, 21*(3), 645–653.

Nielsen, J. (1993). *Usability engineering*. Boston, MA: Academic Press.

Olson, E. (2010). A passive solution to the sensor synchronization problem. *Proceedings of the IEEE/RSJ International Conference on Intelligent Robots and System (IROS)*, Taipei, Taiwan. Available from http://april.eecs.umich.edu/papers/details.php?name = olson2010.

Parsons, T. D., & Courtney, C. G. (2011). Neurocognitive and psychophysiological interfaces for adaptive virtual environments. In C. Röcker & M. Ziefle (Eds.), *Human centered design of e-health technologies* (pp. 208–233). Hershey, PA: IGI Global.

Parsons, T. D., Courtney, C., Cosand, L., Iyer, A., Rizzo, A. A., & Oie, K. (2009). Assessment of psychophysiological differences of west point cadets and civilian control immersed within a virtual environment. *Lecture Notes in Artificial Intelligence, 5638*, 514–523.

Poythress, M., Russel, C., Siegel, S., Tremoulet, P., Craven, P., Berka, C.,… Milham, L. (2006). Correlation between expected workload and EEG indices of cognitive workload and task engagement. In D. D. Schmorrow, K. M. Stanney, & L. M. Reeves (Eds.), *Foundations of augmented cognition* (2nd ed., pp. 32–44). Arlington, VA: Strategic Analysis, Inc.

Sciarini, L. W. (2009). *Noninvasive physiological measures and workload transitions: An investigation of thresholds using multiple synchronized sensors* (Unpublished doctoral dissertation). University of Central Florida, Orlando, FL.

Stanney, K. M., Mollaghasemi, M., Reeves, L., Breaux, R., & Graeber, D. A. (2003). Usability engineering of virtual environments (VEs): Identifying multiple criteria that drive effective VE system design. *International Journal of Human-Computer Studies, 58*(4), 447–481.

Viirre, E., Price, B. J., & Chase, B. (2014). Direct effects of virtual environments on users. In K. S. Hale, & K. M. Stanney (Eds.), *Handbook of virtual environments: Design, implementation, and applications* (2nd ed., pp. 521–530). Boca Raton, FL: CRC Press.

Welch, R. B., & Mohler, B. J. (2014). Adapting to virtual environments. In K. S. Hale, & K. M. Stanney (Eds.), *Handbook of virtual environments: Design, implementation, and applications* (2nd ed., pp. 627–646). Boca Raton, FL: Taylor & Francis Group, Inc.

Section VII

Selected Applications of
Virtual Environments

36 Applications of Virtual Environments

An Overview

Robert J. Stone and Frank P. Hannigan

CONTENTS

36.1 INTRODUCTION

It was Bob Jacobson who, when asked in 1992 about the prospects for commercial uptake of his company Worldesign's services (just before winning its first contract from Evans & Sutherland), said "the companies we're talking to really *do* want to adopt Virtual Reality (VR)… the problem is that they don't want be the *first* to take the risk of doing so" (Jacobson, 1992, personal communication).

And so it was that the VR pioneers of the early 1990s, still slowly recovering from the technological intoxication of the late 1980s, with its head-mounted displays, instrumented gloves, early haptic feedback systems, whole- and part-body spatial tracking systems, and VR gaming units, faced the stark realization that the honeymoon period was over and that real revenues were now essential to survival and growth. But surely, the very fact that VR was appearing on television programs and in prestigious and popular magazines across the globe meant that the so-called *serious* adopters of novel and innovative technologies—those eager to exploit for commercial benefit—would beat a path to the doors of the garage consultancies and overnight academic experts to ensure their business would be the first to benefit from immersion and telepresence. Not so. What confronted the vendors and developers of VR for nearly 15 years was a period of massive uncertainty, especially with regard to the emergence (or lack of emergence) of credible, real-world applications. What also confronted them was an era characterized by expensive, inaccessible, and often unreliable hardware and software technologies, most designed without due attention to the perceptual, motor, and cognitive capabilities and limitations of their target human users.

Yet, despite these issues, there were, worldwide, a few attempts to provide realistic, independent, and evidence-led support to potential industrial adopters of VR through the delivery of case studies and concept demonstrators. In the United Kingdom, for example, one early initiative set out to try and convince both the government and British industry of the value of VR in industry and commerce. Conceived and coordinated by the small VR team within the country's National Advanced Robotics Research Centre (Stone, 1992), the virtual reality and simulation (VRS) initiative came about following a short appearance on the BBC's *9 O'clock News* of a real industrial application, based on the real-time visualization of computer-aided design (CAD) data made available by Rolls-Royce. A small section of Rolls-Royce's Trent 800 civil aero engine, constructed using their Computervision CADDS 4X assets, had, in October 1992, been successfully converted to run on a Division Ltd. *SuperVision* transputer-based system (Figure 36.1, left)—a system that had been commissioned by the Robotics Centre over a year earlier. Although of a highly primitive visual quality and boasting a very low image frame or refresh rate, the demonstration confirmed the feasibility of robust and reliable CAD data conversion and had been well received by Rolls-Royce representatives.

The response to the appearance of this virtual engine during the BBC news feature—the first televised example of a *serious* application of VR in the United Kingdom—was unprecedented. Representatives from a wide range of British industries contacted the Robotics Centre expressing surprise to see that there appeared to be much, much more to VR than the trivial pursuit (much publicized at the time) of shooting virtual pterodactyls at so-called location-based entertainment centers. Building upon this enthusiastic response from industry, the National Advanced Robotics Research Centre team decided to launch the risk-sharing, *try-before-you-buy* VRS initiative in July 1993, bringing together initially 12, later 17, collaborating organizations to fund the world's first wholly industrially funded VR research, development, and exploitation program. The research and

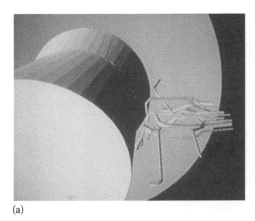 

(a)   (b)

FIGURE 36.1   (a and b) The Rolls-Royce Trent 800 VR Model—the original *SuperVision*-based demonstration (1992; left) and the later Onyx *InfiniteRealityEngine* version (1996; right).

development undertaken by the Robotics Centre team during VRS was to set the standard for all future engagements with industries wishing to exploit the power of VR to revolutionize their business practices and, thus, deliver competitive advantage.

It will come as little surprise that, as one of the founding members of VRS, Rolls-Royce chose to continue to investigate the feasibility of applying VR to aero engine maintenance evaluation. An important aim of this early study was to assess how VR might complement the company's CAD activities, as hinted earlier, especially where there were, at that time, serious limitations in the *intuitive* views one could gain of the CADDS 4X databases (this work was again based on the Trent 800—Figure 36.1, right) and its components. In particular, VR was considered to offer a logical design step between the CAD modeling activities and the costly fabrication of 1:1-scale engine mock-ups (Stone & Angus, 1995). Continuing on the aerospace theme, albeit with a different focus altogether, a small company based at United Kingdom's Manchester Airport, Airline Services, took the initiative to investigate the potential offered by VR in the assessment of new aircraft seat fabrics and cabin furnishings. The virtual model of the cabin interior had to be of a very high quality, both for real-time performance and visual impact—viewers needed to move freely and look very closely at the seat and carpet fabric design (Figure 36.2).

Working in collaboration with Rolls-Royce was another VRS founding member, Vickers Shipbuilding & Engineering Limited (VSEL—today part of BAE Systems). As with Rolls-Royce, they too had a need to do away with highly expensive physical models. For example, at the company's Barrow-in-Furness submarine complex, a one-fifth scale wood-and-plastic model of a Trident nuclear submarine was housed within (as one might expect) an enormous warehouse. Male doll figures (such as *GI Joe* or *Action Man*), despite their anthropometric inaccuracies, were used for basic ergonomics evaluations of deck and compartment layouts. The financial and manpower resource required to build a model of this complexity was huge (in excess of $18 million per model in the early 1990s), not to mention the insurance implications of protecting such a facility. Compare, in Figure 36.3, the visual results of the compartment spaces converted from the company's CAD database in the early 1990s (Figure 36.3a) (hosted on a Silicon Graphics Onyx *Infinite Reality Engine*) with the submarine environment developed for the United Kingdom's *SubSafe* project in 2009 (Figure 36.3b) (hosted on a Sony Vaio VGN laptop).

Similar activities took place in other branches of the engineering community. For example, the first VRS project sponsored by British Nuclear Fuels plc (public limited company) focused on the use of VR to prototype and evaluate, from an ergonomics perspective, the central control room (CCR) of the then-planned mixed oxide plant (SMP), to be constructed at the company's

FIGURE 36.2   Aircraft cabin concept furnishing evaluation using VR.

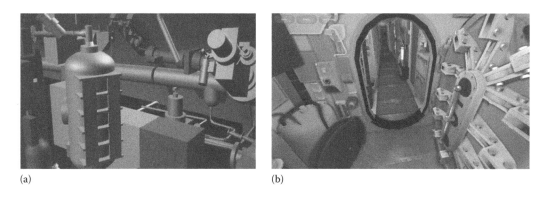

(a)   (b)

FIGURE 36.3   (a) The early VR submarine compartment model (1996; left) and (b) an internal view of the more recent *SubSafe* VR trainer (2008; right).

Sellafield site. The VR *construction* and evaluation process took place using not only the Silicon Graphics machine mentioned earlier but also, and importantly at the time (from a future affordability perspective), a Pentium PC, running Superscape's VR toolkit (VRT), interestingly without any significant degradation in visual and interactive quality (Figure 36.4). This development allowed BNFL representatives to transport the proposed design to different company sites and make desired changes within minutes, rather than the many hours it would have taken them, had they commissioned a large-scale physical model. Representatives of the SMP team were able to move around different versions of the CCR model in 3D and quickly assess preliminary proposals for workstation layout and other features, such as lines of sight across the control room, panel mountings, and light fitting distribution. One encouraging result of the demonstrations was that the virtual CCR helped to remove the technical and language *barriers* so often found when engineers, designers, ergonomists, and managerial personnel from the same company come together for project reviews.

The focus of an exploratory VRS project for Imperial Chemicals Industry (ICI) (once one of the largest manufacturing organizations in the world) involved the production of a detailed VR model

(a)   (b)

FIGURE 36.4   (a) The virtual BNFL CCR: *graphics supercomputer* rendering and (b) PC version.

(a)   (b)

FIGURE 36.5   (a) A view of the virtual ICI SulFerox plant and (b) pump maintenance task.

of selected portions of the company's SulFerox petrochemical plant design for North Humberside. ICI was interested in assessing VR in terms of how the technology might deliver competitive advantage in their future business strategies. A range of *mini demonstrators* were specified as part of the company's VRS project and included *instant change* to important features of the plant, personnel training, green-field construction using a simple virtual crane, and vessel or pump maintenance (Figure 36.5). The United Kingdom's Health & Safety Executive took this project a stage further by adapting the VR model to train personnel in the procedures necessary for undertaking vessel entry and internal cleaning.

Other opportunities that arose during the VRS initiative demanded similar interactive functions to those developed for the engineering and aerospace sectors but were targeted at applications that were, hitherto, not given serious attention throughout the emerging VR *market*. One such application that received considerable press coverage (although more for the fanciful idea of *teleshopping*, as opposed to issues of design and prototyping) was the *virtual supermarket*. The original supermarket effort was sponsored by the Co-operative Wholesale Society (CWS) and was designed to deliver visualization of supermarket layouts, allowing marketing personnel to interact with space planning software, product buyers, and supermarket managers. A similar but more detailed and extensive project was subsequently commissioned by Sainsbury's, the result of which led to the

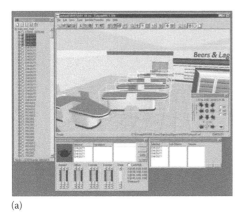

(a)   (b)

FIGURE 36.6 (a) The virtual supermarket: Sainsbury's $Concept_{VR}$ system and (b) product packing evaluation.

development of a PC-based future concept store design system, $Concept_{VR}$, enabling designers to build a virtual supermarket in less than 40 min (Figure 36.6, left). Product-oriented companies, such as Unilever and Procter & Gamble, also joined the consortium to evaluate how VR could make their new products more visible than their competitors on the supermarket shelf (Figure 36.6, right) and more usable in the home (i.e., through the investigation of such features as package opening sequences and storage).

The VRS initiative was influential not only in helping to design the stores in which one shops but also in maintaining the services on and under the streets outside. The first example demonstrator project of this kind was carried out for Welsh Water and focused on a built-up area within the town of Wrexham (Wales). The requirement for VR was to provide an interactive and engaging means of explaining to shop owners and members of the general public why it was necessary to undertake disruptive roadwork in the area (Figure 36.7, left). One popular feature presented with the VR demonstration (using Superscape's VRT) was the augmentation of the area model with digitized photographs of the *actual* sewer condition under a given portion of the virtual pavement (obtained from previous television surveys—see Figure 36.7, right). This information was displayed as a multimedia window insert at the request of the user (by selecting one of a number of small icons distributed along the length of the route concerned).

(a)   (b)

FIGURE 36.7 (a and b) Virtual roadworks awareness-raising demonstration, based on the Welsh town of Wrexham.

(a) (b)

FIGURE 36.8 (a) Virtual business unit in Mid Wales (left) and (b) showing one interior configuration example (right).

Finally, and addressing the marketing sector, the Development Board for Rural Wales selected a small Business Enterprise Park in Newtown, Mid Wales, for their assessment of a concept they described as *VR in a briefcase*. The aim here was to *take* Mid Wales to businesses across the United Kingdom that might be on the verge of relocation. Using Superscape's VRT with a powerful (at the time) multimedia laptop, complete with SpaceMouse and a Virtual IO *i-Glasses* head-mounted display, a model of the business site in Newtown was developed. Not only did this scenario include digitized views of the surrounding countryside, but it also allowed the potential leaseholders to enter a *vacant* business unit and, by using multimedia text and photograph panels, interrogate nearly every feature of the unit architecture and services (e.g., flooring, lighting, double glazing, heating). Then, by depressing a single function key, the bare interior could be changed within seconds to an open plan office environment or an automated factory setup, complete with caged robots (Figure 36.8).

These are just a small selection of applications that helped the VRS initiative demonstrate that, with the appropriate level of engagement with industry and the early provision of sound *non-technology-push* advice and the delivery-focused concept demonstrators, it was possible, given the right resources and skills, to develop a business-led approach to delivering VR services (although nowhere even close to the level of financial return that was being suggested in some of the over-priced VR market surveys that were being published at the time).

But did VRS really make a difference? Are the organizations that took part still exploiting their original investment in VR? Some took VR very seriously indeed. VSEL/BAE Systems, for example, built on the lessons learned as part of their investment in the VRS initiative and established their own digital prototyping and design review theater at Barrow-in-Furness, which was used to support reviews for a number of vessels, both surface and submarine. The company also installed two mini-VR stereoscopic projection facilities inside the submarine construction and assembly hall itself, providing workers with instant access to real-time 3D models of the *Astute* class of submarine. BNFL and Sainsbury's took their facility design prototypes a stage further, but then, due to key members of the organization moving on, the momentum was lost and the VR systems were never developed to their full potential. Many other application demonstrators, too many to list here, suffered similar fates—changes in stakeholders, curtailed research budgets, and a realization that the cost of implementing and maintaining an in-house capability would be extremely prohibitive.

So, from the perspective of potential adopters in application domains such as engineering, defense, and medicine, VR provided a roller coaster ride of achievement and failure throughout the 1990s and early 2000s. It was the belief of the die-hard VR proponents of the time that, come the end of the twentieth century, users of real-time, multisensory computer environments would be wearing head-mounted displays, instrumented gloves, suits, and body-mounted spatial tracking systems or would be found sitting at *Star Trek*™-like consoles with stereoscopic displays or standing within multiwall projection display facilities, such as the *CAVE* (cave automatic virtual environment™). Such a vision of truly *immersive* forms of VR simply did not come to pass and is, even today, many, many years away from a believable implementation.

To use the terminology used to describe *Gartner hype cycles* (Fenn & Linden, 2005), by 1995, VR was already well past its *peak of inflated expectations* and was on course to hit the bottom of the *trough of disillusionment*, where, when reached, it struggled—some even say failed—to make a credible comeback. VR vendors and so-called value-added resellers changed direction (such as Superscape and Division), disappeared altogether (as in the case of Virtuality, VPL and Sense8), or, in the case of one Swedish company, became the subject of a highly publicized bankruptcy and subsequent criminal proceedings. There were also significant failures on the part of so-called academic *centers of excellence* to deliver meaningful and usable intellectual property, let alone evidence of real-world impact. Expensive, unreliable, and costly-to-maintain hardware, an absence of case studies with cost-benefit analyses and a widespread absence of attention to the human factors requirements and limitations of the end users, all took their toll by the end of the 1990s and the early 2000s (Stone, 2004). None of these developments did anything to foster strong industrial confidence in VR, either as a credible technology in its own right or as a market capable of sustaining the year-on-year developmental support and upgrades that the end users would demand.

Yet, despite the failure of VR to deliver during this time, present-day potential adopters, some of whom are described throughout the remainder of this chapter, stand to benefit from a *VR resurrection* of sorts. A resurrection based on the products of a strongly focused, market-driven movement—one that was originally labeled as technically and academically inferior by the proponents and purists of VR. That community is the gaming community. Impressive and engaging entertainment products from gaming companies across the globe, not to mention the availability of real-time rendering engines and software development kits for research and development purposes, have, over recent years, captured the attention of training and education specialists, many of whom were beginning to question whether there would ever be a breakthrough in delivering accessible, modifiable, affordable, and reliable products for serious applications. Exit the *old VR*, enter the *new VR*, the *serious game* (Stone, 2005). There is no one single definition of *serious games* (a term that received as much criticism at its debut as did VR in its early developmental days), although it is widely accepted that they are games *with a purpose*. In other words, they move beyond entertainment per se to deliver engaging interactive media to support learning in its broadest sense. In addition to learning in traditional educational settings, serious games technologies, as can be seen throughout the remainder of this chapter, are also being applied to simulation-based training in defense, healthcare and cultural awareness, in the fields of political and social change, and, slowly, to healthcare and specialized medical and surgical training. Another aspect of serious games—accessibility and affordability—makes their application potential even stronger than their VR predecessors.

Many of the applications summarized in this chapter bear remarkable similarities with those undertaken during the VRS initiative of the early to mid-1990s. The legacy of VRS lives on, and the lessons learned throughout the 5 or 6 years that the initiative existed have proved to be of immense value, both in terms of how to manage the process of effective engagement with industrial end users and in the delivery of independent value-added advice to potential adopters on all manner of issues, from human factors to the procurement of appropriate hardware and software technologies (Stone, 2012). The remainder of the chapter will, by drawing upon just a sample of actual applications, illustrate the use of VR in the defense, medical, and virtual heritage (VH) domains.

36.2 DEFENSE APPLICATIONS

During the 1990s and early 2000s, the defense sector, arguably more so than any other application domain, experienced the highs and lows of trying to adopt VR as a credible, mainstream technology in visualization, training, and systems design. Factors such as commercial naivety on the part of VR companies, significant failures to deliver meaningful and usable intellectual property on the part of so-called academic *centers of excellence*, expensive and unreliable hardware, an absence of case studies with cost-benefit analyses, and a widespread absence of attention to the requirements and limitations of the end users all took their toll by the middle of the new century.

However, long before VR broke free from its NASA and Department of Defense birthplaces to fulfill the insatiable technology needs of an evolving IT commercial market, the future potential of computer games to solve the accessibility and affordability problems of modeling and rendering tools for *serious* interactive 3D applications had already been recognized. In the early 1980s, basic developments in modifiable 3D games technologies for exploitation by communities other than those supporting home entertainment—especially defense—were well under way. For example, Battlezone—a successful 3D wireframe tank game published in 1980 for the Atari—was developed a year later into a *game* to investigate its potential as a training simulator for the US Army's Bradley military vehicle. Much has been written about the future of gaming in defense simulation and training (Stone, 2008a, 2009, 2010). Since 2005, it is fair to say that, while serious games still have some way to go before conclusive statements can be made as to their training efficacy (e.g., promoting positive skills or knowledge transfer and minimizing skill fade), their appearance on the defense training stage in many countries has been met with surprisingly positive receptions. This is despite the fact that the term *serious games* seems to have attracted as many opponents as proponents and, as hinted earlier, there still exists a good population of skeptics who are still recovering from premature investment in costly, unreliable, and overhyped *immersive* display and supercomputing technologies during the VR *era*.

However, even with this early and very positive reception, the exploitation of innovative training technologies for individual and team-based applications is still slow to take off. Largely, the world's armed forces are still training using legacy simulators that, as a result of 1990s hardware and software solutions, are expensive to run, increasingly difficult to maintain, and lack the fidelity that can be achieved with today's serious games development toolkits. Nevertheless, early field deployments of an encouraging number of defense agency-sponsored projects, not to mention the accessibility and affordability of the hardware and software (even though there is still, today, a noticeable reticence within the defense sector to embrace commercial off-the-shelf [COTS] products, despite the potentially huge financial savings COTS technologies offer), have helped to confirm the status of serious games as the *new VR*.

So, what of the evidence? It is fair to say that pockets of evidence are emerging (albeit very slowly) that support the exploitation of *serious games* and VR, particularly in defense training applications. Increasingly, examples of articles presenting laboratory-based experimental results of game-based simulation and training studies are appearing regularly in high-impact academic journals, such as *Nature*. For example, at the most fundamental perceptual–motor and cognitive skills level, experimental evidence suggests that game players tend to show superior performance in a range of tasks (Boot, Kramer, Simons, Fabiani, & Gratton, 2008; Dye, Green, & Bavelier, 2009; Green & Bavelier, 2007), such as those demanding an ability to

- Track fast-moving objects
- Track multiple objects simultaneously
- Filter irrelevant visual information and identify targets in clutter (decision making in a *chaotic* environment) (Roman & Brown, 2008)
- Multitask effectively and switch between multiple tasks
- React to changes in briefly presented visual stimuli
- Mentally rotate 3D objects

On a more general level, and with reference to defense operations at a tactical or strategic level, concrete or definitive evidence is more difficult to find, and relevant, tangible material is often *hidden* in lengthy texts discussing the general pros and cons of game-based simulation or VR. As pointed out by Hays (2005), as part of quite a thorough review of evidence-based learning using gaming technologies, "empirical research on the instructional effectiveness of games is fragmented, filled with ill-defined terms, and plagued with methodological flaws." Hays' review uncovered only 48 out of 270 publications reviewed that contained empirical data on learning effectiveness of game-based instructional tools.

Another important issue here is that good-quality studies are currently few and far between, due in part to the fact that many defense establishments are (unsurprisingly) not receptive to their training régimes being compromised or their classes being split up into control and treatment groups. A related concern is the possible negative impact on the performance and attitudes of control group participants who discover that they may, as a result of the random condition allocation process, have been *deprived* of access to a new and exciting learning technology. Furthermore, measuring the outcomes of VR-based training in naturalistic settings—the military classroom and the training or operational field—is far from straightforward.

Nevertheless, as with the scientific literature, one is witnessing an increase in the publication of results of game-based simulation and training in defense conference proceedings, such as those accompanying the annual I/ITSEC event in the United States. Just one example is the US DARWARS project—an extensive initiative sponsored by the Defense Advanced Research Projects Agency (DARPA) (DARWARS is a shortened version of DARPA WARS) to accelerate the development and uptake of low-cost military training systems. Presented regularly for some time now at the annual I/ITSEC conference, DARWARS exploits PC, gaming, web, and COTS technologies to demonstrate the art of the possible, often driven by specific US requirements and often accompanied by strong evaluation components. For instance, in one particularly impressive study (Wiederholder & Wiederholder, 2006), the effectiveness of desktop training simulators in the teaching of tactical skills (building entry and clearance during simulated urban combat) was demonstrated, along with the practice of stress management (as supported by physiological measure applied during the experiments) and the improvement of performance during real-life combat situations. In addition, the researchers showed that PC-based training improved spatial awareness, reduced the need for frequent communication during a building entry operation, and fostered better anticipation of other team members' movements. One can only hope that, in time, evidence-based reports and papers such as this become the norm rather than the exception.

Having pointed out some of the limitations inherent in the serious games-for-defense arena, it should be said that, despite the limited concrete evidence, there does seem to be a widespread belief and acceptance that game-based simulation will support a more effective transfer of information from the classroom to the real world than conventional methods (e.g., PowerPoint, *chalk-and-talk*, sandbox, or scale model techniques). Under certain circumstances and for some students, such simulation can help facilitate positive attitudinal changes toward the subject being trained (Bredemeier & Greenblat, 1981; Stone, Caird-Daley, & Bessell, 2009a; Stone, Caird-Daley, & Bessell,2009b). Furthermore, game-based learning activities that engage the interest of the learner are often accompanied by more time spent on those activities than would be spent otherwise, leading (Lepper & Malone, 1987) to better learning of the instruction and a more sustained interest in future exposures to the instructional content.

Although the criticisms of Hays and others relating to the lack of empirical data are acknowledged and accepted, the defense training community cannot wait indefinitely for the appearance of experimental evidence to cover every aspect of simulation-based training that may be presented for consideration. In the absence of empirical studies or, indeed, the opportunity to conduct empirical studies (due to some of the barriers described earlier), the community must turn instead to the collation of well-documented case studies, especially those that have benefited from strong stakeholder and end-user input from the outset. One final and important fact to emphasize here is that game-based training will never replace live training. In some cases, it can even be shown that the technology will not even reduce live training (Roman & Brown, 2008). However, this finding is increasingly being challenged, especially in instances where live training is dependent upon the availability of scarce physical assets, as is the case, for example, with submarines or specialist bomb disposal remotely controlled vehicles (RCVs). However, much of the evidence does point to the importance of games and VR as *training multipliers*, making live training more effective (Roman & Brown). Brown (2010) goes further and emphasizes that, although gaming at the military unit level can be considered a training multiplier, the capability of the unit's instructors in administering game-based training plays an important role in determining the effectiveness of the training tool. Just a small selection of the more successful defense applications is presented in the following.

36.2.1 CLOSE-RANGE 20 AND 30 mm WEAPONS SYSTEM TRAINER (ROYAL NAVY: NAVAL RECRUITMENT AND TRAINING AGENCY)

The requirement to close the naval gunnery ranges at HMS Cambridge in South Devon in 2001 was attributed to the increasing costs of live ammunition (at that time around £UK1.5 million p.a.), annual maintenance of the coastal base (unknown), and the expense incurred when flying towed targets (around £UK1.9 million p.a.) (Figure 36.9). A training needs analysis sponsored by the Royal Navy's training board concluded that there was an urgent need to replace the live firing facility with procedural trainers that would allow students to interact realistically with close-range weapons, under the supervision and instruction of weapons directors. The original close-range weapons system (CRWS) installed at the Maritime Warfare School (HMS Collingwood) was the result of a competitive tendering process, won by the primary author's previous company, VR Solutions, designed to elicit *innovative* proposals for a virtual environment trainer for the Royal Navy's 20 and 30 mm weapons, together with a fall-of-shot *spotting trainer* for the 4.5″ Mk8 gun. A later development saw the installation of a basic general-purpose machine gun (GPMG) trainer. The original system described here has, after over a decade of successful service, been replaced with more up-to-date technologies (although it is unclear if the replacement system was the focus of such a comprehensive human-centered design process).

The original human factors observations were conducted during live firing trials at HMS Cambridge, prior to the closure of the base, and at weapons emplacements onboard naval vessels moored at HM Naval Base Devonport. In brief, a weapons director visual (WDV) is located on a raised part of the

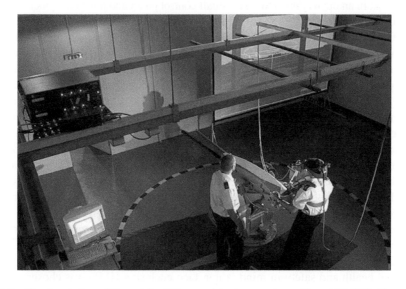

FIGURE 36.9 Overhead view of the original 20 mm virtual gunnery position at HMS Collingwood.

FIGURE 36.10 Own-ship virtual 20 mm location engaging helicopter.

deck (the gunner director's platform, or GDP) and is responsible for relaying target type, bearing, and range data from the ship's operations center to the gunnery team and supervising the final stages of the engagement procedure (using binoculars). The most common close-range deck-mounted weapon in the Royal Navy at the time of the project was the 20 mm *Gambo* (as seen in Figure 36.10), a single-barreled cannon with a range of 2 km, capable of firing 800–900 rounds per minute. The aimer adopts an initial standing position with this weapon and has to rely on his/her weight to provide the momentum to move the weapon in azimuth and elevation. A harness is provided to support the aimer's back, an important feature, especially in some late engagement activities, where the aimer can adopt a backward-leaning posture to elevate the barrel to around 75°–80°. At sea, targets are acquired through an RC35 sight, mounted on the top of the weapon, parallel to the main axis of the barrel. In the case of the 30 mm weapon, aimers sit in an open *cabin* and, via a small control panel and hand controllers, move the gun mounting in azimuth and the barrel in elevation. At sea, targets are engaged via a similar sight to that used on the 20 mm weapon. However, on the 30 mm system, the sight is mounted on a metal bracket that supports it in front of the operator's face. In the case of nonavailability of live ammunition, all of the aforementioned procedures were followed, up to the point when actual firing would normally take place, whereupon the trainees were required to engage the firing mechanism and shout *bang, bang!*

It was obvious from the human factors observations at HMS Cambridge that a completely virtual implementation of the CRWS would not help trainee aimers to acquire the necessary skills to manipulate the weapons in subsequent real-world firing trials and operational settings. It was therefore concluded that the simulation should be built around inert versions of the 20 and 30 mm weapons, so that the motion characteristics and constraints imposed on both seated and standing operators of both weapons, particularly on target acquisition in azimuth and elevation, could be reproduced, thus adding to the realism of the training sessions. In addition, the regular head movements observed between the weapon aimers and the WDV, coupled with the visual scanning of the horizon and the tracking of close-in targets on the sea surface and just above, independent of weapon movement, drove the conclusion that a helmet-mounted display (HMD) and head-tracking solution was desirable in this instance. However, in order to enhance the simulation experience and with operational health and safety in mind, a non-face-enclosing HMD was chosen, affording the end-user peripheral views of their arms and of their immediate real working environment, including the fire control levers on the 20 mm weapon and the control panel on the 30 mm. Furthermore, given

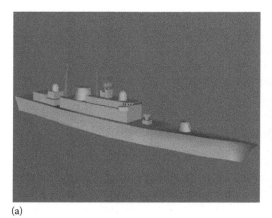

(a)   (b)

FIGURE 36.11   (a and b) Examples of low-fidelity *targets*.

the ranges over which these visual activities were active, not to mention the limitations in HMDs in presenting reliable and consistent stereoscopic images, the need for 3D viewing was judged unnecessary. Consequently, the virtual imagery was presented biocularly (the same rendered image being presented to both eyes). The physical weapon sights were removed from the inert weapons and reproduced virtually, thus allowing aimers to align the virtual reconstructions of their weapons with the virtual targets. The removal of the sights allowed for the mounting of multiaxis tracking systems, so that the movement of the physical weapons could be recorded. Additional trackers were mounted on the aimers' HMDs, supporting head movement independent of weapon movement.

The visual fidelity of the scenes and targets presented to the aimer trainees via the HMD was kept relatively low (see Figure 36.11), due to the distance at which engagements would occur and the high simulated speed with which some of the targets would approach their virtual own ship.

Anecdotal evidence from HMS Cambridge and Collingwood instructors suggested that one of the problems encountered during the training of novice weapon aimers was that of *freezing* just before operating the firing mechanism. It was further suggested that the sound and recoil of the weapons—observed by trainees prior to their own session—might be responsible for this form of *anticipatory response*. However, technical recommendations from the weapons manufacturers suggested that considerable damage could be caused to the inert weapons through the implementation of simulated recoil (such as that provided by a gas blowback system). Consequently, physical recoil was omitted from the final simulation, although attention was given to the quality of the firing sound recordings (captured earlier at Cambridge), to include at least some form of *shock* stimulus during the engagement process. A simulated visual representation of barrel recoil was also provided, although this was not always visible, due to the low resolution and field of view of the HMD used. Motion-based technology was also considered but not included, due mainly to cost constraints but also to the fact that it was felt that the virtual sea state effects implemented would, when viewed through the HMD, generate an adequate sense of own-ship motion. However, the health and safety concerns with the HMDs used meant that, with additional simulated ship motion, even greater attention needed to be paid to adverse effects of the simulation on the user, especially the incidence of postural instability when leaving the CRWS facility.

Just 1 year after the installation of the VR CRWS system (the original cost of which was around $1.13 million), some impressive usage statistics began to emerge. For example,

- Number of courses held: **55**
- Number of simulated rounds fired in first year: approximately **1 million** (cost—**minimal**—simulator power requirements)
- Number of live rounds saved: **39,000** (based on a typical preclosure year at HMS Cambridge; cost about $2.3 million)

- Ab initio student throughput: **329**
- Live aircraft/towed target hours saved: **384 h** (cost—$7666/h = $2.95 million)
- Live downtime avoided due to foul range/bad weather: **33%**
- Instructor feedback from annual Gibraltar live firing trials suggested that those participants who had been exposed to the VR CRWS demonstrated superior marksmanship qualities compared to those who had not

36.2.2 SubSafe (Royal Navy's Submarine Qualification Course, HMS Drake, Devonport)

International submarine fleets are routinely deployed in support of a range of duties, from coastal protection and patrol to the support of scientific research in some of the most inhospitable places on the planet. With such a range of hostile natural and conflict-ridden environments, danger is inevitable and, since the year 2000, there have been a significant number of incidents involving submarines from some of the world's major sea powers (e.g., see http://en.wikipedia.org/wiki/List_of_submarine_incidents_since_2000). Two specific incidents involving submarines of British origin stimulated a reassessment of the way in which submariners are trained, particularly with regard to their spatial knowledge of onboard safety-critical and life-saving items of equipment. The first occurred in October 2004, when the Canadian diesel submarine HMCS *Chicoutimi* (ex-HMS *Upholder*) was struck by a large wave while transiting the North Atlantic. The ensuing fire disabled nine members of the crew due to smoke inhalation, one of whom later died. The second incident occurred onboard the British nuclear submarine HMS *Tireless* in March 2007, while conducting under-ice exercises north of Alaska. During what should have been a routine lighting of a self-contained oxygen generator (SCOG), the unit exploded, resulting in two fatalities.

Innovative technology-based techniques for providing safety training to submariners in a classroom setting, prior to exposing them to a real boat environment, have been under investigation in the United States, United Kingdom, and Australia since the early 1990s (Stone et al., 2009a). From 2006 onward, interest in the virtual submarine concept accelerated, not only with regard to the levels of visual and functional quality that could be achieved through the use of gaming technologies but also with regard to the evaluation of such tools in real classroom settings. In addition, with smaller numbers of more capable submarines becoming the norm in the Royal Navy, vessels alongside for training purposes can no longer be guaranteed. SubSafe was the result of this accelerated interest.

Briefings and observations were conducted within the submarine qualification (SMQ) classrooms at HM Naval Base Devonport (HMS Drake), onboard submarines alongside (when available), and at sea with the crew of HMS *Tireless*. Opportunities were provided to photograph and measure each compartment forward of Bulkhead 59 onboard a selection of *Trafalgar* class vessels and to discuss the interactive qualities of the components contained therein. For example, in many of the accommodation and messing spaces, important valves were located behind seating cushions and bunk cupboard doors. In an interactive style similar to that adopted for the Tornado Avionics Training Facility (ATF) project (described in Section 36.2.8), it was decided that a click-and-remove function would be implemented in the simulation, such that the removal of cushions and opening of doors would be animated to expose underlying objects and a subsequent click would *extract* the object of interest and allow its rotation on-screen.

All objects of relevance to the SMQ training would appear to the trainee when an on-screen circular indicator appeared (the indicator was permanently slaved to the movement of the end user's mouse-controlled view). A simple plan view of the current deck was also deemed to be a useful feature to help with early navigation—recognition of port and starboard, forward and aft, and so on (Figure 36.12a and b). Movement through the submarine (and on the dockside) was to be

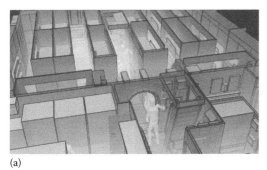

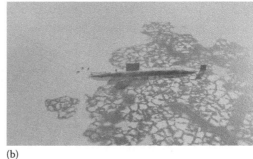

(a) (b)

FIGURE 36.12 (a and b) SubSafe modifications for HMS *Tireless* coroner's court of inquiry animation.

kept as simple as possible. Consequently, a basic keyboard input solution (with mouse movement controlling the direction of view) was implemented, with ladder transits being governed by 3D arrows that appeared on approach to ladders between decks. Mouse-clicking a 3D arrow cue actuated a *teleport* function, making the transition between decks less frustrating than negotiating 3D ladder models (as is often found when playing first-person action games, for instance).

Due to the complexity of a nuclear submarine, it was evident from the outset that to develop an interactive 3D model featuring all vessels, valves, pipes, tanks, controls, and so on would be prohibitively expensive and possibly not beneficial to the SMQ training focus. Consequently, extensive use was made of background textures, imparting a degree of medium fidelity to the submarine as a whole, but maintaining a familiar look and feel for instructors and trainees alike. This included the texturing of pressure vessel walls behind other items of equipment, to convey a sense of the density of pipework evident in many compartments. The geometric nature of the majority of the objects within the submarine was kept as simple as possible, again relying on textures to create immediately recognizable features and locations.

A statistical analysis of knowledge transfer data, collated over a year of experimental trials undertaken in collaboration with the SMQ (south) personnel at HM Naval Base Devonport, revealed that the use of the simulation during classroom training significantly improved the final *walk-round* performance of students (onboard an actual submarine) when compared to that of a control group. Further experiments have investigated the optimum presentation times for SubSafe, comparing a single exposure of the simulation to SMQ students at the end of their 6-week training régime to course-specific exposures during weeks 2–4. Significant attitudinal changes have also been recorded, with participants responding more positively to the notion of using game-based simulation in mainstream RN training after exposure to SubSafe (Stone, 2010; Stone et al., 2009a, 2009b).

In addition to the experimental trials with this first version of the SubSafe trainer, the simulation was modified to provide an animation sequence for a UK coroner's court of inquiry in 2009 relating to the HMS *Tireless* SCOG incident, mentioned earlier (Figure 36.12a). In conjunction with a narrative presented by an RN senior officer, the animation demonstrated the location of the explosion, the subsequent smoke propagation throughout the forward sections of the boat, and the activities of the crew in attempting to rescue casualties within the forward escape compartment. The SubSafe simulation was also used during the courtroom proceedings, to help familiarize attendees with the layout of a *Trafalgar* class submarine.

Finally, and again following the delivery of the first SubSafe demonstrator, information exchanges with the Canadian and Australian navies have supported those countries' attempts to launch their own SubSafe equivalents, for the *Victoria* and *Collins* class SSKs, respectively. In the United Kingdom, consideration is, at the time of publication, being given to exploiting the lessons learned from the SubSafe project to early concept designs, in particular, command space prototyping for future generation submarines.

36.2.3   SUBMARINE RESCUE ("VIRTUAL LR5": THE ROYAL NAVY'S SUBMARINE ESCAPE, RESCUE, AND ABANDONMENT SYSTEMS PROJECT TEAM)

The *Virtual LR5* concept demonstrator study (Figure 36.13) was conducted at the request of the Royal Navy's submarine escape, rescue, and abandonment systems (SMERAS) project team, following earlier exposure to the SubSafe project described earlier. The LR5 manned submersible (Figure 36.14), at the time part of the UK submarine rescue system (UKSRS), was, as with its US counterpart, the deep submergence rescue vehicle (DSRV), designed to rendezvous with a disabled submarine (DISSUB), create a watertight seal between the hatch to the submarine's escape tower and a special transfer *skirt* on the bottom of the submersible, and then transfer survivors to the safety of a support ship on the surface equipped with appropriate hyperbaric and medical facilities. The system was also

FIGURE 36.13   Virtual LR5 submersible.

FIGURE 36.14   LR5 deployed.

designed to provide life support for 16 rescued submariners and 3 crew for, worst case, up to 96 h while underwater. The SMERAS team felt that a cost-effective opportunity existed for a game-based simulator to provide an early (*prewet*) form of pilot training, emphasizing the critical components of approach and mating with a DISSUB under a variety of subsea conditions, including high levels of turbidity, poor lighting, strong currents, and extreme submarine resting angles.

Human factors observations were undertaken onboard the LR5 during a short, 3 h dive in Lamlash Bay, just off the coast of the Scottish Isle of Arran. Additional experiences during a dive undertaken prior to this—onboard the submersible's sister vessel, the LR2—were also referred to during the subsequent simulation design process. On the occasion of the LR5 dive, there was no support submarine present (the UKSRS team often work with Dutch or Norwegian diesel [SSK] submarines, as these are easier to place on the seabed than British nuclear attack submarines [SSNs]). Instead, a large *mating target*—essentially a circular hatch-like plate on four legs—was placed on the seabed. The LR5 was launched by crane from the Royal Maritime Auxiliary Service vessel *Salmoor*, a mooring and salvage platform, and spent a total of three hours submerged. During this time, the submersible conducted two *mock* dockings with the underwater target and the crew conducted simulated evacuation and emergency surface procedures. Recordings of the dive and mating attempts were captured using a small video camera clamped to a panel within the LR5 pilot's forward command module. While most of the footage was too dark to discern any detailed pilot activities, once the submersible's external lights were switched on (just before starting an approach to the docking target), sufficient detail could be seen to enable appropriate external fidelity conditions to be defined for a game-based simulator. However, the key development challenges were (a) the simulation of a view of a DISSUB through the submersible viewing dome (with appropriate visual distortions) and (b) the provision—for the trainee pilot—of additional simulated CCTV views from those external cameras mounted within and behind the transfer skirt. These views are exploited by the submersible pilot once the escape tower hatch (or the mating target in the case of the LR5 exercise witnessed) had passed under the main viewing dome, outside of his immediate field of view.

The subsea environment was based on a scenario created for another game-based simulation project, the *Virtual Scylla*, in which a 3D model of the ex-RN Leander class frigate was developed for the National Marine Aquarium (NMA) in Plymouth, to support climate change and marine ecosystem education (described later in this chapter). However, the LR5 simulation demanded greater attention to visual and functional fidelity as a result of the environment being viewed (in the main) by the virtual submersible pilot in situ, as opposed to via a basic camera mounted onboard a remotely operated vehicle (ROV), as was the case for the *Virtual Scylla*. Underwater effects were achieved firstly by rendering the scene around a 3D *Kilo* class submarine model using a high level of exponentially increasing fog density (the submarine's fin and masts can just be seen beyond the virtual viewing dome in Figure 36.15). Then, a particle effect was linked to the virtual camera position that represented the end user's viewpoint, such that particles were emitted at relative velocities to the camera's movement only. A Gaussian blur filter was added to approximate focusing imperfections through the submersible viewing dome. Another effect simulated the image refraction caused by the Perspex viewing dome, by rendering the first three stages to a texture and applying it to the virtual dome within the environment. Also, and for early illustrative purposes, an additional particle effect was implemented for escaping air bubbles from the casing and a damaged mast well. When in contact with the virtual LR5, these particles endowed the simulated submersible with positive buoyancy. Implementing these effects endowed the DISSUB scenario with a realistic appearance especially given the possible real-world lighting and turbidity conditions submersible pilots might experience.

A low-fidelity submersible interior was created to simulate the pilot's viewpoint. The cockpit contained an additional two simulated CCTV displays, relaying real-time virtual views of the cameras mounted externally (Figure 36.15). For the purposes of this early demonstrator, flying the virtual submersible and looking around the cockpit area were made possible by the use of controls provided on an Xbox gamepad.

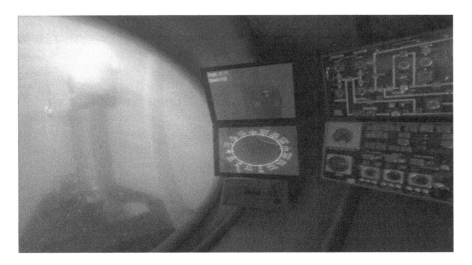

FIGURE 36.15   Simulated LR5 cockpit with dome and CCTV views.

36.2.4   Defense Diving Situational Awareness Trainer (UK Defense Diving School)

From forensic lake diving to the identification and recovery of defense material, personnel, or lost munitions, defense divers (Figure 36.16) find themselves exploring some of the most hazardous environments known to man, often in conditions of severely degraded visibility, strong currents, and low temperatures. Following a presentation to the UK Joint Service Defense Diving School (DDS), interest was expressed in the potential offered by simulation to help train—in a classroom setting— the procedures and recording processes that need to be undertaken during the investigation of artifacts that may be attached to the hull of a vessel or on the seabed. In particular, the challenge was to provide some form of simulation that would help trainees master the process of mentally recording (i.e., without note-taking aids) and reporting such features as artifact size, shape, markings, condition, location, and the nature and condition of the immediate seabed area.

FIGURE 36.16   Trainee divers at the RN Defence Diving School.

The human factors observations were conducted during a very short visit to the DDS, which included a briefing from diving instructors on the procedures undertaken for detecting and reporting on underwater ordnance. Information and training videos were also provided by the DDS. One key issue noted was the lack of awareness training for tasks involving hull inspections on specific vessels. Divers qualifying from the DDS with minimal experience might be expected to dive on new (and unfamiliar) vessel types, including *Astute* class (Figure 36.17, left) submarines, *Type 45* destroyers, and the *Queen Elizabeth* class (Figure 36.17, right) aircraft carrier. Another, perhaps more serious issue was the lack of appropriate methods for analyzing and reporting on underwater artifacts, with DDS personnel commenting on frequent inconsistencies from divers reporting on the same object underwater. Consequently, it became apparent that the simulation solution would not be as *straightforward* as some of the others reported within this review of VR applications. For example, the simulation tool would need to provide instructors with a means by which vessels, artifacts, debris, plants, and other features could be preprogrammed into scenarios and saved for subsequent presentation to trainees. The ability to change scenarios while a simulation was running—alter visibility, for example, or trigger some form of underwater incident—was also necessary, as was a means by which the instructor could monitor a trainee's progress from a different screen. Detailed after-action review (AAR) was also essential, especially for 1:1 reviews between instructor and trainee.

However, it was stressed from the outset that, despite the extent of its capabilities in the classroom, any simulation would not (indeed could not), under any circumstances, be designed to replace wet training. Instead, the goal of the project was to assess whether or not simulation could, during classroom briefing and individual *hands-on* sessions, help improve defense diving trainees' performance when recording and memorizing features of artifacts they find during their underwater investigations and how these features are then relayed back to the dive supervisors for the assessment of threat and the formulation of render-safe procedures (RSPs). An additional feature that was added, again based on early briefings from DDS personnel, was that of a simple *eyeball* (inspection) ROV, based on the commercially available VideoRay system.

The fidelity assessment for this project stimulated much discussion. While it was accepted that many of the activities undertaken by defense diving teams occurred in underwater conditions of high turbidity and low lighting (often forcing divers to focus on tactile exploration rather than visual), it was accepted that there was still some value in providing a simulator that could present underwater scenarios in a reasonably clear form but could also support progressively degradable conditions. The ability to train defense divers to be aware of what features they should take note of during a mission, including prominent hull and seabed visual features and locations of buddy divers, would also be of value. Also, and as has already been emphasized, trying to simulate all conditions associated with underwater diving was not the aim of this project. Apart from being totally unnecessary, a realistic

(a)   (b)

FIGURE 36.17 (a) Virtual *Astute* (left) and (b) *Queen Elizabeth* class aircraft carrier (right) defense diving scenarios.

FIGURE 36.18 DDS trainee (left) and instructor (right) view of seabed artifact for exploration and identification.

diving simulator would be highly costly, if not impossible, given current-generation wearable and haptic feedback technologies. Consequently, a medium fidelity approach was taken, exploiting sub-sea environments of relative simple geometric and texture features and allowing instructors and trainees to view underwater scenarios on separate screens (Figure 36.18). New techniques were necessary to enhance the variable lighting and turbidity effects, in addition to the disturbance of seabed material, by the diver or ROV, or as a result of an underwater incident. Virtual vessels such as mines, nonordnance items, and subsea features were based on modified bought-in assets, supplemented where necessary with bespoke models and textures.

In its most basic form, the simulation presents trainees with two scenarios (additional scenarios are, at the time of writing, being added). Both require early attention to surface preparation procedures before the dive takes place. One scenario is based on an incident in October 2010 when HMS Astute ran aground off the Isle of Skye, triggering fears of damage to sensitive areas of the ship such as the rudder, aft planes, and propulsor. Building on these events, the simulator enables trainees to undertake an underwater search for ordnance, around the propulsor area, within nearby water columns and on the seabed in the immediate vicinity of the submarine. A second fictional scenario involves the Queen Elizabeth class aircraft carrier, which plays host to a hull inspection and search of surrounding seabed for any potential threats to the ship.

Throughout the simulation, the trainee is presented with a range of challenges, each of which requires careful and thorough navigation of the virtual environment and the identification of ordnance. To mimic real-world conditions, the virtual environments contain representations of both real ordnance and *distraction objects*. These include a wide range of objects, including general junk, oil barrels, wooden boxes, small wrecks, and downed aircraft (placed by the instructor using the simulation's dedicated editing toolkit—see Figure 36.19). All objects can be endowed with differing states, such as the degree of silting, deterioration (rusting, leaking), and position in the water column. During each dive, trainees must contend with low visibility and differing levels of turbidity.

The AAR (Figure 36.20) runs in parallel with the simulator and both captures and analyzes the trainee's performance, in preparation for postsession presentation and review. The AAR system captures the trainee's virtual position location and orientation throughout the simulation session, together with dwell times on all underwater artifacts, objects identified as ordnance (and time of identification), and the trainee's on-screen view when identifying an object. These data are then analyzed to extract AAR-relevant measures, such as

- All objects identified as ordnance by the user in the virtual environment
- The percentage of correct versus incorrect objects identified
- The dwell times on all underwater ordnance in the scenario
- The dwell times on all incorrectly identified items
- Deviation from the ideal search path (preset by the instructor)

FIGURE 36.19   Screen image of defense diving simulation subsea editor.

FIGURE 36.20   Example AAR screenshot (basic review data only shown).

At the request of DDS instructors, and unlike other examples of military AAR, the present replay system has not been fully automated. The instructors expressed a preference to have full control over the replay/review process, thus being able to impart their task-specific knowledge to enhance the AAR session. To support this, the system allows—in real time—the selection of one of a number of camera views of the virtual scenario, the highlighting of particular items of interest, and the *scrolling* forward and backward through the scenario at various step speeds using the mouse wheel.

A short pilot study investigating the effect on performance based on the provision of this unique form of AAR was conducted using student and staff participants at the University of Birmingham (Snell, 2011). In summary, simulated diving trials with AAR were associated with a significant

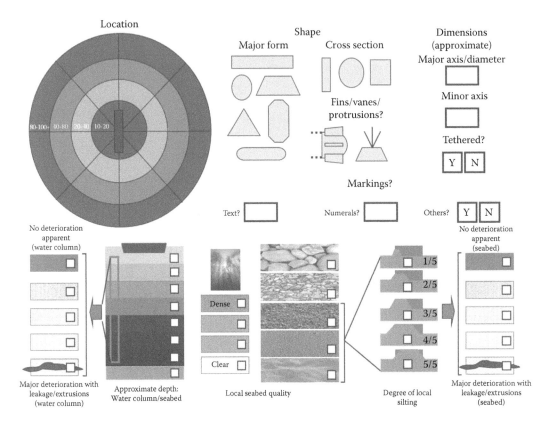

FIGURE 36.21   Recording chart used in simulated diving pilot study.

increase in performance when compared to those without AAR with respect to the number of correct seabed items identified and a decrease in the number of incorrect items identified. Simulated diving trials with AAR were also associated with a significant increase in the reporting scores for underwater ordnance (based on the accuracy with which participants complete a reporting sheet for each correctly identified object in the simulation—see Figure 36.21. Finally, simulated diving trials with AAR were not accompanied with significant increases in search path accuracy (although the results only marginally failed to reach the stated level of significance).

At the time of publication, subsequent investigation and exploitation opportunities are being sought with the DDS and the demolition diving section of the Defense Explosives, Munitions and Search School (DEMSS). Additional demonstrators have been built, including wreck exploration, ROV intervention, and conservation.

36.2.5   Counter-IED Urban Planning Tool (EODSim) (UK Defence Explosive Ordnance, Munitions and Search School)

The United Kingdom's armed forces C-IED (counter-improvised explosive device) subject matter experts (SMEs), based at the DEMSS, conducted a review of their specific needs for classroom-based military aid to the civilian power (MACP) simulation and training and reached two key decisions. The first was that a simulation of a typical homeland scenario would be of benefit to the training of threat awareness, intervention planning, and the formulation of RSPs. However, the success of such a training system would depend on providing instructors with an adequate number of scenarios with which to *expose* trainees to a broad spectrum of potential incidents. The second decision was to play down the dual use of this type of simulation for the training of threat awareness and remotely operated systems skills (driving, manipulation, and remote tool/ weapon use). It was felt that unless (1) the simulated vehicles and subsystems behaved accurately with regard to their physics-based interactions with objects, surfaces, inclines, steps, and so on and (2) the simulated remote system could be operated using the same interface devices as were in use with the real-world vehicles (or closely matched replicas), then there was no need to include them in the urban threat simulator. However, it was accepted that the inclusion of the less realistic vehicle models would be beneficial, but only as a means of illustrating the deployment of a RCV at the end of a threat awareness and RSP process. Consequently, a comprehensive virtual town scenario was constructed that, together with a range of selectable *incident zones* (gas station, hospital, multistory car park, etc.), provided instructors with a dedicated user interface to support prelesson setup of the town and numerous virtual EOD assets (Figure 36.22), together with additional virtual features, such as avatars and different types of vehicles to assist the instructor in developing the lesson in an engaging fashion.

Again, early briefings with DEMSS SMEs were essential in the process of defining the main improvements necessary to take what was being referred to as *EODSim* forward. The development of the new user interface was an iterative process, consulting with DEMSS SMEs at regular intervals. The opening view of the virtual town was exocentric in nature (Figure 36.23), providing an *at-a-glance* appreciation of the town layout and the location of additional zones, as described earlier, that were developed to increase the number of simulated incident contexts. The interface shown in Figure 36.23 is the result of two major iterations, resulting from early classroom deployments of the EODSim tool and feedback from the instructors. Among the functions added included

- *"Rapid transit" point-and-click function.* In order to reduce the time in which the on-screen view could be changed rapidly between exocentric and egocentric (first person) and to support instructors in their locating of virtual assets during the lessons themselves, a mouse-slaved red ball cursor indicated the point to which the instructor could *jump* (via a mouse double left-click). A double right-click would return the instructor to the exocentric viewpoint. Once at the selected location, first-person (*egocentric*) navigation or RCV driving is governed by W-A-S-D or arrow key inputs.

FIGURE 36.22 Scene from EODSim.

FIGURE 36.23 Opening EODSim screen (exocentric town view) and instructor graphical user interface.

- Mouse-controlled pick and place of key assets from drop-down menus (ICP vehicles, RCVs, avatars, devices, and triggers), with the ability to reorient said assets during scenario construction.
- *Camera selection.* The term *camera* refers to the available viewpoints in a given scenario, which include first-person views, the view from an avatar's location, and views from camera positions onboard the RCVs.
- Full slider control over time of day and shadowing.
- A circular cordon range tool—basically a simple red circle, the diameter of which could be set by the instructor (by mouse dragging), depending on the location and *size* of the device under consideration. Familiar yellow-striped cordon *tapes* could also be located when in first-person view, to add to the realism of the incident.
- A range marker (essentially a click-and-drag *rubber band* tool to help trainees estimate distances).
- A red *marker pen* to highlight key features during lessons (the red marks disappear immediately following the instructor's next movement).

In addition to the extended database of new scenarios and objects, new foliage models (trees, plants) were added to increase the level of visual realism throughout the town. The tube train was also animated and could be started and stopped via a single mouse click. This may sound trivial, but ensuring that public transport in the area has been included in the evacuation procedures (and hazards such as electric rails have been isolated) is also key decision point in building up the threat awareness picture. A small selection of avatars was included in this concept demonstrator, although their *presence* simply took the form of animated *placeholders* due to ongoing concerns with the potential negative impact of current-generation avatars when used in real-time training scenarios. At the time of writing, the EODSim system is under evaluation, in parallel with the two further EOD simulation tools described in the following.

36.2.6 AFGHANISTAN MARKET/VILLAGE SCENARIO (UK DEFENCE EXPLOSIVE ORDNANCE DISPOSAL, MUNITIONS, AND SEARCH SCHOOL)

Effective C-IED/EOD training for both operational and homeland security applications is placing considerable pressure on specialist schools to deliver new IED/EOD specialists quickly to front line

(a)   (b)

FIGURE 36.24   (a and b) Two scenes from the virtual Afghanistan village demonstrator.

units. Although the Afghanistan village demonstrator (Figure 36.24) was originally requested to provide urgent resource for a new course to be staged at the improvised explosive device disposal (IEDD) wing of the DEMSS, the project also demonstrated how quickly such a resource could be developed—in less than 4 weeks from the time of the original request. The demonstrator was designed to be used as a predeployment awareness-training tool, including many features—ground signs, building anomalies, threat indicators, unusual objects, and the like—that might be experienced or observed while on a routine patrol in Afghanistan.

The initial human factors assessments for this project were based on briefings from military personnel, with the aim of identifying the scope and specific features that were required to meet the training aims of the new C-IED course. These briefings, together with pictorial and video information culled from web sources and operational reports, enabled a simple plan-view sketch of a generic Afghan village to be developed and used as a *storyboard* to define the typical approach routes and items that were likely to be found while transiting those routes. In addition to the layout of the village, consideration was also given to the technologies necessary to deliver the virtual material. Again, emphasis was placed on instructor delivery, as opposed to 1:1 computer hands-on on the part of the trainees. In order to help the instructor engage with the trainees, while having full control over movement through the virtual world, an Xbox game controller was selected and configured such that it would support walking through the village and, when necessary, *flying*, to enable an overview of the virtual village to be presented for global awareness and discussions relating to insurgent surveillance locations, escape routes, and so on. A simple *binoculars* function was also deemed necessary, so that patrol vulnerability from high or distant insurgent vantage points could be demonstrated. The demonstration was also operable using a mouse and keyboard, if an Xbox controller was unavailable.

One of the key driving features behind the level of fidelity specified for this demonstrator was the need for realistic and effective variable daylighting. When training fine observation skills, time-of-day effects, plus associated ambient lighting and shadowing, are all essential features of a simulation of this nature. These features, when correctly implemented, can help render crucial IED components and cues such as command wires, disturbed earth, and other small, or nearly invisible markers. It is often these near-invisible markers that can make the difference between a safe patrol and one at risk of casualties from IED explosions. Furthermore, strong shadows are capable of concealing even the most obvious (under other lighting conditions) of markers, such as the small rock circle shown adjacent to a distant walkway gate in Figure 36.25. Markers such as these are sometimes placed by parents of Afghan children to indicate *no-play* areas. However, if the walkway shown in Figure 36.25 had to be used during a rush exit from part of the village, the shadowing at a particular time of day might well lead to such a cue being missed altogether.

In addition to the lighting effects, attention was paid to the fidelity of other markers and contexts, such as the poppy and wheat fields. Other markers included small drinks cans, inappropriately located tools (e.g., a spade), discarded cigarette butts (possible evidence of *dicker* activity—observing

(a)

(b)

FIGURE 36.25   (a and b) Effect of time-of-day and shadowing on ground marker visibility.

the movements of previous patrols), wall markings, dead animals, other rock formations, removal of well plank covering, and more. In addition to the main village and market buildings, discarded items, ditches, junk piles, ISO containers, oil cans, and many other 3D features were included within the scenario to quite high levels of fidelity. As well as the ability for instructors to *fly* to different elevated parts of the village, collision detection with all buildings was deactivated, thus allowing them to enter buildings rapidly in order to illustrate views through windows and cracks and other possible vantage points for insurgents.

36.2.7   CUTLASS ROBOT VEHICLE AND MANIPULATOR SKILLS TRAINER (DEFENCE SCIENCE AND TECHNOLOGY LABORATORY)

The MACP (Urban) C-IED demonstrator (EODSim) described earlier included basic remote driving and manipulation examples associated with a range of RCVs. It was generally accepted that these vehicles were included in the simulations for completeness, illustrating the culmination of an RSP decision-making process, rather than as an element supporting the development of remote driving and manipulation skills. The challenge faced with the development of a part-task simulator for the new CUTLASS RCV system (Figure 36.26) was to develop an affordable portable trainer, potentially for multisite deployment, with accurate physics-based representations of remote functions.

One of the major concerns expressed by RCV instructors during observations undertaken at the DEMSS was the regularity with which more recent classes of vehicle (i.e., more recent than and more sophisticated than the long-running UK Wheelbarrow system) are damaged. Many instructors felt that the prime cause of damage was caused by inadequacies in training prior to *hands-on* trials with the vehicles themselves. Current classroom training does not, they feel, equip RCV trainees with adequate skills and system awareness to help them appreciate the capabilities and limitations of the vehicles they are about to operate. Nor does it help trainees to adapt to the notion of controlling vehicles and manipulators remotely, especially in the case of multiaxis manipulator systems (such as that fitted to CUTLASS), where each joystick controls a different function and movement of the manipulator, depending on which of a number of joint-by-joint or multiaxis (*resolved* motion) control modes has been selected. Consequently, early hands-on opportunities can sometimes be characterized by trainees pushing the remote systems to their limits and failing to appreciate those limits when approaching obstacles, such as steep ramps or stairways, narrow corridors, or platforms that may look mountable from a remote camera viewpoint, but in reality are not.

Vehicle damage has a knock-on effect in that a reduction in the size of an RCV fleet means fewer opportunities for hands-on training or significant delays when trainee groups have to share limited resources. Delays also have an impact on skill fade—the fewer the systems (and this is

FIGURE 36.26 CUTLASS RCV.

a particular concern as RCVs become more sophisticated), the greater, potentially, the remote operations skill fades. With future RCVs such as CUTLASS, the cost per vehicle will undoubtedly restrict the number of systems that are made available for training, and the impact of this on basic training, let alone skill fade and refresher training, could be immense. An additional concern, brought about by observations at early CUTLASS trials, related to the remote manipulation behaviors of some of the RCV users. Some—particularly those with Wheelbarrow experience—were seen to extend the CUTLASS manipulator outward, past the front of the vehicle chassis, and leave it in this position, changing the position of the end effector by driving the vehicle forward and backward (mimicking the Wheelbarrow boom) and turning left and right when necessary (a classic case of *old habits die hard*). They would then operate the wrist and gripper functions when in close proximity to the target object. This indicated that the RCV users were not taking full advantage of the nine degrees of freedom the system possesses, thereby compromising its overall performance. The current part-task simulator will allow future CUTLASS users to experience the full range of capabilities of the manipulator system, thereby avoiding this rather restrictive remote manipulation behavior.

The CUTLASS system familiarization trainer (Figure 36.27) also demanded a high level of simulated functional fidelity in order to foster the correct remote operation skillsets. Consequently, detailed attention was given to the design of a virtual environment that would not only provide an adequate introductory training scenario for the end users but would also accurately represent the physics-based qualities of the CUTLASS manipulator, including fine detection of contacts and collisions, grasping, object friction, grasped object slippage, and release (falling under gravity). An early physics-based prototyping simulation was developed for this very purpose, and once refined, the effects were programmed into the virtual environment designed for use with the simulator. The virtual scenario takes the form of a room with a cupboard. A replica (physical) CUTLASS console was also designed and constructed (Figure 36.28). While this console does not possess all of the functions evident with the real system, it contains accurate representations and locations of the key components, including the manipulator mode selection areas on a touch screen.

FIGURE 36.27   Screen shot from CUTLASS RCV simulator.

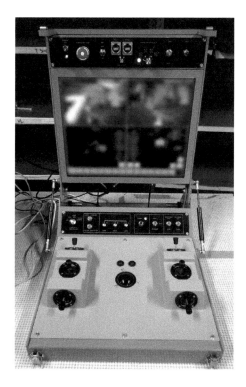

FIGURE 36.28   CUTLASS simulator replica operator interface.

36.2.8 Tornado F3 Avionics Maintenance Trainer (RAF Marham)

The ATF, based at RAF Marham, originally consisted of three key components—a selection of instrumented mock-ups relating to the *Tornado* GR4 strike aircraft, a *Tornado* F3 avionics ground training rig (AGTR), and an F2 rig, modified using F3 and GR4 components, including line-replaceable units (LRUs) and wing pylons. The modified F2 rig was essential for activities such as health and safety training associated with manual weapons loading and the manipulation of heavy LRUs. The ATF replacement project (undertaken by the primary author's previous company, VR Solutions) arose not only as a result of limited access to airframe hardware but also as a requirement to reduce training times and costs. The £UK14 million AGTR facility had, at the time of the project, been in existence for a year and had enabled instructors to reduce course times from 13 to 11 weeks. However, only two students plus one instructor could be present on the rig at any one time. Furthermore, in order to simulate avionics faults, the LRUs provided with the rig had to be removed and inserted into a nearby bench unit, the wiring and functions of which resembled that of an old telephone exchange. As a result of these limitations, each and every student experienced a downtime totaling nearly 3 weeks (i.e., time in which waiting students did nothing).

The virtual *Tornado* F3 was developed on the assumption (confirmed by SMEs) that, by the time technician trainees undertook the avionics simulation course, they would already possess the basic manual skills and knowledge necessary to select and use the correct tools to remove LRU cover panels and the LRUs themselves. It was decided, therefore, that the simulated maintenance task would not benefit from a virtual-reality-like interface that attempted to reproduce the use of spanners, screwdrivers, and other common tools. Not only was this level of technology unacceptable from a cost and reliability perspective, its use would probably focus the trainees' attention more on interfacing with the virtual *Tornado* through what would have been quite cumbersome and unreliable wearable equipment than on performing the task. Consequently, LRU panel opening, extraction, and inspection operations were achieved using *point-and-click* mouse inputs, combined with simple drop-down task sequence menus. Navigation around the external aircraft shell was achieved by mouse motion, coupled with a simple on-screen directional cue.

Figure 36.29 shows the extent of the virtual components and LRUs and simulated test equipment that made up the virtual *Tornado* F3. Each LRU was geometrically accurate and possessed high-definition textures to aid in recognition and fault detection (some even featuring bent connector pins). The external shell of the virtual aircraft was rendered to a medium level of fidelity, with greater detail included only when it was deemed necessary to provide trainees with location recognition cues relating to a specific panel and its underlying LRU. All external moving surfaces were present, including removable and hinged panels, flight control surfaces, and radome. The cockpit areas (pilot and navigator positions) were endowed with the highest level of visual and functional fidelity, including active head-up and navigator displays, together with fully operational controls, actuated by mouse click and/or drag, with directional cues for linear and rotary functions offered by the cursor when over the virtual control in question. Joystick functions were an exception to the point-and-click/drag rule, due to the distribution of the multifunction controls around the joystick handgrip. Consequently, physical pilot and navigator joysticks were provided on each trainee workstation. In order to accommodate the effective learning of virtual and real control inputs and their effect on the aircraft's avionics functions, a triple-screen interface was developed that supported the display of a full-length representation of the virtual aircraft, specific cockpit locations, or combinations of the two.

The ATF system was hosted on a high-specification Windows PC and, uniquely (at the time of installation), featured three screens per workstation, each displaying different working views of the aircraft, avionics bays, over 450 LRUs, and 50 items of virtual test equipment (Figure 36.30). Ten such workstations were produced, fully networked, allowing a minimum of eight students to be trained and supervised by two instructors in basic and advanced *Tornado* F3 avionics maintenance routines, with collaboration between students supported over the local area network as necessary. Soon after its operational debut, the ATF simulator was instrumental in helping Marham instructors reduce training time from 13 to 9 weeks and downtime from 3 to 0 weeks. Indeed, the instructors believed the course could be shortened even further but were reluctant to do so, choosing instead to increase course content, to promote retention through *consolidation breaks*, and to introduce extramural self-paced refresh trials. Further feedback from the instructors suggested that, in contrast to previous courses, ATF students *grasped the concept* (i.e., gained enhanced spatial and procedural knowledge

FIGURE 36.29   Virtual Tornado F3 model showing a selection of LRUs.

FIGURE 36.30   Triple-screen ATF simulator station at RAF Marham.

of the aircraft and LRU distribution) up to 40% faster than achieved by previous non-ATF students. The modified (physical) Tornado F2 rig, as used by the GR4 course students, was retained in order to deliver health and safety training associated with lifting procedures for some of the heavier LRUs. Of particular interest was the cost of the ATF. In total, this amounted to just over one-tenth of the cost of previous non-VR setups (Moltenbrey, 2000; Stone, 2001; Stone & Hendon, 2004).

36.2.9   HELICOPTER VOICE MARSHALING (RAF SHAWBURY, VALLEY, AND ST. MAWGAN)

The term *voice marshaling* refers to the actions of a member of the aircrew of a military or civilian multirole helicopter, whose responsibility is to monitor the position of the aircraft vis-à-vis the external environment through the rear cabin door and to relay important (verbal) flight commands to the pilot. These actions help to expedite an accurate and safe ascent or descent, with or without load, to and from a specific location. The RAF was keen to adopt a low-cost simulation solution in order to improve the quality and efficiency of preflight, ground-based training and to provide a more cost-effective mode of remedial training. Their particular aspiration was to help novice rear-door aircrew master the style and content of their voice interaction with the helicopter pilot. Legacy training techniques, based on scale dioramas and model helicopters, had proven inappropriate for this type of training. Deploying aircraft for remedial training was (and still is) a major problem, due to increased flying restrictions, dwindling personnel resources, and, of course, the actual cost of flying (which, at the time of this project, was in excess of £2500 per flying hour).

The human factors analyses focused on identifying the limitations of current ground and in-flight helicopter voice marshaling (HVM) training methodologies. The analyses were conducted by the primary author of this chapter at RAF Valley (the Search and Rescue Training Unit, SARTU) and RAF Shawbury (the Central Flying School). Both bases possess Bell HT-1 *Griffin* helicopters and the analyses involved short-duration flights in these aircraft over a period of 3 days. A later, similar analysis was conducted onboard a 203(R) Squadron *Sea King* HAR3 helicopter during a short coastal flight out of RAF St. Mawgan.

During reconnaissance of a potential landing or search-and-rescue (SAR) hovering area, both the pilot and voice marshal agree on important features of the terrain (the *5 S's*—size, shape, surround, surface, and slope). They then identify man-made or natural markers for a controlled and safe final approach, including all-important monocular depth/distance cues, such as size and shape constancies, linear perspective, texture gradients, interposition, and parallax. Commonly used markers include telegraph poles, isolated (*NATO Standard 40 ft*) trees and bushes, dwelling features (10 and 20 ft to the tops of a typical house first and second floor windows), and hangars (30 ft to top of hangar door). HVM trainees are expected to be able to *halve* the distance repeatedly between their helicopter and target, thereby ensuring a smooth, rhythmical vocal countdown and, thus, a steady rate of approach on the part of the pilot. When at sea, voice marshals often use breaking wave patterns as markers. As the helicopter approaches the target or landing area, the voice marshals make regular head and torso movements—both outside and inside the aircraft—in order to maintain a strong situational (and, therefore, safety) awareness of the dynamically changing situation.

Of paramount importance to the levels of virtual environment fidelity provided in this simulation were representations of the key markers exploited by voice marshals during the reconnaissance and approach procedures. Given the range at which these features would exist in the real world, highly detailed 3D assets (such as trees with individual branches, ground vehicles and marine vessels with fine structural detail) were considered inappropriate (Figure 36.31). Consider, for example, in the case of training backdrop parallax—instantly judging the altitude of the helicopter based on the extent to which familiar objects, such as trees, are occluded by or visible over other familiar objects, such as buildings. The trees used in the simulation took the form of simple bill boarded texture pairs, the buildings simple flat-shaded structures. The provision of real-time shadowing (highly difficult and expensive to achieve at the time this simulation was developed) had already been discounted, as they are only useful in the judgment of altitude if the helicopter is active over

FIGURE 36.31   (a through c) Examples of virtual environment fidelity provided in the voice marshaling simulator.

artificial surfaces or is involved in load lifting and depositing. A similar decision was made with rotor downdraft effects. Downdraft for SAR applications only provides a useful cue in confined areas (e.g., generating foliage movement) and is influenced by prevailing winds—a 10–15 knot wind will move the downdraft effects aft of the helicopter.

From an interactive fidelity (data input/display technology) perspective, in addition to scene scanning, the regular head and torso movements noted during the in-flight observations were interpreted as a means of obtaining parallax cues, either by using external physical features of the helicopter or by lining up visually with features in the external environment. These behaviors, along with the requirement for the voice marshals to make regular head-up motions to view the horizon and even look under the helicopter on occasions (supervising underslung loads or monitoring very uneven terrain), drove the conclusion that an HMD and head-tracking solution was desirable in this instance. However, in order to enhance the simulation experience and with operational health and safety in mind, a non-face-enclosing HMD was chosen, affording the end-user peripheral views of their arms and of the immediate real working environment. Furthermore, given the ranges over which marshaling target procedures are active, not to mention the limitations in HMDs in presenting reliable and consistent stereoscopic images, the need for 3D viewing was judged unnecessary. Consequently, the virtual imagery was presented binocularly (the same rendered image being presented to both eyes). Finally, the in-flight movement analyses of the aircrew suggested that a simple wooden framework would suffice as a representation of the rear door of the helicopter, provided that the door height was the same as that in the real aircraft and that handholds were located reasonably accurately in parallel with the open doorframes (Figure 36.32).

FIGURE 36.32   The RAF Shawbury voice marshaling trainer.

The HVM facilities have been used by the RAF training teams at Shawbury (2 sets) and Valley (1 set) since their installation in 2002. Their capabilities were extended in 2007, when a dynamic winching simulation was added to complement the rescue scenarios being trained. The RAF has recorded a number of successful training results with the systems, particularly with the remedial training of close-to-failure trainees. The cost of all three simulator systems, including development, hardware, and installation of all three systems equated to roughly 96 flying hours (Stone, 2003; Stone & Mcdonagh, 2002).

36.3 MEDICAL APPLICATIONS

During the late 1980s, many visionaries, notably those at the University of North Carolina and within the Department of Defense in the United States, were developing the notion of the surgeon of the future. Future medical specialists would, they claimed, be equipped with head-mounted displays and other wearable technologies, rehearsing in VR such procedures as detailed inspections of the unborn fetus or gastrointestinal (GI) tract, the accurate targeting of energy in radiation therapy, and even socket fit testing in total joint replacement. For many years, the United States led the field in medical VR. In 1995, one of the leading practical advocates of VR in the United States, Colonel Richard Satava, attempted to categorize achievable applications of VR in medical and surgical domains (Satava, 1995a). He saw developments in the fields of surgical intervention and planning, medical therapy, preventative medicine, medical training and skill enhancement, database visualization, and much more. Satava's original work, sponsored by the Advanced Research Projects Agency (ARPA), focused on large-scale robotic or telepresence surgery systems, which included using VR technologies to recreate the sense of presence for a distant surgeon when operating on, say, a battlefield casualty (Gilbert, Turner, & Machessault, 2005). However, other research efforts began to emerge across the United States (and Europe) using VR in a classic simulator mode to rehearse or plan delicate operations (e.g., as is evident in certain ophthalmic operations).

It was also shown that one could actually use interactive virtual imagery to back up in situ surgical performance through the projection of 3D graphics onto the operative site (augmented reality [AR]) when, as early as 1993, a magnetic resonance image (MRI) had been taken of a patient and overlaid onto a real-time video image of his head (Adam, 1994).

During the late 1980s and early 1990s, the application of VR and associated technologies to the field of medicine and surgery steadily increased, with pioneering (if, at that time, somewhat optimistic) companies such as High Techsplanations (HT Medical) and Cinémed becoming responsible for fueling the obsession with *making surgical simulation real* (Meglan, 1996). By the mid-1990s, advances in computing technology had certainly attempted to provide the means whereby sophisticated and comprehensive anatomical and physiological simulations of the human body could be constructed. From the digital reconstruction of microtomed bodies of executed convicts (e.g., the Visible Human Project, http://www.nlm.nih.gov/research/visible/visible_human.html) to speculative deformable models of various organs and vascular systems, the quest to deliver comprehensive *virtual humans* using dynamic visual, tactile, auditory, and even olfactory data was relentless. Yet, with one or two exceptions, the uptake of these simulations by surgical research and teaching organizations was (and still is) poor. One cannot attribute this failure to a lack of technological appreciation or foresight on the part of individual specialists or administrators within the target user organizations. The poor uptake actually stemmed from an equally poor understanding—sometimes on the part of simulation developers—of the medical needs and human factors requirements of the surgical users and trainees.

The VR community discovered quite quickly that its future could no longer be built on hype and false promises. Many of the early adopters, including those who were convinced they were facing a revolution in surgical and clinical practice, had their hopes dashed as the technology failed time and time again to deliver usable, affordable interactive systems for education, training, and in-theater surgical support. Apart from a small handful of successful products, many of the *high-tech* VR solutions for medicine and surgery were based on highly sophisticated graphics *supercomputers*. It was all too often easy to forget that most medical organizations simply could not justify the excessive initial costs of so-called graphics *supercomputers*—not to mention crippling annual maintenance charges, depreciation, and, in today's rapidly changing IT world, rapid technological redundancy. These systems were not only very expensive; they also suffered from a fundamental problem in terms of training transfer.

It soon became apparent that, despite the impressive supercomputer performance (albeit quoted on marketing material), the goal of *making surgical simulation real*, as mentioned earlier (Meglan, 1996), was unattainable. Although some of the marketing demonstrations were quite impressive, to the keen eye of the surgical user, the virtual bodies were less than *perfect*. The material properties of organs and tissues were not quite right; virtual fluids and gases (e.g., blood, irrigation, and diathermy smoke) behaved in ways that would not be evident in the real patient. Indeed, the anatomical and physiological content of many simulators did little to help foster practical skill acquisition. In fact, the distractions caused by irregularities in the virtual imagery probably led to more examples of *negative transfer* of skills from the virtual to the real.

Currently, many specialists in the world of medical training believe that there has never been a more pressing need to develop effective simulation-based training to fill these gaps, servicing the needs both of individuals and small teams. However, the form that simulation should take is still a constant source of debate, despite years of experience with simulation throughout both defense and civilian medical communities. Should the simulation be real, based on mock scenarios with real actors or instrumented mannequins? Or have virtual or synthetic environment technologies today reached a level of maturity and affordability whereby simulation-based systems can guarantee effective knowledge and/or skills transfer from the point of educational delivery (be that a classroom, the Internet, or via handheld computing platforms) to the real world? Is one form of training more likely to succeed over the other, or should instructors be exploiting both live and virtual contexts? If the latter, how can the balance between the two forms of training be specified?

Many simulation centers in today's *post-VR* era focus their training technology requirements on full- or part-body physical mannequins. Today's instrumented mannequins are able to fulfill the training needs of medical students from a wide range of backgrounds and specialties—nurses, technicians, paramedics, and even pharmacists. Their physiology can be electromechanically controlled in real time, by computer or remote instructor; they can be linked to a variety of typical monitoring devices and are capable of responding to a range of medications, with some systems even delivering symptoms based on the mannequin's *age* and sex. Many of today's mannequin systems also possess impressive *AAR* capabilities, so that events can be replayed and feedback can be provided to trainees on a step-by-step basis, or after a complete scenario has been performed.

36.3.1 V-Xtract

One such system that makes use of physical mannequins and a realistic and reusable training system in a medical domain is V-Xtract, created by Design Interactive, Inc., of Oviedo, Florida. The Vehicle Casualty eXTRACTion (V-Xtract) Trainer is a Phase III SBIR (small business innovation research) used to train first responders on how to extricate casualties from a vehicle in the event of an accident and/or rollover. While V-Xtract is sponsored by the US Army and utilizes a military transport vehicle, it has wide applications for first responders across various domains. V-Xtract includes a configurable vehicle simulator, a performance metric suite, and a training management system (TMS) (Figure 36.33).

The configurable vehicle simulator allows instructors to modify aspects of the scenario that adjust the difficulty of targeted training objectives (e.g., the rotation of the vehicle, activation of combat locks, and jammed restraint systems) while providing a realistic yet safe training environment. Other modifiable characteristics to enhance the realism of the simulation include the following:

- *Environmental effects*: Surround sound speakers replicate enemy fire, explosions, and other sounds; a commercial fog machine replicates smoke; cab lighting replicates day or night conditions; a commercial scent system presents a variety of scents including smoke, burning vehicle, gasoline, and burning flesh.
- *Tilting seating*: Seating within the vehicle moves to provide *trapped* scenarios in which the mannequin is trapped beneath or between seats.
- *Reconfigurable roof hatch*: The military vehicle used can be modified with one of three different roof hatches—a flat roof, a turret, or an MRAP hatch.
- *Space constraints*: Roof panels can be installed and removed to simulate a crushed roof in a vehicle rollover condition.

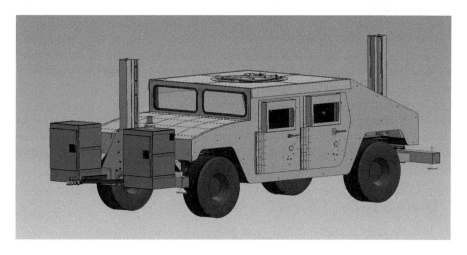

FIGURE 36.33 V-Xtract frame and lift/rotational design.

Over skin

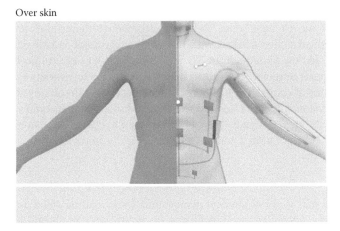

FIGURE 36.34   V-Xtract MMS.

- *Combat locks and door prying/jamming*: Instructors can set door opening mechanisms to numerous states during each scenario requiring a different sequence of actions or different tools to open doors. Each door can also be set independently.
- *Vehicle cutting capability*: V-Xtract uses replaceable hinges and panels that can be cut using extrication tools and replaced for future scenarios.
- *Reconfigurable restraint system*: A custom belt spooling system was designed that allows trainees to cut the seatbelts as they would in a real extrication and for trainers to easily reset and repair the seatbelts for use in future scenarios.

The performance metric and sensor suite provides the capability for instructors to augment low-fidelity mannequins with durable low-cost sensors to record performance on vehicle extraction and medically related tasks. To support the process of collecting performance data on medical tasks, a medical metric suite (MMS) was developed and integrated into the V-Xtract suite. The MMS consists of a suit and integrated sensors that can be placed on a low-fidelity mannequin, (Figure 36.34) such as a Rescue Randy, and software that guides instructors through the process of selecting metrics of interest and setting up the tetherless performance evaluation system. Sensors are hard-coded into the ruggedized suit but are accessible so instructors can replace parts when needed. All data collected with the system are sent to the IOS during the scenario to provide a real-time and AAR review of trainee performance.

The integrated V-Xtract TMS is designed to provide all of the required capabilities to meet the three essential aspects of training vehicle extraction: practice, evaluation, and feedback. The TMS allows instructors to develop and control scenarios by modifying the training environment in real-time while being provided with objective performance metrics in real time.

But do advanced hardware-based developments such as this mean that digital simulation or VR-like solutions are no longer a player in the world of medical training? Certainly not. Since the VR failures of the closing decade of the twentieth century and early years of the twenty-first century, there have been a number of important developments in the digital simulation arena, the most notable being in the realm of *serious games* as previously discussed. The following applications reflect the positive impact VR has had in the medical domain.

36.3.2   MINIMALLY INVASIVE SURGICAL TRAINER

Minimally invasive surgical trainer (MIST) was designed by one of the authors (Stone, 1999a) and evolved from a comprehensive in-theater human factors task analysis (Stone, 2008b; Stone & McCloy, 2004). MIST (marketed until recently by Mentice of Sweden) was a PC-based *keyhole* surgical skills trainer that supports and documents trainees' acquisition of minimally invasive surgical skills in laparoscopic cholecystectomy and gynecology (Gallagher, McClure, McGuigan, Crothers, & Browning, 1998; Taffinder, McManus, Jansen, Russell, & Darzi, 1998). The original MIST system presented the trainees not with high-fidelity 3D human anatomy and physiology but with visually and functionally simplistic objects (Figure 36.35), graphical spheres, cubes, cylinders, and wireframe volumes—all abstracted from an observational task analysis of surgical procedures evident in theater (e.g., clamping, diathermy, tissue sectioning). The analysis made it possible to isolate eight key task sequences common to a wide range of laparoscopic cholecystectomy and gynecological interventions and then to define how those sequences might be modified or constrained by such factors as the type of instrument used, the need for object or tissue transfer between instruments, and the need for extra surgical assistance. By adopting an abstracted-task design process, the MIST simulation content was developed to avoid the potentially distracting effects (and negative skills transfer) evident with poorly implemented virtual humans at the time (Stone & Barker, 2006).

For well over a decade, and with more clinical and experimental evaluation studies than any other simulation-based surgical skills trainer, MIST helped to train the perceptual–motor skills necessary to conduct basic laparoscopic maneuvers involved in cholecystectomy and gynecological minimally invasive interventions. *Believability* was enhanced through the use of realistic (and instrumented) laparoscopic instruments that, together with offline video sequences of actual operations, helped to relate the abstracted-task elements to real surgical interventions. In many respects, the MIST trainer set the standard for a range of subsequent basic surgical skills trainers (including Surgical Science's LapSim and Simbionix's LAP Mentor products). However, the key achievement of MIST was not so much the nature of the final VR product, but the fact that, by adopting a strong human factors approach from the outset, an affordable part-task, surgical simulator could be developed that delivered meaningful and, importantly, measurable interactive training content to medical students (e.g., Gallagher et al., 1998; Taffinder et al., 1998).

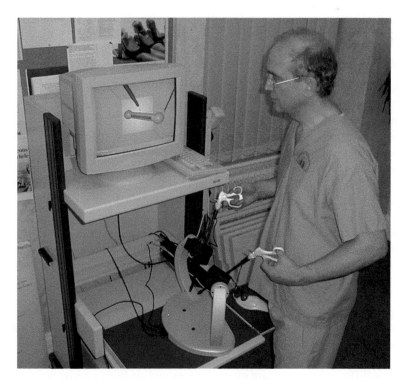

FIGURE 36.35 The minimally invasive surgical trainer (MIST).

36.3.3 University Medical Education in the United States

Northwestern University Feinberg School of Medicine's Center for Simulation Technology and Immersive Learning has, in the past couple years, developed a 1000 ft² VR space complete with six VR trainers. The VR trainers include tutorials on procedures such as bronchoscopy, upper and lower GI endoscopy, ultrasound, and cataract and vitreo retinal surgery. Additionally, the trainers include several tutorials on laparoscopic surgical procedures allowing users to learn or refresh either individual surgical steps or entire procedures as frequently as they desire. Each trainer not only provides the sensitive visual and tactile feedback of an instrument pushing against an organ, but they also provide such detailed feedback as the efficiency of movement of a surgeon's hand. Additionally, students are provided with immediate feedback and recommendations for improvement. Instructors are engaged through AAR capabilities detailing what students practiced and how proficient they were at the task.

According to Dr. Dmitry Pyatetsky, as quoted in the Fall 2011 issue of WardRounds, significant differences in the operating room have been seen in how well prepared first- and second-year residents are after using the VR trainers. Additionally, Dr. Grace Wu observed similar results, remarking that residents are already more experienced with the basic mechanics of surgery and can focus on more complex aspects of surgery earlier in training (Soohoo, 2011).

The University of Illinois–Chicago (UIC) also finds itself on the cutting edge of VR in medicine. The University, with funding from the National Science Foundation's Major Research Instrumentation program and the Department of Energy, has developed CAVE2, an 8 ft tall, 320°, immersive, 3D theater screen. Twenty years ago, UIC developed CAVE, which used four projectors and cost nearly $2 million, to display 3D images. Today's CAVE2 that allows researchers and doctors to view models of their patient's brains at scales "that makes them feel larger than a six-story building or smaller than a molecule," for example, was built using "72 3D, LCD panels with 66 times more brightness, 4176 times more processing power, and 22,500 times the storage capacity" (NSF, 2012). Using CAVE2, neurosurgeons and bioengineers are able to create models of patient's brains and have the resolution and seamless virtual environment required to solve some of the most complex medical problems.

36.3.4 IERAPSI Temporal Bone Intervention Simulator

Integrated environment for rehearsal and planning of surgical interventions (IERAPSI) project (John et al., 2000) was the result of a task analysis that highlighted the need for a simulator to possess higher physical fidelity in one sensory attribute over another. In such a case, it may become necessary to develop, procure, and/or modify special-purpose interfaces in order to ensure the stimuli defined by the analysis as being essential in the development of skills or knowledge are presented to the end user using appropriate technologies. IERAPSI was a medical (mastoidectomy/temporal bone intervention) concept demonstration simulator developed as part of a European-Union (EU)-funded project (Figure 36.36). Here, the task analysis undertaken while observing ear, nose, and throat (ENT) surgeons, together with actual *hands-on* experience using a cadaveric temporal bone, demonstrated that the skills to be trained were mostly perceptual–motor in nature, but the decisions were also complex (and safety-critical), albeit at limited stages of the task. This drove the decision to adopt a hybrid physical fidelity solution to train mastoid drilling and burring skills. The final simulation concept demonstrator was based on

- A low-physical-fidelity visual representation of the temporal bone region (omitting any features relating to the remaining skull areas, middle/inner ear structures, or other structures, such as the sigmoid sinus and facial nerve)
- A high-fidelity software simulation reproducing the physical and volumetric effects of penetrating different layers of hard mastoid cortex and air-filled petrous bone with a high-speed drill

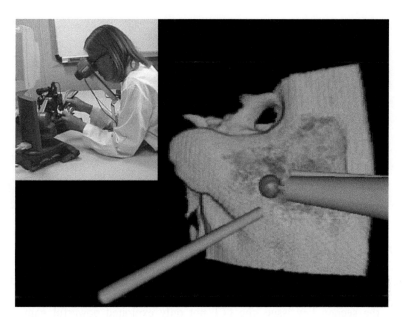

FIGURE 36.36 The IERAPSI temporal bone intervention simulator.

- An interface consisting of a binocular viewing system and two COTS haptic feedback stylus-like hand controllers, capable of reproducing the force and tactile sensations associated with mastoidectomy and the vibration-induced sound effects experienced when drilling through different densities of bone

36.3.5 INTERACTIVE TRAUMA TRAINER

Recent worldwide events have highlighted an urgent need to develop low-cost, distributable training scenarios to help prepare trainee battlefield surgeons to make timely, life-saving decisions and to provide *refresher* training for specialist surgeons. These surgeons (e.g., urologists, pediatric surgeons) may well end up practicing in unfamiliar contexts, such as trauma and emergency medicine, treating military personnel and civilians alike, with the additional stress of the conflict environment. One such application, developed for this specific domain, was a UK concept demonstration sponsored by the MoD, known as the *interactive trauma trainer* (ITT).

The ITT (Stone, 2011; Stone & Barker, 2006) came about following requests from defense surgical teams who expressed a desire to exploit low-cost, part-task simulations of surgical procedures for combat casualty care, especially for refresher or just-in-time training of nontrauma surgeons who might be facing frontline operations for the first time. The ITT was the result of a 6-month proof-of-concept project, the original aim of which was not so much to develop a surgical training prototype but rather to demonstrate two key issues in the field of human factors for synthetic environments. The first issue is the importance of applying human factors/human-centered task analytic techniques early in the trainer development process. The second key issue was to demonstrate the use of game engine technology to deliver useable, affordable, accessible, and distributable real-time interactive 3D training content at a level of interactive fidelity appropriate to the needs of the end user.

The human factors analyses, based on observational analyses and briefings conducted with the United Kingdom's Royal Centre for Defence Medicine and army field hospital specialists, contributed not only to the definition of learning outcomes and performance metrics but to key design features of the simulation and human–computer interface. The task of the user was to make appropriate decisions relating to the urgent treatment of an incoming virtual casualty with a *Zone 1* neck fragmentation wound. Appropriate interventions—oxygen provision, blood sampling, *hands-on*

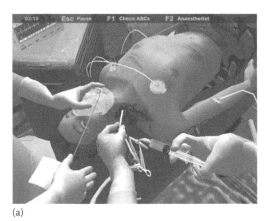

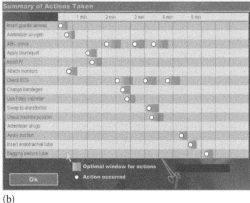

(a)   (b)

FIGURE 36.37   (a) ITT screenshot (left) and (b) performance summary screen (right).

body checks, patient visual and physiological observation, endotracheal intubation, and so on—had to be applied in less than 5 min in order to save the virtual casualty's life.

Addressing the second key issue, the ITT not only exploited powerful game engine software technology, supporting the rendering of high-fidelity models of the virtual casualty, it also exploited a typically simple gaming interface—mouse control for viewpoint change, option selection, and instrument acquisition. The human factors analysis helped to define the end shape and form of the ITT, applying high-fidelity effects only where they would add value to the surgeon's task (Figure 36.37, left). The analysis also ensured that the dexterous tasks (e.g., the use of a laryngoscope, stethoscope, intubation tubes, Foley's Catheter) were committed to clear and meaningful animation sequences, rather than expecting the surgical users to interact with 3D models of instruments via inappropriate control products with limited or no haptic feedback. Finally, users were provided with a feedback screen indicating when actions were taken, whether that action was taken at the most optimal time and an overall score (Figure 36.37, right).

The lessons learned during the execution of the ITT project were also exploited in an advanced interactive simulation system for defense medics called *Pulse!!* (Johnston & Whatley, 2005). Coordinated by Texas A&M University Corpus Christi and funded in 2006 by a $4.3 million federal grant from the Department of the Navy's Office of Naval Research, the *Pulse!! virtual learning space* healthcare initiative was designed to provide an interactive, virtual environment in which civilian and military heath care professionals could practice clinical skills in order to better respond to catastrophic incidents, such as bioterrorism. Using scenarios as diverse as a realistic representation of Bethesda Naval Hospital in Washington DC to a busy clinic populated with patients possibly suffering from anthrax exposure, end users could interact with virtual patients and with other virtual medical personnel to conduct examinations, order tests, and administer medication. The virtual patients were modeled to respond in real time and in an appropriate physiological manner.

36.3.6   VR in Psychology

As well as the early interest demonstrated during the 1990s in developing VR systems to support medical training and interventional planning (Gallagher et al., 2005; Satava, 1995b; Stone & Barker, 2006), many institutions, notably universities in the United States, began to focus on the power of VR to support psychological therapies. VR could, it was argued, effectively present patients with controllable, simulated—sometimes even fantasy-like—realities, while they were present in the safe setting of a therapist's office or clinic. This was in stark contrast to exposing patients to real-world, potentially threatening environments, where a multitude of uncontrollable sights, sounds, smells, and events could *flood* the patient with undesirable experiences. From the early 1990s, through to the early 2000s, and despite the still-primitive and costly nature of the VR equipment available, large numbers of research and clinical projects were undertaken, with applications in treating

- Acrophobia (fear of heights; North, North, & Coble, 1996a; Rothbaum et al., 1995)
- Claustrophobia (fear of confined spaces; Botella, Banos, Villa, & Perpina, 2000)
- Agoraphobia (fear of open spaces; North, North, & Coble, 1996b; North, North, & Coble, 1996c)
- Glossophobia (fear of public speaking; North, North, & Coble, 1998; Pertaub, Slater, & Barker, 2002)
- Arachnophobia (fear of spiders; Carlin, Hoffman, & Weghorst, 1997; Garcia-Palacios, Hoffman, Carlin, Furness, & Botella, 2002)
- Aviophobia (fear of flying; e.g., Krijn et al., 2007; North, North, & Coble, 1997)

36.3.6.1 Distraction Therapy

One of the more powerful examples of applying VR technologies in support of a therapeutic process is that of distraction therapy in pain management and control. Distraction therapy refers to the use of nonpharmacological techniques designed to reduce the onset or severity of an anxiety attack or other stressful/painful situation. The success of distraction therapy is based on the extent to which a patient's limited attentional resources can be *channeled* away from the conscious perception of pain by using alternative forms and optimum levels of sensory stimulation—visual, auditory, haptic (force/touch), proprioceptive (joint position and motion), and even olfactory (smell). The use of VR in this context is appealing, as the technology has always offered the potential to deliver a range of multisensory distraction techniques, courtesy of the numerous (and ever-evolving) forms of interaction and display devices available on the COTS market.

A highly publicized and oft-quoted example of a successful VR distraction therapy system is *SnowWorld* (Figure 36.38). Developed by Hunter Hoffman, David Patterson, and colleagues at the

FIGURE 36.38 SnowWorld (University of Washington HITLab).

University of Washington's Human Interface Technology Laboratory (HITLab) in Seattle, burn victims can, by wearing a head-tracked VR HMD, navigate their way through a simulated *ice canyon* populated with virtual snowmen, penguins, and igloos. Their task is to shoot snowballs at these virtual targets, with the illusion of *cold* being enhanced further by the simulated sound effects, which include the sound of snowballs splashing against the canyon sides or into the representation of an *icy* river. *SnowWorld* is based on the premise that, by capturing patients' attention in a dynamic VE characterized by strong visual simulations of snow, ice, freezing water, and other *visually cold* features, they have less available mental or conscious resource to bring to bear on the processing of pain signals (Hoffman, Doctor, Patterson, Carrougher, & Furness, 2000). Indeed, functional magnetic resonance imaging (fMRI)—a procedure that measures brain activity by detecting associated changes in blood flow—has, in further studies by Hoffman's team, confirmed their hypothesis that *VR analgesia* can, in the case of *SnowWorld* at least, be effective in suppressing pain-related activity in five known regions of the human brain. In some cases, participants (healthy volunteers who were subjected to *tolerable* thermal pain stimuli) showed a greater than 50% reduction in pain-related brain activity while interacting with *SnowWorld* when compared to a condition of no exposure (Hoffman et al., 2004).

Another study addressing the use of VR in burns therapy was conducted by Sharar et al. (2007). In their investigation, 88 patients (aged between 6 and 65) formed the participant group. All required postcutaneous burn passive *range-of-motion physical therapy* (ROM PT) delivered in sessions lasting 3–15 min. ROM PT is undertaken to help preserve the motion capabilities of a patient's limbs, to help regain their strength and endurance, and to prevent pronounced, raised (*hypertrophic*) scarring and the tightening of skin around the wound. Comparisons were made between groups receiving standard pain control (e.g., opioid and/or benzodiazepine analgesics) and those receiving analgesics plus VR distraction using an HMD. Based on self-reported subjective pain ratings (using a scale of 0–100), the addition of VR as a distraction technique resulted in significant reductions in pain ratings for worst pain intensity (20% reduction), pain unpleasantness (26% reduction), and time spent thinking about pain (37% reduction). Eighty-five percent of the participants reported no nausea when using the HMD. Children (75% of the participants, aged between 6 and 18) reported higher sensations of *presence* within and realism of the virtual environment than did adults, but age did not affect the analgesic effects of VR distraction.

36.3.6.2 Imaginal Exposure

The treatment or management of phobias and related psychological conditions will often use imaginal therapy or *imaginal exposure* in the early part of the counseling process. Imaginal exposure is a technique used extensively in cognitive behavioral therapy (CBT, e.g., Rothbaum, Meadows, Resick, & Foy, 2000) and requires the client to describe their traumatic event, step by step and repeatedly, to the therapist or counselor. Alternatively, therapists may actually talk patients through a stress-eliciting scenario. Fear-of-flying sessions held at Royal Air Force (RAF) Brize Norton, for example, have, in the past, used imaginal exposure as part of the treatment with phobic patients (lying down with eyes closed), with the therapist verbally describing the approach to an aircraft from the departure lounge, emphasizing the sights, sounds, and smells they might encounter as they get closer to and actually board the aircraft.

CBT is based on the premise that most psychological problems, including posttraumatic stress disorder (PTSD), are linked to significant negative learning (or cognitive) experiences, affecting one's ideas, mental images, beliefs, and attitudes. Avoiding remembering such experiences—including exposure to actual or similar places, objects, and sensations (even via television, radio, and newspaper coverage)—drastically affects the behavior and day-to-day well-being of the individual concerned, not to mention longer-term health prospects (Collimore, McCabe, Carleton, & Asmundsen, 2008; Galea, Nandi, & Vla, 2005). Of course, not all patients respond to imaginal exposure, despite agreeing to undertake the approach at the outset. Some are simply unable to control their senses and emotions during the imagination process. Unfortunately, research suggests that, without strong emotional engagement, a poor treatment outcome will often result (Jaycox, Foa, & Morral, 1998).

36.3.6.3 Virtual Exposure

One might hypothesize then that by exploiting multisensory VR simulations (typically visual, auditory, haptic, and olfactory), it may be possible to create convincing and engaging simulated contexts to enhance the imaginal exposure process significantly (Hoffman, Garcia-Palacios, Carlin, Furness, & Botella-Arbona, 2003; Hoffman, Hollander, Schroder, Rousseau, & Furness, 1998; Vincelli, 1999). Indeed, in contrast to traditional imaginal exposure and CBT, some research has already suggested that phobia sufferers may be more likely to seek out and complete their courses of therapy if VR is part of the therapeutic process (Garcia-Palacios, Hoffman, Kwong See, Tsai, & Botella, 2001). Difede and Hoffman (2002) report a single-participant case study—a 26-year old female who was treated for acute PTSD and major depression using VR exposure therapy following the attacks on the World Trade Center Towers in New York on September 11, 2001. The patient's avoidance behaviors were quite extreme and included blocking out thoughts of the incident, avoiding newspapers and radio or TV reports, and even refusing to stay at her partner's high-rise apartment. She had not responded to conventional imaginal exposure therapy. Using an HMD to display images and sounds of the Twin Towers being attacked and collapsing (Figure 36.39), the patient attended six VR exposure therapy sessions, rating her severity of distress several times during each session using the Subjective Units of Distress Scale (Williams & Poijula, 2002). Difede and Hoffman (2002) also report that, after the VR treatment sessions, the patient no longer met the criteria for PTSD, major depression, or any other psychiatric disorder, as judged by an independent clinical assessor.

Five years later, Difede et al. (2007) evaluated VR exposure therapy for World Trade Center attack survivors with PTSD by comparing the outcomes of a group of 10 recipient patients with those of a *wait-list control group* of 8 participants. A wait-list control group paradigm is a pseudo-control group of participants who provide an untreated comparison for the active treatment group but are offered the opportunity to obtain the same intervention at a later date. The VR group demonstrated a statistically and clinically significant decrease in scores on the Clinician-Administered PTSD Scale (CAPS) relative to both pretreatment ratings and to those of the wait-list pseudocontrol group. Seven out of ten participants in the VR group no longer carried a PTSD diagnosis, while all of the wait-list controls retained the diagnosis following the waiting period. Five of the 10 VR patients had previously participated in imaginal exposure treatment with no clinical benefit.

Another (highly publicized) application of VR exposure therapy relates to the use of the technology in therapies for treating PTSD in armed forces personnel returning from operational duties.

FIGURE 36.39 Treating acute PTSD following the 9/11 Twin Towers attacks using VR exposure therapy.

Rizzo et al. (2009) report on the use of a well-established VR exposure therapy tool called Virtual Iraq (more recently a similar set of Virtual Afghanistan scenarios have been developed). Virtual Iraq consists of a range of scenarios, including a Middle-Eastern-themed city and desert road environments. The Virtual Iraq system represents quite a *high-tech* solution in the treatment of PTSD, in that, as well as the visual stimuli presented via a COTS HMD, spatial audio, vibrotactile, and olfactory stimuli can also be displayed to the patients. One screenshot from Virtual Iraq is presented in Figure 36.40. The virtual scenarios and stimuli can also be modified in real time, via a clinician-controlled Wizard of Oz interface (in this case, a tablet-like terminal that allows the clinician or experiment to trigger preprogrammed events within that simulation as and when required, such as explosions, enemy fire, appearance of humans, animals, vehicles), while in audio contact with the patient. This interface is instrumental in, as the authors describe it, helping "to foster the anxiety modulation needed for therapeutic habituation."

It is worth noting that the use of VR in the treatment régime described by Rizzo et al. did not occur until the fourth of the 10 sessions with the clinician (held twice weekly, for 90–120 min periods over 5 weeks). The early part of the treatment included clinical interviews, *psychoeducation* relating to trauma and PTSD, general stress management techniques, instruction on the use of the Subjective Units of Distress Scale (as mentioned earlier), and a short (25-min), informal introduction to the Virtual Iraq VE, without any form of trauma trigger stimuli being introduced. In the actual VR sessions, patients were requested to recount their trauma, as if it were happening again, giving as much sensory detail as possible.

In addition to a battery of subjective measures, physiological monitoring (heart rate [HR], galvanic skin response [GSR], and respiration) was also conducted as part of the data collection. Initial analyses of results from the first 20 Virtual Iraq treatment completers (19 male, 1 female, mean age of 28) have indicated positive clinical outcomes. Of the 20 VR treatment completers, 16 no longer met PTSD diagnostic criteria at posttreatment, with only one not maintaining treatment gains at a 3-month follow-up appointment. The authors of this study also reported challenges with dropouts from the sample of active duty participants, seven of whom failed to appear at the first session. In addition, six dropped out prior to the formal VR exposure therapy trials at session four and seven dropped out at various times thereafter.

FIGURE 36.40   Virtual Iraq screenshot.

36.3.6.4 Restorative Environments

Stephen Kaplan, of the University of Michigan, wrote in 1992:

> The difference between nature as an amenity and nature as a human need is underscored by this research. People often say that they like nature; yet they often fail to recognize that they need it … Nature is not merely "nice." It is not just a matter of improving one's mood … rather it is a vital ingredient in healthy human functioning.

Significant global attention is being paid to the relationship between human physical and mental well-being and the availability and status of the urban and natural environments in which they find themselves. Research results suggest that exposure of individuals to these natural or restorative settings can promote stress reduction (Kaplan, 2001; Ulrich, 1981) and assist the recovery of attentional capacity and cognitive function following mental activities or fatigue brought on by *directed attention* (Berman, Jonides, & Kaplan, 2008; Berto, 2005; Kaplan, 1995; Kaplan & Kaplan, 1989).

Restorative environments as simple as window views onto garden-like scenes can also be influential in reducing postoperative recovery periods and analgesic administration. In an oft-cited paper published in *Science*, Ulrich (1984) compared 23 matched pairs of patients who underwent a cholecystectomy (gall bladder surgery). After surgery, patients were randomly assigned to either rooms facing a brick wall or rooms with a view of a natural environment (such as trees or a grassy field). Ulrich found that those facing the natural view had shorter postoperative stays, took fewer analgesics, and rated their hospital stay as more positive than those facing the more *urban* scene.

Research by Tsunetsugu et al. from the Japanese Forestry and Forest Products Research Institute addressed the exposure of subjects to forest and urban settings and the effect that these exposures had on subjective ratings and physiological measures, including blood pressure, heart rate, and salivary cortisol excretion (Park et al., 2007; Tsunetsugu et al., 2007). Their research demonstrated that 15 to 20 min exposures to natural environments, such as a broadleaf forest (the Japanese use the term shinrin-yoku or *forest bathing*), were accompanied by significant lowering of blood pressure, pulse rate, and cortisol levels when compared to similar exposures in a busy city area. Subjective ratings of *calm*, *comfortable*, and *refreshed* were higher in forest conditions than those recorded in the city. More recent results from these research teams have demonstrated a link between forest bathing and significant increases in the number and activity of human natural killer (NK) cells. An important part of the immune system, these cells play a major role in the suppression and eradication of tumors and viral infections (Li et al., 2008).

Li et al. (2008) attribute this effect in part to significant reductions in measured urinary adrenaline (the *stress hormone*) and also to the release by forest trees of phytoncides—active plant substances (e.g., wood essential oils) that can inhibit the growth of, and even kill, bacteria, fungi, and microorganisms. Indeed, these effects appear to persist for up to a week after initial exposure to the forest environment. The Japanese government takes shinrin-yoku very seriously indeed and, at the time of writing, has already invested over $US 4 million in research, with the aim of establishing 100 forest therapy sites within a decade and, thus, promoting a more nondestructive use for the forests that cover nearly 70% of the country.

Other indicators of human response to real-world restorative environments include reductions in blood pressure (Hartig, Evans, Jammer, Davis, & Garling, 2003) and cortisol (van den Berg, 2008). Research also suggests that a reduction of symptoms related to prolonged stress and depression, including those brought about by prolonged adverse weather and annual time changes—a form of *constrained restoration*—can occur as a result of exposure to green spaces and rural outdoor settings (Hartig, Catalono, & Ong, 2007). This may even benefit those who suffer from seasonal affective disorder (*SAD*), a form of seasonal depression characterized by episodes that can reoccur at similar times each year, typically in the winter. The main symptoms of SAD include a low mood and a loss of interest in day-to-day activities, longer sleep periods, and weight gain.

Why do these effects occur? Two theories have been put forward by the early proponents of restorative environments, although neither has the benefit of strong background evidence or experimental support.

Kaplan (1995), for instance, makes reference to attention restoration therapy where, it is claimed, when interacting with a rural environment that is *rich with fascinating* (but subtle) stimuli, attention is *modestly* captured in a *bottom-up, involuntary fashion*, allowing directed attention mechanisms to recover. Urban environments also contain *bottom-up* stimuli (e.g., flashing lights, loud *man-made* sounds, signs, etc.), but these *dramatically* capture attention and require directed attention to overcome the impact of the stimuli.

In contrast, Ulrich (1981) proposes an affective response approach in which sensory patterns within an individual's field of existence prompt automatic and quite dramatic responses. Natural patterns (e.g., from rural environments) lead to a *replenishment of cognitive capacity* by altering the emotional and physiological states of the individual—the initial affective response shapes the cognitive events that follow. Discussions and exchanges relating to the underpinning theory of restorative environments will, no doubt, continue unabated for some time. However, as pointed out by Valtchanov, Barton, and Ellard (2010), "since the emergence of these two theories, the questions of "how" and "why" restorative environments reduce cognitive fatigue, decrease stress levels, and increase an individual's ability to focus have not been thoroughly researched."

Of course, all of the aforementioned examples relate to real-world exposures, be the participants healthy or undergoing a period of hospitalization. But what of individuals who are unable to access and experience real natural environments—patients who present with a variety of psychologically related conditions (e.g., PTSD, depression, attention deficit disorder, pain, and sleep deficit) and who may be confined to sensorially sparse rooms and wards within urban hospitals, within hospices and care homes, or within civilian and military rehabilitation centers? Is it possible that VR technologies could be developed to achieve similar psychological and physiological effects to those described earlier? And could these technologies deliver a new form of therapeutic process that could be used and reused in a variety of passive and dynamic contexts—from recovery in intensive care units to minimizing muscle atrophy prior to the fitting of prosthetic limbs? If so, and to pose a question asked by Wohlhill (1983), "what would it take to simulate a natural environment, one that would in fact be accepted as a satisfactory surrogate?"

36.3.6.5   Image- and Video-Based Restorative Environments

The use of static images as restorative environments includes posters or large-format photographs, murals or wall paintings, and interior decor resembling rural settings (e.g., forest, coastal or lakeside locations, and browsable image portfolios of high-quality natural scenes). A considerable amount of the available literature appears to focus on subjective, *preferential* studies relating to the use of images of natural settings for such applications as marketing/product branding (Levi & Kocher, 2008) or judging views from hotel, apartment, and hospital windows (Bringslimark, Hartig, & Patil, 2011; Mazer, 2010; Simonic 2006; White et al., 2010).

With regard to healthcare issues, as opposed to simple preference assessments, Diette, Lechtzin, Haponik, Devrotes, and Rubin (2003) investigated the use of simple pictures (*murals*) of natural scenes placed at the bedsides of patients due to undergo investigation via flexible bronchoscopy (Figure 36.41). The patients also received taped sounds of nature (such as a babbling brook) before, during, and after the procedure. Patient ratings of pain control (using a five-point scale ranging from poor to excellent) and anxiety showed that, in contrast to a control group, distraction therapy exploiting natural audio and visual stimuli significantly reduced pain but not anxiety.

Nanda, Gaydos, Hathon, and Watkins (2010) presented a review of literature addressing art and posttraumatic stress for war veterans. The review acknowledges the literature relating to the therapeutic effects of using natural images and lists Ulrich and Gilpin's (2003) suggestions for healthcare art, which include waterscapes, landscapes, *positive cultural artifacts*, flowers, and *figurative elements with emotionally positive faces*. However, the authors highlight that this has not been explored in the context of PTSD, specifically of war veterans, and emphasize that careful consideration of veterans' PTSD symptoms is warranted before any visual imagery is selected for environments providing health care to veterans.

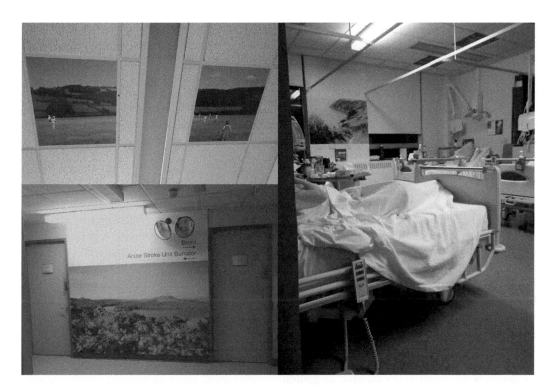

FIGURE 36.41   Static images and murals in a UK hospital environment.

Video representations of restorative environments include background/at-bed/at-chair looping videos presented via projectors or plasma/flat-screen displays (also hospital entertainment screens), as well as closed-circuit television (CCTV) images relayed to the observer from remote external environments. Ulrich and Gilpin (2003) conducted an experiment using over 870 blood donor participants (68% males and 32% females with a mean age of just over 40 years), exposing them to four different environmental conditions, presented using a video player and wall-mounted screen. The conditions were (1) footage of natural environments, (2) footage of urban environments, (3) daytime television, and (4) a blank screen. Using physiological measures (blood pressure and heart rate), the researchers found that stress was lower during the blank screen condition than the daytime television condition and lower during the natural environments condition than the urban environment condition. Throughout, pulse rates were markedly lower during the natural environments condition than the urban.

Friedman et al. (2004) describe two individual participant case studies in which a high-definition television (HDTV) installed in the participants' windowless offices relayed a view from an external camera overlooking a public plaza and fountain area in real time. The results of the studies exposed contrasting opinions. On the one hand, participants found the display psychologically and socially beneficial in terms of an increased sense of *connection*, offering mental breaks. On the other hand, some participants felt that the camera system was distracting and left them with *an immoral sense of surveillance*, specifically relating to the privacy of people whose images were being captured in a public place.

In a later paper, Friedman, Freier, and Kahn (2008) reported an increase in users' connection to the wider social community, connection to the natural world, psychological well-being, and cognitive functioning. Kahn, Severson, and Ruckert (2009) further reported that participants took *mental breaks* to stare at the external views, returning to their office tasks more refreshed. Participants also felt connected to the days passing due to movement of the sun and weather changes.

Abkar, Kamal, Maulan, and Mariapan (2009) performed an experiment in which stress was induced in participants by showing them a video of an accident with excessive blood and gore. To recover from this stress, participants were either shown a video of natural scenes (e.g., trees, moving water, grass) or traffic. Building on the earlier findings of Ulrich and Gilpin (2003) and van den Berg (2008), who also demonstrated a range of positive effects of films of real nature, Abkar et al. (2009) found higher levels of stress recovery in the participants who viewed the natural scenes, as determined by blood pressure, muscle tension, and heart rate. The study showed that the natural setting could elicit responses that included an element of the parasympathetic nervous system linked with the restoration of physical energy.

36.3.6.6 VR-Based Restorative Environments

VR restorative environments use similar at-bed/chair/wheelchair presentation techniques as the video presentations described earlier, but, if implemented appropriately, will support a greater degree of interaction on the part of the user (e.g., virtual environment navigation and exploration, virtual object interrogation, interaction with virtual actors, or avatars) and a more cost-effective means of updating the scenery on a regular basis. One of the surprising discoveries while conducting the literature search was the significant absence of relevant research and experimentation in the field of virtual restorative environments (VREs, i.e., those built using established VR or *serious games* toolkits) for postoperative recovery and their subsequent extension to support rehabilitative procedures. This finding is confirmed in the writings of de Kort and Ijsselstiejn (2006), who point out that the majority of studies have involved the use photographs, slides, or videos in a laboratory setting. Unfortunately, they go on to report the results of three studies that are, themselves, based on passively presented video material of natural environments, where participant *immersion* was defined on the basis of screen size. Of the few studies discovered, only a tiny proportion report relevant experimental findings, most focusing on *what if* issues or the outcome of subjective analyses, many concluding that virtual natural environments *may* have *some degree* of restorative qualities. However, two studies are worthy of mention here.

The first, reported by Waterworth and Waterworth (2004), was based on what the authors described as a virtual *tropical paradise*, developed as part of a EU project called Engaging Media for Mental Health Applications (EMMA). It seems that the EU grant expired before the research team could actually perform any meaningful experiments with their particular VE (despite a funding period of some 8 years), although they (Riva, Waterworth, & Waterworth, 2004) claim to have developed a *biocultural theory of presence*. The EMMA Paradiso VRE, referred to as *Relaxation Island*, is described by the authors as "an archetypical tropical paradise, with lush vegetation, a waterfall and a long beach, all surrounded by mountains ..." Unfortunately, this tends to raise the expectation of the reader, who is then confronted with images of a relatively low-fidelity VE, together with images of an interactive control device resembling a pearl in a seashell. Furthermore, no experimental or controlled therapeutic evaluation of the EMMA system seems to have been conducted. The aim of this VRE, as the authors claim, is to achieve "equal or even greater levels of relaxation without relying on imaginative skill" (e.g., as might be achieved during early stages of hypnosis). While on the island, the participant can choose any of the four zones to learn different relaxation techniques. Two of the zones are beach locations, the third is a waterfall, and the fourth a *cloud zone*. A *calm voice* instructs each participant in the different relaxation exercises at each zone. At the waterfall zone, for example, the therapist is able to enter *worry words* elicited from the participant, each of which appear on the leaves of a nearby plant, before falling into the stream to be carried out to sea.

The second study, by Valtchanov et al. (2010), provides early findings supporting the use of VREs and their similarity in delivering enhanced feelings of well-being as might be achieved using, for example, photographic images or videos (*surrogate nature*). A photorealistic forest covering some 1600 m² was constructed and presented to participants wearing a head-tracked HMD. In addition to the HMD, a *rumble platform* was also used, providing haptic cues to represent taking steps and colliding with objects within the VE. Navigation through the VE was achieved using the buttons of a wireless mouse—moving the participant's viewpoint forward and backward along the current axis of view. Heart rate and skin conductance sensors were mounted onto the participant's nondominant hand, and pre-VE exposure stress levels were induced using a variety of techniques, including self-described stressful experiences accompanied by loud urban noise, followed by the Markus–Peters Arithmetic Test (Peters et al., 1998). Participants were then exposed to the virtual forest environment, allowing them to explore the environment freely for ten minutes (participants in the control condition were exposed to a slideshow comprising ten abstract organic paintings). Skin conductance was found to decrease significantly following exposure to the VRE, although there was no significance in heart inter-beat intervals between the experimental conditions.

36.3.6.7 Virtual Restorative Environment Therapy Project

Virtual restorative environment therapy (VRET) is a project that originated from postgraduate research conducted at the University of Birmingham in the United Kingdom (and is still in progress at the time of writing). The long-term aim of VRET is to develop a range of VREs for the benefit of patients who present with a variety of psychologically related conditions (e.g., PTSD, depression, attention deficit disorder, pain, and sleep deficit) and who may be unable to access and experience real natural environments, including those in hospitals, hospices, civilian and military rehabilitation centers, and care homes (Depledge, Stone, & Bird, 2011). Another aspiration for the VRET project is to conduct fundamental human factors research, using the VEs developed as a test bed, into a range of interactive technologies and software packages and psychophysiological measures of human performance, immersion, and well-being. At the present time, VRET is based around two virtual environments, both representations of real-world locations in the south of Devon— Wembury Bay (Figure 36.42a) and Burrator Reservoir (Figure 36.42b).

In brief (and with additional material and images available at www.virtual.burrator.net and www. virtual-wembury.net), both environments were developed using a variety of 3D modeling, image processing, and run-time tools.

The virtual topography of the environments were based on commercially available digital terrain model (DTM) data—dense fields of digital elevation points at a resolution of 5 m and a vertical accuracy of 1 m. Once the DTM models were converted into a polygon-based mesh, the virtual terrains were of a form suitable for importing into the Unity3D toolkit, where they were flat-shaded and endowed with a high-resolution texture map, itself generated from an aerial photograph of 12.5 cm ground-equivalent resolution. This texture map provided the development team with a visual template that was invaluable in helping to locate key natural and man-made features—trees, large plants, meadows, rocks, streams, buildings, paths, and enclosures. The virtual counterparts of these and many other features were either sourced from the web or *built* from scratch using the 3ds Max or SketchUp Pro modeling tools. A series of photographic, video, and sound surveys were also conducted. Sounds of birdsong, water, wind, and footsteps were then programmed into the

(a)   (b)

FIGURE 36.42   Scenes from Virtual Wembury (a) and Virtual Burrator (b).

FIGURE 36.43 Selection of images showing the interactive display and control elements of the VRET experimental modules.

VE, to create a dynamic soundscape that varies depending on the end user's spatial location. Real time of day (24 h day–night cycle) and simple weather effects were implemented, using the UniSky software system, and particle-based mist and spray effects were also included where it was felt they added a desirable visual quality.

The VREs can be displayed to the end user using a range of devices, from HMDs to plasma screens and projectors. Exploration of, and interaction with, the VE can also be implemented using a range of devices, from basic keyboard and mouse to multifunction hand controllers, such as the Xbox gamepad (Figure 36.43, left, top, and bottom), the low-cost Zeemote JS1 (used for mobile gaming) and Razer Hydra systems (Figure 36.43, right, bottom corner), and the Asus Xtion motion tracking device. The ability to tailor the human–system interface to support appropriate styles of interaction based upon the specific physical conditions presented by the patients (military casualties including amputees, burns victims, etc.) is absolutely essential, if their early and positive engagement (and that of the nursing personnel) is to be successful.

36.4 VIRTUAL HERITAGE

VH may be defined as

> … the use of computer-based interactive technologies to record, preserve, or recreate artifacts, sites and actors of historic, artistic, religious and cultural significance and to deliver the results openly to a global audience in such a way as to provide formative educational experiences through electronic manipulations of time and space. (Stone, 1999b, p. 18; Stone & Ojika, 2000, p. 73)

Ever since the early 1990s, there has been a worldwide interest in the prospect of using VR to recreate historic sites and events for such purposes as education, special project commissions, and showcase features at national and World Heritage visitor centers. The power of VR, as described by Nynex researchers Stuart and Thomas in these formative years, lies with its ability to open up places and things not normally accessible to people from all walks of life and to allow them to explore objects and experience events that could not normally be explored without *alterations of scale or time*. The power of the Internet now supports interaction with remote communities and interaction with virtual (historical) actors. In the context of heritage, VR goes much further, however, in that it offers a means of protecting the fragile state of some sites and can help educate visitors not so much about their history, but in how to explore, interpret, and respect them. For example, VR can display the potential damage caused by the ravages of human intervention and pollution by accelerating the simulated destructive effects to monuments, even large areas of land and sea, over short periods of time (Stone, 1999b).

One can trace the impetus for such interest back even further to an impressive demonstration, staged at the Imagina Conference in Monte Carlo in February 1993. An immersed member of the French clergy was joined by a simple avatar, controlled by an operator in Paris, who provided him with a real-time guided tour of a virtual reconstruction of the Cluny Abbey, a building destroyed in the early nineteenth century (Figure 36.44). This pioneering demonstration proved conclusively that a formative VH experience could be delivered over a standard ISDN communications network. Yet it was not until 1995 that the first VH conference was held at the Assembly Rooms in Bath, United Kingdom, featuring virtual Pompeii (Carnegie Mellon University), virtual Lowry (virtual presence; described further in Section 36.4.1), the Caves of Lascaux (University of Cincinnati), and the Fortress at Buhen, Egypt (ERG Engineering). At the same time, a project instigated by English Heritage was announced to develop a detailed VR model of Stonehenge, complete with a mathematically accurate model of the nighttime sky and a virtual sunrise. Virtual Stonehenge (described in more detail in the following) was presented in June of the following year at the famous London Planetarium, hosted by the equally famous UK astronomer, the late Sir Patrick Moore.

Virtual Stonehenge was also demonstrated live at the second VH conference held in London in December 1996. The 1996 conference was one of computing contrasts. At one end of the spectrum were such projects as virtual Stonehenge, the Battle of Gettysburg (TACS Inc.), and the Tomb of Menna in Thebes (Manchester Metropolitan University), all targeting low-to-medium performance computers. At the other end was the visually stunning *supercomputer*-based works of Infobyte,

FIGURE 36.44   Imagina 1993 Conference Networked VH Demonstration—Cluny Abbey.

notably the widely referenced (at the time) virtual Coliseum project. Another advanced project was announced by Miralab (University of Geneva), which later resulted in an equally stunning virtual recreation and animation of the Terracotta Warriors of Xian.

VH has become one of the few active application domains to survive from the technology-driven VR *era* of the 1990s. Projects throughout the final decade of the twentieth century produced a range of interactive 3D archaeological *exhibits* (Stone, 1999b), with many attracting considerable interest from organizations such as English Heritage and UNESCO. These projects even prompted the launch of an international VH Network. Unfortunately, however, many of the VH demonstrations to date have been very sterile—lacking the dynamic natural features evident in the real world, such as believable environmental effects, both on land and in extreme environments, such as subsea, or remote locations with dynamic seasonal weather changes. Of those that have been developed to a visually and behaviorally rich level of fidelity, most exist in a form that is not accessible to a wide population of beneficiaries—scientists, schoolchildren, students, and even members of the general public.

36.4.1 Virtual Stonehenge and Virtual Lowry

Virtual Stonehenge, a project originally sponsored by English Heritage, set out to deliver a high-fidelity (for the mid-1990s) VR model of the monument and its environs (Figure 36.45). The original aim of the project was to provide end users not only with a real-time exploration capability but also with a means of interacting with certain stone features, thereby exposing historical details, such as Christopher Wren's famous graffiti or axe and dagger marks (Stone, 1998, 1999b). However, the project was *hijacked* to some extent by a well-known computer chipset company, keen to provide sponsorship, but only to demonstrate (and, thus, publicize) the real-time graphics power of its latest processor. As a result, the project emphasis changed from one of strong educational potential to one dominated by technology (and this included the rather distorted display of the VR model using the London Planetarium Dome Theatre, *enhanced* by the use of an inappropriate synthetic odor, allegedly representing *cut grass*).

Fortunately, Virtual Stonehenge has, more recently, been recreated using appropriate *serious gaming* software such as Quest3D and Unity toolkits and the original aspiration of displaying the model's educational potential has been achieved (Figures 36.46 and 36.48).

FIGURE 36.45 Original 1995/1996 *Virtual Stonehenge.*

FIGURE 36.46 *Virtual Stonehenge revisited* using Quest3D and Unity toolkits.

Virtual Lowry (Stone, 1996, 1998), still one of the most popular examples of VR for heritage of its time, was based on the notion of exploiting technology to allow end users to approach and actually enter a famous L.S. Lowry painting—in this case *Coming From The Mill* (circa. 1930). On *reappearing* on the other side of the virtual canvas, the observers were free to explore a 3D reconstruction of Lowry's townscape, complete with animated *matchstalk* figures, all reconstructed using high-definition textures developed from images of actual Lowry paintings (Figure 36.47). This demonstration was commissioned by Salford City Council, as part of their bid for national funding to build the now-established Lowry Centre on Salford Quays in the north of England.

FIGURE 36.47 *Virtual Lowry* (Based on *Coming from the Mill*, circa. 1930).

36.4.2 Human Factors Challenges in Virtual Heritage

Rojas-Sola and Castro-Garcia (2011) emphasized that the two key elements that underpin the generation and acquisition of VH material are *working memory* and *culture*. Working memory, a term more familiar, perhaps, to cognitive psychologists than VH specialists, refers to information drawn from the spoken and written accounts from people who are able and willing to provide their own recollections. Culture, in this context, and again referring to the work of Rojas-Sola and Castro-Garcia, refers to the exploitation of data from ethnographic, anthropological, sociological, and historical sources, including historical and modern maps, local authority archives, and photographic and film collections.

Each of these elements brings with it its own unique human challenges—challenges that are not necessarily made easier with the passage of time nor with the evolution of innovative interactive technologies. There is a key *research question* facing designers and developers of VH, especially where, in the case of industrial heritage, there is still a significant likelihood that the end results will be experienced by surviving and, therefore, highly knowledgeable SMEs. That question is how can the early findings and results of VR or AR recreations be presented and manipulated in such a way as to engage end users, stakeholders, and SMEs and to avoid their alienation, either as a result of errors in historical interpretation or by compromising their experience by exposing them to inappropriate human interface technologies during design and final presentation?

The material from Rojas-Sola and Castro-Garcia's (2011) category *working memory* is absolutely essential to the execution of accurate and educational VH experiences. Each of the *case study* projects described in this part of the present chapter serves to support this statement. Where possible, exposing the owners of *working memories* on a regular basis helps to ensure the accuracy of the virtual sites and artifacts developed. Indeed, early engagement with and iterative exposure of end users or stakeholders are well-established principles in human-centered design processes, as laid down, for example, in international human factors standards such as ISO9241-210 (ISO, 2008). In addition, exposing stakeholders to the evolving VR or AR deliverables could very well help stimulate memories that were not forthcoming during earlier review or recall sessions. However, as time passes, it is a fact of life that the SMEs will diminish in number and there is no guarantee that any knowledge they once possessed will be recorded, archived, or passed into the hands of the generations they leave behind. The attitudes of those generations to the preservation of such knowledge may also be problematic—there is little doubt that, in the past, valuable heritage material will have been lost as a result of "removing the old to make way for the new."

In many respects, material from the *culture* category defined by Rojas-Sola and Castro-Garcia (2011) is a much more complex issue. At one end of a very broad *cultural continuum* of issues is the need to engage closely with the owners of these different cultural material assets and to convince them that their material will be treated carefully and sympathetically. Not only does this refer to care in handling assets such as photographic images, maps, and the like, it also refers to how they are likely to be transformed when implemented in a digital form and how their distribution will be protected (especially if specific assets contain images of personal significance for the owner).

This was demonstrated with the more recent *reincarnation* of virtual Stonehenge mentioned earlier, with its *embedding* of multimedia data into the virtual stones and surrounding terrain (Figure 36.48). Using hyperlinks or *portals* from the main virtual environment to make archived material such as text, images, video, and other virtual objects easily accessible is becoming increasingly popular as an interactive technique in many walks of VR and AR. However, it is important that the digitization of that material does not compromise the owner's expectations or causes disengagement through such practices as image warping, the use of false color or low resolution, highly pixelated images or video sequences, or the inclusion of annotation that may obscure facial or other personal features or characters important to the owner.

Another example is how images sampled from areas within photographs or paintings could be used as textures in the target VR scenario. The virtual Lowry project was an excellent example of this. The *construction* process of the virtual *dreamscape* (a term used by L.S. Lowry himself to

FIGURE 36.48 Information *window* on Sir Christopher Wren activated via a highlighted representation of the architect's graffiti (insert) on one of the stones in *Virtual Stonehenge*.

describe some of his work) demonstrated the importance of engaging with the owners and custodians of the artist's unique works to ensure sympathetic treatment of the painted content. For example, early engagement helped to define acceptable and appropriate levels of fidelity when using scanned area samples of different paintings for texturing buildings, roads, and terrains and for skydome and horizon bill-boarding purposes.

As well as the diminishing number of SMEs in the more recent domain of virtual industrial heritage, detailed and accurate cultural assets are also becoming increasingly hard to find, even with twentieth-century heritage locations. Space does not permit a detailed discussion of the hypothesis that industrial heritage sites and artifacts may be more likely to deteriorate faster than their centuries-old counterparts (due to acts of vandalism, new building projects, accelerated decay of 1990s building materials, etc.). Excellent examples of this particular concern can be seen in books relating to the United Kingdom's *subterranean heritage*, such as Cold War Era nuclear bunkers (Catford, 2010). Even when gaining access to actual physical sites is no longer possible, relevant information and material collections that are temptingly referenced on the web (such as those held by councils and museums) are—frustratingly—often not accessible directly from any online catalogue. To acquire such information may demand significant expenditure on travel to the host locations, and even then, *hands-on* time with the assets themselves may need to be prebooked and of limited duration (not to mention the restrictions that may be in place for digitizing historical material).

Furthermore, where limited or practically nonexistent assets exist (as was the case with the Wembury Commercial Dock project, described later in this chapter), it becomes necessary to extrapolate details from other sources and reference texts and, in many cases, to simply make one's best guess as to the appearance and extent of the virtual environment one is attempting to develop.

FIGURE 36.49 Virtual recreation of Burrator and Sheepstor Halt on the now-abandoned Yelverton-to-Princetown Railway.

Of course, the more unique the site or artifact is, the more one has to expend considerable time and resource in the construction of 3D models that are of the correct design, style, and era. In the United Kingdom, this has proved to be particularly problematic, and the extent to which 3D content has had to be built from scratch has, on numerous occasions, resulted in the distraction of researchers and students from their primary focus of developing interactive and educational VH applications. Recreating a short section of the long-abandoned Yelverton-to-Princetown railway line in South Devon is a particular case in point (Figure 36.49) (Stone, 2012). Here, the absence of specific data relating to detailed features of the small stretch of line being modeled (despite some excellent photographic and video material relating to the Burrator and Sheepstor Halt platform area, as shown in Figure 36.49) has necessitated numerous modeling and remedial modeling activities on the part of researchers. Often, they have to travel to the West Country to meet with SMEs (who do not have Internet access) in order to check the authenticity of their 3D content.

In some cases, it has been possible to acquire such content from such online repositories as Turbosquid, 3D Studio, 3D Cafe, and the like. However, and despite the huge number of 3D models these sites possess, it is often the case that there is little of relevance to the project one happens to be working on at the time. While this situation is slowly improving (with sites such as Trimble/Google SketchUp Warehouse doing much to alleviate the situation), it is still a fact that there are more US-relevant objects and datasets available online than there are for other countries. The temptation is often to download the closest match to the object one requires and to modify the geometry or associated textures. However, this practice runs the risk of attracting significant criticism from, and, potentially, losing the engagement of surviving SMEs.

Finally, and at the other end of the *cultural continuum* from a human factors perspective is the question of how one defines the nature of one's end-user population. What are their informational and educational needs, their current and previous interactive experiences, and their knowledge, skills, and attitudes and how does one use this information to ensure that the VH content is delivered using the most appropriate interactive technologies? Interactive 3D media have to be designed in conjunction with its end users, identifying the skills that need to be trained or the knowledge that has to be imparted and then delivering a solution based on appropriate content, fidelity, and interactive technologies. Furthermore, the solutions must be packaged in a form that can be delivered to the end users in their own working and living environments, as opposed to expecting them to experience the technology in isolated and restricted laboratory-like environments.

The following is a review of a number of recent VR and AR VH projects that serve to emphasize some of the issues discussed earlier. Each project has presented its own challenges, not only in terms of the limitations of the technology during data acquisition in the field and at subsequent demonstrations but also in the collation of relevant historical material and in attempts to engage with a wide range of SMEs, stakeholders, and end users. At the time of writing, the projects are ongoing, although interim demonstrations of the results have been presented.

36.4.3 Project 1: The Wembury Commercial Dock and Railway Proposal of 1909

Wembury is located to in the South Hams district of Devon, to the east of Plymouth and facing out toward Plymouth Sound. Designated as a Special Area of Conservation and a Voluntary Marine Conservation Area (Figure 36.50), the original rationale for constructing a 3D model of this particular coastal region did not evolve from any VH pursuit. Rather, the coastal topography was, as described earlier in this chapter, developed to support research into virtual restorative environments—the exploitation of interactive 3D scenes of natural settings to improve postsurgery recovery of physical and psychological well-being for hospitalized patients (Depledge et al., 2011; Stone & Knight, 2012). It was during the early site surveys for this virtual restorative environment that opportunistic contact with local historians and Wembury Village residents provided the motivation to undertake what became a rather ambitious VH project.

In 1909, a proposal (HL Deb, 1909) was put before the United Kingdom's House of Lords relating to the development of what could have become one of the largest—if not *the* largest—and most successful commercial docks in the country, rivaling other ports at London, Southampton, and Liverpool. Had the proposal not failed, then Wembury Bay would have been changed forever, with the docks, railway, and workers' houses decimating what is, today, one of the most attractive and popular coastal areas in the south west of the United Kingdom.

The port was to consist of breakwaters extending far out into the bay. Two layout proposals were considered, one consisting of a large single continuous dock structure with a railway terminus and the other one boasting four or five *finger* jetties, dry docks, and railway sidings taking passengers directly to and from their ship's berth. The railway would have taken the form of a single-track

FIGURE 36.50 The Real Wembury Bay from the Tower of St. Werburgh's Church (the island in the distance is the Great Mewstone).

branch line from the small town of Plymstock just to the north (with expansion space for a double track branch in the future), offering disembarking passengers a more rapid service to London than that being offered by the competing ports, and even by Millbay Dock in Plymouth itself. Another key issue emphasized by the proposal owners was that the geological nature of Wembury Bay would be ideal for berthing of large-draught vessels, unlike elsewhere in the United Kingdom, where ships had to weigh anchor offshore and then ferry passengers and cargo into the port area.

The proposal failed for a number of reasons, including undercapitalization; reliance on third parties for significant infrastructure elements, such as railway coaching stock; naïve growth and revenue estimates; and the belief that Southampton's expansion plans were already well developed and moving forward. It is also fair to say that parliamentary hostility played a key role in the downfall of the proposal, as did opposition from the London and South Western and the Great Western Railways, for obvious reasons, given their investment into already well-established routes into nearby Plymouth Millbay.

36.4.3.1 Wembury Docks AR and VR Demonstrations

Apart from one paper (Broughton, 1995), one or two simple concept illustrations and plans, limited personal reflections by local historians, some newspaper cuttings (Anonymous, 1908, 1909), and the 1909 hardbound proposal itself (only one copy of which has appeared on eBay in the recent past), there is nothing to convey the magnitude of this engineering project nor the impact it would have had on the Wembury environment and local village residents. Consequently, the cultural assets underpinning the actual 3D models were based on extrapolations from historical British railway and maritime publications. In particular, historical research had to be conducted using references to other UK docks, including Liverpool/Birkenhead, Southampton, and the Port of London, together with Hull, Cardiff, Falmouth, and Bristol, where the current Heritage Dock exhibits proved to be particularly useful.

As well as the topographical model developed using DTM data, described earlier for the VRET Project, the Wembury Dock VH demonstrator exploits a wide range of 3D assets, procured both from online sources and built from scratch. For the AR implementation (and bearing in mind the huge size of the dock), a low-fidelity model was used in conjunction with ARToolWorks' ARToolkit and iOS (Apple Inc.'s mobile operating system) plug-ins for the Unity toolkit, thereby supporting on-site AR visualization using an iPad3 with fiducial markers and a custom-programmed user interface supporting scaling, positioning, and orientation of the virtual images (Figure 36.51).

FIGURE 36.51 Possible multiple wharf layout of the failed Wembury 1909 dock plan visualized using VR.

FIGURE 36.52 One of the Virtual Wembury Dock wharves using the Unity games engine.

To ensure that the final result is viewable by a larger audience than the iPad can support, a higher fidelity VR version of the virtual dock has also been developed (including a virtual model of the ex-Royal Navy Frigate HMS *Amethyst*, the second of the two VH projects described herein (Figure 36.52).

36.4.4 PROJECT 2: HMS AMETHYST'S "FINAL RESTING PLACE"

While researching the history relating to the railway infrastructure in South Devon and Dartmoor for the Wembury Docks project, an opportunity arose to undertake an evaluation of contemporary AR technologies. The results of this have since generated enormous interest, not only in terms of what the project set out to demonstrate, but also in terms of encouraging a new generation of interactive media developers, each with a personal motivation keen to preserve a unique part of British naval history. Information relating to the rail transit of people and raw moorland materials to the ports in South Devon is available from a number of established publications (e.g., Kingdom, 1982, 1991), but one in particular provides an excellent account of the history of the oldest maritime part of the city of Plymouth, namely, Sutton Harbour, within the famous Barbican area. The book *Sutton Harbour*, by Crispin Gill (1997), contained an impressive (if somewhat sad) image of a famous Royal Navy vessel, awaiting the breaker's torch. That image was of HMS *Amethyst*, unceremoniously abandoned in the corner of the harbor, next to the China House—today a popular public house and restaurant (and itself an historic building, dating back to the 1600s).

HMS *Amethyst* was a *modified* Black Swan-class *sloop*, redesignated as a Frigate, pennant number F116, after World War II. In the late 1940s, the ship and her crew made their mark on history—a mark that was to be immortalized in the film *The Yangtze Incident*. While based at Shanghai in 1949, a Civil War was being fought by the communists and the Chinese nationalist party (the Kuomintang). On April 20, 1949, the *Amethyst* was ordered to relieve HMS *Consort*, a ship that was protecting the British Embassy at Nanking on the River Yangtze, and to make preparations to evacuate all British citizens facing the communist advance. While transiting the Yangtze, and despite flying numerous Union Jacks, the communists opened fire, inflicting significant damage and 19 fatalities (including the ship's commanding officer, Lt. Cdr. Skinner), not to mention causing the ship to run aground on a sandbank, at an angle that rendered the firing capabilities of the two forward turrets useless.

While the politicians and media argued about who was to blame for starting the engagement, *Amethyst* was stranded for months in unbearable conditions of heat and an increasing population of

FIGURE 36.53   HMS *Amethyst* at Marrowbone Slip in 1957, minus her bow. (From Anonymous, *The Sphere*, 185–186, May 1957.)

rats and cockroaches. With rapidly dwindling food and fuel supplies, Commander John Kerans, the British naval attaché in China, arrived from Nanking and took command of the ship. On July 30, 1949, Kerans decided to make a nighttime bid for escape. Once again, the ship took heavy fire, but at 05:00 on the thirty-first, the frigate rendezvoused with the destroyer HMS *Concord*. *Amethyst* underwent a refit in the United Kingdom in 1950 and, following additional service in the Far East, returned to Plymouth in 1952, was paid off, and placed in reserve.

Following her final duty, which was to appear as herself in *The Yangtze Incident* film, on January 19, 1957, *Amethyst* was towed into Sutton Harbour in Plymouth, coming to a final stop on Marrowbone Slip (Figure 36.53) (Anonymous, 1957), next to the China House, where she was broken up by Messrs. Demmelweek & Redding. Today, the only physical reminder of the ship's demise is a small commemorative plaque with a single sentence acknowledging her breakup.

36.4.4.1 Augmented Reality Amethyst Demonstrator

The AR *Amethyst* demonstrator project was designed to visualize the ship's final resting place and to draw attention to what must have been a spectacular sight—a 1350-ton, 283 ft long frigate laying silent in a harbor that was, at the time, more used to welcoming small fishing trawlers and sailing ships. The 3D model of the ship was constructed from scratch using 3ds Max and a variety of data sources—from screen grabs of sail-by sequences from *The Yangtze Incident* film and images from the web, to not-very-detailed radio control model plans, and even old reproduction cardboard construction kits from Micromodels (originally costing one shilling and sixpence). As with the Wembury Docks project, the AR software was a commercial product, ARToolkit, with plug-in features for the Unity game development toolkit. The visualization hardware utilized for the in situ demonstration was, again, an iPad3.

The real-world trials of the AR system took place in two locations in the Plymouth area. The first was Marrowbone Slip, described earlier (the present-day context shown in Figure 36.54). For this demonstration, a revised model of the ship had to be used, whereby all significant deck fittings—guns

FIGURE 36.54 Marrowbone Slip and the China House in 2012 with the AR HMS *Amethyst* visualized via the iPad3.

(large and small), life rafts, depth charge launchers, and so on—were removed (as they would have been, prior to delivering the vessel for breaking up). The second location was on the Cornish side of the River Tamar, just south of Her Majesty's naval base at Devonport, where the *Amethyst* made her penultimate journey prior to being paid off and sold to the breakers. Again, the main model had to be modified, with the lower hull region (including propellers and rudder) being removed to create a simple waterline effect (Figure 36.55).

Shortly after the Amethyst AR exercise in Plymouth was completed, a unique opportunity arose to present the findings of the early research and field trials—an invitation to a reunion event for the

FIGURE 36.55 AR HMS *Amethyst* visualized on the River Tamar.

surviving members of the *Amethyst* (including those from the 1949 Yangtze confrontation), together with three generations of family members. One could not have wished for a more critical audience. However, their acceptance and understanding of the significance of the technologies used, plus their recognition of the potential for future educational development, was both surprising and inspiring, especially given the fact that one often hears stories of *technophobia* on the part of the older generation or their unwillingness to accept recreations of past events, be they presented as films, video, or any other form of media. Even the youngest members of the audience, the grandchildren, and great grandchildren of the *Amethyst*'s survivors, expressed a desire to help, by, for example, capturing their ageing relatives' stories on digital video.

36.4.5 Virtual Scylla

Although the Virtual Scylla project set out to demonstrate how a fundamental understanding of how artificial life concepts could be used to drive the development of educational VR simulations depicting the evolution of British coastal marine flora and fauna communities, its impact in demonstrating the *art of the possible* in VH pursuits has been significant. The project focused on marine ecosystems on and around Europe's first artificial reef, the ex-Royal Navy Frigate HMS *Scylla* scuttled in 2004. In very broad terms, artificial life is the scientific study of the behavior of biological organisms and systems in order to simulate how they interact with, and exploit, their natural environments in order to survive, reproduce, colonize, and evolve (or *emerge*).

Artificial reefs, constructed using scuttled vessels, aircraft, surplus military equipment, or modular subsea *building blocks*, are becoming increasingly popular as a means of creating, restoring, or regenerating marine ecosystems, particularly in areas where natural reefs are absent or have been destroyed through pollution, erosion, or catastrophic environmental events such as tsunamis. The artificial reef *movement* is particularly active in the United States (home of the Reef Environmental Education Foundation [REEF]), from locations off the coast of California, South Carolina, Florida, to the Gulf of Mexico, where the toppling of redundant oil rigs to provide marine *safe havens* forms part of the US Minerals Management Service's *rigs to reefs* initiative. New Zealand and Canada are also active, with high-profile sinkings of numerous vessels, including the HMNZS *Wellington* (another ex-Royal Navy Leander Frigate, HMS *Bacchante*), by the organization Canadian Artificial Reef Consulting, featured in the National Geographic 2006 video *The Ship Sinkers*.

On March 27, 2004, Europe's first artificial reef was *launched* by the NMA, as a result of the scuttling, in Whitsand Bay (off the southeast Cornish coast), of the ex-Royal Navy Batch 3 Leander class frigate, HMS *Scylla* (Leece, 2006). HMS *Scylla* (Figure 36.56) was the last frigate to be built at Devonport Dockyard in 1968. During her service in the Royal Navy between 1970 and 2003, she saw action in the Icelandic Cod Wars of the 1970s, engaging in a tit-for-tat ramming session with the Icelandic gunboat Aegir. She missed active duty during the Falklands crisis in 1982, due to an extensive modernization program, as a result of which she was equipped with Exocet and Seawolf missile systems. Much later, in 1991, *Scylla* took part in Operation Desert Storm in the Middle East.

After decommissioning in 1993, *Scylla* was moored in Fareham Creek near Portsmouth, from where she would normally have been removed in due course for dismantling at a commercial scrap yard or sold on to another one of the world's navies. However, some 7 years later, and supported by the South West Regional Development Agency, she was purchased by the NMA for £UK200,000 and towed back to her *birthplace* in Devonport Dockyard for prescuttling stripping, cleaning, and hull modifications to support safe penetration by divers.

Today resting on the sea floor at a keel depth of 24–26 m (depending on tidal conditions, Figure 36.57) and close to another famous West Country wreck—the torpedoed US Liberty vessel James Egan Lane—the *Scylla* provides an excellent opportunity for conducting regular validation and verification studies throughout the course of the Virtual Scylla project. Indeed, at the time of writing, the primary author of this chapter has undertaken four expeditions to the wreck, accompanied by marine and technical specialists from the NMA. The expeditions have yielded a

FIGURE 36.56 HMS *Scylla*, in her operational heyday. (Image courtesy of the NMA.)

considerable amount of information about the declining condition of the vessel and colonization cycles of various forms of marine life, and the use of the Aquarium's VideoRay ROV has been invaluable in this respect.

The development team then set out to obtain as much data as possible about the *Scylla*, up to the point when she was scuttled in 2004, in order to construct as accurate a 3D (3ds Max) model as possible. In the event, and with the exception of images and schematics forthcoming from the NMA, plus access to a scale model built using plans held by Her Majesty's Naval Base at Devonport, very little information was available. Consequently, the early 3D model of the vessel and her new underwater environment had to be constructed from scratch, although (as mentioned earlier) the first VideoRay ROV expedition to the wreck, accompanied by divers from the MBA, helped significantly.

Once the 3ds Max model of the *Scylla* was completed, it was imported into the CryEngine test environment to be scaled and textured. While the *Scylla* model itself looked reasonably convincing, the complete virtual environment was still devoid of an underwater *ambience*. At that point, it was discovered that the CryEngine's underwater fogging capabilities were less accurate than its above-water counter. Therefore, to create an acceptable underwater environment, all virtual water elements were removed from the demonstration and replaced with above-water fogging throughout. In addition, particle effects (simple, semitransparent, sprites) were used to add an illusion of turbidity, as is evident in the real Scylla Reef's setting in Whitsand Bay. Finally, to complete the environment, a basic 3D model of an ROV was added. However, this provided one of the biggest challenges at the time, as the original CryEngine did not support any form of flying vehicle (which, of course, the ROV had now become, given the earlier removal of the limited underwater environment).

FIGURE 36.57 Images from the stern of the Scylla Reef, from a VideoRay ROV survey.

It further transpired that, to achieve realistic ROV motion effects, any software modifications would have to be made deep within the engine's source code. As a result, the first CryEngine-based ROV was endowed with a very limited control system, with the vehicle's forward thrust set to be permanently on.

This problem was overcome when Crytek made their CryEngine 2 product available, supporting not only flying vehicles but gamepad controllers as well (thereby providing multiaxis control of such vehicles—albeit not to the level of realism as was required to simulate the dynamics of a real ROV). Unfortunately, the fidelity of the underwater scenes made possible with CryEngine 2 was also unrealistic as, by trying to achieve a dramatic effect for first-person gamers, the result delivered a constantly modulating, refracted and distorted view of the world (Figure 36.58).

To overcome these limitations (and, in doing so, avoid having to invest considerable sums of money to secure full access to Crytek's software technologies), the decision was taken to investigate the capabilities of a little-known (at the time) real-time 3D environment development tool, Act-3D's Quest3D. The Quest3D tool has many advantages over using commercial games engines, not least the fact that the product supports royalty-free licensing for developed applications. Real-time virtual environments developed using Quest3D can be distributed in the form of stand-alone executables, via a setup program that installs the application or via the web. Versions of the product also provide support for a variety of interactive devices, from joysticks to full CAVE implementations. From the perspective of the Virtual Scylla project, the product provided support for the creation of an accurate and realistic control system for the

FIGURE 36.58   CryEngine rendering of the Virtual Scylla with controllable ROV and static diver.

virtual ROV, implemented using an Xbox gamepad controller. In addition, it was possible to create more realistic lighting and underwater distortion effects.

The final Virtual Scylla environment was constructed in the following stages. Firstly, the vessel and seabed models were rendered using a high level of exponentially increasing fog density. Secondly, a particle effect was linked to the camera position (i.e., the end user's viewpoint) such that particles were emitted at relative velocities to the camera's movement only (unlike the case for both CryEngine products, where particle effects were generated to encompass the entire virtual scene). Finally, a Gaussian blur filter was added to approximate focusing imperfections with the ROV's camera. Another effect supported by Quest3D was remote image refraction caused by the Perspex dome surrounding the ROV camera. The effect was simulated by rendering the first three stages to a texture and applying it to a virtual 3D dome within the environment. This approximated a common effect seen with underwater photography, when there is a tendency for cameras to adjust brightness levels rapidly to compensate for current lighting conditions (high dynamic range processing). Implementing the effect within Quest3D endowed the Virtual Scylla scenario with an almost photorealistic appearance (Figure 36.59) especially given the real-world lighting and turbidity conditions often witnessed on and around the reef itself.

36.4.6   Virtual Plymouth Sound

Less than 3 miles east of the location of the *Scylla* Reef lies the famous Plymouth Sound. From here, Sir Francis Drake set sail in 1588 to defeat the incoming Spanish Armada, allegedly only once he had completed (lost!) his game of bowls on Plymouth Hoe; from here, the *Mayflower* departed in 1652 carrying the Pilgrim Fathers to the New World; and here, today, ships and submarines of the Royal Navy mingle with commercial passenger ferries and pleasure boats of all descriptions. A number of projects undertaken for the British Ministry of Defence and by university students over the past few years are now coming together to fulfill an ambitious scheme to visualize historical events in and around the Sound and to represent them using VR and AR technologies. An above-water 3D model of the coastline, between Whitsand and Wembury Bays, has been developed using DTM data, enhanced with a variety of natural and man-made features (a process similar to that adopted for Virtual Wembury and Virtual Burrator, described earlier). This has been further

FIGURE 36.59 Quest3D rendering of the Virtual Scylla, showing the aft helicopter hangar (with damaged roof—top) and the foredeck area (showing side railings, bridge, and remains of the Exocet launcher rails—bottom).

enhanced with a virtual subsea representation of Plymouth Sound using bathymetric and navigational chart data originally provided by the United Kingdom's Hydrographic Office to develop a multiwindow situational awareness display for mine countermeasures activities (Figure 36.60; Stone, 2012). Using these virtual datasets, it is now possible to represent significant historical events around this part of the British coastline, such as the torpedoing of the US Liberty ship, the SS *James Egan Layne*, by a German U-Boat in 1945 (Figure 36.61, left), and the launch of the United Kingdom's first experimental underwater habitat, the GLAUCUS in 1965 (Figure 36.61, right—a *work in progress* at the time of writing). The ultimate aim is to be able to view and interact these and many other examples as stand-alone VR scenarios or as information vignettes presented as part of a mobile AR application to inform coastal walkers.

36.4.7 Maria

One specific example of Plymouth Sound virtual maritime heritage is based on a unique historical event that took place some 230 years before the deliberate scuttling of the ex-HMS *Scylla*. In 1774, a 31 ft, 50-ton converted wooden sloop called *Maria*, purchased for £UK340, entered the history books as being host to the first of some 65,000 recorded submariner fatalities. The *Maria* was equipped with a wooden-beamed chamber measuring some 12 ft by 9 ft by 8 ft, containing around 75 *hogsheads* of air (24.5 m^3, providing 25 tons of buoyancy and allegedly sufficient for 24 h life support). A topside entry hatch allowed the vessel's only occupant, a shipwright by the name of John

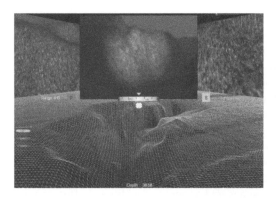
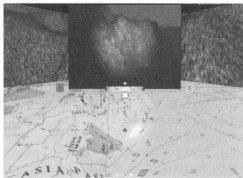

FIGURE 36.60 A concept VR multiwindow situational awareness display for mine countermeasures activities based on Plymouth Sound bathymetric (left) and integrated navigational chart data (right).

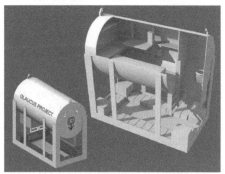

FIGURE 36.61 A VR recreation of the final moments of the US Liberty ship, the SS *James Egan Layne*, (left) and an early 3D model of the UK experimental subsea habitat, the GLAUCUS (right).

FIGURE 36.62 A VR reconstruction of the converted eighteenth-century sloop *Maria*, with cutaway details (left) and chamber interior features (right).

Day, access to the vessel. A system of counterbalance and chains enabled him to close the hatch and render the chamber watertight. Working with specialists from ProMare, a public charity established in 2001 to promote marine research and exploration, an accurate 3D model of the *Maria* has been produced, with the future aim of using VR technologies to reproduce the vessel's final dive from both external and internal perspectives (Figure 36.62).

When designing the *submersible*, Day had also designed the chamber to allow him to communicate with the surface by means of three small releasable markers (white for *I am well*, red for *I am indifferent*, and black for *I am very ill*). The *Maria*'s buoyancy was counteracted by filling the interior with (initially) 10 tons of limestone (although another 20 tons of rocks from local quarries had to be added through hastily-removed deck panels on the day of the dive) and flooding the fore and aft spaces via sluice pipes, opened prior to Day's boarding the vessel. To allow the *Maria* to surface after the experiment, netted groups of 21-ton rocks were to be released by undoing (and quickly plugging!) bolts within the chamber. The *Maria* was towed from Plymouth's Sutton Harbour to Firestone Bay on June 20, 1774, whereupon Day boarded the vessel carrying a hammock, a watch, a candle, a bottle of water, and some biscuits (Figure 36.62, right). With more ballast added, the *Maria submerged* stern first and her disappearance was followed 15 min later by bubbles on the surface, indicative of the escape of a considerable amount of air from an underwater source. After 3 days of effort by 200 dockyard workers, the salvage operation was abandoned. Further extensive salvage attempts were made—one might even say rescue attempts—courtesy of a well-published MD by the name of Falck. Believing Day to be in a state of suspended animation due to the low temperatures of the depths of Firestone Bay, Falck sponsored a month of additional effort to raise the *Maria*. Despite locating the remains of the wreck, his mission came to nothing. As well as the VR reconstruction of the *Maria*'s one and only *dive*, the scenario will be used—together with that of the GLAUCUS mentioned earlier—to deliver educational vignettes for schools and museums relating to underwater physics and human physiology.

36.5　CONCLUDING REMARKS

The applications described earlier only scratch the surface of previous and ongoing VR projects and the topics covered are in no way representative of the potential breadth of the VR applications domain. However, in the view of the author, it is fair to say that, despite many, many years of expectation, there is still no single *killer application* for VR. Of those applications reported in scientific papers, books, trade magazines, and, worst of all, in newsletters from online technology watch sites (many of which seem to contain examples of applications that were *news* many years ago but are being presented as present-day breakthroughs), very, very few indeed have actually made an impact on the real world in terms of industrial or commercial adoption or can be used as a citation supporting the real added value VR can deliver.

There is still a great deal to do if the VR *movement* is ever going to lose its perceived technology-push bias and become a digital pursuit that delivers true impact and competitive advantage for real businesses and real people in the real world undertaking real tasks. Unfortunately, there is no one formula that defines what makes a successful application of VR. However, in the experience of the author, there are a number of key *rules* that will undoubtedly help, especially in the case of the developer community. In brief, these are the following:

1. Every VR project, like interactive digital media in general, *must* be underpinned by a sound and explicit human-centered design process (e.g., Stone 2008a, 2012). This process must be iterative throughout the VR design, development, and evaluation cycle and all lessons learned should be captured an archived for future reuse. Human factors has an indispensable role to play in the delivery of effective and usable interactive media, from the definition of task, context, and interaction fidelities applied to both hardware and software components, through the specification of the *live-synthetic balance* (ensuring that VR is only delivered for those elements of the task where positive skills and knowledge transfer to the real world can be guaranteed), to the evaluation of interim and end results in real-world contexts with real end users and stakeholders.

2. Stakeholder involvement from the outset is essential. Most human-centered design standards emphasize the criticality of early and iterative stakeholder participation, but, with VR, this is all too often ignored in favor of *shoe-horning* end users into more glamorous, high-tech, and PR-worthy (but often unusable) items of technology. It should always be remembered that by the term *stakeholder*, we mean more than just the traditional end user—in the case of VR applications in training and education, the end user also includes

the instructor or educator. Again, all too often, these stakeholders and their requirements are left until the very end of the design and development stages of the project or, worse still, ignored altogether.

3. Unless the project calls for the development of a product, then the use of the word *product* should be avoided altogether and its use discouraged in project review meetings and documentation. The applications described in this chapter have all been *concept capability demonstrator* in nature—not products. If it looks likely that the application demonstrator is destined for use in a real-world training context, then take all steps possible to ensure that the end users are under no illusion that the solution is a prototype and that, if used in such a context (especially in the medical and defense sectors), then the developers cannot accept any responsibility for errors or consequential damage incurred during the use of the hardware and software.

4. A nonprescriptive attitude toward hardware and software solutions is essential, if the VR application is to benefit from all that today's technology has to offer. It *must* be accepted that—despite the marketing hype and rhetoric from some of the world's VR and simulation system vendors—there is not one single product on today's market that will solve 100% of the content design, fidelity, and interactive requirements of at least 80% or so of the applications that present as contractual opportunities. This issue has been well demonstrated in the UK defense sector over the past 3–5 years, where the premature and ill-informed mandating of one particular software product by the UK government has, over time, only served to show that product's incapability of supporting the very broad games-based simulation needs of all three of the main armed forces. Instead, projects have either been canned, before or shortly after entering service, or have required the often costly intervention of third parties—small companies formed around the development of (often unnecessary) *intermediary* software packages made available with the aim of making the government-mandated solution (sometimes, but not always) more usable. The main lesson learned from this episode in the United Kingdom's involvement in VR and games-based simulation is that mandating a single solution for VR or games-based training and visualization projects does not foster meaningful standardization between and within suppliers, does not promote the sharing of common assets, such as 3D models and specialized code, does not guarantee significant cost savings, and does not support the de-risking of military projects.

5. Similar comments apply to the selection, modification, or development of specialized interactive hardware, such as data input and data display devices. It is essential to adopt a *neutral* standpoint and to avoid the premature specification of devices—especially if a client or end user has expressed a desire to include a specific item of interactive hardware in the final solution (often on the basis of a newsletter item they have read or an online video they may have seen). In many respects, today's *kickstarter* era is not that dissimilar from the *garage VR* era of the 1990s. The only major difference between the two is cost, with the items made available to today's prebidders or donators being advertised at significantly cheaper prices than the head-mounted displays and gloves of *yesteryear*. However, just because these products become the focus of intense marketing campaigns and overhyped technology newsletters, with endorsements coming from celebrities across the gaming world (as has been the case for one head-mounted display system and many other interactive gaming products), this doesn't mean that they are necessarily superior to those items of VR equipment that have gone before. Indeed, early experience shows that many of the products still suffer a distinct lack of human factors understanding, and it still remains to be seen how (if at all) they become mainstream interactive technology solutions for real-world applications.

6. Where at all possible, demonstrate an early reuse capability of all model, texture, physics, animation, and other software assets, so that the customer and stakeholders can gain some degree of confidence that their investment will not be wasted, should the initial systems

specified cease to exist or the source company cease trading (as was demonstrated quite clearly in the early days of the VR era, described at the outset of this chapter).

7. Factor in a plan for a change in key personnel or stakeholders before the application development project has finished, especially if the project has military backing. Stakeholder changes are one of the major factors in project cancellation or failure if the work conducted cannot be *sold* appropriately to the incoming personnel, with sound answers to the key questions of cost, value, ongoing support, and end-user engagement. Even if a project does not fail, new stakeholders can contribute significantly to *requirements creep*, by asking for changes or enhancements that enable them to stamp their ownership on what has gone before.

8. Finally, and from a developmental team perspective, remember the 80:20 rule, especially when it comes to delivering the final application demonstrator. What may appear inferior quality to developers (who always take an infinitesimally detailed pride in their graphics, artwork, programming, etc.) will appear as incredible to the customer and end users, many of whom—even today—will not have been exposed to the extraordinary graphics and dynamic qualities of mainstream gaming products. While attention to detail and enthusiasm are not qualities to be suppressed, the effort required to attain that final 20% in quality will almost always eat into a development company's profit line.

If one actually believed the claims of the *purists* throughout the 1990s, VR was, by the end of the twentieth century, destined to have helped computer users abandon the keyboard, mouse, joystick, and computer display in favor of interfaces exploiting a wide range of natural human skills and sensory characteristics. They would be able to interact intuitively with virtual objects, virtual worlds, and virtual actors while *immersed* within a multisensory, 3D computer-generated world. As is evident today, this brave new world simply did not happen. Despite sizeable early investments, national initiatives, expensive (and unexploited) international collaborative projects, and the proliferation of hardware-heavy, so-called centers of *academic excellence*, VR delivered very little of use to the global IT community. A handful of organizations actually adopted VR, but most were deterred from doing so by its complexity and cost. Today's VR supply companies have either passed away or are hanging on by a commercial thread. The academic centers have closed or have been rebranded to fall in line with current research funding initiatives or have simply become expensive technological museums. And the biggest mistake made by the VR community was that it ignored the human factor.

Over a decade on and there is little doubt that games-based learning technologies have the potential to deliver much more than the promises and hype of their VR predecessors—affordability and accessibility in particular. However, to do this, human-centered lessons must be learned. Interactive 3D media has to be designed in conjunction with its end users, identifying the skills that need to be trained or the knowledge that has to be imparted and then delivering a solution based on appropriate content, fidelity, and interactive technologies. Furthermore, the training solutions must be packaged in a form that can be delivered to the end users in the their own working environments, as opposed to expecting them to exploit the technology in isolated and restricted laboratory environments. The solutions must be developed so that the end users can understand and benefit from their contents immediately, supporting their own modifications through simple-to-use shape, texture, and behavioral editors. This is where yesterday's VR failed. This is where today's games-based interactive technologies have the potential to give the interactive 3D community a second chance.

REFERENCES

Abkar, M., Kamal, M. M., Maulan, S., & Mariapan, M. (2009). Influences of viewing nature through windows. *Australian Journal of Basic and Applied Sciences*, *4*(10), 5346–5351.

Adam, J. A. (1994). Medical electronics. *IEEE Spectrum*, *31*(1), 70–73.

Anonymous. (1908, November). Wembury (Plymouth) commercial dock and railway. 8901. *The London Gazette*.

Anonymous. (1909, January). A new English port. *Evening Post*.

Anonymous. (1957, May). The end of a proud record. *The Sphere*, 185–186.

Berman, M. G., Jonides, J., & Kaplan, S. (2008). The cognitive benefits of interacting with nature. *Psychological Science, 19*, 1207–1212.

Berto, R. (2005). Exposure to restorative environments helps restore attentional capacity. *Journal of Environmental Psychology, 25*, 249–259.

Boot, W. R., Kramer, A. F., Simons, D. J., Fabiani, M., & Gratton, G. (2008). The effects of video game playing on attention, memory, and executive control. *Acta Psychologica (Amst), 129*(3), 387–398.

Botella, C., Banos, R. M., Villa, H., & Perpina, C. (2000). Virtual reality in the treatment of claustrophobic fear: A controlled, multiple-baseline design. *Behavior Therapy, 43*(4), 583–595.

Bredemeier, M. E., & Greenblat, C. S. (1981). The educational effectiveness of simulation games: A synthesis of findings. In C. S Greenblat & R. Duke (Eds.), *Principles and practices of gaming-simulation.* Beverly Hills, CA: Sage Publications.

Bringslimark, T., Hartig, T., & Patil, G. G. (2011). Adaptation to windowlessness: Do office workers compensate for a lack of visual access to the outdoors? *Environmental and Behaviour, 43*(4), 469–487.

Broughton, P. W. (1995). *The Wembury docks and railway proposal of 1909. iii.* Wembury, Devon, U.K. Wembury Local History Society Publication.

Brown, B. (2010). *A training transfer study of simulation games* (Unpublished Master's thesis). Naval Postgraduate School, Monterey, CA.

Carlin, A. S., Hoffman, H. G., & Weghorst, S. (1997). Virtual reality and tactile augmentation in the treatment of spider phobia: A case report. *Behavior Research and Therapy, 35*(2), 153–158.

Catford, N. (2010). *Cold War bunkers.* Nottingham, U.K.: Folly Books Ltd.

Collimore, K. C., McCabe, R. E., Carleton, R. N., & Asmundsen, G. J. (2008). Media exposure and dimensions of anxiety sensitivity: Differential associations with PTSD symptom clusters. *Journal of Anxiety Disorders, 22*(6), 1021–1028.

de Kort, Y. A., & Ijsselsteijn, W. A. (2006). Reality check: The role of realism in stress reduction using media technology. *CyberPsychology Behavior, 9*(2), 230–233.

Depledge, M. H., Stone, R. J., & Bird, W. J. (2011). Can natural and virtual environments be used to promote improved human health and well-being? *Environmental Science and Technology, 45*(11), 4659–5064.

Diette, G. B., Lechtzin, N., Haponik, E., Devrotes, A., & Rubin, H. R. (2003). Distraction therapy with nature sights and sounds reduces pain during flexible bronchoscopy: A complementary approach to routine analgesia. *Chest, 123*(3), 941–948.

Difede, J., Cukor, J., Jayasinghe, N., Patt, I., Jedel, S., Spielman, L., … Hoffman, H. G. (2007). Virtual reality exposure therapy for the treatment of posttraumatic stress disorder following September 11, 2001. *Journal of Clinical Psychiatry, 68*, 1639–1647.

Difede, J., & Hoffman, H. G. (2002). Virtual reality exposure therapy for World Trade Center post-traumatic stress disorder: A case report. *CyberPsychology, 5*(6), 529–535.

Dye, M. W., Green, C. S., & Bavelier, D. (2009). The development of attention skills in action video game players. *Neuropsychologia, 47*(8–9), 1780–1789.

Fenn, J., & Linden, A. (2005). *Gartner's hype cycle special report for 2005.* Stamford, CT: Gartner Inc.

Friedman, B., Freier, N. G., & Kahn, P. H. (Jr.) (2004). Office windows of the future? Two case studies of an augmented window. *Poster Presentation—Conference on Human Factors in Computing Systems, CHI 2004.* Vienna, Austria. 1559.

Friedman, B., Freier, N. G., & Kahn, P. H. (2008). Office window of the future? Field-based analyses of a new use of a large display. *International Journal of Human-Computer Studies, 66*, 452–465.

Galea, S., Nandi, A., & Vla, D. (2005). The epidemiology of post-traumatic stress disorder after disasters. *Epidemiologic Reviews, 27*(1), 78–91.

Gallagher, A. G., McClure, N., McGuigan, J., Crothers, I., & Browning, J. (1998). Virtual reality training in laparoscopic surgery: A preliminary assessment of Minimally Invasive Surgical Trainer, Virtual Reality (MISTvr). In J. D. Westwood, H. M. Hoffman, D. Stredney, & S. J. Weghorst (Eds.), *Medicine meets virtual reality.* Amsterdam, the Netherlands: IOS Press.

Gallagher, A. G., Ritter, E. M., Champion, H., Higgins, G., Fried, M. P., Moses, G., … Satava, R. M. (2005). Virtual reality simulation for the operating room: Proficiency-based training as a paradigm shift in surgical skills training. *Annals of Surgery, 241*(2), 364–372.

Garcia-Palacios, A., Hoffman, H., Carlin, A., Furness, T. A., & Botella, C. (2002). Virtual reality in the treatment of spider phobia: A controlled study. *Behaviour Research and Therapy, 40*, 983–993.

Garcia-Palacios, A., Hoffman, H. G., Kwong See, S. K., Tsai, A., & Botella, C. (2001). Redefining therapeutic success with VR exposure therapy. *CyberPsychology Behavior, 4*, 341–348.

Gilbert, G., Turner, T., & Machessault, R. (2005). Army medical robotics research. *International Review of the Armed Forces Medical Services, 78*(2), 105–112.

Gill, C. (1997). *Sutton harbour.* England, U.K.: Devon Books.

Green, C. S., & Bavelier, D. (2007). Action video game experience alters the spatial resolution of vision. *Psychological Science, 18*(1), 88–94.

Hartig, T., Catalano, R., & Ong, M. (2007). Cold summer weather, constrained restoration and the use of anti-depressants in Sweden. *Journal of Environmental Psychology, 27*(2), 107–116.

Hartig, T., Evans, G. W., Jammer, L. D., Davis, D., & Garling, T. (2003). Tracking restoration in natural and urban field settings. *Journal of Experimental Psychology, 23,* 109–123.

Hays, R. T. (2005). *The effectiveness of instructional games: A literature review and discussion* (Naval Air Warfare Center Training Systems Division, Orlando, FL. Technical Report No. 2005-004).

HL Deb. (1909, May 3). Wembury (Plymouth) commercial dock and railway bill [H.L.] c667. 1. London, UK: House of Lords.

Hoffman, H. G., Doctor, J. N., Patterson, D. R., Carrougher, G. J., & Furness, T. A. (2000). Use of virtual reality for adjunctive treatment of adolescent burn pain during wound care: A case report. *Pain, 85*(1–2), 305–309.

Hoffman, H. G., Garcia-Palacios, A., Carlin, C., Furness, T. A., & Botella-Arbona, C. (2003). Interfaces that heal: Coupling real and virtual objects to cure spider phobia. *International Journal of Human-Computer Interaction, 16*(2), 283–300.

Hoffman, H. G., Hollander, A., Schroder, K., Rousseau, S., & Furness, T. A. (1998). Physically touching and tasting virtual objects enhances the realism of virtual experiences. *Virtual Reality: Research, Development and Application, 3,* 226–234.

Hoffman, H. G., Richards, T. L., Coda, B., Bills, A. R., Blough, D., Richards, A. L., & Sharar, S. R. (2004). Modulation of thermal pain-related brain activity with virtual reality: Evidence from fMRI. *Neuroreport, 15*(8), 1245–1248.

ISO. (2008). *ISO 9241-210—Ergonomics of human-system interaction,* Geneva, Switzerland. International Standards Organisation.

Jaycox, L. H., Foa, E. B., & Morral, A. R. (1998). Influence of emotional engagement and habituation on exposure therapy for PTSD. *Journal of Consulting and Clinical Psychology, 66,* 186–192.

John, N. W., Thacker, N. A., Pokric, M., Agus, M., Giachetti, A., Gobbetti, E., … Rubio, F. (2000). *Unpublished IERAPSI Report: Surgical Procedures and Implementation Specification (Deliverable D2)* (EU Project IERAPSI (IST-1999-12175)). University of Manchester, UK.

Johnston, C. L., & Whatley, D. (2005). Pulse!!—A virtual learning space project. In J. D. Westwood, R. S. Haluck, H. M. Hoffman, G. T. Mogel, R. Phillips, R. A. Robb, & K. C. Vosburgh (Eds.), *Medicine meets virtual reality 14: Accelerating change in healthcare: Next medical toolkits* (pp. 240–242). Amsterdam, the Netherlands: IOS Press.

Kahn, P. H., Severson, R. L., & Ruckert, J. H. (2009). Technological nature and the problem when good enough becomes good. In M. Drenthen, J. Keulartz, & J. Proctor (Eds.), *New visions of nature: Complexity and authenticity* (pp. 21–40). Dordrecht, the Netherlands: Springer.

Kaplan, R. (2001). The nature of the view from home: Psychological benefits. *Environment and Behavior, 33*(4), 507–542.

Kaplan, R., & Kaplan, S. (1989). *The experience of nature: A psychological perspective.* New York, NY: Cambridge University Press.

Kaplan, S. (1992). The restorative environment: Nature and human experiences. In D. Relf (Ed.), *The role of horticulture in human well-being and social development* (pp. 134–142). Portland, OR: Timber Press.

Kaplan, S. (1995). The restorative benefits of nature: Toward an integrative framework. *Journal of Environmental Psychology, 15,* 169–182.

Kingdom, A. R. (1982). *The turnchapel branch.* Dorset, U.K.: Oxford Publishing Company.

Kingdom, A. R. (1991). *The Yelverton to Princetown railway.* Devon, U.K.: ARK Publications (Railways).

Krijn, M., Emmelkamp, P., Olafsson, R. P., Bouwman, M., van Gerwen, L. J., Spinhoven, P., … van der Mast, C. (2007). Fear of flying treatment methods: Virtual reality exposure vs. cognitive behavioral therapy. *Aviation, Space, and Environmental Medicine, 78*(2), 121–128.

Leece, M. (2006). Sinking a frigate. *Ingenia, 29,* 27–32.

Lepper, M., & Malone, T. (1987). Intrinsic motivation and instructional effectiveness in computer-based education. In R. E. Snow & M. J. Farr (Eds.), *Aptitude, learning and instruction* (Vol. 3, pp. 255–286). Hillsdale, NJ: Erlbaum.

Levi, D., & Kocher, S. (2008). Virtual nature experiences as emotional benefits in green product consumption: The moderating role of environmental attitudes. *Environment and Behavior, 40,* 818–842.

Li, Q., Morimoto, K., Kobayashi, M., Inagaki, H., Katsumata, M., Hirata, Y., … Krensky, A. M. (2008). Visiting a forest, but not a city, increases human natural killer activity and expression of anti-cancer proteins. *International Journal of Immunopathology and Pharmacology, 21*(1), 117–127.

Mazer, S. E. (2010). *Music and nature at the bedside: Part II of a two-part series.* Retrieved July 2013, from Research Design Connections, Issue 1: http://www.researchdesignconnections.com/pub/2010-issue-1/music-and-nature-bedside-part-ii-two-part-series

Meglan, D. (1996). Making surgical simulation real. *Computer Graphics, 30,* 37–39.

Moltenbrey, K. (2000). Tornado watchers. *Computer Graphics World, 23*(9), 24–28.

Nanda, U., Gaydos, H. L., Hathon, K., & Watkins, N. (2010). Art and posttraumatic stress: A review of the empirical literature on the therapeutic implications of artwork for war veterans with posttraumatic stress disorder. *Environment and Behaviour, 42*(3), 376–390.

National Science Foundation. (2012). *State-of-the-art virtual reality system is key to medical discovery* (Press Release 12-228). Retrieved December 11, 2012. http://www.nsf.gov/news/news_summ.jsp?cntn_id=126209&WT.mc_id=USNSF_51&WT.mc_ev=click

North, M. M., North, S. M., & Coble, J. R. (1996a). Effectiveness of VRT for acrophobia. In M. M. North (Ed.), *Virtual reality therapy. An innovative paradigm* (pp. 68–70). Colorado Springs, CO: IPI Press.

North, M. M., North, S. M., & Coble, J. R. (1996b). Effectiveness of virtual environment desensitization in the treatment of agoraphobia. *Presence: Teleoperators and Virtual Environments, 46,* 346–352.

North, M. M., North, S. M., & Coble, J. R. (1996c). VRT in the treatment of agoraphobia. In M. M. North (Ed.), *Virtual reality therapy. An innovative paradigm* (Vol. 46). Colorado Springs, CO: IPI Press.

North, M. M., North, S. M., & Coble, J. R. (1997). Virtual reality therapy for fear of flying. *American Journal of Psychiatry, 154,* 130.

North, M. M., North, S. M., & Coble, J. R. (1998). Virtual reality therapy: An effective treatment for the fear of public speaking. *International Journal of Virtual Reality, 3*(3), 1–6.

Park, B.-J., Tsunetsugu, Y., Kasetani, T., Hirano, H., Kagawa, T., Sato, M., & Miyazaki, Y. (2007). Physiological effects of Shinrin-Yoku (taking in the atmosphere of the forest) using salivary cortisol and cerebral activity as indicators. *Physiological Anthropology, 26*(2), 123–128.

Pertaub, D. P., Slater, M., & Barker, C. (2002). An experiment on public speaking anxiety in response to three different types of virtual audience. *Presence: Teleoperators and Virtual Environments, 11,* 68–72.

Peters, M. L., Godaert, G. L., Ballieux, R. E., van Vliet, M., Willemsen, J. J., Sweep, F. C., & Heijnen, C. J. (1998). Cardiovascular and endocrine responses to experimental stress: Effects of mental effort and controllability. *Psychoneuroendocrinology, 23*(1), 1–17.

Riva, G., Waterworth, J. A., & Waterworth, E. L. (2004). The layers of presence: A bio-cultural approach to understanding presence in natural and mediated environments. *Cyberpsychology Behavior, 7*(4), 402–416.

Rizzo, A. R., Difede, J., Rothbaum, O., Johnston, S., McLay, R. N., Reger, G., … Pair, J. (2009). Development and clinical results from the Virtual Iraq Exposure Therapy application for PTSD. *Proceedings of the Virtual Rehabilitation International Conference,* June 29–July 2, 2009 (pp. 8–15). Haifa, Israel.

Rojas-Sola, J., & Castro-Garcia, M. (2011). Overview of the treatment of historical industrial heritage in engineering graphics. *Scientific Research and Essays, 6*(33), 6717–6729.

Roman, P. A., & Brown, D. G. (2008). Games: Just how serious are they? *Proceedings of the Interservice/Industry Training, Simulation and Education Conference (I/ITSEC),* December 2008. Orlando, FL.

Rothbaum, B. O., Hodges, L. F., Kooper, R., Opdyke, D., Williford, J. S., & North, M. (1995). Virtual reality graded exposure in the treatment of acrophobia: A case report. *Behavioral Therapy, 26*(3), 547–554.

Rothbaum, B. O., Meadows, E. A., Resick, P., & Foy, D. W. (2000). Cognitive-behavioral therapy. In E. B. Foa, T. M. Keane, & M. J. Friedman (Eds.), *Effective treatments for PTSD: Practice guidelines from the international society for traumatic stress studies* (pp. 320–325). New York, NY: Guilford Press.

Satava, R. M. (1995a). Medicine 2001: The king is dead. In R. M. Satava, K. Morgan, H. B. Sieburg, R. Mattheus, & H. I. Christiansen (Eds.), *Interactive technology and the new paradigm for healthcare* (pp. 334–339). Washington, DC: IOS Press.

Satava, R. M. (1995b). Virtual reality, telesurgery, and the new world order of medicine. *Journal of Image Guided Surgery, 1*(1), 12–16.

Sharar, S. R., Carrougher, G. J., Nakamura, D., Hoffman, H. G., Blough, D. K., & Patterson, D. R. (2007). Factors influencing the efficacy of virtual reality distraction analgesia during postburn physical therapy: Preliminary results from 3 ongoing studies. *Archives of Physical Medicine and Rehabilitation, 88*(12), S43–S49.

Simonic, T. (2006). Urban landscape as a restorative environment: Preferences and design considerations. *Acta Agricultrae Slovenica, 87*(2), 325–332.

Snell, T. (2011). *MCM diver training in virtual environments with AAR system and analysis* (Unpublished final year (MEng) project report). University of Birmingham, Birmingham, U.K.

Soohoo, C. (2011). Welcome to the virtual world: Virtual reality trainers poised to transform graduate medical education. *WardRounds, 28*(3). Northwestern University Feinberg School of Medicine. http://www.wardrounds.northwestern.edu/fall-2011/features/welcome-to-the-virtual-world/

Stone, R. J. (1992). Virtual reality and telepresence. *Robotica, 10*, 461–467.

Stone, R. J. (1996). Virtual Lowry: A world within a world. *Proceedings of the Electronic Imaging and Visual Arts Conference.* London, U.K.

Stone, R. J. (1998). Virtual stonehenge: Sunrise on the new millennium. *Presence: Teleoperators and Virtual Environments, 7*(3), 317–319.

Stone, R. J. (1999a). The opportunities for virtual reality and simulation in the training and assessment of technical surgical skills. *Proceedings of Surgical Competence: Challenges of Assessment in Training and Practice*, November 1999. (pp. 109–125). Royal College of Surgeons and Smith & Nephew Foundation.

Stone, R. J. (1999b, October). Virtual heritage: The willing suspension of disbelief for the moment. *UNESCO World Heritage Review* (pp. 18–27). Paris, France: UNESCO.

Stone, R. J. (2001, April). The importance of a structured human factors approach to the design of avionics maintenance and submarine qualification virtual environment trainers. Proceedings of *ITEC 2001*. Lille, France.

Stone, R. J. (2003, December 1–4). The RAF helicopter voice marshalling simulator: Early experiences and recent enhancements. *Proceedings of I/ITSEC 2003*. Orlando, FL.

Stone, R. J. (2004). Whatever happened to virtual reality? *Information Professional*. Institution of Engineering and Technology, *1*(4), 12–15.

Stone, R. J. (2005). Serious games: Virtual reality's second coming? *Virtual Reality, 8*, 129–130.

Stone, R. J. (2008a). Simulation for defence applications: How serious are serious games? *Defence Management Journal*, 31, 142–144.

Stone, R. J. (2008b). Human factors guidelines for interactive 3D and games-based training systems design: Edition 1. *Human factors integration defence technology centre report*. Downloadable from: http://www.birmingham.ac.uk/stone

Stone, R. J. (2009, August). Exploiting gaming technologies for defence training. *Defence Global* (pp. 90–91). Barclay Media Ltd., Manchester, UK.

Stone, R. J. (2010, March). Serious games: The future of simulation for the Royal Navy? *Review of Naval Engineering, 3*(3), 37–45.

Stone, R. J. (2012). Human factors guidance for designers of interactive 3D and games-based training systems. *Human factors integration defence technology centre booklet*. Downloadable from: http://www.birmingham.ac.uk/stone

Stone, R. J., & Angus, J. (1995, January 26–27). Virtual reality as a design tool in rolls-royce: A business-driven assessment. In J. A. Powell (Ed.), Virtual reality and rapid prototyping for engineering, *Proceedings of an Information Technology Awareness in Engineering Conference (Sponsored by the Engineering & Physical Sciences Research Council).*

Stone, R. J., & Barker, P. (2006). Serious gaming: A new generation of virtual simulation technologies for defence medicine and surgery. *International Review of the Armed Forces Medical Services, 79*(2), 120–128.

Stone, R. J., Caird-Daley, A., & Bessell, K. (2009a). Subsafe: A games-based training system for submarine safety and spatial awareness. *Virtual Reality, 13*(1), 3–12.

Stone, R. J., Caird-Daley, A., & Bessell, K. (2009b). Human factors evaluation of a submarine spatial awareness training tool. *Proceedings of the Human Performance at Sea (HPAS) Conference* (pp. 231–241). Glasgow, U.K.

Stone, R. J., & Hendon, D. (2004, November 3–4). Maintenance training using virtual reality. *Proceedings of the RAeS Conference Simulation of Onboard Systems*. London, U.K..

Stone, R. J., & Knight, J. S. (2012). *Virtual environments for rehabilitation*. Alexandria, VA: Human Factors Integration Defence Technology Centre Publication.

Stone, R. J., & McCloy, R. F. (2004). Ergonomics in medicine and surgery. *British Medical Journal, 328*(7448), 1115–1118.

Stone, R. J., & Mcdonagh, S. (2002, December 3–5). Human-centred development and evaluation of a helicopter voice marshalling simulator. *Proceedings of I/ITSEC 2002*. Orlando, FL.

Stone, R. J., & Ojika, T. (2000). Virtual heritage: What next? *IEEE Multimedia, 7*(2), 73–74.

Taffinder, N., McManus, I., Jansen, J., Russell, R., & Darzi, A. (1998). An objective assessment of surgeons' psychomotor skills: Validation of the MISTvr laparoscopic simulator. *British Journal of Surgery, 85*(Suppl. 1), 75.

Tsunetsugu, Y., Park, B.-J., Ishii, H., Hirano, H., Kagawa, T., & Miyazaki, Y. (2007). Physiological effects of Shinrin-Yoku (taking in the atmosphere of the forest) in an old-growth broadleaf forest in Yamagata Prefecture, Japan. *Physiological Anthropology, 26*(2), 135–142.

Ulrich, R. S. (1981). Natural versus urban scenes: Some psychophysiological effects. *Environment and Behavior, 13*, 523–556.

Ulrich, R. S. (1984). View through a window may influence recovery from surgery. *Science, 224*(4647), 420–421.

Ulrich, R. S., & Gilpin, L. (2003). Healing arts: Nutrition for the soul. In S. B. Frampton, L. Gilpin, & P. Charmel (Eds.), *Putting patients first: Designing and practicing patient-centered care* (pp. 117–146). San Francisco, CA: John Wiley & Sons.

Valtchanov, D., Barton, K. R., & Ellard, C. (2010). Restorative effects of virtual nature settings. *Cyberpsychology, Behavior and Social Networking, 13*(5), 503–512.

van den Berg, A. E. (2008). Restorative effects of nature: Towards a neurobiological approach. In T. Lotus, M. Reitenbach, & J. Molenbroek (Eds.), *Human diversity, design for life* (pp. 132–138). *Proceedings of 9th Congress of Physiological Anthropology*.

Vincelli, F. (1999). From imagination to virtual reality: The future of clinical psychology. *CyberPyschology Behavior, 2*, 241–248.

Waterworth, J. A., & Waterworth, E. L. (2004). *Designing for Life*. Retrieved July 2013, from Relaxation Island: A virtual tropical paradise: http://www8.informatik.umu.se/~jwworth/RelaxIvol2.pdf

White, M., Smith, A., Humphryes, K., Pahl, S., Snelling, D., & Depledge, M. (2010). The importance of water in judgments of natural and built scenes. *Journal of Environmental Psychology, 30*(4), 482–493.

Wiederholder, B. K., & Wiederholder, M. D. (2006). *Physiological monitoring during simulation training and testing*. The Virtual Reality Medical Center. San Diego, CA: Virtual Reality Training Transfer: A DARWARS Study.

Williams, M. B., & Poijula, S. (2002). *The PTSD workbook: Simple, effective techniques for overcoming traumatic stress symptoms*. Oakland, CA: New Harbinger.

Wohlhill, J. F. (1983). The concept of nature: A psychologist's view. In I. Altman and J. F. Wohlhill (Eds.), *Behaviour and the natural environment. Human Behavior and Environment* (Vol. 6, pp. 5–38). Springer.

37 Use of Virtual Worlds in the Military Services as Part of a Blended Learning Strategy

Douglas Maxwell, Tami Griffith, and Neal M. Finkelstein

CONTENTS

37.1 INTRODUCTION

Throughout history, strategy games such as chess and various forms of simulations have aided in military planning and to study potential outcomes of conflict (Lewin, 2012). Early on, complex elements were distilled down to representative pawns. As time and technology continued to evolve, simulations and the worlds they depict have become much richer, giving leader's perspective they might have otherwise overlooked.

The French military and political leader Napoleon Bonaparte once said, "If I always appear prepared, it is because before entering on an undertaking, I have meditated for long and foreseen what may occur" (1831).

As the military services (Army, Navy, and Air Force) look to the future, they are shifting their training plans to a blended learning strategy (BLS) that aids in preparation for the uncertainty and complexity of the global security environment (Air Education and Training Command, 2008; Department of the Army, 2011) (Figures 37.1 and 37.2). Blended learning (BL) is an evolving term frequently defined as online or technology-delivered instruction that may be combined with some face-to-face instruction. It blends the efficiencies and effectiveness of self-paced, technology-delivered instruction with the expert guidance of a facilitator and can include the added social benefit of peer-to-peer interactions. A study by Fletcher and Chatham (2009) reports that it is possible to show a 30% decrease in the time it takes to learn with no decrease in effectiveness when educators develop technology-delivered instruction for appropriate learning content and design instruction according to established learning principles.

Several definitions of BLS are in the lexicon of the training and education community, but one source (Sharma, 2010) provides some relevant definitions of BL for the context of this chapter.

A combination of face-to-face and online teaching—This describes the integrated combination of traditional (classroom or face-to-face) learning with web-based online approaches, such as *Blackboard* or *Moodle* and synchronous (i.e., chat) and asynchronous (i.e., bulletin boards) electronic tools.

A combination of technologies—This could describe distance learning where there are no face-to-face lessons but rather a combination of media and tools are employed (i.e., email and telephone) (Oliver & Trigwell, 2005).

37.2 MILITARY SERVICES SHIFT TO A BLS THAT INCLUDES VIRTUAL WORLDS

In general, the evolution of BLS can encompass many ways to deliver courseware and programs of instruction such as lecture, computer based, distributed, mobile, video conference, and high definition telepresence. Along with some of these more traditional modes of training, one of the technologies gaining a foothold in Service usage is virtual worlds (VW) (Claypole, 2010). This addition to the BLS toolbox adds utility because VWs provide the ability to immerse users physically, mentally, and emotionally while they explore, construct, and manipulate interactive virtual objects in more complex three-dimensional (3D) environments (Dalgarno & Lee, 2010).

As the Services look to a BLS, Dr. Jolly Holden, Chairman Emeritus, United States Distance Learning Association, says that there is increased focus on collaborative tools that facilitate the transfer of learning and adaptive tools used for dynamic content and increased interaction. At a time when the emergence of web-based technologies has radically influenced the ways in which individuals around the world communicate, represent themselves, share ideas, and otherwise

FIGURE 37.1   Air Education and Training Command's vision of training (2008).

interact with one another, these VW environments can influence some of the collaborative goals of the military (Rogers, 2003; Ward & Sonneborn, 2009).

The Air Force made a bold step toward establishing VWs as an official training platform as far back as 2008, when the Air Education and Training Command released their visionary report *On Learning: The Future of Air Force Education and Training* (Air Education and Training Command, 2008) describing their next-generation learning environment equipped with a BLS known as Air Force 2.0. The future learning concept was built on the premise that "Airmen of the future will be able to share their gained knowledge with others, to collaborate, and to operate successfully in and dominate the domains of air, space, and cyberspace. If the Air Force of the 21st century is to be an agile, adaptive, learning organization, it must embrace change, accept risk, cope with reverses, and learn to reinvent itself—constantly" (2008, p. 3).

The center point of the Air Force on Learning concept was that each recruit would receive an avatar upon joining the Air Force that would follow them through their career. This requires that an infrastructure be built that allows the virtual components to link to content management tools that are able to track educational and professional progress. This concept is described as MyBase and is currently in the proof of concept and testing phases. A portion of MyBase is public, with 12 islands at the Air Force's Second Life® (SL) facility. Additional components are behind the Air Force firewall using an authentication system and providing access to resources such as classes and

FIGURE 37.2 The U.S. Army Learning Concept for 2015. (From Department of the Army, *The U.S. Army learning concept for 2015* (Training and Doctrine Command (TRADOC) Pamphlet 525-8-2), U.S. Army Training and Doctrine Command, Fort Monroe, VA, January 20, 2011.)

review materials. The environment allows airmen to predeploy and tour facilities located at their next assignment. This predeployment is expected to reduce ramp-up time and improve productivity as they virtually in-process and complete paperwork.

More recently, the U.S. Army developed the Learning Concept Report for 2015 (Department of the Army, 2011), which is a 66-page document that outlines ways the U.S. Army is seeking to embrace BL for its soldiers worldwide. One goal of the report is to dramatically reduce or eliminate instructor-led slide presentation lectures and begin using a BL approach that incorporates video teleconferencing, simulations, gaming technology, mobile, VWs, or other technology-delivered instructions. The authors of that report discuss that a BLS will be used whenever possible for skill training and performance support when new skills and knowledge requirements are introduced to the class. The underlying impetus of this shift in training focus is that in a climate where every individual has "nearly ubiquitous access to information, the Army cannot risk failure through complacency, lack of imagination, or resistance to change." By embracing BL technology, the Army takes "immediate action to develop a capacity for accelerated learning that extends from organizational levels of learning to the individual Soldier whose knowledge, skills, and abilities are tested in the most unforgiving environments" (Department of the Army, 2011, p. 5).

The Army envisions a 2015 BLS environment that will increasingly employ virtual training. Both resident and nonresident soldiers involved with learning events for individuals, small teams, or large distributed groups will use BL environments that cover a broad range of capabilities, including simulation, simulators, game-based scenarios, mobile, VWs, and others, and may employ additional plug-in technologies as virtual humans, augmented reality, and artificial intelligence to enhance the perception of immersion and realism. The goal of this concept will be that soldiers will have a single online portal where digital learning resources including VWs can be easily found in preferably two, but no more than three clicks. The portal itself may be a 3D VW with natural navigation and interpersonal interactions through avatars, providing access to mentors, peer-based interactions, facilitators, and learning and knowledge content repositories. The portal requires multiple security access levels with ready access to unclassified learning material, and more stringent security requirements for *for official use only*, and secure information.

As with the other two Services, after an Executive Review of Navy Training in 2001, a working group was established to examine Navy training and make substantive recommendations for improving and aligning organizations, incorporating new technologies into Navy training, exploiting opportunities available from the private sector, and developing a continuum of lifelong learning and personal and professional development for sailors (Aplanalp, 2010). The working group's recommendations spawned what came to be known as the *Revolution in Training* (RIT) (U.S. Navy, 2001) (Figures 37.3 and 37.4). A core tenant of the U.S. Navy's RIT was to migrate appropriate tracts of content to a learner directed, web-delivered medium. Their recommendations were followed by a Navy Inspector General Report in 2009 (U.S. Navy, 2009) as well as Department of Navy Instruction in 2010 (U.S. Navy, 2010), which suggests that a blended modality strategy would be the most effective training approach for the following reasons:

- Allows instructors/facilitators to build learner directed content support skills in small increments
- Allows program offices to gradually move small increments of training content to learner directed content as schedule, need, and funds allow
- Provides the opportunity to use the best delivery modality to accomplish the required learning objectives

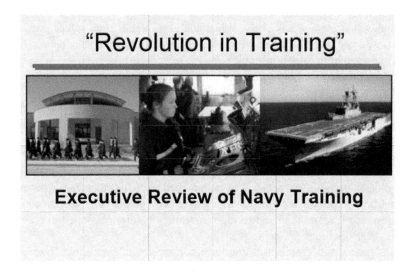

FIGURE 37.3 Revolution in Training. (From U.S. Navy, *Revolution in training: Executive review of Navy training*, U.S. Navy, 2001.)

FIGURE 37.4 A meeting of the Virtual Worlds Education Roundtable (VWER). (Photo courtesy of the VWER: www.vwer.org.)

Taken all together, as suggested in the preceding paragraphs, all the Service components have guidance and documentation that suggests a trend toward moving classroom based and online delivery of training content to a BLS.

37.3 SUGGESTED BENEFITS OF A BLS

As part of the military's push to an overall BLS, the Air Force University in their technical report *Horizons in Learning Innovations* (Air Force Air University, 2011) suggests to get the best *blend* requires ingenious thought and courageous will about how to combine selected components for value and targeted benefits of training and learning. They go on to discuss that by adding serious games, mobile platforms, and VWs in this blend there is the potential to add value to the user by (1) strengthening the affective cognitive linkage through real-life scenarios, (2) strengthening the mind–body linkage using new interface device technology, and (3) providing just in time linkage by accessible modules anytime and anywhere. When analyzing these three values presented by a BL approach, it is postulated that a structured Program of Instruction (POI) may benefit from the addition of serious games, mobile platforms, and VWs.

While serious games is a body of research in itself it has quickly grown in popularity, there are now a large numbers of game engines and platforms supporting game research and Program of Records (PORs) in the Services. This chapter will not attempt to add to that body of knowledge nor will it attempt to tackle the research that is going on in mobile learning. The authors will simply mirror what the authors of the report *Mobile Learning: Transforming the Delivery of Education* (Ally, 2009) discussed in their work, which is that mobile learning, through the use of mobile technology, will allow citizens of the world to access learning materials and information from anywhere and at anytime. With mobile learning, learners will be empowered since they can learn whenever and wherever they want. These attributes of mobile learning have presented an unprecedented momentum in that any well-designed BLS must include a mobile strategy before it is added into any Program of Instruction that is considering a BL approach.

While serious games and mobile learning have quickly found a foothold in the BLS, another component that is receiving a surge in usage throughout the Services is VWs, as noted earlier in this chapter. VWs are computer-simulated environments in which individuals interact with one another via avatars representing their digital selves offering a collaborative tool that has a number of possible uses, from saving money on travel and conference expenses to employee training and soldiers mission rehearsal exercises (Nevo, Nevo, & Carmel, 2011).

Since VWs may have an ability to promote learning based on several recent military studies (Evans, 2010; Martin, 2000; Peachey, Gillen, Livingstone, & Robbins, 2010; Wankel & Kingsley, 2009) which has contributed to their recent increase of use. This increase may be linked to several factors such as the availability of a computer-based simulated environment that is shared by multiple users who interact with each other in a 3D graphical landscape that is always readily available or persistent, the ability to establish personalized avatars, and often user-generated goals that are potentially facilitated by an instructor as opposed to being led by the instructor.

The best way to understand a VW is to experience it. Although reading text on a page describing the experience falls far short of an actual virtual experience, an example from Ender's Game is provided as follows (Figure 37.5):

> His figure on the screen had started out as a little boy. For a while it had changed into a bear. Now it was a large mouse, with long delicate hands. He ran his figure under a lot of large items of furniture depicted on the screen. He had played with the cat a lot, but now it was boring—too easy to dodge, he knew all the furniture.
>
> Not through the mouse hole this time, he told himself. I'm sick of the Giant. It's a dumb game and I can't ever win. Whatever I choose is wrong.
>
> But he went through the mouse hole anyway, and over the small bridge in the garden. He avoided the ducks and the dive-bombing mosquitoes-he had tried playing with them but they were too easy, and if he played with the ducks too long he turned into a fish, which he didn't like ... So, as usual, he found himself going up the rolling hills.
>
> The landslides began. At first he had got caught again and again, crushed in an exaggerated blot of gore oozing out from under a rock pile. Now, though, he had mastered the skill of running up the slopes at an angle to avoid the crush, always seeking higher ground.
>
> And, as always, the landslides finally stopped being jumbles of rock. The face of the hill broke open and instead of shale it was white bread, puffy, rising like dough as the crust broke away and fell. His figure moved more slowly. And when he jumped down off the bread, he was standing on a table. Giant loaf of bread behind him; giant stick of butter beside him. And the Giant himself leaning his chin in his hands, looking at him. Ender's figure was about as tall as the Giant's head from chin to brow.
>
> "I think I'll bite your head off," said the Giant, as he always did.
>
> This time, instead of running away or standing there, Ender walked his figure up to the Giant's face and kicked him in the chin.
>
> The Giant stuck out his tongue and Ender fell to the ground.

Card (1991, pp. 62–63)

This passage of Ender's Game depicts the difficulty that the lead character Ender Wiggin had in distinguishing between what is real and how nonexistent the line between virtual and real can be. It is easy to confuse VWs as game environments, but one major difference is the open-ended exploratory nature of VWs as will be described in a later section. VWs are computer-generated simulated environments that are becoming more realistic with the added capability of technology. Most of the entities in a VW are under direct control of humans; however, that is not a defining factor. These environments can be experienced by multiple users (multiuser) at the same time. A VW is persistent, which means it exists even when there are no users logged into it, and is poised for use 24 h a day, 7 days a week, all the while experiencing and depicting effects from events that occur within the environment (Bartle, 2004).

VWs come in a variety of styles with various purposes and business models. Most current VWs depict geographic spaces where users are represented by avatars who can navigate the spaces. The spaces contain objects that can be created and/or bought via 3D repositories (Thompson, 2011). Taken all together, 3D spaces, avatars, and persistence can make the users be mentally, physically, and emotionally engaged in the experience or, in other words, can make them feel the benefits of immersion and presence.

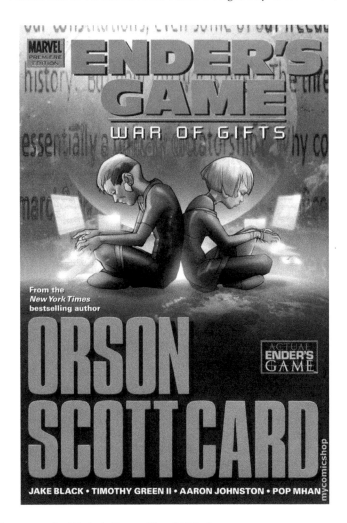

FIGURE 37.5   Cover page for "Ender's Game: War of Gifts."

37.4   BENEFITS OF IMMERSION AND PRESENCE

The learning value of VWs has been the subject of a good amount of research (de Freitas & Veletsianos, 2010; Lee, Ahn, Kim, & Lim, 2006). The premise is that by giving the user the sense of being part of the environment, the experience will have a more realistic and immersive feel. The impression of immersion is related to presence, social presence, and self-presence in a simulated environment. Immersion occurs when the virtual environment (VE) perceptually envelopes an individual such that the individual perceives him or herself to be interacting within that VE rather than within his or her physical surroundings (Blascovich et al., 2002). If a simulated experience is effectively engaging and immersive, the subject will become so completely involved that he or she will experience stress, fear, excitement, or anger as a result of the unfolding events. In other words, there would be minimal physiological, emotional, and physical difference between the immersive environment and the real world.

Presence defines the degree to which a user can actually feel that they themselves exist in the VE. Social presence is the degree to which people feel connected to other people in the VE. Self-presence is the degree to which people believe their own avatar is actually an extension of them (Turk, Bailenson, Beall, Blascovich, & Guidagno, 2004).

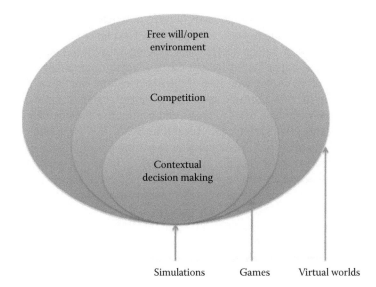

FIGURE 37.6 Simple model of the relationship between simulations, games and virtual worlds. (From Olbrish, K., *Virtual worlds, simulations and serious games, oh my*!, retrieved September 10, 2012, from learningintandem.blogspot.com/2009/05/virtual-worlds-simulations-and-serious.html, May 17, 2009.)

Csikszentmihalyi (1998, 2002) describes *flow* as a state of concentration that is so focused that it results in complete absorption in an activity and results in the achievement of an optimized cognitive state. The premise is that people achieve a positive state if they do something they might succeed at, concentrate on it completely, have clear goals, and know immediately if they have been successful. This theory is a cornerstone to current-day game design and is the focus of a great amount of learning research (Chan & Ahern, 1999; Jackson & Marsh, 1996; Jones, 1998; Pearce, 2005). Although the relationship between serious games for training and VWs is complimentary, there are some significant differences. Where most games exist within a VW or VE, one difference is that VWs are persistent (described previously). VWs are also most often distinguished as supporting multiple users who have the ability to experience the environment simultaneously in a social way. A further factor that tends to distinguish VWs from games can be the extent to which tasks are goal-oriented or open-ended. VWs are more closely linked to open-ended experiences. Pearce (2009) describes games as being more *ludic* or goal-oriented, while VWs tend to be more *padaic* or associated with free will. Another way to consider the relationship between VWs and games is that games or simulations are usually tasks that take place within a VW as is shown in Figure 37.6.

The major benefit of VW technology, and one reason the Services are embracing it, is the ability of these environments to provide a level of adaptability training and critical thinking not associated with a deterministic outcome or scripted POI. As previously mentioned, VWs have the ability for users to behave in a nondeterministic fashion, much like the real world. This spontaneous nature of VW technology provides unique benefits to the user that traditional VEs are lacking. For example, since VWs are inherently social technologies, they support enhanced communications mechanisms allowing for more collaboration opportunities. Common real-time communication mechanisms include text and voice chat, not just person to person, but also person to group.

37.5 DIFFERENT BUSINESS MODEL—SHIFTING COSTS OF CONTENT CREATION TO THE USER

End user control and ability to add to the content pipeline provides a different authoring capability than traditional VEs. Consider the content creation pipeline in traditional simulations, and many video games. More traditional content creation starts with an artist who creates content in a modeling package such as the popular Autodesk® Maya. The artist might then create textures and other artwork in Adobe® Photoshop to apply to the models. Next he or she imports the models and the textures into the VE's *world builder* tool. Lastly, the content is placed, and the world is baked and produced for the users to enjoy. Unfortunately, this is a lengthy and labor-intensive process and changes made to the VE often require a re-bake of the world and the simulation restarted.

One business model in the VW industry has a very different content generation model, demonstrated by SL, which encourages end users to create and manipulate content. Objects are treated as agents that can be independently manipulated in real time by the VW simulation engine, allowing changes to the system to be made without stopping or restarting the simulation engine. The unique nature of this content creation strategy is that the work is done in the environment, while it is running. There is no separate world building application; the world is built in situ. Objects are created and textured in real time and in the presence of other users in the system. Participants in the world function as the world's architects.

One example of how this real-time capability can aid the Services would be with a distributed collaborative design team from all over the world who comes together to design a vehicle. They meet either synchronously or asynchronously in a VW that supports user-created content. The designers could speak to one another or leave notes for one another describing design changes that they are making in real time, such as lowering the center of gravity or changing the shape of the headlights. Copies of the vehicle could be made with variations to assess the best design choices. When the designers have completed their virtual design, they can make copies of it in multiple colors, take it for a test drive, and experience the results of their work virtually. In fact, SL has hosted a range of real-life car manufacturers (e.g., General Motors, Toyota, Nissan, and Mercedes) who have prereleased vehicles to users, often for user feedback and publicity prior to releasing the vehicle in real life (Valdes-Dapena, 2006).

37.6   ADDITIONAL USES OF VWs FOR TRAINING AND EDUCATION

As mentioned previously, one area VWs are gaining attention is in the arena of training and education. VWs Watch estimates that over 95% of UK universities are now using or experimenting with VWs, principally SL (Kirriemuir, 2009). The *SLEducation* Wiki lists over a hundred U.S. educational institutions in SL. Examples include Tulane University, George Washington University, Colorado Tech University, and National Defense University (New Media Consortium, 2012a, 2012b). In addition to increased activity among universities and community colleges, secondary and even primary schools are wading into the use of VWs (New Media Consortium, 2012a, 2012b; Tenkely, 2013). In the private sector, large multinationals like Intel, Raytheon, Lockheed Martin, and IBM (Public Broadcasting Service [PBS], 2010) are using VWs to educate their employees, as well as hospitals (University of North Carolina at Pembroke, 2010; St George's University of London, 2009), governments (Howard, 2010), and the military (e.g., USAF MyBase in SL and the U.S. Army Training and Doctrine Command [TRADOC] in Active Worlds being just two of many examples) are also exploring the use of this technology.

37.7   TOLERANCE AND ACCEPTANCE—INDIVIDUAL DIFFERENCES

One outstanding quality of VWs is their ability to *level-set* the user audience. There are countless stories of individuals who faced racism, sexism, or issues with disabilities who were able to achieve personal success through the use of VWs (Cassidy, 2007). Within VWs the user can establish their identity as they see fit. For example, one literature review (Stendal, 2012, p. 9) shows that individuals with a disability use VWs *for enjoyment, employment, communication, friendship, education, and discussion*. As a result, they gain the sense of independence and empowerment. People with disabilities are able to build social skills and other skills needed for independent living while escaping prejudice. They can explore the world geographically while meeting a wide range of people through virtual field trips. Therefore, VWs may have an appealing quality to help address sociological problems as Gates, Myhrvold, and Rinearson (1996) suggest that people can be treated as equals in the VW and we can use this equality to help address problems that society has yet to solve in the physical world.

Klinger and Coffman (2010) describe how 3D VWs foster social equality in a global economy and discuss issues with ensuring technology is accessible to a diverse population. Given the opportunity to engage with others socially, academically and occupationally, users experience inclusion that reduces the sense of isolation—assuming access to computers, and the necessary computer skills. This is not to say that insensitive activities (racism, sexism, etc.) do not occur in VEs; they do. These worlds are populated by people of all types. The significant difference is that users can define how others see them and even how and if they hear them (using text rather than voice to downplay verbal communications difficulties). Often this may provide the opportunity to accentuate strengths and dispel preconceived notions.

37.8 EXAMPLES OF MILITARY SERVICE VW APPLICATIONS

37.8.1 VIRTUAL WORLD FRAMEWORK

The Office of Secretary of Defense has initiated the Virtual World Framework (VWF), which is an Internet standards–based architecture created with the goal of establishing the groundwork for secured distributed virtual collaboration. The VWF has the goal of aligning the various VW activities throughout the U.S. military. It leverages existing HyperText Markup Language (HTML) 5 standards and is designed to be deployed to a web browser without the need for any third-party executables or installations. The VWF has a client–server specification, which allows for web browsers to behave as VW content viewers with synchronous updates (Lockheed Martin, 2012).

Although the VWF could be used as a lightweight distributed VW, it is not intended solely to be a standalone VW. Rather it is a means to enable various VWs to interoperate. The VWF specification is intended to allow developers to share component content such as video, objects, terrain, and avatars and successfully insert them into specification conformant applications.

The VWF is written primarily in HTML 5, Java script, and Ruby programming language. Although the Department of Defense (DoD) has government use rights to the source code, the code is protected by the Apache 2 open source license and freely accessible by anyone willing to expand its functionality. Apache 2 licensed software allows anyone to use, change, and sell derivative works of the original material as long as the original licensing statements and attributions are included with the distributions. The intended audience is developers of online training systems, simulations, educational content providers, gamers, or anyone interested in learning about the latest advancements in lightweight web-based virtual reality.

37.8.2 AIR FORCE RECRUITING

Another example of how the Services are expanding the role of VWs is by viewing the work of the U.S. Air Force Academy Recruiting (2013) Command, which was one of the early adopters in the use of VWs. At the time of initiation of the MyBase project, more than 15 million accounts worldwide had registered in SL. Air Force officials had hoped that the MyBase area in SL would attract men and women interested in learning more about the Air Force. The site also provides links for enlistment and commissioning information and how to contact the nearest Air Force recruiter. Their vision was to provide an environment that would engage high school students interested in the aspects of the Air Force. LeAnn Nelson, chief of marketing and media at the academy said, "We have to recruit from all 50 states every year, and we wanted a way for kids to be able to get a feel for some of the campus if they can't make it out here to visit." Nelson goes on to say that "it was a way of relating to kids in a way that they already understand" (Zou, 2011, p. 1). During the visit to the online campus visitors can walk around the Academy's campus, buy items from the cadet store and even fly a plane—all without leaving their bedrooms. Potential applicants can view the Academy's facilities and learn important Air Force facts without stepping foot on campus (Zou). Adding to the MyBase recruiting effort the Air Force is taking advantage of a BLS while including social media tools like Facebook and mobile phone applications to reach out to students interested in the Academy.

37.8.3 RESILIENCY

Resiliency has multiple connotations and the military services are using VWs to embrace them all. U.S. Army OneSource provides a Virtual Resiliency Campus to soldiers and their families worldwide. The campus provides an innovative and fun way to build resiliency in the physical, emotional, social, spiritual, and family aspects of their lives. Tools are available on the virtual campus to create and organize an exercise regimen with motivating factors to reach predefined goals. There are resources designed to provide emotional support, engage in social activities, and interact with other families. Spiritual fitness is strengthened through resources and Service directories made available virtually. The campus is available in a private area of the public VW SL (OneSource, 2011).

Additionally, VWs are developing in the area of cognitive therapy and helping with phobias by replicating (or simulating) actual places and events such as giving a speech to a large auditorium. This can aid in building familiarity, putting context into scenarios, or even in reducing panic and phobias. A growing body of research describes the use of VWs for group therapy (Gorini, Gaggioli, Vigna, & Riva, 2008), exposure therapy (Price, Mehta, Tone, & Anderson, 2011; Rothbaum, Garcia Palacios, & Rothbaum, 2012), posttraumatic stress disorder (PTSD) (Freedman et al., 2010), and retraining individuals with brain injuries to reintegrate (Fong et al., 2010).

37.8.4 HISTORICAL FIELD TRIPS

An additional use of VWs has been to travel through time and allow any individual to relive events from the past. One example of this is in the creation of the Twinity Virtuelles Mauer Museum. In 2009, just before the twentieth anniversary of the fall of the Berlin Wall, Twinity recreated a virtual 2 km stretch of the Berlin Wall. The replica invites visitors to time-travel back to 1989 with interactive multimedia content building through key points in history (Twinity, 2009) (see Figure 37.7).

FIGURE 37.7 Berlin Wall in Twinity. (From Twinity, *The Berlin wall in twinity*, retrieved October 12, 2012, from Twinity.com: http://www.twinity.com/en/community/berlin-wall, November 9, 2009.)

37.8.5 MARS EXPEDITION STRATEGY

Showing the expansive capabilities of VWs, the Mars Challenge takes participants into space. The Mars Expedition Strategy Challenge (see Figures 37.8 and 37.9) was created by the U.S. Air Force Air University. It provides an immersive reality challenge to explore strategies for human space-flight beyond low Earth orbit. The Mars Expedition Strategy Challenge begins at the Learning Center with stations on each side that introduce the challenge. The educational environment uses self-assessments, trivia games, and reviews to reinforce learning material. Learners have the opportunity to don a NASA space suit while exploring content videos before embarking onto the space-craft (Stricker, McCrocklin, Holm, & Calongne, 2010).

The simulation features a rocket launch aboard the Orion spacecraft and a flight to into deep space arriving on a Mars Colony where users review the expedition strategy and alternatives.

FIGURE 37.8   Launch Center. (From Stricker, A. et al., *Mars expedition strategy challenge*, retrieved June 18, 2013, from http://marsexpeditionstrategy.blogspot.com/, February 8, 2010.)

FIGURE 37.9   Spacecraft traveling to Mars. (From Stricker, A. et al., *Mars expedition strategy challenge*, retrieved June 18, 2013, from http://marsexpeditionstrategy.blogspot.com/, February 8, 2010.)

Inside the Orion, the learner is seated and a voice describes the experience as the rocket launches, flies upward, and separates into three stages to leave Earth's orbit and approach the deep space explorer. As the flight begins, there is a stunning visual depiction of earth receding as the spacecraft passes the moon. During the flight, the learner visits four consoles and reviews 30 min of videos that explore the strategies for deep space exploration while interacting with the onboard equipment. Upon arrival, the learner disembarks through the hatch to explore the Mars colony. At the end of the simulation, learners participate in a Mars Expedition Strategy Challenge Trivia Game and complete a survey (Stricker et al., 2010).

37.9   ARMY RESEARCH LABORATORY FEDERAL VIRTUAL CHALLENGE

Historically, the military was a significant driver to improvements to the technical capabilities of VEs. Major advancements came about as a result of vision and funding from the DoD (Kang & Roland, 2007). More recently, the commercial game and VW industries have demonstrated better visual realism and shorter development cycles than those available in military simulations (Cohn & Bolton, 2009). Unlike the DoD, the new consumers of VEs are public users and number in the millions, although they do have considerably smaller budgets. Despite a 10% drop in monthly game industry sales in 2011 compared to the previous years (including hardware, software, and accessories), Americans alone were still spending nearly $1 billion each month for game-based entertainment (Greene, 2011). Even during the Cold War, when defense spending was at its apex, the U.S. government could not keep pace with this level of spending. This shifting paradigm has created an opportunity for the military and the modeling and simulation community to become the beneficiary of the rapid pace of improvements made for the entertainment industry (Abramsky, Ghica, Murawski, & Ong, 2004).

One way the military services are leveraging this paradigm shift is by reaching out to developers that have not traditionally been content suppliers to the military through the Federal Virtual Challenge (FVC), previously known as the Federal Virtual Worlds Challenge (FVWC). VW development moves very rapidly within a community of developers who support one another and share resources. The Challenge invites the world's innovators to demonstrate their advancements for recognition and rewards. Each year, the focus areas of the challenge are determined based on significant needs within U.S. government VEs that overlap with active developmental areas in industry. Previous focus areas include training and analysis in VEs, artificial intelligence, engaging training strategies, improved low-cost user interfaces, and providing critical thinking and adaptability training in VEs (U.S. Army Research Lab, Simulation and Training Technology Center, 2013). Aside from the cash award and recognition, winners have the opportunity to demonstrate their innovative solutions to government and industry at a trade conference (Biron, 2013). This is a significant departure to how the U.S. government has explored technology in the past; encouraging greater innovation and more rapid implementation in VEs.

37.10   ENHANCED DYNAMIC GEOSOCIAL ENVIRONMENT

The Enhanced Dynamic GeoSocial Environment (EDGE) prototype is a current research effort at the U.S. Army Research Lab, Human Research an Engineering Directorate Simulation and Training Technology Center (ARL-HRED-STTC) in partnership with the TRADOC G2 (Intelligence) office. It is an effort to leverage commercial game and VW technology to improve training in the operational environment (Dwyer, Griffith, & Maxwell, 2011). Currently, military simulations and training environments are trailing the graphic and feature quality of commercial games and VWs by almost a decade. EDGE was developed to close that gap, providing a graphically rich space that engages learners in training material while being accessible from anywhere in the world with an Internet connection.

The EDGE platform is intended to support a variety of user's mission objectives, meaning the system must be flexible and scalable. This means that terrain, content, behaviors, etc., must be rapidly inserted into the simulation in a cost-effective manner. This flexibility allows for a wide range of scenarios and mission objectives Figure 37.10 shows the arena used to orient the user on how to

FIGURE 37.10   Screen capture of the EDGE training level (June 2013).

navigate the environment prior to using it for training. As the name implies, the intent is that EDGE not only replicate the visual and contextual environment but also has the capability to model complex social networks and communication patterns between them. Actual regions have been built (see Figure 37.11) for use in the environment. They are geospecific areas that are often populated with geotypical buildings generated to support the training purpose. Ultimately, live intelligent feeds can be replicated in the environment allowing a user to experience the environment, its threats and opportunities in safety prior to deployment. Even live television, radio, and weather can be experienced as if the learner had his or her boots on the ground in the specific location (Figure 37.12).

In a combat environment where most casualties occur either in the first months in theater or the last weeks prior to coming home via improvised explosive devices (IEDs), gaining familiarity of the environment as early as possible is critical (Dao & Lehren, 2012). The ability to replicate critical events for learning purposes becomes more meaningful when a trainee is walking in the footsteps, and potentially virtually side by side with others who have been there before them. Social mistakes

Satellite Kuzun   Virtual Kuzun

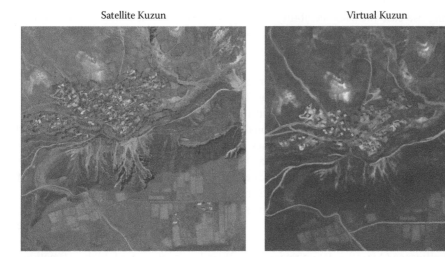

FIGURE 37.11   Comparison of the Azerbaijan town of Kuzun both via Google maps and within the EDGE game (November 2012).

FIGURE 37.12   Screen capture of EDGE within the Kuzun, Azerbaijan terrain (June 2013).

have repercussions that perpetuate over time in an environment that is living and breathing 24 h a day 7 days a week. If the previous unit did a poor job of considering the needs of the native population, the trainee may be faced with the need to repair relationships while ensuring safety. This environment provides the living and breathing patterns of people going to work, tending their flock, and being influenced by their social networks.

Similarly, EDGE is being used to train first-responders in support of the Department of Homeland Security (DHS). The same detailed modeling being used by the military is capable of training police, fire and emergency response teams in dealing with attacks on safety and security in the homeland. This use case has a great amount of flexibility because both the trainees and the opposing forces are live players. This means that the same environment can support very small team training (1–2 individual perpetrators and up to eight first responders) as well as very large team training, with coordinated response including fire/medical rescue, police, dispatch, and command and control alongside a complex coordinated opposing force team.

A developing use case includes the ability to inject new weapons, sensors, and tactics into the environment to support analysis into how these tools might change potential outcomes. Another potential use case is the injection of intelligence data into the immersive 3D space to support actual intelligence fusion and analytics.

EDGE is built on the Unreal game engine to satisfy the need for high fidelity graphics and high-end commercial game functionality, such as the ability to use *sticky* surfaces in the game for cover and concealment. Through the requirements determination process and user evaluations, it became clear that providing soldiers with functionality that was substandard to the capabilities they experienced in popular first-person-shooter games could pose a distraction. Though the geosocial aspect of the environment suggests that if gunplay takes place, something has likely gone awry for the soldier's (or first responder's), the ability to effectively use the tools, such as weapons, at their disposal in a way they are familiar is critical to training.

EDGE's business model is to provide development resources to contributing developers sponsored by the U.S. government. This allows each team to leverage base functionality continuously being developed by various experts and maintain currency with the commercial game industry. Experts in artificial intelligence (AI), after action review (AAR), and user interfaces fold their functionality into the platform's baseline that also provides linkages to various other simulations and command and control systems. The platform capabilities rapidly grow and are shared across the government.

This developmental approach is implemented through a middleware/translation strategy that uses an existing translation tool, the Joint Bus or JBUS, as a means of communicating between various functions and the game engine. Resources have also been devoted to the integration and test of code and coordination with contributing developers. Several design considerations need to be balanced in order to maximize the user experience, these included scalability of the operational environment, scalability of the number of players, frame rate, and complexity of entities.

37.11 EDGE DESIGN CONSIDERATIONS

The EDGE design guidelines provide a framework in which designers of VW systems may be able to maximize the focus on the user experience. For example, the design guideline for scalability of the operational environment can have a big impact on the ability to replicate the size of the operational area whether designing for small urban environments or large regional terrains. EDGE has demonstrated a 24 km^2 at the *2010 Inter-Service Industry Training Simulation Education Conference (I/ITSEC)* when using the Big World Game Engine; however, transitioning to the Unreal game engine reduces the current terrain size to around 10 km^2. Terrain paging strategies are being implemented along with the ability to create high fidelity modeled areas within a larger, lower fidelity map.

The next scale issue is the number of players a system can handle. The current EDGE environment, using the Unreal game engine, is intended to support roughly 64 live players in each shard with the ability to replicate shards automatically based on need. This means that the number of live players being supported by the training environment is limited solely by back-end hardware. EDGE is optimized to run on the Army Common Hardware Platform (CHP), which are typically 3-year-old mid-grade computers. The client load is largely constrained by the number of polygons its hardware can draw at a reasonable frame rate.

The metric of frame rate is highly variable and leads into the remaining scale issue as the differences between client computers may vary greatly. The complexity of each of the entities affects the number a simulation can support. From a client-side perspective, the number of polygons used to draw the world may be reduced to accommodate lower end hardware; however, the sacrifice is often visual fidelity and visual realism. Shown in Figures 37.13 and 37.14 is a side-by-side comparison demonstrating the differences in visual fidelity using two different game engines.

FIGURE 37.13 Combat Observation Post Keating, modeled using the Unity3D Game Engine developed by the Blended Learning Studio artists at the STTC, April 2011.

FIGURE 37.14   Combat Observation Post Keating, modeled by the Blended Learning Studio artists using the CryTek 3D Game Engine, April 2011.

Another strategy used on EDGE to handle entity complexity was to reduce the amount of data passed between the client and the server. To reduce load on the network, each of the entity's state and metadata was minimized. Lastly, the mathematical models at the server side can be reduced to allow for more agents and entities computed in the simulation; however, the scene may become less populated. Another option is to have AI behaviors calculated on the client side, but there is a risk of loss of consistency with what is viewed by all participants in the world. Balancing simulation fidelity with available network and computational resources can be a complex problem when making design consideration decisions in VWs.

37.12   CASE STUDY: MOSES

37.12.1   MOSES: The Military Open Simulator Enterprise Strategy

The ARL-HRED-STTC conducts research into gaming and VW technology for possible application to military training needs. The ARL-HRED-STTC tests and evaluates numerous VW technologies from both industry and the open source community, and it is widely recognized there is still no *one size fits all* solution for all the training and education needs of the military. Some VEs excel at providing classroom focused experiences, other VWs may have better terrain and weather capabilities, still other virtual technologies may have very high accuracy but low graphics quality. The sacrifices needed to support various capabilities are often fidelity in the simulation. What is needed is a flexible virtual training framework that allows for variable fidelity depending on training needs and objectives for the POI that is being considered for VW technology insertion.

One of the VWs investigated was SL, an online 3D VW designed to allow for significant social interactivity. Content in SL is uploaded and presented by the users, called residents, and not created by Linden Lab (the makers of SL). This is a radical departure from the traditional art and content pipelines of competing games, as stated earlier in design and governance.

Objects in SL are interoperable and capable of being scripted to perform various behaviors. This allows for computational steering and represents a major departure from the traditional virtual training environment creation process. These attributes also allow for subject matter experts to directly upload work and training material rather than be completely dependent on artists and modelers.

Initial exploratory efforts by the U.S. military in the SL product revealed a number of useful activities could be accomplished, not only for training but also modeling and simulation. For example, procedural training activities are accomplished through the creation of a virtual representation of a working environment. This allows participants to make mistakes safely and inexpensively. Other great use cases include the preparation for rare or dangerous events. Another useful example is the creation of interactive situations that challenge the user and provide feedback for knowledge training. The simple scripting language in SL allows for a nondeterminant learning environment driven by objectives, not framework limitations.

SL has been established as a useful platform for educators and online learning (Fominykh, 2009). Linden Lab created a profit structure where they benefit from users adding to their framework. However, certain drawbacks to the platform have been exposed. Specifically, the content in the platform is very difficult to export. Since SL makes provisions for all content to have intellectual property rights associated, work created in group settings can become difficult to establish ownership rights, making capture and backup a challenge.

Another drawback is cost. Although it costs nothing for a casual user to register an account and enter SL, individuals or organizations who wish to create content need to purchase space to do so. Private space costs U.S. $1000 setup fee with U.S. $295 per month in maintenance costs (Allison et al., 2011). Since SL is a public forum, certain security and privacy issues arise. Educators often feel the need to obtain this private space so that the students are protected from exposure to the rest of the SL citizens.

Military users need to be able to operate the virtual training environments on networks separate from the Internet. Oftentimes, the training material is sensitive or simulation data have a classified component. Since Linden Lab does not currently offer a version of the SL platform for use behind a corporate or government firewall, the product has limited utility.

In late 2006, Linden Lab made the surprising decision to open the source code to their client code. This allowed enterprising programmers in the open source community to create a reverse engineered software server, called the Open Simulator. Open Simulator is an open source VW server that can be accessed via the same viewer as SL and the developers strive to make the software as closely compatible as possible. In their paper, *The Third Dimension in Open Learning*, Allison et al. (2011) discusses the advantages of using the Open Simulator alternative to SL. The Open Simulator project is an initiative created to address issues such as content portability, security, and cost. Using this open source software, educators can explore the use of a SL environment without making an investment, use the environment in an enclave network to isolate students that may be under age while also being able to backup their work.

MOSES is not a product, but intended to be a best practice strategy for other organizations wishing to deploy an Open Simulator–based VW. The ARL-HRED-STTC and its partners have developed MOSES with the objective of evaluating the Open Simulator and its ability to provide network independent and secure access to a VE. MOSES has several research goals:

- To meet or exceed the service and capabilities provided by the SL product
- To provide a persistent VW in a stable server environment
- To provide a self-serviced voice communications mechanism
- To establish a common base content library with mesh-based content import capability
- To provide in-environment multimedia presentation capability
- To provide online graphical monitoring mechanisms to start and stop server processes remotely
- To provide online graphical user management and setup
- To provide a stable and in-kind SL-like client environment
- To provide guidance to other organizations wishing to replicate MOSES results
- To link to other organizations to expand the MOSES grid

37.12.2   MOSES Deployment 1.0

MOSES was initially deployed in February of 2011 to a limited test group of about a dozen Open Simulator expert users and it debuted to the public on March 22, 2011, at the Defense GameTech Users Conference in Orlando, Florida. It was shown for the first time to a large public crowd at the tutorial, *Virtual Worlds: Advanced Topics with SL—Programming & Management*. The advanced topic was a discussion of the MOSES project and the appropriate approaches to Open Simulator migration with comparisons to Open Simulator versus Linden Research SL products.

The Open Simulator runs as a discrete process that represents 256 m² of virtual terrain. This virtual terrain is called a *sim*, and these sims can be aggregated to create larger operational areas. The amount of processing capability and memory dedicated to each sim are variables left to the administrator and is dependent on resources at hand.

One of the most common questions asked of the MOSES team is, "What kind of hardware is needed to support an Open Simulator grid?" The original MOSES server support consisted of a cluster of nonhomogenous hardware that included 16 decommissioned HP ProLiant BL20p blade servers (graciously provided by another command), two decommissioned Dell PowerEdge servers, and two repurposed SuperMicro servers. Each HP blade server and the PowerEdge servers had two dual core Intel Xeon processors with 4 Gb of random access memory (RAM). The SuperMicro systems had two quad core Intel Xeon processors and 24 Gb of ram.

Gauging load on the various sims is very difficult. Sim load is governed by the amount of content (3D models, sounds, and video), the amount of scripting and the kind of scripting performed, and the number of avatars expected to be supported. For example, a low weight sim may only have terrain and a few buildings for meeting areas. A sim such as this may only need one-fourth of a processor's time and less than 256 Mb of RAM. By contrast, a heavy sim that has over 10,000 individual geometric primitives (prims) and hundreds of scripts running (door openers, greeters, etc.) may need a full processor and over 1 Gb of RAM.

Load presented numerous challenges in the initial deployments of MOSES. MOSES is deployed in a virtualized manner using an open source hypervisor called ProxMox. The ProxMox hypervisor is simply a virtual machine manager that allows for multiple operating system instances to be run on a single server. This is an efficient way to manage large number of software servers and treat hardware as allocated resources. For example, on the Dell PowerEdge servers, a total of eight processors and 8 Gb of ram was available to Open Sim. MOSES was deployed to that hardware on two virtual machines, ran 16 sim instances (two sims per processor) and each sim was allocated 512 Mb of RAM.

The sims deployed to that machine worked well with a sim load of an estimated 5000 prims, moderate script load, and 20 or less avatars logged into it simultaneously. When loads increased, we discovered the deployment was inadequate to support the more aggressive users. Since the hardware in the first deployment was older excess equipment, there was limited control over the configurations. Server load was observed carefully and resources were allocated on an *as-needed* basis to support the heavier users.

37.12.3   MOSES Deployment 2.0

Given the unanticipated level of usage (i.e., as of Summer 2011 there were 200 users) and attention given to MOSES from the academic community, the decision was made to purchase new servers for the project in the summer of 2011. In October, 2011, MOSES was migrated to a new server cluster that will support approximately 20 times the number of current users and simulations. Now that MOSES is relatively stable and has a nuclear community with active projects, the MOSES ARL-HRED-STTC research team can concentrate on higher level questions surrounding VW training. The MOSES team would like to foster an *ecology* of learning that provides context to data, generates repeated exposure of the data to facilitate knowledge transfers and retention, and provides a social network for online mentorship.

37.12.4 MOSES COMMUNITY DEVELOPMENT AND CONTRIBUTIONS

Some participants in the MOSES project are encouraged to develop content on regions given to them for use to advance military training applications in some way (Laboratory, 2013). These regions are granted to participant organizations and the estate managers are required to collect statistics on report their usage of the platform to the MOSES leadership team on a bi-annual basis. Some examples of these estates are documented later.

37.12.4.1 Estate: Virtual Harmony

The Air Force AETC uses MOSES regions called Virtual Harmony Estate. An estate is known as an aggregate of regions used to increase land area. Each region in the estate is organized to support a different set of 3D learning prototypes. These prototypes represent reasoning tools, instructional media examples, and virtual devices for instruction, surveys, and various interactive visualizations.

The Harmony region is the central visitor's center and main teleportation hub for the rest of the estate. Figure 37.15 shows a virtual lecture area in the main hall of the Harmony region.

37.12.4.2 Estate: The Constitution

The Constitution region allows visitors to learn about the issues leading up to the creation of the U.S. Constitution. A person can experience 1790 and investigate the life of Thomas Jefferson, Benjamin Franklin, George Washington, and more. Figure 37.16 shows representative architecture from the period as well as providing an interesting and engaging method for presenting the material inside the buildings.

37.12.4.3 Estate: Tulane School for Continuous Studies

The Tulane School for Continuous Studies maintains four regions in MOSES. Region 1 is used to display educational simulations for economics, finance, and business ethics. Region 2 has photograph galleries used in online teaching. Region 3 has a game theory laboratory and a *Prisoner's Dilemma* simulation. Region 4 has tutorial shops and learning activities. Figure 37.17 shows a mock panel of nonplayer characters used in the Prisoner's Dilemma scenario.

37.12.4.4 Estate: Open Virtual Collaboration Environment

The University of Edinburgh Artificial Intelligence Applications Institute and the U.S. Army Research Lab Human Engineering and Research Directorate are working together on the Open

FIGURE 37.15 Virtual Harmony lecture area in MOSES.

FIGURE 37.16   Air Force AETC Constitution region.

FIGURE 37.17   Tulane University School of Continuing Studies in MOSES.

Virtual Collaboration Environment (VCE) in support of the Dismounted Infantry Collaboration Environment (DICE). The goal of the DICE project is to determine best advice and specialist assistance for when personnel are injured and treated. The Open VCE portion of the project serves many purposes including the creation of virtual synchronous collaboration facilities as well as emergency rehearsal areas. Figure 37.18 shows an outdoor meeting space used for speaking engagements and live video broadcasts.

37.12.4.5   Estate: Air Force Research Lab

The Air Force Research Lab (AFRL) regions are used as a synthetic task environment. This experimentation is designed as a comparative platform for various machine learning techniques. The Cognitive Models and Agents Branch of the Warfighter Readiness Research Division wish to determine the right *mix* of approaches for the use of various adaptive tutors to yield the best results.

FIGURE 37.18   University of Edinburgh Open Virtual Collaboration Environment in MOSES.

37.12.4.6 Estate: Raytheon Missile Systems

The Raytheon Missile Systems regions in MOSES are geared toward technical evaluation and system engineering. Region 1 is used to load the MOSES systems with content and determine the limits at which the simulator can be loaded and still usable. Region 2 is similar to the previous region except the load testing is for scripting rather than 3D geometry. Region 3 is evaluating the MOSES environment for the suitability of hosting collaborative engineering processes. Various engineering collaboration tasks were tested by Arizona State University graduate students. Lastly, Region 4 (as shown in Figure 37.19) is used as a test bed for Concept of Operations (ConOps) visualizations of Command and Control (C2) activities. There are representations of battlefield scenarios in the simulator and are linked to a virtual sand table to allow commanders to have immediate updates of deployed assets.

FIGURE 37.19   Raytheon Missile Systems ConOps and C2 Laboratory in MOSES.

FIGURE 37.20   Air Force Defense Language Institute Sample Training Area by Tech Wizards.

37.12.4.7   Estate: Tech Wizards

Tech Wizards, working under a contract from the Air Force Defense Language Institute, have created a virtual collaboration environment with the purpose of training the students who learn English as a second language on weapon safety and flight briefing procedures in a virtual classroom setting. The simulation uses automated behaviors and artificial intelligences to accomplish its goals. Figure 37.20 shows the entrance to the virtual classroom.

37.12.5   MOSES Future Plans—VW Research into Scalability and Flexibility

In February of 2013, the ARL-HRED entered into a Cooperative Research and Development Agreement with Intel Research, Inc. (Laboratory, 2013). This CRADA specifies a knowledge sharing agreement between the U.S. Army and Industry to conduct research specifically designed to solve critical simulation scalability issues. Properly representing the operational environment for army training needs will require a massive increase in virtual simulation scalability and flexibility. The majority of current game-based VE training applications are only used to train at the small unit level, 40 soldiers or less. The reason for this is the inability for current systems to handle larger numbers of concurrent users in the same place at the same time. This also means there are limited system resources left over for opposing forces and neutral entities. It is believed that VW technology may be used to achieve the goal of full spectrum operations during virtual mission rehearsal exercises.

Current game-based VEs support 10^1 user agents in each level/shard/virtual operational area. Technology research and development efforts are underway to support 10^2 and 10^3 agents, which are representative of the numbers of people in a real operational area. So far, internal research and literature indicates there is no *one* silver bullet solution. A holistic approach to the solution is appropriate.

There are many variables at play when designing and deploying a simulation-based VE using game technology. The first decision the designer must make is how many users the environment must support, usually based on the kinds of training events the system is required to support. For example, a Key Leader Engagement training event may require a village full of users and a couple of platoons. The next decision for the designer is to consider horizontal or vertical scaling to achieve the requirement goal. Typical first-person-shooter games rely on replicated levels or

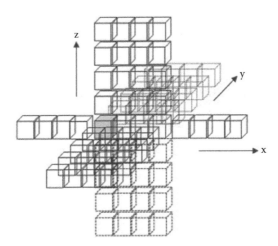

FIGURE 37.21   Horizontal simulator scaling where two or more servers are running several spatial regions and some of the supporting backend services. (From Intel, 2013.)

shards to support large numbers of people. Each group of players are placed in an identical training level; however, the groups do not interact. In order to scale under that scheme, the individual servers must be upgraded with faster processors and more memory. This is known as vertical scaling. There are limits to what a single processor or single server can do and therefore vertical scaling leaves very few options when planning for the total supported numbers of simultaneous users.

The scalability research performed using the MOSES platform is investigating the application of horizontal scaling of the distributed servers. To achieve higher numbers of players in the same training environment at the same time, the MOSES uses a mechanism called the Distributed Scene Graph (DSG). The DSG allows the MOSES load to be distributed across many hardware servers by splitting up the major components of the game engine and allowing them to execute as individual processes. Further, these processes can be distributed across many servers on a network (or over the Internet) to provide additional computational power to each of the virtual training regions as shown in Figure 37.21. Lastly, there is a layer of middleware that performs synchronization of the regions to have a common view to the players. This allows a player logged into one region and a player logged into another region to see and interact with each other in real time.

This prototypical approach to scaled deployments of game-based VE services has demonstrated encouraging results under real-world experimentation. Figure 37.22 shows the targeted configuration used in a scalability exercise performed on March 22, 2013. The head node of the MOSES in Orlando, Florida, housed modified versions of the open simulator grid services. The open simulator services were separated into the physics engine, persistence engine, scripts engine, and client managers. Under normal conditions, all of these services would share a single CPU and memory space. In this exercise, the Open Dynamics Engine (ODE) was replaced with the multithreaded Bullet physics engine and was given 8 processors and 64 Gb of ram. Similarly, the scripts engine was also separated and provided its own dedicated processors and memory. The client manager was replicated and copied to Amazon EC2 server sites located in California, Oregon, Virginia, and Florida.

This dynamic scene graph allows for users logged into any client manager to experience and view any movement or change to the environment made to any of the other client managers. The reason for the distribution of the servers across the United States was to shorten the geographic distance between the end user's client and the nearest available client manager. For example, a user located in Seattle, Washington, would experience less network lag connected to the client manager located in Oregon than if they logged into the client manager located in Florida.

Assuming each client manager can support between 25 and 30 human users each, it is logical to conclude the persistent VW presented in this configuration can support over 100 users and as of

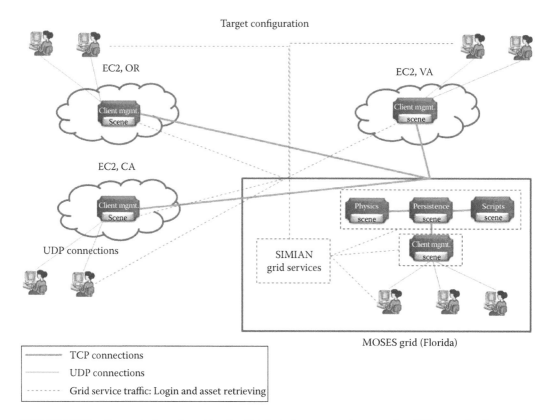

FIGURE 37.22 Conceptual configuration for distributed scene graph supporting 100 users simultaneously in the same virtual space. (From Intel, 2013.)

the latest work demonstrated 60 users. Future tests will determine if this strategy is linearly scalable and if there is an upper limit to the amount of client managers that may be distributed before network saturation prevents further addition of more remote servers.

37.13 LESSONS LEARNED

37.13.1 Costs

As with any technology costs must be considered as a driving factor. With development of VWs for a BLS the landscape can change quite often. Solutions for the military services are no different, not only cost-efficiency must be considered but also how engaging, quantifiable and global the solutions are perceived as matters. VWs lend three options: (1) build from scratch, (2) buy commercial-off-the-shelf (COTS), and (3) modify COTS. There are examples within the military of each strategy with game engines and VWs along with their trade-offs.

The Army began fielding a POR, Virtual Battle Space 2 (VBS2), to fill a training gap for small unit operations in 2009 (King, 2009). VBS2 is a militarized version of a COTS game ArmA, which is currently fielded to over 50 sites around the world for the Army and developed by Bohemia Interactive Studios. Each new customer (i.e., Marine Corp, Australian Army, and UK Army) of Bohemia Interactive receives a custom set of vehicles and features for their scenario development. The vehicles and features that are not sensitive are folded back into the baseline to be shared across the user base. VBS2 brought a great amount of new functionality to Army training at a cost that was well below the cost of developing traditional simulations. This was accomplished by leveraging multiple customers with similar requirements. The Army enterprise purchase of VBS2 was a

$17.7M contract to field 70 gaming systems in 53 locations throughout the United States, Germany, Italy, and South Korea. Each system consists of 52 computers, steering wheels, headsets, and mice (Robson, 2009).

VBS2 provides vast scripting functionality and the ability to edit scenarios on the fly. It was a turn-key procurement of an existing, proprietary solution, or a largely COTS purchase. The U.S. Army did not get access to the proprietary source code, as a COTS purchase, so any future development within the game would need to be provided by the development team at Bohemia Interactive.

As expected the visual fidelity at the time VBS2 was procured was superior to traditional simulations, and the scripting language within VBS2 is very powerful; however, it was not long before users wanted to expand the capabilities of the game beyond what could be done with scripting. At the same time the game industry's visual fidelity continued to improve. As a result, VBS2-2 is coming available with improved functionality and graphics. Keeping current with industry is always a challenge with military procurements and having enterprise license procurement is certainly a cost trade-off that any procurement needs to consider in their market research.

Another example of a game engine that was developed specifically for military use is Delta3D. Delta3D was developed as an alternative to the costly licensing fees charged to the military services for development as well as the end user seat charges. Delta3D is considered an alternative to proprietary solutions that do not provide access to the source code and follow-on development, which must be accomplished by the original provider who now has a monopoly on development or the need to recreating past proprietary work using another platform.

A report on the status of Delta3D (McDowell, Darken, Sullivan, & Johnson, 2006) suggests that 90% of the functionality from one VE to another is the same with minor differences. Delta3D was developed so that that 90% would be complete, available, and free for use without licensing limitations. The design for Delta3D includes a four-part philosophical credo:

- Keep everything open to avoid lock-ins and increase flexibility
- Make it modular allowing the replacement of technologies that mature at different rates
- Make it multigenre
- Build a community so the military is not the only contributor

The Delta3D development team decided that rather than build the engine from scratch, they would use the *best of breed* of existing open source software as building blocks. There is a well-defined, consistent Application Program Interface (API) that defines how the building blocks communicate. New functionality for which an open source solution does not yet exist is either developed by the Delta3D team or one of their users. As stated in the report, "Projects with large user bases are more likely to remain current and state of the art than those with only a small base." Delta3D uses Open Scene Graph (OSG) with Open Dynamics Engine (ODE) as the physics engine. Scripting is accomplished using the Python scripting language in some cases with bindings to connect Python with the C++ Delta3D code. A critical element of Delta3D is the level editor that allows developers to build levels through the graphical user interface (see Figures 37.23 and 37.24).

Since Delta3D is open source and has a knowledgeable user community the costs are relatively low to use the Delta3D engine. Development of content makes use of industry standards and therefore would be comparable with other game engine development. One of the trade-offs for the low cost in the use of Delta3D is that of limited graphic fidelity (see Figure 37.25).

Another example of a VW currently being used is Nexus, which is a VW developed by Engineering and Computer Simulations (ECS) in conjunction with various U.S. Government organizations (National Guard Bureau, U.S. Army, Defense Acquisition University, Department of Homeland Security, and U.S. Joint Organizations). The focus of Nexus is on classroom and virtual field-trip training and exercises. Nexus is moving to a browser-based functionality and focuses on real-world environments. It uses a Joint Knowledge Online (JKO) sign on for access control into specific regions.

FIGURE 37.23 VBS2 image. (From Bohemia Interactive Simulations, *Product*, retrieved June 7, 2013, from Bohemia Interactive Simulations: http://products.bisimulations.com/products/vbs2/overview, 2013.)

FIGURE 37.24 Delta3D level editor (From Navy Post-Graduate School, *Building a world with the Delta3D editor*, available from: http://www.delta3d.org/article.php?story=20050708135927685, viewed July 31, 2013, 2013).

Another example of a platform that is being increasingly used for tight development budgets is Unity. Unity is a cross-platform content viewer. There is a free basic version of Unity available for development. This is useful for basic prototype development, but does not include NavMeshes, path-finding, crowd simulations, Level of Detail (LOD) support, audio filter, video playback and streaming, Inverse Kinematics (IK) rigs, 3D Texture support, real-time shadows, occlusion culling, navmesh, and script access (Unity3D, 2013a, 2013b). Unity Pro, which provides the additional features just described, has a cost to entry of about $5000 per developer seat. It is important to consider that Unity is not strictly considered a game engine because it lacks user management, a game database, inventory, and a physics engine. Adding these elements through a third party expands it into an engine that can be used to support single- and multiuser content (Figures 37.26 and 37.27).

Access to the highest possible graphic quality usually means making use of a first person shooter (FPS) type game engine that has been heavily invested in by industry. These often come with a higher initial licensing cost. FPS engines have been extended through vertical scaling to

FIGURE 37.25   Example of Joint Forward Observer Training Suite–Mobile. (From The MOVES Institute, Naval Postgraduate School, *Joint forward observer training suite-mobile*, 2012.)

FIGURE 37.26   Example of ECS Nexus. (From Engineering and Computer Simulations (ECS), *ECS Orlando Corporate Website*, retrieved June 7, 2013, from Nexus Virtual Worlds: http://www.ecsorl.com/solutions/nexus-virtual-worlds, 2013.)

support large numbers of users with a multiplayer backend that blurs the line between Massively Multi-player Online Games (MMOG) and VWs. These environments, which include the Cry Engine or Unreal Engines, provide advanced development tools. Unreal has a free development kit (UDK) available with release limitations that allow prototyping prior to licensing. Licensing options are often a trade-off with higher cost associated with no end user or developer seat limitations and access to source code. FPS engines are increasingly making their functionality

FIGURE 37.27 Ghost of a Tale Built in Unity. (From Unity3D, *Unity 3D*, retrieved June 7, 2013, from Unity 3D Gallery: http://unity3d.com/gallery, 2013a; Unity3D, *Unity 3D*, retrieved June 7, 2013, from License Comparisons: http://unity3d.com/unity/licenses, 2013b.)

available via browser-based clients and access from multiple platforms. An example of the graphics quality in a FPS game engine is Unreal Engine 4, see Figure 37.28.

There are trade-offs for each development option (buy, build, or modify). Comparing costs of each option is challenging. The development choice should take into consideration cost of licensing as compared to development cost, the ability to maintain currency with industry capabilities and whether the development is a one-time capability or something that is expected to grow and morph

FIGURE 37.28 Screen capture from Epic's Unreal Engine 4 *Infiltrator*. (From Tompkins, B., *1place4tech*, Retrieved June 6, 2013, from Unreal Engine 4 'Infiltrator' & 'Elemental' Demos [Videos]: http://1place4tech. com/2013/03/31/unreal-engine-4-infiltrator-elemental-demos-videos/, March 31, 2013.)

over time (i.e., will source code be needed?). It is important to consider potential license limitations down the road, as some licenses limit the number of user or developer seats. Some licenses may also limit the reuse of 3D assets.

Independent of the strategy used, content is needed to bring life to the environment. Even most proprietary environments are able to consume standard 3D modeling and animation formats. The higher the quality the greater the cost, both in development time and in processing costs. One example of make or buy considerations in this category is best understood when looking at animations. During a development effort, developers consider using downloadable animations at a much lower cost compared to development from scratch. However, looking closely at the license agreement options, the number of potential users and the inability to levy a royalty fee from earnings (because the government is nonprofit), can drive the license cost to an untenable level—and guarantee the animations would not be able to be shared across other government users.

Development of content might include areas such as terrain (elevation), features (buildings, rivers, etc.), characters, animations, props, and vehicles. Not every VE needs all of these elements, but these are common considerations in their development. Each of these areas has the same buy, build, or modify consideration. Terrain generation can be as simple as creating four walls to model a room or as complex as ingesting Digital Terrain Elevation Data (DTED) from the U.S. National Geospatial Intelligence Agency (NGA). Terrain that has been sculpted can then have imagery laid over it or it can be hand-painted. Buildings and structures can be very complex or can be simple blocks with overlaid images. There are many terrain generation tools available on the commercial market as well as government-owned solutions, such as the Rapid Unified Generation of Urban Databases (RUGUD) (Applied Research Associates, Inc., 2013). VWs such as SL have simple terrain morphing tools, and building tools that allow development with little or no training while the current generation game engines require a more rigorous learning curve.

37.13.2 Security

Today, you cannot work in the DoD without receiving daily messages on Information Assurance (IA), perhaps the latest attempt to infiltrate an organization, an alert from the National Security Agency (NSA), Defense Information Systems Agency (DISA), or the U.S. Strategic Command (STRATCOM). The email may be mandating a new IA training requirement, a symposium is being offered, a new pamphlet was developed or flyer for your wall, or a contractor is offering a new workshop or forum. The DoD spends an incalculable amount of funding on IA for travel, conferences, training, labor, and contracts for the development of software, hardware, and information technology support. Army Regulation 25 defines IA as the process for protecting and defending information by ensuring its confidentiality, integrity, and availability (U.S. Army, 2005). Quite often, functionality that is readily available to end users within their homes is not available to government users due to IA lockdowns.

No matter what the BLS or VW being used in a POI, security will be a driver on cost, schedule and performance of the designed and implemented solution. The end user location and the training content will determine the level the BLS environment needs to be secured.

37.13.3 Requirements Definition

The military services' ability to embrace new technology is heavily dependent on their ability to articulate the requirements. Requirements definition challenges and shortcomings in program governance can be attributed to a number of causes that have been repeatedly documented in after action reviews and lessons learned in program management reviews. Some of the causes might be a poor understanding of learner needs, a program's overly schedule-driven approach or DOD's difficulty in overcoming its long-standing cultural resistance to how it trains and learns. Unless these challenges are addressed, there is increased risk that the system will not provide expected capabilities and benefits on time and within budget. This is no different with VWs. Detailing what is to be accomplished, identifying the goals, and using metrics to establish what is required is often accomplished retroactively. Too often technology developers get seduced by requirements creep and the quest for the newest bells and whistles. Often, the use case can be accomplished with a more simple approach. For example, using a VW for a teleconference is much more costly than a simple conference call.

Often the people establishing requirements are experienced in the way things have been done in the past, and may not be intimately familiar with the capabilities that VWs provide. In fact, casual observance has shown that a very small percentage of the instructor and operators of training in the military services have actually spent time exploring a VW capability. It is important that people generating the requirements have the opportunity to *go native* in VWs. This means they need to spend time using and experiencing various VWs to understand how the industry is captivating, engaging, and training users.

Once requirements are set and the environment is established, it is paramount that trainers fully embrace the capabilities of the technology. Research shows that if the instructor is not engaged with the technology, the student will likewise not engage (Armstrong & Yetter-Vassot, 2008; Ertmer, 2005). Instructors who have a full grasp of functionality in the environment can seem almost *magical* in their ability to introduce complex concepts and engage student participation (Boulos, Hetherington, & Wheeler, 2007). Instructors can instantly create complex models, teleport students into locations such as the core of a nuclear reactor or into space to watch the formation of a new star or they can provide real-time exercises and collect and display real-time data.

37.14   SHARED RESOURCES AND SOFTWARE REUSE

Another lessoned learned in the area of VWs is code reuse, also called software reuse, which is the use of existing software, or software knowledge, to build new software (Frakes & Kyo, 2005). One of the most common ideals of software reuse is the potential to have fewer total lines of code written reducing overall development effort. Software reuse can increase productivity, reduce development costs, and minimize schedule overruns. Since reusable software resources should be rigorously tested and verified, reuse also has the potential to improve software quality, reliability, maintainability, and portability. When it comes to a BLS, adopting a model that lets the government focus on achieving its mission objectives through the ability to share assets in order to create content flexibly and cost effective is the ultimate goal. One strategy proving to be valuable is the use of 3D repositories. There are a number of repositories available within the government and work is being done to standardize access across the spectrum of repositories.

One example of a repository is the Milgaming Portal hosted by PEO STRI and the Army Games Studio. Milgaming was launched by the Games For Training (GFT) POR which provides a suite of training software applications installed on PC-based, networked, multiplayer training environment using the COTS/GOTS gaming technology, VBS2, Bilateral Negotiation Trainer (BiLat), Tactical Iraq, Tactical Pashto, and Tactical Dari. GFT enables the army to leverage and influence gaming technologies in order to rapidly deliver relevant training capabilities to support current and future soldier, leader, and collective training and mission rehearsal. GFT has the requirement to develop a web portal to provide comprehensive services to its users and facilitate the management of the program. The top-level requirements for the web portal are tightly integrated and with the authentication and activation of users wanting access.

Another example of a repository is the Office of the Secretary of Defense Advanced Distributed Learning Co-Laboratory (ADL Co-Lab) 3D Repository. The ADL 3D Repository is a website for uploading, finding, and downloading 3D models. Any 3D model may be uploaded, but the system is optimized for certain file types, including .fbx, .dae, .obj, .skp, and .3ds. The 3D Repository provides services for these optimized file types such as extracting polygon count and texture metadata, viewing models in 3D using Flash or O3D plug-ins, and converting models between these file types. One study on initial research with learning repositories was written for the ADL Co-Lab by author Colin Holden suggesting that whatever the exact definition of a repository is, all agree that learning repositories can and should provide access to the increasing supply of digital educational content (Holden, 2003).

37.15 CURRENT TRENDS IN COMMERCIAL VWs

37.15.1 Adoption and Acceptance

VWs have experienced continued growth in recent years. While the rate of additional new users has begun to level off, especially in the 25+ age bracket, there is still growth with the number of users approaching two billion (KZero, 2012a,b). Figure 37.29 shows VW user data from 2012, highlighting the number of users by age groups up in various VWs. Though difficult to see, it is clear that since 2008, the number of users of VWs has significantly increased. A more focused view (Figure 37.30) shows the breakout of adult users. The relative size of the gray dot in the outer most ring shows the number of users in 2012 with L being the largest provider with 31 million registered users. The line projecting out to the gray dot shows the period of growth since 2008 and/or when the specific VW launched (small dot towards center of circle). A data point that is not included in the KZero data is the number of Open Simulator projects started in recent years. As mentioned in the MOSES use case, Open Simulator is a reverse-engineered copy of SL that is freely available to the public with a strong user base and since each user can host their own free and private VW, there is no known database that tracks the number of total users.

Gartner, a technology strategy consultant, describes contextual and social user experiences, such as those in VWs as number three in the top 10 strategic technology trends for 2012 (see Figure 37.31; Gartner, Inc., 2012). Gartner has also developed the concept of the *Gartner Hype Cycle*' (Fenn & Raskino, 2008). The Gartner Hype Cycle graphically describes the predictable path that emerging technologies experience from initial public enthusiasm through disillusionment and ultimately, hopefully, to the plateau of productivity. Clearly not all technology begins and ends in a predictable pattern; however, the Gartner Hype Cycle can provide insight into expected patterns. The five phases of the hype cycle are as follows:

- Technology Trigger: This occurs when the media or word of mouth triggers public interest in a technology breakthrough, though the products themselves may not even yet exist.
- Peak of Inflated Expectation: Early publicity increases expectations with early adopters engaging the technology.

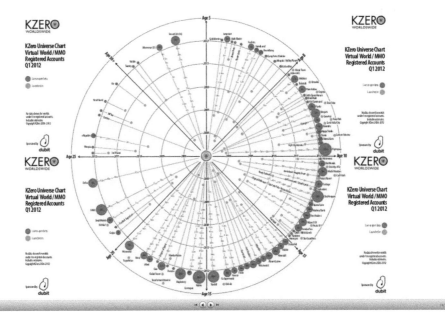

FIGURE 37.29 Virtual world total registered user accounts as of 2012. (From KZero, *Slideshare for Q1 2012 Universe Chart*, available from: http://www.kzero.co.uk/blog/slideshare-q1-2012-universe-chart/, last viewed July 31, 2013, 2012a; KZero, *Virtual World total registered accounts*, 2012b.)

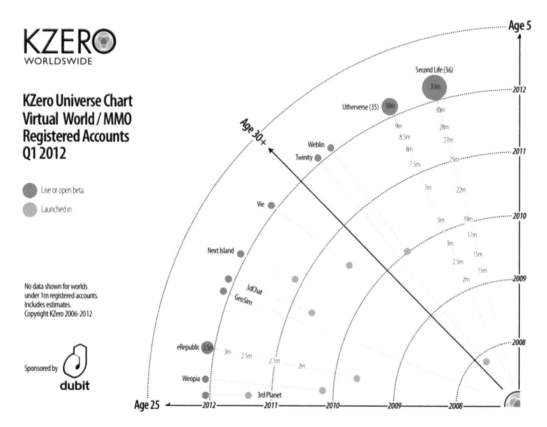

FIGURE 37.30 Detailed breakout of Virtual World/MMO registered accounts for adults. (From KZero, *Slideshare for Q1 2012 Universe Chart*, available from: http://www.kzero.co.uk/blog/slideshare-q1-2012-universe-chart/, last viewed July 31, 2013, 2012a; KZero, *Virtual World total registered accounts*, 2012b.)

Top 10 strategic technology trends for 2012	
Human experience	1. Media tablets and beyond
	2. Mobile-centric applications and interfaces
	3. Contextual and social user experience
Business experience	4. Internet of things
	5. App stores and marketplaces
	6. Next-generation analytics
IT dept. experience	7. Big data
	8. In-memory computing
	9. Extreme low-energy servers
	10. Cloud computing

FIGURE 37.31 How technology trends affect the human, business, and IT experiences. (From Gartner, Inc., *Top 10 technology trends for 2012*, retrieved October 15, 2012, from Gartner.com: http://www.gartner.com/technology/research/top-10-technology-trends/, 2012.)

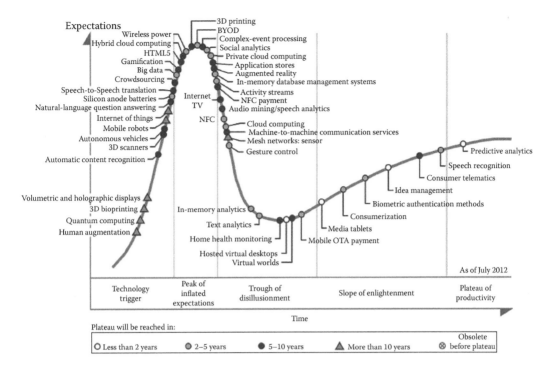

FIGURE 37.32 2012 Gartner Hype Cycle. (From LeHong, H. and Fenn, J., Key trends to watch in gartner 2012 emerging technologies hype cycle, *Forbes*, available from http://www.forbes.com/sites/gartnergroup/2012/09/18/key-trends-to-watch-in-gartner-2012-emerging-technologies-hype-cycle-2/, viewed July 31, 2013, September 18, 2012.)

- Trough of Disillusionment: Public interest wanes as upshoots fail and experiments do not deliver.
- Slope of Enlightenment: Technology matures and benefits become clearer. Second and third generation products appear from a larger base of providers. There is added corporate buy-in but the more conservative companies remain cautious.
- Plateau of Productivity: This is when mainstream adoption is expected. It becomes clearer how to assess provider viability and broad market applicability becomes apparent.

Figure 37.32 shows the Gartner Hype Cycle from July of 2012. The vertical axis shows the relative visibility of the technology while the horizontal axis shows the relative maturity (Fenn & Raskino, 2008). It is clear that VWs are in the trough of disillusionment with an expected 5–10 years before mainstream adoption (Gartner, Inc., 2012). Gartner stresses that "each technology is included because of particularly high levels of hype or because Gartner believes they have the potential for significant impact" (LeHong & Fenn, 2012).

37.15.2 Barriers to Adoption

Barriers vary depending on the user. For example, corporate or government barriers have traditionally been the unusual network requirements of VW technology. IA limits the ability of government or corporate end users to load unique software on their working computers. Those same firewalls may limit message packets or information flowing through *nonstandard* ports, even if the software is loaded.

There are also limitations on the types of information that can be passed through the Internet. Classified, proprietary, or sensitive training material must be thoroughly protected. This is easier to accomplish on secure networks; however, as mentioned earlier, this means that the software must go through a rigorous information assurance review. To date, the number of these environments that have received approval to be used in a distributed manner on government information systems is extremely limited.

Additionally the perceptions, cultural changes and the digital divide also pose challenges to the adoption of new technology being used in a BLS. As of only a few years ago, the use of the word *game* when discussing learning and training was not acceptable. Since then, the commercial market has repeatedly demonstrated the benefits of engaging learners with well designed serious games (Guillen-Nieto & Aleson-Carbonell, 2012). Despite this, there has been a slow adoption rate in both the U.S. government and the corporate world to use VW and game-based learning opportunities. We are still plagued with *death by PowerPoint* for mandatory training. Often, the limitation is not technology but the marriage of instructional designers and game developers who are able to establish engaging learning material with effective scoring mechanisms. There may also be concerns related to people's tendency to *play* in game-based or VW environments and the fear that play does not lead to learning. While play could function as a distraction away from the learning material, well-designed training material takes that play into consideration and uses it to propel the learner forward.

As an example, imagine that you are assigned an immersive VW experience as part of your yearly mandatory prevention of sexual harassment training. Imagine that you are assigned an attractive avatar and that you are provided tasks to do while experiencing jeers and inappropriate comments from your supervisor or peers. This may feel a bit uncomfortable for those who are not used to engaging with an environment through an avatar. However, research shows (Bailenson, 2012) that virtually experiencing the event and making the right choices will influence the user's decision making to a greater extent and over a longer period of time than simply reading about the experience. Further, by placing a user *into the shoes* of another—in this case the recipient of sexual harassment—the user will have a stronger sense of what it feels like to experience inappropriate actions. This was additionally demonstrated through a study (Hershfield et al., 2011) where users were placed in the shoes of a future self, seeing themselves as age-progressed and needing to make better choices for their retirement. The study suggested that having the personal experience of *being older* opened the user's minds and imagination and influenced the users to save more for retirement. Based on this, and similar research, it is not hard to imagine experiencing virtual life with a handicap or to what it might be like to experience racial or gender inequality. The expected outcome is users who are more sensitive to others plights may be more willing to make social change or accommodate others differences.

Another potential barrier to overcome with this technology could be the user interface that can be a distraction or can produce negative training. The means by which the user interacts with the environment is a key design factor with any POI. No one user interface can satisfy all needs and not all trainees are proficient in the use of computers and interface devices. VWs need to support a variety of interface strategies, from the traditional keyboard and mouse to more immersive strategies such as the Brain Computer Interface and Microsoft's Kinect™ as the training need requires. Recently, low-cost COTS interfaces are packaged with software development kits (SDKs) that make it straightforward even for nonprogrammers to link devices. Eye-tracking, facial recognition systems, and brain computer interfaces are making it possible to know what trainees are looking at or what the emotional affect is of the training material. This adds a whole new dimension to training experiences. These interfaces give us insight into subtleties in the user experience that will allow us to individualize training, not just too individual learning preferences, but also based on instantaneous reactions to the material. The ability to bring natural gestures into the environment using motion tracking helps improve social cues and body language limitations that are currently major barriers to natural interactions in VWs. The expectation is that in the near future we will see natural human expressions conveyed into the VW to enhance realism.

37.16   CONCLUSION

VWs, where civilians, soldiers, airman, sailors, and marines can work and interact in a somewhat realistic manner, have great potential to increase the ability of a BL approach to the military's ability to teach and instruct. Although VWs may still be considered games by some, they are rapidly becoming a significant tool where people interact, shop, sell, and make their living. While many reports espouse the potential impact that VWs may have on teaching and learning in the future, there are few empirical studies that inform today's military of the ROI for launching off in this area. The authors recommend additional research in this area.

This chapter explored some aspects and case studies of the military today and how they are using VWs in a BLS. Various business models, design considerations, and research areas were described for consideration. Well-defined requirements and the ability to leverage resource repositories were described as best practices in the development of VWs for learning. Current industry trends and barriers to adoption are considered as they relate to military use cases. Expanding the repertoire of BLS with enhanced presence in 3D environments may significantly promote collaboration and engage learners in everything from recruiting to experiencing the operational environment in which missions will be accomplished. The challenge to overcome the barrier of adoption revolves around cultural acceptance, costs, accessibility, and licensing fees. While security and IA will likely be the primary force holding the military back from taking advantage of this technology.

BIBLIOGRAPHY

Abramsky, S., Ghica, D., Murawski, A., & Ong, C. (2004). *Applying game semantics to compositional software modeling and verification.* Oxford, U.K.: Oxford University Computing Laboratory.

Air Education and Training Command. (2008). *On learning: The future of air force education and training.* Randolph Air Force Base, TX: U.S. Air Force.

Air Force Air University. (2011). *Horizons in learning innovations.* Maxwell AFB, AL: Air University.

Allison, C., Miller, A., Sturgeon, T., Perera, I., & McCaffrey, J. (2011, October). The third dimension in open learning. In *Frontiers in Education Conference (FIE), 2011* (pp. T2E-1). IEEE.

Ally, E. M. (2009). *Mobile learning: Transforming the delivery of education and training.* Edmonton, Canada: AU Press, Athabasca University.

Aplanalp, J. F. (2010). Linking blended learning to navy fleet performance requirements, REAL-TIME. In *Interservice/Industry Training, Simulation and Education Conference (I/ISEC)* (p. 10073). Vol. 2010. No.1, Orlando, FL: NTSA.

Applied Research Associates, Inc. (2013). *ARA—Expanding the realm of possibility.* Retrieved June 6, 2013, from Showcase Project—Rapid Unified Generation of Urban Databases (RUGUD): http://www.ara.com/Projects/p_RapidUnifiedGUD.htm

Armstrong, K. M., & Yetter-Vassot, C. (2008). Transforming teaching through technology. *Foreign Language Annals, 27*(4), 475–486.

Au, W. J. (2008). *The making of second life: Notes from the new world.* New York, NY: HarperCollins.

Bailenson, J. N. (2012). Doppelgangers: A new form of self. *The Psychologist, 25*(1), 36–39.

Bartle, R. A. (2004). *Designing virtual worlds.* Berkeley, CA: New Riders.

Berlin Wall in Twinity. (From Twinity, The Berlin wall in twinity, retrieved October 12, 2012, from Twinity.com: November 9, 2009.)

Biron, L. (2013, April 1). Army picks virtual challenge finalists. *Defense News.* http://www.defensenews.com/article/20130401/TSJ01/304010012/Army-Picks-Virtual-Challenge-Finalists

Blascovich, J., Loomis, J., Beall, A., Swinth, K., Hoyt, C., & Bailenson, J. N. (2002). Immersive virtual environment technology as a methodological tool for social psychology. *Psychological Inquiry, 13*, 103–124.

Bohemia Interactive Simulations. (2013). High fidelity simulation affordable COTS technology. *Product.* Retrieved June 7, 2013, from Bohemia Interactive Simulations: http://products.bisimulations.com/products/vbs2/overview

Boulos, M. N., Hetherington, L., & Wheeler, S. (2007). Second Life: An overview of the potential of 3-D virtual worlds in medical and health education. *Health Information and Libraries Journal, 24*(4), 233–245.

Card, O. S. (1991). *Ender's game.* New York, NY: Tom Doherty Associates, LLC.

Cassidy, M. (2007, May 15). *Flying with disability in second life.* Retrieved October 14, 2012, from Eureka Street: http://www.eurekastreet.com.au/article.aspx?aeid=2787

Chan, T. S., & Ahern, T. C. (1999). The importance of motivation: Integrating flow theory into instructional design. In *Society for Information Technology and Teacher Education International Conference* (pp. 780–782). Vol. 1999. No.1, Chesapeake, VA: AACE.

Claypole, M. (2010). *Controversies in ELT. What you always wanted to know about teaching English but were afraid to ask.* Norderstedt, Germany: LunguaBooks.

Cohn, J., & Bolton, A. (2009). Optimising virtual training systems. *Theoretical Issues in Ergonomics Science, 10*(3), 187–188.

Command, T. D. (2012). PAM 525-8-2 Army Learning Concept. Army.

Csikszentmihalyi, M. (1998). *Finding flow: The psychology of engagement with everyday life.* New York, NY: Basic Books.

Csikszentmihalyi, M. (2002). *Flow: The classic work on how to achieve happiness.* London, England: Rider and Company.

Dalgarno, B., & Lee, M. (2010). What are the learning affordances of 3-D virtual environments. *British Journal of Educational Technology, 41*(1), 10–32.

Dao, J., & Lehren, A. W. (2012, August 21). In toll of 2,000, new portrait of Afghan war. *The New York Times.* http://www.nytimes.com/2012/08/22/us/war-in-afghanistan-claims-2000th-american-life.html?_r=0

de Freitas, S., & Veletsianos, G. (2010). Crossing boundaries: Learning and teaching in virtual worlds. *British Journal of Educational Technology, 41*(1), 3–9.

Department of the Army. (2011, January 20). *The U.S. Army learning concept for 2015* (Training and Doctrine Command (TRADOC) Pamphlet 525-8-2). Fort Monroe, VA: U.S. Army Training and Doctrine Command.

Dwyer, T., Griffith, T., & Maxwell, D. (2011). Rapid simulation development using a game engine-enhanced dynamic geo-social environment (EDGE). *Interservice/Industry Training, Simulation, and Education Conference*, Orlando, FL.

Engineering and Computer Simulations (ECS). (2013). *ECS Orlando Corporate Website.* Retrieved June 7, 2013, from Nexus Virtual Worlds: http://www.ecsorl.com/solutions/nexus-virtual-worlds

Ertmer, P. A. (2005). Teacher pedagogical beliefs: The final frontier in our quest for technology integration? *Educational Technology Research and Development, 53*(4), 25–39.

Evans, A. W. (2010). *Learning for the next: Predicting the usage of synthetic learning environments.* Orlando, FL: University of Central Florida.

Fenn, J., & Raskino, M. (2008). *Mastering the hype cycle: How to choose the right innovation at the right time.* Boston, MA: Gartner, Inc.

Fletcher, J., & Chatham, R. (2009). *Measuring return on investment in training and human performance.* Santa Barbara, CA: Praeger.

Fominykh, M. (2009). *Learning in technology-rich environments: Second life vs. moodle.* Retrieved June 05, 2013, from Academia.Edu: http://www.academia.edu/472272/Learning_in_Technology-Rich_Environments_Second_Life_vs._Moodle

Fong, K. N., Chow, K. Y., Chan, B. C., Lam, K. C., Lee, J. C., Li, T. H., … Wong, A. T. (2010). Usability of a virtual reality environment simulating an automated teller machine for assessing and training persons with acquired brain injury. *Journal of Neuroengineering and Rehabilitation, 7*, 19.

Frakes, W. B., & Kyo, K. (2005). Software reuse research status and future. *IEEE Transactions on Software Engineering, 31*(7), 529–536.

Freedman, S. A., Hoffman, H. G., Garcia-Palacios, G., Weiss, P. L., Avitzour, S., & Josman, N. (2010). Prolonged exposure and virtual reality-enhanced imaginal exposure for PTSD following a terrorist bulldozer attack: A case study. *Cyberpsychology, Behavior, and Social Networking, 13*, 95–101.

Gartner, Inc. (2011, August 10). Gartner's 2011 Hype cycle special report evaluates the maturity of 1,900 technologies. *Press Release.* Stamford, CT.

Gartner, Inc. (2012, October 15). *Top 10 technology trends for 2012.* Retrieved October 15, 2012, from Gartner. com: http://www.gartner.com/technology/research/top-10-technology-trends/

Gates, B., Myhrvold, N., & Rinearson, P. (1996). *The road ahead.* New York, NY: Penguin Books.

Gorini, A., Gaggioli, A., Vigna, C., & Riva, G. (2008). A second life for eHealth: Prospects for the use of 3-D virtual worlds in clinical psychology. *Journal of Medical Internet Research, 10*(3), e21.

Greene, J. (2011, July 14). U.S. Video game industry sales continue slide. *cnet News.* Retrieved August 7, 2011, from U.S. video game industry sales continue to slide: http://news.cnet.com/8301–10797_3–20079621–235/u.s-video-game-industry-sales-continue-slide/

Guillen-Nieto, V., & Aleson-Carbonell, M. (2012). Serious games and learning effectiveness: The case of its a deal! *Computers and Education, 58*(1), 435–448.

Hans, W. (n.d.). *Polzer paramax systems corporation.* Retrieved June 15, 2012, from Government Ownership of Software Boon or bane?: http://webcache.googleusercontent.com/search?q=cache:5P4X-F1bGqwJ:citeseerx.ist.psu.edu/viewdoc/download%3Fdoi%3D10.1.1.54.3714%26rep%3Drep1%26typ e%3Dps+army+software+reuse+why+contractor+do+not+like+it&cd=6&hl=en&ct=clnk&gl=us

Hershfield, H. E., Golodstein, D. G., Sharpe, W. F., Fox, J., Yeykelis, L., Carstensen, L. L., & Bailenson, J. N. (2011). Increasing saving behavior through age-progressed renderings of the future self. *Journal of Marketing Research, 48*, S23–S37.

Holden, C. (2003). *Learning repositories summit: Initial research summary.* Orlando, FL: ADLCo-Lab.

Howard, J. (2010). *Virtual worlds offer real opportunities to learn.* Peoria, IL: Peoria Magazines.

Jackson, S., & Eklund, R. (2002). Assessing flow in physical activity: The Flow State Scale-2 and Dispositional Flow Scale-2. *Journal of Sport & Exercise Psychology, 24*(2), 133–150.

Jackson, S., & Marsh, H. (1996). Development and validation of a scale to measure optimal experience: The flow state scale. *Journal of Sport and Exercise Psychology, 18*(1), 17–35.

Jobs, S. (2006, June 12). Stanford University commencement address. Stanford, CA: Stanford University.

Jones, M. G. (1998). Creating electronic learning environments: Games, flow, and the user interface. *Selected Research and Development Presentations at the National Convention of the Association for Educational Communications and Technology (AECT) Sponsored by the Research and Theory Division.* Vol. 1998. No. 1.

Kang, K., & Roland, R. J. (2007). Military simulations. In J. Banks (Ed.), *Handbook of simulation: Principles, methodology, advances, applications, and practice* (chap. 19). Hoboken, NJ: John Wiley & Sons, Inc.

Kim, S. H., Lee, J., & Thomas, M. K. (2012). Between purpose and method: A review of educational research on 3D virtual worlds. *Journal of Virtual Worlds Research, 5*(1), 1–18.

King, W. (2009, February 26). *HomePage of U.S. army.* Retrieved June 4, 2013, from Virtual Battle Space 2 Army gaming system debuts: http://www.army.mil/article/17502/

Kirriemuir, J. (2009). *Virtual world watch.* Retrieved September 5, 2012, from Silversprite: http://www. silversprite.com/?page_id=353

Klinger, M. B., & Coffman, T. L. (2010). Emphasizing diversity through 3D multi-user virtual worlds. In G. Kurubacak & T. V. Yuzer (Eds.), *Handbook of research on transformative online education and liberation: Models for social equality* (pp. 86–106).

KZero. (2008). *Virtual world competition and Second Life's secret weapon.* London, U.K.: British Journal of Educational Technology. Retrieved September 29, 2012, from http://www.kzero.co.uk/blog/ virtual-world-competition-and-second-lifes-secret-weapon/

KZero. (2012a). *Slideshare for Q1 2012 Universe Chart.* Available from http://www.kzero.co.uk/blog/ virtual-world-registered-accounts-reach-1-7bn-q4-2011/

KZero. (2012b). Virtual World total registered accounts.

Laboratory, A. R. (2013, March 22). *Army research laboratory website.* Retrieved March 23, 2013, from U.S. Army and Intel team up making training virtually a reality: http://www.arl.army.mil/www/default. cfm?page=1779

Lee, S., Ahn, S. C., Kim, H., & Lim, M. (2006). Real-time 3D video avatar in mixed reality: An implementation for immersive telecommunication. *Simulation & Gaming, 37*, 491–506.

LeHong, H., & Fenn, J. (2012, September 18). Key trends to watch in gartner 2012 emerging technologies hype cycle. *Forbes.* Available from http://www.forbes.com/sites/gartnergroup/2012/09/18/key-trends-to-watch-in-gartner-2012-emerging-technologies-hype-cycle-2/, viewed July 31, 2013.

Lewin, C. G. (2012). *War games and their history.* Stroud, England: Fonthill Media.

Lockheed Martin. (2012). *Virtual world framework.* Retrieved October 14, 2012, from http:// virtualworldframework.com/

Martin, G. A. (2000). *Virtual environment technology laboratory research testbed: Project report no. 8 year-two.* Orlando, FL: Institute for Simulation and Training & University of Houston.

McDowell, P., Darken, R., Sullivan, J., & Johnson, E. (2006). Delta3D: A complete open source game and simulation engine for building military training systems. *JDMS—The Society for Modeling and Simulation International, 3*, 143–154.

Merrouche, S. (2011). Learning through MOOing. *Procedia—Social and Behavioral Sciences, 29*, 941–946.

Miller, C. A., Sturgeon, A., Perera, I., & McCaffrey, J. (2011, October 12–15). The third dimension in open learning. *41st ASEE/IEEE Frontiers in Education Conference*, Rapid City, SD.

Nakamura, J., & Csikszentmihalyi, M. (2009). Flow theory and research. In C. R. Shyder & S. J. Lopez (Eds.), *Oxford handbook of positive psychology* (pp. 195–200). New York, NY: Oxford University Press.

Navy Post-Graduate School. (2013). *Building a world with the Delta3D editor.* Available from http://www. delta3d.org/article.php?story=20050708135927685, viewed July 31, 2013.

Nevo, S., Nevo, D., & Carmel, E. (2011). *Unlocking the business potential of virtual worlds.* Boston, MA: MIT Sloan Management Review.

New Media Consortium. (2012a). *Horizon report—2012 higher education edition.* Austin, TX: New Media Consortium.

New Media Consortium. (2012b). *Horizon report—2012 K-12 edition.* Austin, TX: The New Media Consortium.

Olbrish, K. (2009, May 17). *Virtual worlds, simulations and serious games, oh my*! Retrieved September 10, 2012, from learningintandem.blogspot.com/2009/05/virtual-worlds-simulations-and-serious.html

Oliver, M., & Trigwell, K. (2005). Can "Blended Learning" be redeemed? *E-Learning, 2*(1), 17–26.

OneSource, A. (2011). *My Army OneSource.* Retrieved October 1, 2012, from Army OneSource: http://www. myarmyonesource.com/

Peachey, A., Gillen, J., Livingstone, D., & Robbins, S. (2010). *Researching learning in virtual worlds.* London, U.K.: Springer.

Pearce, C. (2009). *Communities of play: Emergent cultures in multiplayer games and virtual worlds.* Cambridge, MA: MIT Press.

Pearce, J. (2005). Engaging the learner: How can the flow experience support e-learning? In *World Conference on E-Learning in Corporate, Government, Healthcare, and Higher Education* (pp. 2288–2295). Vol. 2005. No.1.

Price, M., Mehta, N., Tone, E. B., & Anderson, P. L. (2011). Does engagement with exposure yield better outcomes? Components of presence as a predictor of treatment response for virtual reality exposure therapy for social phobia. *Journal for Anxiety Disorders, 25*(6), 763–770.

Public Broadcasting Service (PBS). (2010). *Frontline—Digital nation: Life on the virtual frontier.* http://www. pbs.org/wgbh/pages/frontline/digitalnation/us/

Robson, S. (2009, January 6). Army paying $17.7M for training game. *Stars and stripes.* Available from http:// www.stripes.com/news/army-paying-17-7m-for-training-game-1.86770, viewed July 31, 2013.

Rogers, E.M. (2003). *Diffusion of Innovation*, New York: Free Press.

Rothbaum, B. O., Garcia Palacios, A., & Rothbaum, A. (2012). Treating anxiety disorders with virtual reality exposure therapy. *Revista de Psiquiatria y Salud Mental, 5*, 67–70.

Sharma, P. (2010). Blended learning. *ELT Journal, 64*(4), 456–458.

St. George's University of London. (2009). St. George's e-learning project wins outstanding ICT initiative of the year 2009. *News archive.* Retrieved June 22, 2013, from St. George's e-learning project wins Outstanding ICT Initiative of the Year 2009: http://www.sgul.ac.uk/media/news-archive/2009/ st-george2019s-e-learning-project-wins-outstanding-ict-initiative-of-the-year-2009

Stendal, K. (2012). How do people with disability use and experience virtual worlds and ICT: A literature review. *Journal of Virtual World Research, 5*(1), 1–17.

Stricker, A., McCrocklin, M., Holm, J., & Calongne, C. (2010, February 8). *Mars expedition strategy challenge.* Retrieved June 18, 2013, from http://marsexpeditionstrategy.blogspot.com/

Tenkely, K. (2013). *Virtual learning world for elementary students.* Retrieved June 26, 2013, from Teaching Community: http://teaching.monster.com/training/articles/10191-virtual-learning-world-for-elementary-students

The MOVES Institute, Naval Postgraduate School. (2012). *Joint forward observer training suite-mobile.*

Thompson, C. W. (2011). Virtual world architectures. *IEEE Internet Computing, 15*, 11–14.

Tompkins, B. (2013, March 31). *1place4tech.* Retrieved June 6, 2013, from Unreal Engine 4 'Infiltrator' & 'Elemental' Demos [Videos]: http://1place4tech.com/2013/03/31/unreal-engine-4-infiltrator-elemental-demos-videos/

Tschang, F. T., & Comas, J. (2010). Developing virtual worlds: The interplay of design, communities and rationality. *First Monday, 15*(5).

Turk, M., Bailenson, J. N., Beall, A. C., Blascovich, J., & Guidagno, R. (2004). Multimodal transformed social interaction. *Proceedings of the ACM Sixth International Conference on Multimodal Interfaces (ICMI)*, State College, PA.

Twinity. (2009, November 9). *The Berlin wall in Twinity.* Retrieved October 12, 2012, from Twinity.com: http:// www.twinity.com/en/community/berlin-wall

U.S. Air Force Academy. (2013, March 3). *U.S. Air Force academy (admissions).* Retrieved June 10, 2013, from Facebook: https://www.facebook.com/AcademyAdmissions/wall

U.S. Army. (2005). *Army knowledge management and information technology.* Washington, DC: Headquarters Department of the Army.

U.S. Army Research Lab, Simulation and Training Technology Center. (2013, April 13). *Federal virtual challenge*. Retrieved June 2, 2013, from http://fvc.army.mil

U.S. Navy. (2001). *Revolution in training: Executive review of navy training.* U.S. Navy. Washington, DC.

U.S. Navy. (2009). *Computer based training.* Navy Inspector General Report to the Secretary of the Navy.

U.S. Navy. (2010). *Naval training systems requirements, acquisition, and management.* Department of Navy, Office of the Chief of Naval Operations Instruction 1500.76B.

Unity3D. (2013a). *Unity 3D.* Retrieved June 7, 2013, from Unity 3D Gallery: http://unity3d.com/gallery

Unity3D. (2013b). *Unity 3D.* Retrieved June 7, 2013, from License Comparisons: http://unity3d.com/unity/licenses

University of North Carolina at Pembroke. (2010). *Hospital for nursing.* Retrieved June 22, 2013, from UNCP in Second Life: Hospital for Nursing—Professional education and clinicals in the virtual world: http://www.uncp.edu/home/acurtis/NewMedia/SecondLife/HospitalForNursing.html

Valdes-Dapena, P. (2006, November 18). Real cars drive into Second Life. *CNN.com.*

Ward, T., & Sonneborn, M. (2009). Creative expression in virtual worlds: Imitation, imagination, and individualized collaboration. *Psychology of Aesthetics, Creativity, and the Arts, 3*(4), 211–221.

Wankel, C., & Kingsley, J. (Eds.). (2009). *Higher education in virtual worlds: Teaching and learning in Second Life.* Emerald Group Publishing.

Weirauch, C. (2011, May). Fidelity and flexibility. *MS&T Magazine*, pp. 10–12.

Zou, J. J. (2011, June 28). *U.S. Air Force Academy recruits caets with a dose of virtual reality.* Retrieved October 12, 2012, from Wired Campus: http://chronicle.com/blogs/wiredcampus/u-s-air-force-academy-recruits-cadets-with-a-dose-of-virtual-reality/32015

38 Team Training in Virtual Environments

A Dual Approach

Tripp Driskell, Eduardo Salas, and William B. Vessey

CONTENTS

In a previous review, Salas, Oser, Cannon-Bowers, and Daskarolis (2002, p. 873) concluded that "As technological capabilities and the complexity of tasks and environments have broadened, likewise, the need for effective synchronicity and coordination of activities among members of crews, groups, teams, and collectives has increased." Over a decade later, this statement still reads true and undoubtedly will be a decade from now. In order to keep up with changes in technology and task environments, organizations have placed a heavy reliance on teams. Several years prior to the first edition of this chapter, Devine, Clayton, Philips, Dunford, and Melner (1999) surveyed US organizations and found that almost half of these organizations leveraged teams to execute organizational outcomes. In a more recent survey of 185 professionals representing 185 organizations, DiazGranados et al. (2008) found that 94% of professionals stated that their organizations employed teams to achieve organizational outcomes. Additionally, there has been an increase in the use of virtual teams—"a team or group whose members are mediated by time, distance, or technology" (Driskell, Radtke, & Salas, 2003, p. 297). Due to the value of teams, considerable resources (e.g., time, personnel, and money) have been allocated to enhancing their effectiveness through training. Much has been learned over the past several decades about the science of training, leading researchers to conclude that training "produces clear benefits for individuals and teams, organizations, and society" (cf. Aguinis & Kraiger, 2009, p. 452; Salas & Cannon-Bowers, 2001).

Although technology is partly responsible for a rapidly changing workplace (Sitzmann, Ely, Bell, & Bauer, 2010), it is also responsible for advancements in training technologies. Virtual environments (VEs) are an example of such a technology that has shown considerable potential utility for team training. One of the major advantages of VEs is its ability to accommodate multiple trainee configurations from individual to team training using both real and/or virtual trainees (Stiles et al., 1996; van Digglelen, Muller, & van der Bosch, 2010). Moreover, technology-based team training represents a cost-effective and flexible means of training delivery (Mathieu, Maynard, Rapp, & Gilson, 2008). This is significant as US organizations spend in excess of $100 billion per year on employee training and development (Green & McGill, 2011).

The appeal of utilizing VEs for training is alluring (e.g., the promise of VE technology is to deliver better training to more people in a more cost-effective manner); however, as with any training intervention, the science of training should be followed. In other words, organizations should not assume that VEs would be successful in and of themselves for enhancing core competencies (i.e., knowledge, skills, and attitudes [KSAs]). A robust theoretical framework for supporting learning within VEs is required (Salas, Oser et al., 2002). A strong theory-based framework will increase the chances that team training in VEs facilitates learning and positive training transfer to the job (Grossman & Salas, 2011; Salas & Cannon-Bowers, 1997).

This chapter adopts the following outline. First, we present a whirlwind tour of team training in VEs covering the past decade. Second, relevant team, training, and technology concepts are defined. Third, a focused review of team training in VEs is provided. This review will enable the reader to develop an understanding of how the VE team training literature has evolved and where future steps need to be taken. Fourth, the chapter outlines the components that comprise an effective learning environment. These components can be used to provide guidance when designing and implementing team training in VEs. And fifth, we advance a dual approach to team training in VEs, specifically focusing on demonstration-based training (DBT; Rosen et al., 2010) and an event-based approach to training (EBAT; Fowlkes & Burke, 2005a). Each approach is reviewed and then discussed in relation to VE team training. Specifically, we provide insight into how DBT and EBAT approaches can be implemented into VEs and propose guidelines for the use of VEs for team training.

38.1   OVERVIEW

The design, implementation, and applications of VEs have been extensively discussed through this book. As a means to deliver training, VEs are unique in several respects (cf. Heinrichs, Youngblood, Harter, & Dev, 2008). First, VEs allow training to be conducted in layouts and locations that directly replicate the operational environment. Second, they afford the ability to suspend trainee disbelief through immersion in realistic VEs. Third, VE architecture can integrate a diversity of training situations and conditions, including those considered too hazardous for physical training. Fourth, VEs can collect, assess, and provide immediate feedback on process and performance outcomes. And lastly, VEs are not constrained by the drawbacks of traditional training (i.e., training teams together, in one location, at the same time). It is also valuable to identify what is unique about *team training in VEs* above and beyond the features noted earlier. Perhaps most significant for team training is the capability of VEs to accurately model the intricacies of teamwork found in real-world settings.

38.2   WHIRLWIND REVIEW OF THE PAST DECADE

In recent decades, advancements in the science of teams and training, in addition to technological advancements, have changed the landscape of how teams are trained. Coupled with the recent push to develop safe, affordable, adaptable, and widely distributed team training, it is unsurprising that after over a decade, the field is almost unrecognizable. VEs are currently being used to train teams in a myriad of domains including business, health care, education, aviation, law enforcement, and the military. Although the use of VEs for training is widespread, they have recently been most widely adopted for team training purposes in the military and health-care domains (cf. Bonk & Dennen, 2005).

In part due to technological constraints, past applications of VE team training have necessarily focused primarily on small team performance. However, with the arrival of *Simulator Networking* (*SIMNET*)—a Defense Advanced Research Projects Agency (DARPA)-sponsored large-scale, real-time, human-in-the-loop system for team training—and the later Serious Games Initiative in 2002, VE team training interventions could be delivered to teams both small (e.g., 3-person teams) and large (100+ person teams). Moreover, team training could be delivered to both collocated and distributed teams. Additionally, team training is currently not limited to government-run organizations, which traditionally had a monopoly on the field. Now, VE team training programs are readily available in over-the-counter and online formats.

Although over the past decade, VE team training has grown by leaps and bounds, questions regarding the efficacy of these interventions remain. Specifically, does VE team training work? And if so, why? In answering the first question, the extant research suggests that it does, in fact, work. The general empirical evidence regarding the efficacy of VEs has been demonstrated in varying domains. Despite the beneficial effects of training in VEs, a definitive answer for *why* it works remains elusive (Coultas, Grossman, & Salas, 2012). Explanations generally are grounded in the empirical link between motivation, engagement, play, and flow and performance outcomes (see Csikszentmihalyi, 2000; Garris, Ahlers, & Driskell, 2002; Pavlas, Jentsch, Salas, Fiore, & Sims, 2012; Ryan & Deci, 2000a).

In reviewing the literature spanning the past decade, we have noticed several trends that are interesting to point out. First is the general shift from real-world training to virtual training. This shift is driven in part by the desire to keep up with advancements in technology. However, it is also in part due to an attempt to cut costs and provide training to a global audience. The second trend that emerges from the literature is the shift within VEs from individual to team training. This is undoubtedly a result of the importance of teams in all facets of society (e.g., health care, business, military). Despite these recent trends, we caution against the blind use of VEs for team training without the necessary application of the growing body of science behind teams and training.

38.3 ASPECTS OF TEAM TRAINING

38.3.1 WHAT MAKES UP A TEAM (AND TEAMWORK)?

In order to discuss team training in VE, it is necessary to first define several key terms. In this section, we will define teams and teamwork, as well as identify several of their key characteristics.

At a conceptual level, a *team* can be defined as two or more individuals working together toward a common goal. Although this definition likely fits the layperson's notion of a team, research dictates that a more detailed definition be established for operational purposes. As is commonplace in scientific literature, a myriad of definitions for *team* exist. In this chapter, we define a team as "a distinguishable set of two or more people who interact, dynamically, interdependently, and adaptively toward a common and valued goal/objective/mission, who have been assigned specific roles or functions to perform, and who have a limited life-span of membership" (Salas et al., 1992, p. 4). An important distinction within this definition is the concept of interdependency, which has been defined as conditions where "each person's actions have an impact on others' outcomes and that individuals are more dependent to the extent that they cannot unilaterally guarantee themselves good outcomes" (Hollingshead, 2001, p. 1081). In brief, interdependency means that a task cannot be accomplished without the input of each team member.

Salas et al. (2002) state that "teams and the individual members that make up teams are characterized by the level of competencies that they have, relative to what would be required to successfully complete a task or a goal" (p. 874). The competencies that characterize teams fall into three categories: *knowledge*, *skills*, and *attitudes*. *Knowledge* represents an understanding of the core concepts, facts, and principles underlying effective team performance (Cannon-Bowers, Tannenbaum, Salas, & Volpe, 1995). *Skills* represent the behavioral and cognitive capabilities required to successfully perform a team task (Cannon-Bowers & Salas, 1997). Lastly, *attitudes* reflect an affective component in reference to team members' attitudes about themselves and the team (Cannon-Bowers et al., 1995).

According to Bowers, Braun, and Morgan (1997), team performance consists of three facets: teamwork, taskwork, and the requirement to time-share taskwork and teamwork. Marks, Mathieu, and Zaccaro (2001) make a critical distinction between taskwork and teamwork where the former describes what teams are doing and the latter describes how they do it. Specifically, *taskwork* refers to the individual tasks the team members partake in during task performance. For the reason that the focus of this section is teamwork, taskwork will not be discussed further. *Teamwork* can be defined as team member coordinated action and consists of three key components: attitudes (e.g., collective efficacy, cohesion), behaviors (e.g., mutual performance monitoring, backup behaviors), and cognitions (e.g., shared mental models, shared situation awareness). Known as the ABCs of teamwork, *attitudes* are a team's internal state that impacts their capacity to interact effectively, *behaviors* are the skills and processes requisite of teamwork, and *cognitions* are the knowledge and experiences of the team that facilitates effective teamwork (Shuffler, DiazGranados, & Salas, 2011).

38.3.2 WHAT IS (AND ISN'T) TEAM TRAINING?

Team training is often misrepresented in both academic and practitioner circles. Thus, it is necessary to explicitly define what we consider to be team training. First, *training* can be defined as a pedagogical approach to enhancing the systematic acquisition of KSAs by means of instructional delivery methods (i.e., information, demonstration, practice) with the aim of improving performance. Extending this to the team level, *team training* can be defined as "a set of instructional strategies and tools aimed at enhancing teamwork knowledge, skills, processes, and performance" (Tannenbaum, Salas, & Cannon-Bowers, 1996, p. 516).

Team training should be considered an active, as opposed to a passive, endeavor, whereby trainees are actively engaged throughout the training lifecycle. The anecdotal and empirical evidence supporting the efficacy of team training for enhancing team processes and outcomes is overwhelming (e.g., Salas, Nichols, & Driskell, 2007; Salas, Rozell, Mullen, & Driskell, 1999; Salas, Wilson, Burke, & Wrightman, 2006; Salas et al., 2008). Although *training* and *team training* are distinct types of interventions, it is important to note that this distinction is arbitrary, in that each follows the same logical order of design, delivery, and evaluation but varies in respect to their primary concentration. For instance, a training intervention that permits two or more individuals to complete a task together is not (necessarily) by our definition team training. Specifically, a training program is team training if and only if it focuses on enhancing team-level competencies. A list of teamwork KSAs that may be enhanced during team training is presented in Table 38.1.

38.3.3 WHAT ARE VEs?

VEs are referred to by a corpus of different terms, among which are virtual reality, virtual worlds, modeling, simulation, and serious games. The usages and definitions of these terms vary by research arena and publication; however, they generally refer to the immersion of trainees into simulated environments (Shebilske, 1993). Although a solitary accepted definition of VE does not exist, for this chapter we will adopt Dorsey, Campbell, and Russell's (2009) definition as training systems that "immerse users in a three-dimensional (3D) world, allowing for real-time interaction with a synthetic environment and objects within it" (p. 204). Considering this chapter's focus on team training, we would expand this definition to meet this emphasis. Thus, this definition may read *VE team training systems'* "immerse users in a three-dimensional (3D) world, allowing for real-time interaction with a synthetic environment, objects within it, and potentially with other system users with the aim of enhancing teamwork knowledge, skills, processes, and performance." Trainee-synthetic environment interactions are accomplished via sensory response (e.g., haptic, auditory, visual) and computer technology for collaboration and knowledge sharing purposes (Salas, Oser et al., 2002). Moreover, Dorsey et al. (2009) suggest that three basic tasks are accomplished by system users in VEs: navigation, object selection, and object manipulation. The basic tasks should be considered when developing a VE team training intervention.

38.3.4 WHAT ARE VTMs?

Virtual team member (VTM) can be defined as "multifunctional autonomous agents (Stiles et al., 1996) simulated images of humans within a VE that function in the role for which they are programmed" (Salas, Oser et al., 2002, p. 875). In addition to employing communication technologies to permit distributed team training, VEs may also include VTMs to supplement individual or team training, thus increasing the flexibility of VE team training programs. For example, a VTM can be used as a standby team member if a team member is missing or unavailable. Moreover, VTMs can function as instructors, mentors, colleagues, and leaders (Salas, Oser et al., 2002). Figure 38.1 depicts the range of potential VTM and real team member (RTM) configurations. Specifically, the involvement of VTMs in teams can range from one to all but one, depending on training need.

In VEs, VTMs act as agents promoting the facilitation of the necessary competencies required for trainees. With current technological capabilities, VTMs can realistically imitate human responses and group dynamics in settings representative of the real world. This allows trainees to receive a reliable and valid learning experience.

TABLE 38.1
Teamwork KSAs

Nature of Team Competency	Description of Team Competency	Knowledge	Skills	Attitudes
Context-driven	Team-specific	Cue–strategy associations	Task organization	Team orientation (morale)
	Task-specific	Task-specific teammate characteristics	Mutual performance monitoring	Collective efficacy
		Team-specific role responsibilities	Shared problem-model development	Shared vision
		Shared task models	Flexibility	
		Team mission, objectives, norms, resources	Compensatory behavior	
			Information exchange	
			Dynamic reallocation of functions	
			Mission analysis	
			Task structuring	
			Task interaction	
			Motivation of others	
Team-contingent	Team-specific	Teammate characteristics	Conflict resolution	Team cohesion
	Task-generic	Team mission, objectives, norms, resources	Motivation of others	Interpersonal relations
		Relationship to larger organization	Information exchange	Mutual trust
			Intrateam feedback	
			Compensatory behavior	
			Assertiveness	
			Planning	
			Flexibility	
			Morale building	
			Cooperation	
Task-contingent	Team-specific	Task-specific role, responsibilities	Task structuring	Task-specific teamwork attitudes
	Task-specific	Task sequencing	Mission analysis	
			Mutual performance monitoring	
		Team-role interaction patterns	Compensatory behavior	
			Information exchange	
		Procedures for task accomplishment	Intrateam feedback	
			Assertiveness	
		Accurate task models	Flexibility	
		Accurate problem models	Planning	
		Boundary-spanning role	Task interaction	
		Cue–strategy associations	Situational awareness	

TABLE 38.1 (continued)
Teamwork KSAs

Nature of Team Competency	Description of Team Competency	Knowledge	Skills	Attitudes
Transportable	Team-specific Task-generic	Teamwork skills	Morale building	Collective orientation
			Conflict resolution	Belief in importance
			Information exchange	of teamwork
			Task motivation	
			Cooperation	
			Consulting with others	
			Assertiveness	

Source: Adapted from Cannon-Bowers, J.A. et al., Defining team competencies and establishing team training requirements, in R. Guzzo, E. Salas, and Associates (Eds.), *Team Effectiveness and Decision Making in Organizations*, Jossey-Bass, San Francisco, CA, pp. 333–380, 1995.

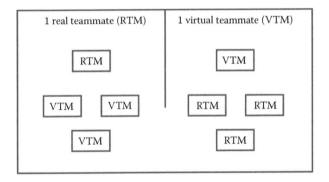

FIGURE 38.1 Range of potential VTM and RTM configurations.

38.4 TEAM TRAINING IN VE: STATE OF THE SCIENCE

As we mentioned, VE team training has been most widely adopted within the military and health-care communities (cf. Bonk & Dennen, 2005). In addition to these communities, VE team training has also been extensively adopted in business, especially at the college level. Business games, for example, have been used for decades to train targeted team behaviors (Tannenbaum & Yukl, 1992). Moreover, with the current science, technology, engineering, and math (STEM) education reform, VEs have also been used for classroom-based training (e.g., K-12, colleges, and universities).

The following section will outline extant literature on team training in VEs ranging from personal computer (PC)-based simulations to massive multiplayer online games (MMOGs) to full-blown simulations. Specifically, the following section will review VE team training in military, business, and health-care domains.

38.4.1 TEAM TRAINING IN THE MILITARY

By and large, the military has been the largest contributor to, and adopter of, VE team training. Computer-based simulations have been used for decades both by the military and by military-sponsored researchers for the purposes of team training. Moreover, full-blown simulators have been used to train aviator skills. However, as previously mentioned, the inception of SIMNET sparked a proliferation of VE team training, especially within the domain of serious games. Recent evolutions in military operations (e.g., asymmetric warfare, distributed warfare) have led the militaries to broaden the use of VEs to include multiplayer games (MPGs), MMOGs, and massive multiplayer online role-playing games (MMORPGs). These VEs allow for both collocated and distributed trainings of military teams that generally represent combat operations. Moreover, VEs have made training more available than at any previous point in history. In reviewing the literature, there appears to be a general trend toward the adoption of larger online-based simulations. In the following section, we will review several of the prominent VEs used by the military for team training.

SIMNET is considered the first large-scale, real-time, human-in-the-loop system for team training (Miller & Thorpe, 1995). Specifically, SIMNET was developed to training tactical team performance including manned vehicle simulations, command postsimulations, and semiautomated forces. Despite a dearth of readily available empirical evidence demonstrating the efficacy of SIMNET for team training, it has undeniably demonstrated its utility in the military (Alluisi, 1991). In the wake of SIMNET, the military has introduced a number of notable VEs designed to train teams. Undoubtedly the most well known of these VEs, falling under the auspices of serious games, is *America's Army (AA)*. AA, developed for commercial use in 2002 by the US Army, is a first-person shooter game initially aimed at increasing recruitment (van der Graaf & Nieborg, 2003). However, the draw of AA as a training tool was quickly recognized. According to Nieborg (2004), besides a tool for recruitment, AA can be used as a test bed and team training tool for the US Army. As a test bed, it has been used to study constructs such as immersion and emotion in VEs (Nieborg, 2004). As a training tool, AA is typically used to simulate soldier combat operations.

In contrast to AA, *DARWARS Ambush!* (Chatham, 2007; Diller, Roberts, Blankenship, & Nielsen, 2004), developed by Pandemic Studios, is a VE developed for noncommercial use to train team convey operations. DARWARS Ambush! has been widely used throughout the US military including with the 1st Brigade, 25th Infantry Division, Stryker Brigade Combat Team, and at the Fort Lewis Mission Support Training Facility (Hussain et al., 2008). In addition to the specific aim of training convoys to better react to ambushes and improvised explosive divisions, DARWARS Ambush! trains specific teamwork competencies (e.g., situation awareness, team adaptability; Raytheon BBN Technologies, 2012). A toolkit, termed DARCCAT, developed for DARWARS Ambush! has demonstrated its feasibility for "automatically detecting critical incidents, identifying performance changes, and evaluating team performance in both live and virtual training environments" (LaVoie et al., 2008, p. 7).

Gorman's Gambit (Hussain et al., 2005) is an example of a VE developed with the specific intent of training teamwork skills. The teamwork skills trained and evaluated in this VE were derived from in-game observations, postdebrief observations, postgame questionnaires, and structured after action review (AAR) exercises. From these methods, the researchers identified 10 skills that drive team performance (Hussain et al., 2008). These skills were extracted from the literature on teams, training, and team training and include leadership, monitoring, backup behavior, adaptability, team orientation, closed-loop communication, team mental models, coordination, communication push (e.g., sharing information), and communication pull (e.g., seeking information).

Additional VEs used by the military include *Virtual Battlespace 1 (VBS1)* and *Virtual Battlespace 2 (VBS2)*. Developed for the US Army by Bohemia Interactive Australia, VBS1 and VBS2 are PC-based systems that train military team tactical operations (e.g., convoy missions). Evaluation of VBS2 demonstrated that trainees held positive attitudes toward the training and felt more prepared to plan convoy operations after training (Ratwani, Orvis, & Kerr, 2010). Moreover, Ratwani et al. (2010) found that training generally enhanced individual- and team-level outcomes. Specifically, team training was found to increase unit process, defined as "soldier belief about how well the unit worked together during the training in terms of different teamwork skills (e.g., communication, monitoring progress towards goals)" (p. 7), and unit cohesion, defined as "soldier attraction to the unit, including their tasks (task cohesion) and other group members (interpersonal cohesion)" (p. 7).

Despite the fact that VEs are widely used to train teams in the military (see additional training games in Table 38.2), there is a relative paucity of empirical evidence assessing their value

TABLE 38.2
Additional Military Training Games

Game	Military Branch
Full Spectrum Warrior	US Army
Quickstrike Time-Sensitive Targeting Trainer	US Air Force
SOCOM: US Navy Seals	US Navy
Close Combat Marines	US Marines
Joint Force Employment	US Joint Forces Command
Modular Semiautomated Forces (ModSAF)	DARPA/STRICOM

beyond the attitudinal level. In fact, Ratwani et al. (2010) state that "Unfortunately, the empirical research linking game-based training to training effectiveness is limited and does not extend much beyond extolling the motivational benefits of games in training" (p. V). Nevertheless, the overall evidence suggests that VE team training in military applications is beneficial, especially under the qualification that the VE team training is grounded in the science of teams and training. Gorman's Gambit provides a good example, as the team training was grounded upon sound training principles.

38.4.2 TEAM TRAINING IN BUSINESS

Business simulations have been used by educators, researchers, and practitioners for over half a century. However, the widespread production of PC-based business simulation games did not occur until the mid-1980s (Faria, Hutchinson, Wellington, & Gold, 2008). According to Wolfe (1993), business simulation games "create environments within which learning and behavioral changes can occur" (p. 447). Despite the implication that business games aim to promote learning and behavioral change, we must first make an important qualification. Per our definition of team training, not all business simulation games should be considered team training. That is, not all business simulations focus on training team competencies. However, although team competencies may not be the focal point of every business simulation, researchers have identified training teamwork as one of five major reasons business games are used (Faria et al., 2008). Specifically, business games often focus on teamwork behaviors (e.g., coordination, communication, conflict management), team attitudes (e.g., cohesion), and moderating factors of team performance (e.g., group size, Anderson & Lawton, 2008, Faria et al., 2008).

Owing to over 50 years of research and application, the extant business gaming literature is expansive (for reviews, see Anderson & Lawton, 2008; Faria, 1990; Faria et al., 2008; Gosen & Washbush, 2004; Greenlaw & Wyman, 1973; Hsu, 1989; Keys & Briggs, 1990; Wolfe, 1985, 1990, 1993, 1997). Despite early uncertainty regarding the efficacy of business games as training tools (Greenlaw & Wyman, 1973), the validity of business games for team training has since been overwhelmingly demonstrated. Reviewing 25 years of research, Faria (2001) concluded that business games are an effective means of training teams and that team characteristics (e.g., cohesion, degree of planning, decision-making organization) are better predictors of performance than individual characteristics. Moreover, Faria (2001) recognized numerous correlates of business game simulation performance. Included in these correlates are the findings that teams outperform individuals (e.g., Nielson, 1975), small teams outperform large teams (e.g., Hoover, 1976), and cohesion (e.g., Wolfe & Box, 1988), team planning and team debriefs (e.g., Hodgetts & Kreitner, 1975; Hornday & Curran, 1988), and positive attitudes toward the simulation increase performance.

Overall, the literature on business games suggests that (1) they work and (2) trainees enjoy them (Faria, 2001; Wolfe, 1997). Although these claims generally hold, Anderson and Lawton (2008) note that business simulation games are least effective in teaching terminology, factual information, basic concepts, or principles. It is our view that the findings from the business gaming literature can be leveraged in the development of VE team training interventions in other domains.

38.4.3 TEAM TRAINING IN HEALTH CARE

According to Wiehagen (2008, p. 1), "Medical simulation and training covers the full spectrum of health care including devices for initial training of combat medics and first responders, devices for training surgical procedures, virtual reality systems for diagnosis and therapy, and team training systems for crisis or incident management." Compared to the military and business, the application of VEs for team training is relatively new in health care. However, following the Institute of Medicine (IOM) report claiming that up to 98,000 medical-related deaths a year are preventable (Kohn, Corrigan, & Donaldson, 1999), health care has been one of the leading adopters of training. Because the IOM identified the need to enhance teamwork (Baker, Day, & Salas, 2006), team training has been at the forefront of mitigating medical errors. Similar to other domains, VEs in health care attempt to match both physically and psychologically the environment in which teams perform. According to Heinrichs, Youngblood, Harter, and Dev (2008), in a typical health-care VE, team members use PCs to interact together with a computer-modeled virtual patient. Heinrichs et al. (2008) note that the primary advantages of medical VEs are their ability to suspend disbelief and model the intricacies of teamwork found in real-world settings. Additionally, VEs in health care are not constrained by the drawbacks of traditional training (i.e., training teams together, in one location, at the same time). The following section will review current research on VE team training in health care.

Despite being a relative newcomer, there has been a dedicated corpus of laboratories and researchers aimed at enhancing the science of virtual team training in health care. For instance, for almost two decades, the Stanford University Medical Media and Information Technologies (SUMMIT; 1990–2008) group developed and tested games for training medical teams (Heinrichs et al., 2008), specifically *Virtual ED I* and *Virtual ED II*.

Virtual ED I is an emergency response acute-care team training program that focuses on team leadership, performance monitoring and backup behaviors, and team communication skills. *Virtual ED I* allows emergency department (ED) staff members the ability to care for a virtual patient that can demonstrate six different trauma cases. The VE team training adopts a cross-training approach in which team members rotate roles in order to developed shared knowledge and understanding of team member roles and responsibilities. Research examining the efficacy of *Virtual ED I* demonstrated that trainees felt immersed in the VE and believed that the training would greatly benefit ED teams (Heinrichs et al., 2008).

The descendant of *Virtual ED I*, *Virtual ED II*, was designed to train ED staff on how to execute a *code triage* for a rapid response to a chemical, biologic, radiologic, nuclear, or highly explosive (CBRNE) incident. Additionally, this team training program aimed to enhance team communication, coordination, and performance monitoring behaviors. Similar to its precursor, trainees in *Virtual ED II* were required to rotate positions. Results examining *Virtual ED II* demonstrated that trainees felt immersed in the VE, became more confident in their ability to respond to a CBRNE incident, and thought that the VE team training would be valuable in training teamwork skills (Heinrichs et al., 2008).

In addition to the approach outlined previously, health care has recently adopted interactive virtual workbenches for team training purposes. The virtual workbenches can project 3D images of a virtual patient or a particular medical situation. Moreover, the projected 3D images can be manipulated (i.e., interacted with) by means of handheld wands (Dorsey et al., 2009).

While the majority of team training in health care is conducted through the use of mannequin-based simulations, VE team training programs do exist. Moreover, there is a recent trend toward the adoption of VEs in health-care team training. In line with the general advantages of VEs for training, Heinrichs et al. (2008) identify seven key advantages of VEs in respect to team training in health care. These include the ability to (1) replicate exact layouts and locations, (2) train distributed teams, (3) provide training at all hours, (4) train a multitude of situations and conditions, (5) include dangerous conditions, (6) easily provide retraining, and (7) assess team performance.

38.5 ESTABLISHING AN EFFECTIVE LEARNING ENVIRONMENT

Team training in VEs has many potential advantages over traditional training delivery (e.g., cost, accessibility, flexibility). However, training must be scientifically grounded in order to facilitate learning and transfer to the operational environment (Salas & Cannon-Bowers, 2001). Unsurprisingly, this adage applies equally to traditional training environments (e.g., classroom-based) as it does to technology-based training environments. Thus, it is important to describe the components of an effective learning environment that can be used when developing team training in VEs. The components that comprise an effective learning environment include the targeted training audience, task requirements, training environment, training delivery, and associated interrelated components (Salas, Oser et al., 2002). These components are represented in the following in a conceptual model of effective learning environments (Figure 38.2). The core components of the model will be discussed in turn.

38.5.1 TRAINING AUDIENCE: FOR WHOM IS THE TRAINING?

One of the primary steps in delivering a team training program is conducting a team needs analysis (Arthur, Edwards, Bell, Villado, & Bennett, 2005). The aim of a team needs analysis is to determine who should be trained, what should be trained, and how the training should be delivered. These objectives require a comprehensive assessment of person, task, and organizational needs (Arthur, Bennett, Edens, & Bell, 2003). The *person analysis* is responsible for delineating who should be trained and identifying the existing characteristics of the team and its individual team members. The types of questions answered in this analysis include the following: What teams need to be trained? What types of teams are these? What are the characteristics of the team? And what are the characteristics of the team members? (Gregory, Feitosa, Driskell, Salas, & Vessey, 2013).

Important team characteristics to consider include "the degree to which team members hold a shared understanding of tasks and their teammates; the team's organizational structure; and the proximity of team members to each other" (Salas, Oser et al., 2002, p. 876). For example, recent research has provided considerable support for the notion that team member's shared understanding (i.e., shared mental model) improves team process and performance (DeChurch & Mesmer-Magnus, 2010; Marks, Sabella, Burke, & Zaccaro, 2002; Mathieu, Heffner, Goodwin, Salas, & Cannon-Bowers, 2000; Smith-Jentsch, Mathieu, & Kraiger, 2005). A team's structure plays a central role in team training design and delivery. As Salas et al. (2002) note, "The composition of team members within the team is a determining factor in how the learning environment should be designed" (p. 877). For instance, hierarchical or horizontal team structures may have differing team dynamics, thus requiring different training emphases (e.g., feedback mechanisms). Lastly, whether a team is collocated or distributed influences the design of the learning environment.

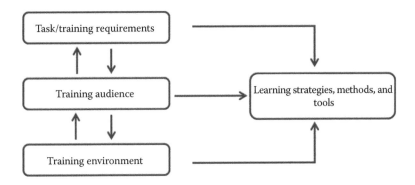

FIGURE 38.2 A conceptual model of a learning environment.

Important individual characteristics to consider include goal orientation, cognitive ability, self-efficacy, age, and conscientiousness (Colquitt, LePine, & Noe, 2000; Towler & Dipboye, 2001). These characteristics have been shown to influence trainee motivation, especially intrinsic motivation, which has been invaluable for effective learning environments (Ryan & Deci, 2000b). For instance, trainee motivation has been linked to positive attitudinal change, enhanced learning acquisition and retention, and transfer of training (cf. Salas & Cannon-Bowers, 2001, p. 479).

38.5.2 TASK REQUIREMENTS: WHAT IS BEING TRAINED?

The *what* of a team training program is uncovered by conducting a team task analysis (Baker, Salas, & Prince, 1991). A team task analysis reveals the specific tasks that a team performs that are vital for teamwork. Specifically, this type of analysis identifies the team competencies that can be advanced during training. An effective learning environment should include the relevant team processes required of successful team performance.

It is useful to note that teams perform under markedly different circumstances. For example, teams may be required to perform under conditions ranging from benign and banal to hazardous and novel. It is thus important to delineate and match the conditions that teams are expected to perform in. The overall fidelity of a simulated environment or VE heavily influences effective learning environments. Generally, general researchers consider three types of fidelity relevant to VE: physical, functional, and psychological fidelity (Bowers & Jentsch, 2001). Physical fidelity represents the degree to which the simulation matches the physical environment of the task. Functional fidelity represents how similar the functions performed during a task in a simulation match the real-world task. And psychological fidelity represents the degree to which the psychological properties of a real-world task are matched in a VE. Each type of fidelity should be considered for team training in VEs.

38.5.2.1 Taskwork versus Teamwork Competencies

It is also important to delineate what is being trained in terms of taskwork versus teamwork competencies. As previously mentioned, taskwork involves individual-level tasks that are needed for successful task completion, but do not require team members to work together as a team, whereas teamwork consists of actions that require more than one individual to complete and reflect team competencies. In order to effectively train teamwork, it is first necessary for individual team members to be proficient in the taskwork required during task performance. Specifically, proficiency in the work task should be attained prior to engagement in team training (Salas, Burke, & Cannon-Bowers, 2002; Salas & Cannon-Bowers, 2000). The team training program should focus on training teamwork competencies, while also promoting the acquisition of individual KSAs.

In respect to taskwork versus teamwork, taskwork competencies do not require interaction with a VTM as they are individual in nature. However, if a VE training program incorporates individual training in order to train team members to a desired level of task proficiency, then that individual training can be delivered by a VTM. Within a VE, for instance, a VTM can present demonstrations or opportunities to practice to the individual trainees. This would allow the trainees the opportunity to become comfortable with the VTM. On the other hand, teamwork competencies require coordinated team member action. In VE team training interventions, VTMs can replace actual team members allowing training to be conducted without the need to have all members present. Again, VTMs can deliver training to the team via information, demonstration, and practice opportunities within the VE.

38.5.3 TRAINING ENVIRONMENT: UNDER WHAT CONDITIONS?

The conditions of the training environment also influence the design of an effective learning environment. The conditions that should be considered include the frequency of training, the length of the training lifecycle, and the number of training sites (Salas, Oser et al., 2002).

First, it may be necessary for training to take place on more than one occasion. This type of decision would likely be made from the information gleaned from the team needs analysis. Second, the duration of a team training program (i.e., from start to finish) should be sufficient in length to produce the desired results and not so long as to negatively impact learning. Lastly, advancements in communications and virtual technologies allow for teams to engage in training from different locations. As a result, considering whether or not training is collocated or distributed is important when designing learning environments.

38.5.4 TRAINING DELIVERY: HOW ARE THEY BEING TRAINED?

It is also important to identify how the training is to be delivered in VEs in order to develop a successful learning environment. Following the science of training (Salas & Cannon-Bowers, 2001), a comprehensive team training program delivers training to the team using three methods: information, demonstration, and practice; each of which can be readily implemented within VEs.

The methods of training delivery are grounded in learning theory (e.g., Bandura, 1977) and are the outlets by which team competencies are acquired. Information-based delivery methods contain stereotypical learning materials (e.g., slide presentation, training manual) that are passively received by the team members. Demonstration-based methods present trainees with visual representations of team competencies. Importantly, both good and bad examples of team competencies should be delivered during training. This allows trainees to develop a more comprehensive understanding of the teamwork KSAs being trained. As the name suggests, practice-based methods afford trainees the opportunity to practice learned teamwork KSAs. Although practice-based methods are an indispensable tool for team training, it is important to note that practice should be guided. This assures that teams practice the relevant team competencies that have been identified as critical (Salas, Oser et al., 2002).

38.6 DUAL APPROACH FOR TEAM TRAINING IN VEs

The following sections will outline the dual approach to team training in VEs mentioned at the beginning of this chapter, centering on DBT (Rosen et al., 2010) and an EBAT (Fowlkes & Burke, 2005a). These descriptions will delimit the individual components that make up each team training approach. Following each approach, a description of how VEs can apply each approach for team training is provided.

38.6.1 APPROACH 1: DEMONSTRATION-BASED TRAINING

DBT is a method whereby demonstrations are used to convey learning content. In essence, DBT is an extension of behavior modeling training (Sorcher & Goldstein, 1972) and social learning theory (Bandura, 1977), which have been empirically validated (Taylor, Russ-Eft, & Chan, 2005). This approach can readily be implemented into VEs via virtual training demonstrations. For example, Dyck, Pinelle, Brown, and Gutwin (2003) identify the potential advantages of learning by observing 3D avatars in VE. The main advantages of DBT are as follows: (1) it represents a flexible training technique and (2) it is cost-effective. Additionally, demonstrations can be combined with other VE training instructional features in order to enhance training. This section will outline the instructional features present in DBT. The instructional features are presented in the following DBT framework (Figure 38.3), based on Rosen and colleagues' (2010) five-component taxonomy including passive guidance/support, preparatory activities, concurrent activities, retrospective activities, and prospective activities. Each of these components is individually described below (Sections 38.6.1.1–38.6.1.5).

38.6.1.1 Passive Guidance/Support

Passive guidance/support represents a method of providing trainees with supplemental and/or organizational information without the trainees engaging in further activities. This method simply

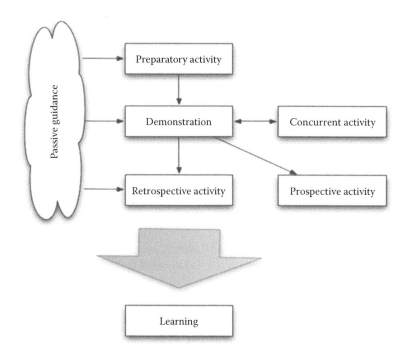

FIGURE 38.3 Instructional features in DBT.

requires the trainees to read or view content before, during, and after training. Examples of passive guidance/support include organizers and summaries, instructional narratives, trainer-provided rule codes, attentional cueing, and passive motivation inducement. Implementing these features in a team training program requires that the information adopted a team-level focus. For instance, attentional cueing during demonstrations should avert trainee focus to the example of critical teamwork behaviors. Moreover, passive motivation inducement—a passive effort to the increase motivation through information provision (e.g., information aimed at increasing trainee's belief in the benefits of team training)—should attempt to enhance the motivation at both individual and team levels.

38.6.1.2 Preparatory Activities/Tasks

Preparatory activities or tasks are those that learners partake in prior to a demonstration (Rosen et al., 2010). These types of activities are unique from passive guidance/support in that they are active (i.e., they require learners to engage in additional activities beyond simple viewing or reading of information). The general aim of these activities is to increase trainee preparedness for the information delivered by the demonstrations. These activities include guided discussions of learning objectives, training aimed to enhance learner understandability, or interventions intended to increase learner motivation.

Observational learning training is an example of an intervention that has been shown to prepare learners for a demonstration's informational content (Browder, Schoen, & Lentz, 1986). Examples of observational learning training include hierarchical encoding training (Hard, Lozano, & Tversky, 2006)—a training aimed at preparing learners to be able to interpret observed behaviors and translate them into component actions—and metacognitive calibration training—a training aimed at increasing an individual's metacognitive processes and avoiding metacognitive miscalibration (Smith, Moores, & Chang, 2005).

In addition to passive motivation inducement, training can include active motivation inducement. Active motivation inducement techniques require trainees to engage in an activity geared to enhance their personal motivation toward the demonstration and associated learning content. These activities, for example, may take the form of exercises focusing on particular teamwork competencies.

38.6.1.3   Concurrent Activities

Concurrent activities occur while the demonstration is already underway (Rosen et al., 2010). These activities are included during training to engage trainees in active learning and have the potential to enhance DBT through augmenting trainee engagement, guiding trainee attention, and promoting information processing. Concurrent activities include perspective taking, guided note-taking, and imitation, each of which can be translated to VE team training.

Perspective taking simply requires trainees to actively adopt the point of view of a model in a demonstration. This is done in order to develop a trainee's understanding of how the model in the demonstration thinks and feels about a situation and why they behave in a certain fashion (Sessa, 1996). Furthermore, perspective taking is intended to permit trainees to reflect on how they would think, feel, and behave in the same situation. In a team training program, perspective taking should focus on team-relevant behaviors, such as coordination or communication. Guided note-taking is a systematic approach where trainers provide trainees with partial note formats as a means to help guide their note-taking (Katayama, 1997). This approach also highlights what information in the demonstration is important to acquire (e.g., teamwork competencies) and provides trainees with valuable after-training resources. In a VE team training program, imitation would likely take the form of mental imitation—as opposed to physical—whereby the trainee imagines themselves performing trained behaviors displayed in the demonstration. Mental imitation may help trainees develop better schemas and encode information at a deeper level.

38.6.1.4   Retrospective Activities

Retrospective activities occur after the demonstration and are intended to have the learner revisit the content presented during the demonstration. Examples of retrospective activities include learner-generated rule codes, group discussion, and imagery exercises.

In postdemonstration exercises, trainees can be called upon to generate their own rule codes (i.e., propositional statements that specify when, where, and how to demonstrate the behaviors viewed during the demonstration; Rosen et al., 2010). This type of exercise can facilitate the content retention (Hogan, Hakel, & Decker, 1986). Group discussions can also facilitate the acquisition and retention of learning content displayed during the demonstrations. This strategy allows trainees to get advice from other trainees and formulate social comparisons, and it helps increase trainee motivation (Prislin, Jordan, Worchel, Semmer, & Shebilske, 1996). Lastly, mental imagery exercise may help knowledge retention and transfer to the operational environment. These types of exercise have been shown to enhance performance in a variety of settings (Driskell, Cooper, & Moran, 1994).

38.6.1.5   Prospective Activities

Prospective activities are executed after training and are aimed at facilitating the transfer of core competencies to novel environments. In contrast to retrospective activities, prospective activities focus on transferring learned competencies to operational environments. Prospective activities include various types of practice, scenario creation, and goal setting.

Practice-based activities are central to effective training interventions and have continuously been identified as an indispensable component of training (Salas & Cannon-Bowers, 2001). Although the axiom "practice makes perfect" may not be entirely correct (i.e., practice should be deliberate and guided; Ericsson, Krampe, & Tesch-Römer, 1993; Salas, Burke et al., 2002), the extant research on the effects of practice on learning is unequivocal—practice improves learning. Practice activities come in many shapes and sizes. One type of practice activity germane to both DBT and VE team trainings is simulation. Simulations allow trainees to practice learned behaviors in environments or situations designed to simulate the real world. In VEs, this may entail trainees watching virtual demonstrations and practicing the behaviors demonstrated using avatars in a simulation environment.

Practice scenario creation and goal setting are prospective activities aimed at increasing trainee buy-in, motivation, and transfer of learning. Scenario creation involves permitting trainees to design practice and simulation scenarios. Druckman and Ebner (2008) note that this may enhance learning by "encouraging active involvement with the material, ... seeing connections between ideas and processes, ... [and] generating analytical questions" (p. 489). Additionally, research demonstrates that trainee-generated scenarios enhance trainee self-efficacy and training transfer (Taylor et al., 2005).

Goal setting requires trainees to develop specific and challenging goals (Locke & Latham, 2002). Researchers have also used the mnemonic SMART to train and describe the model goal (Wade, 2009). Despite variations in the definition of SMART, the most common representation recognizes SMART as specific, meaningful, attainable, relevant, and time bound (Rubin, 2002). According to Locke and Latham (2002), a goal can be defined as an "object or aim of an action, for example, to attain a specific standard of proficiency, usually within a specified time limit" (p. 705). The impetus of goal setting in DBT is to formally require individuals to partake in goal-setting activities in order to increase the likelihood that the learned behaviors will be applied in real-world settings. Extant research has demonstrated that goal setting leads to higher motivation, performance, and training transfer (Locke & Latham, 1990, 2002).

38.6.2 IMPLEMENTING DBT INTO VEs

VEs represent a flexible means of training delivery. In combination with the flexibility of DBT, the instructional features present in DBT can readily be implementing into a VE team training program. The follow section describes how each feature can be integrated with VEs.

Each feature of *passive guidance/support* can easily be implemented into VE team training. These types of training materials are likely easier to apply within VE team training interventions than tradition training interventions (e.g., classroom-based) because they are more easily adaptable. In other words, team training content can be readily updated to represent changes in, for instance, training focus. Furthermore, VE architecture allows trainers to adjust attentional cueing support to meet training demands. For example, where, when, and how a trainee's attention is directed during a demonstration can be altered to focus on different teamwork behaviors present in the demonstration. VTMs could also be used to present passive guidance/support material.

Preparatory activities/tasks aimed at increasing trainee preparedness can be implemented as an introductory training component in a VE team training program. Pretraining procedures may be particularly important to prepare and orient new trainees to a VE. Research indicates that preparatory information can lessen negative reactions to adverse environments (Inzana, Driskell, Salas, & Johnson, 1996). Durlach and Mavor (1995) note that one threat to the effective use of VE is discomfort—feelings of distress or uneasiness that may be related to motion sickness, poor ergonomics, the sopite syndrome, or other factors. At the least, naive users may be expected to experience some initial feelings of unfamiliarity, disorientation, or discomforting feelings in adapting to a VE. These feelings of discomfort are distracting and may detract from training effectiveness. At the worst, these feelings may result in avoidance or limited use of VE training systems for some users. Inzana et al. (1996) found that those who received preparatory information prior to performing a novel task reported less anxiety, were more confident in their ability to perform the task, and made fewer performance errors than controls.

Preparatory interventions may take the form of an instructional presentation preceding the training demonstrations. With the use of communications technologies, teams can engage in these tasks either individually or as a team, from both collocated and distributed settings. Moreover, VEs allow for multiple components to be integrated into a system. We envision a general introductory training component comprised of multiple preparatory tasks and activities. The adaptability of a VE team training intervention would allow researchers and/or practitioners to select which preparatory activities are included in the team training intervention. This has several significant advantages. For example, this would allow trainees to be provided more or less introductory training depending on their level of task proficiency. Additionally, this represents an adaptable method for providing refresher training.

Concurrent activities occur during the presentation of demonstrations. In VEs, the demonstrations would likely be either video demonstrations representing real teams or virtual demonstration of avatar teams. Within VE team training, concurrent activities can be integrated in two primary ways. First, for instance, guided note-taking can be integrated by adding a note-taking section within the system interface. This would allow trainees to take notes in the same technological space as the demonstrations. Second, trainees can be probed to engage in concurrent activities such as perspective taking and mental imitation. Demonstrations could be frozen to allow for probes, or probes could be presented between demonstrations.

Retrospective activities including learner-generated rule codes, imagery exercises, and group discussion are implementable in VEs. Learner-generated rule codes and imagery exercises can be integrated as either individual or team activities. As an individual-level activity, trainees can be proved to develop rule codes and engage in imagery exercises. However, at a team-level, technology-mediated communications may be required to facilitate group discussions. In distributed teams, these group discussions could be conducted, for instance, via telepresence. Applying VTMs in group discussions is impractical as the VTM would need to be able to understand, register, and appropriately respond to a RTM's communications in real time.

Prospective activities, such as practice-based activities, have the most relevance for VE team training and the use of VTMs. Practice-based activities can be entirely conducted within VEs using avatars. As previously mentioned, VEs can permit teams to practice learned behaviors in environments or situations designed to simulate the real world by watching virtual demonstrations and practicing the behaviors demonstrated. As an illustrative example, one of the authors observed an Alcohol, Tobacco, Firearms, and Explosives (ATF) Special Response Team conduct a tactical training exercise using a serious game. A team of about eight members controlling individual avatars practiced room clearing operations. In DBT, these types of exercises can be conducted following demonstrations of how to perform such operations. These demonstrations can, and should be, designed and developed using subject matter experts. Following this example, practice scenario creation can be integrated into DBT to allow trainees to change the virtual location and/or layout of training in order to complete tactical exercises in different domains. VTMs could also be integrated within practice scenarios to supplement or replace team members. It is important, however, that the VTMs be programmed to respond to RTM commands.

38.6.3 Approach 2: EBAT

One method for structuring in a learning environment that is particularly germane to VEs is an EBAT (Dorsey et al., 2009). The EBAT methodology has experienced success in a wide variety of domains including both virtual and physical environments (e.g., health care, Rosen et al., 2008; aviation, Fowlkes, Lane, Salas, Franz, & Oser, 1994; law enforcement, Driskell, Salas, Johnston, & Wollert, 2008; command and control environments, Johnston, Cannon-Bowers, & Smith-Jentsch, 1995; multiservice environments, Dwyer, Oser, Salas, & Fowlkes, 1999; and VE, Brannick, Prince, & Park, 2005). As the name suggests, the EBAT approach is structured around events and aims to deliver principles and guidelines for developing, deigning, and implementing simulation- and experiential-based training (cf. Fowlkes, Dwyer, Oser, & Salas, 1998). Fowlkes et al. (1998) state that "Event-based techniques create training opportunities by systematically identifying and introducing events within training exercise that provide known opportunities to observe specific behaviors of interest" (p. 210). Thus, EBAT affords trainees the opportunities to practice critical teamwork competencies, while at the same time allowing for direct measurement of these observed behaviors. An approach to measurement that has been successfully demonstrated within the event-based framework and is also pertinent to VEs is the targeted acceptable responses to generated events or tasks (TARGET) approach (Fowlkes & Burke, 2005b). TARGET comprises a behavioral checklist developed by SMEs that is used to rate an event based on the appearance or absence of identified *targeted* behaviors.

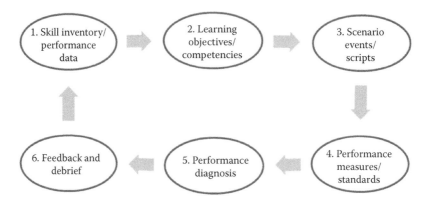

FIGURE 38.4   Components of the EBAT cycle.

As opposed to the development of a formal curriculum for training, scenario-based approaches use the scenario itself as the training curriculum (Cannon-Bowers, Burns, Salas, & Pruitt, 1998). As a consequence, scenario development is central to the success of the training program. In the development of a scenario-based approach to training or EBAT, Cannon-Bowers et al. (1998) adopted a cyclical process (Figure 38.4). Specifically, this cyclical process comprises the following steps: skill inventory/performance data, learning objectives/competencies, scenario events/scripts, performance measures/standards, performance diagnosis, and feedback and debrief. The EBAT methodology is reviewed in this section (cf., Fowlkes et al., 1994; Fowlkes & Burke, 2005a, 2005b).

38.6.3.1   Skill Inventory/Performance Data

A fundamental requirement in the development of any training program, both individual and team, is the identification of critical KSAs relevant to task performance. The critical KSAs in a team training program are determined via a team needs analysis, which is comprised of a team task analysis, person analysis, and organizational analysis. Together, these elements determine what needs to be trained, who needs to be trained, and how they should be trained. Additional techniques that may be leverage to identify core team competencies include knowledge elicitation procedures (e.g., structured interviews) and existing trainee records (e.g., past performance data). These techniques can help determine trainee skill levels, which is valuable when distinguishing what should be trained.

38.6.3.2   Learning Objectives/Competencies

Inherent in cyclical processes, the stages in the EBAT cycle are dependent upon one another. In other words, the information gleaned in stage 1 is used in stage 2, the information in stage 2 is used in stage 3, and so on. For instance, information learned in the skill inventory/performance data stage is used to derive and develop the learning objectives and the core competencies in stage 2. The learning objectives and competencies drive the focus of training. As these elements are based on the team needs analysis and existing performance data, they can range from general to specific in nature. In other words, learning objectives may center on acquiring a specific skill requisite of performance or on a general skill that is transferable across task domains.

38.6.3.3   Scenario Scripts/Trigger Events

As previously mentioned, the success of an event-based training program rests on the quality of the scenarios and trigger events developed for training. The information gained in the initial two stages of the EBAT cycle is leveraged in order to drive the development of these scenarios and trigger events. The number of scripted events in a team training program is dependent on what is going to be trained, who is going to be trained, and how they are going to be trained. Moreover, the trigger events may also range in difficulty. According to Salas et al. (2002), trigger events need to be modifiable in real time based on trainee responses. This helps ensure that the training remains realistic and sufficiently difficult and allows for precise measurement of performance data. The overall aim of the scenario scripts/trigger events is to elicit measurable behaviors of the critical teamwork competencies identified in stage 1.

38.6.3.4 Performance Measures/Standards

Training teamwork competencies is at the very core of every scientifically rooted training program. However, in order to determine the efficacy of any training program, performance must be able to be measured. Thus, performance measures and standards must be developed in order to answer the following question: Did training work? More specifically, to what extent did trainees master the targeted competencies? The strategies and tools for measuring performance in EBATs vary as a function of the identified competencies and the objectives of the training program. Such strategies and tools may include one or a mixture of the following: behavioral observations, semi-automated measurement, or fully automated techniques. Moreover, measurement strategies should aim to assess both process and performance outcomes. Team process represents "members' interdependent acts that convert inputs to outcomes through cognitive, verbal, and behavioral activities directed toward organizing taskwork to achieve collective goals" (Marks, Mathieu, & Zaccaro, 2001, p. 357). The ability to measure process and performance is invaluable in the provision of feedback to trainees.

38.6.3.5 Performance Diagnosis

Performance is diagnosed from information obtained from the performance measures and standards. This information is used to evaluate the efficacy of the team training program (e.g., did trainees learn core competencies). If processes are measured, the causes of performance problems can also be diagnosed.

38.6.3.6 Feedback and Debrief

Based on the diagnosis of individual and team performance, accurate, constructive (both positive and negative feedbacks; Smith-Jentsch, Cannon-Bowers, Tannenbaum, & Salas, 2008), and specific feedback need be provided in a timely manner to trainees (e.g., directly after task performance) (Salas, Burke et al., 2002). Feedback should be focused on team processes and outcomes (i.e., what teams did and the result of those processes). Additionally, feedback provided to trainees should focus on both taskwork and teamwork competencies. Furthermore, feedback should also be provided at both the individual and team levels when possible. This allows trainees to hone in on where and how to improve performance (Salas, Oser et al., 2002). If feedback fails to be accurate, constructive, and/or specific, it may have a deleterious effect on learning. Specifically, bad feedback may negatively reinforce undesirable behaviors and, as a result, stymie the acquisition of taskwork and teamwork competencies.

38.6.4 IMPLEMENTING EBAT INTO VEs

The following section outlines how VEs can implement an EBAT. Although each step of the EBAT approach cannot be directly implemented into VEs, each step influences the design and development of VE team training.

The first step in the EBAT cycle, *skill inventory/performance data*, is aimed at identifying the team competencies developed during training. If previous VE team training data exist, it can be leveraged to determine what needs to be trained, who needs to be trained, and how they should be trained. A unique advantage of VE technology is the capability to reliably retain preexisting performance data over long periods of time. The information gleaned from this step can be used as a focus for designing VEs and VTMs (Salas Oser et al., 2002).

The *learning objectives/competencies* developed in step 2 drive the focus of team training. Learning objectives and competencies can be targeted and programmed into VEs (Salas, Oser et al., 2002). Learning objectives and competencies can be programmed through the use of VTMs. Specifically, "VTMs could be programmed to teach, test, and give feedback to the user based on the objectives" (Salas et al., 2002, p. 885). Additionally, the ability to modify team configurations with VTMs allows training to be adapted to focus on different team competencies and dynamics. VEs also allow teams to easily alternate positions, a team training strategy called cross-training (Cannon-Bowers, Salas, Blickensderfer, & Bowers, 1998).

The execution of *scenario events/scripts* in VEs offers many benefits for team training. VEs are capable of enhancing trainee immersion, or presence, in the virtual world and accurately modeling complex task environments. This affords a substantial degree of flexibility when delivering team training. In VEs, the scenarios can represent hazardous or novel tasks that are too dangerous, logistically difficult, or expensive to conduct in real-world settings. Moreover, one of the more important instructional features of VE, its reconfigurability, allows events to be changed to adapt to different users. As previously mentioned, VEs are capable of delivering better training to more people in a more cost-effective manner.

VEs also allow scenarios and events to be delivered as phased training. Regian, Shebilske, and Monk (1992) claimed that it is not necessarily true that higher fidelity always leads to better training and that many training strategies reduce fidelity early in training to reduce complexity. Friedland and Keinan (1986) have found evidence to support the effectiveness of phased training as an approach to manage training for complex environments. Based on the assumption that a high degree of complexity in the training environment may interfere with initial skill acquisition, phased training is an approach to maximize training effectiveness by partitioning training into separate phases: during initial training, trainees learn basic skills in a relatively low-fidelity or low-complexity environment, and latter stages of training incorporate greater degrees of complexity or realism.

Development of effective *performance measures/standards* is essential for team training diagnosis and feedback. Performance measures and standards can be directly programmed into VE systems. The potential reliability and objectivity of VE technology can allow for consistent and accurate measurement of team process and performance data across time. However, the reliability and objectivity of VE technology only becomes an advantage if performance measures and standards are sound.

VE and VTM technologies can facilitate individual and team process and *performance diagnosis*. This is a result of the reliability and objectivity inherent in VE technology. Salas et al. (2002) note that performance diagnosis in VEs can also be used to operate VTMs in an adaptive manner. For instance, VTMs can be dynamically allocated to either reduce or increase task difficulty based on the attained level of team member and team expertise.

Lastly, VE and VTM technologies can significantly enhance the degree and quality of *feedback and debriefs* provided to trainees. Because performance measures and standards are programmed into VEs, feedback can be provided instantaneously during or directly after training. As mentioned, an important consideration for delivering feedback is that it is provided in a timely manner. Moreover, VTMs can be used to deliver feedback and/or debriefing to trainees. For instance, VTMs can adopt the role of instructor, mentor, or colleague during team debrief sessions (Salas, Oser et al., 2002). Similar to demonstrations in DBT, VTMs could demonstrate correct behaviors when giving feedback.

38.7 VE AND VTM CONTRIBUTIONS FOR TEAM TRAINING

The following section outlines how VEs and VTMs can contribute to enhancing teamwork competencies via team training interventions. VEs offer enhanced flexibility and accessibility, in addition to a reliable platform to deliver team training. These advantages afford trainees the opportunity to learn and practice core competencies in a safe and reliable environment, complete multiple training exercises with both VTM and RTM, and engage in remedial training activities, if necessary. These opportunities support the facilitation of individual and teamwork competencies (e.g., expertise, shared mental models). This proposition is grounded in research on expert performance and decision making (Ericsson, Charness, Feltovich, & Hoffman, 2006; Shanteau, 1992a), as well as early work in human judgment (Brunswik, 1956). For instance, research demonstrates that experts hold deeper mental models than novices, which allow them to attend to more diagnostic environmental cues and subsequently make better judgments and decisions (Shanteau, 1992b). Moreover, these findings extend to the team level (Cannon-Bowers, Salas, & Converse, 1993; Mathieu et al., 2000).

Additionally, information organization and knowledge acquisition is supported by schema theory (similar to mental models), which posits that knowledge is structured in organizational units (Anderson & Pearson, 1984; Mandler, 1984). The structure of organizational units is developed and fined-tuned with repeated guided practices. Thus, repeated practice in VEs for individuals and teams should enhance information organization and knowledge acquisition (Salas, Oser et al., 2002). However, it should be noted that this supposition is under the qualification that training is science driven.

The characteristics of VEs and VTMs can facilitate situation assessment (SA) (Salas, Oser et al., 2002). SA is the ability to *size up* a given situation (Lipshitz, 1993; Noble, 1989) by recognizing relevant environmental cues and patterns (Glaser, 1986) and is the process by which situation awareness is achieved (Endsley, 1995). At a team level, SA represents the degree in which each team member possesses the SA required to perform their responsibilities (Endsley, 1989). At both an individual and team level, SA has been identified as a driver of individual and team performance (Salas, Cooke, & Rosen, 2008). Additionally, within the team cognition literature, shared SA and shared mental models have been acknowledged to be closely related constructs (Burke, Stagl, Salas, Pierce, & Kendall, 2006).

Characteristics of VEs also afford quick and easy modification of training complexity through, for example, imposing temporal constraints or manipulating task characteristics. Modifying the complexity of the training environment can have several beneficial effects. First, scaffolding of task difficulty can help to reduce boredom and increase or maintain motivation by keeping the training challenging. As Garris et al. (2002) note, "we are challenged by activities that are neither too easy nor too difficult to perform" (p. 450). Thus, for instance, as task performance increases during training, so too could task difficulty. Second, exposing trainees to stressful condition (i.e., stress exposure training) can attenuate the effects of stress during real-world performance (Driskell, Johnston, & Salas, 2001). Third, modifying the conditions under which individuals and teams perform can promote adaptability.

As previously mentioned, feedback is an integral component in team training programs. Accurate and reliable feedback promotes the acquisition of teamwork KSAs and enhances training transfer (Gregory et al., 2013). Salas et al. (2002) suggest that VEs and VTMs add to high interobserver reliability, internal consistency, and interexercise correlation. As a result, VEs are particularly well suited to enhance trainee outcomes.

Perhaps the greatest contribution of VEs and VTMs for team training is the ability to circumvent the drawbacks of traditional team training (i.e., the need to train teams together, in one location, at the same time). Specifically, VEs and VTMs permit training teams in different locations, at different times, with any number of team members, and as often as needed. Moreover, many of today's teams perform in complex, dangerous environments where training in real-world settings is impractical (e.g., high-reliability teams). VEs and VTMs allow individuals and teams to train in environments that they would otherwise be unable to train in. Coupled with the capability to accurately model the task environment, psychological fidelity of the task (e.g., stress), and team dynamics, VEs possess significant benefits for team training.

38.8 GENERAL GUIDELINES FOR USING DBT AND EBAT

The following section advances recommendations for integrating DBT and EBAT, learning environment methodology, and VE technology. These guidelines are based on, and expand upon, Salas et al.'s (2002) general guidelines (Table 38.3).

TABLE 38.3

General Guidelines for VE Team Training

VE Characteristics

- A conceptual framework should be used to develop effective learning environments in VEs.
- VE team training should apply sound training principles based on the science of training.
- Structured training methodologies such as DBT and EBAT can enhance training quality.
- Preparatory information can enhance learning outcomes in VEs.
- VEs should allow team members to explore the VE prior to task performance.
- Appropriate feedback of trainees' performances can be provided immediately/real time or delayed until the end of the exercise.
- Using predefined events, VTMs provide opportunities for pinpointing a training audience's deficiencies and proficiencies in learning objectives and for practicing new skills or maintaining old skills.
- VEs should be adaptable and allow for team performance in different simulated environments.
- VEs should accurately mimic the psychological fidelity of the task environment.
- VE scenario and demonstrations should be adaptable to allow for phased training (e.g., increased complexity).
- Team members should be able to unobtrusively communicate from distributed locations.

VTM Characteristics

- VTMs should be designed to demonstrate and build critical team competencies.
- VTMs should be able to demonstrate both taskwork and teamwork competencies.
- VTMs should have the ability to hold different ranking positions within a team and should be able to represent various levels of expertise.
- VTMs should be able to match the psychological fidelity of RTMs.
- The movements and actions of VTMs should be transparent to RTMs.
- VTMs should not negatively intrude on team member task and team performance.
- VTMs should be able to act as substitutes for missing RTMs.
- VTMs should be able to react in real time to changes in the VE caused by RTMs.
- If task proficiency is required prior to team training, VTMs should be able to be used to deliver individual training.
- VTMs should be programmed to accurately perform or demonstrate team competencies.

Source: Adapted from Salas, E. et al., Team training in virtual environments: An event-based approach. In K. M. Stanney (Ed.), *Handbook of Virtual Environments: Design, Implementation, and Applications*, Mahwah, NJ, Lawrence Erlbaum Associates, pp. 873–892, 2002.

38.9 CONCLUDING REMARKS

More teams are being trained today than at any other point in history. These teams come in many different shapes and sizes, with many different needs and objectives. Current advancements in the science of teams and training, in addition to technological advancements, have changed the landscape of how teams are trained. Specifically, a current shift is underway from training teams in real-world settings to training teams in VEs. This chapter set out to capture this change and present a review of team training in VEs covering the past decade.

This chapter also outlined the core components that comprise an effective learning environment. We believe that developing an effective learning environment within VEs is a requirement if team training interventions wish to be successful. Although VE technology may offer the *potential* to deliver more effective training more cheaply, research has only just begun to examine the application of VE technology for training. Data substantiating the efficacy of team training demonstrates that team training is an effective means for enhancing team process and performance (Salas et al., 2007, 2008). As a consequence, we stress the need to adopt sound training principles and reject the notion that an elaborate VE without sound training principles is adequate for training. Specifically, for team training programs, VEs must focus on enhancing team-level competencies. The adoption of a scientifically rooted, structured training methodology can assist in the development of sound VE team training interventions. We envisioned and advanced two approaches we believe to be suitable for team training in VEs: DBT (Rosen et al., 2010) and EBAT (Fowlkes & Burke, 2005a). We believe that the advantages of VE and VTMs (e.g., capability to simulate dangerous and stressful events), coupled with effective learning environments and structured training methodologies, will deliver effective approaches for training teams.

ACKNOWLEDGMENTS

The work presented here was supported by funding from the National Aeronautics and Space Administration (NASA; Grant# NNX09AK48G). The views expressed in this work are solely those of the authors and not those of NASA or the University of Central Florida.

This chapter was partially supported by SBIR Phase 1 NASA contract No. NNX12CE85P awarded to Stottler-Henke Associates. The views expressed within are those of the authors and not the official word of the sponsoring agency or the author's affiliations.

REFERENCES

Aguinis, H., & Kraiger, K. (2009). Benefits of training and development for individuals and teams, organizations, and societies. *Annual Review of Psychology*, *60*, 541–474. doi:10.1146/annurev. psych.60.110707.163505

Alluisi, E. A. (1991). The development of technology for collective training: SIMNET, a case history. *Human Factors*, *33*(3), 343–362.

Anderson, P. H., & Lawton, L. (2008). Business simulations and cognitive learning: Developments, desires, and future directions. *Simulation & Gaming*, *40*(2), 193–216. doi:10.1177/1046878108321624

Anderson, R. C., & Pearson, P. D. (1984). A schema-theoretic view of basic processes in reading. In P. D. Pearson (Ed.), *Handbook of reading research* (pp. 255–292). New York, NY: Longman.

Arthur, W., Edwards, B. D., Bell, S. T., Villado, A. J., & Bennett, W. (2005). Team task analysis: Identifying tasks and jobs that are team based. *Human Factors*, *47*, 654–669.

Arthur, W. R., Bennett, W. R., Edens, P. S., & Bell, S. T. (2003). Effectiveness of training in organizations: A meta-analysis of design and evaluation features. *Journal of Applied Psychology*, *88*(2), 234–245.

Baker, D. P., Day, R., & Salas, E. (2006). Teamwork as an essential component of high-reliability organizations. *Health Services Research*, *41*(4p2), 1576–1598.

Baker, D. P., Salas, E., & Prince, C. (1991). *Team task importance: Implications for conducting team task analysis*. Paper presented at the Sixth Annual Meeting of the Society for Industrial and Organizational Psychology, St. Louis, MO.

Bandura, A. (1977). Self-efficacy: Toward a unifying theory of behavioral change. *Psychological Review*, *84*, 191–215.

Bonk, C. J. & Dennen, V. P. (2005). *Massive multiplayer online gaming: A research framework for military training and education*. Technical Report 2005–1, Office of the Under Secretary of Defense (Personnel and Readiness), Readiness and Training Directorate, Advanced Distributed Learning Initiative, Washington, DC.

Bowers, C. A., Braun, C. C., & Morgan, B. B. (1997). Team workload: Its meaning and measurement. In M. T. Brannick, E. Salas, and C. Prince (Eds.), *Team performance assessment and measurement: Theory, methods, and applications* (pp. 85–108). Mahwah, NJ: Erlbaum.

Bowers, C. A., & Jentsch, F. (2001). Use of commercial, off-the-shelf, simulations for team research. In C. Bowers and E. Salas (Eds.), *Advances in human performance and cognitive engineering research* (pp. 293–317). Mahwah, NJ: Lawrence Erlbaum.

Brannick, M. T., Prince, C., & Park, W. (2005). Can PC-based systems enhance teamwork in the cockpit? *The International Journal of Aviation Psychology*, *15*(2), 173–187.

Browder, D. M., Schoen, S. F., & Lentz, F. E. (1986). Learning to learn through observation. *The Journal of Special Education*, *20*(4), 447–461. doi:10.1177/002246698602000406.

Brunswik, E. (1956). *Perception and the representative design of psychological experiments* (2nd ed.). Berkeley, CA: University of California Press.

Burke, C. S., Stagl, K. C., Salas, E., Pierce, L., & Kendall, D. (2006). Understanding team adaptation: A conceptual analysis and model. *The Journal of Applied Psychology*, *91*(6), 1189–1207. doi:10.1037/0021–9010.91.6.1189.

Cannon-Bowers, J., Salas, E., & Converse, S. (1993). Shared mental models in expert team decision making. In N. J. Castellan Jr. (Ed.), *Individual and group decision making: Current issues* (pp. 221–246). Hillsdale, NJ: Lawrence Erlbaum Associates.

Cannon-Bowers, J. A., Burns, J. J., Salas, E., & Pruitt, J. S. (1998). Advanced technology in scenario-based training. In J. A. Cannon-Bowers and E. Salas (Eds.), *Making decisions under stress* (pp. 365–374). Washington, DC: American Psychological Association.

Cannon-Bowers, J. A., & Salas, E. (1997). Teamwork competencies: The interaction of team member knowledge, skills, and attitudes. In H. F. O'Neil (Ed.), *Workforce readiness: Competencies and assessment* (pp. 151–174). Mahwah, NJ: Lawrence Erlbaum Associates.

Cannon-Bowers, J. A., Salas, E., Blickensderfer, E., & Bowers, C. A. (1998). The impact of cross-training and workload on team functioning: A replication and extension of initial findings. *Human Factors, 40,* 92–101.

Cannon-Bowers, J. A., Tannenbaum, S. I., Salas, E., & Volpe, C. E. (1995). Defining team competencies and establishing team training requirements. In R. Guzzo, E. Salas, and Associates (Eds.), *Team effectiveness and decision making in organizations* (pp. 333–380). San Francisco, CA: Jossey-Bass.

Chatham, R. E. (2007). Games for training. *Communication of the ACM, 50*(7), 37–43.

Colquitt, J. A., LePine, J. A., & Noe, R. A. (2000). Toward an integrative theory of training motivation: A meta-analytic path analysis of 20 years of research. *Journal of Applied Psychology, 85*(5), 678–707. doi:10.1037/0021-9010.85.5.678

Coultas, C. W., Grossman, R., & Salas, E. (2012). Design, delivery, evaluation, and transfer of training systems. In G. Salvendy (Ed.), *Handbook of human factors and ergonomics* (4th ed.) (pp. 490–533). Hoboken, NJ: John Wiley & Sons.

Csikszentmihalyi, M. (2000). *Beyond boredom and anxiety.* San Francisco, CA: Jossey-Bass.

DeChurch, L. A., & Mesmer-Magnus, J. R. (2010). Measuring shared team mental models: A meta-analysis. *Group Dynamics: Theory, Research, and Practice, 14*(1), 1–14.

Devine, D. J., Clayton, L. D., Philips, J. L., Dunford, B. B., & Melner, S. B. (1999). Teams in organizations: Prevalence, characteristics, and effectiveness. *Small Group Research, 30*(6), 678–711.

DiazGranados, D., Klein, C., Lyons, R., Salas, E., Bedwell, W. L., & Weaver, S. J. (2008). *Investigating the prevalence, characteristics and effectiveness of teams: A U.S. sample surveyed.* Paper presented at the INGRoup: Interdisciplinary Network for Group Research, Kansas City, MO, July 17–19.

Diller, D., Roberts, B., Blankenship, S., & Nielsen, D. (2004). *DARWARS Ambush!—Authoring lessons learned in a training game.* Proceedings of the 2004 Interservice/Industry Training, Simulation and Education Conference (I/ITSEC). Orlando, FL.

Dorsey, D., Campbell, G., & Russell, S. (2009). Adopting the instructional science paradigm to encompass training in virtual environments. *Theoretical Issues in Ergonomics Science, 10*(3), 197–215.

Driskell, J. E., Copper, C., & Moran, A. (1994). Does mental practice enhance performance? *Journal of Applied Psychology, 79,* 481–492. doi:10.1037/0021-9010.79.4.481

Driskell, J. E., Johnston, J. H., & Salas, E. (2001). Does stress training generalize to novel setting? *Human Factors, 43*(1), 99–110.

Driskell, J. E., Radtke, P. H., & Salas, E. (2003). Virtual teams: Effects of technological mediation on team performance. *Group Dynamics: Theory, Research, and Practice, 7*(4), 297–323.

Driskell, J. E., Salas, E., & Johnston, J. H. (2006). Decision making and performance under stress. In A. B. Adler, C. A. Castro, & T. W. Britt (Eds.), *Minds in the military: The psychology of serving in peace and conflict* (Vol. 1: Military Performance, pp. 128–154). Westport, CT: Praeger Security International.

Driskell, J. E., Salas, E., Johnston, J. H., & Wollert, T. N. (2008). Stress exposure training: An event-based approach. In P.A. Hancock and J.L. Szalma (Eds.), *Performance under stress* (pp. 271–286). London, U.K.: Ashgate.

Druckman, D., & Ebner, N. (2008). Onstage or behind the scenes? Relative learning benefits of simulation role-play and design. *Simulation and Gaming, 39,* 465–497. doi:10.1177/1046878107311377

Durlach, N. I., & Mavor, A. S. (1995). *Virtual reality: Scientific and technological challenges.* Washington, DC: National Academy Press.

Dwyer, D. J., Oser, R. L., Salas, E., & Fowlkes, J. E. (1999). Performance measurement in distributed environments: Initial results and implications for training. *Military Psychology, 11*(2), 189–215.

Dyck, J., Pinelle, D., Brown, B., & Gutwin, C. (2003). *Learning from games: HCI design innovations in entertainment software.* In Proceedings of Graphical Interface 2003.

Endsley, M. (1989). *Final report: Situation awareness in an advanced strategic mission (NOR DOC 89–32).* Hawthorne, CA: Northrop Corp.

Endsley, M. R. (1995). Toward a theory of situation awareness in dynamic systems. *Human Factors: The Journal of the Human Factors and Ergonomics Society, 37*(1), 32–64.

Ericsson, A. K., Charness, N., Feltovich, P., & Hoffman, R. R. (Eds.). (2006). *Cambridge handbook on expertise and expert performance*. Cambridge, U.K.: Cambridge University Press.

Ericsson, K. A., Krampe, R. T., & Tesch-Römer, C. (1993). The role of deliberate practice in the acquisition of expert performance. *Psychological Review, 100*(3), 363–406.

Faria, A. J. (1990). Business simulation games after thirty years: Current usage levels. In James W. Gentry (Ed.), *Guide to business gaming and experiential learning* (pp. 36–47). East Brunswick, NJ: Nichols/GP.

Faria, A. J. (2001). The changing nature of business simulation/gaming research: A brief history. *Simulation & Gaming, 32*(1), 97–110. doi:10.1177/104687810103200108

Faria, A. J., Hutchinson, D., Wellington, W. J., & Gold, S. (2008). Developments in business gaming: A review of the past 40 years. *Simulation & Gaming, 40*(4), 464–487. doi:10.1177/1046878108327585

Fowlkes, J., Dwyer, D. J., Oser, R. L., & Salas, E. (1998). Event-based approach to training (EBAT). *The International Journal of Aviation Psychology, 8*(3), 209–221.

Fowlkes, J. E., & Burke, C. S. (2005a). Event-based approach to training (EBAT). In N. Stanton, H. Hendrick, S. Konz, K. Parsons, & E. Salas (Eds.), *Handbook of human factors and ergonomics methods* (pp. 47-1–47-5). London, U.K.: Taylor & Francis.

Fowlkes, J. E., & Burke, C. S. (2005b). Targeted acceptable responses to generated events of tasks (TARGETs). In N. Stanton, A. Hedge, K. Brookhuis, E. Salas, & H. Hendrick (Eds.), *Handbook of human factors and ergonomics methods* (pp. 53.1–53.6). Baca Raton, FL: CRC Press.

Fowlkes, J. E., Lane, N. E., Salas, E., Franz, T., & Oser, R. (1994). Improving the measurement of team performance: The TARGETs methodology. *Military Psychology, 6*, 47–61.

Friedland, N., & Keinan, G. (1986). Stressors and tasks: How and when should stressors be introduced during training for task performance in stressful situations? *Journal of Human Stress, 12*(2), 71–76.

Garris, R., Ahlers, R., & Driskell, J. E. (2002). Games, motivation, and learning: A research and practice model. *Simulation & Gaming, 33*(4), 441–467.

Glaser, R. (1986). Training expert apprentices. In I. Goldstein, R. Gagne, R. Glaser, J. Royer, T. Shuell, and D. Payne (Eds.), *Learning research laboratory: Proposed research issues* (AFHRL-TP-85-54). Brooks Air Force Base, TX: Air Force Research Laboratory, Manpower and Personnel Division.

Gosen, J., & Washbush, J. (2004). A review of scholarship on assessing experiential learning effectiveness. *Simulation & Gaming, 35*(2), 270–293. doi:10.1177/1046878104263544

Green, M., & McGill, E. (2011). *State of the industry, 2011: ASTD's annual review of workplace learning and development data*. Alexandria, VA: American Society for Training & Development.

Greenlaw, P. S., & Wyman, F. P. (1973). The teaching effectiveness of games in collegiate business courses. *Simulation & Gaming, 4*, 259–294.

Gregory, M. E., Feitosa, J., Driskell, T., Salas, E. & Vessey, W. B. (2013). Designing, delivering, and evaluating team training in organizations: Principles that work. In E. Salas, S. I. Tannenbaum, D. Cohen, & G. Latham (Eds.), *Developing and enhancing high-performance teams: Evidence-based practices and advice*. San Francisco, CA: Jossey-Bass.

Grossman, R., & Salas, E. (2011). The transfer of training: What really matters. *International Journal of Training and Development, 15*(2), 103–120.

Hard, B. M., Lozano, S. C., & Tversky, B. (2006). Hierarchical encoding of behavior: Translating perception into action. *Journal of Experimental Psychology: General, 135*, 588–608. doi:10.1037/0096–3445.135.4.588; 10.1037/0096–3445.135.4.588.supp

Heinrichs, W. L., Youngblood, P., Harter, P. M., & Dev, P. (2008). Simulation for team training and assessment: Case studies of online training with virtual worlds. *World Journal of Surgery, 32*(2), 161–170. doi:10.1007/s00268-007-9354-2.

Hodgetts, R., & Kreitner, R. (1975). Motivating simulation game performance and satisfaction with performance-contingent consequences. *Simulation Games and Experiential Learning in Action, 2*, 151–156.

Hogan, P. M., Hakel, M. D., & Decker, P. J. (1986). Effects of trainee-generated versus trainer-provided rule codes on generalization in behavior-modeling training. *Journal of Applied Psychology, 71*(3), 469–473.

Hollingshead, A. B. (2001). Cognitive interdependence and convergent expectations in transactive memory. *Journal of Personality and Social Psychology, 81*(6), 1080–1089. doi:10.1037/0022–3514.81.6.1080

Hoover, D. (1976). An experimental examination of group size effects. *Computer Simulation and Learning Theory, 3*, 426–432.

Hornaday, R. W., & Curran, K. (1988). Formal planning, simulation team performance, and satisfaction: A replication. *Developments in Business Simulation & Experiential Exercises, 15*, 138–141.

Hsu, E. (1989). Role-event gaming simulation in management education: A conceptual framework and review. *Simulation & Games: An International Journal, 4*(1), 409–438.

Hussain, T. S., Weil, S. A., Brunyé, T. T., Diedrich, F., Entin, E. E., Ferguson, W., … Roberts, B. (2005). *Gorman's gambit: Assessing the potential of massive multi-player games as tools for military training.* Final project report, BBNT Solutions, LLC & Aptima.

Hussain, T. S., Weil, S. A., Brunyé, T., Sidman, J., Ferguson, W., & Alexander, A. L. (2008). Eliciting and evaluating teamwork within a multi-player game-based training environment. In H. F. O'Niel & R. S. Perez (Eds.), *Computer games and team and individual learning* (pp. 77–104). Amsterdam, the Netherlands: Elsevier.

Inzana, C. M., Driskell, J. E., Salas, E., & Johnston, J. H. (1996). Effects of preparatory information on enhancing performance under stress. *The Journal of Applied Psychology, 81*(4), 429–435.

Johnston, J. H., Cannon-Bowers, J. A., & Smith-Jentsch, K. A. (1995). *Event-based performance measurement system for shipboard command teams.* Proceedings of the First International Symposium on Command and Control Research and Technology, Washington, DC, pp. 274–276.

Katayama, A. D. (1997). *Getting students involved in note taking: Why partial notes benefit learners more than complete notes.* Paper presented at the Annual Meeting of the Mid-South Educational Research Association, Memphis, TN.

Keys, B., & Biggs, W. (1990). A review of business games. In J. W. Gentry (Ed.), *Guide to business gaming and experiential learning* (pp. 48–73). East Brunswick, NJ: Nichols/GP.

Kohn, L. T., Corrigan, J. M., & Donaldson, M. S. (1999). *To err is human: Building a safer health care system.* Washington, DC: National Academy Press.

LaVoie, N., Foltz, P., Rosenstein, M., Oberbreckling, R., Chatham, R., & Psotka, J. (2008). *Automating convoy training assessment to improve Soldier performance.* Proceedings of the 26th Army Science Conference, Orlando, FL.

Lipshitz, R. (1993). Converging themes in the study of decision making in realistic settings. In G. A. Klein, J. Orasanu, R. Calderwood, & C. E. Zsambok (Eds.), *Decision making in action: Models and methods* (pp. 103–137). Norwood, NJ: Ablex.

Locke, E. A., & Latham, G. P. (1990). Work motivation and satisfaction: Light at the end of the tunnel. *Psychological Science, 1*(4), 240–246.

Locke, E. A., & Latham, G. P. (2002). Building a practically useful theory of goal setting and task motivation: A 35-year odyssey. *American Psychologist, 57*(9), 705–717. doi:10.1037/0003–066X.57.9.705

Mandler, J.M. (1984). *Stories, scripts and scenes: Aspects of schema theory.* Hillsdale, NJ: Lawrence Erlbaum.

Marks, M. A., Mathieu, J. E., & Zaccaro, S. J. (2001). A temporally based framework and taxonomy of team processes. *Academy of Management Review, 26*, 356–376.

Marks, M. A., Sabella, M. J., Burke, C., & Zaccaro, S. J. (2002). The impact of cross-training on team effectiveness. *Journal of Applied Psychology, 87*(1), 3–13. doi:10.1037/0021–9010.87.1.3

Mathieu, J., Maynard, M. T., Rapp, T., & Gilson, L. (2008). Team effectiveness 1997–2007: A review of recent advancements and a glimpse into the future. *Journal of Management, 34*(3), 410–476. doi:10.1177/0149206308316061

Mathieu, J. E., Heffner, T. S., Goodwin, G. F., Salas, E., & Cannon-Bowers, J. A. (2000). The influence of shared mental models on team process and performance. *The Journal of Applied Psychology, 85*(2), 273–283.

Miller, D. C., & Thorpe, J. A. (1995). SIMNET: The advent of simulator networking. *Proceedings of the IEEE, 83*(8), 1114–1123.

Nieborg, D. B. (2004). *America's Army: More than a game?* Proceedings of 35th Annual Conference of the International Simulation and Gaming Association (ISAGA) and Conjoint Conference of SAGSAGA. Retrieved from http://www.gamespace.nl/content/ISAGA_Nieborg.PDF

Nielsen, C. (1975). Player performance under differing player configurations in The Investment Game: Some preliminary findings. *Simulation Games and Experiential Learning in Action, 2*, 111–115.

Noble, D. F. (1989). Schema-based knowledge elicitation for planning and situation assessment aids. *IEEE Transactions on Systems, Man, and Cybernetics, 19*(3), 473–482.

Pavlas, D., Jentsch, F., Salas, E., Fiore, S. M., & Sims, V. (2012). The play experience scale development and validation of a measure of play. *Human Factors: The Journal of the Human Factors and Ergonomics Society, 54*(2), 214–225.

Prislin, R., Jordan, J. A., Worshel, S., Tschan Semmer, F., & Shebilske, W. L. (1996). Effects of group discussion on acquisition of complex skills. *Human Factors, 38*, 404–416. doi:10.1518/001872096778701999

Ratwani, K. L., Orvis, K., & Knerr, B. W. (2010). *Game-based training effectiveness evaluation in an operational setting. Research Report* (655803). Arlington, VA: U.S. Army Research Institute for the Behavioral and Social Sciences.

Raytheon BBN Technologies. (2012). *DARWARS Ambush!* Retrieved from http://www.bbn.com/technology/immersive_learning_technologies/darwars_ambush

Regian, J., Shebilske, W. L., & Monk, J. M. (1992). Virtual reality: An instructional medium for visual-spatial tasks. *Journal of Communication, 42*(4), 136–149.

Rosen, M. A., Salas, E., Pavlas, D., Jensen, R., Fu, D., & Lampton, D. (2010). Demonstration-based training: A review of instructional features. *Human Factors: The Journal of the Human Factors and Ergonomics Society, 52*(5), 596–609.

Rosen, M. A., Salas, E., Tannenbaum, S. I., Pronovost, P. J., & King, H. B. (2012). Simulation-based training for teams in health care: Designing scenarios, measuring performance, and providing feedback. In P. Carayon (Ed.), *Handbook of human factors and ergonomics in health care and patient safety* (pp. 573–594). Boca Raton, FL: CRC Press.

Rosen, M. A., Salas, E., Wu, T. S., Silvestri, S., Lazzara, E. H., Lyons, R., …King, H. B. (2008). Promoting teamwork: An event-based approach to simulation-based teamwork training for emergency medicine residents. *Academic Emergency Medicine, 15*(11), 1190–1198.

Rubin, R. S. (2002). Will the real SMART goals please stand up. *The Industrial-Organizational Psychologist, 39*(4), 26–27.

Ryan, R. M., & Deci, E. L. (2000a). Self-determination theory and the facilitation of intrinsic motivation, social development, and well-being. *American Psychologist, 55*(1), 68.

Ryan, R. M., & Deci, E. L. (2000b). Intrinsic and extrinsic motivations: Classic definitions and new directions. *Contemporary Educational Psychology, 25,* 54–67. doi:10.1006/ceps.1999.1020

Salas, E., Burke, C. S., & Cannon-Bowers, J. A. (2002). What we know about designing and delivering team training: Tips and guidelines. In K. Kraiger (Ed.), *Creating, implementing, and managing effective training and development: State-of-the-art lessons for practice* (pp. 234–259). San Francisco, CA: Jossey-Bass.

Salas, E., & Cannon-Bowers, J. A. (1997). Methods, tools, and strategies for team training. In M. A. Quinones and A. Ehrenstein (Eds.), *Training for a rapidly changing workplace: Applications of psychological research* (pp. 249–279), Washington, DC: American Psychological Association.

Salas, E., & Cannon-Bowers, J. A. (2000). The anatomy of team training. In S. Tobias and J. D. Fletcher (Eds.), *Training and retraining* (pp. 312–338). New York, NY: Macmillan.

Salas, E., & Cannon-Bowers, J. A. (2001). The science of training: A decade of progress. *Annual Review of Psychology, 52,* 471–499. doi:10.1146/annurev.psych.52.1.471

Salas, E., Cooke, N. J., & Rosen, M. A. (2008). On teams, teamwork, and team performance: Discoveries and developments. *Human Factors, 50*(3), 540–547.

Salas, E., DiazGranados, D., Klein, C., Burke, C. S., Stagl, K. C., Goodwin, G. F., & Halpin, S. M. (2008). Does team training improve team performance? A meta-analysis. *Human Factors, 50,* 903–933.

Salas, E., Dickinson, T. L., Converse, S., & Tannenbaum, S. I. (1992). Toward an understanding of team performance and training. In R. W. Swezey & E. Salas (Eds.), *Teams: Their training and performance* (pp. 3–29). Norwood, NJ: Ablex.

Salas, E., Nichols, D. R., & Driskell, J. E. (2007). Testing three team training strategies in intact teams: A meta-analysis. *Small Group Research, 38*(4), 471–488.

Salas, E., Oser, R. L., Cannon-Bowers, J. A., & Daskarolis, E. (2002). Team training in virtual environments: An event-based approach. In K. M. Stanney (Ed.), *Handbook of virtual environments: Design, implementation, and applications* (pp. 873–892). Mahwah, NJ: Lawrence Erlbaum Associates.

Salas, E., Rozell, D., Mullen, B., & Driskell, J. E. (1999). The effect of team building on performance. *Small Group Research, 30*(3), 309–330.

Salas, E., Wilson, K. A., Burke, C. S., & Wrightman, D. C. (2006). Does crew resource management training work? An update, an extension, and some critical needs. *Human Factors, 48,* 392–412.

Sessa, V. I. (1996). Using perspective taking to manage conflict and affect in teams. *The Journal of Applied Behavioral Science, 32,* 101–115. doi:10.1177/0021886396321007

Shanteau, J. (1992a). Competence in experts: The role of task characteristics. *Organizational Behavior and Human Decision Processes, 53*(2), 252–266. doi:10.1016/0749–5978(92)90064-E

Shanteau, J. (1992b). How much information does an expert use? Is it relevant? *Acta Psychologica, 81,* 75–86.

Shebilske, W. L. (1993). Visuomotor modularity, ontogeny, and training high-performance skills with spatial instruments. In S. R. Ellis, M. K. Kaiser, & A. J. Grunwald (Eds.), *Pictorial communication in virtual and real environments* (pp. 304–315). Washington, DC: Taylor & Francis.

Shuffler, M. L., DiazGranados, D., & Salas, E. (2011). There's a science for that: Team development interventions in organizations. *Current Directions in Psychological Science, 20*(6), 365–372.

Sitzmann, T., Ely, K., Bell, B. S., & Bauer, K. N. (2010). The effects of technical difficulties on learning and attrition during online training. *Journal of Experimental Psychology: Applied, 16*(3), 281–292.

Smith, D. K., Moores, T., & Chang, J. (2005). Prepare your mind for learning. *Communications of the ACM*, *48*(9), 115–118. doi:10.1145/1081992.1081999

Smith-Jentsch, K. A., Cannon-Bowers, J. A., Tannenbaum, S. I., & Salas, E. (2008). Guided team self-correction: Impacts on team mental models, processes, and effectiveness. *Small Group Research*, *39*, 303–327.

Smith-Jentsch, K. A., Mathieu, J. E., & Kraiger, K. (2005). Investigating linear and interactive effects of shared mental models on safety and efficiency in a field setting. *Journal of Applied Psychology*, 90(3), 523–535.

Sorcher, M., & Goldstein, A. P. (1972). A behavior modeling approach in training. *Personnel Administration*, *35*, 35–41.

Stiles, R., McCarthy, L., Munro, A., Pizzini, Q., Johnson, L., & Rickel, J. (1996). *Virtual environments for shipboard training*. Proceedings of the 1996 Intelligent Ships Symposium. Philadelphia, PA, Philadelphia American Society of Naval Engineers.

Tannenbaum, S. I., Salas, E., & Cannon-Bowers, J. A. (1996). Promoting team effectiveness. In M. West (Ed.), *Handbook of work group psychology* (pp. 503–529). Sussex, England: John Wiley & Sons.

Tannenbaum, S. I., & Yukl, G. (1992). Training and development in work organizations. *Annual Review of Psychology*, *43*(1), 399–441.

Taylor, P. J., Russ-Eft, D. F., & Chan, D. W. L. (2005). A meta-analytic review of behavior modeling training. *Journal of Applied Psychology*, *90*, 692–709.

Towler, A. J., & Dipboye, R. L. (2001). Effects of trainer expressiveness, organization, and trainee goal orientation on training outcomes. *Journal of Applied Psychology*, *86*(4), 664–673. doi:10.1037/0021–9010.86.4.664

van der Graaf, S., & Nieborg, D. B. (2003). Together we brand: America's Army. In M. Copier and J. Raessens (Eds.), *Level up: Digital games research conference* (pp. 324–338). Utrecht, Holland: Universiteit Utrecht.

van Digglelen, J., Muller, T., & van der Bosch, K. (2010). Using artificial team members for team training in virtual environments. *Lecture Notes in Computer Science*, *6356*, 28–34.

Wade, D. T. (2009). Goal setting in rehabilitation: An overview of what, why and how. *Clinical Rehabilitation*, *23*, 291–295.

Wiehagen, G. B. (2008). *Medical simulation and training – Breakthroughs and barriers*. SISO European Simulation Interoperability Workshop, Edinburgh, Scotland.

Wolfe, J., & Box, T. M. (1988). Team cohesion effects on business game performance. *Simulation and Games*, *19*(1), 82–98.

Wolfe, J. (1985). The teaching effectiveness of games in collegiate business courses: A 1973–1983 update. *Simulation & Gaming*, *16*, 251–288.

Wolfe, J. (1990). The evaluation of computer-based business games: Methodology, findings, and future needs. In James W. Gentry (Ed.), *Guide to business gaming and experiential learning* (pp. 279–300). East Brunswick, NJ: Nichols/GP.

Wolfe, J. (1993). A history of business teaching games in English-speaking and post-socialist countries: The origination and diffusion of a management education and development technology. *Simulation & Gaming*, *24*(4), 446–463. doi:10.1177/1046878193244003

Wolfe, J. (1997). The effectiveness of business games in strategic management course work. *Simulation & Gaming: An Interdisciplinary Journal*, *28*(4), 360–376.

39 Visual Perceptual Skills Training in Virtual Environments

Meredith Carroll, Matthew Johnston, and Kelly S. Hale

CONTENTS

39.1 INTRODUCTION

The prevalence of the use of camera and video technology in numerous domains and occupations has led to increased accessibility to and a proliferation of visual data for human analysis. The need for humans to review visual data in industries such as medicine, military, intelligence, and security and convert this into information that can be acted upon requires training in adaptive perceptual skill sets, that is, training search strategies that enable the effective search and recognition of relevant cues to support detection of threats or anomalies under any number of technological, environmental, cultural, and situational conditions. This includes the ability of a radiologist to detect an anomalous mass from a large volume of CT imagery, an airport baggage screener detecting possible threats in carry-on luggage, an intelligence analyst detecting a potential threat from satellite imagery, and the determination of enemy size and location from a military observation post.

The goal of this chapter is to explore how advanced technology, specifically eye tracking, can be used to increase understanding of visual perceptual processes such as search and detection and provide tools that can be used to train search skills using virtual training solutions. This chapter first provides an overview of visual search and visual search strategies and common metrics for visual search diagnosis. Finally, case studies are presented demonstrating integration of metrics for perceptual skill diagnosis in adaptive threat detection training systems.

39.2   VISUAL SEARCH

The visual search process is primarily a perceptual task requiring attention that typically involves an active scan of the visual environment for a particular object or feature (i.e., the threat) among other objects or features (i.e., distracters). Visual search is a combination of the overall scanning of the image or scene using both peripheral vision and clearly focused, or foveal, vision. Attentional focus has been compared to a spotlight, where information contained within the spotlight (i.e., fixation point that captures foveal vision) is actively captured by attention (Posner, 1980). Items outside the spotlight are processed to a certain level, but recognition occurs within the spotlight area where conscious attention is focused. The remaining peripheral information may guide future fixations based on characteristics/features of interest. However, without the focal attention, the information is not adequately perceived and considered. Attentional visual search is theorized to have two components: (1) an effortless component in which stimulus is processed preattentively (Sireteanu & Rettenbach, 2000), essentially *popping out* at the observer, and (2) an effortful component in which attention must be serially allocated to objects in the environment (Treisman & Souther, 1985). These components are influenced, respectively, by (1) characteristics of the visual scene (e.g., saliency of targets and distracters) and (2) learned search strategies. Neskovic and Cooper (2005) promote that fixations are initially driven by stimulus features, while subsequent fixations are constrained/focused by cognitive expectations during the recognition process. More recent summaries have proposed a *guiding representation* that guides attention but is not itself part of the perceptual pathway (Wolfe, Vo, Evans, & Greene, 2011). Within the guiding representation, a number of attributes have been identified that guide attention without reference to specific pathways (serial or parallel, preattentive). Learned search strategies can range from a very structured systematic search in a regular pattern (e.g., alternate up and down scan moving from left to right; Wang, Lin, & Drury, 1997) to a less structured strategy such as searching by area or searching by component/object (Chabukswar, Gramopadhye, Melloy, & Grimes, 2003).

Treisman and Gelade's (1980) feature integration theory model and Chun, Golomb, and Turk-Browne's (2011) taxonomy of internal and external attention point to two key search strategies relevant for visual search: exogenous search and endogenous search. Endogenous search is further subdivided into stimulus-driven endogenous search types—feature-based, position-based, and scene-based—and goal-directed endogenous search.

39.3   VISUAL SEARCH STRATEGIES

39.3.1   Exogenous Visual Search

Exogenous visual search is driven by external attention, where properties of the visual scene itself *pop out* based on basic human visual processing neural pathways, as the nervous system is structured to respond to certain stimuli preferentially (Trick, Enns, Mills, & Vavrik, 2004). Key features of a scene/image are captured automatically by unique receptors for color, motion, and edge in a quick, transient preattentive process (Montagna, Pestilli, & Carrasco, 2009; Treisman & Souther, 1985; Treisman, 2006; van Zoest & Donk, 2004). This process requires little effort, and as it is unintentional, it is difficult to bring under deliberate control (Trick et al., 2004). This search strategy is influenced by saliency or sensory conspicuity of features such as color, motion, orientation, and size suggested to undoubtedly guide attention by Wolfe and Horowitz (2004) who suggest attributes such as luminance, stereoscopic depth/tilt, shape, and spatial location as probable influencers.

39.3.2 ENDOGENOUS VISUAL SEARCH

Endogenous, top-down attentional orienting during visual search occurs when attention is directed in a voluntary way according to our goals and intentions (Mulckhuyse & Theeuwes, 2010). This type of intentional control is often contrasted with exogenous or bottom-up control, in which one's attention is directed in an involuntary way, irrespective of one's goals and intentions (Egeth & Yantis, 1997; Theeuwes, 2004, 2010). These types of orienting are different in several ways as shown in Table 39.1.

Endogenous visual search can be subdivided into feature-, position-, scene-, and goal-based endogenous search.

39.3.2.1 Endogenous Feature-Based Search

Endogenous feature-based visual search is an automatic endogenous process that is habitual, meaning that a particular goal or intention has been repeatedly carried out and reflexes are common to all based on a given individual's specific learning experiences (Trick et al., 2004; Wilder, Mozer, and Wickens, 2010). An example of this type of search would include an intelligence analyst searching for a target that they have learned typically that appears on a rooftop or road. The automaticity of such habits will depend on the frequency of which they are practiced, and such habits can be formed at any time and/or be replaced or fade at any time due to lack of practice or new learning (Trick et al., 2004). This search process is quick and effortlessly captures features, can be evoked unintentionally, and therefore may be more difficult to bring under deliberate control. A benefit of endogenous feature-based search is that saliency maps based on feature saliency are weighted by relevance based on experience or training. Thus, it is thought that this search, while still considered *automatic*, is more refined compared to exogenous search. Specifically, during endogenous feature-based visual search, implicit information about the target stimulus (i.e., overlearned associations) gathered from past experience results in priming of pop-out (Kristjánsson, Wang, & Nakayama, 2002) based on this past knowledge as opposed to the characteristics of the stimulus. Such implicit learning *tunes* the sensitivity of the visual system to previously learned and extracted regularities (Chun & Jiang, 1998).

39.3.2.2 Endogenous Position-Based Search

Attention in position-based search is oriented toward the location at which a target stimulus is anticipated to appear, and this in turn facilitates both detection and discriminative decisions about its perceptual properties. Thus, implicit position information about the potential location of a target stimulus based on past experience is another source of endogenous search, which is a positional analogue of priming of pop-out known as spatial contextual cuing (Chun & Jiang, 1998). An example of this search would include the association of particular anomalies to anatomical structures that are most likely to appear in the same location of medical imagery. Unlike priming, it is long lasting (Chun & Jiang, 2003), is robust to divided attention, and requires little to no excess attentional and working memory capacity (Vickery, Sussman, & Jiang, 2010). Contextual cueing is, however, modulated by selective attention and thus is primarily driven by the attended (vs. unattended) context (Jiang & Chun, 2001; Jiang & Leung, 2005). In endogenous position-based visual search, observers direct their attention to a location in space, and in turn, abrupt onsets elsewhere in the visual fields will no longer capture attention (Theeuwes, 1991; Yantis & Jonides, 1990).

TABLE 39.1

Comparison of Endogenous and Exogenous Visual Search

Endogenous	Exogenous
Relatively slow (300–500 ms)	Fast (100–120 ms, decay by 250 ms)
Sustained	Fleeting
Observer driven	Stimulus driven
Directed by priming, cueing, or learning	Directed by salient features such as color and shape

---

39.3.2.3 Endogenous Goal-Based Search

This type of search involves a cognitively driven (i.e., goal-directed, noncued) willful shift in attention based on top-down object recognition and perceptual *set* processes (Vecera, Behrmann, & Filapek, 2001). People actively search for information relevant to specific goals and perform these tasks in ways that are consistent with expectations and previous learning (Trick et al., 2004). Endogenous, goal-directed visual search utilizes internal attention resources such as knowledge (long-term episodic scene knowledge, short-term episodic scene knowledge, scene schema knowledge, task knowledge), expectations, and current goals (Corbetta & Shulman, 2002) to process what is visually present within attentional focus and decide on a course of action (e.g., when area is relevant to search, identify; when area is not relevant to search, move onto another location). Object representations stored in long-term visual memory represent familiar objects, which provide top-down feedback during visual search tasks and generally lead to the selection of familiar objects (e.g., upright object) more rapidly than less-familiar objects (e.g., rotated object; Vecera & Farah, 1997). Perceptual sets are those expectancies or task goals held by an observer (i.e., task-relevant instructions), a form of *target template* that specifies the types of information that observers can use to guide their behavior. The observer is thought to maintain a template for the target being searched for in working memory and then use it to compare with the currently attended stimulus. However, scanning effectiveness tends to be unaffected by practice and target familiarity (McCarley et al., 2004). To enhance scanning, McCarley (2009) suggests that gaze information could be used to provide feedback on uninspected regions (Dickinson & Zelinsky, 2005) and attentional cuing could be used to highlight potential target items (Wiegmann & McCarley, 2006). In addition, gaze blocking (i.e., masking previously well-examined regions of an image) may be beneficial (Qvarfordt, Biehl, Golovchinsky, & Dunningan, 2010) in further guiding search and focusing goal-directed search strategies.

39.4 VISUAL SEARCH AND THREAT DETECTION

Threat detection has been studied across a range of domains from military aviation to airport security to radiography. Threat detection has been defined in aviation as the process of evaluating relevant cues in the vicinity of one's environment and determining how much of a threat they represent by gathering and reviewing relevant information and deciding on what actions to take (Smith, Johnston, & Paris, 2004), while Fiore, Scielzo, and Jentsch (2004) define it within airport baggage screening as consisting of the ability to rapidly recognize cues in the environment and interpret the meaning and importance of these cues. The radiographic interpretation of the threat detection task can be broken down more granularly with respect to the perceptual components, describing the task as consisting of a search for an abnormality, the recognition of an abnormality, and the decision made regarding the abnormality (Nodine, Mello-Thoms, Kundel, & Weinstein, 2002).

In military observation posts, threats are searched for amid an urban backdrop bustling with distracters, while in baggage screening, threats from a learned banned item list are searched amid complex x-ray images of baggage while radiologists have the benefit of learned anatomical structures to support threat or anomaly detection. It is necessary to develop effective search strategies to successfully detect threats. Search strategies may consist of a quick search of the environment, typically from near to far in military observation post-use cases, focusing on high- and low-priority areas during which time a potential threat may halt the systematic search. Attention is then directed to the new element, evaluation is performed, and if not a threat, search continues. Preattentive search is relied upon during the initial quick search where it is hoped obvious threats would pop out from the scene. Experts will turn, however, to developed search strategies when performing more detailed searches as they attempt to detect more subtle threats. Novices do not have the requisite experience to judge high- from low-priority areas and therefore do not scan in a systematic sequence (Burgert et al., 2007; Jarodzka, Scheiter, Gerjets, & van Gog, 2009; Kasarskis, Stehwien, Hickox, Aretz, & Wickens, 2001) and therefore spend less time than experts attending to relevant aspects of the stimulus (Jarodzka et al., 2009). This lack of structure in their scan pattern is likely influenced most by bottom-up processes, forcing attention to the most salient features in the scene, such as the undoubted attributes mentioned previously (Jarodzka et al.).

39.5 TRAINING STRATEGIES FOR VISUAL SEARCH

A variety of visual search training strategies have proven successful in past research and applications and are critical to effective training (Cannon-Bowers, Rhodenizer, Salas, & Bowers, 1998). Trainees not only must have the opportunity to practice these skills, but relevant training strategies must be incorporated to ensure learning. Training strategies can be employed in practice environments to optimize learning, transfer, and retention (Cannon-Bowers et al., 1998) and they require four principles to be met to be effective:

1. Present relevant information and concepts
2. Demonstrate knowledge, skills, and abilities (KSA) to be learned
3. Create opportunities for trainees to practice the skills
4. Provide feedback to trainees regarding practice (Salas & Cannon-Bowers, 2001)

Visual search training interventions such as presentation of expert scan (Nalanagula, Greenstein, & Gramopadhye, 2006; Sadasivan, Greenstein, Gramopadhye, & Duchowski, 2005), metacognitive strategies (Chapman, Underwood, & Roberts, 2002; Nodine et al., 2002), and attentional weighting strategies (Hagemann, Strauss, & Canal-Bruland, 2006; Williams, Ward, Knowles, & Smeeton, 2002) have proven successful in improving search performance. Two examples of perceptual skills trainers that incorporate these strategies are presented at the end of this chapter. Common practice in visual search training is to provide feedback consisting of knowledge of correct response (KCR) or knowledge of results (KR). This form of purely outcome feedback has been shown successful in some domains (e.g., teacher in service training) (Leach & Conto, 1999); however, in other domains, outcome feedback may not be at a granular enough level to facilitate trainees identifying and improving process-level skills showing performance decrements (Davis, Carson, Ammeter, & Treadway, 2005; Goodman, Wood, & Hendrickx, 2004) and build retention. To address specific performance deficiencies, the appropriate training intervention should be selected. Visual search training strategies primarily fall into six categories, which are described in the following. Of these, process-level feedback, attentional weighting, metacognitive strategies, and expert performance models show the most promise.

39.5.1 PERFORMANCE FEEDBACK

Using a circuit board inspection task, Wang et al. (1997) examined whether search strategy was trainable and whether systematic, natural, or random search strategies led to better defect detection. To train systematic search, subjects were instructed to move their eyes in a regular pattern across the board alternating up and down from left to right, with specific fixation positions defined based on the size of the circuit board. Trainees practiced their scans and were given feedback with respect to whether they had followed instructions or not (i.e., KR). Similar treatment was given for a random search group. Findings indicated that practice with KR performance feedback could significantly change search strategy, both for the better with systematic search and for the worse with random search. KR performance feedback is the feedback method currently used in the field and is thus the current standard.

39.5.2 PROCESS FEEDBACK

Chabukswar et al. (2003) explored the effects of online process plus performance feedback compared to performance (outcome) feedback for a visual circuit board inspection task. Process feedback included both statistical (e.g., percent of area covered) and graphical (i.e., graphical representation of area covered), and performance feedback included items such as number of defaults detected and time to detect. The process plus performance group not only detected more defaults but seemed to develop a more systematic search strategy than the performance feedback group. Nodine, Krupinski, and Kundel (1990) developed a training strategy for radiographic interpretation in which scanned areas with dwell times greater than 1000 ms and for which no indication of lesions occurred were interpreted as detection without recognition and were fed back for reanalysis resulting in increased recognition of initial misses. This strategy is based on findings such as those reported in Nodine et al. (2002) that approximately 70% of lesions that are not reported in mammogram reading attract visual attention, as measured by the amount of visual dwell in the location of the lesion, implying that such misses are covert negative decisions.

39.5.3 ATTENTIONAL WEIGHTING

Attentional weighting focuses on targeting aspects of important cues to attend. Exogenous orienting or highlighting is a technique that has been used to train the use of attentional weighting search strategies. Hagemann, Strauss, and Cañal-Bruland (2006) found that highlighting relevant cues such as areas of the trunk, arm, and racket at the critical times during badminton training led to significant increase in test performance. Williams et al. (2002) used a freeze frame and slow-motion video playback to highlight critical cues to attend in anticipating the direction of tennis strokes. Critical cues were derived from expert performance data extracted via eye tracking. Williams et al. found significant performance improvements that also transferred to subsequent field exercises resulted from instruction that included (1) explicit instruction of critical cues and their associated outcome and (2) guided discovery with the use of verbal probes encouraging trainees to look at a certain area of the body and draw conclusions about the relation of cues to outcomes. Crowley, Medvedeva, and Jukic (2003) developed a perceptual intelligent tutoring system for pathology diagnosis that incorporated a training strategy, which includes *visual hints* by moving a viewer position to an area of interest (AOI), highlighting critical features and providing textual information about the type of feature. The researchers found this method led to improved diagnosis performance.

39.5.4 DIFFICULTY VARIATION

Level of difficulty during training may affect the development of effective search strategies by facilitating development of a more generalizable search strategy. Doane, Alderton, Sohn, and Pellegrino (1996) explored the effect of discrimination difficulty in a simple polygon discrimination task and found that more difficult stimulus tasks to discriminate lead to the development of more effective and more global search strategy and hence better transfer than with easier stimulus tasks to discriminate. Schmidt and Bjork (1992) suggested that operational performance enhancements may be better facilitated by more challenging and diverse training conditions that result in degraded speed and accuracy during skill acquisition (Schmidt & Bjork, 1992). Schmidt and Bjork found that relative to standard practice conditions, three practice conditions, namely, random practice, infrequent or faded feedback, and variation in practice, slowed the rate of improvement during training, resulting in lower training performance at the end of practice, but resulted in enhanced posttraining performance.

39.5.5 METACOGNITIVE STRATEGIES

Chapman et al. (2002) developed a training intervention aimed at improving visual search associated with the driving task, which incorporated both elements of metacognition and expert performance data. Through a series of five training modules, trainees would

1. Practice visual search for driving hazards, indicate hazards, and commentate on what they are looking at in their search
2. Explore in slow motion areas indicated as hazards, commentate on why they are hazards, and listen to an expert commentary on why these areas are hazards

3. Practice visual search while being prompted during pauses to indicate what just happened or what happened next
4. Reexplore in full-speed motion areas indicated as hazards and commentate on why they are hazards
5. Practice visual search for driving hazards, indicate hazards, and commentate on what they are looking at in their search

Results showed not only significant immediate effects on visual search, but some of these effects remained for 3–6 months and many of them transferred to an actual driving task. Nodine et al. (2002) found that in a mammography, diagnosis task that prolonged dwell time on a potential lesion did not notably increase the number of lesions discovered and did increase the error rate for lesion detection. These researchers suggest mentor-guided feedback with instructions to trust only the more confident and early decisions and to quit searching when unsure to improve detection performance in search task.

39.5.6 EXPERT PERFORMANCE MODELS

Expert search performance as illustrated through expert scan paths has been used to train visual search in multiple domains. Sadasivan et al. (2005) examined the effect of a feedforward of expert scan pattern training strategy in an airframe inspection task. In this study, expert scan paths as well as indications of fixation duration as collected via an eye tracker were presented as a static overlay to an airframe image to trainees. This training strategy resulted in 30% greater performance improvements in defect detection accuracy than a practice condition. Similar research (Mehta, Sadasivan, Greenstein, Gramopadhye, & Duchowski, 2005) compared the training effectiveness of different types of expert feedforward training strategies and found that presentation of expert scan data with a decaying trace (fixations remaining on the screen for a brief period of time before disappearing) resulted in a mean gain in number of defects detected after training five times greater than a practice condition. Eccles, Walsh, and Ingledew (2006) performed a study examining expert-novice differences in visual attention allocation in an orienteering task in which significant differences were found (Eccles et al., 2006). Based on these findings, the researchers propose presenting expert models of attention allocation to show trainees how to allocate attention properly to relevant cues, including verbal guidance on when and how to allocate visual attention to relevant cues in the environment.

Presentation of expert scan paths provides strong support for influencing trainee scan strategy, including both location to search and sequence of search. Metacognitive strategies incorporating trainee scan paths could allow trainees to explore their own performance and reflect on areas in which their performance differed from intentions or planned strategies. Attentional weighting strategies that utilize highlighting can be leveraged to direct trainee attention to high-priority areas or areas in which their performance differed from expected. Additionally, attentional weighting strategies that provide background information about features could be used to elaborate on why an area should be searched, targeting the conceptual aspect of search. Process feedback, both graphical and statistical, could be used to provide trainees information with regard to how to allocate their attention both spatially and temporally.

39.5.7 TAILORED FEEDBACK

The strategies identified previously could be made more effective by tailoring them to specific performance decrements. The tailoring or individualizing of feedback to address specific trainee performance decrements has the ability to positively impact performance. Providing trainees with information relevant to what performance areas are in need of improvement allows them to focus on these areas during future performance. Bloom (1984) defined what is referred to as the *2 sigma problem* in which trainees who received one-on-one instruction or tutoring perform two standard deviations above those receiving traditional classroom (i.e., group) training. Bloom believed that through the tutoring process (i.e., one-on-one instruction), all students have the potential to reach these levels. Training scientists have responded with the tailoring of feedback to target-specific performance decrements.

Tailoring of feedback can happen at two levels: (1) tailor to the type of skill and (2) tailor to a specific error. With respect to the first level, different feedback strategies can more effectively target certain types of knowledge or skills as different types of learning tasks require different instructional strategies and methods (Mory, 2004). Utilizing a feedback strategy that can most effectively impact the target skill allows feedback to be tailored to the specific skill decrement. For instance, it may be most effective to incorporate a strategy such as those discussed earlier that have been shown effective in improving visual search. With respect to the second level, as the main function of feedback lies in the correction of errors (Mory, 2004), tailoring the feedback to specific errors allows trainees a better understanding of deficiencies in their performance and how to improve upon these. Incorporating elements of trainee search in comparison to expert search would allow the feedback to be tailored to specific trainee decrements.

As a result, the aforementioned strategies could potentially be combined to provide a powerful training solution that supports trainees in extracting information from both expert scan paths and their own scan paths to guide improvements in search strategy. This level of tailored feedback, however, demands error analysis (i.e., what type of error was made); therefore, a necessary component of tailored feedback is performance measurement and diagnosis.

39.6 MEASURING VISUAL SEARCH PERFORMANCE

Perceptual processes are not easily observable and therefore not easily measured. Head movement can provide gross indication of scanning behavior (Itti & Koch, 2001; Treisman & Souther, 1985) but cannot be used to assess visual attention allocation. Eye tracking can provide insight into subtle physical behaviors such as scanning patterns or internal perceptual processes such as detection that can be used to diagnose root cause of performance breakdowns that are not currently accessible via behavioral metrics (Wang & Winslow, 2014, see Chapter 8). Eye tracking has been used to measure perceptual processes such as visual attention in a driving task (Underwood, Chapman, Brocklehurst, Underwood, & Crundall, 2003) and visual search in a mammogram diagnosis task (Mello-Thoms, Dunn, Nodine, Kundel, & Weinstein, 2002) and integrated with intelligent tutoring systems (ITSs) to monitor attention and interests (Wang, Chignell, & Ishizuka, 2006). Eye tracking may provide invaluable data in understanding how performance unfolded and facilitate the measurement of perceptual performance at a very granular level. However, it is necessary to transform this detailed data into meaningful and actionable performance diagnoses.

39.7 DIAGNOSING VISUAL SEARCH PERFORMANCE

Event-based methods such as event-based knowledge elicitation (Fowlkes, Salas, Baker, Cannon-Bowers, & Stout, 2000) and event-based approach to training (Fowlkes, Dwyer, Oser, & Salas, 1998) have been successful in eliciting knowledge or procedural responses. Such diagnosis methods equate to measurement of behavioral actions and communications to verify if all steps in the procedures were followed and which steps were omitted or performed incorrectly. With respect to the threat detection task, diagnosis has previously been limited by the performance measurement technology, but eye tracking technology can reveal information about a person's perceptual state with several researchers having utilized eye tracking to diagnose perceptual performance deficiencies in studying radiographic interpretation to diagnose where in the perceptual process (i.e., search, recognition, decision) errors occur. Nodine et al. (2002) make distinctions in the classification of errors in misdiagnosis in radiographic interpretation based on fixation duration. In this case, the lack of fixation is interpreted as a searching error, fixation for less than 1000 ms and lack of indication of abnormality are interpreted as a recognition error, and fixation for greater than 1000 ms and failure to indicate as abnormality are a decision-making error. A visual dwell time of 1000 ms is equated to detection as it is considered a significant allocation of visual attention (Nodine et al., 2002), and recognition of an abnormality depends on higher-order cognitive processes (Manning, Leach, & Bunting, 2000). Researchers have proposed that eye fixations can *reveal the sequence of mental operations* during a mental rotation task (Just & Carpenter, 1976, p. 459), which is precisely the goal of using such data in perceptual performance diagnosis.

39.7.1 Case Study 1: Screen Adapt

The Screener's Auto-Diagnostic Adaptive Precision Training (ScreenADAPT) is designed to enhance the detection and recognition skills of baggage screeners by integrating eye tracking metrics and behavioral performance to diagnose root cause of error. ScreenADAPT incorporates a variety of the training strategies for visual search mentioned previously to tailor training content to the performance deficiencies of the individual trainee (Winslow et al., 2013).

As mentioned previously, outcome feedback alone may not be granular enough to identify and improve trainee skills as it cannot provide knowledge of the cause of errors in performance. ScreenADAPT goes beyond simple performance feedback and incorporates five of the six strategies mentioned in the previous section. ScreenADAPT provides performance feedback at intervals during training typically associated with performance results on a pretest that determines the insertion point into the training curriculum and when performance dictates a training intervention is required or when more difficult content is to be introduced. Performance feedback is also provided within the training interventions employed in the form of an after action review (AAR) (Winslow et al., 2013).

When a negative trend in performance is observed, a trainee is assigned to one of two intervention strategies: (1) exposure training, where they are shown serial presentations of threats in bags with characteristics aligned with the negative performance trend, or (2) discrimination training in which the trainee must determine if a threat is the same or different in a parallel presentation of baggage images. In both cases, the user is presented with a bounding box that shows where the threat is located after they make their selection and their success or lack thereof is communicated.

Process-level feedback is provided in the form of representations of scan patterns employed on baggage imagery that was either a miss or false alarm. The trainee is able to see how they scanned a bag where he or she failed to detect or recognize a threat or deemed a threat present when a distractor was present. ScreenADAPT also varies the difficulty of the imagery presented based on characteristics such as the amount of clutter or the orientation of the threat presented. Performance of the trainee can then be related to the characteristics of an image, not the image itself; this allows for generation, in real time, of imagery that progressively increases difficulty as the user improves in performance. Metacognitive strategies are also used by asking the trainee how well they believe they are performing and providing a contrast of their perception with their actual results in an AAR (Winslow et al., 2013).

Empirical results from an investigation into the benefits of ScreenADAPT approach to adaptive training suggest a general improvement in baggage screening performance. Specifically, participants demonstrated a reduction in the number of threat items missed and a reduced response time. Traditional training methods by comparison did show a significantly higher sensitivity and lower false-alarm rate; however, the false-alarm rate was trending higher, while the adaptive training was trending downward after multiple training sessions (Hale et al., 2012). Future research is planned to individually validate specific training interventions as well as add the expert performance model training strategy, which is discussed in the following section.

39.7.2 Case Study 2: ADAPT-AAR

Expert performance models are intended to provide a trainee knowledge of how an expert performs a given task. Typically, this is implemented through presentation of process feedback such as scanning behavior. Auto-diagnostic adaptive precision trainer after action review (ADAPT-AAR) is a perceptual skills trainer, which allows trainees to practice search and detection skills while leveraging eye-tracking technology to assess trainee perceptual skills, demonstrate expert scan patterns

FIGURE 39.1 **(See color insert.)** Expert versus trainee fixations.

and performance, and allow a trainee to compare their scan with an expert scan. In this system, an expert is shown an image and is asked to search for a threat. Their scan pattern and behavioral performance is stored. The expert is also asked to classify areas of the image as high, medium, or low priority. Although a threat may not be present, the trainee should scan areas of high priority. After trainee performance, they are then presented a playback of their scan path and that of the expert (Figure 39.1). The scan path is overlaid on the image and color coded to suggest when they have scanned areas of high or low priority. Figure 39.1 shows an example of expert fixations compared to trainee fixations with red bounding boxes showing AOIs the trainee should have scanned, but did not.

A study was completed to explore the training effectiveness of the expert performance model strategy utilizing eye tracking technology to support both expert demonstration and trainee assessment (Carroll et al., 2013). Eye tracking data was utilized to measure trainee scan data, diagnose performance deficiencies, and provide tailored feedback allowing the trainee to compare their scan with an expert scan. A training effectiveness evaluation (TEE) comparing the ADAPT-AAR method to traditional training was performed within the context of a Combat Hunter training course at the United States Marine Corps (USMC) School of Infantry East (SOI-E). The expert performance model approach demonstrated significant improvements in search strategies compared to traditional instructor-based feedback. Specifically, after training, the trainee's search strategy more resembled that of expert scan strategies, less time was spent in areas of lower priority, and scan paths were more efficient. ADAPT-AAR also demonstrated greater transfer to practical application exercises suggesting this tool would lead to improved field performance (Carroll et al., 2013).

39.8 FUTURE DIRECTIONS

Many perceptual skills training strategies from the training science literature have not reached the field because of the multiple challenges associated with training perceptual skills such as search. The primary challenge is the ability to adequately measure and diagnose search performance to facilitate process-level feedback. In order to be able to debrief at a subtask/process level, it is necessary to be able to distinctly measure a subprocess (i.e., search) and effectively discriminate performance on separate subprocesses (e.g., search and detection). Diagnosis at this level determines where in the perceptual process the process-level error that led to an outcome error occurred and can facilitate feedback to target this specific process. A second challenge is in obtaining the ability to demonstrate search skills. Perceptual skills such as search and detection are subtle or internal processes that are unobservable. As a result, many training strategies consisting of demonstration are infeasible and will instead typically consist of verbal description of skill performance, not actual demonstration. A promising solution to these challenges is the use of eye tracking technology to measure visual performance. With the advancements in eye tracking technology, information about a person's visual attention, once inaccessible, is becoming more attainable. Visual attention can provide important insights to the information used in task performance, such as the importance of various features or cues (Raab & Johnson, 2007). Several studies (Jarodzka et al., 2009; Mello-Thoms et al., 2008; Raab & Johnson, 2007; White, Hutson, & Hutchinson, 1997) have used eye tracking to extract information about scan strategies. These studies have demonstrated that eye tracking can aid in the assessment of perception through measurement of visual attention during observation via gaze, scan path, and fixation data. These metrics can identify which AOIs were gazed upon and the amount of time the AOIs were gazed upon. With respect to the challenge of measuring and diagnosing search performance, researchers studying radiographic interpretation have used eye tracking data to diagnose where in the perceptual–cognitive process (i.e., search, recognition, decision) errors occur. If effective fixation duration thresholds could be established for discriminating the different perceptual processes (i.e., search, detection, recognition), this method could be used in the diagnosis of process-level threat detection errors and discrimination of search versus detection errors. This would facilitate the process-level feedback needed to effectively target the perceptual root cause of performance deficiencies.

With respect to the challenge of utilizing training strategies that consist of demonstration, eye tracking technology provides the ability to capture and present expert search demonstrations. What remains is the need to determine how to effectively present this granular data in order to facilitate effective training of all aspects of search skills. One consideration in targeting search location is that there are two unique aspects of *location* that warrant consideration. The first is the perceptual aspect of the location or what the location visually looks like, which allows identification of a specific area. The second aspect of the location is the conceptual aspect of why the area is a target area (i.e., high-priority area that should be searched), an understanding of which is necessary to abstract specific locations to a higher level of categorization that would facilitate generalization to new environments and situations. There is an opportunity to leverage training strategies from the training science literature that have been proven successful in improving search skills and extend them to address all critical aspects of search performance.

Eye tracking technology is under rapid development with the goal of improving accuracy and reliability and reducing cost. There have been attempts to integrate remote eye tracking with laptops resulting from recent progress in hardware design and implementation. Aside from hardware improvements, software support is also critical toward true gaze-based interaction. Providing a framework that handles low-level eye tracking details and a clean interface for application programmers will be key to promoting gaze-controlled applications. Nontraditional input modalities such as phones and tablets combined with eye tracking technique are also becoming more popular. With the success of smartphones, tablets, and game consoles, it is possible that the eye tracking will be brought into mainstream products to facilitate interactions (Miluzzo, Wang, & Campbell, 2010), thus enhancing the ability to measure and quantify visual perceptual skills for identifying and mitigating deficiencies/inefficiencies and enhancing adaptive training system design.

REFERENCES

Bloom, B. S. (1984). The 2 sigma problem: The search for methods of group instruction as effective as one-to-one tutoring. *Educational Researcher, 13*(6), 4–16.

Burgert, O., Orn, V., Velichkovsky, B. M., Gessat, M., Joos, M., Strauss, G., ... Hertel, I. (2007). *Evaluation of perception performance in neck dissection planning using eye tracking and attention landscapes.* Paper presented at the Medical Imaging 2007: Image Perception, Observer Performance, and Technology Assessment, San Diego, CA.

Cannon-Bowers, J. A., Rhodenizer, L., Salas, E., & Bowers, C. A. (1998). A framework for understanding pre-practice conditions and their impact on learning. *Personnel Psychology, 51*(2), 291–320.

Carroll, M., Kokini, C. and Moss, J. (2013). Training effectiveness of eye tracking-based feedback at improving visual search skills, *International Journal of Learning Technology*, *8*(2), 147–168.

Chabukswar, S., Gramopadhye, A. K., Melloy, B. J., & Grimes, L. W. (2003). Use of aiding and feedback in improving visual search performance for an inspection task. *Human Factors and Ergonomics in Manufacturing*, *13*(2), 115–136.

Chapman, P., Underwood, G., & Roberts, K. (2002). Visual search patterns in trained and untrained novice drivers. *Transportation Research Part F: Psychology and Behaviour*, *5*(2), 157–167.

Chun, M. M., Golomb, J. D., & Turk-Browne, N. B. (2011). A taxonomy of external and internal attention. *Annual review of Psychology, 62*, 73–101.

Chun, M. M., & Jiang, Y. (1998). Top-down attentional guidance based on implicit learning of visual covariation. *Psychological Science*, *10*(4), 360–365.

Chun, M. M., & Jiang, Y. (2003). Implicit, long-term spatial context memory. *Journal of Experimental Psychology: Learning, Memory, Cognition, 29*, 224–234.

Corbetta, M., & Shulman, G.L. (2002). Control of goal-directed and stimulus-driven attention in the brain. *Nature Reviews, 3*, 201–216.

Crowley, R., Medvedeva, O., & Jukic, D. (2003). SlideTutor: A model-tracing Intelligent Tutoring System for teaching microscopic diagnosis. *IOS Press: Proceedings of the 11th International Conference on Artificial Intelligence in Education*. Sydney, New South Wales, Australia, 2003.

Davis, W. D., Carson, C. M., Ammeter, A. P., & Treadway, D. C. (2005). The interactive effects of goal orientation and feedback specificity on task performance. *Human Performance*, *18*(4), 409–426.

Dickinson, C. A., & Zelinsky, G. (2005). Marking rejected distractors: A gaze-contingent technique for measuring memory during search. *Psychonomic Bulletin & Review*, *12*, 1120–1126.

Doane, S. M., Alderton, D. L., Sohn, Y. W., & Pellegrino, J. W. (1996). Acquisition and transfer of skilled performance: Are visual discrimination skills stimulus specific? *Journal of Experimental Psychology: Human Perception and Performance*, *22*(5), 1218–1248.

Eccles, D. W., Walsh, S. E., & Ingledew, D. K. (2006). Visual attention in orienteers at different levels of experience. *Journal of Sports Sciences*, *24*(1), 77–87.

Egeth, H. E., & Yantis, S. (1997). Visual attention: Control, representation, and time course. *Annual Review of Psychology, 48*, 269–297.

Fiore, S. M., Scielzo, S., & Jentsch, F. (2004). Stimulus competition during perceptual learning: Training and aptitude considerations in the x-ray security screening process. *International Journal of Cognitive Technology*, *9*(2), 34–39.

Fowlkes, J., Dwyer, D. J., Oser, R. L., & Salas, E. (1998). Event-based approach to training (EBAT). *International Journal of Aviation Psychology*, *8*(3), 209–221.

Fowlkes, J. E., Salas, E., Baker, D. P., Cannon-Bowers, J. A., & Stout, R. J. (2000). The utility of event-based knowledge elicitation. *Human Factors: The Journal of the Human Factors and Ergonomics Society*, *42*(1), 24–35.

Goodman, J. S., Wood, R. E., & Hendrickx, M. (2004). Feedback specificity, exploration, and learning. *Journal of Applied Psychology*, *89*(2), 248–262.

Hagemann, N., Strauss, B., & Canal-Bruland, R. (2006). Training perceptual skill by orienting visual attention. *Journal of Sport & Exercise Psychology*, *28*(2), 143–158.

Hale, K.S., Carpenter, A., Johnston, M. R., Costello, J., Flint, J., & Fiore, S. M. (2012). Adaptive Training for Visual Search. *Proceedings of the Interservice/Industry Training, Simulation & Education Conference (I/ITSEC 2012*. Orlando, FL. Paper # 12144.

Itti, L., & Koch, C. (2001). Computational modelling of visual attention. *Nature Reviews Neuroscience*, *2*(3), 194–203.

Jarodzka, H., Scheiter, K., Gerjets, P., & van Gog, T. (2009). In the eyes of the beholder: How experts and novices interpret dynamic stimuli. *Learning and Instruction*, *20*(2), 146–154.

Jiang, Y., & Chun, M. M. (2001). The influence of temporal selection on spatial selection and distractor interference: An attentional blink study. *Journal of Experimental Psychology: Human Perception & Performance*, *27*, 664–679.

Jiang, Y., & Leung, A. W. (2005). Implicit learning of ignored visual context. *Psychonomic Bulletin & Review*, *12*, 100–106.

Just, M. A., & Carpenter, P. A. (1976). Eye fixations and cognitive processes. *Cognitive Psychology*, *8*(4), 441–480.

Kasarskis, P., Stehwien, J., Hickox, J., Aretz, A., & Wickens, C. (2001). *Comparison of expert and novice scan behaviors during VFR flight*. Paper presented at the 11th International Symposium on Aviation Psychology. Columbus, OH: The Ohio State University.

Kristjánsson, Á., Wang, D., & Nakayama, K. (2002). The role of priming in conjunctive visual search. *Cognition, 85,* 37–52.

Leach, D. J., & Conto, H. (1999). The additional effects of process and outcome feedback following brief in-service teacher training. *Educational Psychology, 19*(4), 441–462.

Manning, D. J., Leach, J., & Bunting, S. (2000). A comparison of expert and novice performance in the detection of simulated pulmonary nodules. *Radiography, 6,* 111–116.

McCarley, J. S. (2009). Effects of speed–accuracy instructions on oculomotor scanning and target recognition in a simulated baggage X-ray screening task. *Ergonomics, 52*(3), 325–333.

McCarley, J. S., Kramer, A. F., Wickens, C. D., Vidoni, E. D., & Boot, W. R. (2004). Visual skills in airport-security screening. *Psychological Science, 15*(5), 302–306.

Mehta, P., Sadasivan, S., Greenstein, J. S., Gramopadhye, A. K., & Duchowski, A. T. (2005). Evaluating different display techniques for communicating search strategy training in a collaborative virtual aircraft inspection environment. *Human Factors and Ergonomics Society Annual Meeting Proceedings, 49,* 2244–2248.

Mello-Thoms, C., Dunn, S., Nodine, C. F., Kundel, H. L., & Weinstein, S. P. (2002). The perception of breast cancer: What differentiates missed from reported cancers in mammography. *Academic Radiology, 9*(9), 1004–1012.

Mello-Thoms, C., Ganott, M., Sumkin, J., Hakim, C., Britton, C., Wallace, L., & Hardesty, L. (2008). Different search patterns and similar decision outcomes: how can experts agree in the decisions they make when reading digital mammograms? *Digital Mammography, 5116,* 212–219.

Miluzzo, E., Wang, T., & Campbell, A. T. (2010). *EyePhone: Activating mobile phones with your eyes.* Proceedings of the Second ACM SIGCOMM Workshop on Networking, Systems, and Applications on Mobile Handhelds. New Delhi, India: ACM, pp. 15–20.

Montagna, B., Pestilli, F., & Carrasco, M. (2009). Attention trades off spatial acuity. *Vision Research, 49,* 735–745.

Mory, E. H. (2004). Feedback research revisited. In D. H. Jonassen (Ed.), *Handbook of research on educational communications and technology* (2nd ed., pp. 745–783). Mahwah, NJ: Lawrence Erlbaum Associates Publishers.

Mulckhuyse, M., & Theeuwes, J. (2010). Unconscious attentional orienting to exogenous cues: A review of the literature. *Acta Psychologica, 134,* 299–309.

Nalanagula, D., Greenstein, J. S., & Gramopadhye, A. K. (2006). Evaluation of the effect of feedforward training displays of search strategy on visual search performance. *International Journal of Industrial Ergonomics, 36*(4), 289–300.

Neskovic, P., & Cooper, L. N. (2005). Visual search for object features. In L. Wang, K. Chen, & Y.S. Ong (Eds.), *Advances in natural computation* (ICNC 2005, LNCS 3610, pp. 877–887). Springer-Verlag: Berlin, Germany.

Nodine, C. F., Krupinski, E. A., & Kundel, H. L. (1990). A perceptually-based algorithm provides effective visual feedback to radiologists searching for lung nodules. *Proceedings of the First Conference on Visualization in Biomedical Computing, 1990.* Atlanta, Georgia, pp. 202–207.

Nodine, C. F., Mello-Thoms, C., Kundel, H. L., & Weinstein, S. P. (2002). Time course of perception and decision making during mammographic interpretation. *American Journal of Roentgenology, 179,* 917–923.

Posner, M. I. (1980). Orienting of attention. *Quarterly Journal of Experimental Psychology, 32,* 3–25.

Qvarfordt, P., Biehl, J. T., Golovchinsky, G., & Dunningan, T. (2010). Understanding the benefits of gaze enhanced visual search. *Eye Tracking Research & Applications* (ETRA 2010; 2010 March 22–24, Austin, TX), 283–290.

Raab, M., & Johnson, J. G. (2007). Expertise-based differences in search and option-generation strategies. *Journal of Experimental Psychology: Applied, 13*(3), 158–170.

Sadasivan, S., Greenstein, J. S., Gramopadhye, A. K., & Duchowski, A. T. (2005). *Use of eye movements as feedforward training for a synthetic aircraft inspection task.* Paper presented at the Proceedings of the SIGCHI Conference on Human Factors in Computing Systems. Portland, Oregon.

Salas, E., & Cannon-Bowers, J. A. (2001). The science of training: A decade of progress. *Annual Review of Psychology, 52,* 471–499.

Schmidt, R. A., & Bjork, R. A. (1992). New conceptualizations of practice: Common principles in three paradigms suggest new concepts for training. *Psychological Science, 3*(4), 207–217.

Sireteanu, R., & Rettenbach, R. (2000). Perceptual learning in visual search generalizes over tasks, locations, and eyes. *Vision Research, 40*(21), 2925–2949.

Smith, C. A. P., Johnston, J., & Paris, C. (2004). Decision support for air warfare: Detection of deceptive threats. *Group Decision and Negotiation, 13*(2), 129–148.

Theeuwes, J. (1991). Exogenous and endogenous control of attention: The effect of visual onsets and offsets. *Perception & Psychophysics*, *49*, 83–90.

Theeuwes, J. (2004). Top-down search strategies cannot override attentional capture. *Psychonomic Bulletin & Review*, *11*(1), 65–70.

Theeuwes, J. (2010). Top-down and bottom-up control of visual selection. *Acta Psychologica*, *135*, 77–99.

Treisman, A. (2006). How the deployment of attention determines what we see. *Visual Cognition*, *14*(4/5/6/7/8), 411–443.

Treisman, A., & Gelade, G. (1980). A feature-integration theory of attention. *Cognitive Psychology*, *12*(1), 97–136.

Treisman, A., & Souther, J. (1985). Search asymmetry: A diagnostic for preattentive processing of separable features. *Journal of Experimental Psychology: General*, *114*(3), 285–310.

Trick, L. M., Enns, J. T., Mills, J., & Vavrik, J. (2004). Paying attention behind the wheel: A framework for studying the role of attention in driving. *Theoretical Issues in Ergonomics Science*, *5*(5), 385–424.

Underwood, G., Chapman, P., Brocklehurst, N., Underwood, J., & Crundall, D. (2003). Visual attention while driving: Sequences of eye fixations made by experienced and novice drivers. *Ergonomics*, *46*(6), 629–646.

van Zoest, W., & Donk, M. (2004). Bottom-up and top-down control in visual search. *Perception*, *33*(8), 927–937.

Vecera, S. P., Behrmann, M., & Filapek, J. C. (2001). Attending to the parts of a single object: Part-based selection limitations. *Perception & Psychophysics*, *63*, 308–321

Vecera, S. P., & Farah, M. J. (1997). Is visual image segmentation a bottom-up or an interactive process? *Perceptual Psychophysics*, *59*(8), 1280–1296.

Vickery, T. J., Sussman, R. S., & Jiang, Y. V. (2010). Spatial context learning survives interference from working memory load. *Journal of Experimental Psychology: Human Perception & Performance*, *36*, 1358–1371.

Wang, H., Chignell, M., & Ishizuka, M. (2006). *Empathic multiple tutoring agents for multiple learner interface*. Paper presented at the 2006 IEEE/WIC/ACM International Conference on Web Intelligence and Intelligent Agent Technology. Hong Kong, China.

Wang, M. J., Lin, S., & Drury, C. G. (1997). Training for strategy in visual search. *International Journal of Industrial Ergonomics*, *20*(2), 101–108.

Wang, X., & Winslow, B. (2014). Eye tracking in virtual environments. In K. S. Hale. & K. M. Stanney (Eds.), *Handbook of virtual environments: Design, implementation, and applications* (2nd ed., pp. 197–210). Boca Raton, FL: Taylor & Francis Group, Inc.

White, K. P., Jr., Hutson, T. L., & Hutchinson, T. E. (1997). Modeling human eye behavior during mammographic scanning: Preliminary results. *IEEE Transactions on Systems, Man, & Cybernetics Part A: Systems & Humans*, *27*(4), 494–505.

Wiegmann, D., & McCarley, J. (2006). Age and automation interact to influence performance of a simulated luggage screening task. *Aviation, Space, and Environmental Medicine*, *77*, 825–831.

Wilder, M. H., Mozer, M. C., & Wickens, C. D. (2010). An integrative, experience-based theory of attentional control. *Journal of Vision*, *11*(2), 1–30.

Williams, A. M., Ward, P., Knowles, J. M., & Smeeton, N. J. (2002). Anticipation skill in a real-world task: Measurement, training, and transfer in tennis. *Journal of Experimental Psychology: Applied*, *8*(4), 259–270.

Winslow, B., Carpenter, A., Flint, J., Wang, X., Tomasetti, D., Johnston, M., & Hale, K. (2013). Combining EEG and eye tracking: Using fixation-locked potentials in visual search. *Journal of Eye Movement Research*, *6*(4):5, 1–11.

Wolfe, J. M., & Horowitz, T. S. (2004). What attributes guide the deployment of visual attention and how do they do it? *Nature Reviews Neuroscience*, *5*(6), 495–501.

Wolfe, J. M., Vo, M. L.-H., Evans, K. K., & Greene, M. R. (2011). Visual search in scenes involves selective and nonselective pathways. *Trends in Cognitive Sciences*, *15*(2), 77–84.

Yantis, S., & Jonides, J. (1990). Abrupt visual onsets and selective attention: Voluntary versus automatic allocation. *Journal of Experimental Psychology: Human Perception & Performance*, *16*, 121–134.

40 Virtual Environments as a Tool for Conceptual Learning

Robb Lindgren, J. Michael Moshell, and Charles E. Hughes

CONTENTS

40.1 INTRODUCTION

In this chapter, we discuss the design of virtual environments (VEs) for conceptual learning: understanding the important facts, principles, and relationships across many academic domains. The use of VEs in both formal and informal educational settings has exploded over the last decade. VEs have been used frequently by educators to connect students to a larger community of learners and as a way to appeal to a generation of children growing up with video games, online social networks, and other sophisticated interactive technologies. But while much of the rationale for using VEs in education is social and motivational in nature, these environments also have special properties for facilitating concept development and shaping student cognition in constructive ways. Rather than a comprehensive review of VEs for learning (for a more exhaustive survey, see Lopez, Hughes, Mapes, & Dieker, 2012; Mikropoulos & Natsis, 2011), this chapter will focus on five strategies for promoting conceptual learning in VEs: *exploration, invention, inquiry, perceptual tuning,* and *action modification.* Examples of each of these strategies used in educational technology research projects will be presented. Before describing these strategies in detail, we first define the space of educational VEs and describe their relevance to some of the predominant theories of how people learn.

40.2 VEs IN EDUCATION

40.2.1 DEFINING VEs IN EDUCATION

Broadly defined, a VE is a computer-generated simulation of a real or imaginary environment. In educational contexts, a student typically interacts with a VE via some form of analog control and within a spatial frame of reference. This basic definition is often extended to include multimodal stimuli (e.g., visual, sound, tactile feedback), immersive displays (e.g., encapsulating screens and head-mounted displays), the blending of real assets with virtual content, and the involvement of multiple users in a shared simulation. For education, the defining features of a VE often reside in the ways in which a learner interacts with the environment—symbolic manipulation through a mouse or keyboard, through an embodied avatar, through direct physical manipulation, etc. VEs constitute numerous types of computer applications that currently occupy the landscape of educational technologies, including serious games, browser- or mobile-based apps and simulations, multiuser virtual environments (MUVEs), and virtual and mixed reality.

40.2.2 AFFORDANCES OF VEs FOR LEARNING

Much of the excitement that has been generated around VEs in education comes from several perceived affordances of these environments for learning. Specific instructional approaches in VEs that are aligned with learning theory will be discussed shortly. First, we review some of the basic characteristics of VEs that have initially led educational practitioners and researchers to explore their applications.

Likely the most cited reason for using VEs in education is that they are engaging, fun, and novel platforms for learning that often have little resemblance to school. Studies have shown that, irrespective of other effects, students consistently rate learning experiences in virtual worlds such as MUVEs as more enjoyable than other educational platforms with comparable content (e.g., Sullivan et al., 2011). The importance of this attribute of VEs should not be underestimated given the challenges of keeping students attuned to their academic development amidst compelling and sophisticated media technologies designed primarily for entertainment. Another significant benefit of VEs for learning is their ability to simulate experiences that may be impractical or impossible to access in the real world, such as exploring the ocean floor or navigating through a field of ozone molecules. VEs can cheaply provide experience with otherwise expensive equipment, and they make it possible to make mistakes while practicing delicate procedures that can have costly or even deadly consequences in the real world. Virtual field trips provide context that is hard to replicate in a lab while affording experimental controls and replicability that is hard to achieve in the field (Fiore, Harrison, Hughes, & Rutström, 2009). Moreover, virtual excursions consume no jet fuel; virtual laboratories do not explode; virtual dissection kills no animals. These environments can also be used to provide safe access to everyday social interactions for individuals with disabilities such as autism (Parsons, Leonard, & Mitchell, 2006).

Given their spatial nature and their reliance on navigation metaphors, VEs also have unique affordances for developing a student's spatial knowledge representations of a domain (Dalgarno & Lee, 2010). Representing complex concepts and ideas spatially has special properties for learning that can be exploited with interactive simulations and virtual worlds (Lindgren & Schwartz, 2009), and studies have shown that spatial perspectives can be manipulated in VEs in ways that enhance knowledge (Luo, Luo, Wickens, & Chen, 2010). There are also advantages to the common practice of situating students in VEs as embodied actors or avatars. Carrying out learning interactions through avatars provides a means of externalizing and reflecting upon one's knowledge that can facilitate higher-order thinking (Falloon, 2010).

Although contemporary uses of VEs in education often fail to go beyond replications of real-world activity in virtual spaces, a particularly valuable feature of VEs is the ability to *transform*

everyday interactions such that they provide new insights and reveal misconceptions about important phenomena. Dede (2009) describes how immersive simulations give new perspectives and present opportunities for visualizing complicated data systems. In their studies of learning in virtual
reality, Bailenson et al. (2008) describe the effects of what they call transformed social interactions
(TSIs) where synthetic constructions of the social environment, such as positioning a student in a
classroom where virtual colearners exhibit positive behaviors (listening, taking notes, etc.), benefit
student learning. Advances in augmented reality (AR) technologies make it possible to overlay these
kinds of activity transformations directly on top of real-world scenes, such as with AR tablet applications that place data and other digital elements within the device's camera stream. A particularly
important form of augmentation that VEs can offer students is the delivery of targeted and intelligent feedback in real time. Effective feedback systems in these environments leverage features of
video games that scaffold user performance through gradually increasing challenges, perceptual
supports, and finely tuned scoring systems that measure progress toward achieving important objectives (Charles, Charles, McNeill, Bustard, & Black, 2011; Honey & Hilton, 2011). Hand in hand with
the ability to deliver rich and targeted feedback is the ability to record detailed accounts of learner
actions and decisions within the VE via logfiles and other recording capabilities. These data can
be fed back to the learner for reflection—an activity commonly referred to in the simulation and
training literature as *after-action review*, a process that can help a user identify weaknesses in performance or strategy (Lopez et al., 2012). The data that can be collected about learner behavior in
these environments also provide an excellent foundation for the emerging field of learning analytics
(see Long, Siemens, Conole, & Gasevic, 2011).

A final set of affordances of VEs in education pertain to their ability to convey realistic settings and situations upon which to anchor learning. Current technologies allow for the construction
of high-fidelity simulations of human activity that possess both graphical and structural realism.
Although there is still a dearth of empirical studies demonstrating the direct benefits of VE fidelity,
there is strong sentiment in the field that accurately depicting knowledge application scenarios in
the real world increases the possibility of successful learning transfer (Dalgarno & Lee, 2010). Not
only can VEs capture physical realism, but some have argued that they also are more successful at
preserving the social and cultural authenticity of the learning domain (Warburton, 2009). Finally,
while collaborative learning is not the primary focus of this chapter, there has been a considerable
amount of research devoted to demonstrations of VEs as an effective platform for team work and
community building (e.g., Burton & Martin, 2010).

40.2.3 OBSTACLES TO THE EFFECTIVE USE OF VES IN EDUCATION

Despite the numerous affordances for education detailed earlier, it is certainly not the case that VEs
have replaced the traditional classroom. There still exist many challenges to effectively using simulations, MUVEs, and virtual reality technologies in education, particularly at the scale of broad
implementation (Honey & Hilton, 2011). One reason is cost. While setting up a VE using one of
the many publically available platforms may be relatively cheap, populating the environment with
effective content and maintaining that content can be expensive. Additionally, the use of VEs that
go beyond ones that can be run on desktop computers, such as tablet-based environments and mixed
reality platforms, can quickly exceed the resources of a typical brick and mortar school. It can be
especially difficult to justify the cost of these environments if effective learning in VEs is dependent
on individual differences, as some research has suggested (Waller, 2000).

Another challenge is that VEs are complex, and so orchestrating the outcomes that educators
would like to see from these environments is nontrivial. For example, although there is great potential to facilitate learning transfer with VEs, matching the relevant characteristics of the virtual task
with the context features of the target real-world task is an extremely nuanced and precarious undertaking. Likewise, many contemporary VEs admirably seek to support collaboration, but the addition
of multiple people into the learning space only increases the complexity. A further encumbrance is

that even if a VE does have a significant effect on student learning, the types of experimental and discovery-oriented activities that are performed in these environments are frequently incommensurate with the traditional forms of assessments used to measure learning (Schwartz, Lindgren, & Lewis, 2009). Thus, designers of VEs must frequently invent new forms of assessment to effectively capture the particular learning affordances of the environment.

A final challenge of using VEs in education that is the main target of this chapter is the fact that designers and practitioners will often utilize VEs in an effort to generate new learning without a coherent pedagogical strategy or the use of sound theories of learning. The next section describes a few such theories that are particularly well suited for enactment within VEs, followed by a set of specific approaches to cultivating conceptual learning.

40.3 APPLICABLE LEARNING THEORY

40.3.1 CONSTRUCTIVISM

Constructivist learning theory is rooted in Dewey's (1966) idea that education is driven by experience. Piaget (1997) put this notion in cognitive terms by describing how a child's knowledge structures are constructed through exploratory interactions with the world around them. Constructivism has been deeply influential in educational design over the last several decades, pushing instructors to create authentic interactions with the world that are consistent with the knowledge students are expected to develop (Duffy & Jonassen, 1992). VEs can provide opportunities for learners to be participants in goal-driven, authentic activities, and the interactions that they have with objects, systems, and people in these environments facilitate the construction of knowledge about these activities (Dede, 1995; Winn, 1999). Dede (2009) describes the immersive qualities of VEs that permit students to *suspend disbelief* that their actions are taking place inside a digital environment rather than the real world. In their review of educational VEs, Mikropoulos and Natsis (2011) assert that the vast majority are based on the theory of constructivism and that constructivist principles can be used to evaluate their effectiveness. While VEs are widely seen as effective for learning because they provide good approximations of real-world interactions, there are some who have argued that VEs can be superior constructivist learning environments because of their capacity for control and reflection (e.g., Chittaro & Ranon, 2007). While actions in the real world may only practically be completed a fixed number of times, actions in VEs are repeatable and contextual factors (wind, friction, social climate, etc.) are changeable, allowing for efficiency in experimentation.

40.3.2 SITUATED LEARNING

The idea that knowledge is inseverably tied to the situation in which it is acquired is referred to as situated learning (Brown, Collins, & Duguid, 1989). Situated learning theory and *cognitive apprenticeship*—the associated pedagogical approach where students are enculturated into expert practices through authentic activity and social interaction—challenge the long-standing premise that conceptual knowledge can be abstracted from concrete situations and taught in a compartmentalized manner. VEs are increasingly seen as effective spaces for cultivating situated learning given their capacity to accurately simulate not only physical contexts but social and cultural contexts as well. For example, Barab, Thomas, Dodge, Carteaux, and Tuzun (2005) describe MUVEs as "persistent social and material environments, universes with their own culture and discourses" (p. 90). Researchers have shown that VEs, especially those employing embodied avatars, can engender high degrees of social presence (Annetta & Holmes, 2006) and that people interact socially in VEs in ways that largely mimic people's behavior in the real world (Yee, Bailenson, Urbanek, Chang, & Merget, 2007). VEs can also convey culture, such as the culture of conducting chemistry experiments through interactions in a virtual laboratory. A student's avatar, for example, could be positioned at a realistic, perhaps scarred and beat-up, countertop, where they are free to move around the lab. Perhaps the user has to find the Erlenmeyer flask in a drawer, set it up properly, use correct safety procedures with acid and basic solutions, and run the risk of breaking equipment if it is handled carelessly. In short, the virtual laboratory (like a real laboratory) is designed to teach the culture of the experience—the collection of constraints, risks, and procedures that a real chemist must consider.

Researchers investigating the relationship between video games and learning have, in particular, invoked situated learning theory by arguing that the immersive and complex environments available in video games facilitate situated learning by requiring context-relevant action and the immediate application of new knowledge (Gee, 2004; Shaffer, 2006; Squire, 2006). In fact, Gee (2007) describes the actions, tools, and representations of a virtual world contained within a video game as a *situated learning matrix*. Positioning a user within a situated learning matrix allows them to develop an *identity* as a competent actor within a learning domain (Foster, 2008; Gee, 2007; Shaffer, Squire, Halverson, & Gee, 2005). Just like video games effectively position players as warriors, pilots, etc., educational VEs have great potential to situate learners as scientists, engineers, and historians such that they are equipped with proper tools and given the appropriate perspectives to facilitate conceptual knowledge development in the target domain.

40.3.3 PERCEPTUAL AND EMBODIED LEARNING

A final type of learning theory that can be leveraged by VEs centers on the properties of people as physically embodied with a perceptual system that is closely linked with their capacities for thinking and learning. In their seminal work on perceptual learning, Gibson and Gibson (1955) showed that people's ability to make key visual distinctions between stimuli got better with practice, even without explicit knowledge of the features of the stimuli that distinguished them. The implication was that rather than a process of continued abstraction and getting farther away from the physical world, learning means getting closer to the world through the tuning and refinement of one's perceptual system. Helping people learn in this way is something that VEs are well positioned to do through controlled and augmented presentation of a simulated world. For example, one approach to instruction that is grounded in our perceptual abilities is the use of *contrasting cases* (Bransford, Franks, Vye, & Sherwood, 1989). Exposing learners, either visually or through some other sensory modality, to similar instances that vary only on one or two dimensions—such as viewing the paintings of the same artist over different stages of her career—exposes the critical features that differentiate the cases. With VEs, it is possible to show contrasting cases embedded within natural contexts while still maintaining the efficiency of presentation capable in computer environments. Another approach is to use the abilities of VEs to explicitly highlight and guide learners to see the important and defining characteristics of a domain, what anthropologist Charles Goodwin calls *professional vision* (Goodwin, 1994). Desktop simulations that use *virtual highlighting* or AR environments that use digital overlays to make aspects of a real-world scene more salient are both strategies that VEs can use to develop more expert ways of seeing.

An extension of the argument that learning is perception-based is that our physical bodies present specific constraints on how we think and make sense of the world around us. Embodied cognition is the idea that humans' cognitive processes are rooted in the body's interactions with the physical environment (Wilson, 2002), and recently, researchers have described how physically enacting concepts and moving one's body in new ways can engender learning (Alibali & Nathan, 2012; Glenberg, Gutierrez, Levin, Japuntich, & Kaschak, 2004). Educational VEs have been rapidly increasing their ability to support various forms of embodied interaction, ranging from using human avatars in desktop-based virtual worlds (e.g., Falloon, 2010) to virtual reality environments where people can interact naturally in immersive spaces using head-mounted displays (e.g., Bailenson et al., 2008) to mixed reality environments where people interact with digital elements while moving around in the physical world (Birchfield & Johnson-Glenberg, 2010).

40.4   APPROACHES TO PROMOTING CONCEPTUAL LEARNING IN VEs

The preceding section described broad theories of learning that have the potential to be enacted using the special properties of VEs. In this section, we describe five specific pedagogical strategies for cultivating conceptual learning in VEs. As appropriate, we offer examples from the research literature where these strategies are being employed or where there is a strong potential for the strategy to be used.

40.4.1   EXPLORATION

In many cases, new learning begins with the opportunity to organically experience the important events and relationships present within a domain, without the cognitive burdens entailed by formal instruction. Through exploration, a learner is driven to fill gaps in their knowledge by their own questions and curiosity, and it frequently leads to a more intuitive and grounded understanding of the target system or phenomena. Exploration also "aids the learner to find new boundaries, to push back on what they know and to help them to engage socially and conceptually with others" (de Freitas & Neumann, 2009, p. 346).

As described previously, VEs present rich simulations of systems and events that can be traversed in a multitude of ways depending on the interests and inquiries of a particular learner. Research in educational technology has shown that allowing students to explore simulations prior to receiving formal instruction improves learning outcomes (e.g., Brant, Hooper, & Sugrue, 1991; Lavoie & Good, 1988). There is also evidence that virtual exploration can have advantages over physical exploration. A study by Yuan, Lee, and Wang (2010) showed, for example, that a virtual manipulative environment for working with polyominoes was found to be easier to use and elicited more critical discussion amongst participants than a physical version of the same activity.

Exploration was the key motivation for the design of the microworlds that emerged in the 1980s—Logo-driven simulations where learners had access to a *subset of reality* such that they could experiment freely with relationships in math and science (Papert, 1980; Rieber, 1992). The same rationale of exploration exists today with more contemporary VEs. For example, the *PhET* simulations (Wieman, Adams, & Perkins, 2008) provide a suite of interactive applications where learners control the parameters affecting a particular event (e.g., launching objects out of a cannon, rubbing a balloon on a sweater), observing its outcome, and then running the simulation again with new parameters. Many of these simulations are playful—encouraging the learner to be liberal in their modification of parameters, sometimes to humorous effect.

Exploration has also been a strong component of learning in more immersive VEs, such as screen-based 3D virtual worlds. In these environments, a learner will typically control an avatar that has free reign to walk (and sometimes fly) through the world that is frequently populated with other avatars with whom the learner can interact. The tendency of people to explore multiuser VEs can itself be exploited for learning purposes, as was the case with a research study involving *Whyville*, a popular virtual world targeted at teenagers. Kafai (2008) introduced a virtual epidemic called *Whypox* into the community where an infected avatar developed red spots on their face. Investigations of how the disease was spreading through the community led to the elucidation and transformation of misconceptions about the spread of computer viruses.

Informal learning environments, such as museum exhibits or web-based applications that can be accessed by after-school clubs or at home, put a special emphasis on exploration (Bell, Lewenstein, Shouse, & Feder, 2009). These environments are typically more conducive to exploratory behavior because learners are less constrained in terms of time or curriculum. Informal learning environments also present opportunities for more innovative types of exploration, such as using one's body to explore a domain of conceptual knowledge. One such example of this type of exploratory learning VE is the *MEteor* simulation game—an immersive mixed reality simulation where middle school students take on the role of an asteroid moving through space (Lindgren & Moshell, 2011). *MEteor* is a 30 ft by 10 ft simulation environment with interactive floor projections and a tracking system that allows for real-time feedback on student predictions about where the asteroid will move as it travels past planets and other objects in space. Interactive VEs are increasingly capable of accepting body-based input, and the premise of *MEteor* is that giving students the opportunity to involve one's body in exploring the conceptual space of mechanics and planetary astronomy will enhance their intuitions and better prepare them for formal learning. See Birchfield and Johnson-Glenberg (2010) for a similar example of a VE that encourages exploration through full-body movement and metaphor-based interaction.

40.4.2 INVENTION

A more constrained and focused form of exploration is invention, where a learner is tasked with constructing a specific design that meets some objective or fulfills a desired function. Through the process of invention, a learner can come to understand important pragmatic constraints and how abstract principles can be operationalized within a concrete design. Like exploration, research has shown that invention activities, such as creating a statistical measure to solve an applied problem (e.g., determining the accuracy of a pitching machine), improve reasoning and better prepare student for future learning interventions (Schwartz, Chase, Oppezzo, & Chin, 2011; Schwartz & Martin, 2004).

Providing opportunities for invention and creation in digital environments was a central thrust of constructionist learning theory—an extension of constructivism that places an emphasis on tangible interaction and the construction of objects in the service of cognitive development and learning (Kafai & Resnick, 1996; Papert, 1980). VEs have special affordances for invention in that new designs can be specified and evaluated without the initial need for costly resources. For example, the *RockSim* (Apogee, 2012) allows users to create new designs for model rockets and subject the designs to a physics engine so as to evaluate the likely success of its launch. With their ability to deliver instantaneous feedback and near-infinite parameters for visual design, VEs are extremely well suited for processes of *tinkering* and rapid prototyping, processes that can advance a learner's conceptual understanding of the design space (engineering constraints, aesthetic considerations, etc.).

A somewhat broader notion of invention—the creation of unique social, biological, economic, and architectural scenarios—is nearly ubiquitous in contemporary VEs such as video games and online virtual worlds. So-called God games such as *SimCity* and *Spore* allow users to invent systems and characters and test their vitality in the face of environmental conditions such as consumer demand, predators, and warfare. While these design- and invention-driven VEs are clearly calibrated for entertainment, there is an argument to be made that users learn, at least implicitly, the underlying structures and principles that ultimately determine if one's creation will be successful. In other immersive VEs, such as *Second Life* or the currently popular *Minecraft* game world, the environment itself is subject to design—users are given free reign to build new objects and structures that persist in the environment. These VEs provide custom interfaces for building and *crafting* that often resembles software and design processes used for real-world environments.

A good example of an invention platform that outputs original digital media artifacts is the visual programming environment *Scratch* (Maloney, Resnick, Rusk, Silverman, & Eastmond, 2010). *Scratch* is an accessible programming environment that allows children to create original games and interactive stories, and it is one of many emerging VEs contributing to a growing *participatory culture*, where youth can be media *creators* rather than simply media consumers (Jenkins, 2006). It is important to note that media creation activities can be put in service of conceptual learning. Baytak and Land (2011), for example, describe a project where students were tasked with creating games in *Scratch* to reinforce lessons in an environmental science curriculum.

40.4.3 INQUIRY AND HYPOTHESIS TESTING

Inquiry is a learning activity intended to acclimate students to authentic scientific practices (e.g., conducting experiments and constructing evidence-based arguments) (Chinn & Malhotra, 2002). Inquiry certainly has both exploration and invention components, but it follows a more regimented process, namely, the scientific method. While following this process most often results in conceptual learning about science, inquiry is comprised of skills that can be of service to a range of disciplines including the social sciences, humanities, and even the arts. Designs of all types (study designs, interface designs, artistic designs, etc.) have associated hypotheses—about the causes of earthly events, human responses to stimuli—that can be evaluated through a process of inquiry.

One way that inquiry can be supported using VEs is to provide structural supports and appropriate toolsets for carrying out optimal investigations—guiding a learner through the steps of constructing a hypothesis, controlling extraneous variables, interpreting data, etc. An early online VE designed specifically to support classroom-based inquiry is the Web-based Inquiry Science Environment (*WISE*) (Linn, Davis, & Bell, 2004). *WISE* was developed to scaffold student thinking and encourage reflection; it is also extensible in that it allows teachers to add new learning modules that follow the general inquiry structure. Another way that VEs can support inquiry is to simply provide controlled and interesting contexts in which scientific investigations can occur. An intriguing example of this was a project described by Nordine (2011) where students used projectile events in the original *Super Mario Bros.* video game to conduct sophisticated investigations of kinematics concepts.

A particular type of VE that has been cited for its affordances for supporting inquiry is the MUVE. These expansive virtual worlds not only possess structure and interesting phenomena for scientific study but can also provide authentic social contexts and opportunities for situated learning (Qian, 2009). The two most ambitious and significant projects utilizing a MUVE for supporting inquiry learning and conceptual development are *River City* and *Quest Atlantis*. The *River City* project (Dede, 2009) is a self-contained standards-based science curriculum built around a virtual nineteenth-century town where disease has run rampant. Small teams of student avatars work together to investigate the causes of these health issues and communicate a remedial course of action to the community. Studies of student learning in the *River City* environment have shown that students indeed acquire inquiry skills and that learning these skills in a VE had advantages over physical tasks (Ketelhut, Dede, Clarke, Nelson, & Bowman, 2008; Ketelhut & Nelson, 2010). *Quest Atlantis* (Barab et al., 2005) is an educational VE that is farther reaching than *River City* in that it strives to present a platform for learning in many different disciplines: math, science, literacy, etc. *Quest Atlantis* poses various academic challenges in the context of a fictional world in crisis, and students respond to these challenges and experience the consequences of their response. *Quest Atlantis* researchers argue that their environment leads to higher conceptual understanding by putting students in a state of *conceptual play* (Barab, Ingram-Goble, & Warren, 2009). Numerous other VEs aim to develop inquiry skills, and as the graphics and intelligence capabilities of these environments continue to advance, the ability to foster immersive experiences that cultivate learners' identities as nascent scientists will only get better.

40.4.4 PERCEPTUAL TUNING AND DIFFERENTIATION

In Section 40.3.3, we discussed the affordances of VEs for perceptual learning. At the most basic level, perceptual learning occurs because VEs and simulations are typically simplified versions of real-world systems, thus they allow users to focus on the important features of the target stimulus or environment (Winn et al., 2006). However, VEs can also offer more explicit strategies for tuning the perceptual systems of learners such that they are more in line with the target concepts and the knowledge held by disciplinary experts. Lindgren and Schwartz (2009) describe different ways that VEs can support perceptual differentiation—seeing, hearing, or feeling the important distinctions that define a domain. VEs based on modeling and experimentation, for example, typically allow for rapid trials or iterations that can be juxtaposed in ways that reveal the important distinctions. This approach of presenting learners with contrasting cases, described earlier, is well suited for VEs and has been frequently employed in digital learning applications. Jacobson (2008), for example, describes the use of a hypermedia environment for providing contrasting cases relevant to the understanding of complex systems. Another project utilizing a game-based VE engages learners with solving environmental problems on an alien planet by making comparisons to cases of similar problems on Earth (Williams, Ma, Feist, Richard, & Prejean, 2007). The contrasting cases approach presents learners with opportunities to make their differentiation knowledge explicit and in some cases prompts users to articulate this knowledge verbally.

With 3D and immersive VEs, it is possible to tune and refine a learner's perspective in more implicit ways. These environments have the added feature of permitting a near unbounded number of perspectives on an object or situation (e.g., the structure of a molecule, an event of historical significance). Having the opportunity to experience phenomena from multiple perspectives in VEs has been shown to benefit learning (Chittaro & Ranon, 2007; Dede, 2009). For example, Limniou, Roberts, and Papadopoulos (2008) found that having students view chemical reactions in an immersive CAVE environment compared to a 2D simulation of the same reactions leads to increased comprehension of molecular structure, as well as higher engagement and interest in the learning content. There is also the strong potential to achieve perceptual tuning in immersive VEs by manipulating a user's visual field and giving them a specific perspective designed to draw out the important knowledge in the environment and allowing a learner to see the domain as if they themselves were an expert. Lindgren (2012) describes a study where participants either viewed a walk-through of a training simulation from the perspective of an avatar that was controlled by an expert in that environment or viewed the exact same events from a disembodied *camera-on-the-wall* perspective. Results of a postassessment showed that participants who viewed the environment through the eyes of an expert were better able to identify the critical components and potential hazards in the simulation space.

40.4.5 ACTION AND SKILL MODIFICATION

On the surface, this section does not sound like it would have much to do with conceptual knowledge—many psychologists and practitioners see skill development as a qualitatively different type of knowing than possessing the concepts of a science or a history curriculum. However, as the preceding discussion on embodiment suggests, current perspectives on thinking and learning portray a tight coupling between a person's actions and aptitudes for navigating the physical world and their capacities for reasoning and problem solving (Abrahamson & Lindgren, in press; Alibali & Nathan, 2012; Glenberg et al., 2004). Thus, it could be the case, for example, that proficiency with procedures involved in conducting a petri dish experiment in the biological sciences makes a learner more receptive and better prepared to synthesize the knowledge that comes from such an experiment. Likewise, dexterity with using the instruments of an engineer or an archeologist may make it more likely for a person to acquire new insights and perceive the regularities of the domain.

VEs for skill and action modification are especially prevalent in the medical fields and in the military, but in recent years, there have also been attempts to utilize more training-focused applications with younger learners and in schools. One such project used a MUVE environment and embedded pedagogical agents to help a learner to develop students' writing skills (Warren, Stein, Dondlinger, & Barab, 2009). Another project called the *Virtual Chemistry Laboratory* was designed to familiarize distance learning students with the spatial organization, procedures, and professional culture of a chemistry lab prior to these students coming to the actual laboratory to perform physical experiments (Dalgarno, Bishop, Adlong, & Bedgood, 2009). The study results indicated that students who used the virtual laboratory were more prepared for the orientation, assembly, and operation tasks they were faced with in the physical laboratory (for a similar finding for a biological dissection task, see Akpan & Andre, 1999).

A final example of a recently developed and novel VE targeting skill in the area of teacher education is the *TeachLive* platform (Hughes & Mapes, in press). In this environment, a novice teacher or one wishing to improve their skills stands in front of a large display screen showing several children in a simulated classroom. The movement and behaviors of the children in the classroom are controlled by a blend of agency (programmatically determined behaviors) and a single *live* actor who remotely controls the group of avatars using a sophisticated control interface. This system supports teacher rehearsal by using a virtual classroom with distinct student personalities and learning styles. The intent is to help teachers practice pedagogy, classroom management, and content delivery, without involving real students where damage might occur to the learning processes of both children and the inexperienced teachers. The environment provides opportunities for tutoring and reflection, and it is currently being used by numerous universities as part of their teacher preparation programs (Dieker, Lingnugaris-Kraft, Hynes, & Hughes, in press).

40.5 FUTURE OF VEs FOR LEARNING AND EDUCATION

The application of new technologies in education is often greeted with skepticism, and yet Gamage, Tretiakov, and Crump (2011) make the observation that teacher perceptions of the potential of VEs for learning are relatively high. This positive outlook presents an opportunity for designers and researchers to continue innovating and keep pushing the potential of these environments for improving the way people learn both formally and informally. It is also important that this community continue to develop new forms of assessment that are capable of detecting the particular kinds of learning that manifest from experiences using VEs, so that the momentum for building these environments for a myriad of learning applications does not dissipate.

As for the design of the VEs themselves, we predict that more sophisticated graphics engines and the accessibility of development kits will push forward more immersive and physically interactive VEs into more mainstream use in classrooms, at home, etc. These advances will likely elevate the pedagogical strategies described in Sections 40.4.4 and 40.4.5. For example, lightweight goggle-based VR displays are likely to become prevalent and substantially cheaper in the next decade, making it possible to deliver immersive displays that transmit expert perspectives, perhaps even blending expert knowledge with a user's natural viewpoint of the real world via digital overlays. Generally speaking, the boundaries that separate VEs and the physical world will get fuzzier, with elements of VEs comingling with objects and landmarks in the real world, such as with GPS-based mobile applications and activities fueled by augmented reality technologies.

We also predict that the emphasis on VE learning designs will start to shift substantially toward collaborative systems and environments where students can work together effectively in teams. With massively multiplayer online games such as *World of Warcraft,* researchers have observed the emergence of highly sophisticated and effective instances of coordinated action (Williams & Kirschner, 2012). It is important to understand these VE phenomena and the conditions under which they occur so that equally effective collaboration environments can be created within serious games and online educational platforms.

What constitutes a VE for learning has changed tremendously over the last decade, and it is hard to predict exactly what form they will assume going forward. Nevertheless, we believe the future of VEs in education is bright. As the technology evolves, we will undoubtedly discover new affordances that will expand our list of pedagogical strategies. What is important is that we continue to exploit these affordances, working to create environments that are optimally suited to the educational context, to the target knowledge domain, and to the learner.

ACKNOWLEDGMENTS

The funding for the projects described in this chapter was provided by the NSF (DRL1114621, CNS1051067, IIS1116615), ONR Code 30 (N000141210052), and the Bill & Melinda Gates Foundation. Any opinions, findings, and conclusions or recommendations expressed in this material are those of the authors and do not necessarily reflect the views of the funding institutions.

REFERENCES

Abrahamson, D., & Lindgren, R. (in press). Embodiment and embodied design. In R. K. Sawyer (Ed.), *The Cambridge handbook of the learning sciences* (2nd ed.). Cambridge, MA: Cambridge University Press.

Akpan, J. P., & Andre, T. (1999). The effect of a prior dissection simulation on middle school students' dissection performance and understanding of the anatomy and morphology of the frog. *Journal of Science Education and Technology, 8*(2), 107–121.

Alibali, M. W., & Nathan, M. J. (2012). Embodiment in mathematics teaching and learning: Evidence from learners' and teachers' gestures. *Journal of the Learning Sciences, 21*(2), 247–286.

Annetta, L. A., & Holmes, S. (2006). Creating presence and community in a synchronous virtual learning environment using avatars. *International Journal of Instructional Technology and Distance Learning, 3*(8), 27–43.

Apogee. (2012). RockSim. Retrieved from http://www.apogeerockets.com/Rocksim/Rocksim_information. Accessed on March 23, 2013.

Bailenson, J. N., Yee, N., Blascovich, J., Beall, A. C., Lundblad, N., & Jin, M. (2008). The use of immersive virtual reality in the learning sciences: Digital transformations of teachers, students, and social context. *Journal of the Learning Sciences, 17*(1), 102–141.

Barab, S., Ingram-Goble, A., & Warren, S. (2009). Conceptual play spaces. *Handbook of research on effective electronic gaming in education* (pp. 989–1009). New York, NY: IGI Global.

Barab, S., Thomas, M., Dodge, T., Carteaux, R., & Tuzun, H. (2005). Making learning fun: Quest Atlantis, a game without guns. *Educational Technology Research and Development, 53*(1), 86–107.

Baytak, A., & Land, S. M. (2011). An investigation of the artifacts and process of constructing computers games about environmental science in a fifth grade classroom. *Educational Technology Research and Development, 59*(6), 765–782.

Bell, P., Lewenstein, B., Shouse, A. W., & Feder, M. A. (Eds.). (2009). *Learning science in informal environments: People, places, and pursuits.* Washington, DC: National Academies Press.

Birchfield, D., & Johnson-Glenberg, M. (2010). A next gen interface for embodied learning: SMALLab and the geological layer cake. *International Journal of Gaming and Computer-Mediated Simulations, 2*(1), 49.

Bransford, J. D., Franks, J. J., Vye, N. J., & Sherwood, R. D. (1989). New approaches to instruction: Because wisdom can't be told. In S. Vosniadou & A. Ortony (Eds.), *Similarity and analogical reasoning* (pp. 470–497). New York, NY: Cambridge University Press.

Brant, G., Hooper, E., & Sugrue, B. (1991). Which comes first the simulation or the lecture? *Journal of Educational Computing Research, 7*(4), 469–481.

Brown, J. S., Collins, A., & Duguid, P. (1989). Situated cognition and the culture of learning. *Educational Researcher, 18*(1), 32–42.

Burton, B. G., & Martin, B. N. (2010). Learning in 3D virtual environments: Collaboration and knowledge spirals. *Journal of Educational Computing Research, 43*(2), 259–273.

Charles, D., Charles, T., McNeill, M., Bustard, D., & Black, M. (2011). Game-based feedback for educational multi-user virtual environments. *British Journal of Educational Technology, 42*(4), 638–654.

Chinn, C. A., & Malhotra, B. A. (2002). Epistemologically authentic inquiry in schools: A theoretical framework for evaluating inquiry tasks. *Science Education, 86*(2), 175–218.

Chittaro, L., & Ranon, R. (2007). Web3D technologies in learning, education and training: Motivations, issues, opportunities. *Computers & Education, 49*, 3–18.

Dalgarno, B., Bishop, A. G., Adlong, W., & Bedgood, D. R., Jr. (2009). Effectiveness of a virtual laboratory as a preparatory resource for distance education chemistry students. *Computers & Education, 53*(3), 853–865.

Dalgarno, B., & Lee, M. J. W. (2010). What are the learning affordances of 3-D virtual environments? *British Journal of Educational Technology, 41*(1), 10–32.

Dede, C. (1995). The evolution of constructivist learning environments: Immersion in distributed, virtual worlds. *Educational Technology, 35*, 46–52.

Dede, C. (2009). Immersive interfaces for engagement and learning. *Science, 323*(5910), 66–69.

de Freitas, S., & Neumann, T. (2009). The use of "exploratory learning" for supporting immersive learning in virtual environments. *Computers & Education, 52*(2), 343–352.

Dewey, J. (1966). *Democracy and education: An introduction to the philosophy of education.* New York, NY: The Free Press.

Dieker, L. A., Lingnugaris-Kraft, B., Hynes, M., & Hughes, C. E. (in press). Mixed reality environments in teacher education: Development and future applications. In B. Collins & B. Ludlow (Eds.), *American council for rural special educators.*

Duffy, T. M., & Jonassen, D. H. (Eds.). (1992). *Constructivism and the technology of instruction: A conversation*. Hillsdale, NJ: Lawrence Erlbaum Associates Publishers.

Falloon, G. (2010). Using avatars and virtual environments in learning: What do they have to offer? *British Journal of Educational Technology*, *41*(1), 108–122.

Fiore, S. M., Harrison, G. W., Hughes, C. E., & Rutström, E. E. (2009). Virtual experiments and environmental policy. *Journal of Environmental Economics and Management*, *57*(1), 65–86.

Foster, A. (2008). Games and motivation to learn science: Personal identity, applicability, relevance and meaningfulness. *Journal of Interactive Learning Research*, *19*(4), 597–614.

Gamage, V., Tretiakov, A., & Crump, B. (2011). Teacher perceptions of learning affordances of multi-user virtual environments. *Computers & Education*, *57*(4), 2406–2413.

Gee, J. P. (2004). *What video games have to teach us about learning and literacy*. New York, NY: Palgrave Macmillan.

Gee, J. P. (2007). Learning and games. *The John D. and Catherine T. MacArthur Foundation Series on Digital Media and Learning*, 21–40.

Gibson, J. J., & Gibson, E. J. (1955). Perceptual learning: Differentiation or enrichment? *Psychological Review*, *62*(1), 32–41.

Glenberg, A. M., Gutierrez, T., Levin, J. R., Japuntich, S., & Kaschak, M. P. (2004). Activity and imagined activity can enhance young children's reading comprehension. *Journal of Educational Psychology*, *96*, 424–436.

Goodwin, C. (1994). Professional vision. *American Anthropologist*, *96*, 28.

Honey, M., & Hilton, M. L. (2011). *Learning science through computer games and simulations*. Washington, DC: National Academies Press.

Hughes, C. E. & Mapes, D. P. (in press). Mediated dialogues through multiple networked avatars. *Journal of Immersive Education*.

Jacobson, M. (2008). A design framework for educational hypermedia systems: Theory, research, and learning emerging scientific conceptual perspectives. *Educational Technology Research and Development*, *56*(1), 5–28.

Jenkins, H. (2006). *Fans, bloggers, and gamers: Exploring participatory culture*. New York, NY: NYU Press.

Kafai, Y. B. (2008). Understanding virtual epidemics: Children's folk conceptions of a computer virus. *Journal of Science Education and Technology*, *17*(6), 523–529.

Kafai, Y. B., & Resnick, M. (1996). *Constructionism in practice: Designing, thinking, and learning in a digital world*. Mahwah, NJ: Lawrence Erlbaum.

Ketelhut, D. J., Dede, C., Clarke, J., Nelson, B., & Bowman, C. (2008). Studying situated learning in a multi-user virtual environment *Assessment of problem solving using simulations* (pp. 37–58). Mahweh, NJ: Lawrence Erlbaum.

Ketelhut, D. J., & Nelson, B. C. (2010). Designing for real-world scientific inquiry in virtual environments. *Educational Research*, *52*(2), 151–167.

Lavoie, D. R., & Good, R. (1988). The nature and use of prediction skills in a biological computer simulation. *Journal of Research in Science Teaching*, *25*(5), 335–360.

Limniou, M., Roberts, D., & Papadopoulos, N. (2008). Full immersive virtual environment CAVE in chemistry education. *Computers & Education*, *51*(2), 584–593.

Lindgren, R. (2012). Generating a learning stance through perspective-taking in a virtual environment. *Computers in Human Behavior*, *28*, 1130–1139.

Lindgren, R., & Moshell, J. M. (2011). Supporting children's learning with body-based metaphors in a mixed reality environment. *Proceedings of the Interaction Design and Children Conference*, 177–180.

Lindgren, R., & Schwartz, D. L. (2009). Spatial learning and computer simulations in science. *International Journal of Science Education*, *31*(3), 419–438.

Linn, M. C., Davis, E. A., & Bell, P. (2004). *Internet environments for science education*. Mahwah, NJ: Lawrence Erlbaum Associates.

Long, P., Siemens, G., Conole, G., & Gasevic, D. (Eds.). (2011). *Proceedings of the First International Conference on Learning Analytics and Knowledge*. Banff, Alberta, Canada: ACM.

Lopez, A. L., Hughes, C. E., Mapes, D. P., & Dieker, L. A. (2012). Cross cultural training through digital puppetry. In D. M. Nicholson (Ed.), *Advances in design for cross-cultural activities part I* (pp. 247–256). Boca Raton, FL: CRC Press.

Luo, Z., Luo, W., Wickens, C. D., & Chen, I. M. (2010). Spatial learning in a virtual multilevel building: Evaluating three exocentric view aids. *International Journal of Human-Computer Studies*, *68*(10), 746–759.

Maloney, J., Resnick, M., Rusk, N., Silverman, B., & Eastmond, E. (2010). The scratch programming language and environment. *CM Transactions on Computing Education (TOCE)*, *10*(4), 1–15.

Mikropoulos, T. A., & Natsis, A. (2011). Educational virtual environments: a ten-year review of empirical research (1999–2009). *Computers & Education, 56*(3), 769–780.

Nordine, J. C. (2011). Motivating calculus-based kinematics instruction with Super Mario Bros. *Physics Teacher, 49*(6), 380–382.

Papert, S. (1980). *Mindstorms: Children, computers, and powerful ideas.* New York, NY: Basic Books.

Parsons, S., Leonard, A., & Mitchell, P. (2006). Virtual environments for social skills training: comments from two adolescents with autistic spectrum disorder. *Computers & Education, 47*(2), 186–206.

Piaget, J. (1997). *The language and thought of the child* (3rd ed.). London, U.K.: Routledge.

Qian, Y. (2009). 3D multi-user virtual environments: Promising directions for science education. *Science Educator, 18*(2), 25–29.

Rieber, L. P. (1992). Computer-based microworlds: A bridge between constructivism and direct instruction. *Educational Technology Research and Development, 40*(1), 93–106.

Schwartz, D. L., Chase, C. C., Oppezzo, M. A., & Chin, D. B. (2011). Practicing versus inventing with contrasting cases: The effects of telling first on learning and transfer. *Journal of Educational Psychology, 103*(4), 759–775.

Schwartz, D. L., Lindgren, R., & Lewis, S. (2009). Constructivism in an age of non-constructivist assessments. In S. Tobias & T. M. Duffy (Eds.), *Constructivist instruction: Success or failure?* (pp. 34–61). New York, NY: Routledge/Taylor & Francis Group.

Schwartz, D. L., & Martin, T. (2004). Inventing to prepare for future learning: The hidden efficiency of encouraging original student production in statistics instruction. *Cognition and Instruction, 22*(2), 129–184.

Shaffer, D. W. (2006). *How computer games help children learn.* New York, NY: Palgrave Macmillan.

Shaffer, D. W., Squire, K. R., Halverson, R., & Gee, J. P. (2005). Video games and the future of learning. *Phi Delta Kappan, 87*(2), 104–111.

Squire, K. (2006). From content to context: Videogames as designed experience. *Educational Researcher, 35*(8), 19–29.

Sullivan, F. R., Hamilton, C. E., Allessio, D. A., Boit, R. J., Deschamps, A. D., Sindelar, T., … Zhu, Y. (2011). Representational guidance and student engagement: Examining designs for collaboration in online synchronous environments. *Educational Technology Research and Development, 59*(5), 619–644.

Waller, D. (2000). Individual differences in spatial learning from computer-simulated environments. *Journal of Experimental Psychology: Applied, 6*(4), 307–321.

Warburton, S. (2009). Second Life in higher education: Assessing the potential for and the barriers to deploying virtual worlds in learning and teaching. *British Journal of Educational Technology, 40*(3), 414–426.

Warren, S. J., Stein, R. A., Dondlinger, M. J., & Barab, S. A. (2009). A look inside a MUVE design process: Blending instructional design and game principles to target writing skills. *Journal of Educational Computing Research, 40*(3), 295–321.

Wieman, C. E., Adams, W. K., & Perkins, K. K. (2008). PhET: Simulations that enhance learning. *Science, 322*(5902), 682–683.

Williams, D., Ma, Y., Feist, S., Richard, C. E., & Prejean, L. (2007). The design of an analogical encoding tool for game-based virtual learning environments. *British Journal of Educational Technology, 38*(3), 429–437.

Williams, J. P., & Kirschner, D. (2012). Coordinated action in the massively multiplayer online game World of Warcraft. *Symbolic Interaction, 35*(3), 340–367.

Wilson, M. (2002). Six views of embodied cognition. *Psychonomic Bulletin & Review, 9*(4), 625–636.

Winn, W. (1999). Learning in virtual environments: A theoretical framework and considerations for design. *Educational Media International, 36*(4), 271–279.

Winn, W., Stahr, F., Sarason, C., Fruland, R., Oppenheimer, P., & Lee, Y.-L. (2006). Learning oceanography from a computer simulation compared with direct experience at sea. *Journal of Research in Science Teaching, 43*(1), 25–42.

Yee, N., Bailenson, J. N., Urbanek, M., Chang, F., & Merget, D. (2007). The unbearable likeness of being digital: The persistence of nonverbal social norms in online virtual environments. *Cyber Psychology & Behavior, 10*(1), 115–121.

Yuan, Y., Lee, C. Y., & Wang, C. H. (2010). A comparison study of polyominoes explorations in a physical and virtual manipulative environment. *Journal of Computer Assisted Learning, 26*(4), 307–316.

41 Applications of Virtual Environments in Experiential, STEM, and Health Science Education

Angelos Barmpoutis, Benjamin DeVane, and James C. Oliverio

CONTENTS

41.1 INTRODUCTION

This chapter presents examples of utilizing virtual environments for experiential learning and training purposes, with applications to several areas in the science, technology, engineering, and mathematics (STEM) and health sciences. The chapter starts with a general introduction to experiential learning, followed by a presentation of various technologies for enhancing the experience in virtual environments. The rest of the chapter is organized into two sections that discuss specific examples of virtual environments for experiential learning and therapeutic medical applications, respectively. More specifically, the examples will demonstrate the following: the use of low-cost haptic devices in virtual environments for learning nanotechnology, experiential learning environments for forest education, the use of virtual reality theaters as an educational tool in the arts and the humanities, virtual environments for therapeutic solutions, interactive tools for treating motor disabilities using BCI for interaction with virtual environments, and experiential learning applications for microsurgical training using mixed-reality and haptic feedback. The chapter concludes with a final section that discusses future research directions.

41.2 EXPERIENTIAL LEARNING THROUGH VIRTUAL REALITY

Virtual environments are *designed experiences*—informational and intentional contexts of activity that are designed to create a particular user experience. Designed experiences are digital structures of participation that help a person *learn through doing*—building knowledge through meaningful action in a simulated context. Virtual environments can offer "designed experiences, in which participants learn through a grammar of doing and being" (Squire, 2006; p. 24). These structured experiences guide learners as they confront exciting virtual problems in new knowledge domains, introduce them to new context-dependent skills and practices, and present them with problem-solving opportunities that mobilize consequential, just-in-time feedback informing users about the limits of their new capabilities.

Virtual environments are potentially powerful learning tools, as they provide learners with opportunities to experience trying out new knowledge and skills in a *safe* space and with a structure for reflecting on those experiences that makes them meaningful and memorable. Experiential learning in these spaces is a result of action and reflection in a consequential and purpose-driven context. Such a framework for learning is very different from behaviorist and information processing models.

41.2.1 FOUNDATIONS OF EXPERIENTIAL LEARNING

The theoretical roots of *designed experiences* come out of the experiential learning theory of John Dewey (Squire, 2006). Dewey was a prominent American pragmatist, philosopher, and educator in the late nineteenth and early twentieth century, who brought the experiential theory of learning to prominence in his famous experimental Laboratory School in Chicago (Kolb, Boyatzis, & Mainemelis, 2001). In 1896, Dewey, concerned with the rigid and disciplinary pedagogical model in most American schools, founded the laboratory school as a philosophy laboratory in which students were taught through methods that acknowledged the unity of knowing and doing. Students, in other words, would learn by doing.

Directed by Dewey while he was still chair of the philosophy department at the University of Chicago, the laboratory school sought to teach students by connecting what they were learning to real world, everyday actions and occupational practices, which often took the form of trade skills and crafts. Instead of placing students in highly disciplined classroom environments where instruction was rote, Dewey sought to use occupational and trade skills to help students connect knowledge learned in school to the everyday, real world that they encountered outside the classroom. In his popular 1896 book *The School and Society*, which launched progressive education on a wide scale in the United States, Dewey articulated the rationale underlying his pedagogical practice:

> In critical moments we realize that the only discipline that stands by us, the only training that becomes intuition, is that got through life itself. That we learn from experience, and from the books and sayings of others only as they are related to experience, are not mere phrases. But the school has been so set apart, so isolated from the ordinary conditions and motives of life, that the place that children are sent for discipline is the one place in the world that is most difficult to get experience—the mother of all discipline worth the name.
>
> **Dewey (1899, p. 15)**

In Dewey's vision, learning and school activities should be joined with the ordinary habits and practices of social and professional life. In both his writings and his work in the laboratory school, Dewey laid out a framework for experiential learning that would have far-reaching implications for American education (Dewey, 1916). This experiential learning model was constructed around three *unities*:

1. The unity of abstract knowledge and doing in the real world
2. The unity of action and reflection
3. The unity of the individual with the community

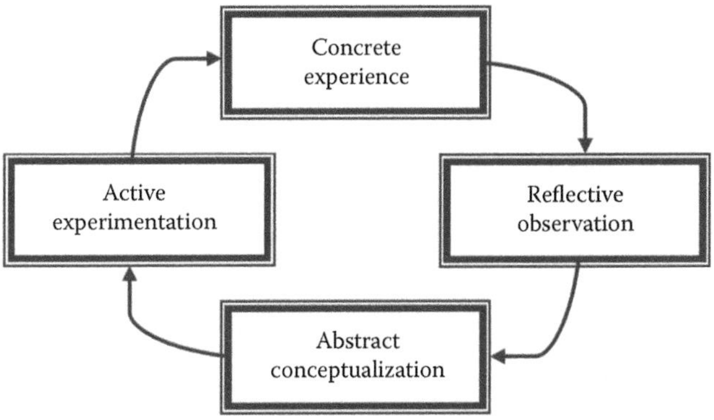

FIGURE 41.1   David Kolb's learning cycle.

Learners build their knowledge through doing, develop deeper understandings by acting and reflecting on their action, and build better understandings of the subject matter and themselves by doing so in a social environment. Based on Dewey's unities, David Kolb (Kolb, 1984) characterized experience as one of the most essential stages of learning (Figure 41.1). According to Kolb, the learning process is a four-stage cycle that begins with concrete experience, which leads to reflection and action through experimentation. By studying Kolb's model, it becomes clear that it is founded on the first two unities of Dewey's framework. These two unities of experiential learning become critically important when we turn our attention to understanding digital games and simulations as designed experiences.

41.2.2   DESIGNED EXPERIENCES IN GAME-BASED AND VIRTUAL LEARNING ENVIRONMENTS

The question posed then is this: How can learning be experiential, but also structured so that the knowledge internalized is needed, relevant, and useful? The strength of the experiential learning paradigm is its ability to foster profound knowledge building and reflection in learners. But the struggle historically for experiential learning has been ensuring that learners build the *right* knowledge for any particular domain. Well-crafted virtual learning environments solve this puzzle. Good learning games are designed experiences because they allow players to learn by acting and experiencing in a well-ordered, problem-solving space. Players of these games and simulations "learn through a grammar of doing and being" (Squire, 2006, p. 19) that is organized around a functional epistemology, a system of knowing that is based on doing. As such, the designed experiences found in virtual learning games and simulations grow out of the foundations of Dewey's experiential learning: action and reflection (Gee, 2007; Squire, 2006). A player confronts a problem in a digital space, thinks about the solution to the problem, acts to try to solve that problem, receives feedback on the efficacy of that action, and then reflects on that feedback (see Gee, 2005). These action–reflection cycles are pervasive throughout well-designed learning games and simulations.

Good learning games and simulations structure experiential learning so that action occurs in a well-ordered content domain. In other words, the shape of game-based experiential learning is specified beforehand by the game designer. While decisions taken by the players carry them into different trajectories of play (and even into emergent, unintended circumstances), the overarching structure of a player experience is a product of a fixed design. But well-designed games and simulations often give the player an authentic feeling of experience and empowerment, allowing them to choose from meaningful trajectories of the play (DeVane & Squire, 2008). In this way, well-designed games and simulations solve some of the core problems of experiential learning and reflective practice. Learning through experience in games and simulations is not free-form and uncontrolled, but rather well-ordered and specified. At the same time, the experience players have in virtual spaces is immersive, rich, and empowering. Learning games and simulations accomplish this feat because of the following characteristics:

1. Well-ordered problem solving. Good learning games and simulations present problems for the player to solve that are appropriate for their skill level. These problems provide the player with a skill-based situation in which they can test their knowledge themselves by engaging in knowledgeable action (Gee, 2007). Problems encourage players to engage in both knowledgeable action and reflection.
2. Meaningful goals. Goals are clearly defined and their rationale articulated by a narrative. Overarching goals and their narrative rationale provide players with a meaningful context that motivates them to achieve success in problem solving. Upon achievement of goals, players are often presented with rewards. Meaningful and authentic contexts for action encourage players to master and reflect on the content they are learning (Squire, 2006).
3. Possibility spaces. Many games and simulations present players with a multitude of choices, decisions, and possibilities for action, letting them formulate an individualized solution to a problem, or a custom strategy to accomplish a goal. These virtual worlds are play spaces where players can experiment with their knowledge, beliefs, values, and identities (Squire, 2006; Squire & Jenkins, 2002).
4. Feedback and information. Additional information helps players reflect on their action—both during and after the action. This information most often takes the form of feedback and just-in-time information. Feedback is information given to the players after they have completed a problem-solving action or series of actions. Just-in-time information is information that is delivered just as the players confront a problem where they will use that information. This way of delivering information stimulates reflection in action. Players understand the relevance of this information to their goals and reflect on the relevance of the information to their action (Gee, 2005). Both feedback and just-in-time information stimulate players to both reflect in and reflect on their action (Steinkuehler & Chmiel, 2006).
5. Empowerment and reward. Players get a sense of empowerment and accomplishment from their achievements and real sense of their action's capacity to have an impact in the virtual space. This sense of empowerment heightens the player's appreciation of their experience (Gee, 2005). Players are varyingly rewarded with new abilities, goods, badges, and decorations that show their accomplishments. The rewards given to them for their accomplishment heightens the pleasure players feel at their achievement.

41.3 AUGMENTING EXPERIENCE USING NATURAL USER INTERFACES

Experience of living organisms is based on their sensory input, which is a lifelong sequence of input signals generated by the surrounding environment including the organisms themselves. These signals are processed by a set of neural systems that produce the result of perception. The traditionally recognized methods of perception in human beings are sight, hearing, touch, smell, and taste.

In virtual environments, the user's experience is formed as if it were in real environments, since the virtual reality systems (usually a set of computer-based devices) are part of our physical environment and produce stimuli that trigger our sensory responses. Depending on the particular application of a virtual reality system, a small subset of the user's senses may be used in order to form his or her perception of the virtual environment. For example, typical home-based commercial virtual environments generate visual projections and simulate sounds. Furthermore, technological limitations may narrow the modalities of the generated stimuli, which usually take the form of a visual field, sound waves, and/or haptic feedback. Olfaction and gustation are considered in many applications less important senses, due to the lack of variations in the sensory input of the former in many real-world environments, as well as the occluded physical location and the small sensitivity range of the human organ of the latter. Nevertheless, technologies for arbitrary odor or flavor composition are currently under research (Krueger, 1996) and have many potential applications in virtual environments, especially in fields where odors have important role, such as chemistry, pharmacology, and medicine.

In the following sections, various technologies for augmenting the experience in virtual environments are discussed, with emphasis to visual perspective, hearing, and touch.

41.3.1 Sight

The sensory input from vision takes the form of two ordered projections of the light in the field of view of our eyes. This primary input signal is processed by our brain to form higher-level information that we experience as the stereoscopic view of the light-emitting or light-reflecting environment in our field of view. The depth of the objects in our field of view is another type of high-level information that we experience as part of our vision and helps us understand the tridimensional structure of our surrounding environment. Although we experience depth better using both of our eyes, human brain can generate the depth information from a monoscopic view based on the motion of the objects, their shading, our focus, and other visual patterns. All the aforementioned processes should be taken under consideration, in order to render a visual representation of a virtual environment that will be experienced by the users in a virtual 3D space.

In several educational applications, virtual environments more often than not are visualized in single flat 2D screens, due to their relatively low cost and their availability in the majority of households in various forms of electronic devices, such as desktop and laptop computers, mobile phones, tablets, game consoles, and TV sets ("Report on teens, video games and civics," 2008). Concave curved or polygonal sets of 2D screens (Cruz-Neira, Sandin, DeFanti, Kenyon, & Hart, 1992; Cruz-Neira, Sandin, & DeFanti, 1993) have been used in immersive virtual reality environments, such as flight or car simulators and other applications related to learning or training (see Section 41.5.3). In the earlier cases, due to the lack of stereoscopic visual sensory input, the user cannot experience the depth of the virtual scene unless the depicted objects move in the virtual space, or their perspective in the rendered 2D projection changes with respect to the position of the user. The latter requires the use of a tracking device, such as a depth camera, that can track accurately and in real time the position of the head of the user in the real-world space. The response of the overall visualization system that includes user tracking and rendering should be fast enough, in order to generate at least 25 frames/s for a smooth user–computer interaction, which corresponds to the average frame rate that can be processed by our sensory input system for vision.

The level of immersion can be significantly enhanced by employing stereoscopic 2D rendering and feeding each image as a sensory input to the corresponding eye. The two images in a stereoscopic projection can be displayed using either a pair of head-mounted displays or a single 2D display for rendering both images simultaneously by polarizing differently the light in the two projections. In the case of head-mounted displays, an accelerometer should be used in addition to the tracking mechanism in order to track in real time the orientation and the location of the user's head.

Depending on its nature, an educational application may involve group viewing and interactions between users located in the same physical space, such as in virtual flight simulators with a physical room that simulates an actual cockpit, or it may involve single person's viewing and interactions in the virtual environment with objects and potentially other users located in different physical spaces, such as in surgical simulations. Group experiences using the same visual stimuli limit significantly the level of immersion, since there are no personalized projections of the virtual environment (even in the case of stereoscopy) and it is assumed that the users' head is positioned at a fixed location and orientation in the 3D space. On the other hand, the direct interactions between users in the same physical space may have significant educational value in certain applications, while the level of immersion may not play an important role in the learning process.

TABLE 41.1
Comparison of Visualization Technologies

Technology	Stimulation	Limitation
2D screens	Single eye	Fixed perspective
2D screens with head tracking	Single eye	Single-person experience
Stereographic screens	Both eyes	Fixed perspective
Stereographic screens with head tracking	Both eyes	Single-person experience
Stereographic head-mounted displays with head tracking	Both eyes	Single-person experience

Highly immersive virtual environments are employed in applications that require from the users to fully understand the tridimensional structure of the virtual space, orient themselves, and interact with virtual objects. Such environments can be implemented by triggering the human sensory input for vision, using stereoscopic head-mounted displays in conjunction with real-time tracking of the location and orientation of the head.

From the aforementioned categorization of the systems for visualizing virtual environments based on the type of educational applications, it was made clear that body tracking is closely related not only to human–computer interaction but also to vision. Real-time tracking can be performed using depth cameras, such as Microsoft Kinect sensor, and accelerometers attached on head-mounted displays. Such devices allow for natural user interaction with the virtual environment, which is one of the key characteristics of immersive environments. Natural user interaction can be associated with other human senses besides vision, such as audition and tactile stimulation, which are discussed in the next sections. Finally, Table 41.1 provides a comparison of the aforementioned visualization technologies with respect to the type of generated stimuli and their limitations.

41.3.2 HEARING

The auditory sensory input contains smaller amount of information compared to the input from vision, which can be represented as a pair of waveforms produced by a directional sound. An audio wave generated by a sound source located at a particular point in the 3D space is received differently from the ears of different individuals, due to the variations on the shape of the earlobes as well as the shape of the head, volume of hair, and the presence of other objects (such as glasses), which as a whole reflect/absorb differently the sound waves.

In virtual reality applications, it is possible to simulate these processes by using head-related transfer functions (Blauert, 1997), which model how a particular earlobe receives a sound wave generated at a given distance and orientation from the ear. There are several libraries of head-related transfer functions that correspond to different shapes of head, styles of hair, and other facial features (Muller & Massarani, 2001) that can be used in order to simulate realistically the auditory sensory input using a stereo headset. In addition, one needs to take into account how the virtual objects reflect or absorb the waves generated by virtual sound sources. As a result, the generated auditory input depends on the orientation and position of the head that can be tracked using an accelerometer and a depth camera, respectively, as it was discussed in Section 41.3.1.

Audio plays important role in many educational applications where sound can extend the ability of the users to understand the 3D virtual environment, even when part of it is not within the field of view. For instance, in a car simulation, the user should be able to understand the presence of nearby vehicles and their approximate location from their engine's noise. Other applications may not produce at all visual stimuli, especially when they target visually impaired audience or when their main goal is associated with hearing.

The aforementioned applications were based on stereo audio wave simulation using head tracking, which is more suitable for personal rather than group experiences. An array of speakers placed in different locations of a physical space may be used to generate an approximation of sound waves in a 3D space with many users who interact directly with each other. In this case, each user receives a personalized auditory input based on his position in the physical space without the need of a tracking mechanism.

Finally, in some virtual reality applications, 3D audio simulation may not offer significant educational value. In such case it can be approximated by a simplistic audio generation mechanism that is properly panning prerecorded stereo audio waves based on the distance from the user.

41.3.3 TOUCH

Besides sensory input from vision and audition, tactile stimulation increases significantly the level of immersion in virtual environments, allowing the users to touch virtual objects. By touch, one can feel the texture, as well as any forces applied to the virtual objects, such as gravity and acceleration, which are important in many applications for experiential learning. For instance, experiments in physics can be performed in a virtual environment by interacting with solid objects with different textures and shapes, spring forces, magnetic forces, and liquids with different viscosities (see examples in Sections 41.5.1 and 41.6.3).

Tactile stimulation can also be used to experience collision with virtual objects, while one navigates a virtual environment. In such applications, the users can have a better understanding of the tridimensional space and the distances between the virtual objects. Furthermore, multiple users may interact with each other by feeling their presence by touch and collaborate in various ways.

There are many haptic devices commercially available that can be employed in educational applications, such as haptic mice, haptic gloves, and touch screens with haptic feedback. Each of the aforementioned natural user interface devices has different range of functionalities that can serve different needs of experiential learning systems. A haptic mouse can be employed for realistic tactile stimulation with large forces ideal for physics simulations; it requires, however, that the user is located near the device, which is usually placed on the top of a desk. Less realistic tactile stimulation is generated by haptic gloves and screens, which generate small vibrations in order to inform the user of the presence of virtual objects and hence enhance the interaction with the environment without restricting the location of the user. Finally, haptic screens have many applications in educational systems for visually impaired users.

The next section discusses how to design educational applications and covers the traditional methodologies and theories that have been used during the design of many of the projects presented in Sections 41.5 and 41.6.

41.4 DESIGNING VIRTUAL SYSTEMS FOR LEARNING THROUGH AGILE DEVELOPMENT

More and more, developers of game-based and virtual environments are utilizing agile development as an organizing production principle. Agile development frameworks emphasize the inherently uncertain and dynamic nature of software development and look toward incremental and iterative organizational solutions to deal with that uncertainty (Rajlich, 2006). Actually, a family of different software models in design practice, agile development, emphasizes the following (Beck et al., 2001):

1. Individuals and interactions over processes and tools
2. Working software over comprehensive documentation
3. Customer collaboration over contract negotiation
4. Responding to change over following a plan

Agile development is rapidly transforming software development paradigms in industry, but it has been slow to penetrate academic and research institutions (Rajlich, 2006). At the same time, however, agile methods present solutions to endemic problems in academic software development: software bloat, unclear milestones, and lacking communication between developers and stakeholders.

Most academic research projects still use the traditional *waterfall* paradigm of software development by default. The waterfall paradigm focuses on anticipating all costs and challenges by planning out every facet and every process of a piece of software's development lifetime and following that plan to the letter in a linear manner. Critics point to many problems they say are associated with the traditional *waterfall* development processes, especially when implemented in research settings (Rajlich, 2006). However, software development is full of change—client needs change, financial resources change, and technological system changes. In both game development and software development, waterfall paradigms have lead to low rates of project completion, lower rates of project success, and increasing cost overruns (Highsmith & Cockburn, 2002).

41.5   INTERACTIVE VIRTUAL WORLDS FOR EXPERIENTIAL LEARNING

In this section, three applications of virtual reality for experiential learning are presented focusing on K-12 and undergraduate education. A virtual environment for learning nanotechnology using devices for tactile stimulation is discussed in Section 41.5.1; an experiential learning virtual forest is reviewed in Section 41.5.2 and finally a virtual theater application for teaching drama in Section 41.5.3.

41.5.1   Use of Haptics in a Virtual Environment for Learning Nanotechnology

There are many examples in literature that report use of haptics in environments for experiential learning (Jackson, Pawluck, & Taylor, 2011; Park et al., 2010; Pawluk, Hoffman, McClintock, & Taylor 2009; Williams, Chen, & Seaton, 2003; Williams, He, Franklin, & Wang, 2007). Due to the nature of tactile stimulation, as it was discussed in Section 41.3.3, most of the applications are related to STEM areas, such as physics and engineering, in which dynamic systems of objects can be simulated by defining their physical properties as well as the forces that are applied to them.

In all these educational virtual environments, the main question that is being raised is, are there any notable improvements on the student's learning outcome when haptic feedback is employed along with the visual stimuli? Although the reported level of engagement in the majority of these applications was relatively high, which is an important factor in learning process, there has not been a clear answer to the aforementioned question. In some cases (Park et al., 2010), there was no significant difference between environments with visual simulation only and those with visual and haptic simulation. This result may be due to possible lack of realism in the way that the tactile stimulation was generated according to Jackson et al. (2011). This leaves room for further evaluation of the educational value of the current available haptic devices, and it also opens directions for improvements on these technologies.

Simulation of haptic feedback has been utilized by many researchers for experiencing scaled textures and forces from the nanoscale world, either for research (Sitti & Hashimoto, 2003) or for teaching (Jackson et al., 2011; Jones, Andre, Superfine, & Taylor, 2003; Pawluk et al., 2009). Figure 41.2 shows a screenshot from the haptic nanoscale application that was presented by Curtis Taylor and his collaborative developing team at the University of Florida Digital Worlds Institute in Jackson et al. (2011).

This project, dubbed *HapNan*, is an educational tool geared toward middle school students, which incorporates the use of a haptic mouse and virtual reality to create an interactive learning environment that has the potential to contribute to the traditional teaching methodologies for sciences within a classroom setting. Students who tested the HapNan technology were able to explore many scientific concepts (such as Van der Waals forces) through sight and touch (Figure 41.2). Low-cost commercially available haptic mouse was used in this environment, and it was programmed to simulate various surface textures and different types of forces (such as gravity, attraction, repulsion,

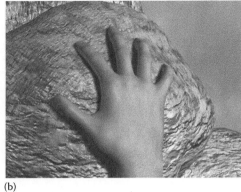

(a)   (b)

FIGURE 41.2 **(See color insert.)** (a) Middle school students exploring various textures and forces using virtual environment with haptic simulation. (b) A screenshot showing the simulated virtual environment.

and resistance) thus making these concepts tangible. The key goal of this tool is to help young students to better understand concepts, such as the covalent bonds within a water molecule and the relationship between gravity and inertia at a variety of physical scales.

The HapNan project is comprised of 11 gamelike levels that cover physical scales ranging from the planetary through cellular down to the nano and atomic domains. A quiz system is built in to the program that allows individual instructors the ability to gauge and assess how much information each student is retaining after exploring each level and potentially help scientists evaluate the effect of tactile stimulations on the learning process.

41.5.2 Virtual Learning Forest

Virtual environments for experiential learning have many applications in sciences with laboratory-based training sessions. Some of the key advantages of such tools are (1) cost-efficiency compared to real laboratories, (2) safety (the students can familiarize themselves with the proper laboratory safety procedures in a virtual setting before they go to an actual laboratory), and (3) accessibility, which is also important for distance education.

For forestry education, experiential learning systems show much potential for offering immersive experiences that simulate environmental areas including forests (Beck, 2009), as well as virtual laboratory tools and processes (Jung, Beck, Bannister, Cropper, & Staudhammer, 2010) that simulate traditional real-world teaching and training methodologies. Such an educational interactive virtual reality system, named *Virtual Learning Forest*, was presented in Jung et al. (2010) and offers visual simulation of longleaf pine ecosystems.

The target audience for Virtual Learning Forest is mainly undergraduate natural resource students. More specifically, the system has been tested by undergraduate forestry students in sampling, mensuration, and silviculture classes at the University of Florida and Virginia Polytechnic Institute and State University. The students were involved with testing and evaluating the virtual environment during the development of various phases of this product. The fully implemented version of a 3D Virtual Learning Forest based on the longleaf pine ecosystem was presented in Jung et al. (2010). In addition to the tridimensional simulation of this ecosystem, the virtual environment simulates several traditional tools and instruments for measuring the diameter of the trees (Figure 41.3), the distance of the student from a specific tree, and the height of the longleaf pine trees. Finally, the system is connected to a database that records the progress of the students in various virtual laboratory exercises that simulate traditional laboratory methodologies and measure how students react to the virtual system and how this impacts their learning experience.

FIGURE 41.3   A screenshot from the Virtual Learning Forest that shows the diameter tape tool, which simulates the process for measuring the diameter of a longleaf pine tree in a real-world forest.

41.5.3   Virtual Reality Theater for Teaching Classical Drama

In addition to the STEM areas, virtual reality has many applications in the humanities (Barmpoutis, 2013) and the performing arts (Masura, 2007). In this section, an application of virtual reality to classical drama is briefly presented as a tool for experiential learning.

The teaching methodologies in higher education for classical drama thus far have been limited to the study of the plot from the original text in conjunction with secondary bibliography that provides information regarding the circumstances of performance based on archaeological and literary evidence. This method of studying and analysis of the text, however, has left gaps in knowledge and room for subjective, or even incorrect interpretations, especially when spatiotemporal interactions of the portrayed characters play significant role in the study of the original text.

Virtual reality provides solutions that enhance the existing traditional teaching methodologies in the humanities, both in classroom and in distance learning (Barmpoutis, 2013). The classroom projector or computer screen can be used as a *magic mirror* in which the users can see themselves standing in the middle of a virtual theater, which is a real-life digital replica of the well-known ancient theater of Epidaurus. This augmented-reality system utilizes a low-cost depth/video camera that is connected with the computer as a regular USB web camera. Such depth cameras are available by many companies that produce devices for natural user interaction such as gaming engines (Microsoft Kinect) and peripheral devices (LeapMotion). One of the key advantages of these devices is that they do not require any technological skill by the users; on the contrary, the users can use their natural body motion and everyday life gestures to give instructions to the computer. Students are found to be familiar with such natural user interface platforms that already exist in the majority of households due to their affordable price range, based on national reports ("Report on teens, video games and civics," 2008). The prototype system presented in Barmpoutis (2013) uses the information acquired from the depth sensor to reconstruct instantly the 3D body image of the user, which is displayed as a real-time 3D holographic video stream in the middle of the virtual theater. The camera follows the motion of the user, who can move, walk, act, and dance, and the corresponding motion is instantly transferred to his or her 3D image in the virtual scene (Figure 41.4).

The students using this prototype system can intuitively understand the size of the stage and the structure of the theater by simply walking in the virtual space and also by visually comparing the

FIGURE 41.4 A screenshot from the virtual reality theater that uses Microsoft Kinect as a tool to understand different aspects of performance in the classical world.

size of their holographic body with the size of the depicted virtual elements of the theater, which is an automatic process of our brain that is triggered any time we visit a new space (Glenberg, Semin, & Smith, 2008). The student can also choose between first and third person's view, hold virtual props, and replicate his or her body many times in the virtual space in various different arrangements. Such an experiential learning environment can be used either by the instructor as a novel teaching tool or by the students who will ultimately understand better the circumstances of performance in the Greco-Roman world by personally interacting with virtual objects.

41.6   EXPERIENTIAL LEARNING IN HEALTH SCIENCE APPLICATIONS

This chapter presents three medical applications using virtual reality. In all of our examples, the virtual environments offer to their users experiences that can be used either as therapeutic solutions (Sections 41.6.1 and 41.6.2) or as tools for medical training (Section 41.6.3).

41.6.1   Virtual Environments for Therapeutic Solutions

Virtual environments are easily accessible, intuitive, and effective forms of virtual reality that provide secure real-time interaction between multiple users with many applications to real-time experiential learning, training, and therapeutic treatment modalities (Josman et al., 2006; Levy et al., 2009; Rose, Brooks, & Rizzo, 2005; Rothbaum, Hodges, Ready, Graap, & Alarcon, 2001). These readily accessible immersive environments provide both psychosocial and psychoeducational interventions and vocational training for service-disabled veterans suffering from posttraumatic stress disorder (PTSD) and mild traumatic brain injury (MTBI).

The prevalence of PTSD and MTBI in returning combat veterans is estimated at 20%, and the Defense and Veterans Brain Injury Coalition estimates that roughly 40% of injured warriors suffer from MTBI ("The neurological burden of the war in Iraq and Afghanistan," 2006). Often the most troubling symptoms of PTSD and MTBI are behavioral: mood changes, depression, anxiety, impulsiveness, emotional outbursts, intolerance of crowds, hypervigilance, or inappropriate laughter. Other symptoms include disturbances in attention and memory as well as delayed reaction time during problem solving. Behavioral therapy, a mainstay of successful rehabilitation, generally requires that veterans be able to organize and negotiate daily activities, travel considerable distances, and interact with individuals in public to meet scheduled appointments—the very areas where impairments may be most profound. An additional barrier to rehabilitation for many returning combat soldiers is generational. Many younger soldiers feel uncomfortable entering veterans' administration medical centers, where both the patients and health-care providers are older and thus may be perceived as unsympathetic or unable to understand the issues of younger veterans.

Virtual environments hold great potential to solve many of the problems that hinder effective treatment, support, and reintegration of wounded combat veterans into their families and into society. There are many examples showing that the use of virtual reality can contribute greatly to cognitive and affective rehabilitation and recovery (Josman et al. 2006, Levy et al., 2009; Rose et al., 2005; Rothbaum et al., 2001).

Figure 41.5 shows a screenshot from the experiential learning virtual reality system for veteran rehabilitation that was presented in Levy et al. (2009). The system simulates a grocery store scenario that offers computational, navigational, and memory challenges. The content of the interaction can be carefully structured, controlled, and monitored by medical professionals through the Internet. This prototype environment allows (1) the therapist and veteran to navigate the grocery store together each occupying distinct avatars, (2) the therapist to switch to other avatars with various roles (e.g., the cashier), and (3) the therapist to set the parameters of the interaction and the environment, such as money in virtual wallets, number of other autonomous avatar shoppers, ambient noise level, and collision scenarios with other shoppers.

Such systems embody capabilities to constantly enhance their efficacy based on user input, including user-designed content, clinical outcomes, and general use patterns. Virtual reality interventions can be accessed not only during traditional sessions in the offices of health-care providers but also from a computer in the privacy of the patient's homes. In many cases, this can reduce or even eliminate the need for recurring travel to a distant medical facility. They provide opportunities for interaction with others in fully immersive, more controlled, and less threatening virtual environments and build on preexisting technological skill sets possessed by many combat veterans in their 20s and 30s. Thus, virtual environments in general show great potential to augment conventional treatment of PTSD/MTBI in returning combat veterans.

Therapeutic simulations of ordinary social and vocational challenges allow veterans to practice problem solving in virtual real-life scenarios. Participants are able to review the consequences of decisions and get direct and immediate feedback regarding choices and responses. Unlike real life, participants have multiple opportunities to learn to deal with challenging situations with no penalty for failure through an interactive experiential learning mechanism.

FIGURE 41.5 Simulation of a grocery store scenario for returning combat veteran rehabilitation through an experiential learning environment.

41.6.2 INTERACTIVE TOOLS FOR TREATING MOTOR DISABILITIES

Brain–computer interfaces (BCI) offer tremendous promise as assistive systems for motor-impaired people, as new paradigms for natural user interaction with virtual environments, and as vehicles for the discovery and promotion of new computational principles for autonomous and intelligent systems (Berger et al., 2007; Sanchez & Principe, 2007; Vaughan & Wolpaw, 2006). At the core of a motor brain–machine interface, there is an embedded computer with information processing models of the motor system to dialogue directly and in real time with the user's action–perception cycle derived directly from the nervous system. Thus, the level of immersion in a virtual reality system can be enhanced using BCI, and it can be employed in applications that assist patients in learning new motor skills and transition from a disabled state back to a fully functioning state in real-world scenarios.

Virtual environments offer numerous advantages over the real world for BCI training and experiential learning. Primarily, virtual reality offers a flexible training environment that has the complexity of the physical world, but the kind of complexity that can be introduced to the user in a controlled manner to stimulate brain function and increased motor performance. Additionally, such environments can be effective enabling mechanisms to move away from the reductionist approach that has been used traditionally and facilitate the measurement of natural movement within interactive complex environments that mimic those encountered in daily life.

Figure 41.6 shows an example of a virtual environment controlled using BCI hardware. The visual appearance of the environment could be reskinned to simulate various daily life processes, such as driving a car, catching a ball, or others that involve a motor process performed using a traditional joystick and enhanced by the BCI device. This environment has been utilized to perform various experiments using two BCI hardware devices, a MindWave device (manufactured by NeuroSky) and an EPOC device (manufactured by Emotiv). The number of BCI sensors is 14 for EPOC and 1 for MindWave, and they communicate with the developed computer software using a regular USB port.

The software performs training and testing of the pattern recognition algorithm in real time, using a gaming environment that motivates the user to perform brain activities related to turning left or right. The input signal is captured during each of the critical turning points of the game and converted into a list of features that are provided to a machine-learning algorithm. This algorithm finds distinctive characteristics and associations between the features that correspond to a *left turn* and those that correspond to a *right turn*. An intuitive scoring mechanism is used to evaluate the performance of the users by counting the percentage of the successful turns during a game session. In a manner similar to that of a personal trainer at an athletic club, a virtual therapist is available

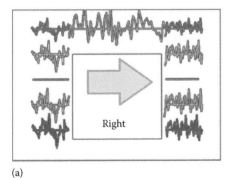
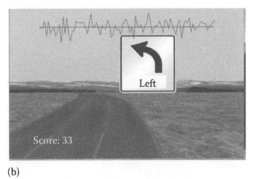

(a)   (b)

FIGURE 41.6   An example of a brain-controlled virtual environment. (a) A training mechanism associates the acquired brain signals with the directional visual stimuli. (b) The user can navigate within the virtual environment using brain signals.

in this system to guide in a program of neuro-rehab therapy tailored to the individual patient. The ultimate goal of such environments is to assist disabled patients to relearn motor skills by gradually transitioning the control of the environment from BCI to a joystick device, which is fully utilized in more advanced training stages.

41.6.3 VIRTUAL BRAIN FOR MICROSURGICAL TRAINING

Mastery of the neurosurgical skill set involves many hours of supervised intraoperative training since the brain is exquisitely sensitive to touch, anoxia, and derangements of its internal environment. A recent step toward a more technologically advanced neurosurgical training has been to capture high-definition images in stereoscopic format and make them available on smartphones, laptops, and 3D television and for stereoscopic projection for large groups (Rhoton, 2007). Furthermore, surgical simulation and skill training offer an opportunity to teach and practice advanced techniques before practicing them on patients. Simulation training can be as straightforward as using real instruments and video equipment to manipulate simulated tissue in a box trainer. More advanced experiential learning virtual reality simulators are now available and ready for widespread use (Alaraj et al., 2011).

Such experiential learning systems use real-time haptic feedback mechanisms to perform realistic neurosurgical simulations that reproduce the operative experience, not restricted by time and patient safety constraints. Computer-based, virtual, or mixed-reality platforms offer just such a possibility (Lemole et al., 2007).

An example is the existing UF Department of Neurosurgery ventriculostomy simulator, which utilizes some of these technologies. Ventriculostomy is a common operative procedure performed by neurosurgeons. The procedure consists of drilling a small hole in the skull through which a small catheter (~2.0 mm diameter) is passed to enter one of the brain ventricles. The indications for placement of a ventriculostomy may be therapeutic, for example, to drain excessive fluid accumulation within the ventricle; diagnostic, for example, to measure the pressure within the ventricle; or sometimes to accomplish both drainage of fluid and pressure measurement simultaneously (Figure 41.7).

The UF Department of Neurosurgery ventriculostomy simulator uses mixed reality to simulate the operative procedure that combines the use of a 3D interactive virtual brain model rendered in real time and an actual physical skull model with a tracking sensor and haptic feedback. The student using the UF simulator experiences the reality of drilling through an acrylic polymer-simulated skull created with a 3D printer. Next, the student is able to pass a catheter through the constructed

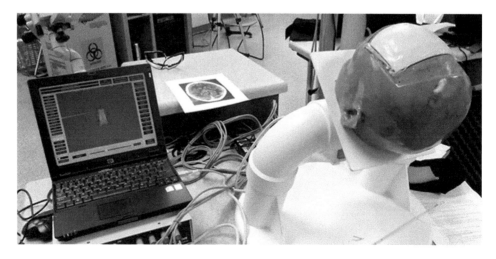

FIGURE 41.7 (**See color insert.**) Picture of the UF Department of Neurosurgery experiential learning mixed-reality ventriculostomy simulator with haptic feedback.

skull opening and into a substance that closely simulates the turgor and resistance of the human brain. The catheter is equipped with electrical sensors that can be tracked and linked via computer interface to a computer simulation of a human brain, complete with ventricles and adjacent critical brain structures that must be avoided in carrying out the operative procedure safely.

Another example of a 3D tool for teaching human neuroanatomy is the work done by the department of anatomy and neurobiology at Boston University School of Medicine (Estevez, Lindgren, & Bergethon, 2010). This tool also uses acrylic polymer models designed from real brain specimens; however, there is no virtual component attached to the models and the interaction is limited to the haptic feedback from the polymer material. A virtual reality alternative was developed by the department of neurosurgery at the University of Illinois, Chicago, and was primarily used as an educational tool for neuroanatomy (Lemole et al., 2007).

41.7 CONCLUSIONS AND FUTURE DIRECTIONS

Educational or *serious games* are one of the most emerging and at the same time constantly evolving form of experiential learning. Some of the examples presented in Sections 41.5 and 41.6 contain intuitive game rules and scoring mechanisms in order to motivate the learners and inform the educators about the efficacy of the applied teaching methodologies. Converting the learning process into the form of a game does not only make learning more enjoyable but also improve its outcomes while increasing the level of engagement.

Platforms for visualizing virtual reality environments for experiential learning are now available to different types of devices ranging from handheld tablets and smartphones to desktop computers and TV sets. This scalability increases significantly the available audience and offers to the academic institutes' new ways for delivering educational material. However, there is still room for technological improvements especially in the areas of low-cost stereoscopic rendering in scalable platforms, haptic feedback devices, and fully immersive simulations that include olfaction and gustation, which are important in many medical and scientific educational applications.

REFERENCES

Alaraj, A., Lemole, M. G., Finkle, J. H., Yudkowsky, R., Wallace, A., ...Charbel, F. T. (2011). Virtual reality training in neurosurgery: Review of current status and future applications. *Surgical Neurological International*, 2(1), 52.

Anonymous. (2006). The neurological burden of the war in Iraq and Afghanistan. *Annals of Neurology*, 60(4), A13–A15.

Anonymous. (2008). Report on teens, video games and civics. *Pew Internet & American Life Project*. Received from http://www.pewinternet.org/

Barmpoutis, A. (2013). Tensor body: Real-time reconstruction of the human body and avatar synthesis from RGB-D. *IEEE Transactions on Cybernetics, 43*(5), pp. 1347–1356.

Beck, H. W. (2009). *Lyra virtual world environment (Lyra VWE): Educational applications in agriculture and natural resources*. Seventh World Congress on Computers in Agriculture and Natural Resources Conference Proceedings. Reno, NV.

Beck, K., Beedle, M., Van Bennekum, A., Cockburn, A., Fowler, W.C.M., Grenning, J., ... Kern, J. (2001). Manifesto for agile software development. Agile Alliance. http://agilemanifesto.org/

Berger, T. W., Chapin, J. K., Gerhardt, G. A., McFarland, D. J., Principe, J. C., Soussou, W. V., ... Tresco, P. A. (2007). *International assessment of research and development in brain-computer interfaces*. Baltimore, MD: Study Report by World Technology Evaluation Center.

Blauert, J. (1997). *Spatial hearing, the psychophysics of human sound localization*. Cambridge, MA: MIT Press.

Cruz-Neira, C., Sandin, D. J.. & DeFanti, T. A. (1993) *Surround-screen projection-based virtual reality: The design and implementation of the cave*. SIGGRAPH'93: Proceedings of the 20th Annual Conference on Computer Graphics and Interactive Techniques. Anaheim, CA, pp. 135–142.

Cruz-Neira, C., Sandin, D. J., DeFanti, T. A., Kenyon, R. V., & Hart, J. C. (1992). The cave: Audio visual experience automatic virtual environment. *Communications of the ACM, 35*(6), 64–72.

DeVane, B., & Squire, K. D. (2008). The meaning of race and violence in Grand Theft Auto. *Games and Culture*, *3*(3-4), 264–285.

Dewey, J. (1899). *The school and society: Being three lectures by John Dewey*. Chicago, IL: University of Chicago Press.

Dewey, J. (1916). *Democracy and education*. New York, NY: MacMillan Company.

Estevez, M. E., Lindgren, K. A., & Bergethon, P. R. (2010). A novel three-dimensional tool for teaching human neuroanatomy. *Anatomical Sciences Education*, *3*(6), 309–317.

Gee, J. P. (2005). Learning by design: Good video games as learning machines. *E-Learning*, *2*(1), 5–16.

Gee, J. P. (2007). *What video games have to teach us about learning and literacy* (2nd ed.). New York, NY: Palgrave Macmillan.

Glenberg, A. M., Semin, G. R., & Smith, E. R. (2008). *Embodied grounding: Social, cognitive, affective, and neuroscientific approaches*. Cambridge, MA: Cambridge University Press.

Highsmith, J., & Cockburn, A. (2002). Agile software development: The business of innovation, *Computer*, *34*(9), 120–127.

Jackson, D., Pawluk, D., & Taylor, C. R. (2011). *Development of haptic virtual reality gaming environments for teaching nanotechnology*. American Society for Engineering Education (ASEE) Annual Conference & Exposition. Vancouver, British Columbia, Canada.

Jones, M. G., Andre, T., Superfine, R., & Taylor, R. (2003). Learning at the nanoscale: The impact of students' use of remote microscopy on concepts of viruses, scale, and microscopy. *Journal of Research in Science Teaching*, *40*, 303–322.

Josman, N., Somer, E., Reisberg, A., Weiss, P. L., Garcia-Palacios, A., & Hoffman, H. (2006). Busworld: designing a virtual environment for post-traumatic stress disorder in Israel: A protocol. *Cyberpsychology, Behavior, and Social Networking*, *9*(2), 241–244.

Jung, Y., Beck, H. W., Bannister, M, Cropper, W. P., & Staudhammer, C. (2010). *Virtual learning environment: An application to the longleaf pine ecosystem*. Proceedings of the ASABE Annual International Meeting. Dallas, TX.

Kolb, D. A. (1984). *Experiential learning: Experience as the source of learning and development*. Englewood Cliffs, NJ: Prentice Hall.

Kolb, D. A., Boyatzis, R. E., & Mainemelis, C. (2001). Experiential learning theory: Previous research and new directions. In Sternberg, R. J., & Zhang, L., *Perspectives on thinking, learning, and cognitive styles* (pp. 227–247). Mahwah, NJ: Lawrence Erlbaum.

Krueger, M. W. (1996). *Addition of olfactory stimuli to virtual reality simulations for medical training applications*. Defense Technical Information Center, Annual Report.

Lemole, G. M., Jr., Bannerjee, P. P., Luciano, C., Neckrysh, S., & Charbel, F. T. (2007). Virtual reality in neurosurgical education: Part-task ventriculostomy simulation with dynamic visual and haptic feedback. *Neurosurgery*, *61*, 142–148.

Lenhart, A., Kahne, J., Middaugh, E., Macgill, A. R., Evans, C., & Vitak , J. (September 15, 2008). Teens, video games, and civics. Pew Internet & American Life Project. http://www.pewinternet.org/files/old-media/Files/Reports/2008/PIP_Teens_Games_and_Civics_Report_FINAL.pdf.pdf

Levy, C. E., Oliverio, J., Sonke, J., Hundersmarck, T., Demery, J., Tassin, C., … Omura, D. (2009). Virtual environments for cognitive and affective dysfunction in MTBI and PTSD: Development of a 21st century treatment platform. *Military Health Research Forum (MHRF)*.

Masura, N. (2007). *Digital theatre: A "live" and mediated art form expanding perceptions of body, place, and community* (Dissertation), University of Maryland, College Park, MD, August 31–September 3, 2009, Kansas City, MO.

Muller, S., & Massarani, P. (2001). Transfer function measurement with sweeps. *Journal of Audio Engineering Society*, *49*(6), 443–447.

Park, J., Kim, K., Tan, H.Z., Reifenberger, R., Bertoline, G., Hoberman, T., & Bennett, D. (2010). *An initial study of visuohaptic simulation of point-charge interactions*. IEEE Haptics Symposium. Waltham, MA, pp. 425–430.

Pawluk, D., Hoffman, M., McClintock, M., & Taylor, C. R. (2009). *Development of a nanoscale virtual environment haptic interface for teaching nanotechnology to individuals who are visually impaired*. American Society for Engineering Education (ASEE) Annual Conference & Exposition. Austin, TX.

Rajlich, V. (2006). Changing the paradigm of software engineering. *Communications of the ACM*, *49*(8), 67–70.

Rhoton, A. L., Jr. (2007). 3D anatomy and surgical approaches of the temporal bone and adjacent areas. *Neurosurgery*, *51s*, 1–250.

Rose, F. D., Brooks, B. M., & Rizzo, A. A. (2005). Virtual reality in brain damage rehabilitation: Review. *CyberPsychology and Behavior*, *8*(3), 241–262.

Rothbaum, B. O., Hodges, L. F., Ready, D., Graap, K., & Alarcon, R. D. (2001). Virtual reality exposure therapy for vietnam veterans with posttraumatic stress disorder. *Journal of Clinical Psychiatry*, *62*(8), 617–622.

Sanchez, J. C., & Principe, J. C. (2007). *Brain machine interface engineering*. San Rafael, CA: Morgan and Claypool.

Sitti, M., & Hashimoto, H. (2003). Teleoperated touch feedback from the surfaces at the nanoscale: Modelling and experiments. *IEEE/ASME Transactions on Mechatronics*, *8*(2), 287–298.

Squire, K. (2006). From content to context: Videogames as designed experience. *Educational Researcher*, *35*(8), 19–29.

Squire, K. D., & Jenkins, H. (2002). The art of contested spaces. In King, L. & Bain, C. (Eds.), *Game On* (pp. 62–75). London, U.K.: Barbarican.

Steinkuehler, C., & Chmiel, M. (2006). Fostering scientific habits of mind in the context of online play. In S.A. Barab, K.E. Hay, N.B. Songer, & D.T. Hickey (Eds.), *Proceedings of the International Conference of the Learning Sciences* (pp. 723–729). Mahwah NJ: Erlbuam.

Vaughan, T. M., & Wolpaw, J. R. (2006). The third international meeting on brain-computer interface technology: Making a difference. *IEEE Transactions on Neural Systems and Rehabilitation Engineering*, *14*, 126–127.

Williams, R. L. II, Chen, M. Y., & Seaton, J. M. (2003). Haptics- augmented simple-machine educational tools. *Journal of Science Education and Technology*, *12*(1), 1–12.

Williams, R. L. II, He, X., Franklin, T., & Wang, S. (2007). Haptics- augmented engineering mechanics educational tools. *World Transactions on Engineering and Technology Education*, *6*(1), 1–4.

42 Design and Development of 3D Interactive Environments for Special Educational Needs

Sue Cobb, Tessa Hawkins, Laura Millen, and John R. Wilson

CONTENTS

42.1 INTRODUCTION

In the previous edition, we presented a structured approach to the design, development, and evaluation of virtual environments (VEs) used for learning or, as we called them, *virtual learning environments* (VLEs[*]), with close liaison with the representative users and professionals at all stages of development to ensure adequate design of VEs to support learning. This approach had been developed

---

[*] Virtual learning environments is a term that has been adopted in education to include any remote learning context (i.e., learning taking place virtually rather than physically), including video conferencing and e-learning. Thus, in this chapter we refer to virtual environments (VEs) for learning and no longer use the term VLE.

through a series of case studies in which we established a framework for planning and designing VE construction (the VEDS framework, see Eastgate, Wilson, & D'Cruz, 2014, Chapter 15) and a procedure for the involvement of schools, teachers, and pupils in the design and development of VEs for learning (Brown, Standen, & Cobb, 1998; Neale, Brown, Cobb, & Wilson, 1999). This approach is still applicable and has been used successfully in further projects. However, we have also learned the need to be flexible and adaptable so that we can facilitate the involvement of a broader, and more diverse, design team. This is particularly important as, since the last edition, there has been a much wider application of VEs for learning in a variety of contexts, and our research group has focused more on special educational needs (SEN). We have extended our design methods to facilitate end-user (children with SEN) involvement in participatory design activities. This chapter provides an update to the previous edition and will show how different stakeholders contribute to the design and evaluation of VEs for SEN using a case example from the COSPATIAL project.

42.2 VEs FOR SPECIAL EDUCATIONAL NEEDS

Over the last decade, the use of technology in schools has dramatically increased. Typical classrooms now use interactive whiteboards and displays, access the Internet, and use educational software in most subject areas; and the number of computers, laptops, and tablets available to students is increasing (Wastiau et al., 2013). The use of information and communication technologies (ICT) is now seen as an educational objective in itself as technology proficiency becomes increasingly important in everyday life and work. The definition of virtual reality (VR) has broadened in this time and can include 3D interactive games and gesture-based control input. Many widely available commercial and gaming applications now make use of these features. Despite the growth in the application and use of educational technology, there are still few guidelines for the design and development of 3D interactive environments for learning (Minocha, Kear, Mount, & Priestnall, 2008).

While early educational computing allowed users to practice repetitive tasks (Solomon, 1986), VEs offered the potential for a different kind of arena for learning, enabling the facility for experiential learning by exploring and interacting with 3D simulations that could provide visualization of concepts that cannot be seen in the physical world (Bricken, 1991; Dede, 1995; Winn, 1993) or rehearsal of tasks and activities that may not be practical in the real world, such as machine operation (Chen & Toh, 2005; Lapointe & Robert, 2000). The key advantage of using VEs for learning is that they allow a greater degree of self-paced and determined exploration as well as the practice of naturalistic tasks in realistic settings (Brown, Cobb, & Eastgate, 1995; Brown et al., 1998; Cobb, Brown, Eastgate, &Wilson, 1993). VEs can be used in education to explore concepts, objects, or systems that cannot otherwise be physically experienced (Mikropoulos & Natsis, 2011), providing potential benefits in *representation* (e.g., showing an abstract concept or showing the spatial or temporal scale and relationships of objects and systems that are beyond human perception), *interaction* (where this is not possible with the real-world equivalent model or environment, e.g., where there is a safety, security, or ecological concern), and *access* (VEs can provide remote access to virtual models where access to real-world equivalents is just not possible for practical reasons). VEs can therefore provide unique opportunities to support a wide variety of educational objectives, from the practical (such as learning how to use equipment in a laboratory—Dalgarno, Bishop, Adlong, & Bedgood, 2009) to the conceptual (such as learning the structure of an atom—Kontogeorgiou, Bellou, & Mikropoulos, 2008).

Constructivism has been advocated as a learning theory upon which to build an instructional technology (Duffy & Jonassen, 1992; Papert, 1988; Perkins, 1996; Youngblut, 1998) and is still the most reported education theory supporting the development of VEs for education (see review by Mikropoulos & Natsis, 2011). An early study conducted by the authors (Neale et al., 1999) investigated the extent to which VEs used for special educational learning met constructivist principles (using principles developed from Jonassen, 1994). It was found that the applications supported moderate rather than extreme forms of constructivism. Different learning environments supported distinct learning principles in varying ways. This was mainly due to differences in the objectives of the VEs. VEs that aimed to promote communication skills were found to support collaboration and social negotiation but contained more abstract visual content and did not represent the natural complexity of the real world accurately. Conversely, VEs that aimed to teach real-world life skills much better represented the natural complexity of the real world and presented more authentic tasks.

VEs have a number of characteristics that may promote certain behaviors that enhance special needs education, such as self-directed activity, motivation, and naturalistic learning (Neale et al., 1999). These features of VEs go some way to supporting learning according to a constructivist perspective. In all of the VEs examined, it seemed necessary to include some form of behaviorist/objective learning. In the cases explored, teachers and students with a learning disability collaborated *shoulder to shoulder* using the VE. Teachers always provided some directive or instructional role; this meant that the students were more likely to take in important points of focus and understand the meaning of their action.

Over recent years, there has been an increase in the development of technology and technology-based interventions specifically for children with special needs and children with autism (Bölte, Golan, Goodwin, & Zwaigenbaum, 2010; Hourcade, Bullock-Rest, & Hansen, 2011; Williams, Wright, Callaghan, & Coughlan, 2002). This increase may be attributed to a large number of reports (both research related and anecdotal) that indicate that adults and children with autism are drawn to and motivated by computer-mediated activities (Battocchi et al., 2008; Colby, 1973; Goldsmith & LeBlanc, 2004; Hardy, Odgen, Newman, & Cooper, 2002).

42.3   DESIGN AND DEVELOPMENT OF VEs FOR LEARNING

There are three general phases in the development of VEs for learning: specification of the application and its content, programming, and evaluation. While these phases are distinct and progressive in themselves, the development of the final product requires a good deal of iteration (see Figure 42.1); thus, although evaluation is shown here as a separate phase, formative evaluation must be integrated throughout the development stage. A user-centered approach taken in the specification and design of VEs will result in a more useful and appropriate program.

Case examples were presented in the previous edition of this chapter (Cobb, Neale, Crosier, & Wilson, 2002) in order to illustrate how we involved schools directly in this process in both mainstream education (Virtual Radlab) and special needs education (the Virtual life skills project) settings. At the end of the chapter, we present a model showing how the project development team and associated stakeholders contributed to the process of VE design and evaluation (see Figure 42.1).

The Life Skills Education project was a community-based research project intended for students with learning disabilities aged 16+ to learn about and practice skills needed for independent living (Brown et al., 1998). A virtual city comprising four main elements was constructed: a virtual house, virtual supermarket, and virtual café linked together by a virtual transport system. Students were given a task scenario that required them to make decisions and perform tasks in the virtual world, including making a shopping list and planning a bus journey to the supermarket to buy food. The objective was to see whether rehearsal of everyday tasks such as catching a bus to go to the supermarket in a simulated VE would facilitate familiarity leading to increased confidence in conducting similar tasks in real life.

The project team considered it very important that the tasks developed in the virtual world were relevant and meaningful to the students. One of the key features of this project was that the end users, young people with specific learning disabilities, themselves were directly involved in the design process to inform decisions concerning content and the VEs (e.g., the choice of learning scenarios) and the evaluation of the VE interface (Meakin et al., 1998). The students formed a user group that met with the project researchers and technology development team on a regular basis to agree on design decisions and to review the prototypes of the virtual world as it was developed. Meetings were held once a month either at the university or at the school and followed a semistructured plan

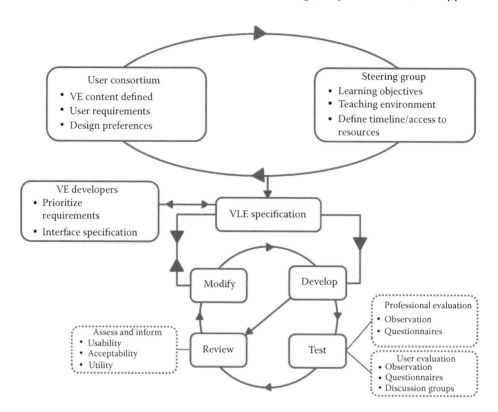

FIGURE 42.1 User-centered design and evaluation model. (Cobb, S.V. et al., Development and evaluation of virtual environments for education, In K.M. Stanney (Ed.), *Handbook of Virtual Environments*, Lawrence Erlbaum Associates, Mahwah, NJ, pp. 353–378, 2002.)

in which the team first held a discussion about what the skills the students wanted to learn about in the virtual world. This was facilitated by one of the teachers from the school, and notes were written on flipcharts. At the end of the discussion, the group voted on any issues that required a decision. Food was provided and then the students spent time exploring the technology prototypes in small groups. This part of the session was informal, and the students talked to the researchers about what they liked and disliked in the virtual world. A group decision was made about what aspects in the virtual world to change or further develop over the next month.

The use of an inclusive participatory design approach was very successful for this project. Students appreciated being asked about what they wanted to learn (Meakin et al., 1998) and commented on how much they enjoyed the social aspects of the user group meetings, "I love coming to the User Group—it's good fun" (p. 7), and the value of their contribution to the project, "I know I'm useful here" (p. 7). Over the period of time, teachers commented that many of the students had grown in confidence and become more self-assured. The user group model was taken forward into new projects that were taking place at the school with some of the students remaining members for several years; "I liked helping in the User Group and I want to help people again now" (p. 8). User group involvement in informing the design of the prototypes meant that the technology was developed in response to the actual needs of people with learning difficulties, not the assumed needs, represented by proxy. Evaluation of the project outcomes found positive results for the usability and enjoyment of performing the virtual tasks and evidence of increased confidence and awareness of what to do in real-world situations (Cobb, Neale, & Reynolds, 1998).

The participatory design approach developed in the Virtual Life Skills project was continued in VE development projects designed for use with children and young people on the autism spectrum, although adapted to accommodate different requirements for social interaction and communication inherent in this type of user group. Two projects examined the use of VEs to support the learning and rehearsal of social communication skills in young people with autism: Asperger's syndrome (AS) Interactive and COSPATIAL. An overview of methods for the development of these case studies with specific focus on how the direct involvement of schools in the project facilitated *student voice* is given by Parsons and Cobb (2013). Here, we use these case studies to show, in more detail, how the process of school involvement in the participatory design of VEs was further developed and adapted specifically for SEN applications. We use the VEDS framework (Eastgate et al., 2014; Wilson, Eastgate, & D'Cruz, 2002) as the basis for our design process.

42.3.1 SPECIFICATION PHASE

In this section, we emphasize the involvement of user groups in the early conceptualization stages. At each stage, representatives of users and professionals should be involved to ensure that concepts are relevant and interfaces are accessible. Figure 42.2 shows how users and professionals can influence the VE development, described in the following sections.

42.3.1.1 Initial Contact with Schools/Organizations

The first step is to make contact with the schools or organizations that may use the VE. Initial contact may arise through several routes:

1. Cold contact with schools to recruit interest in the project
2. User representatives approaching the developers asking them to meet a specific need
3. General interest from potential users after seeing a demonstration or publicity

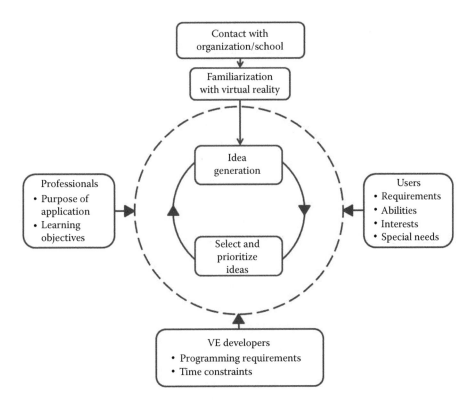

FIGURE 42.2   User-centered VE specification.

It is important that VEs are developed with a clear view of the value of VR as an aid for learning. The user-centered design approach involves teachers/professionals in the first initiation of the VE design. They will determine the purpose of the VE, as an additional teaching aid to support existing methods or as a specific program aimed at providing learning in a unique way. In addition, they inform VE developers about the practicalities of computer-based teaching within their organization (e.g., resources available, staff interest and motivation to use the program, and acceptability).

42.3.1.2 VR Familiarization

After initial contact, the next step is to familiarize the teaching professionals or support workers with VR technology and features of VEs. It is important to give a realistic view of VR and VEs in accordance with the users' needs and level of available resources.

The involvement of the schoolteachers in the design of the VEs developed would help to ensure its relevance and applicability as a classroom tool. The educator has an important role to play in the design of learning environments.

We have found it useful to start with a teacher workshop to introduce the teachers to VR and collect feedback on their first impressions of working with this technology. This also provides an opportunity for the teachers to identify possible application areas for VR in the relevant education domain and any barriers to its implementation and use. Consideration needs to be given as to where to hold the workshop: after school, during school, or off-site. This decision will be based on a number of factors including the availability of teachers and the ease with which technology can be transported for demonstration in schools.

42.3.1.3 Concept Generation

Once teachers and support workers have been made aware of VR technology and its capabilities, they are then in a better position to be able to comment on how it might be useful to them. A brainstorming session can be held to generate ideas for specific applications of interest. The teachers and support workers will have the responsibility for using the technology on a day-to-day basis and its integration within existing teaching methods. It can also be useful to involve users at this stage, where permissible. Users can give an insight into areas that they are interested in learning about, issues they want to more fully understand, and where current teaching methods are inadequate. In this way, the VE produced should be derived from an authentic learning need expressed by the users themselves or the professionals responsible for teaching them these skills/issues (Crosier & Wilson, 1998).

Technical and practical information regarding what ideas are feasible in terms of time and technology capabilities also needs to be provided by researchers and VE developers. VE scenarios that maximize VR capabilities should be developed, so consideration needs to be given to other technologies or methods that may be equally useful in teaching the subject area so as not to *reinvent the wheel.*

At the beginning stages of the design process, the team should not focus on one idea and modifications thereof; it is better to generate a range of ideas during this phase, without spending too much time on each. Some ideas may be quickly discarded, but the process of discarding ideas itself can bring to light requirements that may not have previously been explicitly stated. Beginning with multiple design ideas decreases the chances that an inappropriate or unworkable idea will be pursued and also tends to generate more effective feedback from the design participants (Tohidi, Buxton, Baecker, & Sellen, 2006).

42.3.1.4 Selection and Prioritization of Ideas

The early stages of idea generation frequently end up with more possible VEs than can practically be developed. Prioritization of these ideas should involve a number of professional teachers or carers, who will have a sense of what learning or development objectives are most important for the target user group. In the case of mainstream education, decisions may be governed by curriculum restrictions.

Various methods can be used to facilitate the selection process. We have used the Delphi technique to facilitate the teacher prioritization of curriculum topic that VE could be most beneficial for (Crosier, Cobb, & Wilson, 2002), theme-based content analysis to consolidate ideas generated by focus group activities (Neale and Nichols, 2001), and, in the COSPATIAL project, principles of cognitive behavior therapy to define the activities that could be provided within the VE to support student learning (Zancanaro et al., 2010).

Involvement of relevant groups of professionals at this stage of VE development is critical; it gives them an insight into *why* VE is developed, and it also gives them an opportunity to contribute information for the VE to be relevant to their students or clients. Just as important, their involvement in the early stages of development means that the host or client organizations will be much more understanding and helpful when their time and cooperation is needed in the later stages of VE development. Since VEs must not be one-off initiatives but should be embedded into schools and other institution's general operations, early two-way dialogue with the organization will clear the path for this.

42.3.2   VE DEVELOPMENT PHASE

There is no single correct approach to building a VE. In practice, the approach taken depends upon what the VE is intended for, how well formalized the idea is, how specific the user is in their requirements, and the experience and programming style of individual VR programmers involved. What is most important is that the communication channel between the user (or user champion) and the programmer works well. In some cases, this may not be a direct one-to-one communication channel but involve several links between different people such as teachers, support workers, researchers, and programming managers.

Development of VEs for education is an iterative process. In this, and the previous edition of this book, our colleagues Wilson et al. (2002) and Eastgate et al. (2013) detail a general framework for the development of VEs. Initial brainstorming with teachers or support workers provides the context for the learning environment, the specific learning objectives, and an outline of the layout and visual representation of learning scenarios. During these discussions, sketch-style storyboards are drawn up, which are then presented to the VE programmers, who examine them and make recommendations back to the review or steering group about what they think is feasible within the timescale allocated. This may require the review group to prioritize the features of the VE or to identify particular learning objectives that should be the main focus of the VE. In our experience, we have found that initial user requirements are often based on what they already know about VE capabilities; having seen the familiarization demonstrations, users will ask for similar scenario.

Therefore, it is useful at this stage to have a discussion between programmers and the review group at which the programmer suggests alternative ways to achieve the required effects or ways to reduce programming effort, but does not tell the user what their objectives are. It is important that user objectives for a new VE are not derived from what they have already seen, but that they are guided into stating their actual requirements. The level of programmer involvement in initial contact with users and professionals will affect their level of understanding of user needs. There is often a different level of communication required between the programmer and the user, for example, more detailed questions about the design of a layout or a task within a VE. If the VE developers are involved from the start of the process, then they will have a greater understanding of the messages that the VE should put across and how this should be done.

The design of any computer program may be directed by the use of relevant guidelines. There are a number of human–computer interaction (HCI) principles and rules, which may be used to guide interface design, and a number of high-level and widely applicable principles can be useful to VE developers. General principles that should be followed by VE designers include the following: know the user population, reduce cognitive load, engineer for errors, and maintain consistency and clarity (Preece, 1994). An in-depth knowledge of the user population is essential to follow all of the other principles. For example, the level of cognitive load that is appropriate will be dependent on the user (children, adult with special needs, teachers, etc.).

Design rules that are more specific instructions intended to be followed by the designer can be found in HCI. However, many design rules developed within general HCI are not relevant to VE design because of the different features and objectives of VEs as compared with graphical user interfaces (GUIs) or more traditional computer interfaces. For example, a user of a VE can navigate in 3D space and interact in a more *naturalistic* way with a representative of their real-world counterparts. Kaur, Sutcliffe, and Maiden (1998) have found that conventional interface guidelines are only partially applicable for the designers of VEs. Designing for user populations who do not conform to general user assumptions compounds this situation.

There are a number of general guidelines and observations about the design of VEs, a few of which include the representation of objects (Slater & Usoh, 1995), qualities of a virtual human (Thalman & Thalman, 1994), and communicating depth (Pimental & Teixeira, 1993). A number of techniques for the design of user interactions have emerged: support guiding the user (Wilson, Brown, Cobb, D'Cruz, & Eastgate, 1995) and support for object alignment (Buck, Grebner, & Katterfeldt, 1996). However, little experimental work or evaluation has been done, and there are few comparisons or established standards (Kaur, Maiden, & Sutcliffe, 1999).

Low-fidelity prototyping plays an important role in HCI, allowing the exploration of design ideas before they are implemented and allowing for multiple designs to be considered. In VE design, *low-fidelity* prototypes often refer to software prototypes that, although quicker to produce than the final product, do not fulfill two of the most critical aspects of this design stage: very rapid and inexpensive iteration and the ability for all design partners to contribute equally. Partners who have no technical background (which often includes teachers and children) are unable to provide input into software prototypes without going through an intermediary (Beaudouin-Lafon & Mackay, 2012), thus reducing their role at this critical stage. It is preferable to begin with rapid offline prototyping (such as interface sketches, written descriptions, storyboards, and low-fi 3D models) and progressively increase fidelity as the design process progresses.

42.3.3 FORMATIVE EVALUATION

Formative evaluation and iterative development of VEs should ensure a more usable and useful product. All evaluation activities seek to explore how successful a design is (or how successful a design idea might be once implemented), but formative evaluation should also serve to find ways to improve the design. Formative evaluation should take place regularly throughout the design process and feeds back into development. The original project plans, system specifications, and existing design ideas should be revisited and updated as a result of formative evaluation activities. In order to identify and plan appropriate evaluation activities, it is essential to consider who will be involved, where and how the activity will take place, and what is being evaluated.

In the case of early design concepts, we might consider the following:

- Are the designs technically feasible to implement, given the time, resource, and practical constraints of the project?
- Are the designs appropriate for the target user group, and can they be developed to cater for the specific needs, abilities, and disabilities of those users?
- Will the students enjoy or be motivated by the activities suggested?
- Will the applications support learning and meet the learning objectives?

As designs progress, we would also look to assess content, functionality, ease of use, and educational effectiveness.

Heuristic evaluation and other expert-based methods are often applied in evaluation activities but are less useful in VEs. Few design heuristics exist specifically for VE design, and VEs are often both novel and complex, meaning that predictive evaluation methods generally do not provide an accurate assessment (Bowman, Gabbard, & Hix, 2002). Furthermore, heuristic inspections tend to have limited value in terms of informing the continued design process, highlighting potential issues without really showing better solutions. In VE design, there is currently no sufficient substitute for formatively evaluating with users. Qualitative data, collected from observation studies, interviews, and questionnaires, can be particularly illuminating in formative evaluation, as they are used to describe and interpret what is going on.

The design team should be regularly reviewing the designs as they progress, thus ensuring input from the relevant informants at all stages in the process. It is, however, necessary to carry out wider evaluation activities at key points in the development process, involving an extended representation of teaching professionals and students. Observing students using the system—in context, where possible (e.g., in the classroom and with the same support or facilitation that they would usually have)—remains the best way to discover how usable and effective the system is and how it could be improved. It is possible to run formative evaluations with end users even when the designs are still conceptual; see Section 42.6 for further information.

42.3.4 OUTCOMES: SUMMATIVE EVALUATION AND DEPLOYMENT

There are generally one or two of two broad objectives for educational software projects, whether they are research-based or commercial: either to investigate the use or effectiveness of software or to develop something that can be used in future. It is important to be clear about these final goals from the outset, as final evaluation and deployment activities can be anything from a small-scale task to a very substantial undertaking worthy of a project in their own right.

Summative evaluation activities may be carried out toward the end of the project. Summative evaluation is not intended to feed into the development process (as formative evaluation is) but rather to provide a judgment on the outcome. If an iterative, user-centered design process has been followed, it is unlikely that there will be any substantial changes to be made at this stage (and it is generally too late to do so), although minor modifications may be made in light of summative evaluation outcomes, and it may further inform the deployment and dissemination processes. In projects particularly aimed at SEN, rigorously planned testing of abilities may be required during recruitment and pre- and posttechnology intervention.

Determining the measures for summative evaluation in VEs for education requires careful consideration. Task performance metrics, particularly, are not necessarily the same in this application area as in others. Typically, in interface evaluations, we measure speed and accuracy to assess performance. In this case, however, they may not be appropriate: being able to complete a task quickly and without error does not mean the application is educationally effective. When assessing the interface, we therefore need to take care to separate errors in *using the interface* (e.g., performing the wrong action to interact with an object) from errors in *the educational task* (e.g., choosing the wrong chemical in a virtual science experiment, because the student has not understood the concept). The former indicates usability issues with the interface, while the latter may be acceptable or even preferable, provided that the student is supported in identifying and rectifying that mistake.

It is always advisable during the final stages of a project to contact all of the schools and organizations who have participated or contributed in some way, and they should be updated on the status and outcomes of the project. This is an important aspect of long-term relationships between the researchers, developers, and schools, and it will make it more likely that the schools will consider participating in future projects. If working software is available at the end of the project, it should be offered to participating schools where possible. If the software is being deployed to other organizations, it can be provided to participating schools in advance: they will appreciate the opportunity to have first access and may provide additional feedback about, for example, the installation and setup process or presentation of materials, tutorials, and user manuals.

42.4   INCLUSIVE DESIGN TOOLBOX FOR VLE DEVELOPMENT

The AS Interactive project examined the usefulness of VEs to support the learning of social communication skills for teenagers with Asperger's Syndrome (AS) (Cobb et al., 2002). Although the technology advocated was the same as that applied in the previously mentioned Life Skills Education project, the purpose of using VEs for student learning was very different. The Life Skills Education project had examined the use of VEs to support the learning of procedural tasks and activities, allowing students to practice the steps and sequence of activities needed to complete a given task and to make decisions at appropriate stages. For example, taking a bus from a virtual house to a virtual café or supermarket required them to plan what time to leave the house and decide whether they would need to take money or could use a bus pass at the given time. They needed to plan the route to take to get to the bus stop and then select the correct bus that would take them to the required destination (see Brown et al., 1998 and Cobb, 2007 for further examples). The AS Interactive project examined whether VEs could be used to support and enhance social awareness and social skills among adults with Autism Spectrum Disorders (ASDs). The research questions addressed included the following: How could we replicate social situations in VE? How do users understand and interpret these VEs? How could we use VEs to encourage social interaction? How could we support and measure the generalizations of skills? While the composition of the stakeholder team engaged in the project was similar to that in the Life Skills Education project (e.g., computer scientists, researchers, education specialists, teachers, and pupils), the difference in the type of educational objectives for which the VEs were intended required these stakeholder groups to provide different types of information to inform the design of the VEs.

Throughout the project, a toolbox of methods was applied that allowed stakeholders to contribute to VE design and review at different stages of development (Neale, Cobb, & Kerr, 2003). The stages of design development were similar to the previous project: First, the learning objectives were defined, then decisions were made about the layout of objects in the VE and the types of user interactions required (Stage 1). The visual appearance of the VE content and interface was planned and revised, arriving at a complete description of the activities that a user would be able to do in the VE (Stage 2). Utility and usability of the design was then tested and refined through an iterative review of VE prototypes (Stage 3). Table 42.1 shows the variety of prototyping methods used to inform different design elements throughout these stages of the VE development process, and which stakeholder groups were involved.

The VE scenarios developed in the AS Interactive project and findings from formative and summative evaluation studies are described by Cobb (2007), and comment on the importance of involving students with autism spectrum conditions (ASC) in the process of informing VE design is made by Parsons and Cobb (2013).

42.5   CASE EXAMPLE: COSPATIAL

The COSPATIAL project examined the use of collaborative technologies to engage children with ASC in learning social skills. While the AS Interactive project had demonstrated the successful use of VEs for student learning of appropriate behavior and responses in a specific social context, COSPATIAL explored the use of collaborative VEs (CVEs)* as a means of engaging children in social communication with each other through a medium with which they are comfortable (a computer interface). Three applications that emphasized different social skills were developed: one collaborative puzzle game; and the other application that allowed structured, facilitated rehearsal of social conversation with a peer; and a final application that used mixed reality (video avatars in a virtual room) with the aim of providing an interim step toward being able to transfer these social skills into real-world practice. An overview of the COSPATIAL development process is shown in Figure 42.3.

---

* Collaborative virtual environments (CVEs) allow several participants to share and interact with the same virtual environment even though they may be physically located in separate places. Participants are usually represented in the CVE by avatars which they can move around the virtual environment. Participants may be able to jointly interact with objects in the CVE and can communicate directly with each using any combination of microphones, video or text.

TABLE 42.1

Prototyping Methods and Their Contribution to VE Design Specification

Prototyping Method	User Involvement and Stage Applied	Example of Use	Design Element Informed
Written storyboards	P, V Stage 1	Establish the learning objectives and define tasks to be performed	Function
Written storyboards	P, V Stage 1	Describe interactions in the VE, e.g., "The VE initiates the audio…"	Interface
Pictorial storyboards	U,[a] F, V Stage 1	Define layout of objects in VE and describe interaction activities	Content, layout, activities
Hand-drawn-picture storyboards	P, U,[a] F, V Stage 2	Visual representation of VE (presentation and interaction methods)	Content, interface
Computer screen shots	P, U,[a] F, V Stage 2	Review of VE content, layout, interaction methods, and user viewpoints	Content, interface
3D low-tech models	P, U,[a] F, V Stage 2	Mapping 3D layout of VE and movement of users within it	Content, layout
3D low-tech models	P, U,[a] F, V Stage 2	*Determination of navigation*/interaction requirements to perform tasks	Function, interface
Acting/role-play	U, F, V[b] Stage 2	Efficacy of VE layout and need for auto-viewpoints at key points of interaction	Content, layout
VE prototype review	U, F, V[b] Stage 3	Review all aspects of VE (content, layout, graphics, colors, complexity, interaction methods, feedback, etc.)	Function, usability, interface, and implementation
Online walkthrough (e.g., multiple activity analysis)	U, F, V[b] Stage 3	Describes how/where user should get support from teacher or facilitator, e.g., "Verbal prompt is required to encourage the user to discuss what they should do."	Usability, implementation

Source: Neale, H. et al., An inclusive design toolbox for development of educational virtual environments, Presented at *Include 2003*, March 25–28, Royal College of Art, London, U.K., 2003.

Key: User type: P (domain professionals), U (users), F (user group facilitators), V (VE developers).

Stage of development: 1 (initial design specification), 2 (early development), 3 (formative review).

[a] Depending upon users' ability to cope with abstract representations.

[b] VE developers access this information directly if appropriate or via video or feedback from facilitators (consent required).

42.5.1 INITIAL CONTACT WITH SCHOOLS/ORGANIZATIONS

Contacting educational professionals to discuss participation in the project was the first activity in COSPATIAL, and it was undertaken even before a plan of work had been composed. This is effective for several reasons. First, it avoids the risk of not being able to recruit design partners once the project is underway, and the recruitment process does not consume project time and thus allows the project to be planned more efficiently. Secondly, it satisfies project funders or stakeholders that the project is viable, and the professionals can endorse the project and help to shape a case for the impact and importance of the work. And perhaps most importantly, it also allows education representatives to be involved from the very beginning—including input into the planning stage itself—thus adhering to the principles of true participatory design. In addition to the benefit of using their expertise and experience in shaping the project plan (e.g., in this case, the teachers had unique insights into how design and evaluation activities might be best planned so as to accommodate children with ASC and their teachers/therapists), this also emphasizes the role of those professionals as equal design partners and fosters a feeling of ownership of the project.

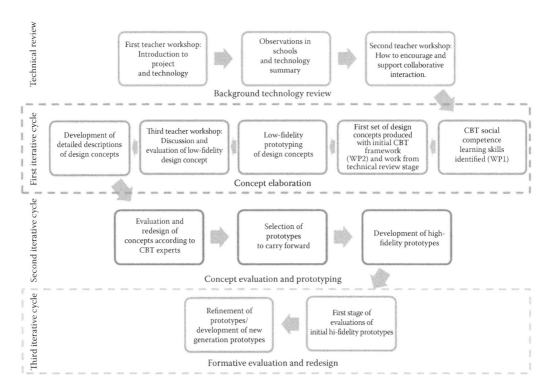

FIGURE 42.3   COSPATIAL development process overview.

We initially contacted schools and individuals we had already worked with (in the AS Interactive and related projects and activities) and formed a core design team, including representatives from three schools: two specialist schools for children with ASC and one mainstream school, selected on the grounds that the software was aimed at using in both of these contexts.

During the planning stage, the researchers involved in the core design team visited the schools to develop a practical understanding of the research questions that the project would aim to address. Through observations and discussions with teaching staff, the needs of the children were assessed, and the target user group was further defined. *Autism* is an umbrella term for a broad spectrum of conditions, and children with autism can have highly varied needs, abilities, and disabilities. A specific technology could not effectively support every child on that spectrum. The educational needs of the various potential users of the technology were considered, and it was decided that the CVE technology best afforded supporting the social skills of mid- to high-functioning children with at least some existing verbal skill. We were looking to support students working at level P6 and above in personal, social, health and economic education (PSHE) according to the UK National Curriculum at that time.

We established that it would be necessary to involve more organizations in the project. While other professionals would not form part of the core design team, teachers and students would be involved in specific design and evaluation activities, and this would require school support. We made contact with several schools through participating in an education and technology exhibition (the BETT show) and followed up with those who registered an interest, taking the opportunity to discuss how and when they could be involved. It was helpful to discuss the mutual benefits of being involved in the project; emphasizing the benefits to the school and their students and also how this would help to create effective software for children with ASC produced a very positive response, and schools were enthusiastic about participating in the project. This was an excellent channel for contact as the attendees of the show already had an interest in the use of ICT for teaching and were therefore perhaps more receptive to novel technologies than an average member of teaching staff.

Further contacts were made by directly cold-contacting schools with letters, emails, and phone calls; this was a more time-consuming process, and a very limited number of schools cold-contacted went on to participate in the project. We found greater success when we were able to identify a contact responsible for SEN coordination within their school and contact them directly. For all schools recruited we offered to travel to the school for an initial meeting in which we presented the project aims and discussed how they might be able to participate. Participation in research projects is a considerable commitment for schools, and it is important that they are informed about expectations, benefits, and required resources (e.g., time from teaching staff and students; space and technology requirements for design and evaluation activities) before making that commitment.

It takes considerable time and effort to establish contact with (and furthermore effective working relationships with) school partners. The value of maintaining existing relationships should therefore not be underestimated; involving the school partners right through to the dissemination stages of a project—and beyond—is highly recommended. In the longer term, these continuing relationships benefit researchers, schools, and students alike (Parsons et al., 2013).

42.5.2 VR FAMILIARIZATION

Following the planning and contact stage, an initial workshop was held with teachers, researchers, and VE developers to discuss the project objectives and approach. For COSPATIAL, this involved discussions around the nature of collaboration, what skills the students should be learning and how, and what key features of the technology should be.

In addition to the initial teacher workshop, the researchers and developers carried out a number of school visits and observations. These visits were designed to be as unobtrusive as possible, so as not to disrupt the routine of the students with autism. Researchers acted as participant observers. This enabled the students to become familiar with the researchers gradually. These early sessions were not recorded as it was decided that this may make the children feel uncomfortable; instead, the researchers wrote *experience reports* immediately after each session. The school visits provided a highly valuable method of understanding the needs of children with autism in the context of the schools in which they were based.

The school visits provided the researchers and developers with good background knowledge about the educational context, the needs (both educational and practical) of children with autism, and the use of existing technology in schools.

The school visits and initial teacher workshop combined therefore served to transfer knowledge between the design teams: the education professionals learned about the technology and the design process, while the research and development teams learned about children with autism and their educational needs. This knowledge transfer was a critical stage in the design process, yet it is important to stress that this only provides a basic background and cannot substitute the substantial experience of domain professionals: the teachers were still not technology experts and the developers were not autism or education experts. It is therefore key to successful design that the participatory design (PD) team comprising representatives of each different role continue to collaborate throughout the project.

42.5.3 CONCEPT GENERATION

A second teacher workshop was held in COSPATIAL, this time focusing more on specific research questions, ideas, and issues to be addressed and generating design concepts. The affordances and constraints of VE technologies were discussed in relation to the project goals, allowing the participatory design team to establish what type of technology to use and how the features of the technology could support the educational objectives. In particular, the following features were highlighted:

- VE technologies could be highly motivating for children with ASC, encouraging them to spend time interacting with peers (which was to be encouraged for this user group).
- The ability to record and replay VE sessions could provide a valuable opportunity for students to reflect on the sessions with a teacher or therapist.
- The ability to take on different perspectives (either during a session or during replay) could enable children to understand other people's viewpoints.
- The ability to provide control over the environment and stimuli was critical to meet the needs of the user group.

This session also served as a starting point for the generation of conceptual ideas and designs, and a number of ideas were discussed. Following this workshop, 12 design ideas were defined, using the information obtained from a number of site visits. A textual description of each concept was recorded in addition to examples of how this might be used.

42.5.4 Selection and Prioritization of Ideas

The 12 activity ideas were all reviewed by therapists, who assessed how they could be suitably implemented to support learning within the educational framework that had been selected for this project (which was based on a cognitive behavioral therapy approach; Bauminger, 2002, 2007a, 2007b). The PD team considered this input along with objectives, suitability for the end users, and feasibility and narrowed the list to five; these ideas would be carried forward for further exploration. Storyboards and low-fidelity prototypes were created for each of these five design ideas. A further teacher workshop was held to agree upon design ideas moving forward. This workshop consisted of a short presentation followed by discussion of the storyboards for each game. Feedback from all participants was recorded on post-it notes, which were placed around the scenario in question (Figure 42.4).

The CVE technology lends itself to supporting different educational objectives in different ways, and it was decided that more than one activity would eventually be integrated into the CVE in order to take advantage of these different technological affordances. Two examples are shown in Table 42.2.

FIGURE 42.4 Teacher workshop sessions.

TABLE 42.2
Primary Educational Objectives and CVE Affordances Identified
for the COSPATIAL Games

Activity	Primary Educational Objectives	Primary Use of CVE Affordances
Collaborative building game	• Understand the importance of and rehearse task-based collaboration skills • Sharing information • Understand another person's visual perspective	• Shared virtual space with separate individual displays (can be used to constrain the information visible to each player) • First-person viewpoints in 3D space
Social conversation activity	• Understand the importance of and rehearse social conversation skills • Develop relationships with peers	• Ability to be remotely located and therefore reduce the anxiety that some children experience with face-to-face communication, which can prevent social interaction in the real world • Option to customize in-game displays for particular children (e.g., topic stimuli, symbols) • Ability to replay and review CVE session

It was considered that two or three of these subactivities would be implemented. It was decided not to commit absolutely to specific ideas at this stage, as decisions might later be influenced by the outcomes of design and evaluation activities in developing the first activity. A decision on which activity would be most suitable for initial development took into consideration the following:

- Suitability for the target student group (informed by teacher partners)
- Likely educational effectiveness (informed by teacher partners and educational psychology researchers)
- Effective use of the CBT framework (informed by the CBT therapists/researchers)
- Motivation for the students (informed by design activities with the students; see Section 42.6 for further information)
- Technical feasibility (informed by VE developers)
- Effective use of affordances (informed by VE developers and HCI experts)
- Suitability for evaluation in relation to the project objective of exploring whether the use of CVE technology could improve real-world social competence skills (informed by research team and educational psychologists)

42.5.5   VE DESIGN AND DEVELOPMENT

We followed a highly iterative process in COSPATIAL, obtaining feedback from all those in the core participatory design team regularly (Figure 42.5).

We progressed from a range of high-level conceptual ideas, to 12 more tangible design concepts, and to 5 ideas with a more detailed specification and eventually went on to develop 3 applications:

1. *Block Challenge*, a collaborative puzzle-style game in which pairs of children work together to build a tower of colored blocks
2. Talk2U, an application for students to learn about and rehearse social conversation skills with their peers
3. Face2Face, a mixed-reality application in which students have video avatars in a virtual room

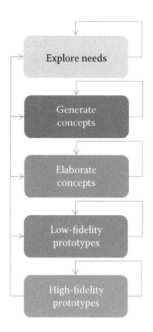

FIGURE 42.5 Iterative stages of the COSPATIAL project.

As the project progressed, the number of design options decreased while the level of detail and fidelity for each design increased (with both the selection and design processes informed by needs and objectives as earlier described).

A range of methods were used to represent design ideas to enable input and feedback during the design process. It is important that all members of the design team, including those with no technical background, can understand the design ideas, provide useful feedback at appropriate stages, and create or make modifications to the designs. To promote high-level feedback, we initially created comic strip-style storyboards, supplemented by short written descriptions (Figure 42.6). These were

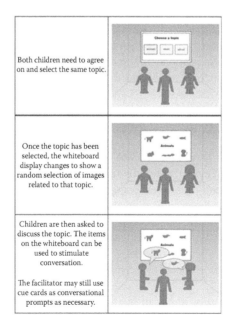

FIGURE 42.6 COSPATIAL Talk2U application storyboards.

easy for the teacher partners to understand and were quick and easy to create and modify, allowing rapid iterations of the design ideas based on feedback. In some cases, the designs would be modified during a workshop or meeting.

We explored a number of low-fidelity prototyping methods and found that intentionally omitting noncritical details provided the most appropriate feedback. This also ensured that there were no misinterpretations of the designs; when the storyboards were more visually detailed or had a more polished look-and-feel, the design partners often expected that the final program would look that way (when, in fact, the visual design had not yet been explored). Producing basic design storyboards makes clear to the participants that the design is unfinished, and this also allows evaluation participants to suggest more radical changes to the designs. As the design ideas were evaluated, modified, and elaborated, more frames were added to the storyboards, and more detail added to the frames.

It should be noted that the storyboards can effectively show a single sequential activity in the VE, but when there are multiple possible events and actions at any given time (as is frequently the case in 3D VEs that can be freely navigated), the storyboards alone are insufficient to detail all of the possible options. As the design process continued, we generated task-flow diagrams showing the range of options possible to a user at each stage in the activity (Figure 42.7).

The interface design, following on from the storyboarding phase, was also highly iterative. Individual aspects of the design were mocked up (using both static and interactive low-fidelity prototypes) for evaluation before being implemented.

Figure 42.8 shows the task concept for the COSPATIAL *Block Challenge* game in which two players are required to collaboratively select and rotate multicolored blocks to build a tower

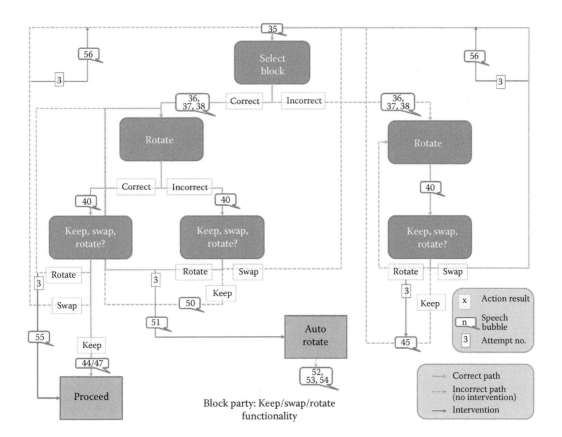

FIGURE 42.7   *Block Challenge* block selection flow diagram.

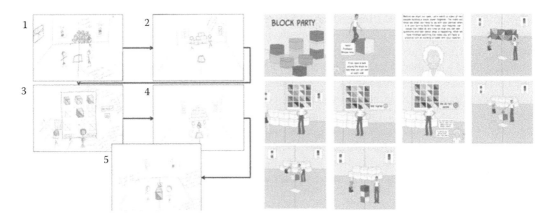

FIGURE 42.8   *Block Challenge* low-fidelity prototypes.

with different colors on each side. In this concept design, the VE was divided into two sections, with a player in each half of the virtual room. A partition down the center of the room would prevent each player from entering the other player's section. This was considered important as each player was to be given their own *target pattern* of colors that were needed for the tower block and these would be displayed on the wall in their section of the room but should not be visible to the other player. A primary objective of the game was that the targets would be different for each player and therefore players would need to talk to each other to find out what colors were needed.

Figure 42.9 shows low-fidelity mock-ups of the user interface for the COSPATIAL *Block Challenge* game. A graphic artist was recruited to develop the design of the interface, and these designs were iteratively reviewed through the user-centered design process. Figure 42.10 shows a low-fidelity mock-up of the COSPATIAL Talk2U program. The screen interface was overlaid onto the PowerPoint illustration of the CVE, and this enabled the participatory design team to review the impact of on-screen information on visibility and interaction with characters and information within the VE.

Once the key design decisions regarding task objectives, activity sequence, and facilitation of users for each scenario had been agreed by the participatory design team, high-fidelity prototypes were created. These were constructed using the same virtual reality development software that would be implemented for the final program (DEMON—3D environment for multiuser simulation). As with all previous stages of the PD process, the development of the high-fidelity prototypes also went through several iterative review cycles. Initial prototypes focused on the placement of 3D objects in the virtual room and the use of the computer interface controls for the movement of the user around the 3D space and manipulation of virtual objects. Figure 42.11 shows the first version of the *Block Challenge* training activity and early

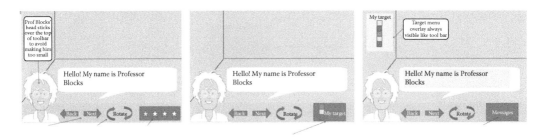

FIGURE 42.9   *Block Challenge* 2D overlay prototypes.

FIGURE 42.10   Talk2U low-fidelity prototype.

FIGURE 42.11   Development of the *Block Challenge* interface.

development of options for selection and rotation of the multicolored blocks. Each prototype was reviewed by the participatory design team.

Walkthroughs of the CVE prototype were carried out as a method of obtaining feedback from the teachers in terms of both usability and learning potential. This was a useful method and provided an in-depth analysis of the program, so it was important to record all feedback in an effective manner. To do this, we used screenshots of key stages within the program on large sheets of paper that could then be annotated by those present during the walkthrough and during postwalkthrough discussions. This enabled the team to record feedback relating to specific aspects of design by annotating screenshots quickly and clearly. The annotations produced in the session were then transferred onto a computer-drawn sheet to produce a clearer record of feedback for the developer. Figure 42.12 shows an annotated screenshot for the introduction stage of the *Block Challenge* training scenario, which includes introductions to Professor Blocks, the navigation tools, and the virtual room.

Annotated screenshots were also used to highlight important interface design changes made by the team through review meetings and discussions. For example, Figure 42.12 documents the changes made to one part of the *Block Challenge* training activity following a review by the participatory design team. Specifically, the "A" button was added to the interface to notify the program to continue to the next section replacing the "Back" and "Next" arrows that were found to be too confusing during the CVE walkthrough with the teachers. In addition, for the perspective-taking

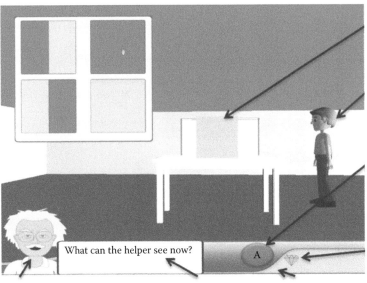

Lighting in CVE room adjusted to allow color matching activity.

Helper changed to a cartoon-style, male character.

"A" button to continue—used to allow potential compatibility with other interaction devices in the future.

User collects jewels when a question is answered correctly. The jewel is then shown in the tool bar.

Professor Blocks in bottom left corner of the screen.

Speech bubble moved to toolbar to allow better visibility of blocks.

Toolbar along bottom of screen.

FIGURE 42.12 Annotated screenshot showing the *Block Challenge* training activity.

activity, the stars from the previous activity were replaced by jewels. The number of jewels collected represents the number of correct color selections made during the perspective-taking task.

Formative evaluation and redesign of the prototypes were conducted throughout the PD process. There were four main revisions of the COSPATIAL *Block Challenge* training activity as shown in Figure 42.13.

In this example, the same stage of the training task is presented—where the user is asked to select the color that the avatar positioned on the other side of the block can see. Each version of software development was reviewed by the research team, teachers, and children through the user-centered design process, and recommendations were fed into alteration requirements for the next version. Version 1 shows the first high-fidelity development of the *Block Challenge* training room with Professor Blocks providing user instructions at the bottom right of the screen. Users are requested to select the color from the panel that matches the Block face as viewed by the virtual helper character. In version 2, a ceiling has been placed on the room and the Professor Blocks character has been modified. The Professor has been moved to the left-hand side of the screen and a toolbar has been added. Navigation buttons to move forward and back through the program and reward feedback for successful collection of stars are displayed in the toolbar. In version 3, the toolbar buttons have been changed to "A," "B," and "R" to facilitate the use of game controllers for interaction with the program. The Block platform has been changed to a table so that the block is positioned higher up. The virtual helper character has been moved so that she is in full view. In version 4, the toolbar buttons "B" and "R" were removed as they would not be used. The virtual helper was changed to a male character and repositioned behind the block (as this would be the true position to view that face of the block). Professor Block's speech bubble was moved onto the toolbar so that it did not obscure objects in the VE. Lighting in the room was also adjusted.

Progressively increasing the fidelity of the designs with regular feedback from the whole participatory design team was an effective and efficient process, but it should be noted that this process is not always suitable for obtaining feedback from children with special needs themselves. We discuss how methods can be adapted to include children with special needs in the design process in Section 42.5.

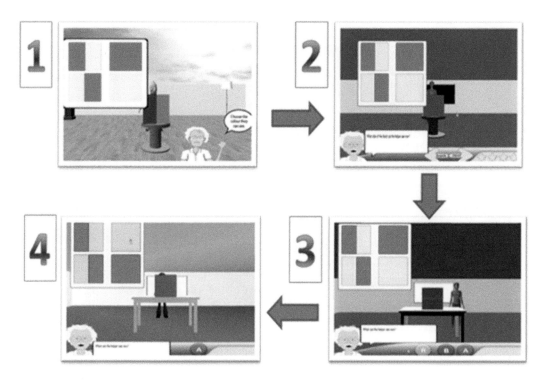

FIGURE 42.13   Development of the *Block Challenge* activity.

42.5.6   FORMATIVE REVIEW AND EVALUATION

While our general participatory design approach ensured regular evaluation, in COSPATIAL, we conducted a number of planned formative evaluation activities to incorporate a wider group of teaching staff and students.

As the development process progresses, the designs become more detailed and of a higher fidelity; they are therefore less abstract, which in turn makes it easier to involve users with SENs who may struggle with abstract representations. This is the case with children with autism, and therefore, at earlier stages in the project, it was necessary to increase the involvement of teaching staff as user advocates and to adapt our design representations and evaluation methodologies to enable children with autism to participate meaningfully (see Section 42.6 for further detail).

Our formative evaluations included a number of workshops, interactive demonstration sessions, and usability testing sessions and involved teachers, teaching assistants, speech and language therapists, children with autism, and typically developing children. Figure 42.14 shows key stages of the design progression informed by formative evaluation activities.

42.5.7   OUTCOMES

Block Challenge is a collaborative puzzle activity in which two children work together to build a tower of blocks in the VE. Children each control an avatar in a shared virtual room, and they both have to choose and move the blocks together.

Successful completion of the task requires the following:

* Collaboration with peers to achieve a common goal
* Task-focused communication
* Thinking about other people's perspectives

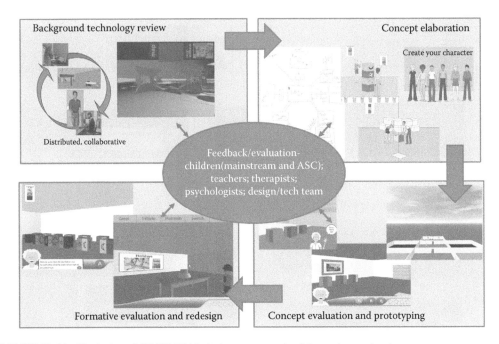

FIGURE 42.14 Evolution of COSPATIAL designs as a result of formative evaluations.

Talk2U is a program designed to allow children to practice social conversations. Two children and a facilitator each control an avatar in a shared virtual room and can talk to each other using microphone headsets. The children agree on conversation topics, and the facilitator uses visual cue cards to provide prompts to each player (Figure 42.15).

Face2Face is a flexible application that uses mixed reality to support social interactions. The users each have video avatars in a virtual room. Face2Face was designed to be a step in between the Talk2U VE and real-world interaction. While Talk2U is a more structured activity and focuses solely on conversation, the video avatars in Face2Face introduce additional social cues such as facial expressions and body language.

A large-scale formal intervention study was conducted in the final year of the COSPATIAL project to assess the technology's effectiveness as an educational tool. Since the technology and moreover its applications in this project were very novel, a pilot study initially involved 14 children,

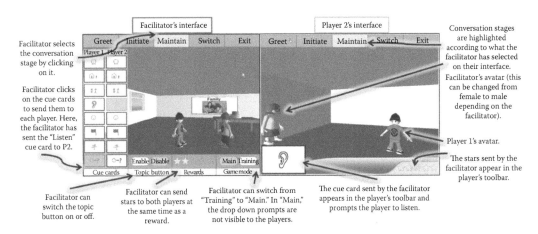

FIGURE 42.15 Setup interface design storyboard.

paired according to diagnosis and level of language ability (according to the British Picture Vocabulary Scale, BPVS), and examined whether the CVE applications were suitable for the target population and for the intervention study. The students carried out CVE activities in their pairs with a facilitator (typically a teacher or teaching assistant from the students' schools). The sessions were video-recorded for analysis, and we administered questionnaires for both students and facilitators.

The main CVE intervention study involved 22 children and took place over a period of approximately four months. The main analyses for the study included standardized measures (for language, IQ, executive functions, and autism characteristics), cognitive and behavioral tasks, and interactions between children while completing the CVE tasks. The measures were taken at baseline and following the interventions. Extensive analyses on the video-recordings from the sessions were carried out. The children showed positive collaboration and communication behaviors in the CVE sessions, and moreover, the findings from the study suggest that the CVE intervention had a positive effect on children's real-world social conversation skills. Further information about the intervention study and outcomes can be found in the work of Parsons et al. (2012).

The COSPATIAL software was initially designed purely to investigate the research question of whether CVE technology could potentially be used to support the learning of social competence skills in children with ASC, and this was addressed in the intervention study. However, following the intervention study, we decided to provide the software to schools that had been involved in the project initially and to make the software available to other schools and organizations if that was successful.

In order to make the software accessible for schools, we developed a user-friendly setup interface for teachers, allowing them to configure the network setup even if they had no technical experience or support and allowing them to customize the games to suit the needs of their students. The setup interface (Figures 42.16 and 42.17), although a smaller development project, was, again, developed iteratively with input from teachers. Since this setup interface was only for use by teaching staff, it was not necessary to include children in this process.

We also created a simple installation package including the necessary components to run the games. Some of the schools that had been involved in the project offered to act as test sites for the software, installing, setting up, and running sessions and providing feedback on the process and any issues they encountered. It was important to run a trial on the software in schools with their own hardware, and we made the decision to allow the schools to do this trial independently and without our support to ensure that they were able to use everything successfully. Final modifications were then made on the basis of school network setups, and finally a user manual was created.

42.6 INVOLVEMENT OF END USERS

It has been commonly accepted as good practice to involve users in the design of technology for many years (Scaife, Rogers, Aldrich, & Davies, 1997). Over the last 20 years, this has extended to cases where children are the end users of the systems; it is now widely acknowledged that children have an important role in the design (Bekker, Beusmans, Keyson, & Lloyd, 2003), and considerably more published research now reports involvement of child end users in the design (e.g., Bekker et al., 2003; Druin, 1999; Garzotto, 2008). Involvement of children with special needs remains limited, however (Benton, Johnson, Brosnan, Ashwin, & Grawemeyer, 2011; Parsons & Mitchell, 2002). A resulting lack of guidance for involving children with special needs in the design (Parsons, Millen, Garib-Penna, & Cobb, 2011) deters or makes it difficult for other researchers and developers to create effective technology for those users.

Often, adult proxies or typically developing children are involved to substitute the participation of the actual end users in the design process. The involvement of advocates or representatives is, itself, highly valuable; they may be better able to communicate some needs of the users and in some cases may be users (or secondary users) of the systems themselves; for example, teachers and therapists may be required to facilitate the use of educational software. However, the representatives may have their own agenda for the design or may make incorrect assumptions about end users' needs and capabilities.

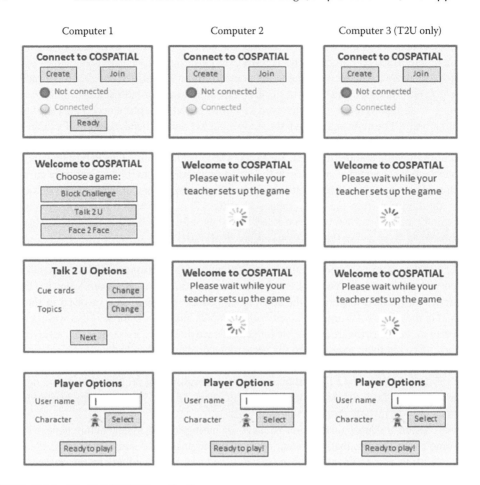

FIGURE 42.16   The Talk2U CVE application.

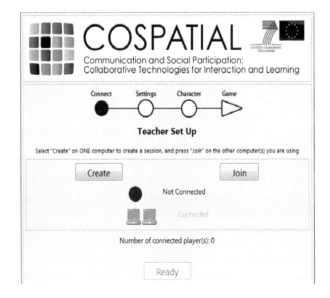

FIGURE 42.17   COSPATIAL setup application networking screen.

According to Hall, Woods, and Aylett (2006), teachers contribute to the design process by providing information about curriculum, learning requirements, and a classroom perspective. In the case of special needs education, we feel that the role of teaching staff is much broader. The teacher acts as a user representative, and their experience with the target student group makes them uniquely placed to express the needs of those students, particularly in cases where the students are unable to explain their own needs. The teachers themselves have experience in how to adapt tuition to suit the needs of their students and thus have important insights into how the educational technology should be designed for those students. Furthermore, the teachers' experience can be drawn upon to plan and carry out effective design sessions, since it is not just the technology itself but also the PD methodology that should be adapted for students with special needs, to ensure that they are able to contribute effectively (in addition to ensuring an ethical approach in which students feel comfortable participating in the activities). It is important to continue to explore methods to involve users with disabilities in the design process—in addition to their teachers—to ensure suitability of the systems and to give the users a voice and sense of ownership.

There is no single approach suitable for involving all users in developing VEs for children with SEN. All conditions are different; and the specific needs, disabilities, and abilities of the particular user group will determine the best approach. However, in all cases of VEs for children with SEN, we need to adapt our methodology to allow the end users to contribute effectively to the design process.

After making contact with schools to discuss participation and obtaining consent from all the relevant parents, carers, and staff, we also seek consent from the children themselves, with carefully adapted participant information sheets and forms. Consent forms may need to be provided some time in advance in order to have time to process them, and it may be necessary to have an adult to help them read and understand the consent. In order to establish an appropriate plan to involve children with SEN in the design, the first and most important step is to gain an understanding of the target user group.

Gaining an understanding of the user group involves spending time with the students, becoming familiar with them, and allowing them to become familiar with you. For some children with SEN, it can be more difficult to interact with strangers than it would be for typically developing children or adults, and it is important that they are comfortable with you before you embark on any design activities. Try to explore their likes and dislikes as well as just their needs and abilities. It can also be valuable to talk to their teachers, support staff, and parents.

It is important to help the child to understand their involvement in the process. Many children, and particularly those with SEN, are accustomed to learning from adults but not to adults wanting to learn from them. The children need to understand that their contribution is genuinely valuable and be clear that if we ask them questions, it is not to test them but to find out what they really think, to help them understand what you want to get out of the sessions and what you want from them.

In COSPATIAL, we followed a broadly similar approach with children with autism as with other design participants: they participated initially in VE familiarization sessions, carried out concept generation sessions, reviewed a selection of design concepts, provided feedback as the designs were elaborated and developed, and carried out usability testing sessions. Although the approach is similar, the planning and conduct of the activity sessions were very specifically tailored to our participants' needs.

Generally, all design and evaluation activities involving children with SEN will benefit from careful planning and a more defined structure than may usually be the case. This is particularly true if, as with autism, the children are most comfortable with the routine and structure. Planning ahead also allows teachers to feed back on the activity planning to ensure it is suitable and is likely to work well. It may be advisable to send a timetable for the session in advance, or at least to present it at the beginning of the session. For children with autism, we used visual timetables to help us explain what would happen in sessions. We then pinned the timetable to the wall and crossed off activities as they were completed. Although structure is often beneficial, it is important to be flexible and to be prepared to adapt plans if necessary.

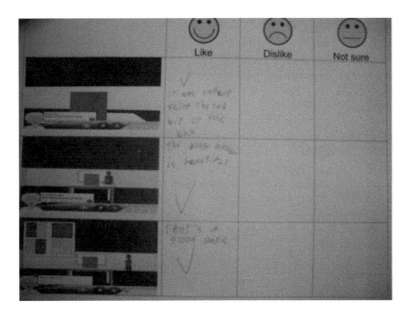

FIGURE 42.18   Student feedback sheet.

Our design sessions are supported by plenty of visual aids. This helps to convey meaning, particularly if the children have limited language capabilities. Words can be replaced or supplemented by pictures and symbols (see, e.g., Figure 42.18, which shows a series of images from along with "like," dislike," and "not sure" symbols, providing an easy way for students to show what they did and did not like about a design idea).

For children with autism—and for some other groups of children with SEN—dealing with abstract concepts can be difficult. As designs are developed and their level of fidelity and detail increases, the level of abstraction decreases and they become more tangible. Thus, it is usually much easier to involve children in the review of more advanced design ideas, and their involvement tends to increase as a project progresses. Nonetheless, it is important for the children to inform the design process at much earlier stages, and there are several strategies to enable this. These strategies are often simple; even minor modifications to activities can make them more tangible and easier to understand. For example, in a *design a game* activity session (for relatively high-functioning children with autism), simply adding a picture of a computer screen as a frame for their design concepts helped them to understand the activity; see Figure 42.19.

It can be helpful to break tasks down into smaller components and focus on the constrained aspects of design rather than to ask children generally to give us ideas. For example, we might ask them "what should your character look like?" or "what happens if you win the game?" We can also provide examples or cues to help the children get started and to know what to do. This has proven to be very effective when students seem reluctant to actively put ideas forward. The risk in such cases is that in doing so, we may influence the child's input, and therefore, such an approach should be used cautiously.

For children with autism specifically, we made a number of additional adaptations to our usual PD methodology:

- Many children with autism use symbols in addition to, or instead of, written words. We used symbol sets that were already used in schools and familiar to the students when we created materials for our design and evaluation sessions.
- We avoided any hypothetical questions and asking students to imagine a situation or a solution, since children with autism often have impairments in imaginative thinking. Instead, we framed activities such that they drew on real experiences.

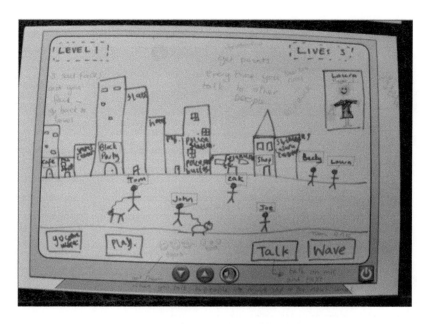

FIGURE 42.19   Concept generation with children with high-functioning autism.

Some children with autism take things literally and may be unable to cope with things that do not *fit* with their existing expectations, so we avoided, for example, portraying unrealistic situations and characters.

We had considerable success in involving children with autism in the design process. The involvement of children in the design process directly influenced the design, from fundamental to minor aspects. We also made steps toward facilitating a proactive contribution from children with SEN, in addition to the more limited reactive role (e.g., as a tester) that is generally the more common role for children with disabilities in the design (Guha, Druin, & Fails, 2008).

We cannot, however, claim to have involved the students as truly equal participatory design partners; they participated regularly throughout the process but were not involved in every activity. We also acknowledge that the requirement to provide clear structure, examples, and instruction also puts the researchers in a position of more influence. It may be impractical to attempt to create truly *equal* roles, when the diversity of roles is in fact the strength of a PD team, but we continue to explore ways to include children with SEN in the design process to improve the effectiveness, usability, and adoption of technologies for those users. Recent work has explored the use of VE technology in design activities; this has initially proven effective for children with autism (Millen, Cobb, Patel, & Glover, 2014).

42.7   DISCUSSION

This chapter has discussed the importance of user involvement at all stages of the design, development, and evaluation of VLEs. As a culmination of a series of VLE development projects in which professionals from mainstream and special needs education worked with VE developers at VIRART (Brown et al., 1995; Cobb et al., 1993; Crosier, Cobb, & Wilson, 1999; Neale et al., 1999), the Virtual Life Skills project provided the opportunity for a broad range of professionals and users within the community to collaborate in VLE development (Brown, Kerr, & Bayon, 1988; Brown, Neale, Cobb, & Reynolds, 1999; Cobb et al., 1998; Meakin et al., 1998). An initial user-centered design model, highlighting this, was given in the previous book (Cobb et al., 2002) and shown in Figure 42.1. Further exploration of this model through the AS Interactive and COSPATIAL projects has shown that we can extend the participatory role of design partners—including the

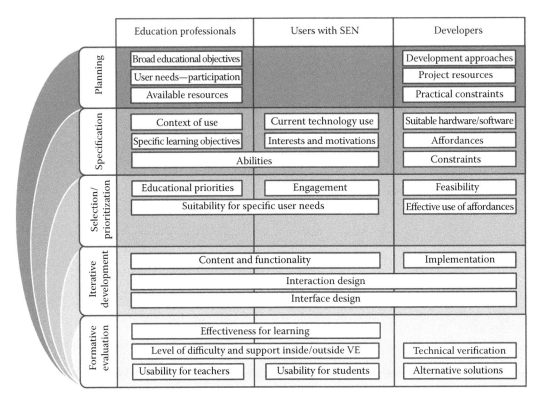

FIGURE 42.20   Elements of design informed by participants in the development process.

development team, teaching professionals, and students—and that adapting our methodology to involve those participants improves both the VE development process and the VE implementation for education. Participants with unique roles can inform the development process in different ways at different stages, and it is therefore important to involve representatives of each relevant role throughout the design process. Figure 42.20 shows examples of the information provided by each group, which steers the design process.

It should be noted that each project will have a unique design team and participants will have varying and sometimes overlapping roles. We use *education professionals* in this model broadly to include any participants who can inform educational objectives and approaches and who have experience with the target student group. In our projects, this has included class teachers, head teachers, teaching assistants, school ICT coordinators, SEN coordinators, and therapists. *Developers* may include academic researchers, HCI or interaction design experts, and graphic designers in addition to programmers. Users with SEN refer most importantly to a representative sample of the target user group themselves but, in some cases, may include user champions, typically developing children, or other children with matched abilities where appropriate. In some projects, some users may fulfill more than one role (Wilson et al., 2002). For example, a teacher may act as an educational informant and a user advocate, or a developer may also be a teacher. In these cases, it is important to encourage those participants to consider their input from the perspective of each role. Equally, it becomes even more important to involve further participants from outside of the core design team.

In addition to extending the role of participants in the design process, this updated model now reflects the involvement of teachers through the planning process. Schools should also be involved where possible throughout and beyond the dissemination stages of a project.

While a project typically begins with planning and progresses through development and evaluation activities, the stages are highly interdependent and the process is always iterative.

It can be seen that there are a number of iterative loops within the process. At the outset of the project, the interests of end users and the professionals who will apply the VE are gathered (via the user consortium and steering group). These two groups work together to define the objectives and content for the VE. During this process, storyboards are drawn up, and the user consortium and steering group work with the VE developers adapting the storyboards until they provide a specification that is both technically feasible and still meets the requirements of the user consortium and steering group.

Once the VE specification has been agreed, the VE developers start the creation of the VE modules. It is important that there is a facility for the regular review of VE development to ensure that the developers have interpreted the storyboards appropriately. It is recommended that review meetings should be held at least once every month, but the more frequently project members can review VE development, the better. It is often the case that, when members of the user consortium and steering group actually see their ideas put into effect in VE development, they change and/or add to their original requirements. This may be because the VE is not developing as they had envisaged or because they get new ideas as they see what can be done. They may also become more confident in expressing their views as they become more familiar with VR technology. This phase of VE development can therefore be extremely creative, and it is worthwhile taking time to complete this phase fully.

The process of VE design and development requires several iterative cycles through which representatives of the user consortium and steering group review various prototypes and inform modifications until the specification requirements are met. This forum allows what may be the first opportunity for project members to discuss the specific features of VE design, and decisions made at this stage can have a huge impact on the usability and acceptance of the final VE. This chapter has indicated that existing HCI design guidelines are limited in their application for VE design. By involving users and professionals in regular informal review and evaluation sessions can help to highlight and resolve potential barriers to usability.

When the internal process of VE development is complete, formal evaluations can take place. These evaluations feed into VE review, which can then be used to provide recommendations for additional design modifications. Early evaluation studies provide feedback on the general usability and acceptance of the VE. If this takes place during project development, the user consortium and steering group will be able to comment on the results and may be able to offer suggestions for design modifications. It is not advisable to leave all of the formal evaluations until the end of the project as there may be insufficient time and resources to make any necessary amendments.

Evaluation studies may include a number of measures in order to capture information from a number of sources to address questions related to usability, enjoyment, and educational effectiveness. It is important to take context into account as this may greatly impact how the technology is used, for example, in terms of the numbers of users sharing a computer and the length of time users have to explore in a classroom situation as compared with home study.

42.8 CONCLUSIONS

The model/framework presented in the previous chapter has allowed us to develop useful and usable VEs. However, it will be constantly updated as we refine and improve our ways of working. For example, we plan to design and evaluate new VEs with increased numbers of user groups from different organizations, cultures, and countries. This may require additional iterations with the model to update or refine for different user groups.

Only with careful design, development, and evaluation of software can we begin to understand the impact of well-designed, appropriate, and useful VEs in the general field of education. Both content and its representation in the VE, taking into account the qualities of VR technology and its intended users, need to be considered first. In this chapter, we have introduced a framework for this to be carried out. Only when evaluations have shown that the content covered is both appropriate and well presented and the VE is usable, then we can begin to comprehend the true value of VEs as a learning tool.

ACKNOWLEDGMENTS

The authors would like to thank all of the users and professionals who have worked with them in the development of the VEs described in this chapter. The *AS Interactive* project was funded by the Shirley Foundation (2000–2003). The COSPATIAL project was funded under the ICT programme (Grant Agreement no. 231266; 2009–2012). Some of the images appear in the COSPATIAL project deliverable reports and are reprinted here courtesy of the COSPATIAL consortium.

REFERENCES

Battocchi, A., Gal, E., Sasson, B., Pianesi, F., Venuti, P., Zancanaro, M., & Weiss, P. L. (2008). Collaborative puzzle game – an interface for studying collaboration and social interaction for children who are typically developed or who have Autistic Spectrum Disorder. *Proceedings of the Seventh ICDVRAT with ArtAbilitation*, Reading, U.K., pp. 127–134.

Bauminger, N. (2002). The facilitation of social-emotional understanding and social interaction in high-functioning children with autism: Intervention outcomes. *Journal of Autism and Developmental Disorders*, *32*, 283–298.

Bauminger, N. (2007a). Brief Report: Individual social-multi-modal intervention for HFASD. *Journal of Autism and Developmental Disorders*, *37*, 1593–1604.

Bauminger, N. (2007b). Brief Report: Group social-multimodal intervention for HFASD. *Journal of Autism and Developmental Disorders*, *37*, 1605–1615.

Beaudouin-Lafon, M., & Mackay, W. E. (2012). Prototyping tools and techniques. In J. A. Jacko (Ed.), *The human-computer interaction handbook* (3rd ed., pp. 1081–1104). Boca Raton, FL: CRC Press.

Bekker, M., Beusmans, J., Keyson, D., & Lloyd, P. (2003). Kidreporter: A user requirements gathering technique for designing with children. *Interacting with Computers*, *15*, 187–202.

Benton, L., Johnson, H., Brosnan, M., Ashwin, E., & Grawemeyer, B. (2011). IDEAS: An interface design experience for the autistic spectrum. *Proceedings CHI'11 extended abstracts on human factors in computing systems.* Vancouver, British Columbia, Canada: ACM Press.

Bölte, S., Golan, O., Goodwin, M. S., & Zwaigenbaum, L. (2010). What can innovative technologies do for Autism Spectrum Disorders? *Autism*, *14*, 155–159.

Bowman, D., Gabbard, J. L., & Hix, D. (2002). A survey of usability evaluation in virtual environments: classification and comparison of methods. *Presence: Teleoperators and Virtual Environments*, *11*(4), 404–424.

Bricken, M. (1991). Virtual reality learning environments: Potentials and challenges. *Computer Graphics*, *25*(3), 178–184.

Brown, D. J., Cobb, S. V., & Eastgate, R. M. (1995). Learning in virtual environments (LIVE). In R. A. Earnshaw, J. A. Vince, & H. Jones (Eds.), *Virtual reality applications* (pp. 245–252). London, U.K.: Academic Press.

Brown, D. J., Kerr, S. J., & Bayon, V. (1988). The development of the virtual city. In P. Sharkey, D. Rose, & J.I. Lindstrom (Eds.), *The second European conference on disability, virtual reality and associated technologies* (pp. 89–98) September 10–11, Skovde, Sweden.

Brown, D. J., Neale, H. R., Cobb, S. V. G., & Reynolds, H. R. (1999). Development and evaluation of the virtual city. *The International Journal of Virtual Reality*, *4*(1), 28–41.

Brown, D. J., Standen, P. J., & Cobb, S. V. (1998) Virtual environments, special needs and evaluative methods. In G. Riva, B. K. Wiederhold, & E. Molinari (Eds.), *Virtual environments in clinical psychology: Scientific and technological challenges in advanced patient-therapist interaction* (pp. 91–103). Amsterdam, the Netherlands: IOS Press.

Buck, M., Grebner, K., & Katterfeldt, H. (1996). Modelling and interaction tools for virtual environments. In *Proceedings of the virtual reality World'96 conference*. Stuttgart, Germany: IDG Conference & Seminars.

Chen, C. J., & Toh, S. C. (2005). A feasible instructional development model for virtual reality (VR)-based learning environments: Its efficacy in the novice car driver instruction of Malaysia. *Educational Technology Research and Development*, *53*(1), 111–123.

Cobb, S. V., Brown, D. J., Eastgate, R. M., & Wilson, J. R. (1993). Learning in virtual environments (LIVE). In *Proceedings of the 'Science for Life' Conference*, University of Keele, U.K., September 2.

Cobb, S. V., Neale, H., Crosier, J., & Wilson, J. R. (2002). Development and evaluation of virtual environments for education. In K.M. Stanney (Ed.), *Handbook of virtual environments* (pp. 353–378). Mahwah, NJ: Lawrence Erlbaum Associates.

Cobb, S. V. G. (2007). Virtual environments supporting learning and communication in special needs education. *Topics in Language Disorders*, *27*(3), 211–225.

Cobb, S. V. G., Neale, H. R., & Reynolds, H. (1998). Evaluation of virtual learning environments. In P. Sharkey, D. Rose, & J.I. Lindstrom (Eds.), *The second European conference on disability, virtual reality and associated technologies* (pp. 17–23). September 10–11, Skovde, Sweden.

Colby, K. M. (1973). The rationale for computer-based treatment of language difficulties in nonspeaking autistic children. *Journal of Autism and Developmental Disorders*, *3*, 254–260.

Crosier, J. K., Cobb, S., & Wilson, J. R. (1999). *Virtual reality in secondary science education – establishing teachers' priorities*. Presented at CAL'99, London, U.K.

Crosier, J. K., Cobb, S., & Wilson, J. R. (2002). Key lessons for the design and integration of virtual environments in secondary science. *Computers and Education*, *38*, 77–94.

Crosier, J. K., & Wilson, J. R. (1998). *Teachers' priorities for virtual learning environments in secondary science*, Presented at VRET'98, London, U.K.

Dalgarno, B., Bishop, A. G., Adlong, W., & Bedgood Jr., D. R. (2009). Effectiveness of a Virtual Laboratory as a preparatory resource for Distance Education chemistry students. *Computers & Education*, *53*(3), 853–865.

Dede, C. (1995). The evolution of constructivist learning environments: Immersion in distributed, virtual worlds. *Educational Technology*, *35*(5), 46–52.

Druin, A. (1999). Cooperative inquiry: Developing new technologies for children with children. *Proceedings of the SIGCHI conference on human factors in computing systems: The CHI is the limit*. Pittsburgh, PA: ACM.

Duffy, T. M., & Jonassen, D. H. (1992). Constructivism: New implications for instructional technology. In T. M. Duffy, & D.H. Jonassen (Eds.), *Constructivism and the technology of instruction* (pp. 1–16). Hillsdale, NJ: Lawrence Erlbaum Associates.

Eastgate, R. M., Wilson, J. R., & D'Cruz, M. (2014). Structured Development of Virtual Environments. In K. S. Hale, & K. M. Stanney (Eds.), *Handbook of virtual environments: Design, implementation, and applications* (2nd ed., pp. 353–390). Boca Raton, FL: Taylor & Francis Group, Inc.

Garzotto, F. (2008). Broadening children's involvement as design partners: From technology to "Experience." *Proceedings of the Seventh International Conference on Interaction Design and Children*. Chicago, IL: ACM.

Goldsmith, T. R., & LeBlanc, L. A. (2004). Use of technology in interventions for children with autism. *Journal of Early and Intensive Behavior Intervention*, *1*(2), 166–178.

Guha, M. L., Druin, A., & Fails, J. A. (2008). Designing with and for children with special needs: an inclusionary model. *Proceedings of the Seventh International Conference on Interaction Design and Children*. Chicago, IL: ACM.

Hall, L., Woods, S., & Aylett, R. (2006). FearNot! involving children in the design of a virtual learning environment. *International Journal Artificial Intelligence Education*, *16*, 327–351.

Hardy, C., Ogden, J., Newman, J., & Cooper, S. (2002). *Autism and ICT: A guide for teachers and parents*. London, U.K.: David Fulton Publishers.

Hourcade, J., Bullock-Rest, N., & Hansen, T. (2011). Multitouch tablet applications and activities to enhance the social skills of children with autism spectrum disorders. *Personal and Ubiquitous Computing*, *16*, 1–12.

Jonassen, D. H. (1994). Thinking technology: Toward a constructivist design model. *Educational Technology*, *34*(4), 34–37.

Kaur, K., Maiden, N., & Sutcliffe, A. (1999). Interacting with virtual environments: An evaluation of a model of interaction. *Interacting with Computers*, *11*, 403–426.

Kaur, K., Sutcliffe, A., & Maiden, N. (1998). *Applying interaction modelling to inform usability guidance for virtual environments*, Position Paper for The First International Workshop on Usability Evaluation for Virtual Environments, December 1998, De Montfort University, Leicester, U.K.

Kontogeorgiou, A. M., Bellou, J., & Mikropoulos, A. T. (2008). Being inside the quantum atom. *PsychNology Journal*, *6*(1), 83–98.

Lapointe, J. F., & Robert, J. M. (2000). Using VR for efficient training of forestry machine operators. *Education and Information Technologies*, *5*(4), 237–250.

Meakin, L., Wilkins, L., Gent, C., Brown, S., Moreledge, D., Gretton, C., ... Mallett, A. (1998). *User group involvement in the development of a virtual city*, Paper presented at the Second European Conference on Disability, Virtual Reality and Associated Technologies, Skovde, Sweden.

Mikropoulos, T. A., & Natsis, A. (2011). Educational virtual environments: A ten-year review of empirical research (1999–2009). *Computers & Education, 56*, 769–780.

Millen, L., Cobb, S., Patel, H., & Glover, T. (2014). A collaborative virtual environment for conducting design sessions with students with autism spectrum conditions. *International Journal on Disability and Human Development, 13*(3).

Minocha, S., Kear, K., Mount, N., & Priestnall, G. (2008). *Design of learning spaces in 3D virtual environments.* In: Researching Learning in Virtual Environments International Conference (RELIVE08), November 20–21, 2008, Open University, Milton Keynes, U.K.

Neale, H., Cobb, S., & Kerr, S. (2003). *An inclusive design toolbox for development of educational virtual environments*, Presented at Include 2003, March 25–28, Royal College of Art, London, U.K.

Neale, H., & Nichols, S. (2001). Theme-based content analysis: A flexible method for virtual environment evaluation. *International Journal of Human-Computer Studies, 55*(2), 167–189.

Neale, H. R., Brown, D. J., Cobb, S. V. C., & Wilson, J. R. (1999). Structured evaluation of virtual environments for special needs education. *Presence: Teleoperators and Virtual Environments, 8*(3), 264–282.

Papert, S. (1988). The conservation of Piaget: The computer as grist for the constructivist mill. In G. Foreman & P.B. Pufall (Eds.), *Constructivism in the computer age* (pp. 3–14). Hillsdale, NJ: Lawrence Erlbaum Associates.

Parsons, S., Charman, T., Faulkner, R., Ragan, J., Wallace, S., & Wittemeyer, K. (2013). Commentary – bridging the research and practice gap in autism: The importance of creating research partnerships with schools. *Autism, 17*(3), 268–280.

Parsons, S., & Cobb, S. (2013) Who chooses what I need? Child voice and user-involvement in the development of learning technologies for children with autism. EPSRC Observatory for Responsible Innovation in ICT. http://responsible-innovation.org.uk/torrii/page-resource/Case%20Studies. Accessed April 8, 2014.

Parsons, S., Garib-Penna, S., Rietdijk, W., Millen, L., Hawkins, T., Cobb, S., ... Alessandrini, A. (2012). COSPATIAL Deliverable 5.2: Final report on evaluation of prototypes. http://cospatial.fbk.eu/sites/cospatial.fbk.eu/files/COSPATIAL%20D5.2.pdf. Last accessed June 14, 2013.

Parsons, S., Millen, L., Garib-Penna, S., & Cobb, S. (2011). Participatory design in the development of innovative technologies for children and young people on the autism spectrum: the COSPATIAL project. *Journal of Assistive Technologies, 5*, 29–34.

Parsons, S., & Mitchell, P. (2002). The potential of virtual reality in social skills training for people with autistic spectrum disorders. *Journal of Intellectual Disability Research, 46*(5), 430–443.

Perkins, D. N. (1996). Minds in the 'Hood. In B.G. Wilson (Ed.), *Constructivist learning environments: Case studies in instructional design* (pp. v–viii). Englewood Cliffs, NJ: Educational Technology Publications.

Pimental, K., & Teixeira, K. (1993). *Virtual reality: Through the new looking glass.* New York, NY: McGraw-Hill.

Preece, J. (1994). *Human-computer interaction.* Boston, MA: Addison-Wesley Longman.

Scaife, M., Rogers, Y., Aldrich, F., & Davies, M. (1997). Designing for or designing with? Informant design for interactive learning environments. *Proceedings of the ACM SIGCHI Conference on Human Factors in Computing Systems (CHI'97)* (pp. 343–350). Atlanta, GA: ACM Press.

Slater, M., & Usoh, M. (1995). Modelling in immersive virtual environments: A case for the science of VR. In R.A. Earnshaw, J.A. Vince, & H. Jones (Eds.), *Virtual reality applications.* London, U.K.: Academic Press.

Solomon, C. (1986). *Computer environments for children - A reflection on theories of learning and education.* Boston, MA: Massachusetts Institute of Technology.

Thalmann, N. M., & Thalmann, D. (1994). Introduction: creating artificial life in virtual reality. In N.M. Thalmann, & D. Thalmann (Eds.), *Artificial life and virtual reality.* Chichester, U.K.: John Wiley & Sons.

Tohidi, M., Buxton, W., Baecker, R., & Sellen, A. (2006). Getting the right design and the design right: Testing many is better than one. *Proceedings of the ACM SIGCHI Conference on Human Factors in Computing Systems (CHI'06)* (pp. 1243–1252). New York, NY: ACM Press.

Wastiau, P., Blamire, R., Kearney, C., Quittre, V., Van de Gaer, E., & Monseur, C. (2013). The use of ICT in education: A survey of schools in Europe. *European Journal of Education, 48*, 11–27.

Williams, C., Wright, B., Callaghan, G., & Coughlan, B. (2002). Do children with autism learn to read more readily by computer assisted instruction or traditional book methods? *Autism, 6*, 71–91.

Wilson, J. R., Brown, D. J., Cobb, S. V., D'Cruz, M. D., & Eastgate, R. M. (1995). Manufacturing operations in virtual environments (MOVE). *Presence: Teleoperators and Virtual Environments, 4*(3), 306–317.

Wilson, J. R., Eastgate, R. M., & D'Cruz, M. (2002). Structured development of virtual environments. In K.M. Stanney (Ed.), *Handbook of virtual environments* (pp. 353–378). Mahwah, NJ: Lawrence Erlbaum Associates.

Winn, W. (1993). *A conceptual basis for educational applications of virtual reality* (Technical Report No. HITL-TR-93-9). Human Interface Technology Laboratory, University of Washington, Seattle, WA.

Youngblut, C. (1998). *Educational uses of virtual reality technology* (Technical Report No. D-2128). Institute for Defense Analyses, Alexandria, VA.

Zancanaro, M., Weiss, P. L., Gal, E., Bauminger, N., Parsons, S., & Cobb, S. (2010). Teaching social competence: In search of design patterns. In *Proceedings of the ACM conference on interaction design for children - IDC2010*. Barcelona, Spain: ACM Press.

43 Virtual Environment–Assisted Teleoperation

Abderrahmane Kheddar, Ryad Chellali, and Philippe Coiffet

CONTENTS

43.1 INTRODUCTION

Teleoperation is the technology of robotic remote control. Teleoperation systems are aiming to optimize the synergy between humans and machines toward achieving tasks within hazardous and inaccessible environments. This human–machine association is hybrid and differentiates teleoperation systems from pure robotic solutions. Human operators are used for their cognition and decision-making abilities, while robots perform the physical actual interactions within uncertain sometimes unknown, and dynamically changing remote worlds.

Historically, the first teleoperation systems were built after the Second World War for needs in nuclear activities. These early systems used a master–slave concept with two symmetrical arms. The operator handles the master arm, while the slave replicates the operator's motions at the location where the task is being performed. In the earliest systems, master and slave were mainly mechanically connected. Later systems were electrically powered, affording the possibility of any distance between master and slave. In the 1980s, computers were introduced as control systems opening the way to computer-aided teleoperation (CAT). Contemporary CAT utilization has been deeply modified by the emergence of virtual environment (VE) technology.

In early systems, the absence of sophisticated electronics and computers obliged a symmetrical mechanical device to correctly transfer motions and energy from the operator to the slave device. Nevertheless, a great aid to manipulation came from the integration of force feedback, allowing the operator to *feel* what he or she manipulated remotely. The introduction of computers led to further enhancements in the design of master and slave devices. Whereas the slave kept its mechanical structure, the master could be reduced in size, as well as transformed into a joystick with force feedback or into a set of different minisystems that were easy for the operator to move. However, these new master interfaces, although better adapted to the human, did not resolve the general difficulty of correctly performing remote tasks. Indeed, teleoperation tends to mix two sensory–motor systems that are different in nature. Remote operations require visual access to the activities being conducted in the slave environment. Classic cameras, even stereoscopic ones, did not succeed in adequately informing the operator. The operator thus often became disoriented (see Darken & Peterson, 2014, Chapter 19), lost sight of the interest, and became afflicted with motion sickness-like symptoms (see Lawson, 2014a, Chapter 23) after several minutes of work. VEs, with their ability to provide partial or total immersion (see Chertoff & Schatz, 2014, Chapter 34), play a central role in improving this situation in reducing the discrepancies between the two agents. The great interest in VE-enhanced teleoperation is based on expected improvement in teleoperation efficiency and effectiveness: from gains in both ergonomics and user-friendliness (see Gabbard, 2014, Chapter 28), as well as an information feedback point of view.

43.2 TELEOPERATION ISSUES

Teleoperation has matured from both technical and conceptual point of views. Yet there are still many issues that may be solved in order to achieve satisfactory or convincing solutions. These problems are summarized in the following:

- *The unreachable ideal transparency*: The ideal master–slave system is the one where the slave reproduces exactly the aims of the master and reports exactly about the working environment state. In addition to stability (i.e., in terms of a control theory point of view, stability means that a system will remain close to its original motion trajectory even if it is slightly perturbed away from it), one of the most important teleoperator characteristics is transparency. The latter can be seen as the unbiased action/sensing fidelity, which can be evaluated through a measurement index (see Laurence in IEEE TRA, 1993). Although trade-offs between stability and transparency have been found, such solutions have largely been derived from a pure control theory. It is an established fact that ideal transparency can never be achieved by conventional bilateral control unless it is considered under other criteria or conceived differently. This limitation comes mainly from (1) the operator's action and the feedback being conveyed through the master–slave chain before reaching the target task and the operator channels, respectively, and (2) this chain including dynamics that cannot be neglected or compensated for without compromises in stability or operator safety. Challenging bilateral coupling seems to be one way to deal with this problem.

Within a unified formalism, teleoperation issues can be stated as solving master–slave sensory–motor discrepancies and distortions in four main dimensions: (1) time (time-delay issues), (2) Euclidean space (scale changes), (3) sensing (limited force and tactile feedback), and (4) motoric (when hand controls complex motions such as a walking humanoid robot). VE helps toward the ideal transparency by providing synthetic features, cues, and mechanisms that supplement and/or correct the natural sensory–motor channels.

- *Time delay in control of remote systems*: For long-range teleoperation (in space or for intercontinental operations), communications are delayed, and resulting time delays between master and slave constitute one of the most crucial problems. Indeed, time delay affects both transparency and stability of the system. Operators' actions and sensory feedback are delayed leading to mismatches between felt and actual situations. Consequently, an inherent and structural instability of the slave occurs. In conventional control solutions (those without VE techniques), constant and variable time-delay-based solutions have been proposed by Anderson and Spong (1988) by means of a strategy resulting in a smart *damping* design via low-level control. This method has been more elegantly formulated by Niemeyer and Slotine (1991) and astutely adapted for the case of nonstationary time delay. Kosuge et al. (1996) and Oboe and Fiorini (1998) also proposed solutions to deal with variable time delay. These basic approaches were recently much more sophisticated and applied to VE force feedback devices using Internet communication media (see Hirche & Buss, 2012). Nevertheless, it is a utopia to try to rely on force feedback from the remote environment when delays are of the order of seconds (IEEE TRA, 1993).

- *Taking into account human factors*: Industrial robotics is sometimes opposed to teleoperation solutions with regard to flexibility; industrial robots are designed to be autonomous in taking decision in very controlled and known tasks (preprogrammed behaviors); telerobots are versatile enough and can cope with various unpredictable tasks, thanks to human's online support. Unfortunately, this flexibility has a cost; in many ways, it is related to operators' adaptation to the teleoperation system. Indeed, to perform a task, the operator must be trained. This training aims at allowing operators to create mental schemes and motor mechanisms to lower sensory–motor discrepancies with the goal to reduce the mental/cognitive workloads toward achieving safely and effectively the targeted tasks. VE provides such simulation and mission preparation tools, including time delays, scale changes, and poor and distorted sensory feedback. Despite this, simulation solutions are only approximations of the reality not covering all possible situations. In practice, teleoperation systems designers integrate and quantify the adaptation and the easiness of handling such systems. This issue known also as human factors and ergonomics is an old well-stated problem that still has no satisfactory evaluation methodology, yet means of assessing VE usability are under development (see Gabbard, 2014, Chapter 28).

- *Slave autonomous behavior and human-machine shared control*: Using more autonomous robots was a way to reduce interfacing effects. Many tentative solutions concerning abstracted human–machine cooperation, that is to say, shared control architectures and semiautonomous teleoperators, have been proposed. This class of approaches aims at optimizing the full system performances by relying on two pillars: (1) the robustness and the reliability in executing tasks for robots and (2) the effectiveness of decision-making skills of human operators. Within such a framework, autonomy sharing, modeling, and quantification appear as narrowly linked to the degree of robot autonomy and to the level of operator intervention, which can range from full manual to symbolic (supervisory) control (Sheridan, 1992). In fact, shared-autonomy schemes shift the classical issues to new ones dealing unanswered questions about the relation between man's intervention methods and robot autonomy level.

- *Reliability and safety*: Most of the time, teleoperation is involved in critical tasks where safety and reliability are key issues. Unlike computers, the manipulator arms, if used improperly, have the physical capability of destroying themselves and their environments. In some contexts, operators work very often in the vicinity of powered axes, where dangers in addressing system integrity, losses of functionality, and device or sensor degradations exist (Struges, 1996). Powered manipulation arms are inherently slow devices; no quick or jerky motion should be attempted in using them. The use of high-powered master–slave devices is unquestionably as dangerous as today's industrial robots. For telesurgery, similar issues (and perhaps more ethical) are also present. VE as a training tool or as an executive platform can help improve safety in teleoperated system. Augmented reality solutions, for instance, are used for navigation or for displaying safety limits and alerts.

43.3   VE-BASED TELEOPERATION

Nowadays, VE reached a maturity stage and offers a wide spectrum of techniques and tools to generate synthetic but realistic multimodal stimuli. This allows creating tailored solutions to address teleoperation problems. However, this richness makes difficult inferring standard ways VE could assist teleoperation. This difficulty, paradoxically, constitutes the strength and cleverness of VE contributions. Therefore, the remainder of this chapter concentrates on the state of the art in VE-based teleoperation, including examples of the capabilities of some teleoperator systems and current challenges in VE-assisted teleoperation schemes. Topics covered include coping with time delay, enhancing CAT, improving bilateral and shared control, human-centered architectures, enhancing sensory feedback to the human operator, and improving operator safety with human factors issues.

43.3.1   Master–Slave Systems and the Importance of VEs

Basically, a master–slave system can be seen as the assembly of four components: the human controller, the human sensor, the telerobot controller, and the telerobot sensor. The human controller generates commands to be executed by the robot controller. The robot sensory system captures the status of its environment and displays it to the human sensory system.

In early teleoperation, the human–telerobot cross-links were direct. Physically made of cables, the links transmit actions and reactions between slaves and operators. As well, operators have direct sight and complete visual access to the remote working environment. These primary systems evolved to account for safety and feasibility constraints, namely, isolating and introducing distance between operators and working environments. This distance implied the use of indirect and synthetic links. For the sensory part, cameras and screens replaced direct visual inspection. In the same vein, contact forces sensed by the remote robot were displayed visually to the operator as a vector representing both the direction and the intensity the felt force. For the motor part, remotely controlled and electrically actuated robots replaced the cables.

In the general case, the flows of sensory–motor data exchanged between partners are mainly synthetic, from sensing to display and from motor action to control. In addition, the flows are not necessarily addressing equivalent sensory–motor channels. VE, and its ability to handle synthetic worlds, appears as the natural candidate to interface operators and telerobots; it allows to convey the inherently hybrid bidirectional information with the aim to guarantee an optimal synergy between partners by making the information available and intelligible to both sides.

43.3.2   Coping with Time Delay

Predictive displays offer a solution to dealing with time delay inducing instability in bilateral force reflection teleoperation. Predictive display development was pioneered by the efforts of the Massachusetts Institute of Technology's Human–Machine Laboratory. Noyes and Sheridan (1984; see Sheridan's review in IEEE TRA, 1993) developed a graphical predictive display where the manipulator was displayed as a stick graphic figure overlaid on the delayed video image of the manipulator. Experimental results have proven the efficiency of the proposed strategy. Bejczy, Kim, and Venema (1990, quoted in Kim, 1996) improved the stick figure with a high-fidelity graphic predictive display that they called the phantom robot. Two graphic models (wire frame and solid) accurately depicting the actual slave robot can be overlaid on the delayed video feedback. The operator chooses, according to the context, one of the two graphic robot representations (wireframe or solid) to act as a predictor. Since the graphic representation of the actual slave robot is faithfully reproduced, accurate virtual to video superimposition on a common display window was made feasible, thanks to a VE calibration method thoroughly discussed in Kim (1996).

Figure 43.1 shows a teleoperation instance in a space operation simulation case. The predictive *phantom robot* reacts in real time and smoothly to the operator commands. When subsequent

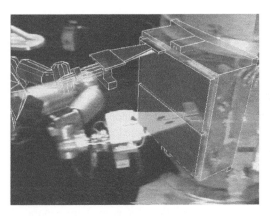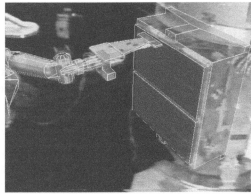

FIGURE 43.1 Predictive display by means of augmented reality techniques for space teleoperation. (Courtesy of Kim [1996], Jet Propulsion Laboratory, NASA, Pasadena, CA.)

predicted actions are considered satisfactory, trajectory commands are sent to the actual robot. This method, although enhancing operator performance, still uses a move-and-wait strategy in which moving concerns the virtual robot and waiting concerns actual and virtual robot matching. It must be noted that force display is obviously prohibited because only the model of the robot is available. To maintain force feedback, however, Kotoku proposed using a VE representation of the entire remote location and its components (including the slave robot) as a predictive display (Kotoku, 1992). In this case, the system consists of the master force feedback arm; the VE, including a virtual robot actually coupled to the master arm, the actual slave robot, and its environment; and communication media linking the VE to the actual robot controller. The virtual robot is operated within the VE in a similar manner as if the virtual robot were real. Indeed, force feedback is artificial, generated from the VE computer simulation, and flows from estimated interaction forces between the virtual robot and the VE. Since object geometric modeling is based on a polyhedral representation, the artificial force feedback algorithm is based on polyhedron interpenetration calculations, including collision detection and other well-known features used in computer graphics simulation (Anderson, 1994). In fact, this system splits the bilateral loop into two local loops, the master bilateral loop coupling the operator to an estimated remote environment (VE) and the slave loop managing actual robot interaction with its real environment provided that the interrelated (i.e., embedded local autonomy, necessary degrees of freedom, and suitable sensors instrumentation) are convenient.

In the previously cited predictors, distant robot control is achieved according to the online predictive planned trajectory. An alternative clever way to perform teleoperation in the presence of considerable time delay is teleprogramming. As shown in Figure 43.2, the principle of teleprogramming is nearly similar to VE predictive teleoperation. Indeed, no bilateral coupling is achieved

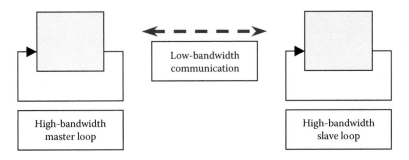

FIGURE 43.2 A high-level view of teleprogramming. (Adapted from Funda, J. et al., *Presence Teleop. Virt. Environ.*, 1(1), 29, 1992.)

between the master arm and slave robot. The master site is set up with a VE coupled with the master arm. This fact allows high-bandwidth virtual robot operation with immediate visual and haptic feedback. The slave arm is also a high-bandwidth closed loop between the remote robot and its environment and concerns the classical sensory feedback needed for the controller.

Compared with the predictive display method, teleprogramming is peculiarly different. Indeed, on the one hand, robot instructions are symbolic and somehow flexible rather than low level and accurate. Subsequently, this requires some robot autonomy. It also makes teleoperation nicely continuous (i.e., not move and wait). On the other hand, feedback from the remote location to the master site matters to the execution status. In addition to providing real-time interaction, the virtual representation software continuously monitors the slave robot and any object in its grasp for collision detection or contacts with the VE. Subsequent macrocommands are generated by integrated software, based on a priori task knowledge and a predefined command language. To prevent execution errors, due to virtual model uncertainties, guarded commands (motion, force, etc.) are generated rather than absolute or static commands. This teleprogramming architecture, together with experimental results, is thoroughly described in Funda, Lindsay, and Paul (1992).

Another variance of predictive display and teleprogramming was developed at the German Aerospace Center (DLR). The Hirzinger team conducted the first actual space experiments on a space robot technology called ROTEX, which flew with the space-shuttle COLUMBIA from April 26 to May 6, 1993 (see Hirzinger et al. in IEEE TRA, 1993). Among the four operational modes of the ROTEX are teleoperation from ground using a predictive display and telesensor programming. Both operational modes are based on a virtual simulation of the ROTEX with its remote space laboratory features (see Figure 43.3). The predictive display contains a model of the up and down communication link delay as well as a model of the actual states of the real robot and its environment features, more peculiarly moving objects. Measured object poses are compared with estimates as issued by an extended Kalman filter. This filter predicts, and graphically displays, the situation that will occur after the up-link delay has elapsed and allows the performed action loop to be closed by operator control, via shared control, or purely in an autonomous mode. This kind of prediction has been made possible due to a nearly perfect world model together with knowledge of the dynamics of objects under zero gravity. The telesensor-programming operational mode is a kind of teleprogramming but differs in that it is nearly teaching by showing, as applied to robot off-line programming. In the proposed telesensor programming, complex tasks are split up into elemental moves for which a certain constraint-frame and simulated sensor-type configuration holds. This provides the actual remote robot with simulated sensory data that refer to relative positions between its sensorized gripper and the environment. This compensates for any kind of inaccuracies in the absolute positions of robot and real world. These operational modes were successfully tested for multiple space laboratory tasks, including grasping of floating objects and assembly tasks.

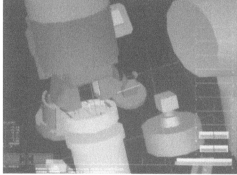

FIGURE 43.3 The DLR MARCO system: overview of the telesensor-programming interface (left), the ROTEX experiment VE setup, and the gripper sensor simulation (right). (Courtesy of Hirzinger, DLR, Munich, Germany.)

The DLR conducted research toward a unified concept for a flexible and highly interactive programming teleoperation station. The system, called MARCO, is designed as a 2-in-2 layer concept representing a hierarchical control structure ranging from the planning to executive layer. User layers are comprised of the task and operation modules, whereas the robotic expert layers' module concerns elementary operations, actuator commands, and sensor phase. In the frame of a DLR–NASDA (Japanese Aerospace Agency) joint program, called German ETS-VII Experiment (for GETEX), the aforementioned modular task-directed programming scheme (MARCO) was successfully involved (April 1999) in the recent unmanned teleoperation of NASDA's ETS-VII free-floating space robot (ETS-VII, 1999).

Other ETS-VII teleoperation experiments have been conducted from the ground (ETS-VII, 1999). The purpose of the ETS-VII mission is to examine the ability of canonical tasks to support and confirm future space investigations such as building and operation of the international space station, inspection and repair of orbiting satellites, and planetary exploration. Teleoperation tasks include an onboard satellite antenna assembling experiment, truss structure teleoperation, servicing, orbital unit replacement, add-on tool exchange, free-floating target satellite capture, and visual inspection. ETS-VII ground robot control systems use operator aids based on VE techniques to assist telemanipulation. These include predictive computer graphics, shared control capabilities, imaginary guide planes (a kind of a core-shaped virtual wall) to guide the robot arm motion to a desired position and to inhibit undesirable motions, various force controllers (for local compensation of geometric discrepancies), visualization and verification of motion commands using a motion simulator, multimodal interfaces, a visual aid system for direct teleoperation using a predictive force method, and teleoperation through virtual force reflection (including potential force fields, virtual force as physical constraints, adaptive virtual force by probing environment; ETS-VII, 1999).

Figure 43.4 shows screen snapshots related to the ETS-VII teleoperation experiments. In conclusion, predictive displays and teleprogramming seem to be an attractive solution to deal with time-delayed teleoperation. Many other advanced teleoperation applications, such as mobile robots (an earlier application domain of teleprogramming initiated by a joint effort of the Laboratoire d'Automatique et d'Analyse des Systèmes [LAAS] at Toulouse, France, and the GRASP Laboratory at the University of Pennsylvania; see also the mobile robot teleoperation section in this chapter), subsea robots, and even flying robots, also use teleprogramming and graphic predictive displays. In terms of pure control theory, predictive displays and teleprogramming could likely be seen as an implementation, in the frame of teleoperation, of the well-known Smith predictor controller, proposed in 1957.

43.3.3 Enhancing CAT

CAT (known also under the name of teleassistance) can benefit from the modern human–machine interfaces associated with VE technology. Resourceful developments and solutions of VE-aided teleoperation that have potential for enhancing CAT include virtual fixtures, active guides, and graphical programming metaphors.

One of the applications of VE techniques in CAT is directed toward fitting abstract perceptual information within the human–machine interface. The work proposed by Rosenberg (1992) explores the design and implementation of computer-generated entities known as *virtual fixtures*. Virtual fixtures are defined by Rosenberg (1992, p. 4) as "abstract sensory information overlaid on top on the reflected sensory feedback from the workspace which is completely independent of all sensory information from the workspace." Although they functionally embody fixtures in the real world, there are many benefits inherent to virtual fixtures, as compared to real physical ones, because they are computer generated. Indeed, virtual fixtures can be extended to include other modalities, such as visual (see Badcock, Palmisano, & May, 2014, Chapter 3), auditory (see Vorlander & Shinn-Cunningham, 2014, Chapter 4), and even tactile (see Dindar, Tekalp, & Basdogan, 2014, Chapter 5) sensations used alone or in cross-modal combination. Rosenberg highlights these advantages by an instance consisting of plotting a straight line. Obviously, the use of

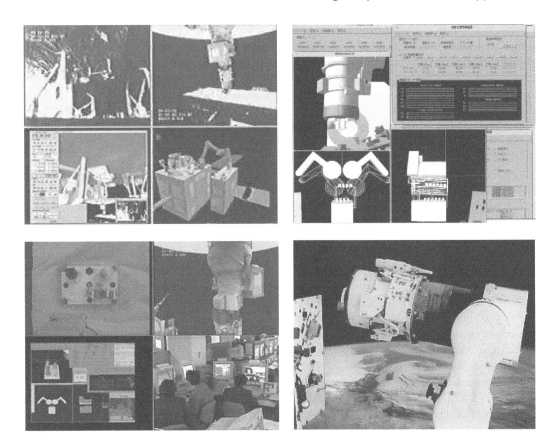

FIGURE 43.4 The ETS-VII unmanned space teleoperation experiment. Clockwise: target satellite handling experiment, displayed CG image of simulation (predictive simulation system), an image from an onboard TV camera, and the whole setup. (Courtesy of NASDA, Arlington, VA, http://oss1.tksc.nasda.go.jp/ets-7.)

a ruler enhances human operator performance in plotting a straight line as compared to when no ruler is used. This may be seen as what has been proposed earlier in the frame of CAT, namely, the possibility to freeze some of the robot degrees of freedom while constraining the operator to control the slave robot in the remaining ones. Virtual fixtures are different, since they offer a more powerful and more flexible tool, which does not act on the slave robot directly. Indeed, more attention is given to operator assistance rather than the robot, through artificial sensory feedback overlaid on top of the sensory feedback from a remote teleoperation work site, which serves as a perceptual aid for task performance. How to make virtual percepts using perceptual rather than physical parameters is the main issue being investigated in using virtual fixtures. How does it work? Simply, as the operator interacts with virtual fixtures (via haptic or other defined modalities), appropriate reactive sensations are computed and fed back to the user. According to Rosenberg, abstraction can be used to operate on virtual fixtures, thereby enhancing interactivity (see Section 43.3.5). Rosenberg has demonstrated that performance could be enhanced by 70% in telerobotic tasks that use virtual fixtures. Thus, virtual fixtures might be used to reduce operator load, facilitate supervisory control, ease the control of complex tasks, and even to *compensate* for performance degradations due to time delay.

Sayers (1999) used virtual fixtures to automatically generate low-level robot commands appropriate to the tasks to be performed. Although Sayers' team gave an implementation within a teleprogramming context, the developed strategies can be used in an augmented reality context as well. In their experiment, synthetic fixtures present the operator with task-dependent and context-sensitive

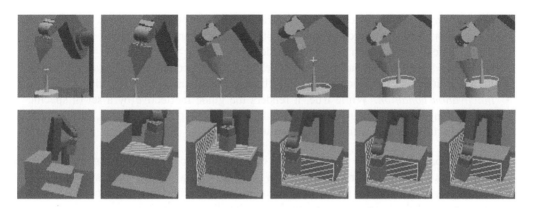

FIGURE 43.5 The use of synthetic fixture in teleprogramming-based teleoperation. (Courtesy of Sayers [1999], the GRASP Laboratory, University of Pennsylvania, Philadelphia, PA.)

visual and force cues. No attempt was made to provide realistic force feedback. Instead, the intention was to supply the operator with force and visual cues that can best aid task performance.

In Figure 43.5, screen snapshots show an instance highlighting the use of virtual fixtures in a VE teleprogramming application. In the upper row of Figure 43.5, a single-point fixture (represented as a cross) is used to help the operator bring the end point of the robot-held tool to a point in space. When the telerobot tool is operated near the desired location, the virtual cross fixture is activated and pulls the robot-held tool toward the cross-fixture location. In this case, the virtual fixture has not been defined to completely constrain operator (thus robot) motion. The system anticipates the operator's next desired action. In the same instance, another fixture allowing the robot to achieve circular motions is activated just after the operator decides to leave the cross-fixture location. In the lower row of Figure 43.5, a set of fixtures were defined to aid the operator in achieving flat face-to-face contacts. Each time the operator moves near the surface, the appropriate face-to-face virtual fixture is activated.

Kheddar (2001) attempted to give a unified formalism to the virtual fixture metaphor and named it *active virtual guide* in the frame of a hidden robot teleoperator (see Section 43.3.4). Under this metaphor, virtual fixtures are classified in three categories:

1. *Pure operator assistance:* This group includes the well-known graphic metaphors used in CAT (e.g., sensory substitutions), in which virtual guides are not directly linked to robot control. Their role is mainly focused on assisting the operator to intuitively perform desired tasks.
2. *Pure remote robot control assistance:* This group includes virtual mechanism concepts, or any other virtual metaphor, induced in the low-level control necessary for strict execution of a real task by the actual robot.
3. *Operator and robot shared assistance:* This group includes those fixtures that are dedicated at the same time to operator and robot assistance in terms of robot autonomy sharing. Actual task completion results are issued from a combination of the virtual task designed by the operator and an autonomous module linked to robotic tasks. Hence, this category is split into three subclasses: autonomous function, semiautonomous function, and collaborative function.

Virtual guides lead then to a unified structure composed of the following items:

- *Attachment:* A particular spot associated to the virtual guide. The virtual guide can either be statically attached (to any frame within the VE, object, robot controller part, etc.) or dynamically attached, appearing on a specific event (collision detection between objects or robot interaction with its environment, etc.) in a specific spot.

- *Effect zone:* A virtual guide may be associated with an effect area (volume, surface, part of the robot reach space, etc.), which may play the role of an action zone or an attractive/repulsive field within which the virtual guide acts.
- *Activation condition:* For each virtual guide, an activation condition is allocated.
- *Function:* It defines the functionality, thus the reason for existence of the guide.
- *Inactivation condition:* While true, it renders the guide ineffective by a set of specific actions. Any whole structured virtual guide may be completely removed, attached, or replaced off-line or online (during teleoperation).

In Figure 43.6, a teleoperated task consisting of grasping an object has been simulated with a cylindrical guide. The same geometry guide can serve as a collaborative (in this case, instance), semiautonomous, or autonomous guide. In an autonomous style, the virtual object can be assumed grasped when one of the operator fingers touches the handle (unrealistic but easy operator grasp), which triggers an autonomous robotic grasp. In a semiautonomous instance, the virtual object operator grasp is based upon physical realistic laws. When the virtual operator grasp is stable, the robot triggers an autonomous grasp. In a collaborative mode, operator grasping is realistic, the guide effect being used just to align the robot axis according to the object handle frame, thus freezing the guide robot gripper orientation to that of the handle and leaving the robot controlled according to the operator translational hand movement.

Graphic programming by means of VE technique is one of the most attractive paradigms used in advanced teleoperation architectures. On the one hand, graphic programming may be seen as one of the potential solutions to the well-known operator—telerobot shared control and autonomy problem. On the other hand, it might be seen as the ultimate way to fulfill both user-friendly and human factor performance with high ergonomic characteristics. Some instances are presented in the following to give a better idea of how graphic programming can carry out the previously cited points.

Figure 43.7 represents the task line and motion guide paradigm, which is an original form of teleoperation proposed by Backes, Peter, Phan, and Tso (1998). Motion guides consist of fashioning paths intuitively in the robot VE. Then, the actual telerobot is constrained to follow these paths when this is possible. Subsequent continuous commands sent to the actual telerobot are only 1D: forward, back, or halt along the motion guide. Motion guides are graphically represented using paths (curves and lines), and

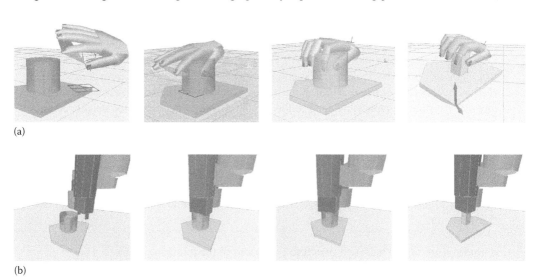

(a)

(b)

FIGURE 43.6 The use of collaborative active virtual guides in a grasping assist task for both the operator (a) and the slave robot (b) in the frame of the hidden robot concept–based teleoperation. (From Kheddar, A., *IEEE Trans. Syst. Man Cybern. A*, 31(1), 1, 2001.)

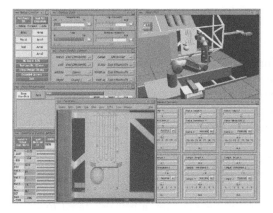

FIGURE 43.7 Task line and motion guide: a graphical programming paradigm for space teleoperation. A surface inspection task (left). The whole teleoperation test-bed system (right). (Courtesy of Backes, Jet Propulsion Laboratory, NASA, Pasadena, CA.)

the motion direction is specified using arrow icons. Specification and modification of motion guides is achieved interactively off-line or online (during teleoperation). Task line is a metaphor to design subtasks attached to motion guides, which is represented in the VE by icons. A task line is a collection of command sequences that are executed when the telerobot reaches the task line location. Among the general commands found in a task line, reactive sequences can be included to be performed upon specific events related by means of internal or external robot sensory capabilities (see Figure 43.7).

At the Sandia National Laboratories, graphical programming tools based on 3D graphics models (called Sancho; see Figure 43.8) are being developed to be used as an intuitive operator interface to program and control complex robotics systems, including teleoperation for nuclear waste cleanup

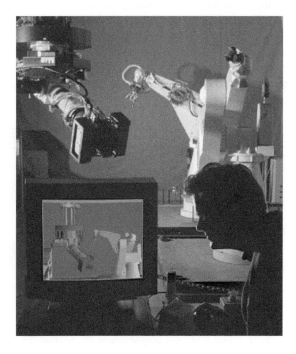

FIGURE 43.8 The Sancho graphic-based robot programming experimental system. (Courtesy of the Sandia National Laboratories, Albuquerque, NM.)

(Small & McDonald, 1997). Sancho uses a general set of tools to implement task and operational behavior and works in four major steps:

1. The tasks are defined, by the user, by means of menus and graphical picks in the simulation interface (define/select step). In this step, a 3D world model is used to simplify task definition; indeed, direct object selection is a primary method for defining task work location and to control task execution. At this stage, task definitions are based essentially on two paradigms: (1) operation based, by which robot tasks are subsequently derived from a set of individual sequence executions that consist of the task to be accomplished, and (2) task based, less flexible than the first and in which case, the task results from a set of predefined useful goals (predefined subtasks).
2. The system plans and tests the adopted solutions under operator monitoring and supervision (plan/simulate step).
3. If the adopted plans are considered satisfactory by the operator, a network-based robot program of the approved planning is sent to the robot controller (approve/download step).
4. Finally, the Sancho system monitors the robot and updates the simulator's environment with the data fed back from the actual robot environment (see Figure 43.8).

Another graphic programming system case, called TAO 2000, was developed by the French Nuclear Center (see Figure 43.9). TAO's graphic programming interface foundation is hierarchical mission decomposition (i.e., a functional objective decomposition) into a set of processes or methods. A process finds an expression in a set of operations (abstract functions); each operation is then decomposed into a set of tasks (i.e., generic functions) and each task is described by a set of actions (i.e., physical functions), which are easily mapped into telerobot executions (physical forms). The whole process constitutes an exploitable generic skill database. For a given mission (for instance, gate inspection), the operator selects a set of processes from a predefined skill database (e.g., an unscrewing process) to achieve each stage of the mission (the unscrewing step) and maps it onto the concerned object (one of the gate screws). The mapping (map unscrewing skill to the pointed gate screw) constitutes an operation that will be executed by the telerobot agent. The operation consists of a set of field containing selected symbolic tools, such as positioning area, an active virtual guide, recovery procedure, calibrating procedure, and agent behavior mode (autonomous, semiautonomous, or full manual teleoperation; Fournier, Gravez, Foncuberta, & Volle, 1997).

The TAO 2000 system is equipped with the PYRAMIDE 3D modeling system. As shown in Figure 43.9, only the functional objects (i.e., those involved in the specific mission—gate inspection) are rendered to unburden operator mission preparation.

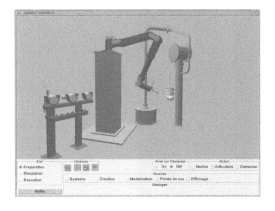
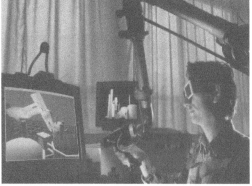

FIGURE 43.9   TAO 2000: a VE-based graphical programming for nuclear maintenance (gate inspection mission case) (left); the whole system with the MA23 master force feedback arm (right). (Courtesy of Gravez and Fournier, Advanced Teleoperation Service, CEA, Paris, France.)

43.3.4 IMPROVING BILATERAL AND SHARED CONTROL

Using a virtual intermediary world between the master and slave robot offers additional paths in control architecture design. All the same, the human operator can gain assistance from low-level control. VE techniques contribute subtly to low-level control as well as provide a potential way to efficiently carry out shared control. From the remote robot control point of view, the great advantage of reflex control is the absence of high-level reasoning before an action takes place. It is a direct link between information and action without passing through a decision-making stage. Reflex behaviors are necessary to take advantage of robot capabilities and to make robot control easier for the operator. Moreover, when time delays prohibit bilateral coupling between the master and slave, reflex behavior becomes an intrinsic property to allow robot autonomy. Nevertheless, because decision is built into the information sensed, this inclusion of information is only possible if the environment is well defined. Therefore, it is necessary to know beforehand, by some other means, the characteristics of the environment components involved in the teleoperation tasks. This unsolved problem relates to sensing and analysis of data. The use of VE may significantly improve this situation. Although the environment is unstructured, it is, generally, not totally unknown. Thus, applying VE to this problem means that two categories of virtual objects may be created. The first category refers to the minimal meddling of what is known about the environment. An example is that of a robot pushing against a surface of unknown hardness. If the robot is guided under force control and the surface is actually very soft, then the system becomes unstable. Even if the surface is very hard, this may also lead to instability. Traditionally, adding a mechanical compliance at the end of the robot arm absorbs energy. But if, instead, real-environment hardness K_e is replaced by a virtual hardness K_v, which is much larger than K_e and is given a priori, then the robot, which applies a force F_c on the surface, becomes stable. Hence, an unknown characteristic of the real environment can be replaced by a known virtual characteristic, which assures a good interaction with the contacted surface (Fraisse, Pierrot, & Dauchez, 1993). The same principle could probably be extended to most characteristics of unstructured environments.

The second category of virtual objects is the previously presented virtual fixtures or active virtual guides. While the concept has been previously applied in computer-assisted teleoperation, it can be extended to enhance robot autonomy and to improve operator assistance through low-level control. As stated, the principle consists of introducing geometrical artifacts along which the robot is constrained to move. This has been expressed through the virtual mechanism principle.

Kosuge et al. (1995) proposed an alternative control algorithm for telerobotic systems. For a given task, a task-oriented passive virtual tool is designed so that its dynamic behavior matches that of an ideal actual tool to be used as an interface between the remote robot and the object involved by the given task. Indeed, both robot maneuverability and stability are improved, while indirectly making bilateral control easier for the operator. As part of the TAO 2000 teleoperator project (see Section 43.3.2 and Figure 43.9), Joly and Andriot (1995) derived bilateral control laws from a simulated virtual ideal mechanism. Compared to Kosuge et al.'s (1995) work, the virtual mechanisms suggested here are connected to both the master and slave arms via springs and dampers. Using this, it is possible to impose any motion constraint to the teleoperator, including nonlinear constraints (complex surfaces and shapes) involving coupling between translations and rotations. This idea is being applied to most haptic interface controllers (see Dindar et al., 2014, Chapter 5) interacting with VEs.

In a recent work, Chellali and Pham (2011) used a shared control-based multirobot system to perform a video surveillance application (Khelifa et al., 2012). Assuming that the working environment is known, the operator can specify the full task to achieve in different ways. Indeed, the operator can just design the task to perform, namely, specify the limits of the environment to be covered by the fleet of mobile robots, letting the latter handle the low-level and the midlevel tasks such as local path planning, determine the global path for each robot according to other robots, and manage the embedded sensing acquisitions. The operator can as well control completely and individually each robots. In the first case, the operator acts as a supervisor of the full system, thanks to a virtual representation of the working environment. In the second case, the operator handles low-level controls to solve any blocking situation or to correct trajectories if needed. Such a system has a tunable shared control ranging from the full autonomy of the robots to the full control.

The research previously cited shows that VE techniques are not limited to operator assistance but could also serve as an intermediary in achieving remote tasks by replacing problems related to human–robot–environment interactions with human–robot–virtual mechanism–environment interactions.

43.3.5 HUMAN-CENTERED ARCHITECTURES

The purpose of any teleoperator is not the perfection of a master design, or an adequate executing machine, or the control architecture, or even the present remote environment state. The purpose is rather a future environment state expressed through its transformation. Indeed, only this transformation is of interest. On the other hand, almost any complex task to be remotely performed can be broken down into subtasks that can be expressed by a set of elementary moves for which a certain sensory-based state holds (i.e., a set of relations between the motion space and sensors space). An important remark can also be formulated to justify the design of human-centered telerobotic architectures; the way human operators perform tasks is in most cases not the most suited way for robots and vice versa (Kheddar, 2001). This basic idea constitutes the foundation of various so-called human-centered architectures.

In the *hidden robot* telerobotic architecture proposed by Kheddar (2001), the goal of an ideal teleoperation system is defined by the possibility to build an intermediate world keeping only a functional copy of the real remote environment adapted to the desired task transformation. The part of the system devoted to the execution of the task must involve additional transformations implicating the intermediate world as a real one. Thus, the proposed teleoperation scheme leads to the design of two separate subsystems and the development of the link layer necessary for their connection. Therefore, to the operator, a representation of the real environment (with the possibility of changing object location, shape, etc.) is made to suit his or her ergonomic requirements. Moreover, this representation handles what is necessary to be adapted to operator skill and dexterity, free from any constraint or transparency compromise inherent to bilateral (real or virtual) robot control. Following the previously cited considerations, it is easy to see that in any intermediate functional representation, the first object to be eliminated will be the *picture* of the remote system moving the real operational tool. There are at least three advantages of such an approach compared to the use of a classical VE-based representation. First, the master station components (VE plus haptic device and sensory feedback) are adapted to direct task performance by the operator without any interference due to robot control. Second, the slave station is adapted to the task achievement by the robot without any direct interference due to the operator. Finally, bilateral data exchange is performed according to clever cross-modality-based task transformations. This concept was revisited and extended in Berman, Friedman, and Flash (2005) to associate task affordance to objects that are telemanipulated.

At the Institute of Robotics Research (IRF), Germany, a clever VE system was developed for teleoperation. It allows for intuitive control of a multirobot system and several other automation devices with a standardized, easy-to-understand user interface. The general aim of the development was to create a universal frame for projective virtual reality in which it is possible to *project* tasks, carried out by the user in the virtual world, into the real world with the help of robots or other automation components. In a projective virtual world, the user can act as in the real world. Therefore, this kind of human–machine interface reaches the previously not achieved standard of intuitive operability, even for complex automation devices. The less the user has to know about the real automation device, the better the design of the human–machine interface.

In the projective virtual reality approach, with the help of robots, changes made in the virtual world are *projected* in the real world. Thus, this approach builds a bridge between VE and robotic automation technology by providing techniques for the connection of these two fields (Freund & Rossmann, 1999, see Figure 43.10).

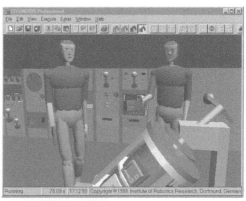
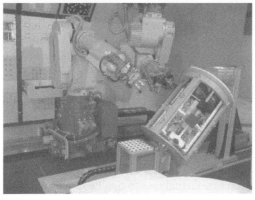

FIGURE 43.10   Projective virtual reality concept for space and complex task telecontrol. Top left: the operator interface (an immersion-based interfacing), top right: the operator virtual projection into virtual avatars, and bottom: the actual robots mapping the virtual avatars' actions into actual ones within the remote location. (Courtesy of Freund and Rossmann, the Institute of Robotics Research, Dortmund, Germany.)

43.3.6   TELEPRESENCE AND BRAIN–COMPUTER INTERFACE PHYSICAL EMBODIMENT

One of the current human-centered architecture technologies is telepresence or telesymbiosis (Vertut & Coiffet, 1985). When coupled to other ways of reproducing natural human actions at remote places, telepresence provides users with the possibility of creating, with different degrees of realism, the notion of presence (see Chertoff & Schatz, 2014, Chapter 34). Telepresence has been defined by Sheridan (1992, p. 6) as a human–machine system in which the human operator receives "sufficient information about the teleoperator and the task environment, displayed in a sufficiently natural way, that the operator feels physically present at the remote site." Very similar to immersion-based VEs, telepresence strives to achieve an actual feeling of presence at a remote real location. The end goals of both telepresence and immersion-based VE are fundamentally the same—a human interface that allows a user to take advantage of natural human abilities when interacting with an environment other than one's direct surroundings. In the case of teleoperation in real environments, this is achieved by projecting the operator's skill and dexterity while reflecting sensory feedback so realistically that the operator feels present at the remote site. Indeed, telepresence borrows a lot from teleoperation technology because a remote physical device is needed to act on the remote environment while feeding back sensory information to the operator. The experience of being fully present at a real-world location remote from one's own physical location is obviously afforded by a high degree of transparency and realism. Someone experiencing transparent telepresence would therefore be able to behave, and receive stimuli, as though at the remote site. As for teleoperation, for any telepresence system, there are three essential components: (1) the operator site technology, (2) the communication link, and (3) the remote site technology. Therefore, video (see Badcock et al., 2014, Chapter 3), audio (see Vorlander & Shinn-Cunningham, 2014, Chapter 4), and haptic (see Dindar et al., 2014, Chapter 5) display systems, such as head-mounted displays (HMDs), autostereoscopic display screens, stereo headphones, and gloves or other devices equipped with touch feedback, may all be used by the operator. Control also needs to be exercised by the human operator, and devices such as head and body tracking, joysticks, master hands and arms in the form of gloves and exoskeleton structures, and other application-specific controllers are used. Contrary to teleoperation systems, telepresence VE techniques could be seen as being more concerned with interfacing technology and not interfacing strategy. If realistic and fully transparent feedback is the top consideration, illusion-based artifacts may be useless.

Tachi's team started studies on tele-existence, (Tachi, 1998). Tele-existence is an advanced type of telepresence system that enables an operator to perform remote manipulation tasks dexterously with the feeling that he or she exists in the remote environment where the robot is working. He or she can *tele-exist* in the VE, which a computer generates, and be able to operate the real environment through the virtual space. One of the distinguishing characteristics of the system is the use of VE to allow operator self-projection. Two self-projection characteristics can be defined. The first one, defined by the authors, concerns the drawing of an operator's legs, body, arms, hands, and so on from which the operator feels as if he or she is self-projected onto the virtual human in the VE. The operator *mapped* to the virtual robot can freely move around in the building, though the real robot can move only in restricted areas (see Figure 43.11).

The second self-projection function that might be understood from using a VE tele-existence is the possibility to allow the operator, in an actual environment telepresence frame, to switch from a realistic representation to a virtual one so that he or she can plan strategy to exercise telepresence tasks, thanks to a better understanding of the real environment from one's virtual representation. In this case, the operator is split into two operators: the one who can explore (*what-if* strategies) the remote location without an actual, but rather a virtual, physical action ported on it (obviously a virtual representation is needed) and the one who can actually act on the real environment (a virtual representation is not necessarily needed in this case). Recent advances in virtual reality technology make it possible to substantially improve this tele-existence system from its original version to the so-called TELESAR V having advanced multimodal sensing/feedback technologies embedded within a more dexterous telemanipulator (Fernando et al., 2012) used as a human surrogate.

Beyond telepresence, *physical embodiment* is a very recent concept, but it differs from telepresence in that it challenges the feeling of ownership of a different body (the robot), which the user *wears* and adopts as own body through which actions and perception are then achieved. Recently, Gergondet et al. (2012) were able to steer a humanoid robot as a human surrogate using brain–computer interfaces (BCI) solely. The purpose of the research is to provide solutions by which a human user feels entirely embodied in a physical humanoid robot. Good illustrative examples are the movies *Avatar* and *Surrogates*. The idea is to control the humanoid by thought alone (see Figure 43.12). In order to do so, robot embedded vision is fed back to user's screen (ideally worn). The video flow is then segmented online, and learned recognized objects appearing within the field of view will be superposed under their 3D models, which will flicker at different frequencies. Steady-state visually evoked potential pattern indicates which object user's intention is focused on. When determined, the object task affordance will trigger a robotic action that shall be executed as closely as possible (in terms of motion) to user's preferences. Complex tasks are not achieved using this approach. In another recent work, VEs are used to train the classifier to identify brain activity patterns related to basic motion task. Cohen et al. (2012) investigated using fMRI as a BCI to steer a humanoid robot. Here, a VE is used in several training stages to direct a virtual avatar by thinking of motion of his left or right hand and legs. The scanner measures

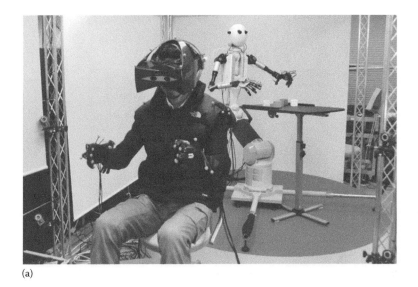

(a)

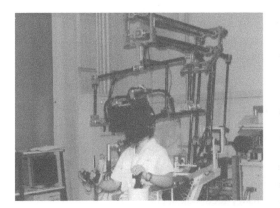

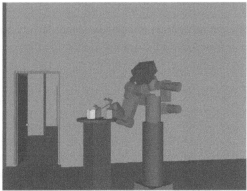

(b)

FIGURE 43.11 **(See color insert for Figure 43.11a.)** (a) TELESAR (tele-existence slave anthropomorphic robot) system. (b) From left to right: The master station, VE tele-existence, and real-environment tele-existence and the TELESAR V. (Courtesy of Professor Tachi, Keio University, Tokyo, Japan.)

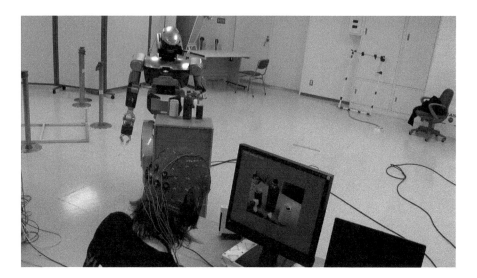

FIGURE 43.12   Physical embodiment experiment: using BCI in AR to steer a humanoid robot HRP-4.

the changes in blood flow to the primary motor cortex to identify characteristic pattern of each thought movement that is then sent via Internet to the humanoid robot. VE can then be used in BCI-based control schemes as intermediary for control or user training for brain activity identification prior to real execution.

43.3.7   ENHANCING SENSORY FEEDBACK TO THE HUMAN OPERATOR

Exact sensory feedbacks in teleoperation are inherently impossible with existing technologies. As mentioned earlier, both controls and sensing information are conveyed through interfaces having latencies, which create synthetic stimuli more or less equivalent to those that operators may feel in direct interactions. These stimuli are usually incomplete or/and distorted. Thus, using VE as a means to compensate for lacking and distorted sensory feedback seems to be natural. That is to say, replacing, superimposing, or combining artificial sensory feedback to the real one is what most systems implement to achieve effective operations.

The most common enhancements are related to the visual channel. It deals with time-delay issues, with cluttered environments (with a lot of occlusions) or degraded visual conditions (with poor lightning or foggy environments) or simply to address nonvisual information (haptic information displayed graphically such as force/torque vectors).

In the presence of time delay, VEs were either used as an overlay to actual feedback (predictors' case) or used as a whole virtual intermediary to allow local feedback with respect to operator sensory bandwidths (for instance, teleprogramming). Using VE as an overlay to real feedback is through augmented or mixed realities. These techniques are also used to improve corrupted or lacking visual sensory feedback. For instance, when vision feedback is defective (undersea boisterousness or cloudiness, smoky working area, or any poor-visibility environment), an augmented or VE display will help the operator in driving the remote robot. The effectiveness of this kind of sensory feedback support has been proven in experiments by Oyama, Tsunemoto, Tachi, and Inoue (1993) and Mallem, Chavand, and Colle (1992). The virtual fixtures introduced by Rosenberg are also powerful tools to improve sensory feedback. Virtual fixtures are essentially sensory overlays to actual feedback allowing operators to increase performances in terms of safety and accuracy. Moreover, fixtures can be invisible if the operator gains no benefits from visual cues, they can be viewed as a synthetic solid virtual object if rich visual cues are useful for the task, and they can even be turned into a transparent glassy solid if visual cues are important and the operator does not want an occluded workspace. Virtual fixtures can be conceived as selective visual filters to block particular distraction, enhance contrast, provide depth cues, and even magnify a part of the workspace. Indeed, one could imagine the huge number of possible combinations that could be achieved by overlapping real sensory feedback in order to improve actual sensory feedback.

Sensory feedback enhancement can also stand out through different sensory substitutions. For instance, the combination of visual fixtures, artificial sound, and tactile stimuli could substitute for a lack of actual sensory feedback from the remote robot location.

In addition, using a whole virtual representation in the case of teleprogramming could be seen as a kind of sensory improvement because it maintains force feedback allowing more intuitive task achievement (even if this is within a VE). As far as a virtual representation is adopted, the number of sensory feedback modalities could be increased and combined in different ways, thanks to various I/O interface technologies. For instance, the ROTEX experiment lacks force feedback, though a VE was used to perform space robot teleoperation. Instead, a SpaceBall was used for the virtual/real robot telesensor programming, and virtual interaction forces were monitored and substituted into visual cues to be displayed to the operator on a screen. Hence, force feedback is not always necessary if adequate substitution over the operator interface is conceived to tackle this shortcoming (Lécuyer, Coquillart, Kheddar, Richard, & Coiffet, 2000).

43.3.8 Improving Safety and Human Factors

In most teleoperator schemes using VE as an additional intermediary, operator safety could be addressed in two ways:

1. The first is preventive and seems to imply the possibility to affix safety functionality integrated within the VE by means of computer programming.
2. The second way deals with the possibility to fit the VE within human sensory capabilities so that teleoperation tasks could be achieved, thanks to less risky, user-friendly, I/O devices derived from VE technology (3D mouse, space balls, sensing gloves, desktop force feedback devices, etc.).

Obviously, both ways could be combined, and additional features may be added to the remote site, namely, by exploiting robot autonomy and enhancing its perceptual issues.

There might be various strategies to act on the VE by adding functions to improve operator safety. One of them is straightforward; the operator would have the opportunity to simulate a teleoperation task before actually performing it. This is a kind of a *what-if* strategy. Another more transparent strategy is to prevent eventual collisions by means of clever active collision avoidance in the VE, as has been proposed for the previously cited TAO 2000 system (see Section 43.3.2 and Figure 43.9). Extending this last principle to more general tasks, the VE can be fashioned to include functionally a kind of *intelligent* filter. The latter may guide or prevent situations that may compromise the operator's safety (see Figure 43.13).

In general, for direct bilateral master–slave coupling, the safety risk is the aggregate likelihood of master, slave, and communication media damage (or functionality loss). Since an intermediate VE is used, (1) the antagonistic well-known transparency–stability problem is subsequently shifted to a local human–VE transparency problem without compromising any of the slave stability; thus, (2) one can state that safety risk is reduced of master dysfunction or a VE crash. The first could be diminished by using more user-friendly interfaces. Subsequently, additional mapping functions are added to derive telerobotic commands (as is the case of the hidden robot architecture). In this case, the VE must offer rich and astute sensory feedback modalities, however, potentially restricting experiential telepresence, because this could be counterproductive. Indeed, in a telerobotic system, a certain detachment may be desirable to keep the operator from becoming totally immersed in a manual control phase such that he might resist returning to higher-level monitoring (this prohibits

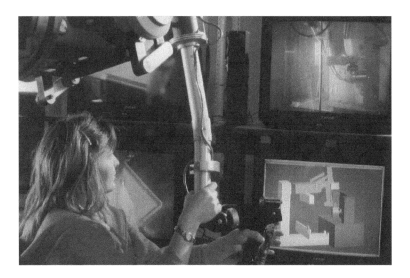

FIGURE 43.13   Improving both visual sensory feedback and safety for the operator: The VE representation enhances visual cues and acts as a filter using an active collision avoidance algorithm developed for teleoperation purposes. (Courtesy of Fournier, TAO 2000 Project, the French Nuclear Center, Saclay, France.)

the use of HMDs in many actual teleoperator systems). As far as a VE crash is concerned, this problem connects software engineering (i.e., one must foresee recovery procedures to restart teleoperation from the crash state) with debugging facilities (e.g., frequent VE state saving). While safety and sensory feedback improvement receives special attention, no generic performance taxonomy exists for classical teleoperator evaluation (though a VE usability taxonomy does exist, see Gabbard, 2014, Chapter 28). Yet one can state that in essence, VE-assisted teleoperation is a move toward user-friendly, more refined control, rich multimodal feedback, but it is still difficult to give a qualitative and a quantitative measurement to these improvements. Concepts for teleoperation can be derived from human factor evaluation obtained from more general application results (Stanney, Mourant, & Kennedy, 1998).

43.4   VEs AS A POWERFUL TOOL FOR SPECIAL PURPOSE TELEOPERATORS

There are many special teleoperation applications among which VE plays a considerable role, either because without VE techniques, these applications would not be feasible or because VE brings a considerable improvement and/or contribution.

43.4.1   TELESURGERY

Improvement in the precision and reliability of robotic systems has won the trust of physicians and medical personnel. Indeed, applications of robotics in the medical field relate to surgery, rehabilitation, and general services including laboratory and prosthetics and orthotics. Likewise, VE technology is a part of many medical applications. This section focuses on VE-based teleoperation aspects as applied to medicine and more specifically to telesurgery. Since the essence of surgery is precision and motion coordination of surgeons, it is not surprising that robotics—the technology of controlled motion—is investigated and widely used in operating wings. Because of the special safety needed, surgical robots are more *assistive* than *active*. This means that in terms of robot control, the surgeon may always be in the control loop.

In minimally invasive surgery (MIS; e.g., laparoscopy, thoracoscopy, arthroscopy), robots serve as telemanipulators used to guide microinstruments. An operation is performed with instruments and viewing equipment inserted into the body through small incisions made by the surgeon, in contrast to open surgery, which uses large incisions. Like arthroscopes used in orthopedics, MIS requires endoscopes. Endoscopic surgery has many common points with teleoperation, because the surgical environment could be seen as *remote*, with sensing and manipulation projected via the endoscope and other long instruments. Nevertheless, endoscopes resemble mechanically coupled master–slave systems, that is, action/feedback is directly mapped from the surgeon to the patient's organs. The well-known limitations of master–slave systems are hampering surgeon's abilities (Tendick, Jennings, Tharp, & Stark, 1993). In this mechanical *bilateral control* scheme, VE techniques enhance tactile sensory capabilities (Guiatni, Riboulet, Duriez, Kheddar, & Cotin, 2013) or augment visual feedback. Eventual use of robots as an intermediary for handling surgical tools is called telesurgery. Therefore, telesurgery is based on existing solutions from computerized teleoperation, and VE applications likely contribute in many ways to fulfill reduced dexterity, workspace, and sensory input and feedback. Also, because tools are manipulated through medical robotic systems, the patient could actually be at a remote location from the surgeon (Marescaux & Rubino, 2003). The other advantage gained from telesurgery is the use of a robot in a semiautonomous mode. Indeed, robots can move with a very low speed and can find less constrained paths, thanks to their own haptic sensory or haptic sensory information (see Dindar et al., 2014, Chapter 5) fed back to the surgeon. This implies less healthy tissue damage for patients, resulting in shorter recovery time and reducing surgeon stress.

Teleoperation technology has also been investigated in the development of emerging microsurgery systems. An instance is the robot-assisted microsurgery (RAMS) system, developed at JPL/NASA (see Figure 43.14, Charles et al., 1997). The RAMS system is a six-degree-of-freedom master–slave telemanipulator with

1. Different control schemes including direct telemanipulation, which includes task-frame referenced haptic feedback and shared automated control of robot trajectories
2. Facilities such as physical scale of state-of-the-art microsurgical procedures
3. Features to enhance manual positioning (e.g., procedures such as manual positioning and tracking in the face of myoclonic jerk and tremor that limit most surgeons' fine-motion skills)

Another example demonstrating the VE contribution in telemicrosurgery is the prototype system developed by Hunter et al. (1993), for eye surgery applications. This system includes two force-reflecting interfaces (a shaft shaped like microsurgical scalpel) to control the left and right limbs of the slave microsurgical robot. A stereo camera system is used to relay visual feedback, on a worn

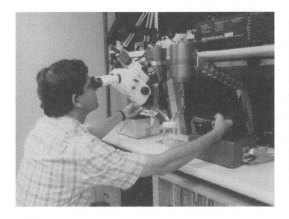

FIGURE 43.14 Master–slave microtelesurgery system. (Courtesy of the RAMS Project at the Jet Propulsion Laboratory, Pasadena, CA.)

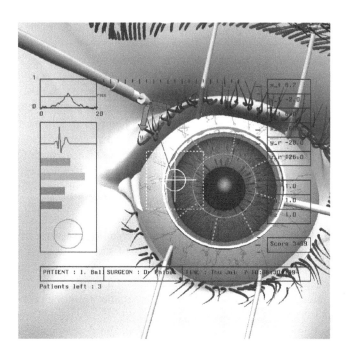

FIGURE 43.15   Screen snapshot of a VE eye representation overlaid with additional information being used for online telesurgery as well as for training and simulation. (Courtesy of Auckland University, Auckland, New Zealand.)

helmet, to the surgeon. The camera position is controlled through the surgeon's head movements, which lead to interactive changes of the camera point of view (see Figure 43.15).

There are many applications where surgical assistance robots can be considered as VE/computer-guided robots through picture processing and augmented reality. The latter technique contributes to surgeon's operation strategy improvement. Enhancement of surgical environments with image overlay has been proposed for online use in neurosurgery (stereotactic brain surgery), orthopedics (like hip replacement), microsurgery, obstetrics, plastic surgery, and other specialties. In these applications, the physician can view medical images or computer-generated graphics overlaid on and registered with the patient. For example, in neurosurgery, a rendering of a brain tumor can be displayed inside the patient's head during surgery, providing localization and guidance to a surgeon. Three-dimensional image overlay capabilities can be used in place of, or to complement, telerobotic systems.

There are also many investigations to use telepresence and/or teleoperation techniques allowing a surgeon to operate at a location remote from his real physical location. But because of ethical and safety reasons, even if telerobots were used clinically, human assistants at the remote site would certainly be needed.

43.4.2   TELEOPERATION AT MICRO- AND NANOSCALE

There is an important demand to reveal the nature of much smaller worlds. Micro and nano (abridged μn) system technology, including μn-robotics, is becoming a challenging area of research because of its potential applications. Interrelated applications concern industry (μn: assembly, sensors, actuators, and mechanics, miniaturization); information technology (disk storage with high density, memory, semiconductor, and integrated circuits); biotechnology, biomedical, and genetic sciences (genes, biological particles, and DNA manipulation; repairing or understanding mechanisms, cell handling, noninvasive eye and plastic surgery); and chemistry and materials (fabrication of μn-structures and man-made materials, study of related quantum effect devices).

Consideration of μn-specific problems, in addition to task application, tools, and interconnection technologies' specific requirements, leads obviously to many flexible μn-manipulation concepts: purely manual teleoperation, automated, and robotics by means of flexible and cooperating μn-robots. Up-to-date, scanning tunneling microscopy (STM), scanning electron microscopy (SEM), and atomic force microscopy (AFM) seem to be the common tools for scanning and manipulating at the μn-scale. Each of the aforementioned microscopes has a range of applications linked to the remote environment nature and the kind of desired μn-tasks. Each has specific limitations while being used in an actual μn-manipulation. For instance, biological samples, such as cells, cannot be visualized using an SEM because their electrical properties can change during envisaged μn-manipulations.

Ideal performance requirements are such that a human operator manipulates in the normal-size world μn-parts and performs tasks (such as cutting, grasping, transportation, assembly, scratching, digging, and stretching), which have a direct similar mapping at the μn-world. Indeed, construction of the μn-manufacturing world is dependent on solutions adhering to the following constraints:

- The working environment must be perceivable by the operator, and information in the processing scene must be transmitted accurately to the operator. As far as μn-tasks are concerned, tools must be arranged in the observing area (colocality), bilateral magnification must be stable and fully transparent, and direct and natural perception is required with 3D movements, dynamic images, sound, and haptics.
- The grasping, release, or assembly of μn-items in the μn-world need perspicacious procedures, which are completely different from those commonly used in the macroworld. Gripping with forceps may not be adequate in many cases. Indeed, vacuum-assisted gripping by electric power or fluid and release under vibration, the use of μn-forces and adhesion, the use of plunger mechanism, and other astute techniques are mostly used. Moreover, the success and performance of μn-grasping, release, or assembly functionality together with μn-manipulation are, in many cases, dependent on operator skill.
- The operational remote environment is actually a hostile environment to humans because μn-world components' behavior is very complex to understand and to manipulate. Moreover, since humans operate based on macroworld model physics, tasks cannot be easily executed by the operator. In the μn-world, mass and inertial forces are negligible, μn-interaction forces are nonlinear, and resultant forces from van der Waals, capillary, electrostatic, pull-off, rubbing phenomena, and even radiation forces of light exceed the gravity.
- Task taxonomies might include operational sequences (e.g., positioning, assembling, grip, release, adjust, fix-in-place, push, pull) and processing steps (e.g., cutting, soldering, gluing, drilling, twisting, bending).

To link the macroworld to the μn, an interface that can match the two physical worlds and compensate for human operator inaccuracy is indispensable.

From the vision feedback point of view, VE can bring a considerable contribution. Indeed, the field of view or scanning is in most cases restricted to a small area, and the distance between μn-objects and the lens or the probe is very small. Moreover, on the one hand, scanning (which is on the order of seconds to minutes in some cases) does not allow online imaging. On the other hand, since the same single probe is used for both functions, scanning and μn-manipulating, these cannot physically be achieved in parallel whatever the scanning speed. Finally, μn-operation is executed, in general, within the field of view. For these reasons, a 3D VE topology can be built and displayed to the user. A 3D virtual μn-world can be a global view of the real μn-world or restricted to the actual working area, according to the operator needs. Static or intuitively manipulated multiple views are then allowed in real time. As well, remote features may be augmented with multiple contrasting colors to present data in a comprehensive and easily interpreted manner.

The lack of direct 3D visual feedback from the μn-world on the one hand and the fragility of the manipulated μn-objects on the other hand make force feedback an essential—even unavoidable—component of the macro-/μn-world interface. Indeed, it is primordial to understand well the condition of the probe during operation. An excessive force applied on a μn-object may lead to a nonnegligible degree of probe or object deformation, destroy the μn-object, or make the μn-object flip away. As well known, μn-interaction is not reproducible enough to automate μn-procedures. Hence, teleoperation control mode seems to be attractive; however, the traditional bilateral control through master–slave coupling has to face many severe problems: understanding of μn-dynamics together with reliable modeling, the effect of various nonlinear forces is completely different from the macroworld (attractive, repulsive, and adhesion forces), bilateral stable, transparent, robust scale mapping, and others. Thus, it is not always possible to adapt easily conventional bilateral methods; moreover, monitoring small forces of 1 μN to 1 nN range needs very accurate sensors. Thus, solutions using a VE-based intuitive interface, which hides the details of performing complex tasks using SPM in combination with 3D topography, seem to be an attractive solution. As for assisted teleoperation at the macroworld, the interface would include virtual tools or virtual effective probes (Finch et al., 1995) together with a 3D representation used as a functional intermediary to map operator's actions to the μn-world (rough to fine methodology) and vice versa. Then, the degree of abstraction of the commands would be determined by the capabilities of the control system. Hereafter, some developed prototype systems are mentioned.

At the University of Tokyo, Sato's team developed one of the most advanced μn-teleoperators (Sato, 1996). As a haptic interface, the system utilizes a touch sensor screen and specially designed, pencil-shaped master manipulator to enable sensitive bilateral μn-teleoperation, thanks to its lightweight and small inertia during movements. A μ-handling result (see Figure 43.16 at right, courtesy of Professor Sato, the University of Tokyo) is a micropyramid constructed with a μ-handling robot inside an SEM. This system is currently being enhanced by a more user-friendly interface based on bilateral behavior media for μn-teleoperation.

Other instances are from the Nanomanipulator Research group at the University of North Carolina (Finch et al., 1995). Their system is using advanced VE techniques as an interface enhanced with virtual tools (grabbing, scaling, flying, etc.), virtual measuring fixtures (marked mesh, etc.), standard VCR functions (replay, save, etc.), and virtual modification tools. The main applications are concerned with biological studies and virus manipulations (see Figure 43.17); the PHANToM is a force feedback desktop mechanism for VE applications (http://www.sensable.com).

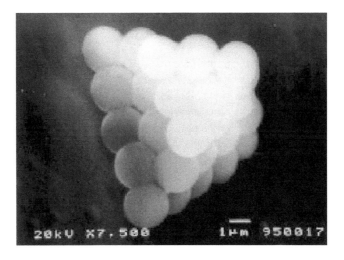

FIGURE 43.16   Micropyramid constructed with a μ-handling robot.

FIGURE 43.17 Different *n*-manipulators configurations (from left to right): PHANToM-based desktop configuration, a workbench together with the PHANToM force feedback stylus, and the Argonne robotic master arm (right). (Courtesy of the Nanomanipulator Research group, University of North Carolina, Wilmington, NC.)

FIGURE 43.18 Molecular docking using advanced VE interfaces. (Courtesy of the IDock research group, University of North Carolina, Wilmington, NC.)

At the same university, studies are being conducted in the frame of molecular docking. Indeed, today's computer-aided molecular design software (Sibyll™, MolMol™, etc.) lacks user-friendly interfaces, and algorithmic search of the configuration space is extremely costly. In return for advanced VE techniques and an *intelligence augmentation*, human–machine system (see Figure 43.18), chemists, and drug designers would gain in efficiency and time (Ouh-young, Pique, Hughes, Srinivasan, & Brooks, 1988).

Another tele-nano-manipulation system is being developed in the Hashimoto laboratory at the University of Tokyo. This system is using an AFM for both scanning and manipulating at the nanoscale. As a user-friendly interface, the tele-nano-manipulator is dotted with a one-degree-of-freedom haptic interface and a 3D topology builder for nanomanipulations (Sitti & Hashimoto, 1999). The controller is actually a bilateral mode based on impedance shaping to allow the operator to feel the forces from the nanoworld. Different control strategies utilizing the haptic interface have been tested, and results are thoroughly described in Sitti and Hashimoto (1999; see Figure 43.19.)

43.4.3 MOBILE ROBOT TELEOPERATION

Since long ago, mobile robotics had the connotation of autonomy and artificial intelligence and was not concerned with teleoperation. Early mobile robots were designed with a *teleoperation* mode so that a human operator could unilaterally control them during transportation and setup phases.

(a)

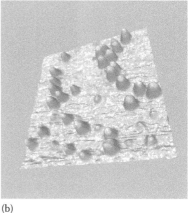

(b)

FIGURE 43.19 Tele-nano-manipulation using one-degree-of-freedom haptic interface (a) and scanned synthetic converted AFM picture (b). (Courtesy of Sitti and Hashimoto, the University of Tokyo, Tokyo, Japan.)

Difficulties in achieving completely autonomous behavior in terms of autodecision and of a mobile robot unilateral teleoperation mode turn out not to be simple when the robots are conceived as a complex structure (subsea and flying robots), with specific locomotion (such as many legs) or propelling mechanisms, or when refined feedback using the operator haptic channel is attempted. Pioneering work in the latter was done by Clement, Fournier, Gravez, and Morillon (1988). Teleoperation mode can be considered as an intrinsic characteristic of advanced autonomous mobile robots. We emphasize this proposition with a simple example—the conceiver of the well-known Honda Humanoid robot reported: "… this robot is completely autonomous" and with insistence declared, "But, it can be teleoperated as well." We stress that this specificity tends to draw closer a classical teleoperation field researchers utterance: "… this robot is teleoperated, but, it can be autonomous as well." As far as mobile robot teleoperation is concerned, VE techniques can bring similar advantages as for classical robots. From the related literature, many VE-based mobile robot architectures have been proposed. Keywords that seem to be a standard in the way VE techniques are used include off-line mission planning, highly interactive environment for abstract programming at various levels, simulation and prediction of complex missions with a virtual teleoperated vehicle, robot computer-assisted design, and training, safety, and user-friendly control through augmented reality (superposition of virtual and real environments may reveal discrepancies or sensor defects).

In the work of Komoriya and Tani (1990), an actual robot together with a VE set around it constitutes a simulation system considered to lie between the full computer simulation and the full actual system. This hybrid system provides an efficient testbed for developed control laws and planning algorithms, without any damage risks for the robot.

Another VE technique for mobile robot and humanoid teleoperation has been proposed by Kheddar, Neo, Tadakuma, and Yokoi (2007), which allows prediction and bilateral robot teleoperation control with multimodal feedback, including haptics. In this proposed scheme, operator commands will involve desired robot reorientation, speed, and acceleration; embedded tool(s) control; global inertia redistribution; and other factors. It has been suggested that pertinent parameters that the operator may feel include stability margins, inertial forces, applied torque, joint limit margins, contact points and forces, related obstacle distance, and robot attitude. All these parameters are computed locally (i.e., based on the virtual model of the robot and its environment) and fed back to the operator locally, thanks to a suitable sensory substitution. The appropriate sensory substitution is chosen to stimulate the operator sensory channel so that subsequent telerobotic actions are reflex generated. In this case, the VE constitutes a filter to intuitively control safely the robot since actual robot controls are sent only if some selected parameters fit within allowed margins. Nevertheless, bilateral control is permitted only if the communication time delay between the master and robot is small enough (less than 1 s). In the case of larger time delays, the VE is functionally used similarly to arm teleoperation (i.e., as a predictive or a teleprogramming interface). This is thoroughly explained in the following cases.

The exceptional instance highlighting the use of VE techniques in mobile robot teleoperation is indubitably the Mars Pathfinder mission (a NASA Discovery exploring Mars planet). Indeed, a VE-technology-based supervision and control workstation was designed to remotely command the *Sojourner* rover, which landed on Mars by July 4, 1997. A graphical user interface supplied numerous available rover commands (i.e., macrooperations with respective various parameters, such as Calibrate Heading with Sun, Capture Images, Drive, GoTo WayPoint). Off-line, a 3D terrain model is processed from both previous mission accumulated stereo images obtained from an embedded IMR camera (a stereoscopic imager) and the partial VE already constructed from previous missions. Hence, using a simple SpaceBall as an I/O device together with a stereo rendering of the virtual working environment and a virtual model of the Sojourner rover, the operator designs plans and actions to be achieved by means of a set of graphic programming metaphors. Subsequently, when the simulation results are considered to be satisfactory, a control code is generated and transmitted, by means of the Deep Space Network (up to 6–11 s time delay, thus prohibiting manual teleoperation), to the actual Sojourner rover on Mars (see Figure 43.20).

FIGURE 43.20 Clockwise: elements of the VE interface before assembly, main rover control workstation program window, the actual Sojourner rover at work on Mars, and driver's VE interface with waypoints shown. (Courtesy of the Jet Propulsion Laboratory, NASA, Pasadena, CA, http://robotics.jpl.nasa.gov/tasks/scirover/homepage.html.)

More enhanced interfaces and telerobotic technology are being developed for future unmanned exploration of the *red planet* to provide scientists with a telepresence interface for real-time interaction with, and interpretation of, the returned geophysical data (Backes, Kam, & Tharp, 1999).

In the frame of subsea robotics, Paul's team at the GRASP Laboratory applied the teleprogramming concept they developed in the frame of a subsea remotely operated vehicle (ROV) equipped with a robotic arm (Sayers, 1999). An experiment has been conducted with the Deep Submergence Laboratory of the Woods Hole Oceanographic Institution (WHOI laboratory). Bandwidth of some Kbits/s and round-trip communication delays of 7 s are typical of the subsea acoustic transmission mean. The operator disposes of a VE remote site wherein control or supervision tasks can be achieved. Remote slave actions are governed by those resulting from the master station. Eventual discrepancies are resolved from a continuous comparison between the virtual (simulated) and the real internal and external embedded sensors. Subsequent mismatches and diagnosis are deduced by the operator that is responsible for the suitable recovery procedures to be taken (see Figure 43.21).

At the French Institute of Research and Exploitation of the Sea (IFREMER), the control of robots for intervention, reliability of these systems, and dynamic stabilization of the machines at low speed are the aspects under investigation. A VE technique named Virtual Environment for Subsea Vehicles (VESUVE) has been developed for 3D subsea scene simulation and visualization.

VESUVE offers a general framework for creation, visualization, real-time animation, and interactive manipulation of any kind of virtual subsea scene involving deep sea vehicles. The main objective of VESUVE is to study, implement, and evaluate VE technologies for subsea vehicles. Its potential areas of application are engineering of subsea vehicles (visual simulation for integration, test, and validation of the system; real-time visualization of CAD data; visual simulation for pilot training or scientist familiarization); operation of subsea vehicles (3D piloting aid for ROV or towed vehicles, 3D navigation aid, telepresence, visual simulation for mission rehearsal or debriefing, teleprogramming for autonomous or semiautonomous vehicles); and subsea robotics R&D (virtual lab for experimental subsea robots in research fields like control, mission programming, navigation, fault detection diagnosis and recovery, and real-time presentation of simulation results; see Figure 43.22).

Obviously, there are still many other remote mobile robotic teleoperation systems using telepresence and VE interfaces, such as the VEVI (a distributed Virtual Environment Vehicle Interface) developed at the NASA Ames Research Center's Intelligent Mechanisms Group (Piguet et al., 1995;

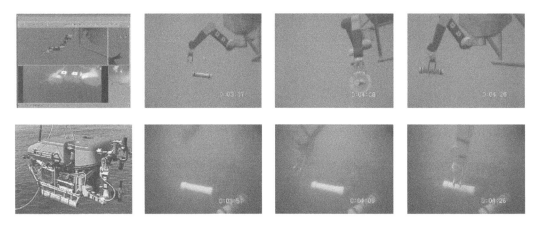

FIGURE 43.21 Teleprogramming as a mean for subsea ROV teleoperation. Upper row pictures design of the master station; lower row pictures show snapshots of the video feedback from the slave (JASON ROV) environment. (Courtesy of Sayers, the GRASP laboratory, University of Pennsylvania, Philadelphia, PA and the WHOI Laboratory, Woods Hole, MA.)

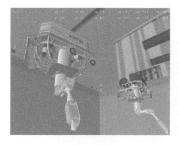  

FIGURE 43.22 VISUVE case studies: cooperative ROVs and control simulation, offshore planning, and the *Titanic* mission simulation, an aid for the actual remote control of the VICTOR 6000 ROV. (Courtesy of the IFREMER, La Tremblade, France, http://www.ifrem.fr/ditidsiw3_uk/engins/vesuve.html.)

Hine et al., 1995), adopted to the Dante mission at Alaska, Antarctica, and other locations, and the advanced computer project to assess Chernobyl damage. We do hope that these few instances being used on actual hazardous environment missions have shown the benefits that VE brings to advanced teleoperator schemes in the special mobile robot teleoperation context.

43.4.4 WEB-BASED TELEOPERATION

Some of the most challenging purposes for today's multimedia systems are to include haptic data exchange (see Dindar et al., 2014, Chapter 5) and to allow physical teleworking, through the network. Currently, various remote robots can be controlled through the Internet network by means of any Internet browser.

Control interfaces are using Java applets with different available command buttons allowing any user connected to the Internet to experience actual remote control of robots (see Figure 43.23)—arms, mobile cameras, telescopes, and even toy trains! Tasks are generally concerned with object assembly, exploration (of museums), among others. The user may see the result of his actions via continuous video feedback (in general, the refresh rate is too slow due to network traffic).

Many laboratory experiments involving teleprogramming through the Internet and diverse other network protocols have been conducted. Many of these, such as the ROTEX experiment and subsea teleoperation, have already been mentioned. An advanced ATM network has been used by the Fukuda team in Japan for a multimedia telesurgery context. The satellite network has also been used in a telesurgery experiment between the Bejczy team at the Jet Propulsion Laboratory and Rovetta team at the Politecnico di Milano, Italy (references are quoted in Kheddar, Tzafestas, Coiffet, Kotoku, & Tanie, 1998).

Other experiments have been conducted between the Laboratoire de Robotique de Paris and the Mechanical Engineering Laboratory involving the control of parallel multirobots through a single operator and VE interface based on the hidden robot concept (Kheddar et al., 1998). With different robots controlled in parallel, a common intermediary functional representation is imperative and enhances the interest of the proposed concept.

As shown in Figure 43.24, the teleworking experiment consisted of a four-piece puzzle assembly within a fence on a table. The real remote assembly operation was to be performed by slave robots (one situated in Japan and three in France; for the first experiment, one in France for the experiment described herein). The operator performs the virtual puzzle assembly using his own hand, skill, and naturalness. Visual and haptic feedback is local and concerns only the graphic representation of the remote task features without any remote robot. Operator–VE interaction parameters are sent to another workstation in order to derive robot actions (graphically represented for software check and result visualization) and do not involve direct operator's action/perception. Video feedback was kept for safety and error recovery purposes.

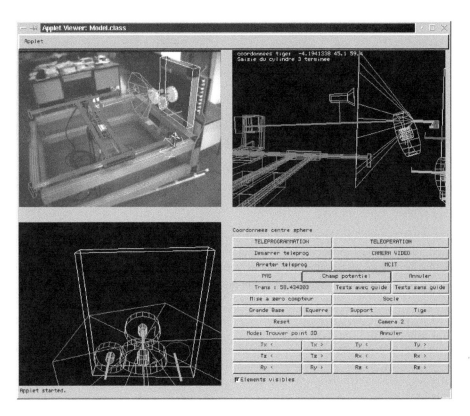

FIGURE 43.23   An Internet web-based teleoperation interface using augmented reality and online assistance by means of virtual fixtures. (Courtesy of Otmane, the Complex Systems Laboratory, Evry University, Evry, France.)

Web-based teleoperation has a direct impact on industry; possible applications include telemaintenance, telemonitoring, and telesupervision. As an instance, a European project for maintenance system based on telepresence for remote operators was launched in September 1996 (Bisson & Conan, 1998). The resulting technology enabled remote users to train themselves to deal with maintenance tasks by connecting to a *virtual showroom* where they could learn maintenance procedures through computer-augmented video-based telepresence and augmented reality techniques.

The teleoperation environment consisted of (1) an extended prototype of the equipment to be maintained, that is, the target equipment, customized to allow remote control and extensive diagnostic, and (2) auxiliary devices such as manipulators and vision systems, used both for equipment servicing and for enhancing user interoperability. Marking techniques allow registration of a virtual model of the video designed equipment provided by the remote camera. The virtual model is annotated with links to a multimedia database containing specific information (e.g., operating and maintenance instructions, functional data describing regular or faulty conditions). Speech recognition is used as well to allow spoken commands. This system is dedicated to the training of complex maintenance and installation scenarios for industrial companies that cannot afford on-site complex training equipment (see Figure 43.25).

NASA's Jet Propulsion Laboratory and Ames Research Center are developing a next-generation ground system called RoCS for (Rover Control Station) for use on planetary rover missions. The RoCS architecture includes a module called WITS for web interface for telescience (Backes et al., 1999). The latter serves two purposes: (1) an interface for scientists to use from their home institutions to view and download data and generate commands for the remote rover and (2) to make the interface available to the public domain (http://robotics.jpl.nasa.gov/tasks/scirover). Thus, any

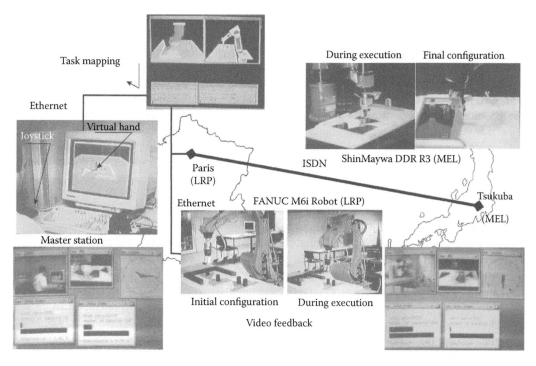

FIGURE 43.24 Parallel long-distance multirobot teleoperation. (Courtesy of the Labaratoire de Robotique de Paris and the robotics department of the Mechanical Engineering Laboratory, Paris, France.)

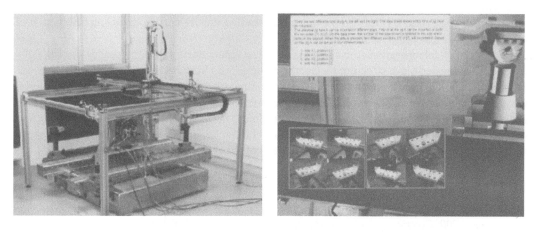

FIGURE 43.25 Teleoperation environment. An XYZ table plus yaw pitch camera to explore the maintenance equipment (left). Output (AR picture) from the MAESTRO, answer to query: "What are the different jig types?" (Courtesy of Bisson, Thomson CSF, France.)

person connected to the server could downlink images and generate rover waypoints, science targets, and commands, which, however, cannot be sent to the actual or true rover.

43.5 PROGNOSIS AND CONCLUSION

Undoubtedly, there is still much to say concerning the prosperous manner in which robotics can benefit from VE techniques (Burdea, 1999; see Burdea & Coiffet in Nof, 1999). It is impressive to note that a cleverly conceived yet *simple* VE intermediary representation contributes to solving the time-delay problem, offers ingenious metaphors for both operator assistance and robot autonomy sharing problems, enhances operator sensory feedback through multiple sensory modality admixtures, enhances operator safety, offers a huge possible combination of strategies for remote control and data feedback, shifts the well-known antagonistic transparency/stability problem into an operator/VE transparency one without compromising the slave stability, offers the possibility to enhance—in terms of pure control theory—remote robot controllers, allows new human-centered teleoperation schemes, permits the production of advanced user-friendly teleoperation interfaces, and makes possible the remote control of actual complex systems, such as mobile robots, μn-robots, and surgery robots. Subsequently, VE techniques seem to provide the ultimate magic solution solving all the hard teleoperation problems at a glance. Nevertheless, this considerable opening gives rise to additional new problems.

One problem deals with VE modeling techniques (see Kessler, 2014, Chapter 11). There is a grand demand for robust software, which can reproduce the VE with high fidelity and sufficiently rich realism to allow both real-time user-friendly operator interaction and easy implementation of necessary metaphors, such as virtual guides, that allow operator assistance and enhance local sensory feedback and robot autonomy sharing in an intuitive way (i.e., without setting the operator at a determined control layer). Nevertheless, in some cases, namely, when remotely controlled systems are complex, there is a need for a virtual representation that cleverly keeps only functional aspects of the complexity of the remote environment, thereby unburdening the operator and facilitating control or supervision. This aspect is only possible if a clever bilateral mapping exists between real environments and their virtual functional representations.

The other hard problem with using VE technology is the error detection and recovery inherent to VE—real-environment discrepancies. This problem has received little attention from the teleoperation research community. It could be tackled from two ways: a high-fidelity error detection strategy (eventually involving the operator) that matches what is expected to be done in the VE and what is actually done in the real environment and a strategy for recovery based on robot autonomy and local sensory robot perception. If this problem can be solved, a merely partial virtual representation of the real environment could be acceptable.

In conclusion, teleoperation efficiency was almost stagnant during several decades since the discovery of the master–slave structure by Goertz in the 1950s (Vertut & Coiffet, 1985), and that was in spite of technical improvements and the use of computers. It seems that VE brings total renewal through the development of telepresence or tele-existence concepts. Now, the desired efficiency could be a reality, not only allowing a clear improvement in traditional applications such as nuclear or space activities but opening numerous new fields of interest. Thanks to VE technology, teleoperation/telepresence could soon represent a major percentage of robotics applications.

REFERENCES

Anderson, R. J. (1994). Teleoperation with virtual force feedback. *Lecture notes in control and information sciences 200: Experimental robotics III* (pp. 366–375). New York, NY: Springer-Verlag.

Anderson, R. J., & Spong, M. (1988). Bilateral control of teleoperators with time delay. *Proceedings of the IEEE Conference on Decision and Control* (pp. 167–173). Austin, TX.

Backes, P. G., Kam, S. T., & Tharp, G. K. (1999). The Web interface for telescience. *Presence: Teleoperators and Virtual Environments, 8*(5), 531–539.

Backes, P. G., Peters, S. F., Phan, L., & Tso, K. S. (1998). Task lines and motion guides. *Presence: Teleoperators and Virtual Environments, 7*(5), 494–502. Retrieved from http://robotics.jpl.nasa.gov/people/backes/homepage.html

Badcock, D. R., Palmisano, S., & May, J. G. (2014). Vision and virtual environments. In K. S. Hale, & K. M. Stanney (Eds.), *Handbook of virtual environments: Design, implementation, and applications* (2nd ed., pp. 39–86). Boca Raton, FL: Taylor & Francis Group, Inc.

Bejczy, A. K., Kim, W. S., & Venema, S. (1990). The Phantom robot: Predictive displays for teleoperation with time delay. *Proceedings of the IEEE International Conference on Robotics and Automation* (pp. 546–551). Cincinnati, OH, May 1990.

Berman, S., Friedman, J., and Flash, T. (2005). Object-action abstraction for teleoperation. *Proceedings of the IEEE International Conference on Systems, Man and Cybernetics* (pp. 2631–2636), 3, Waikoloa, HI.

Bisson, P., & Conan, V. (1998). The MAESTRO Project: An augmented reality environment for telemaintenance. *Proceedings of the 6 éme Journées de travail du GT réalité virtuelle.* Issy-les-mounlinaux, France.

Burdea, G. C. (1999). Invited review: The synergy between virtual reality and robotics. *IEEE Transactions on Robotics and Automation, 15*(3), 411–422. Retrieved from http://www.caip.rutgers.edu

Burdea, G. C., & Coiffet, P. H. (1999). Virtual reality and robotics. In S. Y. Noy (Ed.), *Handbook of industrial robotics* (pp. 325–333). New York, NY: John Wiley & Sons.

Charles, S., Das, H., Ohm, T., Boswell, C., Rodriguez, G., Steele, R., & Istrate, D. (1997). Dexterity-enhanced telerobotic microsurgery. *Proceedings of the International Conference on Advanced Robotics* (pp. 5–10). Monterey, CA. Retrieved from http://robotics.jpl.nasa.gov/tasks/rams/homepage.html.

Chellali, R., & Pham, H. (2011). Frequency modulation based vibrotactile feedback vs. visual feedback in a multimodal interface for 3D pointing tasks in teleoperation. *Proceedings of the IEEE International Conference on Robotics and Biomimetics* (pp. 14–19). Pukhet, Thailand.

Chertoff, D., & Schatz, S. (2014). Beyond presence. In K. S. Hale, & K. M. Stanney (Eds.), *Handbook of virtual environments: Design, implementation, and applications* (2nd ed., pp. 855–870). Boca Raton, FL: Taylor & Francis Group, Inc.

Clement, G., Fournier, R., Gravez, Ph., & Morillon, J. (1988). Computer-aided teleoperation: From arm to vehicle control. *Proceedings of the IEEE International Conference on Robotics and Automation* (pp. 590–592). Philadelphia, PA.

Cohen, O., Druon, S., Lengagne, S., Mendelsohn, A., Malach, R., Kheddar, A., & Friedman, D. (2012). fMRI based robotic embodiment: A pilot study. *Proceedings of the IEEE/RAS-EMBS International Conference on Biomedical Robotics and Biomechatronics (BioRob)* (pp. 314–319). Rome, Italy, June 24–28.

Darken, R. P., & Peterson, B. (2014). Spatial orientation, wayfinding, and representation. In K. S. Hale, & K. M. Stanney (Eds.), *Handbook of virtual environments: Design, implementation, and applications* (2nd ed., pp. 467–492). Boca Raton, FL: Taylor & Francis Group, Inc.

Dindar, N., Tekalp, A. M., & Basdogan, C. (2013). Dynamic haptic interaction with video. In K. S. Hale, & K. M. Stanney (Eds.), *Handbook of virtual environments: Design, implementation, and applications* (2nd ed., pp. 115–130). Boca Raton, FL: Taylor & Francis Group, Inc.

ETS-VII. (1999). Space Robot sessions on ETS-VII experiments and results. *Proceedings of the International Conference on Advanced Robotics* (pp. 243–273, 329–360, 409–436). Tokyo, Japan, TA1-A TP1-A, TP2-A.

Fernando, C. L., Furukawa, M., Kurogi, T., Hirota, K., Kamuro, S., Sato, K., …Tachi, S. (2012). *TELESAR V: TELExistence surrogate anthropomorphic robot, ACM SIGGRAPH 2012.* Los Angeles, CA: Emerging Technologies.

Finch, M., Chi, V. L., Taylor, II, R. M., Falvo, M., Washburn, S., & Superfine, R. (1995). Surface modification tools in a virtual environment interface to a scanning probe microscope. *Proceedings of ACM Symposium on Interactive 3D Graphics* (pp. 13–19). New York, NY. Retrieved from http://www.cs.unc.edu/nano/etc/www/nanopage.html.

Fournier, R., Gravez, Ph., Foncuberta, P., & Volle, C. (1997). MAESTRO hydraulic manipulator and its TAO-2000 control system. *Proceedings of the Seventh ANS Topical Meeting on Robotics and Remote Systems* (pp. 840–847). Augusta, GA.

Fraisse, P., Pierrot, F., & Dauchez, P. (1993). Virtual environment for robot force control. *Proceedings of the IEEE International Conference on Robotics and Automation* (pp. 219–224). Atlanta, GA: IEEE Press.

Freund, E., & Rossmann J. (1999). Projective virtual reality: Bridging the gap between virtual reality and robotics. *IEEE Transactions on Robotics and Automation, 15*(3), 411–422. Retrieved from http://www.irf.de/welc_eng.htm

Funda, J., Lindsay, T. S., & Paul, R. P. (1992). Teleprogramming: Towards delay-invariant remote manipulation. *Presence: Teleoperators and Virtual Environments, 1*(1), 29–44.

Gabbard, J. L. (2014). Usability engineering of virtual environments. In K. S. Hale, & K. M. Stanney (Eds.), *Handbook of virtual environments: Design, implementation, and applications* (2nd ed., pp. 721–748). Boca Raton, FL: Taylor & Francis Group, Inc.

Gergondet, P., Kheddar, A., Hintermuller, C., Guger, C., & Slater, M. (2012). *Multitask humanoid control with a brain-computer interface: User experiment with HRP-2.* 13th International Symposium on Experimental Robotics. Québec City, Quebec, Canada.

Guiatni, M., Riboulet, V., Duriez, C., Kheddar, A., & Cotin, S. (2013). A combined force and thermal feedback interface for minimally-invasive procedure simulation, *IEEE/ASME Transactions on Mechatronics*, *18*(3), 1170–1181.

Hirche, S., & Buss, M. (2012). Human-oriented control for haptic teleoperation. *Proceedings of the IEEE*, *100*(3), 623–647.

Hunter, I. W., Doukoglou, T. D., Lafontaine, S. R., Charrette, P. G., Jones, L. A., Sagar, M. A., … Hunter, P. J. (1993). A teleoperated microsurgical robot and associated virtual environment for eye surgery. *Presence: Teleoperators and Virtual Environments*, *2*(4), 265–280. Retrieved from http://biorobotics.mit.edu and http://www.esc.auckland.ac.nz/Groups/Bioengineering

IEEE TRA. (1993). *IEEE Transactions on Robotics and Automation* [Special issue], *9*(5).

Joly, L. D., & Andriot, C. (1995). Imposing motion constraints to a force reflecting telerobot through real-time simulation of a virtual mechanism. *Proceedings of the IEEE International Conference on Robotics and Automation* (pp. 357–362). Nagoya, Japan.

Kessler, K. D. (2013). Virtual environment models. In K. S. Hale, & K. M. Stanney (Eds.), *Handbook of virtual environments: Design, implementation, and applications* (2nd ed., pp. 259–284). Boca Raton, FL: Taylor & Francis Group, Inc.

Kheddar, A. (2001). Teleoperation based on the hidden robot concept. *IEEE Transactions on Systems Man and Cybernetics, Part A*, *31*(1), 1–13.

Kheddar, A., Neo, E., Tadakuma, R., & Yokoi, K. (2007). Enhanced teleoperation through virtual reality techniques. In Ferre, M., Buss, M., Aracil, R., Melchiorri, C., & Balaguer, C. (Eds.), *Advances in telerobotics, volume 31 of Springer tracts in advanced robotics* (pp. 139–159). Springer-Verlag.

Kheddar, A., Tzafestas, K. S., Coiffet, Ph., Kotoku, T., & Tanie, K. (1998). Multi-robot teleoperation using direct human hand actions. *Advanced Robotics*, *11*(8), 799–825.

Khelifa, B., Chellali, R., & Hauptman, T. (2012). ViRAT: An advanced multi-robots platform. *Proceedings of the IEEE ICIEA* (pp. 551–556). Singapore.

Kim, W. S. (1996). Virtual reality calibration and preview/predictive display for telerobotics. *Presence: Teleoperators and Virtual Environments*, *5*(2), 173–190. Retrieved from http://robotics.jpl.nasa.gov/people/kim/csv/homepage.html

Komoriya, K., & Tani, K. (1990). Utilization of virtual environment system for autonomous control of mobile robots. *Proceedings of the IEEE International Workshop on Intelligent Motion Control* (pp. 439–444). Istanbul, Turkey: IEEE Press.

Kosuge, K., Itoh, T., Fukuda, T., & Otsuka, M. (1995). Tele-manipulation system based on task-oriented virtual tool. *Proceedings of the IEEE International Conference on Robotics and Automation* (pp. 351–356). Nagoya, Japan: IEEE Press.

Kosuge, K., Murayama, H., & Takeo, T. (1996). *Bilateral feedback control manipulation system with transmission time delay*. Proceedings of the International Conference on Intelligent Robots and Systems, Vol. 3 (pp. 1380–1385). Osaka, Japan.

Kotoku, T. (1992). A predictive display with force feedback and its application to remote manipulation system with transmission time delay. *Proceedings of the IEEE/RSJ International Conference on Intelligent Robots and Systems* (pp. 239–246). Raleigh, NC: IEEE Press. Retrieved from http://www.mel.go.jp

Lécuyer, A., Coquillart, S., Kheddar, A., Richard, P., & Coiffet, P. (2000). *Pseudo-haptic feedback: Can isometric input devices simulate force feedback?* Proceedings of the IEEE International Conference on Virtual Reality. New Brunswick, NJ, pp. 83–90.

Mallem, M., Chavand, F., & Colle, E. (1992). Computer-assisted visual perception in teleoperated robotics. *Robotica*, *10*, 93–103. Retrieved from http://www.univ-evry.fr/labos/cemif/index.html

Marescaux, J., & Rubino, F. (2003). Telesurgery, telementoring, virtual surgery, and telerobotics. *Current Urology Reports*, *4*(2), 109–113.

Niemeyer, G., & Slotine, J.-J. (1991). Stable adaptive teleoperation. *The IEEE Journal of Oceanic Engineering*, *16*(1), 152–162.

Noyes, M. V., & Sheridan, T. B. (1984). A novel predictor for telemanipulation through time delay. *Proceedings of the Annual Conference on Manual Control*. Moffelt Field, CA: NASA Ames Research Center.

Ouh-young, M., Pique, M., Hughes, J., Srinivasan, N., & Brooks Jr., F. P. (1988). Using a manipulator for force display in molecular docking. *Proceedings of the IEEE International Conference on Robotics and Automation* (pp. 1824–1829). Philadelphia: IEEE Press.

Oyama, E., Tsunemoto, N., Tachi, S., & Inoue, T. (1993). Experimental study on remote manipulation using virtual reality. *Presence: Teleoperators and Virtual Environments*, *2*(2), 112–124.

Piguet, L., Nygren, E., & Kilne, A. (1995). *VEVI: A virtual environment teleoperations interface for planetary exploration*. Proceedings of the SAE 25th International Conference on Environmental Systems (preprints). San Diego, CA. Retrieved from http://img.arc.nasa.gov/VEVI

Rosenberg, L. B. (1992). *The use of virtual fixtures as perceptual overlays to enhance operator performance in remote environments* (Tech. Rep. No. AL-TR-1992-XXX). Wright Patterson Air Force Base, OH: U.S.A.F Armstrong Laboratory.

Sato, M. (1996). *Micro/nano manipulation world*. Proceedings of the IEEE/RS J International Conference on Intelligent Robotics and Systems. Osaka, Japan, pp. 834–841. Retrieved from http://www.lssl.rcast.u-tokyo.ac.jp:80/~tomo

Sayers, C. (1999). *Remote control robotics* [online]. New York: Springer Verlag Ed. Retrieved from http://www.cis.upenn.edu/~sayers

Sheridan, T. (1992). *Telerobotics: Automation and human supervisory control*. Cambridge, MA: MIT Press.

Sitti, M., & Hashimoto, H. (1999). Teleoperated nano-scale object manipulation. In R. Jablonski, M. Turkowski, & R. Szewczyk, R. (Eds). *Recent advances on mechatronics*. New York, NY: Springer-Verlag.

Small, D. E., & McDonald, M. J. (1997). *Graphical programming of telerobotic tasks*. Proceedings of the Seventh Topical Meeting on Robotics and Remote System. Augusta, GA, pp. 3–7. Retrieved from http://www.sandia.gov/LabNews/LN10–25–96/robot.html

Stanney, K. M., Mourant, R. R., & Kennedy, R. S. (1998). Human factors issues in virtual environments: A review of the literature. *Presence: Teleoperators and Virtual Environments*, 7(4), 327–351.

Struges, R. H. (1996). A review of teleoperator safety. *International Journal of Robotics and Automation*, 9(4), 175–187.

Tachi, S. (1998). Real-time remote robotics—Towards networked telexistence. *IEEE Computer Graphics and Applications*, 18(6), 6–9. Retrieved from http://www.star.t.u-tokyo.ac.jp

Tendick, F., Jennings, R. W., Tharp, G., & Stark, L. (1993). Sensing and manipulation problems in endoscopic surgery: Experiment, analysis, and observation. *Presence: Teleoperators and Virtual Environments*, 2(1), 66–81. Retrieved from http://robotics.eecs.berkeley.edu/mcenk/medical

Vertut, J., & Coiffet, Ph. (1985). *Teleoperation and robotics: Applications and technology*. Englewood Cliffs, NJ: Prentice-Hall.

Vorländer, M., & Shinn-Cunningham, B. (2014). Virtual auditory displays. In K. S. Hale, & K. M. Stanney (Eds.), *Handbook of virtual environments: Design, implementation, and applications* (2nd ed., pp. 87–114). Boca Raton, FL: Taylor & Francis Group, Inc.

44 Evolving Human–Robot Communication through VE-Based Research and Development

Stephanie Lackey, Daniel Barber, Lauren Reinerman-Jones,
Eric Ortiz, and Joseph R. Fanfarelli

CONTENTS

44.1 INTRODUCTION

Advances in robotics, automation, artificial intelligence, and perceptual capabilities facilitate moving beyond traditional human–robot interaction (HRI) (e.g., teleoperation) to human–robot communication (HRC) (e.g., speech, gesture, tactile). HRC provides the underlying capabilities required for collaboration between human and robot team members, such as natural interfaces, bidirectional transactions, and flexible modality applications. Evolving the role of robots, from tools to teammates, is emerging as a critical area of interest within the dismounted infantry domain and represents one of the greatest returns on investment opportunities for future fighting forces.

This chapter describes the typical implementation of robots and unmanned systems within the operational environment, and then turns to the state of the art in HRI highlighting the associated challenges. Next, foundational human–human communication theories impacting the emerging science of HRC are presented, followed by descriptions of research and development efforts using virtual environments (VEs) to explore and fill the gaps inhibiting human–robot collaboration and teaming. Finally, the chapter concludes with descriptions of two important VE application areas: interface design and training.

44.2 ROBOTS IN COMPLEX MILITARY DOMAINS

The emergence of robots in manufacturing and healthcare testifies to their proficiency in task areas requiring repetition and precision. One of the most notable domains in which robots play a daily role is the United States Armed Forces (USAF). In 2001, the U.S. Congress mandated one-third of the operational/ground combat vehicles be replaced by unmanned or remotely controlled robots (U.S. Congress, 2001). As a result, the USAF, consistently at the forefront of robot research, development, and implementation, widened its unmanned systems focus to include advancing HRI.

Military teams typically employ robots for search and rescue tasks, ordnance disposal, mine clearing, and remote targeting. For the purpose of this discussion, it is important to recognize that the term *robot* encompasses unmanned and remotely controlled systems, including Unmanned Aerial Systems (UAS) and Unmanned Ground Systems (UGS).

Traditionally, military operators manually control (e.g., teleoperate) unmanned assets and/or observe autonomous task performance (Reinerman-Jones, Taylor, Sprouse, Barber, & Hudson, 2011). One example, the Predator UAS, performs reconnaissance and surveillance tasks, and, when required, deploys hellfire missiles to hostile targets (see Figure 44.1).

Another example, the Threat and Local Observation Notice (TALON) UGS, provides remote disposal and detonation of improvised explosive devices (IEDs). Prior to the implementation of robots for explosive ordnance disposal (EOD), and in their absence, specialists approached an explosive on foot to manually disarm it. Disarming an IED in this manner opens the specialists to several dangers, including those posed by the explosive being disarmed, secondary explosives, and hostile snipers. Having saved an estimated 788 lives at the time of this writing (RSJPO, 2012), the TALON removes humans from direct interaction with such threats as it can be operated from a safer location—in cover and away from the IED (Figure 44.2).

44.3 STATE OF THE ART IN HRI

Teleoperation refers to the explicit operation of a robot through a human–computer interface (Chen, 2010) and serves as the contemporary standard for HRI. This often takes the form of a human operator providing continuous control of a robot asset via manual manipulation of a joystick while monitoring its progress through one (e.g., TALON) or more (e.g., Predator) visual displays.

As teleoperated tools, robots provide many benefits by executing tasks that would otherwise need to be performed by humans (Ogreten, Lackey, & Nicholson, 2010). Such tasks are frequently performed in dangerous environments as described earlier or with high-risk materials (Barber, 2012).

(a)　　　　　　　　　　　　　　　　　(b)

FIGURE 44.1 (a) Predator UAS in flight (From Lopez, E., Untitled photograph of a predator UAS, Retrieved October 2012, from http://www.defenseimagery.mil, 2012); (b) teleoperation of a UAS. (From Allen, D.R., Untitled photograph of UAV teleoperation, Retrieved September 2012, from http://www.defenseimagery.mil, 2011.)

(a)   (b)

FIGURE 44.2   (a) TALON UGS manipulating a disarmed IED. (From Showalter, J., Untitled photograph of a TALON UGS, Retrieved September 2012, from http://www.defenseimagery.mil, 2008); (b) teleoperation of a UGS. (From Takada, K.G., Untitled photograph of TALON teleoperation, Retrieved September 2012, from http://www.defenseimagery.mil, 2006.).

Once a potential threat (e.g., IED) is identified during an EOD mission, the operator deploys a robot (e.g., TALON) and navigates it to the area of interest by watching a video feed from the robot, controlling its movements via joystick manipulation. Upon the robot's arrival to the area of interest, the operator visually inspects the IED based upon the video provided by the robot. The operator then remotely controls the robot to perform fine-grain movements to detonate or deactivate the explosive components. When the IED is neutralized, the operator remotely *drives* the robot to its home station.

As shown in the example given earlier, robots are extremely effective at entering environments and completing tasks that present safety concerns for humans. The ability to repair or replace damaged robots mitigates the negative repercussions of an unsuccessful event. An IED detonation in the vicinity of a TALON will likely mean the damage of electrical or otherwise nonorganic components. Such a detonation within the vicinity of a human would be catastrophic. Therefore, one of the greatest benefits of robots is improved soldier safety.

Robots are also used when confined boundaries physically restrict a human from entering an area of interest. This may be the case in a search and rescue task within a collapsed building. In this situation, the operator may maneuver a smaller robot (such as the PackBot in Figure 44.3) through narrow openings in order to locate survivors. Through teleoperation, the operator can manipulate a joystick to make very small, precise, and controlled movements in order to effectively navigate a constricted environment. Locating survivors in this manner allows rescue personnel to increase effectiveness while maintaining safe working conditions (Casper & Murphy, 2003).

44.3.1 Challenges with Current Robot Use

The aforementioned IED example illustrates limitations of the current HRI paradigm and presents some key challenges to evolving the role of robots from tools to teammates. Teleoperation limits the flexibility and type of interaction between a human and a robot, thus reducing the efficiency and effectiveness of HRI. During the execution phase of a mission, unidirectional communication occurs as the operator gives a specific command to the robot such as *turn left*. Barring malfunction, the robot obeys, executing the command. This marks the end of that interaction. At this point, another command may be issued such as *send charge* to detonate an explosive. If no command is provided, the robot remains static. In some instances, bidirectional communication may occur, but it is typically limited to the relay of basic sensor information such as video feed or GPS data, which may be used by the operator to analyze the environment before giving another command.

FIGURE 44.3 Compact form of the PackBot makes it a suitable replacement for human exploration of confined spaces. (From Contreras, M.A., Untitled photograph of a PackBot, Retrieved October 2012, from http://www.defenseimagery.mil, 2010.)

From the operator's perspective, this paradigm negatively impacts situation awareness and multitasking—two key elements of military operations. Teleoperation requires continuous control of a robot and thus occupies a majority of the operator's attentional resources. For example, since an operator's visual resources must be allocated to the robot's video display, the ability to survey the surrounding environment is compromised, subsequently reducing situation awareness (Barber, 2012). Additionally, continuous joystick manipulation reduces the operator's ability to efficiently use a weapon in response to real or potential threats in his/her immediate surroundings. In a hazardous environment, these issues diminish threat detection capabilities and require additional personnel to maintain security.

The unidirectional nature of teleoperation is beneficial for performing low-level, fine-grain commands. However, this type of master–slave relationship inhibits multitasking. Research shows that teleoperation negatively impacts workload and reduces the operator's ability to successfully execute secondary tasks (Chen & Joyner, 2009; Cosenzo, Chen, Reinerman-Jones, Barnes, & Nicholson, 2010; Reinerman-Jones et al., 2011). Several studies report a reduction in performance, an increase in error rate (Chen & Joyner, 2009; Pettitt, Redden, Pacis, & Carstens, 2010), and an increase in communication complexity due to additional processes (Cosenzo, Capstick, Pomranky, Dungrami, & Johnson, 2009) during teleoperation.

Remote operation of a robot via mechanical controls and a static display lacks intuitive cues for human operators since nowhere in nature is a similar phenomenon present. Teleoperation is a learned behavior requiring comprehensive training to acquire proficiency. Collaborative, bidirectional interaction based upon natural, intuitive, and flexible communication modalities represents a critical gap. Addressing this gap advances the state of the art in robot operation in three areas critical to shifting the HRI paradigm from using robots as tools to integrating robots as team members: natural and intuitive interfaces, automation applications, and shared mental models (SMM).

44.4 EVOLVING HRC

The strategic application of multimodal HRC capabilities mitigates the risks inherent to the current HRI paradigm. A brief summary of how humans communicate with one another provides context for the solutions proposed.

44.4.1 HUMAN–HUMAN INTERACTION

Human–human interaction offers a foundation for identifying the modalities that can or should be explored for mixed-initiative teams. Moving beyond the overarching concepts of interaction and focusing on communication leads to the investigation of verbal (e.g., speaking) and nonverbal (e.g., gesturing) communication methods. Previous work in multimodal communication (MMC) presents a compelling argument for the inclusion of a variety of communication methods on the continuum of explicit and implicit communication.

Explicit communication refers to any form of communication that is intentionally conveyed through sound, sight, touch, etc. (Lackey, Barber, Reinerman, Badler, & Hudson, 2011). Examples include audibly speaking to give a command, pointing to an object to show location, or a tap on the shoulder to get a person's attention (see Figure 44.4).

FIGURE 44.4 Examples of explicit communication modalities: (a) speaking (sound) (From Berner, A., Untitled photograph of soldiers speaking, Retrieved October 2012, from http://www.defenseimagery.mil, 2012); (b) pointing to a location (sight) (From Wade, S., Untitled photograph of a soldier pointing, Retrieved October 2012, from http://www.defenseimagery.mil, 2009); (c) *come here* gesture (sight) (From Katzenberger, M., Untitled photograph of a combat medic signaling, Retrieved October 2012, from http://www.defenseimagery.mil, 2012); (d) tapping shoulder to signal readiness (touch). (From Hineline, T., Untitled photograph of a marine signalling readiness, Retrieved October 2012, from http://www.defenseimagery.mil, 2010.)

TABLE 44.1
Communication Modalities and Examples

Modality	Delivery	Explicit	Implicit
Auditory	Speech, sounds	Language	Tone, rate, pitch
Visual	Gesture, posture, facial expression, gait, social distance	Intentional pointing, hand signals	Unintentional body language, intensity, eye contact, talking with hands, emotions
Tactile	Belt, vest	Intentional touching, patterns	Pressure, patterns, shakiness

Source: Adapted from Lackey, S.J. et al., Defining next generation multi-modal communication in human robot interaction, *Proceedings of the Human Factors and Ergonomics Society 55th Annual Meeting*, Las Vegas, NV, pp. 461–464, 2011.

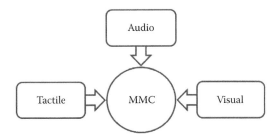

FIGURE 44.5   MMC incorporates multiple channels.

Alternatively, implicit communication denotes unintentional conveyance of contextual information, including emotion (Lackey et al., 2011). It serves as a complementary layer to explicit communication through observable behaviors such as rate of speech, tone of voice, and inadvertent body posture or gestures. Table 44.1 summarizes explicit and implicit communication examples that occur during human–human interaction that represent potential opportunities to improve HRC.

Incorporating and integrating elements from explicit and/or implicit communication methods signifies MMC (Lackey et al., 2011). MMC blends audio, visual, and tactile communication modes to simultaneously and/or redundantly transmit information (see Figure 44.5).

Figure 44.6 illustrates a typical human–human MMC scenario. A person audibly speaks the words *move forward* (explicit audio) and consciously motions to a location (explicit visual). The phrase is inadvertently spoken very loudly and quickly, expressing urgency (implicit auditory), and accompanied by rapid arm gestures (implicit visual). MMC represents the natural ways humans communicate with one another to facilitate SMMs, collaborate, and ultimately enhance team performance.

Human–human MMC methods serve as the foundation for HRC. Identifying the most beneficial application of individual and blended modalities is of primary importance to HRC and directly supported by the existing body of work in human–human communication. Similarly, human–human lexicons drive the development of HRC vocabulary sets and syntax guidelines (Barber, 2012).

44.4.2   Innovative Solutions

Implementation of natural interaction methods between humans and robots represents the logical progression and evolution of HRC. Traditional robot control methods resulted from initial requirements for developing remotely operated vehicles, planes, and boats.

Enabling capabilities, such as automation, continue to advance. In fact, automated advanced route-planning minimizes distances travelled and maximizes fuel resources for multiple unmanned systems more effectively than human planning (Chen, Barnes, & Qu, 2010; McLain & Beard, 2000). This and other capabilities offer an opportunity to offload tasks previously requiring human

FIGURE 44.6   Example of MMC. An army sergeant gives the order to move forward, simultaneously using vocal commands and arm gestures. (From Thomas, T., Untitled photograph of army sergeant giving orders, Retrieved October 2012, from http://www.defenseimagery.mil, 2012.)

intervention to a robot. Furthermore, automation provides a foundation to implement mixed-initiative teaming and allocate tasking more effectively among human and robot team members (Hearst, Allen, Guinn, & Horvitz, 1999).

Successful teams communicate well, often due to their SMMs of the given situation and surrounding environment. This is especially critical within dismounted operations where team members may or may not be colocated and must coordinate quickly. Timely information sharing in dynamic dismounted environments directly impacts the development of SMMs. Mental models serve as *internal representations* of the environment, elements within the environment (e.g., people and machines), and the relationships between elements (Phillips, Ososky, Grove, & Jentsch, 2011).

Facilitating the development of SMMs between human and robot teammates provides improved mutual understanding of roles, responsibilities, and tasks. From this understanding, a team member may begin to anticipate the needs of his or her teammates (Schuster et al., 2011). If Teammate One observes that Teammate Two's workload exceeds their ability to attend to critical secondary tasks (e.g., identify potential threats), then Teammate One may assume part of the secondary task responsibilities. By anticipating a need to offload task responsibility, Teammate One takes on that role to optimize team performance and improve safety. This represents fundamental logic in terms of human–human interaction but is less transparent when a robot performs such reasoning and back-up behavior.

By leveraging human–human communication models in addition to emerging advancements in artificial intelligence and perceptual sensors, opportunities to evolve HRC arise and allow for true human–robot collaboration. Critical issues moving forward include identifying interface designs that optimize human–robot collaboration and developing tools for familiarization and maintenance training. The following section explores the evolution of HRC and leads to a discussion of the role VEs play in addressing the challenges presented.

44.5   VE-BASED RESEARCH CONTRIBUTING TO THE ADVANCEMENT OF HRC

Investigating the challenges presented earlier is imperative to transitioning robots from tools to teammates. However, the current state of the art in robot autonomy is limited in terms of robot perception and intelligence required for accurately classifying and reasoning about physical and human terrain. For example, today's robots are capable of identifying obstacles impeding navigation but cannot reliably distinguish between walls or buildings let alone a human teammate versus a civilian. VEs offer cost-effective research tools capable of compensating for such limitations by simulating next-generation robot perception and intelligence and providing opportunities to test novel HRC systems, interfaces, and training methods prior to full-scale development and deployment. The following section describes methods of communicating with future ground robots using modalities (auditory, visual, tactile) people use in human-to-human communication, the state of the art in technologies to implement these modalities, and examples of how VEs are used to support testing and development of HRC interfaces. Next, potential applications of VE technologies to introductory and maintenance training are presented. The proposed HRC VE training examples described offer a glimpse into the possibilities.

44.5.1 DESIGNING INTERFACES

Historically, auditory communication is the most studied modality for MMC. Efforts to develop Natural Language Processing (NLP) date back to the first demonstration of a working system in the 1950s (Hutchins, 2004). NLP aims to eliminate the need for a specific vocabulary or syntax to enable interactions analogous to human-to-human conversation. Existing NLP commercial systems are capable of converting speech into text with some variability in word classification accuracy, susceptible to noisy environments, and possess limited semantic understanding (Anusuya & Katti, 2009; Duan & Cruz, 2011; Elmer-DeWitt, 2012). Some systems exhibit limited interpretation and understanding of the complete human vocabulary. However, research into discrete commands using a limited lexicon for HRC reduces the complexity of the problem while providing a solution more suited for military operations (Pettitt, Redden, & Carstens, 2009; Tellex et al., 2011). Speech recognition provides a natural method for humans to communicate to robots; however, voice synthesis from robot-to-human requires additional considerations. For example, the type of voice a robot uses (male, female, and synthesized) impacts approach distances to the robot and perception of intelligence (Powers & Kiesler, 2006; Walters, Syrdal, Koay, Dautenhahn, & te Boekhorst, 2008). Research investigating nonvoice audio feedback for robot-to-human communication is more limited than voice-based research and has primarily focused on mounted applications. Research into audio cueing using specialized three-dimensional (3D) spatial audio displays as localization feedback determined participants are able to effectively convey information in moving vehicles with high levels of vibration and respond more quickly when audio accompanies visual signals (Haas, 2007).

VEs facilitate development of vocabularies for, and creation of speech recognition systems within, robot tasks. To test the results of speech recognition in a robot task, it is inherent that the robot is able to execute commands given predictably. For example, Tellex and Roy (2006) developed a simulated maze and mobile robot commanded by using a speech interface. The simulation provided a robot capable of obstacle avoidance and navigation without issues related to sensor reliability and hardware failure. This VE provided consistent robot performance enabling researchers to focus their investigations on lexicon development and speech interface performance navigation tasks (Tellex & Roy, 2006) prior to building physical platforms. Without this type of consistency, it would be unclear whether task performance was impacted by robot navigation or the speech interface. VEs also support development of synthesized speech systems for HRC. Simulated faces representing humans or robots assist research in understanding the role of visual cues for improving speech synthesis and perception (Morishima, 1993; Nijhold & Hulstijn, 2006). Physical representations of such faces would be costly to produce in terms of equipment and labor, and may not be able to emulate specific visual cues under investigation.

Visual signaling using arm-and-hand gestures is another natural method for communication that humans employ on a daily basis (Wexelblat, 1995). Gestures, posture, and other visual cues are helpful in distinguishing the person (or robot) being addressed when working within a team (Stiefelhagen et al., 2007). Gestures are population dependent (Pettitt et al., 2009), so guidelines addressing each population needs are important. For example, the U.S. Army Field Manual for Visual Signals defines arm-and-hand signals for combat formations and vehicle drivers and/or crews in daytime and nighttime conditions (U.S. Army, 1987). This provides a starting point for developing standardized signals and gesture recognition tools for human-to-robot communication in military operations. Emerging technologies provide new methods of gesture recognition without requiring direct line of site. This capability increases the robustness of visual signals for communication to a robot in environments containing a large number of visual obstructions. Accelerometer-based input devices, using wireless communication, demonstrate this capability through classification of multiple arm-and-hand gestures from the U.S. Army Field Manual for command of a robot vehicle (Varcholik, Barber, & Nicholson, 2008). Electromyography combined with an inertial measurement unit incorporated within a hands-free sleeve is another emerging method of hand-and-arm signal classification with accuracies of 90% (Stoica et al., 2012). Similar to auditory communication, visual signaling from robots to humans for enabling bidirectional communication is still limited. Research to date on robot visual signaling is constrained to anthropomorphic robots in academic environments, not military platforms, using basic manipulators and body movement (Bauer, Wollherr, & Buss, 2008).

Similar to research in speech interfaces, VEs play a role in the development of visual classification systems. Mentioned previously with speech, it is important to ensure malfunctions in robot asset performance do not affect metrics for evaluating an interface. To develop a lexicon of gestures and test an accelerometer-based gesture glove for HRC, the authors of this chapter developed a virtual robot. This virtual robot could travel to different locations within a large geo-typical Middle Eastern terrain and respond to start, stop, and other direct manipulation commands while staying on roads and avoiding simulated people. These capabilities allowed developers to focus on gestures to use to command a robot through a reconnaissance and surveillance task without focusing on requirements to make a real robot find roads and see people. Using this approach, the authors evaluated the performance of the gesture glove independently of robot performance. Transitioning of the gesture glove for use within live environments occurred upon completion of this verification and validation stage using a VE.

A novel method of robot-to-human communication currently under investigation is tactile displays. Tactile displays enable communication via the tactile modality, by manipulating the skin on key body locations, typically around the abdomen (White, 2010). Tactons, or tactile icons, describe combinations of vibro-tactile stimulation using a range of parameters, including frequency, amplitude, body location, duration, and rhythm (Brown, Brewster, & Purchase, 2006). Tactile displays support hands-free navigation tasks due to the egocentric nature of directional tactons when located on the abdomen, reducing navigation time (Elliot, Duistermaat, Redden, & van Erp, 2007). Gilson, Redden, and Elliot (2007) developed tactons, matching visual signals from the U.S. Army Field Manual with participants able to classify with high accuracy and minimal training. Matching tactons to visual signals reduced response time to audio cueing when combined with tactile alerts (Haas, 2007). Tactile displays enable an additional modality for layering audio and visual messages for developing robust MMC given the observed effectiveness and perception accuracy in noisy environments (Merlo et al., 2006).

VEs have provided a key motivation for the development of tactile displays to improve spatial awareness and immersion. In the early stages of tactile belt design, key questions regarding body locations, sensory congruency, cross-modal interaction, perceptual illusions, and attention were still largely unanswered (van Erp, 2000). Studies cataloging vibro-tactile spatial resolution around the torso demonstrated improved navigation within VEs and help answer these questions (van Erp, 2000; van Erp & Werkhoven, 1999). Results from this and related efforts directly influenced the design of the tactile belts used for hands-free navigation and tacton representations (Elliot et al., 2007; Gilson et al., 2007). Furthermore, Varcholik and Merlo (2008) used a VE to test a commander's ability to guide multiple users to a destination point using a tactile belt presenting direction tactons.

VEs directly support integration and development of auditory, visual, and tactile communication modalities. By providing a consistent environment with varying levels of complexity, developers can avoid the restrictions of live robot platforms and focus on the communication content and the underlying technology delivering and interpreting messages. Future interfaces, flexibly combining different modalities, will facilitate seamless integration of robots with manned systems. However, steps are required to prepare users for these interactions within advanced mixed-initiative teams. The next section addresses one of the most important steps to bring robots into everyday life: training.

44.5.2 TRAINING

As a new era of HRC emerges, the importance of preparing users for sweeping technology innovations, such as a robot collaborator, becomes evident. Training will serve as a key risk mitigation strategy during the migration from traditional HRI to mixed-initiative HRC. Virtual training tools offer cost-effective options for identifying and developing (Redden & Elliot, 2010) HRC familiarization and maintenance training solutions. Furthermore, as mixed-initiative team communication methods evolve from single modes (e.g., voice, gesture, and tactile) to blended modes, user training must also evolve. The following examples illustrate the future of VE training implementations.

An ISR mission comprising one soldier and one robot, requiring voice and gesture communication from the soldier and tactile feedback from the robot, could be trained in a variety of ways using VEs. Familiarization training for issuing voice commands and performing arm gestures could be presented using laptop-based simulations and a smartphone. The trainee could learn and practice voice commands using speech recognition software on the laptop while a simulated robot would respond (e.g., visual and/or auditory feedback) based upon the input received. Arm gestures could similarly be trained by using accelerometer-based data from the smartphone when a gesture is performed. Interpreting and understanding tactile messages from a robot could be trained in an analogous way on a low-fidelity laptop simulator.

Team training for a mixed-initiative Fire Team comprising three humans and one robot could be presented in a high-fidelity 360° VE. As part of the overall training, communication skills learned in the laptop VE would be leveraged to accomplish mission objectives in an immersive environment. The human trainees would use voice and gestures to communicate with each other and to the robot. The simulated robot within the VE would acknowledge commands via tactile messages and send video to the smartphones of the human trainees. Such a scenario may also occur in a lower-fidelity, laptop- or PC-based virtual simulator, where a trainee queries a virtual robot for navigational information about a person of interest (POI) via voice communication. The virtual robot would return the requested information via tactile signals (e.g., alerts and acknowledgements) and video stream (e.g., visual information about POI's movements). The trainee would then be required to interpret the signals and navigate an avatar to the corresponding location in the simulation environment.

While familiarization training aids in the acquisition of new abilities, maintenance training reduces performance degradation over time. Next-generation HRC will similarly require maintenance training to facilitate consistent HRC performance. A similar scenario to the one described previously may be modified for this purpose. However, instead of a formal, large-scale VE, use of a smartphone application could provide mobile refresher training. Scenario details could be presented through the phone's visual display and audio channel. Performance could be evaluated by speech and accelerometer-based recognition software. This solution offers greater portability, allowing users to have VE-based maintenance training capabilities, regardless of location.

The inherent flexibility of simulation-based training technologies, specifically VEs, compared to traditional classroom methods offers significant advantages. Emerging research in the field of scenario-generation and automated-adaptation points to increasing opportunities for individualized training (Martin & Hughes, 2010; Martin et al., 2009). Adaptation, when properly implemented, can be beneficial to training. For example, adapting difficulty level to an individual trainee enhances training effectiveness (Choi, Qi, Gordon, & Schweighofer, 2008). The previous training examples can be extended to demonstrate adaptive difficulty in training. Here, the robot's tactile signals may be presented with long breaks in between, at first, giving the trainee plenty of time to process the signals. As the trainee's response time decreases and accuracy increases, the signals may be presented in more rapid succession. If the trainee's response time increases beyond expectation, or accuracy falls below a predetermined threshold, then the signals may be presented more slowly. Thus, the simulation adapts to the trainee's present ability, manipulating signal interarrival time to provide adaptive difficulty. As a trainee progresses from familiarization to maintenance, or from procedural to higher order cognitive skills, VEs become increasingly capable of tailoring specific aspects of a scenario to optimize the trainee's performance.

Existing robot infrastructure and protocols facilitate training in VEs. Communication protocols such as the Joint Architecture for Unmanned Systems (JAUS) can be used equally well in live and simulated environments. Modifications to a live robot's interaction capabilities can be directly ported to a virtual representation of that platform within a VE (Barber et al., 2008), thus providing seamless integration of real robot capabilities within virtual training environments. Previous research efforts demonstrate this capability. One instance is the architecture of the Mixed Initiative eXperimental (MIX) testbed, which is designed to support multiple Department of Defense (DoD) standards, for seamless integration within live and virtual environments. The MIX testbed's Operator Control Unit (OCU) is able to communicate with any platform implementing JAUS. This capability enables developers to test novel designs with virtual entities and later transition to live platforms without any additional costs due to software and hardware changes (Barber et al., 2008). HRC VE training can exploit the same capability by directly transitioning operational robot software upgrades to simulation entity models.

VEs provide immersive experiences well suited to dynamic environments incorporating robots as teammates, such as collaborative soldier–robot field operations. Additionally, such technologies provide an ideal, safe environment for users to experience interacting with a virtual robot, prior to working with a real device. VEs offer the opportunity to expose users to the capabilities and limitations of emerging robot devices, thus managing user expectations and improving user confidence via familiarization and maintenance training. By capitalizing upon VE strengths, important human system integration challenges can be addressed. Any disruptive technology, such as a robot teammate, encounters adoption and acceptance issues, but VEs are primed to be part of the human–robot teaming solution.

44.6 FUTURE OF HRC

The evolution of robots, from tools to team members, is in its early stages. Great strides have been made in terms of establishing a theory for which to conduct research. That foundation comes from human–human interaction theory, particularly the field of communication. The definitions of multimodal, explicit, and implicit communication have been modified to reflect application to HRI. Building on this framework, similar modalities and language rules are now applied to HRC. The result is novel uses of existing displays, adaptive displays, and development of VEs for experimentation and training. Experimentation employing VEs is founded on transformed communication theories and HCI principles. The intersection of the fields of communication, robotics, HCI, and VE training still has many gaps, both identified and yet to be discovered. However, the theories that once existed independently are forever connected for a future of scientific investigation and refinement.

The transformative impact of HRC evolution is equally significant for application. Military operations have been the driving force behind this emerging field but are not the sole beneficiaries. Applications extend to other industries that use, or soon will employ, robots for task completion. Local police, disarming bombs or searching for missing persons, benefit from the current HRC research. Healthcare, particularly in the instance of the robotic surgical system known as *the Da Vinci*, is primed for implementing HRC research. In-home care is moving toward automation that is likely to feature robots. These real-world applications will mandate a high level of human–robot cohesiveness, which will only be attained through continued HRC research and development.

Effective HRC will provide a foundation to bring human–robot teaming to the wider marketplace. Ultimately, the efforts presented will assist in the design and implementation of revolutionary robot capabilities. Introducing such disruptive technologies will also require training. VEs play a vital role in both areas.

REFERENCES

Allen, D. R. (2011). Untitled photograph of UAV teleoperation. Retrieved September 2012, from http://www. defenseimagery.mil

Anusuya, M. A., & Katti, S. K. (2009). Speech recognition by machine, a review. *International Journal of Computer Science and Information Security, IJCSIS, 6*(3), 181–205.

Barber, D. J. (2012). *Investigation of tactile displays for robot to human communication* (Unpublished doctoral dissertation), University of Central Florida, Orlando, FL.

Barber, D. J., Leontyev, S., Sun, B., Davis, L., Nicholson, D., & Chen, J. (2008). The mixed-initiative experimental testbed for collaborative human robot interactions. *Symposium on Collaborative Technologies and Systems* (pp. 483–489). Irvine, CA.

Bauer, A., Wollherr, D., & Buss, M. (2008). Human-robot collaboration: A survey. *International Journal of Humanoid Robotics, 5*(1), 47–66.

Berner, A. (2012). Untitled photograph of soldiers speaking. Retrieved October 2012, from http://www. defenseimagery.mil

Brown, L. M., Brewster, S. A., & Purchase, H. C. (2006). Multidimensional tactons for non-visual information presentation in mobile devices. *Eighth Conference on Human-Computer Interaction With Mobile Devices and Services, 159*, pp. 231–238. Helsinki, Finland.

Casper, J., & Murphy, R. R. (2003). Human-robot interactions during the robot-assisted urban search and rescue response at the World Trade Center. *IEEE Transactions on Systems, Man, and Cybernetics, Part B: Cybernetics, 33*(3), 367–385.

Chen, J. (2010). Robotics operator performance in a multi-tasking environment. In M. Barnes, & F. Jentsch (Eds), *Human-robot interactions in future military operations* (pp. 293–314). Burlington, VT: Ashgate.

Chen, J., Barnes, M. J., & Qu, Z. (2010). RoboLeader: An agent for supervisory control of multiple robots. *Proceedings of the Fifth ACM/IEEE International Conference on Human-Robot Interaction (HRI '10)* (pp. 81–82). Nara, Japan.

Chen, J., & Joyner, C. (2009). Concurrent performance in gunner's and robotic tasks and effects of cueing in a simulated multi-tasking environment. *Proceedings of the Human Factors and Ergonomics Society 52nd Annual Meeting* (pp. 237–241). New York, NY.

Choi, Y., Qi, F., Gordon, J., & Schweighofer, N. (2008). Performance-based adaptive schedules enhance motor learning. *Journal of Motor Behavior, 40*(4), 273–280.

Contreras, M. A. (2010). Untitled photograph of a PackBot. Retrieved October 2012, from http://www. defenseimagery.mil

Cosenzo, K., Capstick, E., Pomranky, R., Dungrami, S., & Johnson, T. (2009). *Soldier machine interface for vehicle formations: Interface design and an approach evaluation and experimentation (ARL-TR-4678)*. Aberdeen, MD: Aberdeen Proving Ground, U.S. Army Research Laboratory.

Cosenzo, K., Chen, J., Reinerman-Jones, L., Barnes, M., & Nicholson, D. (2010). Adaptive automation effects on operator performance during a reconnaissance mission with an unmanned ground vehicle. *Proceeding of the Human Factors and Ergonomics Society Annual Meeting, 54*(25), 2135–2139. San Francisco, CA.

Duan, Y., & Cruz, C. (2011). Formalizing semantic of natural language through conceptualization from existence. *International Journal of Innovation, Management and Technology, 2*(1), 37–42.

Elliot, L. R., Duistermaat, M., Redden, E., & Van Erp, J. (2007). *Multimodal guidance for land navigation*. Aberdeen, MD: Aberdeen Proving Ground, U.S. Army Research Laboratory.

Elmer-DeWitt, P. (2012). Minneapolis Street Test: Google gets a B+, Apple's Siri gets a D. Retrieved 8 October 2012, from CNN Money: http://tech.fortune.cnn.com/2012/06/29/minneapolis-street-test-google-gets-a-b-apples-siri-gets-a-d/

Gilson, R. D., Redden, E. S., & Elliott, L. R. (2007). *Remote tactile displays for future soldiers*. Aberdeen, MD: Aberdeen Proving Ground: U.S. Army Research Laboratory.

Haas, E. C. (2007). Integrating auditory warnings with tactile cues in multimodal displays for challenging environments. *13th International Conference on Auditory Displays*, pp. 127–131. Montreal, Quebec, Canada.

Hearst, M., Allen, J., Guinn, C., & Horvitz, E. (1999). Mixed-initiative interaction: Trends & controversies. *IEEE Intelligent Systems, 14*, 14–23.

Hineline, T. (2010). Untitled photograph of a marine signalling readiness. Retrieved October 2012, from http://www.defenseimagery.mil

Hutchins, J. W. (2004). *The Georgetown-IBM experiment demonstrated in January 1954*. Sixth Conference of the Association for Machine Translation in the Americas, pp. 102–114. Washington, DC.

Katzenberger, M. (2012). Untitled photograph of a combat medic signalling. Retrieved October 2012, from http://www.defenseimagery.mil

Lackey, S. J., Barber, D., Reinerman, L., Badler, N. I., & Hudson, I. (2011). Defining next generation multimodal communication in human robot interaction. *Proceedings of the Human Factors and Ergonomics Society 55th Annual Meeting* (pp. 461–464). Las Vegas, NV.

Lopez, E. (2012). Untitled photograph of a predator UAS. Retrieved October 2012, from http://www.defenseimagery.mil

Martin, G., & Hughes, C. (2010). A scenario generation framework for automating instructional support in scenario-based training. *Proceedings of the 2010 Spring Simulation Multiconference* (pp. 1–6). Orlando, FL.

Martin, G., Schatz, S., Bowers, C., Highes, C. E., Fowlkes, J., & Nicholson, D. (2009). Automatic scenario generation through procedural modeling for scenario-based training. *Proceedings of the Human Factors and Ergonomics Society 53rd Annual Meeting* (pp. 1949–1953). San Antonio, TX.

McLain, T., & Beard, R. (2000). Trajectory planning for coordinated rendezvous of unmanned air vehicles. *Proceedings of AIAA Guidance, Navigation, and Control Conference*. Denver, CO: AIAA.

Merlo, J. L., Terrence, P. I., Stafford, S., Gilson, R., Hancock, P. A., Redden, E. S., … White, T. (2006). Communicating through the use of vibrotactile displays for dismounted and mounted soldiers. *25th Army Science Conference*. Orlando, FL.

Morishima, S. (1993). Facial expression synthesis based on natural voice for virtual face-to-face communication with machine. *Virtual Reality Annual International Symposium* (pp. 486–491). Seattle, WA.

Nijhold, A., & Hulstijn, J. (2006). Multimodal interactions with agents in virtual worlds. In N. Kasabov (Ed.), *Future directions for intelligent systems and information sciences: The future of speech and image technologies, brain computers, www, and bioinformatics (studies in fuzziness and soft computing)* (1 ed., pp. 148–173). Heidelberg, Germany: Physica.

Ogreten, S., Lackey, S. J., & Nicholson, D. M. (2010). Recommended roles for uninhabited team members within mixed-initiative combat teams. *The 2010 International Symposium on Collaborative Technology Systems* (pp. 531–536). Chicago, IL.

Pettitt, R., Redden, E., & Carstens, C. (2009). *Scalability of robotic controllers: Speech-based robotic controller evaluation*. Aberdeen, MD: Aberdeen Proving Ground, U.S. Army Research Laboratory.

Pettitt, R. A., Redden, E. S., Pacis, E., & Carstens, C. B. (2010). *Scalability of robotic controllers: Effects of progressive levels of autonomy on robotic reconnaissance tasks* (No. ARL-TR-5258). Army Research Lab Aberdeen Proving Ground Md.

Phillips, E., Ososky, S., Grove, J., & Jentsch, F. (2011). From tools to teammates: Toward the development of appropriate mental models for intelligent robots. *Proceedings of the Human Factors and Ergonomics Society 55th Annual Meeting* (pp. 1491–1495). Las Vegas, NV.

Powers, A., & Kiesler, S. (2006). The advisor robot: Tracing people's mental model from a robot's physical attributes. *Proceedings of the First ACM SIGCHI/SIGART Conference on Human-Robot Interaction* (pp. 218–225). Salt Lake City, UT.

Redden, E., & Elliot, L. (2010). Robotic control systems for dismounted soldiers. In F. G. Jentsch, & M. J. Barnes (Eds.), *Human-robot interactions in future military operations* (pp. 335–352). Burlington, VT: Ashgate.

Reinerman-Jones, L., Taylor, G., Sprouse, K., Barber, D. J., & Hudson, I. (2011). Adaptive automation as a task switching and task congruence challenge. *Proceedings of the Human Factors and Ergonomics Society Annual Meeting*, 55(1), 197–201. Las Vegas, NV.

RSJPO. (2012). Robotic systems joint project office. Retrieved September 2012, from http://www.rsjpo.army.mil/index.html

Schuster, D., Ososky, S., Jentsch, F., Phillips, E., Lebiere, C., & Evans, A. W. (2011). A research approach to shared mental models and situation assessment in future robot teams. *Proceedings of the Human Factors and Ergonomics Society 55th Annual Meeting* (pp. 456–460). Las Vegas, NV.

Showalter, J. (2008). Untitled photograph of a TALON UGS. Retrieved September 2012, from http://www.defenseimagery.mil

Stiefelhagen, R., Ekenel, H., Fügen, C., Gieselmann, P., Holzapfel, H., Kraft, F., … Waibel, A. (2007). Enabling multimodal human-robot interaction for the Karlsruhe humanoid robot. *IEEE Transactions on Robotics*, 23(5), 840–851.

Stoica, A., Assad, C., Wolf, M., Sung You, K., Pavone, M., Huntsberger, T., … Iwashita, Y. (2012). *Using arm and hand gestures to command robots during stealth operations*. Proceedings of Multisensor, Multisource Information Fusion: Architectures, Algorithms, and Applications 2012. 8407. Baltimore, MD.

Takada, K. G. (2006). Untitled photograph of TALON teleoperation. Retrieved September 2012, from http://www.defenseimagery.mil

Tellex, S., Kollar, T., Dickerson, S., Walter, M. R., Banerjee, A. G., Teller, S., … Nicholas, R. (2011). Understanding natural language commands for robotic navigation and mobile manipulation. *Proceedings of the National Conference on Artificial Intelligence (AAAI 2011)*. San Francisco, CA.

Tellex, S., & Roy, D. (2006). *Spatial routines for a simulated speech-controlled vehicle*. Human Robot Interaction (HRI) 2006, pp. 156–163. Salt Lake City, UT: ACM.

Thomas, T. (2012). Untitled photograph of army sergeant giving orders. Retrieved October 2012, from http://www.defenseimagery.mil

U.S. Army. (1987). Visual signals. *Visual Signals: FM 21–60*. Washington, DC: U.S. Army.

U.S. Congress (2001). National defense authorization act for fiscal year 2001. *National Defense Authorization Act for Fiscal Year 2001*. Washington, DC.

van Erp, J. B. (2000). *Tactile displays in virtual environments*. RTO HFM Workshop on "What Is Essential for Virtual Reality Systems to Meet Military Human Performance Goals?". The Hague, the Netherlands: RTO MP-058.

van Erp, J. B., & Werkhoven, P. (1999). *Spatial characteristics of vibro-tactile perception on the torso (Report TM-99-B007)*. Soesterberg, the Netherlands: TNO Human Factors.

Varcholik, P., Barber, D. J., & Nicholson, D. (2008). *Interactions and training with unmanned systems and the Nintendo wiimote*. Interservice/Industry Training, Simulation, and Education Conference. Orlando, FL.

Varcholik, P., & Merlo, J. (2008). *Gestural communication with accelerometer-based input devices and tactile displays*. Proceedings of the 26th Army Science Conference. Orlando, FL.

Wade, S. (2009). Untitled photograph of a soldier pointing. Retrieved October 2012, from http://www.defenseimagery.mil

Walters, M., Syrdal, D. S., Koay, K. L., Dautenhahn, K., & te Boekhorst, R. (2008). *Human approach distances to a mechanical-looking robot with different robot voice styles*. The 17th IEEE International Symposium on Robot and Human Interactive Communication, pp. 707–712. Munich, Germany.

Wexelblat, A. (1995). An approach to natural gesture in virtual environments. *ACM Transaction on Computer-Human Interaction*, 2(3), 179–200.

White, T. (2010). *Suitable body locations and vibrotactile cueing types for dismounted soldiers*. Aberdeen, MD: Aberdeen Proving Grounds, U.S. Army Research Laboratory.

45 Clinical Virtual Reality

Albert "Skip" Rizzo, Belinda Lange, and Sebastian Koenig

CONTENTS

45.1 INTRODUCTION

Virtual reality (VR) technology offers new opportunities for the development of innovative clinical research, assessment, and intervention tools. VR-based testing, training, and treatment approaches that would be difficult, if not impossible, to deliver using traditional methods are now being developed that take advantage of the assets that are available with VR technology. As research evidence continues to indicate clinical efficacy, VR applications are being increasingly regarded as providing innovative options for targeting the cognitive, psychological, motor, and functional impairments that result from various clinical health conditions. VR allows for the precise presentation and control of stimuli within dynamic multisensory 3D computer-generated environments, as well as providing advanced methods for capturing and quantifying behavioral responses. These characteristics serve as the basis for the rationale for the use of VR applications in the clinical assessment, intervention, and training domains. This chapter begins with a brief review of the history and rationale for the use of VR with clinical populations. This chapter then focuses on reviewing four fundamental areas where clinical VR has shown significant potential to enhance clinical practice and research (exposure therapy, neuropsychological assessment [NA] cognitive/physical rehabilitation, and the use of virtual human [VH] agents). At the end of each of these sections, a detailed use case is presented. The goal of this chapter is to present a clear rationale for VR use across diverse areas of clinical practice and present examples of how this has been done successfully. While significant work has been done in other areas of clinical VR (e.g., pain distraction, eating disorders, social skills training), a full treatment of such a broad literature is beyond the scope of this chapter. Thus, we have opted to provide more depth on specific clinical areas where VR has been applied to address anxiety disorders with exposure therapy, NA, cognitive/physical rehabilitation, and emerging work developing VHs to serve the role of virtual patients for clinical training and as health-care agent to promote access.

45.2 HISTORY AND RATIONALE FOR CLINICAL VR

VR has undergone a transition in the past few years that has taken it out of the realm of expensive toy and into that of functional technology. Over the last 15 years, a virtual revolution has taken place in the use of VR simulation technology for clinical purposes. Although media hype may have oversold VR's potential during the early stages of the technology's development, a uniquely suited match exists between the assets available with VR technology and applications in the clinical sciences. The capacity of VR technology to create controllable, multisensory, interactive 3D stimulus environments, within which human behavior can be motivated and measured, offers clinical assessment and intervention options that were not possible using previously available approaches. The unique match between VR technology assets and the needs of various clinical application areas has been recognized by a determined and expanding cadre of researchers and clinicians who not only have recognized the potential impact of VR technology but have now generated a significant research literature that documents the many clinical and research targets where VR can add value over traditional assessment and intervention methods (Glantz, Rizzo, & Graap, 2003; Holden, 2005; Lange et al., 2012; Parsons & Rizzo, 2008; Parsons, Rizzo, Rogers, & York, 2009; Powers & Emmelkamp, 2008; Riva, 2011; Rizzo et al., 1997; Rizzo, Schultheis, Kerns, & Mateer, 2004; Rizzo et al., 2013; Rizzo & Kim, 2005; Rizzo, Lange, Buckwalter et al., 2011; Rizzo, Lange, Suma, & Bolas, 2011; Rizzo, Parsons et al., 2011; Rose, Brooks, & Rizzo, 2005). Based on these reports, there is a growing consensus that VR has now emerged as a promising tool in many domains of clinical care and research.

Virtual environments (VEs) have been developed that are now demonstrating effectiveness in a number of areas in clinical psychology, neuropsychology, and both cognitive and motor rehabilitation. A short list of areas where clinical VR has been usefully applied includes fear reduction in persons with simple phobias (Parsons & Rizzo, 2008; Powers & Emmelkamp, 2008), treatment for posttraumatic stress disorder (PTSD) (Difede & Hoffman, 2002; Difede et al., 2007; Rizzo, 2010; Rizzo, Difede, Rothbaum, & Reger, 2010; Rizzo, Parsons et al., 2011; Rizzo et al. 2013; Rothbaum, Hodges, Ready, Graap, & Alarcon, 2001), stress management in cancer patients (Schneider, Kisby, & Flint, 2010), acute pain reduction during wound care and physical therapy with burn patients (Hoffman et al., 2011) and in other painful procedures (Gold, Kim, Kant, Joseph, & Rizzo, 2006), body image disturbances in patients with eating disorders (Riva, 2011), navigation and spatial training in children and adults with motor impairments (Rizzo, Schultheis, Kerns, & Mateer, 2004; Stanton, Foreman, & Wilson, 1998), functional skill training and motor rehabilitation with patients having central nervous system (CNS) dysfunction (e.g., stroke, TBI, SCI, cerebral palsy, multiple sclerosis [MS]) (Holden, 2005; Merians et al., 2010), and for the assessment and rehabilitation of attention, memory, spatial skills, and other cognitive functions in both clinical and unimpaired populations (Brooks et al., 1999; Brown, Kerr, & Bayon, 1998; Matheis et al., 2006; Parsons et al., 2009; Pugnetti et al., 1995; Rizzo et al., 2006; Rose et al., 2005). To do this, VR scientists have constructed virtual airplanes, skyscrapers, spiders, battlefields, social settings, beaches, fantasy worlds, and the mundane (but highly relevant) functional environments of the schoolroom, office, home, street, and supermarket. Emerging research and development is also producing artificially intelligent (AI) VH patients that are being used to train clinical skills to health professionals (Kenny, Parsons, & Garrity, 2010; Lok et al., 2007; Parsons et al., 2008; Rizzo, Parsons, Buckwalter, & Kenny, in press).

In essence, clinicians can now create simulated environments that mimic the outside world and use them in clinical settings to immerse patients in simulations that support the aims and mechanics of a specific assessment or therapeutic approach. And this state of affairs now stands to transform the vision of future clinical practice and research in the disciplines of psychology, medicine, neuroscience, and physical and occupational therapy and in the many allied health fields that address the therapeutic needs of children and adults with health-care issues and clinical disorders. As well, the clinical and research targets chosen for these applications reflect an informed appreciation for the assets that are available with VR technology (Rizzo Strickland, & Bouchard, 2004) by clinicians/developers initially designing and using systems in this area. These initiatives give hope that in the twenty-first century, new and useful tools will be developed that will advance clinical areas that have long been mired in the methods of the past.

By its nature, VR simulation technology is well-suited to simulate the challenges that people face in naturalistic environments and consequently can provide objective simulations that can be useful for clinical assessment and intervention purposes. Within these environments, researchers and clinicians can present ecologically relevant stimuli embedded in a meaningful and familiar context. From this, VR offers the potential to create systematic human testing, training, and treatment environments that allow for the precise control of complex, immersive, dynamic 3D stimulus presentations, within which sophisticated interaction, behavioral tracking, and performance recording are possible. Much like an aircraft simulator serves to test and train piloting ability under a variety of controlled conditions, VR can be used to create relevant simulated environments where assessment and treatment of cognitive, emotional, and motor problems can take place under a range of stimulus conditions that are not easily deliverable and controllable in the real world.

When combining these assets within the context of functionally relevant, ecologically enhanced VEs, a fundamental advancement could emerge in how human assessment and intervention can be addressed in many clinical and research disciplines. For example, instead of relying solely on unverifiable imagery processes in clients with anxiety disorders to produce the therapeutic effects of habituation, graduated exposure to feared or trauma-relevant stimuli can be delivered systematically in VR. As well, rather than trying to predict real-world functional performance from a decontextualized measure of attention when assessing children suspected of having attention deficit hyperactivity disorder (ADHD), one can look at the effects of systematically increasing ecologically relevant attentional demands in a VE, such as a classroom, social setting, or home. These examples illustrate how VR technology can be used to provide exquisite timing and control over context-relevant imagery and stimulus load/complexity, all of which can be manipulated in a dynamic fashion contingent on the needs and responses of the client or research participant. Within such VEs, human performance can be digitally captured in real time in support of a precise and detailed analysis of relevant responses in relation to systematic stimulus presentations. In this regard, VR can be seen as capable of producing the *ultimate Skinner box* for conducting human research, assessment, and intervention.

This trajectory is expected to continue supported by the revolutionary advances in the underlying VR enabling technologies (i.e., computation speed and power, graphics and image rendering software, display systems, interface devices, immersive audio, haptic tools, wireless tracking, voice recognition, intelligent agents, and authoring software) that allow for the creation of low-cost and usable VR systems capable of running on a commodity-level personal computer. Such advances in technological *prowess* and accessibility have provided the hardware platforms needed for the conduct of human research and clinical intervention within more usable and useful VR scenarios. This convergence of the exponential advances in underlying VR enabling technologies with a growing body of clinical research and experience has fueled the evolution of the discipline of *clinical VR*. And this has now supported the emergence of accessible VR systems that can uniquely target a wide range of psychological, cognitive and physical clinical targets, training objectives, and research questions.

This is in sharp contrast to what was possible in the mid-1990s when discussion of the potential for VR applications in the clinical assessment and intervention domains first emerged (Pugnetti et al., 1995; Rizzo, 1994; Rose, Attree, & Johnson, 1996). At that point in time, the technology to deliver on the anticipated VR *vision* was not in place. Consequently, during these early years, VR suffered from a somewhat imbalanced *expectation-to-delivery* ratio, as most users who eagerly lined up to try such systems during that time will attest. The *real* thing never quite measured up to expectations generated by some of the initial media hype, as delivered, for example, in the films *The Lawnmower Man* and *Disclosure*! Yet the idea of producing simulated VEs that allowed for the systematic delivery of ecologically relevant challenges was compelling and made intuitive sense. As well, the long and rich history of encouraging findings from the predecessor literature in aviation simulation (Hays, Jacobs, Prince, & Salas, 1992) lent support to the concept that testing, training, and treatment in highly proceduralized VR simulation environments would be a useful direction for clinical disciplines to explore (Johnston, 1995; Rizzo, 1994). Within this context, a small group of innovative clinicians and researchers also began the initial work of exploring the use of VR technology for applications designed to treat simple phobias (Hodges et al., 1995; Lamson, 1994; Rothbaum, Hodges, & Kooper, 1995), while others addressed cognitive/functional performance in populations with CNS dysfunction (Brown et al., 1998; Pugnetti et al., 1995; Rizzo, 1994; Rose et al., 1996). While a good deal of this early work employed the costly, cumbersome, low-resolution VR head-mounted displays (HMDs) that were available at the time or simply used flat-screen monitors or stereoscopic projection approaches, these systems began to produce encouraging results (Cromby, Standen, Newman, & Tasker, 1996; Rizzo et al., 1998; Rose, Brooks, & Attree, 2000; Stanton et al., 1998). From these nascent efforts, findings emerged that began to demonstrate the unique value of the technology, served to inform ideas for future applications and created a grassroots level of enthusiasm for using VR that has continued to grow and be supported into the present day. However, now instead of having to rely on $200,000 graphic workstations (and other expensive peripheral technologies) that were required back in 1990s, clinicians and researchers in the twenty-first century can now create and deliver compelling virtual worlds using a standard laptop and a $1500 HMD or stereo television. The technology has now caught up with the vision, and such exponential advances are expected to continue to advance the science and practice in the discipline of clinical VR.

This chapter is designed to present a general overview that will detail the history, rationales, and key research for clinical VR applications in four areas: (1) exposure therapy for anxiety disorders and PTSD, (2) NA for CNS dysfunction, (3) motor rehabilitation for neurological impairment and injury, and (4) VHs for clinical training and promoting health-care access. Sections 45.2 through 45.4 were selected based on the consistent evolution of the research and the growing clinical adoption of applications in these areas. The VH area, although in a nascent stage of development, was selected based on its estimated potential for future growth and clinical impact by the authors. At the end of each section, use cases are presented from work in our lab that illustrates the process for design, development, and evaluation of a system in each of the larger areas. While significant VR research and development activity is ongoing in the areas of substance abuse, eating disorders, pain distraction, social skills training, etc., it was necessary to constrain the scope of what could be presented while still providing sufficient detail on these core areas within the page limitations for this chapter.

45.3 VR EXPOSURE THERAPY

The use of VR to address psychological disorders began in the mid-1990s with its use as a tool to deliver prolonged exposure (PE) therapy targeting anxiety disorders, primarily for specific phobias (e.g., heights, flying, spiders, enclosed spaces). PE is a form of individual psychotherapy based on the Foa and Kozak (1986) emotional processing theory, which posits that phobic disorders and PTSD involve pathological fear structures that are activated when information represented in the structures is encountered. Emotional processing theory purports that fear memories include information about stimuli, responses, and meaning (Foa & Kozak, 1986; Foa, Skeketee, & Rothbaum, 1989) and that fear structures are composed of harmless stimuli that have been associated with danger and are reflected in the belief that the world is a dangerous place. This belief then manifests itself in cognitive and behavioral avoidance strategies that limit exposure to potentially corrective information that could be incorporated into and alter the fear structure. As escape and avoidance from feared situations are intrinsically (albeit, temporarily) rewarding, phobic disorders can perpetuate without treatment. Consequently, several theorists have proposed that conditioning processes are involved in the etiology and maintenance of anxiety disorders. These theorists invoke Mowrer's (1960) two-factor theory, which posits that both Pavlovian and instrumental conditioning are involved in the acquisition of fear and avoidance behavior. Successful treatment requires emotional processing of the fear structures in order to modify their pathological elements so that the stimuli no longer invoke fear, and any method capable of activating the fear structure and modifying it would be predicted to improve symptoms of anxiety.

Imaginal PE entails engaging mentally with the fear structure through repeatedly revisiting the feared or traumatic event in a safe environment. The proposed mechanisms for symptom reduction involves activation and emotional processing, extinction/habituation of the anxiety, cognitive reprocessing of pathogenic meanings, the learning of new responses to previously feared stimuli, and ultimately an integration of corrective nonpathological information into the fear structure (Bryant, Moulds, Guthrie, & Nixon, 2003; Foa & Hearst-Ikeda, 1996). Thus, VR was seen early on to be a potential tool for the treatment of anxiety disorders; if an individual can become immersed in a feared VE, activation and modification of the fear structure were possible. From this, the use of VR to deliver PE was the first psychological treatment area to gain traction clinically, perhaps in part due to the intuitive match between what the technology could deliver and the theoretical requirement of PE to systematically expose/engage users to progressively more challenging stimuli needed to activate the fear structure.

Moreover, even during the early days of VR, this was not so technically challenging to achieve. VEs could be created that required little complex user interaction beyond simple navigation within a simulation that presented users with scenarios that represented key elements of the targeted fear structure that could be made progressively more provocative (views from tall buildings, aircraft interiors, spiders in kitchens, etc.). And even with the limited graphic realism available at the time, phobic patients were observed to be *primed* to suspend disbelief and react emotionally to virtual content that represented what they feared. In general, the phenomenon that users of VR could become immersed in VE's provided a potentially powerful tool for activating relevant fears in the PE treatment of specific phobias in the service of therapeutic exposure.

From this starting point, a body of literature evolved that suggested that the use of virtual reality exposure therapy (VRET) was effective. Case studies in the 1990s initially documented the successful use of VR in the treatment of fear of flying (Rizzo et al., 1997; Rothbaum, Hodges, Watson, Kessler, & Opdyke, 1996; Smith, Rothbaum, & Hodges, 1999), claustrophobia (Botella et al., 1998), acrophobia (Rothbaum et al., 1995), and spider phobia (Carlin, Hoffman, & Weghorst, 1997). For example, in an early wait-list (WL)-controlled study, VRET was used to treat the fear of heights, exposing patients to virtual footbridges, virtual balconies, and a virtual elevator (Rothbaum et al., 1995). Patients were encouraged to spend as much time in each situation as needed for their anxiety to decrease and were allowed to progress at their own pace. The therapist saw on a computer monitor what the participant saw in the VE and therefore was able to comment appropriately. Results showed that anxiety, avoidance, and distress decreased significantly from pre- to posttreatment for the VRE group but not for the WL control group. Examination of attitude ratings on a semantic differential scale revealed positive attitudes toward heights for the VRE group and negative attitudes toward heights for the WL group. The average anxiety ratings decreased steadily across sessions, indicating habituation for those participants in treatment. Furthermore, 7 of the 10 VRE treatment completers exposed themselves to height situations in real life during treatment although they were not specifically instructed to do. These exposures appeared to be meaningful, including riding 72 floors in a glass elevator and intentionally parking at the edge of the top floor of a parking deck.

This research group then compared VRET to both an in vivo PE therapy condition and a WL control in the treatment of the fear of flying (Rothbaum, Hodges, Smith, Lee, & Price, 2000). Treatment consisted of eight individual therapy sessions conducted over 6 weeks, with four sessions of anxiety management training followed either by exposure to a virtual airplane (VRET) or exposure to an *actual airplane* at the airport (PE). For participants in the VRE group, exposure in the virtual airplane included sitting in the virtual airplane, taxi, takeoff, landing, and flying in both calm and turbulent weather according to a treatment manual (Rothbaum, Hodges, & Smith, 1999). For PE sessions, in vivo exposure was conducted at the airport during sessions 5–8. Immediately following the treatment or WL period, all patients were asked to participate in a behavioral avoidance test consisting of a commercial round-trip flight. The results indicated that each active treatment was superior to WL and that there were no differences between VRET and in vivo PE. For WL participants, there were no differences between pre- and post-self-report measures of anxiety and avoidance, and only one of the 15 WL participants completed the graduation flight. In contrast, participants receiving VRET or in vivo PE showed substantial improvement, as measured by self-report questionnaires, willingness to participate in the graduation flight, self-report levels of anxiety on the flight, and self-ratings of improvement. There were no differences between the two treatments on any measures of improvement. Comparison of posttreatment to the 6-month follow-up data for the primary outcome measures for the two treatment groups indicated no significant differences, indicating that treated participants maintained their treatment gains. By the 6-month follow-up, 93% of treated participants had flown since completing treatment. Since that time, an evolved body of literature of controlled studies has emerged, and two recent meta-analyses of the available literature (Parsons & Rizzo, 2008; Powers & Emmelkamp, 2008) concurred with the finding that VR is an efficacious approach for delivering PE, that it outperformed imaginal PE and was as effective as in vivo exposure. A newer meta-analysis and a systematic review of this literature expand on the findings in this area (Opris et al., 2012; Scozzari & Gamberini, 2011) and further confirm that VR can improve outcomes when delivering evidence-based cognitive behavioral therapy (CBT).

In the late 1990s, researchers began to test the use of VRET for the treatment of PTSD by immersing users in simulations of trauma-relevant environments in which the emotional intensity of the scenes can be precisely controlled by the clinician in collaboration with the patients' wishes. Traditional PE typically involves the graded and repeated imaginal reliving of the traumatic event within the therapeutic setting. However, while the efficacy of imaginal exposure has been established in multiple studies with diverse trauma populations (Bryant, 2005; Rothbaum & Schwartz, 2002; Van Etten & Taylor, 1998), many patients are unwilling or unable to effectively visualize the traumatic event. This is a crucial concern since avoidance of cues and reminders of the trauma is one of the cardinal symptoms of the DSM-IV-r (American Psychiatric Association, 2000) diagnosis of PTSD. In fact, research on this aspect of PTSD treatment suggests that the inability to emotionally engage (*in imagination*) is a predictor for negative treatment outcomes (Jaycox, Foa, & Morral, 1998). Thus, VRET offers a way to circumvent the natural avoidance tendency by directly delivering multisensory and context-relevant cues that aid in the confrontation and processing of traumatic memories without demanding that the patient actively try to access his or her experience through effortful memory retrieval. Similar to PE for specific phobias, this approach is believed to provide a low-threat context where the patient can begin to therapeutically process the emotions that are relevant to the traumatic event as well as decondition the learning cycle of the disorder via a habituation/extinction process. The rationale for this approach was bolstered following the start of the wars in Iraq and Afghanistan with specific committee reports indicating that CBT with trauma-focused exposure has the highest level of research evidence in support of its therapeutic efficacy (IOM, 2007, 2012).

The first effort to apply VRET began in 1997 when researchers at Georgia Tech and Emory University began testing the *Virtual Vietnam* VR scenario with Vietnam veterans diagnosed with PTSD (Rothbaum et al., 2001). This occurred over 20 years after the end of the Vietnam War. During those intervening years, in spite of valiant efforts to develop and apply traditional psychotherapeutic and pharmacological treatment approaches to PTSD, the progression of the disorder in some veterans significantly impacted their psychological well-being, functional abilities, and quality of life, as well as that of their families and friends. This initial effort yielded encouraging results in a case study of a 50-year-old, male Vietnam veteran meeting DSM-IV-r criteria for PTSD (Rothbaum et al., 1999).

Results indicated posttreatment improvement on all measures of PTSD and maintenance of these gains at a 6-month follow-up, with a 34% decrease in clinician-rated symptoms of PTSD and a 45% decrease on self-reported symptoms of PTSD. This case study was followed by an open clinical trial with Vietnam veterans (Rothbaum et al., 2001). In this study, 16 male veterans with PTSD were exposed to two HMD-delivered VEs, a virtual clearing surrounded by jungle scenery and a virtual

Huey helicopter, in which the therapist controlled various visual and auditory effects (e.g., rockets, explosions, day/night, shouting). After an average of 13 exposure therapy sessions over 5–7 weeks, there was a significant reduction in PTSD and related symptoms. *For more information, see the 9-minute virtual Vietnam documentary video at* http://www.youtube.com/watch?v=C_2ZkvAMih8.

Similar positive results were reported by Difede and Hoffman (2002) for PTSD that resulted from the attack on the World Trade Center in a case study using VRET with a patient who had failed to improve with traditional imaginal exposure therapy. This group later reported positive results from a WL-controlled study using the same World Trade Center VR application (Difede et al., 2007). The VR group demonstrated statistically and clinically significant decreases on the *gold standard* clinician-administered PTSD scale (CAPS) relative to both pretreatment and WL control group with a between-groups posttreatment effect size of 1.54. Seven of ten people in the VR group no longer carried the diagnosis of PTSD, while all of the WL controls retained the diagnosis following the waiting period and treatment gains were maintained at 6-month follow-up. Also noteworthy was the finding that five of the ten VR patients had previously participated in imaginal exposure treatment with no clinical benefit. Such initial results are encouraging and suggest that VR may be a useful component within a comprehensive treatment approach for persons with combat/terrorist attack-related PTSD. *For more information, see the virtual World Trade Center video at* http://www.youtube.com/watch?v=XAR9QDwBILc.

45.3.1 USE CASE: THE VIRTUAL IRAQ/AFGHANISTAN PTSD EXPOSURE THERAPY PROJECT

War is perhaps one of the most challenging situations that a human being can experience. The physical, emotional, cognitive, and psychological demands of a combat environment place enormous stress on even the best-prepared military personnel. The stressful experiences that are characteristic of the Operation Enduring Freedom (OEF) and Operation Iraqi Freedom (OIF) warfighting environments have produced significant numbers of returning service members (SMs) at risk for developing PTSD and other psychosocial health conditions. In the first systematic study of OEF/OIF mental health problems, the results indicated that "… The percentage of study subjects whose responses met the screening criteria for major depression, generalized anxiety, or PTSD was significantly higher after duty in Iraq (15.6 to 17.1 percent) than after duty in Afghanistan (11.2 percent) or before deployment to Iraq (9.3 percent)" (p. 13) (Hoge et al., 2004). Reports since that time on OEF/OIF PTSD and psychosocial disorder rates suggest even higher incidence rates (Fischer, 2010; Seal, Bertenthal, Nuber, Sen, & Marmar, 2007; Tanielian et al., 2008). For example, as of 2010, the Military Health System recorded 66,934 active duty patients who have been diagnosed with PTSD (Fischer, 2010), and the Rand analysis (Tanielian et al., 2008) estimated that at a 1.5 million deployment level, more than 300,000 active duty and discharged veterans will suffer from the symptoms of PTSD and major depression. With total deployment numbers now having increased to over 2.2 million, the Rand analysis likely underestimates the current number of SMs who may require (and could benefit from) clinical attention upon the return home. These findings make a compelling case for a continued focus on developing and enhancing the availability of evidence-based treatments to address a mental health-care challenge that has had a significant impact on the lives of our SMs, veterans, and their significant others, who deserve our best efforts to provide optimal care.

With this history in mind, the University of Southern California (USC) Institute for Creative Technologies (ICT) created an immersive VRET system for combat-related PTSD. The treatment environment was initially based on recycling virtual assets that were built for the commercially successful Xbox game and tactical training simulation scenario, *Full Spectrum Warrior*. Over the years, other existing and newly created assets developed at the ICT have been integrated into this continually evolving application. The *Virtual Iraq/Afghanistan* application now consists of a series of 14 virtual scenarios designed to resemble the general contexts that most SMs experience during an OEF/OIF deployment, including Middle Eastern-themed city, village, and roadway environments (see Figure 45.1). For example, the Iraq and Afghan City settings have a variety of elements

(a)   (b)

FIGURE 45.1   Scene from Virtual Iraq/Afghanistan city and roadway scenarios. (a) Screen shot of city scenario. (b) Screen shot of Humvee interior from roadway scenario.

including a marketplace, desolate streets, checkpoints, ramshackle buildings, warehouses, mosques, shops, and dirt lots strewn with junk. Access to building interiors and rooftops is available, and the backdrop surrounding the navigable exposure zones creates the illusion of being embedded within a section of a sprawling densely populated mountainous or desert city.

Vehicles are active in streets and animated virtual pedestrians (civilian and military) can be added or eliminated from the scenes. The software has been designed such that users can be *teleported* to specific locations within the city, based on a determination as to which components of the environment most closely match the patient's needs, relevant to their individual trauma-related experiences. The roadway scenarios consist of a travel on paved or dirt roads through expansive mountainous/desert areas with intact and broken down structures, villages, occasional areas of vegetation, bridges, battle wreckage, a checkpoint, debris, and VH figures. The user is positioned inside of a Humvee or MRAP vehicle that supports the perception of travel within a convoy or as a lone vehicle with selectable positions as a driver or passenger or from the more exposed turret position above the roof of the vehicle. The number of soldiers in the cab can also be varied as well as their capacity to become wounded during certain attack scenarios (e.g., IEDs, rooftop/bridge attacks).

All of the city, village, and roadway scenarios are adjustable for time of day or night, weather conditions, illumination, night vision, and ambient sound (wind, motors, city noise, prayer call, etc.). Users can navigate in all scenarios via the use of a standard gamepad controller or by way of a replica M4 weapon with a *thumb-mouse* controller that supports movement during dismounted foot patrol contexts. The mock weapon interface option was based on results from our ongoing user-centered design process with experienced SMs who provided frank feedback indicating that to walk within such a setting without a weapon in hand was completely unnatural and distracting. However, there is no option for firing the weapon within the VR scenarios. It is our firm belief that the principles of exposure therapy are incompatible with the cathartic acting out of a revenge fantasy that a responsive weapon might encourage.

In addition to the visual stimuli presented in the HMD, directional 3D audio, vibrotactile and olfactory stimuli can be delivered into the *Virtual Iraq/Afghanistan* scenarios in real time by the clinician. The presentation of additive, combat-relevant stimuli into the VR scenarios can be controlled in real time via a separate *Wizard of Oz* clinician's interface, while the clinician is in full audio contact with the patient. The clinician's interface is a key feature that provides a clinician with the capacity to customize the therapy experience to the individual needs of the patient. This interface allows a clinician to place the patient in VR scenario locations that resemble the setting in which the trauma-relevant events occurred and ambient light and sound conditions can be modified to match the patients description of their experience. The clinician can then gradually introduce and control real-time trigger stimuli (visual, auditory, olfactory, and tactile), via the clinician's interface, as required to foster the anxiety modulation needed for therapeutic habituation and emotional processing in a customized fashion according to the patient's past experience and treatment progress.

The clinician's interface options have been designed with the aid of feedback from clinicians with the goal to provide a usable and flexible control panel system for conducting thoughtfully administered exposure therapy that can be readily customized to address the individual needs of the patient. Such options for real-time stimulus delivery flexibility and user experience customization are key elements for these types of VRET applications.

The specification, creation, and addition of trigger stimulus options into the *Virtual Iraq/Afghanistan* system have been an evolving process throughout the development of the application based on continually solicited patient and clinician feedback. This part of the design process began by including options that have been reported to be relevant by returning soldiers and military subject matter experts. For example, Hoge et al. (2004) presented a listing of emotionally challenging combat-related events that were commonly reported by their Iraq/Afghanistan SM sample. These events provided a useful starting point for conceptualizing how relevant trigger stimuli could be presented in a VR environment. Such commonly reported events included "Being attacked or ambushed … receiving incoming artillery, rocket, or mortar fire … being shot at or receiving small-arms fire … seeing dead bodies or human remains …" (p. 18). From this and other sources, we began our initial effort to conceptualize what was both functionally relevant and technically possible to include as trigger stimuli.

Currently, the system offers a variety of auditory trigger stimuli (e.g., incoming mortars, weapons fire, voices, wind) that are actuated by the clinician via mouse clicks on the clinician's interface. Clinicians can also similarly trigger dynamic audiovisual events such as helicopter flyovers, bridge attacks, exploding vehicles, and IEDs. The creation of more complex events that can be intuitively delivered in *Virtual Iraq/Afghanistan* from the clinician's interface while providing a patient with options to interact or respond in a meaningful manner is one of the ongoing focuses in this project. However, such trigger options require not only interface design expertise but also clinical wisdom as to how much and what type of exposure is needed to produce a positive clinical effect. These issues have been keenly attended to in initial nonclinical user-centered tests with Iraq-experienced SMs and in the current clinical trials with patients. This expert feedback is essential for informed VR combat scenario design and goes beyond what is possible to imagine from the *ivory tower* of the academic world.

Olfactory and tactile stimuli can also be delivered into the simulation to further augment the experience of the environment. Olfactory stimuli are produced by the *Enviroscent, Inc. Scent Palette*. This is a USB-driven device that contains eight pressurized chambers, within which individual smell cartridges can be inserted, a series of fans and a small air compressor to propel the customized scents to participants. The scent delivery is controlled by mouse clicks on the clinician's interface. Scents may be employed as direct stimuli (e.g., scent of smoke as a user walks by a burning vehicle) or as cues to help immerse users in the world (e.g., ethnic food cooking). The scents selected for this application include burning rubber, cordite, garbage, body odor, smoke, diesel fuel, Iraqi food spices, and gunpowder. Vibration is also used as an additional user sensory input. Vibration is generated through the use of a *Logitech* force-feedback game control pad and through low-cost (<$120) audio-tactile sound transducers from *Aura Sound Inc.* located beneath the patient's floor platform and seat. Audio files are customized to provide vibration consistent with relevant visual and audio stimuli in the scenario. For example, in the roadway scenarios, the user experiences engine vibrations as the vehicle moves across the virtual terrain and a shaking floor can accompany explosions. This package of controllable multisensory stimulus options was included in the design of *Virtual Iraq/Afghanistan* to allow a clinician the flexibility to engage users across a wide range of unique and highly customizable levels of exposure intensity. As well, these same features have broadened its applicability as a research tool for studies that require systematic control of stimulus presentation within combat-relevant environments (Rizzo et al., 2012). A direct link to a *YouTube* channel with videos that illustrate features of this system and of former patients discussing their experience with the VRET approach can be found at http://www.youtube.com/user/AlbertSkipRizzo.

The *Virtual Iraq/Afghanistan* system was designed and built from a user-centered design process that involved feedback from active duty SMs and veterans that began with solicited responses to the initial prototype. User-centered design feedback needed to iteratively evolve the system was gathered from a system deployed in Iraq with an Army combat stress control team and from returning OEF/OIF veterans and patients in the United States. Thus, leading up to the first clinical group test of treatment effectiveness, initial usability studies and case reports were published with positive findings in terms of SMs acceptance, interest in the treatment, and clinical successes (Gerardi, Rothbaum, Ressler, Heekin, & Rizzo, 2008; Reger et al., 2008; Reger, Gahm, Rizzo, Swanson, & Duma, 2009; Wilson, Onorati, Mishkind, Reger, & Gahm, 2008).

The Office of Naval Research, the organization that funded the initial system development of *Virtual Iraq/Afghanistan*, also supported an initial open clinical trial to evaluate the feasibility of using VRET with active duty participants (Rizzo et al., 2010). The study participants were recently redeployed from Iraq/Afghanistan at the Naval Medical Center San Diego and at Camp Pendleton and had engaged in previous PTSD treatments (e.g., group counseling, EMDR, medication) without benefit. The standard treatment protocol consisted of 2× weekly, 90–120 min sessions over 5 weeks. The VRET exposure exercises followed the principles of PE therapy (Foa, Davidson, & Frances, 1999) and the pace was individualized and patient driven. Physiological monitoring (heart rate, galvanic skin response, and respiration) was used to provide additional user state information to the clinician to help inform the pacing of the VRET.

The first VRET session consisted of a clinical interview that identified the index trauma and provided psychoeducation on trauma and PTSD and instruction on a deep breathing technique for general stress management purposes. The second session provided instruction on the use of subjective units of distress (SUDS), the rationale for PE, including imaginal exposure and in vivo exposure. The participants also engaged in their first experience of imaginal exposure of the index trauma and an in vivo hierarchical exposure list was constructed, with the first item assigned as homework. Session 3 introduced the rationale for VRET, and the participant experienced the VR environment without recounting the index trauma narrative for approximately 25 min with no provocative trigger stimuli introduced. Sessions 4–10 focused on the participant engaging in the VR while recounting the trauma narrative.

Generally, participants were instructed that they would be asked to recount their trauma in the first person, as if it were happening again with as much attention to sensory detail as they could provide. Using clinical judgment, the therapist might prompt the patient with questions about their experience or provide encouraging remarks as deemed necessary to facilitate the recounting of the trauma narrative. The treatment included homework, such as requesting the participant to listen to the audiotape of their exposure narrative from the most recent session as a form of continual exposure for processing the index trauma to further enhance the probability for habituation to occur. Self-report measures were obtained at baseline and prior to sessions 3, 5, 7, 9, and 10 and 1 week and 3 months posttreatment to assess in-treatment and follow-up symptom status. The measures used were the PTSD Checklist–Military Version (PCL-M) (Blanchard, Jones-Alexander, Buckley, & Forneris, 1996), Beck Anxiety Inventory (BAI) (Beck, Epstein, Brown, & Steer, 1988), and Patient Health Questionnaire–Depression (PHQ-9) (Kroenke & Spitzer, 2002).

Analyses of the first 20 active duty SMs to complete treatment (19 male, 1 female, mean age = 28, age range 21–51 years) produced positive clinical outcomes. For this sample, mean pre-/post-PCL-M scores decreased in a statistical and clinically meaningful fashion: 54.4 (SD = 9.7) to 35.6 (SD = 17.4). Paired pre-/post-t-test analysis showed these differences to be significant ($t = 5.99$, df = 19, $p < .001$). Correcting for the PCL-M no-symptom baseline of 17 indicated a greater than 50% decrease in symptoms, and 16 of the 20 completers no longer met DSM-IV-r criteria for PTSD at posttreatment. Five participants in this group with PTSD diagnoses had pretreatment baseline scores below the conservative cutoff value of 50 (prescores = 49, 46, 42, 36, 38) and reported decreased values at posttreatment (postscores = 23, 19, 22, 22, 24, respectively). Individual participant PCL-M scores at baseline, posttreatment, and 3-month follow-up are in Figure 45.2.

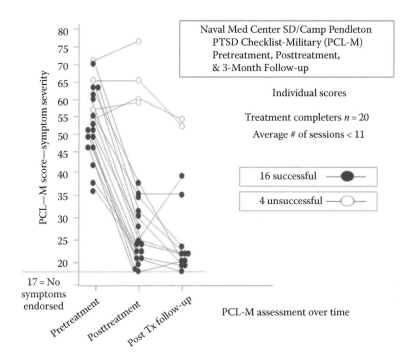

PCL—M score—symptom severity

80
75
70
65
60
55
50
45
40
35
30
25
20

17 = No
symptoms
endorsed

Naval Med Center SD/Camp Pendleton
PTSD Checklist-Military (PCL-M)
Pretreatment, Posttreatment,
& 3-Month Follow-up

Individual scores

Treatment completers *n* = 20

Average # of sessions < 11

16 successful

4 unsuccessful

Pretreatment   Posttreatment   Post Tx follow-up

PCL-M assessment over time

FIGURE 45.2 PCL-M scores across treatment.

Mean BAI scores significantly decreased 33% from 18.6 (SD = 9.5) to 11.9 (SD = 13.6) (t = 3.37, df = 19, p < .003), and mean PHQ-9 (depression) scores decreased 49% from 13.3 (SD = 5.4) to 7.1 (SD = 6.7) (t = 3.68, df = 19, p < .002) (see Figure 45.3). The average number of sessions for this sample was just under 11. Also, two of the successful treatment completers had documented mild and moderate TBIs, which provide an early indication that this form of exposure therapy can be useful (and beneficial) for this population. Results from uncontrolled open trials are difficult to generalize from and we are cautious not to make excessive claims based on these early results. However, using an accepted military-relevant diagnostic screening measure (PCL-M), 80% of the treatment completers in the initial VRET sample showed both statistically and clinically meaningful reductions in PTSD, anxiety, and depression symptoms, and anecdotal evidence from patient reports suggested that they saw improvements in their everyday life. These improvements were also maintained at 3-month posttreatment follow-up.

Other studies have also reported positive outcomes. Two early case studies reported positive results using this system (Gerardi et al., 2008; Reger & Gahm, 2008). Following those, another open clinical trial with active duty soldiers (n = 24) produced significant pre-/postreductions in PCL-M scores and a large treatment effect size (Cohen's d = 1.17) (Reger et al., 2011). After an average of seven sessions, 45% of those treated no longer screened positive for PTSD and 62% had reliably improved. In a small preliminary quasi-randomized controlled trial (RCT) (McLay et al., 2011), 7 of 10 participants with PTSD showed a 30% or greater improvement with VR, while only 1 of 9 participants in a *treatment-as-usual* group showed similar improvement. The results are limited by small size, lack of blinding, a single therapist, and comparison to a set relatively uncontrolled usual care conditions, but it did add to the incremental evidence suggesting VR to be a safe and effective treatment for combat-related PTSD. Finally, at the recent 2012 American Psychiatric Association Convention, McLay (2012) presented data from a comparison of VRET with the traditional, evidence-based PE approach in active duty SMs. The results showed significantly better maintenance of positive treatment outcomes at 3-month follow-up for *Virtual Iraq/Afghanistan* system compared to traditional PE (McLay, 2012). The overall trend of these positive findings (in the absence of any

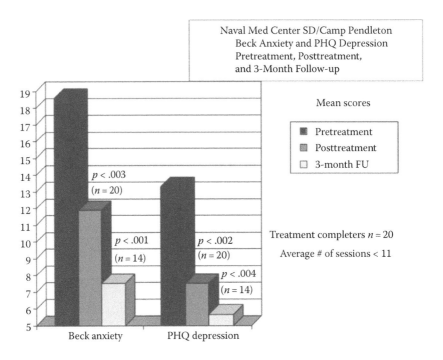

FIGURE 45.3 BAI and PHQ depression scores.

reports of negative findings) is encouraging for the view that VRET is safe and may be an effective approach for delivering an evidence-based treatment (PE) for PTSD.

Four RCTs are ongoing with the *Virtual Iraq/Afghanistan* system with active duty and veteran populations. Two RCTs are focusing on comparisons of treatment efficacy between VRET and pro-longed imaginal exposure (PE) (Reger & Gahm, 2010, 2011), and another is testing VRET compared with VRET + a supplemental care approach (Beidel, Frueh, & Uhde, 2010). A fourth RCT (Difede, Rothbaum, & Rizzo, 2010–2013) is investigating the additive value of supplementing VRET and imaginal PE with a cognitive enhancer called D-cycloserine (DCS). DCS, an *N*-methyl-D-aspartate partial agonist, has been shown to facilitate extinction learning in laboratory animals when infused bilaterally within the amygdala prior to extinction training (Walker, Ressler, Lu, & Davis, 2002). The first clinical test in humans that combined orally administered DCS with VRET was performed by Ressler et al. (2004) with participants diagnosed with acrophobia ($n = 28$). Participants who received DCS + VRET experienced significant decreases in fear within the VE 1 week and 3 months posttreatment and reported significantly more improvement than the placebo group in their over-all acrophobic symptoms at 3-month follow-up. The DCS group also achieved lower scores on a psychophysiological measure of anxiety than the placebo group. The current multisite PTSD RCT (NICoE, Cornell-Weill, and the Long Beach VAMC) is testing the effect of DCS vs. placebo when added to VRET and PE with active duty and veteran samples ($n = 300$). The Department of Defense (DOD) funding support for these RCTs underscores the interest that the DOD/Veterans Affairs (VA) has in exploring this innovative approach for delivering exposure therapy using VR.

While RCTs are the gold standard for emerging treatment approaches to gain wide acceptance by the scientific community, it should be noted that at its core, the therapeutic model/principle that underlies VRET (CBT with exposure) is in fact evidence based. VRET is simply the delivery of this evidence-based treatment in a format that may serve to engage a wider range of patients in the necessary confrontation and processing of traumatic memories or *fear structures* (see Foa et al., 1999) needed for positive clinical outcomes. Thus, even equivalent positive results with PE in these RCTs would validate its use as another safe and evidence-based therapeutic option. As well, the

VRET approach could serve to draw SMs and veterans into treatment, many of whom have grown up *digital* and may be more likely to seek care in this format compared to what they perceive as traditional talk therapy. This is important since numerous reports from both military and civilian blue-ribbon panels underscore the importance of breaking down *barriers to care* for improving the awareness, availability, accessibility, and acceptance of behavioral health care in the military DOD Mental Health Task Force (DOD, 2007), Institute of Medicine (IOM, 2007, 2012), Dole-Shalala Commission Report (Dole et al., 2007), the Rand report (Tanielian et al., 2008), and American Psychological Association (APA) (APA, 2007).

This research has been supported by the relatively quick adoption of the VRET approach by approximately 55 military, VA, and university clinic sites over the last four years. Based on the outcomes from our initial open clinical trial and similar positive results from other research groups, we are encouraged by these early successes and continue to gather feedback from patients regarding the therapy and the *Virtual Iraq/Afghanistan* treatment environments. Patient feedback was particularly relevant for the *Virtual Iraq/Afghanistan* project since the recent full rebuild of the system (using *Unity 3D* software) aimed to provide more diversity of content and added functionality to meet the needs of the target user group. As well, the new system was designed to facilitate the development, exploration, and testing of hypotheses relevant to improving PTSD treatment, in addition to the simulation for other purposes including PTSD and neurocognitive assessment and for the creation of a stress resilience training system (Rizzo, Parsons et al., 2011; Rizzo et al., 2012).

Interest in VR technology to create tools for enhancing exposure therapy practice and research has grown in recent years as initial positive outcomes have been reported with its implementation. The enthusiasm that is common among proponents of the use of VR for exposure-based treatment partly derives from the view that VR technology provides the capacity for clinicians to deliver specific, consistent, and controllable trauma-relevant stimulus environments that do not rely exclusively on the hidden world and variable nature of a patient's imagination. Moreover, if one reviews the history of the impact of war on advances in clinical care, it could be suggested that VR is an idea whose time has come. For example, during World War I, the Army Alpha/Beta test emerged from the need for better cognitive ability assessment and that development later set the stage for the civilian intelligence testing movement during the mid-twentieth century. As well, the birth of clinical psychology as a treatment-oriented profession was borne out of the need to provide care to the many veterans returning from World War II with *shell shock*. Based on these examples, one of the outcomes of the OEF/OIF conflicts could be the military's support for research and development in the area of clinical VR that could potentially drive increase recognition and use in the civilian sector. However, this will only occur if positive cost-effective outcomes are produced with military VRET applications. It should also be noted that any rush to adopt VRET should not disregard principles of evidence-based and ethical clinical practice. While novel VR systems can extend the skills of a well-trained clinician, they are not intended to be automated treatment protocols that are administered in a *self-help* format. The presentation of such emotionally evocative VR combat-related scenarios, while providing treatment options not possible until recently, will most likely produce therapeutic benefits when administered within the context of appropriate care via a thoughtful professional appreciation of the complexity and impact of this behavioral health challenge.

45.4 NEUROPSYCHOLOGICAL VR APPLICATIONS

In the broadest sense, neuropsychology is an applied science that evaluates how specific activities in the brain are expressed in observable behaviors (Lezak, 1995). Effective NA is a prerequisite for both the scientific analysis and treatment of CNS-based cognitive/functional impairments as well as for research investigating normal functioning. The NA of persons with CNS disorders using psychometric evaluation tools serves a number of functions. These include the determination of a diagnosis, the provision of normative data on the status of impaired cognitive and functional abilities, the production of information for the design of rehabilitative strategies, and the measurement of treatment efficacy. NA also serves to create data for the scientific understanding of brain functioning through the examination of measurable sequelae that occur following brain damage or dysfunction.

Thus, treatment and rehabilitation of the cognitive, psychological, and motoric sequelae of CNS dysfunction often rely on assessment devices to inform diagnosis and to track changes in clinical status. Typically, these assessments employ paper-and-pencil psychometrics, hands-on analog tests, computer-delivered continuous performance tests, and observation/rating of behavior in real-world functional environments or within the context of physical mock-ups. On one end of the spectrum, traditional neuropsychological methods applied to impairment assessment and rehabilitation commonly use paper-and-pencil-based psychometric tests. Although these approaches provide highly systematic control and delivery of constrained performance challenges, they have also been criticized as limited in the area of *ecological* validity, that is, the degree of relevance or similarity that a test or training system has relative to the *real* world, and in its value for predicting or improving *everyday* functioning (Neisser, 1978). Adherents of this view challenge the usefulness of constrained paper-and-pencil tests and analog tasks for addressing the complex integrated functioning that is required for successful performance in the real world. On the other end of the spectrum, a common method applied in the occupational sciences discipline to assess and rehabilitate functional abilities employs behavioral observation and ratings of human performance in the *real world* or via physical mock-ups of functional environments (Weiss & Jessel, 1998). Mock-ups of daily living environments (i.e., kitchens, bathrooms) and workspaces (i.e., offices, factory settings) are typically built within which persons with cognitive impairments are observed while their performance is evaluated.

Aside from the real economic costs to physically build the environments and to provide human resources to conduct such evaluations, this approach is limited in the systematic control of real-world stimulus challenges and in its capacity to provide detailed performance data capture. As well, many functional environments in everyday life do not easily lend themselves to mock-ups, as is readily apparent in the domain of driving skill assessment and training. In this regard, *behind-the-wheel* driving assessments, considered to be the gold standard in this area, are often conducted in only the safest possible conditions (i.e., good weather, low-traffic roadways) and actually provide a limited *window* into how driving performance would fare under more realistic (and often unpredictable) conditions.

Neuropsychology is an area where VR stands to have significant impact. While the last 20 years have seen the majority of VR research and clinical application occur in the area of exposure therapy, the considerable potential of VR for the study, assessment, and rehabilitation of human cognitive and functional processes has also been recognized (Pugnetti et al., 1995; Rizzo, 1994; Rizzo, Buckwalter, & van der Zaag, 2002; Rizzo, Schultheis, Kerns, & Mateer, 2004; Rose et al., 1996, 2000, 2005). Indeed, in a US National Institute of Health report of the National Advisory Mental Health Council (1995), the impact of VR environments on cognition was specifically cited with the recommendation that "Research is needed to understand both the positive and negative effects of such participation on children's and adults' perceptual and cognitive skills ..." (p. 51). One area where the potential for both *positive and negative effects* exists is in the application of VR for NA, rehabilitation, and research. In this regard, VR could serve to advance the study of brain–behavior relationships as well as produce innovative evaluation and intervention options that are unavailable with traditional methods.

What makes VR application development in this area so distinctively important is that it represents the potential for more than a simple linear extension of existing computer technology for human use. This was recognized early on in a visionary article (*The Experience Society*) by VR pioneer, Myron Krueger (1993), in his prophetic statement that "... Virtual Reality arrives at a moment when computer technology in general is moving from automating the paradigms of the past, to creating new ones for the future" (p. 163). In this comment, Krueger encapsulated what had also been so limited in neuropsychology's approach to using computer and information technology at that time and opened a conceptual door to VR's potential to advance neuropsychological research and practice. Indeed, neuropsychology's use of technology up to that time could be characterized as mainly translating existing traditional paper-and-pencil tools directly into computer-delivered formats. In its defense, neuropsychology has been increasingly integrating advanced neural imaging technology tools (i.e., fMRI, DTI, SPECT, QUEEG, CT) in its quest for a better accounting of the structure and processes underlying brain–behavior relationships. However, while these advances in *response* measurement have led to new findings and conceptualizations, the *stimulus* delivery end of the equation has lagged behind. By way of VRs capacity to place a person within an immersive, interactive computer-generated simulation environment, new possibilities exist that go well beyond simply automating the delivery of existing paper-and-pencil testing and training tools on a personal computer. VR offers the potential to fundamentally advance this area with innovative applications that leverage the immersive, involving, and interactive assets available in VEs to deliver quantifiable analog-like stimulus protocols within the context of functionally relevant (and controllable) environments. Until now, these features have not been pragmatically available with existing methods in neuropsychology and thus VR may have plenty to offer in these vital and challenging areas.

A primary strength that VR offers for assessment is in the creation of simulated *everyday* environments in which performance can be tested in systematic fashion. By designing VEs that not only *look like* the real world but actually incorporate challenges that require integrated functional behaviors, the ecological validity of assessment methods could be enhanced. As well, within a VE, the experimental control required for rigorous scientific analysis and replication can still be maintained within simulated contexts that embody the complex challenges found in naturalistic settings. Thus, on a theoretical level, VR-derived results could have greater predictive validity and clinical relevance for the challenges that patients face in everyday life compared to paper-and-pencil testing. On a more pragmatic level, rather than relying on costly physical mock-ups of functional assessment and rehabilitation environments, VR offers the option to produce and distribute identical *standard* simulation environments. Within such digital assessment (and rehabilitation) scenarios, normative data can be accumulated for performance comparisons in an automated fashion, needed for assessment/diagnosis and for treatment/rehabilitation purposes.

Thus, it is possible that the use of VR technology could revolutionize our approach to NA. However, while encouraging on a theoretical level, the value of this technology for neuropsychology still needs to be substantiated via systematic empirical research with normal and clinical populations that can be replicated by others. To accomplish this first requires specification as to the real assets that VR offers that add value over existing assessment methodologies as well as further exploration of its current limitations. The current status of VR technology applied to clinical populations, while provocative, is still limited by the small (but growing) number of controlled studies in this area. This is to be expected, considering the technology's relatively recent development and the lack of familiarity with VR technology by established researchers employing the traditional tools and tactics of their fields. In spite of this, a nascent body of work has emerged that can provide knowledge for guiding future research efforts.

In the mid-1990s, using graphic imagery that would be considered primitive by today's standards, Pugnetti et al. (1995, 1998) developed an HMD-delivered VR scenario that embodied the cognitive challenges that characterize the Wisconsin Card Sorting Test (WCST). The scenario consisted of a virtual building within which users were required to use environmental clues to aid in the correct selection of appropriate doorways needed to pass from room to room through the structure. The doorway choices varied according to the categories of shape, color, and number of portholes. Similar to the WCST, the correct choice criteria were changed after a fixed number of successful trials, and the user was then required to shift cognitive set, look for clues, and devise a new choice strategy in order to successfully pass into the next room. In one study, Pugnetti et al. (1998) compared a mixed group of neurological patients (MS, stroke, and traumatic brain injury [TBI]) with normals' performance both on the WCST and on this HMD-based executive function system. Results indicated that the VR results mirrored previous anecdotal observations by family members of everyday performance deficits in the patient populations. Though the psychometric properties of the VE task were comparable to the WCST in terms of gross differentiation of patients and controls, weak correlations between the two methods suggested that the methods measured different aspects of these functions. A detailed analysis of the VR task data indicated that specific preservative errors appeared earlier in the test sequence compared to the WCST. The authors suggested that "... this finding depends on the more complex (and complete) cognitive demands of the VE setting at the beginning of the test when perceptuomotor, visuospatial (orientation), memory, and conceptual aspects of the task need to be fully integrated into an efficient routine" (p. 160). The detection of these early *integrative* difficulties for this complex cognitive function may be particularly relevant for the task of predicting real-world capabilities from test results. This was further evidenced in a detailed single subject case study of a stroke patient using this system. In this report (Mendozzi, Motta, Barbieri, Alpini, & Pugnetti, 1998), results indicated that the VR system was more accurate in identifying executive function deficits in a highly educated patient two years poststroke, who had a normal WCST performance. The VR system, although using graphic imagery that would never be mistaken for the real world, was successful in detecting deficits that had been reported to be limiting the patient's everyday performance, yet were missed using existing NP tests. These results are in line with the observation that patients with executive disorders often perform relatively well on traditional NP tests of *frontal lobe function* yet show marked impairment in controlling and monitoring behavior in real-life situations (Shallice & Burgess, 1991).

Similar findings were reported by McGeorge et al. (2001) in a study comparing real-world and virtual-world *errand running* performance in five TBI patients and five matched normal controls. The selection of the patient sample for this study was based on staff ratings that indicated poor planning skills. However, the patient and control groups did not differ significantly from normative values on the behavioural assessment of the dysexecutive syndrome (BADS) battery (Wilson, Alderman, Burgess, Emslie, & Evans, 1996). Videotaped performance of subjects was coded and compared while performing a series of errands in the University of Aberdeen psychology department (real world) and within a flat-screen VR scenario modeled after this environment. Performance in both the real and VE, defined as the number of errands completed in a 20 min period, was highly correlated ($r = .79$; $p < .01$). Interestingly, while the groups did not differ on age-corrected standardized scores on the BADS, significant differences *were* found between the groups in both the real world and virtual testing. This finding suggests several things. First, performance in the real and virtual world was functionally similar; second, patient and control groups could be discriminated equally using real and virtual tests while this discrimination was not picked up by standardized testing with the BADS; and third, both measures of real and virtual world performance showed concordance with staff observations of planning skills. That these results support the view that VR testing may possess higher ecological value is in line with the observation by Shallice and Burgess (1991) that traditional neuropsychological tests do not demand the planning of behavior over more than a few minutes or the prioritization of competing subtasks and may result in less effective prediction of real-world performance.

Other research has also demonstrated VR's usefulness for NA, particularly for visuospatial processes (Kaufmann, Steinbügl, Dünser, & Glück, 2005; Koenig, Crucian, Dalrymple-Alford, & Dünser, 2009, 2010; Koenig, Crucian, Dünser, Bartneck, & Dalrymple-Alford, 2011; McGee et al., 2000; Parsons, Rizzo, van der Zaag, McGee, & Buckwalter, 2005; Rizzo et al., 2001). For example, VR systems have been used to assess mental rotation, a cognitive function whereby a person needs to visualize the movement and organization of objects in 3D space (Shepard & Metzler, 1971). Mental rotation is important for everyday tasks such as driving, organizing items in a limited space, and any activity that relies on dynamic imagery for prediction of object movement. In the normal population, men outperform women on the mental rotation task. However, *hands-on* performance on a VR spatial rotation task that mimicked the task structure of the original mental rotation task showed no gender difference, and mental rotation was shown to be dramatically improved in both women and low-performing men after a brief 10–15 min period of interaction with the hands-on VR interaction task (Rizzo et al., 2001). Along similar lines, the Morris water maze test of spatial navigation and place learning, commonly used with rodents, has also been simulated in a VE as a test for humans (Astur, Taylor, Mamelak, Philpott, & Sutherland, 2002, Astur, Tropp, Sava, Constable, & Markus, 2004). In this application, the person being tested must use visual cues in the surrounding environment to help guide navigation to a hidden platform. Used in conjunction with fMRI, the test can be applied to determine whether a person has decreased hippocampal activity that might be indicative of Alzheimer's disease (Shipman & Astur, 2008) or schizophrenia (Folley, Astur, Jagannathan, Calhoun, & Pearlson, 2010). Such integration of VR as a complex stimulus delivery tool with advanced brain imaging and psychophysiological techniques may allow neuropsychology to reach its stated purpose, that of determining unequivocal brain–behavior relationships.

Along with this seminal work with adults, other researchers have followed the aircraft simulation metaphor with the creation of VR worlds designed to assess cognitive and functional performance in children with a range of CNS-related disease or injury conditions. For example, researchers have created virtual homes, classrooms, public spaces, and traffic-filled streets to test and train children with autistic spectrum disorder (ASD) and other developmental disorders on activities relevant to fire safety (Rizzo, Strickland, & Bouchard, 2004), social skills (Parsons et al., 2000), functional attention (Rizzo et al., 2006), street crossing (Bart, Katz, Weiss, & Josman, 2008; Strickland, Marcus, Mesibov, & Hogan, 1996), and earthquake safety (Raloff, 2006). Even a virtual obstacle course was created for determining if a child is capable of using a motorized wheelchair safely and effectively and to support training if they are not quite ready yet (Inman, Loge, & Leavens, 1997). For more details on clinical VR research in children, see Parsons et al. (2009) and Rizzo, Lange et al. (2011).

45.5 USE CASE: THE VIRTUAL CLASSROOM ATTENTION ASSESSMENT PROJECT

The original virtual classroom project began in 1999 as part of a basic research application program at the University of Southern California aimed at developing VR technology applications to improve our capacity to understand, measure, and treat the cognitive/functional impairments commonly found in clinical populations with CNS dysfunction as well as advance the scientific study of normal processes and function. The virtual classroom was designed as an HMD VR system for the assessment of attention processes in children. Efforts to target this cognitive process were supported by the widespread occurrence and relative significance of attention impairments seen in a variety of clinical conditions that affect children. Notable examples of childhood clinical conditions where attention difficulties are seen include ADHDs, TBI, and fetal alcohol syndrome. With these clinical conditions, VR technology provides specific assets for assessing attention that are not available using existing methods. For example, HMDs that serve to occlude the distractions of the outside world are well suited for these types of cognitive assessment applications. Within an HMD, researchers and clinicians can provide a controlled stimulus environment where attention (and other cognitive) challenges can be presented along with the precise delivery and control of *distracting* auditory and visual stimuli within the VE. This level of experimental control allows for the development of attention assessment/rehabilitation tasks that are more similar to what is found in the real world and, when delivered in the context of a relevant functional VE, could improve on the ecological validity of measurement and treatment in this area. The first project with the virtual classroom focused on attention assessment in children with ADHD. The heterogeneous features of ADHD, a behavioral disorder marked by inattention, impulsivity, and/or hyperactivity, have made consensus regarding its diagnosis difficult. Furthermore, traditional methods for assessing ADHD in children have been questioned regarding issues of reliability and validity. Popular behavioral checklists have been criticized as biased and not a consistent predictor of ADHD, and correlations between concordant measures of ADHD, such as parent and teacher ratings of hyperactivity, have been repeatedly shown to be modest at best and frequently low or absent (Barkley, 1990; Colegrove, Homayounjam, Williams, Hanken, & Horton, 1999). Due to the complexity of the disorder and the limitations of traditional assessment techniques, diagnostic information is required from multiple types of ADHD measures and a variety of sources in order for the diagnosis to be given (American Psychiatric Association, 2000; Greenhill, 1998). Thus, in the area of ADHD assessment where traditional diagnostic techniques have been plagued by subjectivities and inconsistencies, it was believed that an objective and reliable VR strategy might add value over existing approaches and methods.

The initial research version of the system was run on a standard Pentium 3 processor with the nVIDIA G2 graphics card. The HMD used in this study was the V8 model from virtual research. Tracking of the head, arm, and leg used three 6-degree-of-freedom magnetic *flock-of-birds* trackers from Ascension Technology Corp. In addition to tracking head movement in real time to update the graphics display in the HMD, the tracking system also served to capture body movement metrics from the tracked locations. This provided concurrent data on the hyperactivity component that is a commonly observed feature of ADHD. The research version of the virtual classroom scenario consisted of a standard rectangular classroom environment containing desks, a female teacher, a blackboard across the front wall, a side wall with a large window looking out onto a playground and street with moving vehicles, and, on each end of the opposite wall, a pair of doorways through which activity occurred (see Figure 45.4). Within this scenario, children's attention performance was assessed while a series of common classroom distracters (i.e., ambient classroom noise, activity occurring outside the window) were systematically controlled and manipulated within the VE. The child sat at a virtual desk within the virtual classroom, and on-task attention was measured in terms of reaction time performance and error profiles on a variety of attention challenge tasks that were delivered visually using the blackboard or auditorily via a virtual teacher's voice.

Prior to any clinical tests with the virtual classroom, we applied a user-centered design methodology to insure that the application was usable and safe for children. User-centered approaches

(a)

(b)

(c)

FIGURE 45.4   Scenes from the virtual classroom. (a) Screen shot of a bus driving past window (common classroom distracter), (b) screen shot of classroom scenario with other students and teacher present, and (c) screenshot of a person entering the room (common classroom distracter).

generally require the involvement of the targeted user group in the early design and development phase of scenario development. This involves a series of tight, short heuristic, and formative evaluation cycles conducted on basic components of the system. Consideration of user characteristics in this fashion has increasingly become standard practice in VR development (Hix & Gabbard, 2002). In the virtual classroom's user-centered design evaluation phase, 20 nondiagnosed children (ages 6–12 years) tried various evolving forms of the system over the first year of development, and their performance was observed while trying out a variety of basic selective and alternating attention tasks. One such task involved having users recite the letters that appeared on the blackboard while naming the color of a virtual-chapter airplane that passed across the classroom at random intervals. We also solicited feedback pertaining to aesthetics and usability of the VE and incorporated some of this feedback into the iterative design–evaluate–redesign cycle. Overall, these initial results indicated little difficulty in adapting to the use of the HMD, no self-reported occurrence of side effects as determined by posttest interviews using the simulator sickness questionnaire (SSQ) (Kennedy, Lane, Berbaum, & Lilienthal, 1993), and excellent performance on the stimulus tracking challenges.

Following this user-centered design phase, we conducted a clinical trial that compared eight physician-referred ADHD males (age 6–12 years) with 10 nondiagnosed male children. The groups did not significantly differ in mean age, grade level, ethnicity, or handedness, and all ADHD-diagnosed children were currently taking stimulant medication as treatment for their condition. However, ADHD participants were off medication during the testing period with all testing occurring between 9 and 11 a.m. prior to normal medication ingestion. ADHD participants were excluded from the study if they presented with comorbid autism, mental retardation, full-scale IQ score < 85, or head injury with loss of consciousness greater than 30 min. These same exclusion criteria were applied to the normal control group. Research participants were instructed to view a series of letters presented on the blackboard and to hit the response button only after he or she viewed the letter "X" preceded by an "A" (successive discrimination task). This AX continuous performance task consisted of the letters A, B, C, D, E, F, G, H, J, L, and X. The letters were white on a gray background (the virtual blackboard) and presented in a fixed position directly in front of the participant. The stimuli remained on the screen for 150 ms, with a fixed interstimulus interval of 1350 ms. The target letter X (correct hit stimuli) and the letter X without the A (incorrect hit stimuli) each appeared with equal probability of 10%. The letters A and H both appeared with a frequency of 20%. The remaining eight letters occurred with 5% probability.

Participants were instructed to press the mouse button as quickly and accurately as possible (with their dominant hand) upon detection of an X after an A (correct hit stimuli) and withhold their response to any other sequence of letters. Four hundred stimuli were presented during each of two 10 min conditions. The two 10 min conditions consisted of one without distraction and one with distractions (pure audio-classroom noises, pure visual-chapter airplane flying across the visual field, and mixed audiovisual—a car *rumbling* by the window and a person walking into the classroom with hall sounds occurring when the door to the room was opened). Distracters were each displayed for 5 s and presented in randomly assigned equally appearing intervals of 10, 15, or 25 s, and 36 distracters (9 of each) were included in the 10 min condition. As well, 6-degree-of-freedom tracking from the head, arm, and leg was used to produce movement metrics needed to analyze the motor hyperactivity component in conjunction with the cognitive performance. VR performance was also compared with results from standard neuropsychological testing. The following results summarize the outcomes of this study:

- No significant side effects were observed in either group based on pre- and post-VR SSQ testing.
- Children with ADHD had *slower* correct hit reaction time compared with normal controls on the distraction condition (760 ms vs. 610 ms; $t (1,16) = -2.76, p < .03$).
- Children with ADHD had higher reaction time *variability* on correct hits compared with normal controls on both the no-distraction (SD = 220 ms vs. 160 ms; $t (1,16) = -2.22$, $p < .05$) and distraction conditions (SD = 250 ms vs. 170 ms; $t (1,16) = -2.52, p < .03$).

- Children with ADHD made more omission errors (missed targets) compared with normal controls on both the no-distraction (14 vs. 4.4; $t\,(1,16) = -4.37$, $p < .01$) and distraction conditions (21 vs. 7.2; $t\,(1,16) = -4.15$, $p < .01$).
- Children with ADHD made more commission errors (impulsive responding in the absence of a target) compared with normal controls on both the no-distraction (16 vs. 3.7; $t\,(1,16) = -3.15$, $p < .01$) and distraction conditions (12.1 vs. 4.2; $t\,(1,16) = -3.22$, $p < .01$).
- Children with ADHD made more omission errors in the distraction condition compared to the no-distraction condition (21 vs. 14; $t\,(1,16) = -3.50$, $p < .01$). No such differences on omission and commission errors were found with the nondiagnosed children across no-distraction and distraction conditions.
- Exploratory analysis of motor movement in children with ADHD (tracked from the head, arm, and leg) indicated higher activity levels on all metrics compared to nondiagnosed children across both conditions.
- Exploratory analysis of motor movement in children with ADHD also indicated higher activity levels on all metrics in the distraction condition compared to the no-distraction condition. This difference was not found with the normal control children.
- Exploratory analysis using a neural net algorithm (support vector machine analysis) trained to recognize a stereotypic leg movement on the first five participants in each group was able to accurately discriminate the remaining subjects to groups at 100%.

These data suggested that the virtual classroom had good potential as an efficient, cost-effective, and scalable tool for conducting attention performance measurement beyond what existed using traditional methodologies. The system allowed for controlled performance assessment within an ecologically valid environment and appeared to parse out significant effects due to the presence of distraction stimuli. Additionally, the capacity to integrate measures of movement via the tracking technology further added value to this form of assessment when compared to traditional analog tests and rating scales. In this regard, an HMD appeared to be the optimal display format. Although one of the common criticisms of HMD technology concerns field of view (FOV) limitations, in this application, the limited FOV fostered head movement to supplant eye movement as the primary method for scanning the virtual classroom. This type of *poor man's* tracking of behavioral attention within the controlled stimulus environment obtained in the HMD allowed for ongoing documentation as to where the user was *looking* during test stimulus content delivery. For example, a child missing a target while directly looking at the blackboard is illustrating an attentional error that is fundamentally different from the occurrence of a missed target due to the child looking out the window at a distraction. The documentation provided by head tracking in an HMD can be used to produce metrics of percent time on task during stimulus *hit* trials as well as allowing for a recreation of a naturalistic behavioral performance record for later review. In the current research, ADHD children were found to miss targets due to looking away from the blackboard during 25% of the *hit* trials as opposed to normal subjects who were documented to be looking away at less than 1% of the time. This form of integrated cognitive/behavioral performance record of attention performance during delivery of systematic distraction is simply not obtainable using other methods. More detailed information on the rationale, methodology, other studies, and long-term vision for this project can be found in Bioulac et al. (2012); Gilboa et al. (2011); Parsons, Bowerly, Buckwalter, and Rizzo (2007); Pollak, Barhoum-Shomaly, Weiss, Rizzo, and Gross-Tsur (2010); and Rizzo et al. (2006). *For more information, see a virtual classroom video at* http://www.youtube.com/watch?v=JBIhey7sjzg and http://www.youtube.com/watch?v=daUu3iXyWWY.

45.6   USE CASE: THE ASSESSIM OFFICE PROJECT

The Assessim Office (AO) project is a complex cognitive assessment in a virtual office environment inspired by the virtual office object memory assessment (Matheis et al., 2007). The AO consists of a large office room containing virtual characters and a range of interactive objects (e.g., printer,

FIGURE 45.5   Assessim Office VR cognitive assessment.

monitor, phone, remote control) that enable researchers and clinicians to simulate complex real-life scenarios that closely reflect the cognitive demands of everyday occupational tasks (Figure 45.5). In collaboration with the USC ICT and the Kessler Foundation Research Center, the AO was developed to assess divided/selective attention, complex problem solving, and prospective memory (Koenig et al., 2012; Krch et al., 2013). The application systematically exposes the user to a complex set of stimuli that vary in cognitive demands, cognitive domains, and priorities. The temporal overlap between stimuli requires the user to prioritize tasks and manage a scenario that is very similar to real-world requirements in a busy office environment. Despite its complexity, the AO offers precise control over stimulus timing, frequency, distracters, and instructions, thus providing the foundation for high ecological validity and potentially a distinct advantage in predictive validity over traditional paper-and-pencil NAs. The application was developed in an iterative, user-centered process that involved clinicians, researchers, and software engineers from the USC ICT and Kessler Foundation Research Center.

In more than 20 iterations, the application's input scheme, instructions, task difficulty, and data collection algorithms were tested and improved to best accommodate the needs of clinicians and patients (Koenig et al., 2012). As a result of this iterative development approach, it is possible to quickly make changes to any of the application's components, which in turn produced a system with significant flexibility for adjusting the application's functionality to different patient populations or virtual task demands. Thus, with very little development effort, the AO can accommodate a wide variety of virtual assessment and training scenarios and reduce development costs at a time when research and health-care budgets are decreasing. When the application's flexibility is combined with a robust database infrastructure for data storage and analysis, such a VR cognitive assessment and training system could provide the ultimate clinical toolbox for working with a wide range of patient populations. Since its first clinical use as an executive function assessment, the application has been extended to quantify attention performance in children with ADHD and adults with sleep disorders (Figure 45.6).

The AO has been used in a pilot trial to assess executive functions in seven adults with TBI, five adults with MS, and a control group of seven healthy adults (Krch et al., in press). Participants were placed at a virtual desk in the environment and had the option to interact with the keyboard, printer, a projector remote control, and several other items in the virtual scene. Moreover, as part of a more complex decision-making task, participants were given two sets of criteria for accepting and printing incoming real estate offers. Ringing phones and ad

FIGURE 45.6   Assessim Office extension for attention assessment.

displays on the computer screen provided distractions to the virtual tasks. Each AO session was identical for each participant and lasted approximately 15 min. The key outcomes of this trial are summarized as follows:

- All experimental groups were able to use the AO without experiencing difficulties in user input or comprehension of the VE. All participants were successful in navigating through the virtual office using PC mouse inputs only.
- The AO was able to simulate a range of cognitive tasks that placed high cognitive demands on the participants. As a consequence, MS and TBI patients experienced difficulties in comprehending the requirements of the decision-making tasks (i.e., printing real estate offers). Patients mostly ignored these tasks and thus made significantly less errors than the healthy participants who attempted to solve the task, but committed several errors.
- The AO was able to differentiate TBI patients from healthy controls through significantly less correct real estate decisions in the decision-making task ($U = 4.50$, $p = .007$, $r = .69$).
- MS patients committed significantly more errors during the decision-making task ($U = 4.00$, $p = .030$, $r = .64$). Further, the MS group more often failed to successfully turn the projector light back on, which indicated impaired divided attention.
- Only few of the VR cognitive tasks showed significant correlations to traditional paper-and-pencil NA. For TBI patients, correct email responses significantly correlated with the D-KEFS color–word inhibition/switching test ($p = .037$). Errors for the decision-making tasks and the divided attention task correlated significantly with the D-KEFS Trail Making Test Condition 4 ($p = .030$ and $p = .005$). For MS patients, only one of several decision-making tasks correlated significantly with the WASI matrix reasoning test.

These results only provide little evidence for a clear relationship between VR cognitive tasks and their pencil-and-paper counterparts that assess similar cognitive constructs. While these pilot results show promise for the feasibility of the tested VR tasks, it is currently unclear whether a strong relationship with traditional NA is actually necessary. Instead, the results of several pilot studies (Koenig, 2012; Krch et al., 2013) suggest that the more complex nature of VR tasks could purposefully deviate from pencil-and-paper assessments and in turn provide value through increased predictive validity for activities of daily living (ADLs). However, the relationship between cognitive performance in this VE and real-world scenarios is still unclear. A systematic assessment of the AO (or similar VR systems) in relationship to real-world functional tasks is now needed to provide evidence for the utility of the complex measurements that are possible within a VR cognitive assessments.

45.7   GAME-BASED REHABILITATION

Evidence-based interventions to improve motor function stem from an ever-changing literature in the understanding of neuroplasticity, recovery processes, and the requirements for treatment dosing, timing, and context. The World Health Organization's International Classification of Functioning, Disability and Health (ICF) (WHO, 2001) expands traditional concepts of physical and occupational therapy by including the impact of identified deficits to the patient's unique social and vocational contexts. Clinicians base treatment plans on the ICF framework for assessing the level of pathology (e.g., asymmetric peripheral vestibular function), impairment (e.g., gaze or gait instability), functional limitations (e.g., ability to perform job-related tasks), and disability (e.g., inability to function in one's chosen vocation) (Scherer & Schubert, 2009). For the care of individuals with neurological impairment or injury, the ICF model provides a rehabilitation-focused complement to traditional medical, diagnosis-driven, or mechanism-of-injury-driven models of clinical management (Rauch et al., 2008). Individualized treatments are often based on existing methods and tools and combined in meaningful ways to address the individual's capacity and needs. Previous research has identified the process of treating and tracking progress of persons with neurological impairments as variable, complex, and difficult (Saarela & Lindmark, 2008; Thornton et al., 2005; Weightman et al., 2010).

Rehabilitation providers can address limitations in function and participation and improve performance in all areas by providing interventions to improve physiological function, motor function, and cognitive performance. Optimal rehabilitation therapy involves high doses of skilled repetitive exercises. High demands are placed on the clinician's time, professional knowledge, and experience to find and work out the most effective and appropriate methods of treatment for the individual's rehabilitation requirements and goals. Resources are limited, and it can be difficult for clinicians to provide these exercises during one-on-one time in the clinic setting. Repetitive skilled practice is crucial for rehabilitation, often requiring individuals to perform exercise programs as *homework* outside of standard therapy sessions (Weightman et al., 2010). Furthermore, once outpatient treatment programs are completed, few options exist to provide continuation of care for individuals that still require much needed exercise programs and monitoring. Home-based and clinic-based exercise programs are often provided to clients in the form of a list of exercises with pictures and instructions. However, the individual is responsible for completing the exercises accurately and reporting progress back to the clinician. Adherence to self-guided exercise programs in the clinic and home settings is notoriously low and very difficult to quantify due to the reliance on an individual's subjective feedback and reliance on the maintenance of accurate recording of their exercise sessions in the form of exercise diaries. Providing clients with tools that can motivate them to practice home-based exercise activities in a safe, structured, and well-monitored way is a large gap in the area of rehabilitation as a whole and particularly in the area of neurorehabilitation. Current home exercise tools and activities are often lacking the capability to quantitatively track performance and provide progress feedback to the client and clinician within and across sessions. These tools often are limited to one area of impairment, such as upper limb mobility or dynamic balance.

The use of interactive technologies such as video games and VR systems for exercise and rehabilitation has expanded rapidly over the past fifteen years. Early VR research and development efforts that encourage people to exercise and assist in the relearning of movements have been successfully applied to a wide range of functional deficits (Adamovich et al., 2004; Henderson et al., 2007; Merians et al., 2002; Stewart et al., 2007). Interactive technologies, such as VR and video games, demand focus and attention, can motivate the user to move, and provide the user with a sense of achievement, even if they cannot perform that task in the *real world* (Rizzo et al., 2005). Some of the first clinical VR systems were specifically developed for motor rehabilitation in the

1990s. Initially, these VR applications required expensive computing and tracking hardware, which limited potential access to clinicians and patients in the clinic or home setting due to cost, space, and maintenance requirements. In recent years, there has been a shift from high-cost VR and robotic systems to low-cost, easily accessible video game systems. Video game technology and tools are improving in a way that allow for the development of low-cost game-based applications that are accessible for everyday use in clinic and home settings. The worldwide acceptance and enjoyment of physically interactive gaming consoles such as the Nintendo® Wii™, Nintendo® Wii Fit™, Sony Playstation®2 EyeToy™, Sony Playstation®3 MOVE™, and Microsoft® Xbox360 Kinect™ have demonstrated a shift toward exercise programs that can distract clients and provide long-term motivation. The underlying motion sensing and 3D graphics technologies that are used in these commercial game systems allow the user to engage in entertaining, physically interactive games using gross body movements that are not bound by the limits of a mouse, joystick, or gamepad interface. A number of researchers have treated small samples of neurologically impaired participants using video game systems such as the Sony PlayStation2 EyeToy (Flynn et al., 2007; Haik et al., 2006; Neil, Ens, Pelletier, Jarus, & Rand, 2013; Rand et al., 2008) and Nintendo® Wii™ (Deutsch et al., 2008; Pompeu et al., 2012; Saposnik et al., 2010), with promising results.

However, off-the-shelf video games for consoles such as the Nintendo Wii and Microsoft Xbox should be used with caution in the rehabilitation setting. User testing and exploratory studies of off-the-shelf video games in the clinical setting have demonstrated that the games are not safe or suitable for all patient populations without direct supervision. The tasks and game actions required of the player in these off-the-shelf video games are not designed for rehabilitation; they are designed for entertainment. Therefore, the level of challenge, progression of task difficulty, and required movement patterns are not developed to align with specific rehabilitation goals and principles of motor learning. While suitable for some patient populations, the level of challenge is often too high for people with physical impairments. Provision of feedback to the player on specific components of performance is not available in off-the-shelf video games. The feedback after game sessions is often negative, with terms such as *you failed* and *unbalanced*, thus providing the player with potentially frustrating feedback that could reduce motivation and sense of accomplishment. Clinicians have identified issues with off-the-shelf games, indicating that they do not have control over game task parameters, including the type, number, location, and speed of game objects/targets. Many people with neurological impairment or injury are generally unable to perform all of the required movements and actions in the game. Research and anecdotal evidence in the clinical setting have demonstrated that patients with neurological injury can develop *cheat* moves or use compensatory body movements to accomplish game tasks, thus defeating any rehabilitative benefit of the game (Lange, Flynn, & Rizzo, 2009). Providing options to change the game objects and requirements could potentially customize the games to patients' individual requirements and reduce the potential for compensatory and poor movement patterns. While game technology has the potential to provide quantitative data that could be used to track progress, the video games often do not record meaningful information for tracking player progress. Most games only provide a general game score. Scores are very ambiguous measures of player performance and assume that the player has full mental and physical abilities. These issues with off-the-shelf video games have led to a number of researchers beginning to use existing video game hardware to develop customizable software that addresses specific rehabilitation goals.

Larger clinical trials of off-the-shelf games and customized VR and video game interventions have been conducted in the past few years. Laver et al. (2011) completed a Cochrane review to evaluate the effects of VR and interactive video games on upper limb, lower limb, and global motor function after stroke. Laver et al. (2011) used a very broad definition of VR and video game intervention; therefore, a wide range of interventions with largely varied rehabilitation goals, complexity, cost, and hardware/software requirements were included ranging from driving simulators, assistive devices, robotic devices, off-the-shelf video games (Wii), and treadmill walking to visual–perceptual training. The review reported positive effects of VR on upper limb function and ADL performance, but it highlights a larger issue of the lack of cohesive and structured direction in the field of VR rehabilitation as a whole (Laver et al., 2011). Similar issues were highlighted in a meta-analysis focused on VR for upper limb rehabilitation following stroke (Saposnik et al., 2011). The 12 studies included in the analysis used widely varied interventions ranging from off-the-shelf games, haptic systems, and instrumented gloves to virtual teacher avatars. The variety of intervention methodologies provides some explanation for the mixed results of this meta-analysis (Saposnik et al., 2011). In conclusion, there appears to be little consensus on the effectiveness of VR and video game interventions in the rehabilitation setting. This lack of consensus stems from several issues including (1) a broad and variable definition of VR; (2) heterogeneity of developed interventions and tools, in particular the use of standard tools vs. customizable tools; (3) variation in rehabilitation goals and research questions; and (4) a lack of consensus on appropriate study design and research methodology. Well-developed trials with clearly defined interventions and comparators are a required next step toward the evaluation of VR technologies in the rehabilitation setting.

45.7.1 Use Case: Jewel Mine Application

Our initial work at the USC ICT has focused on addressing the challenge for creating low-cost, home-based VR systems for motor and cognitive assessment and rehabilitation. In an attempt to overcome the previously mentioned discrepancies in research methodology and VR system design, the focus of our research has been threefold: (1) assess the usability of off-the-shelf games and consoles across a range of user populations (TBI, stroke, spinal cord injury, and amputation and older adults at risk of falls), (2) using the initial feedback to repurpose or develop low-cost interaction devices that are appropriate for use within the rehabilitation setting, and (3) design, develop, and test original games specifically focused on rehabilitation tasks. Our initial observational studies brought attention to the need to test off-the-shelf video game technologies and raised awareness of potential issues with using these systems in neurological rehabilitation (Flynn & Lange, 2010; Lange et al., 2009). Through a series of ongoing focus groups, surveys, and user tests with clinicians, patient groups, and caregivers, we identified future directions of research and development (Lange, Flynn, Proffitt, Chang, & Rizzo, 2010; Lange et al., 2009, 2011; Proffitt & Lange, 2013). Based on these initial observations, we developed an original VR rehabilitation tool called Jewel Mine that directly addresses the needs specified by patient and clinician feedback. Specifically, Jewel Mine allows the user (i.e., patient or clinician) without computer programming skills to modify stimulus delivery parameters and to extract and view performance data following interaction with the game system.

Jewel Mine is a rehabilitation therapy tool designed to motivate people with orthopedic and neurological injury or impairments, including stroke, TBI, spinal cord injury, and balance disorders associated with aging. The Microsoft Kinect is connected to a Windows PC that is running the Microsoft Kinect SDK and a custom software application. In Jewel Mine, a player takes on the task of gathering jewels from a mine shaft by reaching out and touching each jewel individually (Figure 45.7a). The jewels are located in 3D space in front of the user and placed at an appropriate distance away from the player based on their level of ability (Figure 45.7b). The VE and associated tasks can be changed easily during game play. For example, the scene can be switched instantly to a meadow where the user must reach out to gather flowers (Figure 45.7c). The level of challenge and type of task can be customized within the software through simple menus (Figures 45.7d and e). Task performance and movement data based on the tracking data from the Microsoft Kinect can be saved and analyzed (Figure 45.7f). This analysis provides the clinician with information about player performance within and across sessions, data that usually are not captured with traditional clinical interventions, especially in the home setting.

The system is currently undergoing user testing and iterative development in collaboration with 10 clinics across the United States and internationally with a range of different patient populations. Formal user testing sessions involve semistructured interviews around the particular prototype we are testing and how it potentially can be used in the clinic or home setting (Lange et al., 2009, 2010, 2011).

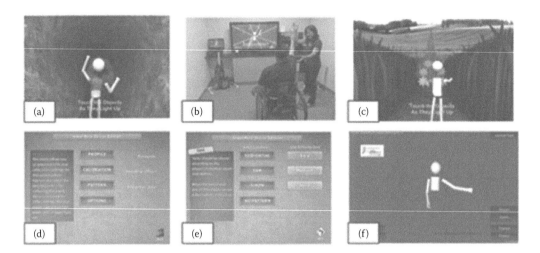

FIGURE 45.7 Screen shots representing features of the VR rehabilitation tool. (a) Screen shot of player represented on screen within 'Jewel Mine' scenario. (b) User interacting with VR tool in clinical setting. (c) Screen shot of player represented on screen within 'Field of flowers' scenario. (d) Screen shot of menu that enables clinician to set difficulty level. (e) Screen shot of menu that enables clinician to customize the task for the user. (f) Screen shot of player represented on screen during replay of previous session.

A large number of participants have identified the potential for this game-based rehabilitation technology to be used in the home setting to assist with providing feedback and to improve adherence to exercise protocols. Participants in our focus groups and user testing sessions identified the potential for this technology to improve physical and cognitive functions, improve or prevent dependence on others for assistance with ADLs, and provide support and feedback on progress with exercise protocols (Lange et al., 2009, 2010; Proffitt & Lange, 2013).

The Jewel Mine application has also been the focus of case studies, feasibility, and initial efficacy studies in a variety of settings with different clinical populations. In a recent case study, a 69-year-old woman with Parkinson's disease and a history of frequent falls performed 1 h of balance activities using the Jewel Mine program two times a week for 8 weeks. During the intervention, the use of in- and out-of-game modifications assisted in challenging the player while reaching, turning, and dual tasking. Gait and balance performance was assessed before and after the intervention, and the participant demonstrated significant and clinically meaningful improvement in forward functional reach (pre, 6 in.; post, 9.5 in.). This study demonstrated that customized VR using full-body movement tracking can be feasibly used during balance training in the clinical setting to enhance key components of rehabilitation tasks (Rademaker et al., 2013).

A study was recently completed as part of a Rehabilitation Engineering Research Center (OPTT-RERC, funded by NIDRR) that was designed to assess the feasibility and initial efficacy of using the Jewel Mine game as a balance training tool with the older adult population (Proffitt et al., 2012; Requejo et al., 2012; Wade et al., 2012). Thirty older adults interacted with the Jewel Mine game's virtual reaching tasks and a comparable real-world task of similar motor demands. Participants enjoyed the Jewel Mine game and showed greater levels of absorption and presence compared to the real task. They had less of an internal locus of attention during the game and focused more on the VE (Proffitt et al., 2012). The VE also impacted the motor performance of the older adults as they showed differences in the use of planning motor actions and online feedback between the virtual and real environments (Wade et al., 2012).

An individual with stroke completed a case study using the VR Jewel Mine game as a 6-week in-home exercise intervention. The system was easily set up in the home environment after some adjustments to the relative locations of each piece of technology (computer, monitor, Kinect sensor, etc.). The intervention was easily integrated into the daily routine of the participant. The individual was already very active in their day-to-day activities and frequently completed thier own in-home stretching and strengthening exercises. In 6 weeks, the participant completed an average of 185 min/week (approximately 37 min/day over a 5-day week). The individual's mood and energy levels before and after playing were generally positive but fluctuated throughout the study and depended on family and personal schedule demands. The participant showed trends toward improvement in functional mobility, self-rated balance confidence, and self-rated occupational performance. With regard to usability, the participant rated the system positively and provided suggestions for future improvements during an in-depth interview postintervention (Proffitt & Lange, 2013).

The key advantage of designing customizable VR tools is to provide the clinician and/or patient with the ability to alter elements of game play in order to tailor treatment tasks for individual users and expand the use of these tasks to a wider range of level of ability. The use of VR rehabilitation tools should always complement and reinforce the goals of existing therapies while improving motivation to perform therapeutic exercise programs. There is still much to learn about the design, development, and implementation of these technologies. More structure and consensus is required by researchers and clinicians in this field to define which VR and video game tools are appropriate for different populations and rehabilitation situations.

45.8   VH AGENTS

Recent shifts in the social and scientific landscape have now set the stage for the next major movement in clinical VR with the *birth* of intelligent VHs. With advances in the enabling technologies allowing for the design of ever more believable context-relevant *structural* VR environments (e.g., combat scenes, homes, classrooms, offices, markets), the next important challenge will involve *populating* these environments with VH representations that are capable of fostering believable interaction with *real* VR users. This is not to say that representations of human forms have not usefully appeared in clinical VR scenarios. In fact, since the mid-1990s, VR applications have routinely employed VHs to serve as stimulus elements to enhance the realism of a virtual world simply by their static presence.

For example, VRET applications have targeted simple phobias such as fear of public speaking and social phobia using virtual social settings inhabited by *still-life* graphics-based characters or 2D photographic sprites (i.e., static full-body photo images of a person with background content removed to allow for relatively seamless insertion into a virtual world) (Anderson, Zimand, Hodges, & Rothbaum, 2005; Klinger, 2005; Pertaub, Slater, & Barker, 2002). By simply adjusting the number and location of these VH representations, the intensity of these anxiety-provoking VR contexts could be systematically manipulated with the aim to gradually habituate phobic patients and improve their functioning in the real world. Other clinical applications have also used animated graphic VHs as stimulus entities to support and train social and safety skills in persons with high-functioning autism (Padgett, Strickland, & Coles, 2006; Rutten et al., 2003) and as distracter stimuli for attention assessments conducted in a virtual classroom (Parsons et al., 2007; Rizzo et al., 2006). Additionally, VHs have been used effectively for the conduct of social psychology experiments, essentially replicating and extending findings from studies on social influence, conformity, racial bias, and social proxemics conducted with real humans (Bailenson & Beall, 2006; Blascovich et al., 2002; McCall, Blascovich, Young, & Persky, 2009).

In an effort to further increase the pictorial realism of such VHs, Virtually Better Inc. (www. virtuallybetter.com) began incorporating whole video clips of crowds into graphic VR fear of public speaking scenarios. They later advanced the technique by using blue-screen-captured video sprites of individual humans inserted into graphics-based VR social settings for social phobia and cue exposure substance abuse treatment and research applications. The sprites were drawn from a large library of blue-screen-captured videos of actors behaving or speaking with varying degrees of provocation. These video sprites could then be strategically inserted into the scenario with the aim to modulate the emotional state of the patient by fostering encounters with these 2D video VH representations.

The continued quest for even more realistic simulated human interaction contexts led other researchers to the use of panoramic video capture (Macedonio, Parsons, Wiederhold, & Rizzo,

2007; Rizzo, Ghahremani, Pryor, & Gardner, 2003) of a real-world office space inhabited by hostile coworkers and supervisors to produce VR scenarios for anger management research. With this approach, the VR scenarios were created using a 360° panoramic camera that was placed in the position of a worker at a desk, and then actors walked into the workspace, addressed the camera (as if it was the targeted user at work), and proceeded to verbally threaten and abuse the camera, vis-à-vis the worker. Within such photorealistic scenarios, VH video stimuli could deliver intense emotional expressions and challenges with the aim of the research being to determine if this method would produce emotional reactions in test participants and if it could engage anger management patients to role-play a more appropriate set of coping responses.

However, working with such fixed video content to foster this form of *faux* interaction or exposure has significant limitations. For example, it requires the capture of a large catalog of possible verbal and behavioral clips that can be tactically presented to the user to meet the requirements of a given therapeutic approach. As well, this fixed content cannot be readily updated in a dynamic fashion to meet the challenge of creating credible real-time interactions with a VH, with the exception of only very constrained social interactions. This process can only work for clinical applications where the only requirement is for the VH character to deliver an open-ended statement or question that the user can react to, but is lacking in any truly fluid and believable interchange following a response by the user. Consequently, the absence of dynamic interaction with these virtual representations without a live person behind the *screen* actuating new clips in response to the user's behavior is a significant limiting factor for this approach. This has led some researchers to consider the use of AI VH agents as entities for simulating human-to-human interaction.

Clinical interest in AI VH agents designed for interaction with humans can trace its roots to the work of MIT AI researcher, Joe Weizenbaum. In 1966, he wrote a language analysis program called ELIZA that was designed to imitate a Rogerian therapist. The system allowed a computer user to interact with a virtual therapist by typing simple sentence responses to the computerized therapist's questions. Weizenbaum reasoned that simulating a nondirectional psychotherapist was one of the easiest ways of simulating human verbal interactions and it was a compelling simulation that worked well on teletype computers (and is even instantiated on the Internet today, http://www-ai.ijs.si/eliza-cgi-bin/eliza_script). In spite of the fact that the illusion of ELIZA's intelligence soon disappears due to its inability to handle complexity or nuance, Weizenbaum was reportedly shocked upon learning how seriously people took the ELIZA program (Howell & Muller, 2000). And this led him to conclude that it would be immoral to substitute a computer for human functions that "… involves interpersonal respect, understanding, and love" (Weizenbaum, 1976).

More recently, seminal research and development has appeared in the creation of highly interactive, AI, and natural-language-capable VH agents. No longer at the level of a prop to add context or minimal faux interaction in a virtual world, these agents are designed to perceive and act in a 3D virtual world and engage in face-to-face spoken dialogs with real users (and other VHs), and in some cases, they are capable of exhibiting humanlike emotional reactions. Previous classic work on VHs in the computer graphics community focused on perception and action in 3D worlds but largely ignored dialog and emotions. This has now changed. Intelligent VH agents can now be created that control computer-generated bodies and can interact with users through speech and gesture in VEs (Gratch et al., 2002, 2013). Advanced VHs can engage in rich conversations (Traum, Marsella, Gratch, Lee, & Hartholt, 2008), recognize nonverbal cues (Morency, de Kok, & Gratch, 2008; Scherer et al., 2013), reason about social and emotional factors (Gratch & Marsella, 2004), and synthesize human communication and nonverbal expressions (Thiebaux et al., 2008). Such fully embodied conversational characters have been around since the early 1990s (Bickmore & Cassell, 2005), and there has been much work on full systems that have been designed and used for training (Kenny, Hartholt et al., 2007; Prendinger & Ishizuka, 2004; Rickel, Gratch, Hill, Marsella, & Swartout, 2001), intelligent kiosks (McCauley & D'Mello, 2006), and virtual receptionists (Babu, Schmugge, Barnes, & Hodges, 2006). Both in appearance and behavior, VHs have now passed through *infancy* and are ready for service in a variety of clinical and research applications.

These advances in VH technology have now supported developments for clinical VR applications in two key domains: (1) the creation of virtual patients that can be used for training novice clinician care providers and (2) VH support agents to serve as online guides for promoting anonymous access to psychological health-care information and self-help activities. These areas are detailed later.

45.9 USE CASES: VH FOR CLINICAL TRAINING AND FOR HEALTH-CARE INFORMATION ACCESS

45.9.1 VIRTUAL STANDARDIZED PATIENTS

An integral part of medical and psychological clinical education involves training in interviewing skills, symptom/ability assessment, diagnosis, and interpersonal communication. In the medical field, students initially learn these skills through a mixture of classroom lectures, observation, and role-playing practice with *standardized patients*—persons recruited and trained to take on the characteristics of a real patient, thereby affording medical students a realistic opportunity to practice and be evaluated in a simulated clinical environment. This method of clinical training was first attempted in 1963, when Dr. Howard Barrows at the University of Southern California trained the first human standardized patient (HSP) (Barrows & Abrahamson, 1964). Since that time, the use of live actors has long been considered to be the gold standard medical education experience for both learning and evaluation purposes (Adamo, 2004; Jack et al., 2009). HSPs are paid actors who pretend to be patients for educational interviews and provide the most realistic and challenging experience for those learning the practice of medicine because they most closely approximate a genuine patient encounter. HSPs are also a key component in medical licensing examinations. For example, the United States Medical Licensing Examination (USMLE) step 2 clinical skills exam uses SPs and is mandatory for obtaining medical licensure in the United States (cf. http://www.usmle.org/). HSP encounters engage a number of clinical skill domains such as social skills, communication skills, judgment, and diagnostic acumen in a real-time setting. All other kinds of practice encounters fall short of this because either they do not force the learner to combine clinical skill domains or they spoon-feed data to the student with the practice case that turns the learning more into a pattern recognition exercise, rather than a realistic clinical problem solving experience. The HSP is the only type of encounter where it is up to the learner to naturalistically pose questions to obtain data and information about the case that then needs to be integrated for the formulation of a diagnostic hypothesis and/or treatment plan.

Despite the well-known superiority of HSPs to other instructional methods (Berkhof, Van Rijssen, Schellart, Anema, & Van der Beek, 2011; Howley, Szauter, Perkowski, Clifton, & McNaughton, 2008), they are employed sparingly. The reason for this limited use is primarily due to the very high costs to hire, train, and maintain a diverse group of patient actors. Moreover, despite the expense of standardized patient programs, the standardized patients themselves are typically low-skilled actors and administrators face constant turnover resulting in considerable challenges for maintaining the consistency of diverse patient portrayals for training students. This limits the value of this approach for producing realistic and valid interactions needed for the reliable evaluation and training of novice clinicians. Thus, the diversity of clinical conditions that standardized patients can characterize is limited by availability of human actors and their skills. HSPs that are hired may provide suboptimal variation control and are limited to healthy-appearing adult encounters. This is even a greater problem when the actor needs to be a child, adolescent, elder, person with a disability, or in the portrayal of nuanced or complex symptom presentations. The situation is even more challenging in the training of psychology/social work and other allied health professional students. Rarely are live standardized patients used in such clinical training. Most direct patient interaction skills are acquired via role-playing with supervising clinicians and fellow graduate students, with closely supervised *on-the-job* training providing the brunt of experiential training. While one-way mirrors provide a window for the direct observation of trainees, audio and video recordings of clinical sessions are the more common method of providing supervisors with information on the clinical skills of trainees. However, the imposition of recording has been reported to have demonstrable effects on the therapeutic process that may confound the end goal of clinical training (Bogolub, 1986), and the supervisor review of raw recordings is a time-consuming process that imposes a significant drain on resources.

In this regard, virtual patients can fulfill the role of HSPs by simulating diverse varieties of clinical presentations with a high degree of consistency and sufficient realism (Stevens et al., 2005), as well as being always available for *anytime–anywhere* training. Similar to the compelling case made over the years for clinical VR generally, VP applications can likewise enable the precise stimulus presentation and control (dynamic behavior, conversational dialog and interaction) needed for rigorous laboratory research yet embedded within the context of an ecologically relevant simulated environment. Toward this end, there is a growing literature on the use of VPs in the testing and training of bioethics, basic patient communication, interactive conversations, history taking, clinical assessment, and clinical decision making, and initial results suggest that VPs can provide valid and reliable representations of live patients (Beutler & Harwood, 2004; Bickmore & Giorgino, 2006; Bickmore, Pfeifer, & Paasche-Orlow, 2007; Kenny, Rizzo, Parsons, Gratch, & Swartout, 2007, 2010; Lok et al., 2007; Parsons et al., 2008; Rizzo et al., in press; Triola et al., 2006; Wendling et al., 2011).

The USC ICT began work in this area in 2007 with an initial project that involved the creation of a virtual patient, named *Justin*. Justin portrayed a 16-year-old male with a conduct disorder who is being forced to participate in therapy by his family. The system was designed to allow novice clinicians to practice asking interview questions, to attempt to create a positive therapeutic alliance and to gather clinical information from this very challenging VP. Justin was designed as a first step in our research. At the time, the project was unfunded and thus required our lab to take the economically inspired route of recycling a virtual character from a military negotiation-training scenario to play the part of Justin. The research group agreed that this sort of patient was one that could be convincingly created within the limits of the technology (and funding) available to us at the time. For example, such resistant patients typically respond slowly to therapist questions and often use a limited and highly stereotyped vocabulary. This allowed us to create a believable VP within limited resources for dialog development. As well, novice clinicians have been typically observed to have a difficult time learning the value of *waiting out* periods of silence and nonparticipation with these patients. We initially collected user interaction and dialog data from a small sample of psychiatric residents and psychology graduate students as part of our iterative design process to evolve this application area. The project produced a successful proof of concept demonstrator that then led to the acquisition of funding that currently supports our research in this area.

Following our successful Justin proof of concept, our second VP project involved the creation of a female sexual assault victim, *Justina*. The aim of this work was twofold: (1) explore the potential for creating a system for use as a clinical interview trainer for promoting sensitive and effective clinical interviewing skills with a VP that had experienced significant personal trauma and (2) create a system whereby the dialog content could be manipulated to create multiple versions of Justina to provide a test of whether novice clinicians would ask the appropriate questions to assess whether Justina met the criteria for the DSM-4-r diagnosis of PTSD based on symptoms reported during the clinical interview.

For the PTSD content domain, 459 questions were created that mapped roughly 4 to 1 to a set of 116 responses. The aim was to build an initial language domain corpus generated from subject matter experts and then capture novel questions from a pilot group of users (psychiatry residents) during interviews with Justina. The novel questions that were generated could then be fed into the system in order to iteratively build the language corpus. We also focused on how well subjects asked questions that covered the six major symptom clusters that can characterize PTSD following a traumatic event. While this approach did not give the Justina character a lot of depth, it did provide more breadth for PTSD-related responses, which for initial testing seemed prudent for generating a wide variety of questions for the next Justina iteration. In the initial test, a total of 15 psychiatry residents (6 females, 9 males; mean age = 29.80, SD 3.67) participated in the study and were asked to perform a 15 min interaction with the VP to take an initial history and determine a preliminary diagnosis based on this brief interaction with the character. The participants were asked to talk normally, as they would to a standardized patient, but were informed that the system was a research prototype that uses an experimental speech recognition system that would sometimes not understand them. They were instructed that they were free to ask any kind of question and the system would try to respond appropriately, but if it did not, they could ask the same question in a different way.

From postquestionnaire ratings on a seven-point Likert scale, the average subject rating for believability of the system was 4.5. Subjects reported their ability to understand the patient at an average of 5.1 but rated the system at 5.3 as frustrating to talk to, due to speech recognition problems, out-of-domain answers, or inappropriate responses. However, most of the participants left favorable comments that they thought this technology will be useful in the future and that they enjoyed the experience of trying different ways to talk to the character in order to elicit a relevant response to a complex question. When the patient responded back appropriately to a question, test subjects informally reported that the experience was very satisfying. Analysis of concordance between user questions and VP response pairs indicated moderate effect sizes for trauma inquiries ($r = .45$), reexperiencing symptoms ($r = .55$), avoidance ($r = .35$), and the non-PTSD general communication category ($r = .56$), but only small effects were found for arousal/hypervigilance ($r = .13$) and life impact ($r = .13$). These relationships between questions asked by a novice clinician and concordant replies from the VP suggest that a fluid interaction was sometimes present in terms of rapport, discussion of the traumatic event, the experience of intrusive recollections, and discussion related to the issue of avoidance. Low concordance rates on the arousal and life impact criteria indicated that a larger domain of possible questions and answers for these areas was not adequately modeled in this pilot effort, and this is now being addressed in our next-generation VH research and development.

We are currently collaborating with the USC School of Social Work, Center for Innovation in Research (CIR), which essentially is an MSW program with an emphasis on military social work. The current project with CIR focuses on the creation of military VPs that will allow social work trainees to gain practical training experiences with VHs that portray behavior more relevant to military culture and common clinical conditions (Figure 45.8). A sample video of the military VPs being interviewed by a social work trainee can be found here: http://www.youtube.com/watch?v=PPbcl8Z-8Ec.

Follow-on work to these VP projects has been funded to develop a toolkit that allows clinical educators to author VPs for clinical training. One of the aims of the system is to build an interface

(a)   (b)

FIGURE 45.8 Military male virtual patient characters (versions 1 and 2). (a) Early version of male virtual patient character and (b) refined version of male virtual patient characters.

that allows clinical educators to create a virtual patient with the same ease as creating a Powerpoint presentation. Such virtual patients, authored by clinical professionals, would then become available to an open-source community in order to broaden the opportunities for diverse clinical training experiences.

If this exploratory work continues to show promise, we intend to address a longer-term vision—that of creating a comprehensive DSM diagnostic trainer that has a diverse library of VPs modeled after each diagnostic category. The VPs would be created to represent a wide range of age, gender, and ethnic backgrounds and could be interchangeably loaded with the language and emotional models defined by the criteria specified in any of the DSM disorders. We believe this vision will also afford many research opportunities for investigating the functional and ethical issues involved in the process of creating and interacting with VHs and patients. While ethical challenges may be more intuitively appreciated in cases where the target user is a patient with a clinical condition seeking a virtual clinician, the training of clinicians with VPs will also require a full appreciation of how this form of training impacts clinical performance with *real* patients. These are not trivial concerns and will require careful, ethical, and scientific consideration. The capacity to conduct clinical training within simulations that provide access to credible VH patients where novice clinicians can gain exposure to the presentation of a variety of clinical conditions will soon provide a safe and effective means for learning skills before actual training with real patients and for supplementing continuing education throughout the professional life span. And as the underlying enabling technologies continue to advance, significant opportunities will emerge that will reshape the clinical training landscape.

Thus far, we have presented an example of the creation and use of VH characters to serve the role of digital *standardized patients* for training clinical skills, in both psychological and medical care domains. This initial effort and other VP project we have in progress have also led to new opportunities for exploring the use of VHs to serve as online mental health-care guides or coaches (Rizzo, Lange, Buckwalter et al., 2011).

45.9.2 SimCoach: An Online VH Health-Care Guide for Breaking Down Barriers to Care

Research suggests that there is an urgent need to reduce the stigma of seeking mental health treatment in SM and veteran populations. One of the more foreboding findings in an early report by Hoge et al. (2004) was the observation that among Iraq/Afghanistan War veterans, "… those whose responses were positive for a mental disorder, only 23%–40% sought mental health care. Those whose responses were positive for a mental disorder were twice as likely as those whose responses were negative to report concern about possible stigmatization and other barriers to seeking mental health care" (p. 13). While US military training methodology has better prepared soldiers for combat in recent years, such hesitancy to seek treatment for difficulties that emerge upon return from combat, especially by those who may need it most, suggests an area of military mental health care that is in need of attention. Moreover, the dissemination of health-care information to military SMs, veterans, and their significant others is a persistent and growing challenge. Although medical information is increasingly available over the web, users can find the process of accessing it to be overwhelming, contradictory, and impersonal.

In spite of a Herculean effort on the part of the US DOD to produce and disseminate behavioral health programs for military personnel and their families, the complexity of the issues involved continues to challenge the best efforts of military mental health-care experts, administrators, and providers. Since 2004, numerous blue-ribbon panels of experts have attempted to assess the current DOD and VA health-care delivery system and provide recommendations for improvement (DOD Mental Health Task Force [DOD, 2007], National Academies of Science Institute of Medicine

[IOM, 2007], Dole-Shalala Commission Report [2007], the Rand report [Tanielian et al., 2008], APA [2007]). Most of these reports cite two major areas in need of improvement:

1. Support for RCTs that test the efficacy of treatment methodologies, leading to wider dissemination of evidence-based approaches
2. Identification and implementation of ways to enhance the health-care dissemination/delivery system for military personnel and their families in a fashion that provides better awareness and access to care while reducing the stigma of help seeking

For example, the APA Presidential Task Force on Military Deployment Services for Youth, Families and SMs (APA, 2007) presented their preliminary report in February of 2007 that poignantly stated that they were "… not able to find any evidence of a well-coordinated or well-disseminated approach to providing behavioral health care to SMs and their families." The APA report also went on to describe three primary barriers to military mental health treatment: *availability, acceptability, and accessibility.* More specifically

1. Well-trained mental health specialists are not in adequate supply (*availability*).
2. The military culture needs to be modified such that mental health services are more *accepted* and less stigmatized.
3. Even if providers were available and seeking treatment was deemed acceptable, appropriate mental health services are often not readily *accessible* due to a variety of factors (e.g., long waiting lists, limited clinic hours, a poor referral process, and geographical location).

The overarching goal reported from this and other reports is to provide better awareness and access to existing care while concurrently reducing the complexity and stigma in seeking psychological help. In essence, new methods are needed to reduce such barriers to care.

The SimCoach project aims to address this challenge by supporting users in their efforts to anonymously seek health-care information and advice by way of online interaction with an intelligent, interactive, embodied VH health-care guide. The primary goal of the SimCoach project is to break down barriers to care (e.g., stigma, unawareness, complexity) by providing military SMs, veterans, and their significant others with confidential help in exploring and accessing health-care content and, if needed, for encouraging and supporting the initiation of care with a live provider. Rather than being a traditional web portal, SimCoach allows users to initiate and engage in a dialog about their health-care concerns with an interactive VH. Generally, these intelligent graphical characters are designed to use speech, gesture, and emotion to introduce the capabilities of the system, solicit basic anonymous background information about the user's history and clinical/psychosocial concerns, provide advice and support, present the user with relevant online content, and potentially facilitate the process of seeking appropriate care with a live clinical provider. An implicit motive of the SimCoach project is that of supporting users who are determined to be in need, to make the decision to take the first step toward initiating psychological or medical care with a live provider.

It is not the goal of SimCoach to break down all of the barriers to care or to provide diagnostic or therapeutic services that are best delivered by a live clinical provider. Rather, SimCoach was designed to foster comfort and confidence by promoting users' private and anonymous efforts to understand their situations better, to explore available options and initiate treatment when appropriate. Coordinating this experience is a VH SimCoach, selected by the user from a variety of archetypical character options (see Figure 45.9), who can answer direct questions and/or guide the user through a sequence of user-specific questions, exercises, and assessments. This interaction between the VH and the user provides the system with the information needed to guide users to the appropriate next step of engagement with the system or with encouragement to initiate contact with a live provider. Again, the SimCoach project is not conceived as a replacement for human clinical providers and experts. Instead, SimCoach aims to start the process of engaging the user by providing

FIGURE 45.9 SimCoach archetypes— (a) retired sergeant major, (b) female civilian, (c) female aviator, and (d) battle buddy.

support and encouragement, increasing awareness of their situation and treatment options, and assisting individuals who may otherwise be initially uncomfortable talking to a live care provider.

Users can flexibly interact with a SimCoach character by typing text and clicking on character generated menu options. Since SimCoach was designed to be an easily accessible web-based application that requires no downloadable software, it was felt that voice recognition was not at a state where it could be reliably used at the start of the project in 2010. The feasibility of providing the option for full-spoken natural language dialog interaction is currently being explored to determine if off-the-shelf voice recognition programs are sufficiently accurate to maintain an engaged interaction between a SimCoach and a user. The options for a SimCoach's appearance, behavior, and dialog have been designed to maximize user comfort and satisfaction but also to facilitate fluid and truthful disclosure of clinically relevant information.

Focus groups, Wizard of Oz studies, and iterative formative tests of the system were employed with a diverse cross section of our targeted user group to create options for SimCoach interaction that would be both engaging and useful for this population's needs. Results from these user tests indicated some key areas that were determined to be important including user choice of character archetypes across gender and age ranges, informal dialog interaction, and, interestingly, a preference for characters that were not in uniform. Also, interspersed within the program are options that allow the user to respond to simple screening instruments, such as the PCL-M (PTSD symptom checklist) that are delivered in a conversational format with results fed back to the user in a supportive fashion. These screening results serve to inform the SimCoach's creation of a model of the user to enhance the reliability and accuracy of the SimCoach output to the user, to support user self-awareness via feedback, and to better guide the delivery of relevant information based on these self-report data. Moreover, an enhancement in user engagement with a SimCoach may be produced if a more accurate assessment of the user's needs is derived from this process to inform the relevancy of the interaction.

Engagement is also supported by insuring that the specific health-care content that a SimCoach can deliver to users is relevant to persons with a military background (and of course their significant others). This was addressed by leveraging content assets that were originally created for established DOD and VA websites specifically designed to address the needs of this user group (e.g., Afterdeployment, Military OneSource, National Center for PTSD). Our early research with this user group indicated a hesitancy to directly access these sites when users sought behavioral health information with a common complaint being that there was a fear that their use of those sites may be monitored and might jeopardize advancement in their military careers or later applications for disability benefits. In spite of significant efforts by the DOD and VA to dispel the idea that user tracking was employed on these sites, the prevailing suspicion led many of the users in our samples to conduct such health-care queries using Google, Yahoo, and Medscape. To address this user concern, supplemental content presented by the SimCoach (e.g., video, self-assessment questionnaires, resource links) is typically *pulled* into the site, rather than directing users away *to* those sites.

As the system evolves, it is our view that engagement would be enhanced if the user was able to interact with the SimCoach repeatedly over time. Ideally, users could progress at their own pace over days or even weeks as they perhaps develop a *relationship* with a SimCoach character as a *go-to* source of health-care information and feedback. However, this option for evolving the SimCoach comfort zone with users over time would require significant database resources to render the SimCoach capable of *remembering* the information acquired from previous visits and to build on that information in similar fashion to that of a growing human relationship. Moreover, the persistence of a SimCoach memory for previous sessions would also require the user to sign into the system with a user name and password. This would necessitate the SimCoach system to *reside* on a high security server, such that content from previous visits could be stored and accessed with subsequent visits. Such functionality might be a double-edged sword as anonymity is a hallmark feature to draw in users who may be hesitant to know that their interactions are being stored, even if it resulted in a more relevant, less redundant and perhaps more meaningful interaction with a SimCoach over time. Likely, this would necessarily have to be a clearly stated *opt-in* function, as the technology may support this in the future. Users also have the option to print out a PDF summary of the SimCoach session. This is important for later personal review and for the access to links that the SimCoach provided in the session to relevant web content or to bring with them when seeking clinical care to enhance their comfort level, armed with knowledge, when dealing with human clinical care providers and experts. We have also created software authoring tools that allows other clinical professionals to create SimCoach content to enhance the likelihood that the program will evolve based on other care perspectives and emerging needs in the future.

The current version of SimCoach is presently undergoing beta-testing with a limited group of test-site users. Results from this user-centered testing will serve to advance the development of a SimCoach system that is expected to undergo a wider release in 2013. Although this project represents an early effort in this area, it is our view that the clinical aims selected can still be usefully addressed within the limits of current technology. However, we expect that SimCoach will continue to evolve over time based on data collected from ongoing user interactions with the system and advances in technology, particularly with improved voice recognition. Along the way, this work will afford many research opportunities for investigating the functional and ethical issues involved in the process of creating and interacting with VHs in a clinical or health-care support context. While the ethical challenges may be more intuitively appreciated, the functional technology challenges are also significant. As advances in computing power, graphics and animation, artificial intelligence, speech recognition, and natural language processing continue to develop at current rates, we expect that the creation of highly interactive, intelligent VHs for such clinical purposes is not only possible, but probable.

The systematic use of intelligent VHs in clinical VR applications is still clearly in its infancy. But the days of limited use of VH's as simple props or static elements to add realism or context to a clinical VR application are clearly in the past. The birth of this field has already happened; the next step is to insure that it has a healthy upbringing. *For more information, see a virtual patient and SimCoach video at* http://www.youtube.com/watch?v=PPbcl8Z-8Ec and http://www.youtube.com/watch?v=JiHlTioZktc.

45.10 CONCLUSIONS

The use of computer-based VR simulation technology will play an increasing role in how clinical care (and training) and scientific research is conducted in the future. Advances in the underlying enabling technologies and continuing cost reductions in system hardware are expected to make it possible for VR to shortly become a mainstream tool in this area. With this view in mind, this chapter has aimed to provide a detailed specification of rationales, research, and development across four major areas of clinical VR.

However, VR is not a panacea for all the challenges that occur in clinical care. By presenting the history, rationale and brief reviews of key literature, along with use cases for the application of VR in each section, it is hoped that clinicians and researchers may use this information to understand the basis for when VR is well-matched to the needs of clinical care in certain areas and when it is technological overkill in others. As well, the task of building really good VR systems that are both usable and useful is a challenging endeavor that requires an interdisciplinary mix of domain-specific knowledge. This sort of collaboration can be supported by providing clinicians with an informed view of what is possible with the technology and why, with what makes the most sense from a clinical perspective. Armed with this information, clinical experts can begin to partner with scientists in more technical disciplines for the thoughtful development of clinical VR applications that meet the needs of clinical populations of interest. Finally, one of the guiding principles in the clinical VR area concerns how novel and innovative VR approaches can extend the skills of a well-trained clinician. As long as VR technology is viewed in the context of its value as a *tool* for extending clinical skill rather than a replacement for it, we believe that effective VR system development and application will bring the benefits of the information age to clinical populations in ways that go beyond the limits of the past.

REFERENCES

Adamo, G. (2004). Simulated and standardised patients in OSCEs: Achievements and challenges 1992–2003. *Medical Teaching, 25*(3), 262–270.

Adamovich, S. V., Merians, A. S., Boian, R., Tremaine, M., Burdea, G. S., Recce, M., & Poizner, H. (2004, September). A virtual reality based exercise system for hand rehabilitation post-stroke: transfer to function. In Engineering in Medicine and Biology Society, 2004. IEMBS'04. 26th Annual International Conference of the IEEE (Vol. 2, pp. 4936–4939). IEEE.

American Psychiatric Association. (2000). *Diagnostic and statistical manual of mental disorders* (revised 4th ed.). Washington, DC: American Psychiatric Association.

American Psychological Association. (2007). *Presidential task force on military deployment services for youth, families and service members. The psychological needs of U.S. Military Service Members and their families: A preliminary report.* Retrieved April 18, 2007, from http://www.apa.org/releases/MilitaryDeploymentTaskForceReport.pdf

Anderson, P. L., Zimand, E., Hodges, L. F., & Rothbaum, B. O. (2005). Cognitive behavioral therapy for public-speaking anxiety using virtual reality for exposure. *Depression and Anxiety, 22*(3), 156–158.

Astur, R. S., Taylor, L. B., Mamelak, A. N., Philpott, L., & Sutherland, R. J. (2002). Humans with hippocampus damage display severe spatial memory impairments in a virtual Morris water task. *Behavioural Brain Research, 132*, 77–84.

Astur, R. S., Tropp, J., Sava, S., Constable, R. T., & Markus, E. T. (2004). Sex differences in a virtual Morris water task and a virtual eight-arm maze. *Behavioural Brain Research, 151*(1–2), 103–115.

Babu, S., Schmugge, S., Barnes, T., & Hodges, L. (2006). What would you like to talk about? An evaluation of social conversations with a virtual receptionist. In J. Gratch et al. (Eds.), *IVA 2006*, LNAI 4133 (pp. 169–180). Berlin, Germany: Springer-Verlag.

Bailenson, J. N., & Beall, A. C. (2006). Transformed social interaction: Exploring the digital plasticity of avatars. In R. Schroeder & A. Axelsson (Eds.), *Avatars at work and play: Collaboration and interaction in shared virtual environments* (pp. 1–16). Springer-Verlag.

Barkley, R. A. (1990). *Attention deficit hyperactivity disorder: A handbook for diagnosis and treatment.* New York, NY: Guilford Press.

Barrows, H. S., & Abrahamson, S. (1964). The programmed patient: A technique for appraising student performance in clinical neurology. *Journal of Medical Education, 39*, 802–805.

Bart, O., Katz, N., Weiss, P. T., & Josman N. (2008). Street crossing by typically developed children in real and virtual environments. *Occupation, Participation & Health, 28*(2), 89–96.

Beck, A. T., Epstein, N., Brown, G., & Steer, R. A. (1988). An inventory for measuring clinical anxiety: Psychometric properties. *Journal of Consulting and Clinical Psychology, 56*, 893–897.

Beidel, D. C., Frueh, B. C., & Uhde, T. W. (2010). *Trauma management therapy for OIF/OEF veterans.* Department of Defense United States Army Military Operational Medical Research Program. Retrieved from http://www.psych.ucf.edu/faculty_beidel.php

Berkhof, M., Van Rijssen, H. J., Schellart, A., Anema, J., & Van der Beek, A. (2011). Effective training strategies for teaching communication skills to physicians: An overview of systematic reviews. *Patient Education and Counseling, 84*(2), 152–162.

Beutler, L. E., & Harwood, T. M. (2004). Virtual reality in psychotherapy training. *Journal of Clinical Psychology, 60*, 317–330.

Bickmore, T., & Cassell, J. (2005). Social dialogue with embodied conversational agents. In J. van Kuppevelt, L. Dybkjaer, & N. Bernsen (Eds.), *Advances in natural, multimodal dialogue systems.* New York, NY: Kluwer Academic.

Bickmore, T., & Giorgino, T. (2006). Health dialog systems for patients and consumers. *Journal of Biomedical Informatics, 39*(5), 556–571.

Bickmore, T., Pfeifer, L., & Paasche-Orlow, M. (2007). Health document explanation by virtual agents. In *Intelligent Virtual Agents. Lecture Notes in Computer Science, 4722*, 183–196.

Bioulac, S., Lallemand, S., Rizzo, A., Philip, P., Fabrigoule, C., & Bouvard, M. P. (2012). Impact of time on task on ADHD patient's performances in a virtual classroom. *European Journal of Paediatric Neurology, 16*(5), 514–521.

Blanchard, E. B., Jones-Alexander, J., Buckley, T. C., & Forneris, C. A. (1996). Psychometric properties of the PTSD Checklist (PCL). *Behaviour Research and Therapy, 34*(8), 669–673.

Blascovich, J., Loomis, J., Beall, A., Swinth, K., Hoyt, C., & Bailenson, J. (2002). Immersive virtual environment technology: Not just another research tool for social psychology. *Psychological Inquiry, 13*, 103–124.

Bogolub, E. B. (1986). Tape recorders in clinical sessions: Deliberate and fortuitous effects. *Clinical Social Work Journal, 14*(4), 349–360.

Botella, C., Banos, R. M., Perpina, C., Villa, H., Alcaniz, M., & Rey, A. (1998). Virtual reality treatment of claustrophobia: A case report. *Behaviour Research and Therapy, 36*, 239–246.

Brooks, B. M., McNeil, J. E., Rose, F. D., Greenwood, R. J., Attree, E. A., & Leadbetter, A. G. (1999). Route learning in a case of amnesia: A preliminary investigation into the efficacy of training in a virtual environment. *Neuropsychological Rehabilitation, 9*, 63–76.

Brown, D. J., Kerr, S. J., & Bayon, V. (1998). The development of the Virtual City: A user centred approach. In P. Sharkey, D. Rose, & J. Lindstrom (Eds.), *Proceedings of the 2nd European Conference on Disability, Virtual Reality and Associated Technologies (ECDVRAT)* (pp. 11–16). Reading, U.K.: University of Reading.

Bryant, R. A. (2005) Psychosocial approaches of acute stress reactions. *CNS Spectrums, 10*(2), 116–122.

Bryant, R. A., Moulds, M. L., Guthrie, R., & Nixon, R. D. (2003). Imaginal exposure alone and imaginal exposure with cognitive restructuring in treatment of posttraumatic stress disorder. *Journal of Consulting and Clinical Psychology, 71*(4), 706–712.

Carlin, A. S., Hoffman, H. G., & Weghorst, S. (1997). Virtual reality and tactile augmentation in the treatment of spider phobia: A case report. *Behavior Research and Therapy, 35*, 153–158.

Colegrove, R., Homayounjam, H., Williams, J., Hanken, J., & Horton, N. L. (1999, August). *The problem of overreliance upon behavioral checklists in the diagnosis of ADHD.* Chapter presented at the 107th Annual Conference of the American Psychological Association, Washington, DC.

Cromby, J., Standen, P., Newman, J., & Tasker, H. (1996). Successful transfer to the real world of skills practiced in a virtual environment by student with severe learning disabilities. In P. M. Sharkey (Ed.), *Proceedings of the 1st European Conference on Disability, Virtual Reality and Associated Technologies* (pp. 305–313). Reading, U.K.: University of Reading.

Department of Defense Task Force on Mental Health. (2007). An achievable vision: Report of the Department of Defense Task Force on Mental Health. Falls Church, VA: Defense Health Board.

Deutsch, J. E., Borbely, M., Filler, J., Huhn, K., & Guarrera-Bowlby, P. (2008). Use of a low-cost, commercially available gaming console (Wii) for rehabilitation of an adolescent with cerebral palsy. *Physical Therapy, 88*(10), 1196–1207.

Difede, J., & Hoffman, H. G. (2002). Virtual reality exposure therapy for World Trade Center post-traumatic stress disorder: A case report. *Cyberpsychology and Behavior, 5*, 529–535.

Difede, J., Cukor, J., Jayasinghe, N., Patt, I., Jedel, S., Spielman, L., … Hoffman, H. G. (2007). Virtual reality exposure therapy for the treatment of posttraumatic stress disorder following September 11, 2001. *Journal of Clinical Psychiatry, 68*, 1639–1647.

Difede, J., Rothbaum, B. O., & Rizzo, A. (2010–2013). *Enhancing exposure therapy for PTSD: Virtual reality and imaginal exposure with a cognitive enhancer. Randomized controlled trial.* Retrieved from http://clinicaltrials.gov/ct2/show/NCT01352637

Dole-Shalala Commission. (2007). *Serve, support, simplify: Report of the President's Commission on Care for America's Returning Wounded Warriors.* Downloaded on November 21, 2011 at: http://www.nyshealth-foundation.org/content/document/detail/1782/

Fischer, H. (2010, December 4). *United States military casualty statistics: Operation New Dawn, Operation Iraqi Freedom, and Operation Enduring Freedom, Congressional Research Service 7-5700:RS22452.* Retrieved from http://opencrs.com/document/RS22452/

Flynn, S. M., & Lange, B. S. (2010). Games for rehabilitation: the voice of the players. In Proceedings of the 8th International Conference on Disability, Virtual Reality and Associated Technologies (ICDVRAT), Chile (pp. 194–195). Indiana University.

Flynn, S., Palma, P., & Bender, A. (2007). Feasibility of using the Sony PlayStation 2 gaming platform for an individual poststroke: A case report. *Journal of Neurologic Physical Therapy, 31*(4), 180–189.

Foa, E. B., & Hearst-Ikeda, D. (1996). Emotional dissociation in response to trauma: An information-processing approach. In L. K. Michelson & W. J. Ray (Eds.), *Handbook of dissociation: Theoretical and clinical perspectives* (pp. 207–222). New York, NY: Plenum Press.

Foa, E. B., & Kozak, M. J. (1986). Emotional processing of fear: Exposure to corrective information. *Psychological Bulletin, 99*, 20–35.

Foa, E. B., Davidson, R. T., & Frances, A. (1999). Expert consensus guideline series: Treatment of posttraumatic stress disorder. *American Journal of Clinical Psychiatry, 60*, 5–76.

Foa, E. B., Steketee, G., & Rothbaum, B. (1989). Behavioral/cognitive conceptualizations of post-traumatic stress disorder. *Behavior Therapy, 20*, 155–176.

Folley, B. S., Astur, R. S., Jagannathan, K., Calhoun, V. D., & Pearlson, G. D. (2010). Anomalous neural circuit function in schizophrenia during a virtual Morris water task. *Neuroimage, 49*(4), 3373–3384.

Gerardi, M., Rothbaum, B. O., Ressler, K., Heekin, M., & Rizzo, A. A. (2008). Virtual reality exposure therapy using a virtual Iraq: Case report. *Journal of Traumatic Stress, 21*(2), 209–213.

Gilboa, Y., Rosenblum, S., Fattal-Valevski, A., Toledano-Alhadef, H., Rizzo, A., & Josman, N. (2011). Describing the attention deficit profile of children with neurofibromatosis type 1 using a virtual classroom environment. *Research in Developmental Disabilities, 32*(6), 2608–2613.

Glantz, K., Rizzo, A. A., & Graap, K. (2003). Virtual reality for psychotherapy: Current reality and future possibilities. *Psychotherapy: Theory, Research, Practice, Training, 40*(1), 55–67.

Gold, J. I., Kim, S. H., Kant, A. J., Joseph, M. H., & Rizzo, A. A. (2006). Effectiveness of virtual reality for pediatric pain distraction during IV placement. *CyberPsychology and Behavior, 9*(2), 207–213.

Gratch, J., & Marsella, S. (2004). A domain independent framework for modeling emotion. *Journal of Cognitive Systems Research, 5*(4), 269–306.

Gratch, J., Morency, L. P., Scherer, S., Stratou, G., Boberg, J., Koenig, S., … Rizzo, A. (2013). User-state sensing for virtual health agents and telehealth applications. *Proceedings of the 20th Annual Medicine Meets Virtual Reality Conference*, San Diego, CA.

Gratch, J., Rickel, J., Andre, E., Cassell, J., Petajan, E., & Badler, N. (2002, July/August). Creating interactive virtual humans: Some assembly required. *IEEE Intelligent Systems, 17*(4), 54–61.

Greenhill, L. L. (1998). Diagnosing attention-deficit/hyperactivity disorder in children. *Journal of Clinical Psychiatry, 59*(Suppl. 7), 31–41.

Haik, J., Tessone, A., Nota, A., Mendes, D., Raz, L., Goldan, O., ... & Hollombe, I. (2006). The use of video capture virtual reality in burn rehabilitation: The possibilities. *Journal of Burn Care & Research, 27*(2), 195–197.

Hays, R. T., Jacobs, J. W., Prince, C., & Salas, E. (1992). Requirements for future research in flight simulation training: Guidance based on a meta-analytic review. *International Journal of Aviation Psychology, 2,* 143–158.

Henderson, A., Korner-Bitensky, N., & Levin, M. (2007). Virtual reality in stroke rehabilitation: A systematic review of its effectiveness for upper limb motor recovery. *Topics in Stroke Rehabilitation, 14*(2), 52–61.

Hix, D., & Gabbard, J. L. (2002). Usability engineering of virtual environments. In K. Stanney (Ed.), *Handbook of virtual environments* (pp. 681–700). New York, NY: L.A. Erlbaum.

Hodges, L. F., Kooper, R., Meyer, T., Rothbaum, B. O., Opdyke, D., de Graaff, J. J., ... North, M. M. (1995). Virtual environments for treating the fear of heights. *Computer, 28*(7), 27–34.

Hoffman, H. G., Chambers, G. T., Meyer, W. J., Araceneaux, L. L., Russell, W. J., Seibel, E. J., ... Patterson, D. R. (2011). Virtual Reality as an adjunctive non-pharmacologic analgesic for acute burn pain during medical procedures. *Annals of Behavioral Medicine, 41*(2), 183–191.

Hoge, C. W., Castro, C. A., Messer, S. C., McGurk, D., Cotting, D. I., & Koffman, R. L. (2004). Combat duty in Iraq and Afghanistan, mental health problems, and barriers to care. *New England Journal of Medicine, 351*(1), 13–22.

Holden, M. K. (2005). Virtual environments for motor rehabilitation: Review. *CyberPsychology and Behavior, 8*(3), 187–211.

Howell, S. R., & Muller, R. (2000). *Computers in psychotherapy: A new prescription.* Chapter downloaded on April 23, 2011 at: http://www.psychology.mcmaster.ca/beckerlab/showell/ComputerTherapy.PDF

Howley, L., Szauter, K., Perkowski, L., Clifton, M., & McNaughton, N. (2008). Quality of standardised patient research reports in the medical education literature: Review and recommendations. *Medical Education, 42,* 350–358.

Inman, D., Loge, K., & Leavens, J. (1997). VR education and rehabilitation. *Communications of the ACM, 40*(8), 53–58.

IOM (Institute of Medicine). (2007). *Treatment of posttraumatic stress disorder: An assessment of the evidence.* Washington, DC: The National Academies Press. Downloaded on October 24, 2007 from http://www.nap.edu/catalog/11955.html

IOM (Institute of Medicine). (2012). *Treatment for posttraumatic stress disorder in military and veteran populations: Initial assessment.* Washington, DC: The National Academies Press. Downloaded on July 15, 2012 from http://www.iom.edu/Reports/2012/Treatment-for-Posttraumatic-Stress-Disorder-in-Military-and-Veteran-Populations-Initial-Assessment.aspx

Jack, B., Chetty, V., Anthony, D., Greenwald, J., Sanchez, G., Johnson, A., ... Culpepper, L. (2009). A reengineered hospital discharge program to decrease rehospitalization: A randomized trial. *Annals of Internal Medicine, 150*(3), 178–187.

Jaycox, L. H., Foa, E. B., & Morral, A. R. (1998). Influence of emotional engagement and habituation on exposure therapy for PTSD. *Journal of Consulting and Clinical Psychology, 66,* 186–192.

Johnston, R. (1995). Is it live or is it memorized? *Virtual Reality Special Report, 2,* 53–56.

Kaufmann, H., Steinbügl, K., Dünser, A., & Glück, J. (2005). General training of spatial abilities by geometry education in augmented reality. *Annual Review of CyberTherapy and Telemedicine: A Decade of VR, 3,* 65–77.

Kennedy, R. S., Lane, N. E., Berbaum, K. S., & Lilienthal, M. G. (1993). Simulator sickness questionnaire: An enhanced method for quantifying simulator sickness. *International Journal of Aviation Psychology, 3*(3), 203–220.

Kenny, P., Hartholt, A., Gratch, J., Swartout, W., Traum, D., Marsella, S., & Piepol, D. (2007). Building interactive virtual humans for training environments. *Proceedings of the Interservice/Industry Training, Simulation and Education Conference (I/ITSEC),* Orlando, FL.

Kenny, P., Parsons, T., & Garrity, P. (2010). Emerging concepts and innovative technologies: Virtual patients for virtual sick call medical training. *Proceedings of the Interservice/Industry Training, Simulation, and Education Conference 2010,* Orlando, FL.

Kenny, P., Rizzo, A. A., Parsons, T., Gratch, J., & Swartout W. (2007). A virtual human agent for training clinical interviewing skills to novice therapists. *Annual Review of Cybertherapy and Telemedicine 2007, 5,* 81–89.

Klinger, E. (2005). *Virtual reality therapy for social phobia: Its efficacy through a control study.* Paper presented at Cybertherapy 2005, Basal, Switzerland.

Koenig, S., Crucian, G., Dünser, A., Bartneck, C., & Dalrymple-Alford, J. (2011). Development of a spatial memory task in realistic virtual environments. *Proceedings of the Virtual Reality International Conference*, Laval, France.

Koenig, S. T., Crucian, G. P., Dalrymple-Alford, J. C., & Dünser, A. (2009). Virtual reality rehabilitation of spatial abilities after brain damage. *Annual Review of CyberTherapy and Telemedicine, 7*, 105–107. Reading, UK: The University of Reading.

Koenig, S. T., Crucian, G. P., Dalrymple-Alford, J. C., & Dünser, A. (2010). Assessing navigation in real and virtual environments: A validation study. In P. M. Sharkey & J. Sánchez (Eds.), *Proceedings of 8th International Conference on Disability, Virtual Reality and Associated Technologies* (pp. 7–16). Viña del Mar/Valparaíso, Chile. Reading, UK: The University of Reading.

Koenig, S. T., Krch, D., Chiaravalloti, N., Lengenfelder, J., Nikelshpur, O., Lange, B. S., ... & Rizzo, A. A. (2012, September). User-centered development of a virtual reality cognitive assessment. In International Conference Series on Virtual Reality and Associated Technologies, Laval, France.

Koenig, S. T., Krch, D., Chiaravalloti, N., Lengenfelder, J., Nikelshpur, O., Lange, B. S., ... Rizzo, A. A. (2012). User-centered development of a virtual reality cognitive assessment. *Proceedings of the 9th International Conference on Disability, Virtual Reality and Associated Technologies*, Laval, France.

Krch, D., Nikelshpur, O., Lavrador, S., Chiaravalloti, N. D., Koenig, S., & Rizzo, A. (2013, August). Pilot results from a virtual reality executive function task. In Virtual Rehabilitation (ICVR), 2013 International Conference on (pp. 15–21). IEEE.

Krch, D., Nikelshpur, O., Lavrador, S., Chiaravalloti, N.D., Koenig, S., & Rizzo, A., Pilot results from a virtual reality executive function task, *2013 International Conference on Virtual Rehabilitation (ICVR)*, pp. 15–21, 26–29 Aug. 2013. doi: 10.1109/ICVR.2013.6662092

Kroenke, K., & Spitzer, R. L. (2002). The PHQ-9: A new depression and diagnostic severity measure. *Psychiatric Annals, 32*, 509–521.

Krueger, M. W. (1993). The experience society. *Presence: Teleoperators and Virtual Environments, 2*(2), 162–168.

Lamson, R. (1994). Virtual therapy of anxiety disorders. *CyberEdge Journal, 4*(2), 6–8.

Lange, B., Flynn, B., & Rizzo, A. A. (2009). Initial usability assessment of off-the-shelf video game consoles for clinical game-based motor rehabilitation. *Physical Therapy Reviews, 14*(5), 355–363.

Lange, B., Flynn, S., Chang, C. Y., Liang, W., Si, Y., Nanavati, C., ... & Rizzo, A. S. (2011). Development of an interactive stepping game to reduce falls in older adults. International *Journal on Disability and Human Development, 10*(4), 331–335.

Lange, B., Koenig, S., Chang, C.-Y., McConnell, E., Suma, E., Bolas, M., & Rizzo, A.A. (2012). Designing informed game-based rehabilitation tasks leveraging advances in virtual reality. *Disability and Rehabilitation, 34*(22), 1863–1870.

Lange, B. S., Flynn, S. M., Proffitt, R., Chang, C. Y., & Rizzo, A. A. (2010). Development of an interactive game-based rehabilitation tool for dynamic balance training. *Topics in Stroke Rehabilitation, 17*(5), 345–352.

Laver, K., George, S., Thomas, S., Deutsch, J. E., & Crotty, M. (2011). Virtual reality for stroke rehabilitation. Cochrane Database of Systematic Reviews, 9, CD008349.

Lezak, M. D. (1995). *Neuropsychological assessment*. New York, NY: Oxford University Press.

Lok, B., Ferdig, R. E., Raij, A., Johnson, K., Dickerson, R., Coutts, J., & Lind, D. S. (2007). Applying virtual reality in medical communication education: Current findings and potential teaching and learning benefits of immersive virtual patients. *Journal of Virtual Reality, 10*(3–4), 185–195.

Macedonio, M. F., Parsons, T., Wiederhold, B., & Rizzo, A. A. (2007). Immersiveness and physiological arousal within panoramic video-based virtual reality. *CyberPsychology and Behavior, 10*(4), 508–515.

Matheis, R., Schultheis, M., Tierksy, L., DeLuca, J., Millis, J., & Rizzo, A. (2006). Is learning and memory in a virtual environment different? *The Clinical Neuropsychologist, 21*, 146–157.

Matheis, R., Schultheis, M. T., Tiersky, L. A., DeLuca, J., Mills, S. R., & Rizzo, A. A. (2007). Is learning and memory different in a virtual environment? *The Clinical Neuropsychologist, 21*, 146–161.

McCall, C., Blascovich, J., Young, A., & Persky, S. (2009). Proxemic behaviors as predictors of aggression towards Black (but not White) males in an immersive virtual environment. *Social Influence, 4*(2), 138–154.

McCauley, L., & D'Mello, S. (2006). A speech enabled intelligent Kiosk. In J. Gratch et al. (Eds.), *IVA 2006*, LNAI 4133 (pp. 132–144). Berlin, Germany: Springer-Verlag.

McGee, J. S., van der Zaag, C., Rizzo, A. A., Buckwalter, J. G., Neumann, U., & Thiebaux, M. (2000). Issues for the assessment of visuospatial skills in older adults using virtual environment technology. *CyberPsychology and Behavior, 3*(3), 469–482.

McGeorge, P., Phillips, L. H., Crawford, J. R., Garden, S. E., Della Sala, S., Milne, A. B., ... Callander, J. (2001). Using virtual environments in the assessment of executive dysfunction. *Presence: Teleoperators and Virtual Environments, 10*, 375–383.

McLay, R. N., Wood, D. P., Webb-Murphy, J. A., Spira, J. L., Weiderhold, M. D., Pyne, J. M., & Weiderhold, B. K. (2011). A randomized, controlled trial of virtual reality exposure therapy for post-traumatic stress disorder in active duty service members with combat-related post-traumatic stress disorder. *Cyberpsychology, Behavior, and Social Networking, 14*, 223–229.

McLay R. N. (2012, May 8). *New technology to treat post traumatic stress disorder.* Paper presented at the American Psychiatric Association Convention, Philadelphia, PA.

Mendozzi, L., Motta, A., Barbieri, E., Alpini, D., & Pugnetti, L. (1998). The application of virtual reality to document coping deficits after a stroke: Report of a case. *Cyberpsychology and Behavior, 1*, 79–91.

Merians, A. S., Fluet, G. G., Qiu, Q., Saleh, S., Lafond, I., & Adamovich, S. V. (2010). Integrated arm and hand training using adaptive robotics and virtual reality simulations. In *Proceedings of the 2010 International Conference on Disability, Virtual Reality and Associated Technology* (pp. 213–222). Gothenberg, Sweden.

Merians, A. S., Jack, D., Boian, R., Tremaine, M., Burdea, G. C., Adamovich, S. V., ... & Poizner, H. (2002). Virtual reality–augmented rehabilitation for patients following stroke. *Physical Therapy, 82*(9), 898–915.

Morency, L.-P., de Kok, I., & Gratch, J. (2008). Context-based recognition during human interactions: Automatic feature selection and encoding dictionary. In Digalakis, V., Potamianos, A., Turk., Pieraccini, R., & Ivanov, Y. (Eds). *Proceedings of the 10th International Conference on Multimodal Interfaces.* Chania, Greece: IEEE. pp. 181–188. ISBN 978-1-60558-198-9.

Mowrer, O. A. (1960). *Learning and behavior.* New York, NY: Wiley.

National Institutes of Health. (1995). *Basic behavioral science research for mental health: A national investment. A report of the U.S. National Advisory Mental Health Council* (NIH Publication No. 95-3682). Rockville, MD: Author.

Neil, A., Ens, S., Pelletier, R., Jarus, T., & Rand, D. (2013). Sony PlayStation EyeToy elicits higher levels of movement than Nintendo Wii: Implications for stroke rehabilitation. *European Journal of Physical Rehabilitation Medicine, 49*(1), 12–21.

Neisser, U. (1978). Memory: What are the important questions? In M. M. Gruneberg, P. E. Morris, & R. N. Sykes (Eds.), *Practical aspects of memory* (pp. 3–24). London, U.K.: Academic Press.

Opris, D., Pintea, S., García-Palacios, A., Botella, C., Szamosko, S., & David, D. (2012). Virtual reality exposure therapy in anxiety disorders: A quantitative meta-analysis. *Depression and Anxiety, 29*, 85–93.

Padgett, L., Strickland, D., & Coles, C. (2006). Case study: Using a virtual reality computer game to teach fire safety skills to children diagnosed with Fetal Alcohol Syndrome (FAS). *Journal of Pediatric Psychology, 31*(1), 65–70.

Parsons, S., Beardon, L., Neale, H. R., Reynard, G., Eastgate, R., Wilson, J. R., ... Hopkins, E. (2000, September 23–25). Development of social skills amongst adults with Asperger's syndrome using virtual environments: The 'AS Interactive' project. In P. M. Sharkey, A. Cesarani, L. Pugnetti, & S. Rizzo, (Eds.), *Proceedings of 3rd ICDVRAT Conference* (pp. 163–170). Alghero, Sardinia, Italy. Reading, U.K.: University of Reading.

Parsons, T., Bowerly, T., Buckwalter, J. G., & Rizzo, A. A. (2007). A controlled clinical comparison of attention performance in children with ADHD in a virtual reality classroom compared to standard neuropsychological methods. *Child Neuropsychology, 13*, 363–381.

Parsons, T. D., & Rizzo, A. A. (2008). Affective outcomes of virtual reality exposure therapy for anxiety and specific phobias: A meta-analysis. *Journal of Behavior Therapy and Experimental Psychiatry, 39*, 250–261.

Parsons, T. D., Kenny, P., Ntuen, C. A., Pataki, C. S., Pato, M. T., Rizzo, A., & Sugar, J. (2008). Objective structured clinical interview training using a virtual human patient. *Studies in Health Technology and Informatics, 132*, 357–362.

Parsons, T. D., Rizzo, A. A., Rogers, S., & York, P. (2009). Virtual reality in paediatric rehabilitation: A review. *Developmental Neurorehabilitation, 12*(4), 224–238.

Parsons, T. D., Rizzo, A. A., van der Zaag, C., McGee, J. S., & Buckwalter, J. G. (2005). Gender differences and cognition among older adults. *Aging, Neuropsychology, and Cognition, 12*(1), 78–88.

Pertaub, D.-P., Slater, M., & Barker, C. (2002). An experiment on public speaking anxiety in response to three different types of virtual audience. *Presence, 11*(1), 68–78.

Pollak, Y., Barhoum-Shomaly, H., Weiss, P. L., Rizzo, A. A., & Gross-Tsur, V. (2010). Methylphenidate effect in children with ADHD can be measured by an ecologically valid continuous performance test embedded in virtual reality. *CNS Spectrums, 15*, 125–130.

Pompeu, J. E., Mendes, F. A., Silva, K. G., Lobo, A. M., Oliveira, T. P., Zomignani, A. P., & Piemonte, M. E. (2012). Effect of Nintendo Wii™-based motor and cognitive training on activities of daily living in patients with Parkinson's disease: A randomised clinical trial. *Physiotherapy, 98*(3), 196–204.

Powers, M., & Emmelkamp, P. M. G. (2008). Virtual reality exposure therapy for anxiety disorders: A meta-analysis. *Journal of Anxiety Disorders, 22*, 561–569.

Prendinger, H., & Ishizuka, M. (2004). *Life-like characters: Tools, affective functions, and applications.* Berlin, Germany: Springer.

Proffitt, R., & Lange, B. (2013). User centered design and development of a game for exercise in older adults. *International Journal of Technology, Knowledge & Society, 8*(5), 95–112.

Proffitt, R., Lange, B., & Rose, D. (2012). User-centered design and feasibility of a game for balance training in older adults. Part of symposium titled: Exploring dynamic balance disability and cognitive load using novel technologies in healthy elders. 65th Annual Scientific Meeting of the Gerontological Society of America, San Diego, CA, November 14–18.

Pugnetti, L., Mendozzi, L., Attree, E., Barbieri, E., Brooks, B. M., Cazzullo, C. L., … Rose, F. D. (1998). Probing memory and executive functions with virtual reality: Past and present studies. *CyberPsychology and Behavior, 1,* 151–162.

Pugnetti, L., Mendozzi, L., Motta, A., Cattaneo, A., Barbieri, E., & Brancotti, S. (1995). Evaluation and retraining of adults' cognitive impairments: Which role for virtual reality technology? *Computers in Biology and Medicine, 25,* 213–227.

Rademaker, M., Fox, E. J., Lange, B., Thigpen, M., & Fox, C. (2013). Made for Rehab: Feasibility and effect of a clinician adapted video game using the Kinect™ on balance and falls in a woman with parkinsonism. American Physical Therapy Association, Combined Sections Meeting. San Diego, CA. January.

Raloff, J. (2006). Virtual reality for earthquake fears. *Science News* [Internet]. August 5 [cited October 30, 2010]. Available from http://www.thefreelibrary.com/Virtual+reality+for+ earthquake +fears-a0151188408

Rand, D., Kizony, R., & Weiss, P. T. L. (2008). The Sony PlayStation II EyeToy: Low-cost virtual reality for use in rehabilitation. *Journal of Neurologic Physical Therapy, 32*(4), 155–163.

Rauch, A., Cieza A., & Stucki, G. (2008). How to apply the International Classification of Functioning, Disability and Health (ICF) for rehabilitation management in clinical practice. *Eur J Phys Rheabil Med, 44*(3), 329–342.

Reger, G., & Gahm, G. (2008). Virtual reality exposure therapy for active duty soldiers. *Journal of Clinical Psychology, 64,* 940–946.

Reger, G., & Gahm, G. (2010). *Comparing virtual reality exposure therapy to prolonged exposure (VRPE extension). Randomized controlled trial.* Retrieved from http://clinicaltrials.gov/ct2/show/NCT01193725?term=Reger&rank=2

Reger, G., & Gahm, G. (2011). *Comparing virtual reality exposure therapy to prolonged exposure (VRPE extension). Randomized controlled trial.* Retrieved from http://clinicaltrials.gov/ct2/show/NCT01352637

Reger, G. M., Gahm, G. A., Rizzo, A. A., Swanson, R. A., & Duma, S. (2009). Soldier evaluation of the virtual reality Iraq. *Telemedicine and e-Health Journal, 15,* 100–103.

Reger, G. M., Holloway, K. M., Rothbaum, B. O, Difede, J., Rizzo, A. A., & Gahm, G. A. (2011). Effectiveness of virtual reality exposure therapy for active duty soldiers in a military mental health clinic. *Journal of Traumatic Stress, 24*(1), 93–96.

Requejo, P., Wade, E., Proffitt, R., McConnell, E., Lange, B., Liu, J., & Mulroy, S. (2012). Reaching performance in a virtual and real-world task. Part of symposium titled: Exploring dynamic balance disability and cognitive load using novel technologies in healthy elders. 65th Annual Scientific Meeting of the Gerontological Society of America, San Diego, CA, November 14–18.

Ressler, K. J., Rothbaum, B. O., Tannenbaum, L., Anderson, P., Zimand, E., Hodges, L., & Davis, M. (2004). Facilitation of psychotherapy with D-cycloserine, a putative cognitive enhancer. *Archives of General Psychiatry, 61,* 1136–1144.

Rickel, J., Gratch, J., Hill, R., Marsella, S., & Swartout, W. (2001). Towards a New Generation of Virtual Humans for Interactive Experiences, in *IEEE Intelligent Systems July/August 2002,* pp. 32–38.

Riva, G. (2011). The key to unlocking the virtual body: Virtual reality in the treatment of obesity and eating disorders. *Journal of Diabetes Science and Technology, 5*(2), 283–292.

Rizzo, A. (2010). Virtual Iraq/Afghanistan and how it is helping some troops and vets with PTSD. *Veterans Today.* Available from http://www.veteranstoday.com/2010/07/29/virtual-iraqafghanistan-and-how-it-is-helping-some-troops-and-vets-with-ptsd/

Rizzo, A., Difede, J., Rothbaum, B. O., & Reger, G. (2010). Virtual Iraq/Afghanistan: Development and early evaluation of a virtual reality exposure therapy system for combat-related PTSD. *Annals of the New York Academy of Sciences (NYAS), 1208,* 114–125.

Rizzo, A., Difede, J., Rothbaum, B. O., Reger, G., Spitalnick, J., Cukor, J., & Mclay, R. (2010). Development and early evaluation of the Virtual Iraq/Afghanistan exposure therapy system for combat-related PTSD. *Annals of the New York Academy of Sciences, 1208*(1), 114–125.

Rizzo, A., Lange, B., Buckwalter, J. G., Forbell, E., Kim, J., Sagae, K., ... Kenny, P. (2011). An intelligent virtual human system for providing healthcare information and support. In J. D. Westwood et al. (Eds.), *Technology and informatics* (pp. 503–509). Amsterdam, the Netherlands: IOS Press.

Rizzo, A., Lange, B., Suma, E., & Bolas, M. (2011). Virtual reality and interactive digital game technology: New tools to address childhood obesity and diabetes. *The Journal of Diabetes Science and Technology, 5*(2), 256–264.

Rizzo, A., Parsons, T. D., Lange, B., Kenny, P., Buckwalter, J. G., Rothbaum, B. O., ... Reger, G. (2011). Virtual reality goes to war: A brief review of the future of military behavioral healthcare. *Journal of Clinical Psychology in Medical Settings, 18*(2), 176–187.

Rizzo, A. A. (1994). Virtual reality applications for the cognitive rehabilitation of persons with traumatic head injuries. In H. J. Murphy (Ed.), *Proceedings of the 2nd International Conference on Virtual Reality and Persons with Disabilities*. Northridge, CA: CSUN.

Rizzo, A. A., & Kim, G. (2005). A SWOT analysis of the field of virtual rehabilitation and therapy. *Presence: Teleoperators and Virtual Environments, 14*(2), 1–28.

Rizzo, A. S., & Kim, G. J. (2005). A SWOT analysis of the field of virtual reality rehabilitation and therapy. Presence: *Teleoperators & Virtual Environments, 14*(2), 119–146.

Rizzo, A. A., Bowerly, T., Buckwater, J. G., Klimchuk, D., Mitura, R., & Parsons, R. D. (2006). A virtual reality scenario for all seasons: The virtual classroom. *CNS Spectrums, 11*(1), 35–44.

Rizzo, A. A., Buckwalter, J. G., & Neumann, U. (1997). Virtual reality and cognitive rehabilitation: A brief review of the future. *The Journal of Head Trauma Rehabilitation, 12*(6), 1–15.

Rizzo, A. A., Buckwalter, J. G., & van der Zaag, C. (2002). Virtual environment applications for neuro-psychological assessment and rehabilitation. In K. Stanney (Ed.), *Handbook of virtual environments* (pp. 1027–1064). New York, NY: L.A. Erlbaum.

Rizzo, A. A., Buckwalter, J. G., Bowerly, T., McGee, J., van Rooyen, A., van der Zaag, C., ... Chua, C. (2001). Virtual environments for assessing and rehabilitating cognitive/functional performance: A review of project's at the USC Integrated Media Systems Center. *Presence: Teleoperators and Virtual Environments, 10*(4), 359–374.

Rizzo, A. A., Buckwalter, J. G., Forbell, E. Difede, J., Rothbaum, B. O., Lange, B., ... Talbot, B. (2013). Virtual reality applications to address the wounds of war. *Psychiatric Annals, 43*(3), 123–138.

Rizzo, A. A., Buckwalter, J. G., John, B., Newman, B., Parsons, T., Kenny, P., & Williams, J. (2012). STRIVE: Stress resilience in virtual environments: A pre-deployment VR system for training emotional coping skills and assessing chronic and acute stress responses. In J. D. Westwood et al. (Eds.), *Technology and informatics*. Amsterdam, the Netherlands: IOS Press.

Rizzo, A. A., Buckwalter, J. G., Neumann, U., Kesselman, C., Thiebaux, M., Larson, P., & Van Rooyan, A. (1998). The virtual reality mental rotation/spatial skills project: Preliminary findings. *CyberPsychology and Behavior, 1*(2), 107–113.

Rizzo, A. A., Ghahremani, K., Pryor, L., & Gardner, S. (2003). Immersive 360-degree panoramic video environments. In J. Jacko & C. Stephanidis (Eds.), *Human–computer interaction: Theory and practice* (Vol. 1, pp. 1233–1237). New York, NY: L.A. Erlbaum.

Rizzo, A. A., Parsons, T., Buckwalter, J. G., & Kenny, P. (in press). The birth of intelligent virtual patients in clinical training. *American Behavioral Scientist.*

Rizzo, A. A., Schultheis, M. T., Kerns, K., & Mateer, C. (2004). Analysis of assets for virtual reality applications in neuropsychology. *Neuropsychological Rehabilitation, 14*(1/2) 207–239.

Rizzo, A. A., Strickland, D., & Bouchard, S. (2004). Issues and challenges for using virtual environments in telerehabilitation. *Telemedicine Journal and e-Health, 10*(2), 184–195.

Rose, D., Brooks, B. M., & Attree, E. A. (2000). Virtual reality in vocational training of people with learning disabilities. In P. Sharkey, A. Cesarani, L. Pugnetti, & A. Rizzo (Eds.), *Proceedings of the 3rd International Conference on Disability, Virtual Reality, and Associated Technology* (pp. 129–136). Reading, U.K.: University of Reading.

Rose, F. D., Attree, E. A., & Johnson, D. A. (1996). Virtual reality: An assistive technology in neurological rehabilitation. *Current Opinion in Neurology, 9*, 461–467.

Rose, F. D., Brooks, B. M., & Rizzo, A. A. (2005). Virtual reality in brain damage rehabilitation: Review. *CyberPsychology and Behavior, 8*(3), 241–262.

Rothbaum, B. O., & Hodges, L. F. (1999). The use of virtual reality exposure in the treatment of anxiety disorders. *Behavior Modification, 23*, 507–525.

Rothbaum, B. O., Hodges, L. F., & Kooper, R. (1995). Effectiveness of virtual reality graded exposure in the treatment of acrophobia. *Behavior Therapy, 26*, 547–554.

Rothbaum, B. O., Hodges, L., Ready, D., Graap, K., & Alarcon, R. (2001). Virtual reality exposure therapy for Vietnam veterans with posttraumatic stress disorder. *Journal of Clinical Psychiatry, 62*, 617–622.

Rothbaum, B. O., Hodges, L. F., & Smith, S. (1999). Virtual reality exposure therapy abbreviated treatment manual: Fear of flying application. *Cognitive and Behavioral Practice, 6*, 234–244.

Rothbaum, B. O., Hodges, L. F., Smith, S., Lee, J. H., & Price, L. (2000). A controlled study of virtual reality exposure therapy for the fear of flying. *Journal of Consulting and Clinical Psychology, 68*, 1020–1026.

Rothbaum, B. O., Hodges, L., Watson, B. A., Kessler, G. D., & Opdyke, D. (1996). Virtual reality exposure therapy in the treatment of fear of flying: A case report. *Behaviour Research and Therapy, 34*, 477–481.

Rothbaum, B. O., & Schwartz, A. (2002). Exposure therapy for posttraumatic stress disorder. *American Journal of Psychotherapy, 56*, 59–75.

Rutten, A., Cobb, S., Neale, H., Kerr, S. Leonard, A., Parsons, S., & Mitchell, P. (2003). The AS interactive project: Single-user and collaborative virtual environments for people with high-functioning autistic spectrum disorders. *Journal of Visualization and Computer Animation, 14*(5), 233–241.

Saarela Holmberg, T., & Lindmark, B. (2008). How do physiotherapists treat patients with traumatic brain injury? *Advances in Physiotherapy, 10*(3), 138–145.

Saposnik, G., Levin, M., for the Stroke Outcome Research Canada (SORCan) Working Group (2011). Virtual reality in stroke rehabilitation: A meta-analysis and implications for clinicians. *Stroke, 42*, 1380–1386.

Saposnik, G., Teasell, R., Mamdani, M., Hall, J., McIlroy, W., Cheung, D., ... & Bayley, M. (2010). Effectiveness of virtual reality using Wii gaming technology in stroke rehabilitation a pilot randomized clinical trial and proof of principle. *Stroke, 41*(7), 1477–1484.

Scherer, M. R., & Schubert, M. C. (2009). Traumatic brain injury and vestibular pathology as a comorbidity after blast exposure. *Physical therapy, 89*(9), 980–992.

Scherer, S., Gratch, J., Morency, L. P., Stratou, G., Boberg, J., Koenig, S., ... Rizzo, A.A. (2013). Automatic behavior descriptors for psychological disorder analysis. *Proceedings of the 10th IEEE International Conference on Automatic Face and Gesture Recognition*, Shanghai, China.

Schneider, S. M., Kisby, C. K., & Flint, E. P. (2010, December 10). *Effect of virtual reality on time perception in patients receiving chemotherapy. Supportive care in cancer.* Retrieved from http://www.springerlink.com/content/?k = (au%3a(Susan+Schneider)+OR+ed%3a(Susan+Schneider))+pub%3a(Supportive+Cancer+Care)

Scozzari, S., & Gamberini, L. (2011). Virtual reality as a tool for cognitive behavioral therapy: A review. In S. Brahnam & L. C. Jain (Eds.), *Advances in computer intelligence. Paradigms in healthcare 6*, SCI 337 (pp. 63–108). Berlin, Germany: Springer-Verlag.

Seal, K. H., Bertenthal, D., Nuber, C. R., Sen, S., & Marmar, C. (2007). Bringing the war back home: Mental health disorders among 103,788 US veterans returning from Iraq and Afghanistan seen at Department of Veterans Affairs facilities. *Archives of Internal Medicine, 167*, 476–482.

Shallice, T., & Burgess, P. (1991). Higher order cognitive impairments and frontal lobe lesions in man. In H. S. Levin, H. M. Eisenberg, & A. L. Benton (Eds.), *Frontal lobe function and dysfunction* (pp. 125–138). New York, NY: Oxford University Press.

Shepard, R. N., & Metzler, J. (1971). Mental rotation of three-dimensional objects. *Science, 171*, 701–703.

Shipman, S., & Astur, R. (2008). Factors affecting the hippocampal BOLD response during spatial memory. *Behavioural Brain Research, 187*(2), 433–441.

Smith, S., Rothbaum, B. O., & Hodges, L. F. (1999). Treatment of fear of flying using virtual reality exposure therapy: A single case study. *The Behavior Therapist, 22*, 154–158.

Stanton, D., Foreman, N., & Wilson, P. (1998). Uses of virtual reality in clinical training: Developing the spatial skills of children with mobility impairments. In G. Riva, B. Wiederhold, & E. Molinari (Eds.), *Virtual reality in clinical psychology and neuroscience* (pp. 219–232). Amsterdam, the Netherlands: IOS Press.

Stevens, A., Hernandez, J., Johnsen, K., Dickerson, R., Raij, A., Harrison, C.,... Lind, D. S. (2005). The use of virtual patients to teach medical students communication skills. *The Association for Surgical Education Annual Meeting*, New York, NY.

Stewart, J. C., Yeh, S. C., Jung, Y., Yoon, H., Whitford, M., Chen, S. Y., ... & Winstein, C. J. (2007). Intervention to enhance skilled arm and hand movements after stroke: A feasibility study using a new virtual reality system. *Journal of NeuroEngineering and Rehabilitation, 4*, 21.

Strickland, D., Marcus, L. M., Mesibov, G. B., & Hogan, K. (1996). Brief report: Two case studies using virtual reality as a learning tool for autistic children. *Journal of Autism and Developmental Disorders, 26*(6), 651–659.

Tanielian, T., Jaycox, L. H., Schell, T. L., Marshall, G. N., Burnam, M. A., Eibner, C., ... Vaiana, M. E. (2008). *Invisible wounds of war: Summary and recommendations for addressing psychological and cognitive injuries. Rand report.* Retrieved from http://veterans.rand.org/

Thiebaux, M., Marshall, A., Marsella, S., Fast, E., Hill, A., Kallmann, M., ... Lee, J. (2008). SmartBody: Behavior realization for embodied conversational agents. *Proceedings of the International Conference on Autonomous Agents and Multi-Agent Systems (AAMAS)*, Estoril, Portugal.

Thornton, M., Marshall, S., McComas, J., Finestone, H., McCormick, A., & Sveistrup, H. (2005). Benefits of activity and virtual reality based balance exercise programmes for adults with traumatic brain injury: Perceptions of participants and their caregivers. *Brain Injury, 19*(12), 989–1000.

Traum, D., Marsella, S., Gratch, J., Lee, J., & Hartholt, A. (2008). Multi-party, multi-issue, multi-strategy negotiation for multi-modal virtual agents. In *8th International Conference on Intelligent Virtual Agents*. Tokyo, Japan: Springer.

Triola, M., Feldman, H., Kalet, A. L., Zabar, S. Kachur, E. K., Gillespie, C., ... Lipkin, M. (2006). A randomized trial of teaching clinical skills using virtual and live standardized patients. *Journal of General Internal Medicine, 21*, 424–429.

Van Etten, M. L., & Taylor, S. (1998). Comparative efficacy of treatments of posttraumatic stress disorder: An empirical review. *Journal of the American Medical Association, 268*, 633–638.

Wade, E., Proffitt, R., Requejo, P., Mulroy, S., & Winstein, C. (2012). Visuomotor researching behavior to virtual and real targets depends on postural requirements in healthy elders. American Congress of Rehabilitation Medicine Annual Conference, Vancouver, Canada, October 9–13.

Walker, D. L., Ressler, K. J., Lu, K. T., & Davis, M. (2002). Facilitation of conditioned fear extinction by systemic administration or intra-amygdala infusions of D-cycloserine as assessed with fear-potentiated startle in rats. *Journal of Neuroscience, 22*, 2343–2351.

Weightman, M. M., Bolgla, R., McCulloch, K. L., & Peterson, M. D. (2010). Physical therapy recommendations for service members with mild traumatic brain injury. *The Journal of Head Trauma Rehabilitation, 25*(3), 206–218.

Weiss, P., & Jessel, A. S. (1998). Virtual reality applications to work. *Work, 11*, 277–293.

Weizenbaum, J. (1976). *Computer power and human reason*. San Francisco, CA: W.H. Freeman.

Wendling, A., Halan, S., Tighe, P., Le, L., Euliano, T., & Lok, B. (2011). Virtual humans versus standardized patients in presenting abnormal physical findings. *Academic Medicine, 86*(3), 384–388.

Wilson, B. A., Alderman, N., Burgess, P. W., Emslie, H., & Evans, J. J. (1996). *BADS: behavioural assessment of the dysexecutive syndrome*. Bury St. Edmunds, U.K.: Thames Valley Test Company.

Wilson, J., Onorati, K., Mishkind, M., Reger, M., & Gahm, G. A. (2008). Soldier attitudes about technology-based approaches to mental healthcare. *Cyberpsychology and Behavior, 11*, 767–769.

World Health Organization. (2001). The World health report: 2001: Mental health: new understanding, new hope.

46 Modeling and Simulation for Cultural Training

Past, Present, and Future Challenges

Kathleen M. Bartlett, Denise Nicholson,
Margaret Nolan, and Brenna Kelly

CONTENTS

46.1 INTRODUCTION: DEMAND SIGNAL FOR CULTURAL TRAINING

The need for cultural training for military, diplomatic, and business persons who work among members of diverse societies outside their own personal experience has been established by numerous sources. Noting that cultural knowledge and warfare are inextricably bound, some experts believe that cultural training is a national security priority (McFate, 2005). According to General David Petraeus, knowledge of the cultural terrain can be as important as, or even more important than, knowledge of geographical terrain, leading to the suggestion that we must study culture in the same way that we study geography (Jager, 2007).

It has also been argued that a rich understanding of the sociocultural environment is a key ingredient for success in counterinsurgency (COIN) activities: "In order to maximize the prospect of success, the joint force must understand the population and operating environment, including the complex historical, political, socio-cultural, religious, economic, and other causes of violent conflict" (United States Department of Defense, 2010). However, culture-specific training prepares forces for engagements within a targeted culture; to meet the needs of diverse cultural engagements, efforts that focus on culture-general (or transcultural) training are needed. As noted by researchers at the Florida Institute of Technology, although culture-specific knowledge is sometimes conducive to mission success, effective performance across a variety of cultural settings requires an emphasis on cultural learning and adaptation (Wildman, Skiba, Armon, & Moukarzel, 2012). Given the urgency of this demand signal, examination of the intersections of culture, operational needs, and training, particularly modeling and simulation (M&S) solutions past, present, and future, are presented in this chapter.

46.1.1 ROLE OF CULTURE IN OPERATIONS AND TACTICS

Military goals have recently shifted from an aggressive military tactics approach to COIN (advocated during the Donald Rumsfeld era) to winning the hearts and minds of civilians (Jager, 2007). Current Pentagon leaders advocate an approach that emphasizes cultural knowledge and ethnographic intelligence as the foundation of COIN doctrine (Jager). In light of that thrust, and as operations become more distributed, forces need more than surface knowledge of how individuals speak, look, and act within a given culture.

For example, military operations frequently require leaders to anticipate the actions of, interact with, and influence individuals and groups from diverse cultural contexts that are very different from their own (Abbe, Gulick, & Herman, 2007). These operations demand a broader cultural capability that enables leaders and squad members to adapt to any cultural setting, and training for this capability must provide culture-general knowledge and skills as a complement to language skills and region-specific knowledge (Abbe et al., 2007). Training that provides a series of isolated cultural details alone (e.g., women do not leave the home, chiefs control small villages) cannot prepare forces to solve problems and make good decisions when events occur outside the realm of specific cultural descriptions; ongoing and future operations will require training that enables outsiders to recognize and respond to cultural patterns, even within an unfamiliar cultural setting.

The recent focus on cultural knowledge in operations and tactics has allowed field commanders in Iraq and Afghanistan, for example, to radically reassess the challenges (and failures) of COIN efforts in those environments. Strategies need to be identified that will link culture-specific and culture-general knowledge in order to train squad leaders and members of all forces to analyze and respond to unexpected, poorly defined, complex cultural problems and to make critical, time-sensitive, effective decisions. The kinds of cultural knowledge required at the tactical level (e.g., the cultural knowledge of specific customs) are distinct from the kinds of cultural knowledge required to define an overarching strategy on COIN (Jager, 2007). Training that provides both detailed culture-specific information and generalized cultural philosophy can better equip forces to operate in a variety of diverse environments.

46.1.2 Cultural Markers

Cultural patterns and markers consist of behaviors, attitudes, and norms that are learned as a member of a societal system. The cultural values in any social system often determine (or at least influence) the behaviors of the members of the group. For example, achievement-driven, individualistic societies are inhabited by persons with different attitudes and behaviors than those within a social, collectivist society where members are expected to uphold the interests of the group.

Cultural markers include social hierarchies that influence participation and status of its members (e.g., women and minorities), that establish frameworks for cultural cooperation and communication, that enforce norms of behavior (e.g., rituals and dress), and that define the role of institutions (e.g., government, religion, military, and education). Cultural membership also shapes the language, thinking, and perceptions (Boroditsky, 2011) of those who experience it firsthand, and in order to instill that attitudinal knowledge in users, immersive training environments seek to replicate cultural norms and behaviors, as well as geographic terrain.

46.1.2.1 Perception of and Response to Cultural Differences

When presented with unfamiliar environments or when acclimating to a foreign culture, individuals progress through a series of stages (Lane & Ogan, 2009), which have been defined as follows: (1) identifying new perspectives, (2) understanding differences, (3) operating within the foreign system, (4) managing or being able to communicate among the differences, and (5) adopting foreign behaviors as familiar (Caligiuri, Noe, Nolan, Ryan, & Drasgow, 2011, p. 10). When confronted by perplexing cultural situations, individuals often first assess dissimilar behaviors by comparing them to familiar practices or by relating them to their own culture (Schatzki, 2003).

Regardless of the method of introduction to another culture (e.g., travel, literature, film, or formal training), exposure provides an individual with experiences and observations that allow for informed comparisons of cultural markers or cues. Dress, language, and nonverbal gestures (e.g., greetings, proximity of personal space) provide obvious indicators for comparison. Deeper patterns exist within institutional levels, such as religious practices, educational methods, and government or military roles, but the complexity of variations in social practices and social roles often leads to misinterpretation and misunderstanding of unfamiliar cultural behaviors. In addition, these patterns and behaviors do not exist as static or unchanging practices, but are composed of dynamic and sometimes unpredictable emotions and events (Pavon, Arroyo, Hassan, & Sansores, 2008), which also complicate attempts to understand and to function effectively among members of diverse cultures.

Individuals often respond to cultural differences by making comparisons to inherent stereotypes derived from an outsider perspective (Schatzki, 2003). In the past, discomfort among expatriates and detrimental misunderstandings between members of differing cultures created a need for cross-cultural training, which typically included overviews highlighting specific cultural attributes and behaviors (Bhawuk & Brislin, 2000; Littrell & Salas, 2005). Cross-cultural training ideally instills an elevated level of cross-cultural and social competence among persons, regardless of societal membership, and minimizes reliance on stereotypical thinking (i.e., an *us versus them* mentality), but this goal has rarely been met.

46.1.3 How Has Culture Been Trained?

While definitions of cross-cultural competence vary among disciplines, it is generally described as a skill set that encompasses an individual's ability to assess and comprehend differing patterns of social behavior (Abbe et al., 2007; Rose-Krasnor, 1997). Even though identifying and understanding these differences remains a critical component of cross-cultural proficiency, the skill begins not with the comprehension of unfamiliar cultures, but with an understanding of one's own culture (Wildman et al., 2012). In other words, to eliminate the influence of cultural stereotypes, individuals must understand the ethnographic nature of their own comparisons (Schatzki, 2003).

Cultural education has been implemented for decades in an effort to prepare members of the military, diplomats, and business people to work and live abroad. Conventional methods involve attribution and cultural awareness training, behavior modification and interaction training, didactic and language training, and experiential training (Littrell & Salas, 2005). Emphasis varies between culture-general approaches, the teaching of self-awareness and cultural sensitivity to culture-specific training, and the teaching of elements of a particular culture (Bhawuk & Brislin, 2000; Littrell & Salas, 2005). Specific methods may include didactic classroom instruction, the viewing of educational videos, and/or the use of consultants or previous expatriates who possess firsthand knowledge of the culture of focus (Bhawuk, 2001). M&S systems that replicate cultural experiences have also recently emerged as a training option to familiarize users with norms and behaviors of, and to practice interactions with, realistically depicted members of specific cultural groups within specific geographical regions. However, in these systems, the cultural markers generally serve only as a backdrop for other training goals.

46.1.3.1 Cross-Cultural Assimilators

Cultural training methods have frequently employed cultural assimilators (vignettes) designed to present descriptions of perplexing situations that may be encountered in a foreign culture. The student evaluates the scenario and proposes a reason for a potential misunderstanding. Assimilators are intended to help students acquire cultural sensitivity, to enhance students' abilities to make isomorphic attributions (Bhawuk, 2001), and to familiarize students with the values of other cultures.

Additionally, assimilator training allows for reversal of intrinsic stereotypes developed by individuals over the course of a lifetime (Bhawuk, 2001). Assimilators are generally credited with building cultural awareness and knowledge among both civilian and military users (Abbe et al., 2007). However, since many assimilators are based on the personal experiences and limited knowledge of their designers and are not empirically tested, training effectiveness may be limited (Abbe et al., 2007; Bhawuk & Brislin, 2000).

46.1.3.2 Smart Cards

A cultural training method developed specifically for the military (Selmeski, 2007) is the cultural training tool known as the smart card, a pocket-sized, laminated, 16-panel, folding pamphlet, compiled by the army and marines for servicemen and women in the field (Davis, 2010). While the smart card was created to supplement prior training by providing a portable and accessible reference guide for recognizing attributes in a specific foreign culture, the fixed structure and unsupported descriptions of the cultures represented have resulted in a negative perception of their value (Davis). When interviewed about the effectiveness of the cultural training they had received, which included smart cards, a majority of marines stated that the training was not useful because it focused on static instructions and lacked contextual relevance (Davis).

46.1.3.3 Online and Blended Instruction

In recognition of the evolving need for transferable cultural education, the US Air Force added a culture-general component to their existing culture-specific course work in the Community College of the Air Force program (Wildman et al., 2012). Lessons, video demonstrations, and examinations are provided in a self-paced, online format. The lessons aim to instill an understanding of culture by emphasizing patterns that exist in all cultures (e.g., relationships, government, sports; Wildman et al.) as well as "nonverbal communication, paralanguage, cross-cultural conflict communication styles, active listening, and interaction skills" (AFCLC, 2012, para. 1). While these lessons demonstrate an increased awareness of the importance of developing a basic understanding of culture, they rely on conventional teaching methods using direct presentation of information. Blended culture training that extends beyond informational sessions and provides an opportunity for practice may prove more beneficial (Wildman et al.).

One example of that type of blended culture training is provided by the US Army Human Terrain System (HTS), which was developed based on situations that occurred in Iraq and Afghanistan during

2005–2006, when operations conducted without local social and cultural knowledge and produced negative affects among the local populations. As a result of increased demand for sociocultural information in the area of operations, the US Department of Defense (DoD) validated the urgent need for sociocultural support (human terrain concept) to combat commanders in Iraq and Afghanistan, and funding was provided through the Joint Improvised Explosive Device Defeat Organization (JIEDDO) as part of their organizational goals. In early 2006, the US Army Training and Doctrine Command G-2, supported by JIEDDO, responded to the operational need by developing a concept to provide social science support to military commanders in the form of Human Terrain Teams (HTTs). The HTS conducts a 40-day training course to train and certify the teams. HTS also provides interactive online training and tutorials for the training of the HTS elements, as well as a repository of reach-back materials for access no matter where they are deployed in the world (see www.humanterrainsystem.army.mil).

46.2   ROLE OF MODELING AND SIMULATION

Indigenous population role players provide the most accurate cultural, lingual, emotional, and physical interaction available, but are very expensive to use for training and practice. With the current economic environment and budgetary restraints, the US Marine Corps (USMC) and US Army have been investigating courses of action and alternatives to reduce the number of live role players used on an annual basis. Gen. James Mattis, who spearheaded the creation of the infantry trainer before coming to the US Joint Forces Command (USJFCOM), testified recently before the US Senate Armed Services Committee, where he called the development of combat simulators a national priority, saying "dramatic advances in immersive simulation, artificial intelligence and gaming technology must now be harnessed to bring state-of-the-art simulation to small infantry units" (USJFCOM, 2010, para. 13).

These types of technologically advanced methods for cultural training based on immersive experiences were originally utilized for tactical training such as flight simulations (Brawner, Holden, Goldberg, & Sottilare, 2011). Today, however, 3D immersion employing synthetic terrain and virtual characters with limited artificial intelligence (AI) has become a prominent part of military cultural training. Immersion offers the user a way to step inside a virtual world, allowing for the inclusion of sensory features that deepen the level of a user's engagement (Dede, 2009).

Simulation and serious games offer a secure environment in which dangerous cultural interactions may be practiced (Zielke et al., 2009) and provide an important supplement to didactic classroom settings and live training scenarios (Belanich, Mullin, & Dressel, 2004). Through the use of AI scripting, characters situated within simulations react to human users in an active and spontaneous manner (Zielke et al.). Replicating cultural situations via simulation provides an opportunity for users to learn from those representative reactions in order to develop cognitive and observational skills (Raybourn, 2005; Zielke et al., 2009) and practice tactics, techniques, and procedures (TTPs). An added benefit of computer-based immersive environments for cultural training is the ability of the software to capture quantitative data regarding user reactions and outcomes in order to empirically retest (Zielke et al.). Like other cultural training methods, the architecture of M&S has had to adjust (at the data level) to the recent shift in training goals, from warfighting maneuvers to civilian communications and distributed operations.

46.2.1   CURRENT M&S APPLICATIONS

Current simulations offer a wide range of live, virtual, and constructive (LVC) platforms to provide scenarios with culturally realistic conditions for kinetic and nonkinetic decision-based training, some of which have cultural competence as a goal, but many (perhaps most) provide cultural cues and markers as scenario-based conditions for other types of training (e.g., weapons employment training). Some systems have the capability to operate in real time as well as longitudinally, offering the additional benefit of cumulative practice.

For instance, in a game-based system using 3D advanced digital animation techniques (3D ADAT), poor etiquette on the part of the human user in real time may ignite gossip among avatars representing unfamiliar cultures, which in turn decreases the user's credibility as the game progresses. Training of communication skills may involve soliciting information from members of an outsider culture to test the cultural competency of the human user. Many serious games and simulations pertaining to culture require the game player to tactfully engage in conversation with a variety of avatars in order to accomplish the mission or goal of the game. Generating a high level of trust in the game player among members of the cultural group represented is the way to *win* the game or to accomplish the goals of the scenario. Additionally, simulations are being developed with options that allow users, even individuals who do not have advanced knowledge of computers or programming skills, to tailor events represented to achieve a variety of training goals (see Section 46.2.2.1). Many systems are also designed for use by multiple human players rather than existing as a single-player game.

46.2.2 EXAMPLES OF TECHNOLOGY-ASSISTED CULTURAL TRAINING

Multiple products have been developed to offer solutions to changing military cultural training needs. While technologically advanced in comparison to former training methods, most of these products only represent cultural aspects of a single geographic domain (e.g., Iraq or Afghanistan). However, ALELO developers offer multiple options for cultural and language training via 3D simulation and serious games. Their applications make use of voice recognition technology and AI as supplemental teaching methods. Some of the ALELO systems that educate soldiers on foreign language and culture include the Tactical Language Training System (TLTS), the Tactical Language and Cultural Training System (TLCTS), and the Operational Language and Cultural Training System (OLCTS) (Johnson et al., 2004; Miller, Wu, Funk, Johnson, & Vilhjalmsson, 2007). Users cannot succeed in these systems without appropriate language skills; however, avoiding offensive or ambiguous nonverbal gestures also marks successful cultural interactions (Table 46.1).

Another common game scenario requires the human user to elicit information from nonplayer characters (NPCs) in order to accomplish a mission. Performance measures include analysis of user decisions about which NPCs to approach, and user speech and conduct toward the NPCs (Johnson & Valente, 2009; Miller et al., 2007). For example, the TLTS combines game design principles and development tools with learner modeling, pedagogical agents, and pedagogical dramas/scenario design elements. Trainees practice missions in a simulated game world, interacting with NPCs. A scenario can be developed to include a virtual aide (NPCs) to assist the trainees who are acting as operational characters to give them *guidance* and contextual performance feedback. AI can be used to control the behavior of the NPCs as well as to manipulate events to increase or decrease the friction points and drama. While cultural skills are necessary for mission success in TLTS and TLCTS, the main objective involves acquiring lingual skills specific to a certain region. ALELO's OLCTS places a greater emphasis on intercultural communication (Johnson & Friedland, 2010). In the ALELO products, intelligent tutoring systems (ITS) provide user guidance and feedback to improve performance.

To instill general cultural awareness, the Virtual Cultural Awareness Trainer (VCAT) South Africa, also developed by ALELO, begins with an educational component defining culture and its importance. The program delivers scenarios set in a specific region, with an emphasis on situational awareness and etiquette rather than language skills (Johnson, 2010). VCAT Afghanistan, however, includes a more language-based curriculum in response to demands from the field (Heinatz, 2011). USJFCOM's Joint Knowledge Online (JKO) uses VCAT Afghanistan to train interpreters in extending culturally appropriate greetings and building rapport with local populations. The program is broken into five modules that teach students how to address local grievances, coordinate humanitarian assistance projects, set up checkpoint manning, conduct training with Afghan National Security Forces, and perform home searches. In an assessment of the ALELO serious game Tactical Iraqi, McCarthy (2009) recommends a precondition of self-awareness training before embarking on the Tactical Iraqi journey so that users are able to appropriately recognize their cultural biases.

TABLE 46.1

Military Applications of M&S Cultural Training Programs

Name of Program	Method of Use	Training Focus	Scenario	Emphasis	Evaluation Method	References
Elect BiLAT	PC Authorable Game-based simulation	Culture-specific	Mission/story-based (user is an army officer with mission, i.e., rebuild town) Assess problem, plan, and conduct meeting	Negotiation skills Culture-specific bilateral engagement Cultural sensitivity	Intelligent tutoring Socratic AAR	Hill et al. (2006)
TLCTS	PC/handheld devices Authorable Game-based simulation	Culture-specific	Mission/story-based User practices specific language and cultural skills with AI character	Language and speech with appropriate cultural cues (communication skill-builder lessons)	Games or quizzes Intelligent tutoring Dialogue performance Task assessment	Johnson and Valente (2009)
3D ADAT	PC Game-based simulation	Culture-specific	Mission/scenario-based Completed by correctly communicating with and assessing NPC mood	Cross-cultural competence with Afghan culture	Analysis of interactions within the community Assessment of NPC emotional state	Zielke et al. (2009)
VCAT	Web-based course Game-based simulation	Culture-specific	Mission/scenario-based User moves through story by communicating with NPC and speaking language	Cultural awareness Language Communication skills	Virtual coach Intelligent tutor	Heinatz (2011)
FPCT	PC Game-based simulation	Culture-specific	Mission/story-based User views prologue and goals and must assess and communicate with NPCs Analyze, plan, act, and gather intelligence	Cross-cultural decision making Situational awareness	User's assessment of NPC moods, attainment of trust, and quality of information gathered	University of Texas Dallas Arts and Technology (2011)

(continued)

TABLE 46.1 (continued)
Military Applications of M&S Cultural Training Programs

Name of Program	Method of Use	Training Focus	Scenario	Emphasis	Evaluation Method	References
VECTOR	PC Authorable Game-based simulation	Culture-specific	Mission-based Peacekeeping carried out by communications within community	Cross-cultural sensitivity Negotiation Language	Intelligent tutor AAR Feedback messages displayed on screen	Deaton et al. (2005)
Adaptive Thinking and Leadership System	PC network Game-based simulation	Culture-specific	Scenario-based User is Soldier, native Iraqi, or invisible evaluator Perspective taking	Teamwork Intercultural communication Adaptive thinking	Evaluation tests given by (human) instructor AAR	Lane and Ogan (2009)

That same study also notes that while *tangibles of culture* (i.e., scenery, gestures, and language) are well represented, *intangibles* such as hierarchies in Iraqi culture, group behaviors, and representations of power are lacking in the Tactical Iraqi cultural training model (McCarthy, p. 39).

ALELO has recently developed a modeling system that addresses components of the whole society approach missing from their simulations. VRP 2.0 builds on previous ALELO products through the inclusion of macro-, micro-, and mesosocial components (Johnson, Sagae, & Friedland, 2012, p. 1) meant to encourage a more complete range of engagement. The VRP 2.0 system has the capability to run longitudinally (i.e., a user decision made early on will affect mission success later) and to contextually shift depending on scenario incidences (e.g., an IED explosion or disruption of general daily activities) (Johnson et al., 2012).

A system known as Enhanced Learning Environments with Creative Technologies for Bilateral Negotiations (ELECT BiLAT) also offers a culture-specific approach to cultural training. The ELECT BiLAT serious game requires the user to navigate social situations inside Iraq. The game familiarizes the user with situations that involve gathering intelligence, connecting to resources, and organizing meetings, negotiations, and follow-up activities to achieve mission success (Hill, Belanich, Lane, & Core, 2006). The game includes feedback through the application of a virtual tutor to improve social interactions and negotiation skills (McCarthy, 2009; Zielke et al., 2009).

Similarly, 3D ADAT, Virtual Environment Cultural Training for Operational Readiness (VECTOR), and the First Person Cultural Trainer (FPCT) involve navigating cultural obstacles (e.g., customs, religions, traditions) in a specific culture. Winning the games consists of successfully gaining NPC trust, eliciting information, apprehending desired subjects, or coexisting in the community by exhibiting culturally appropriate manners (ATEC, 2012; Deaton et al., 2005; Zielke et al., 2009). These games provide experiences intended to develop the skills required by US forces to understand and operate within other cultures, including knowledge of norms based on gender, religion, and status. User responses to interpersonal space (i.e., proxemics), interactions, and emotional or personality tendencies and predispositions are included and evaluated. Importance is placed on the need for forces to understand perceptions of and attitudes toward American culture in order to instill awareness and sensitivity. The overarching goal is to identify and appropriately respond to divergences from American norms.

46.2.2.1 Fielded Training Systems

The virtual gaming system Virtual Battle Space 2 (VBS2), a 3D, first-person shooter, games-for-training platform used by the USMC, army, and other joint forces for individual or collective squad-level training, includes an application programming interface (API) tailor-made for connecting external hardware or applications to the VBS2 simulation engine to expand a training footprint by incorporating other combat elements to exercise communication and coordination TTPs, for example, surrogate call-for-fire devices, externally controlled AI, and custom user interfaces. The system also contains software tools to create scenario content and allows the trained user to import real-world terrain areas and to create and configure new 3D models. The development aspect of VBS2 supports both training and simulation. For example, trainers can use VBS2 to develop scenarios and import geo-specific terrain, while integrators might develop custom user interfaces in VBS2 for a specific simulator, for example, a Javelin trainer or a vehicle driving simulator. The 3D models are built using doctrinally correct parametric data to drive the *behavioral* aspects such as vehicle speed, munitions effects, or destructibility (Table 46.2).

VBS2 "provides realistic semi-immersive environments, dynamic terrain areas, hundreds of simulated military and civilian entities, and a range of geo-typical (generic) as well as actual geo-specific terrains" (Brown, 2010) where trainees and characters operate in a shared, immersive environment that supports mounted and dismounted operations, multiple combat platforms, small arms, and vehicle-mounted weapons. Applications feature US Army, USMC, and multinational equipment and weapons platforms, and more than 100 users can participate in the same networked exercise. The VBS2 simulation platform is potentially robust for culture training, but development

TABLE 46.2
Fielded Training Systems

Name of Program	Method of Use	Training Focus	Scenario	Emphasis	Evaluation Method	References
VBS2	PC and/or laptop network Authorable Simulation game engine	Individual, collective culture-specific	Integrates synthetic terrain, virtual characters, modeled munitions, weapons systems, levels of damage/destruction, and visual effects Radios and tracking systems used in training	Small unit TTPs First-person shooter, games-for-training platform	Resident AAR and conducted AAR Replay available	Brown (2010)
IIT	Mixed-reality training environment Simulated avatar images on mock-up buildings/rooms	Collective, culture-specific	Stress exposure Trainees navigate environment complete with auditory, visual, and olfactory cues Virtual characters and live role players	Decision-making skills Mental resiliency TTPs Application of training Kinetic and nonkinetic skills, urban TTPs, culture TTPs, MEDEVAC, stress inoculation	Instrumented for video and audio capture and recording of shots for AAR conducted by controllers	Henderson and Feiner (2009); Muller et al. (2004, 2008)

of cultural aspects during training depends on fidelity requirements to support training objectives, scenario authoring decisions, and the availability of desired synthetic and model behaviors and characteristics. VBS2 supports a 3D scenario editor and terrain importer tool and offers an after-action review (AAR) capability. VBS2 also provides integration capability with LVC architectures (Brown, 2010) and can be federated and networked for distributed operations, but it is primarily configured as a suite of laptops used to train small units in kinetic and nonkinetic TTPs.

VBS2 is fielded and used in daily training by the US Army, USMC, NATO, and ABCA (American, British, Canadian, and Australian) nations and supports multiple users and multiple training options. For example, a convoy trainer hosted on VBS2, a personal computer-based, first-person shooter game, can be networked to involve many players at once. The game offers simple drag-and-drop controls to build a scenario on one of the various realistic terrain databases available in the system's menu (Brown, 2010). Players can drive armed trucks through the desert; conduct dismounted patrols in an urban environment; work with tanks, artillery, and aircraft in the open field; or interact (on a limited basis) with civilians and coalition partners (Brown, p. 32). While various games may outperform VBS2, the game is readily available for use by individual marines on their personal computers or battalion laptops (Brown). However, the cultural aspects of the system are limited to those that support the primary training experience and may not be representative of the entire cultural milieu.

In November 2008, I Marine Expeditionary Force (I MEF) generated an Urgent Universal Needs Statement (UUNS) articulating the need for a high-end training venue that sufficiently replicated the complexities and stresses of the modern battlefield, including the cultural terrain, in order to expose small unit leaders to cues, conditions, and stressors and to train decision-making processes (MROC Decision Memorandum, 2009). To meet that need, the Infantry Immersive Trainer (IIT), a mixed-reality training environment, was designed and developed by an integrated product team (IPT) composed of USMC stakeholders, engineers, and training specialists to hone small unit collective skills in a safe, realistic, and rapidly repeatable venue (see Henderson & Feiner, 2009; Muller, Cohn, & Nicholson, 2004; Muller, Schmorrow, & Buscemi, 2008). The IIT uses four critical means to support this purpose: hyperrealism (immersion), scenario-based training design, exposure to stress, and an after-action assessment. An IIT footprint is fielded for training at I MEF, Camp Pendleton, CA; II MEF, Camp Lejeune, NC; and III MEF, Hawaii. An experimental IIT, colocated at I MEF, is being used as a test bed for development and evaluation of technologies to determine future human performance training requirements and capabilities.

The IIT footprint provides realism by closely replicating the contemporary operational environment (COE), a realistic urban environment—a living, working village—complete with high-fidelity visual, structural, auditory, olfactory, and tactile characteristics (i.e., sounds, smell, and feel), as well as interaction with a live role-playing population, to immerse the training audience in a chosen environment and create a sense of realism. Detailed scenarios are defined to meet unit-determined mission essential tasks (MET) and drive *population* and *opposing force* actions. These actions create situations that force the small unit leader to exercise judgment, make decisions, and apply training skills within a specific set of circumstances and environmental conditions that closely replicate the lowest and highest levels of stress and complexity found in theater. The exposure to realism and stress is intended to develop *mental resiliency* that enables small unit leaders and squad members to better handle emotional and traumatic situations and to increase unit proficiency and confidence in decision making under demanding circumstances. The AAR and assessment capabilities facilitate effective learning and provide the unit chain of command with a tool to evaluate unit performance under the most demanding situations and to determine preparedness for deployment.

The focus of this training venue helps refine squad leader decision making and allows the squad to integrate various individual and collective skills and to employ unit standard operating procedures (SOPs) in a complex environment. Within the small unit training continuum, and considering the crawl–walk–run paradigm, this event is at the high end of the *run* spectrum when integrated into unit predeployment training plans. Location within the unit training plan considers training and education skill levels necessary for optimal exercise benefit.

The IIT immersive training environment includes the following critical components and assets that are used to develop training scenarios:

- Training facility with environmental representation to impact human senses (sight, sound, smell, feel) including (1) visual stimuli, via geographically similar landscape, people, animals, building structures, vehicles, weapons, explosives, building furnishings, and decorations (atmospherics); (2) scent stimuli to replicate smells that will be experienced in theater as well as smells associated with threat indicators and warnings; (3) audible stimuli to replicate normal environmental activity, battlefield, and weapons effects; (4) and physical stimulus to replicate weather (when feasible) and weapons effects (concussion for explosions and sense of pain when struck by direct fire training systems and simulators).
- Cultural, ethnic, and linguistic representation to include virtual and live role-player speech, community activities, and individual actions and reactions accurately depicting current theater of operation. During execution, unscripted unit actions and role-player reactions (within cultural norms) are also included.
- Operational support across the full range of military operations (ROMO) to allow application of kinetic and nonkinetic responses, including key leader engagements, combat profiling, nonlethal fires and simulated direct fire, indirect fire, and IED effects. Flexible training levels provide the best possible training benefit to units with varying military occupational specialties and degrees of competence.
- Interactive training entities (people, avatars, high-fidelity virtual entities, systems) acting in accordance with training objectives and threat scenarios to *draw* the marine into a realistic situation. Mixed-reality entities, incorporating live and virtual actors, act/react consistently with cultural norms in a cause-and-effect manner to replicate complex situations, rapidly dictate changes across the ROMO, support human terrain understanding, and exercise moral, legal, and tactical decision making.
- Training support tools to enable personnel to receive training objectives from units and to associate them to METs with all necessary supporting information, that is, role-player biographies, social network diagrams, intelligence reports, etc., to enable development of historical context and a coherent *story line* and to create master scenario event lists (MSELs). It also includes tactical support equipment and minor training devices to accurately replicate TTPs.

46.2.3 Shortcomings of Current M&S Cultural Training

Despite these technological innovations, many M&S systems include cultural flaws and misrepresentations similar to the culture-specific training methods discussed earlier. A lack of input from social scientists often results in scenarios based on opinions and/or cultural interpretations that are situation-based and narrowly emulated (Belanich et al., 2004). Similarly, while credited with improving the user's cultural awareness, an absence of empirical analysis has made cultural representations difficult to validate (Abbe et al., 2007; Brown, 2010). In many systems, the models are not adjustable or editable by trainers, resulting in scenarios that do not coincide with evolving goals (Belanich et al., 2004). Additionally, simulations are not meant to be a sole means of cultural training. Supplementary lessons and feedback by subject matter experts (SMEs) are important variables for successful cultural training (Belanich et al., 2004).

In recent years, scientists have made dramatic improvements in creating models that provide players with geographically and culturally realistic simulations, as well as user-friendly software systems. However, one of the biggest shortcomings to achieving *cultural fidelity* in virtual training systems is the state of the art in AI and human behavior modeling in synthetic characters/

avatars. Current systems have not yet achieved the necessary advancements in speech recognition/ generation, facial and hand gestures, spontaneous decision making, and reactions, but ongoing research in this area is being supported by the Office of Naval Research (ONR), the services, and many universities (see Schmorrow, 2011).

Empirical analysis on the effectiveness of simulated cultural training methods, however, is still in the beginning stages, and a common ontology regarding cultural representations and interactions has yet to be established (Tolk et al., 2010).While the use of such training tools is commonplace and constant, they are often integrated in training with little to no scientific support for their effectiveness (Abbe et al., 2007; Caligiuri et al., 2011). As noted by Abbe et al. (2007), the USMC Center for Advanced Operational Culture Learning (CAOCL) has adopted the term operational culture, which embodies the goal of teaching cross-cultural knowledge. However, "the content has not yet been empirically tied to outcomes, and no measures have been developed to assess this knowledge" (p. 37).

In addition, while a representation of society that encompasses all components (e.g., geography, social systems, political systems, religious practices, cultural norms) is recognized as advantageous, many of the programs concentrating on the inclusion of these components are not yet fully developed. While attention to cultural detail in simulations remains of high importance, designers of such representations may overstimulate the user. Exaggerated or overly decorated atmospheres in a given scenario can distract the user from the intended lesson (Caligiuri et al., 2011). Although culturally accurate representations remain important when emulating a specific culture, educational goals of instruction remain the priority.

An absence of standards for researching and assigning cultural attributes to simulations often leaves determination of methods for data collection on culture and the application of what's culturally relevant to the discretion of the software developer or consulting SME. These representations have the potential to be obsolete or inaccurate opinions of the *experts* consulted (Schatz, Folsom-Kovarik, Barlett, Wray, & Solina, 2012). For example, HTTs have been utilized to collect observational and survey data among foreign cultures, which has then been analyzed by SMEs and incorporated into simulations (Albro, 2010). However, these representations are time sensitive and allow for personal bias among data collectors and SMEs, making it possible for the simulated material to be outdated, irrelevant, or conducive to poor decision making (Albro, 2010). Reliable, accurate methods to assess representation of cultural factors in delivery systems and their effect on learner performance need to be developed.

46.3 METHODS FOR MEASURING CULTURAL COMPETENCE AND TRANSFER OF TRAINING

While numerous projects have focused on development of cultural instruction, additional work is needed to assess its impact on learning and performance (Abbe & Gouge, 2012, p. 16). To effectively measure a construct or competence in any given area, a clear definition of what is to be measured must be established (Deardorff, 2006; Hardison et al., 2009; Ross, Thornson, McDonald, & Arrastia, 2009). Given the lack of consensus among academics, professionals, and SMEs about the qualities encompassed by culture, it is not surprising that this ambiguity extends into the definition of what it means to exhibit cultural competence (Ross et al., 2009). Without a clear definition of what is to be evaluated, measurement methods and results are inconsistent. Methods vary from self-assessments, to third-party observation, to evaluations designed to elicit qualitative data (e.g., interviews or focus groups).

Overwhelmingly, questionnaires are utilized to assess cross-cultural competence. Unfortunately, many of these techniques lack robust reasoning or provide irrelevant results. For instance, measures of cultural competence within the health-care profession generally involve self-assessment inventories that often lack consistency and sometimes revolve around race and ethnicity (Kumas-Tan, Beagan, Loppie, MacLeod, & Frank, 2007). Research has revealed the inclination of assessments in the medical field to stem from a Western viewpoint, encouraging a White, English-speaking population to develop tolerance of *different* cultural traditions (Kumas-Tan et al., 2007).

As this example illustrates, the reliance on comparisons of *our* culture versus *their* culture (Kumas-Tan et al., 2007) allows for an ethnographic bias, which, if not addressed in cultural training, may contaminate the effectiveness of the training program. A few examples of evaluations cited in this study include the Cross-Cultural Adaptability Inventory (CCAI), the Multicultural Counseling Inventory (MCI), and the Cultural Competence Self-Assessment Questionnaire (CCSAQ). Each of these is noted by the authors to define culture in terms of ethnicity and race and to define cultural incompetence as a lack of knowledge of other cultures. Cultural competence in this case is referred to as comfort with oneself and people of other cultures (Kumas-Tan et al., 2007). Additional methods of measurement (as described in Caligiuri et al., 2011) include situational judgment tests, behavioral ratings, and knowledge assessments (pp. 38–39). Other ongoing longitudinal studies are being conducted by social scientists at CAOCL and elsewhere (see https://www.tecom.usmc.mil/caocl/Pages/CAOCL-Dispatch-November-2011.aspx and https://www.tecom.usmc.mil/caocl/Pages/CAOCL-Dispatch-November-2011.aspx).

Developing the proficiency to operate in unfamiliar cultures clearly involves more than the acquisition of cultural facts or the ability to overcome cultural bias; it requires the application of a *cultural agility* that enables the trainee to integrate learned skills (Caligiuri et al., 2011). Therefore, measuring cultural competence by collecting data from self-assessments with a limited scope of what culture encompasses is not likely to provide pragmatic results. Development of an assessment with a clear definition of what is to be measured, along with descriptions of specific results desired from training and sufficient data for statistically significant analysis, represents a better approach (Caligiuri et al.).

46.3.1 Evaluating Cultural Competence

In an analysis of existing cultural competence evaluations performed by Abbe et al. (2007), the researchers indicate that assessments vary depending on how culture has been interpreted. The items of measurement in assessments chosen by Abbe et al. (2007) evaluate different individual characteristics, depending on whether cultural competence is thought of as multidimensional (including cognitions, behaviors, and motivations), developmental, behavioral, trait-based, or some other entity (Abbe et al., 2007). The authors note that while many of these measurement instruments are supported by empirical evaluation, none of them have been developed specifically for the military and few have been analyzed after use with a military audience.

In their analysis of this research gap, Abbe, Geller, and Everett (2010) applied several measurements to a sample of army active duty members. They examined the Multicultural Personality Questionnaire (MPQ), the Intercultural Development Inventory (IDI), and the Cultural Intelligence Scale (CQS), each of which corresponds with general cross-cultural skills, not regionally specific cultural knowledge. Researchers determined that these measurements might be useful for assessing motivational aspects of cross-cultural competence and the ability to adapt; however, the tools were less valuable for assessment of performance-related training outcomes (Abbe et al., 2010). The authors did not include analysis of assessments created for a military population (Navy Overseas Assignment Inventory, Cross-Cultural Interaction Inventory, Cross-Cultural Adaptability Scale, and the Cross-Cultural Competence Inventory [CCCI]) due to the lack of evidence of their effectiveness (Abbe et al., 2010). An additional challenge relates to the ambiguous nature of what is to be measured (i.e., indicators of cross-cultural effectiveness), compounded by the lack of an ontology to define training strategies and goals. Future work will need to establish an inventory of culture-general skills that can be used for accurate assessment of training outcomes and to guide instructional goals, as well as instructional delivery.

Currently, CAOCL provides training to marines headed to Afghanistan, for example, that is tailored to the mission and area to which the marines are embarking (e.g., training given to a marine going to the Kabul area will be different than a marine going to Helmand province). The standard training package is approximately 2 days and generally consists of five lessons in five broad categories: how the population interacts with their physical environment; typical appropriate and acceptable social activities; what the population's belief and value system is based on, that is, religion, folklore, legends; the current political system and how it affects individuals and their interactions with that system; the impact of the economy and social structure; and some tactical language and nonverbal communication skills for interactions with the local population.

Most of the training takes place in the classroom with some marine-on-marine role playing. After classroom training, the marines may have opportunities to practice skills in practical applications and simulation events, culminating close to marines' deployment date with an evaluation exercise where they are assessed in mission-based, scenario-driven exercises with indigenous role players at the MOUT facilities or in the IIT. Once marines leave a classroom-based environment, the opportunity to conduct collective exercises as a team or unit is impacted by many variables, including timing, availability, and changing missions. Outcome-based *effectiveness* is a by-product of how well each marine in their role is able to employ these skills in theatre, where he or she is assessed by unit leaders.

46.3.2 CHALLENGES OF PERFORMANCE-BASED M&S MEASURES FOR MILITARY APPLICATIONS

Many factors drive the increasing demand for accurate evaluation, measurement, and assessment of cross-cultural competence among members of the military, particularly the need to determine effectiveness and to justify the costs of training programs (Caligiuri et al., 2011). Assessments often influence placement for specific assignments, diagnosis of training needs, individual feedback for further development, evaluation of training, and readiness assessment (Caligiuri et al.). However, evaluation of individual learning styles and program effectiveness (Abbe et al., 2010; Caligiuri et al., 2011) remains an important consideration as a *one size fits all* training approach may not be the most appropriate method for military cultural training.

Additionally, the measurement of individual or group efficacy of communication in a dynamic system such as culture compounds the challenge of defining accurate evaluation methods (Brown, 2010). Brown also notes that "Performance measurement is a science of its own, and problems are exacerbated when reviewing performance of teams instead of performance of individuals" (Brown, p. 42). The author suggests that measurement efforts should target both levels (individual and team/unit). These challenges, in addition to the need for swift program implementation, may skew evaluation of cultural training effectiveness, placing creators of training scenarios in a position to analyze the effectiveness of their programs without adequate time for empirical analysis.

In spite of obstacles to measurement, efforts to analyze existing programs that utilize M&S have been implemented. For example, Kobus and Viklund (2012) used situational judgment tests to analyze the Future Immersive Training Environment (FITE) and IIT programs. In these examples, participant outcomes are compared to SME outcomes in order to determine program effectiveness. Therefore, a student user who demonstrates cultural skills or communication strategies that differ from the SMEs' portrayal of appropriate behaviors may be regarded as unsuccessful. While this type of experiential testing before and after program implementation may be seen as progress, it exemplifies the inconsistent nature of standards that define success in cultural training programs, since outcomes of individual SMEs may vary, and other potentially successful student outcomes may be rejected. In some instances, the SMEs are placed in both designer and evaluator positions. Currently, the Training Support Centers (TSC) at USMC Training and Education Command (TECOM) are defining metrics and a protocol to be used for training effectiveness evaluations and to further refine the concept of operations for IITs. The MEFs and TECOM TSC have been tasked with the *official* assessments/evaluations, but that process is also ongoing.

A military-specific, universally available evaluation method may help fill the gap in measuring cross-cultural competence within the armed forces. Most cultural assessments for civilian training use scales, surveys, inventories, or questionnaires; a pragmatic approach would be to establish a similar instrument to serve as a baseline for assessing cultural competence in the military (Ross et al., 2009). Comparable measures of cross-cultural competence for the military are being developed by the Cognitive Performance Group (3CI, C-CAT, 3CLO), the RAND Corporation (Hardison et al., 2009), and other teams of researchers (Ross et al.). Ross et al., for example, created a CCCI to test for individual attributes that may predict success in military missions based on cross-cultural communication. While this assessment identifies individuals with the potential to exhibit an elevated level of cross-cultural competence that might lead to success in military missions, it does not evaluate performance outcomes. Researchers Hardison et al. recommend creating performance standards in order to match cross-cultural competence skills and abilities to measurement tools that vary depending on the behavior and skill level to be measured.

Overall, review of the literature points to a general agreement on the importance of training evaluation. A definition and description of measurement criteria for cultural competence is needed to initiate the assessment process. Awareness of individual versus team measurements and comparisons for performance outcomes would improve overall analysis (Brown, 2010). In addition, the use of multiple assessment methods (i.e., qualitative and quantitative), when time permits, would provide more robust data for evaluation of training effectiveness (Abbe et al., 2007, 2010; Davis, 2012; Lane & Ogan, 2009).

46.3.3 M&S MEASUREMENTS: FEEDBACK

Feedback incorporated into simulations offers a means of measuring the cultural skills of student players based on real-time performance rather than the reliance on self- or third-party assessment (Sims et al., 2012). For example, ITS mimic live instructors by recognizing errors, acknowledging improvement, and providing responses dependent on the actions of specific players (Billings, 2012). Implementation of individualized instruction like ITS should be theoretically based and empirically documented since the goal of the ITS is to re-create the individual attention that a student would experience with a private tutor (Billings). While useful in supporting instruction of factual information, ITS often lack the ability to engage the student in scenarios requiring cognitive reasoning that advance his or her analytical skills (Billings, 2012; Nicholson, Fiore, Vogel-Walcutt, & Schatz, 2009). Situated tutors modeled within a simulation's software, on the other hand, provide users with the opportunity to employ deeper reasoning techniques by "combining features of intelligent tutoring systems with simulated environments" (Schatz, Oaks, Folsom-Kovarik, & Dolletski-Lazar, 2012, p. 166). Most serious games emphasizing cultural training utilize ITS, which may improve transfer of training.

46.3.4 TRAINING TRANSFER

Training transfer is described as "the extent of retention and application of knowledge, skills and attitudes from the training environment to the workplace environment" (Bossard, Kermarrec, Buche, & Tisseau, 2008, cited in Brown, 2010, p. 37). The degree of transfer depends on the level of application involved. Vertical transfer refers to the transfer of skills to circumstances resembling those used in training (Brown, 2010), and horizontal transfer describes an instance in which the learner is able to apply training throughout various situations. By internalizing the training lessons, the student reaches a skill level that is innate and applicable to scenarios beyond that of the training context (Bossard et al., 2008; Brown, 2010). Given the variety of multicultural operational environments that military members must navigate, horizontal transfer and the ability to adapt training skills to different circumstances represent essential goals (Bossard et al., 2008). Unfortunately, measuring the effects of training remains a complex process made more challenging by the notion that "there are no absolute right or wrong answers in dealing with socio-cultural encounters ... only what does or does not work within a given cultural context" (Sims et al., 2012, p. 1477).

Currently, common methods of evaluating the efficacy of cultural training involve student self-assessments, academic examinations testing knowledge of subjects, SME or instructor reviews of student performance, or comparison of performance to a certain standard (Brown, 2010, p. 51). With no generalizable standard for cross-cultural competence, a systematic means of measuring training transfer supported by scientific methods is lacking (Brown). As a result, some M&S programs have been implemented without sound reasoning and causal connections (Wildman et al., 2012) and/or lacking identification of outside variables that may affect the effectiveness of training (Brown). For instance, backgrounds and learning styles of individual trainees, differing methods of training and feedback, and fluctuating measurement methods all cloud results of perceived transfer of training (Abbe et al., 2010; Brown, 2010; Caligiuri et al., 2011). Many assessments are based on individual perceptions or performances, ignoring the fact that military units are composed of groups of people (Brown). As mentioned by Sims et al. (2012), evaluations of culture through individual points of view ironically support a Western orientation. Marines, for instance, are given an opportunity to *transfer* skills and knowledge in the IIT. However, since that is predeployment training, they are *evaluated* by their unit leaders, not in a controlled, empirical study.

For the DoD, validation is the process of determining the degree to which the model or simulation provides an accurate representation of the real world from the perspective of the intended user (Davis, 2012, p. 22), which may not be the best predictor of training effectiveness or training transfer. While the debate over the importance of visual fidelity and scenario realism in M&S systems continues, it has been established that a believable story relating to learning goals enhances training transfer more than a game that demonstrates superior graphics (Hill et al., 2006). Perhaps "every commander yearns for the level of training [that is] so real that one cannot tell the difference from actual combat" (Brown, 2010, p. 10), but basing simulations on realistic cultural narratives that involve the user's imagination and create a suspension of disbelief may be a more important goal.

46.4   CULTURE-GENERAL TRAINING

While culture-specific programs emphasizing language training and factual knowledge and/or narratives based on geographically specific areas have dominated cultural awareness education in the past (Abbe & Haplin, 2010; Lane & Ogan, 2009; Wildman et al., 2012), the curriculum has primarily consisted of identification of behaviors that deployed forces should practice or avoid in order to diplomatically coexist with members of a foreign culture (Abbe & Haplin, 2010; Deaton et al., 2005). Knowledge of appropriate and inappropriate actions may result, but this form of training does not offer a broad understanding of why these actions are advantageous or deleterious.

Additionally, although culture-specific training provides accessible information for the region of deployment, the material may contain flaws that can be potentially damaging to observation and communication skills (Abbe & Haplin, 2010). Culture-specific data sources, while sometimes regionally accurate, also allow for invalidated personal opinions or ethnographic generalizations (Abbe & Haplin, 2010; Wildman et al., 2012). For instance, the information may only apply to a small part of an entire region and not extend to neighboring subcultures within the same region. More importantly, the information provided through culture-specific training frequently does not apply to other, dissimilar cultures, thereby rendering the cost of development, and the investment of students' time and effort to learn the material, useless when missions shift to other regions (Abbe & Haplin, 2010; Deaton et al., 2005; Schatz, Folsom-Kovarik et al., 2012).

Culture-specific training offering authorable structures of M&S systems allows for a variety of scenarios and outcomes, but that same feature can contribute to inaccuracy and negative training. In many cases, instructors and SMEs possess the ability to alter scenarios based on what they deem to be important cultural aspects. They also may control encounters between the student user and NPC that they consider relevant, which can lead to misrepresentations or inaccurate, stereotypical behaviors.

Rather than offer a narrow, sometimes skewed representation based on a designer's interpretation of a specific culture, culture-general training offers lessons that apply to multifarious regions and cultures. Scenarios depicting unfamiliar cultural cues without specific details about existing cultures may provide better training opportunities for a more universal cultural approach (Deaton et al., 2005). While culture-specific methods treat culture as a static entity that may be learned with an instructional list of what to do and what not to do, culture-general training broadly emphasizes cultural awareness and sensitivity, and recognition of cultural similarities as well as cultural differences, which may aid in relationship building (Abbe & Haplin, 2010).

Training cultural knowledge requires more than enumerating a set of behavioral rules to a trainee. To some extent, the trainee needs to internalize a mindset; a specific set of cultural encounters in a training scenario will never cover all possible experiences that a soldier may encounter in real life, but needs to be designed to generalize into a broader understanding. Significant academic work establishes the validity of this kind of culture-general approach (Deaton et al., 2005).

46.4.1 Benefits of Culture-General Training

Culture-general training may enable military forces to adapt to multiple cultures by instilling *intercultural effectiveness* (Watson, 2010, p. 94) and *cultural agility* (Caligiuri et al., 2011, p. 7). Rather than narrowing the cultural focus to impart detailed facts about a specific region, culture-general training emphasizes development of the cognitive skills necessary to appropriately assess a foreign environment (Abbe & Haplin, 2010; Deaton et al., 2005; Wildman et al., 2012) and to decide "what, in a given situation, is actually worth paying attention to and why" (Caligiuri et al., 2011, p. 11). To produce cultural sensitivity and reduce bias, culture-general training promotes an awareness of one's own culture and helps trainees control impulsive reactions to unfamiliar stimuli (Wildman et al., 2012, p. 9).

Ideally, culture-general training results in an individual with the observational and perceptual skills to adapt to a foreign atmosphere with or without knowledge of the specific culture (Abbe et al., 2007; Abbe & Haplin, 2010; Deaton et al., 2005; Wildman et al., 2012). Deployed forces receiving this kind of training would better understand *both how to communicate and who to communicate with* (Wildman et al., p. 6).

In recognition of the importance of transferable cultural skills, the army and the air force place culture-general training at a higher level of importance than culture-specific training (Watson, 2010). Current culture-general frameworks include Geert Hofstede's five values dimensions (Hofstede, 2001), the Global Leadership and Organizational Behavior Effectiveness program's nine dimensions of values and practices (Housea, Javidanb, Hangesc, & Dorfmand, 2002), and Gary Klein's cognitive dimensions (Klein, 1999). These approaches possess broad relevance as any culture can be characterized in terms of where it falls along the continuum of a particular dimension (Abbe & Haplin, 2010). In future efforts, culture-specific training might best be used as a supplement to training of culture-general competence or to meet mission specialization needs.

46.4.2 Culture-General Training: Archetypal Patterns of Life

Across societies, cultures, and geographic regions, reoccurring patterns of generalizable human behavior emerge, and within a given sociocultural context, recognizable and reasonably stable patterns of life (PoL) can be observed, enabling identification of a *baseline* of normal activity. Under the Perceptual Training System and Tools (PercePTs) Program sponsored by the ONR, researchers are exploring how, in conjunction with culture-specific and culture-general knowledge, understanding these archetypal (or universal) PoL may better prepare personnel to operate in kinetic and nonkinetic cross-cultural settings (Schatz, Folsom-Kovarik et al., 2012). In the context of cultural training, an archetype is a prototypical pattern, symbol, or concept. It is a general template from which unique derivatives arise (Schatz, Folsom-Kovarik et al., 2012).

Building on this definition, PoL might be described as the archetypal emergent properties of a complex sociocultural system (Schatz, Folsom-Kovarik et al., 2012, p. 4) based on universal behaviors. For example, universal human emotions include fear, sadness, and frustration, and universal practices include cooking food, sleeping in individual or group quarters, joining mates via rituals, identifying interactions based on kinship, exchanging greetings, wearing clothing or wraps, dividing labor, organizing relations hierarchically, making music, creating nonlinguistic symbols, and participating in death rites (Schatzki, 2003, p. 8). Many universals are collective rather than individual: music and dance are found in all societies, for example, but not all individuals participate (Brown, 2004). In terms of behavior, human universals include aggression, gestures, gossip, and certain facial expressions; in the realm of the mind, universals include emotions, dichotomous thinking, empathy, and psychological defense mechanisms; many universals have distinctive neural underpinnings (Brown, 2004).

Additionally, a handful of basic and archetypal literary patterns, often called *narrative universals*, account for two-thirds of the plots in all narrative tradition (Hogan, 2006). These foundational themes—patterns of literary life—are common across all cultures. Similarly, social scientists have articulated the features of universal social roles (Masolo et al., 2004), which are defined by a number of elements: these *patterns of relationships* (Sowa, 2000, cited in Masolo et al., 2004, Section 2.1) are observable across societies and cultures (Schatz, Folsom-Kovarik et al., 2012, p. 4) and include *ethnocentrism, play, exchange, cooperation, and reciprocity* (Brown, 2004, p. 47).

Applied to the current topic, universal PoL fall into archetypal configurations. That is, emergent properties can be organized into categories that describe classes of PoL that share similar general features, and it is theorized that these classes will manifest (likely with different nuanced characteristics) across all societies and cultures (Schatz, Folsom-Kovarik et al., 2012, p. 5). For example, archetypes within COIN operations have been identified, such as the patron–client network, the opportunistic insurgent, and the accidental guerrilla (Turnley, Henscheid, Koehler, Mulutzie, & Tivnen, 2012). Further, these generalizations transcend specific cultures (e.g., Iraq, Afghanistan); they describe and predict emergent PoL across all insurgent groups (Jager, 2007; Turnley et al., 2012). Another example of archetypal (i.e., universally occurring) patterns of observable behavior might be characterized as the presence of key informants. In any given group a "key informant is someone who (1) knows important things, (2) understands why you are asking about these things, and (3) is willing to tell you. The ability to identify and work cooperatively with key informants can speed cultural learning" (Caligiuri et al., 2011, p. 13).

46.4.2.1 Benefits of PoL for Culture-General Training

Understanding archetypal PoL will allow personnel to intuitively recognize emergent sociocultural patterns within any context and use their skills to support sociocultural sensemaking, the development of mental baselines of normality, anomaly detection, and communication of sociocultural cues (Schatz, Folsom-Kovarik et al., 2012, p. 6). Culture-general training that includes study of PoL may help personnel integrate static cultural knowledge and theoretical cross-cultural awareness and actively leverage their knowledge, skills, and attitudes to support sociocultural sensemaking (Schatz, Wray et al., 2012). The work of theorists who have argued for the importance of sensemaking processes in cross-cultural adjustment (e.g., Abbe, Gulick, & Herman, 2008; Black, Mendenhall, & Oddou, 1991; Osland & Bird, 2000) and other training methods to enhance personnel's *metacultural awareness* (Johnson & Friedland, 2010) supports this culture-general approach, as do anthropological theories noting that core behaviors occur transculturally and correspond with predictable human universals (Tomas, 1998, cited in Schatz, Folsom-Kovarik et al., 2012; Bhawuk & Brislin, 2000).

46.4.2.2 Culture-General Training: Recognition of PoL and Anomalies

Recent developments in culture-general training suggest that keen cognitive and observational skills can be developed that allow individuals to analyze their surroundings regardless of specific cultural qualities and permit a swift, accurate evaluation of a baseline of cultural norms and the perception of cultural anomalies. The human mind's pattern recognition abilities (when properly trained) can sift, connect, and draw meaning out of what seems to be random noise (Schatz, Folsom-Kovarik et al., 2012, p. 5). Proficiency in anomaly detection is transferable to all social scenarios, regardless of culture.

For example, the Combat Hunter training program serves as the primary USMC program of instruction for learning sustained observation of social PoL and for enhancing personnel's social, cultural, and behavioral sensemaking skills. Through the Combat Hunter course, personnel learn how to evaluate people's biometric signs, read human behavior, identify geographic indicators (like footprints), and objectively analyze the *atmosphere* of a locale. Broadly speaking, Combat Hunter trains situation awareness, sensemaking, mental simulation, and dynamic decision making for urban operational environments (Schatz & Nicholson, 2012, pp. 2–3; Schatz, Reitz, Nicholson, & Fautua, 2010).

The idea of referencing human universals (or archetypes) as an aid to understanding other cultures has long been suspect, since many patterns of behavior were derived exclusively from a Western viewpoint (Schatzki, 2003). However, due partly to a recent, more respected multidisciplinary approach to the study of human patterns, an acceptance of the plausibility of human universals has emerged. It is possible that persons from unlike societies use a similar process to assess other cultures; individuals may attempt to understand those who are thought to be *different* in their own culture in the same manner that they assess members of a *different* culture (Schatzki, 2003), possibly via pattern recognition, comparison, and analysis.

46.5 M&S INNOVATIONS SUPPORTING CULTURE-GENERAL PoL TRAINING

Affordable and scalable M&S training systems that can replicate archetypal PoL rely on software that adds lifelike human characteristics to real-time visual simulations and enables the easy creation of crowds and individuals who are autonomous and terrain-aware and who react intelligently to ongoing events (Blank, Broadbent, Crane, & Pasternak, 2009). To model human networks (i.e., organized activities requiring multiple agents performing in differing roles over a length of time resulting in measurable output), large numbers of reactive civilians demonstrating PoL need to be incorporated; thus, "characters must be reactive to warring stimuli" (Blank, Pasternak, Crane, & Broadbent, 2011, p. 7) and reinforce the training narrative. Systems need to be able to model a complex network and to depict how countermeasures might disrupt or destroy a network (Blank et al., 2011). Recognizing patterns of behavior among members of networks can enhance warfighters' ability to perform COIN, antiterrorism, peacekeeping, and IED detection activities.

Recently developed platforms for these types of systems, such as DI-Guy, host options that provide PoL depicted by believable, humanlike characters, since poorly represented human characters (or an absence of humans) may distract users from the training goals of the immersive environment (Blank et al., 2009). These programs rely on a technical approach of nonlinear, hierarchal software programs and a combination of AI Minds, SmartObject, and SmartBuilding frameworks. Used by all branches of the US Armed Forces, these software systems also provide the ability to author human PoL activity outside of the dedicated graphical user interface (GUI) (Blank et al.). DI-Guy Scenario and AI functionality is supported with the Unity game engine (a developmental platform). The following developments demonstrate the potential for M&S to support PoL culture-general training:

- DI-Guy SDK is a software library that allows authors to quickly integrate humans into real-time visual application.
- DI-Guy Scenario is an interactive 3D visual application for authoring human performances.
- DI-Guy AI adds AI to the characters, enabling them to autonomously navigate and react to their changing environment.
- DI-Guy Roleplayer and Lifeform Server stations are authorable, human-centric training solutions designed for both single-player use and distributed simulation populations.

To support PoL AI, and to model the most basic PoL concepts, scheduling techniques for loosely specifying behavior are required. For example, the prepackaged farm SmartBuilding contains one or more farm buildings, designated fields surrounding the farm, and members (agents) of the farming family. In the simulation, everyone on the farm wakes at dawn, across a range of times. The children, a subset of people on the farm, go to school at 7 a.m., but they do not all leave at exactly at the same time. GUI support for scheduling is provided to the author for easy creation, review, and editing of the schedules of the agents. The method typically designates a crowd or a subset of a crowd, the time for the goal/state change, and a +/− time duration during which the system will command the transition (Blank et al., 2011).

In Platoon Leader, an ECO Sim system used in preparation for USMC deployment to Afghanistan (Marine Corps Tactic and Operations Group), DI-Guy AI provides critical PoL activity to simulate real-world challenges faced when combating insurgencies embedded within the general populace (Blank et al., 2011). Marines command and control a company of platoons, squads, and individuals in distributed simulations involving hundreds of marines, insurgent fighters, and Afghani civilians. Platoon Leader supports interaction at all levels.

Other systems have recently been developed with similar capabilities. In July 2012, for example, VT MÄK released VR-Forces 4.0.4 and B-HAVE (Brains for Human Activities in Virtual Environments) 2.0.4, both of which feature PoL modeling. Powered by Autodesk Kynapse, B-HAVE uses advanced AI technology to provide complex and realistic behaviors and background traffic within MÄK's VR-Forces simulation environment. Using B-HAVE, VR-Forces' entities can analyze terrain topology, intelligently navigate complex urban environments, automatically plan and follow paths through 3D building interiors, dynamically avoid collisions with obstacles or other entities, and flee from threats. B-HAVE enables VR-Forces' users to create complex PoL, where streams of people and vehicles follow context-sensitive patterns. These patterns add realistic background traffic to VR-Forces' scenarios, giving depth to simulations (Kogler, 2011).

Users can quickly create crowds of people and vehicle traffic that behave in natural, lifelike patterns, such as children on their way to school in the morning or workers' vehicles leaving buildings during lunch. The patterns are developed without programming: Using only the GUI, users can compose patterns to model human behavior, taking simulation conditions (like time of day) into account. A B-HAVE-controlled entity is able to perform dynamic collision avoidance with other moving entities, even as the entity moves along its chosen path through the terrain. B-HAVE entities will avoid other entities in the scenario, whether those other entities are controlled by B-HAVE or not. B-HAVE's algorithm keeps the *goal point* in mind as it moves to avoid collisions so that it can always choose a direction that makes sense in the context of the entity's primary task (VT MÄK, 2012).

The Organizational and Cultural Criteria for Adversary Modeling (OCCAM) project models *soft* aspects of human behavior to meet the need for training US forces to predict behaviors of opposing forces in different cultural environments. Simulated environments are modeled to present a practice environment for trainees, but often the models fail to effectively address some of the *soft* representational PoL issues, including modeling the influence of culture on groups and individuals, the ways organizations influence individual behavior and vice versa, the quantifiable differences between individuals, and the influence of situational factors. The difficult problem of behavior prediction is exacerbated by a general lack of specific data about individuals and groups, and the uncertainty and unreliability of available information that is inherent to asymmetric warfare (Charles River Analytics, 2004).

To meet these needs, the OCCAM project focuses on the development of a software-based decision aid that incorporates organizational and cultural influences on individual and group PoL behaviors. The OCCAM decision aid is able to infer additional characteristics of, and linkages among, individuals, groups, and events, even when given sparse data. The OCCAM tool also integrates modern psychological profiling approaches and cultural anthropology methods with traditional social network analysis techniques. Users of this software can evaluate the impact of social, cultural, organizational, and individual factors on the behavior of an individual or group, leading to improved mission planning and execution (Charles River Analytics, 2004).

46.5.1 ADAPTIVE CULTURE-GENERAL TRAINING IN A VIRTUAL ENVIRONMENT: PERCEPTS

The renewed emphasis on culture-general training for distributed operations has prompted development of numerous culturally based serious games and simulation training systems, software, and platforms supporting accurate representation of PoL, like those discussed earlier; consequently, new goals for cultural scenarios in military training have emerged. First, cultural cues need to be developed and displayed at a depth beyond surface level. Rather than presenting an image representing a stereotypical version of a software designer's interpretation of culturally relevant details (e.g., headdress apparel or traditional attire), social patterns are being researched and incorporated into simulations like the PercePTs virtual observation platform. These archetypal patterns may involve multiple levels of observable physical interactions, communications, and routines that occur among members of social groups, and training can involve recognition of these cultural patterns from a third-person perspective (e.g., watching interactions between genders, observing shopping behaviors). Identification of these patterns (or archetypes) of cultural behaviors (i.e., PoL) will allow an observer to monitor societal norms to establish baselines of day-to-day activities, ultimately enabling a user to detect anomalies from baselines and develop cultural analytical skills (Schatz & Nicholson, 2010; Schatz et al., 2012).

The PercePTs virtual observation platform is being developed to provide this type of culture-general training using an immersive situated tutor designed to instruct perceptual–cognitive skills and related communication competencies within the context of sustained observation. USMC personnel will observe a virtual location (e.g., a small town) from a combat outpost located between 300 and 1000 m away. Using virtual reality optics, such as visually injected binoculars and night-vision goggles, marines will observe the PoL within the virtual village. They will have to make sense of the patterns of activity to establish a mental *baseline* of normal activities, identify anomalies, and, ultimately, predict deleterious events before they occur (i.e., *left of bang*) (Schatz et al., 2012, p. 6). Learning objectives include development of sociocultural sensemaking and mental baselines, identification of anomalies, and development of effective communication skills (Schatz, 2012, p. 2; Schatz & Nicholson, 2010). An anomaly occurs when something above or below the baseline happens. An above-the-baseline anomaly reflects the presence of a new event or object in the environment (e.g., a new vehicle in a neighborhood). Below-the-baseline anomalies reflect the absence of something (e.g., many fewer people in a town square) (Schatz et al., 2012).

Within this system, the virtual squad will model effective communication, demonstrate the impacts of poor communication, and give trainees hints or feedback, based on the scenario's intrinsic narrative (Schatz et al., 2012). After the trainees have observed activities from the virtual observation platform and established a mental baseline, the instructor will trigger a sequence of events (e.g., activities leading up to a bombing in the village marketplace). For example, an event might involve the delivery of bomb-making supplies. The trainees would observe the terrorist cell leader meeting a pickup truck and bags of fertilizer being moved into a home. This represents an obvious anomalous cue, because these chemicals would not normally be stored in a residence. If novice trainees miss observing the anomaly, the instructor can choose to scaffold them by triggering an event (e.g., squawking chickens) intended to draw their attention to the delivery. For advanced trainees, the same kind of cuing event occurring in a distant area can provide a distraction. In this way, manipulating intrinsic cue quantity or cue misinformation can scaffold or challenge a perceptual skill (Schatz et al., 2012).

The architecture for the virtual observation tower will include a control agent system that can monitor progress within scenarios, estimate trainees' proficiency as scenarios evolve, and invoke tailoring strategies (Schatz et al., 2012, p. 3). In support of PercePTs, Wray and colleagues (2009) have developed a general dynamic tailoring architecture that monitors student behavior during simulation-based practice and attempts to scaffold, to challenge, or to engage the learner based on the learning context (Schatz et al., 2012, p. 2). Like dedicated human tutors, these adaptive instructional technologies tailor learning content, delivery, and/or context to the unique needs of the learners (Schatz et al., 2012, p. 1).

46.6 RECOMMENDATIONS AND FUTURE WORK

The best long-term solution for building and sustaining cultural capability will include a blend of culture-specific and culture-general training approaches, including M&S games and scenarios developed specifically to support cross-cultural competence. Culture-specific training will be of greater usefulness in a broad range of diverse regional and cultural settings when training is not based solely on static knowledge of cultural differences, but includes culture-general study that provides a framework for understanding recognizable PoL (i.e., how people in general behave) in diverse social interactions and institutions, leading to recognition of cultural commonalities and minimizing stereotypical cultural biases.

In order to maximize the cultural readiness of deployed forces, cultural training should be employed not only as a part of a predeployment culture-specific curriculum but also as ongoing culture-general education. Explicit, advanced instruction pertaining to specific entities within a culture could supplement culture-general knowledge. Developing higher levels of understanding and possessing a strategic perspective on culture takes time; identifying and screening for foundational attitudes, knowledge, and skills that are prerequisites for developing cultural competence could lead to the development of cultural training methods that target individual learning needs.

Future work will also need to develop reliable, accurate methods to assess representation of cultural factors in M&S systems and their effect on learner performance. An ontology or inventory of culture-general skills is needed to define training strategies and goals, and a clear definition of performance outcomes and specific results desired from training should be developed. This framework should define cross-cultural effectiveness and guide development of instructional goals and delivery. Performance standards are needed to match cross-cultural competence skills and abilities to measurement tools that vary depending on the behavior and skill level to be measured (and individual/group learning styles). In addition, a military-specific, universally available evaluation method is needed for cross-cultural competence.

Finally, a hybrid approach to cultural training should be developed that combines culture-specific didactic instruction and guided culture-general practice supported by individually mentored, active M&S cross-cultural experiences. Incorporating M&S experiences based on archetypal PoL into culture-general training will allow personnel to intuitively recognize emergent sociocultural patterns within any context and use their skills to support sociocultural sensemaking, the development of mental baselines of normality, anomaly detection, and communication of sociocultural cues. M&S systems for culture-general training that include PoL instructional delivery, supported by AI or ITS, may help integrate static, culture-specific knowledge and theoretical cross-cultural awareness for improved transfer of training and increased operational effectiveness in a broad range of unfamiliar cultural settings.

REFERENCES

Abbe, A., Geller, D., & Everett, S. (2010). *Measuring cross-cultural competence in soldiers and cadets: A comparison of existing measurements* (Technical Report 1276, pp. 1–58). Arlington, VA: United States Army Research Institute for the Behavioral and Social Sciences. http://www.dtic.mil/cgi-bin/GetTRDoc?AD=ADA533441

Abbe, A., & Gouge, M. (2012, July–August). Cultural training for military personnel, revisiting the Vietnam era. *Military Review, 97*(4) 9–17. http://usacac.army.mil/CAC2/MilitaryReview/Archives/English/MilitaryReview_20120831_art005.pdf

Abbe, A., Gulick, L. M. V., & Herman, J. (2007). Cross-cultural competence in army leaders: A conceptual and empirical foundation. A report prepared by the United States Army Research Institute for the Behavioral and Social Sciences, Arlington, VA.

Abbe, A., & Haplin, S. (2010, Winter). The cultural imperative for professional military education and leader development. *Parameters, 39*(4), 20–31. http://www.carlisle.army.mil/usawc/parameters/Articles/09winter/abbe%20and%20halpin.pdf

Air Force Culture & Language Center (AFCLC). (2012). *Cross-cultural communication.* http://www.culture.af.mil/culture_intro3cclass.aspx

Albro, R. (2010). Anthropology and the military: AFRICOM, 'culture' and future of human terrain analysis. *Anthropology Today, 26*(1), 22–24.

Arts and Technology (ATEC) at University of Texas Dallas. (2011). *The first person cultural trainer white paper.* http://www.utdallas.edu/~maz031000/res/FPCT_White_Paper.pdf

Belanich, J., Mullin, L., & Dressel, J. D. (2004). *Symposium on PC-based simulations and gaming for military training* (ARI Research Product 2005-01, pp. 1–10). Arlington, VA: Advanced Training Methods Research Unit, U.S. Army Research Institute for the Behavioral Social Sciences.

Bhawuk, D. S., & Brislin, R. (2000). Cross-cultural training: A review. *Applied Psychology: An International Review, 49*(1), 162–191. http://dbr.shtr.org/v_1n1/dbrv1n1b.pdf

Bhawuk, D. P. S. (2001). Evolution of culture based assimilators: Towards a theory based assimilator. *International Journal of Intercultural Relations, 25*, 141–163. http://thestrategist.in/wp-content/uploads/2012/03/Evolution-of-culture-assimilators.pdf

Billings, D. R. (2012). Efficacy of adaptive feedback strategies in simulation-based training. *Military Psychology, 24*(2), 114–133.

Black, J., Mendenhall, M., & Oddou, G. (1991). Toward a comprehensive model of international adjustment: An integration of multiple theoretical perspectives. *Academy of Management Review, 16*(2), 291–317.

Blank, B., Broadbent, A., Crane, A., & Pasternak, G. (2009, July 13–16). *Defeating the authoring bottleneck: Techniques for quickly and efficiently populating simulated environments.* Paper presented at the IMAGE Conference, St. Louis, MO. http://www.diguy.com/diguy/files/IMAG-Authoring_Bottleneck.pdf

Blank, B., Pasternak, G., Crane, A., & Broadbent, A. (2011, November 28–December 1). *Smart objects and agents for autonomous high population simulated environments.* Paper presented at the Interservice/Industry Training, Simulation, and Education Conference (I/ITSEC), Orlando, FL. http://www.diguy.com/diguy/files/IITSEC- SmartObjects.pdf

Boroditsky, L. (2011, February). How language shapes thought. *Scientific American*, 304, 62–65. Published online, January 18, 2011.

Bossard, C., Kermarrec, G., Buche, C., & Tisseau, J. (2008). Transfer of learning in virtual environments: A new challenge? *Virtual Reality, 12*, 151–161.

Brawner, K. W., Holden, H. K., Goldberg, B. S., & Sottilare, R. A. (2011, November 28–December 1). *Understanding the impact of intelligent tutoring agents on real-time training simulations.* Paper presented at the Interservice/Industry Training, Simulation, and Education Conference (I/ITSEC), Orlando, FL. https://litelab.arl.army.mil/system/files/IITSEC2011_Brawner_etal_Understanding%20the%20Impact%20of%20Intelligent%20Tutoring%20Agents%20-%2011008_0.pdf

Brown, B. (2010). *A training transfer study of simulation games* (Master's thesis). Naval Post Graduate School, Monterey, CA. http://www.dtic.mil/cgi-bin/GetTRDoc?Location=U2&doc=GetTRDoc.pdf&AD=ADA518353

Brown, D. E. (2004). Human universals, human nature, & human culture. *Daedalus, 133*(4), 47–54.

Caligiuri, P., Noe, R., Nolan, R., Ryan, A. M., & Drasgow, F. (2011). *Training, developing, and assessing cross-cultural competence in military personnel* (Technical Report 1284, pp. 1–67)). Arlington, VA: United States Army Research Institute for the Behavioral and Social Sciences. http://www.dtic.mil/cgi-bin/GetTRDoc?Location=U2&doc=GetTRDoc.pdf&AD=ADA559500

Charles River Analytics. (2004). *Organizational and cultural criteria for adversary modeling (OCCAM): Modeling "soft" aspects of human behavior.* https://www.cra.com/about-us/news.asp?display=detail&id=89

Davis, P. K. (2012). Fundamental in empirical validation of an analysis with social science models. In D. M. Nicholson & D. D. Schmorrow (Eds.), *Advances in design for cross-cultural activities part II* (pp. 87–98). Boca Raton, FL: CRC Press.

Davis, R. (2010). *Culture as a weapon* (Middle East Report 255). Washington, DC: Middle East Research and Information Project. http://www.merip.org/mer/mer255/culture-weapon

Deardorff, D. K. (2006). Identification and assessment of intercultural competence as a student outcome of internationalization. *Journal of Studies in International Education, 10*, 241–268.

Deaton, J. E., Barba, C., Santarelli, T., Rosenzweig, L., Sounders, V., McCollum, C., … Singer, M. (2005). Virtual environment cultural training for operational readiness (VECTOR). *Virtual Reality, 8*(3), 156–167.

Dede, C. (2009). Immersive interfaces for engagement and learning. *Science, 323*(5910), 66–68.

Hardison, C. M., Sims, C. S., Ali, F., Villamizar, A., Mundell, B., & Howe, P. (2009). Cross Cultural Skills for Deployed Air Force Personnel. The RAND Corporation, Project Air Force (pp. 1–184). http://www.rand.org/pubs/monographs/2009/RAND_MG811.pdf

Heinatz, S. (2011). *New JKO Program Promotes Afghanistan language, cultural awareness.* United States Joint Forces Command Joint Warfighting Center Norfolk, VA. http://www.jfcom.mil/newslink/storyarchive/2011/pa020111.html

Henderson, S. J., & Feiner, S. (2009). Mixed and augmented reality for training. In D. M. Nicholson, D. Schmorrow, & J. Cohn (Eds.), *The PSI handbook of virtual environments for training and education: Developments for the military and beyond* (pp. 134–156). Westport, CT: Greenwood Publishing Group.

Hill, R. W., Belanich, J. H., Lane, C. H., & Core, M. (2006, November 27–30). *Pedagogically structured game based training: Development of the elect Bilat simulation.* Paper presented at the 25th Army Science Conference, Orlando, FL.

Hofstede, G. (2001). *Culture's consequences: Comparing values, behaviors, institutions and organizations across nations*. Thousand Oaks, CA: Corwin Press.

Hogan, P. C. (2006). Narrative universals, heroic tragi-comedy, and Shakespeare's political ambivalence. *College Literature, 31*(1), 34–66.

Housea, R., Javidanb, M., Hangesc, P., & Dorfmand, P. (2002). Understanding cultures and implicit leadership theories across the globe: An introduction to Project GLOBE. *Journal of World Business, 37*(1), 3–10.

Jager, S. M. (2007). On the uses of cultural knowledge (Working paper). Carlisle, PA: Strategic Studies Institute. http://www.strategicstudiesinstitute.army.mil/pdffiles/pub817.pdf

Johnson, L. W. (2010). *A simulation-based approach to training operational cultural competence*. Retrieved from http://alelo.co.uk/files/ModSim_Johnson_final.pdf

Johnson, L. W., Beal, C., Fowles-Winkler, A., Lauper, U., Marsella, S., Narayanan, S., ... Vilhjalmsson, H. (2004). Tactical language training system: An interim report. In J. C. Lester, R. M. Vicari, & F. Paraguacu (Eds.), *Intelligent tutoring systems 2004* (pp. 336–345). Berlin, Germany: Springer-Verlag.

Johnson, L. W., & Friedland, L. (2010). *Integrating cross-cultural decision making skills into military training*. Retrieved from http://www.alelo.com/files/Johnson_WL_370.pdf

Johnson, L. W., Sagae, A., & Friedland, L. (2012). VRP 2.0: Cross-cultural training with a hybrid modeling architecture. In D. M. Nicholson (Ed.), *Advances in design for cross-cultural activities part I* (pp. 307–316). Boca Raton, FL: CRC Press.

Johnson, L. W., & Valente, A. (2009). Tactical language and cultural training systems: Using AI to teach foreign language and cultures. *AI Magazine, 30*(2), 72–83.

Klein, G. (1999). *Sources of power: How people make decisions*. Cambridge, MA: The MIT Press.

Kobus, D. A., & Viklund, E. P. (2012, July 21–25). *Assessing changes in decision making as a result of training*. Paper presented at the Applied Human Factors and Ergonomics Conference (AFHE), San Francisco, CA.

Kogler, J. (2011). Patterns of Life with B-HAVE 2.0. VT Forces Trackback Blog, August 29.

Kumas-Tan, Z., Beagan, B., Loppie, C., MacLeod, A., & Frank, B. (2007). Measures of cultural competence: Examining hidden assumptions. *American Medicine, 82*(6), 547–547.

Lane, H. C., & Ogan, A. E. (2009, July 6–9). *Virtual environments for cultural learning*. Paper presented in the Culturally-Aware Tutoring Systems 14th International Conference on Artificial Intelligence in Education, Brighton, U.K. http://people.ict.usc.edu/~lane/papers/Lane-Ogan-CATS09-VirtualEnvironmentsForCulturalLearning.pdf

Littrell, L. N., & Salas, E. (2005). A review of cross-cultural training: Best practices, guidelines, and research needs. *Human Resource Development Review, 4*(3), 305–334.

Masolo, C., Vieu, L., Bottazzi, E., Catenacci, C., Ferrario, R., Gangemi, A., & Guarino, N. (2004, June 2–5). *Social roles and their descriptions*. Paper presented at the 9th International Conference on the Principles of Knowledge Representation and Reasoning (KR04), Whistler, Canada.

McCarthy, M. (2009). *Developing cultural competence at the tactical level: The art of the possible* (Master's thesis). U.S. Army Command and General Staff College, Leavenworth, KS. http://www.dtic.mil/cgi-bin/GetTRDoc?AD=ADA512487

McFate, M. (2005). The military utility of understanding adversary culture. *Joint Force Quarterly, 38*, 42–48. http://www.dtic.mil/doctrine/jel/jfq_pubs/1038.pdf

Miller, C., Wu, P., Funk, H., Johnson, L., & Vilhjalmsson, H. (2007). *A computational approach to etiquette and politeness: An "Etiquette Engine™" for cultural interaction training* (Report for Defense Advanced Research Projects Agency and U.S. Army Aviation and Missile Command. Contract Number W31P4Q-04-C-R221). Arlington, VA: Defense Advanced Research Projects Agency. http://hq.sift.info/sites/default/files/documents/2007/MWFJV-BRIMS07-vfin.pdf

MROC Decision Memorandum (DM). (2009). I MEF Urgent Universal Needs Statement. The Pentagon, VA: United States Marine Corps.

Muller, P., Cohn, J., & Nicholson, D. M. (2004, December 6–9). *Where's the Holodeck?* Paper presented at the interservice/industry training, simulation, and education conference (I/ITSEC), Orlando, FL.

Muller, P., Schmorrow, D., & Buscemi, T. (2008, September). The infantry immersion trainer: Today's Holodeck. *Marine Corps Gazette*, 14–18. Paper no. 1773.

Nicholson, D., Fiore, S., Vogel-Walcutt, J., & Schatz, S. (2009). Advancing the science of training in simulation-based training. *Proceedings of the Human Factors and Ergonomics Society Annual Meeting, 53*(26), 1932–1934.

Osland, J. S., & Bird, A. (2000). Beyond sophisticated stereotyping: Cultural sensemaking in context. *Academy of Management Executive, 14*, 65–77.

Pavon, J., Arroyo, M., Hassan, S., & Sansores, C. (2008). Agent-based modelling and simulation for the analysis of social patterns. *Pattern Recognition Letters, 29*, 1039–1048.

Raybourn, E. M. (2005). Adaptive thinking and leadership training for cultural awareness and communication. *Interactive Technology and Smart Education, 2*(2), 131–134.

Rose-Krasnor, L. (1997). The nature of social competence: A theoretical review. *Social Development, 6*(1), 111–135.

Ross, K. G., Thornson, C. A., McDonald, D. P., & Arrastia, M. C. (2009). *The development of the CCI: The cross-cultural competence inventory* (Report). Patrick Air Force Base, FL: Defense Equal Opportunity Management Institute. http://www.deomi.org/EOEEOResources/documents/Development_of_the_CCCI-Ross.pdf

Schatz, S., Folsom-Kovarik, J. T., Barlett, K., Wray, R. E., & Solina, D. (2012, December 3–6). *Archetypal patterns of life for military training simulations.* Paper presented at the Interservice/Industry Training, Simulation, and Education Conference (I/ITSEC), Orlando, FL.

Schatz, S., & Nicholson, D. (2012). Perceptual training for cross-cultural decision making. In D. M. Nicholson (Ed.), *Advances for design in cross-cultural activities part I* (pp. 3–12). Boca Raton, FL: CRC Press.

Schatz, S., Oaks, C., Folsom-Kovarik, J. T., & Dolletski-Lazar, R. (2012). ITS + SBT: A review of operational situated tutors. *Military Psychology, 24*, 166–193.

Schatz, S., Reitz, E. A., Nicholson, D. M., & Fautua, D. (2010, November 29–December 2). *Expanding the Combat Hunter: The science and metrics of Border Hunter.* Paper presented at the Interservice/Industry Training, Simulation, and Education Conference (I/ITSEC), Orlando, FL.

Schatz, S., Wray, R., Folsom-Kovarik, J. T., & Nicholson, D. (2012, December 3–6). *Adaptive perceptual training in a virtual environment.* Paper presented at the Interservice/Industry Training, Simulation, and Education Conference (I/ITSEC), Orlando, FL.

Schatzki, T. R. (2003). Human universals and understanding a different socioculture. *Human Studies, 26*, 1–20.

Schmorrow, D. (2011). *Sociocultural behavior research and engineering in the Department of Defense Context* (DTIC Accession Number: ADA549230). Washington, DC: Office of the Secretary of Defense (Research and Engineering). http://www.dtic.mil

Selmeski, B. R. (2007). Military cross-cultural competence: Core concepts and individual development (Air Force Culture & Language Center (AFCLA) Contract Report 2007-01, pp. 1–45). Centre for Security, Armed Forces & Society Royal Militia College of Canada. Department of Politics and Economics at the Royal Military College of Canada.

Sims, E., Glover, G., Friedman, H., Culhane, E., Guest, M., & Van Driel, M. (2012, July 21–25). *Developing a performance-based cross-cultural competence assessment system.* Paper presented at the Applied Human Factors and Ergonomics Conference (AHFE) (pp. 1474–1483), San Francisco, CA.

Tolk, A., Davis, P. K., Huiskamp, W., Klein, G. L., Schaub, H., & Wall, J. A. (2010, December 5–8). *Towards methodological approaches to meet the challenges of human, social, cultural, and behavioral (HSCB) modeling.* Paper presented at the Winter Simulation Conference, Baltimore, MD.

Turnley, J., Henscheid, Z., Koehler, M., Mulutzie, S., & Tivnen, B. (2012, February 28). COIN of the socio-cultural realm: Emphasizing socio-cultural interactions in counterinsurgency modeling and simulation. *Small Wars Journal.* http://smallwarsjournal.com/jrnl/art/coin-of-the-socio-cultural-realm

United States Department of Defense. (2010). *Irregular Warfare: Countering Irregular Threats.* Joint Operating Concept Version 2.0. http://www.dtic.mil/futurejointwarfare/concepts/iw_joc2_0.pdf

United States Joint Forces Command (USJFCOM). (2010). Statement of General James M. Mattis, USMC Commander, United States Joint Forces Command. Statement before the Senate Armed Services Committee, March 9. http://www.jfcom.mil/newslink/storyarchive/2010/sp030910.html

VT MÄK. (2012). Artificial intelligence behavior modeling—B-HAVE. http://www.mak.com/products/simulate/artificial-intelligence-behavior-modeling.htm

Watson, J. R. (2010). Language and culture training: Separate paths? *Military Review, 90*(2), 93–97.

Wildman, J. L., Skiba, T., Armon, B., & Moukarzel, R. (2012, December 3–6). *A paradigm shift in cultural training: Culture-general characteristics of culturally competent forces.* Paper presented at the Interservice/Industry Training, Simulation, and Educational Conference (I/ITSEC), Orlando, FL.

Wray, R. E., Lane, H. C., Strensrud, B., Core, M., Hamel, L., & Forbell, E. (2009). *Pedagogical experience manipulation for cultural learning.* Paper presented at the Workshop of Culturally Aware Tutoring Systems, 14th International Conference on Artificial Intelligence in Education. http://people.ict.usc.edu/~core/papers/AIED09-CATS_PedExpManip_WrayLane-Final.pdf

Zielke, M. A., Evans, M. J., Dufour, F., Christopher, T. V., Donahue, J. K., Johnson, P., … Flores, R. (2009). Serious games for immersive cultural training: Creating a living world. *IEEE Computer Graphics and Applications, 29*(2), 49–60.

47 Immersive Visualization for the Geological Sciences

William R. Sherman, Gary L. Kinsland,
Christoph W. Borst, Eric Whiting, Jurgen P. Schulze,
Philip Weber, Albert Y.M. Lin, Aashish Chaudhary,
Simon Su, and Daniel S. Coming

CONTENTS

47.1 INTRODUCTION

One characteristic of geologists is their ability to think in 3D. Whether this ability is developed as they advance as geologists or they were drawn to geology because they possessed this ability is an open question. However, experienced geologists possess this ability to such a degree that two or more geologists may verbally communicate about a complex 3D, real-world problem—each of them building and modifying a 3D image in their mind as they receive information from the others. They can verbally query the image of another and as a group may actually change perspective in the virtual/mental 3D image. That is, they are capable of agreeing to look at the 3D problem from different positions in space … or time. They often need to trace the development of a particular geologic situation through geologic time. The group is capable of coming to consensus and running a very similar 3D *movie* in their individual minds. This *movie* is tweaked by evidential or opinion input from those involved, developing a cause and effect–based plausible explanation of the present geological situation. Similarly, the group may run this movie into the future to make predictions. Using immersive tools for visualizing the data is, therefore, both natural and powerful for geologists. Immersive 3D computer graphics technologies are tools that combine an intuitive 3D viewing environment with easily learned interaction mechanisms. These advanced 3D tools represent a natural operating platform for geologists who of necessity must think in 3D.

The following statement may be a tautology; however, it is worth specifically stating for its relevance to the use of immersive visualization: "All interpretation occurs in the mind." The specific point is that before interpretation can proceed, information must get into the mind. In the example earlier, the information is communicated verbally. In many other cases, information enters the mind visually. Often, this information is in the form of 2D images. In the mind, 2D images from several different perspectives and at different scales are often (in the case of geologists … very often) constructed into 3D spaces. These 3D spaces, then, are the basis for making interpretations. While geologists are quite good at assembling 3D spaces from various input formats, it is more efficient to have the information enter the mind as already correctly constructed 3D images. This is one great advantage of immersive visualization for geologists.

Another advantage is the ability of one or several geologists to freely move about in the 3D volume. The effect on the geologist of being immersed in their data is similar to being in the field where curiosity drives movement, inspection, and analysis. To geoscientists, this type of interaction is often intuitive—if, while in the field, the geoscientist wants to look at some feature that has aroused curiosity, he or she simply walks over there and looks at it (as part of their education and experience, most geoscientists have had some field work). Contrast this with what happens in many other environments wherein the investigator must make decisions, consciously or unconsciously, related to effort versus reward. That is, decisions must be made as to whether investigation of the curious feature is worth the effort of gathering the additional data necessary to reach a conclusion. Of course, in this process, the conclusion is estimated, perhaps incorrectly, and compared to the cost. One member of our team (Kinsland) experienced this firsthand: he had written and published a paper using 2D representations of aerial topographic light detection and ranging (LIDAR) data to interpret the evolution of a particular river. Later, he investigated the LIDAR data in an immersive 3D system, allowing him to move, interact, and continuously query as to latitude, longitude, and elevation. Within just a few minutes of *mindlessly* but curiously moving about in the data he said, "I have to write another paper. I have found an alternative explanation." He had poured over the 2D data for many months before he wrote the original paper. Less than 20 min of walking within the

FIGURE 47.1   (See color insert.) Immersive technologies (aka VR interfaces) enable thecreation of new methods of interaction useful for geologists to inspect and interact with their data. Here Gary Kinsland and a student from the University of Louisiana at Lafayette collaboratively explore the Chicxulub crater on a large immersive projection system. (See Section 47.4.1 for more details.)

data, as if he were in the field, resulted in the basis for another paper, which was also published. Another example comes from the work done with data over the Chicxulub impact crater (the one that probably killed the dinosaurs) (Figure 47.1). For several years, Kinsland made 2D contour maps of the topographic data, making pseudo-3D maps of the data with color-coded elevation representations. All were useful in building the mental 3D image. However, in discussing the day he was reintroduced to his data in a head-mounted display (HMD) and allowed to walk about within his data, he relays that "I was totally enthralled and stayed in the HMD longer than I later was told I should have. To this day, years later, my mental 3D image of the topography over the crater is largely that which I saw that day in the HMD."

One of the weaknesses of the human mind when dealing with multiple 3D datasets from the same geographical area is in getting the georeferencing done correctly. That is, it is difficult to mentally compare two overlapping 3D datasets such as topography and magnetic intensity. The interpreter is quite capable of inspecting and interpreting each 3D mental image from various perspectives and at various scales. However, it is often useful to query the correlation of features between the two. For example, does a valley in the topographic data correlate spatially with a particular anomaly (outstanding feature) in the magnetic data? *Georeferencing* of datasets is well handled by computers, which can blend the data into a single coordinate system. The two images may then be immersively viewed, compared, and interpreted simultaneously using any of several techniques described in this chapter.

In short, the world is 3D, and geologists think in 3D. Why reduce the 3D world to 2D images for communication into minds where 3D images must then be reconstructed and interpreted? Immersive environments are the key to accurate 3D thinking and offer improved understanding of almost all geology-related datasets.

47.2   COMMON GEOLOGY DATA TYPES

Data for the geological sciences, of course, are centered on the Earth and other planetary bodies. Nevertheless, there are still many different forms that geological data can take. Geology data can be 1D, 2D, or 3D; it can be static or time varying; and it might be spaced at regular intervals or be entirely unstructured.

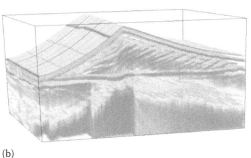

(a)   (b)

FIGURE 47.2 (a) Geologists collecting GPR data over a sand dune and (b) the associated visualization. (See Section 47.4.4 for more details.)

In this section, we briefly describe sources of geologic data and the forms those data take in terms of traditional visualization techniques (Figure 47.2).

47.2.1 TERRAIN ELEVATION DATA

A fundamental data type for geology is the height field of a terrain. Specifically, the term *terrain* (or *bare earth model*) generally refers to the actual ground surface of the planet, whereas *surface* often includes natural (e.g., forests) and man-made (e.g., buildings) objects protruding above the ground. The expressions *digital terrain model* (DTM) and *digital surface model* (DSM) are often used to explicitly indicate what has been measured. However, digital elevation model (DEM) data are ambiguous and may refer to either.

Terrain data are often collected using remote sensing technologies such as RADAR (collected via satellite or space shuttle) or other data types (collected via airborne platforms). Terrain data may also be collected using classical surveying techniques or even by GPS tracking. For the most part, the measuring technologies will capture the height of many points on the surface, but these points are irregularly spaced. Often, these points are processed into a regular, 2D (aka *raster*) pattern. The DEM format as specified by the US Geological Survey (USGS) is a 2D *raster* data format with elevation values specified in a regular grid. Modern sources of DEM data include the Shuttle Radar Topography Mission (SRTM) data collected in 2000 (Farr & Kobrick, 2000; Farr et al., 2007) and the Advanced Spaceborne Thermal Emission and Reflection Radiometer (ASTER) data released in 2011 (Yamaguchi, Kahle, Tsu, Kawakami, & Pniel, 1998) (Figures 47.3 and 47.4).

47.2.2 SATELLITE IMAGERY

Images captured by orbiting satellites constitute another quintessential data type for geology. In the cases of imagery, the data are generally captured as regular 2D *raster* fields of spectral response. Captured spectral fields generally span from the infrared, through the visible colors and into the ultraviolet regions of the electromagnetic (EM) spectrum.

Due to the nature of an orbiting object scanning a curved surface, the image data require reprojection to *warp* the image to match either a Cartesian or polar mapping.

While we will see that there are many opportunities for registering and viewing different sources of data into a single representation, it is natural to begin by combining imagery data with elevation data to create a basic reconstruction of the real world. Even with no other data added, this in itself can be a good tool for reconnoitering geologic expeditions into the field to prepare a better plan for where to gather data (Figure 47.5).

FIGURE 47.3 Geophysical logs of wells are *hung* from a surface generated from SRTM terrain data from North Louisiana. In this system, the virtual *wand* may be used to highlight and toggle a well icon located on the terrain. Toggling turns the well log representation on/off underneath the icon of the well. In the illuminated well data here, yellow is SP and blue is the resistivity. Though not yet implemented, correlation of features in these logs while within the 3D environment promises to facilitate and improve the interpretation of such subsurface data when compared to 2D methods of correlation. (See Section 47.4.1 for more details.)

FIGURE 47.4 A stereo pair representing terrain contours near Mount Rainier. Using a stereoscope, viewers can see the relief of the terrain elevations.

47.2.3 2D FIELDS

In addition to the obvious terrain-height data, or even spectral imagery, that correspond with latitude and longitude, there are other fields that map to the 2D surface of the planet. These data may be collected as an input stream recorded by a moving craft such as an airplane or boat, or they may be collected on the surface by moving measurement equipment to areas of interest. Two fields of particular interest to geology are gravity and magnetism.

FIGURE 47.5 Here, a user modifies an Esri shapefile from within the virtual environment, presented in a four-sided CAVE system.

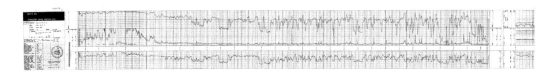

FIGURE 47.6 Well log. This is a log of a well as typically utilized in the petroleum industry to investigate the properties of the subsurface. The header, at the far left, contains information about the location of the well, date logged, well owner, etc. The log extends from left to right from the shallowest to the deepest data and is usually viewed and interpreted vertically (rotated 90° clockwise). The measurements represented by the *squiggly lines* are identified at the ends of the lines as are the scales for the values. The values in this log are (from bottom to top in this image) SP, resistivity, and conductivity (calculated from resistivity).

Magnetic data are typically collected through airborne instruments flown in a pattern over the region of interest (such as the Chicxulub crater presented earlier). Airborne collection of gravity data, however, lacks sufficient precision to be useful for detailed analysis. Thus, gravity data are more typically collected by laboriously moving measurement instruments on the surface from location to location. The irregularly sampled data may then be interpolated into a regular *raster* format more convenient for some visualization techniques—for example, using the Kriging interpolation technique as done in the case study in Section 47.4.1.

Each of these, and other, fields (e.g., radioactivity, resistivity, chemical, heat flow) is interpreted based on knowledge of the regional geology of the survey and the likely types of subsurface materials, which would lead to *anomalies* in the data. *Anomalies* are areas of the data that have values distinct from their surroundings—implying that in the subsurface, there are materials that are distinct from their surroundings (Figure 47.6).

47.2.4 3D Material Property Data

Tomography techniques can be used to determine a property of material within a volume by passing a signal through the volume. There are several signal types that can be used to reveal different properties (and which have different effective ranges and resolutions). In medical imaging, we generally think of CT scans that use x-ray signals. CT scans can also be performed on soils and minerals to analyze properties such as porosity. On a larger scale, seismic tomography is performed with sound waves (sent from well to well). RADAR and many other wave sources can also be used to produce 3D tomogram volumes. For example, Muon absorption tomography is being used to image the interior structures of volcanoes (Tanaka et al., 2007).

Seismic volumes are tomograms collected using seismic reflection surveying, whereby signal sources and receivers are located on the surface with the signal reflecting off horizons in the subsurface. Processing the received signals yields representations of the distribution of various seismic properties within the subsurface. The most common property represented is the *reflection coefficient*—a measure of differences in velocity and density between adjacent rock layers.

47.2.5 Ground-Penetrating Radar (GPR) Data

Specialized radar devices use transmit and receive antennas separated on the order of 1 m from one another sending and receiving a signal aimed downward into the ground and measuring the returned/reflected time-response signal caused by changes in the dielectric constants encountered at subsurface material boundaries. Often, the contrasts in the dielectric constant will correlate with the acoustic impedance contrasts of a seismic reflection survey.

The result of a single sounding is a 1D signal response measured over time, where that time correlates with depth, though not necessarily in any specific ratio. The general means of visualization is to show 2D vertical images (e.g., *fence diagrams*) comprises of regularly spaced soundings of the data (e.g., where that data may be processed to remove low-frequency effects). Thus, the real-time signal from the receiver is plotted against distance along the surface line. A recent trend in some domains has been to create a 3D tomogram volume constructed by performing a 2D survey over a surface, respectively, versus a 1D survey for the fence diagrams.

Ground-penetrating radar (GPR) is used as a tool for a variety of fields such as archaeology, environmental site characterization, hydrology, concrete and road inspection, sedimentology, and glaciology. It is a nondestructive subsurface measurement technique that, in addition to preserving the site, is less difficult and expensive than excavation or drilling (Bristow, Duller, & Lancaster, 2007).

47.2.6 Well Log Data

Data collected while a well hole is bored may be captured. This may be during an active process of accessing a reserve of natural resources, or the collection of the data may be the entire goal of the boring operation (such as when searching for the increased presence of a material such as gold). These *well logs* or perhaps *multivariate well logs of physical properties down a borehole* (Fröhlich, Barrass, Zehner, Plate, & Göbel, 1999) are 1D in space but sample many different properties along the path. It should be noted that with modern drilling technology, these wells are not necessarily in a straight-line path, but can be directed in different directions along the length of the well path.

Similar to other *ground-penetrating* techniques, an accumulation of data from multiple wells can be used to produce a regional layout of the subsurface structure or as control points to constrain and improve models of subsurface volumes.

47.2.7 Vector Geometry Objects with Attributes

There are many geological applications that make use of geometric data specified on a 2D manifold (i.e., surface) using points, edges/lines, and polygons/polylines to indicate places and regions on the surface. There also are many applications of these data that are less than scientific in nature, such as for political or engineering purposes. Attributes are typically assigned to the points, lines, and space within a polygon. Additionally, as the original 2D manifold (e.g., the surface of the Earth) is frequently not planar, *metadata* (data about the data) are kept of properties such as which projections were used in mapping the data from the physical 3D space onto a flat *map*. Vector geometry data for geology might include features such as topographic levels represented as contour lines, along with natural and man-made features such as rivers, roads, forests, and dams.

Among the common formats of *geospatial vector data* is the Esri *shapefile*. Although Esri products are commercial, the format of the shapefile is broadly available, and many other commercial as well as free software tools are able to read and write them. There are other common formats as well, including the WKT/WKB formats (aka well-known text/binary). There are also web markup languages with vector geometry designed as geospatial data representations such as Geography Markup Language (GML) and Keyhole Markup Language (KML).

47.2.8 Breadcrumb Trail

Collecting location data as a subject moves while performing a survey or collecting samples along a pathway or just capturing the path itself are examples of survey traverses. In many respects, a breadcrumb trail can be considered a form of vector geometry (aka *geospatial vector data*) where the attributes are the samples collected at points along the path. One additional attribute is that of *time*. The time attribute can then be used to determine rate of movement along the path.

Typically, a GPS system is used to capture and log the trail. An autonomous differential GPS can be used to achieve 1 m vertical resolution and 1 m horizontal resolution.

47.2.9 Point Cloud Data

Point cloud data are a collection of specific locations in space (vertices) generally with additional information for each recorded point, such as intensity. Frequently, the point locations exist where there is solid or liquid matter. This is a consequence of the data typically being captured by scanning devices that report the location of surfaces struck by the scanning beam. A further consequence is that for scans of solid objects, the points will all be on the surface of that object and thus can be used to derive a geometric shape from the points.

Features of point clouds are that they are irregular, and not constrained to a 2D manifold as vector geometries typically are. Each point can be multivariate. That is, each point can have many attributes such as intensity, color, temperature, as well as categorical data such as *vegetation* or *road*. Point cloud data are most effective when they are rendered in great numbers.

A common method of capturing point cloud data is through LIDAR technology. A LIDAR scan consists of radially captured data as a laser mounted on a rotating platform with a spinning mirror sends and receives the light signal. The speeds of the mirror and platform rotation affect the resolution of the data collected. LIDAR scans easily capture millions to billions of point data in a single scan.

The LAS (American Society for Photogrammetry and Remote Sensing, 2013) format is an effort by the Imaging and Geospatial Information Society (IGIS) to try to facilitate a standard format for LIDAR data exchange between users.

47.3 GEOLOGY VISUALIZATION

Many scientists work with specific datasets collected simultaneously, or simulated as a group, the result of all being in the same coordinate space. This scenario is atypical for geologists, who generally gather data in a variety of ways and at different times, resulting in data that can be difficult to correlate and thus properly visualize. The goal might be to model and characterize a reservoir of natural resources or to "determine the stratigraphy from a drill hole" (Fröhlich et al., 1999). In this section, we discuss the relationship of geological data to standard and specialized visualization techniques, leading up to techniques specific to immersive interface technologies.

47.3.1 RELEVANT ISSUES IN GEOLOGY VISUALIZATION

Geology has particular needs when it comes to visualizing data for analysis. This is not to imply that the techniques used to meet these needs are necessarily unique to geology, but there are particular issues of great importance to geologists, and there are particular visualization methods that address them better.

A major issue with geology analysis is: "where is that?" For example, an anomaly might be evident in the data, but determining where that anomaly is in the physical world (*in the field*) can be difficult. Also, geologists and particularly geophysicists work with physical phenomena that occur hidden from view—under the surface of the earth—and thus are more difficult to measure and more difficult to *see*. The limited ability to take subsurface measurements also leads to irregular collections of data—not only spaced irregularly, but data with holes in it, and collected at varying times.

Perhaps the most significant issue is registering data collected by different technologies, at different times, and from different locations such that comparisons of attributes can be made with datasets aligned to one another. Bringing multiple geology datasets such as mineral content, water content, and magnetism together is an important means for comparing datasets in search of features linked between them. Also, the differing frames of reference will often be mapped into Cartesian space using different mathematical projections. Thus, the data must be reprojected into a common projection as well as translated/scaled/rotated to be coregistered. We can take this one step further and align them to a standard earth-based frame of reference, which is referred to as *georeferencing* the data.

47.3.2 APPLYING STANDARD VISUALIZATION TECHNIQUES TO GEOLOGY

The sciences share many attributes, such as the use of mathematics to explain (and predict) observed phenomena. Thus, it is not surprising that there are many *standard* scientific visualization techniques that work well for analyzing and explaining geological data.

A common technique for geology is the use of height-field maps (or warping) whereby a 3D surface is created from scalar values used to displace an otherwise flat plane proportionate to the data field. The displacement is performed orthogonal to the plane in either positive or negative directions. The most intuitive use of this is to map terrain-height data to height mapping, but any 2D scalar field can be represented this way in order to quickly identify peaks, valleys, and gradients. Another 2D technique used with scalar values is to apply a mapping from data into colors. In some cases, the colors may come from actual photographic capture and represent the physical appearance of the terrain. This need not be the case of course, and mapping an arbitrary selection of colors onto a representation is thus referred to as *pseudocoloring*. Further, a 2D coloring of one data field can be draped over a 2D height-mapped field of another dataset, allowing the researcher to more easily compare the relationship between the two values.

For 3D scalar fields, such as those created through tomographic techniques or building up a volume from 1D signals such as GPR, there are three common visualization techniques: isosurfaces, volume rendering, and slices.

Isosurfaces are an extension of the isocontour line technique that produces a surface within a 3D volume over which all the values are constant. As with contour lines, there may be many disjoint surfaces of the given value. Also, multiple values can be selected to show more of the internal structure of the data, as is done with terrain contour lines. Isosurfaces can be used to show boundaries between subsurface layers or perhaps to show the shape of a subsurface reservoir.

Volume rendering is an alternative 3D visualization technique that can be used to see multiple features within a volume. As opposed to isosurfaces, where the 3D data are converted into solid structures within the volume, volume rendering treats the data as a nebulous cloud. The view through the *cloud* contains colors based on the values along the line of sight. Colors as well as opacities are mapped onto specific data values. This is referred to as the *transfer function*. Through the use of partial opaqueness (i.e., partial transparency), particular values within the data can be highlighted to differing degrees, which thus lead to the nebulous-style rendering.

Reducing the volume to specific *slices* through it is another way to peer into a 3D volume. A common visualization method is to slice through one of the cardinal planes (i.e., orthogonal to one of the primary axes). A slightly more complex algorithm can be used to slice through the data at any angle. In either case, in interactive visualizations, the user is generally given the opportunity to move the slice around, giving a better perception of the entire dataset. A method more specific to geology visualization is to take a slice along a traverse (see Section 47.3.3).

Point cloud data can be converted into geometric shapes with techniques such as *alpha shape and Delaunay triangulation* (Edelsbrunner & Mucke, 1994). But with modern hardware rendering capabilities, a more common method has become to simply render all, or a large subset of, the points themselves. Each point may be colored by any of the associated multivariate data (actual color, time of flight, classification, etc.).

The coloring technique discussed earlier with the 2D visualization techniques can equally be applied to 1D and 3D scalar fields as well. The choice of how to map values into colors is a very important one, though all too often, a default *rainbow* color mapping is used. Poor choice on *color mapping* can disguise important features or create the appearance of an interesting phenomenon that doesn't exist (Rogowitz & Trenish, 1998).

The representation of vector data such as from flow or gradient fields can be done in a handful of ways, affected by whether the field is in steady state or changes over time. A common and easy to comprehend flow field representation is *moving particles* that are released from strategic locations and advected through the field. *Streaklines* and *streamlines* represent the paths that particles follow over time or paths of particles that are infinitely fast. *Line interval convolution* (LIC) is a technique that shows the entirety of a steady-state flow field by warping a white noise texture. The LIC technique works best as a 2D technique (perhaps on a 2D slice through a 3D field), whereas the other techniques work equally well for 3D data.

47.3.3 VISUALIZATION TECHNIQUES FOR GEOLOGY

The nature of geological phenomenon taking place below the planet's surface (though sometimes made visible through geologic or human activity) is primarily reflected by means of data collection and then subsequently how to link subsurface visualizations to recognizable surface landmarks.

Given that some features of a planet's geology are reflected in the topography of the surface, an early visualization technique was the use of *isocontour lines* drawn at regular intervals to give both a quantitative measure of the height of the terrain as well as a qualitative view of the gradient of the surface. This was also extended to be an early use of stereoscopic data analysis through the creation of stereo-pair topographic contour maps (Blee, 1940; Gay, 1971) and before that aerial photography (Bagley, 1917; MacLeod, 1919). (Although the use of stereo to view terrain features has since become commonplace, the use of standard [*flat*] *topographic contour and terrain maps* remains prevalent.)

The *vector geometry objects with attribute* data discussed in Section 47.2.7 also lead to specific visualization techniques that can be thought of as cartographic in nature. Different polygonal *shape* regions might be colored differently based on vegetation, land use, or geologic era.

The nature of well log or GPR data (at least when looking at a single transmission and return) maps easily into representations of the data as a signal similar in nature to the recording of a single seismic trace—a *1D needle vibration movement plotted over time*. Of course, for well log or GPR data, the data are sensory responses plotted over depth. The 3D nature of the measured object begins to become apparent through multiple plots of well logs or GPR recordings.

Fence diagrams, which are 2D data slices taken along a surface traverse, are a geological visualization technique that is used to view subsurface data that cuts across an important terrain feature. As mentioned earlier in Section 47.3.2, fence diagrams are similar to some standard *slice* visualization techniques used in other fields of research. Because we know that data on either side of the slice will generally be similar to the slice itself, it allows the researcher to make overall conclusions regarding the neighboring geologic material.

Techniques that *cutaway* part of the solid representation of the planet's surface are clearly very important when working with data measured from beneath the terrain surface. Similar to how some medical visualizations might show an otherwise complete external view of a person, with certain layers of skin, muscle, etc., removed to show the internal organs in a particular location, the same can be done for the Earth or other celestial body. We can strip away the topsoil to see what's beneath. Often the edges of the *hole* created by the cutaway will themselves show subsurface data, using the slice techniques described earlier. Then internally, techniques such as isosurfaces or volume rendering might show reservoir formations or mineral content.

47.3.4 TEMPORAL DATA

While it's easy to jump to the conclusion that the time scale of geologic processes limits the need for datasets involving more than a single time step, this is of course not the case. For starters, there are geologic events that happen in human-scale time: active volcanic activity, fault shifts, landslides, as well as slower erosive activities to name four. Furthermore, there are unbounded areas in which geologic processes might be computationally simulated, allowing for the prediction of how land masses, etc., change in much larger time scales. Small simulations might have hundreds of time steps, and given sufficient compute resources, thousands or more time steps might be produced.

We think of *time-lapse* movies as ones where slowly changing phenomena are brought to life through long-interval photography presented back in rapid succession. In this way, we can more easily see behavior and flow that was difficult to perceive at real time. Likewise, geologic and other scientific data can be collected over time with large time intervals between collections, and these data can be viewed in rapid presentation. Two areas where this is more common in the geologic sciences are in *time-lapse seismic survey data* and multiple LIDAR captures used for detecting and measuring changes such as before and after earthquakes and landslides. The application in the case study in Section 47.4.2 (*LidarViewer*) is frequently used with just this type of data.

But in geology, even static data represent change over time. Deposited sediments built up over long periods of time forming underground rock layers and other phenomena. Working through the layers, geologists draw conclusions, which then allow the creation of temporal models of geological activity. One such example is highlighted in the case study in Section 47.4.4 where the changes in sand dunes are analyzed through an application with an immersive interface and tools that specifically make use of the 6-DOF (degree of freedom) inputs afforded by that interface—versus a 2-DOF device such as a computer mouse. Similarly, archaeologists generally correlate the depth in which an artifact is found with increase in age as exemplified in Section 47.4.3.

47.3.5 BENEFITS TO GEOLOGIC VISUALIZATION FROM THE IMMERSIVE INTERFACE

There are both general and specific ways in which immersive technologies can benefit the visualization of scientific data. Some of the general benefits derive merely from the accidental properties of a general immersive system: substantial compute power and display size and resolution, whereas the benefit of perspective rendering is specifically part of the nature of immersive systems.

Perspective rendering is what makes immersive systems (virtual reality [VR]) unique. It is the means by which, when an immersed viewer moves their head, the world responds as though it were actually there before them. This feature provides improved perception of the relationship between objects based on that ability of moving one's head to see the world from a new perspective.

Most immersive display systems have a larger *field of regard* than a typical desktop or laptop display. The field of regard is a measure of how many directions you can look and see the virtual world. Larger screens, multiple screens, and screens that follow you are typical ways of increasing the field. Increased field of regard benefits visualization tools by providing more real estate to place the data. Geologically scaled simulations can often generate large datasets, and having the ability to put as much data in *space* as possible can be helpful. Related to that, when focused on a particular phenomenon, it can be helpful to see it in its larger context. The large screen *space* allows for the larger context while placing important details front and center, especially for higher-resolution systems. On the other hand, when surrounding the user with data, it can become difficult for them to find the user interface. Some systems therefore allow the user to control the placement of the interface, as with the LidarViewer application in Section 47.4.2. Another solution is to use an interface that is always front and center when summoned, such as the radial menu discussed in Section 47.4.1.

The other *accidental* benefit of using immersive systems is the increased computation and rendering capabilities that are often included as part of the integrated solution. Thus, more data can be held in memory, processed, and rendered than with a typical laptop or desktop computer.

There are a myriad specific user interface features that can be added to an immersive application. At this point, there has been only minimal coalescence around common user interface techniques or libraries. This stew of interfaces is not necessarily a good thing in the long run, but as the still quite nascent technology evolves, this flexibility provides opportunities for new and quite useful innovations. Some of these innovative user interfaces are discussed in the case studies of Section 47.4. In Section 47.4.1, the use of *3D magic lenses* is demonstrated as a means to correlate information from differing but spatially overlapping datasets. In Section 47.4.2, a handheld *painting* operation is provided in order to quickly, yet specifically, select points that belong to a particular grouping. In Section 47.4.4, a *virtual Brunton compass* was created to mimic the utility of the real article but applied within the virtual world.

47.4 EXEMPLAR APPLICATIONS AND LESSONS LEARNED

As immersive technologies began maturing in the late 1990s, opportunities for visualization researchers to work with geologists began to materialize. Two good examples were reported at the 1999 conference on visualization hosted by the IEEE (Institute of Electrical and Electronics Engineers). These examples include work by Fröhlich et al. (1999), which provided a unique interface for geologists to view seismic volumes, and well logs using isosurface, and slice-plane techniques controlled by a newly created 6-DOF input device dubbed the Cube-Mouse. The other related project from the 1999 visualization conference is by Winkler, Bosquet, Cavin, and Paul (1999) where they extended an existing geoscience visualization program to work in an immersive display system. Experimentation continued in the first decade of the new millennium as exemplified by a borehole drilling planning application (Dorn, Touysinhthiphonexay, Bradley, & Jamieson, 2001), as well as the use of GPR to construct an underground ant colony network based on the boundary between soil and air under the surface (LaFayette, Parke, Pierce, Nakamura, & Simpson, 2008).

In the subsections that follow, we will take a quick look at four applications that make use of immersive technologies to assist in the process of gaining insight into data above and below the planetary surface. These applications were written in collaboration with research geologists who were seeking better ways to understand their data. In general, we'll give a little background information, discuss the visualization techniques used, present the specific ways immersive technologies were used, and discuss some of the benefits and other outcomes of the project.

FIGURE 47.7 Users viewing Chicxulub crater data. Two displays are shown: a large projection display at the Louisiana Immersive Technologies Enterprise (LITE) and a *fishtank* display (left) that allows a user to reach into a 3D volume with a haptic stylus under a mirror. The two systems were networked for a demonstration at the Gulf Coast Association of Geological Societies (GCAGS) 2006 geology convention. Two movable boxes are visible in the main display: one selects a source region of the main mesh, and the other shows a secondary dataset from that source region. A radial menu can be seen near the top center. Annotative marks of various colors appear, arranged along concentric arcs near the standing users.

47.4.1 APPLICATIONS FOR EXPLORATION OF TERRAINS, ASSOCIATED GEOPHYSICAL DATA, AND WELL LOGS AT THE UNIVERSITY OF LOUISIANA AT LAFAYETTE

Borst and Kinsland applied VR environments to explore geoscience datasets including (1) a collection of 3D datasets over the Chicxulub impact crater—for example, Borst and Kinsland (2005); (2) data for insight into water drainage in the Lafayette, Louisiana region (Kinsland, Borst, Best, & Baiyya, 2007); and (3) well log data from the North Louisiana Coalbed Methane database—for example, Kinsland, Borst, Tiesel, and Bishop (2008). They visualized these datasets in HMDs, single-wall stereoscopic projection displays with tracked viewpoint, a CAVE-like display, a *reach-in*-style *fishtank* display (a desktop display that provides a haptic stylus tool in a fishtank-sized display volume), and a low-cost 3DTV-based environment. The systems have been used both for geological investigations by Kinsland and colleagues as well as for educational purposes in geology classes (see Figure 47.7).

47.4.1.1 Rendering and Interaction for Terrain Interpretation

Graphical rendering and display of these datasets for VR was relatively simple compared to supporting user interaction. Early versions of the systems achieved real-time rendering with mesh tiling and culling, careful use of graphics card features, and reduced-resolution meshes textured with colors and shading from higher-resolution versions. Toolkits such as OpenSceneGraph (Wang & Qian, 2010) and VR Juggler (Bierbaum et al., 2001) provided common graphics operations and access to standard VR devices in a portable manner (with extensions necessary for the fishtank devices). Toolkits for 3D interaction and 3D interaction techniques, in general, are less developed, leading to trade-offs between interface quality and development time. For general system control, a menu toolkit was developed as a compromise between naïve interfaces (e.g., functions mapped to many buttons) and specialized 3D interaction techniques (e.g., direct or widget-based manipulation of objects). User studies of the ray-based menu interaction support the use of large radial layouts, contextual location of menus at objects of interest, and automatic scaling for constant projected size (Das & Borst, 2010a, 2010b). Contextual location can cause menu placement behind other objects, so the menus were rendered in a way that reveals them through occluding objects, with reduced intensity.

The ability to directly move through the data with natural navigation techniques is of prime importance for data exploration. Kinsland notes a personal affinity for navigation with natural movement (head motion and walking) in HMD and CAVE displays. This is augmented with wand-based flying, *grab the world*, and scaling interactions to provide a greater range of movement. Unconstrained navigation can lead to uncomfortable views, especially when objects are too close to the viewer, where stereoscopic images are difficult to fuse. Thus, when the user wants to see a location on the terrain in more detail, it can be better to scale up the scene about the point of interest than to fly closer. The difference is not always clear to, or adequately controlled by, users—especially those not accustomed to stereoscopic viewing. In single-sided displays, ranging from small-scale fishtank displays to larger projection walls, good distance and dataset orientation can be maintained with *center of workspace*–style constraints (Ware et al., 2001). In Borst and Kinsland's systems, this is achieved in part by a constrained grab where translations are constrained to lateral sliding in a workspace-centered dataset-aligned coordinate frame and rotations are constrained to fit an azimuth elevation model for the centered frame.

With a good navigational scheme in place, the next feature of importance is the ability to place and adjust annotative markers. In the impact crater and LIDAR terrain systems, users place and edit markers on surfaces to annotate interesting features. The approach varied depending on display type. In a single-wall display, the marking and editing usually involved ray-based pointing at targets. In earlier CAVE and HMD implementations, users moved their hand (wand) directly over targets to drop markers in place. The ray-based manipulation requires less navigation, as the user can be further from the target, but it is less precise at a given dataset scale due to fundamental problems of the imprecision of long-range ray pointing. In the fishtank display configuration, annotations can be more sophisticated and precisely placed: its pen-like tool has stable mechanical tracking and can be used to draw directly on datasets. The pen also supports force feedback that allows users to feel virtual surfaces. It was shown that the force feedback improves performance of users tracing paths on terrains, particularly when paths follow terrain features rather than cross over bumps (Raghupathy & Borst, 2012). Researchers previously suggested various ways for force feedback to aid interaction in geoscience-type applications—for example, Harding, Kakadiaris, Casey, and Loftin (2002) and Komerska and Ware (2004). For example, force cues could support point or terrain editing, menu selection, and interactive 3D *widgets*. Others have discussed various benefits of fishtank configurations over fully immersive displays (Demiralp, Jackson, Karelitz, Zhang, & Laidlaw, 2006; Mulder & Boschker, 2004). They have relatively low-cost and space requirements, high angular resolution, and (for the *reach-in* approach) a focal plane suited to hand-based interactions with visualized objects.

47.4.1.2 Well Log Visualization

In the well log visualization system, users view vertical plots of resistivity and spontaneous potential (SP) under a terrain and can select a plot or well to move markers along its axis. All or any subset of the well logs may be displayed simultaneously. Their vertical scale is adjustable so that the user might investigate the signatures of the individual logs or of a selection of logs at various scales to highlight features. Correlation markers that were determined for each well in a separate 2D system are also displayed in this system.

This system, as it is, serves to convey 3D relationships of the wells in a way that pseudo-3D renderings may not. When inside the data, a geologist feels more as if they are viewing and interpreting

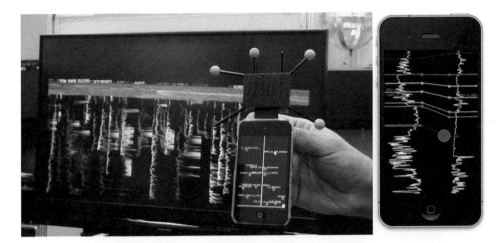

FIGURE 47.8 Low-cost well log visualization system (Mitsubishi 3D DLP TV and iPod Touch with markers for OptiTrack camera-based tracking) showing well logs (curves) hanging underneath a terrain generated from SRTM data. The iPod Touch presents an overview of the well log scene that resolves occlusion in the main view and supports rapid touch-based selection. A middle vertical line represents a virtual ray in the main view, which is locked during a selection step after coarse ray pointing. Right: (Constructed conceptual illustration) Well log *picks* illustrated as horizontal lines with associated depth and text annotation. A highlighted pick on the left log is being associated with a pick on the right log by a drag gesture.

from within the earth. If many well logs are visible simultaneously at various depths, the stereoscopic visuals may help disambiguate overlapping plots. Being able to overview multiple well logs can be helpful for determining subsurface structures, lithologies, fluid contents, faults, etc., based on plot features across well logs. In the future, segments of multiple wells could be selected, moved, and scaled for comparison (correlation). Correlation points marked on multiple wells could influence a computer-generated 3D surface or volume interpolated/extrapolated throughout the dataset to aid further investigation (this type of interpreter/computer interaction is common in packages that aid interpretation of seismic data volumes). VR would help preserve, correct, and maintain constant visibility between geospatial relationships of features, reducing mental transformations between 2D physical images and 3D mental images.

To get an overview of underground features, a user can view many plots at once. This presents a problem for ray-based selection when desired targets are partially or fully occluded by other plots. To aid this selection, the researchers developed *Handymap*, which uses thumb motion on the touch surface of a tracked iPod Touch to allow more precise targeting and selection depth control (Prachyabrued, Ducrest, & Borst, 2011). This approach is additionally motivated by its benefits for low-cost VR environments with reduced tracking accuracy: it does not require very precise or stable tracking (Figure 47.8). The touch surface could additionally be useful for tasks such as menu selection or symbolic input (text entry).

47.4.1.3 3D Lens Immersive Interface Technique

While developing the systems described earlier, Borst and Kinsland explored 3D lens-based techniques for managing multiple dataset views and filtering options in VR systems for interpreting geoscience data. In 2D desktop interfaces, windowing systems are a well-established standard for organizing multiple views, for example, showing a dataset in multiple views or with multiple visualization options, multiple related or colocated datasets, collaborative views that communicate information from remote collaborators during networked operation, or multiple temporal views (different time points or animated views). This motivated various researchers to incorporate related mechanisms into 3D interfaces. Plumlee and Ware (2006) explained how such windowing can reduce errors related to visual working memory limits when compared to alternative navigation techniques such as zooming.

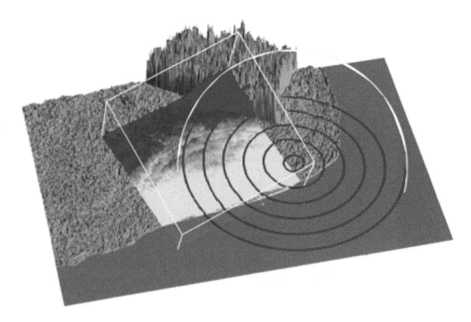

FIGURE 47.9   Two lenses apply effects in constrained regions on a terrain. A box-shaped lens applies a colormap, and a distance tool lens adds concentric contours about its center. For clarity, the lenses are large and the main view is zoomed out.

Some basic approaches to windowing in stereoscopic 3D environments include placing 2D-style windows into 3D spaces—for example, *Task Gallery* (Robertson, et al., 2000)—or using portal-like windows into other 3D scenes or views as with the *SEAMS* project (Schmalstieg & Schaufler, 1999) or tunnel windows (Kiyokawa & Takemura, 2005). Another approach is Itoh et al.'s 3D *WorldBottles*, which are containers for separate scenes that are each drawn on their container surface (Itoh, Ohigashi, & Tanaka, 2006). WorldBottles supports operations such as transporting objects between bottles and transporting users into bottles.

In contrast, Borst and Kinsland extended rendering methods and application of volumetric (3D) lenses, originally introduced by Viega, Conway, Williams, and Pausch (1996) as a 3D version of 2D lens techniques (Bier, Stone, Fishkin, Buxton, & Baudel, 1994; Perlin & Fox, 1993). Described simply, these 3D lenses are boxes containing 3D objects different from the main scene rendering. A *lens* effect is apparent when the chosen content appears to modify the surrounding main view, for example, when the lens contains a version of the main dataset with a different colormap applied (Figure 47.9). The most recent rendering techniques for such lenses in 3D include mechanisms for composing effects when multiple lenses intersect, for example, combining a colormap with a magnifying effect. A comparison of lens rendering techniques is given in Borst, Tiesel, and Best (2010).

In addition to lens-type effects, 3D lenses can be used more generally as containers for secondary datasets or views, leading Borst et al. to use the term *volumetric windows* (Borst, Baiyya, Best, & Kinsland, 2007). A similar term, *volume windows*, was previously used to refer to a 3D windowing system proposed for holographic displays (Kurtenbach, Fitzmaurice, & Balakrishnan, 2002). Volumetric lenses can also be used to *cut out* portions of environments, resembling clipping volumes—for example, Weiskopf, Engle, and Ertl (2003). Lenses configured to contain miniaturized versions of a 3D environment resemble world-in-miniature techniques (Stoakley, Conway, & Pausch, 1995).

Compared to other secondary view techniques, volumetric lenses may better preserve viewing and interaction for 3D environments. Stereoscopic and motion parallax effects are present, unlike for 2D-in-3D windows. Compared to portals or flat lenses in 3D, a volumetric lens view can be more stable because the same dataset region is affected by (contained in) the lens regardless of viewpoint. Raghupathy and Borst (2011) explained that this simplifies placement of secondary views in a terrain-marking application. They demonstrated performance and subjective improvements with a box view compared to portal- or mirror-type views in a small-scale VR environment for drawing on terrains. Their evaluation also underscores the importance of direct interaction in secondary views: users should be able to annotate visualized features directly in secondary views rather than use them only as a visual guide while annotating in a main view.

Borst and Kinsland initially applied 3D lenses to deal with multiple datasets over a single region. Kinsland's geology research task was to mark extrema and ridges (e.g., low points from limestone sinkholes) on a terrain and check for corresponding features in a gravity data surface. Interpretive marks placed on one dataset were projected to the other to help check consistency of features. Prior to incorporating multiple views, much time was spent switching between datasets, manipulating viewpoints, and toggling filtering options. For example, to examine minima, Kinsland applied large-magnitude negative scale or flipped a mesh upside down to view the bottom. Long and repeated sequences of such manipulations occurred as Kinsland sought the best marker placement. A lens system was added that allowed users to *see through* one dataset to another with a lens, grab and move copies of dataset regions in boxes to arrange multiple views in the scene, change filtering and view options in lenses individually, and reach into any of the lenses to place or adjust annotative marks. Based on informal observations, there was some time spent setting up the lens views, but during consistency checking, this subsequently reduced the switching and manipulation of datasets. The ability to fine-tune annotative marks was enhanced because the user could immediately reach into any of multiple views to adjust markers. Kinsland notes that *seeing through* one mesh to another in a movable lens supports comparison between datasets with little overall distraction or focus shift: the lens box can be moved back and forth while the user remains focused on the area of interest and the surrounding context is largely preserved.

Other tasks handled by windows of conventional 2D visualization systems may be addressed in 3D by 3D lens systems. For example, 3D lenses may be useful for communicating a remote collaborator's filtering options or view of a dataset. However, differences between local and remote viewing geometry in head-tracked immersive visualization make it difficult to replicate a remote user's view precisely while still providing stereoscopic viewing and perspective appropriate to the local viewer. Recently, Borst et al. have rendered *time warp* lenses that navigate time-varying datasets by allowing a user to select time, relative time offset, or animated playback inside of constrained regions selected by lens volumes (Borst, Tiesel, Habib, & Das, 2011) (Figure 47.10). Future work can improve 3D lens and window systems with better interaction and arrangement techniques and integration into scene graph tools.

FIGURE 47.10   A user in a projection environment uses ray-based interaction to set up a relative time warp lens. The color in this particular lens indicates increased rainfall 6 h beyond the context.

47.4.2   LidarViewer: An Immersive Point Cloud Visualization Tool

The *LidarViewer* application (Kreylos, Bawden, & Kellogg, 2008) is a LIDAR data visualization tool that runs on a wide range of hardware platforms supported by the Vrui VR integration library (Kreylos, 2008). LidarViewer depends on the Vrui VR integration toolkit for its immersive functionality support, which integrates display configuration with 3D interaction devices. When properly configured, the Vrui toolkit layer allows the LidarViewer application to provide the same functionality on either a desktop platform or VR platform (CAVE-style or head-based display) (Figure 47.11).

LidarViewer was developed under the auspices of the KeckCAVES lab at the University of California, Davis, Department of Geology. Not surprisingly, the KeckCAVES lab is dedicated to the visualization and analysis of geologic data through the use of immersive interfaces and other advanced techniques.

47.4.2.1   Visualization Techniques

LIDAR capture devices generate a large set of points—a *point cloud*—with the 3D location of the optical signal return along with other data such as intensity and time of flight. Often, color data are captured through a secondary sensor and assigned as an attribute to a location data point. LidarViewer then presents these point clouds using an optimized rendering technique that builds an octree structure from the dataset bringing points into the scene as needed to present the full resolution within the nearby field of view (Figure 47.12).

With the data loaded and visible in the 3D immersive environment, the user can select a subset of data points, determine the distance between points or planes, as well as perform real-time 3D navigation through the dataset. These data interaction and visualization techniques provide the user with a true immersive experience. The user visualizes the data as though actually present in the location where the data were collected, thereby providing the user with insight into their data not otherwise possible.

In preparation for using the immersive data visualization tool, LidarViewer provides a pair of data conversion and preprocessing utilities. The *LidarPreprocessor* utility comes with dozens of

FIGURE 47.11   Data from an airborne LIDAR scan capturing the path of a powerline grid are presented using LidarViewer in a CAVE display.

FIGURE 47.12   Using LidarViewer in a CAVE system to view changes in terrain from natural events, seasonal changes, and human activity. Airborne LIDAR scans from two time frames are represented, with points from the initial scan in red (darker) and points from the subsequent scan in green (lighter).

optional directives and converts a standard LIDAR data format into a binary format suitable for the octree rendering algorithm. Data processed into LidarViewer format can be further enhanced with the *LidarIlluminator* utility. The primary enhancement performed by LidarIlluminator is the calculation of normals for each of the points in the cloud. The *normal* data can then be used by the LidarViewer rendering algorithm to produce improved lighting effects.

Menus within LidarViewer enable the user to control shading and other rendering options. Shading options include rendering points at their given intensity level, using any RGB values that may have been assigned to the points or a pseudosun reflection model that makes it easier to see surface features such as roughness. Points can also be colored based on their distance above an assigned *ground* plane. Additionally, the number and size of the points can be controlled in order to balance between rendering quality and rendering speed.

47.4.2.2   Immersive Interface Techniques

Written with the Vrui VR library, LidarViewer inherits several standard Vrui interaction modes, including a built-in menuing system, which includes a choice of methods for navigating through the world. The Vrui menus and widgets are designed to work reasonably well in configurations ranging from a desktop workstation up to large-scale CAVE immersive systems. LidarViewer uses the menu system to bring up dialog controls for shading and other effects, but LidarViewer also has interaction specific tools that make use of the immersive interface.

A key feature of LidarViewer is the ability to select points using 3D painting and selection gestures. All points within a certain region of the handheld controller are highlighted as they are added to the selected group of points. The size of the selection region can be adjusted through a menu dialog. It is also possible to remove points from the selection as desired. Once an acceptable set of points have been selected, a mathematical description of the points can be calculated. The description is based on a set of simple primitive shapes such as a plane, a sphere, or a cylinder. The mathematical description can then be used to make measurements or other annotations within the data.

47.4.2.3 Idaho National Laboratory LidarViewer Workflow

As an open-source project, LidarViewer is also used at other immersive visualization centers. One such group is the Center for Advanced Energy Studies (CAES) in Idaho Falls, Idaho, who have successfully used LIDAR data and immersive environments to better understand and quantify geological change. One successful method involves overlaying georeferenced LIDAR datasets acquired at different times, coloring the time-series point clouds to highlight changes over time.

The LidarViewer application creates a dramatic immersive scene allowing viewers to gain insights for a variety of geologic and natural resources including the following:

1. Land mass change due to natural earth movement such as landslides, earthquakes, erosion, or similar changes (Glenn, Streutker, Chadwick, Thackray, & Dorsch, 2006)
2. Mechanical man-induced changes such as excavation, construction, or road building
3. Seasonal vegetation changes such as those experienced in agriculture and wildland fire
4. Short- and long-term changes in vegetation canopy of a given forest or rangeland due to natural growth, fire, beetle infestation, or forestry operations
5. Changes in water levels such as lakes, ponds, rivers, or snowfields

These changes have been observed in time-series datasets from airborne LIDAR, as well as terrestrial laser scanning (TLS) data. These datasets can also be fused to provide both high-resolution (mm to cm) point clouds (from the TLS), coupled with the large geographic extents (from airborne LIDAR) (Murgoitio, Shrestha, Glenn, & Spaete, 2013). Additionally, LIDAR datasets can be fused with high-resolution digital imagery, including hyperspectral data to provide both true and false-color perspectives of the point cloud (Moore et al., 2011; Olsoy et al., 2012). The unique capability of immersive environments to view this 3D point cloud data in its native 3D format makes this representation very intuitive for exploring, understanding, and quantifying change (Figure 47.13).

The capability of demonstrating time variance using traditional geographically oriented LIDAR datasets leads to an exciting extension of this technology in the change detection of buildings, infrastructure,

FIGURE 47.13 A user explores a wooded area captured by high-resolution LIDAR scans presented in a CAVE with LidarViewer.

bridges, and other important features. The ability of LIDAR to capture highly accurate data regarding current state and then capture feature evolution through subsequent scans opens the door to a whole new world of very important historical preservation activities. The CAES has acquired a very-high-fidelity scan of the nearby Experimental Breeder Reactor No. 1 National Historical Landmark. The first scan will serve as the baseline, while subsequent scans will be studied for change detection activities.

Researchers affiliated with CAES have successfully combined LIDAR data from airborne sensors with terrestrial-based sensors, thereby creating an impressive multiresolution representation of a forest. The air-based scans give the overall perspective and orient the viewer to the surrounding area, while the ground-based scan provides a unique view of details in areas where details are required.

47.4.2.4 Outcomes and Lessons

The experience at Idaho National Laboratory and Idaho State University is just one site that makes use of the LidarViewer application to interact with point cloud data. Researchers at KeckCAVES and the USGS have analyzed and measured phenomenon such as the shift caused by earthquake activity or the movement of structures whose foundations have been compromised by landslide activity.

In the case of earthquake analysis, a site along the San Andreas Fault for which LIDAR scans had been taken both before and after a large earthquake allowed scientists to calculate before and after measurements of various surface features (Gold, Cowgill, Kreylos, & Gold, 2012). In the landslide usage case, a house with much of the foundation washed away was measured from the same location on a daily basis, and by determining the mathematical description of the near wall, the movement of the house could be determined to millimeter accuracy (Kreylos et al., 2006).

The ability to visit, or revisit, sites of interest in and of itself is a great benefit to the geologist whose *laboratory* can be anywhere on the planet, including sites that can be expensive and otherwise impractical to visit on a recurring basis. Thus, the ability to virtually visit these sites with considerable detail is a great boon. And while measurements can be directly obtained while in the field, the ability to take them in the immersive environment is far easier (given that it is possible to fly or rescale the world) and also can be more accurate through the assistance of basing measurements on a large collection of points. Consequently, LidarViewer is not just for revisiting geological and other sites, but can be better than those visits by allowing the visitor to take more measurements, explore the site faster, and leave behind annotations of intriguing discoveries.

47.4.3 Visualization of Multimodal Geophysical Survey Data

Recognizing that geophysical exploration produces large amounts of data, generated by a variety of acquisition devices, a team of archaeologists and computer scientists at the California Institute for Telecommunications and Information Technology (CALIT2) created a project to fuse aerial and satellite imagery and elevation; infrared, magnetic, and EM surface data; GPR; photographs of artifacts and man-made structures; and a database with information about found artifacts (such as 3D location, type, radiocarbon dates, and pictures; Lin et al., 2011). Some of the acquired data were further processed into derived data products: extensive image sets of 3D structures on the ground have been processed with a structure-from-motion (SfM) algorithm (University of Washington's Bundler; Snavely, Seitz, & Szeliski, 2006) to derive a colored 3D point cloud of the structure, and 3D models were created manually for structures that no longer exist but are assumed to have existed in the past.

Archaeologists can often verify their hypotheses by digging up what they believe to be buried somewhere. This principle does not hold for CALIT2's archaeologists working in Mongolia to find Genghis Khan's tomb and other historical sites, because the Mongolian government does not allow them to excavate. This means that they have to rely entirely on what is readily accessible from the surface and data that can be acquired about subsurface objects.

The StarCAVE (DeFanti et al., 2009) at CALIT2 is a 34-megapixel immersive VR system and a unique tool for archaeologists, because it allows them to not only visit a site of interest on another continent without traveling to it but also to show all the available data about the site in its geospatially correct locations. The goal of this effort is to explore the use of the StarCAVE to enable noninvasive *virtual excavation* through the 3D VR reconstruction of geophysical survey data of an archaeological site that has been investigated since July 2010 as a component of the Valley of the Khans project, a noninvasive remote sensing survey for burial sites in Northern Mongolia.

For the archaeologists, working with this VR tool gives them the unique ability to view not only a spatially correct visualization of their data but also one that fuses all the data types they generate. This is because no off-the-shelf tools exist that solve this problem for them. Understanding spatial correlations between the multitude of data types the archaeologists collect is key to making new discoveries.

There are no off-the-shelf solutions for VR systems, which can display all of the data types this system supports. AEGIS Easy 3D (Aegis Instruments, 2013) supports GPR data but does not run in a CAVE. Existing GPR visualization software for VR, such as Billen et al. (2008), does not allow point visualization with the data points following the terrain under which they are located (needed for a precise analysis of the data). Commercial tools such as EON's software suite (EON Reality, 2013) or VTK (Kitware, Inc. VTK, 2013) are not flexible enough to support all the data types and visualization modes, with the required scalability due to the size of the data.

47.4.3.1 Visualization Techniques

The general idea of the visualization tool is to display all the data in their respective locations on a virtual reconstruction of the site of interest. In order to display geospatial data correctly on a terrain map, the archaeologists had to get accurate terrain height. This was done by processing the height data from a Trimble VX Spatial Station to tessellate a terrain surface. This provided a terrain elevation as the missing coordinate for the 3D display of any dataset arrays. One visualization mode (*Hill*) shows the terrain with a height map on it, where a terrain height is color coded onto the surface.

Magnetic (Aitken, 1958; Becker, 1995) and EM (Fröhlich & Lancaster, 1986; Tabbagh, 1986) conductivity are 2D data types, as they result in surface maps of variations of their respective physical properties. A magnetic survey is a passive detection of contrasts in the magnetic properties of differing materials, whereas EM measures the conductivity and magnetic susceptibility of soil. A derived data type is based on the density of artifacts found in 3×3 m grid cells around the surveyed site, defined by the locations of measurement probes, and displayed as a false colormap on the terrain. In each case, a suitable color gradient is used to map the data values to colors.

GPR (Goodman, Nishimura, & Rogers, 1995; Novo, Grasmueck, Viggiano, & Lorenzo, 2008) and electrical resistivity tomography (ERT) (ITSR, 2000) generate 3D data arrays. GPR transmits an EM pulse and measures a reflected signal that is dependent upon the dielectric properties of subsurface material. ERT derives subsurface structures of resistive materials from electrical measurements made by electrodes. The archaeologists use a 1 m grid for the electrodes. Note that the depth information in both data arrays is for distance from the surface, which means that neighboring data samples with the same depth value are not normally on the same plane but follow the curvature of the terrain.

Another important data type used in this work is photographs. Pictures of the probes and their surrounding areas are displayed on the ground in their respective locations. Locations of the various artifacts that were found in the area are represented as blue or red cubes. The red cubes represent artifacts that have been radiocarbon dated (with dates shown above the cube). Both types of cubes can be clicked on, which will display a picture of the artifact in a window floating in front of the user.

Two of the objects at the site were digitized with an SfM approach: many overlapping photographs were taken from different viewing angles, and then the Bundler algorithm analyzed the pictures and computed the camera positions they were taken from in 3D space (Snavely et al., 2006). This allowed the PMVS2 algorithm (Furukawa & Ponce, 2009) to calculate a dense, colored point cloud indicating the geometry of the digitized object. One of these objects is a collection of bricks found on the ground; the other is a modern shrine built by local residents. The latter serves more as a landmark than to understand history.

Finally, a 3D modeling tool was used to create a typical temple from the Yuan Dynasty, which existed at the site around the time the radiocarbon dates indicate. The temple was placed and oriented on a particularly dense collection of bricks and subsurface material to indicate that this is the likely location of such a structure.

47.4.3.2 Immersive Visualization Application

This immersive visualization tool was implemented as a plug-in for the VR middleware software CalVR (Schulze, Prudhomme, Weber, & Defanti, 2013). CalVR is based on the OpenSceneGraph API (Wang & Qian, 2010) and runs on Linux-based visualization clusters.

Figure 47.14 shows the application displayed in CALIT2's StarCAVE. The image on the left shows the site with the magnetic data surface layer (mostly green, with some yellow lighter and red darker areas), the 3D temple model (white), and the probe site images. It can be seen that the probes were arranged on a 3 × 3 m grid. The picture on the right shows the point cloud of a modern shrine on the site, which serves as a reference point for the archaeologists. This point cloud was not tessellated in order to stay as true to the acquired data as possible.

Figure 47.15 shows two views of the subsurface data. Both show GPR points. Every sphere corresponds to a sampled data point. The colors indicate value, normalized to a color range from green to yellow to red. The image on the left was taken from a viewpoint on the surface, showing only the

(a) (b)

FIGURE 47.14 (a) StarCAVE user high above the archaeological site and (b) looking at the shrine model.

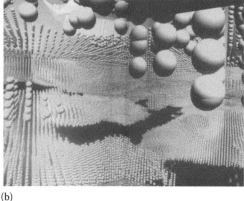

(a) (b)

FIGURE 47.15 Ground-penetrating data visualized as spheres (a). Lighter colors indicate higher measured density (b).

(a)   (b)

FIGURE 47.16   SfM and radiocarbon dates (a), images of the magnetic probes, and artifact photograph (b).

GPR data points at a certain depth (set with a 3D menu). The image on the right shows all available data points from a camera position below the ground.

An artifact location cube indicating a carbon dated artifact, along with its dated time range, is shown in the left picture of Figure 47.16. This picture also shows the 3D digitized ground structure, done with the Bundler algorithm. In the picture on the right, one can see the photograph of a probe location on the ground and the photograph of the selected artifact in the top left corner of the image. Artifacts are selected by pointing and clicking on them. Selecting an artifact updates the picture in the fixed frame at the top left. The frame itself can be moved around by clicking and dragging it with the input device. The picture on the right also shows what a translucent ground texture with underlying GPR spheres looks like. The transparency level can be set in the menu with a slider.

The application's VR menu is shown in Figure 47.17. The three subimages show three different submenus of the application. The leftmost picture shows the Model submenu, the middle the Surface submenu, and on the right the Subsurface submenu. In each submenu, the user can toggle the display of the respective visual elements as well as set their parameters. In the Surface menu, the transparency level for the surface texture can be set. The Subsurface menu allows selecting a layer of spheres to display or to show all spheres. The Density option in this menu allows setting a cutoff point, which hides all spheres with a density below that value. Rendering all spheres creates a very densely packed grid, because our data consist of about 20 million GPR and 400,000 ERT data points.

All menu adjustments result in immediate visual feedback. The application loads all required data into memory at startup and, thanks to optimized rendering algorithms, do not suffer from loading lag or other delays once the data have been loaded. This is possible by downsampling the photographs to sizes they typically are displayed at and a fast, shader-based point rendering algorithm for the GPR data. The terrain is not complex enough to require sophisticated terrain rendering algorithms and can be loaded into memory in its entirety.

The menu allows selecting one of the four surface types and one of the two subsurface types, and it allows displaying any number of 3D models and pictures desired. This is because the latter can coexist without occlusion, whereas the former two data types would occlude one another.

47.4.3.3   Benefits Derived

This successful VR software application for geologists and archaeologists fuses 12 different data types in one interactive visualization tool. The ability to see all these data types georeferenced on the same terrain and get an idea of the age of the found artifacts as well as their 3D locations is new for the domain scientists. It is very valuable for them to be able to spatially correlate all these different data types, which is more intuitive through the immersive interface than with their existing

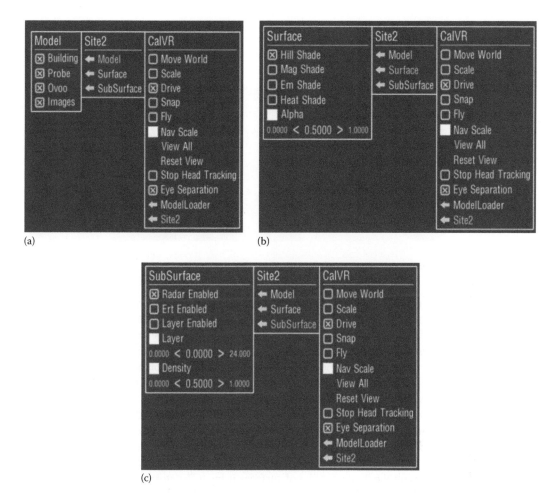

FIGURE 47.17   (a–c) Three submenus of the VR application.

desktop software tools. Additionally, the ability to see the data *to scale* is very useful, as is the ability to navigate through the data space with a 3D input device. Both geophysicists and archaeologists like using the application in CALIT2's StarCAVE, which can accommodate up to 10 people who can collaboratively explore the dataset.

Future goals for this application are to support additional data types, such as LIDAR point data, and datasets that are larger than the size of the computers' RAM. This will require integrating existing out-of-core approaches, which could perhaps be improved by optimizing spatial access for multiple different data types.

47.4.4   DRI Lancaster Sand Dune Layers

Sgambati et al. also applied immersive visualization and analysis to GPR data (Sgambati, Koepnick, Coming, Lancaster, & Harris, 2011)—in this case, data capturing the subsurface layers within sand dunes. They began with the immersive tool Toirt Samhlaigh (O'Leary, 2010) to volume render stacks of GPR images in four-sided CAVE and six-sided CAVE-style displays and on a low-cost immersive display. During their collaboration with geologist Nick Lancaster, they introduced an immersive widget called a virtual Brunton compass (Figure 47.18) for measuring dip and strike (orientation) of features inside the volume and also visualized these measurements and the ground surface in context with the volume.

FIGURE 47.18 A real Brunton compass and the virtual version in use.

Gathering data with GPR works well on sand dunes because sand has a high resistivity, allowing for good penetration of EM energy. This enables researchers to look deep into the large-scale sedimentary structures below the surface (Bristow et al., 2007). Analysis and annotation of the underground structures helps researchers understand and present how given sand dunes evolved over time. Building on this data, researchers can extrapolate information on past climates and wind directions over the region. Deposits of ancient sand dunes also occur in the rock record. Many of these ancient Aeolian sandstones are important reservoirs for hydrocarbons.

Characterizing the sediments of modern sand dunes in order to understand the conditions under which they were formed benefits from taking measurements of the angle and direction of dip of primary and secondary sedimentary structures. Airflow over the dunes affects their shape, and thus by measuring the shape of ancient dunes, the wind flow patterns of the past can be deduced. Working in the field, geologists measure the dip and strike of surfaces using a Brunton compass. Unfortunately, most desktop GPR visualization tools do not offer a good user interface allowing the geologist, accomplished in the use of a real Brunton compass, to transfer this skill to using a virtual version of the tool.

47.4.4.1 Visualization Techniques

The project team leveraged Toirt Samhlaigh for its immersive volume visualization capabilities. They created a 3D tomogram volume of the data from the stack of 2D subsurface cross-section images from the GPR (see also Figure 47.19 and Section 47.2.5). Working with the 3D volume, the geologist (Lancaster) manipulated the visualization controls to highlight features of scientific significance—that is, he manipulated the transfer functions, lighting parameters, and slice plane's positions to best view the data. A semitransparent ground surface provided context. Finally, the virtual Brunton compass tool was added to allow measurement and annotation of the dip/strike across the data. These markings then provided a visualization of the nonuniform vector field, indicating direction with a cone pointing in the direction of the steepest gradient.

47.4.4.2 Immersive Interface Techniques

Leveraging the Toirt Samhlaigh immersive volume-rendering tool, the project team had a firm footing on which to add features specifically for visualization of GPR-sourced data volumes. Addressing the shortcoming of most desktop GPR visualization tools, the virtual Brunton compass was added to enable users to efficiently and precisely measure dip and strike of features. The interface comprises a virtual plane, with a dynamic dip line, attached to a 6-DOF wand that the user aligns with features to measure orientation. With one click and a small hand motion, the user aligns the next

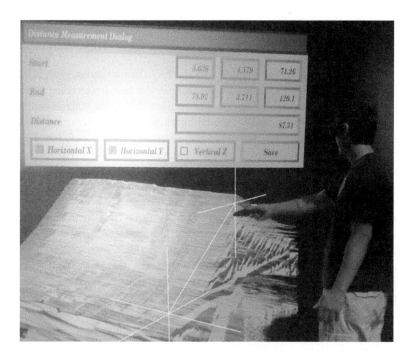

FIGURE 47.19 Using the measurement feature of Toirt Samhlaigh, a researcher simultaneously measures the length, height, and width of the sand dune slope.

measurement and, with a second click, saves that measurement. The collection of all the dip and strike measurements is then saved for off-line (on the desktop) analysis.

47.4.4.3 Outcomes and Lessons

Lancaster, the geologist on the project, stated that this "application improves current tools used by researchers or practitioners who are interested in these datasets." He tested this application in a high-resolution (1920 × 1920) six-sided CAVE-like display, on a standard resolution (1050 × 1050) four-sided CAVE, on a low-cost 3DTV-based immersive display (67″ with 1080p using checkerboard stereo) similar to the IQ-Station (Sherman, O'Leary, Whiting, Grover, & Wernert, 2010), and on a standard nonimmersive (mouse and keyboard only) computer. All were usable for basic visualization and data exploration. The 6-DOF interaction offered by the immersive displays demonstrated an advantage over the mouse/keyboard interface in the interactive use of the virtual Brunton compass and virtual ruler for taking dip, strike, and distance measurements. Lancaster noted major positives that included the ability to quickly view the dataset from different viewpoints in addition to being able to intuitively collect quantitative information on dip and strike.

Specifically, Lancaster noted that interactive tools "enable extraction of quantitative information from dataset" and that "Compared to field investigations, working in the CAVE is a lot easier and quicker." In comparing the taking of measurements in the real versus virtual worlds, he said: "to perform actual measurements of strike and dip on the beds and surfaces imaged by these data, we would have to excavate the dune and expose these features, which would be logistically difficult and not feasible in most cases." While using the immersive tools, he would frequently change viewpoints to confirm and refine placement of the Brunton compass markers. During his sessions in the various immersive systems, the visualization scientists on the team noted the differences in how much he was able to take advantage of their unique qualities. In particular, they found that the wider field of view of the CAVE-style displays over the low-cost 3DTV-based systems afforded the scientist the ability to place more user interface elements in the world without interfering with the visualization, whereas the 3DTV-based display became increasingly cumbersome to use as the smaller screen became cluttered.

Lancaster also pointed out some deficiencies in the tool. In particular, determining the scale of the data was difficult, which led to some *uncertainty in dip and strike measurements*. Perhaps not surprisingly, taking angle measurements on a nonuniformly scaled volume can be disconcerting. Even if calculations are correct, results are nonintuitive. This shortcoming could perhaps be reduced through greater practice or by improving the interface. Another deficiency emanates from the use of 2D menus and controls (widgets) contained within a 3D world. This problem is exacerbated when attempting to control small features with shaky position tracking. One possible solution is to move the 2D widgets onto a tablet, matching the 2D nature of the touch interface, though this could then add a fatigue factor from prolonged holding of the tablet.

Back on the positive side, basic computer graphics lighting also turned out to be important in understanding the structure of the data. The ethereal nature of volume rendering is often difficult to fuse and reduces the effectiveness of stereoscopic rendering. Motion parallax is still helpful, but through the use of GPU lighting effects, features within the volume can come to life more, providing some hooks that the brain can connect with, and make the 3D object more compelling. Also, the manipulation of lighting parameters in and of itself can provide structural cues as shading on surfaces changes.

Contrasted against the difficulty of manipulating 2D widgets in a 3D space, using the 6-DOF position tracking to place and orient the virtual Brunton compass proved to be a very big win. But even some of the 2D interfaces provided benefits not available in other tools. In particular, the Gaussian transfer function editor of Toirt Samhlaigh was much easier to use than the standard piecewise linear transfer function editor. This benefit results from the unique Gaussian curve editor that requires fewer precise selection actions by the user to obtain a desired curve. Sliding the Gaussian around by its center provides a quick way to explore a new dataset for salient features.

One feature of the immersive GPR visualization tool that is not splashy, but crucial, is the basic ability to save and restore the state of the visualization. Not only does this allow the researcher to return to work where they left off, but this also allows them to save a particularly interesting scenario to share with colleagues.

Overall, the geologist found great benefit from the GPR visualization tool, which was based on a volume visualization application with specific extensions for geology visualization, including a virtual Brunton compass widget with natural 6-DOF manipulations. In anecdotal comparisons of various display types, the visualization researchers note how the increased real estate of a four- or six-sided CAVE display provided the space needed to place controls without interfering with the data representations.

47.5 BENEFITS

While not universal, there are demonstrable benefits from the use of immersive technology in many areas, including the visualization and analysis of scientific data from the geological sciences. But demonstrable benefits are not a sufficient criterion in determining whether a new immersive visualization program or tool will pay dividends. Even the most proven tool must overcome the hurdle of breaking into the regular workflow of the researcher. For larger immersive displays, this problem is exacerbated by the distance the researcher must travel to make use of the system. Smaller immersive systems can be available in the lab, and while their benefits may fall between desktop and location-based immersive systems, proximity to the research can make them much more valuable.

A single favorable anecdote indicates that an immersive system *can* be useful, and there are several such anecdotes, some discussed in Section 47.4. For example, in Section 47.4.1, within 20 min in an immersive projection system, geologist Kinsland discovered something new in data he had been poring over. In Section 47.4.2, researchers were able to explore that which could not otherwise be explored—hidden data under the surface at a site for which digging was prohibited. Also they found that seeing their data *at scale* was much more useful than viewing a small model on a computer desktop. In Section 47.4.3, the LidarViewer application has been shown to be valuable for several researchers in its ability to naturally select a group of points and then measure them precisely. And in Section 47.4.4, the immersive interface was helpful through the natural placement techniques used to annotate the dip/strike within the subsurface data, which also was from a site that would have been too big (and deep) to dig.

But beyond anecdotes, there is also experimental evidence that demonstrates advantages in using immersive technologies for scientific visualization. In the realm of geosciences, Gruchalla et al. demonstrated an improved ability in navigating through a complex well-path landscape (Gruchalla, 2004). In the wider field of immersive visualization, Prabhat's experiment on analyzing fluid flow through a heart found immersive technologies provided increased performance over desktop tools (Prabhat et al., 2008), and Brady et al. described tasks (following indistinct fibers through volume data) found to be nearly impossible to perform on the desktop were doable in a CAVE display (Brady et al., 1995).

Of course, there are fields related to the geosciences in their need to visualize data within the context of the physical world that also benefit from these and similar immersive tools. Archaeology has already been mentioned, plus other areas such as community planning, which has been used at the CAES facility using the same LidarViewer tool described in Section 47.4.2.

The benefits are not automatic, however—purchasing a new CAVE is not sufficient to guarantee great results. Acquiring the hardware is only part of the process. There is good software available, but not to an overwhelming degree, and expertise is required to get the software installed, configured, and put to use. In some cases, there won't be a software package that directly meets the scientists' needs, and so custom software will be needed, requiring more expertise as well as more time. And then once again, only if the hurdles to use are low enough, will scientists be fully engaged.

The observation of Marshall McLuhan "*I wouldn't have seen it if I hadn't believed it*" serves as another cautionary consideration. The ability to vary scales (e.g., human versus continental) or select how data are filtered offers the possibility to see features that aren't obvious in the field or when observing the unfiltered data. But to avoid perceiving what isn't there, it is also important to go back to the field or to the unfiltered data. For example, a subtle fault whose areal extent isn't visible in the field becomes identifiable as a fault by its horizontal form and extent in LIDAR data viewed immersively. Back in the field, the subtle topography can be confirmed to clearly identify a fault.

In the end, VR is a medium that can assist our search for details revealing the nature of our planet. As we pointed out in the introduction, most geology is intrinsically 3D. Interpreting and communicating about 3D data is thus most effectively (quickest, most accurate, etc.) done in 3D, physically immersive, viewing. While geologists have learned how to interpret 2D representations into 3D, by presenting the data directly in three living dimensions, it becomes easier for the mind to ingest. We also largely avoid the pitfalls of preconceptions that can unintentionally cloud the interpretation.

47.6   FUTURE AND CONCLUSION

Interest and excitement around immersive technologies—*Virtual Reality!* —ebbs and flows over the course of time. In general, it has followed the typical pattern of the Gartner's technology hype curve (Fenn, 1995) in that after an initial spark, there was huge interest and media exposure, surpassing the accomplished merit of the time. For VR, this *peak of inflated expectations* occurred in the mid-1990s. Following that, during the *trough of disillusionment*, the pendulum generally swings too far to the negative, and many give up on the technology. Those who continue and others who gradually join the effort may persevere through the *slope of enlightenment*, perhaps to eventually reach the *plateau of productivity.*

Presently, we can consider the technology of immersive interfaces (VR) to be somewhere on the *slope of enlightenment* with value gradually being demonstrated to more and more users. That does not mean that there are not those who test the technology and find it insufficient for their needs, for that is the nature of all technology. But we are seeing an increasing number of scientific researchers who have found value in immersive technologies, several in fields related to geologic research.

One trend bolstering immersive technologies is the increasing availability of needed technologies through consumer channels. Most prominently, the availability of 3D television displays has made the prospects of building a low-cost immersive system possible. The ubiquitous nature of 3D acceleration in just about every computing device enables interesting virtual worlds, and the last major piece of the puzzle is position tracking, which can range from home brew using Wii™ remote controls or PlayStation Move™ controllers to professional tracking systems costing one or two orders of magnitude more. But even professional systems are significantly less expensive and perform better than what was available a decade ago. Integrating a low-cost immersive system can be done on an individual or small group basis, but there are benefits to be gained by working together in a community (Figure 47.20). The IQ-Station (Sherman et al., 2010) project is one such effort whereby an open recipe for the hardware is provided and refined, taking into account the changing availability of consumer products (IQ-Station Consortium, 2013).

Another precursor to wider usage of immersive technologies for visualization as well as other uses comes from the growth of end-user software that is quickly usable. This growth can be fostered in two ways. One way is through the adoption of immersive, or VR, options to existing visualization software—particularly software that is already in use by the user community. Some examples of visualization tools that provide this option include the commercial packages Avizo (Visualization Sciences Group, 2013) and Ensight (Computational Engineering International, Inc., 2013), as well as open tools such as ParaView (Kitware, Inc. ParaView, 2013).

The other primary way to spur growth is to provide a user interface that is not only easy to use but widely adopted. When there is inconsistency between how to use immersive software tools, then more effort is required for end users to become familiar and fluent with each tool. Consistency

FIGURE 47.20 A user interacting with a volume dataset using the Toirt Samhlaigh application running on a portable and low-cost IQ-Station.

allows users to quickly move between applications, and when tools become available with needed features, users can readily jump to the new (and probably improved) tool. This ability to keep on the forefront of software technology will better hold the interest of those who use the technology because it helps them improve their research.

The ebb and flow in the amount of usage and attention garnered by immersive technologies will likely continue. But even now, overall, the evidence suggests that there is value in immersive systems benefiting many in the scientific communities, including the geological sciences. We anticipate that as proliferation of the technology continues to spread, the audience will grow steadily, the hardware and software systems will improve, and some will be available ready-to-use right from the time of purchase.

All these advances, coupled with the ascension of new scientists with technology usage, engrained in their lives, looking to take advantage of any technology that provides dividends, and we can expect usage of immersive technologies in the geosciences over the coming years to be even more fruitful.

REFERENCES

Aegis Instruments. (2013). *Easy 3D-GPR visualization software.* Retrieved August 1, 2013, from http://www. aegis-instruments.com/products/brochures/easy-3d-gpr.html

Aitken, M. J. (1958). Magnetic prospecting: I—The water Newton survey. *Archaeometry, 1,* 24–29.

American Society for Photogrammetry and Remote Sensing (ASPRS). (2013). *LASer (LAS) file format exchange activities.* Retrieved August 1, 2013, from http://asprs.org/Committee-General/LASer-LAS-File-Format-Exchange-Activities.html

Bagley, J. W. (1917). *The use of the panoramic camera in topographic surveying: With notes on the application of photogrammetry to aerial surveys* (U.S. Geological Survey Bulletin #657). Washington, DC: Government Printing Office.

Becker, H. (1995). From nanotesla to picotesla—A new window for magnetic prospecting in archaeology. *Archaeological Prospection, 2,* 217–228.

Bier, E. A., Stone, M. C., Fishkin, K., Buxton, W., & Baudel, T. (1994). A taxonomy of see-through tools. In *ACM CHI* (pp. 358–364). New York, NY: ACM

Bierbaum, A., Just, C., Hartling, P., Meinert, K., Baker, A., & Cruz-Neira, C. (2001). VR juggler: A virtual platform for virtual reality application development. In *VR '01 Proceedings of the Virtual Reality 2001 Conference* (p. 89). Washington, DC: IEEE Computer Society.

Billen, M., Kreylos, O., Hamann, B., Jadamec, M., Kellogg, L. Staadt, O., & Sumner, D. (2008, September). A geoscience perspective on immersive 3D gridded data visualization. *Computers & Geosciences, 34*(9), 1056–1072.

Blee, H. H. (1940). Third dimension maps. *Military Engineer, 32,* 187–190.

Borst, C. W., Baiyya, V. B., Best, C. M., & Kinsland, G. L. (2007). Volumetric windows: Application to interpretation of scientific data, shader-based rendering method, and performance evaluation. In *The 2007 International Conference on Computer Graphics and Virtual Reality* (pp. 72–78). CSREA Press.

Borst, C. W. & Kinsland, G. L. (2005). *Visualization and interpretation of 3-D geological and geophysical data in heterogeneous virtual reality displays: Examples from the Chicxulub impact crater.* Paper presented at the Transactions: Gulf Coast Association of Geological Societies (Vol. 55), New Orleans, LA.

Borst, C. W., Tiesel, J.-P., & Best, C. M. (2010, May/June). Real-time rendering method and performance evaluation of composable 3D lenses for interactive VR. *IEEE Transactions on Visualization and Computer Graphics, 16*(3), 394–410.

Borst, C. W., Tiesel, J.-P., Habib, E., & Das, K. (2011). Single-pass composable 3D lens rendering and spatiotemporal 3D lenses. *IEEE Transactions on Visualization and Computer Graphics, 17*(9), 1259–1272.

Brady, R., Pixton, J., Baxter, G., Moran, P., Potter, C. S., Carragher, B., & Belmont, A. (1995, October). Crumbs: A virtual environment tracking tool for biological imaging. In *1995 Proceedings of IEEE Biomedical Visualization* (pp. 18–25). Copyright@1995, IEEE.

Bristow, C., Duller, G., & Lancaster, N. (2007). Age and dynamics of linear dunes in the Namib Desert. *Geology, 6*(35), 555–558.

Computational Engineering International, Inc. (CEI). (2013). *Ensight virtual reality.* Retrieved August 1, 2013, from http://www.ceisoftware.com/virtual-reality-vr

Das, K., & Borst, C. W. (2010a). An evaluation of menu properties and pointing techniques in a projection-based VR environment. In *IEEE 3D User Interfaces (3DUI)* (pp. 47–50). Washington, DC: IEEE Computer Society.

Das, K., & Borst, C. W. (2010b). VR menus: Investigation of distance, size, auto-scale, and ray-casting vs. pointer-attached-to-menu. In *ISVC (Visual Computing)* (pp. 719–728). Berlin Heidelberg: Springer-Verlag.

DeFanti, T. A., Dawe, G., Sandin, D. J., Schulze, J. P., Otto, P., Girado, J., … Rao, R. (2009, February). TheStarCAVE, a third-generation CAVE and virtual reality OptIPortal. *Future Generation Computer Systems, 25*(2), 169–178, Elsevier.

Demiralp, C., Jackson, C. D., Karelitz, D. B., Zhang, S., & Laidlaw, D. H. (2006). CAVE and Fishtank virtual-reality displays: A qualitative and quantitative comparison. *IEEE Transactions on Visualization and Computer Graphics, 12*(3), 323–330.

Dorn, G., Touysinhthiphonexay, K., Bradley, J., & Jamieson, A. (2001). Immersive 3-D visualization applied to drilling planning. *The Leading Edge, 12*(20), 1389–1392.

Edelsbrunner, H., & Mucke, E. P. (1994). Three-dimensional alpha shapes. *ACM Transactions on Graphics (TOG), 13*(1), 43–72.

EON Reality, Inc. (2013). *Experience more with virtual reality.* Retrieved August 1, 2013, from http://www.eonreality

Farr, T. G., & Kobrick, M. (2000). Shuttle Radar Topography Mission produces a wealth of data. *American Geophysical Union EOS, 81*, 583–585.

Farr, T. G., Rosen, P. A., Caro, E., Crippen, R., Duren, R., Hensley, S., … Alsdorf, D. (2007). The Shuttle Radar Topography Mission. *Reviews of Geophysics, 45*, RG2004. doi:10.1029/2005RG000183.

Fenn, J. (1995). *When to leap on the hype cycle* (Gartner ID: SPA-ATA-305). Gartner Group RAS Services.

Fröhlich, B., Barrass, S., Zehner, B., Plate, J., & Göbel, M. (1999). Exploring geo-scientific data in virtual environments. In *Proceedings of IEEE Visualization '99* (pp. 169–173). San Francisco, CA: IEEE Computer Society Press.

Fröhlich, B., & Lancaster, A. (1986). Electromagnetic surveying in current Middle Eastern archaeology: Application and evaluation. *Geophysics, 51*, 1414–1425.

Furukawa, Y., & Ponce, J. (2009). Accurate, dense, and robust multi-view stereopsis. In *IEEE Transactions on Pattern Analysis and Machine Intelligence, 32*(8), 1362–1376.

Gay, S. P. (1971). Morphological study of geophysical maps by viewing in three dimensions. *Geophysics, 36*(2), 396–414, Society of Exploration Geophysicists, 04/1971. ISSN (print): 0016-8033, ISSN (online): 1942–2156.

Glenn, N., Streutker, D., Chadwick, D., Thackray, G., & Dorsch, S. (2006). Analysis of LiDAR-derived topographic information for characterizing and differentiating landslide morphology and activity. *Geomorphology, 73*(1), 131–148.

Gold, P. O., Cowgill, E., Kreylos, O., & Gold, R. D. (2012). A terrestrial lidar-based workflow for determining three-dimensional slip vectors and associated uncertainties. *Geosphere, 8*(2), 431–442.

Goodman, D., Nishimura, Y., & Rogers, J. (1995). GPR time slices in archaeological prospection. *Archaeological Prospection, 2*, 85–89.

Gruchalla, K. (2004, March). Immersive well-path editing: Investigating the added value of immersion. In *IEEE Proceedings of 2004 Virtual Reality* (pp. 157–164). Washington, DC: IEEE Computer Society.

Harding, C., Kakadiaris, I. A., Casey, J. F., & Loftin, R. B. (2002). A multi-sensory system for the investigation of geoscientific data. *Computers & Graphics, 26*(2), 259–269.

IQ-Station Consortium. (2013). *Open-recipe low-cost immersive display.* Retrieved August 1, 2013, from http://iq-station.org

Itoh, M., Ohigashi, M., & Tanaka, Y. (2006). WorldMirror and WorldBottle: Components for interaction between multiple spaces in a 3D virtual environment. In *IEEE Tenth International Conference on Information Visualization, 2006 (IV, 2006)* (pp. 53–61). Washington, DC: IEEE Computer Society.

Kinsland, G. L., Borst, C. W., Best, C. M., & Baiyya, V. B. (2007). *Geomorphology and Holocene Fluvial Depositional History in the Mississippi River Valley near Lafayette, Louisiana: Interpretations of LIDAR data performed in 3D virtual reality.* Paper presented at the Gulf Coast Association of Geological Societies 2007 Convention. Houston, TX.

Kinsland, G. L., Borst, C. W., Tiesel, J. P., & Bishop, C. E. (2008). Imaging digital well logs in 3D virtual reality: Investigation of northern Louisiana Wilcox fluvial/coal strata for coalbed natural gas. *Gulf Coast Association of Geological Societies Transactions, 58*, 517–524.

Kitware, Inc. (2013). *ParaView/users guide/CAVE display.* Retrieved August 1, 2013, from http://www.paraview.org/Wiki/ParaView/Users_Guide/CAVE_Display

Kitware, Inc. (2013). *"VTK—The visualization toolkit" software.* Retrieved August 1, 2013, from http://www.vtk.org

Kiyokawa, K., & Takemura, H. (2005). A tunnel window and its variations: Seamless teleportation techniques in a virtual environment. *HCI International*. Las Vegas, NV.

Komerska, R., & Ware, C. (2004). Haptic state–surface interactions. *IEEE Computer Graphics and Applications, 24*(6), 52–59.

Kreylos, O. (2008). Environment-independent VR development. In Bebis, G. et al. (Eds.), *Lecture Notes in Computer Science (LNCS): Vol. 5358. 2008 Proceedings of the International Symposium on Visual Computing (ISVC), Part I* (p. 901–902). Berlin, Germany: Springer-Verlag.

Kreylos, O., Bawden, G., Bernardin, T., Billen, M. I., Cowgill, E. S., Gold, R. D., ... Sumner, D.Y. (2006). Enabling scientific workflows in virtual reality. In *Proceedings of the 2006 ACM International Conference on Virtual Reality Continuum and Its Applications* (pp. 155–162). New York, NY: ACM.

Kreylos, O., Bawden, G. W., & Kellogg, L. H. (2008). Immersive visualization and analysis of LiDAR data. In Bebis, G. et al. (Eds.), *Lecture Notes in Computer Science (LNCS): Vol. 5358. 2008 Proceedings of the International Symposium on Visual Computing (ISVC), Part I* (pp. 846–855). Berlin, Germany: Springer-Verlag .

Kurtenbach, G., Fitzmaurice, G., & Balakrishnan, R. (2002). Volume management system for volumetric displays. U.S. Patent application 10/183,966.

LaFayette, C., Parke, F. I., Pierce, C. J., Nakamura, T., & Simpson, L. (2008). Atta texana leaf cutting ant colony: A view underground. In *ACM SIGGRAPH 2008 Talks* (Vol. 53, p. 1). Article No. 53. ACM: New York, NY.

Lin, A. Y.-M., Novo, A., Weber, P. P., Morelli, G., Goodman, D., & Schulze, J. P. (2011). A virtual excavation: Combining 3D immersive virtual reality and geophysical surveying. In Bebis, G. et al. (Eds.), *Lecture Notes in Computer Science (LNCS): Vol. 6939. 2011 Proceedings of the International Symposium on Visual Computing (ISVC), Part II* (pp. 229–238). Berlin, Germany: Springer-Verlag .

MacLeod, M. N. (1919, June). Mapping from air photographs. *The Geographical Journal, 53*(6), 382–396. http://www.jstor.org/stable/1780414

Moore, C., Gertman, V., Olsoy, P., Mitchell, J., Glenn, N., Joshi, A., ... Lee, R. (2011, December 5–9). *Discovering new methods of data fusion, visualization, and analysis in 3D immersive environments for hyperspectral and laser altimetry data*. San Francisco, CA: American Geophysical Union.

Mulder, J. D., & Boschker, B. R. (2004). A modular system for collaborative desktop VR/AR with a shared workspace. In *IEEE Proceedings of 2004 Virtual Reality* (pp. 75–82). Washington, DC: IEEE Computer Society.

Murgoitio, J., Shrestha, R., Glenn, N., & Spaete, L. (2013). Airborne LiDAR and terrestrial laser scanning derived vegetation obstruction factors for visibility models. *Transactions in GIS, 18*(1), 147–160. John Wiley & Sons Ltd. doi:10.1111/tgis.12022

Novo, A., Grasmueck, M., Viggiano, D., & Lorenzo, H. (2008). 3D GPR in archaeology: What can be gained from dense data acquisition and processing. *Twelfth International Conference on Ground Penetrating Radar*, Birmingham, U.K.

O'Leary, P. (2010). *"Toirt-Samhlaigh" software*. Retrieved from http://code.google.com/p/toirt-samhlaigh/

Olsoy, P., Gertman, V., Glenn, N., Joshi, A., Mitchell, J., Whiting, E., ... Spaete, L. (2012, March). Interactive exploration and data fusion of multimodal remote sensing data. In *ASPRS Annual Conference* (pp. 19–23). Curran Associates, Inc. Sacramento, CA.

Perlin, K., & Fox, D. (1993). Pad: An alternative approach to the computer interface. In *Proceedings of SIGGRAPH 1993* (pp. 57–64). New York, NY: ACM.

Plumlee, M., & Ware, C. (2006). Zooming versus multiple window interfaces: Cognitive costs of visual comparisons. *ACM Transactions on Computer-Human Interaction, 13*(2), 179–209.

Prabhat, Forsberg, A., Katzourin, M., Wharton, K., & Slater, M. (2008). A comparative study of desktop, fishtank, and cave systems for the exploration of volume rendered confocal data sets. *IEEE Transactions on Visualization and Computer Graphics, 14*, 551–563.

Prachyabrued, M., Ducrest, D., & Borst, C. W. (2011). Handymap: A selection interface for cluttered VR environments using a tracked hand-held touch device. In Bebis, G. et al. (Eds.), *Lecture Notes in Computer Science (LNCS): Vol. 6939. 2011 Proceedings of the International Symposium on Visual Computing (ISVC), Part II* (pp. 45–54). Berlin, Germany: Springer-Verlag.

Raghupathy, P. B., & Borst, C. W. (2011). Investigation of secondary views in a multimodal VR environment: 3D lenses, windows, and mirrors. In Bebis, G. et al. (Eds.), *Lecture Notes in Computer Science (LNCS): Vol. 6939. 2011 Proceedings of the International Symposium on Visual Computing (ISVC), Part II* (pp. 180–189). Berlin, Germany: Springer-Verlag .

Raghupathy, P. B., & Borst, C. W. (2012). Force feedback and visual constraint for drawing on a terrain: Path type, view complexity, and pseudohaptic effect. In *3DUI 2012* (pp. 157–158). Washington, DC: IEEE Computer Society.

Robertson, G., van Dantzich, M., Robbins, D., Czerwinki, M., Hinckley, K., Risden, K., ... Gorokhovsky, V. (2000). The task gallery: A 3D window manager. In *CHI* (pp. 494–501). New York, NY: ACM.

Rogowitz, B. E., & Trenish, L. A. (1998). *Why should engineers and scientists be worried about color?* Retrieved from http://www.research.ibm.com/people/lloydt/color/color.HTM

Schmalstieg, D., & Schaufler, G. (1999). Sewing worlds together with SEAMs: A mechanism to construct complex virtual environments. *Presence: Teleoperators and Virtual Environments, 8*(4), 449–461.

Sgambati, M. R., Koepnick, S., Coming, D. S., Lancaster, N., & Harris, F. C. Jr. (2011). Immersive visualization and analysis of ground penetrating radar data. In Bebis, G. et al. (Eds.), *Lecture Notes in Computer Science (LNCS): Vol. 6939. 2011 Proceedings of the International Symposium on Visual Computing (ISVC), Part II* (pp. 33–44). Berlin, Germany: Springer-Verlag.

Sherman, W. R., O'Leary, P., Whiting, E. T., Grover, S., & Wernert, E. A. (2010). IQ-Station: A low cost portable immersive environment. In Bebis, G. et al. (Eds.), *Lecture Notes in Computer Science (LNCS): Vol. 6454. 2010 Proceedings of the International Symposium on Visual Computing (ISVC), Part II* (pp. 361–372). Berlin, Germany: Springer-Verlag.

Schulze, J. P., Prudhomme, A., Weber, P., & DeFanti, T. A. (2013, March 4). CalVR: An advanced open source virtual reality software framework. M. Dolinsky & I. E. McDowall (Eds.), In *Proceedings of SPIE 8649, The Engineering Reality of Virtual Reality 2013* (pp. 864–902). Burlingame, CA. doi:10.1117/12.2005241

Snavely, N., Seitz, S. M., & Szeliski, R. (2006). Photo tourism: Exploring image collections in 3D. *ACM Transactions on Graphics (Proceedings of SIGGRAPH 2006)*.

Stoakley, R., Conway, M., & Pausch, R. (1995). Virtual reality on a WIM: Interactive worlds in miniature. In *ACM CHI* (pp. 265–272). New York, NY: ACM Press/Addison-Wesley Publishing Co.

Tabbagh, A. (1986). Applications and advantages of the slingram electromagnetic method for archaeological prospecting. *Geophysics, 51*, 576–584.

Tanaka, H. K. M., Nakano, T., Takahashi, S., Yoshida, J., Takeo, M., Oikawa, J., ... Niwa, K. (2007, November). High resolution imaging in the inhomogeneous crust with cosmic-ray muon radiography: The density structure below the volcanic crater floor of Mt. Asama, Japan, *Earth and Planetary Science Letters, 263*(1–2), 15, 104–113. http://dx.doi.org/10.1016/j.epsl.2007.09.001

TSR. 2000. Electrical Resistance Tomography for Subsurface Imaging. Innovative Technology Summary Report. June 2000. OST/TMS ID 17. Characterization, Monitoring, and Sensor Technology Crosscutting Program and Subsurface Contaminants, Focus Area. DOE-EM-0538. Available at http://www.osti.gov/scitech/biblio/768919. doi:10.2172/768919

Viega, J., Conway, M. J., Williams, G., & Pausch, R. (1996). 3D magic lenses. In *ACM UIST* (pp. 51–58). New York, NY: ACM.

Visualization Sciences Group (VSG). (2013). *Avizo extensions.* Retrieved August 1, 2013, from http://www.vsg3d.com/avizo/extensions

Wang, R., & Qian, X. (2010, December). *OpenSceneGraph 3.0: Beginner's Guide.* Birmingham, U.K. Packt Publishing, ISBN: 1849512825.

Ware, C., Plumlee, M., Arsenault, R., Mayer, L. A., Smith, S., & House, D. (2001). GeoZui3d: Data fusion for interpreting oceanographic data. In *Oceans 2001* (pp. 1960–1964). Honolulu, HI: MTS/IEEE Conference and Exhibition (Vol. 3).

Weiskopf, D., Engel, K., & Ertl, T. (2003). Interactive clipping techniques for texture-based volume visualization and volume shading. *IEEE Transactions on Visualization and Computer Graphics, 9*(3), 298–312.

Winkler, C., Bosquet, F., Cavin, X., & Paul, J. C. (1999, October). Design and implementation of an immersive geoscience toolkit (case study). In C. Hansen, D. Silver, & L. Treinish (Eds.). *IEEE Proceedings of Visualization '99.* San Francisco, CA (pp. 429–432). IEEE Computer Society Press.

Yamaguchi, Y., Kahle, A. B., Tsu, H., Kawakami, T., & Pniel, M. (1998). Overview of Advanced Spaceborne Thermal Emission and Reflection Radiometer (ASTER). *IEEE Transactions on Geoscience and Remote Sensing, 36*(4), 1062–1071.

48 Information Visualization in Virtual Environments
Trade-Offs and Guidelines

Nicholas F. Polys

CONTENTS

48.1 INTRODUCTION

Imagine yourself a hyperintelligent being from another galaxy. You are a luminous egg. You are exploring the universe in search of molecular resources and other life. As you warp into the Milky Way, your dashboard alights, detailing the presence of numerous electromagnetic signals and valuable atomic elements in the solar system. Turning your ship toward a small yellow star, you consider the attributes of each planet in the system: size, mass, chemical compositions, and environmental qualities. Patterns in these attributes generate hypotheses about the history and habitability of the planets in this system. Of the top planets you examine closer, you find the third stone from the sun is a *Blue Marble*—a rare gem in this entropic universe. You fly in to take a closer look….

There are several reasons to motivate this chapter with this thought experiment. The first is to examine our assumptions about other minds. The second is that this narrative includes some technology that augments the cognition of our alien pilot to help them make sense of the situation, evaluate alternatives, and arrive at an actionable decision. Third, it illustrates the prototypical cycle of information visualization activities: overview, filter/zoom, and details on demand.

If you were able to *suspend your disbelief* and go with the story, you created your own virtual reality. It is indeed amazing that this rich process of mental activation was achieved with only 127 words! No doubt, each of you readers filled in some unique details in your own mind, and these will vary widely. When presented with explicit visuals (about your alien race, the look of the dashboard, the ship), the variance is reduced and we are closer to a shared understanding in our minds. A picture tells a thousand words (Larkin & Simon, 1987) and an experience can explain a lifetime. But how does this happen? How do our minds make sense out of the noisy and incomplete data fed to them by the senses?

Certainly, things like biological survival, reproduction, and quality of life occupy most of our time; however, many have argued that an understanding of the nature of our minds is our greatest quest. Religion, philosophy, and anthropology have been the historic proving grounds for a deeper understanding of our understanding. More recently, however, cognitive psychology and neuroscience are providing startling discoveries and discussions from the mechanics of perception and action to the nature of emotion, knowledge, and cognition.

We believe the brain to be the seat of our consciousness, though there is no one locus of control. Once upon a time, when Descartes' *dualist* philosophies held sway, the pineal gland was thought to be the controller (metaphorical joystick) of the homunculus driving our actions on the physical plane. This has proven an inadequate explanation. Similarly, we cannot say that the periaqueductal gray is the locus of transcendental bliss or euphoria. In addition, there is no one cell in your brain that activates when you recognize your grandmother—memories and associations are distributed throughout the vast network of the brain. While amazingly complex, we are beginning to understand the computational processes embodied by these vast networks of neurons; this is especially true in the visual system.

48.1.1 OBSERVER AND OBSERVED

The visual system accounts for approximately 3 out of 8 lb in the human brain. We know that our visual system has evolved sensitivities specific to attributes of the environment that impact our survival: high-resolution sensing of luminance for nighttime activities, color sensitivity to distinguish food from poison and to break camouflage, stereo vision to judge trajectories and distances, and retinotopically varied sensitivities (such as different subsystems for focus and periphery) to label objects and locate them in space.

We also know that the mind fills in details with what it expects to see: optical and illusions and Gestalt effects, prototypical qualities of an object or event, L3++3rs in a word are all examples. So clearly, our perception is not just a bottom-up process constricted purely from our sensory data. Our minds use abstractions and assumptions to get by.

The mind is adaptable and plastic. It is reconfigurable by culture and technology—those imperatives that move too fast for biological evolution. It is a good thing, too, since the emergent properties of the variety of societies around our global village would challenge the most advanced protocol droid. While each of us inhabit a unique and personal reality, we engage in linguistic communication and collective hallucination effortlessly every day. We can also recover some mental functions after damage such as stroke or trauma when our brains successfully remap processing functions from the damaged area to new areas and networks.

We may be brains in a vat. We may be butterflies in Hawaii dreaming we are men or imagining a perfect storm. We have serious reasons to be skeptical of our sensory inputs, and this is not new news. Indeed, Plato's *Analogy of the Cave* was an original inspiration for *Virtual Reality*; this analogy explored the nature of perception and knowledge and reality and asked the fundamental question of "Why do you believe your senses?" Two millennia later, the philosophers Hume and Kant pushed the limits of this reasoning. Hume's skepticism and problems with inductive inferences and epistemology have infected generations of philosophers. Kant's metaphysical challenge with knowledge existing independent of experience (famously explored in Critique of Pure Reason) points to this same quandary of existence for a priori or disembodied concepts.

For our purposes, we will adopt a pragmatic stance by considering our mind as an embodied organism, which is the product of eons of biological evolution on this earth. As such, its sensory system and physical musculature are tuned to function in *this* electromagnetic spectrum, *this* gravity field, and *this* atmosphere. Because the mind is embodied, we have the possibility for immersive technologies to substitute and supplement the senses. All sorts of creative gadgets and systems for sensory substitution have been invented and developed (Bowman, Kruijff, LaViola, & Poupyre, 2004); for a historical and particularly expansive vision, readers should consult the book *Virtual Reality* (Pimentel & Teixeira, 1993). Through the present day, this equipment is built with the goal of improving the *immersion* of the user—the objective degree of sensory substitution provided. This should not be confused with the concept of *presence*, which is the subjective rating of the user to *being there* in the virtual environment (VE). Now, more than ever in history, we can deliver synthetic sensory stimuli that are interactive and of high fidelity.

48.1.2 MEDIA AND MESSAGE

The cognitive scientist looks not just on the nature of the sensory transducers and actuators but also at the representations and information processing required to get from stimulus to response. Cognitive psychology takes a computational perspective on mental representations and their processing—it attempts to understand mental activity in terms of the computational work required to encode an input, transform and manipulate that representation, and create an output or behavior. Thus, cognitive psychology considers the complexity, kinds, components, and scale of processing between stimulus and response. Beginning with the nervous system of worms and *Planaria*, scientists have worked their way up the evolutionary ladder mapping more and more of the complex neural circuits that build our reality.

Considering the human observer and the evolutionary development of its perceptual and cognitive systems, we can examine the qualities of stimuli that communicate information. Information about the state of the environment can be received through several sensory modalities of course, and humans are demonstrably *tuned* to certain visual wavelengths, audio frequencies, chemicals (taste and smell), and physical contacts (haptics). These are all prime candidates to off-load information rendering to. However, each has limitations in terms of the information types, ranges, and dynamics of the values they can accurately represent. For example, in both haptics and sonification, humans have different sensitivities to frequency and amplitude making quantitative values difficult to portray. In addition, these sensory modalities are quickly habituated, meaning that over time they become less sensitive to signal (what does it take for you to feel your clothes?).

Perhaps not explicitly stated, much of this work is concerned with understanding how we humans can solve such difficult problems on a daily basis when our working memory is so limited (Miller, 1956). Visualization can amplify cognition (Card, Mackinlay, & Shneiderman, 1999) in several ways, but especially by providing an external data store and processing service for costly mental operations (Zhang & Norman, 1994). Visual displays provide key input to visual working memory (Baddeley, 2003; Baddeley & Logie, 1999; Logie, 1995). Thus, we can optimize our stimuli for pre-attentive processing and still establish the crucial, shared representational correspondence between the concepts in the mind and the evidence in the data.

This approach has born much fruit, which feeds our understanding in areas from mental rotation (Sheppard, 2004), line graph reading (Lohse, 1991), and robotics to expert systems in chess, medical diagnosis, and jeopardy. Taken in full, the experimental literature shows that there are clear patterns in human abilities and competencies: spatial visualization (Sorby, 2009), chunking strategies, and the art of memory are all skills that can be learned and improved. There are also clear biases in human reasoning including confirmation bias, fundamental attribution error, blind-spot bias, anchoring bias, projection bias, and representativeness bias (Kahneman & Tversky, 1979). Interactive, decision-making tools can help mitigate and alleviate common mistakes due to these biases (i.e., Evans, 1989). Similarly, our alien explorer uses an information display dashboard to augment their reasoning and decision making.

48.2 BACKGROUND

Readers should be aware of a notional distinction between *scientific visualization* and *information visualization*. Traditionally, scientific visualization refers to a situation where there is some natural spatial mapping for the data (i.e., an airplane wing, a globe). In contrast, information visualizations can map data values into any spatial domain (i.e., a scatterplot, a pie chart). However, many of the same perceptual and cognitive challenges exist between these fields. In addition, when scientists are concerned with gene expression data or high-dimensional patterns in a nuclear reaction, this distinction is less useful. *Visual analytics* (Thomas & Cook, 2006) gets around this distinction by focusing on the inclusion of interactions and statistical tools to augment the user's reasoning and sensemaking process. As we will see in this chapter, VEs can be used with advantage for all these brands of visualization.

48.2.1 GRAPHICAL INFORMATION

Primary factors in visualization design, which we will describe in the following, concern both the data (its dimensionality, type, scale, range, and attributes of interest) and human factors (the user's purpose and expertise). Different data types and tasks require different representation and interaction techniques. How users construct knowledge about what a graphic *means* is also of inherent interest to visualization applications. For users to understand and interpret images, higher-level cognitive processes are usually needed. A number of authors have enumerated design strategies and representation parameters for rendering signifieds in graphics (Bertin, 1983; Tufte, 1983, 1990), and there are effects from both the kind of data and the kind of task (Shneiderman, 1996).

Card, Mackinlay, and Scheiderman (1999) have examined a variety of graphical forms and critically compared visual cues in scatterplot, cone trees, hyperbolic trees, tree maps, point of interest, and perspective wall renderings. As we shall see, their work is important since any of these 2D visualizations may be embedded inside, or manifested as, a VE. Interactive computer graphics present another level of complication for designers to convey meaning as they are responsive and dynamic and may take diverse forms. There are challenges both on the input medium to the user and on the action medium for the user. These are known as the gulf of evaluation and the gulf of execution, respectively (Norman, 1986). Typically in the literature, visualizations are described and categorized per-user task such as exploring, finding, comparing, and recognizing (patterns). These tasks are common in interactive 3D worlds as well. By Norman, information objects should be depicted with affordances for such activities.

48.2.1.1 Visual Markers

The nature of visual perception is obviously a crucial factor in the design of effective graphics. The challenge is to understand human perceptual dimensions and map data to its display in order that dependent variables can be instantly perceived and processed preconsciously and in parallel (Friedhoff & Peercy, 2000). Such properties of the visual system have been described (i.e., sensitivity to texture, color, motion, depth), and graphical presentation models have been formulated to exploit these properties, such as preattentive processing (Pickett, Grinstein, Levkowitz, & Smith, 1995; Treisman & Gormican, 1988) and visual cues and perception (Keller, 1993).

TABLE 48.1
Accuracy Rankings for Visual Markers by General Data Type

Data Type	Quantitative	Ordinal	Nominal
Graphical representation	Position	Position	Position
	Length	Density	Color
	Angle/slope	Color	Texture
	Area	Texture	Connection
	Volume	Connection	Containment
	Color/density	Containment	Density
	(Cleveland & McGill, 1984)	Length	Shape
		Angle	Length
		Slope	Angle
		Area	Slope
		Volume	Area
		(Mackinlay, 1986)	Volume
			(Mackinlay, 1986)

General types of data can be described as quantitative (numerical), ordinal, and nominal (or categorical). Visualization design requires the mapping of data attributes to *visual markers* (the graphical representations of those attributes). Information mappings to visualizations must be computable (they must be able to be generated by a computer), and they must be comprehensible by the user (the user must understand the rules that govern the mapping in order to interpret the visualization). The employment of various visual markers can be defined by the visualization designer or defined by the user.

Tools such as Spotfire (Ahlberg & Wistrand, 1995) and Snap (North & Schneiderman, 2000) are good examples of expanding this interactive user control over the mapping, display, and coordination of views. Open source tools such as ParaView (Squillacote & Ahrens, 2006) provide capable user control over virtually every aspect of the pipeline through a graphical user interface (GUI) and a Python-scriptable interface. In addition, a set of *modes* of interaction have been proposed for exploratory data visualizations, which attempt to account for user feedback and control in a runtime display (Hibbard, Levkowitz, Haswell, Rheingans, & Schoeder, 1995). For the publishing of interactive 3D information visualizations, Polys described several paradigms for mapping data attributes to the scene graph through Extensible Markup Language (XML) and Extensible 3D (X3D) (Polys, 2003, 2005). Table 48.1 summarizes the ordering of visual markers by accuracy for the general data types. These rankings lay a foundation for identifying parameters that increase the information bandwidth between visual stimuli and user.

48.2.2 ATTENTION

Treisman and Gormican (1988) noted extensive human processing of unattended sources of information and articulated a robust theory called *preattentive processing theory*. *Preattentive perceptual processing* is involuntary, parallel, and efficient. The efficiency advantages of automatic processes make them a desirable target for certain aspects of visualization. By leveraging the preattentive processes of perception, we can make some tasks, such as outlier detection in visual search, much easier (e.g., Ware, 2000, 2003). However, some aspects of complex task performance should not be automatized in order to guarantee the user's sensitivity and flexibility to novel situations. These aspects of performance should remain controlled and receive proper attentional resources. In contrast to automatized processes, controlled processes can be characterized as declarative, serial, and explicitly managed by trainable conscious or *top-down* strategies (Gopher, 1996).

Management of attentional resources can be determined by the environment but also by the *user strategy*. The striking effects of the contrast between automatic attentional processes and those guided by top-down or instructional processes are clear in Simons work on attentional capture and inattentional blindness (Simons, 2000). When instructed or given one kind of stimuli, another unexpected type may go unnoticed. In retrospect or under different instruction, the same unexpected stimuli are obvious. The perceptual system can be high-jacked by top-down control of attention, sometimes resulting in the phenomena described as *attentional blink* (Rensink, 2000). The human perceptual system can also be primed for detection of spatial and linguistic stimuli nonconsciously (at a presemantic level) (Tulving & Schacter, 1990).

As Green and Bavelier have shown (Green & Bavelier, 2003), a minimal *practice* period of 10×1 hour sessions on first-person, 3D-action video games (e.g., *Medal of Honor*) can significantly improve user performance in attentional enumeration as measured by tests of *useful field of view* (UFOV) and attentional blink. Interestingly, this effect was not observed in subjects trained with Tetris (an exocentric, 2D spatial puzzle game). Clearly, there are rich dynamics between bottom-up and top-down processing, and these have a direct impact to a user's experience and the usability of an application.

48.2.3 SCIENTIFIC VISUALIZATION

The ScienceSpace Project showed that *conceptual learning* can be aided by features of immersive VEs such as their spatial, 3D aspect, their support for users to change their frames of reference, and the inclusion of multisensory cues (Salzman Dede, Loftin, & Chen, 1999). The curriculum modules included learning about dynamics and interactions for the physics of Newton, Maxwell, and Pauling. It seems likely that this advantage would also transfer to desktop courseware and applications. Indeed, education researchers have shown improved student performance by augmenting science lectures with desktop VEs including the *Virtual Cell* environment for teaching biology and the processes of cellular respiration (McClean, Saini-Eidukat, Schwert, Slator, & White, 2001; Saini-Eidukat, Schwert, & Slator, 1999; White, McClean, & Slator, 1999).

This is compelling evidence for the value of VEs as learning tools and for concept acquisition during the development of a user's mental model. For example, the New York University (NYU) School of Medicine (Bogart, Nachbar, Kirov, & McNeil, 2001) published a number of anatomy courseware modules in Virtual Reality Markup Language (VRML) that provide an information-rich interface to detailed models of the human head. The value and need for such tools have long been recognized (Farrell & Zappulla, 1989; Kling-Petersen, Pascher, & Rydmark, 1999). Recently in this vein, Google Inc. (Google, 2010) presented an interactive anatomy VE natively in the HTML 5 web browser using WebGL as the rendering engine.

In computational science and simulation, structure and function are often related over time; several visualization techniques are common. For example, mapping variable ranges to color scales, drawing isosurfaces (contours) at various thresholds in the data, tracing streamlines in the flow field, and animating these properties over time. Figure 48.1 shows an example of a scientific visualization in a VE using many of these techniques.

Glyphs are visual markers that can represent multiple variables as attributes on a single object. The preattentive power of Chernoff-face glyphs leverages human sensitivity to patterns in faces to show up to 27 separable attributes (Chernoff, 1973). Glyphs, as other shapes, have been applied to high-dimensional data sets including tensors representing material properties (Hashash, Yao, & Wotrin, 2003), brain activity circuits (Ropinski, Oeltze, & Preim, 2011), or fluid properties such as pressure and velocity. Figure 48.2 shows that a scientific visualization of a heated rotating disk temperature is mapped to color and the velocity field is represented by streamlines and glyphs (from the ParaView tutorial) exported to X3D and rendered in the VT Visionarium VisCube.

There has been much research into *volume* rendering techniques since the emergence of the field, most of which is outside of the scope of this chapter. For a general survey of volume rendering

(a)

(b)

FIGURE 48.1 **(See color insert.)** HVAC simulation for a green energy retrofit composed and rendered with ParaView, at (a) time step = 15 and (b) time step = 81. (Image: Drs. Burns, Borggaard, Herdman, Cliff, Polys [2012]; Courtesy of US Department of Energy, Washington, DC.)

techniques, see Kaufman and Mueller (2004); for a perceptual evaluation approach, see Boucheny, Bonneau, Droulez, Thibault, and Ploix (2009). Volume data are generated and used in several domains: geophysics, medical imaging, and noninvasive sensing of objects as varied as bridges, fossils, and luggage. There are a number of visualization techniques that can be applied to voxel data. For example, the ISO/IEC standard X3D 3.3 specifies several *render styles* that can be used to assign appearances to voxel data (Web3D, 2012). These render styles define the mapping of values in the volume to their visual representation; mappings can be based on transfer functions, specific material definitions, and lighting functions. Typically, volume data sets are segmented, meaning that voxels are marked as belonging to a specific group (a region of interest such as bone or kidney); groups can then be assigned with different render styles (Figure 48.3). As an additional step, the values in the volume may be used to compute an explicit mesh or surface at a given threshold (e.g., using algorithms such as marching tetrahedrons or marching cubes [Lorensen & Cline, 1987]).

FIGURE 48.2 Cone glyphs on streamlines representing fluid temperature and velocity. (Image: Nicholas Polys & Patrick Shinpaugh, 2010.)

48.2.4 Information Visualization

Card, Mackinlay, and Schneiderman have defined information visualization as "The use of computer-supported, interactive, visual representations of abstract data to amplify cognition" (Card et al., 1999, p. 7). This definition provides us with a simple starting point to describe visualization techniques as it distinguishes abstract data from other types of data that directly describe physical reality or are inherently spatial (i.e., anatomy or molecular structure). Abstract data include things like financial reports, collections of documents, web traffic records, or derived statistics, such as the distribution of genetic features per category (e.g., Figure 48.4). Abstract data do not have obvious spatial mappings or natural visible forms; thus, the challenge is to determine effective visual representations and interaction schemes for human analysis, decision making, and discovery. For a good review of information visualization applications and techniques, see Chen (2004) and Spense (2007).

The archetypal visualization pipeline has three important steps (see Figure 48.5). First, data are extracted and transformed from some raw sources into tables that contain the objects and attributes of interest. Second, these objects and attributes are mapped to some visual marker. Third, the resulting markers and structures are rendered to some camera or view. If the visualization is a high-resolution print, the process may stop there. In interactive visualization systems, the user may have control over any step in this pipeline.

Network data are a rich source for visualization challenges and breakthroughs. Nodes and edges can be used to represent the relationships and dynamics of many systems. Each kind of structure (hierarchical tree or graph), relationships (directed or undirected, containing weighted or hyper-edges, etc.), and task has a set of *inappropriate* techniques. The challenge is to find appropriate visual mappings and layout techniques that make the topology understandable and avoid the notorious *hair-ball* views. For example, network security (Fink, Muessig, & North, 2005) used a novel visualization tool to identify anomalies and patterns in router and server traffic. For hierarchies and trees, 3D cone trees (Robertson, Mackinlay, & Card, 1991), tree maps (Schneiderman, 1992), and hyperbolic distortions (Lamping & Rao, 1996) are now well-established techniques.

Chen (1999) showed that an individual's spatial cognitive aptitudes significantly determined their performance when analyzing a semantic citation network where the network was 2D and search

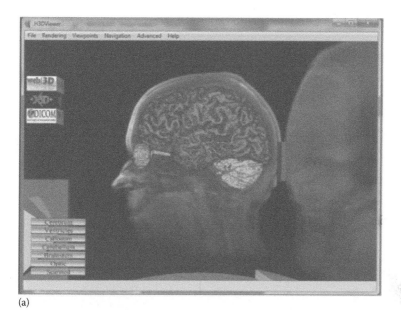

(a)

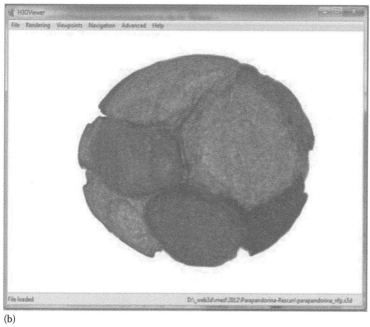

(b)

FIGURE 48.3 Volume rendering in X3D: (a) a brain MRI segmented and interactively composed (Andy Wood & Nicholas Polys, 2012) and (b) a segmented CT scan of the Parapandorina fossil (Andy Wood, Nicholas Polys, Shuhai Xiao, 2012).

relevance was shown as a bar graph on each node (the StarWalker application). Later, Geroimenko and Chen (2006) demonstrated several ontology visualization techniques with data from the semantic web. The unification of graph visualization and analysis tools into one user interface environment has also generated insightful analyses of cell-signaling pathways (Hossain, Akbar, & Polys, 2009, 2012). Figure 48.6 shows some example views of graph data in a VE (a constrained force-directed layout [left] and a radial tree layout [right]).

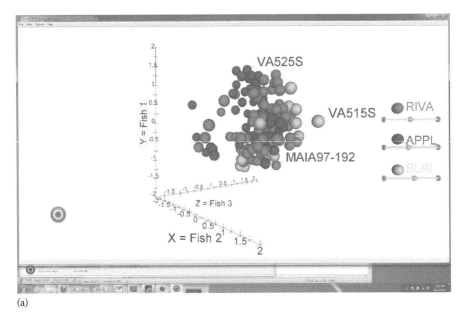

(a)

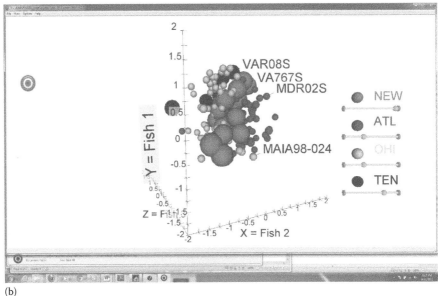

(b)

FIGURE 48.4 Interactive scatterplots rendered in X3D 'examine' mode; here users can manipulate the HUD sliders to adjust the radius of a category; (a) shows fish gene expression variables by geographical region, (b) shows it by watershed.

Geospatial data share many properties with scientific data, in that the data attribute map natural to a spatial basis—they belong to a point or region in real 3D space. Geospatially based data present several unique cognitive and usability challenges (i.e., Slocum et al., 2001). Monmonier (1990) proposed some early strategies for the visualization of geographic time-series data. As web services and application programming interfaces (APIs) become standardized for 3D georeferenced content (e.g., terrain, imagery, and other layers with features), we will see a rapid expansion of mirror worlds and *mash-up* environments such as 3D Blacksburg (Tilden, Singh, Polys, & Sforza, 2011); see Figure 48.7.

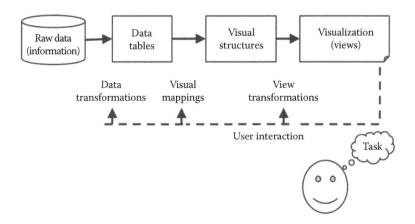

FIGURE 48.5 The archetypical visualization pipeline. (Adapted from Card, S. et al., *Information Visualization: Using Vision to Think*, Morgan Kaufmann, San Francisco, CA, 1999.)

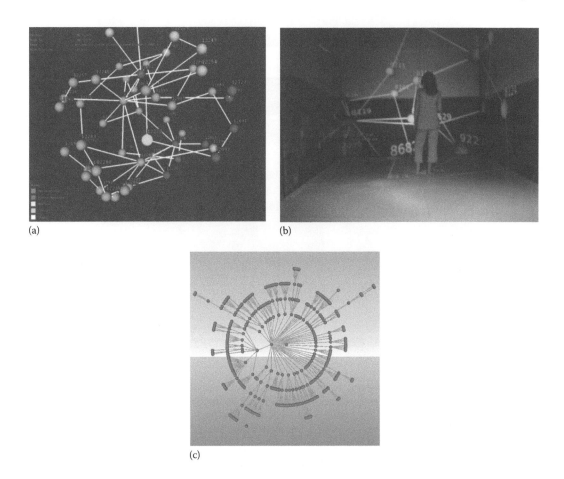

FIGURE 48.6 Network data presents special challenges since space and layout are flexible, but representing topology clearly is crucial in cell-signalling data: on a desktop (a) and in the Virginia Tech CAVE (b) (Images from Henry & Polys, 2010) (c) a radical-tree layout of an industrial product ontology (Image by Radics & Polys 2011).

(a)

(b)

(c)

FIGURE 48.7 X3D Blacksburg leveraging multiple data sources and displays: (a) Level of Detail (LOD) 1 from geospatial building footprints, (b) downtown LOD 3 with textured architectural features, and (c) coordinated multi-touch and immersive Viscube environments (Virginia Tech).

48.2.4.1 Multiple Views

A growing body of work is leveraging object-oriented software design to provide users with multiple linked views or renderings of data. North and Shneiderman (2000) showed that users can construct and operate their own coordinated multiview layouts. North (2001) has also described a taxonomy of tightly coupled views and experimental evidence supporting user performance advantages with multiple coordinated views such as a significant speedup on overview and detail tasks. Multiple visualizations can be coordinated by simple event communication where (1) selecting data items in one view <—> selects those items in other views, (2) navigating to a data item in a view <—> navigates to that data item in other views, and (3) selecting data items in one view <—> navigates the user to that data item in other views, for example. This *visualization schema* approach allows users to easily and reliably build their own coordinated multi-view visualizations (North, Conklin, Idukuri, & Saini, 2002).

Such simple coordination events allow multiple views to be customized and composed in a structured way. Roberts (1999) and Boukhelifa, Roberts, and Rodgers (2003) have described additional models for coordinating multiple views for exploratory visualization including 2D and 3D views. In Roberts' Waltz system, for example, multiform 2D and 3D visualizations of data are displayed and coordinated as users explore sets and subsets of the data. As we shall see, VEs can be coordinated with other views and this concept can be extended to visualizations embedded inside the environment. Hochheiser and Shneiderman (2004) developed and tested *TimeSearcher*, an interactive visualization tool that allows analysts to examine abstract time-series data, such as stock market or census data, with advanced queries and filters.

48.2.5 VIRTUAL ENVIRONMENTS

A VE is a synthetic, 3D or 4D world rendered in real time in response to user input. The first three dimensions are spatial and typically described in Euclidean coordinates x, y, and z. The fourth dimension is time; objects in the VE may change properties over time, for example, animating position, size, or color according to some user interaction, clock, or timeline. The data structure used to represent a VE is called a *scene graph* and consists of two parts: the *transformation hierarchy*, which describes the spatial relationship of objects and groups, and the *behavior graph*, which describes the flow of events (connections) between nodes during runtime. At each time step, the scene graph is traversed for rendering. The organization and semantics of a particular scene graph can have a direct effect on what is visible (rendered), what is interactive, and how application logic is structured. For example, the application of lighting equations to geometric shapes in the scene is often scoped to only effect siblings and their children in the scene graph; similarly, pointing and drag sensors are active over siblings and their children. Many flavors of scene graphs and their APIs have come and gone over the years.

Most notable for their durability, portability, and interoperability are the ISO/IEC standards of VRML (Web3D, 1997) and X3D (Web3D, 2012). These scene graphs are web aware and platform agnostic and have been used with a huge range of applications over the years including multiuser persistent environments (i.e., Figure 48.8). Since the mid-1990s, ISO scene graphs have taken advantage of the exponential growth of graphics and rendering hardware—not only can they still be run by many software tools (at higher frame rates than ever), they can also be run across a wide range of hardware platforms from mobile to immersive. This *spectrum of immersion* is a reality for 3D user interface (3DUI) designers. X3D provides a powerful level of abstraction for the development and deployment of virtual worlds in a network-aware (web) environment (Brutzman & Daly, 2007; Polys, Brutzman, Steed, & Behr, 2008).

While both immersive and desktop platforms may render at different resolutions and may provide stereoscopy, the common setup is that desktops are monoscopic and can support higher resolutions. There is a significant ongoing research thrust to understand the differences between VE platforms and to understand what design parameters should be changed when migrating content and applications (and why) (Bowman & McMahan, 2007). In general, 3DUIs in VEs consist of three activities—navigation, selection, and manipulation (Bowman, Kruijff, Joseph, LaViola, & Poupyrev, 2001)—and desktop and immersive systems require different sets of interaction techniques (Bowman et al., 2004). For example, desktop input devices (mice) are not tracked and have fewer degrees of freedom than those typically used in immersive settings. Without windows and icons and menus and pointers (WIMPs), designers have a new creative freedom to integrate the scene graph with novel display and input devices (i.e., Behr & Reiners, 2008).

FIGURE 48.8 An architect's model of the Virginia Tech Center for the Arts, processed to X3D and published as persistent multiuser world. (Image: Nicholas Polys & Dane Webster, 2011.)

Design principles for interaction techniques in VEs have been described in terms of performance and naturalism (Bowman, 2012). In spatial navigation, for example, the travel technique should impose minimal cognitive load on the user and be learned easily so that it can be automatized and used *second nature* (Bowman et al., 2004). Pierce et al. (1997) first leveraged user perspective and proprioception in demonstrating image plane interaction techniques for selection, manipulation, and navigation in immersive environments. A number of combinations of 3DUI interaction techniques and levels of immersion can bring about the sensation of *presence* (the subjective feeling of *being there*) in the user (Witmer & Singer, 1998). In this chapter, we will dig deeper into the research on design principles for information and interaction techniques across desktop to immersive systems.

48.2.5.1 Information-Rich Virtual Environments

Information-rich virtual environments (IRVEs) seek to unify the presentation of perceptual and abstract information displays in a natural way (i.e., Bolter, Hodges, Meyer, & Nicho, 1995). In IRVEs, the virtual space serves as a context for the methods of VEs and information visualization to be combined and so enables a unified interface for exploring the relationships between objects, space, and information. There was early evidence as to the value of enhancing perceptual/spatial information with visualizations of abstract and temporal data (e.g., Bowman, Hodges, & Bolter, 1998; Bowman, Hodges, & Allison, 1999). The theory, tools, and research agenda behind IRVEs were first formalized in Bowman et al. (2003). Each spatial item, which we call a *referent* (an object, a location, a group, people, and place), may have a variety of time-varying attributes or properties, that is, abstract and temporal information corresponding to it.

The challenge is to support users in analyzing such heterogeneous environments and understanding the relationships and patterns both *within* information types and *between* information types. IRVEs aim to render clear views of complex systems. The challenges and design space for IRVEs in desktop VEs were described in Polys and Bowman (2004) and include early techniques for visual attributes, layout attributes, and aggregation attributes that determine how abstract visualizations/annotations can be rendered with respect to their spatial referents. PathSim is an agent-based simulation to simulate the interaction of the immune system to Epstein–Barr virus

(Duca et al., 2007; Shapiro et al., 2008). In PathSim, agents travel and interact on a network approximating the physiology; thus, a visualization service framework was devised to portray the time-series simulation data in the context of the body (Polys, Bowman, North, Laubenbacher, & Duca, 2004; Polys, Kim, & Bowman, 2007). PathSim visualizer allows users to navigate a multiscale 3D environment with organs and tissue; users can view spatialized agent concentrations as numeric counts, histograms for the current time, and line graphs showing agent concentrations over time.

48.2.5.2 Display: Sizes and Resolution

Both resolution and physical size of a display play an important role in determining how much information can or should be displayed on a screen (Wei, Silva, Koutsofios, Krishnan, & North, 2000). Swaminathan and Sato (1997) examined the advantages and disadvantages of large displays with various interface settings and found that for applications where information needs to be carefully studied or modified, *desktop* settings are useful, but for collaborative, shared view and nonsustained and nondetailed work, a *distance* setting is more useful. Tan, Gergle, Scupelli, and Pausch (2003) found evidence that physically large displays aid user's performance due to increased visual immersion; Mackinlay and Heer (2004) proposed seam-aware techniques to perceptually compensate for the bezels between tiled monitors.

While a number of studies have examined the hardware and the display's (physical) field of view (e.g., Dsharp display [Czerwinski et al., 2003]), less is known about the performance benefits related with the software field of view (SFOV) and VEs. However, Draper, Viirre, Furness, and Gawron (2001) studied the effects of the horizontal field of view ratios and simulator sickness in head-coupled VEs and found that 1:1 ratios were less disruptive than those that were far off. There is also a good body of work on SFOV in the information visualization literature, typically with the goal of overcoming the limitations of small 2D display spaces. Furnas, for example, introduced generalized fisheye views (Furnas, 1981, 1986) as a technique that allows users to navigate data sets with *focus plus context*. Gutwin's recent study (2003) showed that fisheye views are better for large steering tasks even though they provide distortion at the periphery.

48.3 ACTIVITY DESIGN

Ben Shneiderman (1996) outlined a task- and data-type taxonomy for interactive information visualization. Top-level tasks for navigating and comprehending abstract information are enumerated as overview, zoom, filter, details on demand, relate, history, and extract. Overview refers to a top level or global view of the information space. Zoom, filter, and details on demand refer to the capability to *drill down* to items of interest and inspect more details (of their attributes). History refers to the *undo* capability (i.e., returning to a previous state or view), and extract is visualizing subsets of the data. Enumerated data types are 1D, 2D, 3D, multidimensional, temporal, tree, and network. Since each of these can be part of a VE, we will refer to these distinctions throughout the remainder of the proposal.

Generally, image information is best to display structure, detail, and links of entities and groups; text is better for procedural information, logical conditions, and abstract concepts (Ware, 2000). From an activity design perspective, a tool in the design process is a task-knowledge structure analysis (Diaper, 1989; Sutcliffe, 2003; Sutcliffe & Faraday, 1994), which concentrates on user tasks and the required information resource to formalize an entity-relationship model. This model enables the effective design of multimedia interfaces and information presentation by formalizing what media resources the user needs access to and when. This is an important technique for the design of IRVEs, as it intends to formally identify items that need user attention and to minimize perceptual overload and interference per task. Such an analysis can also help to identify familiar *chunks* of information that can improve cognitive and task efficiency.

Munro, Breaux, Patrey, and Sheldon, (2002) outlined the cognitive processing issues in VEs by the type of information they convey (Table 48.2). In reviewing VE presentations and tutoring systems, the authors note that VEs are especially appropriate for navigation and locomotion in complex

TABLE 48.2
Taxonomy of Knowledge Types for VE Presentations
(per Munro et al., 2002)

Location Knowledge	Structural Knowledge
• Relative position	• Part–whole
• Navigation	• Support dependent
• *How to view* (an object)	(i.e., gravity)
• *How to use* (an object access	• Containment
and manipulation)	
• Affordances (e.g., a door)	
Behavioral Knowledge	**Procedural Knowledge**
• Cause and effect	• Task prerequisite
• Function	• Goal hierarchy
• Systemic behavior	• Action sequence

environments, manipulation of complex objects and devices in 3D space, learning abstract concepts with spatial characteristics, complex data analysis, and decision making.

Ultimately, the visual analytic process of perception, interpretation, and making sense drives some actionable outcome or decision. However, there are clear patterns of human irrationality when it comes to reasoning about probabilities, seeking disproving evidence, and evaluating syllogisms, among others (Kahneman, 2003; Kahneman & Tversky, 1979). Taking the experimental psychology and cognitive science evidence in sum, there are two main sources of these cognitive biases: (1) *wrong heuristics* and (2) *lazy evaluation. Wrong heuristics* deals with the strategies users subconsciously adopt to reason under uncertainty. Typically, these constructs of the algorithmic mind are incomplete or simply erroneous. While Evans (2003) pioneered the dualist model, Stanovich (2011) has proposed a tripartite theory that distinguishes these errors as a result of *mindware gaps* or *contaminated mindware.* Several early cognitive science researchers have noted the potential for interactive computer visualization tools to mitigate these biases by replacing wrong heuristics with correct ones (e.g., Evans, 1989). Secondly, *lazy evaluation* refers to the mind's default to *miserly* information processing. For example, in several predictable situations, there are clear tendencies to ignore relevant information or to favor the most convenient explanation. One strategy to combat this is the integration of checklist tools that encourage users to seek disproving evidence for their hypotheses (i.e., Heuer, 1999). In our alien pilot scenario, we might imagine it as a supremely rational being (unlike us) or that the ship dashboard and controls are built with an advanced intelligence that can reason through complicated probabilities, chains of events, contingencies, and constraints and only give the pilot the top *reasonable* options.

48.4 INFORMATION DESIGN GUIDELINES

Through VE technology (the scene graph and interactive 3D rendering), there is a vast design space for information visualization and visual analytics. Visual markers, multiple views, immersive displays, and interaction modalities all provide a rich palette for the transformation of data to information. Facing this enormous challenge, designers *should* be intimidated. For developers and users, there are many risks but also many rewards. In this section, we will look at the latest research into the trade-offs of applying information visualization techniques for insight in VE applications.

There are two crucial properties of mappings from data to visual information: The first is that they must be computable, that is, able to be calculated by the renderer in a reasonable time; and, second, they must be comprehensible, meaning that the user must be able to invert the mapping function to determine the properties of the data that produced the visual representation. This last step is greatly aided by color legends, for example. There is also a third property that should not be underestimated: creativity. Creativity in the mapping process can lead to novel views that can further drive new insights.

48.4.1 3D RENDERING

Trade-off 1: (+) Perspective rendering can provide preattentive cues for depth and distance judgments.

(−) Perspective rendering can introduce distortions of space that make traditional methods of measurement difficult.

Guideline 1.1: If measuring X, Y, or Z position, include axis-aligned orthographic camera positions and background grids.

Rationale: Computer-aided design (CAD) and digital content creation (DCC) modeling tools do this all the time—effectively making multiple 2D views from one 3D scene. Orthogonal cameras provide a view frustum with parallel sides; therefore, objects are not drawn with perspective rendering.

Guideline 1.2: If additional accuracy is required with perspective renderings, introduce additional rendering tools such as stereoscopy or pop-out textual/numeric labels (details on demand by selection).

Rationale: Stereoscopy (binocular disparity) is a strong depth cue in *personal space* (within a few meters) and loses its effectiveness linearly with distance through the *actions space* to the *vista space* (Cutting & Vishton, 1995). Numeric labels explicitly represent the data value but require that these data values are available and can be converted to strings for display and that there is display space for the labels.

Guideline 1.3: If additional user flexibility is required with perspective renderings, introduce additional interactive tools such as head tracking, axis-aligned measuring planes, or 3D tape measure widgets.

Rationale: Head tracking (below as Guidelines 5.1 and 7.3) can provide strong, somatically congruent cues for the sense of motion parallax. Provide interactive widgets, such as 3D measuring tape, to give the user the ability to designate arbitrary objects to measure (Hagedorn & Dunkers, 2007).

Guideline 1.4: For search tasks, use a higher SFOV; for comparison tasks, use a lower SFOV.

Rationale: In general, matching the SFOV to the field of regard (FOR) is a good idea. Higher SFOVs act like a fisheye lens, rendering more of the scene but distorting the space at the periphery. Low SFOVs provide more of telescoping effect toward the focal ray. Polys, Kim, and Bowman (2005) showed that overall, users can be more accurate and faster on IRVE tasks with a higher SFOV; this is true especially on search tasks. Heads-up-displays (HUDs) or visualizations in viewport space (HUDs) can compensate for low SFOVs on crowded environments.

48.4.2 COLOR AND LIGHTING

Trade-off 2: (+) Scene graphs provide expressive parameters to define the appearance of 3D objects in an illuminated space.

(−) Real-time rendering includes shading surfaces, which can introduce perceptual artifacts distorting understanding of the underlying data.

Guideline 2.1: Avoid the rainbow color scale. For nominal and categorical data, select opponent color channels first (red, green, yellow, blue, black, and white) and then perceptually distinct colors (pink, cyan, gray, orange, brown, and purple).

Rationale: The perceptually distinct color choices listed here come from Colin Ware's excellent book *Information Visualization: Perception for Design* (2004). Borland and Taylor summarized the preattentive properties of color and luminance scales especially regarding linear perceptual ordering and sensitivity to spatial frequencies (Borland & Taylor, 2007).

Guideline 2.2: Use perceptually linear color scales for mapping. For high-frequency ordinal data, use blackbody radiation color scale (black to red to yellow to white). For interval and ratio data, consider a scale with equally sized bands of ordered colors.

Rationale: Ordering and judging distance in color spaces is something our visual systems are naturally tuned to, thus the perceptual linearity requirement. While the luminance channel is especially attuned to details, it is also sensitive to context and field effects of brightness. This is the reason why x-rays are still gray scale and why radiology reading rooms have such controlled lighting. The blackbody radiation scale was originally described in the area of ultrasound imaging (Pizer & Zimmerman, 1983) to keep this sensitivity to high spatial frequencies despite changes in the brightness of the context/surround. The X3D scene graph, for example, provides interpolation through HSV space between the RGB keyframes of the colorInterpolator node.

Guideline 2.3: If the accurate presentation of a 3D object's shape or surface is important, use an isoluminant color scale that varies in saturation, for example, from red to gray to green. Use diverging color maps to generate double-ended color maps, for example, from blue to white to red.

Rationale: Luminance information is heavily used by the visual system to judge the 3D properties of objects such as shape and curvature. Thus, Ware (2004) recommends that color scales with a constant luminance value be employed for 3D objects. The diverging color scale method as formalized by Moreland (2009) provides a good balance among the requirements for perceptual linearity, color-blind readers, and low impact on 3D object shading. These color scales are built into the ParaView toolkit.

48.4.3 MULTIPLE VIEWS

Trade-off 3: (+) Scene graphs provide additional dimensions to represent high-dimensional spaces and complex relationships among multiple variables.

(−) Working memory is limited, so relationships between views must be explicit.

Guideline 3.1: Provide clear visual feedback on how the views are related, for example, by coloring highlighted (or active) items across views consistently. In the case of multiple-view scientific visualizations, coordinate the camera positions and orientations relative to the 3D scene.

Rationale: Visual correspondences between related items across multiple views such as shared color, blinking, or motion leverage the preattentive powers of perception and load on the Gestalt association cues of similarity (color) and common fate (items across the views change in synchrony from one user action—selection). These techniques are generally known as *brushing and linking* interaction. Automatic synchronization of the virtual camera viewpoints for users avoids requiring the users to duplicate interaction or perform and maintain these 3D transformations (mental rotations) in their head.

Guideline 3.2: Evaluate multiple-view layouts by the following criteria: diversity, complementarity, parsimony, and decomposition. In addition, consider the task-knowledge structure and Gestalt principles in choosing a layout.

Rationale: Baldonado, Woodruff, and Kuchinsky (2000) originally proposed these four criteria. They also put forward four additional criteria for the presentation and interaction design of multiple-view visualization applications: space/time resource optimization, self-evidence, consistency, and attention management. Recent empirical research supports these guidelines (Convertino, Chen, Yost, Young-Sam, & North, 2003) and methodologies for designing multiple views should evaluate their design according to these criteria. Here, the task-knowledge structure (Sutcliffe & Faraday, 1994, 2003), the Gestalt principles, and the *squint test* (van der Geest & Loorbach, 2005) can be useful methods to apply.

48.4.4　Information-Rich Virtual Environments

Trade-off 4: (+) Information visualizations can be spatially referenced to 3D objects.

　　　　　　　(–) Tighter graphic associations cause more occlusion between objects and views of their properties.

Guideline 4.1: In general, the following should be applied:

A. Choose visibility over occlusion and association.

B. Increase the proximity between annotations and their referents.

C. Minimize the relocation of annotations.

D. Display global and group attributes in a visible, screen-aligned display space.

E. For speed, choose legibility of annotations; for accuracy, choose relative size annotations.

F. For search tasks, choose strong connectedness; for comparison tasks, choose minimal connectedness.

Rationale: A set of experiments was conducted to understand the dynamics of the principal main trade-offs of IRVEs: the occlusion–association trade-off and the legibility–relative size trade-off (Polys, 2006; Polys, Bowman, & North, 2011). The first trade-off addresses the interaction of depth cues and Gestalt cues in IRVEs where the objects in a 3D space are augmented by additional visualizations, which we generally refer to as *annotations*. For example, a 3D part in a machine may be annotated by a text label of its name, a visualization of its temperature status, its pressure tolerances, or a time-series prediction of its failure profile under current conditions. Annotations in our view may be textual or graphic and may be corendered with their referent in any of several coordinate systems (world, object, user, viewport, and display).

In general, the evidence shows that insuring visibility of both spatial and abstract information types is one of the most important design concerns. By the empirical data, we have shown that advantageous user performance can be achieved with very few cues (the less association) in an IRVE. There are, however, particular circumstances where visual configurations of high association and high occlusion can be advantageous, specifically cases where the depth cue of occlusion and the Gestalt cue of proximity can be beneficial, for example, on large displays, high SFOVs, and tasks that require accuracy in comparisons.

The second IRVE trade-off of legibility–relative size speaks specifically to the problem of rendering the annotation and its referent with consistent depth cues versus rendering the annotation with a guaranteed scale to be legible. This work shows that overall, the legibility of annotations is more important than the depth cue of relative size. It also presents a classic user interface trade-off of speed and accuracy. The results show that when annotations are scaled for legibility, users complete the tasks faster but also less accurately than when they are rendered with the consistent depth cue of relative size. This also suggests that users can gain valuable spatial information simply by the act of navigation (to achieve legibility).

48.4.5　Platforms

Trade-off 5: (+) Increased immersion can improve spatial awareness and task performance.

　　　　　　　(–) Increased immersion can be disorienting or otherwise taxing to some users.

Guideline 5.1: Choose immersive rendering platforms with special attention to task requirements and tailor the information design to the platform.

Rationale: User task performance can be improved through effective use of various components of immersive technology, for example, by increasing the FOR (FOR = screen surround) or adding stereo rendering and 6 degree-of-freedom (DOF) tracking. Prabhat, Forsberg, Katzourin, Wharton, and Slater (2008) tested a range of immersive platforms including desktop with head tracking, desktop with stereo with head tracking, and stereo with head tracking in a four-wall CAVE; subjects were asked questions about the spatial relationships among items in volume renderings of microscopy data sets. While not tested as independent components, they found a similar pattern across data sets that increased immersion was a significant contributor to improved performance.

Schuchardt and Bowman (2007) showed that the performance on spatial search and comparison tasks regarding natural underground structures was significantly better by time and accuracy with a higher level of immersion: four walls with stereo and head tracking outperformed one wall with monoscopic rendering and no head tracking. A follow-up study was conducted, which independently varied these three components of immersion for small-scale spatial judgment tasks (Ragan, Kooper, Schuchardt, Bowman, 2013). This study showed that accuracy is improved by increasing the FOR or including head tracking; stereo and head tracking together are advantageous for time. Finally, the highest immersion condition (with all three components of immersion) had the fewest errors, while the lowest immersion condition took the longest time.

Trade-off 6: (+) Greater screen size and resolution can present more information.
　　　　　(−) Legibility and sense making requires proper management of overview and detail and focus plus context.
Guideline 6.1: On large displays, increase the proximity of visualizations/annotations to their referents.
Rationale: Yost and North (2006) evaluated the perceptual scalability of attribute-centric or space-centric visualizations on large, high-resolution displays. They found that large displays can support faster access to more attributes across several glyph types and that embedded space-centric views yielded faster performance on large displays. In Ball and North (2007), the authors note that larger displays with head and input device tracking were advantageous by speed for reading 2D maps with overlaid information visualizations and that this was due to user's preference for physical navigation over virtual navigation.

Andrews, Endert, Yost, and North (2011) provided a good overview of the challenges and opportunities with large, high-resolution displays. Summarizing their research, the authors describe some guidelines for applying information visualization design techniques to large displays, for example, balancing physical and virtual navigation, exploiting multiscale aggregations, and using embedded visualizations and legends. Evidence from the information design of IRVEs also lends support to this guideline for the tighter spatial coupling of annotations and referents on larger screens (Polys et al., 2005; Polys, Shapiro, & Duca, 2007). In addition, increased display size and resolution have been shown to significantly improve user performance in navigation, search, and comparison tasks in IRVEs (Ni, Bowman, & Chen, 2006).

48.5　INTERACTION DESIGN GUIDELINES

In recent years, the spectrum of immersion has expanded as high-end trackers become more multi-modal, more sensitive, and more accurate; on the low end, the interaction capabilities of commodity gaming and handheld devices are also growing. In the vast area between these extremes, there is a huge design space for user interaction. In this section, we will consider how VE technology can be leveraged to enable interactive information visualization applications.

48.5.1 Navigation

Trade-off 7: (+) VEs provide a natural 3D spatial basis for visual representation.

(−) Users can still get disoriented and confused in the virtual space.

Guideline 7.1

A. Include clearly named viewpoints in the scene; include an easy way for users to reset their viewpoint.

B. Consider using invisible walls to guide users toward targets.

C. Provide navigation aids such as a compass, maps, and worlds in miniature.

D. Consider magic metaphors and combinations of ego- and exocentric navigations.

E. Provide an easy means for expert users to switch between navigation modes as needed.

> **Rationale**: In an experimental evaluation of travel techniques in a maze environment, all subjects commonly travelled to higher elevations to get a bird's-eye (survey) view of their location in the environment (Bowman et al., 1999). Several metaphors and techniques have been developed over the years; a comprehensive survey of 3DUI technologies and techniques can be found in the book *3D User Interfaces: Theory and Practice* (Bowman et al., 2004).

Guideline 7.2: Strive to balance navigation techniques with the structure of the environment and the task.

> **Rationale**: There is a demonstrable interaction effect of navigation metaphors on user performance depending on the IRVE layout technique used. For example, in a crowded environment, Chen, Pyla, and Bowman (2004) compared Homer versus Go-Go navigation techniques and object versus viewport (HUD) space for annotation rendering. The HUD labels provided significantly better performance overall, and there was a significant advantage to the combination of Go-Go and HUD (Chen et al., 2004). In an experiment to assess spatial understanding of network graphs as measured by landmark, route, and survey knowledge, researchers independently varied FOR (one vs. four screens) and navigation paradigm (ego- vs. exocentric) (Henry & Polys, 2010). They found that for survey tasks, such as how many nodes of type T are in the graph, egocentric flying was significantly more accurate than exocentric orbiting across FOR. However, for route tasks, such as counting the number of nodes between nodes X and Y, the exocentric orbiting technique provided comparable accuracy on the low FOR condition.

Guideline 7.3: Use head tracking and stereoscopy to promote naturalism in perspective rendering on large-screen immersive setups.

> **Rationale**: In an early VE experiment, travel in an immersive VE was evaluated with tracked head-mounted display (HMD) and showed that view direction should be separable from travel direction (Bowman, Koller, & Hodges, 1997). Raja, Bowman, Lucas, and North (2004) and Raja (2006) evaluated immersive platforms for information visualization tasks using scatterplots and surface plots (Figure 48.9). They found speed, accuracy, and subjective trends favoring head tracking for answering questions about scatterplots and surface plots. Stereo rendering showed a similar speed improvement with scatterplots.

There have been a number of studies over the years that have evaluated the benefits of stereo and/or head tracking on human performance. Tracking technologies continue to improve in accuracy and reduce in latency and cost; new stereo rendering technologies are rapidly improving with stereo being used to deliver compelling content from live sports events to home gaming systems to feature Hollywood films! A recent study with volume analysis tasks (Laha, Schiffbauer, & Bowman, 2012) found that each of these immersive technologies (FOR, stereo, and head tracking) contributed a statistically significant positive effect on performance. It is worth noting the interaction effects

(a)

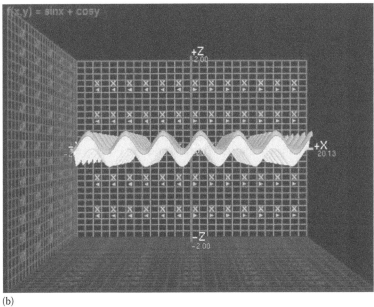

(b)

FIGURE 48.9 Images from Raja's studies—(a) immersive scatterplot and (b) equation viewer screenshot.

that were observed: it seems that for complex visual and spatial searches, FOR and head tracking together cause highly significantly more accuracy and user confidence, less time to completion, and lower difficulty ratings.

Bowman and McMahan (2007) pose the potent question "how much immersion is enough?" McMahan, Bowman, Zielinski, and Brady (2012) worked to separate the effects of display fidelity and interaction fidelity in a virtual reality (VR)-target-based shooter game. They found that certain tasks benefit more or less from immersive technologies and that mixed levels of display and interaction fidelity can be detrimental to performance. Therefore, visualization designers must consider the nuances of their application and platform when developing 3DUIs.

48.5.2 SELECTION

Trade-off 8: (+) VEs provide flexible algorithms and techniques for real-time picking.

(−) Odd-shaped, distant, and sparse objects are difficult to select or group in 3D.

Guideline 8.1: Consider magic techniques in developing 3DUIs.

Rationale: The virtual world offers the opportunity to execute our tasks more easily and efficiently than we can in the real world, where we are constrained by physics. A number of techniques are worthy of note: selection by image plane (Pierce et al., 1997), spotlights (Liang & Green, 1994), dynamic aperture spotlights (Forsberg, Herndon, & Zeleznik, 1996), and other *3D cursors* (Zhai, Buxton, & Milgram, 1994). Peck, North, and Bowman (2009) developed and tested multiscale interaction techniques on a large, high-resolution display for a hierarchical puzzle completion task. Similar to the Forsberg aperture technique, this technique scaled the size of the cursor based on the user's physical distance to the display.

Dykstra (1994) showed that VEs could be coordinated with embedded 2D windowing (X11) spaces. Snap2Diverse (Polys, North et al., 2004) used the snap-together visualization toolkit to coordinate multiple views of a molecule database where interactive 2D views were displayed on one wall and other walls displayed the coordinated 3D view. These 2D *application textures* embedded in 3D space provide quick federation and prototyping of VEs with other visualization tools and GUIs.

Guideline 8.2: Consider alternative metaphors for distant selection and for selection in crowded environs.

Rationale: In some early research on how to improve selection techniques, Wingrave, Tintner, Walker, Bowman, and Hodges (2005) examined the role of individual differences among users and found that magic techniques can be detrimental if users are highly automatized in *overlearned* tasks such as reaching. General mathematical models of 3D distal pointing have been developed and described based on experimental evidence (Kopper, Bowman, Silva, & McMahan, 2010). Pointers that can curve around obstacles have been proposed (Olwal & Feiner, 2003). *Flavors* have been proposed that extend or nuance existing techniques for some advantage, for example, raycasting selection and SQUAD selection to handle dense environments (Cashion, Wingrave, & La Viola, 2012).

48.5.3 MANIPULATION

Trade-off 9: (+) 3D input devices offer additional degrees of freedom to define natural user interactions.

(−) The metaphors and usability for 3DUIs are still an evolving research area.

Guideline 9.1: Explicitly map your device degrees of freedom; balance learnability and efficiency per application.

Rationale: Wilkes and Bowman (2008) looked at extending the *Homer* technique; with the same device and degrees of freedom, they mapped a more natural interface for selection and manipulation at a distance. The 2010 3D User Interface Contest held in association with the IEEE VR and the Symposium on 3D User Interfaces (Figueroa et al., 2010) has demonstrated several innovative approaches to the task of collecting items in a structured and crowded VE. Deep consideration of user's task seems to be the most important criterion for a successful 3DUI: *naturalism* and magic in 3DUI techniques are clearly an area where this applies (i.e., Bowman, McMahan, & Ragan, 2012).

48.6 REFLECTIONS ON *THE NEXT REALITY*

Scene graphs and 3D graphics libraries bring real expressive power to visualization design. The additional dimension of 3D in real-time spatial views provides a rich palette of visual representations for data. As VE content becomes more prevalent and scene graphs are supported natively across the web, we will continue to see development of innovative and dynamic infographics and dashboards. Building these new interfaces and applications, we must remember to *test early and often*. Pilot studies and *discount* usability methods can support quick design iterations with large gains.

Computing technology is becoming ubiquitous and is infiltrating everyday aspects of our lives from our education, research, and work to our entertainment. Mobile devices are rendering hardware-accelerated interactive 3D worlds today. Relatedly, 2D multitouch and 3D gesture interfaces are becoming more refined and widespread. It is an exciting prospect as these display and interaction devices begin to communicate in a user-centered information ecology. The WILD room (Beaudouin-Lafon et al., 2012) is an exciting example where this transparent sharing and coordination of views across display devices can bring new insights. Like sonification and visualization, we will continue to build and evaluate coordinated multimodal interfaces including haptic rendering and feedback. In addition, new perspectives such as affective computing and brain–computer interfaces (BCIs) will continue to evolve and offer value for information visualization and VE designers.

Considering the state of the art of networked devices, sensors, and VEs, the imminent horizon is the fusion of real and virtual information spaces. There are several examples of *mirror worlds* out there today: persistent virtual worlds that, through various sensors and services, reflect the objects and events in the real world. As the technology and techniques underlying augmented reality improve, it will be increasingly possible to register and composite VE graphics with objects in the real world. Indeed, we can expect to see end-user applications becoming smarter as they integrate sensor information from local and online sources to augment our cognition and situational awareness.

Technology churns with market forces and despite the clear potential, several rounds of software and companies have been born with great fanfare and then died. Across all these reinvented flavors of the wheel, the scene graph is the true enabler. Readers are encouraged to examine the flexible and expressive scene graphs such as VRML, humanoid animation (H-Anim), and X3D (Web3D, 1997, 2013). As ISO standards, this content has shown an amazing durability; virtual worlds built in 1997 are still being run across the latest hardware and operating systems with the most modern input sensors and displays.

Like the alien pilot presented in the opening example, human users browse, search, compare, look for patterns, and evaluate alternatives through visual analytic environments. Through further research into the perceptual and cognitive capacities of humans, perhaps one day the interfaces to these environments will be transparent. As we better learn how to take advantage of the natural preattentive and pattern-recognition skills of the human, we can increase the transfer of information between the computer and the mind. Thus, in my view, the grand challenge is ultimately an optimization problem: across all these channels of communication (aka the bandwidth of the senses), we are looking to increase the throughput of information across the real *last mile*—the distance between data and insight.

REFERENCES

Ahlberg, C., & Wistrand, E. (1995). *IVEE: An Information Visualization and Exploration Environment.* IEEE InfoVis, (Spotfire): www.spotfire.com.

Andrews, C., Endert, A., Yost, B., & North, C. (2011). Information visualization on large, high-resolution displays: Issues, challenges, and opportunities. *Information Visualization, 10*(4), 341–355.

Baddeley, A. (2003). Working memory: Looking back and looking forward. *Nature Reviews Neuroscience, 4*, 829–839.

Baddeley, A., & Logie, R. (1999). Working Memory: The multiple component model. In A. Miyake & P. Shah (Eds.), *Models of working memory: Mechanisms of active maintenance and executive control* (pp. 28–61). New York, NY: Cambridge University Press.

Baldonado, M., Woodruff, A., & Kuchinsky, A. (2000). Guidelines for using multiple views in information visualization. In V. D. Gesu, S. Levialdi, & Tarantino, L. (Eds.), *Proceedings of the Working Conference on Advanced Visual Interfaces (AVI)* (pp. 110–119). New York, NY: ACM.

Ball, R., & North, C. (2007). Realizing embodied interaction for visual analytics through large displays. *Computers and Graphics, 31*(3), 380–400.

Beaudouin-Lafon, M., Huot, S., Nancel, M., Mackay, W., Pietriga, E., Primet, R., ... Klokmose, C. (2012). Multisurface interaction in the WILD room. *Computer, 45*(4), 48–56.

Behr, J., & Reiners, D. (2008). Class notes: Don't be a WIMP: (http://www.not-for-wimps.org). In *ACM SIGGRAPH 2008 Classes* (pp. 1–170). Los Angeles, CA: ACM.

Bertin, J. (1983). *Semiology of graphics.* Madison, WI: University of Wisconsin.

Bogart, B., Nachbar, M., Kirov, M., & McNeil, D. (2001). VR Anatomy Courseware. New York, NY: NYU School of Medicine, NYU School of Medicine, http://www.wiki.csoft.at/vrml/skull/info.html

Bolter, J., Hodges, L. F., Meyer, T., & Nichols, A. (1995). Integrating perceptual and symbolic information in VR. *IEEE Computer Graphics and Applications, 15*(4), 8–11.

Borland, D., & Taylor, R. M. (2007). Rainbow color map (still) considered harmful. *IEEE Computer Graphics and Applications, 27*(2), 14.

Boucheny, C., Bonneau, G.-P., Droulez, J., Thibault, G., & Ploix, S. (2009). A perceptive evaluation of volume rendering techniques. *ACM Transactions on Applied Perception, 5*(23), 1–24.

Boukhelifa, N., Roberts, J. C., & Rodgers, P. (2003). A coordination model for exploratory multi-view visualization. In J. C. Roberts (ed.), *International Conference on Coordinated and Multiple Views in Exploratory Visualization (CMV 2003).* Los Alamitos, CA: IEEE.

Bowman, D., Davis, E. T., Hodges, L. F., & Badre, A. N. (1999). Maintaining spatial orientation during travel in an immersive virtual environment. *Presence: Teleoperators and Virtual Environments, 8*(6), 618–631.

Bowman, D., Hodges, L., & Bolter, J. (1998). The virtual venue: User-computer interaction in information-rich virtual environments. *Presence: Teleoperators and Virtual Environments, 7*(5), 478–493.

Bowman, D., Koller, D., & Hodges, L. F. (1997). Travel in immersive virtual environments: An evaluation of viewpoint motion control techniques. In *Proceedings of the IEEE Virtual Reality Annual International Symposium (VRAIS'97).* Washington, DC: IEEE Press.

Bowman, D., Kruijff, E., Joseph, J., LaViola, J., & Poupyrev, I. (2001). An introduction to 3-D user interface design. *Presence: Teleoperators and Virtual Environments, 10*(1), 96–108.

Bowman, D., Kruijff, E., LaViola, J., & Poupyrev, I. (2004). *3D user interfaces: Theory and practice.* Boston, MA: Addison-Wesley.

Bowman, D., North, C., Chen, J., Polys, N., Pyla, P., & Yilmaz, U. (2003). Information-rich virtual environments: Theory, tools, and research agenda. In *Proceedings of ACM Virtual Reality Software and Technology.* Osaka, Japan: ACM SIGGRAPH.

Bowman, D. A., & McMahan, P. (2007). Virtual reality: How much immersion is enough? *Computer, 40*(7), 36–43.

Bowman, D. A., McMahan, R. P., & Ragan, E. D. (2012). Questioning naturalism in 3D user interfaces. *Communications of the ACM, 55*(9), 78–88.

Bowman, D. W., Hodges, L., & Allison, D. (1999). The educational value of an information-rich virtual environment. *Presence: Teleoperators and Virtual Environments, 8*(3), 317–331.

Brutzman, D., & Daly, l. (2007). *X3D: Extensible 3D graphics for web authors.* San Francisco, CA: Morgan Kaufmann Publishers Inc.

Card, S., Mackinlay, J., & Shneiderman, B. (1999). *Information visualization: Using vision to think.* San Francisco, CA: Morgan Kaufmann.

Cashion, J., Wingrave, C., & La Viola, J. J. (2012). Dense and dynamic 3D selection for game-based virtual environments. *IEEE Transactions on Visualization and Computer Graphics, 18*(4), 634–642.

Chen, C. (1999). *Information visualization and virtual environments.* New York, NY: Springer.

Chen, C. (2004). *Information visualization: Beyond the horizon.* New York, NY: Springer.

Chen, J., Pyla, P., & Bowman, D. A. (2004). Testbed evaluation of navigation and text display techniques in an information-rich virtual environment. In *Virtual Reality.* Chicago, IL: IEEE.

Chernoff, H. (1973). The use of faces to represent Points in k-dimensional space graphically. *Journal of the American Statistical Association, 68*, 361–368.

Cleveland, W. S., & McGill, R. (1984). Graphical perception: Theory, experimentation and application to the development of graphical methods. *Journal of the American Statistical Association, 79*(387), 531–554.

Convertino, G., Chen, J., Yost, B., Young-Sam, R., & North, C. (2003). Exploring context switching and cognition in dual-view coordinated visualizations. In J. C. Roberts (ed.), *International Conference on Coordinated and Multiple Views in Exploratory Visualization*. Washington, DC: IEEE Computer Society.

Cutting, J. E., & Vishton, P. M. (1995). Perceiving layout: The integration, relative dominance, and contextual use of different information about depth. In W. Epstein & S. J. Rogers (Eds.), *Perception of space and motion: Handbook of perception and cognition*. San Diego, CA: Academic Press, Zurich, Switzerland.

Czerwinski, M., Smith, G., Regan, T., Meyers, B., Robertson, G., & Starkweather, G. (2003). Toward characterizing the productivity benefits of very large displays. In M. Rauterberg, M. Menozzi, & J. Wesson (Eds.), *INTERACT*. CA: IOS Press.

Diaper, D. (1989). Task analysis for knowledge-based descriptions (TAKD); The method and an example. In D. Diaper (Ed.), *Task analysis for human-computer interaction* (pp. 108–159). Chichester, U.K.: Ellis-Horwood.

Draper, M. H., Viirre, E. S., Furness, T. A., & Gawron, V. J. (2001). Effects of image scale and system time delay on simulator sickness within head-coupled virtual environments. *Human Factors, 43*(1), 129–146.

Duca, K. A., Shapiro, M., Delgado-Eckert, E., Hadinoto, V., Jarrah, A. S., Laubenbacher, R., … Thorley-Lawson, D. A. (2007). A virtual look at Epstein-Barr virus infection: Biological interpretations. *PLOS Pathogens, 3*(10), 1388–1400.

Dykstra, P. (1994). X11 in virtual environments: Combining computer interaction methodologies. *j-X-RESOURCE, 9*(1), 195–204.

Evans, J. (1989). *Bias in human reasoning: Causes and consequences*. Hillsdale, NJ: Lawrence Erlbaum and Associates.

Evans, J. S. B. T. (2003). In two minds: Dual processing accounts of reasoning. *Trends in Cognitive Sciences, 7*, 454–459.

Farrell, E. J., & Zappulla, R. A. (1989). Three-dimensional data visualization and biomedical applications. *Critical Reviews Biomedical Engineering, 16*(4), 323–326.

Figueroa, P., Kitamura, Y., Kuntz, S., Vanacken, L., Maesen, S., De Weyer, T., … Bowman, D. A. (2010). 3DUI 2010 contest grand prize winners. *IEEE Computer Graphics and Applications, 30*(6), 86–96, c83.

Fink, G. A., Muessig, P., & North, C. (2005). Visual correlation of host processes and network traffic. In *VIZSEC '05: IEEE Workshops on Visualization for Computer Security*. Washington, DC: IEEE Computer Society.

Forsberg, A., Herndon, K., & Zeleznik, R. (1996). Aperture based selection for immersive virtual environments. In *Proceedings of the Ninth Annual ACM Symposium on User Interface Software and Technology* (pp. 95–96). Seattle, WA: ACM.

Friedhoff, R., & Peercy, M. (2000). *Visual computing*. New York, NY: Scientific American Library.

Furnas, G. W. (1981). *The FISHEYE view: A new look at structured files*. Murray Hill, NJ: AT&T Bell Laboratories Technical Memorandum.

Furnas, G. W. (1986). Generalized fisheye views: Visualizing complex information spaces. In *ACM Proceedings of Computer Human Interaction* (pp. 16–23).

Geroimenko, V., & Chen, C. (2006). *Visualizing the semantic web: XML-based internet and information visualization*. New York, NY: Springer.

Google (2010). Body browser. http://www.zygotebody.com. Accessed on 4/7/2014.

Gopher, D. (1996). Attention control: Explorations of the work of an executive controller. *Cognitive Brain Research, 5*, 23–38.

Green, C., & Bavelier, D. (2003). Action videogame modifies visual selection attention. *Nature Reviews Neuroscience, 423*, 534–537.

Gutwin, C., & Skopik, A. (2003, April 5–10). Fisheye views are good for large steering tasks. *CHI 2003*, Ft. Lauderdale, FL.

Hagedorn, J. G., & Dunkers, J. P. (2007). Measurement tools for the immersive visualization environment: Steps toward the virtual laboratory. *Journal of Research of the National Institute of Standard & Technology, 112*(5), 257.

Hashash, Y. M. A., Yao, J. I.-C., & Wotring, D. C. (2003). Glyph and hyperstreamline representation of stress and strain tensors and material constitutive response. *International Journal for Numerical and Analytical Methods in Geomechanics, 27*, 603–626.

Henry, J. A. G., & Polys, N. F. (2010). The effects of immersion and navigation on the acquisition of spatial knowledge of abstract data networks. *Procedia Computer Science, 1*(1), 1737–1746.

Heuer, R. J. (1999). Pherson Associates LLC; 2nd Edition (2007), Military Bookshop (April 30, 2010).

Hibbard, W., Levkowitz, H., Haswell, J., Rheingans, P., & Schoeder, F. (1995). Interaction in perceptually-based visualization. In G. Grinstein & H. Levkoitz (Eds.), *Perceptual issues in visualization*. New York, NY: Springer.

Hochheiser, H., & Shneiderman, B. (2004). Dynamic query tools for time series data sets, timebox widgets for interactive exploration. *Information Visualization, Palgrave-Macmillan, 3*(1), 1–18.

Hossain, M. S., Akbar, M., & Polys, N. F. (2009). Storytelling and clustering for cellular signaling pathways. *International Conference on Information and Knowledge Engineering*, Las Vegas, NV.

Hossain, S., Akbar, M., & Polys, N. (2012). Narratives in the network: Interactive methods for mining cell signaling networks. *Journal of Computational Biology, 19*(9), 1043–1059.

Kahneman, D., & Tversky, A. (1979). Prospect theory: An analysis of decision under risk. *Econometrica, 47*, 263–291.

Kahneman, D. A. (2003). Perspectives on judgment and choice: Mapping bounded rationality. *American Psychologist, 58*, 697–720.

Kaufman, A., & Mueller, K. (2004). Overview of volume rendering. In D. C. Hansen & C. R. Johnson (Eds.), *The visualization handbook* (pp. 127–174). Academic Press, 2005. New York, NY: Elsevier.

Keller, P. R. (1993). *Visual cues: Practical data visualization*. Piscataway, NJ: IEEE Computer Society Press.

Kling-Petersen, T., Pascher, R., & Rydmark, M. (1999). Virtual reality on the web: The potential of different methodologies and visualization techniques for scientific research and medical education. *Studies in Health Technology and Informatics, 62*, 181–186.

Kopper, R., Bowman, D. A., Silva, M. G., & McMahan, R. P. (2010). A human motor behavior model for distal pointing tasks. *International Journal of Human-Computer Studies, 68*(10), 603–615.

Laha, S., Schiffbauer, J. D., & Bowman, D. A. (2012). Effects of immersion on visual analysis of volume data. *IEEE Transactions on Visualization and Computer Graphics, 18*(4), 597–606.

Lamping, J., & Rao, R. (1996). The hyperbolic browser: A focus + context technique based on hyperbolic geometry for visualizing large hierarchies. *Journal of Visual Languages and Computing, 7*(1), 33–55.

Larkin, J. H., & Simon, H. A. (1987). Why a diagram is (sometimes) worth ten thousand words. *Cognitive Science, 11*, 65–99.

Liang, J., & Green, M. (1994). JDCAD: A highly interactive3D modeling system. *Computers and Graphics, 18*(4), 499–506.

Logie, R. H. (1995). *Visuo-spatial working memory*. Hove, U.K.: Psychology Press.

Lohse, J. (1991). A cognitive model for the perception and understanding of graphs. In S. P. Robertson, G. M. Olson & J. S. Olson (eds.), *Proceedings of the SIGCHI Conference on Human Factors in Computing Systems: Reaching through Technology*. New Orleans, LA: ACM.

Lorensen, W. E., & Cline, H. E. (1987). Marching cubes: A high resolution 3D surface construction algorithm. *Computer Graphics, 21*(4), 163–169.

Mackinlay, J., & Heer, J. (2004). Wideband displays: Mitigating multiple monitor seams. In E. D. Erickson & M. Tscheligi (Eds.), *CHI*. Vienna, Austria: ACM.

Mackinlay, J. D. (1986). Automating the design of graphical presentations. *ACM Transactions on Graphics (TOG), 5*(2), 110–141.

McClean, P., Saini-Eidukat, B., Schwert, D., Slator, B., & White, A. (2001). Virtual worlds in large enrollment biology and geology classes significantly improve authentic learning. In J. A. Chambers (Ed.), *Selected Papers from the 12th International Conference on College Teaching and Learning (ICCTL-01)* (pp. 111–118). Jacksonville, FL: Center for the Advancement of Teaching and Learning.

McMahan, R. P., Bowman, D. A., Zielinski, D. J., & Brady, R. B. (2012). Evaluating display fidelity and interaction fidelity in a virtual reality game. *IEEE Transactions on Visualization and Computer Graphics, 18*(4), 626–633.

Miller, G. A. (1956). The magic number seven, plus or minus two: Some limits on our capacity to process information. *Psychological Review, 63*, 81–93.

Monmonier, M. (1990). Strategies for the visualization of geographic time-series data. *Cartographica, 27*(1), 30–45.

Moreland, K. (2009). Diverging color maps for scientific visualization. In D. Hutchinson, T. Kanade, J. Kittler, J. M. Kleinberg, F. Mattern, J. C. Mitchell, … G. Weikum (Eds.), *Advances in visual computing: Lecture notes in computer science* (Vol. 5876, pp. 92–103). Berlin, Germany: Springer.

Munro, A., Breaux, R., Patrey, J., & Sheldon, B. (2002). Cognitive aspects of virtual environment design. In Kay M. Stanney (ed.), *Handbook of virtual environments: Design, implementation, and applications*. Mahwah, NJ: Lawrence Erlbaum Associates.

Ni, T., Bowman, D. A., & Chen, J. (2006). Increased display size and resolution improve task performance in information-rich virtual environments. In C. Gutwin & S. Mann (Eds.), *Proceedings of Graphics Interface* (pp. 139–146). Toronto, Canada: Canadian Information Processing Society.

Norman, D. A. (1986). Cognitive engineering. In D. A. Norman & S. D. Draper, (Eds.), *User centered system design* (pp. 31–61). Hillsdale, NJ: Lawrence Erlbaum Associates.

North, C. (2001). Multiple views and tight coupling in visualization: A language, taxonomy, and system. *CSREA CISST Workshop of Fundamental Issues in Visualization*.

North, C., Conklin, N., Idukuri, K., & Saini, V. (2002). Visualization schemas and a web-based architecture for custom multiple-view visualization of multiple-table databases. *Information Visualization, 1*(3/4), 211–228.

North, C., & Shneiderman, B. (2000). Snap-together visualization: Can users construct and operate coordinated views? *International Journal of Human-Computer Studies, 53*(5), 715–739.

Olwal, A., & Feiner, S. (2003). The flexible pointer—An interaction technique for selection in augmented and virtual reality. In *Conference Supplement of ACM Symposium on User Interface Software and Technology (UIST '03)* (pp. 81–82). Vancouver, Canada: ACM.

Peck, S. M., North, C., & Bowman, D. (2009). A multiscale interaction technique for large, high-resolution displays. *IEEE Symposium on 3D User Interfaces, 2009 (3DUI 2009)*. Lafayette, LA.

Pickett, R. M., Grinstein, G., Levkowitz, H., & Smith, S. (1995). Harnessing preattentive perceptual processes in visualization. In G. Grinstein & H. Levkoitz (Eds.), *Perceptual issues in visualization*. New York, NY: Springer.

Pierce, J. S., Forsberg, A. S., Conway, M. J., Hong, S., Zeleznik, R. C., & Mine, M. R. (1997). Image plane interaction techniques in 3D immersive environments. In Andy van Dam (ed.), *Interactive 3D Graphics*. ACM.

Pimentel, K., & Teixeira, K. (1993). *Virtual reality: Through the new looking glass*. New York, NY: Windcrest Books/McGraw-Hill Inc.

Pizer, S. M., & Zimmerman, J. B. (1983). Color display in ultrasonography. *Ultrasound in Medicine and Biology, 9*(4), 331–345.

Polys, N., Bowman, D., North, C., Laubenbacher, R., & Duca, K. (2004). PathSim visualizer: An information-rich virtual environment for systems biology. In *Web3D Symposium*. Monterey, CA: ACM Press.

Polys, N., Brutzman, D., Steed, A., & Behr, J. (2008). Future standards for immersive VR: Report on the IEEE VR 2007 workshop. *IEEE Computer Graphics and Applications, 28*(2), 94–99.

Polys, N., Kim, S., & Bowman, D. (2007). Effects of information layout, screen size, and field of view on user performance in information-rich virtual environments. *Computer Animation and Virtual Worlds, 18*(1), 19–38.

Polys, N., North, C., Bowman, D. A., Ray, A., Moldenhauer, M., & Chetan, D. (2004). Snap2Diverse: Coordinating information visualizations and virtual environments. *SPIE Conference on Visualization and Data Analysis (VDA)*, San Jose, CA.

Polys, N. F. (2003). Stylesheet transformations for interactive visualization: Towards Web3D chemistry curricula. In C. Bouville (Ed.), *Web3D Symposium*. St. Malo, France: ACM Press.

Polys, N. F. (2005). Publishing paradigms with X3D. In V. G. Chanomei Chen (Ed.), *Information visualization with SVG and X3D*. Springer-Verlag.

Polys, N. F. (2006). Display techniques in information-rich virtual environments. *Computer Science and Applications*. Blacksburg, VA. http://scholar.lib.vt.edu/theses/available/etd-06152006-024611/Virginia Polytechnic Institute and State University. PhD: 176.

Polys, N. F., & Bowman, D. A. (2004). Desktop information-rich virtual environments: Challenges and techniques. *Virtual Reality, 8*(1), 41–54.

Polys, N. F., Bowman, D. A., & North, C. (2011). The role of depth and gestalt cues in information-rich virtual environments. *International Journal of Human-Computer Studies, 69*(1–2), 30–51.

Polys, N. F., Kim, S., & Bowman, D. A. (2005). Effects of information layout, screen size, and field of view on user performance in information-rich virtual environments. In G. Singh, R. W. H. Lau, Y. Chrysanthou, & R. P. Darken (Eds.), *Proceedings of ACM Virtual Reality Software and Technology*. Monterey, CA: ACM SIGGRAPH.

Polys, N. F., Shapiro, M., & Duca, K. (2007). IRVE-serve: A visualization framework for spatially-registered time series data. In O. Gervasi & D. P. Brutzman (Eds.), *Proceedings of the Twelfth International Conference on 3D Web Technology*, Web 3D 2007, Perugia, Italy, April 15–18, 2007. ACM 2007, 137–145.

Prabhat, Forsberg, A., Katzourin, M., Wharton, K., & Slater, M. (2008). A comparative study of desktop, fishtank, and cave systems for the exploration of volume rendered confocal data sets. *IEEE Transactions on Visualization and Computer Graphics, 14*(3): 551–563.

Ragan, E., Kopper, R., Schuchardt, P., & Bowman, D. A. (2013). Studying the effects of stereo, head tracking, and field of regard on a small-scale spatial judgment task. *IEEE Transactions on Visualization and Computer Graphics, 19*(5), 886–996.

Raja, D. (2006). The effects of immersion on 3D information visualization. *Computer science*. Blacksburg, VA: Virginia Tech.

Raja, D., Bowman, D. A., Lucas, J., & North, C. (2004). Exploring the benefits of immersion in abstract information visualization. *Eighth International Immersive Projection Technology Workshop*, Iowa State University, Ames, IA.

Rensink, R. A. (2000). Seeing, sensing, and scrutinizing. *Vision Research, 40*, 1469–1487.

Roberts, J. C. (1999). On encouraging coupled views for visualization exploration. *Visual Data Exploration and Analysis VI, Proceedings of SPIE, IS&T and SPIE*, San Jose, CA.

Robertson, G. G., Mackinlay, J. D., & Card, S. K. (1991). Cone trees: Animated 3D visualizations of hierarchical information. In S. P. Robertson, G. M. Olson, & J. S. Olson (Eds.), *CHI* (pp. 189–194). New York, NY: ACM.

Ropinski, T., Oeltze, S., & Preim, B. (2011). Survey of glyph-based visualization techniques for spatial multivariate medical data. *Computers & Graphics, 35*(2), 392–401.

Saini-Eidukat, B., Schwert, D. P., & Slator, B. M. (1999). Designing, building, and assessing a virtual world for science education. *International Conference on Computers and Their Applications*, Cancun, Mexico.

Salzman, M. C., Dede, C., Loftin, B. R., & Chen, J. (1999). A model for understanding how virtual reality aids complex conceptual learning. *Presence: Teleoperators and Virtual Environments, 8*(3), 293–316.

Schneiderman, B. (1992). Tree visualization with tree-maps: A 2-dimensional space filling approach. *ACM Transactions on Graphics, 11*(1), 92–99.

Schuchardt, P., & Bowman, D. A. (2007). The benefits of immersion for spatial understanding of complex underground cave systems. In A. Majumder, L. F. Hodges, D. C. Or, & S. N. Spencer (Eds.), *Proceedings of the 2007 ACM Symposium on Virtual Reality Software and Technology* (pp. 121–124). Newport Beach, CA: ACM.

Shapiro, M., Duca, K. A., Lee, K., Delgado-Eckert, E., Hawkins, J., Jarrah, A. S., … Thorley-Lawson, D. A. (2008). A virtual look at Epstein-Barr virus infection: Simulation mechanism. *Journal of Theoretical Biology, 252*(4), 633–648.

Sheppard, L. M. (2004, January/February). Virtual building for construction projects. *IEEE Computer Graphics and Applications, 24*, 6–12.

Shneiderman, B. (1996). The eyes have it: A task by data type taxonomy for information visualizations. *Proceedings of IEEE Visual Languages*, Boulder, CO.

Simons, D. J. (2000). Attentional capture and inattentional blindness. *Trends in Cognitive Sciences, 4*, 147–155.

Slocum, T. A., Blok, C., Jiang, B., Koussoulakou, A., Montello, D. R., Fuhrmann, S., & Hedley, N. R. (2001). Cognitive and usability issues in geovisualization. *Cartography and Geographic Information Society Journal, 28*(1), 61–75.

Sorby, S. A. (2009). Educational research in developing 3-D spatial skills for engineering students. *International Journal of Science Education. Special Issue: Visual and Spatial Modes in Science Learning, 31*(3), 459–480.

Spense, S. (2007). *Information visualization: Design for interaction.* New York, NY: Pearson.

Squillacote, A. H., & Ahrens, J. (2006). *The ParaView guide.* Clifton Park, NY: Kitware.

Stanovich, K. E. (2011). *Rationality and the reflective mind.* New York, NY: Oxford University Press.

Sutcliffe, A. (2003). *Multimedia and virtual reality: Designing multisensory user interfaces.* Mahwah, NJ: Lawrence Erlbaum and Association.

Sutcliffe, A., & Faraday, P. (1994). *Designing presentation in multimedia interfaces.* New York, NY: ACM Press.

Swaminathan, K., & Sato, S. (1997, January). Interaction design for large displays. *ACM Interactions, 4*, 15–24.

Tan, D., Gergle, D., Scupelli, P., & Pausch, R. (2003). With similar visual angles, larger displays improve performance. In *Proceedings of the SIGCHI Conference on Human Factors in Computing Systems* (pp. 217–224). New York, NY: ACM.

Thomas, J. J., & Cook, K. A. (2006, January/February). A visual analytics agenda. *IEEE Computer Graphics and Applications, 26*, 10–13.

Tilden, D., Singh, A., Polys, N. F., & Sforza, P. (2011). Multimedia mashups for mirror worlds. In J. Royan, M. Preda, T. Boubekeur, & N. F. Polys (Eds.), *Proceedings of the 16th International Conference on 3D Web Technology (Web3D '11)*. Paris, France: ACM SIGGRAPH.

Treisman, A., & Gormican, S. (1988). Feature analysis in early vision: Evidence from search asymmetries. *Psychological Review, 95*(1), 15–48.

Tufte, E. (1983). *The visual display of quantitative information.* Cheshire, CT: Graphics Press.

Tufte, E. (1990). *Envisioning information.* Cheshire, CT: Graphics Press.

Tulving, E., & Schacter, D. L. (1990). Priming and human memory systems. *Science, 247*(4940), 301–306.

van der Geest, T., & Loorbach, N. (2005). Testing the visual consistency of web sites. *Technical Communication, 52*(1), 27–36.

Ware, C. (2000). *Information visualization: Perception for design.* New York, NY: Morgan Kauffman.

Ware, C. (2003). Design as applied perception. In J. M. Carroll (Ed.), *HCI models, theories, and frameworks: Towards a multidisciplinary science* (pp. 10–26). San Francisco, CA: Morgan-Kaufmann.

Ware, C. (2004). *Information visualization: Perception for design*. San Francisco, CA: Morgan Kaufmann.

Web3D, C. (1997). Virtual Reality Modeling Language (VRML). http://www.web3d.org/realtime-3d/specification/all (retrieved 4/14/204) as ISO/IEC 14772.

Web3D, C. (2013). Extensible 3D Graphics (X3D). http://www.web3d.org/realtime-3d/specification/all (retrieved 4/14/2014) as ISO/IEC 19775, 19776, 19777.

Wei, B., Silva, C., Koutsofios, E., Krishnan, S., & North, S. (2000). Visualization research with large displays. *IEEE Computer Graphics and Applications, 20*(4), 50–54.

White, A. R., McClean, P. E., & Slator, B. M. (1999). The virtual cell: An interactive, virtual environment for cell biology. *World Conference on Educational Media, Hypermedia and Telecommunications (ED-MEDIA 99)*, Seattle, WA.

Wilkes, C., & Bowman, D. A. (2008). Advantages of velocity-based scaling for distant 3D manipulation. In S. Feiner, D. Thalmann, P. Guitton, B. Frohlich, E. Kruijff, & M. Hachet (Eds.), *Proceedings of the 2008 ACM Symposium on Virtual Reality Software and Technology* (pp. 23–29). Bordeaux, France: ACM.

Wingrave, C. A., Tintner, R., Walker, B. N., Bowman, D. A., & Hodges, L. F. (2005). Exploring individual differences in raybased selection: Strategies and traits. In *Proceedings of the Virtual Reality 2005, VR 2005*. IEEE.

Witmer, B. G., & Singer, M. J. (1998). Measuring presence in virtual environments: A presence questionnaire. *Presence: Teleoperators and Virtual Environments, 7*(3), 225–240.

Yost, B., & North, C. (2006). The perceptual scalability of visualization. *IEEE Transactions on Visualization and Computer Graphics, 12*(5), 837–844.

Zhai, S., Buxton, W., & Milgram, P. (1994). The "silk cursor": Investigating transparency for 3d target acquisition. In B. Adelson, S. T. Dumair, & J. S. Olson (Eds.), *Proceedings of the SIGCHI Conference on Human Factors in Computing Systems* (pp. 459–464). New York, NY: ACM.

Zhang, J., & Norman, D. A. (1994). Representations in distributed cognitive tasks. *Cognitive Science, 18*, 87–122.

49 Entertainment Applications of Virtual Environments

Adams Greenwood-Ericksen, Robert C. Kennedy, and Shawn Stafford

CONTENTS

49.1 INTRODUCTION

> Virtual environment (VE): three-dimensional data set describing an environment based on real world
> or abstract objects and data. Usually virtual environment and virtual reality are used synonymously, but
> some authors reserve VE for an artificial environment that the user interacts with (Blade & Padgett, 2002).

Many researchers in the area of virtual environments (VEs) might be surprised to find that according
to most definitions of *VE*, the first video game, *Tennis for Two* (Higinbotham, 1958), may well also
have been the first digital VE. It had many of the characteristics we associate with VEs today. It was
artificial in nature, but represented a physical 3D real-world space (a tennis court) using a 2D display
(an oscilloscope) in which two users could interact with the VE and each other simultaneously by
means of crude input devices and a rudimentary graphical user interface.

49.1.1 WHAT'S IN A NAME?

Obviously, we see very few technological similarities between the thousand-pound apparatus
that nuclear physicist William Higinbotham constructed on the campus of Brookhaven National
Laboratory and the latest 3D full-immersion VR systems or current-generation first-person perspective video games. Nevertheless, video games and VEs have been closely linked throughout
their mutual history, and it may well still remain the case that the majority of VE systems in the
world are video games. This is due at least in part to the fact that much of the technology that drives
VEs and video games has been shared throughout their mutual history, and indeed, it is often hard
to tell the difference between the two. As graduate students and researchers, two of the authors of
this chapter routinely built VEs for research purposes that were based on commercial off-the-shelf
video games.

Indeed, it has been argued that in many cases, the *purpose* of an interactive device is the only
characteristic that distinguishes a *serious* VE from a video game and, even then, that may change
periodically with the needs and intent of the user (Prensky, 2001). For the sake of brevity, then,
it may possibly be more valuable to dispense with unhelpful semantic debates and simply view
video games (or most video games, at any rate) as being a partially overlapping subset of VE
intended by their developers to produce an entertaining, or at least engaging, interactive experience for the user.

Practitioners and researchers in game development, for their part, do not seem to have been
particularly worried by definitional issues as far as VEs are concerned. Game designers have
invested much time and toil in pursuit of new and previously unrealized applications of VE technologies to drive novel and entertaining experiences. Ultimately, from the perspective of game
designers and developers, it probably does not really matter whether a product is *called* a VE or
a video game so long as it achieves its intended purpose of entertaining (or at least engaging)
its users.

49.1.2 VIDEO GAMES AS ENTERTAINMENT-DIRECTED VIRTUAL ENVIRONMENTS

It is clearly the case that video games represent an intensely lucrative application of VE technology.
The Entertainment Software Association (ESA), the leading trade organization for the video game
development industry, estimated that industry revenue generated in 2011 exceeded $25 billion
(Entertainment Software Association, 2012) and that the industry has been growing at an average
of 10% per year since they began collecting economic growth data in 2005. By comparison, the
overall US economy grew at less than 2% per year during this period (Siwek, 2010). Moreover, with
thousands of professionally developed games currently on the market, it could well also be the case
that video games represent the most common application of VE technology as well.

Bearing this in mind, it appears that the pertinent question is not how to draw a distinction between
VEs and video games but rather how to design VEs that are intended for entertainment purposes so

that they achieve their objective. In essence, how do we build VEs so that they function, to a greater or lesser degree, as video games, regardless of any secondary purposes for which they might be intended?

If it is the intent of the developer in applying VE technology to develop a product whose primary purpose is to entertain, then it seems that the overall goal is to keep the player engaged, active, and immersed in the VE, regardless of other secondary considerations such as exposure to learning material or progress toward training objectives.

The authors have noted anecdotally that it appears that research appearing in simulation and VE-themed publications is often highly focused on hardware development and configuration and deals with software primarily in the context of how it works with the hardware. By contrast, game developers and designers tend to focus far more on the characteristics and features of the software and leave consideration of the hardware up to the user, often offering products that work on many different hardware configurations and architectures. Previous VE research into the design features of video games has broadly identified a number of key elements that characterize successful games, such as fun, play, rules, and goals (Mitchell & Savill-Smith, 2004). However, these individual elements tend to be scattered and only weakly supported.

Similarly and simultaneously, researchers and practitioners in the game industry and academia have been working in parallel to develop an understanding of what makes video games *fun* and to identify the factors that would drive successful designs to achieve this. Among the earliest researchers in the field whose ideas achieved broad recognition and acceptance was Thomas Malone, who in 1981 identified three broad factors in successful video games: challenge, fantasy, and curiosity (Malone, 1981). This tripartite model of game success remains influential and fits well with well-known academic models of immersive experience that are heavily used in the study of VEs and other immersive simulations, including the popular SCI model (Ermi & Mayra, 2005).

These components provide a reference point for addressing the less tangible elements of design for entertainment-focused VEs. They are especially valuable in that they allow us to link research related to existing constructs from the behavioral sciences to the design of successful VE games. Broadly speaking, these three factors correspond to three important constructs drawn from the behavioral sciences and will form the basis of an effectively designed entertainment VE.

First, the element of challenge is associated with gameplay, which typically is an interactive experience composed of repetitive task elements of various levels of difficulty. As with any repetitive task, achieving both high performance and motivation is a function of training. Techniques drawn from simple and operant conditioning approaches can be usefully applied here to draw in and maintain user engagement in the tasks that make up the game (Grimes, 2011).

Second, the experience that Malone refers to as *fantasy* is the induction in the user of a sense of physical presence or imaginative immersion in the artificial environment. A great deal of literature deals with the areas of presence and immersion, and much of this research is relevant to us in achieving and maintaining both user presence (Weibel & Wissmath, 2011) and the impression of control (Maguire, van Lent, Prensky, & Tarr, 2002).

Finally, creating and sustaining motivation to remain engaged with the VE is critical (BECTA, 2006). Leveraging the element of curiosity through continuous delivery of new and targeted VE content to users is a necessary ongoing process in generating and sustaining motivation, and identifying the specific needs and interests of users is critical if designers are to successfully achieve the objective of immersing users in the environment (Bartle, 1996).

The present work, therefore, will focus on how to generate the three key requirements for designing successful entertainment-oriented VE games:

1. Developing challenging and compelling gameplay based on behaviorist principles
2. Fostering subjective immersion through both the setting and presentation of the VE game
3. Supporting and maintaining player motivation while interacting with the VE game, including the role of individual preferences and personality differences

By addressing these three areas, it should be possible to leverage existing findings in the behavioral sciences to provide practitioners and researchers with a basic overview of the requirements to design and develop successful VE games.

49.2 BEHAVIORAL CONDITIONING AND GAMES

By studying human behavior at its most basic level, the behaviorist school of psychology sought insight into the ways in which the human animal responds to elementary stimuli in neutral, positive, or negative ways and how these interactions can be leveraged to create desirable behaviors. In the context of games, a great deal of the compelling nature of gameplay can be explained through the workings of classical and operant conditioning.

49.2.1 CLASSICAL CONDITIONING

In 1899, Petrovich Pavlov began to develop a set of principles that would later be formalized as classical conditioning (Pavlov, 1927, 1955). Classical conditioning is a staple of behavioral and conceptual learning for players in video games and VEs. The premise of classical conditioning is that it is possible to manipulate the time and location in which an unconditioned stimulus (US) occurs if it is naturally paired with an unconditioned response (UR) and furthermore that this connection is biological in nature (see Figure 49.1) (Table 49.1). This relationship is biological, thus natural, and no training is needed. That is, a US produces a reflexive UR.

Although the focus of this chapter is not on the physiological elements of behavior, it is important to acknowledge the role of the human hypothalamus in creating the footprint for the US–UR

FIGURE 49.1 Basic US to UR relationship.

TABLE 49.1

Guidelines to Support the Use of Classical Conditioning in VE Games

1. US and CS should be paired on multiple occasions that are close in time to enhance the possibility that conditioning will occur.
2. US and CS relationships produce stronger URs when the US and CS are centered in human attention so that focused conditioning occurs. A peripheral CS might not ever produce a CR because it might never be noticed or attended to.
3. The critical nature of human discrimination may prevent a CR from being associated with a CS in some situations. This may occur if the original environment surrounding the initial pairing of US and CS is too different from the subsequent environment where the CS is presented alone (without the US) or if the user is not aware of the CS or interprets the CS as a different stimulus.
4. The human cognitive habit of generalization may also cause an unintended CS–CR relationship to occur. If the CS is very similar to other stimuli in the game, it is possible that a neutral or opposite stimulus may produce an unintended emotional response in the player.
5. Weakly conditioned US and CS relationships may cause a process called extinction, whereby the CR ceases to manifest entirely. The beginning of extinction can be measured by observing a weakening of the magnitude of the CR over time.
6. Classical conditioning is considered a permanent change in human memory. Even after extinction of the CR, it is possible that a CR may spontaneously reoccur in the future. This effect is termed *spontaneous recovery*. Spontaneous recovery can be triggered by environmental cues or other cues that activate the original US and CS pairing memory trace (a neural connection that remains in memory even if the associated behavioral link is no longer observable).

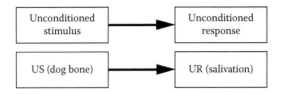

FIGURE 49.2 Typical example given for US to UR relationship.

relationship within humans. Fighting, fear, food, and sex can all create natural US and UR pairings. The classic textbook example of a US and UR pairing is that of a dog bone (the US) that causes a dog to salivate (the UR) (see Figure 49.2). Unconditioned (U) simply means that without training the dog, it is possible to present a stimulus (dog bone, S) and expect a response (salivation, R).

Humans are certainly more complex than animals when it comes to the process of classical conditioning. Humans can explicitly train themselves to become sensitized to food types, for instance, thus resisting a measurable behavioral change like salivation. Though some food recognition pathways in the brain might become active, little to no impact on behavior will be measured in a scientific experiment. However, we do know that fighting, fear, and sexual presentations will generally produce high levels of arousal in a positive or negative valence vector. Simply put, natural US components in nature are powerful biological tools that can be used to make the user of a VE game (the player) feel some particular level of arousal in conjunction with a directional component related to a positive or negative subjective mood state (see Figure 49.3). These items are often measured in game research using, but not limited to, such techniques as galvanic skin response (GSR) and electromyography (EMG) (Carnagey, Anderson, & Bushman, 2006; Gilleade & Allanson, 2005; Nacke & Lindley, 2008; Sykes & Brown, 2003). Figure 49.3 shows the relationship between common subjective mood states and the associated levels of positive or negative arousal magnitude and valence.

Imagine that we measure or have knowledge that a video game player is feeling something (the UR) as the result of fearful virtual scene (the UC): for instance, that the scene is dark, low frequency noises are present, and the space is occupied by strangers. We can then employ our classical conditioning techniques by allowing this experience to continue for an optimum amount of time to reinforce the salience of the effect (see Figure 49.4).

In the classic experiment by Pavlov (1955), he paired the ringing of a bell with the presentation of the dog bone to the dog. Memory is encoded in human and animal brains such that two things seen together are physically connected via neural pathways. The more often this dog bone (the US) and bell are paired

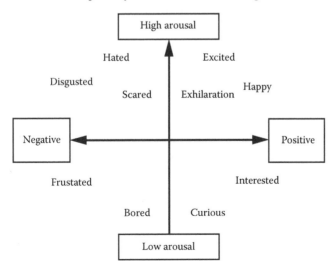

FIGURE 49.3 Arousal and valence (negative/positive).

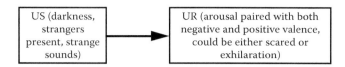

FIGURE 49.4   Creating a feeling in a game player.

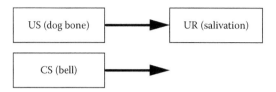

FIGURE 49.5   Natural relationship between a dog bone, a bell, and a dog's salivary response.

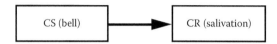

FIGURE 49.6   Relationship after conditioning: salivation would still occur in the absence of a bone.

together, the more likely it becomes that we can remove the dog bone in our experiment and still observe drool produced in the dog's mouth. We have vast amounts of information about why this happens and would direct the reader to any physiological psychology textbook for further research on the memory connections produced by classical conditioning (for instance, see Carlson, 2012). The bell in Pavlov's experiment was named the conditioned stimulus (CS) (see Figure 49.5). Only when the dog bone (the US) was removed after many simultaneous pairings with the bell (the CS) was Pavlov able to confirm that the dog still displayed the UR (salivation) despite the absence of the US (the bone). This new salivation (which biologically is still the same salivation that would result from a dog viewing a presumably tasty bone) as a result of hearing a bell would be considered a conditioned response (CR) (see Figure 49.6).

Thus, initially, we would observe the following natural relationships:

When the bell stimulus (the CS) is presented in conjunction with the dog bone (the US) for many trials, a link is formed behaviorally between the CS and the US. Eventually, it is possible to remove the dog bone and observe the following relationship instead:

Thus, we have associated the original UR with a new stimulus (the CS). This CS and CR relationship is the result of memory trace pairings within the brain.

The US in a video game is not as intuitive to understand as are real stimuli in the real world. However, if we agree that humans immerse themselves in games by projecting reality into the virtual world (see Section 49.3), then we can readily argue that a virtual US can be as impactful on a human as a real-world US allowing for conditioning to occur. This is exactly what we see in video games.

Recently, while playing a video game, I came across one of many salient examples of classical conditioning. The scene presented the player with a dimly lit environment in a train station. Throughout the train station could be found sick strangers who spoke incomprehensible phrases to walls and to themselves (picture escaped inmates from a mental institution who have lost the ability to do anything other than simply wander about muttering to themselves). Low-frequency noises with the occasional high-frequency metal against metal dropped crowbar sound populated the environment. Against the wall was a campaign poster for the arch villain of the game. This train station scene lasted for several minutes with several overtly placed posters throughout. A very simple way to connect a negative feeling with a face (later to be identified as the arch villain) is by presenting the villain at the same time and place with very basic unconditioned stimuli that populated the train station. The result is that the player feels some type of negative emotion (the UR) when the image of the arch villain (the CS) is seen in the subsequent levels of the game. This is a useful tool for VE game designers to induce desired emotional responses in users to further gameplay and create story.

49.2.2 Recent Research in Classical Conditioning

Recent research in classical conditioning is shedding light on CRs such as the feeling (CR) the player gets after being negatively conditioned to a poster (CS). We have had knowledge for many years that viewers are often somewhat conditioned even when they are fully aware of the intended manipulation (Razran, 1940) such as in the case of a commercial jewelry company that is employing a psychological manipulation technique pairing beautiful happy people with an expensive ring that they appear to be purchasing. However, humans are generally quite intelligent and are able to easily monitor media such as television commercials for the overt classical conditioning techniques used by advertisers. For instance, most viewers immediately realize that a particular ring does not automatically bring feelings of happiness even though that couple in the commercial look like they are having a really great time buying it. As such, we must look at the intention of the humans and their willingness to suspend their critical nature (sometimes called discrimination) in each and every situation they encounter throughout their day. In general, humans enjoy games and they want to experience emotional feelings that are both good (love, fun, excitement) and bad (afraid, shocked, awe) in nature. If we can account for factors such as anxiety and willingness to experience a particular emotion, we might be able to determine why later reactions to an arch villain poster are much stronger than the initial fear of being in a scary place (the train station). Stress, fear, and the player's need to process the story may combine in an active memory reinforcement system where the player is reliving the initial pairing of the US–UR/CS presentation. This would occur in what we call active imagination (what the player is thinking) and is quite possibly associated with high levels of arousal and negative and or positive valence, thus producing a player-induced and powerful CR response. However, we also need to acknowledge the vast amount of research into the biology of classical conditioning and remember that conditioning can be permanent (Bouton, 1994; Rumpel, LeDoux, Zador, & Malinow, 2005) and therefore potentially confusing during memory recall as new, potentially contradictory, conditioning memory traces are added to existing relationships over time.

Classical conditioning is likely to remain a very powerful technique in games and VEs regardless of whether a designer has knowledge that they are employing the technique or not. However, some basic principles that are drawn from classical conditioning and can be used to influence users of a VE game are included in the summary at the end of this chapter.

49.2.3 Operant Conditioning

Although some scholars may disagree (Rescorla, 1988), classical conditioning is commonly viewed as a natural and passive form of conditioning that only requires time and space, whereas operant conditioning is considered to be active in nature and generally requires manipulation by outside forces. Whether shaping behavior or producing habits, the process of operant conditioning is a simple and powerful system of manipulations. Using these techniques, it is possible to reinforce or punish behaviors (see Figures 49.7 and 49.8). A reinforcer is something, usually an experience, which is fulfilling or satisfying. The reinforcer can fill a deep psychological need (social rewards, emotionally gratifying rewards) or can be a simpler reward such as food, money, or sex. By contrast, a punishment is something that is unpleasant such as a spanking or negative social comment. It is obvious that both of these terms only become relevant when we can determine what is fulfilling or satisfying in a

FIGURE 49.7   Reinforcers improve emotional states.

FIGURE 49.8   Punishers decrease pleasant states.

given population that we are trying to manipulate. Before we move to punishment, let us first discuss reinforcement and its relation to games and VEs as informed through basic research (Table 49.2).

Historically, basic research into operant conditioning is drawn primarily from work on rats. If a rat produces a behavior that we want to encourage, then we provide the rat with cheese immediately after the rat produces the said behavior (see Figure 49.9). If a rat produces a behavior we want to discourage, then we shock the rat immediately after the rat produces the behavior (see Figure 49.10). The implications are obvious when we need to shape behavior in games or VEs.

TABLE 49.2
Guidelines to Support the Use of Operant Conditioning in VE Games

1. Only when the behavior and reward relationship is obvious do we encourage a behavior. If people do not know why they get a score or benefit, then we will not be able to properly influence their behavior.
2. Rewards are interpreted by each person differently, so we should vary rewards based on player types, profiles, skill levels, and experience.
3. Rewards are always assigned a level of value by the humans getting the reward. In general, as we use a reward, we will decrease its value.
4. We should seek to understand the deeper reward/behavior relationships in a game. There is a difference between (1) reward relationships based on arbitrary points or experience leveling points/behaviors and (2) rewards that satisfy a player's psychological needs (psychological rewards/behaviors). The two reward relationships may be psychometrically the same in one game, but different in another game, either intentionally (to devalue kill stealing) or unintentionally, thus causing serious problems for the retention rate of the game. For example, a player likes the feeling of seeing their direct action result in an object blowing up, but is not rewarded with points equivalent to that feeling. No matter how well rationalized by the designer, this conditioning approach will be difficult to employ effectively within a game because of the inherent disconnect between the two rewards ("I feel great because I blew that up, but this game is broken because I did not get the points!").
5. Variable ratio schedules are the most predicative schedules we can use with humans in encouraging a behavior.
6. A severe initial punishment is more effective than a series of weaker smaller punishments.
7. Punishment is only effective when the punishment/behavior relationship is obvious and the punishment fits the crime. If the punishment/behavior relationship is not obvious or the punishment is too severe, then we can unintentionally produce resentful demoralization (the feeling that no matter what we do, we will be punished).
8. Classic punishment is next to impossible in games and VEs. It is extremely difficult to measure the actual feeling a player gets when their avatar is punished. It is more likely that we are giving a player a time-out, which is less effective in shaping behavior.
9. The change in the positive or negative value of the user's state will determine the effectiveness of the punishment. Time-outs are neutral states. If the initial state was not highly valued by the user, then the time-out will be less effective.

FIGURE 49.9   Encouraging behavior.

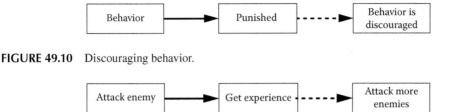

FIGURE 49.10   Discouraging behavior.

FIGURE 49.11   Model for leveling used in many video games.

If we want to encourage players to mindlessly grind 80 levels in a video game, then we should provide a complement of reinforcers across many dimensions to support this behavior (see Figure 49.11). For instance, we may grant a reward such as experience points, currency, or an item for carrying out the behavior we want (defeating an enemy, for instance), provide appealing sensory input, or allow the player to advance to another level or another part of a narrative.

The example earlier describes a fixed ratio schedule. However, humans and animals do not operate efficiently under one-to-one behavior reinforcer schedules. It may very well be that this type of schedule provides very little challenge to humans.

49.2.4   SCHEDULES OF REINFORCEMENT

A fixed ratio schedule would suggest that every time the player performs an action we want to encourage, we immediately provide a reward. However, research shows that while this schedule is effective in animals, it is not as effective in humans. As much as we would like to stereotype humans as lazy, it appears that we do appreciate some type of challenge and at least a bit of uncertainty in task outcomes. In fact, a variable ratio schedule, where rewards are given only after several behavioral actions are executed, can be much more predicative of long-term behavioral influence and learning.

Therefore, fixed ratio reinforcement schedules follow the following pattern:
For every behavior (attacking the enemy), I get a reinforcer (experience points); therefore, I attack more enemies.

By contrast, variable ratio reinforcement schedules look like this:
For every second or third behavior (attacking the enemy), I get a reinforcer (experience); therefore, I attack more enemies.

Recently, some games have switched from fixed ratio to modified variable ratio schedules. In these games, while each enemy killed provides some experience, every second or third enemy killed will provide some bonus experience or unique treasure. Variable ratio schedules are often considered the most powerful schedule of reinforcement for both animals and humans, in part because behavior persists long after it has ceased to be reinforced.

Variable interval and fixed interval schedules produce the weakest impact on behavior. While some games use interval schedules as a basis for reinforcers, this does pose a few obvious problems. It has been well documented in the literature that both animals and humans can quickly identify an interval schedule resulting in slowing or extinction of expression of the behavior. The idea behind interval schedules is that the subject gets a reward-based randomized (variable interval) or set (fixed interval) points in time. In a variable interval schedule, the player would get a reward every second or third minute regardless of behaviors produced in a game. In a fixed interval schedule, the player would get a reward every minute. In both of the interval schedules, a player need only exist within the environment to get rewards. Humans and animals tend to catch on quickly that reinforcers are connected to time schedules and not actions, thus reducing the willingness in the player or animal to produce that behavior when compared to ratio schedules. We recently reviewed a game that rewards bonus experience based on fixed time schedules. A survey of attitudes toward this set time mechanic suggested that while players like the extra experience, it actually devalues any play time that does not get bonus experience, perhaps

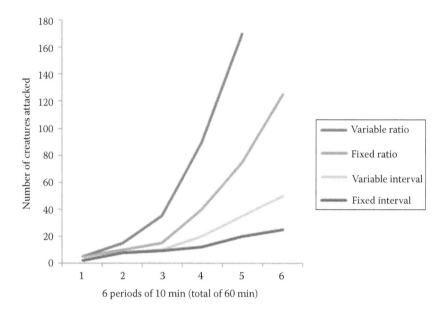

FIGURE 49.12 Number of creatures attacked over time: four basic reinforcement schedules.

reducing overall game value. We also recently identified a handful of games that have been accused by parents as causing overstimulation and overplay in children. Some of these companies now actually use time-based reward schedules that enhance ratio schedules with the specific intent of reducing play time. Players quickly understand that value is greatest only at specific points in time. Additional research is needed before we can fully know the impact of such schedules on player behavior (see Figure 49.12).

Therefore, fixed interval reinforcement schedules follow the following pattern:
While doing behavior (attacking the enemy), I get a reinforcer every minute (experience); therefore, I attack more enemies.

By contrast, variable interval reinforcement schedules follow the following pattern:
While doing behavior (attacking the enemy), I get a reinforcer every second or third minute (experience); therefore, I attack more enemies.

Obviously, these techniques can be useful to the designers of VE games who wish to control the expression of certain behaviors in players.

49.2.5 PUNISHMENT AND NEGATIVE REINFORCEMENT

Creating a negative state in a human or animal is a valid but less powerful way to shape behavior when compared to positive reinforcement. Many scholars agree that positive reinforcement is advantageous when behavior can be shaped over an extended period of time and where long-term learning is required. Raising a child over many years will allow for many thousands of behavior and reinforcer connections to be made by that child. A caregiver can shape behavior by rewarding only those behaviors that are seen as positive. Creating a negative state can be extremely effective when time is constrained or when the behavior is hazardous to the immediate well-being of the player (or child!). However, we may just want to discourage behavior by creating a negative state. Contrary to the views of some scientists as to the actual psychological impact on the subject, the typical opinion of many researchers is that negative reinforcement is the more humane way to reduce or shape a behavior. By removing something of value from the player, we can shift the player from a positive state to negative state. For instance, if a player engages in action X, we might remove item Y, which has some value to them. Behaviors are always followed by a removal. However, humans are excellent at creating negative reinforcement avoidance strategies. For instance, players can remove or hide items that hold value such that no negative state can be created. Players can also learn to avoid any situation where negative reinforcement can be applied. A demoralized player may even purposefully destroy an item of value seen to be at risk so as to control the situation, thus reducing the opportunity for negative reinforcement.

Punishment is a very controversial topic when it comes to shaping behavior. Two camps of scientists usually form when discussing punishment. One camp of scientists may suggest punishment should be avoided altogether or alternatively reduced to a minimal presence in any environment dealing with humans. Another group of scientists would suggest that punishment, like conditioning, should be evaluated on a case-by-case basis as the situation, environment, and behaviors vary. The controversy surrounding punishment results from the inherent ability of punishment to cause long-term physical or emotional harm to the subject. The classic form of punishment is an electrical shock (Gupta & Nagpal, 1978), 100 dB white noise for 10 s (Gupta & Shukla 1989), or aversive physical action causing harm such as spanking or requiring an individual to hold their body in push-up position for some extended period of time. The classic form takes a subject from a positive state (some type of positive state behavior that needs to be reduced) to a negative state (pain, discomfort, or other negative emotional state) that a subject would normally want to avoid.

We have a very obvious problem when it comes to punishment in a video game or VE. All physical punishment in a video game is experienced vicariously. The punishment is actually being experienced by a player's digital avatar, rather than the player themselves, thus reducing its visceral impact. As such, punishment is relevant only to the degree that we can measure psychological components such as empathy and projection within the player toward an avatar. Most punishments in games or VEs are comparable to time-outs such as a parent might give a child. One such approach might be to physically punish a player's avatar (which may have little to no impact on the player) and then make the player wait for 30 s before they can engage in play again. A time-out is a form of punishment that is distinctly different from classic punishment. A time-out is seen as going from a positive state (some type of positive state behavior that needs to be reduced) to a slightly negative but neutral state (boredom) that a subject would normally want to avoid. First-person shooters (FPSs) are one of the best examples of time-out over implementation and misuse. For example, an FPS player will not change behavior when given a time-out if they do not care about the game and/or do not enjoy the game. If the game is multiplayer, those who do care about the game will be receiving a punishment because a teammate has been removed from play. In essence, the punishment punishes others and not the intended target. A time delay punishment has little to no effect on these players and will not encourage better gameplay. Ultimately, the relative change in state predicts the power of punishment on shaping behavior. In some cases, a punishment (such as withholding access to gameplay) will have little to no effect on shaping behavior.

Behavior = extremely fun—punishment will have a powerful effect on shaping behavior.

Behavior = moderately fun—punishment will have a powerful effect on shaping behavior.

Behavior = extremely fun—time-out will be perceived as powerful and will have a powerful effect on shaping behavior.

Behavior = moderately fun—time-out will not have a powerful effect on shaping behavior because it is not perceived as a significantly large enough change in state.

49.2.6 Recent Research in Operant Conditioning, Punishment, and Negative Reinforcement

Much of the current research on punishment is focusing on the saliency of the punishment behavior relationship (Fischer, Greitemeyer, Kastenmuller, Frey, & Oswald, 2007). The obvious, perceived, and continuous connection between punishment and behavior will predict the punishment's influence on shaping behavior. In operant conditioning, much of the research is being produced by a number of *monetized game* research groups, including those focused on the traditional gambling or *gaming* industry. The gambling industry has a huge stake in understanding the equation that encompasses the encouragement of behavior while offering minimal rewards that ensure profit margins are met. Can we then foresee a generally agreed upon equation for player return on investment thresholds in the game industry that would identify the perfect return on investment (rewards) needed to produce long-term behavior without causing a player to become bored or quit? Probably not. Because most game genres require different behaviors and offer distinct rewards, it is more likely that we will see the development of clusters of weighted variables in equations that look at and weigh the reward types and behavior types used by clusters within specific game genres.

For example, likelihood of play (g) = behavior (a) (type) × reward (type [b] and schedule [c]) or g = a × (bc), where a ranges from 0 to 1 depending on internal value of behavior outside of reward; b is 4(VR), 3(FR), 2(VI), and 1(FI); and c ranges from 0 to 1 depending on the value of the reward.

49.2.7 Design Implications of Behavioral Conditioning

Operant conditioning, punishment, and negative reinforcement therefore are and will remain very powerful techniques in games and VEs regardless of whether a designer has knowledge that they are employing the technique or not. Some basic components of these techniques are presented in the summary section at the end of this chapter.

49.3 IMMERSION IN VIRTUAL ENVIRONMENT GAMES

One of the critical issues to be considered in the design of entertainment VEs is how to support and maintain user engagement and immersion in the environment. Unfortunately, when discussing immersion and engagement, one must consider that there are multiple competing models and definitions that describe these constructs (Adams, 2004; Arsenault, 2005; Ermi & Mayra, 2005; Sweetser & Wyeth, 2005; Witmer & Singer, 1998). Therefore, while there is general agreement throughout the industry that immersion is a key goal of interactive design, the industry is not in similar agreement on how to measure immersion or even on how precisely to define it (Jennett et al., 2008; Qin, Rau, & Salvendy, 2009) (Table 49.3).

49.3.1 Defining Immersion

For purposes of this work, we will adopt the broadly accepted Witmer and Singer (1998) definition of immersion: "a psychological state characterized by perceiving oneself to be enveloped by, included in, and interacting with an environment that provides a continuous stream of stimuli and experiences" (p. 227). Although it enjoys broad, but not universal acceptance (see Slater, 1999), this definition is useful because it does not focus solely on the physical trappings of immersive experience—rather, it specifically defines immersion as psychological in nature and emphasizes the role of interaction and presentation of novel stimuli and experiences. Accordingly, to paraphrase this definition, in order to create an immersive VE game experience, it is necessary to present an artificial experience that (1) represents a coherent artificial environment that is (2) compelling and interactive and (3) driven by a continuous series of stimuli, which presumably present some kind of meaning to the user.

Although much academic research in the simulation and training domain regarding VEs focuses on the technological hardware-based antecedents of immersion, game designers typically focus on the software environment, in large part because this is generally the only factor of the experience over

---

TABLE 49.3

Comparison of Differing Existing Frameworks for Describing Immersion

Malone (1981)	Ermi and Mayra (2005)	Adams (2004)	Arsenault (2005)
Challenge	Challenge	Tactical	Systemic
Fantasy	Imaginative	Narrative	Fictional
Curiosity (sensory/cognitive)	Sensory	Strategic	Sensory

which they have control. In the video game industry, games are typically marketed either to specifically designed and standardized consoles, over which the game developers themselves have no control, or to a wide diversity of handheld, laptop, or desktop devices with wildly varying hardware and display capabilities, the specifications of which are equally out of the hands of developers. In either case, designers and developers have access only to the software components and have therefore concentrated extensively on utilizing this half of the VE system to achieve their goal of creating an immersive experience.

Not all designers of VEs labor under these restrictions; however, the lessons learned by designers operating in this context are equally valuable to the community of VE designers as a whole. Thus, determining how to achieve a successful and compelling immersive experience in VE games can be seen in large part to be a function of software design. Although there are different perspectives on defining immersion and there as yet exists no general agreement on a universal model in any domain, developers of entertainment-oriented VEs may draw on insights from all of these sources to help inform their design decisions.

49.3.2 FLOW AND THE GAMEFLOW MODEL

One construct that is closely related to immersion in interactive VE games in general and in particular to models of immersion such as the SCI model is the construct of flow (Csikszentmihalyi, 1990; McClure, 2009). Flow is a concept that many game designers find to be important to the gameplay experience (Chen, 2006), but does not describe immersion directly. While a full explanation of flow is beyond the scope of this work (see Csikszentmihalyi, 1990), in general, flow theory suggests that there exists a zone of optimal experience that is determined by an ideal interaction between the challenge of a task and the skill of the operator performing the task such that when the two elements are correctly matched, a subjectively pleasurable experience is achieved wherein performance may also be enhanced. Although entertainment applications (VE or otherwise) do not necessarily concern themselves with the absolute level of performance achieved by users, the induction of a positively affective experiential state is of course of interest to anyone attempting to create a successful entertainment-themed VE.

Moreover, a model of successful design in games has arisen based explicitly on Csikszentmihalyi's flow hypothesis. The gameflow model (Sweetser & Wyeth, 2005) posits that good games create immersion by inducing a flow state in which players

1. Experience an altered perception of time
2. Experience an emotional connection to the game
3. Experience an increasing detachment from their actual surroundings
4. Experience a lowering in self-awareness
5. Experience a *visceral* connection to the game

Unfortunately, the gameflow model tends to be descriptive with regard to immersion rather than proscriptive; that is, it concerns itself more with identifying and describing the effects of immersion rather than how to induce it in the first place. It is, however, important in that it suggests to us that users who are immersed in a VE game increasingly accept the reality presented by the game as their own, at least temporarily. A poignant illustration of this can be seen in the terminology commonly used by gamers when the player's avatar suffers death. Players almost never say "he/she/it killed my avatar" or "… killed my character." Instead, they almost invariably say "he killed *me*."

49.3.3 SCI MODEL OF IMMERSION

In contrast to the gameflow model, Ermi and Mayra (2005) have proposed a model in which factors critical to generating and sustaining immersion are grouped into three key areas designated *sensory*, *challenge*, and *imaginative*. According to this model, stimuli presented to players should be intended to create an esthetically appealing sensory experience (van den Hoogen, IJsselsteijn, & de Kort, 2009), create ongoing meaningful tasks that will challenge them (Becta, 2001; Csikszentmihalyi, 1990; Maguire et al., 2002), or fire their imagination (Brooks, 2003; Bizzocchi, 2007). One can easily observe how well these three groups of immersive factors line up with the design principles for supporting immersion games as identified by Malone (1981) and discussed in the introduction to this chapter. However, it is important to note that Malone broke up his construct of curiosity into sensory and cognitive curiosity, which does complicate the apparent congruency. According to Malone, sensory curiosity is about "the attention-attracting value of changes in the light, sound, of other sensory stimuli of the environment (p. 363)," whereas cognitive curiosity, by contrast, focuses on the user's drive to enhance their understanding of the rules and nature of the presented domain area (in this case, the virtual world). Accordingly, Ermi and Mayra's model can be seen as incorporating a broader definition of the construct of *sensory immersion* than Malone's.

Additionally, Arsenault (2005) has proposed changes to the SCI model based on specific criticisms of its limitations. The argument has been made that *fictional immersion* is a more accurate term to describe the construct than *imaginative immersion*, as it makes it clear that the user is involved in the fictional story or construct presented in the game rather than using their imagination to create a story or context and that *systemic immersion* is a better term than *challenge immersion*, as *challenge* implies active interactive struggle between the user and system, which may not be present in all games.

Accordingly, there are a number of design implications that creators of VE games can derive from the SCI model and the concept of flow. First, the user should be constantly engaged in interactive decision making. This helps to support the psychological flow state and therefore enhances the subjective experience of the user. It is therefore also critical that the level of difficulty experienced by users in moment-to-moment interaction must be carefully calibrated to be high enough to cause the user to exert themselves to the utmost while still remaining low enough to allow them to be successful. Whenever possible, the environment should be designed so as to draw the user in by presenting a novel, intricate, or esthetic sensory experience (note this does not necessarily have to be *positive* esthetic experience—battlefields, for instance, are compelling but not necessarily beautiful).

In the body of literature surrounding game design, a tripartite framework reminiscent of the models presented earlier exists, although it does not fit quite as cleanly into the categories identified by Malone and echoed by the SCI model. It lacks, as well, the process-based emphasis of the Brown and Cairns model. Nevertheless, it deals more explicitly with the problem-solving and exploratory nature of interactive entertainment and as such may provide helpful insights to those designing VEs for entertainment purposes.

49.3.4 TACTICAL, STRATEGIC, AND NARRATIVE IMMERSION MODEL

Game designer Ernest Adams proposed yet another three-level model of immersion: tactical immersion, strategic immersion, and narrative immersion, which appear to correspond broadly, but not entirely, with the constructs identified both in Ermi and Mayra (2005) and in Malone (1981). The model was first proposed in Adams (2004) and elaborated subsequently in Adams (2005).

Tactical immersion is seen as the *moment-by-moment act of playing the game* and is *physical and immediate* (Adams, 2004). It seems to correspond to both the immersive construct of *curiosity*, in that it includes a constant stream of experience, and the construct of *challenge*, in that it focuses on a continuous and repetitive series of decisions that determine the moment-by-moment status of the game. Tactical immersion is supported by seamless interaction paradigms, including user interfaces that are accessible, efficient, and reliable. Additionally, tactical consistency is important—gameplay mechanics should not change drastically throughout the course of the interactive experience.

Strategic immersion, by contrast, is defined as the player's longer-term engagement with the game, and while it may include elements of challenge, it is more objective focused. In this sense, it corresponds well to Malone's subconstruct of cognitive curiosity (how can I understand and control this environment better?) but less clearly to the *sensory curiosity* subconstruct or to Ermi and

Mayra's *sensory* construct. It is important that users feel that they are able to understand and affect reasonably long-term outcomes in the VE game. This means that while chance is a useful way to vary event outcomes, it should not be overused or the experience may become frustrating to the user. Note that although both depend greatly on randomized outcomes, flipping a coin is not a good game, but poker is.

Narrative immersion hinges on the quality of the presentation of the story in the VE. Note that *story* does not necessarily require a formalized narrative, but may refer to the implicit story told by events. A man sneezing and falling out of a chair is a story, with identifiable plot elements (itchy nose, attempt to avoid sneezing, sneezing, losing balance, flailing, falling over, landing on the floor), a character (the man in the chair), and a resolution (lying on the floor). In games, a great deal of narrative value derives from smaller elements of story such as emotion, setting, and micronarrative (Bizzocchi, 2007). Good quality interactive stories include elements of temporal and causal flow, physical and psychological context, and user participation (Brooks, 2003). This last element reflects game designer Sid Meier's famous observation that "a game is a series of interesting choices" (Meier, 2012, as quoted in Meier, 2012). It is also important to maintain the *suspension of disbelief* (Coleridge, 1817) by avoiding reminders or references that violate the *4th wall* of the VE (Adams, 2004).

So we see a theoretical convergence of ideas suggesting that in order to maximize immersion in VEs intended for purposes of entertainment, the goal should be to maximize the user experience of these three general construct areas. This simple framework, in combination with related findings from the simulation and training and game design literatures (Madigan, 2010; Maguire et al., 2002; Qin et al., 2009), allows some specific recommendations to be advanced.

49.3.5 Designing VE Games to Support Challenge-Related Immersion

Challenge is a feature that appears over and over in the literature in connection to immersion. Csikszentmihalyi's concept of flow (Czikszentmihalyi, 1990) explicitly calls out challenge as a key feature of the *effortful success* on which the creation of a flow state depends. After all, if the task is too easy, there is no need to generate the intense focus necessary to feel *in the zone*. Equally, if the task is too difficult, successful performance is too rare or too hard to achieve, and the intrinsic reward element that helps to sustain the flow state is lost. The moment-by-moment experience of interacting with a game is commonly referred to in the game industry as *gameplay* (Crawford, 2003). The game design literature also regularly addresses this issue under the broad heading of *game balance*. It is generally understood that the game must be hard enough to challenge the player, without being so hard as to become discouraging.

In the game industry, balance is typically assessed and refined through the use of empirical trials using individuals drawn from the potential target demographic for a game. These procedures are commonly known as *playtests* and are in some ways reminiscent of participatory analysis techniques like *naturalistic observation* or *think-aloud* protocols in usability (Nielsen, 1993). Some games use alternative means to ensure a consistent challenge level. One example of this is the use of adaptive difficulty algorithms, which use some kind of coarse measure of user performance or stress to drive the difficulty level of the repetitive gameplay tasks (McClure, 2009).

Implicit in the notion of *balancing* the challenge of VE gameplay, as well as the overall concept of *game*, is the notion that some sort of *winning* or a state of successful performance must be achievable by the player. It is important that the player be aware of the difference between success and failure. Good feedback has the four characteristics of being understandable by the user, timely, specific, and presented in such as way as to enable adaptive behavior (Tighe & O'Connor, 2005). Ensuring that the player is able to successfully engage the ongoing gameplay component therefore requires that players be provided with constant feedback that meets those requirements. For example, if players move into a certain area of a VE game and are blocked by an obstacle, they should immediately receive a clear signal that they cannot proceed in this direction as well as an indication why and how to proceed. For instance, an audible grunt and camera shake effect triggered by contact with a wall could be effective feedback in a case like this, especially if the shake effect automatically turns the camera in a direction in which movement is possible.

Another key factor in maintaining immersion through challenging gameplay is allowing for continuous tactical decision making (Adams, 2004; Maguire et al., 2002), where the player is constantly weighing the benefits and costs of any given course of action. For instance, in a VE game focused on automobile racing, the user must constantly be judging the line their vehicle should take and how that will affect their speed, location on the track, and spatial relationship with other vehicles and objects in the environment. This helps to keep the users' cognitive resources engaged in the ongoing moment-to-moment experience, which is an important factor in maintaining immersion (Qin et al., 2009).

49.3.6 Designing VE Games to Support Fantasy or Imaginative Immersion

Presenting and maintaining a detailed and believable artificial world is critical to maintaining user engagement with a VE game. This does not refer solely to the physical environment itself but also to the design of those elements that encourage the player to consider themselves a part of the world represented in the VE (Malone, 1981). However, it is certainly the case that part of the goal is to achieve what has been termed *spatial immersion*, which is the degree to which the users feel themselves to be physically present in the simulation. While many models of *presence* exist, Wirth et al. (2007) have proposed a model that has been applied to the assessment of game presence in the industry in the past (Madigan, 2010). The model argues that spatial immersion relies upon a number of factors, notably a high level of involvement, a suspension of disbelief, and a strong and consistent mental model of the spatial situation. Therefore, this suggests some specific design guidelines.

First, maintaining the cohesion of the environment and narrative is important to maintaining the user's immersion. One key goal in this process is to maintain the user in a narrative bubble in which the sensory information they receive reinforces and is consistent with their understanding of the workings of the world the VE presents. Inconsistencies between narrative and environment (*it says there's a mountain there, but I see a lake*) remind the user of the artificiality of the environment and of their physical existence outside of the VE. It is also critical that the environment and behavior of computer-controlled entities be consistent. Nonplayer characters (NPCs) should appear to behave in reasonable ways, for instance, not getting *hung up* on elements of the environment and running stupidly in place trying to escape. Additionally, it is important that if narrative is present, it be consistent and constructed according to the conventions of good storytelling. This means using traditional narrative structures whenever possible and avoiding awkward dialog, weak voice acting, poorly drawn characters, and plot devices that do not make sense or do not fit into the larger narrative. For this reason, most modern large budget video games generally employ professional writers to develop their narrative arc.

49.3.7 Designing VE Games to Support Sensory or Curiosity Immersion

Malone (1981) posited two subcomponents to the construct of *curiosity*, a sensory component and a cognitive component, and argued that sensory curiosity can be enhanced by presenting a rich sensory environment that engages as many sensory modalities as possible, although, due to technical limitations, at the present date, these are typically most often auditory and visual in nature. Moreover, he suggested that sensory input could be effective as decoration, to enhance fantasy, as a reward, or to represent some feature of the artificial world. Therefore, the stimulation of as many sensory modalities as possible would seem to be desirable in helping to inculcate a state of sensory immersion. However, more research is probably needed in this area to determine exactly which modalities and how many in combination create an optimal experience for the user in different types of VE applications.

Cognitive curiosity, by contrast, is aroused by presenting the user with a new world to explore, complete with complex rules and knowledge structures to understand. Games often create cognitive curiosity by presenting players with a world where their existing expectations of reality are challenged or do not apply. The game typically does not explain this to the player directly, but rather forces them to discover how to be successful through trial and error. This engages the user's cognitive curiosity and draws them into the experience. Therefore, designers of entertainment-focused VEs should carefully consider how players are introduced to the virtual worlds they create in order to maximize the role of cognitive curiosity in engaging the user.

49.3.8 Involvement Model of Immersion

Brown and Cairns (2004) have proposed an alternative model that focuses not on describing the elements of immersion, but rather on tracing the process by which immersion occurs. In the model, the first stage of immersion is seen as the phase where the user develops an initial impression of their likely level of interest in interacting with the game and determines the level of time, effort, and attention they are willing to allocate to learning to use it. This phase is termed *engagement*. Once the user has decided to engage with the game, the next step is to become further involved with the game such that they become emotionally invested in it. In this stage, the player is less aware of the outside world and begins to focus on the game and neglect their surroundings. This is associated with the suspension of disbelief, in that players increasingly locate their *self* in the VE and begin to accept the alternate reality presented by the artificial world. This second stage is termed *engrossment*. Finally, *total immersion* occurs when the user experiences a sense of *presence* in the VE. In this stage, users are fully immersed in the VE and are either not aware of or totally disregard the outside world. To achieve this state, games must present a multimodal experience that stimulates auditory, visual, and mental processes in the user so that they feel that they are actually present in the virtual world.

49.3.9 Design Implications of the Involvement Model

To enhance engagement, the first of the three steps, designers should focus on lowering barriers to access for users. This means focusing on creating an intuitive and efficient user interface and considering user preferences in things like environment design, thematic focus, and subject domain. Users who like sports are more likely to engage with VEs that incorporate or reflect themes or locations associated with competitive sports. Some of these will vary from user to user, and so procedures that carefully identify what kinds of people will be using the VE and what their individual preferences are will be helpful. Early and frequent usability and human factor testing is also critical to ensure a seamless interaction between the user and the interface. The same emphasis on interactive decision making and repetitive tasks that appeared in the Malone (1981) and SCI (Ermi & Mayra, 2005) models is seen here as well, as well as the emphasis on providing a compelling narrative to provide meaning and context to gameplay. Finally, providing maximal sensory fidelity using as many senses as possible is recommended, although this assumes that the designer has the ability to determine the specifications and capabilities of the hardware component of the VE.

49.3.10 Emerging Input Devices

A number of recent hardware developments have begun to place input technologies historically seen primarily in high-end *serious* VEs into the hands of the developers and users of entertainment-themed VEs. At the time of this writing, infrared, inertial, and video and audio tracking systems to allow some limited haptic and vocal input from players are included or widely available for most current-generation game consoles such as the Nintendo Wii, XBOX 360, and Playstation 3. For instance, some video games now allow the player to give verbal or gestural commands to control in-game avatars or other characters. It can be expected that future console and, increasingly, PC-based systems will begin to use these technologies more frequently, which may allow future designers of VE games to leverage these immersion-enhancing products.

At the present time, the capabilities of these systems are in constant and rapid flux. Currently, the XBOX 360 Kinect (Microsoft, initially released in February 2010) as of the 1.5 update incorporates not only gestural and vocal command recognition but also facial expression recognition. The Playstation Move (Sony, released September 2010) incorporates 3D motion tracking using accelerometers as well as magnetometric location tracking within the earth's magnetic field and some simple object recognition. The Nintendo Wii (Nintendo, initially released November 2006) incorporates accelerometers and infrared detection systems to determine location and movement as well as an available weight distribution sensor and an optional pulse oximeter (Pigna, 2009).

Note that the same principles and guidelines that applied to system output to the user should be followed when designing input systems. For instance, gestural interfaces should probably be consistent with gestures the player uses in everyday life or with those used by computer-generated entities in the VE game. Further, similar devices have been used in the past in nonentertainment applications, allowing future designers of VE game systems to benefit from that existing body of literature.

49.3.11 Guidelines for the Design of Virtual Environment Games to Maximize Immersion

In conclusion, a number of general design recommendations can be advanced based on the existing literature that might be expected to enhance user immersion in entertainment-themed VEs. These conclusions are grouped by Arsenault's (2005) revised model into factors that support fictional (Table 49.4), sensory (Table 49.5), and systemic immersion (Table 49.6).

TABLE 49.4
Guidelines to Support Fictional Immersion in VE Games

1. Identify and, whenever possible, try to cater to user preferences in environment, subject domain, and core interaction paradigm.
2. Avoid violating the 4th wall or otherwise acknowledging the user's status outside of the VE—this breaks the suspension of disbelief and damages immersion.
3. Be careful to avoid problems with narrative (consistency, plausibility, etc.).
4. Ensure that the story provides context, logical event progression, and strong interactive elements that allow the user to influence the outcome (or feel as though they influenced the outcome of the game).
5. Follow conventions of good storytelling (use traditional narrative structures when possible and avoid clumsy dialog, uninteresting or poorly drawn characters, and unrealistic or illogical plot devices).

TABLE 49.5
Guidelines to Support Sensory Immersion in VE Games

1. Be sure that the environment and entities within the environment are internally consistent—that is to say that they act in reasonable, predictable, and realistic ways.
2. High visual fidelity encourages engrossment and helps players become more deeply involved with the VE.
3. Provide detailed stimulation to as many senses as possible. This helps the user to become totally immersed in the environment.
4. Whenever possible, ensure that the environment itself draws the user in by presenting a novel, intricate, or esthetic sensory experience.

TABLE 49.6

Guidelines to Support Systemic Immersion in VE Games

1. Avoid awkward or illogical interactive tasks. Computer-controlled or computer-generated entities should behave in logical or natural-seeming ways.
2. Avoid excessive reliance on chance-based outcome decisions where the user has little control over the outcome.
3. Core interaction mechanics must be consistent throughout the entirety of the interactive experience.
4. Support engaging gameplay (repetitive but challenging and interesting tasks that make up the moment-to-moment experience of interacting with the VE game) with a compelling narrative to provide meaning and context.
5. The level of difficulty in the core gameplay interaction should be carefully balanced as to allow success while still posing a high degree of challenge.
6. The user should constantly be engaged in interactive decision making.
7. To enhance engagement, everything possible should be done to ensure users have access to the VE. This includes identifying and removing barriers such as poor or intimidating user interfaces through iterative testing and development.
8. User interface (UI) and control devices must be accessible, reliable, and user friendly.

49.4 MOTIVATIONAL THEORY IN GAME DESIGN

Why do people play video games? And what makes them continue to play? The concepts that answer these questions are not unrelated to those that explain why people behave in any setting and specifically in VEs. People do what they are motivated to do. Human motivation can be described as that force that guides human behavior and can be represented in terms of directionality, intensity, and persistence. That is, motivation has properties that direct one's energy toward certain behaviors, the intensity or effort level exhibited in such behaviors, and those that cause sustained effort over some period of time. People are motivated to play games to the extent that they draw or direct energy and effort toward play behavior. The motivation to play has intensity in that one plays with higher or lower levels of effort, and motivation to play will be temporal in that at some point the behavior will extinguish. There are numerous taxonomies and theories that attempt to classify how players are motivated to play games.

49.4.1 GAME PLAYER TYPES

Bartle (1996) is often attributed with the development of the first gamer classification taxonomy. In his work, he focused on classifying the types of humans that play multiuser dungeon (MUD) video games. He suggests that there are four player types based on the things that are enjoyable about these types of video games: achievers, explorers, socializers, and killers. Achievers enjoy accumulating points, scores, wins, and status in the game environment. Explorers are more interested in finding out the nuances of the game world, such that they believe that experiences and leveling are important, but value them primarily in that they enable them to reach the expansive limits of the game world. Socializers are obviously more focused on the interaction with other players. The gameplay emphasis for these users tends to be the setting for social events, and the interest in the gameplay strategy and exploration are merely tools that facilitate relationship building and experiencing. Killers are focused on imposing their will on other players and play to experience the perception of power. Bartle's work was based less on psychology or personality theory and was instead primarily empirically derived. Despite the conceptual limitations of this work, it is still somewhat popular among game designers when considering player types, especially for MUDs and other massively multiplayer online role-playing games.

Later, researchers built on the Bartle (1996) types in an attempt to validate a more theory-based set of player types. Yee (2006) conducted a factor analytical study in which he developed a 39-item survey based in part on the Bartle (1996) types. The study identified three factors rather than four, which were called *achievement*, *social*, and *immersion*. Additionally, 10 subcategories loaded relatively well onto the three factors, which are listed in Table 49.7.

TABLE 49.7
Player Motivation Factors and Associated Subcategories

Achievement	Social	Immersion
Advancement	Socializing	Discovery
Mechanics	Relationship	Role playing
Competition	Teamwork	Customization
		Escapism

Source: Yee, N., *Cyberpsychol. Behav.*, 9(6), 772, 2006.

This work was important in that it was an early attempt at identifying game player motivations based on empirical research. However, as the authors acknowledge in their subsequent work (Yee, Ducheneaut, & Nelson, 2012), there were significant limitations to the resulting taxonomy. First, the 39-item survey measured the 10 categories, but they were not specifically developed to measure the apparent factors. Second, the scale was in English, which likely limits its validity in other non-English-speaking cultures. Finally, the survey had not demonstrated the ability to predict behavior. In their most recent work (Yee et al., 2012), they address these three issues by conducting item analysis studies in order to identify which of the items provide the most information for their hypothesized respective factors. Their results were encouraging as they demonstrated relatively robust loadings, and they yielded a 12-item scale that was based on a more theoretical approach. The same publication reports a validation study in which the 12 items are translated to Chinese, the results of which suggested that the factors and dimensions generalize across the populations of both language speakers. Yee et al. also reported on additional work that demonstrated predictive validity with self-reported game behaviors in the massively multiplayer online role-playing game World of Warcraft (Blizzard Entertainment, 2004).

Others have made progress in identifying the motivation or drive to play. Jen and Teng (2008) attempted to correlate five of the dimensions (e.g., discovery, role playing, teamwork, advancement, and escapism) identified by Yee (2006) and Yee and colleagues (2012) with the dimensions of the better known Big 5 (Goldberg, 1992), which include extroversion, openness, agreeableness, conscientiousness, and neuroticism. They were able to employ a shorter version of the Big 5 validated by Saucier (1994) and found that most of their hypotheses were supported. For example, openness was correlated with discovery and role playing; conscientiousness related to escapism; extraversion related to teamwork; agreeableness related to advancement; and neuroticism related to teamwork.

There are other taxonomies that classify gamer motivations, such as Kim and Ross's (2006) seven motivation dimensions of sports gamers: *knowledge application, identification with sport, fantasy, competition, entertainment, social interaction,* and *diversion.* King, Delfabbro, and Griffiths (2010) described five structural characteristics of video games in terms of psychological concepts including *social, manipulation/control, narrative/identity, reward/punishment,* and *presentation.* Westwood and Griffiths (2010) used the King et al. (2010) structural elements as the basis of their Q-sort and subsequent factor analysis to identify six gamer types that they labeled *story-driven solo gamers, social gamers, solo limited gamers, hardcore online gamers, solo control/identity gamers,* and *casual gamers.*

There are other approaches and resultant taxonomies (e.g., Bijvank, Konijn, & Bushman, 2012; Jansz & Martens, 2005; Sherry, Lucas, Greenberg, & Lachlan, 2006), and although the earlier referenced efforts have been helpful in classifying gamers to some extent, it is still necessary to understand the psychological processes involved with video gameplay and what the actual motivational mechanisms are that drive and sustain play behaviors. These taxonomies are useful, but other work has targeted the motivational processes at play in the mind of the gamer.

49.4.2 SELF-DETERMINATION THEORY

Taxonomies are useful for categorizing and classifying humans, including game players, which is helpful in understanding who plays games and that there are different types of people that play games with different behavioral tendencies. However, personality and other individual differences do not address the psychological mechanisms that actually drive one to engage in the behaviors associated with video games and other forms of entertainment in VEs. Early ludologists suggested that at its simplest form, gameplay can be classified as one or more of the following: competition, chance, imitation, or shock (Caillois, 1958). What then would motivate behavior to experience these or other aspects of play? As mentioned previously, motivation is that force that drives or draws one to behave in ways that satisfy needs. It can be very strong or less so. It directs attention and effort toward a myriad of behavioral choices based on the extent the behavioral choice satisfies the most salient needs. This motivational force has properties that will sustain behavior until the salient need is satisfied or until another need becomes more salient and the associated motivation redirects effort. Gameplay is one set of behaviors that competes for attention as would any other.

One motivation theory specifically related to video games that has received a good amount of attention over the last decade, specifically in the area of video game motivation, is self-determina-tion theory (SDT) (Ryan & Deci, 2000). SDT suggests that motivation can be classified into *intrin-sic* motivation and *extrinsic* motivation. Intrinsic motivation is the "inherent tendency to seek out novelty and challenges, to extend and exercise one's capacities, to explore and to learn," (Ryan & Deci, 2000, p. 70). Extrinsic motivation would then be defined as that which drives behavior "of an activity in order to attain some separable outcome," (p. 71). Thus, the drive to behave can either be to experience the behavior itself (intrinsic) or in order to attain some discrete outcome that results from the behavior (extrinsic). This is important because it has been well established that humans perform differently based on whether they are motivated internally or externally. For example, Deci and Ryan (1991) found that intrinsic motivators tend to result in higher levels of performance on the same task, and when the behavior itself is seen as the reward, performers tend to exhibit more creativity and stick with the task longer (e.g., Deci & Ryan, 1991). In fact, when external outcomes (reward or punishment) are provided for behaviors that are intrinsically satisfying to people, they actually tend to shift their type of motivation from intrinsic to extrinsic, which in turn tends to reduce the effectiveness of that behavior (see Deci, Koestner, and Ryan [1999] for a meta-analysis on this topic).

To describe motivations as either intrinsic or extrinsic is overly simplistic. It is likely that some aspects of gameplay are intrinsically motivating, some are extrinsically motivating, and many are likely to share some intrinsic and some extrinsic characteristics. SDT suggests that motivation varies on a continuum from amotivation (disinterest) to extrinsic motivation and finally intrinsic motivation. One regulates one's behavior based on the motivation such that amotivation results in nonregulation and intrinsic motivation results in complete self-regulation. However, extrinsic motivators are regulated in several ways. They can be externally regulated at their extreme, which involves rewards and punishments. They may also be regulated through introjection, which involves some degree of self-control as well as rewards and punishments that are internal to the behaver. Identified regulation is still externally motivated, but the rewards are personally important and are valued. Finally, integrated regulation occurs with extrinsic motivation when the external reward is in complete synthesis with self. That is, when a reward is external, but in complete congruence with personal values and goals, it is still considered extrinsic, but on the continuum from amotiva-tion to intrinsic motivation, it is the closest to intrinsic motivation. An example of this might be an employee who loves his or her job but still expects to be well paid for doing it. Ryan and Deci (2000) also suggest that the benefits of improved effectiveness can be maximized the more regulated the extrinsic motivation is. They suggest behaviors are intrinsically motivating to the extent that they satisfy three primary needs: relatedness, competence, and autonomy. Regulation of extrinsic moti-vators occurs to the extent that they are integrated into satisfying these three needs (see Ryan and Deci [2000] for a more comprehensive discussion on this topic).

TABLE 49.8
Guidelines to Support User Motivation in VE Games

1. Designers should be mindful of what motivates human behavior in general and specifically while participating in entertainment-based VEs.
2. Gameplay should always be designed with the consideration of which of the three primary needs are meant to be satisfied: relatedness, competence, and/or autonomy.
3. For game behaviors that tend to be external reward based, the design should consider the four levels of extrinsic self-regulation (i.e., external, introjected, identified, and integrated) and consider possible content that will facilitate integration.
4. When behaviors within the game are likely to be intrinsically motivated (behavior is the reward), designers should take measures to avoid attaching extrinsic rewards to those behaviors.

This approach to explaining motivation is quite relevant in explaining video game and entertainment virtual reality behaviors in several ways. Some gameplay involves behaviors that are intrinsically satisfying, and some behaviors are instrumental in getting to some external outcome or reward. When engaged in an multiplayer online role playing game (MMORPG), many of the behaviors involve assuming some virtual identity and then acting to kill monsters or other characters. The experience of killing a monster is intrinsically motivating or satisfying to the extent that the actual behavior is the reward. However, the experience is extrinsically motivating to the extent that the drive behind the behavior is not (just) the experience of killing a monster, but the prize or coins or recognition that results. As in real life, behavior in a virtual or game life tends to take on similar motivational properties.

Recent studies have provided valuable insight into the self-regulation of video game players. One particular series of studies were done by Ryan, Rigby, and Przybylski (2006), which investigated the motivation effects of video gameplay and the extent to which they satisfy the basic needs of relatedness, competence, and autonomy. They found through self-report that game enjoyment tended to relate to perceived in-game autonomy and the feelings of competence within the game. Their study also demonstrated that players' perceptions of relatedness, competence, and autonomy also each predicted future play on associated games. In a later study, they found that in-game violence was associated with autonomy and competence satisfaction (Przybylski, Ryan, & Rigby, 2009). In their most recent publication, they report that the most intrinsically motivating games allowed for players to experience their *virtual self* most consistently with their reported perceived *ideal self* (Przybylski, Weinstein, Murayama, Lynch, & Ryan, 2012).

49.4.3 DESIGN RECOMMENDATIONS TO SUPPORT PLAYER MOTIVATION

It is well supported that humans tend to try harder and longer when they are behaving in a manner that is intrinsically motivated. When the behavior is the reward, rather than some external outcome that is separate from the behavior, humans experience more enjoyment and are more satisfied. As with everyday life, some behaviors are not particularly enjoyable or personally satisfying, so extrinsic rewards make those behaviors more attractive. Therefore, game design should consider the motivational properties of various aspects of their game and consider these concepts during the process.

Based on the recent literature and practice outlined earlier, we proposed several design considerations to enhance the play experience through application of intrinsic and extrinsic motivation theory, summarized in Table 49.8.

49.5 CONCLUSIONS

It is likely that video games and other entertainment applications will remain a very lucrative and important application of VE technologies in the future. By approaching the process of design with the aim of creating a product that leverages principles drawn from an understanding of behavioral conditioning, immersion, and motivation theory, developers of VE games should be able to achieve higher levels of user interest, utilization, and satisfaction. This might reasonably be expected to improve the practical, critical, and financial value of such systems.

REFERENCES

Adams, E. (2004, July 9). Postmodernism and the three types of immersion. *Gamasutra*. Retrieved 12/26/2007, from http://designersnotebook.com/Columns/063_Postmodernism/063_post-modernism.htm

Adams, E. (2005). *Letting the audience onto the stage: The potential of VR drama*. Paper presented at the Virtual Storytelling '05 Conference, Strasburg, France. Retrieved 8/9/2010, from http://www.designers-notebook.com/Lectures/VRDRAMA.pdf

Arsenault, D. (2005). *Dark waters: Spotlight on immersion*. Paper presented at Game on North America 2005 International Conference Proceedings, Ghent, Belgium. Retrieved from https://www.academia.edu/224757/Dark_Waters_Spotlight_on_Immersion

Bartle, R. (1996). *Hearts, clubs, diamonds, spades: Players who suit MUDs*. Retrieved 8/15/2012, from http://www.mud.co.uk/richard/hcds.htm

Bijvank, M. N., Konijn, E. A., & Bushman, B. J. (2012). We don't need no education: Video game preferences, video game motivations, and aggressiveness among adolescent boys of different educational ability levels. *Journal of Adolescence, 35*, 153–162.

Bizzocchi, J. (2007). Games and narrative: An analytical framework. *Loading—The Journal of the Canadian Games Studies Association, 1*(1), 5–10.

Blizzard Entertainment. (2004). *World of Warcraft*. [Video Game]. Irvine, CA: Blizzard Entertainment.

Blade, R. A., & Padgett, M. L. (2002). Virtual Environments Standards and Terminology. In K. Stanney, *Handbook of Virtual Environments*. Erlbaum, Mahwah.

Bouton, M. E. (1994). Context, ambiguity, and classical conditioning. *Current Directions in Psychological Science, 3*, 49–53.

Brooks, K. (2003). *There is nothing virtual about immersion: Narrative immersion for VR and other interfaces*. Retrieved 8/6/2012, from http://www.immersive-medien.de/sites/default/files/biblio/Immersive_Not_Virtual.pdf

Brown, E., & Cairns, P. (2004). *A grounded investigation of immersion in games*. In E. Dykstra-Erickson, and M. Tscheligi, (Eds.), Extended Abstracts of the 2004 Conference on Human Factors and Computing Systems CHI 04, Computings (pp. 1297–1300). Retrieved from http://discovery.ucl.ac.uk/55390/

British Educational Communications and Technology Agency (BECTA). (2006). *Engagement and motivation in games development processes*. Retrieved from website: http://dera.ioe.ac.uk/id/eprint/1677.

Caillois, R. (1958). *Les jeux et les hommes*. Paris, France: Gallimard.

Carnagey, N. L., Anderson, C. A., & Bushman, B. J. (2006). The effect of video game violence on physiological desensitization of real-life violence. *Journal of Experimental Social Psychology, 43*, 489–496.

Carlson, N. R. (2012). *Physiology of behavior* (11th ed.). Upper Saddle River, NJ: Prentice Hall.

Chen, J. (2006). *Flow in games*. (MFA Thesis) available from http://www.jenovachen.com/flowingames/Flow_in_games_final.pdf. Retrieved 4/2/2014, from http://jenovachen.com/flowingames/thesis.htm

Coleridge, S. T. (1817). *Biographia literaria*. Cambridge, MA: Harvard.

Condie, R., Munro, B., Seagraves, L., & Kenesson, S., British Educational Communications and Technology Agency (BECTA). (2001). *The impact of ICT in schools—a landscape review*. Retrieved from website: http://www.teindia.nic.in/e9-tm/Files/ICT_Documents/ImpactICT_Becta.pdf

Crawford, C. (2003). *Chris crawford on game design*. Thousand Oaks, CA: New Riders.

Csikszentmihalyi, M. (1990). *Flow: The Psychology of Optimal Experience*. New York: Harper and Row.

Deci, E. L., Koestner, R., & Ryan, R. M. (1999). A meta-analytic review of experiments examining the effects of extrinsic rewards on intrinsic motivation. *Psychological Bulletin, 125*, 627–668.

Deci, E. L., & Ryan, R. M. (1991). A motivational approach to self: Integration in personality. In R. Dienstbier (Ed.), *Nebraska symposium on motivation: Vol. 38. Perspectives on motivation* (pp. 237–288). Lincoln, NE: University of Nebraska Press.

Deci, E. L., & Ryan, R. M. (2000). The 'what' and 'why' of goal pursuits: Human needs and the self-determination of behavior. *Psychological Inquiry, 11*, 227–268.

Entertainment Software Association. (2012). *Sales and genre data*. Retrieved 7/29/2012, from http://www.theesa.com/facts/econdata.asp

Ermi, L., & Mayra, F. (2005). Fundamental Components of the Gameplay Experience: Analysing Immersion. In S. Castell & J. Jensen (Eds.) *Worlds in Play: International Perspectives on Digital Games Research*, vol. 37, pp. 37–53. New York, NY: Peter Lang International Academic Publishers.

Fischer, P., Greitemeyer, T., Kastenmuller, A., Frey, D., & Oswald, S. (2007). Terror salience and punishment: Does terror salience induce threat to social order. *Journal of Experimental Social Psychology, 43*(6), 964–971.

Gilleade, M. K., & Allanson, J. (2005). Affective videogames and modes of affective gaming: Assist me, challenge me, emote me. *Digital Games Research Association Changing Views: Words in Play*, Vancouver, Canada. Retrieved from http://www.digra.org/dl/db/06278.55257.pdf

Goldberg, L. R. (1992). The development of markers for the Big Five factor structure. *Psychological Assessment, 4*, 26–42.

Grimes, M. L. (2011). *Operant conditioning in MMORPGs* (Masters thesis). Retrieved 7/17/2011, from http://www.gamecareerguide.com/features/975/operant_conditioning_in_.php

Gupta, B. S., & Nagpal, M. (1978, May). Impulsivity/sociability and reinforcement in verbal operant conditioning. *British Journal of Psychology, 69*(2), 203–206.

Gupta, B. S., & Shukla, A. P. (1989, February). Verbal operant conditioning as a function of extraversion and reinforcement. *British Journal of Psychology, 80*(1), 39–44.

Higinbotham, W. (1958). *Tennis for two* [computer software]. Upton, NY: Brookhaven National Laboratory.

Jansz, J., & Martens, L. (2005). Gaming at a LAN event: The social context of playing video games. *New Media & Society, 7*(3), 333–355.

Jen, S., & Teng, C. (2008). Personality and motivations for playing online games. *Social Behavior and Personality, 36*(8), 1053–1060.

Jennett, C., Cox, A. L., Cairns, P., Dhoparee, S., Epps, A., Tijs, T., & Walton, A. (2008). Measuring and defining the experience of immersion in games. *International Journal of Human-Computer Studies, 66*(9), 641–661. doi:10.1016/j.ijhcs.2008.04.004

Kim, Y., & Ross, S. (2006). An exploration of motives in sport video gaming. *International Journal of Sports Marketing & Sponsorship, 8*, 34–46.

King, D., Delfabbro, P., & Griffiths, M. D. (2010). Video game structural characteristics: A new psychological taxonomy. *International Journal of Mental Health and Addiction, 8*, 90–106.

Madigan, J. (2010, June 27). The psychology of immersion in computer games. *Blog Post*. Retrieved 8/8/2012, from http://www.psychologyofgames.com/2010/07/the-psychology-of-immersion-in-video-games/

Maguire, F., van Lent, M., Prensky, M., & Tarr, R. (2002). *Defense combat sim olympics: Methodologies incorporating the cybergaming culture*. Paper presented at I/ITSEC Conference, Orlando, FL.

Malone, T. W. (1981). Toward a theory of intrinsically motivating instruction. *Cognitive Science, 4*, 339–369.

McClure, D. (2009, December 4). Adaptive difficulty. *Game Career Guide Feature*. Retrieved 8/8/2012, from http://www.gamecareerguide.com/features/805/adaptive_.php

Meier, S. (2012). *Interesting decisions*. Presentation at the 2012 Game Developers Conference, San Francisco, CA.

Mitchell, A., & Savill-Smith, C. (2004). *The use of computer and video games for learning: A review of the literature*. Retrieved from website: http://www.lsda.org.uk/files/pdf/1529.pdf

Nacke, L., & Lindley, C. A. (2008, November 3–5). Flow and immersion in first-person shooters: Measuring the player's gameplay experience. *FuturePlay*, Toronto, Canada.

Nielsen, J. (1993). *Usability engineering*. San Francisco, CA: Morgan Kaufmann.

Pavlov, I. P. (1927). *Conditioned reflexes*. Oxford, England: Oxford University Press.

Pavlov, I. P. (1955). *Selected works*. New York, NY: Foreign Languages.

Pigna, K. (2009, June 2). Satoru Iwata announces Wii vitality sensor. *1UP News Feature*. Retrieved 8/14/2012, from http://www.1up.com/news/satoru-iwata-announces-wii-vitality

Prensky, M. (2001). *Digital game based learning*. New York, NY: McGraw Hill.

Przybylski, A. K., Ryan, R. M., & Rigby, C. S. (2009). The motivating role of violence in video games. *Personality and Social Psychology Bulletin, 35*, 243–259.

Przybylski, A. K., Weinstein, N., Murayama, K., Lynch, M. F., & Ryan, R. M. (2012). The ideal self at play: The appeal of video games that let you be all you can be. *Psychological Science, 23*(1), 69–76.

Qin, H., Rau, P. P., & Salvendy, G. (2009). Measuring Player Immersion in the Computer Game Narrative. *International Journal of Human Computer Interaction, 25*, 107–133.

Razran, G. H. S. (1940). Conditioned response changes in ratings and appraising sociopolitical slogans. *Psychological Bulletin, 37*, 481.

Rescorla, R. A. (1988). Pavlovian conditioning: It's not what you think it is. *American Psychologist, 42*, 151–160.

Rumpel, S., LeDoux, J., Zador, A., & Malinow, R. (2005). Postsynaptic receptor trafficking underlying form of associative learning. *Science, 308*, 83–88.

Ryan, R. M., Rigby, C. S., & Przybylski, A. K. (2006). The motivational pull of videogames: A self-determination theory approach. *Motivation and Emotion, 30*, 347–364.

Saucier, G. (1994). Mini-markers: A brief version of Goldberg's unipolar Big-Five markers. *Journal of Personality Assessment, 63*(3), 506–516.

Sherry, J., Lucas, K., Greenberg, B. S., & Lachlan, K. (2006). Video game uses and gratifications as predictors of use and game preference. In P. Vorderer & J. Bryant (Eds.), *Playing video games. Motives, responses, and consequences* (pp. 213–224). Mahwah, NJ: Lawrence Erlbaum Associates.

Siwek, S. E. (2010). *Video games in the 21st century: The 2010 report.* Retrieved 7/29/2012, from http://www.theesa.com/facts/pdfs/VideoGames21stCentury_2010.pdf

Slater, M. (1999). Measuring presence: A response to the Witmer and Singer presence questionnaire. *Presence: Teleoperators and Virtual Environments, 8*(5), 560–565. doi:10.1162/105474699566477

Sweetser, P., & Wyeth, P. (2005). GameFlow: A model for evaluating player enjoyment in games. *Computers in Entertainment, 3*(3), 1–24. doi: 10.1145/1077246.1077253

Sykes, J., & Brown, S. (2003, April 5–10). Affective gaming: Measuring emotion through the gamepad. *CHI*, Ft. Lauderdale, FL.

Tighe, J., & O'Connor, K. (2005). Seven practices for effective learning. *Educational Leadership, 63*(3), 10–17.

Weibel, D., & Weissmath, B. (2011). Immersion in computer games: The role of spatial presence and flow. *International Journal of Computer Games Technology*, 2011. Retrieved from website: http://dx.doi.org/10.1155/2011/282345

Westwood, D., & Griffiths, M. D. (2010). The role of structural characteristics in video game play motivation: A q-methodology study. *Cyberpsychology, Behavior and Social Networking*, 13, 581–585.

Wirth, W., Hartmann, T., Bocking, S., Vorderer, P., Klimmt, C., Holger, S., ... Jancke, P. (2007). A process model for the formation of spatial presence experiences. *Media Psychology, 9*, 493–525.

Witmer, B. G., & Singer, M. J. (1998). Measuring presence in virtual environments: A presence questionnaire. *Presence: Teleoperators and Virtual Environments, 7*(3), 225–240.

Wouter M. van den Hoogen, Wijnand, A., IJsselsteijn, & Yvonne, A. W. de Kort. (2009). Effects of sensory immersion on behavioural indicators of player experience: Movement synchrony and controller pressure. In 2009 *Digital Games Research Association Conference: Breaking New Ground: Innovation in Games, Play, Practice and Theory.* Retrieved from website: http://www.digra.org/wp-content/uploads/digital-library/09287.18127.pdf.

Yee, N. (2006). Motivation for play in online games. *Cyberpsychology and Behavior, 9*(6), 772–775.

Yee, N., Ducheneaut, N., & Nelson, L. (2012). Online gaming motivations scale: Development and validation. In J.A. Konstan, E. H. Chi, & K. Hook (Eds.), *Proceedings of the ACM International Conference on Human Factors in Computing Systems* (pp. 2803–2806.) New York: Association for Computing Machinery.

Section VIII

Conclusion

50 Virtual Environments
History and Profession

Richard A. Blade, Mary Lou Padgett,
Mark Billinghurst, and Robert W. Lindeman

CONTENTS

50.1 INTRODUCTION

Although this handbook largely focuses on the current status and future vision of virtual environment (VE) technology, it would be unjust to develop such a work without providing acknowledgment to the pioneers whose vision and innovativeness laid the groundwork for the science, technology, and applications described here. This chapter provides a very brief historical overview of some of the key people and major milestones that led to the current state of the art and describes recent trends over the past decade that are likely to continue in the near future. In addition, readers of this handbook may be left wondering where they can gain more information about the VE profession. This chapter provides a number of such references, including websites, associated periodicals, major conferences and trade shows, major organizational players, and general and historical references.

50.2 BRIEF HISTORY OF VIRTUAL ENVIRONMENTS

Not surprisingly, the science fiction literature envisioned the generation of artificial or illusory environments well before there was any idea of the technology to accomplish such a feat. For example, a famous 1950 science fiction story by Ray Bradbury (Bradbury, 1976) called *The Veldt* depicted a playroom where children experienced an African landscape. The animals ultimately ate the parents. The first attempts at developing such a technology used electronics, but not computers, which at that time were much too primitive. In 1956, Morton Heilig, a filmmaker, developed *Sensorama* (US Patent No. 3,050,870), a mechanical virtual display device (see Figure 50.1). Perhaps the first

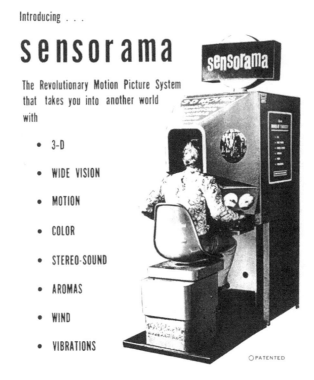

FIGURE 50.1 Poster advertising Heilig's Sensorama.

head-mounted display (HMD) was developed by Philco in 1961, which permitted remote viewing via a video camera (Kalawsky, 1993).

In the early 1960s, computer graphic technology was beginning to be developed. The technology for computer-generated VEs probably began in 1963, when Dr. Ivan Sutherland developed the first interactive computer graphics system, called *Sketchpad*, while a graduate student at the Massachusetts Institute of Technology (MIT; Sutherland, 1965). Sketchpad created highly precise engineering drawings that could be manipulated, duplicated, and stored. In 1968, Sutherland joined with David Evans to build an HMD at the University of Utah (Sutherland, 1970). Shortly thereafter, Thomas Furness, working at Wright-Patterson Air Force Base, developed an HMD that he called a *visually coupled system* (Furness, 1969). Furness went on to develop a virtual cockpit flight simulator in 1981 (Furness, 1988), and then founded the Human Interface Technology Laboratory (HIT Lab) at the University of Washington, an early leading academic research center in virtual reality (VR). The early HMD developments of Sutherland and Furness were continued by various government organizations and private companies, such as Honeywell and Hughes to produce a wide variety of designs.

While Sutherland and Furness were working on the technical aspects, Myron Krueger was studying the artistic and psychological aspects of VEs at the University of Wisconsin. Krueger first concocted *GlowFlow* in 1969, a kind of artistic light and sound show that was controlled by a computer but involved no computer graphics (Krueger, 1977). In *GlowFlow*, audiences actively participated in the show by moving around on pressure-sensitive plates embedded in the floor, though most were unaware of the controlling mechanism. The following year Krueger developed the much more elaborate *Metaplay* with a grant from the National Science Foundation and the university's computer science department. This system involved 800 pressure-sensitive switches and an

8 × 10 f rear projection screen on which people viewed video images of themselves from a camera, superimposed on computer graphics generated by a minicomputer. During the next several years, Krueger developed ever more elaborate artistic works that provided interactive experiences to audiences, such as *Videoplace*, which immersed an individual into a computer-generated world inhabited by other human and virtual participants where the laws of cause and effect could be changed from moment to moment (Krueger, 1985). Krueger is credited with coining the term *artificial reality* in 1973 (Krueger, 1985).

In 1977, with the TRS-80, Apple II and PET 2001, the first affordable personal computers were introduced. That same year, the first glove device for controlling a computer, the Sayre glove, was developed at Electronic Visualization Lab in Chicago. Also in 1977, Kit Galloway and Sherri Rabinowitz created virtual space with the Satellite Arts Project, implementing the vision of Arthur C. Clark and Marshall McLuhan's global village, interconnecting the people of the world via electronic communication. In what Galloway and Rabinowitz called *Hole in Space*, they set up a large video screen and camera in public spaces in New York City and Los Angeles to effectively videoconference in a group setting. A decade later, they went on to found the *Electronic Café* in Santa Monica, California, allowing videoconferencing on an individual basis.

By 1979, researchers at the Architecture Machine Group at MIT began developing spatial data management systems, and in the early 1980s, they began interactive control of video playback with Apple II computers for use in computer-aided instruction (CAI). In 1981 those researchers, including Scott Fisher, produced the *Aspen Movie Map*, in which a person could navigate, using a touch-sensitive display screen, along 20 miles of streets in Aspen, Colorado (Fisher, 1982). This did not use an HMD; rather, it was an elaborate *CAVE*-type (CAVE automatic virtual environment) system that projected scenes on the surrounding walls of a room.

Also in the early 1980s, the Defense Advanced Research Projects Agency (DARPA) funded a global war game simulator called *simulation networking*, or *SIMNET*, that broke new ground in large-scale networks. Using SIMNET and its later upgrade, Distributed Interactive Simulation (DIS), hundreds of soldiers all over the world could sit in tank, helicopter, and fighter bomber simulators, testing their combat skills against one another in real time, much like a video game, but much more realistic. The cost savings and safety of such simulators will undoubtedly continue to drive the development of military battle simulators ever more in the future (see Maxwell, Griffith, & Finkelstein, 2014, Chapter 37).

In 1984, the movie *The Last Starfighter*, written by Jonathan R. Betuel and produced by Edward Denault and Gary Edelson, merged live action with computer graphics to save millions of dollars in production costs. Also in 1984, William Gibson wrote a science fiction novel, *Neuromancer*, in which he introduced the now-ubiquitous term *cyberspace*. In 1986, Lucasfilm, a firm founded from the revenues of the 1977 *Star Wars* movie series, began developing computer movies. Since that time, the movie industry has steadily increased its use of computer graphics to reduce costs and create astounding visual effects.

Inspired by the research of Tom Furness, and on a limited budget, in 1981, the National Aeronautics and Space Agency (NASA) built their own HMD from liquid-crystal displays (LCDs) out of cheap Sony Watchman TVs. This was called the Virtual Visual Environment Display (VIVED) and was connected to a magnetic tracking system and graphics computer to create a fully immersive VE. Scott Fisher joined the project in 1985 and worked on adding spatial sound and glove-based gesture input for a more compelling VR experience. This became the VIEW (Virtual Interface Environment Workstation) project and was used to explore possible user interface for astronauts and immersive visualization.

Also in 1985, a team under Dr. Frederick P. Brooks, Jr., began experimenting with 3D perception of molecules at the University of North Carolina. In the next decade, that project developed into the *Nanomanipulator*, where a person wearing an HMD could manipulate single atoms or small groups of atoms by a force-feedback arm electronically connected to a scanning tunneling electron microscope (Taylor et al., 1993; see Stone, Hannigan, & Murphy, 2014, Chapter 36). The Nanomanipulator project was one of the first academic VR efforts and helped establish University of North Carolina as a leader in the field. Since that time, they have gone on to develop many essential VR technologies such as fast graphics hardware, tracking systems, HMDs, and the widely used Virtual Reality Peripheral Network (VRPN) framework, which provides seamless, efficient support for VR hardware.

The glove input device used in the VIEW system was VPL Research's *DataGlove*, which was invented by Thomas Zimmerman (US Patent No. 4,542,291) and operated by a programming language developed by Jaron Lanier (Rheingold, 1991). In 1985, Jaron Lanier and Jean-Jacques Grimaud founded VPL Research, Inc., to produce state-of-the-art human interface devices (Kalawsky, 1993). Lanier, who coined the term *virtual reality*, was an interesting, creative, and eccentric character that the public came to identify so closely with the technology that many incorrectly viewed him as the founder of VEs. In the middle and late 1980s, VPL Research became famous for making what was intended to be the first consumer-grade VE hardware: the *DataGlove* glove and the *EyePhone* HMD. Together with the Body Electric software, they were able to provide the first complete commercial VR system. Unfortunately, after only a few years, VPL Research came on hard financial times and filed for bankruptcy in 1990, with the assets finally being acquired by Sun in 1999.

In 1988, Chris Gentile, of Abrams/Gentile Entertainment, developed the *PowerGlove* for Nintendo Home Entertainment System (see Figure 50.2; Gardner, 1989). Licensed and marketed by Mattel, the PowerGlove became a best-selling toy in 1989 and 1990 but was discontinued at the end of 1991 because it failed to produce expected revenues. That was probably the first truly consumer-grade VE hardware. In the 1990s, the PowerGlove remained a prized item for VE professionals and amateur enthusiasts alike enabling them to add inexpensive gesture input to their VR systems.

Another company to be created from the NASA VIEW project was Fakespace Labs, which was founded in 1990 and developed the *BOOM** (Binocular Omni Orientation Monitor), a stereoscopic display mounted on a boom for tracking position (see Figure 50.3). This provided a very wide field of view display combined with button input for interaction with the virtual scenes. Fakespace went on to develop the PinchGlove glove-based input system and a wide variety of other VE display technology and peripherals.

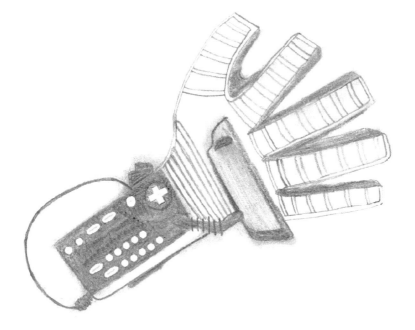

FIGURE 50.2   Mattel's PowerGlove. (Drawing courtesy of Kay Stanney.)

---

* BOOM is a registered trademark of Fakespace Labs, Inc.

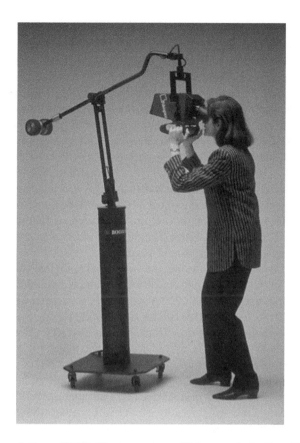

FIGURE 50.3   Fakespace's Boom HMD. (Photo courtesy of Fakespace Labs, Inc., Mountain View, CA.)

Around the demise of VPL, other companies also began to sell VE software and integrated systems. Division PLC was founded in the United Kingdom in 1989 and produced the dVS operating environment, graphics hardware, and HMDs and input devices. At the time, they were the only company providing a complete professional VR system. They were also the first VR company to become publically listed in 1993. In 1991, another UK company, W-Industries began to sell *Virtuality*, a complete VE system for interactive, interpersonal gaming. They provided several different games for arcades the theme parks through the mid-1990s. In the United States, in 1991, Sense8 Corporation developed the WorldToolKit software for building VEs, and AutoDesk demonstrated their Cyberspace PC-based VE CAD system at *SIGGRAPH 89*.

Although the VR industry originated in the United States, significant research work was also occurring in Japan. One of the most unique VR facilities was founded in 1986 in Kyoto, Japan, as a joint effort between the Japanese government and Nippon Telephone & Telegraph (NTT). The Advanced Telecommunications Research (ATR) Institute International gave equal weight to engineers and artists and hosted a very large number of international researchers. This cross-disciplinary environment produced some of the most interesting works of the day and helped jump-start Japanese VR efforts. Some of the significant research efforts there included work in VR teleconferencing, gesture input, virtual humans, locomotion, and haptic interfaces.

In 1989 and the early 1990s, the general public became aware of the concept of VEs and their potential, and enough people began working in the field that the first professional VE conferences were held. The Interfaces for Real and Virtual Worlds conference was held in France in March 1992 and followed a few months later by Medicine Meets Virtual Reality in San Diego, a conference that is still the most popular conference for medical applications of VR, and the Meckler Conference on

VR in September. In September 1993, the HIT Lab at the University of Washington worked with the Institute of Electrical and Electronics Engineers (IEEE) to organize the *Virtual Reality Annual International Symposium (VRAIS)*, which became the *IEEE VR Conference*, the leading academic conference for VR research.

Advances were also being made in novel input and display systems. Through most of the 1990s, and deep into the 2000s, Prof. Hiroo Iwata from the University of Tsukuba, Japan, pioneered one inventive approach to VR locomotion after another. Iwata's team put together an impressive string of accepted demos at *SIGGRAPH (1994–2007)*, including robotic footpads that follow the step of the user (Iwata, Yano, & Nakaizumi, 2001), to *magic* shoes that slide on the floor during walking (Iwata, Yano, & Tomioka, 2006), to the CirculaFloor (Iwata, Yano, Fukushima, & Noma, 2005), consisting of a swarm of coordinating smart floor tiles that move around to catch the footfalls of the user. Dr. Susumu Tachi started one of the earliest VR labs in Japan, and his research has covered several novel display devices, such as the retroreflective camouflage head-mounted projection display (HMPD) (Inami et al., 2000) and Twister (Kunita et al., 2001), which is a cylindrical surround display that provides visuals using persistence of vision.

Japanese researchers have also been leading the way on research into olfactory (smell) displays. Broadly speaking, research into delivering olfactory cues can be divided into two main problems, creating the smell and delivering the smell. Dr. Michitaka Hirose created one of the first delivery systems using a mask (Hirose, Tanikawa, & Ishida, 1997), while Dr. Yasuyuki Yanagida developed a projection-based system, relieving the user from wearing anything (Yanagida, Kawato, Noma, Tomono, & Tetsutani, 2004).

In 1993, SGI announced their *Reality Engine*, a computer capable of running significant VE applications, beginning the current era of rapidly multiplying applications. Significant computer power has since developed on relatively inexpensive personal computers. In 1996, VE climbed onto the Internet, with the creation of Virtual Reality Modeling Language (VRML); however, though a plethora of VRML tools were available, it never really caught on widely as *the* standard for 3D on the web. Finally, in a well-publicized application of VE, astronauts were trained to do repairs on the Hubble space telescope using a full-scale VE simulator constructed at Goddard Space Flight Center (see Stone et al., 2014, Chapter 36).

Thus, by the mid-1990s, all of the components for developing VR systems were commercially available, and there was growing research in academic and industrial laboratories around the world. Many people were predicting a commercial boom and the widespread use of HMDs and personal VR systems within the next decade or so. However, this was not to be the case. In November 1994, Nintendo announced the Virtual Boy low-cost 3D gaming display that was mounted on a tripod and provided an immersive gaming experience. Despite a significant advertising campaign, the product never took off and disappeared from the market in late 1996. Similarly, Virtual i-O released a consumer HMD in 1995, but although well designed, this struggled in the market and the company was bankrupt by 1997. The late 1990s was a period of consolidation for the industry and become known as the *VR winter* as many start-ups went bankrupt or were acquired. For example, W-Industries was sold to Cybermind in 1997, Sense8 was acquired by EAI in 1998, and Division was brought by PTC in 1999.

In 1994, Fred Brooks from UNC famously surveyed the field of VR in a talk entitled *Is There Any Real Virtue in Virtual Reality?* and came to the conclusion that VR almost worked. There were lots of prototypes, but except for simulation and entertainment, VR was not yet capable of doing real work. Five years later, he repeated the same assessment and came to the conclusion that VR barely worked but that it has crossed the threshold of usefulness and was now really real. By the early 2000s and beyond, there was a resurgence in the VR industry and in companies using VE technology for a wide variety of applications. Although it was still difficult for companies providing traditional HMD-based VR systems, key elements of the technology have been commercialized through other means, such as input devices for game consoles, desktop 3D graphics applications, complex scientific visualization systems, and location-based theme park experiences.

Today, applications of VE technology include architectural design with walk-throughs, VE fantasy games, VE flight simulations, and educational materials, including human anatomy walk-throughs and map fly-arounds (see Stone et al., 2014, Chapter 36). NASA plans virtual exploration of the solar system. The entertainment industry not only creates many movie effects with computer graphics but provides VEs in the form of passive and interactive movie rides that include motion simulation (see Greenwood-Ericksen, Kennedy, & Stafford, 2014, Chapter 50). The military continues to develop simulators for all aspects of training and is working on telerobotics for battlefield medical treatment (see Maxwell et al., 2014, Chapter 37). The medical industry trains surgeons using virtual patients and *telesurgery* is expected to be prevalent some time in the future. Psychologists use VEs to desensitize against phobias, for pain treatment, and for post-traumatic stress disorders. Disabled persons use VEs to train wheelchair operation as well as a host of other rehabilitative applications. Financial professionals use VEs to visualize stock market and other financial trends.

In Fred Brook's (1999) paper, he identified four key technologies for VR systems: (1) the visual, aural, and haptic displays that immerse the user in the virtual world; (2) graphics rendering; (3) tracking systems; and (4) systems for building and maintaining detailed and realistic models of the virtual world. In the last decade, significant advances have been made in each of those areas, but there are still important research challenges that need to be addressed.

In terms of immersive displays, interesting research is being conducted on autostereoscopic VE displays that allow people to see 3D graphics without wearing any glasses on their head. For example, the 360° light field display research of Jones, McDowall, Yamada, Bolas, and Debevec (2007) uses a rotating mirror to create a 3D volumetric display that can be seen without glasses. There is also a need for research in the area of multiuser 3D displays. Projected VR systems such as CAVE displays can typically only support a single viewpoint, and so are of limited use for groups of people. However, using high-speed projectors could allow time slicing of the VR content so that several people could have their own views. Froehlich's C1x6 work (Kulik et al., 2011) allows up to six people to look at a projected VR screen, but each has independent views in the VR space. Aside from graphics, there is also work needed for haptic rendering and force-feedback devices.

Graphics rendering has advanced significantly to the point where modern cell phones have more graphics power than the SGI RealityEngine computers that were used to generate early virtual worlds. In order to continue this development, there is more work needed on mobile graphics and especially GPU hardware and algorithms. Graphics rendering has also moved away from fixed function pipelines to programmable graphics processing unit (GPU) rendering pipelines and the use of shader languages. The first OpenGL Shading Language (GLSL) appeared in 2004 and defined graphics operations that could be executed in the graphics pipeline to control the hue, saturation, brightness, and contrast of pixels used to construct a final image. Shader programming enables the creation of very realistic cloth simulation, advanced lighting effects such as depth of focus blur, and photorealistic rendering, among others. However, more research is needed on how to use shader languages to develop compelling VE experiences.

Research should also be conducted on the next generation of scenegraph software. Since 1991, scenegraphs have been used to provide a node-based structure defining the graphics scene, the objects within it, and the relationship between them. However, current scenegraphs do have some limitations, such as not being designed to support multiuser online graphics applications or the difficulty of coordinating changes in distributed graphics systems. So research is being conducted on new types of scenegraphs that can simultaneously optimize for spatial, state, semantic, and central processing unit (CPU) considerations.

A variety of tracking technology has been developed to capture user input for VR, but there is still more work that needs to be done in this area. Current tracking systems use optical, magnetic, acoustic, mechanical, or inertial technology or other methods to monitor user head or hand positions. However, these can have a limited range, low accuracy, poor latency, or other problems. Welch and Foxlin (2002) provide a good summary of the various tracking approaches, coming to the conclusion that no single method is suitable for all conditions, but different technologies will work well for specific situations. Recently, researchers have developed large-area high-resolution optical tracking methods (Welch et al., 2001), hybrid techniques that can track users on a city scales (Reitmayr & Drummond, 2006), and low-cost full-body tracking systems (Suma, 2011). This is an active area of significant ongoing research as new technologies become available.

Aside from the graphics, display, and tracking technologies, creating the virtual world content and maintaining the model database is also a significant challenge. Early VEs were mostly constructed by hand in a time-consuming process, but in recent years, there have been a range of techniques developed to speed up the process. Computer vision methods have been developed that can generate realistic 3D models from photographs or video sequences. Researchers have even been able to successfully synthesize entire cities in a day from large numbers of photographs (Agarwal, Snavely, Simon, Seitz, & Szeliski, 2009). However, in these cases, the models are all rigid, so research still needs to be done on how to capture deformable objects or for automatic generation of content such as terrain models, vegetation, and buildings. Tools that enable users to rapidly create 3D worlds are also needed, based on technologies such as sketching, freehand input, or multimodal interaction.

By far, the most important recent trend within the VR community is what could be called the democratization of VR, characterized by mass-market, low-cost, high-performance hardware, sophisticated game engine software, and established content creation pipelines. Most of the advances in these areas can be traced to the video game boom that has prevailed for the past two decades, which has brought us expressive input devices, such as the Wii Remote, Sony MOVE, and the Microsoft Kinect. Emerging devices such as brain–machine interfaces, the Leap Motion finger tracking system, and the Razer Hydra magnetic tracker (Figure 50.4) can now each be had for about US$100.

In terms of displays, there has been a steady stream of low-cost HMDs from various manufacturers over the past two decades, each bringing out a device that costs about US$500, but ultimately failing for lack of a market beyond the 200 (or so) VR research labs. It remains to be seen if the newest player in this space, the Oculus Rift (Figure 50.5), will fare better than its predecessors and make inroads into the tough-to-crack video-game market.

The general availability of VR hardware and software has led to some interesting phenomena. Using the Kinect as an example, its initial release led to an avalanche of *Kinect Hacks* produced by people outside the *traditional* VR community. But while much of this work had a high coolness factor, interaction with the Kinect eventually led people to many of the same problems that had

FIGURE 50.4 Razer Hydra low-cost magnetic tracker.

FIGURE 50.5 Oculus Rift HMD.

been encountered by researchers using older technology, namely, problems with fatigue (e.g., people would rather sit on a couch and rest their hands on their thighs than be forced to hold them up in front of them in order to control a system), noise in the resulting data, computational limits, and environmental issues, such as operating in bright light. Still, many groundbreaking techniques have emerged, such as KinectFusion (Newcombe et al., 2011).

Another aspect that the game industry has given us is support for massive numbers of concurrent users. The popularity of massively multiplayer online role-playing games (MMORPGs), such as World of Warcraft, as well as other multiplayer game genres, in addition to the more generally connected lives of the average person, has given rise to hardware and software infrastructures that support low-latency, high-bandwidth applications. Online environments such as Second Life (Figure 50.6) have shown that persistent VEs can be developed that support hundreds of thousands of simultaneous users.

In summary, over the last 50 years, VR has evolved from science fiction and crude graphics to a thriving research community with many commercial applications. There are important research problems still to be solved, but there are a large number of institutions working in these areas. Clearly, no one questions the potential and ultimate development of VE, including applications not yet conceived.

50.3 MAJOR APPLICATION AREAS

Many VE applications have developed over the past decade (see Chapters 36 through 50, this book). Provided in the following is a short list of some typical projects and resources from the main application areas and associated reference sites. Links to websites for many VR applications and technologies can be found at VResources, the VR resource page, http://vresources.org/.

50.3.1 Scientific Visualization

- The CAVE, Electronic Visualization Lab, University of Illinois at Chicago, http://evlweb. eecs.uic.edu/
- The Nanomanipulator, Department of Computer Science, University of North Carolina at Chapel Hill, http://cismm.cs.unc.edu/2001/08/nanomanipulator/

FIGURE 50.6 Second Life educational space.

- Virtual Reality Room Exhibits, Electronic Visualization Lab, University of Illinois at Chicago, http://www.evl.uic.edu/EVL/VROOM/HTML/OTHER/HomePage.html
- The Scientific Visualization and Virtual Reality (SVVR) group at the University of Amsterdam, http://www.science.uva.nl/research/scs/visualization/
- Medical scientific visualization and virtual reality from University of Iowa, http://www.uiowa.edu/mihpclab/jni/visualization/visualization.html
- Visionair European visualization facilities, http://www.infra-visionair.eu/index.php

50.3.2 ARCHITECTURE AND DESIGN

- HIT Lab at the University of Washington knowledge base on Architecture and Virtual Reality, http://www.hitl.washington.edu/projects/knowledge_base/VRArch/
- The Virtual Reality Design Lab (VRDL), in the School of Architecture at the University of Minnesota, http://vr.design.umn.edu/
- VTT Virtual Reality Applications for Building Construction, http://cic.vtt.fi/4d/

50.3.3 EDUCATION AND TRAINING

- Learning in Virtual Environments, Human Interface Technology Lab, University of Washington, http://www.hitl.washington.edu/projects/learnve/
- The Virtual Reality and Education Laboratory at East Carolina University, http://vr.coe.ecu.edu/

- Umea University online survey of Virtual Reality in Education; http://www8.informatik. umu.se/~dfallman/projects/vrie/
- HIT Lab On the Net resource for Virtual Reality and Education, http://www.hitl. washington.edu/projects/knowledge_base/education.html

50.3.4 ENTERTAINMENT

- Activeworlds.com, Inc.—Developers of 3D virtual worlds to be distributed over the Internet, http://www.activeworlds.com/
- Second Life—the largest online 3D virtual world, http://secondlife.com/
- HIT Lab On the Net resource for Virtual Reality and Entertainment, http://www.hitl. washington.edu/projects/knowledge_base/games.html

50.3.5 MANUFACTURING

- VR use in car manufacturing at Ford, http://www.vr-news.com/2013/05/17/ virtual-reality-at-ford-3d-glasses-virtual-projections-and-3d-printing/
- Project Vega—Virtual Reality in Product Design, http://www.project-vega.ro/
- Boeing's Advanced Manufacturing Research Centre at the University of Sheffield, http:// www.amrc.co.uk/research/support/vr/

50.3.6 MEDICINE

- Treatment of Parkinson's disease, Human Interface Technology Lab, University of Washington, http://www.hitl.washington.edu/research/parkinsons/
- The Medical Virtual Reality laboratory at the University of Southern California Institute for Creative Technologies, http://ict.usc.edu/groups/medical-vr/
- The Virtual Medical Environments Laboratory at the National Capital Area Medical Simulation Center, http://simcen.org/about.html
- HIT Lab On the Net resource for Virtual Reality and Medicine, http://www.hitl.washington. edu/projects/knowledge_base/medapps.html

50.4 PERIODICALS FOR PROFESSIONALS

There are a handful of journals and other publications that focus on VE science, technology, and applications. Provided in the following is a list of the main publications and associated reference sites:

- *The International Journal of Human-Computer Studies*, published by Elsevier, is a monthly journal that publishes original research over the whole spectrum of work relevant to the theory and practice of innovative interactive systems, including VR and other graphics interfaces; http://www.journals.elsevier.com/international-journal-of-human-computer-studies/
- *The International Journal of Human–Computer Interaction*, published monthly by Taylor & Francis. This academic journal addresses the cognitive, social, health, and ergonomics aspects of work with computers and emphasizes both the human and computer science aspects of the effective design and use of computer interactive systems; http://www.tandfonline.com/loi/hihc20

- *The International Journal of Virtual Reality*, a journal of VR research and applications, published irregularly but nominally four times per year. Though published in English, the journal emphasizes international involvement and cooperation; http://www.ijvr.org/
- *Presence: Teleoperators and Virtual Environments*, published by MIT Press, is an academic journal published bimonthly and provides a scientific forum for current research and advanced ideas on teleoperators and VEs; http://www.mitpressjournals.org/loi/pres
- *Virtual Reality,* published by Springer, is a quarterly journal published since 1995 that aims to disseminate research and provoke discussion in the area of VR. Submissions are welcomed on a wide range of topics relevant to VR, including VR technology, applications, assessment, and philosophical and ethical issues; http://www.springer.com/computer/image+processing/journal/10055
- *The Journal of Virtual Reality and Broadcasting* is an open-access e-journal published since 2004 covering advanced media technology for the integration of human–computer interaction and modern information systems; http://www.jvrb.org/

50.5 MAJOR CONFERENCES AND TRADE SHOWS

There are a number of conferences and trade shows that focus on VE science, technology, and applications. Provided in the following is a list of many of these forums:

- *ACM Conference on Human Factors in Computing Systems (CHI)*; http://www.acm.org/sigchi/
- *ACM Multimedia Conference*; http://www.acm.org/sigmm/
- *Annual IASTED International Conference on Computer Graphics and Imaging (CGIM)*; http://www.iasted.com
- *Games Developers Conference (CGDC)*; http://www.gdconf.com/
- *Electronic Entertainment Expo (E3 EXPO)*: a trade show for new multimedia, entertainment, and educational products; http://www.e3expo.com/
- *Eurographics Workshop on Virtual Environments (EGVE)*
- *Human–Computer Interaction International*; http://www.hci-international.org/
- *ICAT*: the oldest international conference on virtual environments, sponsored by the Virtual Reality Society of Japan and the University of Tokyo; http://www.ic-at.org/
- *IEEE Virtual Reality Conference (IEEE VR)*; http://www.ieeevr.org
- *International Conference in Central Europe on Computer Graphics and Visualization (WSCG)*; http://www.wscg.eu/
- *International Conference on Disability, Virtual Reality and Associated Technologies (ICDVRAT)*; http://www.icdvrat.reading.ac.uk/
- *International Immersive Projection Technology Workshop*
- *Interservice/Industry Training, Simulation, and Education Conference (I/ITSEC)*, Orlando, FL; http://www.iitsec.org
- *ITEC*: a major conference on equipment and simulation for education and training; http://www.itec.co.uk/
- *IVR*: a comprehensive VR exhibition in Japan; http://www.ivr.jp/en/about/
- *Medicine Meets Virtual Reality (MMVR)*: the foremost conference on virtual reality and health care; http://nextmed.com/
- *Modeling and Simulation (MS)*; http://www.iasted.org/conferences/home-802.html
- *SIGGRAPH*: a very large exhibition and conference on all aspects of computer graphics, http://www.siggraph.org/attend/annual-conferences/
- *Society for Information Display Symposium (SID)*; http://www.sid.org/
- *SPIE Annual Meeting and Exhibition*, http://spie.org/

- *Symposium on Virtual Reality Software and Technology (VRST)*; http://www.vrst.org/
- *Trade Show of the International Association of Amusement Parks and Attractions (IAAPA)*: the latest VE and simulation rides, http://www.iaapa.org
- *United Kingdom Virtual Reality Special Interest Group Conference (UKVRSIG)*: http://www.crg.cs.nott.ac.uk/groups/ukvrsig/groups.html/

50.6 RESEARCH LABORATORIES

There are a number of research laboratories that focus on VE science, technology, and applications. Provided in the following is a list of some of the primary endeavors:

- Advanced Interfaces Group, University of Manchester Department of Computer Science, England; http://aig.cs.man.ac.uk/
- Army Research Institute; http://www-ari.army.mil/
- Computer Graphics and User Interfaces Lab, Columbia University; http://www.cs.columbia.edu/graphics/
- Department of Computer Science, University of North Carolina at Chapel Hill; http://www.cs.unc.edu/
- Electronic Visualization Lab, University of Illinois at Chicago; http://evlweb.eecs.uic.edu/
- Graphics Visualization Center, Georgia Institute of Technology; http://www.gvu.gatech.edu/
- Human Interface Technology Lab, University of Washington; http://www.hitl.washington.edu/
- Institute for Simulation and Training Visual Systems Lab, University of Central Florida, http://www.ist.ucf.edu/
- Iwata Lab, University of Tsukuba, Japan; http://intron.kz.tsukuba.ac.jp/
- Studio for Creative Inquiry, Carnegie Mellon University; http://studioforcreativeinquiry.org/
- Virtual Reality and Additive Manufacturing Lab at the Missouri University of Science and Technology—http://web.mst.edu/~vram/
- Virtual Reality Applications Center, Iowa State University, http://www.vrac.iastate.edu/index.php
- Virtual Reality Lab, National Center for Supercomputing Applications; http://www.ncsa.uiuc.edu/
- Virtual Reality Lab, University of Michigan, http://www-vrl.umich.edu/
- Virtual Reality and Computer Integrated Manufacturing Laboratory at Washington State University, http://vrcim.wsu.edu/
- Vision and Autonomous System Center (VASC), Carnegie Mellon University, http://www.vasc.ri.cmu.edu/

50.7 ORGANIZATIONS AND RESOURCES

There are a number of societies and professional organizations that have been established to promote VR and also some valuable online resources, including the following:

- EuroVR—the European Association for Virtual Reality and Augmented Reality, http://www.eurovr-association.org/
- The International Society for Presence Research; http://ispr.info/
- Michael Louka's collection of online Virtual Reality resources, http://www.ia.hiof.no/~michaell/home/vr/VirtualReality.html
- Road to VR online VR resources, http://www.roadtovr.com/

- The UK Virtual Reality Society, http://www.vrs.org.uk/
- The Virtual Reality Society of Japan, http://www.vrsj.org/english/about/
- VR News and Resources by New World Architects, http://vroot.org/
- VR News, http://www.vr-news.com/
- The UK Virtual Reality Society, http://www.vrs.org.uk/

REFERENCES

Agarwal, S., Snavely, N., Simon, I., Seitz, S. M., & Szeliski, R. (2009, September 29–October 2). Building Rome in a day. *2009 IEEE 12th International Conference on Computer Vision*, Kyoto, Japan.

Bradbury, R. (1976). The veldt. *The Illustrated Man*. New York, NY: Bantam Doubleday Dell.

Brooks, F. P. (1994, November 30). Is there any real virtue in virtual reality? Public lecture cosponsored by the *Royal Academy of Engineering and the British Computer Society*, London, U.K. http://www.cs.unc.edu/~brooks

Brooks, Jr., F. P. (1999). What's real about virtual reality? *IEEE Computer Graphics and Applications, 19*(6), 16–27.

Fisher, S. (1982). Viewpoint dependent imaging: An interactive stereoscopic display. In S. Benton (Ed.), *Processing and Display of Three-Dimensional Data* (*Proceedings of SPIE*) (Vol. 41, p. 367). April 8, 1983.

Furness, III, T. A. (1969). The application of head-mounted displays to airborne reconnaissance and weapon delivery. *Proceedings of Symposium for Image Display and Recording* (Technical Report No. TR-69-241). Wright-Patterson Air Force Bose, OH: U.S. Air Force Avionics Laboratory.

Furness, III, T. A. (1988). Harnessing virtual space. *Society for Information Display Digest, 16*, 4–7.

Gardner, D. L. (1989). The PowerGlove. *Design News, 45*(23), 63–68.

Gibson, W. (1984). *Neuromancer*. New York, NY: Simon & Schuster.

Greenwood-Ericksen, A., Kennedy, R. C., & Stafford, S. (2014). Entertainment applications of virtual environments. In K. S. Hale & K. M. Stanney (Eds.), *Handbook of virtual environments: Design, implementation, and applications* (2nd ed., pp. 1291–1316). Boca Raton, FL: Taylor & Francis Group, Inc.

Hirose, M., Tanikawa, T. K., & Ishida, K. (1997). A study of olfactory display. In *Proceedings of the Virtual Reality Society of Japan 2nd Annual Conference* (in Japanese pp. 155–158).

Inami, M., Kawakami, N., Sekiguchi, D., Yanagida, Y., Maeda, T., & Tachi, S. (2000). Visuo-haptic display using head-mounted projector. In S. Feiner & D. Thalman (Eds.), *Proceedings of IEEE Virtual Reality 2000* (pp. 233–240). Los Alamitos, CA: IEEE.

Iwata, H., Yano, H., Fukushima, H., & Noma, H., (2005) CirculaFloor. *IEEE Computer Graphics and Applications, 25*(1), 64–67.

Iwata, H., Yano, H., & Nakaizumi, F. (2001). Gait master: A versatile locomotion interface for uneven virtual terrain. In *Proceedings of the Virtual Reality 2001 Conference (VR'01)* (p. 131). Washington, DC: IEEE Computer Society.

Iwata, H., Yano, H., & Tomioka, H., (2006). Powered shoes. *SIGGRAPH 2006 Conference DVD*, New York, NY.

Jones, A., McDowall, I., Yamada, H., Bolas, M., & Debevec, P. (2007). Rendering for an interactive 360 light field display. *ACM Transactions on Graphics (TOG), 26*(3), 40.

Kalawsky, R. S. (1993). *The science of virtual reality and virtual environments*. Wokingham, England: Addison-Wesley.

Krueger, M. (1977). Responsive environments. In *Proceedings of the National Computer Conference* (pp. 423–433). Dallas, TX: AFIPS Press.

Krueger, M. (1983). *Artificial reality*. Reading, MA: Addison-Wesley.

Krueger, M. (1985). VIDEOPLACE: A report from the Artificial Reality Laboratory. *Leonardo, 18*(3), 145–151.

Kulik, A., Kunert, A., Beck, S., Reichel, R., Blach, R., Zink, A., & Froehlich, B. (2011). C1x6: A stereoscopic six-user display for co-located collaboration in shared virtual environments. *ACM Transactions on Graphics, 30*(6), 188–200.

Kunita, Y., Ogawa, N., Sakuma, A., Inami, M., Maeda, T., & Tachi, S. (2001) Immersive autostereoscopic display for mutual telexistence: TWISTER I (Telexistence Wide-angle Immersive STEReoscope model I). In H. Takemura & K. Kiyokawa (Eds.), *Proceedings of IEEE Virtual Reality 2001* (pp. 31–36). Los Alamitos, CA: IEEE.

Maxwell, D., Griffith,T., & Finkelstein, N. (2014). Use of virtual worlds in the military services as part of a blended learning strategy. In K. S. Hale & K. M. Stanney (Eds.), *Handbook of virtual environments: Design, implementation, and applications* (2nd ed., pp. 957–998). Boca Raton, FL: Taylor & Francis Group, Inc.

Newcombe, R. A., Izadi, S., Hilliges, O., Molyneaux, D., Kim, D., Davison, A. J., … Fitzgibbon, A. (2011). KinectFusion: Real-time dense surface mapping and tracking. In *Proceedings of the 10th IEEE International Symposium on Mixed and Augmented Reality 2011 (ISMAR '11)* (pp. 127–136). Piscataway, NJ: IEEE.

Reitmayr, G., & Drummond, T. W. (2006). Going out: Robust model-based tracking for outdoor augmented reality. In *IEEE/ACM International Symposium on Mixed and Augmented Reality 2006 (ISMAR 2006)* (pp. 109–118). Piscataway, NJ: IEEE.

Rheingold, H. (1991). *Virtual reality*. London, U.K.: Secker & Warburg.

Stone, R. J. & Hannigan, F. P. (2014). Applications of virtual environment.. In K. S. Hale & K. M. Stanney (Eds.), *Handbook of virtual environments: Design, implementation, and applications* (2nd ed., pp. 881–956). Boca Raton, FL: Taylor & Francis Group, Inc.

Suma, E. A., Lange, B., Rizzo, A., Krum, D. M., & Bolas M. T. (2011). FAAST: The flexible action and articulated skeleton toolkit. *IEEE, Virtual Reality*, pp. 247–248. Singapore.

Sutherland, I. E. (1965, May 24–29). The ultimate display. *Proceedings of the IFIP Congress* (Vol. 65, No. 2, pp. 506–508). New York, NY.

Sutherland, I. E. (1970). Computer displays. *Scientific American, 222*(6), 56–81.

Taylor II, R. M., Robinett, W., Chi, V. L., Brooks, F. P., Jr., Wright, W. V., Williams, R. S., & Snyder, E. J. (1993, August). The nanomanipulator: A virtual-reality interface for a scanning tunneling microscope. In *Computer Graphics (Proceedings of SIGGRAPH'93)* (pp. 127–134). ACM.

Welch, G., Bishop, G., Vicci, L., Brumback, S., Keller, K., & Colucci, D. N. (2001). High-performance wide-area optical tracking: The hiball tracking system. *Presence: Teleoperators and Virtual Environments, 10*(1), 1–21.

Welch, G., & Foxlin, E. (2002). Motion tracking: No silver bullet, but a respectable arsenal. *IEEE Computer Graphics and Applications, 22*(6), 24–38.

Yanagida, Y., Kawato, S., Noma, H., Tomono, A., & Tetsutani, N. (2004). Projection-based olfactory display with nose tracking. In S. Coquillart, K. Kiyokawa, J. E. Swan II, & D. Bowman (Eds.), *Proceedings of IEEE Virtual Reality 2004* (pp. 43–50). Piscatsway, NJ: IEEE.

Author Index

Subject Index